Human – Computer Interaction:

Theory and Practice (Part II)

Volume 2

Human Factors and Ergonomics
Gavriel Salvendy, Series Editor

Bullinger, H.-J., and Ziegler, J. (Eds.) : Human–Computer Interaction: Ergonomics and User Interfaces, *Volume 1 of the Proceedings of the 8th International Conference on Human–Computer Interaction*

Bullinger, H.-J., and Ziegler, J. (Eds.) : Human–Computer Interaction: Communication, Cooperation, and Application Design, *Volume 2 of the Proceedings of the 8th International Conference on Human–Computer Interaction*

Hollnagel, E. (Ed.) : *Handbook of Cognitive Task Design*

Meister, D. (Ed.) : *Conceptual Foundations of Human Factors Measurement*

Meister, D., and Enderwick, T. (Eds.) : *Human Factors in System Design, Development, and Testing*

Smith, M. J., Salvendy, F., Harris, D., and Koubeck, R. J. (Eds.) : *Usability Evaluation and Interface Design: Cognitive engineering, Intelligent Agents and Virtual Reality*

Smith, M. J., and Salvendy, G. (Eds.) : *Systems, Social and Internationalization Design of Human–Computer Interaction*

Stephanidis, C. (Ed.) : *Universal Access in HCI: Towards an Information Society for All*

Stephanidis, C. (Ed.) : *User Interfaces for All: Concepts, Methods, and Tools*

Ye, N. (Ed.) : *The Handbook of Data Mining*

Jacko, J., and Stephanidis, C. (Eds.) : *Human-Computer Interaction: Theory and Practice (Part I)*

Stephanidis, C., and Jacko, J. (Eds.) : *Human-Computer Interaction: Theory and Practice (Part II)*

Harris, D., Duffy, V., Smith, M., and Stephanidis, C. (Eds.) : *Human-Centred Computing: Cognitive, Social and Ergonomic Aspects*

Stephanidis, C. (Ed.) : *Universal Access in HCI: Inclusive Design in the Information Society*

Human – Computer Interaction:

Theory and Practice (Part II)

Volume 2 of the Proceedings of HCI International 2003
10th International Conference on Human - Computer Interaction
Symposium on Human Interface (Japan) 2003
5th International Conference on Engineering Psychology and Cognitive Ergonomics
2nd International Conference on Universal Access in Human - Computer Interaction
22 – 27 June 2003, Crete, Greece

Edited by

Constantine Stephanidis
ICS-FORTH and University of Crete

Julie Jacko
Georgia Institute of Technology

LEA

2003 LAWRENCE ERLBAUM ASSOCIATES, PUBLISHERS
Mahwah, New Jersey London

Lawrence Erlbaum Associates, Inc., Publishers
10 Industrial Avenue
Mahwah, New Jersey 07430

Human -Computer Interaction : Theory and Practice (Part II) / edited by Constantine Stephanidis and Julie Jacko.

p. cm.

Includes bibliographical references and index.
ISBN 0-8058-4931-9 (cloth : alk. paper) (Volume 2)

ISBN 0-8058-4930-0 (Volume 1)
ISBN 0-8058-4932-7 (Volume 3)
ISBN 0-8058-4933-5 (Volume 4)
ISBN 0-8058-4934-3 (Set)

2003

Books published by Lawrence Erlbaum Associates are printed on acid-free paper, and their bindings are chosen for strength and durability.

10 9 8 7 6 5 4 3 2 1

Preface

The 10th International Conference on Human-Computer Interaction, HCI International 2003, is held in Crete, Greece, 22-27 June 2003, jointly with the Symposium on Human Interface (Japan) 2003, the 5th International Conference on Engineering Psychology and Cognitive Ergonomics, and the 2nd International Conference on Universal Access in Human-Computer Interaction. A total of 2986 individuals from industry, academia, research institutes, and governmental agencies from 59 countries submitted their work for presentation, and only those submittals that were judged to be of high scientific quality were included in the program. These papers address the latest research and development efforts and highlight the human aspects of design and use of computing systems. The papers accepted for presentation thoroughly cover the entire field of human-computer interaction, including the cognitive, social, ergonomic, and health aspects of work with computers. These papers also address major advances in knowledge and effective use of computers in a variety of diversified application areas, including offices, financial institutions, manufacturing, electronic publishing, construction, health care, disabled and elderly people, etc.

We are most grateful to the following cooperating organizations:

- Chinese Academy of Sciences
- Japan Management Association
- Japan Ergonomics Society
- Human Interface Society (Japan)
- Swedish Interdisciplinary Interest Group for Human Computer Interaction - STIMDI
- Asociación Interacción Persona Ordenador - AIPO (Spain)
- Gesellschaft für Informatik e.V - GI (Germany)
- European Research Consortium for Information and Mathematics - ERCIM

The 282 papers contributing to this volume (Vol. 2) cover the following areas:

- Mobile and Ubiquitous Computing
- User and Context-awareness
- Agents and Avatars
- Interaction Techniques and Modalities
- Supporting Access to Information, Communication and Cooperation
- Visualization and Simulation
- HCI Applications and Services
- Human Factors and Ergonomics

The selected papers on other HCI topics are presented in the accompanying three volumes: Volume 1 edited by J. Jacko and C. Stephanidis, Volume 3 by D. Harris, V. Duffy, M. J. Smith and C. Stephanidis, and Volume 4 by C. Stephanidis.

We wish to thank the Board members, listed below, who so diligently contributed to the overall success of the conference and to the selection of papers constituting the content of the four volumes.

Eduardo Salas, *USA*
Dirk Schaefer, *France*
Neville A. Stanton, *UK*

Universal Access in Human-Computer Interaction
Julio Abascal, *Spain*
Demosthenes Akoumianakis, *Greece*
Elizabeth Andre, *Germany*
David Benyon, *UK*
Noelle Carbonell, *France*
Pier Luigi Emiliani, *Italy*
Michael C. Fairhurst, *UK*
Gerhard Fischer, *USA*
Ephraim Glinert, *USA*
Jon Gunderson, *USA*
Ilias Iakovidis, *EU*
Arthur I. Karshmer, *USA*
Alfred Kobsa, *USA*
Mark Maybury, *USA*
Michael Pieper, *Germany*
Angel R. Puerta, *USA*
Anthony Savidis, *Greece*
Christian Stary, *Austria*
Hirotada Ueda, *Japan*
Jean Vanderdonckt, *Belgium*
Gregg C. Vanderheiden, *USA*
Annika Waern, *Sweden*
Gerhard Weber, *Germany*
Harald Weber, *Germany*
Michael D. Wilson, *UK*
Toshiki Yamaoka, *Japan*

We also wish to thank the following external reviewers:

Chrisoula Alexandraki, *Greece*
Margherita Antona, *Greece*
Ioannis Basdekis, *Greece*
Boris De Ruyter, *Netherlands*
Babak Farschian, *Norway*
Panagiotis Karampelas, *Greece*
Leta Karefilaki, *Greece*
Elizabeth Longmate, *UK*
Fabrizia Mantovani, *Italy*
Panos Markopoulos, *Netherlands*
Yannis Pachoulakis, *Greece*
Zacharias Protogeros, *Greece*
Vassilios Zarikas, *Greece*

This conference could not have been held without the diligent work and outstanding efforts of Stella Vourou, the Registration Chair, Maria Pitsoulaki, the Program Administrator, Maria Papadopoulou, the Conference Administrator, and George Papatzanis, the Student Volunteer Chair. Also, special thanks to Manolis Verigakis, Zacharoula Petoussi, Antonis Natsis, Erasmia Piperaki, Peggy Karaviti and Sifis Klironomos for their help towards the organization of the Conference. Finally recognition and acknowledgement is due to all members of the HCI Laboratory of ICS-FORTH.

Constantine Stephanidis
ICS-FORTH and University of Crete, GREECE

Julie A. Jacko
Georgia Institute of Technology, USA

Don Harris
Cranfield University, UK

Vincent G. Duffy
Mississippi State University, USA

Michael J. Smith
University of Wisconsin-Madison, USA

June 2003

HCI International 2005

The 11th International Conference on Human-Computer Interaction, HCI International 2005, will take place jointly with:

Symposium on Human Interface (Japan) 2005
6th International Conference on Engineering Psychology and Cognitive Ergonomics
3rd International Conference on Universal Access in Human-Computer Interaction
1st International Conference on Virtual Reality
1st International Conference on Usability and Internationalization

The conference will be held in Las Vegas, Nevada, 22-27 July 2005. The conference will cover a broad spectrum of HCI-related themes, including theoretical issues, methods, tools and processes for HCI design, new interface techniques and applications. The conference will offer a pre-conference program with tutorials and workshops, parallel paper sessions, panels, posters and exhibitions. For more information please visit the URL address: http://hcii2005.engr.wisc.edu

General Chair:

Gavriel Salvendy
Purdue University
School of Industrial Engineering
West Lafayette, IN 47907-2023 USA
Telephone: +1 (765)494-5426 Fax: +1 (765) 494-0874
Email: salvendy@ecn.purdue.edu
http://gilbreth.ecn.purdue.edu/~salvendy
and
Department of Industrial Engineering
Tsinghua University, P.R. China

The proceedings will be published by Lawrence Erlbaum and Associates.

Table of Contents

Section 2. User and Context-awareness

Section 3. Agents and Avatars

Section 4. Interaction Techniques and Modalities

Section 5. Supporting Access to Information, Communication and Cooperation

Section 8. Human Factors and Ergonomics

Section 1

Mobile and Ubiquitous Computing

MobiGuiding, a European Multimodal and Multilingual System for Ubiquitous Access to Leisure and Cultural Contents

Carlo Aliprandi
Synthema S.r.l.
Lungarno Mediceo 40
56127 Pisa - Italy
aliprandi@synthema.it

Michel Athénour
Cityvox S.A.S.
BP 65
13303 Marseille – France.
michel@cityvox.com

Sara Carro Martinez
Telefónica I+D S.A.
Parque Tecnológico de Boecillo,
47151 Boecillo, Valladolid – Spain.
scm@tid.es

Nikos Patsis
VoiceWeb S.A.
40 Ag. Konstantinou St.,
GR-151 24 Maroussi – Greece.
npatsis@voiceweb.gr

Abstract

The MobiGuiding project is aimed at building an European interactive guide network, on a common innovative platform, in which contributors will compute the information relating to leisure and cultural events in their locations. This guide will be available on all Internet and mobile devices, the focus being on the 3rd generation mobile phones. MobiGuiding will be an adaptive interactive interface, helpful for end-users, event organisers and content suppliers. It will feature the most developed multimedia and multimodal technologies, including speech servers, natural language processing and localisation.

1 Introduction

MobiGuiding will exploit Ambient Intelligence for leisure and travelling purposes, thereby disseminating the benefit of adopting new interaction experiences with state-of-the art standards, contributing to the building of a common European culture, through sharing cultural information at a cross cultural level. MobiGuiding will contribute to the actual success of UMTS, which must rely on adapted contents, building a strong European network extendable to outside countries.
The ultimate goal of the project consists in implementing a Network, a Model and a System, which will enable the publishing of content and the provision of services in the most up-to-date way, especially on mobile devices: the System will be a real application for mass access to existing contents, that will finalise research and development efforts coming from consortium partners in highly specialised information technology areas.
Providing leisure and cultural information for many languages and countries in Europe, MobiGuiding will face a challenging effort. This paper presents the early results publicly available at the time of writing of this paper, showing how the transfer of research technologies of mobile

Internet, natural language processing and speech servers will enable the development of the MobiGuiding System.
In Section 2 we present some state of the art of currently available systems for mobile tourism; in Section 3 the MobiGuiding System architecture and some distinguishing features are presented. In Section 4 we describe conclusions and attended advantages of our approach.

2 State of the Art

It is nowadays commonly accepted that future IT applications will be inherently mobile, in such a pervasive manner that many aspects of our everyday life will be transformed. Current mobile technologies, like GSM and WAP have provided a first answer to this need, and it is envisaged that the new 3rd and 4th Generation mobile systems will make another revolution.
The travel and tourism market is reflecting these changes and the type of services offered is rapidly evolving. Thus we are assisting at an increasing importance of services provided through wireless devices with new applications now emerging and soon appearing in the market.
Many of these services are coming from research projects, as PALIO (Andreadis et al., 2001), GUIDE (Cheverst et al, 2000) and BUSMAN (Izquierdo, 2002).
PALIO will design an open communication and information architecture to support tourists in using services in a transparent way. This project is mainly based on providing new and free tourism services through WAP protocol over GSM-SMS and GPRS. GUIDE has developed a prototype hand-held computer based city guide basically on webcentric data. MobiGuiding, with respect to these, will be different in the kind of content available and wideness of its update, and also in the extension of mobile devices support. BUSMAN is similar in wideness of content to be indexed and delivered: it is aimed at developing a secure and efficient multiterminal system for the delivery and access to multimedia content. It will be mainly dedicated to video processing, whereas MobiGuiding will be more general in the kind of media supported.
Others features of MobiGuiding, like contents translation and localisation, vocal access and advanced additional services, will in general distinguish it from other service providers, in order to attract customers.

3 The MobiGuiding System

MobiGuiding will implement a Network, a Model and a System, which will enable the publishing of content and the provision of services in the most up-to-date way.
What MobiGuiding will bring is the ability, for content producers (both public institutions and private companies joining the Network), to feed the platform in their own way, whatever the format. Content producers will be able to make their data accessible to all mobile users, accessing specific back-office services which will be organised and optimised for accepting all major electronic versions of content.
MobiGuiding will develop a powerful content management System for all the major mobile devices: Web, Wap & Wap 2.0, iMode, PDAs, SMS alerts, speech services, and, later on, MMS and 3G devices.
Among the many content services that the MobiGuiding system will offer this paper is focused on the main one, which can be presented by the following scenario:
Jane is a woman, who has just arrived to represent her company in Greece; she is invited to a dinner by a Greek potential client. She doesn't know anything about the Greek's habits, when would be polite to speak about business issues, or what to dress or even what she would be discourteous to say. She turns on her mobile device and while she dresses up she asks for information on Greek food and habits. All the online travel guides of different editors are selected.

Two of them are in her native language, one has a special chapter about meals. She asks the mobile device to read that chapter and she finds the relevant information about Greek habits. Just in case, she asks for business habits, so when the dining time arrives she does perfectly, and doesn't try to pay her part of the bill to avoid offending her hosts.
This has been identified by the 'Report on Services to Offer' deliverable (Martinez et al., 2002) as the main required service: to know what to do, where to go and what will be found there, users turn to tourist guides.
We will provide the same mobile online content as tourist guides do, about:

- Events: culture, habits, places to go for leisure or entertainment, celebrations
- Monuments and Museums, with historical and artistic information
- Gastronomic: restaurants and typical dishes, with their history and tradition
- Tourist tours: available organised tours to visit the main places

3.1 System Architecture

The MobiGuiding System architecture stems upon a decomposition into 3 main components: the Contents Management Platform, the Localisation Platform and the Vocalisation Platform.
The Contents Management Platform will play the important role of data collecting, checking, XML transforming (Bray et al., 2000) and storing on multi-entry databases. Employing a 3-tier architecture (as showed in Figure 1) collected data will be accessed by most devices and requests, simply adapting the transformation layer to a particular new device format.

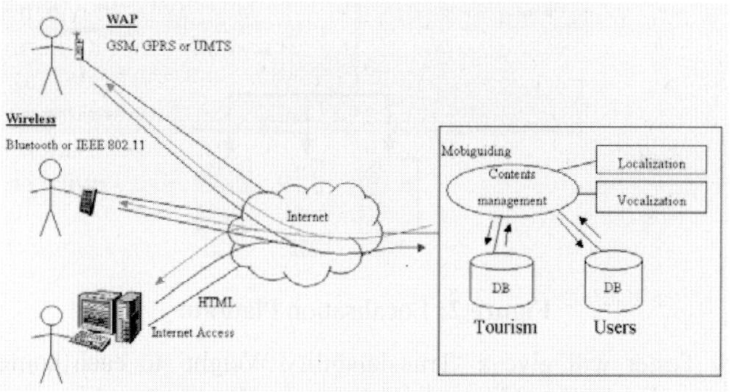

Figure 1: MobiGuiding Architecture

Although most of the technical specifications are under development, some constraints have been identified; for example the Contents Management Platform will rely on specific developments of an existing and available commercial product based on SQL for content storing and on XML-XSL for data dispatching (Adler et al., 2001). The application logic layer will be separated from the content and the presentation layer: this will be based on standards as HTML for PDAs and WML for Wap that will be extended to cHTML or Basic XHTML for iMode and VoiceXML for speech. The Localisation and the Vocalisation Platforms are presented in the next sections.

3.2 Multilingual Features: the Localisation Platform

Being not aware of existing services providing leisure and cultural information in so many languages and countries in Europe, the consortium is facing a challenging effort.
Our innovative Localisation Platform will primarily address technology innovation, but will also try to shift content language adaption from a pure technology oriented-process to a workflow-

oriented process. Key factors as shifting the translation responsibility among all the content production chain and giving support to the whole workflow of content localisation in a business oriented perspective will be taken into account.
State of the art technologies for Automatic Translation (AT) will be made available, as described in (Aliprandi, Neri & Priamo, 2002) and in (Bernth & McCord, 2000), with the ultimate goals of improving the standard quality of translation and its existing level of coverage, addressing an increasing number of covered languages (starting from the specific touristic domain).
Depending on the content adaption demand that will be required by the content providers and by the end users, it will be defined the core of the Localisation Platform: the Translatability Tester (see Figure 2). It will be a Decision Support Tool that will assess the machine translatability of contents ensuring the most appropriate and cost effective mix between AT solutions (Machine Translation, Computer Aided Translation, Computer Assisted Translation) and Human Revision.

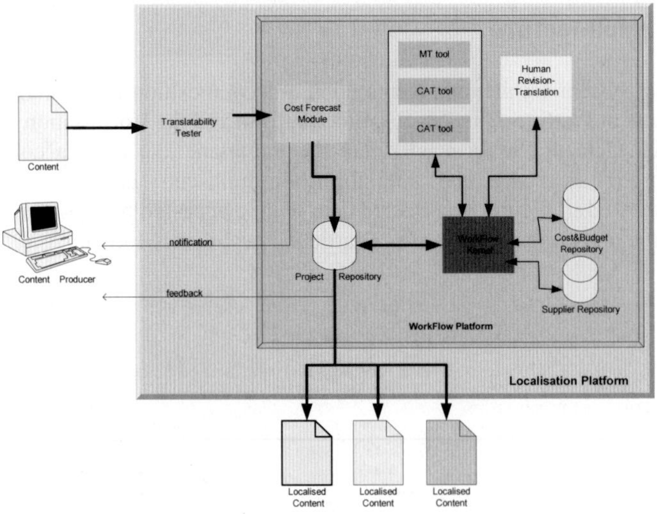

Figure 2: Localisation Platform

The Translatability Tester will give a 'Translatability Weight' to each content, assessing key factors as the Changing Frequency (CF) and the Accessing Frequency (AF).
Once the content has been given a Translatability Weight and has been submitted to the Localisation Platform, a notification is given to the content producer; then the WorkFlow Kernel will manage the whole process of localisation and translation, granting task scheduling, cost budgeting and, when requested, scheduling resources (translator/revisionist suppliers).

3.3 Natural Speech Features: the Vocalisation Platform

MobiGuiding will employ modern ASR technology and effective grammar design allowing for the creation of intuitive Voice User Interfaces not limited to constrained interactions. Well-designed dialogues and Nuance GSL grammars (Nuance Communications, 2001) will cater for the acceptance and recognition of a wide range of expected phrases, with high accuracy rate. The user will be able to speak naturally without having to be restricted to small vocabularies. One more important facility will be multiple semantic extraction, where a user's phrase conveys more than a simple command or information item, thus eliminating the need of multiple questioning to solicit user's information.

The main advantages of MobiGuiding ASR technology are:

- Usage adapted to users' abilities and their natural environment
- Friendly & natural interfaces, even for users without familiarisation
- Freedom from strict tree-like menu application dialogues
- Seamless selectability from large range of options (eg. city events)

Designing a Voice User Interface involves designing of the dialogues, the flow and possible turns within the application, along with the allowed commands at any given point, reflected in the grammars. All these components will be developed with usability as a central concern. MobiGuiding will allow mixed-initiative voice commands where needed, to aid usability. For instance, consider a train timetable application. The system will ask in sequence the information items: depart station, destination station and date. The user can answer those questions one by one, or can alternatively issue a command for the whole journey: *"I would like to go to Madrid from Barcelona tomorrow"*. Using advanced ASR grammar, the information items can be located anywhere in the phrase, only the missing items will be prompted for by the system.

4 Conclusions

This paper has presented MobiGuiding, an European Multimodal and Multilingual System for Ubiquitous Access to Leisure and Cultural Contents. After a brief survey of state of the art of similar systems, we have provided an overview of its architecture and services.
The MobiGuiding approach has differences and potential advantages since, shifting the focus from a pure-technology oriented model to a networked business model it will bring benefits both to content providers and users. Content providers will benefit from costs reduction, market increment, sharing of benefits; users will benefit from contents and devices wideness, customisations as multilingual localisation and natural speech access.
In conclusion, even if we have a service that is actually offered by similar systems, MobiGuiding will differ in such a way that we trust it will be attractive, increasing the importance of touristic services provided through wireless devices.

References

Aliprandi, C., Neri, F., & Priamo, A., (2002). Automatic Translation Tools for Powerful Multilanguage Localisation. Synthema Lexical Systems Lab. internal report. Pisa, Italy.

Adler, S., et al. (2001). XSL 1.0. W3C. Retrieved from http://www.w3.org/TR/xsl

Andreadis, A., et al., (2001). Personalized Access to Distributed Data Sources. In proceedings of the Workshop on Information Presentation and Natural Multimodal Dialogue. Verona, Italy.

Bernth, A., & McCord, M., (2000). The Effect of Source Analysis on Translation Confidence. In proceedings of the 4th Conference of the Association for Machine Translation in the Americas. Cuernavaca, Mexico.

Bray, T., et al., (2000). XML 1.0. W3C. Retrieved from http://www.w3.org/TR/REC-xml

Cheverst, K., et al., (2000). Developing a Context-aware Electronic Tourist Guide: Some Issues and Experiences. In proceedings of CHI 2000. Netherlands.

Izquierdo, E., (2002). BUSMAN. Retrieved from http://www.elec.qmul.ac.uk/busman/index.htm

Nuance Communications, (2001). Nuance Speech Recognition System Ver 7.0, Nuance Grammar Developer's Guide.

Martinez, S., et al., (2002). Report on Services to Offer. MobiGuiding internal report.

Advanced User Interface for the SAFEGUARD Professional Driver Seat

Angelos Amditis
Ioannis Karaseitanidis

National Technical University
of Athens – ICCS, Greece
{angelos,gkara}@esd.ece.ntua.gr

Oliver Stefani

Fraunhofer – IAO
Germany
Oliver.Stefani@iao.fhg.de

Simon Sartor

ISRINGHAUSEN
Germany
Simon.Sartor@isri.de

Abstract

Low back pain is more frequently present in patients involved in occupations that are related to driving automobiles, motorcycles, buses, tractors, trucks and heavy construction and agriculture machinery. Furthermore, low back pain is the leading major cause of disability in those younger than 45 years. Health and comfort issues emanating from prolonged seating and long-time exposure to vibrations are critical for the majority of professional drivers. The EU co-funded project SAFEGUARD aims at developing a new seat that will improve comfort, while decreasing the risk for health damage. In addition, it aims at developing in a broader sense a methodology, to evaluate seating comfort. The solution proposed within SAFEGUARD is to provide automotive companies with advanced seat features that enhance driver's safety and comfort. Within this paper the functionalities proposed for the SAFEGUARD seat will be explained and their expected impact on driver's health and safety will be briefly presented. The methodology used to reach the final result on the User Interface will be explained and the way all functionalities have been incorporated will be presented.

1 Introduction

Seating motionless for long period while driving and prolonged exposure to vibrations have severe implications to the health of professional drivers. To deal with this problem SAFEGUARD moved towards the development of a new seat system where the functional combination of newly developed or updated sensors and actuators results in decreased risks for driver's health while enhancing their comfort feeling. In addition as a step towards future developments a new integrated methodology to assess safety and comfort related parameters in prototype professional seats composed by a number of modules has been developed.

SAFEGUARD is based on a step-to-step methodology [1] leading to the development of an advanced prototype seat, the functionality and operations of which are controlled by an advanced User Interface. Based on an etiological analysis of health related professional drivers' problems, both from a clinical and from a traffic safety point of view a set of evaluation criteria was pre-selected which were measured through a driver/seat interaction model, VE tests and physiological measurements of the driver. The technological gaps and seat inefficiencies specified have been met by design and application of innovative sensors and actuators. The realization of enhanced ergonomical aspects and the development of an innovative User Interface has taken place. The required demonstrators are already built, namely the VR mock-ups and two prototype "intelligent" seats (one for combine harvester and one for truck) have been already integrated in two demonstrator vehicles. The proposed evaluation criteria, methodologies and seat concepts will be validated in a series of Verification Pilots, including laboratory Pilots and on-site Pilots.

All seat functionalities are controlled through an innovative User Interface. For its development all current guidelines and regulations have been taken into account (including the layout of the menus on the display, color combinations, size and type of fonts, symbols and menu structures) and a state of the art of relevant products has taken place. The aforementioned led to the identification of different aspects to be considered within the development of the user interface in order to meet the needs of the professional driver.

2 Design process

The different functions to be implemented on the SAFEGUARD system were a result of the etiological analysis of the health related problems of professional drivers. The suggested functions have been implemented through a series of sensors and actuators, which had to be controlled by a User Interface that would enable user to have control over them in a way that would not distract driver's attention to the primary driving task. It must be noted that different functions have been implemented concerning the two different SAFEGUARD demonstrators, a truck and a tractor. This paper will focus on the truck user interface, which has been developed to include all existing SAFEGUARD functionalities and possible future system expansions. Thus the list of functions included are:

- Posture Control: headrest adjustment, backrest inclination adjustment, horizontal position adjustment, height adjustment, seat pan inclination adjustment, seat cushion depth adjustment, integrated Pneumatic System, memory function, Active Moving Seat, pressure distribution control device, weight scale, free Swivelling Seat, free Swivelling Saddle Seats
- Vibration isolation: adjustment vertical vibration damper / semi active damping, draw down, horizontal damping
- Climate control: electronic climate control
- Other functions: Audio feedback
- Optional, non-seat functions: mirror adjustment, seat belt adjustment, cabin heating, window operation, radio operation, telephone operation, on-board computer, navigation system

It must be noted that although all the functions can be supported not all of them have been implemented to the demonstrator.

Basic features of the User Interface were pre-selected according to background experience, including:

- Rotary knob on armrest, either with push/pull function and no further buttons or with push function and 1-2 further buttons.
- TFT-display (either on armrest or on dashboard) with menu selection of UI-functions and graphic display of current adjustment.
- A voice control, allowing the driver to choose the desired function by voice, without looking at the display, and then adjusting this function with the rotary knob

The first step was an overview of the state of the art of user interfaces and other seat-controls in different types of vehicles. From this SoA analysis conclusions were reached about the final adoption of UI elements. In detail the fact that the more functions a seat has, the more difficult it is for the driver to operate it correctly, because of the large number of knobs and levers used, was addressed [2]. In addition, a multifunctional operating element can be even more difficult to operate because of the multitude of degrees of freedom with a different meaning for each possible direction.

The fact that the driver can have an overview over the functionalities of the vehicle seat without the need to move far from it with the use of the rotary knob was the dominant reason for implementing such one. Additionally a small screen was chosen as the main visual display. A larger screen would bare the disadvantage of being fixed on the dashboard, which is probably the only point in the vehicle offering enough space for its implementation. This would bring the control panel over the seat settings far from the driver while seated, which contradicts the initial condition that the UI should be easily reached by the driver. Finally voice input was rejected since it would be based upon a user-dependent (limited vocabulary) requiring a significant learning time, or a user-independent (extended vocabulary) that would not guarantee its application in case the user has a non-standard voice even though no learning phase is required in this case. This makes the use of a voice input system very indecisive.

When all I/O modalities have been defined conclusions from User Needs (UN) analysis were introduced to the design process. The findings of the UN – which was conducted with bibliographical survey and relevant questionnaires to end-users – refined the design process. Through a step down process recommendations could be withdrawn from general remarks addressed to all UI-related parameters. Through this survey became obvious that the aim of an optimum UI is to meet the needs of professional drivers, by:

- The use of a meaningful and practical_UI, which will provide a clear visual feedback on the adjustments allowed by the SAFEGUARD seat
- A pleasant and adaptable to the users' characteristics design, which will affect highly the acceptance and willingness to use

Among the principles [3] to be accomplished by the UI in order to meet the user needs were indicated the following:

- **Safety and Health**
- **Easiness to use (Design and efficiency)**
- **Manoeuvrability**
- **Purchase and Maintenance cost**

Recommendations on how to achieve all aforementioned goals were provided.

Moving further the aforementioned findings were filtered through a series of guidelines. The guidelines target at a wider range of users / drivers, including elderly drivers, users with reduced visual acuity or with colour blindness. Although guidelines found were not tested for the specific needs and goals of SAFEGUARD seat system, which reduces their reliability, the provision of a full set of guidelines was considered useful as, in any case, the validity of such "empirical" guidelines is only further established by cross reference to the strategies followed by existing products and prototypes, which are not available at present. So guidelines concerning including the layout of the menus on the display, color combinations, size and type of fonts, symbols and menu structures were investigated and used when applicable.

The actual design of the user interface was performed in three steps. First, considerations and proposals were made concerning the layout of the user interface. Then preliminary tests were conducted using a driving simulator, to examine among other things the effects of different screen positions and screen sizes, and to compare different proposed icons, fonts and colours. On the basis of the guidelines, the proposed design and the preliminary tests, the design of the UI was then finalised.

Within the preliminary tests several parameters were under investigation including usefulness, willingness to have, pleasantness, usability. The subjects were asked to make cross-comparisons between symbols and icons and between the SAFEGUARD and pre-existing UIs. Additionally objective parameters were measured with non-intrusive techniques (i.e. workload, time to perform, user error count. Finally through comprehension interviews the subjects were able to summarise their main conclusions and provide the designers with useful feedback. In conclusion the key points can be summarised as follows:

- Turn-Push device is preferred to push device
- Time to perform adjustment tasks is slightly quicker with Turn-Push device.
- Grey is preferred over colour.
- Dashboard integrated display is preferred over seat mounted.
- Large display is preferred over small display.
- Without driving: time to perform seat adjustment is quicker with dashboard integrated display.
- While driving: time to perform seat adjustment is quicker with seat-mounted display (occlusion through steering wheel while driving around curves).
- Size of display does not explicitly affect time to perform seat adjustments

3 Result

Through the aforementioned process SAFEGUARD UI designers reached final conclusions regarding the controller (display in combination with a force feedback operating device), the final seat adjustments controlled by the UI, the display characteristics and the UI layout itself. The following screenshots and pictures demonstrate the final result [4].

4 Conclusions

SAFEGUARD project addresses health, safety and comfort issues emanating from prolonged seating and exposure to vibrations while driving. To compensate with this a new seat system was developed which is controlled by an innovative UI. The design process of the UI and the way relevant guidelines and User needs are taken into consideration is presented. To verify the effectiveness of this final UI-design, SAFEGUARD Consortium is and will be performing comparison tests, where the prototype seat with the new UI is compared to a reference seat with conventional seat controls. These tests are expected to demonstrate the ergonomic and high-usability characteristics of the developed User Interface.

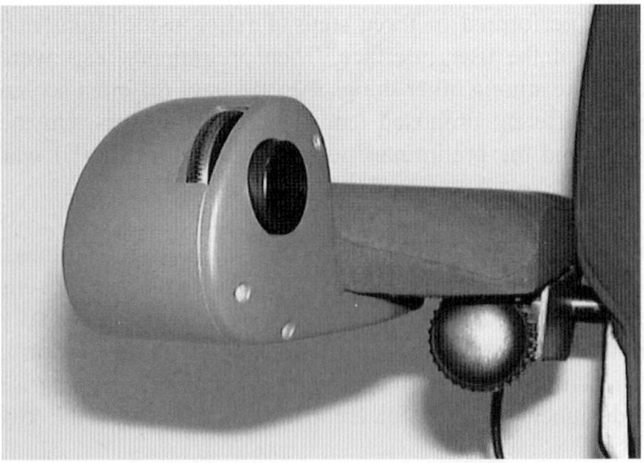

Figure 1: Operating Element – armrest front end

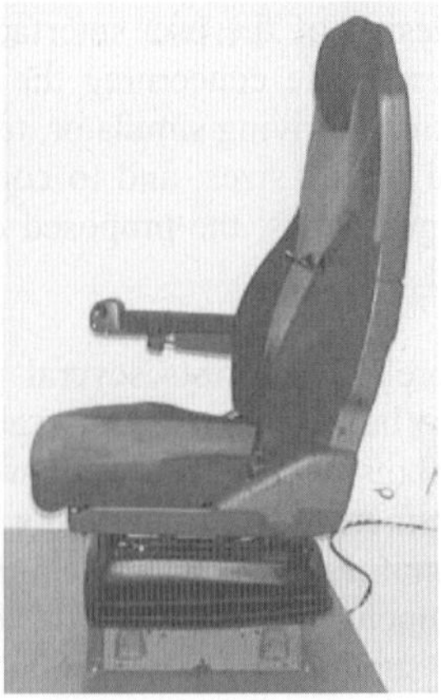

Figure 2: Prototype Seat

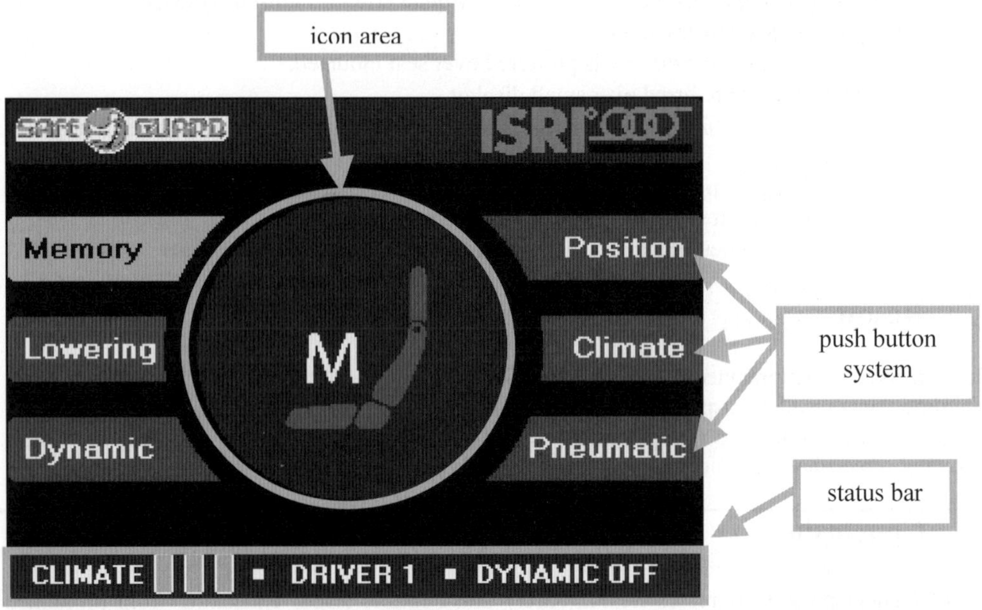

Figure 3: Actual UI - Main Menu - Memory

References

[1] "Innovative seat system for professional drivers – the SAFEGUARD project", Due to GfA/ISOES Conference Proceedings, Munich, May 2003

[2] "Human Factors aspects of manual computer input devices", Greenstein S. J., Arnaut Y. L.,. In Salvendy G. (Ed.). Handbook of Human Factors, 1450-1489, New York: Wiley, 1987

[3] ISO TC159 Ergonomics. "SC1 Ergonomic guiding principles-WG on visual display requirements"

[4] SAFEGUARD project, "Optimum User Interface", April 2002

Key Issues in Automotive HMI for Elderly and Disabled Drivers - The CONSENSUS Approach

Guido Baten

Belgian Road Safety Institute
Brussels
guido.baten@bivv.be

Maria Panou

Transeuropean Consulting Unit
of Thessaloniki, Greece
mpanou@truth.com.gr

Abstract

The need for self-depending transport being a most basic need for every human being, highlights the necessity to provide to the group of elderly and disabled (E&D) the necessary support and to secure their safety and success in reaching their destination. Driving a car is of high significance to their mobility, as the car is their second most popular transportation mean.

On the HMI level, whatever is good for all is usually good also for Elderly & Disabled people, but the level of gain is disproportionate. Differences in the eccentricity of the information provision or the part of the visual field where the information is presented makes a much greater difference for Elderly & Disabled drivers [Klein, 1991].

The introduction of telematic aids and Advanced Driver Assistance Systems (ADAS) has initially provided a new obstacle to the drivability of E&D people. Initial systems have inflexible characteristics (operation elements, controls location, system actuation and feedback times) that were not appropriate for many E&D drivers. Into the bargain, several ADAS control elements are placed in areas incompatible with driver adaptations and aids for E&D drivers. This paper presents such considerations from Pilots conducted within projects TRAVELGUIDE, EDDIT, TELAID, and TELSCAN. If an appropriate HMI is selected, ADAS may reduce the workload and in fact enhance the drivability of E&D drivers. Relevant tests conducted within TELAID project show an enhancement in parking task performance and ACC functions when appropriately designed ADAS are used.

CONSENSUS, a Thematic Network promoting the driving ability of E&D through common methodologies and normative tools, assesses the appropriateness of ADAS functionalities and their HMI to the residual abilities of the target group, and proposes new functionalities and interaction principles for their benefit. Preliminary findings highlight the importance of a T-junction aid, lane recognition and maintenance and blind spot avoidance functions, as well as, parking support for E&D drivers. Interface modalities considered focus on combined use of audio and visual messages, as well as, the use of haptic messages of adaptable frequency, intensity, and sequence.

1. Introduction

There is a strong need for disabled people to drive a private car for their transportation, since the car is the second most popular transportation mean used by them, after the metro or tram (Figure 0), which is not available for those living in smaller cities or villages. These results arise from the TELSCAN project survey that was conducted on September 1999 and involved 137 E&D users (75% of them being wheelchair users). In addition, driving is a modality more user friendly for E&D people than public transportation, according to a number of experts, as many of them are not able to walk the required distance, stand for a long time and have the overall physical endurance to use public transportation means.

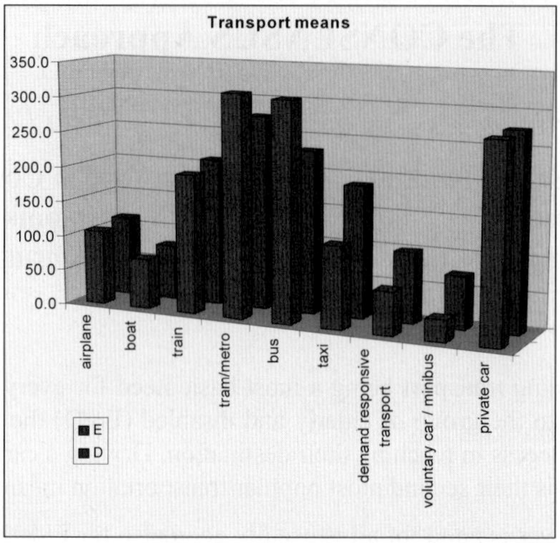

Figure 0: Transport means used by Elderly and Disabled group surveyed (where E stands for Elderly and D for Disabled) within TELSCAN project. The y-axis denotes the average use: 0 (never), 100 (very occasionally), 200 (yearly), 300 (monthly), 400 (weekly), 500 (daily).

Indeed, over 25% of elderly Americans are not able to use public transportation means [Walter and Goo, 1969]. There is no analytical data on the involvement of disabled drivers in traffic accidents. A relevant study estimates that wheelchair users are involved in 3% of traffic accidents, which is quite reasonable for a group summing up to around 4% of the overall population and with somewhat restricted traffic patterns. Still, the low percentage of E&D drivers and the significant problems self-reported by them (Final Report of CONSENSUS feasibility study) denote their need for further support. The above, indicate the strong need for the existence of systems that will actually assist the E&D in the driving task. Such systems could of course include simple driving aids but also high technology systems, as the Advanced Driver Assistance Systems (ADAS).

2. CONSENSUS Project

CONSENSUS is a **Network of Excellence,** to systematically exchange information on driving ability assessment of Disabled people, promote relevant technology transfer within EU and access to expertise and resources of highly specialised Centres to other less specialised country authorities, using state of the art Telematics tools and procedures and experimenting new IT support tools (such as a database and an expert knowledge tool).

A Thematic Group has been formed within CONSENSUS, where the involvement of car adaptation and ADAS manufacturers is targeted, in order to correlate the assessment procedure with existing and prospective telematic aids and vehicle design issues. In this way, guidelines will be extracted for the optimal in-vehicle HMI aids design in order to be usable by disabled people. Moreover, the high experience of involved experts will contribute to the formulation of common evaluation methodologies and scientific, legal and industrial consensus on vehicle control by disabled drivers and to the specification of the necessary in-vehicle platforms to further promote their driving ability, taking into account the recent developments of ADAS and in-vehicle Information Systems (IVIS).

In addition, ADAS and IVIS of appropriate HMI are considered within CONSENSUS as possible solutions to enhance certain disabled drivers residual abilities and help them become or remain safe drivers. Below, the preliminary findings of this Group are discussed.

3. Specific problems of the disabled related to automotive systems HMI

Many of the measures that must be applied in the traffic system to make it accessible to E&D drivers are very costly. To establish priorities and the order of precedence for new measures, a feedback system informing operators about common problems is required. More specific problems, related to the use of in-vehicle HMI include [Panou et al., CONSENSUS, 2002]:

- The location of the devices can cause problems for persons with impaired lower limb mobility, but usually not the system itself. Most common problems are height and location of controls.
- Using the control elements of an in-car system is the most usual problem for people with limited arm movement. Car drivers can get help by reaching and communicating with equipment in the car that provides information during the trip, monitors the vehicle, and pays parking fees and tolls with an integrated system. Such integrated solutions may consist of remote controlled systems for opening doors, paying parking fees, etc.
- People with problems on the upper body suffer from a limited field of vision (LFV) because they cannot twist around and encounter problems in reaching equipment placed at certain angles. The location of different automotive systems is thus important. Equipment must not be placed too high up, but within the natural LFV. In cars, for example, information can be reflected in the windshield. Keypads and keyboards for communicating with traffic information office should be easy to reach. The problem of LFV can be relieved by systems that extend the field of vision and show the surroundings when driving/parking a car.
- People with co-ordination or dexterity problems entail difficulties in using or reaching input, activation/deactivation buttons. To provide accessibility, they must be able to communicate with the system without the need for physical contact. The solution can be found in voice-activated systems.

4. Towards an automotive HMI appropriate for the disabled

Results arisen from TRAVELGUIDE project [TRAVELGUIDE Del.5] provide some qualitative insight to their preferences. Tests were conducted with 6 disabled people versus different user groups in the city of Thessaloniki, aiming at evaluating a driver information service in actual conditions. The systems included Internet, SMS, RDS, VMS and radio as information provision means and supported different interfaces for each. The behavioural effects of the systems per driver type and system are depicted in Figure 1 below:

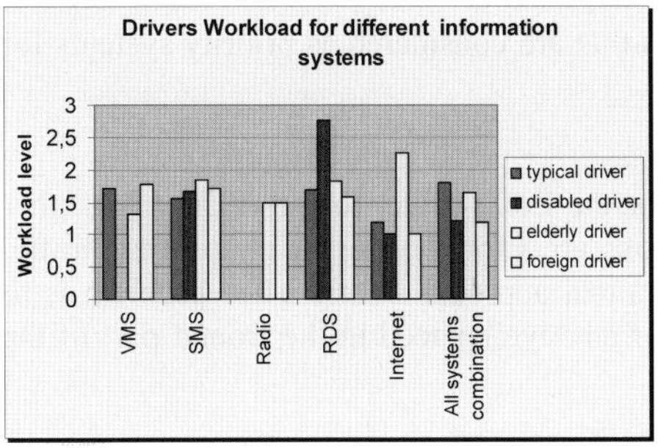

Figure 1 Drivers opinion regarding the workload increase level with the use of information systems.

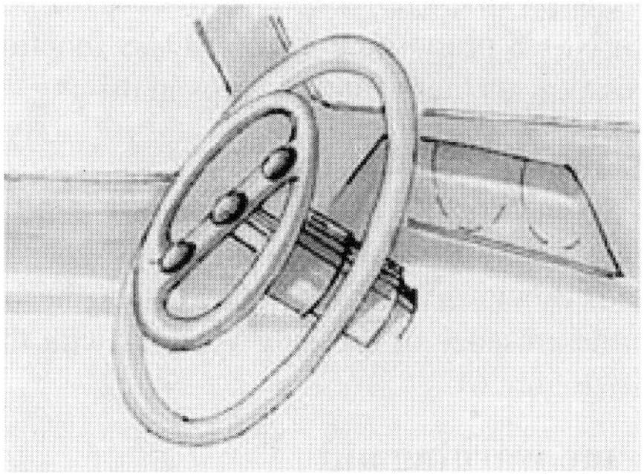

Figure 2 Appropriate alternative placing of ACC control buttons, operable by disabled drivers.

Regarding the control functions of ADAS, they need to be flexible to be used by drivers of adapted vehicles. As an example, if the ACC controls are placed on an acceleration lever or accelerator ring (Figure 2) for drivers requiring hand controls, then the driver could activate the ACC at the same time as controlling the speed.

Disabled drivers felt that there was low workload when interfacing VMS and radio, while they felt a high workload in RDS operation. For the remaining systems, the workload level was in accordance with that of all other drivers. It was felt by the test team that the identified workload problems had not much to do with the information mean itself, but with its HMI. Especially short RDS text format is a good example of non-functional HMI for E&D users.

Regarding the design of each mean itself, a good design for disabled drivers needs to be relevant in content and brief in text, with big characters. The use case below comes from a relevant VMS test performed within TELSCAN project (TR 1108):

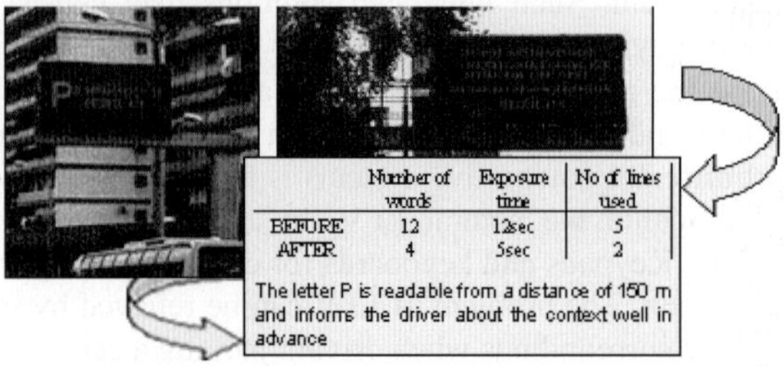

Number of words
Exposure time
No of lines used
BEFORE 12 12sec 5
AFTER 4 5sec 2
The letter P is readable from a distance of 150 m and informs the driver about the context well in advance

Figure 3 Example of an appropriate (left-hand) and an inappropriate VMS text for disabled drivers [TESCAN Annual Review Report, 2000].

In particular, the combination of kinesthetic (tactile) and audible warnings has been found to be of particular usefulness to E&D drivers. From a TELSCAN collaborative test with the AC-ASSIST project, it is recommended that either kinesthetic (brake jerk) and/or audible (spoken) warnings are given at about 4 sec. time-to-collision, on the assumption that speed difference is constant. However, if there is a change in speed difference, it should be reflected in the calculation of a safe time distance. This design of a collision warning considers the needs of elderly drivers, and would allow for alternative warnings in case of, e.g. a hearing impairment. A warning should be followed if necessary by automatic intervention (emergency bake).

4. High priority ADAS functions for the disabled

CONSENSUS proposed a list of ADAS functionalities that are considered as priority systems for E&D drivers, as these fill their gaps in driving.

4.1 T-junction aid

When a vehicle stops at a T-junction, it is difficult for the E&D driver to see if there are other vehicles coming from the crossing lane. These situations are strongly present in an urban traffic scenario. Previous tests have shown that the best technical approach to solve the problem is to detect the presence of crossing vehicles with the aid of an "eye" placed in the frontal part of the vehicle (Figure 6).

4.2 Lane warning aid

This system was developed in the European PROMETHEUS project and further elaborated in project LACOS. Highway and extra urban roads are the typical environments where the system is useful: using a CCD camera and a processing device, observing the road ahead, the system detects car position in the lane and warns the driver when lane markings are being crossed. A larger effort is necessary to choose the right HMI approach for E&D drivers, which is the key point for system acceptability. Whether such a system could decisively help such drivers in lane keeping is to be investigated in the future (Figure 4 and Figure 5).

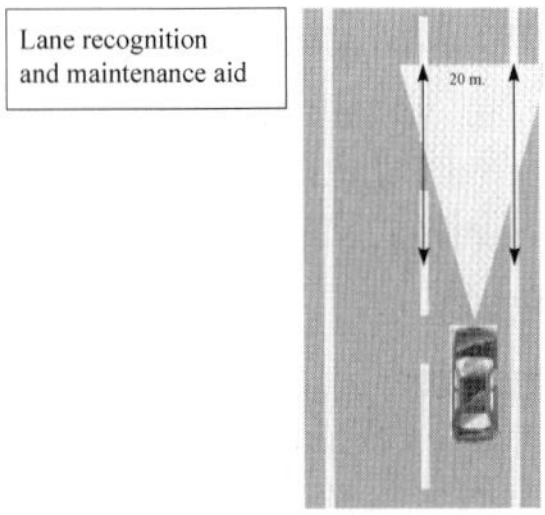

Figure 4 Representation of lane warning aid functionality.

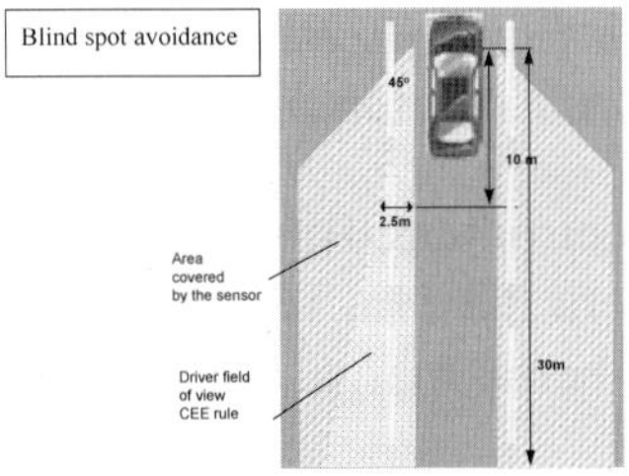

Figure 5 Representation of blind spot avoidance aid functionality.

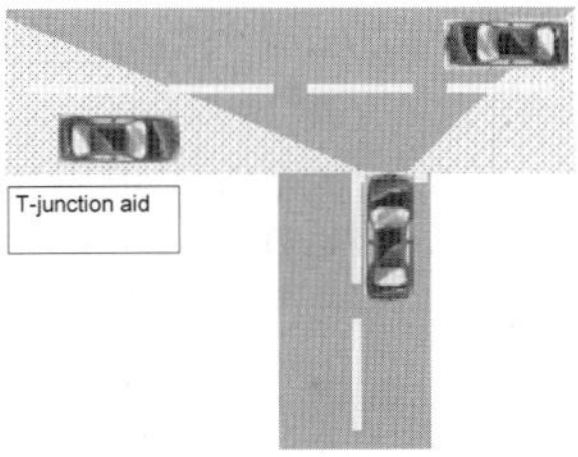

Figure 6 Representation of T-junction aid functionality.

4.3 Blind spot avoidance aid

Many E&D related road accidents are caused by lane changing manoeuvres while overtaking is in progress. Such situations can be detected by a dedicated sensor system able to produce the appropriate warning information to the driver. The presence of "blind spots", areas in the lateral and backside zone not perceivable by the driver through the lateral mirrors, makes a device to detect the presence of overtaking vehicles being useful for car drivers, especially for those drivers, whose level of attention decreases dangerously when driving time increases.

5. Conclusions

Until today, ADAS and IVIS as well as standard HMI has not been designed with the E&D users in mind, nor have the functionalities that he/she requires for information and support been adequately researched. Within the CONSENSUS Network, the relevant initiatives and evaluations performed so far are gathered and analysed, with the aim to collect guidelines and select priority functionalities on automotive HMI design for E&D drivers. The project will conclude to a roadmap for further research work to be conducted and will recognise the major gaps in the knowledge domain of designing in-car systems for all users.

References

Andreone, L. et al. (February 2002), TRAVELGUIDE project (GRD-1999-10041) Deliverable 5.

Barham, P., Alexander, J., Ayala, B., and Oxley, P., (Dec. 1994), Evaluation of reversing aids. Part II : Rear parcelshelf-mounted system used by Rover. CEC DRIVE II EDDIT Project V2031, Del.19A.

Brouwer, W.H., Ponds, R.W.H.M. (1994), "Driving Competence in older persons", Disability and Rehabilitation Journal, Vol. 16.

Klein, R. (1991), Age-related eye disease, visual impairment, and driving in the elderly. Human Factors, Vol. 33(5), p..521-525.

Naniopoulos, A., Bekiaris, E. and Nicolle, C. (eds.), (January 1994), Design Guidelines for Aids for DSN, CEC DRIVE II Project V2132 TELAID Del.7.

Panou at al. (April 2002), CONSENSUS Internal Report: Information on Telematic aids and ADAS for PSN drivers as part of the User acceptance enhancement.

TESCAN project, Annual Review Report, 2000.

Waller, J.A. & Goo, J.T. (1969), "Highway crash and citation patterns and chronic medical conditions", Journal of Safety Research, Vol.1.

Developing in-Car PDA-Based Tour Guides

Francesco Bellotti, Riccardo Berta, Massimiliano Margarone and Alessandro De Gloria

D.I.B.E.- Department of Electronic & Biophysical Engineering – Univ. of Genoa
Via Opera Pia 11a, 16145 Genova (Italy)
{franz, berta, marga, adg}@dibe.unige.it

Abstract

Mobile computing and communication technologies can have a great impact in improving modalities of fruition of the heritage by providing tourists with high-quality multimedia information and services that can be seamlessly accessed in car and outside.

The implementation of such applications requires facing challenging Human-Computer Interaction (HCI) issues, that emerge from the peculiarities of the new technological medium. For this reason, we have designed a toolkit for the development of mobile applications for tourists.

Such a toolkit, called Mobile Application Development Environment (MADE), manages low level multimedia programming aspects, allowing the developer to focus on the high level HCI problems.

We have implemented a Tourist Digital Assistant (TDA), which is a commercial hand-held computer loaded with tourist applications built using MADE (fig.1). TDA samples are currently undergoing an advanced experimentation stage in various European sites.

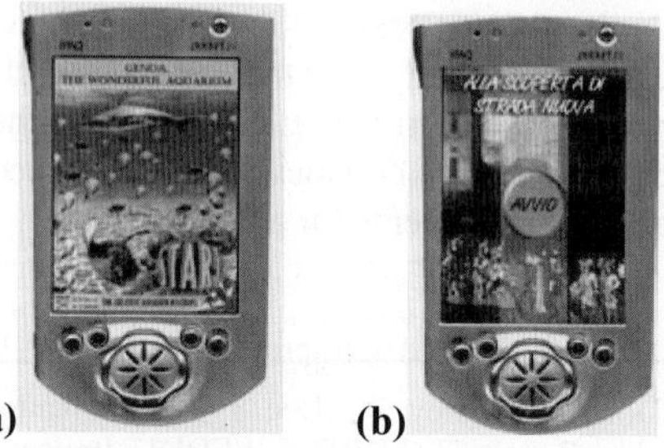

(a) **(b)**

Figure. 1 Snapshots of the TDA developed with MADE. (a) Genoa Aquarium, and (b) Genoa Strada Nuova.

1 Introduction

The automotive market of the next few years will be characterized by a wide number of services offered to drivers and passengers [1, 2]. The services will range from safety systems to personal communication, from entertainment to provision of location-aware information.

The car will become an intelligent environment able to meet the information and entertainment needs of driver and passengers. Wireless network connections will support real-time communication with the external world. Traveling by car will be no longer considered as a waste of time, since mobile connectivity will allow seamless fruition of contents and services.

This upcoming scenario can have a great impact in improving modalities of fruition of the heritage by providing tourists with high-quality multimedia information and services. In particular, handheld devices can become a sort of "hardware portals" through which information can be accessed both in car and outside, following the user all along her/his path (Fig. 2).

Figure. 2 Snapshot from a car cockpit integrating a palmtop computer.

In order to fulfill such expectations, new challenging issues have to be tackled. These issues concern in particular HCI, which is particularly critical for such small-size computers, that are to be used by the general public [3, 4].

Part of the research work described in this paper has been developed in the context of the E-TOUR project. E-TOUR has been co-funded by the European Commission through the European Union's Fifth RTD Framework Program (1998-2002).

In this paper we analyze these issues and present a toolkit ad-hoc designed to enhance development of tourist applications. Finally, we show an application sample and the first results of user tests.

2 HCI issues

A key factor to the success of a tourist guide is given by the ability to provide interaction modality able to support use by general tourists, who are not computer experts and may even have never used a computer.

Standard HCI solutions (e.g. the desktop metaphor of the Windows OS) may not be successfully applied to the new tourist application. In fact, the computational tool (the hand-held device) and the users (tourists) are significantly different from the personal computer and the usual desktop computer users.

2.1 Choice of the hand-held computer

In order to implement the TDA, we have carefully examined and tested hand-held devices available on the market, evaluating their suitability for the tourist site environment through the criteria such as: pocket size, high quality visual and audio interface, and support for multimedia. Based on these criteria, we chose the WinCE based PocketPC Compaq iPAQ as the most suitable platform for the developed requirements.

2.2 Characterization of the user

A tourist is a "distracted" user, completely involved in the real environment of the visit, which demands his full attention.

We have identified three main tourist's characteristics that have to be carefully considered in the development of the HCI: limited learning time, limited attention, and discretion. All these features suggest a HCI design approach supporting easy interaction so that the user can focus on her/his main tourist activity.

2.3 Model of the interface

On the basis of the user needs analysis and of the constraints of the mobile platform (e.g. limited screen size), we have developed the following specification for the interface.
The TDA's interface provides the contents and the controls to navigate between the various multimedia contents (e.g. images, video, audio).
The interface controls are all immediately visible on the screen and easily comprehensible. This is achieved by labelling every button with an explanatory name and with a metaphoric icon, selected among those more largely known (e.g. borrowing the CD-player interface icons from consumer electronics). This prevents the necessity for providing instruction-sheets or long explanations when users rent the TDA.
Buttons are sized so that they can be pressed with fingers. This reduces intrusiveness of technology in the visit, avoiding the need for styluses, which may be of a hindrance to users.
In order to support a synergistic experience, images and videos are always synchronized with the speech comment and with the real subjects of the visit.
Images representing the site map help the tourist to orient herself/himself within the environment and to plan the visit according to her/his preferences. Since the limited size of the display prevents a full visualization of the map, only the portion of the map close to the current user position is displayed. Then, tourist can move through the map using directional buttons. The map features also some active areas, that can be clicked to go to specific information about the selected attraction or about important services.
The TDA application also features an animated assistant that helps users to navigate through menus. The animated assistant is an animated icon, always visible on the interface, that allows users to move among the TDA sections (e.g. map, items, history, etc).

3 Toolkit

In order to implement the TDA, we have developed the Mobile Application Development Environment (MADE) toolkit, which supports efficient development of HCI modalities on hand-held computers.
The MADE toolkit is composed of an Object Oriented language, called Micromultimedia Scripting Language (MSL) and a software layer, called MiroMultiMedia Player (MMMPlayer), that interprets MSL scripts.

3.1 MSL

MSL is an Object Oriented language that supports development of interactive multimedia programs. The language specifies components encapsulating multimedia information. Such a high-level tool allows designers to concentrate on the analysis of HCI methodologies, exploring several solutions, without caring about low-level aspects of writing code for multimedia management and event handling. MSL enhances design productivity by employing the Object Oriented technology, exploiting code-reuse and program extensibility.

To the best of our knowledge, MSL is the first scripting language ad-hoc designed to support efficient development of applications on hand-held platforms featuring high-quality multimedia and wide usability by the general public, in particular in the tourism field.
MSL is extensible, allowing incremental addition of new types of components and functionalities as they emerge from customers' needs. At present, MSL supports the following functionalities:

- Multimedia compressed formats (jpeg, mpeg, mp3)
- Database access (local and remote)
- GPS
- Internet connection (wired and wireless)
- W-LAN (IEEE 802.11) and PAN (Bluettoth) connection
- Interactive maps based on vectorial graphics. Interactive maps offer added value services such as layered information (e.g. restaurant and hotel layers), location aware services (e.g. find-me-the-nearest...) and shortest path search.
- Intelligent tour-planning

3.2 MMMPlayer

MMMPlayer is the software engine that manages user presentation by interpreting applications written in the MSL language.
MSL is completely platform independent: MSL files can be deployed on any kind of platform hosting the MMMPlayer. The player is partly platform independent (the part that manages the logic of the multimedia presentation) and partly platform dependent (the low level drivers to access multimedia hardware at high performance). This enhances portability, since porting involves only the low-level part.
Extensibility of MMMPlayer (i.e. the possibility of easily including new functionalities) is supported, since adding new service class does not modify the structure of the system, but just adds new messages that are exchanged between the core system and the service components.
These features make MMMPlayer an innovative tool in order to support efficient multimedia production on hand-held devices.

4 A Sample application: the tour guide for castellon

The tour guide for the Spanish region of Castellon is a simple example of an MSL application. The guide involves 4 main sections, accessible from the main page. The first section features the whole map of the region, which can be zoomed and tapped to obtain detailed information about places and resorts (Fig. 4).

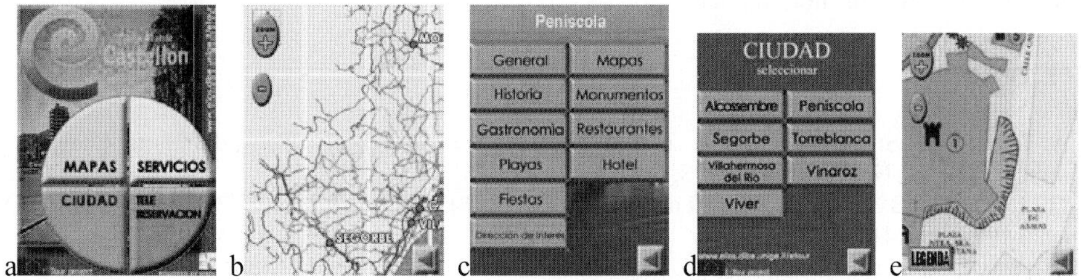

Figure. 4 The main page of the tour guide (a). The regional interactive map (b) and the list of information contents for the Peniscola town (c). The city menu list (d) and a snapshot from the city map of Pensicola (e)

The city contents (second section) can also be directly accessed through the "ciudad" link from the main page (Fig. 1d and e).
The third section provides access to database information, which is queried through user-friendly wizard-like access patterns, rather than through the OS-standard user interface elements.
The last section hosts a hotel tele-reservation service developed by TravelsoftIT, a partner of the E-Tour consortium.

5 Conclusions

We have used MADE to develop TDAs for some European tourist sites, such as the the Genoa historical city centre, the Bohusläns Museum in Uddevalla, the Spanish region of Castellon and Genoa's Costa Aquarium.
Results coming from the field tests have revealed high acceptance by the general public. Usability of the tool is a necessary condition for the success of the tool. Other key factors to success are quality of contents and style of presentation.
The most significant result is the substantial preference of visitors for multimedia presentations. The main motivation reported by users consists in the possibility offered by multimedia systems of easily correlating the argument presented in the description with the direct experience.
The most useful suggestions coming from the domain experts include: integration of the multimedia guide in the digital offer package of the tourist site (i.e. web site, CDs) and integration of all the information services useful for the tourist in the hand-held computer. In this way she/he can use just the palmtop, avoiding resorting to different tools (e.g. paper maps, books, phones, etc).

Next generation cars are likely to become a sort of telematic centers connecting drivers and passengers with the external world. Such an intelligent environment will feature several terminals. Some of them will be fixed (e.g. configurable dashboards, control panels, rear-seat screens) other will be portable, able to follow the user in her/his activities also outside the car.
Among such terminals, palmtop computers will have ever more relevance, thanks to their ability to join high computing power and communication capabilities in a small-size device. Thus, they represent an ideal bridge to connect the user with the car, home and office environments and can provide ubiquitous services, such as ubiquitous and seamless navigation and orientation support.
Thus, developing contents and interfaces for such platforms in a rapid and efficient way becomes a key-factor in the provision of added-value services, also for automotive clients.
In this paper we have shown the main issues concerning development of services for general tourists and briefly described a toolkit, MADE, ad-hoc designed to support efficient development of tour-guides on hand-held devices.

References

[1]A. Jameel, M. Stuempfle, D.Jiang and A. Fuchs, "Web on Wheels: Toward Internet-Enabled Cars", Computer, Vol. 31, No. 1, January 1998, pp. 69-76.
[2] T. Lewis and B. C. Fuller, "Fast-Lane Browsers Put the Web on Wheels", Computer, Vol. 32, No. 1, January 1999, pp. 141-144.
[3] J. Pascoe, N. Ryan and D. Morse Using While Moving: HCI Issues in Fieldwork Environments. ACM Transactions on Computer-Human Interaction, Vol. 7, No. 3, September 2000.
[4] G. Bieber and M. Gierish, Personal mobile navigation systems-design considerations and experiences, Computer & Graphics, August 2001, Vol 25, No. 4, pp.563-570.

Providing Traffic and Route Guidance Information to Tourists

Sascha M. Breker

Leibniz Research Centre for Working Environment and Human Factors, Dortmund, Germany
breker@ifado.de

Karel A. Brookhuis

University of Groningen, Department of Psychology
Groningen, The Netherlands
k.a.brookhuis@ppsw.rug.nl

Pirkko Rama

VTT Building and Transport, Helsinki, Finland
pirkko.rama@vtt.fi

Abstract

Traffic and route guidance information is highly relevant for tourists driving in unfamiliar traffic networks. Advanced traffic technology enables adaptation of driver support to both the current situation on the roads and information preferences of road users. In a stepwise procedure it was examined in what way route guidance information should be provided and displayed in an international traffic corridor. Native and non-native drivers as well as traffic experts from several European countries rated content and layout of 12 different versions of a Full Colour Information Panel (FCIP). Four optimised FCIP versions and their effect on driver behaviour were then tested in a driving simulator experiment. This article summarises the experimental phases and the guidelines that were extracted from the results.

1 Introduction

Navigating in an unfamiliar traffic network is one of the most demanding driving tasks. Way finding becomes even more difficult for non-native drivers who are unfamiliar with local traffic regulations and the local language. Comparisons between traffic injury patterns of international and national Greek tourists on the island Corfu indicate such an excess risk among foreign travellers: Traffic-related accidents are more frequent and more serious among foreign tourists (Petridou, Dessypiris, Skalkidou & Trichopoulos, 1999). Advanced traffic information systems like in-vehicle navigation systems or roadside variable message signs enable adaptation of driver support to both the current situation on the roads and information preferences of different driver subgroups. A better understanding of the information needs and preferences of drivers is thus a necessary precondition for taking full advantage of the opportunities that are offered by new traffic technology (Molnar & Eby, 1997).

The experiments described in this article were conducted both as a part of the EU sponsored project TRAVELGUIDE[1] and for the municipality of the Dutch city of The Hague. Their main

[1] TRAVELler and traffic information systems: GUIDElines for the enhancement of integrated information provision services, GRD1-1999-10041. For an overview, see Breker et al., 2001.

objective was to develop design guidelines with respect to information provision via new roadside traffic information systems. The secondary objective is related to the municipality of The Hague, which plans to place a new Full Colour Information Panel (FCIP) over a motorway in the direction of the adjacent city of Scheveningen. This beach resort is very popular among both Dutch and foreign pleasure travellers.
A FCIP consists of a matrix of red, green and blue Light Emitting Diodes. Its working is based on the addition of these three additive primary colours, comparable to that of an ordinary colour television, yielding the possibility to depict virtually any image. The screen content of the FCIP can be programmed with a normal computer. The computer also integrates feedback signals that might dynamically act upon the screen content, enabling a change of the information displayed on the panel in quick response to the current situation on the roads. Optimal employment of the new technical possibilities is, however, not straightforward. If the FCIPs are not designed properly, it will not only be difficult for drivers to understand the information. Traffic safety may additionally be endangered by distracting the driver's attention. Optimisation of the FCIP design should therefore result in a flexible traffic information system that is effective as well as efficient in terms of the drivers' information processing resources. Focus of the experiments was thus on tailoring the design of the FCIP to the drivers' information preferences as well as to more general aspects of human information processing. By reducing the number of information elements and by organising the remaining elements in a meaningful way cognitive demands placed on the driver by the new roadside information panel should be kept to a minimum.

2 Method

This article is primarily an overview of the FCIP study in general, and not so much a detailed description of the experiments that were carried out. The design of the experiments is thus described at a general level only. Figure 1 shows a schematic overview of the steps taken to optimise FCIP content and layout.

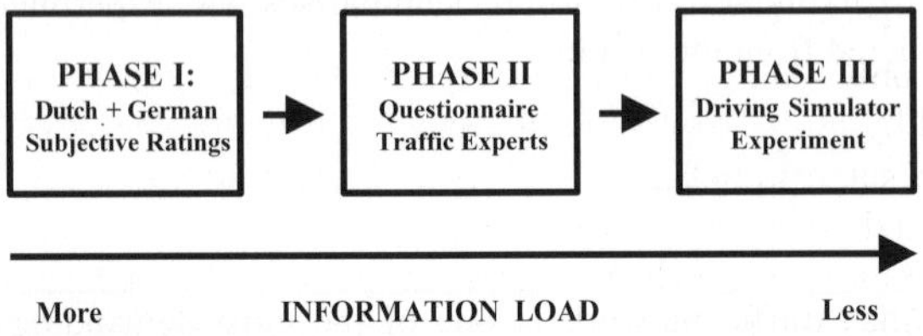

Figure 1: Schematic overview of the conducted experiments.

Following an initial literature survey on human-machine interface design factors, 12 different FCIP versions were tested in *phase one* (see 'Results' section for FCIP examples, figures 2-4). A total of 52 and 30 subjects participated at the Dutch and the German test site, respectively. The study was carried out in a classroom setting. Various designs of FCIPs and the information categories displayed (advisory route choice, travel time, route alternatives, parking facilities, park & ride information) were systematically evaluated via subjective ratings and a card sorting procedure. In *phase two* 18 traffic experts from five different European countries were asked to rank those four FCIP versions favoured in phase one on the basis of their knowledge, and to complete a short questionnaire requesting them to comment on the different versions. *Phase three* consisted of a driving simulator experiment with 20 participants. The four optimised FCIPs were designed in a way that differed with regard to the way the message was presented (textual vs. graphical), the type of symbols indicating the optimal route (arrow vs. 'smiley') and the type of information displayed at the top part of the sign (destination name vs. travel time). The simulator was fixed based, consisting of a car with original controls linked to a Silicon Graphics computer that recorded driver behaviour and computed the environment at 30 Hz. The simulated road was an abstracted copy of the motorway where the FCIP will be installed. The environment was projected on a screen that covered 165° angle horizontal view, and 45° vertical view (for more details on the simulator see Van Wolffelaar & Van Winsum, 1995). The participants were instructed to 'drive to Scheveningen to spend a day at the beach'. After they passed the FCIP the participants had to choose between three route alternatives. Once a route was chosen they were interviewed about their motives, including an assessment of their acceptance of the FCIP. At the end of the session the participants were asked to rank the FCIPs based on their personal preference. During the ride speed and lateral position of the car were registered at 10 Hz.

3 Results

Selected results will be reported according to the three-phased structure of the study. In figure 2, the FCIPs preferred by most Dutch and German subjects *in phase one* are shown.

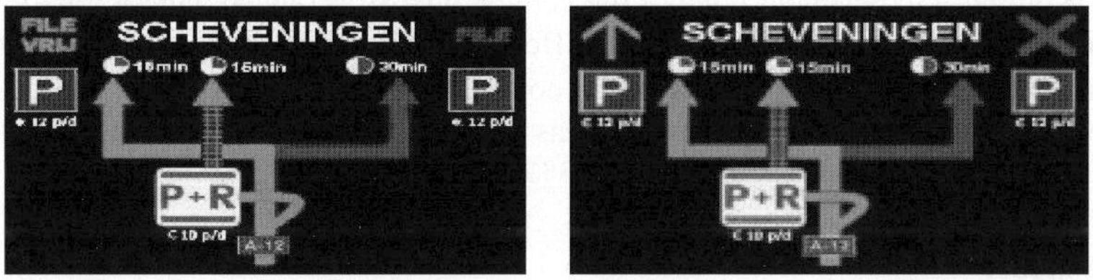

Figure 2: FCIPs mostly preferred in phase one - Dutch (left) and German (right) subjects. Right-hand arrow is red, left-hand arrows are green, icons in colour, text in white.

Out of the 12 initial FCIP versions, the most popular alternative of the Dutch test site (18% first place rankings) ranked only fifth place at the German test site (9% first place rankings). Despite that, the FCIP favoured by most German participants (18% first place rankings) deviates from the one preferred by Dutch subjects in only one respect: Instead of the Dutch words 'file' and 'file vrij' for 'congestion' and 'congestion free' language-independent icons symbolise the current situation on the roads. Both in the Netherlands and in Germany Travel Comfort (41% on average), Travel Time (43%) and Travel Cost (33%) were rated as the most important information categories for route choice .

Figure 3 shows the FCIP version favoured in *phase 2* by most international traffic experts (31% first place rankings). The traffic experts clearly preferred a version displaying no language except for the name of the destination on top of the panel.

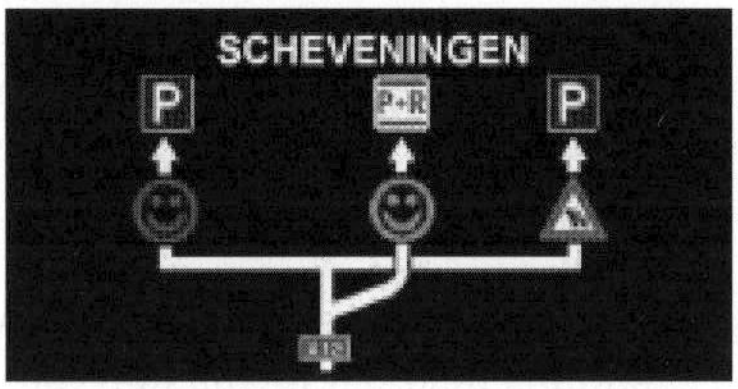

Figure 3: FCIP preferred in phase two - traffic experts. All arrows are white, icons in colour, text in white.

Additional comments of the experts highlighted more detailed aspects that have to be taken into account when designing a traffic information panel for an international corridor. The park-and-ride (P&R) concept is, for example, established in the Netherlands and Germany, while the concept is not well known in France.
Figure 4 shows the FCIP that was preferred in phase three (driving simulator experiment, 32% first place rankings).

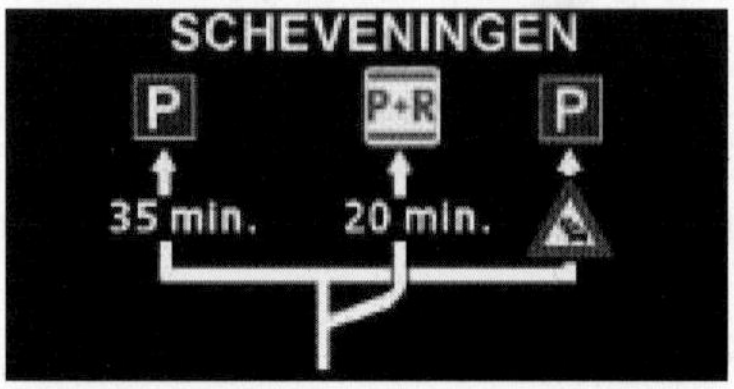

Figure 4: FCIP preferred in phase three – driving simulator experiment.
All arrows are white, icons in colour, text in white.

In the driving simulator experiment none of the vehicle measures (speed, lateral position, and their derivates) differed significantly between the different FCIP conditions. All FCIP versions thus produced similar effects on driver behaviour. In contrast to the previous phase, the least preferred FCIP in the driving simulator experiments was the version using the 'smiley' as the icon symbolising the current situation on the roads (13% first place rankings).

4 Discussion

Already during the classroom experiment with native and non-native drivers it became clear that too much information might result in the opposite as objected: hardly any information transfer at all at the cost of increased risk. The input from traffic experts emphasised the necessity to keep content and layout of the FCIPs as simple as possible by using language-independent icons that are easy to understand. The rankings and subjective evaluations in the first two phases lead to the development of four optimised panels, which were then tested in a driving simulator experiment in a setting copying the traffic environment where the panel will be installed. Most notable change of the FCIP design is the simplification of the road structure during the course of the study. Main result of the final phase is that most participants preferred a basic 'road skeleton' when actively engaged in driving in a simulated traffic environment.
Ergonomically correct, well-legible and comprehensible traffic information provision enforces thus both qualitative and quantitative design constraints, which are mutually depending on each other. The better the quality of the layout, the more information can be provided to the driver. Clear information elements that are easy to understand contribute vice versa to the quality of the layout. Following recommendations can now be formulated for traffic information provision to tourist drivers in international traffic corridors:

- A diagrammatic layout is always preferable to a textual one. The road structure should be a simple abstracted version of the actual road structure. Deviations from the actual road structure are only acceptable if they contribute to the understanding of the driver.
- The use of colour coding should be kept to a minimum. Visual chunking by colour may be used in order to subdivide the total image into separate units.
- Pictorial information is preferable against textual information. It is inherently language-free and has thus the potential to enhance comprehensibility especially in

the case of non-native drivers. Specific findings concerning individual icons show that a red x-shaped symbol is associated with a closed road. A 'smiley' is an incongruent symbol in a traffic environment, and the P&R concept is not known in some European countries.

- Apart from symbol-like statements like 'OK', and the indication of destination (city) names, language use should be kept to a minimum.

However, even with a layout of highest quality, there is always the risk for quantitative information overload. The problem here is the difficulty to define what is an isolated information element in a given context. For example, is a parking sign made of a blue background and the letter P in front processed as one single information unit? Or are the background, the letter, and maybe even the features of the letter processed separately from each other? When evaluating new methods of traffic information provision the only effective way to deal with this problem is practical testing in order to find an optimal design solution.

5 Conclusion

Guiding principle for the design of effective traffic information provision systems for drivers in general, and tourist drivers in particular should be parsimony. Simplify the representation of the road structure, show only simple icons, and use only a limited number of information elements in order to avoid information overload. Exploiting ergonomic design aspects adds at least some degrees of freedom to this parsimony: a clear layout allows a larger number of information elements to be displayed. Habits and familiarities of the target group have to be taken into account as well. If information is to be provided in an international traffic corridor, information should be not only language-independent, but the icons and symbols that are displayed also have to be well known to foreign drivers.

6 References

Petridou, E., Dessypiris, N. Skalkidou, A. & Trichopoulos, D. (1999). Are traffic injuries disproportionally more common among tourists in Greece? Struggling with incomplete data. *Accident Analysis and Prevention*, 31, 611-615.

Breker, S. M., Verwey, W. B., Naniopoulos, A., Bekiaris, E., Lilli, F., Wevers, K. & Brookhuis, K. A. (2001). Adapting advanced traffic information provision to road user needs in Travelguide - a progress report. In *Proceedings of the 8th World Congress on Intelligent Transport Systems.* Sydney: ITS Australia.

Luoma, J. and Rämä, P. (2001). Comprehension of pictograms for variable message signs. *Traffic Engineering and Control.* February 2001, 53-58.

Molnar, L. J. & Eby, D. W. (1997). Preliminary guidelines for the development of advanced traveler information systems for the driving tourist: Route guidance features. In *Proceedings of the 30th ISATA International Symposium on Automotive Technology and Automation.* Croydon: Automotive Automation Ltd.

Van Wolffelaar, P. C. & Van Winsum, W. (1995). Traffic simulation and driving simulation - an integrated approach. In *Proceedings of the Driving Simulator Conference (DSC '95).* Toulouse: Teknea.

Delivery of Services on Any Device
From Java Code to User Interface

Davide Carboni, Andrea Piras, Stefano Sanna, Gavino Paddeu

CRS4 – Center for Advanced Studies, Research and Development in Sardinia
VI strada Ovest, Z.I. Macchiareddu,
09010 Uta (Cagliari), Italy
{dadaista, piras, gerda, gavino}@crs4.it

Sylvain Giroux

Dept. of Mathematics and Computer Science,
University of Sherbrooke
Sherbrooke, Canada
sylvain.giroux@dmi.usherb.ca

Abstract

The design and the implementation of software for mobile computers and for pervasive computing environments entail several issues and lead to new requirements. Applications must handle at run-time the heterogeneity of delivery contexts in terms of devices features, network bandwidth, operating systems and so forth. The issues related to the access pertain both to delivery (can the device run the service?) and to user interfaces (how the user interact with the service through device?). This paper describes, on one hand, an approach where user interfaces are generated at run-time, and, on the other hand, a deployment strategy that, taking charge of distributing the code according to the device capacities, leads to the "application apportioning".

1 Introduction

The design and the implementation of software for mobile computers and for pervasive computing environments [Weiser, 1993] entail several issues. Enhancing physical surroundings with computing capabilities leads to new requirements: *"A device is a portal into an application/data space, not a repository of custom software managed by the user. An application is a mean by which a user performs a task, not a piece of software that is written to exploit a device's capabilities. The computing environment is the user's information-enhanced physical surroundings, not a virtual space that exists to store and run software."* [Banavar et al., 2000] Thus, applications must handle at run-time the heterogeneity of delivery contexts in terms of devices features, network bandwidth, operating systems and so forth. In fact, a user may want to access his personal data/applications from different devices in different context of use: from desktop PC to cellular phones. The issues related to the access belong both to delivery (can the device run the service?) and to the user interface (how the user interact with the service through device?). Delivery must cope with network bandwidth, processing power, operating systems, etc. Human-computer interaction must deal with screen size (if any), colors, and input/output modalities. The design and the implementation of a brand-new interface for each category of devices are tedious and error-prone tasks. Moreover, when applications get mobile or pervasive, one even does not know beforehand what devices will be involved. This paper describes an approach where user interfaces are generated at run-time in a mobile or pervasive context. On the one hand, there are rendering engines that, given a category of device, generate the appropriate user interfaces (UI). These engines exploit reflection on Java code in order to understand which data can be accessed by the user and which actions are permitted on that data. On the other hand, there is a multi-device deployment strategy that, taking charge of distributing the code according to the device capacities, leads to the "application apportioning" [Banavar et al., 2000].

2 Target Devices

The access to an application from unpredictable delivery contexts involves the automatic UI generation at run-time. To support it, a categorization of devices according to their features is needed. In addition to devices physical features, our categorization is also based on the availability or not of a Java Virtual Machine (JVM) since Java is our privileged language for implementation of the aimed distributed applications. The many features already available in Java have been proven invaluable. Furthermore Java is widely used to develop network-based applications. Three main categories were sorted out: fat clients, thin clients, and Web-like clients. Fat clients groups together personal computers (PC), Unix workstations, and laptops. All of them are capable to run a standard Java Virtual Machine, to download mobile Java code, and are provided with WIMP (Windows Icons Mouse Pointing) input/output modalities. A part or all of the application code can be moved across the network to the target "fat client" device. For their part, thin clients can only run a limited JVM, for instance the Java Micro-Edition. They can run only a limited subset of Java and they cannot download nor execute mobile code. As a consequence, Java classes must be loaded from the local disc. Since the application code cannot reach the device, a *service viewer* written in Java has to be pre-installed on the device. This service viewer connects to the application and exchange data using an XML protocol germane to the Web-Service protocol. A service viewer generates an appropriate user interface according to the XML data received, and interacts with the remote object invoking its methods. Thin clients category groups together personal digital assistants, smart phones, and set-top boxes. The last category is Web-like clients. These devices support no JVM at all but they are equipped with a third-party browser (WAP-enabled phones, VoiceXML [Voice, 2000], and HTML browsers). They can connect via HTTP to a Web server and download documents compliant to their specific mark-up languages. For instance, a WAP phone will work only with WML documents [Wireless, 2001], while a voice browser will work only with VoiceXML documents.

3 Rationale

The rationale behind our approach is that the code of an interactive system can be decomposed in two main parts: models and views. Such a decomposition is borrowed from the object oriented programming practice in which the well known Model-View Control, (see Observer pattern in [Gamma et al., 1995]), design pattern has been formalized and widely used in the design of graphical user interfaces. Such a pattern organizes code for a clear separation of the logic (the Model) and the UI (the View-Controller). In object oriented systems models and views are objects, and objects are reactive systems that can be activated by the invocation the methods in their public interface (the set of operations that can be activated by a client). A model is independent from the delivery context while views are strictly tied to it. Let's take a pocket calculator application as an illustration. A calculator can be implemented as a model object which is activated whenever the user triggers a method that appends a new digit to the display or a method that selects the mathematical operation to perform. The model part is independent from the graphical user interface. With respect to the model part, the only thing to adjust at run-time is "where" such an object is loaded and executed. In the case of fat clients, the model moves across the network, is loaded by the client JVM and executed on the user's device. In the case of thin and

Web-like clients, an intermediate remote tier (remote with respect to the user's device), performs code loading, code execution, and communicates with the user's device using an ad-hoc protocol. On the contrary, a calculator view —that is the user interface— is strictly dependant on the user' device. In the case of fat and thin clients, the view part of the application is generated at run-time by means of graphical components such as buttons, text fields, dialogs and so forth. Indeed, views are concrete "interactors" and must be defined by composing concrete widgets belonging to a specific toolkit. In the case of Web-like clients, view objects are only a mark-up presentation of the model: HTML for Web browsers, WML for WAP browser and so forth. So, in the latter case, the final presentation is delegated to a pre-installed third-party browser.

4 From Java Code to User Interfaces

The issue turns into thus taking Java code of the model part of an application to generate a user device for a given class of devices. A "rendering engine" is an application that performs this "rendering process". To address this issue we have developed MORE (Multi-platform Rendering Engine). MORE is on one hand a framework for defining and composing models, and on the other hand, a set of rendering engines that generates the view part of an application. At run-time, the rendering engine builds on-the-fly the right view by transforming models into structures of concrete and interactive objects. Widgets such as text fields, radio buttons, and more sophisticated controls, are deployed as placeholders for the data they refer to: strings, numbers, booleans and more complex data types like object arrays and collections. The rendering process uses reflection to extract directly from the code the "displayable" information, that is public accessors of fields and public methods. Complex models are viewed as composite graphical objects and the rendering process is applied recursively till atomic interactors are obtained. The advantage of this approach is that the same model can be accessed from any device for which has been implemented a rendering engine. The process of inspecting composite models to generate a graphical user interface can be summarized by the following algorithm:

```
Render(a model object) :  a view object
Create a container for the actual platform //e.g. a JFrame in Java Swing toolkit
       For each "displayable" field of the model object:
              Look-up for a run-time editor for the field.
              If found: add the editor to the container.
              Else: add a link, e.g. a button, to reapply this rendering
                   algorithm recursively to the value of the field
       For each method of the object:
              Add a trigger, e.g. a button, to execute the corresponding method.
Return the container
```

A run-time editor is a component aimed to write/read the value of an object field. For instance, if our model contains a field of type `java.lang.String`, the above algorithm looks up a run-time editor for such a type, that will likely be a text field graphic component. To foster the development of new applications, MORE provides run-time editors for main basic Java types: `java.lang.String`, `java.util.Number`, `java.lang.Boolean`, and `java.util.Calendar`. Besides, it provides run-time editors for composite types like `java.util.Collection`, and `java.lang.Object[]`. For those types, the editor allow users to select, display, modify, remove, and add elements to the collection or to the object array. Finally, during the development of multi-media and geo-referenced applications, emerged the need for some new Java types: `Situated` and `MultimediaResource`. `Situated` is Java interface that provides information about the physical position of an object either in a Cartesian format (latitude and longitude) or in a topological format (city, street, number, etc.). An `Situated[]` component is displayed as a map centered on the barycentre of the array, in which the user can zoom in, zoom out, select an object and display it in a separate view. The last type we added is the `MultimediaResource` type which is basically a wrapper to a multimedia file of type among JPEG, GIF, MP3, wav, and mpeg.

5 A Case Study : a Museum Guide for Mobile Users

A museum wants to put on-line information on its collection of paintings. The information should be available from any device. So tourists may use their PCs or laptops at home to prepare their visit to the museum or they may use their handheld devices or their cellular phones in the museum to obtain a complement of information. To build and deploy this application, solely the model part has to be implemented in Java. The core classes are `Museum` and `Picture` (Fig. 1). A `Museum` instance contains a collection of `Picture` and the `String` member field `city` holds the name of the city. In both classes, read and write accessors methods are defined in compliance with the name pattern `get<field>()` and `set<field>()`. For each category of devices, an appropriate rendering engine will 1) use the method `toString()` returns a text description of the whole object, 2) for basic type member field, generate text fields to access and to modify their value if allowed (`set<field>()` is `public`), 3) for aggregate data type, generate a set of widgets enabling to browse, select and view their content, and 4)generate "buttons" for other public methods. Then for any devices, the resulting generated user interface allows to browse the list of paintings, select one, and eventually invoke the operation `export()`. The rendering engine for fat clients yields to an interface based on Java Swing (Fig. 4) where browsing the collection and viewing items are performed at the same time in a single composite frame, while the rendering engines for handheld devices (Fig. 3) and WAP-enabled devices (Fig. 2), browsing the collection and viewing items are performed in two separate steps due to the severe screen size constraints.

6 Current Implementation

Rendering engines were implemented for all the run-time editors described in §4 on the following platforms: fat clients such as laptops and workstations; thin clients such as palm and iPaq handheld computers, interactive TV, and smart phones; Web-like clients such as Web browsers, WAP browsers, and VoiceXML browsers. In the case of Web-like clients, the result of the rendering process is an intermediate XML representation which is then translated in a mark-up language suitable for the actual browser. MORE uses a XSLT processor and an XSL style sheets to perform these transformations. However information obtained by reflection is not always sufficient. On one hand, even if the fields may have names compliant with Java naming and coding conventions, the extracted name may still be difficult to read or understand. For instance, a field `departureDate` should be visualized as "Departure date". Such a transformation could be done automatically decomposing strings according to the position of the capital letters. On the other hand, fields and methods of an object are not ordered in the code following precise criteria, thus the view is generated laying out components with a unpredictable position. Since views are automatically generated at run-time, programmers do not have complete control of the presentation. To address these issues, our rendering engine takes as input additional information, in form of a text file, which specifies aliases for names (allowing also the internationalization of the user interface), and how to layout fields and methods in the container.

7 Conclusion

The framework we described provides immediate and significant advantages. MORE makes "design once, display anywhere" closer to reality. A programmer solely has to write the model in

Java. Reflection of Java code allows to extract the structure of Java types and our system uses this structure to generate a user interface. The model then can be deployed on any device for which there is a rendering engine. MORE is easily extensible as new devices categories emerge. So as technology evolves, it is not necessary to rewrite the UI but simply to adapt the rendering engine to this device category. This approach is more germane to Silica [Rao, 1991] and plug-ins than to model-based ones [Paterno, 2000].

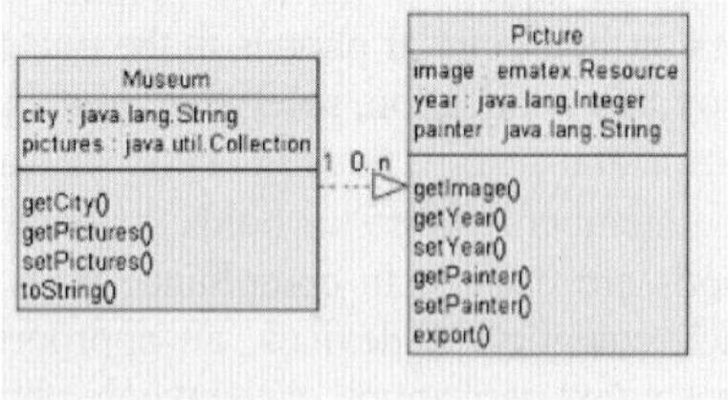

Figure 1: UML Diagram for a simplified version of Museum information system.

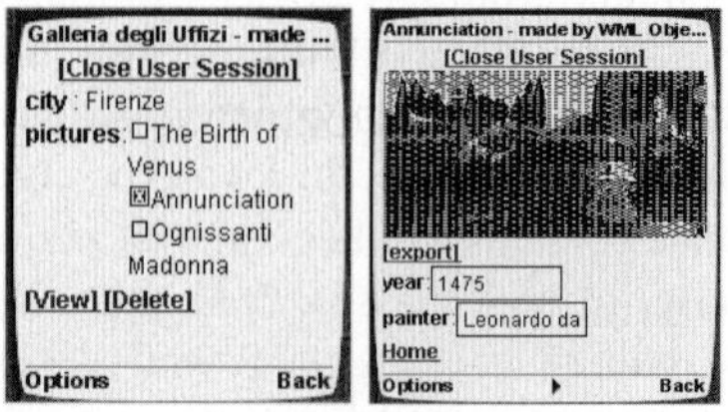

Figure 2: The WML Interface generated by the rendering engine for web-like client.

Figure 3: The HTML Interface generated by the rendering engine for thin client (iPaq).

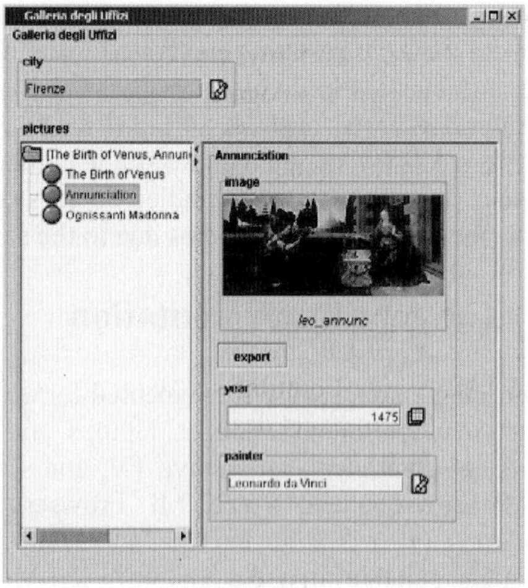

Figure 4: The Java Interface generated by the rendering engine for fat clients.

8 References

G. Banavar et al. (2000) Challenges: An Application Model for Pervasive Computing. *MobiCom 2000*, pp. 266-274.

Gamma, E., Helm, R., Johnson, R., & Vlissides, J. (1995) Design Patterns, Addison-Wesley.

Paterno, F. (2000) Model-based desing and evaluation of interactive applications, Springer.

Rao, R. (1991) Implementational Reflection in Silica, *ECOOP 1991*, LNCS 512, Springer-Verlag, pp. 251-267.

Voice eXtensible Markup Language (VoiceXML) v1.0 (2000), http://www.w3.org/TR/voicexml.

Wireless Markup Language v2.0, Wireless Application Protocol WAP-238-WML-20010911-a, (2001), http://www1.wapforum.org/tech/documents/WAP-238-WML-20010911-a.pdf.

Weiser, M. (1993) "Some Computer Science Problems in Ubiquitous Computing," *Communications of the ACM*, pp.75-84.

Function Analysis and Control Panel Design of in-Car Computer Systems (E-Car)

Jun-Kai Chiu

National Tsing Hua University
Hsinchu, Taiwan, R.O.C., 300
d917818@oz.nthu.edu.tw

Sheue-Ling Hwang

National Tsing Hua University
Hsinchu, Taiwan, R.O.C., 300
slhwang@ie.nthu.edu.tw

Abstract

A Variety of computer subsystems have been proposed as efficient tools for drivers nowadays. For example, electronics map, communication and multimedia systems have become standard equipments in the car. Through the use of these advanced technologies, drivers can enjoy high-quality life such as receiving real-time information, driving with amusements, quick to reach the destination, etc. However, driving is a mental-driven and highly information-processing task which requires more cognitive ability and less physical skill. It is necessary to estimate human mental workload and physical limit during driving. The objective of this study is to develop a safe and useful control panel which can offer drivers appropriate and useful information during driving. The results of questionnaire study were the priority of E-Car functions in three specific situations. The results can be used to consider E-Car necessary functions in the subsequent experiment. The results of virtual reality experiment revealed that with fewer numbers of buttons on the control panel, drivers feel more comfortable, effective, convenient and safe during driving and manipulating E-Car system at the same time. The result also showed that to converse nine subsystems with 3 buttons substantially better than with five or nine buttons. It can be considered on E-Car control panel design in the future.

1 Introduction & background

As surroundings keep up with E-tide, it is important about the following in-car technologies. Several subsystems such as Mobil Phone, Electronics Map, AM/FM radio, CD/Tape, Video display in reverse gear, Guard against burglary & Accident handle, Internet, Video telephone, VCD/DVD movie player…etc, have been integrated in automotive in automotive industry.
Radio operation and telephone dialing during driving had deeply influence on the driving performance, but did not affect the driving pathway. (Hayes, et al.,1989 ;Green, et al.,1993; Tijerina, et al.,1995). Another research of actual car driving experiment shows that radio operation has high correlation with car accidents (Wierwille, Tijerina,1994). According to these results, to simplify the control panel of multiple E-Car function is necessary for convenient and safe driving.
Previous research has ever demonstrated that driving simulators can be used for the collection of driver behavior (Koutsopoulos, Polydoropoulou and Ben-Akiva, 1995). The advantages of driving simulation is on building-up a safer, controlable, and relatively inexpensive research enviroment. Thus, the evaluation of different control panel designs can be conducted in virtual driving enviroment before producing and installing the E-car system into vehicles.

2 Method

This research included two parts, function requirement survey and control panel design experiment.

2.1 Function requirement survey

2.1.1 Questionnaire design

The contents are collected from several resources: automobile industry, motor repair shops, offices of motor vehicles, car drivers and passengers, E-Car websites…etc. After collected of E-car Data, it is helpful to weave a formal questionnaire. The questionnaire included 5 questions:

Questions1-3: Three situations were simulated, working, travel, and rest.

Question4: Driver's favourite position of LCD information Display.

Question5: Driver's favourite control type of E-Car system.

Six personal information (sex, age, vocation, driving experiment, reside area, education) were recorded for relevant analysis.

2.1.2 Sampling

The questionnaire participators were selected from population of drivers in Taiwan, and must have the driving licenses. According to monthly report of Ministry of Communications in October, 2001, the total number of drivers approaches to 9587487. It approaches a normal distribution. In accordance with statistics sampling conditions, 95% dependent route, 6% sample error, the minimum is sampled to represent the entire population. The formula is:

$n \geqq \{[\sqrt{P(1-P)}\ /e]*Z\alpha/2\}2$, where $P=1/2, \alpha=0.05, e=0.06$

Finally, we obtain $n \geqq 1067.11$, and round up 1068.

2.1.3 Procedure

The questionnaire was surveyed in Taiwan, include eastern, western, south, and northern area. The duration of questionnaire survey was from November, 2001 to December, 2001, totally two months. Among 1200 original questionnaires, 1135 were valid questionnaires which satisfies the minimum sampling number.

2.2 Control panel design experiment

2.2.1 Test materials and equipment

A virtual reality experiment was conducted to evaluate the effect of control panels which were designed basing on the results of functional requirement survey. We rebuilt three new control panels from existing control panels of car audio (as shown in figure 1), and used the components of car to simulate hardware environment of car driving (as shown in figure 2). A computer program was developed by using a VR software "3D Webmaster" (as shown in figure3), and all the components were connected with a Personal computer.

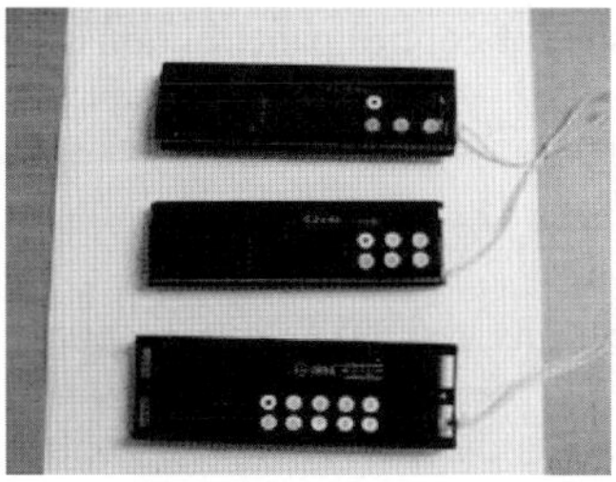

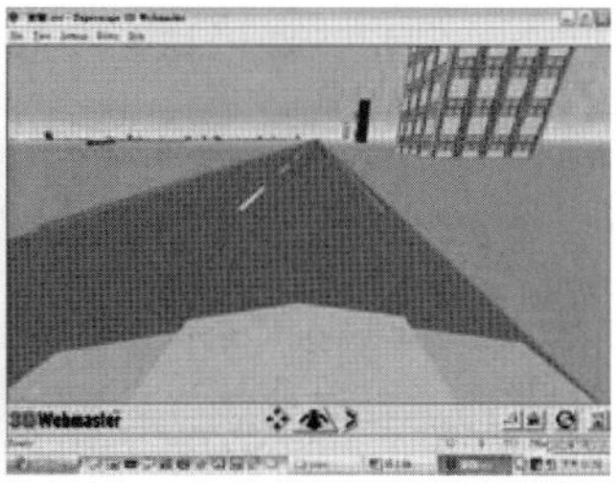

Figure 1: Control panels **Figure 2:** Hardware structure **Figure 3:** Virtual reality software

2.2.2 Experimental procedure

Twenty-seven subjects participated in the experiment. Participators were drawn in Virtual reality Environment, and were asked to drive as well as possible. The secondary task was to select the functions of the E-Car system randomly. The independent variable of experiment was types of control panels (with 3, 5, 9 buttons). The dependent variables included driving performance and E-Car system control performance.

3 Results

3.1 Results of Questionnaire survey

The priorities of E-Car functions in three specific situations (working, travel, and take rest.) were analyzed by T-test as shown in table 1.

Table 1: The priority of E-Car functions in three specific situations

Priority situations	**High**	**Middle**	**Low**
working	Mobil Phone, Electronics Map, AM/FM radio	CD/Tape, Video display in reverse gear, Guard against burglary & Accident handle	Internet, Video telephone, VCD/DVD movie player
Travel	Electronics Map, Mobil Phone, AM/FM radio	Internet, CD/Tape, Guard against burglary & Accident handle	Video telephone, VCD/DVD movie player, Video display in reverse gear
Rest	VCD/DVD movie play, CD/Tape, AM/FM radio	Mobil Phone, Internet, Electronics Map,	Video telephone, Video display in reverse gear, Guard against burglary & Accident handle

Among the six personal information (sex, age, vocation, driving experiment, reside area, education), there was no statistical significance in ANOVA table. This result implied taht personal characters did not affect the priority of E-Car functions.

3.2 Experimental results

Analyses of variance (ANOVA) were conducted on the dependent measures of driving performance and control panel performance. The major findings were discussed in the following sections and could be referred as empirical evidences to select the best control panel type.

Meanwhile, the result of this study might not only facilitate the design of control panel buttons but also minimize disturbances in driving control.

3.2.1 Control panel performance

The control panel performance include three parts, number of correct hits, redundant hits, response time. There was no statistical significant
Ce in ANOVA table. This result implied that diffenent control panels did not affect the control performance.

3.2.2 Driving performance

The driving performance include two parts, frequency of accidents, lateral deviation. First, about the frequency of accidents, the ANOVA table of major dependent variable (different control panel) were summarized in table 2. The main effect in between control panel was significant on frequency of accidents (F=3.504, p=.0461). This result implied that different control panel affects the driving performance. In order to analyze the effect of three different control panel. The Tukey statisical method was summarized in table 3. The effect between panel 1 and panel 3 was significant on frequency of accidents (p=.048). panel 1 cause more wrong hits than panel 3. This result implied that 3 buttons in use, was better than 9 buttons.

Table 2: ANOVA table of panels in frequency of accidents

	Sum squares	df	Mean square	F-value	P-value
Within	84.222	2	42.111	3.504	.046
Between	288.444	24	12.019		
Total	372.667	26			

Table 3: Tukey statistical analysis of panels in frequency of accidents

(I)panel	(J)panel	Average difference (I-J)	Standard Error	P
1	2	3.2222	1.6343	.141
	3	4.1111*	1.6343	.048
2	1	-3.2222	1.6343	.141
	3	.8889	1.6343	.851
3	1	-4.1111*	1.6343	.048
	2	-.8889	1.6343	.851

** $P \leq .05$

Second, about the lateral deviation, there was no statistical significance in the ANOVA table of major dependent variable (different control panel). This result implied that diffenent control panel did not affect the lateral deviation.

4 Discussion

In questionnaire survey, we sampled enough subjects to find the priority of E-Car functions in three specific situations. The result can be used in appropriate control panel design and the E-Car design of automobile manufacturing industry.

In control panel design experiment, the results of statistical analysis revealed that :

1. The control panel involved many functions, but did not influence the E-Car control panel performance.
2. Reducing numbers of buttons on control panel improves the driving performance. The 3 buttons control panel causes fewer accident hits than 9 buttons.

Virtual reality experiment brings advantages of less cost and more safety, and the Virtual reality software (3D webmaster) was easily to be used in experiment. However only 256 colors can be used to present the simulated driving enviroment in 3D Webmaster. It is much different from the colors in the real world.

5 Conclusion and Further Study

In this study, functional requirement and importance of E-Car functions were obtained by questionnaire survey. The result of control panel design experiment shows that 3 buttons design could maintain the driving performance and safety. It can be considered on E-Car control panel design in the future.

As the multi channel of human cognition, human can accomplish several goals in the same time. Beside the control panel, several different control types can be used to control E-Car system, such as vision, voice, touch screen...etc. For further research, one may compare the performance of different control modalities.

6 Acknowledgement

Partial of this study was supported by National Science Council under number NSC 90-2218-E007-031.

References

Koutsopoulos, H., A. Polydoropoulou and M. Ben-Akiva, "Travel Simulators for Data Collection in the Presence of Information," Transportation Research C, Vol. 3, No. 3, pp. 143-159.

Green, P., Hoekstra, E., & Williams, M. (1993, November). Further on-the-road tests of driver interfaces: examination of a route guidance system and a car phone (Report No. UMTRI-93-35). Ann Arbor, MI: University of Michigan Transportation Research Institute

Hayes, B.C., Kurokawa, K., & Wierwille, W.W. (1989). Age-related decrements in automobile instrument panel task performance. Proceedings of the Human Factors Society 33rd Annual Meeting, pp. 159-163.

Tijerina, L, Kiger, S., Rockwell, T. H., & Tornow, C. E. (1995a). Final report - Workload assessment of in-cab text message system and cellular phone use by heavy vehicle drivers on the road. (Contract No. DTNH22-91-07003). DOT HS 808 467 (7A), Washington, DC: U.S. Department of Transportation, NHTSA.

Tijerina, L., Kiger, S. M., Rockwell, T. H., & Tornow, C. (1995b). Workload assessment of in-cab text message system and cellular phone use by heavy vehicle drivers on the road. Proceedings of the Human Factors and Ergonomics Society 39th Annual Meeting, pp. 1117 - 1121.

Wierwille, W. W. & Tijerina, L. (1995). An analysis of driving accident narratives as a means of determining problems caused by in-vehicle visual allocation and visual workload. Paper presented at the Fifth International Conference on Vision in Vehicles, Glasgow, Scotland, September, 1993. (Conference proceedings, North Holland-Elsevier Press, Amsterdam, 1996.)

Architecture for a full-dynamical Interaction in Pervasive Computing

Michèle Courant, Sergio Maffioletti, Béat Hirsbrunner

Department of Informatics
University of Fribourg
Chemin du Musée 3
CH 1700 Fribourg, Switzerland
michele.courant@unifr.ch

Stéphane Le Peutrec

Department of Computer Science
University of Applied Sciences
of the State of Vaud
Rte de Cheseaux 1
CH 1004 Yverdon, Switzerland
stephane.lepeutrec@smile.ch

Abstract

This paper presents an interaction model which takes place in a multi-layer approach of interaction for pervasive computing. Our focus is put on the *context-awareness* layer, where a full-dynamical interaction with applications is made possible. This layer is composed of entangled levels addressing different types of the time-space-knowledge coordination involved in user-application interaction. The core component implementing this architecture, an external and customizable application interface called *application image*, is finally described.

1 Introduction

In the forecoming years, Internet technology will be strongly affected by the generalized deployment of applications onto ubiquitous devices, and the generalization of resource sharing through network computing. Adaptive components, openness and interoperability of software entities, dynamic plugging and software composition are the new requirements brought on by this context onto HCI design. The present work is a contribution to these requirements. It aims at providing concepts, techniques, and architectures allowing to interface *a priori* unknown applications dynamically, during runtime, with minimal application disturbance and user cognitive overload, while promoting a user-centred interaction schema. Application domains for this model encompass unpredictable interaction situations, as those resulting from the high heterogeneity of devices and services prevailing in pervasive network computing, especially complex multi-user applications, like large-scale shared simulations and smart environments.

The here described interaction model takes place in a multi-layer approach of interaction for pervasive computing. The paper therefore starts with an overview of this approach, before focussing on the aspects of the model where the full-dynamicity of interaction is specifically tackled. Section 3 hence describes the architecture proposed for the the *context-awareness* layer, while the core component of the model implementation is presented in section 4. Section 5 is a brief positioning of the work. The conclusion summarizes the main features of the model, and ends up by discussing the relevance of the adopted approach regarding medium-term goals where it should be considered as one step towards universal interfaces for pervasive network applications.

2 A Multi-layer Approach of Interaction

The here described HCI model takes place in a multi-layer approach of interaction for pervasive computing. In this approach (cf figure 1), the first layer, the *physical layer*, manages the available physical devices (PDA, desktop computer, mobile robot, mobile phone, or network device). The second layer, the *ubiquitous access layer*, is responsible for service and resource management: in the descending sense, it handles problems like the adaptation of a basic software resource to the

available displays, and reciprocally offers wrapping structures to the upper layer (Maffioletti 2002). The third layer, called *context-awareness layer*, appears globally as a control layer, where intelligence and context-awareness of the user interaction may be addressed in terms of knowledge-space-time coordination with the application. The fourth layer is the *application layer*. (Note that in our model, the user is considered as the ultimate entity involved in the interaction process. As such it could correspond to the lowest –zero– level, that is in direct contact with the devices).
Most of works on pervasive computing up to now concern the low layers, which seem at the first glance to be the more specific to this highly technology-dependant field. Our focus here is however put on the third layer, where adequate architectures with rather new requirements are still needed. The role of the context-awareness layer is basically to build semantic correspondences between the user's and the application's universes, which, once implemented –by synchronisation and binding mechanisms– become the support of a full-dynamical interaction. Our claim is that, despite a change in complexity essentially due to a semantic closeness with the user in this layer, strong similarities with the lower layers exist, namely with the ubiquitous access layer, and consequently that homogeneous ways of handling problems may be found.

Application Layer
Context–awareness Layer *Events – Structures*
Ubiquitous Access Layer *Services – Ressources*
Physical Layer *I/O Devices – Network Devices*

Figure 1: **A multi-layer approach of interaction for pervasive computing**

3 Architecture for the Context-awareness Layer

Context-awareness is an emerging concept of pervasive computing, that has been up to now mainly used for denoting low-level context-sensitivity of applications in this field, typically time and location. Our position however, is that both these low-level parameters, and the high level parameters coming from the user gain to be merged in one context-awareness layer, as far as from a user-centred perspective, they affect –together and undistinguishly- the application dynamics perceived by the user. The role of this layer consists in integrating the ingredients of different levels influencing the user-application interface: as well "external" events, like time and location belonging to the user's environment, than "internal" events belonging to the application universe, but also the specific cognitive bridges that can be established by the user with the application.
For tackling this integration problem, which requires interwoven levels of interaction dynamics, we propose an interaction model including several components. Its prerequisites regarding applications are minimal: they only encompass an object-oriented implementation, the delivering of a *conceptual interface* namely containing an application ontology, and a registration about an interconnection server. The *interconnection server* constitues a framework of the model: it is a broker in charge of managing application announcement and discovery. The *instrument generator* is then in charge of building the *ad hoc* user-application interface, which is called *virtual instrument*. An *ontology matching unit*, here defined as an interactive interface design unit, assists the instrument generator in being responsible for connecting the two ontologies, the one coming from the application, and the other coming from the user. Globally, we have then three functional levels: (1) *interconnection,* where the interconnection server is used, (2) *instrumentation,* aiming at the user interface building and adaptation, and (3) *interaction*, where the final exchanges between user and application actually take place (cf figure 2). The interaction dedicated

components of the application are respectively the declarative interface (*conceptual interface*), and two kinds of operational interfaces: the *port server*, in charge of the meta-interaction with the interconnection server and the instrument generator, and the *dedicated ports*, modelling the information flux handling with the user-dedicated virtual instruments.

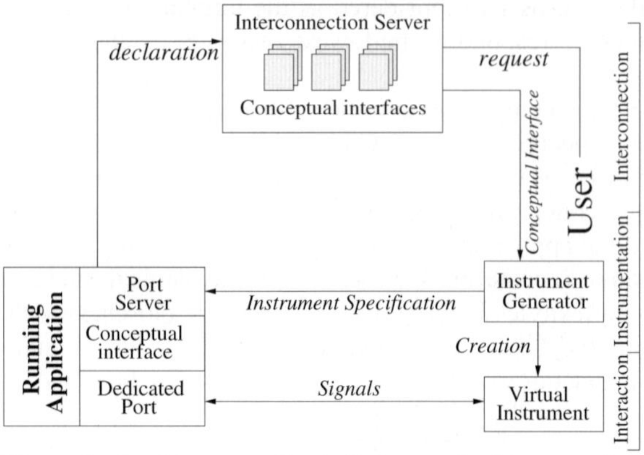

Figure 2: **Architecture for full-dynamical interaction**

4 Implementation Model: Application Image

4.1 Objectives

As far as the interaction model is oriented towards pervasive networks, that is distributed environments, its implementation must entail a minimal degradation of the application behaviour, namely regarding the scheduling of its internal events. Second, it must support concurrent interaction and dynamic interface plugging by users who want to interact simultaneously. Finally, it must work efficiently under the hypothesis of a cognitive scoping of the application based on the user's preferences and perceptual capabilities.

For similar reasons than those invoked in (Ponnekanti, 2001) and (Courant, 2002), we then proposes a user-centred external interface with a weak anchoring, to which the application delegates all the interaction tasks related to a given user. Such an interface uses an avatar of the application, which is permanently reflecting its current state, and is tailored to the users' needs. Such avatars are called *images* of the application. Images are the core components of virtual instrument implementation, the "lenses" reflecting the application behaviour, which concentrate the light and re-send it through the lower layers to the user's eyes. But conversely also, they may act on the application universe in case of interactive application. That the reason why we will now focus on this component (further readings on the other components and the virtual instrument metaphor can namely be found in (Le Peutrec, 2000) and (Courant, 2002)).

4.2 Structure and behaviour

The application image first extracts information from the application, and transforms it according to the user's needs, and conversely propagates user's initiatives to the application. Conceptually, it contains two types of information: (1) information originally existing in the application, and simply translated in terms of concepts belonging to the user's ontology, and (2) synthesized information, that does not correspond exactly to application notions: such information is secondary information derived from application concepts. It is hence clear that different images of a given application coupled to different users might have nothing in common, either between them, nor with the application proper terminology.

Images have been developed in Java on the virtual instrument model. As shown in figure 3, an image object is composed of five parts. The first part (*Kernel*) contains attributes and methods, which have their correspondants in the application. The second part (*Private State*) consists in additional attributes and methods augmenting the kernel, and corresponding to synthetized information. The third part (*Kernel Interface*) is a set of methods dedicated to the kernel – and eventually to the private state– updating. Theses methods are invoked when updated data are coming from the application. The fourth part (*Output Interface*) groups methods invoked by the user to access the state of the image object. The fifth part (*Input Interface*) groups methods invoked by the user for modifying the state.

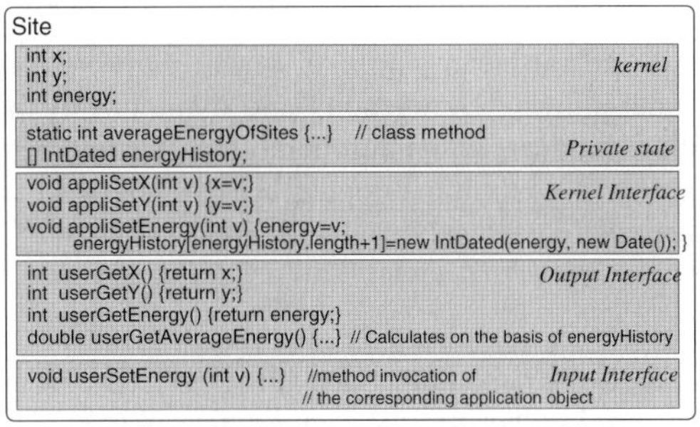

Figure 3: **Structure of an Image Class**

5 Evaluation and related works

The presented model, especially the core concept of image grounding its implementation– shares several objectives and methods with many other works within the user interface research community. It namely joins the CSCW domain, by the need to provide shared working universes, multi-user application interfaces, remote action possibilities, and multimodal interaction (Bentley 1994), together with the need to face synchronization and consistency problems resulting from the management of views. Our work mainly differs in supporting distinct user and application ontologies through images: our focus is put onto a correspondence between the user and application semantic universes, whereas the focus of CSCW relies mainly in communication, coordination, and cooperation between users.

The image concept has also strong similarities with certain works realized within the domain of monitoring of applications (Eisenhauer 1998, Muralidhar 2000, Ribler 1998). However, these works generally address the situation of application debugging, and consequently concentrate themselves on a direct and real-time access to the application objects. Our model then differs by the conceptual flexibility offered, and a higher interfacing level. Yet, the way theses systems take into account the distributed nature of applications let them be an interesting source of knowledge transfer for our further developments.

The multi-layer approach of interaction proposed as a framework for our model has also convergence and complementarity with other works, within the ubiquitous computing and web technology development fields. The UbiDev middleware (Maffioletti 2002), and the COCA model, a work handling ontologies with automatic classification (Schubiger 2002) are respectively constituting the ubiquitous acces layer and the physical layers, which are complementary of our work in this approach (Pai 2003).

Most of the infrastructures for pervasive computing, which aim at a certain level of abstraction, however put the focus onto the physical devices (Schmidt 1999), while the application side has remained quite neglected up to now. However the convergence of ubiquitous networks with web technology in a wider sense largely confirm the basic principles of our model, especially concerning heterogeneity management, resource description and late binding between software components (Adjie-Winoto 1999, agentcities 2002, W3C 2002).

6 Conclusion

With many specific needs, like application sharing, heterogeneous and dynamical environments, the domain of ubiquitous networks brings on new constraints on HCI. We have then described an interaction model for a multi-layer approach of interaction for pervasive computing. In this model, the third – *context-awareness* –layer is in charge of a user-centred handling of knowledge-time-space dependancies involved in user-application interaction. The core component of this layer, called *application image*, is an external application interface, entailing minimal perturbation on the application behaviour. Such an interface is customizable to user's needs, and can be automatically derived and updated using a declarative component, called *conceptual interface*, provided by the application. The whole model allows heterogeneity and dynamics of the underlying devices and resources, and assures full-dynamics, and interoperability management at the user-application frontier. Regarding medium-term goals, the adopted approach should then be considered as one step towards universal interfaces for pervasive network applications.

7 References

Adjie-Winoto W. Schwarz E., Balakrishnan H., Liley J. (1999): The design and implementation of an intentional naming system. In *Operating System Review* 34 (5) : 186-201.

AgentCities (2002): http://www.agentcities.org/

Bentley, R., Rodden, T, Sawyer, P. Sommerville, I.: Architectural Support for Cooperative Multi-user Interfaces, IEEE Computer 5(27), 1994, 37-46.

Courant, M., Le Peutrec, S. (2002): Towards Autonomous Application Interfaces. *In Kybernetes: The International Journal of Systems & Cybernetics,* Vol. 31 n°9/10, 1306-1312.

Eisenhauer, G., Schwan, K. (1998): An Object-Based Infrastructure for Program Monitoring and Steering. *In Proc. of the 2nd SIGMETRICS Symposium on Parallel and Distr. Tools*, 10-20.

Le Peutrec, S., Courant, M. (2000): Revisiting HCI for Networking Computers: A First Breakthrough for Artificial Biology. *In Proc. of the International ICSC Symposium on Biologically Inspired Systems* (BIS'2000). Wollongong, Australia.

Le Peutrec, S., Courant, M. (2002): Dynamic Interfaces for Autonomous Applications. *In Proc. of the 6th World Multi-Conf. on Systemics, Cybernetics and Informatics*. Orlando, U.S.A.

Maffioletti, S.& al., (2002): Ubidev: An Homogeneous Environment for Ubiquitous Interactive Devices. *In Short paper Proc. of the 1st Int. Conf. Pervasive 2002*, Zürich (Switzerland), 28-38.

Muralidhar, R., Kaur, S., Parashar, M. (2000): An Architecture for Web-Based Interaction and Steering of Adaptive Parallel/Distributed Applications. In Proc. of the European Conference on Parallel Processing, 1332-1339.

Pai Group (2003): Welcome Project Leaflet 2003. http://diuf.unifr.ch/pai/research.

Ponnekanti, S., Lee, B., Fox, A., Hanrahan, P., Winograd, T. (2001): ICrafter: A Service Framework for Ubiquitous Computing Environments. *In Proc. of Ubicomp 2001*, 56-75.

Ribler, R., Vetter, J., Simitci, H., Reed, D. (1998): Autopilot: Adaptive Control of Distributed Applications.. *In Proc. of the Seventh IEEE International Symposium on HPDC,* 172-179.

Schmidt A., Aidoo K.A., Takaluoma A., Tuomela U., van Laerhoven K., van de Velde W. Advanced Interaction in Context. *1th Int. Symposium on Handheld and Ubiquitous Computing*, Karlsruhe (Germany), 1999 & Lecture notes in computer science; Vol 1707, Springer: 89-101.

Schubiger, S. (2002): A Model for Software Configuration in Ubiquitous Computing Environments. 181-194. *In F. Mattern, M. Naghshineh (Eds): Pervasive Computing,* Springer.

W3C (2002): http://www.w3.org/Consortium/Points/

Effective Warning of a Drowsy Driver - the AWAKE Experience

Manfred Dangelmaier

Fraunhofer IAO
Nobelstr. 12
70569 Stuttgart
manfred.dangelmaier@iao.fhg.de

Dieter Spath

Fraunhofer IAO
Nobelstr. 12
70569 Stuttgart
dieter.spath@iao.fhg.de

Angelos Bekiaris

Hellenic Institute of Transport
6th km. Charilaou - Thermi Road,
57001 Thermi, Thessaloniki, Greece
abek@certh.gr

Claus Marberger

University of Stuttgart IAT
Nobelstr. 12
70569 Stuttgart
claus.marberger@iao.fhg.de

Abstract

The European AWAKE project aims at an "effective assessment of driver vigilance and warning according to traffic risk estimation". This paper deals with the Driver Warning System of AWAKE. It describes the development of the warning concept and the resulting warning devices. In particular, the issues of human-machine interaction for drowsy drivers are addressed.

1 The drowsy driving problem

Drowsy driving is an important factor in road accidents in Europe and worldwide. Approximately one fourth to one third of road accidents are related to driver drowsiness. The AWAKE project (IST-2000-28062) is part of the Information Society Technologies Programme of the Fifth Framework of European research activities and aims at the real-time diagnosis of driver drowsiness and warning measures to avoid those kinds of accidents in the future.

AWAKE develops a system comprising:

- a hypovigilance detection module using non-obtrusive driver- and traffic-oriented sensor technology,
- a traffic risk estimation module based on behavioural and traffic sensors as well as map data and
- a multi-modal driver warning system (DWS), which provides the driver with an appropriate feedback of her or his vigilance state, adapted to the current traffic risk level.

This paper focuses on the latter subsystem. The DWS warns the driver in order to provide an appropriate feedback to avoid micro-sleep and related traffic risks. Other potential options are the automatic interventions such as emergency calls and automatic driving capabilities. The conceptualisation of the human-machine interface (HMI) of such a system is described by Bekiaris and Dangelmaier (2000). AWAKE however focuses on drowsiness and related driver warnings, leaving the responsibility for appropriate actions to the driver.The system interaction of the AWAKE system with a drowsy driver is a special case of human-computer interaction with embedded systems which differs considerably from, for example, an office application. It can be characterised by following issues:

- Interaction is a complementary task to the primary driving task.
- Interaction takes place mainly at hypovigilant driver states.
- Driver drowsiness detection and warning systems require only infrequent interaction.
- Interaction might occur in critical traffic situations at high risk levels, caused by drowsiness.

These boundary conditions elucidate that designing the interaction for drowsy drivers is a challenging task and requires more care than desktop IT solutions.

2 Design approach

The development process of the AWAKE driver warning system follows a user-centred approach with three steps:

1. Conceptualisation
2. Virtual prototyping and testing (VP&T)
3. Physical prototype implementation and testing (P&T)

However, the design process is not strictly linear. Several iterations are needed, which include alternate design and evaluation phases. The methodological and technological approach for the iterative virtual prototyping and testing of the HMI of in-vehicle systems is e. g. shown by Dangelmaier et al. (2002).

Figure 1 shows an overview of the conceptualisation phase and its outcomes. General guidelines for in-vehicle systems (e. g. European Commission, 2000) and specific guidelines for the design of warning devices (e. g. Lerner et al., 1996) are used together with the pre-defined system architecture to derive a basic warning concept. This establishes both the "philosophy" and "style" of the warning device and HMI, respectively.

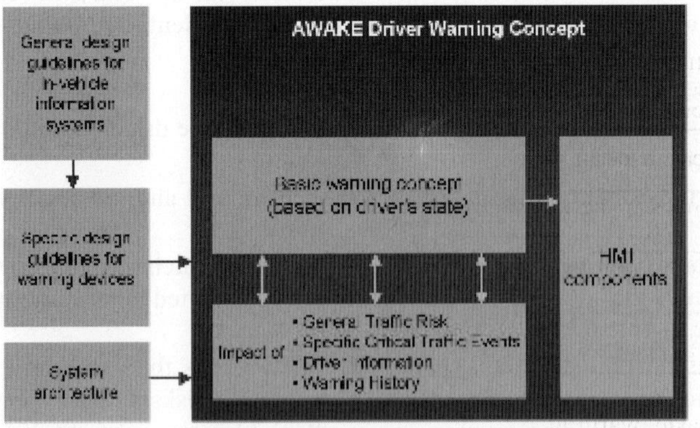

Figure 1: Iterative development process with virtual prototyping and testing

3 Basic warning concept

The basic warning concept follows four elementary requirements formulated by Wickens et al. (1998): the warning must be noticed; the warning must be read / heard; the warning must be understood; the warning must be accepted.

Applying these criteria and subsequent requirements to the special conditions of human-machine interaction leads to a warning strategy of three steps.

1. Attracting the driver's attention;
2. Informing the driver
   a. by a feedback of the hypovigilance diagnosis and
   b. on preferable behavioural options;
3. Alertness raising and keeping measures.

Drawing the attention of the drowsy driver requires suitable stimuli of sufficient intensity. Audible and tactile warnings are well suited for this purpose, because they will be perceived if they are designed accordingly. Visual warnings however are directional, require open eyes and correct head position and bear a certain risk to be missed.

Driver warnings can be expected to occur infrequently. This means that the driver experiences untrained interaction. Thus abbreviated information such as icons are not suitable. Speech messages seem to be more appropriate to inform the driver on his/her hypovigilance and behavioural options. In contrast to static visual displays acoustic messages are not persistent. A "repeat" function can help to recall missed information. Acceptance of speech warnings is influenced by factors such as frequency of alarms and frequency of false alarms. In case of reliable hypovigilance warning systems, there is already some experimental evidence that speech warnings are well accepted. (Bekiaris et al., 2000).

The third phase of the basic warning concept suggests short-term alertness raising and keeping measures to maintain the driver vigilance in post-warning stages. Table 1 shows that besides expected benefits vigilance maintenance by auditory and haptic output and vigilance keeping tasks in a post-warning situation will be considered for AWAKE.

Table 1: Arguments for and against alertness raising and keeping measures for drowsy drivers

Positive arguments	Negative arguments
The user will expect to exploit the potentials.	The user might misuse this functionality for prolonged drowsy driving (behavioural adaptation).
The user often cannot stop the car immediately to take a rest.	The effect of alertness raising measures might be overestimated.
There is an ethical obligation to support the driver if possible.	Additional research efforts on effectiveness are required.
The variable additional costs for simple alertness keeping measures are low.	Higher efforts for effective and optimized systems can be expected.

The AWAKE basic warning concept also takes advantage of the concept of cautionary and imminent warnings (e. g. Lerner et al., 1996). Imminent warnings, per definition., require immediate actions whereas cautionary warnings do not. Cautionary warnings are communicated to the driver in case the AWAKE sensor system detects abnormal driving behaviour, probably due to hypovigilance. Imminent warnings are given as soon as there is physiological evidence of driver drowsiness, e. g. by blink rates and blink duration. Even if in the latter case there is no current imminent collision risk, there exists a permanent probability of micro-sleep which then immediately could lead to critical traffic situations and possible collisions.

4 Warning devices

Based on the underlying considerations of the basic warning concept, the warning devices can be specified:

- Acoustic module for alerting sounds and driver information using the car speakers;
- Vibration device, for example at the belt lock for haptic warnings;
- Visual warning and state indication device;
- Buttons mounted at the steering wheel for repeat and confirm actions (alertness raising tasks);
- Card reader for personalisation of the AWAKE system.

The locations of the visible AWAKE HMI components for passenger cars are presented in Figure 2. In order to follow the ergonomic principle of functional grouping most of the interface components are integrated at the rear view mirror.

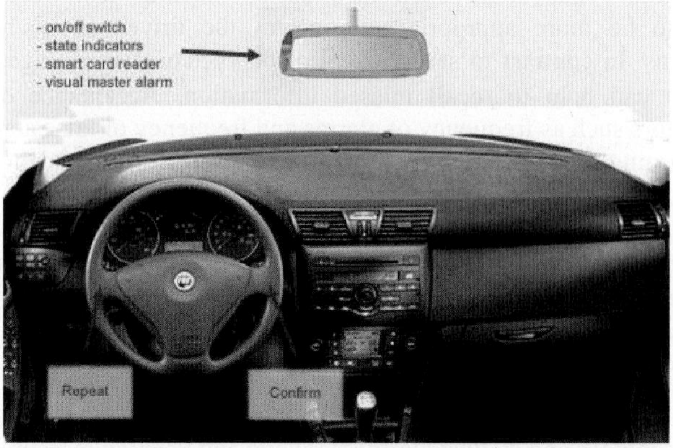

Figure 2: HMI layout (not visible: vibration device mounted at the belt lock)

Figure 3 shows the layout of the mirror device with red, yellow and green state indicators at the left upper edge, the on-off-button below them and the slot for the chip card behind them. The red visual flashlight alert is located behind the mirror glass at the left hand side and covers a larger area of the mirror to improve perception. A disadvantage is the remote reach zone, which led to the decision to put the repeat and confirm buttons, which are operated while driving, at the steering wheel.

Personalisation is an important issue for an HMI. ISO 9241-10 formulates "suitability for individualisation" as one of the principles to design a dialogue. A software is suitable for individualisation if the user is able to adapt the settings according to her/his individual needs. This principle can be applied to the AWAKE system as well. To deal with the issues "untrained user" and "distraction from the primary driving task", a pre-programmed chip card is used for personalisation together with a self-adaptation of the driver warning system, taking into account:

- differences in the native language of drivers
- impairments related to the output modalities of the system such as visual and hearing impairments
- disorders and habits which result in a higher frequency of interaction due to occurrence of warning situations

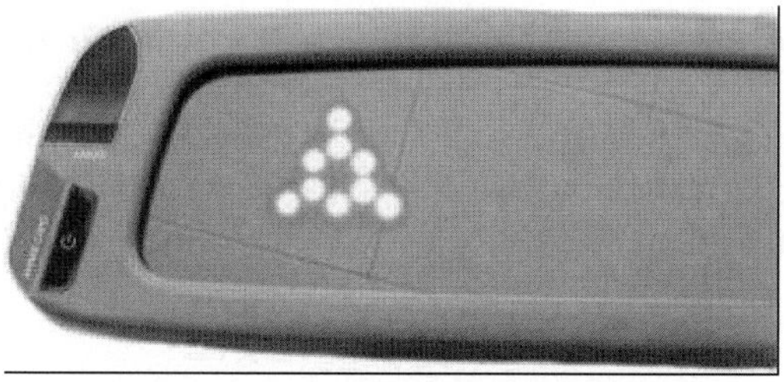

Figure 3: Mirror device for passenger cars

5 Outlook

In summary it can be said that a homogeneous and adaptable concept for a drowsy driver warning system has been developed. By personalisation it is able to compensate for the needs of specific driver cohorts such as elderly and handicapped drivers. It is also adaptable for the requirements of sleep disorder patients, who might use the AWAKE system as a technical aid for their special needs. The subsequent steps comprise the construction of a mixed virtual and physical prototype of the AWAKE driver warning system and usability testing in a driving simulator. Test results will be used within the iterative design process to further optimise the warning strategy, the layout and the warning device design.

6 References

Bekiaris, E., & Dangelmaier, M. (2000). Conceptualisation of the Human-Machine Interface of an Integrated Driver Monitoring and Emergency Handling System. In: *ITS Journal* 5, 279-291.

Dangelmaier, M. Widlroither, H., Wenzel, G., & Haberhauer, M. (2002). Evaluation of Interactive In-Car Systems in Immersive Environments. In: *DSC 2002, Driving Simulator Conference, Paris, France, September 11,12 & 13, 2002*, pp 227-237. Paris : INRETS-Renault.

EN ISO 9241-10: Ergonomic Requirements for Office Work with Visual Display Terminals (VDTs), Part 10: Dialogue Principles, 1996.

European Commission (2000). Commission Recommendation of 21 December 1999 on Safe and Efficient In-vehicle Information and Communication systems: A European Statement of Principles on Human Machine Interface. In: *Official Journal of the European Communities* L19, pp 65-68.

Lerner, N.D., Kotwal, B.M., Lyons, R.D., & Gardner-Bonneau, D.J. (1996). Preliminary Human Factors Guidelines for Crash Avoidance Warning Devices. Springfield: National Technical Information Service.

Wickens, C. D., Gordon, S. E., & Liu, Y. (1998). *An introduction to human factors engineering.* New York: Addison Wesley Longman.

Pen-based Ubiquitous Computing System for Visually Impaired Person

Nobuo Ezaki

Toba National College of Maritime Tech.
Mie 517-8501, Japan
ezaki@toba-cmt.ac.jp

Kimiyasu Kiyota

Kumamoto National College of Tech.
Kumamoto 861-1102, Japan
kkiyota@tc.knct.ac.jp

Hotaka Takizawa

Toyohashi University of Tech.
Aichi 441-8580, Japan
takizawa@parl.tutkie.tut.ac.jp

Shinji Yamamoto

Toyohashi University of Tech.
Aichi 441-8580, Japan
yamamoto@parl.tutkie.tut.ac.jp

Abstract

We propose the pen-based ubiquitous computing system for visually impaired person. In general the software keyboard is used for the character input at PDA. However, a visually impaired person cannot use the software keyboard directly because s/he cannot select the character key-button on the screen monitor. As a solution to these problems, we adopt an on-line character input system using handwritten character recognition technology instead of a keyboard for visually impaired person. Our character input system is installed on pen-based PDA system with two buttons that are able to control for all command operations. In order to investigate the performance of the proposed system, we have addressed the issue of using the PDA to use in sending, receiving and managing electronic mail by wireless LAN. However, since many characters having a small number of strokes (such as Hiragana and Katakana) were contained in the mail documents, the character recognition accuracy was significantly decreased in our recognition system. So we also developed an improvement method of the recognition accuracy by combining the voting method with our original recognition method and two conventional methods. In addition, a phrase unit order recognition method based on a tri-gram probable language model was added to the system. By applying these additional improvement methods, the recognition accuracy was improved to 97.7% from 66.2%, and we confirmed the required performance for utilization.

1 Introduction

In recent years, visually impaired persons are increasing by eye diseases and traffic accidents. There are about 200,000 persons with acquired blindness in Japan. Information society in our country, it is very important to use a personal computer. Therefore computer application support for the visually impaired person has become an important theme. In general, the user inputs a character using a keyboard. However, since the visually impaired user cannot see the key position, so keyboard operation is very cumbersome. As the computer support for a blind person, Braille typewriter has been developed in commercial use. However, only 10, 000 acquired visually impaired persons know Braille points in Japan. Therefore, keyboard input operation is not effective method for the visually impaired users in Japan.

As one solution of these problems, we have proposed an on-line Japanese character input system for a visually impaired person (Ezaki et al., 2000). This recognition method recognizes the handwriting character written without visual information, and inputs it into a computer. Our previous system is a desktop type model, so it is composed of a personal computer and a hand controller. However, the opportunity to use a computer in various situations is increasing in recent years. Therefore we examine applying this on-line Japanese character input system for the visually disabled person to PDA. First, we arrange substituting the function of the hand controller in desktop computer for the jog-dial button on PDA. The system can recognize 3,216 characters including JIS First Level characters. However a recognition accuracy of a character having a small number of strokes is decreased for fewer features. Then we also investigated the improvement method of the recognition accuracy for the character having fewer features. Furthermore we consider using voting method with our previous method and two other conventional recognition methods in which the different feature was used. In addition, character recognition of a phrase unit order was also extended based on a tri-gram probable language model as an error correction algorithm.

2 Computing system on PDA

2.1 System configuration

Figure 1 shows the system configuration of a pen-based Japanese character input system on PDA for visually impaired person. This system is composed of the PDA and voice synthesizer software. The user can acquire various screen information by the voice synthesizer. The system also includes character recognition and an error correction function.

The user writes some characters to the touch display using a stylus pen. Then the computing system begins to recognize each character and to generate candidate phrases using an error correction method. The first candidate phrase is output by the voice synthesizer. Then the user selects a correct phrase using a jog-dial as shown in Fig.3.

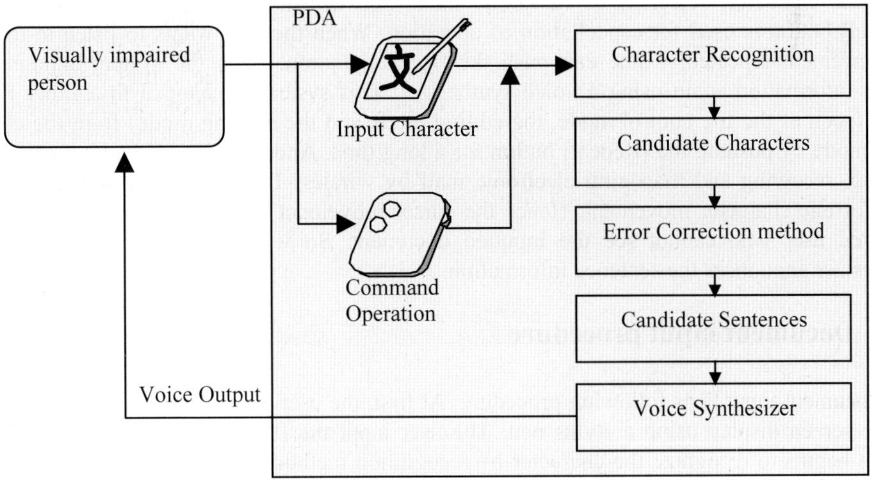

Figure 1: System configuration of PDA

2.2 Button arrangement and Function

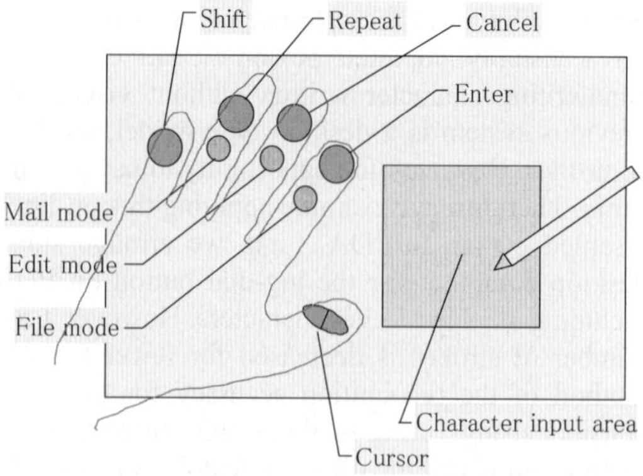

Figure 2: Control board for a desktop computer

We have developed the control board of desktop-type computer for command operation and character input (Fig.2). It consists of an electronic tablet for character input and 9 buttons for command operation. The control board is used for all computer command operations without using a keyboard.

We have implemented to PDA these command functions. The PDA has a tablet function on the screen display. Therefore the user can write a character to the display directly by his finger or a stylus pen. Furthermore there is a step in a character input area to easy description for visually impaired user.

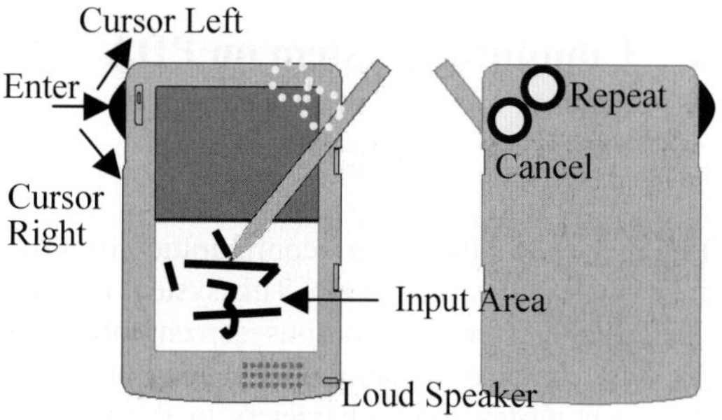

Figure 3: Button arrangement on PDA

In general, right-handed user holds the PDA by left hand and writes character by right hand. Therefore we designed existing jog-dial and two new buttons for system operation. The jog-dial is used for [Enter] and cursor movement. And we have installed two buttons in the backside of the PDA as the [Repeat] function and the [Cancel] function. The [Enter] button is used for operation determination, such as decision of a candidate character, selection of sub-mode menu and decision of a one-character input. The [Cancel] button is used for cancellation of operation. When the user wants to listen to the voice announcement repeatedly, s/he can push the [Repeat] button. Then the system announces the screen information again using a voice synthesizer. This system is changed to another function mode (such as the file control mode, the editing mode and the mailing mode) from the character input mode by pushing the [Repeat] button for a long time. And we have also addressed to use in sending, receiving and managing electronic mail by wireless LAN. Here the [cursor] button is used for one character movement. Hence the cursor movement is a serious problem for visually impaired user who cannot see the inputted document. So when the cursor is moved, voice synthesizer announces the sentence information.

2.3 Document input procedure

The document input is as following procedure. At first, the user writes one character to input area on the screen display using a stylus pen. The user input the [Enter] key by jog-dial. Then, the system begins to recognize the character by recognition method. Next the system changes to the character write mode automatically. The user can make a document by repeating this procedure. After one phrase is inputted, the [Enter] button is pushed again. Then the error correction function starts automatically, and the voice synthesizer announces the first candidate phrase. If a correct answer is announced, the user pushes the [Enter] button. If a wrong answer is announced, the user selects other candidate phrases by the [cursor] button. When there is no correct phrase in the candidates, the user selects one of the candidate phrases. After that, the user re-inputs a miss - recognition character one by one.

3 Improvement of character recognition method

When our system is applied to the application of E-mail, since many characters having a small number of strokes (such as Hiragana and Katakana) were contained in the mail documents, the character recognition accuracy was significantly decreased in our recognition system. It was also difficult to distinguish similar characters for one-character order recognition algorithm.
Then we investigate the new recognition method for improving recognition accuracy. These are the voting method for one-character recognition algorithm and the error correction method based on tri-gram probabilities as a language model.

3.1 A voting method

We propose a voting method for high accuracy character recognition by using three unique on-line character recognition algorithms. These systems are our developed recognition algorithm for a visually impaired person (Toyohashi Method) and other two recognition algorithms for the commercial use system (Recognition method A and B). In voting method, a score is given according to the ranking of ten candidate characters outputted from each algorithm. For example, the system gives 10pt for 1st candidate character's similarity, 9pt for 2nd candidate and 1pt for 10th candidate. The total score of each candidate is calculated as voting similarity (Fig.4) (Nishimura et al., 1999).

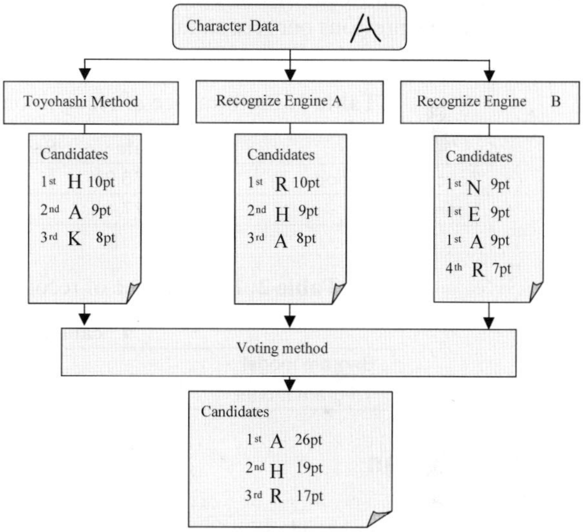

Figure 4: Voting method

3.2 Error correction method by using tri-gram

We also use the tri-gram model for error correction method. The input pattern sequence $X = X_1 X_2 \cdots X_n$ to be the candidate character sequence $C = C_1 C_2 \cdots C_n$ is expressed by the formula (1).

$$\log P(C \mid X) = \log \sum_{i=1}^{n} P(X_i \mid C_i) + \omega \log P(C_1^n) \quad (1)$$

$P(X_i \mid C_i)$ is pattern similarity between X_i and C_i. ω is weight. $P(C_1^n)$ is approximated by formula (2).

$$P(C_1^n) = \prod_{i=1}^{n} P(C_i \mid C_{i-1} C_{i-2}) \quad (2)$$

The tri-grams $P(C_1^n)$ have been calculated from about 50,000,000 characters of text in a newspaper. Largest $\log P(C \mid X)$ is realized by the Viterbi algorithm (Nakagawa et al., 1996).

3.3 Performance of recognition accuracy

We tested the performance of the proposed improvement recognition algorithm for the non-learning 85 mail document samples including about 16,000 characters that were written by blindfold person. The recognition accuracy was 66.2% in the 1st candidate by using the Toyohashi method (the recognition method for one-character order). By using voting method, the recognition accuracy improves to 77.1% (Table 1).
The recognition accuracy of the 1st candidate phrase was 97.1% by tri-gram model (Table 2). Within the best three candidate phrases, the recognition accuracy was improved to 98.5%. Therefore we confirmed that the improvement method corrected miss-recognized Japanese characters in the previous our recognition algorithm.

Table 1: Performance of recognition (One-character order)

	1st candidate	**Best 3 candidates**	**Best 10 candidates**
Toyohashi Method	66.2%	94.0%	98.2%
Voting Method	77.1%	94.8%	99.5%

Table 2: Performance of recognition (Phrase order)

	1st candidate	**Best 3 candidates**
Bi-gram model	96.8%	98.3%
Tri-gram model	97.1%	98.5%

4 Conclusion

We proposed the pen-based ubiquitous computing system for the visually impaired person in the PDA. The advantage of the system is it works to the information devices as mobile computer. By this study for user interface and the improvement of recognition processing in PDA, the new information system was offered to the visually impaired person. Our future work is the experimental evaluation by using this system in a school for the blind and the social welfare organization.

References

Ezaki N., Hikichi T., Kiyota K. & Yamamoto S. (2000), A pen-based Japanese Character Input system for the Blind Person, Proc. of 15th ICPR, Vol.4, 372-375.

Nishimura H., Kobayashi M., Maruyama M. & Nakano Y. (1999), Character Recognition Using HMM by Multiple Directional Feature Extraction and Voting with Bagging Algorithm, IEICE, Vol.J82-D-II, No.9, 1429-1434.(in Japanese)

Nakagawa M., Akiyama K., Le Van Tu, Homma A. & Higashiyama T. (1996), Robust and Highly Customizable Recognition of On-line Handwritten Japanese Characters, Proc. of the 13th ICPR, 269-273.

Design Principles For A Collaborative Hypervideo User Interface Concept In Mobile Environments

Matthias Finke, Matthias Grimm, Mohammad-Reza Tazari

Computer Graphics Center
Fraunhoferstr. 5
D-64283 Darmstadt
{Matthias.Finke, Matthias.Grimm, Saied.Tazari}@zgdv.de

Abstract

Hypervideo content can be understood as a video-based hypermedia information structure providing a user the option to exert influence on the way the video-related content is presented. It offers a user the opportunity to take an active part in the presentation. It has been proven that the combination of interactivity, video content and multimedia documents forming a hypermedia structure is a powerful way to present information. Collaborative knowledge construction within communities on the Internet is a hot topic today, providing support for exchanging and retrieving multimedia content based on group conversation. With this paper, we announce a first approach to a hypervideo collaborative learning interface concept for mobile environments. This approach enables users to extend the interactive video content with multimedia documents during a presentation in order to exchange and share experiences and views with other members within a mobile community.

1 Introduction

The rapid development of new technologies and the huge distribution of information content force the society into a so-called lifelong learning process. As a result, the validity of information today is decreasing and people have to reconsider their standard of knowledge continuously.
Knowledge construction is a process performed often by a single person, for instance, by browsing the WWW. A different approach to creating knowledge is exercised within communities. A community provides the basis for an information exchange and information retrieval between users. These two abstract functions fulfill the precondition for any community conversation and thus the transfer and access to information resources. These information resources within a community can be modeled by a *dynamic information space*. The information space is dynamic, since it is modified and extended by user contributions from the community. The figure below shows an abstract model of a community and the associated dynamic information space.

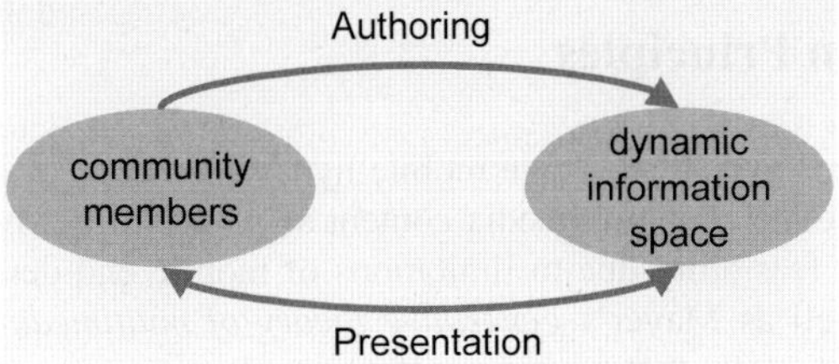

Figure 1: Model of a community

Communities can have a number of different objective targets, for instance, to share experiences or to work together on complex problems, which might be very difficult to solve by a single individual. Newsgroups, where community members do have read and write access to information resources, are one concise example within this context. Knowledge that is constructed through community discussions is denoted here as collaborative learning. With the development of new technologies, communities today have the advantage of being mobile. Mobile networks and PDAs, capable of managing rich multimedia content, like images, video, graphics, etc., are available for public use. This offers communities with the goal of knowledge construction a high degree of freedom and supports a more flexible mode of learning.
Video content has been proven to be very efficient in collaborative learning environments. Hypervideo also offers the advantages of video content and extends them towards an interactive distributed presentation, allowing the user to exert influence (Locatis, Charuhas, and Banvard, 1990; Sawhney 1996). Hypervideo is based on the Dexter Hypertext Reference Model by Halasz and Schwartz (1994), including the notation of *links* and *nodes*. With this model, objects in a video sequence, which represents *video nodes*, can be combined with information elements via a *video link*. Here, the information elements are defined as additional information units supporting the actual video content by providing, for instance, further descriptions of objects in a video sequence. The advantage of combining objects with additional information instead of the whole video frame is a higher degree of resolution. In this way, more than one object can be annotated with additional information in the same video frame and thus the same sequence. The figure below shows the relation between a video node and an associated document.

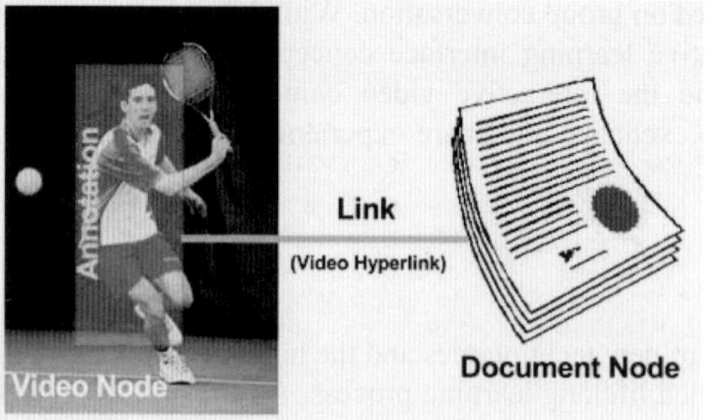

Figure 2: Hypervideo structure

Hypervideo used by distributed mobile communities defines a dynamic information space that consists of the digital video itself and the additional information created by user contributions. The implementation of a hypervideo system within a mobile environment used by distributed communities for the purpose of collaborative knowledge construction requires new concepts regarding the user interface for such an application. With this paper, a user interface concept is presented with the focus on minimizing the cognitive load (Sweller, 1999) of a user.

2 General Design Principles

Our user interface concept for collaborative mobile hypervideo content is dependent on both the instructional design guidelines for multimedia content as proposed by Sweller (1999) and Mayer (2001) and the technical feasibility due to limitations of mobile end-devices. The *cognitive load theory* by Sweller, as well as Mayer's *generative theory of multimedia learning,* are based on experimental results and state that a more effective learning process can be achieved by minimizing the cognitive load of a learner's working memory resources. The working memory modelled by Baddeley (1986) consists of an independent auditory and visual working memory, which is limited regarding its capacity. It is comprehensible that the available cognitive resources should be applied to the learning process, avoiding irrelevant information content in order to save the working memory from overloading. Based on this consideration, a major requirement of the mobile user interface concept presented here is to reduce the cognitive load of a user.

The presentation of complex illustrations or complex coherences is not always preventable and it is comprehensible that the cognitive load will increase. As a result, *user disorientation* can occur and the learning efficiency is decreased. For complex interactive presentations like collaborative hypervideo, Schnots and Zink (1997) propose to provide the user with an overview of the overall media structure and the learning target. The benefits of providing an overview of hypervideo structures are carefully worded by Guimarães, Chambel, and Bidarra (2000). They state that with the disclosure of the hypervideo structure, a user can obtain an overview by navigating within the information space, preventing the problem of disorientation. In addition, the navigation within an information space provides the user with a summary about the different sub-topics, which form the whole presentation. This is especially helpful for non-linear narrative hypervideo content where the users determine by their interactivity the presentation flow and tempo. Since the reduced display size of mobile end-devices does not allow the presentation of large complex illustrations or complex coherences at the same time, such content has to be separated. Therefore, the visualization of the hypervideo structure in a mobile environment has to be established to prevent user disorientation resulting in a cognitive overload.

Cognitive load can also be reduced if the so-called *split attention effect* is prevented. The split-attention effect as defined by Sweller (1999) or the *contiguity principle* as defined by Mayer (2001) state that information elements that refer to each other should be arranged as close to each other as possible so that the users do not have to split their attention between the different information resources. This relates to both a spatial and temporal arrangement. In addition, this indicates that an information element that is irrelevant to the current learning process should not be presented in order to keep the focus of a user on the learning material. An implication of this consideration for a mobile user interface concept is that it is preferable to apply a full-screen modus for the video content, the additional information elements and the visualization of the hypervideo structure. To arrange related elements as close as possible, hyperlink techniques are applied in order to combine them spatially and temporally.

Within a collaborative learning community, cognitive overload can arise due to the lack of social presence between participants. One major problem in this context is the generation of common background knowledge as described by Hesse, Garsoffky and Hron (1997). The common background knowledge of a community is an important aspect participants have to be aware of before formulating their contribution. Insufficient background knowledge increases the cognitive load when accessing inappropriate not appropriate contributions. Palme (1992) proposes to provide a *message history* and a *contribution archive* that enables the user to obtain information about former contributions and their authors. If a community accesses the same content and modifies it like the dynamic information, inconsistencies have to be prevented; otherwise, contradictory activities might occur. Ellis, Gibbs and Rein (1991) suggest a *concurrency control* so that, for instance, only one participant at a time has access to an information unit. Even here, a contribution archive might be applicable. Ahern (1993) states that in collaborative environments, information elements that are related to each other should be linked spatially to each other. A fast access to related information units enables a user to gain a better understanding of the topic and thus reduces cognitive load. This consideration supports the already defined requirement of hypervideo structure visualization, as discussed above. The overall design principles for *groupware-interfaces* described by Hewitt and Gilbert (1993) have to be considered, as well. These design principles address issues concerning *self-explanatory capability*, *software ergonomics*, *error robustness* and *controlling*.

3 Realization

Based on the *general design principles* in chapter 2, our design of the user interface concept starts with the identification of the main processes for the collaborative hypervideo presentation. Participants in a community can extend the dynamic information space by their contribution (*authoring component*) or extract data out of it (*presentation component*). We can conclude five main processes, which determine the user interface framework. These processes are defined as *video presentation, additional information presentation, hypervideo structure visualization, video node generation* and *video link definition* (see figure below). The authoring component is required to enable the user to *generate video nodes* and *define video links* (for a detailed discussion, refer to Finke, Balfanz, Jung and Wichert, 2001). The presentation component extracts data from the dynamic information space based on user interactions. The data can be the digital video content, the additional information units or information about the hypervideo structure itself.

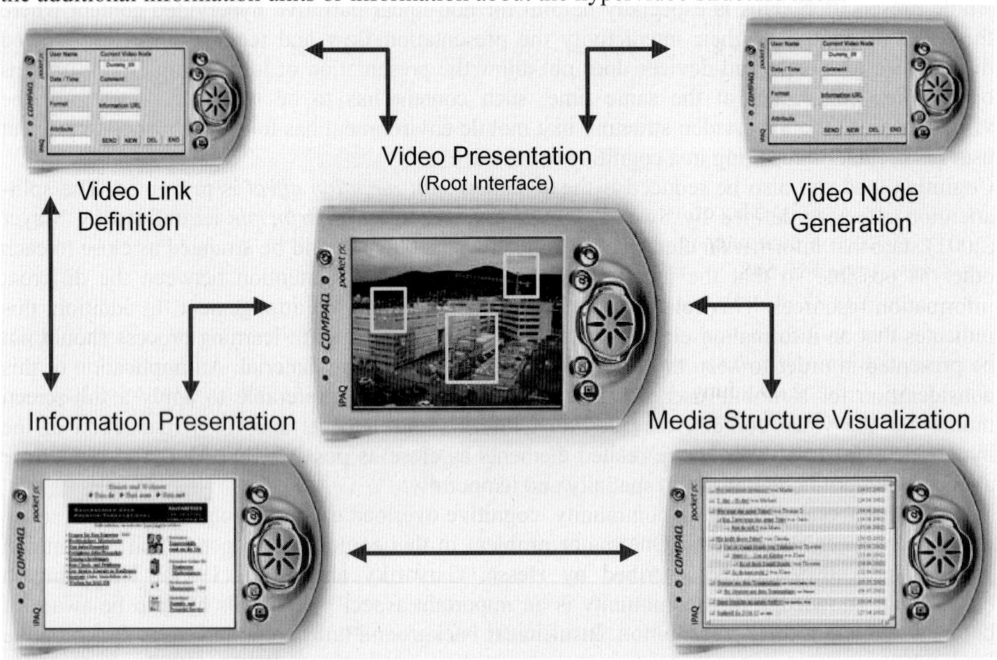

Figure 3: User interface concept

Even if these processes are associated with each other, they remain autonomous processes. Based on the display limitation and the discussion in chapter 2, the identified five main processes will be treated as stand-alone full-screen user interfaces. This implies that only one process will be accessible by the user at a time. The advantage of this approach is a reduction of the cognitive load due to the described requirements. Since the concept will end up with five sub-user interfaces, a switchover from one interface to another has to be comprehensible for the user, according to the requirement of the split-attention effect (Sweller, 1999 and Mayer, 2001).

The video presentation sub-interface can trigger all other sub-user interfaces and can be understood as the root interface of the application. Every interface enables the user to return back to the root interface. The hypervideo structure visualization interface can trigger the video presentation or the additional information presentation per/by user request only. The additional information interface enables the user to return back to the video presentation or to enter the video link definition interface. From the two interfaces for video node generation and video link definition, a user can only return back to the video presentation. This arrangement of the sub-user interfaces provides users with a recognizable presentation structure.

4 Conclusion and future work

With this paper, we introduce an approach regarding design principles and concepts of a user interface for collaborative mobile hypervideo content for the purpose of distributed knowledge construction. Design principles have been investigated with the focus on the reduction of the learner's cognitive load. It has been shown that a cognitive overload will decrease learning efficiency. The concept of the user interface presented here focuses on the reduction of the cognitive load due to the limited capacity of the human's working memory. Major results are the video and information content separation on small end-devices, the use of full-screen modus and the importance of the previously described navigation principles due to user orientation requirements. Future work will focus on the refinement of the presented concept and a usability field test.

5 References

Ahern, T.C. (1993). The effect of a graphic interface on participation, interaction and student achievement in a computer-mediated small-group discussion. *Journal of Educational Computing Research*, 9(4), 535-548.

Baddeley, A.D. (1986). Working memory. Oxford, England: Oxford University Press.

Ellis, C.A., Gibbs, S.J. & Rein, G.L. (1991). Groupware: Some issues and experiences. *Communication of the ACM*, 34 (1), 38-58.

Finke, M., Balfanz, D., Jung, C., Wichert, R. (2001). An interactive Video System supporting E-Commerce Product Placement, Multimedia, Internet, Video Technologies (*MIV 2001*), Malta.

Guimarães, N., Chambel, T. & Bidarra, J. (2000). From cognitive maps to hypervideo: Supporting flexible and rich learner-centred environments. *Interactive Multimedia Electronic Journal of Computer-Enhanced Learning, 2* (2).

Hesse, F.W., Garsoffky, B. & Hron, A. (1997). Interface-Design für computerunterstütztes cooperatives Lernen. Information und Lernen mit Multimedia. 2. Aufl. Beltz-Verlag. 253-266.

Hewitt, B. & Gilbert, G.N. (1993). Groupware interfaces. In D. Diaper & C. Sanger (Eds.), Mindweave: Communication, computers and distance education (pp. 50-62). Oxford: Pergamon Press.

Locatis, C., Charuhas, J. & Banvard, R. (1990). Hypervideo. Educational Technology Research and Development, 38 (2), 41-49.

Mayer, R.E. (2001). Multimedia Learning. Cambridge: Cambridge University Press.

Palme, J. (1992). Computer conferencing functions and standards. In A.R. Kaye (Ed.), Collaborative Learning through computer conferencing. The Najaden papers (pp.225-245). Berlin: Springer Verlag.

Sawhney, N. (1996). HyperCafe: Narrative and Aesthetic Properties of Hypervideo, *Proceedings of the Seventh ACM Conference on Hypertext*. New York: Association for Computing Machinery.

Schnotz, W. & Zink, T. (1997). Informationssuche und Kohärenzbildung beim Wissenserwerb mit Hypertext. *Zeitschrift für Pädagogische Psychologie/German Journal of Educational Psychology*, 11(2), 95-108.

Sweller, J. (1999). Instructional design in technical areas. Camberwell, Australia: ACER Press.

Developing a Ubiquitous Reception-Hall Using the User-Centred Design Usability Engineering Process Mode

Granollers, T.; Lorés, J.; Solà, J.; Rubió, X.

GRIHO (http://www.griho.net). University of Lleida (Spain) (http://www.udl.es)
{tonig, jesus}@griho.net

Abstract

In this paper we study the validity of our Model Process, which enables the design of interactive applications centred on the user by integrating Usability Engineering with Software Engineering, applying it to a ubiquitous paradigm project with an uninitiated user. We explain a project carried out with the aim of converting the reception-hall of a company into a ubiquitous environment where the possible customers and visitors are surprised by projections of audio-visual material offered by the receptionist. We wish to emphasize that the main user of the application has gone, without the need for specific formation later on, from being a user with no knowledge at all of graphical interface systems to using a system that offers multimodal interaction.

1 The Problem

The management team of a medium-sized company, in their desire for leadership, constant customer attention and quality innovation, decided to redesign the visitor's reception-hall in a singular way. They singled out a main objective: to transform the above-mentioned reception-hall into a singular environment where the person who enters is surprised by means of multimedia items related to him or her visualized on a large, high-definition wall screen. In this way the hall is personalized towards the visitor's characteristics.

2 Interactive System Description

The surprise mentioned in the problem section, that which is attempted to offer the visitor, will consist of multimedia resources formed by photos, videos, presentations –photo and video compositions– and personalized texts according to the above-mentioned person's profile (professional, interests, hobbies, etc). This personalisation can even be guided by the current conversation that the visitor is holding with a third party.

The information will be digitalized and stored in the system and this will have to allow the receptionist to search for and present some visual information of interest for the customer, with the objective to surprise him/her during the short time that they remain in the hall.

Therefore, designing a system capable of recognizing the peculiarities of all the persons who may be located in the hall is not possible, so the receptionist's role has a vital relevancy; she must be able to decide which contents will be most suitable for each occasion. The selected person to develop this role –as a main user of the application– is a woman who has never used an interactive system based on graphical environments. In spite of this she is very suitable for this project because of her knowledge acquired during the whole working life in the same work place, which has provided her with important information about the peculiarities, uses and hobbies of most of the clients. Thus, this user's mental model (Norman, 1983) will condition the application.

3 General requirements

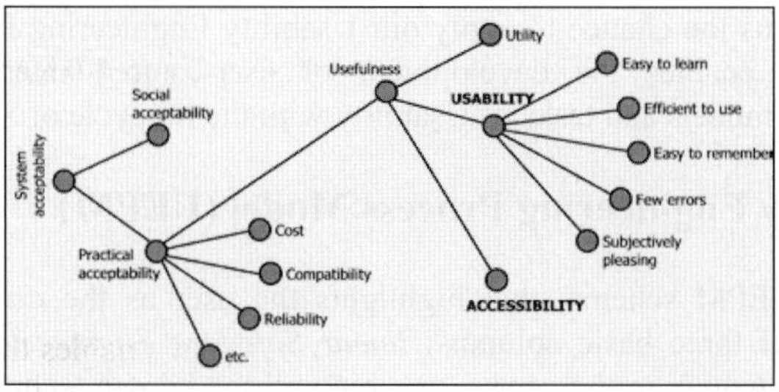

Figure 1: A model of the attributes of system acceptability (Nielsen, 1994).

Figure 1 manifests that "*the overall acceptability of a computer system is a combination of its social acceptability and its practical acceptability*" (Nielsen, 1994), and during the implementation of the project here described, all the attributes that Nielsen refers to have been conveniently considered, although due to the project's peculiarities we highlight:

- *Easy to learn and remember*, as we have indicated, the receptionist had never used graphical interfaces.
- Within the *efficiency to use* we focused on the fast response required as a result of the short time that the visitor will remain in the hall waiting for someone to attend him/her. During this time the system's user must be able to show something of interest on the wall screen.
- Due to the "public" scope of the application results, *social acceptance* becomes a critical factor, so without this the rest of the work doesn't have any significance.

On the other hand due to the work environment and the main user characteristics the application must be provided with a *multimodal interaction style*, enabling the use of the most convenient style for every interaction, according to the action to be carried out. This multimodality is offered by means of:

- the *voice*: with a microphone and related software the user can fulfil all searching options with only his/her voice (Figure 2 – left).
- *touch*: a flat tactile screen facilitates all the user's interactions. It is important to bear in mind that the user has never used a mouse device before (Figure 2 – right).
- and a *wide variety of media input* resources such as scanner, CD-ROM, DVD, digital camera, etc.

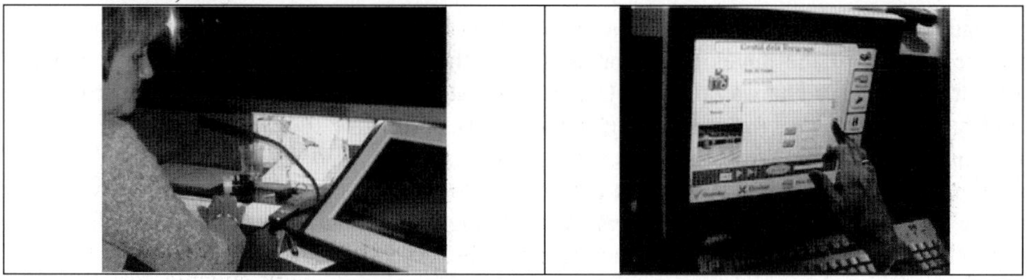

Figure 2: The user interacting by voice and by touch.

During the analysis requirements it was observed that in the future some functionality would be exported to other company's rooms, therefore we decided to use distributed technology with a Web interface style.

4 Developing the Interactive System

This project has given us the chance to apply our Usability Engineering Process Model (UEPM) (Lorés, 2002), which enables the development of user-centred interactive system design, integrating the specific models and tasks of Usability with the lifecycle of Software Engineering.

4.1 The Usability Engineering Process Model (UEPM)

Figure 3 shows the UEPM scheme, that highlights the user as the centre of the process. In summary, it consists of three basic columns: *linear Software Engineering lifecycle* (Pressman, 2001) on the left, *prototyping*, in the centre, grouping all the known techniques about prototypes, and *evaluation*, on the right, uniting all the validation methods.
The diagram distinguishes, furthermore, two kinds of arrows, the "narrow" ones correspond to the Software Engineering (SE) model and the "wide" ones enable the first tenet of Usability Engineering "to test early and often" (Redish et. al., 2002), providing a constant iteration at every development state implying the user as much as needed.

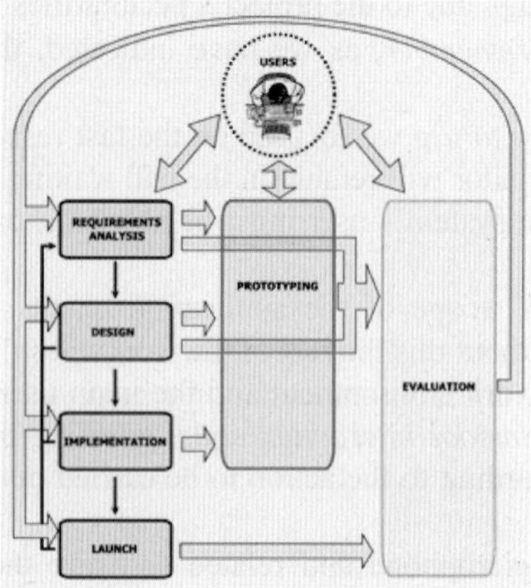

Figure 3: Usability Engineering Process Model (UEPM).

4.2 UEPM methodology applied to the project

As already mentioned, UEPM integrates SE with UE. In that way we use SE protection activities that give support to the development process with the main purpose of obtaining a product of demonstrable quality. The way that our UEPM integrates with SE is by focusing the latter towards Usability using *Configuration Management* (CM), *Software Quality Assurance* and *Risk Management* activities.
Concerning CM, as a mechanism that allows "one to manage the change" throughout all the software's service life, we use a *Configuration Management Work-Sheet* (CMWS) reflecting in a chronological order all the activities done indicating the corresponding UEPM phase.
In figure 4 we can observe the great effort carried out in prototype and evaluation phases as well as their regular distribution along the process. This leads intuitively to the idea that a valid user centered design is being used from the very beginning.

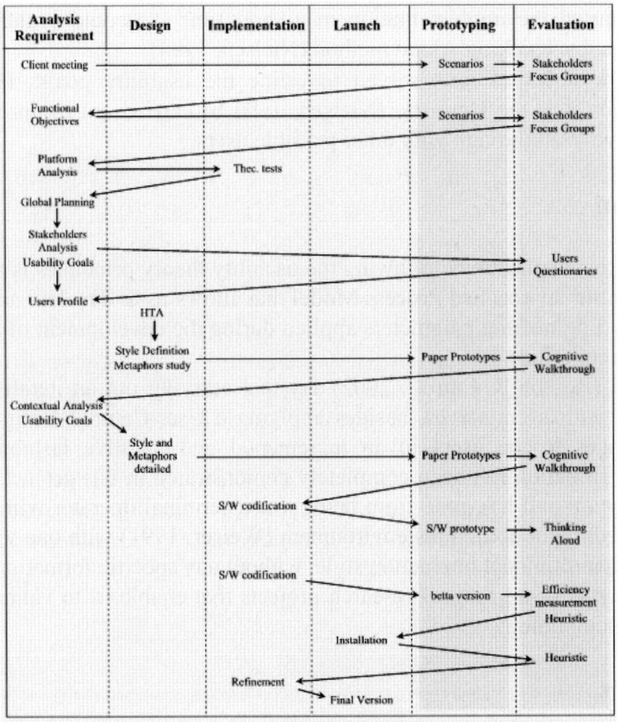

Figure 4. Configuration Management Work-Sheet corresponding to the current project.

Let us explain in more detail how the project was developed:

- After a first client-developer meeting, that pointed out the singularity of the project with the need of a remarkable main-user centred design and important stakeholders (Bevan, 2001; Poulouidi, 1999; Pouloudi, 1997) influencing the functional goals, the first prototypes were implemented. Those were carried out by means of Textual Story Boards (Lorés, 2002) reflecting possible scenarios in the reception-hall using the new system. In their evaluation the main user and stakeholders were involved.
- This first evaluation mainly served to detail the functional objectives. A second scenario and evaluation was carried out.
- From the results of evaluations the development team started with a first platform analysis and some technological tests enabling the first planning.
- Next step was a detailed and consistent system requirements. New stakeholders were identified –technological partners and other department heads joined the design team– with their particular objectives. Also the user's profile was determined by means of an inquiry. And, during this phase main usability goals were determined.
- Once in design phase we proceeded with a Hierarchical Task Analysis (Annet & Duncan, 1983), and with the user interface style definition (several workshops between the main-user and the design team were carried out).
- After that a first Paper Prototype emerged which was evaluated by the Cognitive Walkthrough method (Wharton et al., 1994).
- At this point a new iteration started. Requirement Analysis, Design, Paper Prototype and Evaluation where carried out again overcoming the main functional aspects and those referred to style and interface details.

- Once started the implementation in a short time we could evaluate a first software prototype by Thinking Aloud method (Nielsen, 1994).
- Finally a series of tests served to refine the usability goals. The most important evaluation was the Heuristic (Nielsen, 1994 b) carried out by three experts and new Cognitive Walkthrough with other possible users.

5 Conclusion

To develop interactive applications following the usability theory principles is not a trivial task. In the paper we apply our approached Process Model that allows us to develop an interactive system under the Usability Engineering parameters applied during the development of an application with some significant peculiarities.
Rosson and Carrol (Rosson & Carroll, 2002) say that carrying out an iterative prototyping and evaluation process with users serves, besides applying a User-Centred Design, to introduce the technological changes to the users in an incremental and iterative fashion, facilitating their knowledge. This affirmation has been completely demonstrated in this project, thus the person in charge of the reception-hall has gone from being a text-terminal operator with a textual interface to completely controlling a ubiquitous environment (Weiser, 1991) with one web-based graphical interface and with a multimodal interaction style, without any specific formation.
This project forms part of a group of research projects that enable us to validate and modify the above mentioned methodology

6 References

Bevan N. (2001). Research Manager at Serco Usability Services. Information from http://www.usability.serco.com/trump/methods/recommended/stakeholder.htm

Annett J. & Duncan K. (1967). Task analysis and training in design. In Occupational Psychology, Num. 41.

Lorés J. et al (2002). Curso de Introducción a la Interaccion Persona-Ordenador. CD-ROM 84-607-2255-4

Nielsen J. (1994). Usability Engineering. Academic Press. ISBN 0-12-518406-9

Nielsen, J. (1994 b). Heuristic evaluation. In Usability Inspection Methods (Nielsen J. y Mack R. L. eds.). John Wiley & Sons, New York.

Norman D. A. (1983). Some observations on mental models. D. Gentner & A. Stevens, eds., Lawrence Erlbaum, Hillsdale, NJ

Pouloudi, A. & Whitley, E.A.(1997). Stakeholder identification in inter-organisational systems: gaining insights for drug use management systems. European Journal of Information Systems, 6 p.1-14

Poulouidi, A. (1999). Stakeholder Analysis as a Front-End to Knowledge Elicitation. At & Society, 11 p.122-137

Pressman, R. (2001). Software Engineering: A Practitioner's Approach. (5th Ed.). Mc Graw-Hill

Redish J., Bias R.G., Bailey R., Molich R., Dumas J. & Spool J.M. (2002). Usability in Practice: Formative Usability Evaluations – Evolution and revolution. Proceedings of CHI2002

Rosson M.B. & Carroll J.M. (2002), Usability Engineering: scenario-based development of HCI. Morgan Kaufmann

Weiser M. (1991). The Computer of the twenty-first century. Scientific American, p. 94-104.

Wharton C. et al.(1994). The cognitive walkthrough method: a practitioner's guide. In Usability Inspection Methods (Nielsen J. y Mack R. L. eds.). John Wiley & Sons, New York.

User Interface Techniques for Mobile Agents

Matthias Grimm *Mohammad-Reza Tazari* *Matthias Finke*

Computer Graphics Center (ZGDV) e.V.
Fraunhoferstr. 5, 64283 Darmstadt, Germany
{Matthias.Grimm, Saied.Tazari, Matthias.Finke}@zgdv.de

Abstract

In mobile environments, the usage of mobile agents is very common. After being configured by the user, mobile agents can perform tasks autonomously and present their results when they have finished their task. In most cases, communication between user and agent is only necessary at the time of configuration and at the time of presentation, when the agent has accomplished its task. Our research activities deal with agent-based applications for mobile users with mobile devices of low processing power and without Java capabilities, which yields the question of how user interfaces can be implemented. In this paper, different approaches for the communication between users and agents are summarized and classified with respect to our applications. An agent platform gateway is introduced as our approach to user interface implementation and some services of the gateway are described in detail.

1 Introduction

Mobile agents are a very commonly used paradigm in mobile environments. They are software modules that are able to move through the network autonomously in order to accomplish a task they were given by their owner. In mobile environments, this ability can be used in order to reduce network traffic in online negotiations and thus reduce network costs, provided that the mobile device supports both the execution and the transport of agents by means of Java code. In this case, agents can migrate to the mobile device, be configured by the user for a specific task, and can migrate back to the network. There, they can carry out their task by negotiating with other agents on various platforms on distributed hosts, and finally, they can either migrate back to the device and present the results or can send a message to the user containing the results.

Mobile systems are characterized by manifold mobile devices with capabilities that vary over a wide range. Some typical restrictions are poor bandwidth, small displays, different operating systems and different Java capabilities. Designing and implementing software for these kinds of devices implies many considerations concerning the distribution of the software (client-side vs. server-side), programming language, and support of different platforms and capabilities.

The rest of the paper is organized as follows: In section 2, we discuss a couple of existing approaches to user-agent communication with their advantages and disadvantages. In section 3, we introduce the mobile agent framework we developed and, in section 4, our approach to HTML-based agent interfaces is described in detail.

In this paper, the concept 'agent' means a unit of Java software that is able to travel across the network between servers and which runs on an agent platform. By 'mobile devices', we mean PDAs. Laptops are mobile by means of being easily transportable, but they are not very handy to use, as they have to be booted, require considerable power and are quite heavy. We restrict our consideration of devices to those that are only of limited Java capability, i.e. we do not have a full-blown Java VM on our devices.

2 Existing Approaches

Before discussing different approaches of user-agent communication, we want to present three different acts of communication as introduced in (Mihailescu, Gamage & Kendall, 2001).
The *initial interaction* enables the user to create an agent and set its instructions and start parameters. The *in-progress interaction* enables the agent to communicate with its creator during the processing of a given task. The agent might, for example, ask for some additional information, as the task might not be accomplished with the initial parameters. Finally, with the *completion interaction*, the agent can present the results of the completed task to the user.
We want to start our classification of approaches with respect to the location of the agent the user communicates with. Two cases are obvious: the agent might migrate to the user's device or the device is not capable of accepting and executing agent code. In the latter case, the agent can migrate across different servers in order to accomplish a given task for the user, but a different protocol has to be used for the communication between the user and the agent. We will discuss these two main cases in detail in the following subsection.

2.1 Methods of GUI Representation

2.1.1 Agent located on mobile device

When the device is capable of executing Java code, the standard methods for displaying a Java GUI – based on Swing or AWT – can be used. One problem of this approach is the code size of the agent, as it includes the code for displaying the user interface and the appropriate event handling. Since our focus targets mobile devices connected to the Internet via a radio network, code size is expensive. The other problem is that the devices and platforms we focus on are not capable of executing the agent platform and thus do not support an efficient secure migration of agents. And if the device was able to execute the Java code, the agent had to move to the device anytime communication with the user is necessary, which is again expensive due to the radio network. Therefore, this approach is not suitable for our system.
When agents migrate onto the user's device for communicating, the payload of the information the agent takes along with it must be large or the dialogue between the agent and the user must be complex. If both are not the case, the network costs are not worthwhile.

2.1.2 Agent located on server

When an agent can't migrate to the user's end-device, a different communication protocol has to be chosen for the communication between the user and the agent, and different methods are needed for the agent to display a user interface remotely on the mobile device.

Email One common approach is the presentation of task results as an email. This approach has numerous immediate benefits (La Corte, Puliafito & Tomarchio, 1999). The user can disconnect from the platform and gets a message when new information is present. Almost all devices support email and mail management software is cheap and widespread.
But, there are also disadvantages. The user is not able to interact with the agent when the agent has moved from its home-platform. One example of end-user interaction based on Active Mail (mail including executable code) is presented by (Schirmer, J. & Hayn, R., 1998).

HTML One major advantage of HTML as the user interface description and HTTP as the communication protocol is the platform independence. A web browser exists for almost every device, and as HTML is pure text, it can be compressed very efficiently. Therefore, agents can take along the compressed HTML code without growing too big in code size.
One problem is that mobile agents are difficult to address due to their mobility. When an agent carries a Java servlet with it, its URL changes with every migration. Therefore, something like a home environment is needed, which has a fixed URL. Since our agents communicate using the agent communication language FIPA-ACL, a conversion between HTTP requests, FIPA messages and HTML pages is needed. Another disadvantage is the lack of a mechanism for information *push*. When the agent has accomplished its task, it might want to push some task results to the user's mobile device. Since HTML only supports *pull* communication, an alternative solution is needed here.
Applets Very often UIs of Java mobile agents are realized as Java applets. (Da Silva, da Silva & Romao) present the combination of an agent, an applet and an HTML page as the *conventional model* of UI generation. They propose an integrated model offered by AgentSpace, where specific applets for each agent class are not needed. The idea is to use a rather generic applet that supports the management and presentation of the interfaces provided specifically by each agent. The authors differentiate between an *initialization interface* and a *management interface*, both provided as callbacks by the agent. When a user wants to interact with an agent, the generic applet is downloaded to the client's web browser. Then, it calls the management interface and the UI is displayed as a specific agent frame in the applet.

3 Our Mobile Agent Framework

Our framework is distributed across several logical units. As both users and agents can be mobile, there is the need for a personalized agent platform the users and the agents know about. We call it the user's 'homebase', and its main purpose is to facilitate homogenous access to the user's resources, quite similar to the *virtual home environment* (VHE) presented in (3GPP). The homebase consists primarily of the agent platform and some personalized services. As agent platform, we chose SeMoA (Secure Mobile Agents) developed by Fraunhofer IGD (Roth, V. & Peters, J.). SeMoA addresses several security aspects, such as protecting agents against malicious hosts and preventing agents from being changed by other users.
A context management service keeps track of the user's current location, as well as the location's characteristics – such as temperature, network access, and people present. As another part of the context, the device being used by the user and its characteristics – such as color and sound capabilities, display size, applications, as well as personal preferences of the user (static preferences, as well as conditional preferences) are provided by the context manager. Agents can use this context information in order to generate their dynamic user interfaces.
The agents running on the SeMoA platform can communicate using FIPA-ACL. As content languages, we use primarily XML and RDF. Agents have the ability to asynchronously send and receive messages or to send and receive messages synchronously, just like RPC. The agent tracking and locating service (ATLAS) of the platform facilitates the message routing between agents residing on different platforms on physically distributed systems.
As the platform is based on the Java programming language, the system on which agents should be executed has to supply support for the JDK 1.4., which is not the case for PDAs and Windows CE-based Pocket PCs. Since these devices are our target system platform, our mobile agents are not able to migrate to the end-devices in order to display their user interfaces or embed applets.

4 The Thin Client Gateway

Although the performance of mobile devices is evolving fast, there are technical restrictions, such as a missing Java platform that supports serialization, which is crucial for agent migration.
On the agent platform (homebase), there are a couple of agents that perform specific tasks. The agents communicate using FIPA-ACL. As the web-browser on the end-device does not understand

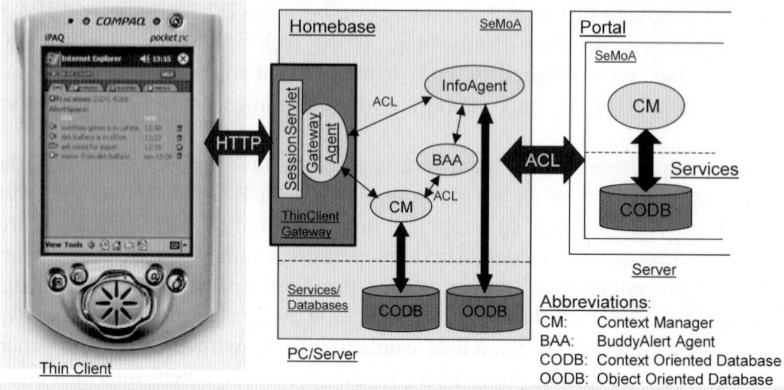

Figure 1 The homebase with the agent platform and the thin client gateway

ACL, there needs to be a translation service that translates between FIPA-ACL and HTTP. This service is the *thin client gateway*, serving as the entry point to the user's homebase (Figure 1).
The main components of the client gateway are a Java servlet (*SessionServlet*) as interface to the mobile device, and an agent (*Gateway Agent*) as interface to the agents located on the platform. Agents that provide a service for the user can register at the Gateway Agent, providing an HTTP request that invokes the service. This HTTP request contains an ACL message, which is interpreted by the gateway and sent to the specific agent when the user activates the link. As the receiver of the ACL message is a mobile agent, it might be located on a different platform than the homebase. The agent tracking service of the SeMoA platform can locate the agent and send the message to it. With this mechanism, we created a location-invariant communication protocol, which tunnels ACL messages through HTML and HTTP and vice versa. The homebase of the user does not change its location, thus, any agent can create HTML pages and send them to the user's device, enabling the user to send messages back to the agent, regardless of its location.

4.1 Services

Before an agent can use the thin client gateway for displaying the user interface, it has to register itself as a service. In order to register, an agent has to send an ACL message to the Gateway Agent where it provides a URL. The URL has to contain the web address of the SessionServlet and the ACL message that has to be sent in order to invoke the agent's service. When the user connects to the homebase, a list of all registered services is displayed, each entry as a link. Activating the link by clicking on it, the user makes the gateway send the encoded ACL message to the agent, waiting for a response. The agent can create an HTML page from a template, and return it as the content of a reply message. This page is returned to the browser by the servlet of the gateway.
The most important service for our applications is the user notification or *push* service. It enables a mobile agent to display a notification on a user's end-device. In most cases, this notification is the result of a transaction of the agent or it signals the change of the agent's state. The push service is

realized as a socket server on the end-device that listens on a specific port. When an agent needs to push a page to the end-device, the servlet sends its URL to that port, and the server causes the client's web browser to open the URL. If the device is not connected at the time the push event occurs, it is stored in the homebase and sent as soon as the device reconnects.

5 Conclusions

In this paper, we introduced some issues regarding user interface implementation for mobile agents. A couple of different approaches were discussed and evaluated for use with our system. Since our system deals with mobile devices of medium capabilities, the design and implementation of HTML interfaces turned out to be the most practicable approach.

We described our approach to an agent gateway service, which connects the agents' FIPA-ACL world with the widespread HTTP/HTML world, namely the *Thin Client Gateway*. With the gateway, the locations of the mobile agents do not change from the user's point of view, and the communication between user and agents is not restricted by the agents' capability to migrate across different platforms. The gateway could successfully be applied within the map project.

The static character of HTML is still a problem. A page cannot reload specific parts, but only a new page as a whole, which is a problem in mobile environments where communication is expensive. Furthermore, animations as result of user interaction are not possible. For these reasons, we are going to evaluate Flash as a UI language in the future within the *mummy* project.

References

3GPP Technical Specification 22.121 v4.1.0: "The Virtual Home Environment (Release 4)", March 2001.

La Corte, A., Pulifiafito, A., Tomarchio, O. (1999). An Agent-based framework for Mobile Users. 3rd European Research Seminar On Advances In Distributed Systems (ERSADS'99), Portugal, April 1999. Retrieved from http://citeseer.nj.nec.com/lacorte99agent.html

Da Silva, A.R., da Silva, M.M., Romao, A. (1999). User Interfaces with Java Mobile Agents: The AgentSpace Case Study. *First International Symposium on Agent Systems and Applications Third International Symposium on Mobile Agents*, Palm Springs, California, October 1999, retrieved January, 07, 2003, from http://berlin.inesc.pt/alb/papers/1999/asa99-asilva.pdf

FIPA (2002). FIPA ACL Message Structure Specification, Retrieved from http://www.fipa.org/specs/

Mihailescu, P., Gamage, C., Kendall, E. A. (2001). Mobile Agent to User Interaction (MAUI). *Proceedings of the 34th Hawaii International Conference On System Sciences (HICSS-34),* retrieved from http://www.pscit.monash.edu.au/~patrikm/maui.pdf

Lingnau, A., Drobnik, O. (1998). Agent-User Communications: Requests, Results, Interaction. In *Proceedings Second International Workshop, MA'98*, Stuttgart, Germany, September 1998, p.209ff

Roth, V., and Jalali, M. (2001). Concepts and architecture of a security-centric mobile agent server. In *Proc. Fifth International Symposium on Autonomous Decentralized Systems (ISADS 2001)*, Dallas, Texas, U.S.A., March 2001, 435-442.

Schirmer, J., Hayn, R. (1998): Multimedia authoring and synchronization within ActiveM3 - a system for composing active email messages as subtypes of mobile agents. *SYBEN'98, Interactive Multimedia Service and Equipment*, Zurich, Switzerland, 18-22 May '98, 456-465.

Ambient Interfaces: Design Challenges and Recommendations

Tom Gross
Fraunhofer Institute for Applied IT—FIT
Schloss Birlinghoven, 53754 St. Augustin, Germany
tom.gross@fit.fraunhofer.de

Abstract

Ambient interfaces go beyond the classical graphical user interface and use the whole environment of the user for the interaction between the user and the system. In this paper we describe the design and implementation of ambient interfaces to facilitate the easy and smooth cooperation and coordination among geographically dispersed team members. We report on results of user involvement in the design and implementation of these ambient interfaces and user feedback from evaluation. We discuss consequences for the design of ambient interfaces and derive design guidelines.

1 Introduction

Ambient interfaces use the whole environment of the user for the interaction between the user and the system. They present digital information through subtle changes in the user's physical environment such as variations of light, sounds, or movements. They capture natural interactions of the user with physical devices such as switches, buttons, or wheels and translate them into digital commands (Gross, 2002; Wisneski et al., 1998). Ambient interfaces go beyond the classical graphical user interface and do not consume real estate on the computer screen; they make user interaction with the system easier and more intuitive. Their properties of a calm technology (Weiser & Brown, 1996) are particular useful for situations, in which users want and need permanent background information without being disrupted in their foreground tasks.

Ambient interfaces can be used for any setting or system, where the subtle presentation of information is vital. This is particularly the case for awareness information environments. Awareness information environments aim to facilitate the easy and smooth cooperation and coordination among geographically dispersed team members by providing them with group awareness. Typically, this group awareness comprises information about co-workers such as their presence and availability as well as about shared artefacts such as the creation or alteration of documents (Dourish & Belotti, 1992). Awareness information environments face the trade-off that on the one hand users need up-to-the-moment and detailed group awareness information and on the other hand users want to perform their foreground task without frequent disruptions (Hudson & Smith, 1996). Ambient interfaces have the potential to considerably reduce this trade-off.

In this paper we describe the design and implementation of ambient interfaces that were used to augment an awareness information environment. We report on results of user involvement in the design and implementation of these ambient interfaces and user feedback from the evaluation. We discuss consequences for the design of ambient interfaces and derive design guidelines.

2 Ambient Interfaces

The ambient interfaces were designed and developed to augment the Theatre of Work Enabling Relationships (TOWER) open awareness information environment. TOWER aims to support mutual awareness and chance encounters among geographically dispersed users with sensors that capture information and various indicators that present the information on the users' computer desktop with pop-up windows, tickertapes, and a 3D multi-user environment. This 3D environment presents shared documents as buildings and users as animated avatars wandering

through the virtual cities. TOWER consists of an event and notification infrastructure that captures and processes awareness information, a space module that dynamically creates the 3D space according to the changes to shared documents, a symbolic acting module that creates and animates the avatars according to the users' actions, and a docudrama module that can replay past states of the 3D multi-user environment.
Several ambient interfaces were designed, developed, and integrated into TOWER; they can communicate directly with the event and notification infrastructure. In the following we present multimodal ambient interfaces and AwareBots.

2.1 Multimodal Ambient Interfaces

Multimodal interfaces address multiple human sensory modalities and multiple channels, of the same or different modalities (Buxton, 1994). Some examples of multimodal ambient interfaces that we developed are a fan, a lamp, and a fish tank. The fan addresses the haptic sense of the user—it blows air into the face of the user. The desktop lamp addresses the visual sense—it points to the ceiling of an office room and illuminates the ceiling, its intensity can be changed in a subtle way. The fish tank addresses the visual and auditory sense of the user—it can release bubbles in different intensities, these bubbles can not only be seen, but also heard, if their frequency goes beyond a certain threshold (cf. Figure 1a).

2.2 AwareBots

AwareBots are ambient interfaces presenting awareness information in the shape of robots. Several AwareBots were developed with the LEGO Mindstorms Robotics Invention System (The LEGO Group, 2002). LEGO offers several advantages for the participatory design of ambient interfaces: users can easily build and change their AwareBots, users can easily personalise existing AwareBots, and the AwareBots are aesthetically pleasing. For instance, the RoboDeNiro AwareBot (cf. Figure 1b) can lift its hat when another user logs in; it can rotate its body when new email has arrived; and the user can press its arm in order to log into the system.
Details about the multimodal ambient interfaces and the AwareBots as well as about other ambient interfaces developed in our group can be found in (Gross, 2002); some mobile interfaces that were also integrated into TOWER were described in (Gross, 2001).

3 Evaluation

The TOWER environment as well as the ambient interfaces were conceptualised and implemented in a multi-national project, which was partly funded by the European Union and started in January 2000 and ended in July 2002. Besides several research partners from academia and industry, two companies participated as application partners.

3.1 Application Partners

Aixonix is a small enterprise with a staff of 25 people working at two sites in Germany and the U.S. Aixonix develops, operates and sells Web-based systems for the transfer of technical as well as scientific knowledge, and provides consulting services in information management. At Aixonix the TOWER environment and the ambient interfaces

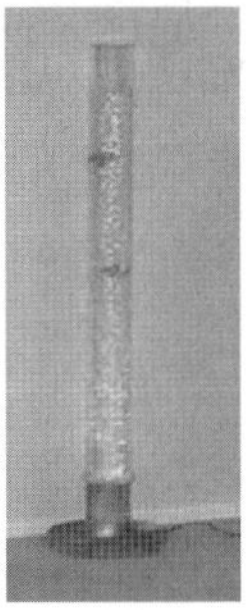
(a)

(b)

Figure 1. a) fish tank; b) RoboDeNiro AwareBot.

were used to inform the users at the different sites about each other.

WSAtkins is a global enterprise with thousands of employees in over 90 offices throughout the UK and several other offices in more than 25 countries. The company is among the world's leading providers of professional, technologically based consultancy, and support services. The motivation of WSAtkins to use the TOWER environment and its ambient interfaces was based on a worldwide reorganisation and a newly created concept of a 'one company - one team' culture in the year 1999. The TOWER environment and its ambient interfaces were used to inform employees from the different sites about each other and to stimulate group cohesion.

3.2 User Involvement

The participatory design and development as well as the evaluation of the TOWER environment and the ambient interfaces can be characterised by two iterative circles.

In the *outer circle* users participated in the requirements analysis and in the evaluation of the prototypes. In expert workshops, which started in January 2001 at Aixonix in Aachen, Germany, and at WSAtkins' main office in Epsom, U.K., the prospective users were interviewed about their work practice and gave feedback on early design ideas, mock-ups, and first ambient interface prototypes. Throughout the rest of the project the environment and the ambient interfaces were deployed incrementally and discussed in regular user workshops. Furthermore, we produced and analysed log files for the email correspondence (only with email headers such as sender, recipient, subject, date and time), for the activities in shared workspaces, and for logins and logouts to and from the TOWER environment.

In the *inner circle* the TOWER environment was used among the design team within the project. New features and new ambient interfaces were first introduced at the different sites of the design team. In fact, several members of the design team had their own LEGO Mindstorms Robotics Invention System packages and produced their own AwareBots and defined their own mappings between the input and output of the AwareBots and the represented information and actions. Various application scenarios were tried out to explore the usability of the features and to check whether they provide sufficient benefits for users to justify their introduction into work practice. Upon acceptance the features were demonstrated and discussed in the user workshops.

On a whole we primarily used qualitative measures of user involvement and evaluation, since they are often more adequate for empirical studies in cooperative settings; the evaluation of the log files was the only quantitative analysis.

4 Guidelines for the Design of Ambient Interfaces

The guidelines (cf. Table 1) originate from various sources: basically, the application partners' input from the participatory design activities and the empirical results of the evaluations were combined with findings from literature on usability goals and principles for GUI-based systems from Preece et al. (2002), on interaction design for multimodal consumer products from Bergman (2000), and on guidelines for the design of applications for mobile devices from Weiss (2002).

Ambient interfaces should be *effective*.	Quality in terms of how good the ambient interfaces does what they are supposed to do. For instance, one basic goal for ambient interfaces was to make use of users' peripheral awareness, not disturbing the user from performing the foreground task.
Ambient interfaces should be *efficient*.	Way an ambient interface supports users in carrying out their tasks. For instance, pressing the arm of RoboDeNiro for logging in to the TOWER environment relieves users from typing user names and passwords.
Ambient interfaces should be *safe*.	Protection of the user from dangerous and undesirable situations. For the PC a considerable body of knowledge on hardware ergonomics exists. For ambient interfaces, especially if they are multimodal, little knowledge on ergonomics exists.
Ambient interfaces should have good *utility*.	Right kind of functionality so that users can do what they want and need to do. Ambient interfaces should be used for easy input for simple actions or for subtle presentation of simple information. They are not well suited for complex information.
Ambient interfaces should be easy to *learn* and *remember*.	Ease for the users to learn a system and to remember how to interact with the system. For ambient interfaces this is particularly challenging, because users do not have experience yet; novel metaphors are used; and traditional help systems are not available.
The functionality of ambient interfaces should be *visible*.	Clear communication to the user at any time which choice she has and what the system is expecting from her. Visibility can be easily achieved through the physical affordances of the ambient interfaces.
Ambient interfaces should give the users adequate *feedback*.	Information to the users to tell her that her input was received and analysed properly, and that the corresponding actions have been or will be performed. The feedback for the login function of the RoboDeNiro is a negative example: as pressing the arm can mean a login or a logout, the users often were not sure if they logged in or out.
Ambient interfaces should provide *constraints*.	System awareness of the user's current situation and possible next steps and appropriate actions of a user. For the ambient interfaces we built constraints play a minor role, because the functionality of input and output in quite simple and easy to handle and does not require complex interaction or multi-step interaction.
Ambient interfaces should provide an adequate *mapping*.	Mapping between controls and their effects should be adequate. This is a particular challenge for a multimodal, distributed multi-client system. In TOWER, the effect of the input often cannot be seen on the ambient interface *per se*, but rather on other indicators
Ambient interfaces should provide *consistent* functionality.	Similar operations and similar control elements should be used for achieving similar tasks. In a distributed participatory design and development process, consistency can only be achieved by frequent exchange among the different sites.
Ambient interfaces should be *adequate* for the target domain.	Adequacy for the target domain such as the environment, in which the ambient interfaces are installed; the users, who will use the system; and the tasks that will be performed on the ambient interfaces. For homogeneous target domains such as in TOWER, adequacy can be achieved easily.
Design and development should be *participatory*.	Stimulation of users to contribute to the design of the ambient interfaces at very early stages. The experience from the TOWER project has clearly shown that users somehow liked the ambient interfaces designed and developed by others, but that they had much more fun when developing their own ambient interfaces.

Table 1. Design guidelines for ambient interfaces.

5 Conclusions

The design, development, and evaluation of the ambient interfaces as well as the design guidelines derived are only a start into a more systematic exploration of the systematic design for ambient interfaces. The whole area of ambient interfaces is rather new and to some extent still in its infancy. The results and guidelines provided should, therefore, be interpreted as a first attempt in combining knowledge from literature on design and usability of computing artefacts and initial experiences with the design, development, and use of actual ambient interfaces. And we are probably far from a usability engineering theory and practice that is already established for classical graphical user interfaces.

Acknowledgments

The research presented here was carried out and financed by the IST-10846 project TOWER, partly funded by the EC. I would like to thank all my colleagues from the TOWER team. I would also like to thank the anonymous reviewers for their valuable comments on the earlier version of this paper.

References

Bergman, E. (Ed.). (2000). Information Appliances and Beyond: Interaction Design for Customer Products. London, UK: Academic Press.

Buxton, W. A. S. (1994). Human Skills in Interface Design. In L. W. MacDonald & J. Vince (Eds.), Interacting with Virtual Environments, New York: Wiley, 1-12.

Dourish, P., & Belotti, V. (1992, Oct. 31-Nov. 4). Awareness and Coordination in Shared Workspaces. Proceedings of the Conference on Computer-Supported Cooperative Work - CSCW'92, Toronto, Canada. 107-114.

Gross, T. (2001, Feb. 7-9). Towards Ubiquitous Awareness: The PRAVTA Prototype. Ninth Euromicro Workshop on Parallel and Distributed Processing - PDP 2001, Mantova, Italy. 139-146.

Gross, T. (2002, Jan. 9-11). Ambient Interfaces in a Web-Based Theatre of Work. Proceedings of the Tenth Euromicro Workshop on Parallel, Distributed, and Network-Based Processing - PDP 2002, Gran Canaria, Spain. 55-62.

Hudson, S. E., & Smith, I. (1996, Nov. 16-20). Techniques for Addressing Fundamental Privacy and Disruption Tradeoffs in Awareness Support Systems. Proceedings of the ACM 1996 Conference on Computer-Supported Cooperative Work - CSCW'96, Boston, MA. 248-257.

Preece, J., Rogers, Y., & Sharp, H. (2002). Interaction Design: Beyond Human-Computer Interaction. N.Y.: Wiley.

The LEGO Group. (2002). LEGO MINDSTORMS, http://mindstorms.lego.com/. (Accessed 18/1/2003).

Weiser, M., & Brown, J. S. (1996). Designing Calm Technology, http://www.ubiq.com/hypertext/weiser/calmtech/calmtech.htm. (Accessed 18/1/2003).

Weiss, S. (2002). Handheld Usability. N.Y.: Wiley.

Wisneski, C., Ishii, H., Dahley, A., Gorbet, M., Brave, S., Ullmer, B., & Yarin, P. (1998, Feb. 25-26). Ambient Displays: Turning Architectural Space into an Interface between People and Digital Information. Proceedings of the First International Workshop on Cooperative Buildings: Integrating Information, Organisation, and Architecture Workshop - CoBuild'98, Darmstadt, Germany.

Interacting with Mobile Intelligence

Lynne Hall
University of Sunderland
Computing and Technology,
Sunderland, SR6 0DD
lynne.hall@sunderland.ac.uk

Adrian Gordon, Russell James
Mimosa-Wireless Ltd.
13 Moorlands, Consett,
Co. Durham, DH8 0JP
mimosa@mimosa-wireless.co.uk

Lynne Newall
University of Northumbria
School of Informatics,
Newcastle, NE1 8ST
lynne.newall@unn.ac.uk

Abstract

The mobile device market offers considerable potential for software products, however, currently there are few compelling applications for hand-helds. We identify the potential for mobile intelligent systems and briefly describe the software that we have developed to enable mobile intelligent system construction. We outline the requirements for a mobile intelligent system for the corporate sector and discuss our attempts to create a demonstrator. The demonstrator is evaluated with some success. Future work is briefly discussed.

1 Introduction

The impact of mobile devices has been considerable over the past year and it is widely expected that the market for wireless data services will continue to grow significantly (Wong & Jesty, 2002). Market research identifies that the era of mobile computing has arrived. However, there has been a significant downturn in the mobile market (Reuters, 2002) seen not only through the limited purchases of new devices but also in a lack of upgrades occurring for hand-helds.

The response to this downturn has been to add functionality to mobile devices, particularly phones. The focus of the market in the winter period of 2002/3 has been to take one of the most successful mobile device applications, SMS (Harmer & Friel, 2001), and upgrade this with the potential to send and receive multimedia short messages (MSM). This has resulted in the sale of a vast number of units.

For the smart phones and Personal Digital Assistants (PDAs) there is no obvious salvation such as the photo-phone and it can be suggested that a significant factor in the downturn of the mobile market (particularly for handhelds) relates to the lack of use-worthy (usable and useful) and compelling software for these devices. Whilst application development has focused on the communication aspects of mobile devices, potential also exists for the exploitation of off-line applications. However, there are few compelling off-line applications for mobile devices, with scaled down versions of office software, retro-games and personal organisation providing the main product offerings. This lack of applications hampers the acceptance and use of these devices, particularly at the corporate level.

This paper discusses the approach used with an Application Development Environment (MADE) and Execution Environment (MEE) that we have developed for the creation and deployment of intelligent applications for mobile devices. Earlier experimentation using toy AI problems has ensured that applications created with MADE and deployed using MEE function at an acceptable

rate on a range of mobile devices. Here, our aim was to determine the translation of user and domain requirements into a use-worthy application that is usable by the intended user population and useful for this group.
Section 2 briefly considers the mobile sector, focusing on corporate use of mobile devices and the trends seen in software development for this market. Section 3 discusses our approach to the development of innovative software for mobile devices, using artificial intelligence techniques to enable access to tailored, focused, expert knowledge without requiring costly communications. Section 4 discusses the development of an mobile intelligent system demonstrator for the corporate sector. Section 5 considers our approach suggests some directions for future work.

2 Corporate Use of Mobile Devices

Apart from communication, the most successful corporate application for mobile devices continues to be the organiser, replacing the time manager or filofax. However, except for calendars, address books, diaries and other essentially administrative support applications there is typically little appropriate use made of the mobile in the work environment.
In an attempt to make the mobile device more useful to mobile workers the recent focus of software for mobile devices such as PDAs and smart phones focuses on mimicking the functionality offered by broadband PCs. However, this approach has had limited success due to the constrained nature of mobile devices. For example, applications that require intensive input and display of large amounts of data (e.g. office software) are not easily usable on constrained devices without the addition of peripherals, such as keyboards. The utility of mobile devices is further reduced by the problems of accessing information resources through such tiny displays (Rist & Brandmeier, 2001). Users are also forced to perform numerous operations by selecting very small icons. The problems of physically manipulating miniaturized versions of "standard" keyboards and pointing devices further reduces the utility of mobile devices.
Activities are often supported through the use of horizontal applications such as wireless e-mail and messaging, workgroup applications, or applications for corporate information access and financial transactions. The dominant approach to providing functions beyond these to wireless computing devices has been by means of so-called microbrowser technologies, such as WAP, and i-Mode. However, the microbrowser model is not suited to the delivery of data services to all wireless devices (Evans & Baughan, 2000):

- It is not suitable for highly interactive and business critical applications because of its dependence on the transmission network - as the network becomes unavailable or slows down, so does the application.
- It is not cost-effective for many users, because of the (sometimes significant) costs of sending data over cellular networks

Information rich environments (e.g. web sites) are also negatively effected by the display capabilities of mobile devices, and are typically viewed in an impoverished format, with a reduced set of information available. Further, user expectations of tools to enable high bandwidth activities (e.g. video downloads) are rarely met (Charny, 2001), with the devices and the infrastructure unable to cope with such applications (Sherman, 2001), typically resulting in user frustration.
The characteristic of mobiles that has been most widely exploited is the communications potential of these devices. However, for the PDA and smart phone market, this potential should only be one part of the product offering not necessarily the dominating force. It can be suggested that the potential offered by mobile devices may be best exploited by off-line applications, with the focus

on portability rather than connectedness, at least until infrastructural and content issues are more firmly resolved.

3 Intelligent applications for mobile devices

Intelligent applications, developed using Artificial Intelligence techniques, provide expert knowledge and advice to users, and have characteristics that make them relevant for the mobile platform. The application of such techniques has been successfully applied to significant problems in a range of domains. Though the technologies for implementing intelligent applications are well established, they have not yet been adapted for delivery via mobile computing devices, because such devices are limited in terms of their memory size and processing power.
Portable intelligent systems could offer compelling user experiences, see table 1. For example an intelligent system operating on a mobile device can offer an employee immediate access to tailored, focused, expert knowledge, without requiring costly communications. The advantages of making an intelligent system mobile are readily apparent. For example, a horticulturist would clearly want to be able to monitor the health of a crop whilst in the field; Health and Safety audits would clearly benefit from being at least partly conducted on site; a consumer would want dietary and nutritional advice whilst out shopping.

Table 1: Examples of Intelligent Systems

Domain	Typical Tasks / Functions
Agriculture	diagnose and propose treatments for plant disorders advise on plant care schedule fertilization and irrigation
Health and Safety	interview users about buildings, work practices and policies conduct detailed compliance reviews on such things as fire regulations
Finance	help independent financial advisers to propose packages of financial products tailor products to individual clients support investment decisions
Consumer market	advise shoppers on diet and nutrition plan and monitor an exercise regime help the amateur gardener choose and care for suitable plants

The awareness of the lack of intelligent applications for mobile devices has resulted in our creation of a proof of concept prototype that allows the development and deployment of intelligent systems on mobile devices. Intelligent system creation requires a powerful development environment, which offers facilities for rapidly constructing, modifying, and testing intelligent systems. To enable this intelligent system to then function on mobile devices the executable has to be dramatically compressed, so that in deployment it will use the minimum amount of memory and processing power possible. This has resulted in the construction of:

- Mimosa Application Development Environment (MADE): which allows intelligent systems to be built and tested on a powerful desktop computer
- Mimosa Execution Environment (MEE): which allows such systems to be deployed on a variety of mobile computing devices, ranging from mobile phones to Personal Digital Assistants (PDAs). MEE compresses the knowledge base into a sufficiently small application for the limited memory and power of typical mobile platforms.

The intelligent system is developed using MADE, which provides an extensible language for developing rule based applications, together with an environment for executing, testing and debugging such applications. This runs on a conventional personal computer and permits the creation of "write once, run anywhere" intelligent systems for mobile devices. Once developed, an intelligent system is deployed on a mobile or wireless device using MEE, which is capable of executing intelligent systems effectively and efficiently on a range of mobile computing devices.
Solving the technical requirements of MADE and MEE has resulted in an environment with potential for developing mobile intelligent systems. To explore the possibilities offered, we are in the process of creating a number of demonstrators, focusing on the commercial, corporate and consumer sectors. These demonstrators will enable the assessment of the feasibility of a range of mobile computing hardware and software configurations for delivering intelligent systems using the Mimosa architecture. Here, we discuss the use of the Mimosa architecture to create a demonstrator for the corporate sector.

4 Corporate Demonstrator

Mobile devices could offer an ergonomic and usable approach to the requirements of mobile workers (i.e. any employee who regularly operates in a non-PC-supported environment or who works outside of the workplace, even if that location is home). These requirements relate to identifying a satisfying task structure that fits within an efficient, enjoyable user experience, based within the context of the user having to be mobile and with the device being optimal.
Sector and user analysis identified a number of specific requirements that aid in determining whether applications would be appropriate for mobile devices:

- Rapid, effective resolution of tasks
- Narrow focus of application goals: *thinware*
- Users will have widely differing levels of device experience and computer literacy (this will often initially be novice and low)
- High pragmatic user expectations (Ralph & Shephard, 2001), based on use of social and recreational software, i.e. no toleration of poor interaction design
- Use of medium must add value / reward the user, this may be in improved worker satisfaction as the total user experience includes social style
- Activity where application to be used must require mobility and use of the application should add value to the business process and worker it supports (Sacher & Loudon, 2001)

For the corporate sector a demonstrator application is being constructed that supports the sales staff of a Call Centre operation. The sales staff spend most of their time on-the-road visiting clients and ideally would like to make instant decisions in relation to whether or not they will accept the client's business. The domain selected was appropriate, as workers are mobile and have little desire to support their work (which is largely face-to-face) with a lap top, but find the idea of digital support tempting. The intelligent system makes recommendations about the data provided by the sales staff based on the application of a set of heuristics, for example relating to manning levels and call pattern type. An early prototype of the Call Centre application was evaluated by sales staff with mainly positive results. The evaluation criteria for the demonstrator were based on the requirements identified above.
The task structure requires only a limited amount of information, most of which can be categorised and then selected by pointing rather than text entry, for example call patterns are represented graphically. Task structure is based on the business model created with expert knowledge and

results in rapid task resolution. The value to the sales staff relates to the knowledge encoded within the system and this requires regular updating (quarterly). The reward to the sales staff relates to the social style, with staff identifying the demonstrator, or more appropriately the device itself, as cool. Users identified that it would support their work functions in a portable, discrete and fashionable manner. However, issues related to monitoring and acceptance of demonstrator consultations still need to be resolved and its integration into the already existing work system.

5 Conclusions

In this paper, we have provided an overview of the development environment MADE and the deployment achieved through MEE. We have discussed the construction of a demonstrator application for the corporate sector. This Call Centre demonstrator matched the requirements of the domain, sector and user and suggests the potential of an intelligent mobile system to provide tailored, focused support in a timely, accurate and appropriate manner.

Future work focuses on further development of this and other demonstrators, focused on the commercial and consumer sectors. For the commercial sector, a hotel demonstrator is under development that incorporates the use of an electronic marketplace populated by intelligent agents. For the consumer sector, a poker tutor is under development. Each of these demonstrators highlights significant issues both for the requirements of use and the design of the interaction.

There are few examples of intelligent systems being developed for mobile devices publicly available. This sector is one which offers considerable potential, however, few applications of any calibre are available. The creation of innovative products will enable a significant step forward.for the sector enabling enable mobile devices to graduate from trite applications such as games and to-do lists to useful, intelligent systems that can provide informed responses to complex user queries.

References

Charny, B. (2001,). Are smart phones too smart? *CNET News.com.*

Evans, B. G., & Baughan, B. (2000). Visions of 4G. *IEE Electronics and Communication Engineering Journal, 12*(6), 293-303.

Harmer, J. A., & Friel, C. D. (2001). 3G Products - What will the Technology Enable? *BT Technology Journal, 19*(1), 24 - 31.

Ralph, D., & Shephard, C. (2001). Services via Mobility Portals. *IEE Electronics and Communication Engineering Journal, 13*(4), 148-154.

Reuters. (2002). *Cell phone industry on shaky ground.* CNet News.com. Available: http://news.com.com/2100-1033-823197.html

Rist, T., & Brandmeier, P. (2001). *Customizing Graphics for Tiny Displays of Mobile Devices. Paper presented at Mobile HCI 2001. Available http://www.cs.strath.ac.uk/~mdd/mobilehci01/procs/rist_cr.pdf*

Sacher, H., & Loudon, G. (2001). Uncovering the New Wireless Interaction Paradigm. *ACM interactions, January/February.*

Sherman, E. (2001). Little Big Screen. *Technology Review.*

Wong, R., & Jesty, R. (2002). *Wireless Internet: Applications, Technology and Market Strategies* : ARC Group.

Embedded versus Portable Interfaces for Personalizing Shared Ubiquitous Devices

David M. Hilbert and Jonathan Trevor

FX Palo Alto Laboratory
3400 Hillview Avenue
Palo Alto, CA 94304 USA
{hilbert, trevor}@fxpal.com

Abstract

Everywhere we go, we are surrounded by shared devices: TVs, stereos, and appliances in the home; copiers, fax machines, and projectors in the office; phones and vending machines in public. Because these devices don't know who we are, they provide the same user interface and functionality to everyone. This paper describes a system for personalizing workplace document devices— projectors, public displays, and multi-function copiers—that has been in use for over two years in our organization. We compare user interfaces that are *embedded* (i.e., integrated or co-located with the shared device) versus *portable* (i.e., accessible via portable devices such as mobile phones or PDAs). We summarize lessons learned for others designing interfaces for shared ubiquitous devices.

1 Personalization and everyday devices

In the Web's infancy, no matter who you were, you saw the same Web pages as everyone else. Today, Web sites remember who you are and tailor their content to match your needs. Amazon.com uses personalization to expedite ordering and to suggest products you may want to buy, based on past purchases and purchases of others with similar interests. Instead of attending to the details of entering billing and shipping information and locating products, these tasks recede into the background so you can focus on *shopping.*

Indeed, this trend is beginning to affect *everyday* shared devices, such as cars and televisions. The BMW 7 Series car remembers drivers' seat, mirror, and steering wheel settings and recalls them automatically when family members use their unique keys to enter the car. Rather than fussing with seats, mirrors, and steering wheel, you just *drive.* TiVo digital video recorders learn your TV viewing habits and automatically record shows for you. Instead of searching through listings and manually setting-up recordings, you just turn on your TV and *watch*. Our work is situated in this tradition, but with a focus on workplace document devices, such as photocopiers.

2 Embedded versus portable interfaces

In designing a system to personalize workplace document devices, we immediately faced a fundamental question in ubiquitous computing: whether to use embedded or portable interfaces?

Rather than picking one approach and simply "validating' it, we wanted to perform a comparative evaluation to gain more general insights to inform future designs. Would users feel uncomfortable accessing personal resources through an interface embedded in a public device? Would they feel better accessing their data via their own cell phone? What if users forget their cell phones or wireless connectivity is unreliable? Are larger, embedded user interfaces inherently more usable than tiny, portable interfaces provided by cell phones and PDAs? While we had numerous intuitions, we wanted to compare embedded versus portable interfaces to observe their relative strengths and weaknesses in practice.

3 Personalizing workplace document devices

We began our exploration of shared device personalization by observing our own colleagues interacting with three document devices in our workplace: the projector in our formal conference room, a large plasma display in our brainstorming room, and the multi-function copier in our mailroom. In each case we noticed users engaging in peripheral activities that took their attention away from presenting, brainstorming, and printing. In each case we identified ways in which personalization could help.

For the projector in our conference room, we envisioned a personal interface for seamlessly locating and opening users' recently edited presentations. This would allow users to focus on presenting instead of spending time locating their documents on the network using the podium PC. For the plasma display in our brainstorming room we envisioned a personal interface for streamlining the activity of locating and opening users' working documents. Thus, users could focus on brainstorming instead of copying documents from their office PCs to laptops and then connecting laptops to the plasma display. Finally, for the multi-function device (MFD) in our mailroom we envisioned a personal interface so users could print personal documents while standing in front of the MFD. This way people wouldn't always need to be in their offices to start print jobs, and they could access other personal features such as: "scan to my desktop" instead of scanning to a public network folder; "fax to my contacts" instead of looking-up the fax number and copying and entering it into the MFD; and re-using their personal copier preferences and fax history.

3.1 The Personal Interaction Points system

The Personal Interaction Points (PIPs) system is a web-based application that generates embedded and portable interfaces for personalizing the projector in our conference room, plasma display in our brainstorming room, and MFD in our mailroom. We immediately thought of numerous personalization features, but wanted to focus our effort on a small set of functionality with a large payoff for users. This was particularly important since we wanted to secure real use in order to observe the effects of personalization on real users. Thus, we decided that integrating personal computer file access with a shared device's functions would be a good starting point. PIPs embodies this by giving "smart" access to people's file history at shared devices, just like the Windows recent "Documents" menu gives access at desktop computers. The "smart" part of the system is that it matches file types to the function types of devices. So for a conference room projector, the system automatically selects PowerPoint presentations from the user's file history as the documents the user is most likely to want to present.

The main distinguishing feature of PIPs is combining users' networked resources—or "personal information clouds"—with device-specific interfaces for performing common device tasks. So instead of remotely accessing your desktop (Richardson et al. 1998) to locate and start a presentation on the presentation PC, you use a special interface for showing presentations that links directly to your recently edited presentations. We now briefly describe the embedded and portable interfaces. For more details see (Trevor et al. 2002).

3.1.1 The embedded interface

Each embedded PIP consists of a touch screen on or near the shared device and a Radio Frequency Identification (RFID) card reader. The embedded interface is activated when the user approaches the shared device and swipes their ID card over the card reader. The system reads the users' encrypted password from the card and starts an NT authenticated process that runs as the user. Users who forget their cards can also login by entering their username and password.

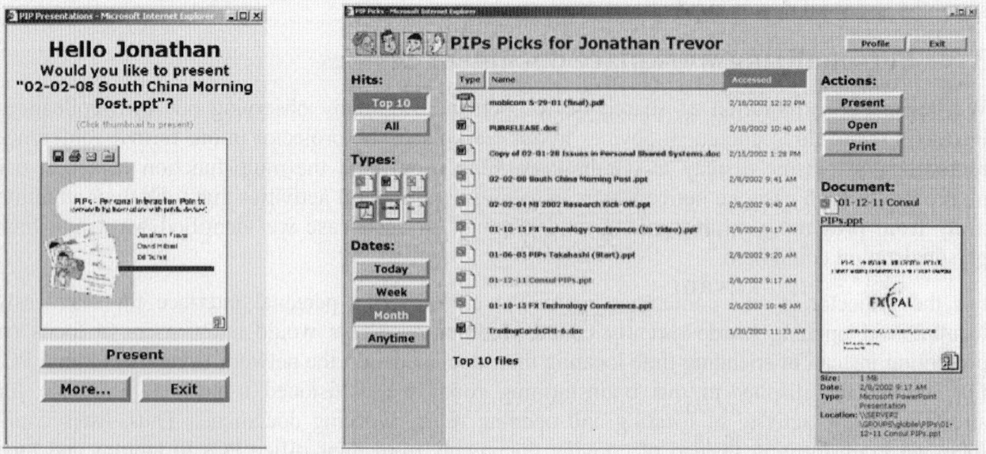

Figure 1. The "best pick" interface suggests the document (or documents in the case of the printer) the user is most likely to want to use on the PIP-enhanced device (left). Selecting the thumbnail causes a device-specific action to be performed on the document (present for the projector, open for the plasma display, and print for the printer). The user may also select "More" to explore other resources in their file history (right). The left frame allows users to filter the file list in the middle frame. Selecting columns in the middle frame sorts the list and selecting a file causes the right frame to display actions applicable to the selected file, its name and thumbnail, and other file details.

The PIP web application then generates the personal interface by fetching and resolving the shortcuts stored in the user's recent file list on their PC. The PIP presents a "best pick" interface with the recent file (or files) the user is most likely to want to use at the PIP-enhanced device (Figure 1 left). The user may then perform a default action (such as present, open, or print), by selecting the document's thumbnail using the touch-screen provided by the embedded PIP. Files are accessed over the network from their original locations, so users needn't plan ahead or copy files anywhere. If the best pick interface does not contain the user's desired document, the user can press the "More" button to bring up the "full" interface (Figure 1 right). This allows the user to access virtually any document (via the device) that they have ever accessed on their office PC.

3.1.2 The portable interface

For the portable interface, users point their portable Web browsers at the PIPs homepage. Selecting a PIP-enhanced device from the homepage activates the portable PIP for that device. Standard browser authentication is used to login users. Once authenticated, the PIP application fetches and resolves the user's recent file list in the same way it does for the embedded interface.

While we strove to keep the embedded and portable interfaces as similar as possible, we were forced to miniaturize the interface and make other minor modifications for portable devices with small displays, such as Pocket PCs. For instance, the "full" interface (Figure 1 right) was split into two pages: one for the main file list and another to show selected file details. However, the most notable difference comes after a file has been selected for presentation, brainstorming, or printing. Users of the embedded interface can use the touch screen, keyboard, and mouse attached to the device to perform subsequent actions on the device. However, in the portable case, the user may not be close enough to the shared device to control it directly, so we created a simple "remote control" interface to emulate the functions available on the shared device.

4 Lessons Learned

Over two years have passed since we deployed the PIPs system. Today, about three quarters of our approximately thirty-person research staff actively uses it, and so far no one who has used it has subsequently stopped using it. The presentation PIP is used for over half the presentations given in our formal conference room. The brainstorming PIP is used for nearly all documents accessed in our brainstorming room. The printer PIP is rarely used and has since been decommissioned.[1] In the following subsections we summarize lessons learned regarding embedded versus portable interfaces for personalizing shared ubiquitous devices.

4.1 Embedded advantages

In our experience, embedded interfaces are more *usable, available,* and *simpler to implement* than portable interfaces.

Embedded interfaces are more **usable** than portable interfaces, due primarily to their larger displays (15" and up in our case) and flexible input mechanisms (touch screen and optional keyboard and mouse). Usability issues were most noticeable for complex tasks, such as document editing in our brainstorming room. The portable interfaces also confused users and altered the user experience, particularly when users were accustomed to interacting directly with the shared device. One of our users asked: "Do I need to load my presentation onto the Pocket PC before presenting in the conference room?" Apparently the separation of the personal interface from the underlying shared device obscured the fact that users' documents are always opened over the network, regardless of whether they use the embedded or portable interface. Another user reported that with the embedded interface, "you feel you have a real relationship with the device, but with the portable [interface], you feel you have a relationship with the portable device rather than the actual device." He concluded: "I think of them as two completely separate applications."

Embedded interfaces were also more **available** than their portable counterparts. We observed the portable interfaces suffering from nearly every imaginable availability issue ranging from batteries dying and wireless network failures, to users forgetting their devices in their offices.

Finally, embedded interfaces were slightly **simpler to implement** than the portable. In both cases, we faced challenges integrating personalization with existing device hardware. We worked around this by developing PC "proxy" interfaces to drive each device (projector, plasma display, and MFD). In the embedded case, once users select a file they can continue to control the device using the device's existing interface. But since portable users may not be close enough, portable interfaces must provide additional remote controls. These remote interfaces are typically more difficult to use than the hardware they are emulating and may require significant additional effort to develop.

These observations taken together suggest designers should consider incorporating personalization into already existing embedded interfaces—to the degree possible—rather than creating new portable interfaces, especially when shared devices support complex tasks.

4.2 Portable advantages

On the other hand, portable interfaces have an edge in terms of *remote control* and *privacy.*

---

[1] People in our organization didn't use the printer PIP often since they were typically in their office, or not far from it, when they needed to print. A personalized MFD would clearly be more useful in public locations or in large organizations where users aren't always within a few steps of their office PCs.

Users found the **remote control** capabilities of the portable interfaces to be quite useful, particularly for simple and on-going tasks in large spaces, such as advancing through slides in our conference room. However, they found remote control to be far less compelling for complex tasks (such as document editing) and single-shot interactions (such as printing) in smaller spaces.

Similarly, users appreciated the **privacy** of the portable interfaces, particularly in the conference room setting. This is because they could access their information clouds via a small private display before presenting, instead of using a larger display that others in the room could see. But again, this was only noticed in the conference room where the mood is more formal and speakers are presenting to colleagues from other projects, the whole lab, or visitors. Privacy was not an issue in informal settings, such as our brainstorming room, or for quick interactions in low-traffic areas, such as printing in our mailroom.

The observed advantages of embedded and portable interfaces taken together suggest designers should consider *hybrid solutions* that allow users to interact with portable interfaces for remote control and highly sensitive tasks, and embedded interfaces for more complex tasks. For instance, users could use a portable interface (e.g., a mobile phone) to select and transmit personal resources to a public device (e.g., an MFD), at which point they could switch to the public device's embedded interface to complete their tasks (e.g., to adjust printing and output options).

5 Conclusions

A little personalization can go a long way toward improving the user friendliness, efficiency, and capabilities of shared document devices. We transformed the user experience of three shared document devices in our lab. Now when we use the presentation PC, we no longer see it as a general PC for finding and opening presentations. We view it as a specialized device that allows us to swipe our ID card to begin presenting. The former tasks of locating and opening documents have receded into the background so we can focus on *presenting.*

Ubiquitous computing researchers typically implement either wholly embedded or wholly portable interfaces. Since we were unsure about which approach to take, we embodied both alternatives in our system and deployed it in a variety of situations to compare the two. Our technique led to a system that is still in use more than two years after its introduction. Today our embedded interfaces are far more popular than their portable counterparts. This could change as wireless devices and networks become more dependable and pervasive, and as users become more accustomed to using them. However, our experience suggests that embedded (or hybrid) interfaces may be inherently more suitable for particular shared devices, namely, those that support complex tasks requiring complex interfaces. If we had simply decided to go with a portable approach, our system would not have achieved the usage it did, and therefore we would not have gained the knowledge we have. We believe our experimental technique is applicable to other HCI, CSCW, and ubiquitous systems research, and encourage others to decide for themselves (Trevor et al. 2002).

6 Acknowledgements

We would like to thank Bill Schilit for his inspiration and guidance.

References

Tristan Richardson, Quentin Stafford-Fraser, Kenneth R. Wood, Andy Hopper (1998). Virtual Network Computing. *IEEE Internet Computing,* 2(1), 33-38.

Jonathan Trevor, David M. Hilbert, and Bill N. Schilit (2002). Issues in personalizing shared ubiquitous devices. *Proceedings of the 4th International Conference on Ubiquitous Computing (Ubicomp 2002).*

Supporting Ubiquitous Information on Very Small Devices is Harder than you Think

David M. Hilbert, Jonathan Trevor

FX Palo Alto Laboratory
3400 Hillview Avenue
Palo Alto, CA 94304 USA
{hilbert, trevor}@fxpal.com

Bill Schilit

Intel Research
2200 Mission College Blvd.
Santa Clara, CA 95052 USA
bill.schilit@intel.com

Abstract

A basic objective of ubiquitous computing research is ubiquitous information: the ability to utilize any content or service, using devices that are always at hand, over networks that don't tie us down. Although much progress has been made, the ideal remains elusive. This paper reflects on the interrelations among three dimensions of ubiquitous information: *content, devices,* and *networks.* We use our understanding of these dimensions to motivate our own attempt to create a ubiquitous information system by combining unlimited World Wide Web content with mobile phones and mobile phone networks. We briefly describe a middleware proxy system we developed to increase the usefulness of very small devices as Internet terminals. We conclude with a post-mortem analysis highlighting lessons learned for others interested in information systems for very small devices.

1 Dimensions of ubiquitous information

While people have made inroads toward ubiquitous information, the ideal remains elusive. One reason is that the very notion of "any information, anytime, anywhere" places conflicting requirements on the *content*, *devices*, and *networks* that make up information systems.

Consider, for instance, laptop computers and local area networks. The combination of powerful processors, flexible user interfaces, high-resolution displays, speakers, and high-bandwidth networking make it easy to interact with rich and interactive content. However, these same characteristics place practical limits on device portability and network mobility. Mobile phones and networks, on the other hand, make the opposite trade-off. They provide extreme device portability and network mobility at the cost of greatly reduced content capabilities, due to limited user interfaces and lower bandwidth networking. Palm-sized computers with wireless WAN cards provide a middle ground: they are more portable than laptops, and more usable and content-capable than mobile phones, but far less pervasive than phones. It seems you can't have it all.

It gets worse. Current trends suggest these tensions will grow in the future. Desktop computers and wired networks keep getting faster and more powerful, meaning ever-richer content will flourish on the Web. At the same time mobile phones keep getting smaller and lighter, meaning tiny user interfaces, processors, and wireless networking will continue to be hard-pressed to handle the content. While wireless devices and networks are increasing in power too, the smallest and lightest devices will be the most ubiquitous, ensuring that usability and interactivity will continue to be problematic. In essence, we face *natural tensions* whenever we attempt to maximize "ubiquity" along all dimensions at once.

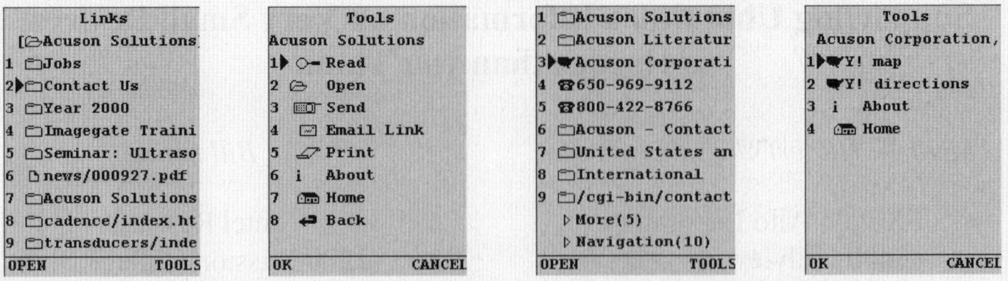

a b c d

Figure 1: M-Links on a Neopoint 1000 web phone. M-Links retrieves requested Web pages and returns native format (WML, HDML, CHTML) screens. In this example, M-Links presents the user with a list of links to navigate the homepage of Acuson, a medical equipment manufacturer (Figure 1a). The open folder indicates the page the user is currently navigating. The closed folders indicate links to other pages. File icons indicate links to non-HTML content, such as PDF or multimedia. When the user presses the "TOOLS" soft key, the interface switches to a list of actions the user can perform on the current link (Figure 1b). For instance, pressing "Read" would allow the user to read all the text on the page. Pressing "CANCEL" returns the user to navigation mode (Figure 1a) again. Selecting "Contact Us" navigates to Acuson's contact information page (Figure 1c). M-Links displays not only the links but other useful information such as phone numbers and addresses. The "Navigation" item (bottom of screen) collects links that repeat across many Web pages. The "More" item (second from bottom) shows the remaining links on the Web page that would not fit on the current screen. In (Figure 1c), the user selects Acuson's street address and presses the "TOOLS" soft key to show a list of actions appropriate for addresses (Figure 1d). For instance, if the user selects "directions", m-links passes the street address to Yahoo! Maps and returns the directions to the user in text form. The "About" and "Home" options allow the user to get more information about the current link and to return to the m-Links home page respectively.

2 A real world ubiquitous information system

We wanted to create a real world ubiquitous information system using content, devices, and networks available today. We knew that in the future, wearable displays and novel input devices would remove *some* of the usability limitations of portable devices, and publicly accessible embedded computers with usable displays and interfaces will someday be more ubiquitous. However, these future possibilities aren't helpful in creating a real system now. So we decided to build upon Wireless Web phone technologies because of their trend towards ubiquitous devices and networks.

2.1 The M-Links middleware proxy

While the current Web browsing model works well for desktop and laptop computers, it is less well suited to phone-tops. Cell phones typically accommodate only three to twelve lines of text, and their design emphasizes portability and features such as battery life, audio clarity, and ease of selecting names from a phonebook. Web interaction has, so far, been a secondary concern. Thus, we developed M-Links, a middleware transducing proxy, to help mobile phone users access and actually *do* things with a wider range of Web content than before. M-Links differs from other Web transducers (Brooks et al., 1995) in that it factors Web browsing into two separate interfaces: one for navigating links and the other for performing actions on links. See Figure 1.

When users access a Web page using m-Links, they see a list of links from that page and can dig through the list in the same way they dig through folders on their desktop to locate files. When they find a link, they may invoke services, analogous to right clicking on a document and using the context menu on a desktop interface. Although users can't do much directly on the phone with content such as large PDF documents or MPEGs, m-Links users can always do *something* with any content they find, even when their device is not equipped to handle the content itself. For instance, once users have located content of interest, e.g., a digital movie or novel, they can always send the content (or URL) via email for later use on the desktop or other device by simply selecting the link and applying the e-mail service. Alternatively, users can use M-Links to invoke third-party content translation services, for instance, to convert PDF documents into text formated for the phone, once these services become available.

M-Links is both simple to use and powerful. Its simplicity comes from the *navigation interface* with its server-side data-detectors for bubbling-up useful bits of information, such as phone numbers and addresses: separating links from page content makes navigation a matter of selecting a link from a list. Its power comes from the *action interface* with its open systems architecture for incorporating new network-based services, similar to browser plug-ins: because users can apply various Web-based services to any link, they can do more with content than simply read it on their phones. For more detailed discussion of the M-Links infrastructure and user interface, see (Schilit et al., 2001) and (Trevor et al., 2001) respectively.

3 Lessons Learned

Our research approach differs from traditional approaches in that we were teamed with business people to develop a working prototype in conjunction with a business plan that we could show to potential users, investors, and business partners. Feedback from all these people helped identify areas in which M-Links could be improved.

3.1 Technical Issues

We quickly realized **the need for speed** could not be overestimated. A basic assumption all along was that we needed to optimize the interface for devices with limited displays and input mechanisms. This led to our separating links from content to speed navigation, and our use of server-side data detectors to bubble-up interesting bits of information such as phone numbers and email addresses. However, people still found it painful to navigate to find content. They said things like "I wish I could find all the phone numbers on this site at once" or "I know what I'm looking for is a PDF file, but I can't remember where it is!" This led us to develop yet more data detectors (e.g., for street addresses) and suggested we might also increase speed by providing a *filtering* interface, that would instantly bubble-up all the links on a site of a given type specified by the user.

Another lesson was that **no device is an island**. People felt frustrated having to register using their email address via the phone keypad, and wanted to know why they couldn't set up their account from their desktop PC. This suggested a substantial support infrastructure at the desktop might greatly increase the chances of user acceptance. We began developing a Web site for account management and have begun exploring other mechanisms for leveraging desktop interactions to facilitate phone-top interactions. For instance, instead of specifying URLs using key entry at the phone, users might select from a list of "favorites" or "recently visited" sites gleaned from their desktop Web usage. Another issue is that people often had a hard time re-locating content on the phone that they had already located at the desktop before. This led to the idea of adding another filtering feature: namely, "show me all the links I've accessed from this site before".

As time passed, we also began to appreciate the importance of **hands-free use**. We repeatedly heard that U.S. mobile phone users, in contrast to Japanese users, use their phones while otherwise engaged, such as while driving. In Japan, Wireless Web use is pervasive on public transportation, as users are free to direct their attention to their phones. This led us to investigate a voice interface for M-Links. However, poor integration between the voice and data capabilities of mobile phones at the time made developing such an interface problematic. We expect this limitation to decrease in the future with 3G mobile phone networks.

Finally, we learned that **infrastructure limitations** were more significant than we had expected. The idea that users could instantly access the Wireless Web simply wasn't true. It often took several seconds to establish a data connection. Furthermore, the reliability and performance of the networks left users frustrated by dropped connections and slow page retrievals. Finally, while users always had their phones with them, phone batteries didn't live up to the promise of "anytime" use. While all of these are limitations of the devices and networks upon which M-Links was built, and not M-Links itself, we underestimated the power of frustrated expectations to reduce users' willingness to consider new capabilities. Again, many of these problems will likely be improved with future advances in mobile phone service.

3.2 Non-Technical Issues

In addition to technical issues, numerous non-technical issues also inhibited our success.

First, there were significant **business-related challenges**. Substantial investment (in time and capital) is required to move research prototypes to commercial products. In some cases, a large marketing campaign may be required to draw attention to, or promote, a lifestyle in order to sell a product, as in the case of AT&T's recent mLife campaign. Furthermore, the lack of a micro-payment system for Wireless Web services (like the DoCoMo iMode model in Japan) means there is no simple mechanism for third-party service providers (such as ourselves) to charge small amounts of money per use of their service without arranging profit-sharing agreements with the big players. Thus, reaching a large number of customers requires relationships with telephony service providers (such as Sprint PCS and AT&T), applications providers (such as Yahoo), or infrastructure providers (such as Inktomi). We were beginning to establish relationships with key partners in all of these areas when the Internet bubble burst, leaving us in an inhospitable business climate.

Second, we now realize that limited **marketplace maturity** and **customer readiness** also impeded our progress. In a nutshell, our vision depended on people actually using their Wireless Web phones to access the Wireless Web. While this condition was met in Japan and other countries, the U.S. has continued to lag: people use their phones to place calls, not access information. This makes it hard for "add-on" information services such as M-Links to take off, and continues to fuel the desire for ever-smaller phones, a trend at odds with phones being used as information devices.

3.3 The dimensions reconsidered

In choosing a ubiquitous information platform, we now realize we should have asked ourselves: **ubiquitous for what?** In the U.S., mobile phones are used ubiquitously as *communications* devices, but not as *information* devices. We were seduced into thinking our platform would include all those devices people were already using as communications devices, when in fact our platform was really only those devices already being used as information devices—a far smaller pool. Given this perspective, another approach would have been to pick devices used less pervasively in general, but more pervasively *as* information devices, such as palmtop computers.

We also should have asked ourselves: **ubiquitous for whom?** By deciding up-front that we wanted to support ubiquitous information for everyone, instead of selected groups, we were forced into providing generic services for a horizontal market, instead of targeted services for vertical markets. A problem with generic services is they often imply lower value, which in turn implies the need for very low adoption costs. This intensified the pressures on infrastructure issues including network speed, reliability, and performance—attributes beyond our control. Another strategy would have been to focus on specific applications with higher value for particular types of customers, thus increasing their willingness to tolerate current infrastructure limitations. After achieving success in vertical markets, we might then have spread out into adjacent markets to increase ubiquity.

In short, we believe there was **a cultural catch-22** at work that we didn't fully appreciate at the time. Since our parent company is Japanese, we were acutely aware of (and motivated by) the huge success of mobile phone information services in Japan. Additionally, analysts in the U.S. were saying a key reason American's weren't satisfied with the Wireless Web was that they, to a far greater extent than Japanese, were accustomed to accessing WWW content on their desktops, and thus felt frustrated they couldn't access the same content on their mobile phones. So there appeared to be a great opportunity: combine the runaway success of Wireless Web usage (as demonstrated in Japan) with the ability to access a much wider range of content on the WWW (as is the custom in the United States). Ironically, while Japanese are enthusiastic about using the Wireless Web, they are far less interested than Americans in accessing WWW content. At the same time, while Americans are enthusiastic about accessing WWW content, they are far less willing than Japanese to use mobile phones to do it. In other words, we envisioned a community of users as enthusiastic about Wireless Web usage as Japanese, and as enthusiastic about accessing WWW content as Americans: a community that doesn't currently exist—at least not in the U.S.

4 Conclusions

Despite its limitations, M-Links achieved many of the goals we set out for it. Namely, it substantially increases the content capabilities of highly portable (ubiquitous) devices operating over very wide area wireless (ubiquitous) networks. However, if we were to do it over again, we would consider retargeting M-Links to take advantage of devices that are more pervasively used as information devices today (such as palmtop computers), and perhaps for specific markets (such as mobile sales and repair professionals) in which users may be more willing to tolerate limitations in current infrastructure technologies. Nonetheless, by leveraging extensible network-side services, our approach still offers substantial value to small Internet device owners by allowing them to exploit the computing resources and network connectivity of larger, more powerful devices to increase the ways in which they can display, share, and otherwise manipulate Web content using very small devices.

5 Acknowledgements

We would like to thank T.K. Koh for her development support and Debra Go and Patrick Ahern for their business expertise.

References

Charles Brooks, Murray S. Mazer, Scott Meeks, and Jim Miller (1995). Application-specific proxy servers as HTTP stream transducers. *Proceedings of the 4th International World Wide Web Conference (WWW '95).*

Bill N. Schilit, Jonathan Trevor, David M. Hilbert, and Tzu Khiau Koh (2001). M-Links: An infrastructure for very small Internet devices. *Proceedings of the 7th Annual International Conference on Mobile Computing and Networking (MobiCom 2001).*

Jonathan Trevor, David M. Hilbert, Bill N. Schilit, and Tzu Khiau Koh (2001). M-Links: From desktop to phonetop: A UI for Web interaction on very small devices". *Proceedings of the 14th Annual Symposium on User Interface Software and Technology (UIST 2001).*

COMUNICAR: Subjective Mental Effort when driving with an Information Management System

Marika Hoedemaeker

TNO Human Factors
Kampweg 5, 3769 DE Soesterberg
Hoedemaeker@tm.tno.nl

Roland Schindhelm, Christhard Gelau[1], Francesco Belotti[2], Angelos Amditis[3], Roberto Montanari[4], Stefan Mattes[5]

Abstract

A fixed-base driving simulator experiment was carried out to test an in-vehicle Information Management system which was developed within the COMUNICAR project. The subjective data on workload (RSME: Rating Scale Mental Effort) provided only weak evidence in favour of information management. Considering the special situation in this experimental set-up learning effects and (implicit) memory effects could be found as explanation for these results.

1 Introduction

COMUNICAR (COmunication Multimedia UNit Inside CAR) is a project in the IST programme of the European Commission. COMUNICAR's main goal is to design, develop and test an easy-to-use in-vehicle multimedia Human-Machine Interface (HMI). One basic element of this HMI is an Information Management system that defines how, when, where and in which format to give the messages to the driver, thereby taking into account his/her level of workload and the driving conditions. When more than one information message arrives to the driver simultaneously, the Information Management system prioritises the information. This results in a serial presentation of the messages instead of presenting all messages to the driver at the same time, as soon as they come in. The hypothesis is that driving with such an Information Management system leaves driver workload largely unaffected and that therefore driving performance will not suffer from an overload of information.

The testing strategy adopted by the COMUNICAR project is an user-centred iterative design and evaluation approach employing different test settings (laboratory, simulator, on-road). This paper describes the simulator study and part of its results. The objective of the simulator study is to assess the impact of Information Management (IM) on driving performance (hence safety) and on

[1] BASt, Federal Highway Research Institute, Germany
[2] DIBE, University of Genova, Italy
[3] ICCS, University of Athens, Greece
[4] CRF, FIAT Research Centre, Italy
[5] DaimlerChrysler AG, Stuttgart, Germany

subjectively experienced workload. The driving simulator allowed us to put the principles of IM to a critical test because risky and potential dangerous situations can be presented to the participants (i.e. the presence of obstacles, conditions of poor visibility) to measure participant's performance (e.g. lane keeping) under these circumstances. Comparative evaluation was used to determine the effects on driving safety and workload with the COMUNICAR IM concept as compared to driving without such a system.

The driving simulator experiment was set up in such a way that drivers encountered several events on a route without IM and on a route with IM. For example, events could be overtaking a slow car, driving through fog etc. while additionally the driver received low-priority information like an SMS messsage or a phone call. Without IM the low-priority information was presented simultaneously to the road event. With IM the low-priority information was delayed and presented later, when the situation was rated as being not dangerous anymore. It was expected that – if IM is effective – safety relevant parameters as well as subjective workload ratings should be better in the condition with IM as compared to the condition without IM.

2 Method

A total of 36 subjects (18 male, 18 female, average age 32) drove the DaimlerChrysler fixed base driving simulator with and without the COMUNICAR IM system activated. On both routes they encountered the same events, like passing a stationary object or overtaking a slow vehicle. During these events a message from one of the information systems was presented, like an incoming phone call or traffic information. At the top of the centre console there was a screen especially designed by the COMUNICAR project to display the outputs of these information systems (see Figure 1). The combination of an event and a message resulted in a total of 12 different scenarios that the subjects encountered during the route. Table 1 presents the complete list of these scenarios.

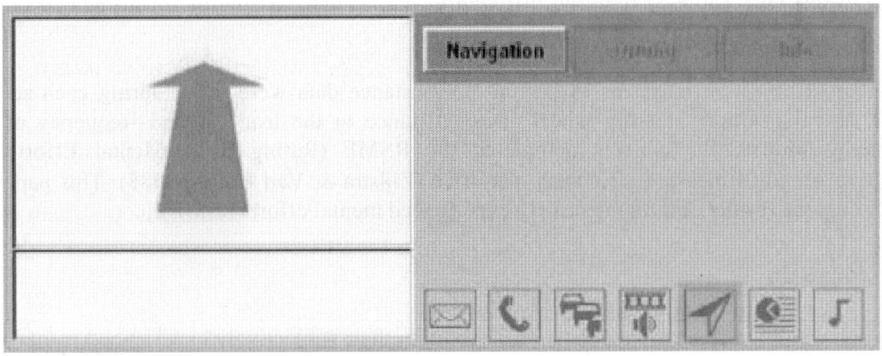

Figure 1: Display of information at the top of the centre console.

Table 1: Overview of the 12 different scenarios (a combination of a road event and an information message) that the subjects encountered during the ride in the driving simulator.

	Road event	Information message
1	Fog area	Navigation message
2	Fog area	Phone call
3	Rain area	Traffic information
4	Rain area	SMS
5	Rain area	Navigation message
6	Roadworks obstacle	Navigation message
7	Bus obstacle	Traffic information
8	Overtake slow vehicle	Navigation message
9	Overtake slow vehicle	Phone call
10	Vehicle in front brakes	Phone call
11	Vehicle in front brakes	Traffic information
12	Cross intersection	Traffic information

Without the Information Management system the message was presented exactly during the event, while with IM the message was postponed to right after the event. Under a third baseline condition the subjects just drove the route with the events, but no information messages were presented. The order of these three conditions was balanced over subjects to prevent memory and order effects. To make sure the participants took notice of the provided information, they had to answer questions about the information after finishing each drive (condition).

The simulated road was a two-lane road, one for each direction, with a width of 3.25 m for each lane. A few hundred meters after the start the driver approached a vehicle in front. The instruction was to follow this vehicle. The speed of the leading vehicle was constant at 80 km/h unless the driving events (see Table 1) required differently (e.g. at the crossing or for a planned braking manoeuvre).

A number of objective measures of driving performance data were taken during each scenario (speed, lateral position, steering wheel angle, distance to the lead car and frequency of lane warnings). Additionally subjects completed the RSME (Rating Scale Mental Effort) as a subjective workload measure after each test drive (Zijlstra & Van Doorn, 1985). This paper will only discuss the results of the subjectively experienced mental effort (RSME).

3 Results

After each condition the participants gave a rating on their subjectively experienced mental effort on the RSME (value between 0 and 150). Mean ratings were 37.8 (with IM), 36.4 (without IM) and 31.6 (reference without any information). The main effect of condition was slightly significant ($F(2,54)=2.36$, $p<0.10$) (see Figure 2). Separate t-tests showed that the condition with IM and without IM were not significantly different from each other ($t(32)=0.4$). Both the condition with IM and without IM showed a tendency to produce higher workload ratings as compared to the reference condition, ($t(32)=1.8$, $p<0.08$).

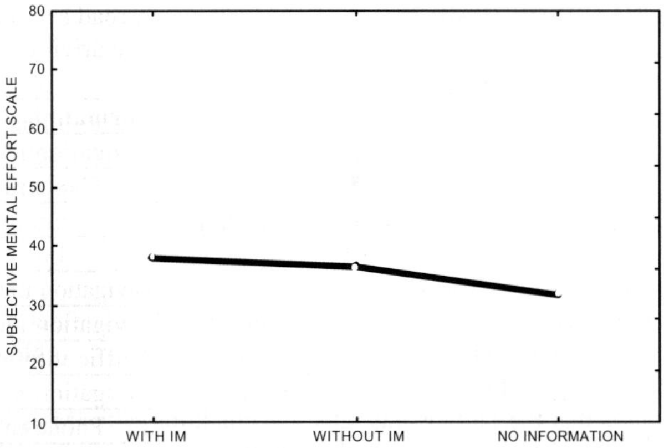

Figure 2: Main effect of condition (with and without IM and the reference condition)

The results also show an interaction effect of condition with the order in which they were presented to the subjects (F(10,54)=2.36, p<0.02). There were 6 different orders in which the three conditions were presented. Figure 3 shows that the participants that started with IM rated this condition as the one with the highest workload, while the participants that started without IM rated this condition as the one with the highest workload. When starting with the reference condition, subjects indicate no difference in workload between the conditions.

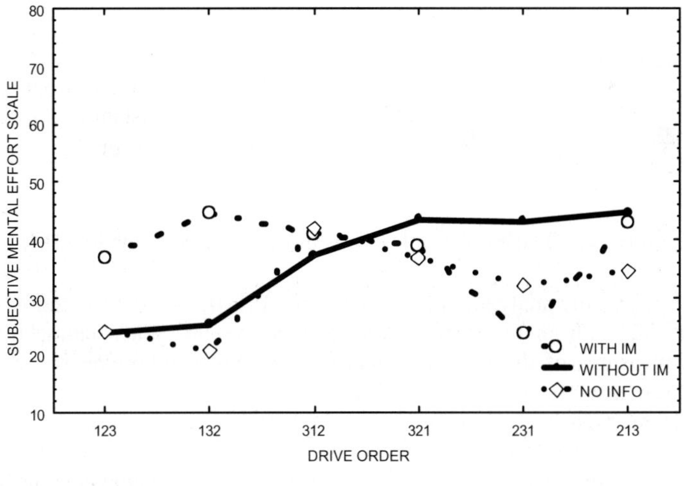

Figure 3: Interaction effect of condition and the order in which the conditions were presented (driver order). 1=drive with IM, 2=drive without IM; 3=drive without INFO

4 Discussion and conclusions

At first sight the results of the RSME are contrary to expectation. It was expected that driving with IM would decrease experienced mental workload compared to driving without IM when messages are presented during difficult events on the road. Because this should create a so called "overload situation". However, this is not clearly the case. On the contrary, driving with IM is on average rated as leading to more workload compared to driving without IM.

There are two explanations possible for this result:
The first one is that the effect of IM is overruled by the interaction effect with what condition participants start out. This condition is rated as the one with the highest workload, whether without or with IM. So there could be a kind of learning effect.
The second explanation is that the participants rated the condition with IM as being more effortful because they used their (implicit) memory on the number of events to give the rating. Figure 4 illustrates that the number of events, as experienced by the participants, is larger for the condition with IM. Consequently, participants who employ such a frequency-based "availability heuristic" when they are asked to give a rating on mental effort might tend to rate the condition with information management as being more effortful.

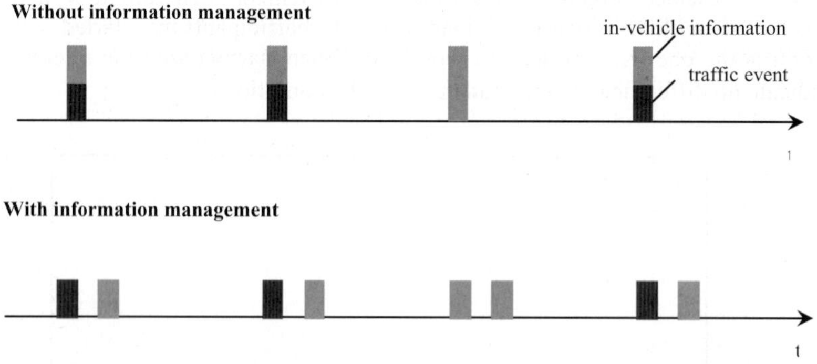

Figure 4: Distribution of loading "events" without and with IM.

It is concluded that, by only analyzing the results of the RSME, we only found weak evidence of the effects of driving with an Information Management system. When taking into account the other dependent measures of driving performance data, maybe more specific evidence can be found to show the advantages of driving with IM.
This paper leads at least to an important methodological conclusion: be careful with the interpretation of data coming from the RSME. We suggest that future research should further explore the validity of especially the second explanation as this would provide further insights into the role of (possibly biasing) memory processes on ratings of subjective workload.

References

Zijlstra, F.R.H. & Van Doorn, L. (1985). *The construction of a scale to measure perceived effort.* Delft, The Netherlands: Department of Philosophy and Social Sciences, Delft University of Technology.

Destination Entry While Driving: The Benefit of Constrained Options to Act in Multitask Situations Illustrated by Two Route Guidance Systems

Georg Jahn[1], *Andreas Keinath*[1], *Josef F. Krems*[1], *Christhard Gelau*[2]

[1]Chemnitz University of Technology
georg.jahn@phil.tu-chemnitz.de, andreas.keinath@phil.tu-chemnitz.de, krems@phil.tu-chemnitz.de

Federal Highway Research Institute
gelau@bast.de

Abstract

In-vehicle information systems have to support time-sharing of visual attention and interrupted operation. We argue that these design needs are met by constrained options in interface dialogues. Task analyses and empirical data for manual destination entry functions of two route guidance systems are presented. The two systems differ with respect to the demands for visual attention and controlled processing. With one system, users have to decide among options during the entry dialogue, whereas with the other system destination entry proceeds in a single line of action. Data from a driving study and from three groups of different ages in training studies confirm the task analyses and show the benefit of constrained options.

1 Using Information Technology While Driving

Drivers perform a dynamic task that requires continuous attention to the road ahead to avoid accidents. Sometimes driving is easy going and drivers can engage in parallel activities, like when driving on a straight road at moderate speed in little traffic. Driving sometimes demands full attention, leaving no spare mental capacity for additional tasks, like when driving in an unfamiliar city and having to decide which turns to take. In such situations the driving task requires controlled perceptual sampling, controlled processing of information and conscious decisions about options to act. If the driving task requires controlled processing, drivers usually cancel parallel tasks and concentrate on safe driving.

Tasks performed parallel to driving vary in demands as does the driving task itself. Some are simple like listening to the radio, others demand controlled processing similar to having to decide which turns to take. In-vehicle tasks that require controlled processing are difficult to perform as a secondary task while driving and are vulnerable to interruptions (Miyata & Norman, 1986). They may even create safety hazards by delaying responses to sudden incidents and by impairing drivers situational awareness. In this paper, we will look at different destination entry methods with regard to the amount of controlled processing they need.

---

[1] Department of Psychology, Chemnitz University of Technology, 09107 Chemnitz, Germany
[2] Federal Highway Research Institute (BASt), Brüderstraße 53, 51427 Bergisch Gladbach, Germany

1.1 Strategic distribution of attention

While the vehicle is in motion, drivers are acting in a multitask situation. At the very least they have to control the vehicle and they have to monitor the traffic situation for possible incidents that require responses. Driving also includes navigating familiar or unfamiliar routes, maintaining awareness of traffic regulations and interacting with other traffic participants. Drivers strategically manage their attentional resources and foreground activities in adaptation to situational demands. If necessary, they slow down in order to reduce situational demands.

The driving tasks that must not be suspended, vehicle control and monitoring for incidents, require continuous visual attention. Even if driving is reduced to these basic tasks, visual attention can be directed away from the road only for short intervals. There is a characteristic pattern for time-sharing of visual attention between driving and a parallel in-vehicle task that requires visual attention. Drivers glance to the in-vehicle task for 1 to 3 s alternating with shorter control glances to the road ahead. The duration of glances to the in-vehicle task is task-dependent, but the mean glance duration usually does not exceed 1.6 s (Lansdown, 2001). Empirical studies also show that in-vehicle tasks normally are performed as continuously as possible with the driving task backgrounded. If the driving task demands controlled actions, the in-vehicle task is suspended and continued later on. Drivers try to schedule in-vehicle tasks in low demand intervals to avoid interruptions.

1.2 Manual destination entry with route guidance systems

One example for a visually demanding in-vehicle task is manually entering the destination into a route guidance system. In some systems, destination entry is only accessible while the vehicle is parked, however, with many other systems it is accessible and can be used while driving. If a destination is not preprogrammed and selected from a list, destination entry includes spelling. Repeated manual entries are needed and system responses to these inputs have to be verified visually. The visual and manual demands of destination entry are comparable to those of other tasks performed with in-vehicle information and communication systems. Similar text entry and list selection occurs while using phones, PDAs and information services.

2 Two Design Solutions for Destination Entry

In a driving study (Jahn, Krems, & Gelau, 2002) that focused on the acquisition of skills in doing in-vehicle tasks, two route guidance systems with comparable functions were used. Both are operated with manual controls, nonetheless they differ in interesting ways. In this paper, we only highlight features that affect the facilitation of use under time-sharing conditions.

The first system (System A) is available as an add-on accessory. It has a colour monitor and is operated by a hand-held remote control with four "arrow" keys for navigating the display and an "OK" key. The second system (System B) is a built-in system, it too has a colour display, but is operated by a single dial button next to the display in the console.

2.1 Layout of the spelling screens

After choosing the destination entry function, with both systems the city and the street name of a destination have to be spelled by entering at least some characters. The spelling screens of both systems are schematically shown in Figure 1. The screen of System A is divided in three parts, one showing the alphabet for character selection, one showing the characters entered so far, and one showing four lines as a window on the database of cities or street names. The window on the respective database is adjusted dynamically in response to the string of entered characters. Switching to the database is necessary to select a city or street name. This is also possible if the intended string is not yet visible. The user has to decide when to quit entering characters to initiate a switch to the database. Monitoring the often very similar four lines is visually demanding.

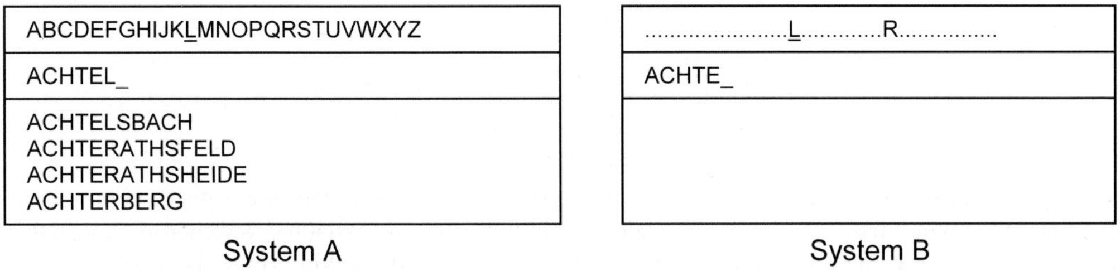

Figure 1: Schematized screens of the two route guidance systems

In contrast, the spelling screen of System B only displays characters for selection and characters already entered. System B employs an intelligent speller. It refers to the respective database offering only those characters for selection that would arrive at a match in the database. Additionally it automatically adds characters to the current string that are already disambiguated. Entering letters continues until the speller indicates a match with an entry in the database.

2.2 NGOMSL models of the spelling procedures

We present fragments of NGOMSL models (Kieras, 1996) to clarify the differing entry methods of the two systems and to specify the respective demands for controlled processing. The top-level method for destination entry is the same for both systems:

```
Method for goal: enter destination.
  Step 1. Accomplish goal: enter string <CITY>.
  Step 2. Accomplish goal: enter string <STREET>.
  Step 3. Accomplish goal: start route guidance.
  Step 4. Return with goal accomplished.
```

The model for System A contains a selection rule set for deciding when to switch to database selection that users encounter inevitably while accomplishing the goal of destination entry:

```
Method for goal: enter string <STRING>.
  Step 1. Retain that string is <STRING>.
  Step 2. Accomplish goal: proceed with entering.
  Step 3. Decide: If spelling screen is displayed, Then goto 2.
  Step 4. Press ok button and forget that string is <STRING>.
  Step 5: Return with goal accomplished.
Selection rule set for goal: proceed with entering.
  If <ENTERED CHARS> is short,
    Then accomplish goal: enter character <NEXT CHAR>.
  If an entry in <DATABASE WINDOW> matches <STRING>,
    Then accomplish goal: select in database.
  If entries in <DATABASE WINDOW> are similar to <STRING>,
    Then accomplish goal: select in database.
  If <ENTERED CHARS> is not short and entries in <DATABASE WINDOW> are
    not similar to <STRING>,
    Then accomplish goal: enter character <NEXT CHAR>.
  Return with goal accomplished.
```

In contrast, System B's destination entry is accomplished without a selection rule set:

```
Method for goal: enter string <STRING>.
  Step 1. Retain that string is <STRING>.
  Step 2. Accomplish goal: enter characters.
```

```
    Step 3. Verify that <STRING> is displayed.
    Step 4. Press dial button and forget that string is <STRING>.
    Step 5. Return with goal accomplished.
  Method for goal: enter characters.
    Step 1. Accomplish goal: enter character <FIRST CHAR>.
    Step 2. Read <ENTERED CHARS> from screen.
    Step 3. Decide: If <ENTERED CHARS> match <STRING>, Then goto 5.
    Step 4. Accomplish goal: enter character <NEXT CHAR> and goto 2.
    Step 5. Return with goal accomplished.
```

The dynamically adjusted database window of System A is permanently present as a reminder of the option to switch from spelling to list selection in the database. Deciding whether to choose this option requires perceptual sampling to test the conditions of the selection rules. The selection rule set and the frequency of this decision might differ between individual users. However, this decision is visually demanding and requires controlled processing. In contrast, spelling with System B is straightforward and the display provides clear feedback on progress in a single line.

3 The Benefit of Constrained Options To Act

The higher demands for controlled processing and visual attention affect destination entry performance in System A compared to System B. In a driving study we trained six drivers aged 35 to 55 with 68 destination entry tasks on System A and six different drivers with 100 tasks on System B (there was insufficient time for 100 tasks with System A). In three further studies we trained participants of three different age groups who all held valid driving licenses. Each participant performed 100 tasks with one of the two systems in a parked car. One group (named *parking* in Figure 2) consisted of 44 participants aged 35 to 55 as in the driving study, an *old* age group consisted of 16 participants aged 65 or older, and a *young* age group consisted of 16 participants aged 18 to 25. Figure 2 shows Total Task Times (TTT) for the three age groups trained in the parked car and Total Glance Times (TGT) for the *driving* study group. TGT is the sum of the durations of display glances that occurred during a destination entry. It was determined from video recordings. Mean TTT and mean TGT are displayed for three training levels, they are averaged over four destination entry trials each (1-4, 65-68, 97-100).

The mean TGT in the driving study (filled symbols) was higher for System A than for System B. This difference partly resulted from the differing controls. The remote control of System A required training. However, even after considerable training there remained an advantage with System B that is best explained by the void of decision making about options. Also, subjective ratings of the required mental effort were scored lower for System B. Subjective ratings of mental effort were sensitive to the specific design differences only under multitask conditions of the driving study, not in the training studies in the parked car.

The TTT for the middle age group in the parked car was similar to TGT in the driving study for System A. TTT and TGT in the middle age groups are lower for System B, however TTT was higher than TGT. System B had longer delays between entry steps. Participants in the driving study used these delays for glances to the road and therefore the delays are not completely reflected in TGT. Participants in the old age group using System B showed an advantage over those using System A. This was maintained throughout the training even after substantial practice. Participants in this group differed in many respects from participants in the other groups including prior experience with computers, perceptual speed and speed of controlled processing.

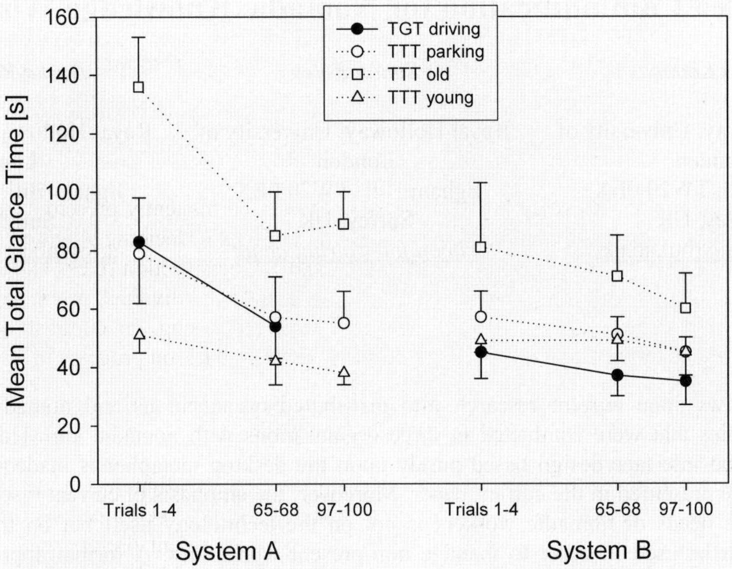

Figure 2: Mean Total Glance Times (TGT) in a driving study and mean Total Task Times (TTT) for three age groups (parking 35-55, old > 65, young 18-25) in training studies at three training levels (two for System A in the driving study); error bars denote *SD*, *N* is given in the text.

With System B age group differences in TTT were less pronounced. Young participants had no difficulties with System A, they even were faster with System A, because young participants who used System B were slowed down by the longer delays of System B.

As a conclusion, these results of two design examples illustrate that interfaces used in multitask situations under time-sharing of visual attention should minimize visual demands and demands for controlled processing. Both design needs are met by constrained options, because deciding among options to act also requires perceptual sampling. Alternative options might be kept accessible for the advanced user, but should be reduced in salience or hidden. A single line of action is most effective to minimize controlled processing that is challenging in multitask situations and that is vulnerable to interruptions.

References

Jahn, G., Krems, J. F., & Gelau, C. (2002). Skill-development when interacting with in-vehicle information systems: A training study on the learnability of different MMI concepts. In D. de Waard, K. A. Brookhuis, J. Mooral, & A. Toffetti (Eds.), *Human factors in transportation, communication, health, and the workplace* (pp. 35-48). Maastricht, NL: Shaker Publishing.

Kieras, D. E. (1997). A guide to GOMS model usability evaluation using NGOMSL. In M. Helander, T. Landauer, & P. Prabhu (Eds.), *The handbook of human-computer interaction* (2nd ed.) (pp. 733-766). Amsterdam: North-Holland.

Lansdown, T. C. (2001). Causes, measures, and effects of driver visual workload. In P.A. Hancock & P.A. Desmond (Eds.), *Stress, workload, and fatigue* (pp. 351-369). Mahwah, NJ: Erlbaum.

Miyata, Y., & Norman, D. (1986). Psychological issues in support of multiple activities. In D. A. Norman & S. W. Draper (Eds.), *User centered systems design: New perspectives on human-computer interaction* (pp. 265-284). Hillsdale, NJ: Erlbaum.

Interface or Interspace? Mediated Communication for Nomadic Knowledge Workers

Stavros Kammas

Royal Holloway, University of London
Egham Hill, TW20 0EX, Surrey, UK
s.kammas@rhul.ac.uk

Simon Foley

Royal Holloway, University of London
Egham Hill, TW20 0EX, Surrey, UK
simon.foley@rhul.ac.uk

Duska Rosenberg

Royal Holloway, University of London
Egham Hill, TW20 0EX, Surrey, UK
d.rosenberg@rhul.ac.uk

Abstract

The paper draws upon current research into distributed organisations and nomadic working. Empirical studies that were conducted in three organisations with nomadic knowledge workers demonstrate that interface design based purely upon the desktop metaphor is inadequate for the reasons that are presented in the current paper. Moreover, the emphasis of current research for the communicative needs of nomadic workers is not on the technology itself but on the way that technology can be used in order to manage non-present workforces. A formal approach to the dynamic "interspace" where nomadic workers communicate reveals the complexity of the situation. In this paper, the authors present a model of the requirements for mediated communication across synchronous and asynchronous spaces.

1 Introduction

In the new economy, knowledge is the main asset in the business context (Davenport & Prusak, 1997). There is a continuous challenge for knowledge to 'be there' and moreover to 'be there on time' (Nonaka, 1994). Under these competitive circumstances contemporary organisations, in order to assure their commitment to knowledge performance, need to make advances in the way they are coping with work time and work load (Morris, Meed, & Svensen, 1996). Moreover, the global competitive environment forces modern enterprises to become involved in international and innovative projects, which demand the employment of large numbers of people in different locations. Therefore, the main challenge for these organisations is to ensure fast and easy access to informational and human resources.

However, in the context of multidisciplinary projects - tasks, business activities and processes often occur in dispersed locations according to the needs of the project and the availability of the resources. Multidisciplinary members are often mobile, repositioning themselves either as a team or as sub-teams or as individuals. To some extent this lack of permanency in the physical work context - a feature of nomadic working (Skyrme, 1999) in multidisciplinary teams, becomes problematic with regard to collaborative working. For example, organisations might run several different projects that occupy a number of people: sometimes the same person might work in more than one project; sometimes there might be partnerships of more than one organisation in each project. Therefore in many circumstances, it might not always be possible or necessary for the project to have a physical space – making any definition of the project space as whole, ambiguous.

However, when people work together, studies show that they prefer to work at the same place with a view to using a common collaborative space. The absence of a permanent space of interaction can, in some circumstances, destabilise existing networks of people and reduce opportunities for collaborative working. The provision of permanent spaces of interaction in mediated environments might reduce some of the ambiguity within the project space and promote additional opportunities for collaboration. This paper therefore examines the mediating factors that either hinder or facilitate multidisciplinary project teams in gaining rapid and unproblematic access to informational and human resources.

Dix, Finlay, Abowd, & Beale, 1998) outline different modes of collaboration: people can work:

- Face to face - *synchronous collocated.*
- At the same time but from different places - *synchronous non-collocated.*
- At the same place but at different times - *asynchronous collocated.*
- At different times and from different places - *asynchronous non-collocated.*

Depending on the constraints of the situation or even what people find more convenient at the specific time, they choose one mode over another when working with others. After describing their investigations into how people in multidisciplinary teams decide on these working modes and select appropriate technology to communicate with others, the authors of this paper propose a framework for mediated communication. This framework addresses how a sense of permanency in mediated environments might be introduced into the dynamic "interspace" of information; an interspace rather than an interface, where nomadic workers in multidisciplinary teams, might secure fast and easy access to informational and human resources.

2 Empirical Research

To explore the concept of interspace, investigations were carried out in five major work sites; Office, Home, Client, Commuter, and Public Site. Nine semi-structured interviews, plus nine role-play interviews based on user-specific scenarios, have been conducted in three organisations in the UK and Spain with nomadic knowledge workers. The emphasis of the empirical research was on how nomadic knowledge workers organise their workspace, define their work tasks, use resources including technology at work and interact with colleagues at work.

The limitations of the media and specifically the interface, were the main focus of the empirical studies. In such highly distributed environments, where co-workers need to have a shared understanding and common ground needs to be established between the different mobile users, the interface that follows desktop metaphor is inadequate. The metaphor fails to address the complex requirements of knowledge workers who need direct access to their colleagues' current project space if they are to collaborate effectively. The diffusion of time, space, people and resources creates new collaborative rules of which knowledge workers, software and workplace designers need to be aware.

For example, investigations into Office Site Working demonstrate that the primary concern in the establishment of technology enabled spaces is the maintenance of permanent spaces of interaction through the provision of informational spaces. Some of the mediating factors which informants believed contribute to structural stability are displayed as a causal network in Fig. 1. That means a mediated space that can support evolving business processes whilst preserving the emotional and contextual awareness can contribute to effective communicative and collaborative activity.

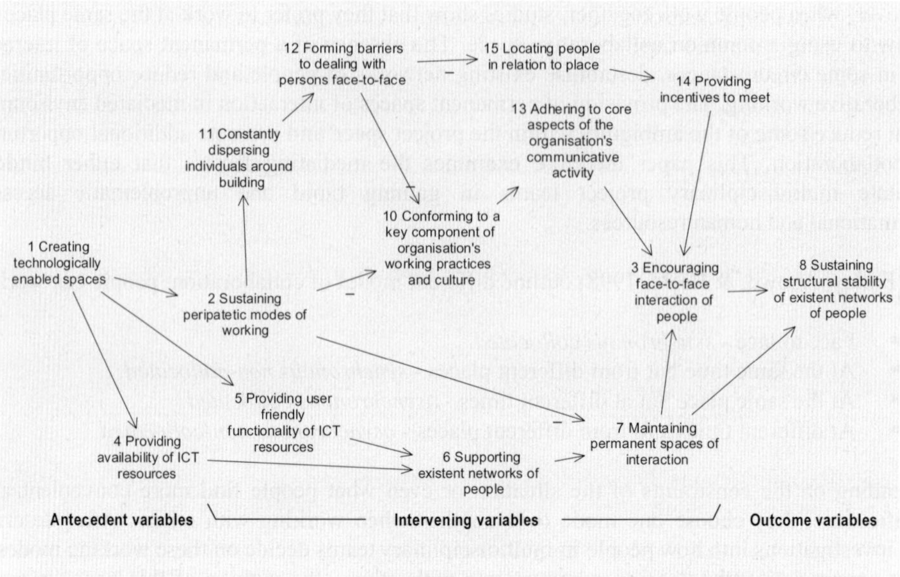

Fig. 1. Technology Enabled Spaces for Office Site Working

Similarly for Client Site Working, the primary concern in designing technological interfaces should not only be to enable access to people and resources, but also to sustain emotional support and interaction between colleagues at different sites and locations. In other words, interfaces must allow individuals to identify the constraints to communication from their physical environment as well as the possibilities for communication in mediated spaces. For example, interfaces must allow users to emphasise and account for geographical distance. Moreover, they should represent the existence of those physical boundaries across space, be it physical, mediated or hybrid. We therefore conclude that the global village idea that the new technology has made distance irrelevant, or obsolete, does not hold, since distance is conveyed not only in physical terms, but also in organisational and social.

In the case of Home Site Working, if knowledge workers are to achieve a work/home life balance that enables them to maximise productivity and organise their time effectively, then the primary concern in designing interfaces is to help users who wish to manage distractions to identify optimal measures, times and spaces for working uninterrupted on concentrated activities and to correspondingly communicate their availability to others. In other words, it is necessary to provide nomadic workers with the means to organise their work in relation to their domestic and familial obligations. To restore the boundaries that have been merged, blurred or even lost in new ways of working to a degree of representation that allows individuals working within collaborative networks to organise and focus their activities accordingly.

Finally, for the Commuter and Public Site Working, if knowledge workers want to make optimal use of commuting time to maximise productivity then the primary concern in designing interfaces to this end is to provide users with the means to restore and sustain distinct boundaries between the various working locations. In other words, interfaces should enable users, not only to engage in many active and passive productive activities whilst travelling, but also to coordinate conflicting demands on their time and energy. For example, consideration should be given to the wider application of existing interfaces that provide asynchronous information on optimal travelling times to avoid traffic; train delays caused by engineering works; to the availability of quite train carriages for concentrated preparation of design or documents; to online modification to flight departure times etc.

3 The Framework

To capture the complexity of the relationships between nomadic knowledge workers, the analysis employs the concept of communication frames that systematically relate information from various sources. Frames can also be a formal basis for integrating heterogeneous elements into a coherent structure, thus providing key parameters for the development of a Model for Mediated Communication of Nonadic Knowledge Workers. The communication frames capture the key attributes of participants in a communicative event, for example, who they are, where they are located, what activities they are engaged in. Furthermore communication frames need not always capture information about human participants, but can also represent resource mediators within technology configurations that enable communication in hybrid - that is, integrated physical and mediated – work environments.

The communication frames verify the parameters of an exclusive communicative situation (agent, topic, activity, setting, resources). They represent a particular set of communicative constraints (or mediating factors) on an individual knowledge worker's activities and assigned roles within a business process.

In the Model for Mediated Communication of Nomadic Knowledge Workers, we address the aforementioned research questions and show how the information from a variety of sources is integrated into common ground, that is, how various communication channels are co-coordinated. We also show what function individual channels have in this context and how they relate to the use of shared artifacts. Finally, how the resulting shared knowledge is retained in the common ground of different kinds of participant (speakers/listeners, over-hearers, trackers). In other words, we introduce the central concepts of the Model (resource mediators: gatekeepers, active, passive and asynchronous interfaces).

For example, the Model in Fig.2 represents how the parameters of combined communication frames, when regarded in hybrid (synchronous and asynchronous) environments, identify possible configurations of various communicative situations according to formal rules - rules which reflect and articulate the needs of knowledge workers (people) and how they relate to the collaborative features (i.e. resource mediators: gatekeepers, active, passive and asynchronous interfaces) that constrain and expand the work context (people, process, place and technology (tools).

The central concepts described within the Model were constructed from the theoretical base of the (Extended) Common Ground Approach (Clark, 1996, 1991, 1989) and the key themes that emerged from our empirical studies. The Model enables researchers to analyse and interpret the dynamic aspects of communication and awareness issues and their potential impact on interaction design. Some of these issues relating to interface and boundary control emanated from our empirical studies.

In summary, the Extended Common Ground Approach, incorporated into this model, looks at the communicative event and takes into account the extra-linguistic context of communication such as shared resources, shared space, shared objects and knowledge worker activities, the degree of participation in joint activities and tasks, interaction in hybrid environments and the role of the workspace in terms of shared articulations, shared artefacts and enabling technologies. The Extended Common Ground thus integrates the resource(s), object(s) or representation(s) as mediating factors in the process of creating shared, mutual or common knowledge of the people involved in work activities.

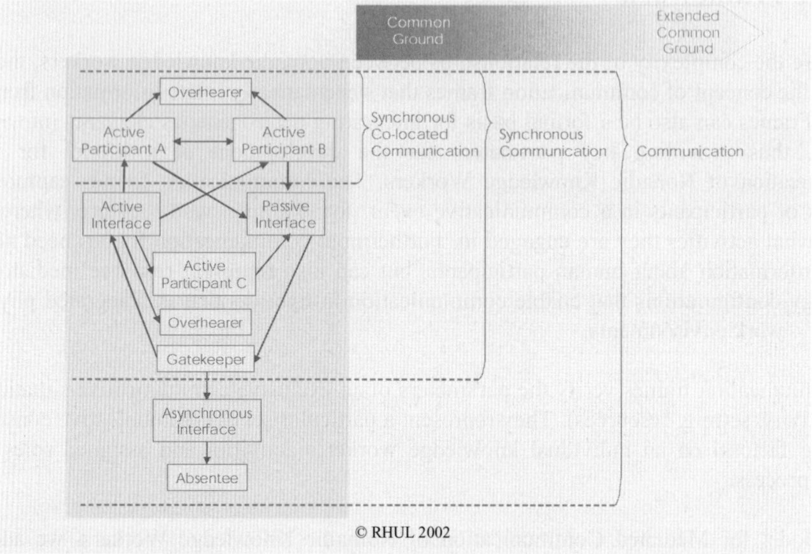

Fig. 2 Model for Mediated Communication of Nomadic Knowledge Workers

The current research has been conducted as part of the Sustainable Accommodation in the New Economy Project (SANE IST 2000-25257).

4 References

Clark, H. (1996). Using Language. Cambridge University Press.

Clark H., & Brennan S. (1991). Grounding in communication, 127-149 in Resnick L.B., Levine J.M., and Teasley S.D., (eds) *Perspectives on socially shared cognition.* American Psychological Association.

Clark H., and Schaefer E. (1989). Contributing to discourse. *Cognitive Science*, 13, 259-292.

Davenport, T. & Prusak, L. (1997). Working Knowledge: How Organizations Manage What They Know. Harvard Business School Press.

Dix, A., Finlay J., Abowd G., & Beale, R. (1998). Human Computer Interaction. Prentice Hall Europe.

Morris, S., Meed, J., & Svensen N. (1996). The Knowledge Manager: Adding Value in the Information Age. Pitman Publishing.

Nonaka, I. (1994). A Dynamic Theory of Organisational Knowledge Creation. *Organisational Science*, 5(1), 14-37

Skyrme, D.J. (1999). Knowledge Networking: Creating the Collaborating Enterprise. Massachusetts: Butterworth-Heinemann.

Wurman, R. S. (2001). Information Anxiety 2. Indianapolis, IN: QUE.

Telemurals: Catalytic Connections for Remote Spaces

Karrie Karahalios and Judith Donath

MIT Media Lab, Sociable Media Group
20 Ames St. E15-468
{kkarahal, judith}@media.mit.edu

Abstract

Mediated communication between remote social spaces is a relatively new concept. One current example of this interaction is video conferencing among people within the same organization. Large-scale video conferencing walls have begun to appear in public or semi-public areas such as workplace lobbies and kitchens. These connections provide a link via audio and/or video to another space within the organization. When placed in these spaces, they are often designed for casual encounters among people within that community. Thus far, communicating via these systems has not met expectations. We are exploring a different approach to linking spaces through the use of what we are defining as a social catalyst. These catalysts are incorporated into our installation, *Telemurals*. An ethnography study of *Telemurals* will be conducted between two graduate dormitories to see how the catalysts affect interaction.

1 Introduction

In this work, we are creating an audio-video communication link between remote spaces for sociable and casual interaction. Some drawbacks to current systems that have been studied include lack of privacy, gaze ambiguity, spatial incongruity, and fear of appearing too social in a work environment [5]. We believe that many of these problems stem from designing interfaces that directly map to face-to-face interaction.

With this work, we are diverging from the approaches of current audio-video connections and focusing on encouraging social interaction by designing a series of social catalysts. We are not creating a substitute for face-to-face interaction, but rather new modes of conversational and physical interaction within the spaces.

2 Social Catalysts

The main idea of the social catalyst is to initiate and create mutual involvement for people to engage in conversation. For example, in a public space, it is not customary to initiate conversation with random strangers. However, there are events that act as catalysts and connect people who would not otherwise be communicating with each other.

Such a catalyst may be an experience, a common object like a sculpture or map, or a dramatic event such as a street performer. Sociologist William Whyte terms this phenomena triangulation:

"*A sign of a great place is triangulation. This is the process by which some external stimulus provides a linkage between people and prompts strangers to talk to each other as if they were not.*" [6]

Our hypothesis is that the creation of a social catalyst as an integral part of the social environment will aid mediated communication between spaces by providing a spark to initiate conversation and the interest to sustain it.
The social catalysts of our installation extend Whyte's triangulation principle into the display and interface of the connected space. The form of our catalyst is abstract. It alters the space and communicative cues between the two spaces. One such catalyst is a connection where current conversation of the users appears as graffiti in the environment. This allows the occupants to see they are affecting the space and might encourage them to alter it. While the possibilities are infinite, the challenge is determining which agents on the interface are effective as social catalysts and why.

In our linking of two spaces in the *Telemurals* installation, we are augmenting the appearance of the familiar audio-video wall interface with stimuli that are initiated at either end of the connection. The wall is intended to be not only a display, but an event in itself; the system becomes both medium and catalyst.

3 Telemurals

Telemurals is an audio-video connection that abstractly blends two remote spaces. The initial setup is straightforward. Two disjoint spaces are connected through an audio-video wall. Video and audio from each space is captured. The two images are then rendered, blended together, and projected onto the wall of their respective space.

Duplex audio is transmitted between the two locations. To provide feedback and comic relief, the audio is passed to a speech recognition algorithm. The algorithm returns text of the closest matching words in its dictionary. This text is then rendered on the shared wall of the two spaces. The goal here is to make it clear that the users' words are affecting the space without necessarily requiring 100% accuracy of the speech recognition system.

A figure of the current implementation of *Telemurals* is shown in Figure 1. Silhouettes of the participants in the local space are rendered in orange. The participants at the remote end are rendered in red. When they overlap, that region becomes yellow. The aim of this cartoon-like rendering is to transmit certain cues such as number of participants and activity level without initially revealing the identity of the participants.

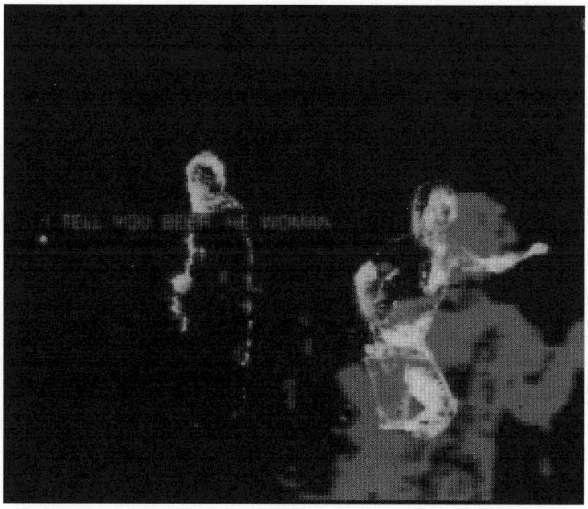

Figure 1. Current Telemurals implementation.

Participation is required for this communication space to work. To reinforce a sense of involvement, we provide the system with some intelligence to modify its space according to certain movements and speech patterns. That is, the more conversation and movement between the two spaces, the more image detail will be revealed to the participants at each end. The silhouettes slightly fade to become more photo-realistic (see Figure 2). This prompts the participants to move closer into the space to see. If conversation stops, the images fade back to their silhouette rendering. We want the participants to choose their own level of commitment in this shared space [4]. The more effort they exert, the more they see of both spaces.

Figure 2. Example of fading from silhouette towards more photo-realistic.

Much thought has been given to the design of the renderings in *Telemurals*. We wanted to maintain the benefits of video in their simplest form. Adding video to a communication channel improves the capacity for showing understanding, attention, forecasting responses, and expressing attitudes [3]. A simple nodding of the head can express agreement or disagreement in a conversation. Gestures can convey concepts that aren't easily expressed in words; they can express non-rational emotions, non-verbal experiences. Yet these cues are not always properly transmitted. There may be dropped frames, audio glitches. Lack of synchronicity between image and audio can influence perceptions and trust of the speaker at the other end. Other challenges include equipment placement. For example, camera placement has long been a reason of ambiguous eye gaze in audio-video links. A large camera offset gives the impression that the person you are speaking to is constantly looking elsewhere.

With *Telemurals*, we are creating an environment where rendered video maintains subtle cues of expression such as posture and hand motion, yet also enhances other cues. For example, changes in volume alter the style of the rendered video. By adding another layer of abstraction into the video stream, we can enhance cues in a manner that is not possible in straight video streams.

In this project, the abstraction of person, the blending of participants, the graffiti conversation, and the fading from abstract to photo-realistic are the social catalysts for the experience. This new wall created by filtering creates an icebreaker, a common ground for interaction, and an object for experimentation.

Telemurals is currently installed in a common area within the Media Lab. The first public *Telemural* installation will connect two recreation halls of MIT graduate dormitories. The connection will run for two months during a social hour at the two dormitories.

4 Evaluation

The field for this observation study is the semi-public space within the two chosen dormitories. We expect the participants to be graduate students who live in the respective dormitory and their friends. We are primarily interested in seeing, (1) how people use *Telemurals*, (2) if the catalysts attract them, and (3) how we can improve the system.

4.1 Methodology

The *Telemurals* observation will take place in March and April of 2003. *Telemurals* will run for two hours each Wednesday and Sunday night in conjunction with a coffee hour/study break. Signage will be placed in the entryways of both spaces to describe what is being transmitted and the privacy concerns of the project.

Video cameras will record people in front of both *Telemurals* spaces. The footage from these tapes will be used to annotate patterns of use for this study and will then be discarded. There will also be a person at each *Telemurals* site to mitigate technical difficulties. The videotape and notes will then be annotated. Initially, we are interested in observing:

- How long people speak using *Telemurals*
- The number of people using the system at any one time
- The number of people present but not interacting
- The number of unique users (if possible)
- The number of repeat users (if possible)
- The number of times and the duration that people use *Telemurals* in one space only
- The number of times and the duration that *Telemurals* is used in photo-realistic mode
- Repeated patterns of interaction: gestures, kicks, jumps, screams

These are factors that we believe are indicative of levels of interaction. However, one must always be open to the unexpected and attempt to find other underlying patterns as well in studying the social catalysts.

4.2 Privacy

When running such a project and study, it would be irresponsible to ignore privacy concerns. The audio and video transmitted in the *Telemurals* interface is not saved or stored in any way. We hope to mitigate this problem with proper signage.

References

1. Bly, S. and Irwin, S. Media Spaces: Bringing people together in a video, audio and computing environment. *Comm. ACM36,1*, 28-47.
2. Galloway, K. and Rabinowitz S. Hole in Space. Available at http://www.ecafe.com/getty/HIS/
3. Isaacs E. and Tang J. What Video Can and Can't do for Collaboration: A Case Study. *Multimedia '93*.
4. Jacobs, J. *The Death and Life of Great American Cities*. New York: The Modern Library. 1961.
5. Jancke, G., Venolia, G., Grudin, J., Cadia, J., and Gupta, A. Linking Public Spaces: Technical and Social Issues. Proceedings *of CHI2001*.
6. Whyte, W.H. *City: Rediscovering the Center*. New York: Doubleday. 1988.

Visual Interfaces for Mobile Handhelds

Bernd Karstens — University of Rostock, 18055 Rostock/Germany, karstens@informatik.uni-rostock.de

René Rosenbaum — University of Rostock, 18055 Rostock/Germany, rrosen@informatik.uni-rostock.de

Heidrun Schumann — University of Rostock, 18055 Rostock/Germany, schumann@informatik.uni-rostock.de

Abstract

Mobile handhelds become smaller and more handy which leads to new challenges in human computer interaction. This effects in particular the design of the visual interface. New paradigms must be developed, especially due to the limited display size and computing power.
In this paper we present techniques to concrete problems in the display of large images and browsing the World Wide Web via mobile handhelds. The proposed techniques have been evaluated in practical tests. The given results show the improved support in navigation, orientation and interaction.

1 Introduction

Small mobile devices have become more powerful and popular in recent years, and are used in different application areas. A typical example are personal mobile navigation systems. However, since mobile handhelds suffer from limited resources, like screen space, interaction facilities and computational power, new paradigms for information presentation and especially navigation are needed.

In information visualization special presentation techniques for efficient use of screen space have been developed. So called FOCUS & CONTEXT-TECHNIQUES are a popular example of this approach (see [Leung et al., 1994][Keahey, 1998]). These techniques combine a focus view, which shows a part of the layout at a high degree of detail, and a context view, which presents the whole information in lower detail to provide an overview. These concepts can be used for information presentation on mobile handhelds, as well [Björk et. al, 1999][Fishers et al., 1997][Buyukokten et al., 2000]. Besides special presentation techniques we need concepts to reduce the amount of represented information in an appropriate way. A first approach of information hiding is the FILTER FISHEYE VIEW [Furnas, 1981], where information is hidden based on the distance to a point of interest. Another example is the EVENT HORIZON [Taivalsaari, 1999], which can be considered as an abstraction of a certain set of information objects. Here, the visibility of information is controlled by a radial movement of objects into or out of the EVENT HORIZON. Further examples for information hiding are SECTION OUTLINING based on text filtering and generation of abstracts [Brown et al., 1996], or POWER BROWSER [Buyukokten et al., 2000] with image removal or scaling. However, information hiding also requires suitable interaction mechanisms to explore the hidden information.

This paper shows, how FOCUS & CONTEXT-TECHNIQUES can be used for navigation and presentation of large raster images on mobile handhelds (see section 2). Furthermore, we propose an approach for navigating throughout the Web on mobile devices using information hiding (section 3). Future work and conclusions closing our contribution in section 4.

2 Exploration and navigation of large images using mobile handhelds

In mobile environments, the size of large images often exceeds the display area of the users output device. There are too many pixels which have to be displayed eligibly. This problem can be solved by using FOCUS & CONTEXT-TECHNIQUES from the field of information visualization [Karstens et al., 2002]. Here, it is important to provide a fast response of the system to enable the user to develop a *feeling* for the presentation. Hence, during interaction, speed is more important than quality.

An established technique for displaying large raster images on small displays is the RECTANGULAR FISHEYE-VIEW [Rauschenbach, 1999][Rauschenbach et al., 2001]. Here, a focus area is displayed in a distorted and therefore reduced context area. The context is divided into belts. Near the focus belts are less distorted than near the image borders (Fig. 1a). The focus area itself is displayed without any distortion, which makes it rather easy to perceive image content and to navigate within the image.
All tasks regarding interaction are executed by manipulating the focus. This eases the rapid interactive exploration on small devices by providing only a few defined interaction points on the display which can be altered by the user. All other transformations are implemented transparently to the user. To achieve real-time interaction, a belt-sequentially screen redrawing is used. First, the new focus region is drawn, than the first context belt and so on.

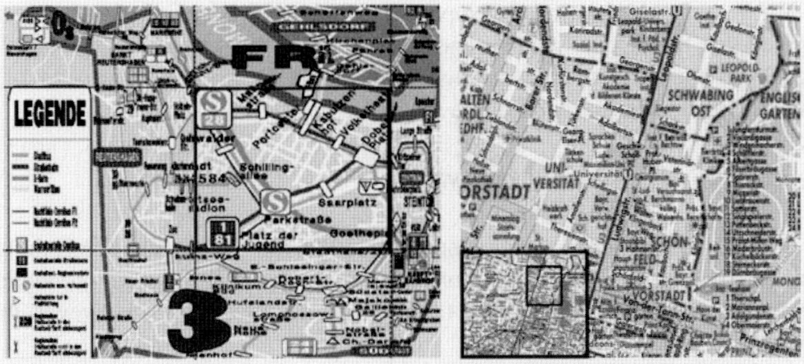

1a) Rectangular FishEye-View 1b) Large Focus-Display

Figure 1: Presentation techniques for large images on small devices.

In many applications the user is more interested in a specific detail than in the context. To satisfy this demand, the LARGE FOCUS-DISPLAY (Fig. 1b) uses a rather different strategy to combine focus and context without to change the distortion within the context. Most of the available display area is used for the focus. Only a small part is used for the context to provide orientation during interaction. It is a downscaled version of the image, where the currently displayed focus region is drawn in. This allows a panning-like display in combination with an additional view of the whole content. This compensates many disadvantages of ordinary panning. All defined user interactions are executed using the content view only. Nevertheless, the content region hides some of the focus area. By allowing to place the content rectangle somewhere on the screen, we solve this problem via interaction. Due to their demands to processing power, the image representation is quite simple, which make it rather fast during interaction and useful especially for mobile palm-size devices.

With regard to the presented two display techniques, the following direct interactions are defined:

Move Focus: This function allows changes of focus position within the context. Moving will usually be exploited if the point of interest moves slowly into the neighbourhood of the focus and the user wants to explore the image *continuously*. Moreover, the action **Move Context** of the LARGE FOCUS DISPLAY enables a user to move the content view's borders within the focus view and allows to explore the complete focus. Thus, a user can always determine, where the context view has to be positioned.

Jump: Using this function, the focus can be set instantaneously to a new position possibly far away from the old one by specifying the centre position of the new focus. This function is usually exploited, if there is no interest to explore spatial relationships between them. Due to the nature of such a *discrete* step, there is a hash change in the focus view. Nevertheless, the user has still all the orientation since inside the context only the focus position has been updated.

Resize Focus: This function allows the user to adjust the size of the focus region, e.g. the user controls the size of the area of interest Additionally, the user is able to determine the actual importance of the context. So, he can shrink or completely hide the context if it is currently of little interest, or maximize to the focus size if it is temporarily important. Therefore, we have to adapt the whole representation.

The FISHEYE-VIEW saves screen space by supporting focus and context, but it is nevertheless not always guaranteed that the display is large enough to display the whole image. Therefore another action has been provided especially for this view. With **Zoom&Pan** the size of the whole representation can be adjusted by specifying a zoom factor.

3 A case study – Browsing the web

If we display sites in the Internet Explorer running on a PDA, the user must scroll the site to reach whole information. If using the Internet Explorer one might use *–Fit to screen-* to avoid horizontal scrolling of text parts or adapt the *-Text size-* to downsize displayed fonts. However, these options have well known drawbacks.

Therefore, special adaptations are necessary to present WWW sites on mobile handhelds effectively. Thus, we distinguish 4 approaches [Karstens et al., 2002].
Device based adaptation: Special web pages are available for different types of devices. Examples are web pages using WML.
Layout based adaptation: Different layout classes are defined based on certain criteria, like degree of abstraction. These classes are applied to web pages.
Client based adaptation: This approaches like FOCUS&CONTEXT (Flip Zooming [Björk et. al, 1999] or CZ Web [Fishers et al., 1997]) use original information and change the presentation directly on the device.

Document based adaptation: The structure or content of the web page is transformed. Examples are section outlining (Zippers [Brown et al., 1996] or scaling (Power Browser [Buyukokten et al., 2000]).

In our solution we combine a client based adaptation with principles from the document based adaptation. The aim was not to develop a browser with the whole functionality, but to demonstrate the concepts by using documents containing typical HTML elements. Furthermore, we want to allow access to all information without any adaptation of the web server. The core of our technology is a special tree structure called WEBGRAPH used to represent the layout of a web page. The graph is created while parsing the page, whereas the tree is presented during display time. The root of a Webgraph is a frame or a paragraph node, representing the whole document. A leaf node represents a single web element like text, illustration, link, or heading. All other interior nodes contain composite elements like frames, tables or paragraphs. They have at least one child node. Edges specify relationships between nodes. Hence, children of a paragraph node can be a paragraph, table, text, illustration, link or heading node. Nodes in every layer are aligned, e.g. a heading node is the first child node in a paragraph. For table nodes, the table content nodes are ordered regarding row and column number. The WEBGRAPH allows a different treatment of node classes, sub graphs or single nodes. Based on the WEBGRAPH, three concepts (SECTION FOLDING, RELATIVE SIZE and KEYWORD OVERVIEW) have been developed.

SECTION FOLDING is an enhancement of section outlining, but it operates directly on the WEBGRAPH. Every node or sub graph of the Webgraph can be easily hidden in the presentation or can be replaced by an info-bar. Clicking an info-bar element presents hidden information. Thus, the whole information can be explored incrementally by keeping the context. The label for info-bar elements are generated from associated text information like headings or a beginning sentence. Pictures are replaced by a downscaled version. After clicking, they are displayed in original size or using the RECTANGULAR FISHEYE-VIEW (see section 2). Regarding the initial state of the view, an automatic decision between folded and fully represented information is necessary. For this, we use the FILTER FISHEYE VIEW [Furnas, 1981]. Therefore a degree of interest is determined by starting from the current point of interest node in the Webgraph and evaluating a distance function and certain levels of detail. Uninteresting information can be hidden. Thus, the point of interest and its local environment are fully presented.

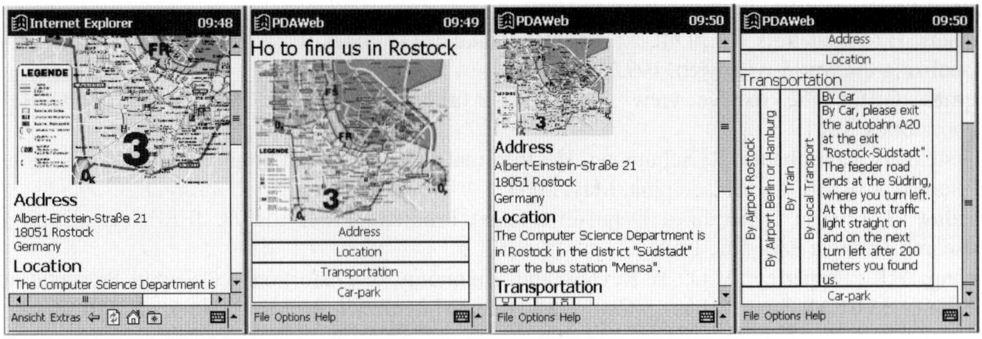

Figure 2: HTML-Page displayed with PDAWeb.

With the concept RELATIVE SIZE the extent of elements within a web page is adapted to the display on a mobile device. This technique is similar to already offered options, which avoid scrolling on textual elements. However, for structural components further techniques are necessary. Several options are possible, e.g. the size can be obtained by absolute or relative size compared to parent or sister nodes or a collection of child nodes.

Figure 2 shows an ordinary Web site, displayed using the Internet Explorer (left) and three different presentations of this site using our proposed concepts. In second picture all information is folded and the picture is scaled. The third picture shows a paragraph of interest and the local environment unfolded, whereas the last picture shows a table with one unfolded table row.
Fast navigation on web sites is also possible using the approach of KEYWORD OVERVIEW, which uses techniques from text filtering. Keywords are listed in an additional window. Thus, it is possible to jump to different nodes of the Webgraph. The keywords are arranged either according to object type or position in the web document. Thus, the keyword overview eases the orientation and navigation especially in unstructured sites.

4 Conclusion

Presentation and interaction on mobile handhelds requires a special treatment to consider the limited resources of these devices like screen space, interaction facilities and computational power. In this paper, we have described how to display large images and how to browse the WWW in such environments. We implemented our concepts on PocketPC-handhelds. We have achieved proper presentations and nearly real-time interaction. Further work will concentrate on usability tests. Moreover, we want to enhance our work on graph drawing to show Webgraphs in other representations and to allow an appropriate navigation on complex web structures.

5 References

Björk,S.; Holmquist, L.E.; Redström, J.; Bretan, I. Danielson, R.; Karlgren, J.; Franzén, K.: West: a Web Browser for small terminals. Proc. ACM Conference on User Interface Software an Technology 1999, ACM Press, 1999.

Buyukokten, O.; Molina, H.G.; Paepcke, A. Winograd, T.: Power Browser: Efficient web browsing for pdas. Technical report, Stanford University, Stanford, 2000.

Brown, M.H.; Weihl, W.E. : Zippers: A focus+context display of web pages. Technical Report SRC-140, Digital SRC Research Report, 1996.

Fishers, B.; Agedelidis, G..; Dill, J.; Tan, P.; Collaud, G.; Jones, C.: Czweb: Fisheye views for visualizing the world-wide web, 1997.

Furnas, G.W.: The fisheye view: a new look at structured files. Technical Report, Bell Laboratories, 1981.

Karstens, B; Rosenbaum, R; Schumann, H,: Information presentation on mobile handhelds, to be presented at IRMA'03, Philadelphia, May, 2003

Keahey, T.A.: The Generalized Detail-In-Context Problem. Proc. Vis'98, IEEE Visualization, Research Triangle Park NC, 1998.

Leung, Y.K.; Apperley, M.D.. A Review and Taxonomy of Distortion – Oriented Presentation Techniques. ACM Transactions on Computer-Human-Interaction, Vol. 1, 2, 1994, pp. 126-160.

Rauschenbach, U.: The Rectangular FishEye View as an efficient method for the transmission and display of large images. Proceedings IEEE ICIP'99, Kobe, Japan, Oct.25-28, 1999

Rauschenbach, U.; Jeschke, S.; Schumann, H.: General Rectangular FishEye Views for 2D Graphics; in: Computers and Graphics, 25(4),2001, S. 609-617

Taivalsaari, A.: The Event Horizon Interface Model for Small Devices, Technical Report TR-99-74, SUN Microsystem Laboratories, 1999

Collaborative Visual Jockey using Mobile Phones

Haruhiro Katayose *Tsuyoshi Miyamichi* *Naruki Mitsuda*[1]

Kwansei Gakuin University
Sanda, 669-133 Japan
http://ist.ksc.kwansei.ac.jp/~katayose/

Wakayama University
Wakayama, 640-8510, Japan
miyamichi@rinku.zaq.ne.jp

Abstract

VJ "Visual Jockey" is the latest remarkable entertainment watching a series of impressive projected images to eyesight, fitting the sophisticated music. This paper presents a VJ system which allows multiple audience to take part in the VJ operation, using mobile phones. We introduce the system architecture based on a server/client, and the overview of trial experimental performances using the proposed system.

1 Introduction

VJ "Visual Jockey" is the latest remarkable entertainment watching a series of strongly impressive projected images to eyesight, fitting the sophisticated music. VJ is an equivalent term to DJ. DJ "Disk Jockey" is the entertainment of music operations; on the other hand VJ adds images to elements of DJ. VJ has been spreading out as the principal highlight of the clubs, especially of the underground culture of young generation (Katayose, 2000).

The VJ performance is performed using a VJ software, which provides the VJ operators real-time operations of image and video. There are some commercial software products, such as digistage's "motion dive[2]" or Steinberg[3]'s "x<>pose". "Motion dive" is a de facto standard product in Japan. Most of the commercial products are designed to be used by a solo operator. There is no software that allows multiple users to take part in the VJ operation. The style that the audience enjoys the VJ performance is attractive itself. Furthermore, we got interested in what may happen, when the audience directly participate in VJ performance (Miyamichi, 2000).

We have to prepare input means for plenty audience to contribute in VJ procedures, for such a goal. We decided to utilize the mobile phones. The number of mobile phone users exceeded 1 billion in April 2002, according to EMC[4]. Most of young generations possess their own mobile phones. Young people use them as internet-terminals, and global wireless telecommunications carriers are building up the internet functions more and more. Recent models are equipped with cameras and can transmit the video messages via e-mail.

We would like to investigate this theme, from the artistic interest, and also from the sociological interest regarding impact of the corroborative process by multiple people at a place. In section 2, we are going to introduce the system overview. In section 3, we will describe access methods and

---

[1] manda@sys.wakayama-u.ac.jp
[2] http://www.digitalstage.co.jp/motiondive/
[3] http://www.steinberg.net/en/
[4] http://www.emc-database.com/

some types of the VJ performances. Finally, in section 4, we are going to shows the actual performances that we have held and to discuss the impact of corroborative performance.

2 System

The goal of the VJ system presented here is to enables audience who possess mobile phones to participate in the VJ performance. We had a plan to hold a performance at any place from the beginning of this project. The performance site may not always be a place where internet facility is available. It was also supposed that we are going to perform the VJ, outdoors. Therefore, we designed the system as we can set up with mobile facilities as shown in Figure 1; the system consists of one or two note-PCs, mobile phones for PPP connection, a LCD projector, and some speakers. We set up the message server at our laboratory in Wakayama University.

We adopted a server/client architecture for the implementation of the network application, which is compatible with the use of public internet services provided by carrier companies. The technical layout is shown in Figure 2. The participants of video mixing send their messages to the remote

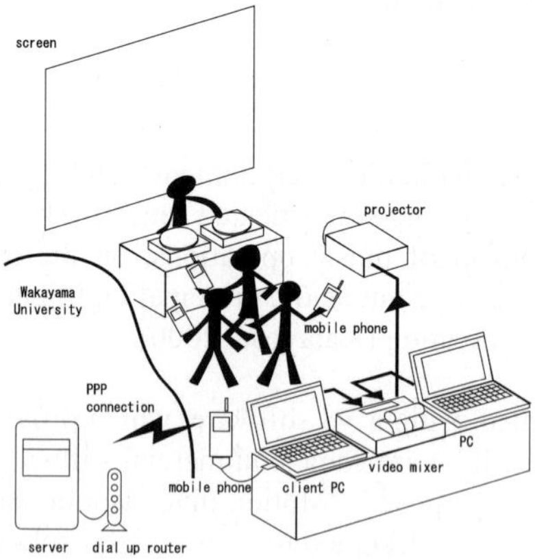

Figure 1: A Setup of collaborative visual jockey using mobile phones

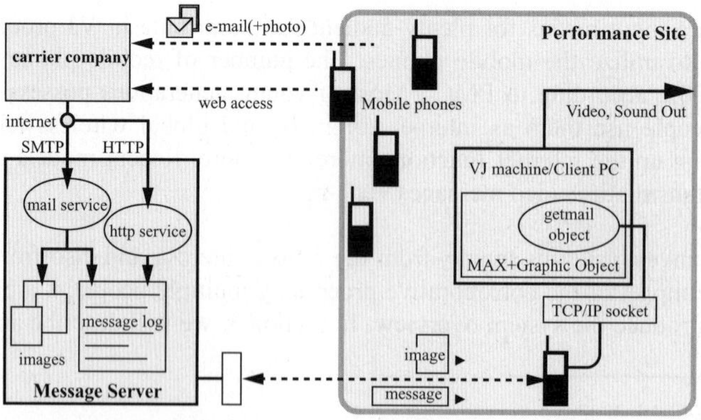

Figure 2: A technical layout of the collaborative VJ system

message-server, with the e-mail function or web browser of their mobile phones.

The message-server is connected to the VJ machine, on which the visual mixing is executed. The connection between the message server and the VJ machine may be ethernet or wireless-LAN or PPP connection. The massages from the participants are transmitted to the VJ machine via the massage server.

The latest mobile phones are equipped with a digital camera. Using this camera, the participants can send the e-mail that a photograph was attached. When the photo-attached e-mail is sent to the message server, it separates the message and image and preserves them. In the process of forwarding this message to the VJ machine, the message server notifies that an image is attached to the message and forwards the image according to the demand of the VJ machine.

The messages sent from the participants consist of a control tag followed by texts or image data. The participants can specify the visual effect using the control tag. Figure 3 shows an example of web form for message entree. Using this form, the participants can send a message with a simple operation.

3 VJ Programs

The visual mixing program executed on the VJ machine is implemented with Max[5], a visual program environment for multimedia. We also used a video effect package for called nato and jittar. This environment enables even amateur programmers to program audio/visual applications. We wrote a general authoring interface, which assists a VJ operator to write her/his own video mixing programs, and a socket API for TCP/IP communications.

The basic functions available on the VJ machine are 1) getting message from the server, 2) design assistance of video featuring the received massages, and 3) mixing with other video materials. The simplest application of collaborative VJ is to let the audience control all of the thing. We assume a style, that a VJ operator takes part in and navigates the way to process the message sent by the audience, is more attractive.

VJ designers can author VJ programs according to their artistic requirement, beforehand, as shown in Figure 5. The prescribed program is very helpful for a VJ operator to select and process the audiences' messages as they might best fit at the live situation.

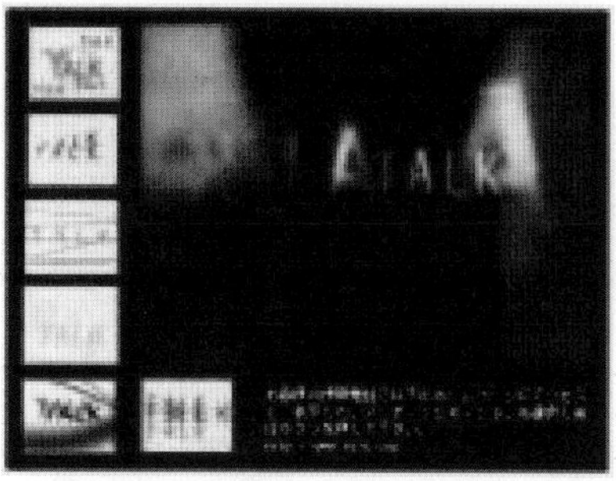

Figure 3: An message Entree web Form

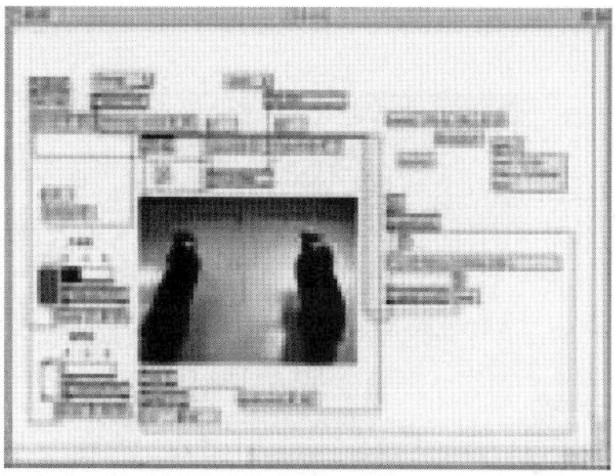

Figure 4: An authoring user-interface on Max

---

[5] http://www.cycling74.com/products/maxmsp.html

We can suppose various interesting presentations for collaborative VJ performance. One of the interesting implementations is giving the majority vote as to what they do next, to the audiences.

4 Holding Performances

4. 1 Message Transfer Response

It was supposed that we are obliged to connect the site and the message server with ppp connection, for the preliminary event. We measured the response time of the message transfer before the events, in using ppp connection (9600 bps) between the VJ machine and the message server. Japanese carriers provide two types of e-mail transfer, *general service* and *quick-mail.* When the message was sent using the general e-mail service, it took more than ten seconds to transfer messages from the mobile phones to the VJ machine. The time delay varied a lot, depending on the network situation. When using the quick-mail service and the web entrée form, the message was sent within a couple of seconds. We decided to use the web entrée form, for the preliminary events.

4.2 Performances

We held four trial VJ events in Japan so far; at a club PEPO NA MALAIKA indoor (2002, Feb), at Tannowa beach outdoor (2002, Sept.), at inter-college of computer music at Kurashiki-Sakuyo University(2002, Dec.), and at the demonstration session of a symposium on Entertainment Computing(2003, Jan.).

Figure 5 shows some samples of visual effects used at the trial events. Figure 6 and Figure 7 is a scene from the event at a club PEPO NA MALAIKA, and a scene at Tannowa beach, respectively. Generally, audience of VJ-performance dance to the music. The projected video plays the role of stage lighting, which is synchronized with the music. In the experimental performance which we held in a indoor club, participants used the screen as a electronic notice board. Some audience took their own pictures when the message they had sent appeared on the screen. We observed extraordinary tension, compared with general VJ events.

The picture shown in Figure 7 is that of the outdoor event. The numbers of the participants is relatively small. The audience enjoyed the word association game using the projected video. This game seemed to connect unfamiliar people. Young participants were sending terms which seems to be catchy in their own society. It accelerated the excitement of the participants.

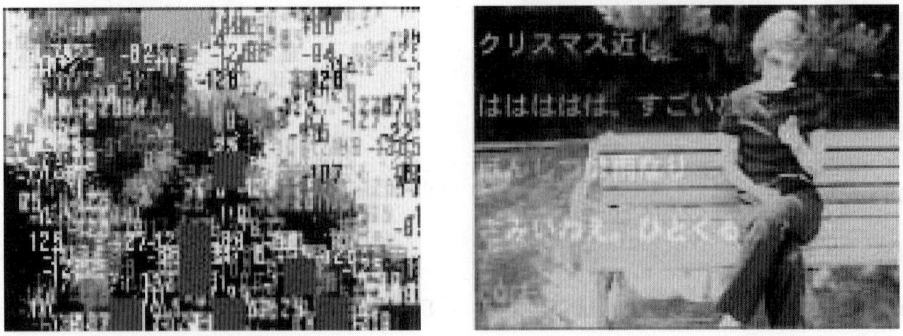

Figure 5: Sample of visual effects used at the trial events

Figure 6: Performance at a club, indoor

Figure 7: Performance at a beach, outdoor

When the other people noticed that excitement was shared by a specific group, it threw a wet blanket over the whole atmosphere.

5 Concluding Remark

We have developed an environment for collaborative visual jockey and have executed trial performances, with the goal of opening a new artistic field and to investigate the impact when many people participate in the collaboration using mobile phones. This is an ongoing activity. At this point, we have not gathered sufficient evidences to insist on an academic assertion regarding the collaborative VJ art. However, we have observed some interesting participants' behaviors, which might be the next research targets.

Recent progress of communication technology is a following wind for our activity. The number of access points which supports wireless LAN is increasing. The key industries of the mobile phone have been sifting into the types equipped with cameras. If the participants of the VJ event can send the pictures, the VJ event would be more striking. Image transmission is technically possible, for the current version of our system. We are going to execute the experiment using the high-speed wireless-LAN, for the next performance. We would like to show the latest experimental result at the conference. The latest report is also available from our web site[6].

Acknowledgement

This work was supported in part by Japan Society for the Promotion of Science under grant JSPS-RFTF 99P01404.

References

Katayose, H. (2000). Using Multimedia for Performance, Y. Anzai (Ed.), *Multimedeia Informatics "Self Expression"*, pp. 67-113, Iwanami-Shoten (in Japanese)

Miyamichi, T., Katayose, H., & Mitsuda, N. (2000). Video Mixing by Mobile Phone: Communication, *Interaction2002*, pp. 59-60 (in Japanese)

[6] http://ist.ksc.kwansei.ac.jp/~katayose/cvj/

A Circular Fashion Menu System Based on Human Motor Control Knowledge for the Pen-based Computer

Kimiyasu Kiyota

Kumamoto National College of Technology
Kumamoto 861-1102, Japan
kkiyota@tc.knct.ac.jp

Nobuo Ezaki

Toba National College of Maritime Technology
Mie 517-8501, Japan
ezaki@toba-cmt.ac.jp

Hotaka Takizawa

Shinji Yamamoto

Toyohashi University of Technology
Aichi 441-8580, Japan
takizawa@parl.tutkie.tut.ac.jp *yamamoto@parl.tutkie.tut.ac.jp*

Abstract

This paper proposes a new type of dynamic menu system that was designed based on human motor control knowledge for a pen-based computer. This system is named FMS (Floating Menu System). Proposed new system is a kind of a pie menu system, so the menu command options organized around the current cursor position. The advantage of the pie menu is that the user can select the command option only by control of the movement direction, without need for accurate control of movement amplitude. The FMS was experimentally compared with the conventional drop-down menu system in young and adult subjects by using a stylus pen or a standard mouse as a pointing device. The experimental results revealed that total selection time was significantly reduced in the proposed menu system, especially in the elderly subjects. The reduced movement time using the FMS can be explained in terms of reduced movement complexity. We are investigating whether FMS is applicable as a pen-based computing system for a blind person.

1 Introduction

The effectiveness and user acceptance of an application are determined primarily by the design of the user interface (Marcus et al., 1995). Menus are an important component of the user interface. These provide an easy-to-use visual interface that allows the user to browse and select an option from a list of command options. There are several types of menus for the window system, such as a fixed, a scrolling, a drop-down, a pop-up menu, etc. Each of these menus can be cascading into submenus. In recent window system, the cascading drop-down menu system is used as a standard menu. However, this standard menu system requires exact selection of a user. This is the factor that the novice user makes the menu hard to use. Future menu systems should be designed using the motor control knowledge about drawing and hand movement by using equipment such as a mouse and a stylus pen. (Plamondon et al., 1990). Based on this concept, we propose a new type of dynamic menu that is arranged in a circular fashion. The user can select to the required command option by control of movement direction only, without need for accurate control of movement amplitude in the new system. The new concept for reducing move time very much is explained by the Fitts' law (Fitts, 1954). The feature of our new designed menu system is the use of the direction ingredient of gestures to select menu options.

2 A conventional drop-down menu system

The various menu systems are used in the computing system in recent years (Rogers et al., 1994). The drop-down menu is used most frequently in the application of the window system. This menu appears when the user selects a command option from the menu bar at the top of an application window. The user clicks using the stylus pen on a command option target. Subsequently, the user can move the stylus pen along the linear submenu, and click on one of the command option. A second-level drop-down submenu expands when the top of the stylus pen remains across that option. The advantage of such a submenu is that the user is not overloaded with a large number of menu options at a time. The movement pattern required to select a submenu option is a rather complex movement contorting of vertical and horizontal sequences as shown in a Fig.1. The movement time towards a menu option is proportional with $\log_2((2W)/S)$, where S is the size of the command option and W is the distance from the current position of the stylus pen to the target option. In addition, to enter the submenu, the user needs the exact pen movement to vertical / horizontal direction. In this way user has to look at a screen and has to perform this operation correctly. Exact pen movement in this system becomes difficult for the elderly person, and a visual disabled person cannot access any longer.

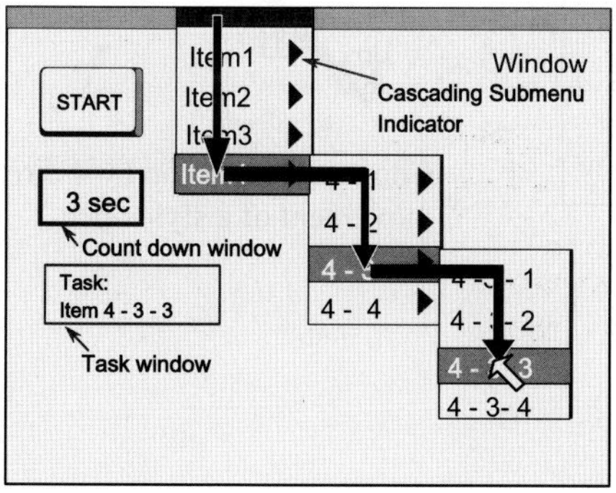

Figure 1: A conventional drop-down menu system

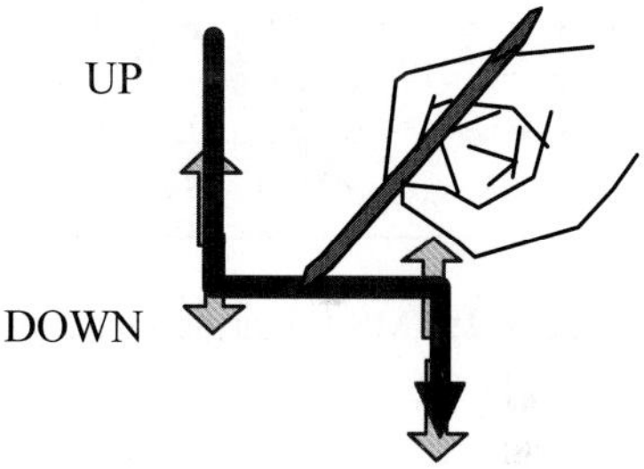

Exact pen movement to vertical / horizontal direction is required

3 New concept of the Floating Menu System (FMS)

Figure 2 shows the new concept menu system where the command options are organized in a circular fashion. This is named FMS (Floating Menu System). The FMS is a kind of a pie menu system, so the menu items are arranged in a circular fashion. The advantage of the pie menu is that the user can select the item only by control of the movement direction, without need for accurate control of movement amplitude. However, the conventional pie menu decreases the selection efficiency, when the number of command options increases. So we are able to apply this circular fashion menu for elderly or visually disabled person, after investigating these problems. As the solution of the above-mentioned problem, our proposed FMS has two concepts. One is arranging only four options on one menu window, and the other is the sub menu window of a layered structure in order to correspond to the many command options. The operation of this system is as following. The first-level circular menu window appears in the center of a screen when the user pushes the side button of a stylus pen. Next s/he selects the one command option from 4 command options on the first-level menu window. The user moves a stylus pen from the center circle to one of the 4 directions then the second-level circular submenu window appears. Furthermore, s/he moves towards one of the 4 command options on the second-level circular submenu window using the stylus pen from the current position again. Finally a third-level circular submenu window appears and the user clicks a side-button of the stylus pen on the target command option. Since all of these selections are mainly determined by a direction, we think that these operating methods are useful to visually impaired person. And we also hypothesise that this menu system greatly reduces the movement time. Furthermore, since top of stylus pen does not travel across unwanted menu options, the fact that the pen moves onto a menu option can be interpreted as selecting that menu option. The advantage of FMS is that user can produce frequently occurring submenu choices very quickly by producing a multi-stroke pen gesture consisting of a sequence of pen movements with little movement amplitude. In this way, accurate movement demands are less severe and the frequent changes of direction are no longer necessary.

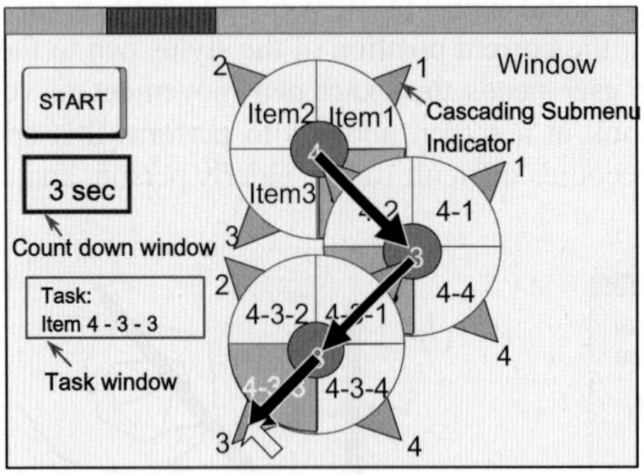

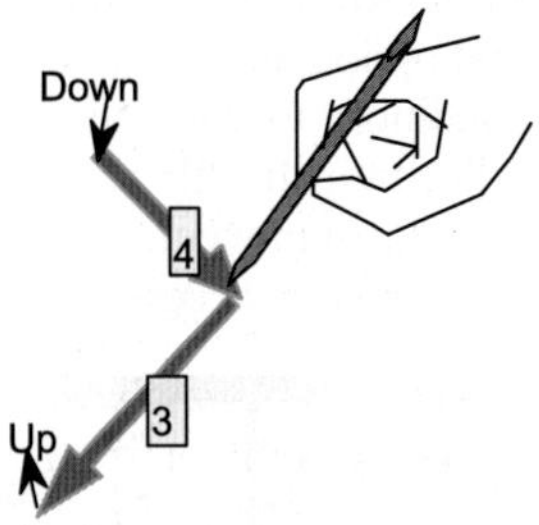

Command selection by 4 directional movement of a stylus pen

Figure 2: FMS (Floating menu system)

4. Experiment

4.1 Subjects and apparatus

Eight subjects participated in the experiment. Their age and gender can be found in Table 2. Subjects were divided into two groups, four subjects were adult users (2 males, 2 females; ages 41-61 years), and four subjects were young children (2 males, 2 females; ages 10-17 years). Both experienced and inexperienced users participated, but all had some experience with a mouse-based computer interface. All were right handed except subject No.7 and were without any known visual or motor impairment. A notebook type PC was used with an 11- inch screen (800 x 600 dots SVGA LCD; a size of 18.5cm x 24.5cm). A linear drop-down menu system and FMS were implemented in Microsoft Window 98^{TM}. Examples of experimental menu system are shown in Figures 1 and 2, respectively. Each menu system has 3 layers of menus with 4 command options. Therefore the third-level menu allows selection of 64 command options. We tested 2 control-pointing devices, which are a mouse (Microsoft ^{TM}; a standard mouse) and a stylus pen on a digitiser (WACOM ^{TM}; KT-0405RN; 128mm x 96mm).

4.2 Procedure

Subjects were explained about tasks for each menu system by experimenter. Testing took place in four sessions (2 menu systems and 2 pointing devices), and the sequence of the sessions was randomised. Each trial with in a block was done as follows:

(1) The subject clicked a start button by the pointing device

(2) After three seconds (the screen showed a count down; steps of 1 sec), the target command option were shown on the screen

(3) The subject had to start to select into the three-level submenus the target option by using a mouse / a stylus pen

Sub-sub-submenu options to be selected (64-choice) were randomised. A total movement time is measured from the time the command appeared until the user has selected the correct item. In this experimental trial, the mean total movement time (MT) and its standard deviation (SD) are measured for each subject on each task.

4.3 Results

Table 1 shows the comparison between FMS and a drop-down menu system by adult and young children group, and Table 2 shows the experimental results for each subject respectively. We compared MT and SD of the two menu systems and the two pointing devices. Table 2 shows that MT was faster for the FMS than for a drop-down menu system except in the case of subject No.6 using a mouse [$F(1,6)=116$, $p < 0.001$]. This effect was strongest for the elder subjects [$F(1,6)=18.0$, $p < 0.01$]. In addition SD was smaller for the FMS than for a drop-down menu system [$F(1,6)=16.5$, $p < 0.01$]. In respect of the pointing device, it was also found that MT was faster for a mouse than a stylus pen [$F(1,6)=6.9$, $p < 0.05$]. This might be due to the fact that all subjects were unaccustomed to operate a stylus pen.

Table 1: The comparison between adult and young groups

	F M S	Drop-down
Adult Group	4.67 (Sec)	8.45 (Sec)
Young Group	2.51 (Sec)	5.00 (Sec)

Table 2: The mean total movement times for each subject per task, and pointing device

			Mouse		Stylus Pen	
Subject number	Sex	Age (years)	FMS Mean± SD(sec)	Drop-down Mean± SD (sec)	FMS Mean± SD (sec)	Drop-down Mean± SD (sec)
1	M	61	2.51± 0.30	4.95± 0.58	5.31± 2.14	8.60± 4.12
2	F	55	3.74± 0.74	7.67± 2.88	5.36± 2.20	8.60± 4.70
3	M	45	2.57± 0.46	7.20± 4.21	3.78± 1.42	5.05± 0.65
4	F	41	3.54± 0.88	5.51± 2.53	4.22± 1.00	11.54± 8.31
5	M	17	1.72± 0.45	2.94± 1.65	1.83± 0.26	3.06± 0.57
6	F	15	2.53± 1.23	2.45± 0.94	1.67± 0.24	3.90± 1.00
7	M	12	2.29± 1.05	2.79± 0.95	3.37± 0.91	6.75± 5.95
8	F	10	3.56± 0.43	4.21± 3.22	3.17± 0.85	6.29± 2.83

SD: Standard Deviation

5 Application for a blind user

We have developed the pen-based input system for the blind person (Ezaki et al., 2001). Using this system, they can easily input the Japanese character without well -training. In general, the drop-down menu or the button menu system is used to do the menu selection using stylus pen in the pen-based computing system. However the present our system uses the hand controller (the desktop computer) or jog-dial (PDA type) for menu selection. Because they are not able to see the cursor position and the menu buttons, the blind person is fairly difficult to operate a command option directly. However FMS can be selected the command option by using only 4 directions determination, without relying on exact movement amplitude. Since a command option is uniquely chosen by not distance but the direction, this menu system is effective in a blind person. However, the voice output function of a menu item is needed in this menu system as shown in Fig. 3.

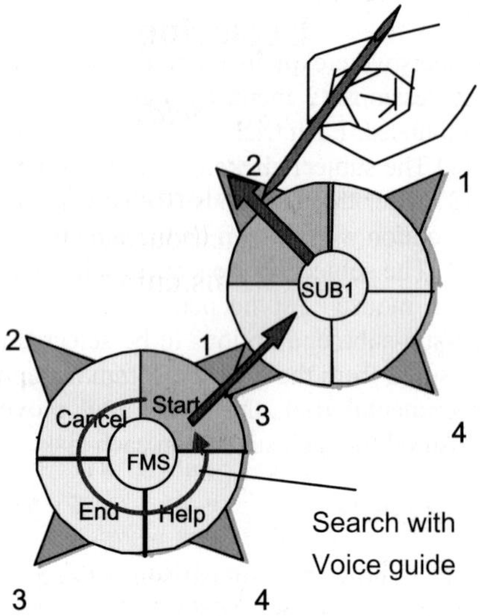

Figure 3: FMS with voice assistance for blind person

6 Conclusion

This paper proposed a new menu system, named the FMS was organized in a circular fashion. This system shows the pop-up menu options organized around the current cursor position. We examined the total movement time on the newly designed FMS and compared with a conventional drop-down menu system for eight subjects. The results revealed that total movement time was significantly reduced in the new menu system, especially in the older subjects. The reduced movement time using the FMS can be explained in terms of movement efficiency rules. It was confirmed that users could produce frequently occurring submenu selects very quickly by producing a multi-stroke gesture consisting of a sequence of cursor movements. Now we are developing this system with voice guide of a command for the blind user.

References

Marcus A., Smilonich N. & Thompson L. (1995). The cross-GUI Handbook for Multi-platform User Interface Design. *Addison-Wesley Publishing Company*.

Ren X. & Moriya S. (1997). The best among six strategies for selecting a minute target and the determination of the minute maximum size of the targets on pen-based computer. Human-computer interaction , 7, 85-92.

Fitts, P. M. (1954). The information capacity of the human motor system in controlling the amplitude of movement. *Journal of Experimental Psychology*, 47, 381-391.

Rogers Y., Sharp H., Benyon D., Holland S. & Carey T. (1994). Human-Computer Interaction, *Addison-Wesley Publishing Company*, 285-296.

Ezaki, N., Kiyota, K. & Yamamoto, S. (2001). Pen-based electronic mail systemfor the blind, Proc. of 9th International conference on human-computer interaction, Vol.1, 450-454.

Evaluating the Usability of Mobile Systems: Exploring Different Laboratory Approaches

Jesper Kjeldskov

Department of Information Systems
University of Melbourne, Australia
jesperk@staff.dis.unimelb.edu.au

Mikael B. Skov

Department of Computer Science
Aalborg University, Denmark
dubois@cs.auc.dk

Abstract

This paper addresses mobile system usability evaluations. Three different think-aloud evaluations of the same mobile system were conducted for the purpose of comparing the results by each approach. The evaluations spanned the use of test subjects with and without domain specific knowledge and the use of simple laboratory setups as well as high-fidelity simulations of the use context. The results show some significant differences between the results produced by the three approaches. However, it is indicated that while the recreation of a highly realistic use context resulted in a number of unique usability problems being identified, a large percentage of the usability problems found in total were identified already in the simple laboratory setups.

1 Introduction

Evaluating the usability of mobile systems constitute a potential challenge since their use is typically closely related to activities in their physical surroundings and often requires a high level of domain-specific knowledge (Nielsen, 1998). This can be difficult to recreate in a usability laboratory. Thus moving the evaluation into the real world may seem like an appealing approach. However, conducting usability studies in the field is also problematic. Access to real users and realistic settings can be difficult, data collection is complicated and means of control are limited.

This study is motivated by the need for evaluating the usability of a highly specialized mobile collaborative system supporting the coordination of safety critical work tasks on large container vessels (Kjeldskov and Stage, 2002). Evaluating this application was a challenge for a number of reasons. First of all, real users from the container vessels were not available for usability evaluations. Secondly, evaluating the system in the real world was not possible due to safety issues. Thus, the evaluation had to be done without going into the field and with limited access to prospective users. This challenged our ability to create a realistic laboratory setting.

2 Method

Three different evaluations of the mobile prototype were conducted for the purpose of comparing different approaches to creating realistic laboratory settings for mobile system evaluations.

2.1 Standard Laboratory with Non-Domain Subjects

Our first evaluation was conducted in a standard usability laboratory facilitating the observation of two physically separated subject rooms from a central control room through one-way mirrors.

Figure 1: Usability laboratory setup

Figure 2: Video from laboratory evaluation

Three two-subject teams, all computer science students, were given the task of coordinating a work task on board a fictive container vessel, communicating exclusively by means of textual commands on their mobile devices. The test subjects received a 15-minute introduction to the use context of the prototype application: the overall operation supported by the system, the basic concepts and maritime notions involved, the distribution of work tasks and present procedures of communication and coordination. Afterwards, one test subject was asked to act as captain on the bridge in one subject room while the other acted as officer on the fore mooring deck in the other room. The test subjects were asked to think-aloud during the evaluation. An evaluator located in each subject room observed the test subjects and asked them about their actions.

The laboratory setup consisted of two Compaq iPAQs connected through a wireless network displaying the interfaces for the officer on the fore mooring deck and the captain on the bridge respectively. Two A4 handouts depicted standard patterns of mooring and explained 10 basic concepts of the maritime context for quick reference. The test subjects were seated at a desk with the mobile device located in front of them (figure 1). Cameras mounted in the ceiling captured high quality video images of the evaluation sessions: overall views of the test subjects and close up views of the mobile devices. The video signals were merged into one composite signal and recorded digitally (figure 2).

2.2 Standard Laboratory with Domain Subjects

Our second evaluation applied the same laboratory setup, introductory procedure and tasks as described above. However, for the purpose of increasing the realism of the evaluation, we altered the experiment in two ways. First, we brought in prospective users from the nearby Skagen Maritime College. All test subjects were thus skilled sailors with practical experience with the operation of large vessels. Secondly, we introduced a simple paper mock-up of a ship in harbor and central instruments on the bridge. Apart from introducing a more realistic context of use, the purpose of this mockup was to supply the test subjects with a tool for explaining their strategies and actions. The test subjects acting as captains were thus asked to operate the controls of the mockup as they would operate the controls on the bridge in the real world and use the model of the ship to illustrate the process as it developed over time.

2.3 Advanced Laboratory with Domain Subjects

Our third evaluation took place in a temporal usability laboratory at the simulation division of Svendborg International Maritime Academy and used their state-of-the art ship simulator for creating a highly realistic but yet safe and controllable experimental setup. The ship simulator consisted of two separate rooms: an operator room (also resembling the fore mooring deck) and a simulated bridge fully equipped with realistic controls and instruments (figure 3). The simulator was set up to imitate the operation of a large vessel in challenging weather and traffic conditions corresponding to a real world situation. The academy provided test subjects with practical experience on the operation of large commercial and military vessels. Three teams of two test subjects were given the introduction and overall task described above. During the evaluation, the test subject acting as captain had to consider all aspects of maneuvering the ship as well as communicating with personnel, harbor traffic control etc. and taking into consideration the movements of other vessels. Two evaluators located on the simulated bridge and operator room respectively observed the test subjects and asked questions for clarification. As in the standard laboratory studies the prototype setup consisted of two Compaq iPAQs and four high quality video images of the evaluation sessions were recorded digitally (figure 4).

Figure 3: The high-fidelity ship simulator

Figure 4: Video from evaluation in simulator

2.4 Data Analysis

The data from the evaluation sessions consisted of three video recordings with the total of 18 test subjects. The analysis of this data aimed at creating three lists of usability problems experienced by the test subjects: one list for each of the three studies. The videotapes were analysed in three steps. First, problems experienced on deck were identified by examining the videos while listening to audio from one subject room only. Secondly, the same was done for identifying problems experienced on the bridge. Finally, the videotapes were examined listening to audio from both subject rooms simultaneously. This analysis was done in a collaborative effort between the two authors allowing an immediate discussion of each identified problem.

3 Results

Figure 5 outlines the distribution of the problems identified in the three evaluations. The three blocks signify the standard laboratory with non-domain subjects (#1), standard laboratory with domain subjects (#2), and advanced laboratory with domain subjects (#3) studies, each divided into problems identified on the bridge and on the deck respectively. Each column represents a unique usability problem and a black box means that the problem was identified by that approach.

Totally, 58 unique usability problems were identified in the three evaluations. 7 problems were identified in all three evaluations and on both the bridge and on the deck. Some of these problems were related to interaction issues e.g. finding out which elements on the screen to interact with. Another general problem found was that many test subjects did not see all relevant state changes

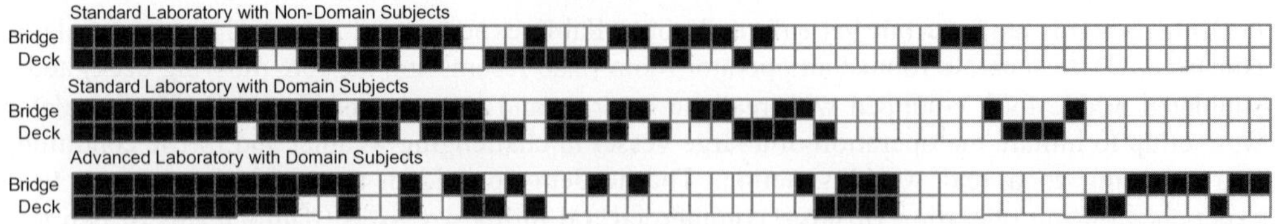

Figure 5: Distribution of identified usability problems in the three evaluations

in the system, which sometimes caused them to miss commands or confirmations. Other general problems were related to the correlation between the representation of the ship and ongoing activities in the system and real activities on the ship.

The study revealed no significant difference in the number of identified problems for the three evaluations. The standard laboratory with non-domain subjects in total identified 37 usability problems, where the standard laboratory with domain subjects identified 40, and the advanced laboratory with domain subjects identified 36. However, the number of unique problems identified was significantly higher for the advanced laboratory study, identifying 12 unique problems of which 3 were discovered both on the bridge and on the deck. In comparison, the standard laboratory with non-domain subjects identified 5 unique problems and the standard laboratory with domain subjects 6. On an overall level, it is noticeable that 35% of the total number of problems was identified by domain subjects only, stressing the importance of including prospective users. On the importance of a realistic use context, however, it is noticeable that 80% of the total number of problems was actually identified outside the simulator. On the other hand, almost 40% of the total number of problems was not encountered by domain users in the simulator. This is an interesting issue as it is difficult to say whether these problems are simply not relevant since they were not discovered in the most realistic setup or if the simulator setup infused so much complexity on the session that some shortcomings might have ignored.

4 Discussion

The key literature on usability testing and engineering typically states that usability evaluators should minimize their influence on the conduction of the test and on the data analysis, e.g. (Nielsen, 1993; Rubin, 1994). This usually means assigning real users to the test in order to identify realistic and relevant problems, and carefully elaborating assignments in order to focus on relevant aspects of the system. However, several studies have shown that various aspects of an evaluation influence the results, e.g. the evaluator-effect (Jacobson, Hertzum, and John, 1998), the number of test subjects (Lewis, 1994), and the level of investigator intervention (Held and Biers, 1992). In our study, we have deliberately explored the influence of changing different settings of the evaluation: varying the test subjects by including non-domain users and applying different levels of contextual realism to the laboratory setup.

While no significant difference was found regarding the number of problems identified in the three studies, analyzing our results qualitatively, a number of interesting differences emerge. First of all, the character of the unique problems identified in each study was different. While the 5 unique problems identified by non-domain users could generally be related to lack of knowledge about the use context, many of the 18 unique problems identified by domain subjects in the standard and advanced laboratory studies combined were concerned with highly relevant issues such as the representation of the task in the system and lack of flexibility in coordination and communication. E.g. some of the domain subjects wanted to specify commands more detailed than supported by the system. Some of the 12 unique problems identified in the advanced laboratory were furthermore related to critical issues such as not being able to cancel commands when changed conditions in the simulation required this. This turned out to be a critical problem since the captain had to apply different means of communication in order to achieve his goal, which resulted in further problems regarding the representation of the real world in the system. No such situations occurred in the standard laboratory and none of the non-domain subjects wanted to cancel commands or expressed that not being able to do so could be a problem. Secondly, the realism of the environment of the advanced laboratory implied that the test subjects had to operate other systems and consider other information apart from the evaluated mobile prototype. Hence, the test subjects' attention towards the mobile device in the advanced laboratory was lesser than in the standard laboratory studies. This resulted in test subjects often missing updates or changes on the display of their mobile device in advanced laboratory sessions while this was not a significant problem to users in the standard laboratory. Also the realism of the context in terms of weather and traffic conditions induced by the simulator made an impact on the results of the study causing some of the test subjects to apply approaches to the operation and request procedures, which were not supported by the system. None of the test subjects in the standard laboratory evaluations experienced that problem.

Our study indicates that central usability problems of a mobile system can be identified in laboratory settings. While it is shown that including prospective users and recreating realistic contexts support the identification of qualitatively different problems, it is also shown that a large number of a mobile system's usability problems can be found in simpler laboratory approaches. The results of our study are limited in a number ways. First, the number of test subjects applied in each evaluation limits their general validity. In order to be able to make general recommendations, the study needs to be replicated with more subjects and probably varying the system. Secondly, the test subjects were not required to be as mobile in the evaluations as they would have been in a corresponding real world situation. Finally, we have not at this point assessed the severity of the identified problems. This means that we cannot conclude on the implications of the discovered problems and also limits our comparison of the three studies. This analysis is forthcoming.

References

Held, J. E. and Biers, D. W. (1992) Software Usability Testing: Do Evaluator Intervention and Task Structure Make Any Difference? *Proceedings of the Human Factors Society 36th Annual Meeting*. Santa Monica, HFS, pp. 1215 – 1219

Jacobson, N. E., Hertzum, M., and John, B. (1998) The Evaluator Effect in Usability Studies: Problem Detection and Severity Judgments. *Proceedings of the Human Factors and Ergonomics Society 42nd Annual Meeting*, pp. 1336 – 1340

Kjeldskov, J. and Stage, J. (2002) Designing the User Interface of a Handheld Device for Communication in a High-Risk Environment. *Adjunct Proceedings of the 7th ERCIM Workshop on User Interfaces for All*, Paris, France

Lewis, J. R. (1994) Sample Sizes for Usability Studies: Additional Considerations. *Human Factors*, 36(2), pp. 368 – 378

Nielsen, C. (1998) Testing in the Field. Proceedings of the third Asia Pacific Computer Human Interaction Conference. Werner, B. (ed.), *IEEE Computer* Society, California, pp. 285-290

Nielsen, J. (1993) *Usability Engineering*. Boston, Academic Press

Rubin, J. (1994) *Handbook of Usability Testing – How to Plan, Design, and Conduct Effective Tests.* John Wiley & Sons

Towards a Model for an Internet Content pre-Caching Agent for Small Computing Devices

Andreas Komninos

University of Strathclyde
Livingstone Tower, 26 Richmond st.,
Glasgow, G1 1XH, UK
andreas@cis.strath.ac.uk

Mark D. Dunlop

University of Strathclyde
Livingstone Tower, 26 Richmond st.,
Glasgow, G1 1XH, UK
mark.dunlop@cis.strath.ac.uk

Abstract

In the near future Internet access will be commonly available in two forms to most palmtops: wide bandwidth of a low-cost land-based connection while docked, and limited capacity expensive wireless connections while mobile. The coming 3G networks will help the bandwidth but the disparity of connection cost and bandwidth between docked and undocked devices will continue. This paper proposes a model for automatically pre-loading (or pre-caching) palmtop computers with websites likely to be of use in order to support the user's daily activities. To achieve this, the model extracts information from a user's diary and feeds this into a predictive system.

1 Introduction

Small personal computing devices are available to almost everyone today. These are presently distinguished between those used primarily for communication, such as mobile phones, and those known as Personal Digital Assistants (PDAs). However, the distinctive line between these two categories slowly starts to disappear.

A characteristic, which signifies this blending of devices, is the incorporation of PIM functions into many of today's mobile phones. Most importantly, the typical calendar application, which is central to PDAs, is now available in most modern mobile phones. Another application which is currently available on PDAs and is slowly appearing on mobile phones is the Web Browser. Until recently, surfing the web on a mobile phone has been restricted to WAP purpose-built sites, mainly due to the slow speeds available through the GSM network (~9.2Kbps) and the limited display capabilities of mobile phones. With the advent of wider bandwidth communication standards (GPRS), at least part of the problem is being addressed.

The distinction line between PDA and mobile phone is expected to disappear with the coming of 3G networks and 3G mobile devices, which will provide several communication services currently limited to desktop computers. Perhaps the most important of these is broadband connection to the Internet and its vast amount of resources. There is however some concern regarding the costs of this new service. The 2.5-G GPRS transmission standard has been around for a while, but it is still very expensive to use. The same is expected for 3G networks, once they become widely available, with analysts estimating that users will be reluctant to accept the high costs of 3G-network usage.[i]

2 Predicting content needs

It would be possible for a user to pre-fetch all of the data that would be required through their low-cost, land-based (and often broadband) internet connection and feed it into their mobile device. The problem with this scenario would be that users would have to spend a considerable amount of time browsing for and saving websites and then transferring to the mobile device. Furthermore, documents might need extra processing in order to appear properly in the small device.

An intelligent software agent could examine the user's electronic calendar and try to estimate the kind of activities they will perform and, possibly, what kind of internet content they may need, in order to support those. The web content downloaded through this predictive system, would be stored in the user's desktop, processed and then transferred over to the mobile device.

The predictive system can be enhanced over time to learn and remember a users' preference for different categories of information. Such a predictive system should be able to obtain information from the user directly and indirectly. The majority of information should be obtained through indirect means, in order to minimise interference with the user's other activities. However, the system should maintain its ability to directly interact with the user, in order to resolve any possible uncertainties, which will be essential to the initial stages of the system's training process.

3 Agent structure

Words such as "learn", "guess", "remember" have been previously used in our theory regarding the way the predictive system should work. These prompt an insight of the human memory as a starting point to model the pre-fetching system. Cognitive psychology can offer paradigms for a flexible, adaptable and interconnected "knowledge-base" structure.

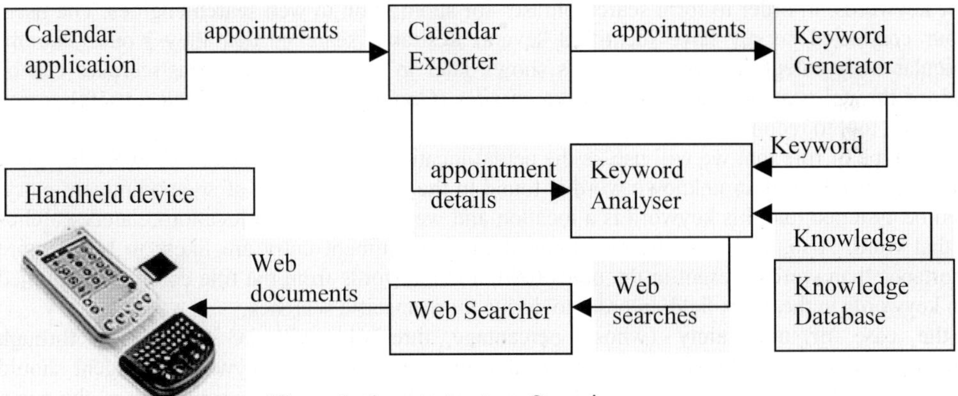

Figure1: Agent structure Overview

In the human brain, various tasks, such as storing information into working memory, are performed with the simultaneous work of various centres. A paradigm of this is the central executive, which uses other agent-type centres, like the rehearsal loop, in order to process and store information in the working memory area. This means that the central executive, which is the "main processor" for the system can offload results from previous tasks to other, smaller capacity agents, so it is free to engage in other difficult processing tasks[ii]. From this paradigm, we adopt a modular form for the agent's structure, such as depicted in figure 1.

The advantages of such a structure are obvious. Instead of relying on one main executive to perform tasks in a serial manner, it is possible to have several "sub-agents" working collaboratively in order to perform tasks with enhanced efficiency, in order to minimise the total running time of the retrieval process.

4 Prediction management

The agent should have at least some knowledge of the world it operates in, in order to perform effectively and produce meaningful results. Ideally the agent should be able to augment its knowledge level. This can be achieved either by allowing the agent to add further knowledge to what it already possesses, or by allowing the agent to obtain knowledge about the knowledge it already holds. We shall call the latter meta-knowledge and refer to it later on. Also the agent should hold some knowledge about how to extract keywords from the appointment entries. This is an important aspect of the agent, since, without proper keywords, there can be no satisfactory web searches.

We conducted a study in the way users write in their calendars, by obtaining actual calendar entries and analysing their content. From this, it has been determined that users typically fall into three categories (hasty, meticulous and average), according to their style of input. From this study, we can formulate some rules for inferring additional information from hastily written entries (e.g. it has been found that when an entry's subject is the name of a person, it is 90% certain that the user will be meeting that person) and for determining which words and in which context can become keyword candidates. Keywords will be compared to a database of known items to identify their significance. In his work "A calendar with common sense"[iii], Erik T. Mueller has describes a potentially useful database for common-sense knowledge items, which could be useful starting point. Information particular to the user, such as names of friends or colleagues from the user's phone book, where she lives, etc, should also be used as part of the rules.

Once a keyword has been identified as such, the agent should be able to associate it with several other keywords, in order to form search phrases for submission to web search engines. The users in our conducted survey have identified several desirable searches that they would like for particular entry categories (an example is shown later in table 1). Other items/searches can be obtained through statistics held by web search sites (Google, for example, have a useful search suggestion tool to recommend related keywords[iv]).

Another type of rule that we will use, is the actual location, where a particular keyword has been found. For example, if an unknown word is found in the "location" field of an appointment entry, it can be assumed that this keyword is a location and we should conduct location-related searches for that word. Also, words in the title field of the appointment entry are likely to be of more importance than words present in the notes field, and keywords from the title could be combined with keywords in the notes field, in order to derive more specific searches.

In the case of uncertainty (whose percentage threshold will be determined through experimentation) during the attempt to apply the rules in a particular keyword, the agent should ask for clarification from the user. This knowledge will be kept for future reference, as the agent learns about items relative to the user.

The success of the conducted searches will be measured through relevance feedback (section 5.2), and this will in itself form another rule, which will be remembered for later use.

5 Memory Organisation

Through the conducted user survey, it was been noted that appointment items tend to fall within a few category types, the most frequent of which are meetings, tasks and reminders. This evidence is further supported by the fact that most modern electronic calendars provide facilities for categorising appointments. Early work into calendar usage by Kelley & Chapanis[v] and Kincaid et.al[vi], shows similar results to our findings. It should be therefore possible, instead of providing the agent a list of related search terms for each possible keyword, to cluster keywords into categories and associate search terms with each category. The use of clustering with keywords should mean that the knowledge database is somehow smaller in size, while remaining just as efficient.

These category clusters will consist of a category descriptor (against which the keywords will be compared) and a list of related search terms. In [2], it is explained that when reading words in a text, the human brain does not actually read the words letter by letter. Instead, some of the letters of the word are read and then the brain "guesses" what letters should come between the read letter and the next one. From relevant experiments, it is apparent, the brain appears to be using a ranking mechanism for the knowledge it holds, and uses the most frequently used (higher scoring) items when guessing.

Table 1: Sample Keyword category

Category Name	Category descriptor	Related Terms	Weight
"Location"	Travel, fly, population, city, town, train, bus	Map	0.7
		Hotel	0.8
		Airport	0.6
		Car Hire	0.4
		Museum	0.3

Following this ranking paradigm from the brain system, the agent should be able to make distinctions between search terms that are important to the user and those which are rarely used. This can be determined using relevance feedback techniques, such as Rocchio's algorithm[vii]. Initially, search terms will have a score (weight) associated with them, which will be pre-defined, something that can be achieved with the usage of the TF/IDF (term frequency/inverse document frequency) algorithm through the analysis of user statements of desired searches and keyword appearance in related web documents. As the users continue to make use of the agent, these weights will be adjusted according to the relevance feedback that will be obtained from the user.

Explicit feedback should be obtained rarely and with discretion, so as not to impede the user or interfere in their activities. It is expected that the majority of information will be collected through implicit feedback, which is unobtrusive and also can provide accurate results.

There are several heuristics[viii], which can be used to implicitly measure relevance. Some examples may be whether a user has clicked a document title to see the document, the amount of time they have spent reading the document or whether they have followed any hyperlinks from that document. Each of these heuristics should have its own weight, as it is for example obvious that when one spends some time reading a document obviously indicates much more interest than simply having opened the document.[ix]

By combining this relevance feedback information with observation of the user behaviour, it is possible to gain the meta-knowledge that we talked about previously. Meta-knowledge is important because it can allow the agent to choose which part of its knowledge it will use, in order to provide results that are useful to the user.

6 Summary

In this paper we have described a model for the management and organisation of knowledge for a pre-caching agent, based on paradigms from the human working memory system. The agent will initially be equipped with some general knowledge that is necessary, in order to support its

function. However, some of this knowledge will be "forgotten" and new knowledge will be "learned", as the agent adapts to the user's preferences and needs.
The heuristics for the evaluation and estimation of the users needs are based not only on findings of previous research, but also on the study of actual calendar entries and their users. These heuristics will provide a realistic rulebase, which will support the agent's evolutionary process.
The agent's structure is such that will allow a flexible approach to the user's needs, by evaluating and adapting to their style of input and personal preferences. This will be achieved through the constant evaluation and weighting of the appropriate factors, using relevance feedback techniques.
.

7 Future work

Currently, the model has not yet been fully implemented into a piece of executable software, so the overall functionality of the model remains to be examined. In the course of this examination, it will be interesting to observe not only the functionality of the proposed system, but also the behaviour of its users and the evolution of their interactions with this system.
In our study of calendar users, it was indicated by the majority of the users that they would be willing to change their input styles, in order to help the program achieve better results. Will they adhere to this statement? Will there be changes in the way detailed entries are made? Will people begin to insert keywords, or even search phrases in their calendars?
The flexible capacities of such a model could be utilised in other scenarios, where the pre-caching of documents is desirable. For example, this system could be enhanced to allow the inclusion of items such as recent emails or text messages and documents that may be relative to a situation, such as a meeting. We believe that the system could prove a useful aid for the increasing number of people who use their mobile computing devices to support their daily activities.

8 References

---

[i] CLAIRE WOFFENDEN (2000), Users 'will not pay for high 3G costs', http://www.vnunet.com/News/1112409

[ii] DANIEL REISBERG, Cognition – exploring the science of the mind, 2nd edition, W.W. Norton & co, New York. p. 15-17

[iii] ERIK T. MUELLER (2000), A calendar with common sense, ACM Intelligent User Interfaces, p.p. 198-201

[iv] GOOGLE AdWord suggestion tool, https://adwords.google.com/select/main?cmd=KeywordSandbox

[v] KELLEY, J. F., & CHAPANIS, A. (1982). How Professional Persons Keep their Calendars: Implications for Computerization. *Journal of Occupational Psychology*, 55, pp. 241-256.

[vi] KINCAID, C. M., DUPONT, P. D., & KAYE, A. R. (1985). Electronic Calendars in the Office: An Assessment of User Needs and Current Technology. *ACM Transactions on Office Information Systems*, 3(1), pp. 89-102.

[vii] J.J. Rocchio (1971), Relevance feedback in information retrieval, in the SMART retrieval system, Prentice Hall,

[viii] YOUNG-WOO SEO & BYOUNG-TAK ZHANG (2000), Learning users' preferences by analyzing web-browsing behaviours, ACM Agents

[ix] YOUNG-WOO SEO & BYOUNG-TAK ZHANG (2000), A reinforcement learning agent for personal information filtering, ACM Agents

Adaptive Smart Home System

Vidas Lauruska, Paulius Serafinavicius

Siauliai University
Vilniaus st. 141, LT-5400 Siauliai, Lithuania
E-mail: vidas.lauruska@tf.su.lt

Abstract

This paper describes the developing system for disabled users with physical and speech impairment. It would let the disabled to work with the personal computer, to use today's means of communication (e-mail, SMS, phone) and, especially, to take a full control of home environment (lighting, blinds, door, windows, heating, ventilation, air conditioning, security system, etc.) The aim of the project is to develop a modular communication and control system which would be controllable using single input device – any on-off switch and capability to adapt the system according to user's need.

1 Introduction

The system we are developing is designed for physically disabled (can't move arms, legs) with speech and language impairments. The system has been designed for severe disabled persons. We choose this group of disabled people because there are many people suffering from such disability and this group is more closed and unsafe in the society.

The main problem for disabled people is that they face some difficulties using the traditional environment control and communication systems or they can't use it at all. The Smart Home system can help them and can increase the quality of their life (Kubilinskas, Lauruska, 2001). The system we are developing makes more accessible environment control and communication for the elderly and disabled people.

Environment control and predictive communication system suggests new capabilities for physically disabled people with speech and language impairment. Many systems like this one are expensive or not enough adaptable for persons with severe disabilities.

The system we have designed enables access for disabled people to control domestic electric appliances, to send an e-mail and SMS, to type and print a text. This will help them to communicate and feel more independent from other persons.

There are many equipments and software applications designed for the disabled like special programs for text typing, computer control, home environment control, etc. These equipments and applications are mostly separate and they don't have a possibility working together as one system. This makes many inconveniences when installing a Smart Home system for disabled and also it results in decreased system's functionality. Because of personal needs the system would have a possibility of easiest adaptation according to them.

2 Methodology

For the maximum efficiency of various devices and equipments control, they would communicate with each other. That means all of them must be connected to a network. So there could be used any home automation network. But there are some requirements to the home automation network making it suitable for the disabled:

- All the controllable devices, actuators and sensors must be connected to the common data exchange network.
- The system's automation devices must be independent as more as possible of user's intervention due to his physical or mental disability.
- The system must be very reliable and safe. All the network devices must be maximum independent so if one is out of order; this will not effect the main system's functioning.
- Communication and control must be available from one device – remote control (ON-OFF switch) the type of which depends on person's disability.
- Remote control must be wireless due to mobility of the disabled user

There are two medium types we have used to develop the system:

- Wired medium - twisted pair cable (TP).
- Wireless medium - radio frequency (RF), infrared rays (IR).

The wired medium (TP) has been used to establish communication between non-portable devices (home environment control). The IR medium – for the control of domestic electric appliances (having IR remote control capability). The RF medium has been used for communication between non-portable system part and wireless remote control. There are many home automation networks such as X-10, HES, CEBus, LonTalk, EIB etc. Most of them are getting to the same key problems:

- Compatibility imperfections between different manufacturers' products.
- No support for wireless medium at all (or not for all devices).

It's very important to select the most suitable home automation network for the Smart Home implementation. After detailed analysis of them we've chosen EIB home automation network with twisted pair (TP) medium as a basis for the Smart Home system because:

- The EIB network is decentralised.
- There is used CSMA/CA (Carrier Sense Multiply Access / Collision Avoidance) medium access method.
- There can be used RF and IR for data transfers between devices if necessary.
- Possibility of EIB connection to Ethernet, Internet, ISDN.
- Easy installation and device connection.

For a RF wireless medium we've selected a Wireless Local Area Network (WLAN) operating on 2,4GHz frequency. This allows designing the Smart Home system with different modifications depending on disabled user home environment and needs. The main function of WLAN is to transfer data from wireless remote control to other system modules (usually non-portable).

The system's general block diagram is shown in figure 1. There has been used server-client method. The server realizes all the control tasks in response to the clients' requests, and also sends data, for example, audio and video signals, when they are required.

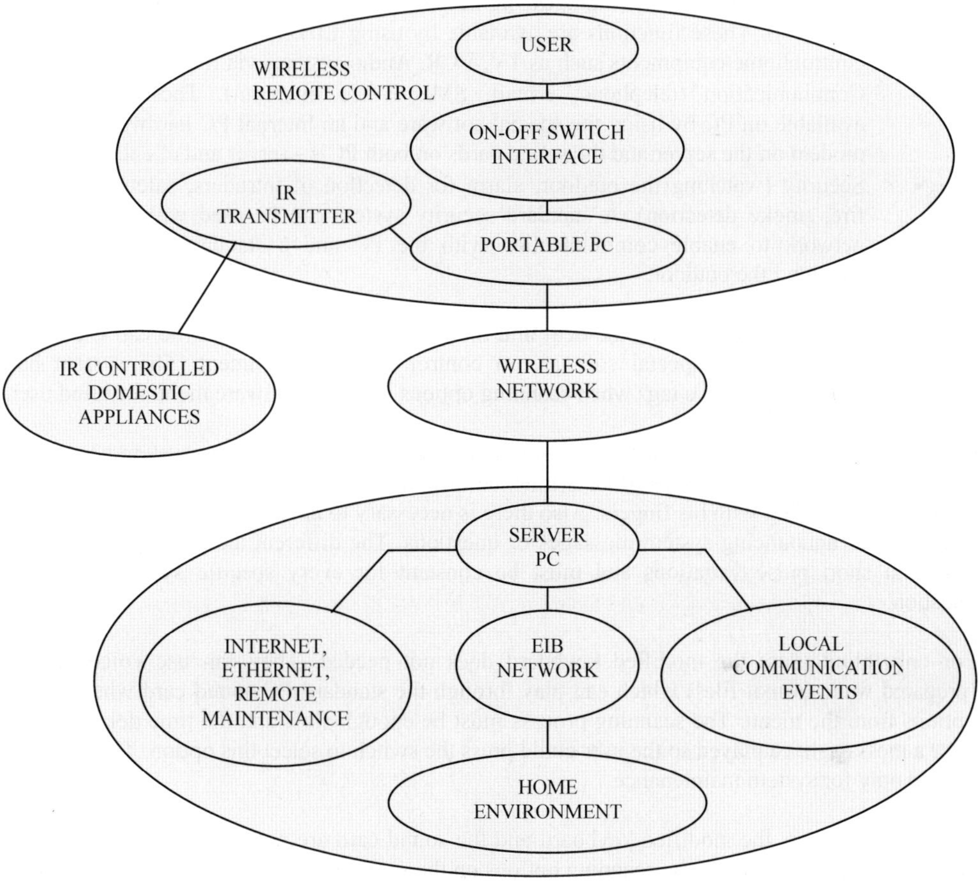

Figure 1. General block diagram of the Smart Home system for the disabled

The principle of system's functionality is based on scanning control buttons (elements) on monitor screen (virtual keyboard), and waiting for control signal at the same time. The virtual button is active on the screen for a short time. The time can be set up programmably. The next button is activated if there is no control signal, etc. The user sends control commands by ON-OFF switch. The switch shapes the binary signal which comes to the portable PC (client) and is used as input data for the scanning software interface. The main function of the remote control PC is to scan menus with control and communication options. When the user presses the switch, the portable PC sends data to the server PC via wireless network. There is connected RS232/EIB adapter on the server. It converts RS232 signals to EIB telegrams and vice versa. When EIB device gets appropriate telegram, it does its job and after has finished, it sends a response telegram to the server PC and the client PC displays an appropriate message according to this telegram.

3 Results and Discussion

This system allows the integration of the functions that the disabled must have access to:

- Environment control (includes lighting, heating, shutters, motorised doors and windows). These functions are available by using EIB network. Also the system can control home equipments such as TV, VCR, Audio systems via IR transmitter.
- Communication (telephone, e-mail, SMS, voice intercom). These functions are available on PC by using the special software and an internal PC hardware - the voice modem on the server and the sound cards on both PC's - server and client.
- Security (watching the outdoor, alarm for detection of intruders, safety alarms: gas, fire, smoke detection). A standard security system can be used with RS232 or EIB network to enable communication with the PC and a standard video camera for watching the outdoor.

The system can be accessible for the deaf and the blind. If the user is deaf, he can use text and graphic messages in the special software to control home environment. The system has a possibility to announce voice tags when scanning options from the software menu for blind user.

There is necessary to have an additional modified keyboard for making the system accessible to deaf and the blind. The surfaces of keys must be different so the blind and deaf user could make out them by touching with his fingers. Also there is necessary to have vibrating alert in this special keyboard for announcing system messages or questions. The different alerting signals consist of long and short pulse variations and must be constant for every specific system message or question.

For only blind users the modified keyboard does not needed. They can use voice messages (prepared wave format files) which can play through the standard PC sound card when scanning options from the menu. The scanning process must be enough slow and the time delay is needed after a message have played so the user could press the switch to select this option. The PC display is used only for system maintenance.

For only deaf users the modified keyboard and the sound card are not needed. User can read the system messages and watch the scanning options on the screen (usual LCD or CRT monitor).

4 Comparison with related systems and advantages

There are many similar systems developed for disabled (Vera, Jiménez, Roca, 2001), (Panek, Zagler, Beck, Seisenbacher, 2001). There is still not available any similar market solution because most of them offer only single equipments or assistive software applications (Hitech Systems Inc.), (EnableMart). Usually it's impossible to implement them as a single system.
The main advantages of our developing system are:

- Integration of environment control and today's communication events.
- Optimized virtual keyboard controlled by a single switch.
- Capability to adapt the system for severe disabled users.
- Multilanguage support (future development).

- High flexibility and reliability.
- Easy installation and adaptation.
- Relatively low price (depending on user needs).

5 Conclusion and future developments

We have chosen to use modular structure in our developing system's software and hardware. It allows more flexible adaptation to individual user abilities and needs. Also there is a possibility of easy system upgrade by changing or adding new home environmental equipments and communication events when necessary. The usage of standard and certified devices shows the system's reliability, flexibility and safety.

The system's cost is relatively cheap and depends how many EIB network devices are being installed accordingly the software modules also including PC and wireless network hardware prices.

Now we are developing a predictive software application module, which allows more comfortable and faster access to various communication events, especially text typing using single ON-OFF switch. The usage of today's communication technologies allows adding new events such as video conference and telecare to the Smart Home system.

6 Acknowledgements

This research was carried out during the project PACS of the Eureka program with the co-financing of Lithuanian Research and Studies Fund.

References

Kubilinskas E., Lauruska V. (2001). Smart home system for physically disabled persons with verbal communication difficulties. Assistive Technology – Added Value to the Quality of Life. AAATE'01. ISSN 1383-813X. IOS Press; 300-304.

Jose A. Vera, Manuel Jiménez, Joaquin Roca.(2001). EIB Bus as a Key Technology for Integrating People with Disabilities. A Case Study.

Paul Panek, Wolfgang L. Zagler, Christian Beck, Gottfried Seisenbacher. (2001). Smart Home Applications for disabled Persons - Experiences and Perspectives.

European Installation Bus Association. http://www.eiba.com

Hitech Systems Inc. http://www.hitech.com/

EnableMart. Assistive technology for vision, hearing, mobility impaired people. http://www.enablemart.com/

Effects of Chinese Character Font and Size on Visual Performance between Different Age Groups

Tzai-Zang Lee

Kun Shan University of Technology
No. 949 Da-Wan Road Yung Kung City
Tainan, Taiwan 710
leetz@mail.ksut.edu.tw

Jian-Zhe Huang

National Cheng Kung University
No 1 Ta-Shueh Road, Tainan,
Taiwan 701
r3688401@ccmail.ncku.edu.tw

Abstract

The use of computer is getting more and more popular. In order to catch up with the trend, the elderly as well as other age groups people have to use computers more frequently. From literature on vision researches, visual capabilities start deteriorating after 40 years of age. This research intended to investigate how do elderly people do on VDT compare to younger generation. Three types of most frequently used Chinese character font (Thin-ming, Zhong-yuan, and Biao-kai) with three levels of character size (25, 40, and 55 minutes of visual angle) were chosen to test visual performance and preferences .The results showed that character font and size both had significant effects on visual performance. Subjects performed best with Thin-ming font and worst with Biao-kai. The bigger the character size, the better the visual performance. Subjective preference was also affected by font and size. Zhong-yuan was the most preferred font and Biao-kai the least. 55 minutes of visual angle resulted in the greatest subjective preference, while 25 minutes the worst. Young subjects performed better than elderly people did

1 Introduction

When using computers, there are several factors that influence reading performance. Factors can be categorized into two broad categories: hardware and information presentation. Hardwares set the basics for information presentation. Parameters of information presentation, such as character size, word spacing, line spacing, and word font will in turn influence reading performance.

The use of computer is getting more and more popular. In order to catch up with the trend, the elderly as well as other age groups people have to use computers more frequently. From literature on vision researches, visual capabilities start deteriorating after 40 years of age. This research intended to investigate how do elderly people do on VDT compare to younger generation.

2 Literature Review

In this section, the authors reviewed related researches have been done on the variables: elderly, font size, and character font.

2.1 Elderly people

According to World Health Organization (WHO), when a person reaches 65 years of age or older is considered as an elderly people. Owing to improving living conditions, nutrition, and medical care, people are living longer and longer. Up to 1998, the elderly people made up 8.26% of the total population in Taiwan (Yeh, 2000). The figure would go up in a steady trend, which means that there will be more usability issues should be considered in daily life..

2.1.1 Visual characteristics of elderly people

Retina is where vision captured. Rods and cones in retina transformed light and color into nervous activities. Lens serve the function to focus the images onto retina, a process called accommodation.

Pelosi and Blumhardt (1999) pointed out that visual capabilities as well as auditory capabilities, mental abilities, and reaction time will deteriorate as people getting older. When people get older, usually after 40 years of age, vitreous and aqueous humour will be blurred, lights will become diffused in the eyeballs. Images of figures become difficult to be focused on the retina. (Sanders and McCormick, 1993)

2.1.2 Visual design for elderly people

On the way of aging, visual and hearing capabilities degrade gradually. Signals have to be bigger or louder to be well perceived. Transgenerational design stresses on the usability of products for different age groups, especially for elderly people. Generally speaking, designs fit for elderly generation will be also fit for younger generation, but not vice versa (Chen, 1996).

Pirkle and Babic (1988) pointed out that there were several suggestions for visual design for elderly people:

a. Take age difference in mind and provide proper lighting and contrast.

b. Provide only necessary information, avoid fancy but unrelated noises.

c. Use unsaturated instead of saturated colours.

d. Provide visual displays with high contrast.

e. Avoid glare, either direct or reflected.

2.2 Fonts of type and font size

From Pirkle and Babic's (1988) suggestions, we know that for elderly people, the bigger the visual displays, the better the visual performance. But Kingery and Furuta (1997) also pointed out that the bigger the font size, the lesser information would be transmitted in a single presentation. There should be an optimal trade-off point. Now, let us look at some related researches on this regard.

2.2.1 Western letters

According to Sanders and McCormick (1993), there are more than 30,000 fonts of type (or typefaces) in western letters. They can be classified into 4 classes: Roman, Gothic (sans serif), Script, and Block letter. Roman types are the most commonly used for text printing. The other classes of typeface can be used for other purposes.

Tullis et al (1995) investigated the effects of typeface and size of letter on reading speed, correct response rate and preference. They found that the effects were significant with bigger and bold Arial (8.25, 9.0, 9.75) and Ms Sans Serif (8.25, 9.75) had better and fast reading performance. Kingery and Furuta (1997) kept subjects in a constant 50 cm viewing distance to read messages in Times New Roman, Airal, Book Antiqua, and Century Gothic typefaces in 27, 39, and 47 minutes of visual angles (VA). They found that in 27 and 47 minutes of VA, there were no significant performance differences among the four fonts. But in 39 minutes of VA, Times New Roman had the best effects on reading performance.

2.2.2 Chinese characters

Chinese Characters are more complex than Western letters. There are also different fonts of characters, some are more artistic, some are more commonly used for hard copy printings. Shieh et al (1997) compared the reading performance on Thin-ming and Biao-kai fonts on VDT. They found that Thin-ming font had better effects . In the same year, Yu (1997) asked subjects to read message on different fonts and sizes on hard copies and found that: a. preferred fonts were not necessarily better for reading performance, b. with 40 minutes of VA, characters with thin stokes (e.g.: Thin-ming, and Soong) were perceived better, c. with 80 minutes of VA, characters with thicker strokes (e.g.: Biao-kai, and round type) were perceived better. Both studies used college students as subjects. How elderly people will do?

3 Methods

The current study intended to investigate the effects of typefaces and font size of Chinese characters on elderly and young people. Three most popular Chinese typefaces were chosen:

Thin-ming, Biao-kai, and Zhong-yuan. Each typeface was shown to subjects in three sizes: 25, 40, and 55 minutes of viewing angles. 24 elderly and 24 young subjects were asked to participate in the experiment. Each subject was to look at the center of CRT from a fixed distance of 50 cm. Characters were shown on screen with 50 milliseconds duration. There was a signal 1.5 seconds before each presentation. Subjects were asked to write down characters been presented. There were 9 combinations of typefaces and font sizes. 5 characters in each condition were presented.

The second part of the experiment was to test the preference of typefaces and font sizes. A short statement was presented in the 9 conditions on the screen. Each subject rate one's own preference for each of the 9 conditions. Data were collected and analyzed.

4 Results

4.1 Legibility

An ANOVA (Analysis Of VAriances) was conducted to test the effects on legibility. The results showed that font size had very significant effects on legibility ($F(2,91)=148.77$, $P<.01$). The bigger the viewing angle of characters, the better the subjects could read correctly. Typefaces also had very significant effects on legibility ($F(2,91)=9.15$, $P<.01$). There was no significant difference between Thin-ming and Zhong-yuan, but both had significant difference with Biao-kai. The effects of age was also very significant ($F(1,46)=262.93$, $P<.01$). Young subjects performed better than elderly people did.

There were significant interaction effects of font size and age ($F(2,91)=38.65$, $P<.01$). For young subjects, the size effects leveled off after 40 minutes of VA (3.88, 4.57, 4.79) but elderly people kept steady improvement (1.44, 3.03, 4.29). There were also significant interaction effects of typefaces and age ($F(2,91)=3.35$, $P<.05$). For young subjects, they performed Biao-kai the least well (4.15), improved on Thin-ming (4.49) and Zhong-yuan (4.60). For elderly people, Biao-kai (2.69) also the least performed font, but improved on Zhong-yuan (2.82) and Thin-ming (3.25).

4.2 Preferences

Another ANOVA was performed to see the effects on preferences. Font size had very significant effects on preferences ($F(2,91)=106.46$, $P<.01$). Subjects liked bigger characters, the bigger the better. Typefaces also had very significant effects on preferences ($F(2,91)=11.27$, $P<.01$). All subjects liked Zhong-yuan best (6.63), then Thin-ming (6.2) and Biao-kai (5.69). For age differences, young subjects had significant preferences for all reading materials. There were no other significant effects on preferences/

5 Conclusion

There are several conclusions can be drawn upon the results:
(1). The bigger the characters, the better the performance or preference. Subjects preferred and performed better on bigger characters. Tullis et al (1995) had the similar findings.
(2) Biao-kai font was the least preferred Chinese font. Subjects also performed worst on VDT by using Biao-kai font characters.

References

Chen, J.J. (1996) Transgenerational Design for Man-Machine Interface, *Journal of Mingchi Institute of Technology*, 28, pp. 169-176.

Kingery, D., and Furuta, R. (1997) Skimming Electronic Newspaper Headlines: A study of typeface, point size, screen resolution, and monitor size, *Information Processing & Management*, 33, pp.685-696

Pelosi, L., and Blumhardt, L.D., (1999) Effect of Age on Working Memory: an Event-related Potential Study, *Cognitive Brain Research,* 7, pp.321-334.

Pirkle, J.J., and Babic, A.L. (1988) Guidelines and Strategies for Design Transgenerational Products: An Instructors' Manual, Syracuse University, New York

Sanders, M.S. and McCormick, E.J. (1993) Human Factors in Engineering and Design, Singapore: McGraw-Hill,

Shieh, K.K, Jung, J.H., and Chen, M.T (1997) The Effects of Characteristics and Working Conditions of Computer Operation on Visual Performance and Fatigue. *Journal of the Chinese Institute of Industrial Engineers,* 14(3), pp. 237-245. (In Chinese)

Tullis, T.S., Boynton, J.L. and Hersh, H. (1995) Readability of Fonts in the Windows Environment, *Fidelity Investments*, ACM.

Yeh, Z.C. (2000) Facing the Aging Society. Journal of Open University, 21, pp. 48-52. (In Chinese)

Yu, F. (1997) A Study of Readability and Imagery of Printing Fonts of Chinese Characters. Yunlin University of Technology. (In Chinese)

Pitfalls in International User Testing, and How to Avoid (some of) Them – A Case Study

Magnus Lif

Enea Redina AB
Smedsgränd 9, SE-753 20 Uppsala, Sweden
magnus.lif@enea.se

Abstract

Performing international user testing can be very powerful. However, conducting tests in several countries in parallel increases the risk of adding bias and costs to the study, e.g. delays, mistakes and errors can prove to be more costly then when testing in one country only.

This case study presents an international usability evaluation performed with users in the UK, France, Germany and Sweden. The studies were conducted in a lab setting where 48 users were observed performing a predefined set of tasks while "Thinking Aloud". Based on our experience from this and other international evaluations a number of pitfalls are highlighted and ways to avoid them are discussed.

1 Introduction

Most international companies acknowledge the importance of taking cultural differences into account when designing a product or service for a global market. Some companies realises the same holds for a website aimed at an international audience. Marcus & Gould (2000) claim that "companies that want to do international business on the web should consider the impact of culture on the understanding and use of Web-based communication, content, and tools (p. 34)". Cultural differences may have impact on the way users interact with, their expectations on and perceptions of the web site. Mrazek & Baldacchini (1997) distinguish between "locale differences" which arise due to country and region such as language, legal regulations, paper formats and responses to marketing campaigns, and "non-locale differences". They claim that non-locale differences in user needs – such as size and colour – are not related to country but due to variations in user type and segmentation. Their view is that "what are categorised as cultural differences do not impact the human-computer interaction model (p. 20)". Instead they state that evaluators should be careful when interpreting a result as being caused by culture. The aim of this case study is not to identify potential cultural differences. Instead the focus is on how to reduce bias in international user testing which in turn will help identifying true differences between countries.

Evaluating international websites can be done with or without users. A typical method not involving users is Heuristic evaluation where Human-Computer Interaction (HCI) experts inspect the system while logging potential usability issues (Nielsen & Molich, 1990). The issues are categorised according to a set of heuristics. For an international site, to increase the likelihood of identifying differences between countries, HCI experts from the targeted countries should conduct the inspection. Not involving users in the process will keep the costs down, however several and severe issues may not be discovered due to lack of domain knowledge.

Involving users is of course an advantage. To reduce travel costs, an interesting option is remote usability testing which can be defined as "usability evaluation wherein the evaluator, performing observation and analysis, is separated in space and/or time from the user" (Hartson, Castillo, Kelso & Neale, 1996, p. 228). A typical method for remote testing is on-line questionnaires. These can be distributed to a large number of users at a low cost and are most useful for gathering the users subjective views. It is more difficult to capture quantitative behavioural data or qualitative data concerning specific usability problems. By not observing the user performing the scenarios, finding out where, how and why users go wrong can be difficult. Hartson et al. (1996) describes a relatively successful method in a case study where video conferencing software was used to transfer a scan converted view of the users monitor from a remote location. This together with an audio connection made it possible for the evaluator to run a usability test remotely although the quality of the audio/video sometimes made it difficult for the evaluator to follow the users actions and comments. Apart from the downside of having little visual feedback as to what the user is doing remote evaluation usually requires the user to install and run quite unfamiliar utilities (Nielsen, 2000).

Remote evaluation can sometimes be a good option. However, one requirement for this international study was that the evaluators should fully understand the users' culture and language. A better option was therefore to involve some of the HCI consultants from our local offices.

2 Method

This case study describes an international user evaluation of an e-commerce website targeting 16 European countries. The aim was to identify potential usability problems with the site. The evaluation involved users and HCI consultants from four different countries. It was planned and coordinated from our UK office and was conducted in the UK, France, Germany and Sweden using a beta version of the site. These countries were selected because they had the largest number of customers and the largest share of Internet users. In each of the countries we had an office with experienced HCI consultants.

The first set of tests was run in the UK. The same study was then conducted in parallel in the other countries the following week. The method was based on the "Thinking Aloud" method where users are asked to perform a predefined set of tasks while thinking out loud (Lewis, 1982). To get more quantitative and comparable results, completion of and the time to perform certain tasks were measured. The users' faces and the computer screen were recorded on video and the client was able to follow each session from a room next door.

From each country 12 users participated. These were categorised by domain knowledge and experience from shopping online. In total 48 users were involved in the study. This would allow us not only to compare differences between countries but also between users with different domain knowledge as well as users with different level of on-line shopping experience. All participants were spending at least 1.5 hour per week on the Internet. Also, they reflected the target audience in terms of age and gender.

3 Results

While planning and conducting an international user evaluation several pitfalls generally occur and each mistake or error is likely to be costly due to the number of countries and people involved. Also, the risk to add bias increases. Under the device "if anything can go wrong, it will" the results are here presented according to where typical pitfalls occur.

3.1 The cloned evaluator

Getting comparable results when working with four evaluators is difficult enough when consultants from the same team are involved. When working with colleagues from offices in different countries, with different background, speaking different languages and potentially using different techniques it is even more likely that bias will be introduced. To reduce bias it is important that all evaluators know exactly how to perform the studies, what information to collect, how to collect the data, and in what format to deliver the results. Therefore, great effort was made to make sure the set-up of and conditions for the studies were as similar as possible in the four countries. Detailed guides and templates for data collection and documentation of the results were developed and translated into the local language. To make sure the tests were run in the same way by all evaluators an HCI consultant from the UK office, well informed about this particular study, were present to guide and help the local evaluator. This turned out to be very useful.

3.2 It is all about the logistics

Setting a date for the actual study is of course crucial. There is a lot of planning involved. Users need to be recruited, usability labs have to be booked and the evaluators need to be allocated. Ideally, you want to set a date well in advance to allow as much time as possible to find the people you need. However, there is a dilemma since dates do have a tendency to slip and slipping dates can turn out to be costly when several evaluations are done in parallel.

For this study the participating users were recruited locally by an external agency. We did inform the agency about the dates for the studies early on in the project to allow them enough time to find appropriate users. We also informed the evaluators and booked the usability labs. However, during the process the client changed the date three times due to delays in the project. This, of course, made it difficult both for the recruiters, the evaluators and for us to plan our time. For one of the date changes the client even had to pay the recruitment agency extra for the added amount of work.

3.3 The unpredictable user

Another area of concern is the quality of the users. Working with users from four different countries increases the risk that not all users will fit the predefined user profile. In our case we worked with a recruitment agency used on a regular basis by our client. This agency had local offices or partners in each of the selected countries, which made the process easy and also lowered the risk for misunderstandings. Also, we made sure screeners were designed and translated well before starting to recruit.

To double check that the users fitted our profile control questions were asked in the beginning of each evaluation. It turned out that a few users did not fulfil our requirements and therefore had to be excluded. Also, there were a couple of users that did not show up at all. To deal with these kind of issues we recruited 15 instead of 12 users in each country, i.e. three more users than we needed, which turned out to be enough.

3.4 The unreliable application

When running the test application from one server, if the system goes down it will happen in all countries at the same time. This can have great impact on the study, especially if it takes time to fix the problem. Ideally, the system should be stable while performing the evaluation. However, since we usually want to evaluate the application at a relatively early stage (to allow time for redesign) it is likely that problems will occur.

In our case we were running the user scenarios on the system beforehand to identify and correct issues with the site. We also confirmed with the project manager that no one was allowed to touch the test environment during the evaluation period and we had immediate access to technical expertise in case something went wrong.

During the evaluation the application went down a couple of times. One time because the test environment was taken down, even though they were not allowed to. Because of this one user test had to be cancelled, i.e. loosing 3 users in total. Luckily we had recruited enough extra users to cover them.

3.5 The localised application

Finally, it is important that the test application is localised before the evaluation. All content should be translated to the local language and design elements such as images should be swapped to local versions when applicable. If this is not done properly it is likely that the result of the evaluation will be affected.

In our case almost the whole application was localised before the test. However, there were a couple of buttons and content areas that were not translated. Luckily this had only a very limited effect on the study.

3.6 The next stage

The data from each individual user test was analysed by the local HCI consultant and the results were delivered to our UK office in the predefined format. All results were then analysed and compared in the UK. Based on the findings recommendations for redesign were made. The result was presented to the client both verbally and in written format well before the launch of the site to allow time for redesign.

Several interesting and important issues were identified during this evaluation. Most of the issues were discovered in all countries. There were no significant differences between users with different domain knowledge or experience from shopping online. There were indications of differences between countries such as:

- In Germany, France and Sweden the users were not willing to pay with credit card. They preferred other payment methods. In the UK this was not an issue.
- Format for address and date-of-birth differed between the countries.
- In Sweden title is regarded as old fashioned and is no longer used. It therefore did not make sense to the users. This was not a problem in the other countries.

However, an important result here was that the number of issues related to differences between the countries was low. This may be a result of the countries being similar culturally.

4 Discussion

International user testing can be very powerful. However, to identify true differences between countries bias needs to be kept to a minimum. There are a number of pitfalls that should be avoided both to reduce bias and to keep the costs down. Below are some recommendations based on what we have learnt from this and other similar projects.

Before the evaluation

- Inform key people early on about the date for the study. Also, tell them if it is likely that the date will slip.
- Use a reliable recruitment agency, preferably with offices in the affected countries, and make sure screeners are properly translated to avoid users not fitting the profile.
- It is likely that users will not show up or have the wrong profile. Always over recruit or arrange to get quick replacements if needed.
- To get comparable results careful planning is absolutely key. Make sure that documents, guides, etc are translated and delivered to the local evaluators well in advance of the tests.
- The evaluators should be fluent in both the local language and your mutual language.
- Make sure that the application is translated and localised before testing it.

During the test

- Always ask control questions when users arrive to ensure they fit the profile.
- Involving an HCI person from the coordinating team can help to reduce bias.
- Running tests in parallel on the same server is a risk. If the application goes down it happens in all countries.
- Be clear about not touching the test environment during the evaluation.
- Someone needs to be at hand to quickly solve technical issues with the application.

Finally, allocate enough time after the test for redesign and do not forget that even smaller changes to the site may be time consuming when translation and localisation are needed.

References

Hartson, H. R., Castillo, J. C., Kelso, J., & Neale, W. C. (1996). Remote Evaluation: The Network as an Extension of the Usability Laboratory. In J. T. Tauber (Ed.), Proceedings of Human Factors in Computer Systems CHI'96 (pp. 228–235). New York: ACM.

Lewis, C. (1982). Using the "thinking-aloud" method in cognitive interface design. (*IBM Research Report RC 9265, 2/17/82*). Yorktown Heigts, NY:IBM T. J. Watson Research Center.

Marcus, A. & Gould, E. W. (2000). Crosscurrents. Cultural Dimensions and Global Web User-Interface Design. *Interactions*, 7 (4), 32-46.

Mrazek, D. & Baldacchini, C. (1997). Avoiding Cultural False Positives. *Interactions*, 4 (4), 19-24.

Nielsen, J. (2000). *Designing Web Usability*. USA: New Riders Publishing.

Nielsen, J. & Molich, R. (1990). Heuristic evaluation of user interfaces. In J. C. Chew & J. Whiteside (Eds.), *Proceedings of Human Factors in Computer Systems CHI'90* (pp. 249–256). New York, NJ: ACM.

Browsing and Visualisation of Recorded Collaborative Meetings on Small Devices

Saturnino Luz

University of Dublin
Trinity College, Dublin, Ireland
luzs@cs.tcd.ie

Masood Masoodian *Gary Weng*

The University of Waikato
Hamilton, New Zealand
{M.Masoodian,cw41}@cs.waikato.ac.nz

Abstract

Small handheld computing and communication devices such as Personal Digital Assistants (PDA) have experienced great increases in processing power as well as popularity. The enhanced multimedia capabilities of these devices has opened new perspectives for user interface design. This paper presents techniques that exploit such capabilities for visualising and browsing multimedia meeting memories. These techniques have been embodied in a prototype Handheld Meeting Browser which supports browsing of synchronous audio and text produced in on-line collaborative meetings.

1 Introduction

In today's work environment it is very common for people to attend conventional or on-line meetings in which issues are discussed, decisions made, and responsibilities allocated. Afterwards, people generally receive the minutes, which they often combine with their own personal notes to create some kind of *meeting memory*. In scenarios where meeting activity is mediated or otherwise supported by computers, it has become possible (and increasingly common) to record the full content of meetings on synchronous tracks of text, audio, and sometimes graphics and video. While recording the whole meeting guarantees that no relevant content is lost, locating and retrieving relevant information from the recording becomes a great challenge (Moran et al., 1997).

In order to address this issue, we have developed a system called COMAP (Masoodian & Luz, 2001) which supports on-line meetings over audio and text channels, allows recording, and creates indexed links between recorded audio and shared textual documents. COMAP implements various forms of visualisation to provide fast and effective access to different parts of text and, most importantly, audio. However, the fact that COMAP has been designed for desktop computers with conventional displays and full input-output capabilities limits the scope of its usefulness as a meeting memory tool. Analysis of users use of low-tech meeting memories suggest that portability is an important factor in their success, as well as user's continued reliance on them.

In this paper we present a Handheld Meeting Browser (HANMER) which has been designed to combine the convenience of conventional meeting minutes with the power of full meeting recording. HANMER represents a natural extension of COMAP by allowing users of to download meeting memories onto their devices for convenient access while they are on the move. This is done by exploring those COMAP mappings between synchronised audio and textual streams which are most adequate for presentation on small screens. Meetings are originally conducted on desktop computers where participants enjoy access to full-duplex audio and a shared text editor. The recorded data then undergo further processing prior to downloading by HANMER. In what follows we describe a set of lightweight indexing and visualisation techniques which have been used in the prototype.

2 Meeting recording and indexing

The overall architecture of HANMER and related components is shown in Figure 1. The components on the top half of the diagram are part of COMAP. They consist of a meeting recorder and a preprocessor. The recorder includes functionality to allow two or more users of desktop computers to communicate synchronously over an audio channel while working collaboratively on a shared text editor (Masoodian & Luz, 2001). The system makes extensive use of the Real Time Protocol (RTP) which supports synchronous multi-party communication over multicast-enabled networks (Schulzrine, Casner, Frederick, & Jacobson, 1999). Recording is started as soon as the first user joins the session and continues until the last user leaves it. RTP packets are stored as they arrive in order to preserve time-stamping, source, synchronisation and other control data. For scenarios in which the mobile devices have access to a wireless LAN or similar connectivity, a direct connection between mobile devices and the recording server could be considered to enable mobile clients to request streaming of the desired audio and text segments. However, we have not exploited this possibility so far. At the moment, a synchronisation server is used.

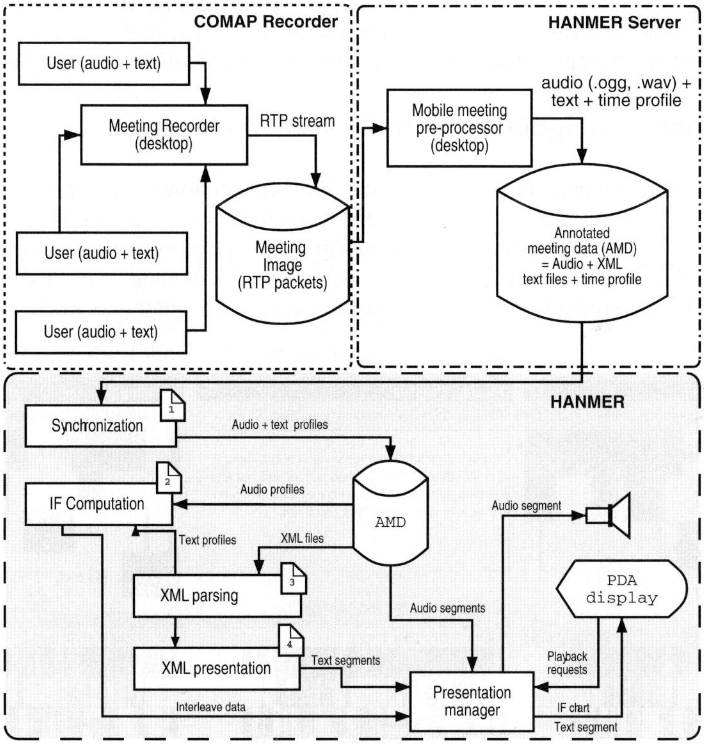

Figure 1: Overall system architecture

Preprocessing consists of keeping time stamps, extracting relevant features from textual input and determining *neighbourhoods* of text and speech segments (Masoodian & Luz, 2001). The preprocessing modules also convert audio and text data back to standard encoding. Text is converted into XML-formatted documents which contain annotations describing user activity over periods of time. Audio payloads are also transcoded into a compressed format in order to minimise memory requirements on the PDA.

HANMER is based on the concept of content mapping between temporal and contextual neighbourhoods (Luz & Masoodian, 2000). The main assumptions behind the idea of neighbourhoods are that recorded text and speech can be clustered into natural segments, and that text acts as a focus for meeting activities, therefore providing an intuitive starting point for memory access. Different levels of analysis determine different types of text and audio segments, varying from superficial to deep segmentation. Examples of superficial analysis include text segmentation into formatting units as well as speech segmentation into silence-delimited audio intervals. Deep segmentation would include automatic determination of discourse structure, turn-taking and speech acts (Stolcke et al., 2000). Although our prototype currently deals exclusively with superficial segmentation, the theoretical framework (neighbourhood mapping) is general enough to accommodate deep segmentation as well.

We define temporal and contextual neighbourhoods as follows. A segment of audio is a temporal neighbour of a text segment if that audio segment was recorded while the text segment was being modified or discussed by the participants, or if it is in a temporal neighbourhood of a related text segment. There could be multiple audio segments in the temporal neighbourhood of a document section, each corresponding to different time intervals during which that section was active. A segment of audio is a contextual neighbour of a text segment when it shares features (e.g. keywords) with that segment. Text segments may also have multiple contextual audio neighbours.

3 From content neighbourhoods to user interfaces

HANMER uses temporal neighbourhoods to guide information flow on the browser. The user may start browsing meeting records by reading a text segment produced during the meeting and, for instance, continue by playing back audio segments in the temporal neighbourhood of that segment. Since typically there will be many such audio neighbours, a method is needed which supports visualisation of these segments and their relationships with other contents as well as the activities of other participants.

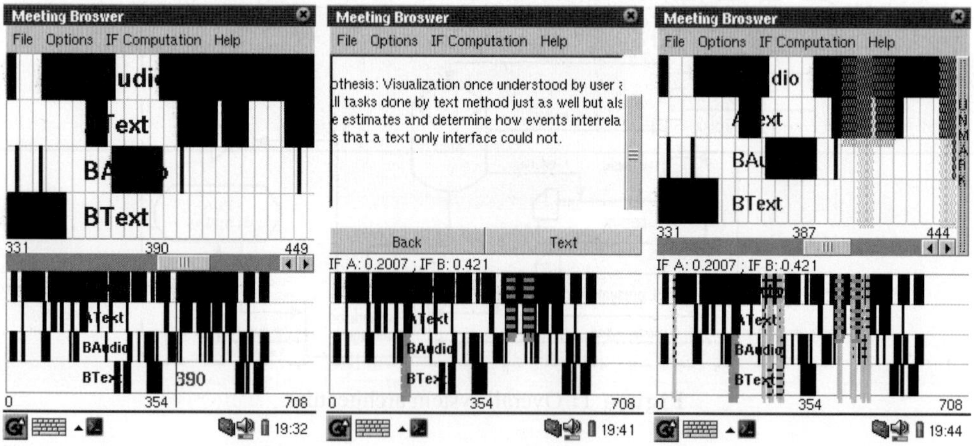

Figure 2: HANMER on a Zaurus SL-5500 PDA

The HANMER visualisation interface consists essentially of a time series of audio and text events plotted as a step function whose values range from 0 to the number of participants times the number of media streams. Initially, the user is presented with a bird's eye view of the meeting (Figure 2, left). The user can navigate by selecting (on the lower half of the screen) sections to be examined in more detail (on the upper half of the screen), and requesting retrieval of text segments, or audio playback. Usually, navigation will be guided by a text component (Figure 2, middle). If the user is interested in a subject referring to a particular text segment, he may select it, thereby causing all intervals of the meeting in which that subject was active to be highlighted on the meeting profile (Figure 2, right). This narrows the search down to a few segments. However, since browsing also involves audio, listening to even a small number of segments might turn out to be a very time-consuming activity. We address this issue by ranking the candidate segments according to a novel relevance metric.

3.1 Interleave Factor

In order to rank intervals by relevance we introduce a metric that quantifies the extent to which an action conveyed by speech is accompanied by another in text. We call this metric *Interleave Factor* (IF) as it is based on the hypothesis that concurrent events will often be semantically related (Luz, 2002). Although intervals with the highest levels of activity are naturally highlighted by the step function, given the limited screen real-estate these patterns tend to become very hard to discern in longer meeting profiles. IF helps identify the relevant intervals unambiguously, regardless of the amount of information contained in a profile.

The IF metric is founded on a continuous probability distribution: the probability of a speech event a in a meeting m is given by equation (1), where $bt(.)$ denotes the time at which the event started and $et(.)$ the time at which it ended. Similar probabilities will also be estimated for text events (denoted here as t, with or without subscripts).

$$P(a) = \frac{et(a) - bt(a)}{et(m) - bt(m)} \qquad (1)$$

The basic idea behind IF is that the degree of interleaving is described by the probability that a speech and a text event overlap within a certain time interval. In order to estimate this, we split the meeting into small sub-intervals, estimate partial IF's for those intervals separately, and then combine these partial estimates to arrive at the final figure. Let's assume a meeting of total duration M seconds split into L-second long intervals $I_1, ..., I_i$, where $1 \leq i \leq M/L$. For each of these intervals we then calculate the probability of overlap between speech and text events (respectively S_i and T_i) in the interval considered, as shown in equation (2).

$$P(S_i \cap T_i | I_i) = \frac{\sum_{j=1}^{k} a_{ij} \sum_{j=1}^{m} t_{ij}}{L^2} \qquad (2)$$

In equation (2), k is the number of speech segments and m the number of text segments. Speech and text segments are represented by a_{ij} and t_{ij}. For two or more participants, intervals of the same media must be merged before the sum is performed. IF is then defined as the sum of these conditional probabilities normalised by the number of intervals, as shown in equation (3).

$$IF = \frac{\sum_{i=1}^{M/L} P(S_i \cap T_i | I_i)}{M/L} \qquad (3)$$

In our tests we have assumed $L = 10s$, which seems to produce IF scores which agree with the visual feedback provided by the step function plot.

Once IF scores are calculated for all sub-intervals, the ones assigned the highest scores are highlighted so as to indicate the points most likely to contain relevant information.

3.2 Implementation

The prototype has been implemented in Java and tested on both Embedded Linux and Microsoft Windows platforms running the Jeode Java Virtual Machine. The code is available under an open-source license (GPL) at the author's web site.

4 Conclusion

We have conducted preliminary studies in order to evaluate the validity of the ideas that guided the implementation of HANMER, namely content neighbourhoods and interleave factor. Various student-supervisor meetings were recorded in which participants worked on shared documents (Luz & Masoodian, 2002). These recordings were then examined using the prototype. Although further testing is needed to better assess the usability of the system, data collected so far indicate that even superficial segmentation techniques such as tiling, information extraction, and silence suppression can be quite effective in helping users structure and access meeting data, particularly speech. They also suggest that IF provides a reliable indicator of segment relevance. Unlike existing meeting browsers which rely essentially on automatic speech recognition, HANMER uses lightweight indexing techniques coupled with a highly compact visualisation interface. These features make it particularly suitable to small devices.

References

Luz, S. (2002). Interleave factor and multimedia information visualisation. In H. Sharp, P. Chalk, J. LePeuple, & J. Rosbottom (Eds.), *Proceedings of Human Computer Interaction 2002* (Vol. 2, pp. 142–146). London.

Luz, S., & Masoodian, M. (2000). Mapping collaborative text and audio communication. In *Proceedings of Webnet 2000: World conference on the WWW and internet* (pp. 769–770). San Antonio, Texas.

Luz, S., & Masoodian, M. (2002). Supporting online meetings between research students and supervisors. In *Proceedings of e-learn 2002, conference on e-learning in corporate, government, healthcare, and higher education* (pp. 1870–1873). Montreal, Canada: AACE Press.

Masoodian, M., & Luz, S. (2001). COMAP: A content mapper for audio-mediated collaborative writing. In M. J. Smith, G. Savendy, D. Harris, & R. J. Koubek (Eds.), *Usability evaluation and interface design* (Vol. 1, pp. 208–212). New Orleans, LA, USA: Lawrence Erlbaum.

Moran, T. P., Palen, L., Harrison, S., Chiu, P., Kimber, D., Minneman, S., Melle, W. van, & Zellweger, P. (1997). "I'll get that off the audio": A case study of salvaging multimedia meeting records. In *Proceedings of ACM CHI 97 conference on human factors in computing systems* (Vol. 1, pp. 202–209).

Schulzrine, H., Casner, S., Frederick, R., & Jacobson, V. (1999, February). *RTP: A transport protocol for real-time applications.* IETF Internet Draft draft-ietf-avt-rtp-new-04.

Stolcke, A., Ries, K., N, C., Shriberg, E., Bates, R., Jurafsky, D., Taylor, P., Martin, R., Ess-Dykema, C. V., & Meteer, M. (2000). Dialogue act modeling for automatic tagging and recognition of conversational speech. *Computational Linguistics*, *26*(3), 339–373.

User-Interface Design and Culture

Aaron Marcus, Valentina-Johanna Baumgartner, Eugene Chen

Aaron Marcus and Associates, Inc. (AM+A)
{Aaron, Valentina, Eugene}@AMandA.com

Abstract: By using examples from corporate Websites from several different countries, this analysis compares user-interface components (metaphors, mental models, navigation, interaction, and appearance) with Hofstede's cultural dimensions (power distance, individualism-collectivism, gender roles, uncertainty avoidance, and long-term time orientation). One can observe several typical patterns.

1 Introduction

We seek to determine to what extent corporate designs exhibit cultural differences and to demonstrate how culture theory, *e.g.,* Geert Hofstede's [Hofstede], is helpful for such research. By combining Hofstede's five dimensions (power distance, individualism-collectivism, gender roles, uncertainty avoidance, and long-term time orientation) with five components of UI design (metaphors, mental model, navigation, interaction, and appearance) [Marcus, 1997], we examine 25 ways a Website may be affected by culture.

2 Method

When combining Hofstede's cultural dimensions and UI design components, we use a five-by-five matrix that allows for 25 fields of interest (Fig 1). To find out how Websites are affected by culture we analyzed "B2B and B2C" Websites from three different continents (USA, Europe, and Asia), as shown in Fig. 2.

	PD	IDV	MAS	UA	LTO
Metaphor					
Mental Model					
Navigation					
Interaction					
Appearance					

Figure 1: Conceptual Matrix

	US	EU	Asia
Business	Sapient (S)	Siemens (SIE)	Hitachi (HIT)
	PeopleSoft (PEO)	SAP (SAP)	
Consumer	McDonalds (MCD)	IKEA (IKE)	Sony (SON)
	Coca Cola (COC)	Mercedes (MER)	Mazda (MAZ)

Figure 2: Companies whose websites were studied

2.1 Key findings

	PD	IDV	MAS	UA	LTO
Metaphors	SIE HIT	SIE		S SIE SAP IKE MER	
Mental Model	HIT	S SIE PEO MER	SIE	SIE	SIE
Navigation	S		SIE	SIE McD	
Interaction	COC		SIE McD MER		HIT
Appearance	SIE PEO		McD COC MER	SIE McD	SIE

Figure 3: Matrix of Websites within the mapping of culture to UI component dimensions

3 Analysis of culture dimensions and UI components

The following section discusses Hofstede's culture dimensions and within them user-interface components.

3.1 Power Distance (PD)

Metaphors: According to Hofstede, countries with high PD focus on expertise, authority, and/or experts. Applied to the field of UI design one can assume that visual metaphors in such high PD countries would show buildings or objects with a clear hierarchy. On the Siemens Website we see the Netherlands (low PD) uses the eye-level portion of a person's face as a metaphor for the home "button", whereas Malaysia (high PD) uses a city's skyline. The Netherland's picture is an "equal" (level) look into someone's eyes (see [Kress and van Leeuwen]); Malaysia's skyline view shows official buildings. (Fig. 4)

Netherlands (PD 38)	Malaysia (PD 104)
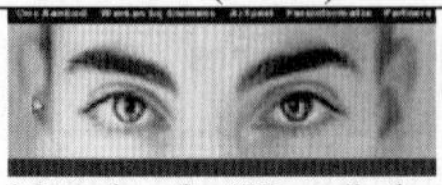 Metaphor for "Home": the face / the eyes of a person	 Metaphor for "Home": an official building

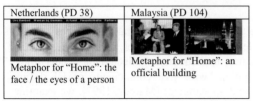

Figure 4: Siemens: Personal Images vs. Official Buildings

Italy (PD 50)	Singapore (PD 74)
 Images of **people**	 Images of **leaders, official Websites**

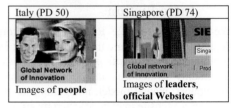

Figure 5: Siemens: People vs. Leaders

Mental Model: It seems likely countries with a high PD prefer complex, highly organized structures and reference data with little or no relevancy ranking. Countries with a low PD might prefer simple, informally organized structures, and less structured data with some or much relevancy. The Hitachi Website shows a contact page in Canada (low PD) that offers limited, but well-structured contact data. The Hitachi Website in Singapore (slightly higher PD) offers much contact information on one page. As opposed to the Canadian contact page, the information on the Singaporean contact page is highly categorized.

Navigation: Following Hofstede's theory, we assume low PD countries prefer open access and multiple options; but high PD countries use more authentication and prefer restricted choices. For example, the careers frequently-asked questions (FAQ) page from the German Sapient Website (low PD) offers a variety of possibilities for how to apply for a job but the Indian Website (high PD) describes only one way to apply: applicants must use a standardized process and apply via a Web form.

Interaction: Interaction refers to input and output sequences, including feedback for the user. The Coca Cola Website provides a good example that feedback in low PD countries can mean "supportive error messages", while feedback in high PD countries contains severe error messages. When one tries to log in to the members' section on the Denmark site with an incorrect password, the error message politely (using words like "please…") summarizes what went wrong with possible solutions. The opposite occurs for the Malaysian Website: The expression "Bzzzzt!" seems impolite and does not explain what went wrong. The error message "wrong password!" seems like a scolding, and no possible solutions are presented.

Appearance: We assume countries with low PD prefer Websites that use "normal" people, symbols, and colors. Countries with high PD might use images of leaders; national themes, insignia, and colors. Supporting examples are found on Italian and Singaporean Siemens Websites. The image used as a Home button in the upper-left corner shows a man and a woman in the Italian version, while the Singaporean Website uses the picture of an official monument surrounded by skyscrapers. (Fig. 5)

3.2 Collectivism vs. Individualism (IDV)

Metaphors: Per Hofstede, we assume metaphors used in low IDV countries might be relationship- and content-oriented, while those in high IDV countries might be action- or tool-oriented. Compare Brazil's (low IDV) McDonalds Website with the US's (high IDV) (Fig. 6). In the US version, more individualism appears with an image of one man who represents the company. The Brazilian Website uses group images.

Brazil (IDV 38)	United States (IDV 91)
 Images of **groups** an organizations that should visualize the section "McDonalds in Brazil"	 Images of a **single person** to visualize the "Corporate" section of McDonalds USA.

Figure 6: Brazil vs. US: group vs. single-person images

Mental Model: For high IDV countries, we assume the individual is most important, and they might use very product- or task-oriented mental models, in which personal achievement is maximized. Collectivist countries might emphasize role-oriented models and underplay personal achievement. For individualist and collectivist approaches within text, we compare localized PeopleSoft Websites. Compare the significant difference regarding emphasizing personal achievement in the Singaporean (low IDV) versus the German (high IDV) "About PeopleSoft" sections. Singapore's Website speaks about the role the company plays in the world's economy and mentions the employees and partners. The German Website simply mentions the company's founding date and location, but emphasizes the CEO, who is mentioned by name.

3.3 Femininity vs. Masculinity (MAS)

Metaphors: Comparing the Finnish (low MAS) with the Austrian (high MAS) McDonalds Website, we find a metaphor on the front page that supports the idea that low MAS countries focus on family and shopping, whereas high MAS countries prefer sports and competition.

Mental Model: Applying Hofstede's assumptions about MAS to the component of mental models, we assume we shall find social structures in low MAS countries and work/business structures in high MAS countries. We also expect detailed views and relationship-oriented approaches in low MAS countries, whereas we expect high-level views and goal-oriented approaches in high MAS countries. The Siemens Website supports this assertion: Whereas the Norwegian (low MAS) careers page is very relationship-oriented (the main sections are entitled "What we are looking for" and "What we can offer"), the Austrian page (high MAS) emphasizes advanced education possibilities for employees, which seems goal-oriented.

Navigation: The contact page of the Siemens Website offers multiple choices in Sweden (low MAS) but only one possibility to contact the local company in Japan (high MAS). This example supports the assertion that low MAS countries would prefer multiple choices and polychronic approaches, whereas high MAS countries would prefer limited choices and synchronic approaches.

Interaction: We assume high masculinity countries prefer game-, mastery-, and individual-oriented approaches. In countries emphasizing gender differentiation and competitiveness less, we expect these approaches less and more practical, function-oriented approaches. The McDonalds Website is an example that supports this assertion: The Swedish (low MAS) Website focuses on client service by providing many ways to contact the company. On the Austrian (high MAS) Website, it is much easier to find the fun and games section than contact information. The fun section contains technical content such as screensavers and a link to send an e-card. A client-service section is not available on the Austrian Website.

Appearance: In countries with low MAS, we expect harmonious shapes. For example, although the Mercedes-Benz localized Websites are very similar, we find a major difference in the design for Sweden (low MAS) and Germany (high MAS). The visual design approach from Sweden uses soft edges and shapes while the German layout focuses on clear structure and avoids cuteness.

3.4 Uncertainty Avoidance (UA)

Metaphors: Per Hofstede's theory, we assume countries with low UA might prefer, novel, unusual references and abstraction, while cultures with a high UA prefer familiar, clear references to daily life and representation, not abstraction. For IKEA, a European furniture store known for its casual advertisement style and low prices, the Swedish (low UA) Website uses the ambiguous slogan "Nothing is impossible." The French (high UA) Website uses the specific slogan "Design at [a] small [low] price". The same pattern holds for imagery: When comparing the British (low UA) and the Belgian (high UA) Siemens Websites, we find pictures that act as metaphors. The UK Website shows a very dynamic photo of unidentifiable technical objects and the slogan "Welcome to SIEMENS in the UK," *i.e.*, an abstract representation of the company. The Belgian Website shows varied pictures of daily life, which act as representations. (Fig. 7).

<table>
<tr><th>United Kingdom (UA 35)</th><th>Belgium (UA 94)</th></tr>
<tr><td>

Novel, unusual references, abstractions</td><td>
Familiar, clear references to daily life, representations</td></tr>
</table>

Figure 7: Siemens: abstraction/ representation

<table>
<tr><th>United Kingdom (UA 35)</th><th>Belgium (UA 94)</th></tr>
<tr><td>

Ambiguous, varied imagery</td><td>
Simple, clear, consistent imagery</td></tr>
</table>

Figure 8: Siemens: variety vs. consistency

Mental Model and Navigation: We expect tolerance for ambiguousness and complexity in countries with low UA. Conversely, we expect simple, explicit, clear articulation; limited choices; and binary logic in countries with high UA. Because the components of mental model and navigation are closely related (structure and process), they are considered together and are impacted similarly as in the previous description. Both Switzerland and Belgium are multilingual countries. When a user enters the Siemens Website of Switzerland (low UA), it is possible for her/him to choose among languages, but it is also possible to access directly several links. The Belgian Website offers a more binary logic: users always must decide at the beginning in which language they want to explore. Only then can they navigate deeper.

Appearance: We assume low UA countries expect tolerance for more perceptual characteristics involved in purely ornamental or aesthetic use and less redundant coding of perceptual cues. Countries with high UA may prefer simple, clear, and consistent imagery, terminology, and sounds. Users may expect highly redundant coding of perceptual cues. We find an example by comparing the Belgium Siemens Website (high UA) with the UK (low UA). The imagery is more consistent on the Belgian Website. (Fig. 8)

3.5 Long-Term Time Orientation (LTO)

Mental Model: Hofstede's theory seems to imply that long-term time-oriented countries would more actively pursue the long-term perspective. The following example shows the difference in mental model concerning long-term time orientation: Pakistan's (low LTO) Siemens Website states the size and locations of the company. China (high LTO) focuses on the long-lasting history of the company.

Interaction: Regarding interaction in short-term time oriented countries we assume that distance communication is accepted as more efficient; and, therefore, anonymous messages are tolerated more. Inhabitants of long-term time oriented countries may prefer face-to-face communication and harmony. We can find an example of this pattern at the Hitachi Website. The US (low LTO) Website offers a contact page on which the user can find only a Web form to place a message. At the Singaporean (high LTO) Website, we find a Web form as well as personal contact information. **Appearance:** Short-term time-oriented countries seem more likely to focus on achieving goals quickly; hence, they might tend to show fewer things, avoid overly ornamented imagery, and focus on achieving practical goals. Long-term time oriented countries might do just the opposite. Siemens shows the use of imagery in both long- and short-term time-oriented countries. China (high LTO) uses warm, fuzzy images and pictures of groups, whereas Pakistan (low LTO) concentrates on showing tasks or products. (Fig. 9)

Pakistan (LTO 0)	China (LTO 118)

Concentration on showing **tasks** or products	Warm, fuzzy images, pictures of **groups**

Figure 9: Siemens Website: task-oriented vs. group-oriented

4 Conclusion

In this research, we discovered that the matrix-oriented method helps to organize and analyze data collection. Initial observations suggest that cultural habits run deeply and operate even under constraints of global design specifications. In high IDV and low PD countries, variations from standard practice are likely to be most frequently observed. We mention that presenting the examples cited, while useful to illustrate patterns, does not necessarily mean that, *ipso facto*, any particular pattern is the *right* way to design or revise a UI for a particular application or culture. The designer must consider both context and culture. In addition, the UI designer should consider how these patterns may influence cultures and design conventions, which undergo continuous change. One likely result of such research is a "culture-base" with specific conditions and predictable results that would inform a content management system (CMS). However, to draw specific conclusions and to use them in a CMS, more data are needed. Further research could produce quantitative and qualitative results that may feed culture-localization templates and tools.

References

Hofstede, Geert (1997). *Culture and Organizations: Software for the Mind*, New York: McGraw-Hill.

Kress, G. R., and Van Leeuwen, T. (1996). Reading Images: The Grammar of Visual Design. New York: Routledge.

Marcus, Aaron. (2002) "Globalization, Localization, and Cross-Cultural Communication in User-Interface Design," in Jacko, J. and A. Spears, Handbook of HCI, Lawrence Erlbaum, Mahway, pp. 100-150.

Marcus, Aaron, "International and Intercultural User Interfaces," in User Interfaces for All, ed. Dr. Constantine Stephanidis, Lawrence Erlbaum Associates Publishers, New York, 2000, pp. 47-63.

Marcus, Aaron. (1999). "Globalization of User-Interface Design for the Web," in Proc., 1st Internat. Conf. on Internat. of Prod. and Systems (IWIPS), G. Prabhu and E. M. Delgaldo, eds., 20-22 May1999, Backhouse Press, Rochester, NY, USA, 165-172.

Marcus, Aaron (1997)."Graphical User Interfaces," Chapter 19, in Helander, M., Landauer, 0 T.K., and P. Prabhu, P., Eds., Handbook of Human-Computer Interaction, Elsevier Science, B.V., The Hague, Netherlands, 1997, ISBN 0-444-4828-626, pp. 423-44.

Marcus, Aaron, and Emilie W. Gould (2000). "Crosscurrents: Cultural Dimensions and Global Web User-Interface Design," Interactions, ACM Publisher, Vol. 7, No. 4, July/August 2000, pp. 32-46.

Vehicle UI and Information-Visualization Design

Aaron Marcus,

Aaron Marcus and Associates, Inc. (AM+A)
Aaron@AMandA.com

Abstract: As vehicles acquire significant telematics and advanced displays, user-interface and information-visualization design must account for usability and appeal issues radically different from preceding decades. This revolution poses significant design challenges including culture and branding issues.

1 Introduction

Powerful computer hardware, software, networks, and wireless communication capabilities are becoming routinely embedded in vehicles. Vehicle user-interface (UI) and information-visualization (IV) design are significant challenges to vehicle systems manufacturers, who are relatively less experienced in general with all aspects of software development. In particular, UI and IV development contain these essential phases: planning, research, analysis, design, implementation, evaluation, documentation, and training. Beyond traditional performance- and usability-oriented human factors issues, UI and IV developers now must address usefulness and appeal, including issues of branding, cultural diversity, in short, user experience. The UI and IV components [Marcus, 1997] include the following:

Metaphors: Fundamental concepts communicated via words, images, sounds, and tactile experiences.

Mental models: Structures or organizations of data, functions, tasks, roles, and people at work or play.

Navigation: Movement through the mental models, *i.e.*, through content and tools.

Interaction: Input/output techniques, status displays/feedback, plus behavior of humans machines.

Appearance: Essential perceptual attributes, visual, auditory, and tactile characteristics

Of special interest is information visualization, the abstract or representational means for communicating structures and processes. Typically, IV design focuses on tables, forms, charts, maps, and diagrams.

2 New Challenges for Vehicle UI+IV Design

Beyond extremely important human factors issues [Marcus et al, 2002] in successful vehicle UI and IV development, other issues are emerging, just as they are for other platforms, such as desktop/Web, information appliances, and portable mobile devices. Among other issues are the following:

- Intelligent, or "smart" vehicles will have significantly new hardware and software for UI+IV. For example, large and/or multiple LCD display screens enable dynamic, unconventional displays quite different from the electromechanical displays of past vehicle systems. Sensors and wireless Web connectivity imply the possibility of fundamentally new content requiring adequate means for IV.
- In addition to usability, usefulness and appeal will have increased importance. Among other human factors analysts, [Jordan] has introduced four levels of pleasure: physical, psychological, social, and cognitive. In the vehicle realm, these concerns will affect the rider/driver experience.

- As for other platforms, the content for vehicle displays will include not only informational communication, but persuasive and aesthetic. Especially important is the issue of branding, and more specifically, how best to co-brand offerings of content providers.
- With increased commuting times, vehicles will become temporary workplaces and homeplaces for drivers and riders, thereby motivating the use of additional content and tools. Additional *usage spaces* [Marcus and Chen, 2002], to be accounted for in vehicle UI and IV design, may include information about one's own identity, entertainment, relationship-building/maintenance (*e.g.*, communication with others), self-enhancement (including education and health monitoring), access to information resources (e.g. travel information, references, and other databases), and commerce. As complete information, persuasion, and entertainment environments, vehicles may evolve UI and IV interactive displays depicting virtual presence in other spaces; games played with other vehicles' drivers and riders; status or instruction concerning driving/parking manoeuvres, repairs, parts re-ordering; complex loyalty program advertisements; etc.
- Localization of global vehicle products/services must consider cross-cultural communication and other dimensions of differences [Marcus, 2001]. English-speaking countries constitute eight percent of the world's population, but by 2005, approximately 75% of Internet users will be non-English speaking. Cultural factors will need to be considered more frequently. Already, 80% of corporate Websites in Europe offer more than English even though launching multi-language Website portals with a dozen European languages is a significant burden. [Hofstede] has posed five fundamental culture-dimension ranges: power distance (high *vs.* low), individualism *vs.* collectivism, masculinity *vs.* femininity, uncertainty avoidance (high *vs.* low), and time-orientation (long *vs.* short). [Marcus and Gould, 2000] show how these differences affect Web UI+IV design. More flexible, customisable vehicle UI+IV design will exhibit similar differences. For example, drivers with high uncertainty avoidance might desire more detailed displays.

3 Beyond Human Factors and Culture

Vehicle UI+IV developers should also consider these additional issues [Marcus, 2002a].

[Nisbett, Peng, *et al.*] studied how cognition itself is a product of culture. Their research shows Asian cultures seem more relationship-oriented and more comfortable with logically contradictory situations; European and North-Americans seem more object-oriented and more oriented to either-or logical relations. One possible conclusion is that Western assumptions of a universal focus/value on reasoning, categorization, and linear cause-and-effect explanations of situations and events may be biased. Such conclusions differ markedly from classical assumptions that would affect the design of UIs and IVs.

[Cialdini] proposes these dimensions of persuasion, by which people convince others to think or act in a particular way: reciprocation, consistency, social validation, liking, authority, and scarcity. Several of these dimensions combine in any situation to lead to someone's actions, and different cultures typically emphasize one factor over others. His analysis seems to imply that vehicle UI and IV persuasion elements might rely significantly on these differences to affect e-commerce and m-commerce.

[Bailey, et al] has studied the dimensions of trust in Website design and proposed these dimensions: attraction, dynamism, expertness, faith, intentions (revealed objectives and goals), localness, and reliability. It seems likely that these same factors would vary by culture and play a role in future vehicle UI and IV design, especially in regard to persuasive communication content related to m-commerce.

[Gardner] published dimensions of intelligence and argued for their consideration in reforming education. Likewise, these dimensions could affect approaches to vehicle UI and IV design.

4 New Prototype Vehicle UI+IV Displays

As an example of new approaches, the author's firm designed conceptual prototypes of new status displays that might replace conventional vehicle dashboards. The goal was to design new paradigms based on the executive information system (EIS) concept; information might be explored upon demand ("drilling down"). These prototypes are based on lexical, syntactic, semantic, and pragmatic considerations but have not been evaluated yet with users. They are intended to inspire new directions in UI+IV development.

Figure 5 (left) shows a 4 x 3 tabular display with frequently needed/required indicators of speed, fuel level, engine temperature, current driving gear, direction of turning, etc. The default display shows these references in three possible states: satisfactory, needing attention, and unsatisfactory. We assumed drivers will become familiar and comfortable with the position of the display and be able to identify the reference simply by its position and color, much as drivers search for and recognize traffic control lights in the physical environment. Alternatives show identical denotations as color strips and as more detailed charts for those with higher uncertainty avoidance requirements or to promote more attention to speed limits or performance measures. The latter also displays data values of important markers in a range of values as well as a digital time readout. These displays indicate how symbolism, color, and layout, among other visual syntax considerations, might affect the traditional appearance of vehicle status displays and account for legal, culture, cognition, performance, and preference considerations.

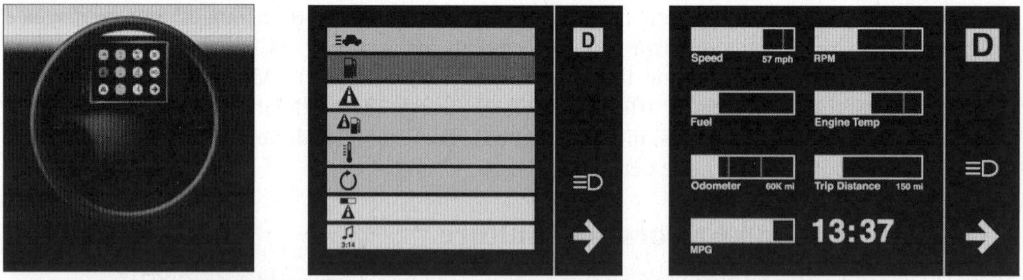

Figure 5: Prototype default vehicle information visualization and alternatives.

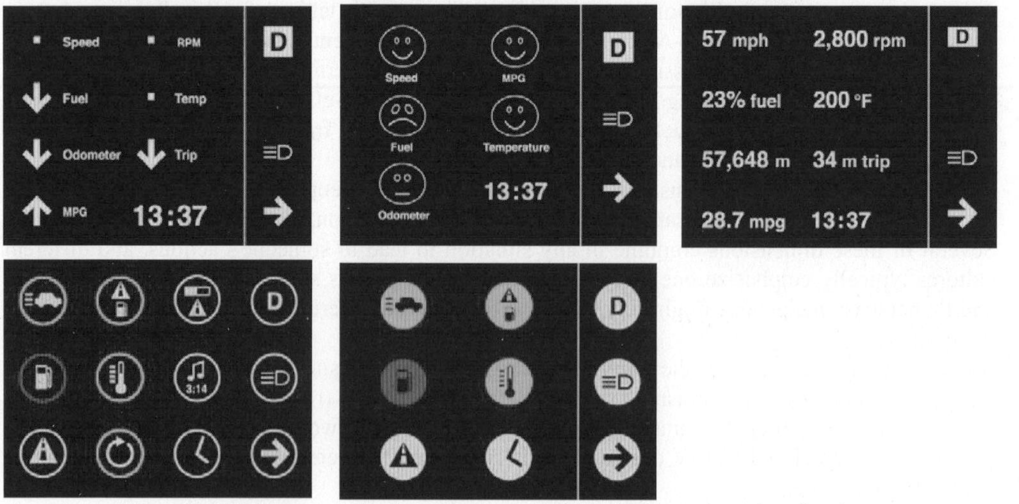

Figure 6: Further syntactic alternatives for vehicle information visualizations.

Figure 6 shows alternative ways of showing tabular organizations of the default display, including the use of arrows, alphanumeric displays, Chernoff faces, symbols depicted with color-coded outlines *vs.* solid disks of color, and alternative choices for displays. Some displays might be required legally, while others might represent personal preferences. Note that social engineering might come into play. For example, drivers might see displays of their performance or safety record, *e.g.*, trends of exceeding the speed limit, in order to encourage proper behavior. This data might be relayed to gasoline pumps. The driver whose record is good might receive a lower price for gasoline, thus turning performance data into an economic benefit. Considerable refinement and evaluation through focus groups and user testing would determine which approaches were worth further investigation, especially preference and performance differences related to culture, age, gender, education, etc. For example, typical culture/language text-reading directions suggest that display-column sequence or individual symbols might be reversed left-to-right for European-language text readers *vs.* right-to-left-reading languages such as Arabic or Hebrew.

Figure 7 shows two alternative situations in which co-branded informational/persuasive communication might be merged and displayed. This alternative shows a presumably required set in the left-most column, a driver-customized set in the second-from-the left column, and a set of "sponsored" signs in the third column. The appearance of these commercial signs indicates the possibility of branding and loyalty systems playing a significant role in what traditionally have been "pure" information displays. For example, when the fuel level reaches a warning or unsatisfactory state, these additional symbols might appear to remind the driver that he/she might also locate the preferred fuel vendor, "refuel" on food, and perhaps seek a preferred roadside rest-stop. Special coupons or offers might appear to encourage certain behaviors. The right column shows a set of required signs indicating the current driving gear, status of forward lights, and turn indicator. Signs would/could be in dynamic blinking, flashing, or other animated states, including pulsing size changes typical of many computer UI and IV designs. The navigation map indicates that sources of food and fuel would be indicated for both generic and sponsored or "branded" signs.

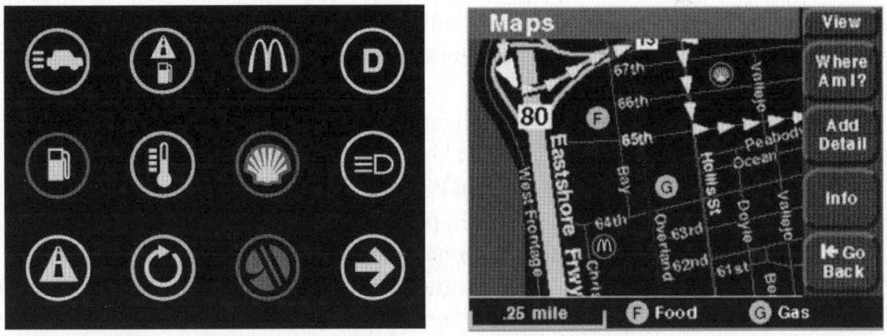

Figure 7: Vehicle information visualizations with co-branded persuasive communication elements

5 Conclusion

Vehicle product/service design requires improved cross-cultural communication in UI and IV development. Large-area, dynamic, relatively low-cost LCD displays make radically different designs possible. Consequently, innovative UI+IV thinking, including cross-cultural analysis and design, and exploration of human factors and visualization issues beyond the traditional concerns of usability must be considered more integrally in all development phases. Developers will need check-lists, templates, and guidelines. Having a better understanding of syntactic possibilities, the mappings of culture dimensions to UI components, as well as to such dimensions as trust or intelligence, will inform designers, enabling them to make better decisions about usability, aesthetics, and emotional experience, and to design better vehicle UIs and IVs.

6 Acknowledgements

The author acknowledges [Marcus, 2002b], on which this paper is based, the assistance of his staff in researching and analyzing the human factors issues of the driver's experience and the support of BMW, Germany, to investigate this topic, and Neil Griffiths, AM+A Designer/Analyst, for his assistance in designing the prototype advanced information-visualization displays.

7 References

Bailey, Gurak,and Konstan (2001). "An Examination of Trust Production in Computer-Mediated Exchange," Human Factors and the Web 2001 Conference, 5 June 2001, http://www.optavia.com/hfweb/

Cialdini, Robert, "The Science of Persuasion," *Scientific American*, Vol. 284, No. 2, Scientific American Inc: New York, February 2001, pp. 76-81 (www.influenceatwork.com).

Gardner, H. (1985). Frames of Mind, The Theory of Multiple Intelligences. Basic Books.

Hofstede, Geert (1997). *Cultures and Organizations: Software of the Mind*, New York: McGraw-Hill

Jordan, Patrick W. (2000). *Designing Pleasurable Products*. Taylor and Francis, London.

Marcus, Aaron (1997)."Graphical User Interfaces," Chapter 19, in Helander, M., Landauer, 0 T.K., and P. Prabhu, P., Eds., Handbook of Human-Computer Interaction, Elsevier Science, B.V., The Hague, Netherlands, 1997, ISBN 0-444-4828-626, pp. 423-44.

Marcus, Aaron. (1999). "Globalization of User-Interface Design for the Web," in *Proc.*, 1st Internat. Conf. on Internat. of Prod. and Systems (IWIPS), G. Prabhu and E. M. Delgaldo, eds., 20-22 May1999, Backhouse Press, Rochester, NY, USA, 165-172.

Marcus, Aaron (2000). "Designing the User Interface for a Vehicle Navigation System: A Case Study," in Bergman, Eric, editor, Information Appliances and Beyond: Interaction Design for Consumer Products, Morgan Kaufmann, San Francisco, ISBN 1-55860-600-9, http:www.mkp.com, pp. 205-255.

Marcus, Aaron. (2000)."International and Intercultural User Interfaces," in User Interfaces for All, Ed. Dr. C. Stephanidis, Lawrence Erlbaum, Mahwah, New Jersey, USA, pp. 47-63.

Marcus, Aaron (2001). "Cross-Cultural User-Interface Design," in Smith, Michael J., and Salvendy, Gavriel, eds., *Proceedings*, Vol. 2, Human-Computer Interface Internat. (HCII) Conf., 5-10 Aug., 2001, New Orleans, LA, USA, Lawrence Erlbaum Associates, Mahwah, NJ, USA, 502-505.

Marcus, A. (2002). "User-Interface Design, Culture, and the Future," in De Marsico, M., Levialdi, S., Panizzi, E., eds., AVI-02, *Proc.*, Working Conf. on Advanced Visual Interfaces, Trento, Italy, ACM: New York, 22-24 May 2002, pp. 15-27.

Marcus, Aaron (2002). "Information-Visualization for Vehicle User-Interface Design," Information Visualization, 09 Sep 2002, 1:2, 95 - 102 .

Marcus, Aaron (2002). "Mapping User-Interface Design to Culture Dimensions," Workshop *Proceedings*, IWIPS 2002, Fourth International Workshop on Internationalization of Products and Systems, (Austin, TX, USA); Product and Systems Internationalization, Inc.: Austin; 2002; 89-100.

Marcus, Aaron. (2002) "Globalization, Localization, and Cross-Cultural Communication in User-Interface Design," in Jacko, J. and A. Spears, *Handbook of HCI*, Lawrence Erlbaum, Mahway, NJ, pp. 100-150.

Marcus, Aaron (2003). "Mapping User-Interfaces to Culture," Chapter 2, in Aykin, Nuray, ed., *International User-Interface Design*, Lawrence Erlbaum Associates, Mahwah, NJ, USA, in press.

Marcus, Aaron, and Emilie W. Gould (2000). "Crosscurrents: Cultural Dimensions and Global Web User-Interface Design," *Interactions*, ACM Publisher, Vol. 7, No. 4, July/August 2000, pp. 32-46.

Marcus, Aaron, and Chen, Eugene (2002). "Designing the PDA of the Future." *Interactions*, ACM Publisher, www.acm.org, 9:1, January/February 2002, 32-44.

Marcus, Aaron, Eugene Chen, Luke Ball, Junghwa Lee, Karen Brown. (2002). "Report on Future HMI Directions" Report by Aaron Marcus and Associates, Inc., to BMW, Inc., 100pp, confidential.

Nisbett, Richard E., Kaipeng Peng, Incheol Choi, and Ara Norenzayan.(2001). "Culture and Systems of Thought: Holistic vs. Analytic Cognition," *Psychological Review*, 108, pp. 291-310.

User Requirements and Customer Benefit Analysis in the Design of a Novel Driver Support System for Night Vision

Michele Mariani, Serena Palmieri
University of Siena
Via dei Termini 6
Siena
Italy
mariani@unisi.it, palmieri@lettere.media.unisi.it

Luisa Andreone, Fabio Tango
Fiat Research Center
Strada Torino 50
Orbassano (Torino)
Italy
l.andreone@crf.it, f.tango@crf.it

Abstract

EDEL is a three years R&D project funded within the European Fifth Framework IST programme for Intelligent Transport Systems. The project main goal resides in the development of an advanced vision enhancement system for night driving based on: near infrared sensors, a novel illumination system, and an innovative human machine interface. EDEL's expected impact is on safety in road transport, in the prevention of accidents and consequent fatalities and injuries. Infrared imaging may be well-suited to night conditions only if the information is displayed to the driver adequately. An appropriate consideration of user requirements in the design phase will increase EDEL benefits and customer acceptance. Authors of this paper devised a focus group concept that was conceived to identify both customer benefit and initial system performances specification. Groups were defined to represent both those people who are most exposed at incidents when driving at night (young and elderly) and the majority of the driving population (middle aged). This approach allowed to come up with a set of unified requirements for the system to be designed.

1 Introduction

The European dimension of the problem of road accidents is known to be on huge numbers of dying or severely injured people every year in the European Community. 30% of road accidents that happen at night involve 50% of people killed on the roads. Darkness is a major risk factor for driving. While drivers travel just 28% of their miles at night, 55% of all motor vehicle fatalities occur at night. There are 10.4 fatal involvement, 3.5 injury involvement, and 9.1 crash involvement per 100 million miles at night, as opposed to only 2.2 fatal involvement, 1.9 injury involvement, and 5.9 crash involvement during the day. Ninety percent of a driver's reaction depends on vision, and vision is severely limited at night. Depth perception, colour recognition, and peripheral vision are compromised after sundown. 62% of pedestrian/motor vehicle deaths occur at night, and visibility is a major factor. In addition, "Motorists fail to appreciate the limitations of their visual functions at night (Leibowitz and Owens, 1986). Accordingly, motorists may routinely behave in a manner that is not supported by their visual ability at night" (Ward and Wilde, 1995). Thus, reduced visibility crashes may be alleviated by systems that compensate for the drivers' inability to see adequately (Tijerina et al. 1995).

2 The EDEL Project

The three-year EDEL project (www.edel-eu.org) is co-funded by the European Commission INFSO within the initiatives of the 5th Framework programme, and gathers a number of companies[1] to develop a fully integrated driver support system for night vision application based on near infrared sensors and on a novel illumination system. The project has started on March 2002 and is expected to come to an end by February 2005. The foreseen activity tackles the following issues:

- the development of an automotive specified CMOS camera with enhanced features;
- the development of a semiconductor light source base NIR illumination system based on multi-element array laser technology (the illuminator will be integrated in a newly designed vehicle projector headlamps);
- the development of a dedicated Human Machine Interface that should extend driver's vision by highlighting the potential obstacles on the images.

Technological innovation and marketing needs are constantly pushing for the introduction of new devices in the car environment. It is however uncertain whether such technologies will really improve the quality of the driving experience and the safety of our roads. To ensure that these two relevant aspects are taken into account along the design of EDEL, a User Centred Approach (Norman and Draper, 1986) has been adopted as a blueprint for the project. As a consequence, the project started from its very beginning with two parallel activities, namely the technological benchmarking (Knoll et al. 2002) and the customer benefit analysis (Mariani et al. 2002). While the first activity is focused on understanding the pros and cons of the current state of the art technologies, the second one is focused on understanding the context of use, users' needs and expectations. The method and the results coming from this second activity will be the topic of the remaining of this paper.

3 Customer benefit analysis

Reduced visibility crashes may well be alleviated by systems that compensate drivers' inability to see adequately· However, from the user's point of view, several problems can affect the degree to which in-vehicle technologies could achieve their potential to enhance safety and mobility. These problems are associated to equipment design and operation that do not consider human factor requirements and expectations of the driver. Whether or not customers will purchase and use the EDEL system to increase their safety and mobility depends on how well their needs are understood from the beginning and incorporated into the system architecture design. To achieve such an understanding and as part of the Workpackage 2 (System functional specification and architectural design), an analysis of customer benefit and system performances specifications was conducted.

3.1 Method

Among the different measurement techniques available in the qualitative research domain, several choices are offered. Among these, focus groups techniques (Edmunds, 1999) have been endorsed by different researchers (Hackos and Redish, 1998; Kuhn, 2000; Caplan 1990) as the most

[1] The EDEL project partners are: Centro Ricerche Fiat, Jaguar Cars Limited, Robert Bosch GmbH, Universität Karlsruhe (TH), Osram Opto Semiconductors GmbH & Co. OHG, Hella KG Hueck & Co., Università degli Studi di Genova, Università degli Studi di Siena

cost-effective ones to specify customer needs for user centred product: "talking with a relatively small number of people in considerable depth as distinct from counting simplified responses from larger, more representative samples of the general population. Such research is a necessary first step towards understanding, public reactions to new products" (Charles River Associates, 1998). Focus groups are an appropriate approach to (adapted from Kuhn, 2000):

- gather requirements about the new system along with their prioritisation;
- gain information about their functions and context of use;
- support the design process with ideas for development;
- elicit perceived expectations, benefits and willingness to pay for it;
- learn about potential user reactions to product concept prior to several systems being available for field testing;
- obtain insights into consumer perceptions of the relative risks of various types of collisions;
- identify possible demand side barriers to the widespread or rapid adoption in new vehicles;
- identify potential market segments of likely early adopters or other groups that have particular interest in night vision systems;
- identify key features or attributes that are most important in consumers' evaluations of the product.

In conclusion, focus groups, being successfully used both in market research and in user centred design, appear to be the best option to specify both end customer benefits and system performance. By performing focus groups with potential customers it has been possible to identify the attributes of acceptance that are personal to an individual user and would probably determine the ultimate decision to purchase/use the system. Focus group discussions and prospective use of multiple subjective measures provided answers to questions alike:

- Which are the contextual conditions under which EDEL should be most effective?
- Which system format (content; mode of presentation; timing) might mostly effectively influence the Vision Enhancement System (VES) clients purchase decisions?
- Which basic design and implementation guidance emerges from discussions with potential consumers?
- Which pricing policy could be more effective?

3.3 EDEL Focus Group Concept

A focus group concept was devised to identify both customer benefit and initial system performances specification. Such concept was defined in strict co-operation with consortium car makers, to be sure that focus groups would have addressed the appropriate design questions. Each session lasted two to three hours, with a pre-determined structure allowing the use of exercises and visual aids in a focused discussion. In each session, two ad hoc questionnaires were administered: (a) pre focus group and (b) post focus group. Questionnaires were set up to collect standardised answers that could be further quantitatively elaborated. Discussions were conducted according to the following plan:

- short introduction to project's aims:
- self introduction of each participant to the discussion;
- filling in the pre focus group questionnaire; structured discussion;
- filling in the post focus group questionnaire;
- debriefing.

The discussion script was used by the moderator in order to keep the group on track. It is made up of three main sections:

- context of use;
- needs, preferences and envisioning;
- marketing issues.

Five focus groups were performed:

- two sessions with young(<25) drivers;
- two sessions with middle aged (25<x<60) drivers;
- one session with older (>60) drivers.

Groups were defined to represent both those people who are most exposed at incidents when driving at night (young and elderly) and the majority of the driving population (middle aged).

3.4 Results

This section summarises the main findings, offering an overview of the main subjective preferences about the system to be designed.

3.4.1 System major expected benefits

The system should be mostly effective in:

- Highlighting unexpected, sudden events;
- Improving visibility of road signs;
- Assisting driving in unknown roads;
- Timely detecting obstacles.

3.4.2 General system architecture

The following, general system architecture attributes have been highlighted:

- Simplified representation (instead of straightforward reproduction of the external road scene) to allow fast detection of hazards and reduce eyes-off-the-road danger;
- Even when a real scene is displayed, highlights (even coloured) on the possible obstacles/dangers should be provided;
- Additional text or icons could be displayed for other relevant information;
- Stand alone, pop-up display located either laterally (young and elderly) or centrally (middle aged) to the driver's line of sight, over the instrument cluster.

3.4.3 System features

The following features were mentioned:

- Display should be big enough to allow an easy comprehension of the road scene (similar to or even bigger than navigation systems);
- Controls (orientation, brightness, contrast, volume, on-off) should be provided. Such controls should be positioned either laterally to the display or on the steering wheel;
- Information on obstacle tracking (speed, angle, distance) should be provided;
- The display should be adjustable to the driver's or passenger's sight, anti glare.

3.4.4 *System preliminary marketing considerations*

- Overall, the majority of subjects were in favour of purchasing an EDEL-like system;
- 'Safe' was the most quoted, while 'easy' and 'simple' were the second most quoted ones;
- The most important single reason for not purchasing an EDEL-like system lies in its novelty.

References

Caplan S. (1990). Using focus group methodology for ergonomic design. *Ergonomics*, 33(5), 527-5313.

Charles River Associates (1998). Consumer acceptance of automotive crash avoidance devices. Interim Report: Boston, Massachusetts.

Edmunds H. (1999). The Focus Group Research Handbook. NTC Business Books: Chicago.

Hackos J.T. & Redish J.C. (1998). User and task analysis for interface design. Wiley and Sons: New York.

Knoll P., Apel U., Beutnagel U., Fiess R., Grimm D., Andreone, L., Shelton R., Heerlein, J., Abel B., Burg M., Bellotti F., Ferretti E. & De Gloria A. (2002). Technological benchmarking of system components. EDEL technical Deliverable 2.2.

Kuhn, K. (2000). Problems and benefits of requirements gathering with focus groups: A case study. *International Journal of Human-Computer Interaction*, 12 (3&4), 3093) 2 5.

Leibowitz H. W. & Owens D. A. (1986). We drive by night. *Psychology Today*, January, pp. 55-58.

Mariani M., Palmieri S., Foschi P., Pisani T., Shelton R., Andreone L. & Vercellino G. (2002). Analysis of customer benefit and system performances specifications. EDEL technical Deliverable 2.1.

Norman D. A. & Draper, S. W. (1986). User Centered System Design. Erlbaum: Hillsdale, NJ.

Tijerina L., Browning, N., Mangold S.J., Madigan E.F. & Pierowicz J.A. (1995) Analysis of Reduced Visibility Crashes and Potential IVHS Countermeasures (NHTSA Technical Report DOT HS 808 201).

ARK: Augmented Reality Kiosk

Nuno Matos, Pedro Pereira

[1]Computer Graphics Centre
Rua Teixeira Pascoais, 596
4800-073 Guimarães, Portugal
{Nuno.Matos, Pedro.Pereira}@ccg.pt

Adérito Marcos[1,2]

[2]University of Minho
Dep. Information Systems
Campus of Azurém, 4800-058 Guimarães, Portugal
marcos@dsi.uminho.pt

Abstract

This paper aims at presenting a very first prototype of an Augmented Reality (AR) system that as been developed in recent months at our research group. The prototype adopts a kiosk format and allows users to directly interact with an AR environment using a conventional data glove. The most relevant feature of this environment is the use of a common monitor to display AR images, instead of employing specific Head-Mounted Displays. By integrating a half-silvered mirror and a black virtual hand, our solution solves the occlusion problem that normally occurs when a user interacts with a virtual environment displayed by a monitor or other projection system.

1 Introduction

The biggest difference noted between Virtual Reality (VR) and Augmented Reality (AR) is the lack of real objects on side of VR. There are two approaches presented so far the research community to develop AR environments: one applies see-through HMD where computer generated images are superimposed over the image of the real world; the other approach employs a rear projection system while mixing the two worlds (Aliaga 1997).

AR environments imply the combination of real objects together with computer generated ones. When these later are presented using a rear projector or a monitor an undesired effect denominated occlusion may arise making difficult to reach a final harmonised mixing of both worlds. In fact, when interacting with the virtual environment, the user's hand and the other real objects can potentially hide the projected image, i.e., the virtual objects, thus turning unfeasible to have a virtual object closer than the user's hand. To solve this problem, it should be possible to project the *virtual* image between the eyes of the viewer and the real objects (Wloka 1995) (Berger 1997) (Breen et al. 1996) (Balcisoy 2000).

In our prototype, we have applied and extended some technology principles behind the Virtual Showcase solution (see (Bimberet al. 2001) for more detail) in order to integrate direct interaction of real and virtual objects, for instance, by using data gloves while implementing a low-cost hardware configuration. Also in ARK we present a solution for the occlusion problem.

In this paper we start by exposing, in detail, the problem of occlusion in AR, then we present some related work in the area. In the next section we describe our own solution – the ARK prototype, in terms of its setup configuration and the results of some initial validation testes. Finally we draw up some conclusions and future work.

2 Occlusion effect

Occlusion occurs between real and the virtual objects in AR environment, typically when the projection of the later ones occludes the view of the real scene. (Wloka 1995) (Berger 1997).

This phenomenon can be described as follows: when interacting with the augmented environment, the user's hand and other real objects can hide the projected image, thus making it impossible to have a virtual object closer than the hand. One potential solution for the problem is to project the image between the eyes of the viewer and the real objects in order to avoid the occlusion problem. For better understanding this, we can imagine two of the following situations that may occur: in the first one the hand or arm of the user is always on front of virtual object, while in the second one, they are always behind of virtual object (see Figure 1). In both cases the user loses eye contact, thus information, with part of the AR scene (virtual or real).

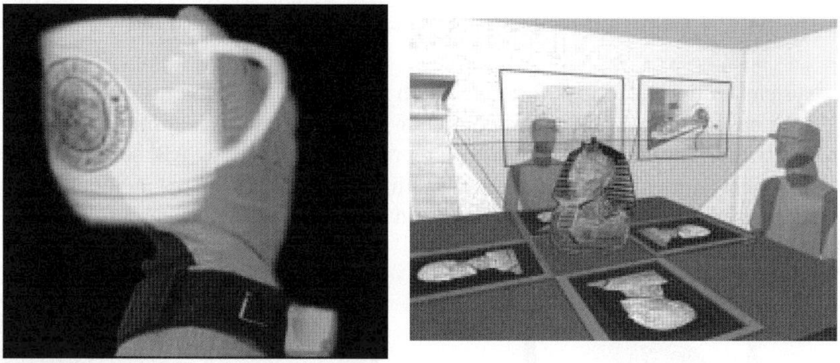

Figure 1: An occlusion example (left); the planar reflection of Virtual Showcases (right) (Courtesy of Virtual Showcase consortium)

By enabling the user to directly interact with the virtual/real content, another problem arises, namely, the computer generated image displayed at the mirror may occlude the hand of the user when the virtual object is behind the hand. These various aspects of the occlusion problem will be discussed later when we present the ARK prototype.

3 Related Work

As mentioned before, the ARK prototype applies and extends some of technology principles behind the Virtual Showcase solution (Bimber et al. 2001) in order to integrate direct interaction of real and virtual objects while implementing a low-cost hardware configuration.

Virtual Showcases are built up from a transparent material (such as glass or Plexiglass) and are laminated with a half-silvered mirror foil. The sides of the Virtual Showcase can simultaneously transmit the image of the real objects as well as reflect displayed computer graphics. Real objects can be superimposed with virtual supplements by displaying stereoscopic 3D, or monoscopic 2D computer graphics that is reflected by the Virtual Showcase (see Figure 1).

Even though VS solution represents an almost perfect solution for combination of real and virtual objects with relative small dimensions, it does not allow interaction with the objects. Users are passive observers of the created scene. The ARK prototype extends this solution by bringing interactivity with the objects recreated, even if only for one user at each time.

(Fuhrmann et al. 1999) have researched occlusion in AR collaboration scenarios, where real objects could be inserted behind the mirror, in order to solve partially the occlusion problem. Also in the project ARCADE (Encarnação et al. 1999), a large stereoscopic back-projection systems such as the Virtual Table, has been used to create VR environments for different application scenarios such as automotive and aircraft design and planning, while the user was able to interact with the objects in immersive mode. The interaction with such a setup is still mostly limited to camera movement and very simple object manipulation. The resulting exclusive use for design review does not exploit the full potential of the Virtual Table. This type of setups are limited to the high-costs involved and are hard to handle due to the dimensions of the devices such as the Virtual Table from ™BARCO. The ARK system allows a simpler, cost-effective solution where the kiosk-format is highly suitable for many type of application scenarios focused on the individual access to AR information that additionally also requires interaction.

4 The ARK system

The ARK prototype applies and extends some of the technology principles and solutions developed in context of the Virtual Showcase. It also adapts reflection and one mirror with interaction. The ARK's projection is obtained by a single monitor instead of by a stereoscopic 3D graphics.

The ARK system adopts a kiosk-format which therefore focuses on the presentation of information both based on real object and computer generated one. Accordingly it does not adopt the see-through HMD based setup but aims at provide a complete AR solution for small to medium sized objects, integrating interaction facilities. The ARK setup includes a monitor as image projection, tracking system, silver glass and shuttle glasses for stereoscopic viewing, as well as data gloves for interaction.

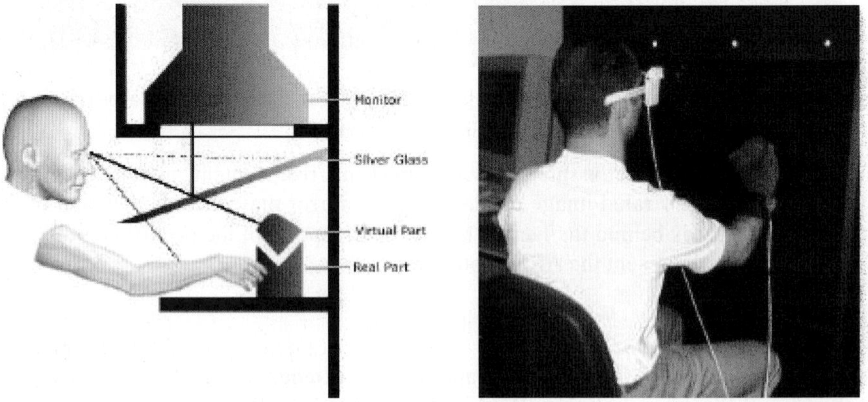

Figure 2: ARK's functional structure (left); aspect of the interaction environment.

The image displayed on the monitor is reflected by the half-silvered mirror, in active stereo, thus creating the illusion that the image space is behind the mirror (see Figure 2).

4.1 Setup of ARK

The current ARK set-up is contained in a wooden structure that supports the monitor face-down and the half-silvered mirror. This wooden structure also reduces the level of outside light that the half-silvered mirror receives, thus allowing a brighter image to be displayed at the mirror. All the ARK structure dimensions are designed to confine a single user alike an information kiosk.

The main structure behind this approach consists of a common 21 inches monitor and a half-silvered mirror. The user interaction occurs directly inside the projection space, by wearing a data glove and by placing the hand beneath the mirror. As mentioned before the occlusion effect arises here when the computer generated image displayed at the mirror occludes the hand of the user when the virtual object is behind the hand. We solve the problem by creating a black virtual hand, which is placed each time at the same spatial position as the hand of the user (see Figure 3).

Because of the real hand is lighter than the surface of the mirror, the user sees actually his hand inside the virtual environment.

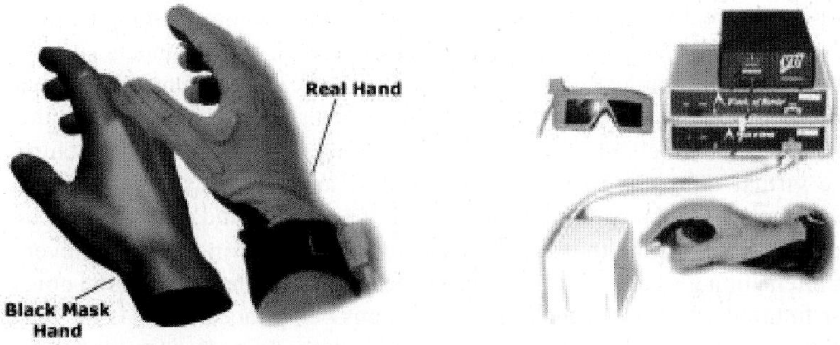

Figure 3: Black virtual and real hand (left); tracking system, shuttle glasses and data glove used in ARK

The input device used in the prototype is a TMCyberGlove with 22 sensors spread by the hand fingers that easily detect any kind of motion in real-time. All finger motions are allowed and, so far, no failures have been detected in order to cause any form of blending. Using this glove, the user can manipulate both, the real and the virtual objects, creating a greater realism effect. The range of the hand is limited (1.00x1.00 meters), however enough to give us the necessary movement freedom inside of the kiosk (Grave et al. 2000). The tracking system in use in the ARK solution is the Tracking Ascension TMFloack of Birds, which integrates two position sensors: one on the glove and the other one on the shutter glasses (see Figure 3).

The position sensors on the shutter glasses and in the glove allows the accurate tracking necessary to project the virtual objects into the augmented scene according to the user's head position, orientation, as well as hand and finger actual position. The computer used is a SGI Octane and the VR modelling software is the TMVirtual Design2 from VRCOM.

The image projection is made through a 21" display (monitor) that lies down over the half-silver mirror at a distance tested and considered sufficient for satisfactory good projection. ARK suits quite well for scenarios of kiosk-based applications as also for small audiences (see Figure 3).

4.2 Initial Validation Tests

ARK initial validations tests, made so far, were of two types: with hybrid objects; and the verification of the occlusion effect. In the first ones we have obtained perfect static hybrid objects, where the blend between the real and virtual worlds was perceptually perfect. However when we tried to test the reaction of hybrid objects with the user head position and orientation, we found some problems: the vertical Y axis presented incorrect values. This occurred because we used only one mirror and the virtual objects were projected in opposite Y axis. Applying a reflection matrix we managed to project the image on the monitor with the wrong axis, but correctly in the silver mirror, thus solving the problem.

Regarding the occlusion tests - even thought the solution was perceptually acceptable for most of scenarios, namely with small and geometrical regular objects such as cans or coups, we have detected some problems arising from the imperfect overlapping effect of the real hand with the virtual black hand. This means the overlapping expected from both objects was not always perfect thus leading to partial occlusion. The solution lies on investing more time on the calibration features which also depend, unfortunately, in many ways on the accuracy of the tracking system adopted (Fuhrmann et al. 2000).

5 Conclusions and Future Work

In this paper, we have presented the ARK prototype - an augmented reality system which allows interaction between real and virtual objects. Many of the technical problems of AR systems are even more challenging than those of self-contained virtual environments. It is necessary to do additional research to obtain realism of virtual images in real-time. On the other hand, we have the occlusion problem. One would expect the user simply employs his hand to grab the object.

If the user puts his hand behind of the projection surface than we have the occlusion problem, but if he uses a virtual interactive glove this problem disappears. Based on this principle we have designed and implemented a solution based on using a virtual hand as a mask for the real hand.

In fact, the user is not aware that his virtual hand is actually in the scene, even knowing he is wearing an interaction glove. From his point of view - his own hand grabs the object.

For the near future work - one of the biggest problems being faced so far is the high complexity of the calibration step. The black virtual hand must be extremely well calibrated with the hand of the user in order to achieve the desired effect.

6 References

(Aliaga 1997) Aliaga, D.G., "Virtual Objects in the Real World", in *Communications of the ACM (CACM)*, Vol. 40. No. 3., March 1997, pp. 49-54.

(Balcisoy et al. 2000(Balcisoy, S., Torre R., Fua, P., Thalmann, D., "Interaction Between Real and Virtual Humans: Playing Checkers", In *Proc. Eurographics Workshop on Virtual Environments,* Amsterdam, Netherlands, 2000.

(Bimber et al. 2001) Bimber, O., Fröhlich, B., Schmalstieg, D., and Encarnação, L.M., "The Virtual Showcase", In *IEEE Computer Graphics & Applications*, vol. 21, no.6, pp. 48-55, 2001. New York: IEEE Computer Society Press ISSN: 0272-1716.

(Berger 1997) M.-O. Berger, "Resolving Occlusion in Augmented Reality: A Contour-Based Approach Without 3D Reconstruction", In *Proceedings of the Conference on Computer Vision and Pattern Recognition, 1997.*

(Breen et al. 1996) Breen D., Whitaker R.T., Rose E., Tuceryan M., "Interactive and Automatic Object Placement for Augmented Reality", in *Computer Graphics Forum (Proceedings of EUROGRAPHICS'96)*, 15(3):C11-C22, 1996.

(Encarnação et al. 1999) L. M. Encarnação, A. Stork, D. Schmalstieg, R. Barton III, "The Virtual Table - A Future CAD Workspace", in *Proceedings of Computer Technology Solutions conference (former Autofact)*, Michigan, Detroit, USA, September 13-19, 1999.

(Fuhrmann et al. 1999) Fuhrmann, A., Hesina, G., Faure, F., Gervantz M.; Occlusion in collaborative augmented environments; Computer & Graphics, Vol. 23, N 6; December 1999 (ISSN 0097-8493).

(Fuhrmann et al. 2000) Fuhrmann, A., Schmalstieg, D., Purgathofer, W., "Pratical Calibration" *in Proc. for Augmented Reality*, Eurographics Virtual Environments 2000; Amsterdam; June 2000.

(Grave et al. 2000) Grave L., Silva A. F., Escaleira C., Marcos A., "Ambientes Virtuais de Treino para Montagem de Cablagens Eléctricas", in *Proc. of 1º Workshop de Sistemas Multimédia Cooperativos e Distribuídos, COOPMEDIA 2000*, vol. 1, pp. 87-96. Portugal, Coimbra, 2000.

(Summers et al. 1999) Summers V., Booth K., Calvert T., Graham E., MacKenzie C.L., ; "Calibration for augmented reality experimental testbeds", in *Proc. of the 1999 Symposium on Interactive 3D Graphics, 1999*, Pages 155 - 162.

(Wloka 1995) Wloka, M. M., Anderson B., „Resolving Occlusion in Augmented Reality", *in Proc. of 1995 Symposium on Interactive 3D Graphics*, Monterey, Association for Computing Machinery: 5-12.

End-User Programming Tools in Ubiquitous Computing Applications

Irene Mavrommati

Research Academic Computer Technology Institute
R. Feraiou 61, Patras, Greece
Irene.Mavrommati@cti.gr

Achilles Kameas

Research Academic Computer Technology Institute
R. Feraiou 61, Patras, Greece
Achilles.Kameas@cti

Abstract

Our future environments will consist of an increasing number of augmented artefacts; not only information appliances, but also ordinary objects enhanced with computing and communication capabilities. People may find themselves thrown into a world consisting of distinct artefacts, yet interconnected via an invisible web of network services. The approach presented in this paper is to enable people make their own applications with 'augmented' artifacts, which are treated as reusable "components", by creating an appropriate infrastructure and tools.

1 Introduction

The paper describes research that has been carried out in "extrovert-Gadgets" (e-Gadgets), a research project funded in the context of EU IST/FET proactive initiative "Disappearing Computer". The Project proposes a generic architectural style, which can be used to describe everyday environments populated with computational artifacts, the Gadgetware Architectural Style (GAS) The overall innovation of the GAS approach lies in viewing the process where people configure and use complex collections of interacting eGadgets (GAS aware artefacts), as having much in common with the process where system builders design software systems out of components. People's environment is therefore seen as being populated with many intercommunicating artifacts, that people can associate in ad-hoc and dynamic ways. This paper presents the approach for the tools that enable people to realize these associations.

2 Basic Concepts

In the GAS approach a vocabulary (Figure 1) acts as a common referent between people, objects and their collections. the eGADGETS capabilities (PLUGS) can be inter-associated (by means of a software/hardware tool, the Editor) with invisible links (SYNAPSES) in many possible ways. A collection of objects functioning together in this way to serve one specific purpose is a GADGETWORLD (GW). The adopted style provides an infrastructure for open applications. The eGadgets are mostly common objects that range from simple (tags, lights, switches, cups) to complex ones (PDAs, stereos) and from small ones (sensors, pens) to large ones (desks, rooms, buildings). Although e-Gadgets have a digital shelf and capabilities, they mostly maintain their original physical shape (no external interfacing/visualization elements (i.e. screens) are added to them). With an external devide, (an Editor), people can supervise e-Gadgets and purposefully create Synapses between them. Anybody can actively shape their environment by associating eGadgets into collaborating collections (the Gadgetworlds). In contrast to the typical system-engineering process, there is no a priori system to be built. Creating a Gadgetworld is done by its "system architecture" by altering associations among participating eGadgets. The Gadgetworlds dynamic nature, needs to be understood by people. Nevertheless people should not need to be engaged in any type of formal "programming" in order to achieve the desired functions.

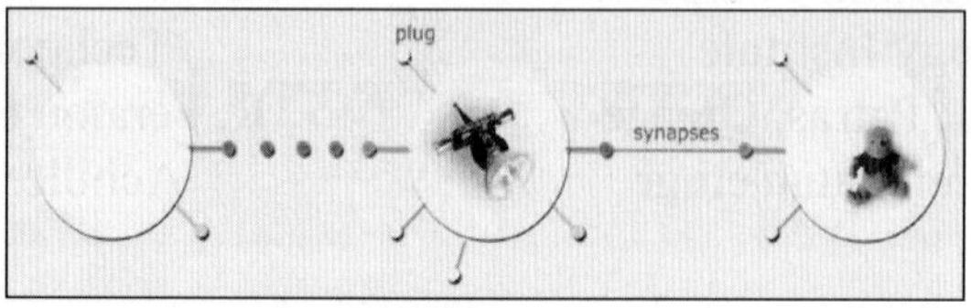

Figure 1: the eGADGETS capabilities (PLUGS) can be inter-associated by means of a software/hardware tool, the Editor.

A desk, a desk and a floor lamp, two books, a clock, a chair, a carpet, an MP3 player, and a Mathmos tumbler light are currently converted into e-Gadgets. This set of 10 sample e-Gadgets has been created (Figure 2), as test implementations of embedding the proposed platform into everyday objects without drastically altering their physical form. Yet, in order to use eGadgets, people have to adapt their task models to include the newly offered possibilities.

GAS-OS, can be considered as a component framework (Schneider & Nierstrasz, 1999). GAS-OS manages resources shared by eGadgets, determines their interfaces and provides the underlying mechanisms that enable communication (interaction) among eGadgets. Thus, it can be considered as a mini-operating system "residing" between the eGadgets intrinsic functions, the hardware and the people's will to create more complex behavior. The proposed concept supports the encapsulation of the internal structure of an eGadget (eGadgets are treated as "black boxes", based on their public interface as manifested by the Plugs), and provides the means for composition of an application, without having to access any code that implements the interface (kameas et al 2002). Thus, this approach provides a clear separation between computational and compositional aspects of an application (Schneider & Nierstrasz, 1999), leaving the second task to end users. The benefit of this approach is that, to a large extent, system design is already done, because the domain and system concepts are specified in the generic architecture (Holmquist et al 2001); all people have to do is realize specific instances of the system. The possible variation is declared in the Plugs, which serve as the primary mechanism for reuse.

The basic assumption is that an eGadget is an autonomous and self-contained artifact, which locally manages its resources (processor, memory, sensors/actuators etc). Some eGadgets will be intelligent, in the sense that they will be able to learn and improve their function by observing the consequences of their actions. On the other hand, Gadgetworld construction by people has some interesting particularities with direct implications on the design of the GAS: In contrast to the typical system engineering process, systems are not built to achieve a pre-specified functionality. It is the intention that people should try out their ideas until a satisfactory Gadgetworld configuration is reached; People should not need to be engaged in any type of formal "programming" in order to achieve the desired functions. eGadgets need to explicitly "advertise" their interconnection capabilities to users in a comprehensible way.

3 Example of Plugs and Synapses:

Lets assume a story involving the use of everyday artefacts: John, 21 is familiar with PC use, (text editing, web-searching). John has recently created his Study Application (a simple ubiquitous computing environment), with a new "extrovert-Gadgets" system he recently bought. He has set up his Study Environment to turn on the light automatically, when he is studying on his desk, since the desk light switch is at the back of the shelves and it is hard to reach it.

The simple story described above can be described as a Gadgetworld consisting of a Desk, a desk-lamp, a book, a chair. The overall Gadgetworld function (Figure 2) can be described as:
When the particular CHAIR is NEAR the DESK AND ANY BOOK is ON the DESK, AND SOMEONE is sitting on the CHAIR AND The BOOK is OPEN THEN ADJUST the LAMP INTENSITY according to the book LUMINOCITY.

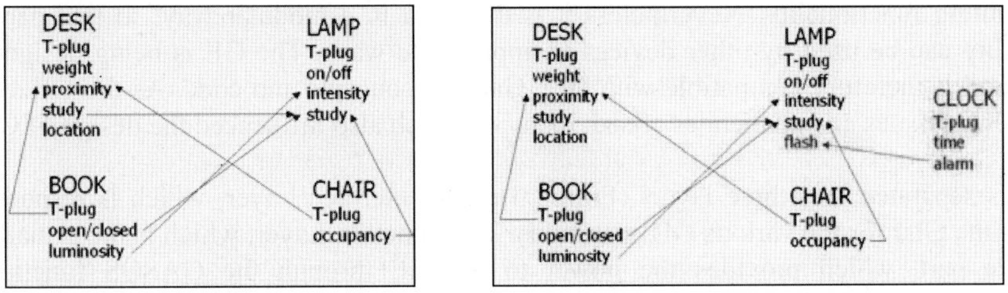

Figure 2: Graphic visualization of two synapse sets (one is a subset of the other). Nevertheless this visualization could be more difficult to grasp as the number of e-Gadgets increases.

4 Editing

A Gadgetworld Editor (GE) is a facilitator device that can be used for Gadgetworld creation and editing. Since a GW can be regarded as a collection of Synapses, the main role of the GE is to allow the user to create and edit Synapses. Since a Synapse is established between two Plugs, all the user has to do is to indicate which two Plugs to associate; the remaining actions can be performed by the GE via the services offered by the GAS-OS.

The purpose of the Editor is threefold: a) to indicate/make visible the available eGadgets and Gadgetworlds b) to form new Gadgetworlds c) to assist with debugging, editing, servicing, etc. (Markopoulos et al, 2003). The editor identifies the eGadgets, as well as the current Gadgetworlds that are available and displays them for supervision. Plugs can also be viewed. Plugs have a direct relation to the sensors / actuators and the functions implemented in the eGadget by its manufacturer. Nevertheless, as Plugs are distinguished to lower or higher level, some of them (especially high-level ones) may not always be obvious to people, apart from via the editor. Synapses, the links between the compatible Plugs of eGadgets, are visualized and manipulated through an editing matrix of the Editor. People can form Synapses between certain Plugs, thus creating a Gadgetworld.

Plug identification and selection is a task that depends heavily on the user expertise. A novice user might not be interested in or understand more than just the description of the Plug and hence base his/her selection on the natural language description provided by the Plug implementer. On the other hand, an experienced user will prefer to have all the technical details to facilitate a proper selection of the Plug. Creation of the required number of Synapses eventually results in the formation of a new Gadgetworld. Once a GW (or a Synapse) is established, the part of GAS-OS that runs on each eGadget will ensure proper working. The GW remains operational as long as required by the user (unless there's technical inability to maintain its functionality) and finally it might be deactivated or broken down by destroying its Synapses. Based on these requirements, the GE should provide the following services to the user:

- Discovery of eGadgets and their Plugs
- Information about the discovered eGadgets, Plugs, their capabilities and offered services.
- Supporting the user in creation, editing and destroying a Synapse
- Creation of Gadgetworlds (as a defined collection of Synapses)
- Gadgetworld operation (activation, deactivation, elimination) and management (i.e. editing)

Many possible devices/interfaces can be considered for the Editor. The Full Portable Gadgetworld Editor (GE) developed, is a software module to run on a PC or Laptop that offers the complete set of GW editing functionality. Nevertheless it is designed in a modular way, so that parts of it's functionality can be used by other devices in appropriate ways. The GE is being designed as an eGadget and is therefore compatible with all e-Gadgets concepts and code. As the GE developed several design issues and challenges arised, many of which also influenced the design of GAS-OS.

The GE is separated into three layers (Figure 3): The Interface Layer, which is responsible for providing the GUI for the various GE operations. The Function Layer, which is the actual GASOS compatible part, which provides the power to the GE through the GASOS functions. The Intermediate Layer, which is a well-defined interface among the Interface and Function layers.
In this design, the GUI is handled at the Interface Layer and all the functions required of the GUI (like discovery of eGadgets, gadgetworld activation, formation etc.) are mapped to the actual GASOS functions in the Function Layer via the Intermediate Layer. Thus, GE implementation on different devices (PC, PDA, Mobiles) requires only the Interface Layer to be changed depending upon the GUI capabilities of the device in question. The rest of the system remains the same.

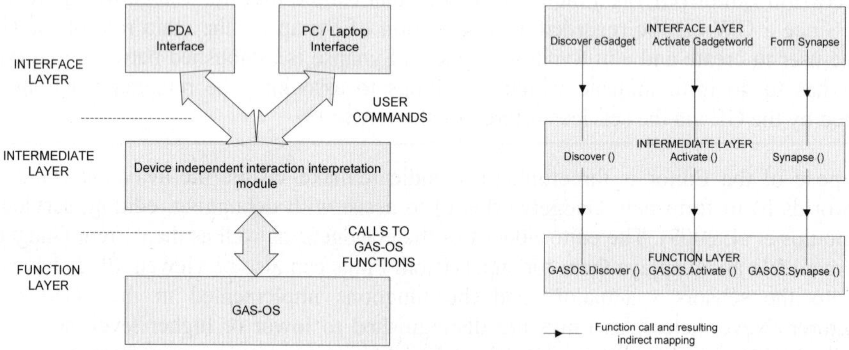

Figure 3: Scematic representation of the GadgetWorld Editor layers.

The GE provides its complex functionality through a simple to use but feature rich graphical user interface (Figure 4). The interface depends on the capabilities of the specific GE device selected. In the first phase, the GE is being developed for the desktop PC environment which allows a large number of GUI features (that allow interaction such as drop down menus, drag & drop, etc). It allows two roundabout working scenarios – the clockwise (from eGadgets to plugs to gadgetworlds) and the counterclockwise (from gadgetworlds to eGadgets to plugs). In the clockwise scenario, the user creates a new gadgetworld from the very basic operation of selecting the eGadgets, plugs and then forming the individual synapses. On the other hand, the counterclockwise scenario is more advance in nature and procedure, which enables description-based gadgetworld formation by searching for already established gadgetworlds or using the description as a recipe to figure out the eGadgets and the synapses required for the gadgetworld.

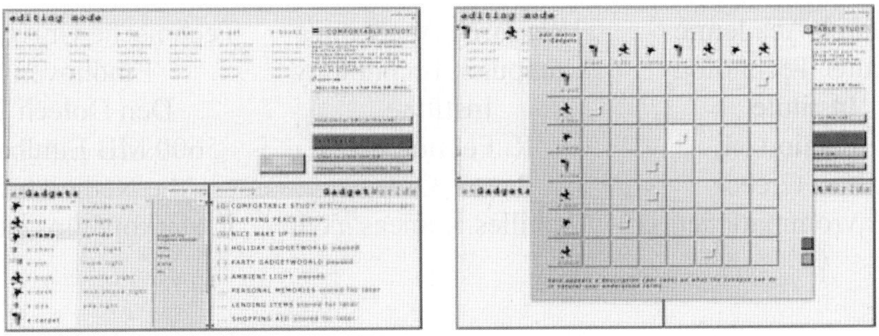

Figure 4: the Graphical User Interface (design draft) of the implemented Gadgetworld Editor.

5 Conclusions

A novelty of this approach is that artifacts are treated as reusable "components". The component architecture is made directly visible and accessible to people, via an Editor, so that it enables them to act as programmers. This end user programming approach may be especially suitable for Ubiquitous computing applications. The possibility to reuse artefacts for several purposes, - not all accounted for-, opens possibilities for emergent uses of ubiquitous artifacts, whereby the emergence occurs from people's use.

6 Acknowledgements

We would like to thank P. Wason, D. Riggas and N. Drossos for creating the first working software of a GadgetWorld Editor, and P. Markopoulos for its evaluation feedback.

7 References

L.E. Holmquist, F. Mattern, B. Schiele, P. Alahuhta, M. Beigl and H.W. Gellersen, "Smart-Its Friends: A Technique for Users to Easily Establish Connections between Smart Artefacts", in Proceedings of UBICOMP 2001, Atlanta, GA, USA, Sept. 2001.

A. Kameas, D. Ringas, I. Mavrommati and P. Wason, "eComP: an Architecture that Supports P2P Networking Among Ubiquitous Computing Devices", in Proceedings of the IEEE P2P 2002 Conference, Linkoping, Sweden, Sept. 2002.

Mavrommati, A. Kameas, P.Markopoulos: Visibility and accessibility of a component based approach for Ubiquitous computing applications: the e-Gadgets case, HCII 2003, Heraklion, Krete

Schneider, J.G. and O. Nierstrasz, Components, scripts and glue, in Software architectures – advances and applications (J. Hall and P. Hall – eds), Springer-Verlag 1999, pp 13-25.

Visibility and Accessibility of a Component-Based Approach for Ubiquitous Computing Applications: the e-Gadgets Case

Irene Mavrommati

Research Academic Computer Technology Institute
R. Feraiou 61,
Patras, Greece
Irene.Mavrommati@cti.gr

Achilles Kameas

Research Academic Computer Technology Institute
R. Feraiou 61,
Patras, Greece
Achilles.Kameas@cti

Panos Markopoulos

Technische Universiteit Eindhoven
Den Dolech 2
5600 MB Eindhoven
the Netherlands
P.Markopoulos@tue.nl

Abstract

The paper firstly presents the concepts and infrastructure developed within the extrovert-Gadgets research project, which enable end-users to realize Ubiquitous Computing applications. Then, it discusses user-acceptance considerations of the proposed concepts based on the outcome of an early evaluation.

1 Introduction

Extrovert-Gadgets (e-Gadgets) is a research project that is part of the EU-funded Disappearing Computer initiative (http://www.disappearing-computer.net). The project extends the notion of component-based software development to the world of tangible objects, thereby transforming objects in peoples' everyday environment into autonomous artefacts (the eGadgets), which can be used as building blocks of larger systems. The ubiquitous computing environments formed by such artefacts are intended to be accessed directly and to be manipulated by untrained end-users.

eGadgets (www.extrovert-gadgets.net) have a tangible self and a software self. They range from simple objects (e.g. lights, switches, cups) to complex ones (e.g., PDAs, stereos) and from small ones (e.g., sensors, pens, books) to large ones (e.g., desks, rooms, buildings). It is intended that a lay person (i.e., who does not have software development skills), can actively shape his/her environment by associating eGadgets into collaborating functional collections (the GadgetWorlds). Such a system can be perceived and analyzed from the standpoints of interaction, which happens at different levels: between eGadgets, between people and eGadgets, or between people and GadgetWorlds. The paper presents the concepts and the infrastructure that has been developed within the e-Gadgets research project, and further it discusses a preliminary assessment of the user acceptance considerations relating to the proposed concepts.

2 Concepts and infrastructure

In the project's approach a vocabulary of basic terms acts as a common referent between people, objects and their collections (Figure 1). These are the following:

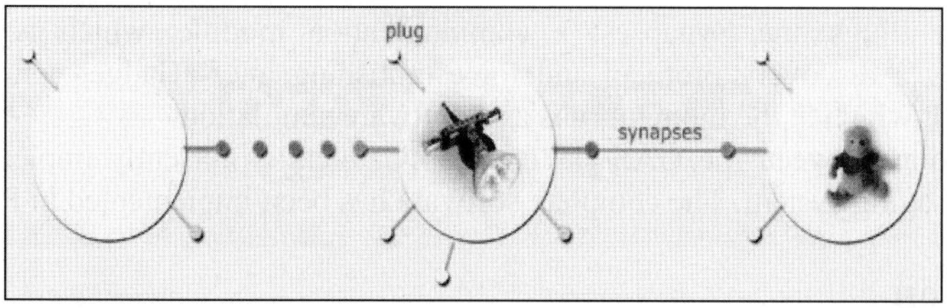

Figure 1: A depiction of the basic vocabulary terms: eGadget, Plug, Synapse, GadgetWorld

- An *eGadget* is defined as an everyday physical object enhanced with sensing, acting, processing and communication abilities. Moreover, processing may lead to "intelligent" behavior, which can be manifested at various levels.
- *Plugs* are software classes that make visible the eGadget's capabilities to people and to other eGadgets.
- *Synapses* are associations between two compatible Plugs. People can utilize Plugs in order to form Synapses (and thus create GadgetWorlds). In this context Synapses can be described as the invisible links between the physical objects.
- A *Gadgetworld* is a functional configuration of associated eGadgets, which collaborate in order to realize a collective function.
- The *Gadgetware Architectural Style* (GAS) constitutes a generic framework, shared by users and designers, for consistently describing, using, reasoning about GadgetWorlds. GAS defines the concepts and mechanisms for people to define GadgetWorlds out of eGadgets.

The design of this novel hardware / software architecture supports computation with limited resources, ad-hoc networking and distributed resource sharing. The e-Gadgets project is developing a set of concepts defined within GAS and GAS-OS, which is the middleware that enables the composition of GadgetWorlds as distributed systems (cf. Kameas et al, 2002). A set of 10 sample eGadgets has been created, as a test implementation of embedding the proposed platform into everyday objects. Further, a software tool, the GadgetWorld Editor, was created to facilitate the composition of GadgetWorlds. People can purposefully associate the Plugs of different eGadgets and synthesize GadgetWorlds as ordered sets of Synapses. As eGadgets can communicate and interact, peoples' environments exhibit a highly dynamic behaviour. Intelligent mechanisms are employed, aiming to learn from people's use of a eGadgetworld and (transparently) optimise it and ease the formation of GadgetWorlds. An eGadget is made of:

- A matrix of sensors and actuators and an FPGA-based board, which implements communication among the sensor board and the processor. It was decided to use the iPAQ as a platform to host the Java Gadget-OS and GAS-OS software because of the small size of these PDA modules. A separate FPGA based PCB was designed to complement the iPAQ and act as a programmable interface between sensors and the iPAQ. This PCB is in compact form 10cm x 10cm and allows up to 200 sensors actuators to be connected to the respective Gadgets. Communication between this board and the computational platform is through RS232. Two computational platforms are used, an IPAQ and a laptop, supporting WinCE. In some case, wireless transmission between the conditioning circuitry and the FPGA board is used in order to maintain user context.
- A processor module (including RAM and wireless module), which is currently served by an iPaq or a Laptop with the necessary wireless cards.
- The GAS-related middleware, which includes the following modules: Gadget-OS, which is specially implemented per every eGadget and used to control its resources; GAS-OS, which

manages the Plugs and Synapses; a communication module, which is responsible for implementing a discovery protocol and for establishing inter-eGadget communication; and eGadget GUI, which at the moment runs as a software simulation

The communication between the GAS-OS of two eGadgets is currently implemented using an XML-based messaging system. The complete software has been implemented in Personal Java.

3 Evaluation

A central research question for the e-Gadgets project is whether the end-user will be inclined and able to use GAS to serve his/her needs. An expert appraisal was carried out for evaluating the proposed interaction concepts and technology, with respect to the end user requirements. At the time of the evaluation, as most technology was still under development, a first version of the Editor was used, together with paper-mock ups of eGadgets. The nature of the evaluation was formative, i.e., it aimed to suggest directions for the next steps of the project, which would ensure that user needs are taken into account. Thus it did not focus on detailed interaction (look and feel) of the proposed GadgetWorld Editors, but on the concepts and on the role that e-Gadgets technology could play in fulfilling user needs. The evaluation evolved around two axes: Comprehensibility of the concepts and interaction and willingness to use such technology.

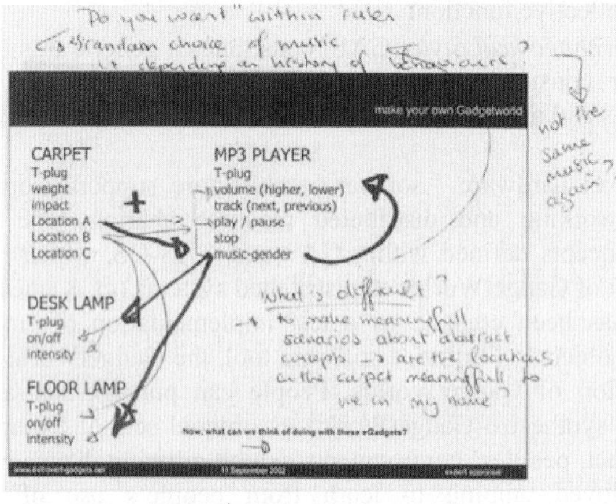

Figure 2. A GadgetWorld configuration made during the problem solving part of the evaluation.

The evaluation was conducted in two phases. First an expert review was conducted in the form of a workshop. Subsequently, the cognitive dimensions framework (Green and Petre, 1996) was applied to assess how well the e-Gadgets concepts support end-users to compose and personalise their own ubiquitous computing environments.

The evaluation workshop involved three experts in user system interaction and two of the authors. The end-user of eGadgets is considered to be a 'technophile' but not a programmer; the three experts matched this profile. The following activities were performed during the workshop.

- One of the authors introduced the basic concepts in a fashion similar to a seminar, in order to familiarize experts.
- A collection of four scenarios was discussed that highlighted different usage/interaction design issues. A small discussion in a focus group format followed each scenario.

- A problem solving exercise was set to gauge the extent to which these experts could build their GadgetWorld and further to reflect on what they consider as problems for the end-user.
- The GadgetWorld Editor interface and some video-prototypes for more advanced interfaces for constructing GadgetWorlds were demonstrated and expert opinions were solicited.
- An open-ended discussion elicited global level feedback for the project.

Some of the most recurring themes during the discussion are noted here:

- Ubiquitous computing technologies embedded in physical objects effectively add hidden behaviour and complexity to them. Problems may arise if this behaviour is not observable, inspectable and predictable for the user.
- Intelligence causes problems of observability and of unpredictability for users. It must be used with caution and this should be reflected in the demonstrations built.
- Constructing and modifying GadgetWorlds is a problem solving activity performed by end-users. As such, it has an algorithmic nature and thus good programming support should be offered[1].

3.1 Cognitive Dimensions

Following the last observation, one of the authors evaluated the GAS using the Cognitive Dimensions framework (Green and Petre, 1996), a broad-brush technique for the evaluation of visual notations or interactive devices. It helps expose trade-offs that are made in the design of such notations with respect to the ability of humans to translate their intentions to sequences of actions (usually implemented as programs) and to manage and comprehend the programs they compose. Broadly, GadgetWorlds can be perceived as applications (that is, complex programs), which are composed in non-textual manner. Thus, this theoretically sound technique can be used to provide insight into selecting between alternative choices with respect to providing tools for GadgetWorld construction. Below, we discuss some of the most interesting points resulting from this evaluation, along the dimensions defined in (Green and Petre 1996):

- Closeness of Mapping. eGadgets implement an architectural abstraction stemming from the nature of software composition, which is also close to the construction of real world objects, such as buildings and machines. However, due to the properties of the digital self of eGadgets, users might conceptualise their tasks in a variety of ways, such as stimulus-desired response, rules, sequences and constraints between entities, etc. In this way there will always be an initial gap between their intentions and the resulting functionality of a GadgetWorld, which people will have to bridge based on the experience they will develop after a trial-and-error process. A GW Editor can shorten this initial gap, by allowing several different ways of expressing the user's goals.
- Diffuseness/Terseness. At this stage e-Gadget appears to be diffuse, i.e., it has few conventions that need to be learnt.
- Hidden Dependencies. They are side effects, an issue well known to programmers: a state change in one component may have non-visible implications on the function of another. In the conceptual diagrams used during the discussion (Figure 2), dependencies were directly visible. However, in the graphical user interface shown in the evaluation, connections and

[1] For example, one of the experts (Figure 2) commented on the gap between concepts that are meaningful for the system (e.g., the definition of locations A,B,C) and how the user identifies concepts meaningful to them (e.g., how does the user understand the concept of a location in the carpet, is it in the centre or periphery, is in left or right). The mapping of concepts from the mental model to the system model is a programming activity (e.g., defining the borders of locations A,B,C) that remains an open issue in terms of end-user programming.

their rules are not shown. Some way of visualising and inspecting such connections needs to be added.

- Role Expressiveness. This dimension relates to the extent to which users can discern the relation between parts of the program and the whole. As an object can belong to several GadgetWorlds, the effect it has on each is not easy to understand from the physical appearance. Future developments of the GadgetWorld Editor will need a way to illustrate to the user how the specification of the parts influences the dynamic behaviour of the GW (similar to debuggers in Object Oriented environments).

3.2 Evaluation results

Currently e-Gadgets seems to be pitching at an abstraction level appropriate for end-users, but much depends on the quality of the tool support foreseen. The feedback to the project, is captured as four tentative design principles:

- The Gadgetworld behavior should not surprise the user, i.e. automation or adaptation actions should be visible and predictable (or at least justifiable).
- Simple tasks should remain simple even in an intelligent Gadgetworld. Intelligence should be applied to simplify complex tasks.
- End-users acting as GW developers should be supported with at least as good tools as programmers have at their disposal, e.g., debuggers, object browsers, help, etc.
- Multiple means to define user intentions should be supported by the graphical editor, as the users tasks tend to be comprehended and expressed in a variety of ways.

Future sessions using working prototypes are planned to evaluate the ease of GadgetWorld composition and debugging by end-users.

4 Concluding remarks

eGadgets is a bold attempt to provide end-users with the ability to construct or modify ubiquitous computing environments. It takes concepts of component based software development and applies them to treat physical objects as components of Ubiquitous Computing environments. This paper has summarized the concepts of the eGadgets project, the current first experimental demonstrator and our attempts to include considerations of end-user acceptance and usability to the formation of the concepts. Currently, a richer and more robust concept demonstrator is being implemented that will support a test with end-users.

5 Acknowledgements

We would like to thank M.M.Bekker, A.Gritsenko and C.Huijnen for participating in the review.

6 References

Green,T.R.G., Petre, M. (1996). Usability analysis of visual programming environments: a cognitive dimensions framework. Journal of Visual Languages and Computing. J. Visual Languages and Computing, 7, 131-174.

A. Kameas, D. Ringas, I. Mavrommati and P. Wason, “eComP: an Architecture that Supports P2P Networking Among Ubiquitous Computing Devices”, in Proceedings of the IEEE P2P 2002 Conference, Linkoping, Sweden, Sept. 2002.

Vehicle Navigation Systems: Case Studies from VDO Dayton

Irene Mavrommati

Aegean University /
Computer Technology Institute
73 Nafsikas street,
Patras, Greece
Irene.Mavrommati@cti.gr

Abstract

This paper draws from the experiences of work for VDO Dayton MS5000 car navigation system, and describes from a design perspective some generic issues applicable in detailed interface design of car navigation systems.

1 Introduction

The paper is a presentation of work done for vehicle navigation systems, and conclusions drawn from it from a design perspective. It draws from the case study of the User Interface Design phase of the VDO Dayton MS5000 car navigation system. Design issues that apply to the design of all vehicle navigation systems are highlighted.

2 UI design and car navigation systems - VDO Dayton MS5000

The initial User Interface design phase of the VDO Dayton MS5000 car navigation system (image 1) took place in 1998. It can be considered one of the early examples of car navigation systems. The first generation of car navigation systems had constraints posed from the technical requirements (for example low screen resolution, memory, use of non proportional fonts etc), as is normal for a first generation of system platforms. Nevertheless some user interface issues that the design team has faced during the design of the VDO Dayton MS5000 apply more generally to the design of vehicle navigation systems.

The main task of a car navigation system is to help drivers getting from one point to another, which they have to define. Route instructions and orientation are given on a map as well as with generated speech. (In the map different scales of abstraction are possible). Other related content is locating petrol shops, restaurants, traffic-jam avoidance, find alternative routes, e.t.c. Two core prime actions in a car navigation system are

a) stating an address in the system, via a keyboard entry and

b) then, upon starting the journey the system directs the driver to the specified address.

In the User Interface Design of a car navigation system, a balance should be made between the interface and:

- the user
- the content
- the control device (the same interaction and interface may have to be designed open for very different input devices, according to the car-manufacturer it is addressed to.
- the environment (lighting conditions, sound level, e.t.c.)
- the technical possibilities and constraints of the platform
- other issues related to upgrades, (expandability to include more functions, different input devices e.t.c)
- branding image to be promoted.

The on-screen design for a car navigation system involves several design challenges (Mavrommati, 2001). The immediacy of the information presented and the time taken to understand it is very important as the users attention is shared between it and the driving task. The positioning of the display is a key factor here, as it is often in the driver's peripheral vision; the driver may need to move his eyes or even his head to read it properly. Another factor is the amount of attention the user can give to the display (and the system) at any one time. A car display should distract the users attention away from the road for an absolute maximum of a couple of seconds. These factors have consequences for the readability, colour, luminance, and contrast of the screen and information & graphics shown on it.

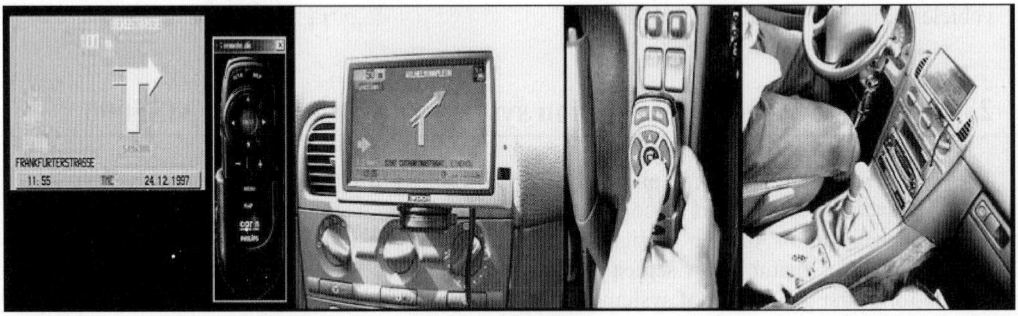

Image1: a sample screen of an interactive prototype based on the PC, and the VDO Dayton MS5000 system, as it was finally produced.

3 Design fine-tuning of the on-screen-user interface

Good user interface and interaction are essential since life and death circumstances are at stake. The driver's attention should remain at operating the vehicle, and not the car navigation system. Therefore immediacy and readability of the information are paramount in the design of the on-screen-information.

There are pragmatic considerations relating to the interface design of a car navigation system that influence the design, such as for example the lighting conditions. While driving, the environmental lighting conditions may change dramatically, ranging from very dark to very bright; this has an influence on reflections on the systems screen and makes it harder to read at times. To compensate for clarity and readability, the screen-brightness has to adapt (image 2), while on-screen graphics should be prepared to maintain their quality during these adaptations. Reflections off the screen can reduce the contrast of text and graphics making it more difficult to read and this can also happen if the graphics are 'too dark' for the environment. If however the screen and graphics are too bright for the environment then this 'glare' can make the screen uncomfortable to look at and will affect the ability of the viewer's eyes to be adjusted for the lower ambient light level. This may be a problem in some situations, for example when using a car navigation system at night where the drivers ability to see outside clearly is vital.

To plan and conduct regular screen tests is crucial in the design process, in order to ensure the readability of the screen designs, under the various possible environmental conditions. Having frequent screen tests helps designers to build up an understanding about which fonts colours and graphics work best in a range of situations. Design awareness about the capabilities of the target screen and its viewing conditions is raised; this is important as the designed is mostly done on a computer screen (different viewing distance, resolution, colour quality than the target screen). A final decision on the screen design should not be reached until the interface is reviewed on the actual screen, as the design there may look different.

4 Visibility for all

In car navigation systems are designed for everybody including the mature and the colour-blind. Mature people need increased brightness and contrast within the screen in order to compensate for age related eyesight problems. Colour-blind people tend to describe red, dark yellow or green as brown and therefore highlighting of important screen elements in these colours or tints may go unnoticed. Adequate luminance contrast between screens, to ensure legibility by all should be catered by the design. A common trick is to test and fine-tune the design contrast by converting the design into greyscale. Yet, nothing compensates having a screen review with some people representing from these target groups, who can verify the overall readability.

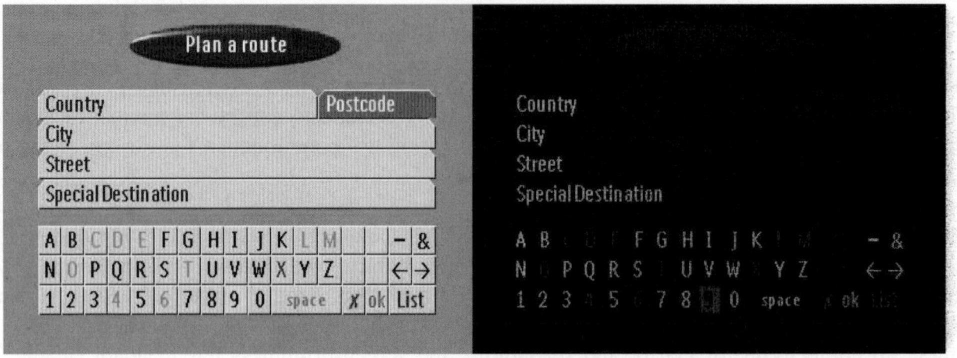

Image2: Screen designs for the keyboard entry.
The same screen is shown for the day and for the night situation. What seems to be completely dark, gives in fact exactly the same result regarding readability in a dark environment.

4.1.1 Detailing and explaining interaction

While flowcharts or other diagrams are invaluable for making a map of navigation, they cannot express the dynamic aspects of how the user experience may be implemented (i.e. animation, sound, e.t.c.). Progressing from concept storyboards to interactive demos to perhaps a real system at the end are natural steps to take. Each time the fidelity is increasing, but the flexibility of it to be changed is decreasing. An interactive prototype is a tool for further development of the application or product. It is also often used as a communication tool for showing developers how certain aspects would work. Finally it can be used as an evaluation tool.

It's important to represent the whole interaction dynamically to give insight on aspects such as:

- The requirements on the user in terms of mental workload and attention.
- The system's readability in relation to interaction (i.e. clicking / touching / distance).
- The effect of sounds, animations and colours.
- The logical (or not) sequence of the screens;
- Other sensorial stimuli.

The experience as a whole is more than the sum of these separate parts, and in bringing the integrated experience to life the demo performs a complementary role to the flowcharts in the design process.

5 Conclusions

In the course of the User Interface Design work done for VDO Dayton, several variations of designs were created; mock-ups of the designs were taken further into creative interactive prototypes (these were done in Macromedia Director). Demo prototypes have not been evaluated with users in working prototypes of the system, (during car-testing sessions for example), during that project; nevertheless certain interaction elements of importance / frequent use, (such as alternative keyboard entry methods and layouts), were demonstrated as interactive prototypes giving a better feeling of the interaction involved to the project team. This process, together with benchmarking against the technological constraints of the system, had lead the selection of particular interaction design styles to be taken further. Screen designs have been reviewed in consultation with target users (color-blind), for ensuring appropriate legibility and clarity. The selected Screen-designs of the User Interface proposals (not in an interactive version) were tested in the final system and improvement actions were taken as a result.

6 References

Mavrommati. Chapter 6, Design of on-screen user interfaces. User interface design for electronic appliances, edited by K. Baumann and B. Thomas, Taylor and Francis, 2001

Off Board Networking of Car Navigation Systems & Services

Frazer McKimm

MdesRCA
CEO DHS Italy Viale.F. Crispi 9. Milan 20121 ITALY
mckimm@dhs-ltd.com

Abstract

Recent advances in technology both outside and inside the vehicle will allow for the design of highly personalized telematic products. New car navigation systems (off and on board) will enable users to customize their user-interface (UI) and access a wide range of new services. DHS has focused on developing a small screen UI template that allows for users to interconnect their mobile phone, residential gateway units and Web-enabled vehicles via a range of standardized voice- and screen-based commands (Fig.1). Hardware suppliers, telecommunication, and content providers will allow users to customize and transplant their UI across devices. Recent work for Fiat Auto has confirmed the prime importance and potential of the automobile in this unbroken communication chain.

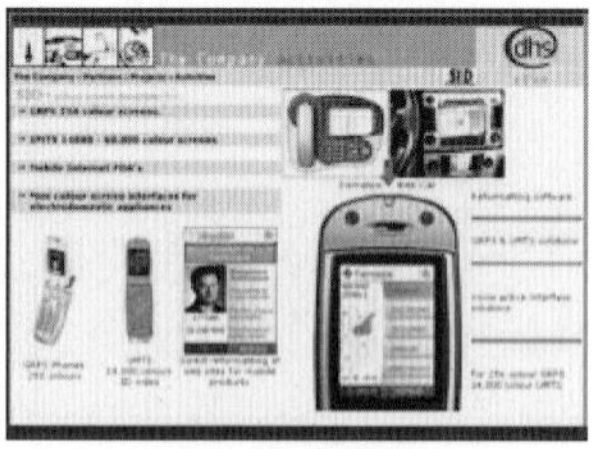

Figure 1

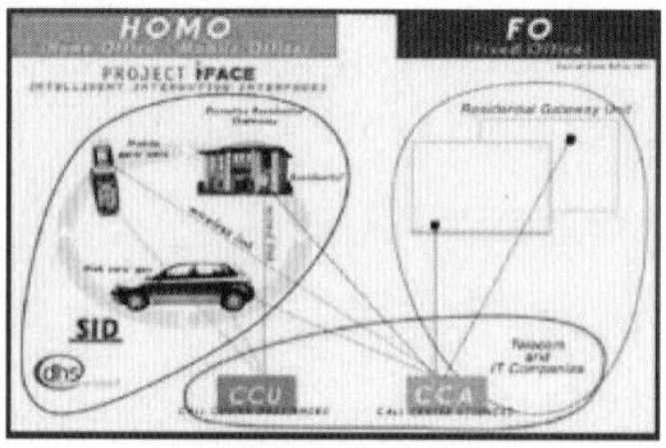

Figure 2

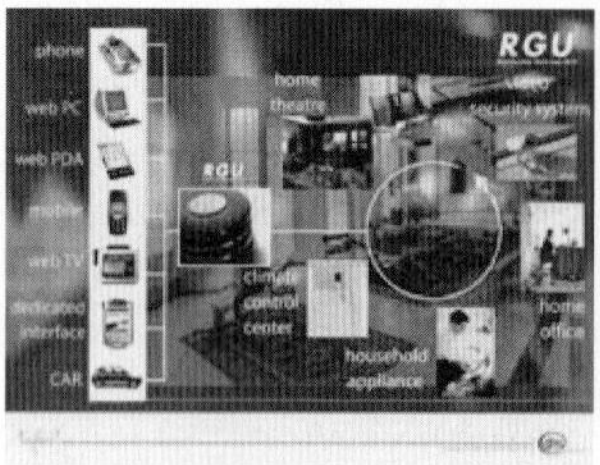

Figure 3

1 Project iFACE TM (Intelligent Interactive User-Interface)

iFACE was established in 2000 as a continuous research project aimed at developing a small screen UI template system, applicable across a broad range of mobile devices. Prime aims are providing a UI road map that progresses from dot matrix, bi-chromatic pictogram-based displays to full color (256+) multimedia displays. The template system will be downloaded onto a user's vehicle, mobile phone, or PDA. Via this UI (voice, text, and icons) users will communicate with a remote customer information/data center that provides subscriber services (Fig. 2).

The next generation of data-center-based information will allow for the access of large quantities of remote devices via broadband ADSL and mobile links. DHS and SPIN (an IBM Global Services partner company) are currently defining the UI service access icons. These will then be embedded in the UI device becoming active once the service options are downloaded onto Web-enabled cars, multimedia mobile phones and fixed-line residential gateway units (RGU) (Fig. 3). A RGU acts as a home server unit, allowing for interconnection of home products via a local area network (LAN) and external wireless devices. Current technology using WAP solutions over GMS/ GPRS data networks enables the sending of SMS commands for the access to other devices.

2 Web Car

The future of UI design for car navigation in the European market includes both on and off board solutions, by which route maps are downloaded onto the car navigation unit via a GSM/GPRS wireless connection or directly via CD ROM. DHS has been studying the potential for off board systems (Fig. 4), in which the interconnectivity lies outside the car. Manufacturers are focusing on simple, easy-to-use features like SMS and downloading MP3 sound files from the Internet. The progress from present off board car systems to full multimedia ones can be defined in four steps.

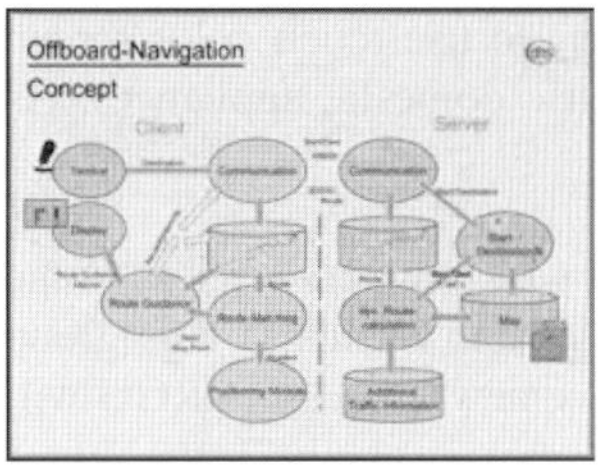

Figure 4

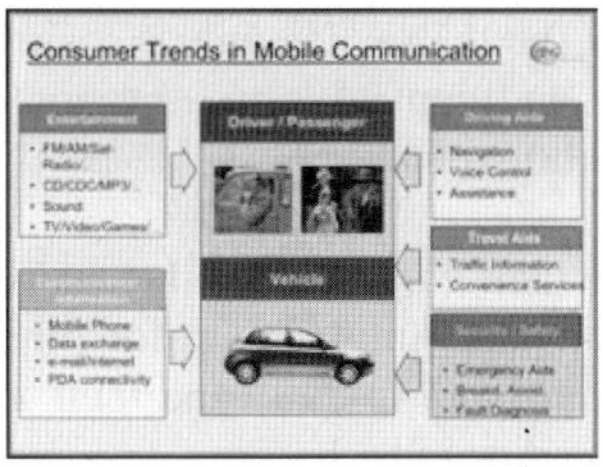

Figure 5a

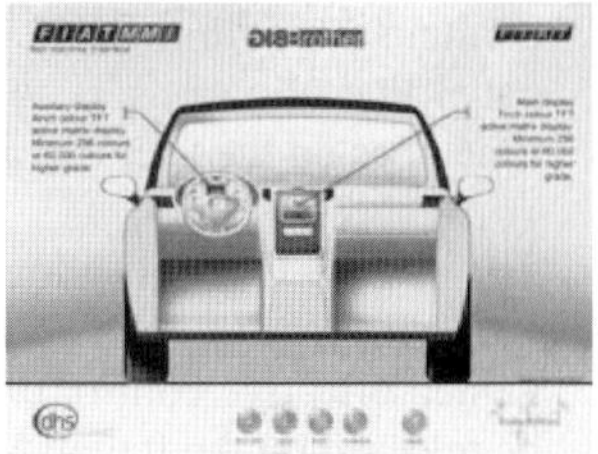

Figure 5b

Step 1: These are currently 1DIN based systems (the German DIN standard specification for size of the car radio box incorporated into the instrumentation panel) that have bi-chromatic dot matrix pictogram displays. Fiat Auto will introduce its new Punto "internet radio" product in May 2003, which will allow for downloading of e-mail, SMS, MP3 files using WAP (Wireless Application Protocol) and access to the voice-based information support of the Fiat Targa service center.

Step 2: Early 2005 will see the introduction of basic diagrammatic maps. 1DIN based systems, with an integrated architecture, i.e. screen and mother unit in one box.

Step 3: Mid 2006 will allow for higher quality diagrammatic maps with 256 colors, the screen will assume a full 1 DIN format.

Step 4: Late 2007 will see the introduction of 2 DIN full multimedia off board systems (Fig.5a and 5b). This next generation of advanced Telematics, processing units will enable downloading and local storage of high-quality interactive bitmap images. Screens will be superimposed on the car template UI system, which will have an optional integrated or distributed screen architecture.

In the context of the car GPS systems, present WAP technology offers many location based service possibilities. The Fiat MMI project discussed later, looks into developing new control concepts and dedicated UI corresponding to "Step 4" advanced outboard car systems.

Current DHS work focuses on system analysis and validation of the present generation Blaupunkt "Connect Nav" systems. The second phase of this work involves working with AM+A in the US to identify the parameters by which the car system can be localized. Current research into these localization parameters will help form a basis for future systems.

Residential Gateway: The focus here is on "domotics" (home automation, Fig. 6) and enabling remote access from a user's car or mobile phone to domestic appliances and home-based Web enabled products (alarm system, heating, Web cams, *etc.*).

Figure 6

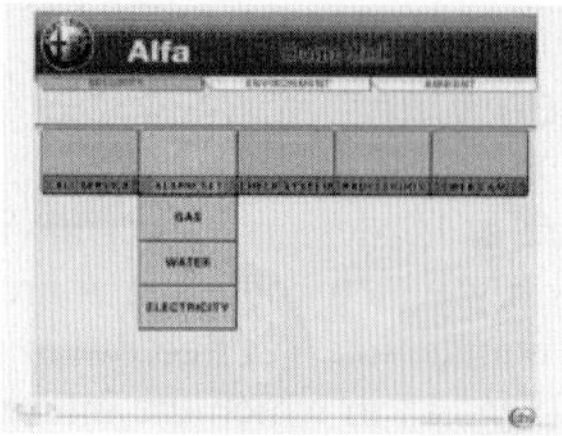

Figure 7

The "DOMOLINK"™ product designed by Studio McKimm for Boffi Kitchens will work in conjunction with a RGU. The DOMOLINK UI unit is a simple, easy to use touch color screen panel that allows users to access and control a broad range of electric domestic appliances. As time goes on the user can add additional features.

The remote UIs will allow access and control of connected devices with a wired/wireless home LAN. Project work in this area has been undertaken for Pirelli Cables and Systems. This group gained effective control of Telecom Italia in 2001 and their innovative new broadband RGU will be released on the Italian Market in April 2003.

3 MMI Fiat Motors: Next Generation

The project ran from September 2001 to January 2002. The Fiat UI simplifies and standardizes the way information is conveyed via careful hardware and screen design and functions as a manufacturer-independent UI system that can be submitted to a supplier as a reference guide during the bidding phase, taking into consideration the varied user groups that Fiat counts as its customers.

The information must be both accessible and adaptable to individual needs as well as be adaptable later to other brands in the Fiat Auto group (Alfa Romeo or Lancia) (Fig. 7). The UI is designed to interact with the FIAT Targa Information services, in which drivers/passengers sends SMS messages to a call center that calls back to the driver to guide him/her.

Phase 1. The FIAT brief was to develop a 2 DIN distributed architecture display and UI system with greater use of multimedia customization and dedicated functions.

Phase 2. DHS was to define a road map for strategic alliances with service, content and other related hard and software suppliers and manufacturers. It was also proposed to design a new dashboard/instrumentation panel to allow for extra functionality.

Phase 3. Study greater interconnectivity with wired or wireless products, especially in relation to residential gateway products for SoHo (Small Office Home Office) and/or the domestic environment.

4 Control Interface Analysis

This defines what areas of internal space on future systems and what kinds of standard controls.

Pen pad: This concept allows for the passengers (and driver when the car is stationary) to write messages, do drawings, select icon or GIF images from a library, digit and type messages.

5 Logic structure

The UI breaks the information into cluster groups that include entertainment, communication and information, driving aids, travel aids, security, and safety aids.

There are two proposed variations to the presentation of services. One is the standard screen-based series: SOS, Car System, and Ambient. The other is the complete user cluster-group of services, which can be accessed by the insertion of a pin-code combined with insertion (or proximity) of a smart-card user. This design allows the system to be profiled and gradually customized to different user requirements, a core concept for the UI, because it allows for molding the vehicle UI skeleton.

6 Screen Structure

Each of the three UI proposals follows a different design concept.

Concept 1: Parallel Path Progression Based on the idea of using a full auxiliary mini screen color display with a high pixel and color definitions (in the absence of which a dot-matrix display can be added). The concept revolves around a parallel path progression through functional layers, so that the need for the driver to look away from the display is reduced to a minimum.

Concept 2: Split Screen Data View This concept allows for the showing of two kinds of information on the screen. At any given time the screen data can be moved by means of a physical sliding button. The information is layered with the previous level shown in a faded tone.

Concept 3: 3D virtual Space Here, the screen information is represented as a road with functions displayed as signposts, the screen UI has changed from layers and levels to a spatial system.

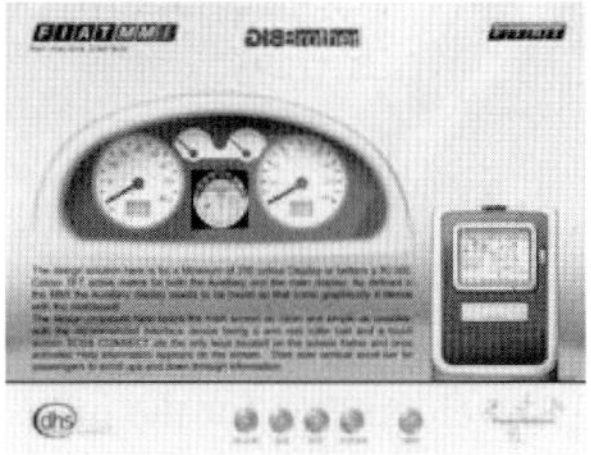

Figure 8

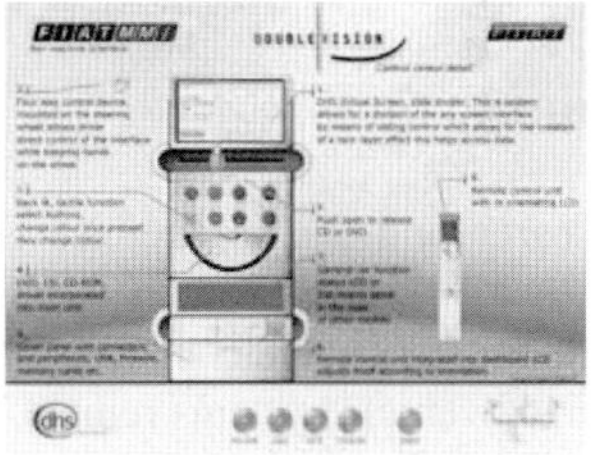

Figure 9

7 Products

The following proposals show how the three UI concepts can be developed as design solutions.

Concept 1: "Big Brother" The design solution here is for a display of minimum 256 colors (4000 colors proposed). TFT active matrix for both the auxiliary and the main display. As defined in the MMI, the auxiliary display needs to be round so those icons graphically blend with the dashboard. The design proposals here keep the main screen as clean and simple as possible, with the recommended UI device being an armrest roller ball and a touch screen. SOS and Tagra CONNECT are the only keys located on the screen frame. Once activated, help information appears on the screen. There are side mounted vertical scroll bars for passengers to scroll up and down through information (Fig. 8).

Concept 2: "Double Vision" In this design solution, the screen is not touch-based and there is a steering-wheel-activated command integrator also with a removable remote control for the passengers. A double screen data display system allows for rapid path flow and combines physical screen UI with a virtual one(Fig. 9). As you slide the control button, a virtual bar moves in parallel. The key design features of this product are:

- DHS Virtual Screen, slide divider that allows for division of screen UI by means of sliding control that allows for the creation of a twin layer effect. This approach helps access data.
- Four-way control device mounted on the steering wheel allows driver direct control of the UI while keeping hands on the wheel.
- Backlit, tactile function select buttons, change color once pressed.
- DVD, CD, CD-ROM driver incorporated into main unit.
- Cover panel with connectors and peripherals, USB, Firewire, memory cards etc.
- Remote control unit integrated into dashboard. LCD adjusts itself according to orientation.
- General car function status LCD or Dot matrix panel in the case of other models.
- Remote control unit with re-orientating LCD.
- For passengers there is a remote control unit integrated into the main unit. The screen is designed to be at the same eye level as the instrument panel with an easy to use UI.
- Push open to release CD or DVD.

Concept 3: "Leonardo" In this design solution the information is displayed in a 3D context and the possible UI access is through a rotational dial which has the main function groupings marked on the outside surface, like a chronograph watch. As you turn the dial you select the main function. The letters then change color. There is an integrated entertainment solution that allows for DVD viewing form a back seat mounted TFT display.

8 Conclusion

As a research and design exercise the Fiat Auto UI gave a very clear indication of the kind of features that would make their system UI stimulating and useful for their customers. Despite the recent gloom in the telecom market, the emerging opportunities for interconnected and highly personalized vehicle systems remain one of the most promising developments for the industry.

9 Explanation of Terms

RGU: Residential Gateway Unit. This device acts as a home server unit, allowing for the interconnection of home products via a local area network (LAN) and external wireless devices
Off Board Navigation: system by which route maps are downloaded onto the car navigation unit via a GSM/GPRS wireless connection
On Board Navigation: system by which route maps are downloaded directly onto the car navigation unit via CD ROM
DIN: German standard specification for the size of the car radio box incorporated into the instrumentation panel

GUI for Graphical Data Retrieval by Means of Semantic Filtering

Zdenek Mikovec

Czech Technical University in Prague
Karlovo n. 13, Praha 2
Czech Republic
xmikovec@fel.cvut.cz

Martin Klima

Czech Technical University in Prague
Karlovo n. 13, Praha 2
Czech Republic
xklima@fel.cvut.cz

Radim Foldyna

Czech Technical University in Prague
Karlovo n. 13, Praha 2
Czech Republic
r.foldyna@sh.cvut.cz

Abstract

The new approach of manipulation with graphical information in mobile environment is discussed. This approach is based on usage of special semantic filters that use the semantic description of the graphical information (e.g. ownership of the graphical object, connection between rooms) to allow faster and more precise specification of user's region of interest. With these filters the user can specify more complex queries that should give more satisfactory results. The retrieval process consists of two steps. First, the user obtains information about the semantic structure of the graphical information and defines the semantic filter. Secondly the interesting graphical data are retrieved only. Software technology used is based on the use of XML/XSLT (W3C, XML) and the semantic description is done in MPEG-7 standard. The first prototype tests are positive.

1 Introduction

Fast progress in the mobile technology has enabled us to run a lot of applications that we can find on the desktop computers and on the mobile (PDA) devices as well. The mobile environment however differs in specific aspects that make it difficult to perform the same task as on the desktop computers. The user interface in mobile environment requires new approaches. One of the most important ones is communication with graphical information by means of a small screen. This situation requires specific methods for handling graphical information. The ability of the user to browse efficiently huge amount of very complex graphical information is limited not only due to the limitations of the mobile device and wireless network but also due to the user's existing situation (e.g. limited time to decision making). Our solution tries to increase the user's efficiency using semantic filters that help the user to find required information by filtering out information that is not necessary in the existing situation. These filters are based on semantic description of the graphical information, which describes relations between objects that are of non-structural nature (e.g. ownership, functional relations, membership in a group).

The goal of this research is to develop a tool that would enable more efficient retrieval of 2D/3D graphics in mobile environment. The tool should help the user to find the relevant information he wants to retrieve by defining filters based on semantic description of the retrieved information.

2 Problem statement

Users working in mobile environment are facing several specific problems. Apart from the limitations of the mobile device (e.g. small screen, low processing power, low memory, slow and unreliable network connection) the most important problem is dynamically changing situation in which the user appears. The user is usually forced to solve tasks immediately and in very short time. This totally different environment (in comparison with an office desktop working place) needs new approaches to user interaction.

We suggest that the solution to this problem could be found in more intensive usage of semantic description of the graphical information in the process of searching important information for the user. This semantic description is analyzed and used to eliminate unnecessary information. Such filters are called semantic filters.

In this solution three main problems arise. The first problem is to define data model for semantic description of graphical information. The second problem is the design of suitable user interface of filter definition dialogue in mobile environment. There are certain problems such as limited data input ability, small display, low memory and processor power. The third problem is generation of the semantic description. The most important semantic information could not be automatically retrieved from the graphical information (the human assistance is needed) and current authoring tools support the semantic description creation insufficiently (Balfanz, 2002).

3 Semantic filters

The efficiency of semantic filters strongly depends on the quality of semantic description provided along with the graphical information. Rich and well-structured description allows the user to specify very efficient semantic filters. The weakness of semantic filters is the current lack of semantic description and insufficient support by authoring tools.

3.1 Data model for semantic description

To define an appropriate model for semantic description, we first have to analyze the graphical information we want to deal with. In our case we are focused on relatively complex graphical information like 2D vector based pictures and 3D scenes. This graphical information is represented by graphical objects (e.g. circle, line, cube) with their description (e.g. colour, material) and relations among them (e.g. hierarchical order).

The same approach can be used to describe the semantic description of the graphical information. We will understand the semantic description in the form of a graph where nodes represent the semantic entities and edges represent relations among these entities.

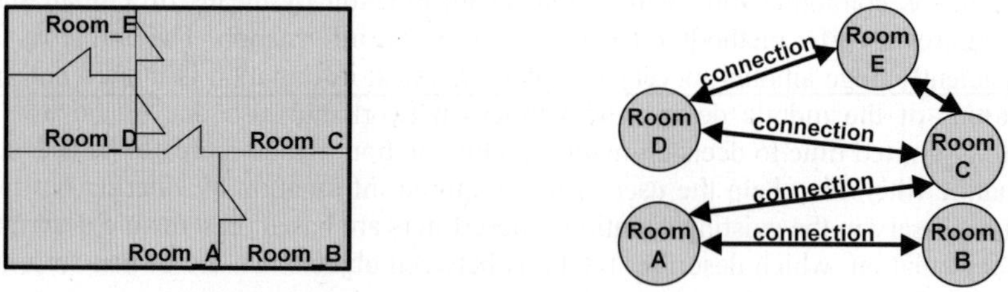

Figure 1: Semantic description of an apartment

For expressing such a model we can utilize the MPEG-7 standard, the general model for description of multimedia information. This rather complex description format consists of following sets (source: MPEG-7 Multimedia description schemes document):

- media (e.g. storage format, encoding)
- creation & production (e.g. title, creator, classification)
- usage (e.g. rights holders, access right)
- structural aspects (e.g. spatial, temporal, spatio-temporal components, colour, texture)
- semantic aspects (e.g. functionality, membership in a group)

For our purposes the last set (semantic aspects) will be used. It allows us to describe the semantics by means of a graph, so it fits our needs.
The semantic aspects could be divided into three categories (following MPEG-7 definition):

- semantic entities (e.g. objects, events, concepts, states, places)
- semantic attributes (i.e. attributes of semantic entities)
- semantic relations (e.g. entity-entity relation, entity-relation relation)

As an example of a semantic description we can show a description of an apartment (see Figure 1). On the left hand side we can see an apartment that consists of five rooms connected together with doors. A very simple semantic description graph can describe these connections between rooms (right hand side of the figure), where nodes represent rooms and edges represent connections between rooms.

3.2 Usage of semantic description in filter definition

The question is how such a semantic description (described in the previous chapter) can be used for information filtering and what are the advantages.
The semantic description provides more precise object classification that can be used for finding objects (e.g. Are there any objects in the apartment owned by Mr. Smith?). This classification can be used to refine the calculation of the level of detail (LOD) for each object.
Next the semantic relations among objects can be analyzed in order to define more precise regions of interest (ROI). This brings totally new definition of ROI that could be described as follows:

ROI definition: (object_A, semantic relation, object_B)

For instance the definition of ROI for rooms that are connected with Room_A looks as follows:

ROI definition: (Room_A, connection , *)

The semantic filter will then select only objects that are in this ROI. The selection is room Room_A, Room_B and Room_C (see Figure 1). Another real example of semantic filtering of 3D scene can be seen on Figure 2.

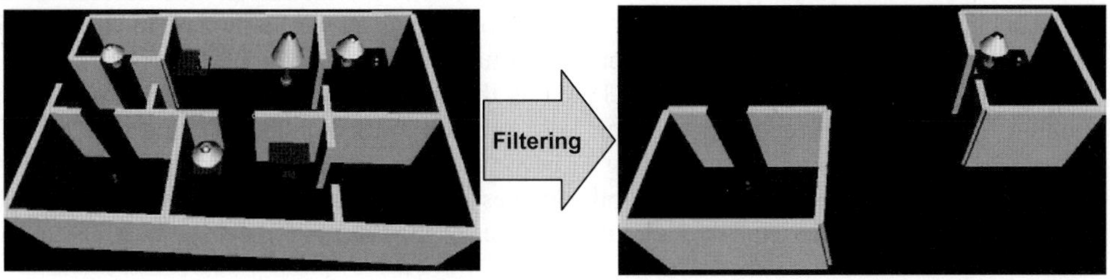

Figure 2: Filtering example

4 Mobile user interface for filter definition

4.1 Mobile scenario

In this research we are introducing a scenario where the user in the role of an inspector is doing inspection of a building site. He is interested in a particular part of the currently visited building and wants to check the reality with the plan stored on the information server, where all graphical information is kept. This scenario shows us that our system must provide tools for defining user's ROI by means of semantic filter specification. The filter specification is sent from the client side to the server side and applied on the selected data (in this case the information about the currently inspected building). This reduction of information complexity minimizes not only the data transfer but also the time the user needs to analyze the retrieved information.

4.2 Special aspects of mobile UI

When designing a UI for mobile environment following aspects were taken into account:

1. Limited device processor power and memory capacity
2. Small display size, small letters
3. Problem with window switching (dialogues)
4. Problematic text input
5. Different operating systems used by PDA's

Based on the limitations of mobile environment we have defined a set of rules that were followed when implementing the user interface:

1. We avoid any animations and any processor intensive rendering tasks. The user interface for filter specification is implemented in a text based form. Due to the significant growth of processor power of the mobile devices in the last couple of years, the problem of running JVM becomes one of the minor problems of UI implementation on PDAs.
2. The user interface must be compact, easily readable and fitting into one screen only. The user must get fast overview and understanding of what is happening with the application without having to scroll or open other windows.
3. The operating systems used on PDA (Windows CE, PalmOS) have very weak tools for context management. Focusing on Windows CE as our current target platform we have recognized that the user can easily get lost in an application that uses several different windows or uses dialog windows to report messages. Due to the limited screen size there is no navigation panel and the task switching mechanisms are not well designed. We use only one active window, which displays all the output and input fields on one screen.
4. The text input is one of the greatest problems while working with PDAs. The devices have a text recognition system which compared to a keyboard input is very slow and uncomfortable. This is why we use lists of choices in all possible situations to reduce direct text input.
5. We have chosen Java Virtual Machine (JVM) as the system platform. The advantage of this solution is its platform independence. The JVM runs on most mobile operating systems (e.g. Windows CE, PalmOS) and is promising to become commonly supported not only on PDAs but also on mobile phones and other mobile devices. The current existing drawback is that the virtual machine consumes relatively a lot of resources of the PDA.

5 System implementation

Our system is based on client/server architecture (see Figure 3). On the server side the MPEG-7 (semantic description) analyzer and graphical data extractor resides. On the client side the semantic filter creator and graphical data viewer resides.

We will demonstrate the functionality of our system on an example where the user wants to analyze part of the VRML model of a building (follow arrows on Figure 3). In the semantic description (MPEG-7) there is for example the following information: of which walls the rooms are composed, connection between rooms, materials of objects, etc. As the first step the semantic description of the model is analyzed on the server side and overview information is sent to the client side. Then the user defines his region of interest by configuring the semantic filter. This is done by specifying the objects and relations among them as triplets [object; relation; object] (e.g. Any room; connected with; room A). This query is sent back to the server side (step 2) and the multimedia extractor extracts the requested data from the VRML model (step 3) according to the query (semantic filter configuration). This reduced model is then sent to the client side and visualized by the multimedia viewer (step 4).

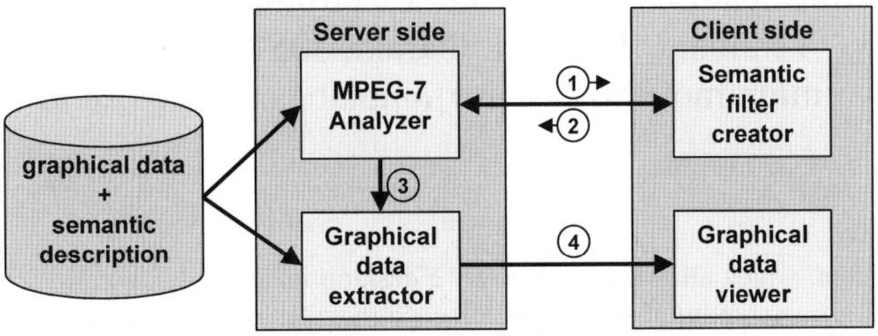

Figure 3: MPEG-7 Analyzer Architecture

6 Conclusion

We have developed a novel system that uses semantic description based filters for retrieval of 2D/3D graphical data in mobile environment. The system helps the user to find the needed information by analyzing the semantic description of the graphical data and configuring the semantic filter. This semantic filter is then used to extract only the needed information from the graphical data. The semantic description is defined in standard MPEG-7 format.

The research is running in the framework of MUMMY project (Mobile knowledge management - using multimedia-rich portals for context-aware information processing with pocket-sized computers in facility management and at construction site) and is funded by the Information Society DG of European Commission (IST-2001-37365). See http://mummy.intranet.gr/.

References

Balfanz D. (2002), Automated Geodata Analysis and Metadata Generation, in Proceedings of SPIE Conference on Visualization and Data Analysis Vol. 4665 (pp. 285-295)

Li J. Z., Ozsu M. T., Szafron D. (1995), Query Languages in Multimedia Database Systems, Technical Report, The University of Alberta Edmonton

Mikovec, Z., Klima, M., & Slavik, P. (2002), Manipulation of Complex 2D/3D Scenes on Mobile Devices, In Proc. of the 2nd IASTED International Conference Visualization, Imaging and Image Processing (pp. 161-166), Anaheim: Acta Press. ISBN 0-88986-354-3

Mikovec, Z., Klima, M., & Slavik, P. (1999), Structural and semantic dialogue filters, In Proc. of the 2nd International Workshop Text, Speech and Dialogue (pp. 280-285), Plzen: Springer-Verlag, ISBN 3-540-66494-7

Metso M., Sauvola J. J. (2002), Media wrapper in adaptation of multimedia content for mobile environments, In Proceedings of SPIE Vol. 4209, Multimedia Systems and Applications III (pp. 132-139)

MPEG-7, from ISO/IEC JTC1/SC29/WG11 2000: http://mpeg-7.com; http://www.cselt.it/mpeg/

W3C Consortium, XML: The Extensible Markup Language, from http://www.w3.org/XML/

A Framework for Transferring Desktop Images and Remote Operations in Multiple Computer Environments

Motoki Miura *Buntarou Shizuki* *Jiro Tanaka*
Institute of Information Sciences and Electronics, University of Tsukuba
Tennodai, Tsukuba, Ibaraki, 305-8573 Japan
{miuramo, shizuki, jiro}@iplab.is.tsukuba.ac.jp

Abstract

A remote display system, which allows users to view a desktop image and to control a remote machine, can be valuable for small group communication in which each member is operating several machines. However, conventional server/client based remote display systems have not been considered for group communication or multi-machines environments. Most of these systems restrict the direction of transferring from server to client. Switching the direction of transferring may require several procedures. Moreover, the settings of the server on each host tend to be troublesome.

We propose a tool for group communication based on remote display system with natural and intuitive interface. We have implemented a system "comDesk," which utilizes P2P (peer to peer) mechanism suitable for group communication. P2P mechanism minimizes the users' setting burden with autonomous configuration and reduces restriction of transferring directions. comDesk provides a visual interface, in which users can specify both the source host and the destination host freely, with simple dragging operations. comDesk will encourage small group communication based on sharing desktop images.

1 Introduction

Many personal computers with network infrastructures are now deployed in group working environments like offices and laboratory workplaces. In these environments, workers usually handle their computers individually. However, they sometimes want to share information and communicate with other workers casually. Email, chat and instant messages can be used for textual communication but they are not suitable for the visual one.

We have been focusing on a remote display system for visual based communication among members of a group. It will be useful for sharing images, instructing usage of an application on a colleague's machine and understanding the current situation of the others. Conventional remote display systems such as VNC (Virtual Network Computing) [Richardson et al., 1998] and some commercial software are becoming popular. Such systems are valuable for personal use of remote operation, however, the following problems arise when we apply these systems to group communication:

(a) the settings of these systems tend to be troublesome.

(b) the interface for recognizing and controlling transferring sessions (connections) is not enough.

2 Framework for Group Communication System

To solve these problems, we propose a framework of remote display system for environments with multiple users and more than two computers.

2.1 Autonomous Configuration

In case that many computers are set, distinction of server/client may increase the possibility of troubles. One of the major troubles is "restriction of transferring direction," which derives from asymmetric configuration. The restriction may cause cognitive overhead from users. The other major trouble is the "setting," specifying server host for each host becomes harder when the user lets many computers join the group.
Therefore we choose a design in which each host has capabilities of both server and client. In this coordination, the user never suffers from any restriction regarding the transferring direction. Also we choose P2P (peer to peer) techniques for autonomous configuration mechanism, which allows users not to select a server when they join the group. Using this mechanism, no administrative server host is required for the group. Tuples of host, user and password are kept by the appropriate host in the group.

2.2 Equality in Controlling Sessions

We think that interfaces for recognizing and controlling transferring sessions are crucial in group communication. "Desktop screen image" may include large quantities of personal information. Thus users may feel uneasy if these interfaces are not provided.
A transferred image is usually shown as a window with session controls in the destination host. Therefore the user of the destination host can easily control the session. However, the owner of the source host has no intervention except for disconnection. We believe that *the owner of the source host should have the right to control the session as much as the user of the destination has*. For example, the owner should understand where and how the transferred screen window is shown by the destination host, and should control the size, resolution and location of the transferring image. Accordingly we provide a visual interface which allows users to control the current transferring sessions at any time. Since the situation can be changed frequently, the view should be updated as often as possible.

3 comDesk : a Communicable Desktop System

We have implemented the system "comDesk," a communicable desktop system based on the above framework. "comDesk" is designed to minimize the users' setting burden with autonomous configuration, and enables the users to specify both the source host and the destination host by using simple dragging operations on the virtual hosts icons. The users can also control the position and the size of the transferred desktop images by similar dragging operations. comDesk is fully implemented in Java with RMI (Remote Method Invocation). Autonomous configuration is enabled introducing P2P technique. Since our system works on almost all platforms without any configuration, the user can casually communicate with the other members.

A typical scenario of comDesk use is (1) more than two computers and users participate in a group (2) each user usually works independently (3) when communication is necessary, a session starts. Each user can join/leave the group at any time.

comDesk consists of two components: comDesk Commander and comDesk Window. comDesk Commander is an interface for controlling image transferring processes. comDesk Window is a holder of remote desktop image, and allows a user to manipulate the source host by transferring operations on the desktop image.

3.1 comDesk Commander

comDesk Commander represents the joining host as an icon (named *HostIcon*). Figure 1 shows an example of comDesk Commander running on host zidane. Figure 1 represents four computers (crescent, zidane, tidus, phobos) that are joined. HostIcon can show a desktop thumbnail image. HostIcon also shows "*WindowIcons*" which represents comDesk Windows displayed on the host. Arrows indicate the transferring direction. In Figure 1, desktop images of crescent, tidus are transferred to zidane. In general, every joining host can start comDesk Commander at any time, and can recognize the current transferring sessions. comDesk Commander shows (1) host name (2) thumbnail (3) owner's name (4) session status, including where comDesk Window is shown.

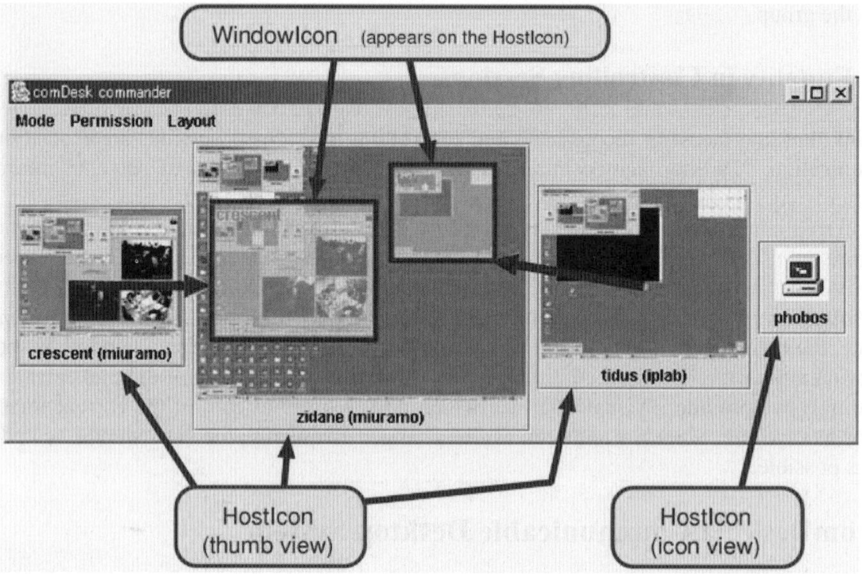

Figure 1: comDesk Commander

A transferring session starts by a drag&drop operation on HostIcon. When the user drags a HostIcon and drops it to another, comDesk Commander creates a transferring session. As a result, a new window (comDesk Window) appears in the dropped host. The dropped point determines the position of the window.

After the session starts, both the owner of the source host and the user of the destination host have permission to control the session by manipulating the WindowIcon shown in the HostIcon.

Dragging a WindowIcon affects the corresponding comDesk Window, that is, the location and size of the comDesk Window can be changed by remote hosts. The lower-right area of WindowIcon is allocated for resizing, and the rest of the area for moving. In addition to this, the owner of the source host can drop the WindowIcon to another HostIcon, which causes "re-transferring of window." The "re-transferring of window" operation is more effective than the procedures of disconnection and re-connection. The common drag and drop interaction for both HostIcon and WindowIcon is intelligible for consistent paradigm of transferring operations to the users.

For the owner of the source host, the "re-transferring of window" is also used to regain the transferring image. Other functions like changing resolution and disconnection are performed using the pop-up menu, which appears by pressing right button on the WindowIcon.

3.2 comDesk Window

comDesk Window is a window that shows a desktop image of the source host at the destination host. The user of the destination host can see the desktop image of the source host. comDesk Window appears at the destination host after the session starts. comDesk Window consists of two parts: remote desktop area and controller. The largest part of a comDesk Window is a remote desktop area. In Figure 2, the desktop of crescent is shown at the remote desktop area. The user of the destination host can change the resolution by "zoom rate combo-box."

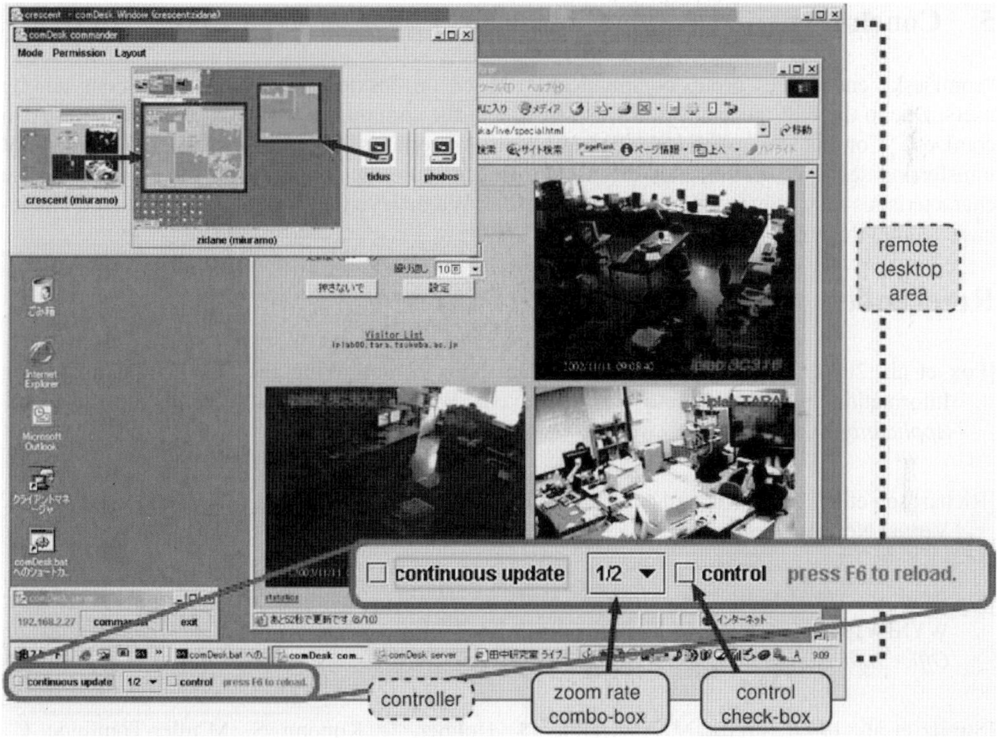

Figure 2: comDesk Window on host zidane. This comDesk Window corresponds to the left WindowIcon in the comDesk Commander (Figure 1). The source host is crescent and the destination host is zidane.

While the "control check-box" is on, mouse and keyboard operations are transferred to the source host. Therefore the user of the destination host can control the source host remotely.
Also the owner of the source host can control the session's properties (location, resolution, controllable option, and disconnection) by selecting pop-up menu from the corresponding WindowIcon on his/her comDesk Commander. If the owner of the source host makes the "control checkbox" disable from the menu, the user of the destination host never operates the source host.

4 Related Work

Colab (Collaboration Laboratory) [Stefik et al., 1987] is a pioneer of large screen meeting rooms at Xerox PARC designed for small working groups of two to six people. It has a facility of multiuser interface, that provides a synchronized view and telepointing. i-land [Streitz et al., 1999] project in GMD (currently Fraunhofer-Gesellschaft) develops several communication devices embedded to a wall (DynaWall), table (InteracTable) and chairs (CommChair) for natural multi-user collaboration. iRoom [Fox et al., 2000] is a large display conference system with a mechanism for accepting heterogeneous devices like PDAs. In iRoom, communication infrastructure based on HTTP is proposed and developed for multi-device coordination. These environments with large display augment face-to-face meeting. comDesk focuses on casual and informal image-based communication among users, generally work independently.

5 Conclusion

"comDesk" encourages group communication based on desktop images and operations among users. Due to the P2P mechanism, the users can join the community with less setting burden. Also comDesk Commander provides efficient visual interfaces to share status information of transferring sessions, and enables the users to reconfigure the properties of the sessions. These characteristics may reduce the anxiety of the users. We believe this tool makes the relaxed image-based communication smoother.

References

[Fox et al., 2000] Fox, A., Johanson, B., Hanrahan, P., and Winograd, T. (2000). Integrating Information Appliances into an Interactive Workspace. *IEEE Computer Graphics & Applications*, 20(3):54–65.

[Richardson et al., 1998] Richardson, T., Stafford-Fraser, Q., Wood, K. R., and Hopper, A. (1998). Virtual Network Computing. *IEEE Internet Computing*, 2(1):33–38.

[Stefik et al., 1987] Stefik, M., Bobrow, D. G., Foster, G., Lanning, S., and Tatar, D. (1987). WYSIWIS Revised: Early Experiences with Multiuser Interfaces. *ACM Transactions on Office Information Systems*, 5(2):147–167.

[Streitz et al., 1999] Streitz, N. A., Geißler, J., Holmer, T., Konomi, S., M¨uller-Tomfelde, C., Reischl,W., Rexroth, P., Seitz, P., and Steinmetz, R. (1999). i-LAND: An interactive Landscape for Creativity and Innovation. In *Proceedings of the CHI 99*, pages 120–127.

The Effects of Display Orientation and Target Position on Target Pointing Tasks on a PDA

Masafumi Ogasawara

Kochi University of Technology
185 Miyanokuchi,
Tosayamada-cho, kochi
782-8502, Japan
030254u@ugs.kochi-tech.ac.jp

Sachi Mizobuchi

Nokia Research Center/
Keio University
2-13-5 Nagata-cho,
Chiyoda, 100-0014
Tokyo, Japan
sachi.mizobuchi@nokia.com

Xiangshi Ren

Kochi University of Technology
185 Miyanokuchi,
Tosayamada-cho, kochi
782-8502, Japan
ren.xiangshi@kochi-tech.ac.jp

Abstract

This study investigates the effects of display orientation and target position on pointing tasks on PDAs (Personal Digital Assistants). Subjects were asked to perform pointing tasks with three PDA orientations (vertical, horizontal 1 and horizontal 2). We examined the best PDA orientation and target position by movement time, error rates, throughput and coordinates (x, y). The results showed that the vertical orientation was the best of the three. There was no significant difference between horizontal 1 and horizontal 2. Regarding the position of the target, the right target position was a little better than left target. In particular, subjects found it easier to tap the top of the target on the right.

1 Introduction

The orientation of most displays on hand-held information devices (e.g. PDAs) is vertical orientation and a stylus is commonly used to input data. There are some studies on the usability of pen-input interfaces for PDAs, e.g., selection strategies, target selection performance [1][2]. However, little research has been reported regarding the physical aspects of PDAs such as display orientation.

Vertical PDA displays and horizontal PDA displays are currently on the market but the choice between these two types of display is usually made without any quantitative consideration.

Kato and Nakagawa (1998) performed an experiment on the effects of target position on target pointing tasks on tablet PCs. According to their study, pointing to the right or lower right target position was slower than other target positions. They concluded that this was due to the fact that a target in these positions is hidden by the right-handed user [3]. However this result has specific reference to a tablet PC and is not necessarily relevant to PDAs. The effect of target position on target pointing tasks on PDAs has not been elucidated yet. This study examines the effect of both display orientation and target position on the pointing task.

2 Experiment

2.1 Subject

Twelve subjects (6 male, 6 female, aged from 20 to 22, all right handed) were tested in the experiment. None of the subjects had previous experience using PDAs.

2.2 Design

The PDA used was NTT docomo G-fort units running Windows CE, 85 mm (width) x 135 mm (height) x 25.5 (thickness). The weight of the PDA was 300 g. The display was 240 x 320 pixels, monochrome, 65536 colours and TFT liquid crystal display (1 mm is about 0.24 pixels). A 16 cm acrylic pen was adopted as the input device. Experimental software was developed with Java.
We prepared three different PDA display orientations (vertical, horizontal 1 and horizontal 2) using this software. In the horizontal 1 orientation, the keys are placed on the left side of the PDA (Figure 1). In the horizontal 2 orientation, the keys are placed on the right side of the PDA (Figure 2).

Figure 1: Horizontal 1

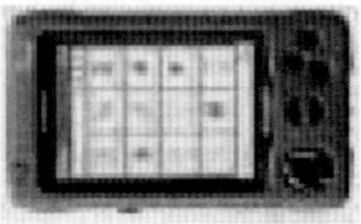

Figure 2: Horizontal 2

The task was the one-direction pointing task defined in ISO 9241-9 (Figure 3). Two rectangles were shown on the display. One was filled and the other was unfilled. Subjects sat down and held the device with their non-dominant hands. They were instructed not to rest their hands on the table or any other objects during the test. Upon contact the rectangles would switch places and the subjects would again attempt to point to the unfilled rectangle. During the task, a beep sound was made whenever the subjects missed the target.
Before the testing, the subjects were asked to point to the unfilled rectangle (called "target" below) with the pen as quickly and accurately as possible. All subjects performed 10 warm-up trials.
The experiment was a 3 x 3 x 3 within-subjects factorial design. The factors and levels were as follows:
Display orientations: Vertical, horizontal 1, horizontal 2
Target widths: 10, 20, 40 pixels (2.4, 4.8, 9.6 mm)
Distances between the center of targets: 100, 150, 200 pixels (24, 36, 48 mm)

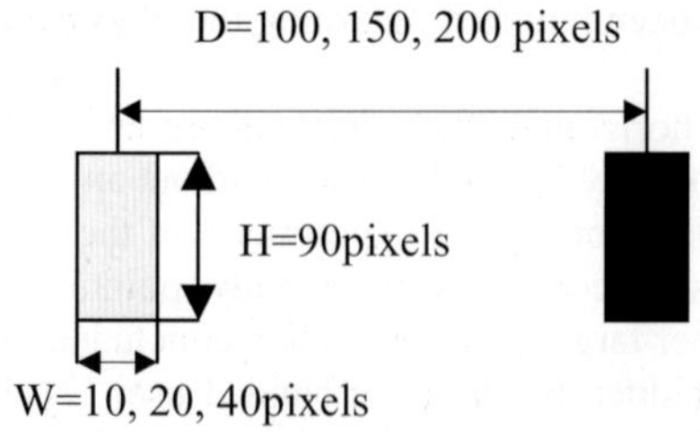

Figure 3: Target condition

Each subject performed the task in 30 trials in each of nine conditions. The height of the targets was 90 pixels in all trials. Targets were presented in different order to the various subjects.

With 12 subjects and 270 trials per display orientation (3 widths x 3 distances x 30 trials), the total number of trials in the experiment was 9,720 (12 subjects x 270 trials x 3 display orientations). Times between each switch and the coordinates for each point of contact were logged on the device.

3 Result

3.1 Effects of display orientation

3.1.1 Movement time

The mean movement times for the vertical, horizontal 1, and horizontal 2 were 393.15 ms, 415.33 ms, 413.72 ms respectively (Figure 4). The result of an analysis of variance (ANOVA) did not reveal any significant differences between three display orientations.

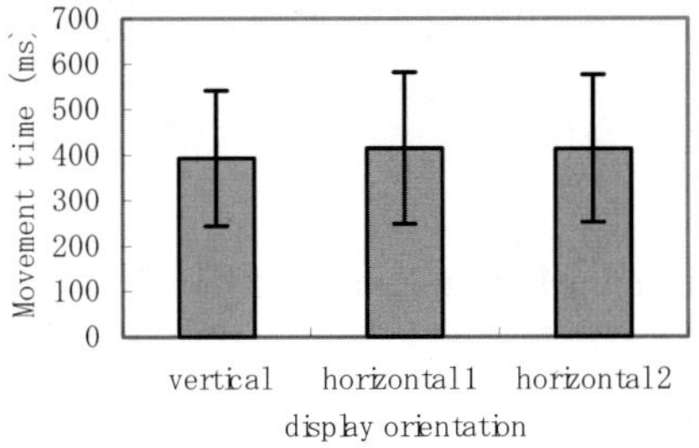

Figure 4: The mean movement time with SDs for display orientation

3.1.2 Error rates

The mean error rates for the vertical, horizontal 1, and horizontal 2 orientations were 12.99%, 16.97%, and 18.94% respectively (Figure 5). The results of nonparametric analysis did not reveal any significant differences between the three display orientations.

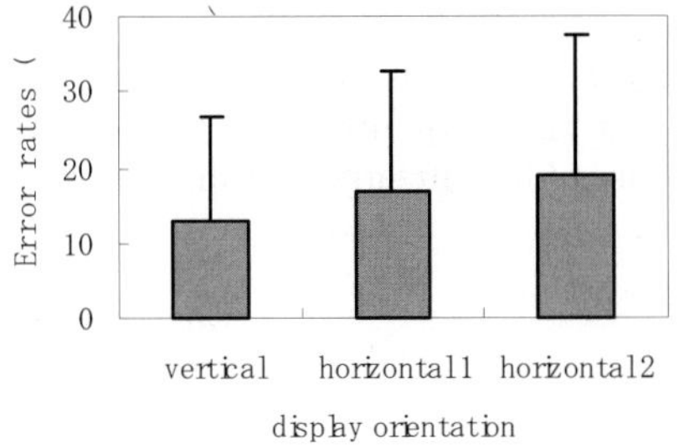

Figure 5: The mean Error rates with SDs for display orientation

3.1.3 Throughput

The mean throughputs for the vertical, horizontal 1, and horizontal 2 orientations were 8.3 bits/s, 7.57 bits/s and 7.52 bits/s respectively (Figure 6). The results of an analysis of variance showed a significant main effect for display orientation ($F2, 321 = 5.56$ $p < 0.01$). The results based on the Bonferoni method of analysis showed that there were significant differences between the vertical and horizontal 1 orientations ($p < 0.01$), and between the vertical and horizontal 2 orientations (p <

0.01), but no significant differences between horizontal 1 and horizontal 2. The vertical was the most efficient display orientation.

Overall, the vertical orientation was the best of the three orientations. The vertical display orientation seems to be the easiest to use, allowing users to hit the target more accurately.

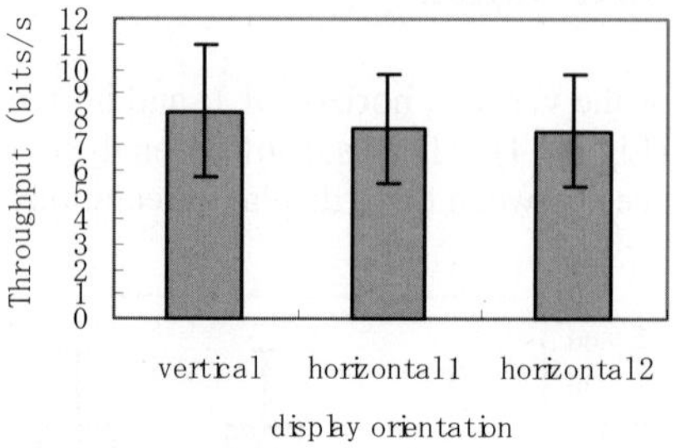

Figure 6: The mean Throughput with SDs for display orientation

3.2 Effects of target position

3.2.1 Movement time

The mean movement time for the target positions were 411.76 ms for the right target, 403.35 ms for left target (Figure 7). There was no significant effect for the target position.

3.2.2 Error rate

The mean error rates for the target position were 15.89% for the left target, 15.78% for the right target. There was no significant difference between the left target and right target.

3.2.3 Throughput

The mean throughputs for the target positions were 7.62 bits/s for the left target, 7.80 bits/s for the right target. There was no significant difference between the left target and right target.

3.2.4 Contact position

Figure 7 shows the coordinates for the points of contact for the horizontal 2 orientation when the target width was 40 pixels and the target distance was 100 pixels. Almost all subjects pointed to the lower part of the target in the case of the left target and the upper part of the target in the case of the right target.

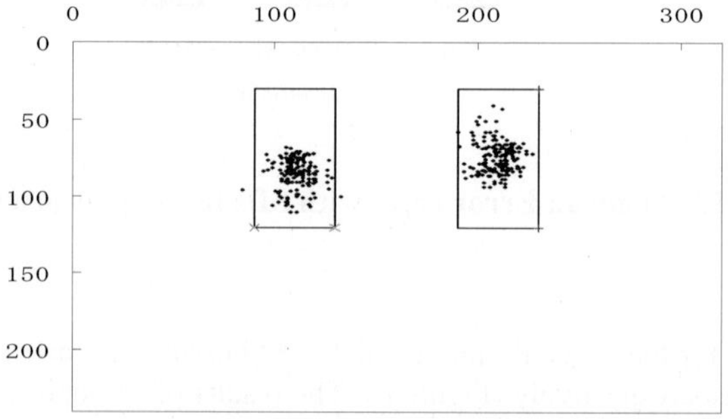

Figure 7: Contact position W=40 D=100 in the horizontal 2

4 Discussion and Conclusion

4.1 The effects of display orientation

According to the effects of display orientations, the results showed that the vertical affected faster target selection than horizontal 1 and horizontal 2. The results also showed the vertical orientation had a lower error rate than horizontal 1 and horizontal 2. We concluded that this was because the device is easier to hold with the vertical orientation than with the horizontal orientation. The vertically oriented PDA is more stable than the horizontal. The subjects performed the tests without resting their hands on a table or any other object, because we wanted to test off-desk of mobile situations. Therefore the subjects tended to be influenced by the size and shape of the apparatus.

4.2 The effects of target position

The results of the effects of target position were different from the research previously mentioned [3] which found that the right target were slower than the other target positions. Two reasons led to this result. First, the targets were fixed in our experiments. The subjects were able to regulate their right hand so that the right target could remain visible throughout the trials. Secondly, the right target could not be easily hidden by the right hand because the perpendicular height of the rectangular target was sufficient to allow constant orientation for the user.

The co-ordinates for the points of contact showed that very selections were affected at the lower part of the target when the target was on the right. We consider that the reason for this is found in the structure of human wrist. The natural path of the right hand as it moves to the left of the display is down. As it moves to the right, the natural path is toward the top, as if the hand were pivoted at the bottom right corner of the device.

Pointing to left side is a little different from pointing to right side on a PDA. However, pointing to right side is a little faster than pointing to left side. It is therefore easier to point to the upper right area when point to the right side of a PDA. Therefore, the best target position is upper right.

5 Summary

In this study, we performed experiments which investigated the effects of display orientation and target position on target pointing tasks on a PDA. The results show that the vertical orientation is the best target position, and the best target position for PDAs is the upper right area of the display.

Acknowledgments

This research was partially supported by a Grant-in-Aid for Scientific Research (Young Researchers (B) No. 14780338), and High-Tech Research Center Development Program (No. 7512002002), both in Japan.

Reference

[1] Ren, X. and Moriya, S.: Improving Selection Performance on Pen-Based Systems: A Study of Pen-Based Interaction for Selection Tasks, *ACM Transactions on Computer-Human Interaction (ToCHI)*, Vol.7, No.3, pp.384-416 (2000).

[2] Mizobuchi, S., Mori, K., Ren X., and Yasumura, M.: An empirical study of the minimum required size and the minimum number of targets for pen input on the small display, In *Proceedings of Mobile HCI 2002*, pp.184-194 (2002).

[3] Kato, N and Nakagawa, M.: An Investigation into the Usability of a Pen for Designing Pen User Interfaces, *Information Processing Society of Japan*, Vol.39, No.5, pp.1536-1546(1998).

[4] ISO9241-9: 2000 Ergonomic design for office work with visual display terminals (VDTs) – Part 9: Requirements for non-keyboard input devices. *International Standardization Organization.*

An Evaluation of Text Entry Methods in a Standing Posture for Application to an Immersive Virtual Environment

Noritaka Osawa

National Institute of Multimedia Education & The Graduate University for Advanced Studies, Japan
osawa@nime.ac.jp

Xiangshi Ren

Department of Information Systems Engineering, Kochi University of Technology, Japan
ren.xiangshi@kochi-tech.ac.jp

Motofumi T. Suzuki

National Institute of Multimedia Education, Japan
motofumi@nime.ac.jp

Abstract

Cooperative work in a distributed immersive virtual environment, which connects virtual reality systems with immersive projection displays such as CAVE, usually requires users to put annotations in the virtual world. Simple 3D symbols or 3D icons are useful in a fixed task; however, simple symbolical annotations are insufficient for more general cooperative work because complex abstract annotations cannot be represented by them. Therefore, we need textual annotations in an immersive projection display system. We conducted a pilot experiment to evaluate text input methods using handheld devices for an immersive virtual environment. A personal digital assistant (PDA) and a mobile phone were used as input devices. They were used by six participants who were standing during the experiment. The experimental results show that the average time to input short Japanese phrases was the shortest for handwritten notes on the PDA, followed by button-input on the mobile phone, soft keyboard input on the PDA, and recognition of handwritten characters on the PDA.

1 Introduction

Text input is one of the most frequent human-computer interaction studies. However, the research has paid little attention to immersive virtual environments. In distributed immersive virtual environments, although manipulation of a spatial position is often attempted, it is sometimes necessary to input annotations that show the meaning of the position in the text. We also need text input in the immersive virtual environment's application. For example, text is necessary for putting the name of the identifier in an immersive programming system (Figure 1). Moreover, text is necessary to make an explanation in an immersive three-dimensional presentation system that would be similar to a presentation graphics tool, such as PowerPoint in two dimensions.

An experiment on text entry in an immersive environment was performed by Osawa and Sugimoto (2002). It measured and compared the performance and user preferences of an integrated method of speech and hand-manipulation, a virtual keyboard, and a method using speech-recognition. To complement this experiment, we performed a small-scale pilot study to investigate methods, features, and tendencies of text input using handheld devices in a standing posture, as they might be used in immersive virtual environments. We are still at a very early stage of exploring text input in immersive virtual environments. Thus, we conducted an experiment to compare five input methods in a real office, rather than in an immersive virtual environment. We asked subjects to perform text input tasks while standing, because users of an immersive environment are usually standing when they use the environment.

MacKenzie and Soukoreff (2002) review mobile text entry methods and optimization techniques, and give a survey of key-based and stylus-based text entry techniques. The key-based techniques include a telephone keypad, other small keyboards (Sugimoto & Takahashi, 1996) and a five-key text entry. The stylus-based techniques include traditional handwriting recognition, unistrokes (Goldberg & Richardson, 1993), and gesture-based text input (Venolia & Neiberg, 1994). However, these studies focus on an alphabetical character set. Our work measured the performance of text entry of the Japanese character set. Japanese text is composed of kana characters (phonograms) and kanji characters (Chinese ideograms). We will call it "mixed text" in this paper. A kana-kanji conversion system is usually used to input a mixed text with an ordinary kana keyboard or QWERTY keyboard. It takes several steps to convert a sequence of kana or alphabetical characters into a Kanji character. In other words, Japanese text entry needs more steps than alphabetical character entry.

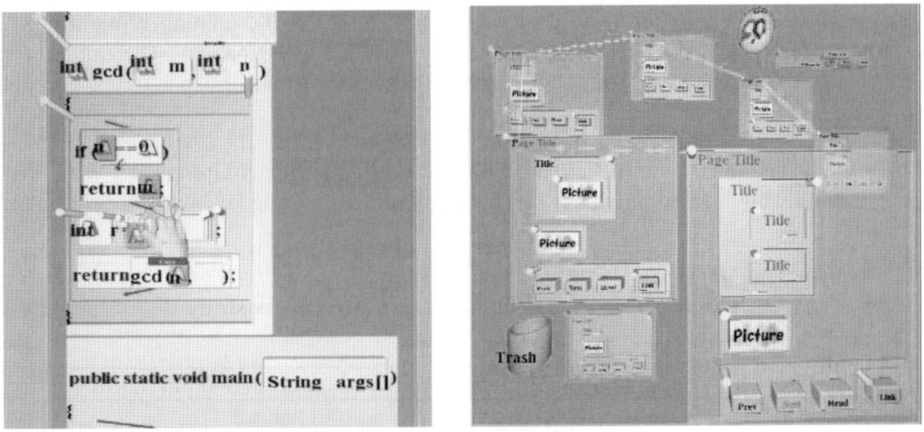

(a) Immersive programming system (b) 3D presentation system

Figure 1: Text in an immersive projection display system

2 Method

2.1 Subjects

Six subjects (4 male, 2 female, all university students) took part in the experiment. They were between 19 and 23 years old (mean age = 20.8 years).

2.2 Design

A personal digital assistant (PDA, Compaq iPAQ PocketPC) and a mobile phone were used as the input devices. We used three functions of the PDA: handwritten recognition, software keyboard, and handwritten entry. We also used a QWERTY keyboard as a baseline for comparison. Thus, we used five input methods in the experiment.

(1) Handwritten recognition on the PDA
The unit's handwritten character recognition function was used.

(2) Software keyboard on the PDA
The unit's software keyboard function was used.

(3) Mobile phone
The subjects used different mobile phones. We asked the subjects to use their own mobile phones if they use them on a daily basis. All subjects had their own mobile phones and used them in the experiment. Thus, the input method was subject dependent. We wanted to know how fast a typical mobile phone user can input text, rather than make a comparison of specific input methods of the mobile phone.

(4) Handwriting on the PDA
"Handwriting" here indicates that the input strokes by pen remain as they are. This is similar to the handwriting on paper.

(5) QWERTY keyboard
The keyboard of a notebook PC (Toshiba Dynabook Satellite 2550X running Windows 98) was used. The subject was asked to input characters on Windows Notepad. The subjects were unfamiliar with the keyboard layout. This may have degraded the performance of text entry using the keyboard.

The tasks were performed in an office, not in an immersive virtual environment. We asked the subjects to stand while performing these tasks, except while using the keyboard (they were seated for this task).

Table 1: The texts in Japanese inputted in the study

Type of task	Text in Japanese	Translation in English
Name	オブジェクト 2	Object 2
Comment 1	デザイン再検討	Review design
Comment 2	綴りを訂正	Correct spelling
Attribute 1	緑を濃く	Deepen green
Attribute 2	直線部分を短く	Shorten a straight part
Structural direction 1	これらを連結する	Connect these
Structural direction 2	5 個に分離する	Separate into 5 pieces
Structural direction 3	要素をグループ化	Group elements
Structural direction 4	上へ移動	Move upward
Structural direction 5	タイトルを中央揃え	Center the title

The tasks required them to input short Japanese phrases or sentences (Table 1). We assumed that multimodal text input would be used for cooperative work in a virtual 3D space. Therefore, we chose short phrases related to the construction and modification of a 3D scene. The phrases were the same as in the previous study (Osawa & Sugimoto, 2002).

2.3 Procedure

First, the experiment was explained to each subject. Each subject was given three practice trials for each input method before the experiment started. The subject was then asked to input the texts by each input method. The time elapsed was between beginning to input and finishing to input each short sentence. They were also asked to rank (on a scale of 0 to 9) the input methods in response to six questions, such as their satisfaction with the method and their desire to use that method. These questions were also the same as in the previous study.

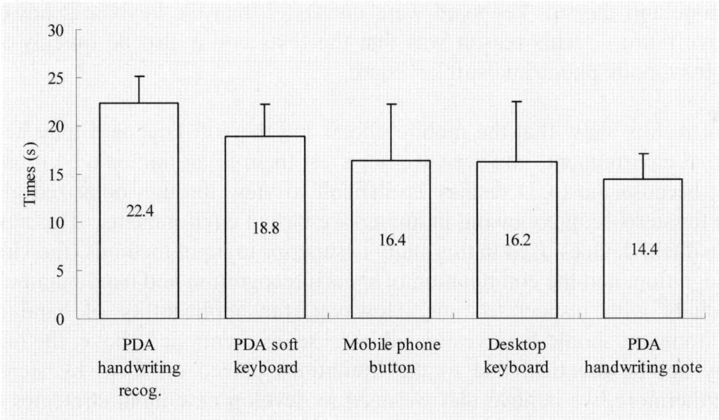

Figure 2: Mean input times with standard deviation error bars

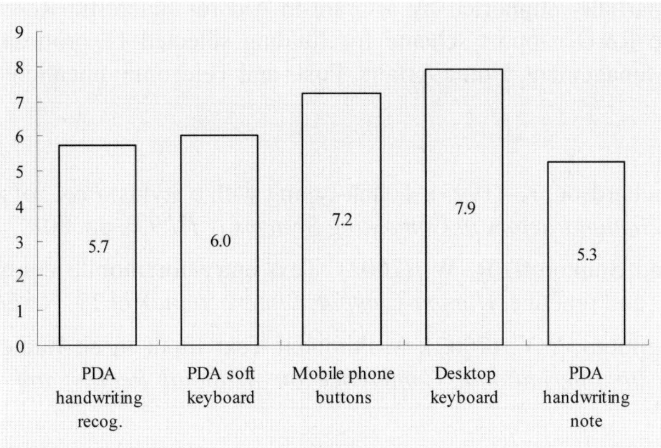

Figure 3: Summary of subjective evaluation for the input methods (0 = lowest preference, 9 = highest preference)

3 Results And Discussions

An ANOVA (analysis of variance) with repeated measures was used to analyze performance in terms of input time and subjective preference. There was a significant interaction in input time between the input methods, $F(4,25) = 2.87$, $p < 0.05$. Figure 2 shows the mean input times. It shows that the mean input times increases in order of handwritten note on the PDA, desktop keyboard input, button-input on mobile phone, soft keyboard input on the PDA, recognition of handwritten characters on the PDA, and speech input. Figure 3 shows a summary of the subjective ratings for the input methods. These ratings were based on the average value of the answers given by the subjects to the six questions. Significant differences were seen among the input methods, $F(4,25) = 24.16$, $p < 0.001$. We do not believe that the desktop keyboard is usable in immersive environments; however, we used it as a baseline. Thus, with the exception of the desktop keyboard, the mobile phone was the most preferred of the other three methods.

The mobile phone and the soft keyboard were not faster than the keyboard; however, they had higher subjective ratings. This reason was that the texts could also be quickly and accurately inputted using the mobile phone and soft keyboard.

Our experiment results show that the mobile phone and the soft keyboard may be applied to an immersive virtual environment; however, we need test them in an immersive virtual environment in future work because the LCD display is difficult to view through polarized glasses or LCD shutter glasses for stereoscopic viewing in immersive virtual environments. This study shows that the input by handheld devices used in this study is superior to input methods (the virtual keyboard, the speech-recognition, and the combination of speech recognition and hand manipulation) used in the previous study with respect to time. However, the applicability of hand devices to an immersive environment should be investigated because, as mentioned above, the methods used in this study may not be practical due to the limitations placed on them by stereoscopic view equipment. Furthermore, we believe that we need to develop new input strategies or devices for text input in an immersive virtual environment.

Acknowledgments

This research was partially supported by a Grant-in-Aid for Scientific Research (14380090) in Japan, and by "The R&D support scheme for funding selected IT proposals" program of the Ministry of Public Management, Home Affairs, Posts and Telecommunications in Japan.

References

Goldberg, D. and Richardson, K. (1993). Touch-typing with a stylus. *Proc. of ACM INTERCHI'93 Conference on Human Factors in Computing Systems (CHI'93)*, pp. 80-87.

MacKenzie, I.S., and Soukoreff, R. W. (2002). Text entry for mobile computing: Models and methods, theory and practice. *Human-Computer Interaction*, Vol.17, No.2&3, pp.147-198.

Osawa, N., and Sugimoto, Y.Y. (2002). Multimodal Text Input in an Immersive Environment. *Proc. of The 12th International Conference on Artificial Reality and Telexistence (ICAT 2002)*, pp.85-92.

Sugimoto, M. and Takahashi, K. (1996). SHK: Single hand key card for mobile devices. *Companion Proc. of CHI 96 Conference on Human Factors in Computing Systems*.

Venolia, D. and Neiberg, F. (1994). T-Cube: A fast, self-disclosing pen-based alphabet. *Proc. of the ACM Conference on Human Factors in Computing Systems (CHI'94)*, pp. 265-270.

A Simple Learning Procedure for Gesture Based Control of Robot Arm Movement

Paulraj M Pandiyan, G. Sainarayanan, R. Nagarajan, Sazali Yaacob

AI Research Group, School of Engineering and Information Technology
Universiti Malaysia Sabah, Kota Kinabalu, Malaysia

Abstract

Gesture recognition has for long been considered as one of the most futuristic options of Artificial Intelligence and user interface. In this paper, a novel statistical method is used to implement Gesture recognition using Artificial Neural Networks (ANN). A conventional back propagation (BP) network has been used for this supervised training scheme. To improve the training time, a simple scheme namely adaptive learning rate method is proposed and the same is implemented and compared with the conventional BP method.

1 Introduction

Gesture-based input is becoming increasingly important because of the increasing need for devices capable of supporting complex, continuous interaction in a range of contexts. Gesture recognition can be of great importance in various applications such as man-machine interaction and surveillance systems where suspect behaviors need to be detected automatically. A great deal of effort has been dedicated recently to the challenging scientific problem of gesture recognition. This paper describes a method for detection and recognition of the palm using ANN which classifies the RGB details of the gesture Further a simple scheme is proposed for the effective training of the neural network. In this scheme, every neuron will have its own learning rate, which is varied adaptively during training the network.

2 Gestures

Gestures can be defined as the use of motions of the limbs or the body as a means of expressing or emphasizing an idea, sentiment or attitude. Hand gesture is an important way for communication between people. On a simple human-machine interface, the hand gesture recognition is important. In a development of systems based on gesture recognition, there are three problems namely static recognition of hands, hand tracking, dynamic gesture recognition. A large number of work has been performed on these types of gesture recognition (Mishra et. al., 1996; Sharma et. al, 2000; Hyeon-Ryulee & Rim, 1999; Davis & Shah, 1994). In this paper, a simple scheme to represent the static gesture using statistical mean is presented. The symbolic representation of the gesture is then used to represent the direction of movement of a robotic arm. These direction gestures are means of indicating that a process has to be done with the movement of a robotic arm. For example, clutched fists with thumb facing upwards (best of luck symbol) could mean move up by a certain distance.

3 Scope of Movement

The whole scope of the movement of the robotic arm is shown in Figure 1 (Duric et. al., 2002). The robotic arm can move through 12 distinct places in three different planes. A simple code language is used to move to the various positions of the robotic arm. In the Figure 1, 'A' represents the upper plane of motion, while 'B' represents the middle plane of motion and 'C' the lower plane. In this, 30 degree steps are used between the discrete points on each plane and the total rotation is restricted to 90 degrees. Hence, in each plane there are a total of 4 points. These 12 points are represented by 12 different gestures of the palm, gesture refereeing upper plane is shown in Figure 2.

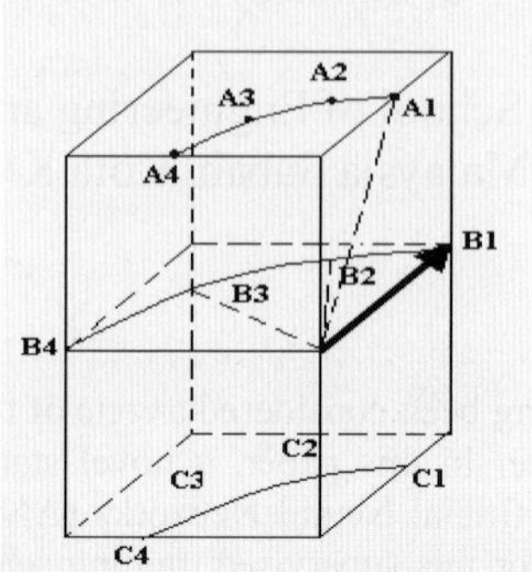

Figure 1: Scope of Movement of the Arm

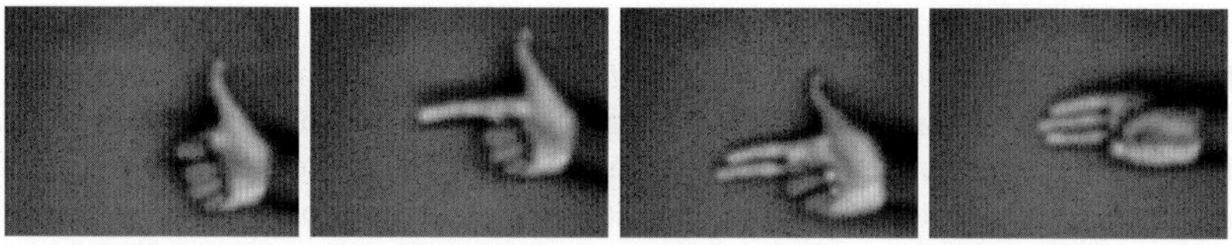

Figure 2: Gestures referring to four points on the Upper plane (A)

Note:
12 sets of similar gestures at various locations, using the same experimental set-up, constant illumination parameters and in the same working area have been taken and used for training.

The constant illumination parameters for the images representing 12 gestures are given in Table 1.

Table 1: Illumination parameters

Background : Black	
Contrast	120
Saturation	130
Brightness	120
Exposure	189
Gain	70
White balance	255

4 Statistical Approach

In the case of true RGB images, the intensities of the red, green, blue components are separated and the mean, median and standard deviation for the whole image are calculated. Along with this, the mean, median and standard deviation for the luminosity of the image are found. Each gesture is represented by a total of 12 parameters.

5 Neural Network Model

BP network (Werbos, 1990) has always been a widely used method in the case of supervised training schemes. Such methods work well with clustering type of problems. In this case, the various gestures are classified into 12 different clusters, representing the 12 different points on the 3-D working plane. This problem requires 12 input neurons and 12 output neurons in the case of RGB images. The hidden neurons are fixed through trial and error method for efficient training. Along with the conventional BP network, a simple scheme is suggested for improving the training of the network.

6 Improvements in BP Network

6.1 The Adaptive Learning Rate Method

In conventional methods (Werbos, 1990), the learning rate for all the neurons are fixed at a particular value and their values are adapted based on the error and the direction of error (Ooyen, & Nienhuies, 1992; Jacobs, 1988; Tollenaere, 1990; Rigler et. al., 1991; Weir, 1991; Eaton & Olivier, 1992). In this paper a simple adaptive learning rate scheme is proposed wherein the learning rate of each and every neuron is different and fixed by a scheme considering the weights connected to that neuron and the output of the same. The learning rate of a neuron not only depends on the error signal but also on the weight that are connected to it.

$$L_{h\_j} = \frac{\sum_i w_{ij}}{\sum_i [w_{ij}]^2} z_j \qquad \ldots (1)$$

$$L_{o\_k} = \frac{\sum_j v_{jk}}{\sum_j [v_{jk}]^2} y_k \qquad \ldots (2)$$

where, $L_{h\_}j$ represents the Learning Rate of the jth hidden neuron.
L_{o_k} represents the Learning Rate of the kth output neuron.
w_{ij} represents the weight connected from i^{th} input neuron to j^{th} hidden neuron.
V_{jk} represents the weight connected from j^{th} hidden neuron to k^{th} output neuron.
z_j represents output of the jth hidden neuron.
y_k represents output of the kth output neuron.

In this scheme, the learning rate is updated only when the Cumulative Error of the present epoch is less than the Cumulative Error of the previous epoch else the learning rates of the previous epoch are retained.

6.1.1 Nomenclature

x : Input training vector $\mathbf{x} = (x_1, x_2, \ldots, x_i, \ldots, x_n)$.
t : Output target vector$\mathbf{t} = (t_1, t_2, \ldots, t_j, \ldots, t_m)$.

Net input to Z_j, $z_{inj} = v_{oj} + \sum_i x_i v_{ij}$ and $z_j = f(z_{inj})$.

Net input to Y_k, $z_{ink} = w_{ok} + \sum_j z_j w_{jk}$ and $y_k = f(y_{ink})$.

6.1.2 Training Algorithm

Step 1 : Initialize the weights.
Initialize the Learning Rate for each neuron.

$$L_{h\_j} = \frac{\sum w_{ij}}{\sum [w_{ij}]^2} \quad (\text{ if } L_{h\_j} > 1,\ L_{h\_j} = 1/L_{h\_j})$$

$$L_{o\_k} = \frac{\sum v_{jk}}{\sum [v_{jk}]^2} \quad (\text{ if } L_{o\_k} > 1,\ L_{o\_k}] = 1/L_{o\_k})$$

Step 2 : While stopping condition is false, do Steps 3 to 10.

Step 3 : For each training pair **x**: **t**, do steps 4 to 9.

Step 4 : Each input unit X_i, i = 1,2, …, n receives the input signal x_i and broadcasts it to the next layer.

Step 5 : For each hidden layer neuron denoted as Z_j, j = 1,2, …, p.

$z_{inj} = v_{oj} + \sum_i x_i v_{ij}$ and $z_j = f(z_{inj})$.; Broadcast z_j to the next layer.

Step 6 : For each output neuron Y_k, k = 1 to m.

$y_{ink} = w_{ok} + \sum_j z_j w_{jk}$ and $y_k = f(y_{ink})$.

Step 7 : Compute δ_k for each output neuron Y_k.

$\delta_k = (t_k - y_k) f'(y_{ink})$.; $\Delta w_{jk} = L_{o\_k}\, \delta_k\, z_j$. $\Delta w_{ok} = L_{o\_k}\, \delta_k$, since $z_0 = 1$.

Step 8 : For each hidden neuron,

$\delta_j = \delta_{inj} f'(z_{inj})$ where $\delta_{inj} = \sum_k \partial_k w_{jk}$ $\Delta v_{ij} = L_{h\_j}\, \delta_j\, x_i$. $\Delta v_{oj} = L_{h\_j}\, \delta_j$, since $x_0 = 1$.

Step 9 : Update weights

$w_{jk}(\text{new}) = w_{jk}(\text{old}) + \Delta w_{jk}$.; $v_{ij}(\text{new}) = v_{ij}(\text{old}) + \Delta v_{ij}$.

Step 10: Test for stopping condition.

If Stopping condition is true, Break
Else
If Cumulative Error < Previous Cumulative Error

$$L_{h\_j} = \frac{\sum w_{ij}}{\sum [w_{ij}]^2} z_j$$

$$L_{o\_k} = \frac{\sum v_{jk}}{\sum [v_{jk}]^2} y_k$$

Else Continue.

6.2 Experimental Results and Discussions

A two layer neural network with 12 input neurons, 12 hidden neurons and 12 output neurons is considered. The neurons are activated by the sigmoidal activation function. Using the proposed scheme, the network is trained for an error tolerance of 0.01 and the results are compared with conventional method and tabulated in Table 2. While training the network with conventional BP, the learning rate and momentum factors are tuned to obtain optimal results. both the conventional and the proposed schemes, the initial weights are randomized between –0.5 to 0.5 and normalized. The experiment is repeated for 50 randomized weights and Table 2 summarizes the mean epoch, minimum epoch, maximum epoch and the standard deviation for both the conventional and the proposed procedure. From the results it can be observed that for proposed procedure the mean epoch, maximum epoch, minimum epoch and the standard deviation are very less when compared to the conventional back propagation procedure.

Table 2: Comparison of Proposed Scheme with Conventional BP

Input Neurons : 12	Hidden Neurons : 12	Output Neuron : 12
Momentum Factor = 0.7	Activation function: Binary Sigmoidal	
Learning Rate = 0.1	Training Tolerance: 0.01	
Parameters	**Conventional BP**	**Proposed BP**
Mean Epoch	15746.833008	4485.919922
Max Epoch	18061.000000	4923.000000
Min Epoch	14161.000000	4019.000000
SD	980.503906	204.222519

7 References

Davis, J. & Shah, M., (1994). Visual Gesture Recognition", IEEE Proceedings - Visual Signal Process, Vol.141, No.2, pp 101-106.

Duric, Z., Li, F., and Wechsler, H., (2002). Recognition of Arm Movements, Proceedings of the Fifth International Conf on Automatic Face Recognition.

Eaton, H. A. C. & Olivier, T. L., (1992). Learning Coefficient Dependence on Training Set Size, Neural Networks, vol.5, pp 283- 288.

Hyeon-Ryulee & Rim, (1999). HMM Based Threshold Model Approach for Gesture Recognition, IEEE on Pattern Analysis and Machine Intelligence, Vol. 21, No.10, pp 961-973.

Jacobs, R. A., (1988). Increased Rates of Convergence through Learning Rate Adaptation, Neural Networks, vol.1, pp 295-307.

Mishra, N., Prasanna, T.V., Singh, M. P., Birla, B. K. & Lal, A., (1996). Gestures as Symbolic Communication, Proceedings of the 31st Annual Convention of the Computer Society of India, Bangalore, pp 22-29.

Ooyen, A.V., & Nienhuies, B., (1992). Improving the Convergence of the Back Propagation Algorithm, Neural Networks, Vol.5, pp 465-471.

Rigler, A. K., Irvine, J.M. & Vogl, T.P., (1991). Rescaling of Variables in Back Propagation Learning, Neural Networks, vol.4, pp 225-229.

Sharma, R., Zeiler & Schulten, (2000). Speech/Gesture Inteface to a Visual-Computing Environment, IEEE Computer Graphics and Applications, pp 29-37.

Tollenaere, T., (1990). SuperSAB: Fast Adaptive Back Propagation with Good Scaling Properties ", Neural Networks, vol.3, pp 561-573.

Weir, M. K, (1991). A Method for Self-Determination of Adaptive Learning Rates in Back Propagation , Neural Networks, vol.4, pp 371-379.

Werbos, P. J., (1990). Back Propagation through Time: What it does and how to do it, Proceedings of IEEE, Vol 78, No.10, pp 1550-1567.

Novice Drivers Training in ADAS HMI - The TRAINER Results

Panou M.[1a], *Bekiaris E.*[1b], *Dolls J.*[2], *Knoll C.*[3], *Falkmer T.*[4]

[1]Aristotle University of Thessaloniki, Greece,
[a]mpanou@certh.gr, [b]abek@certh.gr
[2]Polytechnic University of Valencia, Spain, jdols@mcm.upv.es
[3]Fraunhofer IAO, Germany, christian.knoll@iao.fhg.de
[4]Swedish National Road & Transport Research Institute, Sweden,
torbjorn.falkmer@vti.se

Abstract

TRAINER (GRD1-1999-10024) is an EU co-funded project with a 3-year duration, expected to end on March 2003. The project aims to effectively reduce the traffic accidents of novice drivers through a series of initiatives, leading to a new, improved and yet cost-effective pan-European driver training methodology. TRAINER developed innovative and yet cost-efficient tools, in order not to substitute, but to support the driver's training procedure. A multimedia training tool-MMT (software) and a modular driving simulator have been developed and appropriate curricula for their integration into the training system of 4 countries (namely Belgium, Greece, Spain and Sweden) have been issued. This paper is focused on the MMT part that is related to scenarios on ADAS use. Specifically, the ADAS scenarios cover the areas o of Navigation, Cruise Control and Adaptive Cruise Control systems. Moreover, assessment results of the MMT ADAS scenarios are included, tested by users in 4 European countries (Greece, Sweden, Spain and Belgium).

1. Introduction

While many attempts have been made for the development of relevant theory training tools (videotapes, software, etc.), these are based on generic issues, consisting of random scenarios, not reflecting the actual trainees needs. TRAINER project produced 31 scenarios that are based on the actual needs of the novice driver, concentrating on their frequent errors and accidents. Within them, a key identified need is related to training and information provision on the use, risks and limitation of Advanced Driver Assistance Systems to the drivers, before getting on the road with such systems on their cars. Such training seems to be necessary also for liability reasons related to ADAS implementation (ADVISORS project). In Deliverable 3.2 of ADVISORS project (GRD1-1999-10047) it is proven that there is a need for training and information provision on the use, risks and limitation of Advanced Driver Assistance Systems to the drivers, before getting on the road with such systems on their cars.

2. TRAINER Tools

Both a multimedia training tool-MMT (software) and a modular driving simulator have been developed and appropriate curricula for their integration into the training system of 4 countries (namely Belgium, Greece, Spain and Sweden) have been issued. The tools are available in 8 languages: English, Greek, German, Italian, Spanish, Dutch, French and Swedish. More specifically, the tools developed within TRAINER are listed below:

- Development of a new interactive multimedia training tool. Development of a low cost and enhanced reliability stationary driving simulator.Development of a mean cost and high performance semi-dynamic driving simulator.Design and development of a normative driver behaviour database. There are approximately around 100 scenarios built, that have been scientifically based upon identified frequent and severe problems of novice drivers. These cover all areas of driver training, such as basic control and manoeuvring, hazard identification and perception, visual cues, use of new driver assistance systems (ADAS), driver behaviour under adverse conditions, or influence of alcohol, fatigue, etc., overconfidence to systems such as ABS, etc. (TRAINER Del. 2.1, 2001).

This paper is focused on the MMT that consists of scenarios related to ADAS. Many attempts have been made for the development of relevant theory training tools (videotapes, software, etc.) (Groot at al., 2001). These are based on generic issues, consisting of random scenarios, not reflecting the actual trainees needs. The material covered within the MMT has been selected from experts, according to the main gaps and problems of the trainees as well as the current training material provided by each country. Furthermore, the contents of the MMT are based on the 'GADGET matrix' (Hatakka et al. 1999).[1]

The scenarios included in the multimedia tool address both the training of novice drivers and the re-training of experienced drivers to new in-vehicle aids. They are basically grouped in two categories: Scenarios to support strategic tasks, including the use of new in-vehicle aids (indicatively):

- Economic driving.
- Use of ABS and in-vehicle aids (ADAS).
- Cruise control.
- Visual field and perspectives of drunk drivers.
- Use of safety belt.

- Scenarios to support tactical/manoeuvring tasks (only indicative):
  - Distance keeping.
  - Hazard perception.
  - Interaction with vulnerable road users.
  - Interaction with animals and other moving obstacles. Initially, the user has to select the language he/she wishes and then to login to the software. After that the main course selection screen enables the users to select the desired course/session they wish to be trained about. The sessions are divided into 4 blocks, referring to the categories shown in Figure **2**.

Each session/course follows the same structure, namely:

- Description of the problem that needs to be addressed.
- Description of the aim of the scenario.
- Theoretical issues (presented using various means)User evaluation (Exercise/test)The information included within the MMT, is presented with various multimedia means, such as videos, animations, interactive videos/animations, photos, interactive photos, interactive table, voice messages, sounds.

The MMT has been designed to be used as a complimentary training tool to the actual training defined by the law of each country. This tool is not meant for self-learning and surely does not, by any means, intend to replace the current theoretical training in various countries. The software

---

[1]According the Gadget matrix the driving task involves also decisional and motivational aspects (i.e. strategic and behavioural) as theseplay an important role in the involvement of novice drivers in accidents. Novice drivers can have superior manoeuvring skills and still have many crashes. Teaching scanning and anticipating, as well as self-evaluation skills, appear to be promising ways to reduce accident rates of novice drivers (Hatakka et al. 1999).

will be installed in a driving school classroom, and will be used by the trainees under the supervision of a driving instructor.

Figure 1: Language selection screen.

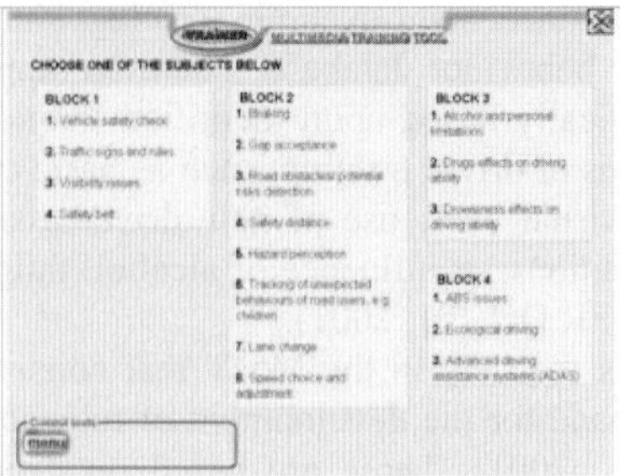

Figure 2: Course selection screen.

3. ADAS scenarios in the MMT

The ADAS scenarios included in the MMT, cover the areas of Navigation, Cruise Control and Adaptive Cruise Control systems. The basic functionalities are presented to the users, with the help of multimedia material, i.e. videos displaying route guidance (as part of the navigation system), pictures of User Interfaces of ADA systems placed on the control panel of cars, and voice messages (Figure 3). After the description of the systems characteristics three tests follow, with seven multiple-choice questions in each and time limitation of one minute (Figure 3). A feedback message informs the user on his/her test performance for each test, when the user presses the button 'Submit'. Finally, explanations are shown for each question, justifying why the correct answer is as indicated.

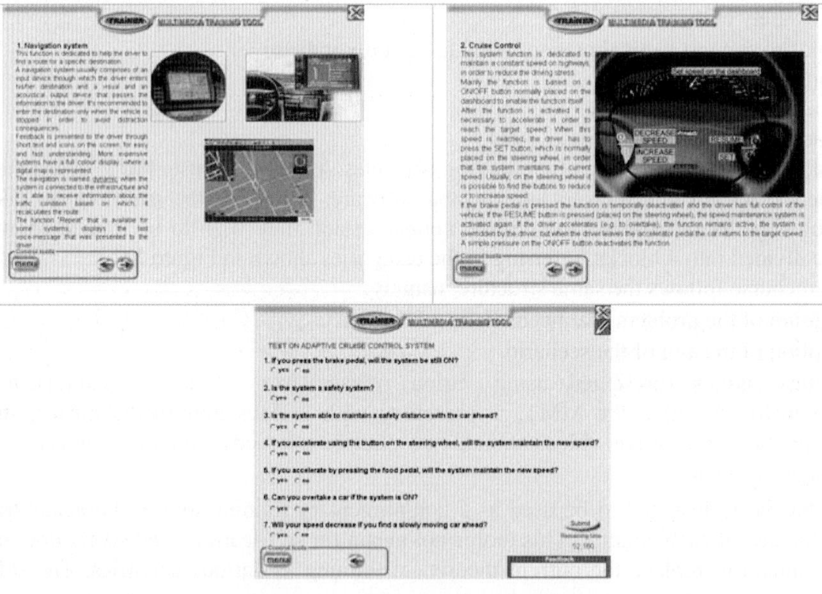

Figure 3: Snapshots of ADAS scenarios (TRAINER Del. 3.1, 2002).

There is a scoring system based on criteria that a trainee has to fulfil in order to pass the assessment provided in the software (TRAINER Del. 5.1, 2002). The educational objectives of each individual scenario differ in general, according to the training curriculum. These variations have been considered in the final scoring system. The specific relevance of a given scenario is expressed as a weight assigned to the scenario in question. The total sum of the scenarios' weights indicates the highest possible score attainable during the MMT test mode. The importance of individual MMT scenarios was judged among members of the TRAINER consortium in order to get a combined mean weight for each scenario. Experts representing different geographical areas (Northern, Central, Southern Europe, etc.) and with different professional backgrounds (traffic safety, driving instruction, driver performance research, etc.) judged the importance of individual scenarios in order to get a combined mean weight for each one. Weights were given for each scenario, with a smaller weight indicating a less important scenario and vice versa (the highest weight given is 7.3 points). The weight of the ADAS session is 2.33 points. This is reasonable, since although this issue is important to be included in the training course, it is not however of the highest priority, compared to other subjects (e.g. speed adjustment, braking on curves, etc.).

4. Tests results

The overall evaluation of TRAINER training tools is carried out in 4 European countries (Greece, Sweden, Spain and Belgium). The evaluation is comprised of two parts: a pre-test evaluation and a final evaluation. As the final valuation is expected to finish by end of March, in this document the pre-evaluation results are presented.
The total number of users that participated in the pre-evaluation was 51 (see Table 1). The average age of the users was 27.8 years, average experience in computer use was 6.5 years and the subjects were half male and half female.

Table 1 Number and type of users that participated in the pre-evaluation of the MMT.

	Germany	**Greece**	**Sweden**	**Spain**	**Total**
Learner	5	10	1	5	21
Novice	5	0	4	5	14
Expert	2	2	9	3	16
Total	12	12	14	13	**51**

Figure 4 highlights the users opinion (expert, novice and learner drivers) relevant to their interaction with the ADAS specific scenarios (TRAINER Del. 3.2, 2002). The ratings are based on a seven-point scale from –3 to +3 (with –3 indicating: not useful, confusing, not at all, etc., and +3 indicating: very useful, clear, absolutely, etc.). The figure shows a general positive feeling of the users with the ADAS training session. The most negative-like results concern the sound, the graphical layout and the colours (not below zero), while the mot positive-like answers have been given for the buttons and control menu on the screen, the text font used, easiness of operation and the level of difficulty of the exercises.

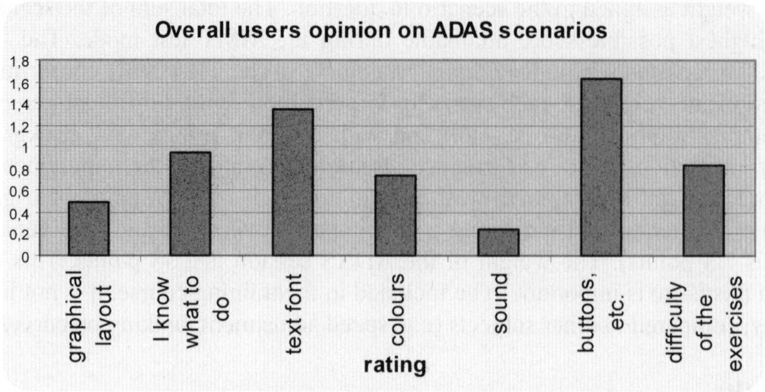

Figure 4: Users opinion towards the ADAS scenarios HMI of the TRAINER MMT.

5. Conclusions

ADA systems are becoming more and more standard features of the vehicle, as it is the case with all new car equipment that has reached the market (e.g. electric lockers, electric windows, airbags, air-conditioning, etc.). This implies a strong need for inclusion of ADAS concepts in the training material of all the countries. The first step that has been realized within TRAINER must be enhanced and optimized by means of theoretical and practical knowledge required. Therefore, the simulators developed for driving training should also be equipped with technical characteristics and software (scenarios with relevant graphical representations) of ADA systems. This applies also to the TRAINER simulators, as an action for future improvement and optimisation.

The pre-pilots results show an overall positive rating regarding the HMI of ADAS scenarios, with some points that should be taken into account for future optimization. However, with the completion of the final pilots conduction, not only more detailed results will be derived, but outcomes will be extracted about the differences of the users relation to ADAS, based on their cultural differences (since the pilots are taking place in 4 different countries covering northern, southern, and central parts of Europe).

References

ADVISORS (GRD1-1999-10047) Deliverable 3.2: 'Framework for Insurance and related Liability issues, gaps and barriers for the implementation and expected Organisational changes', July 2002.

Groot, H.A.M., Vandenberghe, D., Van Aerschot, G., Bekiaris, E. (2001). Survey of existing training methodologies and driving instructor' needs.

Hatakka, M., Keskinen, E., Gregersen, N. P., Glad, A, Hernetkoski, K. (1999), Results of EU-project GADGET, Work Package 3, In. S. Siegrist (ed.): Driver Training, Testing and Licensing – towards theory based management of young drivers' injury risk in road traffic. BFU-report 40, Berne.

TRAINER (GRD1-1999-10024) Deliverable 2.1, 'Inventory of driver training needs and major gaps in the relevant training procedures', January 2001.

TRAINER (GRD1-1999-10024) Deliverable 3.1, 'Interactive Multimedia Tool', September 2002.

TRAINER (GRD1-1999-10024) Deliverable 3.2, 'User Interface of Interactive Multimedia tool', December 2002.

TRAINER (GRD1-1999-10024) Deliverable 5.1, 'TRAINER assessment criteria and methodology', January 2002.

Wireless Input Devices and Their Communication Module for Wearable Computers

Kwang Hyun Park *Jae Wook Jeon*

School of Information and Communication Engineering
Sungkyunkwan University
300 Chunchun-Dong, Jangan-Gu, Suwon, Kyungki-Do, 440-746 Korea
shaia@ece.skku.ac.kr jwjeon@yurim.skku.ac.kr

Abstract

Most input devices for wearable computers have been connected to the CPU board through wires and it may make some problems. Existing wireless devices to get rid of these problems are IR (infra red) and Bluetooth input devices. But these devices have some weak points of restricted operational space and cost, respectively.

In order to have low cost as well as good performance, wireless input devices and their communication module using RF (radio frequency) modules are proposed in this paper. The communication module is connected to the CPU board through the USB port. In order to send some data to the CPU board, each wireless input device can communicate with this communication module through its RF module. Since the frequency range of each wireless input device is the same, the identification (ID) code of each device is assigned and the communication module can distinguish each device by the ID code.

1 Introduction

Recently, many researches about wearable computers have been performed because they can be used to aid inspection, medical, navigation, communication, and cognitive tasks (Brugge & Bennington, 1996. Siegel & Bauer, 1997. Abowd et. al., 1997). Workers can perform more tasks using wearable computers (Greene & Fiske, 1999. Sawhney & Schmandt, 1997. Ross, 2001). Since most input and output devices in wearable computers are connected to the CPU board through wires, these wires do not allow us to put each input device arbitrary location away from the CPU board in a wearable computer. As more input devices are added, the CPU board becomes larger in order to have their corresponding connectors. Furthermore, a wearable computer should be turned off when some input device is being connected or disconnected to the CPU board. Existing wireless input devices to avoid these problems are IR (infra red) and Bluetooth input devices. But, an IR input device has short operational distance and limited direction. Though the performance of a Bluetooth input device is much better than that of an IR input device, the Bluetooth input device is hard to implement and its cost is very high compared with that of an IR input device. Thus, low cost wireless input devices that also have good performance are needed.

In this paper, wireless input devices and their communication module using RF (radio frequency) modules are proposed and implemented for this purpose. In section 2, the proposed devices and communication module are described. Also their communication scheme is described. In section 3, wireless mouse, wireless keyboard, and one communication module are implemented. Section 4 describes the conclusion.

2 Wireless Input Devices and Their Communication Module

Figure 1 shows the proposed wireless input devices and their communication module. The communication module is connected to the main computer through the USB and it can exchange information with wireless input devices by radio frequency (RF) signals ("Universal Serial" 1998). Since each input device uses the same frequency band, it has its own identification (ID) code to avoid the frequency interference. In order to initiate the communication, the communication module sends some information containing one ID code as in Figure 2. If the input device of that ID code responds to send some data, the communication module receives the data and stores them. If there is no or invalid response during the predefined time, the communication module assumes that the called device is off and sends other information containing another ID code. The communication protocol consists of start, ID, data, and end fields as in Figure 3 and Table 1. When the CPU in the main computer asks the data about input devices to the communication module by polling, the data are sent from the communication module to the CPU through the USB. From these data, the CPU can get the information of each input device.

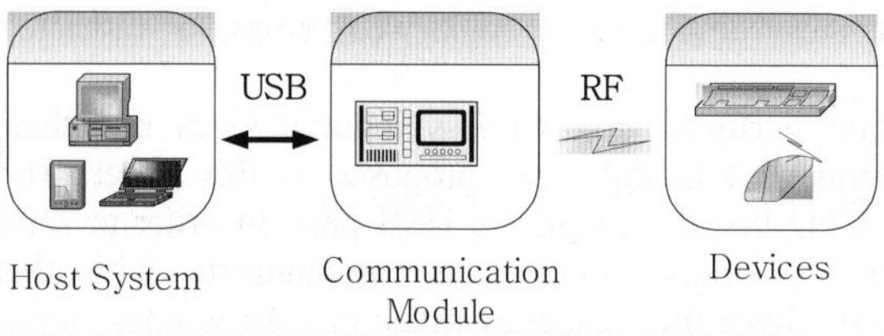

(a): System configuration

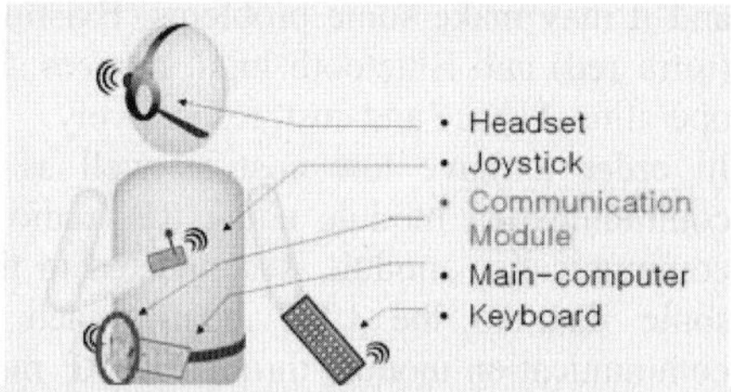

(b) Application in a wearable computer

Figure 1: The proposed system

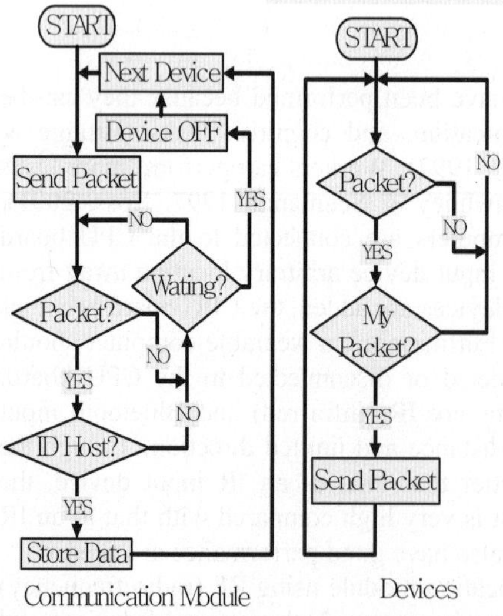

Figure 2: Flow-chart of communication

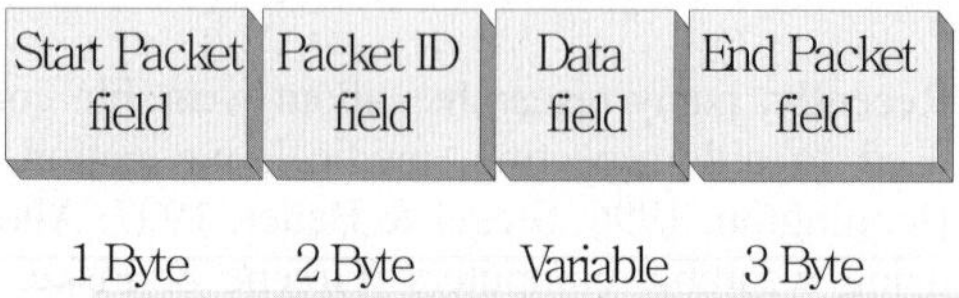

Figure 3: Protocol Packet

Table 1: Packet Field

Offset	Field	Size	Description
1	Start Packet	1	Packet Direction
2	Packet ID	2	ID to receive part (Device or Host)
[*]	Data	Variable	First byte is Data length
[*]+3	End Packet	3	Announcement Packet End

3 Implementation

3.1 Communication module

As in Figure 4, an ATMEL 8-bit RISC controller ATmega163L, a Phillips Semiconductor USB controller PDIUSBD12, and a RadioMetrix RF module BIM-433 are used to implement a communication module ("ATmega163L" 2001, "PDIUSBD12" 1999, "BIM433" 2001). The USB bus can supply all necessary power to drive this communication module.

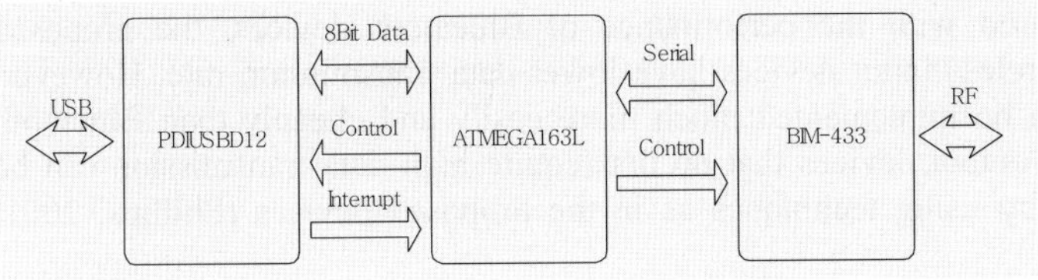

(a): Block diagram

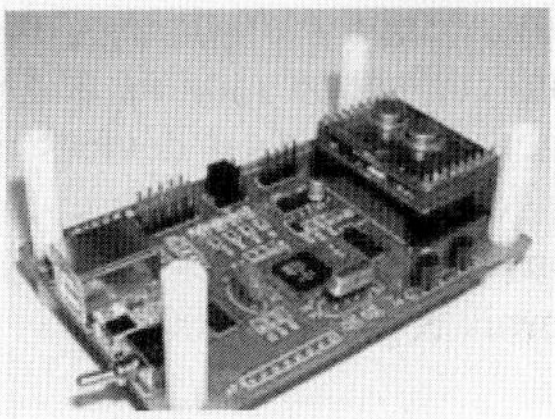

(b): A prototype

Figure 4: Communication module

3.2 Wireless mouse and keyboard

A desktop ball mouse, an 8-bit RISC controller ATmega103, and a RF module BIM-433 are used to implement a wireless mouse as in Figure 5 ("ATmega103L" 2001). Since the motion of ball is represented as an analogue signal by photo-interrupters and slits, this analogue signal should be converted to a digital signal. Analysing this digital signal, the controller can detect the ball mouse motion and its information is sent to the communication module through the RF module.
As in Figure 5, a desktop keyboard, an 8-bit RISC controller Atmega103, and BIM-433 are used to implement a wireless keyboard. Since the keyboard is formed as an *N x M* matrix, the row and column lines are connected to the input and output ports of the RISC controller, respectively. The RISC controller can determine the information of pressed keys by scanning and send this information to the communication module through the RF module.

(a): Wireless Mouse

(b): Wireless keyboard

Figure 5: Wireless devices

3.3 Discussion

Due to the antenna optimisation problem, the maximum rate that one 8-bit RISC controller can transmit to one RF module BIM-433 without any error is 19,200 bps. In order to communicate between two RF modules, each RF module should switch once between transmitting and receiving modes and the stabilization time for switching is needed. The switching time is 3 msec. Thus, the maximum rate that two RF modules can communicate is a little bit less than 19,200 bps. The protocol as in Figure 3 can be changed to accommodate various wireless devices and to improve the reliability of data. Compared with the performance of Bluetooth devices, the proposed communication module and wireless input devices have lower data transmission rate. However, since the proposed devices can be implemented much more easily and cheaply than Bluetooth devices, some wired input and output devices that do not require high data transmission can be converted to wireless devices by using techniques as in the proposed devices (Phillips, 2001. Karnik & Kumar, 2000).

4 Conclusion

Wireless input devices and one communication module for wearable computers are proposed and implemented. Since human cannot send information very fast, most input devices do not require high data transmission rate and they can be easily converted to wireless ones as in this paper. Some output devices of low data transmission can be also converted to wireless ones. Since the communication module is connected the CPU in the main computer through the USB, any wearable computer having the USB host function can use conveniently the proposed wireless input devices.

5 Reference

Abowd, G. D., Atkeson, C. G., Hong, J., Long, S., Kooper, R., & Pinkerton, M. (1997). Cyberguide: A mobile context-aware tour guide. *Wireless Networks*. 3(5), 421-433.

Brugge, B. & Bennington, B. (1996) Application of wireless research to real industrial problems: Applications of mobile computing and communication. *IEEE Personal Communications*. 3(1), 64-71.

Greene, S. R. & Fiske, C. F. (1999). Wearable computers Open A New Era In Support Resource Management. *IEEE Aerospace and Electronics systems Magazine* , 14(11), 33-35.

Karnik, A. & Kumar, A. (2000). Performance analysis of the Bluetooth physical layer. *IEEE International Conference on 2000, personal wireless communications*, 70 -74

Phillips, M. (2001). Reducing the cost of Bluetooth systems. *Electronics & Communication Engineering Journal*, 13(5), 204 –208

Ross, D. A. (2001). Implementing Assistive Technology on Wearable computers. *IEEE Intelligent systems*, 16, 47-53.

Sawhney, M. & Schmandt, C. (1997). Nomadic radio: a spatialized audio environment for wearable computing. *Wearable computers, 1997. Digest of Papers., First International Symposium on , 1997.* 171 -172.

Siegel, J. & Bauer, M. (1997). A field usability evaluation of a wearable system, *Proc. of the The First Int. Sym. On Wearable Computers*, 18-22.

ATmega103L Specification Data sheet. Atmel Co., LTD. Retrieved Jan, 2001, from http://www.atmel.com./

ATmega163L Specification Data sheet. Atmel Co., LTD. Retrieved Jan, 2001, from http://www.atmel.com./

BIM433 Specification Data sheet. Radiometrix Co., LTD. Retrived Jan, 2001, from http://www.radiometrix.com./

PDIUSBD12 Specification Data Sheet. Philips Semiconductors Co., LTD. Retrieved Jan, 1999, from http://www.semiconductors.philips.com/.

Universal Serial Bus Specification Revision 1.1. USB Implementers Forum. Retrieved Sept, 1998, from http://www.usb.org/

Development of Ergonomic Mock-Ups for Usability Testing of In-Vehicle Communicating Systems

Annie Pauzie

French National Institute for Transport and Safety Research
Laboratory of Ergonomics and Cognitive Sciences applied to Transport
INRETS / LESCOT
25 avenue François Mitterrand, case 24, 69675, Bron, France
pauzie@inrets.fr

Abstract

Implemented in-vehicle communicating systems might be a great opportunity to improve road safety. Nevertheless, ergonomics investigation has to ensure that this new situation is not going to disturb the driver. In this context, the development of ergonomic mock ups is a great tool to investigate scenarios and information displays. This approach allows to test modalities of implementation in a concrete and integrated way. It is a way to investigate safety, acceptability and usability of a in-vehicle system without needs of the real database and the real prototype. This paper described the development of various versions of a mock-up displaying alert messages to be tested in real road condition.

1 Context

The vehicle is on the way to change deeply, due to the implementation of telematics technology, allowing to establish bi-directional communication between the driver and the external world. Because of these technologies, the driver is less isolated inside the vehicle and can be contacted to be informed about any kind of immediate or middle term events related to his route.

Obviously, this interactive communication between the driver and the external world can be a relevant assistance for the driver in terms of anticipation, orientation processes and decision making, with positive consequences in terms of road safety.

Nevertheless, the interference due to the display of untimely auditory or complex visual messages with the driving task has to be deeply investigated beforehand, to avoid making the new driving context worse than the original one, rather than better.

The designer is working under time constraint, and has to resolve an important amount of crucial technical problems, in addition to the ergonomics specifications. Furthermore, the designer is aware of the responsibility he has concerning the system liability and safety in the context of the vehicle development. So, according to these constraints, supporting ergonomics data concerning the interface and the dialogue features, set up to make the process easier and more efficient, are usually welcome by designers.

In this context, the approach conducted has been guided by a principle in ergonomics corresponding to the user centred approach, where it is outlined that systems have to match with the logic and the abilities of users, in order to optimise the level of system acceptability and usability. This research aimed at setting up a set of ergonomic guidelines devoted to the system designer use, in order to aid this activity from the ergonomic side, with ergonomic guidelines

defined as "recommendations and advice given to the designers in a form that can be directly applied to the product".
This ergonomics activity is developed in the framework of a French project named "ARCOS". In this project gathering car manufacturers, suppliers and research institutes, it has been planned to develop several functions closely linked to road safety: prevention of secondary accident by means of vehicle-vehicle communication, prevention of dangerous headway (traffic, speed, weather broadcast), prevention of collision with obstacles and prevention of lane departure.
The human factors activity is run in parallel to the work in terms of technical feasibility and reliability. A specific focus has been devoted to the function aiming at "alerting the driver", that is to say informing him/her about a more or less critical event located ahead. The innovative aspect of this function relies in the way this information is activated: manually, by an other vehicle close to the event, and/or automatically, by sensors implemented in vehicle that emitted messages after being crashed in case of an accident happening. The other innovative characteristic concerns the large number of criteria possible to define each alert message: location, type of event, number of vehicles to reach, distance to cover, and so on…

2 Driver's perceptive and cognitive processes

The driver's actions depended upon the type of perceived information and the way this information is cognitively processed. In the road environment, the information can be perceived by the driver without active search ("bottom up" process), or the information can be actively searched by the driver, strategy based upon previous experience of the situation leading to representations of specific visual search strategies ("top down" process) (Neboit, 1980).
Globally, the following steps can be identified:

- perceptive search: set of procedures developed by the driver in order to collect clues necessary to run the driving task
- identification: recognition of clues, that is to say linkage with a category of events. In relation to the level of practice of the task, the number of clues decreases, with a selection of the more relevant clues, in addition to an increase of the number of categories, allowing a better accuracy.
- prediction: anticipation of future events and potential actions based upon perceived clues and use of rules allowing to set up representation of the future state of the system.
- decision: selection and choices associated to prediction mechanisms based upon hypothesis of the driver and allowing to generate the action.

The alert and alarm messages have to be conceived based upon this perceptivo-cognitive mechanism. Then, in case of emergency and critical situation, the signal has to be perceived without requiring active search from the driver. The content, the modality and the timing have to be set up in order to avoid any ambiguity in its identification and meaning, in the prediction in relation to the type of event and in the type of action to develop.

3 Criteria for system evaluation

Design and evaluation processes alternate when developing a system. In the framework of the HMI development of in-vehicle systems, the evaluation phases in realistic environment (driving simulator, or even better, in real road conditions) are crucial to validate the functional characteristics of the system. The tested criteria are: safety, efficiency, usability, acceptability, ease to learn.

The classic methodologies used in this framework consists in driver's behaviour analysis and evaluation of the mental workload while using the system in comparison with a reference situation. Some of the important parameters are visual demand of the implemented screen according to the type of message in various road contexts, understanding and investigation of the planned action the driver intend to do.

The iterative process between design and evaluation requires to have a system interface easy to modify after implementation of improvements according to the first results following a road testing. A real system prototype using important data base and sophisticated on-board technology has several disadvantages: not easily modified, not always available for testing and usually ready at the last stage of the project. In order to overcome all these problems, we propose to develop a mock-up of the HMI which simulated the functions and the interface and interaction modes of the system in real road use. The advantages are to create realistic context of system use in order to collect data about ergonomics before the real achievement of the system prototype, and to make available design recommendations to the developers in advance.

4 Development of a mock-up for alert function

In the case of alert function, if the driver's need is clearly identified, the modalities of information communication still need to be defined: timing in relation to the event, type of message in terms of content and form, type of event relevant to communicate to the driver (highly critical or basic warning), frequency of alert... in order to ensure the maximum of efficiency of the alert.
In order to conduct this approach, several ergonomic mock ups of systems have been developed ; these mock-ups are interactive simulation of in-vehicle communicating systems HMI, presenting the various functions and interfaces that would be available in the final product.

The objective of the mock ups was then :

- to identify and to implement relevant guidelines in the context of HCI for in-vehicle communicating system (alert system, cruise control system, line keeping system, navigation and guidance system)
- to test these identified guidelines in an integrated way during the process of developing various scenarios of ergonomic mock ups
- to validate these guidelines by tests of usability among drivers
- to find new solutions and to set up new HCI design guidelines to answer to drivers requirements.

Development has been made on a portable computer, lately implemented on the dashboard of the experimental vehicle to be tested.

4.1 Examples of mock-ups for alert

The interface has been defined according to the existing design recommendations: strong contrast, use of normalised pictogram, combination of visual and auditory information (Pauzié & Letisserand, 1992, Pauzié & Forzy, 1996, Report FHWA RD-98-057, 1998). The novelty of this approach has been to define non ambiguous messages according to their origin: emitted automatically from sensors after a crash (strong alarm display, figure 1) or emitted manually by a driver witness of a problem on the road (light alert display, figure 2).

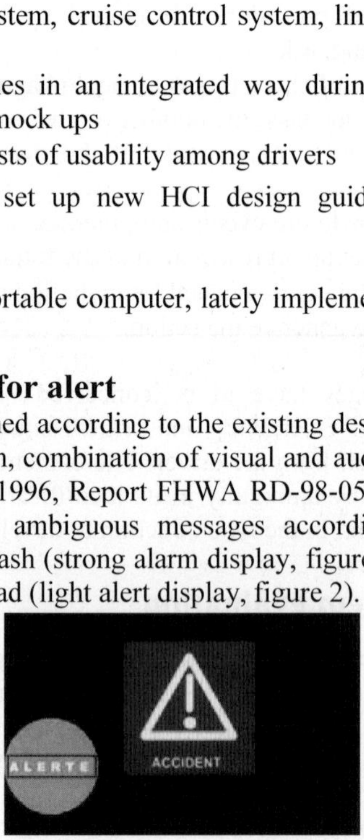

Figure 1: Strong Alarm Display, the pictogram is flashing combined with an intrusive sound

Figure 2: Light Alert Display, textual information about the nature of the event, the accompanying sound is non intrusive.

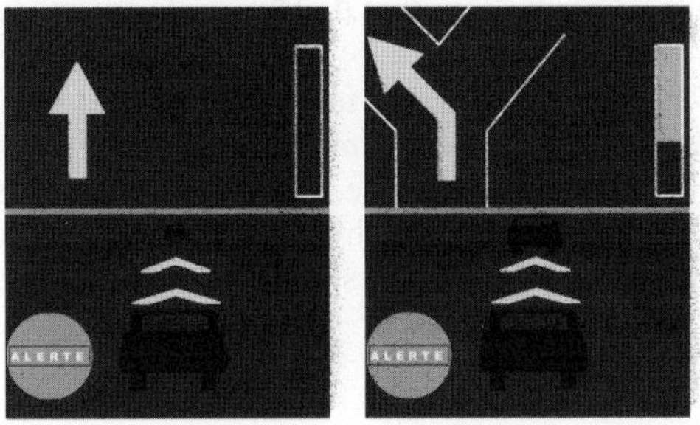

Figure 3: This mode of display allows to integrate other assistance function such as "adaptative cruise control" but does not allow to integrate guidance function which has to be displayed on a separate screen.

4.2 Examples of mock-ups for alert and guidance

In addition to question related to the design of the alert message by itself, the second question was dealing with the modalities of integration of this emergency information in relation to the guidance function. Indeed, due to the widespread of this function in the coming years, and in order to anticipate from an ergonomic point of view about integration, the developed mock-up in this project covered the alternative: alert as isolated function or alert and guidance as combined functions.

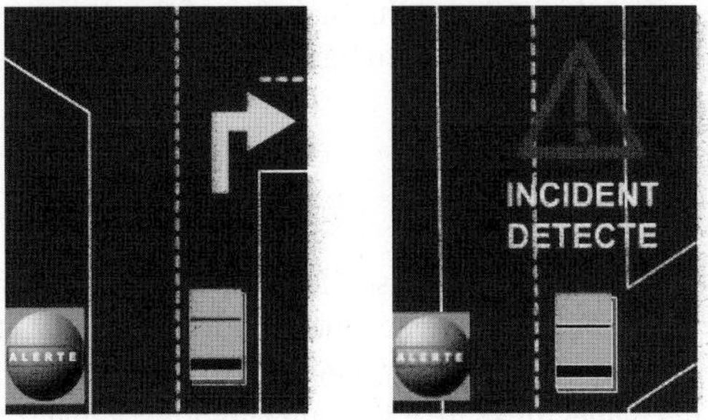

Figure 4: The necessity to integrate both alert and guidance leads to modify totally the visual interface (previous schema).

Figure 5: An intrusive way to inform the driver has to be found when the alert message revealed high critical situation (red flashing background and strong noise).

These two types of mock-ups are going to be tested among drivers in real road conditions, a researcher sited in the back of the experimental car managing the successive dynamic displays according to scenarios defined in advance. The video recordings of driver's behaviour will be analysed afterwards in laboratory frame by frame, in addition to questionnaires, to investigate acceptability and usability of these first versions.

5 Conclusion

This approach allows to test the validity of the design and the functions in a practical and in a flexible way. Indeed, an optimised system design is often a compromise between several choices, and some recommendations might conflict one with the other. Furthermore, the mock-ups development allows to test various ergonomic options among experts, and can be quickly adapted according to the current data coming from comments and feedback. Finally, it allows the validation of the final acceptability and usability of the proposed versions among drivers.
The numerous steps of iterative processes in Design and in Evaluation are time and cost consuming. The use of the ergonomic mock ups makes this stage lighter, as it allows to test many aspects of the product such as interface, interaction, dialogue and available functions, without too much effort, as there is no necessity to establish access to real database whenever there is a modification to implement. So, this effort at the first step of the design concept allows to gain time and cost afterward, as it brings the insurance to have communicating systems well accepted by the drivers' population.

References

Neboit M., 1980, L'exploration visuelle dans l'apprentissage de tâches complexes : l'exemple de la conduite automobile, Thèse de Doctorat, Paris, Université René Descartes.

Pauzié A. & Letisserand D., 1992, Ergonomics of MMI in aid driving systems : approach focusing on elderly visual capacities, in « Gerontechnology », Herman Bouma & Jan A. M. Graafmans (eds.), IOS Press, 329-334.

Pauzié A. & Forzy J-F, 1996, Ergonomic evaluation of guidance and traffic information in the CARMINAT program, III Annual World Congress on Intelligent Transport Systems, Orlando, CDRom.

Report FHWA RD-98-057, Human Factors Design Guidelines for ATIS and CVO, Washington D.C., 1998.

Travel Planning on the Web: A Cross-Cultural Case Study of Where Differences Become Evident Within the Design Process

Sonja Pedell

The University of Melbourne
Department of Information Systems
Victoria 3010, Australia
pedell@acm.org

Helmut Degen

Vodafone Holding GmbH
Global Products & Services
D-40213 Duesseldorf, Germany
helmut.degen@vodafone.com

Kem-Laurin Lubin

Siemens Corporate Research, Inc.
Princeton, NJ 08540,USA
kem-laurin.lubin@scr.siemens.com

Ji Zheng

Siemens Ltd.
China Corporate Technology 7
Beijing 100102, P.R. China
ji.zheng@pek1.siemens.com.cn

Abstract

This case study investigates cultural differences on a ficticious travel website designed in three different cultural contexts (USA, China and Germany), with the same procedure and based on users' input from participants from each country. A set of design factors was suitable to collect requirements in a systematical way as well as to compare the three different websites. Results suggest that cultural differences appear in an early phase of the design process, the requirements gathering phase.

1 Introduction

Recent studies showed that people in different cultures have different requirements and needs regarding the design of user interfaces (Honold, 2000; Prabhu & Harel, 1999; Evers, Kukulska-Hulme & Jones 1999). From an economic perspective, there are two extremes in managing culture oriented design strategies. One is internationalization, by which every country and culture obtains an identical user interface. The other extreme is localization, by which every culture obtains a culture specific user interface. To be economically successful, an appropriate balance of internationalization and localization is necessary.

The key questions in the context of cross-cultural user interface design are: What kind of localization is required for user interfaces that are provided for different cultures? Moreover, resulting from the first question: Where, within the entire design process, should the cross-cultural aspects be taken into consideration?

As a case study, we have chosen a ficticious web-based travel portal for private consumers. The target cultures are those of the USA, Germany, and China. The target user group are "Double Incomes No Kids" (DINKs) living in these countries.

2 Method

The entire design process up to the final design of user interfaces was followed through to investigate the questions proposed above. We focused on use context analysis, requirements gathering, and user interface design according to the usability process of ISO 13407. The processes were kept standardized and were accomplished by local interaction between designers and visual designers. For the use context analysis, questionnaires were used, and for the requirements gathering, the paper prototype procedure NOGAP (Another method for Gathering Design Requirements in E-business User Interface Projects (Degen & Pedell 2003)) was applied. The requirements gathering step bridges the gap between the analysis and the user interface design (Wood 1998). To this extent, it is essential for reasons of efficiency to arrive as close as possible to the desired design of the user interfaces. The sketches contain requirements and design elements. According our experience, the gap between user requirements gathering and user interface design can be bridged very well with the NOGAP method. At the core of this method are six so called "design factors" for the categorization of requirements. These design factors are content, functions, media, wording, layout and linkage.

- Content: Which kind of content do the users expect and how should the content be clustered?
- Function: Which kinds of functions do the users expect? To which content clusters are the functions assigned?
- Media: Which kind of media (e.g. text, sound, picture, movie) is chosen to represent the content or the function?
- Wording: What is an appropriate expression for a function or for a content?
- Layout: How should the single user interface elements be organized and designed on a single webpage.
- Linkage: Which is the expected sequence of single webpages?

During this entire data collection step, the user only works with a sketch form. The user develops and sketches ideas and desirable features (see Figures 1-3) and is guided by the UI designer. The UI designer records the ideas on a requirements form containing the design factors, which he/she also uses as guideline for his/her questions. Based on these ideas and features, the interaction designers created a website concept including the requested content, layout, and "click through". To classify and to compare the differences among the three user interfaces, their characteristics were allocated to the introduced design factors whose quantity and quality allow us to identify where within the design process the cross cultural aspect of user interface design should be taken into consideration.

3 Results

The design factors were used for describing the results of the requirements gathering and the user interface design. As the results show, the design factor "wording" is difficult to compare in different languages. Media is a factor that cannot be investigated by means of single design screens. The final Web design was purely visual. Therefore, we consider only four of the design factors: Content, Function, Layout, Linkage.

3.1 Use Context

The similarities within one country among the participants are extremely high, but some differences among the three countries are evident: While the Americans have most financial freedom for their travel, the German participants have the largest blocks of free time to spend abroad. The Chinese participants spent 15 % of their yearly income on travelling, which was the highest percentage of spent money among those studied. For the Chinese, travelling seems to be a measure of high quality of life.

3.2 Requirements Gathering

Based on the results of the requirements gathering, the main differences between the three countries are related to content and linkage (see Figures 1-3). In terms of content, the Germans and the Chinese participants require comprehensive travel information, whilst the American counterparts do not seem to need this. Also the linkage concept is different between the three countries: Before selecting travel offers, Germans involved in this study expect comprehensive travel information. The Americans prefer a straightforward strategy to travel offerings, without travel information, while the Chinese select first between a "domestic" or "abroad" destination.

3.3 User Interfaces

On the basis of the sketches, which include the requirements of the participants of the respective country for a travel Website, a visual design was created. The designs (see Figures 4-6) clearly illustrate the required contents, functions and basic layout.

3.3.1 Content

The German Website provides lots of travel and country information as well as flight and accommodation information. On the US Website there are no extensive destination pages to browse, but access to concrete flights and hotels, services on site, eg. rental cars, and special travel packages. Other travel options such as cruise and vacation packages are quite salient, but the amount of content is not as high as seen in the German results. The Chinese Website contains a large quantity of information relevant to the target destination. In addition, you can find information which helps the user to decide on a travel destination which can not be found on the German or US travel Website: travel news, links to other travel agencies, Top 10 of most travelled destinations.

3.3.2 Layout and Functions

The overall purpose of the layout of the German Website is to trigger the desire to travel the entire world. The homepage in particular, provides a lot of function-free space to support unrestrained room for the users imagination where to spend the next holiday. The single function filters, content information and offers are clearly distinguished in separate blocks and always stay at the same place while clicking through the site. To support different ways of choosing the travel destination you find different access points (filters of different granularity, the world map, pull-down menus). The US homepage is very compact and gives clear directions of where to enter relevant information. The use of graphics and use of colors is meaningful to support quick entries and to show the relationship and grouping of information in a very structured way.

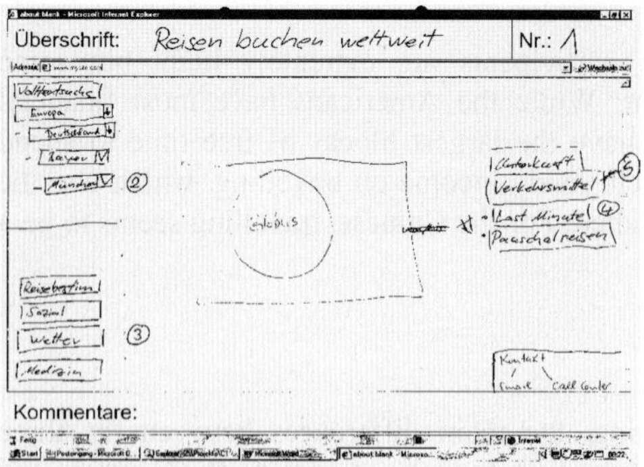

Figure 1: User requirements: Homepage (Germany)

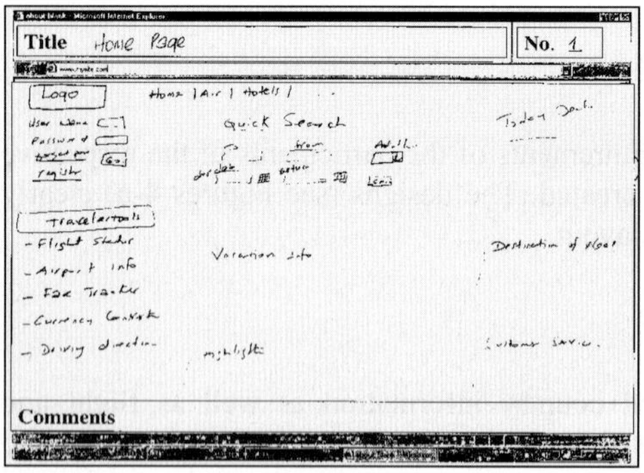

Figure 2: User requirements: Homepage (USA)

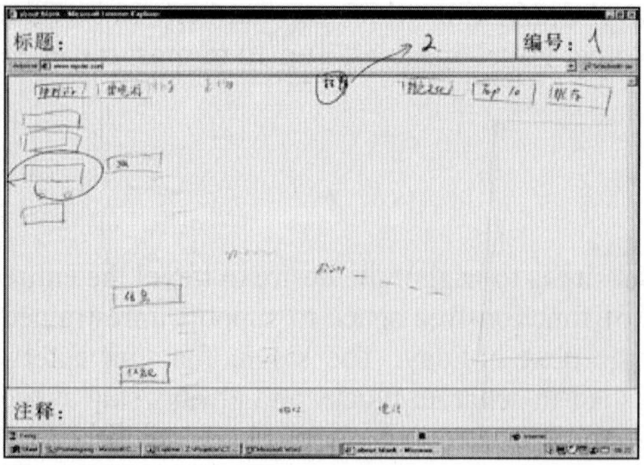

Figure 3: User requirements: Homepage (China)

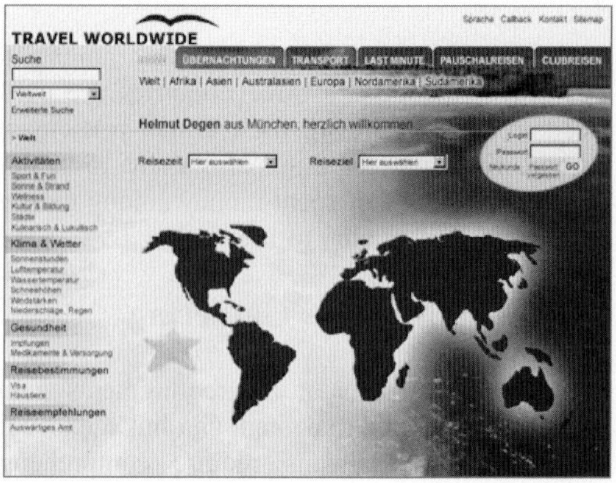

Figure 4: Visual design: Homepage (Germany)

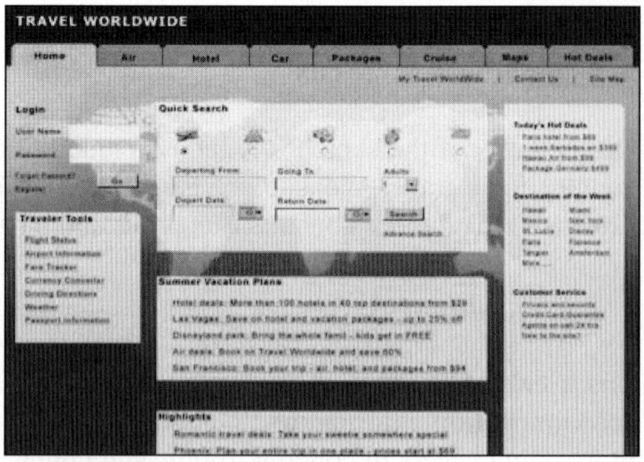

Figure 5: Visual design: Homepage (USA)

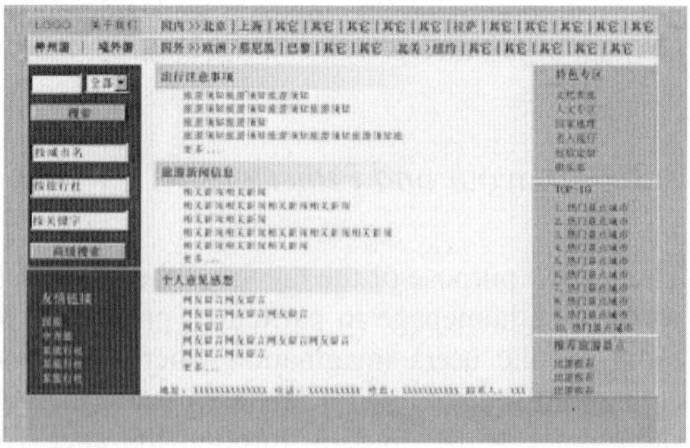

Figure 6: Visual design: Homepage (China)

The contact information is very clearly placed. The Chinese Web Site provides very clear categories of target city information in rows. Regional specific information such as cultural events are highlighted. Photographs of the target information are a central feature of the Chinese Website and take a lot of room.

3.3.3 Linkage

Due to very long vacation periods in Germany, the Germans undertake many trips, always looking for new and exciting travel destinations. Therefore, they browse through the Website to be inspired by information about different travel destinations. Often, Germans merely choose the destination at the Website and book in a travel agency. The Americans have a clear notion where they want to go. Therefore they type in the name of the destination in a Quick Search which leads directly to a booking screen. The procedure is similar to that in a classical internet shop. The intention of Chinese users in visiting the Website is to both browse and book. First, they carry out a regional selection or topic oriented selection, then they focus on the extensive information about the target destination. They book for group travel to the destination and trust that the organization of accommodation and means of transport is dealt with by the provider.

4 Conclusions

The identified differences in the area of travel show a high level of localization: Differences.exist for all investigated design factors in the three culture specific versions. Hence, the concepts for the user interfaces are completely different. According to these results, it is necessary to include user input of different cultures from the very beginning of the design process. Cosmetic changes, eg. changing the colours, carried out late in the process of design, or even after the fact, are insufficient. Managers and user interface designers can see the trend of cultural differences already in an early stage of the design process (after applying the NOGAP method). The final user interface design will incorporate cultural differences established during the requirements gathering phase. Therefore, the effort of a final user interface design to evaluate differences is not necessary.

References

Degen, H. & Pedell, S. (2003). JIET Design Process for E-Business Applications. Appears. In D. Diaper & N. Stanton (eds.), *Handbook of Task Analysis for Human-Computer Interaction.* Unpublished.

Evers, V; Kukulska-Hulme, A. & Jones, A. (1999). Cross-Cultural Understanding of Interface Design: A Cross-Cultural Analysis of Icon recognition. In: Prabhu, G.V.; del Galdo, E. M.(Eds.): *Designing for Global Markets. 1st International Workshop of Internationalization of Products and Systems*, IWIPS 99, May 20-22, Rochester, NY, USA, 1999, S.173-182.

Helfrich, H. (1993). Methodologie kulturvergleichender psychologischer Forschung (pp. 81 – 102). In A. Thomas (ed.), *Kulturvergleichende Psychologie*. Göttingen (u.a.): Hogrefe.

Honold, P. (2000). Interkulturelles Usability Engineering. Düsseldorf: VDI-Verlag.

Wood, Larry E. (ed.) (1998). User Interface Design. Bridging the Gap from User Requirements to Design. Boca Raton, FL: CRC Press.

Warning Strategies Adaptation in a collision avoidance/vision enhancement system

A.Polychronopoulos[1], D.Kempf[2], M.Martinetto[3], A.Amditis[1], H. Widlroither[2], P.C.Cacciabue[3], L.Andreone[4]
[1]Institute of Communications and Computer Systems, Greece
9, Iroon Polytechniou St. 15773, Athens, arisp@mail.ntua.gr
[2]University of Stuttgart, Germany
[3]European Commission – Joint Research Centre, Italy
[4]Centro Ricerche Fiat, Italy

Abstract

In this paper we will focus on the definition of the warning strategies of collision avoidance and vision enhancement systems and their implementation in Human Machine Interaction level. The aim is to achieve the highest balance between a totally supportive and a non-disturbing system. The design follows a user-centered approach starting from a reference cognitive model. A first design of the warning strategies is tested in a static driving simulator; a second design follows, where the HMI is evaluated in a dynamic simulator before the final design and the in-vehicle integration and road tests with subjects. The work described in this paper is part of EUCLIDE[1] project, which aims at developing a driving support system to be helpful in reduced visibility.

1 Cognitive model description

In the design and development of a new support system the first step is to understand the context of use and user behaviour in this context. Therefore, the EUCLIDE project is focused on the driver, as the user interacting with the interface information. The aim is to keep the driver aware and active in the supported driving loop by mean of a trustworthy and friendly human-centred system. In figure 1 it is shown the Reference Model of Cognition (Cacciabue, 1998), including the basic cognitive processes and functions affecting human performance. The focus on human cognition is of primary importance for a support system that produce effective information to the driver for the decision making and the final action.

The external stimuli reach *perception* that is related to the sensory information and affected by subjective expectations. The perceived information is elaborated in the *interpretation* and results in *planning* that is a decision on how to proceed. Finally, the *execution* is the plan implementation, as an action or a new cognitive process. The outcomes of these functions depend also on the *memory/knowledge base* process that works on past experience, knowledge, rules, beliefs etc. and; the *allocation of resources* process that manages the available resources, including knowledge too. This loop describes ideal human behaviours and cognitive performances. However, the information processing mechanism may be interrupted at different levels, according to the human-machine interaction, the socio-technical driving environment, and contextual conditions. The

[1] "Enhanced human machine interface for on vehicle integrated driving support system", funded by the European Commission within the 5th Framework Programme "Competitive and sustainable Growth"

Reference Model of Cognition has to be seen as the starting point in the warning strategies adaptation. In particular, in the case of the EUCLIDE anti-collision system it is essential to give intuitive warnings, so that the time spent by the driver in interpreting warnings and planning emergency actions is reduced to the minimum allowing a quicker reaction to dangers.

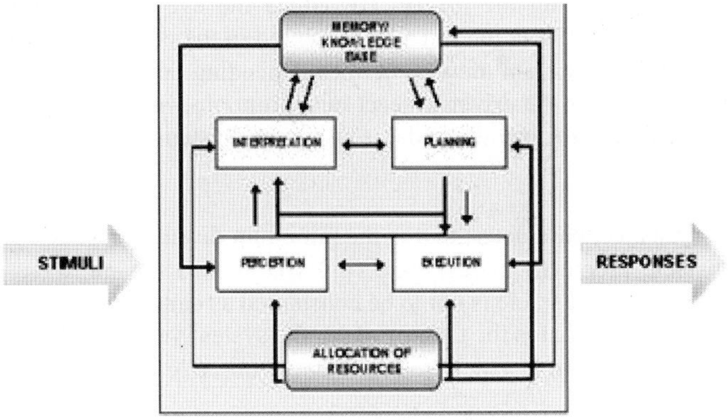

Figure 1: Reference Model of Cognition

2 User-centred approach

According to the state of the art analysis that has been carried out together with the user need analysis there are mainly three types of HMI configurations for vision enhancement systems suitable for the EUCLIDE project: visual head-up displays, visual head-down displays, audio / voice signals and tactile devices. Each of those presents different benefits and drawbacks that are currently under evaluation following the user centred design approach. The methods used in the "User Centred Design" were born during the Eighties (Norman, 1986). They represented a shift in focus from gathering anonymous, statistical data on users (System Centred approaches), to individualize the data through psychological approaches (how users perceive, think, organize and intend during their interaction with technological devices). The adoption of the user centered design ensures that the final product will support the needs, knowledge and skills of the intended users, resulting in an overall improved usability of the system. What is done within the EUCLIDE project is to carry out an iterative design process involving end-users, so that all information collected can be used to modify the system under development. The iterative design is involving the end users at each stage of the development, allowing a fine tuning of all the functional features of the system. Three steps of testing have been defined to allow an iterative user oriented development:

- tests on a static driving simulator using virtual prototypes and a HMI mock-up,
- tests on a driving simulator using virtual prototypes and
- Tests with an integrated solution on demonstrator vehicles in real road scenario.

The different steps of design & test are allowing coming to the most proper definition of the strategies to alert the driver providing acoustical and visual messages combined with the enhanced image of the external scenario: this will represent the first step towards the future scenario of driving supported with synthetic vision combined to direct view.

3 Warning strategies adaptation

The HMI of a collision warning system must immediately indicate a detected dangerous traffic situation and attract the driver's attention. The system must guide the attention of the driver to the dangerous object or road user without distracting the driver. To reduce the eye-off-road to a minimum in the situation of the warning, the information must be presented in the central field of view. The activation of the collision avoidance system must be defined depending on the exposure to a dangerous driving situation and the time for the driver to react in appropriate ways. Based on the data collected with the far infrared camera and the radar sensor, the system detects and calculates the following parameters:

- Object Classification
- Predicted Object Minimum Distance, parameter; that is the minimum distance between a vehicle and a potential obstacle predicted in real time (if PMD=0 then the impact is forecast, if PMD> threshold, then the obstacle is not to be considered as dangerous);
- Time to Predicted Minimum Distance that is the time in which the system vehicle will reach the point of the predicted minimum distances to the potential obstacle.

These parameters are compared with the parameters for collision warning systems mentioned in the literature. The more time a driver has to react and decrease the exposure by changing the driving parameters, the less dangerous the situation is going to be and vice versa. Therefore, the warning of the driver with the collision avoidance system must be activated in varying levels. According to the guidelines for crash avoidance warning devices of the National Highway Traffic Safety Administration (NHTSA), the EUCLIDE system comprises two levels of warnings. The guideline defines the two warning levels as 'imminent crash avoidance warning' and 'cautionary crash avoidance warning'. An imminent crash avoidance situation is one in which the potential for a collision is such that it requires an immediate vehicle control response or modification of a planned response in order to avoid a collision. The warning must be presented at least visually and auditory or visually and tactile. 'A cautionary crash avoidance situation is one in which the potential for a collision requires immediate attention from the driver, and which may require a vehicle manoeuvre, but which does not meet the definition of an imminent crash avoidance situation.' The presentation should also be visual or auditory or tactile but less intrusive than in the imminent case.

4 Driving simulator tests

With the development of collision warning system it is important, to test every functionality after it is determined, in the fields of reliability, effects and potential development. In an early development stage, when a physical prototype didn't exist to run tests with, a virtual prototype was used to evaluate special concepts in a fast and cost-effective way. After this phase it follows the definition of warning systems and major hardware components. On this basis, using the compact driving simulator of IAT, a test about usability and acceptability concerning acoustic and visual warning signals was carried out . Based on the laboratory results the tests series in the full-immersive driving-simulator from Daimler Chrysler in Berlin were planned and defined. These tests will be dedicated for further evaluation on long-term effect of the selected warning-strategies. The tasks of the laboratory tests were the creation of a virtual prototype, the collision warning system EUCLIDE and the implementation of the prototype in the driving-simulator to test the HMI in series of experiments regarding usability and acceptance. The aim was the development of an efficient warning-strategy: an HMI that should help the driver to react as fast as possible to dangerous situations and without the added effect to make him/her nervous or confuse, distract and appeal. The warning strategies should include visual and acoustic warning signals/information for the driver that should be displayed on the head-up (or head-down-displays) and respectively through the speakers for the acoustic signals. To design a test with all possible combination of warning signals it was necessary to accomplish different examinations in the run-up, which should have found out the several parameters for an effective warning-strategy: in particular which combination of the channels of information (visual and acoustical) could be selected as reasonable, effective and less strained. On a head-up-display HMI device the far infrared image of the driving situation was displayed in monochrome. The system warning was presented on the head-up-display. A detected object was tagged with a coloured symbol. The EUCLIDE system with its sensor capabilities is potentially able to distinguish two object categories: humans and vehicles.

The presented symbol was correlated to the category of the detected object. Thereby humans were represented symbolically by a small human figure, cars and trucks by a vehicle and all other non-definable obstacles and objects (i.e. boulder in the street) by a triangle. The size and colour, as well as the position of the warning symbols in relation to the recognized obstacles and objects were tested in the simulator (position of the warning symbol on top of, or beside the recognized obstacles and objects). In the imminent case a classification between different detected obstacles and objects does not exist anymore, only the potential danger was shown in this case. An obstacle or road user that releases the imminent warning requires immediate attention and vehicle control response. The driver was warned by a blinking symbol displayed in the centre of the visual field on the windshield, accompanied by a sound. It was assumed that most obstacles, which cause an imminent warning, are firstly eleasing a cautionary warning. Therefore, these objects were displayed in the head-up-display first, classified by the system and then tagged with a warning symbol. When the brightness was increasing and the threshold of imminent warning was reached, additionally the warning symbol in the windshield was displayed and the warning sound was played. In this case both modes, information and imminent, were active and the content in the head-up-display did not change. Objects and vehicles occurring suddenly and eleasing the imminent mode immediately were also releasing the information mode at the same time. In this case the warnings of information mode and imminent mode were appearing at the same time. As warning symbols a red area, a cross and a warning triangle were selected and tested in the simulator. The used warning symbols should blink for as long as they were shown. The acoustic warning symbols were chosen according to previous studies (Lerner et al. 1996), (Wierville, 1995) with a frequency between 500 and 3000 Hz to attract the attention of the driver. The users classified the different modalities and the results are shown below:

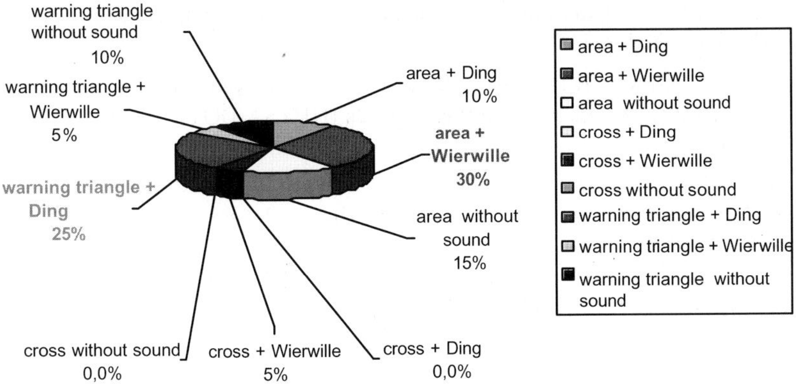

Figure 2: HMI modalities user acceptance

By evaluation of the data from the questionnaire for the head-down-display, only features or content of a warning strategy could be determined, because this test was made statically as mentioned above and the warning strategy only implied visual warning signals. According to the static tests with the head-down-display, for the design of the warning symbols could be made the following preferences:

- 75% of the test persons classified on the basis of a warning triangle as visual warning signal a displayed dangerous situation as potential dangerous
- 45% of the test persons preferred a "big" visual warning signal
- 45% of the test persons would prefer, if the visual signal would cover the obstacle/object
- 47,5% of the test persons preferred the colour red for a visual warning signal

5 Discussion and Conclusions

The study on the results of the tests conducted in the dynamic driving simulator was meant to test the usability and the efficiency of the HMI design (display mode, type of symbols and sounds) which was selected according to the results of the HMI laboratory tests described above. Furthermore we sought to decide among two system variants, one with a simple infrared picture and the other with additional symbols on the same screen for both the imminent and the information mode. On the present results, it turned out that the HMI-solution with the symbols is recommended for further development as it decreases driver workload, produces higher values of TTC, while the driver reduces vehicle's speed in imminent cases.

The design cycle of warning strategies adaptation for the EUCLIDE system was presented. The warning modalities were selected according to user acceptance and were evaluated in a Driving simulator. According to the studies HMI was highly accepted by the subjects. The objective data showed that such a system reduces the risk of an accident and also the subjectively felt workload. It should be emphasized, however, that conclusions on the impact on traffic safety can by no means be drawn from the present results for many reasons (unrealistic high number of events; zero risk in driving simulator, etc.). Therefore, on-road tests are necessary to provide information on these questions.

References

Cacciabue, P.C. (1998). Modelling and simulation of human behaviour in system control. Springer-Verlag, London, UK.

N.D. Lerner, B. M. Kotwal, R.D. Loyns, D. J. Gardner-Bonneau; Preliminary Human Factors Guidelines for Crash Avoidance Warning Devices (NHTSA DOT HS 808 342); Silver Spring (1996): COMSIS

W. Wierville, Rollin J. Fairbanks, Sara E. Fahey, (1995); Research on Vehicle-based Driver Status/Performance Monitoring, Seventh Semi-Annual Research Report, S. Department of Transportation (USA)

Norman, D. A. and Draper, S., User Centered System Design: New Perspectives in Human-Computer Interaction. Hillsdale, NJ: Erlbaum, 1986

A Study of Navigation Support Tools for Mobile Devices

Pei-Luen Patrick Rau

Department of Industrial Engineering
Tsinghua University
Beijing 100084, China
rpl@mail.tsinghua.edu.cn

Yin-Jue Wang

Institute of Communication Studies
Nation Chiao Tung University
Hsinchu 300, Taiwan
mout1024@hotmail.com

Abstract

In this research, the effectiveness of two navigation support tools, history lists and overview diagrams is studied and compared. There are four levels in the independent variable: history lists and overview diagrams, only history lists, only overview diagrams, and none. The results showed that the browsing performance (performance time, step, satisfaction, and orientation) of mobile users who were provided with at least one type of navigation support tools was better than the browsing performance of mobile device users without such navigation support tools. Also, users who were provided with both history lists and overview diagrams, and only overview diagrams were less oriented than the mobile device users who were provided with only history lists.

1 Introduction

Handheld devices are becoming more and more popular. In recent years, the use of wireless in the Internet has rapidly increased. Mobile devices, such as mobile phone, personal digital assistant (PDA), Webpad, handheld PC, and tablet PC play important roles in mobile Web access. With the development of high-speed wireless networking and various types of mobile devices, mobile computing has been widely accepted and applied. Mobile computing and WWW are fundamental technologies for a vision of information access for anyone anytime and anywhere. Many researches have been carried out on the user interface design of the Web and Web usability. However, computers are utilized as the device connected to the Web in many studies. Relatively few studies concentrate on mobile devices and the wireless Web. Moreover, user interface design guidelines for PC and Internet applications do not address the challenges frequently encountered in designing the user interface of mobile devices. Usability principles and standards pertaining to software applications for PCs and electronic appliances are often not applicable to mobile devices even thought the design concept of the user interface of mobile devices user interface emulates that of PC.

The main object of this research is to investigate the effectiveness of navigation support tools on browsing for mobile devices users. Two navigation support tools, history lists and overview diagrams are studied. These two navigation support tools are chosen due to the importance of contextual information during navigation. Structural context information helps forward navigation, whereas temporal context information helps backward navigation.

2 Literature Review

2.1 Navigation on Mobile Devices

Menu based system is very common on small screen devices such as PDAs and cell phones. Han and Kwak (1994) found that searching for menu items on single line displays was three times slower than searching on a conventional display. However, no significant effect was found on hierarchical menu search time with the smaller display and 12 and 24 line windows in another study (Swierenga, 1990).

Jones et al. (1999) studied the effect of small display space on Web-based task completion and showed how reduced displays affected users' approach to Web-based information retrieval. The results showed that 80% of small screen participants indicated the impact of screen size on their ability to complete tasks, whereas it was only 40% in the case of large screen users. However, no difference was found on the average number of forward and return commands for the two screen sizes (30 lines or 1024*768 pixels and 15 lines or 640*480 pixels). Small screen participants did not browse more information compared to their large screen counterparts.

2.2 Contextual Navigation Support Tools

Various navigation support tools have been developed to help hypertext users. Park and Kim (2000) evaluated the effectiveness of the contextual navigation aids in two different types of Web sites, an electronic commerce site with a well-defined structure and a content dissemination system with an ill-defined structure. The results indicated that the structural context information could help users reduce navigation problems by previewing information, and temporal context information may contribute to reducing the navigation problems by reviewing information.

Overview diagrams give a visual representation of the node-link structure of the hypertext and visual cues informing the user (Glenn and Chignell, 1992): where they are relative to other information and where they can move to, from their present location. Parton et al. (1985) found that a tree diagram of the entire menu structure was associated with better performance, better comprehension, and greater satisfaction for novice users. Hammond and Allinson (1989) found that users performed better in both exploratory and directed tasks when they used overview maps. Dee-Lucas and Larkin (1991) showed that the users who were presented with an overview diagram explored more items.

History lists allow the user to directly jump to any of the previously visited nodes. History lists enable users to move to distance nodes that they have visited in the past without stopping at the intermediate locations, and maintain the local context with pages visited staying in the history list (Greenberg, 1993; Park and Kim, 2000). History lists may be presented graphically by showing visited nodes in an iconic form, textually by listing the node names (Nielsen, 1990), or by a combination of both the icons and the node names (Kindersley, 1994). History lists may not be in a strict chronological or reverse chronological order. Several hypertexts have used the backtracking models discussed in the previous section in their implementation of history lists.

3 Methodology

The aim of this research is to empirically examine the effects of two navigation support tools, history lists and overview diagrams, on the browsing performance of mobile devices. Context information in a hypertext document includes structural context and temporal context. The former is an overview of content structure, and the latter is a history list.

3.1 Participants

Thirty-six novice mobile device users aging from 20 to 30 years were engaged in the experiment. Participants were randomly assigned into one of the four groups: no navigation support tools, overview diagrams, history lists, overview diagrams and history lists. Nine participants were assigned to each group, and they were tested individually using a mobile device to perform browsing tasks.

3.2 Experimental Design and Variables

A mixed design was used in this research to test the effect of navigation support tools on different browsing ranges on mobile devices. The independent variable was navigation support. Navigation support has four levels: no navigation support tool, history lists, overview diagrams, both history lists and overview diagrams. Navigation support tool were manipulated between participants. This research used four dependent variables: performance time, step, satisfaction, and disorientation.

Performance time was defined as the total time taken to complete the required browsing task. Performance times were collected for tasks performed by each participant. Time was recorded by the testing hypertext system to the nearest one-thousandth of a second. Step was defined as the total number of nodes that a participant visited for the tasks. Satisfaction was defined as the score obtained through a satisfaction questionnaire, consisting of sixteen questions on the scale of 1 (*lowest satisfaction*) to 7 (*highest satisfaction*). Disorientation was also defined as the score obtained through a satisfaction questionnaire consisting of ten questions (Beasley and Waugh, 1995) on the scale of 1 (*strongly disorientated)* to 5 (*never disoriented*). The lowest score was 10, meaning completely disoriented, and the highest score was 50, meaning not disoriented.

3.3 Tasks

The information browsing tasks on a testing hypertext system were designed for mobile devices. Information was in the second level of the hypertext. Participants were asked to scan and review information relevant to a fixed task. There were six types of search browsing and general purpose browsing tasks for each participant.

3.4 Apparatus

An H3630 iPAQ Personal Digital Assistant with a touch screen of 320*240 pixels was used for the experiment. Each participant was required to perform the tasks independently with no other participants present except for the experimenter. The participants' movements throughout the system were automatically timed to the nearest .001 sec and traced by a testing hypertext system. The testing hypertext system was a well-structured news Web site with a balanced hierarchical structure with four levels.

3.5 Procedure

All participants began to fill out a general information questionnaire concerning their personal characteristics, including age, education, and past computer and Internet experience. Each participant was given a mobile device and on-screen instructions. A brief practice session was then conducted to help the participants understand the operation of the system and the tasks to be performed. Following the practice, each participant performed the information browsing tasks. Participants were instructed to perform the tasks as quickly as possible without sacrificing accuracy. On the completion of the tasks, each participant was given two questionnaires of satisfaction, and disorientation.

4 Results and Discussion

The intention of the experiment was to examine how navigation support tools might influence the browsing performance of mobile device users. Significant difference in performance time (F = 18.33, p = 0.0001) was found, as shown in Table 1. Also, significant differences in step (F = 28.47, p = 0.0001), satisfaction (F = 15.15, p = 0.0001), and disorientation (F = 31.76, p = 0.0001) were also found. According to the results of Student-Newman-Keuls test, the differences between none navigation support tool and overview diagrams, history lists, and both were all significant in performance time, step, satisfaction, and disorientation. Mobile device users provided with navigation support tools were faster, made less moves, and less frustrated and disoriented than those without navigation support tools. Also, significant difference was found in disorientation for providing both types of navigation tools and providing history list. Users provided with both history lists and overview diagrams on mobile devices were less disoriented than those who were provided with only history lists on mobile devices.

Table 1. Results for Navigation Support Tools

Variables	None		Overview Diagrams		History Lists		Both		F	p
	MEAN	SD	MEAN	SD	MEAN	SD	MEAN	SD		
Performance Time (sec)	187.62	14.28	125.42	5.12	120.75	4.30	127.24	7.01	18.33	0.0001
Step	99.22	3.48	77.89	1.44	77.78	4.63	73.67	4.64	28.47	0.0001
Satisfaction	26.8	2.54	32.0	2.50	31.2	1.72	33.7	2.87	15.15	0.0001
Disorientation	33.8	3.27	43.1	1.76	38.0	2.06	43.0	2.24	31.76	0.0001

The aid of temporal context information was not as effective as structural context information on the disorientation for most of the tasks, and time and step for large range fixed search browsing task. As indicated in the literature, providing structural context helps users' forward navigation, whereas providing temporal context helps users' backward navigation (Utting & Yankelovich, 1989). Usually users need help for forward navigation more often than for backward navigation due to the back button design of browsers. Also, Jones et al. (1999) suggested to provide forward navigation aid for browsing hypertext on a small display. And overview diagrams give a visual representation of the node-link structure of the hypertext and visual cues (Glenn and Chignell, 1992), so that users provided with overview diagrams felt less disorientated and performed faster and made less moves than the users provided with history lists.

5 Conclusions

The purpose of this study was to investigate the effectiveness of navigation support tools on browsing for mobile devices users. The results indicate that both the two navigation support tools, history lists and overview diagrams, improve the browsing performance for mobile device users. Also, overview diagrams are more helpful than history lists on browsing with mobile devices in most of the conditions. However, providing both navigation tools does not necessarily mean improvement in browsing performance compared to only providing one of the two navigation tools. According to the results, it is suggested that providing any kind of navigation support tools is expected to improve the browsing performance for mobile device users. Furthermore, providing forward navigation aid would be more effective, especially on disorientation, than providing backward navigation aid. Moreover, forward navigation aid is more effective for large range browsing.

References

Dee-Lucas, D. and Larkin, J. H. (1991). Content map design and knowledge structures with hypertext and traditional text. Poster presented at the ACM Conference on Hypertext, *Hypertext '91*, San Antonio, Texas, USA.

Glenn, B.T. and Chignell, M. H.(1992). Hypermedia: Design for browsing. In H.Rex Hartson and D.Hix(Eds.), *Advances in Human-Computer Interaction*, 3, (pp. 143-183). Norwood, NJ: Ablex publishing Corp.

Greenberg, S. (1993). *The computer user toolsmith: The use, reuse, and organization of computer-based tools*. Cambridge, England: Cambridge University Press.

Hammond, N. and Allison, L. (1989). Extending hypertext for learning: An investigation of access and guidance tools. In Sutcliffe, A., Macaulay, L. (Eds.), *People and Computers*, V, 293-304.

Han, S.H., and Kwahk, J. (1994). Design of a menu for small displays presenting a single item at a time, *Proceedings of Human Factors and Ergonomics society 38th Annual Meeting*, 1, 360-364.

Jones, M., Marsden, G., Mohd-Nasir, N., Boone, K., and Buchman, G. (1999). Improving Web interaction on small displays. *Computer Networks*, 31, 1129-1137.

Kindersley, D. (1994). My First Incredible, Amazing Dictionary. CD-ROM for PC. Dorling Kindersley Publishing. Sunnyvale, CA: MediaMart.

Nielsen, J. (1990). The art of navigating through hypertext. *Communications of the ACM*, 33(3), 296-310.

Parton, D., Huffman, K., Prodgen, P., Norman, K., and Shneiderman, B. (1985). Learning a menu selection tree. *Behavior and Information Technology*, 4(2), 81-91.

Park, J., and Kim, J. (2000). Contextual navigation aids for two World Wide Web systems. *International Journal of Human-Computer interaction*, 12: (2), 193-217.

Swierenga, S. J. (1990). Menuing and scrolling as alternative information access techniques for computer systems: interacting with the user. *Proceedings of Human Factors Society 34th Annual meeting*, 1, 356-359.

Short Span Interaction in Mobile Phone Answering Situations

Lauri Repokari

Helsinki University of Technology
Software Business and Engineering Institute
P.O.Box 9600, FIN-02015 HUT, Finland
Lauri.Repokari@hut.fi

Abstract
The purpose of this paper is to present the concept of short span interaction based on a set of studies where the goal is to understand what happens during the very short moment when the user starts to use his/her mobile phone device.

1 Introduction

During the last ten years, the penetration of mobile phones into the mobile communication device market has grown significantly. Today, almost everyone in Finland has a mobile phone and many of us even own several. The world is getting mobile – at least world communications are getting mobile. Almost everyone in every part of the world would like to own a mobile phone, while numerous people use cell phones and laptops daily and are intent on owning other mobile devices as well. Never before in history has such a remarkably large part of the population carried a communication device with them on a daily basis. This situation has changed mobile phones from luxury devices to mass-market objects. Phone users do not see mobile phones as durables, rather as objects of instant consumption. In Finland, for example, on average, mobile phones are owned for only one and a half years. Consequently, models of mobile phones change rapidly and users must frequently familiarize themselves with new devices and new ways of interacting with them.

In science, and especially in the field of HCI (Human - Computer Interaction), mobile phones and their usability has been a subject of great interest. Technological development has been so rapid that researchers have had to work hard to keep up to date with all the fast-moving developments. Technological development will raise new questions about Human - Device Interaction. Changes and developments in user interfaces and input devices will change the ways in which users interact with the technology. Cognitive requirements and limitations in interaction with new devices will be an interesting challenge for HCI in coming years.

In this paper, a concept of " short span interaction" will be launched. This concept is based on the studies presented below and can be described briefly as: *"interaction is what occurs when the user interacts with the device for the first time and/or without paying attention to the device - user interface functionality. In this situation, the mental model of devices, habits or previous experience will define the decisions the user makes."*

Short span interaction is a concept that helps us to understand this very short moment of human device interaction.

2 Methodology

As this paper consists of four different studies, several kinds of methodology have been used. In the first and second studies, we used observation, while one model of a GSM phone with two kinds of keypad functionality (mirror images) was used as the research tool; in the second study the subjects were interviewed, while in the third we used paper prototypes and the results were statistically analysed. In the fourth study, we programmed laboratory software to enable us to conduct this research; results were analysed statistically. All the studies presented here were conducted by a multidisciplinary research group IERG (Information Ergonomics Research Group) www.soberit.hut.fi/ierg. I am sure that without the multidisciplinary expertise of this group these studies could not have be completed.

3 Theory

The theoretical framework of this paper is engineering science in the field of HCI. Theoretically, these studies are based on the paradigm of cognitive psychology (e.g. Baddelay) and on perceptual psychology (e.g. Traisman) and on engineering science (e.g. Norman). The common element in them all is a willingness to understand human behaviour and technology. This concept of short span interaction simply could not have been presented without input from many different scientific disciplines.

Mobile phones are answered in various situations. Answering should take as little attention as possible so the user can focus simultaneously on whatever s/he is doing, e.g. driving a car, which is a typical situation of divided attention. Actually, mobile phone use situations are nearly always situations of divided attention. To understand the behaviour of users in answering situations, we have to understand the nature of attention itself.

The concept of divided attention means that we share our intentional resources between more than one task at a time. It is difficult, or even impossible, to determine universal attention capacities as they depend on the situation and the abilities of the actor. It is possible to extend one's cognitive capacity with practice performing a certain task (Spelke et al. 1976). The particular modalities used in certain tasks also affect the performance of those tasks.

The concept of attention "spotlight" involves the idea that our attention has two kinds of entity: the sharp spotlight and something that, while not holding our main attention, is nevertheless involved in unconscious information processing. In the typical phone answering situation, the user is usually doing something else consciously, while answering as an automatic function. In this kind of operation, among other things, the ergonomics and keypad layout are important, but attention is still involved in the user's operations. If two tasks are performed using the same modality, action usually becomes more difficult. On the other hand, if the tasks are performed using different modalities (e.g. visual and auditory), the demands they make on us can be coped with more easily (Eysenck and Keane 1990)

In addition to attentional limitations, the user's action is influenced by knowledge provided by the world and internal representations of the world constructed via experiences (Norman 1988). Understanding the behaviour of a physical system and making predictions about its behaviour is based on mental knowledge structures, which represent information about systems. The concept "mental model" is used in multiple meanings in human-computer interaction literature (Ehrlich 1996). In this paper, the term "mental model" refers to the user's knowledge structure of a system. Procedural knowledge concerning the use of a system can be viewed as a functional model of that system (Preece et al. 1994). When a user is not very familiar with the system or the task, attention and moment-to-moment control of action may be necessary. On the other hand, relatively simple, frequently performed, tasks, such as answering the phone, can be done more or less automatically. It has been suggested that well learned, automatic actions differ qualitatively from controlled processing and do not consume attention resources (Schneider & Shiffrin 1977). Even though automatic processing is economic from one point of view, it nevertheless has some disadvantages. Interaction with technical devices is based on perceptual (bottom-up) information available in the interaction situation and also on previous knowledge (top-down information).

The two possibilities when designing a visual mobile phone - user interface are a text-based interface and an icon-based interface. With a correctly designed icon, it is possible to present in compact form the same information as a word. The usefulness of text and icons has been studied at the computer interface (Benbasat & Todd, 1993 Repokari 2002), but there is much less research into smaller displays. In addition, research with icons has mainly been concerned with the semantics: how clear is the meaning of the icon? (Guastello & Traut, 1989; Blankenberger & Hahn, 1991). The visual distinctiveness has not been so intensively studied. The graphical (with colours and animations) user interface will be the future way of showing menu structures (and other things) in mobile phones and other mobile devices.

Visual search is a widely used paradigm in vision research and in studying attention effects in human information processing. In a visual search, the subject has to find a target item among a set of distracter items (for example, Treisman & Gelade, 1980). Visual search has been used in human-computer interaction studies, but not with smaller displays.

In this set of studies, about 24000 interactions occurred between subjects and device. Some of them were very simple, like pointing the paper prototype, but many of them were real answering situations. In every interaction, the action was a part of the mobile phone answering situation. In the laboratory (study four) tests, we used a mobile phone screen size program in which all the icons and terms were from mobile phones on the market.

4 Introduction to the studies

In this chapter, the set of studies on which the concept of short span interaction is based is briefly introduced. The first two studies concentrate on understanding the mental models of users and the effect of the mental model on the usability of the mobile devices. The third study aims to understand the meaning and importance of colour coding in the mobile phone layout. The fourth examines visual search at the mobile phone - user interface from the point of view of cognitive psychology. Full report of these studies: www.soberit.hut.fi/decode/hci.

Study one:
The Effectiveness of Symbol and Colour Coding in the Mobile Phone
This study (ref) concentrated on the mobile phone keys used for answering and ending calls (the SEND and END keys). These keys should indicate their function unambiguously so that fatal errors, like accidentally terminating a call instead of answering it, can be avoided. The de facto standard (DFS) of these keys in Europe is a layout where the SEND key is located on the left and the END key on the right side of the upper part of a mobile phone.

Study two:
The Effect of Mental Models Guiding Users' Actions in Mobile Phone Answering Situations
The goal of this study was to find out how the mental model generated by the use of a certain mobile phone keypad layout affects behaviour in a situation where the user has to answer a call

with an unknown mobile phone. How additional perceptual information influences behaviour was also studied.

Study three:
Influence of Top-down and Bottom-up Information in Mobile Phone Answering Situations
This study continues two previous studies conducted using real mobile phones that indicated top-down information guides the users' action quite strongly. In this study, paper-prototypes of mobile phones were used in order to systematically vary certain keypad properties and to gain more detailed information about the effects of different kinds of information. This study concentrates on the properties of the users' mental models in mobile phone answering situations. What is the role of the mental models and how do the visual features of the phone keypad effect answering? What is the significance of top-down or bottom-up processing?

Study four - Visual search on a mobile phone display
The aim of the study was to compare the search times on a small display when text or icons were used as targets and distracters and to further study the effect of different types of distracters when using icons. We were also interested to see how suitable the traditional visual search paradigm is for human-computer interaction research.

5 Conclusions

As stated above, the purpose of this paper is to present the concept of short span interaction based on a set of studies where the goal is to understand what happens during the very short moment when the user starts to use his/her mobile phone.
In the first and third studies, it was shown that colour is a very strong factor that affects the behaviour of users in mobile phone use (answering) situations. There are also other important factors. Keypad layout should be designed to support the common existing design. In Finland, this means that the answering key should be on the top left of the mobile phone keypad. In the fourth study, we derived strong evidence that icons are an important way of expressing information at the mobile phone - user interface.
An interesting factor in some of the results is that the principles of cognitive psychology do not apply to all areas of short span interaction with small display - user interface devices. In study three, surprisingly, neither the number nor the grouping of the function keys had any effect. Also, in study four, in the Pop-Up search, the set size hadn't the affect that would be expected from very large cognitive psychology research studies.
When we started this set of studies, one interesting topic was our feeling that there was something unique in answering situations that was not explained by traditional psychology and engineering science/HCI. These results are not evidence enough to start to falsify any of the theories in cognitive psychology or HCI, but, nevertheless, our studies and their results are enough to shed some light on the concept of human - machine short span interaction.

6 The Short Span Interaction

Short span interaction consists of a mental model, phenomena like pop-up, perception, attention and other well studied concepts of cognitive psychology. Briefly, short span interaction is not a phenomenon or a theory, rather, it is a use situation where the user is not well prepared to act or accustomed to implementing action, or, as defined above: *"interaction is what occurs when the user interacts with the device for the first time or without paying attention to the device - user interface functionality. In this situation, the mental model of devices, habits or previous experience will define the decisions the user makes."*

What really happens in short span interaction? What do users actually do and what do they assume they are doing? In a nutshell, the user acts and believes s/he is acting as in any normal/usual use situation, where decisions are rational and conscious; in reality, this is not the case. In our studies, this was interesting in two ways, the first of which was that the user did not even recognise that the phone model was changed during the test. In the other study, a proportion of the subjects got stuck in their way of behaviour and could not change their way of acting even if they saw that the phone model was different. These two irrational ways of behaviour can be explained by reference to evidence gained in these studies: users in short span interaction situations do not rely on rational decisions, but, rather, make unconscious decisions.

7 Discussion

While these studies were being conducted, our research team took part in a programme of mobile device concept development. Even if the research questions were taken from actual product development, the studies were designed from the point of view of academic interest. From the start, we wanted to create, test and use methodology that could be adapted to mobile device product development. A deep understanding of the behaviour of users of mobile devices in different use situations provides significant information for mobile device product development teams. We hope that by introducing the concept of short span interaction this paper positively contributes to mobile device product development work. In future research, we aim to create and validate methodology for improving the understanding of the ways users of mobile devices act in different use situations.

References

Patnaik, D. & Beeker, R. 1999. Needfinding: The Why and How of Uncovering the People's Needs. Design Management Journal 1999.

Rumbaugh, J. 1994. Getting started – Using use cases to capture requirements. Journal of Object Oriented Programming, Vol. 11, No.1 September 1994.

Baddeley, A. "Working memory." Science, 1992 255, 556-559.

Benbasat, I., & Todd, P. (1993). An experimental investigation of interface alternatives: Icon vs. text and direct manipulation vs. menus. International Journal of Man-Machine Studies, 38, 369-402.

Blankenberger, S., & Hahn, K. (1991). Effects of icon design on human-computer interaction. International Journal of Man-Machine Studies, 35, 363-377.

Ehrlich, K. (1996) Applied mental models in human-computer interaction. In Oakhill, J. and Garnham, A. (eds.). Mental Models in Cognitive Science, U.K.

Eysenck, M.W., and Keane, M.T. (1990) Cognitive psychology: a student's handbook. Lawrence Erlbaum Associates Ltd., U.K.

Guastello, S. J., & Traut, M. (1989). Verbal versus pictorial representations of objects in a human-computer interface. International Journal of Man-Machine Studies, 31, 99-120.

Nielsen, J. (1993): Usability Engineering. Academic Press. London 1993. ISBN 0-12-518405-0. 358 s.

Norman, D.A. (1988) The psychology of everyday things. Basic Books, Inc., U.S.A.

Preece, J. et al. Human-Computer Interaction. Addison-Wesley Longman Limited, Harlow, UK, 1994.

Schneider, W., and Shiffrin, R.M. (1977) Controlled and automatic human information processing: Detection, Search, and attention. Psychological Review, 84 (1), 1-66.

Spelke E., Hirst W., and Neisser U. (1976) Skills of divided attention. Cognition, 4, 215-230.

Cross Cultural Usability: An International Study on Driver Information Systems

Peter Rößger & Jörg Hofmeister

Harman/Becker Automotive Systems (Becker Division) GmbH
Raiffeisenstr. 34
70794 Filderstadt, Germany
p.roessger@caa.de

Abstract

The interaction between humans and driver information systems has to be as uncomplicated and fast as possible since the main task for the driver is to drive. It is therefore necessary to design human-machine interfaces, that make easy and safe interactions possible. However, the increasing globalisation of product markets make it necessary to take into account cultural differences when developing a human-machine-interface. For this reason, cultural peculiarities and intercultural differences have to be identified and investigated. In the present study we examined if there are any cultural differences between Japan, the USA and Germany with regard to the evaluation of different navigation-systems.

1 Introduction

The technological push in car development leads to growing integration of digital information processing systems in cars. Advanced traffic information systems, automatic speed and distance control, telephones, emergency calls and television are state of the market today. Office applications, internet connections, a variety of telematic information services, and automatic lane keeping are soon to be invented. Often benefits and costs of these new car-systems are not clear.

Due to the fact, that more and more developments are made by suppliers, and most of the suppliers work for more than one manufacturer, knowledge and systems are available for more or less every car producer. Technological developments are not the only criteria for a customer to decide which car to buy. On the other hand customers cognition will be driven by Driver-Information-Systems (DIS), representing high tech inside the vehicle and drawing attention.

2 Driver Information Systems

Driver Information Systems (DIS) in vehicles provide information about the vehicle, traffic, route and more. The information provided gets more complex with the increasing number of functions of DIS. In the near future, systems including audio, navigation, phone, TV, internet, office applications, etc. will be commercially available.

These developments lead to a growing focus on human factors of the driver-car system and DIS in particular. The driver has to be able to perform driving and system use properly. Car manufacturers get the chance to make the difference by realising different car-driver interface philosophies in their products.

3 The Interaction between Drivers and Driver Information Systems

To provide a model that allows industrial human factors research, development and consulting in the area of it is divided into three parts. The three parts are not independent from each other. But they require different research tools and theoretical approaches. That is why it seems to be useful to distinguish between them. Only good solutions in all areas will provide a good human machine interface (HMI). Human factors have an influence on all steps of the development. User-acceptance, meaning the emotional attitude, of the systems is added to human factors work, because many of the tools and research procedures are related to human factors and the results of both are often used in the same departments. In addition, a non accepted system is not ergonomic.

3.1 Anthropometric Human Factors

Anthropometric human factors deal with the turning to the system and how to effect the system. This includes the position of the system, the reach ability for both, driver and drivers mate, and the visibility. These parameters have to fit for all, from short to tall, persons. As well in focus are all input devices, shape, size and position relatively to each other. To provide blind use for expert users these parameters should vary. Regulation powers have to fit the in car environment. Too low powers will lead to wrong inputs, change in regulation powers should give haptic feedback. Anthropometric human factors optimise the information-transfer from the user to the system.

3.2 Cognitive Human Factors

Cognitive human factors analyse the understanding of a system. Consistency and mental models are the main scopes in this area. Consistency includes internal and external consistency. Internal consistent systems are always equal to themselves, input operation always follow the same principles and the structure is in all submenus similar. External consistent systems allow to use dialogue principles know from other technological artefacts, such as computers, cars, or maybe washing machines. Consistency, and even more internal consistency, is the most important factor to influence cognitive human factors.

3.3 Sensoric Human factors

Sensoric human factors look at how information is transported from the system to the user. Main subject is the transport of information via the visual channel. The focus lies on the design of the screen. Font-size and font-types have to fit the needs of the user. Contrasts have to be big enough, colours should not be used for making the screen funnier, cooler or what ever, but to code different system states, to clear transported information, and to structure the output. Other outputs are speech outputs. The speech should be clearly understandable, the timing between a spoken command and its fulfilling has to match the needs of the user. Others than the visual or acoustic channels should be used in addition to transfer information from the system to the driver. Commands for making turn could be transported via the steering wheel, using the tactile or haptic channel.

3.4 Acceptance

Acceptance deals with the emotional setting a user has towards a system. A system might be good from a human factors point of view, but poor from a emotional point of view. Ergonomics lead to systems that are not bad, they do not do any harm to users, are they are easy to use. Acceptance researches on the specials of the systems, the particular look and feel, the difference that might make it desirable. Other points are the prices, the readiness to buy, and the using behaviour.

4 Cross-Cultural Differences

Culture is a persistent situational phenomenon that is manifested through persistent patterns of interaction with the environment. These persistent patterns of behavior include beliefs, values, and mental models. Cultural differences also influence the expectations and handling of technical systems (e.g. Honold, 2000). For example in Japan technical support is considered more important than price. In the USA, however, the price is a driving force in choosing a system. Also, cultural differences in perception caused by phrasing, idioms, symbolic representations or cultural schemas may lead to differences with regard to the level of risk communicated by a warning label (e. g. Zühlke & Röse, 2000).

On the basis of these facts, systems developed for the Western-European market are not naturally useful for non-European markets. Given the increasing globalization of product markets, it becomes more and more important to take into account intercultural differences when developing and designing DIS and human machine interfaces. If companies ignore these cultural specifications, usability and acceptance of a system in a given country may decrease and this may result in economic losses. See also Rößger & Rosendahl, 2002.

5 The Cross-Cultural Study

5.1 Methods

The study was conducted parallel in three major automotive markets, North America (Detroit), Japan (Tokyo) and Germany (Filderstadt). More or less the same DIS where tested with 24-32 subjects in each location (Tab. 1).

Tab. 1: Systems and number of subjects

	USA	Japan	Germany
Rotary-Push Systems (RPS)	32 subjects	24 subjects	24 subjects
Touch Screen System (TS)	32 subjects	24 subjects	24 subjects
1-DIN Radio Navigation System	32 subjects	System n.a.	24 subjects

To collect subjective data, a questionnaire and an interview were used. The questionnaire was developed in a couple of usability studies. Tests on reliability and validity showed good results (Rößger, 2000; Rößger & Smyrek, 1999a,b; Rößger & Smyrek, 2000). The interview was recorded and analyzed later.

To collect behavioral data, videos of the interaction were recorded. They were analyzed for interaction times and mistakes. For a taxonomy of the mistakes see Rößger, Metternich & Smyrek, 2001).

5.2 Results

In this paper only major results of the study will be published. For detailed results refer to Rößger & Rosendahl (2002). The questionnaire has 10 dimensions, due to statistical analysis. Earlier versions of the tool had pretty much the same categories of items (see Rößger & Smyrek, 1999a,b; Rößger & Smyrek, 2000; Rößger, Metternich & Smyrek, 2001). The questionnaire seems to be very stable. So the correlation between Acceptance (one of the dimensions) and the others were calculated.

In Germany the dimensions *Selfdescriptiveness*, *Expectationconformity* and *Appropriateness* correlate highest with Acceptance. This replicates results from earlier studies (Rößger, 2000; Rößger & Smyrek, 1999a; Rößger, Metternich & Smyrek, 2001). In Japan *Expectationconformity* correlates very low with Acceptance. This supports findings of Honold (2000), that Japanese users like to be challenged by technology, so they can work with HMIs, that do not perfectly match their expectations. In the USA the highest score is reached by appropriateness, which supports the North American philosophy of "Keep It Straight And Simple" (KISS; e.g. Foley, 2001).

The only dimension scoring high in all cultures is Consistency. Consistency here means internal consistency. If something works in a particular way in one context or subsystem (even if it might be the second best solution), it has to work the same way in any other context. Another influence on consistency is the "Don't switch between Interaction-languages" – rule, meaning, if the major input device is a rotary-push-device, do not design the interaction in a way, that you have to use soft keys, hard keys, or touch screens in some contexts.

The RPS gets better overall results than the TS in Germany and the USA. The TS is rated better in Japan. Probably the internal structures of the (German) RPS match European and US-American mental models far better, than Japanese mental models. In Japan it is vice versa, some internal features of the system (e.g. making the decision whether to go to a street address, a junction, or a point of interest before typing in the city to go) match the internal models of Asian users.

5.3 Discussion

The results of the study show clearly that intercultural differences between the USA, Japan, and Germany with regard to the evaluation of different navigation-systems do exist. In Germany the RPS gets better results than the TS. In Japan, however, the TS is more accepted than the RPS. The ratings in some dimensions of our usability-questionnaire reveal that menu structure, input device and functionality of the TS satisfies the expectations and behavior of the Japanese participants better than the RPS.

The results of the study show that cultural influences need to be considered when developing information systems. Despite the increasing globalization of product markets, the global user does not exist! This makes great demands on the development of driver-information-systems. In the future it will no longer be sufficient just to "translate" an information system into a different language. Instead, cultural differences with regard to expectations, behavior and handling will be a major aspect to consider when developing technical systems.

Further investigations are therefore necessary to analyze cultural peculiarities. A replication study runs at the moment, avoiding some methodological problems of the described study. Results will be published late 2003.

6 References

Foley, J. (2001). Statement during a presentation at the IIR-Telematics meeting, Feb. 2001, Scottsdale, AZ.

Honold, P. (2000). *Interkulturelles Usability Engineering.* Fortsch.-Ber. VDI Reihe 10, Nr. 647. Düsseldorf: VDI Verlag.

Redelmeier, D.A. & Tibshirani, J.R. (1997). Association between cellular telephone calls and motor vehicle collisions. The New England Journal of Medicine, Vol. 336, No. 7, 453-458.

Rößger, P. (2000). Human Machine Interfaces of Car-Computing Devices. SAE Paper 00LSG29, presented at the SAE World Congress 2000, Detroit, MI.

Rößger, P. & Rosendahl, I. (2002). Intercultural Differences in the Interaction Between drivers and Driver-Information-Systems. SAE Paper 02Annual34, Presented at the SAE World Congress 2002, Detroit, MI.

Rößger, P. & Smyrek, U. (1999a). Bestens bedient? TeleTraffic, NO 11/12 (September/October 1999), 12-15.

Rößger, P. & Smyrek, U. (1999b). Licht und Schatten. TeleTraffic, NO 11/12 (November/December 1999).

Rößger, P. & Smyrek, U. (2000). Reifeprüfung. TeleTraffic, NO 11/12 (November/December 2000), 12-15.

Rößger, P., Metternich, B. & Smyrek, U. (2001). Stunde der Abrechnung. TeleTraffic, NO 1/2 (January/February 2001), 22-25.

Schumacher, R.M. (1995). Ameritech Graphical User Interface Standards and Design Guidelines. Ameritech Cooperation. http://www.ameritec.com, Download at Jan. 12th, 1999.

Zühlke, D. & Röse, K. (2000). Design of global user-interfaces: Living with the challenge, Proceedings of the IEA 2000, Nr. 6, 154 – 157.

Designing a Speech Operated Calendar Application for Mobile Users

Sami Ronkainen

Nokia Mobile Phones
P.O. Box 50, FIN-90571
Oulu, Finland
sami.ronkainen@nokia.com

Juha Kela

VTT Electronics
P.O. Box 1100, FIN-
90571 Oulu, Finland
juha.kela@vtt.fi

Juha Marila

Nokia Research Center
P.O.Box 407, FIN-0004
Nokia Group, Finland
juha.marila@nokia.com

Abstract

In this paper we present the results from the design and evaluation of a speech-operated calendar application for a mobile usage context. The design was evaluated by using Wizard of Oz HTML prototype implementation with high-quality text-to-speech (TTS) prompts, operated by a facilitator instead of automatic speech recognition (ASR) and dialogue manager software. Our evaluation showed that the usability requirements of intuitiveness and learnability were met quite well. However, speed and efficiency needed further work. Overall the users were satisfied with the functionality of the system.

1 Introduction

Since the introduction of mobile phones and PDAs, mobile interaction has grown exponentially. Today, number of different applications known from desktop computer environments including calendar, e-mail and www-browser are used in various mobile contexts. However, mobile interaction sets great challenges to user interface design due to the variety in mobile device platforms on the market with different display and keyboard characteristics. Recent studies have reported severe reduction in data input rate when switching from conventional QWERTY keyboard to a mobile device keyboard alternative (Manaris, MacGyvers & Lagoudakis, 1999). Especially in the mobile context when the user is on the move, the keyboard is hard or in some cases even impossible to use. This applies also to situations where the user might have some parallel tasks or some kind of motoric disabilities. Another issue in mobile usage situations is that it is often difficult to utilize the display of the device. For instance, a user might be walking on street and receive a phone call about a future meeting. Optimally, it should be possible to set up the meeting by checking and editing the user's calendar without having to stop walking.

Speech-operated systems could serve to overcome these obstacles. Although mobile speech recognition is far from perfect, especially in the error correction and recovery, rapidly developing algorithms and evolving hardware solutions can provide a feasible interaction method for mobile devices (Kumagai, 2002; McTear, 2002).

We present here the design principles and results from the evaluation of a calendar speech user interface (SUI) targeted at mobile usage situations. Our aim was to create a speech UI to calendar that had good usability in terms of intuitiveness, learnability and efficiency.

2 The Calendar Speech UI Design

The planned usage context of the calendar was a mobile situation, where the user has a primary task such as walking. Due to that limitation, it was decided that calendar should be very easy to use, even at the cost of reduced functionality. It was assumed that a speech-operated calendar would not be the primary one to be used. In a mobile situation only quick-and-dirty creation of entries and browsing of near future entries would be enough. Entries could be refined later on using a visual calendar, if required.

Earlier internal product development interviews (not published) revealed that people commonly utilize a *minimal information strategy* when creating calendar entries. In other words, only the essential details are commonly stored in a calendar entry. Previous research (Kincaid et al., 1985) and our own interviews further confirmed that the most important targets for calendar usage are: current day, current week, and coming week.

In our design, we introduced three types of possible calendar entries. An *appointment* consists of a date, starting and ending times, an alarm time, a title, and additional details. A *reminder* consists of a date, an alarm time and a title. A *note* contains of only date, a title and additional details. Only a title, date and time-related information were required for each type while other details are optional. The supposedly moderate ASR and TTS resources were compensated for so that simple recording and playback of the user's speech is used for the free-form content, such as titles and additional details.

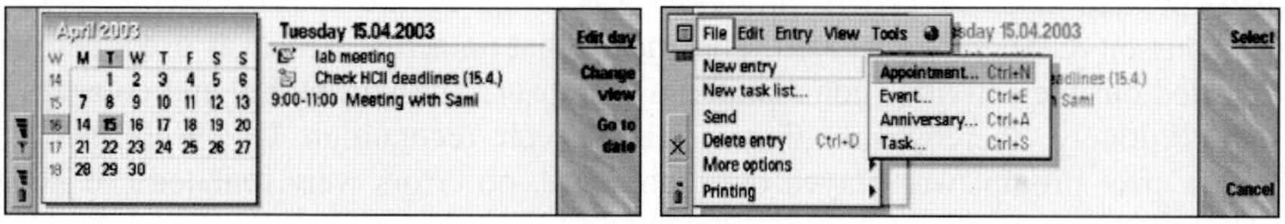

Figure 1: Nokia 9210 calendar view and entry creation

The mode of dialogue in our design is system-initiative. Previous studies demonstrate good performance especially in situations with restricted processing capabilities of the user (McTear, 2002; Walker, Fromer, Di Fabbrizio, Mestel & Hindle, 1998; Oviatt, 1996). However, it requires the most important available commands to be presented in the prompt. Another issue affecting prompt length is the amount of other data to be presented. For instance Lewis (Lewis, 1999) claims that people working as managers had on the average 3.6 calendar entries per day, the average length of an entry being 21 characters. In the 95% percentile, they had a maximum of 9.1 entries per day, with an entry length of 45 characters. Reading out that much text in a single prompt may be time consuming.

A specific problem with speech output is that the information is pushed to the user; it is difficult to skip unwanted information. One way to tackle the problem is to introduce a barge-in capability in the system. In addition to that, we introduced a dialogue where at first only a simple overview of a week or a day was presented. After that the user could start refining the search by selecting the appropriate date and entry type (see Table 1). We assumed that if the user receives moderately short prompts, and can give input frequently, the overall length of the dialogue will not become disturbing.

When creating an entry, a special problem arises. Without a visual presentation, it is hard to make out where the available time slots are, based on the information about existing calendar entries. For instance in Figure 1 one can easily see the available times of the day. However, when the calendar entries are read out, it becomes less obvious. A special feature was developed for searching free time slots when creating a new entry. The user gives the duration of the entry, and the system presents available time slots consecutively. The user can then decide on a suggested time, and the appointment is stored.

System:	*Today, 15th of March. Appointments from 08:00 to 12:00.Two notes. Would you like me to list entry details, create a new entry, find free time, or go to another time?*
User:	List details
System:	*Today, 15th of March. 08:00 to 12:00 <Appointment title>. Note 1 is about <Note title>. Note 2 is about <Note title>. Would you like me to list all details of an entry, modify an entry, create a new entry, find free time, or go to another time?*
User:	List all details
System:	*List all details for an appointment or a note?*
User:	A note

Table 1: Extract of calendar browsing dialogue

3 User Test

The speech-operated calendar HTML prototype was tested by the Wizard-of-Oz (WoZ) method (Dahlbäck, Jönsson & Ahrenberg, 1993). This means that no actual ASR was used, but instead an operator carried out the activities commanded by the user. The vocabulary and command syntax were used accurately, thereby imitating a real speech recognition UI. With incorrect user utterances, an error prompt was played out. Otherwise, no errors were imitated (i.e. recognition rate equaled to 100%). Test equipment consisted of a PC executing the HTML prototype. The test sessions were videotaped for later analysis.

3.1 Test Process

The evaluation consisted of three individual case studies, with users labeled here as A, B and C. User demographics were as follows:

- All female, ages 30/28/26 years
- English skills native/2nd language/daily usage
- Experience with speech UIs incidental
- Calendars used (all): Microsoft Outlook, Nokia 9110 Communicator, Nokia 6210 phone.

The user was initially briefed about the nature and commands of the speech UI they were supposed to use. They were informed that their utterances were to be interpreted by the test facilitator sitting next to them instead of an automatic speech recognition system. Nevertheless, they were instructed to speak and expect responses as if interacting with a machine.

The subjects carried out following tasks with the test prototype: *overview of week and each day's program, insert appointment, insert reminder, search free time for a meeting, edit an appointment.*

In case users felt lost and could not continue usage, they were allowed to ask for help. This type of walkthrough approach enabled us to reveal causes of usability problems as the test went on, and let the users concentrate on following tasks – and us uncover more possible problems – better than an unresponsive observation would.

3.2 Results

3.2.1 Intuitiveness of the Dialogue and Commands

The system was expected to be highly intuitive since users were provided with speech commands from system prompts. Our assumption proved to be correct: the selections were generally made with ease.

The confusions had to with terminology and reference of the prompts. Users A and B often used the utterances in system prompts in their exact format. An example: system prompts *please select the starting time, tell me to list free times or [*other choices*]* – user responds *select the starting time* instead of saying out a time expression. This, however, is in agreement with previous research showing users' strong tendency towards linguistic convergence in speech user interfaces (Oviatt, 1996). User C initially assumed the system to possess advanced language understanding capabilities and used general phrases such as *go back to Monday* which were not supported. Differences in entry terminology – *appointment, reminder, note* – were not clear to users. They all used them interchangeably.

All these intuitiveness problems were apparent with user A: system prompt *do you want to list details for an appointment or a note* elicited response *list details* – that is, the proposition made by the system was understood as a selectable option, and the difference between entry types further dimmed the prompt's meaning.

3.2.2 Subjective Efficiency

System prompts were lengthy, as indicated by all the users in the post-test interview. Likewise, proceeding by the prompts took too long in their opinion. Users B and C wished and tried for a more flexible method of going from one time instance to another than the hierarchic browsing between days, weeks and hours. They wanted to move from within a day view to following week view, or use relative references such as *go back to second note*. The load on users' memory was not excessive, since they asked help from system - by the *repeat* command - only once or twice during usage. Repeat requests were apparently related to the synthetic speech intelligibility more than the prompt content.

3.2.3 Subjective Assessment of Features and Utility

In the post-test interview, users were asked what they liked and disliked about the SUI calendar, and how they would use one in real life.

Most disliked features were: **system's speech was sometimes hard to understand**, **remembering and understanding the selection lists was hard** and **more shortcuts are needed**. Most liked features were: **overview of a day or a week is handy with the system, allows doing something else while browsing calendar (e.g. walking)** and **the find-free-time function**.

Supposed usage in real life was evaluated to be checking free time and week and day overviews.

4 Conclusions

In this article we presented a speech-controlled calendar UI design and evaluation. The design performed well with respect to intuitiveness and learnability, but worse with regards to speed and efficiency. Users regarded the automatic free time search as useful. The design utilized different calendar entry types to simplify entry creation and to speed up browsing. However, the different types were not understood by the test users. Another factor causing problems was the users' strong tendency to imitate system prompts in their utterances, misregarding all system prompts as selection lists. In future designs, offering a list of available selections and querying the user for open-ended data in one prompt should be avoided.

5 Acknowledgements

We would like to thank Mr. Miika Silfverberg from Nokia Research Center for providing valuable information about calendar usage and Ingrid Schembri for providing comments on this article. We would also like to thank all the participants in the test.

References

Dahlbäck, N., Jönsson, A., & Ahrenberg, L. (1993). Wizard of Oz studies: why and how. *Proceedings of the 1st international conference on Intelligent user interfaces,* Orlando, USA, 193-200.

Kincaid, C., Dupont, P., & Kaye, R. (1985). Electronic calendars in the office: an assessment of user needs and current technology. *ACM Transactions on Information System,* 3 (1), 89-102.

Kumagai, J. (2002). Speech Recognition: Talk to the Machine. *IEEE Spectrum*, 39 (9), 60-64.

Lewis, J. (1999). "Information for PDA Application Design: Calendar Entry and Name Length Statistics". *Proceedings of the Human Factors and Ergonomics Society 43rd Annual Meeting*, Santa Monica, USA, 467-470.

Manaris, B., MacGyvers, V., & Lagoudakis, M. (1999). "Universal Access to Mobile Computing Devices through Speech Input". *Proceedings of 12th International Florida AI Research Symposium (FLAIRS-99),* Orlando, USA, 286-292.

McTear, M. (2002). Spoken dialogue technology: enabling the conversational user interface. *ACM Computing Surveys (CSUR),* 34 (1), 90-169.

Norman, D. (1990). The Design of Everyday Things (2nd ed.). Currency/Doubleday.

Oviatt, S. (1996). User-Centered Modeling for Spoken Language and Multimodal Interfaces. *IEEE Multimedia,* Volume: 3 Issue: 3 , Winter 1996, 26–35.

Walker, M., Fromer, J., Di Fabbrizio, G., Mestel, C., & Hindle, D. (1998). What can I say?: evaluating a spoken language interface to Email. *Proceedings of the SIGCHI conference on Human factors in computing systems,* Los Angeles, USA, 582-589.

Yankelovich, N. (1998). "Using Natural Dialogs as the Basis for Speech Interface Design." Submitted to MIT Press as a chapter for the upcoming book, Automated Spoken Dialog Systems, edited by Susann Luperfoy. Retrieved February 7, 2003, from http://research.sun.com/speech/publications/mit-1998/MITPressChapter.v3.html

Basics of Intercultural Engineering: Analysis of User Requirements in Mainland China

Kerstin Röse

Center for Human-Machine-Interaction
User-centered Product Development
University of Kaiserslautern, Germany
PO-Box 3049, D-67653 Kaiserslautern
roese@mv.uni-kl.de

Introduction

For many years, researchers of the social sciences have analyzed cross-cultural differences of interpersonal communication styles and behavior. During the last ten years, usability engineers also have focused on intercultural differences of icon / color coding, navigation, and other human-computer interface components. The time for product development and the time between redesign efforts both are becoming shorter than ever before. Consequently, to prepare effectively interactive products for the global market, we must engineer their intercultural attributes. One basic step for intercultural Engineering is the analysis of user requirements in different cultures. This paper describes the analysis of culture-specific information from users in Mainland China and the application of different methods for different design issues, especially in an intercultural context. Selected results of this analysis will also be presented. The analysis and their results are part of the project Intops-2: Design for the Chinese market, funded by several German companies.

1 Analysis of user requirements in Mainland China

The following three investigation methods are applied in the INTOPS-2 project: test, questionnaire and interview. These methods have been proven by the previous project INTOPS-1 to be quite effective in investigating user-requirements on culture-specific machine design (see also Zühlke et al, 1998). Some new tests and more detailed questionnaire and interview checklist have been developed for the INTOPS 2 project. The analysis was realized by a native speaking Chinese PhD student from the Center for Human-machine-Interaction. Before the on-site investigation different investigation, the analyst prepared materials were prepared in Germany, which include the test materials for a total of eight tests, one questionnaire, and one interview checklist. All these materials were also translated into Chinese.

Table 1: Two tests of the requirement analysis in China

Test	Aim	Material	Subject	Analysis
Concept of grouping	Eliciting the grouping rule and the difference from German one	74 cards with different CNC machine functions	**Only** with experienced CNC machine operators	Preferred structure for grouping, basis for analysis of navigation
Preference to screen layout	Eliciting the familiar screen layout char-acters and difference from German one	Over 20 different cards in form and size, representing screen elements	CNC machine operators	Preferred layout of screen elements

The main categories of questionnaire and interview checklist are the following:

Questionnaire: Basic information about the visited company, information about machine purchasing, requirements on machine service, requirements on machine user-interface, and requirements on machine technical documentation.

Interview: Basic information about the visited company, information about machine purchasing, application situation of imported machines in practice, requirements on machine user interface, requirements on technical documentation, requirements on service, information about work organization and training.

The three investigation methods in this project are complementary to each other. The tests are formulated mainly to verify the hypotheses about user-interface design. The questionnaire is mainly applied to elicit user requirements about general machine design features and functional features. The interview could provide the interviewees with more freedom to express their requirements in a wide range, which could help the investigator to obtain the knowledge of wider and unexpected requirements of the target market. The combination of these three investigation methods could therefore ensure the effect of the investigation. The application of these investigation methods in INTOPS 2 to elicit of design features for different design issues (see also Röse et al 2002) is listed in Table 2:.

Table 2: Application of different methods for different design issues

Design issues		Methods		
		Interview	*Questionnaire*	*Test*
Beyond UI	*Machine functionality*	++	++	
	Technological features	++	++	
	Technical documentation	+	++	+
	Service	++	+	
	General machine design features	+	++	+
Within UI	*Information presentation*	+	+	++
	Dialog and structure	++	+	++
	User support	+	+	+
	Interface physical design	+	+	
	Language issues	+	+	+

The requirement analysis in China was carried out at the end of 2000. During two months 32 Chinese organizations in Shanghai, Beijing, and Chongqing were visited, of which 26 were Chinese industrial enterprises (including Chinese machine tool producers and some machine users). The other 6 organizations included some governmental organizations for import administration and some research institutes for machine user-interface design in China. A total of 42 subjects have participated in the tests, 35 main interviewees were involved and 19 questionnaires were prepared.

2 Chinese user requirements: for screen layout, menu structure, navigation, and help information

2.1 Interface layout

All the six screen layout results are shown in Figure 1. The hypothesized vertical layout of the menu or the function keys on screen has not been demonstrated. Almost all the test results have dominated in laying out the function keys and menu in a horizontal sequence.

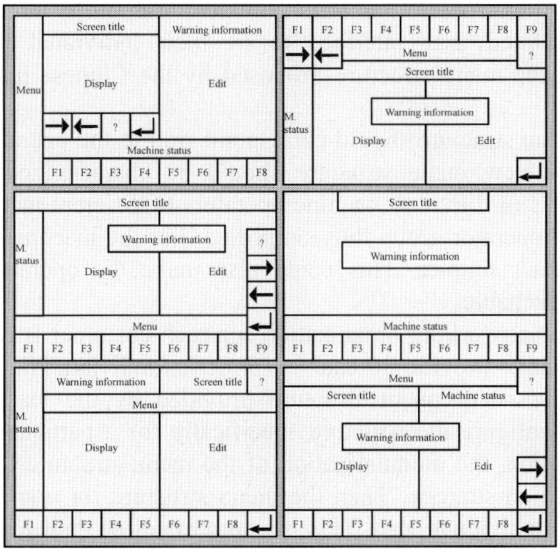

Figure 1: Results of Screen Layout Test

It was also hypothesized that the cultural influences from language (the reading direction) and aesthetic preference should have a large influence on preference for screen layout. However, the screen layout test has indicated that the preference of the Chinese machine operators for machine user-interface screen layout is largely influenced by their actual working experience with some existent machine user-interfaces. This result also means that for the design of a machine user-interface screen layout for the Chinese market, the existent user-interfaces should be investigated thoroughly. Especially the following two points should be considered in screen layout design:

First, the working experience of the Chinese machine operator with the existent user-interface has significant influence on their preference for screen layout. This means that principally all kinds of screen layouts could be adapted well and accepted by the Chinese machine operators in the long run. It is then suggested that the widely applied user-interface screen-layouts in Western countries could be applied in China without significant changes. This superficially "simple" way could be even a better way to facilitate Chinese operators adapting themselves more quickly to the imported machines in the future.

Secondly, Chinese machine operators have very little experience with the operation of a Microsoft Windows user-interface. Therefore, the application of the Windows user-interface for machine operation in China will meet with some problems at the beginning. Furthermore, the more freely structured dialogue provided by the Windows, in comparison to guided dialogue, user-interface also could make the Chinese machine operators unsure about their operations. The results suggested that the application of the Windows user-interface in China is not encouraged for machine operation at the present time.

2.2 Menu Structure

It was assumed before the analysis that the different requirements of Chinese users regarding the menu structure would be influenced by the different thinking patterns of the Chinese in comparison to German users. However, in practice, the most significantly (or most obviously) different requirement in this topic comes from the different working organization. One of the most significant aspects of working organization in China is the more elaborate division of operation tasks. A Chinese operator's task range is narrower than that for a German machine operator. The effectiveness of the German machine user-interface to support the working organization in China is problematic. In fact, this problem has been commented upon by some Chinese interviewees. They require a self-defined user interface to fit their individual production tasks. Some requirements regarding the menu structure proposed by the Chinese machine users include the following:

At first, the menu structure should correspond well to the operation tasks of the Chinese machine operators. The new menu structure should be characterized by the separated (and hierarchical) access rights of different machine operators to different machine functions, with the menu structure for each operator group thus simplified. For the machine operators, the operation functions should be much simpler. This could also make the operation of specific machine function more quickly reachable.

Because the actual working organization for each customer is different, it is impossible for the machine producers to provide for each customer an individual menu structure. In practice, there should be one consistent menu structure provided by the machine producers and the flexibility of users to configure the structure specifically for a particular working organization. Based on this consideration, the modularization of the menu structure to leave room for further adaptation is a good design strategy. Then the menu structure for a specific customer could be configured according to individual need. This freedom to define their own user interface through some openness of the system has been required by many Chinese machine users.

2.3 Dialogue and Navigation

Some direct or indirect information has been obtained from the investigation concerning the design of dialogue forms and navigation system for the user interface. From the problems that the interviewees often identified about their operators and the self-evaluation of the machine operators about their own work, some requirements of the Chinese users concerning machine dialogue and navigation system is revealed. The following points should be helpful for foreign producers to design user interfaces for Chinese operators:

Firstly, Chinese customers have very low assessment of their own machine operators. The operators themselves also have very low self-evaluations. Both groups often remarked that the operators have quite low qualifications and could not understand the complicated machine operations. This status suggests that the machine operators could have potential fear to actively interact with the machine and would prefer to follow the definite operation instructions provided by the system. Consequently the dialogue system must provide more error tolerance for the operation and should have a very clear guide for the operators to enable them follow the operation process.

Secondly, from the questionnaire results it can be determined that the main problem of the Chinese machine operators in machine operation is based on bad understanding of different machine functions and different operation processes. Consequently, the user interface must provide a navigation concept that presents general machine functions and operation processes in an overview, to enable operators to establish one clearer overall image of machine operation more easily.

Because of the general doubt of Chinese machine users about the qualification of their operators, it is required that the possibility to intervene by operators in machine control operation should be low. This means that the machine options and settings should be more standardized and offer minimal possibilities for the operator to change. It therefore seems appropriate that those dialogue forms should be chosen that have only few options.

2.4 Help Information

Chinese machine operators rely very much on help to operate the machines. At present this help is mainly from their colleagues or masters. The machine system, including the technical documents has provided very little help for the Chinese machine operators. In the investigation many Chinese machine operators expressed their requirements for more help on the machine user interface, which includes not only the machine trouble-diagnosis help-information, but also the help for learning normal machine operation. The help information about the machine user interface is in some ways even more important than the accompanying machine operation manuals, because the operators are always trained on the job at the machine, and the operation manuals are not allocated to every operator. Consequently, the provision of more effective help information on user interface must be completed by the machine producers.

3 Conclusion

This requirements analysis is only one small part of the larger project Intops 2: Design for the Chinese Market. The project has constructed one general intercultural human-machine system design structure for the Chinese market. Further research is needed to formulate more detailed design issues and design approaches to conduct the machine localization design in practice. In the near future, preparation of a more detailed design style guide for Chinese UIs is anticipated.

Acknowledgements

The author thanks the sponsors of this project, MAN Roland Druckmaschinen AG, IWKA AG, Bühler AG Amriswil, and Rittal Rudolf Loh GmbH & Co. KG. Also deserving special thanks is PhD-student Long Liu for data collection and analysis in Mainland China.

References

Dong, J. M. and Salvendy, G. (1999). Designing Menus for the Chinese Population: Horizontal or Vertical? In: *Behavior & Information Technology*, Vol. 18, No. 6, P. 467-471.

Röse, K.; Liu, L.; Zühlke, D. (2002): *A Model of Usability Engineering Costs for International User Interface Design.* In: Luczak, H.; Cakir, A.; Cakir, A.: WWDU 2002, Work With Display Units, World Wide Work. Proceedings of the 6th International Scientific Conference on Work with Display Units, Berchtesgaden, May 22-25, 2002, S. 281-284.

Röse, K.; Zühlke, D.; Liu, L. (2003): *INTOPS 2: The Machine User Interface Design for the Chinese Market.* Final Report. Projektabschlußbericht des industriegeförderten Projektes, Design for Mainland China' im Zentrum für Mensch-Maschine-Interaktion, Fortschritt-Bericht pak, Nr. 8, Universität Kaiserslautern, 2003. *im Edit.*

Shih, H. M. and Goonetilleke, R. S. (1998). Effectiveness of Menu Orientation in Chinese. In: *Human Factors*, Vol. 40, No.4, P.569-576.

Zühlke, D., Romberg, M. and Röse, K. (1998). Global demands of Non-European Markets for the design of User-Interfaces. In: *Proceedings of the 7th IFAC/IFIP/IFORS/IEA Symposium, Analysis, Design and Evaluation of Man-Machine Systems.* Kyoto, Japan (1998-09-16-18). Kyoto : IFAC, Hokuto Print : Japan.

Web-Based Applications Using Pen-Based Interfaces and Network-Based on-line Handwriting Recognition

Takeshi Sakurada, Mitsunori Yorifuji, Motoki Onuma, Masaki Nakagawa

Tokyo University of Agriculture & Technology
Naka-cho 2-24-16, Koganei-shi, Tokyo, 184-8588, Japan
takes@hands.ei.tuat.ac.jp

Abstract

This paper presents several applications and their user interfaces that employ pen-based interfaces and network-based on-line handwriting recognition. Pen-based interfaces are again attracting people's attention since several platforms are now available such as Microsoft Tablet PC, Anoto pen and paper, E-pen and many more. We are interested in employing them for WWW applications. We have developed several pen-based applications such as handwritten memo application, questionnaire answering application, crossword puzzle application, etc. All the applications send handwriting to the handwriting recognition server when necessary to have it recognized by the server so that they do not need to employ handwriting recognition themselves. We discuss their merits and the UI design based on the client-server handwriting recognition architecture.

1 Introduction

Pen-based interfaces are attracting people's attention again due to the wide spread of mobile PDAs (Personal Digital Assistant) with pen input, expanding sales of interactive electronic whiteboards, announcement of Anoto Pen from Anoto AB, shipping of Tablet PC of Microsoft Corp., etc.

Pen input has several benefits. It does not require large space. People can use it without training. Pen input on interactive whiteboards can easily attract the attention of an audience. People can express or annotate their ideas without being bothered by how to use it. Thinking is not interrupted by the actions for writing.

There are limitations and restrictions, however, on hardware and systems software (operating systems: OS) on which pen-based applications can be developed by employing a powerful handwriting recognition engine. Moreover, it seems almost impossible to develop consistent applications across the variety of environments employing the same quality or level of handwriting recognition without introducing new software architectures.

This paper presents several applications and their user interfaces (UI) that employ pen-based interfaces based on the network-based software architecture composed of an on-line handwriting recognition server and applications running on a variety of hardware and systems software. These applications can invoke handwriting recognition independent from the environment.

2 Problems of developing applications with handwriting recognition

We have been developing pen-based applications [M,Nakagawa(1999), H.Bandoh(2000),] using our own handwriting recognition engine [O.Velek(2002)]. In order to develop these applications, however, programmers must know how to incorporate the recognition engine. Moreover, sufficient hardware resources of CPU power and memory space are necessary for installing our recognizer. Therefore, it is hard to port the latest recogniser into a small PDA without sacrificing part of the performance.

Another problem is concerned with the software platform used to develop and run pen-based applications. We have been employing Microsoft Windows OS. This is the most powerful platform, but poses a problem when transporting the applications to other operating systems.

Due to these problems, we have chosen to employ a client-server architecture based on a server-side handwriting recognition engine and client applications, and develop applications assuming Web environment as a new common platform.

In the following sections, we begin with Web-based applications, and then present the client-server scheme for handwriting recognition systems through networks. Applications can be developed without including a handwriting recognition engine and executed independent of the problem of systems environment and hardware resources.

3 Prototyping of Web-based applications

We made several pen-based client applications, some of which are Web-based and others are developed in Java. These applications can be executed independent of operation systems. In the subsequent sections, we will show the following applications; Freehand Notepad, Questionnaire Answering, Crossword Puzzle, and Mosaic Kanji Game.

3.1 Freehand Notepad

Fig.1 shows a screenshot of our Freehand Notepad application on which one can take notes freely in handwriting and have the notes be recognized when necessary. The free handwriting recognition requires character segmentation, character recognition, and context processing so that its demand on CPU power is very large. By making the recognition server solve these tasks, clients are freed of any CPU requirement.

3.2 Handwriting Questionnaire Answering

Fig.2 shows a screenshot of our Handwriting Questionnaire Answering system. A writer can tick one of several possible answers to a question with his electric pen, after which the tick is automatically recognized. In case of free text input, the user can write any text, which is then recognized by the handwriting recognition server. This application allows users to answer questionnaires like on a sheet of paper, on a Tablet PC or PDA. The answers are collected and summed up into a report automatically.

Fig.1 Freehand Notepad.

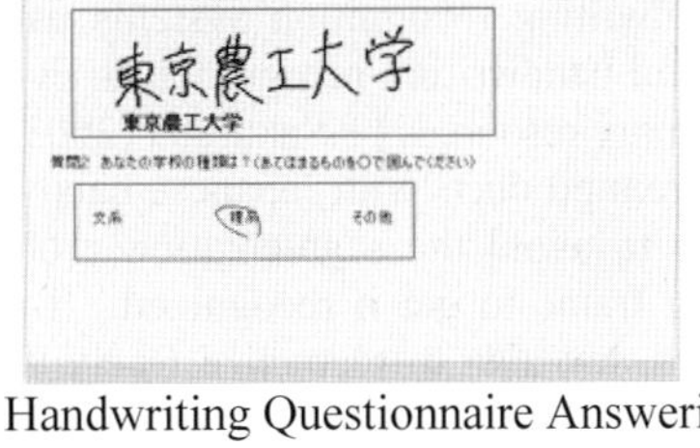

Fig.2 Handwriting Questionnaire Answering.

3.3 Crossword Puzzle

Fig.3 shows a screenshot of our Crossword Puzzle Game. A player can write a letter in a blank square and the software recognizes and displays the printed character. This provides a far better interface than mouse and keyboard, since a player can write a letter by simply moving its hand and writing.

3.4 Mosaic Kanji Game

Fig.4 shows a screenshot of our Mosaic Kanji Game. Although a player may know the rough shape of a kanji (Chinese character) and how to read it, it is often difficult to actually write the character. This game is for learning how to write Kanji correctly.

The player looks at the tessellated kanji on the left side and tries to write the kanji in the right box, guessing from the rough shape. Then, the application tells the player if his/her handwriting is the correct answer by asking the handwriting recognition engine. One's memory of writing is often vague, especially for complicated kanji patterns.

Asian children must be able to write thousands of kanji patterns, so they have to spend some time almost everyday to write and memorize Kanji patterns. This application helps them learn Kanji patterns with fun even on a small PDA.

Fig.3 Crossword Puzzle.

Fig.4 Mosaic Kanji Game.

4 Network-based On-Line Handwriting Recognition System

The above-mentioned applications using handwriting input do not have integrated handwriting recognition. Instead, they access a handwriting recognizer through a network. In the following subsections, we will describe our client-server architecture for handwriting recognition.

4.1 System Requirements

Handwriting recognition systems using networks have been developed to provide the most powerful handwriting recognition to even small and slow systems without sufficient CPU power or memory resources [T.Sakurada (2003)].

Current products with Japanese handwriting recognition on the market employ character writing frames to avoid the segmentation problem where a user writes each character into a separate writing frame to get it recognized. To not only recognize very casually written and deformed patterns, but also to recognize freely written text without writing frames, we have developed a highly reliable recognizer based on a combination of on-line and off-line recognition methods

[O.Velek(2002)] as well as a frame-free, character-orientation-free and line-direction-free handwriting recognition system [M.Onuma(2003),T.Oki(2003)]. The users do not need to care about the input frames, character orientation, line direction or size of characters. They can naturally write and input characters like writing on a sheet of paper with a pen. However, this advanced handwriting recognition engine requires far larger CPU power and memory resources, so it is difficult to run it on a PDA while satisfying real-time constraints.

On the other hand, PHS and wireless LAN make the Internet accessible anytime, anywhere. Therefore, we employed a client-server scheme for handwriting recognition. The very powerful frame-free handwriting recognition is available from any system, requiring only Internet connection capability, so that even mobile terminals with little CPU power and small memory resources can apply sophisticated handwriting recognition by using the network infrastructures.

4.2 Outline of the Client-Server System

Fig.5 shows the overview of the system. The system consists of three parts: client agent, server agent, and recognition engine. The client agent accepts handwriting input and sends it to the server agent. The server agent invokes the recognition engine to get candidates for the recognition result of the inputted pattern and return them to the client agent. The client agent displays them to the user. The user chooses the intended answer from the candidates. If necessary, the user corrects the pattern, which is then resent to the server, and the input process starts again.

In the actual experiment with an 28.8kbps analog modem, we have verified that our system performs input-frame-free character string recognition and returns recognition result in real-time. Moreover, we have also observed that our client-server architecture processes handwriting patterns at much higher speed than many of the clients running their own client-side character recognition.

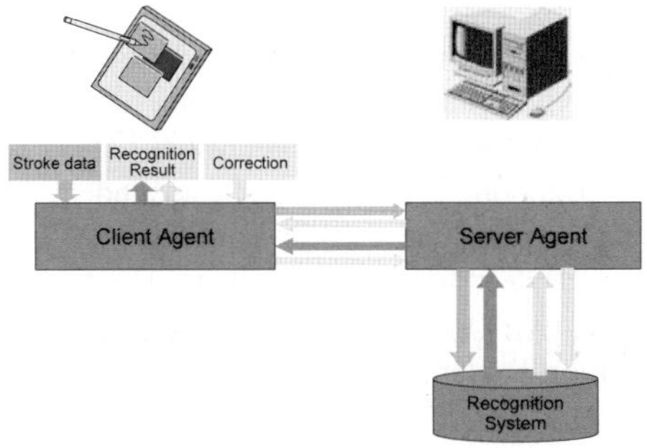

Fig.5 The overview of the system.

4.3 Advantages of the architecture

When this client-server system is employed, it becomes unnecessary to prepare a recognition engine for every environment on the client side. Therefore, the recognition engine imposes no constraints on the client's hardware and software, and the development of recognition engines for different environments becomes very cheap. Moreover, this architecture allows us to develop applications without understanding the details of a handwriting recognition engine.

Improvements or modifications of the handwriting recognizer require only changes on the server side. Once the changes are made, all clients can immediately and simultaneously benefit from these adjustments.

Different client environments can share one handwriting recognition engine, so that a writer can customize all environments simultaneously. For instance, when a person uses two or more PDAs and Tablet PCs, improvements of the recognizer are automatically distributed to all environments rather than requiring the writer to spend time and effort to update all of them independently. Storing all the handwritten patterns and their accepted or corrected recognition results on the server enables an even more powerful customization. In this case, the database of handwritten patterns is automatically expanded allowing continual training of the recognition engine.

5 Conclusion

This paper presented several applications and their user interfaces (UI) that employ pen-based interfaces and network-based on-line handwriting recognition. This architecture frees applications of incorporating a handwriting recognition engine and thus of hardware and software requirements imposed by the engine. We developed client applications assuming the Web environment as a common platform to ensure portability as high as possible. All applications shown here take advantage of the combination of pen interfaces and our client-server architecture for handwriting recognition.

Acknowledgements

This work is being supported by the program "The R&D support scheme for funding selected IT proposals" of the Ministry of Public Management, Home Affairs, Posts and Telecommunications.

References

H. Bandoh, H. Nemoto, S. Sawada, B. Indurkhya and M. Nakagwa: "Development of Educational Software for Whiteboard Environment in a Classroom," Proc. of International Workshop on Advanced Learning Technologies 2000, pp.41-44 (2000.12).

M. Nakagawa, K. Hotta, H. Bandou, T. Oguni, N. Kato and S. Sawada: "A Revised Human Interface and Educational Applications on IdeaBoard," ACM CHI99 Video Proceedings and Video Program and also CHI99 Extended Abstracts pp.15-16 (1999.5).

M. Onuma and M. Nakagawa: "An On-line Writing-box-free Line-direction Free and Character-orientation Free Recognition System for Handwritten Text," (in Japanese) Technical Report for IEICE Japan, PRMU2002-197, Vol. 102, No. 555, pp.49-54 (2003. 1).

O. Velek, S. Jaeger and M. Nakagawa: "A Warping Technique for Normalizing Likelihoods of Multiple Classifiers and its Effectiveness in Combined on-line/off-line Japanese Character Recognition," Proc. 8th International Workshop on Frontiers in Handwriting Recognition, Niagara-on-the-Lake, pp.177-182 (2002.8)

T. Sakurada, M. Yorifuji, M. Onuma and M. Nakagawa: "Design of a Client-Server Scheme for On-Line Handwritten Character Recognition through Networks," (in Japanese) Human Interface Symposium 2002, pp. 687-688 (2002.9).

T. Oki and M. Nakagawa: "Implementation and Evaluation of a Combined Recognition System using Normalization of Likelihood Values," (in Japanese) Technical Report for IEICE Japan, PRMU2002-196, Vol. 102, No. 555, pp.43-48 (2003. 1).

Information Hiding with a Handwritten Message on PDA

Norihisa Segawa *Yuko Murayama* *Masatoshi Miyazaki*

Iwate Prefectural University
152-52,Takizawa-aza-sugo,
Takizawa, Iwate,020-0193, Japan

sega@iwate-pu.ac.jp murayama@iwate-pu.ac.jp miyazaki@iwate-pu.ac.jp

Abstract

In recent years, more attention is paid to the subliminal channel used as one of the technology of security. A subliminal channel is the technique from which the information, that we wants to assure its privacy, is hidden in the normal information. We have proposed the technique of performing a subliminal channel for handwriting information. This paper reports the construction of a handwriting subliminal channel using a personal digital assistant (PDA).

1 Introduction

The Internet has been evolving exponentially those two decades, and various types of communication systems for Computer Supported Cooperative Work (CSCW) are available throughout networks. Some of those systems provide handwriting functions, and handwritten messages are used often for authentication. In other communication systems, however, one needs hiding information. A requirement of a communication system is to provide a subliminal channel for secret communication. This paper reports the construction of handwriting information hiding with a personal digital assistant (PDA). For a PDA with a low power of CPU with a very small memory, the technique of building a handwriting subliminal channel is reported, and the prototype system built according to the technique is reported.

2 Vector Drawing for Handwriting Information

In vector drawing, handwritten information is encoded as a set of points and straight lines (Figure 1) [1]. For sampling we takes two or more points out of a user's handwriting periodically.

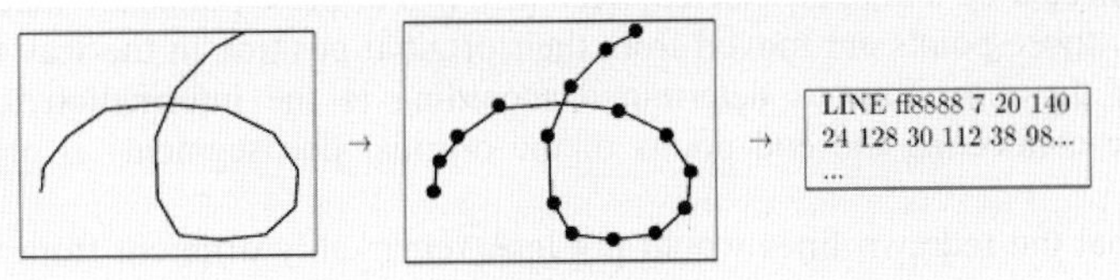

Figure1 Sampling and Encoding in Vector Drawing

When a user writes a stroke during the time between T0 and T1, this stroke can will be expressed as follows:

$$X=F(T) \quad (T0 < T \leq T1)$$

$$Y=G(T) \quad (T0 < T \leq T1)$$

F: a function, which gives the X-coordinate of the trace of a stroke (non-liner) at time T

G:a function which gives the Y-coordinate of the trace of a stroke (non-liner) at time T
T: A lapse of time T of handwriting
X: the X-coordinate of handwriting
Y: the Y-coordinate of handwriting

When the handwriting is sampled in the time interval deltaT, the sampling of a handwritten formula is following.

$$Xs=F(T) \quad (T0 < T \leq T1)$$

$$Ys=G(T) \quad (T0 < T \leq T1)$$

Xs: the X-cooridnate of the sampled point of handwriting
Ys: the Y-coordinate of the sampled point of handwriting
T: A lapse of time T of handwriting

The number of sampling, n, is expressed as follows.

$$n=\frac{(T1-T0)}{deltaT}$$

3 Information Hiding in handwriting function

3.1 Overview

Information hiding is used for an exchanging messages among specific users, which should not be noticed by the others.
We presume that handwriting information is encoded in vector drawing. Our proposal is that the confidential information which is hard for the third party to detect is added to vector drawing. In other words, specific users can exchange the confidential information over a public channel while ordinary users exchange handwritten messages over the same channel without noticing the secret exchange.

3.2 Embedded information

Figure 2 shows the technique of embedding the information shared only by specific users, in handwritten messages which everyone else can read[1]. Firstly a line segment which is composed of two point connected by a line, is selected. Secondly two or more points are taken out of the line segment. Thirdly those points are moved from their original position in the way that the distance and direction of the movement is determined according to the information to be embedded. Fourthly the line connecting the end points of the original line segment is redrawn via those moved points.
It is important that the redrawn lines would not look remarkably different from the original line segment.

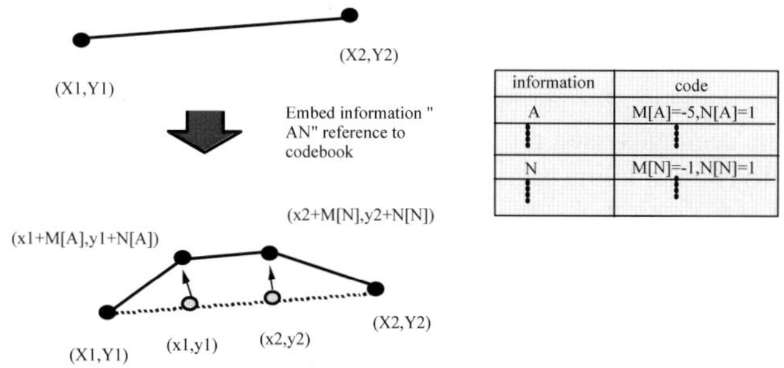

Figure2 Embedded information

A more detailed algorithm is given below:

(1) Divide equally the straight line between A and B into n segments. Figure2 shows an example in which n is 3; we get two more points on that line segment. One of them would be used for embedding information. The shared key information would tell which one to be used.

(2) Produce a codebook, as shown in figure 2, which tells the correspondence between an information item and a code which is composed of the distance and direction of the movement of a point; the direction is expressed in terms of both X-axis and Y-axis. The maximum size of the codebook is:

$$\left|\frac{M}{x}\right| \cdot \left|\frac{N}{y}\right|$$

M: the maximum movement in X-axis
N: the maximum movement in Y-axis
x: the unit length in X-axis
y: the unit length in Y-axis

(3) Embed information into the handwritten message by looking up an information item in the codebook, obtain the corresponded movement information, move the point accordingly, and redraw the line via newly moved point.

3.3 Decoding information

The following procedure is performed in order to extract the embedded information (Figure 3)[1]:

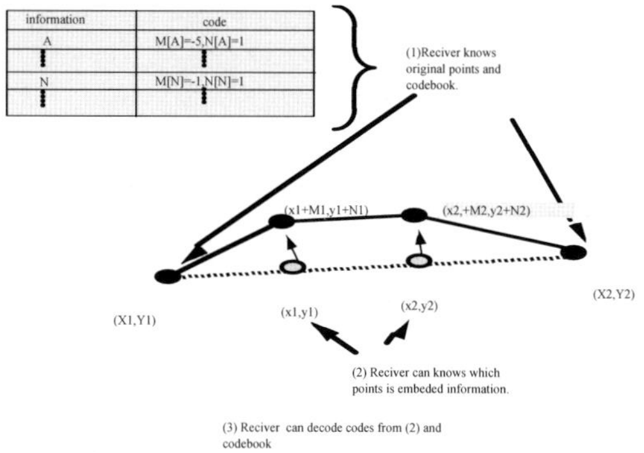

Figure 3 Decoding information

(1) Receive the following key information from the sender who embedded information:
(A) The set of the line segments used for embedding information
(B) The number of division of a line segment
(C) The codebook

(2) Identify the points using which information was embedded, on the basis of key information above.
(3) Extract the embedded information, using the points identified above.
The embedded information is decoded by looking up a code in associated with point movement with the codebook.

4 Information hiding function into PDA

This section describes the technique of operating information hiding stated in section 3 by PDA.

4.1 Overview of PDA

Since the power of electronic products was miniaturized and saved in recent years, various carried type terminals as PDA are developed. A small computer is in a cellular phone and that can be used now also as a computer terminal[2].
PDA has the following features.
(1) It is a very small.
(2) It accepts using a pen.
(4) A mobile communication function can be used.
(4) The capacity of a memory is small.
(5) There is specified OS for PDA

4.2 Construction of a prototype system using a PDA

As section 3.1 described, it is difficult to build a information hiding system on PDA . Since each PDA has specified OS, it is necessary to build a specified system for each PDA. Then the portability be comes necessary , in this research, we chose the Java to assure the portability.

Java for PDA is a subset of Java 2 Standard Edition (J2SE) used conventionally.
Exsamples are given below:
(1) Java 2 Micro Edition (J2ME)
(2) ChaiVM
(3) KaffeVM
(4) Waba
Although (1) is native Java, in this research, we choose (4) Waba. Waba is a language which wabasoft [3,4] developed. Waba can be used with a free license. Waba is designed as a perfect subset of the grammar of a Java language, and the class file and the byte code are also designed as a subset of Java. Therefore, the development environment of the existing Java can be used as it is. Moreover, Waba has the feature that there are classes for making it operate by the small device. Therefore, the system developed this time can be operated by various PDA by using Waba.

4.3 A prototype system for information hiding on PDA

We construct a prototype system for an information hiding explained in Chapter 2 to PDA using Waba.

The prototype system has following functions:
(1) A sender writes handwriting message on PDA
(2) The sender embeds information at handwriting message.
(3) Sender transmits the handwriting message, which embedded information to a receiver by network.
(4)The receiver extracts embedded information
This prototype system is running on a Palm OS. Palm is a system based on a pen input. The system is implemented as Applet of Waba. The function in which a user writes handwriting message, the function to conceal information, and the function that transmit and receive handwriting information are included in this Applet. This Applet size is about 10KB. The size for storing handwriting information is 2KB.
Figure 4 is an example of Applet. Figure 5 is a result of Applet on Palm. A user can write handwriting message freely on this screen. When a sender pushes the [embedded] button in the lower left of Figure 4 after finishing the writing, the information will be embedded by the technique shown in Section 2.The embedded data is sent to the receiver.
When receiver pushes the [extract] button, which is the 2nd one to the left of Figure 4, after displaying the handwriting information sent by the sender, the information currently embedded will appear.

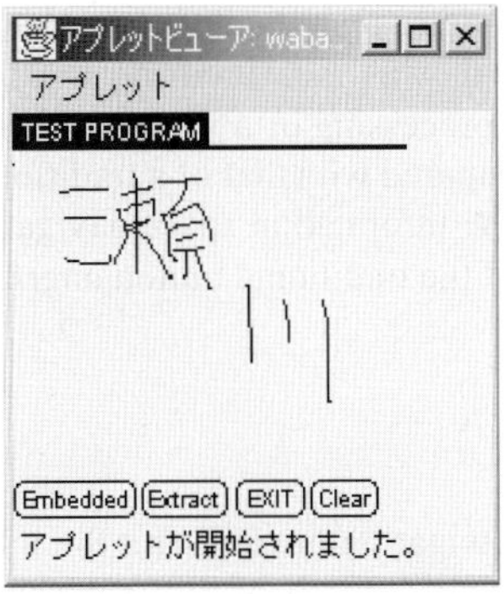

Figure 4 An example of Applet

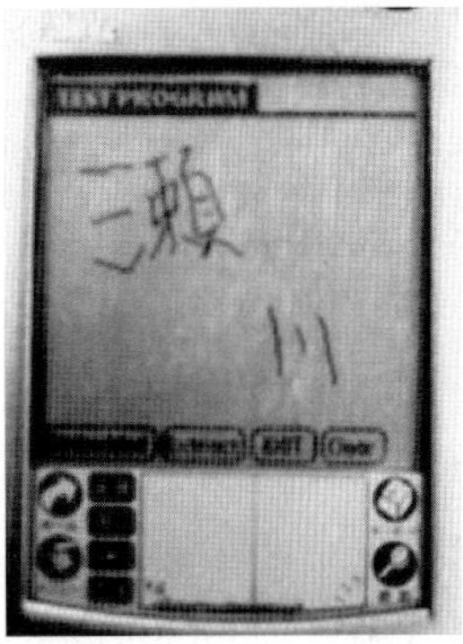

Figure 5 Applet on Palm

5 Conclusion

This paper showed the construction of a handwriting information hiding using the PDA. We considered that handwriting messages is vector drawing.The technique of this information hiding was used and it implemented by using Waba on PDA.
Future work includes the evaluation of the strength of our algorithms as well as implementing the proposed technique using all types of Java for PDA.

References

[1] Norihisa Segawa, Yuko Murayama, Masatoshi Miyazaki: Information Hiding with a Handwritten Message in Vector-drawing Codes, Proceedings of the 35th Hawaii International Conference on System Sciences- 2002, CD-ROM Proceedings, (2002)

[2] Norihisa Segawa, Yuko Murayama, Masatoshi Miyazaki:A Message Board on WWW for On-Door Communication: Proceedings of ACM Multimedia '99 (part 2) , pp187-190, (1999)

[3] wabasoft: http://www.wabasoft.com/ (2002)

[4] Waba World: http://www.cc.yamaguchi-u.ac.jp/~shingo/WabaWorld/ (2002)

An Intuitive Information Space Navigation Method based on the Window Metaphor

Yu Shibuya, Tomoya Narita, Takeshi Yoshida, Itaru Kuramoto, Yoshihiro Tsujino

Kyoto Institute of Technology, Japan
Matsugasaki, Sakyo-ku, Kyoto 606-8585 JAPAN
shibuya@dj.kit.ac.jp

Abstract

In this paper, a new navigation method with a portable information terminal for 3D information space has been proposed. With the method, a user can look around the information space by moving the terminal itself. Furthermore he/she can look into the information space through the display of the terminal as if to look into the outer space through a real window. In order to realize the proposed navigation method, direction of the terminal was measured by a 3D angular accelerometer and user's viewpoint was detected by image processing of a video captured image. Usability of an interface using the proposed method was compared with that of a traditional button interface experimentally. From the subjective evaluations, the information space navigation of the proposed method was easier and more interesting than that of the traditional button interface.

1 Introduction

In our current project, we have been concerning intuitive navigation methods for the information space using a portable information terminal (PIT). There are many kinds of PIT, such as a PDA or a mobile phone. However, most of them have a relatively small display and a few small input devices, so that it is difficult to navigate a huge information space with the PIT.

Fitzmaurica has proposed to use positions and orientation of palmtop computers as input (Fitzmaurica, Zhai, & Chignell, 1993). Rekimoto has used both tilt and buttons to build several interaction techniques of small screen devices (Rekimoto, 1996). Rock'n'Scroll input method (Bartlett, 2000) has used two accelerometers and let users gesture to scroll, select, and command an application without other input devices. In this paper, in order to make the navigation easier with the PIT, we introduce a new navigation method based on the window metaphor and the parallel reality.

2 Parallel Reality and Window Metaphor

Parallel reality differs from both virtual reality and augmented reality. In the virtual reality, users immerse themselves in the information space. In the augmented reality, there is tight connection between the object of the real world and the information. On the other hand, in our proposed parallel reality, there is an information space which is flexibly connected with the real world. That is, the information space is able to be designed independently with the real world but it is easily connected to the real world when the user wants. The user in the real world can look around the information space through a virtual window, a display of the PIT, which connects the real world to the information space.

The user can look around or navigate the information space through the virtual window as he/she does in the real world. For example, if the user wants to look at the left side of the window view, he/she will move his/her head to the right and look into the window. If he/she gets closer to the window, he/she can look the wider view of another side of the window. Furthermore, our virtual window is not fixed in the real world, so the user can hand and move it to the desired direction. These navigation methods are based on the window metaphor, so the user doesn't need any practice.

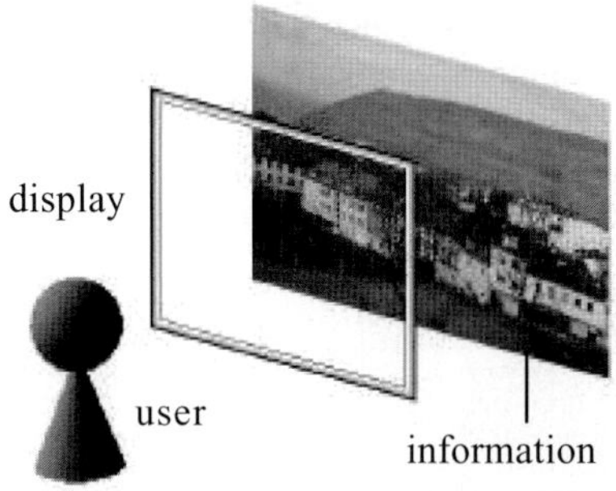

Figure 1: A concept of window metaphor

3 Prototype System

A prototype system with our proposed navigation method is shown in Figure 2. As shown in this figure, the prototype consists of a display, a small video camera, and a 3D angular accelerometer (NEC-TOKIN 3D Motion Sensor). All devices are connected to a PC (Intel Pentium III 933MHz) with the cables. A user's face is captured through the video camera in QVGA (320x240 [pixel]) size and processed by the PC. Depend on the user's view point and the position or direction of the terminal, the PC draws a suitable feedback image on the display.

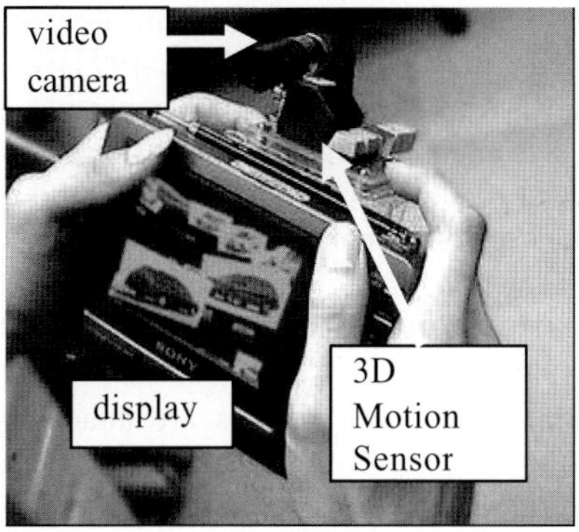

Figure 2: The prototype system

There are three key issues to realize the proposed navigation method. They are detecting and tracking the user's view point, measuring the position and direction of the terminal, and the suitable mapping between the user's action and the information space navigation.

3.1 Detecting and Tracking User's View Point

In order to extract and track the user's eyes, we refer an idea of the circular frequency filter (Kawato & Ohya, 2000). As shown in Figure 2, a small video camera is attached on the PIT and used for capturing the user's face. In the video capture image, suppose that there was a circle and its center was located at the middle of two eyes, and its circumference passes across both eyes (See Figure 3). In such case, there are at least four dark points while checking the brightness through the circumference (See Figure 4). Such dark points indicate the user's eyes and brows. In our system, we prepare many brightness templates of such circle and use them to find the similar pattern from the captured image. We regard the center of the circle as the user's view point. The size of the diameter of the circle depends on the distance between the PIT and the user's eyes. So, the PIT can not only identify the direction to the user's view point but also measure the distance.

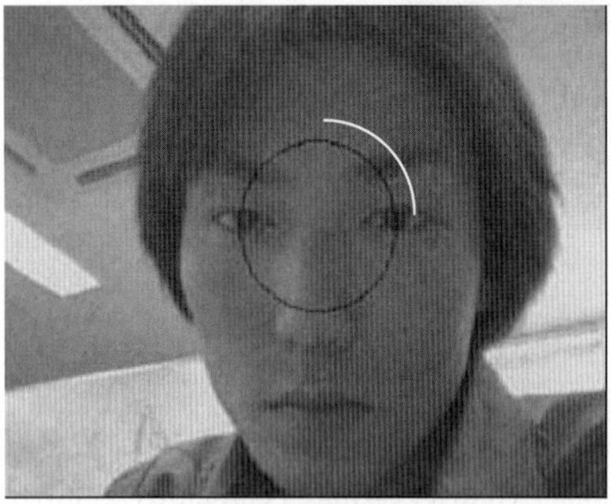

Figure 3: A video captured image for the user's view point detection

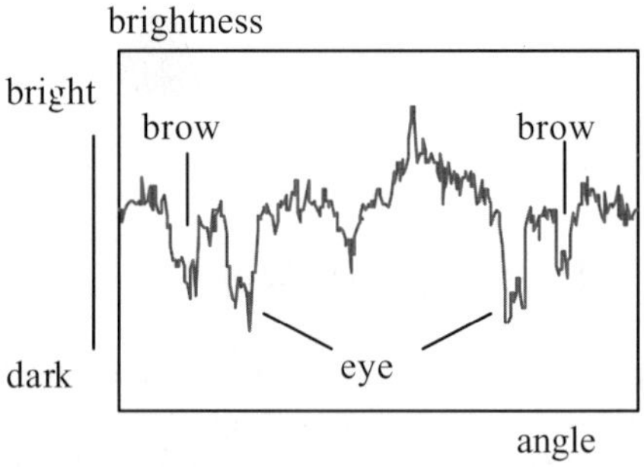

Figure 4: A change of brightness along with the circle of Figure 3

There are many studies that use the head tracking for human-computer interaction. For example, Fish Tank Virtual Reality system (Arthur, Booth, & Ware, 1993) has provided real-time display of 3D scenes using stereopsis and dynamic head-coupled perspective. In the system, the user's head position was measured by the mechanical tracker and his/her movement was restricted by the tracker. Rekimoto has introduced a vision-based head tracker but it used image subtraction method (Rekimoto, 1995). Both methods are not suitable for the PIT.

3.2 Measuring the Position and Direction of the Terminal

For the positioning of the PIT, two small sensors are attached to it. They are a 3D axis accelerometer and a 3D axis angular one. With these accelerometers, the system can measure the position and direction of the terminal and make suitable response to the user. For example, if the terminal turns to left, the display shows the left side out of the current window. If the terminal steps forward, the display shows the deeper part of the information space. However, up to this day, our prototype system uses one accelerometer and can detect only the direction of the terminal.

3.3 Mapping between User's Action and the Navigation

Furthermore, this terminal can also detect the user's action with above sensors. The action is also usable to navigate the information space. For example, if the user pulls the terminal forward to him/her quickly, the system can decide that he/she wants to go close to the object in the display. If he/she pulls the terminal twice quickly, the system can decide that he/she wants to select the object in the display. These mappings between the user's action and the system's reaction are flexible but we should find suitable mapping through our further experiments.

4 Experiment

A prototype using the window metaphor and the parallel reality was implemented and evaluated experimentally. The prototype was compared with the cross direction game pad as a traditional interface.

4.1 Information Space

In the experiment, information was drawn on sheets, like photographs, and they were put in the 3D information space. A screen shot of the information space is shown in Figure 5. The user can look around the information space by moving the PIT itself or by using the cross direction pad. When he/she finds a target information sheet, he/she locates it at the center of the display and pushes the select button. After the selection, a cubic pointer was overlapped on the selected sheet as shown in Figure 5.

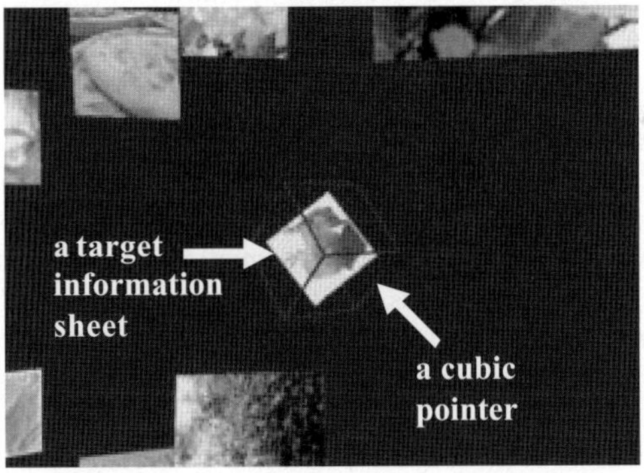

Figure 5: A screen shot of the information space

4.2 Task

There were eleven participants and all of them were students of our university. Thirty information sheets or photographs were allocated in the information space. A participant was asked to select a target information sheet which was rotating. After the selection, next target began to rotate. Each participant was asked to select three of thirty information sheets. As an operation time, elapsed time from the first sheet selection to the third or last sheet selection was measured. The measurement was easier and more accurate than to measure both starting time and ending time of the task. After each task, the participant was asked to answer the questionnaire.

4.3 Result

The averaged operation time is shown in Figure 6. From this figure, the operation time of the prototype was slightly longer than that of the cross direction pad but its difference was not significant. While most participants were not familiar with the proposed interface, they had been already familiar with the traditional cross direction pad. However, the proposed interface was enough fast as the cross directional interface. In other words, the result indicated that the proposed method was as good as the traditional method and it was usable for the information space navigation. Furthermore, from the subjective evaluations, the information space navigation with the proposed method was easier and more interesting than that of the traditional button interface.

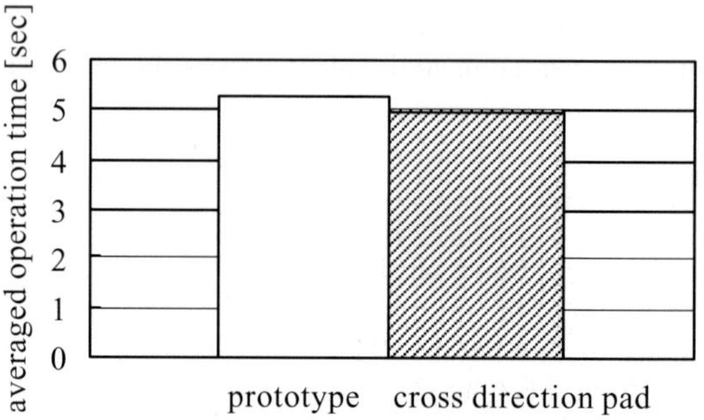

Figure 6: Averaged operation time

5 Conclusion

Our proposed method is suitable for the navigation using the PIT in the huge information space. For example, it might be usable for browsing or searching desired information in the Web space, which consists of huge amount of linked Web pages. While our proposed method offers an intuitive direct access to the object in the 3D information space, such direct access might be difficult by the traditional interface, such as a keyboard or a game pad. Furthermore, it has great advantage of having no input devices explicitly. With this advantage, we can make a new type of the PIT which has a large display with few buttons.

References

Arthur, K. W., Booth, K. S., & Ware, C. (1993). Evaluating 3D task performance for fish tank virtual worlds. *ACM Transaction on Information Systems*, 11 (3), 239-265.

Bartlett, J. F. (2000). Rock'n'Scroll is here to stay. *IEEE Computer Graphics and Applications*, 20 (3), 40-45.

Fitzmaurice, G. W., Zhai, S., & Chignell, M. H. (1993). Virtual reality for palmtop computers. *ACM Transaction on Information Systems*, 11 (3), 197-218.

Kawato, S. & Ohya, J. (2000). Two-step approach for real-time eye tracking with a new filtering technique. *IEEE Int. Conf. on Systems, Man & Cybermetrics 2000 (SMC2000),* 1366-1371.

Rekimoto, J. (1995). A vision-based head tracker for Fish Tank Virtual Reality – VR without head gear -. *Virtual Reality Annual International Symposium 1995 (VRAIS '95)*, 94-100.

Rekimoto, J. (1996). Tilting operations for small screen interface. Proceedings of the 9th annual ACM symposium on User interface software and technology (UIST '96), 167-168.

Optical Stain: Amplifying Vestiges of a Real Environment by Light Projection

Yoshinari Shirai, Tatsuo Owada[1], Koji Kamei and Kazuhiro Kuwabara

NTT Communication Science Laboratories, NTT Corporation
2-4 Hikaridai, Seika-cho, Soraku-gun
Kyoto 619-0237 Japan
{way, kamei, kuwabara}@cslab.kecl.ntt.co.jp

Abstract

Vestiges in the real world convey tacit information that can be clues for understanding activities occurring in one's surroundings. In this paper, we define such information that helps us understand our surroundings as "real-environment-oriented awareness information". However, it takes a long time for some vestiges to emerge, and there may not even be any vestiges remaining, depending on the situation.

To solve these problems, this paper proposes *Optical Stain*, which is a concept for a system that conveys real-environment-oriented awareness information by inconspicuously projecting it as artificial vestiges on real environments. This paper also describes the *Optical Stain*-based bulletin board system that recognizes human activities inside environments with a camera, and superimposes that activity information as artificial vestiges onto the environments with a projector, in addition to outlining the effectiveness of the proposed method that uses this system.

1 Introduction: What is real-environment-oriented awareness information?

The purpose of our research is to amplify "vestiges" in real environments to help people understand his/her surroundings. Vestiges in the real world, including color fading of posters and footprints near a cafe, convey tacit information to a person. For instance, the color fading of posters suggests their age and footprints near the cafe suggests its popularity to people. This tacit information differs from information about objects themselves (e.g., poster or name/type of cafe); it conveys meta-information that can be a clue for understanding some activities occurring in one's surroundings. By paying attention to such kinds of information, one can get a glimpse into those activities, even if the person was not present at the time the activities occurred. We define such types of information as "real-environment-oriented awareness information".

It takes, however, a long time for vestiges to emerge (e.g., fading of posters' color requires a lot of time), and there may not even be any vestiges remaining, depending on the situation (e.g., footprints are not left on a dry day). Therefore, we propose a system to amplify vestiges reflecting people's activities, and whose presentation span is controlled. Furthermore, the system artificially presents vestiges onto real environments by light projection. We call this concept *Optical Stain*. (Hereafter, this paper describes such amplified vestiges as awareness information.)

---

[1] Currently with NTT Publishing Corporation. (owada@nttpub.co.jp)

A variety of awareness support systems that aim to make specified users aware of context in remote places have been researched in the Computer Supported Cooperative Work (CSCW) community (Hudson *et al.*, 1996) in addition to research aimed at calmly conveying others' contexts through physical objects in surroundings (Kuzuoka *et al.*, 1999, Wisneski *et al*, 1998). Our research aims to convey clues for understanding activities in one's surroundings by sharing awareness with unspecified people in the same place.

2 Concept of amplifying vestiges of a real environment

We consider the following three points in designing a system based on the Optical Stain concept that conveys awareness information in a real environment: (1) conveying where the activity occurred, (2) conveying awareness information after the activity occurred, and (3) conveying awareness information with low-intensity stimuli. Noting these points, we propose the following three presentation criteria.

(Criterion-a) Presenting awareness information at the place where the activity occurred
Awareness information should be presented to help people realize the place where the activity occurred. Therefore, we aim to present awareness information at the actual place where the activity occurred, making it easy for people to understand intuitively where the activity actually took place.

(Criterion-b) Presenting awareness information for longer than the actual activities lasted
Awareness information should be conveyed to people who were not present at the scene when the activity occurred. People who didn't see the actual activity can perceive the activity from the awareness information presented, if it is presented for a certain duration after the actual activity had finished. Presenting it for a certain duration can also convey information about the frequency of an activity; people can understand the frequency from the degree of congestion caused by the overlapping of activities that occurred at different times.

(Criterion-c) Changing awareness information presented gradually
Awareness information should be provided without hindering a person's activity. This postulate requires that the awareness information should be presented as a low-intensity stimulus. Since rapid changes attract people's attention, awareness information that accompanies the changes should be presented with gradual changes. Although gradual change is not stimulative, people can perceive it subconsciously.

3 Instance of *Optical Stain*: Real environment that displays changes that occurred in the past

We applied *Optical Stain* to a bulletin board. This system recognizes activities captured in a certain environment with a camera, and superimposes the awareness information onto the environment with a projector (see Figures 1, 2).

We applied the system to a company bulletin board placed in the corridor of the author's laboratory building as a target. In this environment, two characteristic activities were performed on a daily basis on the bulletin board: (1) replacement of bulletins as time goes by, (2) browsing of the bulletin board by people passing by.

A bulletin board with *Optical Stain* presents two activities as vestiges following our three proposed criteria:

(1) After people replaced bulletins (criterion-b), replacement of bulletins was optically presented as bulletin-shapes in their replaced positions (criterion-a) and the images gradually disappeared (criterion-c).
(2) After passers-by read bulletins (criterion-b), browsing of the bulletin board by passers-by was optically presented as human-like-shapes at the bottom of the bulletin board (criterion-a), and these images also gradually disappeared (criterion-c).
Figure 3 shows an example of presenting vestiges on the bulletin board.

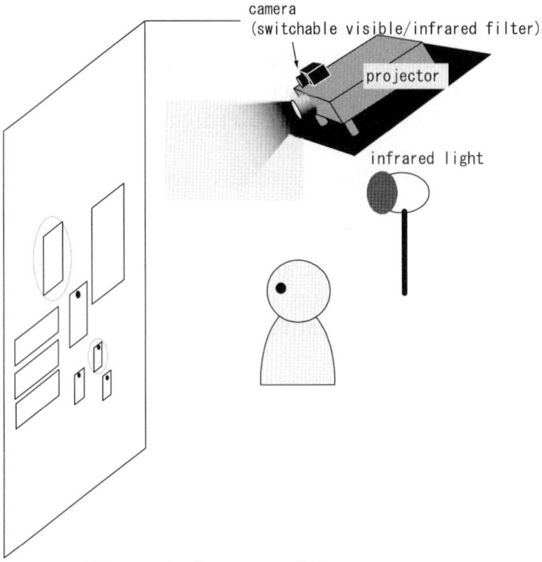

Figure 1: System architecture

Figure 2: System overview

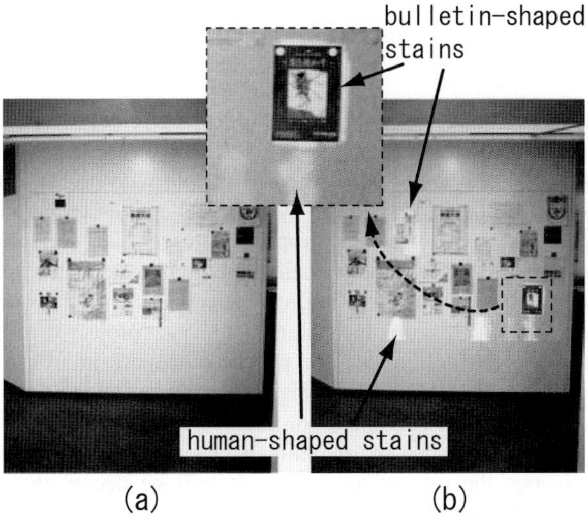

Figure 3: Bulletin board with *Optical Stain* applied
(a): before running the system, and (b): after

4 Experiment

We conducted experiments to confirm the following three points: (1) it influences people passing by the board; (2) people's activities are not disturbed; (3) people understand the meaning of the stains.

To compare the effect of an active system with an inactive one, we split 12 days into 4 terms and activated the system only during the 2nd and 4th terms. We recorded the area in front of the board for one term of this experiment to take some logs.

Because the system was installed in a new place and hence attracted people's attention, we used the last two terms to draw comparisons. That is, we compared the 3rd term (inactive) with the 4th term (active). After the experiment, we also distributed questionnaires to people who passed in front of the bulletin board during the experiment, and obtained the answer from 22 people out of 28.

Table 1 shows the result of video analysis and the following results were obtained from questionnaires.

- 26% of people noticed the stains on the board.
- 61% of people noticed stains in human shapes projected on the lower part of the board.
- 34% of the people answered that they felt bothered by the stains.
- About 20% of the people who noticed the stains correctly identified the meaning of the stains.

Table 1: Result of video analysis (compared 3rd term with 4th term)

	Inactive (3rd term)	Active (4th term)
Gross browsing time (sec)	197	148
Average browsing time (sec)	28.1	13.5
Percentage of people who paused in front of the board (%)	1.07	1.76
Percentage of people who glanced at the bulletin board side when passing by (%)	4.6	13.5

5 Discussion

From the results of this experiment, we obtained the following findings:

(1) It influences people passing by the board

According to the result of video analysis, the rate of people who take a look at the board when passing by increased about threefold when the system was active compared to when the system was inactive. This result suggests that the proposed presentation is recognized for a certain duration when presented, even if the presentation method is not accompanied by impressive animation, like blinking, blasting and librating, etc. Feedback from the questionnaires indicates that about 60% of people noticed stains in human-like shapes projected on the lower part of the board. This result supports the notion that the people who pass in front of the bulletin board recognized the proposed presentation when it was presented for a certain duration. However, less than 30% of people noticed the bulletin shapes on the board. As there is a difference in the rate of awareness between human-like shapes and bulletin shapes, we consider that the visible features of a presented place influence the rate of awareness. It will be necessary to further explore the presentation method that allows this influence in the future.

Moreover, the average board-browsing time when the system is active is almost half that of when the system is inactive. This result can be interpreted as the optical stains on the board fulfilled their function as clues to find primary information more efficiently. That is to say, this result suggests that people can make use of the optical stains when they browse the board.

(2) People's activities are not disturbed
According to questionnaire feedback, only about 30% of the people answered that they became bothered by the stains. This result suggests that our proposed criteria did not disturb people's activities.

(3) People understand the meaning of stains
In a system where real-environment-oriented awareness information is conveyed, not only the awareness information of others, but also our own activities, are presented in those environments. Thus, it appears possible that people gradually understand what types of activities will be presented onto local environments as they spend time within those the environments. Therefore, we expected that people would gradually come to understand the meaning of awareness information presented without special explanation when they often visit the environments. For that reason, we had not informed the subjects of the meaning of stains beforehand, but instead asked the meaning in the questionnaire after the experiment.

Our results show about 20% of all the respondents who noticed stains on notices or on the lower part of the board correctly understood their meaning. Though the percentage was small, the fact that there were people who could understand the meaning may suggest that some people can understand the meaning of awareness information adopted by this *Optical Stain*-based system. As the term of this experiment was short (active term was only 6 days), it is necessary to conduct an experiment over a longer interval to confirm this hypothesis.

As a result this experiment, we have come to the conclusion that vestiges fulfilled presentation criteria and can effectively convey real-environment-oriented awareness information.

6 Concluding remarks

This paper described the concept for a system that artificially leaves some vestiges on real environments. We proposed the three criteria mentioned above for artificially presenting vestiges, and investigated the effectiveness of these presentation criteria by using a bulletin board with *Optical Stain*. We found that presented vestiges fulfilling these criteria do not disturb people's activities and people can recognize these vestiges. We are planning to further develop systems based on the Optical Stain concept to convey real environment-oriented-awareness information.

Acknowledgement

We are most grateful to Mitsunori Matsushita, Takehiko Ohno, Takeshi Ohguro and the members of the Cooperative Systems Research Group for their helpful advice.

References

Hudson, S.E. and Smith, I. (1996). Techniques for Addressing Fundamental Privacy and Disruption Tradeoffs in Awareness Support Systems, *Proc. Computer Supported Cooperative Work (CSCW'96)*, 248-257.

Kuzuoka, H. and Greenberg, S. (1999). Mediating Awareness and Communication Through Digital but Physical Surrogates, *Proc. Human Factors in Computing Systems (CHI'99)*, 11-12.

Wisneski, C., Ishii, H. and Dahley, A. (1998). Ambient Displays: Turning Architectural Space into an Interface between People and Digital Information, *Proc. 1st International Workshop on Cooperative Buildings (CoBuild'98)*, 22-32.

Evaluation of a Text Entry Method for Mobile Devices

Peter Tarasewich

College of Computer and Information Science
Northeastern University
360 Huntington Avenue, 161CN
Boston, MA 02115 USA
tarase@ccs.neu.edu

Abstract

While handheld devices such as cell phones and PDAs continue to proliferate, there is still debate over what methods are best for user interaction with these devices. Text entry, an important part of this interaction, becomes increasingly complex as devices shrink in size. This study tested a PDA text entry method that used a thumbwheel. The character set, which included letters, numbers, punctuation marks, and a space, was implemented as a continuous loop. Turning the wheel upwards or downwards displayed different characters on the screen, and clicking the wheel selected a character. A "snap-to-home" feature was also tested. Thumbwheel-based text entry methods are a viable alternative to keypad, stylus, and voice input on ultra-small mobile devices.

1 Introduction

This research looks at text entry on a personal digital assistant (PDA) using a thumbwheel. As the wheel is turned, different characters are displayed on the PDA's screen. When the thumbwheel is pressed, the character currently displayed is selected. Inspiration for this research came from devices such as the Dymo label maker, but the same principal has also been used on video games and other electronic devices for character entry using knobs or wheels. The movement of a thumbwheel can also be simulated using three keys – two for the clockwise and counterclockwise motions, and one for selection. Research such as MacKenzie (2002) has explored the performance of three key and similar key-based text entry methods. However, our research is unique because:

- the study used text entry that included numbers and punctuation in addition to alphabetic characters;
- the entry technique was evaluated using a realistic mobile device form factor (a PDA) rather than a simulated environment; and
- a thumbwheel was used for text input, rather than a set of keys of buttons.

2 Background

.This section briefly discusses some of the performance testing that has been done with different text entry techniques on mobile devices. A comprehensive review of text entry methods for mobile computing can be found in MacKenzie and Soukoreff (2002). In addition, Tarasewich (2002) provides a discussion of interface design and usability issues for mobile devices.

Recent research has been addressing the usability concern of how well users can perform tasks using the assortment of keypads and keyboards found on many wireless devices. Looking at keypad text entry performance, Silfverberg, MacKenzie, and Korhonsen (2000) created models to predict the entry rates for multi-press, two-key, and linguistic-based keypad text entry methods. Using empirical data, they estimated that expert users could achieve rates of up to 27 words per minute (wpm) using thumb (one-handed) or index-finger (two-handed) input with the multi-press and two-key methods.

MacKenzie (2002) reported the results of an experiment that tested various three-key text entry methods. The keys were left and right arrow keys used to maneuver a cursor over a set of characters (the English alphabet) and a third key to select a highlighted character. One method tested kept the characters in alphabetical order during the selection process. The other method used "linguistic enhancement" to reorder the characters after each selection. This reordering, based on letter pairing probabilities in common English, was an attempt to minimize the cursor distance to the next chosen character. Both methods placed a space character first (before the characters), and "snapped" the cursor home to this position after each entry. Entry rates for each technique were about 9 wpm. Subjects performed slightly better using the enhanced technique than with the characters in a fixed order, but the difference was not statistically significant.

There have been many studies on soft keyboard performance. Those by Lewis, LaLomia, and Kennedy (1999) and MacKenzie and Zhang (1999) found that users could achieve speeds of up to 40 words per minute with a QWERTY layout on a soft keyboard, although speed varied with the devices used, the tasks performed, and the amount of practice. Alternate soft keyboard layouts can produce even higher text entry speeds than the QWERTY configuration, but usually after much experience with the alternate layout (e.g., MacKenzie & Zhang, 1999). A study by Zha and Sears (2001) showed that the size of a PDA soft keyboard did not affect data entry or error rates.

If a stylus is used to write input on the screen of a mobile device (using gesture or handwriting recognition), performance is generally much poorer compared to using any type of keyboard. Studies such as MacKenzie and Chang (1999) found that data entry rates of up to 18 wpm can be achieved using various gesture recognition systems. But these studies did not test performance using handheld devices. More recently, Sears and Arora (2001) compared Jot and Graffiti using Pocket PC and Palm devices, respectively. They used tasks that they felt were more realistic than previous studies, and kept track of data entry times and error rates. Novice data entry rates of 7.37 wpm were obtained for Jot and 4.95 wpm for Graffiti.

Voice recognition technology continues to improve, but there is still the question of how well it works for different applications and tasks. De Vet and Buil (1999) listed some general findings from user studies on the use of voice control compared to entering text data on limited-key devices. User operations that favor voice control included 1) direct addressing of content (e.g., calling out someone's name), 2) menu navigation and option selection, and 3) setting a range (e.g., the starting and stopping times on a VCR). The operation of scrolling through a long list favored the use of cursor keys rather than voice commands for people who were browsing.

3 Methodology

A Sony Clié PEG-S320, with a monochrome screen and the Palm operating system, was used for this study. The device measured 4.6" by 2.9" by 0.6". On the left-hand side of the PDA, near the top, was a thumbwheel (approximately 0.5" in diameter). The standard Palm memo pad application was used as the foundation for a thumbwheel-based text entry application. The memo application provided basic functionality, such as creating, saving, and deleting memos. The applications character set consisted of the space character, the letters A through Z, the numbers 0 through 9, and the punctuation marks { . , ? ' " ( ) ! }. Only upper case letters were implemented. The characters appeared on the PDA's screen. When the wheel was turned downwards, the character set was traversed left-to-right. Turning the wheel upwards traversed the sequence of characters in the opposite direction. Pressing the wheel wrote the selected character on the screen. Three functions, "Done", "Back," and "Enter," were traversed using the wheel as well. Scrolling past "!" got to "Done", and scrolling past "Enter" brought the cursor back to the space character. A "snap-to-home" feature was also implemented. With this feature enabled, the cursor "snapped" back to the "home" character (the space). Otherwise, the cursor remained at the last letter selected.

A study was performed to test the thumbwheel interface design and the methodology described below. Subjects were student volunteers from an information science class. Participants were told they could hold the PDA and use the thumbwheel in any manner they wished. If an error was made, subjects were told to ignore it and continue with the next character (i.e., not to use the back function). Subjects were told to use the "enter" function when they completed each sentence.

Testing consisted of a training session and two task sessions. In the training session, subjects entered a single sentence (see Figure 1 for sentences used in the study). During the training session, subjects saw a tabular representation of the characters on the lower part of the screen. This was done in an attempt to help subjects learn the character order faster. The tabular representation was not visible during any of the subsequent sessions. In each task session, the subjects entered three sentences. The first session consisted of either sentence set one or set two (randomly chosen), and the second session used the other set. The sentences were primarily taken from a set used by MacKenzie (2002), although punctuation and numbers were added for this study. In one of the two sessions (randomly chosen), the snap-to-home feature was enabled; in the other session, it was not. Data was collected on task completion time and error rates for each sentence. After the sessions, subjects were asked for their opinions on the text entry method.

Set	Sentence	Text of sentence
	Training	we are having spaghetti.
1	1A	my watch fell in the water!
	1B	prevailing wind from the east.
	1C	the address is 195 main street.
2	2A	I can see the rings on saturn.
	2B	physics and chemistry are hard?
	2C	he can be reached at extension 482.

Figure 1: Sentences Used in Experiment

4 Results

Three female and seven male subjects (average age = 22 years) were tested. To see if the snap-to-home feature made any difference in entry times or error rates, a Mann-Whitney nonparametric test was performed for each of the sentences (Table 1). For three of the six sentences (1A, 2A, 2B), the entry rate was significantly lower (at the .05 level) with the snap-to-home feature enabled. The entry rate for sentence 2C was also much lower for snap-to-home, but not at the .05

significance level. The error rate was not significantly different with the snap-to-home feature enabled or disabled for any of the sentences.

Table 1: Entry Times and Error Rates With and Without Snap-to-Home Feature						
	Snap-to-Home Enabled		Snap-to-Home Disabled		Significance	
Sentence	Entry Time (seconds)	Errors	Entry Time (seconds)	Errors	Entry Time	Errors
1A	126	0.5	164	0.75	.029	.686
1B	153	2.25	148	1.75	.486	.686
1C	148	2	143	0.5	.914	.114
2A	128	1	172	0.4	.016	.286
2B	141	3	186	2.4	.036	.571
2C	141	0.75	169	1.2	.063	.413

A record was kept on how each subject held the PDA and used the thumbwheel. Given the position of the thumbwheel on the PDA, it was expected that most subjects would hold the PDA in the left hand and use the left thumb to operate the wheel. However, only five (out of ten) people used the device this way. Of these five people, two steadied the device with their right hand during text entry. Of the remaining five subjects, one held the device in their right hand, and used the left thumb. Two subjects held the device in their right hand and used their left index finger. Two subjects held the PDA in their right hands, and used their right index finger. One of these people also used their left hand to steady the device. Some of the subjects switched hands part way through the experiment.

Many subjects commented that a larger thumbwheel, or one that had a smoother motion or smaller turning radius, might have worked better. Many subjects had problems "clicking" the wheel to enter a character; sometimes the wheel would "slip," causing an error. For example, a person would dial to "g" and click the wheel, but the wheel would turn a bit more in the process and actually print "h" on the screen. It may be that this slippage is the cause of many of the errors recorded during the sessions.

Most participants had mixed feelings about the snap-to-home feature. It was viewed positively for two reasons. First, it encouraged downward use of the thumbwheel, which was seen as much easier and more comfortable. Scrolling downwards meant the letters were in alphabetical order and the numbers were in numerical order. One subject commented on passing desired letters when scrolling upwards because they were not used to going backwards through the alphabet. Second, having the cursor return to the space, which was used fairly often in each sentence, was seen as saving time and effort. However, the snap-to-home feature was also viewed negatively because it made it more difficult to type in two letters that were the same or close to each other in the alphabet (e.g., "st").

5 Discussion

At approximately three words per minute, the average entry speeds observed in this study are lower than the speeds found by MacKenzie (2002). But this study is unique in several ways that may explain the increased entry times, including a larger character set and the use of a PDA rather than a simulated environment. With additional training and continued use, entry speeds may significantly increase. The snap-to-home feature seemed to improve text entry speeds, although this may be related to the makeup of the sentences (e.g., number of spaces and repeated characters). Testing with a greater number and variety of sentences is needed. Text entry speeds and error rates could also be affected by the way people held the PDA, something that also needs to be more carefully controlled in future studies. It may be possible to develop more efficient methods for thumbwheel text entry by reducing the total number of keystrokes required. This might be done through "hierarchical" methods. For example, an upper menu may consist simply of the three choices "A", "M", and "0." Clicking on "0" would then present the numeric choices of "0" through "9". An additional click would select one of these lower level choices.

While thumbwheel entry methods may not prove suitable for large amounts of text, they should work well for short notes and messages. Wristwatch-size PDAs are now available (e.g., the onHand PC), and such devices may one day be used for wireless messaging. Their small screen sizes, however, are not conducive to virtual keyboards or to gesture recognition. Voice input is one alternative for these devices, but voice recognition technology is sometimes inappropriate to use due to environmental conditions (e.g., in noisy factories). Thumbwheels may provide one of the few feasible ways of entering text on very small devices in a wide variety of settings.

6 Acknowledgements

The author would like to thank Myra Dideles for programming the PDA used in this study, and for her input into the tasks and methodologies used.

7 References

De Vet, J. & Buil, V. (1999). A personal digital assistant as an advanced remote control for audio/video equipment. In S. A. Brewster & M. D. Dunlop (Eds.), *Proceedings of the Second Workshop on Human Computer Interaction with Mobile Devices.*

Lewis, J. R., LaLomia, M. J., & Kennedy, P. J. (1999). Evaluation of typing key layouts for stylus input. In *Proceedings of the Human Factors and Ergonomics Society 43rd Annual Meeting* (pp. 420-424).

MacKenzie, I. S. (2002). Mobile text entry using three keys. In *Proceedings of the Second Nordic Conference on Human-Computer Interaction* (pp. 27-34).

MacKenzie, I. S. & Chang, L. (1999). A performance comparison of two handwriting recognizers. *Interacting* with Computers, 11, 283-297.

MacKenzie, I. S. & Soukoreff (2002). Text entry for mobile computing: models and methods, theory and practice. *Human-Computer Interaction*, 17, 147-198.

MacKenzie, I. S. & Zhang, S. X. (1999). The design and evaluation of a high-performance soft keyboard. In *Proceedings of the 1999 ACM Conference of Computer-Human Interaction* (pp. 25-31).

Sears, A. & Arora, R. (2001). An evaluation of gesture recognition for PDAs. In *Proceedings of HCI International 2001* (pp. 1-5).

Silfverberg, M., MacKenzie, I. S., & Korhonen, P. (2000). Predicting text entry speed on mobile phones. In *Proceedings of CHI 2000* (pp. 9-16).

Tarasewich, P. (2002). Wireless devices for mobile commerce: User interface design and usability. In B. E. Mennecke and T. J. Strader (Eds.) *Mobile Commerce: Technology, Theory, and Applications*, Hershey, PA: Idea Group Publishing (pp. 26-50).

Zha, Y. & Sears, A. (2001). Data entry for mobile devices using soft keyboards: Understanding the effect of keyboard size. In *Proceedings of HCI International 2001* (pp. 16-20).

Modelling User Context

Mohammad-Reza Tazari *Matthias Grimm* *Matthias Finke*

Computer Graphics Center (ZGDV e.V.), Dept. Mobile Information Visualization
Fraunhoferstr. 5, 64283 Darmstadt, Germany
{Saied.Tazari, Matthias.Grimm, Matthias.Finke}@zgdv.de

Abstract

A major problem in the domain of context-aware and adaptive HCI is the lack of a shared understanding of context. Many of the existing context-aware applications act independently from a shared context management service relying on a curtailed context model not appropriate for interoperability between independently running components. This paper aims at identifying the core of context parameters for a mobile user, followed by a discussion about the classification of these parameters. For each identified class, the most important properties are then determined. The context model presented here is a result of our two major projects from the domain of mobile workplaces and mobile knowledge management (map[1] and *mummy*[2]). Most parts of the model, however, can be applied in arbitrary scenarios and systems.

1 Introduction

A very promising paradigm for adaptive human-computer interaction is introduced by context-aware computing. The basic idea is to adapt the behaviour of computer programs to the user context. In many of the research projects in the area of context-aware computing, however, the idea of context is reduced to user location (see examples in Brown et al., 2000 and Chen & Kotz, 2000). Some researchers have noticed that "there is more to context than location" (Schmidt, Beigl & Gellersen, 1998) but they went no further than analysing some important aspects of context (see in particular Schirmer & Bach, 2000).

Especially in a multi-agent / -component system, where the components run independently and asynchronously, it is crucial to identify the shared context parameters and provide a context model in order to guarantee data consistency and a shared understanding of context. However, providing the context model is not a trivial task, particularly when it aims at a global model that should build the basis to work with arbitrary applications and systems.

J. Schirmer and H. Bach came to the following conclusions when analysing the context (Schirmer & Bach, 2000; see also figure 1): The traditional notion of the context for computer programs was formed of the user running the program and the computing resources and information used by the program. However, when we add the location of the user to the context, we can additionally consider the attributes of the environment. These attributes are grouped in three categories, namely the physical, technical, and social spheres. They also classified context along 3 axes:

- *Change of value*
  - *Static*: hardly changing, e.g. resolution of a display
  - *Dynamic*: changing depending on time and location, e.g. user activity

[1] MAP: Multimedia workplace of the future. See http://www.map21.de/.
[2] MUMMY (IST-2001-37365): Mobile knowledge management. See http://mummy.intranet.gr/.

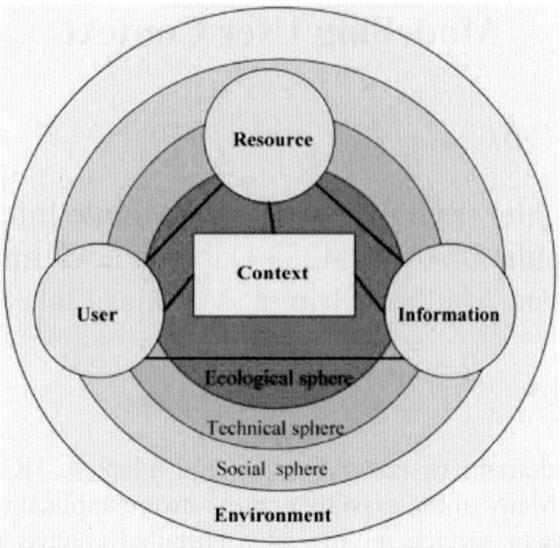

Figure 1: Context and the influence of environment (source: Schirmer & Bach, 2000)

- *Scope*
  - *Local*: "micro world", associated with a single object, e.g. resolution of a display
  - *Global*: "macro world", correlating multiple objects, e.g. location of the user
- *Interpretation*
  - *System-oriented*: unequivocal characteristics of system components
  - *Application-oriented*: interpretation-dependent with different uses

On the basis of these considerations, the next section explores the context of a mobile user. We will try to first identify the most important classes of the user context and then refine them at some level appropriate for the scope of this paper.

2 Mobile User Context

Figure 2 shows how we began to consider the user context. As shown in (Grimm, Tazari & Balfanz, 2002), we could then identify the following groups of context data:

- Profiles of *resources* related to the user context, e.g. available devices, services, documents, etc. Each such profile describes the identity, characteristics, and capabilities of the underlying resource and "knows" about the location and the state of the resource.
- Profiles of *locations*, which describe the identity and the state of the location and list the available resources and the people present at that location. The state of a location results out of the perception of the physical characteristics of the location using sensor data, e.g. temperature, brightness, etc.
- The *current time* in diverse forms, e.g. the absolute time, hour, am / pm, etc.
- *User* profiles consisting of the user's identity, characteristics, capabilities, universal preferences, and the state of the user. User's state includes information about his or her main activity, current terminal, etc., and a reference to a location profile.
- *Application-specific user preferences*.
- Other *application-specific data* that may play a role in the process of determining the user context – especially applications from the domain of personal information management (PIM), e.g. calendar, to-do list, address book, etc. (Satyanarayanan, 2001)

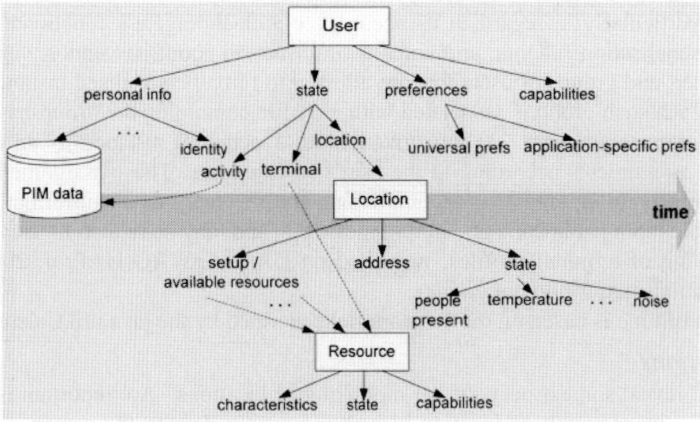

Figure 2: Mental grasp of the user context

2.1 The Context Model

From the above-mentioned groups of context data, we chose the following classes, for they could be modelled more or less independently from applications: documents, agents / services, terminals, locations, users, time, and tasks. Figure 3 summarises some dependencies between these classes. To maintain lucidity, we have left some classes and relations out of this summarised version, among them classes from the domain of personal information management other than the tasks (e.g. appointments and contacts), groups, the relations between two tasks, or between tasks and locations. Also, the class of resources should have been shown as the super class for terminals and services. In the following subsections, we discuss further refinement of the selected classes. Other important aspects of our context model, such as the distribution and inheritance of context parameters, are discussed formerly in (Grimm, Tazari & Balfanz, 2002).

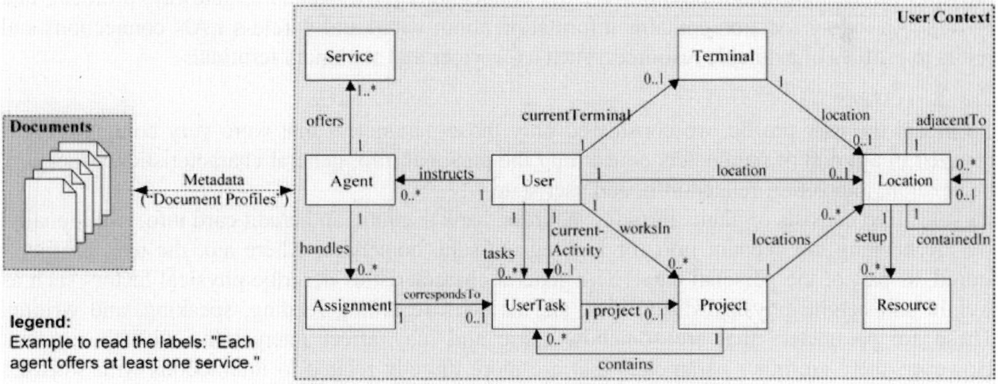

Figure 3: Overview of the context model

2.1.1 Documents, Services, Terminals, and Time

Profiles of documents provide the links between the user context and the information space. Our investigations showed that we can rely totally on PRISM (PRISM, 2002), which is an enhancement of Dublin Core (Kokkelink & Schwänzl, 2002).

For service profiles, we used the service modelling from DARPA Agent Markup Language (DAML-S, 2002) and enhanced the model, adding a new class called "Agent". Each Agent offers some services that are described according to DAML-S, is signed by a third party that guarantees its reliability (particularly in regard to its accordance with the agent's functional specification), provides a personalisation schema, and contains information about its vendor, version, platform, code base, owner, and groups of users that are allowed to run a personalised instance of it on their "homebase". Each such group is associated with a profile containing the group-based defaults for the personalisation parameters. At any given time, a personal instance of an Agent may be handling some assignments according to the user's instructions. Those assignments have a title, refer to the corresponding steps for accomplishing the service, contain information about their state and possible input parameters and output messages, and may be related to user tasks.

For the modelling of terminal profiles, we relied on UAProf (WAG UAProf, 2001) with some minor modifications that we do not go into.

The refinement of time is based on the constant fields defined by the java.util.Calendar class.

2.1.2 Locations

Figures 2 and 3 reveal some aspects of our modelling of locations. An important issue not shown in the figures is the mapping of location-identifying sensor data to the semantic description of locations. Relying on appendix C of the UAProf specification, we expect to see sensor data as an encoding/value pair, e.g. <gps:nmea0183:gll, latitude-longitude-value>. Each such pair forms an ID of the location and each location description may contain an arbitrary number of such IDs.

Generally, it is supposed that a location has an identity and a state. The identity is formed from a label, a type, an address (country, state, region, city, zipCode, street, houseNo, floor, floorArea, roomNo), and a set of IDs. If the profile describes the location of a moving user or object, the identity may have a mobile address component (instead of the fixed one), which refers to two locations as starting and end points of the route, provides the distances between the current location and those points, and specifies the transport means. The state of a location mirrors characteristics like brightness, temperature, noise, users present, and a flag to show if those people are working together or should be considered as individuals not necessarily relating to each other.

A location could be contained in another location and may be adjacent to other locations. An outdoor location extends the state of the location by adding new attributes about the climate and provides information about possible WLAN connections (DNS, domain, gateway, protocol, and SSID). An indoor location contains information about wired and wireless LAN connections and refers to profiles of available resources, such as services and stationary terminals.

2.1.3 Users

To model the user profile, we considered only those parameters that were very common. This resulted in a group of parameters concerning the personal info, general characteristics, education, occupation, interaction-related info, and user state.

Personal info consists of name, birthday, address, bank account, and credit card info. Additionally, we are keeping the account info for logging onto the homebase, where a.o. the user profile is stored, as part of the personal data. The general characteristics describe physical factors such as weight and height, physical disabilities, and abilities regarding reading, speaking, and writing. These are parameters that are somehow static and may affect interactions with the system. However, there are other parameters that are more directly related to interactions with software systems, like expertise, keyboard usage, language, mouse usage, and timeout. Parameters that are of a dynamic nature and describe the state of the user under which interactions take place are grouped in a special component consisting of user's current activity, current terminal, his or her location, motion-state, and orientation. Two groups of user preferences are also considered: (1) session preferences, such as willingness to have voice- or display-based I/O and to disclose his or her location info; (2) browser preferences, like download of applets.

2.1.4 Tasks

To model the user's tasks and goals, we use a hierarchical refinement approach. A task at the highest level can be refined to some network of tasks at the next level. Each of the tasks at the second level can be further refined in a separate plane and this refinement process continues until there are only some atomic tasks. Attributes "parent" and "subtasks" reflect this hierarchical refinement in both directions. Non-atomic tasks are equivalent to goals and sub-goals in our model. On each plane of an arbitrary level, tasks can have other tasks as their prerequisites, i.e. they cannot begin until all of the previous tasks marked as their prerequisites are done. This relation is kept track of by "after" and "before" attributes in both directions. The rest of the modelling uses components describing the identity, state, and scheduling of tasks and the relevant documents, users, and locations.

3 Conclusion and Future Work

Modelling user context is a complex task. Depending on the application scenarios, different aspects of the ambience may be of interest. Providing an overall model causes it to get very large very quickly. Nevertheless, we managed to consider the core of context parameters in a relatively concise model. Although the potential for enhancing this model is obvious, we decided to first examine it and then focus on the privacy issues and the process of setting the session context. The introduced model could successfully be applied within the map project. Now we plan to examine it within the *mummy* project at areas of context-aware authoring (using the creation context as automatic metadata), retrieval (guessing information needs of the user based on the situation), and presentation (adapting the information presentation to the situation).

References

Brown, P., Burleson, W., Lamming, M., Rahlff, O.W., Romano, G., Scholtz, J., & Snowdon, D. (2000). Context-Awareness: Some Compelling Applications. University of Exeter, MIT Media Lab, Xerox Research Center Europe, Motorola Labs, & DARPA. Retrieved January 6, 2003, from http://www.dcs.ex.ac.uk/~pjbrown/papers/acm.html

Chen, G., & Kotz, D. (2000). A Survey of Context-Aware Mobile Computing Research. Retrieved Jan. 6, 2003, from ftp://ftp.cs.dartmouth.edu/TR/TR2000-381.pdf

DAML-S: Semantic Markup for Web Services, Version 0.7, October 2002. The DAML Services Coalition. Retrieved Jan. 7, 2003, from http://www.daml.org/services/daml-s/0.7/daml-s.pdf

Grimm, M., Tazari, M-R., & Balfanz, D. (2002). Towards a Framework for Mobile Knowledge Management. In D. Karagiannis & U. Reimer (Eds.), *Practical Aspects of Knowledge Management* (pp. 326-338). Berlin: Springer (LNAI 2569)

Kokkelink, S., & Schwänzl, R. (2002). Expressing Qualified Dublin Core in RDF / XML. Retrieved Jan. 7, 2003, from http://dublincore.org/documents/dcq-rdf-xml/

PRISM: Publishing Requirements for Industry Standard Metadata (2002). Retrieved Jan. 7, 2003, from http://prismstandard.org/techdev/PRISM Spec-1.2e 09.04.2002.pdf

Satyanarayanan, M. (2001). Pervasive Computing: Vision and Challenges. *IEEE Personal Communications*, vol. 8, no. 4, 10-17.

Schirmer, J., Bach, H. (2000). Context management in an agent-based approach for service assistance in the domain of consumer electronics. *Proceedings of IMC2000*. Retrieved Jan. 7, 2003, from http://www.rostock.igd.fhg.de/veranstaltungen/workshops/imc2000/files/proceedings/05Tazari.pdf

Schmidt, A., Beigl, M., & Gellersen, H-W. (1998). There is more to Context than Location. Retrieved Jan. 7, 2003, from http://citeseer.nj.nec.com/schmidt98there.html

WAG UAProf: User Agent Profile Specification (2001). Retrieved January 6, 2003, from http://www1.wapforum.org/tech/documents/WAP-248-UAProf-20011020-a.pdf

Mobile Contexts of Use: Socio-Spatial Attributes

Kalle Toiskallio and Sakari Tamminen
Information Ergonomics Research Group, SoberIT,
Helsinki University of Technology
P.O. Box 9600, 02015 HUT, Finland
kalle.toiskallio@hut.fi

Abstract

Beside physically based interaction and augmented environments, context-aware computing includes also a viewpoint near this paper. A large qualitative data is used to draw from social processes of everyday life (business and leisure) the most relevant attributes (22) of contexts of use of mobile devices, and experienced mobile contexts in general. Beside the philosophical side of context-aware computing, this paper also contributes the section of "social environment" in ISO 9241-11, since the attributes are classified by ISO-style in main clusters such as "user", "task", and "environment." In the future work this preliminary model of context of use (MCU) will be developed also to support the iteration process of a product development heading to formulate the functional requirement specifications. Mobile settings need specific devices and circumstances, but many aspects of use of them can be generalized. For this reason many of these attributes are applicable beyond mobile settings.

1 Introduction

Elements surrounding a particular action (use context) are widely discussed in the literature dealing with the use of context-aware mobile devices. According to Paul Dourish (2001, 299), "two distinct strands of what we might call context-aware computing within HCI research, although typically conducted and developed in isolation, are in fact two aspects of the same broad program. These two topics are, first, physically-based interaction and augmented environments, and, second, attempts to develop interactive systems around understandings of the generally operative social processes surrounding everyday interaction." First strand is more construction-oriented (e.g. Dey et al., 2001). Although it is highly concerned of social context of use, it doesn't really study it, but rather speculates it analytically. The conclusion usually is that contextual questions are highly complicated and thus almost impossible issues to handle (e.g. Svaneas, 2001). Our paper belongs to the latter line of Dourish's classification. The main idea is that we cannot find any ready-made "frames" around certain activities but the very activities themselves create something we could call social contexts of use. Social factors of context of use are usually shaped by spatial conditions, or at least these two are interwoven together.

As Lucy Suchman (1987) has noted earlier, in the practice users are improvising their interaction with computers all the time, instead of following strictly predescribed guidelines. This makes social contexts of use to change rapidly and proves rigid theoretic-conceptual frameworks too clumsy to be applied in natural environments. However, on the basis of our empirical research we argue that it is possible to define some key attributes of mobile contexts and formulate them as a more general model of context of use (MCU).

2 Theoretical viewpoint and data

Our theoretical standpoints arise from social scientific thoughts of social constructionism: contexts are not understood as some ready-made places that people are only "installed to". Instead, people participating in an occurrence actively create those particular situations and interpretations of them. These finely tuned interpretations then create the human viewpoint on mobile contexts of use. Furthermore, this infinite social constructivity makes us able to present only some terms and concepts that seem to be unavoidable when users create and interpret the contexts around them.

Our empirical data consist of semi-structured interviews (N=52) and observations (N=20) among business and leisure travellers, and different salesmen. We have also done several walk-throughs (N=7) at travel-related semi-public places such as airplanes, trains, and cruise-ferries, including the terminals of each travel mode. Especially the airports have been under our scrutiny.

The data has been analyzed by various qualitative techniques, including discourse and content analysis, ethnomethodological analysis and visual interpretations. On the basis of our analysis we describe 22 attributes that can be used as useful tools when defining the socio-spatial contexts of use, especially those of mobile devices and services.

3 The elements of Model of context of use

ISO 9241-11 mentions social environments – rather passingly – and mixes them with organizational environments. However, sociality, or social factor, is such a transcendent topic that it cannot be simply located in one analytic box, separated from "user" and spatial environment. Sociality is so hard to capture properly in such a simplistic box-illustration precisely because sociality should be in any box that deals with humans – or artefacts done by human. That is why we do not simply offer an additional box of social environment but attributes that are parts of socio-spatial context of use. However, to contribute to the mentioned standard, we have divided the attributes similarly, i.e., into three groups, representing point of *the user and social relations*, *tasks*, and *environmental aspects of the activity and facilities*.

A product development project can take its prototype or a concept as a strainer and draw all the following attributes through it. All the relevant factors will stay in the strainer and should be scrutinised carefully before shipping the product onto market place.

3.1 User and social relations

Although we the modern persons are keen to be autonomous actors, we simultaneously react and construct social circumstances around us. This fact shapes the users' experienced contexts of use. Some very important attributes come from the social relations with surrounding people. Questions to be answered in social context definitions include questions like 1) what is the relation between subject and surrounding people? Are the other people subjects' a) "significant others", i.e., people who are valuable for the subject as themselves, or, are they merely b) "tools" for him/her, that is, valuable because of their certain skills, or, is the subject simply co present with unknown others so that normal social norms are to be observed.

The mobile devices are also easily interpreted as carrying meanings of 2) social comparison. Not just the a) price of it but also the way b) device works, and the way the c) user really can use it. With this kind of resources the user may try to represent a certain 3) social identity. Social identity is one hook that an individual can use when connecting him/herself to 4) desired reference group, or actual member-group. (Advertising uses efficiently this belonging to desired social groups). Within this kind of social meanings the user can have 5) several simultaneous goals, which are influenced by 6) his/her motivation and 7) required intensity of the task. 8) Habitualized task solving processes (=habits) are also a strong factor. 9) The personal goal and especially the motivation can have more or less a) outer or b) inner nature.

3.2 Tasks and mentalities

Dividing task and user strictly is easy only on an analytical level. In practice, attributes describing task are overlapping with the previous class, offering perhaps a slightly different viewpoint.When taking the viewpoint of task, relevant contextual attributes in mobile use is question of 1) preparedness. Many everyday mobile tasks (such as reading papers, or navigating in an unknown environment) cannot be done without at least minimal preparations. Related to habits above, 2) the frequency vs. singular use creates drastically different context of use. Single use is evaluated by one simple time, more frequent act can be also estimated before hand.

One explicitly social and perhaps thus understudied factor is 3) the use in groups vs. individual use. In a group an individual may be allowed to behave as stupidly and as irresponsibly as possible while someone else leading the group. Compared to this, a single user usually behaves rather coherently, carrying all the responsibility of his or her acts alone. If observing for example the navigation in an airport, this difference may be a huge. As a matter of fact, the above mentioned group-member is only concentrating on 4a) multitask because, as usually happens, he or she is simultaneously chatting some one else in the group while the whole group is searching their departure gate, for example. Compared to this, the single traveller needs only to concentrate on 4b) single task, i.e., on his or her own thoughts of finding the needed spot. Finally, 5) the intensity and tempo of the user may vary from total blasé to eager enthusiasm. This variation of intensity of use may of course vary also independently from the actual task.

3.3 Environmental aspects

The environment refers here mostly to audio-visual and spatio-organisational dimensions of use environment. Important questions for the contextual definitions arise from different types of used 1) traffic modes, and episodes of the journey. The rest of attributes include 2) the level of publicity and 3) supposed or "proper" use of place where actions are carried out, which can be indicated in many "official" ways. However, the particular user or even users generally may have certain previous experiences that shape their 4) interpretations of the place. This kind of impressions can be analysed by visual, auditive, and sometimes even haptic effects appearing in that particular environment. 5) The visual environment can deal with the general lightness etc., but also with the visual appearance of a display, for example. 6) The auditive environment can be a) "pluralistic" allowing many different and simultaneous voices, or b) "monoteistic", by insisting that only one kind of auditive effect (or even total silence) is allowed. 7) Haptic effects are not restricted tothe device aone but also other touched surfaces at that moment. Furthermore, in the work situations 8) intra- or inter-organisational tensions can cause – or deny - certain practices. For example, an international mobile phone producer has forbidden its employees to use lap tops at airports. Last but not least, 9) timetables and time shifts in global scale can create inner or outer tensions reflecting in use of mobile devices. Sudden and supposedly dramatic changes in, say, timetables of public transport will rise immediately the mobile phones onto ears of the passengers.

4 Future work

The preliminary gestalt of MCU presented in extremely concise form above must be formulated more carefully in order to be easily used in projects. Currently in one attribute may be hold terms that are analytically on different levels, some of them being loaded by hard social theory, whereas some are mere observations drawn from our data. In addition, some attributes are di- or trichotomies, some are not.

Another challenge is to formulate and validate the MCU as to working model for
a practical method supporting the early stages of product development processes, and perhaps even functional requirement specifications. At the moment, we think this might be possible in five steps:

1. General descriptions of the current work processes and use practices are drawn from collected contextual inquiry data. These can be enriched by earlier experience of designers. In this stage MCU is used as an innovative source and as a checklist.
2. Focused scenario phase, in which MCU is related in the specific case (=the relevant attributes are applied).
3. Consistency and relevancy check for the scenarios. MCU is mostly ruled by "objective" issues. However, on this stage also more subjective attributes checked ones more.
4. Formalized use case descriptions per relevant attribute.
5. Supporting the formulation of functional requirement specifications.

References

Dey, A.K. Abowd, G.D. and Salber, D. A conceptual framework and a toolkit for supporting the rapid prototyping of context-aware applications. In Human-Computer Interaction, 16, 2001, 97 – 100.

Dourish, P. Seeking a Foundation for Context-Aware Computing. In Human-Computer Interaction, 16, 2001, 229 – 241.

ISO (1997) iso 9241-11: Ergonomic requirements for office work with visual display terminals (VDTs). Part 11-guidelines for specifying and measuaring usability. Geneva: International Standards Organization.

Suchman, L. Plans and situated actions. The problem of human machine communication. Cambridge University Press, 1987.

Svaneas, D. Context-Aware Technology: A Phenomenological Perspective. In Human-Computer Interaction, 16, 2001, 379 – 400.

Blind-handwriting Interface for Wearable Computing

Junko Tokuno[*1], *Naoto Akira*[*2], *Mitsuru Nakai*[*1],
Hiroshi Shimodaira[*1], *Shigeki Sagayama*[*3]

*1 Japan Advanced Institute of Science and Technology
{j-tokuno, mit, sim}@jaist.ac.jp
*2 Hitachi, Ltd.
n-akira@crl.hitachi.co.jp
*3 University of Tokyo
sagayama@hil.t.u-tokyo.ac.jp

Abstract

This paper proposes a novel input interface that we call "blind handwriting" for wearable computing. The blind handwriting, which is a word similar to "blind typing" of keyboard, is a particular writing style where the user does not see the pen or the finger movement. Without visual feedback, written characters are distorted, as in the case when the user is blindfolded, and therefore existing on-line handwriting recognition systems fail to recognize them correctly. The sub-stroke based hidden Markov model approach is employed to tackle this problem. When the pen or touch pad is used as an input device, the proposed interface demonstrates a recognition rate of 83% on a test set of 61 people where each person wrote 1016 Japanese Kanji characters.

1 Introduction

As the popularity of personal digital assistants (PDAs) and cellular phones has grown rapidly in recent years, demands for character input interfaces better than keyboards or buttons in mobile computing have increased. One of the extreme cases of the mobile computing is the wearable computing, where computing devices are small enough to be worn on one's body like clothes or accessories. In order for the wearable computing to be effective, a new type of character input interface is needed, since conventional input interfaces such as the keyboard and the mouse are not practical.

Several input methods have been proposed for the wearable computing; for example hand-held keyboard (Twiddler2, BAT) and voice recognition (Furui 2000). Among them, no method exists that is always superior to others in all applications; voice input is not practical in noisy environment or in the case when speaking is not allowed. The hand-held keyboard input requires training to get used to it. The alternative method that we propose in this paper is ``blind handwriting input". The blind handwriting is a particular writing style where the user writes characters without seeing the pen or the finger movement directly or indirectly through the display.

[*1] 1-1 Asahidai, Tatsunokuchi, Ishikawa, 923-1292, JAPAN
[*2] 6, Kanda-Surugadai, 4-Chome, Chiyoda-ku, Tokyo, 101-8010, JAPAN
[*3] 7-3-1, Hongo, Bunkyo-ku, Tokyo, 113-8656, JAPAN

Conventional handwriting input interface, which is widely used in PDAs (Graffiti), assumes implicitly that the users have visual feedback of their handwritings. The assumption of visual feedback turns to a tight restriction in the wearable computing, because a stylus pen and pad should be in appropriate positions so that the user can see them.

On the other hand, the proposed ``blind-handwriting input'' allows the user to wear the devices anywhere, even where they are hard to see. For example, Figure 1 shows an image of "blind handwriting" in a crowded train. A user writes characters on his hip or thigh where a small touch pad or finger movement sensor is attached to. In this position, the user writes characters without seeing his fingers and hence no visual feedback is provided. Instead, the recognition results are fed back to the user as voice through the earphone.

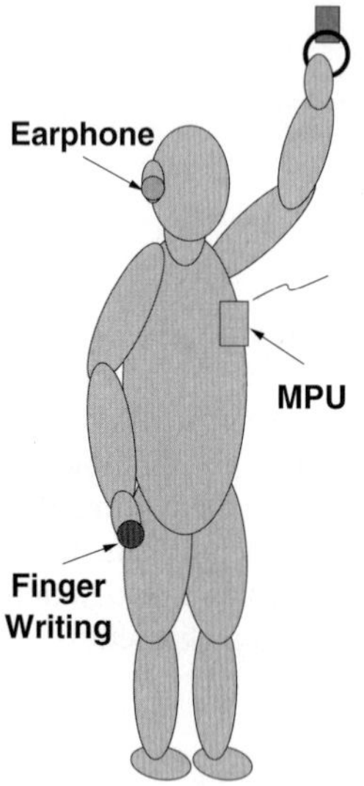

Figure 1: "Blind handwriting" in a crowded train

To realize the ``blind handwriting'' interface, the character recognition engine should be more robust against various distortions than the ones used for ordinary handwriting recognition. This is because the user writes characters without any visual feedback and, therefore, the written characters are seriously distorted when compared with the ones written in the normal writing style with visual feedback. Figure 2 shows an example of blind-handwritten characters that were collected in the experiment described in section3.1. It can be seen from the figure that characters are distorted in various ways, i.e. some of them are slanted, rotated, or with overlapped radicals. These sorts of distortions are enough to make recognition performance deteriorate substantially.

In order to tackle this problem, we employ (1) shift invariant feature, (2) sub-stroke based HMM, and (3) rotation-free decoding.

Figure 2: Samples of blind handwriting dataset: "東", "解", "専", "障", "窓", "推"

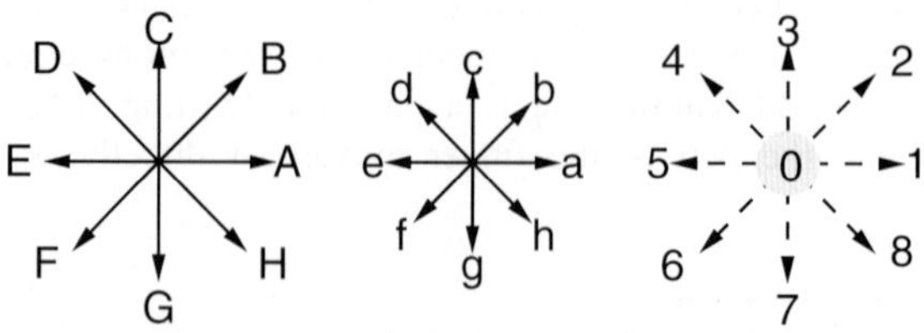

Figure 3: Sub-stroke categories: A-H(a-h) are long (short) sub-strokes with pen down and 0-8 are the direction of pen up.

2 Handwriting Recognition Algorithm Based on Sub-stroke HMMs

The proposed system basically consists of an input feature extractor, sub-stroke models (HMMs), dictionaries and a decoder.

2.1 Input Features

In order to recognize characters whose radicals overlap, shift variant input features are employed. In feature extraction, pen positions (x, y) in the Cartesian coordinate system are sampled at a fixed time interval. Let (dx, dy) be the difference vector between the two consecutive pen-position samples, and (r, θ) be its corresponding feature vector in the polar coordinate system, where r means the norm of the vector, i.e. the Euclidean distance between the two pen-positions ($\sqrt{dx^2 + dy^2}$), and θ represents the angle of the vector. When the pen touches the tablet surface (pen-down), the feature vector (r, θ) represents a velocity vector of the pen at a time instance. When the pen leaves the tablet surface (pen-up), (r, θ) represents a displacement vector between the strokes observed just before and after the pen-up state, because the pen position is not sampled while it is in the air.

2.2 Sub-stroke HMMs

We employ the sub-stroke based hidden Markov model approach (Nakai 2001) to recognize the distorted characters written in the non-visual feedback condition. This approach is effective in mobile use, because the total size of the model is very small and the recognition speed is fast. The details of this approach are as follows.

Based on the knowledge of distinctive features of Kanji characters, we have defined 25 sub-strokes, as shown in Figure 3. Each sub-stroke is modeled by a left-to-right HMM. Three-state continuous-distribution HMM is employed for each pen-down sub-stroke to model the sequence of velocity vectors, while one-state HMM without self-loop probability is used for each pen-up sub-stroke to model the displacement vector. Here, let $\lambda^{(k)} = (A^{(k)}, B^{(k)}, \pi^{(k)})$ be the set of HMM parameters of sub-stroke k, where

$A^{(k)} = \{a_{ij}^{(k)}\}$: The state-transition probability distributions from state S_i to S_j,

$B^{(k)} = \{b_j^{(k)}(\vec{o})\}$: The probability distributions of observation symbols $\vec{o}$ at state S_j,

$\pi^{(k)} = \{\pi_i^{(k)}\}$: The initial state probability distributions.

The observation probability distribution is represented by an M-mixture of Gaussian distributions:

$$b_i^{(k)}(\vec{o}) = \sum_{m=1}^{M} c_{im}^{(k)} \frac{\exp\{-\frac{1}{2}(\vec{o} - \vec{\mu}_{im}^{(k)})^t \Sigma_{im}^{(k)^{-1}} (\vec{o} - \vec{\mu}_{im}^{(k)})\}}{\sqrt{(2\pi)^n |\Sigma_{im}^{(k)}|}},$$

where $\vec{\mu}_{im}^{(k)}$ is the mean vector, $\Sigma_{im}^{(k)}$ is the covariance matrix, and $c_{im}^{(k)}$ is the weighting coefficient. Note that each Gaussian distribution is periodic with a 2π cycle with respect to the polar angle (θ). These model parameters can be trained by the Viterbi training or the Baum-Welch algorithm.

2.3 Recognition

A decoder recognizes an input pattern by referring to the character's sub-stroke sequence, which is obtained by expanding the definition in a hierarchically structured dictionary (Nakai 2001). For example, the definition of the character "二" is 'a 6 A' which means that the two pen-down strokes 'a' and 'A' are connected with the pen-up model '6' in standard stroke order. Similarly, "子" is 'A f 0 G d 4 A' and "字" is 'g 5 g 3 A f 6 A f 0 G d 4 A', where "子" is a partial structure of "字", and both have a common sub-stroke sequence. According to the description in the dictionary, the decoder concatenates the sub-stroke HMMs to generate an HMM of each candidate character, and then calculates the probability that the input pattern be produced from the HMM. This operation is effectively done by the Viterbi search algorithm of a sub-stroke network (Nakai 2001).

2.4 Rotation-free Recognition

In the blind handwriting, many users write slanted and rotated characters, which deteriorate the correct recognition performance of the HMM-based recognition engine. If the angle of slant or rotation is constant and it does not depend on writers, we can cope with the problem in advance. However the angle varies among the users and changes according to writing styles. In the HMM-based approach, if the character rotates, the probability that the input pattern is produced from the character's HMM decreases. The following algorithm is employed to cope with this problem.

[Algorithm for rotated characters]
Let the sequence of feature vectors of an input handwritten character be denoted as $\vec{O} = o_1 o_2 o_3 \cdots o_T$, and $\vec{O}^{\theta_n}$ be the rotated sequence of $\vec{O}$ by an angle of θ_n, where $1 \le n \le N$.

The recognition of the character is done by finding the optimum candidate character $\hat{W}$ and angle $\hat{\theta}$ that satisfies;

$$\{\hat{W}, \hat{\theta}\} = \underset{W_i, \theta_n}{\arg\max}\, P(W_i \mid \vec{O}^{\theta_n}).$$

This optimization problem is effectively solved frame-synchronously by employing the One-Pass decoding algorithm based on the Viterbi algorithm.

3 Experiment

3.1 Handwriting database

In order to investigate the characteristics of the blind handwriting, we created a database of handwriting data. This database, called the JAIST IIPL (Japan Advanced Institute of Science and Technology, Intelligence Information Processing Laboratory) database, consists of several kinds of data sets. Among them, we used the dataset written in 2 kinds of writing style. One of them is the α set, written in visual-feedback conditions and the other is the $\delta - 1a$ set, written in non visual-feedback condition, as shown in Figure 4. The α set covers 1,016 Japanese characters of old and new educational Kanji, containing 109,728 samples collected from 108 writers with standard stroke order. On the other hand, the $\delta - 1a$ set covers 83 Hiragana, 86 Katakana, and 62 Alpha-numeric characters, in addition to the Kanji characters, containing 76,067 samples collected from 61 writers with free stroke order. Character samples of the $\delta-1a$ set are depicted in Figure 2. In addition to the distortion described in Section 1, excess or deficiency of strokes and overlap of separated strokes are observed.

3.2 Experimental Evaluation

In order to evaluate our proposed approach for "blind handwriting", we carried out a recognition task. In the experiment, 47 writers of the α set were used for estimating the HMM parameters, in order to disregard a factor of writer specific difference of stroke order, and 61 writers of the $\delta-1a$ set were used for evaluation. Moreover, we defined the rotated angle for adaptation as 22.5-degree intervals. In other words, the rotated direction is defined in the 16 directions and hence, we can deal with characters which rotate in any direction. The recognition performance is shown in Table1. We can see from the results that the recognition accuracy improved by means of the adaptation for rotated characters. Moreover, from our analysis, we found that in the case of 1,016 Japanese educational Kanji recognition, there are few characters which are mistaken for another character by rotating.

Figure 4: "Blind handwriting"

Table1: Correct recognition rate of blind-handwritten characters [%]

	1-best	10-best
Sub-stroke HMM	72.44	87.59
Rotation-free Sub-stroke HMM	82.86	91.85

4 Conclusion

We proposed a new input interface that we called "blind handwriting" for wearable computing. In order to achieve this interface, we have proposed an on-line handwriting recognition method based on sub-stroke HMMs and adaptation techniques for rotated characters.

References

Twiddler2. http://www.handykey.com

BAT. http://www.onehandkeyboard.com

S. Furui. (2000). Speech recognition technology in the ubiquitous/wearable computing environment. Proc. ICASSP'00, pp. 3735-3738, June, 2000.

Graffiti. http://www.palm.com/products/input

M. Nakai, N. Akira, H. Shimodaira and S. Sagayama. (2001). Sub-stroke Approach to HMM-based On-line Kanji Handwriting Recognition. Proc. ICDAR'01, pp. 491-495, Sept, 2001.

A Run-time System for Context-Aware Multi-Device User Interfaces

Jan Van den Bergh *Kris Luyten* *Karin Coninx*

Limburgs Universitair Centrum
Expertisecentrum Digitale Media[1]
Universitaire Campus
B-3590 Diepenbeek, Belgium
{Jan.VandenBergh, Kris.Luyten, Karin.Coninx}@luc.ac.be

Abstract

In the last few years, the use of mobile computing devices has tremendously increased and expectations are that the growth of mobile computing will continue for many years to come. This evolution creates new challenges for the design of user interfaces. In order to effectively use the new technology, an application will need to be usable on different platforms, in different circumstances and by different users. Thus, the user interface of the application has to adapt to the context in which it will be used. In this paper, we will address the problem of context-aware computing and how a user interface can be adapted according to its context at runtime and still be usable. We will define the concept *context of rendering* in order to effectively incorporate the information provided by the *context of use* in the rendering of a context sensitive user interface. To achieve this, we will combine task models with abstract user interface descriptions and context dependent information. The approach that is taken will be illustrated by the discussion of a proof-of-concept implementation.

1 Introduction

With the future of computing laying in embedded systems and portable devices the current practice of ad hoc design of user interfaces (UIs) can pose serious problems. Even more because the market of these portable devices evolves very fast, which makes it almost impossible to know the properties of the devices on which a product must run when it is designed. These assumptions lead to the following properties of user interfaces for future applications: (1) they should be executable on multiple platforms, (2) their appearance needs to be, at least partially, defined at run-time and (3) they should maintain usability (i.e. be plastic (Calvary, G. et al. 2002)). In this paper we introduce a run-time system that allows for the execution of multi-device user interfaces whose appearance can be defined by both run-time and design time parameters, bundled into a context description. The system builds on the strong foundation of model-based and, more specifically, task-based design.

Several existing methods are also based on task-based design. Some of them use the ConcurTaskTrees (CTT) as notation for the task model. CTT is a hierarchical notation in which an abstract task is refined into more fine-grained abstract tasks, user tasks, interaction tasks and system tasks. Among tasks on the same level, temporal relations are specified such as "enables",

[1] http://www.edm.luc.ac.be

"disables", "choice" and "concurrent". (Souchon, Limbourg & Vanderdonckt, 2002) proposed several methods to allow modeling for multiple contexts of use employing task models in the ConcurTaskTrees notation, while (Calvary et al., 2002) proposed to use translation and reification of different models to generate UIs for multiple contexts of use. Both approaches, however, have in common that they generate different user interfaces for different contexts. (Paternò & Santoro, 2002) proposed the use of a task model in ConcurTaskTrees (CTT) notation as a single base for the design for multiple UIs.

Another approach is taken by (Braubach et al., 2002). They let the user specify the user interface using a domain model, which contains a task model (UML Use Case Diagram), and an object model, a presentation model and a dialogue model. These three models are used in the runtime environment where they are combined with the components from GUI toolkits and domain specific objects. This approach delivers a flexible runtime environment, however the use of the Use Case Diagram implies severe limitations in the specification of the tasks and the runtime environment is rather extensive. In the following sections we will introduce our work, starting with a new definition of context, followed by a discussion of a proof-of-concept implementation.

2 Context

Multiple definitions of context are given in literature. In our approach, we adhere to the definition of context given in (Calvary, et al. 2002). It divides the *context of use* into two parts:

1. The attributes of the physical and software platform, such as the screen size and the available interactors
2. The environmental attributes that describe the physical surroundings of the interaction, such as the location in which the interaction is taking place.

The first part is quite easy to incorporate into the UI rendering mechanism since the properties it describes are directly related to the rendering. The latter part consists mainly of properties that can be measured by software, but need to be interpreted before the information provided by the measurements can be used to adapt the user interface to the context. Since this means that the renderer cannot directly handle the information provided by measurements of the environmental context, we define the *context of rendering* as follows:

1. The attributes of the physical and software platform as in the definition of the context of use
2. The environmental-dependent properties, which is the information provided by interpretation of the environmental attributes, such as the set of available tasks (enabled task set or ETS) that can be influenced by the location where the interaction takes place.

Although both aspects of the context of rendering can describe a lot of aspects, the current implementation is able to incorporate the changes in the tasks that are to be performed, the screen size of the target device and to make decisions about the choice of interactors based on the preferred interactors and the available interactors in a certain context.

3 Modeling and rendering a User interface

In a task-based approach to the design of a user interface, one starts by defining the tasks that need to be performed. We chose the ConcurTaskTrees (CTT) notation (F. Paternò 2000), because it is

one of the most expressive and intuitive task models currently available. In our approach, the designed CTT is refined by adding abstract user interface descriptions (AUIs) to the appropriate leaf nodes of the CTT tree. These abstract user interface descriptions are made in an XML-format. For more details we refer to (Luyten & Coninx, 2001). Because we want to design truly multi-platform and context sensitive user interfaces, we will not require the designer to create platform specific versions of the user interface. Instead, the designer can optionally define different contexts.
Currently, a context of rendering description is limited to a set of possible mapping rules that map abstract interaction objects (such as a choice) to concrete interaction objects (such as an AWT list for the Java platform) and the task set that is to be active in that context. The rest of the contextual information, such as the screen size and the available concrete interactors, is detected at runtime and it is therefore unnecessary to model it explicitly at design time.

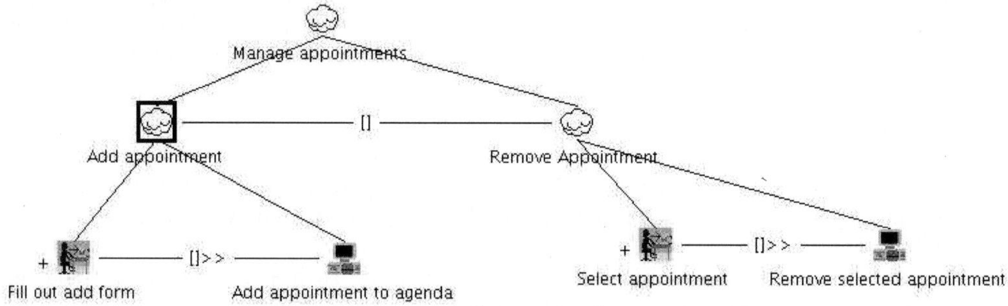

Figure 1 ConcurTaskTrees for the agenda-tool

When the necessary information for a multi-device user interface is modeled, it can be rendered. The rendering is done in three stages:

1. The AUIs that will be rendered are determined based on the information in the context of rendering, the enabled task set[2] that is "active"
2. The context information is read, such as the possible mappings, and determined, such as the screen size
3. The AUIs are translated to final user interfaces and rendered to the screen.

4 Rendering: an example

In this section, we will explain the rendering process as it is done in our current implementation into more detail using a minimal agenda tool as an example. The implementation extends the Dygimes framework, already used in previous work (Luyten & Coninx 2001). The agenda tool has only two interactive functions: adding appointments and deleting them.

Figure 1 shows a CTT-tree for this application. The AUI descriptions for the tasks "Add Appointment" and "Remove Appointment" are linked to the appropriate leafs of the CTT-tree. Listing 1 shows the code for the specification of the month in the former AUI description and the information is translated into an enabled task set description in a compact XML-format. Some mappings are specified in order to adapt the final user interface to the preferences of the user. Two of the specified mapping rules are shown in listing 2.

[2] The enabled task set can be automatically calculated from a CTT model (Paternò, 2000) and modified by changes in the context of use.

The listing shows two mapping rules as XML-tags that specify a mapping of an abstract interaction object (AIO) to a concrete interaction object (CIO). The rules have two mandated child tags in common: *aio* and *cio* which represent the type of AIO the rule applies to and the CIO it will be mapped on using that rule. The first rule has an optional *infocio* tag that specifies the CIO that will be used to display the information specified by the info tag of the abstract interactor in the AUI description. The second rule has an optional *name* tag that limits the application of the mapping rule to the interactors with a name that contains the string specified by the constraint. When there are multiple rules that could apply, the one with the best match is chosen. The absence of a *name* tag is equivalent to an empty *name* tag.

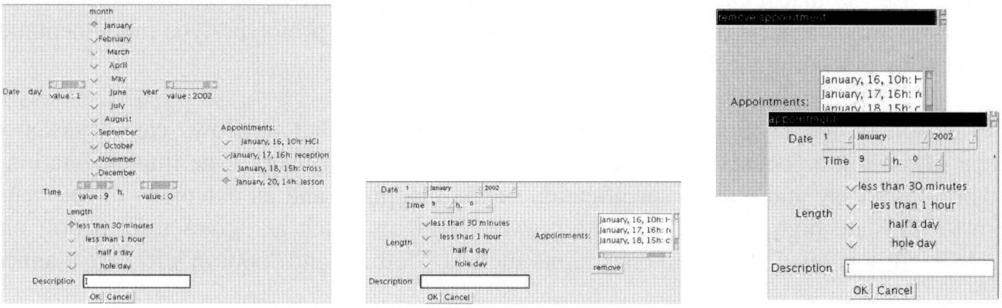

(a) Without mappings (b) With mappings (c) With mappings, small screen

Figure 2: Three different renderings of ConcurTaskTrees in figure 1

When the user interface of the application is rendered, the available tasks are read from the context description. For this simple example, there will be two tasks, except when specified otherwise. When the tasks are read, the applicable mappings are determined and the available screen size is calculated. There might be several rules that could be applied to map an AIO to a CIO, the correct rule is chosen based on the constraint that matches the specific AIO the closest. In our example,

Listing 1: Abstract User interface description

```
<?xml version ="1.0"?>
<ui>
  <title>appointment </title>
  ...
  <interactor>
    <choice name="dateMM">
      <info>month</info>
      <choicetype>single </choicetype>
      <item>January</item>
      ...
      <item>November</item>
      <item>December</item>
    </choice>
  </interactor>
  ...
  </group>
</ui>
```

Listing 2: Two of the specified mapping rules

```
<mapping>
  <aio2cio>
    <aio>choice</aio>
    <cio>awt . VertCheckboxGroup</cio>
    <infocio>awt.Label</infocio>
  </aio2cio>
</mapping>
<mapping>
  <aio2cio>
    <aio>choice</aio>
    <cio>awt.Choice</ cio>
    <name>date</name>
  </aio2cio>
</mapping>
```

both mapping rules in listing 2 could be applied. The first rule has no *name* tag, which means it is the default. The second tag, however, will be used to represent the interactor *date* because it has the most stringent constraint. Figure 2(a) shows the user interface rendered without applying the mapping rules but chosen by a fully automatic mapping engine, figure 2(b) shows the same interface rendered with the mapping rules applied to it. The result of the rendering with the same mapping rules, but in a context where less screen space is available, is shown in figure 2(c).

5 Conclusion and Future Work

We presented a new concept, "context of rendering", that represents a translated and specialized "context of use" so that it can be directly used to adapt the rendering of the user interface taking into account various aspects of the context. We then explained a prototype-type implementation of this concept by extending the Dygimes framework.

The extension of the Dygimes framework in such ways so that it allows easy translation of "context of use" into "context of rendering" is planned as future work. The implementation of a tool that allows easy and guided specification of the mappings and an extension of the mapping system to translate high level constraints between widgets (such as "gives information on" or "cancels") into low level positional constraints such as ("left of" and "bottom left") are planned.

6 Acknowledgements

Our research is partly funded by the Flemish government and European Fund for Regional Development (EFRD). The SEESCOA project IWT 980374 is directly funded by the Institute for the Promotion of Innovation through Science and Technology in Flanders (IWT).

7 References

Braubach, L., Pokahr, A., Moldt, D., Bartelt, A., Lamerdorf, W. (2002). Tool-supported interpreter-based user interface architecture for ubiquitous computing. In P. Forbrig et al. (Eds.), *Interactive Systems*, LNCS 2545 (pp. 89-103). Heidelberg: Springer.

Calvary, G., Coutaz, J., Thevenin, D., Limbourg, Q., Souchon, N., Bouillon, L., Florins, M., & Vanderdonckt, J. (2002). Plasticity of user interfaces: A revised reference framework. In C. Pribeanu & J. Vanderdonckt (Eds.) *Proceedings of TAMODIA 2002* (pp. 127-134). Bucharest: INFOREC Printing House.

Luyten, K. & Coninx, K. (2001). An XML-based runtime user interface description language for mobile computing devices. In C. Johnson (Ed.), *Interactive Systems*, LNCS 2220 (pp. 17-29). Heidelberg: Springer.

Paternò, F. (2000). *Model-Based Design and Evaluation of Interactive Applications*. Heidelberg: Springer.

Paternò, F. & Santoro, C. (2002). One model, many interfaces. In C. Kolski & J. Vanderdonckt (Eds.), *Computer-Aided Design of User interfaces III* (pp. 143-154). Dordrecht: Kluwer Academics Publishers.

Souchon, N., Limbourg, Q. & Vanderdonckt, J. Task modeling in multiple contexts of use. In P. Forbrig et al. (Eds.), *Interactive Systems*, LNCS 2545 (pp. 59-73). Heidelberg: Springer.

Location-Transparent User Interaction for Heterogenous Environments

Chris Vandervelpen, Kris Luyten and Karin Coninx

Expertisecentrum Digitale Media – Limburgs Universitair Centrum
Wetenschapspark 2
B-3590 Diepenbeek – Belgium
{chris.vandervelpen,kris.luyten,karin.coninx}@luc.ac.be

Abstract

Because of the rapid evolution of mobile and embedded computing devices, it is important for a business that its services are accessible for a large group of possible service consumers. These consumers want to gain access to the offered services from heterogeneous environments. In this context, much work has been done on the (semi-) automatic generation of user interfaces (UIs) for services that are mostly targeted toward mobile computing devices. However, most of that work fails to provide actual solutions to the problem of how to provide a suitable interaction mechanism between the generated UI and the application logic that implements the functionality. We will show that by defining the implementation of the application logic as a webservice, we can use webservice technologies to provide two-way communication between the application and its generated UI.

1 Introduction

In a time that large quantities of new networked mobile and embedded devices are shipped to market, it is very important for businesses to make their services available for all these different kinds of appliances (e.g. PDA, cellular phone, set top box, information kiosk, desktop,...) with a short time-to-market. The main problem in achieving this goal is the diversity of all these devices. They all have different limitations such as different operating systems, memory sizes, screen sizes and interaction possibilities (touchscreen, stylus, speech, mouse, keyboard,...). When porting a UI from one environment to another, traditional development techniques require a redesign of the UI in order to take particular constraints into account. To address this problem, much work has been done concerning the (semi-) automatic generation and adaptation of UIs on different hardware and OS platforms (Eisenstein, Vanderdonckt & Puerta, 2001, Nichols et al., 2002). They provide different ways to describe a UI on a platform independent level. This abstract platform independent description then should be instantiated with a concrete platform dependent UI by using a rendering mechanism (e.g. a web browser that renders HTML content). These approaches are often model-based: different models are used that together describe a complete UI. E.g.: A presentation model is used to describe the presentation structure, a domain model can be used to describe problem domain concepts and a platform model is used for describing the possibilities of the target platform. However, considerable less attention has been paid to the way one copes with two-way interaction between the generated UI and the implementation of its functionality which could be either a remote or a local application.

This paper extends previous work (Luyten, Vandervelpen & Coninx, 2002) where a presentation model and the UI rendering framework Dygimes (Dynamic Generation of Interfaces for Mobile and Embedded Systems) are used to generate UIs for heterogeneous computing environments. The Dygimes framework is a platform, device and system independent rendering and interaction system implemented in Java that uses XML-based abstract UI descriptions to represent the presentation model.

In the next section we will discuss the main problems with interaction when dealing with (semi-) automatic generated UIs for local and remote applications. Then, we introduce a solution for these problems based on webservice technologies. In section 5 we discuss benefits and shortcomings of our approach together with some ideas we like to work on in the near future.

2 The need for an interaction model

In general, a UI has to enable the user to interact with a piece of application logic and to let that application logic provide feedback (visual, sound, haptic,...). To enable this kind of two-way communication, UIs are tightly coupled with the application logic. However, when the UI and the application logic are not located on the same device, the tight coupling between both causes problems. Elaborating on a model-based UI design, we propose the usage of an interaction model to overcome the high coupling problem. An interaction model will help the designer to prepare the service or application logic to use location transparent UIs. One can think of this as an "enhanced" Model-View-Controller (MVC) architecture which will be necessary for the next generation of UI toolkits. Such a model is needed to enable the generated interface to be loosely coupled with the application logic.

Until now the Dygimes framework supported only a basic notion of interaction handling where action elements in the abstract UI description could be attached to an interactor. Such an action element denotes which method, from which Java class, would be invoked when the user manipulates that interactor. The problems with this approach were the restrictions to local method invocation and the binding to the Java programming language. Another problem was that there was no notion of datatyping in the system which was problematic when we wanted to pass data between the generated UI and the application logic. It is clear that these shortcomings restricted the interaction model.

A possible way to deal with the aforementioned problems would be to define our own communication protocol between the UI and the logic. For example, (Nichols et al., 2002) uses an XML-based communication protocol that enables two-way communication between appliances and their Personal Universal Controller (PUC). However, they define a static set of self-defined messages and they do not provide a way to describe the interactions and the data flows supported by the system separately. This means that the UI designer still needs a good understanding of the application logic.

We propose a solution that combines existing message passing and communication techniques with out Dygimes framework. This eases the task for the UI designer and allows him to concentrate on designing UIs for multiple platforms (the view) instead of being bothered by the technical details imposed by the implementation and the integration of the application logic (the model). On the other hand, the application logic programmer can implement the services without having to worry about the target devices. Our approach allows a smooth integration of application logic and UIs at runtime, while allowing to separate them during development.

3 Interacting with webservices

The proposed solution is based on the believe that one needs a device and programming language independent description of the possible interactions with the system. This enables the UI designer and the application designer to do their work independently of each other. Such a functionality description supports the generated UI to decide which functionality to invoke as a response to certain user interactions.

To realize this, we define a piece of application logic as being a webservice and the generated UI as being the client that wants to use the webservice. With such an approach we can use existing webservice technologies to achieve our previously stated goals. In this context, we have chosen to use a functionality description based on the Web Services Description Language (WSDL, 2001). WSDL is an XML-based language that enables the description of the operations, messages and datatypes that are supported by a webservice.. Operations consist of request and/or response messages. The language uses some default primitive datatypes, which can be used as parameters and response values for messages (double, integer, boolean,....). It also provides a section to define custom datatypes. By default XMLSchema is used for defining datatypes. We extended this specification by adding a section in which we bind UI interactions to service operations. From now on we will refer to this WSDL-based description as the interaction description.

Once we have an interaction description representing the service functionalities and the bindings with the abstract UI, we can use this information to invoke operations when the user interacts with the generated UI. For this we need a protocol for handling the interactions as described in the interaction description. In our approach we use existing XML messaging protocols to achieve this. One such protocol is the Simple Object Access Protocol (SOAP, 2000). This protocol enables us to invoke functionality on webservices by using Remote Procedure Calling (RPC) based on an XML syntax. Another, more efficient implementation of XML-based RPC, is XML-RPC (XML-RPC, 1999). It is in our intention to evaluate and integrate both implementations into the Dygimes framework.

4 Interaction in Dygimes

With the discussions from the previous sections in mind, we extended the Dygimes framework with a more advanced interaction engine. We now can identify the following parts:

- A piece of application logic (the service) that is annotated with a description of its UI and a WSDL-based interaction description that denotes the functionality. The interaction description also provides a binding between the service and the UI;
- A rendering system which renders the UI description on a particular platform. At the moment, the possible rendering platforms include AWT, Swing, HTML and Applet. The rendering on a cellular phone and a PDA is supported through the J2ME CLDC (Java 2 Micro Edition);
- An interaction engine which uses the interaction description to automatically determine which functionality to invoke on the application logic (which is local or remote) when the user interacts with the generated UI.

We now describe the sequence of events between the different parts when the Dygimes framework is used. To start, the application logic is asked to send its UI and interaction descriptions. When Dygimes receives this XML documemts, it renders the UI and parses the interaction description to build a datastructure in memory. When the user interacts with a particular UI interactor, which has a unique identifier, the following steps are processed by the interaction engine of Dygimes:

- Based on the interaction description, operations triggered by the manipulation of the interactor are determined;
- Data needed for the operation invocation are extracted from the user interface. The datatypes information provided by the interaction description is used to determine the datatypes of the message parameters;
- The interaction engine uses XML-based RPC or Direct Method Invocation (DMI) to invoke the necessary methods of the service;
- The service invokes the method, returns a message back to the UI client and the UI is updated if necessary.

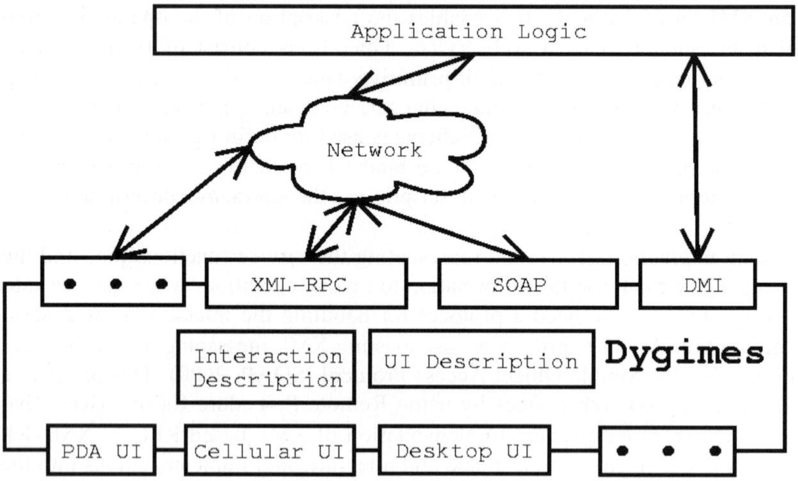

Figure 1: Architecture

Figure 1 gives a closer view of the architecture of the system. Dygimes uses the UI description to render the UI to heterogeneous environments. Interaction occurs through the use of the interaction description together with one of the communication technologies.

5 Conclusions and future work

In this paper we presented an extension of the Dygimes framework which enables the use of an interaction model to handle interaction. Because the proposed approach is based on existing webservices technologies, we can identify the following benefits:

- The system becomes webservices-enabled through the use of SOAP. This will be important in the near future;
- The approach is device and language independent because it is based on XML. If we for example used Java RMI, the system was restricted to the Java programming language;
- The system is usable through firewalls. This would not be possible if we used technologies like CORBA or RMI because they both use a binary communication protocol;
- The system uses common standards: XML and webservices;

- The generation of functional UIs for remote applications becomes much easier;
- The system becomes distributed which enables location-transparency. This means that the location of the application logic is transparent for the user which allows UI migration without the need to reconfigure the UI.

The main problem of this approach could be the overhead caused by the construction of the XML messages. For remote interaction this overhead is minimized by offering the user the possibility to use XML-RPC instead of SOAP. For local interaction we give the user a choice to use XML-based user interaction or DMI which is much faster. The drawback is that with DMI the user is restricted to Java implemented local services.

In the near future we would like to further investigate how webservice and XML technologies can help in achieving effective platform independent and multimodal interactive applications. We could, for example, use the definitions of datatypes in XML to make Dygimes extensible with new and composite interactors.

6 Acknowledgements

The research at the Expertise Centre for Digital Media (LUC) is partly funded by the Flemish government and EFRO (European Fund for Regional Development). The SEESCOA project (Software Engineering for Embedded Systems using a Component-Oriented Approach) IWT 980374 is directly funded by the IWT (Institute for the Promotion of Innovation by Science and Technology in Flanders).

References

Eisenstein J., Vanderdonckt J., & Puerta A. R. (2001). Applying Model-Based Techniques to the Development of UIs for Mobile Computers. In IUI 2001 International Conference on Intelligent User Interfaces, 69-76

Luyten K., Vandervelpen C., & Coninx K. (2002). Migratable User Interface Desciptions in Component-Based Development. In Forbrig P. et al., (Eds.), DSV-IS, volume 2221 of Lecture Notes in Computer Science, Springer.

Java 2 Platform Micro Edition. Sun Microsystems, http://java.sun.com/j2me/.

Nichols J., Myers B. A., Higgins M., Hughes J., Harris T. K., Rosenfeld R., & Pignol M. (2002). Generating remote control interfaces for complex appliances. In User Interface Software and Technology.

World Wide Web Consortium. (2001). Web Services Description Language specification. http://www.w3.org/TR/wsdl

World Wide Web Consortium (2000). Simple Object Access Protocol. http://www.w3.org/2000/xp/Group/

XML-RPC homepage. (1999). http://www.xmlrpc.com

An Event-Based Communication Mechanism to Realize a Mobile Collaborative AR Environment

Reiner Wichert, Matthias Finke

Computer Graphics Center,
Darmstadt, Germany
{Reiner Wichert, Matthias.Finke}@zgdv.de

Mehdi Hamadou

Siemens AG, A&D
Nuremberg, Germany
Mehdi.Hamadou@siemens.com

Abstract

The following work focuses on the requirements and development of a component to provide information for Collaborative Augmented Reality Systems by using Web technologies. In collaborative mobile environments a mechanism is needed to distribute information by event handling on different devices for multiple users. Therefore a component is required to minimise the data traffic which occur during information exchange between all collaborative users. An analysis of useful scenarios and derivation of possible resulting requirements to form the basis for this component will be shown in this paper. Finally the component was successfully integrated in a collaborative AR environment in the industrial field.

1 Introduction

Since humans wants to communicate and collaborate a lot of technologies have been developed. By using the enabling technologies of computer net-working for collaboration a new research field - the Computer Supported Cooperative Work (CSCW) – has been expanded, but soon it was recognised, that traditional teleconferencing and groupware systems has many limitations. Since it became possible to bring real and virtual objects into one common world the research area of Augmented Reality (AR) has emerged. With the idea to combine AR with CSCW, Collaborative AR became possible, where co-located users can experience a shared space that is filled with real and virtual objects (Reitmayr, 2001).

The component-based Collaborative AR System ARVIKA presented in this paper is a situation-oriented and user-centred system in the field of industry. Its primary aim is to test AR in development, production, and servicing. This AR system relieves skilled workers of the planning and monitoring of production and provides support in installation work by directly receiving information on the status of the respective job as a single user system or multiple users in a collaborative AR.

For a co-operation involving multiple users, data for the synchronisation of and communication between components is required. Each individual user can exchange necessary information within a group. In this way, co-operation between multiple users in an AR environment becomes possible, where data must be distributed and updated. For this reason, in Collaborative AR, a component is required to clarify which room the user is in, which machine the user is operating, which work phase a worker is in, how a user gains secured access to his own data, how functionality is distributed or how components communicate with one another. As a result, an analysis of the general architecture of the AR system was carried out. It became clear that an, until now unplanned component for Resource Management was required. The component named ContextManager has the task of optimising the data flow of the components, of storing accumulated information in the system, and of conveying this information to other components.

2 General Approach

Most of the existing AR systems do not support collaborative AR in real time. So, one of the advantages of the component-based AR System will be to support individual users in their simultaneous interaction with augmented objects by using web technologies. Thus, a component was needed to handle the communication between the main components of the base AR system and notify other components or other clients co-operating together when values are changing. With the help of existing base services, the user is locally supplied with information. The information should be pre-filtered for appropriate, timely context. This ContextManager component serves as a central, persistent data manager and communicates with almost every component of the AR system. Components can also convey information beyond their life span via the ContextManager.

The ContextManager (CM) provides all components with a uniform view of information stored in different profiles and conveys them to other components necessary for their function. In the Device Profile, there is hardware-specific data like display resolution, available memory or network connection. In the User Profile, user data, such as name, role, room or position, is stored. Operation-specific data, such as current tasks, work phase, or document currently being used, is entered in the Technical Profile. The optional fourth profile contains data for a multi-user co-operation like the name of the scene or the orientation and the movement of the virtual objects.

Initial, valid application possibilities were found and a definition of useful scenarios was realised. Following an analysis, the component CM must facilitate a scenario, where constant radio contact can be guaranteed all time. The components convey information through the CM. If a value changes, the components that are dependent on the change must be notified. A second scenario was found where Radio contact is lost. In order to guarantee the supply of context-dependent data, the CM must be available on client and server side. The client-sided CM is updated as soon as radio contact is re-established. A last scenario could be where only the server-sided CM is used. By way of a configuration file, it can be determined whether the client-sided CM should be started.

After analysis of the scenarios, possible resulting requirements were deduced. These requirements form the basis for the new component CM, which is integrated into the web-based architecture and is necessary for the both the client and server side, so that context-dependent data can be accessed when radio contact is lost or in a scenario without radio contact. Context information is managed and constantly compared on both sides, in order to guarantee a current presentation of the context. If radio contact is lost, a comparison must be performed immediately after contact is re-established. The components must be notified of a value change by way of an event mechanism. A StorageAdapter ensures a uniform interface between the data management system and the CM, so that any data management system can be implemented. Based on problem analysis, an initial generic architecture was found. In accordance with resulting requirements, the CM was conceived and specified. The demands on the required interfaces were derived from the co-ordinating tuning and analysis work with the components. In a further step, a first version of the CM was implemented and a test environment was also developed as an extension of the CM. After extensive integration and corresponding testing of the new component a client-sided CM was also implemented. A comparison of the data occurs by way of an update from the server.

3 Architecture and Realisation

The web-based client-server architecture of the component-based ARVIKA System (see Fig.1) enables the visual overlaying of computer-generated virtual objects over real objects via AR Technology over the Internet using HTTP protocol. The advantage is that all users have access with less configuration cost through firewalls over large distances and a remote user can be very easily connected to the AR system from any place in the world. The base ARVIKA AR services are implemented by server-side components. Web integration is implemented with a servlet interface

where the output via Internet Explorer is based on Java Server Pages. The runtime environment is the Apache Tomcat Servlet Engine (Wiedenmaier, 2001). The basic components are largely decoupled through the event-based communications of the ContextManager, which stores various context information, while the transparent access to information and its distribution in a wireless and mobile environment is done by the InfoService. The InfoBroker assists the user with a situation-sensitive delivery of information that has previously been described in the product model. A consistent data interface supports the connection of legacy systems for the delivery of process or product data. The WorkflowEngine guides the user through predefined assembly or maintenance steps, offering the proper AR assistance and documents for each step. The Collaboration component enables the integration of remote experts, e. g. a hotline, into AR-assisted work situations, the collaboration of skilled production workers or a teacher with students while assembling a mechanical engine. User Interface Configuration customises the contents and interactions depending on the input / output devices used and the current context of the user (Friedrich, 2001).
The rendering and visualisation component is realised by the AR Browser as a ActiveX component in the web browser on the client side. A video server captures the picture of the camera and provides it to the AR Browser, which is a complete VR System with special AR features. Consequently, the AR Browser can be very easily integrated into every web-based application (Müller, 2001). To permanently synchronise the augmented scene, the viewing direction of each user is tracked continuously and in real-time. The computer vision-based tracking is connected to the AR Browser, and a camera, mounted on the HMD detects a marker by its square arrangement of edges and the matrix on it. Thus, 3D tracking is already possible with only one marker.

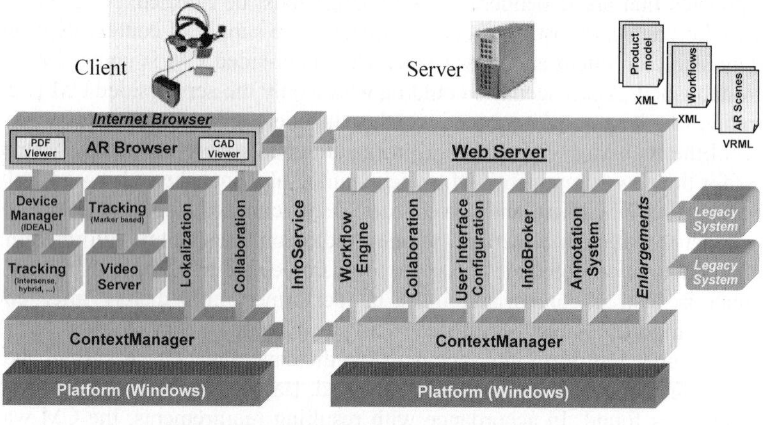

Figure 1: ARVIKA System Architecture

The CM consists of three parts: an event mechanism that assumes the component registration and notification, the StorageAdapter and the ContextInterface to the components. The CM is integrated into the web-based architecture and is present on both the client and server side. The components use the CM on their side. In the event of a request to change a value, the Browser sends a non-ambiguous client identification and informs the servlet by HTTP. This ID is created when the updated context is generated and is transmitted with it from this point on. The ContextServlet that should answer the client request changes the URL into a method call that is then passed on to the CM, which records the value from the associated StorageAdapter and sends it back to the servlet.
The return of the value is also the receipt. The CM contains a chart ContextContainerMap, which links the ID with the corresponding ContextContainer. For each session a new ContextContainer will be built, which stores the used profiles of the belonging session in the chart ProfileMap inside the ContextContainer. To assign the session to the specific ContextContainer an entry, which connects the ID to the according ContextContainer, will be generated for each new client. A compo-

nent can register itself for a value change of a specific property. The changes, for which a component would like to be notified are entered in the CM in the list ListenerMap. If this value is changing a component receives a notification and the value will be stored via the StorageAdapter into the connected data management system by writing. It goes through its chart entries, where the components and other clients have registered themselves to be informed in the event of a change in a certain value and informs them one after another about the change. The client-sided component that requested the change thus receives a receipt for the change. If a receipt is not issued for the change, the component must attempt to make the change at repeated intervals. This can happen, for instance, when radio contact with the server is lost.
The required data for a data exchange within the CM is wrapped in an object called ContextProperty, which is a data structure that is sent through the various classes of the CM. This object is required by the StorageAdapter and the Notification Listener, which is the interface for receiving a notification. The StorageAdapter forms a Java interface between communication protocol and data management system. Therefore, a separate StorageAdapter must be implemented for each data management system. Access to databases or data management systems occurs via an XML-dependent structure. A stored value can have different data types and a time stamp can be set for time-critical values. Conclusions can be derived from the stored values in a ring buffer like a kind of history.

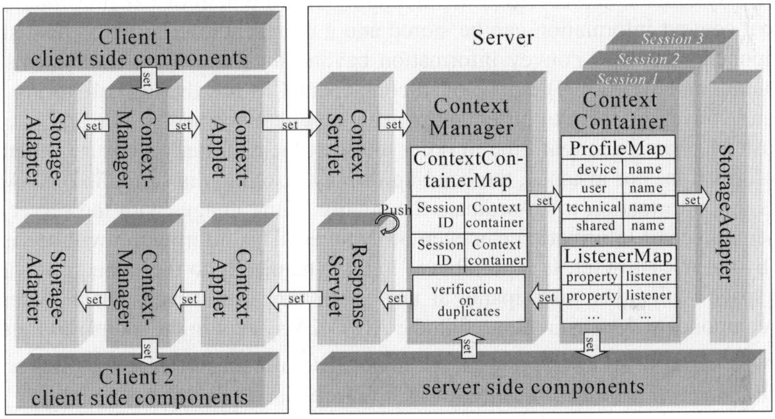

Figure 2: Session-spanned Push Mechanism

The Login Process is handled via Java Server Pages on which, the assignment to the profiles is done. The registration data is sent via the client's registration window to the servlet address on the server. Since this registration data is not yet available, the servlet engine generates a session with a SessionID and manages it. Simultaneously, an instance of the both CMs is generated via the servlet engine upon initial user login. Profile assignment to the appropriate SessionID is intermediately stored in the CM in a chart inside a container. A specific profile value can be called up as needed via the SessionID. Afterwards, the CM is present on the client, as well as the server side.
Via a method call the client-sided profiles are overwritten by the server-sided profiles and the CM start process is complete. To replicate the data between client and server, the various profile data was in the first version synchronised via an update mechanism. Due to the update mechanism, time-critical data were sent to linked clients too late. As a result, a push mechanism was conceptualised to compare the client- and server-sided CMs and notify the components immediately after an event occurs (Broecke, 2001). However, since servlets close the response after an HTTP request, a separation of servlet functionality into two sub-components, the ResponseServlet and the ContextServlet, was both necessary and implemented. Eventually, the test environment was expanded into an authoring tool with which the context-dependent data could be manipulated. As with the up

date mechanism, both CMs are generated the same way and linked with the appropriate profiles. The actual AR start page starts the PushApplet. The ContextServlet then generates two loops to forward events (push), where the first loop in an interval constantly sends the client only a few signs to maintain the HTTP response. Unfortunately, the ContextServlet was blocked for further queries as a result. Hence, this solution also proved unsuitable. For this reason, the functionality was distributed to two servlets on the server side. The ContextServlet receives the client queries and forwards them to the CM. The answer is sent back over the response (that is kept open) via the ResponseServlet. For realisation, both loops described above are started via the new Response-Servlet. In parallel, the PushApplet initialises and starts the ResponseServlet, that then registers itself at the server-sided CM as Listener for all property changes. Should a property be forwarded from the server-sided CM to the ResponseServlet as a callback, this property is first ordered in a queue. The loop gets this property out of the queue and thereby generates an object that is then forwarded to the PushApplet in lieu of the blanks now in real time.

4 Conclusion

In this paper, a concept was developed for a component to provide on-site information for a Collaborative AR System via a web-based architecture. With the help of existing Internet technologies, different context information can be stored and a representation of the system status can be given. Components can also convey information beyond their life span. Subsystems can access data for Reading and Writing. The developed component CM has the task of storing information in profiles and conveying them to other components on client and server side and is compared by way of a Push Mechanism in real time in order to direct their further procedure depending on the incoming information. For instance, the head-up display resolution is stored in the CM. The Info-Service uses this information to co-ordinate the quality of the images with the server or to determine whether a piece of information can even be depicted. This information will be visualised and augmented by the AR Browser. It is possible to login dynamically to the same group and share the augmented space with other participants at any given time from everywhere. After successful testing, the component CM was integrated in the ARVIKA system in the industrial field.

This work was partially sponsored by the German Federal Ministry of Education and Research (BMBF) within the scope of the project ARVIKA (Grant No. 01 IL 903).

5 References

Reitmayr, G., Schmalstieg, D. (2001). Mobile Collaborative Augmented Reality, Technical University of Vienna, *Proccedings of ISAR 2001*, New York, USA.

Wiedenmaier, S., Oehme, O., Schmidt, L., Luczak, H. (2001). AR for Assembly Processes – An Experimental Evaluation, RWTH Aachen, *Proccedings of ISAR 2001*, New York, USA.

Friedrich, W., Jahn, D., Schmidt, L. (2001). ARVIKA-Augmented Reality for Development, Production and Service, *International Status Conference,* pp. 79-89, Saarbrücken, Germany.

Müller, S., Stricker, D., Weidenhausen, J. (2001). An AR-System to support product development, production and maintenance. Retrieved December, 2001, from http://www.inigraphics.net/publications/topics/2001/issue3/3_01a07.pdf

Van den Broecke, Pushlets: Send events from servlets to DHTML client browsers. Retrieved March, 2000, from http://www.javaworld.com/javaworld/jw-03-2000/jw-03-pushlet.html

On Designing Automotive HMIs for Elderly Drivers: the AGILE Initiative

Harald Widlroither, Lorenz Hagenmeyer
Fraunhofer Institute for Industrial Engineering, Stuttgart, Germany
Lorenz.Hagenmeyer@iao.fhg.de

Sascha Breker
Institute for Occupational Physiology
Dortmund, Germany

Mary Panou
Hellenic Institute of Transport
Thessaloniki, Greece

Abstract

The aim of the research presented in this paper is to identify critical issues in the development of automotive human machine interfaces (HMIs) for older drivers by assessing their specific abilities and restrictions. It is shown that the ratio of older drivers will increase significantly and, hence, designing for older drivers will become a critical task. The development of a new tool for the assessment of older people's driving abilities is described. From the results of a literature study and questionnaire based surveys, driving tasks critical for older drivers are identified, and corresponding guidelines for the design of HMIs for older drivers deducted.

1 Introduction

Within the last 20 years the number of controls in a middle-class car has been reduplicated. Constantly more elements, systems and functions are coming into the car in order to increase comfort and to assist the driver. In-vehicle HMIs thus become more and more complex. On the one hand, these functions and systems help to raise comfort and safety, but on the other hand there is an inherent possibility that the same functions and systems might overload the driver with too many information or complex operations. This is especially the case for elderly drivers who may have more problems in processing high loads of information than younger drivers.

Designing HMIs for elderly people is a challenging task that will become even more relevant in the years to come: In the western society the older population is increasing both in absolute and relative terms (TRBNRC, 1988). Moreover, future cohorts of older people will differ from traffic participants today: As in many European countries mass motorization only developed in the 1960s, the number of older drivers[1] will increase significantly. The European project EDDIT estimated that the number of older drivers on European roads is approximately 12% today, and will reach about 20% by year 2010.

Driving as an individual's transport of choice is a key issue in the mobility of the older people. Many of them are not able to walk the required distance, stand for a long time or have the overall physical endurance to use public transportation. Thus, the aim to support older people to drive provides practical, social and personal value, and improves their quality of life significantly by sustaining their independent mobility.

Nevertheless, ageing affects certain capabilities that are important for driving. Both an increase in caution and a decrease in certain functions appear in older drivers' accident statistics. Older persons seem to be aware of their individual deficiencies and do take them into account when driving.

[1] In the literature, 'older' is most often defined as an age over 65. However, drivers in their 60's are, as a group, qualitatively different from those in their 70's and certainly different from those in their 80's due to the fact that accident risk increases nonlinearly with age.

Older drivers have also no particular problem with automatic processes, which place limited demands on their attentional capacity and generally occur under highly predictable conditions of traffic and weather. However, there do exist problems related to non-automatic driving skills especially in demanding traffic situations such as turning, merging, collision avoidance or driving in adverse weather conditions. The most difficult situations seem to be those which demand a rapid processing of large amounts of simultaneously incoming information, like in intersections or inner city driving. As driving is a very complex cognitive activity, it is very important to identify those cognitive functions that are related to driving. In addition to a normal ageing-related cognitive decline, some chronic illnesses may also present aggravating factors, especially if different kinds of medication are involved. Medication could have secondary effects on different cognitive functions which are important for safe driving.

When designing automotive HMIs for older drivers these considerations have to be carefully taken into account. In recent years, intensive work has been carried out in order to develop user-friendly driver support systems for the elderly (Naniopoulos et al. 1994) and relevant guidelines (Naniopoulos et al. 1994, Nicolle et al. 1999). However, a clear identification of the elderly's problems and their capabilities in relation to specific driving tasks and corresponding design guidelines is still missing. The project AGILE[2] takes an integrative approach to these problems: It aims at developing knowledge about needs and deficits of older drivers in order to specify design requirements for this particular target group. The final goal of this process is helping older age groups to continue driving safely because mobility is a significant factor for their well-being.

2 Method

The aim of the AGILE initiative is the development of a new set of training, information, counseling and driving ability assessment and support tools for the elderly, evaluating their full range of abilities. From the data collected, guidelines for the design of HMIs for the older drivers will be deducted. The program is organized in 9 work packages (WPs) as shown in figure 1.

In the first work package (WP1) an etiological analysis of elderly driving related problems was carried out. Literature and accidentological studies, interviews and questionnaire surveys with experts as well as older drivers themselves were conducted. From the analysis of these data, an analytical matrix was composed relating the situation of ageing drivers to different driving tasks. The results were discussed in a pan-European workshop with external experts in the fields of ageing and driving ability assessment.

A state of the art review on existing driving assessment criteria, methodologies and tools is currently performed (WP2). The results will lead to a selection of specific driving assessment parameters and decision criteria for elderly persons. .

On this basis, an elderly driver assessment parameter database will be created (WP3. Additionally, a paper and pencil-based rapid pre-screening tool will be designed in order to help older drivers to make self-assessments. As a preliminary tool for health care professionals it will help them to identify those older drivers who might need a more thorough evaluation.

Existing and newly developed driving assessment tools will be combined to an integrative assessment system, including a neuropsychological test battery, driving simulator scenarios and short on-road tests (WP4). Furthermore, a standardized reference test drive will be developed as a benchmark for the newly created evaluation tools.

The measures described above will be integrated to a standardized and modular driving assessment procedure in combination with a decision and consultation tool (WP5).

---

[2] AGed people Integration, mobility, safety and quality of Life Enhancement through driving, co-funded by the European Union, contract number QLRT - 2001 - 00118

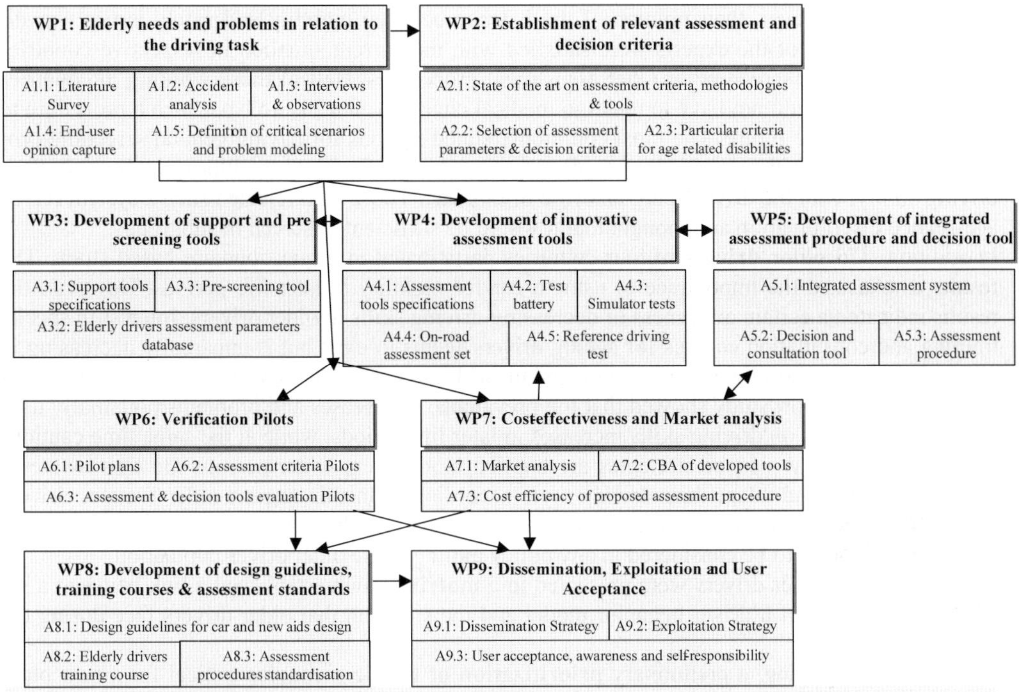

Figure 1: Diagram depicting the structure of the project 'AGILE'

The project results will finally support the development of quantitative and specific guidelines for in-vehicle HMI design to support elderly drivers, a flexible and personalized training course, specifications of new aids to support older drivers and a standardization proposal for a pan-European driving assessment policy (WP8).

3 Preliminary Results

In this section, the main results of the research carried out so far are presented.

The **literature reviewed** constituted that older drivers are in general safe drivers. It was found that the study designs used in most research on the elderly were cross-sectional, but did usually not pay attention to the methodological limitations of this design when generalizing results.

An analysis of several accident databases showed that compared to other drivers, older drivers' accidents occur more often during daylight, on weekdays and on roads that are not affected by snow or ice. According to accident statistics it is obvious, that the elderly choose less demanding conditions when driving - probably a choice younger drivers would make as well if they had the possibility.

The typical older driver accident apparently occurs at intersections, i.e. in situations that require fast processing of information and quick reactions. However, these accident pattern might be less pronounced in the future as today's older drivers differ from the future one's in terms of their driving background. Thus, it is inappropriate to take findings of today and extrapolate them to the older drivers of tomorrow.

A **questionnaire based study** was performed in order to capture attitudes and knowledge among **traffic experts.** Due to a more safety oriented approach the experts were not up-dated on some recent research findings. Most of them were aware of the social importance for the elderly to be

able to drive. However, their opinion of the balance between mobility and safety issues was divided: Almost half of the experts were satisfied with the current situation in respective countries concerning medical screening or other assessment procedures. Half of those working in countries where older drivers do not have to undergo medical checks would like to have such a procedure to be introduced in the future. The most commonly stated reasons for having general screening procedures were to avoid that old drivers are driving while suffering from illnesses affecting their driving ability, with the driver being unaware of this fact. Those who rejected such a system feared age-related discrimination and thought that it would be sufficient to screen in subgroups.
In addition, 473 **older drivers** in four countries participated in a questionnaire based study. The results emphasized the importance of using a car for mobility in later life periods. However, the results indicated a certain awareness of decreasing driving skills in older drivers. Interest in special training and consultation courses for ageing drivers seems to exist but is apparently decreasing in older age groups who might need support the most. In agreement with scientific literature, the results of the self-report study showed that the prevalence of diseases and problems potentially having negative effects on driving skills increases in later life periods, while at the same time cautious compensation and avoidance strategies are applied more often. Older persons are thus aware of potential deficits and adapt their driving style accordingly. In this respect older females drivers seemed to be more cautious or at least more defensive drivers than older male drivers. Therefore, gender aspects have to be considered in any policy aiming at the older driver population.
The problems of older drivers were presented in a **matrix,** which established a link between a sophisticated version of a driver behavior model and the situation that older drivers face during the ageing process.
Based on this mapping, a preliminary **prioritization of traffic scenarios** critical for a new older driver assessment system was achieved. In the order of their priority these scenarios are:

> Intersections, yielding right of way, merging, driving in a complex area, interaction with pedestrians, driving with a secondary task, high informational load, passing and overtaking, adverse weather condition, emergency braking, driving for a longer period, driving under time pressure, way finding in an unfamiliar area, turning in a narrow lane, driving in a narrow lane.

4 Conclusions

It has been identified that the load of information that has to be processed in parallel is critical to the older driver. This situation has two aspects: In addition to direct information about the current traffic situation the driver gets more and more indirect information via new driver support systems. Future car design must therefore consider some kind of work load management in terms of an ergonomic design that helps the older driver in mastering difficult traffic situations such as turning left or merging. An automatic collision detection with an appropriate warning strategy and/or automatic car break might be a solution.
Another issue to be considered is the decline of cognitive abilities by diseases or medications and drugs. Here, a driver monitoring system might represent a solution to warn the driver or even take appropriate action such as reducing speed, altering the steering sensibility and so on.

5 Summary and Perspectives

The ratio of older drivers is steadily increasing so that developing ergonomic HMIs for older drivers will become a more and more critical design task. The results of a literature review and questionnaire based studies of experts and older drivers were presented. It was shown that the opportunity to drive strongly influences the social live of the elderly. Ageing, however, influences older persons' ability to drive. It is concluded that functionally oriented diagnostic instruments testing practical driving skills of individual older drivers are urgently required. Relevant assessment crite-

ria will be identified and implemented in new driving assessment tools. The knowledge acquired in this process will finally be used to define recommendation for in-vehicle HMI design.
It is expected that the adaptation of anti-collision systems and vision support systems to the identified needs and capabilities of older drivers will have great potential to increase traffic safety and driver comfort.

6 Acknowledgements

The authors acknowledge Per Henriksson, Gert Eeckhout, Torbjörn Falkmer, Anu Siren, Liisa Hakamies-Blomqvist, Evangelos Bekiaris and Edna Leue for their significant contribution to the results presented in this report.

7 References

Centraal Bureau voor de statistiek (CBS, 1992). Statistisch Zakboek 1991. Den Haag, the Netherlands.

EDDIT (V2031). Elderly and Disabled Drivers Information Telematics. Technical Annex.

Transportation Research Board of the National Research Council (TRBNRC, 1988). Transportation in an Ageing Society, Vol.1 and 2, Washington DC.

Naniopoulos, A., Bekiaris, E. and Nicolle, C. (1994). Design Guidelines for Aids for DSN. CEC DRIVE II TELAID Project V2132, Del.7.

Nicolle, C., Burnett, G., et al. (1999). TELSCAN Handbook of Design Guidelines for Usability of Systems by Elderly and Disabled Travellers. CEC TAP TELSCAN Project 1999, Del.5.2.

A Human-Computer-Interface Concept for Mobile Devices to support Service & Maintenance Staff in Industrial Domains

Carsten Wittenberg

SIEMENS AG
Corporate Technology
User Interface Design
Otto-Hahn-Ring 6
D-81739 Munich
Germany
carsten.wittenberg@siemens.com

Birgit Otto

SIEMENS AG
Corporate Technology
User Interface Design
Otto-Hahn-Ring 6
D-81739 Munich
Germany
birgit.otto@siemens.com

Abstract

This paper describes the results of a use context and requirement analysis for an user interface concept for mobile devices in the domain of industrial automation. As a result of this analysis advantages for the use of mobile devices in the domain of service and maintenance were recorded. A usability test showed that a user interface on the basis of these requirements is a useful support for the service- and maintenance personnel during their daily work.

1 Introduction

The supervision of modern production systems will be more and more centralised. E.g. autonomous power and heating plants will be spread over the cities and regions. In one central control room a number of such plants will be supervised and controlled. Although these plants are autonomous there is still a need for service & maintenance activities. E.g. for safety reasons different process values has to be controlled onsite regularly (every 24h or 36h) by law. In case of malfunctions the service and maintenance staff has to react very fast and correct to reduce downtimes and costs, and to increase the safety. Although the service and maintenance staff plays a relevant role the working area of service and maintenance staff was not examined under scientific view points until now.

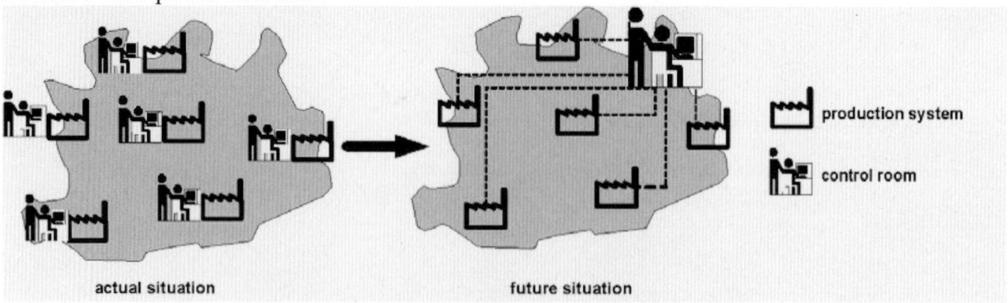

Figure 1: **Future trend in supervisory control**

Except onsite process information the service and maintenance staff has to take every needed information and documentation from a centralised documentation room to the distant plant. The choice of the documentation is made due to assumptions – because of the missing knowledge about the malfunctions. If the staff take the wrong documentation because of incomplete knowledge about the malfunction it is very laborious and time consuming to get the right documentation (as a consequence the downtime increases). In the future this case will become more and more critical because of different heterogeneous decentralised production plants spread over a city or region – each production plant with lots of different process elements form different manufacturers, and with different documentation (Wittenberg, 2002; Wittenberg, Marur & Penzkofer , 2002).

2 Mobile Devices

With the development of new communication technologies like UMTS or WLAN and mobile devices like industrial panel computers enormous possibilities for the onsite support of the service & maintenance staff will be available in the near future. Such panel computers ("web pads") usually have a high-resolution 8" or larger colour display which allows the sharp display of pictures, even under adverse lighting conditions, and are able to access through a wireless network an information server. Fast wireless network protocols like UMTS ensure the necessary data transfer rate. The battery modules become more and more powerful so that these panel computer can be operated alone for a full working shift. The service and maintenance staff can access every needed information and data at every location – even under the most rough industrial environments. This development will change the working situation of the staff. The challenge is to develop an user-centered mobile human-computer interface concept – beginning with a requirement analysis over the concept development to usability tests.

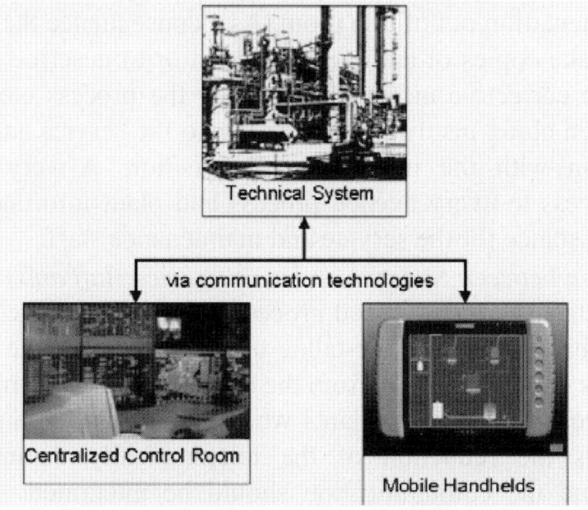

Figure 2: **Communication structure (Wittenberg, 2002)**

3 Requirement & Task Analysis

Following the user-centred development process (DIN EN ISO 13407, 1999; Epstein, Komischke & Wittenberg , 2001) an intensive requirement and task analysis (Kirwain & Ainsworth, 1992) in the domain of power generation was performed. Two onsite analysis and a number of expert interviews with service and maintenance staff were conducted. As a result of this analysis the hierarchical task structure for typical tasks and generic tasks were identified. As a huge potential for optimising the work situation the integration of different information sources like plant documentation, history information and the task schedule into one human-computer interface was identified. Also the acceptance of future technologies like speech in- and output were queried.

3.1 Typical scenarios

Beside this two typical task scenarios were recorded (Penzkofer 2002; Wittenberg et al, 2002). E.g. a common maintenance scenario is the calibration of instruments or drives. The calibration of drives for valves is performed regularly. The process visualisation in the control room shows an impossible process value for a valve (e.g. 112% open, > 100% is impossible). The operator informs the maintenance staff to calibrate this drive. The maintenance staff has to move to the valve and begins the calibration. Due to the missing onsite presentation of the process value of the valve the maintenance staff has to communicate with the operator in the control room to get a feedback about the calibration action. This communication can be very inaudible because of the production noise.

In case of a malfunction of an element of the production process (e.g. a pump) the service staff has to react very fast to keep the safety and reduce downtimes. For a correct diagnosis the staff needs information about the history of the element (message lists), the right documentation (nowadays printed on paper), and ideally a repair instruction. Often spare parts or a special tool are needed which have to be ordered from a different department. An integration of the different information sources and interaction possibilities into one interface is needed.

3.2 Recorded requirements

As a result of the analysis the following optimisation potentials were identified:

- *Information acquisition:* The pick-up of information by the service and maintenance staff is not suitable for the task. Nowadays the complete plant documentation is stored as paper printout in a special documentation room. It is not allowed to take (even parts of) the documentation out of this room. As a consequence the service and maintenance staff has to make copies of the documentation or to describe by hand the needed data, to identify the corrective measures onsite by the process element. If the information which are token along insufficient or wrong so the service and maintenance staff have to begin again with the information acquisition. A very clear requirement is a integrative access to every needed process and plant information which reduces time and walking distance for the service and maintenance staff.
- *Communication between Service- and maintenance staff and operator:* The service and maintenance staff frequently need process information, which is only displayed in the central control room. Via walkie-talkie or mobile phones the service and maintenance staff has to ask this information from the operator in the control room. A problem is the high noise level in production plants which disturb the communication. An important requirement is the reduction of the telephonically communication. The process information and the communication should be integrated in the above mentioned integrative information access.
- *Different data format:* A lot of different data formats and materials (paper, electronically) were found. Starting with an mission list over the plant documentation to the spare parts list for each process element. A unification of the different data formats is desired.
- *Supply of situation-depended interactions and information:* A support of the workflow is required. The service and maintenance staff should get offered situation-depended functions and information, e.g. information for diagnosis or a direct access to the spare part ordering.
- Additionally further detail requirements were recorded.

4 Prototyping & Usability Testing

On the basis of the results of the requirement analysis a widespread interface concept for mobile computing in the service & maintenance domain was developed (Penzkofer, 2002) following the Pattern-Supported Approach (Marur, 2002; Wittenberg et al., 2002). A scenario-based prototype for the above mentioned mobile panel computers with a touch display and five function keys for interaction (no keyboard or mouse) was implemented using Macromedia Flash®. This prototype comprises a task-card concept for the fulfilment of the typical workflow recorded in the requirement analysis. This task-card concepts covers the whole range from the incoming order over the element history and documentation to the spare part ordering. Figure 3 shows the prototype (pump documentation is displayed) on a SIEMENS SIMpad® as it was used in the Usability test.

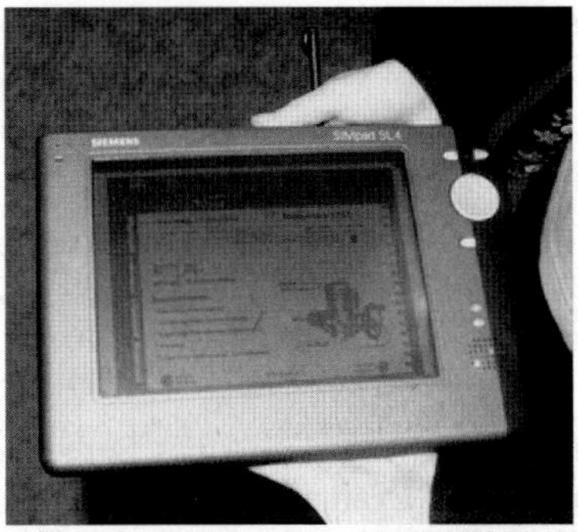

Figure 3: **The prototype on a SIEMENS SIMpad®**

A Usability test with two different user groups was performed. The first user group consisted of experienced service and maintenance staff, the second user group of novices. As a result the acceptance of mobile computers and a widespread interface is very high.

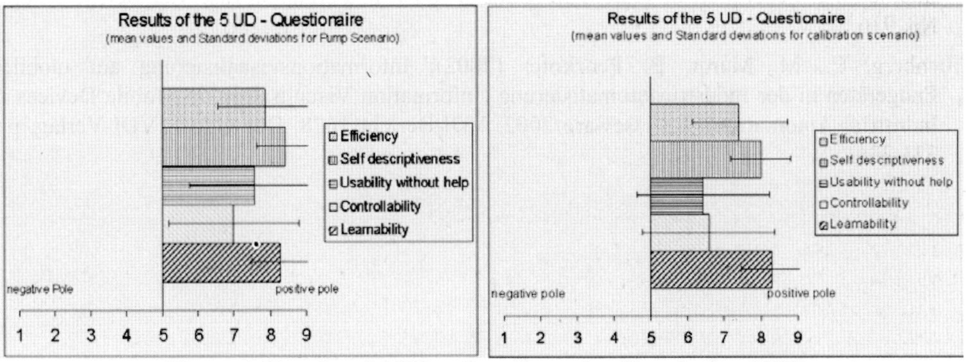

Figure 4: **Results of 5 UD – Questionnaire (Penzkofer, 2002)**

The subjects mentioned consistently that such a concept would support their daily work. The inquiry of the five Usability Dimensions (5UD) shows also a very positive feedback (figure 4) of the user interface concept and the employment of mobile devices in the area of service and maintenance. The continuos high assessment of efficiency, self descriptiveness and learnability allows the conclusion that such applications will be accepted very positive by the end-user. The reason is the provision of all needed information in a mobile portal and the connected reduction of the time-consuming acquisition of information.

References

Epstein, A., T. Komischke, C. Wittenberg (2001): Benutzerorientiertes Design und Usability Engineering in der Industrieautomatisierung (User-centered Design and Usability Engineering in Industrial Automation). Automatisierungstechnische Praxis atp 43 No. 8, pp. 38-45.

DIN EN ISO 13407 (1999): Human-centered design processes for interactive systems. Berlin: Beuth-Verlag.

Kirwan, B., L.K. Ainsworth (1992): A guide to task analysis. London: Taylor & Francis.

Marur, M. (2002): Interaction Design for Mobile Process Control and Maintenance: A Pattern Based Approach. Master Thesis at the Technical University of Eindhoven TU/e, Stan Ackermanns Institute – Center for Technological Design.

Meuser, M, U. Nagel (1991): ExpertInneninterviews – vielfach erprobt, wenig bedacht (Expert Interviews- often tested less considered). In: D. Garz, K. Krimer (Hrsg.): Qualitativ-empirische Sozialforschung: Konzepte, Methoden, Analysen. Opladen: Westdeutscher Verlag.

Penzkofer, B. (2002): Entwicklung und Evaluation eines Bedienkonzeptes für den Einsatz in der Teleautomation (Development and Evaluation of a User Interface concept for the use in Teleautomation). Unpublished Diploma Thesis at the Technical University of Munich, Laboratory for Human-Machine Communikation.

Wittenberg, C. (2002): Bridging the gap between the human operator and the far away technical system by means of pictorial computer graphics and mobile Handhelds. In: E.F. Camacho, L. Basáñez, J.A. de la Puente (Eds.) Preprints of the 15th Triennal World Congress of the International Federation of Automatic Control, Oxford: Pergamon/Elservier Science, Paper No. 916.

Wittenberg, C., M. Marur, B. Penzkofer (2002): Informationsvisualisierung auf mobilen Endgeräten in der Industrieautomatisierung (Information Visualisation on Mobile Devices in Industrial Automation). In: Useware 2002, VDI-Bericht 1678, Düsseldorf: VDI-Verlag, pp. 331-336.

What Tasks are Suitable for Handheld Devices?

Shuang Xu [shuangxu@yahoo.com]
Xiaowen Fang [xfang@cti.depaul.edu]
Susy Chan [schan@cti.depaul.edu]
Jacek Brzezinski [JBrzezinski@cti.depaul.edu]

School of Computer Science, Telecommunication and Information System,
DePaul University, Chicago IL 60604, USA

Abstract

This study investigates the main factors that influence user's preference of tasks to be performed on handheld devices. Thirty-seven participants took part in an observational experiment. Based on the experiment results, the following five factors were identified: perceived ease of use, perceived usefulness, perceived playfulness, task complexity, and perceived security. Discussions of these factors and hypotheses are presented.

1 Introduction

The current state of wireless technology poses many constraints for designing effective user interfaces for wireless applications. Small screen display, limited bandwidth, and the simplistic yet diverse functionality of handheld devices affect usability.

The objective of this study is to address what tasks would be suitable for handheld devices by investigating the main factors that influence user's preference, and perception of tasks performed on handheld devices.

2 Background Literature

Usability research in mobile commerce is a new area. Based on a study of 19 novice wireless phone users who were closely tracked for the first 6 weeks after service acquisition, Palen and Salzman (2002) describe the wireless telephony system as having four socio-technical components: hardware, software, "netware," and "bizware." They indicate that each of these four components has to be designed as user-friendly. Their research suggests a systems-level usability approach. Perry, O'Hare, Sellen, Brown, and Harper (2001) present a study of mobile workers that highlights different facets of accessing remote people and information anytime, anywhere. They identify four key factors in mobile work: the role of planning, working in "dead time," accessing remote technological and informational resources, and monitoring the activities of remote colleagues. In a study trying to understand how web access from a portable appliance changes the way people use Internet at home, McClard and Somers (2000) have investigated how tablet computers would be integrated into household activities and further defined user requirements for such devices. They suggest that an Internet appliance intended for general web access and text-based communication must have three characteristics: 1) Software must contain features that people perceive as useful; 2) Device must be highly portable and comfortable to be used in relaxed positions; 3) Device must have a large enough screen and keyboard to be usable. In the same study, they also identified a list of preferred tablet features on the top of which were surfing the Web, Internet anywhere in the house, and email. An essential goal of m-commerce is to search for mobile values for individual users. Anckar and D'Incau (2002) present a framework that differentiates between the value offered by wireless Internet technology (wireless value) and the value arising from the mobile use of the technology (mobile value). Wireless values are best represented by convenience, cost savings, and cell phones. Services that deliver strong mobile values make m-commerce a dominant channel. These services meet the following five types of user needs: 1) time-critical needs and arrangements; 2) spontaneous needs and decisions, such as auctions, email, and news; 3) entertainment needs; 4) efficiency needs and ambitions; and 5) mobility related needs.

These research findings suggest that a good understanding of the user preference and perceived value of tasks performed on wireless handheld devices is essential for improving usability for mobile tasks.

In this study, an observation experiment was conducted to identify the factors that would affect the preference of suitable tasks to be performed on handheld devices.

3 Method

A wireless web site that provides the function of making advising appointment was developed on the three platforms: Palm, WAP phone and Pocket PC. Similar to the regular web site of School of Computer Science, Telecommunications and Information Systems at DePaul University, these wireless versions allow students to make advising appointments with faculty members. On the server, special scripts were developed to record each user's activities, such as which web page was visited at what time, to a log file. The two tasks employed in this experiment were: 1) Making an advising appointment via an email; and 2) Making an advising appointment via a Web application.

Thirty-seven (37) students from DePaul University participated in the experiment. A pre-experiment questionnaire was used to collect background information of participants. Questions were asked regarding their familiarity with the functionality on the regular web site and their experience with WAP phone, Pocket PC, Palm, and other wireless devices. Among these participants, some used handheld devices before, and some didn't. Most of them had used the advising appointment function on the regular web site before the experiment.

Nineteen (19) participants were assigned to perform the tasks on Palm, 9 participants on WAP phone, and 9 participants on Pocket PC. Each participant was asked to sign a consent form before participating in the experiment. After background information was collected via the pre-experiment questionnaire, each participant was asked to read a one-page instruction about how to use the handheld device employed in the experiment. Then the participant started to perform the two tasks. While the participant was performing the tasks, the experimenter sat aside, observed the participant's performance, and took notes about any unusual events such as mistakes, comments, and complaints. Each participant's activities were also recorded into a log file by the server-side program.

Upon completion of the task, the experimenter interviewed the participant about his/her preference of tasks to be performed on a handheld device, what he/she liked (is this redundant?), and what difficulties he/she had. During the interview, if the participant used any ambiguous terms, follow-up questions were asked for clarification. The interview took about 45 minutes. Each interview was tape-recorded.

4 Results and Discussions

The recorded interviews were transcribed. Transcripts were analyzed and comments from participants were summarized in the following manner:

- Useful comments were extracted from the transcripts and rephrased to reflect similar comments by the researchers based on their professional judgment.
- Each new comment was assigned a unique number and marked with the frequency if a similar or same comment was repeated by the same participant.
- A list of comments for each participant was created.
- Another list was constructed to include comments from all the participants. Similar or identical comments were listed as one.

All the comments relevant to tasks performed on handheld devices were categorized into five factors: perceived ease of use, perceived usefulness, perceived playfulness, task complexity, and perceived security. Table 1 presents some example comments from these five groups. These five factors, which would presumably influence user's preference of tasks to be performed on handheld devices, are discussed as follows.

Table 1: Example Comments for the Five Categories

Category	Example Comments
Perceived Ease of Use	• User would like to use handheld device if the task can be performed without typing. Tapping/clicking on the screen and voice command are preferred.
Perceived Usefulness	• User would like to send/receive emails from wireless devices at anytime and in anywhere. • User would frequently check information about stock, weather, traffic, sports or airline flights.
Perceived Playfulness	• User would frequently play games on handheld devices. • User would frequently download mp3 music on handheld devices.
Task Complexity	• User would not try to do programming or debugging on wireless devices. • User would like simple and straightforward applications on handheld devices.
Perceived Security	• People usually do not feel secure to provide sensitive information through wireless connections because the device is not physically connected to the network.

Perceived ease of use is the extent to which a person believes that using a particular application would be free of effort (Davis, 1989). According to the technology acceptance model (TAM) proposed by Davis (1989), perceived ease of use influences an individual's intention to adopt Information Technology (IT) application. During the interviews of this study, many participants commented about the perceived ease of use. For instance, users prefer to select choices from a dropdown menu than to key in the same information in a textbox on wireless device. If users have to go back to correct some input information in previous pages before submitting, they prefer to have all the previous input saved, instead of having to redo everything. Therefore, perceived ease of use is an important factor affecting user's preference of applications.

Perceived usefulness is the extent to which a person believes that using a particular application would enhance his/her job performance (Davis, 1989). The TAM model suggests that perceived usefulness also influences user's attitude towards usage and intention to use a technology. Based on the findings from a consumer survey, Anckar and D'Incau (2002) point out that wireless services must meet the following five types of user needs: 1) time-critical needs and arrangements; 2) spontaneous needs and decisions, such as auctions, email, and news; 3) entertainment needs; 4) efficiency needs and ambitions; and 5) mobility related needs. The summarized comments from participants show that checking emails, news, weather, stock information were the most frequently performed tasks on wireless devices. Perceived usefulness is no doubt another critical factor.

Perceived playfulness is the extent to which the activity of using a specific system is perceived to be enjoyable in its own right, aside from any performance consequences resulting from system use (Venkatesh, 1999 & 2000). Perceived playfulness is not considered to be a critical factor in regular interface design. But interestingly it turns out that perceived playfulness might affect user's preference of tasks performed on wireless devices. Since a wireless device can be accessed anytime and anywhere, many users prefer to use it to kill time. Many participants mentioned during the interview that they used to play games or listen to mp3 music on wireless devices simply for fun and pleasure. Among the five types of user needs for wireless services identified by Anckar and D'Incau (2002), one is entertainment needs. Prior research on the effectiveness of game-based training has shown that manipulating the level of perceived enjoyment has a significant impact on user's behavior (for technology adoption?)(Venkatesh, 1999 & 2000).

Task complexity is the extent to which a user experiences how complicated it is to successfully accomplish a particular task in an application(citation?). Campbell (1988) proposes a framework that distinguishes the objective complexity of a task from the subjective complexity perceived by a task-doer. Objective complexity is determined by the nature of the task, such as the (?) rate of information changes and information load or diversity. Subjective complexity is influenced by user's familiarity with the task, his/her short-term memory, span of attention, computational efficiency, time constraints, and many other factors. Because wireless devices are used on the move in many cases, tasks performed on these devices would become either secondary tasks or primary tasks with other secondary tasks. In either scenario, different tasks will compete for human cognitive resources. This implicates that users will have significantly less capability of handling complex tasks on wireless devices under mobile conditions than under stationary conditions. During the interviews, many participants indicated that they would prefer simple and straightforward applications on wireless device. Hence, task complexity deserves special attention in the design of a wireless application.

Perceived security is defined as the extent to which a user believes that using a particular application will not expose his/her private information to any unauthorised party (citation?). Previous study has addressed the importance of impersonal trust in online transactions from the consumers (Ba, 2002). Many participants in this study did not feel secure enough to make online payment over the wireless network(s?). Users would not do online shopping on wireless devices unless they were assured that the connection is secure and no personal information, such as credit card number, will be exposed.

Based on above discussions, the following hypotheses are proposed:

Hypothesis 1: Higher perceived ease of use of a task performed on wireless devices will result in higher user preference of the task.
Hypothesis 2: Higher perceived usefulness of a task performed on wireless devices will result in higher user preference of the task.
Hypothesis 3: Higher perceived playfulness of a task performed on wireless devices will result in higher user preference of the task.
Hypothesis 4: Higher complexity of a task performed on wireless devices will result in lower user preference of the task.
Hypothesis 5: Higher perceived security of a task performed on wireless devices will result in higher user preference of the task.

5 Future Research

In this study, five factors that might have affect user preferences of tasks to be performed on wireless devices were derived from an observation experiment. Hypotheses were proposed to explore potential causal (should causal be deleted? In line 8 we only mention correlation analysis) relationships between these five factors and user preferences of tasks. To continue this study, another experiment will be conducted to test these hypotheses. In the next experiment, a set of tasks with different levels and combinations of perceived ease of use, perceived usefulness, perceived playfulness, task complexity, and perceived security will be developed. Participants will be recruited to perform these tasks on wireless devices. Participants will also evaluate the task characteristics and rate their preferences of these tasks. Correlation analyses will be used to find any significant correlations between the five factors and user preferences, and subsequently test the hypotheses.

References

Anckar, B. & D'Incau, D. (2002). Value creation in mobile commerce: Findings from a consumer survey. *Journal of Information Technology Theory & Application*, 4(1), 43-64.

Ba, S. (2002). Evidence of the effect of trust building technology in electronic markets: price premiums and buyer behavior. *MIS Quarterly*, 26(3), 243-268.

Campbell, D. J. (1988). Task complexity: a review and analysis. *Academy of Management Review*, 13(1), 40-52.

Davis, F. D. (1989). Perceived usefulness, perceived ease of use, and user acceptance of information technology. *MIS Quarterly*, 13(3), 319-339.

McClard, A. & Somers, P. (2000). Unleashed: Web tablet integration into the home. *Proceedings of CHI*, 1-6.

Palen, L. & Salzman, M. (2002). Beyond the handset: designing for wireless communications usability. *ACM Transactions on Computer-Human Interaction*, 9(2), 125-151.

Perry, M., O'hara, K., Sellen, A., Brown, B., & Harper, R. (2001). Dealing with mobility: understanding access anytime, anywhere. *ACM Transactions on Computer-Human Interaction*, 8(4), 323-347.

Venkatesh, V. (1999). Creation of favorable user perceptions: exploring the role of intrinsic motivation. *MIS Quarterly*, 23(2), 239-260.

Venkatesh, V. (2000). Determinants of perceived ease of use: Integrating control, intrinsic motivation, and emotion into the technology acceptance model. Information Systems Research, 11(4), 342-365.

SAMIR: A Jack of all Trades Clerk

F. Zambetta, G. Catucci, F. Abbattista, G. Semeraro

Dipartimento di Informatica – Università di Bari
Via E. Orabona, 4 – I-70125
zambetta@di.uniba.it, gracat@email.it, fabio@di.uniba.it, semeraro@di.uniba.it

Abstract

Intelligent web agents, that exhibit a complex behavior, i.e. an autonomous behavior rather than a merely reactive one, are daily gaining popularity as they allow a simpler and more natural interaction between the user and the machine, entertaining him/her and giving to some extent the illusion of interacting with a human-like interface.
In this paper we describe SAMIR, an intelligent web agent satisfying the objectives listed above. It uses: a 3D face animated via a morph-target technique to convey expressions to be communicated to the user, a slightly modified version of the ALICE chatterbot to provide the user with dialoguing capabilities, an XCS classifier system to manage the consistency between conversation and the face expressions. We also show some experimental results obtained applying SAMIR to a virtual bookselling scenario, involving a Web bookstore.

1 Introduction

Intelligent virtual agents are software components designed to act as virtual advisors into applications, especially web ones, where a high level of human computer interaction is required. Indeed, their aim is to substitute the classical WYSIWYG interfaces, which are often difficult to manage by casual users, with reactive and possibly pro-active virtual *ciceros* able to understand users' wishes and converse with them, find information and execute non-trivial tasks usually activated by pressing buttons and choosing menu items. Frequently these systems are coupled with an animated 2D/3D look-and-feel, embodying their intelligence via a face or an entire body. This way it is possible to enhance users trust into these systems simulating a face-to-face dialogue, as reported in [1]. A general observation is that the state-of-the-art systems, though interesting, are often heavy to implement, difficult to port onto different platforms, and usually not embeddable in Web browsers. These reasons lead us to pursue a light solution, SAMIR (Scenographic Agents Mimic Intelligent Reasoning), which turns out portable, easy to implement and fast enough in medium-sized computer environments. SAMIR is a digital assistant where an artificial intelligence based Web agent is integrated with a purely 3D humanoid, robotic, or cartoon-like layout [2]. The remainder of the paper is organized as follows. Section 2 describes the architecture of SAMIR. In Sections 3, 4 and 5, the three main modules of SAMIR are detailed. Some examples of SAMIR in action are given in Section 6. Finally, conclusions are drawn in Section 7.

2 The Architecture of SAMIR

SAMIR (Figure 1) is a client-server application, composed of 3 main sub-systems: The Dialogue Management System (DMS), the Animation Module and the Behavior Manager.

The DMS is responsible for directing the flow of information in our system: When the user issues a request from the web site, an HTTP request is directed to the DMS Server to obtain the HTTP response storing the chatterbot answer. At the same time, based on the events raised by the user on the web site and on his/her requests, a communication between the DMS and the Behavior Manager is set up. This results into the expression the Animation System should assume, coded into a string specifying coefficients for each of the possible morph targets [3] into our system: We use some high-level morph targets corresponding to the 6 fundamental expressions [4] but even low-level ones are a feasible choice in order to preserve full MPEG-4 compliance. After this interpretation step, a key-frame interpolation is performed to animate the current facial expression.

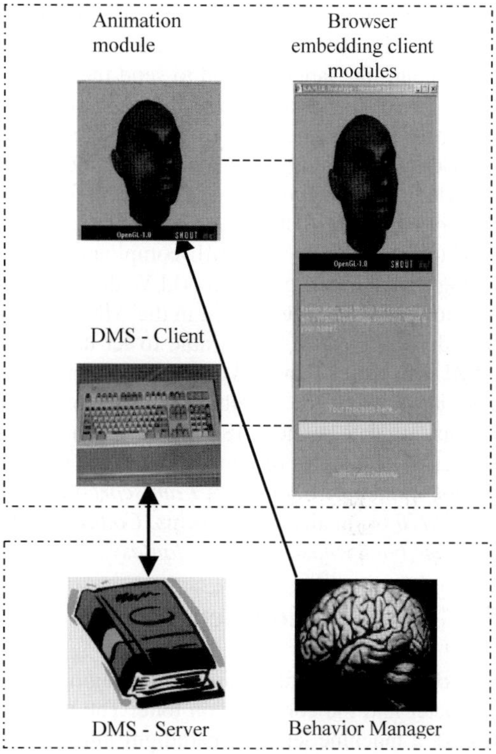

3 The Animation Module

The FACE (Facial Animation Compact Engine) Animation module is an evolution of the Fanky Animation System [5]. We followed the same philosophy introduced in Fanky, that is the implementation of SACs (Standard Anatomic Components). The basic idea underlying them is to define face regions, acting as objects, in an object-oriented sense of the term. The offered services correspond to different low-level deformations such as FAPs (Facial Animation Parameters), used during the animation process, or face sculpting and remodeling. Moreover, SACs made possible to select the numerical method employed to deform the vertices associated to a particular SAC at runtime. We adopt the linear interpolation of a 3D face key-frames because, in our opinion, it represents the best compromise between speed and accuracy. FACE was conceived keeping in mind lightness and performance so that it supports a variable number of morph targets: For example we currently use either 12 high-level ones or the entire "low-level" FAP set, in order to achieve MPEG-4 compliance. Using just a small set of high-level parameters might be extremely useful when debugging the behavior module because it is easier to reason about behavioral patterns in terms of explicit expressions rather than longer sets of facial parameters. An unlimited number of timelines can be used allocating one channel for the stimulus-response expressions, another one for eye-lid non-conscious reflexes, and so on. We are currently developing a custom editor able to perform the same tasks performed by FaceGen but optionally giving more control to the user: This way each user might enjoy the process of creating a new face, tailored to his/her wishes [7].

4 The Dialogue Management System

The Dialogue Management System is responsible for the management of user dialogues and for the extraction of the necessary information for book searching. It can be viewed as a client-server

application composed mainly by two software modules, communicating through the HTTP protocol. The client side application is a Java applet whose main aim is to let user to type requests in a human-like language and to send these ones to the server side application in order to process them. The other important task it is able to perform is retrieving specific information, based on the responses elaborated by the server-side application, on the World Wide Web through the JavaScript technology. On the server side we have the ALICE Server Engine enclosing all the knowledge and the core system services to process user input. ALICE is an open source chatterbot developed by the ALICE AI Foundation and based on the AIML language (Artificial Intelligence Markup Language), an XML-compliant language that gives us the opportunity to exchange dialogues data through the World Wide Web. ALICE has been fully integrated in SAMIR and all its knowledge has been stored in the AIML files, containing all the patterns matching user input.
In order to obtain a system able to let users navigating in a bookshop web site, we wrote some AIML categories finalized to book searching and shopping. Our categories were chosen to cover a very large set of the possible books request: They comprehend the book title, author, publisher, publication date, subject, ISBN code and a general field keyword.
Successful examples of book requests for the Amazon bookshop web site are for example, *I want a book written by Sepulveda*, *I am searching for all books written by Henry Miller and published after 1970* or, in alternative forms, *Could you find some book written by Fernando Pessoa?*, *Look for some book whose subject is fantasy*.

5 The Behavior Generator

The Behavior Generator aims at managing the consistency between the facial expression of the character and the conversation tone. The module is mainly based on Learning Classifier Systems (LCS), a machine learning paradigm introduced by Holland in 1976 [8]. The learning module of SAMIR has been implemented through an XCS [9], a new kind of LCS, which differs in many aspects from the traditional Holland's framework. The most appealing characteristic of this system is that it is strictly related to the Q-learning approach but able to generate task representations which can be more compact than tabular Q-learning [10]. At discrete time intervals, the agent observes a state of the environment, takes an action, observes a new state and receives a reward.
The basic components of an XCS are: The Performance Component, that, on the ground of the detected state of the environment, selects the better action to be performed, the Reinforcement Component, whose aim is to evaluate the reward to be assigned to the system anf the Discovery Component which, in case of degrading performance, is devoted to the evolution of new, more performing rules. Behavior rules are expressed in the classical format ***if*** *<condition>* **then** *<action>*, where *<condition>* (the state of the environment) combines different conversation tones such as: user salutation, user request formulation to the agent, user compliments/insults to the agent, user permanence in the Web page, while *<action>* represents the expression that the Animation System displays during user interaction. In particular, the expression is built as a linear combination of a set of fundamental expressions that includes the basic emotion set proposed by Paul Ekman, namely anger, fear, disgust, sadness, joy, and surprise [4]. Other emotions and many combinations of emotions have been studied but remain unconfirmed as universally distinguishable. However, the basic set of expressions has been extended in order to include some typical human expressions such as bother, disappointment and satisfaction.
In a preliminary experiment, SAMIR has been able to learn some pre-defined rules of behavior and to generalize some new behavioral pattern, updating the initial set of rules [2]. In such a way, SAMIR is comparable with a human assistant that, after a preliminary training, continues to learn new rules of behavior on the ground of experiences and interaction with human customers.

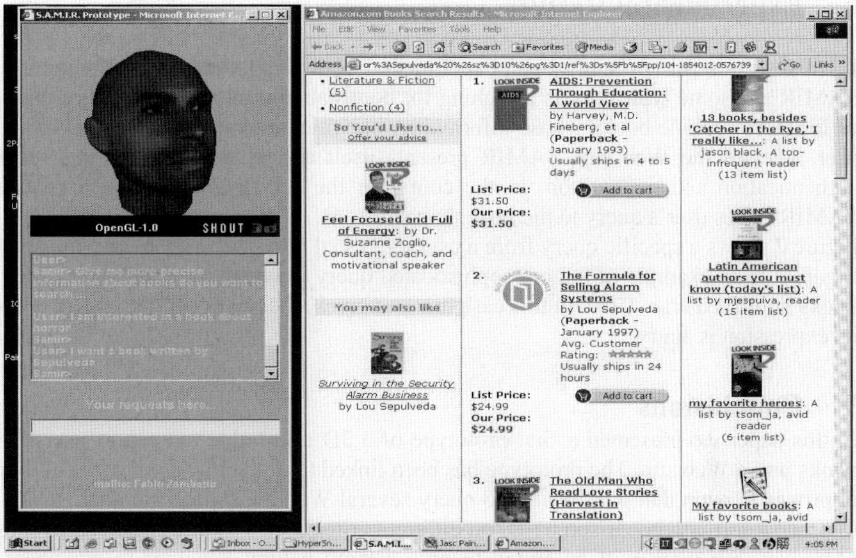

Figure 2: Results about Sepulveda author

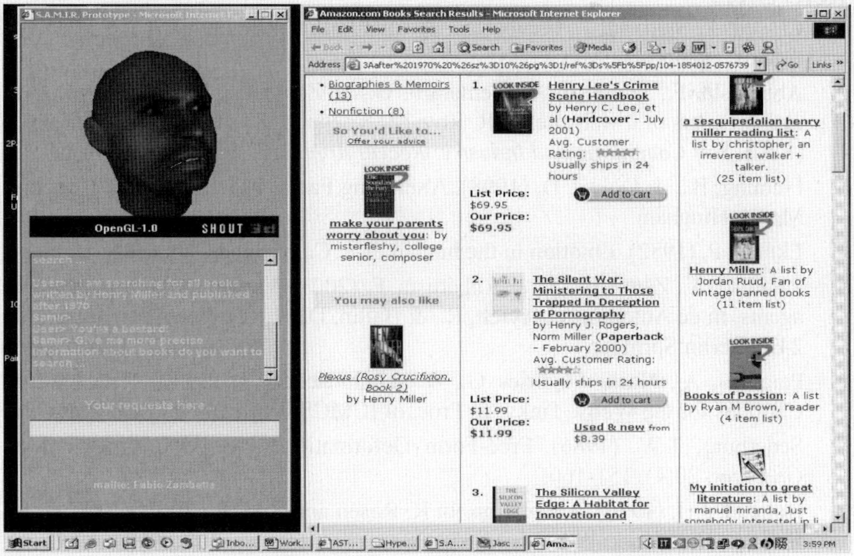

Figure 3: All books by H. Miller published after 1970

6 Experimental Results

In this section we present some experimental results obtained from the interaction between SAMIR and some typical users searching for books about topics like literature, fantasy and horror or for more specific books whose information like title, author and publisher are given. When the user accesses the Web site, SAMIR presents itself and asks to the user his/her name for user authentication and recognition. In the course of the conversation, if the user asks for a book, SAMIR issues user's query to the bookshop site.

Figure 2 shows a specific query from a user interested into a list of books by the author Sepulveda. Figure 3 is an example of a more sophisticated query in which the user requests for Henry Miller books published after 1970. In this case the user gives a heavy insult to SAMIR and, consequently, its expression is angry.

7 Conclusions

In this paper we presented a first prototype of a 3D agent able to support users in searching for books into a Web site. The prototype has been linked to a specific site but we are implementing an improved version that will be able to query several Web bookstores simultaneously and to report, to users, a sort of comparison based on different criteria such as prices, delivery times, and so on.

Moreover, our work will be aimed to give a more natural behavior to our agent. This can be achieved improving dialogues, and eventually, the text processing capabilities of the ALICE chatterbot, and giving the agent a full proactive behavior: The XCS should be able not only to learn new rules to generate facial expressions but also to modify dialogue rules, to suggest interesting links and to supply an effective help during the site navigation.

References

1. Cassell, J., Sullivan, J., Prevost, S., & Churchill, E. (Eds.). (2000). Embodied Conversational Agents. Cambridge:MIT Press.
2. Abbattista, F., Paradiso, A., Semeraro, G., & Zambetta, F. (2002). An agent that learns to support users of a web site. In R. Roy, M. Koeppen, S. Ovaska, T. Furuhashi, & F. Hoffmann (Eds.). *Soft Computing and Industry: Recent Applications*. (pp. 489-496).Berlin:Springer,.
3. Fleming, B., & Dobbs, D. (1998). Animating Facial Features and Expressions. Charles River Media, Hingham.
4. Ekman, P. (1982). Emotion in the human face. Cambridge:Cambridge University Press.
5. Paradiso, A., Zambetta, F., & Abbattista, F. (2001). Fanky: a tool for animating 3D intelligent agents. In de Antonio, A., Aylett, R., & Ballin, D., (Eds.). *Intelligent Virtual Agents*. (pp. 242-243). Berlin:Springer.
6. Paradiso, A., Nack, F., Fries G., & Schuhmacher, K. (1999). The Design of Expressive Cartoons for the Web – Tinky. In Proc. of ICMCS Conference, IEEE Press, 276-281.
7. Sederburg, T.W. (1986). Free-Form Deformation of Solid Geometric Models. *Computer Graphics*, 20(4), 151-160.
8. Holland, J.H. (1976). Adaptation. In R. Rosen and F.M. Snell (Eds.). *Progress in Theoretical Biology*, New York: Plenum.
9. Wilson, S.W. (1995). Classifier Fitness based on Accuracy. *Evolutionary Computation,* 3(2), 149-175.
10. Watkins, C.J.C.H. (1989). Learning from delayed rewards. *PhD thesis*, University of Cambridge, Psychology Department, 1989.

Section 2

User and Context-Awareness

Using Context Information to Generate Dynamic User Interfaces

Xavier Alamán, Rubén Cabello, Francisco Gómez-Arriba, Pablo Haya, Antonio Martinez, Javier Martinez, Germán Montoro

Departamento de Ingeniería Informática
Universidad Autónoma de Madrid
Ctra. Colmenar Viejo, km. 15 Madrid 28049. SPAIN
Xavier.Alaman@ii.uam.es

Abstract

This paper deals with the use of context information to generate dynamic user interfaces. Our framework is a real environment composed of a heterogeneous set of components. The nature of each component can range from a physical device to an abstract concept such as the number of persons in the environment. A middleware, that provides an unified environment model and communicates context changes, is used by two different modal interfaces. This allows to manage environment components without interfering each other.

1 Introduction

Our work is related to the research area known as smart environments or active spaces. A smart environment is a "highly embedded, interactive space that brings computation into the real, physical world". It allows computers "to participate in activities that have never previously involved computation" and people "to interact with computational systems the way they would with other people: via gesture, voice, movement, and context" (Coen, 1998). Thus, these new environments present new challenges (Shafer, 1999) that must be addressed by the research community. Different and highly heterogeneous technologies can be found inside a smart environment, from hardware components, such as sensors, switches, appliance, webcams... to legacy software, such as voice recognizers, multimedia streaming servers, mail agents... On one hand, all of these components have to be seamless integrated and controlled using the same user interface. For instance, a user has to be able to start a broadcasting music server as easily as to turn off the lights. On the other hand, user interaction has to be kept as flexible as possible. It should be based on multiple and distinct modalities, such as web, voice, touch... so that user preferences and capabilities can be considered. Moreover, the environment configuration is highly dynamic and it may change from one environment to another. New components can be added, removed or temporally stopped, and the user interface should be aware of these changes. In our approach, contextual information, gathered from the environment and its components, is used to generate dynamic user interfaces in order to address the problems explained above. This contextual information answers the main questions: who is the user, which of the components can be controlled, and how a user prefers to do it.

A working prototype is described, including: a middleware, that provides an unified context information model and a simple mechanism to communicate context changes, and two different user interfaces, that use it to manage the components of a real environment.

2 Framework description

We have developed a middleware layer. That is the glue between user interfaces and the environment. The interaction between them is based on an event-driven protocol. The contextual changes are published in a common repository, called blackboard (Engelmore and Morgan, 1988), Applications subscribe to the blackboard for the devices changes they are interested. This communication architecture allows to maintain a loose coupling among layers, facilitating the dynamic reconfiguration of environment components and user interfaces. Interfaces employ the information provided by the blackboard to adapt dynamically to the environment configuration and to interact with it. For example, the number of people in the room, the task they are performing and the status of several physical devices (lights, heating, video/audio displays) are represented in the blackboard, and are used by the natural language dialogue-management interface.

The blackboard holds a formal representation of all the environment active components. The nature of each component can range from a physical device to an abstract concept, such as the number of persons in the environment. Each component comprises a set of properties. That can be common or specific. All components of the same type share a set of common properties that describe universal accepted features of a component. Specific properties represent custom application information. Each application can annotate the component representation, so that it can customize the blackboard to its own requirements. The blackboard is not only a set of components but it also stores the relationships among them. A relationship can be of any kind (association, aggregation...) and any direction (unidirectional o bi-directional). It has not a explicit semantic associated, hence a relationship can represent from location information (a device is inside a room) to the flow of multimedia information among the physical devices (microphones, speakers, cameras, displays, etc.).

The blackboard is a graph where each node is a component, and relationships are arcs. In order to navigate this graph we have implemented two independent naming mechanisms. There is a basic namespace in which each component has a unique numerical identifier. Applications can directly access to a component and its properties by this number. Otherwise, a component can be referenced by concatenating the name of all of its parents components. An application can define its own relationship between components and locate a component using its own namespace. Moreover, it is allowed to use wildcards to reference more than one component at the same operation. For instance, it is possible to obtain the state of all the lights inside the environment.

Every environment component publishes a XML description of its features in the central repository. This repository is used as a proxy context information server. Applications may ask the blackboard to obtain information about to the state of any component and change this state. Components descriptions can be added and removed to the blackboard in run-time, and the new information can be reused for the rest of applications. In summary, applications can perform the following operations in the blackboard: retrieve information about the entities and its properties, control the state of components and subscribe to changes in the environment. Every blackboard is a server that can be accessed using client-server TCP/IP protocols. The interaction between applications and the blackboard uses the HTTP protocol, technologically independent of the component nature. HTTP has been chosen as the transport protocol because it is simple and widely spread. XML has been chosen as the lingua franca among layers, because is a standard industry language to interchange information between applications.

3 Applications

3.1 Jeoffrey

Jeoffrey is a web based interface developed to control environment's devices and appliances. It is a custom and partial view of the contextual information stored in the blackboard. Jeoffrey is programmed to be used in a home environment. At start up, it creates a list of the house rooms, and for each room it generates a map that includes the location of the physical devices. Each device is represented by an image. A custom widget is showed when a user clicks on a device image, allowing to control the device. The layout is composed overlapping a fixed image background with each device representation image and is generated every time Jeoffrey is loaded.

The blackboard contains generic information regarding the number of rooms and the devices they host. Each device is represented in the blackboard, and its representation includes the properties required to control it. Thus, Jeoffrey gathers this information to generate rooms views dynamically. This is not Jeoffrey-specific information but it also can be used by any other application. Besides, device representation can include specific information required by Jeoffrey, such as layout coordenates or image representation url. This information is also held by the blackboard but not interferes in other applications performance.

The device control panel is also generated dynamically and it depends on device properties. A set of generic widgets has been defined, such as text areas, buttons, sliders, etc. Therefore, a one-to-one translation between properties and widgets has been established. When device management is required, Jeoffrey reads its properties description from the blackboard, translates properties to widgets and generates a custom control panel composed by the aggregation of simple widgets. Jeoffrey has its own blackboard name-space to facilitate devices naming. The blackboard is a set of devices organized by rooms where each device is located appending the device name to the room name.

When Jeoffrey sets up, it queries the blackboard to obtain all the rooms. For each room, it asks for all the available devices. If a device has been Jeoffrey annotated, the applet reads its particular configuration and generates the web representation and the widgets needed to control it.

Moreover, Jeoffrey uses the blackboard as a proxy to manage the physical devices, like changing the volume speaker, switching the lights, etc., and to receive the changes occurred in the environment. Jeoffrey is subscribed to all the devices events, thus every change in a device state is reflected in the user interface. For instance, a widget named alarm has been defined. If a property has associated a widget alarm, when its value changes, the blackboard will notify to Jeoffrey and it will modify the color of the web image representation. Thus, a Jeoffrey instance can easily coordinate with other applications or Jeoffrey instances.

Left side of Figure 1. shows a Jeoffrey user interface screenshoot. Displayed windows correspond to device control panels. At the right side, a webcam shot shows the experimental environment.

Figure 1: Jeoffrey's Screenshot and webcam laboratory view

3.2 Odisea

An alternative interface to control the environment is provided by means of natural language spoken dialogues. This is called the Odisea system. It uses the contextual information stored in the blackboard to carry on conversations related to device control and user information. Blackboard representation does not have to suffer any modification and it is independent of the Jeoffrey-specific representation.

Odisea is running several environment-specific dialogues that compete to be the one that deals with the current conversation. A dialogue supervisor is in charge of choosing the most accurate dialogue depending on the user input (provided by the speak recogniser) and the contextual information from the blackboard. This supervisor can also activate and deactivate some of the dialogues when they come in or out of scope.

Dialogues implementation is similar to Schank's scripts concepts (Schank and Abelson,1977). A dialogue is formed by a template with gaps that must be filled. The dialogue will guide the user through the script until the template is fulfilled.

Every active dialogue will inform to the supervisor of the state of its template (how much it is completed) after a user utterance. Supervisor collects all this information and gives the control to the most appropriate dialogue.

Every dialogue is focused on a specific task. For instance, the lights dialogue controls the lamps of a room. It has full access to read or modify the values of the lamps stored in the blackboard, as much as other information from the blackboard that can become useful in the dialogue.

Since speak recognition is not completely accurate, context information from the blackboard plays an important rule in the supervisor decisions. When a speaker sentence (or recognizer output) may deal with ambiguities, the supervisor accesses to the information stored in the blackboard to try to solve it. Moreover, it can offer solutions to the user depending on the current context. For instance, if the recognizer output is only the word "lights", the supervisor can check the lights state. If they were off, it could directly offer the user to turn them on.

Odisea dialogues are fully compatible with Jeoffrey's interface. They can be used independently or at the same time. They show two ways of accessing and control to the environment. New interfaces can be added easily and it will not be necessary to modify the blackboard information.

4 Environment

A running prototype has been developed. It consists of a laboratory separated in two rooms. Several devices have been spread out across the rooms. There are two kinds of devices: control and multimedia. Control devices are lighting controls, door mechanism, presence detector, smart-card, etc. Multimedia devices, such as speaker, microphones, TV and an IP video-camera, are accessible through a backbone IP. Control devices are connected to a EIB (EIBA) network, and a gateway joins the two networks. The blackboard that accesses to the physical layer is harmonised through a SMNP (Simple Management Network Protocol) layer that is described elsewhere (Martinez et al., 2003). Both interfaces, Jeoffrey (web-based and automatically generated from the blackboard) and Odisea (natural-language based) can be used to interact with the environment.

5 Conclusions

The user interface required by a smart environment depends on user preference and environment context. Both of them can change over the time, thus the user interface should adapt dynamically. A middleware has been developed in order to facilitate context changes communication. This middleware has been tested in a real environment composed by several devices. Two independent applications allow users to interact with environment devices without interfere between them.

6 Future Work

Our current work is focused on what can be controlled. Future researching is oriented to the other two questions, how and who. For this reason, we will develop new user interfaces using mobile devices, such hand-held PCs, laptop PCs and cell phones. Moreover, user modeling and dialogues management will be improved.

References

Coen, M.H. (1998). Design Principles for Intelligent Environments. In Proceedings of the AAAI Spring Symposium on Intelligent Environments (AAAI98). Stanford University in Palo Alto, California, USA.

EIBA. European Installation Bus Association. http://www.eiba.com.

Engelmore, R. And Morgan, T. (1988). *Blackboard Systems*, Addison-Wesley.

Martinez, A.E., Cabello, R., Gómez, F.J. and Martínez, J. (2003). Interact-DDM: A Solution for the Integration of Domestic Devices on Network Management Platforms. IFIP/IEEE International Symposium on Integrated Network Management Colorado Springs, Colorado, USA.

Schank, R. and Abelson, R. (1977). Scripts, Plans and Goals. Erlbaum, Hillsdale, New Jersey.

Shafer, S. (1999). Ten dimensions of ubiquitous computing. In Proceedgins of 1st International Workshop on Managing Interactions in Smart Environments (MANSE'99). Dublin. Ireland.

Web Site Adaptation: a Model-Based Approach

Nikolaos Avouris, Martha Koutri

Electrical & Comp. Eng. Dept, HCI Group
University of Patras
26500 Rio Patras, Greece
{ N.Avouris, MKoutri }@ ee.upatras.gr,

Sophia Daskalaki

Engineering Sciences Dept.
University of Patras
26500 Rio Patras, Greece
sdask@upatras.gr

Abstract

Adaptation of web sites has been an area of great interest during the last years. In this context, development of techniques for knowledge discovery from web usage data have been developed. In this paper a methodological framework is proposed that exploits knowledge discovery from web usage data in support of a process for off-line adaptation of web sites. The proposed approach is based on a web site model consisting of a set of parameters. These parameters define a multidimensional space in which any web site can be located. The adaptation of the web site is based on combination of usage data and the web site model characteristics.

1 Introduction

The development of new knowledge discovery techniques (Fu, Sandhu & Shih, 1999, Mobasher, Cooley & Srivastava, 1999, Perkowitz & Etzioni, 2000) has contributed to the area of adaptation of interaction between users and web sites during the last years. Web site adaptation can take place either on-line, or more often off-line, when usability problems are identified from usage patterns. Our study is focused in off-line web site adaptation that is achieved by applying web usage mining techniques (Srivastava, Cooley, Deshpande & Tan, 2000) and is accomplished by modifying the web site structure. A crucial step in the process is related to the decision on when the adaptation process should be initiated. The process involves *web usage mining*, which concerns the discovery of users' access patterns from web server logs (Han & Kamber, 2001). The adaptation in our case concerns mostly the *web site structure,* i.e. the placement or removal of hyperlinks in the web site's documents. Processing of web usage data constitutes the first important step during the web site structure adaptation process, since from the navigational behavior of users, possible usability problems can be revealed. The majority of methodologies concerning adaptation of web sites involve a knowledge-discovering phase. In addition, an algorithm for suggesting types of adaptation, needs to be incorporated in the process. In this paper, such an algorithm is proposed, as a part of an integrated methodology, which takes into account the specific characteristics of the web site, in order to suggest the most suitable structure. As discussed in the following, the proposed methodology is based on a model of the web site. This model consists of a set of parameters that define the prevailing structure, the character and expected usage of the site without taking into account traffic analysis data.

A fundamental concept in our study is that of the *web site*. According to the W3C (1999) a website is "a collection of interlinked web documents, including a host page, residing at the same network location". However this definition needs further specification, as is demonstrated in the case of off-line browsers, where users have to determine web site boundaries, since there is not a commonly acceptable definition of the boundaries of a web site (Brunk, 1999). A working definition for the purposes of this research identify a web site as *a set of interlinked web documents including a host page (which defines the website address), residing in the same network location, that contain links to documents of the same set or to external web documents, cover one or more thematic areas, while they are characterized by thematic coherence, have a uniform presentation, including layout of content and links*. All these defining characteristics guide the selection of the appropriate parameters, in order to construct the web site model, used in our approach.

The next section provides an outline of the proposed adaptation methodological framework. Subsequently, the web site model is defined by presenting its structural elements, while section 4 refers to the possible types of adaptation for the modification of web sites structure, and section 5 discusses the way the model parameters affect the procedure of web site adaptation.

2 An extended methodology for off-line web adaptation

The general procedure for web usage mining consists of the following steps (Cooley, Srivastava & Mobasher, 1997, Mobasher, Cooley & Srivastava, 1999): data collection, data preprocessing, patterns discovery, patterns analysis. This explicit sequence of phases adheres to the fundamental principles of Data Mining. However, the special characteristics of WWW adaptation impose a particular treatment regarding *(i) the knowledge discovery from raw web data, (ii) the application of web usage mining results, in order to adapt the interaction of users with a web site*. In terms of knowledge discovery, many researchers propose variations of existing algorithms or develop new methods and techniques. A generic methodology including a decision support module about web site adaptation needs to be established first. In Figure 1, a proposal for an extended methodology for web usage mining, using a flowchart representation, is shown. The web site modeling module, as well as its interactions with the other processes, constitutes the novel component of the process. In particular, if the analysis of usage patterns -resulted by applying web mining techniques in web log data- identifies the need for adaptation of a particular web site, then the modeling module utilizes the usage patterns, as well as the web site documents structure and the knowledge base, in order to make suggestions about the appropriate types of adaptation. So, a generic methodology that associates web mining and web site modeling processes is structured aiming at supporting adaptation decisions.

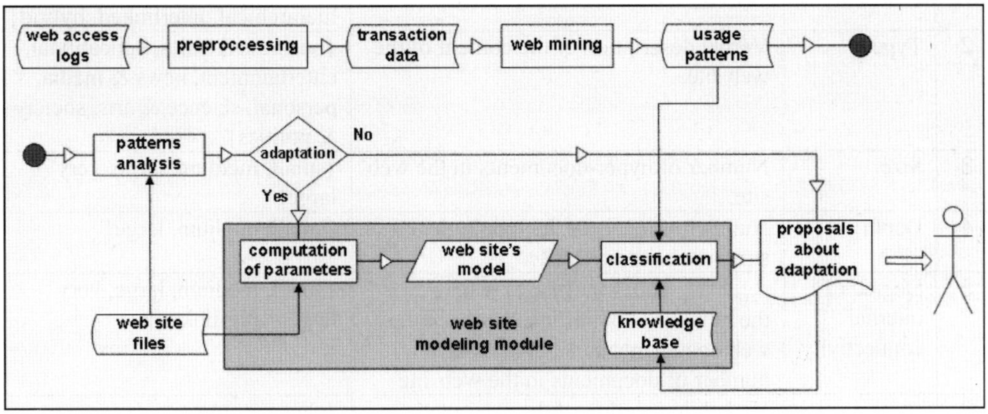

Figure 1. Extended off-line web adaptation methodology

3 A generic web site model

The defined above methodology is based on a web site model. Our attempt to define such a generic web site model in a more formal way yields the definition of such a model as *a descriptive pattern of the web site, which is characterized by the values of some prescribed parameters.* The model parameters need to be defined in such a way that can be related with certain descriptive features of the site. In particular, we propose the following parameters: *address*, *topology*, *type*, *size*, *depth*, *degree of internal connectivity*, and *appearance*. These parameters, relate to the web site working definition, provided in the introduction, and are discussed in the following.

The *Address* concerns the web site's URL specifying the computer that hosts the particular web site and represents its location in the WWW. *Topology* describes the structure of a web site, which results by studying the corresponding web site map. It concerns the order of hyperdocuments in a web site and their inter-connection. *Type* concerns a brief verbal description of the content of a web site. *Size* concerns the number of documents included in the web site. *Depth* measures the number of levels of a given topological structure. *Degree of internal connectivity*, d_{c_in}, is an indicator about the percentage of physically connected web documents in a web site with respect to the total number of documents in the site. In particular: $d_{c\_in} = l_i \cdot 100 / n \cdot (n-1)$ (1), where l_i is the number of hyperlinks between the web documents and n is the total number of documents in the site. *Appearance* concerns a brief verbal description about the presentation of a web site.

An empirical study of model applicability and the range of values of the model parameters has been contacted. In particular, we studied the web sites that have been visited by the users of the Greek Research and Technology Network (GRNet), stored in the proxy server of the Network during a period of 6 months (January-June 2002). The range of the topology, type and appearance parameters have resulted from statistical and empirical study of the values of parameters produced from web site modeling. The range for the rest of the parameters has been produced by the study of the sites, and general experience of interaction with the WWW. We note that address is excluded, because it is actually an identifier rather than a descriptive characteristic. Table 1 represents each parameter of the model, a brief definition, as well as the parameter range of values.

Table 1: Parameters of the web site model

No	Parameter	Definition	Range
1	Topology	The structure of the web site.	{sequential, star-like, hierarchical, interlinked, hybrid}
2	Type	Verbal description of the content of the web site.	{business & sales, educational, entertainment, news & media, personal, science & arts, society & politics}
3	Size	Number of hyper-documents in the web site.	{small, medium, large, very large}
4	Depth	Number of levels of the topological structure of the web site.	{small, medium, large}
5	Degree of internal connectivity	$d_{c\_in} = (l_i \cdot 100) / n \cdot (n-1)$, where l_i is the number of hyperlinks between the web documents and n is the total number of documents in the web site	{small, medium, large, very large, fully-linked}.
6	Appearance	Verbal description of the presentation style of the web site.	{elegant, cute, trendy, rigorous, old-fashioned, ugly, non-artistic}

4 Types of structure adaptation

In this section we present the proposed types of adaptation. A distinction is made between adaptation tasks that concern hyperlink connection and hyperlink formatting. In particular, we describe adaptation of type 1.x concerning hyperlink connection and 2.x for hyperlink formatting.

- *1.1:* Adding new hyperlinks, when some particular web documents are thematically related, while there is not direct physical connection between them. Also bi-directional links, i.e. cross-reference links (Botafogo, Rivlin & Shneiderman, 1992) can be introduced.
- *1.2:* Adding shortcut links, when users visit a particular sequence of web pages, in order to reach a certain target page.
- *1.3:* Removing anchors, when users rarely follow the corresponding hyperlink.
- *1.4:* Totally removing hyperlinks (reference and anchor), when they do not satisfy users needs and desires.

The main types 2.x of adaptation concern formatting of hyperlinks:

- *2.1:* Use of different colors and fonts for highlighting the hyperlink, strongly related with users' needs and desires.
- *2.2:* Use of accompanying icons for recommending or not some particular hyperlinks that interest users or fall outside their scope, respectively.
- *2.3:* Use of a short phrase besides the appropriate hyperlinks for making recommendations - like 2.2 .

As already discussed, the web site model drives the web mining process. This is the subject of the following section.

5 A model-driven adaptation process

The values of the model parameters define the specific web site. The premise of our approach is that decisions of re-structuring the web site should take into account these parameters. In particular, *topology*, d_{c_in}, and *depth* affect connectivity decisions, while *type* and *appearance* affect the formatting of hyperlinks. In addition, we argue that further research needs to be done on the effect that the web site size has on adaptation decisions. So experimentation should be made with various-size web sites, in which evaluation of adaptation decisions should be related to the size of the site. It should be also stressed that the adaptation decisions depend on the combination of parameters for a specific web site. A knowledge base has been constructed after interviews with experienced web-site designers and powerful users. The rules of this knowledge base interrelate the parameters of the website model with adaptation type decisions. An example of this process is given next.

Let us consider an interlinked web site with d_{c_in}=<full linkage>. It has been found that the following 1.x types of adaptation are considered most suitable: 1.2, 1.3, 1.4. Type 1.1 is not considered, because the web site already contains hyperlinks between all its internal documents. In addition, given that depth=<large>, we could concentrate on type 1.2 by adding shortcut links. Besides, if the web site is *educational*, characterized by an *elegant* appearance, then the shortcut links could be highlighted as type 2.1 describes. The application of type 2.2 is not recommended in order to preserve the elegance of the web site. Finally, the educational type of the web site does not lead to application of 2.3 adaptation type. So, we could add a recommendation like "for beginners", in order to help users find their way.

In a similar way, other rules have been derived and a knowledge base has been constructed, which contains a number of possible combinations of values of the model parameters, as well as the corresponding proposals about the types of adaptation. So, the knowledge base contains a set of empirically produced cases, where each one consists of a set of attributes. The attributes are: *topology*, d_{c_in}, *depth*, *appearance* and *type*. Another target attribute, named *recommendation*, is associated to these parameters, containing the recommended types of adaptation for a particular web site. A decision tree corresponding to the given knowledge base was subsequently derived with the use of supervised ID3 algorithm (Quinlan, 1979). The WEKA library of machine learning algorithms in Java (Witten & Frank, 2000) has been used, in order to apply the ID3 classifier into the web site data. This way a generalization of rules that cover other cases, not covered by the experts has been achieved. This knowledge base is combined with usage data for determining the type and place of modification of the web site, as discussed by Koutri & Daskalaki (2003).

6 Conclusions

The described framework has been used in the case of a number of web sites. Evaluation of the effectiveness of this approach is a tedious process, which is still in progress. This involves logging of users' behavior before and after the intervention and collection of subjective user data through questionnaires. A number of modifications in the typology of adaptation types, in the web site model parameters and in the knowledge associating adaptation types to website parameters are expected as a result of this ongoing process. Also various techniques for extraction of patterns of usage need to be tested and used in this process. However one of the main conclusions of the reported research is that off-line web site adaptation needs to be based on models of a generic web site and its structural elements. This model is fully integrated in the web mining process, facilitating decisions and improving traceability of the whole adaptation decision-making process.

7 References

Botafogo, R. A., Rivlin, E., & Shneiderman, B. (1992). Structural Analysis of Hypertexts: Identifying Hierarchies and Useful Metrics. *ACM Trans. on Information Systems* 10 (2), 142-180.

Brunk, D. B. (1999). Overview and Review Tools for Navigating the World Wide Web. SILS Technical Report TR-1999-03.

Cooley, R., Srivastava, J., & Mobasher, B. (1997). Web Mining: Information and Pattern Discovery on the World Wide Web. In Proc. of the *9th ICTAI'97*, Newport Beach: Canada.

Fu, Y., Sandhu, K., & Shih, M. Y. (1999). Clustering of Web Users Based on Access Patterns. In Proc. of the *1999 KDD Workshop on Web Mining*, Springer-Verlag, San Diego: Canada.

Han, J., Kamber, M. (2001). Data Mining: Concepts and Techniques. San Francisco: Morgan Kaufmann.

Koutri M., Daskalaki S., (2003). Improving web-site usability through a clustering approach, Proc. HCII2003, June 22-27, Crete.

Mobasher, B., Cooley, R., & Srivastava, J. (1999). Creating Adaptive Web Sites Through Usage-Based Clustering of URLs. In Proc. of the *KDEX'99*. Chicago: Illinois.

Perkowitz, M., & Etzioni, O. (2000). Towards adaptive Web sites: Conceptual framework and case study. *Artificial Intelligence*, 118, 245-275.

Quinlan, J. R. (1979). Discovering rules by induction from large collection of examples. Michie, D. (Ed.), Expert System in the Micro Electronic Age, Edinburgh University Press, 168-201.

Srivastava, J., Cooley, R., Deshpande, M., & Tan, P. N. (2000). Web Usage Mining: Discovery and Applications of Usage Patterns from Web Data. *ACM SIGKDD Explorations* 1: Issue 2.

Witten, I., & Frank, E. (2000). Data Mining: Practical Machine Learning Tools with Java Implementations. San Mateo: Morgan Kaufmann.

W3C (1999) Web Characterization Terminology & Definitions, www.w3.org/1999/05/WCA-terms

User Modelling Based on Topic Maps

Wolfgang Beinhauer, Franz Koller

Fraunhofer IAO, Nobelstrasse 12, Stuttgart, Germany
User Interface Design GmbH, Teinacher Strasse 38, Ludwigsburg, Germany
wolfgang.beinhauer@iao.fhg.de franz.koller@uidesign.de

Abstract

In large-scaled, multi-user information systems, the key issue is to select the right portions of content for an individual user and to present it in an adequate way. Therefore, the effectiveness of such systems highly depends on the quality of user modelling. In this paper we present a new way of user modelling based on topic maps that holds both the individual interest of a user in specific areas of interest and the structure of knowledge lying behind.

1 Introduction

Extensive context modelling is a key issue in large-scaled information systems that include any kind of personalisation or awareness functionality. In the area of user-centric knowledge management systems or productive environments, the effectiveness of such adaptation techniques depends strongly on the adequacy of a user model. In an open communication platform such as a business community, which is accessed by many users and which holds contributions by a variety of information sources, the user model is the fundamental mechanism for both the selection of suitable content for a single user and for its adequate presentation. Hence, effective user modelling essentially involves knowledge about the preferences of a set of users as well as knowledge about the structure of the content provided for them. Beyond the bare fields of interest, the environmental context, location, time and other external data make up the full user context. In this paper, we will focus on the person-oriented aspects of user modelling as it is required in highly adaptive productive environments and information systems like virtual business communities.

User modelling basically involves the formal description of a user's personal and organizational context. This comprises personal preferences like favourite topics, subjective associations between them and the processes conducted by the user. However, these influencing factors are not static, but may vary over time. For example, with improved knowledge, the focus of interest of a financial consultant might shift from shares to derivatives. Moreover, getting acquainted with derivatives, different types of derivatives will arise at the consultant's horizon such as options and futures. The structure of knowledge becomes finer as new mentions appear to the user's mind. Similarly, the user might adopt best practice guidelines and will change the way he completes his tasks. Hence, an adaptive user model, which represents itself the basis for the adaptation of content and navigation, is subject to changes in granularity, structure and task description. Additionally, it has to be coupled with a workflow engine that recognizes completed tasks by the steps performed and adapts the user interface according to the steps to go. In this paper, we would like to present an integrated dynamic user model, which continuously learns about the preferences of the user.

2 Business Communities

Business communities are a means for knowledge interchange and expertise localisation within one enterprise which is becoming more and more popular. Users of different kinds from different locations can communicate synchronously or asynchronously, have access to large amounts of recent information and are supported during their daily business.

Being a multi-user platform holding large amounts of data, these knowledge management platforms require sophisticated user models for the individualization of content and presentation. However, also role concepts and group concepts have to be included in order to reflect the effect of co-operative learning and progress of parts of the community. On the other hand, the activities not only of a single user can be tracked, but it becomes possible to collect the behaviour of a large number of users. Provided that user tracking happens on a topic level, this enables the dynamic identification of subgroups of users with common interest or similar behaviour. Hence, user modelling in business communities involves more than simple collaborative filtering and content retrieval.

3 Approach

Due to its intermediate position between users and a large amount of available content, a user model serves for several purposes in a business community: firstly, the user interface has to be individualized according to a user's needs and interests. Secondly, also the selection of content has to be personalized, and finally, user modelling is needed for the dynamic identification of subgroups. In order to meet these requirements, a user model needs to be directly coupled to a semantic content retrieval component, a user tracking system and an adaptive user interface. These components are grouped around a domain-specific ontology which provides semantic understanding of the documents delivered and the user's needs.

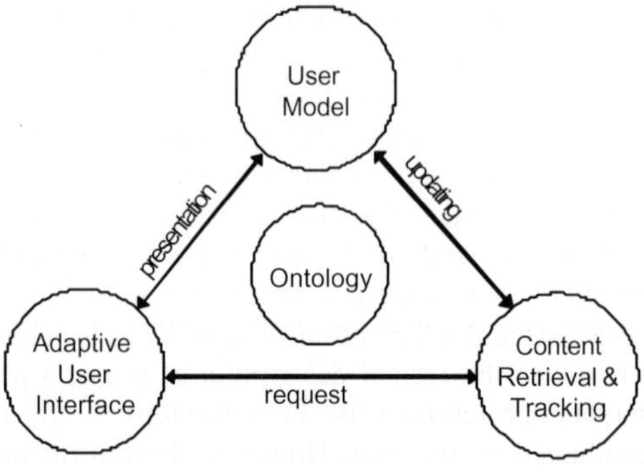

Fig. 1: User Model, User Interface and Content Retrieval

Our user model is based on a domain-specific ontology represented by a topic map. Topic maps have been introduced for describing knowledge structures and associating them with information resources. In 1999, they have been internationally accepted as ISO/IEC standard 13250. Directed graphs as a basis for user modelling have already been employed by several autors, e.g. Baldwin et al. in 2000.

The domain ontology has to be initially created in the course of a dynamic group process, where the topics relevant to the field of application are grouped and linked to each other, making up a directed graph. Each node of the graph represents an item of the application area, while their relations to each other are expressed by the edges. The edges can have name attributes, giving the associations a semantic meaning. This graph consisting of all topics is used as the basis for the modelling of the needs and interests of all single users, groups, roles or even tasks: each topic is weighted according to its relevance for the specified subject, whether it is a role, a person or a task. Physical persons and abstract roles or groups are treated equally with respect to this. Analogously, all edges are weighted, signifying the subjective association strength between the topics. Hence, the overall doma in ontology is the superset of all specific topic maps. In turn, the personal needs and interests of a single user can be reduced to a set of weighting factors for items of the domain ontology. Topics or associations not relevant to one person or task can be eliminated by setting their weight to zero.

Starting from the topic weightings according to the personal preferences, the needs of the group the user is belonging to and the role the user is playing in the application context, a final topic map is calculated by means of a merger algorithm. The algorithm involves a set of rules that are applied to the individual sets of weightings. The topic map finally obtained is directly questioned when building the adaptive navigation structure of the knowledge management system for the business community. Since the model is based on a well-defined ontology, also structural orderings of the suggested topics of interests can be taken into account when creating the navigation structure.

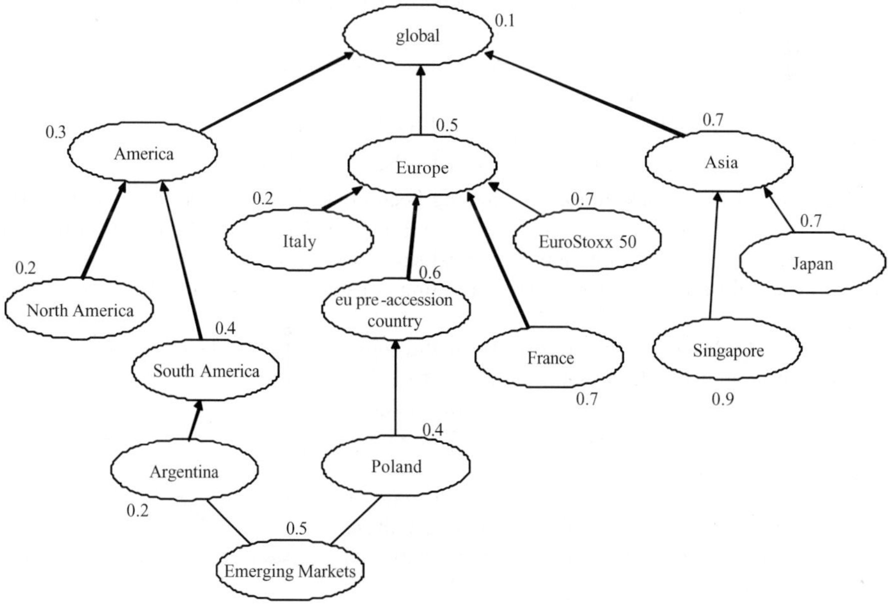

Fig. 2: Excerpt of the weighted topic map (regions for investments)

However, an effective user model has to be essentially dynamic. Therefore, the stored weighting factors both of the topics and the associations are subject to continuous changes. The user model has to be tightly coupled with a tracing component that tracks the user's actions. Each time a user expresses his interest in a topic by clicking on a referring link, his set of weightings gets updated. While personal interests in specific topics may arise or fall quickly, the structure of knowledge is supposed to vary on a larger time scale. Therefore, only the weightings of the topic nodes are subject to changes, not the weightings attached to the association edges.

When tracking the actions performed, the actual documents read by the user are not suitable for the adaptation of the weighting factors of the user model. It is much more interesting to know the topics a resource is dealing with. Therefore, we developed a tracking on the higher topic level and even on the process level. For these purposes, the tracking system must be able to map the contents of a document on the topics included in the ontology. On the process level, single steps have to be recognized in order to guess the task the user is performing. According to the tracking data, the weightings of the topic maps get raised or lowered.

From an algebraic point of view, the representation of the user preferences consists of a topic map *TM*, which in turn contains a set of *N* topics T^N, each of them weighted with a factor w_T^N and a set of *M* associations A^M:

$$TM : \{T^N \times w_T^N \times A^M\}$$

The associations *A* are directed relationships between two topics, attached with an association type Y_A and an association weighting w_A

$$A : \{T \times T \times Y_A \times w_A\}.$$

The value range of both types of weightings is normalized:

$$w_j \in \{0..1\}$$

Finally, a useful set of association types is defined for Y_A:

$$Y_A \in \{is\_a, is\_related\_to, is\_part\_of, \ldots\}$$

One has to be careful since seemingly symmetric associations such as "*is_related_to*" are not automatically bi-directional. Usually, not more than ten different kinds of associations are needed within one domain-specific ontology.

According to the above definitions, the updating function reads as

$$update : TM \times T^K \times n^K \rightarrow TM$$

where T^K is a set of *K* topics with topic T^j having been clicked or arisen n^j times in the documents read since the last call of the *update* function. The transition function of *update* is defined by a set of rules that contain a saturation rule, a normalisation of topic weightings, the topic attraction around heavy-weight topics, a mechanism providing constant overall interest and some non-linear functions.

The saturation rule describes the sensitivity of the topic weight adaptation as a function of the number of clicks. Since a user might just be passing by a topic without profound interest in it, the adaptation curve starts with a low sensitivity. However, if a topic is already heavily weighted, further reading on that topic will not raise it any more: the weighting runs into saturation. In between these flat sections, the weight increases linearly with its occurrence and demand, respectively.

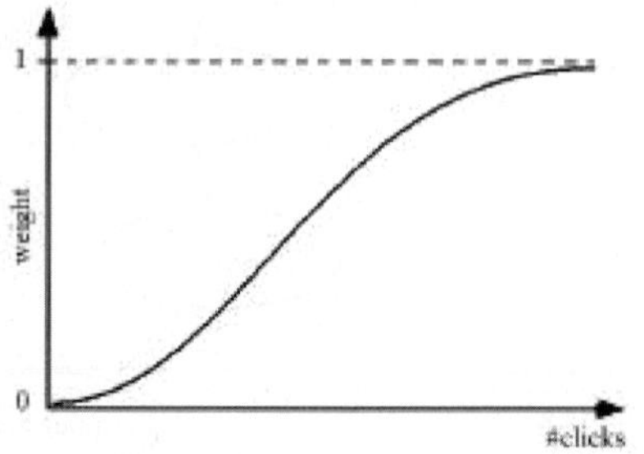

Fig. 3: Saturation curve

In order to guaranty the conservation of the overall interest, all topic weightings undergo a steady renormalization, ensuring a non-inflationary interest of the user. In our mathematical model, this rule reads

$$\sum_{i=1}^{N} w_T^i = const.$$

This rule is an analogy to the energy conservation in physics. Similarly, in order to prevent a concentration of all user interest on a handful of topics, an entropy growth rule has been introduced. Triggered by external sources, a "topic of the day" is suggested for reading and rated with a medium weight.

Finally, analogously to the gravitation potential in physics, heavy-weight topics attract their neighbours in terms of their respective weighting. As a metrics for the distance, the association strength w_A is used.

The set of rules making up the function *update* yield a new set of weightings that represent the user's preferences and interests, which is used as the basis for adaptive content retrieval and user interface adaptation. Additionally, provided that strong interest in one field leads to expertise in it, the user model enables easy expertise location among the users.

4 Results and Future Work

The developed user model has been included and evaluated within a future knowledge management platform for a mutual savings bank. Our first tests of the user model showed that the application of a user model that holds not only interest coefficients but also a knowledge structure is a valuable support for suitable content retrieval and expert finding. If a large amount of information is available, the user model turned out to be very helpful in selecting appropriate content. However, for these purposes, the user model needs to be coupled with a tracking system on a topic level, which requires auto classification of the documents provided if no metadata is available.

By contrast, the adaptive navigation structure based in the weighted topic map did not turn out to be a useful tool in content structuring. The linearization of the network topology of a topic map to a navigation tree was rather confusing to the users and did not lead to productivity increase.

Though the connection to the tracking system is working well, the set of rules for the weightings maintenance needs to be evaluated and should then be further refined. Also, our current auto classification mechanisms are limited to topic-related categorization and will be replaced by more general methods allowing for the recognition of document types.

References

Baldwin, J.F., Martin, T., Tzanavari, A. (2000): User Modelling Using Conceptual Graphs for Intelligent Agents. Proc. 8th Intl. Conference on Conceptual Structures ICCS 2000, Darmstadt pp 193-206

Pepper, S. (2000): the TAO of Topic Maps. Infostream, Oslo, Norway

Kim, J., Oard, D.W., Romanik, K. (2001): User Modeling for Information Filtering Based on Implicit Feedback. Proc. ISKO 2001, Nanterre

Claypool, M., Brown, D., Phong, L., Waseda, M. (2001): Inferring User Interest

Practice of Gathering Requirements with Focus Group in China

Chen Baihong

Legend Corporate R&D
Beijing, P.R.China
chenbh@legend.com

Yang Wanli

Legend Corporate R&D
Beijing, P.R.China
yangwl@legend.com

Abstract

This paper briefly introduced the role of focus groups in the product usability design process. Based on the basic steps of focus group, the differences of recruiting participants in China and in western countries are compared, and the reasons are also analyzed. Lastly the behaviors of typical users in different cities are summarized, and our experiences and methods are also presented in order to share with those would implement focus groups in the usability research in China.

1 Introduction

With the transition from product economy to service economy and then to experience economy, more and more companies begin to strongly sense the importance of transforming the conception of "product-oriented" into "user-centered" in the product design. The academic circles also put up forward some theories and methods about it. From the 80th on, the idea of "user-centered" gradually penetrates through the whole life cycle of product design. The "user-centered" is to concentrate on the users' characteristics and tasks, in order to ensure that product functions meet users' requirements, and to ensure that the operation of products comply with the users' habits in the expected context of use(Beyer & Holtzblatt, 1998). Improving the "quality of use" of products has been the ultimate goal of the usability engineering.

Obviously, user involvement in the product design process has been an important topic. Generally, there are two types of user involvements. One is "user-driven design", that means direct involvement of potential users in the design process. Another is "user-informed design", that means involvement into the context of use by focus group or field studies. The former approach requires considerable user involvement over a long period of time, which is often not applicable to more "user-oriented" products with their relatively short product development times. Moreover previous experiences showed that the user is not always the good designer(Nielsen,1993), so the "user-driven design" is few adopted. At present, the "user-informed design" is often applied in the design process, and focus groups are popular method of obtaining user's requirements (Krueger,1994; Morgan & Krueger, 1998).

Focus group is first used as a qualitative method to prove the hypothesis, or to complement qualitative evaluations or analyses in sociology and society psychology. Focus group is also used in the market survey and product design to gather: new product design ideas; the functions of new products and context of use; the weakness of products, and improvement solutions.

Therefore, focus group could be used as a qualitative interview technique of to collect user's requirements and feelings both before interface design and after use.

A moderator usually guides the implementation of focus group, and collects information through discussion of given topics by several representative users(Fern,1983). In order to obtain certain trends or models, it requires six to eight groups and each group has six to eight users. The information gathered with focus group can be used to support the constant decision making in the product design process. Nevertheless, the expert's suggestion, design decision and detailed solution to the given question cannot be made from focus group. It does differ from other discussion method (e.g., brainstorming), focus group prefers to gather and probe the participants' thoughts and their behaviors in an environment that is as nature as possible. So what the focus group can provide is the prime qualitative data that would help the designer to outline the design goals, describe the tasks and reorganized the workflow. So focus group provides the true data from users instead of design decisions that are based on experiences.

For now, focus group is frequently used at the marketing research. It is still not popularly used in the product design. This paper presents the considerations in making the product design decision by focus group, and discusses the problems of implementation of focus group in China.

2 Elicitation and validation of requirements

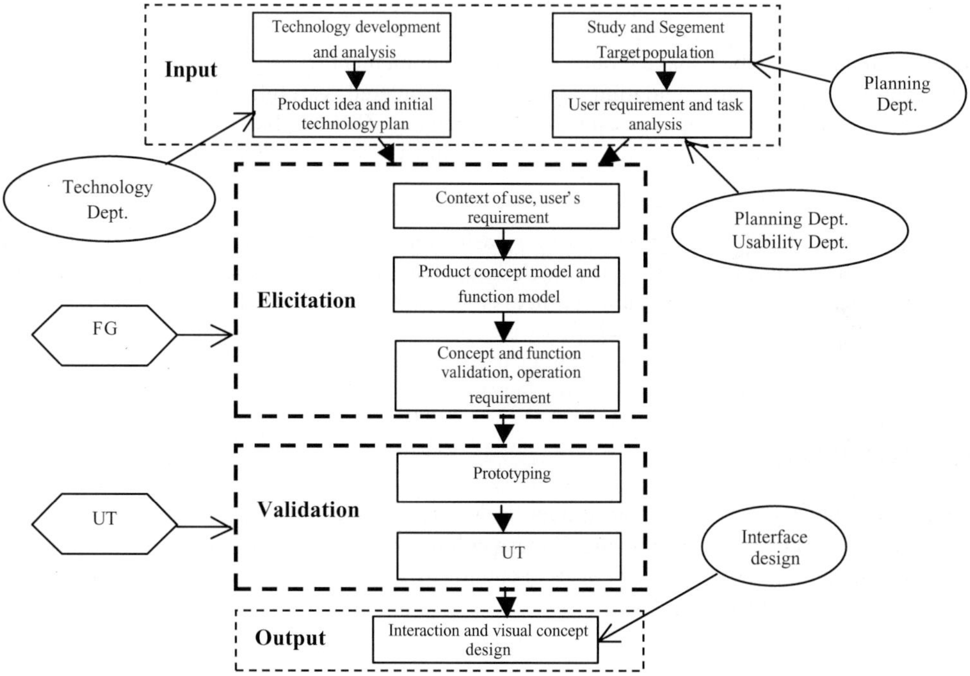

Figure 1: The Process of Elicitation and Validation of Requirements.

Information can be classified as product and technology information, marketing information, and user requirement information. Technique department collects product and technology information, which include product idea and concept, technology trend and technique specification and so on. Planning department and social research department gather the market information, which include company decision and target user segmentation.

Gathering the information such as the user's expected function and context of use must use the technologies of usability engineering. Focus group is the technique to gather the user's requirements. Focus group and usability testing could validate the user's requirement. It can provide constant support for product design decision-making. The usability research department implements the information gathering. The flow of information through the project includes four phases, input, elicitation, validation, and output (Klaus, 2000). Figure 1 shows the process of elicitation and validation of requirements.

In the figure 1, input information includes technology information and marketing information. They are as the input in the process of elicitation and validation of requirements, whereas these two kinds of information would influence with each other. Interaction and visual concept design is the output in the process. It may be required several usability tests and prototypes in the validation phase. The focus group will be used in the phase of requirement gathering and elicitation.

3 Practice in China

Influenced by the Chinese culture, many theories, including the theory of usability research, have to be combined with the Chinese characteristics when applied in China. The employment of focus group is no exception. Since China is a multinational country and covers a vast area, there are many subcultures existing in its culture. Different regions and different nationalities have different dialects and cultures, which make it quite different to implement focus group in China from that in other nations. These differentiations must be considered carefully when collecting user requirements with focus group. For example, the people in Zhejiang like the products with warm-colored outward appearance, while people in Sichuan and Guangdong prefer bright-colored outward appearance.

The implementation of focus group has four basic steps: planning, recruiting, moderating and discussing, and analyzing result. At the beginning, an explicit project plan must be worked out. The basic considerations in the planning process include: the personnel and staffing resource, time plan, the region and the time of each focus group, analysis plan and so on. Chinese subculture must be taken into account in the planning because the subculture will have great influences during the recruiting, moderating and analyzing stage.

3.1 Recruiting

Having the right participants in the groups is key to success. The way to reach the potential participants would be influenced by the Chinese culture and fundamental equipments. It's difficult to use the recruitment experiences of western countries. The table 1 compared effects and costs of different recruitment methods.

Table 1: Effects and costs of different recruitment methods

Recruitment method	China		Western Countries	
	Effect	Cost	Effect	Cost
Advertisement	Low , few responses	About 200$	High	About 15$
Telephone	Low, lack of cooperation	>20$	Medium	<10$
Organization, institution, communities	High	About 120$	Seldom adopted	About 500$

In western countries the usual way of getting participants is to put advertisement on newspapers or Internet. Soon, there are many responses, and the usability experts can choose from them. The private telephone numbers are published and the communication fee is pretty low, so it can find right participants by telephone in some western countries.

But these methods don't fit Chinese situation at all. Not only the cost will be very high, but also the real effect is low. For instance, an advertisement of the size of a name card in a newspaper will cost $200 but it only can attract one or two responses. Advertisement on the Internet is also poor. The effect of telephone is lower. Besides the high fee, most Chinese people are reluctant to talk to a stranger over the phone. They have no trust in the telephone talk. The reserved character of Chinese people prevents them from getting easily involved in the field they are not quite familiar with. Naturally it won't do to try to get the participants through media advertisement or telephone.

However, Chinese people believe more in organization, and they pay more attention to the relationship. Therefore, a method that is suitable for getting participants in China is to establish long-term relationship with some communities and organizations, and to establish long-term cooperation with the selling channels. Getting participants through organizations and relationships greatly reduces the problem of low chances of success due to participant's suspicion, and at the same time it can be guaranteed that the participants can meet our needs.

3.2 Moderating and Discussing

The reserved character of Chinese makes their anxiety easily when they come into the lab. So a comparatively long period of time is needed for foreshadowing and introduction so that they can talk openly and in depth. In order to make participants feel relaxed and comfortable, arrangement of the environment must be noticed. Some snacks and drinks can also smooth over the discussion.

In the discussion process, the participants from different regions have obvious differences. In Shanghai, if there is any suggestion, it will be expressed in detail and even causes dispute among participants. However, in Beijing, the typical participants is talkative regardless of the discussion topic, so there are few useful suggestions you can collected from them. Table 2 shows participants behaviours during the group discussions in typical cities in China and relevant solutions.

Table 2: Typical participant's behaviors in different cities and solutions

Cities	Behaviours	Solution
Beijing	Talkative and negligent of other participants but having few suggestions; Or keeping silent and acting as a "judge".	Reiterate the questions and question insistently. Necessary to pull them back to the theme.
Shanghai	If there is any suggestion, it will be expressed in detail and even causes dispute among participants. They often speak Shanghai dialect.	4 to 5 participants are enough to have a heated discussion.
Guangzhou	Having the tendency to explore the root of the question but the suggestions given may not be their true thoughts. Sometimes they may keep silent if the moderator doesn't speak Cantonese.	Question one by one, pay attention to the expressions and body languages. The local moderator or assistant is valid.
Chengdu	Very friendly but would like to be listeners. Hesitating to speak out his/her own viewpoints. The female is more willing to express her ideas.	Question one by one. Group discussion should be adopted.
Shenyang	Warm-hearted and outspoken. Willing to speak out straightforwardly if they have any ideas and will give many related and unrelated suggestions. Sometimes they also raise awkward questions.	Necessary to make clear the question and the result required. It is just ok not to be opposed against the participants.

From the typical adult participant's behavior, it can be found that moderating focus group in Shanghai and Shenyang is easier. But as far as the teenagers are concerned, if the moderator is older than the participants, the discussion atmosphere is likely to be dull. If select a young moderator or divide the group into two or three small groups, then the participants are willing to express their ideas.

3.3 Analyzing

It is usually necessary to consider comprehensively whether the wording, the background, and the expression are in accordance or not and whether there are some implied meanings before the useful data could be obtained. When focus group is implemented in different cities in China, the superficial data gathered in these cities may sometimes be the same while the implied meanings are not. On the other hand, different wordings may express the same meanings. This phenomenon is greatly influenced by the existence of subcultures in China. This is closely related with the Chinese characteristics of subcultures.

For example, when people judge whether a certain opinion or issue is right or not, the reply made by residents in Shanghai and Guangzhou may represent whether the "product" itself and not the "opinion" itself is good or not. But things are quite different in Shenyang. The answer will aim only at the "opinion" itself. This difference concerns the ways of thinking of connecting, equating, and then comparing a viewpoint and the product itself, which is formed in the consciousness of people during the city development,

4 Conclusion

This paper briefly introduced the role of focus group in the product design process, based on Chinese subcultures and the basic steps of focus group, compared the differences of recruitment, moderating and analysis in China and in western countries, made the following conclusion.

Firstly, It is suitable for getting participants through some communities and organizations in China. The reasons are that Chinese people do not trust strangers and they are reluctant to involve in the field that they are not quite familiar with.

Secondly, the script design and moderating technique should be adjusted according to habits and cultures of different regions in China, because Chinese subcultures would have great influence on the group moderating and result analysis.

5 References

Beyer, H.,& Holtzblatt, K. (1998). Contextual design: Defining customer-centered systems. San Francisco: Morgan Kaufmann.

Fern, E.F. (1983). Focus groups: A review of some contradictory evidence, implications, and suggestions for future research. *Advances in Consumer Research*, 10, 121-126.

Klaus Kuhn. (2000). Problems and Benefits of Requirements Gathering With Focus Groups: A Case Study. *International Journal of Human-Computer Interaction*, 2000, 12, 309-325.

Krueger, R.A. (1994). Focus groups: A practical guide for applied research. Thousand Oaks, CA, Sage.

Morgan, David L, & Krueger, Richard A. (1998). Focus Group Kit, SAGE Publications Ltd.

Nielsen, J. (1993). Usability engineering. San Diego, CA: Academic.

Use-Centered Interface Design for an Adaptable Administration System for Chemical Process Design

Christian Foltz[1], Bernhard Westfechtel[2], LudgerSchmidt[1], Holger Luczak[1]

[1]Institute of Industrial Engineering and Ergonomics
RWTH Aachen University
Bergdriesch 27, D-52062 Aachen
{c.foltz/h.luczak/l.schmidt}@iaw.rwth-aachen.de

[2]Department of Computer Science III
Software Engineering
RWTH Aachen University
Ahornstr. 55, D-52074 Aachen
bernhard@i3.informatik.rwth-aachen.de

Abstract

In this paper, the use-centered interface design for an adaptable administration system for chemical process design is presented. On the basis of the main work activities to be supported with the tool, two different analytical evaluation methods were applied to the prototypical original interface. The derived requirements led to the design and implementation of an alternative interface which was compared with the original one in an experimental study with ten users.

1 Introduction

In general, chemical process design is characterized as a complex, iterative and creative activity typically starting as an ill-defined problem (e.g., Westerberg, Biegler & Grossman, 1997). Within the course of a development project an interdisciplinary team designs and uses different documents or models in various software systems like spreadsheet and flowsheet tools, process simulators, word processors, etc. These data are often stored in document management systems and engineering data management systems, respectively. Additionally, strictly separated software systems address the coordination of the project team supporting classical management functions such as planning, organizing, and controlling by means of project plans.

To permit an integrated view on weakly structured development processes, the above mentioned features of different systems have been combined in one tool called AHEAD (adaptable and human-centered environment for the administration of development processes, Westfechtel, 1999). Within AHEAD three different environments can be distinguished. First, project managers will be supported by the management environment which is concerned with the coordination of teams, i.e. responsibility assignment, monitoring of milestones and deadlines, etc. Second, the developer environment aims to support project team members on the individual level by giving a coherent view on tasks and relevant documents. Third, the modeling environment allows to provide domain specific knowledge, e.g., by defining task types and pre-defined task sequences.

This contribution is concerned with the analysis of the original and the design of an alternative graphical user interface for the developer environment which basically contains two user interfaces. The agenda presents a to-do list showing the developer's tasks with present status and deadline. Choosing one task from the agenda, a new window named work context gives the following detailed information about this task. First, a task net allows to overview related tasks and documents. Second, all documents concerning the task are listed in a document list. Third, with a version graph different document versions can be managed (Figure 1, left side).

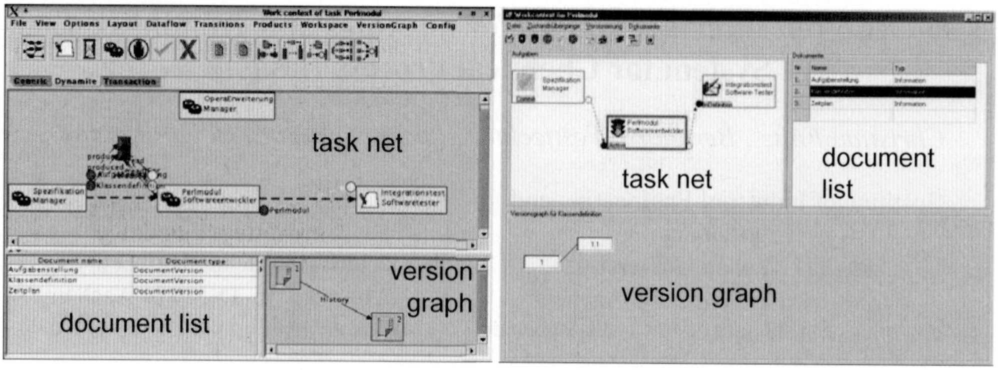

Figure 1: Work context of the AHEAD developer environment (left: original; right: alternative)

2 Analysis and Implementation

The aim of the original, prototypical user interface for the developer environment was mainly to integrate the different software features in one graphical interface. From a computer science perspective this implementation of different conceptual models has been a challenging activity (cf. Westfechtel, 1999).

The design of an alternative interface followed the use-centered design approach. In the use-centered perspective the focus shifts from the interaction between humans and machines to the interaction between humans and work (cf. Flach, Tanabe, Monta, Vicente & Rasmussen, 1998). Consequently, the main work activities which should be supported by the developer environment were identified. In a second step, it was explored which actions and operations have to be performed by an user in order to fulfil these activities using the original interface. For the sake of clarity HTA-diagrams (hierarchical task analysis; Diaper, 1989; Kirwan & Ainsworth, 1992) were used to visualise the necessary interaction steps. Figure 2 reveals that even simple actions like "create a new document" demand a great amount of operations, i.e., browsing through menus and select functions, and mouse movements. To unburden the user from those annoying interactions (Shneiderman, 1998; Johnson, 2000) several single operations were merged into one operation, e.g., when a new document is created an initial version is automatically produced. The grey diagonal bars in Figure 2 denote which user interactions could be saved in the alternative interface without removing functionality.

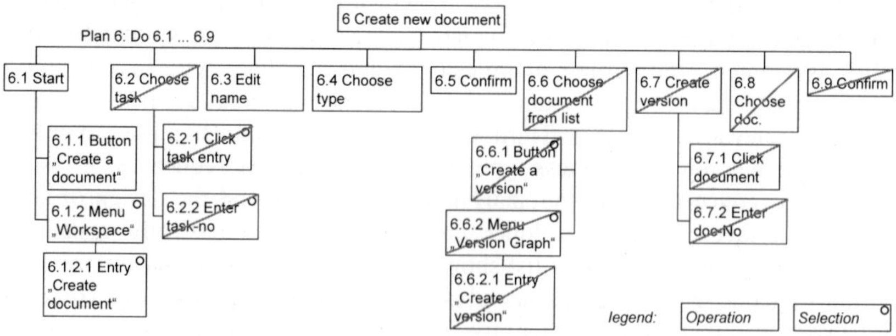

Figure 2: HTA-diagram for the activity "Create new document"

Before prototyping an alternative interface a heuristic evaluation (Nielsen, 1993) was used to uncover further usability problems. Here, different labelling of functions in menu and in button tool tip can be mentioned as well as the arrangement of buttons following neither the importance principle nor the frequency-of-use principle nor the sequence-of-use principle (Sanders & McCormick, 1993).
Based on these insights an alternative interface was designed and implemented with Borland's Delphi. This interface is a mock-up with which all relevant activities can be performed. However, the mock-up cannot be connected and used with the full-functional AHEAD system.

3 Empirical Study

In an empirical study the interfaces have been tested for at least two reasons. First, it should be compared how efficient and effective users can work with both interfaces. Second, it was expected that the empirical study will reveal further usability problems which have not been detected using the analytical evaluation methods.

3.1 Methodology

3.1.1 Independent variables

The two independent variables were the graphical user interfaces (original vs. alternative) and the sequence of use (original-alternative vs. alternative-original).

3.1.2 Dependent variables

The five dependent variables can be distinguished in measures for effectiveness (a. labeling of objects (correct/ false), b. producing states (solved/ unsolved), c. editing tasks (solved/ unsolved)) and efficiency (d. time to produce states, e. time to edit tasks)).

3.1.3 Apparatus

The experiment was performed on a notebook (Pentium III processor with 600 MHz, 128MB RAM) with dual boot option (Windows 98, SuSE Linux 7.3). Due to additional requirements the original prototype could only be used running Linux, although the interface was written with Java. The alternative graphical user interface was developed running Windows 98 and Borland's Delphi. The keyboard, a mouse, and the notebook's touchpad could be used to interact with both software prototypes. The complete experiment was video taped and advised by two investigators. A stopwatch was used to detect the time consumed.

3.1.4 Participants

Ten male participants aged between 25 and 34 years with a mean of 28.8 years (SD = 3.1 years) were recruited for the empirical study. Because it was not possible to recruit experienced chemical engineers, people with experiences in weakly structured engineering processes like software development were chosen. Eight participants worked in the IT sector in positions like system administrator or software engineer. The scenario for testing was build according to this fact. The subject's work experience differed between 0 and 8 years (mean = 3.0 years, SD = 2.4 years). No participant had experience with document and workflow management systems, respectively.

However, some test persons had experiences with project and version management systems. All participants were familiar with at least one version of the Windows operating system and one half is familiar with one of the Unix operating systems including Linux. Participation was voluntary, however, a fee of 50 € each was paid to the participants who were free to abort the experiment at any time.

3.1.5 Procedure

In a preliminary survey the test persons were introduced in the aims of the AHEAD system. Afterwards a questionnaire was used to survey demographic data and information about education and experiences. Each participant used both interfaces whereas one half started with the original interface and proceeded with the alternative one. The other group used the interfaces vice versa. The experiment itself was divided into three parts. First, to test self-descriptiveness and conformity with user expectations (ISO 9241, Part 10, 1996) the participants were asked to label different buttons and screen areas of the graphical user interface. Second, three screenshots representing different states of the software system were presented. The subjects tried to produce this states within 3 minutes. Third, brief written tasks were given to the test persons. These tasks should be solved by the participant within 3 minutes. Finally, parts of the IsoMetrics usability inventory (Gediga, Hamborg & Düntsch, 1999) were used to collect data about the subjective usability estimation of each participant. However, the results of this test will not be presented here. Except for the preliminary survey this procedure was repeated for the second user interface. All in all the experimental study took between 75 and 120 minutes.

3.2 Results and Discussion

In a t-test highly significant differences between the alternative and the original user interface were observed in the dependent variables. These differences were independent of the sequence of using the graphical interfaces. Figure 3 displays exemplary the box plots for the time consumed to solve states and tasks on the left side ($t = -11.485$; $p < 0.01$). On the right side the box plots for the number of solved states and solved tasks are presented ($t = 8.333$; $p < 0.01$). There are great differences in the values for both effectiveness and efficiency measures. Even the best value for the original interface is far away from the worst value for the alternative interface. Interestingly, neither the underlying conceptual model nor the work context's general appearance and arrangement of interface parts has been changed essentially (Figure 1). Instead, the visual appearance of buttons and their arrangement has been changed as well as the sensomotoric effort

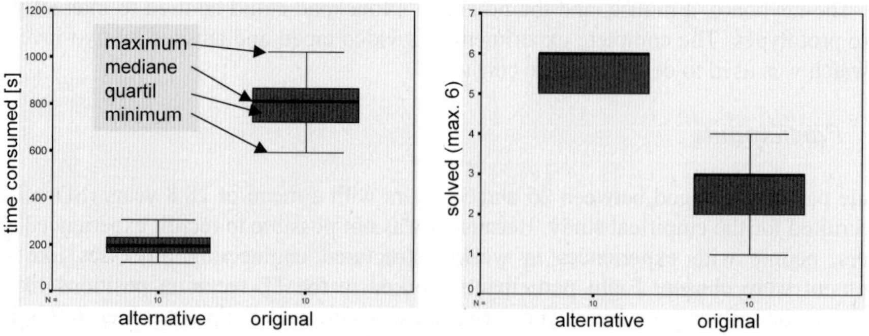

Figure 3: Time consumed (solved states and tasks only) and amount of solved states and tasks using the original and the alternative user interface of AHEAD's developer environment

for using the interface. So far, it cannot be ascertained to what extent the reduction of time consumed is caused by a shortened information processing or by reduced mouse movements. However, the increased number of solved states and tasks using the alternative interface indicates that the information processing has a considerable influence on the time consumed.
Moreover, the empirical study revealed additional usability information. For example, some users remarked that in the document list no hint is given if a document was provided by a third party or created by the user himself. Other participants interpreted the task net view as life cycle of the actual task rather than identifying the actual task and its predecessor(s) and successor(s) and, therefore, had problems in allocating documents. Thus, a further survey should deal with those issues on the conceptual level. In addition, another study should compare the developer environment with "traditional" support tools by measuring the results of realistic work scenarios.

4 Concluding Remarks

Based on an use-centered approach two analytical evaluation methods and ergonomic basics have been applied to the interface design of a novel administration system for chemical process design. This has led to significant differences in both effectiveness and efficiency of use. However, some important shortcomings have not been identified until a test with "real" users, i.e. some questions concerning the conceptual ideas of the tool have not been tackled. Three important conclusions can be drawn from this. First, analysis should be based on the user's work task, rather than on a technology-centered human-computer interaction approach. Second, ergonomics should be integrated into the software development process as early as possible. Therewith, important conceptual questions can be addressed at once. Third, a variety of both analytical and empirical usability methods should be used in all phases of the software design process.

5 Acknowledgement

The authors gratefully acknowledge the financial support of the Deutsche Forschungsgemeinschaft (DFG) within the Collaborative Research Center 476 IMPROVE.

6 References

Diaper, D. (Ed.). (1989). Task Analysis for Human-Computer Interaction, Chichester: Ellis Horwood.
Flach, J.M., Tanabe, F., Monta, K., Vicente, K.J., Rasmussen, J. (1998) An ecological approach to interface design. In HFES (Ed.) *Proceedings of the Human Factors and Ergonomics Society 42nd Annual Meeting* (pp. 295-299). Santa Monica: HFES.
Gediga, G., Hamborg, K.-C., Düntsch, I. (1999) The IsoMetrics usability inventory: an operationalization of ISO 9241-10 supporting summative and formative evaluation of software systems. *Behaviour & Information Technology*, 18 (3), 154-164.
Johnson, J. (2000) GUI Bloopers. San Diego: Academic Press.
Kirwan, B., Ainsworth, L.K. (1992) A guide to task analysis. London: Taylor & Francis.
Sanders, M.S., McCormick, E.J. (1993) Human Factors in Engineering Design (7th ed.) New York: McGraw-Hill.
Shneiderman, B. (1998) Designing the User Interface (3rd ed.). Reading: Addison-Wesley.
Nielsen, J. (1993) Usability Engineering. San Diego: Morgan Kaufman.
Biegler, L.T., Grossmann I.E., & Westerberg, A.W. (1997) Systematic Methods of Chemical Process Design. Upper Saddle River: Prentice Hall PTR.
Westfechtel, B. (1999) Models and Tools for Managing Development Processes. Berlin: Springer.

A Toolkit for Exploring Affective Interface Adaptation in Videogames

Kiel Mark Gilleade, Jen Allanson

Computing Department
Lancaster University
Lancaster
LA1 4YR, UK
{gilleade, allanson}@comp.lancs.ac.uk

Abstract

From its humble beginnings back in the early 1960's the videogame has become one of the most successful form of HCI to date. However if we look more closely at the interactions between the game and gamer it becomes evident little has changed since the advent of *SpaceWar*[1] back in 1961. These interactions are for the most part static and thus predictable, given a particular set of circumstances a game will always react in one particular manner despite anything the player may actually do. Because of this the expected lifespan of a videogame is inherently dependant on the choices the videogame provides; once all possible avenues have been explored the game loses its appeal. In this paper we focus on adapting techniques used in the field of Affective Computing to solve this stagnation in the videogames market. We describe the development of a programming software development kit (SDK) that allows the interactions between man and machine to become dynamic entities during play by means of monitoring the player's physiological condition.

1 Introduction

After the release of *Pong*[1] by Atari back in 1972 the videogame established itself as one of the most popular forms of HCI. In spite of this, research into this particular field of HCI has been relatively slow and only recently has any academic entity attempted to analyse the interactions between the game and gamer. Such examples include AffQuake[2], a modified version of id Software's *Quake II*[3] that used the player's level of arousal to control the games avatar and a variety of affective videogames used to treat neurological problems[4].

However the projects undertaken so far have a very limited area of applicability. The problem inherit in the majority of videogames is that the interactions between the game and gamer are always fixed, for every action the game/player initiates there is always a predetermined reaction by the player/game. Because of this it's relatively easy to predict what will happen given an action initiated by the player or the videogame, which becomes even easier after the player finishes the game. Once all the possible avenues that are open to the player have been explored the game loses its appeal because they're unable to change.

1.1 Interactions as a Dynamic Entity

In order to rectify this problem we believe a videogame must be capable of dynamically changing the ways in which it reacts to the player's presence in the gaming environment. One approach to achieving this is by using the psychological status of the player as a point of reference. Therefore when the player's state of mind changes so do the interactions between the game and the gamer, thus preventing the player from accurately predicting what a given action or reaction will bring about in the videogame. Our premise for the work described here is that by using the player's psychological status, as governed by their level of arousal, to control how the gaming environment reacts to the player's presence, the videogame can become affective. This should in turn enhance the game's appeal and potentially increase its overall life span.

2 Developing an Affective Videogame

To put this theory to the test we decided to carry out some preliminary investigations with a view to creating a programming SDK that could be used by games developers to create affective videogames.

2.1 Suitable Genres

Phase 1 of this project involved the study of the relationships governing the interactions between game and gamer in order to decide which genre of videogame would be suitable for affective interface adaptation. We investigated several of the most common[1] genres of videogame including action, adventure, analytical (puzzle and strategy), role playing games (RPG), simulation and sports and concluded that only those genres that already required a high level of involvement on the player's part would be suitable for this project.

The reasoning behind this is that when developing an affective application of this type we are trying to enhance the interactions that already exist between man and machine, we cannot overtly change the pace of these interactions otherwise we change the nature of the system the user interacts with. And only those genres that require a high level of interaction (player is inundated with objects of interest) are going to be able to invoke a psychological response (whether positive or negative) that we can use to manipulate the gaming environment with.

Suitable genres for incorporation include both the action and adventure genres. Due to their reliance on bombarding the player with a wealth of events which require the player's full and undivided attention in order to prevent failure e.g. a wave of ravaging monsters. The sport genre was also incorporated because the majority of real world sports are highly physical, which is a quality that is inherited by their video game counterparts and as such they operate much like action games. Only the puzzle side of the analytical genre was chosen for incorporation because in the majority of cases with games such as *Tetris*[5] the player needs to respond in rapid succession in order to win. Strategy games like *StarCraft*[6] are often long winded and actions whether initiated by the computer or the player can sometimes take several hours to execute e.g. building a suitable sized force to storm the enemies' base. Because of this the player's level of arousal is unlikely to change by any significant value if at all for the duration of play. The same goes for RPG and

---

[1] Hybrid genres were not analysed e.g. horror-action.

simulation games, the level of interaction is often so slow an emotional response is unlikely to occur. Hence such genres are not suitable for incorporation into our programming platform.

Using this information we decided to build an affective programming platform for the following videogame genres, action, adventure, puzzle, and sports.

2.2 Manipulating the Gaming Environment

Phase 2 of this project involved the identification and isolation of the chosen genres' mutable gaming components. Each component must be capable of being manipulated within the context of the videogame so not to make the player overtly aware any change has been made when one is instrumented. For example a game cannot simply change the stage the player is interacting with because of the way they reacted, as that would spoil the games continuity. However the games difficulty settings could be changed because the resultant effect would blend more transparently within the context of the videogame. An investigation into the chosen genres revealed the following exploitable gaming components to be:

- *Difficulty:* controls the games difficulty setting.
- *Story:* controls the strength of the drama involved in the story line.
- *Music:* controls the music tempo and style.

With these components identified we could begin to develop counter responses to the changes in the player's emotional state.

2.3 Responding to Change

With the identification of the components mentioned above we could move onto phase 3, deciding how we were going to respond to the changes in the players psychological condition during play.

To begin with we needed to decide on how the player's psychological status would be assessed. Videogame developers are often forced to push the hardware their games are run on to the very limit in order to stay competitive. Therefore a complex analysis of the players bio signs in order to assess there psychological condition is not a possibility. Consequently we decided to opt for a simple class based psychological assessment. A physiological aspect (heartbeat rate) of the player is grouped into several classes; where each class represents a different psychological state the player is in. The supported states were as follows, bored (negative state, decreased heart rate), tired, content, excited and ecstatic (positive state, increased heart rate). When the player enters a given class a new set of interactions (unique to that class) will take place in order to prevent the game from losing its appeal. For each class present we allocated a selection of environment changes the game would be able to implement based upon the components it is allowed to manipulate.

With that aside we had to formulate how the gaming experience was going to be enhanced given the player's current psychological state and by how much. In order to do this we needed to take into account what we wanted the game to do given a positive or negative reaction by the player. When the player responds negatively (decreased heart rate) it is reasonable to assume that the player is dissatisfied with the current state of play. Play would therefore need to be invigorated in order to encourage the gamer to play a more active role in the game. This can be achieved by increasing the game's difficulty and overall tempo, as the resultant effect would require the player to interact more with the game in order to remain in play. When the player is responding positively (increased heart rate) to the game we need to retain the current level of interaction to appease the player but we must not over stimulate the player otherwise they may burn out. Therefore during this period we need to increase the level of variety in the gaming environment and slow the overall tempo of the game down using the components available. This can be achieved by changing the games music style, enemy variation and AI capabilities. The scale of the environment change would depend on the degree the player is feeling in either direction e.g. if they're only slightly bored with the current state of play the game tempo is only slightly increased.

3 Implementation and Results

Based upon this analysis we created a software development kit (SDK) for prototyping affective videogames, which we call the Intelligent[2] Gaming System (IGS)[7]. An electrocardiograph (ECG) is employed to assess a player's psychological status. ECG data can be streamed in real time to a prototype videogame developed using the SDK.

In our experimental set up we provided 31 pre-determined game responses to a range of psychological states we envisaged a player might experience during game play for an action-based videogame. Our prototype videogame required the elimination of multiple targets within a set time frame; failure to destroy a target before it escaped the player's field of view would result in the player's health being depleted and eventually death.

The initial evaluation was undertaken by a group of 8 people (2 female, 6 male) aged between 21 and 38. Of the 8, 6 frequently played videogames and 5 of these indicated that they preferred playing action games to any other game genre. For each player a resting baseline ECG measurement was taken. From the baseline we calculated the boundary conditions for each psychological state (bored, tired, etc). This data was then used by the IGS-enabled affective version of the same action game. Within these boundaries streaming ECG data from the player caused the game play to change.

Rather than carrying out an analysis of arousal statistics we wanted to focus on the *subjective* experience. Consequently, for this preliminary study we provided a questionnaire in order to ascertain whether subjects preferred the affective version of the game to the control version (non-affective). All 8 volunteers indicated that they did. In addition the 6 serious gamers indicated that they felt the games life span would be improved by affective means. Of the remaining 2 subjects one was unsure or and the other believed it wouldn't affect the games life span directly.

4 Conclusions and the Future

Through the pilot study described we have demonstrated the principle of affective gaming. We have chosen to do this by providing a platform for the prototyping of ECG-responsive videogames. The responses of our small group of volunteers to an affective adventure game indicate the potentially compelling affect of such adaptive technologies. In addition the gamers among our group agreed with our assertion that this technology has the potential to extend the life span of the game.

---

[2] Machine intelligence can be characterised as the ability to react in light of ongoing events.

The IGS programming platform is still at a developmental stage and we can expect it to be some time yet before the videogame industry is likely to take notice of the benefits of affective videogames. This paper has shown that although there is evidence suggesting affective interface adaptation can lead to an improvement in the gaming experience, further exploration and analysis is required. We hope to carry out further explorations ourselves in the future, and have made the IGS SDK available online in the hope that others will be compelled to use it for their own experiments.

It may be possible to provide a programming platform that allows for responses more in line with the games context, and thus increasing the effect affective interface adaptation would have on the overall gaming experience. For example while playing the psychological horror game *Silent Hill 2*[8], the system could monitor the player's psychological condition and adjust the 'scary' factor according to how the player reacts to the gaming environment created. We could even go as far as recording the player's psychological reactions during key events in the game in order to adjust the next event to something that would have a much greater impact. Such use in hybrid genres such as survival horror could create immensely entertaining environments as the game and gamer are brought closer together.

Although the research described here is relatively limited, its implications can be very far reaching within the domain of videogame research, and the theory behind it can be applied to many other aspects of HCI. By studying how man and machine interact we can identify the relationships that govern those interactions and thus manipulate to make products much easier to use, more fun to interact with or anything else we wish within the context of the devices application. Therefore we feel that the lesson learned from this exploration of affective interface adaptation will be of interest to designers and developers of compelling interactive interfaces in other application domains.

References

1. Herman, L., Horwitz, J., Kent, S. & Miller, S. (2001). History of Videogames. Retrieved January 2003 from http://gamespot.com/gamespot/features/video/hov/

2. AffQuake. Retrieved March 2002 from http://affect.media.mit.edu/AC_listings.html

3. Quake II. Retrieved August 2002 from http://www.planetquake.com/quake2/

4. NASA. (2000). Video Games May Lead to Better Health Through New NASA Technology. Retrieved May 2002 from http://www.gooddealgames.com/articles/NASA_and_Games.html

5. Tetris. Retrieved January 2003 from http://www.tetris.com

6. Starcraft. Retrieved January 2003 from http://www.blizzard.com/starcraft/

7. Gilleade, K. (2002). IGS Programming Platform. Retrieved October 2002 from http://www.comp.lancs.ac.uk/computing/users/gilleade/

8. Silent Hill 2. Retrieved December 2002 from http://www.silenthill2.de

Envisioning Systems Using a Photo-Essay Technique and a Scenario-Based Inquiry

Kentaro Go,[1] *Yasuaki Takamoto,*[2] *John M. Carroll,*[3] *Atsumi Imamiya,*[1] *Hisanori Masuda*[1]

[1]Dept of Comp Sci & Media Eng, University of Yamanashi Kofu, 400-8511, Japan kgo@acm.org

[2]Design Center Fujitsu Limited Kawasaki, 211-8588, Japan takamoto@jp.fujitsu.com

[3]Dept of Comp Sci & Center for HCI Virginia Tech Blacksburg, VA 24060, U.S.A. carroll@cs.vt.edu

Abstract

In this paper, we propose a requirements elicitation method called PRESPE (Participatory Requirements Elicitation using Scenarios and Photo Essays). PRESPE enables participants to reflect upon their personal experiences when using systems and create photo-essays based on this reflection. The participants can then analyze these experiences by forming design concepts, envision scenarios by imagining contexts of use, and create artifacts by sketching these scenarios. We also describe a PRESPE case study that envisions future use of mobile phones.

1 Introduction

PostPet was first released in Japan during the Fall of 1997. Developed by Sony Communication Network, Inc., PostPet opened a new field in e-mail software design by featuring pleasant graphics and animation that contrasted sharply with the various businesslike software programs available. Since that time, over a million PostPet packages have been sold, and PostPet was bundled to more than 7.3 million PCs shipped by December of 2001 (ZDNet Japan, 2001).

This success story demonstrates the undeniable importance of an attractive user interface design. PostPet was not designed to provide any major breakthroughs in email software. Instead, a distinguished designer simply applied an innovative idea to e-mail software so that "a humanized virtual pet delivers our e-mail messages." This idea succeeded in the market in accordance with the potential requirements of the users. Though designers and usability experts constantly seek to create success stories like PostPet, they rarely occur. This illustrates the difficulty of eliciting user's potential requirements prior to design activities and reflecting upon them.

To address this challenge, we focused on design approaches at the early design process. For an integrated approach combining data collection and idea generation, we developed the PRESPE (Participatory Requirements Elicitation using Scenarios and Photo Essays) approach. PRESPE is a scenario-based design approach that creates photo essays and envisions scenarios of system use through the collaborative efforts of users and designers. In this paper, we provide an overview of the PRESPE approach and describe a PRESPE case study to show the various features such as photo-essays and scenarios.

2 The PRESPE Approach

PRESPE is a participatory design approach, in which a group of major stakeholders (including designers and real users) work together to produce and evaluate product designs (Muller, Haslwanter & Dayton, 1997). In the PRESPE approach, there are two roles: coordinators and participants. The coordinators assign a project theme and provide ongoing support for the participants' activities (Figure 1). There are four main activities: (1) reflect, (2) analyze, (3) envision, and (4) translate. For the assigned theme, each participant creates photo-essays to reflect his/her personal experience with existing artifacts.

The participants are divided into several groups; and the rest of the PRESPE activities are conducted as group work. By comparing the individual photo-essays, the participants can analyze shared ideas, identify the concepts behind them, and then develop design concepts. The participants can then use these design concepts as inspiration regarding future uses of the relevant technology when they envision use scenarios and contexts. This activity, called scenario exploration, is a structured brainstorming session with role-playing using scenarios and questions. The participants then translate scenes described in the scenarios into artifacts by making sketches of the scenes (Go & Carroll, 2003).

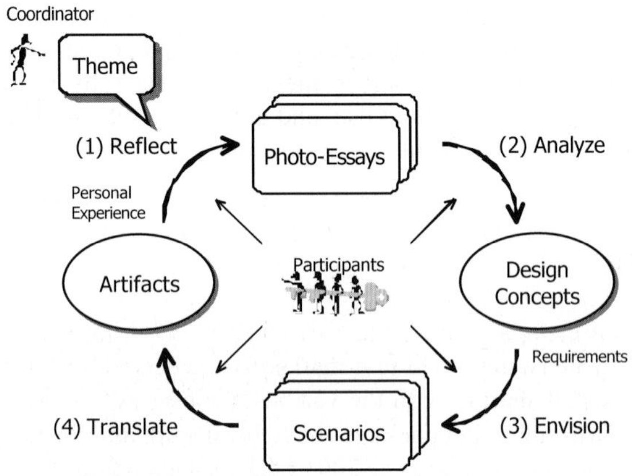

Figure 1: The PRESPE approach

PRESPE employs two devices: photo-essays and scenarios. A photo-essay contains representative photos on an assigned theme and an essay explaining why he/she thinks the photos fit the theme. Photos may be stereograms in order to increase the viewer's sense of reality (Holbrook & Kuwahara, 1998). Scenarios have a textual narrative form (Rosson & Carroll, 2001). During the PRESPE brainstorming session, participants create short scenarios that envision use situations. The participants ask 5W1H (What, Why, Who, When, Where, How) and what-if questions to identify concrete details of various use situations. The answers to the questions are represented as scenarios with detailed information.

3 Case Study: Envisioning Mobile Phones

We conducted a case study of the PRESPE approach with sixteen computer science students. These students possessed some general computer knowledge and basic programming skills, but had no experience in industrial or product design. Box 1 shows the assignment given to the students. Following our PRESPE approach, we asked them to address three themes; "Something I feel happy doing with an IT application," "My favorite IT application," and "Something I usually do with an IT application," by taking a few representative photographs and writing a brief vignette indicating the significance of the photos.

Project Description
The following list contains the themes of this design project. For each theme, do the following:
- Take a pair of photographs (overview and close-up) that describes the theme;
- Write a short essay that explains the meaning of the scene captured in the photographs; and
- Construct a web page that contains the photographs and essay.

Theme
- Something I feel happy doing with an IT application
- My favorite IT application
- Something I usually do with an IT application

Notes
- Consider what each theme means to you.
- Describe the scene in the photographs, and explain why you selected that particular scene.

Box 1: Assignment given to the participants

In less than three weeks, the participants worked part-time and wrote 48 photo essays. They formed five groups, each of which contained three or four participants. The participants extracted 12 keywords from the collective photo essays. Taking the keywords into account, they created 86 scenarios with 49 questions; each participant group created approximately 17 scenarios with 10 questions on average. Most of the scenarios were short (consisting of a sentence or two). Finally, each group created and sketched a future product.

Figure 2 shows photo-essays developed by participants. Figure 2(a) explains an unexpected use of a mobile phone. Figure 2(b) describes channel surfing of television, in which its author assumed televisions are IT products.

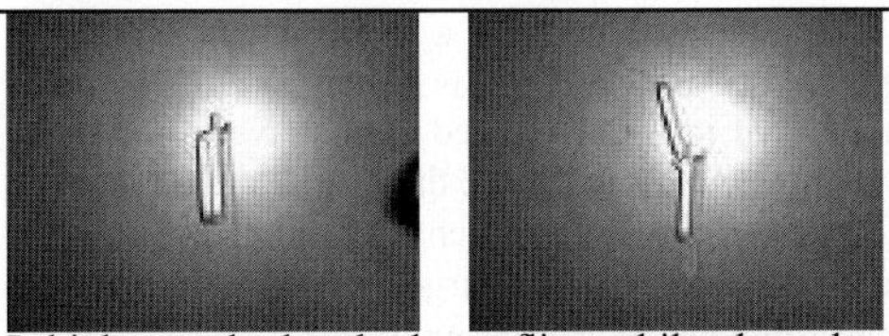

I think everybody who has a flip mobile phone has done this at least once: flipping it open and closing it for no particular reason. This is called clicking. It is obvious to every sensible person that flipping open the phone just for clicking consumes battery power and has no clear benefit, but I love the clicking sound and feeling that I get from clicking, so I do it from force of habit. Because each model of mobile phone offers a slightly different feeling associated with clicking I inadvertently flip open and close my friends' flip phones and display models at mobile phone shops. I feel that using flip phones like this might be a sickness.

(a) Clicking a mobile phone

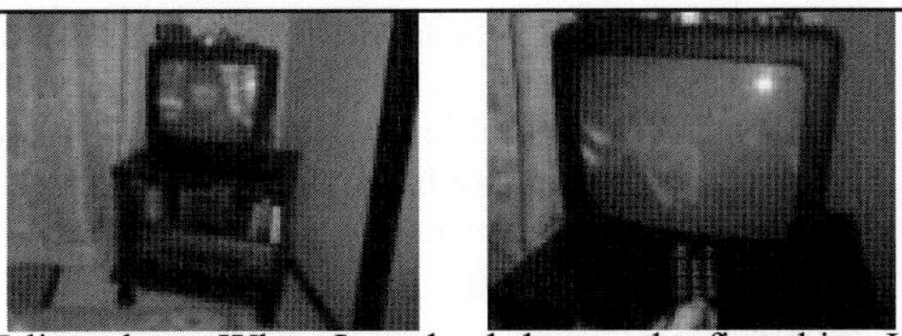

I live alone. When I get back home, the first thing I do is turn on the TV. I guess I might be feeling lonely. I try to find an entertaining program. I watch many kinds of programs, such as variety shows, dramas, and comedies. Since I live alone, I have a habit of channel surfing. Because I don't subscribe to a newspaper, I don't know what TV programs are currently on air. So after turning on the TV, I start channel surfing and stop when I find an entertaining program. During commercial breaks, I start channel surfing again. This is because I don't want to miss any entertaining programs that might be airing at the same time on a different channel. Another reason for this habit is that I am not disturbing anyone since I live by myself. I think that this habit may change depending on my environment.

(b) Channel surfing

Figure 2: Photo-essays written by a participant

During the analysis phase of photo-essays, the participants created keyword descriptions. Box 2 is an example of the keyword descriptions. Box 3 includes a list of keywords extracted as a design concept.

Keyword: Loneliness
Excerpts from the participants' essays on information technology are listed below. The excerpts show that college students may feel loneliness in their lives, and that they might try to forget their loneliness by using information technology.
To me, watching TV covers up my loneliness... (by T)
I live alone. When I get back home, the first thing I do is turn on the TV. I guess I might be feeling lonely... (by W)
In my hometown, I am very socially active, but on campus I am not... I may be feeling lonely... (by H)

Box 2: Keyword description written by participants

Current information
Unconscious action
Detestation of loneliness
Efficient uses of time
Importance of hearing
Is it essential?
Living for the moment
Joy of improvement
Memorable information
Inconsistent activities

Box 3: Keywords selected by the participants

Based on the keyword list, the participants conducted scenario inquiries. Box 4 shows threads of scenario inquiries developed during the scenario envisioning session. The scenarios in the threads evolved through the question-answer process to create and add more details to the original scenarios.

Scenario: One afternoon, I had nothing planned and wanted to hang around with a friend of mine. But I notice that I'd forgotten to bring along my mobile phone.
Question: Why did I forget to bring it?
Scenario: (No apparent reason) I believe that it is convenient to have a mobile phone, but I sometimes forget to take it with me when I go out.
Question: Is it possible to develop a mobile phone that cannot be left behind?
Scenario: I always wear accessories, and I can wear a mobile phone like an accessory.

(a) Mobile phone left behind

Scenario: I'd been talking on my mobile phone, and it shut off because of low battery power.
Question: How much power was left in the mobile phone battery when you went out?
Scenario: I noticed that there wasn't enough power in the battery of my mobile phone when I went out, but I had to leave in a real hurry, so I just grabbed it and went out.
Question: Do I need to make a phone call even though the battery in my mobile phone is dead?
Scenario: I've just remembered that I have to talk to a friend of mine, so I want to use the mobile phone now.
Question: Do I have a battery charger with me?
Scenario: I have a battery charger for my mobile phone, but I can't find any sockets around here.
Question: What if my mobile phone contains a power generator that can charge the battery by itself?

(b) Mobile phone with low battery power

Box 4: Threads of inquiries made and recorded by participants

Figure 3 illustrates some of the product ideas developed in this case study. Figure 3(a) shows a brief sketch of what a future mobile phone might look like. This idea is basically the same as the Whisper concept presented at the ACM CHI 99 conference (Fukumoto & Tonomura, 1999). Figure 3(b) is the concept of mobile phone with a rechargeable power joint. At the final step of the PRESPE approach, the participants conducted a claims analysis (Rosson & Carroll, 2001) on the designed products to enumerate their potential tradeoffs. For example, wearing the ring-and-wristwatch mobile phone may reduce the chances of losing it, but may cause perspiration. This step can be used to evaluate the concepts of designed products.

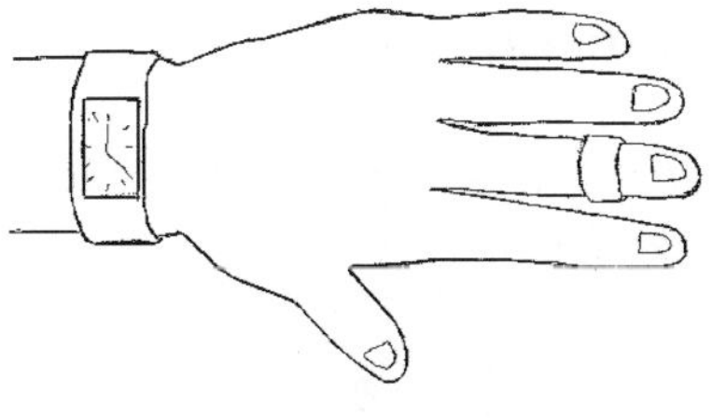

The ring acts as a receiver and the wristwatch contains a microphone; the wristband has a liquid crystal display that identifies the caller, and is able to recognize voice commands (including voice e-mail messages); the wristband is water-resistant to protect it against rain; an alert (a beeping sound, a light going on, and vibration) is issued if the separation between the ring and wristwatch exceeds a specific distance (e.g., 2 or 3 meters); the mobile phone's attractive appearance and style enables it to be a fashion accessory. (by Group 3)

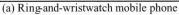

(a) Ring-and-wristwatch mobile phone

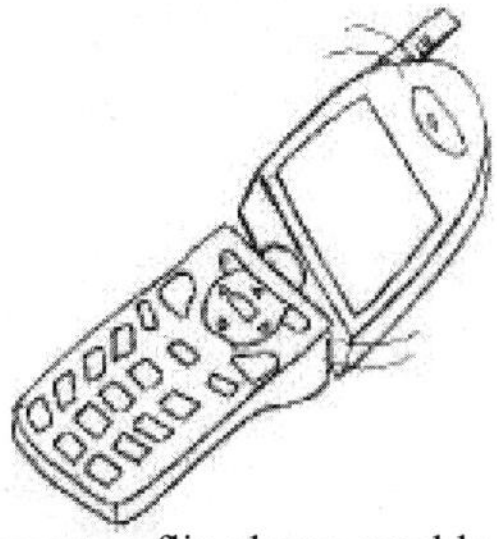

This is, of course, a flip phone capable of receiving messages during battery charge. Rotating the body generates power and recharges the phone battery; and it generates a clicking sound. (by Group 5)

(b) Mobile phone with a rechargeable power joint

Figure 3: Mobile phones envisioned and sketched by participants

4 Conclusion

From the case study, we learned that PRESPE was potentially useful for creating novel ideas regarding system design and requirements. The similarity between the ring-and-wristwatch mobile phone and the Whisper seems more than coincidental. Currently, we are working on a comparative study of the utility and quality of PRESPE between real designers and non-designers. This project would support our current finding that PRESPE enabled participants, even those with no prior design education, to create novel ideas regarding system development.

References

Fukumoto, M. & Tonomura, Y. (1999). Whisper: A Wristwatch Style Wearable Handset. In *Proceedings of CHI 99* (pp. 112-119). ACM Press.

Go, K. & Carroll, J.M. (2003). Scenario-based task analysis. In Dan Diaper and Neville Stanton (eds.), *The Handbook of Task Analysis for Human-Computer Interaction* (in printing).

Holbrook, M.B. & Kuwahara, T. (1998). Collective Stereographic Photo Essays: An Integrated Approach to Probing Consumption Experiences in Depth, *International Journal of Research in Marketing*, 15, 201-221.

Muller, M.J., Haslwanter, J.H., & Dayton, T. (1997). Participatory Practices in the Software Lifecycle. In Helander, Landauer, Prabhu (eds.), *Handbook of Human-Computer Interaction* (2nd ed, pp. 255-297). Elsevier.

Rosson M.B. & Carroll, J.M. (2001). *Usability Engineering: Scenario-Based Development of Human-Computer Interaction*. Morgan Kaufmann.

ZDNet Japan (2001). Retrieved January 27, 2003, from http://www.zdnet.co.jp/news/bursts/0112/10/04.html (in Japanese)

User Learning Modeling in Learnware Design- Case Study with Dynamic Geometry Software

Alex Sandro Gomes[1], *Ana Emilia de Melo Queiroz*[1], *Francisco de Assis Tenório de Carvalho*[1], *Francisco Alves*[2]

[1]UFPE – Centro de Informática
P.B 7851 - 50740.540 - Recife PE Brazil
asg@cin.ufpe.br, aemq@cin.ufpe.br, fatc@cin.ufpe.br

[2]UNIFOR – Universidade Fortaleza
alves@secrel.com.br

Abstract

This article presents a user modeling method, built upon constructivist foundations focusing in process of learning specific concepts. We analysed the real contribution of the use of Cabri Géomètre to the learning of Geometry. The method presented utilises data mining techniques and can be widely used in the evaluation of educational interfaces and in the design of interface agents. It complements traditional approaches for usability tests, adding to the set of tests an approach from the point of view of the learning of concepts.

1 Introduction

Educational softwares for the teaching of Mathematics are of low quality since they do not add information about the learning of concepts (Cuoco & Hoyles, 1996). The advantages offered through direct manipulation (Sedig, Klawe & Westrom, 2001), and calculation facilities, since long heralded as the advantages of computing systems do not effectively contribute to the making of truly different and consistent for the making of in mathematical concepts.

In the process of devising educational software there seems to be a significant difference between the picture of the learning process the designers have in mind, and the real process. Traditionally, the evaluation of educational interfaces is effected through the observation of its constitutive aspects, and the quality of the feedback. The methodology and the paradigm of quality life cycle employed is the same adopted with non-educational software. Interfaces are evaluated descriptively, observing the correct application of principles and guidelines restricted to the creation of interfaces (MacDougall & Squires, 1995). Since educational softwares aim not only at helping in the solving of problems but also in teaching concepts, it is necessary to analyse the learning process inserting it in the evaluation methodology used in educational interfaces. It is necessary a changing of paradigms. We hold the opinion that the learning of concepts takes place through the use of the interface.

This article is organized in the following way: the first section describes the constructivist model, whose users models will be generated through its application in analysis. Then, the process of generating user model data through is described. Finally, some results, and implications to the educational software evaluation and design are presented.

2 Constructivistic Theory for the Analysis of Educational Interfaces

Few studies use constructivist models for the cognitive development to model the interactions of the users with the systems. In the case of educational softwares it is desirable, besides studying the process of learning to use the interface, to analyse the learning of specific concepts that take place when using the software. Therefore, the point is not only learning TO DO something, but learn TO LEARN a concept. There are two advantages in adopting such theoretical standpoint: the existence of a large body of work done according to this theoretical tradition about the learning process (Flavel, 1992), and the existence of explicit definitions of concepts such as instrument and concept.

To engender the theory now described, we adopted the mental scheme model proposed by Vergnaud (1997), which allows the description of the organization of the actions and the identification through inferences of the internal elements, and internal dynamics of such mental schemes. Internal representation of a scheme be composed of information about the reality surrounding the user and the use of artifacts (rules of action), their goals and knowledge identifiable as scientific even though implicit and non-explicitable by users (invariants). The rules are knowledge about: (i) structure aspects of the problems to be solved, (ii) characteristics, functions, and properties of the chosen artifact, (iii) aspects subjacent to the choice of a particular artifact, or (iv) aspects subjacent to a social interaction. The invariants are knowledge that corresponds to concepts or properties of concepts of a reference theory. In the case study we conducted, several invariants related to geometric concepts are put in evidence.

We propose a model that articulates the material and cognitive activity dimensions as happens in the use of interfaces (Gomes, 1999). As such, it is necessary to analyse the actions of the users on the interfaces as being instrumental actions, what implies the adoption of the concept of instrument as defined by the theory of the origin of instruments (Mounoud, 1970; Rabardel, 1995). According to these authors, an instrument exists when a mental scheme organizes the action with an artifact (part of an interface, for instance). Artifacts, material or virtual, are means to achieve a goal. Schemes specialize in the use of particular artifacts. This definition does not take into account the elements of the internal representations of the schemes and its internal dynamics, such as defined by Vergnaud (1997). Since it is not incompatible with the definition of instrument, we have the possibility of redefining the notion of instrument such as proposed by Rabardel (1995).

2.1 Form of Analysis

For this model, the use of educational interfaces is analysed in two steps. First the adaptation of the users to the interface is analysed in a process called instrumental genesis. In this case, the organization of the mental scheme and the aspects of the structure of the artifact are analysed. Second, is analysed the learning that takes place during its use. The instrumental genesis is analysed through following the transformations of the constituent elements: the artifacts and schemes. The analysis of what happens with the artifacts is done from the point of view of the users. Therefore, it is possible to analyse the evolution with time of the functions attributed by the users. It is possible to make a detailed representation of the conceptualization that emerges in relation to the use of the educational interface. See a complete presentation in Gomes (1999).

3 Case Studies

In the case study we analyse the solving of geometric problems with two different systems of instruments (SI): one named "ruler and compasses", and the other a dynamic geometry software named Cabri. Géomètre. We followed the actions of five pairs of fifth grade students. Each pair solved a set of six problems of Geometry. After qualitative analysis, data was prepared in order to allow the extraction of rules about the use of the interface and the learning that is obtained during the actions. It was necessary to make data mining to identify relations between variables. The original database filled with data was treated in an excel file saved as CSv2; to this file an appropriate header was added in order to transform it in ARFF3 file. The Weka system Weka (1999) was the chosen mining tool because it has a GNU public license, is easy to install, and is implemented with JAVA (Java, 2000) what guarantees the portability of the system.

3.1 Data Mining Strategy Analysis

Cross the fields with multivalues (like a Cartesian product) generating new base cases. This case has the advantage of not presenting data loss, however, it introduces the problem of artificially increase the number of cases , what changes some of the metrics used in the evaluation of the extracted (mined) knowledge. In spite of the advent of a new problem introduced by this technique, this approach was maintained. A big obstacle was the great variety of possible values for the categorical attributes and the issue of having a limited set of cases to mine. A grouping strategy was adopted, and it made possible the application of rule-extraction algorithms.
After the use of the strategy described in Table 1, we started the manual selection of attributes. The process of attribute selection gives priority to data that will benefit the extraction of rules, reducing the dimension of the data and thus diminishing the size of the hypothesis space, and allowing the algorithm to operate more efficiently.

Table 1: Preprocessing strategy employed

Strategy	Analysis
Cross the fields with multivalues (Cartesian product) generating new base cases	This case has the advantage of not presenting data loss, however, it introduces the problem of artificially increase the number of cases , what changes some of the metrics used in the evaluation of the extracted (mined) knowledge. In spite of the advent of a new problem introduced by this technique, this approach was maintained.
Grouping of categories	A big obstacle was the great variety of possible values for the catagorical attributes and the issue of having a limited set of cases to mine. A grouping strategy was adopted, and it made possible the application of rule-extraction algorithms.

In some cases, the process of selection of attributes can improve the exactness of the scheme of classification; in other cases, the result can be more compact, and the representation of the learned concept can be easily interpreted. The selection of attributes involves the combination of search and evaluation of the utility of the attributes (Hall, 2000). In this context it is desired to estimate the learning mediated by an instrument, and the mathematical concepts used during the learning process. We present the relevant attributes, and the description of each attribute in Table 2.

Table 2: Attributes and their descriptions

Attribute	Categories
Prod	GOOD, FALSE
Artifact	Compas_Règle, Compas_Crayon, Compas, Crayon, Équerre, I2O, MUA, SD2P
Function	Reporter, Tracer, Trouver
Object	Distance, Droite, Mesure, Milieu, Ouverture du compas, Point d'intersection, OTHER
Theorem	CR, ED, OTHER, MT, SA, SC, TG, TI, VI, MS
Rule	CM-01_(A), CM-02_(A), EQ-02, EQ-02_(A), SD2P-02_(A), RG-07_(A), COR

3.2 Evaluation of the Acquired Knowledge

In this section we shall speak about techniques of analysis of data, and present the results of a benchmarking of classifiers. In this benchmarking, we compared the precision, the coverage of the model, and the percentage of hits. The tests were performed with the Weka© Experiment environment which allows to create, execute, and analyse experiments making possible the ordination of the classifiers based on the chose metrics. The chosen classifiers to take part in the test were, ZeroR, Prism, NaiveBayes, j48.PART.J48 all available in the Weka © system; the motivation for such a choice is that these are categorical prediction classifiers, i. e., predictors that offer results to discreet and categorical domains. In our context, we intend to predict the absolute value Prod, identified as Good or False.

The classifier that scored the best results was J48.PART. The J48.PART (PARtial decision Trees) algorithm is based on two techniques: divide-to-conquer4, and separate-to-conquer5. Aiming at obtaining results coherent with tha analysis of incorrect predictions, we used the J48.PART algorithm associated to the MetaCost classifier (Domingos, 1999).

In the case a better sample, from the point of view of the number of records and the absence of noise, it is possible that there be variations of the metrics (precision, model coverage, and percentage of hits), however, these variations do not invalidate the results obtained given that, besides interpreted rules, process of analysis employed guides the gathering and treating of data.

4 Method for the generation of the decision tree

Concerning the application of the results here obtained in the production of intelligent interfaces for educational software, the rules extracted can guide the development of agents capable of discerning about the adequacy or inadequacy of an action in a specific context. We will demonstrate its contribution for the initial context of quality evaluation of educational software. The fact that the interface allows the masking of operations incorrect from the point of view of the geometric theory should be interpreted as being an intrinsic characteristic of an open interface. What we are looking after is to identify the regularities between the use of certain interface components and the wrong steps made along the process of finding the answer for the problem. These regularities will constitute the model of the user necessary for the interface agents design.

- Object = MESURE: FALSE (43.0/2.0).

This rule relates the use of artifacts of the Cabri Géomètre software to the correction of the user's actions. It informs us that the actions aimed at operating on a measure (MESURE), induced the user to perform an action that deviates from the optimum strategy leading, thus, to a mistake.

- Function = Trace AND Artifact = Compasses AND Theorem = TI: Bon (135.0/2.0)

This rule informs that the use of compasses to solve problems related to isosceles triangles produces a large amount of hits. It tells more: there are aspects of this interface, the compasses, that favours hits in problems involving the concept of isosceles triangle.

- Rule = CM-01_(A) AND Function = Report: Bon (22.0).

This rule is less comprehensive. It states that when compasses are used to report a lenght, the students fare well. It is a good rule, of which an agent makes use to judge if a step of a procedure is correct, and act accordingly.

- Object = Milieu: Bon (29.0).

Similarly, this rule informs us that the location of the median point of segment was correctly determined, independently of the artifact used.

- Theorem = MT: Bon (4.0).

These rules inform us that, when a student makes a decision involving the concept of median, he finds the right answer. It is a simple concept. The first rule additionally informs that when this activity takes place with the aid of the compasses the answer is correct..

- Rule = EQ-02_(A): Bon (20.0).

This rule informs us that the use of compasses implies success. We could replicate the previous reasoning to apply it to the compasses and conclude that there are certain characteristics of the interface square that lead to success, and very likely to a good learning.

- Rule = COR AND Function = Trouver AND Artifact = Compasses AND Theorem = AS: FALSE (20.0).

This rule shows that the same compasses interface, adequate to the correct learning in one situation, is unfit to treat the resolution of problems involving the concept of axial symmetry (AS) directly, without having recourse to another artifact.

5 Conclusions

In this article we strived to demonstrate a theory that could make possible the user modeling for interface agents designs to educational software from the point of view of the learning of concepts. In the case study we analyse the solving of geometric problems with two different systems of instruments (SI): one named "ruler and compasses", and the other a dynamic geometry software named Cabri. Géomètre. What we are looking after is to identify the regularities between the use of certain interface components and the wrong steps made along the process of finding the answer for the problem. These regularities will constitute the model of the user necessary for the interface agents design. The rules obtained are sufficient to guide interface agents for educational software. The same technique and a usability version using the same cognitive model are being used in the design of complete educational application within our research group.

6 Acknowledge

This research was funded CNPq (CNPq/ProTeM-CC Proc. n. 680210/01-6 e n. 477645/2001-1).

7 Bibliographical References

Cuoco A. & Hoyles C. (1996). Software criticism – Helen M. Doerr: Stella ten years later: a review of the literature, *Inter. Journal of Comp. for Mathematical Learning* 1: 201– 224, 1996.

Domingos, P. (1999). A General Method for Making Classifiers Cost-Sensitive". *Artificial Intelligence Group*, Instituto Superior Técnico, Lisboa 1049-001 Portugal 1999.

Flavell J. H. (1992). *A Psicologia do Desenvolvimento de Jean Piaget*, São Paulo: Pioneira.

Gomes A. S. (1999). *Développement conceptuel consécutif a l'activité instrumentée*, Doctoral Thesis, Université Paris V, Paris [Retrieved August 22 from www.cin.ufpe.br/~asg].

Hall M.A. (2000). *Benchmarking attribute selection techniques for discrete class data mining.* Working Paper 00/10, Department of Computer Science, University of Waikato.

Java (2000). *Official Java Page*. Retrieved August 8, 2002, from www.javasoft.com.

MacDougall, A & Squires, D. (1995). A critical examination of the checklist approach in software selection. *Journal of Educational Computing Research*. 12(3), 263-274.

Mitchell, T. (1997). *Machine Learning*. McGraw-Hill, 1997.

Mounoud P. (1970). *Structuration de l'instrument chez l'enfant*, Delachaux: Lausanne.

Rabardel P. (1995). *Les hommes et les technologies Approche cognitive des instruments contemporains*, Paris: Armand Colins.

Sedig K., Klawe M. & Westrom M. (2001). Role of Interface Manipulation Style and Scaffolding on Cognition and Concept Learning in Learnware, *ACM Transactions on CHI*, 8 (1), 34–59.

Shneiderman B. (1998). Designing the user interface: strategies for effective human-computer-interaction, Addison: New Jersey;

Vergnaud G. (1997). The nature of mathematical concepts. In T. Nunes e P. Bryant (Eds.), *Learning and teaching mathematics: An international Perspective*, Psychology: Hove, pp. 5-28.

Weka (1999). WEKA. Retrieved August 8, 2002, from http://www.cs.waikato.ac.nz/~ml/weka/.

Incorporating Adaptivity in User Interface for Computerized Educational Systems

Andrina Granić

Faculty of Natural Sciences, Mathematics and Education, University of Split
Nikole Tesle 12, 21000 Split, Croatia
andrina.granic@pmfst.hr

Vlado Glavinić

Faculty of Electrical Engineering and Computing, University of Zagreb
Unska 3, 10000 Zagreb, Croatia
vlado.glavinic@fer.hr

Abstract

Aside from the great progress in the field of human-computer interaction, witnessed particularly in the last year, in several application areas the need for the development of more suitable and adaptive communication between a human user and the system has been recognized. On the other hand, rapid progress of computer technology made of the computer a strong tool in education by and large, resulting in the development of emulators of human teacher – intelligent tutoring systems, as well as their generators – authoring shells. Our research is concentrated on user-centered, adaptive interfaces for computerized educational systems, and particularly on intelligent computer-aided tools such as authoring shells. In this paper we elaborate the relevant issues and requirements concerning the incorporation of adaptivity in an authoring shell's user interface.

1 Introduction

The aim of human-computer interaction, which nowadays plays an important role in the context of the emerging information society, is to ensure well designed, usable computer system interfaces. Aside from the great progress in the field of human-computer interaction, the need for the development of adaptive communication between a human user and the system has been recognized. On the other hand, contemporary efforts in computer-supported learning and teaching has already introduced a generation of computerized educational systems called intelligent tutoring systems, which could be considered as emulators of human teachers. Moreover, in order to cover different domains of interest, intelligent tutoring systems generators were developed, denoted as authoring shells. Our research is concentrated on user-centered, adaptive user interfaces for intelligent computerized educational systems like authoring shells. In this paper we specifically elaborate on the relevant issues and requirements of incorporating adaptivity in an authoring shell's user interface, along with its design and implementation.

2 Adaptive User Interfaces and Intelligent Tutoring Systems

Adaptive user interfaces are a relatively novel solution to usability problems, which attempt to exploit the capabilities of interactive systems in order to accommodate the great variety of users and/or functionality (Benyon, 1997). Within this context a user interface is called intelligent in the degree it adapts itself and makes communication decisions dynamically, at run-time (Karagiannidis & Stephanidis, 1998). *Intelligent user interfaces* facilitate a more "natural" user-computer interaction, attempting to imitate human-human communication and constitute a major direction in current human-computer interaction research (Maybury, 1999). Several efforts towards the development of adaptive interfaces have been reported in the literature since the early eighties when they appeared, resulting in a number of prototype systems in several application domains like intelligent help, (intelligent) tutoring, information filtering or intelligent multi-media systems (IISIG, 1998).

Rapid progress of computer technology made of the computer a strong tool in education altogether, introducing tutoring systems with a certain level of intelligence. *Intelligent tutoring systems*, ITSs, comprise a recent generation of computerized educational systems, which attempt to mimic the capabilities of the human tutor (Fleischmann, 2000). These systems thus take into consideration the knowledge about what to teach (domain knowledge), the way to teach (teacher knowledge), as well as the relevant information about the student being taught (student knowledge), see Figure 1. As the need to cover a variety of different domains has arisen since, instead of having a number of specialized intelligent tutoring systems, ITS generators were developed, which can be "programmed" for a particular domain by modifying the domain knowledge. Those systems are usually denoted as *authoring shells*, ASs (Barton, 1995).

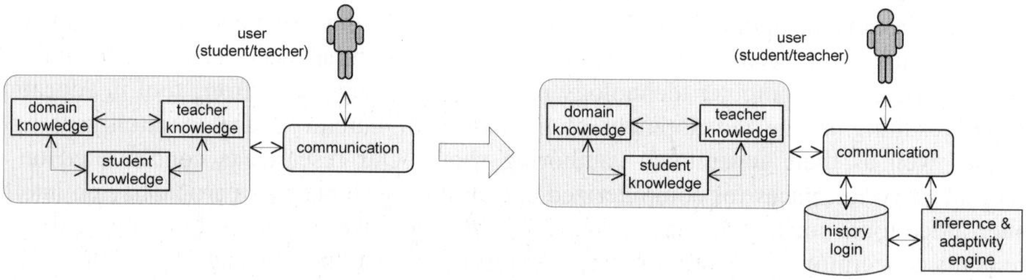

Figure 1: Including adaptivity in ITS/AS user interface

The vast majority of ITSs, as well as existing authoring shells, insure reasonable designs by simply pre-defining the respective interface, providing to and expecting from users (students and teachers) to use one static user interface, i.e. not providing interface design adaptation at all (Murray, 1999). Moreover, when supporting student interaction because of users dealing with concepts – the domain knowledge – yet not understood very well, this communication is inherently complex (Miller, 1988). Additionally, as users have various preferences, experience and knowledge, the need results for supplying suitable alternative ways for different users to communicate with an ITS/AS, requiring its adaptive behavior during run-time. The incorporation of adaptivity into the ITS/AS architecture will be elaborated in the following, based on the assumption that both teachers as well as students do their role better if they have at their disposal a suitable view on the object of their work, i.e. the domain knowledge base. On the one hand, teachers will be processing the domain knowledge base in order to create a knowledge structure that would convey the desired semantics, while on the other hand students will access this domain knowledge and will be tutored by it. Consequently, as a first step towards full interface adaptation that can be successfully used by a wide range of different users, thus creating a powerful alternative to standard static interfaces, we include adaptivity in the ITS/AS user interface design (see Figure 1), starting with domain knowledge base generation.

3 A Domain Knowledge Generator with Adaptive Interface

Adaptive Knowledge Base Builder, *AKBB*, (see Figure 2 for a glimpse of its user interface) is an arbitrary domain knowledge generator with adaptive interface, which provides means for the development of specialized ITSs for particular domains of education (Granić & Glavinić, 2002a),

(Granić, 2002). *AKBB's* adaptive interface is based on knowledge about the user and the interaction session, which is acquired dynamically during run-time. Its design results from several considerations among which the identification of users' individual differences together with their changing knowledge and behavior over time during the system's use. *AKBB's* interface is modeled by following the reference architecture for adaptive systems (Benyon, 1998). *AKBB* is a self-adaptive system based on adaptation of communication rather than adaptation of functionality.

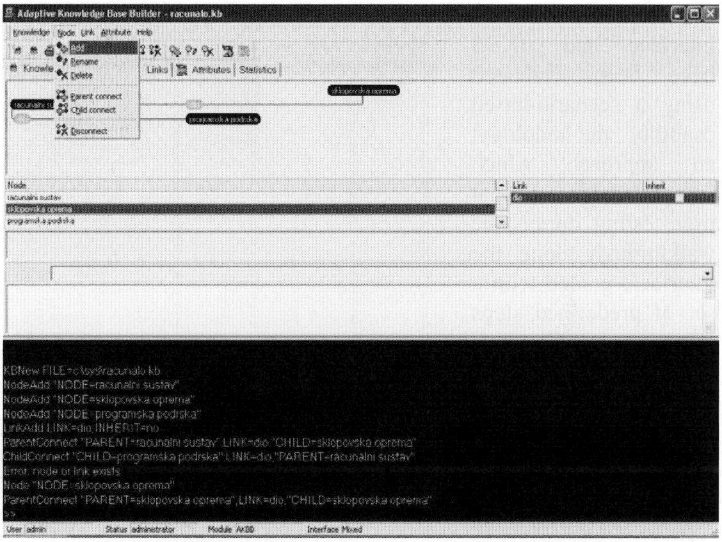

Figure 2: Snapshot of *AKBB* user interface

As the principles for designing human-computer interaction prescribe, novices and infrequent users benefit from using interface dialogue styles that allow users to recognize commands, and are provided with menu selections and fill-in form dialogue styles, in order to reduce their memory load. On the other hand, frequent and more experienced users are provided with a command interface, since they can become irritated with system imposed limitations. Additionally, user individual differences (Egan, 1988) are also taken into account, by considering user *spatial ability*, *experience in command languages*, as well as *incidence of AKBB's usage*. These parameters compose the user model, which is continuously updated in order to record all the relevant aspects of the interaction. According to the users' individual differences and their changing knowledge and behavior during the interaction, the values of these parameters are changing, enabling adequate input for the set of five inferring as well as twelve adaptivity rules. Examples of *AKBB's* adaptivity rules are as follows:

- *Adaptivity rule no. 1*
  if *spatial ability* = high **AND** *experience* = high **AND** *incidence* = high
  then *interface* = successor(*interface*)
- *Adaptivity rule no. 5*
  if *spatial ability* = high **AND** *experience* = low **AND** *incidence* = low
  then *interface* = *interface*
- *Adaptivity rule no. 9*
  if *spatial ability* = low **AND** *experience* = high **AND** *incidence* = no
  then *interface* = predecessor(*interface*)

These rules provide adequate automatic adaptation of *AKBB* user interface, which can be followed through the *Interface* column in *AKBB's Statistics* panel, see Figure 3. Thus, users are offered three different kinds of interfaces with adequate interaction styles:

- a *command interface*, which only enables interaction through the command line,
- a *graphical interface*, which provides an interaction to be performed using combination of direct manipulation and menu selection, enhanced with form fill-ins and mouse right button functions, and finally
- a *mixed interface*, providing the combination of the former two.

Both the AS with the standard static user interface and *AKBB* have been evaluated using a scenario-based usability methodology, an approach which comprehends formal user testing during the walkthrough along the interface, guided with a set of predefined steps (Granić & Glavinić, 2002b). Test users are tested with actual tasks under conditions that are as close to those in the actual system usage. The respective evaluation criteria are expressed in terms of objective performance measures of system use (suitability, adaptivity, learnability, error-rate and memorability), as well as of user subjective assessment (subjective and overall subjective satisfaction). All usability attributes are specified through seven items according to a formal method for specifying operationally defined criteria for success (Whiteside, Bennett & Holtzblatt, 1988). Thus the information is conveyed on the way of determining usability and on whether the current interface design meets the specification criteria and how well it does. Evaluation of quantitative results showed an overall improvement in the range of 10-20% over the non-adaptive interface, thus leading to the conclusion that *AKBB* positively supports a more usable interaction.

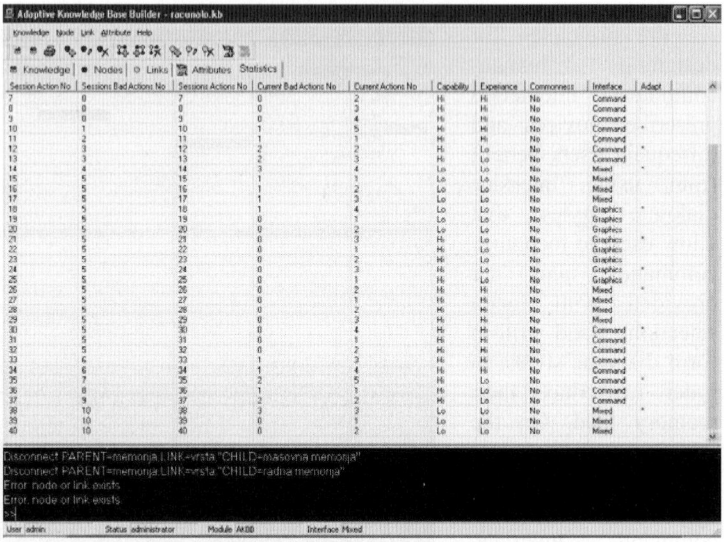

Figure 3: *AKBB Statistics* panel

4 Conclusion

Present day interactive applications make it an ever-increasing requirement to develop user interfaces that exhibit adaptivity. This is even more important for computerized educational systems that are not only to support a more efficient and effective interaction, but also to change the provided interaction styles on the basis of users' individual characteristics. The above reasoning is supported by considering the case of the *Adaptive Knowledge Base Builder*, *AKBB*, an arbitrary domain knowledge generator with adaptive interface. Issues regarding incorporation of adaptivity in *AKBB's* interface, along with its design and implementation are elaborated in this paper. Specifically, the interface is developed by considering the interaction record for each user, resulting in the provision of a selection of interaction styles. Although such mechanism addresses user changing knowledge and behavior over time, in order to augment the "precision" of adaptation the number of input parameters to be monitored is to be enlarged.

5 Acknowledgements

This work has been carried out within project 0036033 *Semantic Web as Information Infrastructure Enabler*, funded by the Ministry of Science and Technology of the Republic of Croatia.

6 References

Barton, M. (1995). Authoring Shells for Intelligent Tutoring Systems. *7th World Conference on Artificial Intelligence in Education*, Washington, USA, August 16-19. Retrieved June 7, 1998, from http://www.pitt.edu/~al/aied/barton.html

Benyon, D. (1998). Employing Intelligence at the Interface. Retrieved September 19, 1999, from http://www.dcs.napier.ac.uk/~dbenyon/IIT.pdf

Benyon, D. (1997). Intelligent Interface Technology. In S. Howard, J. Hammond & G. Lindgaard Eds.), *Human-Computer Interaction: INTERACT'97* (pp. 678-679). Chapman & Hall

Egan, D. (1988). Individual Differences in Human-Computer Interaction. In M. Helander (Ed.), *Handbook of Human-Computer Interaction* (pp. 543-568). Elsevier Science B.V. Publishers

Fleischmann, A. (2000). The Electronic Teacher: The Social Impact of Intelligent Tutoring Systems in Education. Retrieved May 17, 2001, from http://www.student.informatik.tu-darmstadt.de/~andreasf/inhalte/its.html

Granić, A. (2002). *Foundation of Adaptive Interfaces for Computerized Educational Systems.* Ph.D. Faculty of Electrical Engineering and Computing, University of Zagreb, Croatia (in Croatian)

Granić, A., & Glavinić, V. (2002a). User Interface Specification Issues for Computerized Educational Systems. 24th International Conference on *Information Technology Interfaces ITI 2002*, Pula, Croatia, June 24-27, 173-178.

Granić, A., & Glavinić, V. (2002b). Usability Evaluation Issues for Computerized Educational Systems. 11th Mediterranean Electrotechnical Conference *MEleCon 2002,* Cairo, Egypt, May 7-9, 558-562.

Intelligent Interfaces Special Interest Group, IISIG. (1998). Retrieved September 19, 1999, from http://www.dcs.napier.ac.uk/~dbenyon/iisig2.html

Maybury, M. (1999). Intelligent User Interfaces: An Introduction. HCI'99 Tutorial, *8th International Conference on Human-Computer Interaction HCI '99*, Munich, Germany, August 22-27

Miller, J. (1988). The Role of Human-Computer Interaction in Intelligent Tutoring System. In M. Polson & J. Richardson (Eds.), *Foundations of Intelligent Tutoring Systems* (pp. 143-189). Lawrence Erlbaum Associates Publishers, Hillsdale, NJ

Murray, T. (1999). Authoring Intelligent Tutoring Systems: An Analysis of the State of the Art. *International Journal of Artificial Intelligence in Education*, 10, 98-129.

Karagiannidis, Ch., & Stephanidis, C. (1998). Run-Time Adaptation in Intelligent User Interfaces: Automatic, Iterative Decision Making with Feedback. Retrieved September 15, 1999, from http://www.dfki.de/etai/articles/stephanidis-jan-98/paper.html

Whiteside, J., Bennett, J., & Holtzblatt, K. (1988). Usability Engineering: Our Experience and Evolution. In M. Helander (Ed.), *Handbook of Human-Computer Interaction* (pp. 791-817). Elsevier North-Holland, Amsterdam

Model of Intention Inference Using Bayesian Network

Naoki Hatakeyama, Kazuo Furuta and Keiichi Nakata

Dept. of Quantum Engineering and Systems Science, The University of Tokyo
7-3-1 Hongo, Bunkyo-ku, Tokyo 113-8656, Japan
hatake@cse.k.u-tokyo.ac.jp

Abstract

While human ability has not changed within limitations, machine systems have become more complicated in these years. The human therefore has developed automation systems, but automation sometimes causes human errors in some situations. We will need an intent inference method in order to solve this problem, because intent inference is an important element for presenting system information in an appropriate manner. In this paper, we propose an intent inference method, which is based on the whole human cognitive process, and we use a Bayesian network as an inference engine. We made experiments to extract knowledge that is needed to construct the network and compared the result with simulation.

1 Introduction

While machine systems have become more complicated in these years, human ability of recognition, action, and so on have not changed within limitations. The human therefore has developed automation systems in order to control such complicated systems. Human errors, however, are caused by automation, because automatic process is hidden from humans. Development of technique which can properly present information of automatic process will be needed and a method of human intent inference is important for the purpose. In this study we proposed an intent inference model that contains not only the intent inference process of plan recognition but also that of state recognition. Furthermore, we applied a Bayesian network to this model as an inference engine and evaluated how to exploit Bayesian network as a human inference model.

2 Human Cognitive Process

We suppose that somebody observes another person and infers his/her intention. In this case the observer can use the information on perception and action of the observed person in order to infer observed person's intention. Figure 1 indicates the whole process of human cognition. In this figure a person perceives information from the environment, identifies circumstances, decides his/her intention and performs some action. The items that the observed perceived and acted on are the only available information for the observer. We can divide this process into two parts. The process on the left side is called recognition process and that on the right side is called action process. In intention inference these two processes contribute to inference of human cognitive process. In this case , the information that the observer can use is the perception and the action of the observed. The observer infers the person's intention based on these information. We used Bayesian network as an inference engine.

3 Human Knowledge

Next we need to model human knowledge. It is known that an expert has well-designed knowledge composed hierarchically. Therefore we modelled such knowledge by using state hierarchy that is constructed from state nodes and links. Each state node represents an abstract concept and a link between two state nodes represents any relationship. In state hierarchy, the above state stands for a general concept and the below state a specific concept. In other words occurrence of the below state supports occurrence of the above. According to this relation we can get the graph that shows causality of state occurrence. In this graph each state has some symptoms that will be observed when the state occurs. In our model links between some child state and its parent represent OR relation, and those between a symptom and its parents noisy-OR relation. When S and X_i represent a state and a symptom respectively, noisy-OR relation is shown by the next equations.

$$P(-S \mid -X_1, \cdots, -X_n) = 1 \quad , \tag{1}$$

$$P(-S \mid X_1, \cdots, X_i, \cdots, X_n) = \prod_{X_i = true} Q_i \quad , \tag{2}$$

$$P(+S \mid X_1, \cdots, X_n) = 1 - P(-S \mid X_1, \cdots, X_n) \quad , \tag{3}$$

where "+" represents true, and "–" false. Figure 2 shows an example of state hierarchy on the abnormal state identification in a PWR power plant.

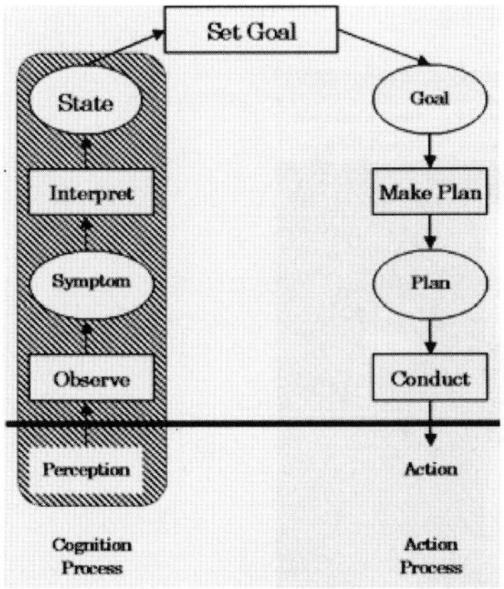

Figure1: Human Cognitive Process

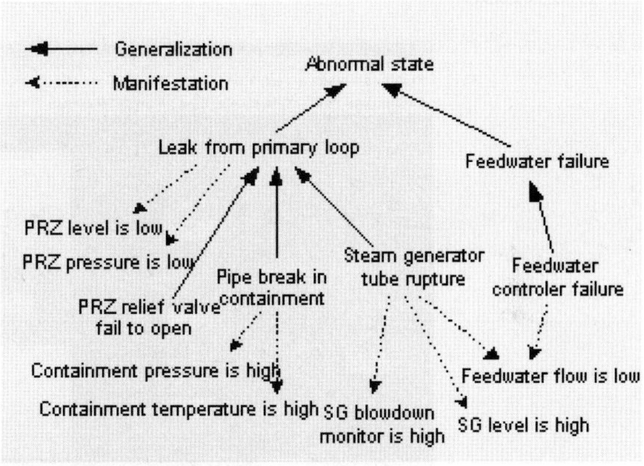

Figure 2: Knowledge for State Recognition of PWR

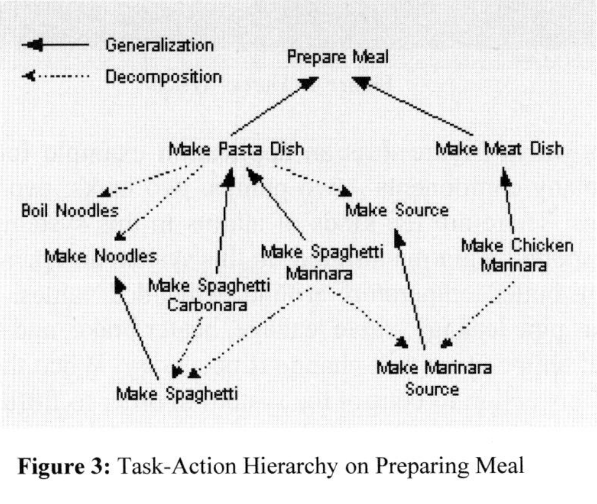

Figure 3: Task-Action Hierarchy on Preparing Meal

Now we consider the knowledge on the action process. Similar to the recognition process, the knowledge on the action process is represented to classify various situations into task hierarchy. Figure 3 shows an example of task hierarchy where a person prepares meal. Here a specific task is classified into more general one to form hierarchy. Moreover each task is related to component actions, which is represented by broken arrow in Figure 3. For simplicity we consider that the link between tasks represents OR relation and the link between task and action represents Noisy-OR relation in this study.

4 Bayesian Network

A Bayesian network is a directed acyclic graph, which consists of nodes and edges connecting between two nodes. In a Bayesian network, each node represents a probabilistic variable, and edge represents the conditional probability between the parent node and its child. The direction of an edge therefore represents the influence from the parent to the child as it is frequently based on causality. As we use a Bayesian network on human cognitive process, we suppose that the value of probability represents the extent of human belief. In this study, we use HUGIN application for calculation of a Bayesian network.

5 Test Case

5.1 DURESS

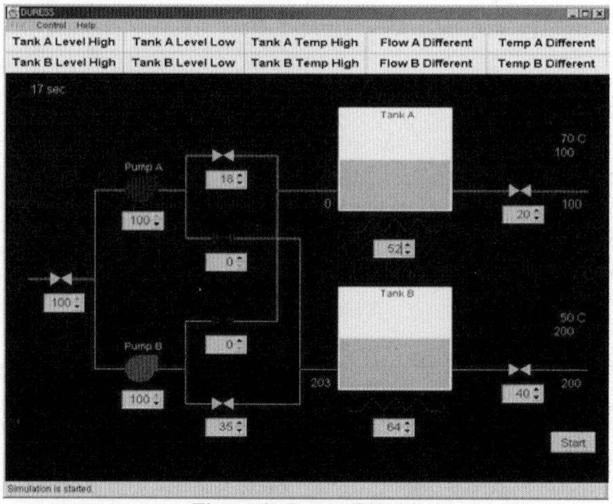

Figure 4: Duress System

We used DURESS shown in Figure 4 as an application example for validation of this model. DURESS consists of many components - two pumps, two tanks, two heaters, some pipes, some valves and some alarms. There are ten kinds of alarms in this system and when some abnormal event occurs, the corresponding alarms turn red. In this system an operator is requested to keep the outlet flow rate and the outlet temperature around demanded values. DURESS is susceptible to abnormal events such as pipe leakage, valve leakage, heater brake, and so on. Settings of the outlet flow rate and the outlet temperature will change if necessary. When the displayed values change, the operator must take any action to control the system in order to fulfil the demands or to identify the system state whether or not any abnormal event occurs. Here plant state identification is the recognition process and control task is the action process in Figure 2. We modelled these processes using a Bayesian network as an inference engine.

5.2 Modelling

We mainly defined four Bayesian networks for representing operator's knowledge: the first one for the normal plant state, the second one for abnormal events, the third one for controlling the outlet flow rate and the outlet temperature, and the last one for operational task and action of DURESS. Former three are represented by state hierarchy and symptoms in operator's knowledge, while the last one by task hierarchy and actions. We defined each of the above knowledge using a Bayesian network.

In terms of intent inference, the state hierarchy should be reconstructed in an appropriate manner. When a person identifies system state, he/she uses symptoms as evidential information of abnormal events. It suggests the way how system states should be classified according to the feature of symptoms as shown in Figure 5. We divide DURESS into three subsystems. The first part contains Valve 0, Pump A, B, and the related pipes. These components affect the whole elements on the downstream of the system. The second part contains Valve 1 to 4, Tank A, B, Heater A, B and the related pipes. Some operation to any component in this subsystem affects the elements of either one of the two outlets. We refer to the stream from Valve 1 or 3 to Valve 5 throughout Tank A as Line A, and the stream from Valve 2 or 4 to Valve 6 throughout Tank B as Line B. The third part contains Valve 5, 6, and the related pipes. This subsystem only affects the outlet flow rate and the temperature on either of Line A or B.

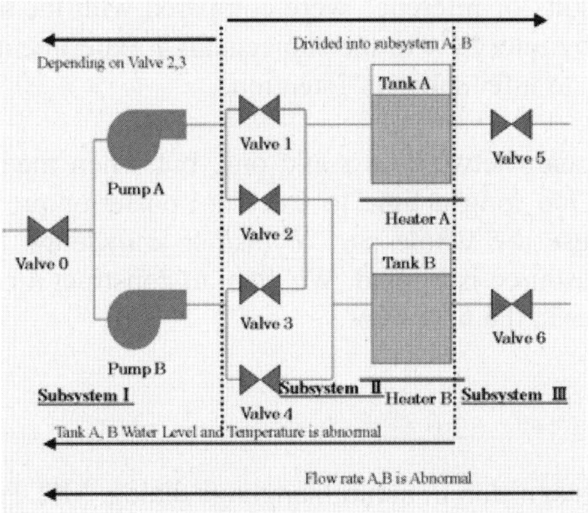

Figure 5: Classification of DURESS According to the Feature of Symptoms

This classification has an advantage that states can be classified according to symptoms. Firstly abnormal states are classified into three subsystems, and then each subsystem into leakage and the other component failures. Finally each type of failure is classified into elements of the corresponding components. Figure 6 shows the resultant Bayesian network of Figure 5.

Two Bayesian networks are constructed each for recognition and action process of a DURESS operator, and we need to combine the both networks into one Bayesian network, because human cognitive process is related between the recognition and action process as in Figure 1. States are connected with the corresponding tasks to get a combined Bayesian network, in which the whole cognitive process in operator's mind is represented. In this Bayesian network, the relation between states is modelled by OR relation, and that between states and symptoms or between tasks and actions by noisy-OR relation. When an observer got information from operator's perception or action, likelihood is set to a symptom node or an action node, and then intent inference is done by calculating posterior probabilities after observation over the whole network.

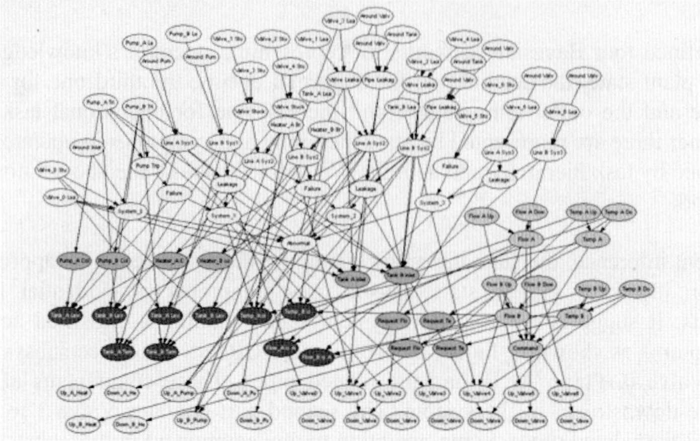

Figure 6: Bayesian Network Corresponding to Figure 5

6 Concluding Remarks

We confirmed the validity of this Bayesian network using some observation examples. Subjects operated DURESS, and results of inference were compared with the actual intention of subjects, which had been identified by interview to the subjects after experimental trials. That results show that this method is effective to infer subjects' intention. .

This time we used a Bayesian network as static one, but when many nodes are involved in a Bayesian network it takes too long a time to calculate posterior probabilities. In human intent inference a person never use the whole part of his/her knowledge but only the relevant part, because human cognitive resource is limited. We want to construct a necessary Bayesian network dynamically in some manner in the next step.

Reference

Carberry, S. (1990). Plan recognition in natural language dialog, MIT Press

Carberry, S., & Pope, W. (1993). Plan recognition strategies for language understanding. *Int. J. Man-Machine Studies*, 39, 529-577.

Hammer, J., & Small, R. (1995). An Intelligent Interface in An Associate System. *Human/ Technology Interaction in Complex System*, 7, 1-44.

Jensen, F. (1996). An Introduction to Bayesian Networks. Springer

Jordan, N. (1990). Allocation of functions between man and machine in automated systems. *Applied Psychology*, 47, 161-165.

Pearl, J. (1988). Probabilistic Reasoning In Intelligent Systems: Networks of Plausible Inference. Morgan Kaufmann.

Vicente, K.J., & Rasmussen, J. (1990). The Ecology of Human-Machine System II: Mediating Direct Perception in Complex Work Domains, *Ecological Psychology*, 2, 207-249.

User Centred Design Utilizing Sensory Analysis

Naotsune Hosono
Oki Consulting Solutions Co., Ltd.
Hosono903@oki.com

Hiromitsu Inoue
Chiba College of Health Science
cal32840@pop07.odn.ne.jp

Yutaka Tomita, Yoshikazu Yamamoto
Keio University
tomita@bio.keio.ac.jp,
yama@ics.keio.ac.jp

Abstract

This paper discusses User-Centred Design (UCD) utilizing a sensory analysis method applied to the initial stage of a new machine designed to meet users requirements. Because of the initial lack of a specific requirement balance, interface designers are required to measure and analyze the value of a user's comfort level. User requirements are observed from an environment aspect in context of use. This paper compared Japanese user requirements with user requirements from other nationalities with varied environmental backgrounds since there are difficulties in defining a general user category group. For the analysis, nine samples were selected as experimental Portable Information Terminals (PIT). The required measurement and analysis was accomplished by combining proposed Marble Method and Correspondence Analysis. In order to validate the analysis results, the correspondence analysis plot charts were shown to expert designers and project managers involved in the design and manufacturing process. It was concluded that the general users group characteristics and proposed method reflect the requirements at the survey and planning step, specification forming step and revision step of the process.

1. Introduction

"Memorable yet Invisible" are the recent user-computer interaction issues, hence the invisible to the user design process must build on meeting diverse user requirements fostering an intuitive, memorable and evolving relationship with the device and the manufacturer. Manufacturers must identify and address user requirements knowing users evolve from novices into experts. The design of a device means it is required to meet the dynamic needs of a user facilitates this invisible process in a memorable manner. In effect, the emergence of new technologies in personal computing allow for the design of smaller and more effective systems and devices that are memorable but not obtrusive with various users' satisfaction [1 & 2] .

2. The initial design with a UCD viewpoint

In June 1999 standards for User-centered design (UCD) were prepared by ISO/TC159 as an International Standard ISO 13407[3]. This standard suggests an approach to interactive system development that focuses specifically on making systems usable under that the product design is further defined in context-of-use scenarios. This paper discusses a sensory analysis method used to determine variable nationality requirements utilizing an experimental Portable Information Terminal (PIT) platform [4 & 5]. Because of generally scarce resources, PIT interface designers must tradeoff various elements of basic requirements. However, in the initial stages of development, designers may have little or no direct experience or reference products to draw upon in developing a specific machine to meet exacting requirements. Given the initial lack of a specific requirement balance, designers must measure and evaluate the value of a user's comfort level while keeping in mind that users gradually and unconsciously evolve from a novice into an expert.

3. Sensory analysis
3.1 Extract samples

The major usability evaluation items are by the reference of PIT features. Required elements of the PIT device include; a compact size, portability, high processing speed, low power consumption, inexpensive, ability to handle various communication needs, et cetera. Additional functions are derived from the standpoint of users specific needs. The eight major components of Display Type, Communications, Battery Life, PCMCIA card slots, Keyboard, Volume, Weight, and Price with their several parameters need to be considered when designing for each application. Nine typical samples were extracted with a combination of these components (Table 1), however Screen Size was fixed at 5-inches throughout the experiment.

Table 1: Extracted nine Samples

Samples	Display	Communication	Battery Life	PCMCIA slots	Keyboard	Volume/Weight	Price (Yen)
Anne	Black/White	No Wireless	20 Hours	1 Slot	On Display only	Cassette Case	50,000
Betty	Black/White	Wireless	20 Hours	1 Slot	On Display only	Cassette Case	62,000
Cathy	Black/White	Wireless	20 Hours	2 Slot	On Display only	Cassette Case	64,000
Dorothy	Black/White	Wireless	60 Hours	2 Slot	On Display only	Soft Cover Book	70,000
Emmy	Colour	Wireless	40 Hours	2 Slot	On Display only	Soft Cover Book	106,000
Francy	Black/White	No Wireless	60 Hours	1 Slot	Attached to Body	Soft Cover Book	62,000
Jane	Colour	No Wireless	40 Hours	2 Slot	Attached to Body	Video Tape Case	100,000
Mary	Black/White	Wireless	60 Hours	2 Slot	Attached to Body	Video Tape Case	76,000
Vicky	Colour	Wireless	40 Hours	2 Slot	Attached to Body	Video Tape Case	112,000

Table 2: Inertia of Japanese

Dimension	Proportion Explained	Cumulative Proportion
1	0.357	0.357
2	0.231	0.588
3	0.163	0.751
4	0.124	0.875
5	0.047	0.922
6	0.034	0.956
7	0.031	0.987
8	0.013	1.000

Table 3: Inertia of Overseas

Dimension	Proportion Explained	Cumulative Proportion
1	0.298	0.298
2	0.232	0.530
3	0.175	0.705
4	0.126	0.831
5	0.078	0.909
6	0.057	0.966
7	0.022	0.988
8	0.012	1.000

3.2 Evaluation Procedures and Data Analysis

The survey assessors consisted of 17 Japanese and 18 East Asian nationalities. In the preparation of this sensory evaluation, the required measurement and analysis are accomplished by utilizing Marble Method [6] and Correspondence Analysis (CA). Each assessor of all nationalities was given 25 marbles (tokens) to put and distribute them among the nine samples. They were permitted to put from zero up to 10 marbles for any of the samples among the nine. The advantage of this method is that it is possible to extract indecisive requirements from users rather than the ranking method or paired comparison method. The assessors were shown three typical mockup models (350g/350cc, 550g/550cc and 850g/850cc) before voting for their reference device. The collected data was analyzed using a statistical algorithm: CA and its SPSS package [7]. The advantage of the combination of Marble Method and CA is easier collection and analysis of data to forecast the relationships between assessors and samples since they can be plotted in the same plane. Additional advantage of this method is that it does not always require a large number of assessors.

4. Results and Considerations

A previous evaluation of the PIT samples in a health care environment, utilizing nurses, therapists and doctors, was structured for a device to be used in hospitals [8 & 9]. This study focuses on general user requirements by comparing Japanese and other nationality cultural viewpoints using sensory analysis. Comparisons of both groups of assessors results show that the first and second inertia are found to account for 53.0% (Table 1) and 58.8% (Table 2) respectively of the assessors preferences. The preferences among the nine models and assessors were plotted using CA (Figure 1 & 2).

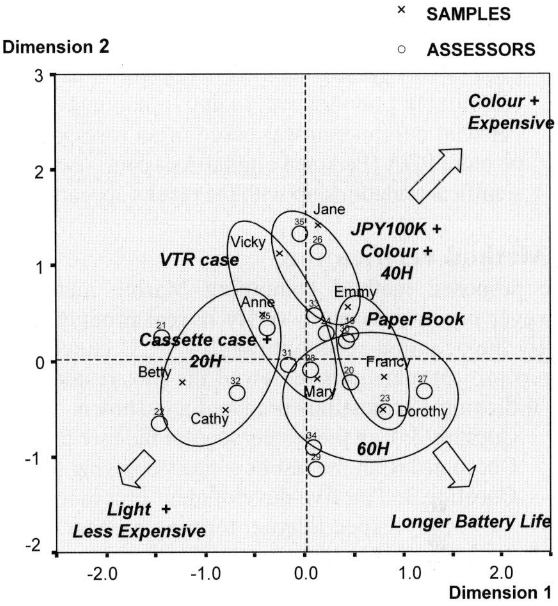

Figure 1: Japanese CA plot

In contrasting the requirements of different cultural backgrounds between the Japanese and the other nationality groups, the followings can be said:

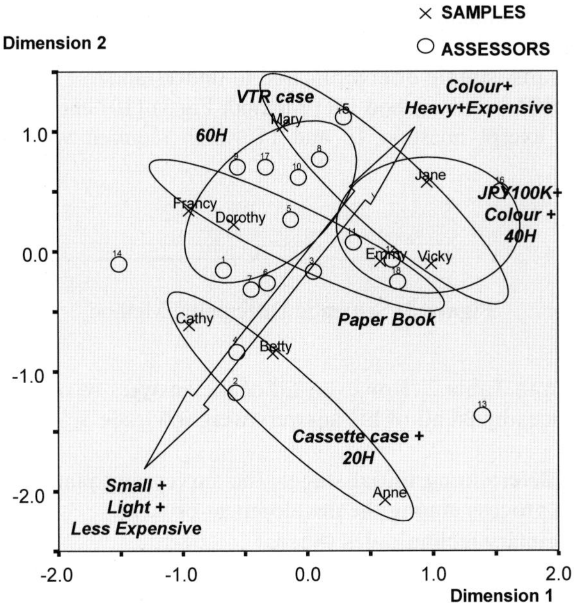

Figure 2: Overseas CA plot

- The evaluation criteria of the Japanese assessors tended to be similarly grouped in the centre of the plot whereas the other nationality assessors choices were scattered throughout the plot and varied individually.
- The Japanese tended to evaluate entire samples. One group focused on the importance of a compact and less expensive sample and others focused on the importance of a colour screen even though it was reasonably expensive. Another attached principal importance to Battery Life.
- Whereas the other nationality assessors are shown to have evaluated the samples by focusing on individual elements such as heavy/light,

large/small and expensive/less expensive considering price most significant.

- All questionnaires were written in the Japanese language for both groups in this survey. Further surveys are deemed necessary to assure equality in understanding the questions, minimizing potential errors due to language issues.
- Personal information was asked of all assessors before the evaluation. There were found to be more PDA (Personal Digital Assistant) users in the other nationalities group. However no significant relationship with the results appears.

5. Method verification

The proposed method combining Marble method and CA method is logically proved however it is necessary to verify in real product development processes. Hence, in order to verify the acquired results through a UCD concept process, three expert designers and four expert project managers reviewed the plot results (Figure 1 and 2). The two plot charts and the following 12 questions prepared and shown.

Question 1: Can the results suggest the user group and user needs?
Question 2: Can the results suggest applying the proprietary technologies (Seeds)?
Question 3: The five development processes are, the survey and planning process, the specification forming process, the experimental production process, the pre mass production process, and the mass production process. Can the results reflect on the survey and planning process?
Question 4: Can the results reflect on the specification forming process?
Question 5: Can the results reflect on the experimental production process?
Question 6: Can the results reflect on the pre mass production process?
Question 7: Can the results reflect on the mass production process?
Question 8: Can the comparison of the two plot charts reflect on the development process?
Question 9: Can the results reflect on the revision stage with feed back results (user requirements) reflecting the market product?
Question 10: Which process applies most suitably?
Question 11: Any other comments on the results?
Question 12: Is this method, and results, simple or complicated to understand?

All the questions are asked by Semantic Differential method (SD Method, Fig.3) [10] choosing from one (minimum) to five (maximum) except question 10 and 11 which required written comment.

The five grades of the SD Methods are:

- Grade 5: The results are quite effective.
- Grade 4: The results are fairly effective.
- Grade 3: Can not be said.
- Grade 2: The results are less effective.
- Grade 1: Can not be used.

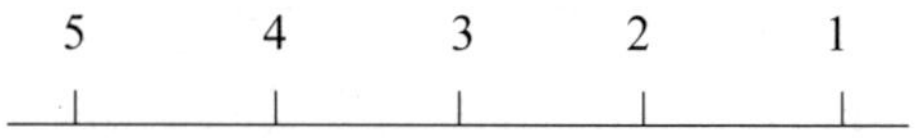

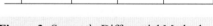

Figure 3: Semantic Differential Method

The expert's assessment results are shown in Table 2. However, all of the answers of one of the experts among seven were omitted since almost all of his answers were in Grade 3.

The summarized views of experts were:

- The results are the most effective in determining initial stages, the survey and planning process and the specification forming process among the development process.
- The results suggest applying the proprietary technologies (Seeds)
- The project managers set a higher valuation on the comparison results of the two plot charts rather than the designers.
- The results effectively apply at the revision stage.
- The results suggest the user group or user needs identification varies among assessors.

- Expressing the plots will require further consideration to allow experts to more easily understandable.

In psychological studies, measurement validity types are construct validity composed from content validity and concurrent validity [11]. Content validity is that the approximate truth of the conclusion that the operational accuracy reflects its construct. Content validity is essentially checking the operation against the relevant content domain for the construct. Concurrent validity is assessing the operational ability to distinguish between groups that it should theoretically be able to distinguish. In this case, the experts assessed the content validity and this leads to construct validity as a consequence.

Table 4: The assessment results by the experts.

	Q1 User group identification	Q2 Necessary seeds identification	Q3 For planning process	Q4 For specification forming process	Q5 For experimental production process	Q6 For pre-mass production process	Q7 For mass production process	Q8 To reflect two plot charts results	Q9 For revision stage	Q12 Understandable
Average	2.833	4.000	4.000	4.167	2.667	1.667	1.167	4.000	4.167	3.000
Standard Deviation	0.983	0.632	0.894	0.408	0.816	0.516	0.408	1.265	0.753	0.632

6. Conclusion

A method was proposed as a guide for designers at the initial stage of a new product design using ISO standards of user-centered design. To derive the general user requirements, Japanese and overseas nationality viewpoints were compared using sensory analysis. The correspondence analysis output results were validated by the experts and the designers and concluded this proposed method was effective in the determining initial and revision stages.

Reference

[1] Norman, D. A. (1999). The Invisible Computer. Cambridge: MIT Press.

[2] Nielsen, J. (1993). Usability Engineering. Cambridge: Academic Press, Inc.

[3] ISO13407. (1999). International Standard: Human-centered design processes for interactive systems. International Organization for Standardization (ISO).

[4] Knaster, B. (1994). Presenting Magic Cap. Reading: Addison-Wesley.

[5] Schneiderman, R. (1994). Wireless Personal Communications. Piscataway: IEEE Press.

[6] Inoue, H. (1993). Item Selection and its Importance. *23rd Sensory Evaluation Symposium.* Union of Japanese Scientists and Engineers.

[7] SPSS. (2000). Statistical Package for Social Science for Windows 11.0J Categories. Tokyo: SPSS Japan.

[8] Hosono, N. Inoue, H. Tomita,Y. (1999). Sensory Evaluation Method for Determining Portable Information Terminal Requirements for Nursing Care Application. *Proceeding of HCI'99,* 1, (pp.953-967).

[9] Hosono, N. Inoue, H. Tomita, Y. Yamamoto. Y. (2002). "Determining User Requirements Using Sensory Analysis". *HCI/EUPA2002.* Vol.16. No.2. (pp.130-133).

[10] Research Committee of Sensory Evaluation. (1973). Sensory Evaluation Handbook. Union of Japanese Scientists and Engineers.

[11] William M.K. (2002). Measurement Validity Types. Retrieved January 24, 2003 from http://trochim.human.cornell.edu/kb/measval.htm

Adaptive Fuzzy Inference Neural Network

Hitoshi Iyatomi and Masafumi Hagiwara

Department of Information and Computer Science, Keio University, Japan
iyatomi@soft.ics.keio.ac.jp

Abstract: An adaptive fuzzy inference neural network (AFINN) is proposed in this paper. It has self-construction ability, parameter estimation ability and rule extraction ability. The structure of AFINN is formed by the following 4 phases: 1) Initial rule creation by self-organized learning, 2) Selection of important input elements, 3) Identification of the network structure by self-organized learning and 4) Parameter estimation using LMS (least mean square) algorithm. When the number of input dimension is large, the conventional fuzzy systems often cannot handle the task correctly because the degree of each rule becomes too small. AFINN solves such a problem by modification of the learning and inference algorithm.

1 Introduction

Studies of fuzzy neural networks that combine both advantages of the fuzzy systems and the learning ability of the neural networks have been carried out.These techniques can alleviate the matter of fuzzy modelling by learning ability of neural networks and have been reported since around the beginning of 1990s (Amano, 1989). Fuzzy neural networks can be applied not only for simple pattern classification but also meaningful fuzzy if-then rules creation; therefore they can be put into practice for various applications. Nishina *et al.* proposed Fuzzy Inference Neural Network (FINN) (Nishna, 1997). FINN is a simple and effective fuzzy neural network that divides input-output data space and extracts fuzzy if-then rules automatically. Since the architecture and behavior of FINN are very applicable, it has been adopted as a basic component for image recognition and interpretation researches. However, its fuzzy modeling for the target task is not always sufficient. A lot of systems which aim at excellent fuzzy modeling and carry out input selection, rule creation and parameter estimation have been proposed (Lin, 1995), (Chiu, 1996), (Linkens, 2001). In these researches Linkens *et al.* reported effective input selection and rule creation method and showed the excellent results on their experiments. However, since these methods use fuzzy c-means (FCM) based algorithm they have to know the number of proper rules in advance. In addition, since the algorithms are complicated and their algorithms and results are presented only for single output system, they have difficulty if they are employed for many applications. By the way, many fuzzy neural network models have common problems derived from their fundamental algorithm. For example, the systems which use gradient decent method sometimes reduce the width of the membership function to negative during the learning. The systems which employ basic fuzzy inference theory make the degree of each rule extremely small and often make it underflow when the dimension of the task is large. In this paper, we propose a new adaptive fuzzy inference neural network (AFINN) that alleviates these shortcomings of the conventional models. AFINN carries out appropriate model construction and solves fuzzy-neuro specific problems by modification of the fuzzy inference and learning algorithm.

2 AFINN – Adaptive Fuzzy Inference Neural Network

To construct appropriate fuzzy model, the following problems should be solved: (1) Identification of the optimum number of rules, (2) Identification of the optimum element of inputs, (3) Identification of the system parameter. The proposed AFINN has the above features based on the simple and versatile FINN architecture.

2.1 Structure of AFINN

Fig.1 shows the structure of AFINN. It consists of two layers. One is the input-output (I/O) layer and another is the rule-layer. The I/O layer consists of the input-part and the output-part. Each node in the rule-layer represents one fuzzy rule. Weights from the input-part to the rule-layer and those from the rule-layer to the output-part are fully connected and they store fuzzy if-then rules. Membership functions as premise part are expressed in the weights.

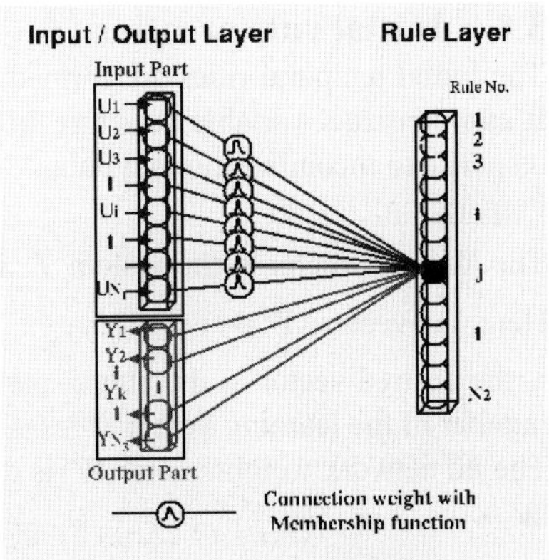

Fig.1. Structure of the AFINN

2.2 Behavior of AFINN

Suppose that the number of neurons in the input-part, which is equal to the dimension of the input data, is N_1, the number of rules is N_2, and the number of neurons in the output-part, which is equal to the dimension of the output data, is N_3.

The input data to the AFINN is expressed as follows: $\boldsymbol{U} = (u_1, u_2 ... u_i ... u_{N_1})^T$. (1)

The subscripts *i, j*, and *k* refer to the nodes in the input-part, those in the rule-layer, and those in the output-part, respectively. The bell shaped membership function represents the if-part of fuzzy rule, which is placed between the *i*th input node and the *j*th node in the rule-layer. The membership function is expressed as: $\mu_{ij} = \exp\left\{-\frac{(u_i - w_{ij})^2}{\sigma^2}\right\}$, (2)

where μ_{ij} is the membership value, w_{ij} is the center value of the membership function and σ_{ij} indicates the width of it. In the rule-layer, many conventional fuzzy systems calculate the degree of the rule by selecting minimum membership value or multiplying them. These calculations, however, often tend to make ρ extremely small and sometimes they cause underflow when the dimension of the task is large. In such a situation, the learning and inference cannot be proceeded correctly. In order to solve such problems, AFINN calculates the degree of the *j*th rule ρ_j as:

$$\rho_j = \Sigma_i^{N1} \mu_{ij}^{1/N_{adj}} . \quad (3)$$

Here, N_{adj} is the compensated factor relating to the input dimension. Then, the inference result of the *k*th node in the output-part is calculated by the following equation.

$$\hat{y}_k = \frac{\Sigma_j^{N2}(w_{jk}\rho_j)}{\Sigma_j^{N2}\rho_j} , \quad (4)$$

where w_{jk} is the weight between the *j*th node in the rule-layer and the *k*th node in the output-part. The w_{jk} corresponds to the estimated value of the *j*th rule for the *k*th node in the output-part.

The logical form of the fuzzy inference if-then rules is given such as:

If u_1 is $\tilde{w}_{1j}$, u_2 is $\tilde{w}_{2j} \ldots u_i$ is $\tilde{w}_{ij} \ldots u_{N_1}$ is $\tilde{w}_{N_1 j}$ **Then** y_k is w_{jk} .

where $\tilde{w}_{ij}$ means the value near w_{ij}. It should be noted here that it depends on the value of σ_{ij} .

3 Model construction of AFINN

AFINN has four phases to achieve appropriate input selection, rule creation and parameter adjustment.

3.1 Initial rule creation

The initial temporal rules are formed in this phase by adaptive self-organized learning algorithm. It can construct variable structure in which the number of rules can be changed dynamically in response to incoming training data.

Preparation:

The *l*th input vector to the system $\boldsymbol{I}^l$ is defined as $\boldsymbol{I}^l = [\boldsymbol{U}^l, \boldsymbol{Y}^l]^T$. (5)

Here, the vector: $\boldsymbol{Y}^l = (y_1{}^l, y_2{}^l, ..., y_k{}^l, ..., y_{N_3}{}^l)^T$ (6)

is the desired vector in the output-part of the I/O layer. The suffix *l* of $\boldsymbol{I}$ indicates the consecutive number of the learning vector $(1 \le l \le L)$.

The *j*th network weight vector $\boldsymbol{W}_j$ is defined to concatenate all the weights relating *j*th rule.

$\boldsymbol{W}_j = (w_{1j}, ...w_{ij}, ...w_{N_1 j}, w_{j1}, ...w_{jk}, ...w_{jN_3})^T$. (7)

The number of current rule N_r is set to 0, the number of updated iterations for the *j*th rule defined as M_j is set to 0.

Step1: The first input vector $\boldsymbol{I}^1$ is used as the first rule $\boldsymbol{W}^1$ ($\boldsymbol{W}^1 = \boldsymbol{I}^1$), N_r is set to 1 and *l*=*l*+1.

Step2: The winner rule j^* is selected from existing N_r rules where Euclidian distance between the *l*th input vector $\boldsymbol{I}^l$ and the *j*th rule $\boldsymbol{W}_j$ is minimum.

Step3: If the distance calculated in step2 is smaller than the threshold ξ_{SELF}, the weights of winner rule $\boldsymbol{W}_j{}^*$ is updated and $M_j{}^* = M_j{}^* + 1$. Otherwise this *l*th input vector is registered as the new weight vector and $N_r = N_r + 1$.

$\boldsymbol{W}_{j^*}(t+1) = \boldsymbol{W}_{j^*}(t) + \varepsilon_{SELF}^{j^*}(\boldsymbol{I}^l - \boldsymbol{W}_j{}^*(t))$ (8)

;where $\left\| \boldsymbol{W}_j{}^* - \boldsymbol{I}^l \right\| \le \xi_{SELF}$, $\quad \boldsymbol{W}_{Nr+1} = \boldsymbol{I}^l$;elsewhere. (9)

Here, $\varepsilon_{SELF}^{j^*}$ is the learning factor defined as $\varepsilon_{SELF}^{j^*} = \varepsilon_{SELFinit}\{(L - M_j)/L\}^2$. (10)

Where $\varepsilon_{SELFinit}$ is the initial learning constant.

Step4: If *l*=*L*: This self-organized learning phase has finished and initial fuzzy rules of AFINN have been created.

else :*l*=*l*+*1*; continue step2 and step3 during *l*<*L*.

3.2 Input selection

It is necessary to select adequate input elements in order to construct suitable fuzzy model. AFINN refers to the procedure of Linkens (Linkens, 2001) that deals with multi-input-single-output system and extends this algorithm for multi-input-multi-output system.

This procedure is composed of two steps.

Step1: In the first step, importance of each input element for the outputs is estimated and elimination of unnecessary inputs is carried out.

First, the *l*th input importance value is calculated using network initial weights.

$$z_{ik}^l = \frac{\sum_j^{N_2} \mu_{ij} w_{ij}}{\sum_n^{N_2} \mu_{ij}}. \quad (11)$$

Where, μ_{ij} is the membership value calculated in equation (2).

Next, the integrated input importance value from the *i*th input to the *k*th output F_{ik} is calculated as

$F_{ik} = R_{ik} / R_k$, (12)

where $R_{ik} = \max_l (z_{ik}^l) - \min_l (z_{ik}^l)$ and $R_k = \max_l (R_{ik})$.

Finally, the syntactic input importance value is calculated as $F_i = \sum_k^{N_3} F_{ik}$. (13)

If this value is smaller than the threshold, this input is regarded as unnecessary and eliminated.

Step2: Then the correlations among remaining inputs are calculated. Based on these values, several groups in which correlations among each input element exceed the threshold each other are created. In each group, one element which has the highest F_i is selected as the representative input.

3.3 Model identification

After the efficient input elements are selected, adaptive self-organized learning explained in **3.1** is carried out again to determine the structure of AFINN such as the number of rules and initial weights.

3.4 Parameter estimation

In this parameter estimation phase, the mean square error between outputs of the network and the desired signals is reduced to adjust the parameters by LMS algorithm. According to the LMS learning principle, the estimation value of the *j*th rule node is updated as:

$$w_{jk}(t+1) = w_{jk}(t) + \varepsilon_{LMS}(y_k - \hat{y}_k)\frac{\rho_j}{\Sigma_n^{N2}\rho_n}. \quad (14)$$

The center value and the width of the *j*th rule are also updated as:

$$w_{ij}(t+1) = w_{ij}(t) + \varepsilon_{LMS}\sum_k^{N3}(y_k - \hat{y}_k) \times\left(\frac{w_{jk}\Sigma_n^{N2}\rho_n - \Sigma_n^{N2}(w_{nk}\rho_n)}{\left(\Sigma_n^{N2}\rho_n\right)^2}\right) \times\frac{1}{N_{adj}}\rho_j\frac{2(u_i - w_{ij})}{\sigma_{ij}^2}, \quad (15)$$

$$\sigma_{ij}(t+1) = \sigma_{ij}(t) + \varepsilon_{LMS}\sum_k^{N3}(y_k - \hat{y}_k) \times\left(\frac{w_{jk}\Sigma_n^{N2}\rho_n - \Sigma_n^{N2}(w_{nk}\rho_n)}{\left(\Sigma_n^{N2}\rho_n\right)^2}\right) \times\frac{1}{N_{adj}}\rho_j\frac{2(u_i - w_{ij})^2}{\sigma_{ij}^3}. \quad (16)$$

These gradient decent based algorithms sometimes make the width of membership function negative during the learning. One solution for this phenomenon is introduction of the minimum threshold for the width of the membership function. However, this sometimes obstructs the learning of the systems in which min-max fuzzy inference algorithm $(\rho_j = \min \mu_{ij})$ is used to calculate the degree of the rule. This algorithm updates only weights associated with the selected rule *j*, so when the *j*th rule meets this threshold, the learning is seriously affected. The systems which use multiplied algorithm $(\rho_j = \prod \mu_{ij})$ hardly meet this problem, however, this makes ρ extremely small and often fail the inference when the dimension of the task is large as mentioned before. AFINN uses equation (3) to carry out learning and inference properly and introduces the limitation $\xi_{\min\sigma}$ to keep the created rules meaningful.

4 Experimental Results

We used the following examples to verify the effectiveness of the proposed AFINN.

4.1 Formula approximation and model construction

We tested a non-linear 5-input-3-output system including one redundant input, u_5, in order to check the model construction ability of AFINN.

$$y_1 = u_1^3 + 3u_2 - 4\sin u_3$$
$$y_2 = \cos^2 u_1 - \tan^3 u_2 + u_4 \quad (17)$$
$$y3 = \sqrt{u_1 + u_2 + u_4}$$

The results of model construction are shown in Table 1. From this table, we can see that the indispensable 4 elements are correctly selected and that redundant input is not selected in all cases. Fig.2 compares the

Table 1. Results of the model construction

ξ_{SELF} (%)	5.0	7.0	10.0	12.0	15.0	20.0
initial # of rule	815	509	208	127	61	23
selected input	u_1, u_2, u_3, u_4					
# of rules (N2)	527	254	80	44	21	8

evaluation results on the proposed AFINN and the conventional models. The proposed learning algorithm shows superior error convergence ability and evaluation results.

4.2 Speech recognition example

Second, we chose speech recognition example to examine the AFINN can handle the task whose input dimension is over hundreds correctly. We used ISOLET (Isolated letter speech recognition) database from UCI Machine Learning Repository. This database can be used as 617-input 26-output classification task. When ξ_{SELF} =12.0%, 296 elements were selected from 617 inputs and 450 rules were created. The classification accuracy for the evaluation data was 93.53%. Even though this accuracy is a little worse than the best result by the conventional back propagation algorithm (95.9%), the proposed system has several merits of fuzzy systems: linguistic rules extraction and their maintenance are possible. Fig.3 compares the error reduction and the evaluation results between the conventional model and the proposed AFINN. The AFINN shows the superior results on error convergence and evaluation. It should be noted here that the system using traditional learning algorithms caused underflow of the degree of rules and could not estimate the system parameters under such high input dimension.

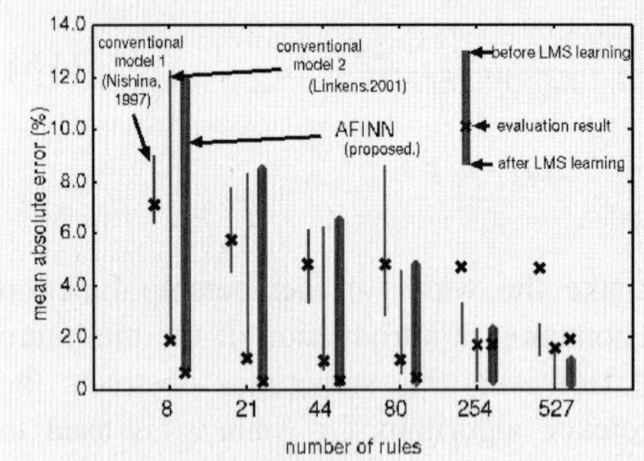

Fig.2 Comparison of the error convergence and evaluation results

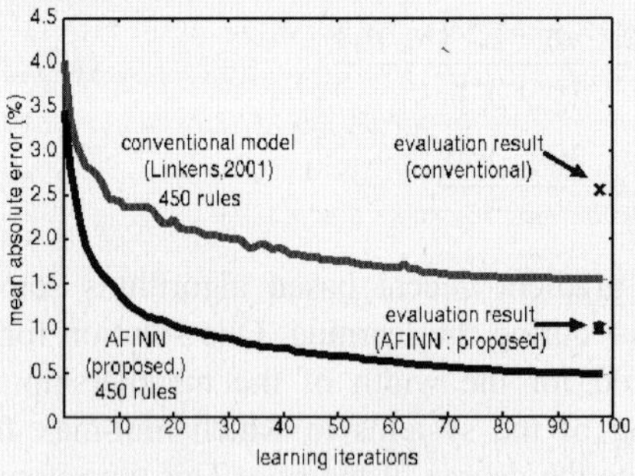

Fig.3 Comparison of the learning and evaluation for speech recognition

5 Conclusions

In this paper we have proposed adaptive fuzzy inference neural network (AFINN). It has many desirable features such as simple structure, adequate self-construction, appropriate parameter adjusting and rule creation ability. AFINN carries out efficient input selection, rule creation and parameter estimation to create appropriate fuzzy model and solves fuzzy-neuro specific problems by modification of the fuzzy inference and learning algorithm. As AFINN alleviates shortcomings of the conventional fuzzy neural network models, it can handle over hundreds dimensional tasks correctly. On account of its simple and applicable structure, AFINN can be used as a basic component for many applications.

References

A.Amano, T.Arisuka (1989). On the use of neural networks and fuzzy logicin speech recognition. *Proc. Int. Joint Conf. Neural Networks. pp.301-305.*

Y.Lin, G.A.Cunningham (1995). A new approach to fuzzy-neural system modeling. *IEEE Trans. Fuzzy Systems, Vol.3, pp.190-197.*

S.L.Chiu (1996). Selecting input variables for fuzzy models. *Journal of Intelligent Fuzzy Systems, Vol.4, pp.243-256.*

T.Nishina, M.Hagiwara (1997). Fuzzy Inference Neural Network. *Neurocomputing 14, pp.223-239.*

M-Y.Chen, D.A.Linkens. (2001). A Systematic Neuro-Fuzzy Modeling Framework With Application to Material Property Prediction. *IEEE Trans. on SMC, Vol.31, No.5, pp.781-790.*

Adaptive Help for e-mail Users

Katerina Kabassi

Department of Informatics,
University of Piraeus
80, Karaoli and Dimitriou St.,
Piraeus 185 34, Greece
kkabassi@unipi.gr

Maria Virvou

Department of Informatics,
University of Piraeus
80, Karaoli and Dimitriou St.,
Piraeus 185 34, Greece
mvirvou@unipi.gr

Abstract

This paper describes a graphical user interface that provides intelligent help to the users of an e-mailing system. The graphical user interface is called Intelligent Mailer (I-Mailer). I-Mailer monitors users while they work; in case a user is believed to have made a mistake, the system intervenes automatically and offers advice. In order to provide individualised help, the system keeps information about each user interacting with the system. This information is maintained centrally on a server based on Web Services. The main characteristic of Web Services is their strong reliance on web standards when they interact with the applications that invoke them. The Server keeps user models for every individual interacting with the client application. In case the Server has not gathered enough information about a particular user, it consults a library of stereotypes and acquires some default assumptions about the user.

1 Introduction

As the number of users that use the Internet in order to process their e-mail increases, the insufficient ability of e-mail clients to satisfy the heterogeneous needs of many users becomes more apparent. Novice users may encounter difficulties due to lack of experience and expert users may face problems due to carelessness or possible tiredness. A remedy for the negative effects of the traditional 'one-size-fits-all' approach is to develop systems with an ability to adapt their behaviour to the goals, tasks, interests and other features of individual users and groups of users (Brusilovsky & Maybury, 2002).

Therefore, a lot of research energy has been put into the development of intelligent user interfaces that could help users organise their mailbox (Payne & Edwards 1997) or help them by performing tasks on their behalf (Lashkari, Metral & Maes, 1994). However, a main drawback of these approaches is that they do not aim at helping users perform the tasks by themselves and, therefore, learn from that experience. This goal however, is addressed by the research area of intelligent help systems. Examples of help systems are UC system (Wilensky et al., 2000), RESCUER (Virvou & du Boulay, 1999), CHORIS (Tyler, Schlossberg, Gargan Jr., Cook & Sullivan, 1991) and Office Assistant (Horvitz, Breese, Heckerman, Hovel & Rommelse, 1998).

For the purposes of helping users organise their mailbox by assisting them perform the tasks by themselves, we have developed a Graphical User Interface (GUI) that offers automatic assistance to users in problematic situations. The system developed is called Intelligent Mailer (I-Mailer). I-Mailer constantly reasons about every user action and provides spontaneous advice in case an action contradicts the user's hypothesised intentions. Hypotheses are generated based on a simulator of human error generations and a limited goal recognition mechanism that is used by the system to improve control. However, the main feature of the system is its ability to adapt its interaction to the needs and characteristics of each individual user.

Many researchers agree that the development of effective assistance systems should be based on a user model which keeps track of what the user is doing (e.g Matthews, Pharr, Biswas & Neelakandan, 2000). However, one problem with user modelling is that the user model of each individual is usually device-dependent. This means that the user model is built and maintained on a particular device in a computer lab. This approach assumes that users always use the same device. However, in a real computer-lab environment users often receive and send messages from different devices. A solution to this problem may be achieved with the incorporation of a user model based on Web Services into the intelligent help system.

2 Intelligent Mailer

Intelligent Mailer (I-Mailer) is an Intelligent Graphical User Interface that works in a similar way as a standard e-mail client but it also incorporates intelligence. I-Mailer's main aim is to help users while sending and receiving e-mail messages and help them organise their mailbox. Therefore, I-Mailer monitors users during their interaction with the system and reasons about all users' actions. In case it suspects that a user may have been mistaken with respect to his/her hypothesised intentions, it provides spontaneous advice. Otherwise, the action is executed normally.

In particular, every time a user issues an action, I-Mailer reasons about it and categorises it in one of four categories, namely "expected", "neutral", "suspect", "erroneous". A command is categorised as expected if it is compatible with the user's hypothesised goals. It is considered suspect if it contradicts the system's hypotheses about the user's goals and erroneous if the command is wrong with respect to the user interface formalities. The command is considered neutral if it cannot be assigned to one of the former categories. Finally, an action is categorised as erroneous if it is wrong with respect to the user interface formalities.

If the action is categorised as expected or neutral, it is executed normally. However, if the action is categorised as suspect or erroneous, the system tries to generate alternative actions that the user may have meant to issue instead of the one issued. Therefore, the action issued is transformed so that similar alternatives can be found, which would not be suspect or erroneous. However, since the transformed action has to fit better in the context of the user's goals, the system reasons about every alternative action generated from the transforms. As a result, each transformed action is categorised in one of the four categories in a similar way as the actual command issued by the user. Finally, only expected actions are selected. In case, the system cannot find any action that is considered better than the action issued by the user, then the user's action is executed normally.

An example of a user's interaction with the system is the following: In an attempt to reorganise his/her mailbox, the user deleted the contents of the folder 'Fifth'. Then s/he accidentally tried to delete the folder 'Fifty' as well. However, the system found the particular action "suspect", because the user had previously stored in that folder a lot of e-mail messages and his/her action would have resulted in losing valuable data. Therefore, the system asked the user whether s/he really meant to delete the folder 'Fifth' instead of "Fifty" for two main reasons:

1. The folder 'Fifth' had been emptied whereas the folder 'Fifty' had not.

2. ‘Fifth’ was very similar to ‘Fifty’, therefore, there could have been a mistake.

3 Web Services for User Modelling

Web services introduced a new model on the Web in which information exchange is conducted much more conveniently, reliably and easily. Web Services interact with the applications that invoke them, using web standards such as WSDL (Web Service Definition Language) (Christensen, Curbera, Meredith & Weerawarana, 2001), SOAP (Simple Object Access Protocol) (Box et al., 2000) and UDDI (Universal Description, Discovery and Integration) (UDDI, 2001). Basing user modelling on web standards has the advantage of enabling the dynamic integration of applications distributed over the Internet, independently of their underlying platforms.

Until recently, user modelling through the Internet has been mainly based on a client-server model. Examples of applications or projects that follow a client-server architecture are Casper (Smyth, Bradley & Rafter, 2002), PACF (Machado, Martins & Paiva, 1999) and Adaptive Information Server (Billsus & Pazzani, 2000). However, in this approach, the developer has to create his/her own communication protocol and the clients may experience problems in receiving data from the Server. For example, if a user works both at home and at work, his/her user model may not function as expected, because the client at work may be behind a firewall that does not allow the user modelling server’s port to pass through. On the other hand, Web Services, follow the XML protocol for sharing data, thus making this data readable via virtually every machine. In addition, Web Services rely on the Hypertext Transfer Protocol and thus gain the advantage of being able to flow through most security systems (Firewalls, Proxy Servers, etc).

The operation of Web Services in I-Mailer is quite simple. I-Mailer incorporates a Server that stores and updates individual user models through Web Services. This Server is called Web Service User Model (WebSUM). The application makes a request to WebSUM and WebSUM returns a string containing the response to the application’s request. Based on this, I-Mailer sends the username and password of the user to WebSUM and WebSUM is responsible for finding the user model and sending this information to the client that requested it. The user model is updated with information gathered locally, through the user’s interaction with the application. In particular, the information, which is acquired locally, is sent to WebSUM so that the user model is updated there. In this way, WebSUM keeps track of intentions and possible confusions of each individual user. This information is available to the application irrespective of the computer where it is running.

Every time the user interacts with I-Mailer, the system requests information from the Web Server that contains the particular user model. WebSUM retrieves the request and returns the result to the client application. WebSUM tries to find the information requested in the individual history of the user. However, there are cases where WebSUM does not have adequate information about the user. In such cases, it consults a library of stereotypes that it maintains. Stereotypes are used to provide default assumptions about users until the user model acquires sufficient information about each individual user. Indeed as Rich (1989) points out a stereotype represents information that enables the system to make a large number of plausible inferences on the basis of a substantially smaller number of observations; these inferences must, however, be treated as defaults, which can be overridden by specific observations. Therefore, stereotypes are used in WebSUM only for capturing the initial impression of a user.

A stereotype is activated during the first interaction of the user with the system. The user has to answer some questions about his/her believed level of expertise, his/her previous experience, his/her knowledge in related topics such as file-store manipulation etc. The users are classified into one of three major classes according to their level of expertise, namely, novice, intermediate and expert. Furthermore, after the user has executed a satisfactory number of commands, the system can also draw inferences about the user's proneness in making mistakes due to carelessness. Therefore, there are two more stereotypes that divide users into two groups, careless and careful. After a stereotype has been activated, the system makes some default assumptions about users' possible errors and can provide some kind of advice.

In the beginning, information is acquired only by the stereotype. However, the system is also constantly collecting information about a particular user's behaviour and errors and informs the individual user model of the user. As the system collects more and more evidence about a user, information is acquired in part by the stereotype and in part from the individual user model. The percentage of information acquired by the stereotype diminishes as the percentage of acquisition by the individual user model increases.

In case a conflict appears, the system always lays more weight on the information acquired from the individual user model. For example, if the stereotype supports that a user's most common error is in the use of a certain command and the individual user model supports that the particular user's most common error is due to a misconception concerning the structure of folders, then the system will favour the view proposed by the individual user model.

4 Conclusions

In this paper we described I-Mailer, an intelligent graphical user interface for an e-mailing program. I-Mailer reasons about every user action and produces advice in cases where the user is having problems with his/her interaction with the system. In order to identify when the user needs help, the system makes hypotheses about every user's intentions. Hypotheses generation is based on a limited goal recognition mechanism.

For the provision of intelligent and individualised help, I-Mailer depends on its WebSUM user modelling component. This component allows for centralised maintenance of information about the user, allowing its access through the Internet from virtually anywhere. The main characteristic of Web Services is that they interact with the applications that invoke them, using web standards. Basing user modelling on web standards has the advantage of enabling the dynamic integration of applications distributed over the Internet, independently of their underlying platforms.

5 References

Billsus D. & Pazzani M.J. (2000). User Modeling for Adaptive News Access. *User Modeling & User-Adapted Interaction*, 10, 147-180.

Box, D., Ehnebuske, D., Kakivaya, G., Layman, A., Mendelsohn, N., Nielsen, H., Thatte, S., & Winer, D. (2000) Simple Object Access Protocol (SOAP) 1.1, *W3C Note*, http://www.w3.org/TR/SOAP.

Brusilovsky, P. & Maybury, M. T. (2002). From Adaptive Hypermedia to the Adaptive Web. *Communications of the ACM*, 45 (5), 31-33.

Christensen, E., Curbera, F., Meredith, G. & Weerawarana S. (eds.) (2001) Web Services Description Language (WSDL) 1.1, *W3C Note*, http://www.w3.org/2001/NOTE-wsdl-20010315

Horvitz, E., Breese, J., Heckerman, D., Hovel D., & Rommelse K. (1998). The Lumiere Project: Bayesian User Modeling for Inferring the Goals and Needs of Software Users. *Proceedings of the fourteenth Conference on Uncertainty in Artificial Intelligence* (pp. 256-265), Morgan Kaufmann: San Francisco.

Lashkari, Y., Metral, M., & Maes, P. (1994). Collaborative Interface Agents. *Proceedings of the 12th National Conference on Artificial Intelligence*, 444-449.

Matchado, I., Martins, A., & Paiva, A. (1999). One for All and All in One - A learner modelling server in a multi-agent platform. In J. Kay (ed.) *Proceedings of the Seventh International Conference on User Modelling* (pp. 211-221). Springer Verlag.

Matthews, M., Pharr, W., Biswas, G., & Neelakandan H. (2000). USCSH: An Active Intelligent Assistance System. In Hegner St.J., Mc.Kevitt P., Norvig P., and Wilensky R. (Eds.) *Artificial Intelligence Review, Intelligent Help Systems For Unix*, 14, 121-141.

Payne, T. R., & Edwards R. (1997). Interface Agents that Learn: An Investigation of Learning Issues in a Mail Agent Interface. *Applied Artificial Intelligence*. 11 (1), 1-32.

Smyth, B., Bradley, K., & Rafter, R. (2002). Personalisation Techniques for Online Recruitment Services. *Communications of ACM*, 45(5), 39-40.

Tyler, S.W., Schlossberg, J.L., Gargan Jr., R.A., Cook, L.K., & Sullivan, J.W. (1991). An Intelligent Interface Architecture For Adaptive Interaction, In *Intelligent User Interface* (pp. 85-109), ACM Press, New York. Addison-Wesley Publishing Company.

Universal Description, Discovery and Integration (UDDI 2001) Version 2.0 Specification, http://uddi.org/specification.html.

Virvou, M., & Du Boulay, B. (1999). Human Plausible Reasoning for Intelligent Help. *User Modeling and User-Adapted Interaction*, 9, 321-375.

Virvou, M. & Kabassi, K. (2002). Reasoning about Users' Actions in a Graphical User Interface. *Human-Computer Interaction*, 17(4), 369-399.

Wilensky, R., Chin, D.N., Luria, M., Martin, J., Mayfield, J., & Wu, D. (2000) The Berkeley UNIX Consultant Project. *Artificial Intelligence Review, Intelligent Help Systems For Unix*, 14 (1/2), 43-88.

A Kansei-based Color Conspicuity Model and Its Application to the Design of Road Signs

Katsuari Kamei — Ritsumeikan University, Kusatsu, 525-8577 Japan, kamei@cs.ritsumei.ac.jp

Eric Cooper — Ritsumeikan University, Kusatsu, 525-8577 Japan, cooper@se.ritsumei.ac.jp

Naoki Fujiiwara — Ritsumeikan University, Kusatsu, 525-8577 Japan

Abstract

Road signs are essential to safety in modern life end their visual design is a key factor to their effectiveness. This paper describes dichromatic color conspicuity based on Kansei modeling. First, we conduct an interactive survey of color conspicuity in which subjects find balanced conspicuity for two figures, one in the middle of the other. The results of the experiment give a measure of equivalent conspicuity for two colors. Next, we develop a model of conspicuity in which the relative area at equivalent balance is the average response from the surveys and this value is drawn on a spline curve. We tested traffic signs for conspicuity based on the model, giving the traffic signs a new quantitative dimension. Our optimal design concept is based on the traffic signs having stronger relative conspicuity for the pictogram of the sign based on the model, and calculated by the relative conspicuity of the two colors in the design. The actual designs shown in this work visually demonstrate the striking power of this model and the difference that such models can make to emotive response.

1 Introduction

Kansei Engineering is a family of methods generally used for product development, such as fashion design, car design, color combination evaluation, and fragrance evaluation. Kansei information, information about the emotive responses of consumers, has become indispensable to the product design process, especially for the products consumers use in their daily lives. This paper considers a slightly different objective for the use of Kansei, the development of road signs. Road sign designers need the tools of Kansei engineering, among other tools, to develop effective and, in some cases perhaps life-saving traffic signs.

Color makes an important contribution to daily life by psychological influences of emotion, image, meaning and symbolism. People can sense two main factors from color. One factor captures the cognitive information of visibility and quantitative color attributes. The other main factor captures the emotive information from imagery associated with the colors. In habitats, two-color designs are preferred for contrast between items that attract attention and items that discourage attention.

This study considers the conspicuity of road signs based on the relative areas of colors. The objective of the study is to propose optimal road sign designs. The target of the study is two-color road signs, in which the relative area of each color is expected to play a large role in making the signs conspicuous.

Colors for road signs are selected to effectively communicate information. Government bodies stipulate the shapes and color combinations of each type of road sign to communicate specific information. This study investigates the consideration of the relative areas of the colors, in addition to the shape and combination.

2 Dichromatic Conspicuity Experiments

This section describes an experiment to investigate the conspicuity of figure and ground color combinations without the pictorial content of the road signs. In visual design, conspicuity may be defined as the relative weight of each part of a design. The objective of the dichromatic color conspicuity survey is to collect Kansei data on human sensibilities about conspicuity of two colors, one within the other, one figure within one ground, on a neutral gray background. The results of the experiment give a measure of equivalent conspicuity for the two colors.

In this color survey experiment, two concentric figures of the same shape appear on a neutral gray background (none of the colors investigated is gray). The *ground* is the large, surrounding shape and the *figure* is the internal shape surrounded by the ground. Subjects manipulate the size of the internal figure using four buttons placed to the right of the figure-ground combination, in the neutral background. Pushing a button marked (+) increases the area of the internal figure and pushing a button marked (-) decreases the area of the internal figure. This color survey interface is shown in Fig. 1, as it appears when the figures are triangles.

The maximum area of the figure is up to the full size of the ground, which results in the ground having zero area. The minimum area of the figure is zero, at which point only the ground is visible. The subject of the experiment adjusts the areas of the figure and the ground until finding the point at which neither figure nor ground stands out more than the other, the point at which figure and ground have equal conspicuity. The subject then clicks a button to record the data and to continue to the next set.

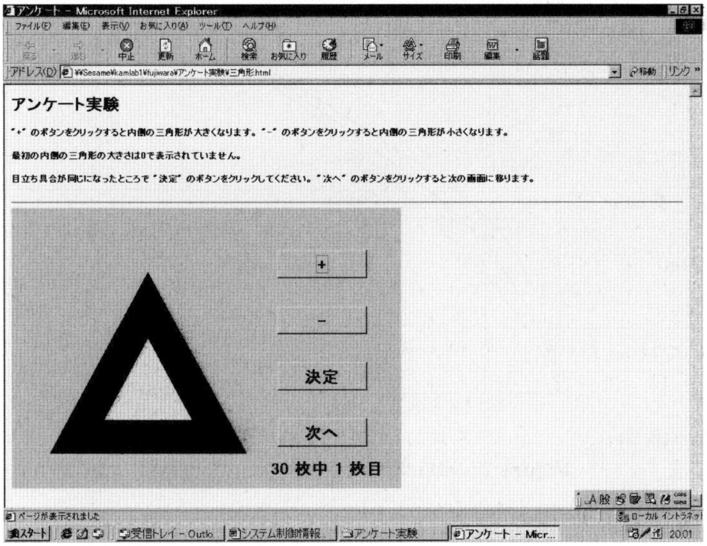

Figure 1: Interactive survey for relative conspicuity of figure and ground.

The data collected is the ratio *p=figure_area/(figure_area+ground_area)*. The experiment collected data on three shapes: circles, squares and equilateral triangles. The colors for figure and ground were six colors (yellow, black, white, red, blue, and green) for a total of thirty figure-ground color combinations. Ten university students completed the thirty color combinations for each of the three shape types.

The results for black or red show the area ratio p is relatively small, indicating higher conspicuity for these colors. When the internal figure is white the area ratio p at equivalent conspicuity is relatively large, indicating a low conspicuity for this color.

3 Dichromatic Conspicuity Model

The two-color conspicuity experiments quantify equivalent conspicuity for a figure and a ground of several different shapes, the area ratio for two different colors of the same shape when one is the ground of the other. The following is a description of a model to generalize the data to non-equivalent conspicuities.

The proposed model defines the conspicuity z (called simply *conspicuity* below) of a figure x relative to the area of the ground, $1-x$, by the relative area of the two. When $x=p$, as in the responses for the conspicuity experiment, $z=0.5$. When $x=0$ the area ratio is zero (no figure color, only the ground color), and when $x=1$ the area ratio is one (no ground color, only the figure color). These three points ($x=0$, $x=p$, $x=1$) are established from the data in the conspicuity experiments. We construct the conspicuity model by drawing a curve through the three points, zero conspicuity for the figure, zero conspicuity for the ground, and equivalent conspicuity for figure and ground. As shown in Fig. 2, we draw this line with a spline curve (Sakurai, 1981), where z is the vertical axis and x as the horizontal axis.

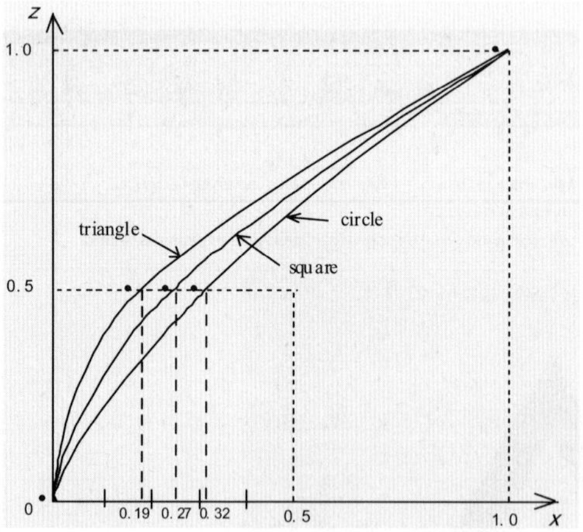

Figure 2: Modeling the conspicuity curve.

The objective of the model is to infer conspicuity of a given figure-ground area ratio. This model is applicable not only to road signs but to all two-color signs with a figure and ground.

This conspicuity model was further investigated by evaluation of nine designs at a time in an interactive survey. The subject in this survey manipulates reduced versions of the dichromatic figure-ground combinations to place them in order of conspicuity. This interactive survey is based on the small multiple ordered evaluation (Cooper & Kamei, 1999), in which subjects evaluate a group of items at once, as opposed to pair-wise comparison. This technique allows comparison of large numbers of stimuli and gives us a more structured model of conspicuity based now on both relative conspicuity at equivalence, and relative conspicuity of each design in the group. The results of this survey showed a good correlation for order of conspicuity of the nine figures in the ordered evaluation and conspicuity determined by the model ($x=p$, p-0.1, and p+0.1 for the circle, triangle, and square shapes, respectively).

4 Application to Traffic Sign Design

Most traffic signs in Japan use pictograms in a dichromatic design to convey information to convey information quickly without requiring the use of legible characters, which are thought to provide quick recognition for a wide variety of drivers and pedestrians (Oota, 1969). Since the pictogram is considered to be the most important part of these designs, we compared the conspicuity of 24 white-on-blue signs, 2 white-on-red sings, and 24 black-on-yellow signs. Nine of the fifty signs are shown in Fig. 3.

p=0.28, z=0.44; p=0.25, z=0.47; p=0.34, z=0.60; p=0.19, z=0.37; p=0.14, z=0.32

p=0.25, z=0.50; p=0.23, z=0.48; p=0.22, z=0.46; p=0.30, z=0.56

Figure 3: Assessments of nine road signs showing *area ratio*, *relative conspicuity*.

For almost every design evaluated, we found that the conspicuity of the pictogram was much lower than that of the ground. In other words, drivers were more likely to see the general color first. In the case of white-on-blue and black-on-yellow signs, this means that drivers and pedestrians would first only know that the signs was one of dozens providing driving information (white-on-blue) or one of dozens providing some type of warning (black-on-yellow). Clearly, the meaning of the pictogram is the most important message in these cases, yet the pictogram is inconspicuous.

We selected nine representative road signs of the fifty road signs tested, those shown in Fig. 3, for simplicity of redesign and for their relatively frequent appearance on the road. We redesigned eight of them to have high conspicuity of the figure, relative conspicuity of 0.60, which was the conspicuity of the highest in the group. These new designs are shown in Fig. 4.

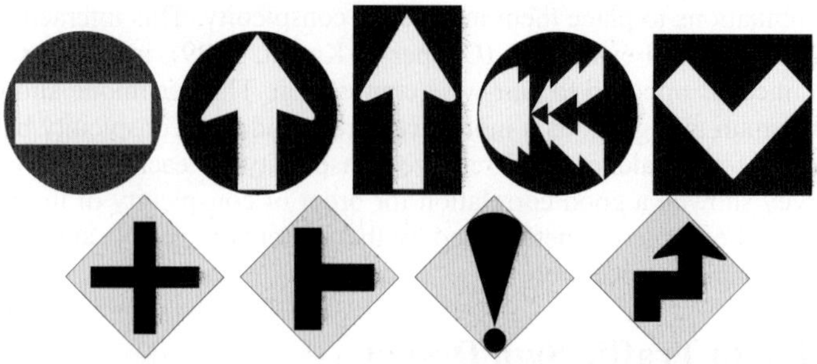

Figure 4: Redesigned road signs, each with relative conspicuity of 0.60.

5 Conclusions

This study graphically shows the possible impact of visual design on our daily lives. The simple conspicuity model suggests that traffic signs may be lacking conspicuity in their key communication channel, the pictogram. Further research is required to show how these designs behave, in other words the responses they evoke in realistic traffic situations. This research would also benefit from further investigation of precisely how the optimums discussed relate to the meanings and use of the signs.

Designs of standard traffic signs in use do not readily change, as the changes require not only replacement of the physical signs themselves but also the reeducation of all drivers, the difficulty of which is determined by the scope of the changes. Conversely, because these changes to the standard designs require huge resources, there is also good reason to exert effort into getting the design of the traffic signs correct. This research offers a quantitative method of supporting such efforts, for an additional dimension and more power to designers and planning teams. While corporations place a huge emphasis on the use of Kansei engineering for design, development and, marketing of new products, civic planners such as designers of traffic signs may also benefit from the tools provided by Kansei engineering.

References

Matsui, A. (1981). *Introduction to Spline Functions*. Tokyo: Tokyo Denki University Press.

Cooper, E., Kamei, K. (1999). An Interactive Survey for Evaluation of Color Placement Support Systems, *Proceedings of Fourth Workshop on Evaluation of Heart and Mind: Heart and Mind '99*, 71-72.

Oota Y. (1969). *Speaking of Pictograms*, Japanese Standards Association.

A design and evaluation of the user authentication system by using characteristics of mouse movements on a soft keyboard

Kentaro Kotani and Ken Horii

Department of Systems Management Engineering, Kansai University
3-3-35, Yamate-cho, Suita, OSAKA 564-8680, JAPAN
kotani@iecs.kansai-u.ac.jp

Abstract

This study addresses the possibility of using mouse movement patterns on a soft keyboard as an authentication system. The objective of this study is to empirically evaluate the proposed authentication by comparing characteristics of parameters extracted by spatio-temporal patterns of mouse motions during password entry on the soft keyboard. The results implied that the authentication system solely by using movement patterns was not a viable methodology. By incorporating with the password entry, the current approach may be useful for adding the security of the computer system.

1 Introduction

Recent development for services using computer technology yielded high demands for network systems in terms of usability and availability. Meanwhile, interests were overwhelmingly shifted to the security of computer systems. To establish advanced safeguards against network crimes, authentication techniques have been developed by using varieties of characteristics. Such characteristics can be categorized into four types of identifiers, i.e., (1) personal identifiers such as keys, PIN numbers, and passports, (2) identifiers based on knowledge such as passwords, (3) identifiers based on human performances such as signatures and behavioural patterns, and (4) biometric identifiers, which are based on physical characteristics such as fingerprints and voices.

Current inventive development for authentication systems in order to access the network by using PCs focuses on biometric-based authentication such as a fingerprint detector incorporated with input devices (Millman, 2000) and an identification system for palm shape detected by matrices of optical sensors assembled with the computer mouse (Sugino, et al., 2001). Advantage of such authentication system is high performance on identification rates. Recent study showed that commercially available fingerprint detectors have capabilities of less than 1% of false rejection rates (FRRs) and less than 0.01 % of false acceptance rates (FARs) on high quality fingerprint databases (Bazen and Gerez, 2002); however, these biometric-based authentication systems were still expensive so that such devices cannot fit as standard equipments for PCs. As Woodward (1999) indicated, there are invasive aspects of the biometric data which may threaten users' privacy so that they may have a resistance to offer personal data like fingerprints to be exposed to the public even though they are treated as strictly confidential data.

This study addresses the possibility of using mouse movement patterns on a soft keyboard as an authentication system. We argue that the proposed system has two benefits, i.e., 1) since many users access the network via PCs, the authentication system for such users should be targeted at the consumer market and, therefore, immediate ease of use and its cost are significantly important. The proposed system does not require anything but standard PC settings. 2) The use of keystroke rhythm has been investigated for authentication by several researchers; however, typing rhythm as an identifier was not sometimes enough to differentiate individuals if the passwords should be stolen. The proposed system can extract parameters not only as temporal patterns of "tapping-on" with a mouse, but also as spatial patterns (i.e. trajectory) by the mouse cursor on the keyboard images that appear on a computer's display. The objective of this study is to empirically evaluate the proposed authentication by comparing characteristics of parameters extracted by spatio-temporal patterns of mouse motions during password entry on the soft keyboard.

2 Methods

A set of experiments was conducted to investigate how individual differences exist in the spatio-temporal patterns by the mouse cursor movements.

2.1 Subject and apparatus

A total of five subjects (21-25 years, 20/20 corrected) were recruited from our department. All subjects were right-handed and had at least a year of experience in PC-use.

A Windows PC controlled the display, measured movement times, and sampled the data including the cursor's horizontal and vertical positions during a trial. Data recorded consisted of a series of x and y coordinates and associated time stamps of each component of the movement. The mouse used in the study was a conventional 2-button serial mouse bundled with the PC, and it was a ball-rotating mouse with no scroll wheel equipped. The size of the mouse was approximately 55 mm in width, 115 mm in length, and 35 mm in height. The weight of the mouse was 90 grams. The Control/Display ratio remained constant, where approximately 4.1 mm of cursor movement occurred per 1mm of mouse movement throughout the experiments. No forearm support was provided to the subjects. The subjects were, however, able to rest their wrist at the rounded edge of the table top. Height adjustment was available for the workplace by adjusting each leg.

2.2 Procedure

At the beginning of the session, the subjects were seated in the relaxed upright position at the workstation at a distance of 35 cm from the 17-inch monitor. The practice session was given until they felt comfortable to use the computer program displaying soft keyboards and other settings in the workstation. The subjects were instructed not to lift the mouse. The session consisted of 10 trials of password entry by pointing-and-clicking each key with the mouse serially. The second session was exactly the same: Only the difference was that the second session was taken place one week after the day of the first session.

2.3 Data analysis

There were four independent variables to be arranged; passwords, layouts of the soft keyboard, trials and sessions. Passwords, which consisted of a certain sequence of characters, are a major factor to determine the total time to complete. Unlike password entry on a keyboard, each key was serially tapped with the mouse on the soft keyboard. Hence, the distances among the keys and the size of each key determine the difficulty to enter passwords. Theoretically, the difficulty for entering a password by using a mouse was determined by accumulating each key-to-key IDs (indices of difficulty) calculated by Fitts' law. A total of four types of strings were tested as passwords. They were 'WATER', 'UJNHK', 'PANEL', and 'QPZMQ', which were determined to see whether any differences in mouse movements can be found in between the meaningful and meaningless strings ('WATER' vs. 'UJNHK', for example) and strings with low and high IDs ('WATER' vs. 'PANEL', for example).

Soft keyboards are defined as keyboard images that appear on a computer's display (MacKenzie, et al., 1999) and typically used for a pen-based system. We designed the system to see whether the mouse cursor movements have individual patterns so as to be used as one of identification parameters. Therefore, the mouse was used for alphanumeric entry in this system. Two types of the soft keyboard were designed; large-key layout and small-key layout (see Figure 1). The large-key layout uses 40 pixels for the size of the keys and the small-key layout uses 20 pixels. The size of the key shown on the monitor was 12X12 mm for the large-key layout and 6X6mm for the small-key layout. The distance between two adjacent keys was 50 pixels (15mm, the center to center distance), and was unchanged between the two layouts. There was no size difference between two soft keyboards. It provides a display region measuring approximately 220X120mm. The distance also affected the time to complete password entry as well. Trials and sessions were also included as dependent variables because they were the key to assure that potential identifiers were consistent over time.

A total of seven parameters were recorded and compared to each other for a potential candidate as an identifier; (1) total time to complete the password entry, (2) average mouse positions on the horizontal direction, (3) average mouse positions on the vertical direction, (4) average velocities of mouse motion, (5) total moving distance, (6) total time that the mouse was not in motion, and (7) average velocities normalized by subtracting the time that the mouse was not in motion from the total time to complete the password entry (i.e., (1) – (6)).

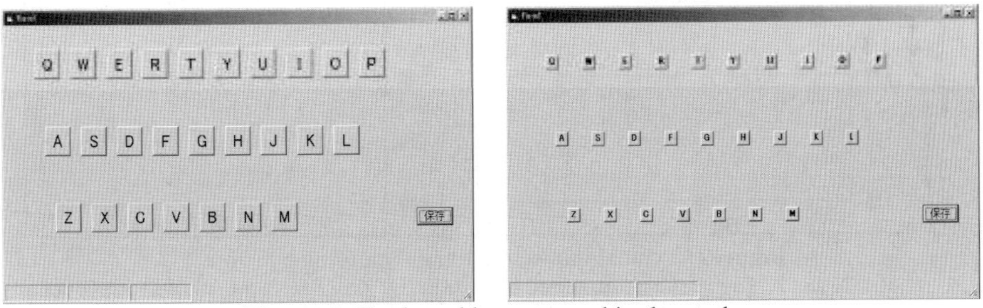

Figure 1: Keyboard layouts used in the study

(Left: large-key layout, Right: small-key layout)

3 Results and discussion

Table 1 summarizes analysis of variances on four independent variables with the subject effect. Passwords and subjects had significant effects on all dependent variables. The effects of trials and sessions, which had a great influence on permanence of identifiers, were not significant for two dependent variables, i.e., average mouse position on horizontal direction and average velocity normalized by subtracting the time the mouse. This implied that these two variables may be the most effective identifiers for user authentication among other parameters. No differences in dependent variables were observed between the meaningful and meaningless strings of passwords.

Table 1: Summary of analysis of variances

	Passwords	Keyboard layout	Trial	Session	Subject
(1)	**	**	NS	**	**
(2)	**	NS	NS	NS	*
(3)	**	NS	*	NS	**
(4)	**	**	NS	**	**
(5)	**	**	NS	*	**
(6)	**	**	NS	**	**
(7)	**	**	NS	NS	**

Note: Numbers in parenthesis indicates independent variables described in the Method section.**:p<.01, *:p<.05, NS: Not significant.

Figure 2 shows superimposed typical mouse movement patterns. Compared with these reaching patterns, it is apparently distinguishable between two, especially due to different shapes in edges.

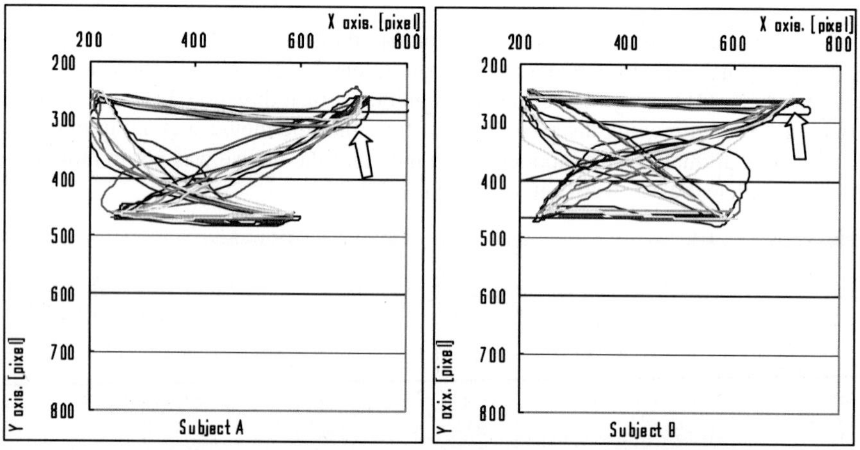

Figure2: Typical mouse movement patterns by subject
(Left: subject A, Right: subject B, tested password: 'QPZMQ' with the small-key layout)
Subject A seemed to perform alignment phase more often than subject B when approaching the key 'P', as marked with arrows at the upper right corner of the trajectories.

Evaluation of a system prototype was performed. In the prototype, all seven variables were used as identifiers. Evaluation session included 20 verification trials and 20 impersonation trials. A total of 20 subjects participated in the session. Based on the results, FRR and FAR were estimated with a system parameter (see Figure 3). The equal error rate was 18%, yielding 36% of combined error rates. This system performance was similar to that by Barrelle, et al. (1996), where they used mouse trajectory in reaching circular targets and showed 38-39% of combined error rates. The results of error rates, however, were not very impressive if compared with signature identification system (Syukri, et al., 1998), where as low as 17% of combined error rates were reported.

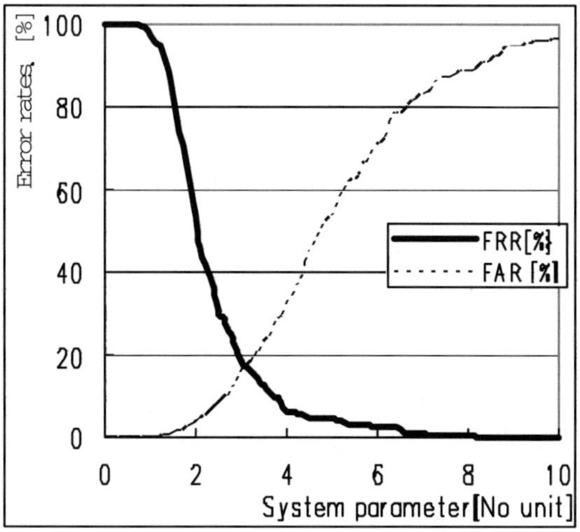

Figure 3: FRR and FAR for the system prototype

4 Conclusion

The results implied that the authentication system solely by using movement patterns was not a viable methodology. By incorporating with the password entry, the current approach may be useful for adding the security of the computer system.

Acknowledgment

This study was supported by the Grant-in-Aid for Scientific Research (Encouragement of Young Scientists (B), 14780334) from JSPS. The authors thank Hiromi Saito for helping the experiment.

References

Barrelle, K., Laverty, W., Henderson, R., Gough, J., Wagner, M., & Hiron, M. (1996). User verification through pointing characteristics: an exploration examination. *International Journal of Human-Computer Studies*, 45, 47-57.

Bazen, A. M., & Gerez, S. H. (2002). Achievements and challenges in fingerprint recognition. In D. D. Zhang (Ed.), *Biometric Solutions for Authentication in an E-World* (pp.23-57). Boston: Kluwer.

MacKengie, I. S., Zhang, S. X. & Soukoreff, R. W. (1999). Text entry using soft keyboards. *Behaviour and Information Technology*, 18 (4), 235-244.

Millman, H. (2000). Give your computer the finger. *Computerworld*, 34(13), 78-79.

Sugino, H., Ito, K., Shimizu, T. & Yasukawa, K. (2001). Continuous biometric personal identification by shape of the hand gripping the mouse, *Proc. of IPSJ (Interaction 2001)*, 41-42. (In Japanese).

Syukri, A. F., Okamoto, E., & Mambo, M. (1998). A user identification system using signature written with mouse. *Lecture Notes in Computer Science*, 1438, 403-414.

Woodward, J. D. (1999). Biometrics: Identifying law and policy concerns. In A. Jain, R. Bolle, & S. Pankanti (Eds.), *Biometrics: Personal Identification in Networked Society* (pp.385-405). Boston: Kluwer.

Field Methods Applied to the Development of e-Learning System

Masaaki Kurosu

National Institute of Multimedia Education
2-12 Wakaba, Mihama-ku, Chiba-shi, Chiba 261-0014 JAPAN
PFD00343@nifty.com

Abstract
For the purpose of obtaining the realistic information about the use of e-learning system, an ethnographic approach was taken. The socio-cultural and historical information about the informants revealed important information on how and why the e-learning system could be used.

1 Introduction

Although the e-learning is expected to be widely used, it is not clearly stated why and how the e-learning should be used. In order to clarify these points and design an effective e-learning courseware, more deliberate approach should be taken with regard to the situational information about the user. The clue to this purpose lies in applying the ethnographic methods before starting the actual design of the courseware.

2 Frameworks on the Design Process

With regard to the design process of interactive systems, many frameworks have been proposed. These include the contextual design by Holtzblatt (Bayer and Holtzblatt 1997), the scenario-based design by Carroll (Carroll 1995, Carrroll 2000, Rosson and Carroll 2001) and the design information framework by Sato (Lim and Sato 2001, Sato 2001). In Fig. 1, these three frameworks are described as a sequence of processes from top to bottom where the contextual design is located in the middle, the scenario-based design on the next to right, and the design information framework on the right.

In the contextual design, the process starts from obtaining the information about the user by applying the contextual inquiry. Then 5 different work models will be created from obtained information. After this process, the work redesigning and making the mock ups follow.

In the scenario-based design, four different scenarios are created of which the problem scenario is the first one where the problems of the user are described in a reasonable story. The activity scenario is a kind of the solution plan. After the activity scenario, the information scenario and the interaction scenario will be created as a concrete design solution. Based on these scenarios, the usability specifications will be summarized.

In the design information framework, the design and the prototyping are conducted based on the information obtained from the user study. During the user study, the observation and the interview are the methods to obtain the information about the user. The final process is the evaluation where the prototype is tested with regard to its validity.

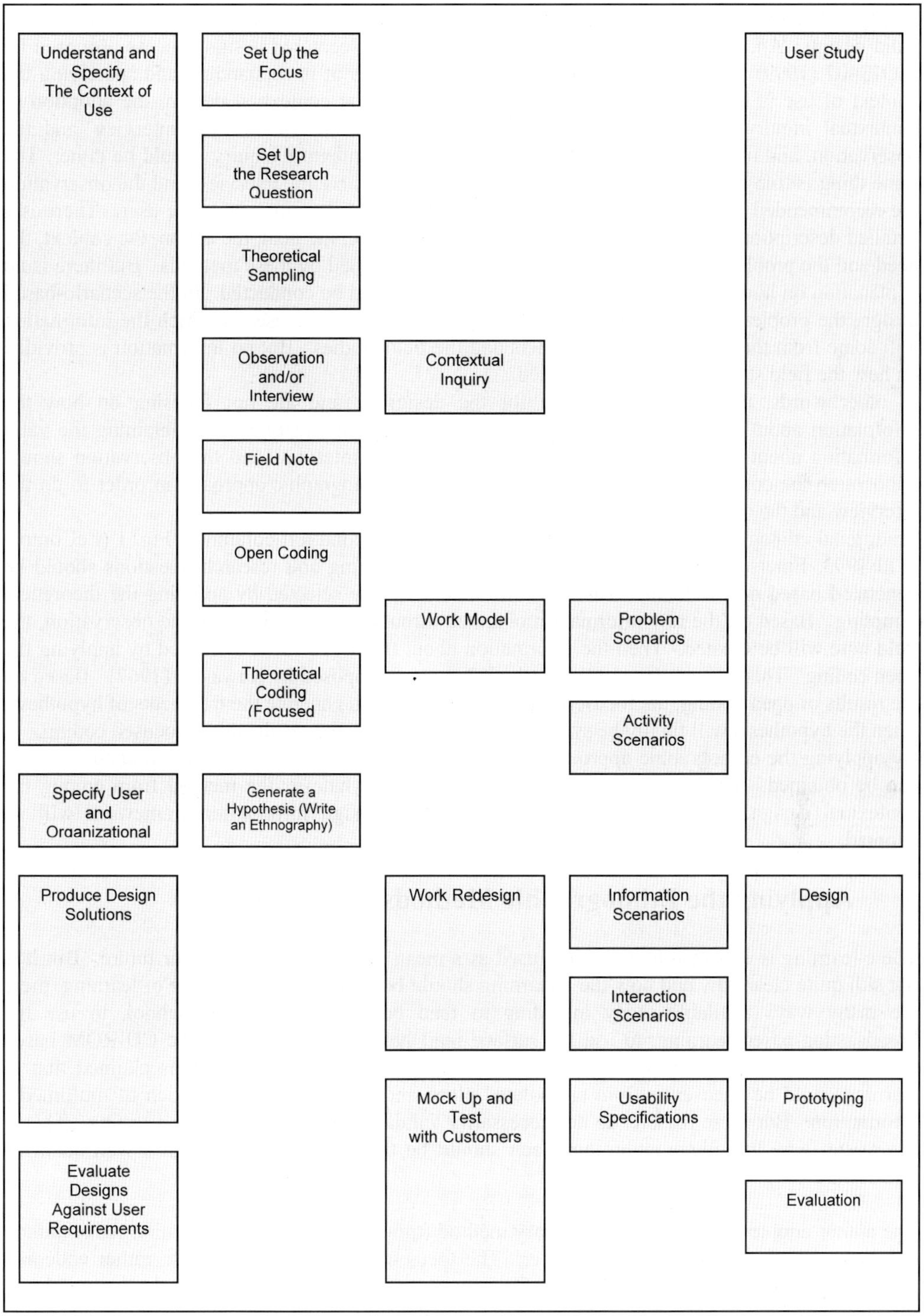

Fig. 1. A comparison of various frameworks in terms of the design process of interactive systems.

All these frameworks refer to the methods that should be applied for the purpose of creating the valid design. And these schemes correspond to the human-centered design process proposed by ISO13407 (1999) on the left of Fig.1. But the first process of understanding and specifying the context of use is not necessarily described in detail. In the contextual design, the adoption of contextual inquiry is recommended. But it is just a combination of the interview and the observation, and no information is provided on how the contextual inquiry should be done. The same thing is true with the design information framework where the interview and the observation are recommended as the basic methods for obtaining the information about the user. There is a detailed description on how the information about the motive, the goal, the action, the context, the need and the problem of the user can be or can not be obtained by these methods. But there is no explanation on how the interview and the observation should be conducted. In the scenario-based design, the problem scenario should focus on the problems of the user of which the information will come from the analysis of stakeholders and the field studies. But no information is provided on how the field study should be conducted.

In other words, these frameworks are for the designing and are not focusing on how the information about the user should be extracted from the real situation. But obtaining the valid information about the user is not straightforward and the interview and the observation should deliberately be conducted. Thus we should revisit the ethnographic approach in order to do the interview and the observation in an adequate manner.

Traditional ethnographic approach is summarized as the second left column of Fig. 1 (LeCompte et al 1993, Emerson 1995). The first process is the focusing and research questions should be generated based on the focus. Then the informants will be selected by applying the theoretical sampling. Based on the raw information obtained through the interview and the observation, the field note will be created. Then the information about the user will be classified by applying the open coding. This is a similar process to the KJ method proposed by Kawakita (1967). Based on the results of open coding, the focused coding will be done to generate the fragments of hypothesis. Then the hypothesis will finally be synthesized by integrating the results of the focused coding.

By applying the ethnographic approach as described above, the valid information about the user can be obtained. This approach should be taken before whatever the method from among the contextual design, the scenario-based design and the design information framework will be adopted.

3 Applying the Ethnographic Methods to e-Learning

The e-learning is expected to be widely used as a mean of education in the near future. But it is not still quite clear why and how the e-learning should be used. For the purpose of learning, there are many ways available today including to read books, to attend the school, to use the broadcasting-based learning, to use the surface mail based learning, to use the CD-ROM based materials, to retrieve information on the web and to use the e-learning. It is claimed that e-learning could be used even from an isolated place and is self-paced with much of multimedia information. But these reasons do not necessarily validate the sole use of the e-learning. This is the reason why the ethnographic approach should be taken in order to design a well-qualified courseware.

The author adopted the interview, the diary method (time chart) and the observation for obtaining the naturalistic information about the user. The focus of interview was, at first, rather wide and included questions about the life history of informants. The life history included the socio-cultural information of the user as well as the historical information. It also revealed the moments when the informants started to use the information devices such as the PC and the Internet. The list of informants is shown as Tab. 1.

Tab. 1 List of informants

Area	Shiga	Osaka	Osaka	Okinawa	Hokkaido	Chiba	Tokyo	Kyoto	Okinawa	Kanagawa
Age	20's	40's	10's	20's	40's	30's	70's	30's	50's	20's
Occupation	Student	Housewife	Student	None	Full time worker	Housewife	Housewife	Housewife	Part time worker	Part time worker
Family	alone	husband and two daughters	father, mother and sister	mother and two sisters	husband, son and daughter	father, mother and husband	alone	husband and three daughters	alone	husband
Diary Method	done	done	done	done	done	done	done	done	done	done
Observation	done	done			done	done	done			
Interview	3 times	once		twice	once	twice	once	twice	twice	

Tab.2 shows an example of the time chart of one informant where it is shown that her time of study was split frequently by the need for childcare. She finally gave up her study because she could not accomplish her study during the pre-determined school term. From this result, it could be said that the e-learning system should not simply be an electronic version of the real university. By taking consideration on the different context of learning of various students, it should not pose a strict school term as it is for the real university. This simple example reveals the validity of applying ethnographic method for understanding the contextual information of the user.

References

Beyer, H. and Holtzblatt, K., 1997, *Contextual Design: Defining Customer-Centered Systems*, Morgan Kaufmann

Carroll, J.M., 1995, *Scenario-Based Design: Envisioning Work and Technology in System Development*, John Wiley and Sons

Carroll, J.M., 2000, *Making Use: Scenario-Based Design of Human-Computer Interactions*, MIT Press

Emerson, R.M., Fretz, R.I., and Shaw, L.L., 1995, Writing Ethnographic Fieldnotes (Chicago Guides to Writing, Editing, and Publishing) (University of Chicago Press)

ISO13407, 1999, *Human-Centred Design Processes for Interactive Systems*

ISO/TR 16982, 2000, *Ergonomics of Human-System Interaction – Usability Methods Supporting Human Centred Design*

Kawakita, J., 1967, Idea Generation Method (in Japanese), Chuo-kouron-sha

LeCompte, M.D., Preissle, J., and Tesch, R., 1993, *Ethnography and Qualitative Design in Educational Researh,* second edition (Academic Press)

Lim, Y. and Sato, K., 2001, Development of Design Information Framework for Interactive Systems Design, *Proceedings of the 5th Asian International Symposium on Design Science*, p.1-7

Rosson, M.R. and Carroll, J.M., 2001, *Usability Engineering: Scenario-Based Development of Human-Computer Interaction*, Morgan Kaufmann

Sato, K., 2001, Creating a New Product Paradigm between Media Space and Physical Space, *Proceedings of International Council of Societies of Industrial Design*, p.362-368

Tab.2 A sample timechart

Time	Place	Event	Time	Place	Event
0:00	Home	Access UCLA	12:00		
0:15		Starts e-Learning	12:15		
0:30			12:30		Lunch
0:45		Submit a homework	12:45		Access UCLA
1:00			13:00		Starts e-Learning
1:15		Visits the discussion	13:15		Visits the discussion
1:30			13:30		
1:45			13:45		Childcare
2:00			14:00		
2:15			14:15		
2:30			14:30		
2:45			14:45		
3:00		Goes to bed	15:00		Access UCLA
3:15			15:15		Starts e-Learning
3:30			15:30		
3:45			15:45		Reads course documents
4:00			16:00		Homework
4:15			16:15		
4:30			16:30		
4:45			16:45		
5:00			17:00		
5:15			17:15		
5:30		Childcare	17:30		
5:45			17:45	Nursery School	Goes to nursery school
6:00			18:00		
6:15		Goes to bed	18:15		
6:30			18:30		
6:45			18:45		
7:00			19:00		
7:15		Childcare	19:15	Home	Childcare
7:30		Housework	19:30		Dinner
7:45			19:45		
8:00			20:00		Childcare
8:15		Childcare	20:15		
8:30			20:30		
8:45	Nursery School	Takes the child to nursery	20:45		
9:00			21:00		
9:15			21:15		
9:30			21:30		Housework
9:45			21:45		
10:00	Home	Comes back	22:00		
10:15		Childcare	22:15		
10:30			22:30		Access the mail
10:45			22:45		
11:00		Housework	23:00		
11:15			23:15		
11:30		Put on the PC	23:30		
11:45		Accesses the mail	23:45		

A Framework for Dynamic Adaptation in Information Systems

Thomas Mandl, Monika Schudnagis & Christa Womser-Hacker

Information Science University of Hildesheim
Marienburger Platz 22 – 31141 Hildesheim - Germany
{mandl, schudnag, womser}@uni-hildesheim.de

Abstract

Adaptation is considered one of the most successful strategies to improve human-computer interaction. This paper shows how adaptation can go beyond common interest based user models. Several systems are presented which allow the adaptation process to significantly alter the systems basic algorithms or knowledge structure. The additional knowledge sources of such models are also presented. Our examples from information retrieval show that individualization may even affect the algorithms or the choice and influence of certain algorithms. A system for library information exemplifies that the whole organizational structure of a system can also be the subject for adaptation. A system for e-learning presents further extensions for user models.

1 Introduction

Recently, significant research has been devoted toward adaptation. Meanwhile many systems have been deployed. Several frameworks for adaptation have been developed. They discuss which parts of information systems can be adapted and how the system is enabled to do so (Kobsa et al. 2000). Three tasks for a system are suggested by Brusilovsky & Maybury (2002):

- Collection of data about the user
- Processing this data to form a user model
- Producing an adaptation effect

Often, adaptation affects the information objects presented based on more or less static interest profiles. In order to adapt information systems to the cognitive preferences of users we need methods which modify the algorithmic reasoning of a system and which alter the system architecture more profoundly. This paper discusses these issues in detail relying mostly on current projects.
In all of them, deep structures of the system are adapted. COSIMIR and MIMOR are examples for the individualization of algorithms while SELIM and MyShelf drastically modify the systems knowledge architecture.

2 Adapting Algorithms in Information Retrieval

Adaptation has often been discussed for information retrieval and many systems have been implemented. In information retrieval systems, adaptation mostly results in the presentation of different result objects. Mostly, these techniques are applied to text documents (e.g. Pazzani &

Billsus 1997). Sometimes, the adaptation of the user interface violates the principle of consistency and as a result, confuses the user. In information retrieval, the results can be modified commonly without negative consequences because the user does not know all information objects available within a corpus.

2.1 Adaptation of Algorithmic Parameters

The COSIMIR (Cognitive similarity learning in information retrieval, Mandl 2000) is completely integrated into the algorithmic core of an retrieval system. The data about the user is collected by recording the user's decisions on relevance feedback. These explicit decisions are directly matched into a learning neural network which implements the matching function. As a consequence, the user model and the system core are implemented within the same module.
A neural network implements an algorithm in which the connection weights determine the outcome. The weights can be interpreted as the long time memory of the system whereas the actual activation of the artificial neurons can be seen as the short time memory.

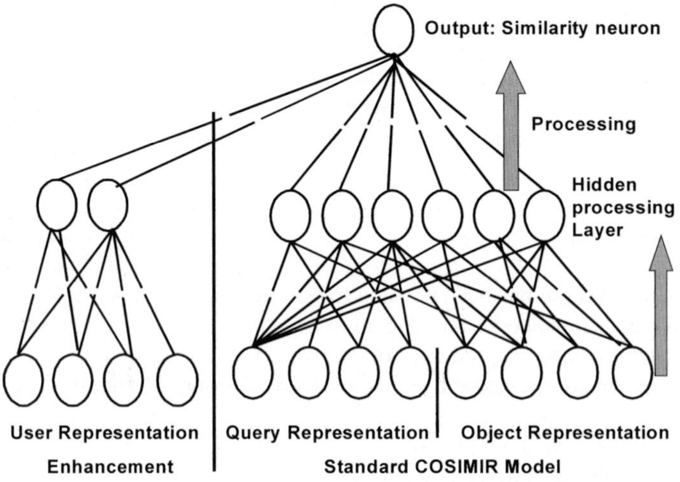

Figure 1: Extended COSIMIR model including user properties

In COSIMIR, the long term memory stores the user models which can contain complex relationships between individual properties of the knowledge items. The activation with the values of certain items initiates processing. In the neural network architecture storing, processing and learning aspects of the user model are intertwined.
Apart from training individual COSIMIR systems for each user, the COSIMIR architecture can be extended and user and context properties can be included. As figure 1 shows, these user properties can be modelled as additional input neurons in the network.

2.2 Adaptation of the Composition of Several Algorithms

Relevance feedback about concrete search results is also the knowledge source for MIMOR (Multiple indexing for the creation of method to object relations, Mandl & Womser-Hacker 2001). MIMOR is a fusion retrieval system which integrates the results from several other search engines in order to improve their performance. Other than most fusion approaches, MIMOR adapts the

linear combination of the retrieval engines. The weights are modified based on the success of the engines in previous searches. An empirical evaluation of MIMOR with data from the Cross Language Evaluation Forum (CLEF) has led to promising results (Hackl et al. 2002).
MIMOR can be individualized by training a model for each user. From a user modeling perspective, MIMOR can be interpreted as a user model coded in the fusion parameters. Each user may receive a completely different result by relying on the same underlying services.

3 Adapting Ontologies in Hierarchical Browsing Systems

Ontologies are a natural and often used concept for the design of information systems. For example, hierarchical internet sites can be seen as instantiation of ontologies. The heterogeneity of cognitive concepts of definitions and their inflexibility are the most serious disadvantages of ontologies.
A typical adaptation approach is presented in Spiliopoulou et al. (2000). It aims at shortening the paths for users. The system analyzes log-files in a relational calculus and uses intelligent prediction to forecast the most typical paths based on the history of the user. For long paths, shortcuts are enabled by inserting links to the predicted target pages at pages visited early during interaction. As a consequence, the software may look different for each user.
However, this approach is unpredictable for the user and can hardly be explained at the user interface level. Furthermore, the inserted link may not be a logical or structural part of the page.
We propose ontology switching as an interaction technique which allows user or system initiated changes of the knowledge structure. MyShelf is based on the model of a virtual library shelf. The advantages of a browsing system are usually accompanied by the disadvantage of inflexibility. As a consequence, MyShelf implements a hierarchical browsing system with several ontologies or taxonomies. At the moment, the user can choose his preferred ordering system. The same information objects are then reorganized and presented in a different hierarchical structure (cf. Mandl & Womser-Hacker 2002). In addition, several automatic adaptation strategies will be implemented without letting the user lose control.
The approach in MyShelf goes further than e.g. the adaptation of links within single web pages to shorten some paths as in Spiliopoulou et al. (2000). By exchanging the basic ontology, almost all potential paths through the hierarchical structure undergo significant change. Even if a path to a desired item may be longer after the adaptation, the interaction may be improved by making orientation easier and by reducing the probability of errors. A longer path may be easier to follow for a user if the path is understandable for the user and if the user searches under the correct categories. The user's cognitive approach toward the domain and its concepts is better reflected by the systems architecture in MyShelf.

4 Adaptation of the Didactic Approach in E-Learning

For many e-learning systems, didactic principles are only implicitly integrated and mainly a behaviorist approach is followed. In SELIM (software ergonomic design for learning with multimedia) we investigate how selecting a certain didactic basis influences the design of the user interface. For that purpose, prototypes based on different learning theories have been implemented for information science content. On the one hand, a combination of behaviorist and cognitivist principles has been implemented (prototype 'bekog'), on the other hand cognitivist and constructivist features have been combined (prototype 'kogkons'). These two approaches differ in the system's structure (linear vs. network structure) and navigational facilities, in the type of exercises presented (reproduction vs. problem-oriented exploration) and in the sequence of instructional texts and exercises (text + exercise vs. exercise + text). User tests showed that learners often give one approach preference over the other. As a consequence, the user should be able to choose between the alternatives as far as this is appropriate for the content of the system (Schudnagis & Womser-Hacker 2002). A further step leads to an adaptation of the didactic model according to the user's preference and learning style. These determining factors may be derived from the user's characteristics which are known and his interaction with the system (Kamentz & Womser-Hacker 2002).

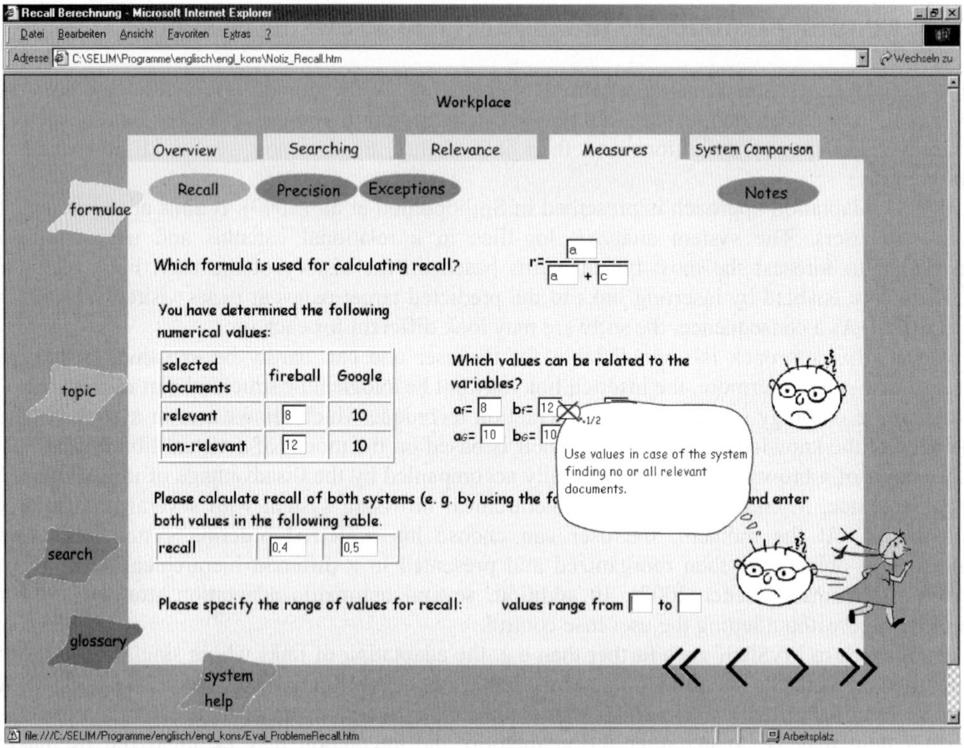

Figure 2: User Interface of SELIM

The individualization affects the entire architecture of the system and will, for example, determine whether the user is offered a linear or a non-linear presentation of the content. Moreover, it has to be decided, whether information is presented first and knowledge is tested in a subsequent step or whether the user works on exercises first in order to explore the topic and is lead to instructional texts afterwards.

Finally, a further example for adaptation is the availability of navigational tools according to the expertise level of the user concerning the topic (e.g. novice who is guided through the most important features) and according to the purpose he uses the system for (e.g. searching for a certain aspect, doing exercises).

Adaptation in SELIM is based on the pragmatics of the modelled learning process. Properties of the cognitive process like didactic approach and order of presentation modify the implemented information system dramatically.

5 Summary

This article discusses several individualized information systems where the adaptation effect reaches beyond modifications of information presentation. In the systems presented, adaptation is understood as a far reaching concept which dramatically alters the algorithms of the system.

6 References

Brusilovsky, P., & Maybury, M. (2002). From Adaptive Hypermedia to the Adaptive Web. *Communications of the ACM* 45 (5). 31-33.

Hackl, R, Kölle, R., Mandl, T., & Womser-Hacker, C. (2002). Domain Specific Retrieval Experiments at the University of Hildesheim with the MIMOR system. In: Peters, C., Braschler, M., Gonzalo, J., & Kluck, M. (eds.). *Evaluation of Cross-Language Information Retrieval Systems. Proceedings of the CLEF 2002 Workshop.* Berlin et al.: Springer [LNCS] .

Kamentz, E., & Womser-Hacker, C. (2002). Cross-Cultural Differences in Academic Styles and Learning Behavior in the Context of the Design of Adaptive Educational Hypermedia. In: Proc. 6th World Multiconference on Systemics, Cybernetics and Informatics (SCI 2002) Orlando, USA pp. 402-407.

Kobsa, A., Koenemann, J., & Pohl, W. (2000). Personalized Hypermedia Presentation Techniques for Improving Online Customer Relationships; St. Augustin: GMD FIT. http://fit.gmd.de/~kobsa/papers/Kobsa-PHPT-draft.doc

Mandl, T. (2000). Tolerant Information Retrieval with Backpropagation Networks. *Neural Computing & Applications. Special Issue on Neural Computing in Human-Computer Interaction.* Vol. 9 (4). 280-289.

Mandl, T., & Womser-Hacker, C. (2001). Probability Based Clustering for Document and User Properties. In: Ojala, T. (ed.): *Infotech Oulo International Workshop on Information Retrieval* (IR 2001). Oulu, Finland. pp. 100-107.

Mandl, T., & Womser-Hacker, C. (2002). Virtual ontologies for browsing interfaces in digital libraries. In: Isaías, P. (ed.): *Proc of the 2nd International Workshop on New Developments in Digital Libraries* (NDDL 2002). April 2, 2002. Ciudad Real, Spain. pp. 39-50.

Pazzani, M., & Billsus, D. (1997). Learning and Revising User Profiles: The Identification of Interesting Web Sites. *Machine Learning* 27. pp. 313-331.

Schudnagis, M. & Womser-Hacker, Ch. (2002). Multimediale Lernsysteme softwareergonomisch gestalten: das Projekt SELIM. Herczeg, M.; Oberquelle, H.; Prinz, W. (eds.). Mensch & Computer 2002. Vom interaktiven Werkzeug zu kooperativen Arbeits- und Lernwelten. Stuttgart et al.: Teubner. pp. 215-224.

Spiliopoulou, M., Pohle, C., Faulstich, L. (2000). Improving the Effectiveness of a Web Site with Web Usage Mining. In: Masand, B., & Spiliopoulou, M.(eds.) Web Usage Analysis and User Profiling. WEBKDD '99. Berlin et al.: Springer [LNAI 1836]

Constructing the user

Lene Nielsen

Copenhagen Business School, Dept. of Informatics
Howitzvej 60, 2000 Frederiksberg, Denmark
ln.inf@cbs.dk

Abstract
Often ethnographic studies are used in the design process to give the designers an understanding of the users. My claim is that when the designers are faced with the ethnographical material they will try to figure out whom the person presented in the material is and they construct a user from the material. In order to make this process successful the presenter of the material must look at this construction process as well as at the material presented.

When we encounter a stranger we have a tendency to see the person as a stereotype. The stereotype is built on knowledge of previous meetings with others, and ordered into categories that form the basis for the stereotype. In this paper I present my experiences with how designers construct a user as a stereotype. In my experience this can happen both when they are presented with field material and when they rely on their own knowledge.

I present how designers negotiated the construction, how the construction went, and which implications it had on the design process. Finally I discuss how a material can evoke sympathy for the user, and how it should be presented. Using inspiration from character building in film and the distinction between empathy and sympathy I discuss how this could influence both field studies and the material presented to the designers.

1 Why design for a single user?

A design process that has a focus on the user as an individual person is part of both participatory design processes and scenario-based design methods. Other methods use individuals as design material, but they are not seen as persons (e.g. use cases). The benefit of a design process that focuses on the individual user is both to see the individual user as a representation of a group of similar users and the user as a person that the designers can engage with. Thereby focusing on the users needs rather than on the designer's whims and ideas. This is achieved by involving actual users in the design process or by creating fictitious characters from research among the users.

Focusing on the single user – the Persona (Cooper, 1999), the Model-user (Nielsen, 2002) - is part of design processes that identifies and breaks down the different groups of users and express these as profiles of single users. Designing for a single user makes it possible for the designers to feel sympathy for the user. But, as I will show later, to feel sympathy craves information that comprises both the user's actions and feelings.

2 The user as a character

Sharrock showed how designers talk about users without ever meeting one, and how, during the talk, the user is constructed as a scenic figure (Sharrock and Anderson, 1994). Users are always present in the designers mind in some form or other in Sharrock's example they are unconsciously part of the design process as somebody the designers speak of without ever meeting one. When designers are faced with ethnographical material they will also try to figure out who the person presented in the material is, what characteristics the person have and they will create a user from the material. This process can end with the designers constructing a stereotype and, as I will argue, stereotypes can be obstacles in a design process that considers users' needs.

In a meeting with a person the initial forming of a category that the person belongs to will eventually be broken when an in-depth knowledge of the person is formed, and the stereotype transforms into a personal character. Designers have often not the possibility to undergo this transformation. They will meet the users in a pre-selected and analysed form. As it is not always possible for the designers to meet the "real" users there will be an act of communicating the fieldwork to them, whether this is video material, interviews or photos. From the material designers will create a user – a fictitious user that do or do not resemble the real person in the ethnographical material. It can be a collective and negotiated process or it can be an individual process. In order to make the process – the move from the stereotype to character – conscious, the presenter of the material must look both on the material presented and at the process of constructing the user - creating a character.

I will distinguish between:

- "The user" who is the person being observed or interviewed.
- "The stereotype" who is a creation done by designers based upon presented material, the designer use their own cultural knowledge to understand in the creation and it has the form of a category.
- "The character" that is based upon presented material and includes the person's sociology, physiology and psychology. The designers create a rounded character that allows them to understand and identify with the person. (Horton, 1999)

3 The workshop

Over the years I have conducted several workshops where the aim was to construct model-users and I have observed how difficult it is for the participant to avoid creating stereotypes (Nielsen and Pedersen, 2002). I see having sympathy for and identification with the user as a design parameter that moves design away from a view on the user as alienated from the designer to a view where the designer understands the users needs and motivations.

The workshop I was invited to observe during my stay at Interactive Institute, Malmoe was different from the workshops I have conducted, as it has a focus on video as a tool to inform the design process (Buur and Binder, 2000). Observations and interviews in an environment are transformed into video snippets. The video snippets represent activities, interactions and processes. The video material is used in a game form that gets the participants to engage in the material, discuss it and use it as an offspring for design. The workshop I observed (Johansson, Fröst et al., 2002) consisted of far more than I use here. The video snippets made it easy for the designers to get a grip of the surroundings and the game form furthered discussions and design ideas. I focus on a partial process I observed in the beginning of the workshop, and not on the workshop as such.

The designers were asked to define Kerstin and her needs in a video game session. They were presented with video snippets of Kerstin and her daily work life and with other persons and daily life. In a game form, the designers were asked to consider Kerstin's needs and understand the presented material from her point of view. During this session the designers both negotiated and constructed Kerstin and interpreted her needs. And they constructed Kerstin as a character. When the first designer was asked to give her impression of Kerstin's needs, she put emphasis on the way Kerstin told about her daily life and mostly on how her environment looked, how neat and well organised it was. This started the construction of Kerstin as a person who seldom came out of her office. In the following example the designer describes Kerstin as competent, a nest-builder, asocial and a bore. Kerstin possessed these features during the whole design process. She was understood in a one-dimensional perspective and stayed so for the rest of the design process.

Designer 1 (D1): "she is the type of person that builds a nest. She is really interested in her job and seeks the information she needs herself. She has a lot of breaks from people visiting her in her office. She is competent but boring. She's a mate, but she's got librarian-looks."
Designer 2 (D2): "(....) She sits with her papers and like to be by her-self and is not very social. Papers must be printed and there must be order."

The construction of Kerstin was of a one-dimensional character. Later in the process a feature from the Kerstin was presented, that she organises yoga-classes for the whole company. The designers included this information and it fitted into their construction of Kerstin of her as alienated from them-selves. Their tone of voice made the audience laugh when the yoga was introduced.

D2 presents Kerstin's view of the designed office to the other designers: "Here I sit because it's very quiet. I have had the yellow screen put up, because I think the room is far too open. (...) In front of me is the person responsible for economics, she sits there because she is so quiet and it's nice and safe to share the space with her. (....) Here behind the wall I usually run some yoga exercises. (Laughter from the audience)

The alienation prevented the designers to engage in Kerstin and identify with her. In this discussion the designer (who also were an employer) sees Kerstin as suffering from stagnation, and she will do so if the designer do not intervene

D2: "I have as an employer a responsibility to get Kerstin to make a move too. I have a responsibility for her not ending in stagnation. I must move the company forward and this implies that I move Kerstin too. So I cannot just fulfil her old needs that she sits in the corner in her little office."

In the group session that followed the designers were asked to design an environment from Kerstin's point of view. Kerstin was quickly put in a corner next to the boss and the environment was designed with features that came from the designer's own needs, discussed during breaks.

Designer 2: "It's common to see somebody bringing a sick child to the office, so we have created a corner with a bed and a sofa. The border between family and work is far to inflexible."

When the designers later presented Kerstin they realised they had created a character they disliked and this furthered a discussion about the users as such and the designers as something other, a very fruitful discussion in the design process. The game format furthered designers to engage in the material, design and discuss this is however not the issue for this paper

The video material holds strong empathic values. It is my claim that there is a need for a move from the empathic feelings to sympathy for the user and in this move there must be an emphasis not only on the environment but also on the user's feelings and emotions.

4 The stereotype

When we encounter a stranger we have a tendency to see the person as a stereotype. The stereotype is built on knowledge of meetings with others and formed as a category. We do not see the person as possessing a unique constellation of characteristics, but add the person to a previously formed category. (Macrae and Bodehausen, 2001)

One definition of stereotypes is that they are "socially constructed representations of categories of people"(Hinton, 2000). A stereotype is very difficult to work with in a design process that focuses on users needs rather than on users goals. It is necessary to get access to the users feelings and knowledge as more than one dimension of the character is needed to raise sympathy.

With inspiration in what it takes to get spectators to feel sympathy for a fictitious character in film I will use Murray Smith's (Smith, 1995) definition of the levels of engagement that comprises the feeling of sympathy. Thereby proposing a framework that can direct the presented fieldwork and enable designer to engage in the user. The different levels of engagement that comprise the structure of sympathy are recognition, alignment and allegiance.

- Recognition is information the enables the designers to construct the character as an individual and continuous human agent.
- Alignment: the process of which the designer is placed in relation to the character's actions, knowledge and feelings.
- Allegiance: the designer's moral evaluation of the character. Allegiance is dependable of the access to the character's state of mind and of the ability to understand the context of actions.

Smith makes a distinction between empathy and sympathy as the two elements comprising engagement. Empathy does not require that we share another person's values, beliefs and goals. It is constituted of emotional simulation and affective mimicry. During the emotional simulation we understand the characters emotions by trying out different emotions that fits the situation. The affective mimicry is an involuntary response to an emotional situation, e.g. the raising of hair in the back of the head when the protagonist feels fear. In order to engage both element are needed and they can disrupt each other and adjust to the understanding of the narrative situation.
To engage in a user not present, the designer engage in the same way as in a fictitious character. What is not presented is negotiated and drawn from the designers' categories of persons. The emotional simulation negotiates the understanding of the user's emotions e.g. tension in a work situation or body language towards a colleague. Recognition is based both on presented material but also on own knowledge. The more presented the less is drawn from own experience. It is the element of feelings, present in alignment and emotional simulation, which becomes the critical factor in the development of engagement from the material.

5 Conclusion

To present field studies to designers is an act of communication that involves choosing both the material and the form of the presented material. And at the same time the presenter must be aware of how the material will be received and interpreted.

Stereotypes are created from common knowledge of what representation certain categories of people have. By adding information that does not support and go against this common knowledge it becomes more difficult to create the stereotype and thereby alienate the person. In choosing the material the presenter should put emphasis on both the usual HCI areas: context, tasks, goals but also give the designers information about who the users are as persons.

The distinction between the rounded character and the flat character can be a guideline for what to look at in the material and also what to look for during field studies. The rounded character consists of actions, emotions and personality (Nielsen, 2002). When the designers get an understanding of the users emotions and personality they understand the users motivation for actions and they do not have to make these up. It is inevitable that we as human beings look for the motivation in order to understand the person in front of us. If we are not presented with clues that gives us an understanding of why, we have to make the "why" up and it is the first and easy pick to find a "why" in the categories

Ethnographic material in HCI is collected, interpreted and communicated. Often the act of communicating the material is overlooked. Focussing on the setting, the work and the situations in the presented material give designers little clue about whom the user is as a person. If the designers are to fully understand the users both ethnographers and researchers are to present material that also consists of and presents the user's emotions and personality.

References

Buur, J. & Binder, T., Brandt, E. (2000). Taking Video beyond 'Hard Data' in User Centred Design. *PDC 2000*, New York.

Cooper, A. (1999). The Inmates Are Running the Asylum. Indianapolis: SAMS.

Hinton, P. R. (2000). Stereotypes, Cognition and Culture. East Sussex: Psychology Press.

Horton, A. (1999). Writing the Character-Centered Screenplay. L.A.: Uni. of California Press.

Johansson, M., P. Fröst, et al. (2002). Partner Engaged Design: New Challenges For Workplace Design. *PDC2002*, Malmö.

Macrae, N. C. & G. V. Bodehausen (2001). "Social Cognition: Categorical person perception." *British Journal of Psychology,* 92, 239-255.

Nielsen, L. (2002). From User to Character. *DIS2002*, London.

Nielsen, L. (2002). Scenarios - a design tool to ensure user-narratives. In Soerensen, Nielsen & Danielsen (Ed.), *Learning and Narrativity in Digital Media* (pp.165-181). Frederiksberg: Samfundslitteratur

Nielsen, L. & G. Pedersen (2002). Understanding users - Merging video card games with model-users and scenarios. *Asian Pacific Computer Human Interaction 2002*, Beijing.

Sharrock, W. & B. Anderson (1994). "The user as a scenic feature of the design space." *Design Studies,* 15 (1), 5-18.

Smith, M. (1995). Engaging Characters: fiction, emotion, and the cinema. Oxford: Clarendon Press.

UISB – The User Interface Specification Browser

Marko, Nieminen, Toni Koskinen, Mikael Johnson

Software Business and Engineering Institute: SoberIT,
Helsinki University of Technology,
P.O.Box 9600, FIN-02015, FINLAND
Firstname.Lastname@hut.fi

Abstract: In this paper, we present the conceptual model and implementation of the User Interface Specification Browser (UISB). The underlying analysis of the requirements for UISB has been presented in our paper presented in HCII2001 (Koskinen & al. 2001). In that previous paper, UISB was outlined to be a solution for the problems that were present in the user interface development environment.

1 Introduction

User interfaces are effective tools for communicating the functionality and structure of a product or an application (software, service) to the intended customers and especially end users. But not only in the external communication, cross-functional internal communication (within the development organisation) of the product features can also benefit from early realisation of user interfaces. The uses of user interface presentations and mock-ups are versatile and in many ways practical. From commercial standpoint this can be seen to be reflected for example in various product advertisements and marketing and brochures as well as in screenshots that make functionality listings more concrete.

The creation and development of user interfaces, however, often takes place in the late stages of development work. During development, developers need to communicate with internal stakeholders as well as customers and intended end-users. In this communication, detailed technical descriptions, abstract models, diagrams, and schemes are used. These development-time deliverables are not always easily understandable by the non-technical stakeholders. During the creation of a revised or completely new product, discussions with a wide range of stakeholders are considered desirable and expected to be fruitful.

Based on this foundation, we can see the need for a development mechanism (working processes and practices) and even a supporting tool that enables the early sketching and creation of user interface prototypes and mock-ups that is tightly integrated with the underlying technical development and documentation. This need and its underlying requirements have been detected in our study that was reported in the HCII'2001 conference (Koskinen & al. 2001).

2 Previous Work and Related Research

The results of the previous study (Koskinen & al. 2001) pointed out several problems related to user interface definition and specification work. Problems were divided into two main groups: 1) problems related to the user interface (UI) documents themselves and 2) problems related to the organisational processes of controlling the UI specifications. The detected problems included: specifications are scattered in several documents that are not always easily found, documentation is cross-linked and contains redundant information but may easily be outdated and therefore contain even contradictory issues, in a distributed development environment technological

constraints (access to network folders, data transfer bandwidth, change notifications) may prevent the utilisation of up-to-date information, not all relevant information from the viewpoint of a single stakeholder is contained in the specifications. These requirements pointed out a need for a tool that is capable of putting together the scattered pieces of information. UISB was outlined to be such a tool.
Additional expectations towards UISB were in making specifications clear and communicative early enough in a development project to relevant stakeholders (UI designers, manual writers, software developers, marketing and management). In practice, this pointed out the need to create UI mock-ups easily and quickly from the specifications. These mock-ups were expected to serve for internal development purposes (clear communication of product features between stakeholders, usability testing) as well as company external purposes (exhibitions, customer probing).

3 Aim and Scope of the Study

The aim of this study was to create and initially evaluate a prototype application UISB that addresses the problems detected in the previous studies. The objective and in many ways aggregated research question in this construction-intensive study is "What kind of tool can enable and support communication about ongoing user interface development work within the development organisation". This question aims at providing support for concurrent visual UI design that is tied closely together with the non-visible technical interaction-related technical development. Additionally, the tool is supposed to provide means for early customer presentations and user testing.

4 UISB: Concept and Prototype

Overview. UISB is a tool for collaborative user interface design that enables different product development stakeholders to review and browse both the current status of user interface and its

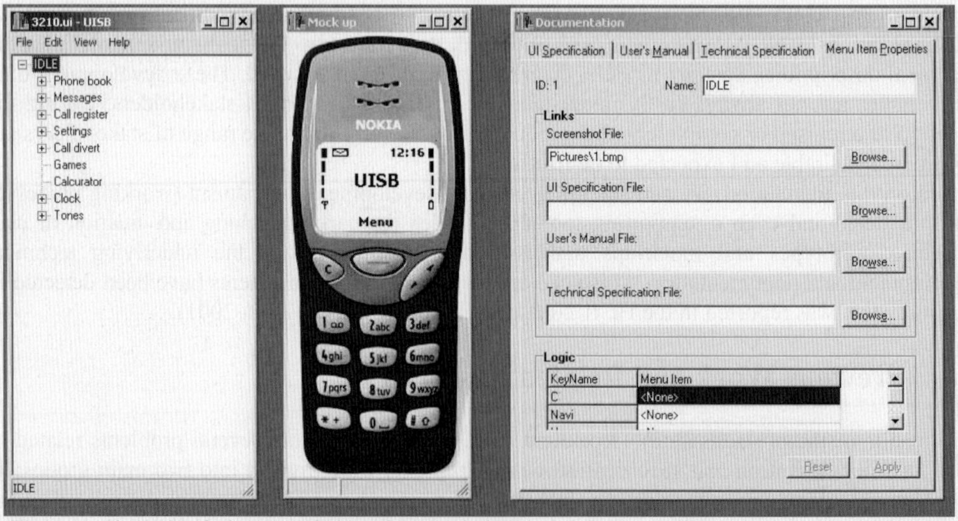

Figure 1. The UISB desktop consisting of three windows: (1, left): a tree-like menu structure of the user interface, (2) a representation of the current state of the visual user interface, and (3) a window presenting the technical aspects (attributes) of the user interface. Additional slips/tabs in window 3 consist of documentation about (a) general UI specification, (b) user's manual, and and (c) UI-related technical documentation.

appearance as well as browse the underlying user interface specification and documentation. The idea of UISB is that the expressive power of information can be increased through interactive multimedia output presentation (Oviatt & Cohen, 2000). The development of UISB has been done in the field of mobile handheld product development. UISB fulfils the basic criteria of a user interface software tool; it helps developers to design and implement the user interface (Myers B., Hudson S., & Pausch R., 2000).

UISB Development Process. UISB has been developed iteratively by applying several different user interface mock-ups in intermediate reviews. The very first prototypes were initiated, developed and presented alongside the original conceptual work.

UISB Usage. In using UISB, the starting point is that the designers have created an initial sketch and a picture of the UI ("UI screen") that will embed other interaction elements (see figure 1). The interaction elements for input (e.g. keys) can be defined in UISB as screen regions (in "UI mockup settings"). The browsing of the UI documentation is done by defining a tree-like menu structure that can be interacted with the input interaction elements ("menu screen; keys"). The UI tree has several properties (in the "documentation screen") that define 1) the visual appearance of the currently viewed screen (screenshot file) but also 2) provide direct (linked) access to the documentation (UI specification file, user's manual file, technical specification file). An additional part the properties defines 3) interaction functionality (logic) for the input elements. These latter properties enable interactive browsing of the user interface. This part of the concept and prototype is expected to make the specifications understandable for a wide range of internal and external stakeholders. They are also expected to make user interface specifications concrete already in the very early stages of the UI specification work for different development stakeholders as well. In this sense UISB resembles user interface toolkits e.g. Hudson, S. & Smith, I. (1997) that allow easy modification of user interface objects.

Basically, the creation of the UI specification elements (documentation files, user interface pictures) has been left outside the UISB environment because different products and even individual designers have powerful domain specific applications for that. As long as these applications provide external access to the data created with them via file/url/database mapping, UISB may be used to browse the data. Some very common file types can even be viewed within UISB but it mainly provides only a browsing/linking front-end.

The UISB prototype has been developed in MS Windows environment with Deplhi and C++builder.

Figure 2. UISB user interface mock-up. The user interface is interactive: it can be navigated by pressing the buttons of the mock-up.

5 Initial Evaluation

5.1 Developer users' evaluation of UISB

Method. The UISB prototype has been evaluated in two sessions. The other evaluation was carried out with the expected "developer users" (UI and software developers, N=6). The evaluation was conducted by following the principles of standard usability test with specific user-originated tasks described as scenarios. The users were asked to create a part of a user interface documentation using UISB and other developer applications. The idea was to combine the different documents and pictures through UISB prototype. The users were encouraged to think-aloud and the evaluation session was videotaped and all problems and improvement suggestions were documented.

Results. Users were able to conduct all test tasks with the UISB prototype. However, the users suggested few improvements to the design. The UISB desktop layout and visible user interface components should be possible to define according to the stakeholder needs. For instance the user interface designer did not want to see the same view as the manual writer. Additionally the UISB should be able to cope with the existing file formats that different stakeholders were using. The most positive comments were given from the interactive a tree-like menu structure. It was considered informative and useful, since it revealed defects in the user interface navigation logic.

5.2 End-users' evaluation of UISB

Method. In the other evaluation, UISB was evaluated as a platform for very early usability testing in a usability laboratory (figure 2). Two different interactive specification was evaluated in an end-user usability testing session (N=16). Users performed test tasks with the two interactive specifications that provided early feedback for the UI developers of mobile instruments.

Results. The distinctions of two interactive specifications became obvious for end-users, since they were able to navigate through the two slightly different versions of user interfaces of mobile instrument. Users were keen to give feedback of differences of the user interface designs. It seemed that UISB provided valuable feedback about the user interface navigational paths and visual design before the actual user interface were constructed. In this sense, UISB can be compared to paper mock-ups. The difference also stated by the end-users is that UISB will allow a realistic navigation through the user interface. In this case the end-user evaluation supported Nielsen's (1990) claims that evaluators using the computer mock-up were more likely to find the major problems than evaluators using the paper mock-up. Obviously, in this case this can be due to the fact that UISB contains more functionality than paper mock-ups in terms of e.g. sound feedback. During the evaluation the users also gave several comments about the visual design of the user interface. Some of them even evaluated the ergonomics of keys. In this sense UISB can slightly mislead the focus of end-users.

6 Conclusions and Discussion

So far, UISB has not been applied in real settings. The early evaluation, however, has indicated the usefulness of UISB for user interface specification work with the possibilities for early evaluation of the interactive user interface. At the same time it visualises the structure of user interface for developer users' it provides a toolkit for end-user usability tests. Like in many studies this study strengthens the claim that the expressive power of information can be increased through interactive multimedia presentation of design.

During the implementation of UISB, we found out that a major issue in the conceptual model for the application is the structure of the contained data. It appears that the data structure (combination of specification elements) should be dynamic instead of fixed, as is the case with the current UISB implementation. This topic is being addressed e.g. in the academic field of "product data management". In our work, this direction has not been studied so far. According to initial discussions with specialists in that research area, it is expected to provide very important insights about more advanced structural model for user interface documentation.

Acknowledgements

This study has been conducted with the support of National Technology Agency of Finland (TEKES) and participating companies in the SMART technology programme. The implementation of the UISB prototype application has been done by Géza Keszei with supporting guidance from the authors.

References

Hudson, S. & Smith, I. (1997). Supporting dynamic downloadable appearances in an extensible user interface toolkit. *Proceedings of the 10th annual ACM symposium on User interface software and technology*. Banff, Alberta, Canada.

Koskinen T., Nieminen M., & Repokari L. (2001). Managing User Interface Specifications in Distributed Development Environment - A Case Study. *Proceedings of HCI International 2001*, New Orleans.

Myers B., Hudson S., & Pausch R., (2000). Past, present, and future of user interface software tools. *ACM Transactions on Computer-Human Interaction (TOCHI)* March 2000, Volume 7 Issue 1.

Nielsen, J. Paper versus Computer Implementations as Mockup Scenarios for Heuristic Evaluation. In Proceedings of IFIP INTERACT'90: Human-Computer Interaction, (1990) pp. 315 - 320.

Oviatt, S., & Cohen, P. (2000). Multimodal Interfaces that Process What Comes Naturally, *Communications of the ACM March 2000/Vol. 43*, No. 3., 45-53.

Development of Radiological Emergency Assistance System "MEASURES"
- Situation Awareness Support Tool for Nuclear Emergency Response Activities -

Shigehiro Nukatsuka *Osami Watanabe*

Mitsubishi Heavy Industries, LTD.
3-1, Minatomirai 3-chome, Nishi-ku, Yokohama 220-8401 JAPAN
shigehiro_nukatsuka@mhi.co.jp osami_watanabe@mhi.co.jp

Abstract

At the time of nuclear emergency, it is important to identify the type and the cause of the accident. Besides with these, it is also important to provide adequate information for the emergency response organization to support proper decision making by predicting and evaluating the development of the event and the influence of the release of radioactivity for the environment. MEASURES (Multiple Radiological Emergency Assistance System for Urgent Response) is a support system which provides not only the current state of the nuclear power plant and the influence of the radioactivity for the environment, but also the future prediction of the accident development. In this paper, the concept and the function of MEASURES are introduced from the view point of SA (Situation Awareness) which is one of the recent topic of Human Factors research.

1 Introduction

At the time of nuclear emergency, it is most important to shut down the nuclear reactor safely, and cope with the accident as soon as possible. On the other hand, for evacuation activities of the residents and the visitors with sufficient time margin, it is important to make adequate decision based upon the accurate prediction of when, how much, and to which direction the radioactivity will be released. If this decision making is inadequate, unnecessary unrest might be stirred up or, to the contrary, it might be resulted in fatal delay in evacuation activities.
Recently, research on "erroneous action which is taken by a belief that it is correct" based on the inadequate situation awareness, as a cause of a human unsafe action, has been studied. Concept of SA and its evaluation method have been advocated. According to Endsley, SA refers to the perception of elements in the environment within a volume of time and space, the comprehension of their meaning, and their projection into the near future (Endsley, 1987).
Until now, many systems have been developed individually. Some systems display current status of a plant, some provide prediction of future status of a plant, and some calculate dispersion of radioactive materials with current value of radiation monitors and relatively simple atmospheric model, for example. However, each system can support just a part of SA in nuclear emergency response activities. In order to support SA in nuclear emergency response activities totally, the diffusion prediction system of the radioactive material which tied up these systems effectively is required.

MEASURES was developed to support appropriate SA in nuclear emergency response activities by combining the plant status display system (perception), the event identification system (comprehension), the inference system for future plant behaviour (projection) and the radioactive material dispersion prediction system with accurate atmospheric model, and by providing information from each system in a rich, familiar and intelligible form for a user.
MEASURES consists of the following four subsystems:

- AIPS - Accident Identification and Processing System
- ACIS - Accident Course Inference System
- ASAS - Accident Simulation Analysis System
- EDPS - Environmental Dose Projection System

Figure 1 shows the system configuration of MEASURES.

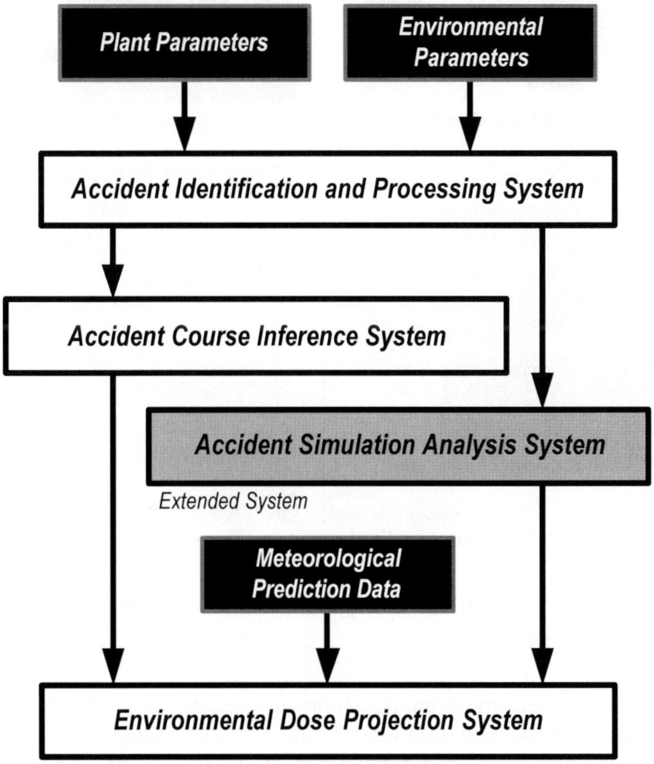

Figure 1: **MEASURES system configuration**

2 Accident Identification and Processing System

AIPS performs the data acquisition, the plant status information display, and the identification of the event. The block diagra m of AIPS is shown in Figure 2. Plant status data via plant computer, environmental parameters such as off-site radiation monitors, wind direction and wind velocity, and the meteorological prediction data such as NOAA (National Oceanic and Atmospheric Administration) are acquired. AIPS displays these data in an intelligible and familiar mimic form. AIPS identifies the event with these plant status data and its diagnostic table. Figure 3 and 4 shows the example of the information displays, and Figure 5 shows the example of the event diagnosis display. By this, AIPS supports grasp of the plant state for the emergency response organization.

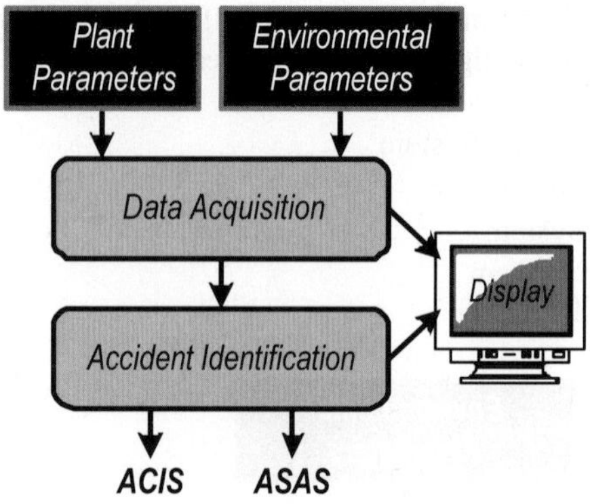

Figure 2: **AIPS block diagram**

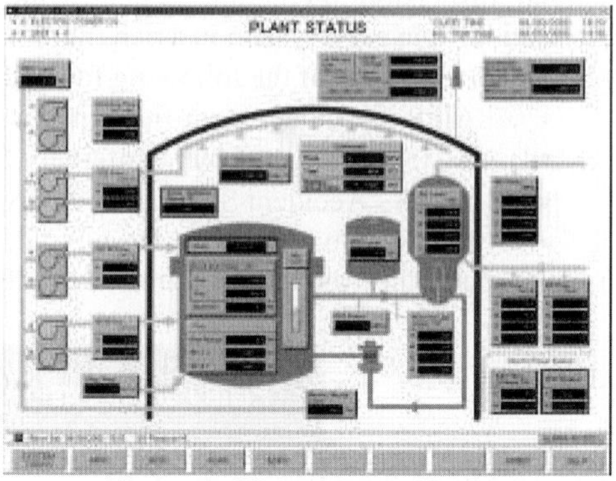

Figure 3: **Plant status display**

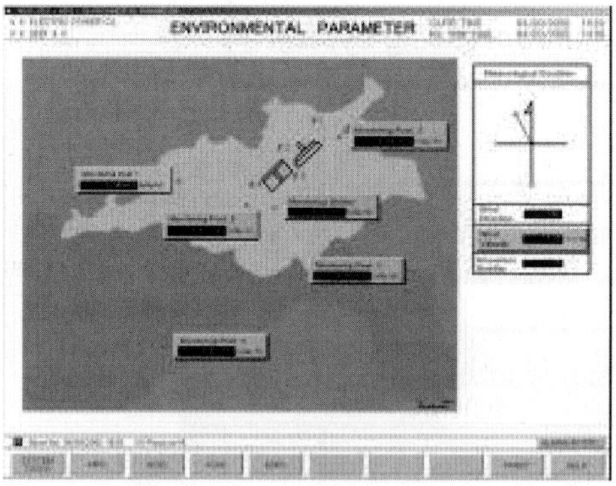

Figure 4: **Environmental parameter display**

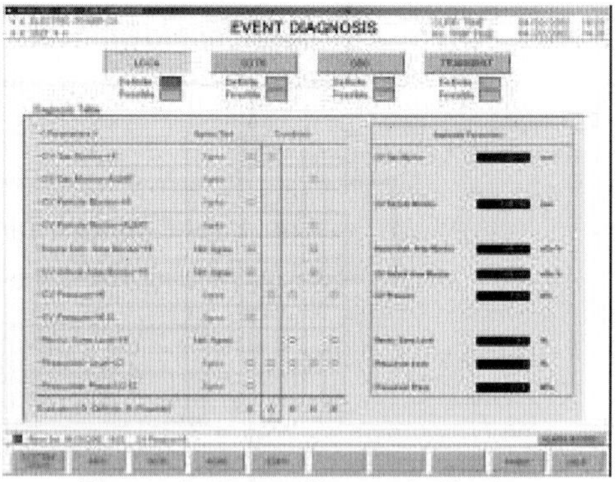

Figure 5: **Event diagnosis display**

3 Accident Course Inference System and Accident Simulation Analysis System

ACIS performs the inference of the event development by retrieving event tree and the estimation of the release time, release rate, and dose rate of radioactivity. Figure 6 shows the block diagram of ACIS. Pre-analyzed data of the event development and the source term data are stored in its database for this inference. ACIS displays the pattern of the event development in the form of event tree. Figure 8 shows the example of the ACIS event tree display. While ACIS infers the event development based on the pre-analyzed data, ASAS can perform the inference of the event development and the sauce term more realistically by means of the severe accident analysis program. Figure 7 shows the block diagram of ASAS. This program covers the analysis from a normal operation mode to a severe accident mode. Figure 9 shows the example of the accident simulation display. With these functions, ACIS or ASAS supports grasp of accident development (the release time, release rate, and dose rate of radioactivity) of the future for the emergency response organization.

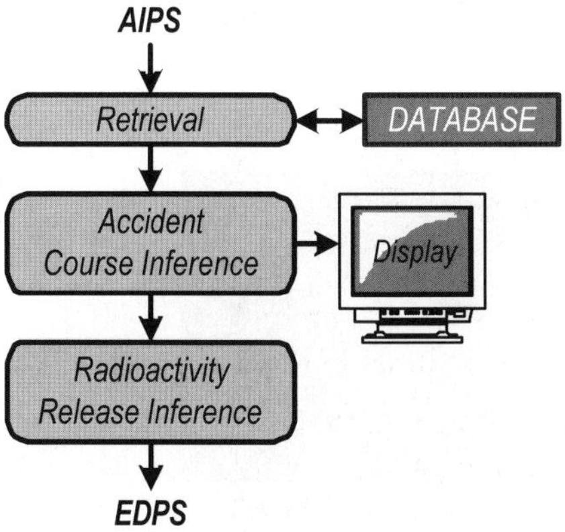

Figure 6: **ACIS block diagram**

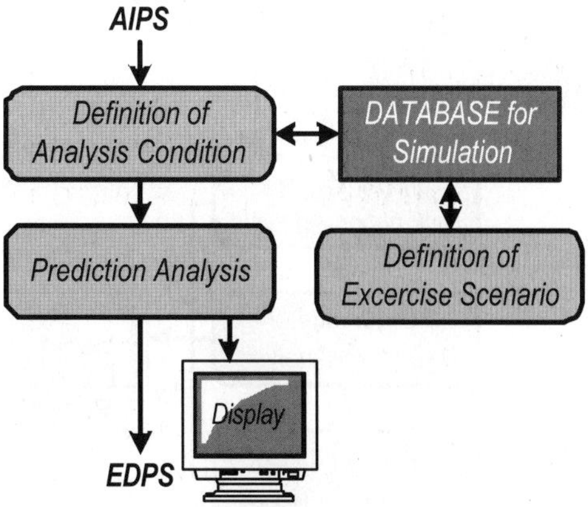

Figure 7: **ASAS block diagram**

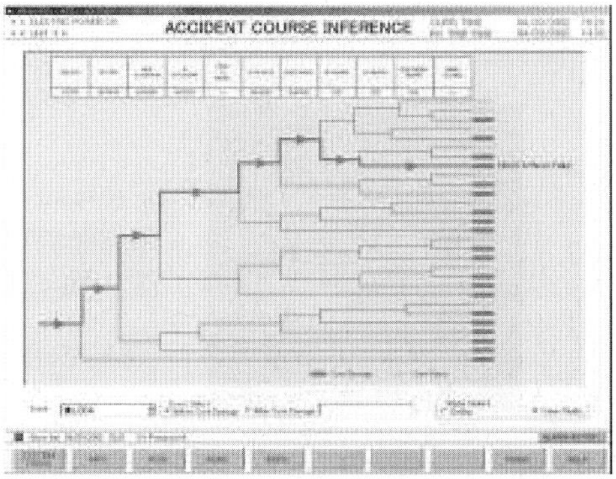

Figure 8: **Accident course inference display**

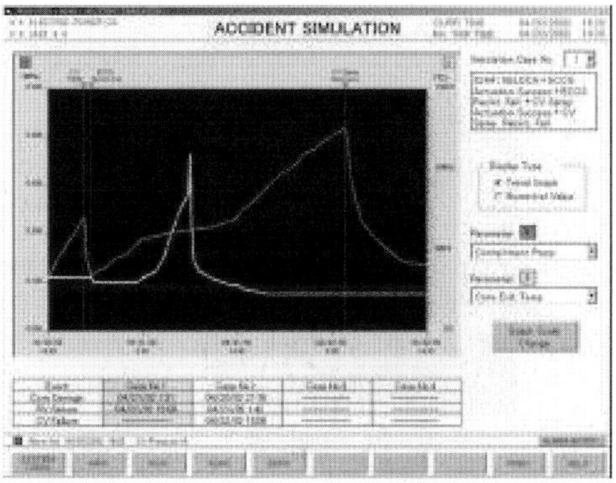

Figure 9: **Accident simulation display**

4 Environmental Dose Projection System

EDPS performs the three dimensional air current distribution analysis, the three dimensional atmospheric dispersion analysis, and the environmental dose projection analysis. Figure 10 shows the block diagram of EDPS. Using the turbulent model, atmospheric air current field is analyzed based on the boundary conditions of wide range meteorological prediction data. Following this analysis, radioactivity dispersion, using the particle method, is analyzed. Then the environmental dose projection analysis is performed to determine distributions of airborne activity concentration, dose rate, and total dose, and to determine external dose by cloud and ground shine, and inhalation dose. EDPS displays those results in the form of wind velocity vector diagram, distribution of airborne concentration, activity distribution deposited on the ground, dose rate distribution, total dose distribution, and animation of plume transfer behaviour. Figure 11 shows the example of the wind vector display, Figure 12 shows the dose distribution display, and Figure 13 shows the plume behaviour display. These diagrams are superimposed with the topographical map around the nuclear power station. Providing environmental dose projection accompanied by the event development in these ways, EDPS supports the decision making of residents/visitors evacuation for the emergency response organization.

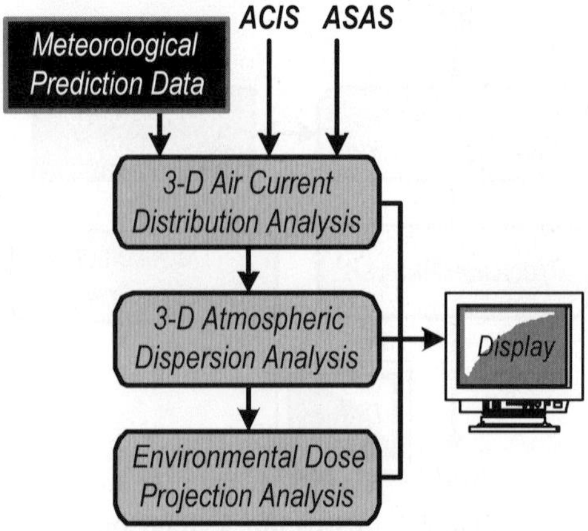

Figure 10: **EDPS block diagram**

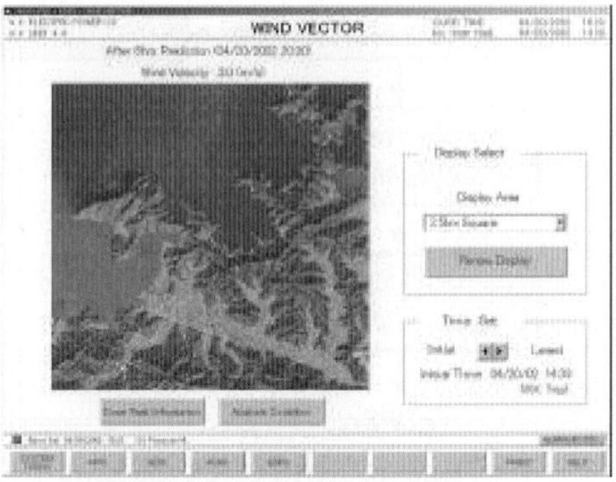

Figure 11: **Wind vector display**

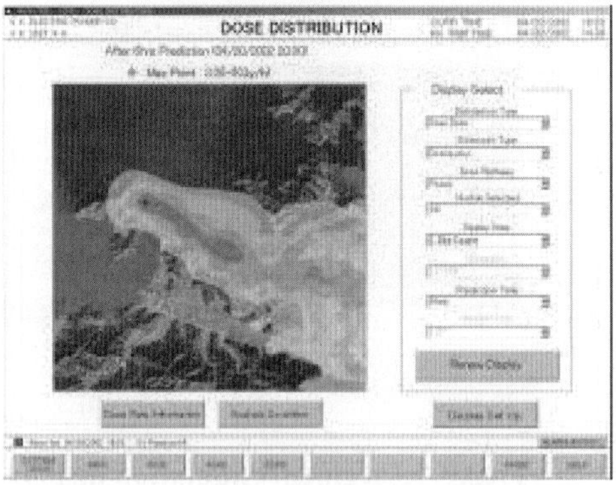

Figure 12: **Dose distribution display**

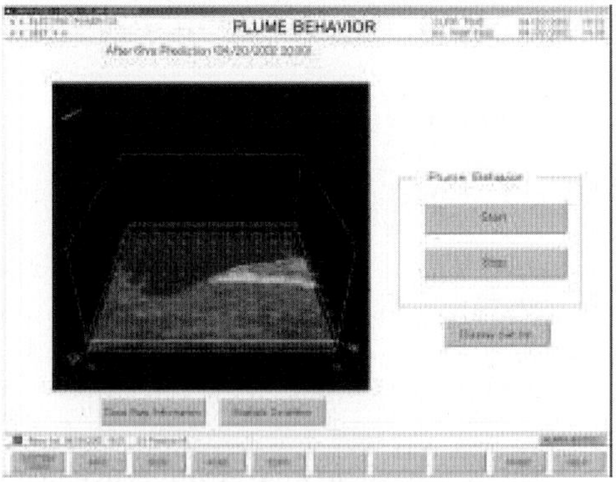

Figure 13: **Plume behavioral display**

5 Summary and Conclusion

MEASURES was developed to support the appropriate SA in nuclear emergency response activities. MEASURES consists of four subsystems, that is, AIPS which displays current plant status and identifies the event, ACIS/ASAS which infer and simulate the future behaviour of the plant, and EDPS which predicts the dispersion of the radioactive material based on an accurate atmospheric model. By combining these four subsystems effectively, MEASURES can support all the aspects of the SA in nuclear emergency response activities.

References

Endsley, M. R. (1987). SAGAT: A methodology for measurement of situation awareness. Hawthorne, CA: Northcorp Corp.

GUI Navigator/Cover: GUI Transformation Systems for PC Novice Users

Hidehiko Okada and Toshiyuki Asahi

Internet Systems Research Laboratories, NEC Corporation
8916-47, Takayama-cho, Ikoma, Nara 630-0101, Japan
h-okada@cq.jp.nec.com, t-asahi@bx.jp.nec.com

Abstract

This paper proposes two systems for transforming the graphical user interfaces (GUIs) of PC applications to enable novice users to use the applications without pre-learning. The essence of the systems is to read GUI screens by using a GUI accessibility method. One system can externally add user-application navigations to application GUIs. The other can replace application GUI screens with more simply designed screens. The effectiveness of the systems has been confirmed by an evaluation with actual PC novices.

1 Introduction

One usability problem of current GUI applications on PCs is that, because of the complexity of their GUIs, interaction sequences are not clear enough for novice PC users to properly complete their tasks without pre-learning the applications. This paper proposes two GUI systems, "GUI Navigator" and "GUI Cover," for solving this problem. The essence of these systems is to read GUI screens, by using a GUI accessibility method, in order to detect what windows and widgets are shown on the current screen.

2 GUI Navigator

A method for solving the problem is to make a computer system navigate user-application interaction sequences for users. In user interactions with GUI applications, users operate GUI widgets such as menu items, buttons and check boxes.. Therefore, an interaction navigation system can navigate by highlighting the widget to be operated in the current screen and displaying a message about the intended operation (e.g., to click on the highlighted menu item, to enter the user name in the highlighted text input field). To achieve such navigation requires the detection of target widget locations on the screen. Our GUI Navigator (GN) detects the locations of widgets on applications' windows without any modifications or extensions in operating systems (OSs) and applications - the widget detection is achieved by reading the current screen. GUI screen reading has been applied to accessibility applications for people with disabilities (Schwerdtfeger, 1991; Kochanek, 1994). For example, a system for blind users reads window titles, widget labels, etc., on the current GUI screen aloud (Mynatt et al., 1994). On Microsoft Windows, the screen reading function can be implemented by using Microsoft Active Accessibility (Microsoft, 1997). The GUI screen information is called the off-screen model (OSM).

Figure 1 shows the system configuration of GN. The system detects a target widget on the current screen according to the pre-specification in an XML-based interaction sequence description file for a user task. Each widget in a sequence is specified by its property values, including the name (i.e., the label), the widget type (e.g., a button), the parent window title, and the widget order on

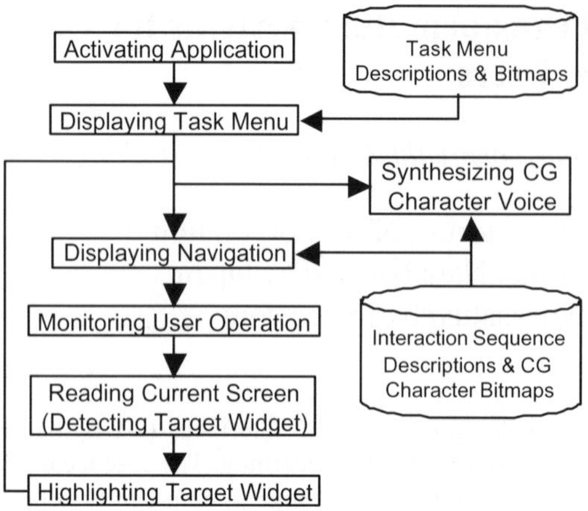

(CG: Computer Graphics)

Figure 1: System Configuration of GUI Navigator

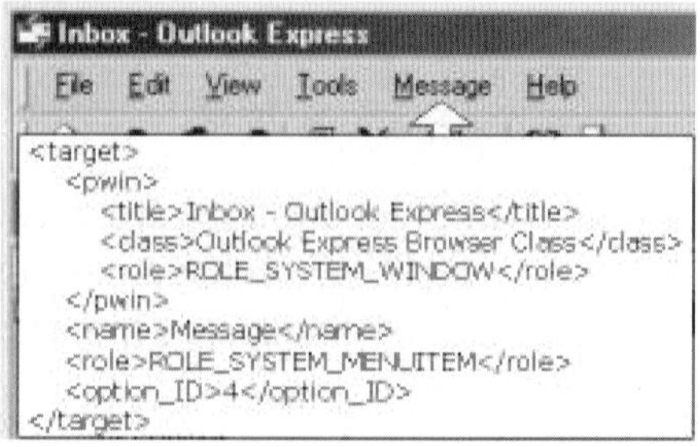

Figure 2: Example of Target Widget Specification in Interaction Sequence Description File

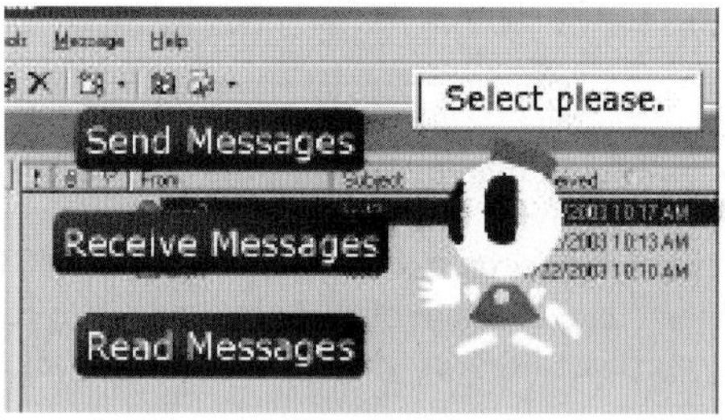

Figure 3: Example Screenshot of Task Menu by GUI Navigator

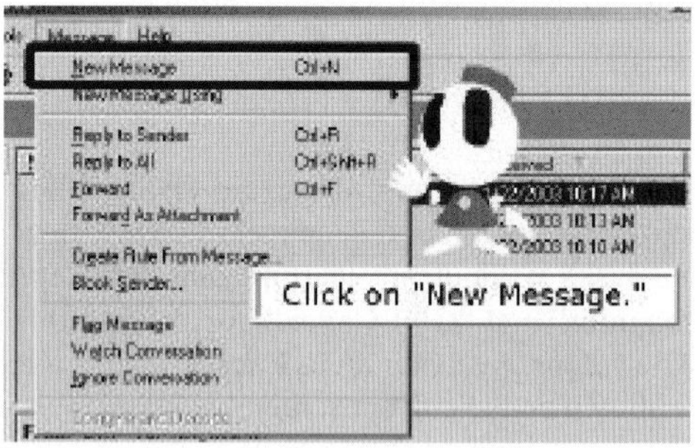

Figure 4: Example Screenshot of Navigation by GUI Navigator

the parent window. Figure 2 shows an example of target widget specification in an interaction sequence description file. In this example, the target widget is the "Messages" menu item on the "Inbox - Outlook Express" window. Figure 3 shows an example screenshot in which GN displays the "Task Menu," a menu of user tasks. After the user selects an item (a task that she/he wants to do) in the Task Menu, GN loads an interaction sequence description file for the selected task. GN then 1) detects the location of the target widget by reading the screen, 2) highlights the detected widget, 3) displays a message about the intended operation, and 4) waits for a user operation to the widget. This cycle of 1)-4) is repeated until the task is completed (i.e., the sequence in the description file is completed). Figure 4 shows an example screenshot in which GN highlights the "New Message" menu item and navigates the user to click on the item. Thus, GN can transform the GUI of any application into one with interaction sequence navigation (i.e., it can externally add the navigation to the GUI of any application).

3 GUI Cover

Another method for solving the problem is to replace the GUI of an application with a simpler one that is more suitable for novices. Our GUI Cover (GC) enables such GUI replacement by covering

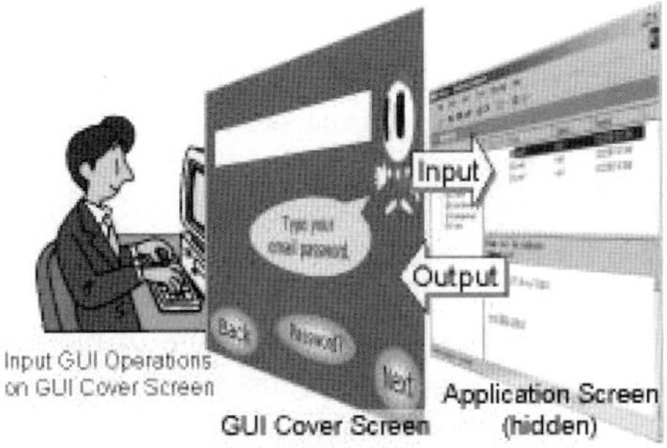

Figure 5: Concept of GUI Screen Replacement by GUI Cover

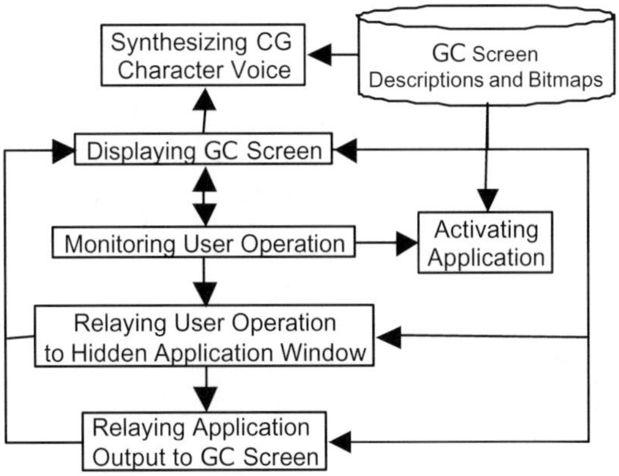

Figure 6: System Configuration of GUI Cover

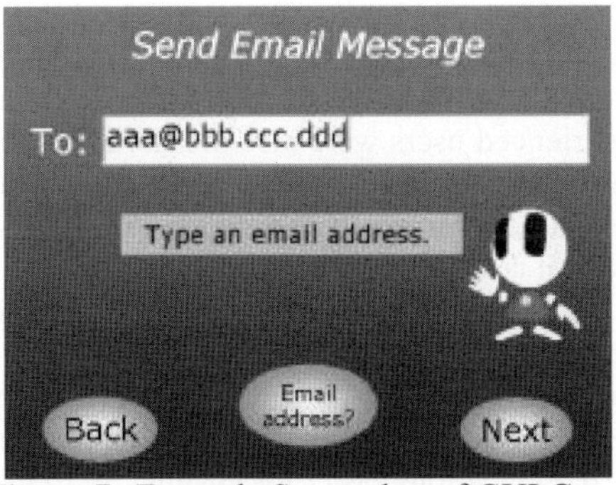

Figure 7: Example Screenshot of GUI Cover

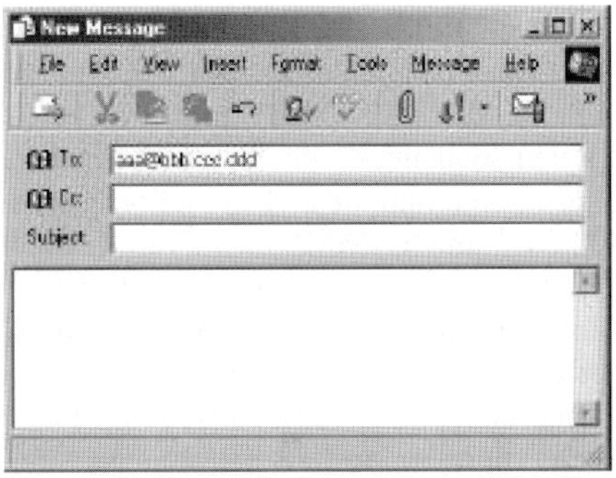

Figure 8: Example Screenshot of Emailer Window Hidden by the Screen in Figure 7

applications' GUI screens with more simplified ones that are designed for novices (Figure 5) - the system displays a window that is always displayed above other windows (thus, application windows are hidden by this topmost window), and shows a simplified screen on the window. Through this simplified screen, users can interact with applications whose windows are hidden. Figure 6 shows the system configuration of GC. The system 1) relays user inputs on the topmost window (i.e., on the GC's simplified screen) to the applications' hidden windows (e.g., an email address entered by a user on the topmost window is copied by GC into the pre-specified email address input field on an emailer window), 2) detects application outputs (shown in the hidden windows) for the user inputs, and 3) relays the detected outputs on the topmost screen. GC shows the cover screen by reading XML-based cover screen description files for user tasks. Tags for specifying widgets in application windows used in description files for GC are the same as in those for GN. Figure 7 shows an example screenshot of the GC screen for entering an email address. The address entered on this screen by a user is relayed to (copied into) the corresponding field (pre-specified in the GC screen description file) on the hidden emailer window (Figure 8).

GC can also be applied without any modifications or extensions in OSs and applications. This is because GC also reads the hidden application windows in order to achieve the input/output relays. Thus, GC can transform the GUI of any application to one with a more simplified screen design.

4 Evaluation

We have evaluated the effectiveness of our systems by an evaluation with real PC novices. Seven subjects participated (27-40 years old, avg. 32.2). They tried to send and receive email messages using an emailer that none of the subjects had used before. The conditions were A) with no GUI transformation (i.e., using the emailer with its own GUI), B) with GN, and C) with GC. If the correct interaction sequences for the tasks are clear enough for the novices, their task performances will be similar to those of experienced users. Therefore, we compared the task performances (task completion time (TCT), operation counts (OC), and total amount of mouse movements between clicks (TAMM)) between novices and experienced users. These task performances were measured by auto-logging user-application interactions with our tool GUITESTER (Okada, H. & Asahi, T., 1999).

Table 1 shows the results of t-tests that test whether TCT*, OC* and TAMM* in condition B and C are significantly smaller than those in condition A, where TCT*, OC* and TAMM* are the task performance gaps in TCT, OC and TAMM between the novice users and experienced users. As described above, small values of TCT*, OC* and TAMM* mean that the task performances of novices are similar to those of experienced users because even novices can complete the tasks without wondering what the next correct operation is or performing a wrong operation. Therefore, the smaller the TCT* OC* and TAMM* values for a condition, the clearer the correct task interaction sequences. The task performances of experienced users were also measured by auto-logging interactions of a user well experienced in using the emailer. As shown in Table 1, the TCT*, OC* and TAMM* mean values for the seven subjects are much smaller in conditions B and C than in condition A. In addition, the t-test results of comparing condition B to A ("B vs. A" in Table 1) shows that condition B is significantly better than condition A in terms of the task completion time, and the t-test results of comparing condition C to A ("C vs. A" in Table 1) shows that condition C is significantly better than condition A in terms of all three task performance measures. Furthermore, all the subjects subjectively preferred conditions B or C to condition A (Table 2). Therefore, our systems are able to make GUI applications easier for PC novices to use.

Table 1: T-test Results for Differences in Condition A-C
(A: with no GUI transformation, B: with GUI Navigator, C: with GUI Cover)
(Task1/2: Sending/Receiving Email Messages)

		TCT*	OC*	TAMM*
A	Task1	66.1	3.7	857
	Task2	47.6	2.71	585
B	Task1	10.6	2	174
	Task2	14.1	1.71	263
C	Task1	10.1	0.7	30
	Task2	8.5	0	145
B vs. A	t-value	2.468 (**p<0.05**)	0.76 (p>0.05)	1.59 (p>0.05)
C vs. A	t-value	2.670 (**p<0.05**)	1.876 (**p<0.05**)	2.053 (**p<0.05**)

Table 2: Novice Users' Subjective Satisfaction

	A	B	C
Most Usable	0 (0%)	3 (43%)	4 (57%)
Least Usable	10 (100%)	0 (0%)	0 (0%)

5 Discussion

We first discuss what types of tasks can be supported by our systems. Both systems navigate user interactions step by step based on *pre-described* interaction sequence files for their tasks. This means that a correct interaction sequence must be pre-specified for each task. Therefore, the systems are applicable to tasks whose correct interaction sequence can be pre-specified, and are not applicable to tasks that do not meet this condition (e.g., creative tasks such as editing Web pages).

We next discuss what type of novice users each system is suited for. In the case of using GN, users see the applications' original GUI screens and input operations on the screens, and thus it is expected that GN users will become accustomed to application interaction manners. Therefore, GN is suitable for novice users who want to study the manners of GUI interactions and become experienced users. In the case of using GC, on the other hand, the applications' original GUI screens are hidden, and thus it is not expected that GC users will become experienced in actual application interaction manners. However, GC users can complete their tasks with more simplified interactions than GN users can. Therefore, GC is suitable for novice users who want to complete their tasks as simply as possible.

6 Conclusion

We have proposed and developed two GUI transformation systems for PC novice users. GUI Navigator transforms application GUIs into those with interaction sequence navigations, and GUI Cover transforms application GUIs to those with more simplified screen designs. These systems can be applied to applications without any modifications or extensions in OSs and applications because they can read GUI screens by using the GUI accessibility method. The evaluation with actual novice users confirmed the effectiveness of our systems.

Future work includes additional evaluations with users of different attributes (e.g., elder novices) and development of user-adaptive GUI transformation methods and systems.

References

Kochanek, D. (1994). Designing an OffScreen model for a GUI. *Lecture Note in Computer Science*, 860, 89-95.

Microsoft Corporation. (1997). Microsoft Active Accessibility. Retrieved January 21, 2003, from http://msdn.microsoft.com/library/en-us/msaa/msaastart_9w2t.asp.

Mynatt, E. & Weber, G. (1994). Nonvisual presentation of graphical user interfaces: contrasting two approaches. *Proceedings of CHI94*, 166-172.

Okada, H. & Asahi, T. (1999). GUITESTER: a log-based usability testing tool for graphical user interfaces. *IEICE Trans. on Information and Systems*, E82-D(6), 1030-1041.

Schwerdtfeger, R.S. (1991). Making the GUI talk. *BYTE*, December, 118-128.

On the Role of User Models and User Modeling in Knowledge Management Systems

Liana Razmerita, Albert Angehrn, Thierry Nabeth

Centre for Advanced Learning Technologies (CALT) INSEAD
Bd. De Constance, F-77300 Fontainebleau France
liana.razmerita@ugal.ro, albert.angehrn@insead.edu, thierry.nabeth@insead.edu

Abstract

The paper elaborates on the role of user models and user modeling for enhanced support in Knowledge Management Systems (KMSs). User models in KMSs, often addressed as user profiles, include user's preferences and are often similar to competency definitions. We extend this view with other characteristics of the users (e.g. level of activity, level of knowledge sharing, type of activity etc.) and we explain the rationale for doing this. The proposed user model is defined as a user ontology based on Information Management System Learner Information Package (IMS LIP) specifications and it is integrated in an ontology-based user modeling system. The whole user modeling module is integrated in a ontology-based KMS called Ontologging.

1 Introduction

The dynamic of change in business environments requires the organizations' knowledge workers to learn continuously and adapt in order to remain competitive. Knowledge is considered the most important asset for organizations and the effective management of knowledge has become an important issue. Knowledge in the context of Knowledge Management Systems (KMSs) consists of experience, know-how and expertise of the people (tacit knowledge) and different information artifacts and data stored in documents, reports available within the organization and outside the organization (explicit knowledge). Nowadays the view of KMSs is often focused on the technology, on the process of capturing, organizing and retrieving information based on notions like databases, documents, query languages and knowledge mining. (Thomas, Kellogg et al. 2001)

In this paper we focus our attention on the role of user models and user modeling for enhanced user support within KMSs. The integration of user models in KMSs opens a large number of research questions. Some of these questions are common to the general objectives of user modeling, some are more specific to the HCI and to KM and some are related to the use of ontologies for representing user models. The problem of integrating user models into KMSs can be broken down into several general questions such as: Why is it important to model users in a KMS? What are the most relevant characteristics of the users in a KMS? What type of users' behavior can be distinguished in a KMS? What type of modeling techniques can be applied in order to track the users' behavior and to maintain user models in a KMS? How can a user model improve the interaction with a KMS or more explicitly: What type of adaptive/personalized services can be provided based on these characteristics? What are the advantages/limitations of applying ontologies in user modeling? What are the perspectives of its use in the context of the Semantic Web? Are security or privacy issues an impediment towards the use of such models and if so how these issues could be overcome?

The aim of this work is to demonstrate the role of user models and user modeling in KMSs so we will address mainly the first three questions within this paper. In the next section we present the rationale for user modeling in KMSs. The third section presents the structure of the user ontology and specific user modeling processes. The fourth section analysis how user model and user modeling processes can enhance the functionality of a KMS. The last section includes conclusions and future research work.

2 User modeling and a next generation of KMSs

Knowledge Management Systems are designed to allow their users to access and utilize the rich sources of data, information and knowledge stored in different forms, but also to support knowledge creation, knowledge transfer and continuous learning for the knowledge workers. Even if traditional KMSs tend to provide more functionality, they are still mainly centered on content manipulation (storing, searching and retrieving). The next generation of KMSs aims to go beyond the mere administration of electronic information; they aim to foster learning processes, knowledge sharing, collaboration between knowledge workers irrespective of their location, etc. In section 4 we will show how user models and user modeling can support these processes and we will emphasize the role of the user model for KMSs. Nabeth and al. (2002) stress that a KMS needs to move from a document-centered approach towards a more user-centered perspective.
Nowadays knowledge management experts see user models or user profiles as the foundation for expertise directories, also known as "people finder systems". Therefore the knowledge workers, the users of KMSs, are usually modeled more from the perspective of skills and competencies similar to the systems for human resources, certain preferences of the users can also be specified.
The research on user modeling is motivated because of two main reasons: 1) differences in individual user's needs and characteristics 2) heterogeneity between different groups of people. Moreover user models and user modeling are the key element for personalized interaction.

3 Building a user ontology for KMSs

The user model is defined as a user ontology describing the different characteristics of a user and the relationships between the different concepts. The definition of the user ontology captures rich metadata about the employee's profiles comprising different characteristics like: name, email, address, competencies, cognitive style, preferences, etc. but also a behavioral profile specific to the user interaction with a KMS. The proposed user model is extending the Information Management System Learner Information Package specification (IMS LIP, 2001).

IMS LIP is structured in eleven groupings in including: Identification, Goal, QCL (Qualifications, Certifications and Licenses), Accessibility, Activity, Competence, Interest, Affiliation, Security Key and Relationship. These groupings are implemented as abstract concepts in the user ontology. The concept Identification contains attributes and other sub concepts that help to identify an individual (name, address, email, etc) within the system. Affiliation includes information on the descriptions of the organizations associated with the user/learner. QCL contains concepts related to the different qualifications, certifications and licenses the user has. Competence contains skills associated with formal or informal training or work history. Activity includes activities related to the education/training work of the user. Accessibility contains concepts related to: user preferences, language information, disabilities etc. The concept Interest contains information on hobbies and other recreational activities. The concept Goal contains learner's/user's goals.

KMSs need to encourage people to codify their experience, to share their knowledge and to be active in the system. For this purpose we extended the IMS LIP groupings with the Behavior concept. The behavior concept describes some characteristics of a user interacting with a KMS: like type_of_activity, level_of_activity, level_of_knowledge_sharing. The Behavior concept and its subconcepts were introduced to "measure" two processes that are important for the effectiveness of a KMS namely knowledge sharing and knowledge creation.

The user modeling systems classifies the users into three categories: readers, writers or lurkers. These categories are properties of the type_of_activity concept. The level_of_activity comprises four attributes that can be associated with the users: very active, active, passive or inactive. The classification of users according to the type_of_activity or level_of_activity is based on heuristics. As knowledge sharing is a critical aspect for the success of a knowledge management system we have introduced a model, which assigns the users to different categories based on their knowledge sharing behavior. Through the level_of_knowledge_sharing we are capturing the level of adoption of knowledge sharing practices. The level of adoption of knowledge sharing is defined using Near's terminology and mapping it into Roger's theory (see Angehrn and Nabeth 1997). The user states in relation to the level of knowledge sharing are: unaware, aware, interested, trial and adopter. The description of the ontology based user-modeling system and its integration into the KMS can be found in Razmerita et al. (2003).

4 Roles of user models and user modeling in KMSs

An important strand of research in user modeling aims to enhance the interaction between the users and the systems. Numerous researchers have reported on: human-agent interaction, how to construct adaptive systems, how to tailor and filter information, how to personalize help and dialogue systems, how to personalize interaction in e-commerce, e-learning etc. (Brusilovsky, 2001; Kobsa et al., 2000; Stephanidis, 2001; Kay, 2001; Andre et al., 2000; Fink and Kobsa, 2000, etc.) All these traditional application areas of user modeling brought us insights on how user modeling can contribute to enhanced features of KMSs.
The integration of user models and of user modeling processes in a KMS can enhance their functionality in several ways, as represented in Figure 1:
Personalization Personalization is an opportunity to provide more "high touch" features for users. Kobsa and al. (2000), studying personalization in the context of online customer relationship, identify the following categories of adaptive features: adaptation of structure, adaptation of content, adaptation of modality and preferences. We use Kobsa's classification of adaptive features but we adapt it to the context of KMSs.
Adaptation of structure comprises different types of personalized views: personalized view based on job title and personalized view based on the domain interest area. "Personalized views are a way to organize an electronic workplace for the users who need an access to a reasonably small part of a hyperspace for their everyday work." (Brusilovsky, 1998). This category of adaptation exploits mainly the properties of the affiliation concept.

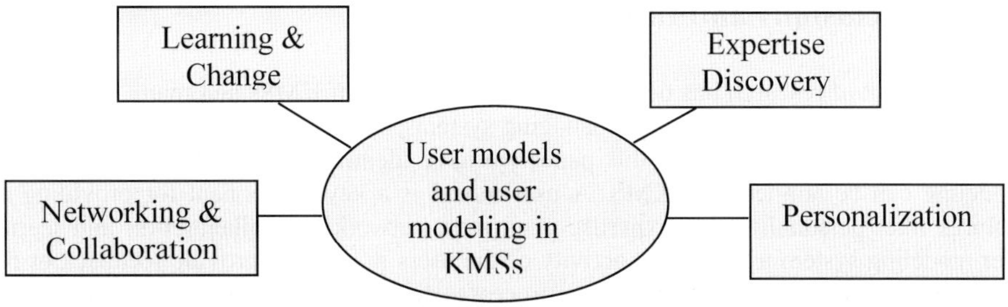

Figure 1: The role of user models and user modeling in knowledge management systems

Adaptation of content could comprise: optional detailed information, personalized recommendations, personalized hints, etc. Different type of agents can be integrated into the system to contribute to a better personalization of knowledge delivery and to provide adapted content (e.g. notification agents, information filtering agents, pedagogical agents, story telling agents etc.) For example, a notification agent could inform a user when new knowledge assets interesting for the user are available to the system. This category of adaptation exploits mainly the properties of the activity and the competency concepts.

Adaptation of presentation and modality can comprise different types of sorting based on various criteria, different types of bookmarks and shortcuts that a user might choose. It empowers the users to choose between different layouts, skins, types of fonts, presence or absence of interface agents, preferred languages, etc. This category of adaptation exploits mainly the properties of the accessibility concept.

Expertise discovery Making the competencies, the qualifications and the domains of interests of the users explicit enables location of domain experts, knowledgeable persons and how to contact them. This is also a valuable option not only for the knowledge workers who need to complete different job related tasks but also for human resource management units especially for big, distributed organizations.

Networking & Collaboration The dichotomy of knowledge as tacit and explicit knowledge implies a requisite for sharing knowledge. The dynamic exchange of tacit knowledge can be facilitated through networking and collaborative tools. Social processes can facilitate networking and collaboration and can be organized to take into account the user's characteristics (hobbies, interests). In certain systems, communities are built based on the user's domain of interests. (Snowdon and Grasso, 2002)

Learning &Change In our view learning is not only a process of acquiring new pieces of knowledge but it often involves a behavioral change for the user at the individual level. We will approach learning from a change management perspective. From this perspective a system can also provide feedback and stimulus for behavioral change at the individual level.

A KMS facilitates storing, searching and retrieving of knowledge assets but a KMS needs also to promote users' participation in knowledge sharing and knowledge creation. Therefore the system tracks a series of "behavioral" characteristics of the user interaction with the system (such as level of activity, level of adoption of knowledge sharing, type of activity etc.). These elements make the user aware of his behavior in the system and are intended to motivate the user to be active in the system. Moreover, based on the identified stages of the users different type of agents can intervene to stimulate and to coach a user towards the adoption of a set of desired behaviors (e.g. adopters of knowledge sharing behavior). More details on the design and the implementation of such a system, called KInCA, are presented in (Angehrn et al., 2001, Roda et al. 2003).

5 Conclusions and future research

Although a lot of research has been conducted in the area of KMSs and many software platforms have been developed as knowledge managing systems, very little work has been done in the field of user modeling for KMSs. In this paper we have identified the different ways in which user modeling can be applied in a KMS. A user model is a key component for providing enhanced features like: personalization, expertise discovery, networking, collaboration and learning. The user modeling system assesses the activity of the users in order to provide feedback or reflection. The user ontology and the user modeling processes described in this paper are part of an ontology-based user modeling system. This user-modeling system is integrated in an ontology-based KMS called Ontologging. Ontologging is an IST EC funded project aiming to implement a next generation of KMSs based on three emerging technologies: ontologies, software agents and user modeling. The evaluation of Ontologging system will provide us further evidence and associated insight to the role of user models and user modeling in KMSs.

References

Andre, E., Klesen, M., Gebhard, O., Rist, T., (2000). 'Exploiting Models of Personality and Emotions to Control the Behavior of Animated Interactive Agents', Fourth International Conference on Autonomous Agents, pp. 3-7, Barcelona.

Angehrn, A., Nabeth, T., (1997) Leveraging Emerging Technologies in Management-Education: Research and Experiences, *European Management Journal*, Elsevier, 15:275-285

Angehrn, A., Nabeth, T., Razmerita, L., Roda, C., (2001). "K-InCA: Using Artificial Agents for Helping People to Learn New Behaviours", Proc. IEEE International Conference on Advanced Learning Technologies (ICALT 2001), August 2001, Madison USA

Brusilovsky, P., (2001). Adaptive Hypermedia, User Modeling and User-Adapted Interaction, Kluwer Academic Publishers, , Printed in the Netherlands, pp. 87-110

Fink, J., Kobsa, A., (2000). A Review and Analysis of Commercial User Modeling Servers for Personalization on the World Wide Web, in *User Modeling and User Adapted Interaction*, Special Issue on Deployed User Modeling, 10, p.204-209,

Kay, J, (2001) Scrutability for personalised interfaces, ERCIM NEWS, Special Theme Issue on Human Computer Interaction, 46, July, 49-50.

Kobsa, A., Koenemann, J., and Pohl, W., (2000). Personalized hypermedia presentation techniques for improving online customer relationships, The Knowledge Engineering Review 16, p111-155

IMS LIP, IMS Learner Information Package http://www.imsproject.org/aboutims.html, 2001

Razmerita, L., Angehrn A., Maedche, A., (2003). Ontology-based user modeling for Knowledge Management Systems, in Proceedings of "UM2003 User Modeling: Proceedings of the Ninth International Conference", USA, to appear

Roda, C., Angehrn, A., Nabeth, T., Razmerita, L., (2003), Using conversational agents to support the adoption of knowledge sharing practices, *Interacting with Computers*, Elsevier, ,

Snowdon, D. Grasso, A., (2002). Diffusing information in organizational settings: learning from experience, Conference on Human Factors and Computing Systems, Minnesota, pp. 331 – 338

Stephanidis, C., (2001). Adaptive Techniques for Universal Access, *User Modeling and User-Adapted Interaction* 11: 159-179, Kluwer Academic Publishers

An Adaptive Human-Computer Design Method Led By Objectives

Charles Santoni

Laboratoire des Sciences de
l'Information et des Systèmes
(UMR CNRS 6168)
Avenue Escadrille Normandie-Niemen.
13397 Marseille Cedex 20 - France
charles.santoni@lsis.org

Pierre Aubert

Laboratoire des Sciences de
l'Information et des Systèmes
(UMR CNRS 6168)
Avenue Escadrille Normandie-Niemen.
13397 Marseille Cedex 20 - France

Abstract

Work presented here, concerns the design of human-computer interaction within the framework of the industrial systems of supervision. In such a context, the user is often considered as a decision-maker [1]. He has to identify different solutions, in order to answer a supervision problem, and choose the most appropriate solution within a complex environment. According to the objectives of the task in progress, the choice can be carried out in various manners by taking into account the characteristics of the user such as, his problem solving ability and his experience level [2]...Thus, for this type of context, it appears important to us to integrate, as soon as the designing phase, adaptivity functions and problem solving functions, in order to generate user-centered [3] interfaces of supervision systems.

The principle of our designing approach makes it possible to consider any session of use as the realization of one or several high level objectives. The designing approach is based on three phases: a conceptual phase allowing the description of the objectives and the definition of the structure of navigation; a logical phase taking into account the presentation aspects of the interface, and an implementation phase allowing the development of the communication interface.

One of the originality of this approach takes place in the concept of the hyperspace [4] used to model the structure and the logical characteristics of a navigation space. In addition, the implementation of the interfacing system is carried out with a multi-agent system making it possible to benefit as much as possible from the distributed aspect of the adaptivity concept.

1 Introduction

The ceaseless progress of the computer systems makes it possible to process a greater number of different types of data. Moreover, in a supervision context, the data flows that the user must apprehend, is becoming more and more large and complex. The only graphics and ergonomics aspects of the current user's interface reduce considerably the knowledge whose user must lay out the system and its interface, and make easier to handle supervision tasks. Nevertheless, it is not enough to face the complexity of the systems. In the framework of supervision systems [5][6], an adaptive human-computer interface could be considered as a result of a compromise between four actors with unceasing and dynamic importance: the users, the tasks, the environment and the computing system [7]. These actors interact to reach a common objective: the control of a

supervised process. The users have to be assisted in order to distinguish the most relevant information in a given situation (in choosing, for instance, the appropriate form of the presentation, the right moment to display it, or the kind of assistance needed). An adaptive human-computer interface enables to simplify complex tasks of an industrial supervision system [8].
The user-centered design method that we are developing, leads toward reducing the efforts of the designer of an interactive supervision system.

2 Design approach

The key point of this approach is to consider any session of use as the realization of one or several high-level objectives. Indeed, we consider that a given objective should be able to be reached by several manners according to the user profile [9]. The identification of these objectives during the designing phase makes it possible to anticipate the individual needs and the preferences of the user and to define directives of navigation in order to satisfy as accurate as possible the objectives.
The development process is composed of three phases:

A conceptual phase which does not take into account the implementation details. It focuses on the navigation structure and content (identification, classification and description of the objectives ; construction of the conceptual diagram of the application domain (process and tasks modeling)).

A logical phase which makes it possible to take into account the presentation and formatting aspects of specific functions of a supervision system (definition of the presentation and the format of the interactive objects).

The implementation phase in the chosen environment (development platform, conversion of the logic diagrams into concrete objects (windows, buttons, lists...)).

3 Models used

The developed method is based on the use of a set of models which aims to guide the interactive system designer throughout the development of the human computer interaction system [8][9] :

The user model, which enables to express the necessary and relevant expectations and characteristics of the different classes of the target users (potential users of the system). It is based on two basic concepts: the *User Role* and the *User Profile*. The concept of *User Role* allows to determine at which category the user belongs to, with regard to the general objective, which should be reached by the use of the interface; the concept of *User Profile* allows to obtain the relevant characteristics of the users.

The user objective model which enables to represent and describe the target of the users' needs, expressed into high-level objective terms. Generally speaking, we can distinguish the informative objectives and the operational objectives (the first type of objectives allows the user to retrieve data, the second one allows to achieve a complex task).

The conceptual model of the domain makes it possible to get an explicit, clear and consistent representation of knowledge and data which concernss the process domain corpus.

The navigation conceptual model allows specifying the structure and the semantic of the navigation within the interface (representing several different views of the interface, each of them integrating the objectives and the specific characteristics of the identified user classes).

The logical level models, which include: the logical model of the users, the hyperspace representation model, the mathematical hyperspace model and the logical model of the domain.

4 The Hyperspace concept

The hyperspace concept [4] represents the logical characteristics of a space of navigation. Each hyperspace refers to a category of operator and can represent either the entire space of navigation offered to the user, or a subset of finer granularity indicated by a unit semantic of navigation, or by a node of navigation or by an element of basic hyperspace. A hyperspace is a static or dynamic composition of elements representing the homogeneous parts of navigation on different abstraction levels. Thus, it makes it possible to represent the same navigation structure in various ways.

This concept allows the designer of human computer interactive systems to concentrate on two additional aspects:

The definition of both the structure and the logical characteristics of the navigation space elements (organization of the various parts of the space),
Specification of the formatting rules for each type of element (physical implementation of each element)

These two aspects being separately processed, some interesting advantages are offered in terms of ease of exploitation and independence with regard to the implementation method.

A hyperspace or a hyperspace element of some level can be decomposed into a number of restricted atoms number, which allows to compute dynamically the space to be presented to the operator, according to his profile. The management of the Hyperspace allows every user to reach a part of this one according to the category to which he belongs.

5 Implementation

Our implementation phase is based on an interface architecture, which is modelled by a multiagent approach. The functions and the characteristics of each actor (user, task, environment and computing system) are performed by agents able to negotiate together in order to compose and display the human computer supervision interface which integrates the constraints established and modelled by each model [10].
The goal of this modelling consists in integrating and benefiting from the distributed nature of the problem of adaptation of human-computer interface, by requiring the active participation of each entity, which composes the system of supervision. Each actor is modelled by a whole set of agents which can control, at their level, the events occurring on the ordered process, the adaptation of the interface (according to the type of user and the nature of the detected events), and to launch the co-operation with the others agents. The advantage is a great simplification of the calculation of adaptation of the user interface.

We distinguish two agent classes within the multiagent organization, [4][11]:

The interface agents associated with different decision levels (synchronizing agents, coordination agents and interaction agents), are directly linked to the implementation of the adaptation calculation of the interface,
The service agents providing a flexible running of the multiagent system, while managing, for instance, the charge distribution into a process of data research (control agents and directory agents).

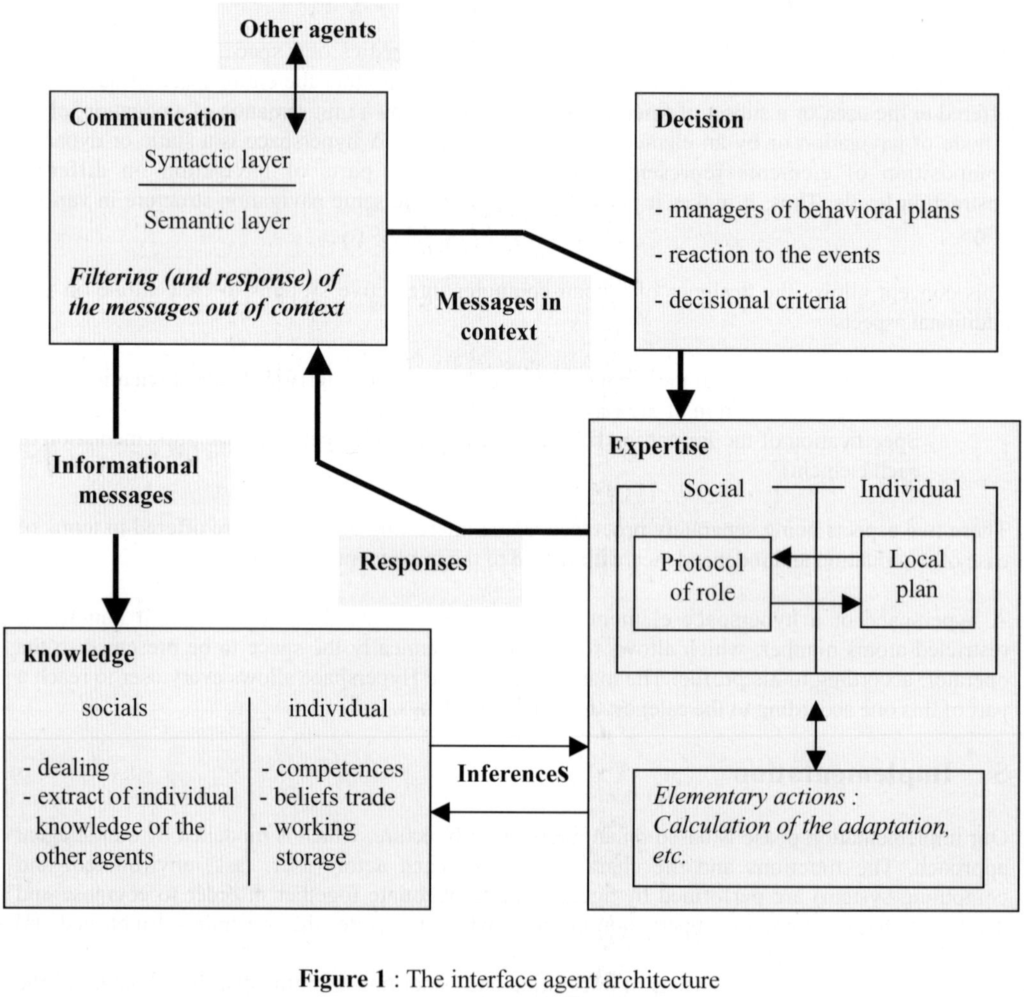

Figure 1 : The interface agent architecture

Each agent is made of several modules, each providing specific and distinct functions: communication, decision, knowledge and expertise (cf. figure 1).
Such architecture of the interface ensures that various profiles of operators (which makes it possible to obtain an adaptation specific to the selection of the commands available and reachable by the operator) as well in tasks of selection of the assistances as in the mode of the data presentation, are taken into account.

6 References

1. P.C. Cacciabue, F. Decortis, B. Drozdowicz, *Modélisation cognitive de l'activité d'un opérateur contrôlant un système complexe.* Symposium sur la psychologie du travail et les nouvelles technologies, Université de Liège. 1990.

2. J. Rasmussen, "Skills, rules, knowledge, signals, signs, symbols and other distinctions in human performance models", *IEEE transactions on Systems, Man and Cybernetics,* Vol SMC 13(3), pp 257-266. 1983.

3. T. Baudel, M. Beaudouin-Lafon, *Outils et Méthodes de construction d'interfaces, Tutoriel IHM'98*, Nantes, France. 1998.

4. P. Aubert, *Etude de la modélisation de l'architecture d'interfaces homme-machine multiagent : lien entre les agents cognitifs et la présentation*, Rapport de DEA, Marseille, France, Juillet 2002.

5. P. Millot, *Supervision de procédés automatisés et ergonomie*, Hermes, *Paris.* 1988

6. J.M. Hoc, *Supervision et contrôle de processus*, La cognition en situation dynamique, Presses Universitaires de Grenoble. 1996.

7. Ch. Santoni, E. Furtado, Ph. François, *Towards adaptive UIMS for supervision systems*, IEEE International Conference on Systems, Man and Cybernetics, Vol. 3, pp. 2598-2603, Vancouver, Canada, October 1995

8. Ch. Santoni, E. Furtado, Ph. François, *« Aid methodology for designing adaptive human computer interfaces for supervision systems »*, HCI International '97, Vol. 2, pp. 501-504, San Fransisco, USA, August 1997.

9. C. Gnaho, *Définition d'un Cadre Méthodologique pour l'Ingénierie des Systèmes d'Information Web Adaptatifs*, Ph-D, Université de Paris I, October 2000.

10. E. Tranvouez, A. Ferrarini, B. Espinasse, *Multiagent modelling and simulation of workshop disruptions management by cooperative rescheduling strategies*, ESS'2001 - 13th European Simulation Symposium and Exibition, Marseille, 2001.

11. B. Espinasse, R. Lapeyre, A. Ferrarini, *A Multi Agents Systems for Modelling and Simulation of Supply Chains*, MLCP'2000, Grenoble, France, 2001.

Mutual Awareness as a Basis for Defining and Assessing Team Situation Awareness in Cooperative Work

Yufei Shu, Kazuo, Furuta, and Keiichi Nakata

Department of Quantum Engineering and Systems Science,
The University of Tokyo
7-3-1 Hongo, Bunkyo-ku, Tokyo 113-8656, Japan
Tel: +81-3-5841-6966 Fax: +81-3-5841-8628
E-mail: shu@cse.k.u-tokyo.ac.jp

Abstract

Our research program is finding out the underlying mechanism of Team Situation Awareness (TSA) and focusing on how to effectively incorporate between human teams and artifacts. In this paper, we argue that the earlier models of TSA, which have been discussed as an intersection of situation awareness (SA) owned by individual team members, are inadequate for study of sophisticated team reciprocal process. We suggest that the definition of TSA is necessary to integrate the notion of individual SA to cooperative team activity. We propose a new notion of TSA, which is reducible to mutual beliefs as well as individual SA in three levels. Then, we develop a feasible TSA inference approach and discuss human competence required to build TSA Through field study, we try to demonstrate how TSA is actively constructed via the inferring practices.

1 Introduction

Situation Awareness (SA) is a key concept that is widely used to assess performance of human-artifact interaction in a number of application domains, such as industrial process control (e.g., nuclear power plant), aviation (e.g., cockpit or air traffic control), etc. In many of these operational contexts, operator crews are embedded in a dynamic environment. Heavily explored by philosophers, psychologists and engineers, Team Situation Awareness (TSA) is becoming more broadly recognized as a critical aspect of team-machine interaction design and assessment. Several studies have already been completed on TSA for these purposes (Salas, 1995). These works mainly discussed TSA as an intersection of SA owned by individual team members. However in a sophisticated team reciprocal process, such intersection is too simplistic in relation to TSA, which has a multi-layered structure of individual SA, mutual awareness, and beliefs among team members. Although individual SA is adequately discussed in various frameworks (Endsley, 1995; Adams, 1995), modeling of TSA involved in a collaborative activity requires a number of additional notions. In particular, understanding of mutual awareness is a necessary element. Furthermore, the ability to infer mutual awareness in TSA is an important aspect to establish collaborative relation between team members or team and machines. The aims of this study are to develop a new notion of TSA, which is reducible to mutual beliefs as well as individual SA, and to propose a TSA inference approach for team-machine cooperative activity

2 Conceptual framework of team situation awareness

We followed the lead of human factors researchers who focused on SA as knowledge created through interaction between a person and her environment and considered TSA as a partly shared and partly distributed understanding of situation among team members. We defined TSA as:
Two or more individuals share the common environment up-to-the-moment understanding of the situation of an environment and another person's interaction with the cooperative task. Firstly, TSA is an awareness of the distributed context and how it changes within a team. Secondly, TSA is an awareness of team members and how they interact within the distributed context. TSA includes two concepts: mutual awareness (MA) and individual situation awareness (ISA). We think that mutual awareness (MA) includes following interrelated elements: mutual responsiveness (Bratman, 1992), knowledge and understanding of each other's work.

3 Definition of TSA

We attempted the use of precise logical formula to express a definition of TSA, which specified a set of heuristic rules. Such rules allow us to reason about TSA, particularly the potential awareness that individuals might have of each other's mental states. As mental phenomenon, we attempt to employ a sort of mental states, such as belief and expectation for other members as evidences to describe TSA. We use modal logic to define TSA, which is similar to that employed in an earlier AI study on joint action (Levesque, 1990). We proposed some formulations to describe the mental processes, such as modal operators, e.g., SA, TSA to interpret awareness, BEL, EBEL and MBEL are used to denote an individual belief, the conjunction of individual beliefs and mutual beliefs respectively and some predicates, e.g., Perceivable, Comprehensible, and Projective to imply mental processes and Hold, State, Symptom to denote system states. The definition of SA in three levels is:

SA1 (m, P) . BEL(m, Hold(P, now)) .Symptom(P)
SA2 (m, P) . BEL(m, Hold(P, now)) .State(P)
SA3 (m, P) . BEL(m, Hold(P, after(now))) .State(P)

The definition of TSA in three levels is:

TSA 1(g, P) . $\cdot_{m,g}$ $\cdot_{Pi,P}$ (SA1(m, Pi).MBEL(g, Hold(Pi, now)))
TSA2 (g, P) . $\cdot_{m,g}$ $\cdot_{Pi,P}$ (SA2(m, Pi).MBEL(g, Hold(Pi, now)))
TSA3 (g, P) . $\cdot_{m,g}$ $\cdot_{Pi,P}$ (SA3(m, Pi).MBEL(g, Hold(Pi, after(now)))).

Where Pi={P|SA (m, Pi)} and P=$\cdot_{m,g}$ Pi.

4 Methodological framework for assessing TSA

From the logic definition, we try to find out the underlying structure of TSA. Considering the dyadic case where the team consists of two members, A and B, TSA(g, p) is simplified as:
TSA(g,P)= SA1(A, Pa).BEL(A,BEL(B, Hold(Pb, t))).BEL(A, BEL(B, BEL(A, Hold(Pa, t)))
holds in A's mind and its counterpart holds in B's mind:
SA1(B, Pb).BEL(B, BEL(A, Hold(Pa, t))).BEL(B,BEL(A, BEL(B, Hold(Pb, t))), where t=now for TSA1 and TSA2, and after(now) for TSA3. It shows that A's TSA consists of three basic components: her own SA (SaPa, or SaCa, SaFa), belief on B's individual SA (BaPb, or BaCb, BaFb) and belief on B's belief on her (A) own SA (BaBbPa, or BaBbCa, BaBbFa) in each of three levels. Here, Pa/Pb is used to denote A's/B's perception, Ca/Cb is used to denote A's/B's comprehension, Fa/Fb is used to denote A's/B's projection, and Ba/Bb is used to denote A's/B's

belief. Later two parts of the formula are corresponding to mutual awareness. This structure is similar to that described in previous philosophical studies (Bratman, 1992; Tuomela and Miller, 1987). Fig.1 shows an illustrative view of the above analysis.

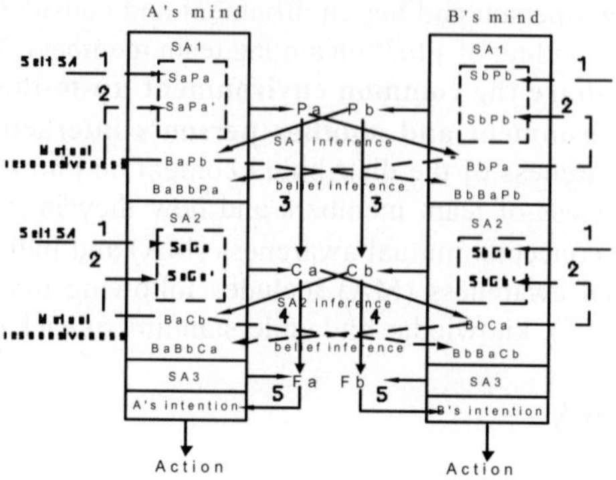

Fig.1 Illustrative structure of TSA and abilities required for its establishment

Since what are in A's mind do not necessarily coincide with what are in B's mind. As for Pa, there are three appearances, and each of them defines a different version of Pa.

Pa={p | SA(A, p)},

Pa'={p | BEL(B, BEL(A, Hold(p, t)))},

Pa"={p | BEL(A, BEL(B, BEL(A, Hold(p, t))))}.

The team has the correct and complete TSA, if and only if the above three versions coincide with each other, i.e., Pa=Pa'=Pa". By assessing the difference among Pa, Pa' and Pa", we can assess the appropriateness of TSA in cooperative activities. Additionally, by comparing the difference between the real TSA and the perfect TSA, we can assess the design of Team-Machine system or team performance.

5 Method for TSA inference

We proposed an inference algorithm of TSA in this study to identify the problem of individual SA and belief inferring. There are two ways to achieve mutual awareness:

1) one is obtained as a result of explicit communication, as in direct face to face verbal exchange or via artifacts;

2) the other one is inferred from implicit verbal communications or from observing the external action of another without explicit verbal exchanges (it is called non-verbal communication).

For communication, we defined communication channels, and considered three communication modes: 1) broadcasting 2) ask and answer 3) instruction.

For the inference method, we identified a set of inferred SA and belief that satisfy the condition SaPa=BbPa=BaBbPa and SaPb=BaPb=BbBaPb for TSA1. It is similar for TSA2 and TSA3. TSA is described as a set of individual SA and mutual beliefs.

5.1 SA inference

The information that is useful for SA inference is what can be observed from the environment, system interface and other team members. In the prescribed system scenario, if we assume that there are two members A and B. The method used in inferring other's perception requires strong mutual knowledge. Firstly, all possible current system state is generated from system symptoms, which are perceived by operator A. Secondly, symptoms that an operator might be expected to pursue are listed from the causality map that is derived by means-ends analysis by consideration of the current system states. Thirdly, procedures to perceive these symptoms are listed in action plan, which prescribes the relation between perceiving actions and all possible perceived symptoms. Then, comparing these perceiving action related symptoms with the expected perceived symptoms, candidates of other's perception are identified. Finally, for each candidate, we use pattern-matching method to calculate its confidence. In this inference method, it is not always the case that one symptom corresponds to one possible state. Under the different system state, the maximum confidence is selected for a certain symptom.

Operator puts the result of perception inference into working memory and uses it as an input of inferring comprehension (SA2). Here, we assume that A believes that B has Pb, also A believes that B has the knowledge to deal with Pb, and A knows this knowledge. Then operator A uses this inferred result (BaPb) to infer Cb with the knowledge as B has.

5.2 Belief inference

B's belief on A's SA (BbPa) is not exchanged explicitly often either. Operator A needs to describe her belief on B's belief (BaBbPa) to infer the B's belief to form her third component. We surmised that B's belief could be inferred from an interpretation of A's own SA, in light of A's beliefs on B's SA that is responsive or in support of A's situation assessment. This corresponds to Bratman's (1992) theory. He thought it is characteristic of shared cooperative activity that each participating agent attempts to be responsive to the intentions and actions of the others, and called this feature "Mutual Responsiveness". We extended this theory so that not only in the intention, or action stage, are the participating agents responsive to each other, but also in the symptom perception and state diagnosis stages. The belief inference procedure is based on ISA inference. Accordingly, the current system state is no longer derived from system symptom perceived by Operator A, but is from the description of B's perceived symptoms, which is formed from A's point of view (BaPb). The inferred belief on perception is an input to infer belief on comprehension. It is also similarity-matching process as previously discussed.

5.3 TSA inference approach

As a result of both ISA and Belief inference, the set of an ISA and beliefs on the other member is obtained, The inference engine then searches for the combinations of them that satisfy the condition: SaPa=BbPa=BaBbPa and SaPb=BaPb=BbBaPb and defines them as TSA.

6 Application of the proposed method

6.1 Plant simulator

The proposed inferring method was applied to operation of DURESS (Dual Reservoir System Simulation) to generate individual SA, belief and test data.

DURESS is a thermal-hydraulic process simulation (Vicente K, and Rasmussen J, 1990). Fig. 2 shows the plant configuration of DURESS. It consists of two redundant feed-water lines that can be configured to supply water to two tanks. On the interface panel, the power of each pump and heater, level of each tank, inlet flow of each tank, outlet flow and temperature, demanded flow and temperature are indicated. All of them are possible perceiving symptoms for Operator A/B to have TSA.

6.2 TSA simulator

Operators' missions are to perceive the symptom, recognize what happened within the simulator plant, and what the other team members perceived, and comprehended. The TSA mo del emulates the operator's information process, situation assessment and interaction with others based upon information received from the environment and supported by an internal mental model. The architecture of the TSA simulation model consists of two distinct but tightly coupled modules: 1) The TSA inference engine, 2) the knowledge base. The data used for the inference (symptom and perceiving action) is put into inference system, then with various kinds of knowledge prescribed in the knowledge base, the inference engine generates its inference result and output are ISA, belief and TSA. The inference engine was developed in a logic programming language (Prolog). Fig. 3 shows the architecture of TSA inference system.

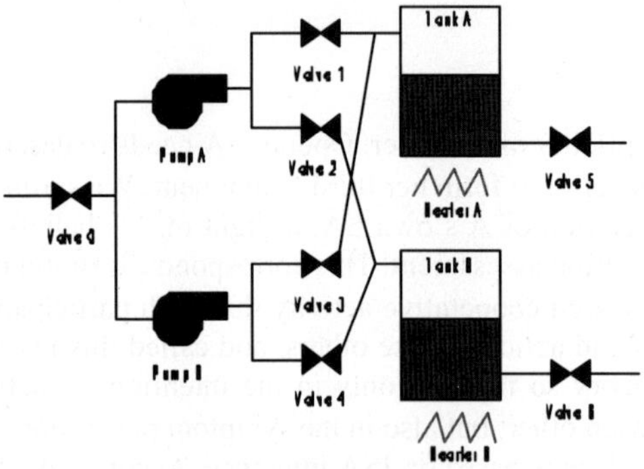

Fig. 2 DURESS configuration

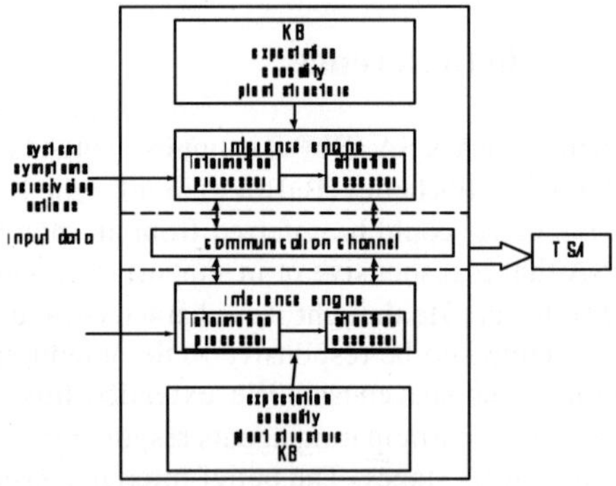

Fig. 3 Architecture of TSA inference system

7 Conclusions

Based upon previous studies of SA, intention, belief, and joint activity in philosophy, AI, and human machine systems, we have proposed a new notion of TSA and supplied its formal definition. We have also proposed a principle to assess appropriateness of TSA and developed concrete methods of inference TSA. It can be used to assess team-machine interaction.

References

1. Adams M., Tenney,Y. (1995). Situation awareness and the cognitive management of complex systems. Human factors, 37(1), 85-194.
2. Bratman M. (1992). Shared cooperative activity. The philosophical review 101(2), P327-341
3. Endsley M.R. (1995). Towards theory of situation awareness in dynamic system. Human factors, 37(1), 32-64
4. Levesque H.J., Cohen P.R. and Nunes J.H.T. (1990). On acting together, Pro. AAAI-1990, P94—99
5. Salas E., Prince C. (1995). Situation awareness in team performance. Human factors, 37(1), 123-136
6. Tuomela R., Miller K. (1988). We intentions. Philosophical studies, 53, P367—398.
7. Vicente K., Rasmussen J. (1990) The Ecology of Human Machine Systems: II. Ecological Psychology, 207-249

The Role of Adaptable Context Representations in Computer Aided Design Activities

Marnix Stellingwerff

Delft University of Technology - Faculty of Architecture
Postbus 5043, 2600 GA Delft, the Netherlands
m.c.stellingwerff@bk.tudelft.nl
and
Hogeschool voor Wetenschap & Kunst, Sint-Lucas Architecture
Paleizenstraat 65-67, B - 1030 Brussels, Belgium
mstel@archb.sintlucas.wenk.be

Abstract

Architectural design is characterized by the important relations between the design object, the user and the urban context. A building cannot move to it's own specific market. Buildings should function in the environment where they stand, for several decades or even for centuries. Therefore, the experience of the urban environment as context for architectural design is an important starting point for most architects.

Design problems are initially ill-defined. A process of exploration, conceptualization, development of alternative solutions and tests of the initial ideas is needed to find interesting and useful solutions. The urban context of a building site can provide hints, inspiration and constrains for various architectural design options. By means of design media, that represent the urban context, the design process can get the form of a constructive conversation between the designer and the applied design media.

This paper will report on an almost finished PhD research for design support systems that visualize the urban environment as a context for new architectural design. The main experiment focused on the constructive conversation while the architect and the researcher 'wandered' through the digitally represented urban environment.

Six architects participated each in two design sessions. During the sessions all actions and reactions were recorded. The virtual reality system showed a 3D-model of a specific urban setting and it allowed to interactively change view-position and to alter various representation types. Thus, the architects were provided with much contextual data for a specific design task. The evaluation of the recordings focused on how the provided data served as information for the architect and subsequently how this information triggered the emergence of their design ideas.

The insights contribute to the current knowledge about the architectural design process, design media and about 3D-models for design. This new knowledge is supposed to lead to more efficient and creative use of existing tools and to developments of new types of design media with new features, which are in accordance with appropriate modes of representation. The gained insights can also lead to a more effective organization of 3D-urban models so that, during the design process, different needs for information can be better supported with accurate visualizations.

This paper will briefly focus on the research framework, the hypotheses and some conclusions of the research.

1 Research framework

In almost every kind of work we have to deal with an increasing amount of information. It is obvious that not all the information can be kept solely in the head. We have to note things down and look things up in agenda's, notebooks, databases and all sorts of documents. The same applies for architects. Architects have the difficult task to deal with the complexity of cities in order to fit a new design into the yet existing urban context. In order to make the fit, it is a necessity to make both the urban environment and the design ideas *'virtually realistic'*. The virtual realism can help to foresee the real future in which design and the context ultimately come together. The superimposed 'image' of the design within its urban setting gives the architect feedback for further development of the design and it allows for having reasonable communication with the client.

The typical aspects for this research are related to information representation and architectural design. The problem, the research goal and benefits can be summarised as follows:

1.1 The problem

- the current urban environment changes rapidly,
- designers need to get and maintain a good view on the actual situation,
- in the design office, the design team depends on collected data and the sometimes cumbersome representation of the data on media,
- the way the collected data is represented has much influence on factual awareness and inspiration during the design process,
- currently available technologies potentially allow designers to get a vivid, changeable, personalised view that instantly answers specific questions about the site, its context and the state of the design,
- however, the new technologies, the designer's preferences, the needs and the effects of different representations are not clear,
- the available media are by far not optimal for providing the right representations, corresponding to the vivid changes in information needs during the design process.

1.2 The goal

- to find a set of visual representation types in perspective projection in accordance with specific information needs of a 'designer in action',
- to find effects of a prototype medium that responds (in a limited way) to the needs of a 'designer in action',
- to be able to advise students and architects on the use of design media and the development of visual representation systems for design and to guide the collection process of useful data for design in a specific context,
- to add new insights on research methods for design and design media research.

1.3 The benefit to the field of architectural design

- the importance of the urban building site as context for design gets renewed attention,
- next to factual realistic information of a building site, there will be new insights on the importance of visual more and less realistic representations in order to foresee the impact

of design in its future context,
- new insights will be gained about inspiration, about getting unexpected thoughts, on the basis of re-presented visual information.

1.4 The benefit to the field of design (media) research

- research of effects and chances for traditional media (e.g. architectural enthescopy and sketching) will get continuation in newer (digital) media, which entail both similar and different characteristics,
- design research will get an example of a 'conversational constructive design study', which partially gains from protocol analyses methods and partially from a more narrative approach that does not exclude 'influences from the research setting' in the discussion of findings.

2 Hypotheses and test-system

In order to make progress in the research, in parallel, theoretical and practical (programming) tracks were followed. The study of numerous theories led to the following set of hypotheses for the research experiment:

1. There is a feeling of situatedness in digital 3D-models that represent a design context.
2. Newly available digital media can provide dynamic views and adaptable representations that lead to different forms of situatedness.
3. Different forms of situatedness stimulate a more vivid reflective conversation, which holds focus, concentration, reframed thoughts and shorter design cycles.
4. Adaptable models OF context provide enough inspiration to become mo dels FOR design.
5. Dynamic views and adaptable representations can be used to evaluate design in its context and stand out against fixed serial visions.
6. The design process (in the office) can benefit from the withdrawal/separation from the real site and all constrains in the built environment as it allows more freedom to think about the (im)possible and (un)desirable alternatives.
7. Combining the 'think aloud method' with intermediate instructions, conversations and direct questions gives more appropriate results.

The first two hypotheses were stated in reference to the work of: (Dibbets, 2002), (Gero, 1997-1998) and (Clancey, 1997).The third hypothesis was stated in reference to the work of: (Schön & Rein, 1994, p.178) and (Schön & Wiggins 1992). The fourth hypothesis was in relation to (Glanville, 1996). The fourth hypothesis was in comparison to (Cullen, 1961). The seventh hypothesis reacts to (Van Someren et.al., 1994) and (Hamel, 1990).

The practical track resulted in a number of programmed prototypes (in VRML and JavaScript) that were finally merged into one system that could be tested by participating architects. These 'test participants' were given tasks related to phases of a design process. According to their 'visual needs' and by means of the prototypes, they could change visual aspects of an urban context-model. They could place their design-model within the context-model and they could also evaluate other designs for the site.

A VRML-model was made of a non-existent neighbourhood. The model consisted of several parts which can be switched on and off in order to get the right view. The view settings can be chosen by means of an icon menu. The object transparency and the atmosphere (fog) in the model were made parametrically adaptable.

While 'wandering' through the virtual model, the architects could express their view-preferences in terms of characteristics (such as: the abstraction or detailed-ness of facades and whether or not there were trees and scale puppets visible), environment characteristics (such as: fog and transparency) and specific viewpoints (such as: eye-level-views or an animated helicopter-view). Several times the architects were questioned about their spatial understanding of the urban site and along the process they could make first sketches and express their design ideas. After the virtual site visit, each architect made a design for the open area in the urban context (see figure 1). In a second research session the design was to be judged in relation to that context. In this second virtual walk, the architects again expressed their preferences and talked about their visual understanding. During the experiment, the model images on the screen were captured and the sound was recorded. This resulted in six illustrated interviews with architects. Thus the research gave many insights in architects' view preferences and the 'design-process-effects' of the views. The flow of the test was arranged by means of a research protocol.

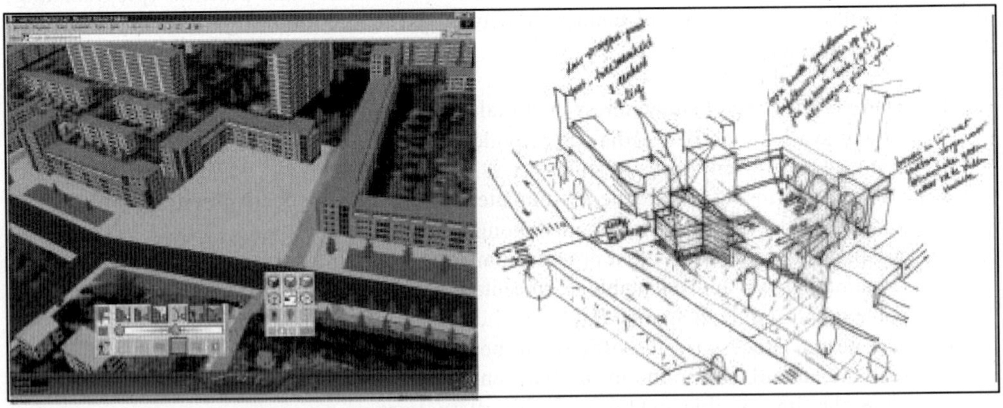

Figure 1: the experiment system and a design sketch by one of the participating architects.

3 Initial research findings

Discourse analyses of the recorded data shows that the virtual representations actually triggered the thoughts of the participating architects. They were situated in the urban context and their ideas were tightly related to what they momentarily saw. Changing viewpoints changed their points of view. Coming back to certain viewpoints activated memories to the same thoughts. However, the lack of enough 'incubation time' for development of design ideas resulted in relatively thin-based design concepts. Overall the experiment provided many outspoken insights of reasoning designers at work.

The research provided many answers and several unexpected insights. In advance of the PhD thesis, the overview in this paper is kept short and just indicative, while it takes many more pages to accurately ground the conclusions. The view-preferences for personal use in the design process differ from the preferences for communication with clients.

As architects are already trained in reading abstract drawings, they often want to see similar abstract views in virtual reality models. The abstractness of form without texture enables them to focus on specific spatial aspects of the design and the site. In the communication with the client, the participating architects preferred to make the design as clear as possible, while the representation of the urban environment (the urban context) could be shown in a less realistic way. In order to achieve this, the use of transparency was an unexpected means to suppress the presence of the environment. Also a black and white representation of the surroundings in combination with a coloured design was preferred. Other interesting results of this research are concerned with the differences between 'overview' (birds' eye view) and 'insight' (eye-level view) and about the emergence of initial design ideas while observing the different model impressions. Most surprising was the way in which the initial design ideas quickly rooted into the architects' narrative.
It can be concluded that such research needs to be continued in order to discover the important role of ever-new design media on the thought processes of architects while they use the new media. Quality of a digital model is too often considered to be equal to features that support realism in the images. Realism is only one quality. Unrealistic, schematic or reduced aspects proved to be other qualities in the representation of a city-model. The provision of environmental context-awareness is possible by means of adaptable model sets in e.g. a VRML presentation. However, it became obvious that the reduction to just one medium impoverishes the perception and understanding of the 'real world' for which architects have to design their buildings. VR can extend the understanding, it cannot replace a site visit. Re-presentation is a powerful aspect of media; the 're' implies adaptability that can be used for alternative views that might trigger alternative thoughts.

4 References

Dibbets, Pauline, (2002). Humans as an animal model? studies on cue interaction, occasion setting, and context dependency. Dissertation, University of Nijmegen.

Clancey, William J., (1997). Situated Cognition; on human knowledge and computer representations. Cambridge University Press, Cambridge.

Cullen, G., (1961). Townscape. London, Architectural Press.

Gero, John S., (1997-1998). Situatedness in design. In: The International Journal of Design Computing, KCDC 1997-1998 volume 1, editors column, see http://www.arch.usyd.edu.au/kcdc/journal/vol1/columns/gero.html

Glanville, R., (1996). The distinctions between models of vs. models for were privately explained after a discussion at the CAD Creativeness Conference 25-27 April 1996, Bialystok, Poland.

Hamel, R. (1990). Over het denken van de architect; een cognitief psychologische beschrijving van het ontwerpproces bij architecten, Publisher Amsterdam : AHA Books.

Schön, D.A. & Wiggins, G. Kinds of seeing and their structures in designing, Design Studies 13(2 1992 135-156.

Schön, Donald A. & Rein, Martin, (1994). Frame Reflection - Toward the Resolution of Intractable Policy Controversies. Basic Books / Harper Collins Publishers, New York.

Someren van M.W., Barnard Y.F. & Sandberg J.A.C., (1994). The think aloud method, a practical guide to modelling cognitive processes, London: Academic Press.

Zeisel, J. (1984), Inquiry by Design: Tools for Environment-Behaviour Research, Monterey, Cal: Brooks/Cole, Cambridge University Press.

Scenario-Based Acceptability Research

Hirotsugu Tahira, Haruhiko Urokohara, U'eyes novas Inc. Japan

8th fl. Fujita INZX Building, 9-5,
Shinsen-cho, Shibuya-ku, Tokyo, 150-0045 Japan
tahira@novas.co.jp, urokohara@novas.co.jp

Abstract

We developed a new quantitative research method using a scenario to obtain information on the context of use, and named "Scenario-Based Acceptability Research." This paper introduces a procedure of this method, how to design an appropriate questionnaire for a test purpose, how to analyze data and evaluate the results, based on our own experience. Various aspects including advantages of this method are also introduced.

1 Introduction

Various attempts to grasp the context of use have done, since analyzing the use of a system by various users has become an essential step to promote a user-centered design approach. A traditional field observation method, however, takes time and cost. While it is difficult to understand users' perception and behavior from conventional questionnaire method, which may lead results ignored the context. Taking the importance of context into account, we developed a new method, called "Scenario-Based Acceptability Research," which makes it possible to conduct a research with an easy procedure, guarantees quantitative results, and encourages users to participate the research more positively and easily. This paper considers the effectiveness of this method, based on our own experience.

2 What is Scenario-Based Acceptability Research?

2.1 Overview

The two faces of "Scenario-Based Acceptability Research" are scenario-based design approach and questionnaires with a checklist. Scenario-based design approach is based on a user interaction scenario, which is a story about people and their activities, using a technology system, in other words, sequences of actions and events to achieve task goals are described. The context from a user's point of view leads to extract problems of the system and design a definite measure to improve it. Questionnaires with a checklist are based on a checklist that scales the importance of some measures to improve the system. Based on the replies from users, the importance of each measure is quantitatively determined to propose an intensive measure, to consider the propriety of each measure, to grasp the difference among users. "Scenario-Based Acceptability Research" focuses on a context oriented approach by a scenario and easy calculation of the scaled data by a checklist. Figure 1 is an example of a questionnaire for all-scenario method of "Scenario-Based Acceptability Research." Actions and events are extracted from the scenario with a definite image of a user and a rating scale is set to each action or event to investigate its acceptability. Test participants who agree with the attributes of the user in the scenario must be screened carefully. Based on the replies from those test participants, acceptability for each action or event is calculated to grasp usability problems, and to lead utility concepts and other various ideas.

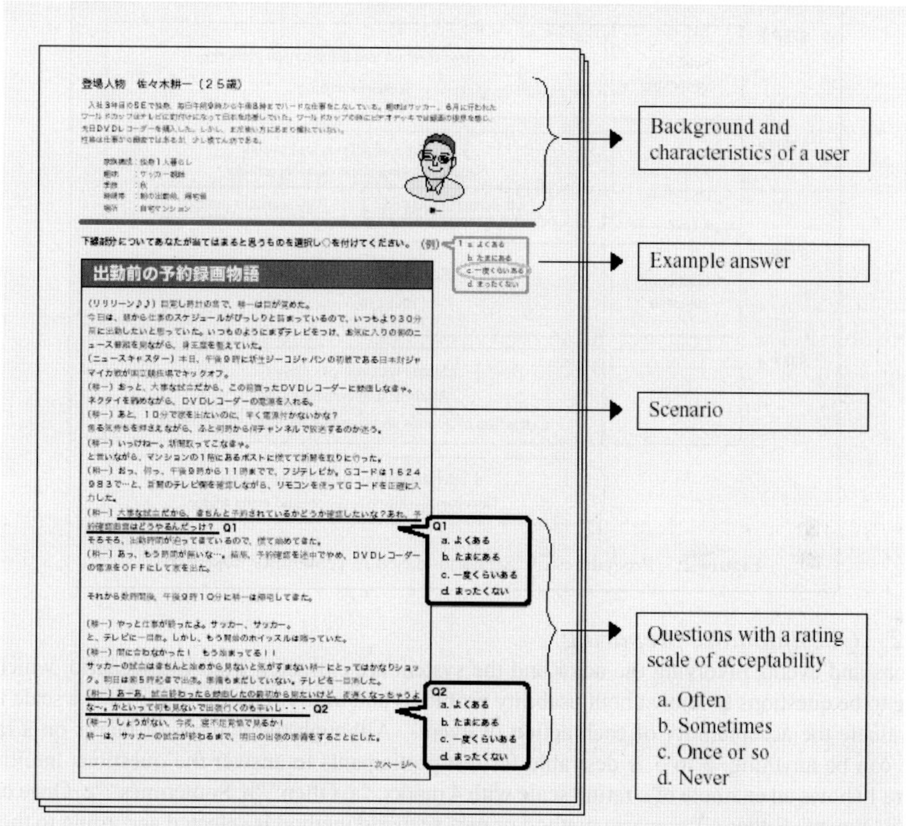

Figure 1: A questionnaire for all-scenario method

2.2 Procedures

2.2.1 Scenario composing

A scenario presupposes a setting where a system is used by an actor in his/her life. The behaviors and thinking of the actor, the appearance and behaviors of the system, and what interactions occur between the actor and the system are described sequentially. The following are tips to make a scenario as rich and realistic as possible:

- Presuppose a setting such as climate, season, time, place, and relationship with other people where the system is used.
- Apart from the purpose of use the system, describe the detailed characteristics of the actor, such as gender, personality, age, family, hobby, and preference.
- Make a story style scenario. The actor's words and thinking should be expressed in natural spoken language.
- Describe the appearance and behaviors of the system in detail.

How well the scenario is composed may impact the replies from the test participants. A scenarist is requested not to compose a scenario based on only his/her experience, but to have a skill to imagine various stories with various kinds of life style. Some systems may require prior investigation and information gathering to compose a scenario with good quality. Figure 2 shows the most general procedure of "Scenario-Based Acceptability Research."

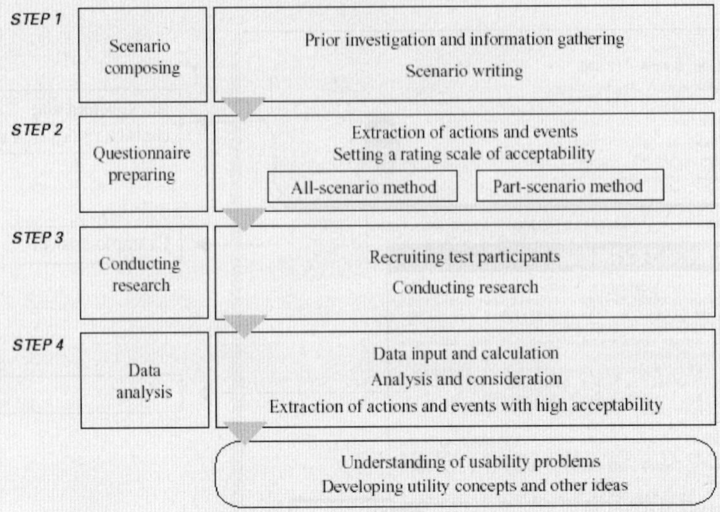

Figure 2: Procedures of "Scenario-Based Acceptability Research"

2.2.2 Questionnaire preparing

Actions and events involving the actor and the system are extracted from the scenario, which are going to be questions to know about usability problems and utility requirements. Then a scale is set to evaluate the acceptability of each action or event. Although the number of marks on a rating scale can be anything, 3 to 5 is desirable for test participants to answer the questions intuitively. Figure 1 shows an example of a rating scale with 4 marks: "a. Often" "b. Sometimes" "c. Once or so" and "d. Never". Either all-scenario method or part-scenario method is selected according to the size and purpose of the research. As shown in figure 1, the whole scenario is given to test participants and questions are underlined or highlighted somehow in all-scenario method. Test participants can easily understand the story and consider the questions more realistically. However, the length of the scenario and the number of questions need to be carefully considered to avoid a burden on test participants. Based on our experience, the appropriate number of questions to each test participant is less than 8. Figure 3 shows an example of a scenario used in part-scenario method. Some parts around questions are taken out of the whole scenario to give to test participants. Larger number of questions can be asked compared to all-scenario method, at the sacrifice of the story. To minimize this sacrifice, the questions need to be written with summarization of the story, which requires extra time and effort to compose.

2.2.3 Conducting research

Test participants who agree with the attributes of the user in the scenario must be screened carefully. If not so, proper data of acceptability may not be gathered. Statistical propriety needs to be considered to decide the number of test participants. This is not different from the case with a general questionnaire. Basically, the research is conducted by distributing the questionnaire to test participants. It becomes more effective if it can be done with surroundings that make test participants easily imagine the context of use. For example, when the research is about car appliances used while driving, drivers who are shopping at retail store of car things or who are at a parking area to take a rest become desirable test participants to gather data with the context of use.

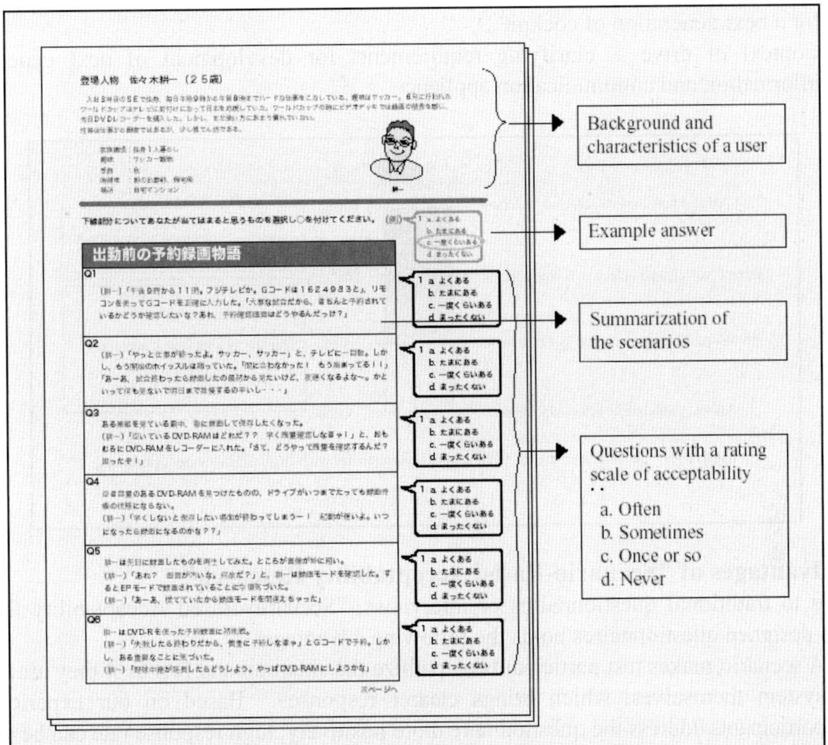

Figure 3: A questionnaire for part-scenario method

2.2.4 Data analysis

The procedure of data analysis is introduced in this section. Any points can be allotted to each mark on a rating scale, apart from the lowest mark 'Never' which should always be allotted 0 point. For example, 3 points to "a. Often", 2 points to "b. Sometimes", 1 point to "c. Once or so", and 0 point to "d. Never". Equation 1 is to calculate the acceptability. X means the point of selected mark, n means the total number of test participants, and k means the highest allotted point among all marks (3 points of "a. Often" in this example). Analysts are to decide a criterion to distinguish questions with high acceptability and low acceptability. For example, assume the analysts decide the case all test participants selected "b. Sometimes" as an important criterion, the calculated acceptability in the case, which is 67 % becomes a borderline as described in figure 4. Based on the acceptability calculated for all questions, problems and requirements are to be prioritized, and then the questions with high acceptability are mainly focused on to lead solutions of usability problems and suggestions of utility concepts.

3 Conclusion

3.1 Case study with "Scenario-Based Acceptability Research"

We have tried to apply this "Scenario-Based Acceptability Research" to various works as follows:

- Creating a concept model of interior elements, and information and communication appliance for a next generation of cockpit for the US
- Context of use research of car navigation systems
- Creating a concept model of information and communication appliance, and future devices

for a next generation of cockpit

- Context of drive ⸾ clarifying requirements for development of next generation of information and communication appliance

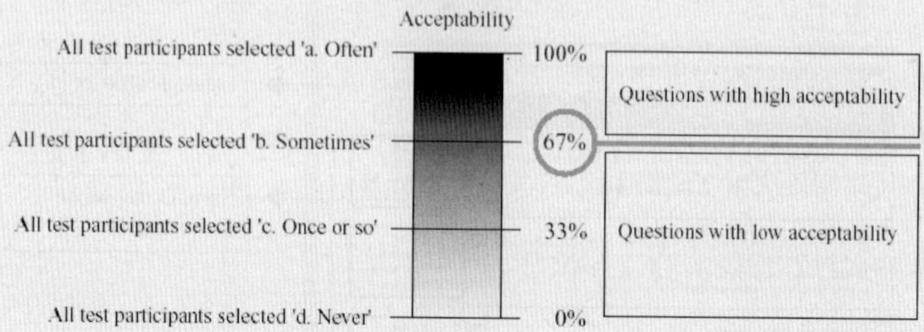

Figure 4: Setting a criterion to distinguish questions with high and low acceptability

3.2 Advantages of "Scenario-Based Acceptability Research"

Compared to traditional questionnaires or interviews, "Scenario-Based Acceptability Research" with well-designed questionnaires holds the following advantages:

- A scenario makes test participants empathize the situation and feel as if they are using the system themselves, which brings clearer responses. Based on our experience, test participants address the questionnaire more positively; high response rate can be expected.
- The priority and importance of usability problems and requirements turns to be quantitatively clear and convincing.
- Problems and requirements from a user's point of view can be extracted out of a scenario, which proposes developers to expand their various ideas toward a definite goal.
- Either all-scenario method or part-scenario method can be selected according to the size and purpose of the research. In other words, flexible approach based on the budget, and available time and effort becomes possible.

These characteristics confirm our belief in "Scenario-Based Acceptability Research" as an effective user-centered design tool. We keep verifying various possibilities to apply this method and its effectiveness.

4 References

Carroll, J.M. (ed.). (1995). Scenario-Based Design Envisioning Work and Technology in System Development. New York: John Wiley & Sons.

Rosson, M.B. & Carroll, J.M. (2002). Usability Engineering Scenario-Based Development of Human-Computer Interaction. Morgan Kaufmann Publishers.

Integrating Machine Learning Methods throughout the Temporal Extent of a Web-based Student Model

Victoria Tsiriga

University of Piraeus, Department of Informatics
80 Karaoli & Dimitriou St, 18534, Piraeus, Greece
vtsir@unipi.gr

Maria Virvou

University of Piraeus, Department of Informatics
80 Karaoli & Dimitriou St, 18534, Piraeus, Greece
mvirvou@unipi.gr

Abstract

In this paper we describe the student modeler of an adaptive and intelligent Web-based algebra tutor, which is called Web-EasyMath. The student modeling process consists of two subsequent phases. The first phase concerns the initialization of the student model, whereas the second phase is responsible for updating the student model based on the observed behavior of the student. Web-EasyMath makes use of techniques inherited from the area of machine learning in both phases of the student modeling process. In particular, the initialization of the model of a new student is based on a novel combination of stereotypes and the distance weighted k-nearest neighbor algorithm. Furthermore, Web-EasyMath maintains the model of each student using a data ageing mechanism as well as the history of the student's actions.

1 Introduction

Web-based educational systems offer platform independence of the application and easy access to it. Students can access Web-enabled tutors from any location, at any time. These advantages ensure that the audience of Web-based applications may be very large and diverse. For this reason, distance learning systems should be highly adaptive to the individual characteristics of each student. However, most existing Web-based educational applications lack the sophistication, interactivity and adaptivity of applications, such as intelligent tutoring systems (Weber and Specht, 1997).

Adapting instruction and feedback to an individual student requires the system to be able to make inferences about the student. Therefore, the student modeling component is very important for Web-based educational applications. The student modeling component of an intelligent tutoring system performs two main functions (Nwana, 1991):

- creates the model of a new student, and
- updates the student model based on the student's interaction with the system.

Although much research work has focused on the identification of efficient methods for updating the student model, the process of the initialization has often been neglected or it has been dealt using trivial techniques (Aïmeur et al., 2002). The initialization of a student model is of great importance and its rationale stems from the fact that it seems unreasonable to assume that every student starts up with the same knowledge and misconceptions about the domain being taught by the educational system. Another important issue that has received little attention in the literature of intelligent tutoring systems is the issue of managing temporal factors of the student models. The management of temporal factors is very important for educational software due to the fact that learning is an evolving dynamic process through which users progress from a situation of unfamiliarity to one of mastery of a knowledge corpus.

Both the creation and the maintenance of student models have been approached using a variety of techniques and methodologies. Recently, machine learning techniques are gaining a lot of popularity in the research area of user modeling. Hence, machine learning has been widely used for modeling the student's knowledge and misconceptions. According to Sison and Shimura (1998), machine learning or machine learning-like techniques have been used (a) to induct a single, consistent student model from multiple observed student behaviors (e.g. Kono et al., 1994; Baffes and Mooney, 1996), and (b) to automatically extend or construct from scratch the bug library of student modelers (e.g. Hoppe, 1994; Sison et al., 2000). Techniques from the area of machine learning have also been used in order to draw inferences concerning student characteristics that are of a higher level of abstraction. For example, Beck and Woolf (2000) have used Linear Regression to determine how likely a student is to answer a problem correctly and how long it will take her/him to generate this response. However, there are no reports on the literature of student modelling to refer to the use of machine learning to address the issue of the initialization of the student model.

In this paper we describe the approach taken for creating and maintaining the model of each student in Web-EasyMath. Web-EasyMath is an adaptive and intelligent Web-based algebra tutor that aims at teaching the domain of algebraic powers. The student modeler of the algebra tutor uses machine learning techniques both to initialize the model of a new student and to update this model based on the observations of the individual student's behavior. In particular, the system creates a model for a new student based on an innovative combination of stereotypes (Rich, 1979) and the distance weighted k-nearest neighbor algorithm (Dudani, 1976; Emde and Wettshereck, 1996). Then, based on the information acquired through the student's interaction with the system, Web-EasyMath maintains the model of each student using a data ageing mechanism (similarly with Webb and Kuzmycz, 1998) as well as the history of the student's actions. Based on the information contained in the student model, the system adapts instruction and feedback to the strengths and weaknesses of each student.

2 The Student Characteristics that are being Modeled

The student model in Web-EasyMath keeps information about which concepts of the domain knowledge are mastered by the student and to what extent. In particular, for each concept in the domain there are two feature-value pairs associated with it in the student model:

- "knowledge level", that takes values within the range [0..1] and indicates the degree of knowledge of the student in this particular concept, and
- "error proneness", ranging from 0 to 1 and which is an estimation of the student's proneness to make mistakes in exercises that test this concept.

Based on these student attributes, Web-EasyMath adapts instruction and interaction to the specific needs of each student. More specifically, the system uses the values of the "knowledge level" and "error proneness" features of the student model in order to support the student when s/he studies theory and solves exercises (Tsiriga and Virvou, 2002a). When the student is studying theory, s/he is supported by the system for the selection of the theory page s/he wishes to study next. There is a combination of two link adaptation techniques to help the student while navigating through the structured theory hyperdocument, namely adaptive link annotation and direct guidance (Brusilovsky, 2001). Furthermore, Web-EasyMath also adapts the presentation of the content of each theory page based on the information contained in the model of the student.

Web-EasyMath constructs dynamically new exercises of different levels of difficulty, based on exercise templates. In order to select the next exercise to present to the student, Web-EasyMath consults the individual student model. Exercises are selected based on the level of knowledge of students concerning the concepts o the domain being taught. After the student has provided her/his solution to an exercise, the system performs analysis of the answer and in case of an error it tries to diagnose its cause. In particular, Web-EasyMath uses information concerning the "error proneness" of the student in each concept in order to disambiguate between possible competing hypotheses about the cause of a mistake.

3 Student Modeling Phases

The student modeling process of Web-EasyMath is partitioned in two different phases. In the first phase the system constructs an initial model for a new student who registers to the system. In the second phase the student modeler of Web-EasyMath updates the student model based on direct observations of the student's interaction performance.

3.1.1 Initial Phase of Student Model Construction

Web-EasyMath uses a novel combination of stereotypes and the distance weighted k-nearest neighbor algorithm to create a model for a new student (Tsiriga and Virvou, 2002b). The student is initially assigned to a stereotype category concerning her/his knowledge level, according to her/his performance on a preliminary test. The system then initializes all aspects of the student model using the distance weighted k-nearest neighbor algorithm among the students that belong to the same stereotype category with the new student. The basic idea is to weigh the contribution of each of the neighbor students according to their similarity with the new student. The similarity between students is estimated taking into account the school class students belong to, their degree of carefulness while solving exercises and their ability in using simple arithmetic operations (addition, subtraction, multiplication and division). The inferences drawn for each new student concern her/his knowledge level and error proneness.

Table 1: Student Models Knowledge Base

Name	Stereotype	Class Code	Degree of Carefulness	Addition	Subtraction	Multiplication	Division
Jim	Beginner	C_1	Careless	0.8	0.8	0.7	0.5
Mary	Beginner	C_3	Averagely Careful	0.9	0.8	0.6	0.7
Sofia	Intermediate	C_1	Careful	0.9	1	0.9	0.8

For example, let us assume the system already has the student records presented in Table 1 in its student models knowledge base. If a new student, after completing the preliminary test was categorized to the beginner stereotype, then the system would initialize her/his student model using information only from students "Jim" and "Mary". Furthermore, if this student was also considered "Careless" and her/his capability in using addition was measured as 0.7, then the information of Jim's student model would have greater contribution to the initialization of the model of the new student. This is due to the fact that Jim's student model is more similar to that of the new student compared to Mary's model.

3.1.2 Maintenance of Student Models

Each time a student issues an action, her/his student model is updated accordingly so as to reflect the current state of the student's progress. More specifically, each time a student visits a theory page that is associated with a specific concept, the feature "knowledge level" of this concept is set to 0.4. This indicates that the student has some knowledge of this concept. However, s/he has to use this concept in exercises before the system infers that the student has satisfactory knowledge of the concept. Every exercise that is associated with a concept is also associated with a difficulty level ranging from 1 (easy) to 3 (difficult). If the student solves exercises of a particular level of difficulty in a percentage greater that a predefined threshold (which is set to 80%), then her/his knowledge level of the associated concept is increased by 0.2 for each level of difficulty. For example, if the student solves the second level of difficulty to a satisfactory percentage, her/his knowledge level is set to 0.8, irrespective of whether s/he has tried to solve exercises of the first level of difficulty.

Furthermore, the data ageing mechanism is used in order to update the student's proneness to make mistakes in each concept of the domain. The data ageing mechanism is rather suitable for reflecting the current knowledge state of a student, due to the fact that it discounts older evidence, placing greater weight on recent ones. In Web-EasyMath, each time a student provides an answer (irrespective of its correctness) to an exercise that evaluates a set of concepts, the error weights of all the evaluated concepts are decreased by a set proportion, for example x. This is done so that the results of the error diagnosis concerning this new answer have more weight. Indeed, according to the error diagnosis results, the system increases the weights of each concept that was not used correctly by a set proportion, y, greater than the one decreased ($x < y$).

4 Conclusions

In this paper we described the student modeling approach of an adaptive and intelligent Web-based algebra tutor. The system makes use of machine learning techniques to create and maintain the student model. Instruction and feedback is then adapted to the strengths and weaknesses of each student. The approach to student modeling that we propose, addresses two main issues that have received little attention in the literature of intelligent tutoring systems. The first is the initialization of the model of a new student and the second concerns the management of temporal factors while updating the student model based on the observed student behavior.

References

Aïmeur, E., Brassard, G., Dufort, H. and Gamps, S. (2002). CLARISSE: a machine learning tool to initialize student models. In: S. A. Cerri, G. Gouardéres and F. Paraguaçu (eds.) *Proceedings of the Sixth International Conference on Intelligent Tutoring Systems, Lecture Notes in Computer Science*, Vol. 2363 (pp. 718-728). Berlin, Heidelberg: Springer-Verlag

Baffes, P. and Mooney, R. (1996). Refinement-based student modeling and automated bug library construction. *Journal of Artificial Intelligence in Education*, 7 (1), 75-116.

Beck, J. and Woolf, B. (2000). High-level student modeling with machine learning. In: G. Gauthier, C. Frasson and K. VanLehn (eds.) *Proceedings of the Fifth International Conference on Intelligent Tutoring Systems, Lecture Notes in Computer Science*, Vol. 1839, (pp. 584-593). Berlin Heidelberg: Springer-Verlag

Brusilovsky, P. (2001). Adaptive hypermedia. *User Modeling and User-Adapted Interaction*, 11 (1/2), 87-110.

Dudani, S. (1976). The distance-weighted k-nearest-neighbor rule. *IEEE Transactions on Systems, Man and Cybernetics*, 6, 325-327.

Emde, W., and Wettshereck, D. (1996). Relational instance-based learning. In L. Saitta (Ed.), *Proceedings of the 13th International Conference on Machine Learning* (pp. 122-130).

Hoppe, U. (1994). Deductive error diagnosis and inductive error generalization for intelligent tutoring systems. *Journal of Artificial Intelligence in Education*, 5 (1), 27-49.

Kono, Y., Ikeda, M. and Mizoguchi, R. (1994). THEMIS: a nonmonotonic inductive student modeling system. *Journal of Artificial Intelligence in Education*, 5 (3), 371-413.

Nwana, H. (1991). User modelling and user adapted interaction in an intelligent tutoring system. *User Modeling and User-Adapted Interaction*, 1 (1), 1-32.

Rich, E. (1979). User modelling via stereotypes. *Cognitive Science*, 3 (4), 329-354.

Sison, R. and Shimura, M. (1998). Student modeling and machine learning. *International Journal of Artificial Intelligence in Education*, 9, 128-158.

Sison, R., Numao, M. and Shimura, M. (2000). Multistrategy discovery and detection of novice programmer errors. *Machine Learning*, 38, 157-180.

Tsiriga, V. and Virvou, M. (2002a). Individualized Assessment in a Web-based Algebra Tutor. In Fernstrom, K. (ed.) *Proceedings of the International Conference on Information Communication Technologies in Education* (pp. 443-448), Athens: National and Kapodistrian University of Athens.

Tsiriga, V. and Virvou, M. (2002b). Initializing the student model using stereotypes and machine learning. *Proceedings of the 2002 IEEE Conference on Systems, Man and Cybernetics.*

Webb, G. and Kuzmycz, M. (1998). Evaluation of data aging: a technique for discounting old data during student modeling, In *Proceedings of the Fourth International Conference on Intelligent Tutoring Systems* (pp. 384-393). Berlin, Heidelberg: Springer-Verlag

Weber, G. and Specht, M. (1997). User Modeling and Adaptive Navigation Support in WWW-based Tutoring Systems. In A. Jameson., C. Paris, and C. Tasso (eds.) *Proceedings of the Sixth International Conference on User Modeling* (pp. 289-300). Vienna, NY: Springer

Relating Error Diagnosis and Performance Characteristics for Affect Perception and Empathy in an Educational Software Application

Maria Virvou, George Katsionis

Department of Informatics,
University of Piraeus,
80 Karaoli & Dimitriou St.
Piraeus 18534, Greece
mvirvou@unipi.gr; gkatsion@singular.gr

Abstract

While the benefits of computer-assisted learning are acknowledged within the educational community, numerous new educational software applications are being developed. Educational software can become very effective if it is adaptive and individualised to the student. However, one important aspect of students that has been overlooked so far in such software applications is affect. This paper describes how system observations of students' behaviour while they interact with an educational application may provide important evidence about students' emotions while they learn. Observations mainly concern the way students respond to assessment questions in terms of the quality and correctness of their answers. The system's inferences about students' emotions are used by the system to adapt interaction to each individual student's needs taking into account their character and mood.

1 Introduction

There is a growing interest for educational software from educational institutions and educational strategic planners as a result of the general acknowledgment of the potential benefits of computer assisted learning. However, to a large extent educational software aims at being used by students without the physical presence of a human instructor. Therefore educational applications have to incorporate as many reasoning abilities as possible. One aspect of students that plays an important role in students' learning and has been overlooked so far is affect (De Vincente & Pain 2002; Kort & Reilly 2002). Indeed, affect has been overlooked by the HCI community in general (Picard & Klein 2002). However, how people feel may play an important role on their cognitive processes as well (Goleman 1995).

In this paper, we present and discuss the student modelling aspects of an educational application that combines evidence from students' errors and detectable performance characteristics in order to recognise important students' emotions and provide appropriate feedback. The educational application that has been used as a test bed for our research is a Virtual Reality game for teaching English. Previous versions of Virtual Reality games did not incorporate such features (e.g. Virvou et al. 2002).

2 Students' emotions in the educational application

The educational game for English has features that are quite common in virtual reality adventure games. Such features include dungeons, dragons, castles, keys etc. In this game the student-player tries to reach the "land of knowledge" and find the treasure, which is hidden there. The difference of the educational game is that one must fight one's way through by using one's knowledge. However, to achieve this, the player has to obtain a good score, which is accumulated while the player navigates through the virtual world and answers questions concerning English spelling. In the game worlds there are animated agents that communicate with the players. There are three types of animated agent, the advisor, the guard of a passage and the student's companion. Animated agents, who act as advisors, lead the student to lessons that s/he has to read. An example of an animated agent who acts as an advisor is the angel, which is illustrated in Fig. 1. On the other hand, guards of passages ask questions to players in order to let them continue their way into the passage and receive more points for their total score. Finally, companions are responsible for showing empathy to the students and help them in managing their emotions while playing and answering questions.

Figure 1: A screenshot of the game.

It is very important that the animated agents may be able to speak to the students. In most of the commercial games there are sound effects that a player hears during the game. However, the effect that makes games more human-like is that of the agents of the game speaking to the player. This feature gives the player a feeling of interplay with the agents, something like the player having a companion in his/her journey. This feature helps the educational game, because the students do not become bored easily.

The game itself may motivate students but it may also cause disappointment and frustration each time a student does not perform so well as s/he would like. Moreover, the testing process where students have to answer questions may cause them anxiety (as exams always do) and thus they may perform worse than they could, if they let their anxiety take over them. On the other hand there are students that may be quite confident and efficient in which case they may only need affective help when they face exceptionally disappointing situations for them (e.g. when they do not remember something correctly and they are not allowed to continue the game).

3 Evidence and inferences

For the purposes of finding out which aspects of the students' emotional state in relation to their performance in the educational game could be modelled, we conducted an empirical study. In this study, computer logging was used to record students' actions while they interacted with the application in a similar way as in (Virvou & Kabassi, 2000) and (De Vincente & Pain 2002). Through computer logging the system may continuously collect objective data for further analysis and interpretation without interfering with users during their interactions with the system (Chou 1999). The collected user protocols were passed on to 5 human experts who were asked to observe students' actions while they played the game and to note down what the students were likely to have felt. As a result, the experts had distinguished between different characters of students and assessed their emotions in relation to the students' characters and correctness of their knowledge.

Taking into account the results of the empirical study, the educational game uses as evidence on students' characters and emotions several actions that relate to typing and mouse movements. Time has played a very important role in our measurements. There are many inferences that can be drawn for the students' feelings and reactions depending on the time they spend before and after making some actions. Some examples of inferences based on observations on time spent for various activities are the following:

- *The time that it takes to the student to answer a question.* This measures the *degree of speed* of the student.
- *Pausing time after a system's response.* The time the computer is left idle after a response to the student is used to measure the *degree of surprise* that the response may have caused to the student.

In addition, certain patterns of actions are used to show aspects of the students' cognitive and emotional state. Some examples of students' actions that are used as evidence are the following:

- *The number of times that a student presses the "backspace" and "delete" button while forming an answer.* This evidence is used to measure the *degree of certainty* of the student concerning a particular answer; the more times the student presses "backspace" and "delete" the less certain s/he is about the answer.
- *Mouse movements without any obvious intent in the Virtual Reality space of the game.* This evidence is mainly connected to the *degree of concentration* or *frustration or intimidation* of the student; the more mouse movements without any obvious intent, the less concentrated or the more frustrated or intimidated the student is. The exact interpretation depends on the context. For example if the mouse movement without any obvious intent occurs some time after the student has been asked a question then it shows frustration since the student does not probably know how to answer.
- *The number of times the student drops out.* This is considered as evidence of *disappointment* or *boredom* depending on the context.

In some cases, inferences are drawn from the combination of two different categories of evidence. For example, the *degree of confidence* is calculated as the means of the degree of speed and the degree of certainty of a student.

The above kind of evidence based on a student's actions is combined with evidence on the student's degree and quality of knowledge of the parts of the lessons that are examined during the game. Therefore for each question asked, the system examines the correctness of the student's answer and if the answer is incorrect it performs error diagnosis. The system also tries to estimate the severity of an error (i.e. whether it was an accidental slip or whether it was due to a persistent misconception).

Examples of some rules that show such kind of combination are the following:

- If a student repeatedly answers questions with a high degree of speed and s/he produces a high degree of incorrectness then this may show *anxiety.*
- If a student repeatedly shows a high degree of confidence irrespective of correctness of his/her answers then this may show *determination* (the student does not give up).
- If a student has given an incorrect answer and then drops out this shows *disappointment.*
- If a student has answered correctly in most questions and drops out before the end of the game then this shows *boredom.*

There are also some more general inferences that can be drawn by using many of the above measurements.

- Degree of efficiency: Depending on the degree of confidence of the student when s/he gives correct answers and his/her degree of concentration, the degree of his/her efficiency is calculated.
- Consolidation: Depending on the percentage of the correct answers and the certainty of the student the system may calculate a degree of consolidation of his/her knowledge.

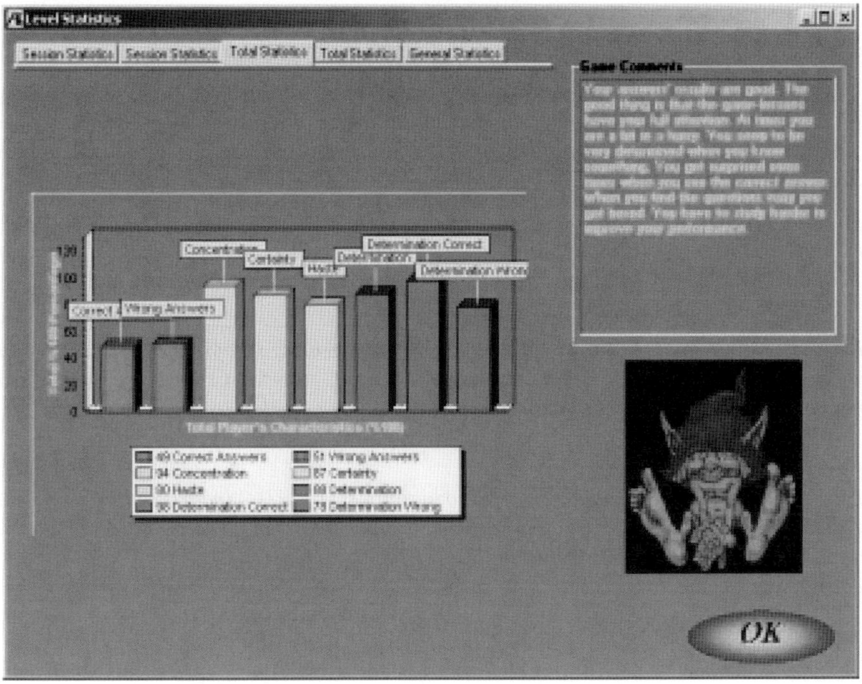

Figure 2: A report of inferences drawn by the system

All kinds of evidence are used by the system to adapt its interaction with the user. Moreover a user may see a report of the inferences drawn by the system. This report is used to show the user a more personal interaction since the system tries to know him/her better. An example of such a report is illustrated in Fig. 2.

4 Conclusions

In this paper we have described how evidence from the students' actions using the keyboard and the mouse may be combined with evidence on the student's knowledge of the domain being taught for drawing inferences for the student's emotional state while interacting with the educational application. Such inferences may be very helpful for adapting the system's advice on the needs of each individual taking into account both his/her knowledge state and character and mood.

References

Chou, C. (1999). Developing CLUE: A Formative Evaluation System for Computer Network Learning Courseware, Journal of Interactive Learning Research, 10(2)

De Vincente A. & Pain H. (2002). "Informing the Detection of the Students' Motivational State: An Empirical Study". In S.A. Cerri, G. Gouarderes and F. Paraguacu (Eds.): Intelligent Tutoring Systems 2002, LNCS 2363, pp. 933-943, Springer-Verlag 2002.

Goleman D. (1995). "Emotional Intelligence". Bantam Books: New York

Kort, B. & Reilly, R. (2002) "An affective module for an intelligent tutoring system". In S.A. Cerri, G. Gouarderes and F. Paraguacu (Eds.): Intelligent Tutoring Systems 2002, LNCS 2363, pp. 955-962, Springer-Verlag 2002.

Picard R. W. & Klein J. (2002) "Computers that recognise and respond to user emotion: theoretical and practical implications" Interacting with Computers 14, pp. 141-169.

Virvou M. & Kabassi K., "An Empirical Study Concerning Graphical User Interfaces that Manipulate Files", Proceedings of ED-MEDIA 2000. World Conference on Educational Multimedia, Hypermedia & Telecommunications, AACE, Charlottesville VA, 2000a. pp. 1724-1726.

Virvou, M., Manos, C., Katsionis, G. & Tourtoglou, K. (2002) "Incorporating the Culture of Virtual Reality Games into Educational Software via an Authoring Tool" In IEEE International Conference on Systems, Man and Cybernetics (SMC 2002).

Acting User Scenario for User Centered Design Team

Kazuhiko Yamazaki

IBM Japan Ltd. and the University of Tokyo
1623-14 Shimotsuruma Yamato city, Kanagawa, 242-8502, Japan
kaz@e.email.ne.jp

Abstract

The purpose of this research is to propose a design approach for information appliances in the age of network computing. This paper focuses on a design method with acting user-scenarios for design collaboration by various types of professionals.

From the user-centered design approach, the collaboration of various types of professionals is important to design information appliances because of complex usage, tools, and systems. One of the difficulties of collaborations among various types of professionals is that it is not easy for them to communicate with each other because their backgrounds, the results of their work, and their technical terms are different. The author has extended the user-scenario method to acting user-scenarios to help the collaboration. The suggested method is that the designer acts out the user-scenario by him.

After proposing the design method, two experiments was performed to evaluate the approach and the results showed that the proposed method helps the designers to design the information appliances. The method supports collaboration among various types of professionals and helps them understand from the user viewpoint.

1 Introduction

This paper focuses on a design method for the creation of information appliances. From the user-centered design approach (IBM, 2000), collaboration with various types of professionals is important to design information appliances (Yamazaki, 1991 and Nielsen, 1993). These professionals include industrial designers, graphic designers, interface designers, usability specialists, and human factors experts as the core of a design team that also extends to planners, engineers, and marketing professionals. One of the difficulties of collaborating with various types of professionals is that it is not easy for them to communicate with each other because their backgrounds, the results of their work, and their technical terms are different.

Another problem for the user-centred design approach is that it is not easy for designers to understand users' feelings and actions, because of the involvement of many resources such as hardware, software, and printed materials such as packages and manuals. Also, the usage is different for each user and the manner of usage changes frequently. To solve these problems, the author extended the user-scenario method to an acting user-scenario that is created by a designer.

2 Problems

There are three problems for design collaboration among various types of professionals: output, motivation, and evaluation.

It is not always easy to understand the output of other professionals. For example, a drawing is a key output from an industrial designer, but for another professional person, it is not easy to imagine a three-dimensional object from a two-dimensional drawing. Icon and screen graphics are key outputs from a GUI designer, but for other professionals, it may be difficult to imagine the

entire software interaction. In order to communicate easily among various designers, common output that most of the professionals will be able to understand is desired.
The next problem is motivation. It is not always easy to understand the working relationships between professionals and the reasons to collaborate with each other. Professionals are individually focused on processing certain parts, and they may not appreciate the importance of collaboration. The method should have capabilities to find the relation with other professionals by each professional.
The third problem is evaluation. When one professional proposes an idea, it may not be easy for other professionals to evaluate the idea, and also it is not easy to understand the relation between the various activities. The design method should provide a capability to easily evaluate others' ideas.

3 Acting User-scenario

Modelling user-scenario is one of the useful methodologies to understand users and share information among designers and related people (Carroll, 1996, Seung-Hun 1999 and Carroll2000). A user-scenario has many purposes such as providing a system vision, a design rationale, usability specifications, functional specifications, a user interface metaphor, prototypes, an object model, a formative evaluation, documentation, training, and help, and an overall evaluation (Carroll, 2000).
The author has extended the user-scenario method to acting user-scenarios to help the collaboration. The suggested method is that the designer acts out the user-scenario by him based on defined user task. For example, in case of software product, designer acts user to pickup user interaction and speak his feelings with draft prototype based on defined user task.
The acting user-scenario is the communication tool for various types of professionals. The design steps are from goals, requirements, creation, evaluation, detail design, and communication to the public. The acting user-scenario can be utilized in any of these steps as a reference for user situations, actions, and requirements.

4 Design Process with Acting User-scenario

As shown in Fig.1, following is the design process with acting user-scenario to design product;

- Define user group: Define several user groups who is targeted user.
- Make basic user-scenario: Make overall user-scenario from getting information of the product, setup, use and upgrade.
- Define user task: Defined step by step user task for one of key user-scenario. User analysis method or observation is good method to help designer. The acting user-scenario method helps designer to define user task and evaluate user task. And also, by sharing this activity by design team, most important information will be shared.
- Make design concept: After defined requirements by user task and user groups, next steps is to make design concept.
- Create Ideas: Based on design concept, several ideas will be prepared. In case of software, it will be better to prepare all sketches for defined user task. Because it will be utilized next phases.
- Evaluate ideas by design team: Before evaluation by user, designer and deign team needs to evaluate ideas by utilize acting user-scenario. Method is designer act user with ideas based on defined user task. Ideas should be visible such as sketch, illustration or draft model. In case of software, there are many specification changes and this method helps designer easy to evaluate.
- Evaluate ideas by user: Ideas should be evaluated by user.

- Decide design: Based on evaluation by design team and user, final design will be decided by design team.

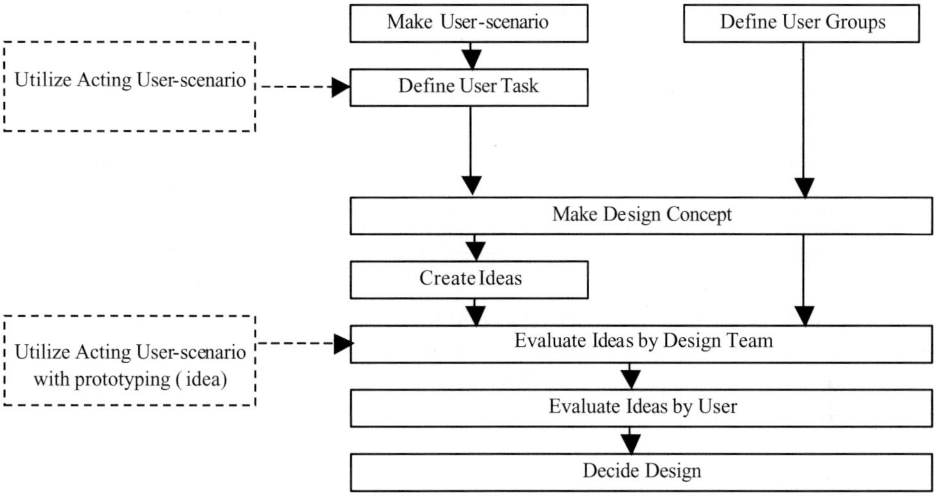

Figure 1: Design process with acting user-scenario

5 Experiment 1

5.1 Purpose and Method

The purpose of the first experiment is to evaluate the proposed design method from designing viewpoint on simple experiment. A total of six volunteers from 24 to 46 years old participated in the experiment, including one woman. All operators are professional designers such as industrial, graphic, and software designers. The participants were instructed how to act out user-scenarios with a written user-scenario as shown in Fig. 2. After each participant made an acting user-scenario, they filled in a questionnaire about the usefulness of acting user-scenarios and their feelings from user viewpoint for each phase.

5.2 Results

As shown in Table 1, these results indicated that the acting user-scenarios are useful in any design phase. All participants indicated acting user-scenario were useful or very useful for communication with various types of professionals.

As shown in Table 2, they indicated it helps designers to experience the user viewpoint by acting user-scenario, except for the evaluation of ideas. The author believes this is because it is not easy for designers to evaluate their ideas, and author confirmed a need to see the actual implementation for evaluation. Overall, all the participants indicated positive or very positive feelings about experiencing the users ' viewpoint by acting out the user-scenario.

Table 1: Usefulness of acting user-scenario for each design phase (number of people)

Design phases	Very useful	Useful	Average	Not useful	Not very useful
Considering concepts	5	1	0	0	0
Creating ideas	3	2	0	1	0
Evaluating ideas	2	2	1	0	0
Making presentations	5	0	1	0	0
Communication with other professionals.	4	2	0	0	0
Overall design	4	2	0	0	0

Table 2: Doing task from user viewpoint for each phase (number of people)

Design phases	Very positive	Positive	Average	Negative	Very negative
Thinking from user viewpoint	2	3	1	0	0
Creating ideas from user viewpoint	2	3	0	1	0
Evaluating ideas from user viewpoint	0	2	4	0	0
Overall (feeling from user viewpoint)	2	4	0	0	0

Background
Situation: Playing golf with a customer at a golf course, Saturday morning
Person: 40 years old businessman who works for a computer company
Tool: Mobile phone with PDA function
Scenario
He is playing golf when a mobile phone alerts him.
He takes the mobile phone out of his pocket and looks at its screen.
The screen indicates emergency e-mail with a confidential mark.
He checks the security status and confirms OK.
He inputs his password and reads the information.
He confirms it is not important.
He inputs his reply, which he will send later.
He takes back the mobile phone into his pocket and continues playing.

Figure 2: Example of user-scenario

6 Experiment 2

6.1 Purpose and Method

The purpose of the next experiment is to evaluate the proposed design method on designing software products. A total of four designer participated in the experiment. All participants are professional designers for software products. The participants were instructed how to act out user-scenarios and design process with acting user scenario. They designed software product by collaboration as one team based on proposed design process with acting user scenario. After each participant designed GUI design for the software, they filled in a questionnaire about the usefulness of acting user-scenarios and their feelings from user viewpoint for each phase.

6.2 Results

As shown in Table 3, these results indicated that the acting user-scenarios are useful in any design phase. All participants indicated acting user-scenario were very useful for evaluating idea by team. The author believes this is because it helps to evaluate detail design, easy to evaluate without user and evaluate in anytime.

Overall, all the participants indicated very useful to feel about experiencing the users' viewpoint by acting out the user-scenario.

Table 3: Usefulness of acting user-scenario for each design phase (number of people)

Design Phase	Very Useful	Useful	Middle	Not Useful	Not Very Useful
Making user task by utilize acting user-scenario	2	2	0	0	0
Creating ideas by utilize acting user-scenario	3	1	0	0	0
Evaluating ideas by utilize acting user-scenario with design team	4	0	0	0	0
Total design by utilize acting user-scenario	3	1	0	0	0
Feel user viewpoint by utilize acting user-scenario	4	0	0	0	0

7 Discussion

The author has proposed a design method for design collaboration using acting user-scenarios. A designer creates this acting user-scenario by himself by acting out the user's task. The major goal of this method is for designers to smoothly collaborate with various professionals and to understand users' feelings and actions. The results of experiments indicate the acting user-scenario helps designer, and most designers feel motivated to use it as a design tool. Also, these results indicated that the acting user-scenarios are useful methods in any design phase, and it helps designers to experience the user's viewpoint during design.

References

Carroll, J.M., 1996, Scenario-based Design – envisioning work and technology in system development, Wiley, US

Carroll, J.M., 2000, *making User-Scenario based Design of human-computer interactions*, The MIT Press, US

IBM, 2000, IBM User Centered Design, http://www.ibm.com/easy

Nielsen, J., 1993, *Usability Engineering*, Academic Press, US

Seung-Hun, Y., Kun-Pyo, L., and Dong-Gun, K., 1999, A Study on the application of scenario based design method for the idea generation, *4th Asian Design Conference*, Nagaoka, Japan

Yamazaki, K., 1991, User interface design for notebook style computer (in Japanese), 7th Human Interface Symposium, Kyoto, Japan, pp143-146

Section 3

Agents and Avatars

Walkable shared virtual space with avatar animation for remote communication

Kinya Fujita and Takashi Shimoji

Tokyo University of Agriculture and Technology
2-24-16 Nakacho, Koganei, Tokyo 184-8588 Japan
kfujita@cc.tuat.ac.jp

Abstract

Human communication over Internet is emphasized by representing the physical image of remote users. A networked virtual reality (VR) system was developed which enables walk-through using a wearable locomotion interface device and the walking avatar animation of remote users. The study consists of two parts of development. The first is a development of a wearable locomotion interface device using walking-in-place, which can be utilized in personal environment. The second is a development of networked virtual reality system, which has functions for walk-through using the developed locomotion interface, communication over Internet and avatar animation of remote users.

1 Introduction

The growth of the computation and computerized graphics power made the virtual space familiar to the public. Internet popularized human communication over such as text chat. On the basis of these backgrounds, networked human communication systems using virtual space are widely studied. Avatar gesture of remote user was implemented for smooth remote communication in virtual space, in a chat communication system (e.g. Cassell 1999).

Walking function is another key to communicate in a shared virtual space because walking allows users moving around voluntarily to control the communication group. As an early study of networked walking, physically synchronized walking between two users has been reported (Yano 2000). In that study, the walking motions of two users were physically linked by synchronizing the two locomotion interface devices over ISDN communication line (64kbps). The purpose of the system was to share the sense of walking as if the two users were a one person, not to control the position of each user in the virtual space. No system has been developed which allows multi-user walk-through in shared virtual space.

In this, study, a wearable locomotion interface device was developed to be utilized at home, and a walkable shared virtual space system with walking avatar animation was developed which enables communication and walking together with other users.

2 Locomotion interface device WARP

Numbers of locomotion interface (LI) devices to allow users walk in a virtual space have been developed. Treadmill-based device is one of the most popular LI. Treadmill allows the users actually walk in the real space, however, it is large and heavy especially in the case the device has a mechanism for free-directional walking, such as Omni Directional Treadmill (Darken 1997). Several types of gesture-based locomotion interface have also been proposed, such as Tilting Disc (Kobayashi 1998). Gesture-based system has advantage in size, however the motion for locomotion control is unnatural because it is different from the actual walking. Therefore, the authors developed a more natural and wearable locomotion interface device using walking-in-place in real space (WARP) (Amemiya 2001). The hardware of WARP, shown in figure 1, consists of two bend sensors for hip joint angle detection, a geomagnetic sensor for body orientation detection, microprocessor PIC16F877 for analog-to-digital conversion and a RS232C communication interface. The walking velocity is calculated using the hip joint angles, and the walking direction is calculated using the body orientation.

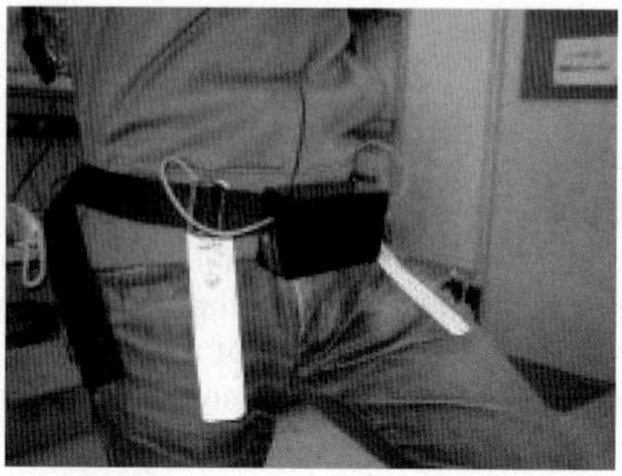

Figure 1: Wearable locomotion interface using walking-in-place in real space (WARP).

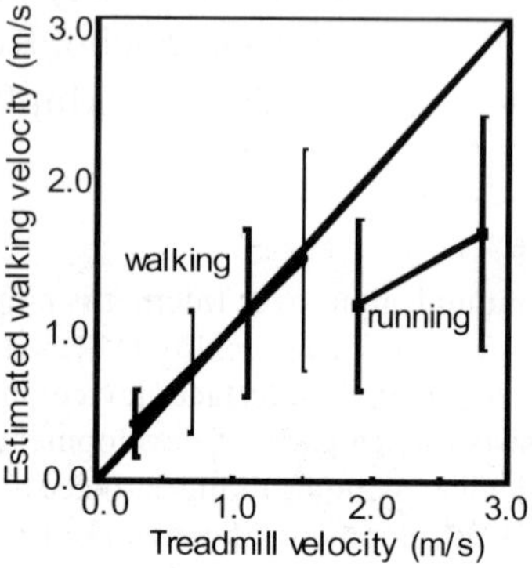

Figure 2: Estimated walking velocity in treadmill walking at various velocities.

By approximating the difference between two hip joint angles during walking to be cosine wave, the walking velocity in virtual space can be calculated from the detected hip joint angle difference θ by using the following equation, where ℓ is the leg length (Amemiya 2001).

$$v = \ell / \pi \sqrt{\left(\frac{d\theta}{dt}\right)^2 + \theta^2 \frac{d\theta}{dt} \Big/ \int \theta dt} \qquad (1)$$

This equation utilizes only the instant values. Therefore, the real-time walking velocity control is possible, which means delay-less walk-through in virtual space. The estimated walking velocity while users walked on a treadmill is shown in figure2. The accurate velocity control was successfully attained. The effect of stride length on estimated velocity was also examined (not shown). It was confirmed that WARP can estimate correct walking velocity at any stride length, because equation 1 reflects both the stride length and stride frequency.

3 Walkable shared virtual space

3.1 Network communication design

Each client needs to communicate over Internet to obtain the information about other users. In the developed system a client/server communication model was utilized as shown in figure3. The server receives the position, velocity vector and curvature from each user, and sends all users status back to each user at constant interval. The suitable communication interval was chosen experimentally as described in section 4, to reduce the network traffic and sever load, and also to minimize the effect of network delay on avatar animation. The prototype software was developed on Windows2000 platform with Visual C and glut library. TCP/IP protocol was utilized for the client/server communication. In the server program, an independent thread was invoked for communication with each client. In the client program, the communication routine was executed in an independent thread to asynchronize the communication with the other calculations.

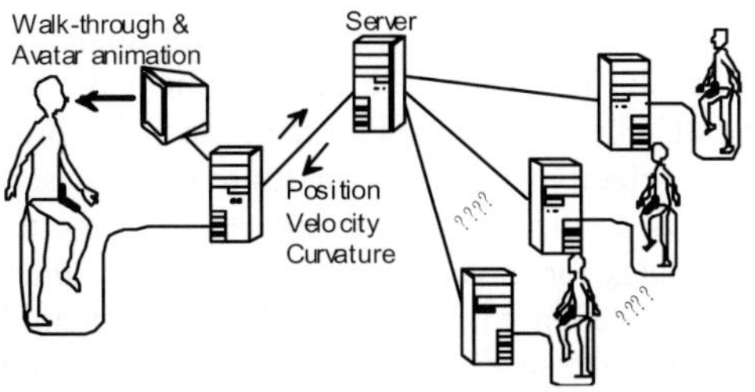

Figure 3: Client/server communication model to share remote-user information

3.2 Prediction and interpolation of remote-user position

For a smooth walking animation of remote-user's avatar with intermittent network communication with inconstant delay, an accurate position prediction and interpolation algorithm is required. The combinations of two prediction methods and three interpolation methods were evaluated. The evaluated prediction methods were the ordinary linear prediction (Macedonia 1994) and the curvature prediction (Wray 1999). The prediction algorithm using curvature is expected to reduce the prediction error in principle, however, the body orientation of the user detected by LI has a fluctuation because of electric noise and mechanical body movement. Therefore, the predicted position contains unexpected position error.

The avatar position needs to be interpolated at each graphical rendering (every 30ms), to move the avatar toward the predicted position. Three interpolation methods, shown in figure 4, were examined. The first is the linear interpolation. It is the most direct method to compensate the error, however, the orientation of the trajectory becomes discontinuous at each communication time. The second is a method to use Hermite curve. It is smooth because the trajectory is continuous, but the unessential orientation change occurs. The third is the proposed "orientation prior method" to reduce the unessential orientation change. It is a method to correct the orientation error by rotating the avatar walking direction. However, the orientation correction does not compensate the position error by itself. Therefore, the position error is reduced gradually by parallel translation. The correction coefficient 0.3 (30 percent of the position error is compensated during one communication interval) was chosen experimentally.

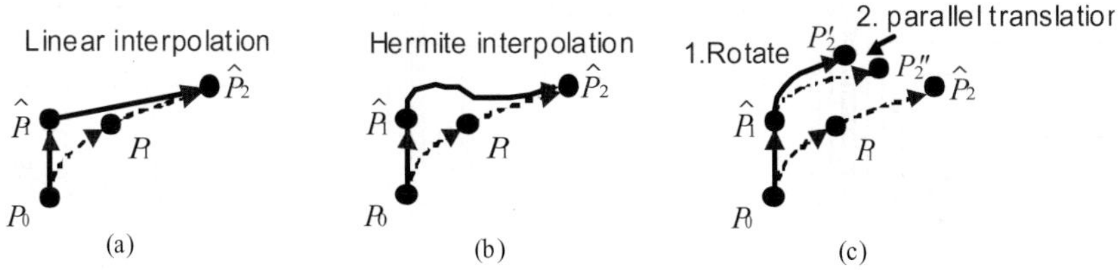

Figure 4: Interpolation methods. $\hat{p}_1$ is the predicted position from the actual position p_0 at t_0 while p_1 is the actual position at t_1. Avatar position was interpolated from $\hat{p}_1$ to predicted position $\hat{p}_2$. (c) is the proposed "orientation prior method".

3.3 Avatar animation

Numbers of works have been performed for generating human walking animation (e.g. Zeltzer 1982, KO 1996). For real-time avatar animation, most conventional rule-based animation method was utilized in this study. The changing patterns of 12 joint angles of the avatar were defined using walking progress index P. The length of each segment of the avatar was obtained by measuring a young male adult. In the developed system, the client program receives the remote-user's gravity center position, velocity vector and curvature from the server at first. The position at the next communication time is predicted and interpolated from the received data as mentioned above. The walking progress index for each rendering time (every 30ms) was calculated using the center of gravity position and stride length. For example, if the current P=0, P will be 0.5 after the remote-user progresses the half of the stride length. The stride length of the avatar was varied with the walking velocity as stride length changes with velocity in actual human walking.

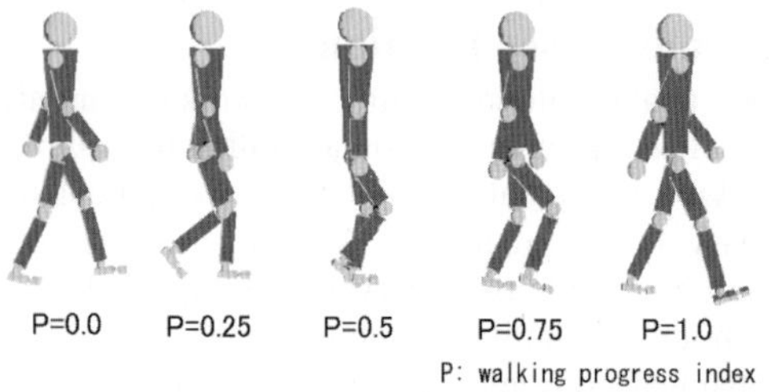

Figure 5: Animation of walking avatar **Figure 6:** Walking in shared virtual space.

4 Evaluation

The example of walking in shared virtual space is shown in figure 6. The system was demonstrated and tested by more than 100 users at the 7th Virtual Reality Society of Japan annual conference. The server performance was examined with 100 client programs running on 10 PCs (10 clients on 1 PC) at 0.8s communication interval. Nominal server response was obtained as designed.

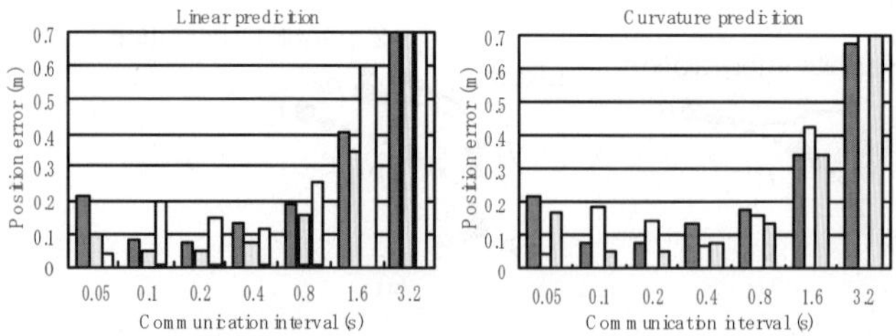

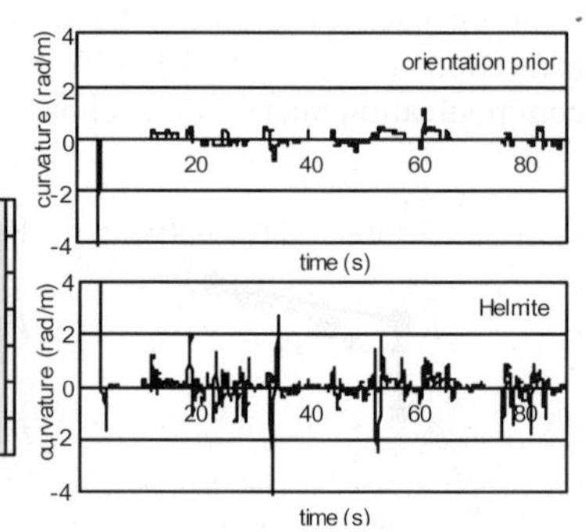

Figure 7: Average position errors. **Figure 8:** Curvature of avatar walking trajectory with two interpolation algorithms.

The prediction error was evaluated with various prediction and interpolation methods by using stored remote user walking trajectory. The results are shown in figure 7. The longer communication interval increased the position error. Especially, the error increased rapidly at longer interval more than 0.8s. Therefore, the developed system utilized communication interval 0.8s. The curvature prediction showed less position error at longer intervals. The interpolation method did not show the significant difference. The effect of interpolation algorithm on orientation change was evaluated by comparing the curvature of avatar trajectory as shown in figure 8. The proposed "orientation prior method" showed smooth walking trajectory than linear and Helmite method. The psychophysical evaluation of naturalness of avatar walking was also performed in 7 young adults using paired-comparison method. The interval scales of the "orientation prior method", Helmite method and linear method were 4.3, -1.1, -.04 respectively. It was confirmed that the reduction of unessential change of avatar walking direction is effective for natural avatar walking animation of remote-user.

5 Conclusions

A locomotion interface using walking-in-place in real space (WARP) was developed as wearable locomotion interface device for at-home use. The real-time velocity control performance of WARP was demonstrated. A walkable shared virtual space with avatar animation of remote users was attained by using WARP and network communication software. The proposed interpolation algorithm was effective for smooth avatar walking animation with intermittent network communication. The developed system is expected for at-home rehabilitation and communication system for elderly people who have difficulty in going out. The voice communication and avatar animation of gestures are currently implemented for more realistic human communication in shared virtual space.

Acknowledgement

This work was partially supported by the program "The R&D support scheme for funding selected IT proposals" of the Ministry of Public Management, HomeAffairs, Posts and Telecommunications.

References

Amemiya,S. Yagi,T. Shiosaki,S. Fujita,K. and Watabe,F. (2001). Development and Evaluation of Locomotion Interface using Walking-in-a-Real-Place (WARP), *Trans. VRSJ*, 6(3), 221-228 (in Japanese).

Cassell,J. and Vilhjalmsson,H. (1999). Fully embodied conversational avatars: making communicative behaviours autonomous. *Autonomous Agents and Multi-Agent Systems*, 2(1), 45-64.

Darken,R.P. Cockayne,W.R. and Carmein,D. (1997). The Omni-Directional Treadmill: A Locomotion Device for Virtual Worlds, *Proc. of UIST*, 7, 213-221.

Ko,H. and Badler,I. (1996). Animating human locomotion with inverse dynamics, *IEEE Computer Graphics and Applications*, 16(2), 50-59.

Kobayashi,M. Shiwa,S. Kitagawa,A. and Ichikawa,T. (1998). Tilting Disc: A Real Scale Interface for Cyberspace, *Proc. of SID 98*, 333-336.

Macedonia,M.R. Zyda,M.J. Pratt,D.R. Barham,P.T. and Zeswitz,S. (1994). NPSNET: A network software architecture for large scale virtual environment, *Presence*, 3(4), 265-287.

Wray,M. and Belrose,V. (1999). Avatars in Living Space, *Proceedings VRML99 Forth symposium on the virtual reality language*, 13-19.

Yano,H. Noma,H. Iwata,H. and Miyasato,T. (2000), Shared Walk Environment Using Locomotion Interface, *Proc. of ACM CSCW2000*.

Zeltzer,D. (1982). Motor control techniques for figure animation, *IEEE Computer Graphics and Applications*, 2(9), 53-59.

"Moving" Avatars: Emotion Synthesis in Virtual Worlds

K. Karpouzis, A. Raouzaiou and S. Kollias

Image, Video and Multimedia Systems Laboratory,
National Technical University of Athens
9, Heroon Politechniou st., 15780 Zographou, Athens, Greece
{kkarpou,araouz}@image.ece.ntua.gr, stefanos@cs.ntua.gr

Abstract

The ability to simulate lifelike interactive characters has many applications. Human faces may act as visual interfaces that help users feel at home when interacting with a computer because they are accepted as the most expressive means for communicating and recognizing emotions. Thus, a lifelike human face can enhance interactive applications by providing straightforward feedback to and from the users and stimulating emotional responses from them. Numerous applications can benefit from employing believable, expressive characters since such features significantly enhance the atmosphere of a virtual world and communicate messages far more vividly than any textual or speech information. In this paper, we present an abstract means of description of expressions, by utilizing concepts included in the MPEG-4 standard. Furthermore, we exploit these concepts to synthesize a wide variety of expressions using a reduced representation, suitable for networked and lightweight applications.

1 Introduction

In everyday life people communicate using speech as well as their face and their body to express their emotions. Research in facial expression analysis and synthesis has mainly concentrated on primary or archetypal emotions. In particular, sadness, anger, joy, fear, disgust and surprise are categories of emotions that attracted most of the interest in human computer interaction environments. Very few studies (Faigin, 1990) have appeared in the computer science literature, which explore non-archetypal emotions. This trend may be due to the great influence of the works of Ekman (Ekman, 1993) who proposed that the archetypal emotions correspond to distinct facial expressions which are supposed to be universally recognizable across cultures. In the contrary psychological researchers have extensively investigated a broader variety of emotions.
Facial and hand gestures and body pose often convey messages in a much more expressive and definite manner than wording, which can be misleading or ambiguous, especially when users are not visible to each other and, therefore, can easily conceal their actual emotional state. While a lot of effort has been invested in examining individually these aspects of human expression, recent research has shown that even this approach can benefit from taking into account multimodal information. As a result, Man-Machine Interaction (MMI) systems that utilize multimodal information about users' current emotional state are presently at the forefront of interest of the computer vision and artificial intelligence community. The real world actions of a human can be transferred into a virtual environment through a representative (avatar), while the virtual world perceives these actions and corresponds through respective system avatars who can express their emotions using human-like expressions and gestures.

In this paper we propose an efficient approach to expression synthesis in networked environments, via the tools provided in the MPEG-4 standard (Preda & Preteux, 2002). More specifically, we describe an approach to synthesize expressions based on real measurements and on universally accepted assumptions of their meaning. These assumptions are based on established psychological studies, as well as empirical analysis of actual video footage from human-computer interaction sessions and human-to-human dialogues. The results of the synthesis process can then be applied to avatars, so as to convey the communicated messages more vividly than plain textual information or simply to make interaction more lifelike (Bates, 1992). Applications of this concept in human-computer interaction are presented, along with actual results of this process, within the MPEG-4 framework.

2 Emotion Representation

The obvious goal for emotion analysis applications is to assign category labels that identify emotional states. However, such labels are poor descriptions, especially since humans use an overwhelming number of labels to describe emotion. Therefore we need to incorporate a mo re transparent, as well as continuous representation, that matches closely our conception of what emotions are or, at least, how they are expressed and perceived.

The activation-emotion space (Whissel, 1989) is a representation that is both simple and capable of capturing a wide range of significant issues in emotion. It rests on a simplified treatment of two key themes:

- *Valence* (Evaluation level): the clearest common element of emotional states is that the person is materially influenced by feelings that are "valenced", i.e. they are centrally concerned with positive or negative evaluations of people or things or events.
- *Activation* level: research has recognized that emotional states involve dispositions to act in certain ways. A basic way of reflecting that theme turns out to be surprisingly useful. States are simply rated in terms of the associated activation level.

The axes of the activation-evaluation space reflect those themes. The vertical axis shows activation level, the horizontal axis evaluation. A basic attraction of that arrangement is that it provides a way of describing emotional states which is more tractable than using words, but which can be translated into and out of verbal descriptions. Translation is possible because emotion-related words can be understood, at least to a first approximation, as referring to positions in activation-emotion space. Perceived full-blown emotions are not evenly distributed in activation-emotion space; instead they tend to form a roughly circular pattern. From that and related evidence, Plutchik (Plutchik, 1980) shows that there is a circular structure inherent in emotionality. An extension is to think of primary or basic emotions as cardinal points on the periphery of an emotion circle. Plutchik has offered a useful formulation of that idea, the "emotion wheel" (see Figure 1).

Figure 1: The Activation – emotion space

3 Facial Expressions

In general, facial expressions and emotions are described by a set of measurements and transformations that can be considered atomic with respect to the MPEG-4 standard; in this way, one can describe both the anatomy of a human face –basically through FDPs, as well as animation parameters, with groups of distinct tokens, eliminating the need for specifying the topology of the underlying geometry. These tokens can then be mapped to automatically detected measurements and indications of motion on a video sequence and, thus, help to approximate a real expression conveyed by the subject by means of a synthetic one.

Modelling facial expressions and underlying emotions through FAPs provides the compatibility of synthetic sequences, created using the proposed methodology, with the MPEG-4 standard. Moreover, archetypal expressions (Parke & Waters, 1996) occur rather infrequently; in most cases, emotions are expressed through variation of a few discrete facial features which are directly related with particular FAPs. Distinct FAPs can be utilized for communication between humans and computers in a paralinguistic form –expressed by facial signs. It is also important that FAPs do not correspond to specific models or topologies, so synthetic expressions can be animated by different (than the one that corresponds to the real subject) models or characters.

An archetypal *expression profile* is a set of FAPs (Tekalp & Ostermann, 2000) accompanied by the corresponding range of variation, which, if animated, produces a visual representation of the corresponding emotion. Typically, a profile of an archetypal expression consists of a subset of the corresponding FAPs' vocabulary coupled with the appropriate ranges of variation.

For our experiments on setting the archetypal expression profiles, we used the face model developed by the European Project *ACTS MoMuSys*, being freely available at the website http://www.iso.ch/ittf. Table 1 shows examples of profiles of the archetypal expression fear (Raouzaiou, Tsapatsoulis, Karpouzis & Kollias, 2002).

Figure 2 shows some examples of animated profiles. Fig. 2(a) shows a particular profile for the archetypal expression *anger*, while Fig. 2(b) and (c) show alternative profiles of the same expression. The difference between them is due to FAP intensities. Difference in FAP intensities is also shown in Figures 2(d) and (e), both illustrating the same profile of expression *surprise*. Finally Figure 2(f) shows an example of a profile of the expression *joy*.

Table 1: Profiles for the Archetypal Expression *Fear*

Profiles	*FAPs and Range of Variation*
Fear ($P_F^{(0)}$)	$F_3 \in [102,480]$, $F_5 \in [83,353]$, $F_{19} \in [118,370]$, $F_{20} \in [121,377]$, $F_{21} \in [118,370]$, $F_{22} \in [121,377]$, $F_{31} \in [35,173]$, $F_{32} \in [39,183]$, $F_{33} \in [14,130]$, $F_{34} \in [15,135]$
$P_F^{(1)}$	$F_3 \in [400,560]$, $F_5 \in [333,373]$, $F_{19} \in [-400,-340]$, $F_{20} \in [-407,-347]$, $F_{21} \in [-400,-340]$, $F_{22} \in [-407,-347]$
$P_F^{(2)}$	$F_3 \in [400,560]$, $F_5 \in [-240,-160]$, $F_{19} \in [-630,-570]$, $F_{20} \in [-630,-570]$, $F_{21} \in [-630,-570]$, $F_{22} \in [-630,-570]$, $F_{31} \in [260,340]$, $F_{32} \in [260,340]$, $F_{33} \in [160,240]$, $F_{34} \in [160,240]$, $F_{35} \in [60,140]$, $F_{36} \in [60,140]$

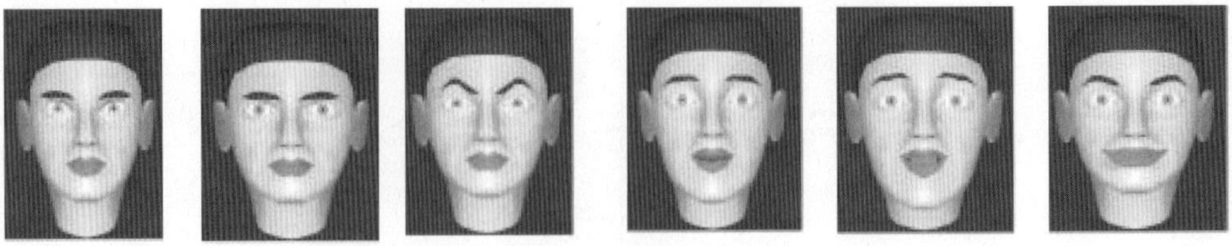

(a) (b) (c) (d) (e) (f)

Figure 2: Examples of animated profile: (a)-(c) Anger, (d)-(e) Surprise, (f) Joy

3.1 Creating Profiles for Intermediate Expressions

3.1.1 Same Universal Emotion Category

As a general rule, one can define six general categories, each characterized by an archetypal emotion; within each of these categories, intermediate expressions are described by different emotional intensities, as well as minor variation in expression details. From the synthetic point of view, emotions belonging to the same category can be rendered by animating the same FAPs using different intensities. In the case of expression profiles, this affect the range of variation of the corresponding FAPs which is appropriately translated; the fuzziness introduced by the varying scale of FAP intensities provides mildly differentiated output in similar situations. This ensures that the synthesis will not render "robot-like" animation, but drastically more realistic results.
For example, the emotion group *fear* also contains *worry* and *terror* (Raouzaiou et al., 2002), synthesized by reducing or increasing the intensities of the employed FAPs, respectively.
Figures 3(a)-(c) show the resulting profiles for the terms *terrified* and *worried* emerged by the one of the profiles of *afraid* (in particular $P_F^{(2)}$). The FAP values that we used are the median ones of the corresponding ranges of variation.

3.1.2 Emotions lying between archetypal ones

Creating profiles for emotions that do not clearly belong to a universal category is not straightforward. Apart from estimating the range of variations for FAPs, one should first define the vocabulary of FAPs for the particular emotion. One is able to synthesize intermediate emotions by combining the FAPs employed for the representation of universal ones. In our approach, FAPs that are common in both emotions are retained during synthesis, while emotions used in only one emotion are averaged with the respective neutral position. In the case of mutually exclusive FAPs, averaging of intensities usually favors the most exaggerated of the emotions that are combined, whereas FAPs with contradicting intensities are cancelled out.
Figure 3(d)-(e) shows the results of creating a profile for the emotion *guilt*, according to the calculated values of Table 2. Plutchik's *angular* measure shows that the emotion term *guilty* (angular measure 102.3 degrees) lies between the archetypal emotion terms *afraid* (angular measure 70.3 degrees) and *sad* (angular measure 108.5 degrees), being closer to the latter.

Table 2: Activation and Angular Measures Used to Create the Profile for the Emotion Guilt

***Afraid* (4.9, 70.3):** $F_3 \in [400,560]$, $F_5 \in [-240,-160]$, $F_{19} \in [-630,-570]$, $F_{20} \in [-630,-570]$, $F_{21} \in [-630,-570]$, $F_{22} \in [-630,-570]$, $F_{31} \in [260,340]$, $F_{32} \in [260,340]$, $F_{33} \in [160,240]$, $F_{34} \in [160,240]$, $F_{35} \in [60,140]$, $F_{36} \in [60,140]$
***Guilty* (4, 102.3):** $F_3 \in [160,230]$, $F_5 \in [-100,-65]$, $F_{19} \in [-110,-310]$, $F_{20} \in [-120,-315]$, $F_{21} \in [-110,-310]$, $F_{22} \in [-120,-315]$, $F_{31} \in [61,167]$, $F_{32} \in [57,160]$, $F_{33} \in [65,100]$, $F_{34} \in [65,100]$, $F_{35} \in [25,60]$, $F_{36} \in [25,60]$
***Sad* (3.9, 108.5):** $F_{19} \in [-265,-41]$, $F_{20} \in [-270,-52]$, $F_{21} \in [-265,-41]$, $F_{22} \in [-270,-52]$, $F_{31} \in [30,140]$, $F_{32} \in [26,134]$

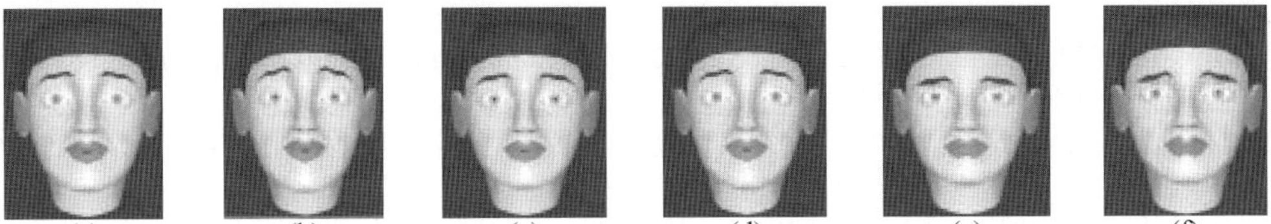

(a) (b) (c) (d) (e) (f)

Figure 3: Animated profiles for (a) afraid, (b) terrified (c) worried, (d) afraid, (e) guilty and (f) sad

4 Gesture Classification

In general, an MPEG body is a collection of nodes. The Body Definition Parameter (BDP) set provides information about body surface, body dimensions and texture, while Body Animation Parameters (BAPs) transform the posture of the body. BAPs describe the topology of the human skeleton.

The low level results of the approach can be extended, taking into account that hand gestures are a powerful expressive means. The expected result is to understand gestural interaction as a higher-level feature and encapsulate it into an original modal, complementing speech and image analysis in an affective MMI system (Wexelblat, 1995). In general, one can classify hand movements with respect to their function as:

- *Semiotic*: these gestures are used to communicate meaningful information or indications
- *Ergotic*: manipulative gestures, usually associated with a particular instrument or job and
- *Epistemic*: again related to specific objects, but also to the reception of tactile feedback.

Semiotic hand gestures are considered to be connected, or even complementary, to speech in order to convey a concept or emotion. Especially two major subcategories, namely *deictic gestures* and *beats*, i.e. gestures that consist of two discrete phases, are usually semantically related to the spoken content and used to emphasize or clarify it (McNeill, 1992).

5 Conclusions

Expression synthesis is a great means of improving HCI applications, since it provides a powerful and universal means of expression and interaction. In this paper we presented a method of synthesizing realistic expressions using lightweight representations. This method employs concepts included in established standards, such as MPEG-4, which are widely supported in modern computers and standalone devices.

References

Bates, J. (1992). The role of emotion in believable agents, *Comm. of the ACM*, 37 (7), 122-125.

Ekman, P. (1993). Facial expression and Emotion. *Am. Psychologist*, 48, 384-392.

Faigin, G. (1990). The Artist's Complete Guide to Facial Expressions. New York: Watson-Guptill.

McNeill, D. (1992). Hand and mind: what gestures reveal about thought. Univ. of Chicago Press.

Parke, F. & Waters, K. (1996). Computer Facial Animation. A K Peters.

Plutchik, R. (1980). Emotion: A psychoevolutionary synthesis, New York: Harper and Row.

Preda, M. & Prêteux, F. (2002). Advanced animation framework for virtual characters within the MPEG-4 standard. *Proc. of the Intl Conference on Image Processing*, Rochester, NY.

Raouzaiou, A., Tsapatsoulis, N., Karpouzis, K. & Kollias, S. (2002). Parameterized facial expression synthesis based on MPEG-4. *EURASIP Journal on Applied Signal Processing*, 10, 1021-1038.

Tekalp, M. & Ostermann, J. (2000). Face and 2-D mesh animation in MPEG-4. *Image Communication Journal*, 15 (4-5), 387-421.

Wexelblat, A. (1995). An approach to natural gesture in virtual environments. *ACM Transactions on Computer-Human Interaction*, 2 (3), 179 – 200.

Whissel, C. M. (1989). The dictionary of affect in language. In R. Plutchnik & H. Kellerman (Eds), *Emotion: Theory, research and experience: vol 4, The measurement of emotions.* New York: Academic Press.

Personality Engineering for Emotional Interactive Avatars

Simon Lock, Jen Allanson, Paul Rayson

Computing Department
Lancaster University, Lancaster, LA1 4YR, UK
{lock, allanson, paul}@comp.lancs.ac.uk

Abstract

This paper describes the design, implementation and deployment of an emotive, interactive avatar. In particular we focus on the process and practice of engineering a suitable personality for an interactive avatar. We begin by introducing an architecture that supports the development of artificial personalities. We then describe the utilisation of this architecture in the construction of a unique personality for an interactive avatar. Following this, we describe the avatar in use in a real world public setting. Along the way, we highlight various factors and considerations concomitant with the engineering of affective avatars, agents and other similar technologies.

1 Introduction

Avatars are most closely associated with virtual reality (VR) applications, where they have traditionally been used to physically represent users within virtual environments. Avatars also occasionally appear in other media as virtual presenters of information [6]. In both of these domains however each avatar's behaviour is dictated directly by the actions of the person whom it represents, they are in effect computer-generated puppets. In recent years we have witnessed an increasing interest in *agents* as representative of users in various domains. The major difference between agents and avatars is that agents can act under their own volition, they learn and display "intelligent" behaviour. In addition agents often do not require a physical manifestation as the focus is on behaviour rather than appearance.

In 1997 Picard first discussed the requirement for human-computer interaction to become engaging on an emotional level [7]. Models of interaction and frameworks for the development of affective agents are now emerging [2, 4, 8]. In addition scripting tools have been explored for the development of agent personalities [9]. This paper describes an alternative method for the construction and operation of emotive, interactive avatars. The avatar whose construction is described was created for a performance art event. The underlying architecture of the avatar was conceived so that we might explore issues of engagement, interaction and conceptual design of interactive, public virtual personalities.

We begin this paper by looking briefly at the role of the avatar in human-machine interaction. Thereafter we introduce our architecture, conceived to support the development of emotionally responsive avatars. A description of the architecture in use for the creation of a particular avatar called "Andrine" follows. We then describe the deployment and use of Andrine. Finally we talk about the wider implications of this work and discuss our plans for future work in this area.

2 Avatar Architecture

We have developed an architecture known as LIA (Light-based Interactive Avatar). LIA is an infrastructure for the construction of reactive video-based avatars. Using the architecture it is possible to prototype emotionally-responsive virtual entities. This in turn enables us to investigate issues of engagement, interaction and representation of information. LIAs are not agents, although they do have agency in that LIA's choose how to respond to a given interaction.

Due to the fact that our first LIA was to be used in a noisy public space, we had to initially limit feedback to visual responses. The vocabulary of emotions currently supported by LIA is based on the expression groups identified by Ekman [3] and includes happy, sad, laughing, angry, annoyed, disgusted. We have also included one generic expression (i.e. open to interpretation). In order to present these emotions, LIA utilises a set of video clips, each of which exhibited a different emotion. The use of these video clips was a conscious decision not to use a computer generated character, as is common in other facial presentation systems [6, 8]. By making LIA presentations video-based we could bypass technical issues of the computer generation of a realistic face and concentrate our effort on behaviour.

The LIA architecture supports a multitude of sensors as potential input sources. One source, which is discussed further here, is short messages service (SMS) text messages.

Text support requires use of the following components from the architecture:

- Message source - provides a supply of textual messages for processing by LIA
- Base Lexicon - classifies the possible senses of a large corpus of common words
- Message interpreter - processes incoming messages from the message source and determines suitable emotions, based on word classification derived from lexicon
- Personality module - a "pluggable" set of components which define an individual personality
- Preference Rules – a set of personal likes and preferences for an individual personality
- Video library - set of video clips for an individual personality, with one for each supported emotion
- Avatar renderer - presents a moving, graphical representation for each of the emotions required by the message interpreter.
- Presentation – large scale projection of the rendered emotion

The base lexicon takes the form of a plain text file containing over 36000 words, each of which is associated with one or more semantic categories. There is a total of 232 different categories with which a particular word may be associated. These include categories such diverse as getting or giving, easy or difficult, safety or danger, anatomy, health and disease, emotional states, food and drink, politics amongst many others. For many categories, positive and negative sub-classes exists. Thus for emotional states, positive sub-classes include calm, happy and confident, whereas negative subclasses include angry, sad and concerned. Of the 232 classifications, 30 were directly relevant for use by LIA. Each of these 30 categories is then mapped to one of the 7 emotions supported by LIA.

The purpose of the message interpreter is to parse the incoming messages and tokenise them into distinct words. The lexicon component of LIA is then used to classify the words in an attempt to derive an abstract "meaning" for the message and thus an appropriate emotion for the avatar to present. Due to the fact that each word can have multiple classifications within the lexicon, a mechanism for ensuring the appropriate selection of the correct classification was introduced. This took advantage of the fact that the classifications of a word in the lexicon were present in general usage probability order.

The avatar renderer produces a display of video representing an appropriate emotion for the stimulus message. Non deterministic and non linear playback of this video is employed to achieve the required presentation of emotions. This involves jumping to particular points in the video and decoding and presenting discrete segments for the display of emotions. This has the effect of presenting a continuous and seamless stream of video, the content of which can be dynamically selected and generated in real-time.

3 Personality engineering

It is possible to engineer any number of different personalities for use with LIA. Each "personality model" is associated with it's own graphical representation as well as a unique set of preferences, likes, dislikes, opinions and so on. For the purposes of this study, an effective personality, known as "Andrine" was constructed on top of the LIA architecture.
An essential element in the construction of an artificial personality is the selection of a suitable graphical representation. Figure 1 shows Andrine's facial expressions for the seven supported emotions.

Figure 1. The seven emotions as portrayed by Andrine

The initial lexicon was derived from previous work in lexical analysis [10, 12] and as such provides an extensive collection of preference neutral background word senses. In order to achieve a realistic artificial personality however, it was essential to inject the personality with a unique set of tastes preferences and opinions. To aid this process, a group of volunteers were asked to provide sets of words which invoked each of the seven supported emotions.

The word-emotion pairs identified in this exercise are referred to as preference rules and can be inserted directly into the lexicon with all of the other entries. These preferences rules differ from the normal contents of the lexicon in that they incorporate the implicit value judgements made by volunteers regarding emotional inferences the words. For example what classification should the word "skiing" be allocated? Such a decision must involve an explicit judgement as to the avatar's preferences, likes and dislikes.

An issue, which results from the manner in which the preference rules were constructed, is the potential for conflicting judgements on the inferences of particular words. Two individuals working on the construction of the preference rules may not agree of the classification and inferences of a particular word or two similar words. To solve this problem, various conflict resolution policies are possible. Automatic consistency checking is difficult due to the need for explicit semantics to defined the relationships between words. Thus to ensure consistency in the development of Andrine, a personality editor was appointed whose job it was to have the final say on the classification of individual words

4 Discussion

Andrine was deployed at the Art-Cels performance and interactive arts event [1]. The Art-Cels event provided us with a large, diverse group of users. Our aim was simply to engage passers by in an interaction (bearing in mind the requirement to SMS Andrine, the users would effectively be paying to interact with her!). Andrine was projected above the bar area (Figure 2) and introduction cards were left on the bar (Figure 3)

Figure 2. The operational environment of LIA

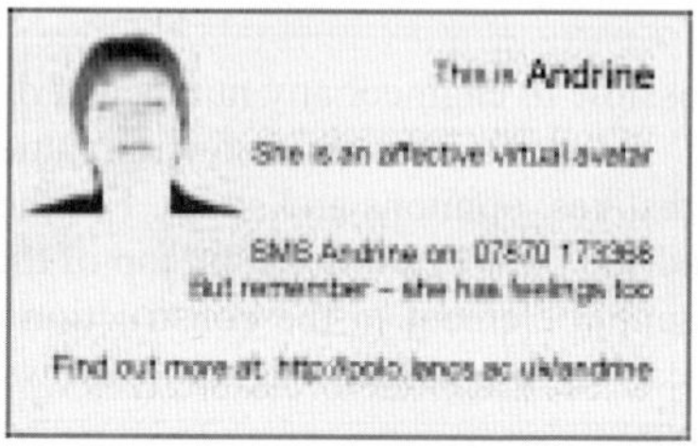

Figure 3. Contact cards for the Andrine personality left on the bar

Over a period of approximately 3 hours a total of 55 text messages were received by the system. The length of the messages received ranged from single word, exploratory test massages through to relatively long sentences (the largest message was 13 words), with the average length of message being 2.9 words. The content of the messages included observations (e.g. "funny lady"), questions (e.g. "is human perception analogue or digital") and requests (e.g. "smile!"). The breakdown of all messages into these three categories is shown in Table 1.

Table 1 Occurrence of message types

Message type	Number	Percentage
Requests	16	29.1%
Observations	13	23.6%
Questions	7	12.7%
Other	19	34.5%

LIA's message interpretation mechanisms, which attempts to overcome errors in the interpretation of single words was seen to perform best with longer messages. Despite the relatively short length of the messages received during the study, this has not had a serious impact on the performance of LIA. Of the 55 messages which were received, 43 (78.2%) were interpreted in a manner that would seem appropriate and only 12 (21.8%) were interpreted in what we considered to be an inappropriate manner.

Although the system did not make information about the sender or the content of messages publicly available, users could not be certain of their anonymity. Due to the fact that interaction took place via a user's phone, tracing messages back to the sender could be possible. This may have effect the freedom and frankness with which the content of messages were composed. On the whole, the messages sent by users were polite, respectful and courteous . There were however a small number of unpleasant, abusive or obscene messages. From the existence of such messages, it is possible to deduce that users viewed the avatar as a non-human entity since the normal social rules controlling acceptable conversation between human participants did not seem to be applied.

5 Conclusions, Contributions and Future Work

Using the LIA avatar creation architecture, we achieved the aim of engaging users and we achieved this through the minimal use of guidelines and rules. Exploration, investigation and discovery on the part of the users can be seen as being the key to the system's success. In addition we were pleased with the high success rate in terms of appropriate responses from the system to the users' input texts. Although the dialogue supported by the system was limited, the potential ability to draw the user into discovering the system's functionality is exciting. In addition we observed user's sharing information with other, would-be users of the system.

An increasing amount of research effort is focussing on issues concerning shared use of public displays [5]. Touch-screens are an obvious means to access these displays. Increasingly interaction with large screens is being realised through the use of personal digital assistants (PDAs). However, PDAs have so far failed to capture the imagination of the wider public. We would argue that the prevalence of mobile phones, coupled with the fact that users' personal phones provide them with a familiar interface, maximises access to interactive public display and information-sharing systems. Issues of charging can be overcome with a little imagination.

We see our work in context of other projects exploring the potential richness of the public information user-experience through clever deployment of technology [11]. We are particularly interested in the potential of entities like Andrine engaging users in a way that encourages them to "discover" the functionality of the system. We need to consider how the dialogue, once initialised, may continue, and in support of what real-world activities? Also, how might we manage multi-user interactions? Answering these questions will provide the focus for future research in this area.

References

[1] Art-cels workshop and performance party homepage, http://www.art-cels.com

[2] Bazzan, A. L. C., Bordini, R. A., Framework for the Simulation of Agent with Emotions: Report on Experiments with the Iterated Prisoner`s Dilemma, proceedings of the 5th Int. Conference on Autonomous Agents, Montreal, pp292-299, 2001.

[3] Ekman, P., Emotion in the Human Face, Cambridge University Press, 1982.

[4] Gratch, J. and Marsella, S., Tears and Fears: Modeling emotions and emotional behaviors in synthetic agents, proceedings of the 5th International Conference on Autonomous Agents, Montreal, Canada, pp278-285, 2001.

[5] Greenberg, S., Boyle, M. and LaBerge, J., PDAs and Shared Public Displays: Making Personal Information Public, and Public Information Personal, Personal Technologies, vol.3(1), March. Elsevier, pp54-64, 1999.

[6] LifeFX homepage, http://www.lifefx.com/

[7] Picard, R., Affective Computing, The MIT Press, 1997.

[8] Predinger, H. and Ishizuka, M., Simulating Affective Communication with Animated Agents, proceedings of the 8th international conference on Human-Computer Interaction (INTERACT 2001), Tokyo, Japan, , pp182-189, 2001.

[9] Prendinger, H. and Ishizuka, M., SCREAM: Scripting Emotion-based Agent Minds, poster session, proceedings of the 1st International Joint Conference on Autonomous Agents and Multi-Agent Systems (AAMAS-02), Bologna, Italy, pp350-351, 2002.

[10] Rayson, P., and Wilson, A., The ACAMRIT semantic tagging system: progress report, In Evett, L. J. and Rose, T. G. (eds.), Language Engineering for Document Analysis and Recognition (LEDAR), AISB96 Workshop proceedings, Brighton, England., pp 13-20, 1996.

[11] Sparacino, F., Larson, K., MacNeil, R., Davenport, G., Pentland, A., Technologies and methods for interactive exhibit design: from wireless object and body tracking to wearable computers, International Conference on Hypertext and Interactive Museums, ICHIM 99, Washington, DC., pp22-26, 1999.

[12] Wilson, A. and Rayson, P., Automatic Content Analysis of Spoken Discourse, In Souter, C. and Atwell, E. (eds), Corpus Based Computational Linguistics, Amsterdam, Rodopi. pp215-226, 1993.

Design of Co-existing space by *Shoji* interface showing Shadow

Yoshiyuki Miwa [*1] *Chikara Ishibiki* [*2] *Takashi Watanabe* [*2] *Shiroh Itai* [*2]
*1:Faculty of Science and Engineering, Waseda University
*2:Graduate School of Science and Engineering, Waseda University
59-319, 3-4-1,Okubo, Shinjuku-ku, Tokyo, Japan
miwa@waseda.jp, c_ishibiki@toki.waseda.jp

Abstract

Since conventional IT media has been designed on a basis of a communication taking place in a space in which the subject (yourself) and object (your communication partner) are spatially separated, a co-existing *Ba* experienced by such as a face-to-face communication is hardly created. In order to solve this challenging problem, present study has been aimed on shadows of each communication participant serving as a communication media to call attention to recognize existence of person in remote distance. Based on evidence that each one of individuals involved in a remote communication can transmit existence of people or materials by virtue of shadows projected on *Shoji* screen, a novel communication system has been developed. With this system, spatial information in which each one is involved can be combined and expanded to a co-existing spatial zone in a non-separatable manner between yourself and your communication partner. Communication experiments using the shadow concept suggested that each embodiment was enhanced and *Ba* was co-shared. Accordingly, it can be considered that this system will serve an effective means to create reliability and secure feeling between individuals involved in a remote communication.

1 Introduction

A creation of a community utilizing IT network has been recently receiving special attentions in various social sectors (Smith, 1999; Ishida, 1998;). However, it is extremely difficult for current IT system to make individuals involved co-share the co-existing feeling (or sharing the same feeling that all involved individuals are present at the location) that a face-to-face communication can easily establish. This can be understood by the fact that one cannot convey his emotional messages such as feeling, timing or silence to his communication partner through an Internet. Because, these emotional feelings are implicit Ba information which cannot be signalled after separating them from the subject (Shimizu, 2000; Dreyfus 2001; Miwa 2001). Hence, with IT communication system, signals are exchanged between a transmitter and receiver, and IT system does not take co-existing Ba design into consideration, which is necessitated to co-share contexts between a signal-sender and signal-receiver. Accordingly, reliability, secure feeling, and creativity in community society cannot be expected to further advance only by promoting current IT system. To break through this dilemma, a design concept on a co-existing virtual space with which each communication partners can co-share Ba has been developing for a remote communication.

How can we develop a virtual space to support for co-sharing Ba between remote locations? Current activities on developing various types of system to achieve this problem have been remarkably advanced, including CAVE, HyperMirror, TWISTER etc., (Cruz-Neira, 1993; Morikawa, 1998; Kunita, 2001). However, since these systems are based on a design of virtual co-sharing space in a separatable manner between the subject (yourself) and the object (your communication partner), a seamless merging between a real space and virtual co-sharing space is hardly accomplished. In order to solve these drawbacks, present study introduces a novel communication system, specifically aiming on shadows of human being or materials reflecting on *Shoji* screen to create Ba. Principle background of such development is based on facts that shadow itself possesses a powerful media function to express its existence and shadows cannot be separated from their origins. Moreover, since *Shoji* screen (Japanese traditional sliding screen made of wooden frameworks and rice paper pasted over the framework) can transmit a light, external activity/situation (on other side of *Shoji* screen) can be easily imagined and judged by internal side (this side of *Shoji* screen) through observing shadows. Peoples living in Japan must have similar experiences. Accordingly, co-existing Ba can be created between remote distances by combing *Shoji* screens and shadows revealing thereon.

Based on the aforementioned principles, a novel communication system has been developed with which spatial information/situation of each individual involved in a communication can be connected and expanded to co-existing space in a non-separatable manner between the subject and the object, by transmitting existences of human being or material in remote distances by virtue of their shadows being projected on the *Shoji* screen.

2 Structure of Communication System

For spatial designing to display shadow's movements onto *Shoji* screen, there could be two unique methods as seen in Figure 1; (a) a face-to-face type by which a silhouette of your communication partner is projected on *Shoji* screen, and (b) mirror surface type by which, being similar to your own shadow, a shadow of your partner is extended from a toe to facing *Shoji* screen. In this research project, a unique communication system has been developed with which your shadow as your other self can go back and forth through *Shoji* screen by exchanging a face-to-face and mirror surface types alternatively. A seen in Figure 2, two rooms (1800 [mm] depth, 1800 [mm] wide and 2000 [mm] high) were prepared remotely. Each room is composed of *Tatami* mattress on floor surface, *Shoji* screen in front face, and both side walls with dark curtains. In the real portion of these rooms, a projector was installed at 2000 [mm] distance from the front *Shoji* screen and 2240 [mm] high with an aiming angle of 12.4 degrees. The projector has two-fold purposes; (i) to project shadow of your communication partner, and (ii) to create your own shadow.

In order to create shadow of your remote communication partner, a infrared thermographic camera was employed for the following reasons; (i) even if two shadows are overlapped, separation of these two can be easily performed, (ii) less sensitive to any possible changes in an optical environment, and (iii) in some cases, an entire body-shadow of your partner is needed to be separated. After a human's thermal image (160 x 120 [pix]) was input to PC, an image processing such as a binarization and noise elimination was followed in order to separate a body image and a background. Then a background was painted with white color and a human body image was colored with black to create a human's shadow image on a real time basis (see Figure 3).

With a mirror surface type, when projecting a your partner's shadow image onto *Shoji* screen facing toward you, it is necessary to re-create a image in such a way that toe location of your partner's shadow image is positioned as the same toe location of your communication partner in remote distances. For this purpose, a taken your partner's shadow image was labeled, its toe position was detected, and standing toe position of your partner is also detected by sensor which was installed in mat on a whole floor. This was accomplished by constructing a touch sensor array, in which 120 touch sensors holding insulating sponge on their both sides are inserted with 150 [mm] pitch between two conductive sheets (see Figure 4). Such sponge has regular pattern of holes (with 25 [mm] in diameter) in every 50 [mm] pitch, so that when a foot is stepping on it, conductive sheets become into contact to detect the toe standing position. The thus detected toe position is transmitted on a real-time basis from a control box, in which H8 microcomputer is installed, to host PC with 9600 [bps]. Then a projected image (i.e., shadow) of your

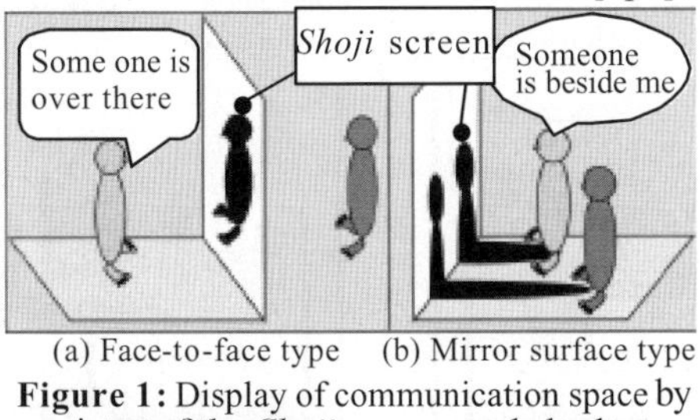

Figure 1: Display of communication space by virtue of the *Shoji* screen and shadow

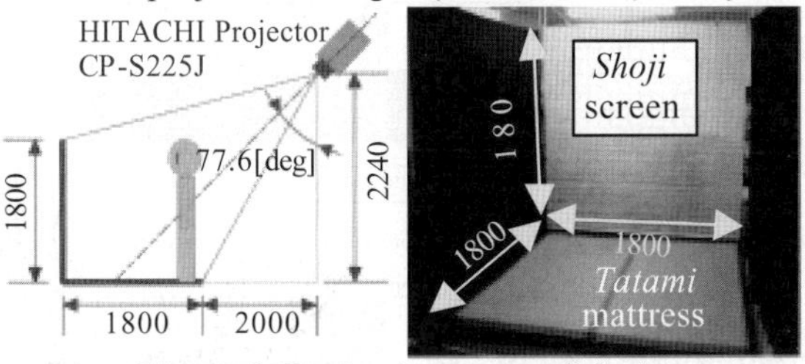

Figure 2: Projective space of shadow

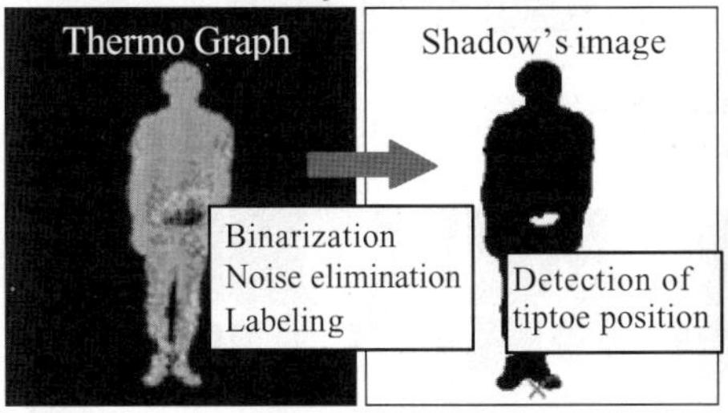

Figure 3: Dissection of shadow by thermo graph

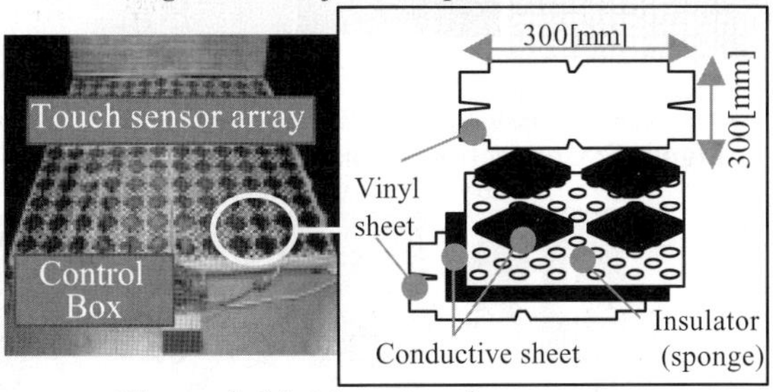

Figure 4: Mat-type touch sensor

communication partner was created by positioning it to toe position of his shadow image's toe position. Furthermore, since shadow's height can be geometrically varied, depending on distance between a rear light source and standing position, for mirror surface type, a human image was manipulated to shorten or enlarge, corresponding to a standing position of your partner. Moreover, in order to connect smoothly a shadow image on *Shoji* screen and shadow projected on a floor, both images projected on respective *Shoji* screen and floor were subjected to the trapezoid compensate using OpenGL. Furthermore, in order to perform a TCP/IP communication of a thus processed and compressed projected images, a two-way animation communication soft (15 [frames/sec], 240 x 120 [pix]) was employed with C++buider 4.

In the next step, a shadow of your communication partner as his other self is stepping over through *Shoji* screen toward your spatial zone. As an alternative way, you can pass through *Shoji* screen to your partner's spatial zone. To achieve these movements, the following expression method was employed. As seen in Figure 5, assuming that there were a fictitious *Shoji* screen at center portion of a floor, each room will be divided into two portions; the first portion (a closer half) to *Shoji* screen is for the face-to-face type and the second portion is for the mirror surface type. If both parties are in a same spatial zone, a shadow is expressed as a mirror surface type. On the other case when one of two parties is passing through a fictitious *Shoji* screen, a shadow of your partner projecting in your own spatial zone is changed from a mirror surface type to a face-to-face type or vice versa.

In this occasion, in order to create a feeling on distance or spatial depth when your communication partner is passing through *Shoji* screen from a remote location, a shadow of yourself projecting on *Shoji* screen is manipulated to become smaller and/or lighter contrast, responding to a longer distance between your partner and a fictitious *Shoji* screen. Similarly, with mirror surface type spatial designing, relative distance is obtained from your position and your partner's position. Depending on the obtained relative distance, contrast of your communication partner's shadow can be varied with 255 steps, or partner's shadow is shrunk or enlarged with factor of approximately 2.

As described in the above, the existence of your partner can be recognized by moving back and forth passing through between your own spatial zone and your partner's spatial zone, resulting in that the co-existing spatial zone will be able to be created. Figure 6 shows an entire structure of this communication system. Figure 7 shows a digested pictures demonstrating sequential movements of your partner's shadow passing through *Shoji* screen toward to you, and changing a facing direction, followed by returning back to behind *Shoji* screen again. From this demonstration, it can be understood that the co-existing spatial zone can be created by moving your partner's shadow passing through *Shoji* screen toward your spatial zone.

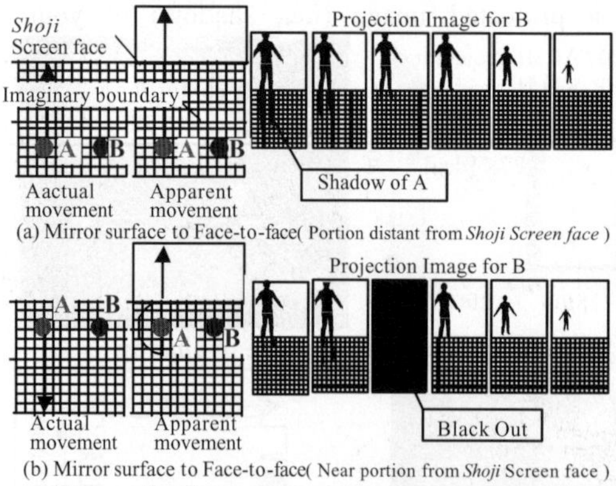

Figure 5: Connection of spaces separated by the *Shoji* screen

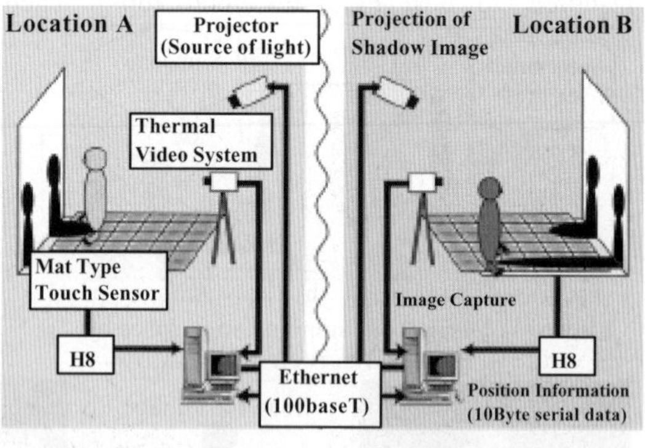

Figure 6: Structure of an entire system

Figure 7: Expression of shadows moving back and forth passing through the *Shoji*

3 Results and Discussion

Two persons participated to investigate this system and each one of them entered to individual room to conduct the communication tests. Two tasks were assigned to them; (1) with a mirror surface type spatial zone setting, they are allowed to talk freely for 3 minutes using their cellular phones, and (2) without any verbal conversation, they should perform their interaction by only observing/judging movements of shadows which they were observing. Twelve test participants were university students with age ranging from 21 to 26 years old. For a control, experimental tests were also conducted that were performed by a normal video chat using a *Shoji* screen. Figure 8 shows several examples of bodily interactions observed during these tests.

Summaries of comments of the test participants are as follows; "I found an easiness to talk.", "I tried to keep some distance from my communication partner.", "When my shadow was touched, I felt as if I was touched directly.", "I had an feeling as if my partner was with me.", or "I felt to move my body." These comments suggest a creation of feeling to co-share the same feeling as their communication partner. On the other hand for a control test using a video chat mechanism, these are comments; "Background was separated.", "I felt as if I was watching the screen.", "Very superficial and flat,", or "I hardly moved." The common factor laying over these comments is that one could not sense the presence of his communication partner. When bodily movements are compared between these two tests, as seen in Figure 9(a), relative positions of both participants did not change with a video chat tests; while, with this system, each participant tried to keep an appropriate distance between them. Small fluctuations in wave pattern were noted. When changes in moving speeds of both participants were investigated, it was found that these fluctuations were rhythmic and a phase-shift relationship between two participants was established, as seen in Figure 9(b). From these tests and comparisons, the present system enables that *Ba* can be co-shared by responding to a shadow's movements, leading to that bodily movements are easily induced and rhythmic synchronization without any recognition can be accomplished. Accordingly, embodiment during a communication can be enhanced by using their shadows and a creation of an entrainment (Condon, 1974) (which has been believed not be impossible with the conventional IT communication) can be realized with this present communication system. Recognition of your partner's presence by shadows can be confirmed by the following experiments. Changes in contrast in shadow images and a shrinkage or enlargement of shadows, depending on the distance between the communication partners, were charged to twelve participants. Results indicated that each participant intended to behave as if they tried to keep an appropriate distance from each other in order to change darkness of your own shadow to be similar to that of your communication partner. Moreover, even when a shrinkage/enlargement of shadows depending on your partner's position was tested, it was observed that each participants behaved to change their standing positions in order to alter your shadow's size to

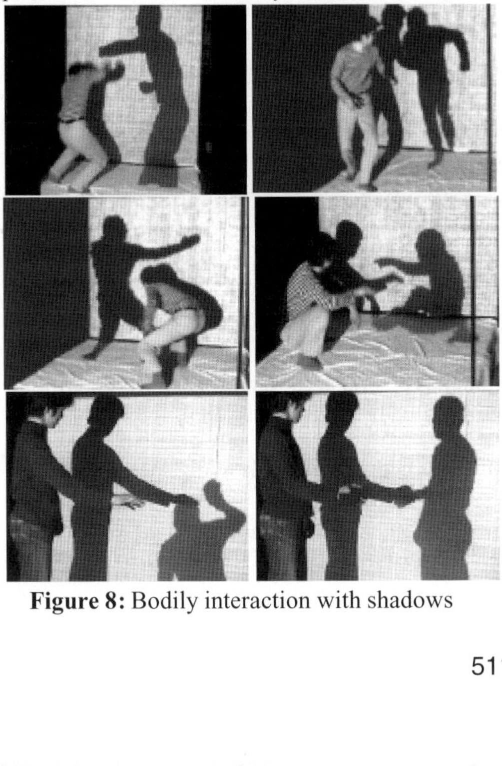

Figure 8: Bodily interaction with shadows

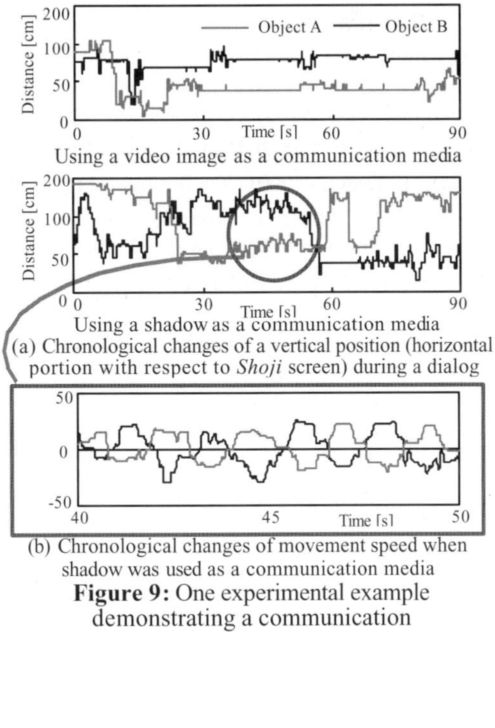

Figure 9: One experimental example demonstrating a communication

shadow's size of your communication's partner, and try to fix your partner's shadow inside the allowed projection screen. These results indicate that the feeling about your partner's presence can be altered, depending on ways how to express shadows. Moreover, if presentation and control of a background are improved, it is expected that a creation of the *Suibokuga* (black and white brush paint) which possesses much higher emotional and spiritual contents and depth can be realized.

As described in the above, present novel communication system is essentially different from a recognizable communication that is achieved by signal operation as done with conventional IT system. This system supports effectively and efficiently co-existing communication that can create more reliability and secure feeling by a mutual recognition of co-sharing the *Ba*. Accordingly, it is also expected, in future, that this system can be applied to create network to compensate conventional IT network system. By the final remarks, although there is a work done on the shadow in a mixed reality field (Naemura, 2002), present study was conducted systematically with a uniqueness of utilizing an existing expression of shadow as a communication media.

4 Conclusions

Summarizing obtained results and findings,

(1) Specifically aiming on *Shoji* screen interface projecting a shadow thereon, a novel remote communication system have been developed to co-share the *Ba* while having a dialogue.

(2) The present system, unlike the subject-object separatable system as a conventional IT technology, is characterized by the fact that different spatial zones are connected in a non-separatable manner between the subject and the object by virtue of the shadows projected on *Shoji* screen.

(3) Since this system strengthens embodiment of individuals involved in communication by connecting spatial zones by movements of a respective shadows, probability of misunderstanding in a net communication will be reduced and adlib co-sharing of scenario can take place easily.

(4) This system is expected to apply for public health service fields including a psychological therapy, rehabilitation, or hospital nursing.

(5) It is further expected that this system can expand its capabilities and performances to convey conditions of remote locations or emotional feelings by continuing investigations on (yet discovered) potential functions of shadows and background (white space).

This study was funded in part by the 2002 Publicly Soliciting Research Program, Japanese Science and Technology Corporation "Chaos and Its Control in Highly Automized Social System" (Principle Investigator: Dr. H. Shimizu). Assistant for developing a system is gratefully appreciated to our undergraduate student, Mr. T. Fumino.

5 References

M.A. Smith, and P. Kollock eds, (1999), "Communities in Cyberspace", Routledge

T.Ishida et al., (1998), " Community Computing: Collaboration over Global Information Networks", John Wiley & Sons

H Shimizu, et al., (2000), Ba and Co-creation (Japanese), NTT publishing co., Ltd., pp.23-177

Hubert L. Dreyfus (2001) :On the Internet (Thinking in Action); Routledge

Miwa, Y. (2001): Communication Technology of "Ba" in Co-Creation [in Japanese]; Systems/Control/Information, 45-11, pp. 638-644

C.Cruz-Neira, T. A. DeFanti et al., (1993), "Surround-screen projection-baced virtual reality:" design and implementation of CAVE",ACM SIGGRAPH 93 Proceedings, pp.132-142

O, Morikawa and T, Maesako, (1998), HyperMirror Toward Pleasant-to-use Video Mediated Communication System, CSCW98 pp.149-158

Yutaka Kunita, Naoko Ogawa et al., (2001), "Immersive Autostereoscopic Display for Mutual Telexistence: TWISTER I (Telexistence Wide-angle Immersive STEReoscope model I)," Proc. IEEE VR 2001, pp.31-36

Condon W. S., et al., (1974), Neonate movement is synchronized with adult speech; Science, No.183, pp.99-101

T. Naemura, T. Nitta, A. Mimura, H Harashima, (2002), Virtual Shadows in Mixed Reality Environment Using Flashlight-like Devices, Trans. Virtual Reality Society of Japan, 7, 2, pp. 227 - 237

Co-creation in Human-Computer Interaction

Yoshihiro Miyake

Tokyo Institute of Technology
Midori, Yokohama 226-8502 JAPAN
miyake@dis.titech.ac.jp

Abstract

The purpose of our research group is to realize a “Co-creation System.” Co-creation means co-emergence of real-time coordination by sharing subjective space between different persons. Human communication with emergent reality like this needs two kinds of processing at the same time. One is explicit communication such as the exchange of messages and the other is implicit embodied interaction such as sympathy and direct experience. Using this dual-processing complementarily, we are developing co-creative man-machine interfaces and interactive media. This new technology will be effective for recovering human linkage, social ethics and mutual-reliability that has been lost in the IT society.

1 Introduction

As the background to this research, we have already pointed out the serious limitation of intelligence realized in artificial systems. Human intelligence can be classified into two different categories: “search” and “emergence.” A searching algorithm based on “completeness” of intelligence has been widely used in IT systems. This kind of intelligence is applicable to definite situations in which every state can be previously defined (Fig.1a). However, in indefinite situations, such as the system including human behaviour and social communication, searching cannot be used. In such unpredictable conditions, another type of intelligence is required and intelligence with emergence is essential for overcoming the limitation of the conventional approach.

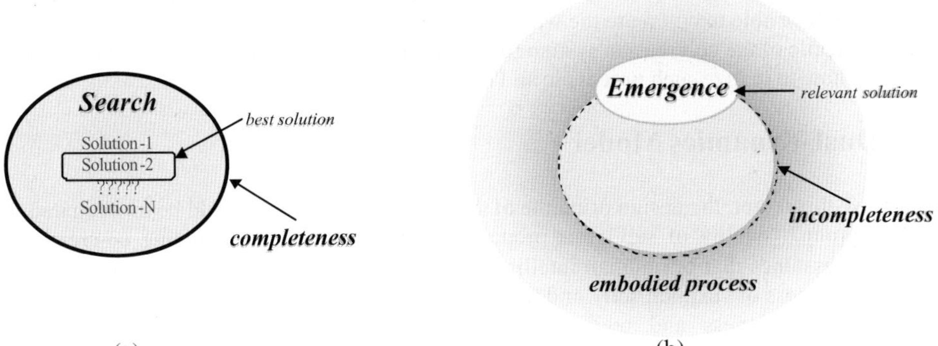

(a) (b)

Figure 1: Search and emergence

Figure 2: Japanese stone garden

2 Incompleteness and Co-creation

In the present IT systems based on the framework of searching, the relationship between humans and artificial systems is a one-sided transfer from the system to human. As a result, the size of the information space prepared in the artificial system becomes extremely large and the human becomes very passive on the other side. This is a problem of design principle based on intelligence with completeness. On the other hand, Japanese culture has very different traditional background, i.e. emergence in real-time. Japanese think that artificial systems should be incomplete, but this is a kind of active "incompleteness" to realize an emergent reality with humans (Fig.1b). Due to this incompleteness, coordination between the system and the human is co-created in real-time.

In this process, embodied interaction plays an essential role in realizing a relevant function with real-time coordination. The stone garden Ryoanji temple in Kyoto is a famous example. There are 15 stones in this garden, however you can never see all stones at the same time (Fig.2). Some stones are always hidden by other stones. This is a kind of active incompleteness. So if you want to get a whole image of the garden, you have to walk around in the garden as part of the garden. Then you will be able to co-create the whole image including you as part of the garden. This is the design principle of the co-creation system that is based on active incompleteness and embodiment. However, how to realize such process in an IT system is still an open question.

3 Dual-dynamics Model

To achieve this target, the design principle of the co-creation system should be investigated. In our research group, "duality of self" was proposed as a hypothesis for realizing co-creation. This hypothesis assumes that our human intelligence is composed of two different processing modes. One is the process of "explicit self" and the other is "implicit self." This explicit self is concerned with self-consciousness and realizes intelligence with completeness. In other words this intelligence is our causal operation in formal logic. On the other hand, the implicit self is concerned with embodiment with active incompleteness. This realizes the interaction between the

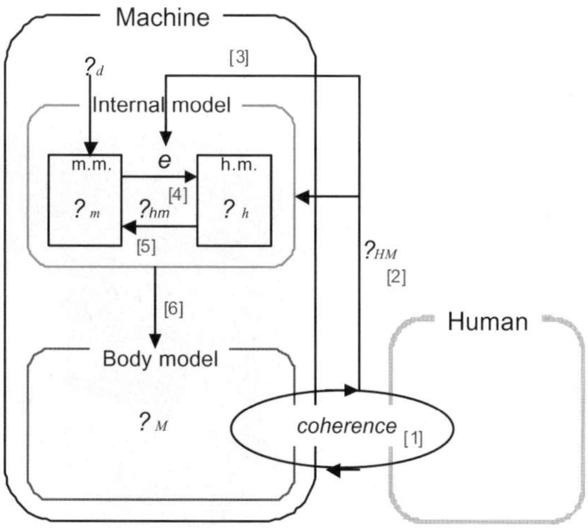

Figure 3: Dual dynamics model

system and the indefinite actual world. Here, the interface between these two processes emerges by "mutual constraint." We regard this emergence as a co-creative process of intelligence.

From this hypothesis, we have already proposed a model of co-creation (Fig.3). This "dual-dynamics model" is composed of two sub-models. One is an internal model to show the explicit process and the other is a body model to represent the implicit process. Since the synchronization phenomenon of the body motion is widely observed as a typical dynamics of the embodied process, a nonlinear oscillator and mutual entrainment are used to show the dynamics of a body model that can be embedded as a part of the indefinite world. This is a kind of mathematical expression of the active incompleteness and temporal coherence can be self-organized in such open space. On the other hand, the internal model is a coupled nonlinear oscillator as closed space to represent the process based on completeness. By the mutual constraint between these two sub-models, an emergent process of intelligence is simulated.

4 Co-creation in Human-Computer Interaction

This model was represented as part of the co-creation process between human and computer. The rehabilitation process of human walking was used as an example because coordinated walking between two persons is widely observed in rehabilitation for elderly people to redevelop their walking ability. Our dual-dynamics model is realized in a personal computer as a virtual walking robot and footsteps are exchanged between the robot and human walker (Fig.4). The footstep of the robot is transmitted to the human by headphones and the footstep of the human is feed back to the robot by a touch sensor. Everyone has had this kind of experience when walking with another person. In such situation, temporal coherence of footsteps between two persons spontaneously appears. This co-creation robot was named "Walk-Mate."

To evaluate the effectiveness of this system, interaction between Walk-Mate and a human with walking impairment was analyzed. As a result, three characteristic properties were observed (Fig.5). The first is mutual adaptation between the human walker and Walk-Mate. The periods of their footsteps mutually coincided with each other after the start of interaction. This is a kind of co-creation dynamics in walking motion. The second is the emergence of global stability including

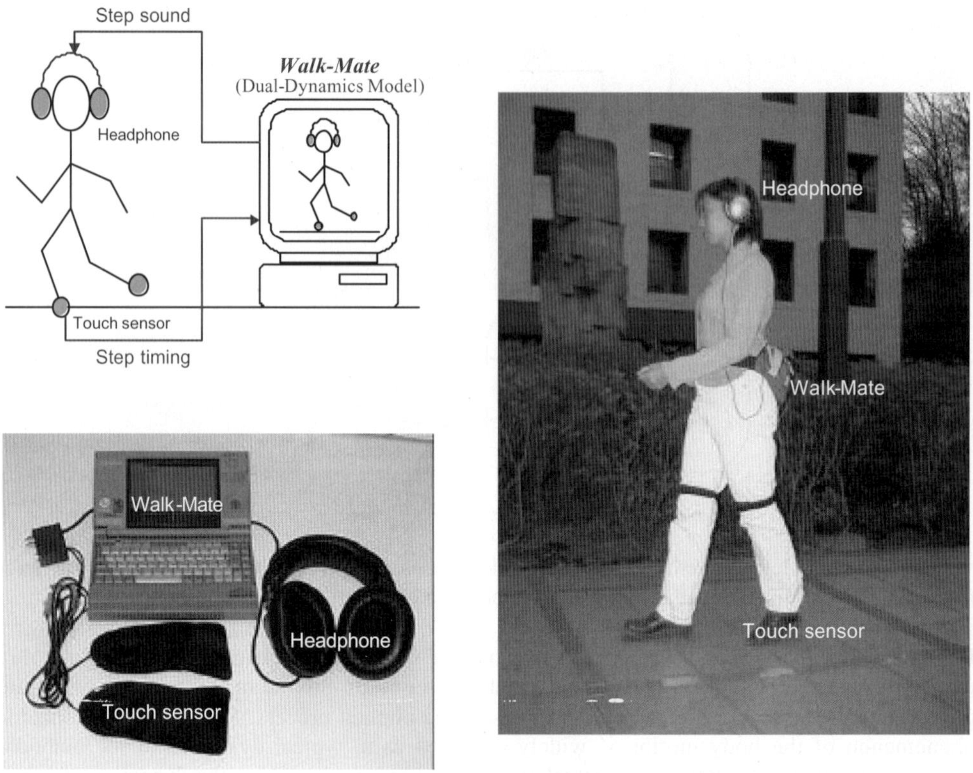

Figure 4: Walk-Mate

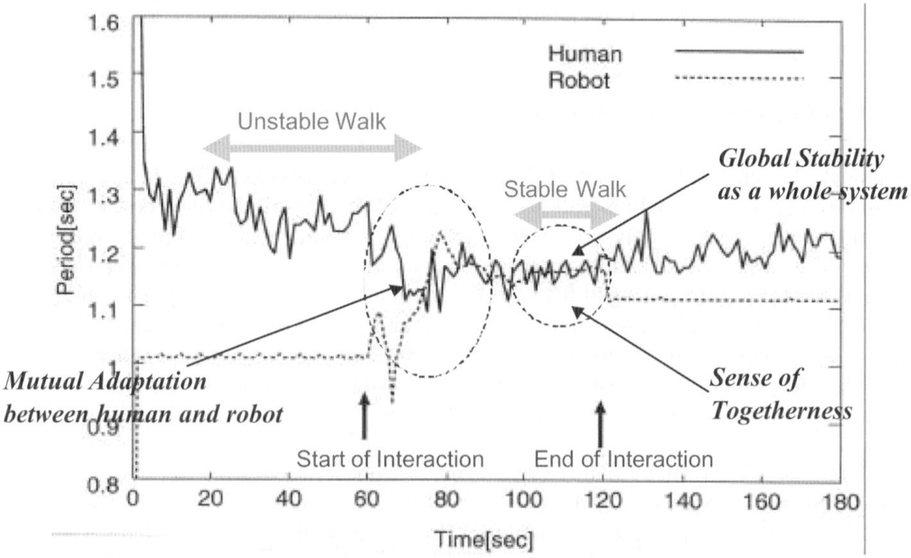

Figure 5: Co-creation process

both the dynamics of the human and the Walk-Mate. Fluctuation of the footstep period due to the walking impairment was significantly decreased in this co-creation process. The third is the sense of togetherness emerging through this process, suggesting this process also realizes the emergence of our mental connection. These results show that our proposed framework could be effective for establishing a co-creation process between humans and artificial systems.

5 Conclusion

In this way, we are developing co-creation systems in a man-machine interface. In the next step, this technology for a co-creative interface should be extended to a network system to support co-creative communication in human linkage. On the Internet, exchange of explicit information is very easy but the implicit process cannot be shared there. This means there is no co-creation on the Internet. As such network system has penetrated into daily life, relation-building problems, such as mutual reliability in a virtual community and togetherness between remote families has become an important subject. This co-creation technology has the potential for overcoming the problems of the modern network society.

References

Miyake, Y. (1997). Design in Life (in Japanese). *Gendai Shisou*, 25(6), 301-317.

Shimizu, H., Kume, T., Miwa, Y. & Miyake, Y. (2000). Ba and Co-creation (in Japanese). Tokyo: NTT Publisher.

Miyake, Y., Miyagawa, T. & Tamura, Y. (2001). Man-machine interaction as co-generation process (in Japanese). *Transaction of SICE*, 37(11), 1087-1096.

Muto, T. & Miyake, Y. (2002). Analysis of co-emergence process on the human-robot cooperation for walk-support (in Japanese). *Transaction of SICE*, 38(3), 316-323.

Yamamoto, T. & Miyake, Y. (2002). Analysis of interaction between player and listener in music live performance (in Japanese). *Transaction of SICE*, 38(9), 800-805.

Takanashi, H. & Miyake, Y. (2003). Co-emergence robot Walk-Mate and its support for elderly people (in Japanese). *Transaction of SICE*, 39(1), 74-81.

Miyake, Y., Onishi, Y. & Pöppel, E. (2002). Two modes of timing anticipation in synchronization tapping (in Japanese). Transaction of SICE, 38(12), 1114-1122.

Miyake, Y. & Miyagawa, T. (1999). Internal observation and co-generative interface. *Proc. of 1999 IEEE International Conference on Systems, Man, and Cybernetics (SMC'99)*, Tokyo, Japan, pp. I-229-I-237.

Anthropomorphic Dialog Agent Development Tool Using Facial Image Synthesis and Lip Synchronization

Shigeo Morishima

Seikei University
3-3-1 Kichijoji-kitamachi Musashino, Tokyo 180-8633 JAPAN
shigeo@ee.seikei.ac.jp

Abstract

In this paper, the basic software for an anthropomorphic spoken dialog agent is introduced. The basic software consists of four modules for speech recognition, speech synthesis, facial image synthesis, and multi-modal dialog integration. This interim report describes the basic design approach and an implementation of the software focusing on efforts to ensure the interactive capability of the spoken dialog agent.

1 Introduction

The basic software introduced here consists of four modules for speech recognition, speech synthesis, facial image synthesis, and multi-modal dialog integration. System control and data management capabilities in a dispersed environment are essential in order for these various modules to interoperate smoothly as a single dialog system, and several systems exhibiting these capabilities have been developed including DARPA's Communicator Program [3] based on MIT's Galaxy-III [2], and the Open Agent Architecture (OAA) developed by SRI [4]. This paper describes the basic design and an implementation of the project software that is intuitive, easy to understand, and ensures fully interactive spoken dialog with the agent.

2 Background

Considering how much faster computers are able to crunch numbers and deal with complex calculations than people, why is that they are incapable of communicating by speech with humans? Just what would it take to enable computers to speak in the same way that people talk among themselves? As the technological foundation for such communications, research so far as focused on such dream technologies as "apprehending speech," "synthesizing speech," and "generating computer graphic representations of real people." Practical success has recently been achieved in applying speech recognition and speech synthesis technologies to synthetic voice reading. It is also now possible as we are already beginning to see in movies to render the realistic movement of actors and actresses with computer graphics. Basic technologies to achieve these sorts of human interfaces have been advanced to a certain level, and research is now concentrating on refining and improving the quality of these capabilities even more. Finally, considerable R&D is also seeking to integrate these technologies to create interfaces and systems that are capable of sustained dialog similar to that between people.

Being actively involved in these developments, the Multi-modal Tool Working Group of the Special Interest Group on Spoken Language Processing of the Information Processing Society of Japan (IPSJ SIG-SLP) over the three-year period from 1998 to 2000 identified anthropomorphic agents as a target of next-generation research, and they have developed a plan to build and make publicly available a research platform through collaborative efforts of researchers. This conception received support of the Information-technology Promotion Agency (IPA), and more than ten research institutes began a cooperative effort to develop the basic software in March 2000.

3 Software Configuration

The anthropomorphic spoken dialog agent that is now under development consists of four basic software modules, all of which will be made available in the form of freeware. By implementing the software as separate modules, this is not only an effective tool for assessing the various constituent technologies, it also provides a versatile R&D platform making it easy to build original dialog systems by simply plugging in different software modules developed independently by the different R&D institutes involved in the project as required.

3.1 Basic Software

New basic software is being developed to integrate and control the dialog component modules and to manage the dialog. Some of the specific projects that are currently under way include (a) an Agent Manager (AM) providing low-level control of the speech recognition, speech synthesis, facial image synthesis, and other modules; (b) the capability to interpret VoiceXML-based high-level dialog descriptions; (c) a Task Manager (TM) for controlling dialog using the functions provided by the AM; and (d) a prototyping tool to provide a GUI environment supporting the setting of parameters and the description and control of scenarios, all things that are necessary to construct dialog systems.

In this paper we will address the issues involved in designing a dialog system from the standpoint of the Agent Manager, and present an implementation

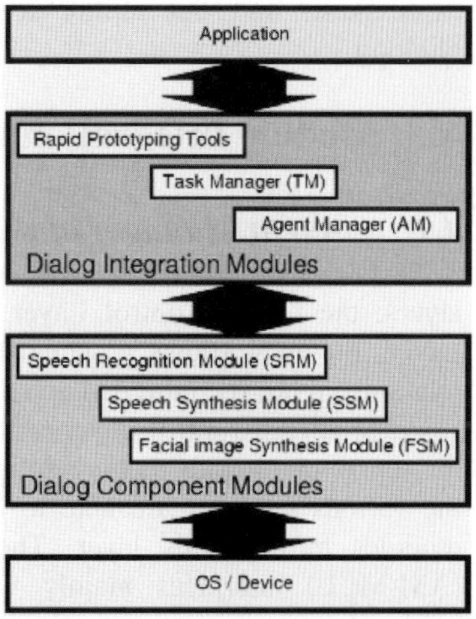

Fig.1 Agent System Platform

3.2 Speech Synthesis

New basic speech synthesis software is being developed that not only clearly reads sentences of mixed kanji and kana (Chinese characters and phonetic script), but also shares data to enable synchronization with a facial image. This enables lip-sync, the synchronization of sound and motion so the facial movements of speech coincide with the sounds. Furthermore, anticipating changes in the nature of the speech to reflect different circumstances or the intent of the speaker, we are also seeking ways to control a range of different emotions and speech rhythms.

3.3 Speech Recognition

Building on the software that came out of the IPA's basic Japanese dictation software development project from April 1997 to March 2000, it should be easy enough to extend the capabilities of that software package to accommodate dialog processing and implement flexible control. Specifically, we are doing away with grammar-based recognition and recognition results, and developing functions that can deal with unnecessary words and poses, and can provide dynamic control of recognition processing.

3.4 *Facial Image Synthesis and Control*[1]

Starting with the IPA's facial image processing system for the human-like kansei agent that was developed from June 1995 to March 1998, we are enhancing the software package to support higher quality agent facial image synthesis, animation control, and precise lip-sync with synthetic and natural speech. Some of the specific enhancements include a GUI able to map standard wire frames to images of heads shot from different angles to easily generate 3D models of human heads, sharing of data with the speech synthesis module, more precise lip-sync, the ability to add any facial expressions, and the ability to control nodding and blinking.

4 Module Integration Processing

4.1 *Agent Manager and Modules*

The Agent Manager (AM) consists of two functional layers: the Direct Control Layer and the Macro Control Layer. Figure 4 shows a schematic representation of the relationship between the AM and the various modules.

The Direct Control Layer (AM-DCL) directly controls the sets of commands that are defined for each module, and the various modules are able to communicate with other modules through this layer. The Macro Control Layer (AM-MCL) interfaces mainly with the Task Manager (TM). By redefining frequently used sequential command sets as macro commands and by taking on inter-module synchronization management and similar low-level module control, the AM-MCL markedly improves operation of the system from the standpoint of the Task Manager.

In principle, the Speech Recognition Module (SRM), and Speech Synthesis Module (SSM), and the Facial Image Synthesis Module (FSM) communicate through the AM-DCL. This means that, in developing a module, one only needs to worry about communication with the AM. This is a major benefit for the present project, because it allows the various modules to be developed at separate locations of the participating R&D firms.

The Task Manager (TM) mainly communicates through the AM-MCL, but if necessary can also communicate with the AM-DCL just like the other modules. All the output from the various modules is supplied to the TM. The TM is thus able selectively adopt whatever data that it needs from among the totality of data that it receives from all the modules.

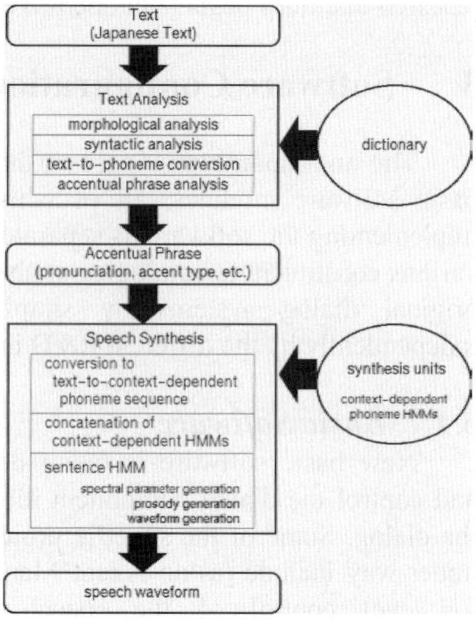

Fig.2 Speech Synthesis Module

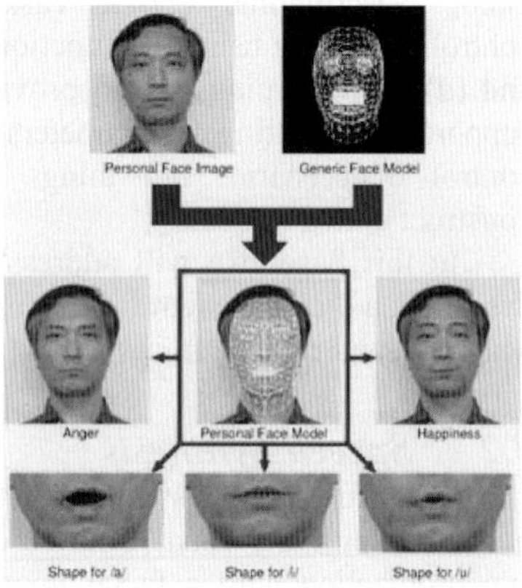

Fig.3 Face Synthesis Module

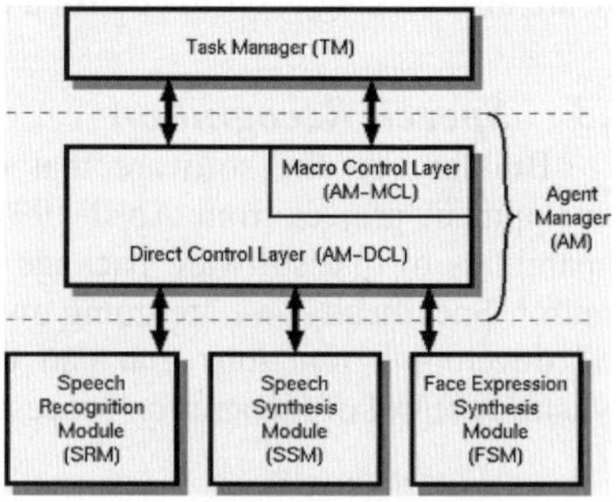

Fig.4 Agent Manager

4.2 *Virtual Machine Models*

In defining the command specifications for communicating between the dialog integration module and the various dialog component modules, the dialog component modules are treated as virtual machine models. For example, Fig. 5 illustrates the relationship between the Agent Manager (AM-DCL) and a virtual machine model. The dialog component modules are managed by defining a parameter slot for each input and output parameter. The macro command definitions in the dialog component modules are managed in a similar way using macro slots. These slots are shared with the AM-DCL, and it is through the slots that the AM-DCL communicates with the dialog component modules.

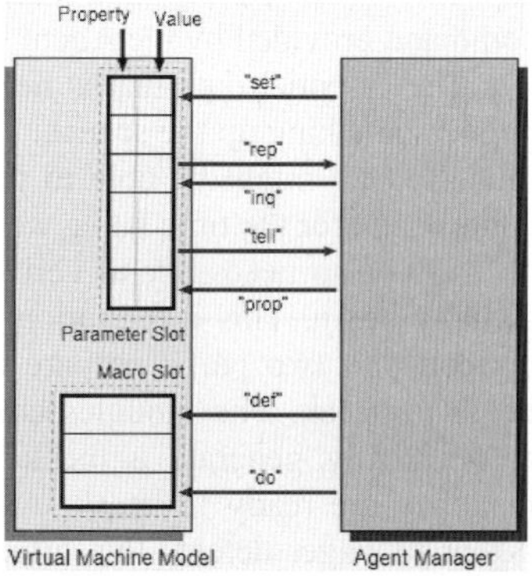

Fig.5 Virtual Machine Model

Each slot possesses a value and a property that are treated like virtual machine model switches, and slots are activated by common commands. The slot values can effect various actions; they can monitor an operating state, direct that an action start or stop, set a particular operating environment, and so forth. Changing the value of slot is immediately reflected in the form of a different action. In other words, changing a slot value is instantly associated with a particular action.

This makes it possible to manipulate all the dialog component modules in a centrally coordinated fashion. Commands are communication specifications that are not dependent on any particular module, while the parameter slots are module-dependent specifications that are not dependent on communication. This means that the difference between dialog component modules is nothing more than the different functions programmed in their parameter slots.

4.3 *Communication Between Modules*

Synchronization between a speech synthesis module and a facial image synthesis module might be achieved (1) by communication over a direct connection that is set up between the two modules, or (2) in the same way as other communication, via the Agent Manager.

Note that while this example only involves synchronization between two modules, the first method would require a synchronization management capability in which every module was aware of every other module. This would make the modules more interdependent while at the same time diminishing their autonomy.

In the second method, synchronization between the two modules is managed by the AM. While this increases the processing costs, it makes it easier to maintain the autonomy of the modules since designers only have to concern themselves with communication between the module and the AM. Because our current priority is to make it easier to ensure the autonomy of modules, we are proceeding on the basis of the second approach. The actual synchronization indicator that defines the exact timing when speech begins and so on is achieved by conveying the system time to the two modules. Very accurate synchronization is achieved using the Network Time Protocol (NTP) that was developed for system time synchronization across networks.

4.4 *Data Sharing Between Modules for Synchronization*

Management of the synchronization between the two modules might be implemented by a higher level module that is separate and distinct from the AM. We are also considering defining and implementing a new type of module that is dedicated exclusively to synchronization. However, considering the importance of the lip-sync capability for agents and how frequently such capability is used in spoken dialog, we have currently implemented this function using a macro command provided by the Agent Manager.

The essential data that is needed for lip-sync when an agent speaks is the durations of each phoneme making up the speech. This information is obtained by interrogating the speech synthesis module. One might be able to think of other kinds of information that would be useful in this context, but for the time being we only use the duration of each phoneme.

It is also necessary to verify that the two modules are ready to speak before speaking can actually begin. This information is obtained by the following procedure. The speech process is divided into two parts: prepare to speak and begin to speak. The modules are designed to automatically generate a message indicating that they are ready to speak. As soon as the Agent Manager detects this information from the two modules, the AM directs that they can actually begin speaking. Figure 6 shows the sequence of commands involved in this process. Note that the commands triggering these sequential processes are actually implemented by the macro command function in the AM-MCL

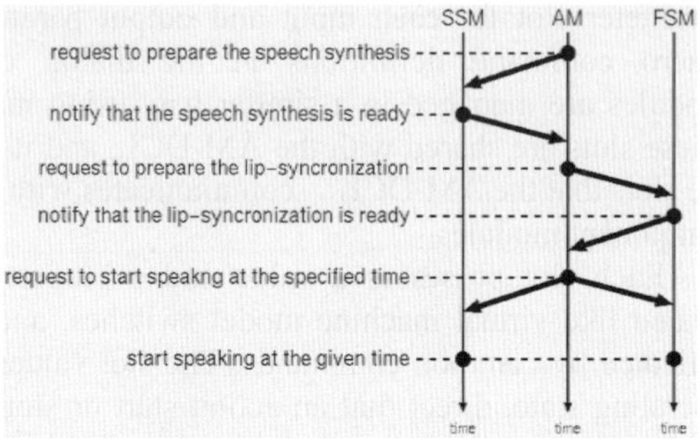

Fig.6 Processing for Lip-synchronization.

5. Conclusions

This paper described the design and an implementation of the basic software for an anthropomorphic spoken dialog agent that ensures interactiveness, a project sponsored by Japan's Information-technology Promotion Agency (IPA).

Considering that this is an interim report on a system that is currently under development, there are a number of unresolved issues that still need to be worked out. As the project unfolds, we will further expand and enhance the functions of the dialog component modules and the multi-modal dialog integration module, and explore the feasibility of incorporating a standard distributed object environment architecture such as CORBA[5].

Acknowledgments

This work is supported so many collaborators in the region of speech recognition, speech synthesis and image synthesis community with their state-of-the-art original technologies.

REFERENCES

1. Tatsuo Yotsukura, Shigeo Morishima, “An Open Source Development Tool for Anthropomorphic Dialog Agent”, -Face Image Synthesis and Lip Synchronization-, Proceedings of IEEE Fifth Workshop on Multimedia Signal Processing, 03_01_05.pdf, 9-11 Dec.2002.
2. Stephenie Seneff, Ed Hurley, Raymond Lau, Christine Pao, Philipp Schmid and Victor Zue: GALAXY-II: A Referece Architecture for Conversational System Development. In ICSLP-1998, pp.931--934, 1998.
3. DARPA Communicator Program, 1998. http://fofoca.mitre.org/.
4. OAA (The Open Agent Architecture). http://www.ai.sri.com/~oaa/.
5. CORBA (The Common Object Request Broker Architecture). http://www.corba.org/.

Developing life-like Agent and Personality that is Familiar with the User-Comic strip for child-

Maki Mushiake

Tsunashima-sou A-101, 36-16 Kamiikedai 4-chome,
Ota-ku, Tokyo 145-0064, Japan.
Tel: +81-3-3728-1426
mushiake@dream.com

Abstract
We produced a familiar character agent and construct comic strip. This is evaluated by its access in a Web setting. This paper describes the method to develop a character. Then I construct comic strip using inter relay chat through learning Japanese sign language. I find character agent to be suited for child learning.

1 Introduction

Recently, human-agent interfaces have been used in social interactions and as an embodied conversational agent, and one proposed application is a personalized pet-like 3D agent that can chat [1]. To develop computer supported collaborative learning for child [2], the service industry, and entertainment, it is crucial that the user be familiar with the agent.

2 Design an agent

2.1 Childlikeness

This section describes animated agents, like the Microsoft chat, and discusses what makes an agent more fascinated and enables it to become familiar. I analyze other research on animated agents [3, 4] and evaluate their familiarity with the user.

2.2 Assisted character

Animal is a typical design of a character to interact with Japanese children (Fig. 1). This animal-like character is sometimes used in childrens' comics or animation. The product designed for children is the same as that for TV animation or comic books. This character can also be useful in learning systems for children. To educate children effectively, the character should encourage each child by learning what information the child requires. For example, the agent can talk to the child and ask what each child needs to learn, and then activate its own behaviour.

2.2.1 Developing a personality familiar to children

The agent's personality has the following components:

- Background knowledge of the agent,
- Animation effects to make the agent appear alive, such as over reaction [5], and the agent's actions and expressions of emotion.

To design an agent that is familiar to children, the agent should be:

- An animal character or a robot character.

Features used to make the character child-like include:

- Facial and body design and[6]
- Gestures.

Figure 1: Typical character agent for child

2.2.2 Environment

To describe this character, I used IRC (Internet Relay Chat). We can use graphical interface, text mode, and more interactive environment, such as 3D avatar chat. In this paper, I used graphical comic chat.

2.3 Setting

MSChat (Microsoft Chat) is an Internet chat program that uses comics in an on-line chat system. It uses cartoon characters to represent the chat participants. These cartoon characters display several poses, for example sit down or jump, and indicate eight emotions automatically.
Already there was Japanese sign-language animation creating system in web page [7].In this system I set up emotional expression and Japanese sign language vocabulary and phrase. I made picture for basic sign language vocabulary. In this system I did not concern the grammar and collocation. Pictures of Japanese sign language were drawn based on Japanese sign language streaming pictures [8],

3 Discussion

3.1 Results

Fig.2 is comic strip of the own character (Fig.2). This first strip they are introduced theirself by sign language and speech. We adapted each pictures to emotional expression and sign language in advance.

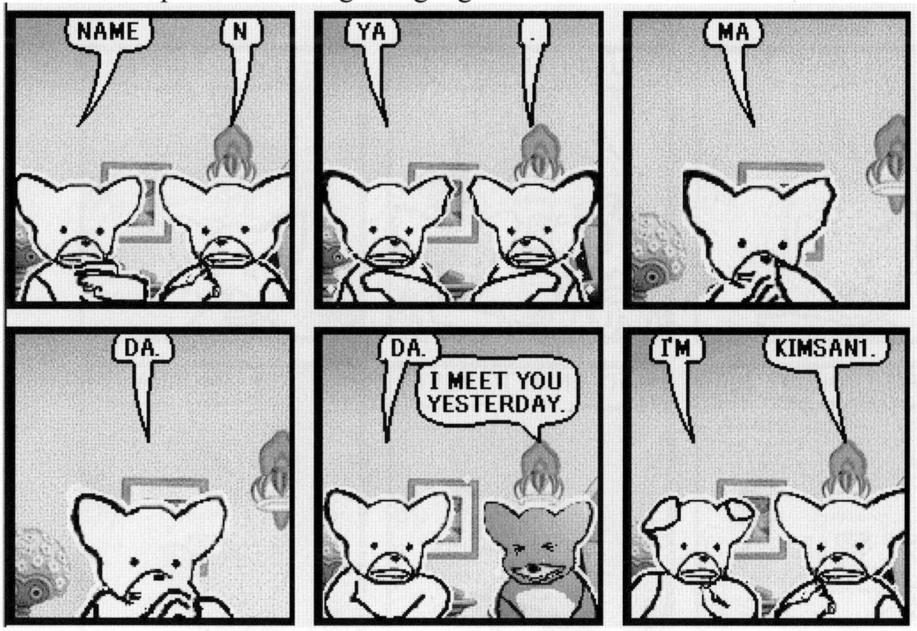

Figure2: Comical chat using the sign language vocabulary and phrase

3.2. Signed language

There are 400,000 deaf people in Japan. Deaf children with normal-hearing parents cannot converse with their parents and develop skills to communicate with others.
Since normal-hearing children learn language through conversation with others, deaf and impaired children need some method to acquire a language system. It is important that language be learned within a critical period. Test results suggest that the emotional expressions of the character can be used as sign language animation.

3.3 Future communication

Many sign language animations have ever developed, for example the sign language dictionary. Therefore it can be easily retrieved by keyword and the hand- position of sign language.
In the future this system is useful for communication tool between deaf or impaired child and normal hearing people with going the research on grammatical analysis in Japanese-to-sign language translation [9].

Figure5: Future communication between deaf people and normal hearing people

3.4 Community assistant

Some life-like agents are already in use to interface person-to-person communication, such as a system that supports new encounters in the meeting place [10], and in future agent-mediated communications and the agent-human Interface will assume a more important role. When using life-like agents as intermediates, harmonious communications are essential among people belonging to diverse cultures with different value systems, to minimize the risk of conflict between groups and communities.

4 Conclusion

To study the design of the product, this article describes animated agents, like the Microsoft agent, and discusses what makes an agent more fascinated by child and enables it to become familiar in the area of Japanese sign language.
Then I analyzed children's familiarity with it and its personality for use with children. After that I contracted comic using sign language vocabulary and phrased. This character is an animal-like agent template product designed for children. Such agents are suited for child learning systems. Using such agents facilitates children's access to computer-mediated learning.

References

Microsoft Chat is a registered trademark of the Microsoft Corporation in the United State.

[1]PAWPAW Web Page from (2003/1):
http://www.so-net.ne.jp/paw/index.html

[2]Ryokai, K. & Cassell. (1999). Computer Support for Children's Collaborative Fantasy Play and Storytelling. *In Proceedings of CSCL '99.*

[3] MS Agent Data by ATR M&C Web Page (2003/1):
http://www2.mic.atr.co.jp/agent/

[4] Tosa, N. (1995).Life-like Believable Communication Agent-NEURO BABY '95,"MIC&MUSE"-. *The Third IEEE International Conference on Multimedia Computing and Systems.*

[5] Lasseter, J. (1987).Principles of Traditional Animation Applied to 3D Computer Animation. *Computer Graphics*, Vol.21, No.4. pp. 35-44

[6]Mushiake, M. (2002). Designing Life-like Agent that is Familiar with the User. SIG-KBS-A203, *.ICS-KBS Joint Workshop in Hakodate*, 73-75. (In Japanese)

[7] Mascot Web Page from: (2003/2)
http://mascot.mis.ous.ac.jp/syuwa/main/frame.htm

[8] Let's learn sign language Web Page from: (2003/2)
http://jurakoubou.cool.ne.jp/syuwa/wmv/index.htm (In Japanese)

[9]Ikeda, R, Iwata, K, & Kurokawa, T (2002).Language Transformation for Japaneseto-Sign Language Translation and Prosessing of the Innflection Rules in It. *Correspndences on Human Interface*, Vol.5, No.1, 19-24.

[10] Sumi, Y. & Mase, K. (2000).Supporting the awareness of shared interests and experiences in communities. *Int. J. Human-Computer Studies,* Vol.56, No.1, 127-146.

The Implementation of RobotPHONE

Dairoku Sekiguchi, Masahiko Inami and Susumu Tachi

Graduate School of Information Science and Technology
The University of Tokyo
7-3-1 Hongo, Bunkyo-ku, Tokyo 113-8656, JAPAN
+81-3-5841-6917
{dairoku, minami, tachi}@star.t.u-tokyo.ac.jp

Abstract

RobotPHONE is a Robotic User Interface (RUI) that uses robots as physical avatars for interpersonal communication. The shape and motion of RobotPHONE is continuously synchronized by a symmetric bilateral control method. Using RobotPHONE, users in remote locations can communicate shapes and motion with each other. In this paper we present the implementation of the RobotPHONE system including Virtual RobotPHONE, which is software for simulating an actual RobotPHONE.

1 Robotic User Interface

The concept of a Graphical User Interface (GUI) that originated from NLS (On-Line System) by Douglas C. Engelbart [1] and Alto by Alan Key [2] changed the way of using computer. GUI simplified the use of a computer and played important role in the spread of computers to the public. However, because GUI is based on the combination of WIMP (Window, Icon, Menu, Pointing Device), recently the interaction method of GUI has been recognized to be limited when interacting with the real world we live in.

NaviCam [3] and Tangible Bits [4] are attempts to overcome such limitations of a GUI by using a physical object that exists in the real world as interface. Many of these attempts use a see-through HMD or a projector to output information to the user. Therefore, considering that direct interaction with the outputted information is still limited, we can state that an output method making use of a real object hasn't been established yet.

On the other hand, personal robots, such as pet robots [5], are a good example of utilizing a real object. Contrary to a CG character on a computer display, these robots have a physical body and that existence attracts people. The robot can be regarded as a computer with a physical body that enables to interact with the real world. Hence, considering its strong capability to interact with real world, a robot will be efficient interface that has input and output method for the real world. In addition, regarding the robot as a general-purpose machine, it is possible to consider that the robot is universal interface. Robotic User Interface (RUI) is the word for this type of interface. That is, the concept of using a robot as an interface between the real world and the information world can be referred to as RUI. An intelligent robot as a physical entity for an Artificial Intelligence agent or a haptic feedback robot arm used in a VR systems is good examples of a RUI.

2 RobotPHONE

RobotPHONE [6] is a RUI system for interpersonal exchange that uses robots as agents for physical communication. The RobotPHONE system employs robots that are called shape-sharing device. The shape and motion of remote shape-sharing devices are always synchronized. Operations to the robot, such as modification of posture, or input of motion, are reflected to the remote robot in real-time. Therefore, users of RobotPHONE can communicate and interact with each other by exchanging the shape and motion of the robot.

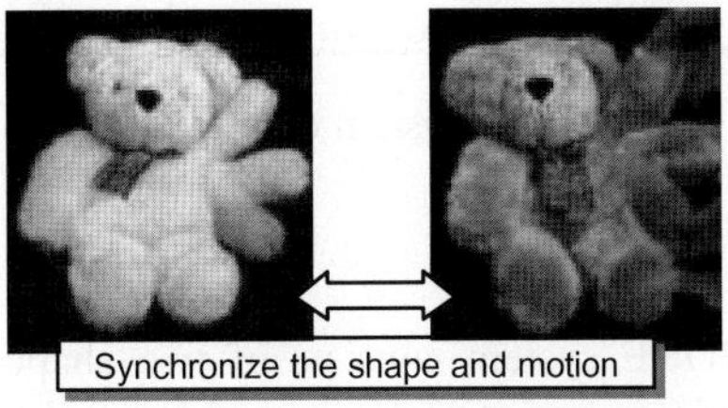

Fig.1 The concept of RobotPHONE

If ordinary telephone is considered to be a device for transferring voice, RobotPHONE is a device for transferring motion. In addition, because RobotPHONE uses physically existing robot as input and output interface, a user of the RobotPHONE system can directly feel the force of a user in the other side of the system. Therefore, users of RobotPHONE can get motion information not only from visual sensation but also from tactile sensation and it is possible to say that RobotPHONE is new type of telephone that can transfer visual, haptic and auditory information in one time. It also can be said that RobotPHONE is a system that makes it possible to have an object exist virtually in a remote place on behalf of user. RobotPHONE transfers the existence of the user not by attempting to transmit the user itself but to transmit the user's substitute.
Consider a mother giving her daughter a stuffed doll to keep her company at night. This is a form of communication aided by a physical entity. RobotPHONE can allow this kind of communication to become possible.

3 System Design

We have been implementing two types of RobotPHONE. The first prototype is a snake-like RobotPHONE. This prototype consists of two six parallel axes snake-like robots as shape-sharing devices. The second RobotPHONE is humanoid type and consists of two humanoid robots, which have the appearance of teddy bears, as shape-sharing devices. In this paper, we will explain the system design and implementation of the humanoid type RobotPHONE.

3.1 System overview

Fig. 2 shows an overview of the RobotPHONE system. Each RobotPHONE consists of a teddy bear-like robot, a controller unit and an IBM-PC/AT compatible computer (PC). The controller unit controls the robot and is connected to the PC via RS-232C. The communication program running on each PC transmits the control data for each robot over a 100Mbps LAN.

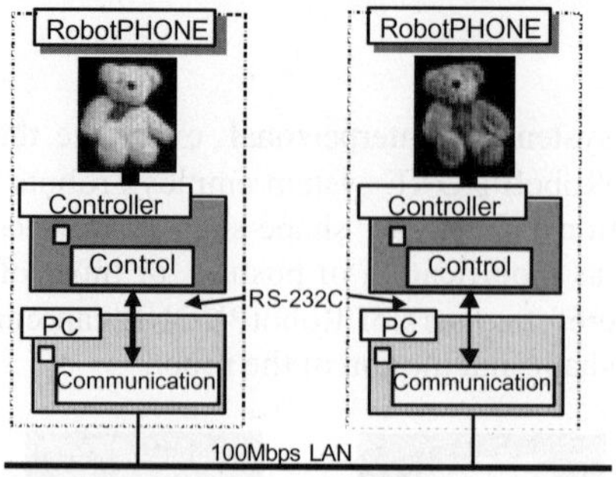

Fig. 2 The system configuration

3.2 The shape-sharing device

The humanoid type RobotPHONE system consists of two shape-sharing devices. The shape-sharing device has 2 degrees of freedom at each arm and leg, 3 degrees of freedom for the head, 11 degrees in total (Fig. 3).

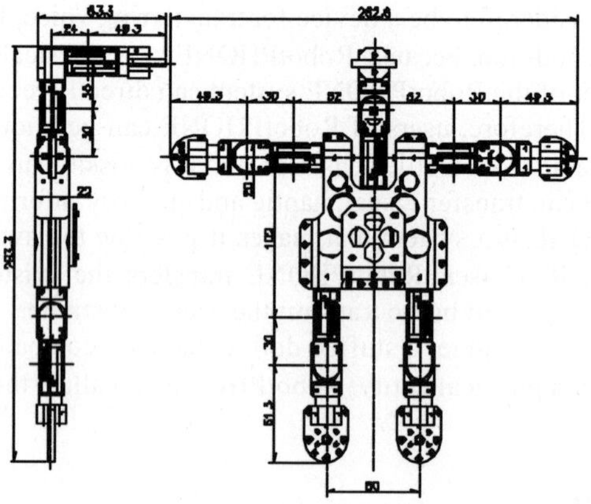

Fig. 3 The mechanical structure of the humanoid-type shape-sharing device

Each joint is composed of a modular unit, which consists of a potentiometer, a 2.6W DC micro-motor and a newly designed 1/60 ultra-compact planetary gearbox. To reduce the weight of the device, we used aluminum and polyacetal resin for body structure. The length of the body is about 30cm.

3.3 The controller unit

All DC motors are controlled by a one-chip microcontroller (AT908535) inside the controller unit. A Pulse Width Modulation (PWM) and a full-bridge driver IC are used to drive the DC motors. The output frequency of the PWM is 0.8[kHz]. The DC motors are controlled by a symmetric bilateral control method, in which the DC motors are controlled to minimize continuously the position difference of each pair. The symmetric bilateral control method is not so suitable for transferring precise force information, but it has the highly desirable merit of simplicity in implementation. By choosing an appropriate servo gain, the torque necessary for bending a joint of the shape-sharing device can be kept easily manageable by a single hand.

3.4 The communication program

To transfer the control data over the network, we used the RCML 2.0 system [7]. The RobotPHONE system is completely symmetrical and there is no distinction of server and client. However, the RCML 2.0 system is basically based on server client models. Therefore, to build an RCML 20 system for the RobotPHONE, we introduced a module called a RCML coordinator (Fig.4).

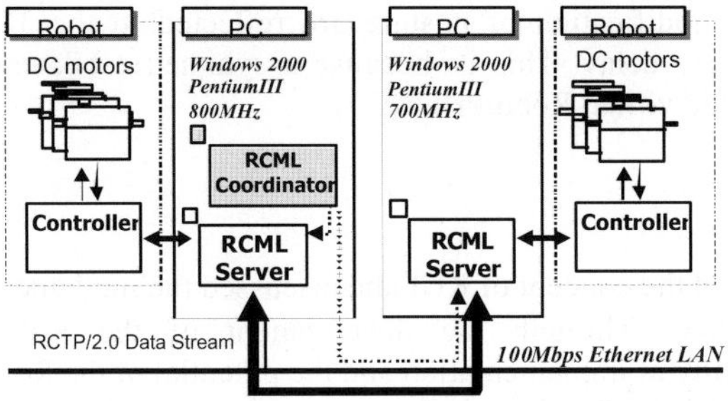

Fig. 4 The RCML 2.0 system for RobotPHONE

As shown in Fig.4, both ends of the system are RCML servers. The RCML coordinator acts as an RCML client for both RCML servers and mediates between the two servers. The first negotiation between the two RCML servers is done through the RCML coordinator. However, the control data is transferred directly by an RCTP/2.0 data stream, which is set up by an RCML coordinator between the two servers.

The control cycle of the system is 21[ms], and the bilateral control is performed in the same way when two controllers are connected directly.

3.5 Virtual RobotPHONE

Virtual RobotPHONE is software that enables the communication between a RobotPHONE user and a non-RobotPHONE user who doesn't have an actual RobotPHONE device (Fig. 5).

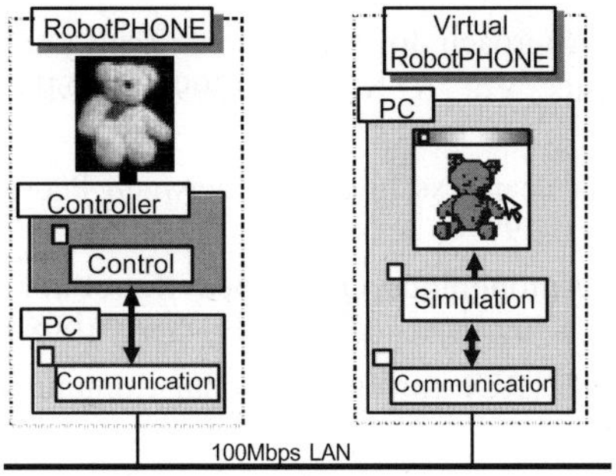

Fig. 5 The configuration of virtual RobotPHONE

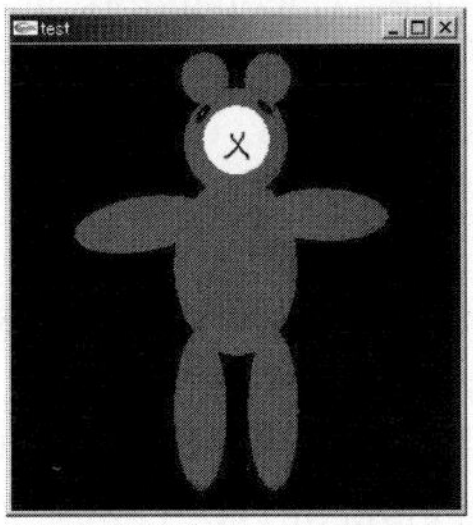

Fig. 6 Screen-shot of virtual RobotPHONE

By simulating the robot used in RobotPHONE, the program constructs a virtual RobotPHONE inside the computer. Virtual RobotPHONE simulates not only the visual information but also the mechanism and the control system of the robot. Therefore, looking from the real RobotPHONE side, the virtual RobotPHONE behaves completely in the same way as a real RobotPHONE including the communication method and control parameters. By simulating the behaviour of a real RobotPHONE, a virtual RobotPHONE can be integrated into the existing RobotPHONE system seamlessly. Virtual RobotPHONE is composed of a 3D rendering part, which displays the 3D CG model of the teddy bear-like robot with OpenGL, and a control simulation part, which simulates the symmetric bilateral control.

Fig. 6 shows the image displayed by virtual RobotPHONE. A user of virtual RobotPHONE can manipulate the real robot by dragging the corresponding part of 3D CG model. Operations on real RobotPHONE such as modification of posture are reflected to the 3D CG model of virtual RobotPHONE without any delay. This is because a symmetric bilateral control is performed between RobotPHONE and virtual RobotPHONE.

4 Conclusion

In this paper we explained the concept of RUI and discussed the implementation of RobotPHONE and virtual RobotPHONE. Through the development of the RobotPHONE system, we demonstrated the feasibility of implementation and the potential of the RobotPHONE concept. We will refine the system design to adapt it for practical use, such as support of long distance communication.

5 Acknowledgments

The part of this work has been supported by CREST of JST (Japan Science and Technology Corporation).

References

1. Douglas C. Engelbart, and William K. English: A Research Center for Augmenting Human Intellect, Conference Proceed-ings of the 1968 Fall Joint Computer Conference, San Francisco, CA, Vol. 33, pp. 395-410, 1968
2. Alan C Key: Microelectronics and the Personal Computer. Scientific American 237, no. 3, pp. 230-244, Sep. 1977
3. J. Rekimoto: NaviCam: A Magnifying Glass Approach to Augmented Reality Systems, Presence: Teleoperators and Virtual Environments, Vol. 6, No. 4 pp.399-412, MIT Press, 1997
4. H. Ishii, and B. Ullmer.: Tangible Bits: Towards Seamless Interfaces between People, Bits and Atoms, in Proceedings of CHI '97, pp. 234-241, 1997
5. M. Fujita and H. Kitano: Development of an Autonomous Quadruped Robot for Robot Entertainment, Autonomous Robots vol.5, pp.7-8, Kluwer Academic Publishers, 1998
6. D. Sekiguchi, M.Inami, S. Tachi: RobotPHONE: RUI for Interpersonal Communication, CHI2001 Extended Abstracts, pp. 277-278, 2001
7. D. Sekiguchi, W. Teng, Y. Yanagida, N Kawakami, S. Tachi: Development of R-Cubed Manipulation Language - The design of RCML 2.0 system, Proc. of ICAT 2000, pp. 44-51, 2000.

2.5D Video Avatar for Networked VRPhoto System[♦]

Youngjung Suh, Dongpyo Hong and Woontack Woo

KJIST U-VR Lab.
Gwangju 500-712, S. Korea
+82-62-970-3157
{ysuh, dhong, wwoo}@kjist.ac.kr

Abstract

In this paper, we propose a novel yet simple way to generate a photo-realistic 2.5D video avatar on the fly for a networked VRPhoto system, which allows users at a distance taking a photo/video int eractively in a shared 3D virtual environment through networks. The avatar is an important medium for users at a distance to feel they seem to be in a same space. The proposed algorithm consis ts of three key steps: (1) 2.5D video avatar generation from a natural background, (2) mesh simplification for efficient transmission of the generated 2.5D video avatar through a limited network bandwidth, (3) real-time augmentation of video avatar into 3D virtual space over network. The pro posed algorithm can be applied to various types of VR applications that require real-time interactions between users at a distance.

1 Introduction

Over last few years, various research activities on avatar generation have been reported. At first, a CG-based avatar has been proposed but it is used in limited applications due to its weakness, i.e., the lack of reality. To overcome the weakness, an image (video)-based avatar has been developed, where the texture of the avatar is segmented from background image and then augmented into virtual world. However, augmenting 2D video avatars onto virtual world is unnatural, since 2D video avatars do not have 3D position information, in general [1]. As a result, it is difficult to allow 2D video avatars to interact with real or virtual objects. To relieve these problems, a 3D video avatar has been developed. However, it usually takes time in generating 3D video avatar and requires extensive computational power in modelling and rendering the avatars [2]. Though 2.5D video avatar has been proposed as a compromise between 2D and 3D video avatars, it also has various limitations since the avatar is generated by exploiting chroma-keying in segmenting a user from the captured image [3].

In this paper, we propose a novel yet simple way to generate a photo-realistic 2.5D video avatar from the natural scene on the fly for its efficient delivery through a network and augmentation into a virtual environment [4]. Figure 1 shows the proposed system block diagram. The proposed algorithm, at first, generates 2.5D video avatar from a natural scene. Then, it performs mesh simplification, which is to represent the mesh model of 2.5D video avatar efficiently for network delivery. At last, the simplified mesh model of 2.5D video avatar is naturally augmented into 3D virtual environment over a network.

[♦] This work is supported by KIST and ICU DML.

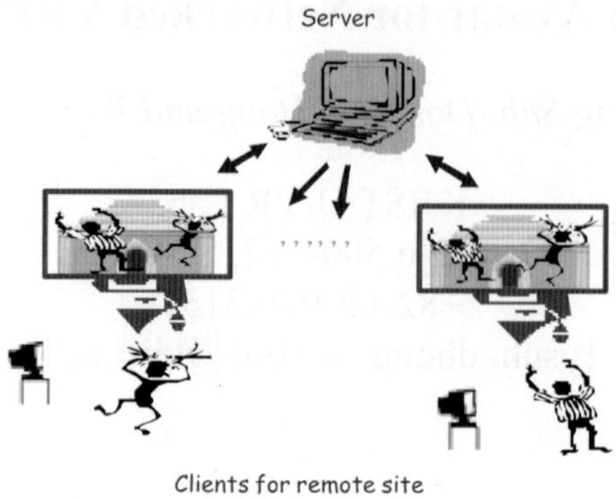

Figure 1: Conceptual diagram of Networked VRPhoto System

The resulting 2.5D video avatar can be a medium for users from different places to provide immersive feeling that they seem to be in a same space. Thus, the avatars can observe existence of the others as well as themselves in a same virtual space, navigating VR space by exploiting their 3D depth information. The proposed algorithm can be widely exploited for VR applications requiring the real-time interactions as well as various collaborations between users at a distance such as teleconference, tele-education, game, and broadcasting.

This paper is organized as follows: in section 2, user segmentation from natural background and real-time mesh model generation are described. In section 3, we explain network delivery and augmentation of the generated 2.5D video avatar. Brief description on the Networked VRPhoto System exploiting the proposed 2.5D video avatar and conclusion are in Section 4 and 5, respectively.

2 2.5D Video Avatar generation

In the proposed algorithm, we adopt a statistical algorithm for detecting moving objects from a natural background scene [5][6][7]. First, we calibrate distorted color values due to the physical limitations of digital cameras. Then we model static background of the compensated image sequences over time. The segmentation algorithm exploits the differences between a trained background image and a current image to segment moving objects from a natural scene. We, first, segment objects in RGB color space by comparing a current image and the modelled reference image. However, RGB color model has weakness in segmenting object without shadow. To segment shadow from the segmented moving objects, we introduce a normalized RGB color model, rgb, with pixelwise dynamic threshold. The rgb color model can detect shadow from the segmented object due to its characteristics. The proposed algorithm overcomes the weakness of the previous background segmentation algorithms by introducing pixel-wised dynamic threshold values considering the characteristics of color channels [8].

For mesh model generation, we first find correspondence between pairs of stereo sequences to estimate 3D image. Then, as shown in Figure 2, we assume that 3D point of each pixel with disparity value might be adjacent to one another also in 3D OpenGL space. Under this assumption, mesh model is generated by triangulation algorithm that connects points according to a specific rule. Also, each vertex that makes up triangular mesh has the texture coordinate with color information. By setting the material property of polygon in model as RGBA array, texture of color image is mapped onto triangular mesh, resulting 2.5D video avatar.

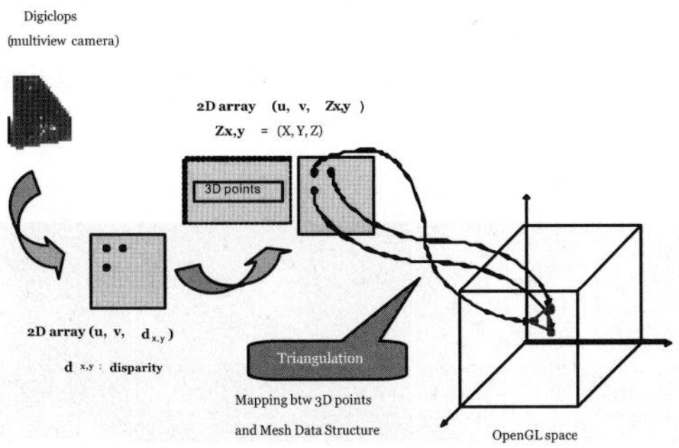

Figure 2: Mesh model generation

3 Augmentation of 2.5D video avatar over networks

Simplifying the generated mesh model of 2.5D video avatar is necessary for efficient transmission and augmentation of 2.5D video avatar through the limited network bandwidth. We can simplify the three components, or vertex, edge, and face of the mesh model. In the proposed algorithm, we simplify the vertices of mesh model using a regular hexahedron. As shown in Figure 3, we form the bounding box which encloses the entire model by exploiting the smallest and largest value of the coordinates of mesh model. After that, the generated bounding box is split into several small cells. Then, if a point of mesh model is included within a specific cell, it is mapped onto the representative point of the cell. Consequently, the number of vertices of mesh model is reduced.

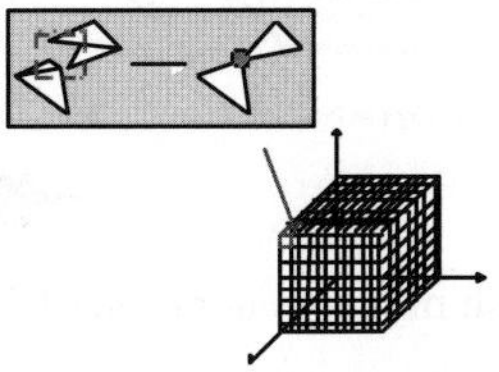

Figure 3: Mesh simplification

The simplified 2.5D video avatar is augmented with virtual cultural heritage considering the camera parameters of real and virtual cameras. The process of the avatar's augmentation over a network is as follows: the environment in which Quanta library is available is set to the clients and a server. A server plays a role in sending data from one client to the others (more than one) simultaneously. Each client sends 3D points and corresponding color values of the avatars, and receives same types of data for one frame from the others to render those.

4 Implementation of VRPhoto System

To prove the usefulness of the proposed algorithm, we implemented "Networked VRPhoto System." In the implemented VRPhoto system, the users from different places can take a photo/video interactively in a shared 3D virtual environment through a network. The system is implemented at Pentium III Dual Xeon 1.0GHz CPU. To obtain images, Digiclops of IEEE 1394 camera is used. Sony UP-DP10 is used as a digital photo printer.

Figure 4 shows the result of the proposed background subtraction. As shown in Figure 4(b), user and its shadows are subtracted from background in RGB color space. However, Figure 4(c) shows only user without shadows, where it is subtracted in the normalized RGB color space. The proposed algorithm can be applied to estimate the light source for VR augmentation applications such as VR studio.

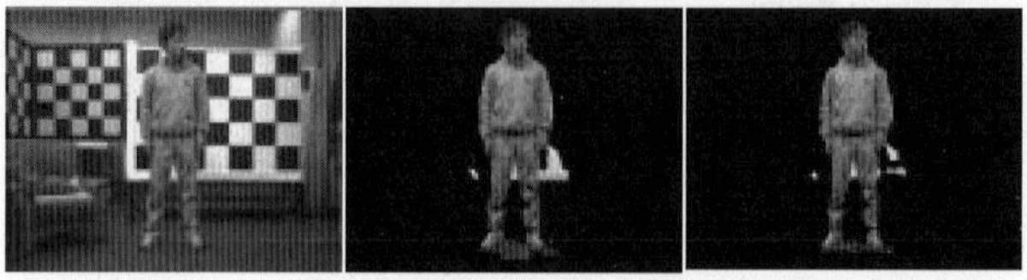

(a) (b) (c)

Figure 4: Background subtraction result. (a) input image, (b) subtraction in RGB color space, (c) subtraction in the normalized RGB color space.

Figure 5 shows the process of constructing the mesh model of 2.5D video avatar exploiting 3D depth information from a 3D camera.

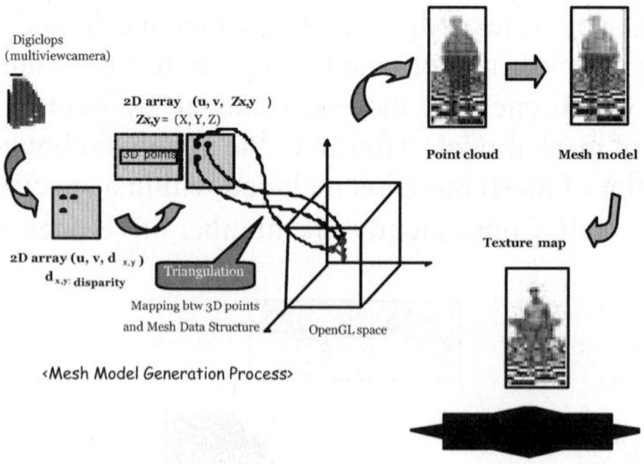

Figure 5: Mesh model generation of 2.5D video avatar

Then, we simplify the polygonal mesh model of generated 2.5D video avatar. Since the delivering cost, as well as rendering cost, of using a mesh model is directly related to the number of polygons, it is useful to efficiently represent the mesh models. Figure 6 shows the results of mesh simplification.

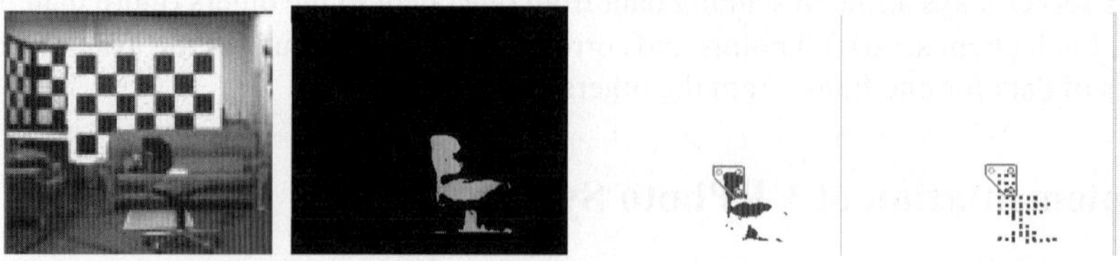

(a) (b) (c) (d)

Figure 6: Mesh simplification. (a) Input image, (b) depth of segmented moving object, (c) original model, (d) simplified model

After all, the simplified mesh models with textures are augmented into a shared virtual space in a server. As shown in Figure 7, the simplified 2.5D video avatar is augmented into 3D virtual environments considering the camera parameters of real and virtual cameras over a network.

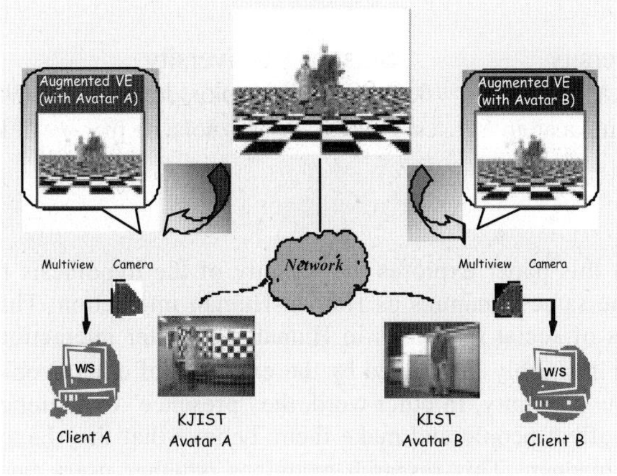

Figure 7: Networked VRPhoto system

5 Conclusion

In this paper, we introduced a novel approach to segment moving objects from natural scene to generate 2.5D video avatar and to augment the avatar naturally into 3D virtual environment through a network using the calibrated camera parameters and depth information of the 2.5D avatar. The proposed algorithm can be applied for VR applications requiring the real-time interactions for various collaborations between users at a distance such as teleconference, tele-education, game, and broadcasting. In addition, without loss of generality the video avatar can be transmitted over the Network by efficiently representing the mesh model.

6 References

[1] "The ALIVE System: Wireless, Full-body Interaction with Autonomous Agents", M.I.T. Media Laboratory Perceptual Computing Technical Report No. 257 http://alive.www.media.mit.edu/projects/alive/

[2] "Virtualized Reality: Constructing Virtual Worlds from Real Scenes", IEEE MultiMedia, vol.4, no.1, Jan. -Mar. (1997),pp.34-47. http://www.cs.cmu.edu/afs/cs/project/VirtualizedR/www/VirtualizedR.html

[3] Hirose, M., Ogi, T., Yamada, T., Tanaka, K., Kuzuoka H., "Communication in Networked Immersive Virtual Environments", 2nd International Immersive Projection Technology Workshop (1998)

[4] Y.Suh, D.Hong, and W.Woo, "2.5D Video Avatar Augmentation for VRPhoto," ICAT'02, pp. 182 -183, Dec. 3-6, (2002).

[5] T. Horprasert, D. Harwood, and L.S. Davis, "A Statistical Approach for Real-time Robust Background Subtraction and Shadow Detection," Proc. IEEE ICCV'99 FRAME-RATE Workshop, Kerkyra, Greece, September (1999)

[6] W. Woo, N. Kim, K. Wong and M. Tadenuma, " Sketch on Dynamic Gesture Tracking and Analysis Exploiting Vision-based 3D Interface," in Proc. SPIE PW-EI-VCIP'01, vol. 4310, pp. 656-666, Jan.(2001).

[7] N. Kim, W. Woo and M. Tadenuma, " Photo-realistic Interactive 3D Virtual Environment Generation Using Multiview Video," in Proc. SPIE PW-EI-VCIP'01, , vol. 4310, pp. 245-254, Jan. (2001).

[8] D.Hong, W.Woo, "Background subtraction for a vision-based interface", ICIP'03 (under review)

Social Influence of Agent's Presence in Desktop Interaction

Yugo Takeuchi — Shizuoka University, 432-8011 Shizuoka, Japan, takeuchi@cs.inf.shizuoka.ac.jp

Keiko Watanabe — Shizuoka University, 432-8011 Shizuoka, Japan, cs8099@cs.inf.shizuoka.ac.jp

Yasuhiro Katagiri — ATR MIS, 619-0288 Kyoto, Japan, katagiri@atr.co.jp

Abstract

The study reported in this paper explores the validity of the hypothesis that Human-Computer interaction displays the same dynamics as Human-Human interaction. The study also addresses the issue of awareness of social responses in Human-Computer interactions. The behaviors and attitudes of people are invariably influenced by the existence of other people even though they do not physically exist one's vicinity. In other word, the "presence" of an imperceptible person has a social power that can affect people and make them believe that they have to respond to it in a socially interpersonal manner. This research examines whether users can recognize individual regions of a desktop as structurally different notional spaces when notional partitions are presented in the form of simple CG images (a CG drawn strip in the center of display screen) on the desktop. The experimental results show that an agent's presence consists not merely in its visual exp ression but in its social effectiveness, which can be simply designed by a CG drawn notional partition. These findings imply the possibility of managing users' perceived presence of an agent by simply presenting notional CG partitions on a desktop screen.

1 Introduction

One of the central issues in Human-Computer interface research has been to instill into computers human-like qualities in terms of both intelligent functionalities and communicative capabilities. Such computers would make it easier for everybody to use and interact with machines. Research on anthropomorphic and believable agents directly addresses this issue, and a number of technologies in artificial intelligence, pattern recognition and multi-modal interface have been devoted to creating "s entience" in computers (Laurel, 1990; Nagao & Takeuchi, 1994; Tosa & Nakatsu, 1996).

Complementary to these technology-oriented approaches, which focus on the design of new types of machines, the problem of human-like qualities in computers has also been investigated from the point of view of the psychology of human reactions to computers: how and to what degree do people personify computers and attribute human-like qualities to them. Particular emphasis has been placed on the social aspects of human responses to computers. Reeves and Nass (Reeves & Nass, 1996) have convincingly demonstrated, through a number of experiments, that human-computer interactions are basically of the same nature as human-human interactions. They found repeatedly in their experiments that people respond to computers, at an unreflective gut level, as though computers were humans. These human behaviors do not result from people's ignorance or from psychological or social dysfunctions but from the fact that social responses are commonplace and easy to generate, even without sophisticated AI or multimedia technologies. Human-computer interactions are fundamentally social and natural.

As described above, human-like agents can strongly induce in their users a perception of social "presence" by how they respond to the users. Therefore, it is necessary to manage adequate human-like responses when users interact with agents. This research examined whether users can recognize individual regions of a desktop as structurally different notional spaces when notional partitions are presented in the form of simple CG images on the desktop. The psychologically perceived presence of an agent affects the social relationship between users and the agent. This helps to establish a relaxed and comfortable computer desktop environment even though the agents are only presented on the desktop screen.

2 Presence

One of the most effective approaches to designing a social environment on a computer desktop is to apply human-like agents. Here, it is expected that the users can interactively and socially achieve their tasks with the agents. In order to use the agents effectively, it is necessary to properly manage the "presence" of each agent. This is because the presence of the agent greatly stimulates users to socially respond to the agent. For example, we illustrated that people are sensitive to social power such as authority, and they may try to conform to those who interact with social power even if they only merely view a theatrical demonstration of the social inter-agent interaction (Takeuchi, Katagiri, & Takahashi, 2001).

Human social acts are strongly influenced not only by the physical existence of others but also by a presence that makes humans mentally conceive of such an existence as fact. At the same time, if others physically exist but a human is not aware of their presence, human social acts would not be influenced by their physical existence.

These observations of different behaviors in human social acts indicate a dependence on the awareness of a presence. Accordingly, the design of social interaction between a human and an agent on a computer desktop screen should thoroughly consider how to actualize a natural and appropriate computing environment. In particular, when plural agents simultaneously appear on the same desktop screen, much consideration should be given to designing social human-agent interaction.

3 Experiment

3.1 Virtual Partition on Desktop Screen

When we present a theatrical play, we tacitly understand that the stage space is virtually divided into two separated areas when a simple board wall or door is built into the stage set. The actors naturally perform just as if the two areas are completely divided by a solid wall, on each side of which one actor is not aware of the presence of another actor side on the other side.

Suppose the concept of the theatrical stage environment is extended to the computer desktop screen environment, then how would the user interpret a CG drawn partition (shown in **Figure 1** as a strip in the center of each desktop screen) between two regions on the screen? If users assume that the CG drawn partition is an actual boundary, in the manner understood on the theatrical stage, we can hypothesize that each region is regarded as a distinct social space. According to this hypothesis, it is predicted that a CG drawn partition can affect the user's awareness of each agent's presence. In order to examine whether the user's awareness of an agent's presence is affected by the CG drawn partition, a psychological experiment was carried out.

3.2 Method

Subjects

43 Japanese university students (24 male and 19 female)

Each subject was randomly assigned to four experimental condition groups.

Materials

Desktop PC with 19-in CRT, Speakers, Vocal Mic

Procedures

In this experiment, each subject was separately interviewed by two independent agents (A1 and A2). The computer desktop screen was divided into two regions by a vertical CG drawn partition in the center.

(1) First, agent A1 at the right side of the desktop screen asked the subject eight questions (**Figure 1**) concerning social morality while maintaining a strict attitude. In this situation, the subject was tacitly required to answer the agent by appearing to conform to desirable morals.

(2) After the first interview, the subject was asked eight questions (shown in **Table 1**: five of the eight were the same questions used in the first interview) about social morality by agent A2 (**Figure 1**), which maintained a friendly attitude. In this situation, the subject did not feel the necessity to answer the agent in conformity to socially endorsed morals because the attitude of the agent A2 induced the subject's self-disclosure. Agent A2 appeared on either side of the desktop screen, depending on the experimental conditions.

Table 1 Five shared questions, from original set of eight, concerning social morality in the two interviews.

(a)	When you want to go somewhere by train, a friend offers to lend you his/her commutation ticket. *Would you accept and use it while pretending to be your friend?*
(b)	When you want to cross a street, you can see a crosswalk a little bit away from your current position. *Would you move to it before crossing the street?*
(c)	When you want to throw your trash in a public trashcan at a park, you see that the trashcan is already full of trash. *Would you give up the idea of throwing your trash away and take it home?*
(d)	*Would you talk with your friend by using a cellular telephone within a closed public space such as a bus or railroad car?*
(e)	On your way home, you can take a shortcut and get home earlier by going across private land. *Would you go across this property without permission?*

Experimental Conditions

[*Separated Condition*] Agent A1 remains on the right side of the desktop screen as agent A2 interviews the subject from the left side of the virtually divided desktop screen.

[*Shared Condition*] Agent A1 remains on the right side of the desktop screen as agent A2 interviews the subject from the same right side.

[*Independent Condition*] Agent A1 entirely disappears from the desktop screen while agent A2 interviews the subject from the left side of the desktop.

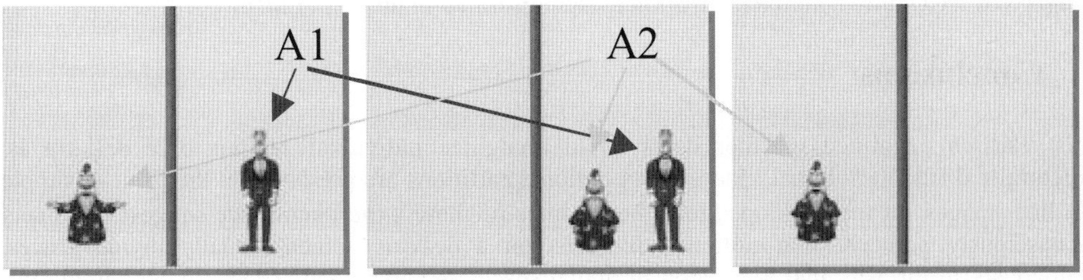

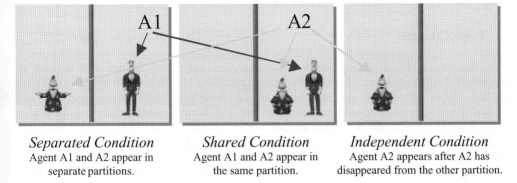

Separated Condition
Agent A1 and A2 appear in separate partitions.

Shared Condition
Agent A1 and A2 appear in the same partition.

Independent Condition
Agent A2 appears after A2 has disappeared from the other partition.

Figure 1 CG drawn partition and appearance position of two agents in each experimental condition.

Predictions

If subjects become aware of the presence of agent A1, they will cautiously answer the questions of agent A2 in the same way as they would answer to agent A1. On the other hand, if subjects were aware of the presence of agent A1, they would not feel it necessary to use caution in disclosing themselves. This is because it is generally considered good social manners to disclose oneself to a non-familiar person in a cautious way without being contradictory.

4 Results and Considerations

The evaluation was based on the number of answers that the subjects changed in their disclosures between the first and second interviews with the agents. **Figure 2** shows the results under each condition. In ANOVA, a main effect was found ($F_{(2,37)}=4.349$, $p<.05$), and there were two significant differences for Separated-Shared ($p<.05$) and Independent-Shared ($p<.05$) conditions.

According to the post-experimental questionnaire, most of the subjects reported that they supposed that they were cautious in answering agent A1 when agent A2 stood on either side of the partition. However, the subjects in fact changed their responses when agent A2 stood on the other side of the partition from agent A1 as well as when agent A2 disappeared from the desktop screen. This result shows that the subjects unconsciously recognized the CG drawn notional partition as a physical wall on the desktop screen. In other words, an agent's presence consists not only in its visual expression but also in its social effectiveness, which can be simply

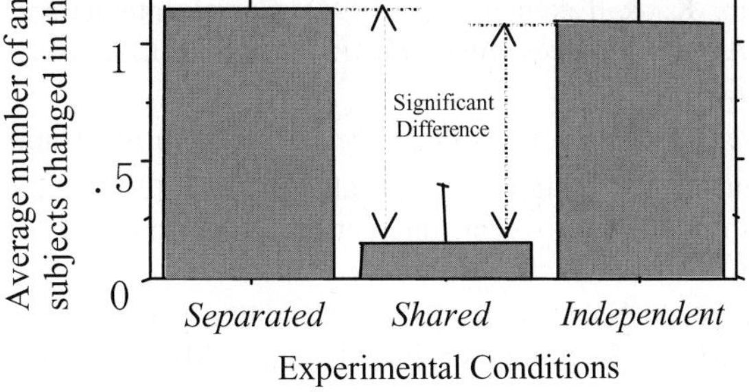

Figure 2 Results of each experimental condition.

designed by a CG drawn notional partition as demonstrated in this experiment.

5 Conclusions

This research examined whether users could recognize individual regions of a desktop as structurally different notional spaces when notional partitions are presented in the form of simple CG line images on the desktop. From the findings of these experiments, we suggest that users unconsciously perceive that notional divisions on a desktop screen actually divide spaces, analogously to the way that a division of actual spaces is understood. These findings imply the possibility of managing users' perceived presence of an agent by simply presenting notional CG partitions on a desktop screen.

Designing a cognitive appropriately desktop interface becomes an even more complex issue when we consider that cyberspace further increases the possibilities through practical network use. Users access cyberspace in order to actualize their requirements, for example, information search using the WWW, network communication, and so on. The many sources of information and activities in cyberspace become increasingly complicated to deal with as network technology advances. Therefore, it is necessary to arrange materials in cyberspace that conform to users' intuition. This might involve classifying two types of material in cyberspace from the viewpoint of their properties. One type of material, which is structurally systematized information such as the contents of Web page or hierarchical semantic data, could be organized into structurally different notional spaces such as Web sites or "folders" respectively. On the other hand, the other type of material, which includes the agents or avatars of users in cyberspace, would be difficult to organize structurally because this type is endowed with human-like characteristics. Such characteristics have a variety of attributes: attitude, affiliation, nationality, age, etc. Therefore, it is best to arrange agents in a social rather than structural space. This kind of social space was investigated in this study as notional partitions.

Social interaction will enhance and improve Human-Agent interaction in intelligent agent applications such as those in CAI systems, navigational guide systems, and expert consulting systems, as well as in virtual environment systems inhabited by avatars and agents that mediate Human-Human collaborations.

References

Laurel, B. (1990). Interface Agents: Metaphors with Character. In B. Laurel (Ed.), *The Art of Human-Computer Interface Design*. Addison-Wesley.

Nagao, K. & Takeuchi, A. (1994). Social Interaction: Multimodal Conversation with Social Agents. *Proceedings of the 12th National Conference on Artificial Intelligence (AAAI-94)*, 1, 22-28.

Reeves, B. & Nass, C. (1996). *The Media Equation*. Cambridge University Press.

Takeuchi, Y, Katagiri, Y. & Takahashi, T. (2001). Learning Enhancement in Web Contents through Inter-Agent Interaction. In Hirose, M. (Ed.), *Human-Computer Interaction --- INTERACT'01* (pp. 480-487). IOS Press.

Tosa, N. & Nakatsu, R. (1996). Life-like Communication Agent: Emotion Sensing Character 'MIC' and Feeling Sensing Character 'MUSE'. Proceedings of the ICMCS96, 12-19.

Construction of Meaning Acquisition Model Using Prosodic Information: Toward a Smooth Human-Agent Interaction

Atsushi Utsunomiya, Takanori Komatsu, Kentaro Suzuki, Kazuhiro Ueda, Kazuo Hiraki

Department of System Sciences,
The University of Tokyo
3-8-1 Komaba, Meguro-ku, Tokyo,
153-8902 JAPAN
au@cs.c.u-tokyo.ac.jp

Natsuki Oka

Matsushita Electric Industrial Co., Ltd
3-4 Hikaridai, Seika, Souraku, Kyoto,
619-0237 JAPAN
oka@mrit.mei.co.jp

Abstract

The purpose of this study is to propose a speech meaning acquisition model, which can utilize prosodic information of her/his speech sound, for realizing adaptive speech interface system. This model was constructed based on a human-human communication establishment process, and as a result of a testing experiment to confirm the competence of this model, we could confirm that this model could learn the meanings of the instructions through actual interaction without utilizing any phoneme information and prepared knowledge about instructions such as dictionary-like database. It is then expected that the constructed meaning acquisition model can be applied for an adaptive interface system which provides a natural interaction environment for its user.

1 Introduction

Recently, many researchers have tried to develop speech interface. Most of the previous studies focused on phoneme information, which can be written by texts, rather than prosodic information. However, this method has at least two technical issues: one is the instability of the speech recognizer, the function of which is to extract phoneme information from actual speech sounds. The other is that a mapping between the particular word and the function of the application must be defined *a priori*. To overcome such problems, some researchers have started utilizing prosodic information for a speech interface. For example, Igarashi & Hughes (2001) developed a speech interface focusing on prosodic information; e. g., when the user says "move up, ahhh" and this interface recognizes the "move up", the screen starts scrolling upwards, while the "ahhh" continues. The user can also control the scrolling velocity by increasing or decreasing her/his voice pitch. The advantage of this method is that a user can control the scrolling distance and velocity interactively with "ahhh" pronunciation instead of using well-defined verbal command, like "move up, 20cm". This interface, however, still requires a prepared mapping between verbal commands and certain functions, like as the traditional way, and users must obey it. For example, users cannot use "roll up" instead of "move up" for this interface.

In this paper, we assume that the desirability of both the user and the interface having the capability to assign meaning to verbal commands as a result of the interaction between them and without using the prepared knowledge about instructions, such as a dictionary-like phoneme database. This communication situation we assumed above is similar to a relationship between a

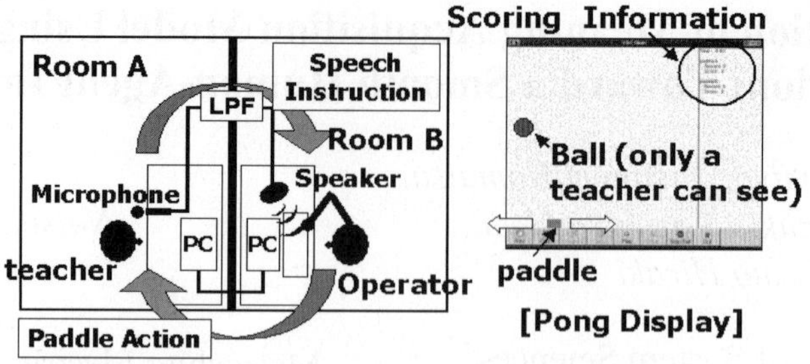

Figure 1: Experimental Environment and Display Setting

speaker (a user) and a listener (an interface), in the case that they do not share the common language. Therefore, to achieve an interactive speech interface system, which can provide smooth and natural communication for users, a requirement is that an interface should have the ability to recognize the intentions of user speech. This ability should be the result of interaction between an interface and a user and of employing prosodic sound features instead of phoneme information.

The purpose of this study is to propose a meaning acquisition model that can recognize a user's intention from prosodic information of her/his speech sound through interaction. Specifically, at first, we carried out a communication experiment in order to observe how people acquire the meanings of unknown utterances when the speech sound is linguistically incomprehensive. We then constructed a meaning acquisition model which can discriminate utterances from salient prosodic features and learn the relationships between these utterances and their corresponding functions, based on the result of the experiment. Finally a testing experiment was carried out to investigate whether the proposed meaning acquisition model could learn the meaning of actual speech sound through actual interaction. It is expected that the result can be a basic methodology for realizing an adaptive speech interface that can smoothly communicate with users.

2 Communication Experiment

2.1 Purpose and Settings

The purpose of this experiment is to clarify how people establish a communicative relationship by acquiring the meanings of utterances in languages they do not understand. Eleven pairs (Japanese, 20-28 years old, 18 men and 4 women) participated in this experiment: they played a "Pong" game. The experimental environment and the display setting are shown in Figure 1. The goal of this squash-like game was to hit a falling ball by a paddle through mutual cooperation. The two subjects are placed in separate rooms: one was a teacher whose task was just to give the operator speech instruction how to move a paddle, while the other was an operator whose task was just to operate the paddle according to the teacher's instruction. Each pair played two consecutive 10 minutes games. The pair was awarded 10 points when the operator hit the ball. To mask phoneme/linguistic information in the teacher's instructions, the teacher's speech instructions were transmitted through a low-pass filter (LPF). The LPF masked the teacher's speech phonemes but did not affect the prosodic features of their speech. In addition, the operator could not see the ball, which was the hitting target, in her/his display. The operator had to recognize the meanings of the teacher's instructions, which were linguistically incomprehensive. (About the detail of this experiment, see the article of Komatsu et al. (2002))

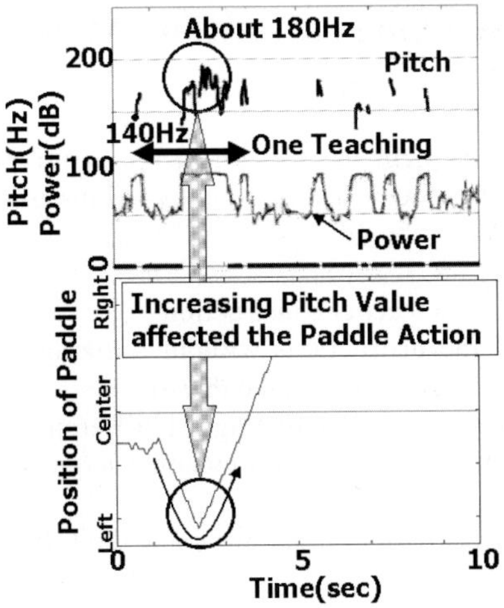

Figure 2: Attention Prosody and Paddle Action

2.2 Results

To evaluate the subject's performance, a value was assigned to a successful action. For each move action, if the operator moved the paddle in the teacher's intended direction, Correct Direction Value (CDV) was assigned one point; if s/he moved it in a different direction, CDV was assigned zero point.

We used a testing statistical hypothesis formed by using binominal distribution to group the subjects. Specifically, we assumed that the subjects recognized the teacher's instructions if the average CDV was more than 0.8. As a result, most pairs were succeeded in understanding the meanings of unknown instructions: Out of the 11 pairs, only two failed to understand any instructions. The nine remaining pairs succeeded in moving the paddle in the teacher's intended direction. In addition, the following phenomena were commonly observed in the successful pairs.

1. The operator recognized the types of given instructions based not on the linguistic information but rather on the salient prosodic features, i.e., the player could recognize that one long utterance was a different instruction than a series of choppy utterances and that there were two types of instructions used by the teacher.
2. The teacher used high-pitched voice to draw the operator's attention. When the operator received such high-pitched voice from the teacher, the operator intuitively recognized that her/his current action was wrong and modified it immediately (see Figure 2). We named such prosodic information "Attention Prosody (AP)". We could confirm that all subject in the experiment used this AP sound features in same way, so that the usage and interpretation of AP seems to be universal.
3. The operator acquired the meaning of an instruction by coupling with her/his action. In this experiment two types of information (reward) were used for this meaning acquisition process. One was the positive reward (hitting the ball), and the other was the negative reward (hearing a high-pitched AP sound). Therefore, it seems that the meaning acquisition process observed in this experiment was reinforcement-learning-like process based on these two types of rewards.

From these results, we assume that this observed process would be the basis of meaning acquisition model.

3 A Proposal of a Meaning Acquisition Model

3.1 Overview

From the result of the previous experiment, we considered that a meaning acquisition model that can recognize the meanings of a user's instructions by means of prosodic sound features must perform the following four tasks:

1. Map the paddle action and the teacher's instruction
2. Acquire the meanings of the instructions without utilizing the prepared knowledge about instructions (e.g. dictionary-like phoneme database)
3. Find critical sound features in speech to distinguish different types of instructions
4. Utilize an AP sound as a negative reward for meaning acquisition process

To realize these requirements in our model, we made the following assumptions:

- When a positive reward is given (i.e. when the paddle hit the ball), the model should recognize that the meaning of the given instruction indicated the current action. Conversely, when a negative reward (i.e. hearing an AP sound) is given, the model should recognize that the meaning of the given instruction did not indicate the current action.
- The instances, i.e., a pair of instruction sound (eight-dimensional sound vector such as pitch, zerocross number, and so on) and paddle action, are stored in internal memory when positive/negative reward is given. Each instance in this memory is assumed to be generated from certain distribution of a mixture of normal distribution, which is used as a mapping between the instruction and the paddle action. Each distribution in the mixture of normal distribution expresses the meaning of certain instruction. The model recognizes the meanings of given instructions based on this mapping.

To understand the meanings of instructions, this model must estimate the parameters (average and variance), of probability distributions to explain the incoming instances. As a basic methodology, we used the EM algorithm (Dempster, 1977). However, this algorithm cannot deal with a negative instance (which means the instance is not generated from certain distribution), so that we developed the extended EM algorithm that could include negative instances for estimating the parameters.

3.2 Testing Experiment

To evaluate the competence of this meaning acquisition model, we carried out a testing experiment to confirm whether this model could learn to recognize the meaning of instructions through interaction with a human instructor. As a testing environment, this meaning acquisition model was incorporated into the paddle component of the software for the "Pong" game that the human operator moved in the communication experiment. This proposed meaning acquisition model does not focus on the phoneme information, so it must learn to recognize the meanings of instructions through prosodic sound features so as to distinguish the different instructions types, regardless of the actual language being spoken. It can do so if there are enough of these prosodic sound features. Therefore, to evaluate the performance of this model, an instructor used the five following types of instructions:

- "Up" and "down" in Japanese ("UE" and "SHITA", respectively)

- “Up” and “down” in English
- High-pitched “ahhh” for “up” and low-pitched “ahhh” for “down”
- Low-pitched “ahhh” for “up” and high-pitched “ahhh” for “down”
- Long “ahhh” for “up” and choppy “ahhh” for “down”

As a result, this model could learn the meaning of all five types of instructions. In all cases, the CDV surpassed 0.8, as well as the human operator of the successful pairs in the communication experiment. From this result, we could confirm that our model has enough ability to learn the meanings of instructions from salient prosodic features of actual speech sounds without any prepared knowledge of instructions, e. g., a dictionary-like database.

4 Discussion and Conclusion

It is generally believed that the implementing the learning mechanism into a speech interface is not appropriate due to consumed times for the learning, so that the instruction-function mapping was defined *a priori* in most previous speech interfaces such as Igarashi & Hughes (2001). Our model, however, could learn to recognize the meanings of instructions during the first two to five minutes of the game. Although some users might complain about the length of learning time, our method has at least two advantages over the normal one. First, our method does not use a traditional speech recognizer, so that this method is free from the problems which most previous speech interface studies faced. Second, our method does not force the user to learn about the interface, e. g., memorizing the name of commands, so that the user can give instructions in her/his way. In short, the interface based on our method might require less user’s cognitive load and could provide more smooth interaction for the users than the previous speech interfaces. However, there is still one issue, that is, how to apply this model for a multi-functional interface. To overcome this issue, an integration this method with other techniques, such as image processing for utilizing gesture information, is expected to be discussed further.

In this paper, we constructed a meaning acquisition model that could recognize the intentions/meanings of user’s instructions through interaction. A testing experiment showed that the constructed model could recognize the meanings of speech instructions based on salient prosodic features. We expect that these results could contribute to achieving an interactive interface. In addition, the results showed that this model could learn to recognize the meanings of instructions regardless of the language spoken by user. This means that our method can be used in speech interface techniques for any language as a “universal interface.”

References

Dempster, A. P., Laird, L., M., & Rubin, D., B. (1977) Maximum likelihood from incomplete data via the EM algorithm. *Journal of the Royal Statistical Society, Series B*, 39(1), 1-38.

Igarashi, T., & Hughes, J. F (2001) Voice as Sound: Using Non-verbal Voice Input for Interactive Control. *Proceedings of CHI 2001*, 77-84.

Komatsu, T., Suzuki K., Ueda, K., Hiraki, K., & Oka, N. (2002) Mutual Adaptive Meaning Acquisition by Paralanguage Information: Experimental Analysis of Communication Establishing Process. *Proceedings of the 24th Annual Meeting of the Cognitive Science Society*, 448-453

Anthropomorphic characteristics of Interface agents

Manuel Velez and Esther Esteban

Dpt. Draw
Facultad de Bellas Artes
University of Granada
18071 Granada, Spain
mvelez@ugr.es, biombo@wanadoo.es

José J. Cañas

Department of Experimental Psycholog
University of Granada
Campus Cartuja
18071 Granada, Spain
delagado@ugr.es

Abstract

The agents are objects presented in many of the interfaces of the computer programs that we are use today. Its inclusion in the interfaces is based on the assumption that they can serve as help for interaction with the interface. However, the empirical investigation on this assumption is offering contradictory results. In the present paper we offer results of an investigation that it being conducted at the University of Granada with the goal of providing support to the hypothesis that says that one possible reason that users have for rejecting the use of agents could be related to their anthropomorphic characteristic.

1 Introduction

Agents are objects present in many of today interfaces and that have been designed with the purpose of helping the users of a computer program to perform their tasks in a more effective way. To be able to do their job, the agents need to know the goal of the users, experience, etc, as well as a good knowledge of the tasks that the user want to perform. The main characteristic of agents is that they can communicate with the users.

Agents are meant to be new forms of interfaces and new ways to interact with them. In the interfaces of direct manipulation the users manipulate inanimate objects that are presented inside the interface. On the contrary, in an interface with an agent the user can communicate with it to look for the help or to carry out a task, separate or jointly. For example, in Internet an agent can be programmed so that it looks for certain type of information during a period of certain time. In this sense, an agent is an automation component of the interface. It can also be programmed to discover when the information that the user is interested in appears. In this case the agent works asynchronously and it does not interrupt the user's work while she/he can be doing something at the same time. But it is also possible that agent's work is carried out synchronously with the user.

Interface agents were introduced to make the interaction easier for the user. It was assumed that agents can facilitate learning and interacting with the computer so that it can be more effective and more pleasant. Therefore, ignoring the concerns that some authors had (for example Norman, 1994; Shneiderman and Maes, 1997; Wilson, 1997), the software designers began to introduce agents in all their programs and now they are present in almost all the interfaces that we use.

However, after several years of having the available agents in almost all the commercial programs, it is not clear that the agents really facilitate the interaction. For example, Dehn and Mulken (2000) have revised recently the empirical investigation conducted to test the effects that agents have on the experience, the behaviour and the performance of user. The conclusions of their review confirmed what many designers and users suspected: there is very little evidence that the interfaces with agents are superior to those that don't have them. In the two analysed aspects, the users' good execution and good attitude toward the interface with agents, only in this last aspect it seems to be certain advantage of the interfaces with agents. However, the users perform their tasks equally well with or without agents.

Consequently, we believe that it is necessary to consider the arguments given against the supposed advantages of the agents and to conduct more empirical investigation to determine how and in what circumstances agents are really effective. For example, it is possible that agents force the user to acquire and reinforce false mental models of the system. If an agent with certain anthropomorphic characteristic behaves somehow as a human being, the users could believe that the agent has other cognitive and emotional abilities that they really lack. That is to say, it is possible that the user expects more than the agent can do, and it would be convenient to eliminate these anthropomorphic characteristic.

According to this line of reasoning, we are conducting a research project to identify the graphical characteristics of interfaces agents. The main goal of the project is to find which of those characteristics could be responsible for the difficulties that users have with them. In this paper we present results from a study in which undergraduate students from Graphical Design, Computer Science and Psychology expressed their preferences about a set of different agents design along the dimension of anthropomorphism. The agents were presented alone or in the context of two popular computer programs (Microsoft Word, PhotoShop).The students made their preference evaluation on an static environment in which the agent was not used to perform any computer tasks. Contrary to previous results, we found that that students prefer the more anthropomorphic agents. We advance an explanation of these results based on the role that agents play on helping users on performing their task.

2 Methods

2.1 Materials

Figure 1 shows the twelve agents that were presented in three experimental conditions: (1) paper without context; (2) Microsoft context; and (3) Phothoshop context.

2.2 Participants

Fifty-three students of Psychology, 53 of Computer Science and 53 of Graphical design participated in the study.

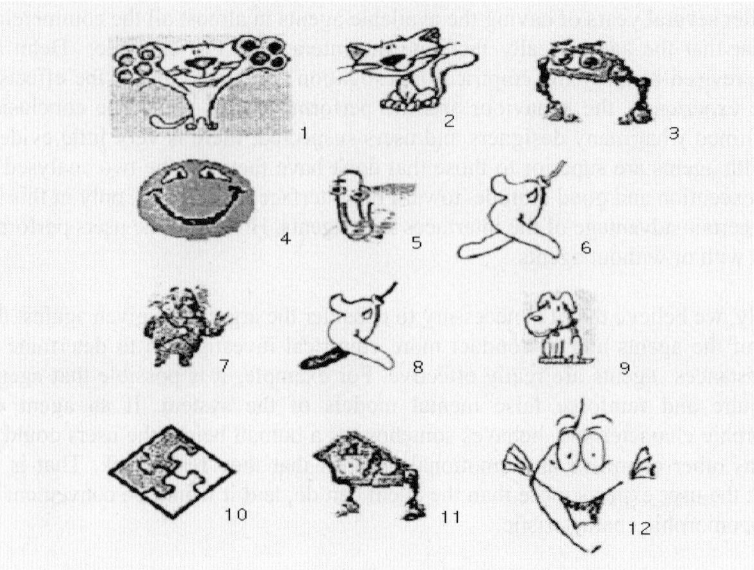

Figure 1. Agents without context

Figure 2. Agents in the WORD and Photoshop contexts

2.3 Procedure

The students were asked to made a preference evaluation of the agents. Students of Graphical Design and Computer Science were presented with the agents in the three conditions. Psychology student made their preference only in the 'paper without context' and 'Photoshop context' conditions. They were instructed to mark the agent that they would like to see in a programme when they make an error and need help to solve it.

3 Results

Figures 2, 4 and 5 show the results for Graphical Design, Computer Science and Psychology students respectively. The data presented in these figures are the number of choices for each agent in each of the presentation condition. As can be seen, all the students preferred agents with anthropomorphic characteristics (1,2 and 7) in all of the presentation conditions. However, agents with design in which these anthropomorphic characteristics are not clear were rejected by the students.

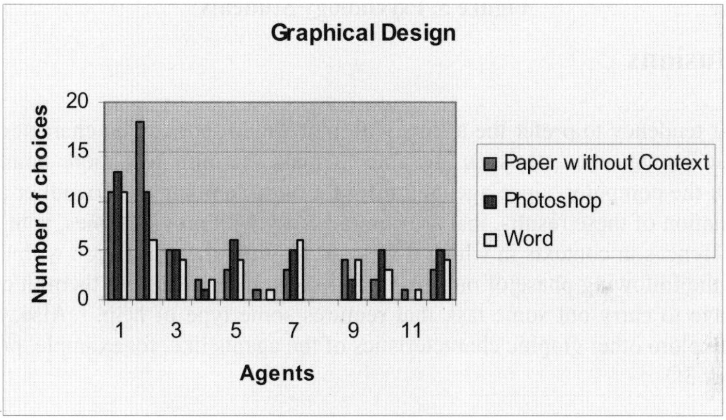

Figure. 3. Graphical design students

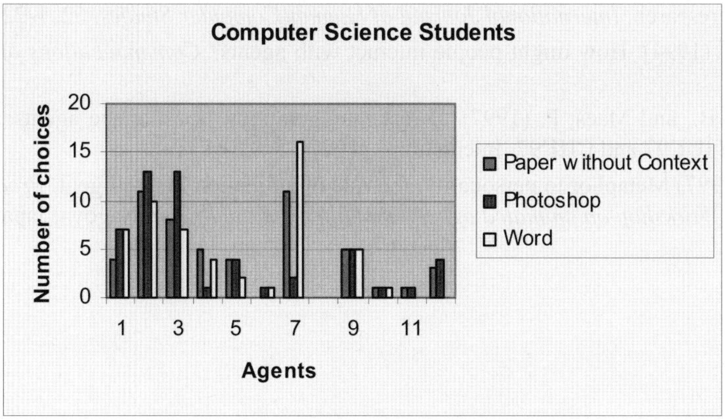

Figure. 4. Computer Science Students

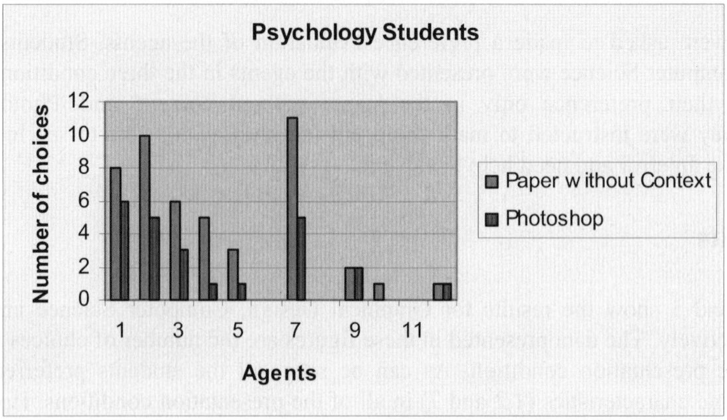

Figure 5. Psychology Students

4 Conclusions

The user's clear tendency to prefer the agents with more anthropomorphic characteristics let us to think that those human characteristics are able to make her/him feel more comfortable in its interaction with the computer, since agents transmit a more familiar and friendlier atmosphere. A possible explanation of these results, that are contradictory with previous ones, is that our students gave their preferences in contexts in which it was not required the interaction with the agents. For this reason, in the following phase of our investigation we plan to present them in contexts where the students have to carry out some task that requires some type of help. Also, in this second phase we will explore other graphic characteristics of the agents like, for example, presenting them in formats 2D or 3D.

References

Dehn, D.M. and van Mulken, S. (2000). The impact of animated interface agents: a review of empirical research. *International Journal of Human-Computer Studies*. 52, 1-22.

Norman, D.A. (1994). How might people interact with agents? *Communications of the ACM*,.37, 68-71.

Shneiderman, B., and Maes, P. (1997). Direct manipulations vs. Interface agents: excerpts from debates at IUI'97 and CHI'97. Interactions, (1997). 4, 42-61.

Wilson, M. (1997) Metaphor to personality: the role of animation in intelligent. *Proceedings of the IJCAI-97 Workshop on animated agents: making them intelligent*. Nagoya, Japan.

AVICE: Evolving Avatar's Movements

Hiromi Wakaki

Graduate School of Frontier Science,
The University of Tokyo
7-3-1 Hongo, Bunkyou-ku, Tokyo,
113-8656, Japan.
wakaki@miv.t.u-tokyo.ac.jp

Hitoshi Iba

Graduate School of Frontier Science,
The University of Tokyo
7-3-1 Hongo, Bunkyou-ku, Tokyo,
113-8656, Japan.
iba@miv.t.u-tokyo.ac.jp

Abstract

In movies and games, lifelike animations expressed in 3D space have frequently been used in recent years. The animations are generally created either by importing the movements of a human being through motion capturing so that the avatar will move in the same manner, or by drawing the movements by human hand so that the avatar will move like a human being. However, there are claims that it is extremely difficult to generate such human movements, unlike drawing 3D pictures. This is attributable to the fact that the creation of movements by hand requires intuition and technology. With this in mind, this study proposes a system that supports the creation of 3D avatars' movements based on interactive evolutionary computation, as a system that facilitates creativity and enables ordinary users with no special skills to generate dynamic animations easily, and as a method of movement description.

1 Introduction

In recent years, the widespread penetration of the Internet has led to the use of diverse expressions over the Web. Among them, many appear to have strong video elements. However, few expressions are based on human beings, with whom we are most familiar. This is deemed to be attributable to the fact that it is not easy to generate human movements.

As a means of expression, 3D-CG description software use inverse kinematics (IK) **(**R. Stuart Ferguson, 2001**)** , which divides the human being into parts and describes its shape accordingly while sustaining the correlation between the joints, so that if its arm is pulled, other joints will also move to make the movement look natural. In practice, however, the detailed description of movements is a manual task in 3D-CG.

In consideration of the above, we realized a system that generates movements through Interactive Evolutionary Computation (IEC) **(**T. Unemi, 1999**)** **(**H. Takagi, 1998**)** based on evaluation by users through H-Anim (See **Sect. 2.1**).

Imagine creating your favorite music or pictures by an automatic system. The desirable output depends on the user's subjectivity, which is generally difficult to be defined in terms of a fitness function. There are many areas as such, not only in the field of art but also in the field of engineering and education. One way to optimize such systems is IEC. IEC is more or less the same as normal genetic programming (GP) **(**Banzhaf et al., 1998**)**, but the user evaluates each individual instead of calculating the fitness value.

With the above consideration, this paper clarifies the following points:

- The interactive EC system called AVICE is proposed for the sake of creating 3D-Avatar motions.

- AVICE supports the user creativity in designing motions, which is empirically shown.
- AVICE's facility is evaluated with some psychological experiment with a traditional system.

2 AVICE System

2.1 Humanoid Animation

The Humanoid Animation Working Group (http://h-anim.org/) specifies how to express standard human bodies in VRML97, which is called Humanoid Animation (hereinafter referred to as "H-Anim"). VRML is deemed to be versatile as it enables viewing over the Internet through a browser.

The name of each body part of a human being and the correlation between joints are specified accordingly. Once the configuration of the body is decided, the rotation and movement of each part can by described in the same manner as ordinary VRML. However, the parameters of rotation and movement are not easy to understand for human beings. For the description of such movements, tools are indispensable for importing the movements of the human body described in H-Anim and moving the parts so as to convert them into VRML.

2.2 Outline of AVICE System

AVICE system integrates Creator, Mixer and Remaker modules (See **Fig. 1**). Each relationship is that Mixer module take in the output of Creator module and other softwares. Creator module makes VRML files suited user's taste from initialization used IEC. Mixer module makes VRML files suited user's taste from inputted files used IEC. If the user wants special remaking, he or her uses Remaker module with VRML files outputted from Creator module or Mixer module. After the above operation, there are the finished VRML files.

In following subsections, we explain each module in details.

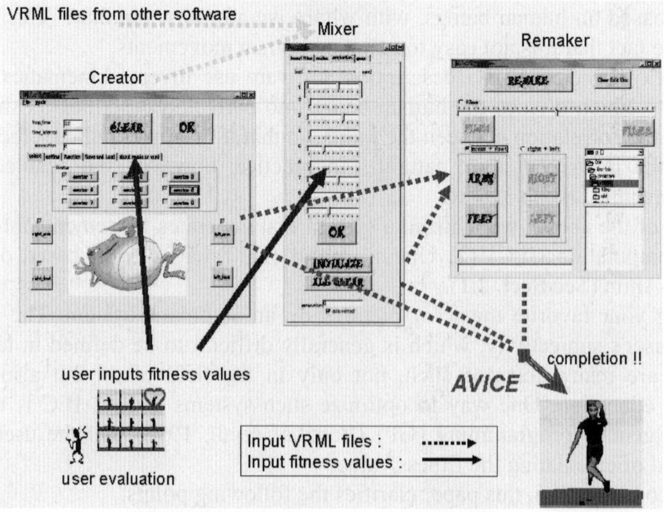

Figure 1: Outline of AVICE system

2.3 Specifications

2.3.1 Creator Module

In this study, we determined the extent to which joints should rotate based on GP genes as a way to describe the movements of the avatar using H-Anim. Evaluation was done by human beings using IEC, and we built a system that can easily and automatically attain smoother movements (H. Wakaki, & H. Iba, 2002). The genes use arithmetic operations, trigonometric functions, min/max, etc. as nodes. The variable is time t. The mechanism is to define the time so that the rotation will be determined with respect to time according to the H-Anim description.

Movements are governed by a function that reads the time and a function describing the rotation (axis and angle) according to the time, as well as a description component that actually makes a certain node rotate joints by combining them.

2.3.2 Mixer Module

The Mixer Module can read both the output of Spazz3D and the VRML of movements of an avatar evolved by IEC. Put differently, it creates animations through IEC based on existing software data. In specific terms, it reads the VRML and imports it into the terminal node of the gene. The read VRML files are archived and the VRML of avatars with a high fitness value is also added to the library, to be used in the event of mutation. Based on GP, it realizes the integration of movements with respect to joints in each node. Further, evaluation by human beings was based on a scale of 1 to 10. There are two fitness values: "the fitness value in each file" (FF) and "the fitness value of each GP individual" (FP). Firstly, a fitness value is defined when the VRML file is registered in the library. The fitness value is 10, unless specified otherwise.

The user looks at the actual movement (PTYPE) expressed by each GP individual based on the VRML file and then defines a fitness value (a value between 1 and 20). Based on the evaluation of the fitness value, the VRML file is registered in the library (**Fig. 2**).

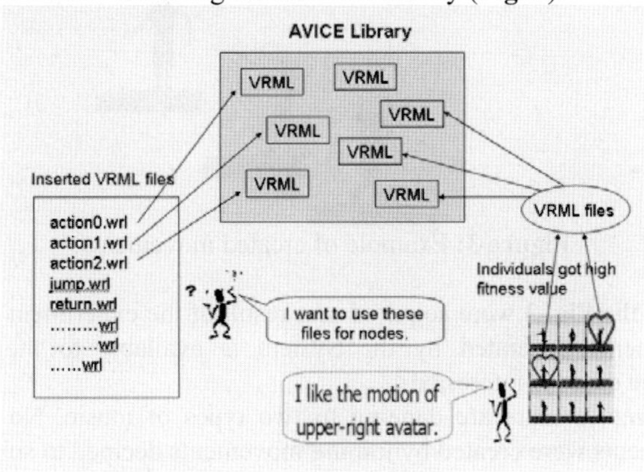

Figure 2: Construction of library.

- The read VRML files are registered in the library.
- VRML files of individuals with a high fitness value are registered in the library.
- Each VRML file in the library has a unique fitness value.
- Those with a higher fitness value are more likely to be imported in the event of mutation, etc.

In the event of crossover, genes of individuals with a high fitness value are chosen.
As explained in the Creator Module, the parts of the avatar and the rotation function with respect to the parts are defined by ROUTE. This section is read, and the Mixer Module gene is activated with respect to the associated functions.

2.3.3 Remaker Module

The Remaker Module has the function to join two VRML files together in the following three aspects.

- Temporally join two files together.
- Join the movement of the upper half of the body with the movement of the lower half thereof.
- Join the movement of the right side of the body with the movement of the left side thereof.

3 Demonstration of Created Movements

Figure 3: Example of created movements.

Movements shown in the **Fig. 3** were acquired as a result of the experiment of AVICE. Of note, a demo of the movements generated by the System is available to the public. Readers are encouraged to view the demo if interested [1].
In this movement demo, avatars are dancing to two types of music. No music data has been imported. The movements were created by joining movements deemed to suitable to the music.
Even if no music data has been imported, it shows that they are in rhythm. The difference between the two also shows that their dance is not only in rhythm to the music but also suits the atmosphere thereof.
It is fair to say that AVICE's merit -i.e., the "ability to select while looking at the movements"- made it easier to create movements suitable to the music.

[1] http://www.miv.t.u-tokyo.ac.jp/~wakaki/avatar/action.html

4 Merits of AVICE

The merits of AVICE are summarized as follows.

- AVICE is effective in assisting the creation of video, because the user can make judgments looking at the movements and the description evolves depending on how good the movements are, unlike scripts that require the user to join stationary states together.
- AVICE facilitates creativity compared to motion capturing and scripts, because the description will evolve based on evaluation as to whether the movements are suitable to the object or not, even if the user does not have a clear image.
- AVICE can be used in collaboration with other software, because movements created by AVICE can be imported into other software and movements created by other software can be imported into AVICE.
- AVICE consists of three modules: the user can employ the suitable function without having to worry about the other functions if he/she wants to create movements randomly (Creator Module), fiddle around with imported movements (Mixer Module) or use optional functions (Remaker Module) (**Fig. 1**) .

5 Conclusion

Few expressions are based on human beings, with whom we are most familiar. This is deemed to be attributable to the fact that it is not easy to generate human movements. If movements of avatars can be created based on VRML, which can be displayed by a browser, it should increase the ways in which individuals express themselves, as they do through paintings and music.

As a tool to achieve that end, this essay described AVICE, a system that automatically generates the avatar's movements to the user's liking. It explained the specification concept, internal configuration and the merits in detail, and revealed its effectiveness based on an experiment by users.

The current version of AVICE is based on the trial specifications of the system used broadly in general. The future issues constitute the objectives of the AVICE system. We firmly believe that new ways of expression will be brought about by its completion.

References

T. Unemi (1999). SBART2.4: Breeding 2D CG images and movies, and creating a type of collage. In *Proceedings of The Third International Conference on Knowledge-based Intelligent Information Engineering Systems* (pp. 288-291). Adelaide: Australia

H. Takagi (1998). Interactive Evolutionary Computation - Cooperation of computational intelligence and human KANSEI. In *Proceeding of 5th International Conference on Soft Computing and Information/Intelligent Systems* (pp. 41-50)

W. Banzhaf, P. Nordin, R. Keller, & F. Francone (1998). Genetic Programming -- An Introduction. Morgan Kaufmann Publishers, Inc.

H. Wakaki, & H. Iba (2002). Motion Design of a 3D-CG Avatar using Interactive Evolutionary Computation. In *Proceedings of 2002 IEEE International Conference on Systems, Man and Cybernetics (SMC02)* .

R. Stuart Ferguson (2001). Practical Algorithms for 3D Computer Graphics. A K Peters Ltd (pp. 211-230).

SAKURA: Voice-Driven Embodied Group-Entrained Communication System

Tomio Watanabe and Masashi Okubo

Faculty of Computer Science and System Engineering
Okayama Prefectural University
111 Kuboki, Soja, Okayama, 719-1197 JAPAN
{watanabe, okubo}@cse.oka-pu.ac.jp

Abstract

This paper proposes the concept of voice-driven embodied entrainment communication system for activating group interaction and communication, and develops the prototype of the system called "SAKURA". SAKURA creates group-entrained interaction in a virtual classroom where voice-driven CG interactive characters called "InterActors" with both functions of speaker and listener are entrained one another as a teacher and some students by generating expressive motions and actions such as nodding, blinking and body motions coherently related to voice input. By using SAKURA, talkers can communicate with the sense of unity through the entrained InterActors including virtual students' InterActors by only voice input through network. The system would be effective in analysis by synthesis for group embodied interaction and communication as well as remote interaction and communication support.

1 Introduction

In human face-to-face conversation, a talker's voice is rhythmically related and synchronized with the listener's movements such as nodding, head and body motions and actions. This synchrony called entrainment in communication generates the sharing of embodiment in human interaction, which plays an important role in essential human interaction and communication (Condon & Sander, 1974; Watanabe, 2001). Hence, the introduction of the entrainment mechanism to human interface is indispensable to the development of human-centered essential interaction and communication systems.

We have developed a voice-driven embodied interactive character called "InterActor" with both functions of speaker and listener for activating human interaction and communication by generating the whole body motion such as nodding, blinking and the actions of head, arms and waist coherently related to voice input. InterActor is the electronic media version of physical interaction robot called InterRobot (Ogawa and Watanabe, 2001), which sets free from the hardware restriction for the human interface of advanced GUI based network communication. By using InterActor, we proposed the concept of voice-driven embodied interaction system for supporting human interaction and communication with the sharing of embodiment by the entrainment in remote communication (Watanabe, Danbara, & Okubo, 2002).

In this present paper, focusing on the human interface of embodied group-mediated communication system, the concept of voice-driven embodied group-entrained communication system is proposed for supporting essential interaction and communication, and the prototype of the system called SAKURA is developed by using InterActors. SAKURA activates group interaction effects of InterActors in the same virtual classroom where InterActors are entrained one another as a teacher and some students. By using SAKURA, talkers can communicate with the sense of unity through the entrained InterActors by only voice input though network. The interaction analysis and sensory evaluation of verbal remote communication demonstrates the effectiveness of the system.

2 SAKURA

2.1 Concept of SAKURA

The concept of SAKURA is shown in Figure 1. Five InterActors as the role of students and one InterActor as a teacher are arranged in a virtual classroom where InterActors are entrained one another on the basis of only voice input through network. From the viewpoint of a teacher, when the talker as the role of a teacher speaks to students' InterActors, InterActors response to the utterance in appropriate timings by means of their entire body motions by nodding, blinking and communicative actions in the manner of listeners. Thus, the talker can talk smoothly and naturally. Then, the voice is transmitted through network to the remote SAKURA system as the role of a student. From the viewpoint of the student, InterActors can effectively transmit the teacher's message to the student by generating body motions in the manner of the speaker for the teacher's InterActor and the listeners for students' InterActors on the basis of the voice, and presenting both the voice and the entrained body motions simultaneously. The student this time transmits his voice through one of students' InterActors in the same way. The only information transmitted or received by this system is voice. It counts that it is human that transmits and receives the information. InterActor just generates the entrained communicative motions and actions on the basis of human voice, and supports the sharing of mutual embodiment in communication. In the system, virtual students' InterActors whose students do not exist are arranged for generating group-entrained interaction effects of InterActors on smooth interaction. SAKURA is an essential communication support system with the function of activating interaction and communication by the group entrainment.

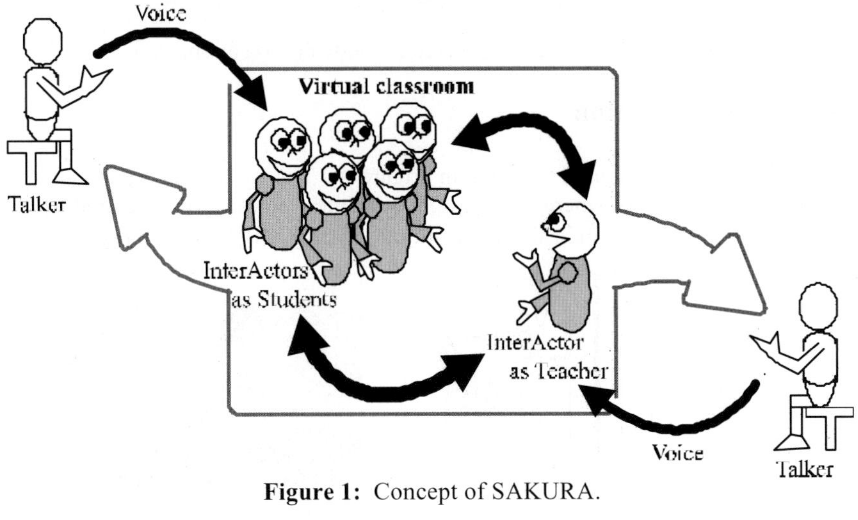

Figure 1: Concept of SAKURA.

2.2 InterActor

An expressive action of InterActor as listener is shown in Figure 2. The body motions were related to the voice input by operating both the neck and one of the wrist, elbow, arm and waist at the timing over a body motion threshold. The threshold was set lower than that of the nodding prediction of Moving-Average (MA) model, expressed as the weighted sum of the binary voice signal to nodding. Five InterActors as students had different values of parameters and their thresholds for group-entrained effects.

An example of the body motion of InterActor as speaker is shown in Figure 3. The mouth operation of InterActor was realized by the switching operation synchronized with the burst-pause of voice. The body actions of InterActor as speaker were also related to the voice input by operating both the neck and one of the other body actions at the timing over a threshold which was estimated by the speaker's action model as its own MA model of the burst-pause of voice to the whole body motion. By introducing both listener and speaker action models into a character, InterActor with both functions of listener and speaker was realized.

Figure 2: Expressive action of InterActor as listener.

Figure 3: Example of InterActor's body motion as speaker.

2.3 System Configuration

The configuration of the system is shown in Figure 4. The system consisted of two PCs, microphones and loudspeakers where a classroom and InterActors were generated by SENSE 8 WorldToolKit Rel. 8, and the voice was transmitted via 100 MB Ethernet. Figure 5 shows an

Figure 4: System configuration.

example of interaction scene from the viewpoints of a teacher and a student respectively. InterActors for attending teacher and students had both function of speaker and listener while virtual InterActors had only the function of listener. Figure 6 shows a scene in which virtual students' InterActors turn round for a student's talk. Their embodied actions such as nodding can activate and assist his/her talk.

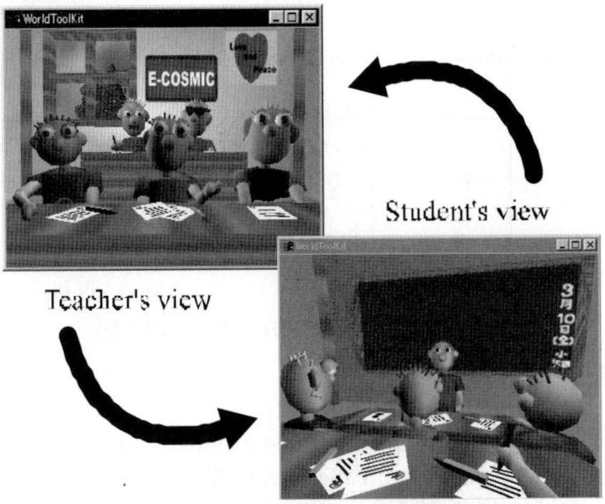

Figure 5: Example of interaction scene.

Figure 6: Virtual students' InterActors turn round for a student's talk.

3 Communication Experiment

Two remote talkers communicated in a separate room by using SAKURA. The experiment was performed in 12 pairs of 24 Japanese students as the role of a teacher and a student respectively under two conditions: one was the entrained mode where InterActors were entrained one another on the basis of voice; the other was the non-entrained mode where students' InterActors had no entrained actions such as nodding. The experiment time was 4 minutes in the non-entrained mode and then 4 minutes in the entrained mode. After that, sensory evaluation for the system was examined with the seven point bipolar rating scales from -3 (not at all) to 3 (extremely) in which the score 0 denotes moderately. Figure 7 shows the result of sensory evaluation for the entrained mode based on the non-entrained mode. In any items except naturalness, the entrained mode was highly evaluated from the viewpoints of a teacher and a student respectively. This demonstrated the effectiveness of SAKURA.

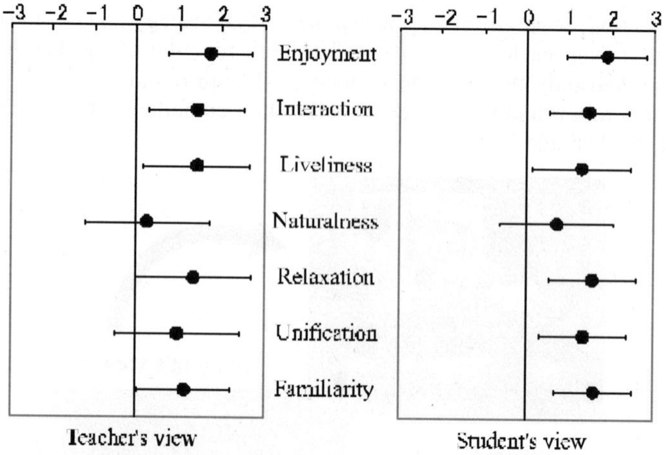

Figure 7: Result of sensory evaluation

4 Conclusion

In this paper, the concept of voice-driven embodied group-entrained communication system was proposed, and the prototype of the system called SAKURA was developed by using voice-driven CG characters called InterActors. SAKURA activates group interaction effects of InterActors in the same virtual classroom where InterActors are entrained one another as a teacher and some students. By using SAKURA, the interaction analysis and sensory evaluation of verbal remote communication demonstrated the effectiveness of the system.

The physical version of SAKURA with InterRobots and InterActor is also developed, which is exhibited in National Museum of Emerging Science and Innovation where visitors come across with dynamic experience of embodied communication. They perceive the effects of group-entrained communication environment intuitively, and recognize the importance of embodied communication.

Acknowledgement

This work under our project E-COSMIC (Embodied Communication System for Mind Connection) has been supported by CREST of JST (Japan Science and Technology).

References

Condon, W.S. & Sander, L.W. (1974). Neonate Movement is Synchronized with Adult Speech: Interactional Participation and Language Acquisition. Science, 183, 99-101.

Ogawa, H. & Watanabe, T. (2001). InterRobot: Speech Driven Embodied Interaction Robot. Advanced Robotics, 15, 3, 371-377.

Watanabe, T. (2001). E-COSMIC: Embodied Communication System for Mind Connection. Usability Evaluation and Interface Design, 1, 253-257.

Watanabe, T., Danbara, R. & Okubo, M. (2002). InterActor: Speech-Driven Embodied Interactive Actor. Proc. of IEEE RO-MAN2002, 430-435.

Intimate virtual communication place supported with networked "lazy Susan"

Shigeru Wesugi, *Kazuaki Ishikawa*

Graduate School of Science and Engineering, Waseda University
59-319, 3-4-1,Okubo, Shinjuku-ku, Tokyo, Japan
wesugi@computer.org
kishikawa@miwa.mech.waseda.ac.jp

Yoshiyuki Miwa

Faculty of Science and Engineering, Waseda University
59-319, 3-4-1,Okubo, Shinjuku-ku, Tokyo, Japan
miwa@waseda.jp

Abstract

Communicating activities, such as education and community meeting, which require bodily interaction at a physically shared place have been supported online between remote locations. Therefore, we have proposed a design framework of sharing a virtual space that reflects mutual bodily action at each remote site in order to create a co-existing space. We focus attention on a communicating situation that people gather around a "Lazy Susan" (a revolving wooden disk), and have developed networked "Lazy Susan" system, that bodily interaction can be shared through rotating the physical disk, and with viewing a virtual hand and virtual disk in the virtual space. This paper describes our experiments indicate that a presence of remote participant is generated at each local place by interacting with real disk and experiencing virtual space. Additionally, some demonstrations about modeling brick on the disk and collaborative drawing indicate that our networked "Lazy Susan" is promising to support a collaborative work.

1 Introduction

Communicating activities, such as education and community meeting, which require bodily interaction at a physically shared place have been supported online between remote locations. Meanwhile it also has been reported that a virtual community doesn't always contribute to a community activity in real life (Dreyfus, 2001). To such a computer-mediated communication (CMC) situation, John Canny pointed out "The Cartesian (and dominant) approach to CMC has broken the interaction into communication channels such as video, audio...in a context-independent way"(Canny, 2000). Consequently, we consider that bodily interaction, such as facial expression and gesture, which serves generating context should be shared directly among participants in remote locales. Our goal for resolving this issue is to create a co-existing space bridging remote locales as a place where mutual bodily interaction is grounded and as a place where remote participants feel a sense of "being collocated". Therefore, we have proposed a design framework of sharing a virtual space that reflects bodily action at each real space in order to share a mutual situation. As one example, we focus attention on a communicating situation that people gather around a "Lazy Susan" (a revolving wooden disk) in Chinese restaurant (Figure 1), and developed networked "Lazy Susan" system so far. This paper describes some experiments on remote interaction indicate an ability for creating a presence of remote participant at local place, and some demonstrations of work in this virtual space show a capability for supporting remote collaborative work.

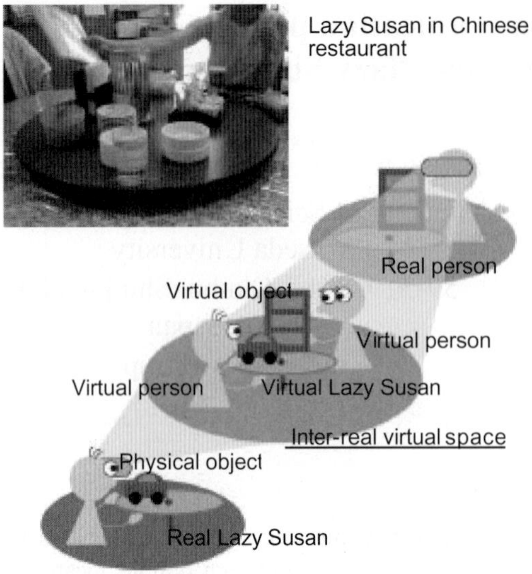

Figure 1: Communicating situation around Lazy Susan in inter-real virtual space

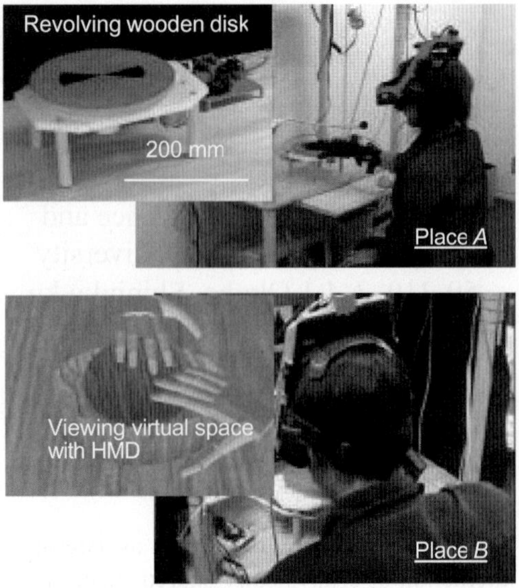

Figure 2: Participants interact with the physical disk through inter-real virtual space

2 Networked Lazy Susan

We have proposed a design framework of sharing a virtual space that reflects mutual bodily action in order to create a co-existing space (Figure 1). Remote participants can share a mutual communicating situation in physically real space, not in a pre-constructed shared virtual space, through the virtual space that represents mutual bodily action at real each site. We refer to this virtual space as an "inter-real virtual space", because the virtual space bridges among remote real spaces as an interface space (Wesugi, 2002). Then the inter-real virtual space can be projected onto the each real place for supporting consistency between real and virtual. In order to implement this idea, we focus attention on a round table including "Lazy Susan" in Chinese restaurant. The reason is that the rotatable disk supports a communicating activity as following three:

- The table is a physical place that people gather around
- Their bodily actions are represented as a rotation of the disk
- They can do a collaborative work on the table while moving and pointing to a real object on the disk.

Based on these items, we developed networked "Lazy Susan" system so far; the rotations of the physical disk at each site are synchronized with the remote disk at other participating sites and with the appearance of the virtual disk in the inter-real virtual space (Wesugi, 2002). The feature of this "Lazy Susan" is that bodily interaction can be shared in two ways; through rotating disk in a physically real space, and with viewing a virtual hand and virtual disk in a virtual space that reflects real situation of both sites. Figure 2 illustrates a situation of both remote participants rotating the disk.

As related line of research, "inTouch", synchronized rotations of remote wooden rollers, was proposed for creating a sense of presence of remote person through haptic interaction (Brave, 1998). Meanwhile, our work is intended to create a co-existing space by focusing on sharing a table that holds spatial dimensions, and to support a collaborative work on the table.

At the next section we evaluate whether sharing an inter-real virtual space creates a presence of remote person, and show some demonstrations about a collaborative work in this virtual space.

3 Communication with Lazy Susan

3.1 Experiments and discussion

Our experiments were designed to investigate the ability of the networked "Lazy Susan" system to support a presence of remote participant at local site. Ten pairs of adult students (16 males, 4 females, 20-28 age), who were unacquainted well with each other, were located in two rooms and interacted through "Lazy Susan" connected by a LAN. They had conversations about a campus life through Microsoft Netmeeting software under the following three conditions for six minutes (Figure 3):

- Condition 1: The participants have a conversation only by voice and don't interact with the disk (Bodily action of remote participant is not represented at local site)(Figure 3a)
- Condition 2: The participants have a conversation and can rotate the disk without experiencing the virtual space (Bodily action of remote participant cannot be visually expressed) (Figure 3b)
- Condition 3: The participants can rotate the disk and see a virtual space representing virtual disk and virtual hands of both his own and the other participant's hand with HMD (Figure 3c).

Under condition 2,3, participants were instructed to be free to touch the table and to rotate the disk. Before the condition 3, participants were asked to wear an optical see-through HMD and calibrate the system by first locating their hands on an origin point and then by using a mouse to adjust the virtual disk to overlay accurately onto the real disk. The order of these conditions was shuffled to each pair. After the experiences, each participant was asked to answer a questionnaire and to write down some comments on a sense of co-presence under each of the three conditions. The questionnaire includes four items in Table 1. Each item is rated by seven scales (0 neutral). Figure 4 illustrates the result of average and standard deviation under four items of the questionnaire. A Wilcoxon, signed-rank test showed a highly significant difference between condition 2,3 (interacting with disk) and condition 1(voice only) under all of four items. However, the score of condition 2 is below zero at item B, C concerning about the place where remote participant exists. This result is consistent with the participants' comments about a sense of sharing the disk. Those comments were evenly split between positive comments indicating a sense of rotating the same disk together, and negative comments such as feeling that the PC was controlling the disk rather than the other participant and that they cannot feel a presence of remote participant well. This indicates that the way of only synchronized rotations of both remote disks supports insufficiently to create a sense of presence of remote person at local site. And it also means a design framework such as "inTouch"(Brave, 1998) which supports only bodily interaction with synchronized moving object cannot create a co-existing space between remote locales.

Concerning about condition 3 (experiencing physical interaction and visual expression), Figure 4

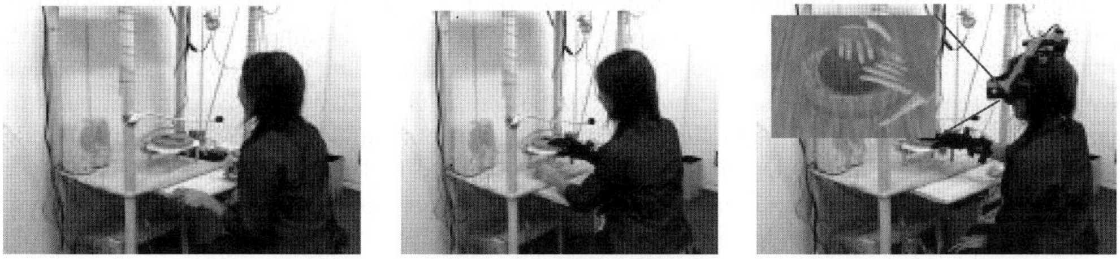

a) Condition 1: Only verbal conversation and no interaction with disk

b) Condition 2: Rotate the disk without virtual image hand / disk

c) Condition 3: Rotate the disk with viewing the virtual disk and virtual hands

Figure 3: Three conditions in experiments

also shows a highly significant difference between condition 3 and condition 2 (without virtual space) under all of four items. Under condition 3, most of the participants also reported a feeling that they and the other person in the pair were working at the same disk together. This feeling of co-presence was particularly strong when a participant put his hand on the disk and 'felt' the other participant rotates the disk; at this point the participants reported a strong impression that the other participant's virtual hand was directly rotating the disk. Additionally, the comments about a feeling that a virtual hand touched my hand, and being situated in face-to-face, also indicate that a presence of remote person was generated at local table.
Above mentioned results of experiments conclude our design framework of sharing a communicating situation through bodily and visually is available to create a presence of remote participant at each local place.

Table 1: Items of questionnaire

A: To which extent if they feel share the same table
B: To which extent if they feel remote one exists beside table
C: To which extent if they feel exist at the same place
D: To which extent if they have a sense of closeness

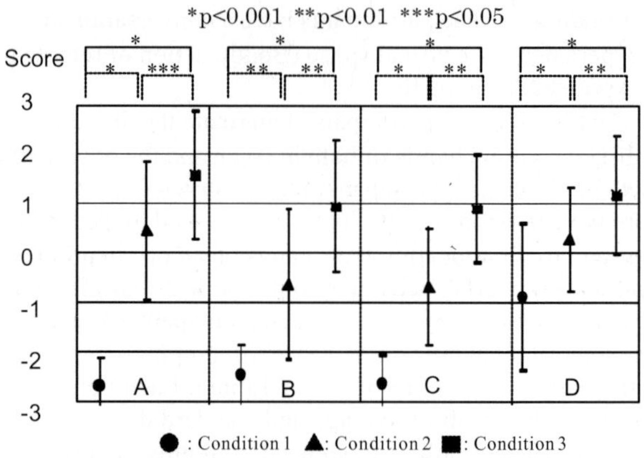

Figure 4: Result of questionnaire on copresence

3.2 Collaborative work

We investigated whether our "Lazy Susan" can support a collaborative work between remote locations. Firstly we integrated this "Lazy Susan" with Modeling Glove system, which was developed so far in order to visually represent modeling process with real brick in virtual space (Wesugi, 2002). This Modeling Glove consists of RFID sensor, 6DOF tracking sensors, and brick attached on RFID tag. Modeling process in real space such as grabbing, moving and releasing bricks is visually represented in a shared virtual space between two remote places, and a remote participant can point to a virtual brick in the virtual space by virtual hand. Figure 5 illustrates that a local participant interacts the remote and virtual brick by rotating the real disk.
Additionally in order to support a collaborative work in physically real space, we constructed tentatively an advanced "Lazy Susan" system, which integrates CCD camera and video projector. Video image capturing remote table is projected to a screen around the local table. Figure 6 shows two or more remote participants talk about the statue on the disk. Remote participant can point to remote object by video image hand and interact with it by rotating the disk back and forth. While this demonstration, a phenomenon was frequently observed that the remote participant almost stretched out to a video image of the falling statue. At that time, the participants reported that they felt as if their bodies expanded to the remote disk.
Furthermore, a collaborative drawing work is also attempted by attaching a white board onto the disk (Figurer7). These trial demonstrations indicate this "Lazy Susan" system is promising to support a remote collaborative work by sharing an inter-real virtual space and by interacting with the disk.

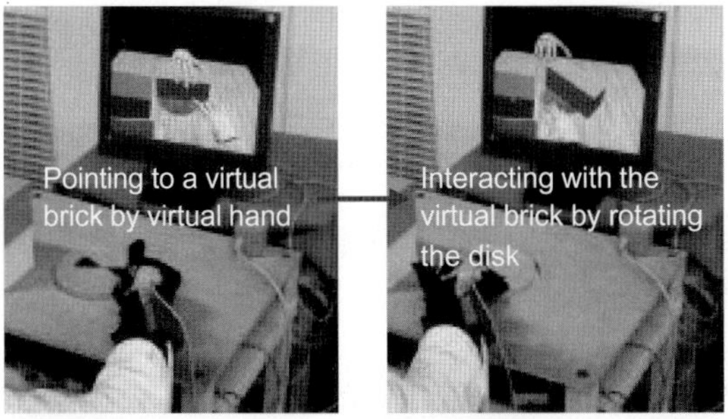

Figure 5: A collaborative modeling on "Lazy Susan"

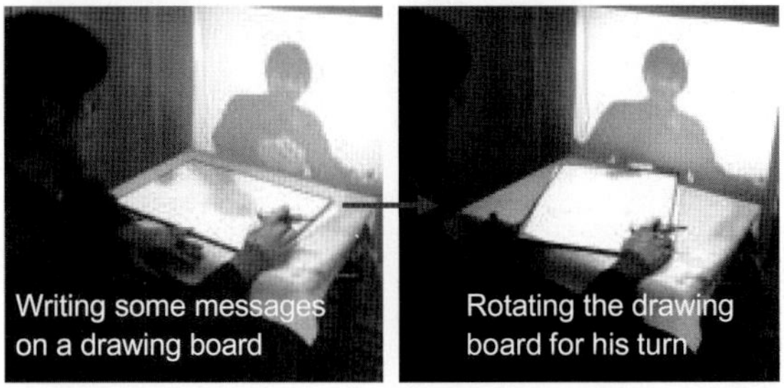

Figure 7: A collaborative drawing on "Lazy Susan"

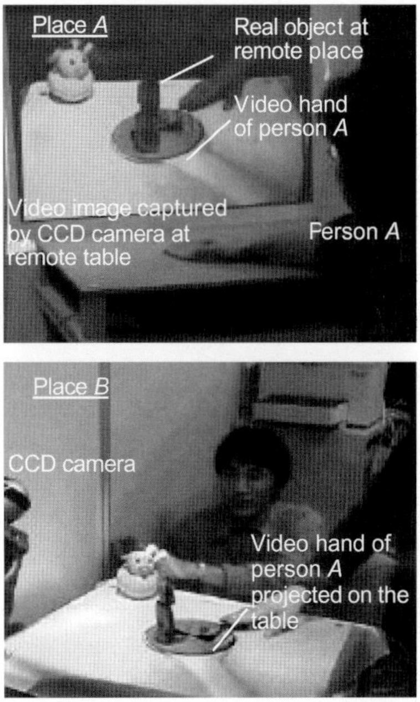

Figure 6: Collaborative work between remote locations

4 Conclusion

We have proposed a design framework of sharing a virtual space that reflects mutual bodily action in order to create a co-existing space, and developed networked "Lazy Susan" system based on this idea. This paper describes our experiments indicate an ability for creating a presence of remote participant at local place by sharing a communicating situation through bodily and visually. Additionally, some pilot demonstrations indicate that our "Lazy Susan" is promising to support a collaborative work by bridging remote locales through inter-real virtual space that reflects bodily action in each real space.

Future works will be that we develop networked "Lazy Susan" for supporting a spatial modeling work and a collaborative drawing work in real space, and evaluate its ability.

References

Brave, S., Ishii, H. and Dahley, A.: Tangible Interfaces for Remote Collaboration and Communication; Proc. of CSCW '98, ACM Press, pp.169-178 (1998).

Canny,J., Paulos,E.: Tele-Embodiment and Shattered Presence: Reconstructing the Body for Online Interaction; The Robot in the Garden: Telerobotics and Telepistemology in the Age of the Internet, MIT Press(2000).

Hubert L. Dreyfus :On the Internet (Thinking in Action); Routledge (2001).

Miwa, Y.: Communication Technology of "Ba" in Co-Creation [in Japanese]; Systems/Control/Information, 45-11, pp. 638-644(2001).

Wesugi, S., Miwa, Y.: Building Brick Interface Supporting for Actual Communication [in Japanese]; Journal of Human Interface Society (2002) (in press).

Wesugi, S., Miwa, Y. :Overlaying a virtual and a real table to create inter-real virtual space; Proc. of SIGCHI-NZ Symposium on Computer-Human Interaction, pp37-42 (2002).

Section 4

Interaction Techniques and Modalities

Keyboard Encoding of Hand Gestures

Nicoletta Adamo-Villani

Department of Computer Graphics Technology, Purdue University-1419 Knoy Hall, West Lafayette, IN 47907
nvillani@tech.purdue.edu

Gerardo Beni

Department of Electrical Engineering University of California at Riverside, Riverside, CA 92521
beni@ee.ucr.edu

Abstract

We present a new human-computer communication method which utilizes the user's typing skills to control the motion of the fingers, arching of the palm, wrist flexion, roll and abduction of a computer generated three-dimensional hand. The idea is based on the realization that a hand gesture path requires the same number (26) of parameters as the letters of the English alphabet. This simplified encoding of hand gesture is valuable in many areas including: (1) teaching fine manipulative skills; (2) teaching dynamic manipulative tasks; (3) representing communicative gestures.

1 Statement of Objective

The purpose of this paper is to show that it is possible to realize a simple human-computer communication of messages encoding hand gestures. The idea is based on the realization that a hand gesture path requires the same number (26) of parameters as the letters of the English alphabet. Since touch typing is an easily acquired and widespread skill, it is possible to conveniently input messages encoding hand gestures if each letter key of the keyboard corresponds to one degree of freedom of the hand

Hand gesture modeling has been the subject of intensive research in the last few years [1][2]. In terms of human computer interaction (HCI), the emphasis of hand gesture research has been on visual interpretation of such gestures in order to achieve a natural man-machine mode of communication. Another area of research in hand gestures for HCI is the development of hardware capable of transmitting the hand gestures to the computer, e.g., the development of sensor-embedded gloves [3][4].

There are several approaches to modeling/representing gestures in either manipulative or communicative (symbolic, pointing or imitative) actions. The main modeling approaches are volumetric and skeletal. The skeletal approach is amenable to clear quantitative description. The human hand, which consists of 27 bones, can be accurately modeled with joint angles totaling 23 degrees of freedom. Most of the joints (especially the carpal ones) have very limited range of motion. Thus the overall task of skeletal description is confined to a parameter space not excessively large in terms, not only of computer representation, but also of human encoding. It is this characteristic that has suggested the approach to hand gesture encoding described below.

A convenient hand configuration encoding would be applicable to many practical tasks: (1) teaching fine manipulative skills, as, e.g., in dentistry, surgery, mechanics, jewellery making, tailoring, sewing; (2) teaching dynamics manipulative tasks as, e.g., musical instrument playing, sport devices handling, cooking/eating tools handling; (3) teaching communicative gestures as, e.g., American Sign Language, HCI visual recognition gestures, and other hand signalling tasks.

A simple encoding of hand gesture configuration would prove very valuable also in storing and communicating hand animation data. A simple set of (23 +3 = 26) component vectors would represent a hand trajectory and could be transmitted with very low bandwidth to animate complex hand models held at the receiver site. However, this simplicity of computer to computer communication of the encoded hand gesture does not become more practical than complex hand gesture representations unless there is a convenient way of human-computer communication of the encoded message.

2 Description of Methods

We introduce a new method which utilizes the user's typing skills to control, with high level of precision, the motion of the fingers (fingers flexion, abduction and thumb crossover), arching of the palm, wrist flexion, roll and abduction of a computer generated three dimensional realistic hand.

The hand has 26 degrees of freedom which can be controlled by the 26 letters of the alphabet. Figure 1 shows the skeletal structure of the hand with its 26 joints and the IK end-effector (represented by the red cross) that controls the positioning of the hand in the 3D environment.

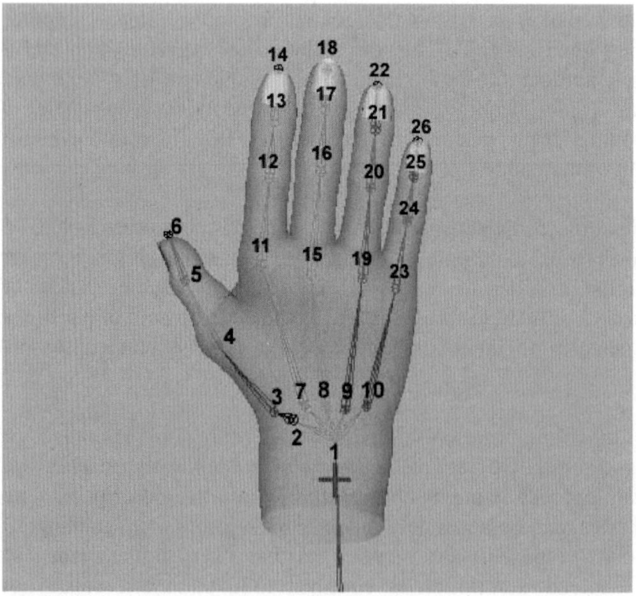

Figure1

Via keyboard input the hand can be positioned in space and manipulated to attain any pose: by touching a letter key the user rotates the corresponding joint a pre-specified number of degrees around a particular axis. The rotation "quantum" induced by each key touch can be easily changed to increase or decrease precision. For specific applications (e.g. fist or single digit action) the number of movable joints can be conveniently reduced.

3 Discussion of Results

The hand that we present was modeled as a continuous polygonal mesh and makes use of a skeletal deformation system animated with both Forward and Inverse Kinematics [5]. The structure of the CG skeleton closely resembles the skeletal structure of a real hand allowing extremely realistic gestures. Using MEL (Maya Encrypted Language) we have created a program that encodes hand gestures by mapping each letter key of the keyboard to a degree of freedom of the hand (Lower case letters induce positive rotations of the joints, upper case letters induce negative rotations of the joints). Figure 2 shows a rendering of the hand with the joints' rotations (23) and IK effector translation parameters (3) mapped to the 26 letters of the alphabet. Table 1 shows the motion output produced by each letter key.

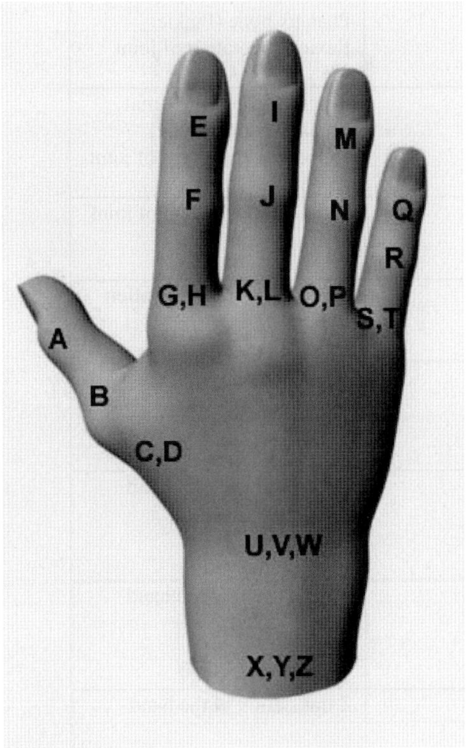

Figure2

Letter Key	Motion output	Letter Key	Motion output
A	Rotation of 1st Distal Phalanx bone (Thumb flexion-z rot of joint 5)	N	Rotation of 4th Middle Phalanx bone (Ring finger flexion- z rotation of joint 20)
B	Rotation of 1st Proximal Phalanx bone (Thumb flexion- z rot of joint 4)	O	Rotation of 4th Proximal Phalanx bone (Ring finger flexion- z rotation of joint 19)
C	Rotation of 1st Metacarpal bone (Thumb abduction-y rotation of joint 3)	P	Rotation of 4th Proximal Phalanx bone (Ring finger abduction- y rotation of joint 19)
D	Rotation of 1st Metacarpal bone (Thumb crossover-z rotation of joint 3)	Q	Rotation of 5th Distal phalanx bone (Pinkie flexion-z rotation of joint 25)
E	Rotation of 2nd Distal phalanx bone (Index flexion-z rotation of joint 13)	R	Rotation of 5th Middle Phalanx bone (Pinkie flexion- z rotation of joint 24)
F	Rotation of 2nd Middle Phalanx bone (Index flexion-z rotation of joint 12)	S	Rotation of 5th Proximal Phalanx bone (Pinkie flexion- z rotation of joint 23)
G	Rotation of 2nd Proximal Phalanx bone (Index flexion-z rotation of joint 11)	T	Rotation of 5th Proximal Phalanx bone (Pinkie abduction- y rotation of joint 23)
H	Rotation of 2nd Proximal Phalanx bone (Index flexion-y rotation of joint 11)	U	Wrist roll (x rotation of joint 1)
I	Rotation of 3rd Distal phalanx bone (Middle finger flexion-z rotation of joint 17)	V	Wrist abduction (y rotation of joint 1)
J	Rotation of 3rd Middle Phalanx bone (Middle finger flexion- z rotation of joint 16)	W	Wrist flexion (z rotation of joint 1)
K	Rotation of 3rd Proximal Phalanx bone (Middle finger flexion- z rotation of joint 15)	X	X translation of the hand
L	Rotation of 3rd Proximal Phalanx bone (Middle finger abduction- y rotation of joint 15)	Y	Y translation of the hand
M	Rotation of 4th Distal phalanx bone (Ring finger flexion-z rotation of joint 21)	Z	Z translation of the hand

Table1

Figure 3 shows an example of keyboard encoding of the "D" handshape of the American Sign Language (ASL) alphabet. Lower case letter keys induce a positive 10 degrees joint rotation.

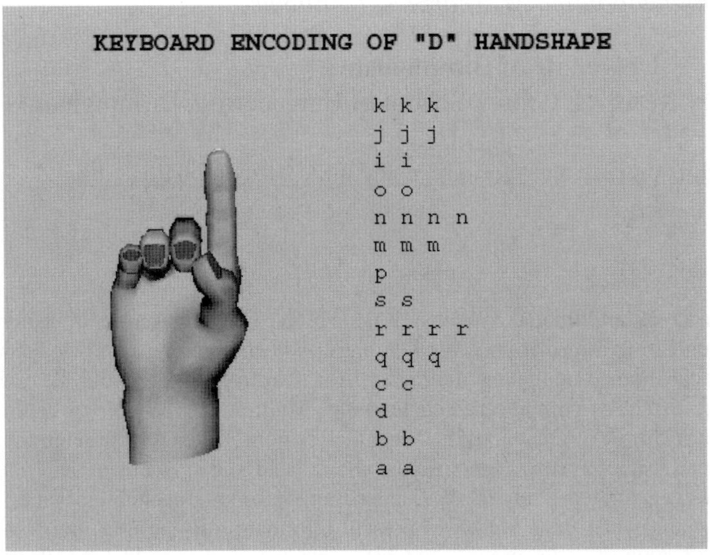

Figure 3

The design of this touch-typing reconfigurable hand can be easily extended to other models. In particular, the design is equally suitable for lower polygonal representation of the modeled hand so that operation outside the Maya environment is possible. In particular we have exported a simplified hand model from Maya to Macromedia Director 8.5 [6]. From such platform the touch-typing hand reconfiguring can be performed on web deliverable interactive application programs. Finally, the design of the touch-typing reconfigurable hand lends itself to easy memorization of the joint-letter relations so that a moderately skilled touch typist can easily acquire dexterity in configuring the modeled hand. Letters can be maintained on the model during the initial phase of acquiring the skill.

References

[1] Huang, T. S. and V. I. Pavlovic. "Hand Gesture Modeling, Analysis and Synthesis". Proc. of Intl. Conf. on Automatic Face and Gesture Recognition, (1995).

[2] Pavlovic, V. I., Sharma, J., and Huang T. S., IEEE Trans. PAMI vol. 19, 7, p. 677-695, (1997).

[3] Data gloves, cybergloves <http://www.metamotion.com/hardware/motion-capture-hardware-gloves-Datagloves.htm>.

[4] Hernandez-Rebollar, J.L., Kyriakopoulos N., and Lindeman R. W."The Acceleglove: A Whole Input Device for Virtual Reality". Conference Abstracts and Applications, Siggraph 2002.

[5]Graft, L., Dwelly B., Riesberg, D., and Mogk C."Character Rigging and Animation", Alias|Wavefront, 2002.

[6] Adamo-Villani, N. and G. Beni. "Teaching Mathematics in Sign Language by 3D Computer Animation". Information Society and Education: Monitoring a Revolution,
International Conference on Information and Communication Technologies, Spain, 2002.

Multimodality and learning: linking science to everyday activities

S. Anastopoulou, M. Sharples, C. Baber

University of Birmingham
Educational Technology Research Group, School of Engineering, Birmingham, B15 2TT, UK
anasto@eee-fs7.bham.ac.uk, m.sharples@bham.ac.uk, c.baber@bham.ac.uk

Abstract

This paper focuses on the role of multimodal systems in education. The importance of using multiple modalities in communicating information while learning has been recently acknowledged by educationalists. Multimodal human-computer interaction for learning tasks could be an alternative path to pioneering effective computer aided learning. Multimodal technology could offer new opportunities to learn from everyday activities in the classroom. The act of moving the hand, for example, could lead to interesting conclusions about hand's displacement and its graphical representation. However, there are several challenges that multimodal technology faces: it is not only a matter of how and when the system will present information to the user, but also where the learner would need support (educational content), what activities would keep them engaged. The paper will report a study aimed to explore these issues empirically.

1 Introduction

A modality can be defined as a means to communicate information, i.e., a sensory channel via which information is passed from or to a person. Modalities can be used individually or in combination. Multimodality refers to the simultaneous or alternate use of more than one modalities to send and receive information. In a multimodal interaction someone may receive information by vision and respond by speech or movement. Human- human interaction, e.g. in a classroom, is basically multimodal: the interaction between visual, actional and linguistic communication can be employed in learning (Kress & Jewitt, 2001). It is argued that use of multiple modalities while learning engages learners' interest and facilitates the process of learning.

Human–computer interaction can be multimodal as well as unimodal. By multimodal human-computer interaction we mean interaction between human and computer that involves interaction devices supporting different response modalities, e.g. pointing and speaking, or supporting the use of at least two sensory modalities, e.g. vision and hearing, or a combination of these (Baber and Mellor 2001; Carbonell 2001). Multimodality could be contrasted to 'unimodality', which is based on the use of only one modality to sense and respond to information. An example of unimodal activity could be watching an animated presentation on a computer and responding only by pressing keys on the keyboard; in this example, the visual-spatial modality is used for both activities.

When technology can support the multimodal interaction in a learning activity, meaning construction can be facilitated. In a highly interactive environment new configurations can be tested by the learner who constructs and negotiates meaning with the aid of the system's feedback (Scaife M., 1996). The integration of different modalities gives the opportunity to configure real-life actions, such as hand movements. Representation of information to each modality is also an issue. If representations such as diagrams or graphs are *easily* produced by the learner, comprehension of the representing concepts is intimately facilitated (Scaife M., 1996). When visual representations are effectively coupled with movements, learners can relate representations to their sense and knowledge about their body and experience body syntonic learning (Papert, 1980). Thus, using movement as a means to record data, which is displayed in graphs and results in correction of the movement is considered as a multimodal learning experience. This paper describes a study that explores whether such an experience is beneficial when learning about kinematics graphs.

2 Learning about graphs

For the students to learn how to interpret a graph, the relation between the movement and the line of the graph is important. Seeing how the graph is plotted by their hand movement and being able to change it as they move about, gives them the ability to test their ideas and discard the problematic ones. When learning about kinematics, for example, pupils often ignore the abstract concept of the graph and think of the graph as a picture of motion, i.e. a line parallel to the time axis could be erroneously assumed to describe a horizontal movement and a line going upwards describes a vertical movement. By using a sensor on their hand to collect data for drawing a graph, the pupils can negotiate their understanding: they relate their physical movement to the appearance of the graph. Looking at the generated graph in real time can also provide 'graphical constraining', that is the real-time graph constrains the inferences that can be made about the underlying represented world (Stenning & Inder, 1995). Thus, the lines are interpreted as movement or lack of it instead of movements along different dimensions. Having a system to generate kinematics graphs from their own data, gives pupils a meaningful situation to consider: an authentic problem that refers to real life situations and which is thus worthwhile to think about.

However, as shown by a pilot study, there are technical difficulties that can arise from the initiation of such an innovative learning experience. Time delays in displaying the graph might weaken the link between the activity and the graph formation. The presentation of the graph also needs to be considered: very condensed graphs can be difficult to relate to the hand movement. An activity can also raise usability issues: a rapid activity does not give enough time to participants to realize how their movement affected the graph. Additionally, the educational focus needs to be related to the students' curriculum for it to be a valid classroom activity.

3 Main Study

The study explores the relation between learners' own hand movements and their graph formation and understanding.

The technology included a position measuring system that sends data to a software that displays the graph in real-time (Figure 1). The learner attaches the sensor to their hand and moves the hand about to see a distance-time graph.

The activity was initially open-ended: the learner could move their hand in any way they wanted. Later, they are asked to generate specific graphs so they had to find out how they should move their hand, aiming at strengthening the link between the activity and the graph. They could move their hand towards any direction but it had to be the same throughout the study.

The learning content was about distance-time graphs, which is an important issue in the science classroom for Key Stage 3 in UK and elsewhere. The pupils need to learn how to identify when the graph shows movement or lack of it; when there is movement, they need to understand its direction and to interpret the slope of a graph in terms of speed of the movement.

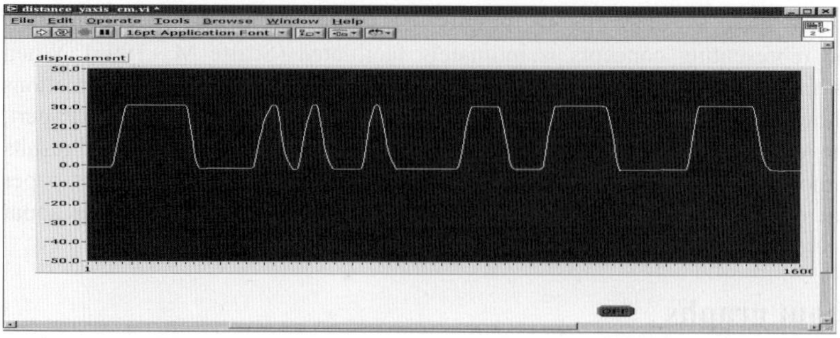

Figure 1: Screenshot of the display

3.1 Method

The study was conducted with 22 students in year 9 (14 years old) of a secondary school in Birmingham, UK. It was expected that the students would know little or nothing about the subject because they had not then been taught distance-time graphs. Students were assigned to the conditions on presentation to the experiment. In each session, there was one student in an empty classroom with the experimenter for about half an hour.

There were two conditions: students who formed graphs as they moved their hand (Doers) and students that thought about their hand movements in order to explain the graphs (Thinkers). The thinkers were also allowed to move their hands, but their movement did not generate a graph plot Both conditions had access to body syntonic learning: they had to imagine the graphs as expressions of their own movements. The experimental condition (Doers), however, had a reinforced experience. They had the chance to correct themselves as they saw the results of their movements on the visual display. It was expected that by relating graphs to hand movements and getting immediate corrective feedback from the display, learners would be more able to understand graphs and will do so more easily and with more interest than those in the control condition.

The feedback from the experimenter was kept to a minimum: the learners received feedback at the beginning of the teaching session where two very simple graphs were explained to them. This was necessary for the thinkers to continue the study. During the teaching session they had to interpret 4 graphs, of varying difficulty. A set of verbal protocols was used, to ensure that all students are given exactly the same instructions. For the teaching session the group was split into the 'Thinkers' and 'Doers'.
The 'Thinkers' were shown specific graphs and they could either say what they would do to generate them or move their hand accordingly. They discussed with the experimenter about the details of the graph, i.e. the name of each axis, the values it would have and what a negative value would mean. Subsequently, students were asked to write down what they said or did.

The 'Doers' had the tracker's sensor attached to their wrist with the aid of a sweatband. They moved their hand about freely to get familiar to the movement and the generation of the graph for approximately 3 minutes. Meanwhile, they discussed the details of the graph with the experimenter. The students tried afterwards to generate specific graphs. After they had caried out the task they wrote down their results.

The third part of the experiment involved the students answering questions in the form of a written test. They could not look back to the previous sheets. Finally, the pupils were asked to reflect on the experience and express their opinion about the study. They completed a short attitude survey, based on a 5 point Likert scale. In particular, they were asked whether they found the session interesting, if they liked it, if the liked watching their own data, how difficult were the questions and whether they felt that they understood the distance-time graphs at the end.

3.2 Results

The results discussed below are based on the sheets completed. The final test results of the two conditions were significantly different (Mann-Whitney test $z = -2.275$, $p<0.05$). Comparison of individual questions shows that 'Doers' were more able to describe the distance-time graph in terms of hand movements and they understood better the meaning of each line on the graph.

From the initial set of questions, it was apparent that none of the students know about distance-time graphs. There was no difference in what the students knew before the study between conditions. From what the students wrote next to the graphs of the teaching session, it appears that the 'Doers' were more able to describe correctly the graphs in terms of their hand movements.

When asked for their opinion about the study, 'Doers' liked it more than 'Thinkers' (Mann-Whitney test, $z=-2.181$, $p<0.05$) and found it more interesting (Mann-Whitney test, $z=-2.355$, $p<0.05$). Most of the students, in all conditions, indicated that they would like to watch their own data. All the participants responded that they understood 'distance-time' graphs.

3.3 Discussion

'Doers' performed better overall than thinkers. In particular, 'Doers' were more able not only to interpret correctly the graphs into general movement but also were able to translate correctly graphs into hand movements. All 'Doers' also mentioned a sensible direction in which the hand would move. This is in contrast to the 'Thinkers' who were less able to translate the movement into a sensible direction. A frequent response was that a straight line expresses movement across and a sloped line expresses a movement diagonally (picture-like effect).

Initially, most of the participants thought that they had to move their hand diagonally in order to draw a diagonal line on the graph. 'Doers' could overcome this problem because they could see that the effect on the graph was not the expected. They had the chance to self-correct themselves via the visual feedback and discover the correct movement. Thus, having access in graph generation resulted in solving common misconceptions: the system constrained their inferences about the underlying represented world.

It was also noticed that 'Doers' tended to have more standardized behaviour than 'Thinkers'. The system and the task triggered their attention and they concentrated on their answers. They stayed focused on the task since they actively constructed knowledge through experimentation. During the study, they negotiated the meaning of the problem at hand and they discovered which concepts where applicable. They were dealing with a real-life situation which gave them an authentic problem which was related to themselves. Conversely, the participants in the control group paid attention to some of the tasks of the session but because of the lack of feedback, they got bored or distracted.

3.3.1 The learning experience: the system, the activity, the educational content

'Doers' were fascinated by the system. Moving their hand about and seeing its distance-time graph on the visual display was an engaging experience. The feedback came from the visual display as soon as they moved their hand and they could alter the pace of their movements to notice the changes on the graph. They were watching the display with interest and they were drawing conclusions about the effect of moving (or staying still) on the graph.

The activity they had to do was not specific. They could move towards any direction as long as they kept it the same throughout the study. They could try to move their hand differently, test new configurations and realise the changes on the graph by the system's feedback. Being able to change the graph as they moved about, they related the graph to their sense and knowledge about their body, thus having access to body syntonic learning.

The study was successful in reaching its educational aims. The main aim was gained by both conditions: they realised that a straight line on a distance-time graph shows no movement and the diagonal line shows movement. However, the ability to interpret the lines of a graph in terms of hand movements with correct direction was much more pronounced for 'Doers'. The third aim of the study was focusing on the slope of the lines: the steeper the line the faster the movement. Students from both conditions were able to answer the relative question correctly. This was expected because it was explicitly explained by the experimenter.

3.4 Conclusion

The study described above showed that introducing multimodal interaction in a learning activity makes learning more interesting and effective: it facilitated students' understanding and engagement. It related students' own hand movement to graph formation and interpretation: instead of focusing on any object of the environment it was thought that the use of their body would interest them more and would trigger them for a better understanding of the graph and what it shows. The integration of different modalities, i.e. kinaesthetic and visual, in a learning activity gives the opportunity to test real-life actions and receive feedback from the system. Visual representations that are effectively coupled with movements, facilitated comprehension of kinematics graphs and related science to their body. Thus, using movement as a means to record data, which is displayed in graphs and results in correction of the movement is considered as a beneficial multimodal learning experience.

Multimodal systems for educational purposes are introducing a new phase in computer aided learning: the aim to develop systems that support learners in ways that enrich the whole learning experience by giving access to information that was previously hard to obtain and visualise.

References

Baber, C., & Mellor, B. (2001). Using critical path analysis to model multimodal human-computer interaction. *International Journal of Human Computer studies, 54*, pp.613-636.

Carbonell, N. (2001). *Recommendations for the design of usable multimodal command languages.* Paper presented at the HCI International, New Orleans, LU.

Kress, G., & Jewitt, C., Ogborn, J., Tsatsarelis, C. (2001). *Multimodal teaching and learning: the rhetorics of the science classroom*. London: Continuum.

Papert, S. (1980). *Mindstorms: Children, Computers and Powerful Ideas*. Brighton : Harvester.

Scaife M., Rogers, Y. (1996). External cognition: how do graphical representations work? *International Journal in Human-Computer Studies, 45*, 185-213.

Stenning, K., & Inder, R. (1995). Applying semantic concepts to analysing media and modalities. In B. Chandrasekaran, J. Glasgow, N. H. Narayanan (eds.) (Ed.), *Diagrammatic reasoning: cognitive and computational perspectives* (pp. pp.303-338). Menlo Park, California: AAAI press.

Making Machines Understand Facial Motion & Expressions Like Humans Do

Ana C. Andrés del Valle & Jean-Luc Dugelay

Multimedia Communications Dpt. Institut Eurécom
2229 route des Crêtes. BP 193. Sophia Antipolis. France
{andres, dugelay}@eurecom.fr

Abstract

Complete interaction amongst humans and machines unavoidably needs computers to understand human emotions. Most emotive information comes from facial motion and expression. This article presents the design of a new procedure for image analysis that is able to understand facial actions on monocular video sequences, without imposing restrictions on the speaker or its environment. The exposed technique follows a global-to-specific analysis approach that tries to imitate the way people analyze face motion: by dividing this analysis in processes of different level of detail.

1 Introduction

Researchers from the Computer Vision, Computer Graphics and Image Processing communities have been studying the difficulties associated with the analysis and synthesis of faces under motion for more than 20 years. The analysis and synthesis techniques being developed can be useful for the definition of low-rate bit image compression algorithms (model-based coding), new cinema technologies as well as for the deployment of virtual reality applications, videoconferencing, to enforce with realism the users' presence in games, etc. As computers evolve towards becoming more human oriented machines, human-computer interfaces, behavior-learning robots and disable adapted computer environments will use face expression analysis to be able to react to human action.

Facial analysis of motion and expression from monocular (single) images is widely investigated and of main interest because non-stereoscopic static images and videos are the most affordable and extensively used in visual media. In this field, research oriented towards the improvement of communications for the hearing impaired, like the video phone application presented by Sarris and Strintzis (2001), prove that understanding facial motion is of great help. We must not forget that "(…) many standard facial expressions, such as shaking the head for 'no' or raising the eyebrows to form a question, are used extensively to convey emotion, emphasis and intensity." (Disability Online, 2001).

Until recently, the analysis of rigid and non-rigid facial motion has been studied separately. Solutions given to deploy general expression analysis, mainly based on only-image processing techniques such as template matching or image Principal/Independent Component Analysis, are often designed to work on faces viewed from a frontal perspective and under controlled environment conditions. Tian, Kanade and Cohn (2000), already present a way to adapt their near-to-front analysis scheme to other head poses. They define "multiple state models", where different facial component models are used for different head poses; this solution proves to be toilsome. Approaches that search for more detailed facial action information need to be well aware of the pose of the head not only in reference to the image plane but also to its physical location. The 3D pose tracking of the head allows better control of the analysis and more freedom of movement to the speaker; we can appreciate these advantages in those approaches that fit a 3D mesh to better track face expressions (Ogata, Murai, Nakamura, & Morishima, 2001). It also permits more efficient image information retrieval by allowing, for instance, the rectification of the analyzed image to a known frontal pose (Chang, Chen, C.-C., Chou, & Chen, Y.-C., 2000). Nevertheless, most solutions end up having their own limitations related to the conditions under which the face is being analyzed. Ideally, proper facial motion analysis should work under any circumstances, regardless of the characteristics of the face and its surroundings. The design of a robust and potentially improvable analysis scheme can be obtained by observing human behaviour at performing this same task.

People can understand facial action even when faces are under very bad lighting or in the presence of disturbing objects over them. This is basically due to the fact that humans are able to automatically reduce the complexity of the analysis into different parts and to do this analysis progressively. First, our sight adapts to the lighting conditions under which the face is and we decide if further understanding is possible; then, we locate the head and get its rigid motion (its pose) and finally, we pay attention to the different details of the face that are interesting to us because they contain expression information. When humans are not able to perform an exhaustive analysis (lighting is very bad, or a significant part of the face is occluded), they make up for the missing information (generally assuming standard human behavior) or they simply accept that they cannot understand the face motion they are observing. In this article we present a complete system that performs facial motion and expression analysis on monocular video sequences trying to simulate this natural and intuitive human conduct. The methods and algorithms presented are designed to work under natural and non-controlled situations: not assuming the use of a calibrated camera, with no need of precise lighting conditions or markers on the person; to be as universal as possible, trying to avoid any system training specific to an individual previous to the analysis; to be as precise as the analysis conditions permit, allowing the user as much freedom of movement as possible and by generating an analysis stratagem permitting potential improvement in its precision; and to potentially work in real-time, thus permitting instant expression understanding from the interpretation of coherent facial animation parameters.

2 Overall Analysis Scheme

We consider face analysis from a video sequence as a function of the general pose of the face on the sequence, the illumination conditions under which the video is recorded and the face expression movements. We believe that synthesizing the analyzed face action on the speaker's clone[1] is a relevant way to check that the extracted motion information is correct. Following such premise, our analysis scheme attempts to obtain facial animation parameters (FAP) allowing as much motion precision as possible during their later synthesis, given the current conditions of analysis.

To obtain animation parameters from video frames in a robust manner, we need to be aware of the lighting on the face as well as of the physical nature of the features we are analyzing; this information will enable our algorithms to work under any lighting conditions and to remain robust throughout the analysis. We first estimate the pose of the face obtaining translation and rotation parameters of the head. And then, we extract some specific features from the face (eyes, eyebrows

[1] Clone: 3D synthetic representation of a person that not only represents realistically its appearance but also has the potential of being animated replicating exactly this person's facial motion

and mouth) and we apply on them some dedicated analysis techniques to obtain face animation parameters.
We utilize a two step process to develop our image analysis algorithms. First, we design image processing techniques to study the features extracted from a frontal point of view of the face. Faces show most of their expression information under this pose and this allows us to verify the correct performance of the image processing involved. Second, we extend our algorithms to analyze features taken at any given pose. This adaptation is possible because the motion models utilized during the analysis can be redefined in 3D space and the accuracy of the retrieved pose parameters is such that enables us to reinterpret the data we obtain from the image analysis in 3D. Controlling the pose permits understanding the evolution of the 2D regions we are analyzing on the image and thus allows us to foresee if doing the analysis over a specific feature will be profitable. It also let us the possibility of designing hierarchical analysis algorithms with different levels of detail depending on the visibility and the quality of the feature. Combining motion information coming from the analysis of different features to deduce the most suitable feature action is one of the best ways to avoid incoherent and unnatural analysis results. The analysis final check should be done in layers and compensate for that information that may be missing from the analysis (e.g. from occluded parts). In our case, we start by making sure that both eyes are behaving in the same way, then, we will introduce the eyebrow motion and finally we will analyze the overall expression when including the mouth. Although this kind of approach limits expression behavior to standard, natural human face motion, it becomes very helpful in situations where face analysis may be difficult.

3 Head-Pose Tracking

To fully understand global head actions in front of a camera, we needed to develop a head tracking algorithm capable of determining accurately the pose in a 3D reference system. Many tracking algorithms tend to work using local 2D image reference systems on which they can only estimate the user's position on the screen (e.g. Bradski's Cam Shift algorithm (1998)). For the sequence of processes to follow: detection of the features to be analyzed, analysis and interpretation of the analyzed results, this information is not accurate enough.
To obtain precise information about the person's location in space, we have developed an algorithm that utilizes a feedback loop inside a Kalman filter. Kalman filtering has been applied to head tracking giving very positive results (Cordea, Petriu, E. M., Georganas, D., Petriu, D. C., & Whalen, T. E., 2001; Ström, 2002) and it enables the prediction of the translation and rotation parameters of the head from the 2D tracking of specific points of the face on the image plane.
The natural drawback of this system is the need of 3D information about the shape of the head we are tracking, that is, we must use a model that provides the 3D coordinates of the points whose projection is tracked on the image and fed to the filter to obtain the prediction of the pose parameters. Very often, a general head model is used; although this apparent drawback can be become a strong advantage if a realistic 3D synthetic representation of the user is available. In (Valente, & Dugelay, 2001), we showed that improvement in the amount of freedom of movement in front of the camera is possible if using the speaker's clone during the tracking. In our approach, the algorithm operating on the image plane extracts the 2D features to be tracked on the video sequence from the synthesized image of the model, onto which the predicted pose parameters have already been applied thus providing an adjusted view of the user in its future pose. Since this approach compares head models and video frames at the image level, models have to be an accurate 3D representation of the speakers, in shape and texture. Furthermore, some lighting compensation must be done on the synthetic world to adapt it to the illumination conditions of the video sequence. Details in how this is performed can be found on the aforementioned reference.

4 Facial Feature Analysis Development for a Frontal Point of View

Designing specific image processing algorithms for each of the facial features being analyzed enables us adapting the design following anatomical and muscular constraints. These constraints take into account the influence of lighting on the appearance of the feature and the restrictions natural human facial movements impose. We have designed motion models that control feature behavior in a frontal position. These models can easily be adapted to scale their complexity and amount of retrieved data depending on the analysis conditions (i.e. size of image, lighting, etc.) This section briefly reviews the main characteristics of the algorithms.

- Eye-state analysis (Andrés del Valle & Dugelay, 2001):

The image processing analysis strategy decomposes the eye tracking actions in two categories: the open-close movement and the eyeball movement. To best exploit the physical characteristics of the eyes, a different algorithm analyzes each action. We define the degree of eye opening as proportional to the inverse of the amount of skin contained within the analyzed feature image. Gaze orientation is related to the position of the pupil on the feature region. Pupils can be defined as the lowest energy points of the eyes. We set a tight cooperation between the two analysis techniques in a temporal state analysis, which allows us to correct possible erroneous results from the algorithms.

- Eyebrow motion analysis (Andrés del Valle & Dugelay, 2002):

To study eyebrow behaviour from video sequences we utilize a new image analysis technique based on an anatomical-mathematical motion model. This technique conceives the eyebrow as a single curved object (arch) that is subject to the deformation due to muscular interactions. The action model defines the simplified 2D (vertical and horizontal) displacements of the arch. Our video analysis algorithm recovers the needed data from the arch representation to deduce the strength of the parameters involved.

- Mouth motion analysis (Andrés del Valle, Perales & Dugelay, 2003):

We have studied and observed the muscular and bone interaction during mouth motion, so not only lips are tracked but also tongue and teeth are taken into account. We have mathematically modeled mouth muscular interaction to deduce which are the minimum needed control points that will permit to synthetically replicate mouth motion through the understanding of its behavior.

5 Coupling Feature Analysis and Pose Tracking

The solution we propose (Andrés del Valle & Dugelay, 2002) defines the feature regions to be analyzed and the parameters of a motion analysis on 3D, over a head model in its frontal position. The complete procedure goes as follows (see Figure 1):

(i) We define and shape the area to be analyzed on the video frame. To do so, we project the 3D-regions of interest (ROIs) defined over the head model on the video image by using the pose parameters obtained from the rigid motion tracking, thus getting the 2D-ROIs.

(ii) We apply each one of the previous image processing techniques on these areas extracting the data required for the motion models.

(iii) We interpret these data from a three-dimensional perspective by inverting the projection and the transformations due to the pose (data pass from 2D to 3D) assuming all data fall on the same image plane. At this point, we can compare the results with the feature analysis parameters already predefined on the neutral head model and decide which has been the feature action.

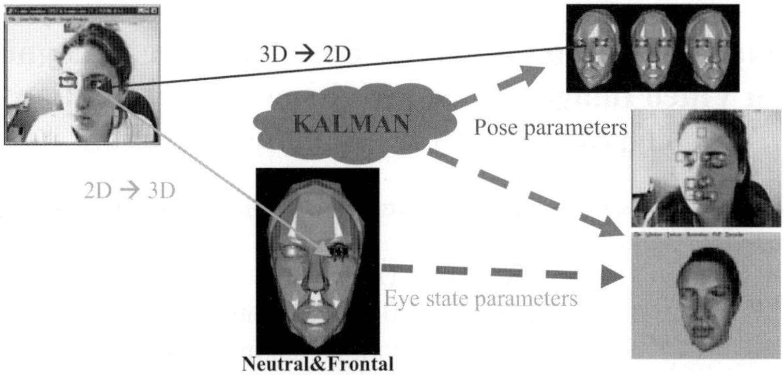

Figure 1: This diagram depicts the different interactions amongst 2D and 3D data during the analysis

6 Conclusions

In this article, we have presented an original approach to tackle the challenging issue of analyzing facial behavior on monocular video input in environments with minimum constraints (without known lighting or specific head pose). In the proposed framework, the complexity of the overall head action understanding is split into the study of several subproblems. Pose and face features are first analyzed independently and then jointly. This allows us to validate the robustness and performance of the algorithms involved because we are able to clearly detect and identify the origin and the nature of unexpected inaccurate results. Therefore it also permits to control the improvement of our algorithm steps while we are developing the different modules.

References

Andrés del Valle, A. C., Perales, F. J. & Dugelay, J.-L. (2003). *Analysis of mouth and lip motion and its coupling with pose tracking*. Technical Report to be published jointly by Eurecom Institute and the Universitat de les Illes Balears.

Andrés del Valle, A. C. & Dugelay, J.-L. (2002) Eyebrow movement analysis over real-time video sequences for synthetic representation. *Lecture Notes in Computer Science*. Springer (Ed.), Vol. 2492, 213–225.

Andrés del Valle, A. C., & Dugelay, J.-L. (2002). Facial expression analysis robust to 3D head pose motion. *Proceedings of the IEEE International Conference on Multimedia and Expo.*

Andrés del Valle, A. C. & Dugelay, J.-L. (2001). Eye state tracking for face cloning. *Proceedings of the IEEE International Conference on Image Processing.*

Bradski, G.R.(1998). Computer vision face tracking as a component of a perceptual user interface. *Workshop on Applications of Computer Vision*, 214–219.

Chang, Y.-J., Chen, C.-C., Chou, J.-C., & Chen, Y.-C. (2000) Virtual Talk: a model-based virtual phone using a layered audio-visual integration. *Proceedings of the IEEE International Conference on Multimedia and Expo.*

Cordea, M. D., Petriu, E. M., Georganas, N. D., Petriu, D. C., & Whalen, T. E. (2001). 3D head pose recovery for interactive virtual reality avatars. *Proceedings of the IEEE Instrumentation and Measurement Technology Conference.*

Disability Online. (2001). Auslan is a sign language. Retrieved January 21, 2003, from http://www.disability.vic.gov.au/dsonline/dsarticles.nsf/pages/Auslan_is_a_sign_language?OpenDocument

Ogata, S., Murai, K., Nakamura, S., & Morishima, S. (2001) Model-based lip synchronization with automatically translated synthetic voice toward a multi-modal translation system. *Proceedings of the IEEE International Conference on Multimedia and Expo.*

Ström, J. (2002, October). Model-based real-time head tracking. *Eurasip Journal on Applied Signal Processing*, 2002 (10), 1039–1052.

Tian, Y., Kanade, T., & Cohn, J. F. (2001, February). Recognizing action units for facial expression analysis. *IEEE Transactions on Pattern Analysis and Machine Intelligence*, 23(2), 97–115.

Valente, S., & Dugelay, J.-L. (2001). A visual analysis/synthesis feedback loop for accurate face tracking. *Signal Processing: Image Communication*, 16(6), 585–608.

Human Information Retrieval Based on Face Recognition in Video Image through Multi-modal Interaction Using Speech and Hand Pointing Action

Yasuo Ariki, Masakiyo Fujimoto, Natsuo Yamamoto and Msahito kumano

Ryukoku University
Seta Otsu, 520-2194, Japan
ariki@rins.ryukoku.ac.jp

Abstract

In this paper, a system is proposed for enquiring human information about famous persons broadcast on television programs such as news or drama. The system is composed of hand-free speech recognition and hand action recognition so that a user can ask about person on the television by speaking in a hand-free mode and pointing the person by his hand. The system can extract and recognize the pointed person and retrieve the human information on the WWW.

1 Introduction

Digital contents are widely broadcast on the television throughout the world. However, it has no function to reply the questions from users such as "who is he?" or "what is this?" about the objects appearing in the contents. Consequently, we can't get the information exactly we want to know from the television interactively. In order to solve this problem, we have to provide the television with an intelligent interaction facility through which we can ask anything about the contents on air.

As the first step to realize this function, we propose in this paper an interactive television system to enquire human information about famous persons broadcast on television programs such as news or drama. In enquiring to the television, a keyboard or mouse is not suitable because we want to focus our attention on the content when watching the television. From this viewpoint, speech and hand action are excellent modalities compared with the keyboard and mouse. In speaking to the television, grasping a microphone or wearing a handset is still cumbersome and unnatural because we want to feel easy when watching the television. From this viewpoint, we employed the hand-free speech recognition using a microphone array.

Speech is a strong tool to communicate with machines. However, speech alone is ambiguous to convey the user's intention to the system. For example, speaking "who is he?" is ambiguous when there are more than two persons on the television. In this case, an unconscious and easiest way is to point out the person by our hand. From this viewpoint, we employed recognition of hand-pointing action using three-dimensional motion capture analysis to understand the user intention.

2 Processing Flow

Figure 1 shows a processing flow in our system. A user is watching the television and when he is interested in someone, he says "who is he?" by pointing the person. The system first gets user's hand-free speech and recognizes it. Simultaneously the system gets the user's hand action and recognizes the position on the television. From the speech and hand action, the system understands the user's intension that he wants to know the name and information about the pointed person on the television. At present, we provide the system with the function of human information enquiring, but it will be expanded to an enquiry like "what is this?" when we prepare the object recognition software.

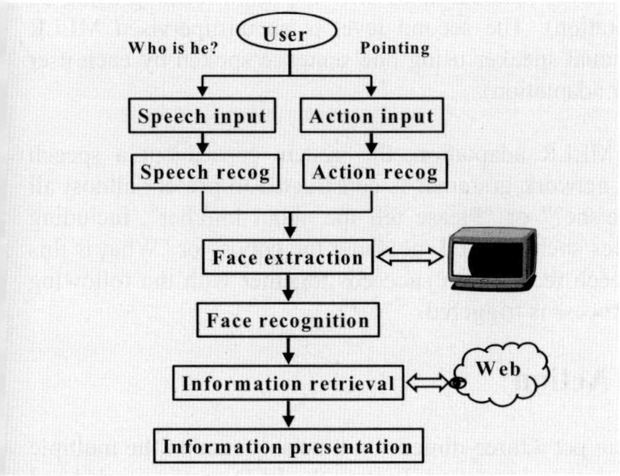

Figure 1: Processing flow in the system

The face of the pointed person is extracted on the television image and then recognized by image processing techniques. Even when the pointed position is out of the face region, the image processing can identify and extract the pointed face. The homepage URL of the pointed person is retrieved through the WWW by using the recognized name. Finally, the homepage of the pointed person is shown on the television and the user can get more information of the person by clicking buttons on the homepage using further hand actions.

3 Hand-free Speech Recognition

3.1 Beam Forming and Estimation of Speaker Direction

Hand-free speech is distorted because of the transfer function between a speaker mouth and a microphone array. It is also noisy because of the superposition of the circumstance noises. In order to solve these problems, beam forming is required to estimate the speaker direction and to carry out noise adaptation. For the beam forming, we employed a delay and sum array method (Flanagan, Jhonston, Zhan, & Elko, 1985).

Figure 2 shows a diagram of the delay and sum array method. Target sound arrives from θ direction and is received at the M microphones placed every d distance in one line. The target sound arriving from θ direction and being received at the n-th microphone is delayed by $(n-1)\,\tau$ for equalization of the phase. All the waveforms with the equalized phase received at M microphones are summed and one waveform is produced with M amplified magnitude. Sounds arriving from other than θ direction produce the waveform with less than M amplified magnitude because the phases are not equalized. Therefore, the virtual microphone sensitive to θ direction can be formed.

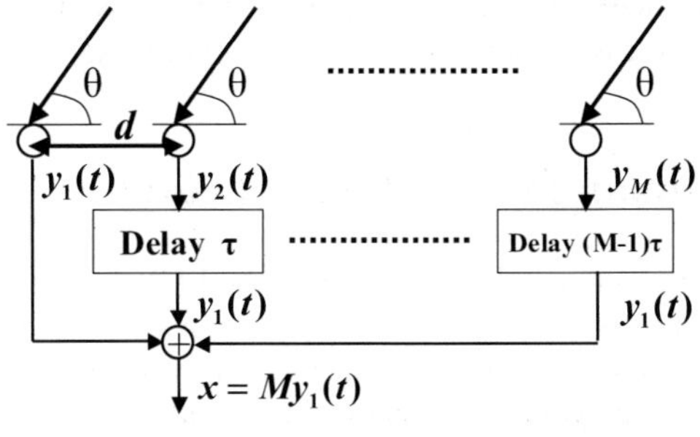

Figure 2: Delay and sum array method

3.2 Noise Adaptation and Speech Recognition

The beam forming described above can reduce sounds other than the target speech. However, the background noise is still superimposed on the waveform after beam forming. To reduce the background noise, we propose here a two levels MLLR (Maximum Likelihood Linear Regression) adaptation. The first level is a supervised MLLR adaptation (Leggetter and Woodland, 1995) of phoneme HMMs (Hidden Markov Models) to the residual noise after the beam forming using 20 sentences spoken by 5 males (noise adaptation). The second level is an unsupervised MLLR adaptation of phoneme HMMs to an individual speaker using one sentence spoken by each user after the first level noise adaptation (speaker adaptation).

After the beam forming and two levels MLLR adaptation, the system carries out a speech recognition using a network grammar. The network grammar is constructed to present almost all the enquiry sentences such as "Who is he/she?" or "Please tell me about him/her", including enquiry about objects other than human faces such as "Tell me about the goods" or "What is this accident?" for further expansion. If the speech recognition succeeds together with the following hand pointing action, the face recognition process is triggered.

4 Analysis of Hand Pointing Action

A three-dimension motion capture device can get a three-dimension time sequence of the multiple LEDs (“Visualeyez,”). Each LED can be uniquely identified. Putting the LED on the user's hand and arm, and connecting the positions of the arm LED and the hand LED, a three dimensional line is computed. When the user points some person on the television by his hand, a crossing point between the three-dimension line and the television display can be computed.

However, it was found that there was a large difference between the user intended point and the computed crossing point. This is because the user intended point is the crossing point between the television display and the line connecting an eye with a hand LED, not the line connecting the arm LED with the hand LED. For the eye position, we put the LED on the user head and estimated the eye position from the head LED position. For more accurate estimation of the crossing point, the finger position is also estimated from the hand LED position.

The crossing point computed is given to the mouse control software so that the user can point anywhere on the television as if his finger is a mouse device. A click action also can be realized by computing the movement of the crossing point. Namely, if the variance of the coordinates of the crossing point is less than some threshold within certain duration, then the crossing point is regarded as a stationary point. This action is recognized as clicking. By this action, the system can discriminate whether the user speech is conversation with his friends or enquiry to the system. If the action is stationary and the user is speaking, then it is recognized as enquiry. Otherwise, it is neglected as his conversation.

5 Human Face Recognition and Human Information Retrieval

For the real time face extraction and recognition, we employed a subspace method (Oja, 1983). This method is based on principal component analysis that can compress dimensions in the observation space by computing the principal axes on the criterion of the maximum variance. The distance or similarity between an input pattern and the category subspaces can be computed in real time.

For the face detection, one facial subspace is created as the facial model. It is created first by collecting many facial images and computing their covariance matrix. Finally it is decomposed by an eigen value decomposition and the facial subspace is created (Ariki, Ishikawa and Sugiyama, 1998). Faces are detected by computing the projection distance between an input image and the facial subspace. If the projection distance is less than some threshold, the input region is identified as a facial region. To find the facial regions with various sizes at several positions, pyramid images are created from the input image by reducing the image size.

For the face recognition, facial subspace is required for each person in advance. It is created using his training data collected by the above described face extraction for each person. Then the projection distance is computed between the extracted facial region and the facial subspace of the respective person. Finally, the pointed person is recognized as the person with the minimum projection distance. By combining the hand-free speech recognition, hand action analysis and face recognition, the name of the person required by the user speech and pointed by his hand on the television is recognized. The human information can be retrieved by searching for the homepage of the recognized person in the WWW.

6 Experimental Result

Five users tried 3 times to enquire about 10 famous persons appearing on news programs broadcast by NHK (Japan Broadcasting Corporation). In total, 150 (5 users x 3 times x 10 persons) evaluations were performed in the hand-free speech recognition, hand pointing recognition and face extraction/recognition. The users were asked to speak the queries while the caster was speaking in the news programs.

Hand-free speech recognition was carried out using the beam forming and two levels MLLR adaptation. The microphone array was composed of 16 microphones, whose distance was 2cm, arranged in a line. A user speaks queries like "Who is he?" while the caster is speaking, at the position two meters apart from the microphone array in a front direction. In the experimental environment, 9 PCs and 4 projectors were working. The total noise level including the caster speech from the TV news was 55 dB at average.

Table 1 shows the conditions of acoustic analysis and HMM structure. The left column of Table 2 shows the hand-free speech recognition result. The table shows the improvement of speech recognition accuracy by the beam forming and two levels MLLR (TL-MLLR) adaptation. The successful 140 trials among 150 were passed to the hand pointing recognition. The total speech recognition time was about 2.5 seconds using the PC with Pentium 4, 1.7GHz x2, 512MB memory. Hand pointing recognition described in section 4 was carried out in the same experimental environment as the hand-free speech recognition. The recognition rate was 100% so that still 140 trials were passed to the face extraction/recognition.

Table 1: Experimental condition

Condition of acoustic analysis		HMM structure	
Sampling frequency	16kHz, 16bit	Number of states	5 states and 3 loops
Feature	12MFCC + power + Δ + $\Delta\Delta$ (39 dim)	Number of mixtures	12
Frame length	20ms	Number of phonemes	41
Frame shift	10ms	HMM type	Left-to-right monophone
Window	Hamming window		

Table 2: Experimental Result

Hand-free speech recognition		Face extraction	Face recognition
Method	Recognition rate (%)	Extraction rate (%)	Recognition rate (%)
Beam forming	67.3 (101/150)	60.0 (84/140)	76.2 (64/84)
Beam forming+TL-MLLR	93.3 (140/150)		

The face extraction result is shown in the middle of Table 2. The extracted faces were judged to be correct manually if the eyes, nose and mouth were all in the extracted face. At present, the extraction rate is low due to the reason that the famous persons were taken from various directions in a real environment. This can be improved by using several facial subspaces with different directions. The extraction time was about one second on the same personal computer.

For the face recognition, we collected in advance 10 facial images as the training data for each famous person by performing the face extraction. The result of the face recognition for 84 trials is shown in the right column of Table 2. The recognition is still not so high and it is attributed to the face direction problem described in the face extraction. The recognition time was also about 1 second on the same personal computer.

7 Conclusion

We proposed in this paper the system for enquiring the human information about famous person broadcast on the television programs such as news or drama. The system is composed of hand-free speech recognition (93.3% recognition rate) and hand-pointing recognition (100% recognition rate) so that the user can ask who the person is on the television by speaking in a hand-free mode and pointing the person by his hand. The system can extract and recognize the pointed person and retrieve the human information through the WWW using the recognized name. This is a sort of an integration of television and WWW using signal processing and pattern recognition. In future, we plan to install the microphone array on the television set.

References

1. Flanagan, J.L., Jhonston, J.D., Zhan, R., and Elko, G.W.(1985). Computer-steered microphone arrays for sound transduction in large rooms. *Journal of Acoust. Soc. Am.*, Vol.78, No.5, pp.1508-1518.
2. Leggetter, C.L. and Woodland, P.C. (1995). Maximum likelihood linear regression for speaker adaptation of the continuous density hidden Markov models. *Computer Speech and Language*, Vol.9, pp.171-185.
3. Visualeyez USER'S MANUAL. *PhoeniX Technologies Incorporated.*
4. Oja, E.(1983). Subspace Methods of Pattern Recognition. Research Studies Press, England.
5. Ariki, Y., Ishikawa, N. and Sugiyama, Y. (1998). Face indexing on video data -extraction, recognition, tracking and modeling-. *Proc. of FG98*, pp.62-69.

Arbitrating Multimodal Outputs: Using Ambient Displays as Interruptions

Ernesto Arroyo

MIT Media Laboratory
20 Ames Street E15-313
Cambridge, MA 02139 USA
earroyo@media.mit.edu

Ted Selker

MIT Media Laboratory
20 Ames Street E15-322
Cambridge, MA 02139 USA
selker@media.mit.edu

Abstract

This work explores the use of ambient displays in the context of interruption. A multimodal interface was created to interrupt users using ambient displays in the form of heat and light. These ambient displays acted as external interruption generators. Experimental results show there are different effects on performance and disruptiveness caused by interruption modalities. Thermal interruptions have a larger detrimental effect than light interruptions on disruptiveness and performance. These results offer guidelines for interface designers to help them select interaction modalities to accommodate people's limitations relative to focus, concentration and interruptions. Furthermore, this work also shows that it is possible to differentiate between modalities and create multimodal interfaces that arbitrate between interruption modalities based on their effectiveness, user's performance, and disruptive effects.

1 Introduction

The use of interruptions is key in the design of human-computer interfaces. In general, current computer environments are becoming more and more complex, with an increasing number of tasks and an increasing number of events computer users have to keep track of (Maes, 1994). Multitasking is useful and natural. Unfortunately people have cognitive limitations that make them susceptible to errors when interrupted (McFarlane, 1999). Researchers have investigated interruptions by looking at how and when to interrupt users in a multitasking environment (McFarlane, 1999 & Czerwinski, 2000). They have also found that users perform slower on an interrupted task than on an uninterrupted task; that is, interruptions are perceived as disruptive. This work identifies key factors that influence the perceived effect of two interrupting modalities and shows the effect of two different interruption modalities on performance and disruptiveness.

Multimodal interfaces provide substantial advantages in efficiency (Oviat & Cohen, 2000). Finger and hand actions with the keyboard and mouse are commonly used to communicate to the computer. Visual and acoustic modalities are the most often used for conveying/presenting information to the user (Srinivasan, 1995 & Tan, Ifung & Pentland, 1997). Current human computer interfaces generally ignore important modalities such as ambient and peripheral visual cues, heat, vibration, smell and the sense of touch. The main focus of multimodal HCI research has been on combining input modalities rather than using multimodal outputs to take advantage of human sensing capabilities. The common and unique characteristics of the human senses allow for the design of computer interfaces that use multiple output modalities, and furthermore, computer interfaces that arbitrate between these modalities.

This work shows that the disruptiveness and effectiveness of interruptions varies with interruption modality. A multimodal interface was created with two ambient displays for interruption: heat and light. User awareness of these ambient displays shifts from the background to the foreground (Wisneski & Ishi, 1998), acting as external interruption generators. Overall, the thermal modality produced a larger decrease in performance and disruptiveness on a task than the visual modality.

2 Approach

This work explores the use of ambient displays in the context of interruption. Ambient displays present information in the modality and form that can be interpreted with a minimal cognitive effort (Wisneski & Ishi, 1998). Ambient displays act as external interruption generators designed to get users' attention away from their current task; they also serve as a media for interruptions.
This paper presents an experiment designed to test the effect of heat and light when used as interruptions. The experiment purpose is to identify the key factors that influence what are the perceived effects for each ambient display.

Interruption involves many subtle low-level mechanisms of human cognition (Bailey, Konstan & Carlis, 2000). In order to study the effects of interruptions, a simplified abstract model of common real world tasks was chosen. The task involves a graphical-textual computer-based game that imposes a high cognitive load (See Figure 1). Examples of people performing this type of tasks are software developers. A debugging task, for example, requires a software engineer to identify and keep track of variable values as they change over the execution of the software. These identification and tracking tasks impose a high cognitive load. Interruptions during this process cause errors, allowing for observations of subjects' responses to be easily broken down into discrete units (Gillie & Broadbent, 1989).

The experiment is set in the context of a computer-based adventure game, similar to online Multi user Dungeon (MUD) games, where the player has to issue commands in order to achieve certain goals. Gillie, et al used a similar approach (Gillie & Broadbent, 1989). A MUD is a network-accessible, multi-participant, user-extensible game in which participants have the appearance of being situated in an artificially constructed place through an entirely textual interface.

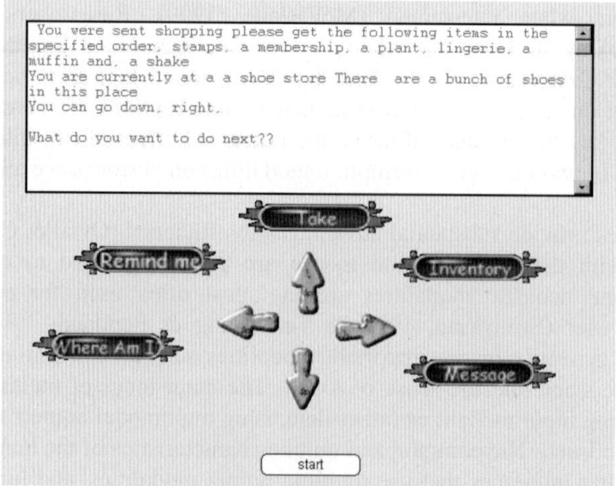

Figure 1: **Graphical-textual computer-based game**

Subject's task is to read directions, memorize a list of items presented to them, explore several locations around a small geographical area, create a mental map about the location and its contents, take objects in the specified order, and decide the next location to go to. This task provides several performance and disruptiveness indicators: score, speed, error rate and overall time. Czerwinski presented a similar experiment where subjects navigated a list of items searching for a book title. The investigator used a memory task to look for effect of disruption (Czerwinski, 2000).

2.1 Participants

23 subjects were randomly recruited and compensated for their time. The sample consisted of 14 males and 9 females with ages ranging from 22 to 34 years.

2.2 Procedure

The experiment has twelve randomly presented trails; each of them contains a fixed-ordered list with six items with in the same category norm (Battig, & Montague, 1969) presented within a plausible story. The list items are distributed randomly in a geographical area contained in a 5x5 matrix where subjects navigate. Once subjects have taken all six objects, the next trial is presented. While subjects perform the primary task, an ambient device attracts their attention by changing temperature or light intensity. Subjects then acknowledge the interruption and read a list of words organized into networks of associated ideas. This dual-task of the experiment is conceptually simple, but difficult to perform due to the high cognitive load.

The order in which the computer presents each problem, the interruption modality to use, as well as the choice of problems to interrupt are randomized. Randomizing keeps subjects from anticipating interruption and balances any novelty effects interrupting modalities may cause. Non-interrupted sessions serve as a baseline for comparison.

3 Results

Heat presented a larger detrimental effect on performance than light. Performance measured by the time to take objects compared against a non-interrupted session, indicates that there is a 24% increase in performance when interrupting with light, $F(1,22)=6.47$, $p<0.019$ and only a 2% increase when interrupting with heat, $F(1,22)=30.89$, $p<0.0005$.

Heat presented a greater disruptive effect than light. Disruptiveness, measured by errors in direction, indicates that light reduces the number of errors by 50%, $F(1,22)=7.23$, $p<0.013$, whereas heat reduces them by only 37%, $F(1,22)=12.757$, $p<0.002$.

Light is more effective for getting user's attention promptly. Light was noticed 42% faster than heat $F(1,22)=7.76$, $p<0.011$.

Subjects were not negatively affected by their non-preferred modality. Based on subjects self-reported preferred modality, there was no main effect of subject's preferred modality in performance $F(1,22)=1.374$, $p>0.254$, neither was an effect in speed observed $F(1,22)=0.006$, $p>0.94$.

Although heat was harder to detect, it was also harder to ignore once it was present. Heat was perceived as a dangerous threat. Light, as opposed to heat, which had an affective component, had no physical interaction with subjects that could be perceived as an invasion their own personal space.

3.1 Discussion

This experiment verifies previous research about interruptions, in that subjects perform slower on an interrupted task than on a non-interrupted task; demonstrating the general effect of interruptions. Furthermore, this experiment also shows that the interruption modality affects performance. The thermal display produced a larger decrease in performance than the visual display. This thermal display also has a greater disruptive effect on the interrupted task than light. Disruptiveness and performance measures agree that heat causes larger of a detrimental effect than light when used as an interruption.

Advances in computer technologies have enabled the creation of systems that allow people to perform multiple activities at the same time. People have cognitive limitations that make them susceptible to errors when interrupted. Unfortunately, interruptions are common to today's multitasking computing user interface experience. Computer interfaces must be designed to accommodate people's limitations relative to focus, concentration and interruptions. These results offer guidelines for interface designers in choosing one modality over the other. Light is more efficient in getting user's attention (42% faster than heat). In contrast, heat takes longer to be noticed but is more disruptive. Heat could be used more reliably in environments where other channels are already saturated overwhelmed with information (i.e., when there are many visual distractions). One advantage of using heat is that users can be interrupted without taking their attention off the screen. With light, users tend to focus their attention to the light source. Additionally, heat is an interruption to a single person; a personalized attention-grabbing device. Unlike ambient lights, which alert all people present at the location where light changes occur, heat can be used to signal messages subtly to a single person, Therefore heat is a personalized attention-grabbing device. By taking these results and applying them to user interface design, a system could maximize the effectiveness of interruptions through proper modality arbitration.

4 Conclusion

This work explored the use of ambient displays in the context of interruption in order to illustrate the use of other perceptual channels in current computer interfaces. A multimodal interface was created to interrupt users using ambient displays in the form of heat and light. These ambient displays acted as external interruption generators. Ambient displays can help orient and situate a person to serve a purpose other than the mere presentation of information—they can serve as a media for creating and changing context about interruptions.

This works contributes to previous research by showing there are different effects on performance and disruptiveness caused by interruption modalities. Thermal interruption has a larger detrimental effect on both disruptiveness and performance. Previous research regarding the general effect of interruptions, in which subjects perform slower on an interrupted task than on a non-interrupted task, was also corroborated.

Human senses differ in their ability to be ignored, precision and speed. The common and unique characteristics of the human senses allow for the design of an interface that uses multiple modalities and, furthermore, of an interface that selects the modality to use based on contextual information. We have shown that it is possible to differentiate between modalities and build multimodal interfaces that select the interruption modality to use based on its effectiveness, user's performance, and disruptive effects. We can now work to improve interfaces that arbitrate between interruption modalities.

We envision utilizing users' physiological responses as feedback to a computer interface, so that the interface could modify the way it communicates with every user by selecting and configuring the adequate modality. Our experiment sets the initial point for understanding how to build interfaces of this type by looking at the effect of different modalities when used as interruptions.

5 References

Bailey, B. P., Konstan, J. A. & Carlis, J. V. (2000). The effect of interruptions on task performance, Annoyance, and Anxiety in the User Interface. *IEEE International Conference on Systems, Man, and Cybernetics.*

Battig, W.F. & Montague, W.E. (1969). Category norms for verbal items in 56 categories: A replication and extension of the Connecticut category norms. *Journal of Experimental Psychology Monograph*, 80, 1-45.

Czerwinski, M., Cutrell, E. & Horvitz, E. (2000). Instant messaging: Effects of relevance and time. *People and Computers XIV: Proceedings of HCI 2000*, Vol. 2, British Computer Society, Eds: S. Turner & P. Turner, 71-76.

Gillie,T. & Broadbent, D. (1989). What makes Interruptions Disruptive? A study of length, Similarity and Complexity. *Psychological Research*, 50, 43-250.

Maes, P. (19940. Agents that Reduce Work and Information Overload. *Communications of the ACM*, Vol. 37, No.7, (July 1994), pp. 31-40, 146.

McFarlane, D. (1999). Interruption of People in Human Computer interaction. *Human Computer Interaction- Interact.*

Oviat, S. L. & Cohen, P.R. (2000). What comes naturally. *Communications of the ACM*, 45-53.

Srinivasan, M. A. (1995). Haptic Interfaces. *Virtual Reality: Scientific and Technical Challenges.* Report of the Committee on Virtual Reality Research and Development, National, Research Council. N. I. Durlach and A. S. Mavor, National Academy Press.

Tan H. Z., Ifung Lu, & Pentland, A. (1997). The chair as a novel haptic user interface. *Proceedings of the Workshop on Perceptual User Interfaces*, 56-57.

Wisneski, C., Ishii, H., and Dahley, A. (1998). Ambient Displays: Turning Architectural Space into an Interface between People and Digital Information. *International Workshop on Cooperative Buildings.*

Parallel Versus Sequential Grammar Systems for Modelling Dialogues[1]

S. Aydin, H. Jürgensen[2]

Institut für Informatik, Universität Potsdam
August-Bebel-Straße 89
14482 Potsdam, Germany
email: aydin@cs.uni-potsdam.de, helmut@uwo.ca

Abstract

A new model of dialogues is proposed, that of parallel grammar systems. In contrast to sequential grammar systems, this model hides the work performed by each participant to evaluate the information, but affords the same performance and has the potential of greater effi ciency. In contrast to other dialogue models proposed in the literature it can capture non-deterministic phenomena in dialogues like change of topic, questions for clarifi cation,etc. We illustrate the approach by a small example.

1 Introduction

We model dialogues, between a human and a computer, using non-standardized spoken language. In several ways our goals are similar to those outlined in (Boye, Hockey & Rayner, 2000). We focus on the structure of dialogues and assume that the linguistic problems have been dealt with. Clearly, this assumption is unrealistic at this point in time. However, it helps to control the problem.

We use grammar systems as the modelling tool. A grammar systems consists of a fi nite set of grammars which co-operate according to a formal protocol pre-selected from a given set of protocols (see e.g. Csuhaj-Varjú, Dassow, Păun & Rozenberg, 1994). This idea is suggested and investigated in the fi rst author's diploma thesis (Aydin, 2001) and was fi rst published in (Aydin, Jürgensen & Robbins, 2001) and then modifi ed in (Aydin & Jürgensen, 2002). That proposal uses a protocol which requires sequential processing of the interactions and memories associated with each grammar. In this paper we consider a parallel protocol. We argue that this leads to a simpler and more natural dialogue model which, moreover, allows for faster information processing. The theory of parallel grammar systems is investigated in (Vaszil, 2000).

In contrast to the usual approaches to dialogue modelling (e.g. Clark, 1996; Cohen, 1991; Cohen, 1997), grammar systems provide a natural and simple way for dealing with spontaneous speech acts, disagreements, misunderstandings, change of opinion, change of topic, etc.

By non-standardized spoken language we mean that the human user of such an interface is not restricted to communicate with the computer by pre-determined commands or keywords; instead, the user may communicate with the computer using his/her own natural (spoken) language expressions,

[1] This research was in part supported by Grant OGP0000243 of the Natural Sciences and Engineering Research Council of Canada.

[2] H. Jürgensen is also at: Department of Computer Science, The University of Western Ontario, London, Ontario, Canada, N6A 5B7.

subject of course to the limitations imposed by the topic of the 'conversation'. As a human-machine interface for spoken language has to deal with various dialogue phenomena there is a need for a stable formal model of dialogues. Grammar systems provide such a model.

We introduce a formal model for dialogues based on parallel communicating grammar systems. Each human participant and computer (or computer window) involved in the dialogue is modelled by a grammar, such that all participants of the dialogue are formalized together as a parallel communicating grammar system.

In our earlier work, a dialogue between a human and a computer is modelled by co-operating distributed grammar systems with memories. These are grammar systems which consist of several grammars working together in a sequential way, that is, each utterance of a participant is modelled and processed one after the other and by taking turns. The memories of the grammars in the system contain private notes of the dialogue participants; in this way, private and public information can be kept separate. As demonstrated in (Aydin et al., 2001; Aydin & Jürgensen, 2002), this model can deal with the variability and unpredictability of dialogues, that is for instance, disagreements between dialogue participants, unexpected changes of opinions or topics, diffi culties concerning comprehension and misunderstandings between the participants.

Our new proposal is to use *parallel* communicating grammar systems (see Csuhaj-Varjú et al., 1994; Dassow, Păun & Rozenberg, 1997) as a model instead. In this way one models that utterances spoken by one participant can be heard and processed by other participants simultaneously. This model is more natural without giving up the advantages of our earlier sequential model and lends itself to faster information processing with speed measured in terms of grammar derivation steps.

The parallel protocol is roughly as follows: The grammars of the system which model the parties involved in the dialogue work on their own sentential forms; the latter represent the utterances and the progression of the turn taking. The sentential forms include the private notes of each participant, which are not known to the other participants of the dialogue, as well as the public information as developed during the dialogue.

For modelling dialogues by parallel communicating grammar systems, a meta-grammar is needed that handles the general conventions of a dialogue for it to be successful (correct and reasonable). This meta-grammar controls and records the dialogue progress and handles the variability and unpredictability of dialogues. The meta-grammar can send queries to the other grammars; in this case, the other grammars have to supply part of their sentential forms, but never their private goals.

From a mathematical point of view, using parallel communicating grammar systems for dialogue modelling rather than sequential ones has the advantage that the rather awkward memories in the sequential system, which make a mathematical treatment diffi cult,can be avoided.

We illustrate, by example, how a typical dialogue with unpredictable utterances and topic changes can be modelled using our approach.

2 Basics

The symbol $\mathbb{N}$ denotes the set of positive integers. An *alphabet* is a fi nitenon-empty set the elements of which are referred to as *symbols*. For background on language theory, see (Salomaa & Rozenberg, 1997).

3 Parallel Communicating Grammar Systems

The formal model we use is that of parallel communicating grammar systems (Dassow et al., 1997; Vaszil, 2000).

Defi nition 3.1 A *parallel communicating grammar system* is a construct

$$\Gamma = (N, K, T, G_1, G_2, \ldots, G_n)$$

with $n \in \mathbb{N}$ with the following properties:

- N is a fi nite non-empty set of *nonterminal symbols,* the *nonterminal alphabet;*
- T is a fi nite non-empty set of *terminal symbols,* the *terminal alphabet;*
- K is a fi nite set of *query symbols,* the *query alphabet,*

  $$K = \{Q_1, Q_2, \ldots, Q_n, \text{Ter}\},$$

  where the index i refers to the i-th grammar G_i of Γ.
- For $i = 1, \ldots, n$ each $G_i = (N \cup K, T, P_i, \omega_i)$ is a grammar with set of rewriting rules P_i and axiom ω_i.

G_1 is said to be the *master grammar* (or just *master*) of Γ.

The formal defi nition of how a parallel communicating grammar systems works can be found in (Dassow et al., 1997; Vaszil, 2000). In the sequel we assume that all grammars G_i are context-free; this implies, in particular that $\omega_i = S_i$ for some symbol $S_i \in N$. Each grammar G_i maintains its own sentential form x_i. Thus, the initial confi guration of the system is the n-tuple $(S_1, \ldots, S_n)$. A confi guration $(x_1, \ldots, x_n)$ is *fi nal* if Ter is the last symbol of x_1. To determine a fi nal confi guration, the sentential forms $x_2, \ldots, x_n$ of the other grammars are not taken into account.

The step-wise behaviour of such a grammar system Γ depends on whether any of the sentential forms in the present confi guration contains a query symbol or not. If no query symbols are present all grammars proceed independently according to their own rules. On the other hand, if a query symbol Q_i is present in any current sentential form, it is replaced by the corresponding sentential form x_i. This is called a *communication step;* whenever a grammar introduces a query symbol the normal re-writing process is halted until all query symbols have been replaced. In this way any grammar can ask for information about the current sentential form of any other other grammar in the system and include that information in subsequent computation steps.

For details and variants of this model see (Dassow et al., 1997; Vaszil, 2000).

To model dialogues, we do not need the full power of parallel communicating grammar systems. It is suffi cient that only the master grammar of Γ can introduce query symbols. Such a grammar system is called *centralized.*

4 A Small Example

Dialogues do not follow simple rules; digression and anaphora problems are common. Rule-based or turn-taking models usually cannot take these phenomena into account. Consider the following rather simple dialogue between a *caller* C and an *answerer* A:

A: Computer Science Department, main offi ce.

C: Is professor Miller in?

A: No, he is back in April. To whom am I talking?

C: This is Peter Smith. Can you tell me how I get to the university by bus?

A: Sorry, but you can fi ndthat information on our website.

C: OK. Is professor Miller teaching theoretical computer science next term?

A: Yes.

C: Thank you. I'll call him April when he is back.

A: I'll leave a message for him saying that you called and that you will call again in April.

C: Thank you. Bye.

Real dialogues are signifi cantlymore complex than this one. This example is only meant to illustrate some of the main features of our approach.

As usual, a dialogue model will distinguish between private goals of each of the participants and public goals for the whole dialogue. Roughly, the public goal in the example is to establish communication between the caller and professor Miller. To achieve this, both A and C need information – their private goals. Syntactically, we separate public and private information in a sentential form by a special symbol.

To model this particular type of dialogue, one would use a grammar system consisting of three grammars: the master G_1; G_A for the answerer; G_C for the caller. The re-writing rules of G_A and G_C reflect the introduction of goals and information. For example, in the fi rststep, G_A provides the information that it is the "Computer Science Department", but also wants to know the "name" of the caller and the "reason" of the call (as private goals); similarly, G_C records that the "Computer Science Department" has been called; the master grammar G_1 records that the dialogue has started. As the dialogue proceeds, G_1 will issue queries whenever new information has been made available by either G_A or G_C; the term of "new information" includes also change of topic, questions concerning comprehension, change of opinion, etc. For example, G_1 will introduce a query symbol after the utterance *"Can you tell me how I get to the university by bus?"* as this changes the topic of the dialogue. The master grammar G_1 will add the symbol Ter when the dialogue is fi nished.

5 Concluding Remarks

Models of dialogue systems for unconstrained natural-language interfaces need to take into account the unpredictability of events in a dialogue. Even for limited domains, the usual models cannot afford this. Grammar systems provide a natural framework for processes like keeping notes, exchanging information, evaluating answers or asking questions. Some of these tasks are performed privately while others require communication. In our model of (Aydin et al., 2001; Aydin & Jürgensen, 2002) the distinction between private and public task was only implicit. The new model using parallel grammar systems achieves this distinction explicitly; moreover, due its parallel operation, it reflects what is going on in a real dialogue more closely. Both our models can deal with spontaneous dialogue phenomena like change-of-topic easily, an issue for which most of the other published dialogue models do not have a systematic solution.

6 References

Aydin, S. (2001). Dialoge als kooperierende Grammatiken. Diplomarbeit, Universität Potsdam.

Aydin, S., & Jürgensen H. (2002). Dialogues modelled by cooperating grammars. In D. Ivanchev, M. D. Todorov (Eds.), *Applications of Mathematics in Engineering and Economics'27* (pp. 613–637). Sofi a:Heron Press.

Aydin, S., Jürgensen, H., & Robbins, L. E. (2001). Dialogues as co-operating grammars. *Journal of Automata, Languages and Combinatorics,* 6, 395–410.

Boye, J., Hockey, B. A., & Rayner, M. (2000). Asynchronous dialogue management: Two case-studies. In D. Traum, M. Poesio (Eds.), *Proceedings of GÖTALOG 2000, Fourth Workshop on the Semantics and Pragmatics of Dialogue.* Gothenburg Papers in Computational Linguistics 00-5.

Clark, H. (1996). Using Language. Cambridge: Cambridge University Press.

Cohen, P., (1997). Dialogue modeling. In R. A. Cole, J. Mariani, H. Uszkoreit, G. B. Varile, A. Zaenen, A. Zampolli (Ed.), *Survey of the State of the Art in Human Language Technology, Lingustica Computazionale,* XII-XIII (pp. 204–210). Cambridge: Cambridge University Press; Pisa: Giardini Editori e Stampatori.

Cohen, P., & Levesque, H. (1991). Teamwork. *Noûs,* 25, 487–512.

Csuhaj-Varjú, E., and Dassow, J., Kelemen, J., & Păun, G. (1994). Grammar Systems: A Grammatical Approach to Distribution and Cooperation. Yverdon: Gordon and Breach.

Dassow, J., Păun, G., & Rozenberg, G. (1997). Grammar systems. In G. Rozenberg & A. Salomaa (Eds.), *Handbook of Formal Languages,* (vol. 2, pp. 155–213). Berlin: Springer-Verlag.

Rozenberg, G. & Salomaa, A. (Eds.) (1997). Handbook of Formal Languages. Berlin: Springer-Verlag.

Vaszil, G. (2000). Investigations on Parallel Communicating Grammar Systems. PhD Thesis, Eötvös University, Budapest, Hungary.

Non Hierarchical Mergeable Dialogs

Eric Blechschmitt

Fraunhofer-IGD
Fraunhoferstrasse 5
64283 Darmstadt
Eric.Blechschmitt@igd.fhg.de

Christoph Strödecke

Fraunhofer-IGD
Fraunhoferstrasse 5
64283 Darmstadt
cstroede@igd.fhg.de

Abstract

We describe a system which combines hypermedia with human computer dialogs to enhance the user interaction. The core of the system consists of an interpreter for a UIML compliant dialog description language. The interpretation of the described dialogs is used to handle user-interaction which is adapted to the available set of modalities provided by the used device. Furthermore, the system is designed to adapt an application as well on mobile and stationary devices allowing different interaction methods and metaphors for the same task. Examples for such adaptation can be performed with WIMP-based desktop systems, voice-menu based mobile phones or Personal Data Assistants (PDA) providing Point and Click metaphor. In the area of mobile end devices the adaptation of user interaction is still a challenging task due to complex applications in use. Control and navigation functions have a strong and demanding influence on the usability of applications. The here presented work shows an approach how navigation and control functions are improved by arranging the dialog steps in a hyper-structure and by assembling dialogs with a merge operation.

1 Introduction

With the increasing performance of mobile computing devices many new applications have become possible and seem to be an interesting field of application. To run such mobile applications there is the need for a standardized access to the functions of all anticipated devices. The functional part of such a standardization can be realized by using a middleware that runs on all operating systems which appear in the application scenario. Such an environment is given by the JAVA Virtual Machine which runs also on stationary and mobile devices. Though all problems seem to be solved, there is one challenge still persisting: the need for the adaptation of the devices' interaction methods which can be designed quite different in particular cases.
With regard to the humans' ability to adapt different situations for their communication we took the approach to use a dialog-oriented language to describe human computer interaction in an abstract manner and developed a basic prototype system (Blechschmitt & Strödecke, 2002). This prototype system has been successfully tested within the MAP-project (Kuhlmann, 2001).
In a natural conversation the exchanged information serves different purposes: beside the content, some information is exchanged to control the information flow. In a face to face situation the participants can use their voice primarily to communicate content information. But the flow of conversation can additionally be controlled by facial expressions and gestures. In fact content and control information can not be fully separated from each other. Facial expressions can both be used to transmit content and to control a conversation. However, no conversation can occur without control. Hence, the less modalities are available, the more control must be mapped into the

remaining modalities. This can be observed in case of a phone call, when content and control is relied on the same channel (i.e. voice).
The control of the information flow is also considered by human computer interaction systems and it appears also as navigation functions. Like the control part in natural communication the navigation allows humans to change the dialog to another topic or to stop or cancel a dialog.
This paper describes our approach to enhance human computer dialogs by control functions with hyperlinked navigation. The realized system is applicable for common user interface types: e.g. graphical user interfaces and text - or speech-based interaction systems.

2 Related Work

There exist a lot of specification for interaction languages for special requirements, such as the VoiceXML specification ("Voice extensible", 2000), the Dialog Description Language (DDL) (Göbel, Buchholz, Ziegert & Schill, 2001), XForms (Dubinko et al., 2001) and the User Interface Markup Language (UIML) ("User Interface", 2000). VoiceXML is made to encode speech-based dialogs without considering other modalities. It is menu-based and supports the inclusion of audio files to increase the impression. Events from other UIs can not be linked into the UI process. DDL describes dialog-based interaction, whereas a dialog is defined as interaction via web-based forms. All inputs are cached until the submission is initiated by the user. There is no further support for multimodal interaction. UIML is an XML-based language to describe UIs in general. The developer has to be aware of the intended devices and must specify a lot of device-dependent attributes. It provides no framework to support multimodal features. XFORMS is comparable to UIML, but contrary to UIML it reduces client-server interactions by direct support of input validation.

3 Dialog description

To encode the dialogs we have developed an XML-based abstract dialog description language which describes interaction from the view of natural human dialogs and which can be mapped to different devices and interaction metaphors.

Structure: Our developed dialog description language defines a dialog as consisting of three sections: The opening section, the middle section and the closing section. The task of the opening section is mainly to establish the communication between human and computer. This includes a salutatory and an introduction to the dialog. The middle section consists of dialog moves which can be composed of further dialog moves. Thus the dialog moves are arranged hierarchically. The tree structure of the dialog moves implies natural traversing strategies of the whole dialog. This was a lack of flexibility. Now, the hierarchy is extended by links among the moves. The links allow to arrange the moves in any desired structure. The main task of the closing part is to terminate the dialog properly and to allow to return into the dialog. This includes to summarize the interactions and to make corrections. It might be possible that some dialog moves are mandatory, but have currently not been processed, these moves are proposed to the user.
The order in which the user navigates through the dialog moves results in a complete traversal of the underlying graph. The user can freely choose among all dialog moves which are linked to the currently focused dialog move. Our system also suggests an automatically chosen dialog move as next step. This is done by graph theoretical algorithms for complete traversal of graphs (cf. traveling salesman, Hamilton paths). By enforcing a sequential navigation through the dialog we

solve the problem of synchronization between modalities with the property of sequential processing and modalities with the property of parallel processing (Nigay, 1995).

The navigation of the user through a dialog results in a path consisting of the visited dialog moves. This path is saved as history of processed moves. Beneath the possibility of navigating back, the history is a source for analyzing a user's behavior, e.g. his preferences, knowledge, desires, believes, interests etc.

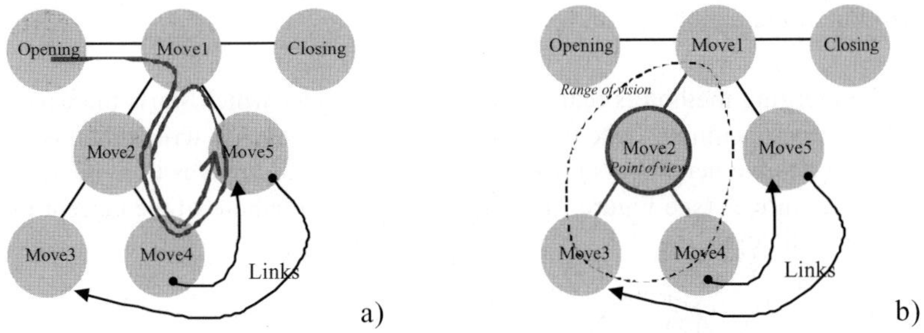

Figure 1: The extended dialog structure can produce cyclic behaviour a) which can be handled by limiting the range of vision from the current move b)

As a result of the arbitrary structure, the interpreter of the dialog must consider possible cycles in the structure of the dialog (see Figure 1a). For example, Graphical User Interfaces rely on a tree structure. We resolve this problem by introducing a *point of view* and a *range of vision* (Figure 1b). A point of view is represented by the current processed dialog move. The range of vision means that not all dialog moves are presented at once. Only the adjacent moves of the current move are presented for the navigation and all other moves are suppressed.

Merging dialogs: The knowledge which is represented by a dialog stems from a particular domain. If two dialogs belong to the same domain then they possibly will contain similar parts of interaction. To reduce the interaction effort we introduced the merge and split operations: two dialogs which result in one new dialog. The merge operation can be revoked by a split operation. The merge operation presumes that the two dialogs have at least one congruent move. The remaining parts of the two dialogs are then connected via the fused moves (see Figure 2).

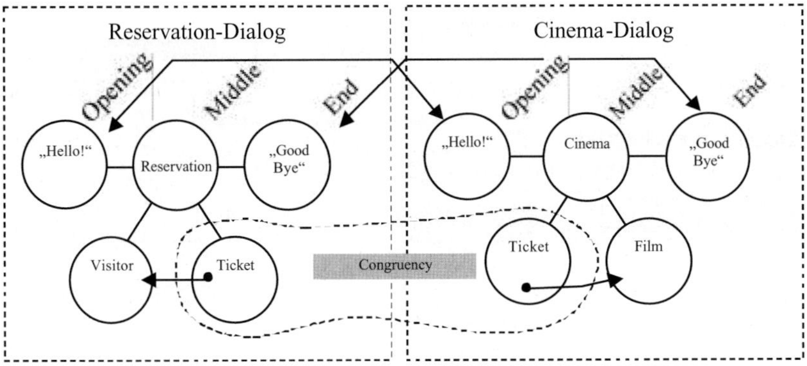

Figure 2: The reservation dialog and the cinema dialog own a congruent move. This ticket move can be used to merge the dialogs

If no congruent moves in the middle section of the two dialogs can be found, the dialogs are only connected via the opening section and the closing section, which are part of every dialog. The base for determining the congruence of particular dialog moves is given by the *is*-attribute which assigns the type of the move.

Sometimes it will be useful that the user can finish the dialog without having visited some dialog moves. E.g. if the unvisited dialog moves are not mandatory or if the user enters the whole dialog for the second time and if he has already provided the referring information previously. Therefore the dialog interpreter tries to complete unvisited dialog moves automatically, when the user enters the closing section of the dialog.

Realization: The merging method is realized by a visitor object which starts the execution of the dialog from the dialog opening. While processing the dialog moves it writes the visited moves to the history list. The visible neighbours are determined by the child moves of the current move and by the links to other moves (see Figure 3). In addition the *is*-attribute of the current move is used to add links to other dialogs.

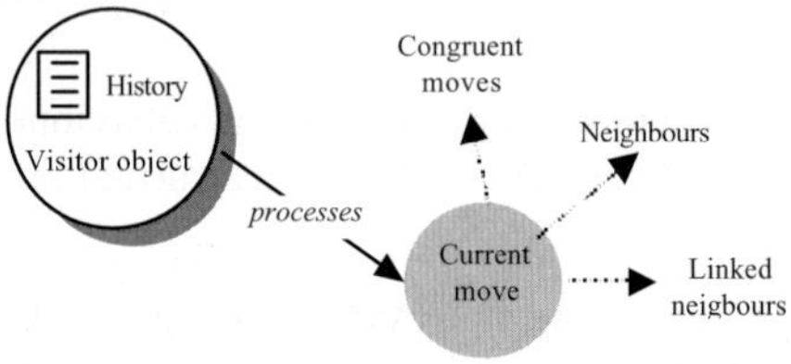

Figure 3: The visitor object processes a dialog move

4 An illustrative example

Figure 4 shows the above mentioned example dialog. This dialog is presented by an UI-Engine, which processes the dialog moves by using common graphic UI elements. From the opening section of the dialog the user can decide to branch into the reservation dialog or into the cinema dialog. Both dialogs provide a link to a ticket move, which is a merged move of both dialogs. It combines the visitor's name with the choice of a film the visitor wants to see.

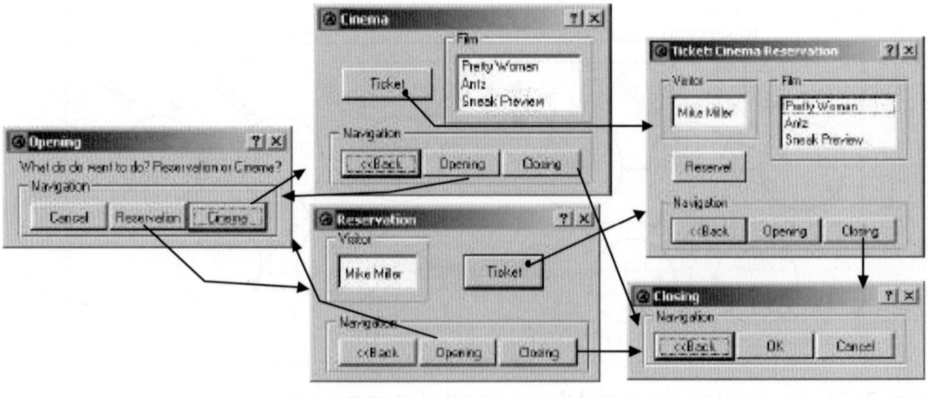

Figure 4: The merged example dialog

5 Conclusions

The presented approach uses aspects of natural conversation as a model for our mobile interaction management system. The model takes advantage of the hyperlinks which are used to support a better navigation. By providing the dialog components with semantic information, different dialogs can be merged together if they contain congruent parts. Even mobile applications can benefit from the presented features since the devices are designed differently in their interaction methods. The prototype shows the advantages of this approach:

- The navigation is enhanced while less dialog elements must be presented at the same time. Therefore the generated user interface needs less display space which is desirable when using mobile devices.
- By the use of an abstract dialog description language the described user interface can automatically be adapted to different devices with different modalities.
- The users' perception is discharged from overloaded user interfaces.
- Merging the dialogs offers an additional way to navigate trough dialogs. It can connect different applications if congruency is found.

The main problem of merging dialogs is the need for an accurate modeling process in the development phase of the dialogs.

References

Blechschmitt, E., Strödecke, C., 2002, An Architecture to provide adaptive, synchronized and multimodal Human Computer Interaction, Proceedings of the ACM Multimedia 2002, Juan les Pins, France

Dubinko, M. et al., 2001, XForms 1.0. W3C. 2001. Working Draft. Retrieved February 10, 2003, from http://www.w3.org/TR/xforms/

Göbel, S., Buchholz, S., Ziegert, T., Schill, A., 2001, Device Independent Representation of Web-based Dialogs and Contents, Proceedings of the IEEE Youth Forum in Computer Science and Engineering (YUFORIC'01). Valencia, Spain

Kuhlmann, H., 2001, Multimedia Workplace of the Future. Office, e-Business and eWork Conference (e2001), Venice, Italy

Nigay, L., and Coutaz, 1995, J. A Generic Platform for Addressing the Multimodal Challenge. Proceedings of the CHI'95, Denver CO

User Interface Markup Language (UIML) Draft Specification. Document Version 17, January 2000. Retrieved February 10, 2003, from http://www.uiml.org/specs/docs/uiml20-17Jan00.pdf

Voice eXtensible Markup Language (VoiceXML) version 1.0. May 2000, Retrieved February 10, 2003, from http://www.w3.org/TR/2000/NOTE-voicexml-20000505/

The Development of 'Hybrid' Multimodal Shopping Systems Within a 'Rapid Ethnographic' Methodology.

Stuart Booth[1], *Steve Westerman*[1], *Karim Khakzar*[2], *Thomas Berger*[3], *Hans-Martin Pohl*[4] & *Katarina Dubravcova*[5]

[1]School of Psychology
University of Leeds
Leeds LS2 9JT
stuartb@psychology.leeds.ac.uk

Abstract

Through the combination of traditional retailing and Internet-based commerce it may be possible to develop 'hybrid' multimodal systems that augment shopping environments in a ubiquitous manner, thereby providing tangible experiences along with flexibility in terms of access to the sales interface. However, to produce innovative designs that are effective, it is necessary to understand the user's perspective. Rapid ethnography provides a useful, time efficient technique through which this can be achieved. The procedure is illustrated through the example of ShopLab, a project aimed at the development of hybrid multimodal systems for traditional city-centre shops. Fieldwork carried out in a made-to-measure shirt shop in Germany is described, along with data analysis methods, including Cognitive Task Analysis (CTA). The results demonstrate the validity of the methodology in permitting an idiocentric understanding of users' needs. Implications for design of hybrid shopping systems are discussed, together with the utility of the rapid ethnographic procedure as a time-deepening approach for HCI in general.

1 Introduction

The growth of Internet shopping has resulted in significant retail change. However, there is potential to go further in integrating the 'virtual' with the 'real' to produce 'hybrid' shopping environments. Multimodal systems could be developed that augment sales interfaces in an interactive, ubiquitous manner. This might blur the boundaries between e-commerce and traditional retailing. 'Augmented commerce' of this nature has been explored by Brody & Gottman (1999), who developed hand-held devices to permit navigation of Internet-based information within physical shops. Customers were able to compare prices of a given store with those of on-line competitors. Future systems could go further through the use of advanced multimodal interfaces, adding a rich sensory quality to shopping experiences, whilst maintaining a high level of utility. To optimise their development, it is essential to gain an in-depth understanding of shopping contexts, including the thoughts and actions of users, together with their interactions. Although there are similarities between different retailers, individual sales

[2] University of Applied Sciences Fulda, Marquandstraße 35, 36039, Fulda, Germany
[3] inter.research, Am Alten Schlachthof 4 (ITZ), 36037 Fulda, Germany
[4] Institute fuer Digitale Medien und Kommunikation GmbH, Leipziger Straße 130, 36039 Fulda, Germany
[5] Houot Agencement, 43 Rue du Centre, BP 4, 88200 Saint-Nabord, France

contexts differ in important respects. It is therefore also important to pay consideration to situational variations. Customisable solutions can be achieved through a combination of ethnographic methods and HCI techniques. Ethnography has gained popularity in marketing due to its ability to provide anthropological insights into customer behaviour (see, for example, Sherry, et al., 2001). It therefore seems useful to adopt similar procedures in the design of hybrid shopping systems.

However, the term 'ethnography' is prone to misuse. Whilst definitions differ, it could be argued that its proper reference is to the development of an idiocentric understanding of human experiences within a given (usually social) context. Unfortunately, the time and resource constraints imposed by the design process mean that fieldworkers are seldom able to conduct the rigorous fieldwork this requires. Data collection, analysis and interpretation often take several months (Bently et al., 1992). In consequence, human factors are often considered relatively late in the design process. As Millen (2000) points out, HCI ethnographers therefore face the challenge of matching short product development cycles with the need to spend time on fieldwork. He advances a solution to this in the guise of 'rapid ethnography'– a time deepening strategy for HCI field research. The methodology comprises three key ideas (Millen, 2000):

- Firstly, the focus of the field research should be narrowed appropriately before initiating data collection in the field. Important activities should be identified and zoomed in upon. This might be achieved through the use of appropriate individuals acting as 'key informants'.
- Secondly, in order to identify exceptional and useful behaviour, multiple interactive observation techniques should be used.
- Thirdly, data analysis should take place through the use of collaborative and computerised data analysis methods.

Millen's procedure provides a time-efficient, yet powerful technique through which an ethnographic understanding might be established. In this paper, the utility of a rapid ethnographic approach is evaluated through the example of ShopLab (ShopLab, 2001), an EC IST-funded project aimed at the development of hybrid multimodal systems for traditional city-centre shops. The project has a particularly challenging design brief. It requires the development of innovative systems that can be applied within a wide variety of city-centre shops across different cultural contexts throughout Europe. The solutions offered must be customisable to support and enhance the distinguishing characteristics of individual shops. Like many projects, ShopLab requires coordination between a geographically dispersed, multidisciplinary and multinational design team. It therefore provides an interesting case for the application of rapid ethnography as a means of developing an understanding of shop-specific system requirements.

2 Method

2.1 Participating Shops

Two traditional city-centre shops were studied using Millen's (2000) rapid ethnographic methodology: a made-to-measure shirt shop in Lauterbach near Fulda, Germany, and a specialist sports shop in Remiremont near Epinal, France. These were selected to represent typical small businesses in Europe, being privately owned and providing personal, specialist services that contrast with the more impersonal approach catered for by their chain-store competition. For reasons of space, the present paper focuses on the work carried out in the shirt shop. However, the procedures used in the sports shop were the same and the results were of a similar nature.

2.2 Procedure

2.2.1 Narrowing the Focus of the Field Research.

Before data collection began the owner of the shirt shop was interviewed to narrow the focus of the research. In Millen's (2000) terms he therefore acted as a 'key informant' who was able to identify important aspects of his businesses that the fieldwork should focus on. The results of the interview indicated that customers often find it difficult to develop clear mental visualisations of shirts before they have been made. Any insights gained in this respect would therefore be useful.

2.2.2 User Observation Through Multiple Interactive Techniques

Having narrowed the research focus, it became possible to develop data collection procedures. Questionnaires were first administered within the store to develop a profile of its customer base. Twenty individuals were asked questions relating to gender, age, occupation, computer-use, etc. Descriptive statistics were compiled and participants were recruited to reflect the resulting profile.

Three fieldwork techniques were used: semi-structured interviews, video observation, and pen-on-paper observation. The interviews focussed on the process of buying a made-to-measure shirt, with emphasis on visual imagery from the customer's perspective. Seventeen individuals took part. For the video observation, digital cameras were installed and positioned to provide good views of product displays, customers and sales staff. An opportunity sample of four customers was videotaped as they went through the process of selecting a shirt. A researcher also made discrete pen-on-paper notes about their behaviour and that of sales staff. This permitted observation from the point at which an individual entered the shop to the point at which he/she left.

2.2.3 Data Analysis

Data were first processed in Fulda. This was necessary to translate materials from German into English. Due to time and resource constraints, the translations covered essential details of the interviews relating to the process of buying a shirt, with emphasis on visual imagery. Written descriptions of the movie recordings were also made, again including only the most important details of the interactions. The pen-on-paper observations were translated in the same way. These materials were transferred electronically to the University of Leeds, along with the raw data.

At Leeds a Cognitive Task Analysis (CTA) was produced in a similar manner to Millen's (2000) use of cognitive mapping techniques. The aim was to describe the typical process through which a business shirt is bought. The data were first analysed to extract themes and stages in the purchase process. Transcripts were broken down individually to form a step-by-step description of the purchase process. They were then combined iteratively in an overall analysis taking account of similarities and differences between individuals' perspectives. Hierarchical task descriptions of behavioural and cognitive operations were then developed. In so doing an ethnographic approach was adopted, emphasising the individual's interpretation of the social and environmental context.

3 Results

Given that the CTA covered the entire process through which a shirt is bought, there is insufficient space within the present paper to provide a full description of the results. However, it is possible to describe part of the analysis to illustrate the types of ethnographic insights gained.

At the top of the CTA hierarchy, Subtask 0 had the title 'Buy Outfit'. Giv en that the task analysis was intended to describe the process through which shirts are bought, it might seem surprising that it did not have the title 'Buy Shirt'. However, in the process of interviewing and observing customers, it became clear that shirts are rarely bought without consideration of the overall outfit they will be worn with. Indeed, when buying formal shirts, most individuals reported following a sequence of actions aimed at selection of an outfit. Within this process the selection of a shirt is dependent upon prior activities. Generally speaking, customers first select a suit (subtask 2) or, if they already possess one, bring an image of this to mind. Having done so, only then are they able to think about which shirts might be appropriate (subtask 3). Once a shirt is selected accessories are chosen, such as ties, handkerchiefs and cufflinks, etc. (subtask 4). It can therefore be seen that purchase of a shirt is a sub-element of a more general process.

4 Discussion

The CTA results have implications for design of hybrid shopping systems. As the example above demonstrates, the rapid ethnographic methodology provides idiocentric insights that might not otherwise be gained. Such understandings are valuable given their potential to encourage innovation. For instance, on the basis of its fieldwork, the ShopLab team is developing designs supporting visual imagery, with an outfit as the starting point, rather than a shirt. Although the shop does not stock suits, the system would allow customers to view them, with opportunities for configuration to match personal clothing items. Shirts can then be chosen to match.

There are many other ways in which a rapid ethnographic approach may improve development of both hybrid shopping systems and more general HCI design. System functions might be tailored to particular circumstances and innovative concepts derived that would not be possible without an understanding of users' personal perspectives. The technique is likely to be particularly useful in the requirements capture phase of development. Given the short time-scales involved, the technique therefore has potential to make ethnographic understandings available to design practitioners, even when operating under tight constraints.

In spite of these advantages, it remains important to better integrate rapid ethnography within the design process. At present there is a danger that the research focus might be narrowed at too early a stage. For example, although key-informants may provide essential insights, their views may not reflect the wider experiences of the general population. As a result, fieldwork might fail to take individual differences into consideration. In the case of ShopLab there is the possibility that the shop owner's views produced an over emphasis on visual imagery. It would therefore be useful to develop a framework in which rapid ethnography is formally integrated within an existing development model, whilst adopting procedures that avoid problems relating to sample bias. However, this might prove difficult given the time constraints development projects typically face.

Other questions relate to the practicalities of participatory design management. HCI practitioners often face problems in terms of acceptance within development projects due to the latter's tendency to focus on engineering concerns. Rapid ethnography may be useful in this respect. It provides a time efficient means through which user experiences can be described and related to design concerns in a way that is understandable for a multidisciplinary audience. When combined with other techniques it can help funnel development from a relatively broad conceptual beginning to more specific concerns as it progresses. As has been demonstrated in the case of ShopLab, representational methods, such as CTA, are useful in this respect. However, the project has also found it necessary to use additional communication tools to engage users in its participatory design process. For example, design storyboards might be useful at a conceptual stage, moving towards semi-functional rapid prototypes later on. Such techniques build upon the base provided by rapid ethnography, ensuring efficient and effective user-centred design.

5 Conclusion

Hybrid multimodal shopping systems have potential to combine the benefits of traditional retailing and Internet-based commerce within augmented shopping environments. However, it is important for developers to generate innovative designs that pay attention to the user's perspective. The rapid ethnographic procedure provides a time-effective technique through which this can be achieved. It has the ability to provide idiocentric insights that might not otherwise be obtained, leading to increased opportunities for successful innovation. Due to the efficiency with which it can be implemented and the value of the results produced, the rapid ethnographic approach might act as a successful time-deepening strategy for HCI practitioners wishing to understand users' personal perspectives, even in the face of ever tightening time constraints. However, the approach might benefit from formal integration with an existing design model, consideration of individual differences and combination with other communication tools for design.

Acknowledgements

Work jointly funded by the European Union 5th Framework and the government of Switzerland.

References

Brody, A. B., & Gottsman, E. J. (1999). Pocket BargainFinder: A Handheld Device for Augmented Commerce. *First International Symposium on Handheld and Ubiquitous Computing (HUC '99)* (Karlsruhe, Germany, 1999).

Bently, R., Hughes, J. A., Randall, D., Rodden, T., Sawyer, P., Shapiro, D., & Somerville, I. (1992). Ethnographically informed systems design for air traffic control. *Proceedings of the Conference on Computer Supported Cooperative Work* (Toronto, November, 1992), 123-129.

Millen, D. R. (2000). Rapid Ethnography: Time Deepening Strategies for HCI Field Research, *Symposium On Designing Interactive Systems* (New York City, 2000), 280-286.

Sherry Jr., J.F., Kozinets, R.V., Storm, D., Duhachek, A., Nuttavuthisit, K., & Deberry-Spence, B. (2001). Being in the Zone: Staging Retail Theater at ESPN Zone Chicago, *Journal of Contemporary Ethnography*, 30 (4), 465-510.

ShopLab (2001). ShopLab homepage. Retrieved February 5, 2003, from http://www.shoplab.info.

A rhetorical model to augment the functionality of adaptive interfaces

Licia Calvi[1]

Dept. of Computer Engineering and Systems Science
University of Pavia
licia.calvi@unipv.it

Abstract

In this paper, we propose a rhetorical model to within-node argumentation that is based on the notion of adaptation to the user and of the construction of an implicit user to adapt to. We show how this can result in a better way of delivering information in personalised Web sites.

1 Introduction

In most current adaptive interfaces, users are modelled according to a set of predefined categories that Brusilovsky (1996) has summarised as being:

- The user's goal
- The user's preferences in terms of interaction style
- The user's background and hyperspace experience
- The user's knowledge level in the content domain of the application.

In a way, the user's goal may be considered as equivalent, or very closed, to the user's motivation why interacting with that particular application, and it is inherent in the nature of the application itself. So, for instance, the user's goal when interacting with an educational application is to learn the content that is modelled in that system, and also the motivation why doing so. In an information retrieval system, instead, the user's goal is to identify and retrieve information, but also her most visible and apparent motivation[2]. Actually, however, such adaptive systems most often limit themselves to modelling the user's knowledge level, which is probably the most straightforward parameter the system can monitor.

If we apply this strategy to Web sites, in particular to those ones offering a personalised information delivery, we soon realise that this way of modelling the user is by far too coarse-grained.

From Jakobson's model of the possible functions a statement may present[3] (1960), several theories[4] have been delevoped that ascribe to Web sites one of the following four functions[5]:

---

[1] Licia Calvi is also affiliated with the University of Parma, Dept. of Italian.

[2] Whereas the most inner and deep reason may be different. In this sense, as for the goal, the user's motivation may be distinguished into a *local* (or shallow) and a *global* (or deep) motivation.

[3] That is: an expressive, a conative (as in B2C sites), an informative, a poetical, a metalinguistical (as in the W3C site), and a contact function.

[4] See, for instance, in (Loeber, 2002).

- To persuade users
- To entertain users
- To instruct users
- To inform users.

A way to augment the functionality of those Web sites, by enabling them to match their users' profile in a more fine-grained way, may therefore be based on a rhetorical model. Such a model foresees a one-to-one correspondence between the function the site performs (among those four mentioned above) and the motivation why the user is exploring that site.

In this paper, we propose such a rhetorical model as applied to the most commonly accepted framework for adaptive interfaces (De Bra et al., 1999). The perspective that we adopt is not really to analyse how the system (i.e., the Web site) should monitor the user's motivation and behave accordingly, rather how it should put together information in a way that matches both the user's motivation and the function the site performs. It will be a proposal for a model for argumentation that can ultimately result in a better way of delivering information in user-adaptive Web-based systems.

The paper is structured as follows: Section 2 introduces the problem of argumentation in hypertext. Section 3 presents our rhetorical model. Section 4 draws some conclusions and indicates the necessary future work.

2 Argumentation in Hypertext

Argumentation in hypertext systems may be approched at two different levels: at the level of the information inside one single node (*within-node argumentation*), and at the level of the information that is spread over several nodes (*between-node argumentation*): this last one clearly corresponds to the navigational path proposed (i.e., enforced) or suggested by the author.

About between-node argumentation, several works already exist[6].

Within-node argumentation seems to be more problematic (unless one intends to resort to the standard argument theory[7]), because of the low argumentative character nodes present, where the content is often limited to simply introducing new notions. To overcome this limitation, we propose a rhetorical model to within-node argumentation that is derived from the New Rhetoric paradigm developed by Perelman and Olbrechts-Tyteca (1969). Within this paradigm, in order to understand a sentence, it is important to take into account the context in which the sentence has been uttered and the implications its interpretation may have as a countereffect on the environment, not simply the sentence taken in isolation. In a way, a sentence is seen as a "mediating structure" in Hutchins's terminology (1995). For the New Rhetoric, then, the relevance of sentence context and the effects of its interpretation on the environment, i.e., a receiver, rely on argumentations. Argumentation implies that the "new" orathor adapts her speech to the receiver in order to be able to have some effect on her, because "argumentation is concerned with what is supposed to be accepted by the hearers" (Perelman and Olbrechts-Tyteca, 1969, p. 65).

Two basic ideas are at the basis of this new rhetoric:

---

[5] Lughi (2001), for instance, develops a taxonomy of Web sites that is based on the empirical observation of their sender identity. Such a taxonomy includes: personal, associative, commercial, and insitutional sites. Although it is possible to ascribe each function illustrated above to each one of these sites (namely, personal sites will most likely perform a poetic function; associative sites an expressive function; commercial sites a conative function; and institutional sites an informative function), many sites might simply be multifunctional.

[6] For instance, (Bernstein, 1998).

[7] As Locke (2000), for instance, does.

1. The speaker must adapt to the audience (in order to have some effect on it).
2. The audience varies depending on the role (or function) it plays.

If we project these principles on Web sites and on the four functions we identified earlier, we recognise a similar functioning. If we consider[8], indeed, a site as a macrotext, whose syntax is based on the notion of standby, we realise that its style strictly depends upon the communication situation it clearly represents (and that is defined by the functions mentioned earlier). In this context, the user plays an active role in the standby syntax enacted by the site, by showing a behavior that is consistent with both the site identity and its function. Although, indeed, there might be multifunctional sites, they all must present what Lughi (2001) calls a *semantic coherence*, i.e., they must deal with only one subject matter. The univocity of its content must be perceived both from the sender side (i.e., *topic*) and from the user's side (i.e., *isotopy*).

So, in synthesis, the model we propose is based on the following reference framework (Fig.1):

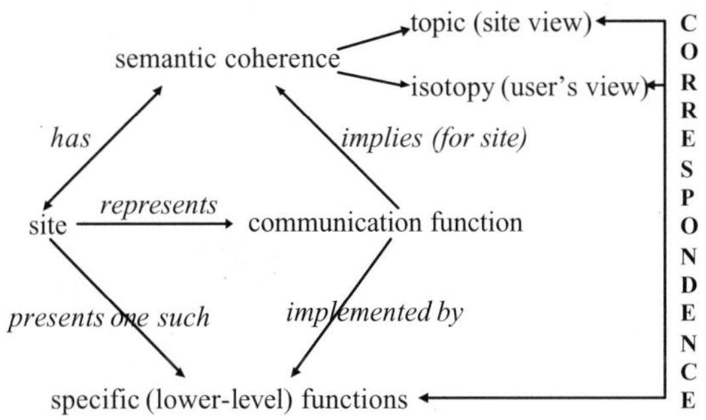

Figure 1: Theoretical reference framework for our model

3 A Proposal for a Rhetorical Model

The rhetorical model we propose relies on the notion of adaptation to the user and of the construction of an implicit user to adapt to, i.e., a user that is implied by the function the site performs and whose implied profile is mirrored in the isotopy. This model specifies how arguments should be built within one single node (i.e., within-node argumentation) in order for the system to match the function the site performs to the user's motivation why visiting that site. Because of the correspondence isotopy-communicative function we mentioned above (Fig.1), we conceive the form the argument may take as a function of the content the site presents.

In what follows, we outline how the four communicative functions mentioned before could be expressed rhetorically, given the context of use and the user's implicit profile (Fig. 2):

- *To persuade:* to fulfil this communicative function, the information in the site should be assembled using rhetorical devices such as demonstration, repetition and anticipation.
- *To entertain:* to fulfil this communicative function, the information in the site should be assembled using rhetorical devices such as allusion, quotation and apostrophe.

[8] As Lughi (2001), for instance, does.

- *To instruct:* to fulfil this communicative function, the information in the site should be assembled using rhetorical devices such as oratorical definition, periphrasis (and other related rhetorical figures like synecdoche or metonymy), interrogation and, as with to persuade, but clearly with a different meaning, repetition.
- *To inform:* it seems more difficult to clearly match this function with rhetorical figures, because it somehow lies at the intersection of the other three functions i.e., at the intersection between persuading and entertaining/instructing. However, this function could be rhetorically realised by means of anticipation and quotation, which assume a different meaning compared to their usage for persuading and for entertaining.

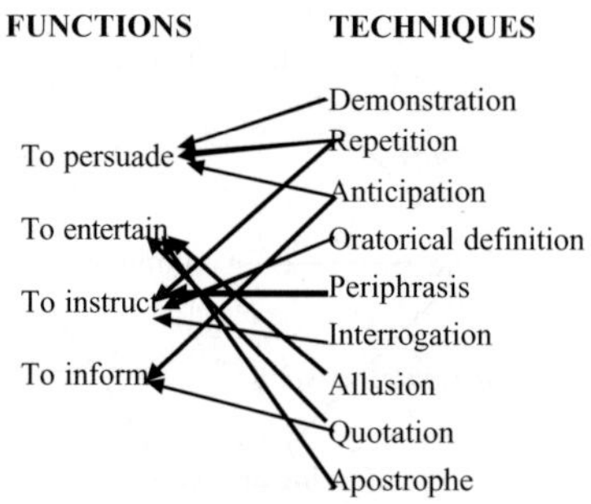

Figure 2: Correspondence between communicative functions and rhetorical techniques

What we claim with this model is that, when assembling information to construct a new Web page, the system must be able to answer the following questions:

- What is the communicative function the page expresses?
- Who is the implicit user that such a function entails? What are her characteristics?
- How do pieces of text need to be assembled in order to comply both with the above mentioned functions and with its implicit user?

Only if the system takes these parameters into account, the resulting Web page will be constructed in a way as to satisfy the user's expectations, as to comply with the motivation why she is visiting that page, i.e., it will be adapted to the user it implies. This will bring about two additional positive effects: it will make the Web site sufficiently attractive for new users (among those with a relevant implicit profile) and it will allow the site to keep them, as long as it will be built in a way as to match their implicit profile, i.e., their motivation.

4 Conclusion and Future Work

In the paper, we have presented a model for information delivery in personalised Web sites that is based on the notion of adaptation to the user and of the construction of an implicit user to adapt to.

We have shown the existence of a correspondence between the communicative function the site expresses and the motivation of the user why visiting that site (i.e., isotopy). This model complies with both these terms by exploiting rhetorical devices, i.e., by assembling information on the same Web page in a way that is consistent with both the site communicative function and the user's isotopy.

Only by answering the questions reported above, the resulting Web page will be constructed in a way as to match the (implicit) user's motivation why visiting that page, and, ultimately, to meet her expectations. So, it will be adapted to the user in a more comprehensive way, because it will be argumentative[9].

This model represents however only the first step towards the realisation of a finer-grained adaptation to the user. In the immediate future, related work will include:

- The definition of a formalism (for instance, a logic) to realise the translation of these rhetorical tools into a self-contained, well-constructed Web page.
- The verification that its semantics is computable.
- The implementation of this formalism in a, possibly already existing, adaptive system.

After this first series of improvements, additional work may consist of:

- The passage from an implicit to an explicit user, by defining a number of devices within the system to monitor the actual user.
- The definition of which characteristics of the user to focus on to implement the rhetorical model and to deal specifically with the user's motivation.

References

Bernstein, M. (1998). Patterns of Hypertexts. *Proceedings of the Ninth ACM Conference on Hypertext '98*, 21-29.

Brusilovsky, P. (1996). Methods and techniques of adaptive hypermedia. *User Modeling and User-Adapted Interaction,* 6 (2-3), 87-129.

De Bra, P., Brusilovsky, P., & Houben, G.-J. (1999). Adaptive Hypermedia: From Systems to Framework. *ACM Computing Surveys*, 31 (4es).

Hutchins, E. (1995). Cognition in the wild. Cambridge: The MIT Press.

Jakobson, R. (1960). Closing Statements: Linguistics and Poetics. In T. A. Sebeok (Ed.), *Style in Language.* Cambridge: The MIT Press.

Locke, M.C. (2000). Arguments in Hypertext: A Rhetorical Approach. *Proceedings of the Eleventh ACM Conference on Hypertext '00*, 85-91.

Loeber, S.G. (2002). Modelling the audience in a Web-based presentation: context-driven, rhetorical role-playing approach. *Proceedings of the Eleventh European Conference on Cognitive Ergonomics* (ECCE11), 251-258.

Lughi, G. (2001). Parole on line. Dall'ipertesto all'editoria multimediale. Guerini e Associati.

Perelman, C. and Olbrechts-Tyteca, L. (1969). The New Rhetoric: A Treatise on Argumentation. Indiana: The University of Notre-Dame Press.

[9] "An efficacious argument is one which succeeds in increasing (this) intensity of adherence among those who hear it in such a way as to set in motion the intended action (...) or at least in creating in the hearers a willingness to act which will appear at the right moment" (Perelman and Olbrechts-Tyteca, 1969, p. 45).

Designing Auditory Spaces: The Role of Expectation

Priscilla Chueng *Phil Marsden*

University of Huddersfield
Queensgate, Huddersfield HD1 3DH, UK
{p.chueng, p.h.marsden}@hud.ac.uk
http://scom.hud.ac.uk/scompc2

Abstract

This paper investigates aspects in designing auditory spaces to support novel forms of interaction in virtual spaces. Initial research on human imagined sounds from places has identified 'expectation' as an important psychological construct. Instead of designing realistic experience, the paper suggests a user's sense of presence as a measure of the user's experience in virtual spaces. The results indicate that using highly expected sounds increases users' sense of presence. As such, it is to propose that designing auditory spaces using expectations as perceived affordance for presence is perhaps a minimal way to engaging users experience.

1 Users Experience in Virtual Spaces

We have a tendency to imitate reality for virtual experience. These virtual environment designs are based on real life metaphors that register our senses as important aspects to optimise the virtual experience. However, perceptual realism attained through accurate perspective projection may not always be the best approach for engaging users' experience. Taking examples from cartography that intentionally distorted to exaggerate features (Monmonier, 1991), the design of virtual experience can be very different from those happening in our everyday world. Moreover, the modern technology affords a novel way of interaction (Gaver, 1991). The concept of 'presence' is has becoming well accepted as the key concept to redefine virtual environments. Presence is defined as the perceptual illusion of "being there" in a mediated environment (Lombard & Ditton, 1997). This illusion of non-mediation occurs when user fails to perceive or acknowledge the existence of a medium in the environment and responds as if the medium were not there.

2 Auditory Spaces

Although vision tends to be the dominant sensory channel, it is reported that auditory cues are important to the establishment of a 'full' sense of presence in virtual environments (Gilkey & Weisenberger, 1995). The ecological approach to auditory perception (Gaver, 1993a & 1993b) notes that although we understand scientific descriptions of auditory perception, phenomenally we don't 'hear' acoustic signals or sound waves but instead we hear events: the sounds of people and things moving, changing, beginning and ending, forever interdependent of the dynamics of the present moment. We 'hear' the sound of silence for example. We hear the semantics of sound producing objects, events and the environment. Ecological approach is the central approach of this research project.

Acknowledging the importance of auditory channels, there has been an increasing activity in the area of auditory display research. This research project focuses on the sound generated from surroundings, which has been termed ecological sound, environment sound, sound-scapes, ambient sound or everyday listening sounds, respectively. We aimed to find out psychological aspects influencing sound scapes in places that is enough to convey characteristics of the place for a sense of presence and engaging users' experience in virtual environments.

3 Empirical Works

The research aims to understand the users' perceptual responses to auditory cues, when they are combined with the understanding of context awareness for interaction in virtual spaces, and how it can be used to support a sense of presence in virtual spaces. The research will provide virtual environment designers with guidelines for designing sound for virtual spaces that will provide engaging users' experience.

3.1 Sounds we imagine to hear in places

3.1.1 Imagination study

We studied user's interpretations of experiencing real life auditory spaces in places. An open interview was carried out to investigate the sounds which people imagine they can hear in four different real life places: a pub, a supermarket, a high street and a park. Results from the initial study (Chueng, 2002) showed participants tended to report sounds according to the context of the place from sound producing objects and events, such as, 'people talking', 'people shouting', the' wind' etc. (Gaver, 1993a & 1993b). From these results, it is to propose that – *expectation* (Chueng, 2002a) can be the important perceived affordances of sound producing objects that contribute to users' experience. *Expectation* defines the extent to which a person will expect to hear a sound in a particular place.

3.1.2 Online Survey

The online survey aimed at gathering quantitatively ratings of the range of highly expected sound producing objects and events reported from the initial study. The place chosen for this study was the 'Pub'. For each sound listed, 20 participants of age 18 to 40 years from the UK rated their expectation and discrimination by answering two questions related to the Pub. (Chueng, 2002c). No time restriction was placed on participant before stating their answers. For each sound event listed, Participants were asked to answer "How frequently do you expect to hear each of these sounds?" and rated in high, moderately high, moderate, moderately low, low, not at all.

We then compared the reported ratings from imagination study to ratings from online survey. T-tests shows that there are no significance differences between expectation ratings from imagination study and online survey ($t=-0.395$, $df=9$, $p= 0.702$); 2). As such, the online survey has proven the validity of the proposed expectation as important roles on people sounds imagined in places from initial study.

3.2 The Effect of Expectation on People's Sense of Presence

Expectation as a dimension for sound design is not new for film sound designers (Bordwell & Thompson, 1996). For example, the dinosaur's foot steps in the movie "Jurassic Park" was created by recording the sound of a heavy weight being dropping from a great height.

From previous studies, expectation is identified as an important construct to be considered in sound design. The aim of this experiment was to investigate if the expectation of sounds in places, influence people's sense of the presence. During the experiment, participants' mental models are prompted by still image of places while listening to audio clip recorded from real life places. Initially, four places from the initial study were chosen for this experiment. : Pub, Supermarket, High Street and Park. However we encountered difficulty in recording clarity of sound from a park. It was therefore decided that train station was chosen to replace the Park and the sound of Park was used for familiarisation of participants at the beginning of each experiment.

3.2.1 Hypotheses

The main hypothesis for this experiment is that a person's sense of presence will increase with high expectation. This is achieved by displaying matching visual images together with matching audio clips. Secondly, correct responses of people's memory task collected from places-related questions of both sounds and visuals are expected to be higher with high expectation. Thirdly, low expectation (mismatched audio-visual) will result people's irritation and hence participants' parasympathetic reading will be higher.

3.2.2 Method

The experiment uses a two factorial within subjects design. The dependent variables are (1) total weight collected from sense of presence questionnaire (rank with the scale of five). (2) Number of correct responses collected from places-related questions of sounds and visuals. 3) Electro dermal reading.

During the experiment, the facilitator uses a Multimedia Pentium PC computer to play back sequence of audio and visual images through Macromedia Director. Images were projected to a white screen in a darken room using a digital projector whilst sound clips were played back using headphones make by Sennheiser model eH2270. To reproduce realistic sound, binaural recording (Gilkey, 1997) methods were used for recording. A pair of mini microphones made by AKG model 417pp was attached to both ears of the sound engineer who stands in the middle of the places to imitate real life hearing of participants. Sound clips with length of ten minutes were digitised to 44khz and 16 bit from a DAT tape to a Cd-ROM. Sound clips were played back using headphones make by Sennheiser model eH2270.

Twenty participants took part in this experiment. They were a mixture of students and staff with ages of 22 to 50 from the School of Computing, Engineering and Music at the University of Huddersfield. None had hearing problems and all either had normal vision or wore prescribed corrective lenses. Participants are exposed to matching or mismatching visual stimuli while listening to sound clips recorded from real life places.

They are asked to 1) record their sense of presence by answering a sound related presence questionnaire, 2) answer place-related questions from each sound listened to and each image viewed. 3) During every audio-visual display, their readings of parasympathetic responses of arousal are recorded using psycho-physiological sensors.

Each participant takes approximately twenty minutes to finish viewing five audio-visual displays. Audio-visual stimuli of the park were used for familiarisation purpose. Following each display, they answer nine sounds-related presence questionnaire taken from SVUP (Larsson, Västfjäll & Kleiner, 2001) with five- scale rating, and places-related questions (three taken from information of sound clip and three taken from information of visual displayed) using paper and pen. Parasympathetic responses of Galvanic Skin Resistance (GSR) and Blood Volume Pulse (BVP) are captured using Datalab 2000 and software Biobench.

3.2.3 Results

The overall mean of presence score is higher for matching audio-visual than mismatching audio-visual. t-test ($t=5.990$, $df=78$, $p<0.00$) shown that presence scores for matching audio-visual is significantly different between matching stimuli and mismatching stimuli. Thus, the user's sense of presence increased with high expectation (matching stimuli). Low expectation with mismatched audio and visual stimuli resulted in a lower sense of presence. Although the overall mean of memory task reponses is higher for higher expectation, the result is however in significance ($t=0.746$, $df=78$, $p=0.458$). Participants' memory tasks are not affected by exposing to mismatching stimuli.

We also takes average of each participants GSR reading for both match and mismatch. Preliminary analysis with t-test shows insignificance result ($t=-0.068$, $df=78$, $p=0.946$). Exposing to mismatched audio-visual stimuli did not caused changes in participants' parasympathetic responses. The preliminary analysis on the physiological data has shown significance result on presence but insignificance results for memory tasks reponses and parasympathetic measurement. This can be resulted from 1) some degree of dependency on types of images and sounds displayed, 2) Memory task questions were too easy, 3) individual differences in GSR readings. Further work in these analysis will be undertaken.

4 Overall Conclusions

The aim of the work described in this paper was to investigate aspects influencing sound that people imagine to hear in real life places, and take the outcome to guide designing sound in virtual environments for an effective users experience.

The imagination study qualitatively investigated what extent people imagine hearing sound from places. Expectation is identified as constructs on how people imagine hearing sound from places. The online survey further investigated and proven the validity of the imagination study. We continued with an experiment investigated the effect of expectation on people's sense of presence. When expectation of a place is met while hearing auditory spaces, people's sense of presence (perceptual illusion of being there) is encouraged. This shows that expectation can be taken as an element to design sound in virtual environment to provoke effective users experience.

5 Discussion

The empirical works have demonstrated the possibility of designing contextually less realistic with expected auditory spaces for an effective user's experience in virtual environments. In fact, sound has been under used as a medium for novel interaction in virtual environments. The current trend of designing sound in virtual environment has two extremes: 1) Imitation of reality by designing high fidelity realistic sound. 2) The use of low resolutions MIDI for network transmission. The former produces sound file digitally big in size but the latter tends to provide less convincing auditory experience. This research aimed to find a solution to designing sounds that is enough for engaging experience without losing audio quality. Taking that we expect to hear certain sound events in places and high expectation contributes to higher sense of presence, it is plausible to employ the extent of 'expectation' as the perceived affordance to be considered while designing sound in a virtual space to achieve an effective user experience.

6 Future Work

Future work involves further investigations on expectation as the perceived affordance to design sound for virtual environments to support users' sense of presence. The investigations will take place in virtual spaces built in active worlds (activeworlds, 2000). An expectation study in virtual environments based on navigation task and presence questionnaire is planned. It is to follow by an evaluation study will conclude the research project by building virtual auditory spaces in virtual environments based on the outcomes gathered from our previous studies and experiments. The research will provide guidelines for future virtual environments developers with regard to designing minimum auditory space to support virtual places for an effective user experience.

7 References

Active Worlds, Online communities website. Retrieved 2000, from http://www.activeworlds.com.

Bordwell, D. & Thompson, K. (1996). Film Art: An Introduction. McGraw-Hill.

Chueng, P. (2002). Designing sound canvas: The role of expectation and discrimination. *Extended proceedings of CHI'012 Conference on Human Factors in Computing Systems*. 848-849. ACM Press.

Chueng, P. (2002). Minimal ecological sound design: the issues of presence in virtual environments. *Extended proceedings of British HCI 2002.* London.

Chueng, P. (2002). Online questionnaire of sounds in pub. From http://www.supersurvey.com/cgi-bin/surveys/s10492.pl

Gaver, W. (1993). How do we hear in the world? Explorations in ecological acoustics. *Ecological Psychology*, 5(4), 285-313.

Gaver, W. (1993). What in the world do we hear? An ecological approach to auditory event perception. *Ecological Psychology*, 5(1), 1-29.

Gaver, W.W. (1991). Technology affordances. *In CHI '91, Human factors in computing systems conference proceedings on Reaching through technology*. 79 – 84. ACM Press.

Gilkey, R. H. & Anderson, T. R. (1997). Binaural and spatial hearing in real and virtual environments. Mahwah, NJ: Lawrence Erlbaum Associates.

Gilkey, R.H. & Weisenberger, J.M. (1995) The sense of presence for the suddenly deafened adult, *Presence*, Vol. 4, No. 4. 357-363.

Larsson P., Västfjäll D. & Kleiner M. (2001). The actor-observer effect in virtual reality presentations. *CyberPsychology & Behavior*, 4(2), 239-246.

Lombard, M., & Ditton, T. (1997). At the heart of it all: The concept of presence. *Journal of Computer Mediated Communication.* 3(2). http://www.ascusc.org/jcmc/vol3/issue2/lombard.html.

Monmonier, M. (1991). How to lie with Maps, Chicago: University of Chicago Press.

Blending Speech and Touch Together to Facilitate Modelling Interactions

Joan De Boeck, Chris Raymaekers, Karin Coninx

Limburgs Universitair Centrum
Expertise center for Digital Media
Universitaire campus, B-3590 Diepenbeek, Belgium
{joan.deboeck,chris.raymaekers,karin.coninx}@luc.ac.be

Abstract

Over the last years, multimodal interfaces are steadily gaining importance. It has been proven that multi-modal interfaces advance the user's behavior in 3D applications such as modeling applications. In this context, force feedback is a modality that already has demonstrated its benefits when directly interacting with virtual objects. On the other hand, voice controlled interfaces recently appear to become more and more popular. It is clear that both modalities, force feedback and speech recognition, have their benefits and their drawbacks. Starting from a representative subset of tasks from a modeling application, this paper describes an interface proposal that allows us to find an appropriate combination of those both modalities. This proof of concept application will be verified by an informal usability study, which also demonstrates the actual user's behavior when more than one modality is available.

1 Introduction and Related Work

It has been proven that multi-modal interfaces improve the users' experience in virtual and 3D environments. Especially, haptic feedback is useful when interacting with virtual objects since it can help to determine whether an object or a user interface element is actually touched or not (Massie & Salisbury, 1994)(Chen, 1999). However, some drawbacks do exist: most haptic devices only provide a limited workspace. Additionally it remains difficult to spatially locate an object or user interface element to interact with. Several solutions already exist to overcome this problem, but these are not always feasible (Giess, Töpfer & Meinzer, 2000) (Raymaekers, De Boeck & Coninx, 2001).

Speech recognition, on the other hand has the potential to free the user's hand by controlling the environment by voice alone. Nevertheless, the mental load of remembering all voice commands, and the requirement to pronounce complex phrases without hesitation, as well as the overall error prone are common weaknesses (Cohen, 1992). Although plenty of work has been done in evaluating multimodal interfaces (Bolt, 1980) (Oviat, 1999) (Sturm, Bakx, Craenen, Terken & Wang, 2002), not much can be found on the combination of force feedback and voice controlled commands in a 3D environment. This paper elaborates on a multimodal interface proposal, concretised on a modelling task, which allows us to find an appropriate combination of both modalities. In this work we restrict the interaction to a *sequential multimodal*[1] interaction. The results of this research should provide us with a sound basis to progress to *simultaneous multimodal*[2] applications.

In a first paragraph, we will explain the experimental set-up; we describe our hardware configuration, our interface proposal and task subset, and finally the additional head tracking. In a next section, we will give more details about the user experiment. Subsequently, the results of the experiment will be discussed in detail. We will end up this paper with our conclusions and future work.

2 Experimental Set-up

Force feedback (and thus direct manipulation) and voice controlled command both have their advantages and disadvantages. In this work, we will try to find an appropriate blend of both modalities in such a manner that both modalities become complementary for a modelling task.
Since most modelling tools draw on a similar set of commands, we have tested only a small subset of common modelling tasks in order to assess the usefulness of both haptics and speech. Although a lot more modelling actions certainly do exist, we believe this command subset will enable us to make a good evaluation of the combination of speech and force feedback.

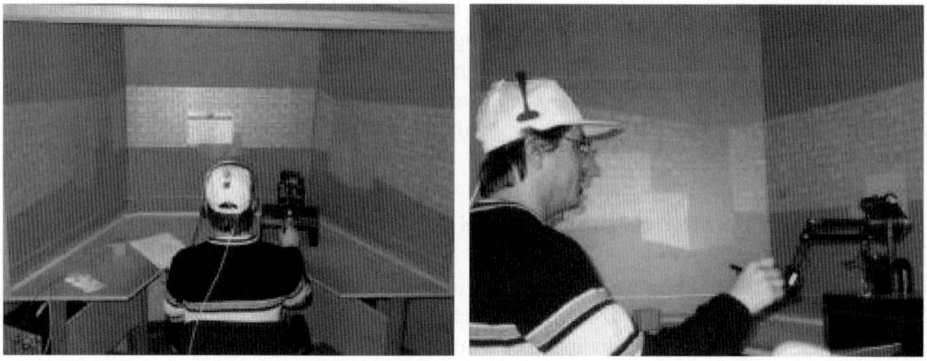

Figure 1: Experimental set-up

2.1 Hardware set-up

The experimental set-up includes a Phantom 1.0 device, mounted in our "Personal Surround Display" (PSD), a personal workspace with 3 synchronized non-stereo projection screens each 90 cm wide and located with an angle of 60 degrees (see Figure 1). The projection transformation in the rendering pipeline of both side screens are adjusted in such a manner that all three screens

[1] In a *sequentially multimodal* interaction, two subsequent actions are executed using different modalities and do not overlap in time (Sturm et al, 2002).
[2] In a S*imultaneous multimodal* interaction, two subsequent actions are executed using different modalities and do overlap in time (Sturm et al, 2002).

have the same projection centre[3]. Working within this PSD results in a wide working space, which provides the user with a semi-immersive environment. Dependent on the task, the PSD's large dimensions force the user to make larger cursor movements with the haptic device, which in this case motivates the addition of speech input.
The speech recognition module uses the Microsoft Speech API and runs on the same machine that supports the force-feedback simulation.

2.2 Task with speech and haptics

The main task in this research application is to select and deselect objects and apply textures to them. To investigate the appropriate blend of both modalities, all commands can be performed by multiple modalities. Commands can be performed by directly picking them from a haptic hybrid 2D/3D menu (Raymaekers & Coninx, 2001) and by pushing against objects in the virtual world. Alternatively, commands also can be given by using voice commands. For instance, an object can be selected by enabling the selection mode from the menu and touching the object. As an alternative, an object (in this example the left cube) can be selected by saying the phrase "Select left object". Combinations of both interaction methods are possible as well: e.g. the menu can be activated by voice, but the commands within the menu can be selected with the haptic device. It is clear that this test application only implements a very limited command set of a modelling tool. Obviously, the smaller this set, the more efficient the speech-modality will turn out. Therefore, we have provided our command set with a sufficient number of textures and a sufficient number of alternative expressions, to achieve a certain load for the subject's short-term memory.

2.3 Head tracking

Since the application, running in our PSD, establishes a large workspace, it is clear that the user has to make relatively large movements with the PHANToM device. To counter this problem and to establish a better cooperation between both modalities, head tracking has been introduced. In our application, a tracking device (a Polhemus Fastrak tracker), mounted on a cap has been used to get the orientation of the user's head. This information is solely used to get the direction in which the user's head is turned. The projection transformation is not affected by the tracker's data. When enabling the menu (invoked by a speech command), the tracker's information is used to show the menu in the user's region of interest. We believe this not only avoids large cursor movements to access the menu, but also limits the user's workload since the menu is brought to the user and must not be sought after.

3 Experimental Task

To determine which part of the task will be performed by which modality, an informal usability test has been conducted. Five subjects, all male with no or little experience with speech recognition systems and force-feedback, were asked to perform a predefined set of tasks on three cubes. Before starting the experiment, each person was allowed to practice for 5 minutes to get used to the environment and to learn the vocabulary. Next, the test consisted of simple tasks (such as selecting an object or setting a texture) and combined tasks (such as setting the texture of a specific object). During the test, each command, succeeded or failed, spoken or pointed, was manually logged. Finally, a questionnaire was filled-up by the test persons.

[3] Note that the walls in Figure 1 appear to be "broken" because the pictures are not taken in the projection centre.

Two hours after the first test, the same subjects were asked to perform another trial, with similar but slightly different assignments. The second test had to be executed without extra practice to minimize short-term memory influence.

4 Results and discussion

4.1 First trial

From the first trial, we can conclude that speech is used as the main input modality. Almost 80% of all commands are vocal. The haptic interaction is rather used as a backup when a speech command is not recognised for a number of times. In our survey, most users reported that speech recognition works comfortable and "surprisingly" adequate, although still 25% of the spoken commands are not recognised. In contrast, direct manipulation has an error rate of only 9%. However, subjects report difficulties accessing the menu which subjectively makes the haptic interaction to appear slower. We also have reported those common problems with haptic interaction in our previous work (Raymaekers & Coninx, 2001) (Raymaekers et. al, 2001).
Although users have extensively practiced the spoken command set, we also can conclude that long commands with more variables are more error prone. The command *"Set left object's texture to wood"* failed in nearly 78%, due to mispronunciations or hesitations. Even the phrase *"Select left object"* failed in 38%. Shorter commands, e.g. to change modes, were much more reliable.

4.2 Second Trial

Two hours after the first trial, subjects were asked to perform another set of instructions, this time without a practicing session. In this second trial we clearly see the effect of the short-time memory: some speech commands are avoided, or fail more often because users hesitate or don't remember the exact wording. The effect is more pronounced with longer commands. The most complicated command *"Set left object's texture to wood"* has only been tried once, without success. Also the command *"Set texture to wood"* was "remembered" in a lot of variations, which resulted in a succeed ratio of only 17%. Hence, haptic interaction has been used more intensively. The haptic menu-command to change the object's texture has been used twice as many times as in the first experiment. From our questionnaire, however, we can conclude that our subjects still prefer verbal interaction, but they report that remembering the command set is a main problem. Therefore, the menu more often is activated as a reminder (20 times against 5 times). In this case the feature to show-up the menu in the region of interest has been evaluated pretty well.
If we take in account the individual behaviour of the subjects between both trials, we see that two users behave in the same way in both experiments. Two users incline to interact more hapticly in the second experiment. Finally one user tries to use more speech commands, but falls back to haptics for some commands, because of too many speech recognition errors.

5 Conclusions and future work

This paper describes an interface proposal, concretised on a modelling task, in which we have sought for an appropriate blend of two modalities, speech and haptics, in such a manner that both modalities become complementary for the task. We can conclude that speech is the preferred modality, although roughly 25% of the commands failed. Haptic interaction was reported to be a valid backup when spoken commands were unsuccessful. When eliminating short-term memory effects, in a second trial, spoken commands become less accurate and users more often substitute on the "backup" modality. This is certainly true for longer commands with more variables. Even then, users still prefer the oral interaction.

Since the size of the menu-items, and the lack of an adequate depth sight, appeared to make the haptic modality slower and thus less popular, we think about enlarging the size of the menu. When activating the menu by voice, we also consider actively move the PHANToM in front of the menu, while still tracking the users' head and showing the menu in the region of interest.

Ultimately, we can conclude that this work can be considered as a promising step towards a multimodal modelling environment with an appropriate cooperation between force feedback and speech input. We expect the benefits of multimodal interaction to be even more pronounced when modalities can be used simultaneously.

6 Acknowledgements

We would like to thank the test persons for their cooperation with the usability test.
Part of the work presented in this paper has been funded by the Flemish Government and EFRO (European Fund for Regional Development).

7 References

Bolt, R.A. (1980) "Put-That-There": voice and Gesture at the Graphics Interface, In SIGGRAPH '80 Proceedings, volume 14, 1980, 262-270

Chen, E. (1999) Six Degree-Of-Freedom Haptic System as a Desktop Virtual Prototyping Interface" Proceedings of First International Workshop on Virtual Reality and Prototyping", June 1999, Laval, FR, 97-106

Cohen, P. R. (1992) The Role of Natural Language in a Multimodal Interface, Proceedings of User Interface Software Technology (UIST'92) Conference, Academic Press, Monterey, California, USA, 143-149

Giess, C., Töpfer, S., & Meinzer, H.P. (2000) Can shadows improve haptic interaction in virtaul environments?, Proceedings of the 2nd PHANToM Users Research Symposium 2000, Selected Readings in Vision and Graphics, vol. 8, July 6-7 2000, Zurich, CH, 1-8

Massie, T.H, & Salisbury, J.K. (1994) The PHANToM Haptic Interface: A Device for Probing Virtual Objects, Proceedings of the ASME winter Annual Meeting, Interfaces for Virtual Environments and Teleoperator Systems, November 1994, Chicago, IL, USA

Oviatt, S. (1999) Ten Myths of Multimodal Interaction, Communications of the ACM, vol. 42(11), November 1999, 72-81

Raymaekers, C., & Coninx, K. (2001) Menu Interactions in a Desktop Haptic Environment" Proceedings of Eurohaptics 2001, July 1-4 2001, Birmingham, UK, 49-53

Raymaekers, C., De Boeck, J., & Coninx, K. (2001) Assessing Head-Tracking in a Desktop Haptic Environment, Proceedings of HCI International 2001, August 6-10 2001, New Orleans, LA, USA, 302-306

Sturm, J., Bakx, I., Cranen, B., Terken, J., & Wang, F. (2002), The Effect of Prolonged Use on Multimodal Interaction, Proceedings of ISCA Workshop on Multimodal Interaction in Mobile Environments, Kloster Irsee, Germany

When Marketing meets HCI : multi-channel customer relationships and multi-modality in the personalization perspective

Alain Derycke, José Rouillard, Vincent Chevrin

Laboratoire Trigone - CUEEP – Bât. B6
Université des Sciences et Technologies de Lille
59655 Villeneuve d'Ascq Cedex - France
{alain.derycke, jose.rouillard}@univ-lille1.fr
v.chevrin@ed.univ-lille1.fr

Yves Bayart

3 Suisses International
La Cité Numérique
245 rue Jean Jaurès
59491Villeneuve d'Ascq
France
ybayart@citenum.com

Abstract

This paper presents a first investigation on new forms of interaction not yet approached broadly by the HCI community. It's the study of the convergence and divergence between the personalizations of interaction seen under two complementary angles: direct marketing, with its multi-channel customer relationships, on the one hand, and HCI view on the other hand. Our aim is to understand this, both theoretically and experimentally in order to design new interactive system for E-Commerce that exhibit a better relationships, such as continuity, between the organisation and the customers. This is achieved by developing a theoretical framework for the analysis and design of such systems and by designing and experimenting a first prototype exploring the potential of integrating voice and Web interactions.

1 Introduction

The multi-modality has been studied since many years in the HCI field, and personalization has been used and studied in marketing for a long time, but seldom, both were studied jointly. However in spite of some similarities between multi-modality, as seen in HCI, and multi-channel, as seen in interactive marketing, there are still differences that are not only due to the viewpoint put of the stakeholder (interactive user or customer) but also due differences into temporal aspects and interaction strategies which are underlying. This field of investigation is relatively new for the HCI community so we decide to start pragmatically by proposing a first framework for the analysis and design of new prototypes using the potential of the multi-modality of the Web technologies such as VoiceXML. This work is rooted into the large experience gained by one of the organisation supporting it, a large Direct Marketing company, 3 Suisses International, which have experimented, or maintained in production, various channels and interactive technologies in its relationships with its customers. This experience have leaded us to study this domain with the goal to have a better understanding of the interaction process through a combination of channels and to derive some new approaches for the future of interactive systems, more integrated and "intelligent".

2 E-commerce, multi-channel customer relationships and personalization

In direct marketing focused on the individual customer, it is not true that development of E-Commerce means the exclusive use of the Web channel. Indeed there are already well established relationships with the customer through a diversity of communications channels (or medium) such as Call Centres for phone, Audiotel and videotext servers, Web, WAP, SMS… Nevertheless Direct Marketing companies have developed also some forms of personalised relations with their customers, and are going toward a One-to-One marketing strategy (Peppers & Rogers, 1997). This have been amplified by the support of the personalization processes into E-Commerce solutions where, thanks to the technologies (Fink & Kobsa, 2000), it is more easy now to apply this kind of personalization, not only based on the direct knowledge of the customers (through her/his passed purchases or actions and from direct filing of some forms) but also indirect knowledge, for example by inferring from user interactions through the HCI. Our view of the E-Commerce reflects the central place of the knowledge about the customer not only as a user of the interface, with his/her preferences and skills, but also as a long term customer relationship, in order to augment or to maintain her/his loyalty or fidelity. We are closed to the definition of the E-Commerce given by (Holsapple & Singh, 2000) "*E-commerce is an approach to achieving business goals in which technology for information exchanges enables or facilitates execution of activities in and across value chain as well as supporting decision making that underlies those activities*". In putting emphasis on the decision process, informed by knowledge learned from the past relations and interactions with the customer, the history, those authors shown also that this knowledge management process is not concerned only by traditional information (*know that*) but also by more procedural knowledge (*know how*) or inferential knowledge (*know why*). It appears to us that this knowledge capture must use all the opportunities of interaction with the customer whatever the channel he/she uses for that purpose. So all the relationships can be seen as a "learning relationship" where both the parties (seller and customer) co-operate, more or less, in order to maintain a mutual understanding of the transaction and to achieve their respective satisfaction.

Our experience, gained from a large base of user relationships with a multi-channel and personalization approach, have shown us, however, two main problems that our research program want to solve:

- One relative to a poor integration of the different channels into a common "*infostructure*", an engineering problem avoiding the good articulation of the channels inside a same transaction;
- And a second relative to the quality of the relationship with the customer, viewed from the marketing viewpoint. Especially it appears that, during a business transaction which can last a long time, the commutation between the different channels (i.e. from Web to phone) can lead to a lost of the business opportunity for the seller and to a frustration for the client. In the HCI words the system is not "*Seamlessness*".

3 A theoretical framework for the analysis of the multi-channel marketing interactions

Our starting point, for our new program of research and the collaboration between HCI and Marketing researchers, is that it exists a proximity between the problem of multi-channel interaction and the multi-modal one, and that the personalization perspective is shared both by the Marketing and HCI communities, even if there are differences either in the aims of this approach or in the temporal grain of the phenomena. The research is done in two directions:

- First, on the elaboration of the theoretical framework in order to analyse and to model multi-channel interactions on a more large view, but still compatible, than those used for the study of multi-modality in HCI. The foundations are taken into the field of the theory of language seen as co-ordination process for the sharing of a common knowledge, and the Grounding concept (Clark & Brennan, 1991). The Grounding provides an analysis framework because it allows characterisation of the various channels in accordance with their temporal and physical constraints. We used also the Activity Theory in order to explain the aforementioned cooperation process between the customer and the selling company, reusing our previous experience for the design of evolving Groupware systems (Bourguin, Derycke & Tarby, 2001).

- Second, an experimental approach which required the set-up of a technological platform and the design of a prototype for exploring the range of problems, both of usability and implementation, of a combination of multi-channel interactions and multi-modal ones. Multi-modal interaction have had a new interest in the Web technology with the advent of natural language interactions favoured by the adoption of the VoiceXML standard and availability of tools to support it.

4 The Prototype of an E-commerce Multi-modal, multi-channel site

Our first prototype already in operation is limited to near synchronous interactions and the available channels are limited to Web and telephone (equivalent to an Audiotel service). The technologies used and how is solved some problems such as synchronisation and articulation of channels are described in (Rouillard, 2002). "*One of the biggest problems in the way of multi-modal progression is synchronisation.*" (Breitenbach et al. 01). We are currently conducting two investigations with this prototype: one relative to an evaluation of the real use with a light personalization derived from users profiles and some elements capture during the session; the other on the extensions of our prototype in order to enlarge the support of others channels (SMS, WAP) and to support more asynchronous articulation of these channels.

We consider that it could be relevant to use HTML and VoiceXML alternatively, within the same application, and why not, synergicly, that is to say, in the same time. We would like to see if multi-channel and multi-modal interactions are useable on E-commerce systems. Indeed, it is illusory to believe that because a system is multi-modal, it will be inevitably used in a multi-modal way by the users (Oviatt, 1999).

The E-commerce application that we propose to develop is a kind of remote shop that sales clothes and others goods, on the Web. The server machine is a PC equipped with Windows NT and IIS, MS-Access, and IBM Websphere (for VoiceXML). The importance of the multi-modality is easy to understand at this point of the interaction. Indeed, if you are not already a client, and you want to become one, you will be asked to give some information such as your first name, surname, address, etc. If a speech recognition system can more or less recognize what you said (particular grammar and vocabulary are used), it is very difficult, nay impossible, to catch what you said if the linguistic domain is not limited. That's why we decided to propose to the new clients to identify themselves in a traditional web page, as we can see on Figure 1.

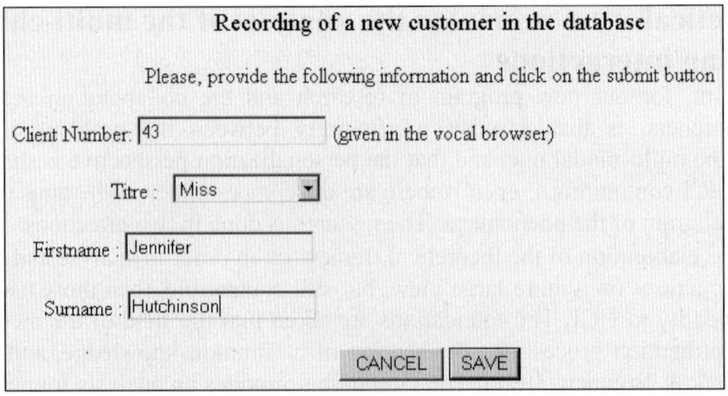
Recording of a new customer in the database

Please, provide the following information and click on the submit button

Client Number: 43 (given in the vocal browser)

Titre : Miss

Firstname : Jennifer

Surname : Hutchinson

CANCEL SAVE

Figure 1 : The information given in this web page will be used by the voice browser

From this moment, the information given by the user on the web page are usable in the entire application, including the phone side. This is possible thanks to dynamic grammars. After the welcome message of the computer, in a first step, the user gives her client number to the system, and if this number matches with a number in the database, the machine asks the user for a password. If it's correct, the user is now identified with a title, a first name, and a surname. The Figure 2 shows an example of oral dialogue between Miss Jennifer Hutchinson (which was unknown a few minutes ago) and the machine. In the oral interaction, the machine asks the user to go to a traditional web page and to complete a form (see Figure 1). At this moment, the synchronization process takes place, and the oral side is waiting for the data coming from the textual and graphic side. If the registration number used by the consumer is correct, the interaction continues on an oral mode.

10:32:15.000 C: Welcome on our web site. Are you already one of our clients?
10:32:26.671 H: no
10:32:26.843 C: We are going to create your client account. In order to register yourself, please give in the Web page value 43 as client number. Did you correctly hear your client number?
10:32:41.468 H: yes
10:32:41.484 C: Ok. I'm waiting while you are completing the Web page.
10:32:41.485 C: (waiting)
10:32:47.093 C: Your registration is correct Miss Jennifer Hutchinson. Would you like to buy something now?
10:32:56.671 H: yes
10:32:56.828 C: What do you want to buy Miss Hutchinson?
10:33:06.640 H: some dresses (…)

Figure 2 : The oral dialogue uses information coming from the textual Web page

In the same way, output synchronization is possible between graphical and vocal interfaces within the same application.

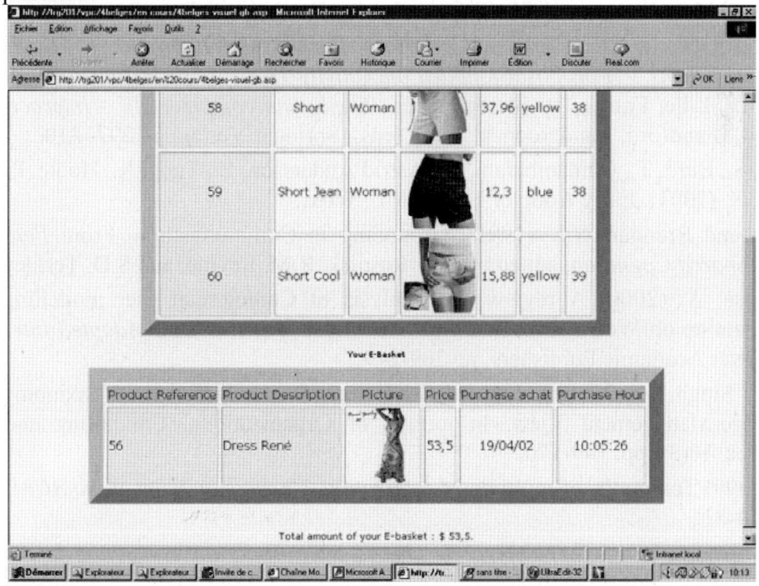

Figure 3 : The web page gives automatically the results of an oral request

The Figure 3 shows how a traditional web page gives the results of an oral user's request. This page is automatically refreshed every 5 seconds (for example), and retrieved the current data useful for the user to choose a product in the virtual shop. So this multi-modal interaction allows seeing on the screen some information that the vocal browser cannot display, like the photos of the goods, in our example.

5 Conclusion

We have defined a theoretical framework that is useful for the classification of the different channel or media used in the interaction in the framework of E-Commerce. We have also conducted a systematic comparison of the potential of combining two channels, for example start with a Web interaction alone followed by a phone interaction (in parallel or in sequence), and determine some combination which are good candidate for a multi-modal like interaction. The first prototype, articulating a natural language interaction, based on the VoiceXML technology, and a Web interaction have shown us that technically, it is possible to couple and synchronize heterogeneous channels in order to give similar or complementary information to the customer: "*In the forthcoming years, two factors will jeopardize the deployment of Web applications: supporting multi-device outputs and one-to-one personalization.*" (Bonifati, Ceri, Fraternali & Maurino, 2000). We believe that a generic architecture will allow strong coupling/decoupling processes. It will be possible to begin on a channel, and to switch seamlessnessly on another one. That's why we are interested in modelling temporal information for this kind of interaction. Our present work is done in two directions: continue to enlarge the model of the channel characterisation in order to support new media such as Multimedia phone, and redesign of the first generation of Audiotel (automatic phone answering machine) in order to support both a better personalization and the possibility to combine this modality with another one such as the Web or the SMS.

6 References

Bonifati, A., Ceri, S., Fraternali, P., Maurino, A. (2000) *Building Multi-device, Content-Centric Applications Using WebML and the W3I3 Tool Suite,* ER2000 International Conference on Conceptual Modeling, pp. 64-75, Springer-Verlag Berlin Heidelberg.

Bourguin, G., Derycke, A., Tarby, J.C. (2001) *Beyond the Interface: Co-Evolution inside Interactive Sytems – A proposal founded on Activity Theory*. Proceedings of IHM-HCI 2001 conference, Lille, France, 10-14 september, *People and computer XV – Interactions without Frontiers*, Blandford, Vanderdonckt, Gray (eds), Springer Verlag, pp 297- 310.

Breitenbach, S., Burd, T., Chidambaram, N., Astrid Andersson, E., Tang, X., Houle, P., Newsome, D., Zhu, X. (2001). Early Adopter VoiceXML, Wrox.

Clark, H.H., and Brennan, S.E. (1991). Grounding in Communication. From *Perspectives on Socially Shared Cognition*, edited by L.B. Resnick, R.M. Levine, and S.D. Teasley.

Fink, J. Kobsa, A. (2000) A review and analysis of Commercial user modelling server for personalization on World Wide Web. In *User Modelling and User-Adapted Interaction*, vol. 10, , Kluwer Academic Publishers, pp 209-249.

Holsapple, C. Singh, M. (2000) Electronic Commerce: from a Definitional Taxinomy Towards a Knowledge-Management View. In journal of Organizational Computing and Electronic Commerce, 10(3), pp. 149-170.

Oviatt, S., (1999) Ten myths of multimodal interaction. *Communications of the ACM*, Vol. 42, N. 11, , p. 74-81.

Peppers, D., Rogers, M. (1997) *Enterprise One-to-one : Tools for competing in the interactive age*. Doubleday Publishing, NY.

Rouillard, J. (2002) *A multimodal E-commerce application coupling HTML and VoiceXML*, The Eleventh International World Wide Web Conference, Waikiki Beach, Honolulu, Hawaii, USA.

Temporal Context and the Recognition of Emotion from Facial Expression

Rana El Kaliouby[1], Peter Robinson[1], Simeon Keates[2]

[1]Computer Laboratory
University of Cambridge
Cambridge CB3 0FD, U.K.
{rana.el-kaliouby, peter.robinson}@cl.cam.ac.uk

[2]Engineering Design Centre
University of Cambridge
Cambridge CB2 1PZ, U.K.
lsk12@cam.ac.uk

Abstract

Facial displays are an important channel for the expression of emotions and are often thought of as projections or "read out" of a person's mental state. While it is generally believed that emotion recognition from facial expression improves with context, there is little literature available quantifying this improvement. This paper describes an experiment in which these effects are measured in a way that is directly applicable to the design of affective user interfaces. These results are being used to inform the design of *emotion spectacles*, an affective user interface based on the analysis of facial expressions.

1 Motivation

Facial displays are an important channel for the expression of emotions and are often thought of as projections or "read out" of a person's mental state (Baron-Cohen et al., 2001). It is therefore not surprising that an increasing number of researchers are working on endowing computing technologies with the ability to make use of this readily available modality in inferring the users' mental states (Colmenarez et al., 1999; Picard and Wexelblat, 2002; Schiano, 2000).

To design the *emotion spectacles*, an emotionally intelligent user interface, we examined the process of emotion recognition in humans. Whereas most existing automated facial expression recognition systems rely solely on the analysis of facial actions, it is theorized that humans make additional use of contextual cues to perform this recognition (Bruce and Young, 1998; Ratner, 1989; Wallbott, 1988). Existing literature, however, falls short of quantifying the type or amount of context needed to improve emotion perception in a form applicable to computer interface design. In this paper, we investigate the effect of temporal facial-expression context on emotion recognition.

2 The Emotion Spectacles and Related Work

The *emotion spectacles* are designed to identify and respond to a user's affective state in a natural human-computing environment. This could be integrated into a wide range of applications ranging from ones that respond to user frustration, ones that gauge the level of user engagement during an online learning task, and virtual salesmen that learn an online shopper's preferences from his/her reaction to the offered goods.

The process with which the *emotion spectacles* infer a user's mental state is two-fold and involves facial action analysis followed by emotion recognition. For the analysis stage, we use dynamic facial action analysis, which identifies and analyzes facial motion in video sequences using feature point tracking. The emotion recognition stage however, has received little attention to date with the exception of Calder et al. (2001) and Colmenarez et al. (1999) who classify facial actions into a number of basic emotions. Our goal in *emotion spectacles* is to automatically infer a wider range of emotions. We propose combining the analyzed facial motion with other contextual information in order to do so.

3 Putting Facial Expressions in Context

A number of studies indicate how humans make considerable use of the contexts in which expressions occur to assist interpretation (Bruce and Young, 1998). Ratner (1989) shows that the reliability of facial expressions as indicators of emotion is significantly improved when they are perceived in relation to contextual events instead of in isolation. Wallbot (1988) found that the same expression perceived in varying contexts was judged to indicate different emotions.

In order to quantify the effect of context, we follow Edwards' (1998) description of an expression of emotion as a sequence of concurrent sets of facial actions, or micro expressions. In addition, we define temporal facial-expression context as the relationship between consecutive micro expressions. It is temporal, because it represents the transition, in time, between two consecutive micro expressions. It is context, because micro expressions are judged to indicate different emotions when perceived with respect to preceding ones, versus in isolation (Edwards, 1998).

In this paper, we examine the effect of temporal facial-expression context on the recognition accuracy of both basic and complex emotions. Basic emotions include happy, sad, angry, afraid, disgusted, surprised and contempt. Emotions such interest, boredom, and confusion, on the other hand, involve attribution of a belief or intention–a cognitive mental state–to the person, and are hence referred to as complex (Baron-Cohen et al., 2001).

4 Experiment

The goal of this experiment was to investigate the effect of temporal facial-expression cues on the recognition accuracy of both basic and complex emotions. We specifically addressed the following questions:

1. To what extent does temporal context have an impact (if any) on recognition of emotions from facial expressions?
2. Is temporal facial-expression context equally effective in improving (if at all) the recognition accuracy of both basic and complex emotions?
3. What relationship exists between the amount of context and degree of improvement, whether this relationship has critical inflection points and whether it tapers off with additional context becoming irrelevant?

4.1 Video Material

Material used throughout the experiment was developed using videos from "Mind Reading", a computer-based interactive guide to emotions (Human Emotions Ltd., 2002). The resource has a total of 412 emotions organized into 24 groups. Six video clips are provided for each emotion showing natural performances by a wide range of people. We picked 16 videos representing 13 complex emotions (such as confused, enthusiastic, and undecided), and 8 videos representing 6 basic ones (such as afraid, angry, and disgusted), for use throughout the experiment. The duration of the videos vary between 3 to 7 seconds (mean= 5.24, SD= 0.45).

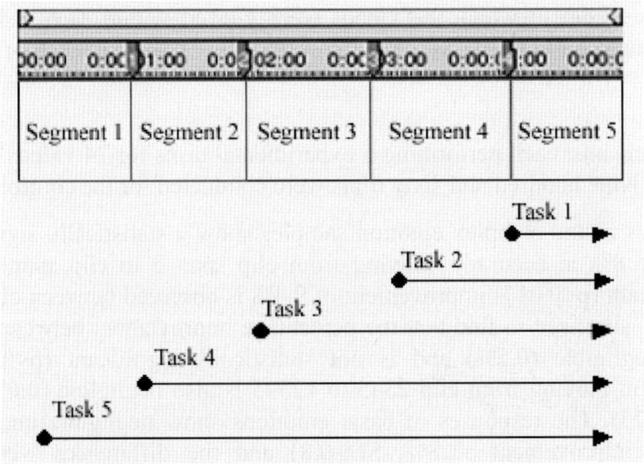

Figure 1 The process of clip construction showing how segments of a video are joined to form the clips used during each of the five experimental tasks.

Each video was divided into five separate segments, such that each segment is composed of essentially different micro expressions. The segments were then joined to construct the clips used during each of the 5 experimental tasks. The process of clip construction is illustrated in Figure 1. The last segment of every video (segment 5) is shown during the 1st task. Clips for the 2nd task are constructed by combining the 4th and 5th segments of a video and so on.

4.2 Experimental Variables

Two independent variables were defined. Clip span (5 conditions) defines the span of a video clip, and emotion category (2 conditions) could either be basic or complex. Accuracy of recognition measured in percentage of correctly identified emotions was the dependent variable.

4.3 Participants

36 participants between the ages of 19 and 65 (mean= 31, SD= 11) took part in the experiment, 21 males and 15 females. Participants were either company employees who covered a wide range of occupations or university research members (mostly computer science). Participants were of varied nationalities, but all had substantial exposure to western culture. All participated on a voluntary basis.

4.4 Experimental Tasks and Procedure

We designed a total of six tasks, five experimental and one control. The experimental tasks were designed to test the effect on recognition accuracy of the last segment of each video, when gradually adding earlier segments. The control task tested that for each emotion, none of the five segments played a greater role in "giving away" an emotion.

During each of the tasks, participants viewed 24 video clips of varying length and were asked to identify the underlying emotion. A forced-choice procedure was adopted, where three foil words were generated for each emotion (for a total of 4 choices on each question). All participants were asked to carry out all five tasks making this a within-subject repeated measures study. This set-up minimizes differences in the task responses that can be attributed to varying emotion-reading abilities of the participants. A typical test took 40 minutes on average to complete, and generated 120 trials per participant. Tasks were carried out in increasing order of clip length to prevent any memory effect. The order in which the videos were viewed within each task was randomized. During the control task, participants were randomly shown segments of each of the 24 videos.

5 Results

Twenty-eight participants, each performing 5 experimental tasks for 24 videos produced a total of 3360 trials in total. Nine hundred and sixty trials were conducted for the control task.

A pair-wise analysis of the complex emotion samples show a statistically significant ($p<0.0001$) improvement of 25.8% in accuracy, moving from clip span 5 to clip span 45. A smaller but statistically significant ($p<0.017$) improvement of 9.4% is observed between clip span 45 and clip span 345. We were surprised to find that the percentage improvement between clip span 345 and 2345 is almost negligible (0.5%) and is not statistically significant ($p<0.9$). Similarly, the improvement seen in moving from clip 2345 to 12345 is also negligible (0.6%) and statistically insignificant ($p<0.75$). The responses of basic emotions show negligible improvement between clip spans (mean improvement =2.8%, SD=0.8) and the differences were not statistically significant. Analysis of the results from the control task showed no statistically significant difference ($p<0.2$) between the recognition accuracy of any of the 5 segments of both the basic and complex emotions.

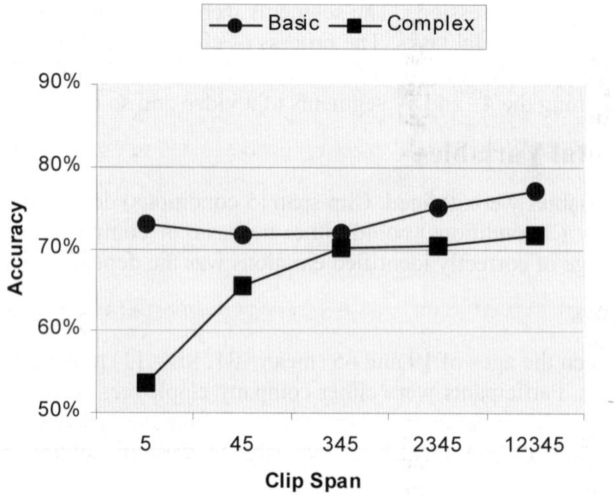

Figure 2: Effect of clip span on recognition accuracy in the case of basic and complex emotions.

The pattern of improvement in accuracy is summarized in Figure 2. In the case of basic emotions, the recognition is nearly constant. In the case of complex emotions, a small amount of context yields a pronounced improvement in recognition accuracy. This correlation however is not linear: as more temporal facial-expression context is added, the percentage of improvement tapers off.

Although we predicted a pronounced improvement in the case of complex emotions, we had also anticipated some impact, even if less obvious, in the case of basic ones. We were somewhat surprised that there was no significant change in accuracy to report during the experiment.

6 Implications on the Emotion Spectacles

Our experimental results have significant implications for the design of affective user interfaces, especially those based on facial affect. To start with, the pronounced improvement of accuracy associated with the addition of temporal facial-expression context suggests integrating context in the design of emotionally intelligent interfaces. Whereas prevalent classification methodologies treat every micro expression in isolation of preceding ones, we suggest that classification should use the result of prior inferences in recognizing the current expression. Our findings in this experiment suggest that the two immediately preceding micro expressions are responsible for all of the statistically significant improvement. This has favourable implications on the complexity from a design point of view.

7 Conclusion

Our work presents a multidisciplinary study on the effect of context on the recognition accuracy of both basic and complex emotions. We show that a relatively small amount of temporal facial-expression context has a pronounced effect on recognition accuracy in the case of complex emotions, but no corresponding effect is seen in the case of basic emotions. The results are used to inform the design of *emotion spectacles*, a facial-expression based affective user interface currently under development, but can also be utilized in the design of embodied conversational agents, avatars, and in computer-mediated communication.

References

Baron-Cohen, S., Wheelwright, S., Hill, J., Raste, Y., and Plumb, I. (2001) The "Reading the Mind in the Eyes" Test Revised Version: A Study with Normal Adults, and Adults with Asperger Syndrome or High-functioning Autism. *Journal of Child Psychology and Psychiatry,* 42 (2), 241-251.

Bruce, V. and Young, A. (1998) In the Eye of the Beholder: The Science of Face Perception. Oxford University Press.

Calder, A.J., Burton, A. M., Miller, P., Young, A.W., Akamatsu, S. (2001) A principal component analysis of facial expressions. *Vision Research,* 41, 1179-1208.

Colmenarez, A., Frey, B., and Huang, T. (1999) Embedded Face and Facial Expression Recognition, *International Conference on Image Processing.*

Edwards, K. (1998) The face of time: Temporal cues in facial expression of emotion. *Psychological Science*, 9, 270-276.

Human Emotions Ltd. (2002) Mind Reading: Interactive Guide to Emotion. http://www.human-emotions.com

Picard, R.W., & Wexelblat, A. (2002) Future Interfaces: Social and Emotional. *Extended Abstracts of The CHI 2002 Conference on Human Factors in Computing Systems*, ACM Press, 698-699.

Ratner, C. (1989) A Social Constructionist Critique of Naturalistic Theories of Emotion. *Journal of Mind and Behavior*, 10, 211-230.

Schiano, D.J., Ehrlich, S., Rahardja, K., and Sheridan, K. (2000) Face to Interface: facial affect in (hu) man machine interaction. *Proceedings of CHI 2000 Conference on Human Factors in Computing Systems,* ACM Press, 193-200.

Wallbott, H.G. (1988) In and out of context: Influences of facial expression and context information on emotion attributions. *British Journal of Social Psychology*, 27, 357-369.

Tablet PC - Using Field Trials to Define Product Design

Evan Feldman	Erik Pennington	Jo Ireland
Microsoft Corporation	Microsoft Corporation	Microsoft Corporation
1 Microsoft Way	1 Microsoft Way	1 Microsoft Way
Redmond, WA 98052	Redmond, WA 98052	Redmond, WA 98052
evanf@microsoft.com	erikpenn@microsoft.com	jireland@microsoft.com

Abstract

This paper outlines some of the key challenges the user research team faced throughout the lengthy product design cycle of Microsoft's Tablet PC. From concept stage through numerous iterations of prototype hardware and software in lab studies, to gathering data from users in the field, the user research team played a key role in defining and designing the end product.

1 Introduction – What is a Tablet PC?

In short, a Tablet PC is a next-generation personal computer optimized for mobility, convenience, and comfort. Tablet PCs are designed to be primary business PCs, providing the full functionality of traditional PCs. These machines support a range of new user scenarios combining various input modes, including keyboard, mouse, pen and voice, together with the convenience of a notepad and the simplicity of a pen-based interface. The Tablet PC is designed to improve and expand the utility of the PC for mobile computer users.

2 Tablet PC User Research Program

In August of 1999, user research for the Tablet PC began in earnest. After reviewing previous pen computing devices (e.g., Windows for Pen Computing, the Apple Newton, etc.), the tablet team set about a two-pronged approach to user research. We began performing interviews and ethnography with a wide variety of potential end users while simultaneously performing prototype testing of new user interface concepts for pen interactions. For this paper we discuss only the ethnographic type research.

2.1 Primary Interviews and Observations

The general approach for initial interviews was to examine how users interact with traditional pen and paper. This naturally led to studying when information workers take notes and the situations in which these notes are taken.

2.1.1 The Meeting Environment

For our participants, meetings provide one of the most natural and prolific note-taking environments. The primary tool we used was to observe as many meetings of different types with as many different participants as possible. The methodology was simple: 30 minutes of pre-meeting interview with the participant, shadow the participant during the meeting, and then up to an hour of post-meeting interview. The pre-meeting interview focused on prepared materials for the meeting, a breakdown of the expected meeting and a description of how the participants "perceived their own note-taking." During the meeting, a team of two or three observers would stand behind the participant to unobtrusively photograph and document the note-taking behavior. After the meeting, we performed an in-depth reconstruction of the notes taken, the manner in which they were written on the page, the use of the pen, and any social interactions that we observed. This was followed by further discussion with the participant about other notes that may have been taken and an indication of actions that would follow, how the notes would be dispatched and archived, and any search or retrieval strategies that might be employed. This provided a strong foundation for the Tablet team as it not only provided the opportunity for hands-on observations of potential end-users but also a strong education on the real needs and environmental constraints of these users.

2.1.2 The Tablet PC Opportunity

From studying this interaction, it became clear that a PC could be valuable to users in meetings but faced many hurdles to acceptance. Many users rightfully recognized that laptop computers as they exist today are not socially accepted in meetings due to several factors: Laptop displays create social barriers, typing is noisy, users are perceived as being unengaged, and more attention is needed to control a laptop. However, while at a meeting, users often find that they need to access documents (both previous notes and computer files) that they don't have at hand, need to update a computer document being discussed, want to check their calendar or contact information, and query others not currently present. The challenge was to enable these activities while creating a device that addressed the shortcomings of using laptops in meetings today. The biggest challenge that we faced was to appropriately test and research whether the Tablet PC solution would meet these needs. For that we needed to branch out into field trials. Over the course of the development of the Tablet PC, we performed four distinct field trials.

2.2 Early Field Trials

Users have a unique way in which they take notes and those notes are integral to their job. Simulating a note-taking meeting environment is extremely difficult. When users were studied in an artificial lab situation, the structure and content of their notes generally did not match those collected in a work environment. Differences occurred because users did not have an investment in organizing and retaining the information. Additionally, the social interactions are hard to simulate. As such, trying to evaluate the tablet software and hardware interaction required us to go farther in testing than traditional lab studies and user interviews.

The field trial methodology was simple: give target users Tablet PC computers for weeks or months and watch how the devices are used, the social interactions that result, and the users' reaction to the device. This methodology shouldn't be considered a "beta test" – the investment on studying the users and the potential reaction to the data far exceeded a traditional beta test which is more focused on finding "bugs" in the product software or hardware. In our case we were looking for how the entire user experience worked and whether the solution we provided would meet the users' needs. Execution for a study of this magnitude was the most difficult part. First we needed to build robust prototypes (both hardware and software) that could last four to eight weeks of daily use, and then we needed to gain the users' agreement to give up their current PC and use the

Tablet as their main device for that period of time. Finally, we needed to create an infrastructure to support the prototypes and users. The methodology is described in more detail in Dray, Siegel, Feldman & Potenza (2001).

The first trial was the shortest, ranging just two weeks with 26 users. In this trial we looked at a very early prototype that had an extremely limited amount of functionality. In fact, this prototype only allowed users to take notes and was not capable of running Microsoft Windows or any other application and thus could not replace the primary PC. Participants in this trial provided us with feedback on the note-taking experience and their ability to use this note-taking device in a real meeting situation. A clear trend among these users was how valuable the note-taking experience was given that they could now integrate their notes with other documents. A second important trend was that the absence of Windows turned out to be a detriment to performing other tasks while mobile.

The second trial was larger in complexity, involving 21 participants with fully functioning Tablet PCs (thus running Windows and standard productivity software) for four weeks. We spent more than three hours with each participant every week following them to meetings, watching them use the device at their desks, and interviewing them about their general usage. We applied many of the same techniques for meeting observations that we used in the initial interviews and observations. This provided one of the few opportunities to determine if the Tablet PC would be a useful, usable, and needed product. During this trial we found that the general value proposition of the Tablet PC was there, but that the implementation of several key pieces was lacking.

Had we not performed this field trial, the Tablet PC would have been launched nearly a year earlier, but would have included functionality that, while more technically advanced, would not have met users' needs and expectations for the tasks that we had envisioned. The users' experiences in the field trial were counter to that of the users during traditional lab tests and internal team members who had been using the tablet up to that point. The internal team members made use of the advanced heuristics and functionality that we had built into the product and enjoyed "playing" with the tablets. The participants in the lab understood and used the functionality we provided with few issues, but as we found out with the field trial this did not match real world usage. The user research in this case clearly pointed out that the tasks that we wanted to enable for the end user were not accomplished with the current functionality and that the internal team members were performing complicated work-arounds to use the pen computer.

2.2.1 Starting Over

While we created a device that could be socially accepted in meetings and allowed the user to drive existing Windows applications with a pen, the ergonomics of the prototype and the utility that we constructed for note-taking had some severe interaction issues as noted above. In the case of the ergonomics, we knew that the prototypes that we built would not necessarily be representative of the final units built by the hardware manufacturers (e.g., Hewlett Packard, Toshiba, Acer, NEC, etc.), however, we discovered several areas where we could definitely influence and change the direction of the hardware built by these manufactures. For the note-taking experience we had built the equivalent of an "ink word processor" whereas users really wanted a "sheet of paper" (this is where the early testers of the device dramatically differed from the end-users in the field trials). In this case, the note-taking utility allowed the user to rotate the page dynamically which would cause the ink on the page to automatically re-flow to best fit the new dimensions of the tablet screen. In addition, the ink on the page would move around based on the action that the user performed – such as inserting a diagram or erasing a word. While the "ink word processor" model was technologically advanced and garnered many accolades, it simply didn't meet with the users' goal of taking freeform notes. Users wanted to lay their ink down anywhere on the page regardless of the lines on the page and have the ink stay there even when they went back to change something. One of the largest insights was that users generally don't want to edit their notes during a meeting and in fact generally don't ever want to edit their notes in the future either. The findings from the second field trial resulted in the Tablet team completely overhauling the note-taking utility.

2.3 Field Trial 3

Over the next six months the user research team went back to basics and tested many prototypes and early code of the revised note-taking utility in the usability labs. We focused on generating a range of note-taking scenarios to look primarily at the interaction model, with the knowledge that the usefulness and real world interaction could be validated thorough investigation during the third field trial. For this trial, we invited 7 participants to use a revised Tablet PC prototype for six weeks. The goal was to determine if the major changes that we had made since the previous trial fit with the user's expected and desired application of this device. In addition, we used this third field trial to explore the overall adoption process of the tablet in more detail. This trial was extremely encouraging as it verified that the major changes to the product did result in a more usable interface for the note-taking utility.

Based on two and half years of research and three different field trials, we finally had found the right combination of technology and implementation to be useful, usable, and needed. The third trial resulted in a resounding success for the dramatic changes that were made to the product. Users were able to accomplish their goal of taking notes in a meeting using a computer in a way that was socially acceptable. While we did achieve our major goal, we also discovered a fair number of additional usability issues that would also need to be addressed.

The remainder of the product cycle until the product shipped was filled with numerous usability studies in the lab, heuristic evaluations, questionnaires, and additional site visits. However, as we ramped down on some of the particular feature testing for the Tablet PC, we started to envision our most elaborate field trial to date.

2.4 Field Trial 4

Beginning in April 2002, the Tablet PC user research team began laying the groundwork for an open-ended, long-term field study. The purpose of this fourth field trial is to observe the complete lifecycle of learning and adoption of this new technology within the context of the users' work environment. Eight participants were recruited to permanently exchange their current PC with close to production quality tablet hardware and software.

Unique to this field trial, the Tablet PCs were delivered to the participants in the original manufacturer's packaging with all peripheral devices and user assistance print materials. This provided the research team with our first opportunity to monitor users in their work environment learning to use a Tablet PC without intervention. In this context we were able to observe the confluence of several situational factors not present in the lab including the participant's need to quickly begin productive work with a new PC, interruptions by co-workers or phone calls during the learning process, and integrating the tablet into their existing workstation.

During the first 12 weeks of the study the researchers met with the participants an average of six times. Consolidating observational and participant self report data allowed the researchers to identify trends in usage and issues that were compiled into a timeline of learning and adoption.

After approximately four months the participants' usage of the tablet had reached a steady state wherein learning and the number of reported issues did not increase notably across participants. The user research team has taken this opportunity to reap additional value from the field trial by periodically replicating lab studies with these participants to compare results of the same tasks performed by novice users (lab) and experienced tablet users (field trial). This process has been very successful in not only validating lab study data but in providing new feature-level insights that are not available from lab participants with limited exposure to the tablet.

Given the steady state of tablet use, regularly scheduled site visits with the field trial participants are no longer necessary. However, the user research team continues to periodically monitor their tablet use and note-taking behaviors as well as solicit feedback on feature ideas and prototypes. The user research team plans to continue to poll these participants as the Tablet PC project evolves.

3 Conclusions

From inception to implementation to planning the next release, the Tablet PC has presented a number of methodological and logistical challenges for usability testing. The challenge began with endeavoring to understand the user's needs and behaviors with regard to electronic note-taking, and continues with identifying new scenarios for information workers to make use of the Tablet PC. Four field trials and numerous lab studies have been employed in an effort to understand how we can both facilitate and enhance productivity with a combination of software and hardware that is simultaneously familiar and innovative. Certainly no single usability research technique could have provided the insight and breadth of data the research group collected and communicated to the development teams.

The field trials provided researchers with observations of contextual behaviors and product interactions that validated or repudiated multiple designs at various stages during the product cycle. Affording participants the time and support to incorporate the tablet into their work life yielded the significant benefit of illuminating both the beneficial and derogatory impacts of the product in a complete end-to-end scenario. Subsequently, issues were uncovered about aspects of the tablet that would not have been the focus of a dedicated lab study and unsuspected interactions between features were discovered. Additionally the field trials provided an invaluable opportunity for members of the Tablet PC development team to witness first-hand users working, and struggling, with their product. Although the financial, time, and resource costs of these field trials were substantial, the studies continued to receive the full support of Tablet PC management because the data consistently translated directly into improvements of the product.

References

Dray, S., Siegel, D., Feldman, E., & Potenza, M. (2001). Why do Version 1 and Not Release it? Interactions: New Visions of Human-Computer Interaction, Vol. 9, No. 2, p.11-16.

Using Confidence Scores to Improve Hands-Free Speech-Based Navigation

Jinjuan Feng, Andrew Sears

Interactive Systems Research Center, Information Systems Department
UMBC, Baltimore, MD 21250, USA

Abstract: Speech recognition systems have improved dramatically in the past two decades, but recent studies confirm that error correction activities still account for as much as 66-75% of the users' time. While researchers have suggested that confidence scores could be useful during the error correction process, the focus is typically on error identification. More importantly, empirical studies have failed to confirm any measurable benefits when confidence scores were used in this way. In this article, we provide data that explains why confidence scores are unlikely to be useful for error identification. Next, we propose a new, confidence score-based, error specification technique and provide results from a simulation that suggests that this new technique is highly effective. These results suggest that confidence scores can be used to support speech-based error correction and lay the foundation for future research.

1 Introduction

Speech recognition systems have improved dramatically in the past two decades. However, users of state-of-the-art speech recognition systems still experience significant difficulty correcting recognition errors (Sears et al., 2001). In speech-based dictation systems, error correction can be viewed as a three-stage process. The first stage is identification: users must locate the errors. The second stage is specification: users must specify which word needs to be corrected. Experienced users can spend as much as one third of their time specifying the words that need to be corrected if speech is the only input method available (Sears et al., 2001). The final stage is actually correcting the error. More efficient error specification and correction techniques are critical if speech recognition systems are to be widely accepted. More efficient hands-free techniques are particularly important when the users' hands are unavailable due to a physical disability or a conflicting task. In this article, we focus on the process of specifying errors. More specifically, we present a novel speech-based technique that utilizes the confidence scores generated by the speech recognition engine to improve speech-based error specification.

2 Related Research

One of the main obstacles that hinder the broad application of speech recognition for dictation-oriented activities is the error correction process. Karat et al. (1999) reported that users with extended use of the speech system still spent 75% of their time correcting errors. As outlined above, these correction activities consist of three stages: identification, specification, and actually correcting the error. Since recognition errors are all correctly spelled words that could be more difficult to be found as compared to typing errors, researchers have attempted to facilitate the error identification process in various ways. Confidence scores are generated by the speech recognition engine and measure of how confident the speech engine is that the output matches what the user actually said. Confidence annotation research typically focuses on using confidence scores to identify speech recognition errors. A significant amount of research has been done in this area. (e.g. Hazen et al., 2002) However, these studies tend to focus on generating information to provide feedback to systems or to assist developers. Few researchers have considered providing feedback to the users of these systems. In contrast, Suhm et al. (2001) investigated a system-initiated recognition error detection mechanism that was based on confidence scores that involved marking words with low confidence scores as possible recognition errors. Since confidence scores are not completely reliable, this process is not precise. Some of the marked words will be incorrect (i.e., correct detections), but others will be correct words that happened to have low confidence scores (i.e., false alarms). Similarly, some recognition errors will have confidence scores greater than the threshold and will be considered likely to be correct (i.e., missed detections). Suhm et al. used classification accuracy to set the threshold, which is the ratio between the number of correctly classified words and the total number of words. Since the number of correct words is typically much larger than the number of recognition errors, using classification accuracy to set the threshold results in a bias toward minimizing false alarms. Unfortunately, results from an empirical study suggested that this approach actually slowed the error correction process. This suggests the reliability of current confidence scores may not be accurate enough to support error identification activities.

Once an error is identified, users must move the curser to that word to correct it. Currently speech-based navigation is accomplished with one of two approaches: target-based navigation or direction-based navigation. Target-based commands move the curser by specifying target word, for example, "Select door". Direction-based commands move the cursor by specifying navigation direction, such as "Move Up". Other more powerful, but more error prone, forms of direction-based commands also exist such as "Move up five lines". Few studies have been conducted on speech-based navigation. Recently, Sears et al. (2002) reported that experienced users of speech recognition systems spent about one-third of their time on navigation activities. Detailed data were reported concerning navigation command failure rates as well as the causes and consequences of failure. The authors proposed several improvements to speech-based navigation. An empirical study indicated that the proposed improvements were effective.

Earlier studies provided insights into speech-based error correction, but research on speech-based navigation is still limited. The attempt to use confidence scores to support error identification is instructive, but the results highlight the need for further exploration.

3 Methods

Our analysis builds on data gathered in an earlier study. In this study, 15 participants created 120 documents using a custom speech recognition application (TkTalk Version 2.0) that employed the Millennium Edition of IBM's ViaVoice speech recognition engine. The resulting documents contained a total of 66950 words. This includes 55495 words that were correctly recognized and 11455 recognition errors (17.1%). In ViaVoice, each word is assigned an integer confidence score. In our study, confidence scores ranged from approximately –15 to +30. The word output by the speech engine as the "most likely" alternative is actually selected based upon a variety of factors including the confidence scores of the individual words as well as language models that consider the surrounding words. In addition, ViaVoice generates alternatives that may be correct if the "most likely" word is not. The first alternative generated is considered the "best alternative" when the "most likely" word is incorrect.

4 Identifying Recognition Errors

Most discussions regarding the use of confidence scores to facilitate error correction have focused on assisting the user in identifying likely recognition errors. Suhm et al. (2001) focused on the raw confidence scores associated with the "most likely" words and used classification accuracy to set the threshold for tagging possible errors. When using raw confidence scores on our data set, classification accuracy is maximized by setting the threshold at -7. Classification accuracy is maximized because over 96% of the correctly recognized words are classified properly, but only 27% of the recognition errors are identified. Clearly, a higher threshold is required to identify a reasonable fraction of the recognition errors. A threshold of -2 increases recall (percentage of recognition errors identified) such that over 50% of the recognition errors are identified, but false alarms also increase substantially. Increasing the threshold to 1 allows approximately two-thirds of the recognition errors being identified, but only 37% of the words identified as recognition errors are actually incorrect. These results suggest that a threshold-based approach that uses raw confidence scores is not likely to be effective if the goal is to facilitate the process of error identification. These results suggest that confidence scores are not likely to prove effective if the goal is to facilitate the identification of recognition errors. Therefore, we shift our attention from identification to specification.

5 Specifying Recognition Errors

5.1 Recognition error sequences

We begin by analyzing the distribution of recognition errors to determine if these errors tend to occur in clusters or separately. We define a recognition error sequence as one or more consecutive words which are recognized incorrectly. In our sample of almost 67000 words, the length of recognition error sequences varied from one to twenty words. As illustrated in Figure 1, less than 30% of the recognition errors occur in isolation. More importantly, this confirms that over 70% of recognition errors appear immediately adjacent to at least one other recognition error. The fact that recognition errors tend to occur in clusters provides new opportunities given the goal of specifying, as opposed to identifying, recognition errors. Given this goal, helping users navigate to a recognition error sequence may be sufficient.

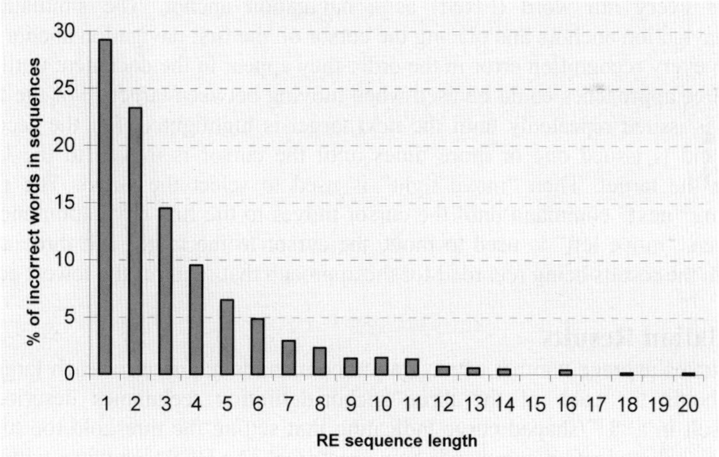

Figure 1. Percentage of incorrect words in recognition error sequences of different lengths.

5.2 Navigation anchors

Our earlier studies of speech-based navigation highlighted the potential benefits and important limitations of simple direction-based navigation (e.g., Move up). Short, fixed, navigation commands minimize errors, but tend to provide limited power. Longer, or constructed, commands provide greater power but are also associated with higher failure rates. Therefore, our goal is to design a speech-based navigation technique that allows for greater efficiency than the simple direction-based commands discussed above while minimizing failure rates through the use of short, fixed, navigation commands. Navigation anchors are central to the proposed solution. In the new system, navigation anchors are strategically selected words within the document. Using two short, fixed, commands (i.e., next and previous) that should rarely fail, users can easily move to the nearest anchor that precedes or follows the current cursor location. Since the navigation anchor may not be the exact word the user needs to correct, we also provide four simple direction-based navigation commands that allow users to move to adjacent words or lines (i.e., move up, move down, move left, and move right). The goal when defining navigation anchors is to facilitate the process of specifying correction targets, not necessarily to identify recognition errors. To analyze the efficacy of a set of navigation anchors, we developed a simulation that determines both which commands and how many commands would be required to navigate to a predefined set of targets.

5.3 Simulation Design

For the simulation, we used the data described in Section 3. While the proposed navigation mechanism includes six commands (i.e., next, previous, move up, move down, move left, move right), only three are used in the simulation: next, move right, and move left. Since we do not anticipate that "move up" or "move down" would be used frequently, and their effects are determined by the screen size and font size, these two commands were not included in the simulation. "Previous" is not used since it only becomes useful after "move down" is used. Three approaches were evaluated for defining navigation anchors. The first method uses the raw confidence score (CS) of the "most likely" word with words having a CS lower than the specific threshold selected as navigation anchors. The second uses the difference between the confidence score (DCS) of the "most likely" word and that of the "best alternative". Again, words with a DCS lower than the threshold are selected as navigation anchors. For comparison purposes, the third approach uses every nth word (Fixed) as a navigation anchor. The simulation begins by highlighting navigation anchors and placing the cursor on the first navigation anchor. The cursor is then moved to every recognition error in the order they appear in the document until it reaches the final error. Three approaches could be used when moving between targets. For the first approach, "move right" is issued repeatedly until the next target is highlighted. For the second approach, "next" command is issued one or more times until the cursor is moved to the last navigation anchor before the target. Then "move right" is used to select the target. The final approach involves issuing "next" command until the cursor moves to the first navigation anchor following the target. Then, "move left" is used to move the cursor to the target. All three approaches are evaluated, with the results being recorded for the approach that requires the fewest commands.

5.4 Simulation Results

Figure 2 illustrates average number of navigation commands required for each target when using various thresholds for each of the three anchor-definition techniques described. All three techniques result in a "U" shaped curve indicating that setting the threshold too high or too low results in additional navigation commands being required. The DCS technique, with a threshold of 1, minimized the number of navigation commands required (2.12 commands on average). As to command composition, "next" accounts for approximately 60% of the commands issued, "move right" accounts for 31% of the commands, and "move left" accounts for only 8% of the commands. Since the failure rate of "next" is expected to be significantly lower, the fact that over 60% of commands are "next" is promising. For comparison purposes, we found that users issued an average of 2.29 direction-based commands to reach a target in our previous study. However, the failure rate of those direction-based commands was nearly 20%. The simulation suggests that the new technique will require approximately the same number of commands for each target.

However, the commands should be easier to issue and less likely to fail, suggesting that this could be an effective and reliable alternative for speech-based navigation.

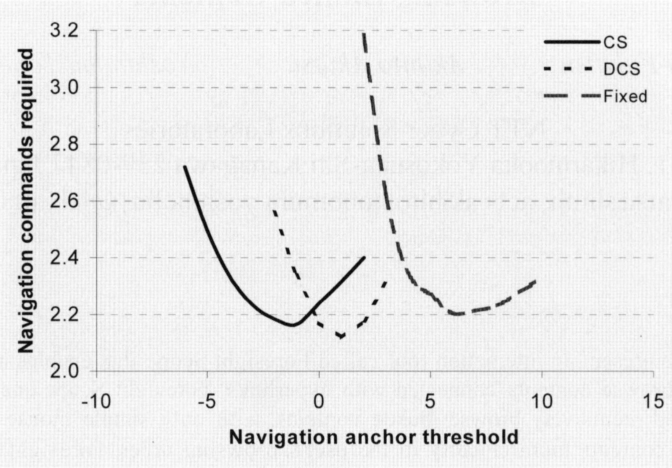

Figure 2. Average number of navigation commands required when using different navigation anchor thresholds and techniques.

6 Conclusions

Existing speech-based error correction techniques tend to be error prone and time consuming. The current article focused explicitly on error specification. Our results suggest that confidence scores are not likely to be effective if the focus is on error identification, but a new navigation technique that is based on confidence scores appears to be both efficient and reliable. The proposed design was evaluated using a simulation with promising results. However, the simulation is limited in that it did not include all of the available commands and did not consider the consequences of failed commands. Therefore, the simulation results may not be representative of actual system usage. Empirical studies are needed to provide a more comprehensive evaluation of the proposed technique. Finally, in addition to the number of commands required and commands composition, other criteria (e.g., number of words highlighted) should be considered when defining navigation anchors.

Acknowledgement: This material is based upon work supported by the National Science Foundation (NSF) under Grant No. IIS-9910607. Any opinions, findings and conclusions or recommendations expressed in this material are those of the authors and do not necessarily reflect the views of the NSF.

7 References

Hazen, T. J., Polifroni, J., and Seneff, S. (2002). Recognition confidence scoring for use in speech understanding systems, *Computer Speech and Language*, Vol. 16, No. 1, pp. 49-67, January, 2002.

Karat, C-M., Halverson, C., Karat, J. and Horn, D. (1999). Patterns of Entry and Correction in Large Vocabulary Continuous Speech Recognition Systems. *Proceedings of CHI 99*, 568-575.

Sears, A., Feng, J., Oseitutu, K., Karat, C-M. (2002). Speech-based navigation during dictation: Difficulties, consequences, and solutions, under review.

Sears, A., Karat, C-M., Oseitutu, K., Karimullah, A., & Feng, J. (2001). Productivity, satisfaction, and interaction strategies of individual with spinal cord injuries and traditional users interacting with speech recognition software. *Universal Access in the Information Society*, 1, 4-15.

Suhm, B., Myers, B. and Wailbel, A. (2001). Multimodal error correction for speech user interfaces. *ACM Transactions on Computer-Human Interaction*, 8(1), 60-98.

Foresight Scope: An Interaction Tool for Quickly and Efficiently Browsing Linked Contents

Shinji Fukatsu *Akihito Akutsu* *Yoshinobu Tonomura*

NTT Cyber Solutions Laboratories
1-1, Hikarinooka Yokosuka-Shi Kanagawa 239-0847 Japan
{fukatsu.shinji, acts.akihito, tonomura.yoshinobu}@lab.ntt.co.jp

Abstract

In this paper, we present an interaction tool called Foresight Scope that enables a user to quickly and efficiently browse contents connected with hyperlinks. Foresight Scope enables the user to continuously and recursively browse linked contents with only simple mouse movement and presents linked contents hierarchically in the user's browsing order. Foresight Scope allows a better and more thorough understanding of the linked contents.

1 Introduction

In many applications, users are faced with the problem of quickly and efficiently browsing contents connected with links (e.g., hyperlink or icon) through a graphical user interface (e.g., web browser or file browser). For example, even though it is inconvenient to move to a new web page and back to a previous browsed web page by clicking each anchor and back button, these actions are frequently invoked in web navigation (Tauscher & Greenberg, 1997). Also, it is troublesome to step up and down deep folder hierarchies by clicking folder icon and back buttons. These problems are caused by the fact that conventional web browsers (or file browsers) are designed for the purpose of enabling a user to read a web page (or folder) and are not designed to enable user operation following hyperlinks (or icons). Nielsen indicated that people rarely read web pages word by word; instead, they scan the page, picking out individual words and sentences (Nielsen, 1997). Therefore, it is essential to develop a system that enables a user to quickly move between contents by following links and to efficiently understand the content of the information at a glance.

In this paper, we present an interaction tool called Foresight Scope that enables a user to quickly and efficiently browse contents connected with hyperlinks (from here on referred to as linked contents). Foresight Scope enables the user to continuously and recursively browse linked contents with only simple mouse movements and presents linked contents hierarchically in the user's browsing order.

2 Related Work

Several approaches have been proposed to enable a user to quickly and efficiently browse linked contents, especially web contents. Visual Preview (Kopetzky & M¨uhlh¨auser, 1999) presentS a preview image of linked contents near an anchor and Zero-Click (Nanno, Saito & Okumura, 2002) presents the linked contents in a pop-up window when the mouse cursor is over the anchor. Fluid Annotations (Bouvin, Zellweger, Grønbæk & Mackinlay, 2002) and InlineLink (Miura,

Shizuki & Tanaka, 2001) enable a user to insert or remove the linked contents near the anchor. These previous works allow a user to quickly browse the linked content of the anchor. However, except for InlineLink, these previous works do not support the continuous and recursive browsing of multiple contents following multiple hyperlinks. In addition, InlineLink makes it possible to insert multiple contents recursively, but in this case the inserted web page becomes far too long and takes too much time to browse. To enable multiple contents following multiple hyperlinks to be browsed efficiently, conventional web browsers present a history list, which is a list of previously visited web pages, that allows a user to visit previously visited pages by using it. This list also allows users to remember the way they navigated previously. However, such conventional history lists are independent from the ordinal operations following hyper links, i.e., clicking an anchor or a back button.

3 Foresight Scope

We have designed an interaction tool called Foresight Scope that enables a user to quickly and efficiently browse contents connected with hyperlinks. Foresight Scope enables the user to continuously and recursively browse linked contents with only simple mouse movements. It also presents linked contents hierarchically in the user's browsing order and enables users to review the way they have navigated previously and the relationship between linked contents (see Figure 1).

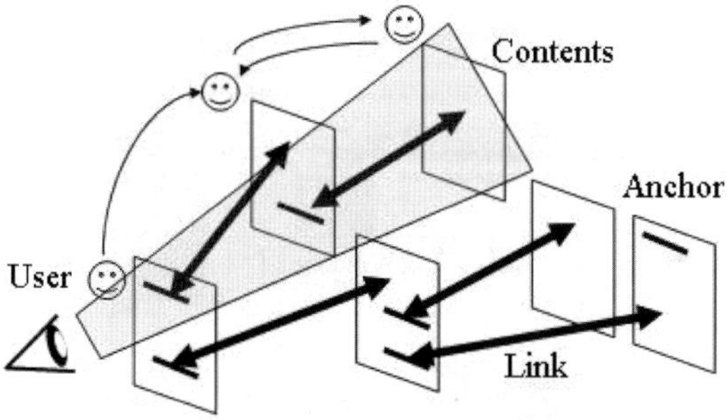

Figure 1: Foresight Scope

3.1 System Interface

With Foresight Scope, when a user puts a mouse cursor on an anchor, the mouse cursor transforms to a pop-up window and the linked content is displayed in the pop-up window (see Figure 2). In this case, the linked content is displayed scaled down according to the size of the pop-up window. And if the user moves the mouse cursor to the pop-up window, the mouse cursor automatically moves into the pop-up window. Furthermore, when the user moves the mouse cursor out of the pop-up window, the pop-up window is closed and the mouse cursor automatically moves back to the anchor. In this case, the pop-up window transforms to the mouse cursor on the anchor. These animations help the user understand the relationship between the anchor and the pop-up window. As a result, the user can browse a linked content following a hyperlink with only simple mouse movements.

Using Foresight Scope, users can browse a linked content in a pop-up window with an ordinal browsing operation. For example, they can scroll the linked content using a wheel mouse and change contents by clicking an anchor in the pop-up window. Also, by double-clicking in the pop-up window, they can display the linked content in a newly opened ordinal web browser.

Figure 2: User's View with Foresight Scope

Furthermore, they can visit previously visited linked contents by moving the mouse cursor into the pop-up windows. As result, they can continuously and recursively browse multiple contents following multiple hyperlinks using Foresight Scope (see Figure 3). Here, the system gradually changes the size of a pop-up window and changes the zoom of a linked content according to the number of following hyperlinks from the beginning (e.g., Window 0). Also, the system recursively enlarges the decision area for closing a pop-up window (e.g., Window 1) to involve pop-up windows (e.g., Windows 2, 3) that are opened from the pop-up window. As a result, the system configures the parent-children relationship between multiple pop-up windows and presents multiple linked contents hierarchically in the user's browsing order. This enables users to review the way they have navigated previously and the relationship between linked contents.

Figure 3: Continuous and Recursive Browsing of Multiple Contents

3.2 System Configuration

Figure 4 shows the Foresight Scope system configuration and the process for presenting a linked content in a pop-up window with it. The system constantly monitors mouse movements on a user's desktop using a mouse event hook Dynamic Link Library (DLL). When the user puts the mouse cursor on the anchor of an HTML document in a web browser, the system receives the relative position of the mouse cursor in the web browser from the mouse event hook DLL. Next, the system utilizes a Component Object Model (COM) to specify the anchor tag under the mouse cursor and get the URL of the anchor tag. Here, the HTML document is parsed based on the Document Object Model (DOM), and the system generates a pop-up window and displays a linked content of the URL in the pop-up window. In this case, the system changes the size of the pop-up window and presents the pop-up window in an appropriate place where it does not occlude the anchor in the web browser. The system also changes the zoom of the linked content by inserting zoom property into the <BODY> tag of the linked content. In this way, Foresight Scope does not need any preparation (i.e., by adding JavaScript) to anchors unlike previous works; it enables a user to enjoy the benefits of Foresight Scope with all of the existing contents. In addition, since

Foresight Scope operates in users' computers, the system can utilize their operational situation or operational history for customization and can cooperate with other applications in their computer.

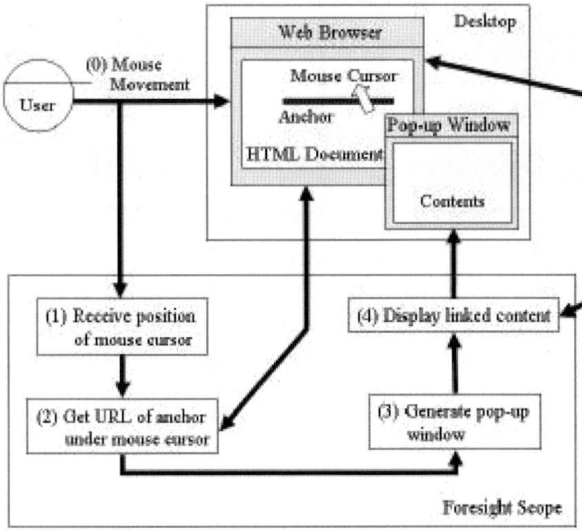

Figure 4: System Configuration

3.3 Extension of Foresight Scope

With Foresight Scope, the linked content is presented scaled down in a pop-up window. It helps a user to efficiently understand information of the linked content at a glance. Here, in order to more efficiently understand information of the linked contents, it is essential to rewrite or remake the linked contents in an appropriate style. For such extension of Foresight Scope, we have introduced a proxy server that rewrites or remakes the linked contents in advance. In this case, a user browses rewritten or remade linked content instead of the original linked contents in a pop-up window. In this section, we use an example to explain the extension for movie contents with Foresight Scope.

When a user accesses a web page that includes anchors to movie contents, the system requests the proxy server to generate thumbnail views of the movie contents. The proxy server then generates thumbnail images of the movie contents at a predefined period and stores the thumbnail images with time code. Also, when the user puts the mouse cursor on the anchor to movie content, thumbnail view of the movie content is presented in a pop-up window from the proxy server (see Figure 5(a)). Moreover, when the user puts the mouse cursor on a thumbnail image, the movie content is play backed from a time code defined by the thumbnail image (see Figure 5(b)). In this case, the movie content is presented from a web server that has the original movie contents. In this way, with the extension for movie contents, a user can understand the overall movie content at a glance and quickly access each scene of the movie contents from the thumbnail images.

4 Conclusion

We have presented an interaction tool called Foresight Scope that enables users to quickly and efficiently browse linked contents, especially those connected with hyperlinks. Foresight Scope enables users to use only simple mouse movements to continuously and recursively browse linked

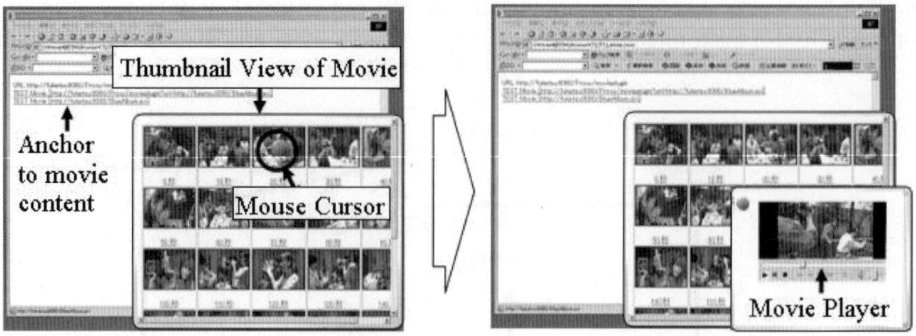

(a) Thumbnail View (b) Playback from Defined Time Code

Figure 5: Extension for Movie Contents

contents and review the way they have navigated them and the relationship between linked contents and the hierarchically displayed linked contents in the user's browsing order. We also described the extension of Foresight Scope using a proxy server and presented an example extension of it for movie contents.

An informal user observation showed that Foresight Scope was useful for browsing linked contents from search engine results, site maps, and image databases that contain many anchors. It was also shown to be useful for browsing discussion boards and FAQs where linked contents are linked hierarchically. In these cases, users could obtain information about contents quickly and efficiently with Foresight Scope. Our future work will include quantitative evaluation of Foresight Scope by comparing it with conventional web browsers, applying it to file browsers, and further extending it by using a proxy server.

References

Bouvin, N., Zellweger, P., Grønbæk, K., & Mackinlay, J. (2002). Fluid annotations through open hypermedia: using and extending emerging web standards. *Proceedings of the 11th International World Wide Web Conference*, 160-171.

Kopetzky, T. & M¨uhlh¨auser, M. (1999). Visual preview for link traversal on the WWW. *Proceedings of the 8th International World Wide Web Conference*, 447-454.

Miura, M., Shizuki, B., & Tanaka, J. (2001). InlineLink: inline expansion link methods in hypertext browsing. *Proceedings of International Conference on Internet Computing*, 2, 653-659.

Nanno, T., Saito, S., & Okumura, M. (2002). Zero-Click: a system to support web browsing. *Proceedings of the 11th International World Wide Web Conference*, 7-11.

Tauscher, L. & Greenberg, S. (1997). How people revisit web pages: empirical findings and implications for the design of history system. *International Journal of Human-Computer Studies*, 47(1), 97–138.

Nielsen, J. (1997). How Users Read on the Web. from http://www.useit.com/alertbox/9710a.html

Magic Pages – Providing Added Value to Electronic Documents

Marcel Götze, Stefan Schlechtweg

Department of Simulation and Graphics, University of Magdeburg
Universitätsplatz 2, D-39106 Magdeburg, Germany
{goetze|stefans}@isg.cs.uni-magdeburg.de

Abstract

Most electronic documents have a predefined structure, and a visualization of a document is adjusted to this structure. In many cases, however, it is desirable to have different visual representations of the document. This becomes especially important if there is additional information available, for instance, annotations. This paper introduces Magic Pages, a user interface technique that supports different views onto an electronic document. Support in this case means to provide an intuitive way of handling the visualization of textual content as well as additional information. Magic Pages are designed based on the user's experience with the handling of paper documents and treat different views as transparent pages set atop the original text.

1 Introduction

Electronic documents are replacing their traditional paper counterparts as a medium for information exchange. Hence, it is of vital importance to provide tools and techniques to work with the information given in electronic documents. This pertains not only to the textual content but also to additional information being attached to the document. Providing "added value" and offering means to handle this added value is the key to the acceptance of electronic documents over their paper equivalents.

Electronic documents can be adaptively visualized and personalized. Personalization in this case is the process of adding information and changing the view onto a document. Hence, it goes hand in hand with an adaptive visualization of the textual content and the additional information. A user (or even multiple users) can add and edit information (e.g., annotations) that is not part of the original document. The text in connection with the annotations should then be visualized based on the user's needs, his or her reading goals (O'Hara, 1996), or the type of reading (Adler & van Doren, 1972). Altogether, personalization enables to deliver the right information to the appropriate place at the desired time and in a form that supports the user's task at hand. The user interface to handle electronic documents should adopt the well known environment from paper documents. One of our main concerns is therefore to provide a familiar environment to the user by exploiting pen-based input as well as paper document metaphors wherever possible to raise the acceptance of electronic documents even more.

In this paper we will present *Magic Pages*, a user interface technique for visualization of and interaction with electronic documents that are enriched with additional information. A Magic Page is a transparent page that is placed on top of a document's text, that performs a certain function such as keyword search, annotation selection, or marking selection, and that finally visualizes the results of these function in the context of the document itself. The proposed user interface is based on the metaphor of a stack of transparent pages and thus handling the Magic Pages resembles handling a stack of sheets of paper.

2 Related Work

The idea presented in this paper draws on research work from quite different areas. In (Erwig, 2000) is stated that there is a very strong need for a user-friendly query interface to XML documents. This is not only the case for structured documents but also for documents that contain rather unstructured, textual data. In order to prevent the user from needing to learn and exploit query languages or difficult query interfaces, a graphical way of specifying and visualizing query results is called for. Magic Lenses as introduced by (Bier, Stone, Pier, Buxton & DeRose, 1993) provide an intuitive way of automatically querying (or transforming) a document and visualizing the result on a per-region basis. A Toolglass widget contains a function that is performed on the objects viewed in a region. The result of this function is directly visualized overlaying the region in question. For textual documents, this technique has been applied in (Phelps & Wilensky, 1998) as part of the Multivalent Document architecture. In both cases, the Toolglass widgets and lenses have to be positioned and resized by the user, adding interaction tasks that are uncommon for paper documents.

Detail-in-context presentations, like Fluid Documents (Chang, Mackinlay, Zellweger & Igarashi, 1998), and focus+context techniques (Document Lense (Robertson & Mackinlay, 1993), Fisheye Views (Furnas, 1986) and Semantic Depth of Field (Kosara, Miksch & Hauser, 2002)) are another source of inspiration. The problem with these techniques is that they visualize information within the document which is not possible when working with paper documents. The mentioned visualization techniques change the document content in a way that the additional information becomes part of the document itself. In contrast to this, the approach presented in this paper supports the visualization of information separately from the document it belongs to.

3 Magic Pages

Adopting the terminology from (Bier et al., 1993), we define a *Magic Page* as a display page together with a filter function on a document and together with a second function that visualizes the result of the filtering. More formally, a Magic Page is a triple $M=(D, f, v)$ where D is (part of) the document model that is displayed on the page, f is a function that returns a selected part of the document model based on some criteria, and v is a function that displays the result of the filter function in a certain way.

3.1 The Document Model

The Magic Pages interface is designed to work on an information enriched textual document. The original electronic document is given as an XML file so that while parsing this file an internal representation can be built that supports the search for specific information. We have decided to use the W3C Document Object Model (DOM) as basis for our work. The Document Object Model defines the logical structure of documents and the way a document is accessed and manipulated (W3C, 2002). Using a DOM parser, the structure of the document is represented as a tree that can be traversed to search for specific elements (nodes) or attributes.

In order to enrich a document with additional information such as annotations or markings, we use the technique presented in (Goetze, Schlechtweg & Strothotte, 2002). Here, the user works with a pen-based interface to "mark up" an electronic document just as he or she would do on paper. Different kinds of pens can be simulated and different types of markings and annotations are supported. All these annotations and markings are stored within the document's XML file and hence become part of the DOM tree. Together with the actual annotation data, information are given that identify the user who added the annotation and the pen type that was used. This is especially useful since readers over time develop a more or less consistent scheme of markings that delivers information which parts of a text are, for example, important, questionable, or unclear. Also, this information enables a distinctive visualization based on different users.

The additional information in the form of annotations and markings may or may not be present in the document. The standard DOM tree, which is built from any HTML or XML document, already contains all necessary information to use the Magic Pages interface. The more additional information is included, however, the more possibilities for filtering and visualizing the documents become available and, hence, the more powerful becomes the Magic Pages interface.

3.2 The Filter Function

The most important part of each Magic Page is the filter function, i.e., the operation on the document model that is performed by the respective page. These functions are implemented by means of the DOM API and here especially using methods that traverse the DOM tree and return a list of all those nodes that are of a certain type, contain a certain text or have a certain attribute.
To describe the various functions more formally, we introduce the following terminology. The set text(D) contains all nodes of the document model that actually contain textual content. The set attr(D) contains all those nodes that hold attribute values which are attached to textual elements.
We can immediately identify several types of functions that yield usable results for the application at hand. These include:

- *document contents selection:* $f(D)$ = text(D)
  This function returns all characters of the document, i.e., the text contents.
- *keyword search:* $f(D) = \{n \mid n \in \text{text}(D); n \text{ contains the keyword as substring}\}$
  This function returns all those text nodes that contain the given keyword.
- *attribute search:* $f(D) = \{n \mid n \in \text{attr}(D); n \text{ contains the given attribute value}\}$
  This function returns the attribute nodes that contain a specified attribute value.

The functions themselves are relatively straightforward and hence their power depends on the parameter values that are given and on the combination of several filter functions. A combination of two filter functions f and h can be performed in two ways. Either we apply h to the result of g which yields $h(g(D))$, i.e., h filters the *already filtered set of nodes or* we apply h to the same domain (document model) as g which yields $g(D) \cup h(D)$.

3.3 The Visualization Function

The second function that is part of each Magic Page works on the results of the filter function and is responsible for presenting the results to the user. Such a presentation dependents heavily on the type of information to be displayed as well as on user preferences. Hence, the visualization function is freely programmable. Some examples include

- *document rendering* to display all of a document's text content by applying either standard formatting instructions or by *following a style sheet*,
- *highlighting keywords* after a keyword search in the context of the whole document by color coding them or by applying other visualization techniques (e.g., Semantic Depth of Field (Kosara et al., 2002)),
- *showing annotations* as they were drawn/written by the user when he or she annotated the document using the pen type and color that was chosen by the user at that time

Even though the visualization is rather specific we can supply some standard Magic Pages where a filter function is coupled with a matching visualization. These pages provide a standard functionality that can be extended by the user.

4 The User Interface

A Magic Page acts as a filter on a page of the document or on other Magic Pages. The document itself and all additional information therefore has to be represented in a way that filtering different information is possible. To add annotations, we built upon the idea of the *Intelligent Pen* described in (Götze et al., 2002). A Magic Page placed on top of a document filters the document page and displays only the information that passes the respective filter. Adding a new Magic Page adds a new filter whose results are displayed on this new page. Depending on the chosen combination rule, it is applied to the document page and all other Magic Pages below and either filters the already filtered results or adds filtered document contents. As an example, given a document that was annotated by “goetze” and “stefans”, a first Magic Page filters all annotations by “goetze”, and a second one filters all annotations done with a blue ballpen. In combination, all blue ballpen annotations by “goetze” are seen through both Magic Pages on top of the document. We can take this further and filter the document contents, for example, by search queries that might result in highlighting all terms matching the query. This aids the reader in seeking for special information.

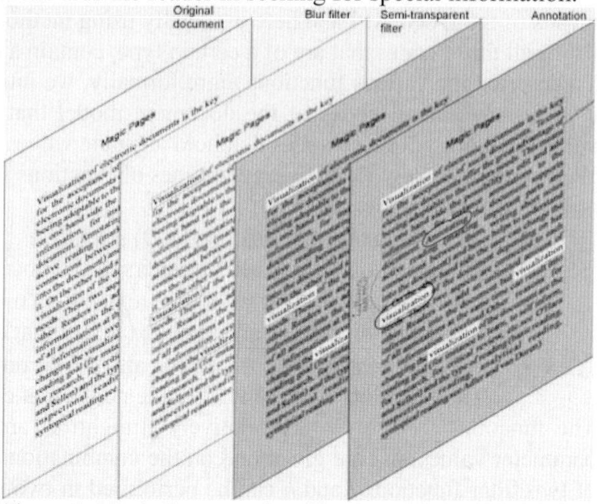

Figure 1: Magic Pages used as filter to guide attention and to add information.

The Magic Pages can be configured to aid the reader having a certain reading goal like “getting an overview” of the document. This can be seen as a more complex query within the electronic document and can, on the hand result in hiding instead of highlighting (possibly large) parts of the document. On the other hand, this can provide additional information as there is the visualization of connections between parts of the document or instructions for further reading. Magic pages can also be configured by the author to provide views onto the document that might be helpful for the user. The “transparency” of a Magic Page ensures that the document's text is always visible in order to provide the context. A schematic overview of the technique can be seen in Figure 1. Besides the possibility to let the user change the order of the Magic Pages, the system itself can reorder them. This is especially necessary in a case where the user adds information that belongs to a special Magic Page, e.g., an annotation. In this case (when the user starts writing) the system will move the desired Magic Page to the top of the stack.

The implementation shows an intuitive way of handling Magic Pages and hence visualizing information in the context of an electronic document. However, it is only a proof of concept, the presented technique offers a general way to structue (additional) information an visualize them in a page structured environment.

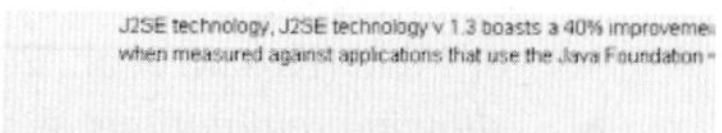

Figure 2: Tabs, used as a handle.

Since the metaphor behind Magic Pages originally came from the use of paper documents we consequently implemented our prototype based on the paper document metaphor. Hence, the user interface consists of different transparent slides laying on top of each other like pages on the desk. The user can add and remove Magic Pages as well as change their order using a simple interface. Each page contains a tab that tells the function of the page and acts as a handle (cf. Figure 2). To change the order the user can simply drag a page to the desired position within the stack. Currently unused Magic Pages can be placed behind the document page where they are inactivated. In most cases, no additional menu is needed for interaction tasks since some predefined pages are supplied. For defining new Magic Pages, i.e., combining filter and visualization functions, a dialogue interface is used so far.

5 Conclusion

In this paper Magic Pages, a new technique for visualizing electronic documents is introduced. For the acceptance of new technologies three supporting principles can be found. The first is to connect new techniques and exisiting (known) habits uf the user. Therefore, the Magic Pages are based on handling page based documents. The second principle is to develop suitable metaphors to simplify the use of new technologies. This is achieved by adopting known aspects of paper documents and transparent slides. The third principle is to adapt the user's behavior to the new technique in a way that he or she realizes that he or she controls the new technique. Obviously, the interaction with a computer application is not the same as interacting with a paper document. Nevertheless, reducing the interface to the known pen and paper interface based on the document metaphor allows it to control the system and, hence, the new technique in a way that complies with the third principle.

With Magic Pages we have provided an intuitive and novel way of dealing with electronic documents containing additional information. An enhanced and selective visualization provides distinctive views and aids the user's understanding of the document. Following the paper document metaphor, the user interface becomes a familiar environment also for electronic documents. Nevertheless, a still pending user study needs to confirm these findings.There are some aspects for further improvement and development. It would be interesting to investigate the use of Magic Pages as an input filter. In the above described system every user input is automatically redirected to the appropriate Magic Page. In contrast to this a Magic Page could also act as a cover and, for instance, prevent a document's page from beeing annotated. Also, since Magic Pages are defined as a rather abstract concept, it should be investigated if different representations of the document itself help to even more improve the usefulness of this technique.

References

Adler, M., & van Doren, C. (1972). How to read a book. New York:Simon & Schuster.

Bier, E.A., Stone, M.C., Pier, K., Buxton, W., & DeRose, T. D. (1993). Toolglass and Magic Lenses: The See-Through Interface. In *Proceedings of SIGGRAPH'93* (pp. 73–80). New York:ACM Press.

Chang, B., Mackinlay, J.D., Zellweger P.T., & Igarashi, T. (1998). A Negotiation Architecture for Fluid Documents. In *Proceedings of UIST'98* (pp. 123-132). New York:ACM Press.

Erwig, M. (2000). XML Queries and Transformations for End Users. In *Proceedings of XML 2000* (pp. 259-269).

Furnas, G.W. (1986). Generalized Fisheye Views. In *Proceedings of CHI'86* (pp. 16-23). New York:ACM Press.

Götze, M., Schlechtweg, S., & Strothotte, T. (2002). The Intelligent Pen—Towards a Uniform Treatment of Electronic Documents. In *Proceedings of the 2nd International Symposium on Smart Graphics* (pp. 129–135). New York:ACM Press.

Kosara, R., Miksch, S., & Hauser, H. (2002). Focus + Context Taken Literally. *IEEE Computer Graphics and Applications*, 22(1), 22–29.

O'Hara, K. (1996). Towards a Typology of Reading Goals. Technical Report EPC-1996-107, Rank Xerox Research Centre.

Phelps, T.A. & Wilensky, R. (1998). Multivalent Documents: A New Model for Digital Documents. Technical Report No. CSD-98-999, Division of Computer Science, University of California at Berkeley.

Robertson, G.G., Mackinlay, J.D. (1993). The Document Lens. In *Proceedings of UIST'93* (pp. 101-108). New York:ACM Press.

W3C DOM Working Group (2002). Document Object Model (DOM) Level 3 Core Specification, Version 1.0. Technical report, World Wide Web Consortium, 2002. http://www.w3.org/DOM/.

Interface Issues for Accessing and Skimming Speech Documents in Context with Recorded Lectures and Presentations

Wolfgang Hürst, Lakshmi Siva Kumar Alapati Venkata

Institut für Informatik, Universität Freiburg
Georges-Köhler-Allee, D-79110 Freiburg, Germany
{huerst, alapati}@informatik.uni-freiburg.de

Abstract

In this paper we discuss different ways to present, skim, and access audio files containing recorded presentations or lectures. Such acoustic signals generally have special characteristics and features that might cause problems when using common techniques for audio access. The purpose of this paper is to evaluate commonly used approaches for presentation and interaction with audio files and to identify their usability and limitations in context with recorded lectures. We discuss different alternatives for presenting audio documents, for example, as a result of an audio-based search engine, as well as methods for accessing and skimming recorded lectures and we identify which techniques are useful (and which are not) in our scenario and therefore should (or shouldn't) be integrated in a final user interface.

1 Introduction

Today, universities commonly use the World Wide Web to distribute teaching and learning material and to make course documents (such as slides from lectures, assignments, as well as other complementary material) accessible to their students. With the advent of multimedia technology these documents are no longer restricted to text or static images but can include continuous media, such as audio or video recordings, as well. At the same time, systems for automatic presentation capturing are becoming more and more popular. For example, many professors at our university record their lectures using commercial presentation recording software or a public domain version of a system called AOF (Authoring on the Fly, *Müller 2000*) that was developed by our research group. The resulting multimedia files, each of which contains one recorded presentation, are made available to the students over the local university network. Therefore, the students do not only have access to the textual material of a course but to the audio recordings, as well.

This paper deals with the usage of these audio files by the students and addresses the question how those documents should be made accessible, how they should be presented, and what user interfaces are most suitable for these purposes. Usage studies with recorded presentations (see *Zupancic 2002*, for example) indicated that the students want to use these files for various tasks, for example, to repeat things they haven't understood completely, to look-up a specific information (e.g., the explanation of an algorithm), to get additional information about a topic, and so on. Therefore, users must be able to get a quick overview of the content of an audio file as well as to find particular parts of interest within a longer document in a fast and easy way. In case of textual documents, aids such as headlines, punctuation, or the layout of the text (such as paragraphs, sections, different font styles and sizes) support users in skimming the data. No such information exists in case of audio because of its dynamic nature. In addition, there is no natural way to skim acoustic data in the same way as a human is able to run over some textual information by just looking at it. As a consequence, there are many research efforts that focus on questions such as how audio files can be presented properly to the users, how they should be made accessible, how the interface can optimally support audio browsing and skimming, and so on. In the following, we review the most important of these approaches in context with recorded lectures. We discuss which of them are useful in our scenario and identify limitations, advantages, and disadvantages of different techniques.

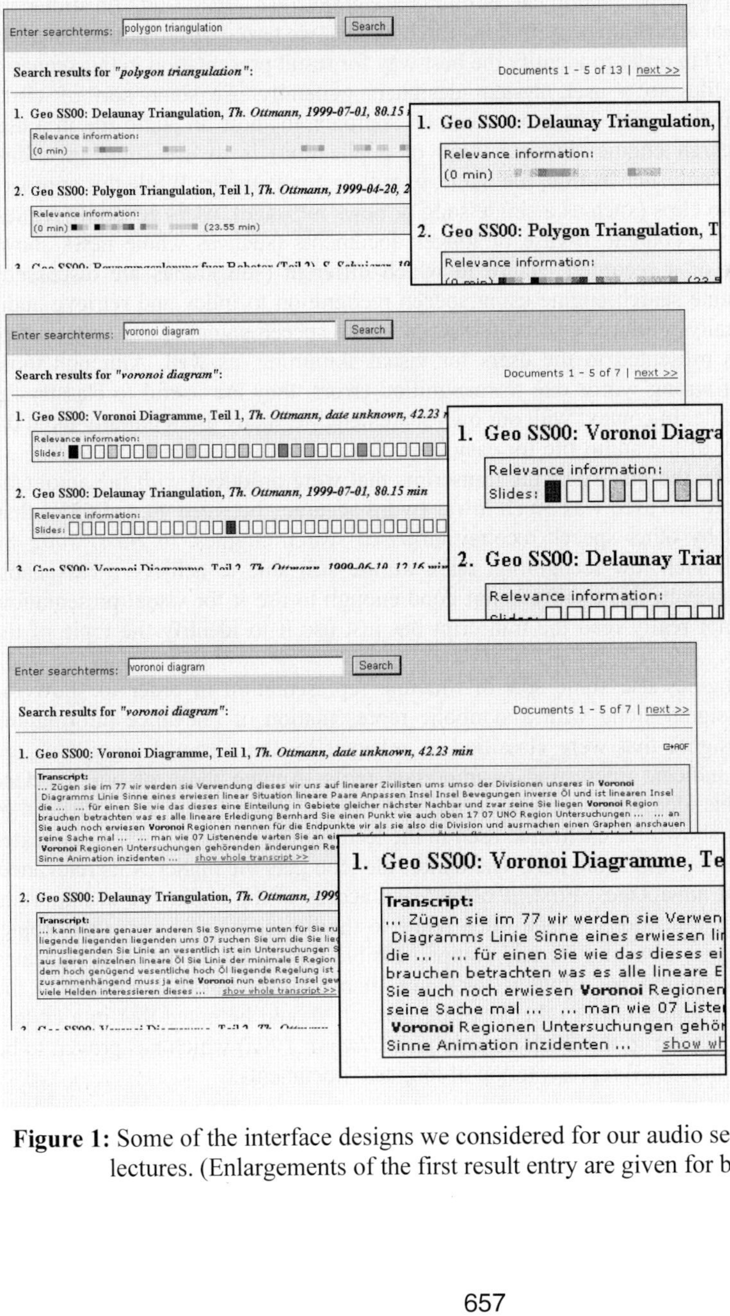

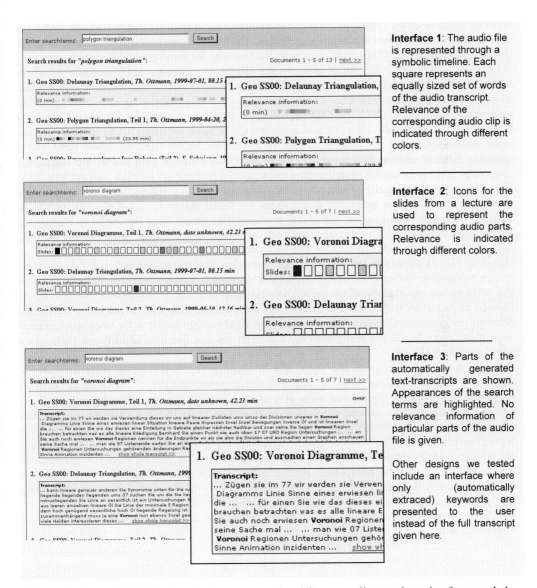

Interface 1: The audio file is represented through a symbolic timeline. Each square represents an equally sized set of words of the audio transcript. Relevance of the corresponding audio clip is indicated through different colors.

Interface 2: Icons for the slides from a lecture are used to represent the corresponding audio parts. Relevance is indicated through different colors.

Interface 3: Parts of the automatically generated text-transcripts are shown. Appearances of the search terms are highlighted. No relevance information of particular parts of the audio file is given.

Other designs we tested include an interface where only (automatically extraced) keywords are presented to the user instead of the full transcript given here.

Figure 1: Some of the interface designs we considered for our audio search engine for recorded lectures. (Enlargements of the first result entry are given for better reading.).

2 Visual Representation of Speech Signals

The question of how an audio document should be initially presented to a user is especially important in context with search engines. Even the best search engines usually do not return a 100% correct result but always deliver documents that have low relevance, as well. Therefore, it is very important to present the retrieval result in a way that allows users to quickly identify the files that are really important for their particular information needs. In our research group, we started to develop a search engine for the audio files of lectures that have been recorded using our AOF software (a demo of the prototype version can be found at *aofSE 2002*; *Hürst 2002* illustrates the used approach and retrieval algorithms). With this search engine, we tested different user interface designs (compare *Figure 1*) in order to identify the best way for result presentation in our scenario. When presenting audio files to a user, system designers generally use some sort of visual representation of the acoustic signal. The most common approach to show an audio file in a user interface (not only in search engines but any kind of application) is to use some meta-data representing the audio content, such as title, file name, or author, for example. While this approach is sufficient for short audio clips (such as a single song or news message), users generally request more information about the content in case of longer documents (such as whole news shows covering different events or a recorded lecture in which different (sub-)topics are discussed). *SpeechBot 2002* is an online search engine using speech recognition to index and retrieve audio files. Here, the automatically produced textual transcript of the speech signal is not only used to search the files, but it is presented to the users for visual skimming, as well. Although those transcripts contain lots of wrong words due to recognition errors, they are useful to classify the overall topic of the audio file. In context with our own search engine, many users initially liked the idea of having a transcript of the audio file for easy browsing and access (compare "*Interface 3*" shown in *Fig. 1*). However, the quality of the transcripts that were produced with the automatic speech recognition software we used was much lower (word accuracy between 40 and 55%) than the performance reported by other speech recognition-based search engines. In *Hürst 2002* we showed in that even with such low recognition rates we are still able to achieve a reasonable retrieval performance. However, this quality is not good enough to use it for visual presentation, even if the users would not really read the transcript but just use it to identify the topic of the corresponding audio file.

A lot of audio search engines use some sort of timeline representation in order to show the duration of the acoustic signal along with a symbolic representation of relevance information. Positions of the speech signal that were classified as relevant by the retrieval algorithms are somehow highlighted to indicate interesting or important parts. Among those of our interface designs which followed this idea, most users preferred "*Interface 2*" from *Fig. 1*. Here, instead of a pure timeline, an icon is shown for each slide used in the lecture. The color of the icon indicates the relevance of the corresponding audio part. The darker the icon gets the higher is its relevance. This approach has several advantages. First, it offers easy access to the audio file at important positions, namely the slide transitions, which often relate to the beginning of a new (sub-)topic. The presented information is very intuitive and comprehensible for the users. It clearly indicates which positions of the recording have higher relevance, if a relevant position appears at the beginning, at the end, or in context with other relevant parts. It should be noted that in a certain way this approach is very similar to the TileBars approach (*Hearst, 1995*) which has proven to be a successful technique for the visual representation of long text documents.

3 Direct Skimming of a Speech Signal

Other approaches to support user browsing of audio files (instead of choosing a visual representation) are based on direct skimming of the acoustic signal. Most of these techniques were originally pioneered by *Arons 1997* and successfully used in other systems, as well. They use, for example, faster replay to enable quick navigation and skimming of speech documents. Studies have proven that humans are still able to classify the topic of a speech conversation even if the file is replayed twice as fast. In addition, those approaches use simple audio analysis techniques for automatic detection of intonation, emphasis, as well as longer parts of silence. Since such characteristics often indicate the beginning of a new topic, they can be used, for example, to enable more comfortable navigation, to shorten the replay, or to automatically generate audio summaries.

Although these techniques work well in various real-world applications (see *Stifelman 2001*, for example), it turned out that when using them in context with our task the same quality of results could not be achieved. We implemented several algorithms similar to *Arons 1997* in order to automatically create a content-based segmentation of the audio signals using simple features such as short-term energy, shot-term zero-crossing-rate, or pitch-detection techniques based on autocorrelation, for example. We evaluated those segmentation algorithms using four lectures from three different speakers that were recorded with our AOF lecture recording software. Using a "perfect", manually generated segmentation for comparison, we calculated precision and recall, two commonly used performance measures in context with information retrieval (see *Baeza-Yates 1999*, Chapter 3, for example), in order to estimate the performance of the segmentation algorithms. We evaluated two different approaches, one based on pauses made by the speakers and one based on automatic emphasis-detection. Overall performance was very low. For example, with pause-based segmentation we got average precision and recall values of 16% and 47%, respectively, meaning that only 47% of all segment borders were found and that only 16% of the segments that were calculated by the algorithm were correct, while the other 84% were wrongly classified. Emphasis-based segmentation was even worse with precision rates as low as 10%.

The most likely reason for such a low performance is that we are dealing with a very different speech signal. First of all, recording is usually done in a classroom or lecture hall (with significant background noise). This and the usage of cheap recording equipment results in a much lower audio quality compared to a high-quality studio recording. In addition, the speech signal produced by the lecturer usually differs from the clear and well-formulated speech used by, for example, professional speakers of news shows. In lecture recording, we are dealing with rather spontaneous and colloquial speech, that has more of a conversational character. Grammatically incorrect sentences, abrupt stops in the middle of a sentence or even within one word, filler words such as "ehm" and "hmm", result in a speech signal with much more variety. Such variations in the signal cause large problems for the automatic segmentation algorithms. In addition, common rules about the characteristics of the usage of human speech (e.g., that long pauses generally indicate the beginning of a new topic) are not always true in case of recorded lectures. For example, by analyzing several recordings from our database we identified significant pauses or other variations in the speech signal that were not motivated by the meaning of the words but by unrelated events. For example, people usually speak much slower when they write on a blackboard and speak at the same time. However, when applying our implementation of the algorithms to recordings of radio news shows we were able to get a performance that was comparable to the one commonly reported in literature by other projects that successfully used these algorithms with clear, high-quality speech signals.

Of all the approaches that are commonly used in order to support a user in browsing and skimming audio documents, the only one that seems to be useful in our scenario is the time compressed replay of audio files. Faster replay of speech files can be used, for example, to quickly skim lecture recordings in order to get an idea of the content (e.g., in order to find the correct results among a set of documents returned from a search engine), or, for example, to find a particular information in a longer document (e.g., the definition of a technical term). However, faster replay usually results in a shifting of the voice, making the presenter sound like a cartoon character. This effect can be avoided using algorithms such as the synchronized overlap add method (SOLA), harmonic compression, or phase vocoding (compare *Arons 1997*). Parts of our current work focus on evaluating different user interface designs that integrate faster audio playback into the AOF player software.

4 Conclusion

In this paper we reviewed commonly used approaches to present and access audio files and discussed their usability in a specific task, namely dealing with recorded presentations and classroom lectures. We argued why those type of documents pose difficulties and the usage of common approaches is not straight-forward and we evaluated the usability of these techniques in our scenario. Our current work focuses on the question if and how the visual information that is sometimes available together with the recorded acoustic signal (such as slides that might have been used in a lecture) can be used in addition to the audio in order to improve skimming of the files. The tests with using faster replay in order to search and skim the audio files indicated that it seems to be a very promising approach to combine audio skimming with the presentation of the available visual information that changes over time, such as the slides used in the lecture.

References

aofSE - Audio Retrieval Demo (2002). *http://ad.informatik.uni-freiburg.de/mmgroup/aofSEaudio/*

Arons, B. (1997). "SpeechSkimmer: A System for Interactively Skimming Recorded Speech." *ACM Transactions on Computer-Human Interaction*, Volume 4, Number 1.

Baeza-Yates, R., Ribeiro-Neto, B. (1999). "Modern Information Retrieval." *ACM Press New York Addison-Wesley.*

Hearst, M.A. (1995). "TileBars: Visualization of Term Distribution Information in Full Text Information Access." *Proceedings of the ACM SIGCHI 1995*, Denver, CO, USA.

Hürst, W., Kreuzer, T., Wiesenhütter, M. (2002). "A qualitative study towards using large vocabulary automatic speech recognition to index recorded presentations for search and access over the Web." *Proceedings of IADIS WWW/Internet 2002 Conference*, Lisboa, Portugal.

Müller, R., Ottmann, T. (2000). "The Authoring on the Fly system for automated recording and replay of (tele-)presentations." *ACM/Springer Multimedia Systems Journal*, Volume 8, Number 3.

SpeechBot – audio search using speech recognition (2002). *http://speechbot.research.compaq.com/*

Stifelman, L., Arons, B., Schmandt, C. (2001). "The Audio Notebook - Paper and Pen Interaction with Structured Speech." *Proceedings of the ACM SIGCHI 2001*, Seattle, WA, USA.

Zupancic, B., Horz, H. (2002). "Lecture Recording and its Use in a Traditional University Course." *Proceedings of ITICSE 2002, 7th Annual Conference on Innovation and Technology in Computer Science Education.* Aarhus, Denmark.

Acknowledgments: This work is supported by the German Research Foundation (Deutsche Forschungsgemeinschaft DFG) as part of the strategic research initiative "V3D2" ("Distributed Processing and Delivery of Digital Documents"). In addition, the authors like to thank Jürgen Dick for valuable input and support.

TOOL DEVICE:
Handy Haptic Feedback Devices Imitating Everyday Tools

Youichi Ikeda, Asako Kimura, Kosuke Sato

Graduate School of Engineering Science, Osaka University
1-3 Machikaneyamacho, Toyonaka, Osaka, Japan
ikeda@inolab.sys.es.osaka-u.ac.jp

Abstract

In this paper, we propose handy haptic devices named Tool Device. Tool Device is designed with metaphor of everyday tools, such as scissors, tweezers and syringe, which have good shape affordance by themselves, and allows users seamless manipulation of multimedia data in an extended information environment. Just as users feel haptic feedback from everyday tools while handling physically, the Tool Devices can also have haptic feedback to show the quantity and freshness of the handling data. We developed two types of the Tool Devices, a Syringe Device and a Tweezers Device, and studied their application to the manipulation of music and text data. Results showed that the Tool Devices were easy to manipulate because of their shape and haptic feedback, and users could use them without any training.

1 Introduction

Because of digital technologies' improvement, users not only can manipulate various types of multimedia data in one machine, but also can transfer or share those data with other machine via high speed network. In this paper, we propose handy input devices named Tool Devices, whose designs imitate shapes and haptic feedbacks of everyday tools. The Tool Devices allow users to manipulate multimedia data, such as text, image and sound, in an extended information environment simply and seamlessly.

D. A. Norman pointed out that users form mental models through experience, training and instruction (Norman, 1990). Since novice users have not undergone any training and instruction, experience is the only trigger for them to estimate the manipulation method. Everyday tools, such as scissors, tweezers and syringe, have good affordance by themselves. Moreover most users have experience of using them. An advantage of the Tool Devices is that, since users already know how to use the original physical tools, even novice users can easily apply those methods to Tool Device manipulation.

In addition, while using an everyday tool, haptic feedback to users' fingers helps them to sense what kind of object they are handling. On the other hand, while using a mouse or a keyboard, there is no haptic feedback that shows how many documents a file includes or when a document is updated. To make intangible digital data touchable, we added force and thermo feedback functions to the Tool Device, which display quantity and freshness of the data.
In this paper, we introduce two types of the Tool Devices: a Tweezers Device and a Syringe Device.

2 Related work

Computer augmented environments incorporate benefit of digital information into real world. For instance, Tangible Bits (Ishii & Ullmer, 1997) feature objects' form and haptic feedback physically to display digital data instead of graphical user interface. Pick-and-Drop (Rekimoto, 1997) uses a physical tool (pen) to transfer digital data directly between information appliances. In these works, user can manipulate digital data by handling objects or tools physically. However both interface systems require users to learn how to use them. The Tool Devices also use objects to manipulate digital data physically, but, since the Tool Devices imitate shapes and haptic feedbacks of everyday tools, users have already known how to use the original tools and can apply those methods to Tool Device manipulation.

Haptic feedback plays an important role in the Tool Devices. To be equipped into a handy Tool Device, the haptic feedback device should be small. Various haptic feedback systems have already been developed, such as PHANTOM ("The PHANTOM"). However, most of them are large systems. Though vibration is used in some small haptic feedback systems, it is a simple signal and cannot display haptic feedback of every day tool manipulation. In our system, we develop small haptic feedback system that displays natural haptic feedback, such as hardness of object.

3 Tool Device

3.1 Tweezers Device

Tweezers is a tool that can clip and release physical objects. When users clip an object, reaction force from object appear to their fingers through the tweezers.

In contrast, the Tweezers Device can clip and release digital data by closing and opening it. When users close the device to clip digital data, the reaction force to open tweezers is displayed to their fingers (Fig.1). The more data users clip, the stronger force is displayed. By opening the device, clipped data is released and the reaction force disappears. This idea is based on the fact that users can recognize size and hardness of object from the reaction force.

The feedback force is created by DC servomotor attached to fulcrums of Tweezers Device and magnitude of the force is controlled by the voltage to the motor. We use a voltage controller and a servo motor controller via RS-232C ports of Windows PC.

3.2 Syringe Device

A syringe is a tool that can suck up and push out liquid. Users can get liquid by pushing and releasing syringe body, and then they can push out the liquid by pushing its body again. The more liquid users get in syringe, the harder its body expands.

In contrast, a Syringe Device enables users to suck up and push out digital data. Firstly users point target data with a head of Syringe Device and push the device body. The depth of user's push controls how much data to be sucked up. When users release the body, the data is copied to the device and the body expands and becomes hard. To paste copied data to other appliances, users must push the device body on them. Then its body becomes soft and returns to its original hardness.

Similar to Tweezers Device, hardness feedback is created by DC servomotor inside a device body. Users can sense how much data has been sucked up into the Syringe Device by the hardness of the device body. Besides, the Syringe Device has another haptic feedback: a thermo feedback that displays freshness of the data. If the sucked data is new, the surface of the Syringe Device becomes hot. The device becomes cold for the old data. This feedback is based on a metaphor of cooking and is controlled by two Peltier elements (thermoelectric modules). A relay controller switches the Peltier elements on/off via PC's RS-232C ports.

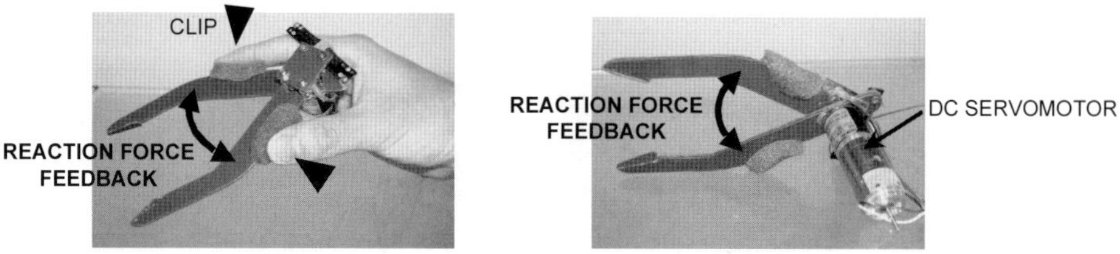

Front Back

Figure 1: Tweezers Device

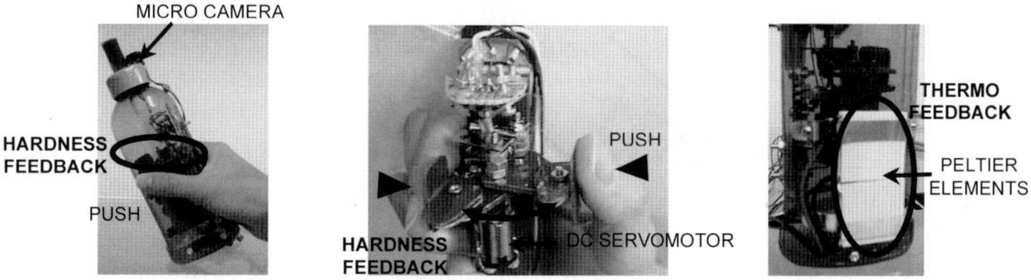

Front Inside Back

Figure 2: Syringe Device

4 Applications

We constructed music player and text editor applications, which enable users to manipulate music data and text data by the Tool Devices simply and seamlessly in an extended information environment. In both applications, users can clip or suck up data between both digital appliances, such as a touch panel display, a speaker and a printer, and real objects, such as CDs and books (Fig.3).

In the music player application, users can select music data from music title lists on PC's touch panel display or CD labels. The selection method using the Tweezers Device is to touch the front and end position of titles and close the tweezers' hands (Fig.4(a)). To play the clipped music, one can release them in a data-loading cup on a speaker by opening the tweezers. There is a touch switch in the data loading cup to detect whether the data are released in the cup or not (Fig.4(b)). With Syringe Device, users can suck up music titles by pointing the head to them (Fig.5(a)) and play it by pushing them out in the loading cup (Fig.5(b)).

Each haptic feedback of the Tool Devices is linked to features of music data. The reaction force and hardness feedback represent the number of selected music titles and thermo feedback shows freshness of data. In addition to such haptic feedback, graphical animation also supports user's "clipping" and "sucking up" manipulation. When users close tweezers, an image of selected data is compressed, then disappears (Fig.4(a)). When users suck data up by syringe, the data image becomes smaller and smaller like sucked liquid (Fig.5(a)).

To clip and suck up data from real objects, such as CD and MD, we attached a micro camera to Tool Devices and identify each object by color code ID markers posted on it. (Fig.6).

In a text editor application, text data can be copied from PC's touch panel or paper to networked PC or printers.

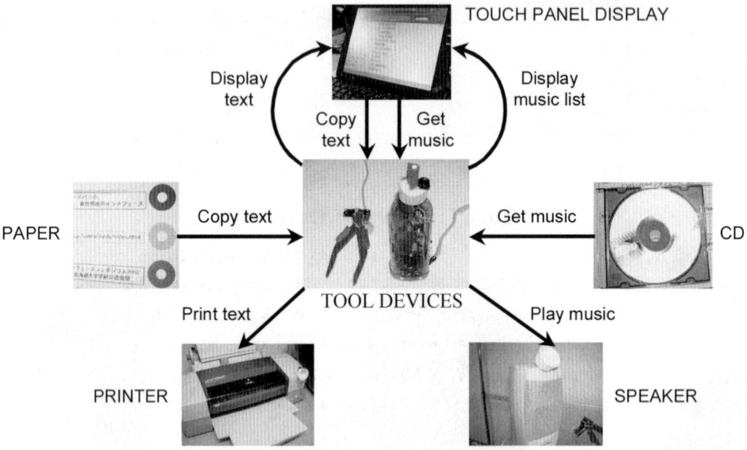

Figure 3: Tool Devices' manipulation in extended information environments

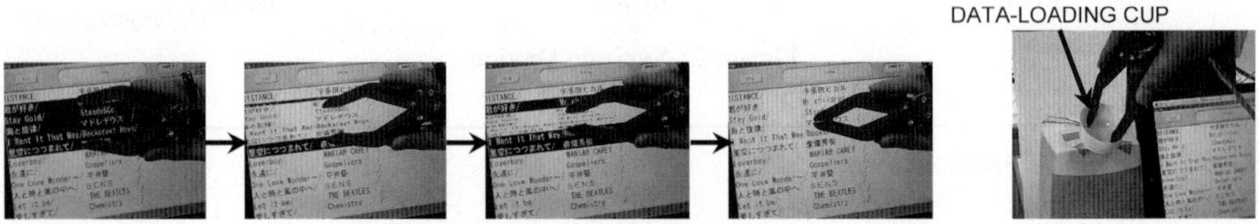

(a)Clipping music data **(b)Releasing music data**

Figure 4: Data manipulation using Tweezers Device

(a)Sucking music data up **(b)Pushing music data out**

Figure 5: Data manipulation using Syringe Device

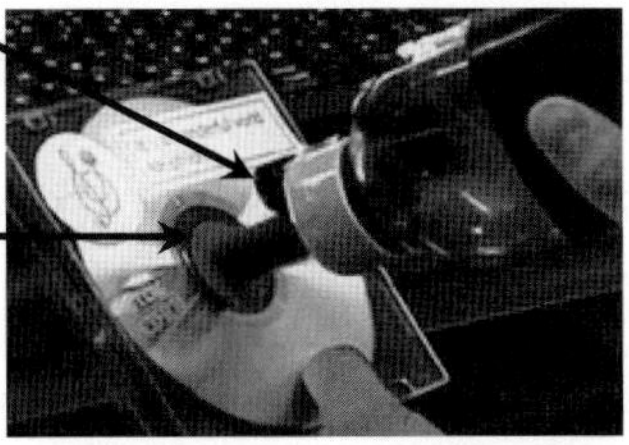

Figure 6: Identification of real objects with color code

5 Evaluation

Experimental task for evaluation was to select some music from music title lists on PC's touch panel display and play them on a speaker. Eight test users, 3 females and 5 males, assessed the easiness of the manipulation using the Tweezers and Syringe Device compared with a mouse and a keyboard. The task for the users was to select some music titles from a music title list on PC's touch panel display and play them on a speaker. They were not instructed or trained before experiments. After experiments, they were interviewed with the impression of the Tool Devises' manipulation and haptic feedback.

The results show that the test users could easily understand how to use Tool Devices and what haptic feedback meant. All test users could finish the task using Tool Devices. All found that the reaction force increased as more music titles were clipped. Seventy five percent of them understood the hardness feedback denoted how many music were sucked up, and 63% found that thermo feedback displayed whether music data were old or new. Most test users commented that the Tool Devices were more enjoyable and entertaining than the mouse or keyboard.

6 Conclusion

We proposed a Tool Device designed as a metaphor of shape and haptic feedback of everyday tools. A Tweezers Device with reaction force feedback and a Syringe Device with hardness and thermo feedback were developed as the examples of the Tool Devices. We also constructed Tool Device applications, where users can manipulate music and text data seamlessly from computer display, CD, or paper to networked computer, speaker or printer. The evaluation experiments showed that the test users could easily understand how to use Tool Devices and what haptic feedback meant.

We believe that a lot of appliances could be targets to be controlled by the Tool Devices. For future work, we are designing other type of the Tool Device and its application.

References

Norman, D. A. (1990). The design of everyday things. Currency Doubleday.

Ishii, H., & Ullmer, B. (1997). Tangible Bits: Towards seamless interfaces between people, bits and Atoms. *Proceedings of Conference on Human Factors in Computing Systems*, 234-241.

Rekimoto, J. (1997). Pick-and-Drop: A direct manipulation technique for multiple computer environments. *Proceedings of UIST'97*, 31-39.

The PHANTOM haptic interface, from http://www.sensable.com/

Tangible Media Player with embedded RF tags*

Seiie Jang, Ning Zhang and Woontack Woo

KJIST U-VR Lab.
Gwangju 500-712, S. Korea
{jangsei, zhangn, wwoo}@kjist.ac.kr

Abstract
We present a new type of media system, nomadic Tangible Media System (*n*-TMS), which allows users to play or control digital media by adopting tangible interface. In order to replace conventional interface such as mouse or keyboard, we introduce two types of RF-enabled objects, i.e. media and control objects. We first attach RF tag to each content object and then retrieve media data using the media object. Using the control object, user plays digital media at any networked computer with RF tag reader. According to experimental results, it offers a simple and intuitive interface with tangible objects for those who are not familiar with computers. The proposed *n*-TMS can be applied to various applications such as interactive education, entertainment, smart toys, etc.

1 Introduction
Rapid development of computer and communication technologies helps people access multimedia over network. Since the concept of "Ubiquitous Computing [1]" was introduced in early 90's, there have been various research activities supporting Weiser's idea, i.e. integrating computers seamlessly into the real world as parts of the environment. Meanwhile, manipulating a media player in a computer is still inconvenient for people who are not familiar with computers even though interactions are largely confined to Graphical User Interface (GUI). Recently, research activities on Tangible User Interface (TUI) have paved a new way to overcome such inconvenience in connecting human and computer. For example, stimulated by Weiser's idea, Ishii et al. looked forward to exploiting physical objects as new interface to access digital information. The "Tangible Bits [2]" allows users to "grasp & manipulate" bits in the user-centric views by coupling the bits with everyday physical objects and architectural surfaces. Tangible bits also enables users to be aware of background bits at the periphery of human perception using ambient display media such as light, sound, airflow, and water movement in an augmented space. Such activities on new interface will provide easy, aesthetically pleasing and emotionally engaging access to digital information even for those who are not familiar with computer.

As explained, Tangible Bits provides intuitive interface for people to control media data by adopting physical objects to access digital information and vice versa. The advantage of Tangible Bits is that they bridge the gaps between cyberspace and the real world, as well as the foreground and background of human activities. For example, the musicBottle [3], genieBottles [4], and MusiCocktail [5] projects are representative applications weaving Tangible Bits into music player. The musicBottle and genieBottles present an interface to access digital information (music or story) using glass bottles as "containers" and "controls". The design gives digital contents to bottles and exploits the human senses of touch and kinaesthesia. While, MusiCocktail allows users to influence certain parameters of a piece of music in the way they mix their beverages. MusiCocktail can enhance the interaction in a social space by providing a new and entertaining

* This work is supported by IITA.

form of interaction with music, by strengthening the sense of community through collaboration. However, all the interactions are focused on the opening and closing of bottles to release virtual contents and the manipulation is taken place only in the special installation table. In addition, the systems also have limited choice of music and stories.

To overcome such disadvantages, we propose a new type of media system, nomadic Tangible Media System (*n*-TMS) with RF-enabled objects. In order to replace conventional interface such as mouse or keyboard, we introduce two types of RF-enabled objects, i.e. media and control objects. We first attach RF tag to each content object (e.g. CD case, poster card, toy, etc.) and then retrieve media data using the media object. The control object plays the retrieved media in the networked computer with RF tag reader. The proposed *n*-TMS offers a simple and intuitive interface with tangible objects for those who are not familiar with computers to easily play or control digital media in computer or over Internet without using mouse and keyboard [6]. The proposed *n*-TMS can be applied to various applications such as interactive education, entertainment, and smart toys.

This paper is organized as follows. In Section 2 we describe the proposed *n*-TMS. The implementation and experimental results are explained in Section 3 and 4, respectively. Finally, the conclusion and future works are discussed in Section 5.

2 Nomadic Tangible Media System

As shown in Figure 1, this system consists of Tangible Object (TO), RF Module (RFM), and Tangible Media Player (TMP). TO contains information about digital media and controls for media player in RFID tags. RFM plays a role in reading or writing the information in a RFID tag, and TMP plays digital media according to the information.

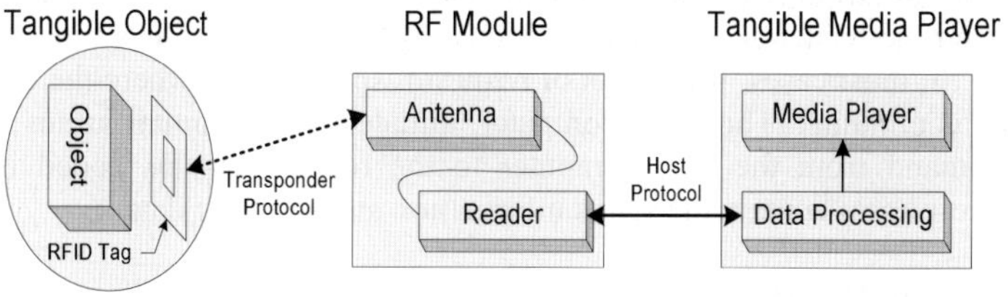

Figure 1: Component of *n*-TMS

2.1 Tangible Object

TO combines cyber space and real world by attaching RFID tags to the objects in real world. RFID tag is a thin and flexible "smart label" which can be laminated between paper and plastic, and contains eight memory blocks, where each block consists of four bytes, as shown in Figure 2. If first block in TO is '00000000', it is a control object, while '00000001' indicates that it is a media object.

As explained, we introduce two types of RF-enabled objects, i.e. media and control objects. As shown in Figure 2(a), Media object (e.g. CD case, poster card, etc.), containing the URL of the multimedia in a RFID tag, consists of number of directory, host address, directory, and file. The number of directory represents the depth of directory of digital media in a remote computer. For example, if the number of directory is 00000010, it means that first two memory blocks and last block, respectively, contain the name of the directories and the file name of the media. Meanwhile, control object (e.g. dice or toy) controls the retrieved media with RFID tags representing 'Play, Stop, Pause, FF, RW, VolumeUp, and VolumeDown' according to the data of control state, as shown in Figure 2 (b).

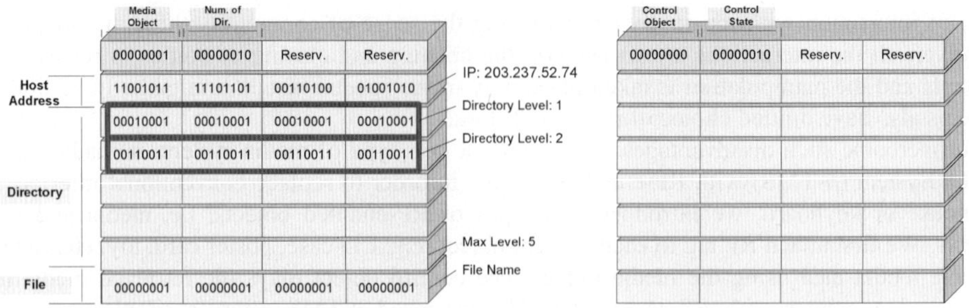

(a) Media Object (b) Control Object

Figure 2: RFID Tag for Media/Control Object

2.2 RF Module

The RFM consists of a tag antenna and a tag reader. To transfer signal from TO to TMP, RFM uses two protocols: transponder protocol and host protocol.

Transponder protocol defines a way to communicate between a tag reader and RFID tags. The transponder protocol is activated when a RFID tag is inside the working area of the tag antenna. Then the antenna transmits the frequency to supply the operating energy and to transmit the data from the tag reader to the tag. The tag derives the operating energy from the transmitted frequency to response to the reader and returns the requested data. The antenna receives the coded signal from the tag and the reader starts to communicate with the host computer.

Host protocol defines a way to communicate between the host system and the tag reader. It is designed for point-to-point, half-duplex communications, with the host controller acting as primary station and the reader as secondary station. Host protocol also is a binary and bytecount-oriented protocol. In most cases, it consists of request/response pairs where the host waits for the response before continuing. The host computer initiates all communications using the host protocol. A command from the host computer to the reader contains coded instructions and parameters. The responses contain the status information and resulting data.

2.3 Tangible Media Player

TMP consists of data processing module and media player. Data processing module transforms the RFM signal in a media object to the URL or RFM signal in a control object to the command of TMP and vice versa. It also allows users to easily make a media object through the 'writing' command. Media player automatically retrieves with the URL and plays the media according to the commands in control object, e.g. Stop (00000000), Pause (00000001), Play (00000010), FF (00000011), RW (00000100), VolumeUp (00000101), VolumeDown (00000110), etc. As shown in Figure 3, the hash table is referred by media player to determine the directory and file name in the media object.

Key	Value
00010001/00010001/00010001/00010001	Demo
00110011/00110011/00110011/00110011	Movie
01110011/01110111/01110011/01110011	Music video

(a) Hash for Directory

Key	Value
00000001/00000001/00000001/00000001	U-VR.meg
00000011/00000011/00000011/00000011	Rainbow.avi
00000111/00000111/00000111/00000111	Myheart.qt

(b) Hash for File

Figure 3: An Example of Hash Table for Media Object

3 Implementation

The proposed n-TMS is implemented in the 'ubiHome', a testbed for ubiComp-enabled home applications [7][8]. The RFM of *n*-TMS is attached under the table in ubiHome, and it is connected to PC where TMP is installed through the RS232 serial port. The resulting media was displayed on TV in ubiHome.

As shown in Figure 4(a), there are two types of TO. A dice is used as a control object by attaching the RFID tags to each side and both CD case and photo of music video are used as media objects by sticking the RFID tag of media object on the objects. Figure 4(b) shows RFM (Texas Instrument's S600 Reader and Antenna Set-RI-K01-320A). TMP, as shown in Figure 4(c), is implemented in JMF (Java Media Framework) which supports some formats of digital media (e.g., .wav, .avi, .mpeg and .qt).

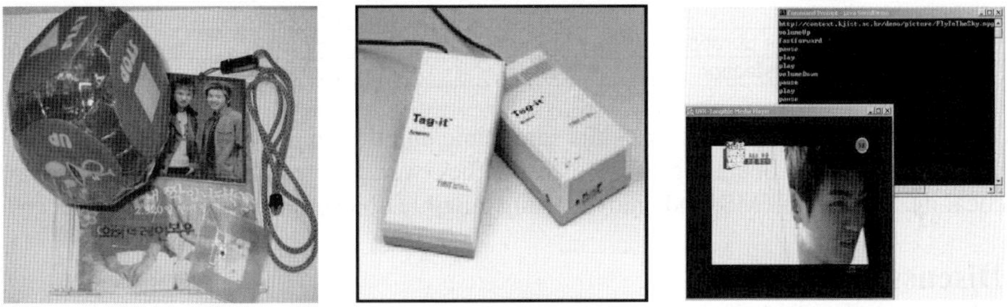

(a) Tangible Object (b) RF Module (c) Tangible Media Player

Figure 4: The Implemented Components of *n*-TMS

Figure 5 and Figure 6, respectively shows the implemented *n*-TMS and the corresponding working area of RFM in ubiHome.

Figure 5: The implemented *n*-TMS in ubiHome

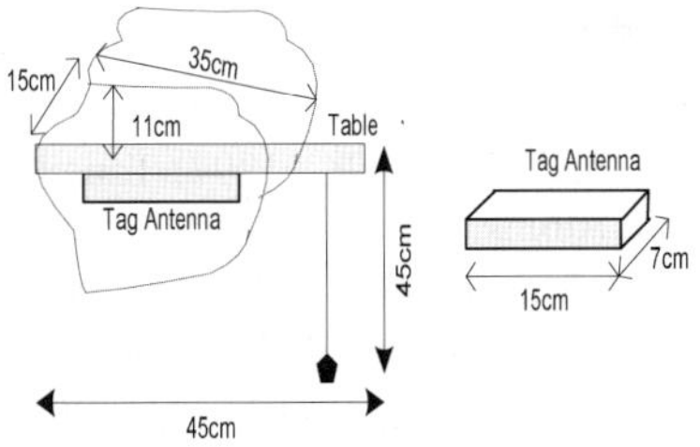

Figure 6: The working area of *n*-TMS

4 Experiments

To show the usefulness of *n*-TMS, we compared TUI-based system (n-TMS) with GUI-based system (WMP: Window Media Player). To perform subjective evaluation, ten volunteers were participated to test both systems. Half of them are above fifties (group A) who are not familiar with computer, while the remaining half (group B) are in twenties with good knowledge of computers. We provided them with the URL information of digital media. We measured the time duration from selecting a digital media to playing it. We also obtained the information about the user's satisfaction after they were given a chance to use with conventional means.

Table 1: The result of Group A

	WMP	*n*-TMS
Time duration	>> 30 min	2~3 min
Satisfaction	Low	High

As shown in Table 1, while most of the group 'A' felt difficulties in operating WMP with mouse and keyboard, they easily controlled *n*-TMS with tangible objects. All of them were satisfied with *n*-TMS as they did not feel any inconvenience to select digital media and control media player.

Table 2: The result of Group B

	WMP	*n*-TMS
Time duration	1~2 min	1~2 min
Satisfaction	Normal	High

As shown in Table 2, group 'B' took the same time duration to use both *n*-TMS WMP because they were familiar with operating PC. However, they were interested in the proposed easy and aesthetical interface of *n*-TMS, and enjoyed using tangible object in the form of a necklace.

5 Discussion

While the conventional Media playing systems use keyboard and mouse as GUI-based interface, the proposed *n*-TMS uses a physical object such as dice, CD case or photo of music video as TUI-based interface. According to some preliminary experimental results and subjective evaluation, we observed that the proposed *n*-TMS allows users an easy, aesthetically pleasing and emotionally engaging access to digital media. In addition, it offers a simple and intuitive interface for those who are not familiar with computers. Without loss of generality, we expect it can be applied to other general applications of PC such as e-mail and web browser.

6 References

[1] Wesier, M. (1993) Hot Topics: Ubiquitous Computing, IEEE Computer.

[2] Ishii, H., and Ullmer, B. (1997) Tangible Bits: Toward Seamless Interfaces between People, Bits and Atoms. In Proc. of UIST, pp. 173-179.

[3] Ishii, H., Mazalek, A., Lee, J. (2001) Bottles as a Minimal Interface to Access Digital Information, CHI Extended Abstract, ACM Press.

[4] Mazalek, A., Wood, A., Ishii, H. (2001) genieBottles: An Interactive Narrative in Bottles, SIGGRAPH.

[5] Mazalek, A., Jehan, Tristan. (2000) Interacting with Music in a Social Setting, CHI. ACM Press.

[6] Zhang, N., Jang, S., Woo, W. (2002) Nomadic Tangible Music Player with RF-enabled Sticker, ICAT, vol. 15, No. 1, pp. 297.

[7] Jang, S., Lee, S., Woo, W. (2001) Research Activities on Smart Environment, IEEK, Magazine, vol. 28, pp. 85-97.

[8] Yoon, J., Lee, S., Suh, Y., Ryu, J., Woo, W. (2002) Information Integration System for User Recognition and Location Awareness in Smart Environment, KHCI.

Multi-Modal Fusion Model, a design based on D.A.I.M, the Decoupled Application Interaction Model

Ing-Marie Jonsson

Royal Institute of Technology/Dejima Inc
160 West Santa Clara Street, San Jose, CA 95113
ingmarie@ansima.com

Abstract

New mobile devices with small screens are becoming widely used. These devices can run applications previously constrained to the desktop, thus creating a demand for multi platform access to all types of applications. Further complications arise from restrictions in interaction modes available. Input is often based on a reduced size keyboard, an on-screen virtual keyboard, or a gesture based text- input system using a stylus and touch-screen.
To address these issues we introduce the Multimodal Fusion Model, as design based on the decomposition strategy of the Decoupled Application Interaction Model. The Multimodal Fusion Model provides support for alternative and fused input and output modalities, and it allows interaction models and user interfaces to be tailor-made for particular devices. The Multimodal Fusion Model is a framework for implementing multimodal interaction and natural language dialogues on multiple platforms. Both user input mechanisms and presented output information can be tailored to a particular device.
We also describe two example implementations of the Multimodal Fusion Model effectively illustrating the robustness and flexibility of the overall approach.

1 Introduction

New mobile, handheld devices with small screens are becoming widely used, both for telephony and for text based communication with people and applications. These devices have wireless Internet access and can run and access applications previously constrained to the desktop. The distinctions between these devices and desktops are becoming blurred, and there is a demand for multi platform access to all type applications (Myers, Hollan & Cruz, 1996). This demand is complicated by the restrictions in interaction models available on handheld devices. Input is often based on a reduced size keyboard, an on-screen virtual keyboard, or a gesture based text- input system using a stylus and touch-screen.

To address these issues we introduce the Multimodal Fusion Model (MFM). The MFM provides support for alternative and fused input and output modalities, and it allows interaction models and user interfaces to be tailor-made for particular devices. The challenge for the MFM is to properly combine interaction technologies to replicate the natural style of Human-Human communication.

2 Overview

The MFM has adopted the decoupling strategy of Decoupled Application Interaction Model (Jonsson, 1999); all functions from input and output modalities to the natural language understanding and dialoguing are implemented as separate modules. However, in addition, MFM introduces a separate Fusion Module (FM) to create application input based on the semantic fusion of multiple input streams. (Bolt 1980, Djenidi, Ramdane-Cherif, Tadj, & Levy, 2002).

The MFM is a generic design utilising frame-based integration and temporal information to generate a multimodal command (Nigay & Coutaz, 1995). This combination allows the MFM to handle overlapping user input, and to decide whether input data should be treated separately, combined, or ignored. Multimodal architectures, such as the MFM, provide at least two advantages over unimodal architectures: 1) better error recovery from recognition problems due to higher probability of (at least) partial command recognition 2) many ways to issue the same command. The result is a more robust and natural interaction model for human computer interaction (Oviatt, 2002).

The MFM design is illustrated in Figure 1. The FM is placed between the input/output modalities, the natural language and dialogue management, the application and finally the presentation module and the access devices.

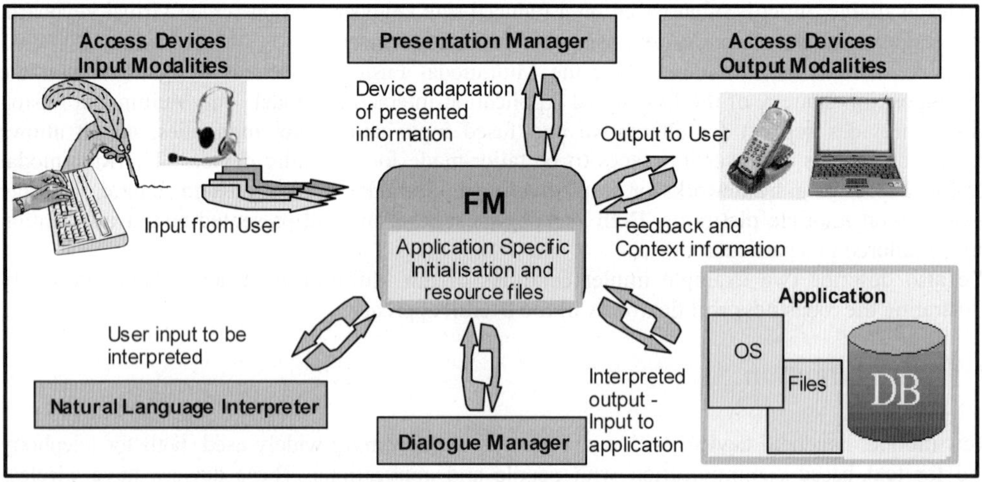

Figure 1: Overview of Multimodal Fusion Model – MFM

The FM is responsible for fusing input from input devices with feedback and context information from the application and the access devices. The fused data is forwarded to the Natural Language Interpreter (NLI). If the fused data is ambiguous or incomplete, the NLI may handover to the Dialogue Manager (DM). The DM can initiate a dialogue with a user and request more information until a viable application command can be discerned and generated by the NLI. The generated command is forwarded to the application. The response from the application is passed to the Presentation Manager that will match the presentation of the response to the access device.

3 Implementation of the MFM

In this section we describe an early and generic MFM that was built for evaluation. The MFM core, i.e. the FM and the inter-module communication are implemented in Java and structured as a group of communicating threads. Each thread is responsible for communication with a particular module of the MFM.

The MFM loads a set of application specific initialisation and resource files during start-up. These files contain information about how the application is started and which resources it utilizes. This includes, for example; connections to databases, operating system and other applications as well as the command vocabulary and fusion style for the application. The MFM core essentially acts as a smart switch that collects user input and feedback from the application and the access devices as incoming events and messages. The messages are matched against the application specific command vocabulary and then combined with previously received messages, ignored, or saved. Saved messages can, for instance, indicate incomplete or ambiguous information and trigger a dialogue with the user to resolve the problem.

The MFM allows for media independent commands and supports incomplete commands (Gourdol, Nigay, Salber, & Coutaz, 1992) through functionality implemented in the FM and the NLI. The FM supports command frames, temporal fusion or a combination thereof. Fused input can essentially be generated in two ways; either by a command frame with all slots filled, or by a timeout event (Djenidi et al, 2002). This means that fused input can be either a complete or an incomplete command when it is sent to the NLI. In the case of incomplete commands, the NLI can generate a partial command for the application or handoff the DM to fill in the missing parts.

3.1 Multimodal System 1

The MFM was used to design a Multimodal Testbed, depicted in Figure 2. The goal of the testbed was to address the problem of user interface design for the 3G wireless networks.

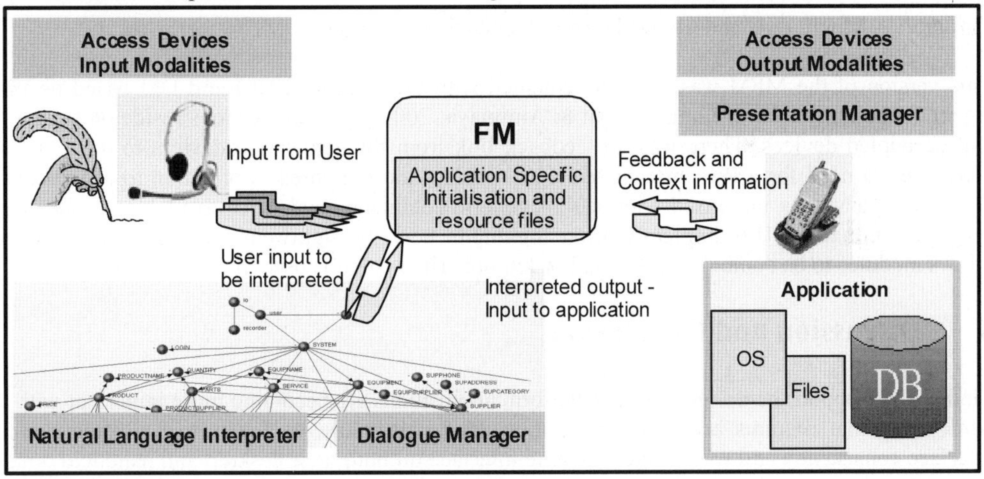

Figure 2: A Multimodal Testbed based on the MFM

The testbed was arranged to use speech and pen input together with text, graphic and speech output on a handheld device. The MFM was implemented with a combined NLI and DM based on an Agent Oriented Platform (Hodjat & Amimaya 2000). The Presentation Manager was integrated with the application, and the FM (based on temporal command frames) was able to handle partially completed commands. A laptop was used to run the Speech Recognizer, the FM and the combined NLI-DM. The Presentation Manager and the application were implemented on a handheld device. This implementation highlights the modular lightweight design of the MFM where component modules are either combined or distributed. This testbed was used in a study with a focus on users' preferred modality and user responses to partially completed commands.

3.2 Multimodal System 2

The MFM was also used to design a system for Multimodal and Multi Platform Applications, depicted in Figure 3. The goal of the system was to provide alternatives to using the keyboard and the mouse to access applications. The system used speech, pen-input, gestures and Morse code as input modalities and speech, audio, text and graphics as output modalities.

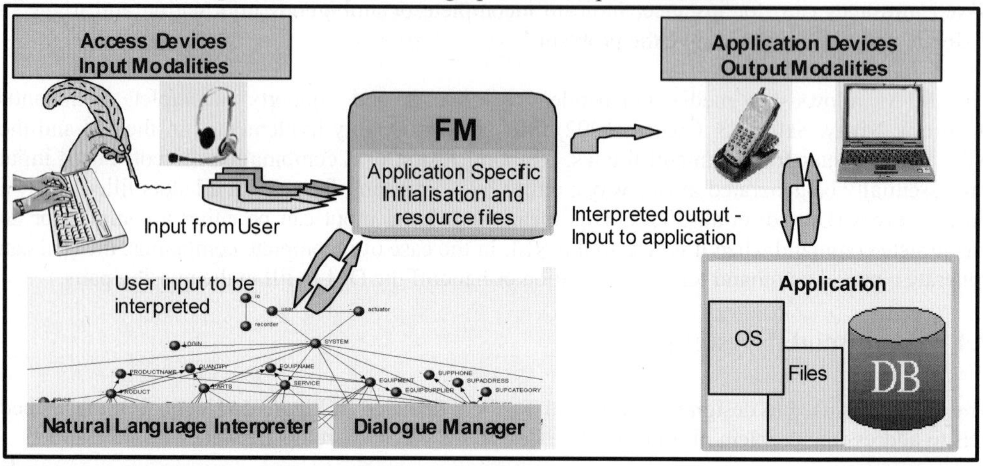

Figure 2: A Multimodal and Multi Platform Application based on the MFM

This version of the MFM was also implemented with the combined NLI and DM based on the Agent Technology from Dejima (Hodjat & Amimaya, 2000). The applications reside on separate and decoupled devices. There was no feedback link from the application devices to the FM, so there was no need for a Presentation Manager to adapt the presented information to a particular device. The FM was initialised with sets of generic commands and when available, application specific commands. This implementation highlights how the lightweight design of the MFM allows modules to be combined, restricted or left out. The system is used in a research lab.

4 Discussion and Conclusion

Models that support the design of multi platform applications are becoming more important with a work force that requires access to applications and data while on the move. People prefer interface consistency and would like the same interface on both their desktop and handheld units. This desire for consistency presents a dilemma to application and interface designers: It would seem optimal to rewrite each application for each device, capitalizing on the various opportunities. However, this approach is time-consuming and expensive, a more economical approach is to design the application only once, and to allow access from multiple platforms. The presented model, the MFM supports this one-time view for the development of multimodal and multi platform applications.

The MFM is a lightweight design model, implemented as communicating processes. This process-based design promotes fault tolerance and robustness. Malfunctions are kept local and individual modules can be updated without halting the application. The lightweight design also enables portability across application domains and a flexibility to combine and integrate modules or to distribute modules. The distribution enables modules to be executed where resources are available. It provides support for alternative and fused input and output modalities, and it allows interaction models and user interfaces to be tailor-made for particular devices. A potential disadvantage is the latency introduced by indirection and communication in the process based architecture.

Future work includes exploring advantages of different command fusion strategies, and how this impacts the generation of partial commands. There is also a need to explore Multimodal output, and strategies for how the presentation of results can be made more natural. Furthermore, there is a need evaluate how user's react and respond to output from multimodal systems, be that partially recognised and executed commands or fused multimodal output.

5 Related Work

Architectures similar to the MFM have been developed at the CAIP center at Rutgers University (Medl et al, 1997). This solution is restricted to sequential signals based on the use of a single demon that issues the appropriate command when all slots in a command frame have been filled. Another related architecture is the Generic Multi-Agent Architectures for Multimedia Multimodal Dialogues (Djenidi et al, 2002). This is an agent-oriented approach using coloured petri-nets to model the fusion and the dialogue. The generic platform for addressing the multimedia challenge developed at LGI-IMAG (Nigay & Coutaz, 1995) is based on adding a fusion engine to the dialogue components of PAC-Amodues, and it provides multi application support similar to the MFM. The difference between the two previously mentioned architectures and the MFMF is the decoupled design where each function (Fusion, Dialogue etc) is handled separately in the MFM.

References

Bolt, R. A., (1980). Put that there: Voice and Gesture at the graphics interface, *ACM Computer Graphics 14,3 262-270.*

Djenidi, H., Ramdane-Cherif, A., Tadj, C., and Levy, N., (2002). Generic Mutli-Agent Architectures for Multimedia Multimodal Dialogs. *Workshop on Modeling of Objects, Components, and Agents*, Aarhus Denmark.

Gourdol, A., Nigay, L., Salber, D., and Coutaz, J., (1992). Two Case Studies of Software Architecture for Multimodal Interactive Systems: VoicePaint and a Voice-enabled Graphical Notebook, *Engineering for human-computer interaction: Ellivuori, North-Holland, 271-284.*

Hodjat, B., and Amamiya, M. (2000) Introducing the Adaptive Agent Oriented Software Architecture and its Application in Natural Language User Interfaces, *In Proceedings of 1st International Workshop on AOSE*, Limerick, Ireland.

Jonsson, I, (1999) The Decoupled Application Interaction Model, D.A.I.M. *In proceedings of Human Computer Interaction International 1999*, Munich, Germany.

Medl, A.,, Shaikh, A., Juth, S., Marsic, I., Kulikowski, C., Flanagan, J.L. (1997). An Architecture for Fusion of Multimodal Information. *Workshop on Perceptual Interfaces*, Alberta Canada.

Myers, M., Hollan, J. and Cruz, I. (1996) Strategic directions in human-computer interaction. *ACM computing surveys, 28* (4). 794-809.

Nigay, L., and Coutaz, j., (1995) A Generic Platform fro Addressing the Multimodal Challenge, *In Proceedings of Computer Human Interaction 95*, Denver, CO, USA.

Oviatt, S.L. Multimodal Interfaces (2002). In the Human-Computer Interaction Handbook: Fundamentals, Evolving Technologies and Emerging Applications, J Jacko and A Sears, Eds. Lawrence Erlbaum Assoc., Mahwah, NJ, 2002, chap. 14, 286-304.

Multi-Session Group Scenarios for Speech Interface Design[1]

Kari Kanto(1), Maria Cheadle(2), Björn Gambäck(2), Preben Hansen(2), Kristiina Jokinen(1), Heikki Keränen(1), Jyrki Rissanen(1)

(1) Media Lab, UIAH - University of Art and Design Helsinki, Finland
(2) SICS - Swedish Institute of Computer Science AB, Kista, Sweden
{firstname}.{lastname}@uiah.fi, dumas@sics.se

Abstract

When developing adaptive speech-based multilingual interaction systems, we need representative data on the user's behaviour. In this paper we focus on a data collection method pertaining to adaptation in the user's interaction with the system. We describe a multi-session group scenario for Wizard of Oz studies with two novel features: firstly, instead of doing solo sessions with a static mailbox, our test users communicated with each other in a group of six, and secondly, the communication took place over several sessions in a period of five to eight days. The paper discusses our data collection studies using the method, concentrating on the usefulness of the method in terms of naturalness of the interaction and long-term developments.

1 Introduction

In the initial design phase of an adaptive speech-based e-mail application, we needed information on how the users would interact with the envisaged system, and data on vocabulary and language use of the future users. The acquisition of such data poses a problem. Human-human dialogue data is unsuitable, because people speak differently to computers than to each other. Furthermore, real e-mail exchange cannot be monitored because of privacy considerations.

The usual solution is to collect the data in a scenario-driven Wizard of Oz setting, and that was our starting point also. However, this method in its traditional form is insufficient for investigation of long-term temporal phenomena, which are an essential factor when designing adaptive applications. Commonly, a scenario comprises a set of tasks that one test user at a time tries to complete during a single session, and in the e-mail domain, tasks involve a mailbox with several pre-generated messages (e.g., Walker 2000). The sessions usually last a few minutes, rarely exceeding fifteen minutes. In our view, some effects of adaptivity need more time to manifest themselves. For example, the formation of shortcuts, choice of new strategies, and other comparable developments of user expertise may appear only after the user is thoroughly accustomed to the system. A longer period of observation may also illuminate certain linguistic phenomena, such as accommodation and convergence, which are important when dealing with inadequate speech recognition (Zoltan-Ford 1991). Finally, longer stretches of time are needed when studying the complicated interplay between an adaptive system and an adaptive user.

---

[1] Work sponsored by the European Union's Information Society Technologies Programme under contract IST-2000-29452, DUMAS (`www.sics.se/dumas`). Thanks to all project participants from KTH and SICS, Sweden; UMIST, UK; ETeX Sprachsynthese AG, Germany; and U. Tampere, U. Art and Design Helsinki, Connexor Oy, and Timehouse Oy, Finland. The wizard interface was developed by Andrew Conroy, UMIST.

Changes in user behaviour over a longer time could perhaps be studied by increasing the number of tasks, but this would soon become tedious for the user and the setting would become increasingly artificial and forced. We have therefore designed a multi-session group scenario method (MSGS) in order to provide a natural setting for observing user behaviour in prolonged system use.

The rest of the paper is laid out as follows: we introduce the MSGS method in Section 2 and the experimental setting in Section 3. In Section 4 we demonstrate the potential of the method with examples from our results. The results are discussed in Section 5, while conclusions are drawn in Section 6.

2 The multi-session group scenario

A Wizard of Oz method with two novel features was conceived: firstly, instead of doing solo sessions with a static mailbox, our test users communicated with each other in groups of six using a simulated speech-based e-mail system via telephone. The e-mail system was controlled by a wizard, and it allowed the subjects to dictate and receive messages, arrange them in folders, etc. Secondly, the communication took place over several sessions in a period of five to eight days during which the subjects played a role defined for them in the scenario. They were advised to call at least twice a day, resulting in a total of around 60 dialogues per experiment.

Having people communicate with each other in the experiment was aimed at creating a more natural and real setting. We wanted the participants to be motivated by the interaction within the group rather than by the interaction with the e-mail system, and to choose their actions based on what happens in the discussion instead of following predestined paths. Methods of dramatic writing (e.g., Egri 1946) were used to prepare a scenario that would incite lively e-mail conversations. Each participant was assigned a fictional character. Five characters in the scenario formed a software development team and the sixth was a customer's representative. All were assigned different tasks and given professional and emotional motives for participating in the e-mail exchange with the others; a tangled web of tensions was woven between them. For example, the customer was the group lead's ex-wife, and the best friend of one of the characters had lost a position in the group to a less competent person, namely the group lead's son. A description of the character assigned to a participant and that character's attitudes towards the others in the group along with some background information and task to solve was presented to each participant at the beginning of the experiment.

The experiments were designed to last for several days in order to gain information on user accommodation and the effects of system adaptivity. The focus was on the development of user expertise and linguistic accommodation.

3 Experimental setup

The six person experiment was duplicated at two sites, UIAH and SICS. The first experiment was conducted in Finnish and the second in Swedish. All participants use computers daily and were 20-35 years old. At both sites three were female and three male. Four participants had previous experience with speech applications, incl. two who were advanced users. One of the participants at UIAH was blind.

The participants were instructed that the speech-based system would basically have the same functionality as an ordinary e-mail program. The participants at SICS were also given an ordinary e-mail account with a web-interface to enable them to review and compose messages in a text modality as well. The users could call from 9 am to 5 pm every day during the experiment.

The telephone calls from the participants were directed through a Dialogic telephony board into a computer where it was recorded and also sent to the loud speakers for the wizards to monitor the call. A wizard interface was used to send the appropriate prompts to the user.

4 Results

To illustrate the potential of the method, we give here examples of different aspects of our results. Generally speaking, the scenario generated active e-mail traffic, even heated discussions, and the subjects seemed committed to playing their characters. A summary of the basic statistics of the two experiments is provided in Table 1.

Table 1: Statistical summary of the experiments

	Finnish participants						
	f_1	f_2	f_3	f_4	f_5	f_6	*average*
number of calls (dialogues)	10	9	16	15	7	4	10.2
messages sent	14	17	19	26	9	3	14.7
messages received	22	19	18	26	11	7	17.2
average call duration (min:sec)	4:28	5:47	5:45	5:24	4:43	4:00	5:01
	Swedish participants						
	s_1	s_2	s_3	s_4	s_5	s_6	*average*
number of calls (dialogues)	9	10	10	16	11	8	10.7
messages sent	18	10	14	9	14	6	11.8
messages received	24	23	26	15	16	16	20.0
average call duration (min:sec)	8:48	5:59	8:04	4:54	3:10	4:32	5:49

The central goal of the method was to study phenomena that occur when a system is used for a prolonged period of time, primarily linguistic developments and changes in the users' level of expertise and in their strategies.

As regards user expertise, one of the participants had a clear curve of development: at the beginning she was hesitant, cancelled actions and didn't always know what to say. After five sessions, she used the system with a highly standardized range of expressions with no signs of hesitation. Another user, who had previous experience with speech systems, picked up the system's vocabulary in her first session and maintained it through the test without exception.

Two users showed other signs of development, by initially restricting themselves to rather short verb phrases when issuing a read-message command, but over time introducing more detailed and specific expansions of the original verb phrase by adding a variety of noun phrases referring to particular messages (Table 2).

Table 2: Developments in phrasing in the Swedish experiment

Initially	After a few sessions
lyssna *listen*	lyssna på det senaste meddelandet från K. *listen to the latest message from K.*
lyssna på meddelandet *listen to the message*	läs båda meddelandena *read both messages*
läs upp *read*	läs upp det första nya meddelandet *read the first new message*
läs upp meddelandet *read the message*	

The shorter phrases were not abandoned but used alongside the more detailed ones throughout the rest of the experiment. The other commands (and some other users, too) show a similar trend, but it is most easily observed in the read-message commands since they are very frequent and were used by all.

Individual interaction strategies were formed, exemplified by one of the participants who developed a habit of calling in once and listening to the new messages, and then calling again a quarter of an hour later to dictate her replies. This pattern becomes evident between her 6th and 7th calls and remains unchanged for the next 8 calls, until the end of the experiment. In her reply to the post-test questionnaire, she mentioned finding it difficult to keep in mind several received messages for reply. Hence, the strategy of separate calls for listening and dictating. This is the sort of development that would not show up in a single-session study.

Another user always accommodated her answers to the system prompts' syntactic form (which becomes easily evident in Finnish as an agglutinative language). For example, when the system asked, "Kenelle haluat lähettää viestisi?" ("To whom do you want to send the message", with the word 'who' in allative case), an accommodating user replied "Katalle" (the name of one of the characters + allative case suffix), whereas an unaccommodating user replied, "Kata" (no case suffix). Some test subjects accommodated only sometimes, usually in sudden situations, and some had no clear tendency towards either.

One of the users, who had previous experience with speech recognisers, never formulated her replies after the model given by the preceding system prompt. She was consistent without exception: for example, when starting to listen to new messages, she said "lue viestit" ("read the messages", plural form) regardless of whether there were many messages or just one.

5 Discussion and future work

The MSGS method seems flexible enough to support observation of several kinds of long-term phenomena. When wizard behaviour is not strictly regulated, the method serves exploration by giving room for unexpected user behaviour. With more strict rules of conduct, the focus moves towards details, for example, testing of particular adaptive functionality. If the experimental system is non-adaptive, users' adaptive behaviour can be studied; otherwise, the interplay between the adaptive system and the adaptive users moves into the forefront.

It is difficult to measure the naturalness and realism of the setting, and presently we must rely on our impressions and the questionnaire that was sent to the test subjects after the experiment. Most participants maintained that they had absorbed or somewhat absorbed their roles, one mentioned that he had not been taken by the scenario at all. All claimed to have been keen to hear what kind of messages they had received. All agreed that they sent fewer and shorter messages in the experiment than they would have if the system had been in real use and the mailbox their own.

The brevity of the messages compared to written e-mails was anticipated due to the modality of the interaction. The fact that we were not able to provide the characters with complete life-long background information is a factor. The relatively small amount of messages, compared to what many users receive in a comparable period of time, is at least partly explained by the small circle of conversants. The researchers "sent" some messages from outside the group and inserted mass mailings, but real-life networks comprise legions of people sending e-mail to each other, which is difficult to simulate.

The rather freeform nature of the setting limits which quantitative measurements can be made. For example, the Dialogue Efficiency Metrics and the Task Success Metrics of the PARADISE evaluation framework (Walker, Litman, Kamm & Abella 1997) are not applicable, because there are no clear-cut tasks for which task success, completion time, etc., could be measured. Dialogue Quality Metrics and User Satisfaction, on the other hand, apply if they are converted to handle frequencies over several sessions instead of single session packages. This increases the importance of the user questionnaire conducted after the test. Comparisons that require Dialogue Efficiency and Task Success Metrics are better served by the traditional single session method.

As mentioned above some users did not accommodate their replies to the system prompts' syntactic features. This occurred most visibly when the user reacted to clearly defined and simple system prompts when preparing to dictate a new message. The system asked first to whom the user wanted to send the message; answering this question without consideration of the question formulation could mean that the user regarded the situation as slot-filling, in a sort of a transfer from graphical e-mail interfaces. This may open some interesting viewpoints to the mental models the users construct when interacting with speech interfaces.

The vocabulary development of some users, establishing more detailed and specific phrasings of certain commands over time, could be attributed to a number of explanations; the participants developing a better understanding of what the simulated system was able to do, or growing more confident and thus, rather than accepting the system's manner of presenting messages, taking charge of the interaction and making sure it delivered what they wanted.

In future studies, we intend to insert a speech recogniser between the user and the wizard. We hope to create better data by replacing as many as possible simulational parts with real ones without forsaking the quick prototyping possibilities of WOZ setups. The data gathered in the Finnish experiment has been used as a basis for a machine learning experiment (Jokinen, Rissanen, Keränen & Kanto 2002).

6 Conclusions

The paper has described a multi-session group scenario as a means to investigate the effects of system adaptivity on user behaviour and also to provide a realistic simulated environment for testing and developing speech applications.

The rather freeform nature of the setting limits which quantitative measurements can be made: there are no clear-cut tasks for which task success, completion time, etc., could be measured. The simulation provided by the scenario is nonetheless reasonably realistic, giving the designer qualitative insights into the research matter. The effects of adaptivity manifest themselves with the method. Functionality requirements can be explored and dialogue/language models extracted with our scenario. Our conclusion is that the MSGS method is useful and will be developed further.

The scenario can also be used with a working prototype, the system need not be WOZ-operated. This is important in the e-mail domain, since the problem of e-mail privacy remains through the whole system development process.

References

Egri, L. (1946). The Art of Dramatic Writing. London: Simon & Schuster, A Touchstone Book.

Jokinen, K., Rissanen, J., Keränen, H. & Kanto K. (2002). Learning interaction patterns for adaptive user interfaces. *Proc. 7th ERCIM Workshop User Interfaces for All*, Paris, France.

Walker, M.A. (2000). An application of Reinforcement Learning to dialogue strategy selection in a spoken dialogue system for email. *J. Artificial Intelligence Research* **12**, pp. 387-416.

Walker, M.A., Litman, D., Kamm, C., Abella, A. (1997). PARADISE: A Framework for Evaluating Spoken Dialogue Agents. *Proc. 35th Annual Meeting of the Association for Computational Linguistics (ACL-97)*, pp. 271-280, Madrid, Spain, July.

Zoltan-Ford, E. (1991). How to get people to say and type what computers can understand. *Int. J. Man-Machine Studies* **34**, pp. 527-547.

Speech-based cursor control: Understanding the effects of variable cursor speed on target selection

Azfar Karimullah *Andrew Sears* *Min Lin* *Rich Goldman*

Interactive Systems Research Center
Information Systems Department
UMBC
1000 Hilltop Circle
Baltimore, MD, 21250
akarim2; asears; mlin4; rich1@umbc.edu

Abstract

Speech-based interactions can be an effective alternative for some people with physical disabilities as well as individuals completing tasks that otherwise occupy their hands. In this article, we investigate one specific component of speech-based interactions: speech-based navigation. Our study deals with speech-based navigation using direction-based navigation commands that result in continuous cursor movements. We compared the performance of two approaches to speech-based cursor control. One uses the standard mouse cursor while the other uses a predictive cursor designed to help the user position the cursor more accurately. Variables such as target size, moving distance, and movement direction were examined for their effect on target selection time, selection accuracy. User satisfaction was also assessed. Target size and direction both had a significant effect on performance, but the predictive cursor did not prove beneficial. The variable speed cursor used in the current study appears to allow for improved performance, especially for small targets, as compared to the fixed speed cursor used in previous studies.

1 Introduction

While state-of-the-art speech-recognition systems typically provide mechanisms for both data entry and cursor control, speech-based interactions continue to be slow when compared to similar keyboard- or mouse-based interactions. While numerous researchers have focused on reducing and correcting recognition errors, few have studied speech-based navigation mechanisms. At the same time, several studies have confirmed that users spend significant time correcting recognition errors (Halverson et al., 1999, Karat et al., 1999, Sears et al. 2001). More importantly, speech-based navigation has been shown to be both time consuming and error prone (Sears et al. 2003).

Two fundamental approaches have been employed for speech-based cursor control. For target-based navigation, the user navigates directly to a desired destination. For example, "Select Friday" could cause the word "Friday" to be highlighted. This approach proves effective when the number of targets is sufficiently limited. However, as the number of targets increases this approach becomes more error prone and it becomes more difficult for the user to remember the name of each target. Further, multiple instances of the same target (e.g., if the word "Friday" appears on the screen several times) may lead to increased difficulties. Speech-based navigation can also be accomplished using direction-based commands. Direction-based navigation is further divided into two types: commands that result in discrete movements and those that initiate continuous movements. For discrete direction-based navigation, the user specifies the direction and distance (e.g. centimeters, inches, characters, or words) the cursor should move. For example, "Move left four words" would move the cursor left four words. For continuous direction-based navigation, which are our focus, the command (e.g., "Move left") causes the cursor to begin moving in the specified direction until the "Stop" command is issued.

In this article, we discuss on method of implementing direction-based navigation that results in continuous movements. We believe continuous movements may prove beneficial for graphical interfaces where the number of possible targets can become quite large. Our previous study (Karimullah & Sears, 2002) provided interesting insights into target selection, cursor speed and command selection, confirming that larger targets and shorter distances resulted in faster target selection, while larger targets also resulted in fewer errors. This earlier study investigated the use of a predictive cursor, but did not identify any significant differences between standard and predictive cursors. However, it did identify areas that needed further research including the effect of cursor speed on target selection. For example, we had employed a cursor moving at constant speed of 20 pixels per second (approximately 53 mm/sec). This speed allowed adequate accuracy for larger targets, but the inability to slow the cursor resulted in high error rates when selecting smaller targets. To select small targets more reliably, we believe the cursor must move more slowly. However, a slower cursor will also increase selection times, which were already substantially longer than when using a mouse. Therefore, to allow small targets to be selected reliably while simultaneously allowing larger targets to be selected quickly, the cursor must be allowed to move at multiple speeds. These observations were explored in a preliminary study designed to evaluate the effect of cursor speed on target selection times and error rates. Results indicated that the speed of the cursor does effect selection times, suggesting that a variable speed cursor may prove beneficial. Further, we concluded that users may benefit from the predictive cursor if the cursor moves at multiple speeds. The current study includes allows the user to control the speed of the cursor while using a redesigned predictive cursor to provide feedback that may simplify the process of accurately selecting targets.

We describe a study that investigates several factors that may affect the efficacy of continuous direction-based navigation using a between-group experimental design. We employed a variable speed cursor thereby giving more control to the user while examining the affects of size, distance and direction of movement on both selection time and errors. The new predictive cursor technique (Sears, Lin, & Karimullah, 2002) was used. This technique provides guidance intended to allow users to more accurately select targets. The actual cursor trails behind a predictive cursor (see Figure 1a). The distance between the cursors is configured for each individual user based on the delay between when the user issues a command and when the cursor actually stops.

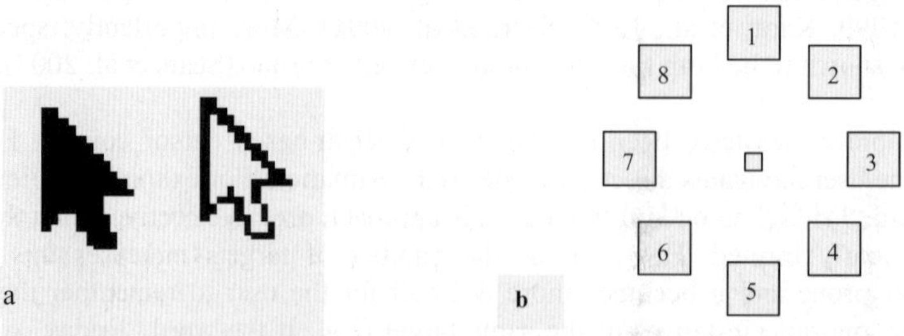

Figure 1: (a) After a "Move Right" command, the predictive cursor (on the right) indicates where the real cursor (on the left) is expected to stop if a "Stop" command is issued at any given time. (b) Targets were positioned in eight directions relative to the home position (the center square).

2 Experiment

Thirty volunteers (9 males and 21 females) participated in the study. They were all native English speakers and had no previous experience using speech recognition technologies. No uncorrected visual impairment or documented hearing, speech or cognitive impairments were reported. Participants were randomly assigned to two groups, one using standard cursor and the other using the predictive cursor. For both groups, the average age of participants was 23 years.

Participants were instructed to use the assigned technique to move a cursor from its home position toward the target and issue a "click" command at the appropriate time to select the target. Targets were positioned around the home position in eight directions (see Figure 1b) at three different distances. The distances were defined as center to center distance from the home position to the target. The three distances were 1.9, 3.8 and 5.7cm (referred to as D1, D2, and D3). There were five target sizes, 0.2, 0.5, 0.7, 1.0 and 1.3cm per side (referred to as S1, S2, S3, S4, and S5). Following a brief training session, each participant selected 120 targets (8 directions x 3 distances x 5 sizes) in random order. The experimental software was implemented using Visual Basic and the Millennium Edition of IBM's ViaVoice speech engine. A desktop PC running Windows NT was used for the experimental sessions. After the target selection task, a questionnaire was used to assess user satisfaction.

The speed at which the cursor moved could be adjusted using the "faster" and "slower" commands. Nine levels of speed were provided. The cursor speed started at the medium level, which corresponded to 20 pps (pixels per second). At any time, issuing a "faster" or "slower" command doubled or halved the current speed. A slide bar was provided to allow users knowing current speed level.

The four independent variables in the study were cursor type (standard or predictive), target size, movement direction (straight or diagonal), and the distance between target and home position. Dependent variables included the time required to select the target, positioning accuracy, and user satisfaction. Five hypotheses regarding the relationship among these variables were tested.

H1: Target size affects selection time and positioning accuracy.
H2: Distance affects selection time.
H3: Movement direction affects selection time and positioning accuracy.
H4: There is interaction between target size and movement direction for both selection time and positioning accuracy.
H5: The predictive cursor will result in better performance and higher satisfaction as compared to the standard cursor.

3 Analysis and Result

No significant difference on target selection time and accuracy was observed between the two mouse control techniques. Consistently, user satisfaction scores were statistically the same between the two techniques. H5 was not supported. Therefore, the data from two techniques were combined in testing other hypotheses (see Table 1). Target size showed a significant effect on selection time [$F(4,880)=5.96$, $p<0.001$] with smaller targets requiring significantly longer than larger targets. Distance and direction also had significant effects on selection times

[F(2,880)=19.24, $p<0.001$; F(1,880)=128.86, $p<0.01$ respectively] (see Figure 2). Longer distances required more time to reach the target as did diagonal movements.

Table 1: Mean target selection times (in seconds) with standard deviations in parentheses

Size \ Distance	Straight			Diagonal			Overall
	D1	D2	D3	D1	D2	D3	
S1	18.88(3.18)	25.48(3.07)	27.93(2.17)	37.41(4.79)	34.60(2.08)	48.16(3.88)	32.08
S2	9.01(0.84)	13.68(0.86)	16.32(0.72)	14.96(0.88)	20.95(1.26)	27.57(1.50)	17.08
S3	7.82(0.75)	10.83(0.83)	14.61(0.93)	11.90(0.92)	19.05(1.44)	23.11(1.06)	14.55
S4	7.09(0.35)	9.44(0.42)	12.28(0.57)	8.77(0.36)	13.59(1.03)	18.48(0.96)	11.61
S5	6.42(0.29)	8.83(0.55)	11.63(0.64)	8.50(0.82)	12.54(0.61)	16.50(1.07)	10.74
Overall	9.84	13.66	16.55	16.31	20.15	26.76	

The accuracy of target selection was measured by counting the number of "click" commands issued outside the target boundaries. Neither distance nor direction showed significant effect on accuracy. However, target size did significantly effect accuracy [F(4,880)=5.48, $p<0.001$] with smaller targets resulting in lower accuracy than larger ones. Therefore, H1 and part of H2 and H3 were supported by the data. Further, significant interactions between size and direction were observed for both selection time and accuracy [F(4,880) =16.77, $p<0.001$; F(4,880)=6.47, $p<0.001$, respectively] (see Figure 2). Therefore, H1, H2, H4, and part of H3 were supported.

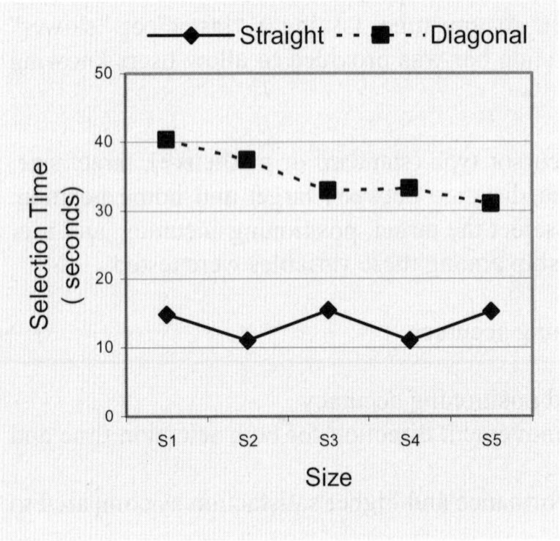

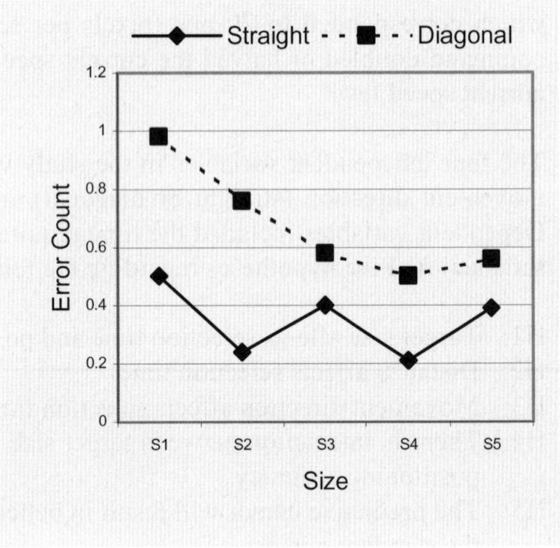

Figure 2: Mean target selection time and number of errors.

4 Conclusions

Our results confirm earlier findings. Smaller targets are more difficult to select than larger targets. Longer distances and diagonal movement both result in longer selection times. Interestingly, our results indicate that the predictive cursor was not beneficial. However, it does appear that a variable speed cursor allows users to select small targets faster and more reliably while simultaneously speeding the selection of larger targets.

Interesting insights regarding the efficacy of a variable speed cursor are possible if we compare the current results to those reported earlier. For example, participants in our previous study needed approximately 54 seconds to select a 0.3cm square target when using a fixed speed cursor. In the current study, participants needed only 24 seconds on a 0.2cm wide target. This suggests that a variable speed cursor may prove beneficial for small targets.

On the other hand, the ability to adjust the cursor speed may tempt users to increase speed when targets appear large enough or far away. While current implementation provides feedback to help users determine when they should slow down the cursor, this feedback may not be effective when the cursor is moving too quickly. Further, this feedback does not help users determine if a faster cursor may be useful. Therefore, it is possible for users to speed up the cursor only to discover that it is moving too fast for the current task. The efficacy of different speeds requires further investigation. For example, it would be useful to determine the maximum speed that the cursor should travel. Based on our current results, we suggest that it is likely to be between 20 and 40pps.

Finally, it appears that users can adapt to the inherent delay between when they issue a command and when it is executed without guidance from the system. Given our results, it is also possible that the reliability of the feedback provided by the predictive cursor may be too low to prove beneficial. Since novices may need additional time to develop effective strategies of speed adjustment, longitudinal studies may prove interesting. Additional studies that focus on the process by which users change the speed of the cursor, the upper and lower bounds for the speed of the cursor, and how user performance changes with experience would also be useful.

5 Acknowledgements

This material is based upon work supported by the National Science Foundation under Grant Nos. IIS-9910607 and IIS-0121570. Any opinions, findings and conclusions or recommendations expressed in this material are those of the authors and do not necessarily reflect the views of the National Science Foundation (NSF).

6 References

Halverson, C., Horn, D., Karat, C-M. and Karat, J. (1999). The beauty of errors: patterns of error correction in desktop speech systems. *Proceedings of INTERACT'99*, 133-140. IOS Press.

Karat, C-M., Halverson, C., Karat, J. and Horn, D. (1999). Patterns of entry and correction in large vocabulary continuous speech recognition systems. *Proceedings of CHI 99*, 568-575.

Sears, A., Feng, J., Oseitutu, K. and Karat, C. (In press). Hands-free speech-based navigation during dictation: Difficulties, consequences, and solutions. To appear: *Human Computer Interaction*.

Sears, A., Karat, C-M., Oseitutu, K., Karimullah, A. and Feng, J. (2001). Productivity, satisfaction, and interaction strategies of individual with spinal cord injuries and traditional users interacting with speech recognition software. *Universal Access in the Information Society*, 1, 4-15.

Karimullah, A. and Sears, A. (2002). Speech based cursor control. *Proceedings of Assets 2002*, 178-185.

Sears, A., Lin, M. and Karimullah, A. (2002). Speech-based cursor control: understanding the effects of target size, cursor speed, and command selection. *Universal Access in the Information Society*, 2, 30-43.

The Optimal Sizes for Pen-Input Character Boxes on PDAs

Taishi Kato *Xiangshi Ren* *Youichi Sakai* *Yoshio Machi*

Kochi University of Technology
185 Miyanokuchi, Tosayamada-cho, Kochi
782-8502, Japan
030256h@ugs.kochi-tech.ac.jp ren.xiangshi@kochi-tech.ac.jp

Tokyo Denki University
2-2 Kanda Nishiki-cho, Chiyoda, 101-0054
Tokyo, Japan
youichi@sakaii.com machi@d.dendai.ac.jp

Abstract

This study seeks to determine the optimal size for pen-input handwritten character boxes on PDAs (Personal Digital Assistants) inside which users can most efficiently write English, Chinese, hiragana, alphanumeric characters, and so on. The results were assessed in terms of high performance factors such as high character recognition rate, minimal stroke protrusions outside the character box (experiment 1); high subjective ratings (experiments 1 and 2), and physiological data such as brain wave information (experiment 2). The last term is a unique evaluation approach which allows us to investigate what kind of character boxes a user can write in while maintaining an implicit relaxed state. The analyses of the results of experiments 1 and 2 show that the optimal size of character boxes for the input of alphanumeric characters is approximately 1.44 x 1.44 cm. Future directions are also discussed.

1 Introduction

This study seeks to determine the optimal size of pen-input handwritten character boxes on PDAs (Personal Digital Assistants) inside which users write English, Chinese, hiragana, alphanumeric characters, and so on. Here we assume that an optimal size exists, and we define it as including the following characteristics: high performance (e.g. a high character recognition rate and a minimum number of strokes protruding from the character box) and a minimal degree of mental and physical fatigue. Thus, our evaluations are based not only on conventional performance data and subjective preferences, but also on physiological data such as brain waves. The latter is a unique approach which allows us to investigate what kind of character boxes a user can write in while maintaining an implicit relaxed state.

Software applications for handwritten character input on PDAs which are currently on the market usually display 2 character boxes, but some have 4 to 8 character boxes. However, the optimal size for handwritten character boxes has not been clearly established. As the number of input boxes increases on the screen, character box size becomes rather small. Users may, however, prefer more input boxes because more characters can be written at once. However, it is assumed usually that if the character box size is large, users may write in more a relaxed state. Furthermore, the size and/or number of character boxes decreases as the amount of information on the PDA screen increases. Obviously, the screen size of PDAs is limited. If designers make boxes too small, handwriting may protrude from the boxes and an incorrect icon or function may be selected. Thus, there is a trade-off between the size of a character box and the number of character boxes. Therefore, if we determine the optimal size of a character box, we believe that it becomes a fundamental factor for the design of PDA screens.

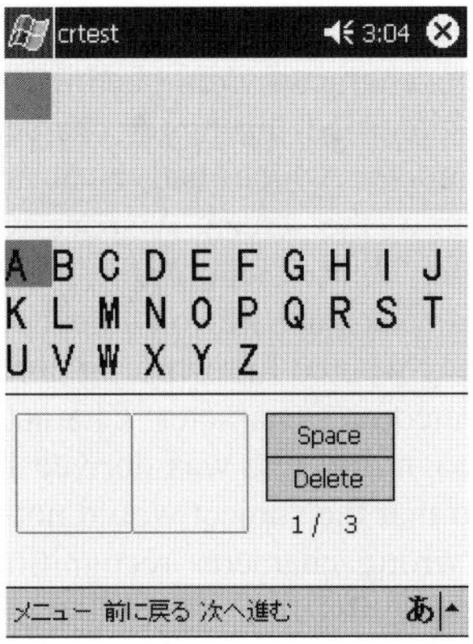

Figure 1: Experiment screen

2 Experiments 1: Performance Evaluation

2.1 Participants

Eleven university students (20 - 22 years old) and 1 university staff member (37 years old), average age 22.8 years old (6 male and 6 female; all right-handed). Two persons had experience in PDA use of about 1 – 1.5 year.

2.2 Apparatus

The hardware we used was a PDA called "iPAQ Pocket PC" running "Windows CE 3.0". The weight was about 190g, and the size was 84 mm (W) x 16 mm (D) x 134 mm (H), the spatial resolution of the screen was 0.24 mm / pixel.

The software for the experiment was developed using Microsoft embedded Visual C++.

2.3 Design

This study tested alphanumeric character input in order to make a simple evaluation of character boxes.

We designed two kinds of character boxes (square and rectangular). The sizes of the squares tested were as follows:

- 0.24 x 0.24 cm
- 0.48 x 0.48 cm
- 0.96 x 0.96 cm
- 1.44 x 1.44 cm
- 1.92 x 1.92 cm (Windows CE standard size)

Moreover, the rectangular input box, 0.6 x 1.18 cm (25 x 49 pixels) referred to the quasi-optimal size of the alphanumeric character input box as previously determined by Ren & Moriya (1995).

2.4 Procedure

First the outline and method of the experiment were explained to each subject. In the practice trails, the subject was asked to input the large and small alphabetic characters (A-Z, a-z) once each, and the number (0-9) twice in order to familiarise them with the experimental environment.

Figure 1 shows the screen on which the subject input the characters.

(1) *The target character was displayed*: The target character, which the subject was to input by hand, was displayed and highlighted in pink in the middle rows. The character input boxes were displayed on the lower part of the screen. The characters actually input into the boxes

by the subjects were displayed in the upper section of the screen. When the character was successfully input, the next target character would be highlighted in pink.

(2) *Character input*: The subject identified the target character and input the character in the boxes with the pen. The character which had been input was then displayed without recognition on the upper section of the screen. A space was inserted between the characters whenever the subject touched the "Space" icon in the lower right of the screen. Touching the "Delete" icon had the effect of a backspace key on a keyboard. Subjects used the "Delete" box icon to remove any character they wanted to rewrite or correct, e.g. if the character which was written by the subject was an incorrect character. Character recognition was not carried out during the experiment. The recognition rate was derived from the data after the experiment. This procedure was followed so that the subject would not develop stress caused by having to rewrite a character when the wrong character recognition result was displayed. Since the purpose was to evaluate the optimal box size, we wanted to eliminate any excessive stress.

(3) Questionnaire: After the input of all characters was completed, we asked the subjects to rate readability, ease of writing, degree of fatigue, box size preference, and overall evaluation on a scale from 1 (worst) to 7 (best).

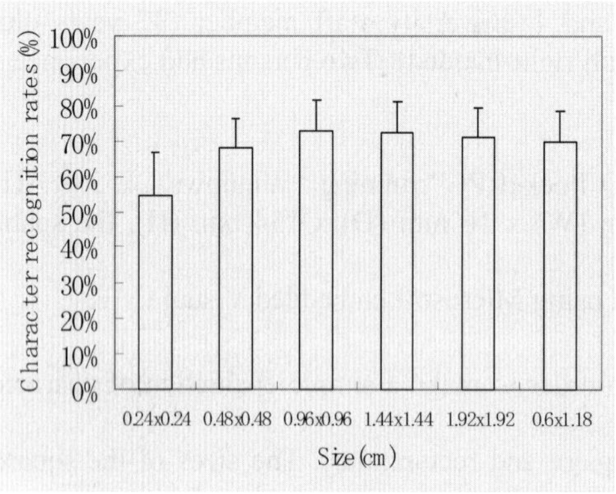

Figure 2: The character recognition rates

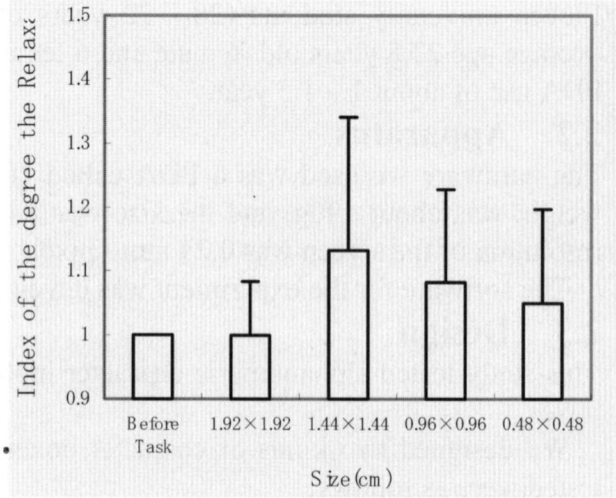

Figure 3: The results of alpha waves

Each subject used the six kinds of box sizes in a different order. A 10 minute break was inserted between tests on each box size.

The total number of characters written for the experiment by the 12 subjects was 5184 characters. At the end of the experiment, stroke data, the number of protruding strokes, the number of error corrections, and the times taken to write each character into the file format were recorded. The number of protruding strokes represented the number of times a stroke protruded out of the box. The number of error corrections was measured by the number of times the "Delete" key was pushed. The writing time for each character was measured as the time from which the subject started to write the character until the instant the character was displayed on the upper section of the screen.

Evaluation indices were: character recognition rates, the number of protruding strokes, the number of error corrections, average writing time for one character, and questionnaires.

2.5 Results and Discussions

No significant difference between the six boxes was found in writing times.

A significant difference between the six boxes was found in the number of protruding strokes, $F(5,66) = 32.96$, $p < .001$. The number of protruding strokes ranged from the rectangular box (mean = 0.25), 1.44 x 1.44 cm box (mean = 0.42), 0.96 x 0.96 cm box (mean = 1.67). The rectangular box had the lowest number of protruding strokes, however, there was no significant difference among these three boxes.

No significant difference between the six boxes was found in the number of error corrections.

A significant difference between the six boxes was found in the character recognition rates (Figure 2), $F(5,66) = 6.56$, $p < .001$. Although the rate of recognition for the 0.96 x 0.96 cm box (mean = 72.80%) was higher than that for the 1.44 x 1.44 cm box (mean = 72.69%), there was no significant difference between the two boxes.

Moreover, the 1.44 x 1.44 cm box received high ratings from the questionnaire.

Based on the above results, we concluded that the optimal character box size was from 0.96 x 0.96 to 1.44 x 1.44 cm box.

3 Experiment 2: Physiological Evaluation

3.1 Participants

Fourteen university students, average age 22.1 years old (all right-handed, 13 male and 1 female). One had experience in PDA use of about 0.5 year.

3.2 Apparatus

The PDA and the software used in the experiment were the same as for Experiment 1 (see Section 2.2). Apparatus used to measure physiological parameters were: an EEG monitor (SYNAFIT 5000,NEC Medical System Corp.), an electrocardiogram (Bio-Biew G (2G54), Data collection System MP100 / EMG100B (BioPAC system Corp.) and AcgKnowledge 3.5.7, ATALA 2.3.

3.3 Design

The number of boxes, and the input posture were the same as Experiment 1. The target characters used were extracted from an English article. We used following four box sizes:

- 0.48 x 0.48 cm
- 0.96 x 0.96 cm
- 1.44 x 1.44 cm
- 1.92 x 1.92 cm (Windows CE standard size)

3.4 Procedure

Explanation to the subjects and practice trials were the same as for Experiment 1 (see Section 2.4). The subjects were asked to input the characters while being assessed according to the TBM (Task Break Monitoring) measuring method as follows. This method was developed by Chen, Ren, Kim & Machi (2000), and was used in this experiment to measure brain waves and R point potential. The EMG value while inputting in each kind of box was expressed as a relative value on the basis of values derived from inputting into the 1.92 x 1.92 cm box.

The following instructions were given to the subjects:

(1) Close your eyes and keep your mind in a relaxed state for three minutes.
(2) Do the input task into the PDA continuously for ten minutes (See Section 2.4 (2)).
(3) Close your eyes and keep your mind in a relaxed state while awake for three minutes.

The following physiological data was recorded simultaneously into the file format.

- EEG (alpha wave): to evaluate the degree of mental fatigue
- ECG (R point potential): to evaluate the degree of mental and physical fatigue

- EMG: to evaluate the degree of physical fatigue

After the test, the subject was asked to fill in a questionnaire in terms of "ease of writing, ease of movement between boxes, degree of fatigue, and overall evaluation".

3.5 Results and Discussions

The alpha waves from tasks using the 1.44 x 1.44 cm box (mean = 1.13, SD = 0.2) were the highest among the boxes (Figure 3). R point potential from the 1.44 x 1.44 cm box (mean = 1.01, SD = 0.02) was also the highest. EMG results from the 1.44 x 1.44 cm box (mean = 0.82, SD = 0.15) were the lowest. Moreover, the results of the questionnaire show that the 1.44 x 1.44 cm box had the high ratings.

The above results show that the 1.44 x 1.44 cm box produces the least mental and physical fatigue.

4 Conclusion

Based on the results of Experiments 1 and 2, we can conclude that the optimal size of a character box for the input of alphanumeric characters is approximately 1.44 x 1.44 cm.

This study shows that the optimal value of a character input box size exists. We also made the following findings from the experiments.

- The larger boxes may not have the higher rate of recognition (Experiment 1).
- The larger boxes may not induce less fatigue (Experiment 2).
- A box size which minimizes fatigue exists (Experiment 2).

These results may be regarded as reflecting universal characteristics of the human use of character boxes. We believe that knowledge of the optimal size of a character input box will be useful when designing the screen interface of PDAs.

We expect to conduct future studies on PDA character input boxes based on these findings. Such studies should examine the effects of variations in the number and placement of character boxes, various kinds of characters (e.g. Chinese characters) and comparisons with input methods that do not include character boxes. The implications may go well beyond PDA usage to PC tablet, and other similar pen input interfaces.

Acknowledgments

This research was partially supported by a Grant-in-Aid for Scientific Research (Young Researchers (B) No. 14780338), and High-Tech Research Center Development Program (No. 7512002002), both in Japan. We wish to thank Hiroshi Tanaka for the valuable comments.

References

Ren, X., Moriya, S. : The Minimal Sizes and the Quasi-optimal Sizes for the Input Square during Pen-input of Characters, Information Processing Society, Japan, Vol.36, No.3, pp.654-657 (1995).

Chen, S., Ren, X., Kim, H., and Machi, Y.: An evaluation of the physiological effects of CRT displays on computer users, in *IEICE Transactions on Fundamentals of Electronics, Communications and Computer Sciences*, Vol. E83-A, No.8, pp.1713-1719 (2000).

A User Study On Advanced Interaction Techniques in the Virtual Dressmaker Application

M. Keckeisen *S. L. Stoev* *M. Wacker* *M. Feurer* *W. Straßer*

WSI/GRIS, University of Tübingen, Tübingen, Germany
keckeisen@gris.uni-tuebingen.de

Abstract

Virtual Environments provide useful tools for the solution of many real world problems. In the last years many Virtual Reality applications have been presented in areas like engineering, medicine, or education. However, it remains important to verify that these tools and the used Virtual Reality hardware are easy to use and improve concrete tasks within the corresponding area. In this work, we present the *Virtual Dressmaker*, an application for interactive physically based simulation of clothing in a Virtual Environment. First, we describe recent technical improvements within our application compared to earlier work. Then, we discuss a recent user study in which we let users compare the usability of our own application with two standard desktop applications: SGI's CosmoWorlds and Maya by Alias|wavefront.

1 Introduction

Research in physically based cloth simulation has achieved major advances in modeling the physical behavior of cloth and developing fast numerical algorithms [1, 2, 3, 4, 8]. Virtual Prototyping for textiles will be a major challenge in the near future. The *Virtual Dressmaker* is an application for interactive physically based simulation of clothes in a Virtual Environment. It provides interaction techniques for assembling pieces of clothes around a chosen figure and seaming them together virtually. Moreover, it allows interaction during the draping process, like selecting and dragging parts of the garments. This scenario allows the user to try on clothes with different sizes and find the appropriate ones. Furthermore, different behavior of the garments can be examined, depending on the material properties of the simulated cloth patterns. Hence, a reduction of the high costs for individual size fitting and custom made-to-measure wear will become achievable. The *Virtual Dressmaker* application enables the user to simulate the physical behavior of textiles made from real CAD patterns. A major demand for the sewing process is to position corresponding seams of the patterns near each other without penetration with other patterns or the figure. For good sewing results and short simulation times good initial placement of the garment patterns is vital. Therefore, our system provides 6DOF interaction tools for setting up an adequate initial condition for the simulation and for manually adjusting parts of the garments during the simulation. First, the user may choose some garment and a 3D character model and then precisely position the patterns around the figure using a set of tools based on 6DOF grabbing and manipulation. After assigning a particular material to the cloth, the patterns can be sewed together. The integrated cloth simulation engine continuously computes the drape of the clothes. The intermediate simulation steps are displayed during the computation, while the user can still navigate freely in the scene, select and drag parts of the clothes. At any time the user can jump back to the pre-simulation state, make corrections, or choose a new or additional garment, set another garment material etc., and trigger the simulation again. Figure 1 shows some snapshots of the application in action.

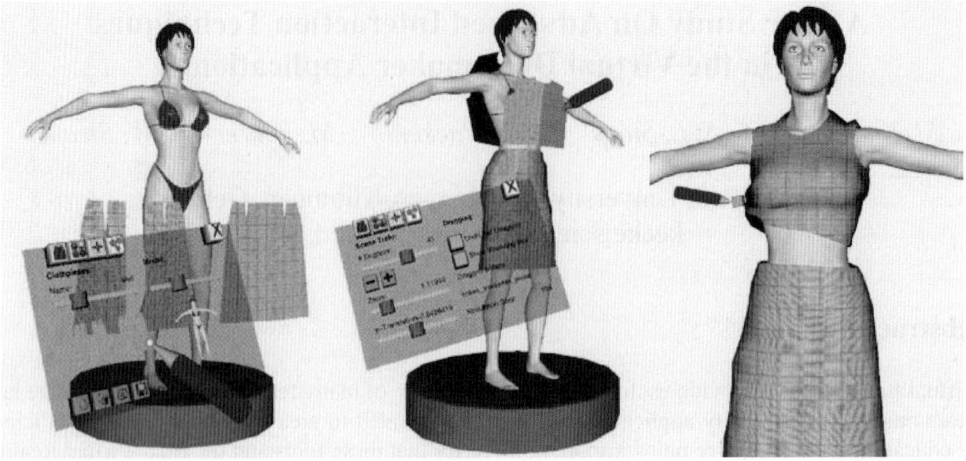

Figure 1. The images show various steps of the try-on process. The left image displays the scene after inserting a set of patterns belonging to a skirt. The result of the simulation of the skirt is shown in the centre image, together with the positioned and simulation-ready parts of a top. The image on the right shows the result of the physically based simulation of both garments, while the user is grabbing a part of the top during the simulation.

One of the most crucial issues of a VR-application is the frame rate. In order to guarantee adequate interaction, we decoupled the application from the simulation module. Our system is realized as a socket-based heterogenous client server system. It consists of two servers responsible for the physically based simulation and the tracking, respectively. The display and the interaction part of the application is implemented as a client, which provides nearly live-size stereo projection combined with 6DOF interaction. For implementing the tracking, we applied an electromagnetic 6DOF tracker (*Ascension, Flock of Birds*) and a stereo table top display (*Barco, Baron*) called the *Virtual Table*. The interaction in our system is based on the *pen* and *palette* paradigm [7] implemented in the *Studierstube* framework [6]. Based upon this framework, we are able to display various virtual tools on the surface of the palette. Depending on the functions of the interface elements, they are grouped into several categories like selection, navigation, and simulation. A detailed description of the *Virtual Dressmaker* application can be found in [5]. In this work, we have presented preliminary usability studies, which have shown that our application allows fast, easy, and precise interaction compared to interaction based on 2D input devices. However, these tests also have shown that some interaction tools had to be improved, especially those for scene navigation and picking. In the following, we first describe technical improvements in our applications compared to the work presented in [5]. Second, we discuss a new usability study that has been done to verify that the new techniques really improve user performance.

2 Improved Interaction Techniques in the Virtual Dressmaker

Based on the results of the preliminary usability studies in [5], the proposed techniques attempt to improve our application's interaction tools, namely scene navigation and picking. In the *Virtual Dressmaker* application, we have implemented two alternative interaction tools for scene navigation. First, there is a navigation sheet on the palette, as can be seen in the centre image of Figure 1. There are sliders for rotation around the vertical axis, zoom, and translation along the vertical axis. While this allows enough degrees of freedom for the given application, users found it tedious to switch to the palette for navigation while they were working with the pen in the scene. Consequently, we implemented a navigation tool within the scene, namely a Navigation Sphere, as it can be seen in the images of Figure 3. The Navigation Sphere supports exactly the same degrees of freedom as the sliders. After clicking on the sphere with the pen, left-right movement relates to rotation, up-down movement to translation, and in-out movement to the zoom functionality. Additionally, we added a Home Button on all sheets of the palette, allowing to reset the camera to it's default position. As to the picking, the patterns are now highlighted, which provides better feedback to the user.

3 Usability Study

In order to verify and classify the new interaction techniques and to obtain a more thorough understanding for the needs of the users in a Virtual Environment, we carried out a new user study with detailed questions about the users' comfort with the application. The experiments consisted in two dress assembly tasks that had to be fulfilled in the *Virtual Dressmaker* environment as well as with the desktop modeling tools CosmoWorlds by SGI and Maya by Alias|wavefront. In all applications we compared the users' time and precision to complete the respective positioning tasks and also the subjective impressions were protocolled. Here we give a preliminary evaluation of the user study which already allows valuable statements on the impact of 6 DOF interaction tools with 3D displays.

3.1 Experiment Setup

The users' task was to go through the complete process of garment assembly twice, preceded by a task for a simple skirt to get used to the interaction tools of the respective application. The recorded tasks consisted in an assembly process for a short skirt (three cloth patterns) and for a pair of trousers (four cloth patterns). The order of applications was randomly chosen such that we could exclude possible training effects. Since most of the users had no experience with garment assembly, we marked the goal positions of the patterns by colored copies of the respective cloth patterns. The participants were asked to place the patterns on the corresponding copies in each application starting with exactly the same initial conditions. During each assembly process the time for completing the task was recorded. Afterwards, the users were asked to fill out a questionnaire concerning the subjective impressions of specific interaction during the experiment such as picking, positioning, rotating of the garments, and scene navigation.

	interaction hardware	display	predefined views	two-handed interaction	keyboard shortcuts	selection of multiple patterns
Virtual Dressmaker	6DOF pen	stereo	yes	yes	no	no
CosmoWorlds	2D mouse	mono	no	no	no	no
Maya	2D mouse	mono	yes	yes	yes	yes

Figure 2. Summary of the differences between the applications compared in the user study.

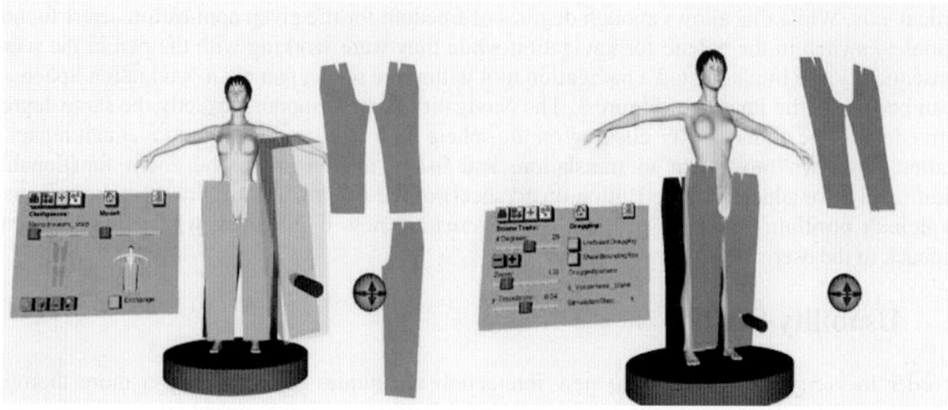

Figure 3. The images show two snapshots of the test scenario in the usability study. In the left image the right front pattern of the pair of trousers is picked. In the right image the left front pattern is moved to the goal position. On the right bottom side of each snapshot the Navigation Sphere is visible.

3.2 Results

Our scenario consisted of 29 users aged between 23 and 35 years (3 female, 26 male) with varying experience with computer modeling software and 6DOF interaction techniques. All of them were users with computer science background, mainly computer science students and Ph.D. students. In order not to measure training effects we mixed the order of the applications for each participant. Moreover, every application and its handling was explained and could be tried out in a simple test scenario afterwards. The results of the time measurements are shown in Figure 4. It is evident that all users completed the positioning tasks faster in the *Virtual Dressmaker* application than in both 2D modeling applications independent of the knowledge of modeling software or 6DOF techniques. We note that better task completion times in the Maya application than in CosmoWorlds result from two handed interaction and better knowledge of the first application. An evaluation of the questionnaire for the rating of the navigation in the scene showed that we could also eliminate prior problems, which were evident in the last user study. The new study was in good correspondence with the old study. Furthermore, we could rise the users' comfort by the Navigation Sphere. 63% of the users rated the Navigation Sphere better applicable than the navigation sliders on the palette.

4 Conclusions

In order to verify and assess the usability of the *Virtual Dressmaker* application compared to current desktop modeling software, we have carried out a set of experimental evaluations. In these trials, the users had to go through the whole process of garment assembly starting from the same initial configuration, however, using different systems for positioning the garment patterns. The results of the usability studies have shown that the users accomplish the given tasks faster in the *Virtual Dressmaker* system than in comparable desktop applications. Moreover, the presented improvements for scene navigation and picking have helped the users to accomplish their tasks. We are currently evaluating the user's subjective impressions concerning 6DOF interaction compared to common 2D user-interfaces. The results will be presented in detail in the near future.

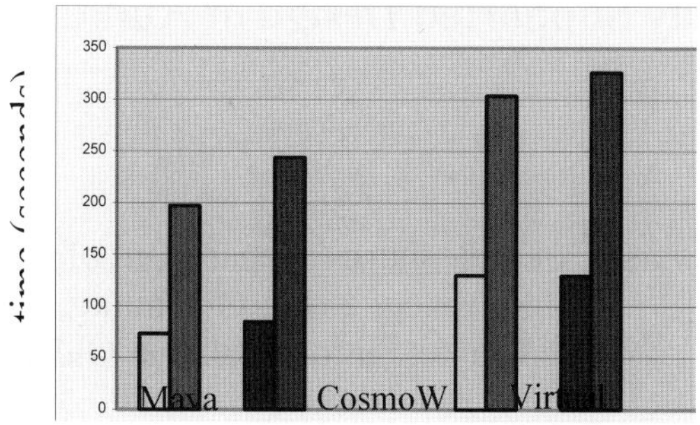

Figure 4. The results: average times for positioning the skirt (nd the trousers ( ) th the respective standard deviation (d)

5 Acknowledgements

We are very grateful to Johannes Hirche who helped us porting the *Studierstube* to Linux. Moreover we thank our Virtual Try-On project partners Hohensteiner Institutes and Frauenhofer IGD Darmstadt for providing 2D garment patterns, seam information, and material parameters for our application. This work was partially supported by the bmb+f Virtual Try-On grant (www.virtualtryon.de).

References

[1] D. Baraff and A. Witkin: Large Steps in Cloth Simulation, SIGGRAPH 98 Conference Proceedings, 1998.

[2] K.-J. Choi and H.-S. Ko: Stable but Responsive Cloth, SIGGRAPH 2002 Conference Proceedings, 2002.

[3] O. Etzmuß, J. Gross, and W. Straßer: Deriving a Particle System from Continuum Mechanics for the Animation of Deformable Objects, IEEE Transactions on Visualization and Computer Graphics, 2002.

[4] M. Meyer, G. Debunne, M. Desbrun, and A. Barr: Interactive Animation of Cloth-like Objects in Virtual Reality, Journal of Visualisation and Computer Animation, 2001.

[5] M. Keckeisen, S. L. Stoev, M. Feurer and W. Straßer: Interactive Cloth Simulation in Virtual Environments, Proceedings of IEEE Virtual Reality, 2003.

[6] D. Schmalstieg, L. M. Encarnação and Z. Szalavari: Studierstube – An Environment for Collaboration in Augmented Reality, Proceedings of the Conference on the 1999 Symposium on 3D Graphics, 1999.

[7] Z. Szalavari and M. Gervautz: The Personal Interaction Panel - a Two-Handed Interface for Augmented Reality, Computer Graphics Forum, 1997.

[8] P. Volino and N. Magnenat-Thalmann: Implementing fast Cloth Simulation with Collision Response, Computer Graphics International Proceedings, 2000.

Visually Supported Design of Auditory User Interfaces

Palle Klante

University of Oldenburg / Department of Computer Science
Uhlhornsweg, 26121 Oldenburg, Germany
Palle.Klante@Informatik.Uni-Oldenburg.De

Abstract

This paper describes the functionality, the used models and an evaluation of the Visual Auditory Interface Design (Visual-AID) prototyping tool which assists the Usability Engineering Process for Auditory User Interfaces (AUI). The process generalizes the experiences made in the INVITE-ZIB-Project, which creates an auditory Web browser for blind users (Donker et al. 2002). The complete user-centered design process defines three main components: guidelines and rules, auditory interaction objects (audIO) and a tool to generate evaluation mock-ups.

1 Introduction

There is a wide range of application possibilities for blind users, but also for sighted users in e.g. mobile scenarios (Shawney & Schmandt, 2000), where only a small or at least no visual display exists as a result of hardware miniaturisation. Future applications are Location Based Services, which nowadays are used with PDAs. The user walks through a museum or a city and dependent on his position he receives information on objects and exhibits in his reach. With current devices the user has to look at a little screen and can't focus on the exhibits. The amount of information is not limited, because the user can actively navigate through the information space. Special attention has to be given on the change between different applications. What happens when the user leaves the museum and get into his car? Actually the interaction objects are changing. The user has to rethink about the used sounds and the behaviour of the objects. A standardization of objects and the use of one specific toolkit of interaction objects for a specific task seems to be useful.

In contrast to the design and the attributes of graphical presentations, there are some differences when designing with acoustic elements. The auditory information processing is primarily time-based and only has a poor resolution in space. A graphical screenshot gives detailed information about the current system status. But when a single sample of peaks is recorded from an audio output, the scene analysis is more difficult. The user has to listen to it for a longer period. These differences and limitations have to be considered in the components of the Usability Engineering Process of AUI.

2 Usability Engineering of Auditory User Interfaces

Usability Engineering supports the idea of design as an intentional and structured process with activity instructions and methods. A detailed description of the final product cannot be given before the actual start of the project. Instead the target should be described via usability goals which should be reached. The product can finally be tested against the completeness of the goals. When designing AUI these design goals have to be adapted and additional criteria have to be considered (e.g. pleasant for a longer period, navigable, easy localization, distinct, aesthetic, joy of use, etc.). These criteria can be categorized in a catalogue of guidelines.

After establishing the usability goals the task for which the application should be build must be identified. The tasks have to be split up in smaller subtasks. A sequence of subtask describes a complete task. These sequences are describing the application flow. To reduce the design complexity early in the design process a disjunction must be done between the model of the user interface and the elements which are used in this model. After finishing the application flow, the elements in each scene must be described. This is partly based on game design. Usually the design team creates a storyboard, then implements the game structure and game idea in one team. Another team implements the figures and actors which can be used in the game structure. It has to be ensure, that every team member of the interdisciplinary design team works with the same design idea. The directions have to be given in a style guide at the beginning of the design process.

For the design of AUIs the scene concept is useful and is the basis for this work. The metaphor of a storyboard is adapted from film production, which also uses time -dependent media and not static elements like in GUIs. The storyboard consists of little stories, which are sequences of scenes. The storyboard is a scene graph. In each scene actors are jumping in, moving in it and finally leaving. Each scene has to be designed for its own, but must also fit to the design of the other scenes.

2.1 Auditory Interaction Objects (audIO)

The design of the actors in a scene are split up in two categories. On the one hand you have to design the information objects. This is the content of the application. The objects which are manipulated by the tools and functions the application offers. The objects are specific to a specific application. A designer has to invent the presentation and behaviour of these objects always from the beginning. It can be described as a picture in an imaging tool. On the other hand there exist interaction objects. The objects are called widgets in GUIs. They offer functions presented in menu entries, buttons to start an action or edit boxes to enter parameters. The audIOs' are based on the concept of abstract interaction objects (Vanderdonckt, 1995). Each object will be described by it`s name, attributes, events, behaviour, sound sources and associated interaction objects. The audIOs should present typical tasks, which are often used.

The basis for the sound design of passive information objects as well as active interaction objects depend on the content of the object. The concept of auditory icons (Gaver, 1989) is used for a limited number of objects which convey a certain meaning. But if objects appear in a large and not limited number, such as the representation of documents, the concept of composed Earcons (Blattner et al. 1989) is used. The finally used concept for presenting audIO are Hearcons (Bölke & Gorny, 1995). Hearcons are permanently hearable objects with a position in space, a varying intensity, an interaction area and a specific sound.

Different sets of audIOs are necessary for different application models. This depends on the used metaphor, and the input-and output devices. The control of the interaction is fixed in the dialogue structure which is controlled by the application. An integration of the audIOs in other applications is easily possible.

3 Visual AID

The prototyping tool Visual-AID is used in different steps of the process. It is a platform for design team members to discuss sound design solutions, for building the above mentioned dialogue models and structures and for evaluating complete prototypes as well. It had shown that paper prototypes are not suitable to perceive the difficulties of an auditory representation. You cannot imagine the actual auditory output just by seeing the objects on a visual display. Therefore it is necessary to present changes in the dialogue model of the AUI already in the development process. The presented tool considers this fact by having an author- and simulation modus. Every single change of the interface definition has a direct impact on the auditory presentation.

In general, GUI-Builders are interaction-based and support a layer-based presentation. AUI are time-based and it is impossible to overlap objects. This is considered by the prototyping tool which consists of four views: The application's core is the **scene and object selection.** It represents the structure of the generated auditory prototype in a tree view and it is technically saved as an XML-document. So it can be exported to other applications and transformed to other user interfaces. First a project must be generated in which different scenes can be integrated. For each scene a main interaction object can be defined. An object can be selected in the icon bar, and will automatically set up in the selected scene or other object. This object will start when the application flow starts the scene. A scene can have only one main interaction object, but this one object can activate various other audIOs within the scene. In the **object property window,** which is connected to the scene and object selection, the attributes for a selected object (the project, interaction space, the scenes and used audIOs with sound sources) can be manipulated. This window is mainly used to define starting conditions for a scene and in what kind of mode it should start.

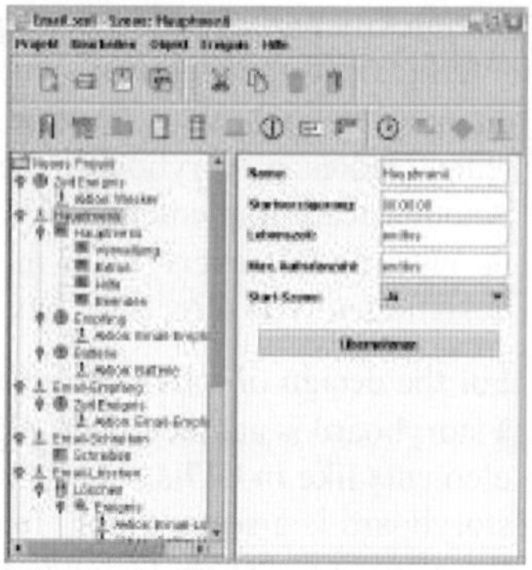

Figure 1: Object selection and property window

The scenes are finally organized in a scene graph. The change from one scene to another is initialised through a user interaction captured by an audIO. The audIO can throw external or internal events. External events have effects on other scenes, while internal events only have effects within the same scene. When an external event is thrown it is possible to define conditional actions. The events will then have different effects. E.g. a scene with the main menu of an application should only be started after a welcome message has finished playing. As well as it is possible to define interaction events from the user, it is also possible to define automatically started time events for a scene. They are used to simulate events from outside the applications. E.g. while selecting E-Mails from a list a message appears after a few seconds to make the user aware, that the battery is low, and the application has to be shut down in a few minutes. The latter scenario shows an important aspect of AUIs: The Designer always has to care about the scenes. Defining a new scene has direct impact of already running scenes. While in GUIs the overlapping or covering of objects has mainly no effects on the correct perception of the actual objects, the audio output will not be recognized correctly from the user, because of interferences. Another problem is considered inside the audIOs. Humans are not able to percept and differentiate more than five to eight auditory signals at once. While the audIOs are never build up with more than five Hearcons, this concept can be broken, when playing more than one scene in the same time. The solution is to change the loudness status for a scene. When the user works in one scene and a dialogue appears in another scene the intensity can be lowered. The underlying

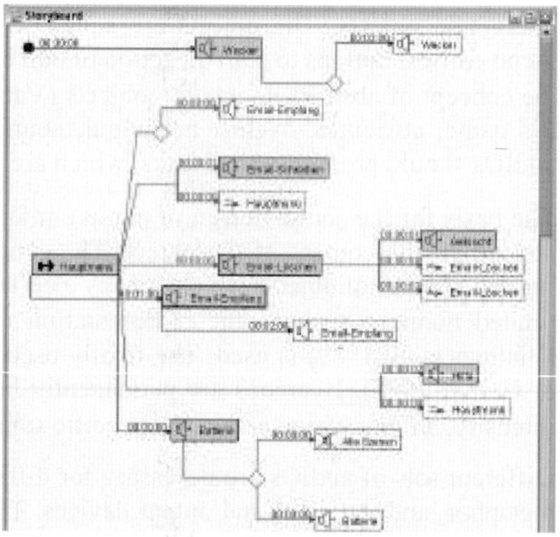

Figure 2: Storyboard and scene graph

building model of AUI will be visible in the **storyboard window.** A special notation contains vertical lines as time events, diagonal lines as interaction events and a junction for conditional events. The knots are scenes which contain audIOs. Each audIO in a scene can release one of the two event types. The storyboard considers the time-based approach of AUI. By clicking on a scene in the storyboard window, a detailed description of the scene appears. The external events are shown, and a timeline for the time-based events is integrated. Each interaction object that throws an external event, has a direction connection to the resulting scene.

By creating a scene and selecting audIOs for it, the changes will be simultaneously shown in the **visual scene description.** It shows the arrangement of the Hearcons used in one scene. The Hearcon attributes (interaction area, position in space, volume and a specific sound) can be manipulated there. The visual scene description can be used as the application window. Possible interaction from the user can be tested. The author has the possibility to view and hear directly a design decision while he makes them and interact with the audIOs.

3.1 Evaluation

Two evaluation phases were conducted to improve the concept and the usability of the design tool. The results of the first evaluation were already implemented and the description in this paper shows the status of the software at the second evaluation. For both evaluations a set of audIOs' was developed for a mobile scenario. The toolkit contained a Menu, MenuItem, Scrollbar, ProgressBar, Separator, 1ofN Selection MofN Selection. The interaction objects were navigable with only four input keys and speech input for hands-free interaction and no a visual display.

The first evaluation has its focus on the question what kind of support the user group needs in the design process. It was an evaluation of the concept. In the first evaluation prototype the possibility to define conditional events was not implemented. The support of the audIO was not finished and finally the ad hoc audio presentation in the scene view doesn't work. The result from this first evaluation was to have a direct feedback and more control over the application flow. The second evaluation had to prove that these advanced concept works. The software was reengineered. At both evaluations ten participants took part and only four participants of the first evaluation also were invited at the second evaluation. Four of them were computer science students, who had already experience in creating auditory user interfaces. One participants was developing graphical user interfaces and had lots of experience with visual tools for GUIs. One person mainly generates composed sounds on the computer. There was one musician, which previously has generated some sound for another project, one physicist with the focus of acoustic, and one psychologist who is working in field of human-environment-relation focused on acoustics.

A detailed task description with context-information was defined. The participants should visually implement an auditory navigation software to manipulate the options and navigate in the content of a mobile MP3-Player. The evaluation task was split up into five subtask. It began by creating a main menu, generating a message when the user enters the application. This first tasks were very easy and had the background to show the participants who to use the interaction objects inside the Visual-AID system The last three tasks were much more complicated. The participants had to use conditional actions and time based events and were ask to simulate a status indicator of a battery, which warns the user every 45sec for a maximum of four times. The participants were asked to think aloud while solving the task. They had to tell what they want to do and what they think where they can find it. The communication with the evaluation leader was rather difficult, because the participants had to hear the output of the system. They must control their design decision and need to know how a scene sounds like. Also the tests were recorded on video. After finishing the test they were interviewed. The whole test last 90 minutes.

All tasks were solved by the participants. One main problem was, that the participants were not involved in the previous concept design and had to edit tasks they did not have invented themselves. The creation of a project was very easy for the participants. The concept was easily understood and internalized. The use of the integrated audIO was also very good. At first it was not obvious (especially for the non computer science group) that the audIO are not part of the application. This is surely a typical behavior of integrated development environments which are mainly used by programmers. So the computer science group had no problem with this integration. The evaluation shows, that there are problems to use the conditional events, as well as the time events. These two features are not normally used everyday and the participants had problems to adapt their ideas to the more formal notation. The participants first have to understand the task and have to think about how to describe it in the software. This was a problem of the lost context. The time events were problematic too, because they can be set on different levels in the prototype. There were global events for the whole applications and local events inside a scene. This in combination with the duration of a scene was hard to manage.

3.2 Conclusion

The direct interaction with the new auditory application was very good accepted. It allows to test the design application ad hoc. It was very easy to test different sound-sources. Mistakes were recognized. It least the main problem during the design process was solved: Not to know, how the final user interface will sound like. The concept evaluation has shown, that the visually supported design of auditory user interfaces helps to test design solution in the design process. Although the prototype user interface is for its own not well designed. There is much work to do, to implement different sets of interaction objects for different usage scenarios. Later it will be necessary to have a connection to the functionality the final application (the user interface prototype is made for) should have. In fact Visual-AID supports a non-formal, flexible and free usage. The tool's architecture takes into consideration to use different sets of independent audIOs'. In the future the tool and the audIO will be made public. Therefore it is necessary to make sure to allow the audio output on a wider sound hardware.

4 References

Bölke, L. & Gorny, P. (1995) Direkte Manipulation akustischer Objekte. In D.Böcker (HG.), Proceedings Software-Ergonomie'95. Darmstadt 1995. Stuttgart Teubner, p 93-106

Blattner, M. M.; Sumikawa, D. A.; Greenberg, R. M. (1989), Earcons and Icons: Their Structure and Common Design Principles. In Human-Computer Interaction, Special Issue on Nonspeech Audio; V4; N1. Lawrence Erlbaum, Seite 11-44

Donker, H.; Klante, P.; Gorny, P. (2002), The Design of Auditory User Interfaces for Blind Users. In Proceedings of NordiCHI 2002, Second Nordic Conference on Human-Computer Interaction October 19-23, 2002, Aarhus, Denmark

Gaver, W. W. (1989) The SonicFinder. An Interface That Uses Auditory Icons. In Human-Computer Interaction, Spec. Issue on Nonspeech Audio; V4 N1. Lawrence Erlbaum, p 67-94

Klante, Palle (2002), Werkzeuggestützte Entwicklung auditiver Benutzungsoberflächen, Presented Paper at Informatiktage 2002, Fachwissenschaftlicher Informatik-Kongreß, 8.-9. November 2002, Bad Schussenried, Deutschland

Sawhney, Nitin; Schmandt, Chris (2000), Nomadic Radio: speech and audio interaction for contextual messaging in nomadic environments. ToCHI: Volume 7, Issue 3 (Sep 2000) Special Issue on human-computer interaction with mobile systems. p 353-383

Vanderdonckt, J. (1995), Tutorial 12: Tools for Working with Guidelines. HCI International'95, 6th International Conference on Human-Computer Interaction, Yokohama, Japan.

VRIO: A Speech Processing Unit for Virtual Reality and Real-World Scenarios - An Experience Report

D. Kranzlmüller[1], A. Ferscha[2], P. Heinzlreiter[1], M. Pitra[2], J. Volkert[1]

GUP[1] and IPI[2], Joh. Kepler University Linz
Altenbergerstr. 69, A-4040 Linz, Austria/Europe
kranzlmueller@gup.jku.at[1] | ferscha@soft.uni-linz.ac.at[2]

Abstract

Human Computer Interaction (HCI) summarizes research and engineering activities related to the communication between human beings and all sorts of "computerized" machines. Within this domain, substantial amount of work is dedicated to the idea of using the human voice as a natural interface for accessing computer systems. The VRIO speech processing unit represents one example of such an interface, where users control the machine via spoken commands. While the application of VRIO was originally intended for Virtual Reality (VR) environments only, a major redesign of VRIO's architecture allows its application to arbitrary scenarios, e.g. within ubiquitous and pervasive environments. This paper describes the revised architecture of VRIO as well as examples of its application in VR environments and for real-world scenarios.

1 Introduction

Comparable to everyday life, communication problems between the human user and the machine may have substantial impact on either of the two communication partners and/or the surrounding environment and are thus not desired. For this reason, many on-going research projects investigate new or improved ways of Human Computer Interaction (HCI). An example is the VRIO prototype, a combination of a software framework and commodity-of-the-shelf hardware, which provides a flexible user interface within arbitrary computing environments.

The original idea of VRIO started in Virtual Reality (VR) environments (Burdea & Coiffet, 1994), in particular the room-sized, 3-D projection-based CAVE Automatic Virtual Environment (Cruz-Neira et al, 1992). While the visual output of the CAVE provides interesting possibilities to experience computer-generated scenarios, the options for user input are often not satisfying. Sophisticated input devices such as the wand, a 3D-like mouse with 6 degrees of freedom (Ware, 1990), require a significant amount of training, while usage of traditional input devices such as a keyboard, is limited by the user's position, posture, and movement in the CAVE.

This example calls for a more human-centred interface design (Landay & Myers, 2001), with possibly natural or intuitive human-computer interaction through multimodal input and output (Ark & Selker, 1999). The deficit of human to computer communication compared to computer to human communication (Damper, 1993) is addressed by a series of ongoing projects in areas such as speech processing and computer vision (Wahlster, 2000).

In this context, the approach of VRIO was to replace or enable parts of the user interaction by voice input. The user, wearing a headset with a microphone, was able to control the VR scenario via spoken commands (Kranzlmüller, Reitinger, Hackl & Volkert, 2001). Due to the invariance of the VR environment, the speech processing was performed on a dedicated workstation located closely by the user.

The application of VRIO in Virtual Reality environments demonstrated some of the limitations of the system but also the feasibility of using speech processing for HCI. Based on this experience, a second prototype of VRIO has been developed. The software framework of VRIO was completely reworked to provide a better suited level of abstraction. The resulting client-server architecture enables interaction between arbitrary command clients - not only speech processing - and the server on the one side, and between the server and arbitrary actuators on the other side.

This paper provides an overview of VRIO's architecture and two exampkes of its application, one within Virtual Reality and one in the real-world. The next section describes the architecture of VRIO and its basic functionality, while example applications are presented in Section 3. A conclusion and an outlook on future work in this project summarizes the paper.

2 Architecture of VRIO

2.1 Overview

The architecture of VRIO resembles the traditional client-server approach. The user interface is provided as a preferably simple and almost invisible device, which receives the commands from the user. The input device transforms the commands in machine-readable form and forwards it via the input interface to the central server. The server analyzes the commands and generates one or more corresponding controls, which initiates the requested actions on connected actuator devices.

An overview of the architecture is provided in Figure 1, with possible command clients on the left side of the server and example actuators on the right side. Of course, it is possible that a command client may also provide actuator functionality.

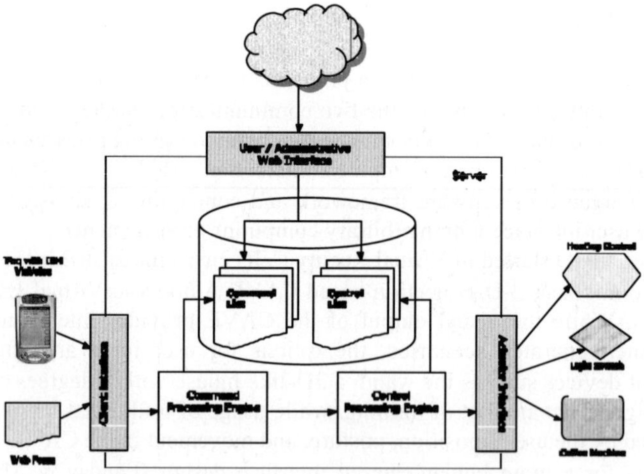

Figure 1: System architecture of VRIO

The communication between the client and the server, as well as between the server and the actuators relies on TCP/IP and HTTP, and can thus be easily integrated in many existing computing infrastructures. In fact, a variety of command clients and actuators, from simple web forms to self-made hardware instruments, have already been tested with VRIO. The mapping between commands and controls is provided by an XML request scheme, which can be easily adapted to different needs and scenarios. In particular, the system is able to dynamically adapt itself to different contexts, depending on the user's location and surroundings. The whole system is supported by an Application Programming Interface (API), which simplifies the usability of the framework for both clients and actuators. The API is (as much as possible) platform-independent in order to facilitate its usage on different devices, from embedded systems to sophisticated high-end installations such as the CAVE.

3 Example

As an practical example of this second generation VRIO, the hardware unit of the command client has been replaced by a personal digital assistant (PDA). Figures 2 and 3 display a VRIO user with a PDA, in this case a Compaq iPAQ. The PDA is equipped with a microphone for processing speech commands and a network interface card for communication with the server. The command client uses IBM's ViaVoice speech processing software (http://www.software.ibm.com/speech) to translate voice commands into corresponding XML requests. The requests are transferred over the network to the server, which maps the command to matching controls as indicated in Figure 1. The controls are then forwarded to the corresponding actuators in order to perform the desired activity.

3.1 Application of VRIO in the CAVE

According to our original intention, VRIO has been utilized within the CAVE Automatic Virtual Environment for different kinds of applications. Within the computational steering environment MoSt, the user controls the execution of a large scale high-performance computing application by navigating through a graphical representation of the program's states during its execution. The user is able to query the systems state, extract behavioural data, or modify parameters to change the program's behaviour.

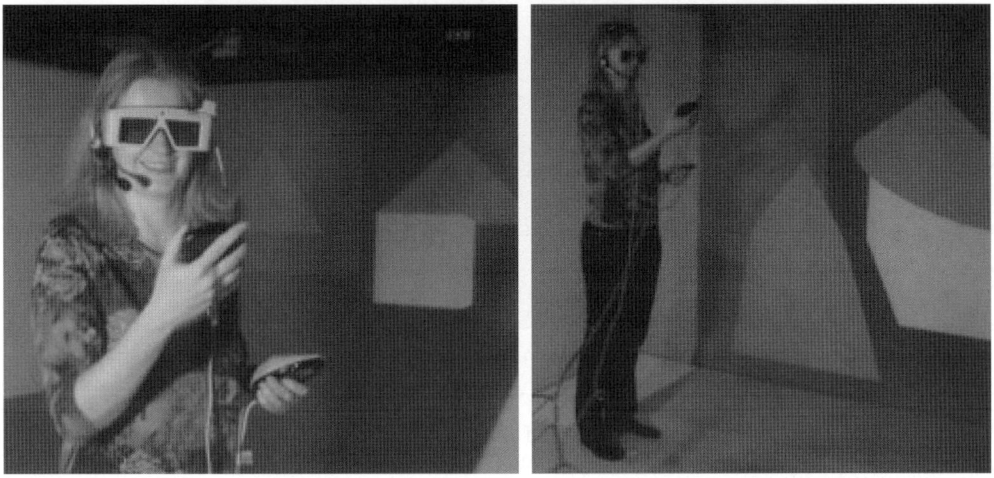

Figure 2: Application of VRIO with the CAVE Holodeck application

Another VR application of VRIO is the Holodeck 3D Editor. In this example, VRIO supports the user when generating arbitrary 3D VR scenarios, which can afterwards be used in artificial worlds. The 3D world is constructed interactively by placing and manipulating 3D objects. A variety of commands for handling 3D objects (e.g. cubes, spheres, cylinders, ...) and other useful items (e.g. light sources) are provided. Objects can be generated, selected and moved within the virtual world. Their graphical representation can be modified by transformation such as scaling and rotation, as well as changing colours of the objects.

An example of the Holodeck 3D editor is given in Figure 2. The user is standing in the virtual world generated by the CAVE. Shutter glasses are required for providing the impression of 3D stereo pictures. The user's position is tracked through the glasses and the 3-D wand, a 6 degree-of-freedoms mouse, which is also used for object movement. The input client of VRIO (a Compaq iPAQ) in the right hand of the user is equipped with a headset to receive the spoken commands. Figure 2 contains some graphical elements of a simple scenario, which have already been positioned and manipulated by the user.

3.2 Application of VRIO in Real-World Scenarios

The redesign of VRIO increases its application domain from Virtual Reality to pervasive computing scenarios (Birnbaum, 1997). One example is a VRIO actuator for a standard interface card, which is used to connect arbitrary electric devices to a computer. With this approach, VRIO is able to switch and manipulate devices, e.g. to control arbitrary gadgets of consumer electronics. Another major effort of integrating this kind of natural user interface is being implemented in the Wireless Campus project, where students will be able to access administrative services of the Johannes Kepler University Linz. Related to this project is the public communication WebWall project developed at the IPI (Ferscha & Vogl, 2002).

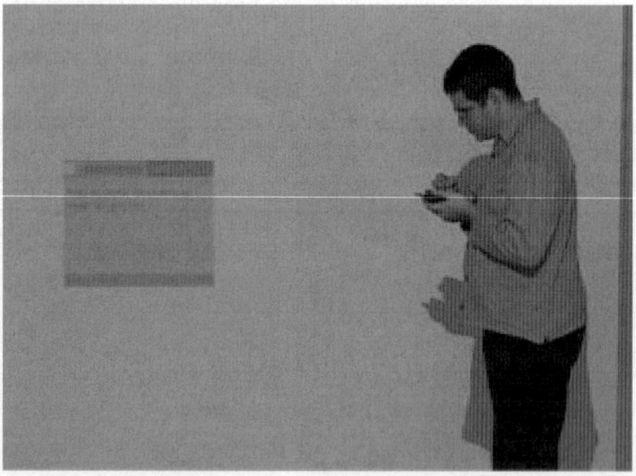

Figure 3: Application of VRIO in front of a public communication WebWall (Ferscha & Vogl, 2002)

The basic concept of the WebWall is based on the provision of large-scale displays at public places, which provides an interface to the World Wide Web (WWW), independent from the available input interface. Through WebWall's connection to the Internet and the cellular phone network, users can utilize the WebWall services as a kind of electronic pinboard for a variety of activities, e.g. placement of small advertisements, event notes, or even multimedia contents.

VRIO represents another sophisticated interface to the WebWall. Instead of typing the messages for the electronic pinboard, users can simply dictate their notes or activate arbitrary actions via spoken commands. An example is shown in Figure 3. The user stands in front of a WebWall, wearing VRIO (in form of a Compaq iPAQ) and speaking to the device. Immediately after the command has been submitted, a new window with the translated note opens on the webwall, containing the message of the user.

4 Conclusions and Future Work

The latest redesign of VRIO transformed the original Virtual Reality Input Output device from the self-contained world of the CAVE to ubiquitous computers and pervasive systems. The new and much more portable approach combined with the adaptable client-server architecture opens VRIO for a set of novel and interesting application areas. First examples in the CAVE and in the real-world have already demonstrated the feasibility of the approach. Several more applications are currently being developed.
An aspect, which has not been sufficiently covered so far, is (linguistic) context, which strongly influences the interaction between communicating human beings. Although the spoken words or gestures of humans are similar in different situations, the context of words or phrases enclosed by other words may trigger different “states” in the communication partners. Thus, the context itself must be considered as a primary source for input within any future human-computer interface.

Acknowledgments

Several colleagues at GUP and IPI have contributed to the development of VRIO. We are most grateful to Ingo Hackl, Bernhard Reitinger, Christoph Anthes, Edith Spiegl, and Simon Vogl.

References

Ark, W.S. and Selker, T. (1999). A Look at Human Interaction with Pervasive Computers. *IBM Systems Journal*, 38 (4) 504-507.

Birnbaum, J. (1997). Pervasive Information Systems. *Communications of the ACM*, 40 (2), 40-41.

Burdea, G., and Coiffet, P. (1994). Virtual Reality Technology. John Wiley & Sons.

Cruz-Neira, C., Sandin, D.J., DeFanti, T.A., Kenyon, R.V., and Hart, J.C. (1992). The CAVE: Audio Visual Experience Automatic Virtual Environment. *Communications of the ACM*, 35 (6), 64–72.

Damper, R.I. (1993). Speech as an Interface Medium: How can it Best be Used?. Proc. Interactive Speech Technology: Human Factors Issues in the Application of Speech Input/Output to Computers, Taylor and Francis, 59-71.

Ferscha A. and Vogl, S. (2002). Pervasive Web Access via Public Communication Walls, Proc. Pervasive 2002, International Conference on Pervasive Computing, Springer-Verlag, 84-97.

Kranzlmüller, D., Reitinger, B., Hackl, I., and Volkert, J. (2001). Voice Controlled Virtual Reality and Its Perspectives for Everyday Life. In: A. Bode, W. Karl (Eds.), Proc. APC 2001 - Arbeitsplatzcomputer, ITG Fachbericht, Vol. 168, Munich, Germany, 101-107.

Landay, J. and Myers, B.A. (2001). Sketching Interfaces: Toward More Human Interface Design. *IEEE Computer*, 34 (3) 56-64.

Wahlster, W. (2000). Pervasive Speech and Language Technology. In: Wilhelm, R. (Ed.), Informatics – 10 Years Back, 10 Years Ahead. Springer Verlag, LNCS, 2000, 274-293.

Ware, C. (1990). Using Hand Position for Virtual Object Placement. *The Visual Computer*, 6 (5), 245-253.

Flow of Action in Mixed Interaction Modalities

Ernst Kruijff *Stefan Conrad* *Arnold Mueller*

Fraunhofer Gesellschaft Sankt Augustin
Institute Media Communication (IMK)
Competence Center Virtual Environments
Schloss Birlinghoven
53754 Sankt Augustin, Germany
Ernst.Kruijff@imk.fraunhofer.de

Abstract

The increasing complexity of immersive virtual environments (VEs) is leading to multivariate interaction. In order to access and display the functionality of the application, developers start mixing interaction modalities by integrating two and three-dimensional input and output devices. Hence, the input structure of an application gets composite, and is difficult to organize. In order to optimize the input structure, and there from forthcoming output structure, this paper focuses at the flow of action within VEs using mixed interaction modalities. We describe the basic concepts behind flow of action in VEs. Successively, we present how these concepts have affected interaction in our software environment AVANGO, by describing aspects of a collaborative virtual engineering application. Overall, we focus on three main areas. The areas, mapping of interaction on devices, allocation of interface elements, and continuous feedback, address major problems of flow of action in VEs.

1 Introduction

The increasing amount of functionality that can be accessed within nowadays VEs poses a distinct need on the employment of flow of action methods. Most VEs are characterized by a wild mixture of interaction techniques. Often, these techniques do not match the tasks, or input devices at hand or impose a severe cognitive load on the user. Furthermore, some of the functionality within a VE are intrinsically of two-dimensional nature, and can be poorly mapped by a three-dimensional interaction technique. In order to match the task characteristics and structure, application builders are starting to mix multiple devices, thereby regularly using several input modalities. Not only do task-specific (specialized) 3D input devices find their way into VEs, developers also make use of 2D I/O devices again.

Flow of action foremost refers to the user's input structure to an application. It takes into account which actions are performed to reach a certain goal, and how these actions can be perceived as a chain of (sub-) actions. The analysis of the flow of action has been discussed in desktop applications. Nevertheless, within immersive applications the focus has rather been on separate actions, and hardly on how these actions can be coupled in a chain of actions.

In this paper, we present a model of flow of action in VEs, that predominantly focuses at the usage of mixed interaction modalities. Coupled to this model, we present general problems that occur

during interaction with a VE. We state how the framework affected our interaction with virtual reality software, AVANGO, by describing a first implementation of concepts.
In this paper we focus at three main areas, we largely overlap with major problems of flow of action in VEs. The areas are mapping of interaction on devices, allocation of interface elements, and continuous feedback.

2 Related Work

Since the rise of desktop applications, quite some research has been made on more general issues of flow of action. This research is foremost based on task analysis and focused at the users and system behavior over time. A good example of research on the syntax of interaction is Buxton's theory on chunking and phrasing (Buxton 1986). More recently, some research has been focused at so called *continuous interaction*. Continuous interaction takes into account the structure of (also) non-conventional control methods like speech interfaces or gestural commands, which form a continuous input flow to an application. Continuous interaction can be seen as the opposite of discrete interaction, in which a user produces more atomic-like actions (Doherty and Massink 1999). The fields of ergonomics and human control also play an important role within the field of flow of action. Control issues, as investigated over a longer period of time in the area of machinery design, show the relevance of dialogue structure controlling (possibly) more complex systems. A general overview of control issues can be found in (Bullinger, Kern et al. 1997).
Within the domain of VEs, flow of action can be regarded as a specific issue for three-dimensional user interfaces (3DUI). Research on 3DUIs has foremost been focused at separable actions like navigation and selection, and not necessarily how actions can be coupled (Bowman, Kruijff et al. 2001). System control in VEs, as being a major focus in this paper, could greatly benefit from a better flow of action in an application.

3 Flow of action in Virtual Environments

In this section, we focus specific factors of flow of action in VEs, especially those that cause problems during interaction with more complex applications. We base our theoretical model on a detailed task analysis performed on a collaborative virtual engineering application, characterized by a high level of functionality. Furthermore, the application exemplifies (also in the interaction framework and case study sections) the usage of mixed interaction modalities by using several 2D and 3D I/O-devices.
The basic issues of flow of action can be traced back to the several stages of information processing (figure 1). In a VE, the interaction with a system is continuous. The user's input device delivers a continuous input stream, and the user's senses receive continuously receive output. Due to the continuity, multiple loops (as shown in figure 1) will be gone though every second. A break in the loop can lead immediately to a disturbance of the flow of action.
The main aim of supporting flow of action in a VE is to allow the user to make use of the (complete) functionality of a system without being burdened by interrupts in the performance of actions or chains of actions. This requires a clear view on how to reach the functionality of a system, how to apply the functionality in a continuous flow, and finally, the communication of the current state of action (interaction mode).

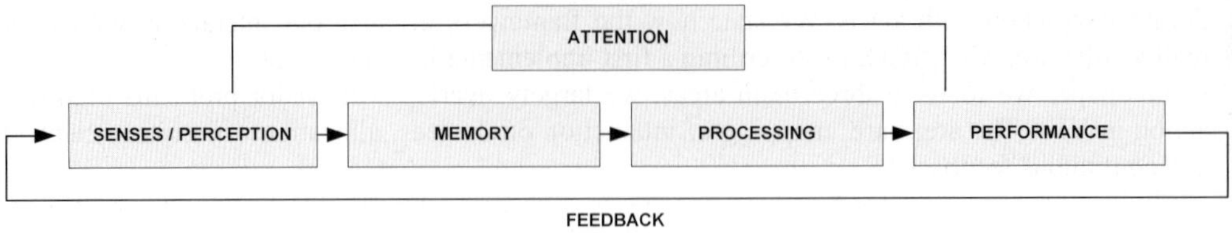

Figure 1: A basic model for flow of action from an information processing perspective

In case of mixed interaction modalities, the input structure to an application gets compound. Even though the mixing of devices can map tasks better, the user's *memory* can easily get overloaded. Many 3DUIs require a minimum level of user experience, since they are regularly not self-explanatory - often functionality is invisible to the user. With the usage of multiple devices, task may be better mapped, but if the user does not know how the functions are mapped to the devices, the task cannot be *performed.* Access of functionality (change of mode of action) is either achieved by pressing a button or by using some kind of system control technique like a flying menu. This change of action is the connector between several tasks or subtasks. Disruption caused by the change of action is a direct disturbance in the flow of action, which can be worsened by failing feedback to the user. Hence, there should be a *continuous feedback* stream to the user, that may be independent of the device. That is, feedback should preferably make use of the same metaphor or technique all the time. This may eventually lead to using one particular sensory channel to supply the user with feedback over time.

Returning to the issue of menus, the inclusion of a desktop-like menu in a VE, or the usage of a PDA or PenPC, mostly leads to the user switching between several focal depths. The menu is two-dimensional and normally in a focal plane close to the user, whereas the actual work area is further away.

Finally, of specific interest for task processing and performance, is the usage of two sensory output channels at once (like gesture and voice (Bolt 1980)), or the application of two-handed interaction techniques, like (Mapes and Moshell 1995). The task is split up in several subtasks being processed by two human output channels simultaneously (parallel processing).

Some general problems can be identified. These problems may seem obvious, but are still regularly occurring pitfalls in many applications.

The *wrong mapping of techniques* on devices is one of the most occurring failures. Developers have to pick from a limited amount of interaction techniques that do not always map the input device in use. Also, input and output devices can be wrongly combined. Overall, wrong mapping leads to performance loss, which directly influences of flow of action in an application. It is a myth that different input devices are capable of transmitting comparable content.

Another regularly occurring problem is *system control overload.* By increasing amounts of functionality, some developers are simply placing more flying menus in the VE. Often these menus have to be large too, for readability issues. The allocation of menus generally overlaps the main work area, leading to attention problems. Rapid eye and head movements of the user are observable, moving from front (menu) to back (work area) focal planes.

Finally, with the increasing complexity of immersive applications, *feedback* to the user is of utmost importance. The user needs to know what is the current mode of interaction, and if a performed action has been succeeded. Often, users need to check multiple times if a task is performed, leading to unnecessary disruptive action loops.

4 Implemented concepts

We have implemented some of the concepts of flow of action in a complex application. As a basis for this application, a collaborative virtual engineering environment, an in-depth task analysis has been performed that specifically looked at task syntax factors affecting flow of action. In the application, a distributed VE, up to 4 remotely connected partners perform design review of an industrial object using direct audiovisual communication. Users have to manipulate the geometry, or design new parts via physical based modeling. During interaction, the users make use of both two-dimensional and three-dimensional data. The system runs at multiple l-shape displays (like the Responsive Workbench^{tm}), using a magnetic tracking system, an adapted stylus, a Cubic Mouse^{tm} (Froehlich and Plate 2000), and a PenPC (Paceblade). The application has been developed in the AVANGO^{tm} software framework (Tramberend 2001).

With respect to the mapping of the functions to the devices, we have analyzed which three-dimensional actions can be performed by the general-task device (the stylus), and which should be performed by the task-specific device (the Cubic Mouse). General task devices often need to make use of some kind of system control technique to change their mode of interaction, whereas task-specific devices regularly integrate the interaction mode into the design of the device. Intrinsically, task-specific devices perform a small amount of actions very well. However, the performance structure of actions performed with the task-specific device regularly include small *repeating interaction loops*. These loops are actions that may not be mapped to the device, like selection of a new object, since performance normally decreases. Nevertheless, since the switching of devices disturbs flow action considerably, we have applied so called *multi-mapping techniques.* Simply said, some actions can be performed by all the devices to avoid device switching. Here the tradeoff between device switching or worse performance is certainly in advantage of the latter, since the repeating interaction loops occur very often.

The two-dimensional tasks, like the session management, are placed on the PenPC. In order to avoid some of the device switching, we added a pen-tip to the stylus, thereby being able to directly control the PenPC with the stylus. The allocation of the graphical user interface elements have particular effects on the attention of the user. The interaction mode changes of the stylus can be achieved by a hand-oriented menu, whereas, as stated before, all two-dimensional actions (including symbolic input) are placed on the PenPC. The PenPC is bidirectional synchronized with the VE, showing the current state of action visualized in the applications scenegraph The PenPC is connected directly in front of the user on the Responsive Workbench. The allocation of the interfaces implies that the user has several focal areas. Foremost, these focal areas are the active work area (the 3D model) and the display of the PenPC. The advantages of using the PenPC are its high readability and low overlap with active work area displayed on the Responsive Workbench. Having two focal areas may have a bad influence on flow of action, since the *focus of attention* may be changing regularly. In our application, we have mapped the functions in such a way, that most of the times, multiple task will be performed serially with one device. In this way, that there is no direct need to switch between a 3D input device and the PenPC in a parallel way. Some exceptions exist, though. When the user is performing symbolic input, the stylus and the PenPC are used in combination. Also, the PenPC can be used for feedback purposes. In order to avoid confusion, having the PenPC in the same place with respect to the user has a big advantage. Due to the fixed position, the focus of attention of the user is always directed to one spot (*directed focus of attention*), thereby minimizing unnecessary search behavior of the user. This behavior is often observed when floating menus are applied, since the user has to check through several focal areas to find the searched widget item. Since the user is switching between devices, continuous feedback needs be taken special care of. This feedback can be attached directly to the input device, for example via a small icon at the stylus tip, or via a feedback method that is independent of the input device in use. Therefor, we have implemented the concept of *cross-device feedback.* First of all, we always display the current interaction mode by a small text-icon, that is placed at the same focal depth as the active work area, in order to avoid switching between focal planes. Nevertheless, for more complex actions, to communicate the active interaction mode is not enough. Therefor, we have applied a scenegraph-oriented interaction mode at the PenPC (Mueller, Conrad et al. 2003). At the PenPC, the currently active node in the scenegraph is displayed, showing detailed data on this node, and the current action performed at the moment. This implies that users can always fall back to the scenegraph-oriented interface, either to receive detailed feedback, or to perform a preferred action directly via discreet input. Even though looking at the PenPC to receive feedback implies a change of focus of attention, it resolves confusion of the user immediately, since the PenPC gives a complete state overview (based on node selection) of the last performed action. The possible attention loss is an affordable tradeoff.

5 Conclusion

In this paper we presented a rough model of flow of action for interaction with a VE using multiple interaction modalities. We identified some major problems of flow of action in VEs, and stated several particular tradeoffs of combining 2D and 3D input devices. We focused at mapping of actions, allocation of interface elements, and feedback issues. We have implemented several concepts in a collaborative virtual engineering environment. Some particular fields of interest are multi-mapping to avoid device switching, directing the focus of attention of the user, and cross-device feedback. In this article several formal tests have been integrated, but more formal evaluation is needed. For example, the analysis of eye and head movements can give new insights in focal area problems.

References

Bolt, R. (1980). "Put-that-there": voice and gesture at the graphics interface. SIGGRAPH'80, ACM.

Bowman, D., E. Kruijff, et al. (2001). "An Introduction to 3D User Interface Design." Presence: Teleoperators and Virtual Environments **10**(1).

Bullinger, H., P. Kern, et al. (1997). Controls. Handbook of Human Factors and Ergonomics. G. Salvendy, John Wiley & Sons.

Buxton, W. (1986). Chunking and phrasing and the design of human-computer dialogues. IFIP World Computer Congres, Dublin.

Doherty, G. and M. Massink (1999). Continuous Interaction and Human Control. European Conference on Human Decision Modeling and Manual Control, Loughborough, Group-D Publications.

Froehlich, B. and J. Plate (2000). The Cubic Mouse, A New Device for Three-Dimensional Input. CHI, ACM.

Mueller, A., S. Conrad, et al. (2003). Multifaceted Interaction with a Virtual Engineering Environment using a Scenegraph--oriented Approach. WSCG, Plzen.

Tramberend, H. (2001). Avango: A Distributed Virtual Reality Framework. Afrigraph'01, ACM.

Clues for the Identification of Implicit Information in Multimodal Referring Actions

Frédéric Landragin

LORIA
Campus scientifique – BP 239
F-54506 Vandœuvre-lès-Nancy CEDEX
FRANCE
Frederic.Landragin@loria.fr

Abstract

The implicit is an imprecise and heterogeneous notion that plays a role, not only in the global comprehension of utterances, but also in the interpretation of reduced phenomena like multimodal referring actions, which combine visual perception, language, and gesture. The identification of the intended referents relies on the correct identification of implicit information that is communicated with the multimodal utterance. The implicit can be linked either to the conjoined use of multiple modalities, to the cognitive status of referents and reference contexts, to the dialogue history, to inferences which occur during the interpretation process, and to interpretative effects for future utterances. For each of these types of the implicit, we focus on the determination of clues that allow the specification of reference domains (structured sub-contexts for reference resolution). We show that this notion of reference domain can integrate all these aspects with computational objectives, thus laying the foundations to a future algorithm for implicit identification.

1 Introduction

With the development of speech and multimodal interfaces, it is easier to imagine to what a spontaneous human-computer interaction may tend towards. If it is pleasant to hear a system react with *"which one?"* to a request like *"remove the blue object"*, and if it is all the more pleasant to see an avatar or a robot pointing out an object and asking *"this one?"*, we want here to claim that such reactions show a good recognition of the user's message and intention but not efficient interpretation abilities. Why? Because in usual communication situations the user's utterance is not intentionally ambiguous but relies on a presumption of success to be interpreted, and because this presumption is based on implicit information (not present in the transmitted code). Following Relevance Theory (Sperber & Wilson, 1995), communication is seen as ostensive and inferential. Inferences have to be made from the ostensive clues to identify the implicit. In man-machine dialogue, the correct interpretation implies the best implicit identification capabilities. What is the nature of the implicit, and how can a system identify it? We focus here on the nature of the implicit that occurs during the resolution of reference in a multimodal context (man-machine dialogue with a visual support). We propose a classification of the implicit and of clues that the system may exploit to identify it. We show how the model of 'reference domains' constitutes an efficient framework for the formalization of all these heterogeneous pieces of information. Then we conclude about the design of multimodal dialogue systems.

2 Implicit and Multimodal Referring Actions

(Grice, 1975) distinguishes two levels of meaning: 'what is said' and 'what is implicated'. The first corresponds to what is explicitly stated in the verbal utterance, and the second to the implicit that is implied from it. The main idea is that the proposition conveyed by the utterance is built from what is said, and that its interpretation requires implicatures. Considering that some information is not said but explicitly communicated, (Sperber & Wilson, 1995) and (Bach, 1994) add a middle level of meaning. This information ('making as if to say') is called impliciture by Bach and completes the proposition conveyed by the utterance. Sperber & Wilson call explicature the resultant enriched proposition. Their cognitive theory of relevance emphasizes the exploitation of ostensive clues for implicit identification, but includes no computational model for that, as we need for man-machine dialogue. With more computational objectives, (Grosz & Sidner, 1986) propose a model of implicit identification through the notions of intentional structure and attentional state. The problem is that these notions are more explicative than operational. The authors say themselves that the two structures are related: the first is a primary factor in determining the second, and the second helps constrain the first. Their computation is by consequence difficult. (Geurts, 1999) formalizes with DRS (Discourse Representation Structure) the implicit linked to presuppositions. Nevertheless, this important extension of the theory of (Kamp & Reyle, 1993) follows the same restriction to linguistic considerations. The lack is in the exploitation of the visual context. In dialogue with visual support, the implicit depends on the verbal utterance as well as on extra-linguistic characteristics. Visual context is taken into account by (Beun & Cremers, 1998) with focus spaces, and by (Kievit, Piwek, Beun & Bunt, 2001) with visual salience. Beun & Cremers present a computational model of identification of focus spaces in dialogue with visual support. One important point is that they determine a strong clue to detect a change of focus space: redundancy. They show that the presence of redundant information in the utterance denotes the user's intention to reconsider the context. Such a model is very near to what we imagine for a dialogue system. It consists in a formalization of the implicit using data structures, and in clues for their identification. To complete this point of view, the DenK system (Kievit et al., 2001) exploits salience as a visual clue to choose between objects when searching for referents. Then salience constitutes a useful implicit information. Another important point of the DenK system is the identification of the implicit in a multimodal context, including speech and gesture. A weak point is that inferences are ignored. We want here to try to put together all these aspects of the implicit in a multimodal context.

The work we pursue in multimodal reference resolution (Landragin, Salmon-Alt & Romary, 2002) shows that each referring action implies the activation of a reference domain. This subset of contextual information can come from visual perception, language or gesture, or can be linked to the dialogue history or the task's constraints. The utterance's components allow to extract the referents from this sub-context and to prepare the interpretation of a future reference. For example, *"the red triangle"* includes two properties that must be discriminative in a reference domain that must include one or more 'not-red triangles'. The use of the referring expression, and particularly the linguistic constraint conveyed by the determiner, is then justified. A further referring expression like *"the other triangles"* may be interpreted in the same domain, denoting a continuity in the reference sequence. On the other hand, the use of a demonstrative determiner (*"this triangle"* associated to a gesture) implies that another triangle is present in the reference domain, in order to justify the spatial contrast denoted by this determiner and the gesture. The same domain will be used for the interpretation of *"the other one"*. Some reference domains may come from perceptual grouping (see (Landragin et al., 2002) for the computation of the Gestalt criteria to build visual domains). Some may come from the user's gesture (Landragin, 2002), others from the task's constraints. All of them are structured in the same way. They include a grouping factor (being in the same referring expression, being in the same perceptual group, etc.), and one or more partitions of elements, each partition being characterized by a differentiation criterion ('red' vs. 'not-red', 'focused' vs. 'not-focused', etc.). This unified framework allows to confront the various contexts, and to model the implicit whatever its origin between perception, speech and gesture.

3 A Classification of Implicit Information

Some implicit information intervenes during the specification of the semantic form of the current utterance, so before its pragmatic interpretation (section 3.1). This implicit may be distinguished from the one used for the pragmatic interpretation, that corresponds to constraints for referents identification (sections 3.2, 3.3 and 3.4). A third kind of implicit intervenes after the pragmatic interpretation and is important for the dialogue continuation (section 3.5).

3.1 Implicit Linked to the Conjoined Use of Multiple Modalities

In order to build the logical form corresponding to the enriched proposition conveyed by the utterance, we have to precise the nature of the implicit link between modalities. First, we consider the presumption that a gesture is associated to the linguistic expression. A simple clue is the presence of a demonstrative determiner. This clue is particularly strong when all anaphora are impossible considering the linguistic context. Another strong clue is a temporal synchronicity between speech and gesture. Second, we consider the implicit link between the pointed objects and the referents. The interpretation may be generic: *"I love these cars"* associated to a gesture pointing out a Ferrari. The main clues are linguistic: the aspect of the predicate (unlike 'love', 'possess' and 'destroy' have a punctual aspect and lead to a specific interpretation); the presence of a numeral or a quantifier (both of them force the specific interpretation). Our idea here is that the presence of a gesture does not change anything to the possible ambiguity between specific and generic interpretation. The very common utterance *"this N"* associated to a gesture pointing out one N can always refer to all N of the corresponding type. Even for a specific interpretation, the pointed object may differ from the referent: *"he has a big head"* associated to a gesture pointing out a hat. The clue here is an incoherence between speech and gesture. Third, the association of speech and gesture is also an implicit association of criteria for searching referents. Following componential semantics, a lot of criteria like category, properties, cardinality, and spatial focusing due to the gesture, work together in the reference resolution process. An example of clue is the association of a definite determiner and a gesture, that denotes the extraction of a particular referent in the domain delimitated by the gesture (Landragin, 2002). The use of reference domains is a way to formalize the implicit link between modalities, because the differentiation criteria can complete each other in order to merge domains built from the gesture and from the linguistic form.

3.2 Implicit Linked to the Cognitive Status of Referents and Domains

The resolution of reference to objects requires to test each object of the situation. Some objects may be more accessible than others. For example, (Gündel, Hedberg & Zacharski, 1993) propose a classification of cognitive status of entities invoked in the interaction. Considering the form of the verbal referring expression, the referent must be focused (*"My neighbor has a dog.* It *kept me awake"*); activated in the low-term memory (*"My neighbor has a dog.* This dog *kept me awake"*); familiar in the long-term memory (*"I couldn't sleep last night.* That dog *kept me awake"*); etc. This implicit has been built from the previous utterances. Another implicit is linked to the visual accessibility of entities. This is the role of salience. The more an object is salient, the more it is accessible for a referring action. A system based on implicit identification must then include a model of salience. Moreover, we said that a referring action implies the activation of an implicit reference domain. The accessibility of such domains is also an important aspect in the interpretation, and constitutes the implicit linked to the cognitive status of the interpretation context. The model of reference domains, that includes a model of salience and the management of a stack of domains, appears as a good way to formalize this kind of the implicit.

3.3 Implicit Linked to the Dialogue History

When following the same referring strategy, the user gives implicitly indications to the system. For example, in the corpus presented in (Wolff, De Angeli & Romary, 1998) and corresponding to a tidying task, two strategies may be identified: tidying guided by the category of the objects, and tidying guided by the visual perception of objects. These strategies correspond to scripts (or frames). If the user always follows one of them, his referring actions may be easier to understand. These scripts can be formalized in reference domains: the grouping factor corresponds to the strategy and the differentiation criteria are linked to the steps.

3.4 Implicit Linked to Inferences for the Interpretation

Following (Ducrot, 1991), the term 'implicitation' groups together all inferences that are computed from the logic form of the utterance and from the context. It includes presuppositions and implicatures. Presuppositions are important in the reference resolution process because they reduce the possibilities. For example, the interpretation of *"tidy these objects"* will only consider the objects that are not tidied yet. Implicatures occur when the user does not follow the co-operative principle of communication (Grice, 1975). For example, he may assert *"these objects have to be tidied"* instead of ordering, with the same communicative intention. The speech act is implicit and has to be identified by the system. In this manner, some utterances may include an ironic intention (or other emotive characteristics) that might be identified, too. The problem is very large because the information needed to understand is boundless. A solution is to take into account only the information linked to the concepts of the utterance and of the visual context. That is what is done in the model of reference domains.

3.5 Implicit Effects for Future Utterances

The last kind of the implicit consists in interpretative effects for future utterances. For example, *"the first"* may be followed by *"the second"*. In French, *"l'un"* will be necessarily followed by *"l'autre"*. These baiting effects are linked to the way information is presented in the utterance (referents are presented taking into account the future possibilities of anaphora). Clues are deduced from the referring expression and also from visual perception or from task constraints. For example, a spatial disposition of objects or a particular sequence of actions due to the task may incite to referring expressions like *"the next one"*. Once more, reference domains provide a structure for the formalization of such heterogeneous constraints.

4 Towards the Computation of Clues

What are the computational consequences of these phenomena? First, the necessity to manage mental representations of the entities that are in the user's mind. Without them, it seems impossible to manage cognitive status and implicit domains corresponding to attentional state. Second, the necessity to clearly define algorithms for the exploitation of the clues that lead to implicit identification. Considering the referring expression, constraints are formulated about the referents and the possible reference domains. These constraints filter the potential referents and allow to identify the correct one, whose cognitive status is then set as focused. A third consequence is methodological. It is very difficult to validate an implicit notion like mental representations. Psycholinguistic experimentations involving sequences of references constitute a solution. The interpretation of expressions like *"the other one"* can prove the existence of an implicit reference domain. The possibilities of such a methodology are numerous.

5 Conclusion

To conclude, we think that a simple microphone associated to algorithms based on implicit identification and on the exploitation of clues like a determiner or characteristics of the visual context will be more efficient than a lot of devices associated to recognition and fusion algorithms. Technical aspects must not mask semantics and pragmatics concerns. The approach of underdetermined semantics and the one characterized by a fine treatment of grammatical words like determiners (as we show with our examples) are just beginning. We hope that the notion of reference domain, that corresponds to unified structures modeling heterogeneous mental representations, will be useful to the design of multimodal systems, first for the interpretation of multimodal referring expressions, and then for the comprehension of the whole utterance.

References

Bach, K. (1994). Conversational Impliciture. *Mind and Language*, 9, 124-162.

Beun, R.-J., & Cremers, A. H. M. (1998). Object Reference in a Shared Domain of Conversation. *Pragmatics and Cognition*, 6 (1/2), 121-152.

Ducrot, O. (1991). Dire et ne pas dire. Paris: Hermann.

Geurts B. (1999). Presuppositions and Pronouns. London: Elsevier.

Grice, H. P. (1975). Logic and Conversation. In: Cole, P., & Morgan, J. (Eds.) *Syntax and Semantics* (vol. 3, pp. 41-58). Academic Press.

Grosz, B. J., & Sidner, C. L. (1986). Attention, Intentions and the Structure of Discourse. *Computational Linguistics*, 12 (3), 175-204.

Gündel, J. K., Hedberg, N., & Zacharski, R. (1993). Cognitive Status and the Form of Referring Expressions in Discourse. *Language*, 69 (2), 274-307.

Kamp, H., & Reyle, U. (1993). From Discourse to Logic. Dordrecht: Kluwer.

Kievit, L., Piwek, P., Beun, R.-J., & Bunt, H. (2001). Multimodal Cooperative Resolution of Referential Expressions in the DenK System. In: Bunt, H., & Beun, R.-J. (Eds.) *Cooperative Multimodal Communication* (pp. 197-214). Berlin & Heidelberg: Springer.

Landragin, F. (2002). The Role of Gesture in Multimodal Referring Actions. In: *Proceedings of the Fourth IEEE International Conference on Multimodal Interfaces*. Pittsburgh.

Landragin, F., Salmon-Alt, S., & Romary, L. (2002). Ancrage référentiel en situation de dialogue. *Traitement Automatique des Langues*, 43 (2), 99-129.

Sperber, D., & Wilson, D. (1995). Relevance. Communication and Cognition. Oxford UK & Cambridge USA: Blackwell.

Wolff, F., De Angeli, A., & Romary, L. (1998). Acting on a Visual World: The Role of Perception in Multimodal HCI. In: *Proceedings of the AAAI Workshop on Multimodal Representation*.

Specifying the User Interface as an Interactive Message

Jair C Leite

Federal University of Rio Grande do Norte
Campus Universitário - Natal, RN - 59072-970 -Brazil
jair@dimap.ufrn.br

Abstract

Semiotic Engineering is a theoretically based approach to HCI in which the computer system is seen as a metacommunication artifact that conveys a message from the designer to the users. In this perspective, the User Interface (UI) is a message that communicates to users what they can do with an application and how they can use it. In a GUI, the designer should use the common repertory of UI widgets as a visual sign system in order to compose the message. However, an interaction designer is not necessarily a good graphic designer and he/she may not know how to use GUI design tools in order to achieve a better communicability. To know what express to users, interaction designers need to formulate the message content in a conceptual level. Our work contributes to UI Design and to the Semiotic Engineering approach by proposing the Interactive Message Modeling Language - IMML - a formalism to the specification of the UI as an interactive message. The IMML provides a semantic model that guides the designer to specify the UI in an abstract and structured way focusing on what he/she really wants to mean to the users. The resulting specification could be mapped on conventional UI widgets.

1 Introduction

Semiotic Engineering is a theoretically based approach to HCI in which the computer system is seen as a metacommunication artifact that conveys a message from the designer to the users (de Souza, 1993). In this thought-provoking perspective, the User Interface (UI) is an interactive message that communicates to users the answer to two fundamental questions: (I) "What kinds of problems is this application prepared to solve?" - the application functionality - and (II) "How can these problems be solved?" - the interaction model. Such message can, in their turn, send and receive messages. This process occurs in a demonstration-like fashion where the designer is saying: "Watch what I mean!"

In the Semiotic Engineering approach, the designers need a formalism that helps them to specify the UI as an interactive message. In this paper we describe our work with the IMML – Interactive Message Modeling Language – a linguistic formalism to the user interface specification. The IMML provides a semantic model that guides the designer to specify the UI in a flexible, abstract and structured way focusing on what he/she really wants to mean to the users. It is a vocabulary for reasoning about both the communicative and interactive purposes of the UI elements. The resulting specification could be mapped on conventional UI widgets. We argue that this communicative perspective provided by the IMML improves the design process and promotes user interface learning.

2 The IMML

The IMML was designed to be used by humans, not by computers, so one of the main concerns is to provide statements that seem like the designer's words. The language has a XML like syntax (eXtensible Markup Language) allowing the composition of nested messages in a very flexible and structure way. Since the IMML is on the conceptual level, the designer focuses only in the essential aspects of the application functionality and interaction without loosing control of its communicability. It considers not only structural aspects of UI screen layout but also the performing and interactive aspects of the user interface medium. The IMML follows a user interface conceptual model composed by a *domain model,* an *interaction model* and a *presentation model.*

2.1 The domain model

The domain model is composed of Domain Objects and Domain Functions. The domain objects refer to database records, document files, e-mail messages and several other objects that users know in their domain. They are represented in computer system as data structures in the programming level and as user interfaces signs (label, icons or widgets) in the user interface level. Using IMML, the designer specify each domain object based on the following template:

```
<Domain-Object name="name" type="type">
```

It is only necessary to specify the name of the object and the type of representation. The IMML has some basic representation types: numeric, string and finite set are examples. These basic types are supposed to cover the main structured data type used in computer systems and they are more appropriate on the abstract level of specification. The following sentences are the specification of two domain objects:

```
<Domain-Object name="Number of copies" type= "numeric">
<Domain-Object name="Printer name" type="finite-set">
```

The second conceptual entity of the domain model is the Domain Function. A domain function refers to a complete process executed by the computer that changes the state of a domain object. From the users point-of-view, a domain function is a computing service that realizes a use case. It is not an internal program function that implements something like a file opening or a window drawing. The specification of a domain function is constructed in IMML as showed in figure 1.

```
<Application-Function name="Printing"
  Operands="File Name, Printer, Number of copies"
  Pre-conditions=" File name must be informed,Printer must be selected,
      Number of copies should be enter or the system print only one copy"
  Post-conditions=" The printer should print the number of copies informed of
    the specified file name."
  Control="Start, Stop, Pause, Resume, Cancel"
  State="Available, Running, Stopped, Finished
```

Figure 1: Application function specification.

The IMML requires the designer to specify the function name and other properties. The operands are the domain object elements that are used or affected by the execution of the function. The pre- and post-conditions define what should be true before and after the execution. These conditions are expressed in IMML informally by using natural language statements. The control properties allow the designer to specify what kind of execution control applies. *Start, Stop, Pause, Resume* and *Cancel* are examples of execution controls. They are used in association with the command element of the interaction model. The state property specifies the states of a function.

2.2 The interaction model

The interaction model represents the interaction process between the user and the application. The basic elements of the interaction model are the *function command, function results and presentation control.*

A Function Command is used to enter information and to control the execution of domain function. It should have construct to provide information to set function operands and to allows the user to control the execution of the function. It allows the user to *start, stop, pause* or *resume* the execution of some function. It is important to note the each function command must be associated with a domain function. A domain function may be associated with several commands.

A function command is a composition of several basic interactions in a structured form. A *basic interaction* refers to the action users do to interact with a UI widget. Examples of basic interactions are a *click on a button, type textual data, click on a checkbox, select data from a list* and *view a result*. The IMML provides the following basic interaction: *entering information, selecting information, activating controls, dragging-and-dropping objects and viewing objects.*

```
<Command name="Print" domain-function="Printing">
 <Join>
   <View> To print a document you must enter the information and …</View>
   <Sequence>
    <Join>
     <Select>
          <Enter Domain-Object="File Name">
          <Select Domain-Object ="File Name">
     </Select>
        <Enter Domain-Object="Number of copies">
   </Join>
   <Select>
    <Activate Control="Start" Domain-Function="Printing">
    <Activate Control="Stop" Domain-Function="Printing">
    <Activate Control="Suspend" Domain-Function="Printing">
    <Activate Control="Continue" Domain-Function="Printing">
   </Select>
  </Sequence>
</Comand>
```

Figure2: Interaction Model Specification.

The IMML provides five structures to create more powerful commands from the basic interaction units. Structures may be nested by other structures. The *sequence* structure is the most common of all and specifies that the user should do the each interaction in an ordered manner. The *repeat* structure is used when the user should do an interaction many times. The *select* structure allows users to choose one of several interactions. The *combine* structure is used when two or more interactions have some kind of dependency. Finally, the *join* structure is used to group interactions that have some relationship but not require a specific order to act. It is modeless. Using IMML, a command should be specified as showed in figure 2.

Function Results are the output of domain function or error messages. They provide the feedback to user about the interaction process. Interaction results are domain object or a direct communication about the interaction itself.

The Presentation Control is an interaction component that is used to allow users to control the presentation of the user interface. Navigation and browsing are examples of this kind of interaction. Pop-up and pull-down menus and the window minimization and maximization controls are also examples of presentation controls. The main difference between function command and presentation control is that the latter does not launch or control a domain function. It only controls the presentation of other interface objects. The presentation control message specification is the statement `<Activate Control="Show" Command="Configure">` in figure 2. It means that the user should activate a widget in order to view a command message. It activates no domain function.

2.3 The presentation model

The presentation model refers to the global message the designer constructs to communicate the domain and interaction message to the user. The main elements are *task environment, command panel* and *domain object display*. The elements are used to organize the presentation of domain and interaction elements.

The task environment contains command panels and domain displays. A task environment may be simple or composite. A composite task environment can contain others task environments. Each task environment is organized by interaction and layout structures.

A command panel organizes and presents the function command to the users. The organization is specified using the presentation and interaction structures. A set of function commands may be organized using a *select, sequence, combine, join* and *repeat* structures. The domain object display is used to represent the states of domain object and function. In most cases, domain object states are the main concerns because it helps users to know about how the things are going on. Windows and Panels are examples of messages that present other messages and the UI objects. Figure 3 shows an example of presentation model specification.

```
<task-environment>
  <command-panel>
    <select>
      <Activate Control="start" domain-function="listUsers">
      <Activate Control="start" domain-function="listFiles">
    </select>
  </command>
  <display>
      <list domain-object="File name">
  </display>
</task-environment>
```

Figure 3: Presentation model specification.

The last component of our presentation model is the Metacommunication Message. A metacommunication message is direct communication that helps users to understand the application. Help and tip statements are common examples of this type of message. They are specified in IMML as statements using the **view** basic interaction as showed in figure 2.

3 Mapping IMML statements into UI signs

To be a practical formalism, the design representation constructed using IMML should be translated into a concrete user interface. The table 1 shows a semantic map the makes a correlation from the IMML keywords and statements to the interface signs. This semantic map may change if the meanings of the interface signs are not in agreement with those known by users. For instance, the *combine* may be mapped into a *frame* widget, whereas the *join* structure may be expressed by a visual grouping of interface signs. The *select text* keyword may be mapped into a *combo box* widget and the *enter text* into a *text box*. The *command button* widget is usually used when there is an *activate* statement.

It is not our intent to provide an automatic translation because we believe that the human talent to creativity and visual communication is fundamental to assure a good communicability. It is the designer that decides what user interface signs should be used. There is no formal meaning assigned to each user interface widget.

Table 1: The semantic correlation between the IMML and UI signs.

IMML Statements	UI Signs (widgets)
`Sequence {...}`	Spatial configuration and setting buttons label color attribute as gray meaning *disable*.
`Combine {..`	Frame widget
`Select {...}`	Spatial configuration (visual grouping)
`Select text`	"Combo box" widget
`Enter text`	"Text box" widget
`Enter number`	"Spin box" widget
`Activate`	Command button

4 Discussion

Our work on the IMML embodies a new perspective on user interface and interaction design. Rather than rely only on the physical interface-centred design activity that occurs through a rapid prototyping process, we propose that the design process is designer-user communicative process, within a semiotic framework that supports it. Specifically, we have created a specification language that might integrate the communication of design intentions with the interaction processes, using interface signs as the unit of communication and interaction. Our research has been guided by the principle that an application's usability is inherently constrained by the lack of good designer-user communication and formalisms that support this design approach.

In sum, the IMML was designed as a language to construct a message that is by itself an interactive medium to human-computer interaction. Our focus is on design situations in which learning and communication is an important goal and the designer is trying (or at least willing) to get beyond merely making a user-friendly interface. We are developing a diagrammatic version of IMML to integrate with the Unified Modeling Language (UML).

Ackonwledgments

Jair C Leite would like to thank to CNPq (Process 470096/2001-2) for supporting his research.

References

de Souza, C.S. (1993) The Semiotic Engineering of user interface language design, in International Journal of Man-machine Studies, 39, 753-773, Academic Press.

Gesture-based Interaction in Digital Museum

Li Zhenbo, Meng Xiangxu, Xiang Hui, Yang Chenglei

College of Computer Science, Shandong University, JiNan, ShanDong, China, 250100
zhenboli@hotmail.com

Abstract

The study of multimodal interaction aims at the development of harmonious Human-Computer Interface. Among many types of input and output modalities, gesture is important. In this paper, we detailed a gesture-based Human-Computer Interface with data-gloves to be used with a virtual digital museum system. Topics covered include hand modeling, three-dimensional interactive primitive definition, gesture motion capture and gesture analysis etc.

1 Introduction

A very important trend of current computer evolution is personating and virtual reality is one of the most important personating technologies. Traditional graphical user interfaces and interactive equipments such as keyboard and mouse are no more suitable nowadays for many applications, e.g. virtual environment (VE) applications. 3D interactive equipments are becoming more and more important. Correspondingly, the design of three-dimensional multimodal Human-Computer Interface (HCI) is necessary. Among many types of input and output modalities including gesture, speech recognition, sensory feedback and vision tracking, gesture becomes important during recent years for HCI studies [1,2,3]. With data gloves we can get all-round gesture information including three-dimensional space motion. And it is easy to implement than computer vision based gesture input [4].

On the other hand, the 24-hour virtual museums become possible with the development of virtual reality technology and the increase of computing power. The advantages of virtual digital museum can be summarized as follows: The visitors can enjoy cultural relics without the restriction of time and place and safety of cultural relics are also guaranteed. The visitors have the chance to see precious cultural relics that would be exhibited in a conventional way because of security considerations; furthermore, with the help of multimodal interaction, the visitors can even "touch" or "manipulate" the relics, which would be important for professionals. Relic restoration and duplicating could also benefit from virtual digital museums. A well-designed virtual museum need convenient interaction methods, which the keyboard and mouse based traditional interactive methods are not relevant. For example, the ordinary cursor has only two degree of freedom; it is not only unnatural but also complicated to accomplish six degrees of freedom input tasks. The gesture based interactive method can make this nature and feasible.

In section 2, a brief description of the history of data gloves is given and information about 5DT data gloves (which are used in our system) are introduced. Topics including hand modeling, three-dimensional interactive primitive definition, gesture motions capture and gesture analysis etc. is covered in Section 3. Implementation results are also given. We then conclude our work and discuss some future directions in Section 4.

2 Data Gloves

Before data gloves appeared, people have made a lot of similar equipments to acquire gesture information. In 1983, Gray Grimes developed the first data gloves to recognize hand gesture at BELL laboratory, and it was used to help the deaf recognize sign language letters. But it did not widely used at that time. In 1987 Thomas Zimmerman etc developed the data gloves possessing ten joints and six degrees of freedom, and this kind of gloves were evolved into the famous VPL data gloves later. It has been widely used until now. Many similar data gloves have appeared after that including Power Gloves developed by Mattel and Nintendo; Cyber Glove developed by James Kramer at Stanford University; 5DT Data gloves developed by 5DT corp. in 1995 etc [5]; These data gloves are more and more convenient, apperceiving more hand motions, imitating hand motion more realistically, and more cheap than before.

3 Gesture-Based Interaction

Using multimodal interaction can strengthen immersion feelings when navigating in virtual museum. Fig.1 describes a multimodal interaction sketch for virtual museum. One important part of that is gesture interaction, one gesture often equals to one user's interactive task, user needn't to partition one interactive task into many subtasks when submitting interactive intention, this will reduce user's interactive burden widely [7]. Gesture-based interaction includes hand modeling; hand analyzing and synthesizing; primary language defining etc.

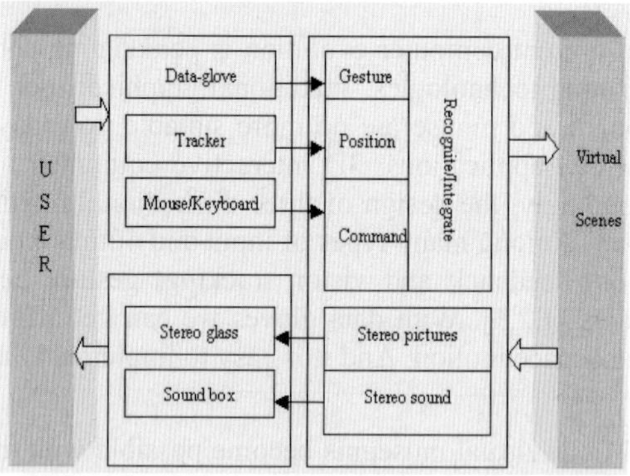

Figure2. Multimodal Interaction Scheme

3.1 Hand Modelling

Hand model is needed as a navigating avatar in our virtual digital museum system. It can help choose navigating paths and manipulate culture relic models. So we have to analyze hand motions and make physical model of the hand available at first.

The human hand is highly articulated. To model the articulation of hand, the characteristics of hand should be analyzed. Fig.3 is the kinetic skeleton model of a right hand, each of the four fingers(Ⅱ ~ Ⅴ) has 4 degrees of freedom, every DIP and PIP joint possesses one degrees of freedom, chiefly being the bending motion, the metacarpophalangeal (MCP) joint has two degrees of freedom due to flexion and abduction. The thumb has a different structure from the other four fingers and has five degrees of freedom, one for the interphalangeal (IP) joint, and two for each of the thumb MCP joint and trapeziometacarpal (TM) joint both due to flexion and abduction. Each of the palm rotational and translation motion has three degrees of freedom. Thus this kinematical model has 27 degrees of freedom [9]. Moreover the human hand is high restriction, and the motion of finger joints are high related, for example $0° \leq \theta_{PIP} \leq 110°$;$0° \leq \theta_{DIP} \leq 90°$, the motion of DIP and PIP joints have approximate such restriction: $\theta_{DIP} = 2/3\theta_{PIP}$ etc [10]. The kinetic model can describe the hand motion property.

Different models are suitable for different HCI applications. For example, in animation application, the shape model should be as realistic as possible and for motion analysis, a simple kinematical model is often more adequate [8]. For the purpose of virtual navigating in our virtual digital museum system, we adapt the kinematics hand model above according to the characteristics of 5DT Data Glove 5:

1. Removed the degrees of TM joint (Thumb), which is regarded as a part of the palm.
2. Removed the rotation degree of the MCP joint and retained the bending degrees of all joints, thus the fingers have 14 degrees of freedom altogether.
3. Retained the pitch and roll degrees and removed the rotation and translation degrees of palm, so whole hand has 16 degrees of freedom. This kind of hand model can well imitate the shape of hand and grabbing actions etc.
4. Moreover, in the navigation the hand is served as an avatar of the user, it also represent the position and direction of user's viewpoint, it will occupy one part of user's sight region all the time. In order to avoid the impact on the navigation and still be able to let the user know the corresponding hand motions, the hand model is set translucent. (Fig. 4):

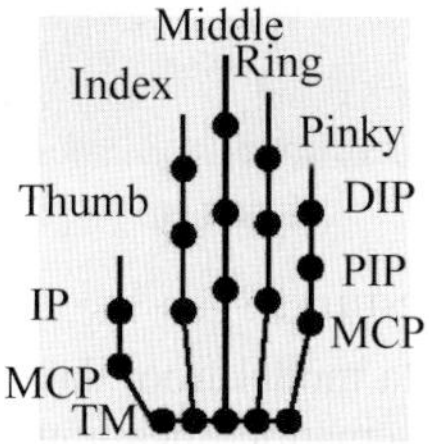

Figure3.Kinematical Hand Structure [9]

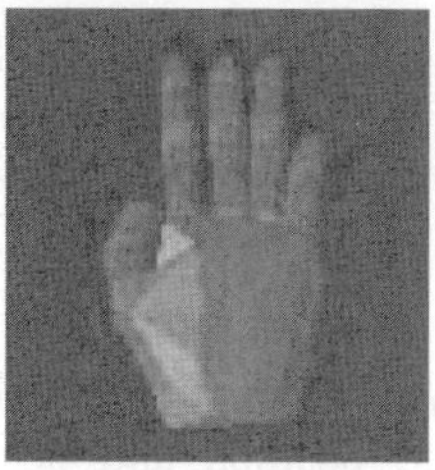

Figure4. The Semitransparent Hand Model

3.2 Analyses and Synthesis of Gesture

The gesture is the physical expression of thoughts. The gesture is used for accomplishing certain interactive tasks: pointing, refusing, grabbing, drawing a flower and shaking hands with somebody etc. The gesture can be divided into the static state gesture and the dynamic gesture. From the motive of the beginning to the aftermost show, the gesture is always followed some fixed motion characteristics on space and time. Kendon described three motion stages of a simple gesture: preparation, stroke and retraction. This kind of explaining has such merit: it can be used to describe any gesture generally and durability. Quek extract 6 rules to distinguish meaning gesture from meaningless movement [11,12]:

1. Movements that comprise a slow initial phase from a rest position proceed with a phase at a rate of speed exceeding some threshold (the stroke), and returns to the resting position are gesture laden.
2. The Configuration of the hand during the stroke is in the form of some recognized symbol.
3. Slow motions from one resting position and resulting in another resting position are not gestures.
4. Hand movements outside some work volume will not be considered as pertinent gesture.
5. The user will be required to hold a static hand gestures for some finite period for them to be recognized.
6. Repetitive movements in the workspace will be deemed gestures to be interpreted.

Huangs etc [14] proposed a method for gesture motion modeling: some joints' movement represents the gesture of static hand. The course preparation, stoke and retraction is used to represent the gesture of dynamic hand. The shape of hand may change during the moving course, but these changes do not contain any gesture information and will be ignored.

According to different application scenarios, gestures can be classified into several types such as conversational gestures, controlling gestures, manipulative gestures and communicative gestures etc [9,13]. For virtual navigation, we consider only the application of static state gesture; the static state gestures would be used in two ways:

1. Controlling gesture used for navigating, the hand model serves as a avatar and the user's viewpoint can be controlled by the position and direction of the model (Fig. 5);
2. Manipulative gesture, which can be used as a natural interactive method with virtual objects such as cultural relic etc.

In order to imitate hand motion, we should synthesize the hand gesture according to the restrictive conditions including gesture rules, state transformation

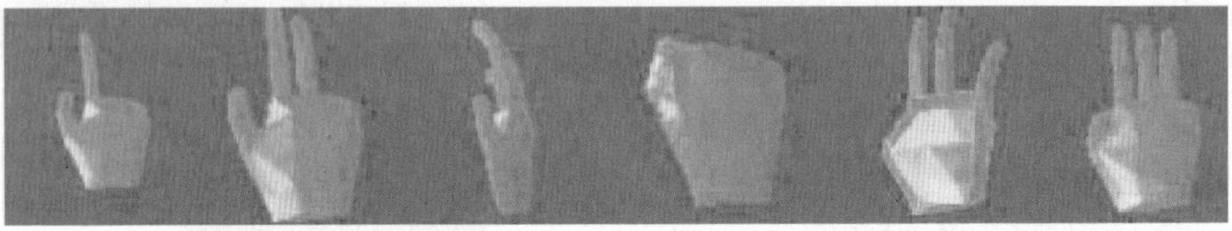

Forward Backward Turn left Restart Look up Look down

Figure5. A Set Of Navigating Gestures

and graphic display processing etc. That is, we build the data model from data gloves inputs and synthesize the hand position and pose with high-level semantic meaning after a series of transformation. Then we would change the hand model accordingly and display it on the screen.

3.3 Primitive Definition and Gesture Recognition

Three-dimensional interactive primitive languages try to understand the user action according to the scene objects or graphic objects. The interactive tasks and primitive are corresponding one to one . When Using gesture interaction at virtual environment, the change of hand model is displayed in the space according to the states transition and restrictions. At the same time, the scene objects should carry out the corresponding tasks according to the gesture. In our virtual digital museum system, the navigating gestures include: Forward, Backward, Turn left, Restart, Look up, Look down etc (Fig.5,6); and the manipulative gestures include: pick, rotate etc (Fig.7). To make computer carry out the predefined actions, we must recognize the gesture first.

At present, there are three methods used for gesture recognition: template matching, NN, statistical analysis. Template-matching technology is used for gesture recognition in our system. A set of template was saved before recognition. When starting recognition, the program will match the data gained by the data gloves' sensors and the template. If they are matching, the corresponding action will be performed.

4 Conclusion and Results

In this paper, we detailed a gesture-based Human-Computer Interface with data-gloves to be used with a virtual digital museum system we designed and implemented. The hand model is used as an avatar for virtual navigation in our immersion systems. Gestures are used to replace the traditional keyboard and mouse to enhance the immersion experience in virtual environments. User can "navigate" in our virtual museum and "manipulate" the virtual objects with gestures. The experimental results are satisfactory and encouraging. For future works, the problem of the recognition of dynamic gesture still needs to be explored and moreover the integration of gesture-based interaction with other interactive channels also is worth further studying.

Figure6. Navigating Gesture

Figure7. Manipulating Gesture

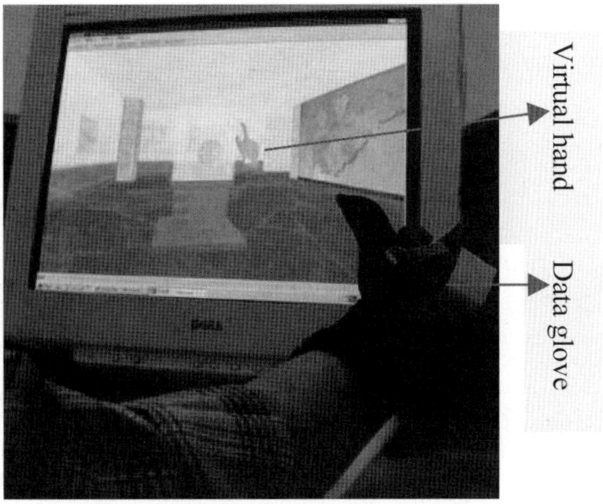

Figure8. Virtual System

5 References

1. J.K.Aggarwal, Q.Cai, "Human Motion Analysis: A Review, " IEEE proc. Nonrigid and Articulated Motion Workshop'97, pp90-102, 1997.
2. R.Kjeldesn, J.Kender, "Toward the Use of Gesture in traditional User Interfaces", IEEE Automatic Face and Gesture Recognition, pp151-156, 1996.
3. Vladimir I. Pavlovic, R.Sharma, T.S.Huang, "Visual Interpretation of Hand Gestures for Human-Computer Interaction: A Review," IEEE PAMI, Vol.19, No.7, July, pp.677-695, 1997.
4. Dong Shihai, Wang Jian, Dai Guozhong.Human-Computer Interaction and Multimodal User Interface [M]. Beijing: Science Press, 1999(in Chinese).
5. D. J. Sturman and D. Zeltzer. A survey of glove-based input. *IEEE Computer Graphics and Applications*, 14(1): 30--39, January 1994.
6. Fifth Dimension Technologies. 5DT Data Glove for the Fifth Dimension User's Manual [Z]. Febrary 2000.
7. Li Yang, et al,"Research on Gesture-Based Human-Computer Interaction", Journal of System Simulation Vol.12, N0.5, September 2000.
8. Ying Wu, Thomas S.Huang. Human Hand Modeling, Analysis and Animation in the Context of HCI In Proc. of Int'l Conf. on Image Processing, Japan, 1999.
9. Ying Wu and Thomas S. Huang, "Human Hand Modeling, Analysis and Animation in the Context of HCI", In Proc. IEEE Int'l Conf. on Image Processing (ICIP'99), Kobe, Japan, 1999.
10. J.Lee, T. Kunii, "Model-based Analysis of Hand Posture", IEEE Computer Graphics and Applications, Sept, pp.77-86, 1995.
11. Francis K.H. Quek, "Toward a Vision-Based hand Gesture Interface", in Proc. of VRST'94, pp.17-34, 1994.
12. ChanSu Lee, et al"The Control Of Avatar Motion Using Hand Gesture" Proceedings of the ACM Symposium on Virtual Reality Software and Technology (VRST'98), pp. 59-65, Nov. 1999, Taipei, Taiwan
13. K.H.Jo, Y.Kuno and Y.Shirai, "Manipulative Hand Gesture Recognition Using Task Knowledge for Human Computer Interaction", Proc. 3rd IEEE International Conference on Face and Gesture Recognition, pp.468-473, 1998.
14. Thomas S. Huang and Vladimir I. Pavlovic, "Hand gesture modeling, analysis, and synthesis," Proc. of International Workshop on Automatic Face-and Gesture-Recognition (IWAFGR), Zurich, Switzerland, June 26-28, 1995.

SIVIT ShopWindow
A Video-based Interaction System

Michael Lützeler, Jens Racky and Hans Röttger

Siemens AG
Corporate Technology, CT IC 5
81730 München, Germany
Jens.Racky@siemens.com

Abstract

At the EuroShop 2002 fair in Düsseldorf, Germany, Siemens exhibited it's Sivit ShopWindow presentation system at the well-known Galeria Kaufhof department store, Düsseldorf, Königsallee for the very first time. This interactive shop window allows the customer to "go shopping" in a virtual department store by controlling an "in-store" multi-media presentation from the outside. This paper presents an overview of the video-based input channel for the presentation system.

1 Introduction

Todays shop-windows require continuous work to attract customers. Often video displays, e.g. video projectors, are installed to enhance attractiveness. Nevertheless, the problem of every sequential presentation remains: The current topic is not necessarily of interest to the current addressee. Adding interaction capabilities overcomes this problem by enabling more intelligent presentation concepts (e.g. avatar-based shop systems) addressing the customer directly.

As we stated in earlier contributions (Maggioni & Röttger, 1999) common input devices are difficult to understand, to learn and to operate. Sometimes, as in a public environment they are even unapt for the task at hand, for example due to their sensitivity to vandalism.

At Siemens Corporate Technology we have developed a solution for an interactive shop window, providing touch screen functionality. In this contribution we focus on the perception module representing the input channel of the implemented system.

2 Environment and Constraints

An interactive system has to satisfy several demands, regarding the interaction concept as well as the sensor technology.

The interaction concept has to be simple to address a broad range of customers. This restricts interaction to touching the shop-window, e.g. for selecting an item, for playing a video or to enter a virtual room.

The sensor module, interpreting the user's actions, has to meet several requirements. Economical aspects include scalability to different display sizes, remove-ability without destruction of valuable components and protection against vandalism. Operational requirements comprise customer independency (e.g. body size, strength, clothing) and avoidance of false-activation.

3 Technical Approaches

Well known are sensitive areas based on photo-resistors. They lack success as they do not integrate seamless with the presentation and due to their technical limitations.

Several other sensor technologies meet the requirements described in the previous section to some extent, like

- Light-barrier-arrays: These provide sufficient accuracy, but require to be mounted at the windows border. This restricts their use to rather small windows. As shop-window sizes are not standardized reusability is doubtful.
- Perception by a laser-range-scanner: This solution offers sufficient resolution and flexible mounting options. Disadvantages are the price for the scanner and arguable health risks due to laser usage.
- Perception by cameras: This class of solutions can cope with the constraints mentioned above, though individual constraints, e.g. regarding illumination conditions, can become serious.

Taking the mentioned requirements and the characteristics of these approaches into account, we decided to develop a video-based perception system as described in the remainder of this paper.

4 System Implementation

4.1 Hardware setup

Our hardware setup, as depicted in figure 1, consists of two processing units, one is dedicated to the multi-media presentation whereas the second performs the recognition task.

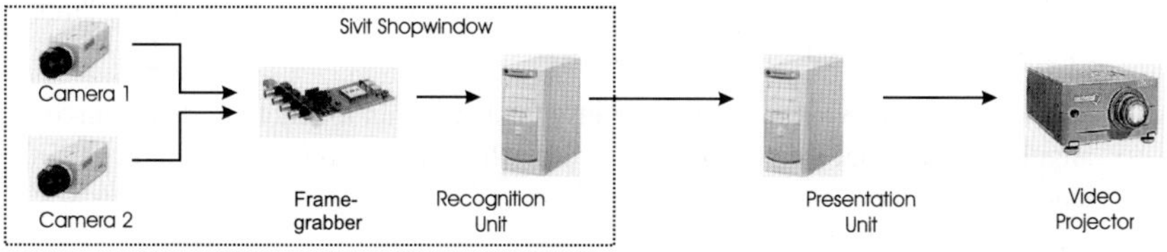

Figure 1. Hardware setup

The recognition unit, a standard desktop computer, is connected to two video cameras in a stereo setup. The cameras are mounted in a way that permits observing the users pointing gestures from inside the shop, see Figure 2. The results of the recognition, e.g. touching the window at a certain position, are sent to the presentation unit via a TCP/IP connection using an XML-based communication protocol.

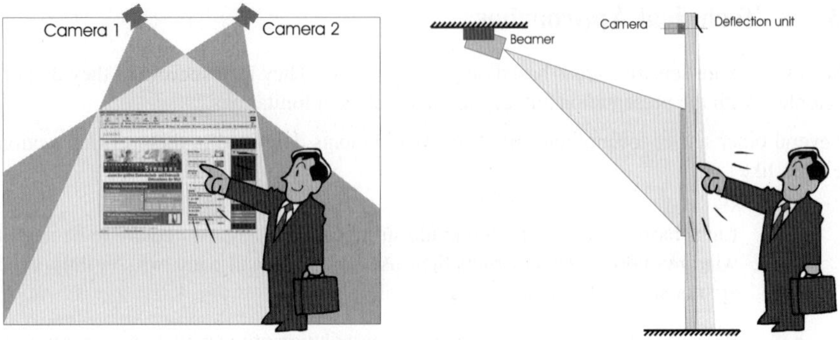

Figure 2. System front- and side-view

4.2 Recognition Unit

The system recognizes the user's touching position on the window. The recognition is based on motion tracking and stereo triangulation of the "touching" position. Our system runs at full camera frame rate (PAL, 25 Hz). Latency is below 100ms.

Figure 3 shows the data flow within this module. As we use a stereo imaging approach, the first four processing steps are present for each sensor.

The processing steps for each sensor are as follows:

1. *Image acquisition*: The interaction area is seen by a PAL color camera via the deflection unit.Cameras are setup with automatic gain and shutter. For our approach it is not necessary to use synchronized cameras.
2. *Motion Detection*: Motion detection is done by a recursive filter, based on a thresholded neighborhood accumulator. For each pixel this filter yields a binary value stating "motion" or "no motion" in the motion–detection–image.
3. *Object Detection*: Motion objects are instantiated within the motion–detection–image as rectangular regions touching one image border, which have sufficient size and contain a sufficient amount of motion.
4. *Object Tracking*: Motion object's bounding box is updated with every frame, until no motion has been detected for a given time, in which case the object is deleted. After updating all existing motion objects, new motion objects are searched for between the existing ones.
5. *Touch Detection*: A motion object is classified as "touching the window" if it's closest border point has exceeded a distance threshold and motion has stopped for a sufficient time.

After processing the individual camera images, both sensors are searched for motion objects in state "touching" to find corresponding pairs of motion objects (object registration).

For registered objects the 3D pointing position is computed using the camera model described in(Bouguet, 1999). The 3D pointing position is then transformed to display coordinates by the well–known Tsai algorithm (Tsai, 1987).

Finally a "touching"–event is generated for the display–position and sent to the presentation unit via the communication interface, where it is by default mapped to a "left-button-click" mouse event.

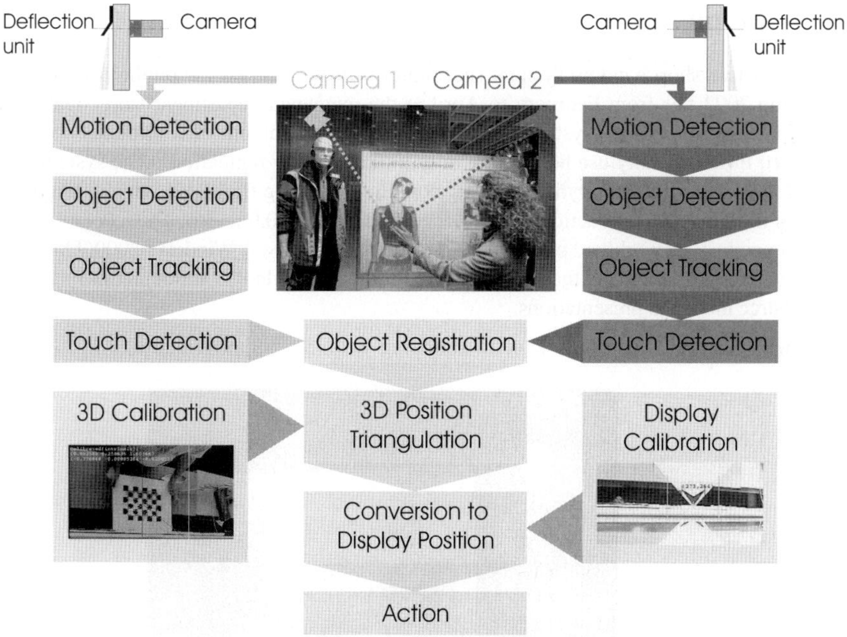

Figure 3. Processing steps and stereo triangulation concept (u.m.).

4.3 System Calibration

Like many touch-screens our system has to be calibrated to relate the users touching position to display coordinates.

The system is calibrated following several steps:

1. *Intrinsic Calibration*: The intrinsic part of the camera model is calibrated before mounting of each camera, using a standard "chess–board"–like calibration tool.

2. *Mechanical Calibration*: Each camera has to be mounted and adjusted to cover the entire interaction area.

3. *World Calibration*: The world related (extrinsic) parameters of the camera models are calibrated jointly for both cameras mounted at their final position with the same tool. By this the system is given the origin of it's 3D-coordinate system.

4. *Display Calibration*: For nine successively displayed calibration points the 3D positions are recorded and then used for a calibration by Tsai's algorithm. For maximum accuracy a dedicated calibration tool shown in Figure 2 is used. The shape has been designed to allow for robust detection of a point under extreme ranges of scaling and distortion.

5 Results

The interactive shop-window system, see Figure 4, was installed for public use during and after EuroShop 2002 fair from February to April at the renowned department store "Galeria Kaufhof" Düsseldorf, Germany. The system provided more than two month of continuous operation from 9 AM to 10 PM. The daily use hours were only limited by city regulations. The system performance proved satisfactory under varying lightning conditions ranging from overcast and bright sunshine to artificial illumination. Interaction was provided on a 1.2m by 0.9m area. The recognition task lead to approximately 10% workload on a PIII 1GHz, running Microsoft Windows 2000(TM). This permits running the presentation and the recognition module on a single PC, although this not recommended for resource intensive presentations.

Customers passing by enjoyed "window shopping" with Sivit-ShopWindow.

Figure 4. Pilot installation (Siemens AG press picture), (SIVIT ShopWindow Product).

6 Conclusions

The presented system is suitable to enhance conventional shop windows in an appealing, flexible and thus cost-effective manner. Future work will concentrate on an indoor variant for compact setups using wide angle cameras.

References

Bouguet, J.-Y. (1999). Visual Methods for Three–Dimensional Modeling. PhD thesis, California Institute of Technology.

Maggioni, C. & Röttger, H. (1999). Virtual Touchscreen - a novel User Interface made of Light - Principles, Metaphors and Experiences. *Proc. of the Eighth Intl. Conference on Human-Computer Interaction*, 301–305, 1999.

SIVIT ShopWindow Product. Siemens AG, I&S MIS1;Contact: Mr. G. Slansky (Gerald.Slansky@siemens.com).

Tsai, R. Y. (1987). A versatile camera calibration technique for high accuracy 3d machine vision metrology using off–the–shelf tv cameras and lenses. *IEEE Journal of Robotics Automation*, 3(4):323–344.

Pointing Gesture Recognition and Indicated Object Detection

Tomohiro MASHITA, Yoshio IWAI, and Masahiko YACHIDA
Graduate School of Engineering Science, Osaka University
Toyonaka, Osaka 560-8531, JAPAN
mashita@yachi-lab.sys.es.osaka-u.ac.jp

Abstract

We propose a method and its application for detecting an object indicated by pointing gestures. The method uses a linear model and the concepts of a cognitive origin and a reference plane. We use a wearable camera system to recognize pointing gestures. Our system consists of an omnidirectional head-mounted camera and methods called random sampling and importance sampling, and the advantages of our system are that the system runs in real-time and that the camera can shoot the arm and an indicated object simultaneously.

1 Introduction

A human being can generate a new idea from others' suggestions by communicating with them. Everyone has had the experience of solving a problem which cannot be addressed alone, but can be solved only by communicating others. It is a fact that communication with others is an effective way to generate a new idea or activate creativity. In such a case, what means and information we use for communication with others is important. What place we exchange communications in is also important.

General means of communication include facial expression, gesture, and speaking. These are performed face to face. Only text and voice are used for communication when people are apart from each other. If a system were developed that could can send and receive realistic facial expressions and gestures, we would be able to communicate with each other more effectively. The important components in such a system are a natural interface and a real-time process.

The operation of pointing and spatial cognition have been studied in the fields of neuroscience and cognitive science for a many years (Brain, 1941; McInTyre, 1997). In that time, the pointing gestureaaa has also been studied in the field of human interfaces and robotics because it is an effective interface for man-machine communication (Kahn, 1996; Waldherr, 2000; Kortenkamp, 1996). We think that pointing gesture is one type of a natural interface for man-machine interaction too. A method proposed by Kahn and Swain (Kahn, 1996) uses positions of head and that of a pointing hand to detect an indicated object. When a human being points to an object with looking at it, the pointing hand doesn't overlap with the object in his/her view. It is easy to understand that the indicated object doesn't fall on the line extending from the eyes or head through the hand. At present, understanding of the pointing operation with respect to reaching distance is increasing, but the knowledge with respect to walking distance is lagging behind, because the relation between an indicated object with respect to walking distance and the pointing gesture is not yet clear. We think that a system becomes more effective at communication if the system can find an indicated object. We therefore propose a system for pointing gesture recognition in real-time and a method for estimating the orientation of an indicated object.

A wearable camera system has several advantages. One of them is that the user wearing the system can move freely because the camera moves with the user if the system's camera has a sufficient field of view. In the case of a fixed camera, the user ensure that his/her position is in the fixed camera's field of view. Another advantage is that it is possible for wearable camera to have the same field of view as the user. Because of these advantages, a head-mounted omnidirectional camera is suitable for our system in order to activate communications.

Our proposed system avoids occlusion problems by using a head-mounted camera and by capturing images from the top. A 3D hand position is estimated from the hand regions by giving the

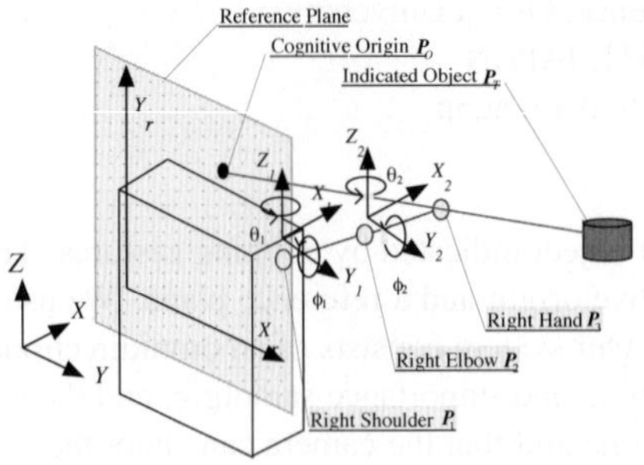

Figure 1: Pointing Gesture Model

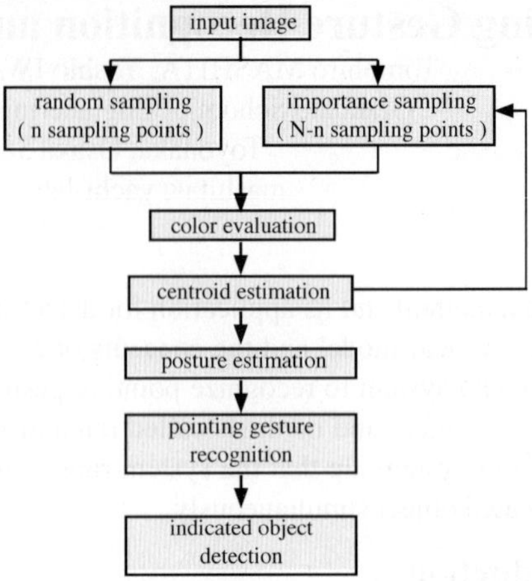

Figure 2: System Flow

lengths of arms in advance. The hand is detected and tracked quickly by random sampling and importance sampling. We use a camera with an omnidirectional sensor (HyperOmni Vision) in order to cover the wide range of hand movement and the indicated object.

In our method for indicated object detection, we use the concepts of the cognitive origin and the reference plane. In these concepts, the direction of an indicated objects is estimated from an estimated cognitive origin, and the range of movement of a cognitive origin is restricted within a plane called the reference plane. We defined the relative position of a cognitive origin in the reference plane and arm posture by linear. Our method can also use linear to detect the direction of the indicated object.

2 Pointing Gesture Model

The pointing gesture model is shown in Figure 1. The cognitive origin, $\boldsymbol{P}_O$, is a point in the 3D space, and we assume that the indicated object, $\boldsymbol{P}_T$, is on the line extending from the cognitive origin to the end of the pointing arm, $\boldsymbol{P}_3$. Parameters θ_1 and ϕ_1 are the directions of the vector from $\boldsymbol{P}_1$ to $\boldsymbol{P}_2$, and θ_2 and ϕ_2 are the directions of the vector from $\boldsymbol{P}_2$ to $\boldsymbol{P}_3$. We define that the cognitive origin, $\boldsymbol{P}_O$, lies in the reference plane. The coordinates of the cognitive origin in the reference plane, (x_r, y_r), are determined from the posture of the pointing arm. We define a pointing gesture as that in which the positions of a user's right shoulder $\boldsymbol{P}_1$, right elbow $\boldsymbol{P}_2$ and right hand $\boldsymbol{P}_3$ are colinear; in other words, the case in which $\theta_2 = 0$ and $\phi_2 = 0$. θ_1 and ϕ_1 are calculated from a vector from shoulder $\boldsymbol{P}_1$ to hand $\boldsymbol{P}_3$. We assume that (θ_1, ϕ_1) and (x_r, y_r) are linear, as expressed by the following equation:

$$\begin{bmatrix} x_r \\ y_r \end{bmatrix} = \begin{bmatrix} \theta_1 & \phi_1 & 1 & 0 & 0 & 0 \\ 0 & 0 & 0 & \theta_1 & \phi_1 & 1 \end{bmatrix} \begin{bmatrix} \alpha_1 \\ \alpha_2 \\ \alpha_3 \\ \alpha_4 \\ \alpha_5 \\ \alpha_6 \end{bmatrix}, \tag{1}$$

where $\alpha_1, \cdots, \alpha_6$ are regression parameters, θ_1 and ϕ_1 are observed parameters estimated from the arm posture.

Next, we explain the learning of the linear model. We presuppose that there are m sets of data of arm postures, $\theta'_1, \cdots, \theta'_m, \phi'_1, \cdots, \phi'_m$ and coordinates of an indicated object, $\boldsymbol{P}_{T1}, \cdots, \boldsymbol{P}_{rm}$ for learning. We can calculate the coordinates of the cognitive origin, (x_{ri}, y_{ri}), from the positions of the right hand and the indicated object because the cognitive origin, the right hand, and an indicated object are colinear. The relation between the cognitive origin and arm posture is expressed by the following equations:

$$\boldsymbol{B} = \boldsymbol{A}\boldsymbol{P},$$

$$\boldsymbol{A} = \begin{bmatrix} \theta'_1 & \phi'_1 & 1 & 0 & 0 & 0 \\ 0 & 0 & 0 & \theta'_1 & \phi'_1 & 1 \\ & & & \vdots & & \\ \theta'_m & \phi'_m & 1 & 0 & 0 & 0 \\ 0 & 0 & 0 & \theta'_m & \phi'_m & 1 \end{bmatrix}, \boldsymbol{B} = \begin{bmatrix} x_{r1} \\ y_{r1} \\ \vdots \\ x_{rm} \\ y_{rm} \end{bmatrix}, \boldsymbol{P} = \begin{bmatrix} \alpha_1 \\ \alpha_2 \\ \alpha_3 \\ \alpha_4 \\ \alpha_5 \\ \alpha_6 \end{bmatrix}. \quad (2)$$

We can calculate the least square solution by using the following equation:

$$\boldsymbol{P} = \left(\boldsymbol{A}^T\boldsymbol{A}\right)^{-1}\boldsymbol{A}^T\boldsymbol{B}. \quad (3)$$

3 Pointing Gesture Recognition System

Our system detects and tracks the targets, the markers on the user's elbow and hand, and the system estimates the directions from the omnidirectional camera's focal point to the tracked targets. The arm posture is analytically calculated from the directions of the tracked targets. An outline of our system is shown in Figure 2. At first, the system receives an omnidirectional image from the omnidirectional head-mounted camera. The camera, HyperOmni Vision, has the same optical characteristics as a common camera. Next, the system decides the sampling points in the image by using random sampling and importance sampling. These sampling methods can reduce the computational cost because they employ sparse sampling. The system evaluates the sampling points by using a color model assumed to be a normal distribution. Then the system estimates the centroids of the targets in the image, if the system can estimate the centroids, it uses them to predict positions for importance sampling in the next frame. Finally, the vectors from the omnidirectional camera's focal point to the tracked targets are calculated from the estimated centroids. The system estimates the arm posture from the vectors, shoulder position and the arm lenth. The shoulder position and the arm length are given in advance. At this point, the system has up to 4 solutions because of sign ambiguity. The correct posture is decided by the pointing gesture recognition module.

4 Experimental Results and Discussion

In experiments, we used a Fastrak tracking system (Polhemus) using electro-magnetic fields to determine the position and orientation of remote objects, and the positions of the head, the right shoulder, the right elbow and the right hand. For learning, test subjects pointed at 75 targets while sitting on a chair. The positions of the targets and a test subject are shown in Figure 3, and the heights of the targets vary between 0 cm, 100 cm, and 155 cm above the floor. The $\alpha_1, \cdots, \alpha_6$ values are different for each test subject because we assume that the parameters can be estimated for each person.

Figure 4 shows the lines extending from the cognitive origins to indicated objects, while also showing the lines in the 3D space projected onto the X-Y plane. In this figure, the difference between each subfigure is the height of the indicated objects. It is clear that the lines in each subfigure in Figure 4 intersect each other at a point between the head and the shoulder, and that the intersection moves from the shoulder to the head in proportion to the height of the indicated object.

Figure 5 shows the relation between each cognitive origin in the reference plane and each θ'_i, ϕ'_i. It can be read from figure 5 that θ is proportional to X_r and ϕ is proportional to Y_r. This result supports our concept of the reference plane.

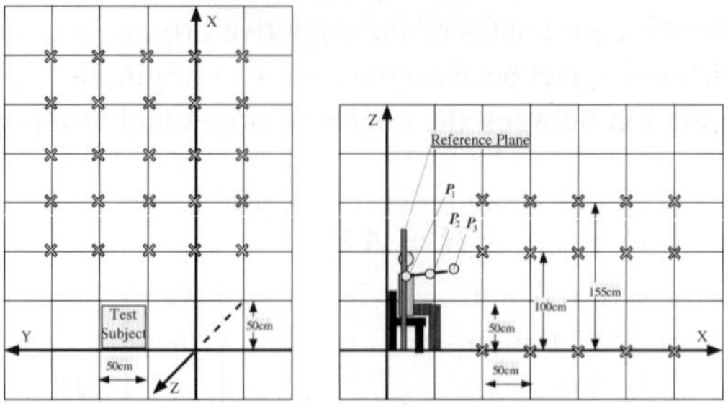

Figure 3: The Positions of The Targets and The Test Subject

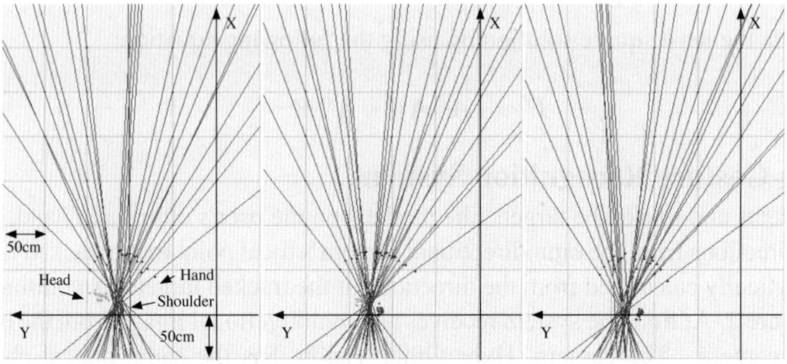

Height of targets (a) 0cm (b) 100cm (c) 155cm

Figure 4: The lines from the cognitive origins to indicated objects

The root-mean-square error of learning datasets at the reference plane is shown in Table 1. In this experiment, the ground truth is the point while the line from each position of the right hand to each position of the indicated object and the reference plane intersect with each other, and the cognitive origin is calculated by Eq. 1. We defined the reference plane as a plane parallel to the $Y - Z$ plane and including a point of the mean of shoulder positions.

The root-mean-square error of another data set in the 3D space is shown in Table 2. In this experiment, each ground truth is the 3D coordinates of each indicated object. The error is calculated from the distance from the ground truth and the line estimated from the cognitive origin and the hand. We think that this error is caused by the error on the reference plane and, is enhanced by the ratio of $||\boldsymbol{P}_3 - \boldsymbol{P}_T||/||\boldsymbol{P}_O - \boldsymbol{P}_3||$.

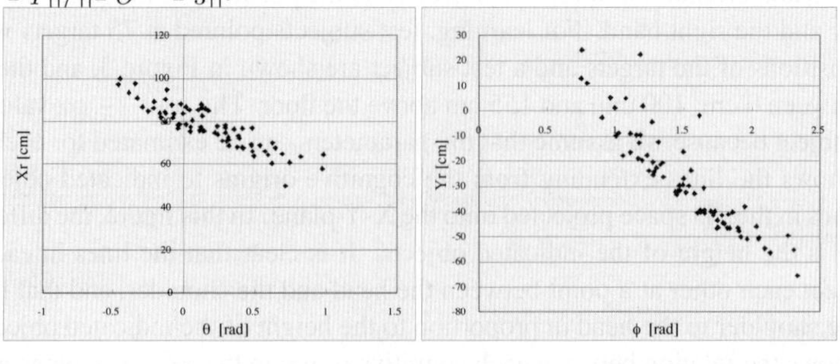

Figure 5: Relation between θ_i', ϕ_i' and X_r, Y_r

Table 1: RMS on The Reference Plane by Learning Dataset

Test subject	RMS [cm]
1	7.123
2	8.991
3	7.151

Table 2: The RMS of The Indicated Objet

Test subject	RMS [cm]
1	27.661
2	45.926

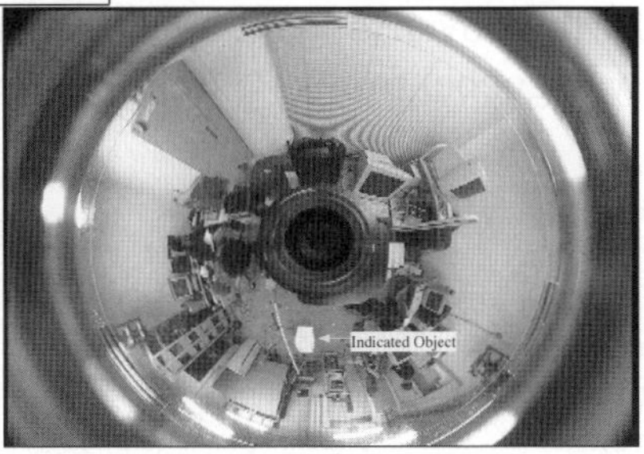

Figure 6: Omnidirectional image and an estimated line

The estimated line projected onto an omnidirectional image is shown in Figure 6. In this figure, the indicated object is a cardboard box. The line maintains a slight distance from the target, though the distance between the target and the line is acceptable because the system searches for the indicated objects around the line.

5 Summary

We proposed a method for detecting the orientation of an indicated object by a pointing gesture. We think that it is acceptable to find an indicated object by searching around the line in an omnidirectional image.

We also proposed a system using a head-mounted omnidirectional camera for pointing gesture recognition. The advantages of our system are that the system runs in real-time and the camera can shoot an arm and an indicated object in the same image. In future work, we intend to improve the system by implementing indicated object detection.

Reference

Brain W. Russell (1941). Visual diorientation with spatial reference to lesions of the right cerebral hemisphere. *Brain*, Vol. 64, pp. 244-272.

Kahn R.E., Swain M.J. Prokopowicz P.N. & Firby R.J. (1996). Gesture Recognition Using Perseus Architecture. *IEEE Computer Vision and Pattern Recognition*, pp. 734-741.

Kortenkamp, D., Huber, E. & R. P. Bonasso, (1996). Recognizing and Interpreting Gestures on a Mobile Robot. *AAAI*, Vol. 2, pp. 915-921.

McInTyre J., Stratta F. & Lacquaniti F. (1998). Viewer-centered frame of reference for pointing to memorized targets in three-dimensional space. *J. NeuroPhysiol.*, Vol. 78, pp. 1601-1618.

Waldherr, S., Romero, R. & Thrun, S., (2000). A Gesture Based Interface for Human-Robot Interaction. *Autonomous Robots*, Vol.9, No. 2, pp. 151-173.

Yamazawa, K., Yagi, Y. & Yachida, M. (1993). Omnidirectional imaging with hyperboloidal projection. *IEEE/RSJ Conference on Intelligent Robots and Systems*,Vol.2, pp. 1029-1034.

Laser Pointer Interaction with Hand Tremor Elimination

Sergey Matveyev *Martin Göbel* *Pavel Frolov*

Fraunhofer Institute for Media Communication
IMK, Schloss Birlinghoven, 53754, Sankt Augustin, Germany
{matveyev,goebel,frolov}@imk.fraunhofer.de

Abstract

The proposed pointing technique realizes a direct control in a graphical presentation environment at a distance with hand tremor elimination. The device for interaction is constructed on the basis of the infrared (IR) laser diode which works in a light diapason invisible for a human eye. The coordinates of the laser spot received from the IR camera are interpreted as a position of the (pointer) cursor after the mathematical processing which is displayed with the help of the projection system and can be interpreted as commands of the application control. The smoothing of the data received from the camera, allows to fix the cursor even on the small objects as for it reduces the effect of the cursor trembling on the screen.

Figure 1: The interaction with graphics application must be as natural as the interaction with the surrounding environment. (as an imitation to M.C. Escher)

1 Introduction

Pointing at a distance is an interaction style for Virtual and Presentation Environments, interactive TV, and other systems where the user is positioned at the outlying distance from the display.
Pointing is the operation of moving an on-screen tracking symbol, such as cursor, by manipulating the input device. Pointing operations are fundamental to human interaction with graphics application [10].
The natural reaction of a person to specify the object located in some distance is to point it out with his arm, a stick pointer, a laser pointer (LP) or simply to look at it and keep an eye on it. Thus at the big distances (up to the projection screen) the use of direct means of controlling of the application on the projection screen based on the LP technique turns out to be easy.
Since the direct interaction method is natural for the distances that exceed the length of the stretched arm more than in several times, then the efforts of different authors to adapt the LP technique and to realize the different methods of interaction with the application with the help of LP seem to be evident.
The use of a laser pointer for interaction on a distance presents various problems.

- As many researchers notice, the problem of the natural hand tremor influences greatly the use of the LP methods[3], [6],[9]. Physiological tremor is up to 8-12 Hz tremor that is of low amplitude. In the paper of Peck [7] the results of the tests of an amplitude of a laser spot jitter on the screen are submitted. The amplitude of fluctuations lays within the limit up to 0.6 degree.

- The next problem that is mentioned by many authors [5],[9] is the impossibility to choose the small objects on the screen. This problem appears to be the direct sequence from the previous one. By solving the "tremor" problem we will solve that one of the "choice".
- The third one consists in the interference of the visible laser spot with the objects that have the same color with the latter [6].

Smoothing of natural hand tremor is topical for direct interaction methods, and particularly for LP technique that is commonly used for presentations and in collaborative work of many users on a large projection wall.

2 Technique

The device for interaction is constructed on the basis of the IR laser diode which works in a light diapason invisible for a human eye and without being an interference to the projection channel of a visual presentation, since there can be no coincidence of the color of the IR laser and the visible object on the screen (see Figure). The LP emits a laser beam which produces a bright seen IR spot in the crossing point with the screen.
The scheme of work is submitted in the following stages:

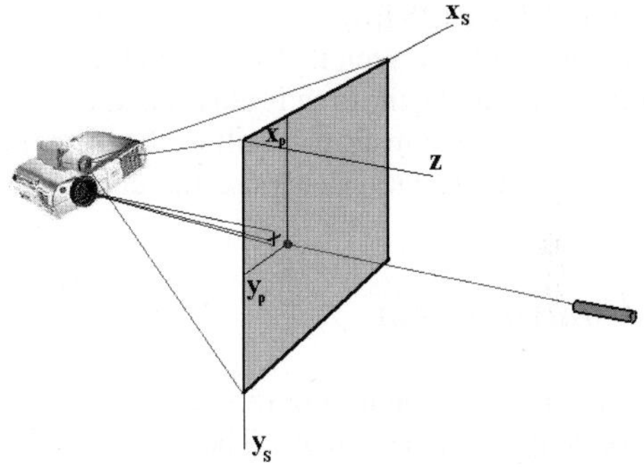

Figure 2: Scheme of interaction. The red point on the screen is the invisible IR laser spot . The blue cross is the corresponding visible cursor (its position can be shifted from the red one)

- The first stage is calibration. At this step we produce mapping from the camera coordinates to the screen coordinates.
- After the capture of the frame the recognition must be made. The recognition is a determination of coordinates and sizes of a laser spot, with the use of hints at about the possible spot position from the stage of data processing.
- The smoothing is produced simultaneously at the stage of processing together with a prediction.

2.1 Calibration

The calibration algorithm allows to construct the mapping from the camera coordinates to the screen coordinates. In our case as compared to the usually used algorithms of calibration the task becomes much simpler, since we use 2D in 2D mapping (see Figure). It can be considered as one of the advantages of our approach.
To take into account the nonlinear distortions we break the screen area into squares. The conformity of the squares vertices in the screen coordinates to points in the camera coordinates is searched automatically without the participation of the user. Thus the degree of nonlinearity

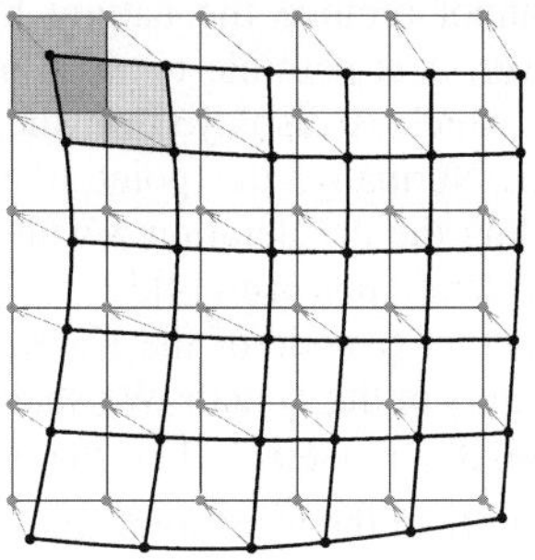

Figure 3: Calibration: 2D in 2D mapping

consideration in our algorithm depends on the amount of subareas into which we break the area of display.

2.2 Data acquisition

We use an operating mode of the IR camera with fields (as contrary to a standard operating mode with frames). The frame consists of two fields, each of which consists of even and odd lines of the frame. Thus, the number of lines in a field is equal to the half number of frame lines, at the equal length of a line. During the period of accumulation of the information on a matrix of the camera we obtain the data either about one frame, or about the two fields. The speed of the information reception from the camera in an operating mode with fields grows twice. For the PAL standard it makes 50 fields/sec (25 fps).
The laser spot on the screen that we should distinguish and identify, represents by itself an area of the elliptical form. As the area is closed and can be represented as a simple geometrical figure, the information contained in a field (half of the frame) is enough to define the coordinates of this area. In our further investigations we use the coordinates of the geometrical centre of the recognized area.

2.3 Control signal

The common Laser Pointer doesn't have a "mouse-button", i.e. the device can only position the laser spot on the screen visually. There is a limitation for the interaction. For example, to send the control signal to an application. The primary task when choosing an object consists in fixing the cursor on the object and transferring the choice of an object signal to a system. The speed of the data reception is 50 measurements per a second. It allows us to use, firstly, the mathematical methods of the data smoothing with a small value of delay of the system reaction on the IR laser pointer moving. Secondly, we use the interruption of an optical signal on 100-200 msec (5-10 fields) as a command to choose an object.
The "loss" of a signal on the time interval defined beforehand can be used as a command to choose an object on which was a cursor at this moment of time. The smoothing of the data received from the camera, allows to fix the cursor even on the small objects as for it reduces the effect of the cursor trembling on the screen.

3 Smoothing

We cannot remove the natural hand tremor without using the additional mechanical devices. However, it is possible to work not with the laser spot on the screen but with its mathematical model which is displayed on the screen by way of the cursor with the help of the projection system. Normally, the point movement is not the movement with constant acceleration. For modelling the acceleration a white noise which puts the stochastic function in concordance with the acceleration is used [4],[2].
The operating mode of the Kalman filter [8] depends at a great extent on the value of the initial parameters of the system. We shall consider the behaviour of Kalman gain in the limiting cases for the initial parameters. Let the measurement error covariance the matrix R_k specifying the uncertainty in the measurements. It means, that with the growth of the error covariance the matrix Kalman filter "trusts" to real measurement z_k ever less and less and smooth an output data (see Figure and Figure). And on the contrary, when the matrix R_k is vanishing to zero the "trust" to the real data increases (see Figure).

Hence, it is possible to get different operating modes of Kalman filter by adjusting diagonal elements values of the matrix R_k. Then, the results of the filter's work in two modes are

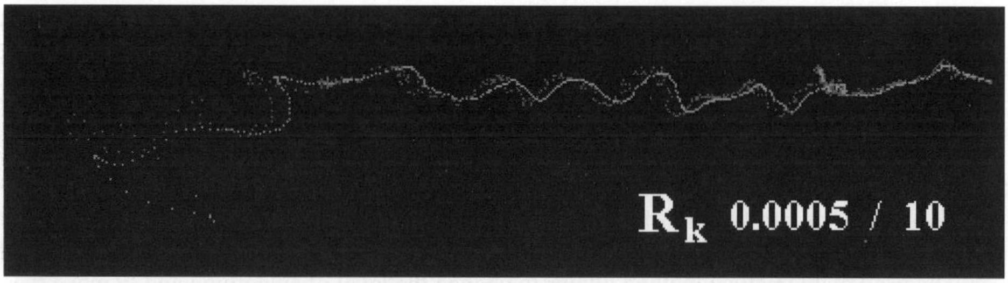

Figure 4: The low level of the data smoothing. The red points are the input data.

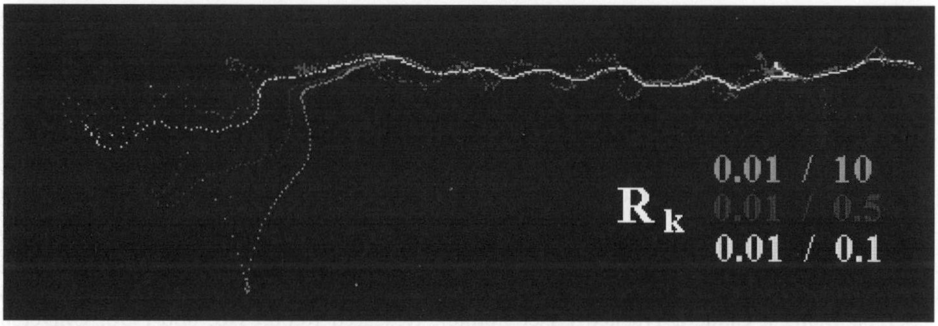

Figure 5: The middle level of the data smoothing. Different time of the Kalman filter stabilization for various initial values of the matrix R_k.

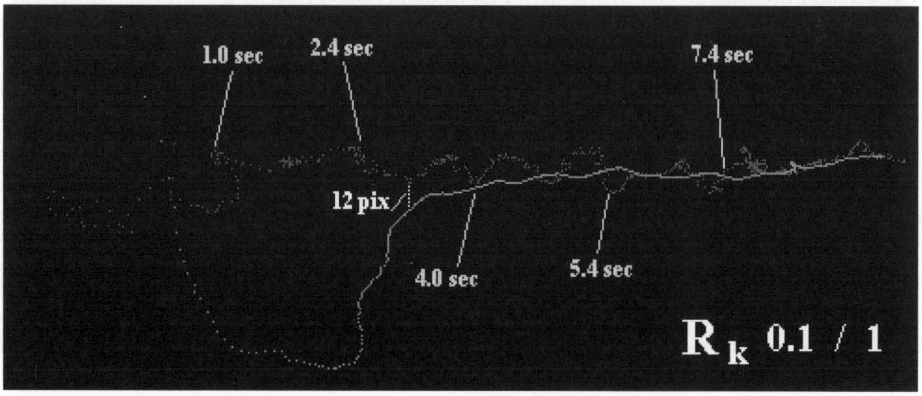

Figure 6: The high level of the data smoothing. Long time of the filter stabilization (~3 sec).

submitted with the same array of the input data. Red points designate the real values z_k, and green ones designate the predicted (or smoothed) values.

As it is seen from the figures Kalman filter faintly reacts to the sharp deviations in a point's movement at the increase of the diagonal elements of the matrix R_k. But at the same time, in this mode filter Kalman badly predicts the following position of a point. This mode corresponds to the smoothing of the input data. Hence, we can use two Kalman filters working in different modes: one for prediction, and the second for smoothing. Thus the user will see the smoothed values so there will be no trembling and he can easily position the pointer on the object. At the same time, the algorithm of recognition, in most cases, will work only with the predicted area of values that allows to reduce the computing expenses in two orders.

4 Conclusion

In the suggested method of the LP interaction with the remote screen the IR diapason is used. It solves the problem of interference of a laser spot on the screen with the objects of the same color. As the point is invisible, then after smoothing its coordinates are used for projection of the visible cursor. The used mathematical method allows to change the level of smoothing and to make the individual adjustments of the LP.

5 References

[1] Azuma, Ronald T. (1997) Course notes on "Registration" and "Correcting for Dynamic Error" from Course Notes #30: Making Direct Manipulation Work in Virtual Reality. ACM SIGGRAPH '97.

[2] Pavel Frolov, Sergey Matveyev, Martin Goebel and Stanislav Klimenko (2002). Using Kalman Filter for Natural Hand Tremor Smoothing during the Interaction with the Projection Screen, in Workshop Proceedings VEonPC'2002, pp. 94-101, Russia.

[3] C. Kirstein, H. Mueller (1998). A System for Human-Computer Interaction with a Projection Screen Using a Camera-Tracked Laser Pointer, Research Report No. 686/1998, Universitat Dortmund FB Informatik LS VII.

[4] Markus Kohler (1997), Using the Kalman Filter to track Human Interactive Motion, Technical Report 629, Informatik VII, University of Dortmund, January 1997.

[5] Brad A. Myers, Rishi Bhatnagar, Jeffrey Nichols, Choon Hong Peck, Dave Kong, Robert Miller and A. Chris Long (2002). Interacting At a Distance: Measuring the Performance of Laser Pointers and Other Devices, Proceedings CHI'2002, Human Factors in Computing Systems, Minneapolis, Minnesota, , pp.33-40.

[6] D.R. Olsen Jr., T. Nielsen (2001). Laser pointer interaction, Proceedings of the SIGCHI conference on Human Factors in Computing Systems, pp.17-22.

[7] Choon Hong Peck (2001). Useful Parameters for the Design of Laser Pointer Interaction Techniques, Proceedings of ACM CHI'2001 Student Posters. Seattle, WA. pp. 461-462.

[8] Greg Welch, Gary Bishop(2000). An Introduction to the Kalman Filter}, Tech Rep. TR95041, Universaty of North Carolina, Dept. of Computer Science

[9] Michael Wissen (2001). Implementation of a Laser-based Interaction Technique for Projection Screens, ERCIM News, No.46, pp.31-32.

[10] MacKenzie, I. S., Jusoh, S. (2001). An evaluation of two input devices for remote pointing. Proceedings EHCI 2001. pp. 235-249. Heidelberg, Germany: Springer-Verlag.

Influential Words: Natural Language in Interactive Storytelling

Steven Mead, Marc Cavazza and Fred Charles

University of Teesside
Middlesbrough, UK, TS1 3BA
{steven.j.mead, m.o.cavazza, f.charles}@tees.ac.uk

Abstract

In this paper, we introduce the use of Natural Language as a paradigm for influence of plans that are used to drive the behaviour of characters in Interactive Storytelling. We briefly introduce our character-centered approach to Interactive Story system is briefly introduced, and the knowledge representation of stories. Using an example based upon a fully implement first prototype, we discuss how the user is able to interfere with a story by issuing advice to the characters, and how the recognized speech acts are mapped onto their plans.

1 Introduction

In this paper, we discuss issues pertaining to the use of Natural Language Processing (NLP) as a paradigm for interaction in Interactive Storytelling. Interactive Storytelling is an increasingly popular area of research, and can be thought of as the convergence of new technologies (Virtual Reality and Computer Games) with traditional storytelling [Cavazza, 2000]. Several paradigms of Interactive Storytelling have been proposed, each with differing perspectives [Mateas, 1997][Szilas, 1999] [Young, 1999].

Our character-centred approach essentially follows proposals by Young [1999], where story is generated from the interaction between different plans. The user is afforded anytime interaction at a physical level, for example taking of objects important to the story, and more interestingly, the user can also make utterances based upon their present understanding of the story. Utterances are processed in real-time and are then intended to influence the development of the story; this can of course be towards both positive and negative conclusions. The example story used throughout the paper is loosely based upon the sit-com television show "Friends", where the premise of the base-line story is that the lead male character ("Ross") is attempting to ask the lead female character ("Rachel") out on a date.

With natural language, you can imagine the possibility of watching a movie and issuing advice to characters – and then observe the advice actually affecting the outcome. Whilst these are the long-term objectives, we are at present we are working with a fully implemented, single-user prototype. There are both practical and theoretical implications that have yet to be considered regarding a multi-user system, for instance how to resolve contradictory advice to the same actor. However, our prototype allows us to explore the possibilities of using natural language in interactive storytelling. Whilst we make no claim of inventing a new approach to NLP, we present results of its use as a natural mo dality of interaction to Interactive Storytelling.

In the next section we briefly introduce the current Interactive Storytelling system (more specifically the knowledge representation aspects), which is used as the platform for testing the natural language system.

2 Interactive Storytelling

We have developed a character-centred approach to Interactive Storytelling. At its core is a planning system that is used to drive the behaviour of autonomous characters. Stories are generated from the interaction between the characters plans, and any particular story instantiation can be considered as the 'cross-product' of these plans, and through encounters with the user. There are other random (non-user) determinants of variability for story generation, such as the random spatial displacement of the actors (and objects), and the resultant temporal effects on action execution.

2.1 Knowledge representation for Interactive Storytelling

Our system uses Hierarchical Task Networks (HTNs) [Erol, 1995], which are suitable because of the knowledge intensive nature of our approach to Interactive Storytelling [Cavazza, 2002]. An HTN describes the behaviours of each character in the story. The lead character's containing the base-line story, whilst the supporting character's HTNs have them behave in a manner appropriate to the base-line story. Figure 1 is a typical HTN.

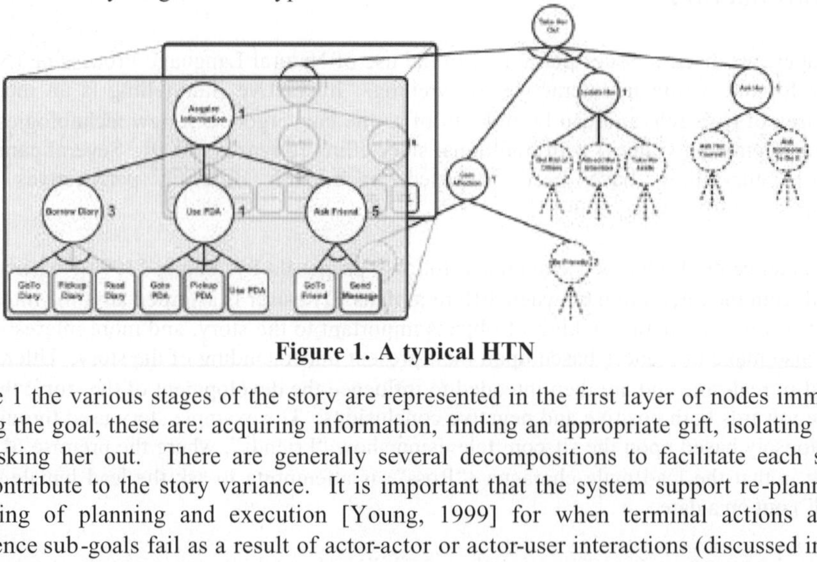

Figure 1. A typical HTN

In Figure 1 the various stages of the story are represented in the first layer of nodes immediately following the goal, these are: acquiring information, finding an appropriate gift, isolating her, and finally asking her out. There are generally several decompositions to facilitate each sub-goal, which contribute to the story variance. It is important that the system support re-planning and interleaving of planning and execution [Young, 1999] for when terminal actions and as a consequence sub-goals fail as a result of actor-actor or actor-user interactions (discussed in greater depth in the next section). As such it is undesirable to compute an entire plan, instead we require local re-planning, to find an alternative solution to the current sub-goal (if possible).

2.2 Implementation

We have used the Unreal engine[1] as a visualisation and development tool for the Interactive Storytelling system. It provides a scripting language (UnrealScript), which is used for the

[1] Epic Games, USA

execution of the low-level actions that are provided by the planner. The Speech Processing sub-system itself is an external application that communicates with the engine via UDP.

3 Interaction in Interactive Storytelling

As the story is dramatised by the characters carrying out their various low-level actions, the user can identify situations that have narrative importance – this is assisted by the use of a text-to-speech system (albeit sounding somewhat robotic). This information is likely and intended to influence the users behaviour, resulting in them interacting with the story[2].

3.1 Physical interaction

Physical interaction in Interactive Storytelling generally is limited to picking up and dropping resources. The impact on the story would rely upon the characters plans and their dependence on the resources existence. For instance, consider "Ross" that is walking towards a flower bouquet. The user can then infer that it has some narrative significance, and take it making it unavailable to "Ross", whose sub-goal will fail invoking re-planning.

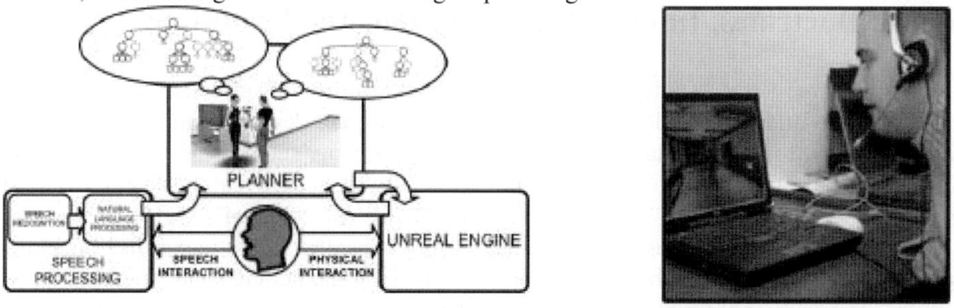

Figure 2. User participation in the development of the story

3.2 Natural Language interaction

However, it is more interesting to consider how the user may interfere with the story by using natural language to affect a characters' plan at a higher level. As previously noted, we envisage future entertainment systems where the user(s) may actively participate in the story development (Figure 2). To make these systems more believable to the user, more natural forms of interaction to enhance their "suspension of disbelief". Natural language for Human Computer Interaction (HCI) is widely researched and developed; here we consider how it can be applied to influence and assistance (or contrast) the development of a story. The user is considered as an "active spectator", however we do not wish to consider the use of language as an alternative to a conventional input device of character control (e.g. ordering an actor to move to a specific spot with the mouse), as this contradicts our notion of influence.

The system has two layers: A Speech Recognition layer and the Natural Language Processing layer. The Speech Recognition layer has been developed using an off-the-shelf package (EAR SDK[3]), which provides tools for developing the surface forms of the recognised sub-language and a C/C++ API to use it. As well as providing utterances for each potential narrative situation, we

2 Of course the user could decide to do nothing!
3 BabelTech, Belgium.

must provide suitable quantity of semantically equivalent utterances to allow the user to speak relatively freely, without needing to learn a strict sub-language – otherwise this could impede their belief of the system. The NLP layer attempts to map the output from the Speech Recognition layer and carries out the actual speech act recognition. In effect, it computes the influence on the characters plan. This is performed in a two-phase template processing system. The first phase attempts to identify the surface form of advice, this includes any selectional constraints on specific slots. Once a match is identified, the second phase will take the semantic information and instantiate a possible speech act. The effect of the speech act would be used to update nodes in the plan that matches its pre-conditions.

The system must also recognise the context in which the utterance is presented, and interpret it accordingly. As observed by Blaylock [2001], "Different planning contexts will result in different interpretations of the same utterance". For example, consider the utterance "Ross, Rachel is in her room", the interpretation is dependant on the stage at which "Ross'" plan has reached. If he is acquiring information by attempting to read "Rachels" diary (which is in her bedroom), then "Ross" should find an alternative solution to acquiring this information. On the other hand, if "Ross" is about to ask her out, then this information can be used to inform him of her location.

4 Advice: Doctrines, Information Provision, and Warnings

We have identified three types of advisory utterance as viable forms of influence in Interactive Storytelling. Doctrine utterances will have a global impact on plan development (of a specified actor), and can be used to avoid certain situations from ever arising, such as the meeting of two actors. Information Provision will allow the user to solve an actors' sub-goal, for instance indicating to "Ross" that "Phoebe" can provide the details of "Rachel's" gift preference. Warnings can be used to inform "Ross" that a situation may arise that could jeopardise his plan. Consider that "Ross" is reading "Rachel's" diary, yet "Rachel" or another character friendly to her is approaching. The user could warn "Ross" of this situation, which would cause a short-term intermediate situated plan to trigger such as making a drink or watching the television. We generally consider these events as situated reasoning [Geib, 1993]. It is also feasible and desirable that the user can accidental or deliberately make incorrect utterances. The result of which would have variable impact on the story, and have potentially disastrous outcomes for "Ross'" plan. For example, lying about "Rachel's" gift preferences.

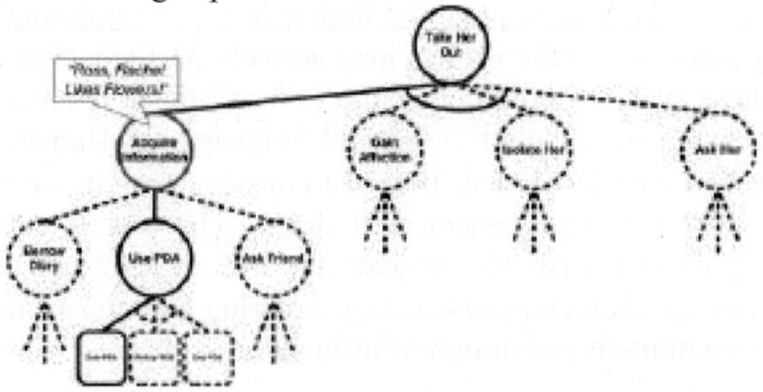

Figure 3. Solving a sub-goal by providing the information.

5 Results

Following on from the previously introduced base-line story, Figure 5 is an actual story instantiation. "Ross" is going towards the "Rachels" PDA, however the user knows that she is using it, and tells "Ross". This causes the closest sub-goal to fail, as it's pre-condition is that the

PDA is free and that there are no other characters around. This triggers re-planning, and speaking to "Phoebe" is computed as an alternative strategy for acquiring information.

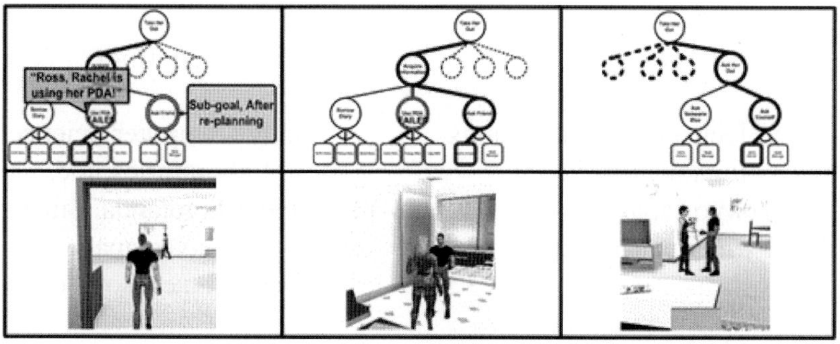

Figure 5. Providing advice to influence a characters plan

6 Conclusions

We have introduced a paradigm for user interaction to influence a story's development in Interactive Storytelling. As interactive entertainment (including Interactive Storytelling) develops, more natural modalities for Human Computer Interaction will be needed, to allow the users to express themselves more naturally. We have presented a first prototype to test the use of natural language that allows a single user to interact with and affect a story. However, we do not use natural language in a traditional dialogue–based context, but rather as a form of influence over the higher-level planning activities of an artificial actor – where speech acts can be used to explicitly modify their plans.

References

Bolter JD, Grusin R. (1998). Remediation: Understanding New Media. Cambridge, Massachussets: MIT Press.

Cavazza M, Charles F, Mead SJ. (2002). Character-Based Interactive Storytelling. *Special Issue on AI in Interactive Entertainment.* IEEE Intelligent Systems. pp. 17-24.

Charles F, Mead SJ, Cavazza M. (2001). User Intervention in Virtual Storytelling. Proceedings of Virtual Reality International Conference. Laval, France.

Erol K, Hendler J, Nau DS, Tsuneto R. (1995). A Critical Look at Critics of HTN Planning. *Proceedings of IJCAI-95*.

Geib C, & Webber B. (1993). A consequence of incorporating intentions in means-end planning. Working Notes – *AAAI Spring Symposium Series: Foundations of Automatic Planning: The Classical Approach and Beyond*. AAAI Press.

M. Mateas. (1997) *An Oz-Centric Review of Interactive Drama and Believable Agents*. Technical Report CMU-CS-97-156, Department of Computer Science, Carnegie Mellon University, Pittsburgh, USA.

Webber B, Badler N, Di Eugenio B, Geib C, Levison L & Moore M (1995). Instructions, Intentions and Expectations. Artificial Intelligence Journal, 73, 253-269.

Young RM. (1999). Notes on the use of Plan Structures in the Creation of Interactive Plot. AAAI Fall Symposium on Narrative Intelligence. AAAI Press.

Interactive Sight: A New Interaction Method for Real World Environment

Yuichi Mitsudo

University of Electro-Communications
1-5-1 Cyofugaoka Cyofu-shi
Tokyo Japan
mitsud-y@hi.is.uec.ac.jp

Ken Mogi

Sony Computer Science Laboratories, Inc.
3-14-13 Higashigotanda Shinagawa-ku
Tokyo Japan
kenmogi@csl.sony.co.jp

Abstract

In this paper, we present a real world interaction system, called Interactive Sight (IS). The user's sight is used as a display, and real world objects in his sight can be treated like icon under IS environment. The Real Eye Communicator, an optical communication system using the user's eye as a medium, plays an instrumental role in constructing the IS environment.

1 Introduction

In a ubiquitous computing[1] environment, there exists a lot of accessible chipsets in the real world. The user can utilize the computing device and access the internet everywhere. In such a situation, it is important to enable the user to select one from a lot of available services.

An approach in the real world computing is to use a position location technology such as the GPS, and provide a rich network of services to the user based on his/her position. Such a system, e.g. the comMotion[3], detects and learns user's behavior based on the time course of his/her geometrical position, and provides an appropriate application to the user by predicting user's behavior. Another approach uses a radio or optical communication system, and provides various services depending on the user's distance from the devise. Electorical Tags[2] is typical examples of this line of research.

In this paper, we present an alternative approach to context-aware computing. In this system, we do not require a position location device or one that measures the physical distance between the user and the device. In stead, we aimed to exploit the user's natural vision of the environment, making use of the user's perceptual process itself to realize an intuitive interactive computational environment.

2 Real Eye Communicator(REC)

Here, we propose a new interactive computing system called the Real Eye Communicator (REC). The Real Eye Communicator is a simple optical communication system. The novel feature of this communicator is using the user's eye itself as a medium for communication. The human eyeball has a high reflective index. The reflected image from the cornea is called the 1st Purkinje image. By using reflected infrared light, it is easy to observe the 1st Purkinje image (figure1).

By observing the 1st Purkinje Image with a photo-diode, the Real Eye Communicator system can receive the optical signal from the optical transmitter in user's sight via the eyeball. The received optical signal is shown in figure4. The unique characteristics of this system is that it is able to receive the information in a parallax-free manner, as the photo diode measures the reflected signal from the cornea coming from the transmitting object in the user's sight. Based on this advantage, we are able to realize a useful interactive computing environment (Interactive Sight), which we explain in the next section.

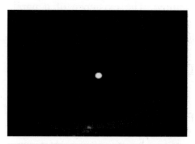

Figure 1: 1st Purkinje Image (Infrared Image)

In addition, the REC has the following useful features.

1) High Speed Data Transfer
2) Simple Structure

By using a simple optical communication system, the REC realize high-speed data transfer from real world ob ject to the user's computer. Specifically, the transmitter can send the fixed size data, such as the device ID in short time, or send larger data to the the receiver via the user's eyeball. At present, we have demonstrated that we can receive pulses of 5 microsec duration. The REC doesn't need a complicated optical system such as a high resolution video, as only the temporal domain of reflected right is required for the communication.

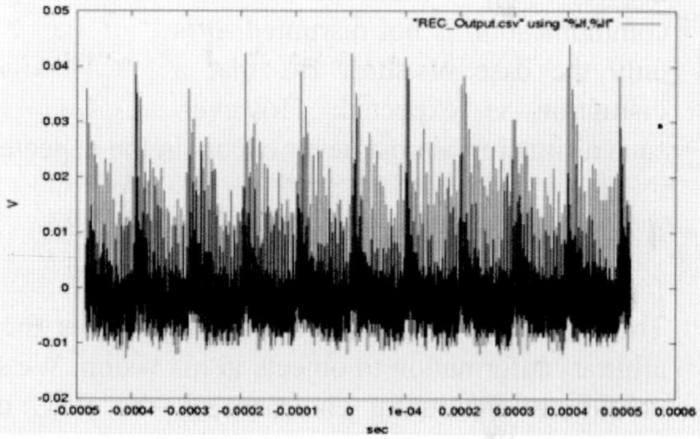

Figure 2: The output of Real Eye Communicator

3 Interactive Sight

By using the REC, we can establish a unique interaction method called Interactive Sight. The user can indicate the intentioned real world object to the system just by pointing at it. We call this action Tele-Click. The Tele-Click is similar to the conventional (physical) click: The user can use his natural sight as a display, and Tele-Click on an object in his/her sight, thus realizing an intuitive method for communication. See Figure3.

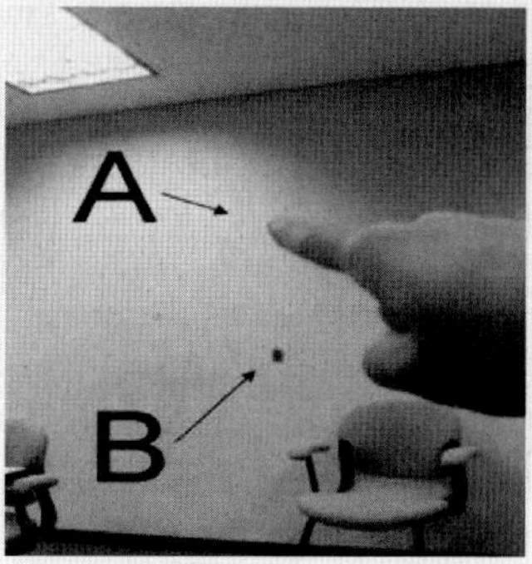

Figure 3: The user excute Tele-Click to transmitter A

The system can specify the Tele-Clicked transmitter, because only the optical signal from Tele-Clicked transmitter is interrupted. To realize such a feature, a parallax-free optical receiver is required. The existence of a parallax would make the accurate implementation of Tele-Click impossible. See Figure5. The Interactive Sight Method requires a parallax-free-receiver, and the REC satisfies this requirement (Figure5).

Light Axes

Figure 4: Only the signal from Tele-Clicked transmitter is interrupted

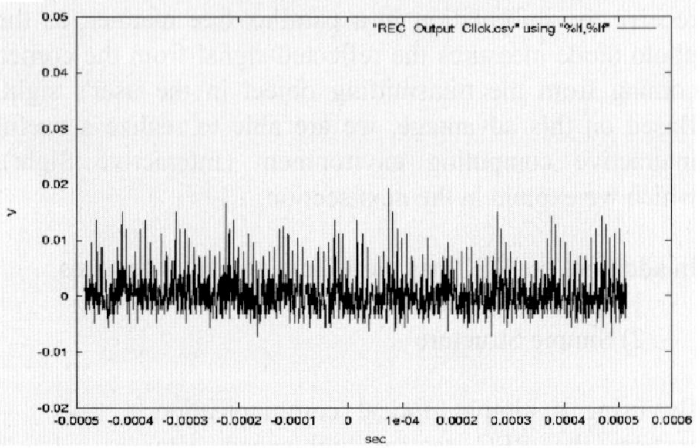

Even a slight difference of light axes can disrupt the accuracy necessary in realizing the Interactive Sight. See figure6. The degree of disruption depends on the length of BC in Figure 6. In most cases that a user encounters, the length of BC is longer than the diameter of the cornea, and the parallax-free communication is not disrupted. It is only the case of short BC that a disruption is expected. However, such a situation almost never occurs in the expected user environment where the REC is applied.

4 Application Model

The Interactive Sight Environment provides the user with a means to attach and manipulate arbitrary information to objects in his vision. We show some basic application models below. In a ubiquitous computing environment, the REC can be used as a medium for perception-ID translation. The transmitter transmits its ID continuously. The receiver to the user receives the ID in the user's vision. If The user tele-clicks a specific object, the system can specify which ID is tele-cliked by observing characteristic interruption time course of the signal. The user's computer can then access relevant information tagged to the ID via the network, and utilize information related to the ID. The "Speaking Description" is one possible application. "Speaking Description" is a kind of audio system where the user can listen to audio data tagged to the tele-clicked object. In the metaphorical sense, the user is able to get the audio data "from his view". It is a way of getting a "description" of the object in sight, in a situation such as the one in a museum. We suggest that such a system is effective for a navigation system for visually challenged people.

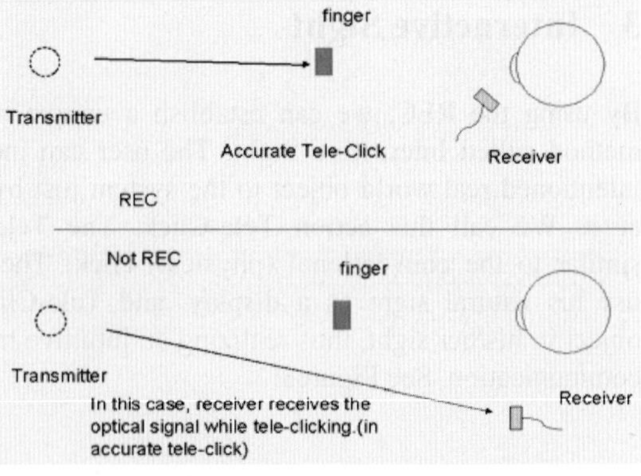

Figure 5: Parallax decrese the accuracy of Tele-Click

Figure 6: Difference of right axes

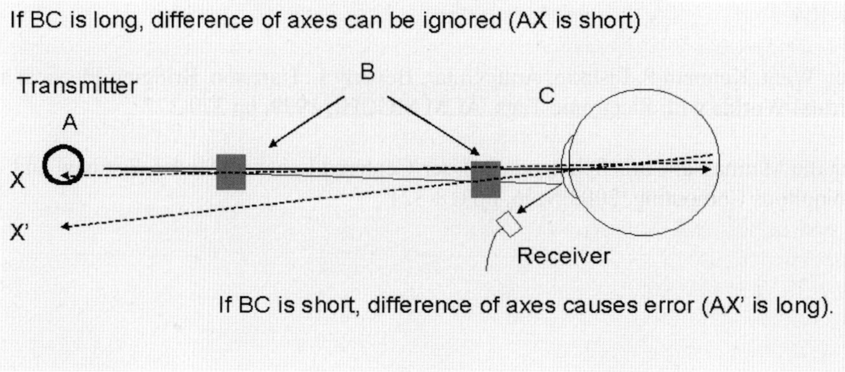
If BC is long, difference of axes can be ignored (AX is short)
Transmitter
A
B
C
X
X'
Receiver
If BC is short, difference of axes causes error (AX' is long).

5 Conclusion and Discussion

In this paper, we presented a novel method for real-world interaction, called the Interactive Sight. We constructed the Real Eye Communicator, which is a optical communication system using user's eye as a medium. This system has a parallax-free communication ability, which is required for implementing the Interactive Sight metaphor. The REC has various merits in visual identification and optical communication. The user can identify real world objects by making use of his natural visual perception, and the parallax-free ability of REC enables to "attach data" to visual perception. This structure enables the system to bypass image processing, and realizes high-speed data transfer in a simplified way compared to a video-based system.

The Interactive Sight Method makes it possible to provide an intuitive interactive method for the user. In Interactive Sight Environment, the user doesn't need to be know details of the network configuration. All that the user have to know is how to "utilize" his/her visual perception has a data in the interactive sight metaphor.

6 Future Work

In the present system, the photo-diode and pre-amplifier circuit is placed in front of the user's eye. The Pre-amplifier system is required to be placed near the photo-diode to avoid the effects of electrical-noise. Using an optical fiber will make it possible to separate physically the optical system and the amplifier system, providing a more natural view to the user.

References

[1] Mark Weiser, The Computer for the Twenty-First Century, Scientific American, 1991, pp. 94-104

[2] Roy Want, Kenneth P. Fishkin, Anuj Gujar, Beverly L. Harrison, Bridging Physical and Virtual Worlds with Electronic Tags, ACM SIGCHI, 1999, pp 370-377

[3] Natalia Marmasse, Chris Schmandt, A User-Centered Location Model, Personal and Ubiquitous Computing, 2002, vol6, pp318-321

Sounds@Work - Auditory Displays for Interaction in Cooperative and Hybrid Environments

Christian Müller-Tomfelde, Norbert A. Streitz

Fraunhofer - Integrated Publication and Information Systems Institute (IPSI)
Dolivostr. 15, D-64293 Darmstadt, Germany
mueller-tomfelde@fhg.ipsi.de, streitz@fhg.ipsi.de

Ralf Steinmetz

Dept. of Electrical Engineering and Information Technology
Darmstadt Univ. of Technology,
Merckstr. 25, D-64283 Darmstadt, Germany
ralf.steinmetz@KOM.tu-darmstadt.de

Abstract

This paper presents the concept and selected realizations of auditory displays and feedback for human-computer interaction in computer-augmented work environments. Our approach is to experiment with sound in order to produce a plausible and coherent audible sensation for users in a collaborative and hybrid environment. This is different from virtual reality environments that use sound to produce an immersive illusion and beyond the "traditional" usage of sound provided by multimedia facilities of desktop computers. Recent developments in information and communication technology indicate a trend that implies the disappearance of the computer as a primary artifact. This will also result in a situation where the sound caused by mechanical parts disappears due to new components or the integration in the environment. This loss of the acoustic aura of computer systems might lead to irritations and uncertainty of users when interacting with computers because they miss familiar sounds as, e.g., caused by a spinning hard disk. Our approach is now to augment the environment with sounds in a controlled way thus providing feedback to users during interaction. In this paper, we present examples of sound feedback for the interaction with so called "roomware" components and demonstrate different possibilities for the reappearance of sounds in work environments. This will not be restricted to recreating the sounds that were lost but we go also beyond by creating new sounds. The sound feedback should help the user to be more engaged in the interaction with real artifacts and their virtual extensions.

1 Introduction

For "traditional" standard desktop computer environments, audio feedback is state of the art and well covered in the literature. Different concepts were developed to map sound effects to user input events enriching the user's interaction with the computer. For example, the concept of Earcons implies the translation of events into short musical sequences of tones. Their semantics have to be acquired by the listener in order to understand their meaning (Blattner et al. 1989). In contrast, the concept of Auditory Icons uses everyday sounds of the environment that are well-known to the user in order to convey information about events and states within a computer system (Gaver 1989). The situation and requirements change when attempting to provide sound feedback in computer supported teamwork situations. This results in new issues because one cannot simply apply the solutions designed and realized for an individual work environment with individual audio feedback to a cooperative teamwork situation. This is due to the specific characteristics of the sound media. Additionally, new advances in computer technology open up new possibilities to apply audio feedback when the computer disappears and further gets increasingly inaudible.

2 Hybrid Environments

In the general domain of computer-based workplace environments, the continuum ranges from traditional desktop computers to immersive, 3D environments. Taking a position in-between, so called "hybrid" environments combine possibilities for direct interaction and manipulation of data represented and displayed by computers with opportunities to manipulate real objects of the environment that are tangible but augmented with information technology. Artifacts in hybrid environments can represent both: objects in the tangible real world as well as in virtual space (Wellner 1993, Ishii & Ullmer 1997). The i-LAND environment (see figure 1) at the Fraunhofer institute IPSI represents our approach to hybrid environments (Streitz et al. 2001). We extended it with additional audio facilities in order to experiment with audio augmentation and feedback for the interaction with hybrid artifacts.

Figure 1: Roomware-components of the i-LAND environment: DynaWall, InteracTable, CommChair and ConnecTable.

In our approach, we used the computer hardware and software of an electronic meeting room and extend the environment with audio facilities. We adapted the existing cooperative software BEACH (Tandler 2001) in order to pass the information of interaction to a dedicated audio system and to form different examples of audio feedback. The aim of this work is not to achieve a realistic and high fidelity audio presentation and feedback, coming along with virtual environments but to provide a plausible auditory display. In the context of group work, the focus is more on usefulness and at the same time unobtrusiveness in every day work situations in combination with new forms of human-computer interaction.

3 Sound in Hybrid Environments

While computer devices are getting increasingly invisible and ubiquitous in hybrid environments, the audible properties and characteristics are getting lost or vanish, too. In some extreme cases, there could be no more a physical reason for the production and emanation of sound. At Fraunhofer IPSI, we built examples of these noise-free computer systems as part of the second generation of the roomware components CommChairs and ConnecTables (Streitz et al. 2001).

This provided us with possibilities to experiment with new sounds for the interaction with computer systems. Either the sounds can imitate real acoustic behavior to ensure the natural quality of the interaction or new sounds can convey additional information to enrich the interaction. In the hybrid environment the audio feedback should support the coherent perception of the real and virtual properties to form one mental representation.

4 Design Guidelines

Some design rules for the auditory displays are suggested to achieve the goal of an integrated sound feedback for hybrid artifacts and to induce a coherent and plausible user impression:

- The loudspeakers should be as invisible as the computer equipment, which is integrated in the artifact. This best solution is the complete integration into the environment or artifact to avoid distraction by the existence of a loudspeaker as an object.
- The sound production should occur at the location where the sound is intended to come from to minimize the problems of virtual positioning and offer to all users at the same time best location cues of the source position.
- The generation of the audio signals should be based on as much data and information about the environment and user's interaction as possible. This helps to avoid monotonous and static signals and to form an unobtrusive and natural sound feedback.
- Immediate and direct feedback for the interaction enables an engaging and multimodal user interface. Therefore, the delay between the control action and the resulting auditory feedback has to be as short as possible.

These guidelines were used to extend the hybrid work environment of the i-LAND project at FhG-IPSI with audio technology to realize auditory displays for different artifacts and work situations in the team room.

5 Sound Feedback at an Electronic Whiteboard

To demonstrate and experiment with audio enhanced interaction for a teamwork situation in a hybrid environment we identified the DynaWall (see background of figure 1). Here it becomes clear, that not all orientation on the display can be managed well by visual cues or feedback. Depending on the distance of the user to the DynaWall his view is restricted but the audio feedback can be used to go 'beyond' these limits. The individual interaction with virtual objects is augmented with an acoustical behavior similar to that of friction between objects on surfaces in real environment. In addition to this feedback which imitates real world behavior, new forms of interaction at the electronic whiteboard, e.g., gesture interaction to open or delete objects are accompanied and embedded in the overall sound schema of the interaction. The sound feedback should be perceived as natural as possible to obtain a plausible multimodal sensation, but without spending too much effort on too realistic simulations. In contrast to a local visual feedback in a multiple computer environment, sound allows to give not only individual feedback but also awareness of the activity to other team members. For example, at the DynaWall the individual interaction with the computer can happen in parallel to or mixed with the speech communication among team members. The full description of the approach can be found in Müller-Tomfelde & Steiner (2001). To prevent the sound feedback from being annoying for the user, we are experimenting with fading sound amplitudes for continuous interaction beyond approx. 3 seconds.

5.1 Interaction with information objects

One of the core features of the BEACH software running on the DynaWall is that it provides a homogenous area for interaction over all three DynaWall segments. To manage the display space at the DynaWall in an appropriate way the software allows moving and throwing of information objects. Holding down your finger or pen on the DynaWall dragging the information objects all over the wall for several meters is not very ergonomic because of the large width of the DynaWall (4.5 m.). So a person A can throw his current selected information object on the electronic whiteboard to person B by simply accelerating the information object in the desired direction. After person A released his finger or pen the information object continues to move in the intended direction but then reduces its speed, just like a real object on a table-top. In our environment the moved and thrown information objects are enriched with sound cues that correspond to the speed and the position of the currently moving information object. Inspired by the sound of an object when it is moved or shuffled over the surface of a tabletop the sound feedback for objects at the DynaWall consists of a seamless looped sound sample with a succeeding processing in real-time corresponding to the users interaction with the object. This approach can be seen as a mixture of parameterized auditory icons and the dynamic data-to-sound mapping, which is used in the context of sonification. Because the interface is enhanced with audio feedback of a sound reaction well known from the physical environment, it is a very plausible and accepted audio augmentation. The audio enhancement of interaction becomes important in a group work situation, in which the peripheral awareness of the partners' activities like annotating documents or shuffling information objects makes the mixed interaction more transparent, safe and understandable.

5.2 Gesture recognition feedback

The interaction with objects at the DynaWall is supported with gestures. We are using an incremental gesture recognition which allows to have a modeless interface. While the user is scribbling the type of input is checked continuously over time. To improve the individual use of these gestures each state of the recognition is passed as an event to the sound system, where the detection envelope, i.e., the tracked states over time can be mapped to feedback sound cues, e.g., to accompany a gesture and to confirm its successful recognition. This envelope stimulates a sequence of sounds like a rhythm pattern when a gesture is drawn on the electronic whiteboard. This 'gesture melody' always begins in the same way but ends different depending on the detected gesture. This pattern could increase also the correct use of gestures. We are working on an implementation, which is adaptive to the habits of the users over time, e.g., the sound feedback fades out the more the user repeats the same interaction or gesture in a short period. This feature could help to avoid redundant information and to spare the resource of attention within a cooperative work situation.

5.3 Sonyfiying pen interaction on virtual surfaces

Usually the qualities of surfaces are named by terms like rough, coarse, soft or smooth. Although these qualities could also correspond to visual properties of the surface areas, they are mainly related and determined by the haptic experiences during the interaction. We suggest an auditory display that offers a cross-modal perception to the users during the interaction, i. e., conveying surface quality information via auditory cues. This feature of a virtual surface is used to give feedback to the user about certain qualities of an object on the surface. Interaction a write-protected document could sound coarse like writing on glass, while "permitted" writing let the interaction sounds smooth like moving a pen on paper. This work is motivated and inspired by the research experiences of the authors with the auditory display of the DynaWall and of smaller pen tablets with integrated graphical displays (Müller-Tomfelde & Münch 2001).

6 Conclusions

In this paper, we presented the use of sound in future collaborative and hybrid work environments. The described examples are part of our research on future and emerging computer environments that will produce no more noise at all. Further examples of audio feedback in the i-LAND environment are realized for the passage mechanism (Streitz et al. 2001) and for the ConnecTables (Müller-Tomfelde 2002). All examples aim at the useful and important audio feedback for the interaction with augmented artifacts in a hybrid environment. The described work is based on the PhD research of the first author at the FhG-IPSI.

7 Acknowledgements

The authors would like to thank the Ambiente-Team and especially Sascha Steiner, Tobias Münch and Steffen Halama for their valuable contributions to the audio extension of the i-LAND project in the AMBIENTE division (http://www.ipsi.fhg.de/ambiente).

8 References

Blattner, M. M., Sumikawa, D. A. & Greenberg, R. M. (1989). Earcons and Icons: Their Structure and Common Design Principles, *Human Computer Interaction*, 4(1), 11-44.

Gaver, W. W. (1989). The SonicFinder: An Interface That Uses Auditory Icons, *Human Computer Interaction*, 4(1), 67-94.

Ishii, H., & Ullmer, B. (1997). Tangible Bits: Towards seamless Interface between People, Bits and Atoms In *Proceedings of the ACM Conference on Human Factors in Computing Systems* (CHI '97), 234 – 241.

Müller-Tomfelde, C., & Steiner, S. (2001). Audio-Enhanced Collaboration at an Interactive Electronic Whiteboard. In *Proceedings of 7th International Conference on Auditory Display, ICAD*, 267-271.

Müller-Tomfelde, C., & Münch, T. (2001). Modeling and Sonifying Pen Strokes on Surfaces, In *Proceedings of the COST G-6 Conference on Digital Audio Effects (DAFX-01)*, 175-179.

Müller-Tomfelde, C. (2002). Sound Effects for a Silent Computer System, In *Proceedings of the COST G-6 Conference on Digital Audio Effects (DAFX-02)*, 227-232.

Streitz, N., Tandler, P., Müller-Tomfelde, C., & Konomi, S. (2001). Roomware: Towards the Next Generation of Human-Computer Interaction based on an Integrated Design of Real and Virtual Worlds, In: J. A.Carroll (Ed.), *Human-Computer Interaction in the New Millennium* (pp. 553-578). New York: Addison Wesley

Tandler, P. (2001). Software Infrastructure for Ubiquitous Computing Environments: Supporting Synchronous Collaboration with Heterogeneous Devices. In *Proceedings of UbiComp 2001: Ubiquitous Computing*. 96-115.

Wellner, P., Mackey, W., & Gold, R. (Eds.) (1993). Computer-augmented environments: Back to the real world. Special Issue of Communications of the ACM, 36(7), 87-96.

Human Cognitive Characteristics in Speech Control for Virtual 3D Simulation on 2D Screen

Miwa Nakanishi, Yusaku Okada

KEIO University, Graduate School of Science & Technology,
Department of Science for Open & Environmental Systems,
Y.OKADA Laboratory (Human Factors & Ergonomics)
Hiyoshi 3-14-1, Kohoku-ku, Yokohama city, Japan, 223-8522
miwa@ae.keio.ac.jp, okada@ae.keio.ac.jp

Abstract

The study attempts to examine the human cognitive characteristics in speech control for the virtual 3D simulation on 2D screens.

To begin with, we conducted a comparative experiment between speech control and manual key control, and analyzed the time, the efficiency and the errors. From the results, we discussed the tendencies of perception, action and emotion. Moreover we illustrated the concept in this paper.

Speech control could be fit for human cognition in virtual 3D simulation in VDT work. And it is expected to enhance the working environment and increase efficiency in industry.

1 Introduction

Today, virtual 3D simulations are being used in more and more applications in industry. Particularly, the one which can be used on standard VDUs is an effective and not so expensive tool in supporting workers' tasks. In general, workers control an object in virtual 3D space displayed on 2D screens with manual controllers suitable for 2D control, such as a keyboard or mouse. In this case, the directions of their actions may not exactly correspond to those of their perception of the object's 3D movement. Alternatively, in the case of speech control, it may be possible for workers to orally control direction in virtual 3D space on the screen. That is to say, speech control is a human-computer-interaction which allows us to exceed the restrictions of 2D.

Therefore we did our research on the basic characteristics of speech control of a virtual 3D simulation on a 2D screen, in particular focusing on human cognition.

Our goal is to suggest a more effective human-computer-interaction in industrial fields.

2 Method

We conducted a comparative experiment between speech control and manual key control. For this experiment, we programmed a simplified simulator. This simulator displayed a cubic image on the screen.

The subjects' task was to move an object to a given target position within the cubic space. Then two kinds of coordinate systems were defined to give the directions where the object was moved. In one coordinate system, the subjects set the origin of the control on the origin of the cube (X:0, Y:0, Z:0), and moved the object along XYZ axis in the following 6 directions: X+, X-, Y+, Y-,

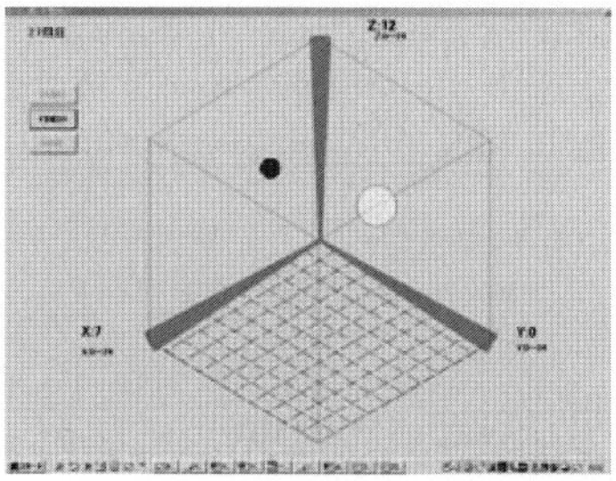

Figure 1
The Experimental Program

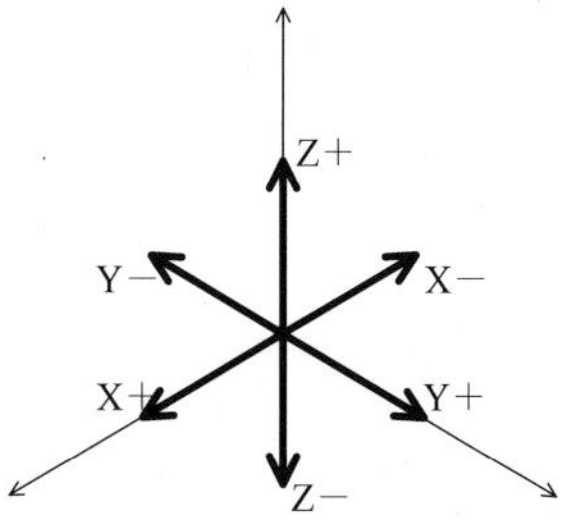

Figure 2-1
The Origin-Based Coordinate

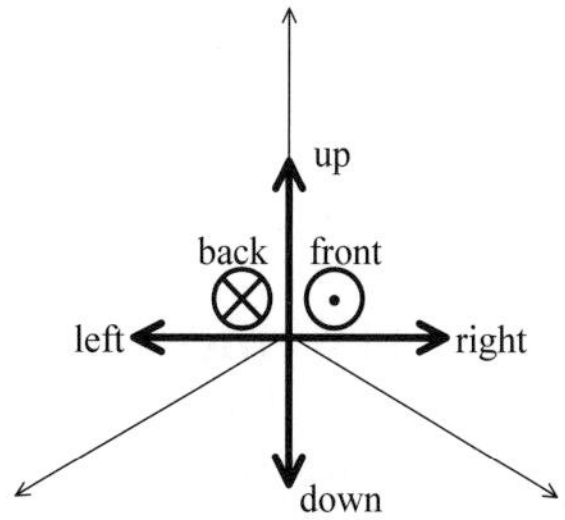

Figure 2-2
The Eyes-Based Coordinate

Z+, Z-. In another coordinate system, they based the origin of control on their own eyes, and moved the object in the following directions: up, down, right, left, front, back. We call the former "the Origin-Based Coordinate", and the latter "the Eyes-Based Coordinate".

The subjects were required to perform the task 30 times in each coordinate system in both speech control and key control. All of the subjects' actions during this experiment were recorded in time sequence and the questionnaires on each experimental condition were obtained from the subjects after the experiment.

3 Result

3.1 The required time for the task

Figure 3-1
The Keys Allocation
in the Origin-Based Coordinate

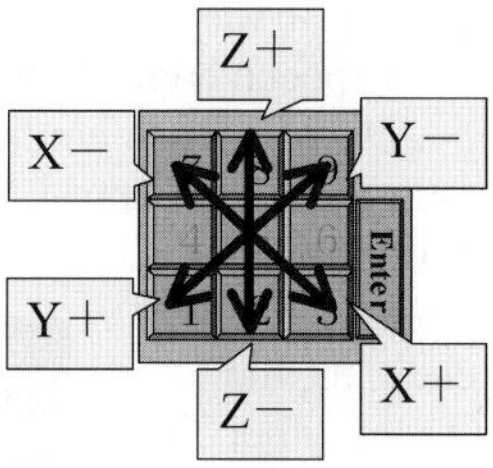

Figure 3-2
The Keys Allocation
in the Eyes-Based Coordinate

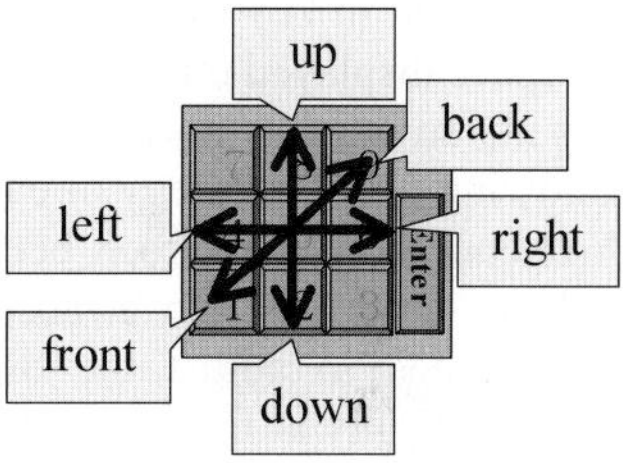

Here, we focused on the time from a control action to the next (Seg-Time), and compared it among 4 experimental conditions. Figure 4 shows the result of the average of Seg-Time and the standard deviation.

The Seg-Time of speech control was longer than that of key control in either coordinate system. It is simply because speaking a word requires more time than pushing a key for each control action. And in both speech control and key control, the Seg-Time of the Eyes-Based Coordinate was shorter than that of the Origin-Based Coordinate. It explains that the Eyes-Based Coordinate enabled the subjects to easily decide the path of the object's movement. The difference between both the coordinate systems was remarkable in speech control because most of words used in the Eyes-Based Coordinate were shorter than ones used in the Origin-Based Coordinate.

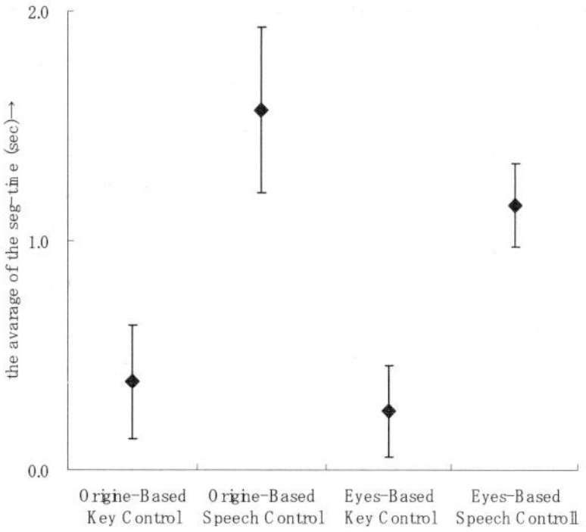

Figure 4
The Time from a Control Action to the Next (Seg-Time)

The standard deviation in the Eyes-Based Control was small in either control. However that in the Origin-Based Coordinate was larger in speech control than in key control. In key control, there was little difficulty with the cognitive flow from intention to action, because the XYZ directions perceived on the display accorded with the directions assigned to the keys to be pushed (see Figure2-1 and 3-1). On the other hand, in speech control, the words to be spoken, that is symbolic XYZ directions, might not accord with the intuitive words in the subjects' mind (Figure 2-2 and 3-2). Accordingly, it was necessary to link intuition to a word, and it caused the large variation of the Seg-Time.

3.2 The efficiency of the object's movement

We defined the following indexes to evaluate the efficiency of the object's movement.

$$Int\_prop\_3DR = \sum^{N} \frac{Dif\_3DR}{Tas\_3DR} \quad \text{(The 3 dimensional efficiency)}$$

N: The number of control actions during a task
Dif_3DR: The 3 dimensional distance between the target and the object by every control action
Tas_3DR: The 3 dimensional distance between the target and the object at the start of a task

$$Int\_prop\_2DR = \sum^{N} \frac{Dif\_2DR}{Tas\_2DR} \quad \text{(The 2 dimensional efficiency)}$$

N: The number of control actions during a task
Dif_2DR: The 2 dimensional distance between the target and the object by every control action
Tas_2DR: The 2 dimensional distance between the target and the object at the start of a task

The earlier the object approached the target position and the smaller the number of control actions was, the smaller the index values are.

The results of *Int_Prop_3DR* and *Int_Prop_2DR* are shown in Figure 5.

First, in case of the Origin-Based Coordinate, *Int_Prop_3DR* was much smaller in speech control than in key control. It means that it was easier for the subjects to image the hypothetical 3D space displayed on the 2D screen in speech control than in key control. However the difference was not so much between *Int_Prop_2DR* of speech control and that of key control. It implies the following typical strategy of each control. In key control, the subjects tended to decrease the visual difference on a 2D screen between the object and the target position at an early stage, and then tended to adjust the difference in the depth tactically at the next stage. On the other hand, in speech control, they tended to perform strategically and sub-consciously at an early stage.

Second, in case of the Eye-Based Coordinate, there was little difference of both *Int_Prop_3DR* and *Int_Prop_2DR* between speech control and key control. As mentioned before, this is because the Eyes-Based Coordinate enabled the subjects to easily find the right path for the object's movement. The reason why the values of both indexes of key control was rather large is because the subjects kept pushing the keys and often moved the object more than their intention. On the other hand, in speech control, such a careless overrun seldom occurred because of slowness of movement.

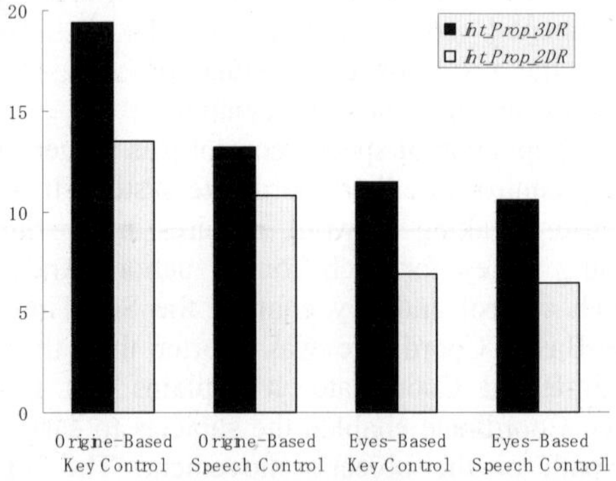

Figure 5
The Index Values
of 3-Dimensional Efficiency (*Int_Prop_3DR*)
and 2-Dimentional Efficiency (*Int_Prop_2DR*)

3.3 Errors during the task

We, moreover, picked up 4 kinds of human errors (Stop_Error, Out_Error, Key_Error and Recog_Error) and analyzed them. The meaning of each error is as follows.

Stop_Error: Stopping the task before the object has not perfectly reached the target position. This error is concerned with perception of 2D.

Out_Error: Moving the object out of the cube. This error is concerned with perception of 3D, in particular, depth.

Key_Error: Using keys which don't correspond to any direction for the object's movement. This error is concerned with action.

Recog_Error: Recognition error. This error is concerned with voice condition, noise, and the recognizer itself.

The results are shown in Figure 6-1, 6-2, and 6-3.

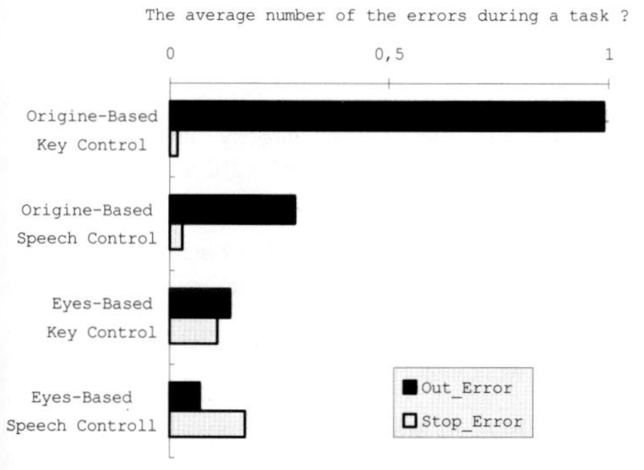

Figure 6-1
The Number of Stop_Error and Out_Error

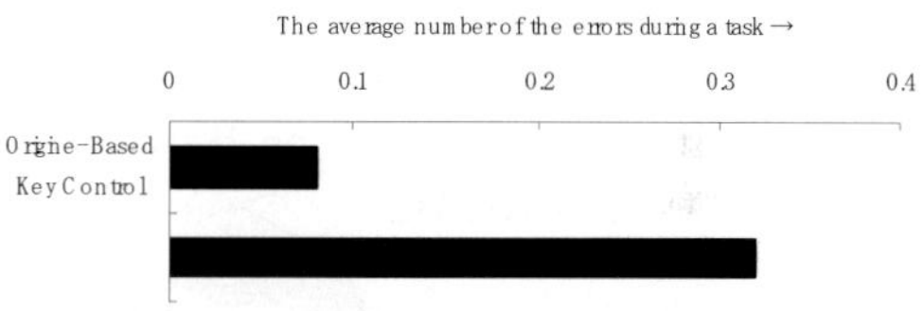

Figure 6-2
The Number of Key-Error in Key Control

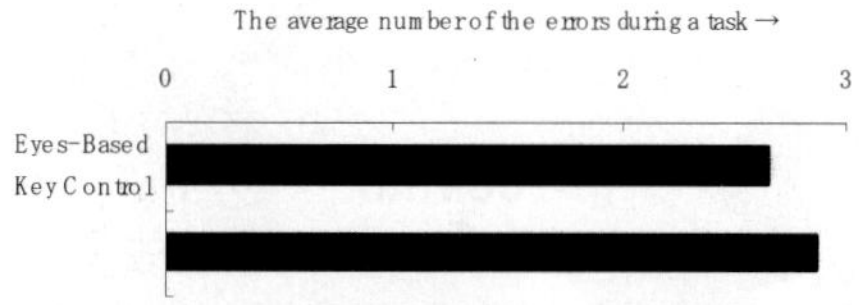

Figure 6-3
The Number of Recog_Error in Speech Control

First, although the difference of Stop_Error between speech control and key control is small in case of both coordinates, Out_Error, especially in the Origin-Based Coordinate, occurred in key control much more often than in speech control. It explains that perception of the hypothetical 3D image tended to work better in speech control than in key control while perception of the 2D image tended to work well in both speech control and key control. Additionally, in key control, as the subjects tended to often hit keys too fast because of its easiness, they moved the object too much out of the given cube.

Second, Key_Error of key control was apt to occur more often in the Eyes-Based Coordinate than in the Origin-Based Coordinate. This is because the Eyes-Based Coordinate involves the depth direction and it was inevitable that the depth was assigned to a diagonal direction on the keyboard (see Figure 3-2).

Third, Recog_Error of speech control occurred about a little less than 3 times per task in any coordinate. It often increased the required time for the task.

4 Discussion

From the above results, the following remarks could be construed as human cognitive characteristics in speech control for virtual 3D simulation on 2D screen.

1) Speech control requires more time than key control. Particularly when a word to be used is not congenial to human intention, it would be necessary for human to call up the word. Contrastively, when the right words can be used to achieve the task, the cognitive problems would be somewhat improved. From another point of view, slowness of speech control could keep away such careless action errors that are often made in key control.
2) It is possible that human cognition is easy to get the sense of 3D directions even if hypothetical 3D space is displayed on the 2D screen.
3) In case that the task is like a virtual 3D simulation on the 2D screen, the Eyes-Based Coordinate would be more suitable for human cognition than the Origin-Based Coordinate. However when the popular manual controllers such as keyboards are used, every 3D direction is forced to be arranged to each 2D direction in case of the Eye-Based Coordinate. It potentially causes the action error. On the other hand, speech control makes the control in any direction much easier as long as the appropriate words are available.
4) The existing popular recognizers cannot recognize all of the words. It does not only cause increase of the required time, but it also makes human feel that the machine does not obey his or her instructions.

From the above discussion, we show the concept of human cognitive characteristics in speech control for virtual 3D simulation on 2D screen in Figure 7.

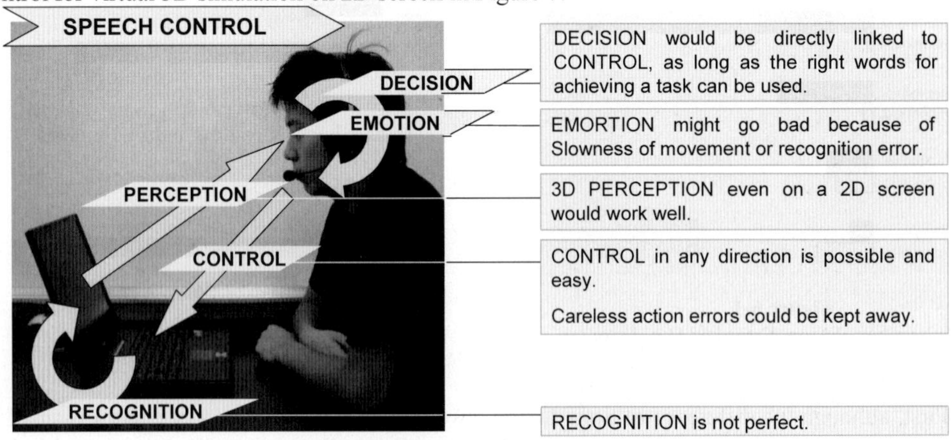

Figure 7 The Concept of Human Cognitive Characteristics in Speech Control
The Number of Recog_Error in Speech Control

5 Conclusion

As a conclusion, the following is suggested for applying speech control to VDT tasks in industry, in particular, virtual 3D simulation. Speech control not only makes it easier for the workers to control an object in any direction by natural language, but also helps them perceive virtual 3D space even on a 2D screen. In other words, it can be a kind of human-computer-interaction which could save human cognitive resources. Hence speech control can work effectively in actual situations dealing with virtual 3D simulation, for example, managing a component warehouse in an industrial plant or sort packages for delivery services and so on. And it is expected to enhance the working environment and increase efficiency in industry.

An XML Based Interactive Multimedia News System

Igor S. Pandzic

Department of Telecommunications, Zagreb University
Igor.Pandzic@fer.hr

Abstract: We present a prototype of an interactive multimedia news system featuring a talking virtual character to present the news on the Web. Talking virtual characters are graphical simulations of real or imaginary persons capable of human-like behavior, most importantly talking and gesturing. In our system a virtual character is used as a newscaster, reading the news on the Web while at the same time presenting images and graphics. The users choose the news topics they want to hear. The content of the news is defined in an XML file, which is automatically processed to create the complete interactive web site featuring the virtual newscaster reading out the news. This allows for very frequent automatic updates of the news. The virtual character is animated on the client using a Java applet, requiring no plug-ins. The bandwidth and CPU requirements are very low and this application is accessible to a majority of today's Internet users without any installation on the end-user computer. We believe that the presented news system combines qualities of other current news delivery systems (TV, radio, web sites) and therefore presents an attractive new alternative for delivering the news.

1 Introduction

We present a prototype of an interactive multimedia news system. The system automatically generates interactive news presentations for the Web, starting with the news text in XML format. The main features of this system are

- Uses a talking virtual character for rich multimedia presentations
- Fully interactive – news on demand
- Accessible to everyone (standard Web browser, modem, average PC)
- Fully automatic content generation from news text

We believe that this kind of system can be used for automatic delivery of up-to-date news on the web, in a TV-style talking presentation format that was previously too difficult to deliver due to production cost and bandwidth constraints. The system architecture (Section 2) uses only the standard Web browser for the end-user delivery. The virtual character is managed by the MPEG-4 Facial Animation Player implemented as a Java applet (Section 3). The modest bandwidth and CPU requirements mean that the content delivered using our system is accessible to the broadest possible audience of Web users – practically anyone who can access the Web can get a reasonable delivery of the news from our system. The newscaster's face is created by an artist and automatically prepared for animation using the Facial Motion Cloning method which allows for fast creation of morph target data necessary for animation. The process of preparing a new newscaster model is described in Section 4. The news are structured into topics and news items in a fairly simple XML format (Section 5). The XML structure attaches appropriate images to news items for later synchronized presentation. An automatic off-line process parses the news, generates the speech and animation for the newscaster, and creates the full directory structure with the news Web site, including the interaction mechanisms for the end-user.

2 System Architecture

The processes involved in producing and presenting the news are schematically presented in
Figure 1. The first step is making the newscaster, i.e. the animatable face model that will present the news on the web. Typically, the newscaster needs to be prepared only once for a new news service. The process involves creating or purchasing a 3D model of a face that will be used as the newscaster, in VRML format. The face model is then prepared for animation using the Facial Motion Cloning method that copies a set of generic morph targets, i.e. the basic facial movements, onto the new model. The face model and the complete set of morph targets are stored in a new VRML file that is ready for animation in the MPEG-4 Facial Animation Player Applet that we describe further on. The whole process of creating the newscaster is

explained in more detail in section 4. The second step in publishing interactive news is preparing the actual news content: making the news (see
Figure 1). The complete news content is first prepared in a structured XML format consisting of topics and news items, and giving reference to separate files with images and graphics. The news processing application processes the input XML file and automatically generates the complete interactive news web site. This involves generating the speech of the newscaster using a text-to-speech system, lip synchronization, and the generation of an appropriate set of web pages organized in a structure that allows interactive news delivery. The process of making the news is presented with more detail in section 5. This process is fully automatic and can be repeated daily, hourly or as needed to ensure that the latest news are always available online.

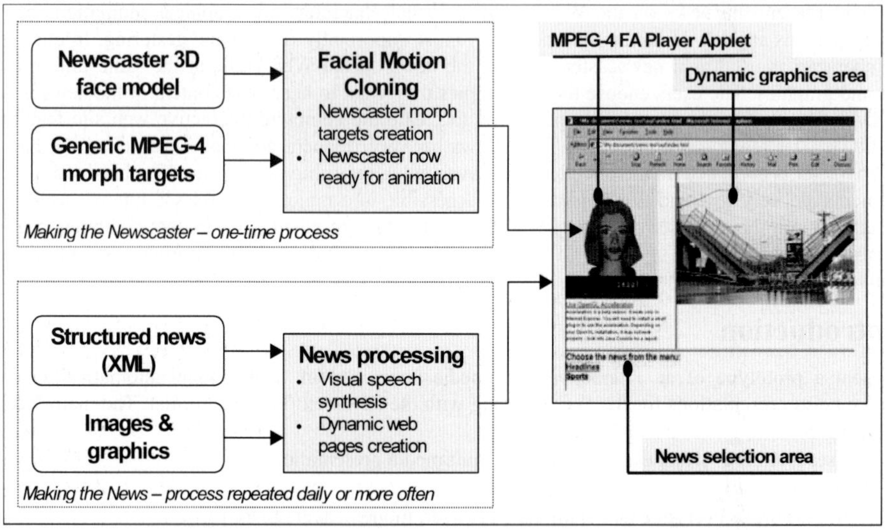

Figure 1: The Multimedia news system architecture

The process of making the news automatically places the whole set of web pages, graphics, speech and lip sync information on the Web site where the interactive news system is available to the public. The final delivery happens entirely at the client. The client is a standard web browser supporting Java (both MSIE and Netscape have been tested). No plug-in is required. This means that the news system is available to a widest possible audience. The actual layout of the pages can of course be modified. The default implementation is shown on the right side of

Figure 1. It consists of a news selection area, where the user chooses the topic of interest (e.g. sports, business…) by clicking on a topic. When the topic is chosen, the newscaster (upper left corner of the web page) delivers the news by speaking. At the same time, appropriate images and graphics are shown in the Dynamic Graphics Area (upper right corner of the web page). The appearance of graphics is synchronized with the speech, giving a full presentation. The newscaster is rendered by the MPEG-4 Facial Animation Player Applet (Section 3) that uses the newscaster face model prepared by the Facial Motion Cloning process (Section 4) and plays the speech sound in sync with the facial animation contained in MPEG-4 FBA bitstreams.

3 The Facial Animation Player

In order to deliver the news on the Web, the player needs to be modest in usage of resources, both CPU and bandwidth. In order to be easily accessible to anyone, it should preferably not require any plug-in or other specific installation on the end-user computer. To allow future portability, the player should be easy to port and adapt to any platform. Following these requirements, the first choice was to make the player MPEG-4 FBA compatible [ISO14496, Pandzic02a]. This choice ensures very low bitrate needs. Because the MPEG-4 FBA decoding process is based on integer arithmetic, its implementation is very compact and it is very modest in CPU usage. MPEG-4 compatibility allows adaptation to a wide variety of facial animation sources.

When the MPEG-4 FAPs are decoded, the player needs to apply them to a face model. Our choice for the facial animation method is interpolation from key positions, essentially the same as the morph target approach widely used in computer animation and the MPEG-4 FAT approach [ISO14496, Pandzic02a]. Interpolation was probably the earliest approach to facial animation and it has been used extensively [Parke74, Arai96]. We prefer it to procedural approaches like [Parke82, Magnenat-Thalmann88, Kalra92], and certainly to the more complex muscle based models like [Platt81, Waters87, Terzopoulos90] for the following reasons:

- It is very simple to implement, and therefore easy to port to various platforms.
- It is modest in CPU time consumption
- The usage of key positions (morph targets) is close to the methodology used by computer animators and should be easily adopted by this community

The way it works is the following. Each FAP (both low- and high-level) is defined as a key position of the face, or *morph target*. To stay consistent with the computer animation terminology, we will use the term morph target throughout the article. Each morph target is described by the relative movement of each vertex with respect to its position in the neutral face, as well as the relative rotation and translation of each transform node in the scene graph of the face. The morph target is defined for a particular value of the FAP. The movement of vertices and transforms for other values of the FAP are then interpolated from the neutral face and the morph target. This can easily be extended to include several morph targets for each FAP and use a piecewise linear interpolation function, like the FAT approach defines. However, current implementations show simple linear interpolation to be sufficient in all situations encountered so far. The vertex and transform movements of the low-level FAPs are added together to produce final facial animation frames. In case of high-level faps, the movements are blended by averaging, rather than added together. Due to its simplicity and low requirements, the Facial Animation Player is easy to implement on a variety of platforms using various programming languages. The implementation we use here is written as a Java applet and based on the Shout3D rendering engine [Shout3D]. It shows performance of 15-40 fps with textured and non-textured face models of up to 3700 polygons on a PIII/600MHz, growing to 24-60 fps on PIII/1000, while the required bandwidth is approx 0.3 kbit/s for face animation 13 kbit/s for speech, 150K download for the applet and aprox. 50K download for an average face model. This performance is satisfactory for today's average PC user connecting to the Internet with a modem. More details on this implementation and performances can be found in [Pandzic02].

4 Making the newscaster

In this section we describe our approach to the production of face models that can be directly animated by the Facial Animation Player described in the previous section. We believe that the most important requirement for achieving high visual quality in an animated face is the openness of the system for visual artists. It should be convenient for them to design face models with the tools they are used to. While numerous algorithmic facial animation systems have been developed, the best-looking animations in current productions are done manually by artists or by facial tracking equipment and performing talent. This manual creation is painstakingly time-consuming, but some aspects can be automated. The concept of morph targets as key building blocks of facial animation is already widely used in the animation community. However, morph targets are commonly used only for high level expressions (visemes, emotional expressions). In our approach we follow the MPEG-4 FAT concept and use morph targets not only for the high level expressions, but also for low-level MPEG-4 FAPs. Once their morph targets are defined, the face is capable of full animation by limitless combinations of low-level FAPs. Furthermore, being MPEG-4 compatible offers access to a growing wealth of content and content sources. Obviously, creating morph targets not only for high level expressions, but also for low-level FAPs is a tedious task. We therefore propose a method to copy the complete range of morph targets, both low- and high-level, from one face to another. This means that an artist could produce one very detailed face with all morph targets, then use it to quickly produce the full set of morph targets for a new face. The automatically produced morph targets can still be edited to achieve final detail. It is conceivable that libraries of facial models with morph targets suitable for copying to new face models will be available commercially. The method we propose for copying the morph targets is called Facial Motion Cloning. Our method is similar in goal to the Expression Cloning [Noh01]. However, our method additionally preserves the MPEG-4 compatibility of cloned facial motion and it treats transforms for eyes, teeth and tongue. It is also substantially different in implementation. Facial Motion Cloning can be schematically represented by Figure 2. The inputs to the method are the source and target face. The source face is available in neutral position (*source face*) as well as in a position containing some motion we want to copy (*animated source face*). The target face exists only as neutral (*target face*). The goal is to obtain the target face with the motion copied from the source face – the *animated target face*.

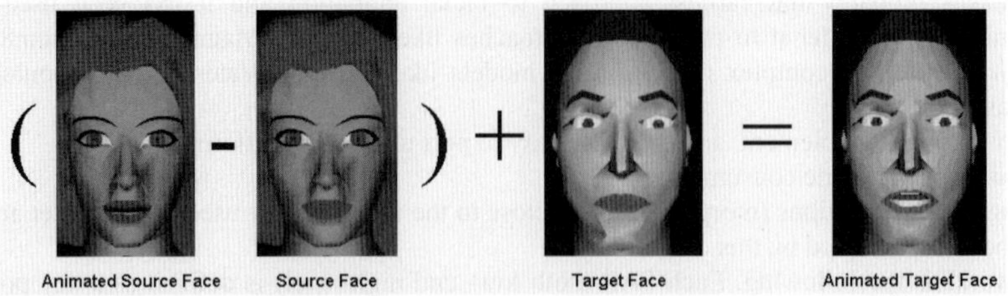

Figure 2: Overview of Facial Motion Cloning

To reach this goal we first obtain *facial motion* as the difference of 3D vertex positions between the animated source face and the neutral source face. The facial motion is then added to the vertex positions of the target face, resulting in the animated target face. In order for this to work, the facial motion must be normalized, which ensures that the scale of the motion is correct. In the *normalized facial space*, we compute facial motion by subtracting vertex positions of the animated and the neutral face. To map the facial motion correctly from one face to another, the faces need to be aligned with respect to the facial features. This is done in the *alignment space*. Once the faces have been aligned, we use interpolation to obtain facial motion vectors for vertices of the target face. The obtained facial motion vectors are applied by adding them to vertex positions, which is possible because we are working in the normalized facial space. Finally, the target face is de-normalized.

5 Making the news

The complete news content is first prepared in a structured XML format consisting of topics and news items, and giving reference to separate files with images and graphics. The news processing application processes the input XML file and automatically generates the complete interactive news web site. This involves generating the speech of the newscaster using a text-to-speech system, lip synchronization, and the generation of an appropriate set of web pages organized in a structure that allows interactive news delivery. This process is fully automatic and can be repeated daily, hourly or as needed to ensure that the latest news are always available. Here is an abbreviated example of an XML file containing an interactive news set:

```
<?xml version="1.0" encoding="ISO-8859-1"?>
<news>
        <logo>newslogo.jpg</logo>
        <introduction>Welcome to the interactive news.</introduction>
        <voice>Mary</voice>
        <topic>
          <name>Headlines</name>
          <item>
                  <text>U.S. airstrike hits the Konduz province
        Sunday.</text>
                  <image>headlines1.jpg</image>
          </item>
          <item>
                  ...
          </item>
          ...
        </topic>
        ...
</news>
```

Each news topic is set as a main menu item that the user can choose. The presentation of a topic consists of a series of news items. Each news item consists of a text to be pronounced by the newscaster, and the image to be displayed simultaneously. The text of each news item is passed to a speech synthesis tool in order to produce speech. The speech synthesis tool (speech engine) is integrated with our software using the SAPI standard, ensuring easy switching between the multitude of available SAPI-compliant speech engines varying in quality and price and supporting different languages. The SAPI-compliant speech engine also provides the phoneme timing information, based on which our news processing application generates the lip sync information and encodes it into an MPEG-4 FBA bit stream. We use the MPEG-4 viseme parameter for this encoding and the viseme blend for a simple co articulation implemented by linear interpolation between neighboring visemes. For each topic, a small program file is generated, containing the order of news items to be played, i.e. presented by the newscaster. When the user chooses a topic, the facial alnimation player plays the news items (i.e. speech and lip-synchronized facial animation) based on this program file, and displays the graphics corresponding to the news items simultaneously in the dynamic graphics area. The process of making the news automatically places the whole set of web pages, graphics, speech and lip sync information on the Web site where the interactive news system is available to the public.

6 Conclusions

We have presented the first implementation of a fully automatic and interactive news system featuring a virtual newscaster. Table 1 presents a comparison of our interactive virtual newscaster system with the current news delivery systems, i.e. the classical news web sites, TV, newspapers and radio).

	News on demand	Speech	Video	Automatic content production	Delivery
Newspaper	NO	NO	NO	NO	PAPER
Radio	NO	YES	NO	NO	RECEIVER
TV	NO	YES	HI QUALITY	NO	TV SET
Standard web site	YES	NO	NO	YES	STANDARD PC
Virtual Newscaster	YES	YES	MEDIUM QUALITY	YES	STANDARD PC

Table 1: Comparison of news delivery systems

We can conclude that the presented news system combines qualities of all other news delivery systems, in particular the low-cost production of news-on-demand typical for the web with the high visual content typical for the TV. At the same time our system does not demand high extra bandwidth. We therefore believe that our system presents an attractive new alternative for delivering the news. A demonstration of the system is available at www.tel.fer.hr/users/ipandzic/frames/Newscaster.

7 Acknowledgements

This research is partly supported by the Croatian Academic and Research Network (CARNet) through the HUMANOID project.

8 References

[Arai96] "Bilinear interpolation for facial expressions and methamrphosis in real-time animation", Kiyoshi Arai, Tsuneya Kurihara, Ken-ichi Anjyo, The Visual Computer, 12:105-116, 1996.

[ISO14496] ISO/IEC 14496 - MPEG-4 International Standard, Moving Picture Experts Group, www.cselt.it/mpeg

[Kalra92] Kalra P., Mangili A., Magnenat-Thalmann N., Thalmann D., Simulation of Facial Muscle Actions based on Rational Free Form Deformation", Proceedings Eurographics 92, pp. 65-69

[Magnenat-Thalmann88] "Abstract muscle actions procedures for human face animation", N. Magnenat-Thalmann, N.E. Primeau, D. Thalmann, Visual Computer, 3(5):290-297, 1988.

[Noh01] "Expression Cloning", Jun-yong Noh, Ulrich Neumann, Proceedings of SIGGRAPH 2001, Los Angeles, USA

[Pandzic02] "Facial Animation Framework for the Web and Mobile Platforms", Igor S. Pandzic, Proc. Web3D Symposium 2002, Tempe, AZ, USA, demonstration at www.tel.fer.hr/users/ipandzic/MpegWeb/index.html

[Pandzic02a] "MPEG-4 Facial Animation - The standard, implementations and applications", Igor S. Pandzic, Robert Forchheimer (editors), John Wiley & Sons, 2002, ISBN 0-470-84465-5

[Parke74] "A Parametric Model for Human Faces", F.I. Parke, PhD Thesis, University of Utah, Salt Lake City, USA, 1974. UTEC-CSc-75-047

[Parke82] "Parametrized models for facial animation", F.I. Parke, IEEE Computer Graphics and Applications, 2(9):61-68, November 1982.

[Platt81] "Animating Facial Expressions", S.M. Platt, N.I. BadlerComputer Graphics, 15(3):245-252, 1981.

[Shout3D] Shout 3D, Eyematic Interfaces Incorporated, http://www.shout3d.com/

[Terzopoulos90] "physically-based facial modeling, analysis and animation", D. Terzopoulos, K. Waters, Journal of Visualization and Computer Animation, 1(4):73-80, 1990.

[Waters87] "A muscle model for animating three-dimensional facial expressions", K. Waters, Computer Graphics (SIGGRAPH'87), 21(4):17-24, 1987.

Speech-based Text Entry for Mobile Devices

Kathleen J. Price and Andrew Sears

Interactive Systems Research Center
Information Systems Department
UMBC
1000 Hilltop Circle
Baltimore, MD 21250, USA
kprice1@umbc.edu asears@umbc.edu

Abstract

As handheld devices become more common and the tasks performed using these devices more diverse, it is necessary to find increasingly efficient data entry techniques. This paper presents an evaluation of a new speech-based data entry technique for handheld devices. TabletTalk supports data entry using server-based, speaker-dependent speech recognition combined with a soft keyboard which can be used to correct errors. Participants were tested in five daily trials to examine the effects of experience. Data analysis reveals minimal errors with a mean text entry rate as high as 25.3 words per minute (wpm) which is substantially higher than the data entry rates obtained with other standard techniques currently supported by handheld devices. Multiple regression indicates that recognition error rates explain 35% of the variance in data entry rates while task length explained 20% and individual differences explained 7.5%. Trial did not have a significant effect on data entry rates.

1 Introduction

Handheld devices (e.g., personal digital assistants or PDAs, pagers, mobile phones) have become a common part of everyday life, supporting a diversity of tasks which require entering varying amounts of data. Standard data entry techniques for mobile devices such as gesture recognition or soft keyboards, are often slow and cumbersome and can be tedious if used to enter large amounts of text. Since mobile phones are used for voice communication, speech-based data entry is a natural adjunct to support activities such as e-mail and Internet access. Speech recognition is already used on some mobile devices to support small vocabulary tasks, such as voice-activated dialing. Due to the computationally intensive nature of speech recognition, greater system capabilities typically result in improved recognition accuracy, especially when dealing with large vocabulary activities. Therefore, this study evaluates the efficacy of speech-based data entry using server-based speech recognition with a soft keyboard to support error correction activities.

2 Related Research

A number of studies have examined data entry techniques for handheld devices. MacKenzie and Zhang (1997) assessed Graffiti for data entry and found that accuracy ranged from 86% to 97%, but they did not investigate data entry rates. Sears and Arora (2002) investigated both Graffiti and Jot, reporting modest data entry rates that ranged from 3.8 to 8.8 words per minute (wpm), depending on the task. Zha and Sears (2001) evaluated the efficacy of small soft keyboards for data entry when used by novices and reported data entry rates ranging from 7 to 12 wpm, depending on the task. Another study using mobile phones with a standard 12 button keypad demonstrated data entry rates of 27 wpm for a single experienced user performing data entry using a multitap technique (Silfverberg et al., 2000).

Discourse on speech and recognition software appeared as early as the 1960's (Holmes et al., 1964) with discussions of speech-based text entry for handheld devices beginning in the early 1990's (Stifelman et al., 1993). Recognition errors have always been a concern, resulting in numerous efforts to develop robust error correction techniques (e.g., Manaris & Harkreader, 1998). Noyes and Frankish (1994) effectively described this topic by stating: "Although the techniques used for recognizing human speech will inevitably become more sophisticated, the problem of errors remains. Total recognition accuracy is an unrealistic and unobtainable goal, one not even obtained in human to human conversation." Researchers have examined user patterns of error correction (e.g., Karat et al., 1999) and the use of multi-modal interactive error correction for speech text entry (e.g., Suhm et al., 2001). A recent study (Sears et al., 2001) reported speech-based data entry rates as high as 15wpm, suggesting that speech has the potential to be an effective solution for mobile data entry as compared to existing gesture-based techniques.

3 Research Objectives

This study examines the efficacy of speech-based entry of short text phrases for handheld devices. Data entry rates (i.e., time to complete each task), accuracy (i.e., number of residual errors after task completion), and user satisfaction ratings (i.e., questionnaire responses) were recorded. The system was evaluated via five daily trials to observe the effects of practice on user performance.

In our previous research, we evaluated the use of both multitap and soft keyboards to correct errors. For novices, we found data entry rates as high as 22wpm when the multitap technique was used to correct errors and 23.5wpm when a soft keyboard was available (Price & Sears, 2002). Given these results, we expect the proposed technique to result in improved text entry rates as compared to existing techniques. As a result, several research questions were examined:

- What data entry and accuracy rates can be achieved using speech text entry when alternative selection and a soft keyboard are provided for error correction?
- How satisfied will users be with this text entry/error correction technique?
- How does practice affect text entry rates?

4 Method

A within-subject design was used, with all participants completing all study trials and tasks. There were five daily trials, four tasks per trial, for a total of twenty tasks. Twelve participants were recruited from the students and staff at UMBC including 4 males and 8 females with an average age of 20. Participants' college majors or career fields varied across a wide range of subjects. All participants had multiple years of computer experience and affirmed daily e-mail use.

The equipment used in this study is intended to simulate the use of a handheld device such as a mobile phone or PDA. A headset with microphone was connected to a laptop computer. Users interacted with a WACOM Cintiq ™ 15X Interactive Pen Display tablet. The software, TabletTalk enables the text entry using dictation, presenting the resulting text and a soft keyboard (measuring 3.5x4 inches) in an area that is approximately the size of a PDA screen (see Figure 1). The keyboard uses a QWERTY layout. The "Dictation" button turns dictation on and off. The "Start" button was used to indicate when the user was ready to begin their task. A stylus was used to place the cursor anywhere within the text. TabletTalk uses the Millennium Edition of IBM's ViaVoice speech recognition engine.

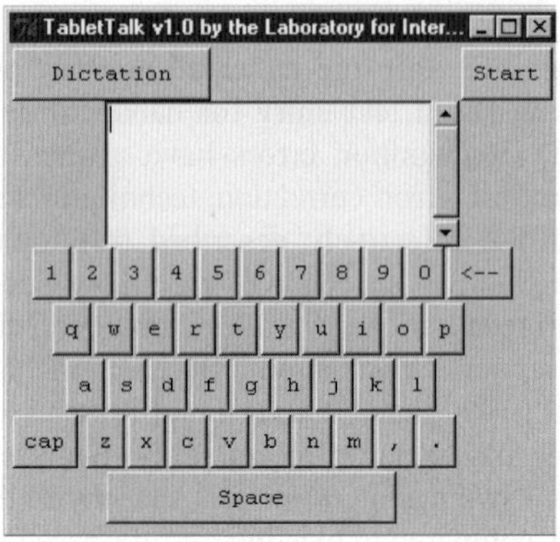

Figure 1. Tablet Talk interface

The software simulates server-based speech recognition and allows both alternate word selection and soft keyboard error correction. Users dictate the desired text, wait for the results, and correct any recognition errors. The speech engine runs locally, but a ten second delay is introduced before displaying results from the speech engine to simulate the network delay that would be experienced with server-based speech recognition. Recognition errors can be corrected by double clicking on an incorrect word using the stylus and selecting the correct word from a list of up to four alternatives that were generated by the speech engine. The correct word is not always available in the list of alternatives. Users can also type the correct word using the soft keyboard.

Participants entered four short text phrases that varied in length during each trial. Mean task lengths were 3.5, 7.5, 14.9, and 34.2 five character words. Different phrases, which represent potential e-mail responses or calendar entries, were used for each trial. The order in which the different tasks were completed was randomized for each participant and trial using a Latin Square design. After screening, a brief introduction, and consent, each participant was asked to complete the standard ViaVoice enrollment process to enable the speech engine to learn their unique speech patterns. Participants were trained on the use of TabletTalk and all study equipment. They were given information about the software, hardware, dictation, phrase navigation, and error correction techniques. Participants were allowed to practice the error correction techniques using four phrases similar in length and nature to the study tasks. When ready to begin, participants were instructed to use the learned techniques to complete the study tasks. The four tasks were presented one at a time in a random order. All performance data (e.g., times, errors, and user actions) were recorded by the software for later analysis. The sessions were also videotaped to provide later clarification if necessary.

Participants returned on four more days to complete the additional trials. On day five, participants were asked to complete a demographics and user satisfaction survey. A ten dollar incentive for participation was provided at the end of each session. The initial study session lasted approximately 45 minutes with follow-up sessions lasting about 15 minutes each.

5 Results

Multiple regression was utilized to analyze the results including data entry rates, accuracy, and subjective satisfaction. Means and standard deviations for text entry rates, measured in words per minute, are shown in Table 1. Multiple regression was used to determine the effect of four variables on the data entry rates: individual differences (i.e., participant), trial, task length, and recognition error rate. Since data entry rates were not normally distributed, a log10 transformation was used to normalize the data. Regression analysis results are summarized in Table 2 (variables are listed in the order they were entered into the regression model).

Table 1: Means/standard deviations of data entry rates (wpm)

	Tasks			
	One	Two	Three	Four
Trial	Mean (Stdev)	Mean (Stdev)	Mean (Stdev)	Mean (Stdev)
1	8.5 (2.5)	9.8 (3.7)	24.5 (3.8)	14.7 (4.0)
2	9.0 (3.2)	10.6 (3.9)	20.4 (7.9)	19.6 (7.0)
3	9.7 (1.8)	16.4 (4.0)	19.2 (7.8)	22.4 (8.2)
4	8.0 (2.7)	15.5 (4.6)	25.3 (8.4)	21.5 (7.9)
5	8.6 (2.2)	18.9 (5.0)	21.1 (8.6)	15.2 (4.3)

Table 2. Multiple regression summary

Variable	Percent Change in Variance	T test
Individual differences	7.5%	4.38 (p<0.001)
Trial	1.3%	NS
Task length	20.0%	8.15 (p<0.001)
Recognition Error Rate	35.0%	-15.09 (p<0.001)

Task length and recognition error rate both significantly influence data entry rates. As expected, text entry rates improve as phrases increase in length. Conversely, data entry rates decrease as recognition error rates increase. Trial does not significantly effect data entry rates. Due to the number of participants, criterion scaling was necessary to evaluate the effect of individual differences on data entry rates. Using the corrected F value, it was confirmed that individual differences did have a significant influence on data entry rates ($F(14, 225)=62.32$, $p<0.001$). Given the sizable effect of recognition errors on data entry rates, it is clear that minimizing recognition errors is an important method of increasing data entry rates. This supports the concept of using server-based speech recognition since servers are likely to provide increased computing power as compared to mobile devices. Satisfaction was assessed by administering a posttest Likert-style satisfaction survey. Overall, satisfaction results were favorable. Suggestions for improvements included: improved speech recognition accuracy (e.g., better recognition of natural speaking patterns), faster processing (i.e., reducing network delays), automatic capitalization (e.g., of proper names), and presenting more alternative words. Accuracy was high with ten residual errors from a total of 240 tasks. Given minimal errors, statistical analysis was not performed.

6 Conclusions

Novices were able to enter text at a rate of up to 25 wpm with minimal content or formatting errors. This, combined with the observation that trial did not have a significant effect on data entry rates, suggests that participants were able to effectively utilize the error correction strategies with minimal experience. Additional studies may include a broader range of participants and tasks as well as improvements designed to address participant concerns about capitalization. While researchers continue to address issues related to recognition accuracy, our own research suggests that the participants' suggestion for the provision of additional alternatives is not likely to prove beneficial with state-of-the-art speech recognition engines. Network delays are beyond the scope of the current project.

7 Acknowledgements

The authors would like to thank Aether Systems, Inc. for their support of this research. This material is based upon work supported by the National Science Foundation under Grant Nos. IIS-9910607 and IIS-0121570. Any opinions, findings and conclusions or recommendations expressed in this material are those of the authors and do not necessarily reflect the views of the National Science Foundation (NSF).

8 References

Holmes, J. M., Mattingly, I. G. & Shearme, J. N. (1964). Speech synthesis by rule. *Language and Speech*, 7, 127-143.

Karat, C-M., Halverson, C., Karat, J., & Horn, D. (1999). Patterns of entry and correction in large vocabulary continuous speech recognition systems. *Proceedings of CHI '99*. pp. 568-575.

MacKenzie, I. S. & Zhang, S. (1997). The immediate usability of Graffiti. *Proceedings of Graphics Interface '97*. pp. 129-137.

Manaris, B. & Harkreader, A. (1998). SUITEKeys: A speech understanding interface for the motor-control challenged. *Proceedings of the 3rd International ACM SIGCAPH Conference on Assistive Technologies (ASSETS '98)*, pp. 108-115.

Noyes, J. M. & Frankish, C. R. (1994). Errors and error correction in automatic speech recognition systems, *Ergonomics*, 37, 1943-1957.

Price, K.J. & Sears, A. (2002). Speech-based data entry for handheld devices: Speed of entry and error correction techniques, Information Systems Department Technical Report, UMBC: Baltimore, MD.

Sears, A. & Arora, R. (2002). Data entry for mobile devices: An empirical comparison of novice performance with Jot and Graffiti. *Interacting with Computers, 14*(5), 413-433.

Sears, A., Karat, C-M., Oseitutu, K., Karimullah, A., & Feng, J. (2001). Productivity, satisfaction, and interaction strategies of individuals with spinal cord injuries and traditional users interacting with speech recognition software. *Universal Access in the Information Society*, 1, 4-15.

Silfverberg, M., MacKenzie, I. S., & Korhonen, P. (2000). Predicting text entry speed on mobile phones. *Proceedings of CHI 2000*. pp. 9-16.

Stifelman, L. J., Arons, B., Schmandt, C., & Hulteen, E. A. (1993). VoiceNotes: A speech interface for a hand-held voice notetaker. *Proceedings of INTERCHI '93*. pp. 179-186.

Suhm, B., Myers, B., & Waibel, A. (2001). Multimodal error correction for speech user interfaces. *ACM Transactions on Computer-Human Interaction, (8)*1, 60-98.

Zha, Y. and Sears, A. (2001) Data entry for mobile devices using soft keyboards: Understanding the effects of keyboard size and user tasks, *Proceedings of HCI International 2001*. pp. 16-20.

Proposal of Grasping Force Interface as Realtime Mickey Ratio Adjuster for Pointing Tasks of Mouse

Sigeru Sato, Muneo Kitajima

National Institute of Advanced Industrial Science and Technology
AIST Tsukuba Central 6, 1-1, Higashi, Tsukuba, Ibaraki, 305-8566, Japan
sato-s@aist.go.jp, kitajima@ni.aist.go.jp

Yukio Fukui

University of Tsukuba, Inst. Information Sciences and Electronics, 1-1-1, Ten-no-dai, Tsukuba, Ibaraki, 305-8573, Japan
fukui@is.tsukuba.ac.jp

Abstract

The paper describes the concept and experimental system of a computer mouse with realtime adaptive Mickey ratio adjustment by grasping force.
Grasping style is adaptive, and it depends not only on the grasped object but also on the purpose of grasping. Even if the same person grasps the same object, when the purpose differs, the grasping style also differs. Therefore, there should be some possibility that the operator's will is detected from grasping style. In pointing tasks, there is a relationship between the motion of mouse and that of the cursor. In usual, Mickey ratio is not adaptive in operation even though it is adjustable before the operation. But this adjustment is a quite difficult function. According to Fitts' law, pointing tasks is easier when the target is larger and the distance is shorter. Large gain means the short distance and small target, small gain means, as opposite to above, long distance and large target. This paper proposes a realtime adaptive Mickey ratio adjustment for an answer of this dilemma. And it is recommended that this adjustment be controlled by operator's grasping force of operation terminal to avoid increase of complexity of operation.

1 Introduction

Grasping style is adaptive, and it depends not only on the grasped object but also on the purpose of grasping. Even if the same subject grasps the same object, when the purpose differs, the grasping style also differs (Napier, 1956). Therefore, there should be some possibility that the operator's will is detected from grasping style.

Fitts' law gives time T that takes an operator for a pointing task on a target of size S at distance of D as:
$T = a + b * \log2 (D/S + 1)$
where "a" and "b" are experimentally determined constants (Fitts, 1954).

This T indicates difficulty of the task, because easier task needs less time. Fitts' law indicates that pointing task is easier when the target is larger and the distance is shorter.

2 Analysis of Pointing Tasks

Considering cursor motion driven and controlled with mouse, Mickey Ratio gives the gain of cursor speed with mouse motion. Mickey Ratio is the gain of mouse to cursor motion. Larger gain gives faster speed. It offers as same effect as smaller target at shorter distance. Smaller gain causes equivalent effect with larger target at longer distance. Constant gain is not effective to decrease D/S of the equation above even if it is adjustable before the tasks.

The most important point of this idea is that during each task, parameters D and S are not simultaneously effective, although they present as D/S in Fitts' law. D is effective when the pointer moves toward the target (in following part of this paper, it is called "approaching phase"), and S is effective after the pointer comes near around the target (in following part of this paper, it is called "positioning phase"). In every task, approaching phase is driven before positioning phase.

If during each task Mickey Ratio were adjusted larger in approaching phase and smaller in positioning phase, the easiness of the task would increase. This is realtime adaptive Mickey Ratio adjustment. And it suggests that pointing tasks include two stages, approaching and positioning.

To adjust gain, new switches or other new gears shouldn't be supplemented, because they increase the complexity of operation system. That has the opposite effect to the purpose of "increase the easiness of remote pointing tasks". The realtime adaptive gain adjustment should be controlled automatically or, at least, naturally.

Some studies are known for icon selecting tasks using computer mouse. Effect of tactile feedback is analyzed by Akamatsu et al. (Akamatsu, Sato & MacKenzie, 1994). Study of Worden et al. shows that sticky icons are effective, especially for older adult people (Worden, Walker, Bharat, & Hudson, 1997). Dennerlein et al. describe advantage of force feedback in some targeting tasks (Dennerlein, Martin & Hasser, 2000). The "sticky icon" is a special variation of realtime adaptive gain adjustment. Because the machine knows where the targets exist in those icon selections, it is easy to distinguish positioning phase from approaching phase.

In general pointing tasks, the target exists only in the user's idea or is created instantly during the pointing task. The target cannot be distinguished by the pointing machine before the user finishes each pointing task. Information of pointer position is no useful to gain adjustment in these general pointing tasks. Approaching to boundary of these two phases is quite a psychological process of the user, some physical or physiological parameter may be influenced. It is only user's operation itself that is available for realtime adaptive gain adjustment function. To provide natural usability of this new mouse, the signal for switching gain (Mickey Ratio) between these two phases should be detected from the user's physiological or physical behaviours. This study proposes that grasping force is available.

3 Method of Mickey Ratio Adjustment

To apply this to pointing tasks, in positioning phase, task requires more precision operation than in approaching phase. It means that muscles are expected more tensioned in positioning phase because more robustness is needed. Positioning phase may be indicated by this tension. It is expected that grasp operation terminal such as joystick switches, levers, computer mouse etc. are grasped naturally stronger in positioning phase than in approaching phase.

Therefore, using grasping force of operation terminal, automatic realtime adaptive gain adjustment can be realized without increase of operation complexity (Sato, Kitajima & Fukui, 2001).

An experimental system is constructed as shown in Figure 1 and 2. To detect grasping force, a thin pressure sensor is set on the mouse surface at the position of thumb. Detected grasping force is processed and assigned to right button of mouse circuit. Experimental tasks can be driven on a

notebook computer. A graphic pointing task program is designed that cursor moves slower when right button signal is active (Sato, Kitajima & Fukui, 2002).
Although experimental system can change only 2 speeds and only special experimental tasks can be realized presently, this concept can be applied to general pointing tasks. Especially, when a mouse signal generate circuit includes grasping force signal processing function, it needs only replacing mouse to use this realtime adaptive Mickey ratio adjuster.

Figure 1: Experimental Sample

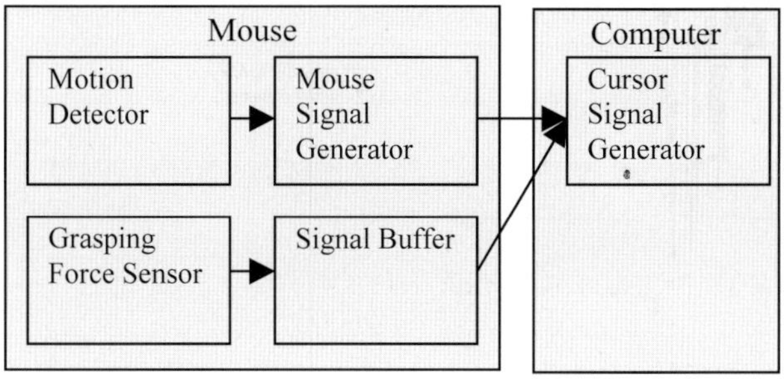

Figure 2: Block Diagram of Experimental Mouse with Realtime Adaptive Mickey Ratio Adjuster

4 Experimental Result

An experiment of drawing task is driven with more than 20 subjects. Figure 3 shows an example of grasping force by one subject. The experimental task is to complete an illustration of Eiffel tower. There are both precision period (the pen is down and a line is drawn) and rough period (the pen is up and only moving the pen position). Grasping force differs these precision and rough periods.

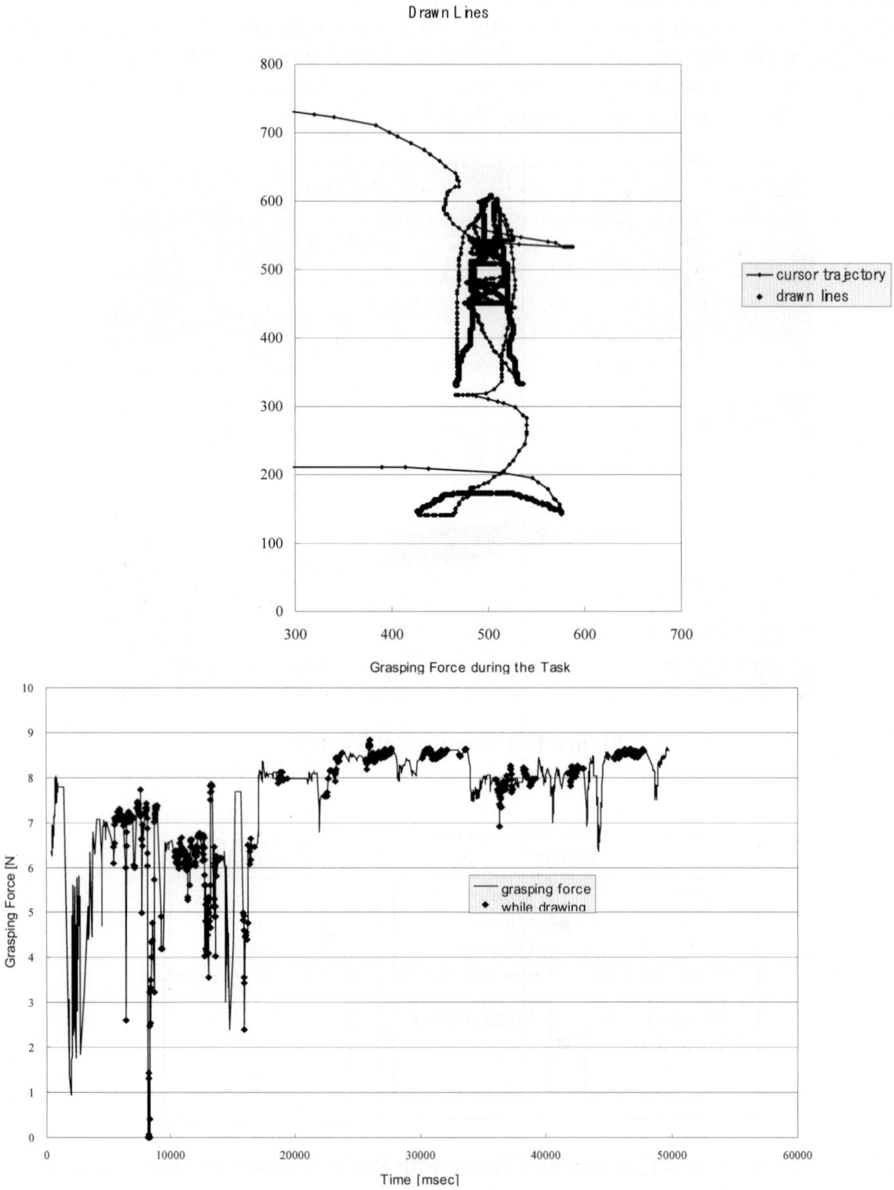

Figure 3: Experimental Result of a Subject

5 Conclusion

A new type of computer mouse that has a function for realtime adaptive mickey ratio adjustment by grasping force is proposed. It is expected to make computer mouse more natural and more comfortable human-computer interface device. When cursor signal generator is built in the mouse, it is more convenient because this function is available only by replacing the mouse.
Remote pointing tasks are separated in approaching and positioning phases. Realtime adaptive Mickey ratio adjustment has advantage that easiness of both of two phases increase. This is caused from the fact that distance to the target is major effective parameter in approaching phase and target size is major effective parameter in positioning phase.
Because the purpose of realtime adaptive gain adjustment is increase easiness of pointing tasks, it must be added without any new complexity.
In general pointing tasks, these two phases cannot be known before the task is started. The boundary between two phases exists only in the operator's mental or psychological process, it is difficult to detect itself. Grasping information interface has a possibility to provide natural usability to realtime adaptive gain adjustment for grasping type operation terminal.

References

Akamatsu, M., Sato, S., & MacKenzie, I. S. (1994), Multimodal Mouse: A Mouse-Type Device with Tactile and Force Display, Presence, 3(1), 73-80.

Dennerlein, J. T., Martin, D. B., & Hasser, C., (2000), Force- Feedback Improves Performance For Steering and Combined Steering-Targeting Tasks, CHI Letters, Amsterdam, 1, 423-429.

Fitts, P. M. (1954), The Information capacity of the human motor system in controlling the amplitude of movement, Journal of Experimental Psychology, 47, 381-391.

Napier, J. R. (1956), The prehensile movement of the human hand, Journal of Bone and Joint Surgery, 38-B, 903-912.

Sato, S. Kitajima, M., & Fukui, Y. (2001), A New Mouse with Cursor Speed Control by Grasping Force, Proc. HCI2001, New Orleans, Poster Sessions, 291-292.

Sato, S. Kitajima, , M., & Fukui, Y. (2002), A Mouse with Realtime Adaptive Mickey Ratio Adjustment by Grasping Force, Proc. UIST2002, Paris, 53-54.

Worden, A., Walker, N., Bharat, K., & Hudson, S., (1997), Making Computers Easier for Older Adults to Use: Area Cursors and Sticky Icons, Proc. CHI 97, 266-271.

Novel Interaction Techniques for Virtual Heritage Applications using Chinese Calligraphy Brush and Virtual Avatar

Meehae Song, Thomas Elias, Wolfgang Müller-Wittig, Tony K.Y. Chan

Centre for Advanced Media Technology (CAMTech)
Nanyang Technological University (NTU)
50 Nanyang Avenue
Singapore 639798
song@camtech.ntu.edu.sg

Abstract

In this paper, we present a digital heritage project incorporating novel interaction techniques in an immersive virtual heritage application. Virtual Reality (VR) technology enables us to digitally recreate immersive and interactive cultural heritage sites that no longer exist or are inaccessible. It is an excellent educational tool providing for a 'hands-on' application suitable for a wide range of audiences. Here we present the development and implementation of the virtual heritage application with early results and future work and specifically focus on the two interaction techniques – the Virtual Tour Guide and the Chinese calligraphy brush.

1 Introduction and Objective

Virtual Reality(VR) technology is widely being used in a diverse range of heritage applications providing a fully immersive 3D environment presenting to the public new possibilities to experience and interact with the cultural heritage sites. Using this VR technology, we have selected to recreate and explore the Peranakans and their culture. The Peranakans are descendants of an early Chinese community that settled in Singapore and Malaysia since the 17th century. They developed a rich culture by blending the Chinese and Malay cultures thus adding another dimension to the Singaporean-Malaysian society. For the interaction techniques, a Peranakan Virtual Tour Guide has been created to guide the visitors through the virtual environment and a Chinese Calligraphy brush has been incorporated providing the visitors with a familiar and easy to use interaction device. These techniques offer the visitors the option to explore and navigate through their specific needs reaching a wide range of audiences.

2 Development and Implementation

The major components of this project are 3D modelling, development of interaction techniques, and high-end visualisation. Sophisticated 3D modelling techniques have been used to create the Virtual Tour Guide and selected cultural material from the Peranakan culture such as furniture, ceramics, silverware, and embroidery and beadwork. These models are capable of real-time rendering using in-house systems for the visualisation. The following sections describe the development and the implementation of the interaction techniques used within the Virtual Environment.

2.1 Virtual Tour Guide

The Virtual Tour Guide we have created for our project interacts with the visitors, shows them around the Virtual Environment, and also provides historic information on the Peranakan culture. He is of Chinese ethnicity and dressed in the traditional Peranakan clothing to fit the context of the Peranakan culture. Subdivision surface modelling technique was used to model the body of the Virtual Tour Guide. The body was then mapped to a base hierarchical skeleton in order to generate animation sets. The Virtual Tour Guide provides information by talking to the visitor. However, he can also show supplemental information in the form of text, images, and video. In order to achieve this, he must be capable of real-time facial animation for speech generation. For the face modelling, a base face was created and viseme face sets were derived from the base head. Visemes are the mouth positions/shapes of a particular sound in speech. The base head uses the different visemes as targets to 'morph' between the different mouth shapes according to the analyzed sound files.

The visitor has the option of whether he/she wants the help of the Virtual Tour Guide. For example, a visitor can explore the Virtual Environment and all its objects by walking through the virtual rooms by him/herself or also ask the Virtual Tour Guide for a complete guided tour. During the tour the visitor follows the Virtual Tour Guide automatically and gets additional information on important exhibits. The visitor can interact with the Virtual Tour Guide at any time. If the visitor wants additional information on an interesting exhibit, he/she simply selects the exhibit and calls the Virtual Tour Guide to trigger for more information.

The Virtual Tour Guide also introduces novel visitors to the Virtual Environment and shows them how they can use the Chinese Calligraphy brush for interaction. He explains the Chinese characters that the visitor can write with the brush as well as their associated functionalities.

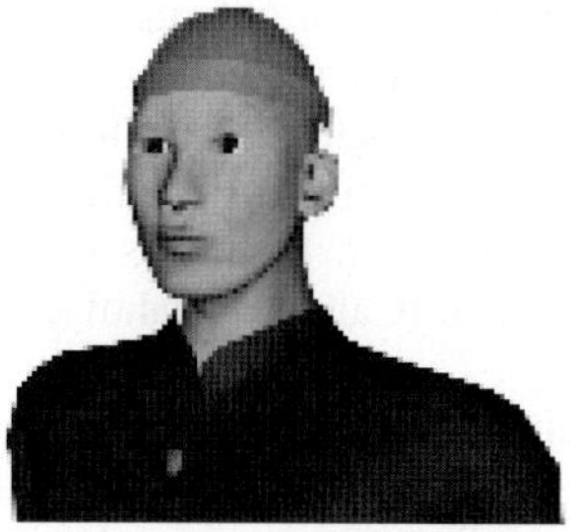

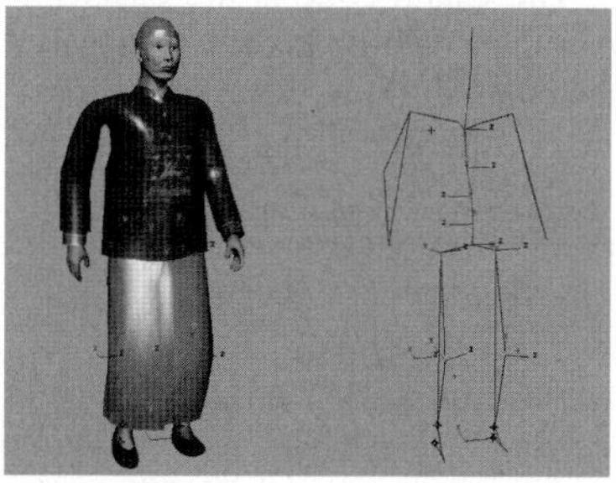

Figure 1: Virtual Tour Guide Setup

2.2 Calligraphy Brush

An important aim of this project is to educate the public and to give an insight to the Peranakans. In order to reach a broader audience - including visitors with little to no experience with VR technology, the interaction method must be intuitive and easy to use. Furthermore, the interaction device should correlate to the context of the application. A good example of this is the Virtual Dunhuang Art Cave from (Lutz & Weintke, 1999). Here, one can use a real flashlight to navigate in the virtual cave. For these reasons, we have chosen to use a Chinese Calligraphy brush as an interaction device. Calligraphy is an art dating back to the earliest day of history and widely practiced throughout China and other parts of the world to this day. It is a very familiar tool to the Asian culture since they are exposed to it at an early age whether it is at school or in other environments. This makes the interaction device an integral part of the culture that the visitor can additionally explore.

2.2.1 Writing with the Calligraphy brush

The visitor can use the brush to write simple Chinese characters that are associated to specific commands the system understands. For this purpose, the visitor holds the brush in one hand and a small board in the other. The visitor writes on the board by moving the brush on the board but not actually using any real ink. The virtual counterparts of the brush and the board are shown as 3D models in the Virtual Environment. They reflect the writing in black ink and provide visual feedback to the visitor. Only simple characters that are composed of a few strokes are used. They are easy to memorise and economical to write making them even suitable for visitors with no knowledge of Chinese. For example, by writing the simple Chinese character "Ren" 人 which means "Person", the visitor calls the Virtual Tour Guide to request help or information.

Writing with the Calligraphy brush is realised by implementing a Chinese character recognition method. A tracking system is used to capture the position and the orientation of the real brush and board. This information is used to update the virtual models of the brush and board to correspond to the real ones. If there was an intersection between the brush and the board, the area of the imprint made by the virtual brush is coloured in black to simulate ink. The captured information is also passed on to the recognition method. An online Chinese character recognition method is used that gathers the relevant data such as the position, time, and pressure of each imprint while the visitor writes. All this data that belongs to one imprint is saved in one element of a Chain-code. After the visitor has written the complete character, the Chain-code contains as many elements as the number of captured imprints. These elements are concatenated and ordered by time. An example for an online recognition method can be found in (Calhoun, Stahovich, Kurtoglu, & Kara, 2002). The Chain-code is the basis for the actual recognition process. This recognition process analyses the Chain-code data to detect the Chinese character. The visitor only uses a few Chinese characters to interact with the system so that a simple recognition method is sufficient. The method used searches and identifies the strokes of the Chinese character. Each Chinese character is composed of different types of single strokes that vary in the position, orientation, and length. The types of strokes and their order identify each Chinese character. Examples for recognition methods that detect strokes are given in (Amin & Sum, 1993) and (Calhoun, Stahovich, Kurtoglu, & Kara, 2002).

The actual recognition process analyses the complete Chain-code and tries to detect the single strokes the Chinese character is composed of. For this purpose, the Chain-code has to be subdivided into parts so that each part only contains elements of one stroke. In order to find these elements, the recognition method searches for higher-than-average changes in the position, speed, and pressure in the Chain-code. These changes occurs when the brush is lifted during the writing of the character to move from one end of a stroke to the beginning of the next. Afterwards, the type of each stroke has to be determined by calculating its length and orientation. The recognition method has a database where the number, type, and order of strokes are stored for each Chinese character. At the end, only the appropriate data that was determined during the recognition process has to be compared to the one stored in the database.

2.2.2 Navigating with the Calligraphy brush

The visitor can also use the calligraphy brush to navigate in the Virtual Environment by pointing with it to a specific direction to indicate where he/she wants to move to. This simple navigation method is sufficient because only walking and turning on the ground level is needed. Pointing with the brush to the left or right side of the screen turns the visitor inside the Virtual Environment. Similarly, pointing the brush at the top or the bottom of the screen moves the visitor forward or backward in the virtual scene. Additionally, the visitor can point at an exhibit to get additional information. A cursor appears on the screen at the position where the visitor points to. In order to give the visitor better feedback, the shape of the cursor changes depending on which side of the screen the cursor is situated at. For example, when the visitor points at the left side of the screen, the cursor shows a left arrow and when he/she points to an exhibit, the cursor is changed to a hand.

In order to implement navigation with the Calligraphy brush, the position and orientation of the brush have to be known. This information is returned by the tracking system as described in section 2.2.1. In order to detect which side of the screen the visitor is pointing to, an invisible plane is defined and placed in front of the screen. The position of the brush and its roll axis define the starting point and direction of a ray. This ray is intersected with the plane in order to find the position of where visitor is pointing at. Figure 2. shows an exemplary setting. It can also be seen that partial areas are defined on the plane. After the position of the cursor is calculated, the partial area the cursor is situated in is determined. If it is one of the four areas at the side of the screen, the cursor is changed into an appropriate arrow and the visitor is rotated or moved respectively within the Virtual Environment. If the cursor is situated in the central area, the system checks if the visitor is pointing at an exhibit by intersecting the ray with the bounding box of all currently visible exhibits and changes the cursor into a hand if an intersection occurs. This action which rotates, moves, or selects an exhibit, does not occur instantly after the cursor changes to a hand. In general, the visitor has to keep the cursor within the defined area for a certain period of time. The visitor thus can cancel the action by moving the cursor to another position before the time elapses. The arrow cursor indicates the elapsing time by changing its colour from yellow to red. The hand cursor indicates it by closing the hand until it becomes a fist. The correct time period is crucial for smooth interaction. A large period will make the interaction slow and a short period can cause false execution of actions not intended by the visitor.

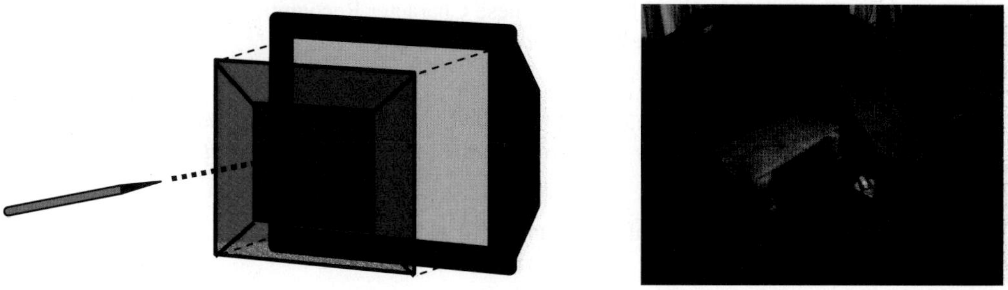

Figure 2: Navigation with the brush

3 Results

The Virtual Tour Guide and all its associated components have been modelled and it is possible to generate facial animations with the appropriate sound files. A selection of Peranakan material culture has been selected and an object library is under work. For the Chinese Calligraphy brush interaction, we developed the online character recognition method that recognizes all the Chinese characters we have defined by detecting the strokes they are composed of. The recognition method is optimised for the defined characters therefore the characters are detected with a high success rate. We also implemented the navigation with the Chinese Calligraphy brush and conducted first in-house user test studies. They showed that the Calligraphy brush increases the edutainment value of the application. Users of Chinese background only have to be told the Chinese characters that the system understands and their functionality. Users of a non-Chinese background need more practice with writing the Chinese characters. The tests also pointed out that the system should display the Chinese characters that can be used as a memory aid. Users who are not familiar with VR systems are able to use the Calligraphy brush to navigate in the 3D environment after observing other users or receiving a short introduction. Some more experienced users prefer to use additional and more powerful interaction devices that offer more degrees of freedom.

4 Conclusions and Future Work

A digital heritage project distinguishes itself by its content. With the help of Virtual Reality technology, we have created a virtual heritage environment on the Peranakans where the visitors can virtually enter and learn new information on the Peranakans and their culture. The main focus of our research work is to provide for interaction techniques that are intuitive and easy to use. For future work, we will work on improving the Virtual Tour Guide by developing more natural facial expressions and animation sets and by making him more "intelligent". To achieve this, we will utilise user profiles, a storyboard in the form of a database that contains prioritised information on all the exhibits, and a learning system that adapts the priorities and presents information that is dependant on who the visitor is and on the visitor's behaviour observed throughout the exhibition. We will also further exploit the possibilities of using the Chinese Calligraphy brush as an interaction device and carry out extensive user test studies on users with different skills, knowledge, and culture to evaluate this interaction method.

5 References

Amin, A., Sum, K. C. (1993) Hand-Printed Chinese Character Recognition. *Inaugural Australian and New Zealand Conference on Intelligent Information Systems*

Calhoun, C., Stahovich, T., Kurtoglu, T. and Kara L. B. (2002) Recognizing multi-stroke symbols. *2002 AAAI Spring Symposium Series – Sketch Understanding, Technical report*

Lutz, B., Weintke, M. (1999) Virtual Dunhuang Art Cave: A Cave within a CAVE. *Proceedings of Eurographics* Milan

DynaGraffiti: Hand-written Annotation System for Interactive and Dynamic Digital Information

Shun'ichi Tano   Daisuke Ogisawa   Mitsuru Iwata   Yusuke Sasaki

Graduate School of Information Systems, University of Electro-Communications
1-5-1 Chofugaoka, Chofu, Tokyo 182-8585, JAPAN
{tano, ogisawa, miwata, yuu}@tlab.is.uec.ac.jp

Abstract

The progress of multimedia and information technology is replacing paper documents by digital documents gradually. We usually annotate on the published materials when we read carefully and understand them. But it is difficult to annotate on the digital documents because of their characteristics that differ from paper documents. In this paper, we propose the handwritten annotation system, called DynaGraffiti, for reading dynamic and interactive digital document and design the architecture to realize it.

1 Introduction

Nowadays the multimedia technology and the network technology enable us freely to access various multimedia information spread in the world. Based on the information technology, many high-tech information systems are developed. However it is gradually well known that such high-tech information systems sometimes do not promote the human's creative thinking but also stop us from thinking [1].

One example is the digital document. According to the rapid progress of information technology, the real papers are being replaced by the digital document. We usually write the handwritten annotation on the published paper when we want to read carefully and try to understand it. Recently we read more the digital information than the ordinary paper-based information. But it is difficult to annotate on the digital documents because of their characteristics that differ from paper documents [2].

In this paper, we analyse the advantage and disadvantage of the digital information and propose the novel handwritten annotation system, called DynaGraffiti, which promotes the advantage and compensates the disadvantage. This work is one of our research activities on the user interface for the intelligent and creative work [1-7]. First we clarify the advantage and disadvantage of the digital information, summarize the requirements for annotation system for digital information readers and design the user interface for the handwritten annotation system, then show the system architecture to realize it.

2 Advantages and Disadvantages of Digital Information

To compare with the ordinary paper-based information, the digital information has two remarkable features. The first is the dynamic nature. It can convey such the dynamic information as the animation, movie, sound, music, speech and so on. The second is the interactive nature. It can respond in accordance with the user's interaction. Due to these two features, the digital information gets popularity and becomes ubiquitous.

On the other hand, the disadvantage is gradually obvious. The most serious disadvantage is that we tend to understand the content of the digital information not deeply but shallowly. Many possible reasons have been discussed. Some research suggests that the information display in multimedia traps us in the experimental cognitive mode and keeps us from the reflective cognitive mode that is indispensable for the creative and intelligent work [1]. We think that the disadvantage comes from the absence of the hand-written annotation system for the digital information.

There are many systems that enable us to write the annotations. The annotation system is a sort of the pen-based system that enables us to write the handwritten drawing on the computer screen. The systems include from the simple drawing software to the annotation system for the multimedia information [8, 9]. None of them cope with the hand-written annotation on the true "Dynamic" and "Interactive" digital information.

We have categorized the problems of current annotation systems and summarize them as two problems.

(i) Must Annotate on Dynamic & Interactive Documents

"Dynamic" and "Interactive" are the most important nature that characterizes the digital document. "Dynamic" means that the digital document contains the moving picture (animation, movie) and sound. "Interactive" means that the digital document accepts the users input and reacts it. Some systems can manage the handwritten annotation that is tagged with sound [8]. But no system can support the handwritten annotation on the "Dynamic" and "Interactive" digital document.

(ii) Must Annotate on Every Digital Document

As pointed out in the above section, some systems can manage the handwritten annotation that is tagged with sound [8] or tagged with movie [9]. But it does not mean that the user can write the annotation at any time and at anywhere on the computer display. The systems work only with the prefixed applications.

3 Requirements for Annotation System for Digital Information Readers

The requirements for the annotation system can be categorized into the following four points.

(1) Annotation on Dynamic Information

The system must enable us to write an annotation on such the dynamic information as the animation, movie, sound and speech as well as the static information. Moreover, since the annotation is given on the dynamic information, the playback function is necessary.

(2) Annotation on Interactive Information

The system must enable us to write an annotation on the interactive information that changes itself in accordance with the user's operation, i.e., pushing buttons and inputting the text. So it is important that the annotation is sometimes treated as the user input to the interactive information.

(3) Simple User Interface

Since the goal of the system is to promote the creative and intelligent work, the user interface must be simple enough not to give us any perceptual load.

(4) Application Independent System Architecture

The system should not be implemented for the specific applications but works with any applications.

4 DynaGraffiti: Hand-written Annotation System for Interactive and Dynamic Digital Information

To meet with the requirements described above, we designed and implemented the prototype system called "DynaGraffiti". To design the user interface, at first, we have made the simple UI principles and then designed the UI to solve two problems summarized in the above section.

4.1 Simple UI Principle

It is very important to design the user interface based on the small number of the operation rules. In DynaGraffiti, everything users have to memorize is *(i) if you write something at the pen point, you can write annotation, (ii) if you write something at the back of pen point, you can delete annotation, and (iii) if you write something while pushing the button of the pen, something will happen.*

The operation *(i)* and *(ii)* are obvious but *(iii)* is not enough to do the intended task. However, when the user fixes the pen at one point while pushing the button, the menu automatically appears. In other words, the user who knows only *(iii)* cannot move the pen but fix at one point, then the menu automatically appears, and the menu teaches how to use DynaGraffiti.

Moreover the gesture commands are simply categorized into "playback" gesture (explained later) in case of writing on something, "command gesture" in case of writing on vacant room, and "menu gesture" in case of staying the pen at one point. Fig 1 shows the summary.

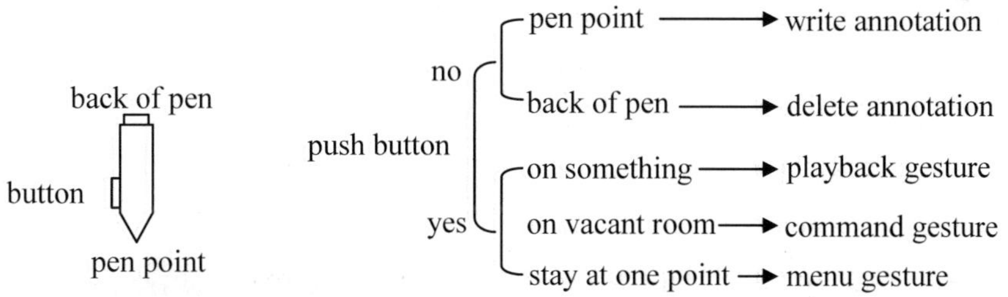

(a) Pen (b) UI Principle

Figure 1: Simple UI Principle

4.2 UI for Dynamic Document

The user can write the annotation even while the movie and sound are being played as shown Fig.2 (a). Note that the sound comes from the speaker and the microphone for the users' voice or the environmental sound. When a user traces on strokes of the annotation, DynaGraffiti plays back the dynamic information that was shown when the specified portion of the annotation was written.

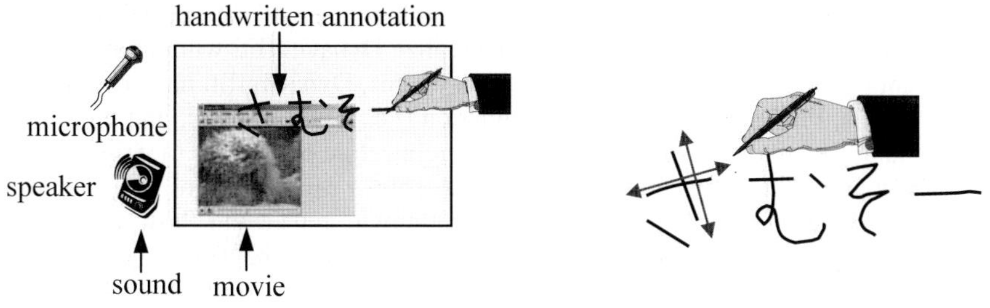

(a) Annotate (b) Playback

Figure 2: UI for Dynamic Document

In other words, the user can look at and hear the past information again, which was displayed at the time of writing the annotation. Changing the tracing speed of the annotation can control the playback speed. It means that the annotation can work as the button of audio-video player. For examples, the user can specify which portion to playback or how fast to playback by trace the written annotation at various speeds as shown Fig.2 (b).

4.3 UI for Interactive Document

The user can interact with the application freely. As summarized in Fig. 1, the drawing while pushing the button of the pen is recognized as the gesture command. There is no difference in writing annotation on the dynamic information or on the interactive information except that some gesture is treated as the mouse operation and some annotation can be used as the input to the interactive information.

The interaction is the mouse operation and the keyboard operation. The mouse click is to draw a check sign "v" as shown in Fig.3 (a). The double click, the right click, and the right double click are similarly assigned to "w", "^", and "M" respectively. The drag is to draw a line that shows the trajectory of the dragging, followed by the click operation ("v").

The keyboard operation is realized by two-step. First the user writes an input string by pen, and then moves it to the input place. The system automatically recognizes handwritten string and input it to the place. In Fig. 3 (b), the user wrote an Internet address by pen and moved it to the address input area of the Internet Explorer. The handwritten address was successfully recognized and pasted.

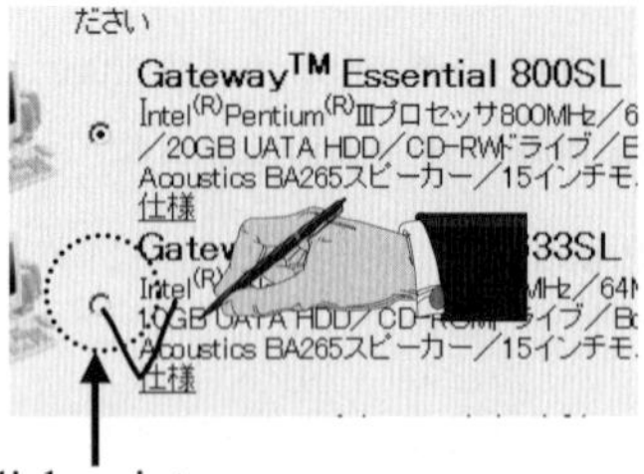

click point

(a) Mouse click

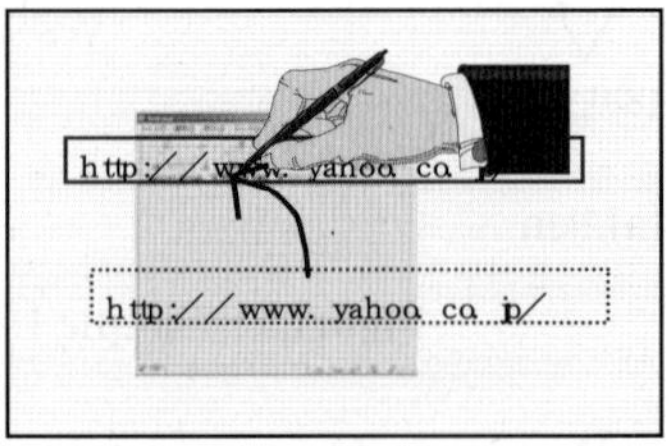

(b) Keyboard operation

Figure 3: UI for Interactive Document

5 Implementation

To meet with the requirement (4), The DynaGraffiti is designed as the special application that mediates the ordinary applications and the user. To realize DynaGraffiti without any modification on the current applications and the current operating systems, DynaGraffiti has the transparent full screen output area and always intercepts the interaction (i.e. display, sound, and mouse) between the application and the user as shown Fig.4 (a). This architecture enables us to write annotation on any application window that plays movie and sound as shown Fig. 4 (b). For example, a user can write annotation on the Internet Explore that plays the movie, music and flash, or the e-learning applications that show the streaming video in which the professor in the distance gives the lecture.

6 Summary

In this paper, first, we pointed out that the dynamic nature and interactive nature characterizes the digital information, and then analysed the advantage and disadvantage of the digital information. Second, we showed that the annotation system for the dynamic and interactive information is indispensable to compensate the disadvantage. Finally we summarized the requirements to the annotation system and showed our DynaGraffiti. At this point, the prototype has been implemented and we are now evaluating by using the multimedia educational contents.

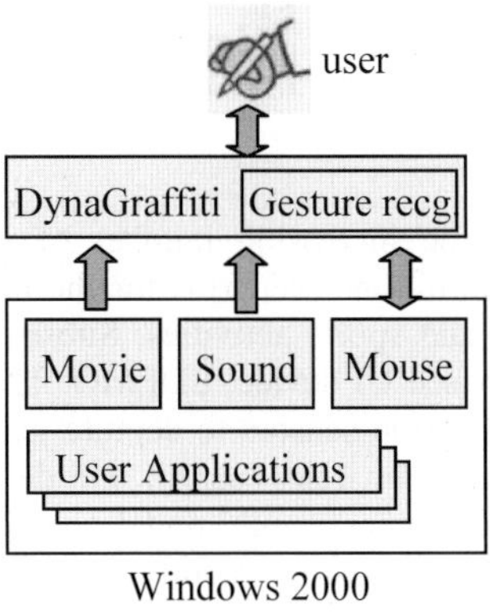

(a) Architecture

(b) Photo

Figure 4: System Architecture and Photo

References

[1] S. Tano: Design Concept to Promote Both Reflective and Experimental Cognition for Creative Design Work, CHI-2003 Extended Abstract, 2003.

[2] D. Ogisawa, M. Iwata, S. Tano: Design and Evaluation of Annotation System on Dynamic and Interactive Digital Information, Technical Report of IEICE, HIP2000-55, pp. 1-8, 2001. (Japanese).

[3] S. Tano, M. Tsukiyama: User Adaptation of the Pen-based User Interface by Reinforcement Learning, HCI-99, pp. 233-237, 1999.

[4] Tano, Sugimoto: Natural Hand Writing in Unstable 3D space with Artificial Surface, CHI-2001 Extended Abstracts, pp. 353-354, 2001.

[5] Tano, Watahiki: User Interface with Seamless Cooperation of Multi-dimensional Media and Application to Design Work, IEEE International Conference on Multimedia and Expo (ICME-2001), pp. 485-488, 2001.

[6] S. Tano, S. Inoue: Overview of Cross-tablet User Interaction for Littery Augmented Paper - LAP - Environment, UIST-2002 companion, pp. 57-58, 2002.

[7] S. Tano, S. Inoue: Cross-tablet User Interaction for Lavishly Augmented Paper - LAP - Environment, PCHI-2002, pp.548-559, 2002.

[8] Lynn D. Wilcox, Bill N. Schilit, Nitin Sawhney: DYNOMITE: A Dynamically Organized Ink And Audio NoteBook, CHI97, pp.186-193, 1997.

[9] Chellury R. Sastry, Darrin P. Lewis, Arturo Pizano: WEBTOUR: A System to Record and Playback Dynamic Multimedia Annotations on WEB Document Content, ACM international conference on Multimedia, pp.175-178, 1999.

EDEMO-Gesture Based Interaction with Future Environments

Jarno Vehmas, Sanna Kallio, Juha Kela, Johan Plomp, Esa Tuulari, Heikki Ailisto

VTT Technical Research Centre of Finland
Kaitoväylä 1, P.O. Box 1100, FIN-90571 Oulu, FINLAND
firstname.lastname@vtt.fi

Abstract

This paper studies the use of gestures for controlling the increasing number of visible and embedded electrical devices in our environment. First, we outline scenarios for the use of small wearable devices for the care of the elderly, home-care and sports activities. Subsequently, we focus on the design and modeling of a wearable gesture based UI device, EDEMO. The EDEMO device is worn on the back of the hand and it uses acceleration sensors to capture gestures. The hardware implementation is based on VTT's SoapBox, a small wireless sensor unit developed for research of ubiquitous computing applications, context awareness, multimodal and remote user interfaces, and low power radio protocols. The use of gesture based interaction with a maze game and for the control of a slide presentation is shown. The results and experiences support our belief that multimodal wearable user interfaces are very useful in pervasive computing environments.

1 Introduction

Common human to human interaction includes a variety of spontaneous and subliminal gestures such as finger, hand, body and head movements, not forgetting facial expressions. However, most of our interaction with computers and devices is carried out with traditional keyboards, mice and remote controls designed mainly for stationary interaction. PDA's, mobile phones and wearable computers provide new possibilities to interact with various applications while moving, but introduce new problems with small displays and minimised input devices. Small wireless gesture based input devices integrated in clothing, wristwatches, jewellery or even mobile terminals could provide intuitive methods to interact with different kinds of devices and environments. These personal interfaces could be used to control e.g. home appliances with simple user definable hand movements. Simple up and down hand movements could be used to control a garage door, the volume of your stereo equipment or your room's lights depending on the user's location or orientation. However, since gestures are unfamiliar for this purpose, research is needed to identify the gestures that people find natural and easy to remember for certain control tasks.

Existing research on gesture recognition can be classified into two main categories; camera-based and sensor-based. The camera-based approach is more suitable for stationary applications, which often require a specific camera setup, lighting conditions and calibration. Recent research demonstrates 3D mice (Hinckley, Sinclair, Hanson, Szeliski & Conway, 1999), interactive movies (Segen & Kumar, 1998) and home appliance control (Fails & Olsen, 2002). In Georgia Tech's Aware Home project researchers developed a camera-based control device called Gesture Pendant (Starner, Auxier, Ashbrook & Gandy, 2000) allowing the wearer to control elements in the home via hand gestures supported by contextual cues such as location, orientation and speech. Sensor-based techniques utilise e.g. tilt, acceleration, pressure, conductivity, or capacitance sensors, to measure gestural movement. An example is GestureWrist, a wristwatch-type hand gesture recognition device using both capacitance and acceleration sensors to detect forearm and hand gestures (Rekimoto, 2001).

Since camera-based gesture recognition has limited support for wearability and mobility, our work focuses on sensor-based solutions. In this paper we introduce a prototype of a wireless gesture based control device called EDEMO, which is attached at the back of the hand by means of an elastic band. In this paper, we survey EDEMO's possibilities to improve the usability of electronic devices in ordinary life.

2 The EDEMO Idea and Use Scenarios

The EDEMO personal user interface is a small portable device that aids making the interaction with electronic devices and services more natural. It captures gestures and other movements and translates them into meaningful commands and observations needed for diagnostic purposes. As such, it can replace or compliment currently prevailing interaction modes with a more natural one.

In order to survey the possibilities of our envisioned EDEMO interface, we attempted to find real-life scenarios that would make use of the capabilities of the interface. After initial idea generation and brainstorming sessions, a small set of use scenarios was chosen for further study; *Personal safety system*, *Personal interface for home care, gaming* and *presentation control*. A use scenario was made for each and the last two scenarios were implemented.

Personal safety system interface: The EDEMO personal user interface is based on gestures, which are detected by the sensors contained in a SoapBox (see chapter 4). The system can be configured to continuously observe our motions. We can utilise this to create a *personal safety system* for example to increase the supervision of elderly people. In case of an emergency, the system detects an unusual situation through monitoring the sensor readings and turns automatically to emergency mode. The system can additionally make a diagnosis of the user's health by means of biosensors (if available), or it simply compares the data of the movements before and after the accident. In emergency mode, the EDEMO personal safety system interface tries to get feedback from the user and if needed calls for help. Potential users of this concept include elderly people, handicapped and maintenance personnel working alone. Also leisure activities can benefit from the EDEMO personal interface device. In general, gestures may provide a very user-friendly interface when the attention of the user is occupied mainly by performing another task simultaneously.

The EDEMO home care system is a conceptual study of the possibilities to use gesture based user interfaces in a household. With a single portable device we are able to make simple measurements of our health. By collecting the data of our own constitution or our children's and comparing that with previous analyses, we are able to make simple diagnoses. For example by analysing the sounds of several coughs the software might be able to tell us if the patient is getting better or worse. A single movement of a hand may be used to initiate the health measurements. The user then touches the patient with the measurement unit and data is stored automatically. Information on the diagnosis would be observed from a small display or a screen nearby.

Gaming and presentation control: The EDEMO device can be used for the control of games, home appliances, car equipment, etc. As such, it can either replace the usual means of control, or it may provide a complementary modality. Specific examples of such use are the control of a maze-game, where tilting the device causes the user's dot to move in a certain direction, and slide presentation control, where previous and next slides are selected through dedicated gestures.

Instead of limiting the use of the EDEMO device to one specific application, it can also be used as a general means of control. The advantage of this is that one needs to learn only one personal interface. An interpreter would take care of translating our gestures into commands for the controlled device. Thus, a single command turn on/off can be used for all of our devices.

3 Design and Modelling the EDEMO Personal User Interface

The aim of the design process was making the EDEMO personal user interface as portable and imperceptible as a watch or a ring. Concept creation began with a brainstorming session, where even the wildest ideas were accepted for the following evaluation rounds. Finally the team selected several acceptable concept ideas, which were illustrated by means of multimedia scenarios. By analysing these six different user interface cases we created the final one – the EDEMO gesture based user interface.

The next task was to determine the right place to wear the EDEMO user interface and to make the design for the shape. For this step the physical size of the SoapBox was the first dilemma. Although the SoapBox is quite a small device (43x31x12 mm), it still puts constraints on its use. When making gestures, hand and especially fingers play a major role. The use of fingers is not feasible, because of the physical dimensions of the SoapBox. The team considered other possible locations for the device and concluded that the most common and natural way to make gestures with a small device is to attach it to the back of the hand. This maximises the readout of movements of a human body, but is still quite unobtrusive and natural.

After some tests the team concluded that attaching the device on the back of a hand could also provide support for other kinds of interactions, for example for detecting movements of fingers (with additional sensors). Tests also indicated that people experienced the EDEMO personal user interface as a natural means for different kinds of gesture based control tasks.

4 Implementation Using SoapBox and HMM for Gesture Recognition

The functional part of the EDEMO UI is based on a SoapBox *(Sensing, Operating and Activating Peripheral Box),* which is a sensor device developed for research activities in ubiquitous computing, context awareness, multi-modal and remote user interfaces, and low power radio protocols (Tuulari & Ylisaukko-oja, 2002) (see figure 1). It is a light, matchbox-sized device with a processor, a versatile set of sensors, and wireless and wired data communications. Because of its small size, wireless communication and battery-powered operation, SoapBox is easy to install into different places, including moving objects. The basic sensor board of SoapBox includes a three axis acceleration sensor, an illumination sensor, a magnetic sensor, an optical proximity sensor and an optional temperature sensor. Sensor information can be utilised in many ways in future intelligent environments and in new mobile user interfaces, both as a source for context information and as an input modality. The SoapBox can be used for example in applications of location information, observation of the user's motion or environmental conditions.

Acceleration sensors can be used to detect dynamic hand gestures. Acceleration-based gesture recognition can use 3D data directly providing more naturalness in generating gestures compared to other approaches. It is used for example in a musical performance control and conducting system (Sawada & Hashimoto, 2000), and a glove-based system for recognition of a subset of German sign language (Hoffman, Heyer & Hommel, 1997). The most used technique for recognising dynamic hand gestures is Discrete Hidden Markov Models (HMM). HMM is a statistical signal modelling technique that can be applied to the analysis of time-series with properties that change over time. HMM is widely used in speech and hand-written character recognition as well as in gesture recognition in video-based and glove-based systems.

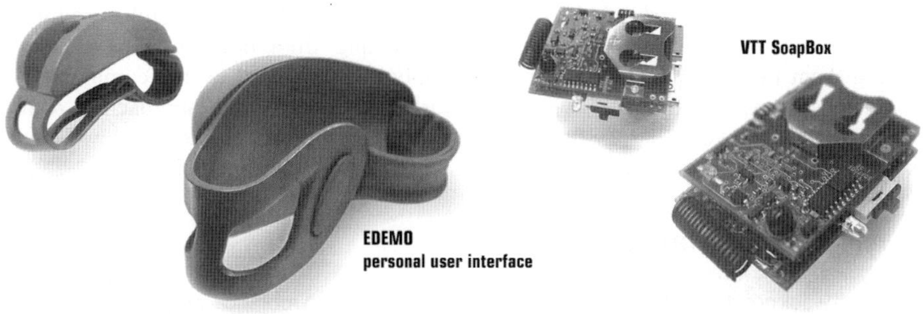

Figure 1: The EDEMO personal user interface (left) and VTT SoapBox sensor device

5 Applications and Experiments

Acceleration sensors are very promising for application in EDEMO-like and future mobile user interfaces. Acceleration sensors measure both dynamic acceleration (e.g. motion of the box) and static acceleration (e.g. tilt of the box). They provide useful information for the recognition of human hand gestures and a new way to add human motion to the human/machine interface. The acceleration sensor capabilities to sense static acceleration can be used to make tilting actuated input devices (Bartlett, 2000). Instead of pressing a button to scroll up or down on the screen, the user can simply tilt the device, enabling more natural one-handed operation.

We have realised two EDEMO applications based on the acceleration sensors of the SoapBox. The maze game is a multi-user game with gestural input. Players navigate trough the maze by tilting a SoapBox (in the EDEMO), causing the "ball" to move. Direction and speed of the ball is calculated from the tilt angle. The gesture based user interface and wireless communication make the playing situation less obtrusive. Players can control the action with natural movements and simultaneously move about freely without limiting cables.

We have also demonstrated natural slide show control, where the most common keyboard commands are replaced by simple dynamic hand gestures. A person can, for example, present a slide show by replacing page up and page down commands with hand gestures and thus avoid walking back and forth on the stage. EDEMO can also be used as a mouse by employing the same tilting interface as for the maze demo and simulating mouse clicks by a hand gesture.

6 Discussion

As our environment gets increasingly enhanced with interconnected devices and embedded services, the challenge to provide a feasible manner of interaction gets ever greater. In order to provide unobtrusive and natural ways of interaction new modalities need to be employed in addition to crucial efforts to personalise and adapt the interaction to the user. In this paper we surveyed the potential of sensor-based gesture interfaces to enhance interaction with such future environments. Several scenarios were presented that showed the potential of the interface not only for control, but also for monitoring. The implementation of the interface involved the design of a novel gesture recognition tool (EDEMO) attached to the back of the hand. The tool also serves as an example of rapid development utilising the SoapBox platform designed in previous projects.

In spite of our promising results, it is unlikely that gestures will be used to replace other means of interaction. Instead, gestures should be seen as an additional modality that can be used in suitable situations or combined with other modalities. Switching from one modality to another, or combining them in a multimodal fashion is currently not supported by the common UI design methodologies. Consequently, we have started work on a UIML-based approach to resolve this problem (Plomp & Mayora-Ibarra, 2002). We also intend to add gestures to this new UI metaphor and study the enhancements needed for a proper support of true multimodality.

7 Acknowledgements

We gratefully acknowledge the support of our partners in the Ambience project and Tekes for the funding. The scenario definition and design work was done in co-operation with the University of Lapland Department of Industrial Design, in particular by Anna Rounaja and Áku Koivukangas.

References

Barlett, J. F. (2000). "Rock 'n' Scroll Is Here to Stay". *IEEE Computer Graphics and Applications,* 20 (3), 40-45.

Fails, J., & Olsen, D. (2002). Light widgets: interacting in every-day spaces. *Proceedings of the 7th international conference on Intelligent user interfaces,* San Francisco, USA, 63-69.

Hinckley, K., Sinclair, M., Hanson, E., Szeliski, R., & Conway, M. (1998). The VideoMouse: a camera-based multi-degree-of-freedom input device. *Proceedings of the 12th annual ACM symposium on User interface software and technology,* Asheville, USA, 103-112.

Hoffman, F., Heyer, P., & Hommel, G. (1997). "Velocity Profile Based Recognition of Dynamic Gestures with Discrete Hidden Markov Models". *Proceedings of Gesture Workshop '97,* Springer Verlag.

Plomp, C. J., & Mayora-Ibarra, O. (2002). A generic widget vocabulary for the generation of graphical and speech-driven user interfaces. *International Journal of Speech Technology 5,* January 2002, Kluwer Academic Publishers, 39-47.

Rekimoto, J. (2001). GestureWrist and GesturePad: Unobtrusive Wearable Interaction Devices. *Proceedings of the Fifth International Symposium on Wearable Computers, ISWC 2001.*

Sawada, H., & Hashimoto, S. (1997). "Gesture Recognition Using an Accelerometer Sensor and Its Application to Musical Performance Control". *Electronics and Communications in Japan Part 3,* 9-17.

Segen, J., & Kumar, S. (1998). Video-based gesture interface to interactive movies. *Proceedings of the sixth ACM international conference on Multimedia,* Bristol, UK, 39-42.

Starner, T., Auxier, J., Ashbrook, D., & Gandy, M. (2000). The gesture pendant: A self-illuminating, wearable, infrared computer vision system for home automation control and medical monitoring. *Proceedings of the Fourth International Symposium on Wearable Computers, ISWC 2000.*

Tuulari, E., & Ylisaukko-oja, A. (2002). "SoapBox: A Platform for Ubiquitous Computing Research and Applications" *First International Conference, Pervasive 02,* Zurich, Switzerland.

Malleable Paper: A User Interface for Reading and Browsing

Jian Wang

Microsoft Research Asia
No. 49 Zhichun Road
Haidian District
Beijing 100080, P. R. China
jianw@microsoft.com

Chengao Jiang

Institute of Software, CAS
No. 4 South 4th street,,
Zhongguancun
Beijing 100080, P. R. China
chinajcg@iel_mail.iscas.ac.cn

Abstract

Reading and browsing of digital documents on screen are becoming increasingly important for users of PC, particularly the Tablet PC. Malleable paper is a user interface designed for reading and browsing of electronic documents. It has two distinctive features: 1) distorted visualization of documents using a paper metaphor; 2) users' tangible interaction with the documents. This interface supports: 1) display of both the reading focus and the context; 2) multi-foci. An empirical evaluation was conducted to compare the performance of malleable paper interface with traditional windowing system. The results suggested that malleable paper interface allows users to read faster and is preferred by users.

1 Introduction

The amount of electronic information is increasing dramatically. People spend more and more time reading and browsing web pages, e-mails, e-books, electronic newspapers and research papers. With recent advancement of wireless technologies, people are getting on-line electronic documents anywhere at any time. However, reading and browsing electronic documents on electronic devices face many challenges from traditional media (Johnson, Jellinek, Klotz, Rao & Card, 1993; O'Hara & Sellen, 1997). Two problems are relevant to the current study: 1) how to make reading easy and fast with appropriate visualization technologies; and 2) how to provide a good reading experience on electronic devices as printed paper does.

Various information visualization techniques (Carpendale, Cowperthwaite & Fracchia, 1995; Carpendale, Tigges, Cowperthwaite & Fracchia, 1998; Furnas, 1986; Lamping, Rao & Pirolli, 1995) have been proposed to help users manage and access large amount of information. Information visualization helps users overcome many difficulties in reading and browsing large-scale information such as navigation and comprehension.

"Overview + detail" and "focus + context" are two effective ways to visualize information. Various techniques, such as cone tree, hyperbolic tree, document lens and 3D pliable surface (Robertson, Mackinlay & Card, 1991; Lamping, Rao & Pirolli, 1995; Carpendale, Tigges, Cowperthwaite & Fracchia, 1998) have been proposed. With the "overview + detail" strategy, the detailed content and the global context are displayed in two separate views. One view contains an overview of the entire document and the other shows details that the user is currently focusing on. With the "focus + context" strategy, details are integrated into the overview of the entire document and displayed in the same view. Recent investigations showed that "overview + detail" can help users read electronic documents (Hornbæk & Frøkjær, 2001).

A natural reading experience is the other problem that needs to be solved (Schilit, Golovchinsky & Price, 1998; Schilit, Price, Golovchinsky, Tanaka & Marshall, 1999). Active reading and a paper-like interface are some of the techniques proposed. Active reading treats annotating, highlighting and scribbling as part of the reading experience. Tangibility is one of the important features of paper-like interfaces and has been shown to enhance navigation (Ishii & Ullmer, 1995; Price, 1998). However, interaction with visualized document has been relatively ignored.

The purpose of this research is to propose a novel reading and browsing user interface to improve the reading experience. The following sections describe the design of a malleable paper interface and an empirical evaluation of the interface.

2 Malleable Paper

Malleable paper interface takes the approach of distorted "focus + context" view to visualize the document. All the information is displayed on a 3D pliable thin surface. The surface has three components: stretched, folded and stacked (see Figure 1a).

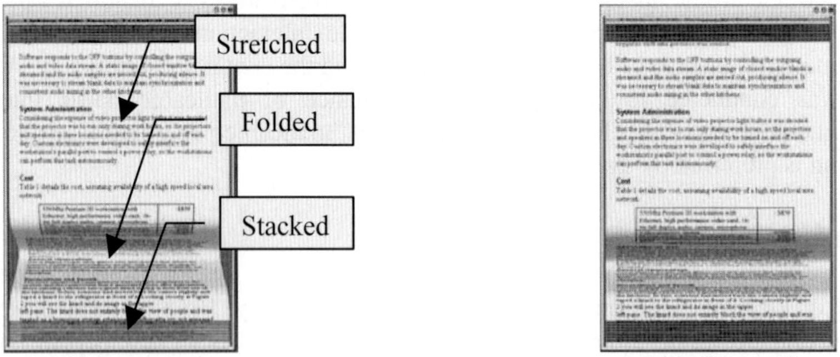

a. Perspective b. Orthogonal

Figure 1: Malleable Paper User Interface

The stretched part is the fully extended part of the malleable electronic paper. It is the focus of the document and has all the detailed information. The folded part is the visually-compressed and partially-readable part of the document. The folded part can be extended fully to be a stretched part or compacted to be stacked. The stacked part has visual cues on how many pages there are before and/or after the current focus. It looks like a pile of "paper." Both the stacked and the folded part provide users with rich context information whereas the stretched part gives users the detailed content. During normal reading, the malleable electronic paper can be fully stretched if users want to see more details. Users can also turn the folded part directly into the stacked part if they want to skip the details. An algorithm is designed to visually simulate the smooth process of stretching, folding and stacking during the interaction.

In addition to the perspective view shown in Figure 1a where both perspective and shading information provide 3D cues for visualization, malleable paper interface can also be displayed in an orthogonal view (see Figure 1b) with shading cues only. The side view as shown in Figure 2a shows how malleable paper interface is stretched, folded and stacked in a 3D space.

No interface gadgets, such as scroll bars or buttons, are used for document interaction. Instead, dragging and press & holding are the two interaction methods for the malleable paper interface.

With dragging, users can interact continuously and directly with the visualized document anywhere as if the document was printed on a piece of flexible paper.

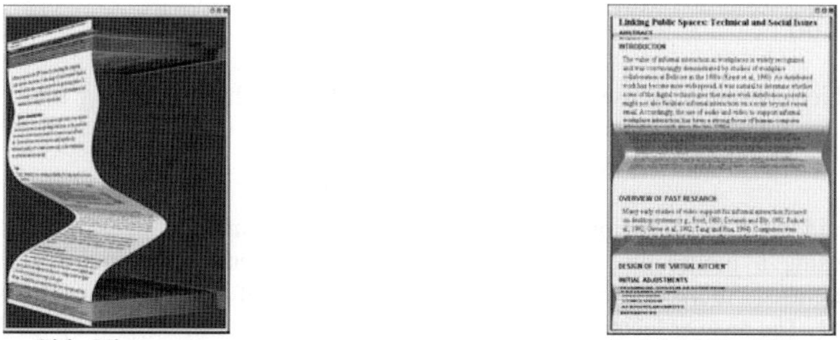

a. Side View b. Multi-Foci View

Figure 2: Multi-Foci and Side View of Malleable Paper

Structured multi-foci malleable paper (see Figure 2b) is designed for documents with hierarchical structures. Structured multi-foci malleable paper can stretch, fold and stack according to the hierarchical structure of the document. With structured malleable paper, lower levels are folded and stacked whereas higher levels are stretched at first. Structured malleable paper can also support multi-foci since users can manually stretch any part of the document at any level.

3 Experiment

3.1 The Tested Interfaces

The two interfaces tested are shown in Figure 3a and 3b. In the windowing interface (as shown in Figure 3a), an "overview + detail" interface recommended in a recent study (Hornbæk & Frøkjær, 2001) was used. The detailed content was shown in the right scrolling pane whereas the hierarchical structure or the overview of the document was shown as a tree view in the left pane. The interface was displayed in full screen.

a. "Overview + Detail" Windowing Interface b. Multi-Foci Malleable Paper Interface

Figure 3: The Tested Interfaces

In the malleable paper interface, the document was displayed in a single view. The size of the view was equal to the size of the right scrolling pane in the "overview + detail" interface. In other words, the malleable paper interface actually took less screen space. The interface was displayed in full screen mode with grey background.

3.2 Method

A within-subject design was used. All the subjects used both the windowing interface and the malleable paper interface. The order in which the subjects used the two interfaces was counter-balanced to minimize carry-over effects.

Thirty-six paid subjects participated in the study. All were undergraduate students and had experience using a mouse and word processing software. Tasks imitating web reading and browsing were used. Subjects were shown various documents and asked to answer questions about the documents. The answer location can be either near (in page 1 or 2) or far (in page 3 or 4) from the beginning of the document. A total of 20 documents were used in the study, 10 for each interface. All the documents were in Chinese, taken from publicly accessible web sites. The documents had similar (but not exactly the same) content and hierarchical structures.

The experiment consisted of two sessions, one for each interface. Each session lasted for approximately an hour. There was a 20 minute break between the sessions. A typical session consisted of training, practicing and test. During training, subjects learned how to use an interface by reading instructions and asking questions. During practicing, subjects performed five trial tasks until they were successful. During test, subjects completed 10 tasks. For each task, subjects read the question from a booklet and pressed the space key to begin. When subjects found an answer, they were instructed to click at the answer position to confirm. If the answer was incorrect, the computer played a warning sound to ask the subjects to correct. The task was completed when the correct answer was found. The task completion time was recorded in a log file along with all the interactions the subjects had with the interface. At the end of each session, subjects were asked to fill out a questionnaire to provide subjective opinion.

3.3 Results and Discussion

Repeated ANOVA with interface type and answer location as independent variables showed that both were significant ($F(1,35)=99.6$, $p<0.01$; $F(1,35)=7.87$, $p<0.01$). The mean task completion time was 41s for the windowing interface and 35s for the paper interface. Interaction between the two was also significant ($F(1,35)=4.27$, $p<0.05$). A post hoc test showed that task completion time was only significantly different when the answer was near (first two pages) the beginning of the document (32s vs. 22s, $t=3.899$, $p<0.01$). Figure 4 shows the interaction effect.

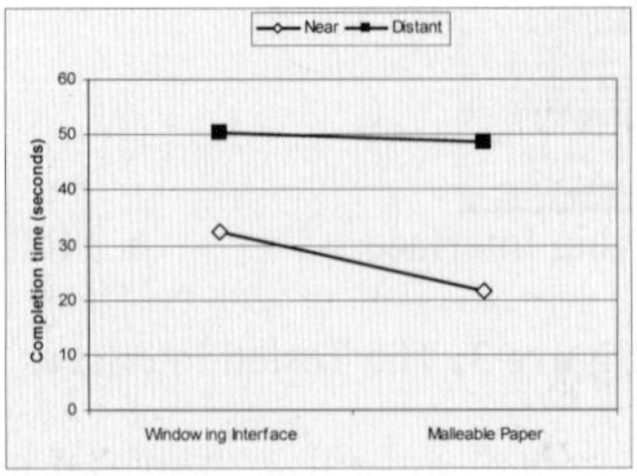

Figure 4: Interaction of Interface Type and Answer Location

The results showed that the malleable paper interface had faster reading and browsing performance when the reading range was within two pages. To understand the interaction effect, log data were analyzed. The analysis showed that it took 10 more seconds to expand the document in the paper interface when the answer was far. This may explain why the speed advantage of the paper interface disappeared when the answer was far from the beginning of the document.

Subjective preference was measured using a 7-point scale, with 1 being "least liked" and 7 being "most liked." The mean was 4.9 for the windowing interface and 5.4 for the malleable paper interface. Paired t-tests showed that the difference was significant ($t=2.14$, $P<0.05$). Subjects had preference of malleable paper interface over windowing interface.

4 Conclusion

This paper describes a novel user interface, malleable electronic paper interface, for reading and browsing of digital documents. Malleable paper interface uses the paper metaphor to provide multi-foci, "context + detail" visualization and tangible interaction. The interaction technique of dragging and press & holding may potentially make the malleable paper interface work better with a pen and a jog dial. An empirical evaluation was also conducted to examine the performance and users' subjective opinion of the interface. The results showed that malleable paper interface has faster reading and browsing performance and is preferred by users.

Reference

Carpendale, M.S.T., Cowperthwaite, D.J., & Fracchia, F. D. (1995). 3D pliable surfaces: For the effective presentation of visual information. In *Proceedings of the ACM Symposium on User Interface Software and Technology*, 217-226.

Carpendale, M.S.T., Tigges, M., Cowperthwaite, D.J., & Fracchia, F.D. (1998). Bringing the advantages of 3d distortion viewing into focus. In *Proceedings of the IEEE Symposium on Information Visualization, IEEE Visualization.*

Furnas, G. W. (1986). Generalized fisheye views. In *Proceedings of CHI'86*, ACM Press, 16-23.

Hornbæk, K., & Frøkjær, E. (2001). Reading of Electronic Documents: The Usability of Linear, Fisheye, and Overview + Detail Interfaces. In Proceedings of CHI'2001 ACM Press, 293-300.

Ishii, H., & Ullmer, B. (1995). Tangible bits: towards seamless interface between people, bits and atoms. In *Proceedings of CHI'95*, ACM Press, 234-241.

Johnson, W., Jellinek, H., Klotz, L., Rao, R., & Card, S. (1993) Bridging the paper and electronic worlds: The paper user interface. In Proceedings of INTERCHI '93, 507-512.

Lamping, J., Rao, R., & Pirolli, P. (1995). A focus + context technique based on hyperbolic geometry for visualizing large hierarchies. In *Proceedings of CHI'95*, ACM Press, 401-408.

O'Hara, K., & Sellen, A. (1997). A Comparison of reading paper and on-line documents. In *Proceedings of CHI'97*, ACM Press, 335-342.

Price, M.N. (1998). Linking by Inking: Trailblazing in a Paper-like Hypertext. In *Proceedings of Hypertext'98*, ACM Press, 30-39.

Robertson, G., Mackinlay, J.D., & Card, S.K. (1991). Cone trees: Animated 3D visualizations of hierarchical information. In *Proceedings of CHI'91*, 189-194.

Schilit, B.N., Golovchinsky,G., & Price, M.N. (1998). Beyond Paper: Supporting Active Reading with Free Form Digital Ink Annotations. In *Proceedings of CHI98*, 249-256.

Schilit, B.N., Price, M.N.,Golovchinsky,G., Tanaka, K., & Marshall, C.C. (1999). As we may read: the reading appliance revolution. In *IEEE Computer 32*, 65-73.

FreeDraw: a Drawing System with Multimodal User Interface[1]

Yue Wang Weining Yue Jizhi Tan Heng Wang Shihai Dong

HCI & Multimedia Lab, Department of Computer Science, Peking University
Beijing 100871, PR. China
wangyue, ywn, tjz, hengwang@graphics.pku.edu.cn
dong@pku.edu.cn

Abstract

In this paper, we present the architecture of FreeDraw, a drawing system with multimodal user interface based on speech. Then the multimodal task design of FreeDraw has been introduced. The experimental results show that using speech modality can improve the performance and intuition of the traditional WIMP, and users are prefer the multimodal interface to traditional mouse and keyboard interface.

1 Introduction

How to design an efficient and natural user interface has become more and more crucial in HCI. Unfortunately, the traditional mono-modality, the bottleneck of interaction efficiency, limits the computer's popularization. To overcome it, multimodal interface (MMI) has become the choice of the efficient, intuitive and mainstream user interface, since multimodal is a normal interaction mode for human-human communication.

Multimodal interface (MMI)[Shihai, Jian & Guozhong, 1999] is a kind of interaction technique based on visual line tracking, speech recognition, hand gesture input, feeling feedback and so on. Users can interact with computers freely by interaction modalities including hand gesture, speech, eye expression and so forth. As the most natural one, speech is a very important interaction modality. It can enhance the usability of the multimodal interface greatly because of the complementary strengths of the traditional graphics user interface (GUI) and speech. Multimodal interfaces are expected to support a wider range of diverse applications, to be usable by a broader spectrum of the average population, and to function more reliably under realistic and challenging usage conditions[Oviatt et al., 2000].

This paper presents a drawing system we have developed named *FreeDraw*. The system has multimodal interface, in which the speech modality plays a very significant role. All the drawing tasks in FreeDraw can be achieved by the integration of the keyboard, mouse and limited speech. Users can combine these three modalities freely as they want. In this paper, we first introduce the architecture of the system and describe the multimodal information processing method in detail.

[1] Grant No. 60033020 from National Natural Science Foundation of China and Microsoft Research Asia supported this research.
Grant No. 2001AA114170 from The National High Technology Research and Development Program of China (863 Program) also supported this research.

Then we give the experimental results of the performance studies, which show that the execution time of most tasks is greatly reduced by the addition of speech modality and users are satisfied with the multimodal interface on a whole.

2 Related Works

Compared to other modalities, speech modality has many remarkable merits. For instance, it frees the hands and is convenient for user. In a multimodal system speech and direct manipulation can compensate for one another[Cohen, 1992] and the addition of natural language can improve the interaction efficiency[Mignot, Claude & Noelle, 1993]. In a experiment designed by Hauptmann[Hauptmann, 1989], more users were likely to combine the speech and hand gesture inpute togather. Nishimoto[Nishimoto, Shida, Kobayashi & Shirai, 1994] verified the benefits of speech input using a multimodal drawing tool. As the results, the multimodal drawing tool with speeh recognition reduced average opration time, the number of command inputs and movement of mouse pointer. The results showed that the system with speech input is beneficial especially on the easiness of learning and reduction of difficulty and drawing task is suitable for the application of speech input.

Many researches have shown speech modality a promising interactive mode, yet few softwares have a speech interface. Two factors weigh heavily against the effective application of speech modality. One is the lack of a mature arithmetic that can integrate speech and other modalitied, the other the low recognition rate and the difficulty of semantics analyse. In our system, the speech modality has been applied well. We make use of limited speech instead of natural language to overcome the low recognition rate; and also use mouse to compensate for speech to solve the recognition error.

3 System Description

3.1 System Interactive Mode

FreeDraw is a drawing system with a multimodal user interface based on speech, mouse and keyboard. Users may utilize any one or combine two or three modalities freely according to their habits to complete a drawing task.

For example, if a user wish to draw a circle, he (or she) may either draw with a mouse like using other drawing tools or just say "draw a circle", or point with mouse while saying "draw a circle". In the second case, system automatically draws a circle. At the third case, system knows what the user wants to draw by processing speech, the user has said while drawing, and don't need to select the *draw circle menu* before drawing. Also, one can say "change it to red", and at this time system will decide which object to select according to the mouse position.

3.2 System Structure

The FreeDraw system consists of three modules. The first is interactive module, which receives and processes user's input. The second is multimodal information process module translating the user inputs into input primitives (IP) and integrating them to generate the drawing tasks. Finally, the drawing module implements the tasks. See Figure 1.

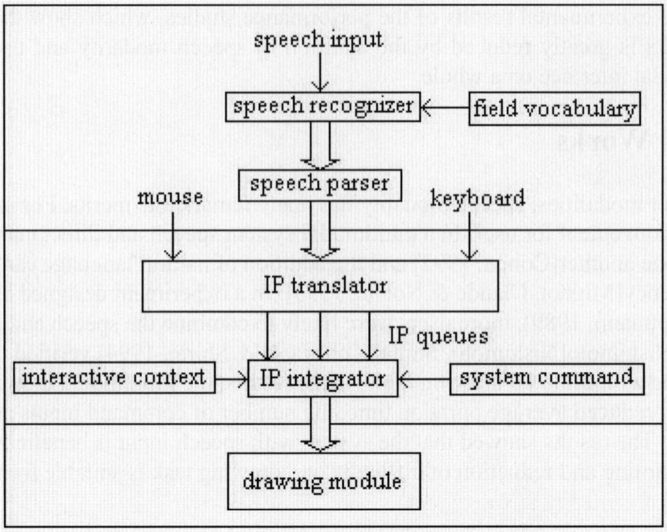

Figure 1: The architecture of FreeDraw

3.2.1 Interactive module

It receives information coming from speech, mouse and keyboard modalities and pre-processes the speech input: recognition and words parsing. FreeDraw only recognizes the speech relative to system command and neglects others. It has been proved that[Bin, Yuquan & Shihai, 2001] it is feasible to limit the vocabulary of natural language in some field, and could improve the efficiency of user interaction.

Since users may use quite different expressions to describe the same task, we established a vocabulary table in order to translate various synonymous words into a standard one. Then the speech information is submitted to the multimodal information processing module along with the input from the mouse and the keyboard.

3.2.2 Multimodal information process module

Information from three modalities is transformed by multimodal information process module into uniform Input Primitives (IP) stored in three IP queues separately. Then the module integrates IPs in three queues into different task based on their time, content and queue number. Finally, send the task to drawing module.

Speech parsing process is related to the requirement of the system. The parser separates words or phrases which are useful to the interaction. For example, if user says "please draw a red line", in this sentence, "draw", "a", "red", "line" are useful. The parser must divide the sentence into words and phrases correctly, then hold the useful words or phrase, and discard others.

3.2.3 Drawing module

This module accepts and completes the integrated task from multimodal information process module and shows the results to the users by screen.

4 Multimodal Information Processing

The multimodal information processing module receives the input information from the three modalities and turns them into an uniform structure -- Input Primitives (IP), as shown in Figure2. Then the IPs, ordered by their time labels, are distributed into three primitive queues according to their source modalities.

User's action	Data information	Modality number	Time label

Figure 2: Structure of the Input Primitive in FreeDraw

The IP integrator processes the primitives in the queues with FIFO order and generates the system tasks with the integrate arithmetic.

FreeDraw has 51 executable tasks, and 30 of them can be implemented with input combined from three modalities, that means users can select and combine speech, mouse and keyboard input freely to complete the tasks. While performing the other 21 tasks, users can use either speech or mouse, and in this case, the mouse modality and the speech modality are interchangeable.

Universal task slot is used to describe the executable tasks, which includes task name, allowable modalities and a list of parameters, as shown in Figure 3.

Task name	Allowable modalities	Is param1 valid	Param1	Is param2 valid	Param2	…

Figure 3: Task slot in FreeDraw

The number of parameter in every parameter table is fixed. Parameter numbers of every task are variable because of the flexible input modalities and their parameters. Some tasks permit various parameter numbers, such as "draw a line" and "draw a red line". There is only one parameter in the former command while two parameters in the latter, so the system will provide a default color for the former case. The default values are determined by the user interactive history and context. This method is proved to be feasible in practice.

5 Performance Studies

The 51 tasks in FreeDraw can be divided into three types: shape and position-related tasks, colour-related tasks and complex tasks (the combination of several simple tasks and are performed as one step). In order to compare the performance of the multimodal interface and the traditional GUI, we carried out a performance experiment. There are 18 tasks in the experiment, six of which belong to shape and position-related tasks, six colour-related tasks, and others complex tasks. We examined the execution time of the three kinds of tasks with multimodal and traditional interface, and the results show that the multimodal user interface with speech is more efficient than traditional GUI (As shown in Table 1).

Table1: Average Execute Time

	Type and position tasks	Colour tasks	Complex tasks
Mouse & keyboard	21.67 (s)	30.69 (s)	37.96 (s)
Multimodal	12.16 (s)	18.77 (s)	22.97 (s)

After the experiment, we invited twenty students who took part in the experiment to evaluate FreeDraw. Here are the users' opinions: First, they approve of the effective and accuracy of multimodal interaction when performing position-related tasks and complex tasks. Second, they are dissatisfied with the low recognize accuracy in a noisy environment, since a high accuracy can be achieved only in quiet place. Third, multimodal user interface with speech modality is accepted gladly for its efficiency and intuition compared with traditional interface. On a whole, sixteen among them preferred to interact with the speech modality because it is more natural.

6 Conclusions

Speech is a promising interactive mode, however it has not been applied pervasively. In this paper, we present the architecture and experiments of multimodal information processing of FreeDraw, a drawing system that can use speech, mouse and keyboard to complete the drawing the tasks. The experimental results show that using speech modality can greatly improve the performance and intuition of the traditional GUI.

7 References

Cohen P.R.(1992). The role of natural language in a multimodal interface. *UIST'92,* pp. 143--149.

Dong Shihai, Wang Jian and Dai Guozhong (1999). Human-Computer Interaction and Multimodal User Interface. Beijing: Science Press

Hauptmann, A. (1989). Speech and Gesture for Graphic Image Manipulation. *In Proceedings of ACM Conference on Human Factors in Computing Systems (CHI'89)*, 241-245.

Mignot C., Claude V. and Noelle C. (1993). An experimental study of future 'natural' multimodal human-computer interaction. In *Proceedings of ACM INTERCHI'93 conference on human factors in computing systems – adjunct proceddings,* 67-68

Nishimoto, T., Shida, N., Kobayashi, T., and Shirai, K. (1994). Multimodal Drawing Tool Using Speech, Mouse and Keyboard. *In Proceedings of International Conference on Spoken Language Pròcessing (ICSLP'94)*, Vol.3, 1287-1290.

S. Oviatt, P. Cohen, L. Wu, J. Vergo, L. Duncan, , B. Suhm, J. Bers, T. Holzman, T. Winograd, J. Landay, J. Larson, and D. Ferro (2000). Designing the user interface for multimodal speech and gesture applications: State-of-the-art systems and research directions. *In Human Computer Interaction*, volume 15, pages 263--322. Addison-Wesley,.

Xiao Bin, Jiang Yuquan, Dong Shihai (2001). A web browser-based multimodal interface for netshopping. *Journal of Computer- aided Design & Computer Graphics* , 2: 168~172.

Enhancing Tangible User Interfaces with Physically Based Modeling

Jens Weidenhausen

Fraunhofer IGD
Darmstadt (Germany)
jens.weidenhausen@igd.fhg.de

Abstract

This paper describes a novel approach to Interaction with Augmented Reality Environments. In Tangible user interfaces (TUI) the user can interact with the virtual parts of the environment through associated real objects. Virtual objects that are not part of the TUI remain unreachable for the user. Integrating a physical simulation can diminish this gap. Every object that is part of the simulation can at least be manipulated indirectly.

1 Introduction

Augmented Reality (AR) can be regarded as a powerful paradigm for enriching a users environment with context sensitive information. A user performing a real world task can be provided with significant information that may help him to perform this task more efficiently (Azuma et al., 2001). One aim is to seamlessly integrate the computer generated information into the real environment. This should apply to the visual appearance and particular to the behaviour of the virtual objects. Tangible user interfaces are a promising way to achieve such a fusion of real and virtual (Ishii & Ullmer, 1997). The user can interact with the virtual parts of the environment through associated real objects. These parts of the scene inherit a physically correct behaviour from their real correspondents. In Augmented Reality Systems where marker based vision tracking is employed this relation can be attained easily by attaching markers to real objects (Slay, Thomas, Vernik, 2001). But what happens to virtual objects that are not part of the TUI?

The associations between real objects and the virtual scene can be classified into three main classes.

- A real object is assigned to a visible virtual object. The real object can be used for direct manipulation of the virtual object. Moving a marker results in a corresponding movement of the virtual object.
- A relation between the real object and a corresponding occlusion geometry. This enables the correct handling of occlusions between real and virtual objects.
- The position or orientation of a marker is mapped to a 3D GUI Element. This can be used to indirectly manipulate scene or object properties.

Typically there is a one to one relationship between real and virtual objects. If the user interacts with one of the dedicated real objects only the assigned virtual object is affected. Some objects will remain passive, i.e. they are not linked to a real object. To overcome this shortcoming we introduced a real-time physical simulation engine to the system that controls the physically correct behaviour of the whole virtual scene. The collision detection provided by the simulation prevents

virtual objects from penetrating each other. Objects that are part of the TUI can be used to move passive objects. As a simple example a virtual ball begins to roll when it's pushed by a real coffee-cup. When the ball reaches the edge of the table it falls to the floor. To achieve this interaction between the real and virtual scene only a coarse model of the real environment is necessary.
Besides the collision detection the most frequently used feature of the physics engine is gravity. It enables the natural placement of virtual objects in the environment. Due to the fact that the simulation can be parameterised to a great extent it can be adopted to the needs of various applications.

2 Overview of Experimental System

The Augmented Reality Core Component developed in the ARVIKA project (Arvika, 2003) called ArBrowser encapsulates the main functionality like tracking, rendering and interaction (Figure 1). While the ArBrowser is implemented as an ActiveX-Control it can be easily integrated into web based applications.

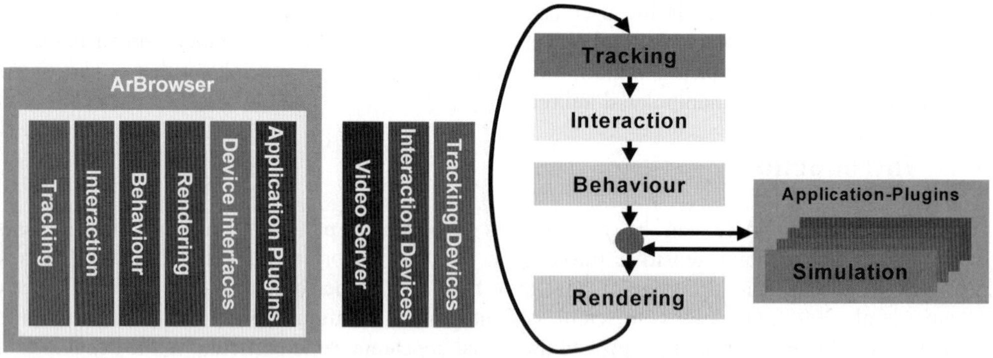

Figure 1: ArBrowser system components (left) and main loop (right)

2.1 Tracking

In the ArBrowser marker based video tracking is employed to determine the users position in the environment. Every marker has a 4x4 pattern for its unique identification. The configuration of the tracking system allows the definition of a world coordinate system for the user tracking as well as the use of local coordinate systems for object tracking. Because all marker coordinates are calculated relative to the camera object coordinates can easily be transformed into the reference coordinate system. This coexistence of all objects in one global coordinate system is strongly needed to perform the physical simulation.

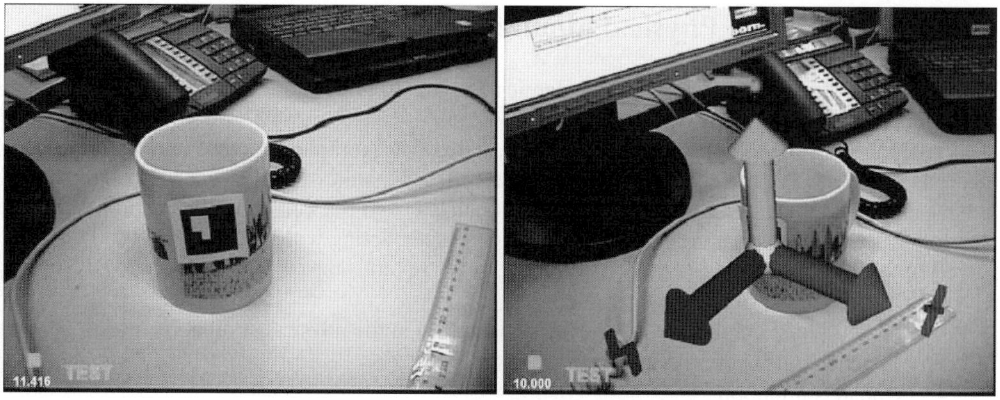

Figure 2: Vision based Tracking using fiducial markers

Like the coffee cup in Figure 2 objects of the real environment can be integrated into the TUI by attaching a fiducial marker.

2.2 Interaction and Behaviour

The Interaction and Behaviour is largely handled by a VR-System developed at the Fraunhofer IGD. It is capable of interpreting the most common subset of VRML97 behaviour descriptions. Through the Device Abstraction Layer IDEAL, (Fröhlich & Roth, 2000) it is possible to connect additional Tracking and Interaction Devices to Scene Nodes of the VR-System.
As renderer the open source scene-graph OpenSG (OpenSG, 2003) is deployed. The scene-graph contains as well the geometry nodes as the behaviour nodes of the VR System.

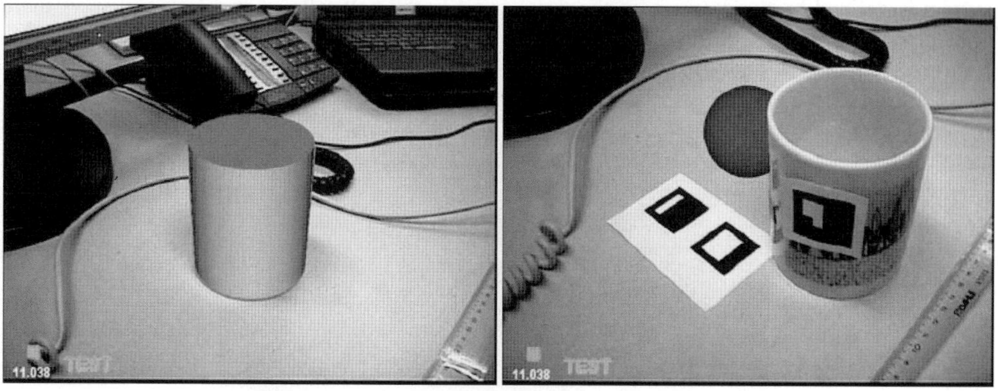

Figure 3: Simple cylinder assigned to fiducial marker (left) and the same geometry used for occlusion handling (right)

The different coordinate spaces of the tracking system are mapped on sub-trees that can be transformed independently. A second scene-graph contains the geometries for occlusion handling. It has the same structure as the main scene-graph, so that every tracked real object can occlude virtual objects (Figure 3). The visual realism can be increased easily.

3 Integration of Physical Simulation

We integrated and adapted the Open Dynamics Engine –ODE (Smith, 2003), a free library for simulating rigid body dynamics for the simulation task.
The integration of the dynamics engine into the ArBrowser is implemented as application plug-in that has full access to the scene graph. At the time the dynamics plug-in receives control in the main loop, the scene graph reflects the resulting changes from tracking, animation and interaction. Thus the user can influence the physical simulation by interacting with tracked objects.
A call to the simulation plug-in can be divided into the several steps where data is passed between simulation and associated scene graph objects:

- Obtain scene graph state of objects participating in the physical simulation.
- Transfer state to simulation objects
- Update simulation
- Pass updated state to scene-graph nodes

Every object in the physical simulation is associated to a geometrical counterpart in the main or the occlusion scene-graph. This geometry is needed to perform the collision detection between the simulation objects. Because in the current version ode can only handle collisions between simple objects, the collisions are computed based on the bounding boxes of the objects. Although this is only an approximation, it seams sufficiently accurate for most cases.
In addition to the pure geometric description, information of material properties and mass are needed.

4 Sample Scenario

A virtual ball placed on a real desk shall be moved with a real pot. The reference coordinate system is defined by the markers on the desk. Initially the ball falls on the desk and stays there. When the bounding box of the pot collides with the ball, it begins to roll according to the defined friction parameters (Figure 4).

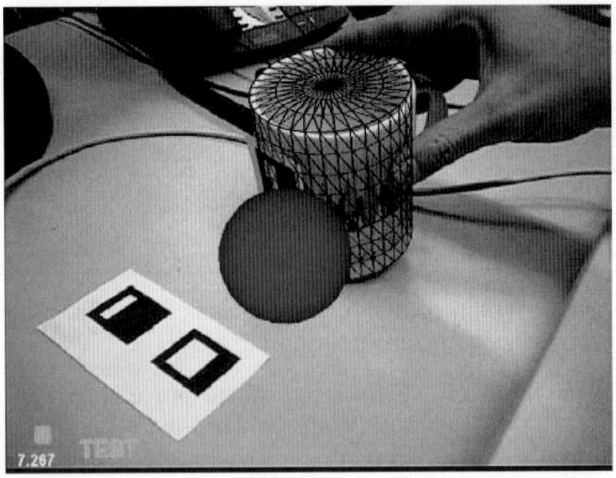

Figure 4: A virtual ball is moved by a real coffee pot on the desk

One problem, when working with the physics engine is, that the real objects must not be moved too fast, because otherwise the collision detection fails or the integration produces errors that will cause the system to explode. When objects are moved slow and not too abrupt the simulation works very stable and produces accurate results.

5 Conclusion

Obviously the system can be used to explain physical phenomena like gravity or friction in mixed reality edutainment demonstrations. In addition to conventional, real demonstrators it has the ability to vary the physical boundary conditions and object properties very easily. The simulated behaviour of objects on mars can be compared to that of real objects in the same mixed reality space.

Besides the additional realism, the physical simulation can be employed for several different applications. The effort of authoring augmented reality content can be reduced drastically even if the resulting scene is static. Virtual objects dropped into the environment will come to rest on the next object underneath them regardless if its real or virtual. By modifying the parameters of the simulation annotations can be placed on the surface of real objects from every direction. By just changing the direction of gravity it is very simple to place annotations on walls or even the ceiling. Due to collision detection and gravity the accurate alignment of objects becomes very easy. In dynamic environments the potential for reducing the engineering effort of creating augmented reality content is much bigger. The behaviour of objects is to the greatest possible extend defined by their physical properties and only additional behaviour must be described by animations or scripts.

Acknoledgement

The presented work is based on the ArBrowser Component that was developed as part of the ARVIKA project, an augmented reality research project funded by the German Federal Ministry for Education and Research (BMBF).

References

Arvika (1999). Arvika Project Home Page, http://www.arvika.de

Azuma, R., Baillot, Y., Behringer, R., Feiner, S., Julier, S., MacIntyre, B. (2001). Recent Advances in Augmented Reality. *IEEE Computer Graphics and Applications 21*, 6 (Nov/Dec 2001), 34-47.

Fröhlich, T., Roth, M.(2000). Integration of multidimensional interaction devices in real-time computer graphics applications. *In Proceedings of Eurographics 2000,Blackwell Publishers, U.K., 2000*

Ishii, H. Ullmer, B. (1997). Tangible Bits: Towards Seamless Interfaces Between People, Bits and Atoms, *In Proceedings of Chi '97*: 234-241

OpenSG (2003), OpenSG Project Home Page, http://www.opensg.org

Slay, H., Thomas, B. Vernik, R. (2002), Tangible User Interaction Using Augmented Reality, Australian Computer Science Communications , ThirdAustralasian conference on User interfaces - Volume 7 January 2002 Volume 24 Issue 4

Smith, R. (2003). Open Dynamics Engine – ODE, http://opende.sourceforge.net/ode.html

Modelling Emphatic Events from Non-Speech Aware Documents in Speech Based User Interfaces

Gerasimos Xydas, Dimitris Spiliotopoulos and Georgios Kouroupetroglou

Department of Informatics and Telecommunications
National and Kapodistrian University of Athens
Panepistimiopolis, Ilisia, GR-15784, Athens, Greece
Tel: +30 210 7275305, Fax: +30 210 6018677
{gxydas, dspiliot, koupe}@di.uoa.gr

Abstract

Most of the every day documents we come across have been composed without any information of how to be rendered in a speech-based user interface. As a result, visual formations that might imply emphasis are being ignored by Text-to-Speech systems or text-adapting applications (screen readers) and furthermore, complex structures, such as tables, are usually being vocalized in a rough linearized form, which leads to a confusing provision of information. In this work we accommodate both cases, by altering segments of emphasis in the content text, leaving a prosodic space for the vocalization of meta-information as well. We present a model for locating emphatic events and assigning to them a custom prosodic behaviour. Events are being divided in implicit and explicit ones. We concluded that the latter requires insertions of text to the linear form of structures in order to be properly realized. A script-based framework (e-TSA Composer) that supports the manipulation of prosodic elements in response of specific meta-information has been used. Finally, a model of table vocalization using our approach shows the significant improvement of the information provision compared to commercial applications.

1 Introduction

The majority of the electronic documents currently composed and viewed are non-speech aware, in the sense that they do not contain information about how to be appropriately vocalized by Text-to-Speech (TtS) systems. This is more conceivable when elements of visual structures have to be spoken in cases of speech-based user interfaces. Even tools for visual impaired people (e.g. screen readers) fail sometimes to deliver a meaningful speech representation of visual structures. Commercial systems do not properly construct a speech format of tables, as cells are parsed in a row, and character formation (bold, italics) are also ignored thought it should cause some prosodic alteration. This problem propagates to all cases of auditory-only interfaces (e.g. telephone Web access, directory services).

The Aural Cascaded Style Sheets (W3C, Aural style sheets) (Lilley & Raman, 1999) is a recommendation of the World Wide Web Consortium (W3C) that concerns the transfer of speech information (mainly prosodic) along with documents. In our work we deal with cases of, by any means, non-speech aware documents. Attempts to deal with the problem of the speech generation of documents have been also made in the past. Raman has developed a system to provide an audio format of (LA)TEX documents, focusing on the vocalisation of complex mathematical formulas (Raman, 1992). To achieve this, he assigned non-speech sounds (to indicate formulas) and prosodic features (to group elements in formulas) to math meta-information. W3C now provides recommendations on speech formatting of mathematics (Ausbrooks et al., 2002) (Kowaliw, 2001).

Realising emphatic events in the speech format of documents serves an augmented and usually meaningful auditory representation of them. Emphasis is a use of language to mark importance or significance, through either intensity of expression or linguistic features such as stress and intonation. Here we focus on the exploitation of speech emphasis in order to achieve an augmented auditory representation of documents with visual meta-information. Section 2 presents the modelling of the so-called *emphatic events*. An application of the model for table vocalization is shown in section 3, while a sort discussion is followed.

2 Modelling emphatic events

In this work we are interested in modelling emphatic events in documents to be used in speech-based user interfaces, as this way we manage to accentuate and, thus, distinguish the actual information (text), while we are able to vocally represent the visual format by using non-emphatic speech elements. Therefore, in order to properly vocalize visual structures, the hierarchy that they represent should be also retained in their speech format. Hierarchies might be in one out of two forms: *list* and *tree*. The first one is almost straightforward to vocalize as all the emphatic events occur in a sequence, without affecting each other. An example of a list is given in Figure 1. This figure presents a Heterogeneous Relation Graph (Taylor et al., 2001) with two relations: the phrase, which carries syntax information about the text on the left, and the cluster, which carries visual meta-information, as has been described in the e-TSA Composer (Xydas & Kouroupetroglou, 2001a), (Xydas & Kouroupetroglou, 2001b). Usually, TtS systems or even screen reader applications parse the phrase relation, hiding any structural or visual information provided by the cluster one (which represents the visual format of the document).

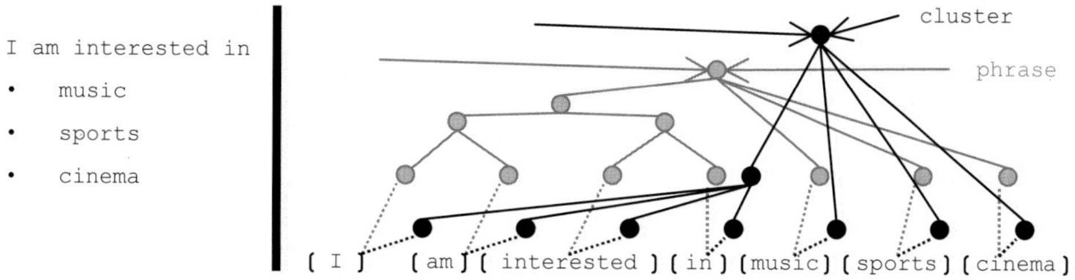

Figure 1

The tree form can be very complicated in vocalization, as it should be clear to the user that the spoken text is part of a specific hierarchy. An example of a tree form is shown in Figure 2. Testing the table on the left with state of the art commercial systems, we got two cases for speech:
A' case : "Table with four rows three columns. Model CC HP. 2000 1800 120 1990 2000 90. 1998 1900 130 table end"
B' case : "Model CC HP 2000 1800 120 1990 2000 90 1998 1900 130."
This however, leads to a misunderstanding of the content of the table.

These examples indicate the need for a more efficient handling of visual structures in Text-to-Speech process. We deal with this by introducing segments of emphasis in the text. We call the

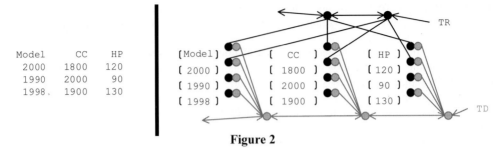

Model	CC	HP
2000	1800	120
1990	2000	90
1998	1900	130

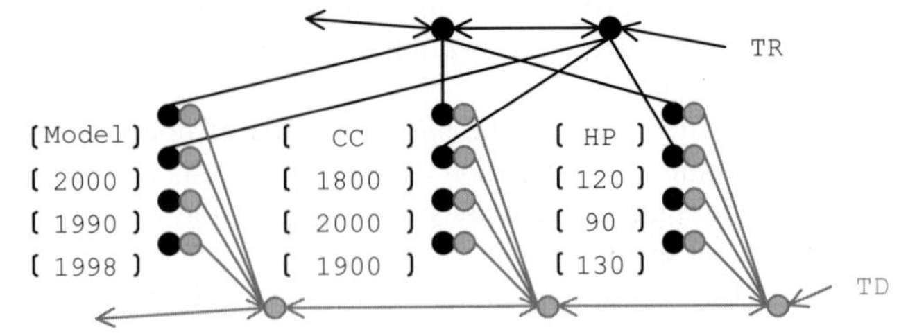

Figure 2

locations and the types of emphasis inside documents *emphatic events*, and these can be divided into two major classes:

- *Explicit events*, which are usually denoted by character special formation (e.g. bold, italics) where the level of emphasis depends on the actual format.
- *Implicit events*, which need to be identified and accessed from the special structures of the document for the structure meaning to be conveyed fully and correctly.

2.1 Explicit emphatic events

Explicit emphatic events are encountered in list form and should accentuate the corresponding cluster of text where they occur. They are conveyed by using text formation such as bold and italics. According to W3C (W3C, Information type elements) (W3C, Document structure) (W3C, 1999), italics is used to denote emphasis, whereas bold is used to denote strong emphasis. Two phrase elements, `<em>`, which indicates emphasis, and `<strong>`, which indicates stronger emphasis, are generally presented by visual agents as italics and bold text respectively. For non-speech aware documents there may be several ways to show explicit emphasis since more than the above may be needed. We classified the cases of explicit emphasis in HTML as:

- 1 or low (italics)
- 5 or medium (bold)
- 10 or high (italics & bold)
- or a value between 1-10 in cases of letters size or other formation.

2.2 Implicit emphatic events

Implicit events are encountered in either list or tree forms. In case of tree, we are using emphasis to denote the text content of structures, by distinguishing text from structural information that should also be vocalized. Implicit emphasis can be identified from structural objects inside documents, for example a row of cells in a table, bulleting, paragraph marking in conjunction with certain headings or capital lettering. In such cases emphatic events can be modelled by processing the structures individually. For example, in case of bullets it is generally required to emphasize the starting word or words of each bullet, and return to normal speech after a comma, if any exists. Nested bulleting reveals a hierarchy that should be taken into consideration, varying the levels of emphasis between the levels of nested bullets (Pitt & Edwards, 1997). Tables can also contain levels of hierarchy, but even when they don't, their complexity is still very high.

2.3 Emphatic events and prosody definition

To identify and classify visual formats of the source document, we use an XSLT-based HTML adapter in the e-TSA framework. This allows us to build the HRG presented in Figure 1 and Figure 2 and combine a hierarchy of visual directives with the traditional linguistic processing of the text. For the proper vocalisation of the documents, anything that is followed by structural meta-information is marked to be emphasised during synthesis, while inserted text representing the structural meta-information is rendered de-emphasised.

In speech, emphasis is delivered through prosody, by raising the tone, making a stressed syllable longer and increasing the loudness. Alternatively changing the prosodical characteristics of function words against content words also emphasizes certain point in sentences. The way that emphasis should be realised relies on the preferences of the user. The e-TSA framework provides a custom pool of Cluster Auditory Definitions (CAD scripts) that can vary the prosodic behaviour of the system, depending on the type of the emphatic event.

3 Vocalizing tables

One of the most common document types, which the on-line community uses in an every day basis, is HTML, which provides visualization meta-information about the text data. We model here the vocalization of one of its most common and quite complex structures; the table. Special recommendations to promote accessibility containing guidelines on how to make the web content accessible to people with disabilities are also provided by the W3C (Chisholm, Vanderheiden & Jacobs, 1999). According to these, the use of `<TH>` (for headers) and `<TD>` (for data cells) is mandatory. The use of `<THEAD>`, `<TFOOT>`, and `<TBODY>` to group rows and `<COL>` and `<COLGROUP>` to group columns is also required to associate data and header cells. This experiment also assumes (according to the guidelines as well) that tables are not used for layout purposes, unless it makes sense when linearised.

Our approach provides a scalable three way modelling of tables in non-speech aware documents, depending on which of the recommendations are present, as well as the content of the tables. In each case, an example of a table that inherits from the one of Figure 2 is provided.

<u>High compliance</u>
Any accessibility oriented text provided is being exploited. The summary for the table is uttered first. That informs the hearer of the information that is going to follow; by all means it is a title. If additional text is provided on the `<TH>` elements it is treated as the starting text for each sentence. Then the value of the corresponding `<TD>` is added. That way one sentence is constructed for each row, and text and cell values added column by column. This would be an ideal situation, however it is not common. The values contained in the data cells are marked as the emphatic parts of the sentences, with high level of emphasis. Example: "This table provides information about cars. Model ***2000***, CC ***1800***, HP ***120***. Model ***1990***, CC ***2000***, HP ***90***. Model ***1998***, CC ***1900***, HP ***130***."

<u>Medium compliance</u>
If the above accessibility oriented text is not provided, the utterances are constructed by alternating the headers and the data values for each row connecting using phrasal patterns like: "header *has* data" or "*for* header *the value is* data". In this case, each pair of header-data is uttered separately. This approach sometimes fails when the tables contain nested tables or type of values with which the added generic text results in unintelligible meaning. Data values are assigned to

high level of emphasis, while headers are also emphasized to distinguish from the inserted phrasal pattern. Example: "For *Model* the value is ***2000***, for *CC* the value is ***1800*** and for *HP* the value is ***120***. For *Model* the value is ***1990***, for *CC* the value is ***2000*** and for *HP* the value is ***90***. For *Model* the value is ***1998***, for *CC* the value is ***1900*** and for *HP* the value is ***130***"

Low compliance

The most generic approach is to model each row as a list of header-data pairs marking the data cell values with high level of emphasis. Example: "Model ***2000***, CC ***1800***, HP ***120***. Model ***1990***, CC ***2000***, HP ***90***. Model ***1998***, CC ***1900***, HP ***130***."

4 Conclusions

We presented a methodology for providing a more understandable speech format of visual elements. We used the e-TSA framework for the identification and classification of the visual meta-information along with emphatic alterations in order to accentuate content text against structured text. In all cases, we achieved a meaningful vocalization of tables compared to that of commercial applications. However, the scalable model presented needs to be supported by a language model in order to provide better phrasal patterns.

References

Chisholm, W., Vanderheiden, G., & Jacobs, I. (1999). Web Content Accessibility Guidelines 1.0, W3C Recommendation, 5 May 1999, *http://www.w3.org/TR/1999/WAI-WEBCONTENT-19990505/*

Lilley, C., & Raman, T.V. (1999). Aural Cascading Style Sheets (ACSS), W3C Working Draft 2 September 1999, *http://www.w3.org/TR/WD-acss*

Kowaliw, T. (2001) Accessible Mathematics on the Web: Towards an audio representation of MATHML Technology and Persons with Disabilities Conference, California North State University Northridge, 2001.

Pitt, I., Edwards, A. (1997). An Improved Auditory Interface for the Exploration of Lists. ACM Multimedia 1997, pp. 51-61

Raman T.V. (1992). An Audio View of (LA)TEX Documents, TUGboat, 13, Number 3, Proceedings of the 1992 Annual Meeting, pp. 372-379

Taylor, P., Black, A., and Caley, R. (2001). Heterogeneous Relation Graphs as a Mechanism for Representing Linguistic Information, Speech Communications 33, pp 153-174

W3C, Aural style sheets, W3C description, *http://www.w3.org/TR/REC-CSS2/aural.html*

W3C, Document structure, W3C description, http://www.w3.org/MarkUp/html-spec/html-spec_5.html

W3C, Information type elements, W3C description, *http://www.w3.org/MarkUp/html3/logical.html*

W3C (1999), HTML 4.01 Specification, W3C Recommendation, *http://www.w3.org/TR/REC-html40*

Ausbrooks, R., Buswell, S., Carlisle, D., Dalmas, S., Devitt, S., Diaz, S., Hunter, R., Ion, P., Miner, R., Poppelier, N., Smith, B., Soiffer, N., Sutor, R., & Watt, S. (2002). Mathematical Markup Language (MathML) Version 2.0 (2nd Edition), W3C Working Draft 19 December 2002, http://www.w3.org/TR/2002/WD-MathML2-20021219/

Xydas, G. & Kouroupetroglou, G. (2001a). Augmented Auditory Representation of e-Texts for Text-to-Speech Systems, in Proceedings of the 4th International Conference on Text, Speech and Dialogue, TSD 2001, Plzen (Pilsen), Czech Republic, September 2001, pp. 134-141

Xydas, G. & Kouroupetrolgou, G. (2001b). Text-to-Speech Scripting Interface for Appropriate Vocalisation of e-Texts, in Proceedings of the 7th European Conference on Speech Communication and Technology, EUROSPEECH 2001, Aalborg, Denmark, September 2001, pp. 2247-2250

A Direct Manipulation Interface with Vision-based Human Figure Control

Satoshi Yonemoto

Kyushu Sangyo University
Matsukadai 2-3-1, Higashi-ku,
Fukuoka 813-8503, JAPAN
yonemoto@is.kyusan-u.ac.jp

Rin-ichiro Taniguchi

Kyushu University
Kasuga-koen 6-1, Fukuoka
816-8580, JAPAN
rin@limu.is.kyushu-u.ac.jp

Abstract

This paper describes a vision based 3D direct manipulation interface which enables realistic avatar motion control and installs virtual camera control by body postures. Our goal is to do seamless mapping of human motion in the real world into virtual environments. With the aim of making computing systems suited for users, we have developed a vision based human motion analysis and synthesis method and have applied it to interactive system. The human motion analysis is unwired approach based on 3D blob tracking, and the motion synthesis is focused on generating realistic motion from a limit number of blobs. To realize smooth interaction among the real world and the virtual environments, we introduce additional mechanism into usual Perceptual User Interface. We assume that virtual objects in virtual environments can afford avatar's action, that is, the virtual environments provide action information for the avatar. Avatar's motion is controlled, based on the idea of affordance extended into the virtual environments. In order to handle interesting virtual objects easily, we also introduce 3rd-person viewpoint control methods coupled with body postures.

1 Introduction

The goal of our research is to develop a vision based 3D direct manipulation interface to realize seamless interaction in virtual environments, which are seamlessly coupled with the real world. Though many human interface devices force users to get used, essentially, systems must be adapted for the users (Weiser, 1993). From this point of view, human motion sensing without physical restrictions, i.e. unwired sensing is the most promising approach to realize seamless coupling between virtual environments in the systems and real world which the user exists. 3D vision systems are applicable for such purpose. In particular, it is very beneficial to estimate 3D body postures, or 3D positions of head, arms, feet, etc. because the postures indicate user's intention. Most related works have focused on classifying states of an user into pre-defined body gesture. However, much less attention has been devoted to acquiring continuous motion of the user and representing natural motion in the virtual scene. We think that realistic presentation of the substitute body is important subject to percept self body in the virtual scene or to communicate with the other existence.

We have already developed two techniques about human motion analysis and synthesis. Stable acquisition of details of human motion is a very difficult problem in computer vision research (C.Wren et al., 1997). Therefore our method is based on motion synthesis from a limited number of perceptual cues, which can be stably estimated by vision process. The first technique is a real-time human motion capture by 3D blob tracking (Yonemoto, 2000). Here, the blobs mean a rough sketch of human body from the images, i.e., a set of coherent image regions, which is mainly used as an effective visual feature. The second technique is a physically constrained human figure motion synthesis to generate natural motion from a limit number of blobs (Yonemoto, 2002). In the virtual environments, basic postures of the avatar can be represented by these methods.

In addition, we address our framework to use virtual scene context as a priori knowledge. We assume that virtual objects can afford avatar's action. In other words, the virtual environments provide action information for the avatar, such as properties of the virtual objects. An important point here is that we can consider scene constraints in the virtual scene to generate more realistic motion beyond the limitation of the real world sensing. Concretely, in object manipulation, scene constraints which lie between the virtual object and users' motion are employed. Every task in the manipulation is strongly related to objects in the virtual environments. Use of a priori knowledge for the virtual objects can make it possible to realize smooth interaction. The idea is based on affordance, extended into virtual environments. We also introduce 3rd-person viewpoint control to handle the virtual scene by body postures of the user.

2 Vision-based Human Figure Motion Control

We briefly show human motion analysis by 3D blob tracking and human figure motion synthesis from the tracked blobs.

2.1 Tracking of 3D blobs

To extract a rough sketch of human body from the image, coherent image regions, or blobs are mainly used as effective visual features. We also employ skin color regions of a face and hands in the image. When a 2D blob is detected in two views, or in multiple views, the 3D position of the blob can be calculated by stereo vision. In blob tracking, precise estimation is not required and, therefore, we have employed a simple stereo calculation. The algorithm of 3D blob position calculation is summarized as follows:

According to camera calibration, for each of the views, a line of sight, or a vector from the origin of the camera coordinate system to the center of mass of the blob, is calculated. Intersecting the lines of sight, the 3D position of each blob is calculated.

2.2 Motion Synthesis from Perception Data

Though many human motion synthesis methods are proposed (J.Kuffner Jr, 1999) (Y.Koga, 1994), we need to develop a method which can reconstruct a human figure model and its motion from observation acquired by vision process. It is difficult to introduce strict dynamics simulation required for the reconstruction which requires the physical properties such as forces and torques, since our system can obtain only positions of head and hands by the vision process. Therefore, as an approximate approach, we have introduced a physically constrained motion synthesis, where only the skeletal model obeys the law of gravity and satisfies point-to-point constraints between adjacent joint points based on Hooke's law. Then, a multiple part model is fitted for the adjacent joint points to estimate the pose parameters such as rotation and translation. Here, we assume that a link between adjacent joint points, or material points, is represented by a strong spring model so as to keep the link length constant. Moreover, we assume that the velocities of the joint points can be approximately calculated from the displacements of the joint positions. All joint positions that was estimated by iteration process keep the balance according to the change of positions of head and hands caused by vision process. The other constraints such as the collision against the other objects can be also added to the iteration process (Yonemoto, 2002).

3 Application to Direct Manipulation Interface

We have applied human motion analysis and synthesis method the above mentioned for 3D inte raction. In order to realize smooth interaction among the virtual objects and the user, in addition, we introduce the following mechanism.

- Avatar motion control based on the idea of affordance in the virtual environments
- Virtual camera control by body postures

3.1 Avatar Control

We assume that each virtual object affords essential information about user's action based on the idea of affordance in the virtual environments. In other words, we assume that interaction among the virtual objects and the user (i.e. avatar) should be properly performed by making each virtual object give rise to *afforded* action. Both the states of the virtual objects and the states of the avatar are decided whether the virtual objects are handled or not, according to the distance between the virtual objects and the user. For simple example, when the user grasps a virtual cup, the *afforded* finger motions to grasp it are generated and the state of the object is changed *static* into *move*. These motions are not acquired by real-world sensing but provided from the virtual object.

In addition, our system can consider scene constraints in the virtual environments to simulate scene events realistically. Since the virtual scene is completely recognized by the system, the context reasoning in the interaction through the virtual environments is not more serious than in the real world sensing.

3.2 Virtual Camera Control

We realize 3rd-person viewpoint control coupled with the body posture. Three virtual camera control methods are employed, based on a setup of virtual camera (see Figure 2).

Type 1: The simplest method is based on relative head position of the user. The movement to right and left direction corresponds to the change of *azimuth*, and the movement to front and back direction corresponds to the change of *distance* r.

Type 2: In object manipulation, the target object in which the user is interested can be specified owing to avatar motion control based on the idea of affordance. To realize smooth interaction, it is desirable to hold the target object centered in the field of view. Turning to the position of the target object, the virtual camera holds it.

Type 3: The third method is based on relative hand position. The hand is used as a substitute for mouse device to control virtual camera. View control by mouse device is often used in CG viewer. The result of grasp identification corresponds to mouse click and in grasping mode, the movement to right and left direction corresponds to the change of *azimuth,* the movement to up and down direction corresponds to the change of *elevation*, and the change to front and back direction corresponds to the change of *distance* r.

4 Results

We have applied our method to desktop-style interaction system, which is implemented on a standard PC. The system installs two firewire cameras and a wide 2D display in front of the user. The system can perform real-time and online from vision process to scene rendering. The user can only monitor projected virtual scene with the 2D display. Figure 1 shows the setup and an example of the virtual scene.

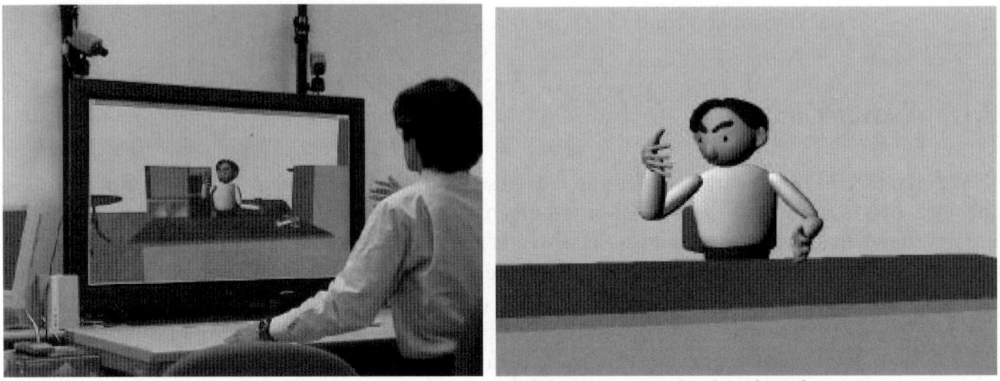

Figure 1: (left) System overview (right) The avatar in the virtual scene

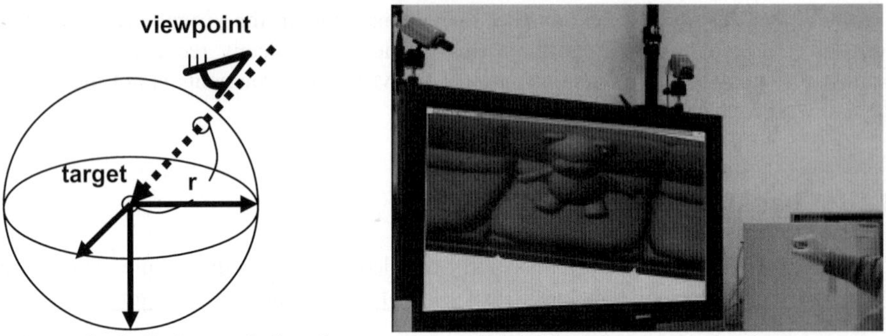

Figure 2: (left) Setup of virtual camera (right) Virtual scene navigation

4.1 Avatar Control

In desktop-style interaction, manipulation tasks such as reaching, grasping and transferring are needed. As perceptual cues, 3D positions of head and both hands, and appearances of hands related to grasping motion are acquired. First, the user performs reaching tasks, and then, according to *afforded* actions, he or she handles interesting virtual objects. The *afforded* actions involve additional motions of manipulation tasks. For example, in grasping, finger motions which cannot be estimated by a limit number of perceptual cues are supplemented to augment the virtual scene. Face direction of the avatar is automatically controlled. In the example of the Figure 1 (right), the avatar tracks his right hand.

4.2 Virtual Camera Control

We demonstrate that the user can smoothly interact with 3D objects by 3rd-person viewpoint control coupled with the body posture. In on-line demonstration (Figure 1), we tested an example of virtual camera control where the avatar is physically represented. First, the user swings the head and directly select favorite viewpoints by Type 1. When right hand approaches a virtual object, the object gives *afforded* motion with finger and face direction, and then the object is held in the center position by Type 2.

Figure 2 (right) shows an example of virtual scene navigation where avatar is not physically represented. Based on Type 3, the user controls virtual camera by right hand, toward an object (teddy) distant from the position of the avatar. The user started to control from the viewpoint of the desk object, and finally, he held the viewpoint of the teddy object on the sofa object.

5 Conclusion

With the aim of making computing systems suited for users, we have developed a vision based human motion analysis and synthesis method, and have applied it to 3D interaction. Our method is based on unwired sensing, the user can easily experience seamless coupling between the real world and the virtual environments. In order to realize smooth interaction among the virtual objects and the user, we have introduced two mechanism: avatar motion control based on the idea of affordance in the virtual environments, and 3rd-person viewpoint control methods coupled with the body postures.

Future works include improvements of virtual camera control by body postures, personal adaptation and application to more complex interaction scene.

Acknowledgments

This work has been partly supported by "Scientific Research on Priority Areas: Informatics Studies for the Foundation of IT Evolution; A03 Understanding Human Information Processing and its Application", one of the programs of the Grant-in-Aid for Scientific Research from the Japan Society for the Promotion of Science (14019071).

References

S.Yonemoto, D.Arita & R.Taniguchi.(2000). Real-Time Human Motion Analysis and IK-based Human Figure Control: *Proceedings of Workshop on Human Motion*, 149–154.

S.Yonemoto & R. Taniguchi. (2002). Vision-based 3D Direct Manipulation Interface for Smart Interaction: *Proceedings of International Conference on Pattern Recognition*, 655-658.

M.Weiser. (1993). Some Computer Science Issues in Ubiquitous Computing: *Communications of the ACM*, 36 (7), 75–84.

C.Wren, A.Azarbayejani, T.Darrell & A.Pentland. (1997). Pfinder: Real-Time Tracking of the Human Body: *IEEE Transactions on Pattern Analysis and Machine Intelligence*, 19 (7), 780–785.

J.Kuffner Jr.(1999). Autonomous Agents for Real-time Animation: *PhD thesis Stanford University*.

Y.Koga. (1994). Planning Motions with Intentions: *Proceedings of SIGGRAPH'94*, 24–29.

Combining Usability with Field Research: Designing an Application for the Tablet PC

Wenli Zhu

Microsoft Research Asia
5F, Beijing Sigma Center
No.49, Zhichun Road, Haidian District,
Beijing 100080,
P. R. China
wenlizhu@microsoft.com

Lori Birtley

Microsoft Corporation
One Microsoft Way
Redmond, WA 98052
USA
loribi@microsoft.com

Nan Burgess-Whitman

Harris Interactive
135 Corporate Woods
Rochester, NY 14623
USA
nburgesswhitman@harrisinteractive.com

Abstract

In designing an application for the Tablet PC, traditional usability methods were used to evaluate and validate the design. However, many of the questions facing the design team, such as how well users can learn and use the product in the long run could not be well addressed by traditional lab studies. A field study was thus conducted and yielded results that validated and expanded the usability study findings.

1 Introduction

Usability testing methods have been well practiced (for example, see Nielsen, 1993) and the effectiveness well documented (Rudisill et al., 1995). However, there are many limitations to the method of usability testing in the lab (Jordan & Thomas, 1994; Bevan & Macleod, 1994). For example, the test environment is artificial. The test tasks are not real. The time spent in the lab is limited. And typically, only users' first-time experience is tested. All of these could make one argue that traditional usability lab testing doesn't take into account users' long-term learning. Usability problems observed in the lab may not correspond to problems users have in the real-world or over the long-run. In recent years, field research and ethnographic methods have been increasingly applied to product and system development (Beyer & Holtzblatt, 1998). These methods advocate observing users "in context," i.e. observing users in their own environment, to understand what they do, why they do it and how they do it.

Field research and ethnographic methods offer an interesting alternative to traditional lab testing. We hypothesize that usability lab testing and field research methods differ in terms of the environment, the tasks, the time spent, the methodology, the experiementer, the cost and the usability problems identified. Table 1 outlines the differences hypothesized.

At Microsoft Corporation, usability professionals are heavily involved in product development (Denning et al., 1990; Kanerva et al., 1997). Iterative lab testing is conducted throughout the product development cycle. This paper discusses how we extended usability lab testing with field research to identify long-term usability problems and understand how users might use a product.

Table 1: Comparison of Usability Lab Testing and Field Research

	Usability Lab Testing	**Field Research**
Environment	Lab environment (a usability lab or a special area set up for testing purposes)	Users' own environment (users' offices or homes or any place where users may do work)
Tasks	Representative tasks that users may use the system for (typically created by the experimenter with domain and system knowledge)	Users' own tasks (real-world problems)
Time	Pre-set, typically 1-2 hours	A longer period of time that may last for days, weeks, months or even years
Methodology	Think out loud (i.e. verbal protocol), experimental design	Ethnographic methods and retrospective verbal protocol (observe as users perform the tasks and review previous experience)
Experimenter	Usually one person (typically a usability or human factors professional)	Usually takes an interdisciplinary team consisted of usability professionals, ethnographic or field researchers, product designers, product developers and marketing
Cost	High	Higher (much higher)
Usability Problems Identified	Typically users' first-time experience. Learning is limited to training and what users can learn during the session.	Users' first-time experience is not as important. Users' longer-term learning and usage patterns are more valuable.

The product evaluated was Microsoft OneNote™. It is a new note-taking and organizational tool designed for the Tablet PC as well as for desktop and laptop PCs. Users can type or hand-write notes (using the Tablet PC's stylus), organize notes into pages, groups of pages, notebooks and folders, search notes, flag notes (for example, mark notes as to-do items) and retrieve the flags. Figure 1 shows a bitmap of Microsoft OneNote. It is still under development and is expected to release in the summer of 2003. The design shown in Figure 1 may change in the final release.

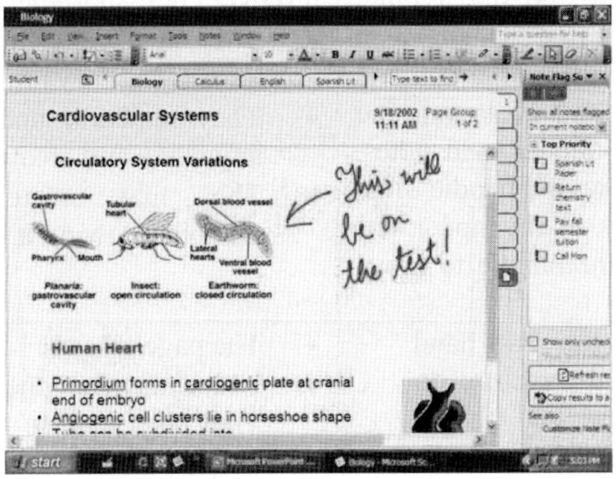

Figure 1: Microsoft OneNote

2 Usability Studies

Iterative usability studies were conducted from product conception to Beta (i.e. the product is fairly functional but is still in test mode).

2.1 Methodology

A typical iterative design and evaluation process was followed, i.e. results from one usability study were used to make design changes and consequently, the new design was tested. Paper prototypes, interactive online prototypes, and real-code were used in the tests. A total of 53 subjects participated in the usability studies. Profile of the subjects was chosen to represent the target user population of the product. Subjects received Microsoft software for their participation. Tasks representative of real-world scenarios were given to the subjects. Most were close-ended tasks, i.e. there was one clear outcome to judge whether the task was completed successfully. Some tasks were open-ended, i.e. there was no clear end state that the subjects had to achieve. Each test session took about 1.5-3 hours. Verbal protocol, success/failure rates, task completion time and errors were collected.

2.2 Results

Table 2 summarizes the major findings from the usability studies.

Table 2: Summary of Major Findings From Usability Studies

Design	**Usability Findings**
Gloms: a feature code-named Glom was designed to guide users write and distinguish between writing and drawing. If the user is writing, glom indicates whether it is a new paragraph or an indented list. The picture below shows a writing glom. 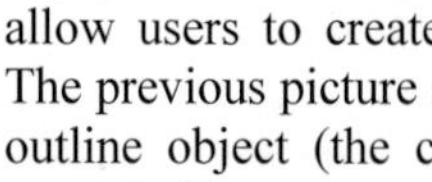	• Users will notice gloms but may not understand or follow them. • Users may not recognize or understand the different gloms for building structured notes such as paragraphs and lists.
Outline objects, handles, click-and-type: users can click any where and start typing. An outline object is created which is an intermediate level between a paragraph and a page. Outline objects and click-and-type allow users to create 2-dimensional notes. The previous picture shows the handle of an outline object (the cross at the upper left corner). Users can use the handle to move the outline object.	Users may not understand outline objects or discover click-and-type. • They may notice the handles but may not know what they are for. • They may not understand they can click anywhere and create a new outline object. • They may create new outline objects without knowing it.
Page tabs are shown on the right hand side of the screen (see the picture on the right). Users can use the tabs to navigate to a page or create a new page (the last tab). 	• The page tabs are hard to discover. • Users may not understand what a page is. They may tend to follow the Word model, i.e. pages are sequential instead of parallel.

3 Field Study

3.1 Methodology

The field study methodology was developed using a combination of qualitative approaches: contextual inquiry, in-depth interviews and focus groups. Each method was designed to elicit particular data about the user experience. A research team comprised of a usability engineer, product planner, research consultant, and program managers conducted the study. The field study was held for eight consecutive weeks. Fifteen participants were recruited. Ten were professionals (knowledge workers) and five were university or graduate students. All subjects were instructed to use OneNote on their Tablet PC in daily situations when they would typically take notes with paper or using keyboard. Each week, the research team would interview and probe the user on his or her experiences. In addition, specific tasks were assigned and the user was asked to report back not only on their challenges in completing the task, but their understanding of feature capabilities and general barriers/enablers within the OneNote application.

3.2 Results

Table 3 summarizes the major findings from the field study.

Table 3: Summary of Major Findings From the Field Study

Design	Field Study Findings
Usage	Subjects successfully entered information using OneNote. Notes were taken in meetings, during phone calls, in conferences, on the bus, in their cars, and during soccer games.
Note organization: pages, notebooks, and folders	Subjects understood the organization. The organization worked well for both students with complex hierarchies and professionals with simple ones.
Discovery of features	Most subjects quickly discovered writing (inking), auto-save, formatting, page tabs, and new page. Many gradually discovered groups of pages, notebooks and folders. Some discovered click-and-type and created 2-dimensional notes. Most did not discover search, note flags or ink highlighter until told.
Gloms	• Most participants ignored gloms. • Most did not notice the difference between the different gloms, or noticed but never understood. • Only two of the fifteen subjects understood gloms and tried to "obey" gloms to create structures such as paragraphs and lists.
Outline objects, handles, click-and-type	• Subjects did not generally understand outline objects. They did not understand the underlying page/object/paragraph hierarchy. • Subjects did not understand they can use the handles to move outline objects. • In "real life" situations, click-and-type is far too likely to create separate objects than to add to existing ones.
Auto-save: OneNote auto-saves every 30 seconds	• Subjects initially panicked when they could not find a way to save. • Once they learned auto-save (after first shut-down and re-open), they appreciated it.

4 Comparison of Findings

A comparison of findings from the usability studies and the field study revealed that:

- *Initial learning:* usability studies found major usability issues with gloms, outline objects, click-and-type and page tabs. Usability studies, however, did not answer questions like whether users will continue to have these problems over time, discover certain features over time or how they might use the product.
- *Learning over a period of time:* the field study found that the lack of understanding of gloms and outline objects persisted over an extended period of time for most users. Difficulties with initial discovery of page tabs and click-and-type were non-issues. Certain features were hard to discover even over an extended period of time.
- *Usage patterns:* usage patterns observed validated the scenarios targeted for OneNote.

Results from the field study further convinced the design team to re-visit some of the difficult design decisions, for example, whether gloms should be visible by default, whether to eliminate outline object handles altogether, and whether click-and-type should bias much more strongly toward adding to existing outline objects than creating new ones. These were all considered before the field study, based on findings from the usability studies. However, due to difficulties in implementing the changes, the design team was unwilling to make the changes right away. The field study provided the strong evidence needed to make the right decisions.

5 Conclusion

Usability lab testing and field research offer unique perspectives and advantages. Designing a brand new application for a brand new platform provided us a great opportunity to combine the two approaches. A comparison of the findings revealed that even though usability studies only tested users' first-time experience, serious usability problems uncovered persisted, even after an extended period of time. Results from usability studies, therefore, should not be freely dismissed or disputed based on the argument that long-term learning did not occur during the limited test time. Serious usability problems are likely to persist even after an extended period of usage.

References

Bevan, N., & Macleod, M. (1994). Usability Measurement in Context. *Behaviour and Information Technology,* 13 (1/2), 132-145.

Beyer, H., & Holtzblatt, K. (1998). Contextual Design. San Francisco, CA: Morgan Kaufmann Publishers, Inc.

Denning, S., Hoiem, D., Simpson, M., & Sullivan, K. (1990). The Value of Thinking-Aloud Protocols in Industry: A Case Study at Microsoft Corporation. In *Proceedings of the Human Factors Society 34th Annual Meeting*, 1285-1289.

Jordan, P. W., & Thomas, D. B. (1994). Ecological Validity in Laboratory Based Usability Evaluations. In *Proceedings of the Human Factors and Ergonomics Society 38th Annual Meeting,* 1128-1130.

Kanerva, A., Keeker, K., Risden, K., Schuh, E. & Czerwinski, M. (1997). Web Usability Research at Microsoft Corporation. In J. Ratner, E. Grosse and C. Forsythe (eds.) *Human Factors for World Wide Web Development*. New York: Lawrence Erlbaum.

Nielsen, J. (1993) Usability Engineering. Chestnut Hill, MA: AP Professional.

Rudisill, M., Lewis, C., Polson, P. G., & McKay, T. (1995) Human-Computer Interface Design: Success Stories, Emerging Methods, and Real-World Context. Morgan Kaufmann Publishers.

A Universal Approach to Multimodal User Interfaces

Pavel Žikovský

Czech Technical
University in Prague
Karlovo n. 13, Praha 2
Czech Republic
xzikovsk@fel.cvut.cz

Zdeněk Mikovec

Czech Technical
University in Prague
Karlovo n. 13, Praha 2
Czech Republic
xmikovec@fel.cvut.cz

Pavel Slavík

Czech Technical
University in Prague
Karlovo n. 13, Praha 2
Czech Republic
slavik@cs.felk.cvut.cz

Abstract

In our present-day information society users want to access their information everywhere, in various life situations, environments and occasions. Using one application in various environments brings up the need of multiple modality user interfaces - there are even cases, where it is impossible to communicate via one modality - driving a car can be a good example (a driver cannot pay attention to visual information). Blind or visually impaired people need these interfaces even more than sighted people. Development of a multimodal user interface is traditionally a difficult task. To make the development of client programmes with multimodal user interfaces easier, we propose a unified Multimodal User Interface Server Architecture, which handles UIML commands and definitions from application-server side and communicates with the client application in (any) given modality. By means of XSL transformations, these XML based user interface definitions are converted on the server to markup languages that can be rendered in various modalities. Optionally, the output can be passed through to modality-render (virtual client), which can render the output directly in the physical modality (e.g. speech). This conversion is done on the fly by the user interface server, which is located logically between the client and application server. The core of the server is based on Cocoon technology.

1 Introduction

The system is based on "client-UI-Application" technology, which is transparent for both the client and server sides. This technology enriches common client-server architecture scheme by adding a user interface server between an application (server in client-server architecture) and the client. We propose the solution, where the server is based on Cocoon technology, which is the only available publication framework based on XML and XSL (Clark J., 1999). As this server can also serve as a user interface for numerous applications, it becomes very close to the "user interface service", i.e. the user interface becomes standalone internet service.
All protocols used for negotiations between the application and UI-server or UI-server-client are based on XML standard, which makes them easy to enrich with possible new formats, properties and modalities. On the way between the client and user interface server, there can also be located a "virtual client" ("modality servlet"), which can render input XML data directly into given (physical) modality. This is meant for thin clients (i.e. phones) who can communicate only in this modality and have no other possibilities to communicate. The following picture explains the scheme more clearly.

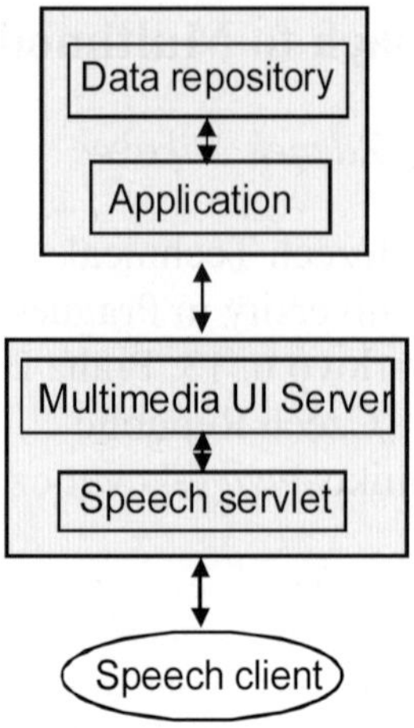

Figure 1: MUI Server with (an example) speech modality servlet

The rectangle around data repository and application indicates the fact, that the multimodal user interface server can act as a user interface rendered for many different applications – there can be more than one application connected to the server. As the application and data repository distance decreases, data-security also decreases.

Like the Internet in general, the client-server hosts the processing power and information on servers but couples that with selected functions and information on each client device (your PC, for instance). This allows us to manage shared information such as vocabulary lists and grammar rules centrally, distributing it to clients as needed.

Good example can be i.e. speech modality, where client-server architecture can handle problems with computational power, as advanced speech services require more memory and processing power than most handheld devices can offer. The server takes care of heavy processing, while the client devices handle a much simpler set of tasks. Therefore, the client device needs less computational power, which means a smaller battery and implies portability.

1.1 UIML

UIML, the User Interface Markup Language (UIML 2.0, 2000) (Abrams et al., 1999) is a language for describing user interfaces via XML. It is intended as generic (therefore also multimodal) description of user interfaces. To create a UI, one writes a UIML document, which includes presentation style appropriate for devices on which the UI will be deployed. UIML is then automatically mapped to a language used by the target device of a particular computing platform, such as HTML, Java, WML, HTML, PalmOS and VoiceXML. Because the language is based on XML, it is easy to write transformations using a tool like XSLT (Clark, 1999) that take the language from one abstract representation to a more concrete representation.

2 "Virtual client" architecture

As in real life we deal with a number of clients with different computational power, available memory and also different network bandwidth used for communication with the server, we have designed an adaptive architecture, which we call "Virtual Client Architecture". The key point of this architecture is that both the client and the server have the same application interface (API)

from the point of view of the sort of "plug-in" which we call "virtual clients. Both of them are written in JAVA (Sun Microsystems, 1997) .
The main difference from common plug-in architecture is that virtual clients can be dynamically used at either the client or server side according to the client configuration. Both the client and server are then "thin", i.e. the input at the client side is raw modality data (sound, video, etc.) and at the server side it is pure XML. The conversion between XML and the raw data[1] is made by certain virtual client. Whether this conversion is made on the client or server side is dynamically decided according to the network bandwidth and client's computational possibilities. This information is sent by the client during communication set-up process.
It is also useful to mention, that virtual clients can be asymmetrical - the conversion XML->Raw data can be done on the client side meanwhile the raw data-XML conversion can be done on the server. This, in fact, means that the conversion in each direction can be represented by one virtual client. Speech can be a good example of such a scheme: Text-to-speech systems are relatively easy tasks, but the speech recognition is much harder to compute and therefore it is not suitable for small mobile devices.
Let us take a more general example: We have a high-end PDA connected to a low-speed network as a client. This device has sufficient computational power and memory, but it suffers from low data bandwidth. Therefore, it is a better idea to perform all the raw data->XML conversion on the client side and transmit only textual XML, which is smaller. So, as the first step, all needed virtual clients are downloaded from the server, initialized and after that the communication within the server stays in pure XML. Then, we connect our PDA to local high-speed network. At that point, there is no need to do all the conversion on the server – all downloaded virtual clients can be released to free some memory (this step is not necessary, but possible – we can just bypass the virtual client and store it for future use) and then the communication is carried out in raw modality data, which can now pass through the high-speed network.

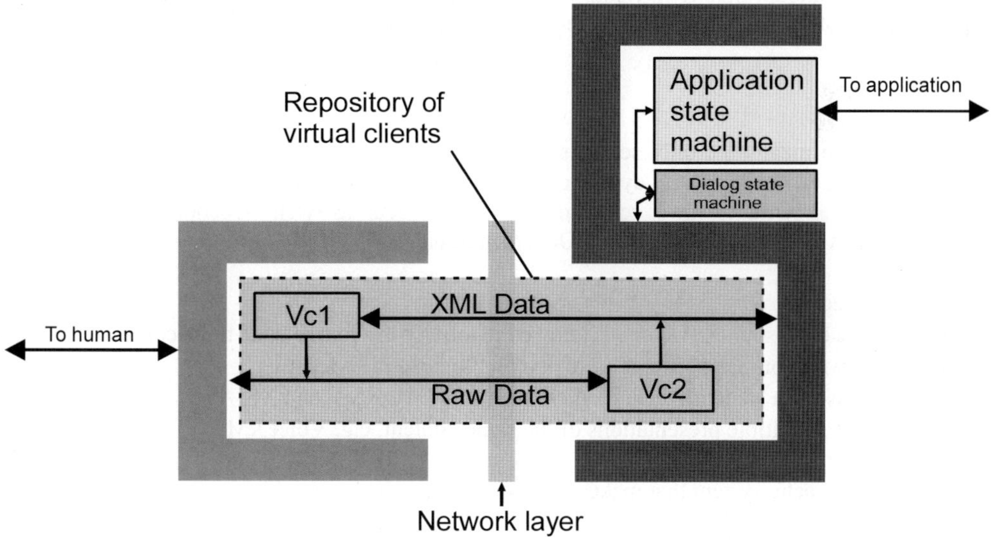

Figure 2: "Virtual client" MMS architecture

[1] Direct data stream, e.g. PCM audio for speech

In Fig. 2, you can see a client (big green "C") and server (big red "S") with virtual clients' repository between then. There are also 2 example virtual clients – Vc1 and Vc2. You can see that the virtual client Vc1 is realized on client side, meanwhile the Vc2 stays on server. For now, disregard the state machines on the top part of "S". They are part of our future work.

2.1 Virtual client types

There are two types of virtual clients according to the direction of the conversion, depending of the format of input, resp. output. In Fig. 3, we call them coder and decoder in the sense of coding to XML. The types can be pure XML and raw modality data.

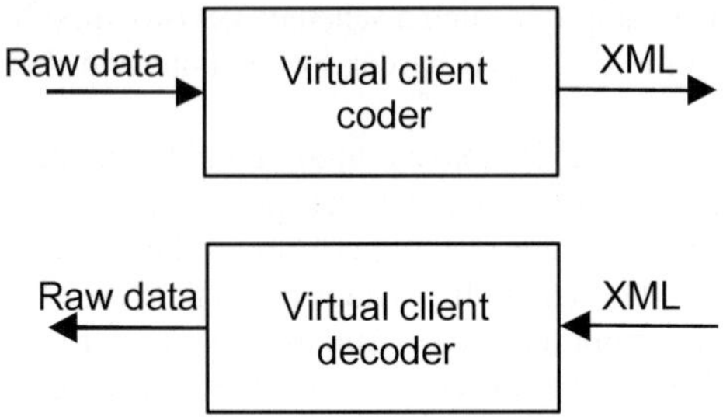

Figure 3: Virtual client types

There are many potential applications for the approach described beginning with applications for visually impaired users and ending with specific multimedia applications in the field of mobile computing. Multimodal mobile navigation system can be a good example of such an application. A pilot version of the system outlined here has been implemented and successfully tested.

3 Cocoon

Cocoon is by no means the only server solution to enable transformations from XML to other markup languages, even with dynamic content generation capabilities. Various software manufacturers, as Oracle, have released similar products as well. Cocoon has still some advantages over other products. (Cocoon Documentation, 2000)

- It is free.
- It is an open source. Anyone is allowed to modify the source code to meet her own needs.
- It allows a complete separation of content, logic, and style.
- It is based on standard technologies, such as DOM, XML and XSLT.
- It allows multiple presentations of the same content, e.g. VoiceXML or WML versions of the original XML documents.
- It has a cache system that makes the server processing more effective.

4 System implementation

As we need the same API on both server and client side, the only possible choose nowadays is JAVA. For our pilot version, we used FreeTTS engine (FreeTTS, 2002), which is free TTS engine written entirely in JAVA. It is based upon Flite TTS developed at Carnegie Mellon University.

Flite is derived from the Festival Speech Synthesis System from the University of Edinburgh and the FestVox project from Carnegie Mellon University. The only disadvantage is that FreeTTS it is written for JAVA 2 and therefore it is unusable for small mobile devices running JAVA Micro Edition. FreeTTS is an isolated class and it acts as the virtual client mentioned before, which moves between the client and server.

5 Conclusions

In this paper we have presented a framework of a system, which can cope with constraints found in multimodal mobile computing devices. UIML is a suitable tool to do this, certainly when it is combined with Java. The usage of XML to generate a user interface description at runtime implies an automatic XSL conversion to various modality dependent formats. The use of Java language on both client and server side together with virtual client architecture can deal with dynamic constraints found at mobile devices.

There are many potential applications for the approach described such as applications for visually impaired users and specific multimedia applications in the field of mobile computing. Multimodal mobile navigation system can be a good example of such an application. A pilot version of the system outlined here has been implemented and successfully tested.

6 Acknowledgements

This work has been performed within the framework of the IST project MUMMY (IST-2001-37365): Mobile knowledge management - using multimedia-rich portals for context-aware information processing with pocket-sized computers in facility management and at construction site. Funded by the Information Society DG of the European Commission. See http://mummy.intranet.gr/.

7 References

Marc Abrams, Constantinos Phanouriou, Alan L. Batongbacal, Stephen M. Williams, Jonathan E. Shuster (1999), "UIML: An Appliance-Independent XML User Interface Language," 8th International World Wide Web Conference, Toronto, May 1999, http://www8.org/w8-papers/5b-hypertext-media/uiml/uiml.html. Also appeared in Computer Networks, Vol. 31, 1999, pp. 1695-1708.

Ali, M. F., M. Abrams (2001), Simplifying Construction of Cross-Platform User Interfaces using UIML. In UIML Europe 2001 proceedings, January 2001, Paris, France. pp. 64-78

Cocoon Documentation (2000), Apache XML Project, The Apache Software Foundation, http://xml.apache.org/cocoon/

Clark, J. (1999) XSL Transformations (Version 1.0), http://www.w3.org/TR/xslt.

Sun Microsystems (1997). Java Remote Method Invocation. ,
http://java.sun.com/products/jdk/rmi/.

UIML 2.0 draft specification (2000). http://www.uiml.org/specs/docs/uiml20-17Jan00.pdf.

VoiceXML, VoiceXMLForum, http://www.voicexml.org/

W3C. XHTML (2000): The Extensible Hyper Text Markup Language, A Reformulation of HTML_4 in XML 1.0,http://www.w3.org/TR/xhtml/

Section 5

Supporting Access to Information, Communication and Cooperation

Evaluating an Online Academic Community: 'Purpose' is the Key

Chadia Abras, Diane Maloney-Krichmar, Jennifer Preece*

Information Systems Department
University of Maryland, Baltimore County
Baltimore, MD 21250, USA
<abras, preece>@umbc.edu, dkrichmar@bowiestate.edu

Abstract

Online communities are fast becoming a major aspect of the daily lives of millions of people around the world. Determining what makes an online community successful is therefore crucial for developers and designers. Why do some succeed and some fail? In order to study these questions, an online academic community was created for students in an interdisciplinary doctoral program at a major university in the United States. Because the program serves working adults with a variety of research interests, it is difficult for members to get together to discuss ideas and socialize. Students expressed the desire for an online community with a discussion board to facilitate communication. A community-centered design approach (Preece, 2000), that involved the students in all aspects of community design and development, was used to create the community. An action research approach, using quantitative and qualitative methods, was used to assess the success of the community. Data was collected through surveys, observations, and interviews and indicated that 25.6% of all members actively participate in discussions, 12.7% never visit the discussion board and 61.7% are occasional readers. Most participants indicate that the purpose of the community should be narrowly focused to fit their particular interest. They emphatically indicate that the discussions should relate to their own specific area, be lively and well moderated. This study reveals that a narrow purpose appears to be the strongest indicator of success for an online community. The purpose brings them in to the community, but lively, interesting discussions keep them coming back.

1 Introduction

The contemporary concept of community encompasses the idea of a group of people that are united by culture, but who do not have to live in the same neighbourhood (Rheingold, 1996; Wellman & Gulia, 1999). Online community can mean different things to different people, in some it might evoke a warm feeling, in others it can lead to thoughts of hate groups and groups catering to people with antisocial behaviour (Preece, 2001.) A community does not have to have physical boundaries; it may be defined as a social network without physical boundaries rather than in terms of space (Rheingold, 1996; Jones, 1998; Wellman & Gulia, 1999). These social networks have loose, more permeable boundaries, interaction is with diverse people, and hierarchies are no longer as well defined as in face to face interactions (Wellman, Boase & Chen, 2002).

For the purpose of this study we will adopt the working definition of an online community established by Preece (2000): an online community consists of, *people* who interact socially as they strive to satisfy their own needs or perform special roles, such as leading and moderating. A shared *purpose* which provides a reason for the community, such as an interest, needs for information, or service. That p*olicies* in the form of tacit assumptions, rituals, protocols, rules, and laws guide people's interactions, and c*omputers systems* that support and mediate social interaction and facilitate a sense of togetherness.

The aim of this research is to establish an academic online community in order to study which usability and sociability heuristics are key to evaluating success in this particular community. The following sections discuss the approach taken in designing the community, the approach taken in the research, and initial findings deduced from questionnaires, observations, and interviews.

2 Establishing the community

The method used to design the online community was a community-centred development approach, where the focus is on the community. This type of design approach is participatory (Preece, 2000) and compliments the action research paradigm adopted as the research approach in this study. In this design, the developers start by assessing the needs of the community, through surveys, observations, and interviews. The design was implemented according to the wishes of the community and the prototype was tested and evaluated. At each step, the designer went back to the community for further assessment. Before initiating the site, the developers tested for usability and sociability in order to assess whether any usability or sociability issues did not work for this particular community.

The online community in this research consists of members of a doctoral program at a major university in the United States. This particular program is interdisciplinary and therefore the interests and aspirations of all the members are varied. The program is a collaboration between eight different departments and the students can specialize in any number of areas depending on their interest. The nature of this program and the fact that the majority of students are working adults makes it hard for members to get together. In essence, after completing the core courses, they hardly see each other again.

The site was launched in 1999, but it remained unused for over a year. Since students in the program continued to express a need for an online community, it was decided to work to revive it (Maloney-Krichmar, Abras & Preece, 2002). As with the first attempt, a community-centred development approach was used (Preece, 2000). When the Web site was introduced again in 2002, the doctoral program had 38 students and had existed for four years. The students were surveyed concerning their wishes and the purpose of the community and the revised site was designed accordingly. The revised site was launched in the spring of 2002 without a discussion board, which was added in fall 2002.

This process was guided by the community-centred development approach, in which the designer refers back to the members of the community, through surveys, questionnaires and interviews, and modifies the design according to the wishes, desires, and needs of the community. The selection of the software to be used was guided by the wishes and needs of the students in the doctoral program. Once the site was designed, it was evaluated by expert reviewers and a group of students that were not part of the community. A second survey was administered to test for usability and sociability. The site was reassessed according to the results while the community was being nurtured by an active moderator. The discussion board was introduced three month later, with many topics for discussions and an active moderator for each topic. Professors and students were asked to moderate. Research shows that when an online community is introduced, it must be nurtured by publicizing it, welcoming members, maintaining a healthy dialogue, and making sure hardware and software are in good working order at all times. The evaluation process never stops and reassessing is continuous (Preece, 2000). This particular community existed in a face to face setting, yet the bonds that held the members together were very weak. It was hoped that in the online setting, the members will be able to find one common purpose that will hold them together, which is an attempt at communication and the desire to stay connected.

3 Action research

3.1.1 Action research approach

Action research seeks to find solutions to problems in the community through a participatory approach. The members of the community and the researchers are involved in all steps of the research and are encouraged to theorize their actions, examine the theory in light of these actions, and change their approach to acting in society in light of these new found theories (Reason & Bradbury, 2001). Habermas' critical theory and cognitive interest are at the base of Kemmis' (1985) action research paradigm, mainly that technical, practical and critical reflections are examined respectively from the viewpoint of the action, as part of a larger context and from a social perspective. Freire (1970) theorizes that people should identify and then analyze their own problems in order to find a solution, meanwhile the researcher takes the role of co-investigator and becomes a part of the research instead of being an objective observer (Selener, 1997). This new role allows the investigator to closely analyze, observe, and reflect on the problems in the community while experiencing them at the same time. The researcher's own reflection on his/her participation in the process can add a rich and thick description to the research. The people become more aware of their environment through participatory research (Reason, 1994).

Action research is cyclical in nature and the methods used in analysis are varied, they may include qualitative and quantitative with an emphasis on community reporting and always going back to the group for further input in the research (Reason, 1994; Reason & Bradbury, 2001). In Information Systems (IS) this method of investigation did not become popular until the 1990s. It was introduced to IS through the works of Mumford, Checkland, and Wood-Harper in the mid 1980s (Baskerville, 1999).

3.1.2 Phases of the study

The participants were divided in three groups, A, B and C. Group A consists of the students who were admitted the first two years of the program. They were familiar with the researchers and had participated in the original study. Group B are students that were admitted after the inception of the original community, they were not familiar with the researchers or the original study. Group C are students admitted after the inception of the second version of the online community. They were admitted when the bulletin board had already been introduced, and their teacher agreed to conduct her class using the bulletin board.

In Phase One of the study only groups A and B participated; the community was established and the potential participants were surveyed in order to obtain a baseline data for computer use, Internet habits, and technological skills. The Webpage was introduced in May 2002, without a bulletin board and participants were observed until September 2002. A second survey was administered to assess the feelings of the participants about the site, whether it is successful, and whether they felt that they belonged to an online community.

In Phase Two the bulletin board was introduced. Group C was admitted and they were administered survey one. For the bulletin board we chose PHPBB (www.phpbb.com), because of its intuitive set up, the streaming design of message layout, and because we were able to add extra features, such as: avatars, smilies, and -email message to a friend.

One student moderator was asked to conduct an active session on the discussion board and one class was conducted on the bulletin board. Two weeks later a new topic was introduced, with a new moderator and a few days later a third discussion board topic was introduced. All three moderators maintained active discussions on several topics in one area of interest for each. Survey three was administered at this point and the data was analyzed and interviews were conducted.

In Phase Three it was decided that the bulletin board had not have enough time to affect the community; therefore, a new topic was introduced, this time it was moderated by a faculty member. A fourth survey was administered in order to assess whether students' feelings about belonging to an online community have changed.

4 Results

When initially surveyed, the members of the program were asked if a Web site in the form of an online community would be useful to them. The results indicated that the site would be useful to the group: 21% were neutral indicating that they would like to see the site before making up their mind, 37% indicated that it would be very useful and 21% said it would be extremely useful. None were negative about the usefulness of such a tool.

After redesign of the online community and analyzing the surveys administered in Phase Two and Three of the study, conducting the interviews, and observing the activity on the discussion board, the survey results were quite different from the initial study outcomes. These results indicated that 25.6% of all members were actively participating in the discussions, 12.7% never visited the board and 61.7% were occasional readers.

Students felt that the community was successful if it were able to maintain their interest, if it provided interesting topics for them, and if they felt connected to the group. They felt that lurking is a form of participation and therefore can lead to success of an online community. They felt that the site should be useful, dynamic not static, and discussions engaging. Only one member indicated that in order for the community to survive, the members have to be committed to the process.

Some of the findings of this study are not new indicators for determining success in online communities, but what is amazing, is that the active users and the lurkers see success as what interests them and not the number of people participating. When a message receives 65 views and only six replies, users perceive this as successful since they can tell how many people are reading. Some members indicated that they would like the discussions to be philosophical and more in depth, while others wanted an environment for light conversation. It is possible to accommodate both views through careful moderation. However, moderation of topics should be in the hands of the members empowering them in the daily operation of the community. The students indicated, in interviews, that the purpose of the online community has to be very narrowly focused to fit their neèds, and since many members expressed different needs, it became apparent that the community needed to be multimodal in order to accommodate all members.

The original purpose, the need to be connected expressed by all was not strong enough to unite them. The data showed a vast divide between what the members indicated on initial survey and what finally materialized. Their desire to have a community is genuine, but they do not want to put an extra effort to assure its survival. The purpose that united the students was not strong enough to make them invest the extra time online. When one expresses need, there should be a measure to how strong that need is, and that would be the single best indicator for the survival of the community. Purpose is the strongest element in sociability heuristics, the others can be achieved to varying degrees. If the purpose is strong enough, the members will come, but of course the developer must not ignore the other aspects of the online community, in order to keep them coming back.

5 Conclusions

The process of designing, implementing and maintaining this online community is cyclical and ongoing. The community emerged very slowly, with very weak-ties between its members. Some felt greater connection than others, but as one member indicated; this community exists face to face and the ties between the members are also very weak in that setting because of their varied interests; therefore, the same types of connection transferred online. The students have varied backgrounds in sociology, literature, education, technology, and languages, which afforded the researchers a rich description of the users' feelings, opinions, and insights.

This was the second attempt to establish this online community. The first resulted in no participation, the second was more successful, but participation and ties are still weak; however, there is a steady growth in the participation of the students with continued commitment and nurturing, the community can grow, become healthy and independent of the researchers' involvement. Communities are dynamic and therefore one should adapt to their ever changing nature (Preece, 2000).

References

Baskerville, R. L. (1999). Investigating information systems with action research. *Communications of the Association for Information Systems, 2*, article 19.

Freire, P. (1970). *Pedagogy of the oppressed.* New York: Herder & Herder.

Jones, S. (1998). Information, Internet, and community: Notes toward an understanding of community in the information age. In S. Jones (Ed.), *Cybersociety: Revisiting CMC and community* (pp. 1-34). Thousand Oaks, CA: Sage Publications.

Kemmis, S. (1985). Action research and the politics of reflection. In D. Boud & R. Keogh & D. Walker (Eds.), *Reflection: Turning experience into learning* (pp. 139-163). London: Kogan.

Maloney-Krichmar, D., Abras, C., & Preece, J. (2002). Revitalizing an online community. *Proceedings of the 2002 International Symposium on Technology and Society* (pp. 13-19). Raleigh, NC. June 6-8.

Preece, J. (2001). Sociability and usability in online communities: Determining and measuring success. *Behaviour & Information Technology, 20, 5,* 347-356.

Preece, J. (2000). *Online communities: Designing usability, supporting sociability.* Chichester, England: John Wiley & Sons.

Reason, P. (Ed.). (1994). *Participation in human inquiry*. London: Sage.

Reason, P., & Bradbury, H. (Eds.). (2001). *Handbook of action research: Participative inquiry and practice*. Thousand Oaks, CA: Sage Publications.

Rheingold, H. (1996). A slice of my life in my virtual community. In P. Ludlow (Ed.), *High noon: On the electronic frontier* (pp. 413-436). Cambridge, Mass: The MIT Press.

Selener, D. (1997). *Participatory action research and social change.* Ithaca: Cornell University Press.

Wellman, B., Boase, J., & Chen, W. (2002). The networked nature of community: Online and offline. *IT&Society, 1*(1), 151-165.

Wellman, B., & Gulia, M. (1999). Virtual communities as communities. In M. A. Smith & P. Kollock (Eds.), *Communities in cyberspace* (pp. 167-194). London: Routledge.

Socio-Technical Evaluation of Computer Supported Work and Learning Systems

Evren Akar, J.H. Erik Andriessen, Jelle Attema and Bige Tunçer[*]

Faculty of Technology, Policy and Management, Delft University of Technology
Jaffalaan 5 2628 BX Delft, Netherlands
e.akar, j.h.t.h.andriessen, j.attema@tbm.tudelft.nl

[*]Faculty of Architecture, Delft University of Technology
Berlaweg 1 2600 GA Delft, Netherlands
b.tuncer@bk.tudelft.nl

Abstract

Although computer supported collaborative work (CSCW) and computer supported collaborative learning (CSCL) are two separate research disciplines, with their own research interests, some recent groupware systems, designed for business settings or educational environments, fall into the research interest of both disciplines. While some recent collaborative work tools are being used as e-learning environments, collaborative learning tools have the features of collaborative work tools and evaluating these tools requires a close look at the findings of CSCW and CSCL disciplines. One important finding of both disciplines is that it is not enough to evaluate technical efficacy of these tools but to evaluate them in a socio-technical context and the evaluation process should cover the design process as well. This work reports on the development of a socio-technical evaluation methodology (STE), which is based on the research findings about collaborative work, computer supported collaborative work, collaborative learning and computer supported collaborative learning.

1 Introduction

Despite the potential of computer supported collaborative learning systems, the research into their use and effectiveness of them is inconclusive. Although there are positive reports, there are also negative reports. The negative outcomes are mainly based on low participation rates and/or varying degrees of disappointing collaboration (Kirschner 2002). Similar outcomes are also valid for computer supported collaborative work (Andriessen 2002). The reasons for these and possibly more negative outcomes may arise as a result of the delicacies in technical and social contexts. Therefore understanding reasons behind this becomes crucial to develop better design and evaluation methodologies. Evaluation helps developers to improve the design processes and the products. However, evaluation of collaborative systems is a complex process (Grudin, 1994). Generally, it is not enough to validate the systems' technical quality and interface usability, their qualities in actual use are the essential criteria. Therefore, it is important to design systems that have qualities in use and to develop evaluation methods designed to evaluate these qualities (Bevan 1995).

A general approach to evaluate a new socio-technical system is developed in the MEGATAQ project (Andriessen et al. 1998) and new methodology is based on those principles that is on: a) evaluation should be user centred, b) evaluation should be integrated in the design process from the very beginning, c) the design and the evaluation process should be iterative, and take place at the concept, prototype and operational stage d) future usage scenarios can support the identification of design requirements in the concept stage and of evaluation criteria. In agreement with Andriessen et al. and based on our other findings from literature we have developed a design-oriented socio-technical evaluation (STE) process which is user-centered, participative and iterative. The evaluation methodology covers concept, prototype and operational phases of the development process. Different from the general approach of MEGATAQ, our evaluation methodology focuses on the learning aspect in particular. We chose a collaborative learning system, INFOBASE, to develop the evaluation approach. INFOBASE is a web-based multimedia-learning environment designed to support group work for architectural design. In the following section we will discuss the relevancy of InfoBase with our approach.

2 Motivation of Evaluation Methodology

The motivation of our evaluation method comes from the findings of two research disciplines. These are Computer Supported Collaborative Work (CSCW) and Computer Supported Collaborative Learning (CSCL). As a starting point we used the models developed in CSCW. One of these models is DGIn model that we will discuss in the subsequent sections. We will also discuss how we adapted DGIn model for learning environments. We give an explanation of CSCW and CSCL, describe their similarities and differences, and discuss their possible inputs for a theoretical framework of our evaluation process below.

2.1 CSCW and CSCL

Computer Supported Collaborative Work (CSCW) and Computer Supported Collaborative Learning (CSCL) are two research disciplines based on the premise that groupware systems can support and facilitate group process and group dynamics. The purpose of CSCW is to facilitate group communication and productivity while CSCL is used to support learners. These systems can support communicating ideas and information, accessing information and documents and providing feedback on problem solving activities. The research interest of CSCW and CSCL covers the techniques of the groupware and their social, psychological, organizational, and learning aspects.

Dynamic Group Interaction Model (DGIn model) is introduced below. The DGIn model was designed using the basic principles of group interaction, integrating group dynamics and related theories, such as media richness theory, activity theory and coordination theory (Andriessen 2002).

2.2 Dynamic Group Interaction Model

The DGIn model (Figure 1) is assumed to be usable for the analysis of all kinds of groups but the importance and role of the factors may differ for various groups. IA distinction is made in the model between input characteristics, processes and outcomes and the context of the functioning group. Three types of outcomes define the success of the group. These outcomes are the function of the group processes and the group processes are determined by the characteristics of the input characteristics. The line of causation may not be always simple, however, the processes may change the input characteristics of the group and the group may go through certain life cycle changes. According to the *Contingency Perspective,* systems should match their environment so that they can be effective. In the case of collaboration technology this idea means that the effectiveness of an application depends on its fit to aspects like the task, the user(group)s and the context in which these user(group)s operate (Andriessen 2002). In the DGIn model, Contingency Perspective poses that the various input characteristics in the model, such as size, task, group composition and group culture, have to be balanced. The better this balance, the higher the chance of adequate functioning of the group.

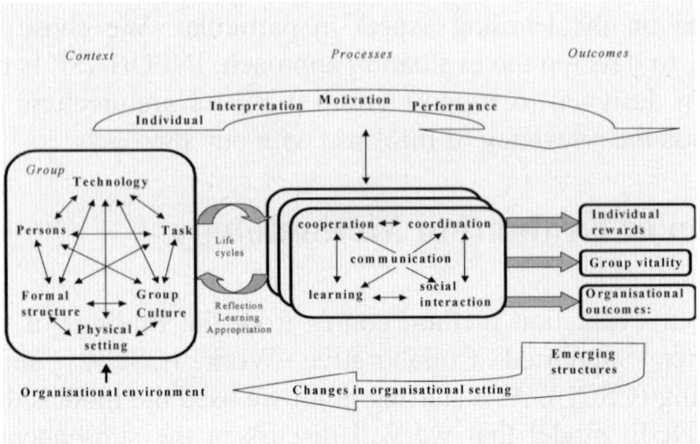

Figure 1. The Dynamic Group Interaction Model (DGIn-model Andriessen, 2002)

Five group processes are defined in the DGIn-model, namely cooperation, coordination, communication, social interaction and learning. According to Andriessen (2002) communication is of a different nature than the other processes. Communication refers to the exchange of signals, to the issue of verbal and non-verbal interaction. It is basic to information sharing, co-ordination, co-operation and social interaction. Andriessen defined co-operation, co-ordination and learning as task-oriented processes. *Co-operation*, i.e. working together, is applying one's knowledge and skills and using appropriate performance strategies. *Co-ordination* is adjusting the work of the group members to fit the goals of the group. *Learning* is exchanging (sharing) and developing information, views, and knowledge. *Social Interaction* is a *group-maintenance oriented* process and it refers to team building to develop trust and cohesion, to drive oriented behavior and to reflection on team activities and team development.

2.3 Learning Aspects

Adapting the DGIn-model into a collaborative learning environment requires particular emphasis on learning processes. Collaborative learning is based on constructivist learning theories. According to Bruner's (constructivist) theoretical framework (Bruner 1966 in Aroyo 2001): "a) learning is an active process in which learners construct new ideas or concepts based upon their current/past knowledge, b)students should be able to continually build upon what they have already learned c) the learner selects and transform information, constructs hypothesis, and make decisions, relying on a cognitive structure (mental models) which provides meaning and organization to experiences. It also allows the students to reflect on the information and create knowledge; d) the instructor should also support and encourage the students to discover links and associations by themselves, e) the instructor should present information to be learned in a form that matches the learner's current state of understanding". In a collaborative learning environment; learners have individual responsibility and accountability, learning interaction takes place in small groups, communication during learning is interactive and dynamic, learners can identify their role

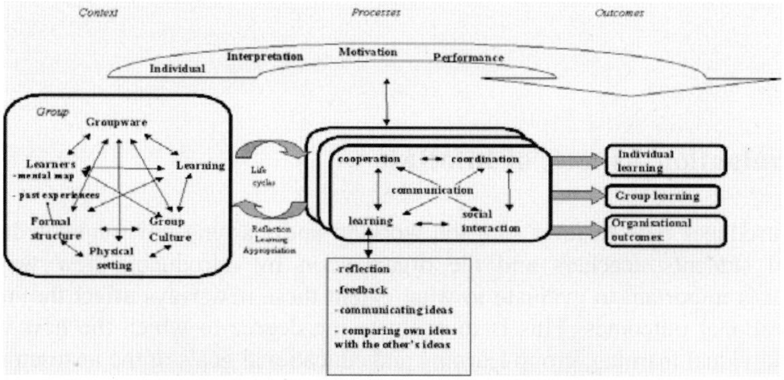

Figure 2 DGIn model adapted for learning situation

in the learning task and participants have a shared understanding within the learning environment. The challenge is to combine the DGIn notions with the notions concerning learning processes to accomplish a comprehensive evaluation. Fig. 2 depicts the new model for learning based on DGIn model. In the next section we will introduce the functionality of InfoBase and discuss them based on the theoretical notions introduced above.

3 InfoBase

The InfoBase learning environment aims to support students to electronically manage and organize their communication and cooperation activities. The functionality of the environment primarily concerns with: a) the storage and management of working documents, b) a powerful and flexible organization of information (Figure 3), c) the presentation, within the group and also to the public (Figure 4) and d) the integration of communication with document management. The environment is interactive and dynamic; flexible, such that the student can adopt the environment independently of any instructor or course; multimedia-oriented, including video, audio, images, text, etc (Figure 5). These features of InfoBase aimed to provide students with features based on the building blocks of a collaborating group, considering relevant theories such as media richness theory, activity theory, coordination theory and constructivist theories.

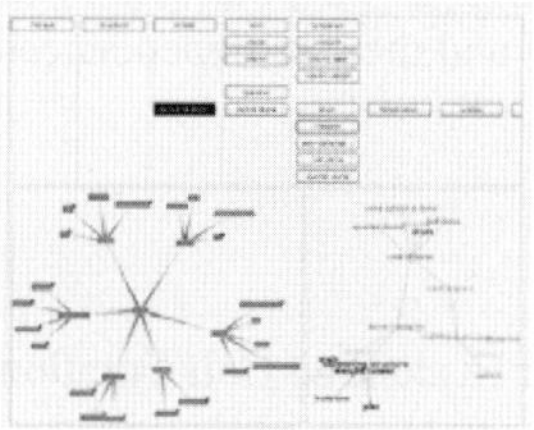

Figure 3. Organization of metadata

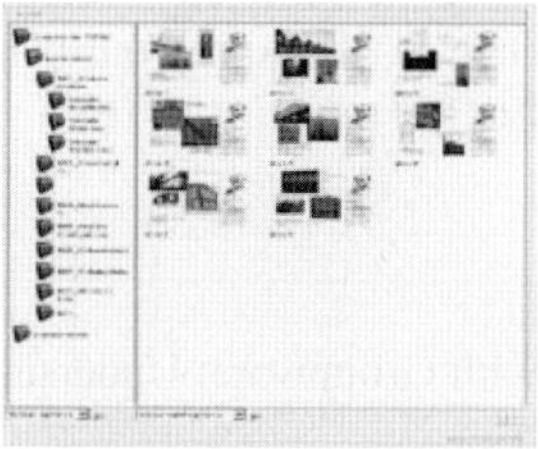

Figure 4. Presentation of group findings

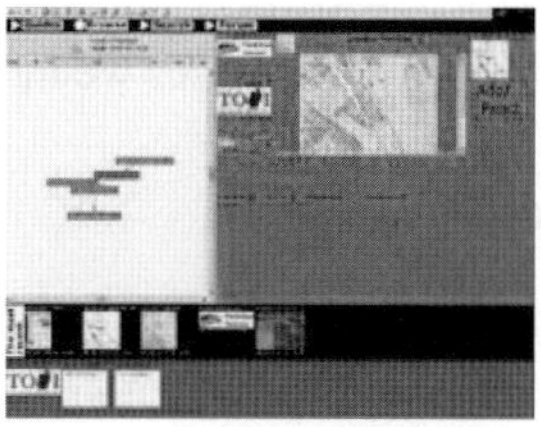

Figure 5. A multimedia environment

INFOBASE is grounded primarily in the field of CSCL and serves as an educational environment, however it is also intended for use in professional environments. Therefore, it is essential to study InfoBase in relation to both CSCW and CSCL. The model we developed by adapting DGIn model is proper to evaluate InfoBase. InfoBase environment provides tools to facilitate group processes and means to support learning processes such as reflection, feedback, communicating ideas and comparing own ideas with the ideas of others. Therefore InfoBase provides us opportunities to test evaluation methodology.

3.1 Evaluation Aspects of InfoBase

InfoBase introduces new features into the working and learning environment. It transforms the activities of students, teachers and the organization by introducing new ways of working. Therefore, it is important to evaluate to what extent these new ways affect the individual, group and organizational outcomes. This is to evaluate the degree to which the new way of working improves individual learning, group learning and educational goals of the instructors.

In addition, the following questions need to be answered: To what extent is individual user or group's action facilitated by the interface? Is the information presented correctly? Can the user or group present their findings using the interface? Is the interface usable and does it guide the user and keep them achieve their task? The question as to what extent the use of media, and what extent the use of media hinders or supports people's interaction; non-verbal communication and context awareness needs to be answered.

4 Future Work

We have developed a socio-technical evaluation approach based on our findings from literature and we aim to develop a firm evaluation methodology based on the results of its application for different tools. We chose InfoBase as the first application tool because it fulfils the requirements arises from both CSCW and CSCL research fields and we were able to apply our methodology starting from the design process to the operational phase of the tool. Covering both design process and operational phase will help us to improve our design oriented approach.

Up to date the concept and prototype phases of InfoBase have been completed. The content and interface of InfoBase is developed based on the findings of workshops where we applied future usage scenarios. The stakeholders were satisfied with the results, and this encouraged us about the relevance of our approach. However, for a complete evaluation we have to wait the results of operational phase. These results will help us to modify our methodology and the interdependences between the processes we introduced in our model.

References

Andriessen, J.H.E. (2002) Working with Groupware: Understanding and Evaluating Collaboration Technology. Springer Verlag London

Andriessen, J.H.E., Koorn, R., & Arnold, A.G. (1998) Multidimensional Evaluation of Generic Aspects of Telematics Applications Quality: The MEGATAQ model. Proceedings of the IFIP Conference, Vienna-Budapest.

Aroyo, L. (2001) Task-oriented approach to information handling support within web-based education. PhD Thesis. University of Twente Press. Enschede, Netherlands.

Bevan, N. (1995). Measuring usability as quality in use. Journal of Software Quality, 4, 115-140.

Grudin, J.(1994). Eight challenges for developers. Communications of the ACM, 37(1), 93-105.

Kirschner, P.(2002). Can we support CSCL? Educational, social and technological affordances for learning. Inaugural Address. Open University, Amsterdam, Netherlands.

FLIRT: Social services for the urban context

David Bell

Philips Research Laboratories
Cross Oak Lane, Redhill, UK
david.bell@philips.com

Ben Hooker, Fiona Raby

Royal College of Art
Kensington Gore, London, UK
b.hooker@rca.ac.uk, fiona@dunneandraby.co.uk

Abstract

The FLIRT project set out to identify new mobile services and enabling technologies that would promote increased use of mobile networks in Helsinki. This paper describes the process we followed to create a number of innovative services to meet the challenge.

1 History

In 1998/1999 the number of text messages (SMS) sent each month was fast approaching the billions that we see today, and data services for information, communication and entertainment were seen as the way to increase revenues from each subscriber. The mobile telecommunications industry created the WAP (Wireless Application Protocol) forum to deliver the standards for mobile data services.

The motivation for the FLIRT project (Flexible Information and Recreation for Mobile Users) was to uncover some fresh ideas for mobile services that would provide drivers for advancement of service-provisioning technologies such as WAP. As described below, many of these services provided a new perspective upon awareness between friends and strangers in the urban setting.

2 Consortium

Perhaps the most significant impact on the approach employed within the project was the make-up of the project consortium. The team from the Royal College of Art (RCA) brought training and experience in Architecture, Multimedia, Graphic Design and Industrial Design; the team from Philips Research Labs (PRL) were familiar with R&D for consumer electronics and could develop much of the infrastructure to deliver tailored services over a mobile network; Philips Consumer Communications provided support for development and integration of software for mobile devices; Infogrames Entertainment brought experience in multiplayer PC gaming over the internet as well as games for portable Platforms; and Elisa Communications provided access to their mobile phone network and urban user community in Helsinki.

3 The Design and Development Process

At a high level, the process followed in the project was an iterative cycle of Requirements, Design, Development and Testing. Within this cycle there was a distinct design activity inspired by the features and constraints of the mobile communications infrastructure (explored further in Raby 2000), and a technical activity driven by the need to realise the emerging service concepts.

3.1 Initial Requirements

The initial technical requirements were elaborated in opening meetings of the consortium in which we exchanged details about current and emerging technologies in mobile communications, mobile phones and mobile gaming platforms. The initial requirements of the target user community were based upon information from the network operator about Helsinki's population and detail about the make-up of the city. The designers needed to know the potential of the technologies and capabilities available to the project such that the final FLIRT services would be innovative yet achievable with the available resources.

3.2 Understanding the Target Community

The first action of the design team of the RCA was to spend several days in Helsinki, undertaking intense observation of people in the city they inhabit, its climate, its culture, its transport, and its media. One way of gathering information was to observe and follow a few subjects though the city. Subjects were mostly identified at the train station, and followed to work or home. Particular attention was made to phone usage and modes of transport. No consent was sought for observation, so this approach is not recommended for the faint-hearted. The foundations of the approach come from work in the 1950s, with Guy Debord's theory of *drifting* through urban spaces to gain new insights (Debord 1958). A similar approach to the study of strangers using their mobile phones outdoors is documented by Weilenmann & Larsson (2001).

From the visit to Helsinki, it was possible to draw some conclusions about the spaces in the city, the roads and squares where many people crossed paths, or the places people stopped to shop or relax. Much material was gathered, including video footage and photos, maps, tickets, magazines and local papers. In the following weeks this material was sorted and sifted and provided inspiration for new services.

3.3 Initial Ideas

Explorations by the RCA yielded a huge diversity of ideas, considering the interface between people and places of Helsinki with real and virtual events each day, and taking an adventurous view of the technological possibilities. Some ideas considered provision of information about the city such as its services and events, at the other end of the spectrum were ideas that dealt with fiction and narrative of game play, in the middle was a rich seam of "daydreaming" ideas where fiction and reality were blurred.

3.4 Interpretation of Technology

The RCA had a number of meetings with Philips and Elisa Communications to better understand the behaviour of mobile technologies and raise questions about technical possibilities that had not been raised before. In this period, the RCA began to elaborate and interpret the features (and flaws) of mobile technologies and saw novel ways to explo it them.

One of the features of cellular communications that sparked the imagination of the RCA was the way in which the "coverage" of the mobile network was apparent as cells of various shapes and sizes across the city. Those cells relate to places of relevance to the population, where they move, where they congregate, and thus reflect natural groupings. Furthermore, hand-off between base-stations (and between cells) is not necessarily at strict boundaries, which encourages the perception that the boundaries between those cells are ambiguous, somehow *elastic* (figure 1.)

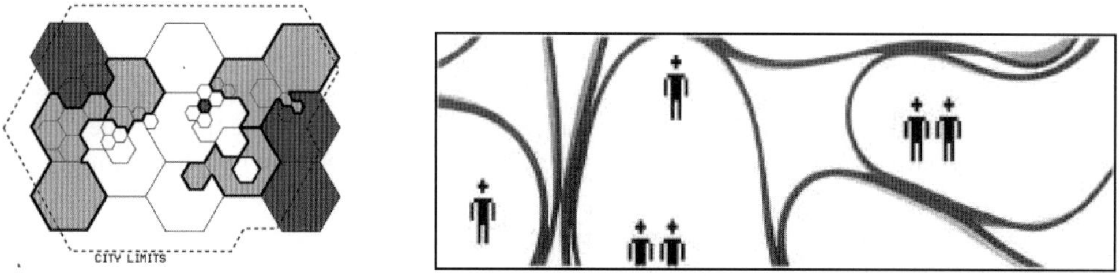

Figure 1: Impression of the cellular network and "elasticity" of cell boundaries

Subscribers give up much information to network operators upon subscription, and by implication permit the network operator to acquire great databases of information about them in relation to calls made, texts sent and their movements throughout the city. The ambiguity of location was therefore seen as an advantage from the user perspective, because it provided a bit of inaccuracy in an otherwise exact system.

One of the key characteristics of the mobile phone is that it is one of the few items which people carry **all** the time, switched on in readiness to receive a call or a text. In its idle mode, the mobile phone is in fact busy exchanging data with base-stations of the mobile network, ready for handover if needed. This makes it a unique device for user interaction. Not only is the phone available to access services, but also available for use by services and to receive *pushed* content. Furthermore, using cell-broadcast it is possible to push content to everyone at a particular place so there are numerous ways to disrupt this idle state.

The constraints of the mobile phone as an interaction device are the small size of keyboard and screen, lack of screen resolution, low processing power, and battery life. There are also then the constraints with the connection the network such as occasional call failures, limited bandwidth, limited size of SMS messages, and variability of SMS delivery times. The constraints of the mobile phone itself were the stimulus to think of the phone as a low-resolution window to the high-resolution events and activities of the surrounding city.

3.5 The FLIRT Infrastructure

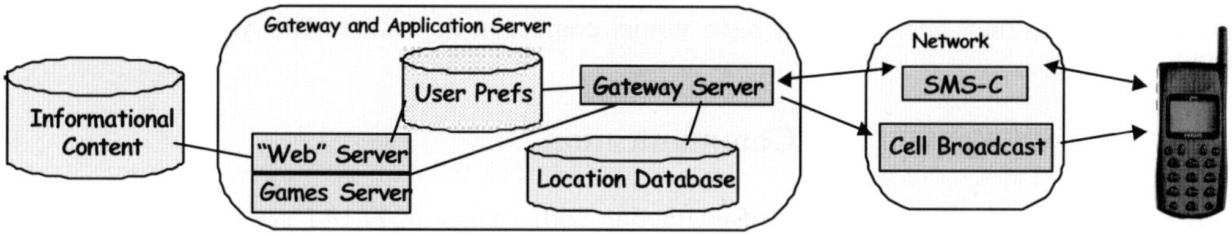

Figure 2: Outline of the FLIRT Service Infrastructure

As the definition of services progressed, the technologies required to realise the required features were prototyped and refined as part of the ongoing development and specification activity. In particular, we needed to be able to locate a user whenever they requested a service, and track them as they moved from cell to cell. We also needed to be able to push services to particular users, or groups of users in a given (cellular) location. The infrastructure to realise this was centred upon a gateway server linked into Elisa's SMS-Centre and Cell Broadcast Centre (figure 2). The gateway was a proxy for exchanges with the mobile phones that contained our micro-browser. This browser was capable of receiving pushed alerts and animated graphics as well as rendering textual content. We built a location database for resolving a user's location from the cell-ID held in their phone, as well as a database for recording user preferences and usage data. The gateway could then retrieve informational content from the Internet via a web server and filter the content according to a user's location. The gateway could also exploit user location to enrich games.

3.6 Services with Social / Emotional benefits

3.6.1 Connection with friends and family – messages in places

The FLIRT infrastructure made it possible to associate content with a particular place as denoted by cell-ID, and users could potentially "post" messages at their current location. One example application is *reminder messages*: at a given location one could post a message for oneself, and subsequent return to that location would trigger receipt of the reminder message. Perhaps greater value comes from the ability to post a message for a friend or relative. For example, a son might use his phone to leave a birthday message for his mother at the train station, knowing that she will pass through later that morning and receive the message on her mobile en route to work. Such located messages have an emotional value beyond SMS because they show effort and thought to physically locate them, which is special to the people involved.

3.6.2 Connection with strangers – pixel kissing

Because we could record the transitions of people from cell to cell, and the times of each transition each day, we could build patterns of movement and use this information as a trigger for services. We considered how we could capture and present information about inhabitants of the city that was relevant to the individual. We considered the interest one might have in the many strangers of the city, wondering if they have similar travel patterns or go to the same destination at a similar time. Maybe one could learn better routes.

Somewhat akin to repeated meetings on a train which might trigger conversation, we devised an application that enabled anonymous interaction between those who repeatedly shared the same routes through the city. Upon receiving such a trigger to play, each player is given a series of questions to which there is a fixed set of answers, the answers given by each player are exchanged and players can then decide if they have virtual contact in the future. This was the essence of "Pixel kissing".

3.7 Idea Refinement and Communication

To better understand the experience of interacting with a range of FLIRT services, a game board was created based upon the map of Helsinki, with models of some of the major sites, models of stereotypical users, and representations of cell boundaries. Scenarios captured a soap story of interactions amongst the characters and the places they frequented. The Helsinki game board was an invaluable tool to play out the FLIRT services in space and time, focusing ideas upon the mechanics of user-location and movement, and the necessary behaviour of the network.

Visual material was key to communication between designers and engineers, to convey the mechanics of the network in relation to mobile subscribers, as well as bringing service concepts to life. Diagrammatic specifications captured times and places for particular interaction events, and web-based animations of services showed the services as they would appear to the user. The animations provided graphics that could be used in handsets as the project progressed.

3.8 Focus Groups

Focus groups were carried out in Helsinki to understand how well the FLIRT services met the desires and needs of the target community. Participants made it clear that they wanted control of the information sent to them, and they liked the idea of receiving News and information about events relevant to their location. There was some concern about "Pixel kissing", maybe there was a risk that personal details would be exposed and abused. The ability to participate and give opinions anonymously, whilst still remaining part of a collective, was regarded as very special.

4 Conclusions

The FLIRT project focused upon design and development of innovative services for the mobile domain, and successfully yielded a number of innovative solutions suitable for realisation in Helsinki. It is possible that services such as Pixel Kissing might be valued in other urban settings, but services providing some sort of social awareness need a density of population and a critical mass of users to make a stimulating impact. Also, such services need to be localised to the user community, but we feel that users have a desire to make an impact on their space (anonymously or otherwise) so there is a need to make sure that "awareness" services can be shaped by the users and exploited in ways that best suit them.

With regard to the process, we broke away from the idea that designers are brought in to style a completed technology and deliberately let the designers and engineers influence each other. We recognised the value of tools for knowledge exchange such as the Helsinki board game and the interactive animations. The design team spent time immersed in the target community, and time entertaining wild ideas before the software team focused in on technical solutions. The eventual solutions were at the limit of what was possible over a mobile network, but were well rooted in technical reality and well in tune with the needs of the user community as we understood them.

Acknowledgements

The FLIRT project (EP26765) was funded under the IT for Mobility theme of ESPRIT Framework IV, and completed in September 2000. Many thanks to the team who participated in the project and all who helped in creating this paper.

References

Debord, G. (1958). Theory of the Dérive. In *Internationale Situationniste #2* (central bulletin of the International Situationist). Retrieved February 12, 2003, from http://situationist.cjb.net/

Raby, F. (2000). Project #26765 FLIRT: Flexible Information and Recreation for Mobile Users. London: RCA CRD Research

Weilenmann, A., & Larsson, C. (2001). Local Use and Sharing of Mobile Phones. In B. Brown, N. Green & R. Harper (eds) *Wireless World: Social and Interactional Aspects of the Mobile Age.* Godalming and Hiedelburg: Springer-Verlag, pp 99-115.

Approaches to the Design and Measurement of Social and Information Awareness in Augmented Reality Systems

Frank A. Biocca[1], Jannick Rolland[3], Geraud Plantegenest,[1] Chandan Reddy,[1] Chad Harms[1], Charles B. Owen[2], Weimin Mou[1], Arthur Tang[1].

M.I.N.D. Lab[1], METLAB[2]
Michigan State University
East Lansing, MI USA

ODALab[3]
University of Central Florida
Orlando, Fl USA

Biocca@msu.edu

Abstract

This paper characterizes the problem of managing focal and peripheral awareness of people, objects, and procedures in augmented reality (AR) systems. We suggest interface design strategies that map the problem to properties of human spatial cognition and briefly introduce how the problem is addressed in three collaborative AR projects.

1 Awareness of augmented reality systems

To perform tasks, navigate, and coordinate behaviour in the physical environment, members of work teams or other coordinated groups must continuously be aware of the state of others, objects, and the environment. Humans possess highly evolved mechanisms for monitoring the location of the body, objects, and the environment in space. Entering into this evolutionarily developed pattern are technologies that can mediate interaction between individuals and dramatically extend the range of other people an individual can be aware of. Networked technologies, especially immersive augmented reality (AR) technologies, are grafted onto the evolved system and mediate the natural relationship between the body and the environment (Biocca, 1997). However, no synchronous technology merely subtracts from the environment; rather it adds new elements to monitor, a virtual environment. More critically in this expanded and potentially limitless virtual environment, the objects, states, and people that one must monitor are outside the physical environment and include distant mediated others such as team members, virtual tools, data objects, and the states of various support technologies such as automobile indicators, etc. Augmented reality promises to bring this world of virtual objects, people, and tools together and to merge them with the busy world of physical objects, people, and tools. In the expanded, parallel, and competing realm of objects and people lies the root of the dilemma of awareness management.

Carefully managing the integration of knowledge in distributed augmented reality systems that support collaborative work is key to providing the right level of interaction and engagement in group communication (Alavi & Tiwana, 2002) But, in awareness systems, the focus is on the substance and the real time resource allocation of attention and short-term memory to the people, information objects, and environments at hand. Awareness information is always required to coordinate group activities, whatever the task domain. A key problem in augmented reality systems is that the virtual and physical objects compete for the attentional resources of the individual, people, and additional information.

Part of this problem of awareness in AR can be initially characterized by a simple two dimensional space depicted in Figure 1. We can posit that for any object or person and any specific step in a task, there is an ideal level of awareness for a tool or person and that this tool or person may inhabit the virtual or physical space or a bit of both, as when real participants are annotated with virtual. This level of awareness can be characterized as a location in this two dimensional space. Focal awareness is defined as the allocation of high levels of attention to and high levels of modelling of the states of physical, virtual, or imaginary people or objects. Focal awareness is likely to use conscious attentional resources. Peripheral awareness is defined as limited consciousness of people or objects, based on largely automatic attentional processes indicating that people or objects are co-present. We will return to the issue of social presence below.

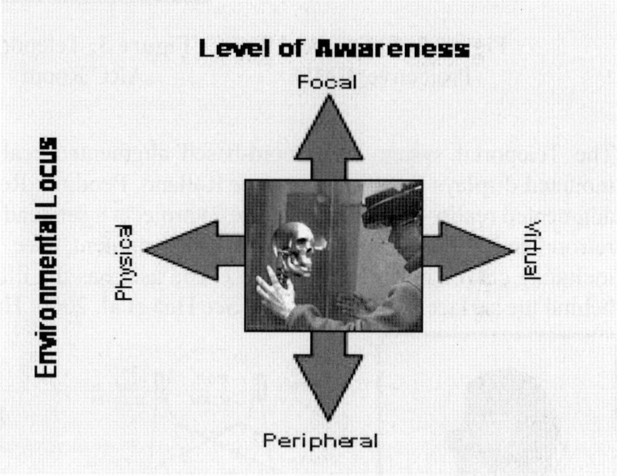

Figure 1: AR awareness continuum

Looking at the design features of awareness systems, most often they have primarily been focused on the awareness of people: their location, behaviour, and states. But in augmented and virtual reality, agency might be attributed to a real physical person mediated by a telecommunication system, an agent representing the system, or tools and environments that may be thought of as having "agency" by the user, for example that a computer system is "aware" of the user's presence. So awareness of others, initially conceived as just awareness of presence and states of office mates or others in a networked environment, might be better conceived as a general awareness of the states of agents, whether they are real people, computer agents, or objects for which a user might attribute some agency. A user may attribute dispositional "agent-like" states in objects such as tools, machines, and data objects. In all cases the key issue that relates people and objects to the user on some common plane is the user's planned, current, or past interaction with the item. A person, tool, or environment might have been, or be the current object of interaction, or is under consideration for interaction in some procedure, be it a conversation, tool manipulation, etc. Thought of this way, similar principles and issues might apply to all information objects and potentially simplify the problem space of awareness management in an augmented reality system.

2 Spatial organization of awareness

A project called Mobile Infospaces works with a fundamental psychological and design property of AR, ***spatial representation***. The project examines the psychological properties of space and its relation to the design of interfaces that manage objects, tools, and people in AR space (for theoretical basis and model see Biocca, Mou, Tang, & Owen, 2003; Mou, Biocca, Tang, & Owen, submitted). The fundamental variable that ties awareness and the augmented reality systems is the use of spatial organization and sensory cues to organize object and social awareness along the continuum of focal to peripheral awareness. As much as half of the brain is dedicated to processing visual-spatial information and guiding spatial movements (Kanwisher, 2001). In our current projects we are exploring various ways in which so-called spatial natural mappings activate primitive or highly learned cognitive responses based on perceptual or motor affordances of objects or agents around the user's body. Many information systems such as graphs and visualizations make extensive use of perceptual and spatial cues. Mapping information organization to the human brain's prodigious capacity for spatial cognition may offer a potential route for managing awareness in mobile, AR interfaces.

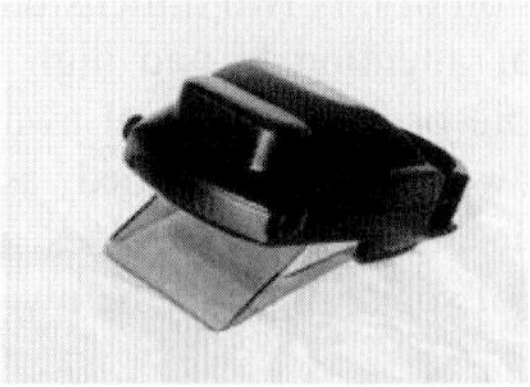

Figure 2: Prototype AR Projective HMD

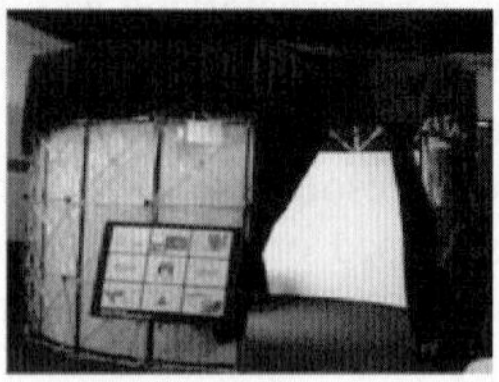

Figure 3: Teleportal ARC Room

Figure 4: Cylindrical volumetric screen with virtual object.

The Teleportal system is a room-based augmented reality system. It uses a projective, head mounted displays (HMD) (Biocca & Rolland, Pending; Rolland et al., submitted). The projective augmented reality HMDs (see Figure 2) projects a set head tracked, stereo images to a strategically retroreflective fabric in the physical environment (See Figure 3) or molded for any surface including curved ones (See Figure 4). The user has the illusion of a 3D object floating in front or behind the surface plane (Figure 4). See Hua et al, 2000; Hua et al., 2003)

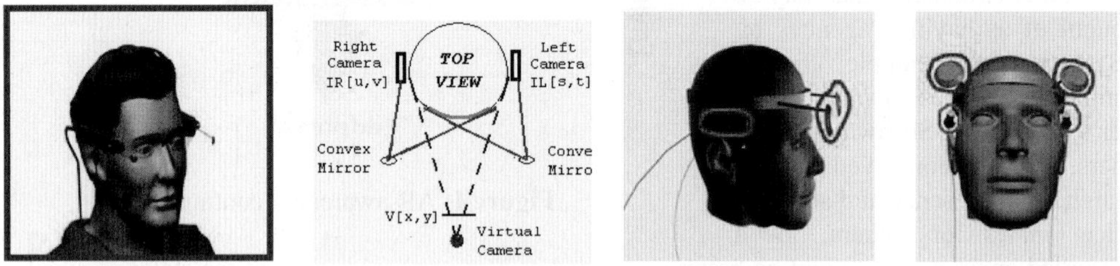

Figure 5: Teleportal Face-to-face capture: (a) concept, (b) schematic, (c, d) current implementation.

In networked teamwork situations others must sometimes move from the peripheral-to-focal awareness. The movement from one to the other is vividly suggested by two English phrases that use spatial metaphors to represent the extremes of the continuum between peripheral and focal: one suggesting peripheral awareness, "he's hardly on my radar"; to one suggesting high focal, almost excessive, awareness and social presence, "this person is in my face." With the 3D face-to-face capture system we are attempting to develop a technology that can use spatial mapping to manage social awareness from the very high focal awareness and high social presence, "he's in my face," to peripheral awareness of the other. The 3D face-to-face system is designed to capture a full 3D image of the face (See Figure 5). It uses head-worn stereo cameras with a set of small convex mirrors. This affords a highly detailed record of the facial expression of the remote user no matter where the individual is looking. We used Structured light mechanism for generating a 2D frontal view of the face and Water's model for generating facial animation. The realism in the 3D face model is dramatically improved by texture mapping. Used as input for an augmented reality system and coupled to data from a head tracker, the technique will allow us to position the 3D head of the remote user in the exact location within the matched networked Teleportal rooms (see Reddy, Stockman, & Biocca, 2003 for details).. This affords (a) awareness of spatially accurate visual 3D head-model of the attention of the remote other in the matched networked Teleportal room, (b) record of non-verbal cues of emotional states (c) high-focal awareness in cases where the other must be "in your face," (d) and allow the computer to be "aware" of facial expressions of the user.

3 Social Presence Measures for Awareness

How does one measure awareness other than within a simple continuum of "aware" or "unaware"? How can we get at the characteristics or qualities of that awareness of others? Finally, how can the general notion of social awareness be extended to not only awareness of the states of mediated others, but to the states of agents and computer tools who take on properties of agents, for example virtual animal assistants? Therefore, in the study of awareness a key issue is the measurement of ***social presence***, especially as mediated agents, virtual agents, and tools with apparent agency move within the continuum between focal and peripheral awareness.

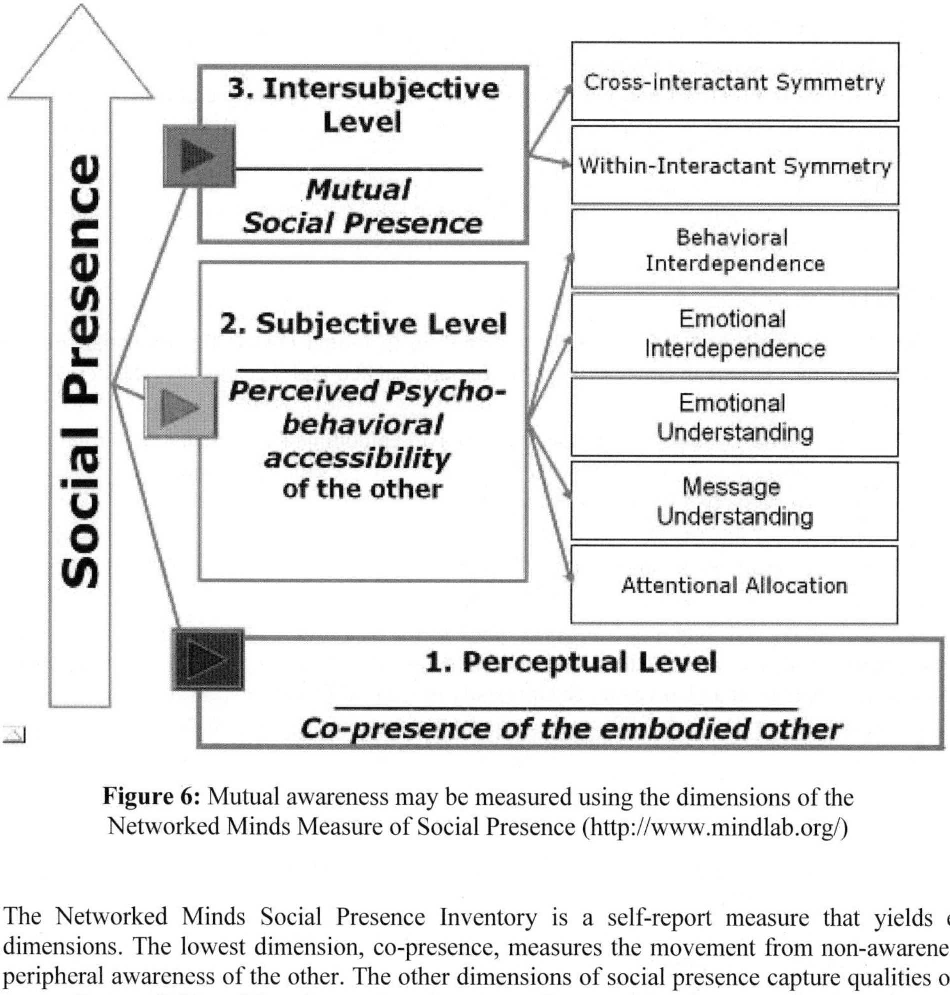

Figure 6: Mutual awareness may be measured using the dimensions of the Networked Minds Measure of Social Presence (http://www.mindlab.org/)

The Networked Minds Social Presence Inventory is a self-report measure that yields eight dimensions. The lowest dimension, co-presence, measures the movement from non-awareness to peripheral awareness of the other. The other dimensions of social presence capture qualities of the sense of accessibility of the other as the other moves from peripheral to more focal awareness, and the qualities of that awareness. The measure includes sets of paired items that assess the degree to which interactants assess themselves, how much they feel socially present, and how much they perceive the other to be socially present. This allows us to derive two other indices:

- **Within-interactant social presence symmetry**: This is an index of the degree to which an individual perceives their sense of social presence to be equal or symmetrical to the other interactants.
- **Cross-interactant social presence symmetry**: This index reports the degree to which each of two or more individuals perceive the other's level of social presence as matching the other's own self-assessed social presence.

These two indices can be used to assess the effects of different interfaces and social interactions on the degree of mutual awareness, measured as the degree to which the others are socially present. The measure might be useful to individuals who study awareness and can be downloaded from http://www.mindlab.org/.

3.1 Summary and conclusion

Awareness issues may be critical to the design of augmented reality systems. We have suggested the use of spatial organization strategy for interface designs. These may assist the user in managing the allocation of focal and peripheral awareness to people, tools, and other objects. A persistent issue is likely to be the competition of the virtual and physical environment for attentional resources, especially when the physical and the virtual are poorly integrated. When poorly integrated, physical presence in one environment may compete with the sense of physical presence in the other. Finally, we have briefly introduced how some of these issues are addressed in the design of a room based and mobile augmented reality systems.

4 References

Alavi, M., & Tiwana, M. (2002). Knowledge integration in virtual teams: The potential role of KMS. Journal of the American Society for Information Science and Technology, 53(12), 1029-1037.

Biocca, F. (1997). The cyborg's dilemma: progressive embodiment in virtual environments. Journal of Computer-Mediated Communication, 3(2), http://www.ascusc.org/jcmc/vol3/issue2/biocca2.html.

Biocca, F., Mou, W., Tang, A., & Owen, C. (2003). Mobile infospaces: A working model of spatial information organization in virtual and augmented reality environments. East Lansing: Media Interface and Network Design Lab (www.mindlab.org).

Biocca, F., & Rolland, J. (Pending). Teleportal Face-to-Face system: Teleconferencing and tele-work augmented reality system, Patent Application: 6550-00048; MSU 99-029, Dec. 22, 2000. United States: Michigan State University & University of Central Florida.

Kanwisher, N. (2001). Faces and places: of central (and peripheral) Interest. Nature Neuroscience, 4, 455-456.

Hua, H., A. Girardot, C. Gao, and J. P. Rolland, "Engineering of head-mounted projective displays, Applied Optics 39(22), 3814-3824 (2000).

Hua, H, Y. Ha, and J.P. Rolland, "Design of an ultra-light and compact projection lens," Applied Optics 42(1), 97-107 (2003).

Mou, W., Biocca, F., Tang, A., & Own, C. (submitted). Spatial cognition and mobile augmented reality systems. Behavior & Information Technology..

Reddy, C., Stockman, G., & Biocca, F. (2003). Face model construction and transmission for telecollaboration. East Lansing: M.I.N.D. Lab.

Rolland, J., Biocca, F., Gao, C., Hua, H., Ha, Y., Harrysson, O., et al. (submitted). Design and prototyping of a teleportal ultralight-weight large field of view head-mounted display. International Journal of Advanced Manufacturing Technology.

An interface for supporting versioning in a cooperative editor

Marcos R. S. Borges
Universidade do Brasil
NCE & IM
mborges@nce.ufrj.br

Alexandre P. Meire
Universidade do Brasil
PPGI - NCE & IM
apmeire@ufrj.br

Jose A. Pino
Universidad de Chile
Dept. of Computer Science
jpino@dcc.uchile.cl

Abstract

A versioning mechanism called mask is introduced. This mechanism allows to keep several versions of diagrams for later reviewing and use. Masks are also useful to provide awareness of the design evolution in a group creative process. Masks have been implemented in a groupware tool developed to support the cooperative edition of UML collaboration diagrams.

1 Introduction

Groupware represents a paradigm shift from computer science, in which human-human rather than human-machine communications are emphasized. To support this new paradigm one of the requirements is to provide mechanisms for awareness, which Dourish & Bellotti (1992) define as an *"understanding of the activities of others, which provides a context for your own activity"*. Without awareness a person cannot build his sense of a group and the human-human paradigm will remain mainly on the intentional level (Sohlenkamp & Chwelos, 1994).

Awareness mechanisms are essential to group support systems in order to transform irregular interactions of group members into a consistent and perceptive performance over time. Awareness mechanisms are important to keep members up-to-date with important events and therefore to contribute to a more conscious acting from their part. This is true either for synchronous or asynchronous interactions (Borges & Pino, 2000).

In a cooperative editor, the use of versions has much to do with awareness. Multiple versions can be used in a cooperative editor to support awareness by means of storing evolution in order to explain the transformation process. Multiple versions can also be used when we want to accommodate different points of view without losing their original context. One of the challenges is the design of a simple and intuitive interface to represent multiple versions of the same object.

A metaphor can be understood as a figure of speech based on some resemblance of a literal to an implied subject (Lakoff & Johnson, 1980). The rationale for the growing number of metaphors is that they provide an intuitive way for users to interact with computers as they were on their own environments (Marcus, 1993). Since different types of software may require different metaphors, the same happens to groupware (Borges & Pino, 1994).

In this work we describe the use of the presentation slide metaphor, called *mask*, to represent multiple versions of diagrams in a cooperative editor. A mask is a graphical element that is virtually placed over another mask to provide a new interpretation of its contents. In a broader usage it may have different shapes, sizes and colors, but in the context of this paper, it is considered to have the size of the diagram area. The mask metaphor can be very useful for designing user interfaces in systems dealing with versioning and annotation.

2 The mask metaphor

Metaphors are ubiquitous in the user interfaces of today's computers. Software designers have been incorporating metaphors into a variety of software from operating systems to information retrieval applications. The motivations for using metaphor in the design of user interfaces are similar to the reasons metaphors have long been popular in education. Many educators have observed that giving students comparisons can help them learn (Smilowitz, 1996).

A metaphor is a visual expression of how a computer system is similar to and works like an entity the user already knows. A metaphor suggests a visual model the user can incorporate into his own model. To be effective both the metaphor and the user's internal model should comprise a cognitive map, images and relationships between images (Carroll, Mack & Kellogg, 1988).

There have been several proposals for the use of metaphors in group support applications. Pino (1996) proposed the used of a stick-on metaphor to deal with multiple versions in co-authoring. Beard, Palanlappan, Humm, Banks, Nair & Shan (1990) proposed the use of the slide metaphor to support group calendaring. The mask is an extension of the slide metaphor, but it also allows elements to be hidden or modified from its underneath mask. A mask has transparent and opaque areas and may contain icons, painted areas (transparent or not) and data in general.

The mask as proposed in this paper allows representing multiple versions generated during the designing of a diagram. Working in a mask means that you can modify the contents of the mask below it. Users represent their ideas by virtually placing a mask on top of an existing diagram and making the desired changes while preserving the previous diagram. Two masks can be compared by placing and removing the mask onto the original one. A version created from another one may be considered a child of it. In this work, we organize the versions as a tree. To keep consistency, a mask cannot be updated once a descendant mask has been generated from it.

A mask can also contain annotations, either general remarks or specific annotations linked to an element of the diagram. In this case, however, a mask keeps the annotation contents of all its predecessor masks, i.e., the user can only add new annotations to existent ones. Only in the case the element is removed from a mask, all associated annotations are also deleted from this mask. A similar scheme is provided for comments made during a chat session.

3 An implementation of masks

We demonstrate the use of the mask metaphor in a groupware tool developed to support the cooperative edition of UML collaboration diagrams. The first requirement is that several users may be able to work on the same diagram. This is shown in Fig. 1: two user windows are overlapped to depict the sharing of the same working space. A WISIWYS interface is provided by the system. The users can make use of telepointers to locate each other's pointing device.

Since different points of view exist during the building of such diagrams, the editing tool should support not only a shared workspace but also multiple versions. This is the aim of the masks mechanism. A user can make changes on a mask without affecting the original diagram.

Suppose a conflict of opinions arises, because two users have different views on how the diagram should look like. One way to solve the problem is to make the users discuss it a priori using some communication device or meeting in person. An alternative way is to let the users present their proposals and have a discussion a posteriori. The advantage of the second approach is that the discussion will probably be more informed and others may participate in it.

In order to follow the second approach, one of the users can simply generate a new mask and make changes in the diagram without disturbing the other users. Figure 2 illustrates the case of two users working on different masks (2 and 4). At any time, however, one user can see what the other is doing or did before by clicking on the corresponding mask in the menu at the right hand side of his window. This menu in Fig. 2 shows, e.g., that mask No. 2 is called “Remove link”, it was created by modifying mask No. 1 and it is closed (marked as X). One may modify the contents of a mask if the creating user leaves it open to contributions.

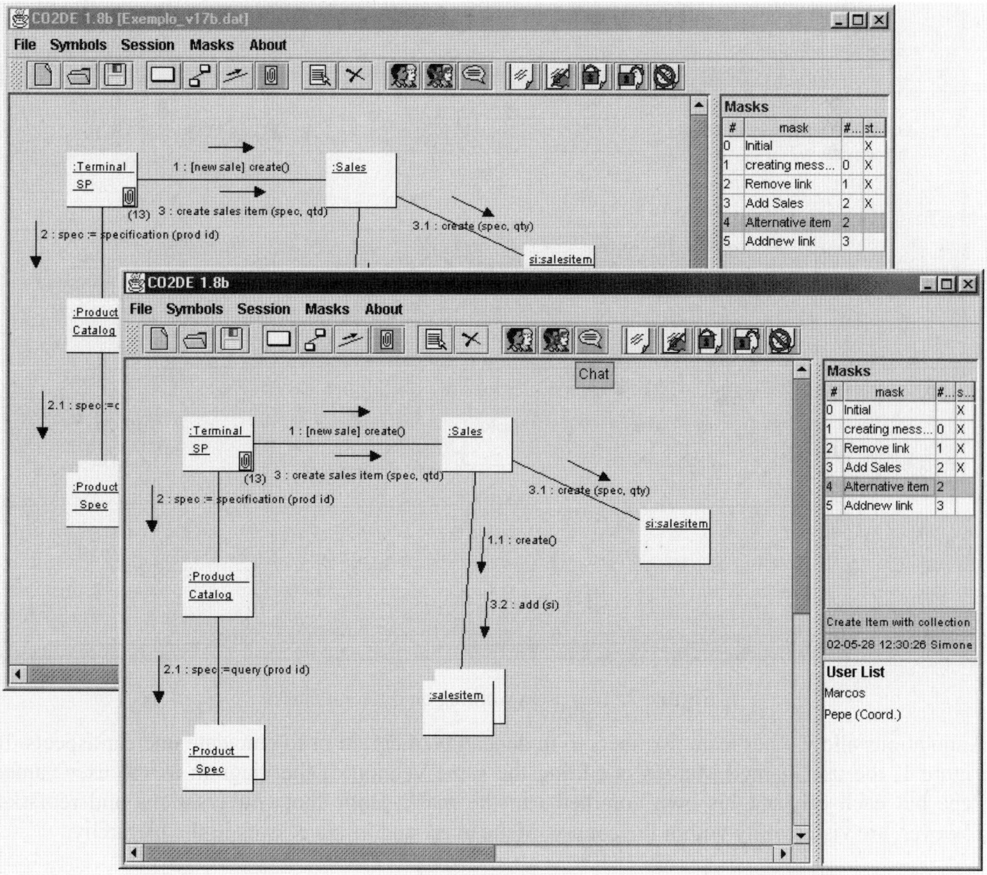

Figure 1. Two users working in the same mask with a WYSIWIS interface

Note that users build a hierarchy of masks in this way, shown as a list at the menu on the right hand side of the window. A mask may be closed, meaning that no more changes are permitted. This option is used to allow users to create successor versions of a mask. A mask can also be blocked (marked by a B), i.e., the mask temporarily accepts changes only from the user who requested the blocking. The mask can be released at any time, but only by the user who blocked it.

The tool may be used by people to make contributions almost independently of other users. However, this is not the goal. The group must work together, which calls for high interaction among its members. Thus, the tool offers several opportunities for this interaction. A chat and an annotation services are examples of interaction support. Also, awareness mechanisms are provided, e.g., the names of users connected at the same time are listed in the user's window.

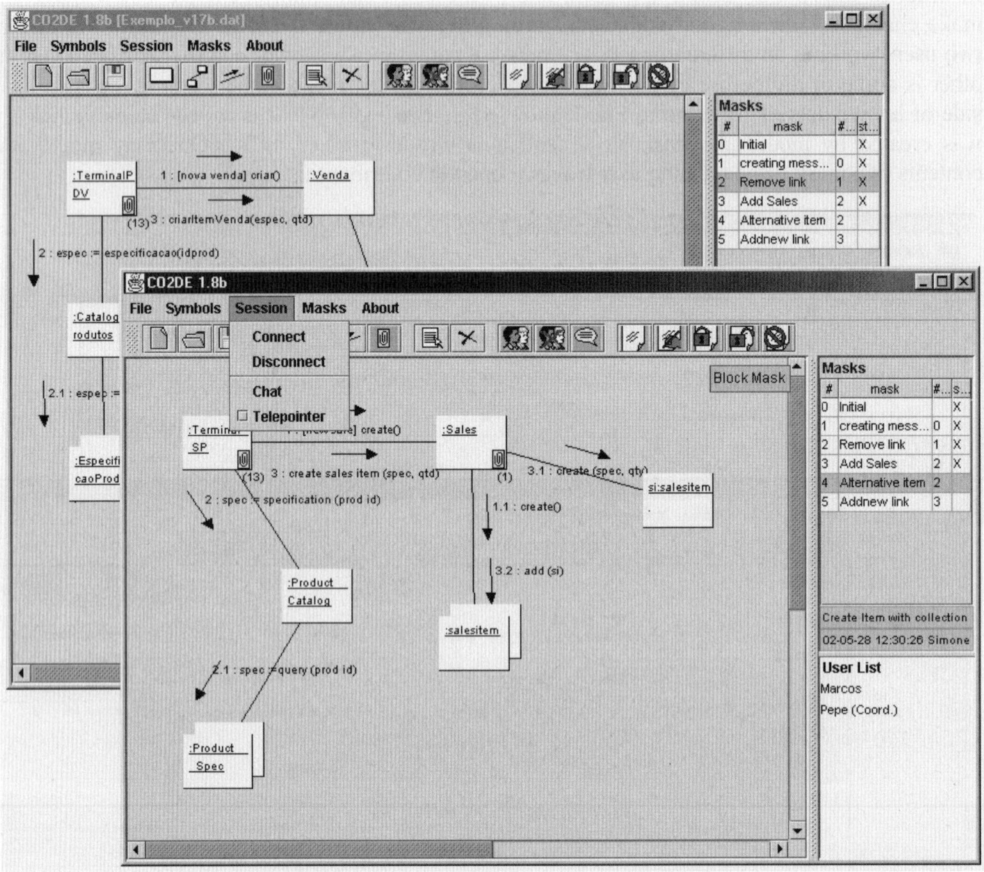

Figure 2. Two users working on different masks

Annotations allow other team members to understand a diagram or to elucidate unclear aspects. In Figure 1, the user named Marcos is editing the same version of the diagram as the user named Pepe but each can add his own comments to the mask. Both chat conversations and remarks, however, are visible only within the context of the mask and its successors in the hierarchy.

4 Conclusions

Versions are frequently needed in group work. They allow the members of the group develop several optional designs before choosing a final one. Indeed, group members may feel free to propose alternative designs to the one(s) already contributed by other participants, because the new proposals do not destroy or delete the previous designs (Pino, 1996). Versions are also useful as holders of the history of the development process.

This paper has presented the masks: version mechanisms for holding several diagrams. It has also presented a specific implementation, intended for the group creation of UML collaboration diagrams. Each mask is presented on screen as a separate window; it also has a representation of the whole versions tree. The mask has contextual information, which may be provided by any participant. This allows *in context* discussion for rich group interaction.

User acceptance of this versioning mechanism will be reported in a forthcoming paper. Initial experiments with three groups in a Software Engineering environment show easy learning and production of several masks for each design project.

Acknowledgments

This work was partially supported by grants from CNPq (Brazil) Prosul AC-62 and Fondecyt (Chile) No. 1000870 and 7000870.

References

Beard, D., Palanlappan, M., Humm, A., Banks, D., Nair, A. & Shan, Y. (1990). A visual calendar for scheduling group meetings. Proceedings of CSCW´90. New York: ACM Press, 279-290.

Borges, M.R.S. & Pino, J.A. (1994). Additions to the card metaphor for designing human-computer interfaces, 4th Workshop on Information Technologies and Systems (WITS '94), Vancouver, Canada, 243-251.

Borges, M.R.S. & Pino , J.A. (2000). Requirements for shared memory in CSCW applications. 10th Annual Workshop on Information Technologies and Systems (WITS´00), Brisbane, Australia, 211-216.

Carroll, J.M., Mack, R.L. & Kellogg, W.A. (1988). Interface metaphors and user interface design. In Helander, M. (Ed.): Handbook of Human-Computer Interaction, Amsterdam: Elsevier Science Publishers B.V., 67-85.

Dourish, P. & Belloti, V. (1992). Awareness and Coordination in Shared Spaces. Proceedings of CSCW '92, New York: ACM Press, 107-114.

Lakoff G. & Johnson M. (1980). METAPHORS we live by. Chicago, IL: University of Chicago Press, 1980.

Marcus, A.(1993). Metaphor design and cultural diversity in Advanced User Interfaces. In M. J. Smith and G. Salvendy (Eds.): Human-Computer Interaction: Applications and Case Studies. Amsterdam: Elsevier Science Publishers B.V., 469-474.

Pino, J.A.(1996). A visual approach to versioning for text co-authoring, *Interacting with Computers* 8 (4), 299-310.

Sohlenkamp, M. & Chwelos, G. (1994). Integrating Communication, Cooperation, and Awareness: The DIVA Virtual Office Environment. Proceedings of CSCW '94. New York: ACM Press, 331-343.

Smilowitz, E. Do Metaphors Make Web Browsers Easier to Use?, Retrieved January 31, 2003, from http://www.baddesigns.com/mswebcnf.htm

Suspenseful User Experiences in Collaborative Virtual Spaces, Enabled by Interactive Narration

Norbert Braun

GRIS, FB Informatik, TU Darmstadt
Fraunhoferstraße 5
64283 Darmstadt,
Germany
NBraun@gris.informatik.tu-darmstadt.de

Oliver Schneider

Digital Storytelling, ZGDV e.V.
Fraunhoferstraße 5
64283 Darmstadt
Germany
Oliver.Schneider@zgdv.de

Abstract

We describe the influence of storytelling on the experience of suspense and immersion in virtual spaces. Following an analysis of the literary and dramatic parentage of interactive story experiences as it relates to German Literary and Drama theory, we derive an approach to automated interactive storytelling with regard to the use of computer support for suspenseful user experiences. An application verifies the discussed approach. The approach provides a method for making virtual environments attractive to users.

1 Interactive Narration Methods

In the field of Virtual Reality and Augmented Reality, we lack a structuring method, which involves the user in the same way traditional content-based media, like movies, television, radio or books, do, namely involvement of the user by giving her an experience of immersion in the content presented by the medium. This immersion is enabled by the feeling of *suspense* - an experience where the audience forgets about the real world and engages totally with the virtual world of their minds, a world stretched by the author or narrator of the story (Ryan 2001).

Traditional content-based media use dramatic structures (stories) to enhance suspense and immersion. Immersion takes place via a number of factors. Ryan (Ryan 2001) distinguishes between spatial, temporal and emotional immersion. Dramatic structures intensify the temporal and emotional immersion: The temporal immersion is provided via the expectations of the audience in regard to the experiences of the protagonist of a story. The emotional immersion is provided via the structure of the Drama itself:

Typically, the first act offers an explanation for the ongoing actions and motivations of the drama's characters. The audience knows both, the goals of the protagonists, as well as what the protagonists will suffer should they not reach those goals.

To define a concept of interactive storytelling for virtual or augmented spaces, we examine the basics of literature and dramaturgy.

Goethe (Goethe 1819) defines three categories (German: Gattungen): Lyric, Epic and Dramatic. The Novella is of the category Epic, but with an internal dramatic structure (exposition, a climax or turning point, a decline and an end. It shows events that are interpreted as turning points in the life of a protagonist and that, therefore, produce suspense for the audience.

For the discussion of the dramatic elements of Interactive Stories, we refer to Staiger's definition of the form category *Dramatic*, (Braak 1996, p.117). He used the term Drama to describe *the playing of characters*. The characteristics of Interactive Stories are classified as Closed One-Place-Drama.

The narrative elements used in Interactive Storytelling are based on Weber (Weber 1998), who defined narration as 'a serial addressing of temporal specified circumstances'.

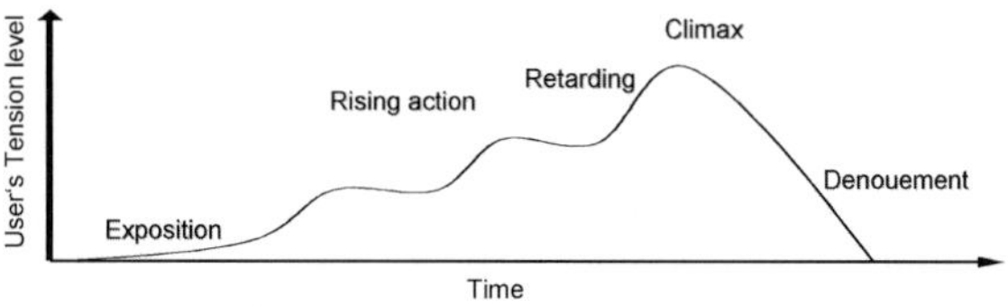

Figure 1: Drama Structure According to Aristotle

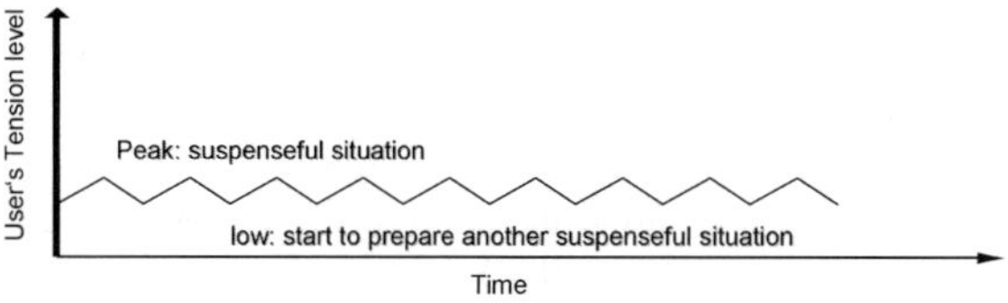

Figure 2: Suspense Structure of a SitCom

Of course, there are several models of dramaturgy to produce suspense, for example the Sitcom (Situation Comedy) TV format, see Figure 2. Those models refer to a time scale of suspense:

Field (Field et al 1987) describes the suspense within a theater play in a three-act structure, with the first act providing an exposition, the second act a confrontation and the third act a denouement of the story. The same holds true for Aristotle: he offered a five-act structure (exposition, rising action, reversal of fortune (crisis), climax, denouement), very close to the Novella structure, see Figure 1. For Hollywood feature films, Fields demands the occurrence of a relevant action every 15 minutes. For the TV format of a Sitcom, the amount of time between two suspense-peaks is much shorter, e.g. every minute. Unfortunately, these dramaturgy models are a bit too unspecific to be used within a computer-processed story narration.

The screenwriting software Dramatica (Screenplay Systems 2003) provides a very detailed specification of suspense produced by personal conflict models. Dramatica is used to describe in detail the conflict of certain persons and the shifting of their mentality, while the actions of the story go on. Unfortunately, Dramatica is a bit too detailed to be used as the basis of our system. A more pragmatic approach is the usage of semiotic -based concepts, like the story morphology of Propp (Propp 1968). Propp's approach is based on characters and their actions within specific parts of a story and can, therefore, be used to track personal conflicts. These conflicts serve as the basic suspense generator for the story's audience.

The next section will provide an overview of our story model. In section 3 the architecture of a storytelling prototype system is outlined, while section 4 and 5 reveal prototype results, a conclusion and an outlook on future work.

2 Approach to Interactive Narration

Our approach to Interactive Narration combines a concept of personal conflict with the reincorporation of user activities. The foundations of our work are based on Audience Participatory Theatre (APT) and a Morphological Function approach to storytelling. The

Morphological Function approach offers the criteria of the interactive mimetic presentation of a Novella; the narrative user interaction aspects are covered by Audience Participatory Theatre.
As found in Wirth [Wirth 1994, p.53], narrative in APT is based on the four general principles of Roles, Pegs, Links and Reincorporation:

- Roles: The definition of characters, their relationships, their objectives and their environments.
- Pegs: A story element within the narrative, somewhat like the smallest piece of a story.
- Links: Tying an unrelated story element into the existing narrative.
- Reincorporation: The technique of linking an element brought up earlier in the story to the current situation.

In accordance with these principles, each user is not simply left alone playing his character, but is carefully, in regard to the ongoing actions and his own experience, integrated into the overall narration of the story.
The general story modelling, the story's structure as demanded by [Wirth 1994, p.53], is found by using the semiotic model of stories founded by [Propp 1968]. Propp defines a story as a set of morphological functions, dependent on the dramatic characters within those functions. Propp showed the possibility of generating new stories by combining functions with regard to a limited number of combination-rules. For a detailed description of Propp's approach and the Interactive Story Engine possibilities, see Braun (Braun 2002).
As the work of Propp is based on the actions of characters with regard to the story as a whole, it can be adapted to the character-based approach of APT to give users a unique and interactive experience of story.
The following section gives an insight into the Scene Engine of such a storytelling system. Within the Scene Engine, the ATP and the Morphological Functions approach get closely connected.

3 Architecture of the Narrative and Collaborative Augmented Reality System

The architecture of our storytelling prototype is adapted to the needs of a collaborative, augmented reality space. This implies several side effects, as a lot of dramaturgical functions used in Virtual Reality or Videos / Movies like camera settings, cuts, etc. are not possible, as the environment is fixed within reality. Authoring details are described by Schneider et al (Schneider et al 2003). Within this section, we provide insight into the possibilities of the combination of ATP and the Morphological Function approach within the scene processing (in the Scene Engine module) of the system - the system is outlined in Figure 3.
A scene consists of a place, a time and the content to be presented, see Field (Field et al 1987). For an augmented reality scene, place and time are defined by the position of the user and the specific time the user uses the system. The user position has to be mapped to a stage (as stored in the Environment Description). The user position, e.g. the user viewpoint, is the camera view of the scene; every virtual object within the scene (given by the Environment Description) is rendered in relation to the user viewpoint.
Every scene of the story is described by a scene script (stored in the Scene Script module). It contains the role of the user, as well as several timing- and content-related instructions for the virtual characters populating the scene. Those instructions are given to the Subject Behaviour module, where they are executed by the virtual characters. The details of the virtual characters' behaviour and speech are predefined within the scene script, but modified by the Subject Behaviour; the local execution is dependent on the user's position and user's actions.
The user interaction possibilities are limited; They are defined in the scene and interpreted by the User Interpretation module. User movements and viewpoints are interpreted as APT links, as the

movements and the viewpoint indicate the objects of interest to the user. Those objects of interest are extracted, are integrated into the scene by using the Environment Description module and are sent to the story engine in order to reincorporate the user's interest in future scenes.

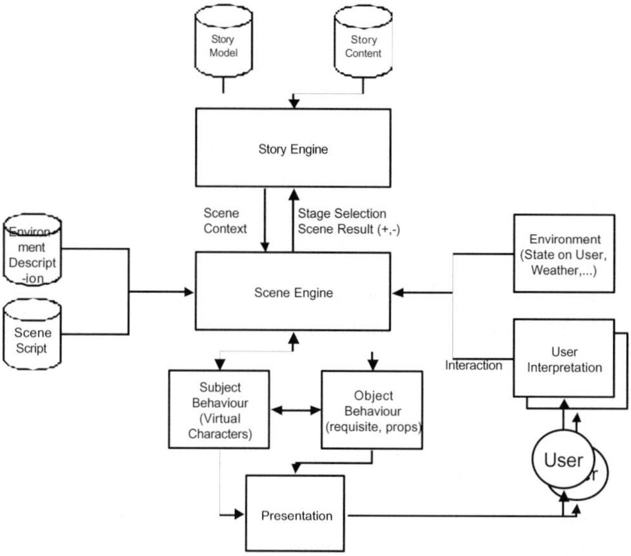

Figure 3: Architecture of the System

By using the predefined scenes of the story engine to handle the virtual characters' behaviour within a scene and by giving scene results, as well as reincorporation, back to the Story Engine to decide which scene to present next, a suspenseful user experience is highly likely.

4 Results

The approach is verified by the Geist project (*Geist*, an AR Environment for the interactive narration of historic stories, (Geist 2003)) of the Centre of Computer Graphics (with the Computer Science Department, University of Darmstadt), Fraunhofer Institute Computer Graphics and European Media Lab. The prototype gives insights into the Thirty Years' War as it unfolded in Heidelberg, Germany.

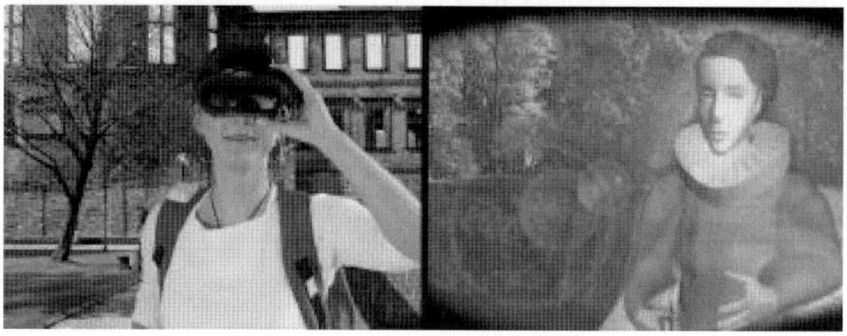

Figure 4: User, Interacting with a Ghost in the *Geist* Augmented Reality Environment

The users (users in the prototype evaluation are kids age 13 to 15) are experiencing a story together, collaboratively playing a role in the story as a unit. The story is structurally predefined in plots and story roles that are adapted to the interactions of the users, e.g. local appearance of the users on different stages, time flow, interactions with virtual characters (see Figure 4) and artefacts, as described in section 3.

5 Conclusion and Future Work

We described a collaborative augmented reality system, enhanced by techniques of storytelling to provide a suspenseful user experience, driven by suspense concepts of interactive narration. We outlined the basics of the narration approach, described the system architecture (with a focus on scene processing) and provided an overview on the resulting project (Geist, an Augmented Reality Environment on the Heidelberg castle). Future work will concentrate on the improvement of the Scene Engine's knowledge of Augmented Reality and on the integration of the APT concepts into the Story Engine, as the user's possibility to integrate pegs (see section 2) is very limited.

6 References

Braak, I. (1969). Poetic in Stichworten: Literaturwissenschaftliche Grundbegriffe. Verlag Ferdinand Hirt, Würzburg, Germany.

Braun, N. (2002). Storytelling & Conversation to Improve the Fun Factor in Software Applications, CHI 2002 *Conference on Human Factors in Computing Systems*, Workshop Funologie, Minneapolis, Minnesota, SIGCHI, ACM, USA.

Braun, N. (2003). Storytelling in Collaborative Augmented Reality Environments, *11th International Conference in Central Europe on Computer Graphics, Visualization and Computer Vision*, Plzen, Czech.

Schneider, O., Braun, N., Habinger, G. (2003). Storylining Suspense: An Authoring Environment for Structuring Non-Linear Interactive Narratives, *11th International Conference in Central Europe on Computer Graphics, Visualization and Computer Vision*, Plzen, Czech.

Screenplay Systems, Inc. (2003). Theory of Dramatica, Retrieved January, 2003, from http://www.dramatica.com.

Field, S., Märthesheimer, P., Längsfeld, W. (1987). Drehbuchschreiben für Film und Fernsehen.

Goethe, J.W. von. (1819) Notes and Essays on the West-Eastern Divan, in *Werke 2 of Goethes Werke* (ed.) Erich Trunz, 14 vols., 1948-60, 126-267.

Geist, Official Geist Project Web Site, Retrieved January, 2003, from http://www.tourgeist.de/.

Propp, V. (1968). Morphology of the Folktale. University of Texas Press.

Ryan, M.L. (2001). Narrative as Virtual Reality. Immersion and Interactivity in *Literature and Electronic Media*, chapter Immersion and Interactivity in Hypertext. Johns Hopkins University Press, Baltimore, USA.

Weber, D. (1998). Erzählliteratur: Schriftwerk, Kunstwerk, Erzählwerk. Vandenhoeck & Ruprecht, Göttingen, Germany.

Wirth, J. (1994). Interactive Acting, Acting, Improvisation, and Interacting for Audience Participatory Theatre, Fall Creek Press, Oregon.

Collaboration and Core Competence in the Virtual Enterprise

Rainer Breite and Hannu Vanharanta

Tampere University of Technology, Pori
Pohjoisranta 11
Pori, Finland
rainer.breite@tut.fi, hannu.vanharanta@tut.fi

Abstract

During the Spring of 2000 and Autumn 2000 and 2001 students with a bachelor's degree in engineering participated in an experiment which was intended to investigate the possibilities to manage a virtual company by means of collaboration. The test subjects developed a virtual company operating on the Internet with suppliers available on the Internet. The test subjects' satisfaction and perception of the experiment were solicited. The task was complex, but taking into consideration the test subjects' background, their degree in engineering, as well as their several years of industrial experience, these students, if any, should have the capabilities to manage the task given and thus provide us with some valuable information about the possibilities of collaboration and virtuality in the Internet.

1 Introduction

In our study the decision-makers are working through several www addresses in order to create a virtual company. Many of the found www addresses are clicked away, while others are brought up for a more thorough inspection. Our main interest is to find out whether collaborative and virtuality is possible on the Internet. The main research question can be divided into the following sub questions: 1) Is it possible to find necessary suppliers on the Internet? 2) Is it possible to evaluate different suppliers on the Internet? 3) Is it possible to create a virtual enterprise on the Internet? 4) Does the Internet environment give support enough for the decision-maker? We assume that the fourth question forms an essential part of the other three questions, as the decision maker needs support for his/her decision making process.

By using the hyperknowledge platform (Chang et al., 1989, 1993, 1994; Vanharanta et al., 1995), we assume that we can deal with structuring and processing of decision problems, which involve creating the virtual enterprise. We start from the idea that our virtual enterprise is an object, which has to be formed on the knowledge base. Thus the decision maker needs support for his/her decision making process, through which s/he forms the virtual enterprise and affects the contracts between the decision maker / the virtual company and the supplier. The decision maker has to evaluate the ability to make the contracts solely by using the information which s/he gets from the computer screen. Thus we address the usability and the utility of the Internet in the context of collaboration by utilizing the hyperknowledge platform. Usability and utility are two equally important factors in the evaluation process. Newell et al. (1972) define utility as "the question of whether the functionality of the system in principle can do what is needed". Usability, in turn, is usually described in terms of criteria like learnability, efficiency of use, memorability, a small number of errors, and subjective satisfaction (Nielsen 1990).

The present study focuses on the two most critical components of the hyperknowledge environment: the user and the contents of the Internet applications. We try to find out whether the Internet applications used are a plausible system for evaluating the decision support activities of business contracts, and what are the advantages and disadvantages of Internet applications from the user's point of view. The validation methods are partially the same that which have been used in expert systems (O'Keefe et al. 1987 and O'Leary 1988) and in the hyperknowledge system evaluations (Vanharanta et al. 1995), i.e. a form of performance validation combined with a questionnaire.

2 Theoretical Framework

2.1 Hyperknowledge and Decision Support System

The concept of hyperknowledge is wider than those of hypertext or hypermedia, although it follows the same general principles. Hyperknowledge is an ideal working and learning environment that holds knowledge and, at the same time, defines the nature of hypertext and hypermedia. The user can navigate freely in this environment (as on the Internet), and widen his/her own knowledge. (Chang et al. 1989, 1993, 1994). The basic goal of this framework is to serve active decision support that enables the decision maker to participate actively throughout the decision process. The framework takes into account that the decision maker cognitively possesses and processes many diverse and interrelated pieces of knowledge e.g. procedural knowledge, descriptive knowledge, reasoning knowledge, etc. The user (i.e. the user's mind) is able to freely deal with and control these different pieces of knowledge. (Chang et al. 1989, 1993, 1994)

A decision support system (DSS) (See Dos Santos et al. 1989, p. 3), in turn, consists of three main components: *The Language and the Presentation System* mediate messages to and from the decision support system. The language System (LS) on the Internet is controlled by using a hypertext transfer protocol (http). The user activates a www page and a www address by using a mouse and keyboard. The presentation System (PS) on the Internet application is everything presented in www page/screen format and which can be printed in paper form. *The Problem Processing System (PPS)* handles all the user requests or responses to and from the various knowledge sources in the system. On the Internet this all means search functions and processes by the Internet application. *The Knowledge System (KS)* contains all the decision support system's knowledge and it stores, in groups, concepts that are related to each other by definition and/or by association. On the Internet applications this means that all available contract information, knowledge and procedures in www addresses, sites, and pages are placed in this system. Thus the decision support systems (DSS) should be a natural extension of the decision maker's internal activities.

2.2 Collaboration in the Internet

In our case collaborative has been defined as follows: "*The ability of two or more people or groups to transfer data and information with the capability of on-line interaction. The distinguishing feature is the ability for many-to-many interactions and information sharing, unlike e-mail where the interaction is one-to-one or one-to-many*" (http://www.collaborate.com). Ideal collaboration in the virtual context means that it is possible to create a virtual enterprise by using the Internet. The virtual enterprise, based on collaborative contracts on the Internet, can then control and evaluate all the needed transactions. This, in turn, means that a company has the ability to form a network, which consists of suitable suppliers and distributors and the members' core competences support the company's core competence.

3 The Empirical Study

We made three separated tests, where we conducted an exploratory study on M. Sc. (Eng.) students' capabilities to create a virtual company on the Internet. The first experiment made in the spring of 2000 (marked with S00), the second in the autumn of 2000 (marked with A00), and the third in the autumn of 2001(marked with A01). Altogether (18 (S00); 30 (A00); 25 (A01) M.Sc.(eng.) students with a bachelor's degree in engineering participated in the experiment. All of the students have industrial engineering as their academic major. The mean age of the participants was (35 (S00), 28 (A00), 29 (A01)) years and the participants had on average (9.2 (S00), 3.6 (A00), 6 (A01)) years of industrial experience. The majority of the test subjects (i.e. 93 %) visit the Internet daily or several times per week. The test subjects are not experts in purchasing management but all are quite well grounded in the theories of purchasing management and they understand several concepts of contracts (cf. User knowledge in a hyperknowledge environment). In every test case the test subjects had already planned their own "case company" and their company had product, where to components were looked. The task given to the test subjects was twofold; (i) to create a virtual company on the Internet and (ii) to report on their experiences and perceptions, especially the problems they faced during encountered during the task. The test subjects represented the knowledge managers and all the contracts were to be made between the knowledge manager and the different suppliers. When choosing the suppliers a requirement was that the transportation costs should be reasonable when taking the geographical distances into account. The test subjects were asked to construct the virtual company organization, i.e. construct a flowchart of different suppliers with their www addresses. This flowchart was to be designed in a logical way and delivered as a report at the end of the experiment. No e-mail contacts were allowed during the experiment between the test subjects and the suppliers. The experiment was conducted in one of the PC-classes at Tampere University of Technology in Pori. The experiment started at 8 a.m. and the test subjects were given 8 hours to complete the given task.

4 Results

Our utility analysis focuses on research constructs, which are based on the hyperknowledge framework, and the validity and utility model for the hyperknowledge environment.(c.f. Vanharanta 1995). In this model the knowledge of the system has been divided into five different types of knowledge: linguistic knowledge e.g. computer explanations, descriptive knowledge e.g. fact data etc., procedural knowledge, reasoning knowledge and presentation knowledge (c.f. Chang et al. 1994). Our research paper is concerned with descriptive knowledge only, thus we examine how the information which is transmitted via the www pages, supports the user's evaluation of business under his/her decision- making. We also refer to the statistical results calculated from the questionnaires. The test subjects' assessments of their own performance was measured on (i) the overall satisfaction rate, (ii) attitudes towards the task and (iii) how experienced they were with the Internet. The attitudes towards Internet as a source for creating a virtual furniture company were measured with the following factors, (i) finding relevant suppliers, (ii) making comparisons between the suppliers, (iii) choosing the suppliers, and (iv) the usability of the suppliers' web sites. The responses were mainly measured using the Likert attitude scales (5-point scales ranging from "strongly agree" (5) to "strongly disagree" (1)).

In this paper the above mentioned research constructs used were divided into three different categories: The first category describes traits and feelings perceived by the test subjects themselves during and after the tests. In our study these constructs are "lost in space", and "cognitive overhead". The second category describes perceived outcomes for the test subjects interacting with the Internet application, e.g. what was the contribution of the virtual environment for the user. In our study these constructions are: "learning", the "creation of comprehensive understanding", "understanding the Internet and its applications", "usefulness in acquisition", and the "utility of the Internet". The third category (satisfaction) describes user satisfaction. (cf. Conklin 1987; Vanharanta et al. 1995)

The task was completed by all the test subjects, however with varying results. The number of suppliers needed for a virtual company was on average 12, the maximum number being 20 and the minimum being 6. The number of visited web sites also varied. On an average the test subjects worked through 9 out of 12 sites when creating the virtual company, the maximum being 13 and the minimum being 5 sites.

4.1 Performance Validity and Utility Assessment

<u>Finding the suppliers on the Internet.</u> According to Vanharanta et al. (1995) one of the problems in hypertext and hypermedia applications is the feeling of being "lost in space". This means that the users loose their "coordinates". In Internet applications, especially www sites, this means that the user is unable to find "the right page" and s/he is not sure, where s/he exactly lays and what the context of the www page is. This means, that the user does not have any familiar framework. (cf. Lost in space in a hyperknowledge environment).In the questionnaire we asked for the subject's opinion by evaluating his/her ability to find "the best supplier" among all the suppliers. On an average the test subjects did not consider finding the suppliers on the Internet as too difficult, the mean value being (2.9 (S00); 3.0 (A00); 2.8(A01).

<u>Making comparisons between the suppliers.</u> We tested learning by asking the test subjects to estimate the ability to make contracts with suppliers. So the test subjects had to evaluate suppliers by using the information which was available on the suppliers' www sites. (cf. Learning and creation of comprehensive understanding in a hyperknowledge environment). Once test subjects find the suppliers they need to make comparisons between them. The comparisons were to be based solely on the information available on the Internet. Intuitively this phase would be more difficult than finding the suppliers on the Internet. The results give some evidence for this but the difference is not significant. We can, however, find without exception that the average values for making comparisons are some cases higher than for finding suppliers on the Internet (mean 3.3 (S00); 2.5 (A00); 2.7(A01)).

<u>The usability and utility of the suppliers' web sites.</u> Our purpose was to indicate how well the user can handle the knowledge which the Internet contains, how aware s/he is of that knowledge, and how close the Internet can be to the ideal hyperknowledge environment. We tested the utility of the Internet by asking about the usability of the suppliers' websites. On an average the test subjects did not see the usability of the suppliers' web sites as too problematic when creating the virtual furniture company, mean value 3.1(S00); 3.2 (A00); 2,9 (A01). This means that the quality of the web sites, the relevant information not provided on the site, such as price and delivery terms, no interactive sites, missing company information which does not give a very reliable picture of the suppliers, just to mention a few.

<u>Satisfaction and expectations.</u> On an average the participants considered the experiment as realistic in comparison to things that individuals and organisations do in a typical or common business situation, the mean value being 2.3 (S00); 3.3 (A00); 3.3 (A01). On an average the students were not very happy with their performance, the mean value being 2.6 (S00); 2.7 (A00); 3.2 (A01). When

asked about it, the majority of the participants 61% (S00); 64% (A00); 56% (A01) thought that it is possible to manage a virtual furniture company, 28% (S00); 28% (A00); 32% (A01) being not sure to the idea. The majority of the students thought, based on their experience from creating the virtual company that it would be possible to agree on co-operation in general and on contracts on either a permanent or a temporary basis. Considering partnership the opinions were divided. In general the test subjects did not find the effort as too demanding, mean value 2.4 (S00); 3.0 A00); 2.9 (A01). The level of interest as well as the challenge of the task were quite good, both scoring 3.05 (S00); 4 (A00); 3.5 (A01) on average.

5 Conclusions

We can conclude, that there were no problems in finding the necessary suppliers on the Internet. The problems start with the evaluation of the suppliers, i.e. comparing and choosing the suppliers. The fact that the participants were not allowed to send e-mails to the suppliers made the task difficult. If the necessary information was not available or not found on the suppliers' web sites then the participants had to live with that and make the comparisons as well as choosing the best suppliers without proper information. According to our findings we can see a slight trend of growing complexity from finding to comparing and to choosing the best suppliers. The differences are, however, not significant.

All the students, with the exception of six, managed in principle to create a virtual company. Whether these companies would in practice be manageable was not tested. The participants were rather positive to the idea of managing a virtual company. Their opinions were divided regarding partnerships but the majority thought that it is possible to agree on co-operation in general and on contracts on either a permanent or a temporary basis. According to the user satisfaction results we can conclude that users seem to need more relevant and supporting information for their decision-making.

References

Chang, A-L. Holsapple, C.W. Whinston, A.B. (1989) A Decision Support System Theory. *Working paper University of Aritzona* Tucson, USA, 53p.
Chang, A-L. Holsapple, C.W. Whinston, A.B. (1993) Model Management Issues and Directions Decision Support System 9(1) pp.19-37
Chang, A-L. Holsapple, C.W. Whinston, A.B. (1994) A Hyperknowledge Framework of Decision Support Systems. *Information Processing & Management*, 30 (4) pp. 473-498
Conklin, J. (1987) Hypertext: An Introduction and Survey. *IEEE Computer* 20(9) pp. 17-41
Dos Santos, B.L. Holsapple, C.W. (1989) A Framework for Designing Adaptive DSS Interfaces *Decision Support Systems*, 5 pp. 1-11
http://www.collaborate.com
Newel, A. Simon, H. (1972) Human Problem Solving Prentice Hall, Enlewood Cliffs, New Jersey
Nielsen, J. (1990) Hypertext and Hypermedia Academic Press Inc., San Diego
O'Keefe, R. Balci, O. Smith, E. (1987) Validating Expert System Performance. *IEEE Expert* 2(4) pp. 81-89
O'Leary, D. (1988) Methods of Validating Expert Systems. *Interfaces* 18(6) pp. 72-79
Vanharanta, H. Käkölä, T. Back, B. (1995) Validity and Utility of Hyperknowledge –Based Financial Bechmarking System. *Proceedings of the 28th Annual Hawaii International Conference on System Sciences*, pp. 221-230

Supporting Operations-Reference Knowledge Development Cycles for Collaborative, Distributed Research

Barrett S. Caldwell
Purdue University
West Lafayette, Indiana
bcaldwel@ecn.purdue.edu

Sudip K. Ghosh
Purdue University
West Lafayette, Indiana
sghosh@ecn.purdue.edu

Abstract

This paper describes information and communication technology (ICT) requirements to support knowledge and resource coordination between distributed collaborative researchers conducting an integrated engineering simulation. The design and research focus is based on a sociotechnical systems engineering model of information flow in dynamic environments. The NASA Specialized Center of Research and Training for Advanced Life Support (NSCORT-ALS) project is the operational setting for this work. NSCORT-ALS requires the coordination of over 20 investigators, with a strong interdisciplinary mix of specialties, spread among multiple research sites. The project will result in ICT capabilities that facilitate and enhance knowledge sharing and resource coordination (including data exchange and presentation), subject to information exchange and uncertainty constraints, task delays, and varying levels of expertise, among teams of research collaborators.

1 Introduction

The NASA Specialized Center of Research and Training in Advanced Life Support (NSCORT-ALS) is a multi-investigator, multi-site research program hosted at the authors' institution. NSCORT-ALS is devoted to the design of an enclosed life support system (ECLSS) capable of supporting a human crew on a mission to Mars. The whole ALS system is divided into many ECLSS sub-systems that give rise to issues of distributed knowledge and expertise. Each of the research collaborators works independently on the sub-system but has to collaborate and coordinate appropriately to make the whole system work as desired. It is not simply a case of people being in different places, but also a question of who knows what, in what knowledge domain and does what task. Given the diversity of the sub-systems, it is impossible for any one collaborator to know all this about the system. There is a need to address these issues of distributed knowledge and expertise and not just storage of data and task lists.

2 NSCORT-ALS and ICT Design

In order to support the development and ongoing collaboration needs of such a collection of researchers, special attention must be given to the design of the supporting information and communication technology (ICT) system. For any distributed expert community, the organization and implementation of ICT provides critical infrastructure and pathwork support to enable knowledge sharing and coordination at teamwork and taskwork levels of analysis (Garrett and Caldwell, 2002a).

The NSCORT-ALS project spans a number of time scales, both shorter (such as a single project meeting or data analysis of a critical experiment) and longer (such as an emerging research direction, or creation of a jointly authored paper). The success of NSCORT-ALS depends on efforts to define user information needs and data exchange requirements, describe information architecture structures for common data exchange capabilities, and evaluate current collaborative ICT implementations. Individual researchers within a distributed research environment will see this effort as part of their own work—defining information needs from other researchers, specifying formats for data exchange, and providing repositories for shared information, results, and emerging results.

One of them is addressing the ICT requirements, but the other part is also addressing the issue of distributed knowledge and expertise in terms of a sociotechnical context. Gaines et al (1997) have addressed some aspects of the ICT requirements of distributed experts. Apart from meeting the functional requirements of the collaborators, it is also important that the ICTs used are easy to learn and flexible for each of the collaborators. This flexibility will not only enhance the adoption by the collaborators, but the ability of the ICT to effectively support the different aspects of data exchange and information flow. Harvey and Koubek (2000) point out some of the sociotechnical aspects of distributed engineering collaboration. They emphasize the use of tele-data and tele-presence to make virtual collaboration follow the group dynamics of the real world as closely as possible. Questions remain regarding whether these ICT implementations provided are sufficiently flexible to support effective collaboration even when technology degradation or limited resources across project groups restrict the theoretical capabilities of the tools.

In a complex and distributed research collaboration environment, researchers will emphasize the development of their own technical capabilities, rather than general ICT design. However, ICT robustness is an important criterion for system development and design. A major ICT requirement maybe availability of high bandwidth, but a lack of it under any uncertain circumstances does not render the other collaboration requirements and needs as redundant. The infrastructure must still be capable of supporting the system, in spite of the lack of required bandwidth, in order to prevent any system failure. The NSCORT-ALS project is in the design and development process and there is a continuous knowledge development and operational reference cycle occurring across various time scales. Such considerations upfront will only ensure that the final system is capable of supporting and sustaining itself. Proper use of information and collaboration tools is essential to enhance the productivity of the group within the defined scope of the NSCORT-ALS projects.

3 Mission Control Center (MCC) Information Flow

The design of the NSCORT-ALS is built on the model of information flow, knowledge sharing, and operational knowledge development cycles in the Mission Control Center (MCC) environment. For NASA, MCC information flow occurs through a variety of communication paths to support operational activity over a range of event types and time scales. Events may describe actions (changes in state of engineering system components, software processes, or human cognitive activities), or interactions (commands to distributed sensors / actuators, network data exchanges, conversations between experts, or human interactive controls of engineering components). Events can also occur over a range of time scales, based on characteristic event cycles of the underlying information flows (see Caldwell and Wang, 2003 for additional information). Multiple information flow and event referencing events share some data exchange requirements and information architecture structures. This distinction is analogous to a telephone / cable jack, which also shares a broadband digital subscriber line (DSL) connection. The DSL connection sits on top of the telephone connection using different line capabilities to support packet sharing at a much higher bandwidth, but flowing through the same physical path as the telephone voice channel.

4 Integrating ICT Information Flows

Based on the MCC model, we distinguish NSCORT-ALS information support needs into the ALS Engineering Simulation System, and the Distributed Coordination Information Flow Network. A third type of information flow is also part of a mature distributed supervisory coordination system—the engineered sociotechnical system itself (see Figure 1). The MCC is a fully operational system, where distinct information flow support is dedicated to maintaining information flows within the space vehicle, and between the MCC and space. However, the NSCORT-ALS is exclusively in the development phase, and without program responsibilities to create an actual operational ALS environment. There are new issues of knowledge development and references to the experiences of the research collaborators and engineering simulations that are unique to this development project. With appropriate ICT design, event referencing for knowledge development in researcher coordination and engineering simulation event cycles can be used in future ALS engineering system operations.

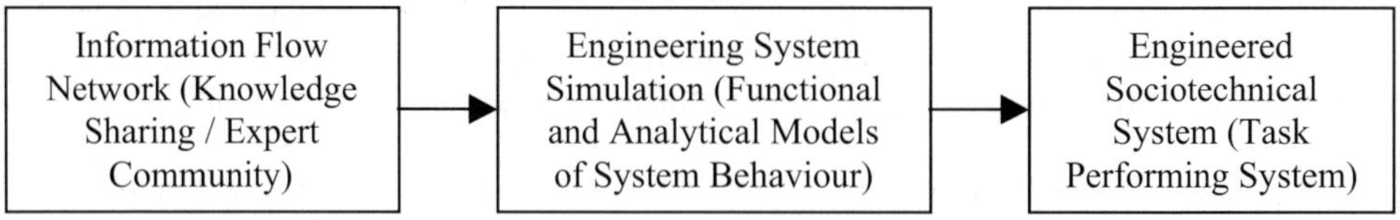

Figure 1: Progress and interaction of linked ICT flows to support information needs in distributed coordination of engineering systems.

Our project effort is being conducted in parallel with researchers focusing on data exchange requirements for the ALS Engineering Simulation System. This parallel project effort emphasizes the development of analytical models of the Engineering Simulation System based on researcher descriptions of functional relationships between ECLSS subsystems. The examination of distinctions between researchers examining specific ECLSS subsystems, and our team focusing on the information flow network between researchers, is in parallel to our understanding of information flow and knowledge coordination in the MCC environment. Communication and information exchange function allocation across the MCC team of system controllers involves engineering system domain expertise, support for maintaining the necessary communication paths, strategic coordination, and management of shared information and knowledge synchronization (Garrett and Caldwell, 2002b).

Our project group also applies the process described in Figure 1 as part of the HCI design effort to create a functional and effective operational ICT system. Prior work in our team has already created interview and survey tools to identify the dimensions of expertise and domains of information needs that researchers require from collaborators, as well as a classification structure for ALS knowledge sharing domains. These elements will form the basis of future ICT development, in much the same way that collaborator interactions will form the basis of future ALS simulation development. The main distinction is at the level of human information flow requirements and event cycles, which comprise different time scales from those of engineering data flow. The question to be answered for supporting ICT event cycles is fundamentally HCI, where a design team intends to support team coordination. These evaluations highlight both user requirements (e.g., ease of knowledge capture, document and data sharing, and task synchronization) as well as system management demands (e.g., server set-up and maintenance functions, security support, and user / task / project permissions management).

However, an examination of Figure 1 immediately generates the question, *what is the system simulation phase for ICT system development*? Our team has recently begun the development of a series of Information Flow Modelling (IFM) system simulations to examine information flow, delay tolerance, and knowledge synchronization dynamics between distributed experts. These simulation models will examine human information exchange activities across operational time scales from minutes to months. The IFM approach emphasizes a combination of classical group dynamics, social information processing, and adaptive feedback control approaches to group-level information flow, knowledge sharing, and task coordination in modelling of knowledge sharing communities. Our development of IFM models can also help identify how knowledge development processes and event cycles at one time scale influence and trigger events at other time scales.

5 Developing IFM Simulation Modules

The IFM model work completed to date is the first in a series of simulations for knowledge sharing communities. The IFM simulation module described here is for a distributed environment consisting of experts and novices. Such a module can simulate small expertise communities like

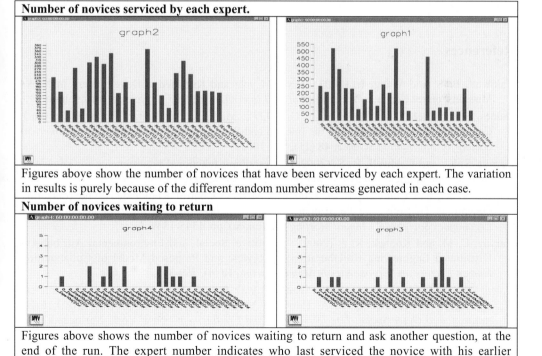

Number of novices serviced by each expert.

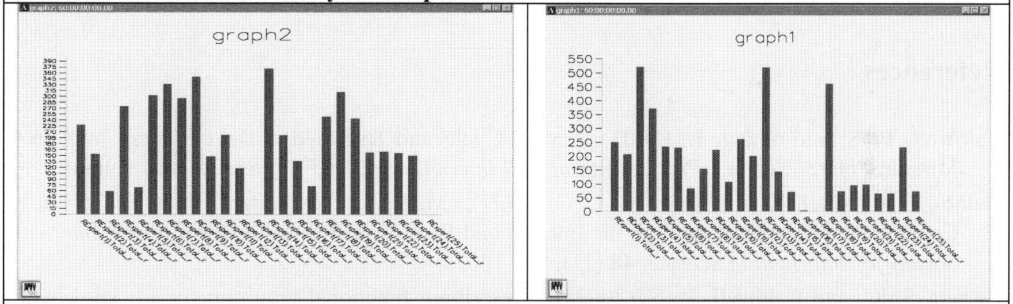

Figures above show the number of novices that have been serviced by each expert. The variation in results is purely because of the different random number streams generated in each case.

Number of novices waiting to return

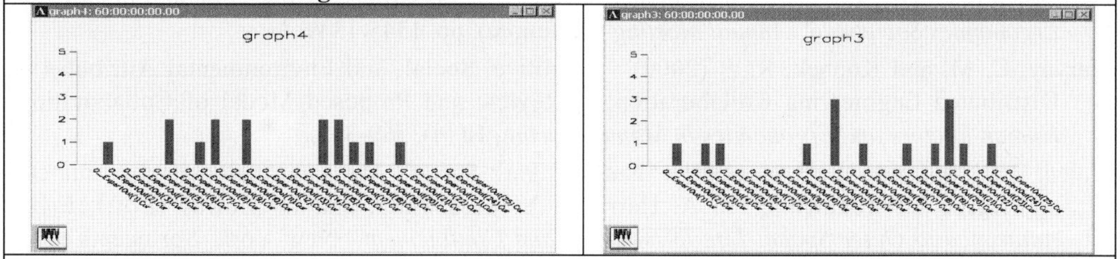

Figures above shows the number of novices waiting to return and ask another question, at the end of the run. The expert number indicates who last serviced the novice with his earlier question.

Figure 2: Figures of IFM dynamics of preliminary "expert query" module.

visitors at an "Ask The Expert" web site, a group of US Park Service rangers working in an isolated visitor centre (Valen & Caldwell, 1991) or those posing questions to the NSCORT-ALS researcher community. Presently, the IFM simulation architecture is event-based and discrete. However, further models will be event-based and more mathematically refined continuous models.

The "expert query" module is divided into three processes: expert initiation, novice creation, and expert-novice interaction. In the expert initialisation process, the 25 experts are defined with expertise ratings in 6 different expertise areas using a knowledge structure matrix with values corresponding to each expert. The second process is novice creation, where various characteristics of the novice, like expertise rating, question area, complexity, etc. are defined. The third process involves the expert and novice interactions. The novice enters the system of experts and chooses an expert according to socially recognizable criteria (such as "awareness" of relative status and who might a reasonable expert to ask). Expert response dynamics also reflect socially recognizable criteria (willingness to answer, annoyance at being asked the same question repeatedly, or simply delays in responding). Depending on whether the novice is satisfied with an answer, they may repeat the process and re-enter the system, or leave the system dissatisfied. Please see Figure 2 for graphical descriptions of results of simulation runs of a preliminary "expert query" module developed by the authors and other members of our research team.

It is still an open question in the human factors, cognitive engineering, and user interaction design fields regarding how dynamic repository, knowledge sharing, and other IFM functions should be supported. In addition to further IFM simulation module development, our research team will also focus on what types of user interfaces are best suited for those functions, and how operational experience is transformed into reference expertise and synchronized knowledge structures across multiple time scales

References

Caldwell, B. S. and Wang, E. (2003). Event Cycle and Knowledge Development in NASA Mission Control Center. *Proceedings of HCI International 2003 Conference*, this volume.

Gaines, B. R., Chen, L. L., Shaw, M. L. G. (1997). Modeling the Human Factors of Scholarly Communities Supported through the Internet and World Wide Web. *Journal of the American Society for Information Science,* **48** (11), 987-1003.

Garrett, S. K., and Caldwell, B. S. (2002a). Describing functional requirements for knowledge sharing communities. *Behaviour and Information Technology*, **21** (5), 359-364.

Garrett, S. K., and Caldwell, B. S. (2002b). Mission Control Knowledge Synchronization: Operations To Reference Performance Cycles. *Proceedings of the Human Factors and Ergonomics Society 46th Annual Meeting* (Baltimore), pp. 1345-1349.

Harvey, C. M. and Koubek, R. J. (2000). Cognitive, Social, and Environmental Attributes of Distributed Engineering Collaboration: A Review and Proposed Model of Collaboration. *Human Factors and Ergonomics in Manufacturing*, **10** (4), 369-393.

Valen, R. J., & Caldwell, B. S. (1991). National Park Service Areas as Analogues for Antarctic and Space Environments. In A. Harrison, Y. Clearwater, and C. McKay (Eds.), *From Antarctica to Outer Space: Life in Isolation and Confinement.* New York: Springer-Verlag, 115-121.

Data Analysis and Visualization for Usability Evaluation for Collaborative Systems

Jeffrey D. Campbell, Enrique Stanziola, Andrew Sears

UMBC
1000 Hilltop Circle, Baltimore, MD 21250 USA
jcampbel, stanziola, asears @umbc.edu

Abstract

GLogger is a data collection, management and analysis tool specifically designed to meet the requirements for usability evaluations of multi-user collaborative applications. It has been used to collect and analyze data from the development of diagrams, collaborative text editing and instant messaging. It provides support for rich meta-data and synchronization needed for multi-user logging. GLogger is distinguished by its flexibility and multiple visualizations of the data. Data views include hierarchical, multimedia, time line, time chart, document and data table.

1 Introduction

Participants in a software usability evaluation can generate a large volume of data. This data can include events recorded by the software being evaluated, contemporaneous and subsequent notes recorded by observers, screen image capture video, audio and/or video of the participants during the evaluation, interview responses, survey question results and post hoc analysis. The evaluation of groupware systems with multiple simultaneous users is more complicated due to the increased number of data files, the importance of recording and analyzing interactions between participants, and the increased difficulty for the observer to monitor multiple participants at once. Many tools have been developed to assist with the analysis of such data, with summaries in Hilbert & Redmiles (2000) and Ivory & Hearst (2001).

This paper describes GLogger, a flexible logging component and analysis system. GLogger addresses the requirement for exploratory sequential data analysis (Sanderson & Fisher, 1994) for software that provides fast access to data, collects data automatically, allows data management, performs calculations and offers data visualizations. GLogger's distinguishing characteristic is support for multiple, flexible visualizations of the usability data including digital audio and video. GLogger records user interaction with a system by recording events, capturing screen image videos, saving contemporaneous observer notes and supporting post hoc analysis. GLogger has been used with three collaborative applications with multiple people working at the same time on different computers in approximately 45-minute sessions. The participants have created diagrams, edited a text document and used instant messaging (IM) to solve a problem. The initial usage of GLogger has recorded keystrokes and mouse clicks in the document along with higher-level events, such as invoking specific menu items. Other higher-level events have been recorded such as people joining or leaving the session. Since the only requirement for using GLogger is that there are event codes to distinguish the events that have occurred, GLogger would be useful for the analysis of many other types of event-driven data.

GLogger was designed for flexibility. Log data in the form of delimited text, XML and database records is supported for input and output. Each event can have different detailed information. For example, a keystroke event would include the key pressed but a "File Save" event could include the file name and format as details. The definition of file formats as well as the analysis functions and viewing formats are defined in text files for easy customization.

The next section discuses several technical requirements for multi-user logging. Section 3 describes the multiple visualizations provided by GLogger. Section 4 concludes with results and directions for continued work.

2 Technical Requirements

The overall requirements for evaluation of computer supported cooperative work applications have been described at length (e.g. Knutilla, Steves, & Allen, 2000). This section focuses on two technical requirements – meta data and time synchronization. It also describes features in GLogger that increase flexibility.

Since there is likely to be event data from each of the participants, GLogger can analyze multiple files together. Similarly, GLogger collects contemporaneous or subsequent annotations in a format that is consistent with the event data, allowing those notes to be analyzed along with the original event data. The meta-data requirement comes from the relatively large number of data and analysis files generated from multiple computers. GLogger maintains data about this data to help manage this increased volume of data compare to single-user software data. GLogger records the IP address, machine name and full file name in each log file to reduce potential confusion after the files have been copied to a common location for analysis. Similarly, it records the version of the application and the logging component within the log file so that the data is interpreted within the proper context if any changes are made to the programs during iterative development. GLogger also provides an easy way to record which participants and observers were using which machines within the log files. GLogger protects the integrity of the original data during analysis since all editing and analysis is recorded in a separate file. All such editing/analysis is saved along with an identifier for the person performing the analysis and the date/time. This allows easy review and revision, if necessary.

Time synchronization is another issue that is more significant for multi-person applications. GLogger has an iterative approach that calculates the offsets between the high-resolution system clock (e.g. the number of milliseconds since the machine started in Microsoft Windows) on the different machines. This feature allows comparison of events logged on different machines and also provides a measure of network latency. A second aspect of time synchronization is matching the data logs to the audio video data. This is currently done by placing a visible/audible mark in each media file that matches a logged event in the data logs.

3 Visualizations

A distinguishing characteristic of GLogger is that it provides a variety of textual and graphical views of the data. There can be multiple instances of each visualization type and each one can display all or a user selected portion of the data sorted chronologically or by any other property. For example, it is often useful to show data for each participant in both individual and group displays. Color-coding and other appropriate formatting is consistent across visualizations. For example, if colors are used to indicate individuals, the same color is used in all views. The different visualizations can be coordinated so that the same time range is displayed in multiple windows. This allows one to see the same events from multiple perspectives, for example, playing the video or screen capture when an interesting portion is found in the log or viewing the log when the media shows a critical incident. The formats provided are Hierarchical, Media, Time Line, Time Chart, Document and Table. Each is described in the following subsections.

3.1 Hierarchical View

The hierarchical view (Figure 1) supports two types of hierarchies. The first displays each event with an event-specific customized line of text with additional details as subordinate items. For example, for text editing and instant message data, it has been useful to define a format that shows the character typed as the main display with the details (time in two formats, user, sequence) only displayed when expanded. The second type of hierarchy is used to display meaningful groups of events. For example, the keystroke events that compose each IM message can be represented as subordinate items to the actual message.

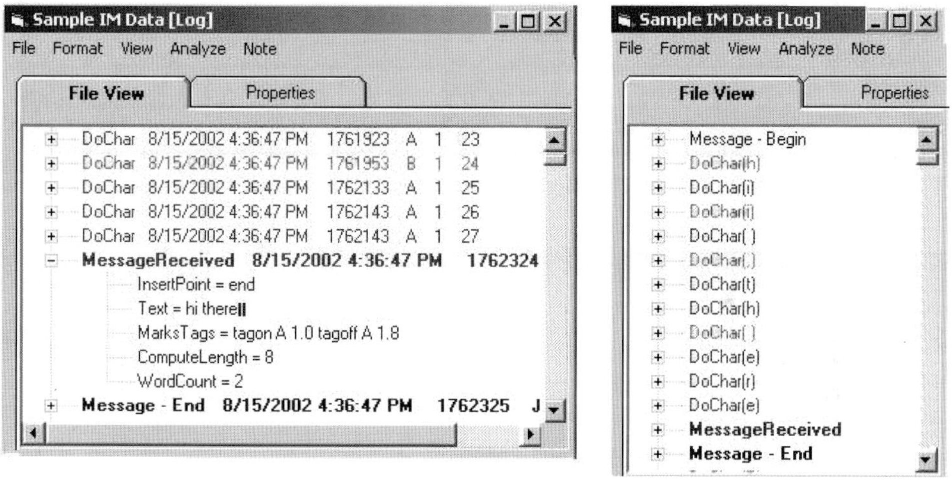

Figure 1 General hierarchical view showing two formats

3.2 Media View

The media view plays digital audio, video or screen capture images. It displays the media time in a format consistent with the event times in the log files. The user can play any portion of the media file. The media file can also be used to coordinate the display of other views to match the time shown in the media. The media view may show the media corresponding to the data displayed in another view. Simultaneous media views are possible, for example to show both a screen capture video and a video of the participants at the same time.

3.3 Time Line View

The time line view is a configurable display showing the occurrence of instantaneous events as short vertical lines and the duration of extended events as horizontal bars. The extended events may be computed using the analysis engine in GLogger. In Figure 2, there is color-coded data for two people (A and B). The upper set of vertical lines (A) is the same color as the upper horizontal lines. In Figure 2, the vertical lines represent keystrokes. The horizontal bars represent the duration of message composition in IM, which is computed as the time between the first keystroke and pressing the Send button. The data shows that person B paused in their typing as they received a message from person A. The time line clearly shows this interesting occurrence what would be difficult to detect using other views. The time line is similar to, but more flexible than, CollabLogger (Morse & Steves, 2000) and the related Flogger (Dyck & Gutwin, 2001).

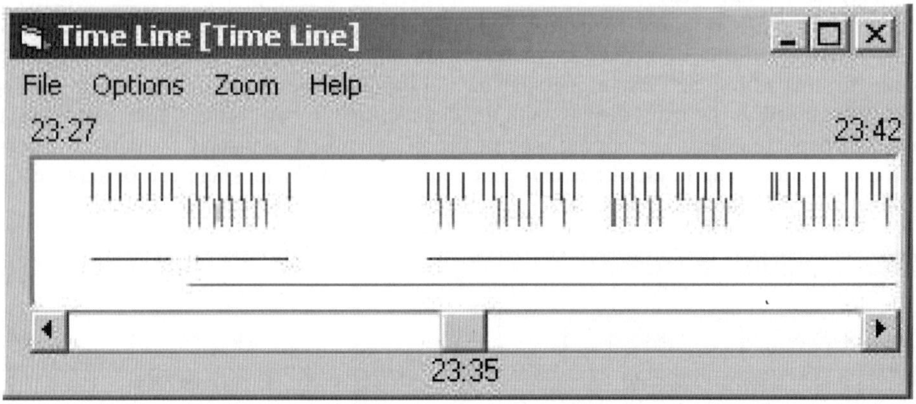

Figure 2 Time line displaying IM keystrokes and messages for two people

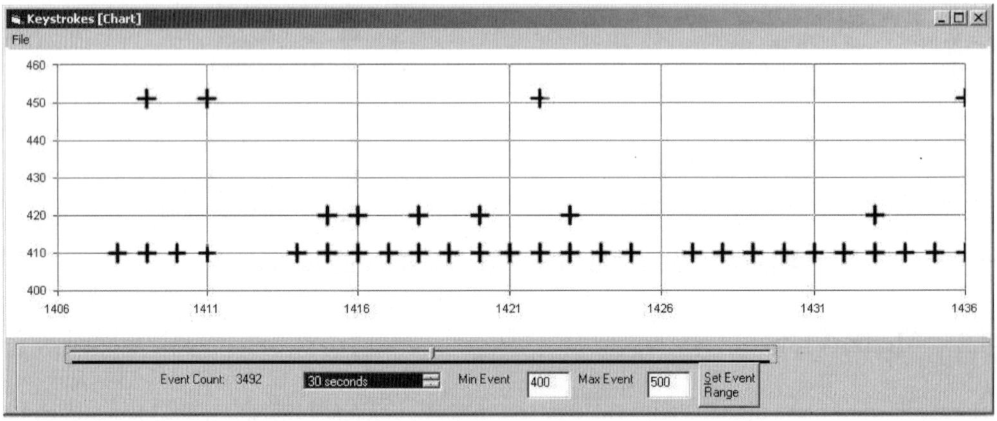

Figure 3 Time chart view showing IM data

3.4 Time Chart View

The time chart view (Figure 3) is a representation of event codes on the vertical axis and time on the horizontal axis. This is similar to Timelines (Owen, Baecker, & Harrison, 1994) that was designed for videotape annotation, but GLogger supports multiple concurrent videos. The purpose is to provide a quick overview of the data patterns without needing to define specific displays for the time line. Figure 3 is a brief portion of IM data showing the pattern of keystrokes (event 410), character deletes (event 420) and sending messages (event 451).

3.5 Document View

Document view displays the final document (diagram, text document, IM messages) to help provide context for the more detailed views. The document view provides an easy link to the more detailed event data by coordinating the other views to match the selected line here.

3.6 Table View

Table view resembles a spreadsheet with each row representing an event and each column one of the data items associated with that event. For example, time and user name are common data items. This view facilitates sorting the data based on any data item and choosing which events to display based on criteria using any of the columns. This view has been useful for quickly finding particular values or ranges of values for coordinated display in another view.

4 Results and Conclusion

Glogger was used to compute detailed statistics and identified message durations for 16 IM sessions containing 63,000 events covering over 6 hours of data in about an hour and a half of analysis time. That included about 10 minutes to specify a new calculation that had not been anticipated while using test data in GLogger.

Analysis of physical sensor data from a mobile computing experiment is pending and will further demonstrate the flexibility of GLogger. Enhancements for multimedia recording and processing are being designed. The full features of GLogger are available in the Microsoft Windows environment. Log creation has also been implemented in Tcl/Tk providing multi-platform support

References

Bayer, S., Damianos, L. E., Kozierok, R., & Mokwa, J. (1999). The MITRE Multi-Modal Logger: Its Use in Evaluation of Collaborative Systems. *ACM Computing Surveys, 31*(2es), Article 17.

Dyck, J., & Gutwin, C. (2001). FLogger: Flexible Log Visualizer v 0.1: University of Saskatchewan, Retrieved February 10, 2003 from http://hci.usask.ca/software/flogger.xml.

Hilbert, D. M., & Redmiles, D. F. (2000). Extracting Usability Information from User Interface Events. *ACM Computing Surveys,* 32(4), 384 - 421.

Ivory, M. Y., & Hearst, M. A. (2001). The State of the Art in Automating Usability Evaluation of User Interfaces. *ACM Computing Surveys*, 33(4), 470-516.

Knutilla, A., Steves, M. P., & Allen, R. (2000, June 2000). *Workshop on Evaluating Collaborative Enterprises - Workshop Report.* Paper presented at the IEEE 9th International Workshops on Enabling Technologies: Infrastructure for Collaborative Enterprises.

Morse, E., & Steves, M. P. (2000, June 2000). *CollabLogger: A Tool for Visualizing Groups at Work.* Paper presented at the IEEE 9th International Workshops on Enabling Technologies: Infrastructure for Collaborative Enterprises.

Owen, R. N., Baecker, R. M., & Harrison, B. (1994, April 24 - 28, 1994). *Timelines, a Tool for Gathering, Coding and Analysis of Temporal HCI Usability Data.* Paper presented in the Conference Companion ACM CHI '94, Boston, MA.

Sanderson, P. M., & Fisher, C. (1994). Exploratory Sequential Data Analysis: Foundations. *Human-Computer Interaction,* 9, 251-317.

Man Machine Cooperation

Bertrand David, Ahmad Skaf

ICTT lab, Ecole Centrale de Lyon
36, avenue Guy de Collongue, 69134 Ecully Cedex, France
{Bertrand.David, Ahmad.Skaf}@ec-lyon.fr

Abstract

Man-machine cooperative systems design is of great interest in a large number of fields and in particular in industrial engineering, where control – command systems are numerous. In this paper we propose an original approach, which leads to dynamic allocation of actions between man and machine. Following a presentation of the problem and comparison with existing work, we describe our proposal based on an original modelling method and an algorithm of distribution which can be used in various contexts. This approach allows in particular dynamic allocation of actions to the man and the machine in relation with the formulated aims. The purpose of designing such systems is to reach global performance in relation with human actor and machine capabilities.

1 Introduction

The concept of cooperation involves the pooling of a variety of competences (human and/or artificial) to achieve a common goal. This can be carried out by a distribution of tasks and responsibilities and by a coordination of actions. Three reasons are at the origin of the study of man-machine cooperation:

1. Low capacity of man to carry out certain actions,
2. The risk for man to deal with certain tasks,
3. Improvement of the performance of the complete system.

The concept of cooperation within the man-machine system implies, for the majority of authors, assistance to be provided to operators (Grosjean, 1999). For Sheridan, for example, collaboration is summarised with a role of permanent assistance allocated to the machine in favour of man. Johannsen (Johannsen, 1986) proposed a vision of collaboration through a decisional structure of the man-machine system while insisting on the importance of the communication between the actors of this system. In addition, it does not consider specificities of the actors taking part in work (man and machine). For him, the machine can theoretically collaborate to complete collective work in any form of organisation valid for a human team. It thus does not distinguish between a machine and a human operator.

Millot (Millot, 1996) developed an approach of collaboration for a system of supervision and monitoring. It considers that man (in a highly automated system) mainly finds his place in the control room where all information on the process is concentrated. The computer controls and supervises the system and the operations. The intervention of man is limited to mode changes and interventions in situations of malfunctions. Millot proposes two types of co-operation: vertical which corresponds to the notion of assistance, and horizontal, more sophisticated, which corresponds to the dynamic allocation of the tasks to be carried out. This allocation is based on an estimate of the workload of the operator (Millot, 1988) and performance of the technical system. Our work is situated in the latter context: the study of shareability of tasks between men and machines in various phases of product life cycle.

Our typology of collaboration between man and machine identifies five categories of activities: (1) automated (carried out by the machine), (2) entirely manual (mandatory carried out by man), (3) those able to be carried out by man with the assistance of the machine (4) or by the machine with human assistance and, the most interesting of them all (5), able to be allocated dynamically to man or to machine. The traditional collaboration approach is limited to static allocation of actions. This approach allocates to each actor the actions to be performed preliminary to the execution of the man-machine system and once and for all. Our purpose of dynamic allocation of actions is to adjust and improve statically defined distribution. For that, it is necessary to take into account indices measuring instantaneously current characteristics of the actors. For each actor, the following indices are required: (1) Capacity to carry out the corresponding action, (2) Performance in this action, (3) Instantaneous availability, (4) Provided cumulated effort.

Each action is allocated to one of the actors in respect of a distribution formula based on selected criteria taken from the following list (which is obviously not complete): duration, cost, effort, reliability and quality. This formula is used for both: local optimisation (task concerned) and global optimisation of all the tasks of the process.

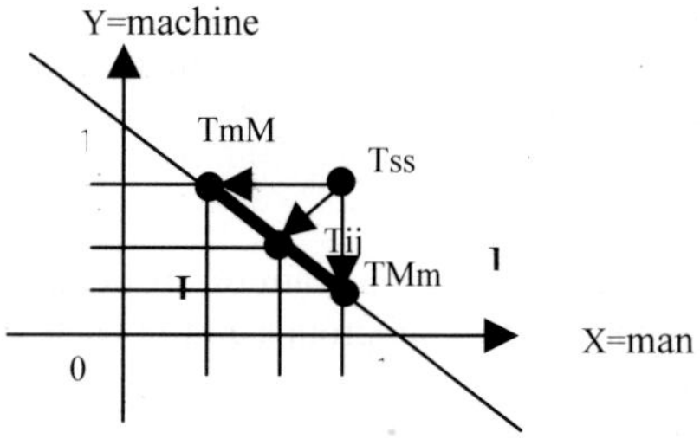

a: over-defined task

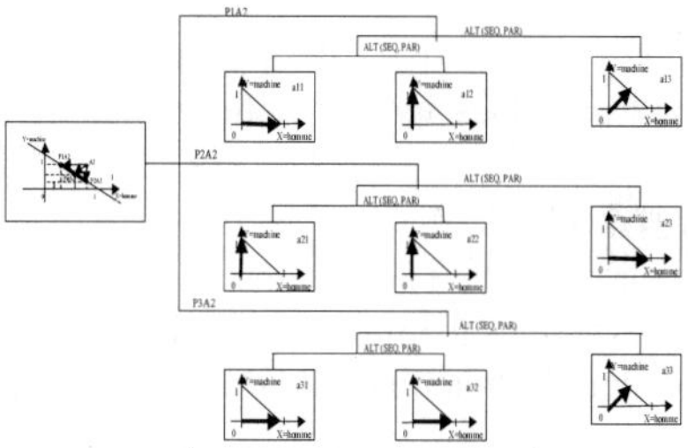
b: MAD tree of tasks

Figure 1: Task modelling in man-machine system

Our work is based on a rigorous work organisation model, based on a set of concepts which are Mission, Task, Activity, Action, divided between all the actors (operational originators, organisers and actors). A mission constitutes the most macroscopic view, the content of which is expressed in the form of tasks to be carried out. It is considered that one or more activities carry out a task. These activities can be divided into actions, considered as elementary in our modelling. From the man-machine point of view, we consider that a mission is composite, i.e. it requires at the same time man and machine interventions. This may also be true with respect to tasks and activities (figure 1). An over-defined task (Tss) can be properly defined by a projection on the segment (0,1 – 1,0), in three different manners TmM, Tij, TMm; the first one maximising machine participation in the task, the last one maximising participation of man and the middle one proposing equitable sharing. At the final level, the action is not composite, i.e. decomposable, but elementary. It can either be individualised, i.e. assigned to a specific actor (man or machine), or joint, requiring the joint intervention of several actors (men, machines or both simultaneously). More explanations about task modelling is given in (Skaf, 2002).

2 Allocation Algorithms

The distributor, whose role is to carry out the allocation, works on the tree of activities which is an enriched MAD tree (Skaf, 2002), on which information on man and machine capabilities is expressed. The interest of the tree depends on the time when the allocation is performed and the allocation strategy used. Its use in the preliminary stage has already been discussed and its disadvantages identified. Its use during the process is undoubtedly more relevant. According to the allocation strategy, it may take into account various aspects of the context in relation with expressed criteria. In particular three strategies appear interesting:

1. Instantaneous allocation: This takes into account the context running - availabilities of human and technical resources and current evaluation of the state of the object.
2. Allocation taking account of the history: This takes account of cumulated context, i.e. actions already carried out and current context. Thus the tiredness of the human actor can be taken into account, because the assessment of his implication in the activities already carried out is available.
3. Overall satisfactory allocation: This is not only a question of taking into account context running and history, but also a more complete view of all the activities. The main goal is to take into account the capacities of man over one period (working day, for example) and to set out again its implication in a rather regular way on the unit of the period.

Next, we will discuss these three strategies from the point of view of their achievement. However we will first introduce two definitions relating to actors and tasks.

2.1 Actors

Assume that we have two groups of actors which represent two teams: a team of artificial actors (or machines) and a team of human actors. Each team is made up of subgroups and in each group the actors (man or machine) are described by a set of attributes.

$$V = \begin{bmatrix} goup\,1 \\ group\,2 \\ \ldots\ldots \\ group\,k \end{bmatrix} \begin{bmatrix} \begin{bmatrix} actor\,1 \\ actor\,2 \\ \ldots \\ actor N\,1 \end{bmatrix} \\ \begin{bmatrix} actor\,1 \\ actor\,2 \\ \ldots \\ actor N\,2 \end{bmatrix} \\ [\ldots\ldots] \\ \begin{bmatrix} actor\,1 \\ actor\,2 \\ \ldots \\ actor Nk \end{bmatrix} \end{bmatrix} \begin{bmatrix} \begin{bmatrix} a11, a12, a13 \ldots\ldots .a1n \\ a21, a22, a23 \ldots\ldots a2n \\ \ldots\ldots \\ aN1, aN2, aN3 \ldots\ldots aN1n \end{bmatrix} \\ \begin{bmatrix} a11, a12, a13 \ldots\ldots .a1n \\ a21, a22, a23 \ldots\ldots a2n \\ \ldots\ldots \\ aN1, aN2, aN3 \ldots\ldots aN2n \end{bmatrix} \\ [\ldots\ldots] \\ \begin{bmatrix} a11, a12, a13 \ldots\ldots .a1n \\ a21, a22, a23 \ldots\ldots a2n \\ \ldots\ldots \\ aN1, aN2, aN3 \ldots\ldots aNkn \end{bmatrix} \end{bmatrix}$$

Assume we have i actors in k groups, each actor with n attributes. To represent all the actors in their groups we use a matrix representation V: V= [group k] [actor i] [attribute n], where for each group k, i varies from 1 to Nk. Nk is the actor's number in the relevant group.

2.2 Tasks

As stated above, the work to be carried out is broken down into missions, tasks, activities and actions. In this classification we specified each identified action by a set of criteria c1, c2, c3... Cn grouped in vector C. The methodology of such a specification is described in (Skaf, 2002). We underline here a very important remark: The aim of action specification is to be able to judge the realization aspects of the action (how?, what effort?, what means, for what price? etc.). This makes it possible to decide which actor can perform it, and also to define the form of assistance, if necessary.

2.3 Allocation

2.3.1 Instantaneous allocation

The process is based on the algorithm sketched in figure 3. Starting from identification of all actions (Skaf, 2001), a flow of actions is processed to dispatch and to determine man's actions, machine's action or man-machine's actions. Each action is identified by a set of attributes. According to the values of these attributes described in a V matrix, an appropriate actor is chosen.

2.3.2 Allocation taking account of history

In this allocation, process historical information on the behaviour of human actors is taken into account. To collect this historical information a register saves continually the action nature n, its execution time t, quality of realisation q and references to actors who have assisted in realisation r. This information is collected in a historical vector h = [n,t,q,r] for each action. Figure 4 shows how historical information is considered in the decision-making process. The allocation process of the action set begins at t=0, for each action carried out by a human actor k, a vector h is added to the historical matrix H. The row number of this matrix is equal to the number of actions carried out by the actor concerned.

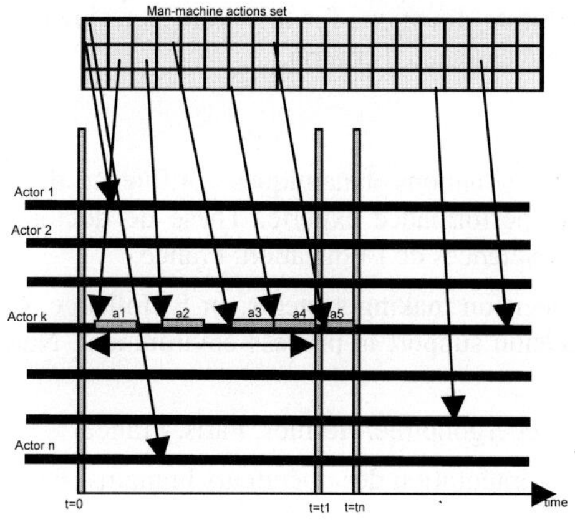

Figure 2: Representation of historical consideration in allocation process

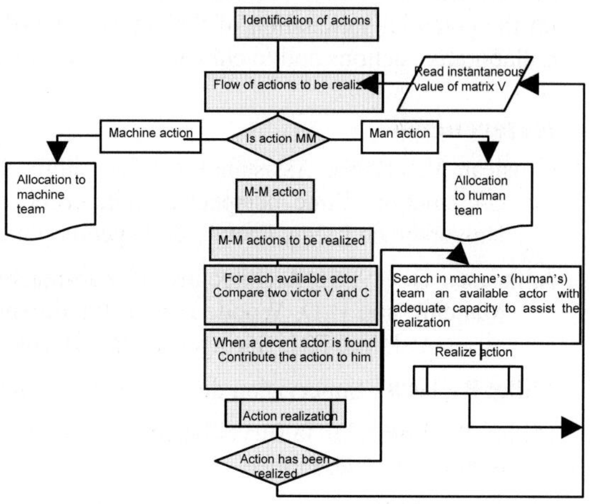

Figure 3 : Instantaneous allocation algorithm

For example in figure 4, actor k carried out 4 actions in interval t = (0;1), so his historical matrix looks like:

$$H = \begin{bmatrix} n1, t1, q1, r1 \\ n2, t2, q2, r2 \\ n3, t3, q3, r3 \\ n4, t4, q4, r4 \end{bmatrix}$$

Thus, to dispatch the next action (action a5) not only matrix V and C will be considered but also matrix H. The algorithm provided above (figure 3) will be modified to take into account the historical matrix of the human actor's behaviour. Each time, before allocating an action, a look at the cumulated context will affect the allocation process.

2.3.3 Overall satisfactory allocation

For overall satisfaction it is not only a question of taking into account context running and history, but also a more complete view of all the activities. The main goal is to take into account the capacities of man over one period (working day, for example) and to set out again its implication in a rather regular way on the unit of the period. This means that a new matrix G related to the human actors has to contain the information on their participation in the mission, in order to take into account this information, and call on them as often as possible if they are generally forgotten or only if it is necessary, if they are very busy.

3 Conclusion

In this paper, task allocation in man-machine systems was discussed. Three allocation algorithms were described in the context of dynamic allocation. Instantaneous allocation represents a mechanism of task sharing based on actor attributes and action specification. Allocation taking into account history gives an answer to the demand to distribute work equally between several human actors. Finally, overall satisfactory allocation tries to integrate the forecasted information on the potential solicitation of the actors. One of the future directions for this work is to identify collaborative actions and to enhance reference model and associated criteria.

References

Grosjean V. (1999). Assistance à la conduite dans les situations dynamiques : influence de la construction d'une perspective temporelle sur la performance experte. Thèse de doctorat, Université de Liège, Faculté de Psychologie et des Sciences de l'Education. France.

Johannsen G. (1986). Architecture of man-machine decision making systems. In E.Hollange, G. Mancini, and D.D. Woode (Eds.), Intelligent decision support in process environment. Nato series ASI, Vol F21. Springer Verlag, Berlin.

Millot P. (1988). Supervision des procédés automatisés et ergonomie. Hermès, Paris, France.

Millot, P. (1996). De la surveillance à la supervision : l'intégration des opérateurs humains, Ecole d'été, Grenoble, France.

SKAF A. (2001) Etude d'un système de supervision et de commande d'un procédé complexe comme un élément de base d'une organisation distribuée comprenant des machines et des hommes. Thèse en Automatique et Productique, Université Joseph-Fourier-Grenoble I. 18 décembre 2001. p. 182.

Skaf A., David B. (2002) Collaboration homme-machine : vers un partage dynamique, IHM 2002, 14e Conférence Francophone sur l'Interaction Homme-Machine, Faculté des Sciences, Poitiers du 26 au 29 novembre 2002.

Capillary CSCW

Bertrand David, René Chalon, Gérald Vaisman, Olivier Delotte

ICTT lab, Ecole Centrale de Lyon
36, avenue Guy de Collongue, 69134 Ecully Cedex, France
{Bertrand.David, Rene.Chalon, Gerald.Vaisman, Olivier.Delotte}@ec-lyon.fr

Abstract

This paper describes the evolution of CSCW towards greater actor availability by taking into account their mobility and the use of portable and handled devices. We call this approach "Capillary CSCW" (by analogy with blood vessels) to express the irrigation of all the actors whatever their location or the devices in their possession.

1 Introduction

The relations between software and their users are increasingly diversified. From a single user interacting with a single application, both the number of applications used simultaneously by a user and the number of users sharing the same applications are constantly increasing. This evolution is at the origin of Computer-Supported Collaborative Work (CSCW). Several classifications have been proposed. We can quote the most famous by (Ellis, Gibbs & Rein, 1991) which classifies the nature of cooperation in a matrix, in regard to synchronous or asynchronous, local or remote aspects of cooperation. Physical distances can vary in length (inside a building, local to a city, regional, national, continental, intercontinental). This classification has been extended subsequently, introducing awareness of cooperation, foreseeability or unpredictability of collaboration and location. The possibility of bringing together geographically distant people is an important contribution of groupware. It is also possible to collaborate without imposing simultaneous work. It can be noted that in all cases, it is a question of managing, via the information processing system, the participation of several people who may be present only virtually (in space and/or time). The first aim of groupware is thus to propose a support for the abolition of space and time distances. Moreover, knowledge and management of the interventions of the multiple participants appears to be necessary. In fact, the participants constitute working groups that have to be organized with respect to working conditions, time and location. The organisation can lead to the definition of different roles, sub-groups and phases of project work.

In-depth analysis of cooperation reveals several dimensions which must be examined, as initially proposed by (Ellis & Wainer, 1994) with the Clover model i.e. a support of production, conversation/communication and coordination between participants. The **production dimension** and more precisely co-production aims at indicating the degree of cooperation. Thus, it implies: structuring and managing production data, identifying shared data and their structure, modeling the support of work activities, identifying sharing characteristics, etc. The purpose of the **coordination aspect** is to describe the organization of the work process coordination and the workflow to ensure a certain dynamics of operation (in push or pull approaches). Thus, it implies identifying the activities (tasks) to be performed, the work phases and their temporal organization (concurrent operations, co-interaction, etc), modeling of the work process, identifying the functional roles of participants, choosing cooperation modes. The purpose of the **conversation/communication aspect** is to define communication media in order to allow exchanges between participants concerning their work. It implies the choice of different conversation modes: synchronous or asynchronous conversation, in text, voice, multimedia manner and tools: textual chat, audio or videoconferencing, and the choice of conversation protocols: point-to-point or multicast, with moderator or free group conversations, etc.

In our study on cooperative design activities, we identified four generic cooperation modes that can be applied to other industrial world cooperative activities. These modes that correspond to finer granularities of interaction are:

- Asynchronous **cooperation:** the various participants interact in the project by exchanging data and working when they can (without co-temporality).
- **In session cooperation:** the various participants work at the same time, but independently. They can communicate (in co-temporality), but cannot visually share the objects of their discussions.
- **In meeting cooperation:** clearly identified participants work and communicate in co-temporality, sharing the objects on which they work and discuss. They have identified roles in relation to the goal of the meeting and their competencies/skills. Their interventions are controlled and allocated by a metaphor of the type "to give the floor to".
- **Close cooperation:** the participants can work, communicate and interact in real time on a subset of shared objects in close collaboration with several other participants. The consequences of their interventions are directly visible to all participants.

2 Nomadism

Significant evolutions have taken place as regards technological infrastructure, among which we can quote networks, communicating mobile objects, light devices of communication (handheld computers), and nomadism, which is defined as the geographic mobility of information and of actors. While informational nomadism is already acquired by network infrastructures and various software architectures like CORBA, nomadism of actors is rapidly evolving. Moreover, we observe significant trends in the market of handheld devices which are mobile, portable, PDA, cellular phones, connected, personalized and "intelligent". In this way, a "handheld computer" is defined as a "treatment" device which accompanies the user in his displacements by providing him with assistance in various situations and for varied tasks.

In mobile communicating object technology, we retain especially three types of characteristics, which we will use:

- autonomous mobile objects: they contain the vital minimum (user interface, network interface, localization mechanisms) and can be used as support with the activities of actors (the PDA for example).
- environment embarked objects: they are not mobile, but can be moved and act as local information sources.
- passive objects: they are not directly connected in a network, but by means of an object such as RFID labels (standard ISO 14443).

The exchanges imagined can take place either between actors via their mobile devices and thanks to the network, or between actors and the objects, or finally between environment embarked objects and the other objects. The concept of attentive environment is obtained by sensors observing the environment and updating the perceived situations. The knowledge of the physical or logic localization of the actors as well as of the objects seems one of the characteristics of mobility (closer contact with the environment). This is not always necessary in all the applications, but can allow new treatments concerning actors, artifacts or tools. For localization outside the buildings, the GPS allows a relatively precise positioning (with a margin of a few meters) anywhere in the world. Inside the buildings, no technology emerges really (Hightower & Boriello, 2001).

Access capacity to information is more or less permanent in relation with the mobility of the devices. Laptop devices are generally fully connected; mobile and handheld devices are to a greater or lesser extent often connected to or disconnected from their sources of information. The term "source of information" must be taken in the broad sense and covers the data bases, the knowledge bases, the websites, etc. The users can disconnect and reconnect easily in both situations (volunteer disconnection related to a particular situation or working context and un-volunteer disconnection related to physical and geographical problems). In these two situations, the synchronization of "information sources" is mandatory to propagate the modifications made on each side of the link.

3 Capillary CSCW

The concept of nomadism (networking, handheld devices, mobile communicating object technology, localization and permanent or non permanent connection) extends the CSCW and allows us to introduce the concept of "capillary" CSCW or Capillary Coooperative Work (CCW). We use this term by analogy with the network of blood vessels. As its name implies, the purpose of the capillary CW is "to extend the capacities provided by co-operative working tools in increasingly finer ramifications, from their use on fixed proprietary workstations to small handheld devices" (Figure 1).

The concept of CCW is tempting because it fits naturally into the continuity of research in the field of co-operative work, by adding the consideration of new mobile devices. In addition to access to information, interaction and collaboration by the virtualization of space and time, it adds mobility of people and, as applicable, the virtualization of their localization or, on the contrary, the consideration of the precise context, of their localization and the localization of the objects on which they are working. The capillary CSCW satisfies a strong need formulated by users who no longer only work in offices in front of permanent stations (generally the issue in work about the CSCW), but in various environments where it is a basic necessity to be able to communicate and collaborate with remote people or systems, in particular for activities of decision-making, expert behaviors or remote repairing situations (Juhlin & Weilenmann, 2001, Schmidt, Lauff & Beigl, 1998). Geographical scattering can concern, and in a non-exclusive way, both the actors (members of a team of project) and the artifacts which are concerned with the activities or the devices which the actors need. The CCW also brings new constraints and new problems to be solved in addition to those which already exist. In relation with the work already carried out in this field, like those of (Sheridan, 1988, Saikali & David, 2001) we can quote mainly:

- management of collaboration and coordination of the mobile actors,
- coherence and validity of the information exchanged between handheld devices which are connected only intermittently to the network and the "group" with the aim of having the most synchronized possible information. In addition, the problem is accentuated by the fact that the handheld devices can communicate separately between the remainder of the network.

- heterogeneity of the communication protocols of the handheld devices,
- constraints of interface and overall capacity of the handheld devices in terms of size of screen, speed transmission, memory, autonomy, as well as the interaction devices,
- consideration of the need for awareness, which is a recurring need in the field for the CSCW, but which should receive specific solutions as proposed by certain experimental systems, among which one can quote "Awarenex" (Tang, Yankelovitch & Begole, 2001). The latter mainly reuses existing visual effects in groupwares such as "Groove" or systems of discussions such as "ICQ" to represent the evolution of the activities of the group.

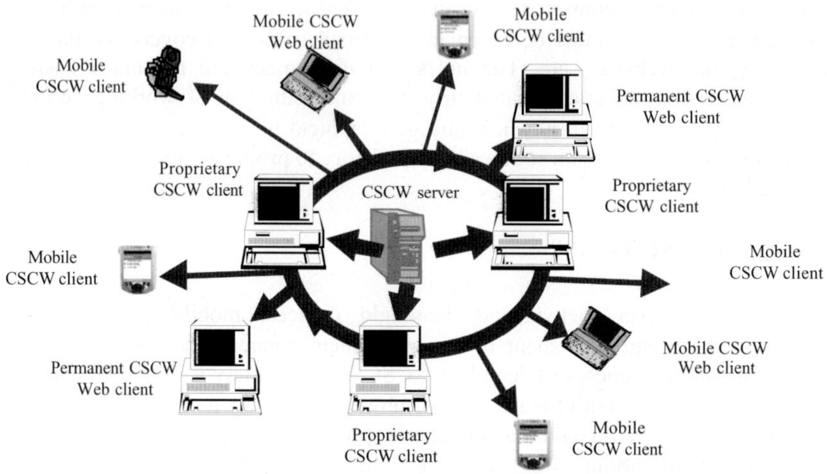

Figure 1: Capillary Cooperative Work

4 Prototypes

A first prototype is a location aware collaborative application. It consists in a localisation and appropriation tool used to exchange documents between fixed actors and mobiles ones with a PDA. The mobile user can also visualise documents on devices that are available in the vicinity (such as printers or data projectors) by dynamic appropriation. For example, a workteam has a meeting but all the members are not present at the office. Some are on other sites of the company, some in another company. To exchange the working documents, each member has a private space on a server reachable by any Internet access. When a member wants to distribute a document, he/she copies it on the working space of other members. Then everyone can work on this document, either by using the usual working environment or, for remote people, by using any device available in his/her surroundings: printer, data projector, etc.. that he/she has previously appropriated dynamically with our application. The application allows users to become aware of the working context of the remote users by knowing which tools and devices are available to them without knowing explicitly their geographical location. This application makes up a component of the capillary collaborative work environment, covering mostly the conversation/communication space of the Clover model.

A second prototype is a WAP application for mobile phones letting people book rooms in the University. It is not designed for long term reservation but for "just in time" use by finding a free room as needed when the user is located near the room which he is asking to occupy. A similar application was designed for online management of student presence at lab works.

5 Conclusion

The prototype on which we are currently working relates to contextual maintenance and crisis interventions for equipment of significant cost and considerable criticality. It combines contextualisation (access to precise information related to each product), collaboration (possibility to work together with different actors located either on the intervention site or elsewhere to group local actors and remote experts when necessary), localization (easy identification by appropriate devices of parts, components or subsets of the machine) and mobility (access using light - handheld devices). By referencing the object on which he/she is working, the user can acquire on his/her device (laptop, PDA, etc..) information about this artefact coming from the enterprise's PDMS (Product Data Management System) (David & Boutros, 2001). In this case, not only the content is adapted to the context but also the human interface which leads to the concept of plasticity of user interfaces (Thevenin & Coutaz, 1999). We thus implement Capillary Cooperative Work principles: cooperation modes, objects and agent mobility, contextualization, localising, interface plasticity and awareness.

References

David B., Boutros N. (2001). DidaRex: Formation et assistance contextualisées. In Proceedings of JIM 2001, Metz, July 2001.

Ellis C.A., Gibbs S.J. & Rein G.L. (1991), Groupware : some issues and experiences. *Communication of ACM*, Vol 34. n°1, 39-58.

Ellis C.A. & Wainer J. (1994). A Conceptual Model of Groupware. In Proceedings of CSCW'94 Conference, p. 79-88, ACM Press.

Hightower J., Boriello G. (2001). Location Systems for Ubiquitous Computing. *Computer*, 34(8), IEEE Computer Society, August 2001, pp. 57-66.

Juhlin O., Weilenmann A. (2001). Decentralizing the Control Room: Mobile Work and Institutional Order. In proceedings of ECSCW 2001, Bonn, Germany, 16-20 Sept. 2001.

Saikali K. David B. (2001) Using Workflow for Coordination in Groupware Applications. In proceedings of IHM -HCI2001. Lille, France. September 2001

Schmidt A., Lauff M., Beigl M. (1998). Handheld CSCW. Workshop on handheld CSCW'98, Seattle USA, 14 november 1998. http://www.teco. edu/hcscw/papers.html.

Sheridan T. B. (1988). Human and computers roles in supervisory control and telerobotics: musing about function, language and hierarchy. In L.P.Goodstein, H.B. Andersen, and S.E. Olsen, (Eds.) *Task, errors and mental models*. Taylor&Francis. London. UK.

Tang J.C., Yankelovich N., Begole J. (2001). ConNexus to Awarenex : Extending awareness to mobile users. In Proceedings of CHI 2001, Seattle, Washington, April 2001.

Thevenin D., Coutaz J. (1999). Plasticity of User Interfaces: Framework and Research Agenda. In Proceedings of Interact'99, Edinburgh, Scotland, pp. 110-117.

Requirements for Intelligent Access to Mankind's Collective Memory in I-Mass

Geert de Haan

Maastricht McLuhan Institute
University of Maastricht
P.O. Box 616
5600 MD Maastricht
The Netherlands
g.dehaan@mmi.unimaas.nl

Abstract

This paper describes a novel way to acquire trustworthy access to cultural heritage information over the internet. Instead of depending on information that is put on the internet on a free and voluntary basis, this approach utilizes the digitization of cultural heritage objects in combination with metadata descriptions generated by experts at museums and libraries.
The paper will present an overview of the I-Mass system, its concepts and the state of the art of its development. It will further describe each of three requirements studies in more detail and compare their results in order to draw conclusions about the relation between costs and efforts and the utility of the methods they employ. Finally, the paper will discuss the results and draw several conclusions about the usefulness of the three methods for requirements analysis for novel information systems.

1 Introduction

I-Mass is a European research and development project in the framework of the Information Society Technology (IST) actions of the European Community. The general problem that I-Mass addresses is providing access to the enormous diversity of cultural heritage information in a useful way. A picture of the Mona Lisa as an illustration for a high school project differs from the highly detailed digital scans used by restoration specialists. Likewise, a professor in renaissance art may want different information about Leonardo Da Vinci than the average tourist.
I-Mass utilizes a multi-agent system, to deal with the variety of description standards and to present information in the proper context to the user, e.g., by adapting the system to the characteristics and preferences of the user by means of dynamically inferring the users' expertise, interests and intentions. For a detailed description of the I-Mass system, see: de Haan (2002) and for more information about the I-Mass project, see: www.i-massweb.org.

In terms of architecture, the I-Mass system consists of five elements or modules (see Figure 1): the digital heritage collections, software agents interfacing to the heritage collections, the knowledge landscape, the user interface software agents, and the user interface and the user model.

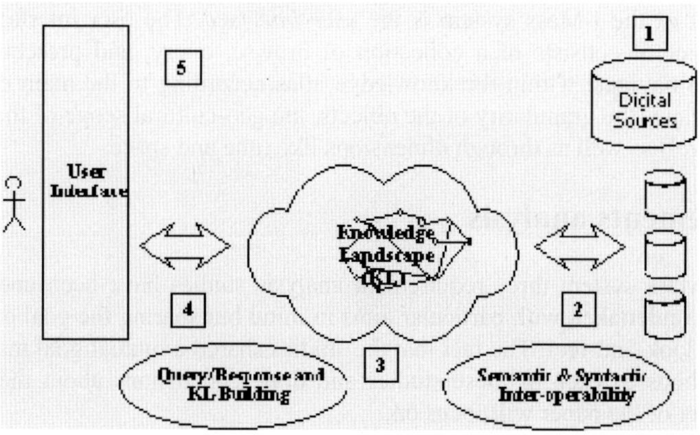

Figure 1: The five main modules of the I-Mass system.

The *digital cultural heritage collections* form the data for I-Mass. Part of the project consists of digitizing and annotating the collections. A number of libraries, museums and archives from Italy and the Belgium-Dutch region of Limburg participate. They provide online access to part of their digital collections as well as a description of the collections according to some metadata standards. The collections are both distributed, since they remain at the site of the content-provider, and they are quite heterogeneous, involving different languages and ranging from archaeology, local history and culture, city and provincial archives, Limburgentia (things having to do with Limburg), books and newspaper articles, organs, Renaissance architecture, 3D models of buildings, and scientific machinery and history.

The *multi-agent system* to interface to the heritage collections forms the second part of the I-Mass system. The role of these agents is threefold: (1) they know what is in their collections and are able to communicate about them, (2) they hide the (metadata) details of their collections from the rest of the system and provide information about collection items in standard formats, and, apart from direct searches, (3) they are responsible for proactively searching for contextual information. As such, these agents provide for syntactic and semantic interoperability. Interoperability is acquired, using information about metadata standards and using information from authority files and reference works, which are also regarded as collections.

The *knowledge landscape* is the third element of I-Mass. The knowledge landscape is a dynamic and abstract representation that is the result of the history of the dialogue with the user, the answers to the user's questions from the digital collections, and the pro-active searching and manipulations by the software agents. The knowledge landscape is a multi-dimensional landscape with dimensions: granularity of the objects (individuals, groups, processes and societies), the geo-cultural scope of the knowledge (local, regional, national and international), and the level of representation (from concepts and definitions to the objects themselves and reconstructions) (Veltman, 1996).

A fourth element, a set of *user interface agents*, is responsible for translating the knowledge landscape into a presentation form or *knowledge atlas* that is appropriate to the user. In the opposite direction, the user interface agent system translates any input from the user, such as information requests and browsing behavior into something that makes sense to the system, and it manages the processing of queries and responses.

The final element of the I-Mass system is the *user interface*. The user interface is not a single element, but, rather, it consists of a collection of browse, query, and presentation modules to enable the user to navigate within the knowledge atlas according to the main dimensions of the knowledge landscape (the granularity of the objects, the geo-cultural scope of the knowledge, and the knowledge level) as well as through dimensions like time and space.

2 Requirements analysis

To design the I-Mass system three requirement analysis studies have been undertaken. Each of these studies was undertaken with particular aims in mind but sharing the goal of defining I-Mass functionality and look-and-feel. The fact that the studies shared a mutual goal makes it possible to compare the methods of each of these studies and draw conclusions about their utility. This is what the remainder of the paper will focus on.

The requirement studies utilized the following methods to gather, analyze and report requirements:

- Study one concerned an interview and observation study, providing results in the form of a general list of requirements.
- Study two concerned a scenario-based approach to derive and validate requirements by means of mock-ups and quick-and-dirty prototypes.
- Study three involved gathering and analyzing interviews with stake holders and presenting the synthesized results as a pictorial vision of the user interface design.

In a *first study*, prospective users of the I-Mass system were observed and interviewed to determine their ways of working and how their work could be supported by a system like I-Mass. A number of scientists, teachers and students staffing a course in Medieval Art, served as subjects. The subject sample, albeit biased towards higher education, forms a good representation of the potential user population of a system like I-Mass. The data were gathered on observatory grids and the results were presented as a list of preliminary user requirements according to several main categories. A further selection of requirements was created using priority assignments by the I-Mass participants until a list of about a hundred preliminary requirements remained. Even though the list of requirements addressed almost every possible aspect of the system, the list simply was much too long to forward design or to inform designers about exactly what to build.

A *second study* was specifically aimed at creating design information. This study used a scenario-based design exercise on the basis of Activity Theory (Nardi, 1996). For a detailed description of the method, see: Rizzo et al. (1997). From in-depth interviews with cultural-heritage experts in the context of work, scenarios were created about how particular tasks were performed, such as selecting the material for a student anthology or writing a thesis. Each scenario was represented in a standard graphical form to enable indicating how I-Mass might support particular steps in the scenario. The results were synthesized in the form of mock-ups and quick-and-dirty prototypes for validation by the experts against the scenario data. Figure 2 presents part of a mock-up design solution for representing and querying resources.
Presenting information about the functionality and look-and-feel of the user interface proved quite worthwhile to inform designers about how to proceed. The use of mock-ups, however, does not guarantee that design proceeds in a proper direction, which is particularly important since an overall vision of the final purpose of the design is still missing. In a subsequent step, the mock-ups were taken together to form a single integrated mock-up. As a rather shallow design solution, this mock-up did not provide the minimally required overall vision of the design solution.

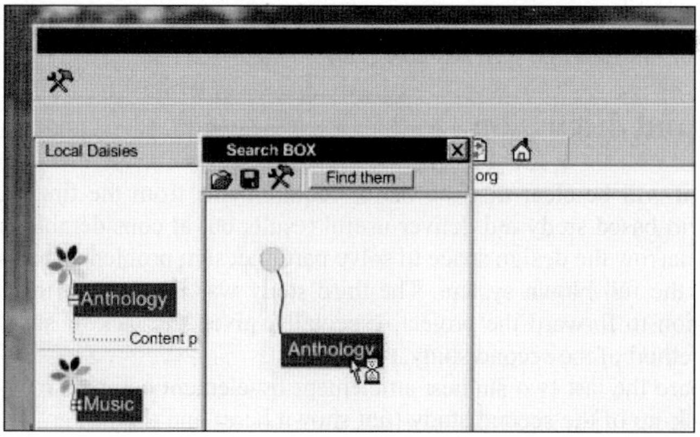

Figure 2: A mockup design solution presenting resources as flowers with attribute leaves.

A *third study* was specifically undertaken to create a representation of what the I-Mass design should eventually look like. Content-providers (museum curators, librarians, etc.) were interviewed assuming that they are know about both, their collections and the prospective I-Mass users. Being presented with a bare search interface such as Google's they were simply asked to present, in text and drawings a so-called *minimal interface*, which provides what is minimally required to perform their tasks (de Greef and Neerincx, 1995). In this study, the results were not presented as a list but rather aggregated into an overall design vision of I-Mass in pictorial form. In a workshop, the results were subsequently fed back for comments and improvements. Figure 2 presents the content-providers' vision of what the system might eventually look like.

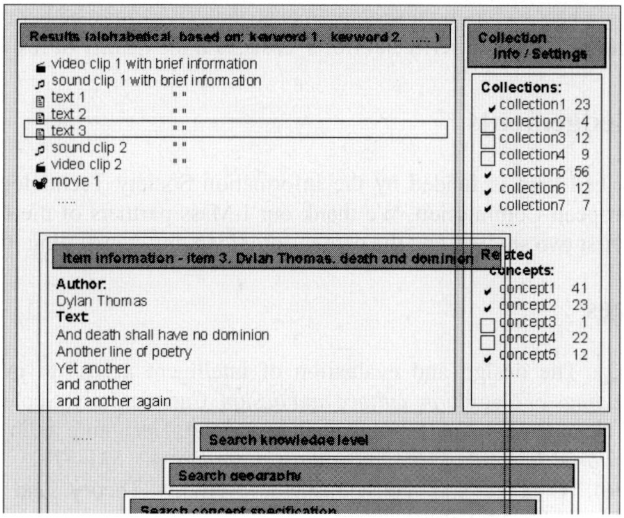

Figure 2: The I-Mass design vision according to the content-providers.

In creating the overall design vision, the third study proved most useful, especially because of the limited amounts of time and effort required. A notable problem, associated with the method of directly asking for the subjects' views was that it proved rather difficult to the content-providers to step out of the *world as we know it*. As a result, the I-Mass system was portrayed as a merely enhanced version of the average web search engine.

3 Results and discussion

From the results, it will be clear that the list of requirements from the first study is simply not useful. The scenario-based study did deliver useful results but at considerable costs. It does seem useful, though, to narrow the design space to solve partial design problems, and only then to widen it again to design the full blown system. The third study was most useful in delivering a sound overall design vision to forward the project, especially, given the lack of such a result from the more expensive method of the second study.

To properly compare the last two studies, an element-by-element comparison was made between the integrated mock-up of the second study (not shown here) and the design vision of the third. It showed a striking similarity in surface characteristics. Both present a search frame, a frame to present and manipulate results, and frames to specify presentation preferences and search constrains. Also, both mock-ups provide facilities to closely inspect individual search as well as facilities to save and reuse results. Unique in the vision mock-up are small icons to tell what each item represents (cf. a picture, a video clip) and a separate frame to choose which collections to search.

Major differences between the mock-ups are found at the deeper characteristics, where the functionality is quite similar but only the integrated mock-up provides facilities to create new description. The design vision mock-up only provides functions for manipulate the results for presentation purposes. In order to compare functionality differences better, more research is needed and a follow-up study is planned to determine whether simple interview and envisioning techniques suffice or that more heavyweight scenario and mock-up techniques should prevail.

Finally, if the comparison of requirements methods shows anything at all, it is that instead of looking for one single best technique for requirements analysis, it may be better to apply methods in parallel to find the best balance between efforts, costs, and the quality of results.

4 Acknowledgements

The research described here is funded by the Information Society Technologies (IST) research program of the European Commission. We thank our I-Mass partners of the University of Siena for conducting the first two studies, and the content-providers for cooperating in the third study.

5 References

de Haan, G. (2002). The design and evaluation of intelligent access to mankind's collective memory. In: *Proceedings of Cognition, culture and design*. Catania, Italy. September 8-11, 47-53.

de Greef, H.P. and Neerincx, M.A. (1995). Cognitive support: Designing aiding to support human knowledge. *International Journal of Human-Computer Studies*, 42, 531-571.

Nardi, B.A. (1996). Context and consciousness. Activity Theory and Human-Computer Interaction. Cambridge, MA, MIT Press.

Rizzo, A., Andreadis, A. and Marchigiani, E. (1997). The AVANTI project. In: *Proceedings of the ACM DIS Conference*, Amsterdam, the Netherlands, August 18-20.

Veltman, K.H. (1996). Universal media searching - SUMS and SUMMA. *Proceedings of EVA '96, Future Strategies and Visions*. London, UK, National Gallery.

Creating social presence through peripheral awareness

de Ruyter Boris *Claire Huijnen, Panos Markopoulos, Wijnand IJsselstein*

Philips Research Eindhoven
Prof. Holstlaan 4
5656 AA Eindhoven
The Netherlands
boris.de.ruyter@philips.com

Technical University of Eindhoven
PO Box 513
5600 MB Eindhoven
The Netherlands
Chuijnen@cuci.nl, P.Markopoulos@tue.nl,
W.A.IJsselstein@tm.tue.nl

Abstract

This paper describes an experimental assessment of affective user benefits that may result from peripheral awareness of a remote friend or group of friends during a shared viewing of a televised event. The experiment suggests that awareness supported through a visual display enhances the level of social presence experienced and increases the attraction individuals feel towards the remote partners, a sign that social interactions can benefit from this type of technology.

1 Introduction

This research explores the potential user benefits from interconnected CE devices. One such benefit is social presence, which refers to the sensation of 'being together' that may be experienced when people interact through a telecommunication medium. As connectivity permeates our daily lives we expect that network infrastructures will become enablers for social interactions. While communication media such as e-mail, telephony, text messaging services for mobile phones, etc., are common, there is more to system-mediated communication than exchanging information. This paper describes research to assess the potential of attaining social presence by maintaining a peripheral (visual) awareness of a connected person or group of persons, outside the context of communication/information exchange tasks. Further, it is assesses the affective benefits that arise out of this interconnection.

2 Social Presence and Group Attraction

The concept of social presence (Short, Williams & Christie, 1976) can be colloquially defined as "the sense of being together" and communicating with someone. While, the main body of research on social presence has concerned its determinants and the measurement of social presence, this study deals with the consequences and the nature of social presence in the context of the home. In social life people are part of many groups that they interact with. The feeling of being a member of such group is called *group attraction*. In general a distinction can be made between two types of groups: primary and secondary groups (Cooley 1990). Primary groups are small, close-knit groups such as families, friendship cliques, children's playgroups, emotionally close peers, etc. Examples of secondary groups are professional associations, business teams, etc. Primary groups are characterized by frequent face-to-face interactions, interdependency and strong group attraction among their members. According to (Cooley 1990) primary groups are "fundamental in forming the social nature and ideas of the individual". Earlier studies have shown that people desire more connections to family and close friends, e.g., see (Hindus, Mainwarin, Leduc, Hagström & Bayley, 2001). It is possible that members of such primary groups are distributed over different locations. With systems that are capable of enhancing social presence, we might be able to connect these different members of the group. We anticipated that increased social presence will improve the relations between group members and that this will be reflected in the attraction individuals feel towards the group of persons they are interconnected with. In the remainder of the paper we report an empirical study that was conducted to validate this conjecture.

3 Empirical Study

An experiment was conducted at the Philips HomeLab, at Eindhoven in the Netherlands. The HomeLab is a test laboratory that looks like a normal house, but thanks to an extensive observation infrastructure provides a 'natural situation' to test the behavior of people at home.

Participants and task. 34 participants, who were all Dutch males, participated in the experiment. They were recruited as groups of 3 friends who enjoy watching soccer games, who have no love relationships, as such relations could bias the measurements of group attraction. The friends were split (2-1) and placed in 2 different rooms (there were 2 groups of 2 persons). During the experiment all participants watched the same soccer game.

Design and conditions. A mixed experimental design was adopted. First 2 kinds of viewers were distinguished: single viewer (only 1 person in a room) and group viewer (2 persons in a room). This was a between subjects condition. Then the information offered about the remote friend(s) was a within subjects conditions, where each subject receives every condition (for approximately 30 minutes). The different conditions are the control condition, a silhouette visualisation and a full video condition. In the control condition all participants watched the same match on TV at remote locations. However, the persons could not see any visualization of the remote participant(s). They were told that their friends were watching the same match simultaneously. This condition is a baseline for comparing the visualizations: It may be that people experience a certain level of social presence when they know about each other that they are engaged in the same activity at a certain moment. The second level of visual information was a black and white (silhouette-like) image of the remote match viewers that is updated in real time when there is a movement of the persons in that room (see figure 1.). The processed visual representation of the persons in the remo te location was projected on the wall behind the screen/on the TV screen people were watching, to provide the image of the remote friends at the periphery of the attention of the test participants. In the third condition participants were shown full live video images of their friends watching the match. In this visualization, more detail is depicted and the people in the visualization are always visible. This is contrary to the silhouette visualization, where people only see silhouettes when there is a change in activity.

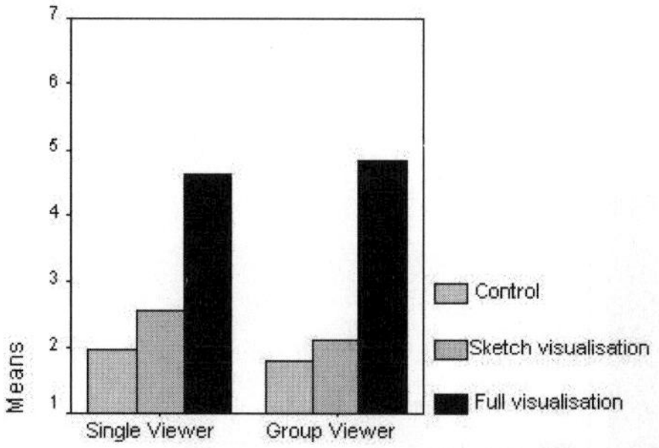

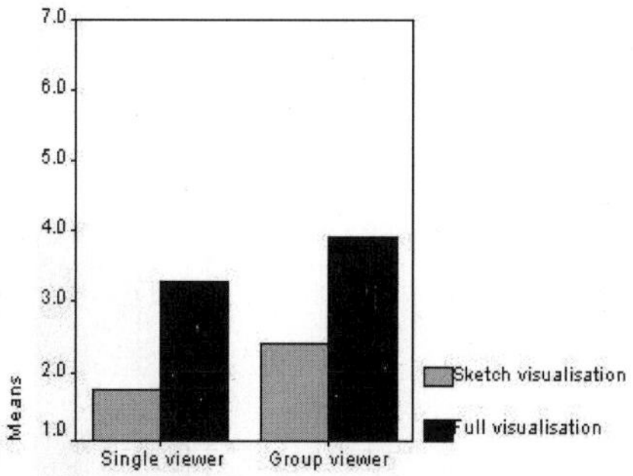

Figure 2. Subjective assessment of the experience by test participants: social presence experienced (left) and the feeling of being watched (right).

The trials were counterbalanced to avoid any potential sequence effects. Every group watched the same match to prevent any effect from differences in games. The match was a classic game for the national Dutch team that can still be exc iting for Dutch soccer enthusiasts, even though it is not a live broadcast. In none of the experimental conditions did we use audio as an experimental condition (audio of the game on television was included in all three conditions). From previous social presence literature it can be expected that adding audio to the visual channel would significantly increase the level of social presence (Short, Williams & Christie, 1976). In this study, however, it was not intended to achieve the maximum level of social presence or to assess the effect of different modalities on social presence. Rather, we were interested to explore minimal and undisruptive means for achieving social presence.

Measures. The independent variables for the experiment are the amount of visual information (none, silhouette or full video) and the group setting (single viewer or 2 persons together). Social presence was measured after each condition by use of the IPO-SPQ (IPO Social Presence Questionnaire) (de Greef and IJsselstein, 2001). The IPO_SPQ makes use of two approaches to measure social presence. It uses the semantic differential items from (Short, et al. 1976) that measures more affective qualities of the medium. Next to these semantic differential items, the IPO-SPQ includes subjective attitude statements about the experience using a 7-point agree-disagree scale. The items from the subjective attitude scale were adapted for the context of this experiment. Another dependent variable is the level of Attraction to Group experienced by the subjects. This was measured by the Group Attitude Scale (Evans and Jarvis, 1986) after each condition. Group Attraction is defined as: 'an individual's desire to identify with and be an accepted member of the group'. The GAS is composed of an equal numb er of positive and negative statements to guard against response set bias. Finally, some questions were included to asses people's attitudes and experiences about the system they were interacting with.

4 Results

On the basis of the data collected with the IPO-SPQ questionnaire, we observe that social presence increases from the control, to the silhouette and the full video condition (see figure 2, left). The difference in social presence between the control and the silhouette conditions is not significant. The difference between both the control and silhouette conditions with the full video condition is significant. In the full video condition people were constantly aware of the activities of their friends, which was not the case in the silhouette condition. There is no difference in the level of experience of social presence between the single and the group viewers. The effect of the silhouette and full visualizations on group attraction is the same for both kinds of viewers (single and group viewers) (see figure 3.). The silhouette and the full video visualization caused a higher

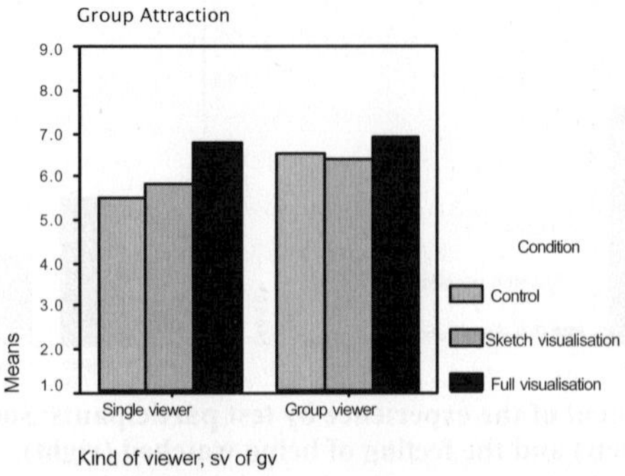

Figure 3. Results for condition on group attraction

group attraction (based on the GAS data) compared to the control. In the control condition there is a difference in the group attraction between the single and the group viewer. The group viewer is more attracted to the group than the single viewer in the control condition. Group attraction increases in the silhouette- and full visualization. The group attraction difference between the single and group viewer disappears when the silhouette or full video representation are displayed.

Some items of the questionnaire gather information about the participant's attitude towards the experimental system. During the full video condition people felt more watched than during both the silhouette and control conditions (see figure 2, right). There was no difference between group or single viewers. People devoted more attention to the television during the silhouette visualization than they did in the full video condition (Figure 4, right) while the silhouette visualization attracted less attention than the full visualization (Figure 4, left). No differences between the single and the group viewer were found.

5 Discussion

The present study investigated if social presence can be established by providing visual display of remote friends that are watching the same television program. By presenting different visualizations of the physical activities in the remote locations in parallel with shared TV content, an experimental condition was created in which the amount of social presence and group attraction could be measured. The results from this study indicate that a low bandwidth visualization of the

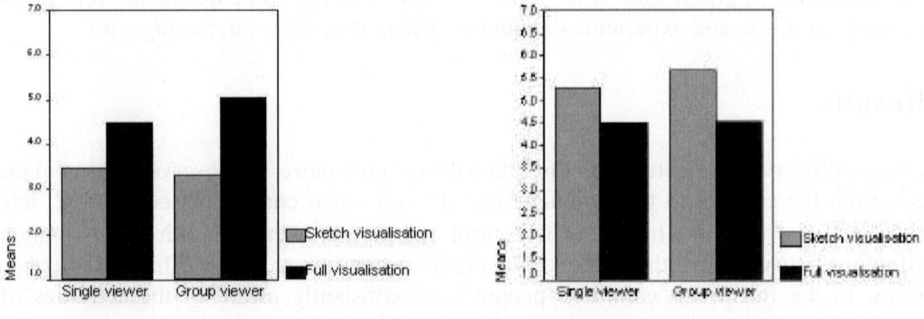

Figure 4: The amount of user attention to the visualisation during the experiment (left) and of user attention on the TV content during the experimental conditions (right).

physical activities from remote locations is capable of establishing a sense of social presence.

Furthermore, the feeling of being part of a group (i.e. group attraction) was increased. When the single viewers did not receive any representation about their friends location, they had a low feeling of belonging to and being an accepted member of the group. The group viewer on the other hand, did have the feeling of belonging to the group in the control condition. When the minimal representation is introduced to the single viewer, this person did experience himself as being an accepted member of the group. The difference between the two kinds of viewers vanished by introducing the visual media.

The silhouette visualization was equally distracting as the full video but it gave participants less the feeling of being observed. This latter aspect of the visualization could be of great importance to create social presence enabling systems for the home environment. Although the full video visualization was stronger in creating the feeling of being together and being part of a group, the low bandwidth visualization using silhouette representations of physical activity is probably more acceptable because it respects user privacy. Further research needs to be conducted for establishing the optimal level of social presence: the present study did not consider whether the amount of social presence created by the experimental visualization was sufficient to be of value for users. Test participants indicated that they would prefer different levels of social presence for different kind of TV programs. People prefer to watch sports and movies in presence of others, whereas they prefer to watch news and documentaries alone. For entertaining programs, viewers enjoy making a cozy atmosphere and experience other person's reactions. More research is needed to investigate the context in which people prefer less or more social presence.

Additional research should assess whether the effects generalise for different relational types. It may be the case that people prefer different levels of social presence depending on the "interaction/communication" partner. Moreover, different kind of activities need to be investigated. It can very well be possible that different kind of activities elicit different levels of social presence. Moreover, it can be interesting to asses the level of enjoyment of the participants during the interaction with different systems. Finally, further research is needed to assess whether differences exist between males and females.

6 References

Cooley, C. H. (1909). *Social organization.* New York: Scribner.

Evans, N.J., & Jarvis, P.A. (1986). The Group Attitude Scale. *Small Group Behavior*, 17(2), 203-216.

Greef, P. de & IJsselsteijn, W. (2001). Social Presence in a Home Tele-Application. *CyberPsychology & Behaviour*, 4(2), 307-315.

Hindus, D., Mainwarin, S. D., Leduc, N., Hagström & Bayley, O. (2001). Casablanca: Designing Social Communication Devices for the Home. *Proc. CHI 2001*, ACM Press, NY, 325 – 332.

IJsselsteijn, W.A., de Ridder, H., Freeman, J., & Avons, S.E. (2000). Presence: Concept, determinants and measurement. *Proceedings of the SPIE*, 3959: 520-529.

Short, J., Williams, E. & Christie, B. (1976). *The Psychology of Telecommunication*, John Wiley & Sons Inc., London, 1976.

Shared displays to increase social presence

Monica Divitini & Babak A. Farshchian•*

* Dpt. Information and Computer Science, NTNU, Trondheim, (N)
•Telenor R&D, Trondheim (N)
Monica.Divitini@idi.ntnu.no; Babak.Farshchian@telenor.com

Abstract

Most systems for promoting virtual presence are focusing on providing awareness information to one user through a personal device, being it a desktop or a mobile device. Shared displays have been lately proposed as a way to provide information, and to promote community building and socialization. Displays situated in public areas allow immersing awareness information into a wider informational and social context. In this paper we discuss the usage of shared visual displays to provide presence information to a working community sharing a common territory.

1 Introduction

In this paper we discuss the possibility to use *shared displays* (e.g. interactive whiteboards or LCD screens located in public areas) combined with mobile devices to provide presence information. Shared displays are widely available in form of, e.g. bulletin boards, whiteboards, postIt, and various printouts in any space inhabited by a community. Shared displays play an important role in supporting the social cohesiveness of a community. We suggest that electronic shared displays that resemble these artifacts, and that are enriched with awareness information, can help promoting social presence. In fact, a groupware tool of this type can provide awareness information in the social context where it belongs, and where it can be shared and commonly used by all the members of the community. In addition, as already pointed out in (Luff & Heath, 1998), the combined usage of shared displays and mobile devices can be used to support different degrees of mobility and a smooth switch from individual to collaborative activities, from public to shared information. In this way the proposed tool can accommodate for mobile and resident users and for various modalities of cooperation. The paper is organized as follows. Section 2 discusses current (paper- based) shared displays as they are used in communities, their limitations, and the requirements for turning these artefacts into electronic shared displays enriched with presence information. In Section 3 we present a system realizing a flexible computer-based shared display, currently in use at Telenor R&D. In Section 4 we discuss how this system can promote social presence and awareness in a community. Section 5 concludes the paper.

2 Requirements

Shared displays play multiple roles in a community and recently there has been a growing interest in creating electronic versions of these artifacts for overcoming the limitations of the paper-based ones (Greenberg & Rounding, 2001). In the following we focus on the role that shared displays play in supporting presence within a community. Figure 1 shows three different displays. (The photo is taken from Telenor R&D's premises.) To the left there is a traditional corkboard with various printouts that can be of interest for the community. In the middle there is a whiteboard where people can mark their availability during the two coming weeks (middle-term availability). To the right there is a printout of a table with the holiday schedule (long-term availability). This simple snapshot clearly points out two ways in which displays promote social presence in a community. First they provide *traces of community life*. A glance at the corkboard is enough to see whether there is any news that one should be aware of, what information is relevant for the specific community, who is expected to do what, and so on. Second, these artifacts act as a *community concierge*, proving information on who is in, who is out, and where each person is.

Figure 1: paper-based shared displays

These artifacts are powerful tools but their capability to support social presence is suffering from some limitations. First, they can quickly become messy, especially when the community life reflected in the artifacts is highly dynamic. Second, there is no direct connection between the information that is provided and the person who is related to that information (and who could provide further explanation if needed). Third, the provided availability information is too coarse-grained. Finally, these artifacts are available only to people in their close physical proximity.

To take advantage of a familiar informational artifact and increase the feeling of presence among the members of a community, we envision an electronic version of shared displays that must:

- Be easy to contribute to: All members of the community (distributed or collocated) should be able to easily contribute to the contents of the display, in order to keep the contents relevant.
- Be interactive: Paper-based displays are mostly passive. We envision a version that allows a higher degree of interactivity between the user and the system and, indirectly, supports interactivity within the community. This requires that users can, e.g. send information to the system, request information from the system, forward information to other community members, and start a communication with someone through the system.
- Support distributed users: In order to support cooperation in geographically distributed communities, the system must allow multiple copies of the same shared display to exist in different locations. To support mobile users the system must support the partial visualization of shared displays and remote contribution to them.
- Support different interaction modalities: In order to support the required level of interactivity as well as remote users, the system must allow the usage of different devices for interacting with the system, e.g. interactive screens, desktop computers, and different handheld devices.

- Be extensible: Since each community has different needs and these needs can evolve with time, it must be possible to extend the system with new services.

3 KOALA: A shared display application

Our ideas are demonstrated in a prototype of a shared display application called KOALA (KOmmunity Awareness with Large Artifacts). KOALA is being developed through a user-centered process. It is in daily use at Telenor R&D, and is being improved based on feedback collected from users. In this section we provide an overview of the main functionality of KOALA, and describe briefly its architecture. The main part of KOALA is its *shared display*, a physical screen hanging in a common area (see Figure 2, to the left). The shared display is where public information posted by the user community is displayed. Users can use dedicated client applications, called *interaction clients*, to post information to the shared display from their desktop PCs, PDAs or mobile phones. For distributed communities, KOALA supports multiple copies of its shared display located in several physical locations. Information posted by the members of the distributed community is displayed on all the copies of the shared display.

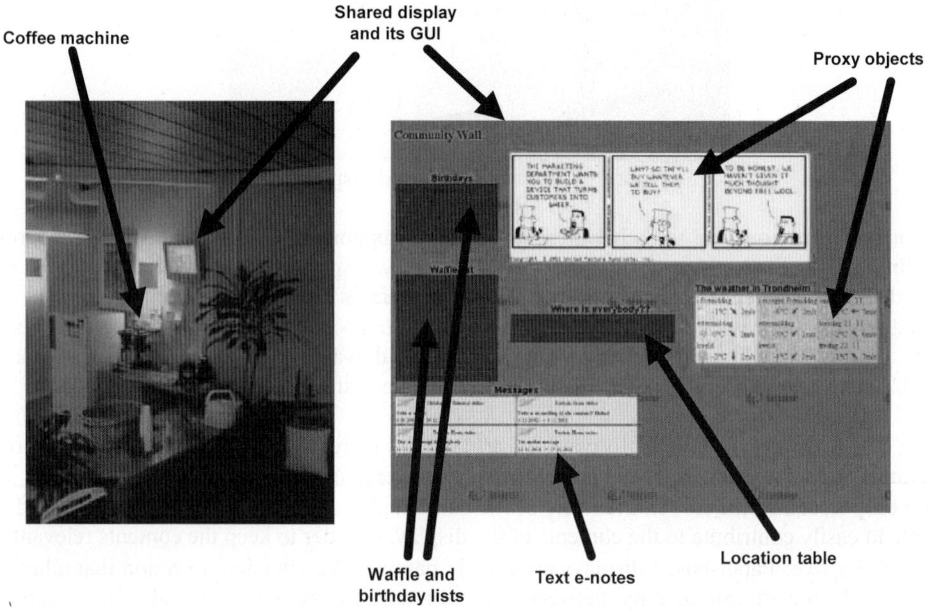

Figure 2: The shared display used at Telenor R&D

3.1 Types of information on a shared display

The right-hand side of Figure 2 shows the contents of a typical KOALA shared display. KOALA supports three generic *informational object* types, which correspond to and improve some of their physical counterparts found in conventional bulletin boards:

- *E-notes*: a multimedia message created by a user and attached to the display. E-notes are similar to conventional notes, printouts, snapshots, etc. that normally hang on bulletin boards.

- *Community objects*: While e-notes are created and owned by a single member of a community, community objects are owned by all the members. An example is the electronic version of a *location table* (see Figure 2) that indicates whether people are in or out of office. Users' locations are detected automatically by tracking their mobile phones.
- *Proxy objects*: A proxy object is used to display some external information, i.e. information not contained within KOALA database. Simple proxy objects show comic strips or weather forecasts (as shown in Figure 2). A more advanced proxy object is a live video stream connected to a video camera showing the cafeteria or any other location.

3.2 User interaction with KOALA

KOALA does not support any direct user interaction with the shared display (e.g. users cannot attach notes to a shared display by simply approaching it, as it is common for conventional bulletin boards). We have chosen to implement user interaction using mobile *interaction clients* instead of advanced display technology such as touch-enabled screens. This choice is motivated by the willingness to support mobile users, and to take advantage of widely available technology. Interaction clients allow users to post different types of e-notes (text messages or snapshots taken with web cameras) or to interact with community objects (e.g. register for making the waffle next Friday). Interaction clients can be used from any Web browser, on a desktop PC or on a mobile device such as a PDA.

Using dedicated interaction clients has two main advantages. First, the shared display itself remains very simple, and viewing does not require any specific effort except glancing at the shared display. This is an important property, allowing KOALA to become a ubiquitous part of an informal environment. Second, using networked interaction clients (on desktop PCs, PDAs or mobile phones) means that users can interact with the shared display from any physical location.

3.3 KOALA design and architecture

KOALA is implemented in a client-server fashion using XML web services architecture (Glæserud & Hoem, 2002). The overall architecture of KOALA consists of KOALA *server* and multiple copies of the *shared display*. KOALA server is in charge of updating the information content of all the existing copies of the shared display. Users who want to interact with KOALA beyond viewing the information on the shared display (e.g. send e-notes) log on to the server using their interaction clients. Modularity and extensibility have been two important requirements when implementing KOALA. All communication is based on XML web services and SOAP (Simple Object Access Protocol) (World Wide Web Consortium, 2002). This makes the architecture highly modular. KOALA server maintains a dynamic directory of services it provides (e.g. informational objects). New types of informational objects can be added relatively easily in form of new web services. Interaction clients are implemented in form of web-based applications, making it possible to use them on both desktop PCs and PDAs. The application running on the shared display is implemented using Java Web Start technology that allows for easy management of changes to the shared display (e.g. when new information object types are added).

4 Support for presence and awareness in KOALA

Although we have not yet performed any formal study of the usage of KOALA, we have done some informal observations of how people use KOALA, and what they use it for. This section provides a summary of these observations. Shared displays are located in common areas, generally within informal surroundings. At Telenor R&D one shared display is located in each coffee corner just above the coffee machine (Figure 2). It is difficult not to see the display when walking to the coffee machine to take a cup of coffee. A glance at the display gives an overview of who is around, latest messages sent to the display, and other information of social character (such as "who has birthday today" or "who will make waffle today"). In this way the display collects traces of the community life, increasing the overall awareness of what is going on. At the same time, the display plays the role of community concierge, at a level of details not possible using the original artifacts (shown in Figure 1). The information presented on the display often triggers informal discussions among people who are around the display at the same time. KOALA aims to support community culture and feelings, by increasing the awareness of users on other community members as well as on the community life. KOALA provides an efficient way of obtaining an overview of social information:

- Awareness of happenings in a community: KOALA provides an overview of the activities in a community. This overview is obtained both through e-notes and through community objects. KOALA encourages people to share information that is of social value, in this way raising overall community awareness.
- Awareness of people: KOALA allows users to contribute with various presence information about themselves, e.g. by sending e-notes (support for multimedia mobile messages, or MMS, is a feature desired by users who own mobile phones with built-in cameras) and by providing their current location. In this way, KOALA contributes to social awareness and presence.

5 Conclusions and future work

In this paper we have discuss the possibility to use displays situated in public areas for supporting the life of a community and for immersing awareness information into a wider informational and social context. We have also presented a system realizing these shared displays. We have briefly discussed the usage of this system for promoting social presence in an existing community. The presented system is currently used at Telenor R&D. This first experimental usage will allow us to better understand the impact of this type of systems on the feeling of presence in the community. The feedback will also be used to refine the system. Using the system in different types of community (e.g. in a family) is an interesting future work. In addition, we plan to study the possibility to enrich the system providing direct interaction with the shared display by using touch screens.

6 References

Glæserud, C. P., & Hoem, T. (2002). Personalized community walls (Student Project Report). Trondheim, Norway: Dept. of Information and Computer Science.

Greenberg, S., & Rounding, M. (2001). The notification collage: posting information to public and personal displays. Paper presented at the Conference on Human Factors and Computing Systems, Seattle, WA, USA.

Luff, P., & Heath, C. (1998). Mobility in collaboration. Paper presented at the Computer Supported Cooperative Work, New York, NY, USA.

World Wide Web Consortium. (2002). Web Services Activity homepage. Retrieved, 2002, from the World Wide Web: http://www.w3.org/2002/ws/

An Interactive Ontology-Based Query Formulation Approach for Exploratory Styles of Interaction

Elena García

Computer Science Department,
University of Alcalá
Ctra. Barcelona km. 33.6 - 28871
Alcalá de Henares (Madrid) Spain
elena.garciab@uah.es

Miguel-Ángel Sicilia, Paloma Díaz, Ignacio Aedo

Computer Science Department, Carlos III
University
Avda. Universidad, 30
28911 Leganés (Madrid) Spain
{msicilia@inf, pdp@inf, aedo@ia}.uc3m.es

Abstract

The use of ontologies enables the development of new Web resource search mechanisms that use metadata instead of lexicographical patterns for query formulation and resolution. As a consequence, new query formulation techniques must be devised and tested. In this paper, an approach to query formulation based on the navigation of the generalization-specialization hierarchy of ontology terms is described as a novel general-purpose approach conceived for users interacting in an exploratory style. The query refinement process also allows the specification of the form of the resources to be retrieved, and query results – that entail the traversal of ontology relationships – can also be used to initiate new query formulation processes. A prototype is described along with required improvements obtained from the initial testing of the approach.

1 Introduction

The process of searching information in the Web is currently carried out by using search engines that essentially take as input a rather simple query string from the user, and then consult indexing structures to find resources in which the given combination of terms appear. These *information retrieval* (IR) methods (Baeza-Yates & Ribeiro-Neto, 1999) are intrinsically constrained by the fact that the current Web is made up of unstructured information in the form of text, videos and other media, which are not intended to be machine-readable but human-readable. With the emergence of the so-called *Semantic Web* technology (Berners-Lee, Hendler & Lassila 2001), Web resources are expected to be associated in some way with *ontologies* (Gruber, 1993), which are large shared consensually-engineered conceptualizations of specific domains, based on logic languages like *description logics* (DL) (Nardi & Brachman, 2002). If the *Semantic Web* vision comes to a reality, Web resources would provide associated *annotations* that connect them with one or several ontologies, providing an explicit semantic characterization. In consequence, a research agenda is needed to design and test new query formulation and resolution strategies for the overall community of Internet users. While the latter area is mainly concerned with the fields of databases and logic queries, query formulation is mainly a matter of *human-computer-interaction* research.

Many interfaces for ontology-based queries like WEBKB (Martin & Eklund, 2000) or ONTOBROKER (Decker, Erdmann, Fensel & Studer, 1999) are oriented towards ontology specialists, which are able to write complex queries (in the WEBKB case) or to formulate them using formalisms oriented towards Frame-Logic syntax (in ONTOBROKER), and therefore are cumbersome for common users. Straightforward ontology query approaches like SHOESEARCH (Helfin & Hendler, 2000) do not provide guidance and help about query refinement to the users. In addition, natural language processing approaches systems like ONTOQUERY (Andreasen, Fischer, & Erdman, 2000) are ultimately hampered by the complexity of human languages, which makes difficult capturing the sense of non-trivial sentences. As a consequence, approaches that study interactive techniques in query formulation have begun to appear, but they are domain-specific, like the one described in (Kapetanios & Groenewoud, 2002), which is tightly bound to domain knowledge on medical information.

In this paper, we describe a domain-independent approach to query formulation using ontologies based on the work described in (Sicilia, García, Díaz & Aedo, 2003), which uses a layered approach for ontology-based descriptions. The ontology contains both a domain ontology and a resource-type ontology, enabling filtered queries that specify that only concrete types of resource (e.g. 'conference papers' or 'Ph.D. Thesis') should be considered. This apparently simple feature significantly improves precision in many situations. The rest of this paper is structured as follows. In Section 2, the query formulation approach and its Web prototype is described. Section 3 briefly sketches the current retrieval mechanism used, along with how new queries can be initiated from query results. Finally, conclusions and future directions are summarized in Section 4.

2 Query Formulation

Our rationale for the interaction design of an ontology-based search system considers two basic guiding principles. The first one is that query formulation must constitute an interactive refinement process that navigates the conceptual schema defined by the ontology, as is done in IR-inspired approaches to database search, e.g., in (Hofstede, Proper & van der Weide, 1996). The second one is that the system should allow for exploratory uses, in which users are not formally trained in the underlying ontology structure, but discover the possibilities of the interface by using it. In addition, it's assumed that the search tool is *thematic* (or multi-thematic), in the sense that the user selects one or several concrete domains before starting the search process. Therefore, the approach is especially devised for a *formal search* mode of interaction, according to the characterization given in (Choo, Detlor & Turnbull, 2000).

As a result of the just described principles, the interface departs from top-level, abstract terms in the ontology (called *entry points*) – that are pre-established by the expert or *ontogroup* that created the ontology –, and proceeds iteratively. The iteration consists in that the user has the option to select one or more of the terms showed, and by clicking a "*refine*" button, he/she obtains the terms that are direct specializations of the selected ones, e.g. if he/she selects the terms 'Usability Evaluation Method' and 'Article', after asking for refinement, he will obtain, between others, the terms 'Journal Article', 'Conference Article', 'Inspection Method' and 'Inquiry Method'. This process may proceed till reaching the level of the instances of these or more concrete terms. It should be noted that all the process is guided by providing a single selection task at each step. An example query on a 'usability evaluation' resource base is showed in Figure 1, in which the user has selected a domain criterion 'Questionnaire' – after refining it from the more general terms 'Inquiry Method' and 'Usability Evaluation Method' – and also a resource type criterion, 'Conference paper'.

As just described, the approach for query formulation uses first an ontology browser that is used to crawl the generalization-specification hierarchy, collecting a collection of ontology terms (in the *T-box*, following DL jargon) in a query refinement process. This allows for the efficient formulation of a conjunctive basic query, since hierarchies are seldom deep, resulting in a low number of user selection actions. Then, the basic query is used to generate a set of instances (extracted from the *A-box*) along with their relations. As relations in ontologies are general-purpose predicates on terms and instances, the user is able to discover specific relationships in the results of the simple query, which may lead her/his to narrow the query expression. In all of the phases, the query is built by navigation and not by writing down strings, which avoids problems with homonymy and more complex forms of ambiguity that are inherent to natural language. Then, the interface allows for iterative definition of queries, since query results are displayed as links that open a new query formulation context taking into account the existing query. This approach is consistent with the guidance and relevance concepts as stated in (Bechhofer, Stevens, Ng, Jacoby, & Goble, 1999).

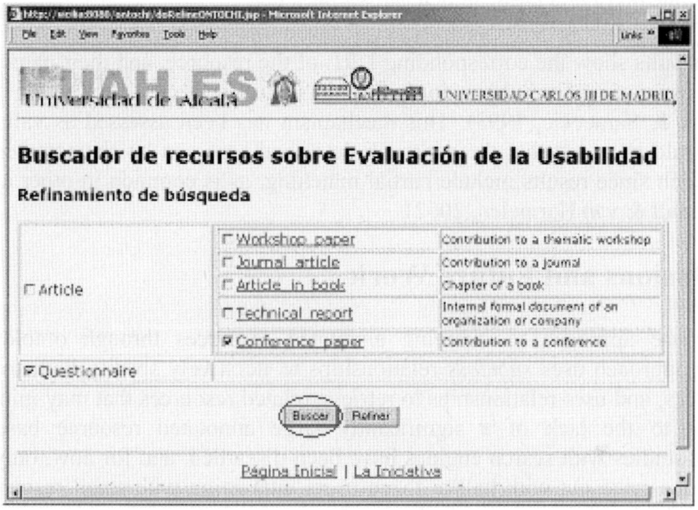

Figure 1: Example query formulation process

After a prototype implementation – developed in Java, using RDF and the JENA toolkit to process it as described in (Sicilia et al., 2003) – and a restricted small test of the just described approach, a number of conclusions have pointed out to further research issues (a thorough user testing phase is left to future work, since we're dealing with a novel concept of querying and further alternatives are required to draw firm conclusions). First, the specific form of resource annotation constraints the interface. In our approach, annotations were based on the reification of resources inside the ontology. This way, resources could be instances of an ontology term, and as such, be related to other instances (resources or not). Other approaches, like associating terms to documents, place constraints on the query formulation, since plain associations can be interpreted in several ways. Second, interactive, guided approaches appear to enable effective searching strategies, and the key aspect of these interfaces is the number, kind and sequence of selections the user should perform to fit its search strategy (possibly adaptive techniques could address differences in user's strategies). And third, it's the exploitation of relationships between terms that provides the advantages over conventional information retrieval schemes. In consequence, that they must be the focus of query building strategies.

3 Results as Query Initiators

Once the query formulation is completed, the system takes the query expressed as a set of terms and proceeds in the retrieval process. Diverse heuristics can be implemented for that purpose, and we chose to implement only those that were directly suggested by potential users. Concretely, two techniques are currently implemented in the prototype:

- First, instances of the terms are retrieved by traversing the *type* relationship in the RDF representation.
- And second, relationships are retrieved according to the following heuristic. First, the extent – i.e. the set of all of the instances, including instances of specializations – of each term in the query is computed. Then, all the extents are merged, and finally, relations involving terms in that set are retrieved.

Both kinds of results show the corresponding URL of the resource, and their names are also links that initiate a new query formulation process that includes them – facilitating term relevance feedback (Spink & Saracevic, 1998). This mechanism has been assessed as valuable from user opinions. It should be noted that the approach described here can be characterized as a flexible querying approach since results include partial matching, as is common in other approaches, e.g. in (Stuckenschmidt & van Harmelen, 2002).

4 Conclusions and Future Work

A general-purpose approach to querying annotated resources through ontologies has been described. The approach uses *subclass* relationships to iteratively specify both search terms and types of resources, and uses relationships to retrieve related resources that may initiate new search processes. Due to the lack of a significantly large annotated resource base, comparative recall/precision studies with search engines have been discarded, and for now, only informal user testing has been carried out with half a dozen users interacting with small resource bases about 'usability evaluation' and 'new technologies in education' (Sicilia et al., 2003). These studies have suggested a number of concrete improvements, including the following:

- Domain-specific concepts and resource types must be clearly differentiated in the interface design since they represent distinct criteria. The latter criterion is known to be complementary to content-related issues (Carvalho, Machado & Barreto, 1999).
- Relationships between terms that are included in the query under formulation are good candidates to refine the query, given the importance of term relevance feedback.
- Novel resource matching techniques may enhance query results, even for the purpose of providing feedback. Some candidates are the traversal of *transitive* relationships and the suggestion of relaxing the query formulation to additional document types.

The described work and results are currently used to develop the second version of the system. In our view, the evaluation of this kind of systems should stress the importance of exploratory control – as advocated in (Warner, 1999) –, so that cognitive analysis may inform their design. In addition, the introduction of additional query specification facilities for the user is considered as desirable due to the fact that automated search has well-known limitations and some users prefer retaining control on strategic search capabilities (Bates, 1990).

Acknowledgements

This work has been supported by the spanish "Dirección General de Investigación del Ministerio de Ciencia y Tecnología", project number TIC2000-0402 (Ariadne).

References

Andreasen,, T., Fischer, J. & Erdman, H. (2000). Ontology-based querying. In H.L. Larsen et al. (Eds.) *Flexible query answering systems, recent advances* (pp. 15-26). Physica-Verlag, Springer.

Baeza-Yates, R. & Ribeiro-Neto, B. (1999). Modern information retrieval. Addison Wesley.

Bates, M.J. (1990). Where should the person stop and the information search interface start?. *Information Processing & Management*, 26, 575-591.

Bechhofer, S., Stevens, R., Ng, G., Jacoby, A. & Goble, C. (1999). Guiding the user: an ontology driven interface. In *Proceedings of the Workshop on use interfaces to data intensive systems* (pp. 158-161). IEEE Computer Society.

Berners-Lee, T., Hendler, J. & Lassila, O. (2001). The Semantic Web. *Scientific American*, May 2001.

Carvalho-Moura, A.M, Machado, M.L. & Barreto, M. (1999). A metadata architecture to represent electronic documents on the Web. In *Proceedings of the 3rd IEEE Metadata Conference.*

Choo, C.W., Detlor, B. & Turnbull, D. (2000). Information seeking on the Web. *First Monday* 5(2).

Decker, S., Erdmann, M., Fensel, D. & Studer, R. (1999). Ontobroker: ontology based access to distributed and semi-structured information. In Meersman et al. (Eds.) *Semantic Issues in Multimedia Systems* (pp. 351-369), Kluwer Academic Publisher, Boston.

Gruber, T. R. (1993). A translation approach to portable ontology specifications. *Knowledge Acquisition*, 5, 199-220.

Heflin, J. & Hendler, J. (2000). Searching the Web with SHOE. In *Artificial Intelligence for Web Search. Papers from the AAAI Workshop* (pp. 35-40). AAAI Press, Menlo Park, CA.

Hofstede, A., Proper, H., van der Weide, T. (1996). Query formulation as an information retrieval problem. *The Computer Journal* 39(4): 255-274.

Kapetanios, E. and Groenewoud, P. (2002). Query construction through meaningful suggestions of terms. In *Proceedings of the Fifth International Conference on Flexible Query Answering Systems* (pp. 226-239). Lecture Notes in Computer Science, 2522 Springer.

Martin, P. & Eklund, P. (2000) Knowledge retrieval and the World Wide Web. IEEE Intelligent Systems, 15(3) 18-25.

Nardi, D. & Brachman, R. J. (2002). An introduction to description logics. In F. Baader, D. Calvanese, D.L. McGuinness, D. Nardi, P.F. Patel-Schneider (Eds.) *The Description Logic Handbook* (pp. 5-44). Cambridge University Press.

Sicilia, M.A., García, E., Aedo, I. & Díaz, P. (2003). A literature-based approach to the annotation and browsing of domain-specific Web resources. *Information Research* 8(2).

Spink, A. & Saracevic, T. (1998). Human-computer interaction in information retrieval: nature and manifestations of feedback. *Interacting with computers*, 10(3) 249-267.

Stuckenschmidt, H. & van Harmelen, F. (2002). Approximating terminological queries. In T. Andreasen, A. Motro, H. Christiansen, H. Legind Larsen (Eds.) *Flexible query answering systems, 5th conference* (pp. 329-343), Springer Lecture Notes in Computer Science 2522.

Warner, J. (1999). "In the catalogue ye go for men": evaluation criteria for information retrieval systems. *Information Research*, 4(4).

Modeling Collaborative Environment

M. Gea[1], F.L. Gutierrez[1], J.l. Garrido[1] AND J.J. Cañas[2]

[1] D. Lenguajes y Sistemas Informáticos
E.T.S. Ingeniería Informática,
University of Granada
<mgea, jgarrido>@ugr.es
18071 Granada, Spain

[2] Dpto. Psicología Experimental,
Facultad de Psicología
University of Granada
delagado@ugr.es
18071 Granada, Spain

Abstract
Nowadays, communications and collaboration activities take an important role in the modern work organization. Although several methodologies are suited to develop software for workgroups, it is necessary to include aspects related with collaboration and group awareness. This paper proposes a conceptual framework to study relevant features of collaborative systems. This methodology focuses on the dynamics aspects of collaborative scenarios and the relationships between groups and their activities. The notation used in this model is an extension of UML, with enhanced expressiveness to represent collaborative human behavior and to relate this analysis to Software Engineering.

1 Collaborative environments

CSCW (Ellis, 1999) is an interdisciplinary discipline in which is analyzed mechanism of communication and collaboration and the systems supporting them. CSCW have been studied and analyzed from different points of view, such as temporal and spatial dimensions (Benford, Greenhalgh, Rodden & Pycock, 2001), economic cost, sociological aspects, etc. However, no importance has been given to the in-depth study of a methodology to cover relevant aspects of such systems in a systematic way. Thus, considerations about the collaboration process itself, and the dynamic context in which is involved, have not been taken into account.

Collaborative environments are highly interactive systems and their design requires the identification and modeling of communication and sharing mechanisms. Another important issue is the representation of the organization in which the users are involved, because it may impose several constraints to the participants (such as rules, strategies, common objectives, social protocols, etc.). Modeling group behavior in collaborative environments involves three important aspects: task analysis, user behavior and group dynamics.

Traditional approaches focus on different aspects of collaborative systems, but they are not suitable to define a relevant feature such as the dynamics. Task analysis (Paternò, 1999) and workflow systems define a static representation of user task and work processes. In real cases, each user plays different roles depending on the organization requirements and the context. In these cases, the static representation of task analysis it is not enough. On the other hand, collaboration implies inherent group knowledge (decision making, protocols, changes in organization, task allocation for participants, etc.) that is convenient to express in the design phase. This information should be obtained from theoretical models of group behavior.

In collaborative environments, the group itself has the cognitive capabilities (perception, memorization, etc.) of the sum of the group members. One theory to explain these processes is the activity theory (Cañas & Waen, 2001). The activity (figure 1) is the smallest meaningful unit for human actions. Activities are embodied to accomplish a goal (objective) using tools within a community, which enables two perspectives: subject and community through institutional rules, and object and community for task allocation (how to do it).

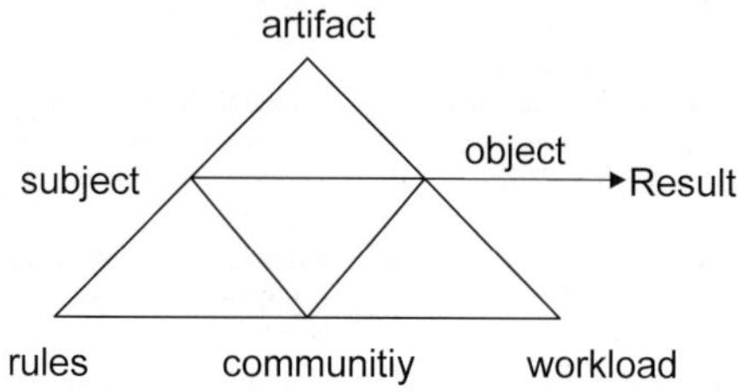

Figure 1: Activity theory

This cognitive view allows us to analyze the necessary information that each user needs to know about the collaborative scenario. First of all, each member must know the scenario and the rules which governs it (rules, activities, artifact…) usually acquired by a learning process. Note that this knowledge may vary on time, so the users must know these changes and adapt their behaviors. Another fact is the group knowledge (group awareness) by their members. In such case, it is important to know the user activities of the group members and the possibilities of collaborate with them (if it is necessary).

Based on the information obtained from the activity theory, we propose a methodology to develop collaborative systems taking into account the representation of the group knowledge and the inherent dynamics of such systems.

2 AMENITIES: A methodological framework proposal

AMENITIES, (acronym for **A MEthodology for aNalyzing and desIgn of cooperaTIve systEmS**) is a methodology based on a description of the group behavior to analyze and design generic cooperative systems (Garrido & Gea, 2001,2002). This framework allows us to achieve a conceptual modeling of a group-centered system that embodies relevant issues of such a structure (organization, evolution, rules, etc) and a smooth transition towards software development (using notation widely used in software engineering). There are three components in AMENITIES: (1) The Requirements Model, (2) the Cooperative Model, and finally, (3) Output of the analysis in the form of a Formal Model and Software Development Models. (Figure 2) shows the relevant aspects of this approach.

The initial phase begins analyzing the system requirements. Several techniques can be used such as ethnographical analysis, formal models (an collaborative extension to PIE model) (Padilla, 2002), or use case diagrams. Each one of these techniques allow us to describe several views of the requirements, that it will be joined in next phase.

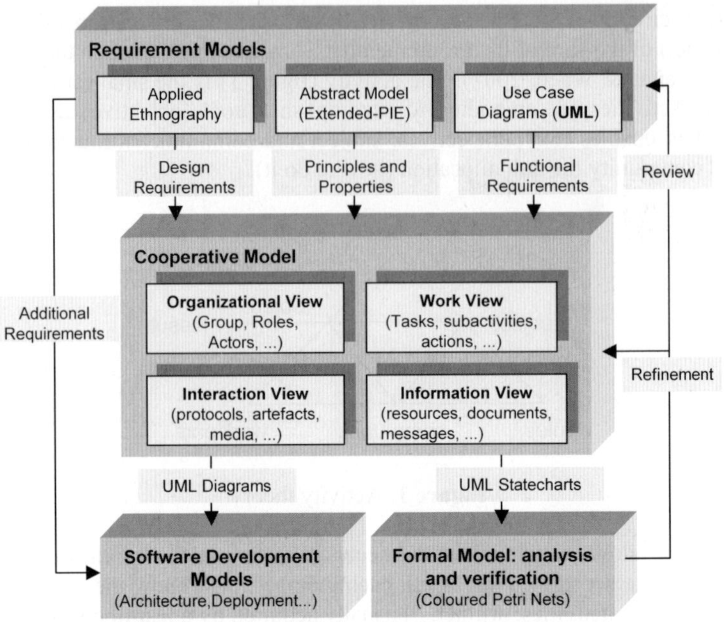

Figure 2: Conceptual framework of Amenities

The cooperative model comprises relevant information about the system. By the nature of the collaborative environment, this analysis is based on the group behavior and related tasks in which the members of the group are involved. AMENITIES organize this information into four views, allowing the representation of different and complementary aspect related to an collaborative environment, that is: organization, work , interactive and the information views.

The organizational view. This view allows us to understand the typology of the organization: the cohesion among participants and their workload. An organization is described as a set of roles and their relationships which they define the group structure and the workload.

The user tasks are conditioned by two kind of relations: capabilities and rules. The capabilities are abilities that the user acquires and allow the user to play a role (usually by a learning process, o by organization imposition). Thus the rules are restrictions carrying out the organization structure. These relations are suited to modeling dynamic situations.

The work view. This vision identifies the knowledge acquisition of each group member in the organizational context. These aspects directly reflect ethical, social and personal considerations. In this view is described the user tasks, assigning the workload to each member of the group. Also it appears relationships among participant in collaborative tasks. This view focuses on user task instead of system process. This is very helpful to analyze cognitive issues (such as the usability, mental models, learning process, etc.)

The interaction view. This view identifies the way in which the task is to be done, the degree of cooperation among group members and protocols. Furthermore, relevant factors concerning the devices used in the interaction process are taken into consideration, since interaction occurs when various artifacts are employed.

The information view. Data and objects are identified and managed with regard to the entire process.

The benefits of this modeling technique can be achieved in several ways. First of all, this conceptual framework can be translated into a **formal model** to analyze and verify properties in a straightforward manner. Thus, the resulting model is suitable to connect with the **software development process** to propose the architectural design of systems.

3 Notation

We have used UML as a modeling tool (Rumbaugh, Jacobson, Booch, 1999). The benefits are evident because is a standard de facto in the industry and it is syntactical and semantically extensible. However we have found difficulties to express notions about group structures and the dynamic aspects involved within. We have chosen statecharts and activity diagrams to specify the group behavior, with minor changes in the semantics of the elements involved. (Table 1) summarizes the main UML elements of interest and how each of these is interpreted in the problem domain.

Table 1. UML elements and their interpretation

UML Diagram	UML Element	Used for / to
Statecharts	State	Role
	Stub state	Set of interruptible substates in a role
	Internal Transition	State or behavior change in an actor playing a role (e.g. starting a new task)
	Transition between states	Role change
	Transition guard	Capability or condition enabling or disabling the behavior change
	History pseudo-state	Remember last execution state of an actor in a particular role
Activities	State	Activity or action
	Branch-merge	Alternative behavior on basis of guards
	Fork-join	Concurrent execution of activities
	Completion transition	Normal execution termination of an activity
	Comments	Add or modify behavior rules, conditions, etc.

The most important diagrams we have used to represent the collaborative model are: the *organization diagram*, the *user roles diagram* and the *task diagram*. Thus, other UML diagrams are used. For example, uses cases diagrams are used to express requirements and class diagram to model the information view.

The *organization diagram* is used to represent all those different roles involved in the collaborative scenario as well as the relationships among them (capabilities to act as another role or rules representing constraints in the organization). This information is represented with an UML state diagram, denoting states as roles and the transitions as rules and/or capabilities. The diagram also describes the dynamics of the scenario by changes of role of the participants due to external or internal (learning process, new abilities of individual) conditions. The *role diagram* summarizes the activities that a member can do playing this role. This information is enriched including triggering events and interruption conditions. This information is represented with an UML activity diagram, where each task is represented as a set of activities. Finally, the *task diagram* describes the individual and group activities, including information about protocols, interaction requirements, and information. Each collaborative task is represented in a single diagram, denoting concurrency and role responsibilities.

4 Conclusion and future research

This paper proposes a methodology to analyze and design collaborative environments based on group behavior. Using a theoretical approach (the activity theory), relevant features are discussed and collected within a conceptual and methodological framework called AMENITIES, in which cognitive issues, workload and group dynamics are addressed.

This notation allows us to describe relevant features of such systems and it can be easily integrated into the Software Engineering process. Future research will consider the development of tools to aid the development process. The management of the shared knowledge of a group is becoming ever more important in various disciplines (business, teaching, etc.), and we consider the proposed framework to be a positive contribution to this goal.

5 References

Benford, S., Greenhalgh, C., Rodden, T., & Pycock J. (2001). Collaborative Virtual Environments. Communications of the ACM July 2001/Vol. 44, No. 79

Cañas, J.J., Waen, Y. (2001). Ergonomía cognitiva, Ed. Panamericana.

Ellis, C. (1999). Workflow Technology. In Computer Supported Cooperative Work. (M. Beaudouin-Lafon, ed.).

Garrido, J.L., Gea, M. (2001). Modelling Dynamic Group Behaviours. In: Johnson, C. (ed.): Interactive System - Design Specification and Verification. LNCS 2220, Springer, (pp 128-143)

Garrido, J.L., Gea, M. (2002). A Coloured Petri Net Formalisation for a UML-based Notation Applied to Cooperative System Modelling. In proc. of IX Workshop on Design Specification and Verification of Interactive System. Rostock

Padilla, N. (2002) Especificación de Sistemas cooperatives. Phd thesis. Universidad de Almería. Spain.

Paternò, F. (1999) Model-Based Design and Evaluation of Interactive Applications. Springer-Verlag,

Rumbaugh, J., Jacobson, I, Booch, G.: The Unified Modeling Language – Reference Manual. Addison Wesley, 1999.

Ontology Based Search for Distributed Agent Platforms

Vlado Glavinic

Faculty of Electrical Engineering and Computing, University of Zagreb
Unska 3, 10000 Zagreb, Croatia
vlado.glavinic@fer.hr

Marko Cupic

Faculty of Electrical Engineering and Computing, University of Zagreb
Unska 3, 10000 Zagreb, Croatia
marko.cupic@fer.hr

Abstract

Information search is a common form of human activity, which certainly represents the predominant activity in an information infrastructure. Its improvement thus substantially contributes to a better human computer interaction. Information search lands itself to augmentation by using a promising new technology basing on agents and ontologies. An agent-based search provides a distributed process and offers the users freedom in accessing the search results. Ontologies, on the other hand, provide context to a search thus improving queries. This paper describes a simple search system based on the previously mentioned principles, with document descriptions expressed using RDF. It consists of three components, including a user interaction one, a mediator and a number of agent environments distributed among information providers.

1 Introduction

Information search is a common form of human activity, which can be augmented by using computers. The computer is a powerful tool for searching, but traditional user interfaces have been a hurdle for novice users and an inadequate tool for expert users (Shneiderman, 1998). As the amount of information nowadays available to users over variety of networks is quite large, there is almost impossible to locate that piece of information relevant to the user, what is commonly known as the "productivity paradox" (Sorensen, O'Riordan & O'Riordan, 1997). An early attempt to solve this problem has envisaged the use of search engines usually equipped with large document bases, which attempt to locate documents possibly relevant to users. Keyword based search engines (Luhn, 1958) are the simplest ones, often resulting with large amounts of documents satisfying the specified keywords but otherwise being contextually irrelevant. Improved versions can use Bayesian networks, which bind documents and keywords basing on statistics, but nevertheless do not enable the specification of the search context (Turtle & Croft, 1991). On the other hand, a new approach would make use of ontologies (Genesereth & Nilsson, 1987), which are explicit specifications of conceptualizations (of world, objects, relations among objects, etc.) (Gruber, 1993) for formal knowledge representations, hence enabling the specification of search contexts, as well as the avoidance of ambiguities like "does a person want to find out information about trees as plants, or trees as a special sort of graphs".

When performing search, it is rather obvious that people intuitively think about objects they want to find ("where is *that document...*") and their relations with a domain of interest ("that *defines a tree structure*?"). I.e. they have a *mental model* (Collins, 1995) comprising of objects (concepts) and relations among them. Since ontologies enable description of this type of relationships, they are strong candidates for usage in the search process. Additionally, private agents could be used to improve the process of searching, thus relieving humans of the process and augmenting the interaction. This approach, which combines ontologies and agents to support search, is described in the paper. The rest of the paper is organized as follows: Section 2 gives the motivation for the use of agents in information search. In Section 3 the reasons that led to and the way of using ontologies is expounded. Section 4 gives a short overview of the search system as well as of some salient implementation details. Section 5 concludes the paper by providing hints for future work.

2 Agent-based Search

Users should be able to perform the search process in an easy and intuitive way. They should be able to specify a search context as well, rather then to enumerate several keywords that may also appear in many documents otherwise irrelevant to them. To support this approach, mobile intelligent agents are utilized to perform the search process on behalf of the user. The agents "understand" the document content, basing on a particular description of documents that can be interpreted by the agents.

In order to additionally improve the interaction through remote connections, users should be given freedom to use both synchronous and asynchronous access. In the former case users would explain to the agent what they want, thus instructing it to visit information providers and send back the matching results. However, the latter case would lead to a more flexible interaction, by enabling users to issue a query and then disconnect, while the agent(s) perform a possibly time consuming search. After some time they could check the search progress and eventually decide to the recall agents and stop search process. In a way this procedure could be compared to a popular metaphor of exchanging typewritten information on the Internet, e.g. the chat facility and e-mail.

Building upon agents to perform the search process brings several advantages: (i) parallel execution – the search process can be executed on many information providers simultaneously, (ii) modification of search algorithms – instead of changing software on every provider, it is sufficient to program a new agent that will use a new search algorithm, and (iii) support for an asynchronous search process – ability to dispatch agents to perform some assignment on behalf of the user.

3 Ontology-based Document Descriptions

Ontologies provide the users the means to express themselves as precisely as possible. Namely, ontologies enable describing all documents without any ambiguities, and effectively provide a way to introduce a search context. E.g. unless some context is defined, when querying about a "tree", it is not clear whether plants or special form of graphs are meant. In the ontological approach every concept has its unique identifier (URI), thus it is not possible to mix-up the above two concepts.

```
<rdfs:Class rdf:about="http://www.fer.hr/#WebPage">
        <rdfs:subClassOf rdf:resource="http://www.fer.hr/#HTMLDocument"/></rdfs:Class>
<rdfs:Class rdf:about="http://www.fer.hr/#WebSite">
        <rdfs:subClassOf rdf:resource="http://www.fer.hr/#WebPage"/></rdfs:Class>
<rdf:Property rdf:about="http://www.fer.hr/#belongsToWebSite">
        <rdfs:label>WebPage X belongs to web site Y</rdfs:label>
        <rdfs:subPropertyOf rdf:resource="http://www.fer.hr/#belongsTo"/>
        <rdfs:domain rdf:resource="http://www.fer.hr/# WebPage"/>
        <rdfs:range rdf:resource="http://www.fer.hr/#WebSite"/></rdf:Property>
```

Figure 1. Part of FERONTO1 ontology representation using RDF

As an illustration a part of the FERONTO1 ontology (Cupic, 2002) represented using RDF (W3C, 1999) is given in Figure 1, where the definition of two concepts (`WebPage` and `WebSite`) and one property (`belongsToWebSite`) can be seen.

Following the ontology in Figure 1 agents can precisely be informed that some Web pages are grouped into Web sites, and subsequently instructed to find all Web pages belonging to some predefined Web site.
The ontological approach, however, has disadvantages as well, the principal one being private ontologies – different authors using their own private ontologies for describing documents. While this is generally a very difficult problem, there are some special cases when inference is possible, and one solution is developed as part of our work.

There are two obvious possibilit ies to store descriptions of documents:

- Links to descriptions or descriptions themselves can be incorporated into the document, what has some important disadvantages. The agent analyzing a document description must visit every single document, be familiar with its structure, parse it and extract descriptions or follow links to them. In real world situations this would mean many agents simultaneously doing the same unnecessary task of parsing large amounts of documents in an attempt to collect descriptions.
- Store descriptions in a description repository that is central for the host. This means that all authors of a document can have access to the repository generating possible problems concerning security and data consistency. Conversely, the authors could describe their documents as above, a single agent updating the repository from time to time (e.g. once a day). This approach is used in the search system described in the following.

4 Search System Overview

We have developed a search system (see Figure 2) consisting of (i) a user interaction component (UIC), (ii) a mediator and (iii) agents and agent environments. The main purpose of UIC is to enable a user to interact with the search system and to control the search process. By dispatching agents and collecting results of their search the mediator enables both synchronous and asynchronous search. The agent environment provides a "living space" for agents. The subdivision of the system components onto hosts is shown in Figure 3. In this scheme information providers are equipped with an appropriate agent environment paving the way for local searches performed by mobile agents.

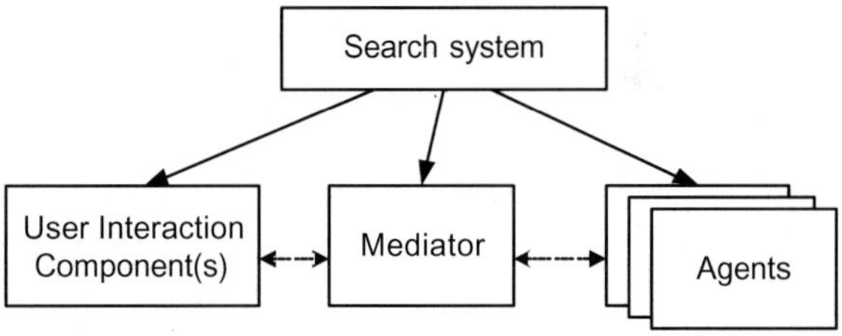

Figure 2. Search system structure

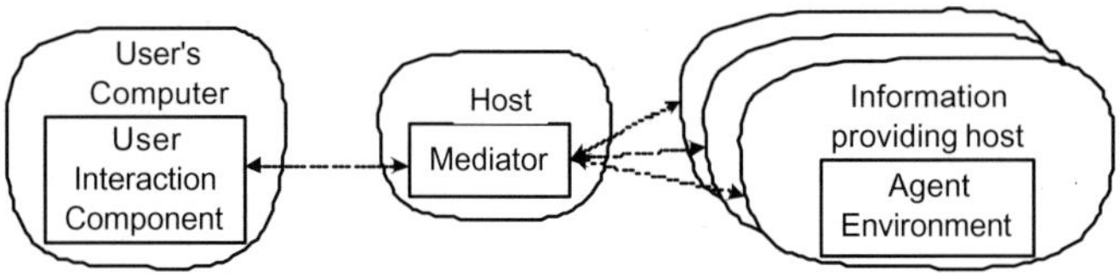

Figure 3. Search system's component placement

The search process goes as follows. The user connects to the search system Web page using a standard Web browser. The search system sends an applet (UIC) by means of which a query can be issued and the search process started. The query is sent to the mediator, which starts an agent and sends it to some information provider, where it performs the search and returns the results. If the agent concluded that there might be some interesting information elsewhere, it clones itself and transports there. In a synchronous search process, the mediator delivers the results to UIC as soon as it gets them. In an asynchronous one, the mediator caches the results, reporting the status and providing the results when the user connects at some later time. This subsequent status and result retrieval as well as termination of the search process can be performed using servlets, too.

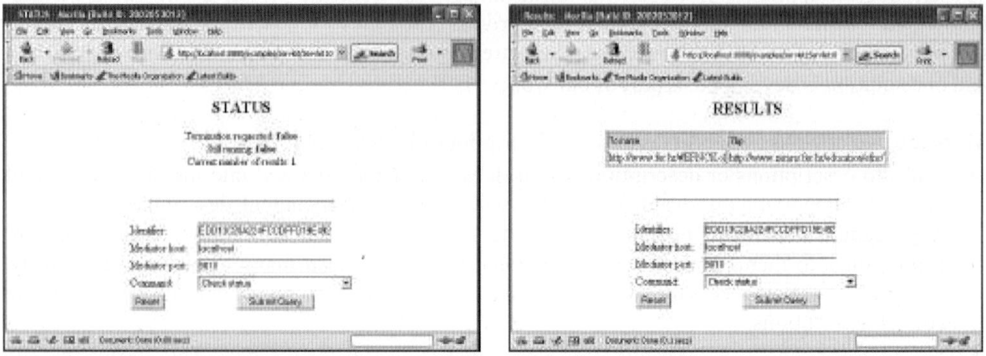

a) asynchronous status checking b) asynchronous result checking

Figure 4. Servlet supported asynchronous operation

4.1 Agents and Agent Infrastructure

The selection of Java as the language supporting agents is a natural one: agents have to be mobile and that is exactly what Java can easily accomplish using its internal mechanisms, what encompass handling TCP connections, capturing a state of an object and serializing it, transferring that information and finally restoring and deserializing it. To enable agents to be mobile, an environment has to be provided to agents for performing their tasks. This supporting infrastructure – agent environments – exhibits the following capabilities: agent acceptance, agent execution, agent cloning, transport of agents to another host, and messaging services for agents (sending/receiving messages to/from agents in the same or different agent environment).

In order to keep the agent body as small as possible, so to minimize bandwidth requirements, the simple search engine for querying RDF triplets is implemented as part of the agent environment instead of being part of the agent itself.

4.2 Search Engine

RDF represents ontology as a collection of simple triplets <subject, predicate, object>, what is very adequate for description storage (e.g. in a relation database) and querying. Our simple search engine supports 4 types of triplets with suitable resolution procedures:

- Hard Triplets: checking whether a triplet exists; performing bindings from existing triplets for variable triplet elements.
- Soft Triplets: usually transformation into a hard triplet.
- Floating Triplets: checking with respect to properties and subproperties.
- Relation Triples: checking semantics that cannot be directly expressed via RDF.

Using agents help the simple search engine can infer even in some cases when private ontologies are met.

5 Conclusion and Further Work

This paper describes a simple search system based on an ontological description of information and on an agent supported search process. By doing so, the search context is insured and distribution of the search process is achieved. The implemented search system uses a rather rudimentary information presentation at the user interface. It is planned to improve the graphical presentation of search results, as well as include a dynamic data completion. E.g. the search for the author of a particular Web page results with that person's unique identifier (URI). However, taking into account that an URI represents a person (described by some ontology), it is possible to automatically retrieve more information about that person, like the full name, e-mail address, of course if the used ontology supports that. This would mean that a search process could be executed in multiple phases – starting with finding some general solutions and then gradually refining them for user automatically, as suggested by (Shneiderman, 1998). We believe that this approach could make information retrieval a little easier then it currently is.

Acknowledgements

This work has been carried out within project 0036033 *Semantic Web as Information Infrastructure Enabler*, funded by the Ministry of Science and Technology of the Republic of Croatia.

References

1. Collins, D. (1995). *Designing Object-Oriented User Interfaces*, Benjamin/Cummings Publishing Company, Inc., Redwood City, CA.
2. Cupic, M. (2002). *Intelligent Search of Information Space within the Information Infrastructure*, Faculty of Electrical Engineering and Computing, University of Zagreb, Zagreb, Croatia (in Croatian).
3. Genesereth, M. R., Nilsson, N. J. (1987). *Logical Foundations of Artificial Intelligence*, Morgan Kaufmann Publishers, San Mateo, CA.
4. Gruber T. R. (1993). *Toward Principles for the Design of Ontologies Used for Knowledge Sharing*, Revision August, 23, http://ksl-web.stanford.edu/KSL_Abstracts/KSL-93-04.html.
5. Luhn H. P. (1958). "The Automatic Creation of Literature Abstracts", *IBM Jour. Res. Dev.* 2(2).
6. W3C (1999). *Resource Description Framework*, Retrieved January 1, 2001, from http://www.w3.org/RDF/
7. Shneiderman, B. (1998). *Designing the User Interface, Strategies for Effective Human-Computer Interaction*, Third Edition, Addison-Wesley, Reading, MA.
8. Sorensen, H., O'Riordan, A., O'Riordan, C. (1997). "Profiling with the INFOrmer Text Filtering Agent", *Journal of Universal Computer Science*, 3(8), 988-1006. http://citeseer.nj.nec.com/sorensen97profiling.html.
9. Turtle, H. R, Croft, W. B. (1991). "Evaluation of an Inference Network Based Retrieval Model", *ACM Trans. Inf. Sys.*, 3.

The Treatment of Collaboration in the Usability Evaluation Models for Collaborative Virtual Environments

Ilona Heldal
Department of Technology and Society
Chalmers University of Technology
SE – 412 96 Gothenburg, Sweden
ilohel@mot.chalmers.se

Abstract

The argument of this paper is that usability evaluations of collaborative virtual environments (CVEs) must consider both technical and social interaction, thus taking into account multiple users. This view is supported by the results from three studies that comparatively examine collaboration and usability issues for different CVE technologies, with different types of applications for co-located users and for distributed use. Three processes are identified that each subject has to undergo in each session working with a CVE application. The first process is the internal attempt to achieve the goal(s) within the particular application. The second and the third are external processes, viz. the need to handle the technology and to pay attention to the partner(s). A model that examines usability problems with special focus on social interaction and collaborative interaction via technology is suggested. This model helps to sort and to identify the usability problems which occur. It can be employed early in the design and thus contribute to developing more usable collaborative applications.

1 Introduction

While the use of various CVE applications is increasing, there are many problems – unresolved aspects regarding networking, devices, interfaces (West & Hubbold 2001), communication modalities and *social* interaction in these environments (Schroeder 2002). So far, much of the effort has been to make the virtual environments (VEs) technologically good enough. The next step in their improvement should result from user requirements. This need is even more urgent in the case of CVEs where a major part (in time, frequency, amount of actions) of the interaction is social.

Usability evaluation models are promising techniques to obtain better applications and more satisfied users. While considerable work is being done on defining usability evaluation models for VEs in general (Stanney 2001; Bowman, Gabbard et al. 2002) and for different collaborative applications for VEs in particular (Hix, Swan et al. 1999; Goebbels, Lalioti et al. 2001; Oliveira 2001), there are currently no usability evaluation methods for CVEs that include extensive collaboration. Steed and Tromp (1998) have defined evaluation methodologies for CVEs by understanding component technologies, and Tromp (2001) has determined systematic usability guidelines that consider changes in group processes during the use of certain applications. Until now, the analysis of collaboration has been performed only on the component level of the evaluation models.

2 Collaboration and usability for CVEs

In CVEs the social interaction and the technical interaction often take place in parallel, or they are interconnected in a non-deterministic way due to the nature of social interaction. There are several factors during the collaboration that can contribute to the effectiveness and enjoyment of the application. While for a single-user system it is easier to determine a user profile, and to follow a user-task analysis, it can be more complex to perform the analysis in the same manner for collaborative applications. It is hard to determine in advance how a group is going to work for a certain application, how much conversation will take place, or how the members will establish a common understanding of the problems.

The hypothesis here is that social interaction without hindrances in CVEs contributes to the effectiveness of the application. The social interaction can be supported to a higher degree by the technology but, to determine exactly how it can be done more knowledge on the usability of collaborative interactions is required. The actual research on usability in CVEs has been reviewed by Heldal (2003) and it is shown that group collaboration, which is analyzed in terms of both the social interaction and the technology that supports interaction, is not included and treated on a general level of the evaluation process.

3 Arguments from three studies

The reason for the design of the three studies was to examine the influence on usability of different technologies and settings. The first study compared two types of systems, a desktop system and an immersive projection technology (IPT) system (Cruz-Neira, Sandin et al. 1993) for visualizing a complex molecular structure for co-located users. For this study 20 focus groups of pairs had to identify different atoms and formations in the molecules (Heldal & Schroeder 2002). The second study compared collaboration in networked IPT systems with collaboration in an IPT system linked to a desktop computer, and with collaboration in the equivalent real-world or face-to-face condition. For each scenario 22 pairs did spatial puzzle-solving, a Rubik's cube-type task (Schroeder, Steed et al. 2001). The third study focused on the main changes during repeated usage of desktop computers for distributed unstructured meetings in VEs. This study examined the collaboration with different tasks in AW (ActiveWorlds, see www.activeworlds.com) for 4 subjects during three months and 10 sessions (Nilsson, Heldal et al. 2002).

The reason for the variation in technology and application is the intention to make more general observations. Collaboration was examined via the participants' own evaluations, by rating and comments and via observation lists, usability in terms of effectiveness (how the subjects reached the goals), efficiency and user satisfaction. Collaboration varied substantially with tasks. The results showed that positive experiences in collaborating contribute to higher effectiveness, independently of the technology used or the applications. Poor collaboration and effectiveness resulted among other things from non-intuitive devices, misunderstandings between collaborators, and a lack of transparency about how the technology supports interaction. Thus the experiences in collaborating in CVEs include both social and technical aspects, and both are crucial in the evaluation of CVEs (Heldal 2003).

The presently available methods for developing usable VEs are not enough for CVEs. In the first of the quoted studies, it appeared that when the subjects were trying to understand, explore, orient themselves and report the complex molecule, the degree of cooperation between them was low. The same result was obtained in the third study for the open-ended explorative tasks. One reason for the low collaboration here was the problematical social interaction and the low feedback from the environment – i.e. the subjects could not be certain whether they followed the same structure or not, and shared the same problems or not.

4 Three cognitive processes

For each study and each session, three cognitive processes have been identified that a user must undergo in order to achieve her/his goals in the CVE:

P1. To find a strategy for achieving the goal(s) within the application, i.e. solving the problems.

P2. To handle the technology.

P3. To pay attention and handle the partner(s), i.e. social interaction.

To do things in the environment, the first two processes – i.e. solving the problems in the application and intuitively handling the devices – should work seamlessly. For solving a puzzle in the distributed settings, this means that one should concentrate on a strategy that leads to the goal (P1) and at the same time handle the real devices in order to interact in a correct way with the virtual environment (P2). Moreover, there often are cases in which the subjects have to manipulate the environment when they speak to each other. Then all three processes must work in parallel.

The importance of the third process, viz. social interaction, is stressed in this paper. Sometimes social interaction determines technical interaction. This was observed on several occasions for the puzzle-solving task, when one subject sitting with a desktop system collaborated with another in an IPT system. When the subjects with the desktop system observed that it was easier to ask their partners about colors of the hidden side of the cubes than to manipulate the cubes themselves, they asked their partners more often. Accordingly, the pair often decided that they should move their avatars in such a way as to keep one in front of the other and thereby have the model between them, thus making it easier to collaborate. In addition, social interaction "saved" unnecessary manipulation activities. The puzzle study shows a higher degree of social interaction (P3) especially in the cases where technology was good enough to be handled intuitively and not as an obstacle to collaboration, as was the case for the networked IPTs and in the real scenarios.

The balance of the three processes differs greatly from application to application. For the molecule visualization study, there appears to be more work in an internal problem-solving process (P1), without visible feedbacks following the steps in problem-solving. Just looking around in the environment, each partner did not know how far advanced in the problem-solving process the other partner was, or which strategy she/he followed. This also contributes to very uncertain social interaction (P3).

These three processes are of different types. The problem-solving is mainly an internal process. The members of the group do not necessarily know how the others solve the problems. Handling the technologies and the social interaction are external processes and the group can follow what a member does. By analyzing the immediate consequences of the processes, additional difficulties that lead directly to uncertainties can be found. Accordingly:

F1. Visible or invisible signs in the VE, depending on the active person, can be the consequence of each step of problem-solving.

F2. Each step of manipulating is an action. The partners can follow these actions, but may misinterpret them in the case of non-symmetrically distributed VEs.

F3. The social interaction. Until now there are very limited ways to follow up these interactions in the VEs.

The presence of the processes for each session in the environment, their contribution to the goals, and the possible difficulties caused by visualizing the feedbacks in them (F1, F2, F3) argue for examining them in connection with problematical activities.

5 A model for examining collaborative usability problems

For CVEs the way of treating work as the result of the three cognitive processes and a switching between them opens a possibility of developing the usability evaluation concept. The quick switching between the processes implies that the work between switches is often small and can be regarded as an element or a fragment. It is then possible to study these fragments carefully in order first to identify hindrances (by technology or application) to positive actions, and secondly to define repeating fragments.

An important aspect that contributes to the usability of CVEs is the clearness of the feedbacks following the interactions. Hindrances often result from misunderstanding the possibilities and the intentions of the partners, as demonstrated in the IPT-desktop scenario for the puzzle-solving task. By following the activities in an experimental set-up, we can understand better what is happening in an environment, regarding important fragments and bottlenecks to activities. In terms of usability, the model examines hindrances, and possible problematical scenarios have to be identified before the hindrances can be examined.
For the subject in the environment, the choice of action is crucial. To visualize the action possibilities of a subject including feedbacks of different kinds, a schematic model is outlined (see Figure 1).

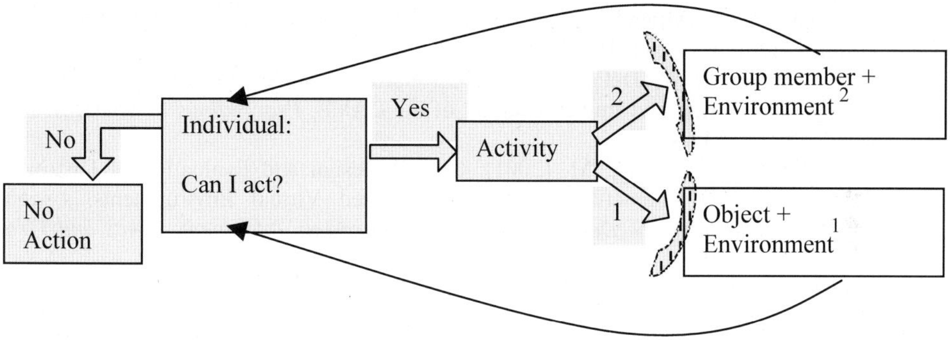

Figure 1. A model for examining the influence of an activity in the collaborative environment.

One way of making CVEs easier to use is to identify and take away technical and social problems, and the disturbances caused by uncertainties during the switches between social and technical interaction. In order to be productive in a CVE, the feedbacks are also very important.

The model is centered on what a member of the group experiences after each social and practical activity in the environment. The key question for an individual is "Can I act?" In order to answer "yes" and thus take action to contribute to the group effort, the individual must experience an understanding of the situation, what to do next, and whether the group accepts the next action. As these experiences change quickly, depending on the activities going on, we must try to break down human activities into discrete components. Even though it is hard to obtain complete results for human activities, certain approximations can be made. For each situation/component, it can be examined whether or not it is possible for the individual to act (take action) depending on technical and social circumstances.
To define the model, the first step is to identify the most important activities that are needed to approach the goal. Here, an activity is the unit of analysis directed at an object, by the subject *or* by some other individual from the same virtual environment (not necessarily real). Each activity has a target. Each activity is composed of social or technical (inter)actions.
The overall examination illustrates the need to follow easier rules in those environments that are more primitive. This requires predefined process structures, clear roles (or description of the roles), ways to handle misunderstandings, and procedures that help the partners in the environment. Another suggestion here is the following: by extending the user-task analysis in the sequential evaluation part of the model defined by Bowman (Bowman, Gabbard et al. 2002) with analysis of social interaction as a multi-user-task analysis, and by analyzing the influences of the three cognitive processes for the application performance metrics, there is a possibility to produce more efficient new guidelines for usability of CVEs.

6 Conclusions

This paper argues that the ways in which usability is considered for the development of single-user VEs are not enough for collaborative environments. Three cognitive processes in CVE work have been identified and, through them, the importance of considering social interaction in the CVE is emphasized. Feedback was identified as an important component. An example of a model has been presented that permits analysis of problematical activities and work-fragments occurring during the interactions.

7 References

Bowman, D., Gabbard, J. & Hix, D. (2002). A Survey of Usability Evaluation in Virtual Environments: Classification and Comparison of Methods. *Presence: Teleoperators and Virtual Environments,* 11(4): 404-24.

Cruz-Neira, C., Sandin, D. & DeFanti, T. (1993). Surround-Screen Projection-Based Virtual Reality: The Design and Implementation of the CAVE. *Proceedings of SIGGRAPH 93*, New York, pp. 135-42.

Goebbels, G., Lalioti, V. & Mack, M. (2001). Guided Design and Evaluation of Distributed, Collaborative 3D Interaction in Projection Based Virtual Environments. *Usability Evaluation and Interface Design: Cognitive Engineering, Intelligent Agents and Virtual Reality*, pp. 26-30.

Heldal, I. (2003). *Usability of Collaborative Virtual Environments. Technical and Social Aspects.* Licentiate thesis, Chalmers Technical University, Gothenburg.

Heldal, I. & Schroeder, R. (2002). Performance and Collaboration in Virtual Environments for Visualizing Large Complex Models: Comparing Immersive and Desktop Systems. *VSMM2002*, Gyeongju, Korea, pp. 208-20.

Hix, D., Swan, II J. E., Gabbard, J. L., McGee, M., Durbin, J. & King, T. (1999). User-centered design and evaluation of a real-time battlefield visualization virtual environment. *Proceedings – Virtual Reality Annual International Symposium*, pp. 96-103.

Nilsson, A., Heldal, I., Schroeder, R. & Axelsson, A. (2002). The Long-Term Uses of Shared Virtual Environments: An Exploratory Study. In Schroeder, R. (Ed.), *The Social Life of Avatars.* Springer, London, pp. 112-27.

Oliveira, J. C. (2001). *Issues in Large Scale Collaborative Virtual Environments*. Ph.D. thesis, University of Ottawa.

Schroeder, R. (2002). Social Interaction in Virtual Environments: Key Issues, Common Themes, and a Framework for Research. In Schroeder, R. (Ed.) *The Social Life of Avatars Presence and Interaction in Shared Virtual Environments.* Springer, London, pp.: 1-19.

Schroeder, R., Steed, A., Axelsson, A. S., Heldal, I., Abelin, A., Wideström, J., Nilsson, A. & Slater, M. (2001). Collaborating in networked immersive spaces: as good as being there together? *Computers & Graphics,* 25(5): 781-88.

Stanney, K. M. (2001). Virtual Reality: Design, Implementation, Usability and Applications. Tutorial at the 9th HCI International Conference, t. A. New Orleans.

Steed, A. & Tromp, J. (1998). Experiences with the Evaluation of CVE Applications. *Collaborative Virtual Environments 1998*, Manchester, UK.

Tromp, J. (2001). Systematic Usability Design and Evaluation for Collaborative Virtual Environments. *Ph.D. thesis*, University of Nottingham, Nottingham.

West, A. & Hubbold, R. (2001). System Challenges for Collaborative Virtual Environments. In Churchill, E.F., Snowdon, D. N. & Munro, A.J. (Ed.) *Collaborative Virtual Environments Digital Places and Spaces for Interaction.* pp.: 44-55.

A Support System for Collaborative Decision Making by People of Different Countries

YunKi Hong[1] *Morio Nagata*[2] *Tai-Suk Kim*[3]

Abstract

As the internationalization of enterprise increasing, more communication between different countries has been needed. At the time of communication between different two country, barriers of decision making are occurred by the fact that the countries differ each other.
In this research we propose new approach in order to solve this problem, and design a prototype system. Our prototype system was tested being compared with the ordinal chat system. In result, the prototype system made decision quickly.

1 Introduction

Determining schedules or decision making at enterprises have been done by people at the same place and the same time. However, the videoconference systems and groupware systems enable us to make decisions in collaboration with people in different places via the internet. Furthermore, the internationalization of businesses forces people of different countries to work together. In case of collaborative decision making of people of different countries, there are many barriers to communicate each other. These barriers are not only languages but also rules, laws, habits and traditions. Thus, it is hard for us to communicate people of different countries. Consequently, it takes much time for people of different countries to make collaborative decisions.
A support system is needed to mediate between people of different countries for collaborative decision making. For communication of people of different countries, the machine translation systems have been used. In order to build up speed and to make qualified decisions, another mediate system is needed. This paper proposes an interactive support system to remove the barriers automatically. We show the effectiveness of our approach by some experiments using our prototype system. Our system can be used with PC, PDA and cellular phone.

2 Mediate System

Our system includes the common knowledge base written in XML[3] for detecting and removing the barriers between people of different countries. Using this knowledge base, our system supports to exchange information for collaborative decision making. The user of this system can input his/her opinions in his/her native language. The input is transferred to the server of our system and inspected by the knowledge base. This system replies to the user in accordance with the following three types of the input. If the input causes no problems, then it will be translated into the partner's language directly. This is the first case. In the second case, our system exchanges information on

[1] Graduate School of Science and Technology, Keio University,3-14-1 Hiyoushi, Kohoku-ku, Yokohama, Kanagawa 223-8522, Japan,ykhong@ae.keio.ac.jp
[2] Faculty of Science and Technology, Keio University, 3-14-1 Hiyoushi, Kohoku-ku, Yokohama, Kanagawa 223-8522, Japan, nagata@ae.keio.ac.jp
[3] Department of Software Engineering, DongEui University, Gaya-Dong, Busanjin-ku, Busan, 617-714, Korea, tskim@hyomin.dongeui.ac.kr

some differences of the countries and translates the input. Exchanging rates of currency and the difference in time are examples of such information. The third case is that the input causes any problem concerning the rules, laws or culture of the partner. In this case, our system asks some questions to the user. For example, the user guesses “Let us meet on the next Thursday”. If the next Thursday is a holiday in the partner’s country, then our system confirms to the user as follows. “The next Thursday is a holiday in the country. Do you reconsider the schedule?”. Another example of the third case is the following. If the user inputs “We would like to purchase the clothes.” ,then our system asks “The tariff of importation depends on the kind of the commodity. Please input the detailed trade name of the clothes”.

2.1 Knowledge Base

In the section, we present the knowledge base which is proposed in this research.
The difference occurring due to country which differs is classified by the country in the knowledge base.
We show that difference is adopted in our prototype system in the section 2.2.

The feature of our knowledge base is described as follows.

- For flexibility, XML is adopted.
- Layered structure.
- Written in the Unicode.
- It is separated by country.

```
<?xml version=”1.0” encoding=”utf-8”?>
<Rule>
    <Korea>
        <Holiday RegExp=”…” Message=”…” Type=”…”>
            .........
        </Holiday>
        ...........
    </Korea>
    <Japan>
        ...........
    </Japan>
</Rule>
```

Figure 1: Structure of Knowledge Base

Figure 1 shows the simple example of the Knowledge base. We describe the structure of knowledge base about figure 1.
There are a Rule Tagged as root element, and Korea and Japan as a child element.
These two elements such as ‘Korea’ and ‘Japan’ are adopted because our prototype system for Korea and Japan is proposed in this research. Using this system for other two countries, we adopt the names of countries instead of the names, Korea and Japan, respectively.

As showed on figure 1, the differences between two countries are embraced in the tags of countries respectively. When contacting with each other from two countries through our prototype system, the differences embraced are utilized for supporting communication. For example, the differences embraced in the 'Korea' tag are for Japanese user and the other differences for Korean.

The XPath[4] expression is used to operate knowledge base remarked above in our system side. When Korean user connecting to our system, the system take information of the descendant elements using XPath " Rule/Japan/* ". In case of Japanese user, the XPath 'Rule/Korea/*' is used for the same purpose.
The expression of '*' corresponds to differences between two countries.

As an example, one of differences is described in the knowledge base in the type as follows.

< NationalHoliday RegExp="…" Message="…" Type="…">

We describe the attributes, RegExp, Message and Type.

The attribute, RegExp is expressed as regular expression for the element embracing itself.
The attribute, Message is the notice which inform user of the meaning of element.
The attribute, Type has three types, Information, Warning and Calculation.

- The type of Information is for reference.
- The type of Warning is for attention.
- The type of Calculation is for exchange rate.

In the result, after our prototype system identifying the user which is connected, the rule of opposite country is selected for the user. After rule was selected, our prototype system inspects the character string which the user inputs by referring the value of RegExp attribute.
If the result which the system inspected is adequate to the rule, namely, there is a part which is applicable, the value of the Type attribute is checked. It is decided to put out warning or to put out reference information.
After that, referring the value of the Message attribute, it draws up the character string as an output and puts out the output to the user.

2.2 Rule

At this version, the rules being applicable in our prototype system are as follows.

Table 1: Rules

	Usual Life	Business
Rules	Unit of money Day of duty Bank hours Time of term of university Meaning of word National holiday	The information regarding the company Patent Premium of stock Delivered item The part where IT enterprise makes the center Circulation structure Liquidation of big business

2.3 Method of Detection

In this section, we show the method of detection which used in our prototype system.
In this research, due to the fact that the rules have some 'Pattern', 'Pattern matching' is adopted as the method of detection.

Here, we describe the rule of 'National Holiday' among the rules presented in the table 1.

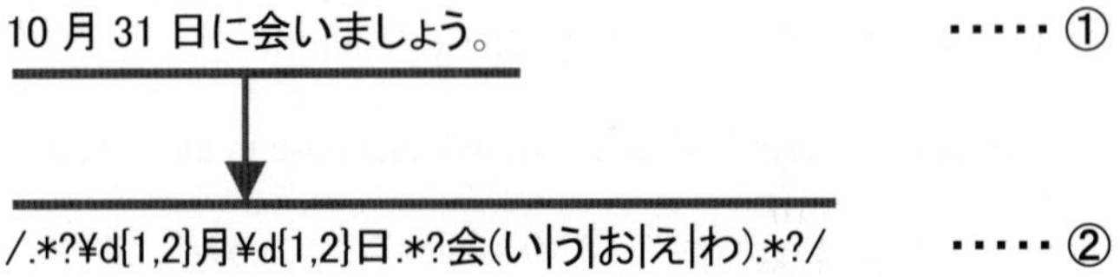

Figure 2 : Pattern of National Holiday

In figure 2, ① shows the character string inputted from user, and is composed the variation of the verb, '会う' [4] and the date.
② is a regular expression corresponding to the pattern of ① inner the system.

3 Our Prototype System

This research aims to make the support of communication between people of all over the world. However, our prototype system is implemented for collaborative decision making on the simple joint project by the Japanese and Korean universities and enterprises. The knowledge base written in XML is constructed for this project. Here, radio communication, the XML and the ASP(the Active Server Page) technologies, Unicode[2] and Regular Expression are used.

Our prototype system has been implemented as a web chat based system. Figure 3 outlines the prototype system. The system consists of the knowledge base, the regular expression module and so on.

In user interface of the system, Japanese people can input his/her opinions in Japanese language. On the other hand, Korean people can input them in Korea a language. To support multi-language is used Unicode.
This server manages and arranges everything. The clients just connect to the web page provided by our system.

---

[4] It mean 'Let's meet' in English

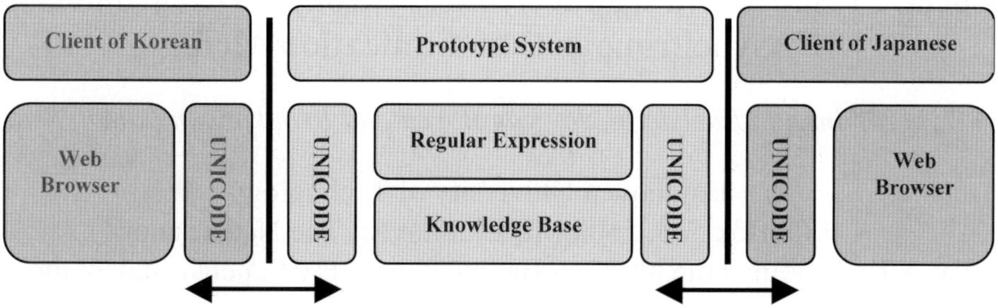

Figure 3: Prototype System

4 Evaluation

Our method was evaluated by using our prototype system. Twelve pairs of Japanese and Korean students tried to make decisions on given problems. Two problems of decision making of simple business games were given to the pairs. Each pair solved one problem by using our prototype system. The pairs solved the other problem by using the ordinal chat system. The average time of the group to solve the problems using our prototype system are compared with the average time of the group using the chat system. Moreover, we used the scores of the results of decision making. The groups using our prototype system made their decisions quickly 1.5 times and the scores are higher than the decisions by the other groups.

5 Concluding Remarks

Our experiment shows that our approach and knowledge base are useful for collaborative decision making by people of different countries. Our prototype system for Japan and Korea will be extended for other countries by replacing the knowledge base and the machine translation system. Furthermore, our system can be applied to experiencing foreign countries in the virtual world.

6 References

[1]Document Object Model (DOM) Level 2 Core Specification http://www.w3.org/TR/DOM-Level-2-Core/

[2]Unicode, http://www.unicode.org

[3]XML(eXtensible Markup Language) 1.0 http://www.w3.org/TR/REC-xml

[4]XPath(XML Path Language) http://www.w3.org/TR/xpath

[5]HATTORI Masakazu, SUEDA Naomichi, 2001, XML Processing Engine Knowledge Factory, Toshiba Review. 56,11, pp. 11-14

[6]Paul McNamee, James Mayfield, 2002, Comparing cross-language query expansion techniques by degrading translation resources, Series-Proceeding-Section-Article, pp. 159-166

Staying in Touch
Social Presence and Connectedness through Synchronous and Asynchronous Communication Media

Wijnand IJsselsteijn, Joy van Baren and Froukje van Lanen

Eindhoven University of Technology
P.O. Box 513, 5600 MB Eindhoven, The Netherlands
W.A.IJsselsteijn@tue.nl, J.K.v.Baren@tue.nl, F.v.Lanen@student.tue.nl

Abstract

The emergence and proliferation of email, mobile communication devices, internet chatrooms, shared virtual environments, advanced tele-conferencing platforms and other telecommunication systems underline the importance of developing measurement methods that are sensitive to the human experience with these systems. In this paper, we discuss the concepts of social presence and connectedness as complementary notions, each relating to a different set of media properties that serve distinct communication needs.

1 Introduction

Human beings are social beings. We have a fundamental need to communicate, to form, maintain and enhance social relationships. This well-known fact is illustrated by the massive success of recent communication media such as email, mobile telephony, and SMS, but the basic insight can be traced back to the days of Aristotle, or even earlier. Maslov's theory of human needs, formulated in the 1950s, illustrates that social interaction is essential to satisfying human needs at several levels, in particular needs for belonging, love, and esteem, although at all other levels the formation of social networks may facilitate the satisfaction of the various needs.

Media technologies have significantly extended our reach across space and time. They enable us to interact with individuals and groups beyond our immediate physical surroundings. An increasing proportion of our daily social interactions is mediated, i.e. occurs with representations of others, with virtual embodiments rather than physical bodies. The extent to which these media interactions can be optimised to be believable, realistic, productive, and satisfying has been the topic of scholarly investigations for several decades – a topic that is only increasing in relevance as new communication media emerge and become ubiquitous.

In order to optimise the range of communication media for different users, contexts (e.g. home, mobile, work), bandwidths, and modalities, we need experiential metrics by which to judge the quality of the appliances, the services, and the interactions they afford. Several measures are already available that allow the assessment of functionality and usability. However, measuring the quality of the mediated interaction itself remains a considerable challenge. How is the mediated interaction experienced? Does it fulfil real communication needs? Does it resonate with the context and the specific user requirements? Does it enable a sense of connectedness, of belonging, of identification? Is it just as good as 'being there' face-to-face, or maybe even better? Or maybe it's a different experience altogether?

We need a theory and measurement methodology that allows us to answer such questions. This clearly requires going beyond technology assessments, and into social psychology, sociology, ethnography, and philosophy of mind (Biocca & Harms, 2002). The concept of social presence, i.e. the sense of being together, provides us with a useful point of departure in this respect, but, as we will argue in this paper, the concept itself is intrinsically limited in its application, such that it cannot account for the whole gamut of experiences associated with communication technologies available today.

2 Social presence and media richness

In their pioneering work, Short, Williams and Christie (1976) conceptualise social presence as a way to analyse mediated communications. Their central hypothesis is that communication media vary in their degree of social presence and that these variations are important in determining the way individuals interact through the medium. Media capacity theories, such as social presence theory and media richness theory, are based on the premise that media have different capacities to carry interpersonal communicative cues. Theorists place the array of audio-visual communication media available to us today along a continuum ranging from face-to-face interaction at the richer, more social end and written communication at the less rich, less social end.

Richer media are traditionally considered to be those that enable the transmission and display of nonverbal communicative cues. In face-to-face communication, the nonverbal channels are continuously attended to and communicate information that is primarily affective in quality and connected with personal relationships. In this respect, the nonverbal channels seem to be less controllable than the verbal channels, i.e. they are more likely to "leak" information about feelings.

Argyle and Dean (1965) argue that interpersonal intimacy is kept at an optimal, equilibrium level through factors as physical distance, smiling, eye contact and personal topics of conversation. Other scholars have added to this list of intimacy behaviours to include factors such as gestures, touching, vocal cues (e.g. tone of voice), turn-taking behaviour in dialogues (e.g. frequency of interruptions), the use of space (e.g. moving towards someone) and verbal expressions directly acknowledging the communicative partner (e.g. 'How did you do that?' or 'I see what you mean'). Wiener and Mehrabian (1968) have applied the concept of immediacy, i.e. the psychological distance a speaker puts between himself and the hearer, to an understanding of speech. They showed that the choice of 'We…' as opposed to 'I…' or 'You…' connote a feeling of closeness and association. Thus, supporting intimacy and immediacy behaviours seems to be particularly relevant for engendering social presence through media.

Biocca and Harms (2002) have made significant advances in developing a more comprehensive theory of social presence. In line with most other definitions, they define social presence as a "sense of being with another in a mediated environment". They continue their shorthand definition by stating that "social presence is the moment-to-moment awareness of co-presence of a mediated body and the sense of accessibility of the other being's psychological, emotional, and intentional states" (p.14). Importantly, they distinguish three distinct levels of social presence. Level 1 is the perceptual level – primarily the detection and awareness of the co-presence of the other's mediated body. The second, or subjective, level entails the sense that the user has of the awareness of the other, and the level of accessibility to the others attentional engagement, emotional state, comprehension, and behavioural interaction. The third level is a dynamic, intersubjective level. It is comprised of the user's sense of the other's sense of social presence of them – i.e. the perceived symmetry of social presence. These theoretical concepts have been translated into a questionnaire measure of social presence that is currently being validated (Biocca, Burgoon, Harms, & Stoner, 2001; Biocca & Harms, 2002).

3 Face-to-face and beyond

The majority of tele-relating studies to date have focussed on audio- and videoconferencing systems in the context of professional, work-related meetings and computer-supported collaborative work (CSCW). Using such systems, participants typically appear in video-windows on a desktop system, or on adjoining monitors, and may work on shared applications that are shown simultaneously on each participant's screen. Examples include the work of Bly, Harrison, & Irwin (1993), Fish et al. (1992), and Gaver et al. (1992). As more bandwidth becomes available (e.g. Internet2), the design ideal that is guiding much of the R&D effort in the telecommunication industry is to mimic face-to-face communication as closely as possible, and to address the challenges associated with supporting non-verbal communication cues such as eye contact, facial expressions and postural movements. These challenges are addressed in projects such as the National Tele-Immersion Initiative (Lanier, 2001), VIRTUE (Kauff, Schäfer & Schreer, 2000), and TELEPORT (Gibbs, Arapis & Breiteneder, 1999), where the aim of such systems is to provide the remotely located participants with a sense of being together .

Complementary to this approach however, is the appreciation and utilisation of the considerable potential of communication media to provide features typically unavailable in face-to-face situations, such as saving the history of interactions, or changing the representation of self and others (Hollan & Stornetta, 1992; Clark & Brennan, 1993; Heeter, 1999). Clark and Brennan (1993) have characterised different properties that communication media may offer (p.229): co-presence (A and B share the same environment), visibility (A and B are visible to each other), audibility (A and B can communicate through speaking to each other), co-temporality (B receives at roughly the same time as A presents, i.e. synchronous communication), simultaneity (A and B can send and receive at once and simultaneously), sequentiality (A's and B's turns cannot get out of sequence as in asynchronous communication), reviewability (B can review A's message), and revisability (A can revise messages for B). Interestingly, these properties indicate that face-to-face communication lacks opportunities offered by some telecommunication media. For example, reviewing a message before sending it proves to be difficult in face-to-face conversation, while email supports such functionality. Inasmuch as these functionalities go beyond mimicking face-to-face encounters, social presence measures need to be complemented to properly account for the user experience in this regard.

4 Staying in Touch

A more recent focus of research in HCI and CSCW, influenced by previous work in Media Spaces (Bly, Harrison & Irwin, 1993) and Portholes (Dourish & Bly, 1992), as well as current trends in ambient intelligence, are *awareness systems* for use in personal settings – either home or mobile. Here, lightweight, emotional, informal forms of communication are being facilitated by systems that help people to effortlessly maintain awareness of each other's whereabouts and activities. Examples include the work by Hindus et al. (2001) and Markopoulos et al. (2003). In line with Marc Weiser's notion of calm computing (Brown & Weiser, 1996), such systems can typically be always-on, yet be very gentle or calm in terms of attentional demands. When attention is asked, it should typically move from background to foreground in an unobtrusive manner. As the attentional demands are kept to a minimum, these systems should blend into the background and are effectively intended to be ignored until the user feels like communicating, i.e. asynchronous communication.

The current generation of awareness systems are all experimental. Field tests have been limited to the workplace, or to Wizard-of-Oz type evaluations of limited functionality prototypes in the home. To date, awareness needs have been served by existing media, such as the telephone. Markopoulos et al. (2003) explored some of the limitations of existing media for awareness purposes, in particular aimed at the elderly as a special interest user group. Reported drawbacks of using existing media for staying in touch included their synchronous nature (not practical for both parties at the same time) and their need to be tied with explicit communication interactions (e.g., needing an excuse to communicate). In addition, with regard to video communication, Bouwhuis (2000) found that it could be perceived to threaten privacy, in particular because the camera would be capturing private information not explicitly intended to be communicated, e.g. certain valuable properties, untidiness of the home or the clothing, presence of visitors, etc.

The aim of awareness systems is often simply to stay in touch, i.e. to be reassured about the well-being of others, to let others share your experiences, to let someone know you're thinking of him/her, or to create opportunities for synchronous communication. In other words, for this type of communication, the informational content of the message is of secondary importance to the emotional, relational content that is being transmitted. What is important to note here is that the concept of social presence may not be the best applicable experiential metric. In effect, when considering the theoretical framework described by Biocca & Harms (2002), outlined previously, the most basic level of perceptual awareness is almost absent. From a media richness point of view, awareness systems may be very poor, and social presence measured along richness dimensions will be low. Yet the sense of connectedness, the feeling of being in touch with the other can be strong and the experience highly appealing. In order to be sensitive to this dimension of human communication, we are currently in the process of developing a *Connectedness Questionnaire* that will not focus on the sense of being together as such (following the face-to-face model), but rather focus on the affective benefits of the awareness systems. Hypothesized affective benefits include a feeling of having company, a stronger group attraction, a feeling of staying in touch, of keeping up-to-date with other people's lives, and a sense of sharing, belonging, and intimacy. In short, awareness systems are not seen as replacing existing communication means, but rather as enriching them, strengthening existing social bonds and enabling new kinds of interactions.

5 Acknowledgements

The authors gratefully acknowledge the support of the IST ASTRA and IST OMNIPRES projects, funded through the European Commission FET (Future and Emerging Technologies) Presence Research Proactive Initiative (http://www.cordis.lu/ist/fetpr.htm).

6 References

Argyle, M., and Dean, J. (1965). Eye contact, distance and affiliation. *Sociometry*, 28: 289-304.

Biocca, F., Burgoon, J., Harms, C., & Stoner, M. (2001). Criteria and scope conditions for a theory and measure of social presence. Paper presented at PRESENCE 2001, Philadelphia, PA, May 21-23 2001.

Biocca, F. & Harms, C. (2002). Defining and measuring social presence: Contribution to the Networked Minds theory and measure, *Proceedings of PRESENCE 2002*: 7-36.

Bly, S.A., Harrison, S.R., & Irwin, S. (1993). Mediaspaces: Bringing people together in video, audio, and computing environments. *Communications of the ACM, 36*: 29-47.

Bouwhuis D.G. (2000). Parts of life: configuring equipment to individual lifestyle. *Ergonomics*, 43(7), 908-919.

Clark, H. H. & Brennan, S. E. (1993). Grounding in communication. In: Baecker, R. M. (ed.), *Readings in Groupware and Computer-Supported Cooperative Work, Assisting Human-Human Collaboration*, pp. 222-233

Dourish, P. & Bly, S. (1992). Portholes: Supporting awareness in a distributed work group. *Proceedings of ACM CHI '92*: 541-547.

Fish, R.S., Kraut, R.E., Root, R.W., & Rice, R.E. (1992). Evaluating video as a technology for informal communication, *Proceedings of ACM CHI '92*: 37-48.

Gaver, W., Moran, T., MacLean, A., Lövstrand, L., Dourish, P., Carter, K., & Buxton, W. (1992). Realizing a video environment: EuroParc's RAVE system, *Proceedings of ACM CHI '92*: 27-35.

Gibbs, S.J., Arapis, C., & Breiteneder, C.J. (1999). TELEPORT – Towards immersive copresence, *Multimedia Systems, 7*: 214-221.

Heeter, C. (1999). Aspects of Presence in Telerelating, *Cyberpsychology and Behavior, 2*: 225-325.

Hindus, D., Mainwaring, S.D., Leduc, N., Hagström, A.E., & Bayley, O. (2001). Casablanca: Designing social communication devices for the home, *Proceedings of ACM CHI'01*: 325-332.

Hollan, J. & Stornetta, S. (1992). Beyond being there. *Proceedings of ACM CHI '92*: 119-125.

Kauff, P., Schäfer, R., & Schreer, O. (2000). Tele-Immersion in Shared Presence Conference Systems, International Broadcasting Convention, Amsterdam, September 2000.

Lanier, J. (2001). Virtually there. *Scientific American*, April 2001, 52-61.

Markopoulos, P., IJsselsteijn, W., Huijnen, C., Romijn, O. & Philopoulos, A. (2003). Supporting social presence through asynchronous awareness systems. In: Davide, F., Riva, G., & IJsselsteijn, W.A., (eds.), *Being There - Concepts, Effects and Measurement of User Presence in Synthetic Environments*. Amsterdam: IOS Press.

Short, J., Williams, E. and Christie, B. (1976). *The Social Psychology of Telecommunications*. London: Wiley.

Weiser, M. and Brown, J.S (1996). Designing Calm Technology, *PowerGrid Journal*, v 1.01, http://powergrid.electriciti.com/1.01

Wiener, M. and Mehrabian, A. (1968). *Language within Language: Immediacy, a Channel in Verbal Communication*. New York: Appleton-Century-Crofts.

Development of an Education System for Surface Mount Work of a Printed Circuit Board

H. Ishii, T. Kobayashi, H. Fujino, Y. Nishimura, H. Shimoda, H. Yoshikawa

Kyoto University
Gokasho, Uji, Kyoto, 611-0011, JAPAN
{hirotake,t-koba,fujino,nishimura,shimoda, yosikawa}@uji.energy.kyoto-u.ac.jp

W. Wu, N. Terashita

Mitsubishi Electric Corporation
8-1-1, Tsukaguchi Honmachi, Amagasaki, Hyogo, 661-8661, JAPAN
{Wu.Wei, Terashita.Naotaka} @wrc.melco.co.jp

Abstract

A part of the surface mount work of printed circuit boards is conducted by human workers and human error resulting from hand working is a serious problem. So the education of workers is very important in order to prevent human errors, but the present education methods using conventional video or paper are not so efficient. The goal of this study is to develop an education system for the surface mount work of printed circuit boards by utilizing a CCD camera, image processing techniques, a laser irradiation device and so on. This paper describes the outline design of the education system, developed functions and the results of their evaluations.

1 Introduction

It is difficult to automate all of the surface mount work of a printed circuit board (PCB), because there are various kinds of electric parts, and their size and shape are quite different. So a part of the surface mount work is conducted by human workers. But human workers may make a mistake and it can be occurred that the assembled PCB cannot be used and it may also become the cause of the accident. Therefore, it is very important to maintain and improve the skill level of the workers. But the workers who work at PCB factories are frequently exchanged, and the education cost of the workers is very high. Especially, in the case that the skilled workers educate the novice workers by explaining the basic knowledge and demonstrating the actual surface mount work, the cost for the education becomes very high. Moreover, it is very difficult to evaluate the skill level of the workers correctly. So it also happens that the workers who have received the education make a mistake frequently at a factory, and they turn out to need to receive the education again. Then in this study, the authors aims at developing an education system which can educate the novice workers so that they can make PCBs by themselves without troubling the other workers.

2 Surface Mount Work of Printed Circuit Boards

2.1 Configuration of surface mount work of printed circuit boards

At a PCB factory, some instruction documents which explain how to mount the electric parts onto PCBs are prepared and they are distributed to the workers in paper form. A PCB is located on a work bench and some parts boxes and instruction documents are located around the work environment. And one kind of electric parts are put into one parts box. The workers mount the electric parts onto the PCBs with their both hands based on the information described in the instruction documents and silk printing printed on the PCBs.

2.2 Present condition and issues of worker education

It is not easy to make PCBs, because it requires special knowledge about electric parts and needs to understand the complicated instruction documents. To be able to make the PCBs, the workers need to understand not only the basic knowledge about electric parts but also the following topics:

- There are various kinds of electric parts, and their size and shape are different. And there are some electric parts which shape is almost same but the kind is different
- There is a polarity in electric parts and the indication of the polarity is different according to the electric parts.
- The efficiency of making PCBs partly depends on the order of mounting electric parts. The best order is not instructed by the instruction documents and the workers are respected to find the best order by themselves.

To educate the novice workers, some strategies can be considered such as using a conventional paper or video material, demonstrating the actual mount work of PCBs and so on. But these strategies requires too much time of the educator so that an education system is desired with which the novice workers can learn the surface mount work by themselves.

3 Education Plan and the Outline Design of the Education System

3.1 Education plan of the surface mount work

The education plan designed in this study is as follows. (Step 1) The education system provides the materials to learn how to make PCBs in the form of conventional video and text. (Step 2) The education system guides the novice worker to conduct the surface mount work one by one. (Step 3) The novice worker conducts the surface mount work. (Step 4) The education system recognizes whether the worker conducted the surface mount work correctly, and judges whether more education is necessary or not. And if more education is necessary, (Step 5) the education system makes a plan to educate the topics the novice worker does not understand yet. From Step 2 to Step 5 are repeated until the novice worker masters the required skill for making PCBs by themselves.

3.2 System requirement

Before designing the education system, the authors determined the required features of the education system as follows:

- The novice workers can experience the surface mount work in the environment similar to the actual work environment. The education system does not restrict the worker's movement.
- The education system can be used easily.
- The skill level of the novice workers can be evaluated quantitatively.
- The education course can be changed according to the skill level of the workers.
- The development cost is not expensive and the burden of the educator is light.

3.3 Outline design of the education system

The functions of the education system can be divided into 2 categories. One is the functions to display the instructions to the workers (display function). The other is the functions to recognize the progress of the surface mount work (recognition function). Concerning to the display functions, not only the function to display the instruction documents, but also the functions to display additional information are necessary such as the kind, location, direction and parts box of the electric parts which should be mounted next. On the other hand, concerning to the recognition functions, not only the function to recognize the kind, location, direction of the mounted electric parts, but also the functions are necessary to recognize what kind of electric parts the worker takes and whether the mount work has been finished. Figure 1 shows the outline design of the education system designed in consideration of the above system requirements. The education system consists of a LCD monitor, a laser irradiation device, a color CCD camera and a personal computer. The surface mount work conducted by the novice workers is captured by the color CCD camera located at the head of the work environment. And the progress of the surface mount work is recognized by image processing techniques. The instructions are displayed by the LCD monitor located in front of the worker and the laser irradiation device located at the head of the work environment. The reason that the color CCD camera is used to recognize the progress of the surface mount work is to avoid to restrict the worker's movement. The reason that the laser irradiation device is used to display the instructions is to make it possible for the workers to reference the guidance of the education system more intuitively by overlapping the instructions over the PCBs. But it is difficult to display all instructions necessary for educating the surface mount work only by using the laser irradiation device, so the LCD monitor is also used.

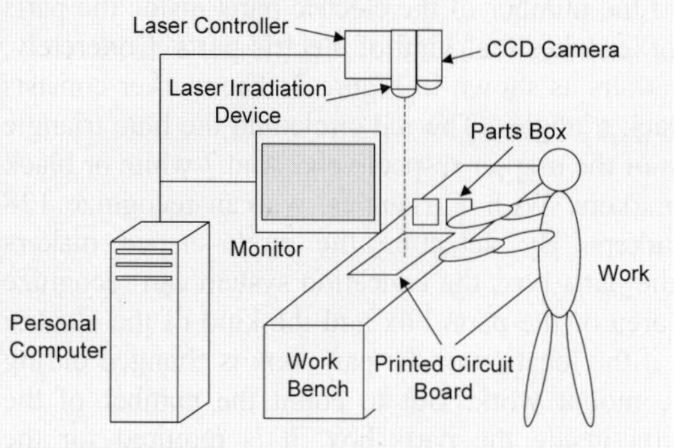

Figure 1: Outline design of the education system.

4 Required Functions of the Education System

To develop the education system, various kinds of functions need to be realised. This section describes the required functions to develop the education system.

4.1 Detect the location of the printed circuit boards

In order to recognize the progress of the surface mount work, it is necessary to detect the accurate location of the PCB, because the other information such as the location of the mounted electric parts can be recognized based on it. In this study, the location of the PCB is detected by using color information as follows. (1) Classify each pixel of the captured image into PCB-area and non PCB-area by comparing with a given threshold in green color component. (2) 4 edge lines of the PCB is detected by using Hough Transform[1]. (3) 4 vertices of the PCB is detected by calculating the intersection of the detected edge lines. (4) The color image of the captured PCB is extracted as a rectangle shaped image by using the information of the 4 vertices and the size of the PCB which information is registered to the education system in advance.

4.2 Detect the location of the worker's hands

In order to detect the location of the worker's hands, the method by using the skin-color information can be considered. But with this method, the location of the worker's hands can not be detected correctly in the case that the worker wears a half-sleeved shirt because the arm is also extracted when extracting the skin-color area. So in this study, the workers are supposed to wear blue color gloves and the location of the worker's hand is detected as the center of the blue area of the captured image.

4.3 Detect the location of the mounted electric parts

The location of the mounted electric part is detected as follows. (1)The differential image is calculated between the captured image of the PCB before and after mounting an electric part. (2) Classify each pixel of the differential image into parts-pixel and non parts-pixel by comparing with a given threshold. (3) The location of the mounted electric part is detected as the center of the largest cluster area of the parts-pixel.

4.4 Detect the kind of the mounted electric parts

There are some electric parts which kind is different but the shape is almost same. So it is very difficult to detect the kind of the mounted electric parts only by using the part's image if the CCD camera is not close enough to the PCB until it becomes the obstacle of the surface mount work. So, in this study, the number of the electric parts inside the parts box is counted instead of detecting the electric parts mounted on the PCB. That is, if the number of the electric parts inside the parts box is decreased, it can be supposed that the worker takes that kind of electric parts. Concretely, the authors have been designed circle-shaped markers as shown in Figure 2. The marker consists of 1 red circle, 1 blue triangle and 7 white or black triangles. The red circle and the blue triangle are used to detect the location and the orientation of the marker respectively. And 7 white or black triangles are used for coding the kind of the markers. With 7 triangles, we can recognize 128 unique markers. By detecting the circle-shaped makers pasted to the parts box, the education system can recognize the inside area of the parts box and the kind of the electric parts even if the location of the parts box is changed during the surface mount work. But to count the number of the electric parts inside the parts box, it is required for the electric parts to be separated each other inside the parts box.

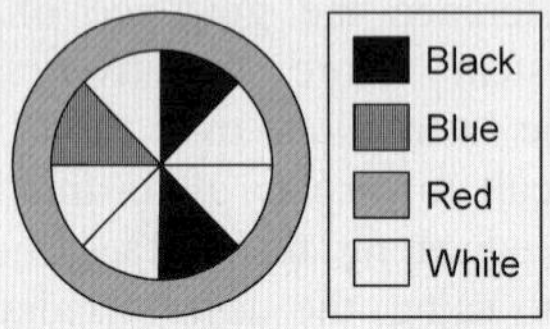

Figure 2: Circle-shaped maker.

4.5 Detect the direction of the mounted electric parts

There are some electric parts which shapes are almost symmetric. So it is difficult to detect the direction of the mounted electric parts only by using the image of them. To make it easy to detect the direction of the electric parts, it may be considered that some makers are added to the surface of the electric parts. But it is not desirable that the worker can detect the added markers for judging the direction of the electric parts, because at the actual factory, these makers will not be added. So in this study, 3 kinds of methods have been developed to recognize the direction of the mounted electric parts. First method is that the image of the mounted electric part is compared with the template images prepared in advance by using pattern matching such as cross-correlation coefficient. Second method is that the surface of the electric part is coated by infrared absorbent which is almost transparency for the workers and when capturing the images, the infrared rays are irradiated on the electric parts. The area coated by infrared absorbent is expected to be darker than the other area when capturing by an infrared CCD camera. Third method is that the surface of the electric parts are marked by color paint complicatedly so that the worker can not distinguish them and the direction of the electric parts are detected by using the color information. At the present stage, the authors have just developed these 3 methods and they will be evaluated to examine which method will be suitable for the education system.

4.6 Function to display the instructions

As mentioned in the subsection 3.3, the instructions to the workers are displayed by using the LCD monitor and the laser irradiation device. The LCD monitor is set as the external monitor of the personal computer and some instruction materials such as the explanation texts and the appearance of the electric parts are displayed on it. The laser direction is controlled by the galvano scanner and it is possible to draw simple figures such as a circle and a rectangle on the arbitrary surface around the work environment. The information for the instructions are stored in the XML format[2] by using special tools developed for constructing the instruction contents.

4.7 Process flow of the education system

The process flow of the education system for educating the novice workers is as follows. (Step1) Read the education scenario from the database. (Step2) Detect the circle-shaped markers pasted to the parts boxes to recognize the kind of the electric parts and their located area. And the location of the PCB is also detected. (Step3) Display the instructions to the worker. (Step4) By watching the location of the worker's hands, judge whether the worker's hands move over the parts box. If the worker's hand moves over the parts box, go to Step 6, otherwise go to Step 5. (Step 5) Detect whether the electric part is mounted on the PCB. If mounted, go to Step 7, otherwise go to Step 4. (Step 6) Count the number of the electric parts inside the parts box. If the number of the electric parts is decreased, judge that the worker takes the electric parts from the parts box and go to Step 4. (Step 7) Extract the area of the mounted electric part. And detect the direction of it. (Step 8) The results detected at Step 5, 6 and 7 are verified with the correct working procedure of the surface mount work. If there is a mistake, display the instruction to fix the mistake. If there remains something to educate, go to the Step 2. Otherwise the education is finished.

5 Evaluation of the Developed Functions

The functions described in the section 4 were developed using Microsoft Visual C++ in Window 2000. Especially, the functions using image processing techniques were developed using Intel Image Processing Library and Open Computer Vision Library. For the evaluation of the developed functions, the authors have developed the temporal work environment. A color 1/3-inch CCD camera which focal length is 8mm was mounted at a height of 50cm from the work bench. The frame grabber has a resolution of 1024 by 768 pixels. The laser irradiation device which can irradiate class 2 laser was also mounted at a height of 50cm from the work bench. The light intensity around the PCB ranged from 600 to 800 lux. Concerning to the function to detect the location of the PCB, the detection error was within 3mm in the condition that the PCB was moved around the work environment. And even if a part of the PCB was hidden by worker's hand, the location of the PCB could be detected correctly. Concerning to the function to detect the location of the mounted electric parts, there were some cases that the location of the electric parts could not be detected correctly. The cause of the failure seems to be the shadow of the electric parts. When the light is irradiated from the side angle, there is a case that the shadow of the electric parts becomes larger than the size of the electric part. In this case, the education system fails to detect the location of the mounted electric parts. This problem will be avoided by locating the light at the head of the work environment. Concerning to the detection of the circle-shaped maker, there were some case that the detection was failed. The cause of the failure seems to be the reflection of the light on the marker. When the light is reflected on the marker, the black color area of the maker is looked as white color area. This problem can be avoided by making the marker from the materials such as suede or cloth which does not reflect the light.

6 Conclusion

The authors aims at developing the education system to educate the surface mount work of the printed circuits boards without troubling the other workers. In this study, several functions necessary for developing the education system have been developed and evaluated. As the future works, the developed functions will be improved and the whole education system will be implemented. After the implementation of the education system, the effectiveness of the education system will be evaluated by conducting the field trial.

References

[1] Illingworth, J. & Kittler, J. (1988). A survey of the Hough transform, *Computer Vision, Graphics, and Image Processing*, 44, 87-116.

[2] Williamson, H. (2001). XML: The Complete Reference, McGraw-Hill Professional Publishing.

Virtual Manufacturing Approach to Collaborative Design and Production for Hard-Tissue Implants

Teruaki Ito

University of Tokushima
Tokushima 770-8506, Japan
ito@me.tokushima-u.ac.jp

Teisuke Sato

University of Tokushima
Tokushima 770-8506, Japan
sato@me.tokushima-u.ac.jp

Abstract

The paper proposes a virtual manufacturing approach for custom-made design and production on hard-tissue implants with porous ceramics. Using digital bone model created from medical data, appropriate shape of implants is designed in collaboration with medical doctor and design engineer. Porous ceramics are regarded as a suitable material for hard tissue implants, but their fragile features are not suitable for mechanical processing to create the complicated shape to fit the damaged portion of bone. Using wax combination method which we propose, production of porous ceramic implants with complicated shape can be available. The paper presents the overview of the idea behind the approach and describes some of the key technologies in our study.

1 Introduction

Porous ceramic implants are used in plastic surgery operation as a replacing material for defected bone tissues because of their characteristic features to absorb bone tissues. Although a variety of product ranges is provided by manufacturers, those hard tissue implants are based on simple geometric shapes to be produced by mass-production. The shape of damaged portion of the bone is normally irregular and very complicated. For custom design of implants, pre-operation has to be made before the main operation. However, multi-stage operation is too much hard for the patients and is not usually conducted. Therefore, one of the most appropriate shapes of implants is selected from the wide range of products, and is applied to the damaged portion of the bone.

In addition to the design issue, material issue of implants is also under discussion. Porous ceramics are regarded as a suitable one for implants, into which the bone tissue grows to become bone after a certain period of time. However, their fragile features are not suitable for mechanical processing to produce the desired shape based on the shape design. Faster time for tissue growth is also required but it is also an open issue.

The objective of the study is to make custom-made design of hard-tissue implants without pre-operation, and to produce the designed implants with porous ceramics. The paper describes the background for custom-made design on hard tissue implants. Then, the paper proposes a virtual manufacturing approach to design and production on hard tissue implants, including digital bone modeling, implant shape design, implant evaluation, and virtual implant production. The paper also covers the mechanical processing on porous ceramics to prepare any complicated shape to fit the damaged portion of bone. Concluding remarks will follow.

2 Backgrounds for Custom-Made Design on Hard Tissue Implants

For plastic surgery of implants, its operation has to be completed within shortest time as possible, satisfying various requirements. In addition to that, it is not usually acceptable to conduct pre-operation to design the shape of implants which should be exactly fit the damaged portion of bone. Pre-operation means a very hard burden on patients, which should normally be avoided. Therefore, one of the most appropriate shapes of implants is selected from the wide range of products, and is applied to the damaged portion of the bone. It may be possible to reshape implants at the time of operation to fit the damaged portion, which is normally very complicated shape. However, to minimize the operation time, it is normally performed to reshape the bone to fit the simple geometric shape of implants.

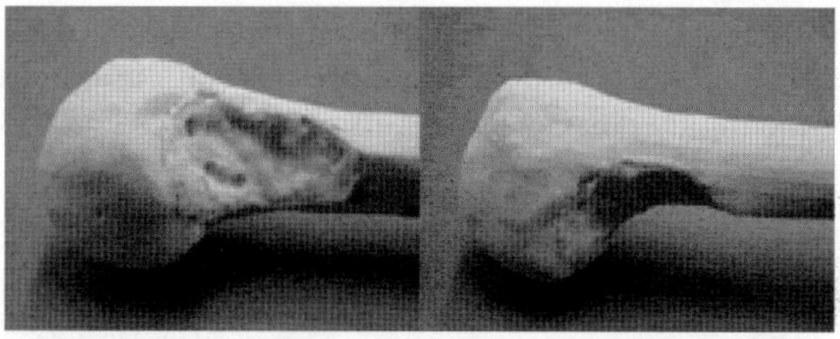

Figure 1: Critical damage near joint portion

Figure 1 shows a critical case of damaged bone near joint portion. If the portion is reshaped to fit a prefabricated implant, then the implant may be applied to the portion. However, with weakened strength of the portion, it may cause the loss of whole joint portion. The risk to lose the whole joint like this can be drastically reduced without reshaping process in this way. The key solution would be associated with the custom-made design on implants which should be exactly fit the damaged shape of the bone, which is normally a very complicated shape. If the custom-made shape design is available, its production can be studied independently. However, two issues have to be considered. One is design issue, or how to design the appropriate shape without preliminary operation. The other one is production issue, or how to produce porous ceramic implants with complicated shape based on the shape design. The study proposes a virtual manufacturing approach to tackle these two issues.

3 Collaborative Design and Production under Virtual Environment

To achieve the goal of study, or custom-design on implants and its production without pre-operation, we have been developing a virtual environment for design and production of implants to evaluate the design, to produce virtual products and to provide the design data to the actual production site. From design to production, several stages are modeled to implement the environment, which includes digital modeling of bone, shape design of implants, rapid prototyping and evaluation, and production.

3.1 Digital Modeling of Bone

For custom-made implant design, digital model of the bone should be created precisely. We use medical data for the modeling, such as CT scan data, and creates digital bone model, from which preparation of physical rapid prototype is also available. The bone in Figure 2, right is a physically rapid-prototyped model based on the digital model shown in Figure 1, left.

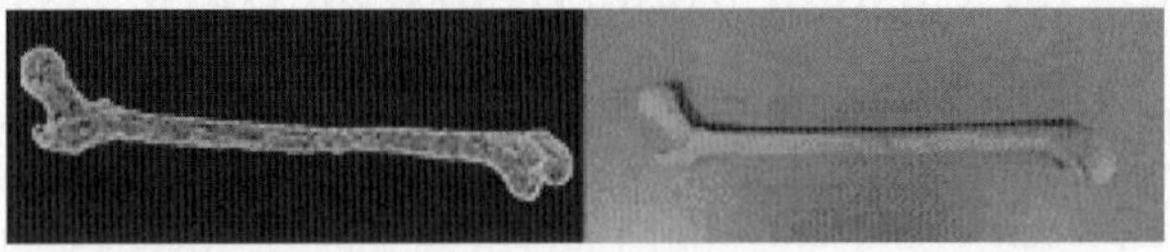

Figure 2: Digital model and its physical prototype

3.2 Shape Design of Implants

Implant shape is designed based on the digital model prepared in the previous stage. Figure 3 shows a sample design of implant, which is designed for the insert implant of damaged bone as shown in the example of Figure 1. Compared to prefabricated implants, the shape of the designed implant is exactly fit to the damaged portion, which means that cutting-off of the normal portion can be minimized. Once the digital design is made, its physical rapid prototype is also available.

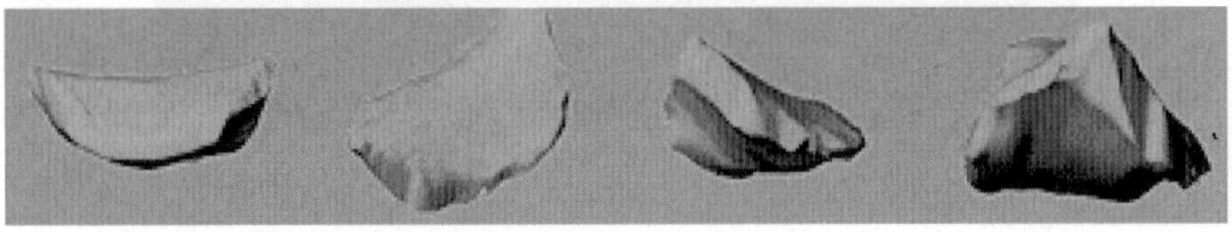

Figure 3: Design of implant

3.3 Evaluation on Implants

Designed implant is evaluated in the virtual environment, which can also be accessed via a network by participating person including medical doctor, design engineer, production engineer, and even patients. Manipulating the implant and the bone, fitness of them is studied, which can also be used for giving information on the scheduled plastic surgery to the patient as an informed consent. The design being provided to production site, the appropriate implant can be manufactured.

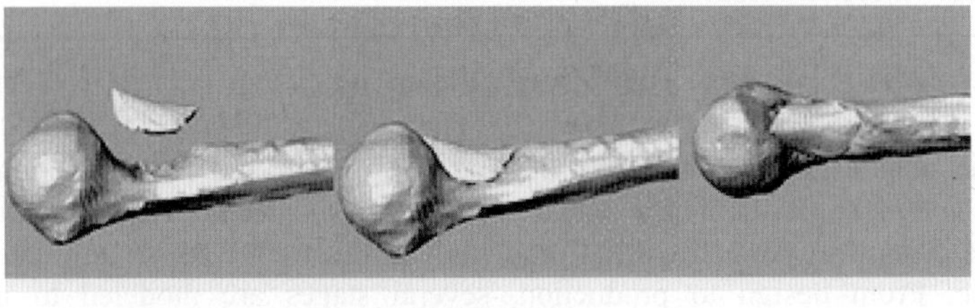

Figure 4: Evaluation of implant in virtual environment

3.4 Virtual Production of Implants

For the production of designed implant, supporting material is also designed in the virtual environment as shown in Figure 5. A collaborative environment is provided to enhance collaboration between medical doctor and design engineer.

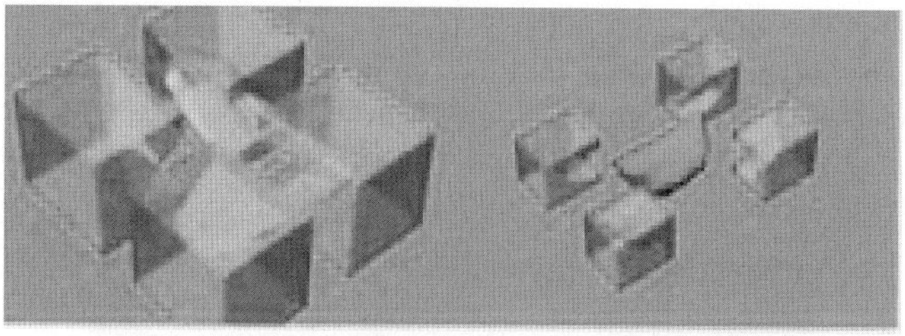

Figure 5: Evaluation and preparation for production in virtual environment

4 Production of Implants with Porous Ceramics

With the design of implant prepared in the virtual environment, its production can be available, if it is made from easily process-able materials, such as metal or hard ceramics. However, it is completely different in the case of porous ceramics.

The potential advantage offered by a porous ceramics implant is its inertness combined with the mechanical stability of the highly convoluted interface developed when bone grows into the pores of the ceramics. Mechanical requirements of the prostheses, however, severely restrict to the use of low-strength porous ceramics to low-load- or non-load- bearing applications. Studies show that, when load bearing is not a primary requirement, nearly inert porous ceramics can provide functional implants. When pore sizes exceed 100micro m, bone will grow within the interconnecting pore channels near the surface and maintain its vascularity and long-term viability. In this manner, the implant serves as a structural bridge and model or scaffold of bone formation. The microstructure of certain corals makes an almost ideal investment material for the casting of structure which highly controlled pore sizes.

To prepare the hard tissue implants using the porous ceramics based on the STL data, we apply a new method using wax combination. Porous ceramics material comprising a special wax is thermally processed to stabilize in its physical shape, and prepared as a starting material for hard tissue implant. On the contrary to normal porous ceramics, which are too fragile to be processed, mechanical processing can be available on this waxed ceramics. After the processing, the unnecessary wax is to be removed by second thermal processing. The picture on the right side of Figure 6 shows an example of sculptured ceramics after the second thermal processing of left one in the figure, which can hardly be processed by a normal method due to fragile features of the porous ceramics.

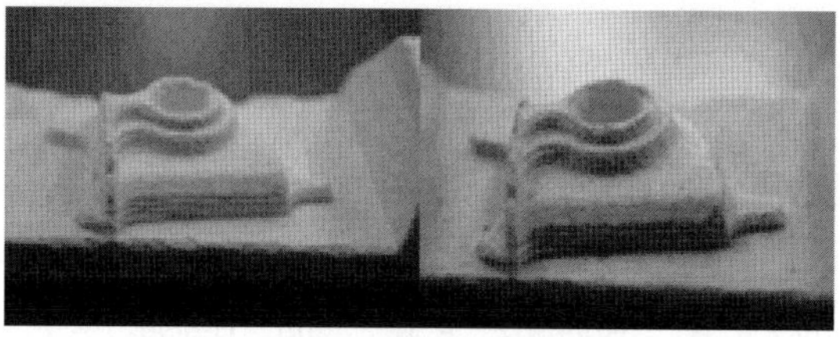

Figure 6: Sculptured porous ceramics by mechanical processing before (left) and after (right) thermal processing

The picture of the left side of Figure 7 shows a starting ceramic bar with special inner structure so that bone tissue can much easily grows into the ceramics. Although the structure is much more fragile than that of normal porous ceramics, mechanical processing to produce any complicated shape can also be possible with our method. This special structure will accelerate the bone formation, however, fragility of the structure may raise another issue which we will report as a separate study.

Figure 7: A starting ceramic bar with special inner structure before (left) and after (right) the thermal processing

5 Concluding Remarks

The paper pointed out the critical issue for requirement on custom-made design and its difficulty in preparation for hard tissue implants. As a new approach to hard tissue implant design and production, the study proposed a custom-made design approach using virtual manufacturing environment. The study focuses on precision process for porous ceramics for hard tissue, and development of a concurrent design system for hard tissue implants. The paper covered the basic idea and picked up some key technologies in the system.

As the goal of the project, medical doctor collaborates with engineers to design the most appropriate shape of hard tissue implants, evaluates it on a virtual environment, and prepares it as a rapid prototyping product. After the evaluation, the implant design will be transferred to the manufacturing site through the network, produce the implant, and deliver it to the doctor in a very timely manner. During the preparation, the doctor can present the patient what sort of material is going to be used for the operation under the virtual environment, and even show the rapid prototype product. Collaborative implant design using virtual manufacturing approach of our study provides a new approach to plastic surgery operation, and a solution to ease the pain of patients.

6 References

Chiroff, R.T., White, E.W., Webber, J.N., and Roy, D.M., (1975). Tissue ingrowth of replamineform implants, J. Biomed. Mater. Res. Symp. 6:29-45.

Hench, L.L. and Ethridge, E.C., (1982). Biomaterials: An interfacial approach, Academic Press, New York.

Holmes, R.E., Mooney, R.W., Buchloz, R.W., and Tencer, A.F., (1984). A coralline hydroxyapatite bone graft substitute, Clin. Orthop. Relat. Res., 188:282-292.

Kesper, B. and Moeller, D.P.F., (2000). Temporal database concept for virtual reality reconstruction, 2000, European Simulation Symposium, pp.369-376.

MetaChart – Using Creativity Methods in a CSCW Environment

Janssen, D., Schlegel, T., Wissen, M., Ziegler, J.

Fraunhofer IAO,
Nobelstr. 12, 70569 Stuttgart, Germany
Doris.Janssen@iao.fhg.de, Thomas.Schlegel@iao.fhg.de,
Michael.Wissen@iao.fhg.de, Juergen.Ziegler@iao.fhg.de

Abstract

An increasing demand for synchronous collaboration with people in different locations has raised the question for adequate tool support of such collaboration activities in the early phases of projects. As work in the early phases is mainly creative and leads from unstructured ideas to structured scenarios, continuity between those steps is essential. We present here a tool-supported method that supports synchronous group sessions as well as asynchronous work in the same environment, based on a generic approach for information generation and structuring.

1 Introduction and State of the Art

Cooperative work remains separated into two areas: Asynchronous cooperative work and synchronous collaborative and creative work. While there already exist numerous technologies for supporting asynchronous cooperative work (e.g. Kortemeyer, Bauer 1999; Shneiderman 2003), support for direct, synchronous collaborative work is still not frequently used in practice (Wissen et al. 2003), although research in this field exists. This is in some cases due to a lack of practical applicability and of support for creative tasks.

For this reason, there is a demand for tools supporting group sessions – especially creative group sessions – performed to develop new ideas, models and products. Lacking computer-supported methods, such a session is often accomplished only by using paper and pencil. The results generated in these meetings often remain unstructured and unused in the further process. Thus, later projects cannot benefit from the results generated in earlier projects. Existing tools for computer-based support of group sessions often cover only the very early stages of a process and are not able to guide and structure the process further on (Prante et. al. 2002). Another aspect is the support for synchronous collaboration in distributed groups. Standalone software is only applicable for group sessions with one central workspace and the group meeting in the same place. Often it is necessary to work with the results continuously in further process steps and to include co-workers at other locations in the meetings. This leads to problems when using conventional methods or standalone software for group sessions.

A computer-based, distributed application for creative work could solve these problems, but other challenges, like usability and acceptance, will arise and have to be handled using up-to-date methods of human-computer interaction.

2 MetaChart: Method and System in Brief

We have developed a tool-based method supporting distributed as well as local and multi-location group sessions. The approach aims at supporting the early to middle stages of group-based project work. These stages consist especially of generating ideas and developing them into semi- or fully-structured project results.

For supporting these highly creative stages of a project, it is necessary to build on a basic modelling concept, which allows an intuitive way of modelling. The MetaChart approach uses a card-based method for this purpose, which is very similar to meta structuring.

It is based on a few, basic elements. A graphical work space, on which the objects can be moved freely, has been developed, using former results (see e.g. Henderson 1986). On this work space, containers can be placed, which are graphically represented as specialized windows. These windows are called MetaCharts and can contain any set of objects. MetaCharts are responsible for ordering, grouping and structuring ideas. Same as windows in GUIs, they can be positioned freely, can be marked with different colours for better mapping and can be given a name and be associated with other MetaCharts.

These two-dimensional structuring abilities are extended by an hierarchical organisation realised by a containment relation between MetaCharts to model hierarchic card systems. This containment allows to insert MetaCharts as childs into other (parent) MetaCharts with unlimited hierarchy depth. Other relations – like associations – can be introduced through the typed relation feature of MetaChart. Typed relations allow a better structuring in phases following the initial partitioning.

MetaCharts can contain different types of cards, like text cards, graphics and various other file types, which can be directly inserted via drag-and-drop from other applications. Unlike material cards, they are underpinned with complex technical features that allow various enhancements and extensions to the basic card metaphor. Every card and MetaChart has content attributes that allow enhancing the model with semantic descriptions.

MetaCharts and cards behave and appear like a specialized window to enhance conformity with user expectations and suitability for learning (see ISO 1998). All objects can be iconified and restored, which makes working with this surface very similar to working with a common PC environment. Once iconified, all objects show their type, while in the de-iconfied state showing their content. All changes, done on any specific item in the workspace, are shared with all group members instantly, while every group member can have the possibility of direct interaction with the system.

3 Creativity Support

What we have described so far is only the support for noting and structuring ideas. Facilitating idea generation means that in addition creativity techniques must be available. Lots of conventional creativity methods have been developed in the past decades to enhance and support the generation of ideas. The MetaChart system has been designed to support the whole process of generating and structuring ideas. Therefore, it supports a couple of well-known creativity methods like, e.g., meta structuring, brainwriting, visual synektik, random stimulation and mindmapping. Other methods, including implementation of the Creativity Cards, are in development. Different system modes help to apply these methods. First, in the meta-structuring mode, it supports the posting of text cards that can be written by any participator. The subsequent arrangement of these cards and a rating ability that gives every participant a certain amount of rating points to allot to text cards are also offered by the system.

4 Intuitive Interaction Techniques in the Group Environment

In order to support creative work in CSCW environments, it is necessary to complement creativity support with intuitive interface technologies as well as with an appropriate working environment. Presentation media can not only be used for the display of the content but also as an interaction media, where team members can work together on the same content. One scenario for MetaChart is the use of a large interactive wall (see figure 1) with additional computing entities. The large interactive wall becomes active using a laser-based system (Wissen 2001). This allows mouse-like interaction while enabling direct pointing, which is important for wall scenarios, because it allows fast and intuitive position changes. Pen gesture recognition and a camera-based hand gesture recognition system enhance the direct interaction experience and allow direct painting on the large interactive wall. Texts can be entered by handwriting on the wall or on tablet PCs, while voice recognition allows speech input of texts and voice commands.

The existing work scenario also includes the use of pen tablets instead of notebooks or desktop PCs. With pen tablets, the interaction is very similar to the use of paper and pencil and, therefore, provides a more direct coupling mode of use. For mobile use, a specially designed interface to MetaChart on a PDA is also available, which allows mobile access to group sessions and will become a more important usage scenario with the faster data transmission methods currently developing.

For better communication between the members of the group, the program includes peer-to-peer and group-based video and audio conferencing components. This way, group members who join a session from other locations can be integrated in a multimodal way as well as group members can establish private peer-to-peer connections to MetaChart participants inside and outside the session group, e.g., for inquries.

figure 1: Environment for creative CSCW work

5 A Four Step Approach

A major application area of the MetaChart system is the area of collaborative content engineering. The MetaChart system provides mechanisms for modelling content. Content can be grouped according to themes, which can then be graphically related to each other using visible associations. Not only can changes be made by generating new objects via menu, but also by drawing objects and their connections freehand.

This method is especially useful when utilizing direct pointing hardware such as pen tablets or tablet PCs. Gesture recognition algorithms identify manually drawn objects and their connections and convert them directly into MetaCharts and their attached associations. The change performed on the model will be instantly made visible to all other group members who, in turn, can perform their changes.

Pure synchronous work would be possible this way, too, but tests show, that it often prevents the session from developing further, if everyone works at the same time with this very open method. Therefore, it will be necessary to find organisational methods for working together or to implement sophisticated tool-based methods for moderation. Both is considered in the ongoing project work.

MetaChart supports an approach composed of four steps from supporting the early stages of idea generating to the late stages of browsing and changing content.

Data Creation: In the first stage, the focus lies mainly on the generation of ideas and data. This stage is usually a highly creative, synchronous stage. Creativity Methods are used for this stage as well as intuitive input methods like gesture recognition and speech input. Usually, in this stage a working group will model the content to be used in the further structuring steps. (see figure 2)

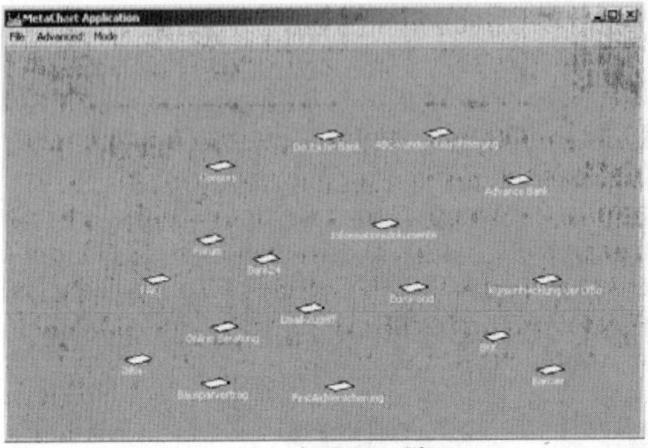

figure 2: Data creation in MetaChart

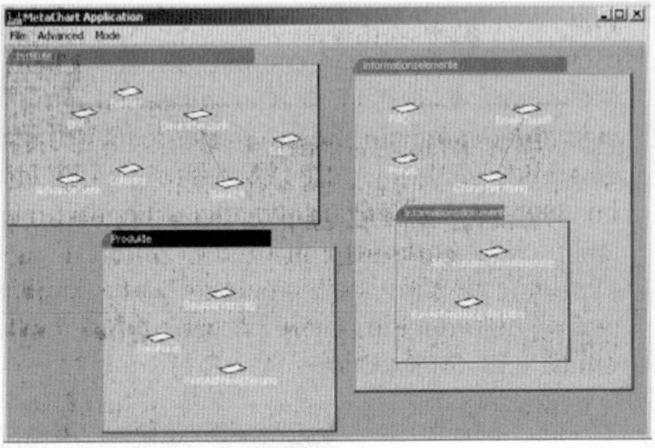

figure 3: Data structuring

Data Structuring: After the initial creation of a normally flat model, which contains an initial collection of ideas, the ideas found have to be structured. This structuring of data is done using associations as well as containment relations. In order to enhance the productivity of structuring, the relations still remain untyped. Containment is modelled by embedding into MetaCharts, which can on their part again be embedded into other (parent) MetaCharts (see figure 3). The MetaChart system offers the possibility to split one session into different new sessions, based on the containment relation. This means that different groups can work independently on different parts of the session in parallel.

Input of Content: In a third step, the data is being further refined, while detailed content and data types are being entered. For this purpose, a tree-based tool named MetaChartExplorer allows easy navigation through all objects in the session as well as easy typing of associations (see figure 4). This stage is usually characterised by delegation and asynchronous group work, although the presentation and discussion of the results may take place in a synchronous manner.

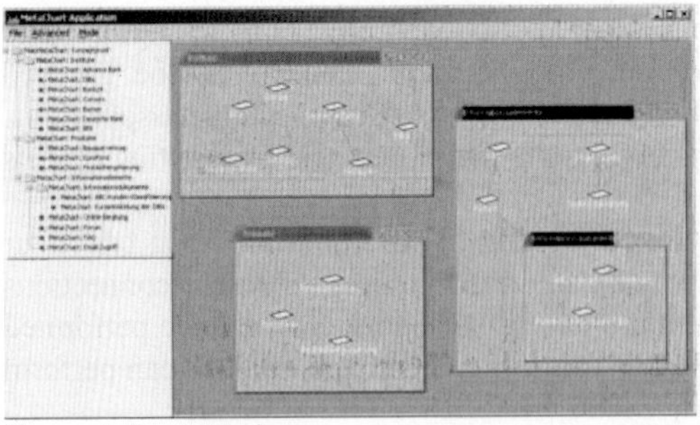

figure 4: Data typing

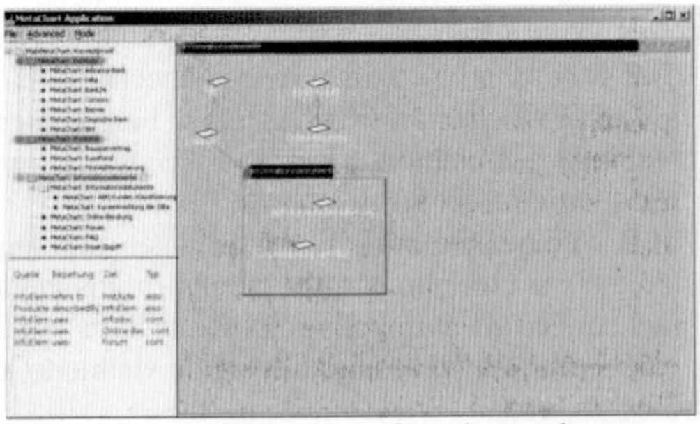

figure 5: Browsing, re-engineering and re-use

Browsing and Data Re-Use: The fourth stage contains methods for browsing the content (see figure 5). The content may also be exported as XML, XTM or RDF. The import and export capabilities allow a connection to other tools (like the ARIS software) and process steps.
Other application scenarios include the development of topic structures and ontologies as well as computer-supported learning. The containment relation allows various other applications that deal with the hierarchic structuring of content.

6 Conclusion

The four stages allow a stepwise creation and refinement of models together with browsing and import/export capabilities. As the MetaChart system is entirely capable of synchronous and asynchronous cooperative work, it allows local as well as distributed groups to join sessions and to go together through the steps of the initial, creative phases. Advanced, multimodal interaction techniques support a fast and intuitive model creation process in group sessions. Application scenarios range from initial idea generation using creativity methods to content modelling and refinement including meta-data description and typed relations.
With its advanced collaboration, modelling and creativity support capabilities, the MetaChart system provides the abilities necessary for IT-based group sessions in the early projects stages.

References

Borghoff, U. and Schlichter, J. (1998): Computer Supported Cooperative Work, Springer-Verlag, Berlin Heidelberg New York

Harel, D. (1987): Statecharts: A Visual Formalism for Complex Systems. Scie. Computer Program, Vol. 8, pp. 514-530

ISO (1998): ISO 9241, Ergonomic requirements for office work with visual display terminals (VDTs), International Standard, International Standards Organisation, Part 10

Henderson, D. A., Card, S. (1986): Rooms: The Use of Multiple Virtual Workspaces to Reduce Space Contention in a Window-Based Graphical User Interface. ACM Transactions on Graphics, 5(3), pp. 211-243.

Kortemeyer, G. and Bauer, W. (1999): Multimedia Collaborative Content Creation (mc^3) – The MSU LectureOnlineSystem, J. of Eng. Educ., 88(4), 421

Prante, Th., Magerkurth, C., Streitz, N. (2002): Developing CSCW Tools for Idea Finding – Empirical Results and Implications for Design, in: Proceedings of the 2002 ACM conference on computer supported cooperative work, ACM Press, New York, pp. 106-115

Shneiderman, Ben (2003): Creativity Support Tools: A Tutorial Overview, in: Proceedings of the fourth conference on Creativity & cognition, ACM Press, New York, pp. 1-2

Wissen, M (2001): Implementation of a Laser-based Interaction Technique for Projection Screens. ERCIM News, No. 46, July 2001, pp. 31-32

Wissen, M. and Ziegler, J. (2001): Creativity support in system and process design. In Proceedings of the 9th Int. Conf. on Human-Computer Interaction (HCI International 2001), Vol. 2 Mahwah, N.J.: Lawrence Erlbaum pp. 119-123

Wissen, M. and Ziegler, J. (2003): Methoden und Werkzeuge für kooperatives Content Engineering, in: Mambrey, P., Pipek, V., Rohde, M.: Wissen und Lernen in virtuellen Organisationen, Physica-Verlag, Heidelberg 2003

Designing Online Communities: Community-Centered Development for Intensively Focused User Groups

Emmanouil Kalaitzakis, Georgios A. Dafoulas, Linda A. Macaulay

Computation Department, UMIST
Manchester, UK
e.kalaitzakis@postgraduate.umist.ac.uk, dafoulas@co.umist.ac.uk, lindam@co.umist.ac.uk

Abstract

In the wake of the 21st century the Internet plays a crucial role in the lives of millions of people. Common use of the Internet includes information retrieval, entertainment and commercial transactions. A significant number of organisations base their existence solely on this relatively new medium for communicating and doing business. In addition, millions of individuals live in a virtual world (cyberspace) where they form relationships and interact with each other by using the Internet. During the past few years, this interaction took the form of online communities, places on the Internet where individuals can communicate with each other using facilities such as e-mail, message boards and real-time chat. This paper describes and models alterations in the development process for online communities and examines how such changes, in a content centered community can increase website traffic and introduction of new members.

1 Introduction

Every day numerous online communities are created, attracting individuals with common interests and same ideas, who want to share their experiences or knowledge with others. Total strangers seem keen enough in exchanging ideas, answering each other's questions or even getting involved in "virtual relationships". A major factor that affects the creation and popularity of Online Communities could be significant events or tragedies. A typical example is the 30 million Americans turned to Online Communities after the terrorist attacks on the World Trade Center and the Pentagon (Preece, 2002). Hence online communities map out the way people behave and interact in an actual societal context. However, "aspects such as individual needs of the members forming the community and their content-complex activities have been overlooked in the study of online communities" (Bogdan, and Cerratto Pargman, 2002). The following sections discuss how the development process was altered to accommodate the specific characteristics and satisfy the special requirements of an online community development project supported by the Department of Trade and Industry (DTI) and Chemical Nutritional Products (CNP) Ltd.

2 Understanding Online Communities

Quite frequently, Internet users tend to believe that by adding a message board in a website, the site is magically transformed into an online community. Also several web developers wrongly believe that when the supporting software is in place and the community is ready to go online, their work is finished and the community can be self-supported. Both views could lead to failure (e.g. limited visitors, short life span). It does not matter how sophisticated the software is when there is no one to use it. "Commercial, non-profit and grassroots organisations have opportunistically viewed these online communities as inexpensive mechanisms for developing customer and donor relationships, with the ultimate goal of increasing revenues" (Hagel and Armstrong, 1997), (Andrews, 2002). However due to lack of trust these online communities do not always achieve their purpose. It is suggested that two steps should be taken after starting an online community in order to tackle this problem (Andrews, 2002): "encouraging early online interaction and moving to a self-sustaining interactive environment." According to Kelly online communities developers must overcome three major obstacles (Kelly, 2002): (i) how to get users to behave well, (ii) how to get users to contribute quality content and (iii) how to get users to return and contribute on an on-going basis. Most of the above concepts were considered during the early stages of this project. The following section provides a detailed description of an online community for bodybuilders that are aware of a specific product range and show special interest to certain athletes and competitions. The online community was created as an extension of an existing "static" website in a primitive state that provided limited information for visitors.

3 Online Communities for Intensively Focused Groups – The CNP Ltd Case Study

In 1998 Kerry Kayes, an English Bodybuilding champion, and the six times Mr Olympia, Dorian Yates founded Chemical Nutritional Products (CNP) Ltd. CNP manufactures and sells nutrition and weight loss products for professional athletes as well as consumers participating in recreational exercise and fitness activities. CNP are a major player in their field. CNP activities include hosting the largest bodybuilding event outside the US, collaborating with key figures in the industry and providing support to many famous athletes. Since January 2002 CNP is involved in a Teaching Company Scheme project (TCS) with the University of Manchester Institute of Science and Technology (UMIST) under the coordination and financial support of the British Department of Trade and Industry (DTI). This project is concerned with creating an Online Community Portal for CNP and further enhancing its e-business solution.

Building a commercially based (Andrews, 2001) online community generated several conflicts regarding the requirements of community users (i.e. what CNPLTD.com visitors want) and the project constraints (i.e. what CNP need). An initial investigation was conducted to understand the main CNP operations and clarify its structure and the underlying business model. Early findings showed rigid distribution network that included several gymnasiums and health centers on the UK and a single distributor for each European country. Some of the resulting constraints included that online sales should not affect sales of existing distributors and that online sales should be avoided outside Europe and especially to the US market which is targeted by CNP's sister company (Dorian Yates Approved). From a user perspective, early findings underlined the imperative need for rich, graphical content, constantly updated news and technical support for training and fitness. This was emphasised by an initial observation of the British Bodybuilding Grand Prix attendants. It is obvious that the specific case study concerns a content centered community for event driven users. Our hypothesis is that in such communities, highly specialised content can increase website traffic and attract more visitors into becoming active community members. To prove this hypothesis a detailed community development process is essential.

4 Community-Centered Development (CCD)

Community-centered development (CCD) is a process that can be used for developing online communities. According to Jennifer Preece, the CCD process involves repeated develop-and-test cycles that can be summarised in the following stages (Preece, 2000): (i) assessing community needs and analysing user tasks, (ii) selecting technology and planning sociability, (iii) designing implementing and testing prototypes, (iv) refining and tuning sociability and usability and (v) welcoming and nurturing the community. The aim of CCD is to focus on the community, by approaching the two interconnected concepts of usability and sociability. "Developing online communities by involving community members in participatory, CCD helps to identify issues, characteristics and idiosyncrasies that address sociability and determine usability" (Preece, 2001).

According to Jennifer Preece, assessing community needs leads to the identification sociability and usability issues while supporting the evolution of an online community (Preece, 2000). Figure 1 shows a proposed model adapting and extending the above concerns based on the initial observation results of the CNP community. Two new sets of issues are identified that relate to the special requirements and behaviour norms of intensively focused groups and to the specific constraints imposed by the underlying business plan of corporate communities. Currently an attempt is made to prove the proposed distinction of issues with a iterative evaluation and redesign process that takes place in three stages. This is further discussed in the following two sections.

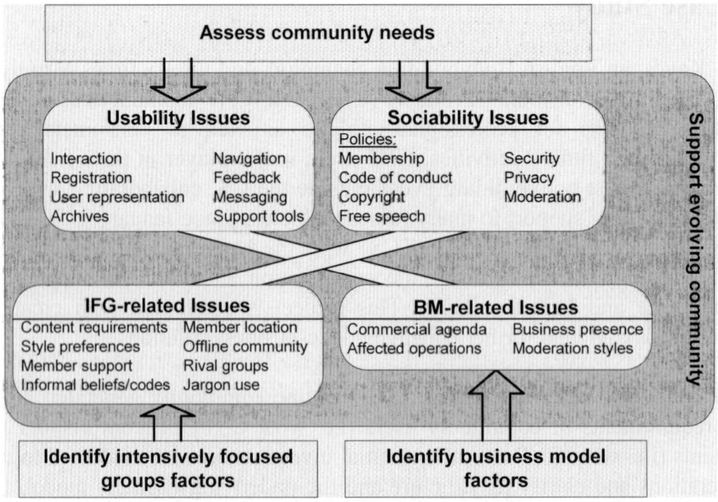

Figure 1: Supporting Online Community Evolution (Adaptation/extension of Preece, 2000)

5 Effects of Introducing an Online Community

A statistical analysis of the community log files in comparison with CNP's previous website log files reveals a number of significant findings. The obvious changes are on daily visitors and page views with a dramatic 100% increase, showing the immediate "attractiveness" effect of introducing an online community. Interestingly enough the user behaviour also shows a tremendous difference. Figure 2 shows the initial web site having an almost constant number of daily visitors. As shown in figure 3, several peaks followed significant community activities with two main peaks of 400% increase during two major bodybuilding events (i.e. Mr Olympia, British Bodybuilding Grand Prix). This observation reveals early evidence that intensively focused user groups, such as the one forming this community, are highly event-driven and require a content-centered community in order to satisfy their needs. Additionally there was a 100% increase on page views per visit and a 6% increase on visitors that spend more than 7 minutes in the community. Finally, a 700% increase on the outgoing traffic per date came as a result to the increased availability of content in the form of competition photos and video streams.

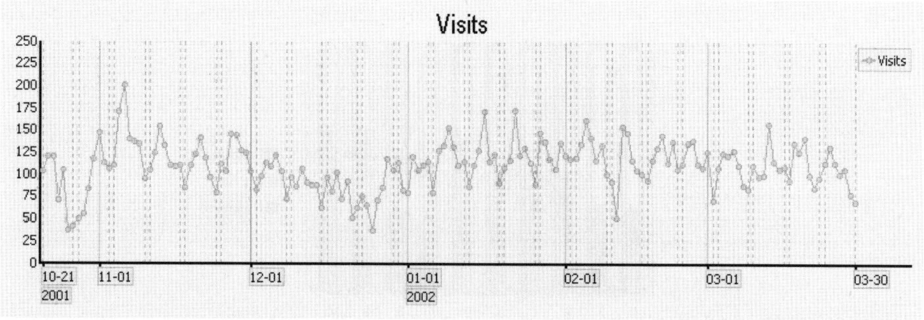

Figure 2: CNP web site visitors (23 week period, Oct 2001 - Mar 2002)

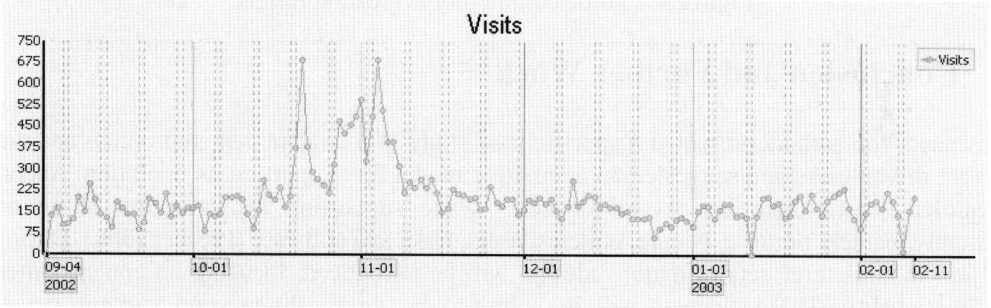

Figure 3: Online community visitors (23 week period, Sep 2002 - Feb 2003)

6 Initial Evaluation Results

During the first six months of introducing the online community an initial evaluation followed the above encouraging results. The evaluation consisted of four parts: (i) a brief questionnaire filled in by community members attending the British Bodybuilding Grand Prix, (ii) a 90-questions usability survey distributed to participants of an electronic commerce course that have used and studied the community for two weeks, (iii) a cooperative evaluation including 1-hour sessions with five of the most active community members and (iv) a statistical analysis of web server logs revealing navigation patterns, numbers and origin of visitors, content popularity, etc. Due to the space limitations for this paper, figure four represents only the summarised results of the first two evaluation parts. Feedback is classified under the following ten categories: (1) accessibility, (2) communication clarity, (3) navigation, (4) consistency, (5) visual presentation, (6) content, (7) privacy/security policies, (8) services, (9) comfort and (10) stickiness. It is clearly shown that the overall response is significantly positive for the first five categories. It is also noticeable that certain community related issues need attention, such as policies and services. Another interesting finding is that the feelings of comfort and 'stickiness' need more development effort. This was mainly due to the high expectations of experienced Internet users that participated to the usability evaluation rather than the core members of the bodybuilding community. Finally, the 'rude' introduction of an online community after a period of two years with an old-fashioned but familiar web site produced a significant percentage of early neutral feelings against the new environment, which are expected to decrease in the following months. With respect to the cooperative evaluation, two key points are made. First, the involvement of CNP provides a trustful environment that makes most members disregarding most policy pages and second, the majority of the community members are novice Internet users and require further online support.

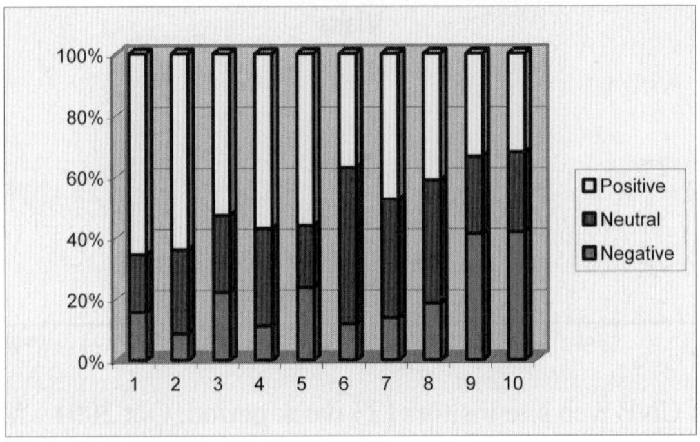

Figure 4: Online community evaluation results

7 Conclusion and Further Work

In this paper the authors presented a specific case study of a content centered online community for event driven users and suggested an alternative model for supporting its evolution. According to Jennifer Preece there are several concerns while supporting the evolution of an online community, namely purpose, people, policies, users, tasks and software (Preece, 2000). During the next four months three sets of major redesign will be introduced, focusing on content, interface and functionality. These changes will be followed by three 100-question surveys structured according to the above six concerns. Further evaluation includes interview and observed discussion sessions with ten of the most active community members, based on the four types of issues identified in the proposed model.

Acknowledgments: Special thanks to the DTI, TCD and CNP and the Hi-Spec research team (Dr Kathy Keeling and David Tomkinson) for their collaboration during the community evaluation.

References

Andrews, D. (2002) Audience-Specific Online Community Design, CACM, Vol. 45, No. 4.

Andrews, D. (2001) Considerations in the Development of Commercially Based Online Communities, in Proceedings of the Seventh Americas Conference on Information Systems (AMCIS 2001), Boston, MA, USA.

Bodgan, C and Cerratto Pargman, T. (2002) Reconsidering support for the members of specialized online communities in Nordic Ergonomics Society's 34th Annual, Congress on Humans in a Complex Environment (NES 2002), Norrköping, Sweden.

Hagel, J. III and Armstrong, A.G. (1997) net.gain: Expanding Markets Through Virtual Communities, Boston: Harvard Business School Press.

Kelly, S.U., Sung, C., Farnham, S. (2002) Designing for Improved Social Responsibility, User Participation and Content in On-Line Communities in Proceedings of the SIGCHI conference on Human factors in computing systems, Minneapolis, Minnesota, USA.

Preece J. (2001) Designing Usability, Supporting Sociability: Questions Participants Ask about Online Communities in Proceedings of the INTERACT '01 conference on Human Computer Interaction, Amsterdam, Holland.

Preece, J. (2000) Online Communities: Designing Usability, Supporting Sociability, John Wiley.

Preece, J. (2002) Supporting Community and Building Social Capital, Communications of the ACM, Vol. 45, No. 4.

Collaborative searching and browsing with a large interactive display

Chris Knowles

Waikato Innovation Centre for electronic Education
University of Waikato
New Zealand
chrisk@waikato.ac.nz

Sally Jo Cunningham

Department of Computer Science
University of Waikato
New Zealand

sallyjo@cs.waikato.ac.nz

Abstract

Observations of collaborative searching/browsing sessions conducted with a whiteboard-sized screen indicate that large screen devices can be useful for supporting collaborative searching, and that they may be particularly suitable for developing group working skills. Usability problems are also identified for the particular large display device used in this study.

1 Background

The need for people to work in multi-disciplinary teams is a mantra commonly rehearsed in many job advertisements and computer science/information systems course outlines. One group skill that has received relatively little attention by researchers is that of gathering information collaboratively, to support later group work. Searching for information—whether conducted via a 'formal' document collection such as a conventional or digital library, or of an unorganized resource such as the WWW—has been regarded as a solitary activity, and information retrieval system interfaces have been designed with a solo searchers in mind. Evidence is accumulating, however, that collaborative searching is a significant event in location such as libraries, where users are physically co-located. These collaborations are mainly unplanned and spontaneous, as searchers ask each other for advice or help [Twidale et al, 1996]; one question, then, is what skills support successful, intentional collaborative information gathering. How collaborative information seeking and retrieving behaviours might be supported through different display devices is also an area that needs further exploration.

The aim of this study is to identify the effects that using a large display device has on the execution of a specific problem solving task, carried out by different user groups, with a view to generating design ideas for further development of the current interface. This is, therefore, an exploratory study that does not seek to provide definitive answers, rather raise further research questions that might indicate fruitful areas for more detailed study.

Figure 1. Using a pen (subject on the right) to interact with the LIDS.

2 Methodology

We observed five groups—three groups of students and two groups of university staff members—engaged in collaboratively searching for information over the World Wide Web. All sessions were recorded on video tape. The searching tasks represented authentic information needs for all groups, motivated by either personal interests or by the demands of a university course. The task had to be relevant to the group in order to provide an incentive to work through the normal behaviours associated with searching for and retrieving information—for example, identifying what was to be searched for, deciding which resources to search, recognizing relevant and appropriate information, and determining when the information need has been satisfied.

When conducting their search, the groups used a LIDS (Large Interactive Display Surface) device [Apperley and Massoodian, 2000; Apperley et al, 2001]. A LIDS device consists of a standard PC connected to a projector, which back-projects the computer screen to a whiteboard-sized surface. Users can enter text through a keyboard or interact directly with the screen by using a pen (in this case, a Mimio ®) as a mouse (see Figure 1).

The two specific target groups in this study, university students and staff, were chosen on the basis of three factors. They had to ; be familiar with personal computers and standard applications associated with internet technology, have used search engines and be familiar with basic interaction devices (keyboard, mouse), and have had some exposure to working in groups, although not necessarily in collaborative information gathering.

The students in each session were members of the same third year computer science course, performing a collaborative search as a preliminary step for a group work course project. The project required the students to locate or design an algorithm for character thinning, and to implement this algorithm. The first step in the assignment was to collaboratively search for information about character thinning—descriptions of algorithms, formally published papers, code sample, or anything else that the members of the group felt would be useful.

The three student groups each included four members. In one group all of the students were non-native English speakers and so their discussions were carried out in their common language, so as to encourage them to discuss the search process without having to first filter their thoughts through the English language. All members of the students groups were under 30 years of age.

The two groups of staff participating in the study were drawn from a research and development group in the University that focuses on online education; collaborative working was an everyday feature of their work. They were older than the student group and were from different disciplines. All the staff were either native English speakers or bi-lingual and all the sessions were carried out in English.

The staff groups were given the choice of what information to look for, with the constraint that it should be a topic of interest to the entire group. The staff had recently returned from a team-building event involving a skiing trip and so the first staff group elected to find out about more adventurous skiing opportunities in the South Island of New Zealand. The second staff group focused on locating sources for food meeting special dietary requirements (these requirements being relevant to a number of members of the unit and their families).

3 Results

The performance of the five groups, when searching, varied considerably in terms of their information behaviours, the amount and type of interpersonal communication, the degree to which the task was completed and perceived to be completed and the extent and manner in which the display device was used to carry out the task and its required activities. These attributes are summarized in Table 1.

Group	Success – unassisted	Success – assisted	Time taken	Key start information behaviours
Student 1			40mins+	
Student 2		√	35	
Student 3		√	35	
Staff 1	√		<10	Goal clarification Evaluation criteria specified
Staff 2	√		10	Goal clarification Evaluation criteria specified

Table 1. Summary of group task completion

The first most obvious outcome was the level of success, assisted or unassisted for each group. Only two of the five groups completed the searching task unassisted—both staff groups. The most interesting elements of this result are the time taken, with one group not managing to gather sufficient information after 40 minutes, whereupon the session was called to a close. The task completion time for the staff groups may have reflected the subject matter, where general context information was better known to them. It may also have reflected another difference with respect to the search approach. Both staff groups made an effort at the beginning of the session to clearly define the task, to agree on what it was and what would constitute a satisfactory answer. The instruction to the groups did not specify this was required—it was done unprompted. Having a clear goal appeared to influence the way in which the groups approached the task.

Interestingly, none of the student groups attempted to clarify the goal at the outset. They were, in effect, participating in a collaborative searching task with only their own interpretations of what was required. In most cases, they did not attempt to arrive at a mutual understanding of the goal until they had spent a considerable amount of time working on the task. The absence of a clearly defined goal was also reflected in their relatively poor intra-group communication.

One advantage of using the LIDS set up for observing shared working is that that the size of the display means that many people comfortably view the screen, preserving the degree of personal space with which they feel most comfortable. However, only one person could use the Mimio pen, and this person had to stand beside the LIDS to use it—thereby dominating the space and becoming the driver of the search. This arrangement literally distanced the other group members from the screen, and so made it difficult for them to actively participate by, for example, pointing out items of interest. The pen holder was also physically isolated from the rest of the group. This is an effect of the device on the execution of the task and the way in which collaboration is affected.

These control and isolation effects are more pronounced with the student groups than the staff groups. The staff are experienced with using shared workspaces and displays. In these situations, there is a device for recording interactions, be it a pen, a mouse, a pointer etc. Each person was used to taking turns with the interaction device. In their use of LIDS, the staff groups maintained this level of turn taking and participation. Individuals come forward to the display to point something out, the person with the pen would then hand it over to enable them to do this. When another group member held the pen, the original pen holder moved back into the group, rather than 'hovering' over the new pen holder. If text entry was required another person, not the pen holder, came forward to do this. The text entry person did not stay at the machine the whole time; once the entry was completed the person moved back to the group. This may be because the keyboard was to the side and, therefore, it was difficult to see the display as the others saw it, so encouraging them to rejoin the group once the entry was completed.

Compare this to the behaviour of the student group and the isolation effect there is much more pronounced. In two of the three student groups, the first person to hold the pen remained the holder throughout the session. The pen holder became a stronger driver of the interaction, partly because of the reluctance of the other group members to take on the role and partly because the level of interaction between the group members was very low. The general level of interaction between the student group members was low. In the first student group all members of the group took an active part in some of the discussions. This group did not, however, succeed in completing the task. In the other student groups the main discussions tended to take place between only two members—in most instances the pen holder and one other person. The 'Wander' strategy, a tactic that involves moving freely between resources and being receptive to new search terms and strategies that are suggested by materials encountered in a search (Twidale et al, 1997), was suppressed. The student groups found it difficult to cue each other that a particular search should be abandoned and a new search begun; the group members tended to politely observe as the pen holder scrolled through lengthy lists of hits, long past the point at which useful-looking results occurred. These student groups experienced a form of search inertia, becoming 'locked in' to the resources, search strings, or strategies. By not refining and agreeing the detail of the task and the criteria that would be used to identify if they had found information to satisfy the task requirements, the isolating effect for the pen holder became more acute.

4 Discussion and conclusions

These observations indicate that the LIDS device can solve a major logistical problem with collaborative searches involving more than two people, by providing a display surface that all members of the team can easily view together. To achieve this visibility, however, the majority of the group members had to sit or stand at least a few feet from the screen, thus reducing their ability to physically indicate their focus of interest (for example, by pointing to a particular spot on the display) or to assume an active role in the search.

The staff groups provided more critical feedback on the design of the LIDS set up. Both groups stated that the use of a pen and keyboard exerted an effect on how well they were able to participate, as the pen holders and keyboard users had to make the effort to move around the room to put themselves back with the group for the discussion. This was something people who are used to working together, or have a lot of experience in working in teaching were comfortable doing. It was noted by one participant that the set up was very useful for small classes in a teaching environment but the dependence on using the pen would partly offset that advantage, e.g. " you would never want to turn your back on children in the classroom". The isolating effect of pen and keyboard could be overcome by using a remote pointer and a keyboard on a long cable. These were seen as ways in which, in teaching scenarios, the teacher could be with the class, have the means to carry out the interaction, be able to designate the interaction role to someone else and still have all the people in the room be able to see the display.

The major problems encountered by the groups—difficulties in discarding infertile resources or search strategies, lack of a clear and negotiated search strategy, awkwardness in turn taking—could only be overcome by training in both search techniques and collaboration skills. Successful collaboration requires a far higher level of communication at the beginning of the search process; the group members must share an understanding of the search goals and come to a consensus on the search strategies that the group will adopt.

References

Apperley, M. and Masoodian, M. (2000): "Supporting collaboration and engagement using a whiteboard-like display", in Shared Environments to Support Face to Face Collaboration: A CSCW Workshop, Philadelphia, Pennsylvania, p 22-26.

Apperley, M. Dahlberg, B., Jeffries, A., Paine, L., Phillips, M. and Rogers, B. (2001): "Lightweight capture of presentations for review", in Volume 12 IHM-HCI Conference on Human Computer Interaction Lille, France, p 41-42.

Brown, M.S., Seales, W.B., Webb, S.B., and Jaynes, C.O. (2001): "Building large-format displays for digital libraries", in Communications of the ACM 44(5) (May), p. 57-59.

Twidale, M.B., and Nichols, D.M.: Collaborative Browsing and Visualisation of the Search Process. *Aslib Proceedings 48(7-8)* (1996), 177-182.

Twidale, M.B., Nichols, D.M, and Paice, C.D.: Browsing is a Collaborative Process. *Information Processing & Management 33(6)* (1997), 761-783

CoVitesse: A Groupware Interface for Collaborative Navigation on the WWW

Yann Laurillau *Laurence Nigay*

CLIPS-IMAG Laboratory, University of Grenoble
Domaine Universitaire, BP 53, 38041 Grenoble Cedex 9, FRANCE
{Yann.Laurillau, Laurence.Nigay}@imag.fr

Abstract

In this paper we present a groupware interface that enables collaborative navigation on the WWW, based on a collaborative navigational task model: the CoVitesse system. The system represents users navigating collaboratively in an information space made of results of a query submitted to a search engine on the WWW. We present the results of an ergonomic evaluation of the interface using heuristics.

1 Introduction

As computers become more and more prevalent, the need for systems that support collaboration, such as peer-to-peer or online games, between and within groups increases markedly. In addition computer users are living in a world of information spaces. One of the most critical needs of users is to be able to efficiently search for information in these large spaces.

These two requirements motivate us to investigate collaborative navigation of information spaces. In this paper, we focus on the synchronous collaboration of users while seeking information on the World Wide Web.

Indeed in everyday life, most information retrieval is based on collaboration between individuals. For example researchers exchange references, standing round the coffee machine ("coffee machine phenomenon"). The existence and significance of collaboration in information seeking have been shown. On the web, navigational behavior often relies on the expertise of other users. One typical social behavior on the web consists of asking a colleague about information we assume the other has the pointer too: such observed social behaviors show that web users are striving for collaboration. For example asynchronous tools, such as email, are commonly used for sharing web pointers.

In this paper, we show a new groupware interface, the CoVitesse system, that enables collaborative navigation in a large information space, the World Wide Web.

The paper is organised in two parts. The first part presents the CoVitesse system and the main features of the interface. In the second part, we present the results of two experimental evaluations.

2 CoVitesse system

The CoVitesse system enables the users to navigate synchronously on the WWW. Four types of navigational tasks are explicitly available to the users: guided tour, opportunistic navigation, shared navigation and coordinated navigation. The four types are fully described in (Laurillau, 1999, p. 308) and in (Laurillau & Nigay, 2000, p. 121). CoVitesse is based on a single-user application, Vitesse, which is described in the following section.

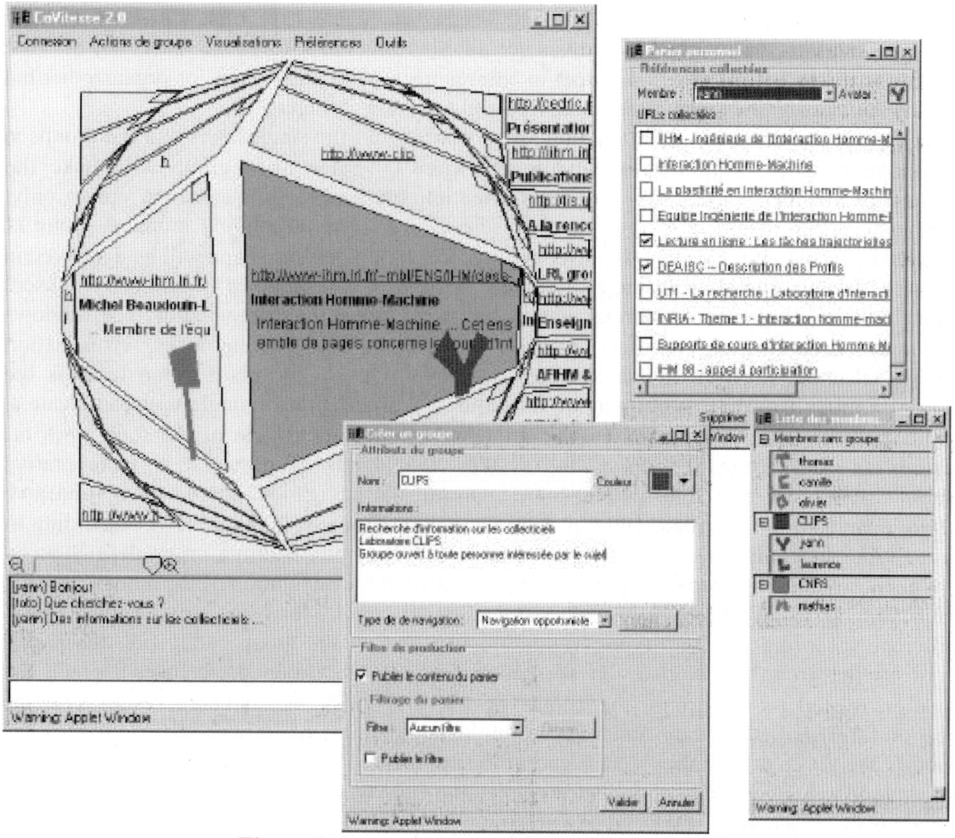

Figure 1: Four windows of the CoVitesse interface.

2.1 The Vitesse system

The Vitesse system visualises the results of a query submitted to a search engine on the WWW. As shown in Figure 1 (top-left window), the overall graph structure of the results is displayed: each retrieved page (node) is displayed. One retrieved page or node is displayed by a polygon. The selection of a node (double click) enables the user to access the web page. We performed a usability study to identify the relevant information to be displayed inside a polygon (Nigay & Vernier, 1998, p. 37). The 2D space is obtained by placing the most relevant retrieved page at the top left of the space.

In Vitesse the user has the choice of the seven visualisation techniques of the result space: birdeye view, polar and cartesian fisheyes. In Figure 1 (top-left window), the current visualisation technique of the information space is the spherical view. At any time the user can freely switch from one visualisation technique to another one (menu "visualisations").

2.2 Covitesse system

When starting a session, a CoVitesse user defines his/her avatar by a shape and a name. The user then either selects an information space or specifies a query that will be sent to a selected search engine. The results of the query define a new information space. The user can then navigate in the information space, observe other users (top-left window in Figure 1, two users are navigating), create or join a group. At any time, a user can see all other users moving in the space, use the chat box (which is below the information space) to communicate with other users, or organize her/his own caddy which contains the marked pages (top-right window in Figure 1).

The user can make visible an additional window containing all the single users and the groups in the information space (bottom-right window in Figure 1). Selecting a group will make the corresponding members observable on the information space. The user can then opt to only observe some of the members of the group by selecting their corresponding icons. A group is represented by a colour and a name. If a user belongs to a group, her/his shape will be displayed in the colour of the corresponding group; else a predefined colour is automatically assigned to a user. Additional windows are available through the menus at the top of the main window. The windows are organised according to five sets of tasks: connection tasks, group tasks, visualisation modalities selection, preferences management and tools. One of the main group tasks is the creation of a group (middle-bottom window in Figure 1): at any time, a single user can create a group, its objective and its style of collaborative work by selecting one of the four types of navigation. In the current version of the system, four kinds of groups, i.e. collaborative navigational tasks, are available: guided tour, opportunistic navigation, shared navigation and cooperative navigation (Laurillau, 1999, p. 308) (Laurillau & Nigay, 2000, p. 121). According to the group type, different functionalities are available. For example, within a group defined as an opportunistic navigation group, a member can take control of the navigation of the other members; such functionality is not possible with a Shared navigation group. Moreover, access rights to data are different according to group types; rights are imposed on the group caddy (i.e. the pages gathered by the group) as well as on the group preferences. Group preferences include the information related to the group, the choice or not to publish the gathered results and the publication filter applied on the results. For example, any member of an opportunistic navigation can modify the group preferences. At the end of the session each user collects the findings of the group gathered in his/her caddy of collected results (top-right window in Figure 1). In addition, the collected results are emailed at the end of a session.

CoVitesse provides persistent access to all the data, modified or produced during a session. These data include information about users and groups such as the avatar shape, the gathered results and the preferences. These data are protected with a simple mechanism of username/ password. Then, when a user starts a new session, she/he recovers her/his own private data.

3 Experimental evaluation

We carried out two complementary experimental evaluations of CoVitesse.

The goal of the first experiment was to evaluate the time response between two users around the World. We wanted to verify if synchronous interaction is possible when users were distant. One user was in the United State, while the other was in France. We selected the guided tour as the appropriate navigational means because the two users are tightly coupled. The results in terms of time response were good: the frame rate was high and we obtained a real time WYSISWIS. Nevertheless this first experiment also stresses the problem of communication between users: a text based chat room was not sufficient enough and the two users quickly established a phone connection that lasted for the duration of the experiment. In addition, we also performed an experiment measuring the crowding impact and the server capacity for managing a large number of users. These observations were confirmed by another experimental evaluation detailed below. We performed an informal experiment with ten robots according to random behaviors (Laurillau, 2002, p. 180). It was easy to build these robots because the CoVitesse system is based on our Clover architecture model (Laurillau & Nigay, 2002, p. 236) and on our Clover platform for groupware development (Laurillau, 2002, p. 129). The experiment was convincing and the server did not collapse. However, it appears that the interaction events exchanged between two remote clients are delayed when robots send events at a rate lower than 200 ms. This may not happen with human users because the Fitts law tells us that expert users interact at a rate of 340 ms. These values were confirmed by the following evaluation. Nevertheless, we are working on the Clover platform in order to obtain better rates.

The second experiment aimed at driving a heuristic evaluation of the interface using ergonomic properties. This experiment has been driven with 10 master's students in computer science using ergonomic rules such as Nielsen's heuristics for single-user interfaces (Nielsen & Molich, 1990, p. 249). A synthesis of the observations dedicated to collaborative activity is shown in Table 1:

Ergonomic properties	Observations	
Observability	(+)	users are observable in the information space
	(+)	users and groups are identified by an avatar, a color and a name
	(+/-)	colors are used to make the difference between members or not of a group. Problems occur if two colors are very close (ex: blue)
	(+)	collected results are stored in a caddy and emailed at the end ofthe session.
	(+/-)	the history of visited web pages is seen as gray polygons in the information space. The information is lost at the end of the session
Published observability	(+)	caddies and preferences are observable
	(+)	published information may be filtered
Consistency	(+/-)	labels are explicit but icons are missing
	(-)	the menu's content for group actions is unstable
Feedback	(+/-)	group modifications are only observable in the member's list window
	(+)	group joining action provides good feedback
WYSIWIS	(+)	strict and relaxed WYSIWIS are mixed (ex: strict WYSIWIS in guided tour)
	(+)	multiple visualisation modalities are available (relaxed-WYSIWIS)
	(+)	group caddy modifications are observable as relaxed-WYSIWIS
	(-)	there is no WYSIWIS for the mouse pointer
Reciprocity	(+)	any avatar is observable in the information space
	(+)	text exchanges are fully observable in chat room
	(-)	preferences and caddies may be filtered
Privacy	(+)	filters may be applied on personal data (cf published observability)
Flexibility	(+)	users and group preferences are available
	(-)	the set of preferences is minimal
Time response	(+)	the interaction time response is very low ("real-time")
	(-)	the system is slow at connection time and no status feedback is given
Task migration	(+)	the navigation control may be given to another user
	(-)	access rights or roles may not be transferred to another user
Reachability	(-)	Other user's work space is not accessible for the shared navigation task

Table 1: Synthesis of observations.

We can extract the following conclusion according to this experimental evaluation. It appears that the use of heuristic rules designed for single-user interactive systems works fine. Indeed, a collaborative interface is also build with elements of single-user interface. For example, the consistency property is not well supported by the CoVitesse system: as show in Table 1, the menu dedicated to group actions must be stabilized because the number of items is different according to the collaborative navigational task. However, the use 'at is' of heuristics designed for single-user applications may lead to contradictions. For example, the reachability property is not well supported by the CoVitesse system when users are member of a group based on a shared navigation: the information space in divided in several parts and each part is assigned to a particular user; in this case, a user is not allowed to navigate in a part assigned to another user. This point is a consequence of the used navigational task model, the shared navigation, which allows strict-WYSIWIS in the information space and relaxed-WYSIWIS locally.

A complementary approach is to evaluate a groupware interface using the extended set of Nielsen's heuristics (Gutwin & al., 2001, p. 123) for groupware. Our evaluation of the CoVitesse system will be completed with the use of these heuristics. However, here are the first observations as shown in Table 2.

Rules	Observations	
Means for intentional and verbal communication	(+/-)	Textual communication only (no audio or video)
Means for intentional and gestural communication	(-)	no means provided
Means for "body" communication	(-)	no means provided
Means for communication through artifacts sharing	(+/-)	moving an avatar in the information space
Protection	(+)	filtering for published observability, concurrency control and role policies when a user is member of a group
Management of coupling	(+)	mixed WYSIWIS defined by our collaborative navigation task model
Coordination of actions allowed	(+) (+/-)	awareness through the avatars, the collaborative tools and the chat negotiation when joining a group or coordination between users according to a kind of collaborative navigation task
Easy user finding and contact making	(+/-)	Users are identified in the information space by their name and avatar. The list of users is given in the user's list window.

Table 2: First observations using extended Nielsen's rules for groupware (Gutwin & al., 2001).

4 Conclusion and further work

In this paper, we focused on collaborative navigation on the web that is a concrete and observed phenomenon although few tools support it. CoVitesse is a tool that supports synchronous collaborative navigation. In addition, we have presented two experimental evaluations of this system. Currently, we are improving the interface based on these experimental studies. Further experimental evaluations of CoVitesse must be carried out, in particular, based on extended Nielsen's heuristics for groupware.

5 References

Gutwin, C., Baker, K. & Greenberg, S. (2001). Heuristic Evaluation of Groupware Based on the Mechanics of Collaboration. In *Proceedings of 8th IFIP International Conference (EHCI)*, 123-140, Springer.

Laurillau Y. (1999). Synchronous Collaborative Navigation on the WWW. In *Extended Abstracts of ACM conference on Human Factors and Computing Systems (CHI)*, 308-309, ACM Press.

Laurillau Y. & Nigay, L. (2000). Modèle de Navigation Collaborative Synchrone pour l'Exploration de Grands Espaces d'Information. In *Proceedings of conférence Francophone Interaction Homme-Machine (IHM)*, 121-128, CRT ILS&ESTIA.

Laurillau, Y. & Nigay, L. (2002). Clover Architecture Model for Groupware. In *Proceedings of ACM conference Computer Supported Cooperative Work (CSCW)*, 236-245, ACM Press.

Laurillau, Y. (2002). Conception et réalisation logicielles de collecticiels centrées sur l'activité de groupe : le modèle et la plate-forme Clover. *Ph.D. dissertation*, 216 p., University of Grenoble, France.

Nigay, L. & Vernier, F. (1998). Design Method of Interaction Techniques for Large Information Spaces. In *Proceedings of Advanced Visualization for Interfaces (AVI)*, 37-46, ACM Press.

Nielsen, J. & Molich, R. (1990). Heuristic Evaluation of User Interfaces. In *Proceedings of ACM conference on Human Factors and Computing Systems (CHI)*, 249-256, ACM Press.

Engineering and Evaluation of Community Support in useworld.net

Sandro Leuchter, Leon Urbas

MoDyS Research Group
Center of Human-Machine-Systems
Technical University Berlin, Germany
Jebensstr. 1, D-10623 Berlin
{leuchter,urbas}@zmms.tu-berlin.de

Kerstin Röse

User-Centered Product Development
Center for Human-Machine Interaction
University of Kaiserslautern, Germany
PO-Box 3049, D-67653 Kaiserslautern
roese@mv.uni-kl.de

Abstract

Design objectives, support services and a development methodology for online communities by means of the case study portal "useworld.net" are presented. The development process that was used emphasizes interdisciplinary requirements analysis and usability testing. As a result special attention was paid to community building and to developing adaptive support which we call socialware. Evaluation results of first prototypes are reported.

1 Introduction

useworld.net is an open user adaptive scientific portal that integrates different information services with collaboration components (see Figure 1). It was jointly developed by a distributed interdisciplinary team at four German universities (Röse et al., 2002). Its purpose is to support information exchange and cooperation within the research area of human-machine-interaction (HMI).

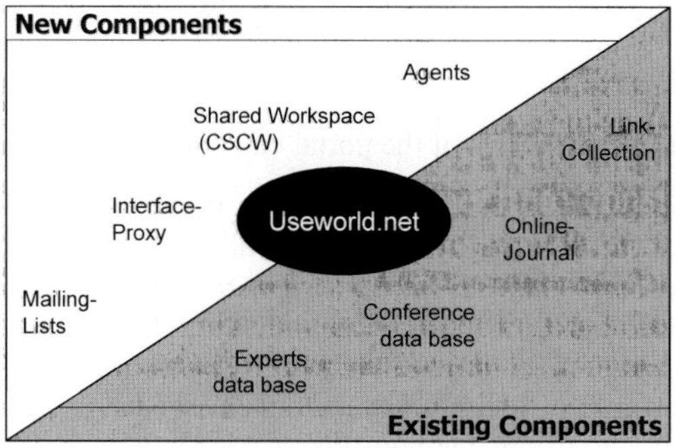

Figure 1: Components providing services in useworld.net

Together with an already existing electronic journal several other information sources are integrated into a browsable web-catalog: conference announcements, link list, job postings, pre-print server, literature references, mailing lists, and expert database. The catalogue and external content is made accessible through a search engine. A shared workspace component enables file-based cooperation in working groups (Künzer & Schmidt, 2002). Registered users can easily form new groups and invite other users into existing workspaces. The operational concept does not include an editorial office (except for the journal) for supplying new content because as a non-profit organization there is no assured and stable way funding it. Thus we apply the idea from several successful online communities that all registered users can act as editors.

2 Community Support

Due to this concept we focused on three main objectives: quality assurance of content, online community building and mediating relevant information on changes to the users providing community awareness.

The community success relies on activity of its users. They only engage if they gain a benefit. Their avail is the content offered by other users so they have to accept it and thus have to trust the content. Quality assurance is central because every registered user is allowed to place new information in public readable sections of the portal. To introduce quality assurance we provide registered users the possibility to rate content. Ratings are used in the portal's catalogue to filter and sort listing views. Thus low rated content will not be displayed at a prominent place. Since the target group of useworld.net is interdisciplinary (psychologists, computer scientists, engineers, graphic designers) the interests and needs of portal users greatly differs. This results in heterogeneous content and ratings. To qualify other users' ratings we apply (user adaptive) relevance information inferred from different sources: use of same workspaces, profile information (in means of catalog categories), activities in the catalogue, and adoption of certain roles within the community.

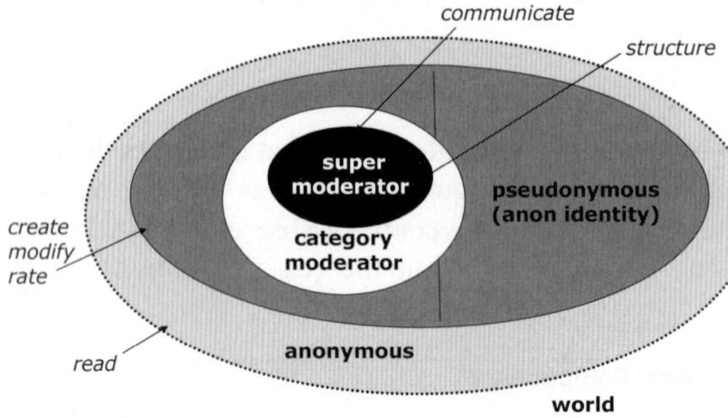

Figure 2: Roles users of the portal can play in the community

While the above objective aimed at consumption of the portals content special efforts had to be made to motivate users to act as content producers. Online community building is addressed by social functions: registered users may adopt moderator roles (see Figure 2); by design: rating is especially easy to fulfil and is graphically emphasized; by technical functions: an interface proxy permits for easy incorporation of new external web-information into the catalogue; by organization: privacy and security are important factors that we paid special attention to.

With useworld.net we offer a system for both direct cooperation in shared workspaces and for indirect cooperation through information sharing, discussing, and producing in the catalogue. An important community support for both is to enable change awareness implemented with agents that collect relevant information for its users and present it in e-mails and personalized portal pages. We adopt the term *socialware* from Hattori, Ohguro, Yokoo, Matsubara & Yoshida (1999) for this technology. Figure 3 shows the principle architecture of the socialware support in useworld.net: Users interact with the information services of the portal ("information and cooperation services"). These activities are monitored by the socialware layer of the portal. Since the same information space of the portal is used by many users in parallel the activities of one users affects other users as well. Examples for such activities are inviting to work groups, cooperation on an object in a shared workspace, discussions, rating of elements in the information space, or creation of new elements. Some of the activities help the socialware layer to form a representation of single users and their interrelations (working groups, interest groups). This user profile information is used by agents to monitor all activities and filter information about changes relevant for single users or groups. Depending on personal settings the personal socialware-agent informs a user on a regular basis by e-mail.

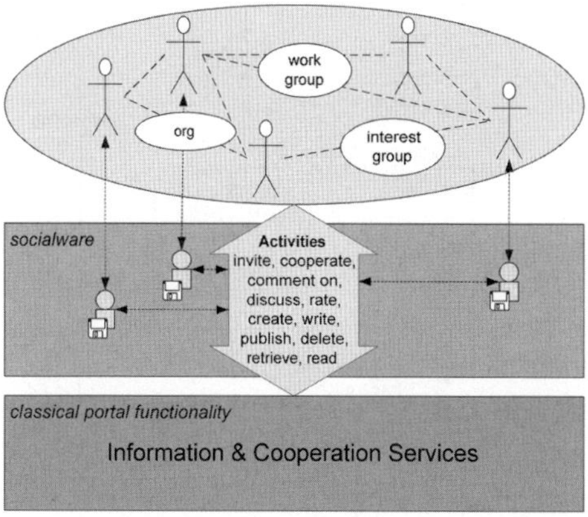

Figure 3: Socialware support in useworld.net through agent support

3 Engineering

Standard software engineering methodologies like the Rational Unified Process are not sufficient for developing innovative collaboration platforms that meet the needs of heterogeneous communities because this is not a straightforward engineering task. Instead this usability engineering problem can be addressed by applying the parallel-iterative engineering approach (PIE, see Figure 4). It suggests user participation in all phases. Technology developers work parallel to human factors specialists in defined phases: After first defining the system's boundaries (e.g. target group and objectives), in the second phase technical functions (features) on one hand and user needs, expectations and abilities on the other are collected to guide the design of function and interaction principles. The result is a specification of use cases and corresponding technical functions. The next parallel step is to implement the technical system and to develop interface designs and organizational embedding. This process phase is accompanied with formative evaluations of mockups and prototypes supporting design decisions in early phases of software development. Finally usability evaluation cycles help to optimize the software. This approach significantly reduces the risks of software development in innovative interaction oriented software projects. The systems engineering view on software development guides usability evaluation: Scenarios are defined on the basis of the use cases. Results are interpreted and weighted with objectives and target group in mind.

PIE defines a universal (macro) process. We have further structured its phases with the agile software engineering methodology "feature driven development" (Coad, Lefebvre & De Luca, 1999) because the notion of features is especially suited to communicate between technical and human factors staff and because features are an appropriate interface between the phases of the macro process.

This process was used to develop the useworld.net portal. During the planning phase a survey was conducted for the human factors side of the requirements analysis (Leuchter, Rothmund & Kindsmüller, 2002). Results showed that the proposed concept "users as editors" is critical. Especially the community building process had to be supported. Thus Kindsmüller, Razi, Leuchter & Urbas (2002) focused a further study on community aspects and user motivation to prepare the design of useworld.net.

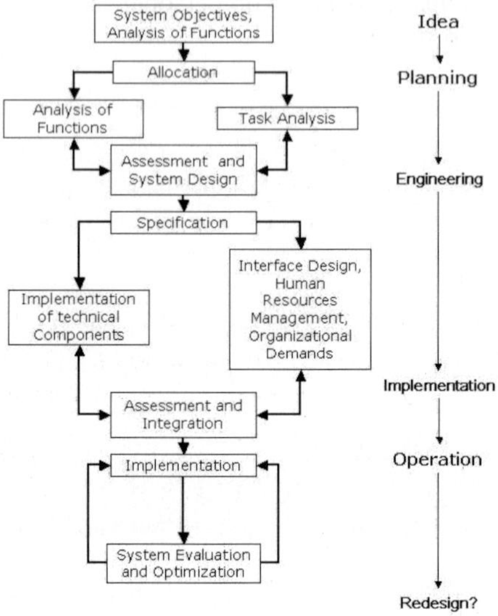

Figure 4: Development process for interactive systems with parallel-iterative engineering (PIE)

4 Evaluation

The first evaluation of the prototype was carried out with 20 subjects from mostly university background (engineers, computer scientists, psychologists). Some background information about the participants:

- *Computer experience/-usage in average:* 12 years, ca. 5 hours daily, mostly professional
- *Internet usage in average:* 6 years, ca. 2 hours daily, mostly professional
- *Search engines in average:* 5 years, ca. 20 minutes daily, mostly professional
- *Online-community-usage:* most less than 5 communities, mostly private

Typical use-situations were tested with a scenario-based-approach. E.g. the task: "Invite the member 'eva' into your workgroup and check the successful sending of your invitation to 'eva'". The participants of the evaluation were expected future user of the useworld.net, but not the same individuals as in the requirement analysis (Leuchter et al., 2002), therefore user on the novice level of useworld.net-usage. The evaluation situation was structured with a short pre-questionnaire about experiences with online-cooperation and internet-usage, followed by a scenario-based testing of selected usage-situations with useworld.net, followed by a post-questionnaire about subjective satisfaction with the usage of the system and finished by an open feedback discussion.

The items of the post-questionnaire were structured into two categories: 'joy of use' and 'ease of use' and provided a rating with five steps. In the remainder, some selected results of this evaluation are reported. Over all, the user-satisfaction was agreeable for a prototype. In total the median for 'joy of use' (used the hedonistic differential) was 3.45 and for 'ease of use' 3.07. These results are not bad but show a potential for redesign. Interesting is the higher score for 'joy of use'.

Finally the participants resumed the evaluation in a focus group feedback discussion:

- useworld.net is an innovative and interesting approach.
- The high degrees of freedom (especially in the CSCW-part) were accentuated.
- They missed a list with all names of useworld.net members.

All users emphasize the good correspondence of navigation (interaction design) between the catalogue- and the CSCW-area. The main actions in both areas were realized with drop-down menus (which include context-dependent main functions).

5 Conclusions

The parallel-iterative engineering development process proved to be useful. Especially the feature-based fine planning scheme made communication in the interdisciplinary team effective. The requirements analysis was very important since it revealed the demand for community aspects consideration. This resulted in the community support for our interdisciplinary scientific target group as detailed above.

The usability test showed that participants were all pleased of the variety of the possibilities the community-portal offered for cooperation. In the scenario-based testing part 90% of all participants have executed the cooperative tasks successfully without any instructions. 77.7 % of the participants could see that other users are also active (logged in) in the community. The offer of global access (independent of location or used system) to the community was for all participants a really important aspect, in combination with the possibility to create the own 'internet-workplace' and a private 'sub-community' integrated in a comfortable portal-community.

We kindly acknowledge financial support by the DFN & BMBF under grant number VA/I-110 within the programme "Einsatz von Netzdiensten im wissenschaftlichen Informationswesen". User adaptive algorithms are jointly developed with the MoDyS Research Group that is sponsored by VolkswagenStiftung within the programme "Junior Research Groups at German Universities".

References

Coad, P., Lefebvre, E., & De Luca, J. (1999). *Java Modeling In Color With UML: Enterprise Components and Process*. Upper Saddle River, NJ, USA: Prentice Hall.

Hattori, F., Ohguro, T., Yokoo, M., Matsubara, S., & Yoshida, S. (1999). Socialware: Multiagent Systems for Supporting Network Communities. *Communication of the ACM*, 42 (3), 55-61,.

Kindsmüller, M.C., Razi, N., Leuchter, S., & Urbas, L. (2002). Zur Realisierung des Konzepts "Nutzer als Redakteure" für einen Online-Dienst zur Unterstützung der MMI-Forschung im deutschsprachigen Raum. [On the realization of the concept „users as editors" for an online-service supporting HMI research in German speaking countries] In GfA (Ed.), *Arbeitswissenschaft im Zeichen gesellschaftlicher Vielfalt. 48. Kongress der Gesellschaft für Arbeitswissenschaft* (pp. 133-137). Düsseldorf, Germany: GfA-Press.

Künzer, A., & Schmidt, L. (2002). An Open Framework for Shared Workspaces to Support Different Cooperation Tasks. In *Proceedings of World Wide Work, Work With Display Units, Berchtesgaden, Germany* (pp 217-219). Berlin, Germany: ERGONOMIC Institut für Arbeits- und Sozialforschung.

Leuchter, S., Rothmund, T., & Kindsmüller, M.C. (2002). Ergebnisse einer Tätigkeitsbefragung zur Vorbereitung der Entwicklung eines Web-Portals für Mensch-Maschine-Interaktion [Results of a task survey for preparation of the development of a web portal for human-machine-interaction]. In GfA (Ed.), *Arbeitswissenschaft im Zeichen gesellschaftlicher Vielfalt. 48. Kongress der Gesellschaft für Arbeitswissenschaft* (pp. 129-132). Düsseldorf, Germany: GfA-Press.

Röse, K., Urbas, L., Gersch, P., Noss, C., Künzer, A.,& Leuchter, S. (2002). Interdisciplinary Development of a Collaborative Portal for a Heterogeneous Scientific Community. In P. Isaias (Ed), *WWW/Internet 2002, Proceedings of the IADIS International Conference, Lisbon, Portugal, 13-15 November 2002* (pp. 385-392). Lisbon, Portugal: IADIS Press.

An End Users Dedicated New Language for Geographical Information Retrieval

Mohamed Limam - Mauro Gaio - Jacques Madelaine

Laboratoire GREYC, CNRS UMR 6072
Universitéde Caen, F 14032 CAEN Cedex
{mouldahm, gaio, madelaine}@info.unicaen.fr

Abstract

After the review of 4 mono-modal query languages, this paper presents a new bi-modal language for geographical information retrieval.

1. Introduction

Geographical Information (GI) can be easily constructed by the use of advanced software applications, thus producing a large amount of this type of information, while the retrieval process is still difficult for end users. Various mono-modal interfaces exist for GI retrieval and access; main categories are: browsing through hyper-links, formal languages, visual language and natural language. All these methods lack of usability for spatial querying. A new hybrid bi-modal approach using sketch and text for GI access is finally proposed.

2. Different existing approaches

The *hyper-media* approach allows navigation trough documents sets by following preexistent hyper-links. Links may also connect part of a map and can be used for geographical acess. This method is very easy to use and needs virtually no training. The Web technology makes it widely available at no cost. The main use of the Web is direct navigation (or exploration) and retrieval with the use of search engines. But search engines have no geographical indexes, making impossible to issue geographical queries (Boursier & Mainguenaud, 1992).

On the contrary, *formal languages* offer the possibility to state queries using artificial languages. We found in this category: GIS programming languages and SQL extensions (Egenhofer, 1994). The main characteristic of these languages is the strict syntax they used to construct a request, leading to a non ambiguous semantic. Their disadvantage is the training needed for their use, making them inadequate for end users who are not database or GIS experts (Egenhofer, 1992).

To deal with these problems, the *visual approach* came to offer an easy and intuitive mean for spatial configuration expression. It uses intuitive visual metaphors allowing to graphically represent spatial objects (e.g. a polygon for a town, a line for a river) (Egenhofer, 1996) (Calcinelli, 1994). As the user can ``draw" its query, it is easy to over-specify a situation leading to some kind of ambiguity (Bonhomme, Aufaure-Portier & Trepied, 2000). Furthermore, the complexity of the implementation increases with the freeness offered to the user; if it is easy to interpret a drawing made with a restricted set of individualized objects (icons, open lines, ...) it becomes very hard to interpret a totally free sketch.

The last method we present is based on *natural language*. It allows users to freely express their queries using text. But end users can find difficult to express the topological relation between three or more objects; it often needs long sentences using many relatives propositions that can be ambiguous. Moreover, the analysis of such sentences may be technically difficult if the user is not constrained. Many problem of interpretation coming from polysemy and anaphora resolution.

Table 1 summarizes the pros and cons of these four methods regarding to the four criteria: user training cost, expressiveness of the language, unambiguous language and technical feasibility of the language implementation. If the use of hyper-media interface and of natural language querying needs nearly no training, visual language are not so directly intuitive and the use of formal languages need a real expertise. The expressivity of the different language are the same if we except the hyper-media technique that does not allow for direct geographical querying. If browsing a document or specifying a formula lead to unambiguous results, a spatial expression in natural language can be ambiguous mainly because of under-specification (Talmy, 1983). Visual languages can be very ambiguous if the interface is not restricted. From the point of view of technical feasibility, hyper-media technique and formal languages are in the best position and many products already exist. This is not the case for visual and natural languages mainly because of the complexity of their interpretation; this complexity increases with the freedom of expression accepted by the system.

Table 1: Comparison summary

	Hypermedia Technique	Formal Language	Visual Language	Natural Language
User Training Cost	++	- -	+	+ +
Expressiveness	-	+	+	+
Lack of Ambiguity	+ +	+ +	-	- -
Technical Feasibility	+ +	+ +	-	- -

One conclusion drawn from the table 1 could be that both visual and natural languages are not usable. But if we analyze deeply the results, we can see that the intrinsic qualities/drawbacks of the two modalities image and text are in fact complementary. This lead us naturally to a bi-modal language that will take advantage of the two modalities to increase its expressiveness. Furthermore, the comparison of the expression of the same query in the two modalities initiate easily an interaction between the system and the end user.

3. The two sides of a geographical query: towards a bi-modal approach

Let's now consider geographical queries as a combination of two components: one component is concerned with the spatial specificities (localization, orientation, etc.) while the second component is concerned with thematic information (names of the entities, quantification of properties) or temporal information (Schlaisich & Egenhofer, 2001).

The basic idea is to use two different modes to express these two components. This leads to a *bi-modal* approach using a *visual language* plus *natural language*. The user will have the ability to specify a spatial configuration using a *sketch* and to precise contextual, thematic or temporal constraints using free *text*.

The text users need to type in, will only be phrases and not full sentences because they will have simply the need to specify a toponym (e.g. city of Caen, region of Basse Normandie) or quantitative, qualitative and temporal observations associated with spatial objects (e.g. number of inhabitants, road quality, river flow, ...).

We propose a new bi-modal language for geographical information retrieval where the spatial specification is expressed with a sketch made of predefined shapes (closed polygons, lines, points) each of them being thematically or temporally defined with a free text. We claim that the combination of these two modes allows an easy expression for the end user while the analysis process is very tractable as the text analysis is restricted to short phrases and the sketch analysis uses only predifined shapes.

These restrictions do not diminish the expressiveness of the system. Sketch mode is well adapted to express the basic geometry of objects such as shape, dimensions, etc., but also to specify the topological relations such as inclusion or intersection. Furthermore it can express orientation and relative positioning of spatial objects. These specifications are not only easy to state but are very precise compared to their counterpart expression in natural language. Take for example the phrase "*A* is at the North-East of *B*"; this expression is rather vague and indeterminate while a sketch can determine the exact localization of the considered objects. Thus, the two modes combination will overcome the opposite drawbacks of under or over specification in the two modes. It will offer an adapted mean to specify geographical information, each mode sketch or text respectively will then be used for the expression of the respective spatial or thematic query component.

4. The prototype

We develloped a prototype of the bi-modal interface. In this prototype, the predefined shapes are points, polysegments for lines, and octagon for spatial regions. After having defined a shape, the user can label it with a phrase in natural language. The figure 1 shows a snapshot for the query about finding highways that pass through a town of over 10,000 inhabitants and along a medieval castle. The castle is represented with a point (small filled circle), the town with an octagon, and the highway with a polysegment. We choose to show the label in a separate window (entitled "informations" on the figure) as it may obscure the sketch in case of lengthy labels.

This example clearly shows that natural language is ambiguous as spatial configurations are under-specified. On the example: is the castle inside the town or not ? This is very precisely specified on the sketch. This is mainly why the system uses natural language in order to give a feedback to the user with paraphrases (showed in the window entitled "Retour"). Each of them is written under a particular focus. The system presents along with each paraphrase a variant of the spatial situation using an extented version of the topological neighborhood concept of (Egenhofer & Mark, 1995) detailed in (Szmurlo & Gaio 1998).

The user may then validate his query or choose a variant that can in turn be reformulated under different focuses.

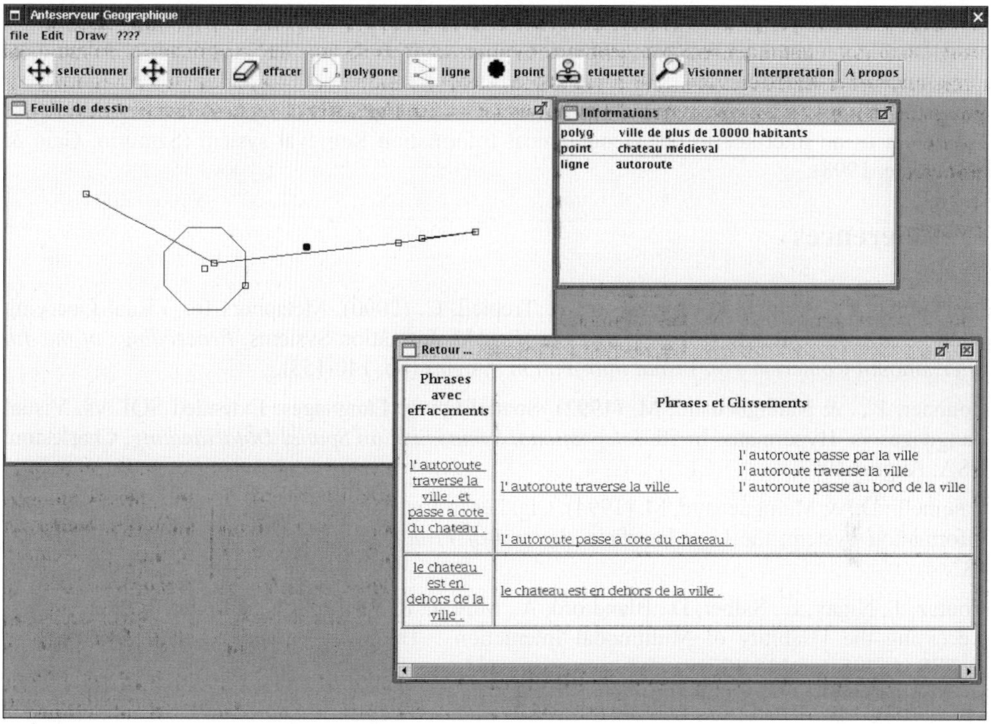

Figure 1: Finding a highway that passes through a town and along a medieval castle.

5. Conclusion

This paper presented a new bi-modal geographical query language. The two modes are clearly complementary (Coutaz et al., 1995). In our case, we use sketch to give a global geometrical vision and text to define the entities by their names or qualities. They can furthermore be used to overcome some drawbacks by using a system/user interaction loop. After the system analysis of the query and its first interpretation, the system paraphrases the situation using *text*. The first aim of the interaction is to allow the user to validate or to modify the query regarding of the system interpretation. The second reason is that sketch may "over-specify" the information (Talmy, 1983) while natural language may be more vague or indeterminate regarding to a drawing. The textual mode has also the advantage to state what are the relevant conditions or restrictions in the query, and will *de facto* rank these conditions because of the inherent linearity of the text.

We can now in turn evaluate our method against the four criteria:

Training cost: the use of the two modes is very easy for all types of users without training as we could verify with some geographers using our prototype.

Expressivity: each mode is used in the domain where it is the most expressive.

Non ambiguity: each mode is used to disambiguate or to drop constraints expressed in the other mode.

Technical Feasibility: we have implemented this bi-modal method in a prototype that allows users to build a query and to interact with the system in order to validate their queries.

Our current implementation is not actually connected to any retrieval system but we plan to use end test it as an interface to our Geographical Information Retrieval system (Szmurlo, Gaio & Madelaine, 1998).

6. References

Bonhomme, C., Aufaure-Portier, M.-A., & Trepied, C. (2000). Metaphors for Visual Querying Spatio-Temporal Databases. In Advances in Visual Information Systems. *Proceedings of the 4th International Conference on Visual Information Systems* (pp. 140-153).

Boursier, P., & Mainguenaud, M. (1992). Spatial Query Languages: Extended SQL vs. Visual Languages vs. Hypermaps. In 5th *International Symposium on Spatial DataHandling*, Charleston, USA, August 1992.

Calcinelli, D., & Mainguenaud, M. (1994). Cigalles: a Visual Query Language for a Geo-graphical Information System: the User Interface. *Journal of Visual Languages and Computing*, 5(2):113-132.

Coutaz, J., Nigay, L., Salber, D., Blandford, A., May, J., & Young, R. (1995). For Easypieces For Accessing the Usability of Multimodal Interaction : The Care Properties. In *Proceedings of INTERACT'95*, (pp 115-120).

Egenhofer, M. J. (1992). Why not SQL! *Int. Journal of Geographical Information Systems*,6(2):71-85.

Egenhofer., M. J. (1994) Spatial SQL: a Query Presentation Language. *IEEE Transactions on Knowledge and Data Engineering*, 6(1):86-94.

Egenhofer, M., & Mark, D., (1995). Modeling Conceptual Neighborhoods of Topological Line-Region Relations. *International Journal of Geographical Information Systems*, 9 (5): 555-565.

Egenhofer., M. J. (1996) Spatial-Query-by-Sketch. In M. Burnett and W. Citrin edts., *VL'96: IEEE Symposium on Visual Languages*, Boulder, CO, (pp. 60-67). September 1996.

Schlaisich, I., & Egenhofer, M. (2001). Multimodal Spatial Querying: What People Sketch and Talk About. In C. Stephanidis, ed., *1st International Conference on Universal Access in Human-Computer Interaction*. New Orleans, LA, pages 732-736, August 2001.

Szmurlo, M., & Gaio, M. (1998). Extended conceptual neighborhoods, *ISPRS Comm. IV International Symposium: GIS - Between Visions and Application*, September 1998, Stuttgart, Germany.

Szmurlo, M., Gaio, M., & Madelaine, J. (1998). The Geographical Anteserver: a Client/Server Architecture for GIS. In *Proc. of EOGEO'98 Workshop*, Salzburg, Austria, Feb. 1998.

Talmy, L. (1983). How Language Structures Space? In Spatial Orientation: Theory,Research and Application. H.Pick and L. Acredolo, Plenum Press.

Study on A Retrieval Method which Reflects Individual Preferences by Preference Analysis with Mediation Variables

Takashi Mitsuishi

Graduate School of Educational Informatics, Tohoku University
Kawauchi, Aoba, Sendai 980-8576, Japan
takashi@ei.tohoku.ac.jp

Abstract

Information retrieval is becoming common in our daily life. However, the nature of available data is extending from simple texts consisting of words and phrases, to data that are related to personal tastes, impression and emotions (*kansei*), such as music, video and foods. The description of such "*kansei*" data by simple words or phrases is difficult. In order to make effective use of database applications containing *kansei* data, we propose a new retrieval method. It analyzes the individual user's preferences of the target *kansei* data through mediation variables. Such mediation variables are related to the target data, so they reflect the estimated preferences on retrieval results, and facilitate users finding data, which best matches his/her preferences. Based on the proposed method, we design an online recipe retrieval system, and show the effectiveness of our method by experimental results.

1 Introduction

Although web based database applications facilitate us access to large-scale databases and get numbers of data, it is not easy to browse or retrieve target data from such databases due to explosion in the amount of data. When a database containing some data, which are difficult to be characterized by some specific words like *kansei* data, it is even more difficult to distinguish or find target data. In order to facilitate users to retrieve target or preferable data from such a database, some effective retrieval methods are required.

2 Retrieval method which reflects individual preferences

2.1 Existing retrieval methods

In order to retrieve the desired data efficiently from some databases, there are many studies on retrieval methods, which enable to query by impressions or reflect preferences on retrieval results.

Some of them provide facilities to query by some words of impression (Yoshida, Kiyoki & Kitagawa, 1998) (Fukuda, Sugita & Shibata, 1998) (Harada, Itoh & Nakatami, 1999) (Ishihara, Ishihara & Nagamachi, 1999). Other studies estimate individual preferences of users to different data, and reflect them on retrieval (Okada, Lee, Kinoshita and Shiratori, 1998) (Kazama,Sato, Shimizu & Kambayashi) (Kawata & Yatsu, 1999).

However, these methods require some specific words in order to characterize the data or estimate preferences of users for data. On the other hand, there exist some databases containing kansei data, for which characteristics are difficult to specify with words. As an example, let's consider the case of a database containing recipes of dishes. The user might be interested in searching by savor and flavor, in order to retrieve recipes of preferable dishes. Although there are many words expressing savor or flavor, it is not easy to specify savor and flavor of different dishes definitely by associating simple or common words. Thus, the existing retrieval methods are not enough to handle this type of databases, because neither we could differentiate one target data from others nor retrieve the intended data.

2.2 Preference analysis and reflection with mediation variables

In order to retrieve target data efficiently from a database containing kansei data, we are proposing a new retrieval method (Mitsuishi, Tada, Sasaki & Funyu, 2001) (Mitsuishi, 2002). For traditional retrieval methods, words, which are able to characterize the target data, must be available. Our proposed method can deal with data for which such characterization words are not available, as we use some other information that are related to the target data. It analyses access histories to a database, estimate individual preferences of users through these related data as mediation variables, and reflects estimated preferences on retrieval results. Details of our proposed method follows.

(1) Preparation: First of all, a mapping function from target data T to mediation variables: M based on the relation between them is defined, as $R(t)$: $\{t \in T\} \rightarrow \{m \in M\}$. When one of target data $t \in T$ is given, $R(t)$ returns one of corresponding mediation variables $m \in M$.

Also a profile $\boldsymbol{p_u}$ of each user u is defined, as a vector consisting of scores on individual mediation variables. When there exist n mediation variables; $m_1, m_2, \ldots, m_n$, $\boldsymbol{p_u}$ consists of n scores; $s_{m_1 u}, s_{m_2 u}, \ldots, s_{m_n u}$ on these mediation values as $\boldsymbol{p_u} = \langle s_{m_1 u}, s_{m_2 u}, \ldots, s_{m_n u} \rangle$. The initial values of such scores are set to 0.

(2) Profiling: Individual preferences of users based on access histories to a database are analysed. When an user u queries data and select one data t from retrieved data, the score $s_{R(t)u}$ in his/her profile $\boldsymbol{p_u}$ is updated, by $s_{R(t)u} + 1 \rightarrow s_{R(t)u}$, where $s_{R(t)u}$ is the score on one of mediation variables $m{=}R(t)$ corresponding selected data t. By repetition of this scoring, preferences of different users for target data can be estimated with mediation variables.

(3) Reflection: When an user retrieves k data; $\{t_1, t_2, \ldots, t_k\}$ by querying to the database, the individual scores of these data are given from his/her profile as $t_1{:}\ s_{R(t_1)u}, t_2{:}\ s_{R(t_2)u}, \ldots, t_k{:}\ s_{R(t_k)u}$. It rearranges the retrieved data list according to these scores, and presents it to him/her.

For example, we could use types of wines as mediation variables for a recipe database. When an user frequently chose some fish dishes, we could estimate that he/her prefers dishes matching white wine than dishes matching red wine. Then, we could recommend him/her chicken dishes matching white wine more than other meats (e.g. pork, beef, and ram).

3 On-line recipe retrieval system

As an application of our proposed method for web-based database application, we develop an on-line recipe retrieval system an example of web based database applications. With this system, the user is able to query by materials of dishes as keys, retrieve a dish list, and get a recipe of a selected dish from a list. When the system presents a list, it rearranges the list in the order of an estimated preference of each user.

3.1 Retrieval method of the on-line recipe retrieval system

In order to rearrange a retrieved dish list according to the preference of each user, this system analyses individual preferences of users with wine types as mediation variables.

At first, we classify wines into 44 types (16 red wine, 18 white wine, 6 rose wine, 4 sparkling wine) by their characteristics based on the book (Tazaki, 1998). For example, red wine could be classified by their taste in volume of body and astringency of tannins (Figure 1). Next, we define a mapping function $R(t)$ from dishes ($t \in T$) to wine types ($m \in M$) according to the compatibilities between dishes and wine types also based on the book (Tazaki, 1998). For example, red wine type A in Figure 1 matches with lightly roasted beef, spicy bean curd, sliced raw tuna, sautéed eggplant.

When an user select one from a retrieved dish list in order to refer a recipe of it, the system regards it as one of preferred dishes for him/her and add a point to a corresponding wine type in his/her profile. This process is repeated, and the more accessed wines get more points than the less accessed wines. In this way, the difference in points indicates the user preferences, and it could be estimated that the dishes corresponding to high-scored wine types are he/her preference. So, the system rearranges a list of retrieved dishes. For example, when a profile of an user had been scored as shown in Figure 2, it shows a type G gets the highest points. By this score, it could be estimated that the dishes matching a type G (e.g. sautéed beef with soy sauce, hamburger steak, and so on) are the most preferable for him/her, and the system could present these dishes at the top in the list of retrieved dishes.

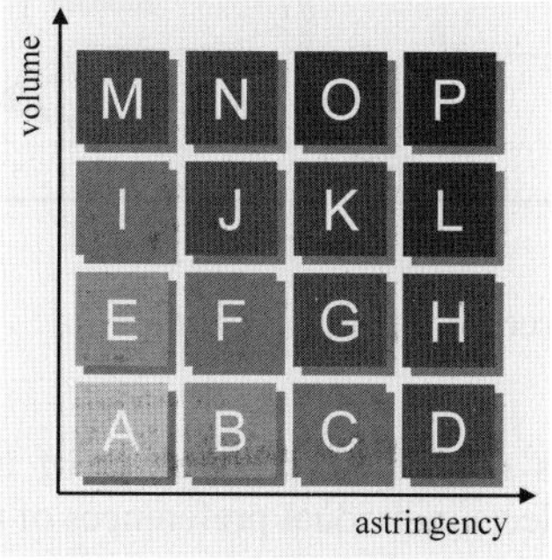

Figure 1: Classification of red wine types

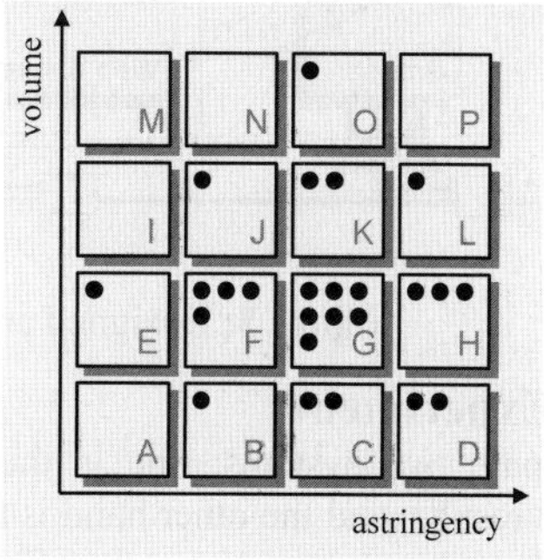

Figure 2: Scoring to red wine types

3.2 System Design and Implementation

The architecture of the system we have designed and implemented is shown in Figure 3. Regarding the system environment, it is based on an UNIX workstation with Solaris2.7, Apache 1.3.12, PostgreSQL 7.0.2 and PHP 3.0.15-i18n-ja.

This is a client-server system, where the server consists of a search engine and a database. The search engine is implemented on a web server and composed from user manager, taste analyser, and result evaluator. It provides facilities to retrieve recipes from database via DBMS, and also analyses individual user's preferences. When an user queries recipes by materials with this system, user manager retrieves a list of dishes from the database. Result evaluator rearranges the retrieved list, and user manager presents it to him/her. When he/she selects one from the list, user manager retrieves a recipe of it, and preference analyzer updates his/her preference by adding a point to a corresponding wine type.

4 Evaluation

In order to confirm the effectiveness of our proposed method, we conducted an experiment with the implemented system.

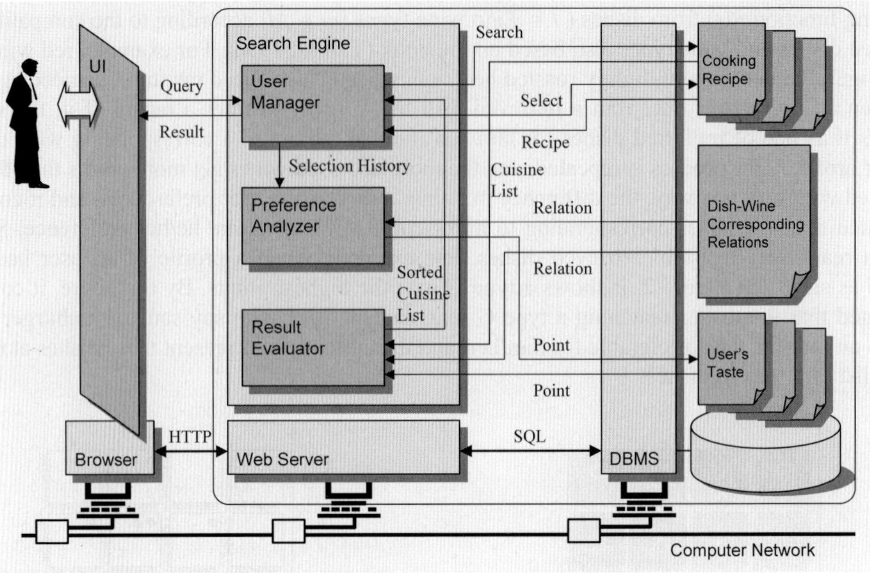

Figure 3: Designed architecture of on-line recipe retrieval system

4.1 Experiment

We prepare two systems; one is the system, which does not reflect preferences of users on retrieval results, and the other hand is the system, which reflects individual preferences of users on retrieval results, and compares the effects between them.

We divide subjects randomly into 2 groups A and B. Group A use the system without our method, and Group B use the system with our method. With these systems, testees prepare daily menus for 6 weeks, 2 menus par a day, total 84 menus. We record position p_{ui}, of which a subject u selects a dish in a retrieved list and the size of the list at ith selection for adding it to menus. And we calculate variance v_{uw} of relative position p_{ui} / n_{ui} in a wth week (14 menus) as follows.

v_{uw} = **Error!Error!Error!**

If the system with our method reflects individual preferences on retrieval results by repetition, it becomes that preferable dishes will be present in higher position in a retrieved list, and testees will select these dishes. As a result, the variance of each testee will get lower.

4.2 Result

As the subjects of our experiments, we have requested the help of 20 undergraduate students. For both groups A and B, we have plotted the variation of the variance values, which are summarized in Figure 4.

Although there are differences between the curves corresponding to each subject, the variance for Group B tends to decrease when compared to the curves of Group A. It means that the on-line recipe retrieval system with our method could reflect individual preferences of subjects, through repeated use of the system. Thus, our proposed method can help users find data, which best match his preferences.

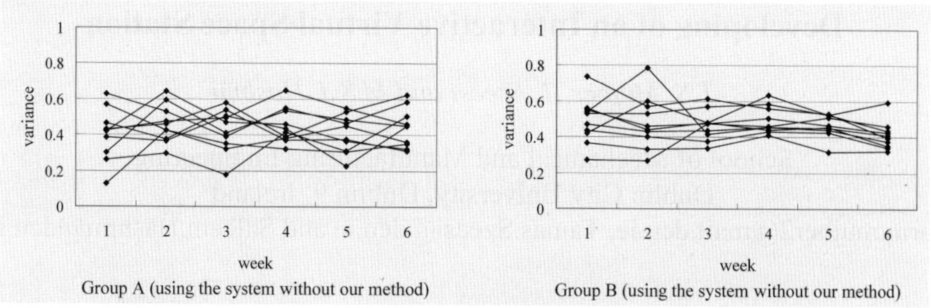

Figure 4: Variation of variance of selected dishes in retrieved dish lists

5 Conclusion

We proposed a new retrieval method, and designed and implemented on-line recipe retrieval system based on the method. This method estimates individual preferences of users by mediation variables, and reflects estimated preferences on retrieval result. We also performed an experiment with the implemented system, and confirmed that it could reflect individual preferences on retrieval results with our method and facilitate users to retrieve preferable data.

References

Fukuda, M., Sugita, K. and Shibata, Y.(1998). Perceptional Retrieving Method for Distributed Design Image Database System, *Trans. IPSJ*, Vol.39(2), pp.158-169.

Harada, S., Itoh, Y. and Nakatani, H.(1999). On Constructing Shape Feature Space for Interpreting Subjective Expression, *Trans. IPSJ*, Vol.40(5), pp.2356-2366. (in Japanese).

Ishihara, S., Ishihara, K. and Nagamachi, M.(1999). Analysis of Individual Differences in Kansei Evaluation Data Based on Cluster Analysis, *KANSEI Engineering International*, Vol.1(1), pp.49-58.

Kazama, K., Sato, S., Shimizu, S. and Kambayashi, T.(1999). HTML Document Correlation Analysis by User's Behavior in World Wide Web Navigation, *Trans. IPSJ*, Vol.40(5), pp.2450-2459. (in Japanese).

Kuwata, Y. and Yatsu, M.(1999). An Automated Follow-up Service for Technical Support Help Desk, *Trans. IPSJ*, Vol.40(11), pp.3896-3905. (in Japanese).

Mitsuishi, T. (2002). A Database Retrieval Method which Reflects Individual Preferences by Taste Analysis with Mediation Variables, *Proc. of Database and Web System Symposium DBWeb2002 (IPSJ Symposium Series)*, Vol.2002(19), pp.291-298. (in Japanese).

Mitsuishi, T., Tada, Y., Sasaki, J. and Funyu, Y.(2001). A Proposal of A Retrieval Method Reflecting Individual Sensibility by Taste Analysis with Mediation Variables, *Poster Sessions: Abridged Proc. of 9th HCII*, pp.151-153.

Okada, R., Lee, E.-S., Kinoshita, T. and Shiratori, N.(1998). A Method for Personalized Web Searching with Hierarchical Document Clustering, *Trans. IPSJ*, Vol.39(4), pp.867-877.

Tazaki, S.(1998). *Shin'ya Tazaki's Everyday Drinkable Wine Selection*, Shin-sei Shuppan-Sha. (in Japanese).

Yoshida, N., Kiyoki, Y. and Kitagawa, T.(1998). An Implementation Method of a Media Information Retrieval System with Semantic Associative Search Function, *Trans. IPS Japan*, Vol.39, No.4, pp.911-922. (in Japanese).

Developing of an Interactive Virtual Space Station

T.S. Mujber, T. Szecsi and M.S.J. Hashmi

School of Mechanical and Manufacturing Engineering
Dublin City University, Dublin 9, Ireland
tariq.mujber2@mail.dcu.ie, Tamas.Szecsi@dcu.ie and Saleem.Hashmi@dcu.ie

Abstract

Virtual Reality allows a user to step through the computer screen into a 3-Dimensional (3D) world. The user can look at, move around, and interact with these worlds as if they were real. This paper is devoted to the development and implementation of a virtual space station by using the Superscape VRT 5.6 package. The virtual world was developed to be interactive where the user can navigate and interact with the object by using a mouse, keyboard and joystick. Some intelligent features have been added to some objects to enhance the interaction and the navigation within the virtual space station. A program written by using Visual Basic 6.0 was developed and interfaced to a program written by using Superscape Control Language (SCL) that enables the user to interact and navigate the virtual world by using voice commands as well as external switches. A circuit was implemented to work as an interface between the external switches and the parallel port of the PC. Head Mounted Display (HMD) has been used as an immersive output device to give a sense of presence in the virtual world. The Head Mounted Display is coupled with a tracker that enables the user to navigate around within the world by moving the user's head.

1 Introduction

Virtual Environments are made up of 3-D graphical images that are generated with the intention of interaction between the user and the objects in that environment. The primary concept behind VR is that of illusion. It focuses on the manifestation of the fantasy world of the mind in computer graphics. High technology is used to convince the user that they are in another reality, experiencing some event that does not physically exist in the world in front of them. It is also a new media for information and knowledge acquisition, and representing concepts of ideas in ways not previously possible. [1] This section discusses designing and creating of an imagination of virtual space station, where the user can travel virtually by spacecraft (see figure 1) to a virtual building in the space. The virtual building consists of reception, lift and two offices, which are located on the 7th and 9th floors. The user can visit these offices by using a virtual lift, which is located at the reception of the virtual building. The reception and the offices are enhanced with different kind of furniture, which include chairs, tables, TVs, and some other furniture. The software that has been used to develop the virtual space station is called Superscape VRT 5.6. Superscape VRT is a complete 3D authoring studio for personal computers that lets users create interactive 3D worlds that can be published on the Internet using Superscape's Viscape, or displayed on the standalone Visualiser platform. Superscape VRT consists of an integrated suite of editors, which are used to create worlds, with two browsers, Visualiser and Viscape, which are used to view the worlds. Textures and sounds can be added to objects in the worlds to make them more realistic, and different lighting setups can also be introduced. Using SCL (Superscape

Control Language), a control language based on the popular "C" language, behaviour to objects in the world can be assigned, and complex actions can also be performed.

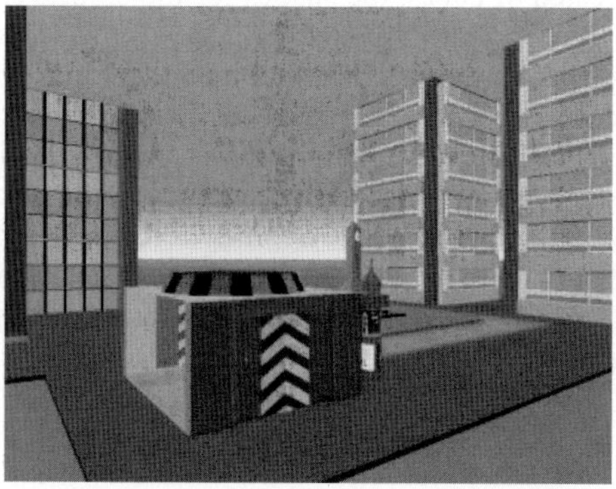

Figure 1: Spacecraft on the Ground

2 Designing and Creating the Virtual Space Station

The steps that have been taken to design and create the virtual space station include: planning the world, building the world, creating shapes, adding attributes to objects and optimising the world.

2.1 Planning the World

Some consideration was taken before the virtual world was built. VRT worlds can be targeted towards either Visualizer for standalone application that can be distributed on CD, or Viscape for 3D Web Pages. Visualizer is used for viewing large complex worlds with many levels of interactions. Viscape worlds need to be as small as possible to minimize their download time. At this stage, all the objects that are going to be included in the world have been decided, and the way of how the user is going to interact with the world. For example, it was decided to build an office with a main door, which can be opened by clicking on it, and to provide the office with some furniture such as a table, chairs, pictures on the wall and some other furniture. A list of objects is created that needs to be included in the world. Creating a list of objects is very useful to check if there are any missing items, which may cause a significant change to the sorting later. For example, the list that has been created for the reception of the virtual building includes: receptionist, table, computer monitor, sofa, TV, telephone, clock, cabinet, plants, wallpapers and main door.

2.2 Building the World

The World Editor of Superscape VRT is used to develop the virtual world by defining the space that each object in the world will occupy. Creating an enclosed environment, such as the office, the first object created is a holding "group" that reflects its size. Series of cubes and groups were created that represent the objects that are going to be built. Some objects were added from the virtual Clip Art libraries. Defining a series of orthogonal cubes at this stage lets us see if there are likely to be any basic sorting problems. This stage is considered as sketching using building blocks where objects do not have to be exact at this stage.

2.3 Creating Shapes

Once the world has been created of cubes, the shapes for the objects can be constructed by using the Shape Editor. The size of each shape's bounding cube can be set according to the size of the object that will use it. A shape consists of the combined facet and point information contained inside a bounding cube. Facets are surfaces created by connecting two or more points together. The smallest component in the VRT world is a point, simply defined as a point in three-dimensional space. VRT uses two kinds of points, which are, Relative and Geometric. Relative is a point, which is given an X, Y, and Z position. Geometric is a point, which is defined in terms of where it is proportionality on a line between two other points. Geometric points were used as possible to create the shapes of the world. They are generally easier to position and are processed more quickly than relative points.

2.4 Adding Attributes to Objects

After that the objects of the virtual space station were created and their shapes were added. Some attributes have been applied to the objects, such as dynamics, colour and some other attributes. For example, the dynamic attribute was given to the spacecraft to make it fly from the ground to space.

2.4.1 Texture and Sounds

Some textures and sounds were added to some objects to enhance the virtual world and to give a better feeling of presence. For example, clock sound was attached to the clock in the virtual world to indicate that the clock is working, as some other sound has been attached to the doors to give a feeling of opening and closing the doors.

2.4.2 Behaviour

Once the objects are created and their attributes are applied, behaviours were assigned by using Superscape Control Language (SCL). SCL is a purpose-built behavioural control language for interactive 3D worlds; it is used to add "intelligence" to objects. For example, the command "ydrive ('Spacecraft') = 50000;" was attached to the spacecraft that gives the Spacecraft a driving force of 50000 units per frame, which makes the Spacecraft flying up.

2.5 Optimising the World

The final step in building the virtual space station is to optimise it so that it runs as fast and smoothly as possible. The speed of animation of the world depends on many items including the number of facets that VRT has to process, the speed of the target processor, and the platform on which is intended to display the world. Some steps were taken while developing the virtual world to optimise the world, which are:

- The objects are grouped together efficiently (in groups of 10 to 15) to make sure that VRT is processing the world in an optimum way.
- The unused shapes, textures and sounds were removed to reduce the file size.
- Timing dialog box in the Visualizer was used to show the elements of the world that take a long time to process and to try to reduce it.

- Adding distance attributes to some objects (replacing a detailed object with a simpler object as it moves further away from the viewpoint) to reduce the processing time, which increases the speed of the world.

3 Navigation of the Virtual Space Station

The viewpoint was attached to a movable object, which made of a holding group object that has dynamics attribute, and a child object that has a rotation attribute. The holding group represents the body of human, and the child object represents its head. The user can navigate the world by changing the position of the object by using a mouse, keyboard, joystick and voice commands (see figure 2). The viewpoint was attached to the object to give a feeling to the user while navigating the world as a walking person "viewing the environment from about 1.6 meters above the ground". The input devices enable the user to interact and navigate the virtual space station by travelling from the earth to the virtual building in the space by the spacecraft (see figure 1). The user then can use the lift to visit the offices on the 7th and 9th floors.

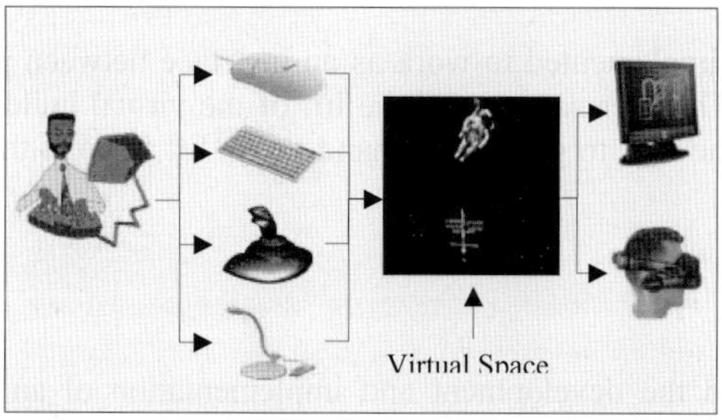

Figure 2: Navigation of the virtual space station

The following figures (3 & 4) show snapshots of virtual offices for the head of school of Mechanical & Manufacturing Engineering and the president of DCU. A Monitor and a Head Mounted Display were used as output devices to view the virtual world. The HMD is coupled with a tracker, which enables the user to navigate around the world by moving his/her head. It can be noticed from the figures below that one of the walls of both of the offices was built to be transparent to show the space view where there is a astronaut diving in the space and two satellites have different kind of movements.

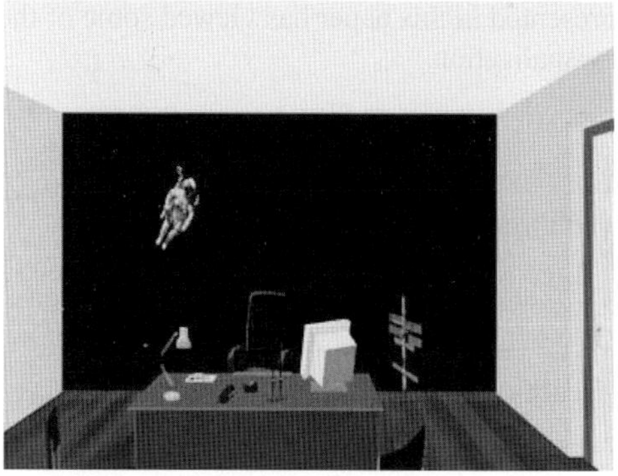

Figure 3: Egocentric View of the Head of Mech. & Manufacturing Engineering Office.

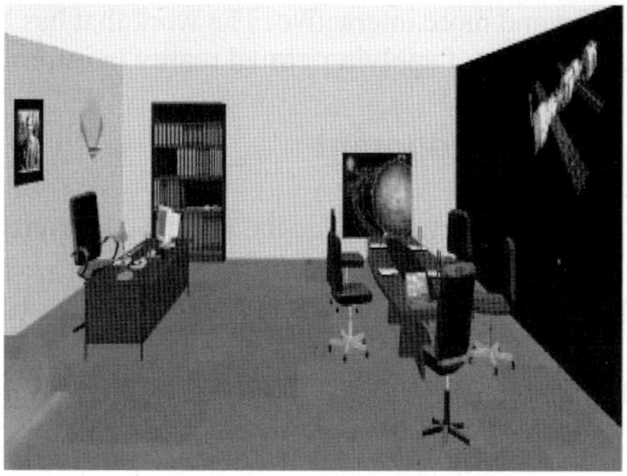

Figure 4: Egocentric View of the President of DCU office.

3.1 Interacting with the World by Voice Commands

A program written in Visual basic 6.0 was developed and interfaced with a program written using Superscape Control Language (SCL) to enable the user to interact with some objects in the virtual world by giving voice commands. For example, when the user gives the voice command *"open spacecraft door"*, the spacecraft's door will be opened, as the voice command, *"fly"* makes the spacecraft fly up, and the voice command *"land"* makes the spacecraft land down. The speech recognition engine that has been used is downloaded from Microsoft Agent Downloads. The speech recognition engine needs to be trained to speech pattern of the users to get better results. Different users (males/females) have tested the voice commands. It was found that the results were much better for the users who have trained their voices than the users who did not. Pronouncing the words correctly and talking slowly gave better response because the speech recognition engine compares what was heard with the phrases in the grammar file.

3.2 Controlling the Lift by External Switches

An interface circuit was implemented to work as an interface between physical switches and the parallel port of the PC. The user can control the lift of the virtual building in the space by three switches, which enable the user to give a command to the lift to go to the 7th floor, 9th floor and the lobby of the building.

4 Conclusions

This paper has discussed the development and implementation of an interactive virtual space station. The virtual space station was created to be interactive, which enables the user to navigate and interact with objects in the virtual world by different ways, which include a mouse, keyboard, joystick, voice commands and external switches that have been connected to the PC by implementing an interface circuit. Head Mounted Display (HMD) was used as an immersive output device to view the virtual world in 3D mode and to give the user the impression of being in the virtual world with the ability to do some navigation by moving the user's head. The virtual space station was developed and implemented to show some of the capabilities of using virtual reality in fantasy that gives the illusion of being in the real world. The main feature of the virtual space station is that it has been developed to be user-friendly to enable the user to interact and navigate the virtual world with more flexibility to make the navigation of the virtual space station easier and more interactive. The work that has been presented in this paper has viewed some of the capabilities of applying virtual reality in information visualization.

Acknowledgment

The authors acknowledge the financial support received through a grant from AMT Ireland.

References

K. Pimentek and K. Teixeira. (1993). Virtual Reality: Through the new looking glass. Windcrest/McGraw-Hill Inc: PA, first edition.

Superscape Inc. (1998). Superscape VRT 5.6 User's Guide. U.K.: Superscape VR plc, third edition.

A Tsunami Hazard Mitigation System from the Viewpoint of Human Interface

Yoshio Nakatani

Public-Use e-Solution Center, Mitsubishi Electric Corporation
2-2, Dojima 2, Kita-ku, 530-8206 Osaka, Japan
E-mail: yoshio.nakatani@melco.co.jp

Abstract

Reducing tsunami hazard is a very challenging problem. This paper discusses requirements and functions of the tsunami hazard mitigation system from the viewpoint of Human Interface. In recent years, these systems increasingly draw attention in relation to the predicted big earthquakes named Tonankai and Nankai in Japan. The characteristics and problems of the systems are discussed by classifying them into the tsunami prediction, the tsunami warning, the tsunami countermeasures, the evacuation preparation, the total information management, the disaster area information and the restoration information systems.

1 Earthquakes and Tsunami Hazard

This paper analyzes the existing tsunami hazard mitigation systems and proposes the near future development themes. Recently in Japan, these systems increasingly draw attention in relation to the predicted big earthquakes named Tonankai (southeast sea) and Nankai (southern sea) earthquakes [1]. Japanese seismologists conclude that there are 50 to 80% probabilities of these earthquakes with magnitude 8.4 or greater, capable of causing widespread damage, striking the western pacific coasts of Japan, in around 2035. Historically earthquakes have occurred eight times in Nankai area and their intervals range from 100 to 150 years. These earthquakes will be the first time for the modern megalopolises to be stricken by the big earthquakes followed by the gigantic tsunamis.

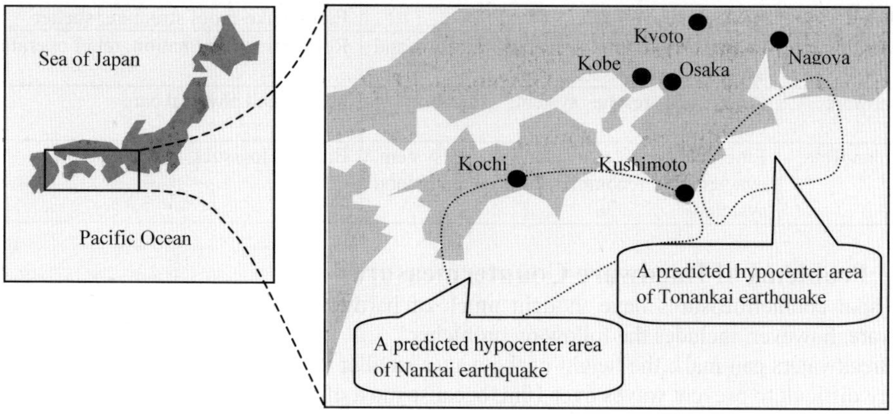

Figure 1: Predicted hypocenters of Tonankai and Nankai Earthquakes

Japanese government strongly promotes the tsunami hazard mitigation program, aiming at its complete equipment in about 20 years, including such systems as the seismological observation systems, the tsunami and tidal wave hazard mitigation stations, and the wide-area hazard information networks.

Tsunami hazard has the characteristics as follows:

(1) A tsunami is a large water wave, generally caused by a magnitude 6.0 or greater undersea earthquake whose hypocenter is shallower than 120km beneath the sea.
(2) One earthquake causes five or six tsunami waves, whose intervals are 50 to 60 minutes. The first wave of the Nankai earthquake will strike Kushimoto in 15 minutes, Kochi in 30 minutes, and Osaka in two hours. The first tsunami is not always biggest.
(3) A propagation speed of a tsunami is decided only by depth on its route. When the depth of water is 4,000m, the speed is 720km/h. It keeps a speed of 36km/h even near the coast. People cannot run away from the tsunami flood. The best countermeasure is to evacuate at once. It is, however, difficult for the people to evacuate just after stricken by an earthquake.
(4) As the speed of tsunami goes slower, its energy concentrates and the wave goes higher. Empirically, the tsunami goes up to the altitude of its double height. The heights of tsunamis in Nankai earthquake are predicted to be 6m or more, and waves may go up to 12m above sea level. The plain of Osaka is wide and flat, and a wide area may be sunk under the water.
(5) The tsunami goes up along the rivers, and climbs over banks.
(6) The tsunami has historically caused significant damage to coastal communities. The Nankai earthquake is predicted to attack many big cities, including Osaka, Kobe, and Wakayama, whose populations sums up to 7 millions.
(7) Small ships may be pushed up to the shores, and causes house collapse and explosions. Even big oil tankers may be broken by clashing into the wharf, which may cause severe oil hazard.

2 Tsunami Hazard Mitigation Systems

To reduce disaster hazard, the following four viewpoints are considered in disaster management: hardware, software, humanware and commandware (table 1)[2].

Table 1: Emergency Management in Disasters

	Emergency Management (Crisis Management)	
	Risk Management (before disaster)	Crisis Management (after disaster)
hardware	Redundancy, failsafe	Earthquake-proof structure, shelter
software	Disaster information, evacuation drills and excercises, antidisaster plan	Restoration information, relief operation
humanware	Relief and rescue system, education of volunteers	Mental and physical care
commandware	Emergency management system, evacuation recommendation, command system	Backup, logistics, restoration plan

2.1 Problems of Hardware Countermeasures

The usual countermeasures have thought much of hardware equipment. The current status of hardware, however, includes the following problems.

(1) Breakwaters can make the height of the waves smaller by 20%. Some breakwaters, however, are difficult to prevent waves over 10m, because most of them are designed for tidal waves.
(2) Most floodgates are equipped by the government, while cities and companies are entrusted to manage middle and small floodgates. There are many floodgates in a port, most of which are

outdated and manually controlled. It is not easy to close all of the floodgates in a short time. For example, the Kochi port has more than 300 floodgates, which have to be closed in 30 minutes after the earthquake origin time. Kochi city has once tried to close them in a typhoon of 2001 to find that it took four hours. The remotely controlled floodgates are increasing, but they cost too much.

(3) The floodgates may not work after earthquakes. There are possibilities that the operators cannot close the gates because they are injured or damaged by the earthquake. The remote controlled floodgates may not be closed due to breaking down of the control network. The radio network is being introduced, but its transfer speed is not enough for sending the real-time camera images to confirm the closing of the floodgates.

Only the hardware countermeasures are not sufficient. Recently rapid evacuation is a most effective countermeasure. To succeed rapid evacuation, software, humanware and commandware are essential.

2.2 Subsystems

The tsunami hazard mitigation system seeks to reduce tsunami hazards, and includes the following seven subsystems.

[Risk management (before crisis)]
- Tsunami prediction system --- predicts the time, the seismic centre and the seismic intensity.
- Tsunami warning system --- predicts the arrival time and the damage.
- Tsunami countermeasures system --- requires the organization to shift to the caution system, to operate the tsunami hazard mitigation stations (to close the floodgates and to announce the tsunami alarm to the residents), and to patrol the cautious areas and facilities.
- Evacuation preparation system --- prepares and announces hazard maps, carries out drills, and educates the residents.

[Crisis Management (after crisis)]
- Total disaster information system --- manages information on disaster and damages, and makes decisions.
- Stricken area information system --- manages information on stricken areas and helps to make a restoration plan.
- Restoration information system --- manages information on progresses of restoration and helps to reexamine a restoration plan.

These systems are used in certain periods of tsunami hazards as shown in Figure 2.

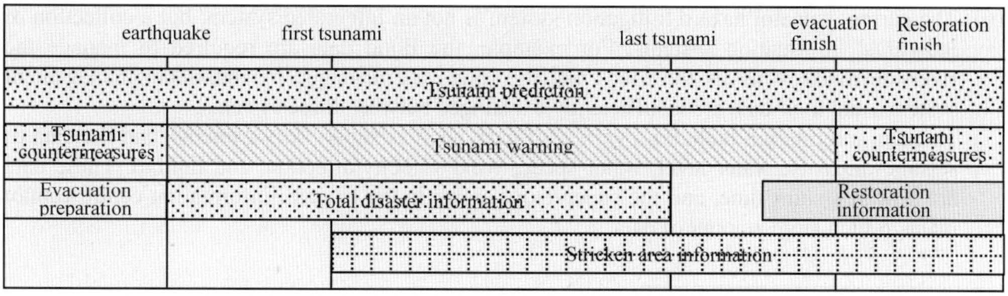

Figure 2: Information Systems in a Tsunami Life Cycle

3 Tsunami Hazard Mitigation Systems

The tsunami hazard mitigation system has several problems from the viewpoint of human interface.

(1) The total hazard mitigation program has to consider the precede earthquake hazard as well as the tsunami hazard. When the tsunami caution is announced, many people want to go back to their homes by train in Japan. Six major railways gather in the centre of Osaka. The trains, however, have to stop when an earthquake occurs. More than one million people are forced to stay in dangerous areas. It is quite possible that buildings are destroyed, roads sink, rivers flood, and many fires break out. The current simulation covers an individual hazard, not multiple hazards. We have to consider the behavior of people when they suffer from earthquakes, tsunamis, and fires.

(2) People forced to stay in the centre of the megalopolis may phone their home by the mobile phones, and the telephone services will be severely overloaded. The Nankai earthquake is the first experience of the big disaster in the mobile phone age. The internet may be out of use because of the excess of access[1]. It is very important to provide them and their families with real-time disaster information (of course in Japanese and other languages). The Japanese government has their own high speed optical fibre networks along the National Roads. By using the network, the disaster information can be provided on the street information boards. To this end, the information systems of the road, the river information, and the port have to be integrated among the government, the prefectures, and the cities.

(3) In Osaka, the big underground shopping centers, linked to the subway stations, are located riverside. The number of passengers is more than one million per day. They may be struck by tsunami flooding after the damage by the earthquake. After the river flooding, they will start to be sunk in about 10 minutes. The tsunami may occur in a storm, and the water is a great menace. We have to know how many people are underground and plan to rescue them in a short time. Some underground shopping centres plan to guide people to the neighbouring earthquake-proof tall buildings. These buildings must not be informationally isolated.

(4) The current tsunami forecast system predicts the arrival time of only the first wave. The first wave is not always biggest. Furthermore, the predicted wave height is the averaged height of the near-coast, not helpful for people assess the risk of their community. So the real-time information from the neighboring areas is can be a great help. For example, tsunami data is required to be automatically transferred from the struck areas to the to-be-struck areas.

(5) If the floodgates are difficult to be closed manually in a short time, the remote closing-operation system using the radio communication technology is required. It is necessary to supervise the closing operation for safety and confirmation. To this end, the high speed radio communication network is required to transfer the live camera images.

(6) The current tsunami hazard mitigation system is not an integrated system, but a collection of individual information systems. For example, the flood data are required to manage the riverside roads, whereas the roads and the rivers are managed by individual management systems, and their data are insufficiently exchanged between these systems.

(7) Tsunami information includes a variety of data, such as supervisory cameras, and the data of seismic intensity, water level, water speed, wind velocity/direction, and rainfall. These data are sent in a short time, and the supervisory crews in the tsunami information centre cannot manage such large amount of data.

---

[1] They say that the homepages provided by the stricken local governments are mainly accessed by the outsiders of the stricken areas, to confirm the safety of their relatives and friends.

(8) In order to solve the problems described above, we have developed the advanced technologies for the disaster supervisory centre, including the following systems:
   - The large-scale multi-screen system that can display the multi camera images, tables, graphs, and web pages (see Figure 2) on the GIS (Geographic Information System), in order to summarize and share the ongoing situation among supervisory crews.
   - The GIS-linked video image retrieve system that collects videos of the cities along with the GPS data, automatically links them to the GIS by using the GPS data, and plays the video images on the GIS (see Figure 3), in order to confirm the situation of the struck area.
   - The tsunami hazard reduction station that remotely supervise and control floodgates, with the image processing subsystem that automatically detects intruders to the floodgates in closing, the floodgate action monitoring subsystem, and so on.

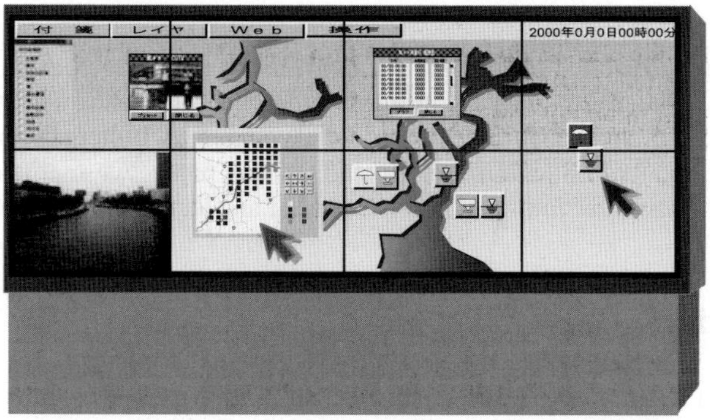

Figure 2: An Example of the large-scale display system

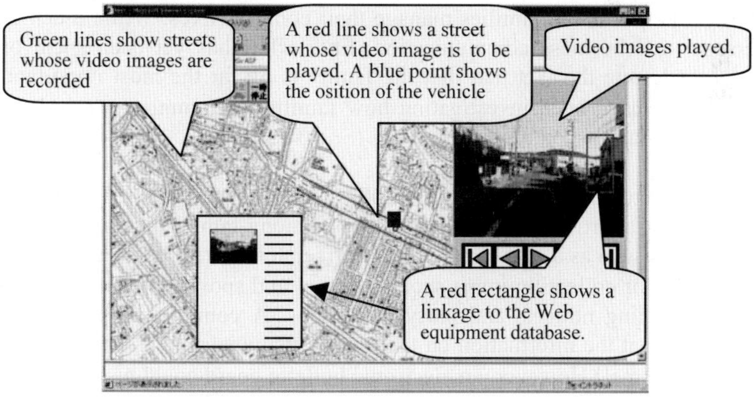

Figure 3: A screen image of the GIS-linked video image retrieve system

References

[1] Y. Kawata: Disaster mitigation due to next nankai earthquake tsunamis occurring in around 2035, Proc. Intl. Tsunami sympo. 2001, 315-329, 2001.
[2] Y. Kawata et. Al.: Urban Tsunami Disaster Mitigation and Humanware Management, Proc. Intl. Conf. Urban Engineering in Asian Cities in the 21st Century, Vol.II, H80-H85, 1996.

Investigating Intra-Family Communication Using Photo Diaries

Hans Nässla

Institutionen för Datavetenskap
Linköping University
SE-581 83 Linköping, Sweden
hanna@ida.liu.se

David A. Carr

Institutionen för Systemteknik
Luleå University of Technology
SE-971 87 Luleå, Sweden
david@sm.luth.se

Abstract

Much effort has been expended on developing applications to manage business and personal time. However, very little has been spent on information technology to manage the modern family's busy, complex life. We have performed a photo-diary of family bulletin boards in order to understand how this common coordination tool is used. In this paper, we report on the results of the photo diaries and discuss their implication for design of an electronic family bulletin board.

1 Introduction

Today's family can be complex; it can consist of children, parents, ex-spouses, and grandparents. Usually, both parents work outside the home. Together, they must supervise children's schools and activities, homework, and cooking. In addition, they must maintain their own personal networks. This combination of activities makes time a scarce resource and suggests that information technology might help modern families manage their complex lives. While much effort has been put into personal and business communication, little effort has yet been spent on intra-family communication – despite the fact that that the family is by far the most important personal network. Therefore, we have been investigating how families communicate in order to design systems to support intra-family communication.

Our early interview studies (Nässla & Carr, 2000; Nässla, 2001) pointed to a bulletin board as being one of the central artifacts for communication among family members. While it may have taken different forms such as a corkboard or the refrigerator door, every family in these interviews used a central place for collecting items such as school and sports schedules, Parent-Teacher Association (PTA) meeting notices, and party invitations. A common problem with these family bulletin boards was that the information on them was often unavailable when needed. (The family member was not at home!) However, an electronic bulletin board could be available from work via the Internet. With the carefully designed software, it could even be available from a mobile phone. But, this design needs to be based how families use their bulletin boards.

Ethnographic field studies may seem to be ideal for investigating how a family manages its information flow, and how it uses a bulletin board. However, even a quick-and-dirty ethnography study tends to include several days of continuous fieldwork, which in another's home is likely to be undesirable if not unacceptable

Methods must be selected to develop an understanding of the phenomenon under investigation in its own terms (O'Brien & Rodden, 1997). The use of diaries for self-documentation may be seen as a tool for research in the home that achieves a relatively high standard of objectivity (Rieman, 1993). Personal interactions are a key part of a successful study, and the participants must be convinced to make a considerable effort. (Indeed, the father in the first family had just purchased a digital camera and was very interested in using it in this study.) It is very imp ortant to let the participants use their own ways to document their lives, when using this kind of self-documentation (Ellegård & Nordell, 1997). The use of photo diaries would allow a longer field study of the family, instead of just occasional visits to their home, while simultaneously providing a fairly non-intrusive way of registering what happens.

In order to understand the type of data on the bulletin board and how it changed over time, we collected detailed information from three families. For two of these families, we were also able to follow changes for 3-4 weeks by collecting a photo diary of their bulletin boards. The families took still photos of their bulletin board whenever new notes appeared or disappeared.

2 Investigating Family Bulletin Boards

2.1 Procedure

We have chosen to use diary studies to investigate the families' use of their bulletin boards. To facilitate this, the families were given a disposable camera, and instructed to take pictures of their bulletin board as soon as something happened on it, whether new notes came up on it or old notes were taken away. The diaries were kept for 3-4 weeks.

After collecting all the photographs and information about the shown notes, we analyzed the photos with regard to what kind of notes were there and in what way the notes were part of the family life. After this analysis, we went back to the families to discuss our findings:

- By whom, and in what way were the notes handled?
- Did the placement in the home affect the use of the bulletin board?
- Were there different areas on the bulletin board used for different kinds of notes?

2.2 The Families

The families, themselves, all consisted of two parents with one or two children. They could be considered “information technology mature” in that they owned at least one home computer with an Internet connection. Table 1 summarizes the families with respect to common aspects. In addition to their bulletin board, families one and two used a family calendar for long-term planning.

Table 1: The families and theirs activities

Aspect \ Family	**1**	**2**	**3**
Children (age)	9	6, 12	12
PCs / Internet access	3 / modem	2 / modem	1 / broad band
First PC (years ago)	20	10	?
Activities/associations	Condo association	Politics, sports	Sports
Bulletin board type	Corkboard	Refrigerator door	Paper clip of notes

2.3 The Bulletin Boards

The bulletin boards were fairly dynamic although the number of notes did not vary greatly. For example, family one began and ended the diary with 13 notes on the bulletin board. However, six new notes appeared during the diary period, and six notes disappeared. In addition, one of the notes disappeared and re-appeared during the three-week study period. The second family had 12 notes when their diary began and 21 when their diary ended. In total, 12 new notes appeared, three notes disappeared, and seven notes went up and down again during the four weeks. The third family had 10 notes collected at the instance we examined them. (See Table 2.)

Table 2: The bulletin boards

Aspect \ Family	**1**	**2**	**3**
Place	Entrance	Refrigerator	Entrance
Notes in beginning	13	12	10
New notes	6	12	-
Notes that disappeared	6	3	-
Notes that appeared and disappeared	1	7	-

All of the notes have a story; some of the notes were temporarily on the bulletin board, while others stayed there for a longer period of time. To illustrate, the first family (Figure 1) had a school schedule (upper left corner), telephone numbers (top part), things to carry away from home (to the right, near the front door), and archival material (bottom part). New notes were placed in the center and moved later toward the edges. The limited area available resulted in notes hanging over the edges. The second family (Figure 2) had a school schedule (top center), recipes and articles (top left and right), school, sport and after-school items (lower left and right), and things to be handled by the parents (just below the school schedule). New notes were placed in the lower center and moved upward with age. The third family collected all notes, such as schedules, sport activities, library information, and lottery tickets in a paper clip at the entrance to their home. This family placed new notes on top.

Regarding the children's use of the bulletin board, we noticed that:

1. The 6-year old son in Family 2 had a note on the refrigerator door with his first written

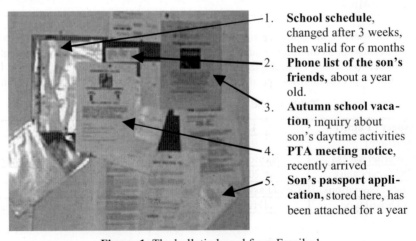

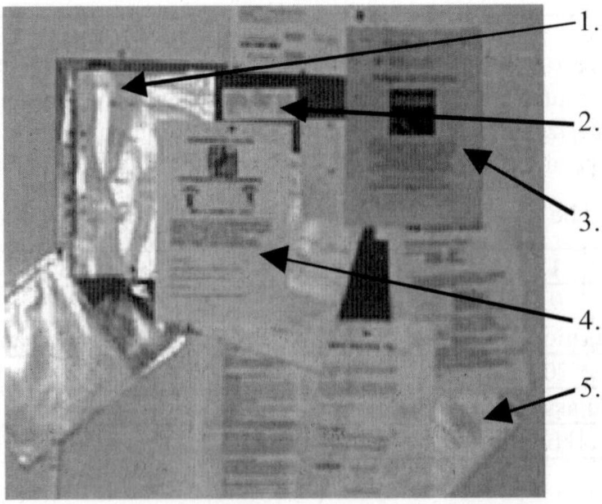

1. **School schedule**, changed after 3 weeks, then valid for 6 months
2. **Phone list of the son's friends,** about a year old.
3. **Autumn school vacation**, inquiry about son's daytime activities
4. **PTA meeting notice**, recently arrived
5. **Son's passport application,** stored here, has been attached for a year

Figure 1: The bulletin board from Family 1.

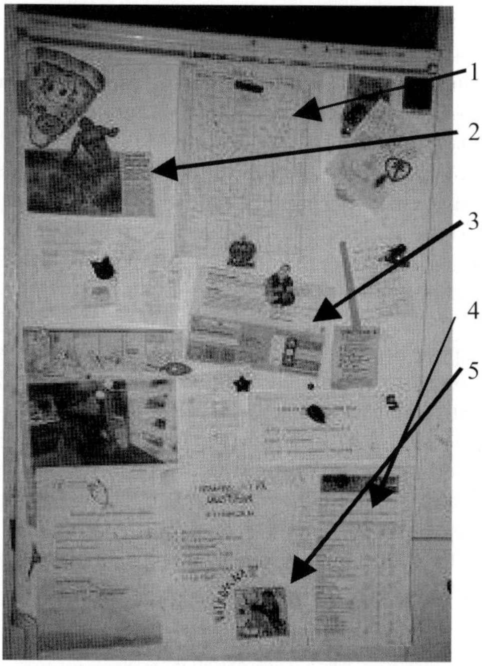
1. **School schedule**, lasts for a semester
2. **An article**, that ought to be read several times
3. **Auto safety inspection**, reminder to schedule a visit
4. **Sport center**, opening hours
5. **PTA meeting**, invitation that has recently arrived

Figure 2: The refrigerator door from Family 2.

sentence. He was not aware of it, but his mother was proud of it.

2. The 9-year old son in Family 1 was not tall enough to reach the school schedule
3. The 12-year old son in Family 2 used the school schedule everyday and also sometimes wrote notes on the refrigerator door
4. The son in Family 3 was also old enough to be responsible for his activities, so he read relevant notes himself, although his parents also reminded him.

The notes on the families' bulletin boards seemed to be of three distinctive characters: short term, long term, and archival. Short-term notes consisted of things like coupons or party invitations with a lifetime of a couple of weeks. Long-term notes would clearly be present for many months. They were things such as children's school schedules. Finally, there were notes such as the electric company's phone number for reporting power outages. This type of note would clearly remain indefinitely. Both families reported that they placed notes "anywhere there was room". However, the photo diaries revealed a clear pattern. The first family placed new notes in the center, and older notes migrated to the edges. The second family placed new notes on the bottom, and notes moved up as they aged. The third family placed all notes in a pile using a paper clip.

3 Conclusions

In this study, the use of three family bulletin boards has been observed by obtaining photo diaries. The diaries have been studied in order to obtain requirements for an electronic bulletin board. Although a limited number of bulletin boards were involved, some preliminary implications for design can be drawn.

- **Paper must be accommodated:** Although many of the notes may be distributed in electronic form in the future, paper will probably still be in use (e.g., rebate coupons and free tickets). That means there will be a need for places to attach paper notes on the electronic bulletin board, perhaps in the form of a corkboard.

- **Note browsing and management must be supplied:** Both families used more space than is likely to be available on a display, yet notes often overlapped and hung over the edges. Clearly, miniaturization, browsing, and organization functions will be needed. However, each family's own note placement method should be accommodated.
- **No special design for children must be created:** Although much of the material concerned the children in the families, it was clearly manipulated by the parents for children younger than about twelve. By this age, the children can read and can be expected to operate the same interface as their parents.

There are also a number of issues that require further investigation. While we are convinced that paper must be accommodated, the question of how remains. Should the bulletin board include a tangible interface and integrate both physical and electronic notes? Or, should physical notes be scanned and then manipulated only in electronic form? In order to make the added complication of a computerized bulletin board pay off, the bulletin board must be accessible from remote locations. How a remote interface is formed also requires research. An Internet based interface can mimic the home interface, but on a PDA or mobile phone the screen may be too small. Therefore, we need to investigate whether it is better to browse over a smaller graphical representation of the bulletin board, or if it is better to use a content-based interface such as a text-based list of notes.

In conclusion, we believe the information obtained in this study can form the basis for expanding the usefulness of family bulletin boards by enabling them to be remotely accessed.

4 Acknowledgments

We are grateful to the families for taking part in this study and allowing our intrusion into their daily life. We would also like to thank Kajsa Ellegård for valuable insights into diary studies and Henrik Artman for ideas at the study's beginning.

5 References

Ellegård, K. & Nordell, K. (1997) Att byta vanmakt mot egenmakt. Självreflektion och förändringsarbete i rehabiliteringsprocesser. (To change powerlessness to empowerment. Reflection and changes in rehabilitation processes. In Swedish) *Metodbok*. Johansson & Skyttemo, Stockholm.

Nässla, H. & Carr, D. (2000) The family connections. *Proceedings of i3 Annual Conference*, Jönköping, Sweden, 13-15 September 2000, 67-71.

Nässla, H. (2001) Grandpa never used to call grandma from work, but now the parents call each other and their kids everyday – family communication on the threshold to the IT society. *Proceedings of HFT 2001 (Human Factors in Telecommunication)*, Bergen, Norway, 5-7 November 2001, 293-296

O'Brien, J. & Rodden, T. (1997) Interactive systems in domestic environments. *Proceeding of DIS '97*, Amsterdam, Netherlands, 18-20 August 1997, 247-259.

Rieman, J. (1993) The diary study: a workplace-oriented research tool to guide laboratory efforts. *Proceedings of InterCHI '93*, Amsterdam, Netherlands, 24-29 April 1993, 321-326.

Teaching Teamwork Online

Lisa Neal
EDS
3 Valley Road
Lexington, MA 02421 USA
lisa@acm.org

Eileen Entin and Fuji Lai
Aptima, Inc.
12 Gill Street
Woburn, MA 01801 USA
ebe@aptima.com, fujilai@aptima.com

Abstract

We report on the development of an innovative program aimed at providing online teamwork training for distributed professionals performing complex, highly interdependent tasks.* Examples of teams for which this training would be appropriate include special purpose (e.g., crisis response) teams within command and control organizations and operations centers, special operations teams, and medical teams. To maximize learning and flexibility, the program couples synchronous and asynchronous approaches to online learning, uses video judiciously to illustrate teamwork in a specific domain, and uses team members as session leaders.

1 Introduction

Teamwork is an essential skill since, without it, even highly skilled professionals lose their effectiveness when they are unable to coordinate their tasks with others. It is challenging to teach teamwork skills well face-to-face, and team bonding is often emphasized over the development of specific teamwork skills. Teamwork training must cater to the demands of professional lives, including busy schedules and diverse locations. Our approach to teamwork training for distributed teams is to use a blended approach to learning, combining synchronous and asynchronous sessions, where synchronous sessions reinforce asynchronous ones; to provide flexible learning that fits participants' schedules; to use video to illustrate good and poor teamwork in a domain; and to use team members as rotating leaders to minimize the demand on experts, increase accountability, and reinforce learning.

We developed the program to provide distributed teamwork training for teams performing complex, highly interdependent tasks. The objective of the program is to provide training in teamwork skills such as communication, coordination, monitoring, and leadership. The underlying assumption is that the team members are already skilled in their own technical or professional area of expertise. In this paper we describe the program we developed and discuss some of the issues and challenges we faced as we developed the program.

* This research is a collaboration between University of Maryland Medical School, Aptima, and EDS, and is supported by the United States Air Force under Contract No. F33615-02-Mp6008 under the supervision of Dr. Peter Crane.

2 The Distributed Teamwork Training Program

To address the need for distributed teamwork training, we developed a web-enabled, scenario-based teamwork skills training program comprised of: (1) information about and examples of teamwork skills; (2) training materials, including video, that are highly relevant to the domain, since the goal of the training is to provide immediately usable training, as opposed to general concepts (3) scenario-based training exercises that provide practice in teamwork skills; (4) guidelines for team-conducted exercise debriefings that do not require the presence of a training instructor; and (5) a leader's manual that helps team leaders conduct web-based training sessions. The program uses web-based asynchronous and synchronous tools that provide capabilities including shared artifacts, audio and text communication, discussion and chat, polling, whiteboards, and archiving. The goal of the program is to have the trainees apply the skills they are learning in the domain in which they work and thereby attain a practical understanding of the teamwork skills, not just a theoretical understanding of teamwork concepts.

The program is comprised of the materials for a self-contained set of scenario-based training sessions, each of which is comprised of three components: (1) a structured pre-brief that explains the training goals for that session and sets forth the initial parameters for a web-based problem-solving scenario; (2) a threaded scenario that provides an initial problem for the team to work through and includes a set of events that refine, complicate, or accelerate the tempo of the teamwork processes; and (3) the structure and resource materials for a post-scenario briefing that uses the threaded events as a framework that enables the team members to analyse their teamwork and derive lessons learned and goals for improvement without the need for a real-time training instructor in the loop.

The initial version of the training program was designed for fellows and residents working in the trauma resuscitation unit of a major medical center. The training program requires eight to ten hours and can be completed within a two week time period. The training is comprised of four modules, and alternates between asynchronous and synchronous modules. Each of the synchronous modules is scheduled at a fixed time while the asynchronous modules are accomplished by trainees as their individual schedules permit, but within a fixed window of time. The synchronous modules allow students to ask questions or seek clarification and receive feedback from the leader on their understanding of concepts. The synchronous sessions are also valuable to build a sense of community and to reinforce what was learned in the asynchronous sessions (Neal, 1997).

The first module introduces and motivates the program and describes the structure of the training. It covers the initial presentation of the concept of teamwork and provides explanations and examples of teamwork skills in both medical and other domains. The module includes both positive and negative examples of teamwork in order to support better generalization and transfer of modelled skills (Baldwin, 1992). The second module motivates participation in the program by providing an opportunity for trainees to interact with one another, and it provides the first opportunity for participants to practice teamwork skills. The third module is comprised of a series of *three sequenced iterations* that provide further opportunity for trainees to apply and practice teamwork skills and to receive feedback from other team members. Each iteration focuses on one or two teamwork skills, and has two phased segments, a *practice phase* in which the trainees apply the concepts they are learning and a *feedback phase* in which they receive feedback on their performance from their peers and leader. In this asynchronous module, trainees go through the practice phase individually, but, in the feedback phase, see how their teammates responded to the exercise and enter their own reactions to the team's performance. The fourth module provides an opportunity for trainees to practice in a synchronous environment. It includes two practice segments focusing on a combination of teamwork skills.

3 Issues and Challenges

3.1 Requirements and Limitations of the Target Population

The initial population for whom this program was developed is emergency medical teams. These teams of highly skilled professionals, each of whom has his or her own specialty, must work together to resuscitate and treat a patient. The purpose of the program is to train them to apply their skills in a team setting. The learner population is well-educated, highly focused and motivated, accustomed to dry, but relevant, presentation of materials, and time-constrained. Due to these factors, presentation of learning materials must: use domain-specific materials that provide examples of, and show the consequences of, poor teamwork; pay little attention to the games and exercises increasingly common in online learning; and accommodate small windows in which to work on coursework; and minimal synchronous interaction due to the difficulty of scheduling it.

3.2 Developing a Blended Approach

A major challenge is to train distributed teams as effectively as co-located teams. By blending synchronous and asynchronous technologies, teamwork skills training takes advantage of the sense of community created by synchronous training, and at the same time affords the scheduling flexibility provided by asynchronous training. The synchronous sessions reinforce and motivate completion of the asynchronous sessions, and their archival means a session can be played back to accommodate scheduling conflicts or improve comprehension or presenter skills. In the asynchronous sessions, team members respond within a limited time period to questions that are posed in a problem situation. A designated leader integrates responses and reports back to the team.

3.3 Implementation and Effective Use of Streaming Video

The effective use of video to provide the problem scenarios is a cornerstone of our training program. Video as a data source to examine work practices has been used in several domains to examine performance of tasks (Mackenzie at al, 1994) and as a training tool (Townsend et al, 1993). Video clips provide rich material for targeted performance review and awareness exercises. The use of video enables setting up of scenarios that are realistic and detailed. In an online course, it is especially important that course materials are authentic and resonate with students in order to keep them engaged (Kearsley and Shneiderman, 1999).

Video is used as stimulus material to facilitate a discourse on teamwork among participants. When judiciously used, video is a valuable tool in training. However, the quality of the video is critical to its effectiveness. For example, if the audio and video are out of sync or there are transmission delays, the viewer may be more annoyed than benefited (Orton, undated). Information in videos that is even slightly out of sync is viewed as less trustworthy (Reeves and Nass, 1996), which would defeat the purpose of using it for specialized training. The use of video is most advantageous when it offers a perspective otherwise unobtainable by the remote viewer or when the interactions of people are captured on video. A major challenge is sharing and making the video clips visible to participants in the session using current synchronous technologies.

3.4 Providing Support and Guidance for the Session Leaders

An online instructor uses many of the same skills as a classroom teacher, but has the added chore of learning to use and manage the use of delivery technology, including how to compensate for lack of visual cues and how to communicate effectively (Neal, 97). The team leader, who is responsible for overseeing each session, receives training materials about how to use web-based technologies, how to conduct a synchronous session, and key points to be covered in feedback and discussion segments of training.

Smith-Jentsch et al. (1998) point out that some teams are not able to engage in effective team self-correction because team members do not know how to provide feedback in a constructive manner. Embedded throughout the program is training in positive and effective team self analysis and correction. Module 2 emphasizes constructive feedback and a team climate that fosters open communication and trust. Each iteration in Module 3 uses a different team member to supervise the training. The use of rotating *peer leaders* provides team members with leadership experience and further invests the team members in the training program.

4 Lessons learned

One difficulty in training teams whose members are geographically distributed is the cost and time required to bring them to a single location. A second is the limited availability of skilled instructors. Our program is a domain relevant, scenario-based *distributed* team-training program that *does not require the presence of a real-time instructor*, thereby surmounting both difficulties. It also addresses another criterion for *successful* teamwork training programs – they must be based in the domain of the trainees. The program meets the need to use domain-relevant materials by incorporating an efficient, streamlined method for conducting a training needs assessment and using this information to generate realistic, engaging scenarios in the target domain.

References

Baldwin, T. T. (1992). Effects of alternative modeling strategies on outcomes of interpersonal-skills training. *Journal of Applied Psychology*, *77*(2), 147-154.

Kearsley, G. and Shneiderman, B. (1999) Engagement Theory, http://home.sprynet.com/~gkearsley/engage.htm.

Mackenzie, C.F., Graig, G.R., Pan, M.J., Hoyt, R., and the Lotas Group (1994). Video analysis of two emergency tracheal intubations identifies flawed decision-making. *Anesthesiology*, 81:763-771.

Neal, L. (1997) Virtual Classrooms and Communities. Proceedings of GROUP '97 Conference, November 16-19, 1997, Phoenix, AZ.

Orton, P. (undated) *Streaming Video – Friend or Foe*, http://www.brandonhall.com/public/pdfs/streaming_video.pdf

Reeves, B. and Nass, C. (1999). *The Media Equation*, CSLI Publications.

Smith-Jentsch, K. A., Zeisig, R. L., Acton, B., and McPherson, J. A. (1998). Team dimensional training: A strategy for guided team self-correction. In J. A. Cannon-Bowers and E. Salas (Eds.), *Making Decisions Under Stress: Implications for Individual and Team Training*, Washington, DC: APA Press.

Townsend, R.N., Clark, R., Ramensofsky, M.L., Diamond, D.O. (1993). ATLS-based videotape trauma resuscitation review. Education and outcome, *J. Trauma*, 34:133-138.

Providing Access to Humour Manipulation for Individuals with Complex Communication Needs

David O'Mara
Annalu Waller
Applied Computing
University of Dundee, UK
domara@computing.dundee.ac.uk

John Todman

Psychology
University of Dundee, UK

Abstract

Children who use computer-based communication aids because of a speech disorder tend to be passive communicators. Taking part in a conversation requires knowledge of the rules of engagement. Effective communicators need to practice their skills. Computer-facilitated humour provides an opportunity for children who use communication aids to play with language and scaffolds developing language skills. High-tech communication aids can provide a platform for the storage, presentation and facilitation of verbally-expressed humour. The unique design challenge of providing interfaces for children with complex communication and physical needs to access and generate humour is presented.

1 Introduction

Physical disabilities and motor co-ordination problems can affect the quality of speech production. The opportunity to use an alternative or augmentative form of communication (AAC) can benefit those individuals (children and adults) who find most difficulty in being understood. This may range from an occasional need to set a frame of reference during conversation through to a reliance on the communication method as a substitute for speech. Computer-based communication aids provide non-speaking individuals with access to the spoken word. However, these individuals tend to be passive communicators who respond to closed questions while seldom initiating conversation (Waller & O'Mara, 2002). Some children who use communication aids may have additional learning difficulties that can impact upon their development of underlying language skills.

Each year, manufacturers of high-tech voice output communication aids (VOCAs) produce smaller and faster equipment with an ever increasing memory and enhanced functions and software designed to make the aid easier and more practical to use. This can mean the user at least has the hardware and software to potentially take a greater part in a conversational episode including enhanced opportunities to control a conversation, to make a speech, to express feelings, and to interrupt to make a timely point.

However, research has reported a rejection rate of around one-third of what would appear well-designed and functional VOCAs (Phillips & Zhao 1993; Riemer-Reiss; 2000). One reason may be the user has never acquired the complex skills needed to engage in conversation, such as; topic initiation; turn-taking; repair; and closure. Therefore, it is important that children and young adults with communication difficulties encounter many natural sources of language practice – facilitating language "play" through humour provides one such encounter.

2 Wordplay and speech disorders

Puns, punning riddles and jokes (verbal wordplay) form a natural part of children's discourse, familiar to every school playground. They provide a structure within which words and sounds are experienced and within which the normal rules of language can be manipulated. The linguist David Crystal suggests it is likely that: "…the greater our ability to play with language, the more we will reinforce the general development of metalinguistic skills, and – ultimately – the more advanced will be our command of language as a whole, in listening, reading, writing and spelling." (1998, p181).

The children who would be expected to gain most from the opportunity to more easily play with language through a voice aid are those whose cognitive skills are least affected and where there is a wide gap between receptive language abilities and expressive abilities. This is commonly seen in children diagnosed with articulation disorders such as developmental apraxia and dysarthria. The specific language impairment apraxia (or dyspraxia) describes a difficulty in sequencing and executing the oral movements necessary for speech, that is there is a difficulty in the motor speech planning. Dysarthria describes the difficulty in coordinating the speech musculature because of damage to the nervous system - although basic language processes are intact, the mechanical production of speech is impaired.

3 Computer-facilitated humour

The regular structures and mechanisms found in much of verbally expressed humour which help to structure wordplay (and enable fun manipulation) make facilitation of some forms of verbal humour through a reasonably high-tech communication aid possible. Researchers at Edinburgh University have implemented a computer program that generates punning riddles from a general-purpose lexicon (Binsted and Ritchie, 1997). The computer program is an implementation of a formal model of punning riddles. Testing with 8-11 year-old children revealed the Joke Analysis and Production Engine (JAPE) produced significantly more joke-like jokes than non-jokes but was less successful when compared with human-generated material. They interestingly conclude their paper with the observation the model could work, probably with more success, with a human lexicon rather than relying on JAPE's inbuilt lexicon.

It is envisaged the voice aid user could become the human lexicon that Binstead and Ritchie speak of - they would be in control of topic initiation and the choise and input of keywords. In this implementation the user inputs some keywords relating to a subject currently being conversed, the software then provides some associations which requires further input from the user who then chooses the joke (actually a punning riddle which due to complex humour rules the system is only designed to generate) they would like to try out. The chosen piece of verbal word-play could be used interactively and narrated through the voice aid. Feedback from conversational partners - smiles or laughs-and, just as importantly stony silences, could be used to help scaffold the user's emerging language and conversational skills. This process closely resembles the trial and error, use (or abuse!), of humour by a typical language user.

However, children with communication difficulties often present with physical and learning difficulties too. Therefore, there needs to be a provision for a simpler interface to the software to play with language. This can be achieved by providing a transparent storage and retrieval interface for favourite jokes which can still be used interactively. A carer, teacher or friend could help at any stage, the important point being that words are being played with, and language rules are being stretched and tested.

4 Interface design

The design goal is to create suitable interfaces that will enable the person using a communication aid to achieve a number of key goals that are specific to personal abilities. The functionality of the humour-play system needs to provide for the following:

1. The transparent storage and retrieval of favourite pieces of verbal humour.
2. A platform for the communication aid user to play with words.
3. The opportunity for the user to create self-generated and tested verbal humour.
4. Ease of output of the chosen material.
5. All areas to be accessible to people with complex communication and physical requirements.

Expert users (teachers and therapists) and end users (individuals with and without disabilities) have been involved in this research from its inception which arose from previous research which introduced the idea of story-based communication to children with complex communication needs who often required a voice aid to facilitate their communication (Waller et al 2001). A number of these children and young adults (one participant's story is told in the above named paper) took great delight in being able to store jokes in their voice aid using the software provided. These jokes would subsequently be told, through the voice aid at opportune moments. Being able to "tell" jokes in such a way often led to the participants being able to introduce a topic and control an audience – something that is difficult to achieve for people with language impairment. The desire to communicate was being encouraged through the success that humour play provided even though the punchline may have been given too quickly or the turntaking procedure not closely followed-thus revealing the limited knowledge of certainly humour and perhaps conversational rules. The software could be successfully used for this purpose but a customized system would offer much more functionality. It could be designed to provide a more transparent interface for the storage, retrieval and output of humorous material requiring as little a cognitive and physical challenge as possible.

The design of interfaces for individuals with complex communication needs provide a unique challenge. Potential users have a range of physical and learning difficulties requiring novel approaches. A further consideration is: do children who have had reduced opportunities to play with language understand verbal humour in a different way to others? More than eighty children with typical and non-typical communication abilities have been playing with words using a task designed to explore their comprehension of humour (O'Mara & Waller, 2003; O'Mara, 2003). This has provided more detailed information that needs to be considered including: difficulties with certain types of jokes and how children handle ambiguous material.

The interface seen in figure 1 provides a simple introduction to a system in which favourite jokes can be simply stored and retrieved under a subject topic chosen by the user. These can be easily adapted for each user. Each icon can be chosen by mouse click and cursor or a scanning procedure. Scanning allows physically disabled users to access an interface using one or more switches. Possible choices are highlighted sequentially until a switch activation indicates that a choice has been made. Scanning tends to be slow and interfaces need to be designed to optimise the order in which options are highlighted.

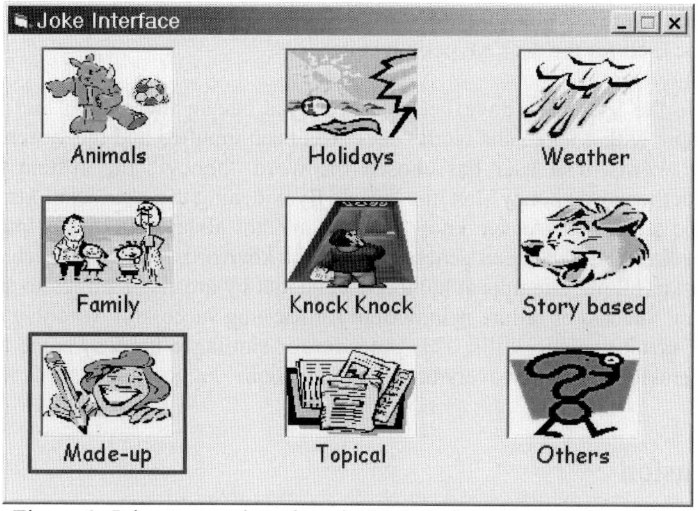

Figure 1. Joke storage interface for high-tech communication aids.

Favourite jokes are stored behind the icon of choice. When clicked the user has access to these jokes and can use them where and when appropriate through the voice output function. The user is encouraged to learn correct delivery of the humour (timing, turntaking, etc) by having a high degree of control of how and when each sentence is delivered. There is also a private area to store jokes which may be suitable for a peer, but not for an aging aunt!

In the example of Figure 1, the icon "Made-up" has been chosen by the scanning procedure. This leads to the computer-facilitated humour function, a composite interface to which is shown in Figure 2 (the steps taken to arrive at the jokes are actually on sequential screens for ease of interaction).

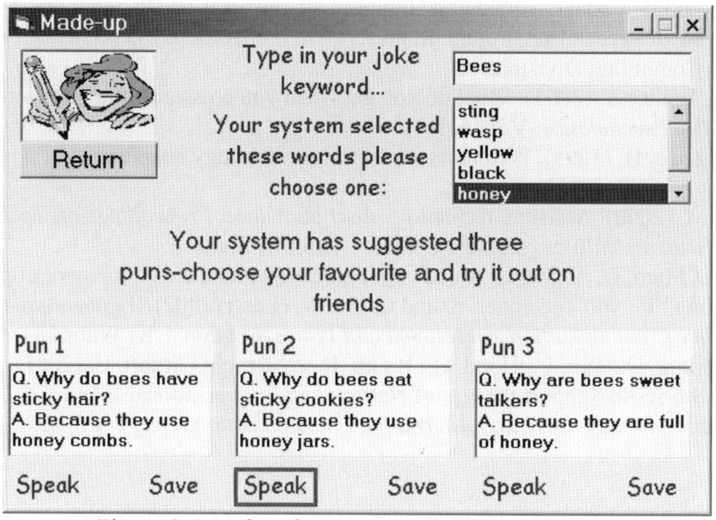

Figure 2. Interface for computer facilitated humour.

This interface has again been designed to reduce cognitive and physical demands - scanning has been introduced wherever possible, options and choices require a mouse click, and word-prediction can be used for the user's type-in text box.

In the example the user may have been chased by a bee in the garden and had decided to introduce the topic with a joke. The word "bees" has been inputted and the system has generated five alternative words. The user has chosen the word "honey"; the system then looks for associations with the word "honey" but not "bees". By widening these associations, several riddles and puns can be generated. As we know humans are not always successful punsters! This is reflected in the jokes that have been generated from the keywords, some of which may be funnier than others! Just as all our attempts at humour can be met by groans or smiles so too can attempts using this system. Success or failure is important for learning successful wordplay and the general development of metalinguistic skills. This procedure is similar to the way some forms of verbal humour are generated and used in everyday communication.

5 Conclusion

Children with complex communication needs do not have the same opportunity to naturally converse and practise their rapidly increasing language skills, which mature from a seemingly chaotic use to a more regular structure. Children with typically developing language have been practising their communication skills from the age they could babble. It is expecting a great deal from a novice aid user with little previous use of language to be a successful conversationalist without similar opportunities to experience and learn how the pragmatics and rules of conversation generally progress. Providing verbal word-play opportunities for the communication aid user may be a stepping stone to the making of a conversationalist!

References

Crystal, D. (1998). *Language Play*. Penguin Books, 80 Strand, London

O'Mara, D.A. (2003). Verbally Expressed Humour: A Conversation Development Approach for Children with Speech Impairment. Manuscript in preparation. University of Dundee, Applied Computing Division.

O'Mara, D. A., Waller A. (2003). What do you get when you cross a communication aid with a riddle? *The Psychologist,* Vol 16, No 2.

Phillips, B., & Zhao, H. (1993). Predictors of assistive technology abandonment. *Assistive Technology, 5*(1), 36-45.

Riemer-Reiss, M. (2000). Assistive technology discontinuance. *Proceedings of: Technology and persons with disabilities conference 2000.* California State

Waller, A., & O'Mara, D. A.(2002). Aided communication and the development of personal storytelling. In, von Tetzchner, S. and Grove, N. (Eds.) (2002). *Augmentative and Alternative Communication: Developmental Issues*. London, UK: Whurr.

Waller, A., O'Mara, D., Tait, L., Hood, H., Booth, L., &. Brophy-Arnott, B. (2001). Using Written Stories to Support the Use of Narrative in Conversational Interaction: An AAC Case Study. *Augmentative and Alternative Communication (Dec, 2001)* p221-232.

GraphSQL: a Visual Query Specification Language for Relational Databases

Marco Porta

Dipartimento di Informatica e Sistemistica – Università di Pavia
Via Ferrata, 1 – 27100 – Pavia - Italy
porta@vision.unipv.it

Abstract

In this paper we present a graphic formalism for visual specification of relational database queries. The purpose is to make the structure of queries clearer, by highlighting the parts of which they are composed and their interrelations. GraphSQL may be thought of as both a visual language to be employed directly for querying relational databases and as a support tool to be used along with "traditional" text-based SQL systems.

1 Introduction

Querying databases is undoubtedly one of those activities for which simplicity and naturalness of interaction assume a fundamental role. Within procedural query languages for relational databases, SQL surely plays a main part, since it is the simplest and most widespread. However, at times, formulations of complex queries may be little intuitive even for the expert, because of the difficulty of identifying and separating the single components of the query, which, put together, represent the logical description of what one wants to obtain from the database. The reason for the possible confusion arises mainly from the textual (and therefore linear) structure characterizing the query, which forces the user to a careful and precise analysis (both syntactic and semantic) of the query string(s). In other words, it is not possible a "quick glance" analysis, except for very simple cases. The field of visual languages shows that in many cases two-dimensional structures allow expressive power to be considerably augmented, both in programming and in database querying (Catarci & Santucci, 1995). For such reason, in this paper we present a graphic formalism whose purpose is to make the structure of SQL queries conceptually more intuitive, by visually highlighting the parts of which they are composed and their interrelations.

Systems for database querying which employ visual formalisms are called *Visual Query Systems* (VQSs). By means of *Visual Query Languages* (VQLs), they exploit the greater bandwidth of the human vision channel to allow the user to better manage large amounts of information. VQSs and VQLs developed are many, in confirmation of their usefulness. Several languages exploit directly the entity-relationship (E-R) conceptual model to specify the query objective (e.g. Angelaccio, Catarci & Santucci, 1990, and Bretan, Nilsson & Hammarström, 1994). Although the transposition of the E-R model, in itself based on a graphic formalism, from conceptual model to visual query language appears a natural choice, there are also alternatives to E-R-like languages, as shown in (Papantonakis & King, 1994) and (Murray, Paton & Goble, 1998). In particular, at the basis of the graphic formalism we will describe in this paper lies the conviction that relational algebra remains essential for expressing queries of a certain complexity in relational databases. Thus, rather than searching for visual representations based on alternative "paradigms", we concentrate on a graphic transposition of SQL able to make the process of specifying query objectives more intuitive, by allowing "global outlooks" which are hardly obtainable through pure textual representations.

2 The Formalism

Figure 1a shows the representation of a basic SQL query. Both database tables (relations) and the attributes we want to obtain as results are represented by means of proper symbols, connected to the central rectangle through lines. The order in which table symbols are arranged does not matter, while the order of attributes, clockwise, determines how they will appear in the result (from left to right). Within the rectangle, which we will call *condition rectangle*, the query condition is specified, i.e. the predicate(s) determining what information will be extracted from the database.

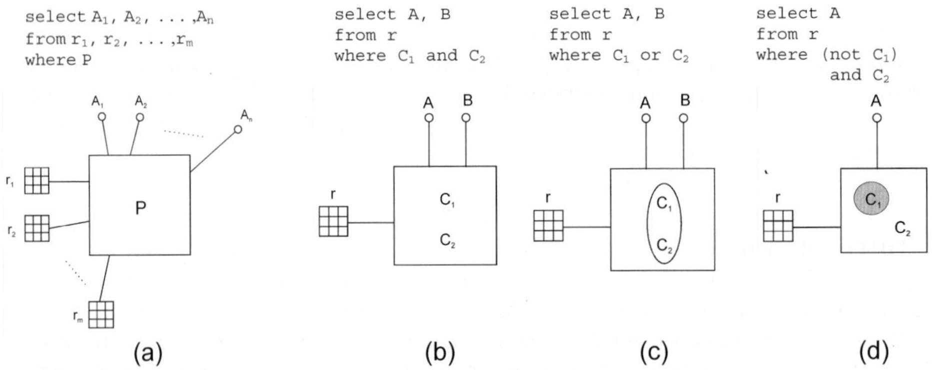

Figure 1: (a) Graphic representation of a basic SQL query; (b) Conditions in AND; (c) Conditions in OR; (d) Negation

2.1 Predicate Composition and Result Selection

Predicates specified within the condition rectangle of a query are implicitly considered in logical AND; if they are in OR, they must be explicitly enclosed in an ellipse. Negation of a predicate, instead, occurs by enclosing it within an ellipse with gray background. The three situations are depicted in Figures 1b, 1c and 1d. Figure 2a shows a query with conditions in AND. The link between table *borrow* and attribute *customer_name* has the purpose to specify that it is just the *customer_name* of relation *borrow* the one we want, not that of relation *customer*. Alternatively, the dotted notation could be used also for the attribute's name. Moreover, as will be clear from the next examples, the first predicate, textually expressed within the condition rectangle (`borrow.customer_name = customer.customer_name`), could also be graphically expressed through an 'equal' operator. Note that the fact that the small circles at the end of attribute lines are solid instead of empty means that, as results, we want distinct values.

Figure 2b shows an example of union between the results of two queries. The attributes (results) on which we want to execute the set operation are enclosed in proper shapes, which can also be arranged vertically and obliquely, to include all what is needed (for set intersection, an ordinary rectangle is used instead of one with round corners).

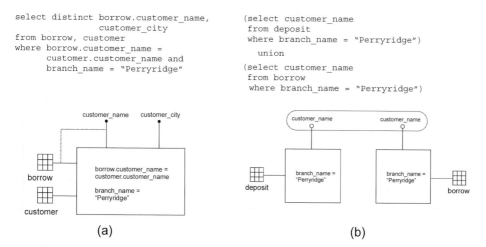

```
select distinct borrow.customer_name,
                customer_city
from borrow, customer
where borrow.customer_name =
      customer.customer_name and
      branch_name = "Perryridge"
```

```
(select customer_name
 from deposit
 where branch_name = "Perryridge")
   union
(select customer_name
 from borrow
 where branch_name = "Perryridge")
```

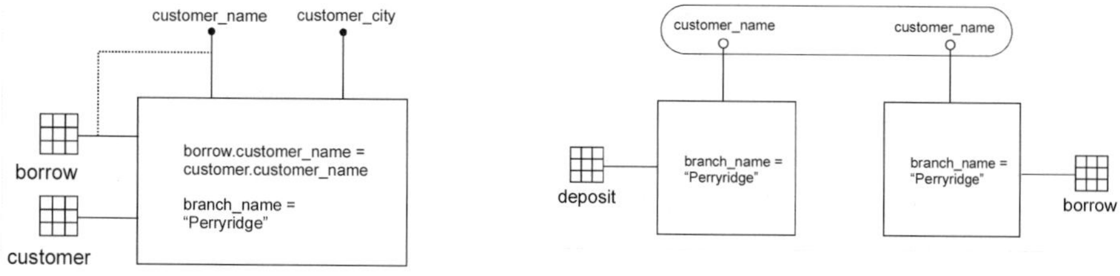

(a) (b)

Figure 2: (a) An example query with conditions in AND; (b) An example of union

2.2 Operators, Tuple Variables and Aggregate Functions

When formulating queries of a certain complexity, it is often necessary to exploit "operators" (set operators, comparison operators, etc.), which make connections between attributes and conditions of different subqueries making up the main query. In the graphic formalism we are proposing, such operators are represented by parallelograms. For example, Figure 3a shows a query which uses the set membership operator (IN). Dotted lines specify which attributes are involved by an operator. Arrows, when present, have the purpose to make the "direction" of application of operators explicit. In this example, the arrow entering the IN operator's parallelogram indicates that *customer_name* attributes obtained as the result of the upper selection (applied to table *borrow*) must be included in *customer_name* attributes of the lower selection (applied to table *deposit*). The continuous line, which we will call *condition line*, starting from the upper condition rectangle and ending at the IN parallelogram states that the fulfilment of the membership is a condition for the selection of the requested *customer_name* attributes in table *borrow*. Note also that the upper rectangle is double: this is to make evident that it refers to the main *select-from-where* construct and that its attributes will be considered as the query result.

Of course, a query can have more than one "parallelogram" operator. Figure 3b shows an example in which two operators are used, '>' and some. Since the two *select-from-where* constructs operate on the same table (*branch*), it is possible to graphically represent it only one time (although this is optional). Attribute *assets* displayed within the main condition rectangle (the upper one) indicates that such attribute is not required as result, but is used only to express a condition. Unary operators can also be used (e.g. the existential operator), still enclosed within a parallelogram but with only one dotted line departing from them.

The formalism we are proposing is also able to describe SQL queries where "tuple variables" are necessary. Consider, for example, the query shown in Figure 4a. Since we need to distinguish between distinct tuples (identified by variables S and T in the SQL query) taken from the same table, we represent it two times. This way, by connecting attribute lines to the proper table, it becomes possible to distinguish between different instances of the same attribute.

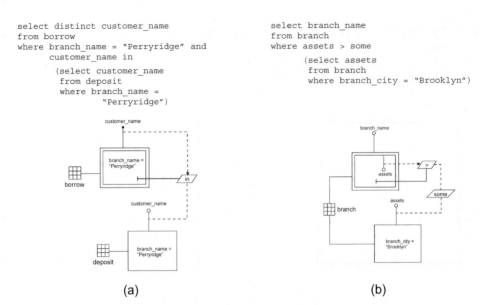

```
select distinct customer_name
from borrow
where branch_name = "Perryridge" and
      customer_name in
        (select customer_name
         from deposit
         where branch_name =
                 "Perryridge")
```

```
select branch_name
from branch
where assets > some
        (select assets
         from branch
         where branch_city = "Brooklyn")
```

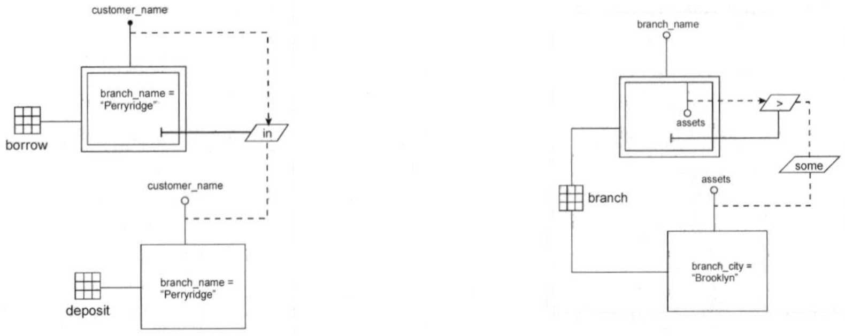

(a) (b)

Figure 3: (a) An example of query with condition depending on a set operator (IN); (b) An example of use of operators '>' and SOME

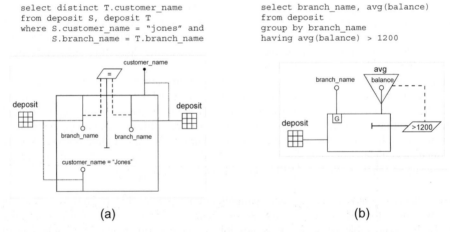

```
select distinct T.customer_name
from deposit S, deposit T
where S.customer_name = "jones" and
      S.branch_name = T.branch_name
```

```
select branch_name, avg(balance)
from deposit
group by branch_name
having avg(balance) > 1200
```

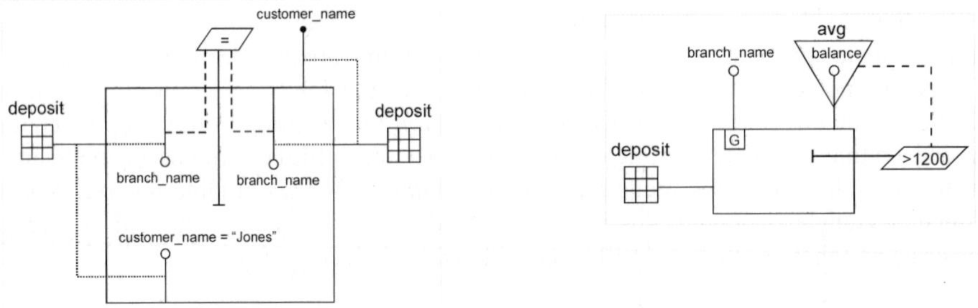

(a) (b)

Figure 4: (a) Graphic transposition of an SQL query using tuple variables; (b) An example of query with a condition which applies to a group of tuples

With SQL it is possible to calculate functions on groups of tuples. In Figure 4b an example is shown where the average of the values of attribute *balance* is calculated, for each value of attribute *branch_name*, and a condition on the average value is then imposed. The fact that the grouping has to be done on the values of *branch_name* is expressed by the small rectangle (with the 'G' inside) at the base of the attribute line. Instead, function 'average' (avg) is represented by the triangle placed on attribute *balance*. We stress again that "what" is external to the condition rectangle (attributes or functions applied to attributes) will constitute the query result. Therefore, the result of this query will be given by the values of attribute *branch_name* and by the results of the application of function avg to *balance* attributes.

3 Comments on the Formalism and Conclusions

The proposed formalism may be thought of as both a visual language to be used directly to query relational databases and as a tool to be used side by side with "traditional" (textual) SQL systems. In the first case, interaction could occur through direct manipulation, by means of drag&drop techniques, and a visual parser would translate graphic queries into SQL, so that they can be submitted to the database. In the second case, GraphSQL would assume a mainly descriptive function, able to graphically highlight the structure of the query under construction. In other words, once a textual SQL query has been specified, its graph would be automatically generated. At last, the two possible uses just described could be integrated to allow both visual construction of query graphs (then translated into SQL) and generation of graphs for textual SQL queries.

Although the visual formalism has not yet been tested through rigorous procedures involving many subjects, informal experiments conducted with people with basic knowledge of SQL have shown that GraphSQL is a promising approach for graphic formulations of relational queries. In particular, its ability to provide a "quick glance" description of the query turns out to be especially useful in case of non-trivial structures, composed of several (and possibly nested) *select-from-where* constructs (like, for example, the one shown in figure 5).

```
select distinct S.customer_name
from deposit S
where (select T.branch_name
       from deposit T
       where S.customer_name = T.customer_name)
    contains
      (select branch_name
       from branch
       where branch_city = "Brooklyn")
```

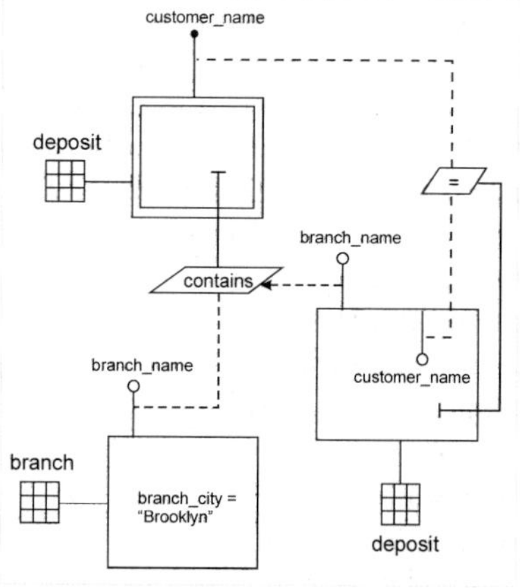

Figure 5: Example of composite query

4 References

Angelaccio, M., Catarci, T., & Santucci, G. (1990). QBD*: A Fully Visual Query System. *Journal of Visual Languages and Computing*, Vol. 1, n. 2, 255-273.

Bretan, I., Nilsson, R., & Hammarström, K. S. (1994). V: A Visual Query Language for a Multimodal Environment. Proceedings of *CHI'94*, Boston, Massachussets, USA, April 24-28.

Catarci, T., & Santucci, G. (1995). Are Visual Query Languages Easier to use than Traditional Ones? An Experiment Proof. Proceedings of the *6th International Conference on Human-Computer Interaction (HCI'95)*, August 29 – September 1, Huddersfield, United Kingdom.

Murray, N., Paton, N., & Goble, C. (1998). Kaleidoquery: a Visual Query Language for Object Databases. Proceedings of the *Workshop on Advanced Visual Interfaces* (*AVI'98*), May 24-27, L'Aquila, Italy.

Papantonakis, A., & King, P. J. H. (1994). Gql, a declarative graphical query language based on the functional data model. Proceedings of the *Workshop on Advanced Visual Interfaces* (*AVI'94*), June 1-4, Bari, Italy, 113-122.

Sociality of an Interface Agent for Sharing Mutual Beliefs in Collaborative Monitoring of Complex Artifact Systems with a Human Operator

Tetsuo Sawaragi, Yukio Horiguchi and *Yasunori Nishimoto*

Dept. of Precision Engineering, Graduate School of Eng., Kyoto University
Yoshida Honamchi, Sakyo, Kyoto 606-8501, Japan.
Email: sawaragi@prec.kyoto-u.ac.jp

Abstract

A view of socially-centred automation view is novel and essential in realizing true human-automation collaborative systems. It is well recognized that collaboration requires the establishment of the following three kinds of relations among the participants; *augmentative* (for mutual incompleteness), *integrative* (of different views), and *debative* (in sharing critics and criticism). In forming these relations, conversation and/or dialog does play an important role in forming and monitoring the status of mutual beliefs between a human operator and an interface agent. In this article, after discussing the aspects of joint activity of conversation, we present a prototype system implementing that for collaborative monitoring complex artifacts.

1 Introduction

Current development of automation technologies has enabled a *human-out-of-the-loop* control system, i.e., *fully-autonomous* control systems. However, for a safety reason, a human is still expected to remain in the loop, where the roles of a human has been shifted from operative tasks to supervisory ones. From a popular human-centred automation view, an essential design issue of a human-machine system is how to let an operator have and maintain a feeling of "commitment", and/or a sense of being engaged in a task. Loss of such feelings may incur over-reliance to the automation and may cause an operator's degradation of vigilance to the status of the plant. In the case of nuclear power plants, there reported many operators' opinion that they are reluctant to a plant operated by a single operator, even if it were guaranteed that technologies operate a plant and fully back up an operator's task. The major reason is because they need some *partners* or *colleagues* who can provide them with another viewpoint and with whom a human operator can talk and consult even when a operator has a strong confidence in his/her own judgment. I think that such an operator's feeling of isolation against the technologies is caused by the design principle of "One-Best way Doctrine" stating that a human is strictly inhibited to deviate from the prescribed procedure, but is forced to correctly trace a tightly-determined sequence of operations. There exists no freedom for an operator to attempt nor explore any creative ideas just because those are not allowed by any sense.

Based upon the above ideas, in this article we propose a novel principle of socially-centred automation, where a human and an automation should be regarded as equivalent partners and we allow them to interact with each other more intimately. We insist that such interactions should be made not only at the system operation levels, but also should be allowed in their creative thinking phases in advance to their executions. Thus, we focus on the design of a human-machine systems at the *conversation* levels in exchanging and sharing their beliefs on the plant status between two actors (i.e., a human operator and an automation). Introducing an interface agent as a partner of a human operator, we propose a dialog model that is embedded within an agent and can manage and control the flow of naturalistic conversations with a human operator just like as a man-to-man conversation.

2 A Novel View of Socially-Centred Automation

Extensive user participation is a key to success in both innovation promotion and the integration of human factors, and "mutual design and implementation" to facilitate user acceptance of computer systems and/or automation is the final goal of the human-machine system design. Toward the final goal of establishing "mutual knowledge of intent" between a human user and automation, currently formal arrangements for incorporating human factors both in system design and system development are sought for, and as a means for that, exploitation of the computer as an aid in system design conjunction with user participation is truly needed. In a word, automation systems should not be designed only from the perspectives of "utility" attained in the final products isolated from the user's participation. Rather, they should be designed from the perspectives of "relations" and "processes" emerging between the artefacts of automation and the human user. In order to keep the users in their active and continuous commitment loops, we have to design the emergence of novel views and perspectives out of the increased and varied interactions between the human user and the artefacts so that the users continuous participation and commitments can be maintained.

In this paper, we focus on the *conversation* activities done between two people. When people engage in a conversation, they typically do so with the intent of making themselves understood. They need to make sure, as they speak, that the other participants are at the same time attending to, hearing, and understanding what they are saying. Since unresolved uncertainties can result in communication failures, people collaborate to establish and maintain the *mutual belief* that their utterances have been understood well enough for current purposes. A conversation is more reminiscent of a *collaborative* effort or *joint activity* than simply a structured sequence of utterances. The process by which participants elegantly coordinate the presentation and acceptance of their utterances to establish, maintain, and confirm mutual understanding has been called *grounding*.

3 Clark's Idea of Conversation as a Joint Activity

A conversation is more than a structured sequence of utterances. People engaged in conversation elegantly coordinate the presentation and acceptance of utterances to achieve and confirm *mutual understanding*. After Clark's idea (Clark, 1996), Paek and Horvitz investigated how people collaboratively contribute to a conversation at successive levels of mutual understanding through *grounding* (Paek and Horvitz, 1999). Based on their idea, we regard a conversation activity as shown in Figure 1.

At the most basic level, which we denote as the *channel* level, a speaker *S* attempts to open a channel of communication by executing behavior such as an utterance or action, for a listener *L*. However, *S* cannot get *L* to perceive his/her behavior without coordination. At the next higher level, the *signal* level, *S* presents something as a signal to *L*, which must be recognized as such by *L*. Hence, *S* and *L* must coordinate on what *S* presents with what *L* identifies (i.e., they have to have a shared attention). At the next higher level of *intention*, *S* signals some proposition *p* for *L*. By focusing on the goals of *S*, the intention level treats the "speaker's meaning" as primary. *S* cannot convey *p* through a representation R(*SL*) without *L* recognizing that *S* intends to use R(*SL*). This again takes coordination. Finally, at the *conversation* level, *S* proposes some joint activity, which *L* considers and takes up. A proposal solicits an expected response defined by that activity. Thus, in short, all four levels require coordination and collaboration in order to achieve mutual understanding.

With respect to this multi-layered conversation model, we would like to stress the following three issues just like a sentential grammar; *syntactic*, *semantic*, and *pragmatic* issues of joint conversation. First syntactic issues deal with a structure of interactions made between *S* and *L*. That is, the roles of reactions to the partner's utterances are very important evidences in monitoring bi-directional flow and grounding status of the conversations. Second semantic issues deal with the consistence of the meanings of the utterances. This requires the sharing of the domain-specific knowledge. Finally, pragmatic issues should deal with the contexts on how a particular utterance is made based upon what preceding utterances, and/or upon what mutual beliefs were obtained so far.

With respect to communication between two parties, Freeman has proposed a new idea referring to interactions between individuals' brains as well as to what are exchanged at the physical interaction levels: Communication between brains requires *reciprocal* construction of representations in accordance with meanings, which elicit other meanings in other brains (Freeman, 1997). A pair of brains can act, sense, and construct in alternation with respect to each other. Freeman illustrates this as a model of two brains *S* and *L*' interaction as shown in Fig.1 *S* has a thought that constitutes some meaning M(*St*). In accordance with this meaning *S* acts to shape a bit of matter in the world to create a representation directed at *L*, R(*LSt*). *L* is impacted by this shaped matter and is induced by thought to create a meaning M(*Lt*). So *L* acts to shape a bit of matter in accordance with M(*Lt*) in a representation R(*LSt*), which impacts on *S* to induce M(*St+1*). By its nature an external representation and/or physical reactions can be used over and over. It cannot be said to contain or carry meaning, since the meanings are located uniquely inside *S* and *L* and not between them.

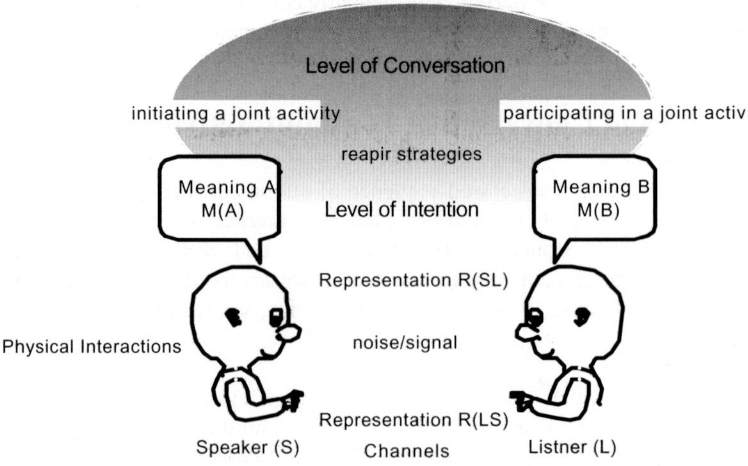

Figure 1: **Conversation as a Joint Activity**

4 Embedding a Conversation Management Mechanism within a Interface Agent

In this article we investigate into a conversation design of an *interface agent*, and discuss about its social relationships with a human operator. An interface agent is a semi-intelligent computer program that can learn by continuously "looking over the shoulder" of the user as he/she is performing actions against some complex artefacts and is expected to be capable of providing the users with adaptive aiding as well as of alternating the activities instead of a human. This is shown in Fig.2. We especially focus on its conversational aspects with a human operator (i.e., upper bi-directional interactions of the figure); how their conversation comes to be grounded based upon what they physically share and exchange at the physical interaction level (i.e., lower part of the figure).

In this work, we first construct a multi-layered task structure for a number of operation procedures as a basis of mutual understanding between a human operator and an agent. We call this structure as a *plan tree*, which is a domain-specific knowledge on how the process should proceeds at a variety of abstraction levels. Note that this is a normative one to be traced by a human operator, but may be different from what he/she actually executes in a descriptive sense. To be a human-friendly partner, an interface agent has to dynamically construct this structure and to manage this in comparing this with what it observes from the plant and the conversation with the operator. Since what is observed by an agent is not always decisive nor deterministic, an agent has to manage the uncertainties of the grounding status of the conversation by gathering many evidences. Those evidences may exist in the available plant data and/or an operator's physical reactions made in reply to an agent's queries to an operator. And the current grounding status may drive an agent to ask a variety of queries to an operator. Dynamically constructed multiple levels form a ladder of co-temporal actions ordered with upward completion. Actions at a given level of dialog are completed bottom up. Furthermore, evidence that one level is complete is also evidence that all levels below it are complete. In other words, evidence flows downward. The control infrastructure provides an environment for exploring the value of intuitions behind upward the notions of completion and downward evidence. We investigate procedures and policies for integrating the results of co-temporal, collaborative inference at multiple levels.

As for a control infrastructure, we develop *conversation stacks*, with which the current focus and mutual understanding shared by a human operator and an agent are monitored and controlled. The dynamics of the conversation is controlled by the system's checking whether a current state of the plant to be monitored is "common ground" or not between an operator and an agent. This is based upon Clark's formulation of "mutual beliefs". In case it is uncertain that a pair of actors is forming a mutual belief, the system lets an agent to make some confirming questions as well as allows an operator to respond to an agent by asking confirming and/or disconfirming a fact. Thus, by allowing intensive interactions between a human operator and an agent, our system handles uncertainty about the status of the joint activity, common ground (a shared knowledge base for dialog), and other higher-level dialog events relevant to the joint activity.

5 Conclusions and Future Perspectives

The details and the implemented examples of the system will be presented. We are analysing the conversational structure dynamically shared and formed in a party of humans and/or a party of a human and a machine. One is the aviation domain dealing with the actual conversation made between a human air traffic controller and a pilot as well as an automated device such as TCAS (i.e., warning device for aircraft collision avoidance). is presented, which is analyzed from the perspectives of our conversation framework. The other one is from a nuclear power plant domain and we demonstrate the conversation between a human operator and an interface agent with respect to the non-normal procedures of the plant anomalies. We will also show how the interactiveness does contribute to the enhancement of the team situation awareness of a human operator and an interface agent.

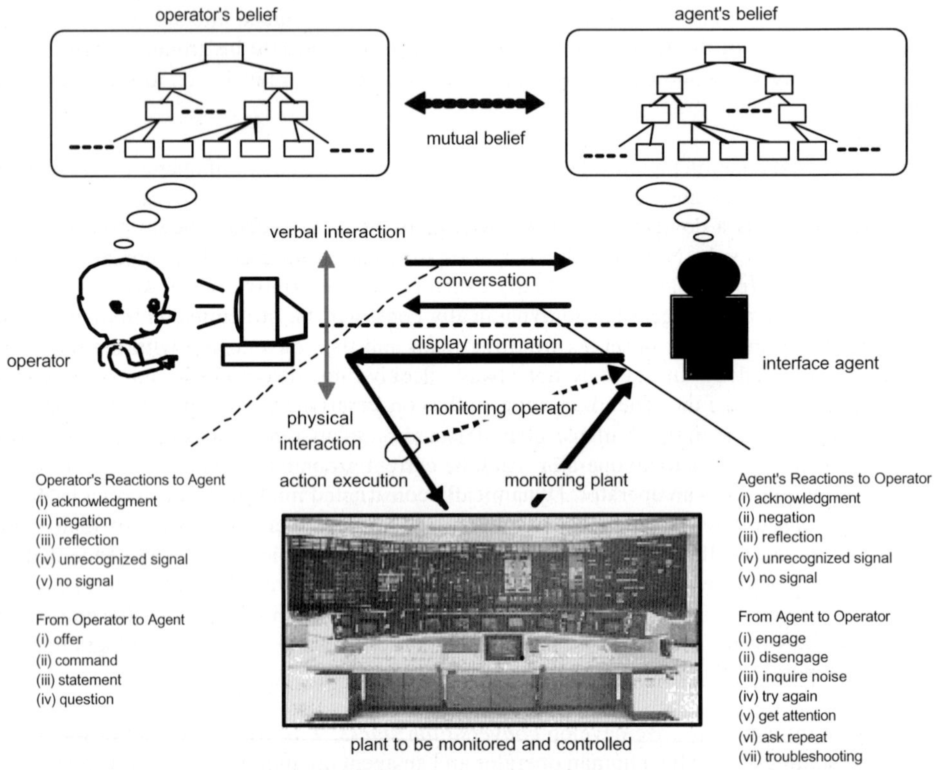

Figure 2: **Mixed-Initiative Sociality Emerging between an Operator and an Agent**

6 References

Clark, H.H. (1996). *Using Language*. Cambridge University Press.

Horvitz, E. & Paek, T. (1999). A computational architecture for conversation. *Proc. of the Seventh International Conference on User Modeling*, 201-210. Springer Wien.

Freeman, W.J. (1997). A Neurobiological Interpretation of Semiotics: Meaning, Representation, and Causality, *Proc. of the 1997 International Conference on Intelligent Systems and Semiotics: A Learning Perspective*, 487-492.

Integration of Heterogeneous Multi-Agent Systems

Hiroki Suguri, Eiichiro Kodama** and Masatoshi Miyazaki***

*Communication Technologies
2-15-28-6F Omachi Aobaku
Sendai 980-0804 Japan
suguri@comtec.co.jp

**Faculty of Software and
Information Science
Iwate Prefectural University
152-52, Takizawa-aza-sugo, Takizawa,
Iwate 020-0173 Japan
{kodama,miyazaki}@iwate-pu.ac.jp

Abstract

Multi-agent systems play important role in human-computer interaction. Agents interact with heterogeneous entities such as humans, other agents, and non-agent software and hardware as media of communications. However, there exist heterogeneous multi-agent systems, which are incompatible with each other. To facilitate the use of agent technology and to promote the agent-based applications and human interface, heterogeneous multi-agent systems themselves must communicate and interoperate with each other. To solve the problem, we take a two-step approach. Firstly, message-level interoperability is achieved by a gateway agent that interconnects heterogeneous multi-agent systems. Secondly, higher-level interoperation of conversations, which consist of bi- (or multi-) directional streams of messages, is governed by interaction protocols. We demonstrate the concept and implementation of dynamic negotiation of interaction protocols. With these techniques, we show how the heterogeneous multi-agent systems have been integrated to enhance future human-computer interaction.

1 Introduction

Multi-agent systems are platform software on which application programs and human interface are deployed. On the platform, agents interact with heterogeneous entities such as humans, other agents, and non-agent software and hardware. Therefore, multi-agent systems play important role in human-computer interaction. Ideally, all multi-agent platforms should be compatible and interoperable with other platforms. That would make the application programs based on a platform be able to communicate and interact with other applications on other platforms.

Unfortunately, that is not the case. There are at least two major multi-agent standards: FIPA and KQML, which are incompatible with each other. Even within the FIPA community, different versions (such as FIPA 97, FIPA 98, Experimental and Standard) make it very hard, if not totally impossible, to assure the interoperability and compatibility between different implementations. The situation is the same (or worse) with KQML where multiple implementations of different specifications are not always interoperable.

To solve the problem, we proposed a gateway agent that interconnects heterogeneous multi-agent systems (Suguri, Kodama & Miyazaki, 2001). In this paper, we report the actual implementation of the gateway. Based on that experience, it became clear that higher level of interoperability was needed beyond the gateway agent that translates single message. We noticed that conversations between agents, which consist of bi- (or multi-) directional streams of messages, are governed by interaction protocols. Thus we came up with the idea of dynamic negotiation of interaction protocols. In addition to the implementation of the gateway agent, we will demonstrate the concept and technique of the dynamic negotiation of the interaction protocols.

2 Understanding single message: the gateway agent

We have identified three architectural elements of the multi-agent systems: (1) structure of mental state, (2) agent communication language (ACL), and (3) message transport and proposed the gateway agent that converts the architectural elements between agent systems (Suguri et al., 2001). Basically, the gateway interprets single message from a sender agent and translates the message to receiver agent. Below we discuss the implementation of the gateway.

2.1 Implementation of the gateway agent

Figure 1 illustrates the internal diagram of the gateway agent.

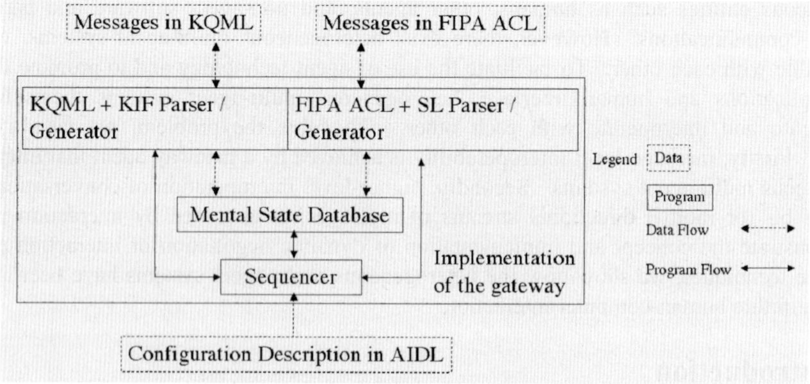

Figure 1: Implementation of the gateway

Configuration Description in AIDL: The file specifies input architecture and output architecture for the sequencer. The file name is passed to the gateway agent when it is invoked from the shell.

Sequencer: At startup, the sequencer reads the input and output architectures from the configuration description in AIDL. It then controls the KQML+KIF Parser/Generator, FIPA ACL+SL Parser/Generator and Mental State Database according to the AIDL description.

Mental State Database: The database stores intermediate messages described in interlingua that are generated when the parser interprets incoming messages. The interlingua record consists of FP, RE and other message parameters. FP stands for Feasibility Precondition, which is a proposition that must be true before the sender issues the message. RE, or Rational Effect, is a proposition which is expected to become true after the receiver accepts the message. The gateway adopts FIPA semantic model of FP and RE (FIPA, 2002a). The database is implemented using PostgreSQL and JDBC.

KQML+KIF Parser/Generator, FIPA ACL + SL Parser/Generator: The parser reads a KQML or FIPA ACL message from the standard input, interprets the message and generates an interlingua expression which corresponds to the incoming message. The interlingua expression is then stored in the mental state database. On the other hand, the generator reads the interlingua expression from the mental state database and generates a KQML or FIPA ACL message, which is written to the standard output. The behavior of the Parser/Generator is controlled by the sequencer.

2.2 Example behavior of the gateway

Let the input be a FIPA ACL message shown below. Basically, this message means that a sender agent (`sender@fipa`) tells a receiver agent (`receiver@kqml`) that a proposition `(= a b)` is false. The gateway has to translate the message into corresponding KQML message.

```
(disconfirm   :sender sender@fipa
              :receiver receiver@kqml
              :in-reply-to irt0
              :reply-with rw0
              :conversation-id cid0
              :language sl1.5
              :ontology ipu-gateway-ontology-v1
              :content (= a b) )
```

When the gateway agent is invoked from the shell, it at first reads in the AIDL configuration file specified in the command line. Since this is a translation from FIPA ACL to KQML, the content of the AIDL is `(:stdin FIPA :stdout KQML)`. Next, the sequencer instructs the FIPA ACL+SL parser to read the original message from the standard input. The parser interprets the communicative act (`disconfirm`), content (`:content (= a b)`), content language (`:language sl1.5`) and ontology (`:ontology ipu-gateway-ontology-v1`) to generate the following pair of propositions (FP and RE) in interlingua.

FP: $B_i \neg\phi \wedge B_i(U_j\phi \vee B_j\phi)$ where ϕ is (= a b)
RE: $B_j \neg\phi$ where ϕ is (= a b)

Next, the sequencer invokes KQML+KIF generator after the completion of the process by FIPA ACL+SL parser. The KQML+KIF generator searches the database for the records that are to be translated into a KQML message. It reads FP and RE propositions, content language and ontology from the database to generate a performative and fill the values for `:content`, `:ontology` and `:language` slots. Other parameters are simply copied from the database to the outgoing KQML message. Final output of the gateway, which is sent out to the standard output, is following.

```
(tell         :sender sender@fipa
              :receiver receiver@kqml
              :in-reply-to irt0
              :reply-with rw0
              :conversation-id cid0
              :language kif
              :ontology ipu-gateway-ontology-v1
              :content (not (= a b)) )
```

In FIPA ACL, `DISCONFIRM` is used to negate the propositional content of the message. Therefore, the KQML message `TELL`s the negation of the content expression in the original FIPA ACL message. A translation from KQML message to FIPA ACL message is processed similarly.

3 Understanding conversation: dynamic negotiation of protocols

In this chapter, it is assumed that communicating agents can understand single message each other by using a gateway agent described in the previous chapter, or by being inside a same community of agents, which means that the agents share a common architecture. Under such an assumption, we discuss the necessity of dynamically negotiating, introducing and interpreting heterogeneous interaction protocols. We propose and demonstrate the concept and technique of dynamic negotiation of interaction protocols using a prototype implementation.

3.1 Merit and demerit of interaction protocols

Usually, a unit of communications between agents in an application scenario, which we call a *conversation*, consists of multiple messages. A simple example of conversation is that an agent A requests something for another agent B by sending a message; and B returns the result of the request to A by sending back another message. Like this request-response conversation, some certain typical forms of conversations can be identified in the interactions between agents, which are called *interaction protocols*.

There are several advantages of utilizing interaction protocols if the participating agents can agree on using them. First, interaction protocols define the sequence of conversations between the communicating agents in advance. Secondly, the interaction protocols are typical flow of conversations that are proven to be useful and practical. Therefore, it is more convenient and effective to simply follow the instructions of the interaction protocols than to dynamically invent the conversations "emergently" by trial and error.

Notwithstanding, some problems exist in utilizing the interaction protocols in real world applications. First, the communicating agents in the application must agree upon the interaction protocols that are used in the conversations prior to the commencement of the interaction. Now, one of the benefits of multi-agent systems is dynamic re-configurability and adaptability. Each agent can dynamically adapt itself as the environment changes to re-configure the system according to the changes that are external and internal to the application. That means all agents in an application must understand all the interaction protocols exactly the same way. As the number of the agents increases in the application, this is getting harder and harder. In addition, if an existing application is modified to use a new interaction protocol, all of the agents constituting the application must be altered to understand the new protocol by hand. It decreases the productivity of the software significantly.

3.2 Concept and technique of dynamic negotiation of interaction protocols

The problems stated above are all due to the fact that current implementation of interaction protocols is internally and statically hard-coded inside each agent. Therefore, we propose to externally describe the behavior of the interaction protocols and dynamically exchange the description between the agents to negotiate, introduce and execute the protocols. The dynamic exchange of the interaction protocols is performed by a regular messaging of agent communications, which is guaranteed to function correctly either by a translation gateway agent or by being inside a same community of architecture.

The technique frees the agents from the problems mentioned in the previous section. Not all agents must understand all interaction protocols beforehand because protocols can be informed to the agents on demand. If new agents, which exploit new interaction protocols, are added to the existing applications, the new agents can teach existing agents how to use the new protocols. Human programmers do not have to modify the code of the agents by hand.

We have designed and implemented the prototype of the dynamic negotiation of interaction protocols based on FIPA interaction protocol library (FIPA, 2002b), its AUML model (Bauer, Mueller & Odell, 2001) and scheme-oriented description of the protocols. Scheme (Kelsey, Clinger & Rees, 1998) is a dialect of lisp that features a small footprint compared to Common Lisp. Even if the agent platforms are programmed in other languages such as C or Java, the scheme interpreter can be easily incorporated. Also, the interpreter is small enough to be embedded even in resource-sensitive application or human-interface agents. The protocol description is passed from the sender agent to the receiver agent as a functional closure that can be evaluated on the receiver. If the receiver is programmed in other languages, proper hooks are required to map the scheme functions to the corresponding routines in the receiver. In that sense, the description of the interaction protocols is not fully implementation-neutral. However, we believe that the prototype has demonstrated the power and usefulness of the idea of dynamic negotiation of the interaction protocols.

4 Conclusion and future work

We have discussed a two-stop approach toward the integration of heterogeneous multi-agent systems. The first phase is interconnecting heterogeneous agent systems by a gateway agent that translates each message used in the communications. We have identified architecture elements of multi-agent systems and built a prototype of the gateway. The second stage is assuring a higher level of conversations between the agents, which consist of bi- (or multi-) directional streams of messages. Based on the first step that guarantees a basic exchange of the messages, we have proposed and demonstrated the concept and technique of dynamic negotiation of interaction protocols using a prototype implementation.

As future work, we are planning to establish measurement criteria of the performance, accuracy, and application productivity for the gateway agent and dynamic negotiation of the interaction protocols based on an application scenario. With the criteria and evaluation results, it is hoped that more concrete usefulness of the approach will be demonstrated.

References

Bauer, B., Mueller, J. P., & Odell, J. (2001). Agent UML: A Formalism for Specifying Multiagent Interaction. In P. Ciancarini & M. Wooldridge (Eds.), *Agent-Oriented Software Engineering* (pp. 91-103). Berlin: Springer

Kelsey, R., Clinger, W., & Rees, J. (Eds.). (1998). Revised^5 Report on the Algorithmic Language Scheme. *Higher-Order and Symbolic Computation*. Vol. 11, No. 1.

FIPA (2002a). FIPA Communicative Act Library Specification, from http://www.fipa.org/specs/fipa00037/

FIPA (2002b). FIPA Interaction Protocol Library Specification, from http://www.fipa.org/specs/fipa00025/

Suguri, H., Kodama, E., & Miyazaki, M. (2001). Gateway that interconnects heterogeneous Multi-Agent Systems. *Abridged Proceedings of the 9th International Conference on Human-Computer Interaction,* 118-120.

Transparent Community: Creating a Novel Community Framework Using P2P Technologies

Hiroshi Tamura[1], *Tetsuji Hidaka*

R&D Division, Hakuhodo Inc.
4-1, Shibaura 3-Chome, Minato-Ku, Tokyo, 108-8088 JAPAN
{hiroshi.tamura, tetsuji.hidaka}@hakuhodo.co.jp

Tetsuya Oishi, Kazuhiro Kikuma

NTT Network Service Systems Laboratories, NTT Corp.
9-11, Midori-Cho 3-Chome, Musashino-Shi, Tokyo, 180-8585 JAPAN
{oishi.tetsuya, kikuma.kazuhiro}@lab.ntt.co.jp

Abstract

In the conventional online communities such as chat systems, bulletin boards and listservers, users who actively provide information are considered valuable, while users who visit online communities just to collect information tend to be regarded as non-contributors. We propose that we should facilitate the latter group of users to turn into active information providers by introducing a novel framework into online communities. We call this new community framework "transparent community." In transparent communities, each user is required to have his/her independent information inventory that can be referred to each other. In this context, we made use of SIONet (Hoshiai et al., 2001), a novel peer-to-peer (P2P) architecture, as an enabling technology to build a communityware prototype which we call "WineDiary." In this paper, we verify the possibility of this novel community framework through the results of user testing on WineDiary.

1 Introduction

Since the mid-90s, online communities including chat systems, bulletin boards and listservers have emerged spontaneously with the rapid popularization of the Internet, and we've come to realize that they were not merely a social phenomenon but also of great economic importance (Armstrong & Hagel III, 1995).

Various methods and tools have been proposed in this field to support users who are willing to be involved in online communities by participating in the communities and consequently contributing toward organizing those communities. There are many users, however, who do not provide information but are solely collecting information in actual online communities (Nonnecke & Preece, 2000). These users have merely been regarded as non-contributors, yet we think they might be potentially great contributors in terms of diversifying and expanding such communities. Furthermore, we might say online communities are getting worthier both socially and economically if non-contributors turn to active contributors to provide information for these communities.

---

[1] The author also belongs to Graduate School of Interdisciplinary Information Studies, The University of Tokyo (3-1, Hongo 7-Chome, Bunkyo-Ku, Tokyo JAPAN).

2 User Classification

Brazelton and Gorry reported there are three user-groups in the online community regarding regional school education (Brazelton & Gorry, 2003). Our criteria of the user classification are similar to theirs. The difference is that they classified the users by observing a particular online community but we classified the users by the strength of their commitment to general online communities and thus we aim to restructure community space taking potential users as well as actual users into consideration.

Table 1: User Classification

User group	Lead users	Silent users	Bargain hunters
Classification criteria	■ Frequently logging in particular chat systems ■ Frequently slipping into particular bulletin boards ■ Often contributing to subscribed listservers	■ Hesitating to log in any chat system ■ Checking particular bulletin boards regularly ■ Subscribing to listservers only for reading	■ Hesitating to log in any chat system ■ Using bulletin boards as information sources when necessary ■ Not subscribing to listservers

Table 1 shows our classification criteria of the users. Lead users are a user group who lead the communities and play vital roles by providing information and by organizing such information in several online communities. Silent users are a user group who regularly collect information in particular online communities and utilize such information for developing their own knowledge, but they rarely provide information for online communities. Bargain hunters[2] are a user group who frequently use the Internet but access online communities only when they need information in there. They are not accustomed to observing any particular community and rarely provide information.

Referring to the classification, we defined the concrete objective of our research was to facilitate silent users to turn into active information providers by proposing a novel framework for online communities and to verify the possibility of the framework.

3 Transparent Community

What is the primary reason that silent users take negative attitudes toward providing information? According to our former survey on frequent users of the Internet, about 90% of the subjects answered that they were interested in providing information but the majority of them answered at the same time that they were feeling uneasy about providing information for anonymous users. We suppose there are psychological barriers in the systems of the conventional online communities.

[2] We suppose that bargain hunters are a very large user group in many online communities, but we think the possibility of their providing information would not be large since they have no intention of regularly accessing online communities by nature. Therefore we excluded them from the subjects of our research. It is still a significant theme how we deal with the user group.

Therefore we could say it would be possible to facilitate silent users to turn into active information providers if we succeed in getting rid of those barriers.

In this context, we propose a novel framework for online communities: "transparent community." In the transparent community, each user has his/her own information inventory, and the inventories exchange information autonomously according to his/her profile and/or context with other inventories (see Figure 1). Therefore users can add and incorporate information to their individual inventory on their own initiative without minding the eyes of other users. By adopting such autonomous distributed models, we can get rid of the prerequisite condition of the conventional communityware that every user should provide information at public space, and consequently can share information in the community simultaneously.

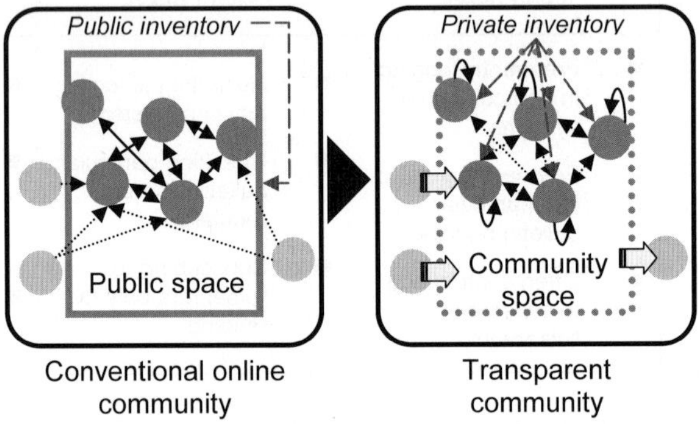

Figure 1: Comparison of the concept models

4 SIONet

As an enabling technology for establishing transparent community, we made use of SIONet (Semantic Information-Oriented Network), which is one of the autonomous distributed network architectures. SIONet works as a platform of P2P network services, having a flexible mechanism for discovering entities based on semantic information, i.e. names, locations, preferences and so on, as a unique feature. This mechanism enables every inventory on SIONet to autonomously discover common attributes and to share actual information with other inventories. The implementation of inventories on SIONet is a set of XML documents, and the definition of the tags for XML description must be done beforehand in each community space. SIONet actually uses these XML tags as the clues to the discovery based on semantic information, and enables us to discover and to retrieve information with fine granularity according to the definition of the tags. See more details about SIONet in (Hoshiai et al., 2001).

5 WineDiary : a Communityware Prototype Using SIONet

As a communityware prototype, we developed "WineDiary" which was aimed at organizing a community of wine-lovers, using SIONet. Users were requested to write diaries about their experiences of wines, and the diaries were sent to peers autonomously. Therefore, the users were motivated to improve their diaries while enjoying information provided by others at the same time.

Basic functions of the application were "diary-issuing", "diary-retrieving" and "P2P-messaging". Diary-issuing was a phase when users described their experiences of wines along with some 90 attributes (see Figure 2), i.e. region of production, brand, grape vintage, grape variety and so on, which were defined by XML tags. Diary-retrieving was a phase when users could see their own diaries and others' in the same window (see Figure 3), and of course they could be retrieved with the combination of attributes. P2P-messaging was an optional phase when users could communicate with each other directly. We provided both synchronous and asynchronous tools.

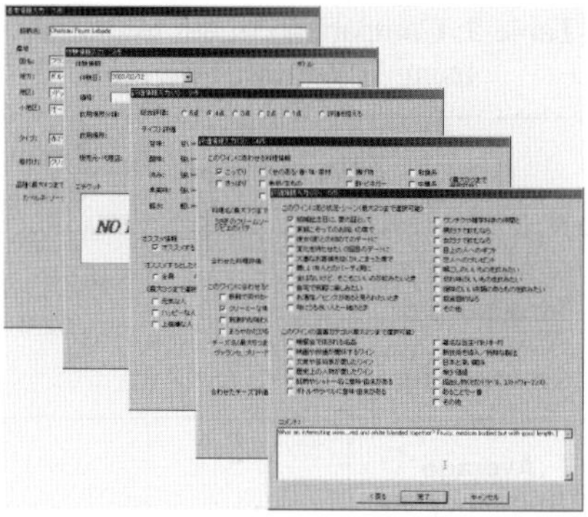

Figure 2: The input forms of diary-issuing phase

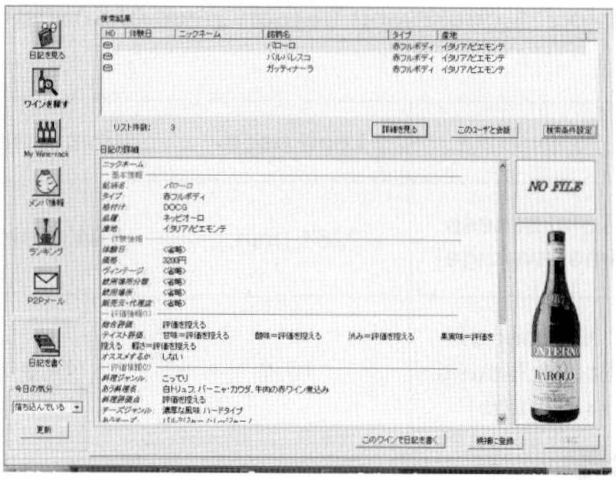

Figure 3: Retrieved result of the diaries

6 Testing Method

We held a user-test last June to July. Some 30 users became the test subjects. The requirement for the recruitment of subjects was that they should be frequent users of the Internet; more specifically, they should browse WWW more than once a day. Besides, we randomized them from offline organizations of the Internet users to avoid lead users' occupying the subject group, resulting in the proportion of subjects in which about 43% were lead users, about 32% were silent users and the rest 25% bargain hunters. We divided the subjects into six groups and appointed one subject out of each group as a representative. We trained the subjects through the representatives for about two weeks prior to the test. During the period of the test period, we left the subjects to their own devices and didn't assign any obligations including setting the minimum access time per day, because our primary objective was to observe the subjects using the application on their own initiative. After the test, we conducted a survey on all the subjects and collected access logs from their PCs. In addition, we held the interviews twice (on the halfway and after the test) with the selected subjects.

7 Results

The result of the log-analysis is briefly indicated at Table 2 and 3. According to the Table 2, silent users, compared to lead users, almost equally contributed to the community from the viewpoint of access time, number of diaries and quality of diaries issued[3]. According to the Table 3, however, silent users didn't use P2P-messaging tools as frequently as lead users did. Related to this topic, several silent users told in the interviews that they felt somewhat uneasy about communicating with anonymous users although they didn't hesitate to open their diaries to them.

[3] The scores were calculated from the number of downloaded files by other subjects.

Wrapping up these results, silent users have great potential to provide information as well as lead users do. And they would be great contributors if they were not forced to have communication at public space where they cannot help minding the eyes of anonymous users. Therefore we can say the transparent community is a very effective framework to establish the online communities where silent users can provide information without feeling constraint.

Table 2: Comparison of the values of contribution

	Lead users	Silent users
Total access time (average)	2281 (min.)	1943 (min.)
Number of diaries issued (average)	16.4	14.8
Quality of diaries issued (average)	54 (points)	50 (points)

Table 3: Comparison of the frequency of using P2P-messaging tools

	Lead users	Silent users
Average frequency of use (synchronous)	128.6	82.6
Average frequency of use (asynchronous)	7.3	5.5

8 Conclusion

In this paper, we proposed transparent community, a novel framework for online communities, which could facilitate silent users, who have long been regarded as non-contributors to the conventional online communities, to turn into active information providers. And we conducted the user-test on WineDiary which was a communityware prototype using SIONet, a P2P network architecture, in order to evaluate the framework. The result was satisfactory and showed the framework working very well.

However, there are many aspects still remaining unverified, i.e. feasibility to general Internet users, other themes, longer periods and so on. We are going to move the test bed of the research from subjects in closed online communities to general Internet users in order to prove the possibilities of the framework more explicitly.

9 References

Armstrong, A., & Hagel III, J. (1995). Real Profits from Virtual Communities. *McKinsey Quarterly*, 1995 Number 3, 126-141.

Brazelton, J. & Gorry, G.A. (2003). Creating a Knowledge-Sharing Community: If You Build It, Will They Come?. *Comm. ACM*, Vol. 46, No.2, 23-25.

Hoshiai, T., Koyanagi, K., Sukhbaatar, B., Kubota, M., Shibata, H. & Sakai, T. (2001). SION Architecture: Semantic Information-Oriented Network Architecture. *IEICE Trans. Commn.*, Vol. J84-B, No.3, 411-424. (in Japanese)

Nonnecke, B. & Preece, J. (2000). Lurker Demographics: Counting the Silent. *Human Factors in Computing Systems*. Paper presented at CHI'00 The Hague, Holland.

A Proposal for Under-the-Door Communications on the Network

Tetsuya Tomita, Yuko Murayama

Graduate School of Software and Information Science
Iwate Prefectural University
152-52,Takizawa-aza-sugo,Takizawa, Iwate Japan
tomita@comm.soft.iwate-pu.ac.jp, murayama@iwate-pu.ac.jp

Abstract

In this research we try and implement communication systems using the metaphor of a door on the World Wide Web (WWW) as a media for informal communications. We call those informal communications through a door “on-door-communications”. This paper introduces a system for the “under-the-door communication” with which a user can slip a message note under the other's door. The system has the following features: 1) communications are asynchronous, 2) if the location of a door homepage is known, anyone can slip in a note any time, and 3) it is only the owner of a homepage who can see the arrival of the message and read the contents. We present the model, design and implementation of our system as well as its operations.

1 Introduction

The “door” of an individual room, such as the one in a student hall of residence, is not only the entrance to the room, but also works as a media for various information exchange with the room resident and the others. In this research, we call informal communications which make use of the metaphor of a door “on-door-communications,” and have implemented systems for such communications on the network (Murayama et al. 2002). The purpose of the research is to explore the possibility of such systems as a communication media.

We have a knock-on-the-door-and-chat system so that a visitor may knock on the door to notify his arrival to the resident of a room and start chatting if she is available. If the resident is not in, the visitor may leave a message making use of our on-door message board system. In some case, the visitor may leave a message under the door. We call the last one “under-the-door communication.” This paper presents the model of our under-the-door communication system, its initial design and the implementation of the prototype system. We also report on its operations.

2 Model of Under the Door Communication

Under-the-door communications are performed between an unspecified number of people and the resident of a room through the door. If the room location is known, anyone can visit there and leave a message under the door. The room could be looked at as an informal communication space. A resident has a key for an access to the room, so that she could see the existence of the message and read it.

The model of under-the-door communication is composed of a door, a visitor, people passing by, the resident of a room, a message, the owner of the room, and the room key (see Figure 1). A door is an interface for the resident to receive a message privately. A visitor submits a message and it will be maintained in the room. None but the resident would notice of the existence of the message. A door enforces an access control for this. People would pass along by and leave the resident a message. The owner lends the resident a room and owns the master key of the door, so that she has an access to the room for maintenance and management.

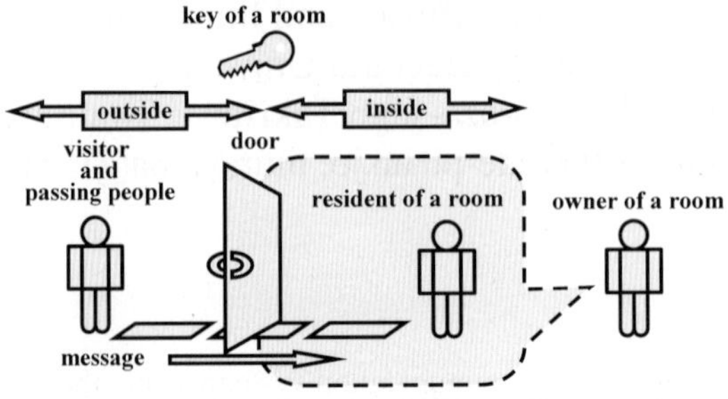

Figure 1: Model of Under the Door Communication

3 Design the Under-the-Door Communication System on the network

3.1 Overview

We consider, the room as a homepage on WWW and a page with a link to the room page as a "door". The resident has a right to have an access for reading messages left in the room page through the door. A message is in the multimedia form, containing text data, a picture, a sound and so on.

Our under-the-door communication system has the client-server structure. The server provides users with an under-the-door communication space which maintains messages. A client functions as a door to the communication space. The owner of the door would be an Application Service Provider (ASP) which provides users with the under-the-door communication service, manages the server, and grants a resident right for use of a door. The resident sets up a link to the under-the-door communication server on her homepage. The resident and visitors acquire the client function through this link. Under-the-door communications are asynchronous and involve the resident and an unspecified number of people.

3.2 Design of User Interface

We design an intuitive user interface so that a user does not need any explanation. A user gets to the service through the link to under-the-door communication on a homepage. A user makes an access to a door by clicking the door page. The interface to the door is presented by a sub window. One can submit a message by dragging and dropping an object on a sub window.

In order to have an access to a message, one has to enter the room. A door has a function for granting a reading access only to the resident. One can enter the room by clicking an authentication icon on the sub window. The authenticated user sees the existence of a message at the door, and reads it by clicking the message icon on the inside-the-door sub window. The message icon shows the existence of a message, and appears if there is any unread message left when the resident is in the room. The icon is displayed in real time so that the resident can see the icon when a message is being inserted by a visitor.

3.3 System Structure

A user such as a resident and a visitor requests for the door function from a web server. Through the door visitor can send a message to the message management server, and the resident can receive a message form the server. Moreover, a user is authenticated and control her access to a message management server.

An under-the-door communication server is composed of a web server and the message management server. The web server sends the function of the door upon a request from a user site. The HTTP protocol was used for transferring the door function and the FTP protocol for message data (see Figure 2).

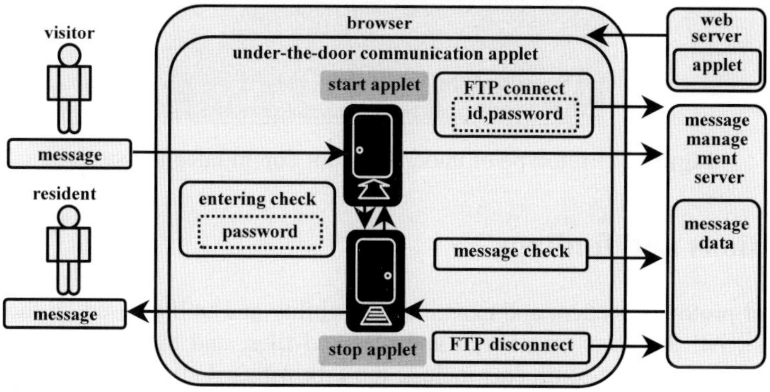

Figure 2: System architecture

4 Implementation of a Prototype System

We used Java for implementation of a prototype system. Java applet was used for a client, and the FTP server was used for management of messages. Table 1 shows the list of software used for implementation.

Table 1: Server Summary

OS	HTTP Daemon	FTP Daemon
Linux version2.4.18	apache-1.3.26	proftpd-1.2.4

A user does not need any installation of client software due to a Java applet. A Java applet is automatically loaded into the user site from a server and it becomes a client. A user requires a Web browser which can run a java applet as well as a connectivity to the Internet, to enjoy our system as a client. The client has the functions to transmit a message to the message management server, to authenticate a user, and to control the user access to a room.

The FTP connection between a client and the server is established when the client applet starts running and is closed when the applet completes. An ID and a password are required for a connection to the FTP server, and they are held within an applet. A user access to the message management server is controlled by the client system. A visitor transmits a message to a message management server from her client system; the client system could control the visitor's writing access to the server, however, our system does not enforce this access control. Any one can transmit a message to the server. On the other hand, a resident at her client system receives a message from a message management server; the client system controls the resident's reading access to the server. Only the authenticated user can have an access to the resident's room. A user can come into the room only when the user password matches with the one kept in the applet. Only for those who pass entrance authentication, the system grants a right of an access for receiving a message from the message management server. Figure 3 shows the display image of our door page.

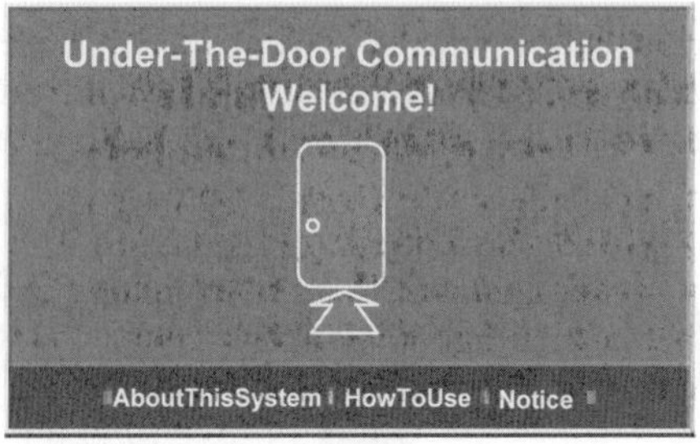

Figure 3: Prototype of under-the-door communication system

5 Operations and findings

We operate our prototype system and examine its usability and utility based on the questionnaire with ten users[1]. All the users transferred and received files, and they reported that they did it without any difficulty. Users comments include the followings: 1) the use of the system was easy because a user does not require any installation of software; 2) they would use it for message exchange with friends as well as for a temporary storage of a file shared by a number of people. We found some problems as well. The current user interface of the prototype is too intuitive and simple for a user to view the total number of messages. Some users raised a security issue that a message could include an information bomb such as virus due to lack of access control over submission. Finally, a management function is needed for a group of people to share the door.

6 Related Work

The Door Awareness System (Nichols, Wobbrock, Gergle & Forlizzi, 2002) has a sensor to detect whether a door in the real world is open or close, and provide this information on WWW. HERMES System (Cheverst et al., 2002) shows the messages from a Web site onto the display installed on a real door. Our system could also have such functions to link to the real world in the future. One of the related work to our system is a report presentation system (Saito et al., 2003), in which a user submits electronic documents through WWW. The system is used for the exchange of somewhat formal information such that one would submit a paper to a conference as well as a student submits a report under the door of her teacher's room. Users wish to be identified in such formal communications, so that they go through registration of the sender identity before they transmits a message. On the contrary, users may stay anonymous in under-the-door communication, and there is no need for registration of a sender. The contents of a message can be anything for informal communication. An electronic mail system provides the similar functions to ours, however, sender's identity is specified in terms of an email address. The contents of an email message can be controlled sometimes due to the size limitation.

[1] The member of Murayama Lab. Faculty of Software and Information Science, Iwate Prefectural University, and Araki Lab. Graduate School of Information Science and Electrical Engineering, Kyushu University.

7 Future work

We consider that a door is in a close state for under-the-door communications. We could improve our system so that the resident can control the door state either in an open state or the close one as in (Buxton, 1997). If the door is open, a visitor could hand in a message directly to the resident. If it is closed, a message would be inserted under the door. When the message is an important and urgent one, a visitor shall knock on the door to see if the resident is available in the room. Such synchronous feature could be considered as an extension function as well. We would like to investigate a link to the real world, as well.

In the real under-the-door communication, when the visitor can be sure of the submission of a message to the resident's room under the door. Moreover, it depends on the physical safety in the real world how difficult it will be for the others to have an access to a message left to the resident under the door. For instance, locking up a room could make it hard for one to steal and tamper a message. With our under-the-door communication system on a network, a client system serves as an entrance door. That is, the network paths between a client system and the server are a part of a room. We need some protection mechanism for submitted message from security threats such as unauthorized disclosure of a message and unauthorized tampering.

8 Conclusions

This paper proposed a new type of communication --- under-the-door communications, and presented the model, design and implementation issues of a system for such communications. With our prototype system, the visitor can submit some multimedia information to a resident --- the owner of a door home page. Future work includes providing more door-related features such as open and close states combining with some real-time communication tools. We need more investigation on security features such as some countermeasures to unauthorized disclosure and tampering.

References

Buxton, W. (1997). Living in Augmented Reality: Ubiquitous Media and Reactive Environments. In K. Finn, A. Sellen & S. Wilber (Ed.), *Video Mediated Communication*. Hillsdale, N.J.: Erlbaum, 363-384.

Cheverst, K., Fitton, D., Dix, A., & Rouncefield, M. (2002). Exploring Situated Interaction with Ubiquitous Office Door Displays. Retrieved February 1, 2003, from http://www.appliancestudio.com/cscw/cheverst.pdf

Murayama, Y., Gondo, H., Nakamoto, Y., Segawa, N., & Miyazaki, M. (2001). A Message Board System on WWW with a Visualizing Time Function for On-Door communication. *Proc. of the 34th Annual Hawaii International Conference on System Sciences (HICSS34)*, 284 -293.

Nichols, J., Wobbrock, J.O., Gergle, D., & Forlizzi, J. (2002). Mediator and Medium: Doors as Interruption Gateways and Aesthetic Displays. *Proc. of the ACM Conference on Designing Interactive Systems (DIS '02)*, 379-386.

Saito, T., Inagaki, T., Shouji, F., Sumiya, T., Nakamura, J., Nonaka, C., & Tanimoto, A. (2001). Report presentation system used PHP. *Proc. of Conference on Education for Information Processing '01*, 523-526.

CMS: A Collaborative Work Environment for the Assurance of Conference Proceedings Quality

David Tuñón Fernández; Sergio Ocio Barriales; Martín González R; Juan Ramón Pérez Pérez;

University of Oviedo
Department of Computer Science, Calvo Sotelo, s/n, 33007 OVIEDO – Spain
david@petra.euitio.uniovi.es, i1652800@petra.euitio.uniovi.es,
martin@lsi.uniovi.es; jrpp@pinon.ccu.uniovi.es

Abstract

There are a set of applications called Groupware that present Internet as a place for cooperative working in spite of the idea of the global database. In this paper we gave an overview of what is collaborative or cooperative work and the requirements of a groupware application. Next we discuss about the conference proceedings and the necessitate of improving a cooperative system to help conference committees and as an approach for conference proceedings quality assurance. Also we argue about why is so important the usability in this kind of systems. Finally we present a web portal as an example of cooperative work to support the conference proceedings.

1 Introduction

The exponential grown of Internet use provides a way to get information, search and in general better access to the information. Although these are important, there is another approach that consider Internet as a way to share information and to support collaborative work of a group of people. This is set of applications are know as groupware.

Hills (Hills, 1997) believes that groupware applications must provide the next five functionalities:

1. Allow cooperative work of a group of people.
2. Allow people to share information and knowledge.
3. Let people to do their job in an automatic way.
4. Must register the work that had been done.
5. Join geography and time, people working at the same time and at the same geography point.

2 Conference workflow

There is no doubt that conference management involves a high amount of tasks. Typically all work depends on organising committee members, they are in charge of promoting the event, handling the submitted papers, controlling the review process, sending notifications to authors and generally all the tedious process concerned.

In the next figure we can see the typical workflow of a conference:

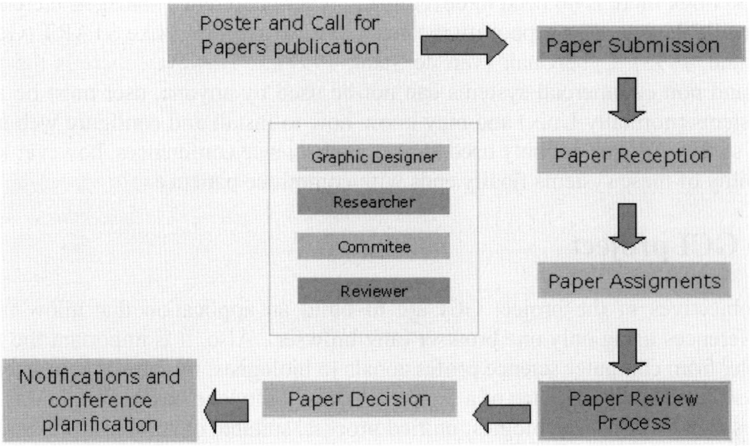

Figure 1. Conference Workflow.

The steps of the conference are the following:

1. Publication of the Poster: A graphic designer prepares a poster for the congress and it is published by the committee members (on a web page done specifically for this congress, on newspapers, …).
2. Publication of the Call for Papers: The committee members send a call for papers asking for participation via e-mail, or put it available on a web page.
3. Paper Submission: Authors, taking care of conference topics, send their papers, usually via e-mail.
4. Handling Submissions: The committee members handle the submitted papers and forward them to the reviewers (in electronic format via e-mail or in printed format).
5. Assignment Papers to Reviewers: The committee do the assignment of the papers to the reviewers concerning to the reviewers skills and biddings.
6. Reviews Submission: The reviewers return their results via e-mail or in printed format.
7. Paper Results: Papers are marked as accepted or rejected.
8. Authors Notification: Authors are informed by the committee. A notification is sent to the author via e-mail or letter.
9. Conference Plan: The conference is planned using the accepted papers.

Generally, the workflow exceed organising committees capacity when they try to start a conference using traditional methods. The amount of papers, normally hundreds, and the needed of keeping people involved in the conference informed are two of more critical aspects. We see the requirement of a management system to run the conference proceedings and assure the quality of conference organization.

3 Conference management systems

We can think there is no need to develop any other conference manager, there are already a set of tools available that offer support to the conference proceedings like START (Gerber, 1998), ConfMan (Lund, 1995), CyberChair (van de Stadt, 1997),… However, there is the fact that this commercial and non commercial systems can not be used by anyone, user must be an expert on operating systems (normally Unix) and may know how to install and configure web servers. That is the reason so that systems are only used at computer science conferences, however later, the low level of usability of these systems finally ends with committee patience.

4 The GCI project

The objectives of the project GCI are to build an application that allow users to start multiple conferences using only one browser (any browser). Also, it is important the fact that any kind of people, from computer science professionals to biologists, must be able to interact with the system, so here it is the significance of a high level of usability and navigation facilities. We reach our objectives using two methodologies: unified process, suitable of an object oriented project and OOHDM (Object Oriented Hypermedia Design Method, (Schawbe, 2002)), used in hypermedia and web projects. Also we use a couple of Java technologies to build dynamic contents from a data base, these are Java Servlets and JSP ("Java Web Site").

4.1 Strong Points

One of the strong points of GCI is its knowledge management. It is able to manage conference organization and on the other hand it collects information about reviewers like paper preferences (often called bids), reviewers knowledge about conference topics and another important information like paper marks and opinions of reviewers. This information collected is organized, stored and the system is able to evaluate it and make paper assignments to the reviewers based on their biddings and knowledge and it is able to decide best papers for the conference.

Another strong point of GCI is the accessible and international interface. Actually Spanish and English languages are supported. Also the application accessibility has been tested using many browsers including text browsers like lynx and links. We use the term accessibility to express that GCI system could be used in any browser that support standard HTML so that in any platform.

4.2 Features

We can summarize the features of GCI as follows:

1. Poster Management: Committee can put the poster available to download.
2. Call for papers Management: Committee can put this document available to authors, reviewers and all the people interested in the conference
3. Submission of Papers: Authors have to be able to send papers to any conference that the application is managing.
4. Review Assignment: Committee can assign papers to reviewers in an automatic way or also doing the assignments at their own.

5. Submission of reviews: It handles with the results submitted by the reviewers. System can put them available to committee members and reviewers.
6. Registration: Users can register their attendance to the conference. This feature is especially useful to the committee members.
7. Agenda: Committee can manage important dates, deadlines. Using this feature you can inform authors about the conference date, the submission deadline, ... Provides a good communication between authors, reviewers and committee members.
8. Links: Committee can manage links to other conferences,...
9. Sponsors Registration: Committee can register sponsors of the conference giving a short description and links to their home pages.
10. Notifications: Automatic and manual notifications about deadlines and other important dates of the conference. A mailing list to inform about deadlines.

5 Inside the system

5.1 Technologies and methodologies

To improve a static web site we just need a HTML editor, but it is not our desire. We want to develop a software application based on web with database access, so we need technologies to generate dynamic pages. We have chosen the following:

- Java Servlets: They solve all the problems of CGI (efficiency), but they mix interface with code, so we need to used it together with the next technology.
- JSP (JavaServer Pages): This technology allows us to create the HTML interface and after this, to add the dynamic sentences. It split interface and code, and furthermore it is a free technology.

To develop GCI we have used the following couple of methodologies of software development: The classic process to build any object oriented application, the unified process. That includes use case diagrams, requirements, class diagrams ... The hypermedia process. A web (hipermedia) application must not be seen only as a traditional application, there are links, navigation,... so we must take care about that, and ensure an efficient and coherent navigation. We reach our purpose using the OOHDM (Object Oriented Hypermedia Design Method). In (Rossi, 1999) authors discuss about the obligation of this hypermedia process.

5.2 Design Patterns

In design process and in the implementation we have used the model view controller (MVC) design pattern. It helps us to separate the system into three independent parts. The model contains application functionality and data. It is represented by the system classes that make the interaction with the data base using JDBC bridge and also it is represented by data repository. View are the representation of data gave to the user. These views are taken from the abstract interface design of the hypermedia process (OOHDM) and are the JSP and HTML pages of our GCI system. The last part is the controller, it handles the user entries and it is represented by the Java servlets of GCI. They capture user data using GET and POST request.

The main advantage of using this design pattern is that it helps us to obtain multiple views of the same model, that is, of the data base. Therefore we have updated views all the time.

This design pattern is critical for the application internationalisation because it aids developer to change views without making any change of model or controller.

6 Usability test

To guarantee the system usability, GCI has been subjected to usability tests with Thinking Aloud techniques, that is, we provide the test users with the product to be tested and a set of task to perform during this test. We ask the user to perform these tasks taking care of the time he or she spends achieving the objective. At the end we ask the user about their impressions of the interaction with the application and try to get a better understanding of the user's mental model and interaction with the product. The results of this test were relevant because they remark the high level of usability of our system, especially analysing the results of the people not used to web technologies and Internet of our usability group.

7 Conclusion and future steps

The only way to reduce work of a conference and make it efficient is applying an automatic and high usable system. In this paper we have discussed the obligation of a management system, the different tools available at the moment and the best technologies available to improve a new system. We also have expressed the reason why it is necessary a new system, as well as the problems of the tools existing and how to fix them.

GCI was used to organize the Third HCI Conference held in Oviedo (Spain) July from 7 to 9, 2002 http://www.di.uniovi.es/%7Emartin/hci/Lab/bulletin.htm. Actually GCI is supporting the International Conference on Web Engineering, ICWE'03 that will be held in Oviedo, Asturias, Spain on July 16-18, 2003 http://www.icwe2003.org/
More information about GCI project can be found at http://www.telecable.es/personales/bbbttt/

References

(Bentley, 1997) Bentley R., Appelt W., Busbach U., Hinrichs E., Kerr D., Sikkel K., Trevor J. and Woetzel G.. Basic Support for Cooperative Work on the World Wide Web. p. 827-846, June 1997, ©Academic Press.

(Gerber, 1998) Gerber , Rich; Hollingsworth, Jeff, & Porter, Adam: "START: Submission Tracking And Review Toolset", 1998. http://www.cs.umd.edu/~rich/start.html

(Hills, 1997) Hills, M. Intranet para groupware. Madrid: Anaya multimedia, 1997, p. 47 (spanish)

("Java Web Site") Java Web Site. Sun Microsystems. http://www.java.sun.com

(Lund, 1995) Lund , Ketil; Halvorsen , Pål & Preuss, Thomas: "ConfMan: Conference Manager", 1995. http://confman.unik.no/~confman/ConfMan/

(Rossi, 1999) Rossi1,3 , Gustavo; Schwabe2, Daniel; Lyardet1 , Fernando "Web Application Models are more than Conceptual Models", 1999. 1 LIFIA, Departamento de Informática, UNLP. Argentina, 2 Dpto. de Informática, PUC-Rio, Brazil, 3 also at CONICET and UNLM

(Snodgrass, 1999) Snodgrass, Rick. Summary of conference management software, 1999 http://www.acm.org/sigs/sgb/summary.html

(Schawbe, 2002) Schawbe, Daniel, Rossi ,Gustavo. The Object Oriented Hypermedia Desingn Model http://www.telemidia.puc-rio.br/oohdm/oohdm.html.

(van de Stadt, 1997) van de Stadt, Richard. "CyberChair", 1997 http://www.cyberchair.org

Collaboration Table: An Alternative Medium for Multi-user Multi-site Cooperation

Hiroyuki Umemuro

Tokyo Institute of Technology
2-12-1 O-okayama, Meguro-ku, Tokyo 152-8552 Japan
umemuro@me.titech.ac.jp

Abstract

In many collaborative tasks such as product design, computers may significantly support cooperation among multiple participants. Many computer tools aimed to support such collaboration, however, allow only one participant to operate at a time, mainly because the computers used have only one pointing device and it may be difficult to identify the participants when the single pointing device is shared. This paper proposes a computer tool to support multi-user collaboration, named Collaboration Table, which allows multiple participants to work on a set of objects in a shared space simultaneously, while identifying and recording the operator of every action. Collaboration Table allows participants to manipulate objects projected on the tabletop directly with their fingers, while a computer traces the movement of each finger separately. Light-emitting diodes (LEDs) are used as participant identifiers and a variation of identification methods is discussed. The tool also supports collaboration among participants in distant locations when connected via the Internet. The effectiveness of the developed Collaboration Table was evaluated by applying it to a product grouping task.

1 Introduction

In many collaborative tasks, computers may significantly support cooperation among multiple participants, whether in the same or in different locations. Computers can provide environments in which participants can work on the same objects and exchange ideas. Such collaboration is essential for creative tasks such as product design.

For computers to support such collaboration, it is essential to record the operations of each participant while allowing multiple participants to work simultaneously in the environment, so that one can see which participant is responsible for a given action afterwards. Many computer tools aimed to support such collaboration, however, allow only one participant to operate at a time. This is mainly because the computers used have only one pointing device and it is difficult to identify the participants when the single pointing device is shared. Alternative approaches include having multiple pointing devices on the same computer, or using multiple computers communicating to each other, although both approaches increase the complexity and the cost of the systems.

The purpose of this paper is to propose a computer tool to support multi-user collaboration, named Collaboration Table, which allows multiple participants to work on a set of objects in a shared space simultaneously, while identifying and recording the operator of every action. The tool also supports the collaboration among participants in distant locations when connected via the Internet. An example of its usage with a product grouping task is given to show its effectiveness.

2 Design of Collaboration Table

2.1 Basic Concept

The basic idea of Collaboration Table is that participants would manipulate objects on the table directly with their fingers, while a computer would trace the movement of each finger separately. The computer is used to generate the objects on the table as well as to identify and track the fingers on the surface of the tabletop.

2.2 Structure of Collaboration Table

As a part of the TRI system, Stappers and colleagues (Keller, Hoeben & Stappers, 2000; Keller, Stappers & Hoeben, 2001; Stappers, Keller & Hoeben, 2000) proposed a sophisticated interaction method using a glass table and a projector. Images are projected downwards onto white tracing paper on the tabletop. When one moves a finger on the surface of the tabletop, its shadow is projected on the paper and it can be captured by a video camera at the bottom of the table. The movement of the finger can be detected using simple image processing techniques and can be used as a pointing device for a computer. When multiple participants work simultaneously on this table, however, it is not easy to identify each participant because all finger shadows look alike and are difficult to distinguish. The Collaboration Table proposed in this paper is based on the original idea of Stappers' group, while it uses light-emitting diodes (LEDs) as active identifiers instead of shadow images, as discussed in **2.3**.

The structure of Collaboration Table is shown in Figure 1. Collaboration Table consists of a table, a computer that works as the collaboration server, participant identifiers, an image projector and an image sensor. The tabletop is used as the workplace for participants. Images of objects generated by the collaboration server are projected by the projector upon the tabletop. The tabletop is made of glass and semi-transparent acrylic board, so that it works as a screen for projected images, while the very strong light on the surface can be detected by the image sensor from the opposite side. An image sensor located at the bottom of the table detects the movements of the participants' fingers and sends this data to the server computer. The computer interprets this information as multiple pointing input signals.

2.3 Participant Identification

To identify who performed a given operation, each participant wears a small LED light at the tip of one finger. When participants move their fingers with attached LEDs on the surface of the tabletop, the light of the LEDs can be detected by a video camera or other image sensors located below the table. Participants can attend to the table from any side, as long as the LEDs on their fingers are within the active area of the tabletop.

There are two methods for identifying participants using LED lights. The first method involves simply using a different colored LED for each participant. This method is easy to implement when color image processing is available, and the resulting response time is relatively short. This method also has an advantage that when images of pointing cursors are projected as location

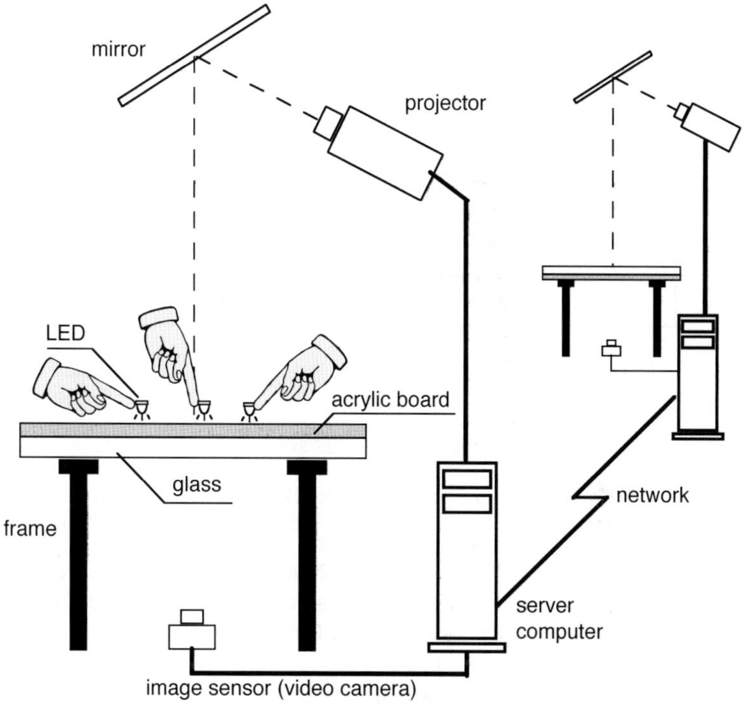

Figure 1. Structure of Collaboration Table

feedback, it is easy to associate cursors with participants' fingers if the same colors are used. The disadvantage of this approach is that the number of participants is limited because the colors used need to be clearly distinguishable. The maximum number of participants per table would be three or four.

The second identification method involves changing the blinking patterns of the LEDs. The LEDs either turn on or off for a unit time (usually 1/F seconds where F is the number of frames the image sensor can capture per second). With this method it is easier to increase the number of participants and color image processing is not necessary.

When the LEDs blink independently, the length of time necessary to identify N participants is equal to (1/F) multiplied by the smallest integer greater than or equal to $\log_2(N)$ + (number of start bits) + (number of stop bits), as shown in panel A of Figure 2. This method has the advantage that the blinking mechanism can be simple and not necessarily connected to the server computer; thus, participants have more freedom to move around. However, the disadvantage of this method would be a rather longer response time as described above and this largely depends on the speed of the image sensor.

The disadvantage above can be compensated to some extent by synchronizing blinks with the collaboration server. In this case, the minimum time length necessary to identify N participants is equal to (1/F) multiplied by the smallest integer greater than or equal to $\log_2(N+1)$ as shown in

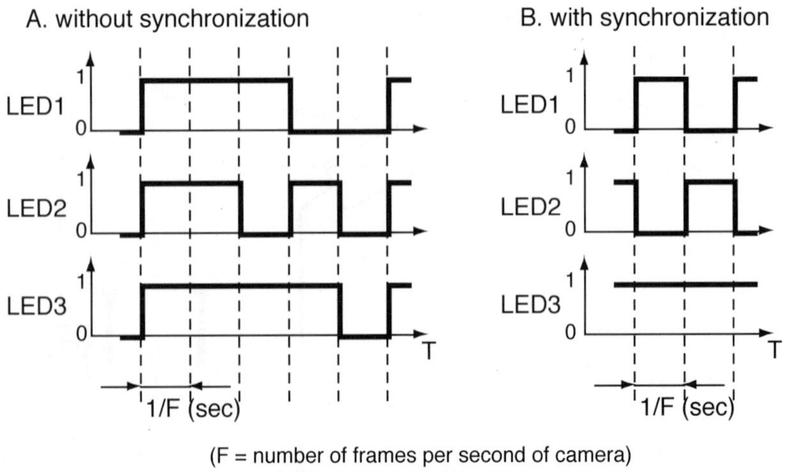

Figure 2. Blinking patterns of LEDs for identification

panel B of Figure 2. In this case, LEDs on the participants' fingers need to be connected to the server computer, limiting the freedom of the participants. Estimating the locations based on historical data may also compensate the possible delay to some extent.

2.4 Interconnection of Collaboration Table

More than one Collaboration Table, each in a distant location, may be connected using an Internet protocol and may exchange the position data of participants. Participants can see the movements of the objects as well as the cursors corresponding to remote participants' operations, while at the same time they can interact with the objects. When combined with voice or character-based communication technology over the Internet, the connected Collaboration Tables allow multiple participants in multiple sites to share the same tabletop workplace and to collaborate simultaneously.

3 Usage Example

The developed Collaboration Table was applied to a product grouping task to demonstrate its effectiveness. In this example, the standalone setting of the Collaboration Table without a network connection was used. Participant identification was performed using color variation of the LEDs.

3.1 Task

The task was to map the similarities of products according to their design attributes in two-dimensional space. Initially small pictures of various cellular phone products were randomly located and projected on the table. Two participants were asked to work together, moving any of the pictures using their fingers with attached LEDs so that the products they thought similar would be closer together while those they thought different would be far from each other.

The pointing operation was defined as follows: when a pointer (LED) was placed on the table where an object existed, the pointer was recognized as having grabbed the object. When the pointer moved after having grabbed an object, it was recognized as dragging the object. When the pointer left the table, it was recognized as having released the object.

3.2 Results

Two participants could work simultaneously, without paying attention to their "turn" to move; they could move objects at any time they liked. None of the participants noticed any significant delay or error in system responses. After ten minutes of cooperation, one two-dimensional map of products was obtained, which represented the consensus of the two participants on the similarity of the products.

In addition, the operations of each participant could be recorded separately. Based on these movement records, an individual similarity matrix of products could also be defined for each participant. Using a multi-dimensional scaling technique, two different mappings could be obtained, illustrating the difference between the attributes that the two participants were making use of to classify the products; although both seemed to have thought the color attribute important for defining the similarity, one seemed to emphasize physical structure while the other seemed to place a greater emphasis on functionality.

4 Discussion

The developed prototype of Collaboration Table highlighted some technical problems. One of the major problems comes from the viewing angle of the image sensor. When the size of the tabletop gets larger and a video camera with a larger view angle lens is used, distortion of the view may be a problem and calibration will require complicated calculation. One possible solution is to make the distance between the tabletop and the camera greater, for instance using mirrors, although such a scheme may be more likely to suffer from noise. Another solution may involve a combination with a large sized touch sensor. If a touch sensor that allows multiple pointing is available, it can be used to give the precise position of the pointer on the tabletop while the image sensor is used to identify participants, though this would make the system more expensive.

References

Keller, A. I., Hoeben, A., & Stappers, P. J. (2000). Aesthetics, interaction, and usability in 'sketchy' design tools. *Exchange Online Journal, 1*(1). Retrieved January 25, 2003, from http://www.media.uwe.ac.uk/exchange_online1/exch1_article4.php3

Keller, A.I., Stappers, P. J., & Hoeben, A. (2001). TRI: Inspiration Support for a design studio environment. *International Journal of Design Computing, 3*, 17. Retrieved January 28, 2003, from http://www.arch.usyd.edu.au/kcdc/journal/vol3/dcnet/keller/

Stappers, P. J., Keller, A. I., & Hoeben, A. (2000). Listen to the noise: 'Sketchy' design tools for ideation. *Exchange Online Journal 1*(1). Retrieved January 25, 2003, from http://www.media.uwe.ac.uk/exchange_online1/exch1_article5.php3

The Design of a Recollection Supporting Device A Study into Triggering Personal Recollections

Elise van den Hoven

Technische Universiteit Eindhoven and Philips Research
Prof. Holstlaan 4, 5656 AA Eindhoven
e.v.d.hoven@tue.nl

Berry Eggen

Philips Research and Technische Universiteit Eindhoven
P.O.Box 513, 5600 MB Eindhoven
j.h.eggen@tue.nl

Abstract

The work in this paper is carried out in the context of the design of a device which supports recollecting personal memories. This device aims to help people recollect or reminisce about their life, together or alone, at home. From the (autobiographical) memory literature we concluded that we needed to fill the device's database with triggers instead of the memories themselves. For building this device we were interested in the following question 'what are the most efficient triggers for recollecting'. We carried out a field-study with media-types as triggers, that could be incorporated in such a device: photos, videos, sounds, smells and physical objects. One of the conclusions from this study is that certain types of triggers in fact reduce the number of memories people write down during recall. Therefore more research on recollection triggers is needed before the design phase of a device which supports recollecting can be finished.

1 Introduction

This paper describes a study preceding the design of a Recollection Supporting Device. By the latter we mean a physical device which can help people recollect memories from their lives, for example when reminiscing with friends or relatives, in the living room on the couch.
The remainder of this paper will give insights into the psychological background of memories and a study will be presented which was aimed at answering the question: what are the most efficient triggers for recollecting personal memories.

2 Memories

Many people feel the need to do something with their memories. They want to share them with others, for example by showing them their photos and talking about their memories. A number of studies have been reported in literature that focus on applications that aim to address this user need (for an overview on digital pictures, see Frohlich et al. 2002). Less knowledge is available on the human experience of remembering personal events in life, and how that can be supported. Therefore we organized a focus group that was conducted to explore the concept of 'memories' or 'recollections' from a user's perspective and to look at what can be done with memories in an application like a Recollection Supporting Device. The most important result was our finding that the ideal 'recollection supporting' device should not contain the *memories* themselves, but the *triggers* to those memories. We looked for support on this finding in human-memory theories.

2.1 Theory

The Constructionist approach to memory (see Guenther, 1998, for an overview) is generally believed to be a valid paradigm of how human memory works. This approach is based on the role of memory in anticipating the future. An important principle is that individual memories change connections between ideas and concepts in the corresponding cognitive systems, which results in the storage of regularities and unexpected deviations. Constructionists believe that remembering is synonymous to reconstructing, based on the information stored. Therefore, forgetting means that too much stored information has changed adaptively, making reconstruction impossible. The Constructionist approach makes clear that, although people are not aware of this, recollections can change over time according to changed knowledge and beliefs.

Both this information as well as the reconstruction process itself support our idea of offering recollection triggers (in a Recollection Supporting Device) instead of the memories themselves. First of all, presenting 'the' memories is not possible, since they are in the owner's head and can change. But if there would be a way of presenting the original memory this could have the same effect as the finding of Conway and Pleydell-Pearce (p. 266, 2000): 'recall of memories that were inconsistent or dissonant with a lifetime period caused strong cognitive reactions'. Therefore, we decided for our experiment not to present 'the truth' in memories but to use triggers instead. In addition, we also did not want to judge the recollections of the participants on their authenticity.

Autobiographical memory is the part of human memory concerning the events of one's life. Conway and Pleydell-Pearce (2000) describe three levels of specificity in autobiographical memories, namely: lifetime periods ('when I was going to primary school'), general events ('every day I would walk home') and event-specific knowledge (ESK), where the ESK represents the smallest unit of a memory or the lowest level of specificity ('but one day I saw this cute puppy').

There is not much literature on triggering of specific autobiographical events, but there is one on autobiographical-picture triggering by Burt et al. (1995). They found that information in photos showing an *activity* brings up most memories followed by *location* and *participants*. The assumed explanation for this order is that most people join in activities mostly with the same people, often in the same location, whereas the activity varies most and is therefore the most unique characteristic. The time and date are also unique but, according to the same study (p. 62), 'time (date) information is not routinely stored in [human] memory'. Another important study by Gee and Pipe (1995) was carried out with physical objects triggering 6- and 9-year old children's recall. (The objects were introduced during an event with a magician and they were later shown to the children during the interviews.) One of their conclusions was that participants interviewed with objects recalled more and more correct information than did participants in the standard interview without objects. Another important finding was that the object triggers had a greater effect on participating children compared to children that observed the magicians activities.

'Compared with memories evoked by photographs or names, memories evoked by odors were reported to be thought of and talked about less often prior to the experiment and were more likely to be reported as never having been thought of or talked about prior to the experiment' (p. 493, Rubin et al., 1984). A recent study into the effect of olfactory and visual triggers (Herz and Schooler, 2002) found that 'naturalistic memories evoked by odors are more emotional than memories evoked by other cues' (p. 21). Since this paper focussed on reported emotionality, the important aspect for the current study is that odors did trigger additional memories after triggering verbally. Based on this overview we decided to use triggers instead of 'memories', to let the participants create their object-triggers and to continue the rest of our idea, since such a study was not done before.

3 Triggering Experiment

We conducted a study to measure which type of recollection trigger is most efficient: photos, sounds, video, smells, or physical objects. This study consisted of two parts. The first part was a one-day unique event and the second part was the completing of two questionnaires: one with and one without a trigger. Before the actual experiment took place a pilot was run with two participants, in order to test the different activities on suitability and the time needed.

3.1 Part One – The Event

68 participants joined in a one-day unique event to Archeon, a history-theme park in the Netherlands. This day the Archeon was only open for us and followed a programme specified and optimized by the authors to ensure that as many experimental requirements as possible were met. Since the participants (to ensure a balanced gender-mix we only allowed adult couples) had to take part in five activities, the most important requirement was that all activities should last equally long, which was decided on 20 minutes. Because the activities could only be done by a maximum number of people at the same time the group of 68 people was divided in 5 subgroups, each group having two guides instructed by the authors before the start of the event. The ten guides were necessary to make sure that each group was in time for their activities at the right location and to collect memory triggers to be used in the second part of the experiment. Each group had the same order of activities, only the start (and thus the end)-activity was different.
The following two (out of five) activities were optimized such that we could gather unique material for each subgroup and for each trigger type:

1. Making a fibula - creating an ancient pin (the predecessor of the safety pin), from copper wire using a hammer, a pair of nippers and a piece of wood, while the room was smelling of vanilla incense,
2. Making felt - turning washed sheep's wool into felt with use of olive soap, and then creating a felt bracelet with it.

We used the following material for triggering the participants' recollections when they were filling out their questionnaires: photos (made by the guides), videos (made by the guides), artifacts (created by the participants themselves during each activity), sounds (recorded by the guides), smells (either copied from the activity [olive soap for making felt] or were applied by the authors [vanilla incense sticks were lit in the room in which fibulas were made]).
The guides of each subgroup were instructed to tape and make photos containing: activity, location and participants (based on Burt et al., 1995).

3.2 Part Two – The Questionnaires

The 68 participants had to fill out two free-recall questionnaires each, one with and one without a trigger. The questionnaires were administered in a living-room setting, 4 or 5 weeks after the Archeon-event. 9 Participants received two questionnaires without trigger. They served as control group to enable testing for order effects and differences in the two types of activities we asked them to recollect ('the making of a fibula' compared to 'making felt'). Furthermore, all six independent variables were balanced over all participants and their questionnaires: gender, activity, order, Archeon-group, days after Archeon they filled out the questionnaires and most important the condition (trigger type). We choose for written accounts compared to verbal accounts for logistic reasons, we now could test more participants in a shorter time frame.

3.3 Data Processing

Since we set-up a unique event which was used for triggering a month later, we could only check the data for ESKs (see Section 2.1). Our method counts the total number of ESKs, scores the type of memory and determines the memories' detailedness, this will be described elsewhere.
This method does not judge the content of memories (this in contrast to other recall studies) it only decides what is definitely not a memory, for example remarks like 'I am not sure about this'.
Two researchers were trained for about 10 hours on the pilot questionnaires before counting independently the actual 136 questionnaires with the newly developed method. Their ESK-data gathered showed a correlation of 0.97. This indicates that if users of this method are trained well, they can count memories in a reliable way.

3.4 Experimental results

Two main results will be presented here. The first one concerns general absolute values scored in this experiment. For example, the average number of ESKs for all participants was 18.5 (St.dev.=9.6), where there was no significant difference between the conditions ($F(df=5,63)=1.88$, $p=0.11$) but NoTrigger and Smell scored highest (Mean=19.6) and Object and Sound lowest (M=15,4). The time it took people to fill out their questionnaires was on average 15.6 minutes (S=6.6), which was significant ($F(5,63)=4.01$, $p=0.003$) and where Smell was faster (M=12.1) than the rest (M=[15.4,17.2]). The average number of details per questionnaire was 127 (S=68), which was not significant ($F(5,63)=2.09$, $p=0.078$) and where most details were generated in the NoTrigger and the Smell condition (M=134) and the least in the Sound-condition (M=106).
The second main result (based on differences between the two questionnaires from each person) is that the NoTrigger condition seems to generate an equal or higher number of ESKs relatively than the triggered conditions (see Figure 1). Actually, we found that women produce significantly less ESKs when they are triggered compared to when they are not ($t(30)=-2.21$, $p=0.035$). This effect was not found for men ($t(27)=-0.71$, $p=0.486$).

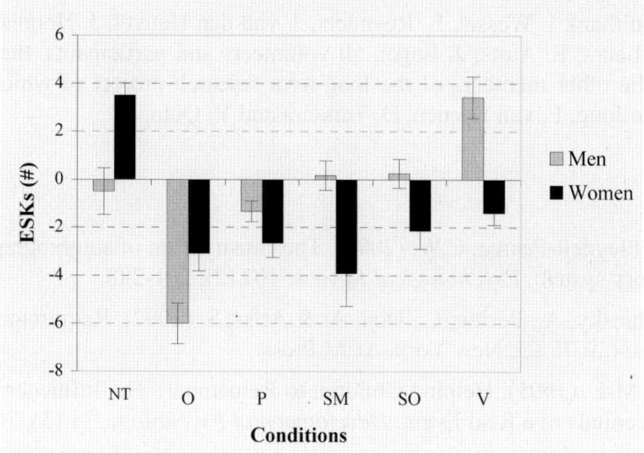

Figure 1. Differences in number of ESKs, NoTrigger-values subtracted from Trigger-values, per gender: NoTrigger (control group), Object, Photo, Smell, Sound and Video.

4 Discussion & Concluding remarks

Based on the theory in Section 2.1 we expected to find more ESKs in certain triggered conditions as opposed to the NoTrigger-condition. But this was not the case. Possible explanations for this finding could be:

a. Filter-effect. From the results described above we got the impression that a trigger has a filtering and restricting effect on the number of memories. Perhaps the more information is given the more constraints are put on the internal search;
b. Speed. It is known from literature that sensory triggers can change mood quickly, but then the memory comes more slowly (Rubin et al., 1984), perhaps this has an influence on the total number of memories recollected for certain triggers;
c. Person-dependency. Triggers might be personal, some people say they have preferences, but this does not prove that those modalities (or combinations) are also internal preferences. Furthermore age could also play a role, which could explain why Gee and Pipe (1995) found that children recalled more in conditions with object-triggers compared to without objects;
d. Long-term effect. We expect certain types of triggers to work better for older events.

We concluded that for our Recollection Supporting Device more research is needed. We do not yet remove triggers from our design requirements, since we believe that triggers might also play an important role in other dimensions of recollecting, for example:

a. Pleasure. We did not test whether certain trigger-types increase the enjoyability of recall;
b. Mood. Olfactory triggers increase emotionality (Herz and Schooler, 2002), perhaps this mood creates a better context and therefore improves other aspects of recall;
c. Intensity. Since we did not look into the content of the memories, there might be a difference in certainty of the reconstructions or intensity of the recollective experience.

Based on the results described above and a planned study we will build a Recollection Supporting Device.

5 Acknowledgements

The authors like to thank I. Wessel, L. Reynders, J. van den Heuvel, J. Hoonhout, G. Hollemans, B. de Ruyter, A. Leurs, E. Aarts, J. Engel, all volunteers and participants, the Archeon for their cooperation and the other members of the long-term research project of which this study was a part: E. Dijk, N. de Jong, E. van Loenen, D. Teixeira and Y. Qian.

References

Conway, M.A. & Pleydell-Pearce, C.W. (2000). The construction of autobiographical memories in the self-memory system. *Psychological Review*, 107 (2), 261-288.

Frohlich, D., Kuchinsky, A., Pering, C., Don, A. & Ariss, S. (2002). Requirements for Photoware. *Proceedings of CSCW'02*, New York, ACM Press.

Gee, S. and Pipe, M-E. (1995). Helping Children to Remember: The Influence of Object Cues on Children's Accounts of a Real Event. *Developmental Psychology*, 31 (5), 746-758.

Guenther, R.K. (1998). Human Cognition. Upper Saddle River, NJ: Prentice Hall.

Herz, R.S. & Schooler, J.W. (2002). A naturalistic study of autobiographical memories evoked by olfactory and visual cues: Testing the Proustian hypothesis. *American Journal of Psychology*, 115 (1), 21-32.

Rubin, D.C., Groth, E. & Goldsmith, D.J. (1984). Olfactory Cuing of Autobiographical Memory, *American Journal of Psychology*, 97, 493-507.

Evaluating technologies in domestic contexts: extending diary techniques with field testing of prototypes

Henriette van Vugt and Panos Markopoulos

Industrial Design, Technical University of Eindhoven
PO Box 513, 5600 MB Eindhoven
henriett@cs.uu.nl and P.Markopoulos@tue.nl

Abstract

This paper discusses the use of diaries in combination with field-testing of prototypes, as a technique for requirements gathering and for user testing. The effectiveness of this approach combined with logging, interviews and questionnaires was assessed with a case study. This case study concerned the design of an awareness system, a computer mediated communication tool that supports related people to feel in touch with loved ones. The results were encouraging, revealing strengths and weaknesses of the adapted diary method.

1 Introduction

1.1 Application domain

Awareness systems are a class of computer mediated communication (CMC) systems that support individuals to maintain a peripheral awareness of each-other's activity, similar to that which we have of our co-worker across the corridor or the neighbour next door to our home. In this paper we focus on the use of an awareness system to help people stay in touch with remote friends and family. Traditionally this need is served by letter writing, telephony and, more recently, e-mail. Special purpose systems have been used experimentally for awareness purposes in working contexts, as for example the early work in Xerox PARC: Portholes (Dourish and Bly, 1992).

The interest in the domestic or leisure use of awareness systems is more recent. Several design concepts have been discussed by the Casablanca project (Hindus, Mainwarin, Leduc, Hagström & Bayley, 2001) and the Interliving project (Beaudouin et al., 2002). A recent study on communication needs of elderly (Romijn, Huinen, Philopoulos & Markopoulos, 2002) showed that communication with remote family members is a high priority for elderly people who live alone. Video communication at the home is, however, not accepted for social communication, one main obstacle being that it captures private information (e.g., clothes, presence of others in the room, untidiness, etc.) in a way that may be threatening or annoying (Bouwhuis, 2001). Consequently, we focus on CMC that does not use video based communication, but rather aims to help people stay in touch by selective, low effort and discrete capture and display of awareness information. Therefore, we digitalise a non-intrusive object commonly found in home contexts and add an intuitive interface which allows the user to show awareness both explicitly (send a short message) and implicitly (presence indication).

1.2 Assessment methods

Studying people's relation to technologies at their homes is challenging, especially when these technologies are still experimental and not used in current households. A crucial goal of user research is to ensure a fit of future technologies into the daily life of the user: Do such systems serve an actual need? Is the interaction suitable for the context? Are they perceived as useful after regular use? Studies of technology-use at home have increasingly endorsed principles of ethnography. By necessity, ethnographic studies of technology use at home must have a short duration, e.g., see (O'Brien, Rodden, Rouncefield, & Hughes, 2000) or (Dray and Mrazek, 1996). Considering home contexts, we note the following limitation to these methods:

- Contextual inquiry and participant observation techniques require the presence of the researcher at the home of the user. This presence interferes with many of the rituals, habits and intimate experiences that are part of domestic life.
- The experience of awareness that is experienced through the day and the week, is hard to assess with short term visits. Instead more prolonged studies are required.
- Ethnographic studies in households are particularly good for analyzing current technology rather than give feedback on new designs.

The need to analyse the experience with envisioned products in context, is partly addressed by experience prototyping (Buchenau and Suri, 2000). This technique relies on role play and using mock-ups in real context. In order to assess the long term use of technology, we combine such an approach with diary studies (Rieman, 1993). Diaries have mainly been used in the requirements gathering phase of product design and in user studies, e.g., (Brown, Sellen, & O'Hara, 2000). This research tool achieves a relatively high standard of objectivity (Brown et al., 2000), and reduces the need for participants to rely on their recollection of past experiences and without the experimenter having to be present during the study. Varying levels of fidelity and functionality of a prototype can serve to simulate the intended usage of a system over longer periods of time and through the day.

2 The case study

The case study approach (Yin, 1994) was used to evaluate the appropriateness and usefulness of the diary used in combination with field testing of a prototype in actual home contexts. Other data collection methods used were questionnaires, log files and debriefing interviews, which will be further described in the methods section. Before going into the prototype design, we first focus on the expectations concerning the usefulness of the (combination of) evaluating techniques. These have been made a priori, based on relevant literature, and described as indications that the method works or that it does not work.

- A diary encourages participants to think about the prototype and make their feelings concrete. Records of intimate or personal thoughts may be considered as a positive indication for the use of the diary technique. Dry, factual reporting of communication taking place could show the inadequacy of the method to capture affective aspects of how users experience this medium for interpersonal connectivity.
- The various techniques should supplement each other. Conflicting or complementary data from the data collection techniques used is considered as a positive indication for their combination. No new insights from the diaries compared to logs, interviews and questionnaires, would be an indication of their inadequacy for our purposes.
- The methods can assess user experiences and opinions concerning the prototype. The content of the diary entries, e.g. whether it incorporates feelings and opinions, and the understanding and involvement shown by the participants are indicators to validate or reject this expectation.

To assess the evaluation methods, a prototype system was created to gauge the reaction of people to awareness systems. Photographs were a central concept in designing the prototype, because they are a typical home decoration, have personal value and can play an important role in social interactions. The system designed was a photo-frame that allows for sharing this intimate awareness over distance. It can further be described as follows, referring to screen prints of the interactive prototype which was written in Macromedia Director:

Two photo-frames are connected to each other over distance, and react directly on input. Each frame shows a photo of the other frame-owner. Two different connections are possible. First, if a frame is looked at, a light is turned on in the other photo-frame (Figure 1). This light fades away in time. Second, greetings can be sent from one frame to the other (Figure 2). Some greetings are pre-recorded, and are textual (e.g. 'Hello!') or pictorial (e.g. a smiling face). One can also type in short, textual messages. Likewise, one can express what is felt when watching the photo. Greetings will fade away in time, just as the light. They can be recalled and retrieved (Figure 3).

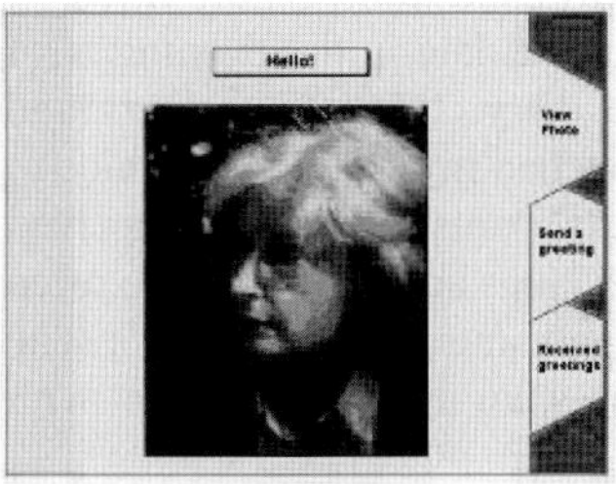

Figure 1: View Photo

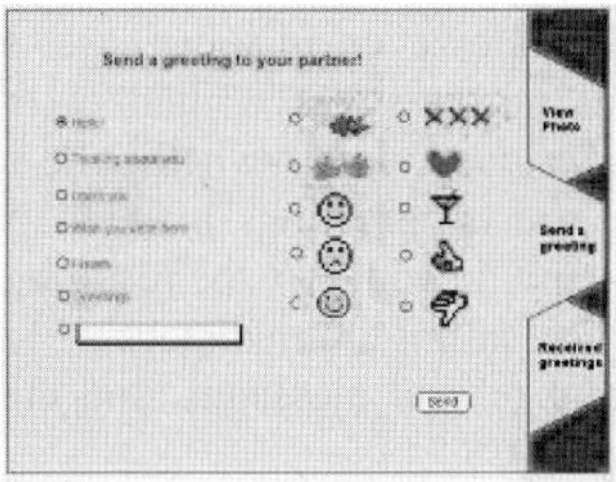

Figure 2: Send a Greeting

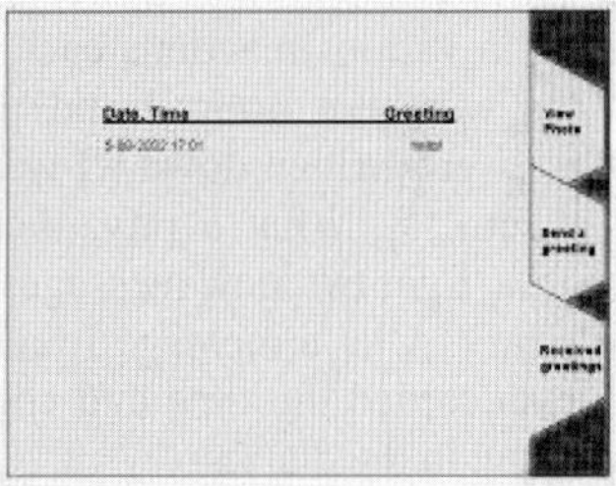

Figure 3: Received Greetings

3 Method

Three couples participated in the study. Each two participants forming a couple lived in different locations in the Netherlands or abroad, over a 90 minutes drive from each other. All the participants had sufficient knowledge of the English language and computers. All three couples (two sisters, a father and daughter and a boyfriend and girlfriend) had frequent telephone contact and saw each other occasionally.

A laptop PC *C-4110 LIFEBOOK* of Fujitsu was used with a standard Microsoft mouse. Windows 98 and Macromedia Director 8.5 were installed. The PC was connected to the mains power supply at the participants' homes and was left constantly on, without interruption, by, for example, screen savers. A prototype personalised to each test participant, based on settings (alarm times, greetings and photograph) chosen by the other person of the couple, was shown on the screen. After personalisation of the prototype with one person of the couple, the experimenter explained the study, the diary and the prototype to the second person of the couple. The portable PC was put in the living room. For the next three days the prototype would produce output based on the alarm times and greetings, and the participant would interact with the system and write in the diary. After three days, the system was disconnected, an in-depth interview was held and the questionnaire filled out.

The diary encouraged the participants to write about their experiences concerning the system. It was semi-structured, guiding participants where to write what, without restricting them in what they should and shouldn't indicate; the first section belonging to a 'day' was an open space to write the time and experiences, and the second section existed of pre-defined questions, e.g. 'What contact did you have today with your partner?'. The questionnaire measured user experience in a quantifiable way and consisted of 15 items with a 1 to 5 rating, e.g., 'It is important to know that my partner thinks about me' or 'A photo of my partner in my home connects me to him/her'. Further, log files of the participants' interaction with the prototype were kept. A debriefing interview was conducted after the diary was completed, to revise things written in the diary, extend on it and detect problems. This is common practice in diary studies (Rieman, 1993).

4 Results

Each data collection method seems to gauge the user experience from a different angle. The result presented here are, though, mainly concerned with the diaries and debriefing interviews. The diary gave us a sense for the participants' experiences when interacting with the prototype. Participants wrote clearly about their experiences and interactions concerning the system and their partner. When participants noticed a greeting or sent one, the diary was always and immediately filled in, resulting in various reactions each day. However, in some cases such as late in the evening, sent greetings were not noticed and thus nothing was written in the diary. Intimate feelings and reactions were written down ('I was curious to know whether he watched my photo', 'It is nice to look at him...' 'What a pity, there are no greetings'). Often participants reacted on received messages ('I got two new messages') and, in addition, the resulting actions and feelings were often made clear ('After a message I feel like calling immediately', 'He gave me a blink. Great!'). The log files recorded the time and content of sent messages, and these were mostly consistent with those found in the diaries.

The debriefing interviews extended on this information which makes it more valuable. The participants reported watching the frame to see whether new messages had arrived ('Handy that the messages can be reviewed'). One participant suggested that the message should remain visible until it is noticed. Further, the interviews aimed at detecting potential problems concerning the diary or prototype, and extended on phrases written in the diary. The use of the prototype with the diaries helped the participants think about the system ('Would be nice if it recognizes who I am', 'The prototype should be reliable', 'I don't want a camera in my house') and formulate ideas ('Message should stay on the screen'). Two participants said that the feelings for their partners were not increased when using the prototype. However, they did indicate in their questionnaire responses that they thought more about the other person. A selected group of people, namely lovers living apart, were seen as the potential user group of devices for social contact.

5 Discussion and Conclusion

This paper focussed on a combination of methods to evaluate a prototype in a participatory, contextual design setting. The photo frame prototype helped to gain a common understanding between participant and experimenter regarding awareness technology. The prototype and the diary let participants experience the system in a limited but effective manner. Further, it made them think about the system, make experiences concrete and formulate (design) ideas and opinions, as data of both diaries and interviews show. Log files and questionnaires, on the other hand, resulted in quantitative data; just as valuable for improving the design, but beyond the scope of this paper. The variety and kind of data obtained suggests that the various techniques indeed seem to supplement each other, as expected. Overlap in data strengthens the evidence, and the variety in data results in a more complete picture of the situation. Therefore, we think that a combination of methods is advisable in studies such as these concerning long term experiences with domestic technology. Such a study is, though, quite demanding for the participants. Having a system in the living-room, having to interact with it and write a diary can not be asked for more than a few days. The evaluation of the prototype gave useful insights, despite the fact that it was only a simulated prototype. Clearly, increasing functionality and realism will enhance the validity of the findings, but it is our contention that useful results can be obtained by testing early prototypes in the field in the manner illustrated.

Several limitations concerning the research should be mentioned. First, only a few participant couples were used. The exact contribute of the various methods is therefore not exactly clear, as the data regarding this subject is only limited. Second, the period of investigation was quite short, in comparison with real-life use. For investigations on the long term, other investigation methods might be needed than the ones described.

6 References

Bouwhuis D.G. (2000). Parts of life: configuring equipment to individual lifestyle. *Ergonomics*, 43(7), 908-919.

Brown, B.A.T., Sellen, A.J. and O'Hara, K.P. (2000). A Diary Study of Information Capture in Working Life. *Proceedings of ACM CHI'00 Conference on Human Factors in Computing Systems*, 438-445

Buchenau, M. and Suri, J. (2000). Experience prototyping. *Proceedings of DIS '00,* 424-433

Dourish, P and Bly, S. (1992). Portholes: Supporting Awareness in a Distributed Work Group. *Proceedings of ACM CHI'92 Conference in Human Factors in Computing Systems*, 541-547

Dray, S and Mrazek, D., (1996). A day in the life of a family: an international ethnographic study. In: D. Wixon and D J. Ramey (Ed.), *Field methods casebook for software design* London: John Wiley & Sons Inc, 145-156.

Hindus, D., Mainwaring, S.D., Leduc, N., Hagström, A.E., & Bayley, O. (2001). Casablanca: Designing social communication devices for the home, *Proceedings of ACM CHI'01 Conference on Human Factors in Computing Systems*, 325-332

Beaudouin-Lafon, M., Bederson, B., Conversey, S. Eiderbäck, B. and Hutchinson, H. Technology Probes for Families. Retrieved July 24, 2002, from http://interliving.kth.se/papers.html

O'Brien, J., Rodden, T., Rouncefield, M., & Hughes, J. (2000). At Home with Technology: An Ethnographic Study of a Set-Top Box Trial. *ACM ToCHI*, 6(3), 282-308.

Rieman, J. (1993). The diary study: A workplace-oriented research tool to guide laboratory efforts. *Proceedings of ACM INTERCHI'93*, 321-326

Romijn, O., Huijnen, C., Philopoulos, A., & Markopoulos, P. (2002) Supporting Relationships with Awareness Systems.. *Proceedings of the 16th British HCI Conference*, volume 2, 78-81

Yin, R.K. (1994). Case study research. Design and Methods (2nd ed.). California: Save Publications.

A Collective of Smart Artefacts Hopes for Collaboration with the Owner

Elena Vildjiounaite, Esko-Juhani Malm, Jouni Kaartinen, Petteri Alahuhta

Technical Research Centre of Finland
Kaitovayla 1, P.O.Box 1100, FIN-90571, Oulu, Finland
{Elena.Vildjiounaite, Esko-Juhani.Malm, Jouni.Kaartinen, Petteri.Alahuhta}@vtt.fi

Abstract

This work studies the interactions between a user and a context-aware system built up from a number of everyday objects augmented with sensing, communicational and computational capabilities, developed for the purpose of helping the user to collect all the personal belongings necessary for a journey, without forgetting anything. We show that carefully planned cooperation between the user and the context-aware system helps to deal with the intrinsic difficulty of reasoning about the user's context and uncertainty in the choice of appropriate actions for the system. The results show that the proposed interaction capabilities, which help to provide services with a relatively simple system configuration, are useful and acceptable to different people.

1 Introduction

The main idea of ubiquitous computing is that computers will be everywhere in the world, together providing more services for the user. At the same time, by taking into account the current user's context, the amount of attention the user has to pay to the system can be reduced. The irresistible appeal of the idea has led to the large family of diverse context-aware applications, each of them functioning on the assumption that context recognition works perfectly. The user's context consists of very many factors, however, and not all of them can be sensed or inferred by a computer system. Dey, Mankoff, Abowd & Carter (2002) propose an architecture for "building more realistic context-aware applications that can handle ambiguous data through mediation" (cited from p.121). The term "mediation" refers to the resolving of ambiguity by involving the user in a dialogue with the computer system. The architecture was evaluated by building three applications: In/Out board, CybreMinder and Word Predictor, in all of which ambiguities were resolved by interaction between the user and the central (or only) computer. The methods of interaction were the following: speech, typing at a keyboard and, in one of the applications, docking a special device called an iButton. In the context-aware tourist guide (Cheverst, Davies, Mitchell & Friday, 2000) the user is involved in solving problems of location estimation by dialogue with a hand-held computer.

Unlike the authors mentioned above, we have developed a context-aware system built through collaboration between a number of smart objects, each of them performing sensing, exchanges of radio messages and reasoning. This approach allows smart spaces to be deployed everywhere in an ad-hoc manner without any need for a powerful central node. The central node is needed mainly for exchange of messages between the user and the smart objects. Accordingly, dialogue between the user and the central node is not enough to resolve ambiguities in such system, so that developers had to design special methods of interaction with the smart artefacts. We describe here the methods of collaboration between the user and the smart objects which were found to be useful in our system.

2 Application Scenario

The constantly decreasing size of microcontrollers and sensors makes it possible to have computation embedded into literally any artefact, moving us towards truly ubiquitous computing (e.g. see the PRINTO project web page). Our application scenario is presented in Figure 1.

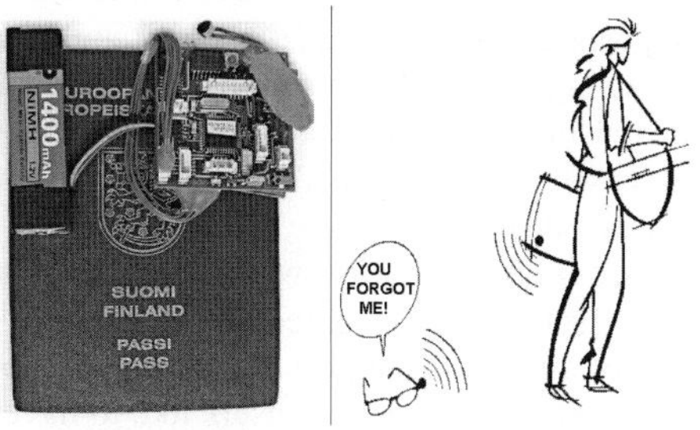

Figure 1: Smart object and application scenario

In the application scenario the computer system helps the user to collect all the things she planned to take with her (or usually takes with her for similar events). Such a system could be useful for luggage handling problems at airports, for the transportation of goods, and in almost every home. Just count how many times we have to pack numerous bags for business, weekend or holiday trips, which include clothes, medicines, books, tools, etc, etc... If all the objects around us were smart and some of them knew that they had to leave for a journey together at a certain time, why should we not shift the burden of writing a long list of items and striking them off one by one?

3 System Description

The system prototype was built on generic hardware developed for research purposes in the Smart-Its project (see Smart-Its project web page). Everyday objects become smart after attaching Smart-Its boards to them (see Figure 1).

Each Smart-It consists of a sensor board and a core board, each containing a PIC microcontroller. The core board is responsible for RF (radio) communication and the sensor board for sensing and for running the application. The sensor board contains a two-directional accelerometer, light, temperature and pressure sensors, a microphone, three LEDs (Light Emitting Diodes) and a loud speaker. The two boards are about 4 x 5 cm in size and are mounted on top of each other.

The user can exchange information with the system via a Compaq iPAQ Pocket PC with Smart-It attached to it via a serial cable. The Pocket PC has a list of journeys and lists of items corresponding to each journey in its memory. After the user has edited or confirmed the journey and the items to be collected, the estimated time of departure and the list of items are broadcast by the radio. The items with IDs on the list become members of the group which must leave on the journey together.

The group members try to detect the current context by exchanging messages, which contain the item's ID, type of movement and information about the RF beacons the item can hear. Each item analyses the messages it has received from the other members every five or six seconds and broadcasts the result of the analysis (whether the item is moving together with the other members or not). The central node (Pocket PC) collects these members' opinions, compares them and presents the information to the user, provided that the confidence level exceeds a certain threshold. It should be noted that, according to our tests, members' opinions differ only when the situation changes (e.g. items start or stop moving), due to the absence of synchronisation between boxes. After a couple of seconds all the boxes come to the same opinion independently. This means that all the members in our system have a good understanding of the overall situation.

4 Resolving of Ambiguities

Despite the relative simplicity of the application, we have found that reasoning about the context and the system's choice of appropriate actions leads to many ambiguous situations. These arise partly from the fact that we are using generic hardware instead of developing a specialised system. We believe, however, that increasing the system's complexity is not likely to guarantee perfect autonomous behaviour in all cases, and thus mediation methods for resolving ambiguity are inevitably needed in context-aware systems.

We argue that involving the user in collaboration with the system makes sense in the following situations: first, when privacy issues are concerned. For example, our sensor boards include microphones, as the processing of audio data improves proximity detection (Schiele & Antifacos, 2001). On the other hand, it is questionable whether users like to have working microphones with them all the time, and whether they are interested in knowing what sensors are included in the system.

Second, when dealing with malfunctioning items. The most certain way of detecting that the user does not have all the necessary things is based on the absence of messages from some items during a given time period. The absence of messages does not necessarily mean that the item is out of range, however, as it might simply be malfunctioning (our prototype is not energy-aware and does not measure the strength of the received RF signal). If the latter is the case, the system probably should not deliver the frequent reminders that the item has been forgotten. Instead, it should let the user mark the item as unreliable and to treat it differently from the others, so as to remind the user to visit a service centre later on, for instance.

Third, when the moment at which an alarm is given is very important. Even if our system were capable of measuring signal strength, this would not help to give an alarm fast enough in the following situation: imagine that several people are collecting their belongings simultaneously in the same place and some things are mixed up. Then they wait for a bus, and one of them jumps into the bus as it is leaving, carrying a misplaced item. If the user wants to prevent such problems, this can be done easily by shaking the assembled pack. The system can easily distinguish between items having similar movement patterns at the same moment and items with a different movement pattern (see Holmquist et al., 2001) and will immediately inform the user.

Fourth, when configuring the placement of beacons. Our system is built in such a way that it can give the user its conclusions about the current state of affairs at different levels of confidence. Many people collect their things by first bringing all of them into one room and only after that packing them into a suitcase (mentioned by our users). The system can detect (with less certainty than on the basis of the absence of messages) that some things are missing by using both beacon information and accelerometer data. We have noticed, however, that its conclusions based on beacon data can be completely wrong, because the communication range is affected by reflection. Sometimes it is enough to move a beacon half a metre to get correct results, and again it is the user who is needed to tune the system.

The last group of situations arises from the flexible and spontaneous nature of human behaviour. Context-aware systems should not act on the assumption that users will plan everything carefully in advance. Concerning our application, this means that the list of items to be taken on the journey

will probably change several times during packing, and some items may be bought on the way to the airport or taken from a refrigerator at the last moment.
After considering all the above issues, we decided to include the following interaction capabilities in our system:
1. Sensor announcements: the system notifies the user which sensors are included, particularly types such as microphones, cameras and fingerprint sensors. The system allows the user to switch these off at any moment (in our current prototype this is simulated.)
2. Explanation abilities: the system is able to acquaint the user with the reasons for its decisions. The user can ask the system's opinion about any item, and it will inform the user whether it considers a given item as having been forgotten or not, at what confidence level, what was the item's movement type and beacon data, and when the last message was received from that item.
3. Error recovery: the ability to recover from mistakes made by either the system or the user. The user can add or remove items from the list and change the "departure time" of any item. If there are problems with beacons, the user can obtain an opinion from the system based only on accelerometer data, without considering the beacon data, or else try to move some of the beacons.
4. Special actions: the system suggests to the user a set of actions which might help it to increase the certainty of context detection. One of the options is to shake an assembled pack or move it back and forth. This interaction technique, as stated by Holmquist et al. (2001), is very easy to use, because it does not matter what kind of movement is carried out. Another option is to move the pack from one room to another, provided that the beacons work reliably.
5. Personalised behaviour: the ability to take the user's preferences into account concerning cases in which everything goes well. In such situations there is always the question of whether the system should remain silent and let the user wonder if it is still functioning, or whether the system should interrupt the user with an "OK" message. (The "special action" of shaking the assembled pack serves this purpose well: i.e. if the user intentionally shakes the pack it means that a message from the system would be desirable.)

5 User Study

Twenty persons (ten men and ten women) of different ages, nationalities and professions took part in the user study. The system was demonstrated in the following scenarios:
Scenario 1: The user packs items into a suitcase and leaves, forgetting something, e.g. a wallet. After the system becomes sure that the wallet is missing, it sends an alarm to the user. This usually happens only after the user has left the apartment.
Scenario 2: The user packs items into suitcase and shakes it. An alarm is sent immediately.
Scenario 3: One of the boards (among the items to be packed or the beacons) gets broken. The system sends an alarm, and the user can find the reason by asking it for an explanation.
After that the users answered the following questions:
1. Would you like to use the system?
2. Would you use the option of switching off sensors such as cameras and microphones?
3. Would you use the "special action" of shaking an assembled pack?
4. Would you use the system's explanation capabilities?
5. Once everything is packed, would you like to receive an "OK" message? (As opposed to receiving a message only if something is missing)

Table 1: Summary of results

Question	1	2	3	4	5
Number of positive answers among 10 men/10 women	8 / 10	7 / 10	5 / 10	7 / 10	4 / 6

A summary of the users' opinions is presented in Table 1. One of the reasons for answering "no" to the first question was the lack of trust in computers, and another was unwillingness to use "memory prosthesis". One of the women was particularly happy with the shaking option and said that she would shake her handbag very frequently, because she was always worried whether she still had all her things with her. Most of the users agreed that this kind of system should also help the user to find forgotten items.

We were also interested in finding out by which medium users preferred to communicate with the system, because the communication device in the current version is a Pocket PC and we are thinking about using a mobile phone for this purpose in the future. Only nine of our users said that they would prefer a phone, however, four suggested a watch, two a device attached to the suitcase, and the others would have preferred a separate device.

We understand that users will not necessarily behave in the same fashion in a real situation as during an experiment conducted with a system prototype, but the results are nevertheless encouraging from the point of view that the people (particularly the women) had nothing against explicit interactions with smart objects.

6 Conclusions

The results of the user study show that people do not object to collaboration with smart personal belongings, in that the prototype context-aware system built up on generic hardware for the purpose of helping a user to collect things for a journey, was in general approved by the users.

The proposed methods of interaction between the user and a number of smart everyday objects, tested on 20 people, allow us to improve the certainty of context recognition, helped us to take personal preferences and privacy issues into account, and facilitated the providing of services with a relatively simple system configuration.

7 Acknowledgements

The Smart-Its project is funded in part by the Commission of the European Union under contract IST-2000-25428, and by the Technical Research Centre of Finland. We would like to thank all the people who took part in the user study for their goodwill.

References

Cheverst, K., Davies, N., Mitchell, K. & Friday A. (2000) Experiences of Developing and Deploying a Context-Aware Tourist Guide: The GUIDE Project. Proceedings of MobiCom 2000, Boston, August 2000, pp. 20-31.

Dey, A., Mankoff, J., Abowd, G. & Carter, S. (2002) Distributed mediation of ambiguous context in aware environments. Proceedings of UIST 2002, pp. 121-130

Holmquist, L.E., Mattern, F., Schiele, B., Alahuhta, P., Beigl, M. & Gellersen, H.-W. (2001) Smart-Its Friends: A Technique for Users to Easily Establish Connections between Smart Artefacts. Proceedings of Ubicomp 2001, Atlanta, GA, USA, Sept. 2001, pp. 116-122

PRINTO project web page. http://akseli.tekes.fi/Resource.phx/tivi/elmo/en/rolling.htx Retrieved February 04, 2003

Schiele, B. & Antifacos, S. (2001) Beyond Position Awareness. Proceedings of the Workshop on Location Modeling at UBICOMP 2001

Smart-Its project web page. http://www.smart-its.org. Retrieved January 22, 2003

Integrated Information System for Supporting Maintenance Activities of Nuclear Power Plants

W. Wu, T. Ohi, Y. Ozaki

Mitsubishi Electric Corporation
8-1-1, Tsukaguchi Honmachi, Amagasaki, Hyogo, 661-8661, JAPAN
{Wu.Wei, Tadashi.Ohi, Yoshihiko.Ozaki} @wrc.melco.co.jp

Y. Zhou, H. Yoshikawa

Kyoto University
Gokasho, Uji, Kyoto, 611-0011, JAPAN
{zhou, yosikawa} @uji.energy.kyoto-u.ac.jp

Abstract

Following the trend of deregulating electricity markets, the issues of cutting down cost of electricity generation is put in front of electricity suppliers. This paper proposed a new approach to decreasing maintenance costs of nuclear power plants (NPPs) in Japan. The current maintenance activities of Japanese NPPs were first reviewed to explain the condition-based maintenance and to describe two types of maintenance activities in current NPPs. An integrated maintenance information system is then proposed for the future maintenance activities coping with cutting cost and also maintaining/increasing the safety and reliability of future NPPs. The configuration and the functions of the integrated information system are explained in details. The integrated maintenance support system is applied to a maintenance scenario of a micro-gas-turbine to show its usability and to evaluate the system functions.

1 Introduction

Nuclear power plant (NPP) is a kind of high-risk systems. To maintain the safety and reliability of an NPP is the most important issue. The maintenance activities in NPPs have contributed to ensure the safety and reliability, and meanwhile, to improve consistently the capacity factor of NPPs in years. However, recently, the well-trained maintenance workers are getting older and would be retired in very near few years. How to maintain the safety and reliability of NPPs in such situations is an urgent subject. On the other hand, following the trend of deregulating electricity markets, the issues of how to cut down cost of electricity generation is put in front of electricity suppliers. It is said that[1,2] high cost of operation and maintenance is one of important aspects that made NPPs have poor competition with other electricity power suppliers.

Generally, maintenance activities can be categorized into following types:

(1) Time-based maintenance (TBM)
Conduct maintenance activities every constant period. In NPPs, it means that stop the plant periodically and inspect, test, replace, and modify the plant equipment so that the reliability is guaranteed to be over the necessary reliability.

(2) Condition-based maintenance (CBM)
Changes maintenance schedule by the condition of the equipment estimated by using various kinds of sensors that monitor the condition of plant equipment constantly. Stop the plant only if the plant reliability is decreasing and approaching to the necessary reliability. Fig.1 shows the difference between TBM and CBM. Compared with TBM, CBM is also referred as the future maintenance style. Moreover, CBM also has different styles according to the maintenance policy, such as reliability-centred maintenance and risk-based maintenance.

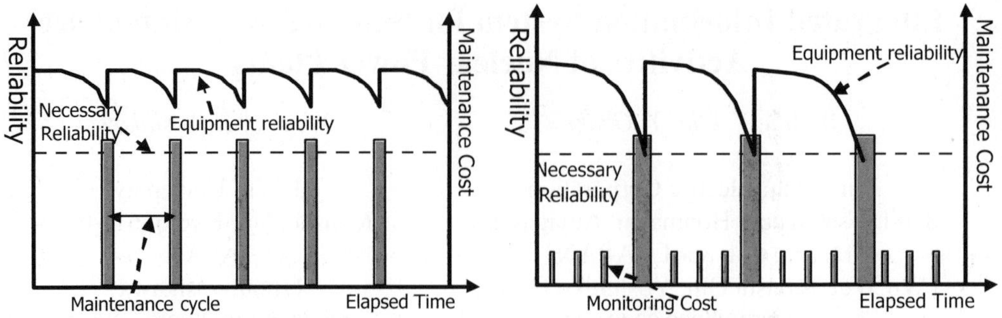

(a) The concept of TBM (b) The concept of CBM

Figure 1: The differences between TBM and CBM [3]

(?) Breakdown maintenance (BDM)
Different from the above, conduct maintenance activities only when the equipment is breakdown. Of course, such breakdown must have no influences on the total plant reliability.

For lots of years, the maintenance activities in Japanese NPPs are mainly carried out with TBM method, typically represented by the 13-months-regulation: Japanese NNPs must stop every 13 months to disassemble, inspect, test and re-assemble plant equipment. Such TBM did contribute to Japanese high-level plant reliability and safety, also in a period did give high-level plant availability. However, the approach required much cost and the further improvement of the capacity factor would be very difficult. Lots of efforts have shown the limitation of TBM. [3]

Considering both the aspects of costs and reliability, it has been suggested to find the best choice and combination of CBM, TBM and BDM based on various kinds of information and operation, maintenance experiences. In this paper, an integrated information system is proposed for supporting maintenance activities of NPPs so that the future maintenance style CBM could be realized, and in the same time, the maintenance cost could be decreased while the reliability and safety of NPPs could be guaranteed.

Basically, there are two types of maintenance activates in NPPs. One is the daily maintenance activity such as patrolling to inspect various kinds of plant equipment, collecting condition parameters of the equipment, identifying the status of the equipments, and so on. The other type of maintenance activities is performed during the refuelling period that is 13-months regulated by Japanese law, such as disassembling, inspecting, testing and re-assembling various kinds of plant equipment. The integrated information system proposed here will provide a common supporting framework for both types of maintenance activities. But in this paper, only the support functions for the daily maintenance activities will be explained since another paper[4] will give a detailed description about the support functions for the maintenance activities during refuelling period of NPPs.

2 Integrated Information System

The concept called as Satellite Operation & Maintenance Centre (SOMC) has been proposed as one of the solutions cutting down the total cost of NPPs[5]. SOMC will integrate the operation & maintenance activities of NPPs. Experts of maintenance will always stay at SOMC, they cooperate with field workers to manage maintenance activities of several plants. Specialists for maintenance would support field workers in a plant site. If necessary, a specialist will be dispatched to the site. It has been anticipated that the efficiency of the equipment and human resources would be increased through such organization. Considering the present regulation and social situation, it is difficult to shift to such integrated form immediately from the present form. In this paper, the integrated information system for supporting maintenance is one of efforts to realized SOMC in maintenance field. From a viewpoint of a human machine interface, the target of the development is strengthening situation-awareness (SA) for a single workers and team-awareness (TA) for a group of workers cooperating with each other.

The integrated information system for supporting maintenance activities of NPPs is consisted of three subsystems, corresponding to the filed workers, experts/well-trained workers in SOMC and the maintenance target equipment. As for the field workers, hands-free is the basic requirement for the support system. So wearable terminal is proposed as the support devices, such as personal digital assistant (PDA), or head-mounted display (HMD), or tablet PC. For the maintenance target equipment, a small-sized device called as Ubiquitous Computing Device (UCD) will be developed to serve as an information gate for various kinds of plant equipment. And for the experts/well-trained workers in SOMC, it is important to grasp whole maintenance tasks progress. So the monitoring and communication functions are the basic functions for supporting the cooperative tasks between the filed and SOMC.

In the following sections, the functions required for such integrated information system are explained in detail in the context of NPP daily maintenance activities, such as patrolling to inspect various kinds of plant equipment, collecting condition parameters of the equipment, identifying the status of the equipments

2.1 Wearable terminal for the field workers

As for the field workers, hands-free is the basic requirement for the support system. So wearable terminal is proposed as the support devices, such as personal digital assistant (PDA), or head-mounted display (HMD) or tablet PC. The required functions for the wearable device are listed below.

(1) Communication function with SOMC

Future maintenance activities in SOMC style may require workers to maintain more than one specific plant. It is impossible for field workers to know all plants so well as now they maintain only one specific plant. The communication functions with SOMC will provide the following information to filed workers;

-Navigation information so that filed workers can arrive at the target equipment certainly.

-Detailed information about maintenance tasks should be conducted

-Task instructions by the experts/well-trained workers in SOMC

(2) Communication function with UCD of maintenance target equipment.

In the context of NPP daily maintenance activities, main tasks are to check the condition of various kinds of plant equipment by collecting condition parameters. The communication with UCD would provide such parameters automatically rather than the current paper-pen-based methods. The filed workers will be notified when abnormal signals were recorded by UCD since last patrolling tasks.

(3) Information presentation function

As a wearable information terminal, it will present various kinds of information to the filed workers;

-The information about the maintenance procedures,

-The information about the conditions of the maintenance target equipment

-The navigation and instruction information from SOMC.

2.2 Support sub-system for the target equipment

In order to realize CBM, it is important to collect large amount of data to estimate the condition of the plant equipments. Furthermore, the data analysis method should be also developed to present meaningful information to the workers rather than just parameter values. In this support system, an industry-use compact computer called as UCD is developed to play such role. The functions of UCD are summarized as follows;

(1) Data collection function
UCD will be installed into the target plant equipment. It connects to various sensors that gather constantly the raw signals from the target equipment. The collected data will be then stored in UCD as historic data of the equipment and will be checked to notify if the equipment are beyond normal conditions.

(2) Communication function with field workers' wearable terminal
Due to the capacity limitation, UCD could not store big amount of historic data of the target equipment over a long time. So it will transfer automatically the data to field workers' wearable terminal when the workers conduct the daily maintenance activities. Also if the historical data are somewhat questionable, UCD will notify the field worker immediately.

(3) Navigation function
Such function is a combination result of the communication functions among UCD, wearable terminal device and SOMC. The location of UCD, i.e. the location of the plant equipment, is known beforehand. So by the wearable terminal device, UCD could inform SOMC indirectly about the location of the moving field workers. SOMC then could provide navigation information to the field workers.

2.3 Support sub-system for SOMC

As for the centralized maintenance station SOMC, it is important to grasp whole maintenance tasks progress, such as where are the field workers, what kind of maintenance tasks is conducted by he filed workers, and what about the progress and so on. So the monitoring and communication functions are necessary to be implemented. Moreover, various kinds of information related with various kinds of maintenance tasks must be prepared and stored in various kinds of databases,

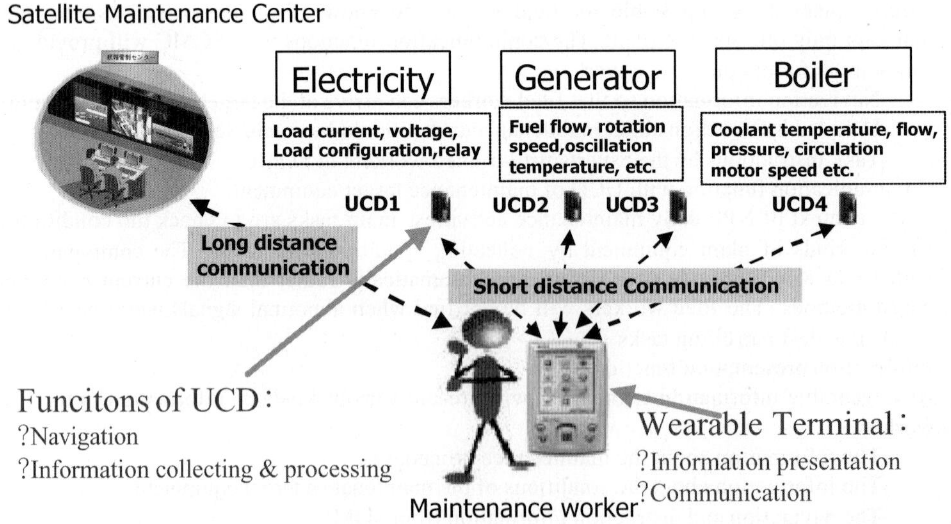

Figure 2: The image of the support system in the context of daily maintenance activities

such as maintenance procedure database, human error exempla database, plant layout database, and so on. Based on the communication and monitoring functions, SOMC will provide the filed workers with various kinds of information such as navigation, maintenance procedures, specific instructions, specific notification related to the past human error exempla and so on. All such support functions are developed to make sure that the maintenance tasks are conducted certainly and efficiently.

3 Application and function verification of the support system

The support system has been applied to an experiment to verify its functions and to review its usability. In the experiment, the support system is configured as follows;

(1) A micro-gas turbine
It is selected as the emulated plant equipment.

(2) A PDA and an HMD
They are used as the filed workers' wearable information terminal device in the context of daily patrolling tasks.

(3) An industry-use compact computer
It is used as UCD so that the data constant collection and communication function could be implemented easily.

(4) Bluetooth technology
This short distance communication technology is used to implement the communication function between UCD and the wearable information terminal device. The features of the technology reflect the concept of ubiquitous computing.

(5) Personal handy-phone system (PHS)
This long distance communication technology is used to implement the communication functions between SOMC and the wearable terminal device since Japanese NPPs has authorized the use of PHS in the plant.

Currently, a patrolling task scenario is designed to verify the functions of the support sub-systems, especially the functions of UCD and wearable terminal devices.

4 Summary

An integrated information system for supporting maintenance activities is proposed to enhance the competition NPPs in the deregulated electricity market. The system is also proposed to enhance the situation awareness and team awareness of maintenance workers in the future SOMC style. Patrolling scenario experiments are carried out to verify the functions and to show the usability of the support system. The results so far obtained show that the basic technology we outlined here is feasible. The further development will focus on how to apply it to the real maintenance tasks.

References

[1] M.Yajima: Energy Security-Theory, Practice, Policy-, Tohyoh Keizai Shinbunsha, 2002

[2] H. Yoshikawa, etc.:How Should We Advance Operation and Maitenance Technologies of NPP for the Age of Deregulated Electricity Market. In Proceedings of International Symposium On the Future I&C for NPP, pp329-336, Nov 2002.

[3] Maintenance Technologies on Nuclear Power Plants, Journal of the Atomic Energy Society of Japan, Vol.44, No.4, 2002 (In Japanese)

[4] T.Ohi, etc.: Development of an Advanced Human-machine Interface System to Enhance Operating Availability of Nuclear Power Plants, to be presented on HCII2003.

[5] T. Nagamatsu, etc: Information Support for Annual Maintenance with Wearable Device, to be presented on HCII2003.

Advanced methods of search query refinement in web environment

Pavel Žikovský
Dept. of Computer Science
CTU Prague
Karlovo nám. 13, 121 35 Prague 2
Czech Republic
xzikovsk@fel.cvut.cz

Pavel Slavík
Dept. of Computer Science
CTU Prague
Karlovo nám. 13, 121 35 Prague 2
Czech Republic
slavik@cs.felk.cvut.cz

Abstract

When we want to retrieve some information from the Internet, we have to know its location (i.e. URL), or we have to use a search engine. Some search engines are provided with a hard-wired list of categories, which can narrow the number of pages to search through. From another point of view, these categories are in fact a set of pages, which correspond to one keyword, most obviously to the name of a category. In this paper, we are describing novel methods and two novel tools for searching the Internet more efficiently. The first tool provides dynamic category evaluation, according to the user's needs (or previous search process) and cluster analysis. It is also capable of an analysis within a set of links at one single web page. The other tool computes and displays efficiency of each category (set of keywords, query). Both of these tools are equipped with 3D data output, which makes presented information more easy to understand and convenient?.

1 Introduction

Anyone who has searched the Internet using search engines extensively knows that one of the biggest problems is to define the query correctly; i.e. to map the properties of the document we are searching for to an appropriate set of keywords. This is not critical when we use a simple query, e.g. two or three keywords. Situation becomes more confusing when the complexity of query increases. When we define a complicated, multi-keyword query, it is very common that we miss the "right page" because there is a keyword in our query, which is "wrong", i.e. it filters our desired page out. This happens because keywords are generated by webmasters and each of them has his own way of defining keywords. One uses a keyword describing a certain fact, another one uses its synonym, etc. . These human aspects make the search process difficult.

Another similar problem is the case when we define a correct query, but the search results in a huge number of pages and we are not sure about another correct keyword, which might refine our query in a right way.

We can deal with these problems with a homogenous set of two tools. One of them, called "CrystalSearch" computes the efficiency of each possible combination of given keywords – it performs a decomposition of a complicated query – and therefore informs us about the efficiency of each combination of keywords. Finally, results are displayed in a 3-D form derived from InfoCrystal™ technology (Spoerri, 1996), which is easy and quick to understand. In comparison with a simple 2-D display form, the 3-D display has more possibilities to place objects while maintaining the ability to distinguish between them on the screen. The other tool, called "PlanetSearch", on the contrary, performs a cluster analysis within keywords in the set of resulting pages and suggests a group of possible keywords, which would most probably enhance the original query in a correct way. The result is displayed as a 3D self-rotating planetary model in VRML environment.

1.1 State of Art

Searching the Internet is one of the most frequent user activities on the web. Most users use simple search engines (i.e. AltaVista™, Yahoo, etc.) for that purpose. These services work as a simple database – they list all pages according to the user query which consist of which contain one or more keywords. Limitations of such simple search services on the Web have led to the introduction of metasearch engines (MetaCrawler) (Dreilinger D. and Howe A.E.,1996). A metasearch engine searches the Web by making requests to multiple search engines such as Google, AltaVista™ or Infoseek. The results from the individual search engines are combined into a single result set. Advantages of metasearch engines include a consistent interface to multiple engines and improved search performance (due to the larger database).

2 CrystalSearch

When we define a multi-keyword query, we are in fact searching for the intersection of n sets, where each represents a query for a single key word from our query. (See Fig. 1).

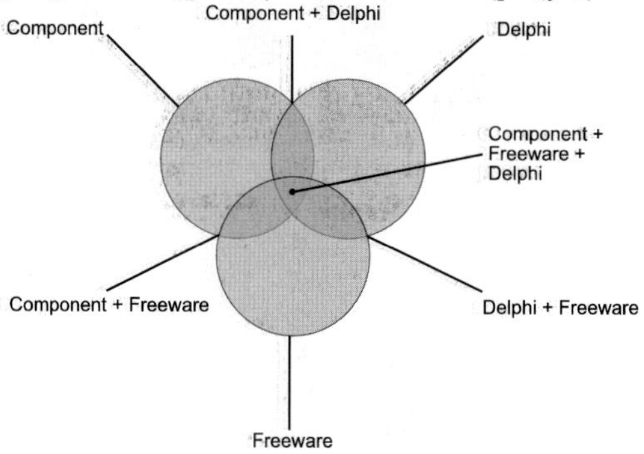

Figure 1: Intersection of three criterion sets

Unfortunately, this way of displaying intersecting sets (Fig. 1), called Venn diagram, is capable to display only intersections of three sets. Four and more sets are not displayable (Andersen, 1992). As we usually cannot limit word count of the query to three, this way of displaying search results would not fulfill our needs. Fortunately, there is a novel way of displaying more than intersecting sets in 2D, called InfoCrystal™. In Fig. 2 there is a scheme of the transformation a Venn diagram into InfoCrystal™. We begin with a common Venn diagram (a). In (b), there is the diagram divided into all its subparts. Finally, in (c), we can see an InfoCrystal™, which represents exactly the first picture (a). This scheme is easily extendable to more than three intersecting sets.

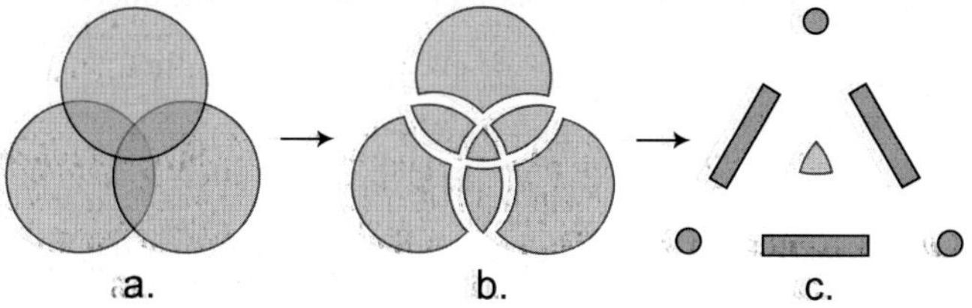

Figure 2: Transforming Venn diagram into InfoCrystal™

2.1 Using CrystalSearch

In our approach, CrystalSearch is directly connected to AltaVista™ search engine. It can be obviously connected to any other search engine. As the user enters a list of keywords (separated by space characters), CrystalSearch application composes a set of queries for all combinations of given keywords and sends it to the AltaVista™ search engine. The result of each of the created queries is then displayed in 3D iconic view. When the user moves a mouse pointer over a crystal, information about the query and its page count is displayed in the bottom of the window. After clicking the crystal, a web browser window with an appropriate AltaVista™ search page is displayed.

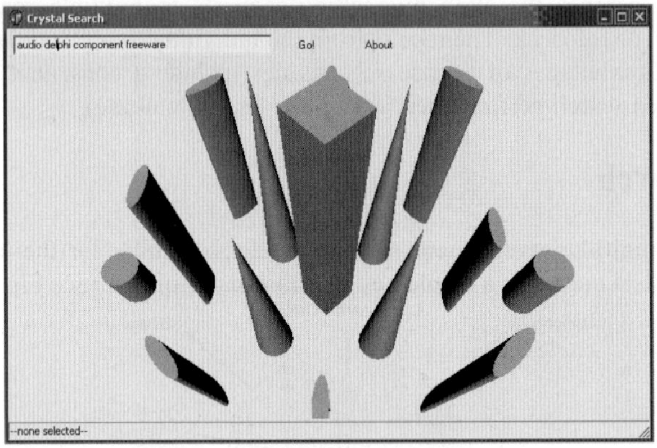

Figure 3: CrystalSearch application window with 3-D crystals

3 PlanetSearch

PlanetSearch is a tool, which helps us to find additional keywords, when we need to refine our query and we are not sure, what the keywords should be. This usually happens, when we obtain too many pages as a result of our query, and we have already used all keywords we know.

Each page indexed by some search engine has a list of keywords affiliated with it. As each query has its corresponding list of pages, which fulfills it, we can perform a cluster analysis (Kaski S., 1997) of the keywords from all the pages. Let us take an example. The first keyword we enter is *music*. Planet Search performs a query with this keyword to AltaVista™ (or another) search engine. It loads the first 20 pages (this value is adjustable and is set to 20 for the response speed purpose) and analyzes the set of keywords within each page. The analyzing process scans the keywords contained in these pages and tries to locate a keywordwhose occurrence is multiple within the given pages. In our example, it found keywords *search*, *people*, *free*, and *download*. Detailed information about the keywords found and their occurrence count follows in Table 1.

Table 1: Keyword count: Cluster analysis results of the query "music"

Keyword	search	people	free	download
Count	16	14	14	14

These new keywords, which have the capacity to refine the query in the best possible way, are displayed as planets in a planetary system with the initial keyword (or a set of keywords; "music" in our example) in the center. Clicking the central ball results in opening the corresponding

AltaVista™ page. Clicking orbital balls will refine central query with its keyword and the program will perform another cluster analysis. The search procedure will be described in the following paragraph.

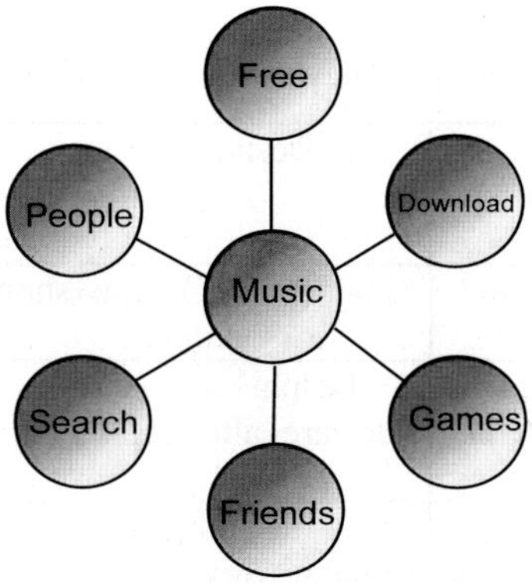

Figure 4: PlanetSearch plain "found keyword" structure

4 Searching the Internet

Our main goal is to increase the usability of web search processes. Both of the tools we provide are meant to be used simultaneously to find the "right query" in the sense of the query, which results in displaying appropriate pages (i.e. pages with the information we have been searching for). Let us take an example: A user searches for Borland Delphi component, which can handle PNG images. Therefore, the initial query would be: "Borland" "Delphi" "component" "PNG". When we examine this query with CrystalSearch, it is clear that the keyword "Borland" does not refine the query well (see Fig. 3). However, the remaining query "Delphi component PNG" still contains a lot of pages. Therefore, we need to extend it with an appropriate keyword. Which one it should be is a good question for the PlanetSearch tool. It analyses pages corresponding to its query and suggests some new "most suitable" keywords. In this case *freeware*, *download* and *graphics*. We want to download the component, so the keyword "download" will be a good refinement. We let the program preform another cluster analysis of the query extended with this keyword. The resulting set of new possible keywords is now *freeware*, *shareware*, *demo* and *shop*. As we want to use our component for free, we choose "freeware". Another set of generated keywords is empty – no keyword cluster has reached minimal count boundary. So we click the central planet to display the list of pages which fulfills our refined query. Other possible outcomes of the search could be:

- Displayed keywords make no sense and/or they are not connected to wanted topic in any way.
- There is already a huge list of keywords and extending the query would make (or already makes) no profit in refining the query.

5 Tests

We conducted a research concerning the efficiency of the keyword searching procedure. Testers were asked to find certain page or information (see Table 2) using a standard search engine (Google) and using our search programs. The queries were taken as "similar", if they contained the same keyword count and when the page we were looking for was a result of all key-words.

The following table contains examples of the test queries. The result is the average number of user interactions while refining the query (mouse clicks and typing). Our test has proved that the help we provide to internet user while searching is meaningful and searching is about two times faster than when using standard search engines.

Table 2: Results of the efficiency tests comparing two searching methods

Page (task) description	Desired query	Iteration count (Google)	Iteration count (PlanetSearch +CrystalSearch)
Find prices of tamagochi toys in Oregon	+tamagochi+Oregon+shop +online	12	6
Find freeware component for parsing AltaVista results for Borland Delphi environment	+Delphi+component+ freeware+altavista+parsing	14	8
Find freeware plug-in for Adobe Photoshop, which performs charcoal effect	+photoshop +plugin+freeware+charcoal	8	3
Find 3D model of some Le Corbusier's villa	+3D +model+LeCorbusier+villa	9	5

6 Conclusions and Future Plans

We have presented two novel tools which help the Internet user to search the internet more easily and efficiently. The help is based on cluster analysis and advanced display techniques.

In the future, we want to implement tracking and saving of keyword groups for each user. This way we will be able to offer "what other people choose" option for each user to optimize the keyword list. Also, we want to unite CrystalSearch and PlanetSearch into one single application, which can perform all presented tasks more user friendly and efficiently. We also plan to implement some sort of data (from search engine) caching, because accessing the search engine for every single query slows down the system performance.

7 References

Andersen E. (1992) The Statistical Analysis of Categorical Data, Springer Verlag, Berlin

Dreilinger D. and Howe. A.E (1996) An information gathering agent for querying web search engines. Technical Report CS-96-11

Dufresne, A., Martial, O., Ramstein (1995), C. Multimodal user interface system for blind and "visually occupied" users: Ergonomic evaluation of the haptic and auditive dimensions. (pp. 163-168). In Proceedings of IFIP International Conference Interactions '95 Lillehammer, Norway.

Kaski S.(1997) Data Exploration Using Self-Organizing Maps, Acta Polytechnica Scandinavica,

MetaCrawler search engine, http://www.metacrawler.com

PlanetSearch engine http://gpv.zde.cz/

Spoerri, A. (1993) "InfoCrystal: a visual tool for information retrieval & management, " Proc. ACM Information & Knowledge Management.

Section 6

Visualisation and Simulation

A Training System for Coronary Stent Implant Simulation

Giovanni Aloisio, Lucio De Paolis, Luciana Provenzano, Massimo Cafaro

Dept. of Innovation Engineering - University of Lecce
Lecce, I - 73100, Italy
{giovanni.aloisio, lucio.depaolis, luciana.provenzano,
massimo.cafaro}@unile.it

Abstract

A coronary stent implant is a therapeutic procedure that requires perfect knowledge of the three dimensional structure of the vessels, hand-eye coordination and manual dexterity in order to safely perform the implantation.
The utilisation of surgical training devices combined with a computer simulation allows avoiding the problems of in vivo practice or the use of animal models. Our aim is to build a tool able to allow medical students to practice extensively this delicate operation avoiding to expose patients to risks. In order to obtain a transparent simulation experience to the user, real-time interactivity is a critical issue because movements of the haptic interface should be produced without delaying the corresponding motions of the simulated instruments.
The system consists of a virtual environment and a haptic interface; it provides both the visualization of the virtual environment, composed of the anatomical coronary arteries, and of the surgical instruments virtual models, and force feedback generated during the interactions of the surgical instruments in the virtual environment.
The interaction between the haptic device and the virtual environment is performed using a collision detection algorithm, in order to determine the contact points, and a collision response module to compute forces to be replicate to the surgeon's hand by means of the haptic interface.
The system has to guarantee accuracy of the simulation and real time interactivity.

1 Virtual surgery

Virtual reality is being applied to a wide range of medical areas, including remote and local surgery, surgery planning, medical education and training, pain reduction and architectural planning of medical facilities. Virtual reality for surgery involves applications of interactive immersive computer technologies to help perform, plan and simulate surgical procedures. Simulation is mostly used in training and it may be used for routine training or to focus on particularly difficult cases and new surgical techniques.
In open surgery, the surgeon cuts the patient's body and uses hands and medical instruments to operate. This is the most invasive form of surgery, with long recovery times. The trend is to minimise the risk for patients using the improved techniques of minimally-invasive surgery.
In this way only a small incisions is executed with more advantages for the patient like less pain, less strain on the organism, and faster recovery. There are also relatively small injuries, and an economic gain arising through shorter illness time.
However, for the surgeon, there are several disadvantages, including restricted vision and mobility, difficult handling of the instruments, difficult hand-eye coordination and no tactile

perception except force feedback. The minimally invasive surgical methods require training, which is different from the traditional techniques; it should be performed frequent practice in a safe environment which mimics the anatomy and physiology of the body as closely as possible to ensure adequate transfer of skills.

The utilisation of a surgical computer simulations allow repeatedly exploring the structures of interest and view them from almost any perspective. This is obviously impossible with real patients, and it is economically infeasible with cadavers which, in any case, have already lost many of the important characteristics of live tissue. Animal experiments are expensive, and of course the anatomy is different.

The advantage is that it provides safe controlled medical environment to practise and to planning surgical procedures reducing the costs of education; different scenarios can be built in the simulator to provide a wide experience that can be repeated as needed. Training can be done anytime and anywhere the equipment is available. Training systems make possible the reduction of operative risks associated with the use of new techniques, reducing surgical mortality.

However, the big challenge is to simulate with sufficient fidelity to minimise the difference between performing the surgical operation with the simulator and on real patients.

Since all of the human organs are not rigid, their shape can change during the interaction. For this reason the realism of the deformation is very important in surgery simulation. Most of the existing training simulations are for applications, where there are not large deformations and mostly manipulation of hard objects. For other applications, deformable models are required to build realistic and efficient simulations. For this reason it is necessary the use of force feedback, modelling of soft tissue and a real-time response to user's action (Gorman, Meier, Kummel, 1999) (Dawson et al., 2000).

Visual realism and real-time interactions are essential in surgery simulation. Real-time interaction requires that each action of the user generates an instantaneous response in the virtual environment, whatever the complexity of its geometry.

The realism can be enhanced using a device that allows integrating the force feedback sensations in order to provide to the surgeon forces as soon as possible close to reality (Zorcolo et al., 1999).

Sometimes it is very important to have images obtained from the actual patient, since life-critical decisions are based on the presentation of patient data.

The aim of this research is to provide a realistic computer-based training system that allows surgeons to practice and evaluate their skills in the coronary stent implant procedure. Our work is part of the HERMES Project of Consorzio CETMA, Brindisi, Italy.

2 Coronary stent implant procedure

A coronary stent implant is a therapeutic cardiac procedure that is used in association with balloon angioplasty to open up a blocked coronary artery. After implantation, the coronary stent becomes a permanent implant to hold the artery open and prevent it from closing back down.

The coronary stent is a small, slotted stainless steel tube that will be passed through the catheter in femoral artery. The doctor initially advances a long, thin guide wire through the sheaths and up across the blockage in the coronary artery. This initial widening of the blockage is necessary to pass the stent catheter through the blockage. Next, the stent catheter is threaded onto the guide wire. The stent is mounted on a deflated balloon. When the stent catheter is positioned in the centre of the blockage, the balloon is inflated and the stent is expanded, spreading the blockage apart. The stent presses slightly into the wall of the coronary artery, keeping it open. The balloon catheter and guide wire are then removed and the stent holds the artery opened reducing the rate of restenosis (see Figure 1).

This intervention requires perfect knowledge of the three-dimensional structures of the vessels, hand-eye coordination and manual dexterity in order to safely perform the implantation.

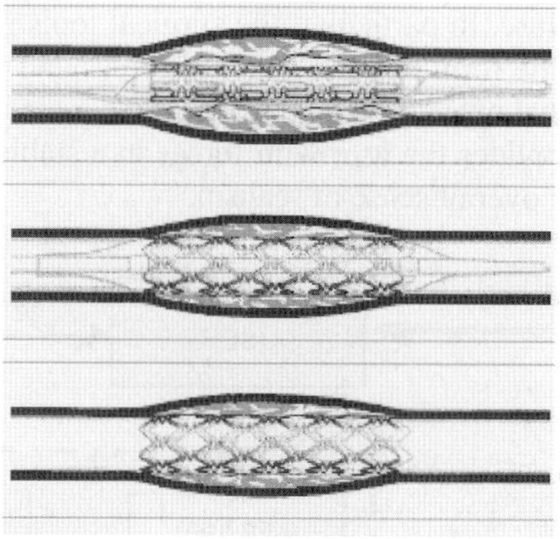

Figure 1: Stent implant procedure

3 Simulator overview

The designed simulator consists of a virtual environment and a haptic interface; in this way both the visual feedback and the force feedback are provided during the interactions of the surgical instruments in the virtual environment.
During the simulation, the user interacts using the haptic interface that likely reproduces the real surgical instruments manipulated by the cardiologist. Position data acquired by the haptic device sensors are used to update the virtual environment and to determine possible collisions between virtual objects.
Collisions between surgical instruments and artery walls produce forces that are replicated on the surgeon's hand by the haptic interface.
The simulation system, depicted in the Figure 2, is composed by:

- a haptic interface able to read the surgical instruments positions and to provide the user with the appropriate forces in the case of collision;
- a virtual environment that is a representation of surgical instruments and part of human body;
- collision detection and response modules able to detect any collision in the virtual scene and to compute forces that are sent to the haptic interface;
- physical modelling that allow describing the mechanical properties of the real body.

The virtual arteries model is built using anatomical models described in medical literature and refined in collaboration with a cardiologist because computer tomography and magnetic resonance cannot be used to build anatomic representation of the coronary arteries due to the heartbeat. In this way, it is possible to have different case studies exhibiting more difficulties (high or not stenosis) and problems due to different arteries geometry.
The haptic interface likely reproduces the real surgical instruments that the cardiologist manipulates during the stent implant operation. Data acquired from haptic device sensors are used to represent the instrument position in the virtual environment and to determine possible collisions between virtual objects. Movements of haptic devices lead to change the virtual environment representation; collisions between virtual objects produce forces to be replicated on the surgeon's hand by the haptic interface.
The haptic interface used to perform the coronary stent implant simulation has been designed and realized by PERCRO Laboratory of Scuola Superiore S. Anna, Pisa, Italy.

This device is able to satisfy the surgeon requirements, so the workspace is not reduced by mechanical constraints. The haptic interface has two degree of freedom controlled by means of motors that produce force and torque resistance; this is enough to replicate the sensations that the surgeon feels in the real procedure. In particular, the surgical tools respond to the following user-applied forces: longitudinal in the form of push and pull forces and torque in the form of twisting of the tool around the longitudinal axis.
An efficient control strategy provides the user with forces in a stable way, without vibrations that would strongly compromise the overall sense of realism.

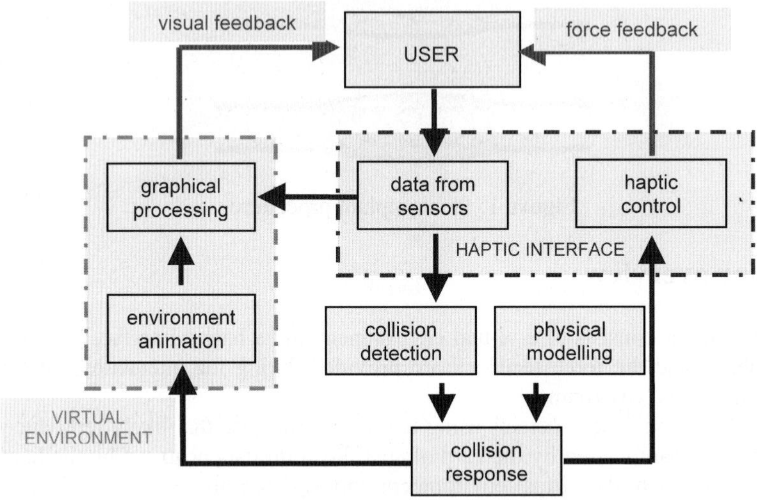

Figure 2: Simulation system

4 Collision detection and response

Physical modelling has the task of determining the dynamic behaviour of virtual objects in order to add realism to the simulation. Simulations based on physically modelling depend highly on the physical interaction between objects in a scene.
Collision detection involves the automatic detection of an interaction of two objects and of the location this interaction takes place. At the moment of impact, the simulation produces a collision response that generates a haptic feedback to the user.
Collision detection is considered as the major computational bottleneck of physically-based animation systems. The problem is still more difficult to solve when the simulated objects deform over time (Lombardo, Cani, Neyret, 1999) (Watt, Policarpo, 2001).
We are interested in methods that detect interpenetrations between polygonal models that are the most convenient for real-time rendering. However, when a collision occurs, the precise knowledge of the intersection region is needed, since it will allow a precise computation of response forces.
Most of the previous work in collision detection between polygonal models has focused on algorithms for convex polyhedra. These algorithms are very efficient, but they are not applicable in the case of a surgery simulation, where the organs are generally non-convex. In this context, it is not convenient to spend time for pre-computing complex bounding volumes, since the objects of virtual environment usually stay in very close configurations and the collision detection check will need to be done at each time step.

Generally, detecting a collision between two objects basically consists in testing if the volume of the first one intersects the second one. We perform dynamic collision detection by testing for an precise intersection between the surgical tool extremity and the triangle mesh that models the artery walls; this computation is performed at each time step. The virtual environment is composed of the coronary arteries anatomical and the surgical instrument models; the artery model is the constraint of the virtual surgical instrument motions. This ensures that the two objects surely collide. The scene is very simple and real-time performances are obtained.
The collision detection algorithm receives in input a short set of triangles relating to the artery section of interest and, in case of collision, produces as output the exact contact point, the incidence angle and the force that the haptic interface has to replicate on the user's hand.
A general description of collision detection and response algorithm follows. At each time step, we check if there is an intersection between the actual position of catheter and each triangle meshes that models a short area of artery walls; if there is a collision the algorithm:

- will find the exact intersection point between a possible triangle and the segment obtained from the previous and actual catheter positions;
- will compute the incidence angle between this segment and plane that include the triangle;
- will compute the force proportional to the penetration distance (between actual position of catheter and contact point in the triangle).

5 Future work

This work represents the first step to realize a simulation of coronary stent implant procedure in order to obtain a computer-based training system. Improvements in medical imaging technologies could be used to build a virtual environment from real patient's images with high diagnostic power. In this way, the simulator could also be used by confirmed surgeons to retrain themselves, maintain their skills for infrequent operations, plan and rehearse the coronary stent implant.
An advanced physical modelling of the artery walls will allow improving the accuracy of the simulation and a new test phase in collaboration with cardiologists will be necessary to guarantee accuracy of the simulation and real time interactivity.

6 References

Dawson S. L., Cotin S., Meglan D., Shaffer D. W., Ferrell M. A. (2000). Designing a Computer-Based Simulator for Interventional Cardiology Training. *Catheterization and Cardiovascular Interventions* 51:522-527, Wiley Press.

Gorman P. J., Meier A. H., Kummel T. M. (1999). Simulation and Virtual Reality in Surgical Education. *ARCH SURG,* vol. 134.

Lombardo J. C., Cani M. P., Neyret F. (1999). Real-time collision detection for virtual surgery, *Computer Animation '9*. Geneva.

Watt A., Policarpo F. (2001). 3D Games – Real-time Rendering and Software Technology (1st ed.). *Collision Detection.* Addison Welsey.

Zorcolo A., Gobbetti E., Pili P., Tuberi M. (1999). Catheter Insertion Simulation with Combined Visual and Haptic Feedback, *Proc. First PHANToM Users Research Symposium (PURS'99)*, Heidelberg, Germany.

prototyping.ppt –
Power Point® for interface-simulation of complex machines

Barbara Bönisch, Jürgen Held and Helmut Krueger

Institute of Hygiene and Applied Physiology
Swiss Federal Institute of Technology, ETH
Clausiusstrasse 25, 8092 Zurich , Switzerland
boenisch@iha.bepr.ethz.ch

Abstract

Prototyping with its spiral circle approach promises iterative development towards users' requirements and allows them to be actively involved in cooperative design. PowerPoint® from Microsoft® Corporation can be used as facade tool and then produces computer-created prototypes of intermediate-fidelity. Part of the interactivity of the real product can be simulated. PowerPoint is mainly used for prototyping web design and trivial interfaces, but was not used for design of anaesthesia respirator machines before. This contribution answers the question if this program is also flexible enough to be suitable for simulating and for prototyping the interface of a machine as intricated as a respirator.

1 Objective and Significance

Designing for a complex work system like anaesthesia, with numerous sophisticated devices all with their own interface concept, needs thorough user participation to intensify integration of their needs into the design of MMI. But users need proper visualization to enable them to fruitful participation. Prototyping with its spiral circle approach promises iterative development towards their requirements and allows users to be actively involved in cooperative design.
"A prototype is an easily changeable draft or simulation of at least part of an interface" (Hackos & Redish, 1998, p. 376). Several tools for prototyping can be distinguished (figure 1):

fidelity	low	intermediate	high
	mock-ups	**facade tools**	**rapid application development**
examples	paper & pencil	Tcl/Tk	Delphi
	sticky notes	Hypercard	Visual Basic
	plastic objects	Powerpoint	NeXT
	whiteboard		UIM/X for X
			C++
			CAD

Figure 1: Prototyping methods and tools

Facade tools allow the creator to specify input behaviour next to drawings and text, something which is not possible with pencil and paper. These prototypes, which look and feel like the actual application, operate on a limited set of artificial data but nonetheless will show users effectively the impact of their actions. PowerPoint® from Microsoft® Corporation can be used as facade tool and then produces computer-created prototypes of intermediate-fidelity. Part of the interactivity of the real product can be simulated. Users can click on the picture of a button and get whatever that button would do in the real product.
PowerPoint is mainly used for prototyping web design and trivial interfaces (f.ex. Pering, 2002), but was not used for design of anaesthesia respirator machines before as far as the authors know. Is this program also flexible enough to be suitable for simulating and for prototyping the interface of a machine as intricated as for example a respirator?

2 Methods

The context of the project is a cooperation between a swiss company for medical devices (manufacturer) and the ETH. Up to now, the project consisted of the following parts (figure 2):

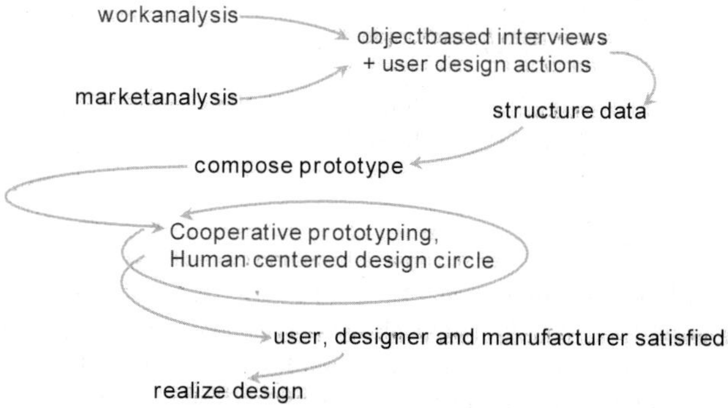

Figure 2: Steps undertaken in the project

In this case study the goal was redesigning the interface of a respirator, a breathing machine, which is used in operation theatres at anaesthesia workplaces (figure 3).

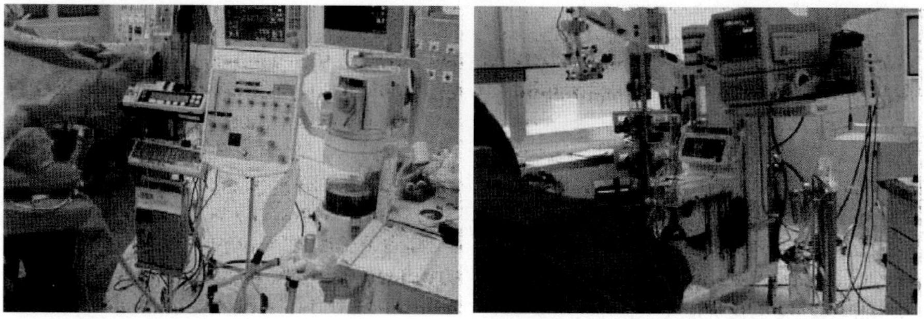

***Figure 3**: Anaesthesia-workplaces with different respirator concepts*

After analyzing work-system and tasks a competitive analysis of the respirator-market was conducted. Furthermore users´ needs were investigated by object-based interviewing. Intending an new interface-solution which combines a touchscreen with a turn&push-knob, a first prototype-version was created with Power Point® from Microsoft® Corporation Version 2000. Information and control parts of the display were designed using the drawing functions, in addition pictures representing knobs were imported. The slides were connected via action settings. Right in the OR-area of the hospital users could try out the design via presentation-mode on a mobile interface simulation system for cooperative prototyping, built with a laptop and a 17′ touchscreen, the very monitor intended for the later respirator (figure 4).

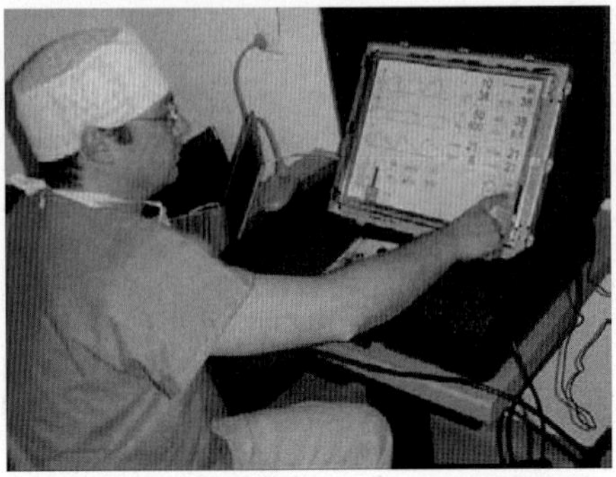

***Figure 4**: A doctor switching between the four different breathing modes*

This design-basis was processed during 14 loops of a human centred design circle. 20 sessions (duration between 15 – 60 min.) were held with 10 doctors, 4 nurses and 3 with the manufacturer.

3 Results

Prototyping with PowerPoint remained possible despite all extensions and additional menus. At the end of the human centered design circle and the interface-simulation could show over 50 elements only for information and the following manipulation possibilities were offered: two functionality-test modes, four patient-specific presettings and one for pre-configuration. "During anaesthesia" four breathing modes with up to eight parameters each and two alarm settings could be changed, additionally five menus offered more than 50 changeable parameters.
The first facade-prototype version started with the main screen and simulated the breathing modes which resulted in five slides, while the end version controlled 103 slides (Figure 5)

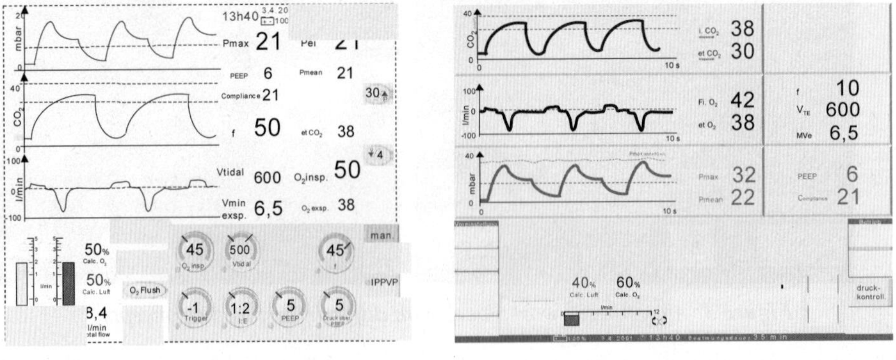

***Figure 5**: Main screens of the same breathing mode compared with 14 redesign loops in between*

Figure 6 shows possibilities of combining information and control in consideration of input via touch screen:

***Figure 6**: Small part of the "evolution of knobs"*

The chosen interface-solution consists of a touchscreen for selecting and of a turn&push-knob for altering the selections. All touchscreen-selection possibilities could be simulated with Power Point. The turn&push-knob and it's effects had to be imagined by the users, which according to expectations proved no problem, as this concept is common in hospital surroundings (figure 7).

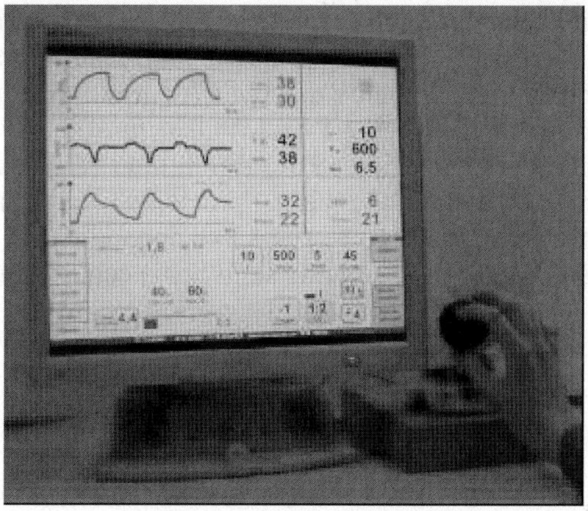

***Figure 7**: Final interface solution together with turn&push-knob and its electrical circuit*

4 Discussion

Without any knowledge of a single programming language a human centered design circle with 14 loops was carried out, controlling a high number of features and parameters via PowerPoint. Comparing this program with other prototyping tools the following conclusion can be drawn:

4.1 Plus

PowerPoint is suitable for parallel design tests and both for horizontal and vertical prototyping. As the program is known so widely, users are not confronted with a blackbox, what Ehn (1993) criticised other prototyping programs for. Adaptations might even be carried out by users themselves during prototyping. Changes can be done quickly, with higher precision than possible if drawn by hand. The realistic presentation and interactive simulation of the users´ path to task accomplishment demands less imagination power from users than paper-pencil-prototypes, resulting in higher face validity. It proved suitable for computer supported cooperative work. As PowerPoint can be found on most computers it allows prototyping at the users´ site even without carrying much equipment. No handling of yet another additional program has to be mastered as with other facade tools (f.ex. Hypercard, Tcl/Tk, Macromedia Flash or Director), and no extra programming skills (f.ex. C++, Visual Basic) are necessary as with interface builders or Rapid

Application Development tools (f.ex. UIM/X, NeXT Interface Builder, Visual Basic, Delphi and C++ Builder).

4.2 Minus

Unlike interface builders PowerPoint does not generate executable code which might be reused in the actual implementation. PowerPoint also does not allow pixel wise positioning of elements. The more slides one is working with, the more a function is missed which would allow to act out changes on a defined number of certain slides.

4.3 Conclusion

Power Point proved handy for Rapid Prototyping of such a complex machine as a respirator. The program fulfils the requirements for good prototyping tools postulated by Szekely 1994: "Ease of use, Fast turn-around, Flexibility, Useful throughout the development cycle, Executable and Version control". Nonetheless it might be essential for most design-projects to use different prototyping tools throughout the development circle to benefit from their respective advantages. In the intermediate phase of the design-process PowerPoint offers a quick way for simulating much more interaction than expected before the project, "providing considerable time and cost savings" (Pering, 2002), as no extra program has to be purchased and mastered (like other facade tools like Hypercard or Macromedia) and no programming skills are necessary (like C++ or Visual Basic).

5. References

Ehn, P. und Kyng, M. (1993). Cardboard Computers: Mocking it up or Hands-on-the-future.
In: J. Greenbaum and M. Kyng (eds.): Design at Work: Cooperative Design of Computing Systems, Hillsdale, New Jersey, p. 169-196.

Hackos, J., & Redosj, J. (1998). User and Task Analysis for Interface Design. New York, Wiley.

Pering, C. (2002). Interaction Design Prototyping of Communicator Devices. *Interactions. New Visions of Human-Computer Interaction.* Volume IX.6 Nov.-Dec., 36-46.

Szekely, P., (1994). "User Interface Prototyping: Tools and Techniques" USC/Information Sciences Institute.

Computer simulations according to different learning theories

Gisela Broder

TAI Research Centre
Helsinki University of Technology, P.O.B. 9555, FIN-02015 HUT
gbroder@tuta.hut.fi

Abstract

This paper investigates the instructors' insights of today's and tomorrow's heating, plumbing and ventilation engineering computer simulations. According to this research today's simulations can be described as supporting behavioristic approach. Expecatiotions about future simulation support constructivistic approach. The main data was collected through qualitatively analysed interviews of ten instructors.

1 Introduction

Simulation is a representation of reality (Gredler 1992, 13 - 16; Ruohomäki 2002, 15). Simulation can be an abstract, simplified or accelerated model of a system, a process or an environment. It includes critical elements of the real-world system. It also simplifies reality emphasizing its most important aspects. The model can be for example a verbal description or a graphical description - mathematical or statistical by nature. (Ruohomäki 2002, 15) This paper investigates instructors insights on heating, plumbing and ventilation engineering computer simulations.

Computer simulations are considered to be a part of computer-assisted instruction (CAI). This paper concentrates mainly on the interaction between the computer and the student in computer-assisted instruction. The pedagogical foundations of computer simulations can be very different in different simulations. In this research I studied how heating, plumbing and ventilation engineering simulations support learning or do they support learning at all. The purpose of this study is to give ideas for future simulations' planning work. The results can be seen in a more general way – not only the way how heating, plumbing and ventilation engineering computer simulation could be build. The learning theories considered give more general aspect to the future planning work.

There are different insights to computer simulations depending on the learning theories. Different learning theories have different outlooks on how learning occurs and how the instructor – in this case simulation - can guide the student. Sometimes these learning theories have something in common. However I introduce them here as separate perspectives of the same issue. In this study I present three different learning theories – behaviorism, experiential learning and constructivism. The purpose of this paper is to ask adult educational centres instructors' viewpoints on today's and tomorrow's computer simulations' methodological aspects analysing viewpoints according to the above mentioned learning perspectives. In conclusion I consider these categorized viewpoints according to the three learning theories moreover. Expectations of future simulations' methodological perspectives are considered according to the learning insights in the conclusion part. First I introduce these three learning theories shortly.

According to behaviorism, learning is understood as receiving information; knowledge is constant. The instructors' role (in this case simulation's role) is to make sure that the learner reacts the way has planned. The instructor leads the whole process without taking different students into account.

(Patrikainen 1997, 75.) For example B. F. Skinner's programmed instruction is based on behaviorism in which knowledge transfer is not questioned.

Instructional simulations can be described as a representation of experiential learning (Brent 1999; Gredler 1992). Simulation constructs a special psychological situation to the participant. The participant is expected to take some kind of role during the simulation session; he/she is expected to empathize in simulated circumstances. (Gredler 1992, 13 - 16.) The experience is occuring in authentic situations during the simulation session should help the students to manage real-world problems.

The experiential learning theory emphasizes the learner as a whole. Cognitive, behavioral and emotional aspects are inseparable (Brent 1999). There are different form of experiential learning (e.g.. Argyris 1982; Boud, Keogh & Walker 1985; Dewey 1938, 1966; Kolb 1984; Schön 1987). During the 1970's the learner was expected to take a more active role than before in the instructional context (Brent 1999). Kolb (1984) saw learning as a process in which knowledge is created through the transformation of experience. This definition includes essential and deviant insights on learning. Learning can not be described as a result as the behaviorists do (Kolb 1984, 25-38).

Constructivism can be described as an umbrella concept to similar approaches. According to the constructivistic approach human beings are active participants who need to understand the world around them. Human beings search for reasons and consequences to different phenomena. Taking this into account, constructivistic learning environment offers problems, tools, advices and support to the participant. Taking constructivistic learning insights into action is complicated because it is very student oriented approach. The student is an active participant, the instructor (in this case the simulation) takes the roles mentioned and acts according to the situational demands. (Rauste-von Wright & von Wright 1994, 133.) Gredler's (1992) insight on computer simulations can be described in terms of constructivism. Gredler (1992) describes computer simulations as interactive, problem-based training programmes. During the simulation session, the participant sees the reaction of his/her action. (Gredler 1992, 13 - 16.)

Computer simulations can be used both in face-to-face and distance education. The need for computer-based simulation arises from a rapid growth of distance education. The need for more interactive computer-based simulations has been recognized, especially because of the growing need of distance education.

There are different basis for implementing computer simulations. The most important base according this research is the way computer simulations affect student's learning. Well formulated educational computer simulations characterize excellently systems' dynamic (Hein & Larna 1992, 106). This can be seen in the way in which student can influence the process. Thereby the student can see the results of his or her action in the process. Computer simulations – and simulations in general – are tools of illustration. In Canada, at Alberta University a computer simulation called Pembina has been created. The purpose of implementing the simulation was to create an authentic work situation to the students. Maynes ja McIntosh (1996) point out that the simulation is not an answer for all challenges in the real world situation. The simulation is based on Kolb's insight on experiential learning. (Maynes ja McIntosh 1996.)

The research problem of this study is the following: *What kind of insight instructors have on today's and tomorrow's computer simulations from a methodological perspective?*

2 Procedure

The data was gathered in two phases. First I collected data with a web-based questionnaire and after that I interviewed instructors of heating, plumbing and ventilation engineering. The purpose of the web-based questionnaire was to identify instructors who have experience of using computer simulations. With the questionnaire I clarified the instructors' experience level of using computer simulations. I chose ten instructors for a thematic interview. The interview data was main data in this study. I analysed the interview data with the qualitative content analysis.

3 Results

All the instructors saw today's computer simulations as part of the learning situation. In that case we can not talk in terms of computer instruction. There are more elements in instruction than just computer simulations, e.g. other equipment such as the Internet and psychosocial factors, such as support given by instructor.

The instructors' insights on simulations as part of the instruction can be characterized to four categories. The categories are: *teaching basic knowledge, support, training* and *distance education*. Learning basic knowledge characterizes what happens before the implementation of simulation. Support describes the instructor's and the program's role affecting learning experiences. The third category, training, characterizes training contains, which happen during and after the simulation session. Distance education describes learning situations, which happen at home or at work.

Firstly, all the instructors emphasized, that students need to get information about the subject before the simulation session because of the nature of the today's simulations available. Today's simulations do not give enough feedback for the user. That is why the instructors saw the feedback given by the instructor to be crucial.

Secondly, support consisted of seven subcategories: demonstration, vocabulary, hints, explanations, feedback and conclusion given by computer simulation. These subcategories follow the categorization made by Allesi (2000). Thirdly, the instructors expected the future simulations to be more training-oriented. According to the instructors learning by doing in different simulated situations enables the real understanding of the subject area.

Finally, today's computer simulations do not support independent learning in simulated sessions. This is also the reason why computer simulations are not in use in distance education. However, distance education was seen as flexible way of learning. Distance education takes also the learner's life situation into account.

4 Conclusion

The instructors' expectations of future simulations' methodological perspectives differ radically from the possibilities given by today's simulations. Today's simulations are more like tools for transferring information from the simulation to the users mind in behavioristic way. This kind of instructional way does not attain the users' real understanding of the subject. Moreover it does not enable independent studying because of the nature of the interaction between the learner and the simulation. Today's simulations do not meet the demands of learners' different capabilities. That is why the instructor's role is essential.

Every instructor pointed out that it is necessary to give to the students information about simulated issues before the simulation session. This instructional context can be described as behavioristic.

Traditionally behavioristic model has been seen suited for giving basic information (Rauste-von Wright & von Wright 1994, 113).

In simulation assisted instruction, in more general CAI, individual working, groupworking and support given by instructor can have different role in the instructional and learning context. Previous studies improves that groupwork betters learning results (e.g. Cohen 1994; Qin, Johnson & Johnson 1995). According to experiential learning approach, learning happens usually in a group context. The supporters of experiential approach expect the learners to become self-directed through the support given by the group. (Rauste-von Wright & von Wright 1994, 141.) The instructors are expected to see themselves as asking the right questions rather than giving the right answers (Chiodo & Flaim 1993, 191 - 193). This refers to the constructivistic as well as to the experiential approach.

According to Day, Hmelo and Nagaraja (2000) the instructor's (in that research named as research assistant) advice during the computer simulation assisted instruction was not as essential for the students who knew about the subject beforehand than for the students who were not familiar with the subject area. It appeared that the high-prior-knowledge group was able to interpret more systematically simulated situations than the low-prior-knowledge group. Students in the high-prior-knowledge group also plans their activities more considerately using the simulation and they are more able to evaluate their progress. The key issue is that they have knowledge beforehand so they had also more meaningful simulation session. The results indicate that it is important to give instruction according to how much students know about the subject beforehand. Students can be seen as the constructors of knowledge. Computer simulations help both groups to organize their activities, but the instructor's role is not as central to the persons who were familiar with the subject. (Day et. al. 2000.) In constructivistic learning environment the instructor (in this case simulation and/or instructor) gives guidance taking students' different knowledge demands into account. By creating a feedback system we can avoid misunderstandings. One of the main findings of this research is the instructor's task to avoid students' misunderstanding of the subject during the simulation session. The instructor's role is crucial because of the inadequacy of simulations' methodology.

Both experiential instruction and constructivism have something in common, for example both of them emphasize the learner's self-reflection. If future simulations would require interaction between the learner and the simulation, the learner could have a more active role. This would enhance learning. Referring to instructors' insight, the simulations available are not interactive in the experiential way according to experiential learning theory. In the Kolb's (1984) experiential learning model, learning as phenomenon has been described through a cycle. The first phase in the cycle is concrete experience, the second phase is considering the experience, the third base is abstract conceptualizing and the fourth is active experiencing. Instructors' insight on today's simulation do not support managing this learning cycle.

The instructors' expectations about the future simulations refer to the constructivistic learning insight which emphasizes the feedback feature as essential for enhancing learning processes. The instructors expect also learning in experiential environment; they expect to have an authentic learning environment where the student can feel the magnificence of discovery.

The model's interactional character is essential also, because it affects knowledge transfer (Klabbers 2000). It has been recognized that the most important indicator of usefulness of the simulation is the learner's capability to solve problems in the real world situations (Vartiainen, Teikari & Pulkkis 1989). Same words can be said discussing about simulation-assisted instruction.

In developing future computer simulations it is worthwhile to create simulations that enable higher level thinking processes and problem solving. Pure behaviorism in traditional form is not enough for students which are active and thirsting for knowledge. Behavioristical approach does not support the competence demands of the new millennium either. Instruction, more concretely – computer simulations – should take the new demands of sustainable development into account.

In reading this paper it is important to notice that the results are not directly related to the types of computer simulations instructors use. More concretely, this paper is not directly related to the methodological characters of the simulations. This paper investigated intructors' insights on heating, plumbing and ventilation engineering computer simulations' methodology. During this research it appeared that instructors have different ideas of the nature of computer simulations. Further research is needed to clarify the students learning results when using computer simulations in instruction. In summary, the meaning of developing simulations' methodological perspectives is to enhance and support learning and working in real-world situations.

References

Allesi, S. (2000). Designing educational support in system-dynamics-based interactive learning environments. Simulation and Gaming 31 (2), 197-229.

Argyris, C. 1982. Reasoning, learning and action. San Francisco: Jossey-Bass.

Boud, D., Keogh, R. & Walker, D. 1985. Reflection: turning experience into learning. London: Kogan Page.

Brent, D. R. 1999. Simulation, Games and experience-based learning: the quest for a new paradigm for teaching and learning. Simulation and Gaming, vol. 30 (4), 498-505.

Chiodo, J. & Flaim, M. 1993. The link between computer simulations and social studies learning: debriefing. Social Studies 84, 119-123.

Cohen, E. G. 1994. Restructuring the classroom: conditions for productive small groups. Review of educational research, 64 (1), 1-15.

Day, R. S., Hmelo, C. E. & Nagarajan, A. 2000. Effects of high and low prior knowledge on construction of a joint problem space. Journal of Experimental Education 69 (1), 36-59.

Dewey, J. 1938. Experience and education. New York: Collier.

Gredler, M. (1992). Designing and evaluating games and simulations. A process approach. London: Kogan Page.

Hein, I. and Larna, R. (ed.). 1992. Near, far, alone, together (in Finnish). Helsingin yliopiston Lahden tutkimus- ja koulutuskeskus. Hakapaino Oy.

Klabbers, J. H. G. 2000. Learning as acquisition and learning as interaction. Simulation & Gaming 31 (3), 380-406.

Kolb, D. A. 1984. Experiantial learning: experience as source of learning and development. Englewood Cliffs, NJ: Prentice-Hall.

Maynes, B. & McIntosh, G. 1996. Compuetr-based simulations of the school principalship: preparation for professional practice. Educational Administration Quarterly 32 (4), 579-596.

Patrikainen, R. 1997. idea of man, idea of knowledge and idea of learning in the class teachers' pedagogical thinking (in Finnish). Joensuun yliopiston kasvatustieteellisiä julkaisuja 36.

Qin, Z, Johnson, D & Johnson, R. 1995. Cooperative versus competitive efforts and problem solving. Review of educational research, 65 (2), 129-145.

Rauste-von Wright, M. & von Wright, J. 1994. Learning and instruction (in Finnish). Porvoo-Helsinki-Juva: Wsoy.

Ruohomäki, V. (2002). Simulation game for organisation development. Development, use and evaluation of the Work Flow Game. Doctoral dissertation. Helsinki University of Technology, Industrial management and work and organizational psychology. Report no 20. Vantaa: Tummavuoren Kirjapaino Oy.

Schön, D. A. 1987. Educating the reflective practitioner. San Francisco: Jossey-Bass.

Vartiainen M. Teikari V. Pulkkis A. 1989. Psychological instruction of work (in Finnish). Helsinki: Otatieto.

Simulation Supported Learning of Soft Computing Models

Bojana Dalbelo Bašic, Vlado Glavinic, Marko Cupic

Faculty of Electrical Engineering and Computing, University of Zagreb
Unska 3, 10000 Zagreb, Croatia
{bojana.dalbelo | vlado.glavinic | marko.cupic}@fer.hr

Abstract

The paper describes a simulator of soft computing models build to assist learning process. It supports various kinds of fuzzy sets operations, fuzzy control systems, processing elements, neural networks, and learning of neural networks using distributed genetic algorithms. The simulator is implemented in Java making it particularly suitable for distance learning.

1 Introduction

During the past years computers have been increasingly used for educational purposes, either as a supplement to the human teacher or as a standalone source of information when the human teacher is unavailable, this latter process being known as asynchronous learning. In this paper we will focus on learning soft computing models. Having in mind that soft computing mo dels have a large number of degrees of freedom, it is very important to have a tool enabling intensive and easy experimenting. To make the learning process as easy as possible, the learning tool should be portable, always available, easy to use, and to facilitate distant learning it should be web-oriented.

In this paper we present a simulator (Dalbelo Bašic, Glavinic and Cupic, 2002), which meets the previously mentioned requirements, along with its use as an extension to classical teaching. The paper is organized as follows. In Section 2 we elaborate on the design of the simulator. In Section 3 an overview of the supported models is given. Section 4 gives an example of simulator usage, while Section 5 concludes the paper.

2 Design of the simulator

It is a known fact the Internet provides a huge amount of educational content (e.g. on-line books, educational communities, etc). However, most of this material is static and does not enable both experimenting with different parameters of the problem under consideration and the dynamic observation of results. To offer students freedom in soft-computing problem exploration, a simulator is developed. The following important guidelines determines the simulator design:

- *Accessibility* – students should be able to learn at a chosen pace and time.
- *Portability* – the simulator should be able to work on many different computer platforms.
- *User-friendly* – the simulator interface should be easy-to-use and intuitive enough to achieve student satisfaction.
- *Easy upgrade* – it should be relatively easy for developers to distribute new versions of the simulator, the whole process being as transparent as possible.
- *Expandability* – it should be easy to expand the functionality of the simulator in order to cope with the practically unlimited number of problems that students could generate.

To meet the above requirements, the simulator is based on Web technology. This additionally permits a continuous simulator development, as well as promptly available upgrades that are transparent to users. In order to enable an easy and secure way for simulator distribution and usage Java is selected as the implementation language. Java programs are executed in an artificial environment, thus ensuring portability. Java applets do not have any access privileges to users' computer file systems, thus providing a controlled environment for simulator execution without any possibility for harmful actions or virus infections. On the other hand, Java applets are executed directly within users' Web browsers, requiring no previous installation.

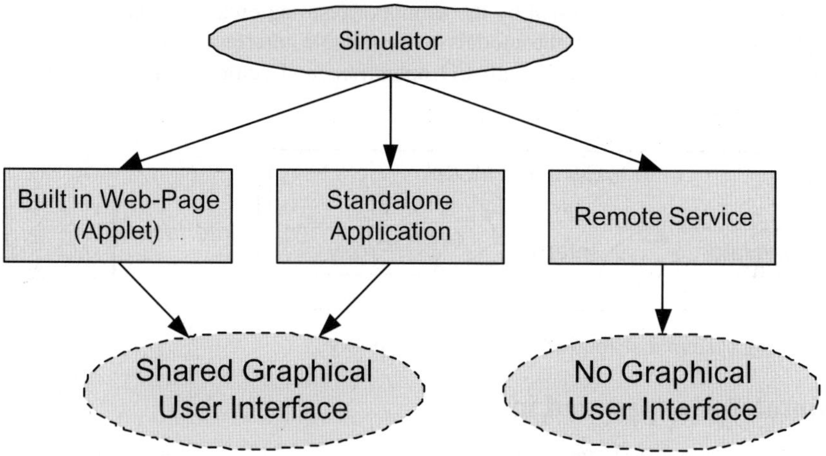

Figure 1. Simulator execution modes and associated interaction styles

Figure 1 presents the simulator execution modes. Along with the normal Web based access the simulator supports also downloads to users' hosts and execution as a standalone application (neither requiring a Web browser nor connection to the Internet). However, the graphical user interface is uniform and consistent, independently whether the simulator runs as an applet or a standalone application. To support more advanced users, the simulator can be run as a server, in which case no graphical user interface is needed.

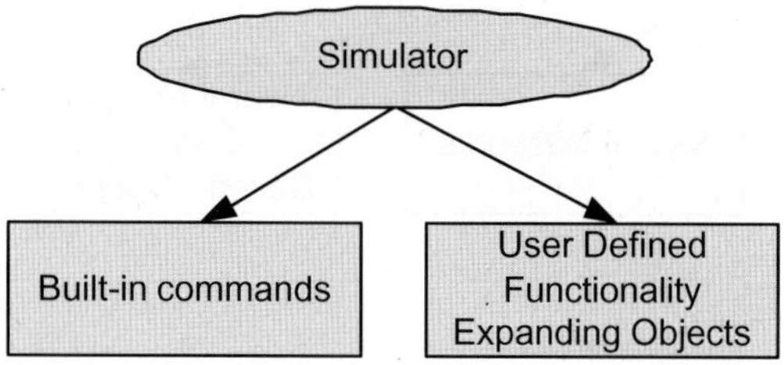

Figure 2. Simulator functionality

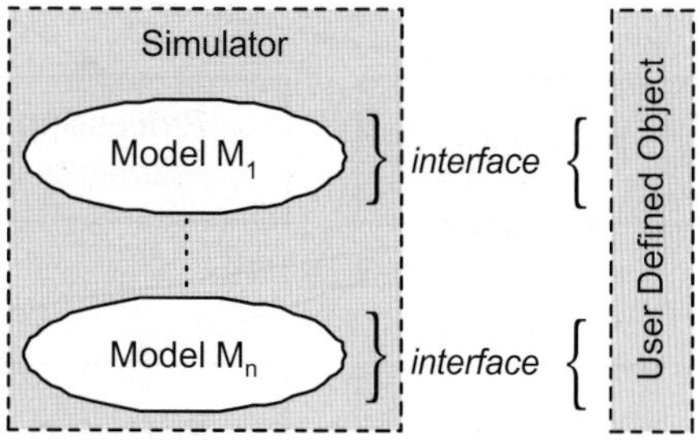

Figure 3. Black-box model

To support a highly abstract thinking about soft-computing models, a black-box paradigm for model representation is utilized. Built-in commands enable the definition of both soft-computing models along with a number of operations to be performed upon them.

Operations not directly supported by the system command set can be custom defined by using Java objects, as indicated in Figure 2. These way users are provided freedom to enrich the simulator functionality and perform experiments with predefined soft-computing models. The latter are then interpreted as black boxes having a predefined interface for input specification, output retrieval and methods invocation.

The simulator capabilities to work with soft-computing models can be incorporated into distributed applications. In this scenario Java applications or even mobile agents can be developed, which reside on a host and utilize black-box soft-computing models, while the actual soft-computing calculations are performed by the simulator residing on another host acting as a server. These applications interact with the models over suitable interfaces as if they were available locally, while a proxy service handles all I/O operations and communication with the simulator across the Internet (see Figure 4).

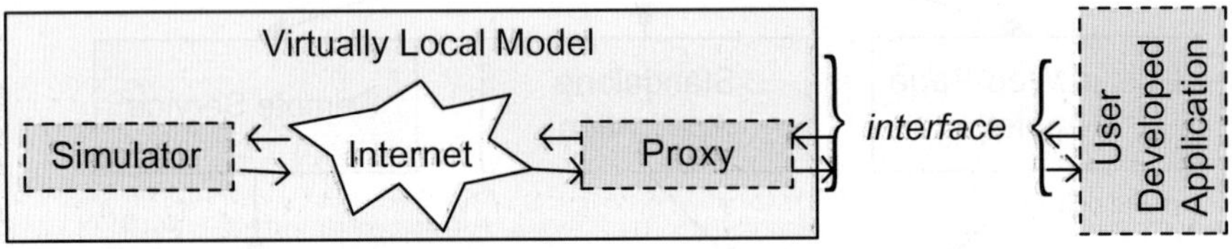

Figure 4. Hiding system distribution through use of proxy

3 Simulator supported models

The simulator covers most of the elementary fuzzy logic models (Zadeh, 1996), simple processing elements and neural networks (Picton, 2000), as well as neural networks training by means of distributed genetic algorithms. This latter is implemented by developing a custom object that utilizes mobile agents and agent environments (Cupic, 2002). Figure 5 gives an overview of simulator-supported models.

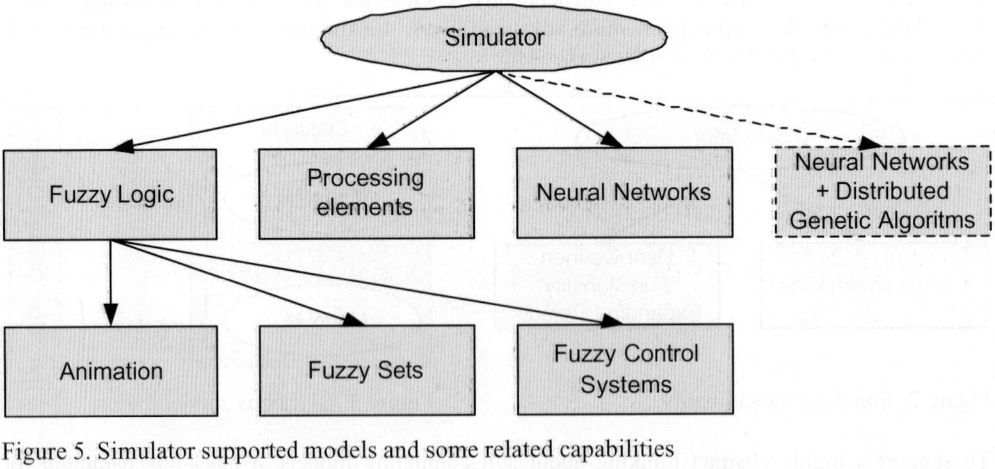

Figure 5. Simulator supported models and some related capabilities

The simulator has a number of visualization capabilities intended to support computer-aided learning. An illustrative example of its use is learning neural networks, where the network knowledge is distributed among all of the weights, and the learning algorithm is rather cryptic (e.g. backpropagation). We strongly believe that an insight in the process of weights changing, as well as total epoch error tracking could significantly improve students' understanding of the network learning process. This comes especially true if having in mind that error tracking is one of the techniques used as the learning stop condition. Figure 6a shows a simple neural network used for learning of the first half period of sin(x), Figure 6b the original function and the result of the learning process, Figure 6c the total error tracking results and Figure 6d the change of weights during the learning process. In order to support a detailed analysis of the learning process, the simulator can also generate a report showing every single calculation made.

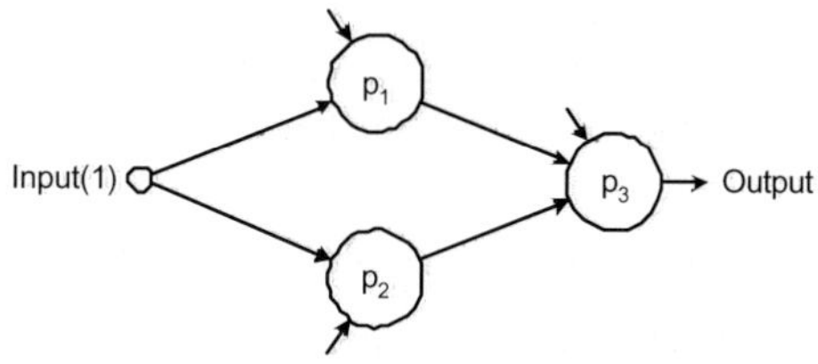

a) Simple neural network

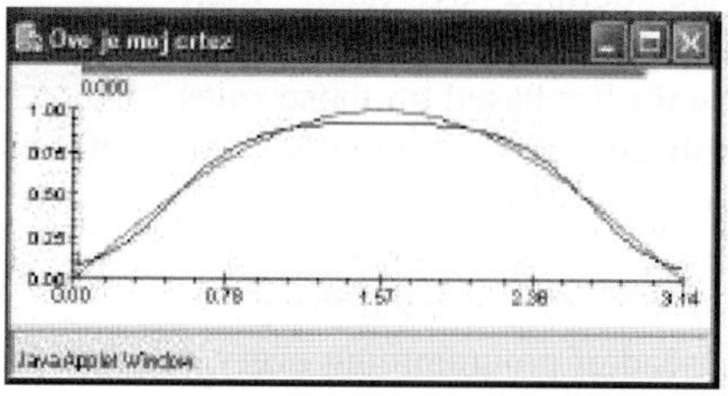

b) Function and results of learning

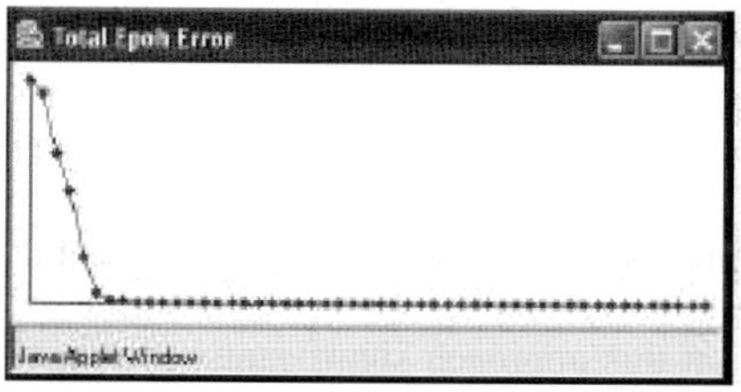

c) Total epoch error

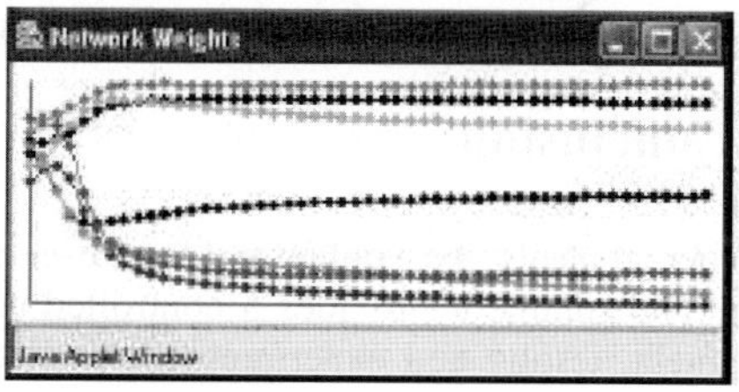

d) Weights tracking

Figure 6. Neural network learning process

The simulator can be used to study the effects of parameterized s- and t-norms and to observe some apparently paradoxical situations. E.g. Figure 7 shows the behavior of Hamacher t-norm applied on two fuzzy sets for two parameter values. Until the definition of the t-norm is consulted it need not be obvious (at least to novices) why there is a peak for value 25 (as shown in Figure 7b). To study effects like this, the simulator supports animated visualizations, such as the one shown in this example. When it is possible, the simulator supports graphical presentations of learning processes for single processing elements, and also detailed table reports explaining every performed step.

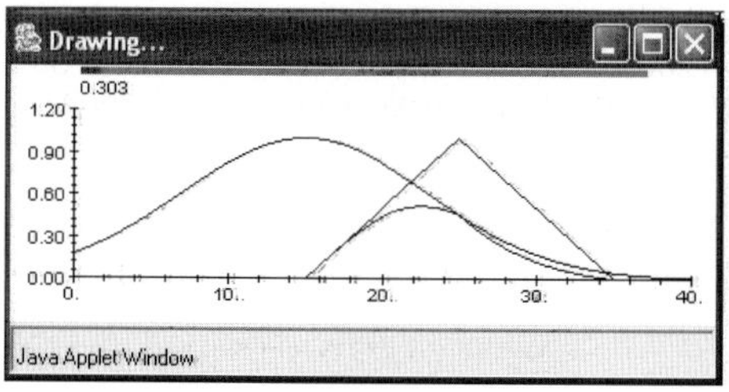

a) Hamacher t-norm for t=0.303

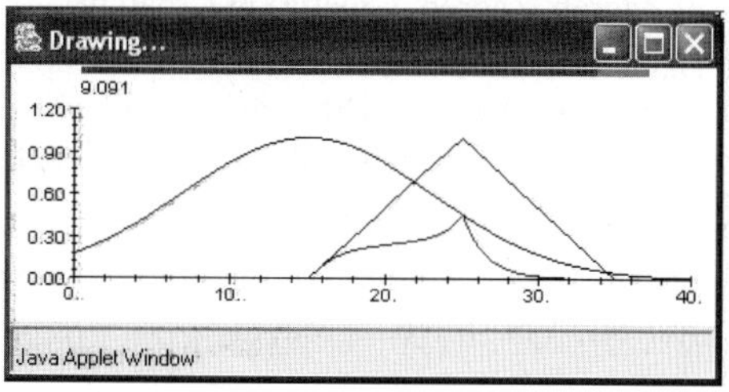

b) Hamacher t-norm for t=9.091

Figure 7. Animated effects of parameter changing for parameterized t-norm on two fuzzy sets

4 Example of Use

To demonstrate the simulator usage in teaching, a simple program developed by one of our students is shown. The vertical movements of a paddle, which has to prevent a ball from leaving the area surrounded by three walls, are defined by fuzzy logic rules. The student developed a Fuzzy Logic Controller based on those rules and utilized the simulator for executing the game. This way he could pay more attention to the logical design of the controller and the game itself, instead of programming every detail of the control system and fuzzy rules. In comparison to the same controllers developed in commercial systems, the results were better. Figure 8 shows a snapshot of the game and automatically generated graphical representation of some input and output linguistic variables defined by the student.

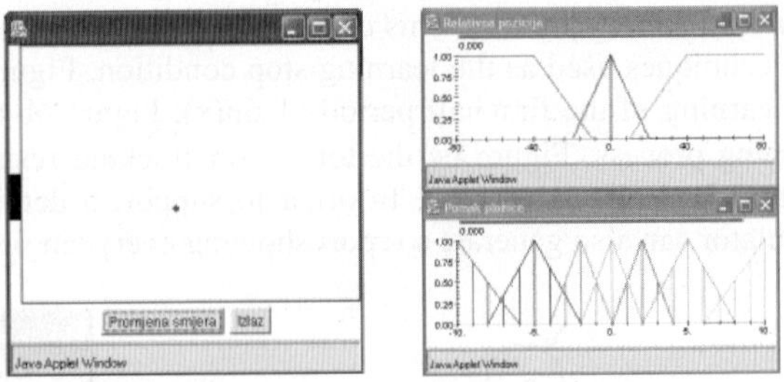

a) Snapshot from the game　　b) Some linguistic variables

Figure 8. Simple game developed using FSIT

5 Conclusion

This paper presents the simulator for soft-computing models developed to assist the learning process. The simulator provides a controlled and secure environment for soft-computing model development, testing and experimentation. We believe that this approach significantly improves this type of human-computer interaction, as discussed in the paper.

Acknowledgements

This work has been carried out within projects 0036033 *Semantic Web as Information Infrastructure Enabler* and 0036025 *Multi-Agent Systems for Dynamic Scene Interpretation* both funded by the Ministry of Science and Technology of the Republic of Croatia.

References

1. Cupic, M. (2002) *Intelligent Search of Information Space within the Information Infrastructure*, Diploma Thesis, Faculty of Electrical Engineering and Computing, University of Zagreb, Zagreb, Croatia (in Croatian).
2. Dalbelo Bašic, B., Glavinic, V., Cupic, M. (2002) Web-oriented Simulator for Soft Computing Models, *Proc. 6th Int'l Conf. on Intelligent Engineering Systems – INES 2002*, A. Lovrencic, I. J. Rudas, Eds., Opatija, Croatia, May 26-28, , 255-260.
3. Picton, P. (2000) *Neural Networks*, Second Edition, Palgrave, 2000.
4. Zadeh, L. A. (1996) Fuzzy Logic = Computing with Words, *IEEE Transactions on fuzzy systems*, Vol. 4, No. 2, May.

Experimental Interfaces for Visual Browsing of Large Collections of Images

Gianluca Demontis, Mauro Mosconi, Marco Porta

Dipartimento di Informatica e Sistemistica – Università di Pavia
Via Ferrata, 1 – 27100 – Pavia - Italy
{demontis, mauro, porta}@vision.unipv.it

Abstract

In this paper we present five experimental interfaces for browsing large collections of images in fast and effective ways, through original interaction paradigms based on new visual metaphors. As it is not always possible to identify precise search criteria for an image database, our methods try to provide the user with a global view of the entire collection, allowing many images to be inspected in short times.

1 Introduction

Within the field of *Information Visualization*, special attention deserves to be given to the so-called *Information Presentation*, whose aim is to provide the user with a global, and, at the same time, detailed, view of some kind of elements.

Approaches followed by existing systems for exploring image databases can be classified into three main categories. The first one includes those systems (very common on the Web) in which the user can manually browse the available images or their previews. Although these approaches allow collections to be analyzed in detail, they often require long times for complete database scans. Methods of the second category, in which every image has textual information associated with it, allow the user to formulate requests in the form of query strings. However, the main limitation of this approach lies in the high subjectivity of descriptive text, as different system administrators are unlikely to choose the same keywords (Enser, 1995). Approaches of the third category, at last, directly exploit intrinsic characteristics of images, such as mathematical measures of color, texture and shape (see for example Flickner, 1995), which are compared with those of the templates to be searched for. Unfortunately, with these methods it is usually very difficult to address the search for complex semantic content, where the identification of different elements is required simultaneously.

In our work, we start from the assumption that it is often necessary to examine image databases to search for something, but without knowing what exactly (for example, to choose an image to be included in a document, for decorative purposes). In such cases, query systems can only be of limited help (because there is no specific query that can be formulated), and the "traditional" browsing techniques are seldom able to fulfill the requirements of the user when dealing with great amounts of images (in the order of thousands). Our approaches aim at providing the user with a global view of the entire collection, by allowing him or her to inspect as many images as possible in the least time possible.

Using Macromedia Flash MX®, we have implemented ICE (*Image Collection Explorer*), a browsing suite made of five different exploration methods, each inspired by a corresponding visual metaphor.

2 System Description

In devising our methods, we drew inspiration from that branch of Information Visualization (called RFIP, from *Rapid-Fire Image Preview*) devoted to studying methods for presenting images in rapid succession. Essentially, RFIP techniques require high exploitation of the available screen area and high visualization speeds. A famous example of RFIP system can be found in the work described in (Wittenburg, Ali-Ahmad, LaLiberte & Lanning, 1998), where multiple images are presented in the service of navigation in Web information spaces.

2.1 Browsing Metaphors

Our system relies upon five metaphors, chosen according to the specific browsing criteria we wanted to obtain and to RFIP methods' basic requirements. The metaphors implemented (all of which allow high levels of abstraction to be achieved) are the following:

Cube. In this method, images are associated with the vertices of one or more cubes, whose rotation allow pictures in the foreground to continually change (Fig. 1). After a complete exploration of the eight images of a cube, other groups can be inspected by pressing a proper button. Magnitude and direction of rotations are controlled by the distance vector between cubes' centers and the mouse pointer (when the pointer is on a cube's center of gravity, the rotation halts).

Snow. Like snow flakes, images in this method rain down from the upper screen area and disappear as the bottom is reached (Fig. 2a). To prevent pictures from continuously overlapping each other while falling, they follow sinusoidal paths with casual periods and amplitudes. The "raining speed" is controlled by the position of the mouse.

Snake. In the snake method, images cover a sinuous path reminding a snake body (Fig. 2b). One after the other, they move along the fixed route, with a perspective effect and without any overlapping. By modifying the amplitude of the sinusoidal path, the user can vary the length of the route; at an extreme (amplitude zero), the snake metaphor can be turned into that of a movie film.

Volcano. Like for a volcano seen from above, in this method images are "erupted" by a central "crater" and, with a perspective effect, slide down laterally along the virtual slopes (Fig. 3a). While the user's focus of attention is centered in the screen midpoint, a global view of the leaving pictures can be always maintained. Since image directions are casual, partial overlappings are possible (the penalty to pay in favor of high browsing speeds and simultaneous display of many images on the screen).

Funnel. Complementary to the Volcano metaphor, the Funnel method exploits the visual effect of a funnel seen from above, where images appear at the screen borders and, with a perspective effect, disappear in the center (Fig. 3b). In this case, the focus of attention may be considered as a "distributed" concept, since there is no single point where pictures originate. Although images may overlap each other (like for the Volcano metaphor), the Funnel method allows very high exploration speeds to be achieved.

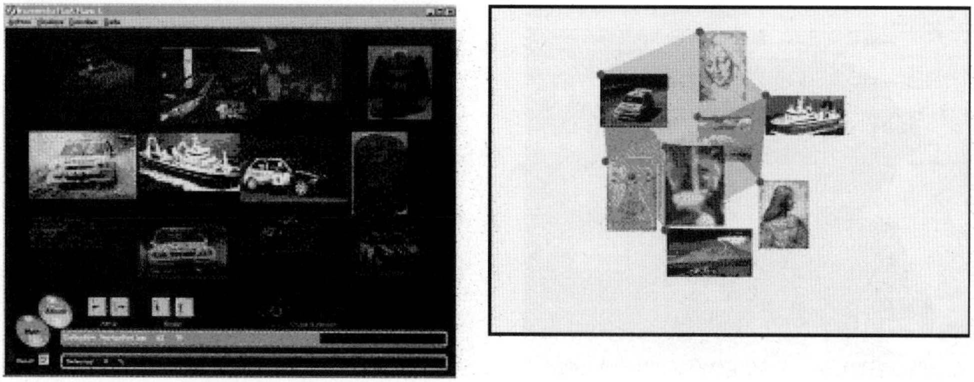

(a) (b)

Figure 1: The Cube method. (a) Images rotate according to the positions of the vertices of two cubes; (b) Metaphor principle

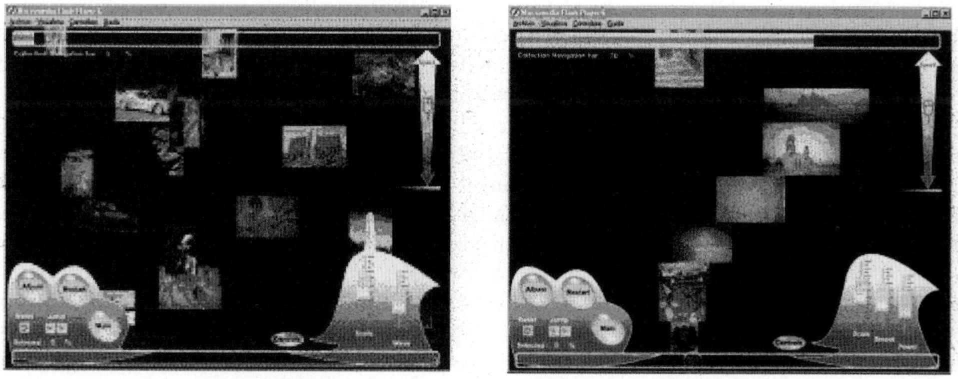

(a) (b)

Figure 2: (a) The Snow method: images rain down following casual sinusoidal trajectories; (b) The Snake method: images follow a fixed sinusoidal path

2.2 Implementation Issues

Although, practically, any programming language/environment allows graphic interface elements to be created in relatively simple ways, Macromedia Flash® is surely one of the most widespread and versatile programs for creating, importing and programming GUI elements. Moreover, its orientation towards the Web made it the ideal choice for our prototype system, aimed at being employed on heterogeneous platforms.

All of the approaches implemented are fully configurable in their main parameters (e.g. scale factors and motion speeds), through proper buttons and sliders. During the browsing, images can be saved for successive detailed inspection, by simply clicking on them.

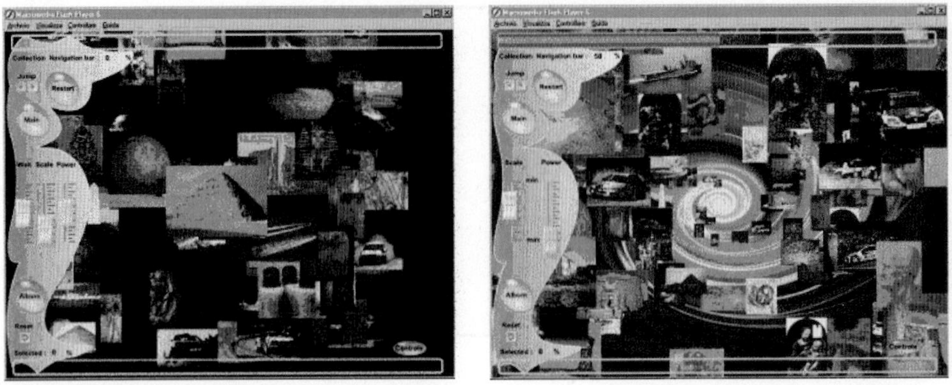

(a) (b)

Figure 3: (a) The Volcano method: images emerge at the center of the screen and disappear at the borders; (b) The Funnel method: images appear at the screen borders and disappear in the center

3 System Testing

To test our browsing methods, two collections of three hundred images each were created, each one containing four distinct themes (ships, animals, landscapes and trains for the first collection, aircrafts, cars, works of art and fractals for the second collection). Each theme was composed of thirty images uniformly distributed in the presentation progression; the other images were selected so as to not create ambiguities with the themes. Fifteen subjects were asked to search for images of a specific theme using the five new techniques devised and two "traditional" methods: the *sequence*, where the user displays images one after the other (by pressing a "next" button), and the *grid*, where images are shown in a tabular manner. To make experiments more balanced (and prevent the user from getting accustomed to a certain group of images), the collection was changed for each consecutive method test, as well as the theme to be searched for. For analogous reasons, the initial training to the system, for each tester, was carried out on a third collection. Considering the high level of personalization of the parameters of each method, single adjustments were maintained independently of the specific user.

The test group was composed of nine men and five women, with ages ranging from 22 to 46 and with different levels of computer knowledge. For each method, and for each tester, the following qualitative/quantitative data were observed: (a) tester's liking, (b) number of images found for the specific theme, (c) number of wrong images selected (i.e. not pertaining to the correct theme), and (d) time necessary for exploring the entire collection.

Figure 4 shows average test results obtained. As you can see, all the five methods developed allow the browsing time to be reduced with respect to traditional techniques. The Volcano and Funnel methods, in particular, are the fastest approaches, although they introduce errors in selecting images pertaining to specific themes. Considering the ratio between the percentage of correct images selected and the browsing duration, the Snake technique seems to be the best of the five, being reliable and fast at the same time.

Although we are aware that further tests (with more testers) are necessary to get a better outline of the methods' potentials and limitations, we think that these first results can be taken as a very encouraging starting point towards new browsing paradigms.

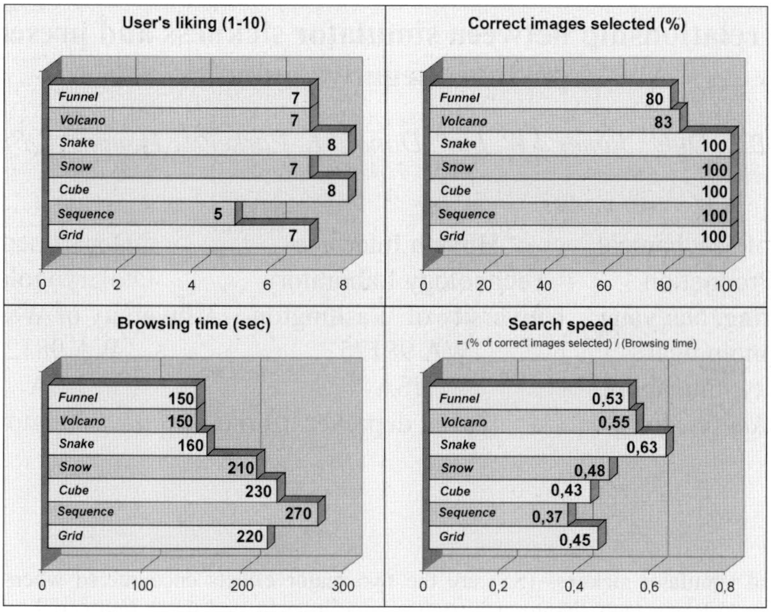

Figure 4: Average test results obtained for the five new methods (Cube, Snow, Snake, Volcano and Funnel) and two "traditional" techniques (Grid and Sequence)

4 Conclusions

In this paper we have presented five new methods for exploring large collections of images. The devised techniques offer manifold approaches to the browsing activity, improving its efficiency and making the user's experience more "pleasant". To evaluate the methods, we carried out usability tests involving several subjects with different computer skills. The experiments have shown that, once the user becomes accustomed to them, and according to the context of the browsing activity, our approaches show interesting advantages with respect to traditional browsing techniques, both in terms of exploration speed and quality of search.

Among the possible developments of the described methods, which we will consider in the near future, there are the use of more complex solids instead of the cube (e.g. a dodecahedron), and the integration of the Cube approach into the Funnel and Volcano techniques (so as to have images moving also in a three-dimensional space).

5 References

Enser, P.G.B. (1995). Pictorial information retrieval. *Journal of Documentation*, Vol. 51, N. 2, 126-170.

Flickner, M. (1995). Query by image and video content: the QBIC system. IEEE Computer, Vol. 28, N.9, 23-32.

Wittenburg, K., Ali-Ahmad, W., LaLiberte, D., & Lanning, T. (1998). Rapid-Fire Image Previews for Information Navigation. Proceedings of *the Working Conference on Advanced Visual Interfaces (AVI '98)*, L'Aquila, Italy, May 24-27, 76-82.

The relationship between simulator sickness and presence: positive, negative, none?

Henry B.L. Duh[1], James J.W. Lin[2], Donald E. Parker[2,3], Thomas A. Furness[2]

[1]School of Mechanical and Production Engineering, Nanyang Technological University, Singapore
mblduh@ntu.edu.sg

[2] Human Interface Technology Laboratory
University of Washington, WA 98125
USA

[3] Department of Otolaryngology,
University of Washington, WA 98125
USA

{jwlin, deparker, tfurness}@u.washington.edu

Abstract

Presence and simulator sickness (SS) are the two major effects encountered when people use virtual reality (VR) systems. Numerous studies have investigated those two variables. However, there is no consistent agreement regarding the association between presence and SS. This paper describes a new method for exploring possible relationships between presence and SS.

1 Introduction

The sense of presence is one of the important factors to consider for virtual environment (VE) design. Presence has been defined as the degree to which participants feel that they are somewhere other than where they are physically located when they experience a computer-generated simulation (see Bystrom, Barfield and Hendrix, 1999). Simulator sickness is one of the significant problems in VR applications. VE users often report nausea, dizziness, disorientation and so on. Users also report post-exposure disturbances / aftereffects. In an ideal VE, users should experience high presence, low SS and few aftereffects. Stanney and Salvendy (1998) noted that knowing possible associations between presence, SS and aftereffects might help VE design. However, although presence, SS and aftereffects have been explored by many studies, there is no general accepted conclusion regarding their relationships.

Some studies reported a positive relationship between SS and presence (Wilson, et al., 1997); other studies showed the negative relationship between SS and presence (Nichols, Haldane & Wilson. 2000; Singer et al, 1997). Thus, the relationship between SS and presence remain unclear? Stanney and Salvendy (1998) suggested that both SS and presence might correlate with intervening variables such as vection. Nichols et al. (2000) agreed.

This study describes an attempt to explore possible relationships between the sense of presence and SS. First, we describe traditional statistical correlation methods to explore the experimental data. Second, we explore a new method to examine those data.

Initial examination of scatterplots, regression models and calculated correlation coefficients may be useful for assessing relationships between the two experimental variables. However, if the variables exhibit no clear relationship or are mediated by other variables, it may not be easy to detect a true association. One of the approaches to analysis in this situation is nonparametric curve estimation. The approach can be used to explore data systematically and to uncover the possible relationships that could be lead to further testing (Fox, 2000). The method has been used in many fields (see Bowman and Azzalini, (1997); Efromovich, (1999); Simonoff (1996); Fox, (2000))

2 Method

Data from the 11 subjects collected in an experiment described by Duh, Parker and Furness (in press) were combined with the data from 20 subjects from Lin et al. (2001). Those experiments used the same apparatus: a Real Drive driving simulator (Illusion Technologies International, Inc). It includes a real Saturn car (General Motor Company), three 230cm X 175 cm screens on which images from three 800 X 600 pixel Sony projectors are displayed. A virtual world was generated by CAVE library software (developed by EVL, University of Illinois at Chicago) using a Silicon Graphic Onyx2 rack. Subjects could wear a shutter glasses – CrystalEyes (StereoGraphics Corp.) and sit in the car. All subjects were 'driven' through the VE along pre-recorded trajectory. The field-of-view could be as wide as 180^0. The SSQ (Kennedy et al., 1993) used for evaluating SS and the E2i questionnaire was used to evaluate presence and enjoyment (Lin et al., 2001).

3 Results and Discussion

Table 1 presents the correlation coefficients relating SS, presence and enjoyment calculated from the combined data. There is a low, non-significant negative correlation between E2i and SSQ scores. There is low, non-significant positive correlation between presence and SSQ scores. However, there is a significant negative correlation between enjoyment and SSQ scores. The presence and enjoyment scores are significantly positively correlated with the E2i scores, as expected.

Table 1. Pearson correlation table of SSQ, E2i, presence and enjoyment score

	SSQ score	E2i score	Presence score	Enjoyment score
SSQ score	1	-.116	.156	-.399**
E2i score		1	.830**	.637**
Presence score			1	.126
Enjoyment score				1

**p<0.01

The results are partly consistent with Nichols' et al. (2000) findings. They used an HMD; a Presence Questionnaire and the SSQ were used to investigate the relationship between SS and presence. 20 subjects were tested. Nichols and his colleagues reported a significant negative correlation between SS and presence. Nichols (1999) also reported a negative correlation between enjoyment and SS. It is interesting that in our study the SSQ scores tended to be positively correlated with the presence scores. This is different from some previous research (see Stanney and Salvendy, 1998) but consistent with reports by Wilson et al. (1997) and Welch (1997).

The combined data from 31 subjects who participated in 3 experiments were further analyzed using a nonparametric smoothing density estimation procedure. The results are shown in Figure 1. As the figures indicate, there is no *linear* correlation between SSQ, presence and enjoyment scores.

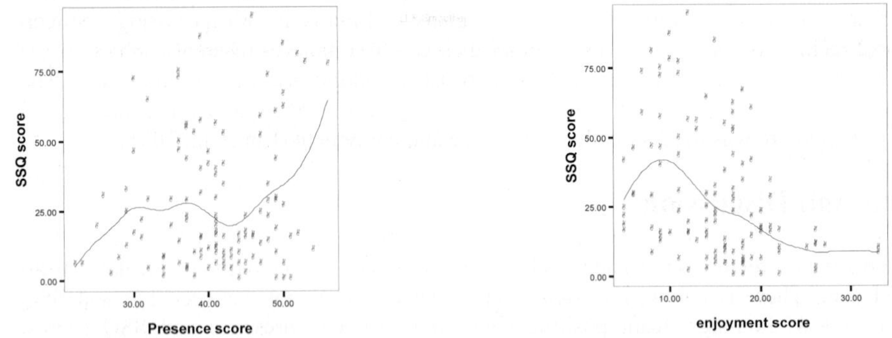

Figure 1. Scatterplot with nonparametric smoothing density estimation of SSQ with presence and enjoyment scores. (31 subjects, 124 observation points)

Regarding the relationship between presence and SSQ scores, for low presence scores, SSQ and presence appear to be positively correlated; with increased presence, SSQ remain at the same level; for higher presence scores, increasing presence is associated with a farther increase SSQ scores. Regarding the relationship between enjoyment and SSQ scores, for low SSQ values, SSQ and enjoyment appear to be positively correlated; however, with farther increases in enjoyment, SSQ scores remain low.

As figure 1 illustrated, the association between presence and SS cannot be described as a simple linear correlation. To obtain certain level of presence in VE, subjects may need to develop some minor SS symptoms. However, with increasing SS symptoms, presence appears to decrease. As Nichols et al. (1997) suggested, SS might detract from the sense of presence. Another interesting finding is that high presence was apparently associated with high SS. The association between enjoyment and SSQ showed the different trend noted above. The implication for VE design from these data is that some minor SS symptoms may facilitate development of presence, especially during the initial period of exposure. However, the VE designer must be able to limit SS to maintain a desired level of enjoyment. This hypothesis needs further examination. Also, the relationship between SS and presence may involve modulating variables such as FOV and stereo.

We need to explore further those intervening variables since presence might be used develop VE interface design criteria.

Acknowledgments

Supported by Eastman Kodak Contract. We thank Habib Abi-Rached for software development.

References

1. Bystorm, K., Barfield, W., & Hendrix, C. (1999). A conceptual model of the sense of presence in virtual environments. *Presence: Teleoperators and Virtual Environments*, 8, 241-244

2. Bowman, A.W., & Azzalini A. (1997). Applied smoothing techniques for data analysis: the kernel approach with s-plus illustrations. NY: Oxford University Press

3. Duh, H.B.L., Parker, D.E., & Furness, T.A. (in press). An independent visual background reduced simulator sickness in a driving simulator. *Presence Teleoperators and Virtual Environments*

4. Efromovich, S. (1999). Nonparametric curve estimation: methods, theory, and applications. NY:Sringer-Verlag.

5. Fox, J. (2000). Nonparametric simple regression: smoothing scatterplots. CA: Sage Publication Inc.

6. Kennedy, R.S., Lane, N.E., Berbaum, K.S., & Lilienthal, M.G. (1993) Simulator sickness questionnaire: an enhanced method for quantifying simulator sickness. *International Journal Aviation Psychology*. 3(3). 203-220

7. Lin, J.J.W., Duh, H.B.L., Parker, D.E., Abi-Rached, H.K., & Furness, T.A. (2002). , Effects of field-of-View on presence, enjoyment, and simulator sickness in a virtual environment. In *Proceedings of IEEE VR 2002*, Orlando, FL, 164-171

8. Nichols, S.C. (1999). Virtual reality induced symptoms and effects (VRISE): Methodological and Theoretical Issue. Ph.D. Thesis, University of Nottingham.

9. Nichols, S.C., Cobb, S.V.G., & Wilson, J.R. (1997). Health and safety implications of virtual environments. *Presence: Teleoperators and Virtual Environments*, 6, 667-675

10. Nichols, S.C., Haldane, C., & Wilson, J.R. (2000). Measurement of presence and its consequences in virtual environments. *International Journal of Human-Computer Studies*, 52, 471-491.

11. Simonoff, J.S. (1996). Smoothing methods in statistics. NY: Springer-Verlag

12. Singer, M.J., & Witmar, B.G. (1997). Presence: Where are we now? In M.Smith, G. Salvendy, & R.Koubek (Eds.), *Design of computing systems: Social and ergonomics considerations* (pp. 885-888). Amsterdam: Elsevier

13. Stanney, K., & Salvendy, G. (1998). Aftereffects and sense of presence in virtual environments: formulation of a research and development agenda. *International Journal of Human-Computer Interaction*, 10(2), 135-187

14. Wilson, J.R., Nichols, S., Haldane, C. (1997). Presence and side effects: complementary or contradictory? In M.Smith, G. Salvendy, & R. Koubek (Eds.), *Design of computing systems: Social and ergonomics considerations* (pp. 889-892). Amsterdam: Elsevier

Synergistic Use of Visualisation Technique and Web Navigation Model for Information Space Exploration

Carla M.D.S. Freitas
Ricardo A. Cava

Federal University of Rio Grande do Sul
Caixa Postal 15064
Porto Alegre, CEP 91501-970, Brazil
[carla, cava]@inf.ufrgs.br

Marco A. A. Winckler
Philippe Palanque

LIIHS-IRIT Université Paul Sabatier
118, route de Narbonne
31062 TOULOUSE cedex 4, France
[palanque, winckler]@irit.fr

Abstract

Problems related to navigation are known as the most frequently reported by users when browsing large Web sites. Users become quickly lost in large collections of pages especially if little contextual information is given to explore the information space. In this paper, we discuss the importance of providing users with a unified representation of structure and navigation information of Web sites as a way to prevent the *lost-in-space phenomena*. We present a notation to support navigation design of Web applications, the StateWebCharts, which explicitly represent structure and navigation in a seamless way. The Bifocal Tree, a technique originally developed for visualizing hierarchical structures, is extended and used to display StateWebCharts specifications, taking advantage of their characteristics to represent both structure and navigation information.

1 Introduction

For Web applications, browsing is one of the main activities carried out by users. However, users' browsing behaviour is poorly understood, and this leads to Web applications showing usability problems (Chi, 2002).

Navigation design is much more than matching pages (Schwabe et al., 2001). It includes: a) design of the classes of pages that determine the *structure* of the web site; b) design of each individual page with links to related pages in the site, which is called here simply *navigation*. Even though notations like WebML (Ceri et al., 2000), OOHDM (Schwabe et al., 2001), and StateWebCharts (SWC) (Winckler et al., 2001) represent both hierarchical and topological structure, they are intended to be used by Web designers and not by end users.

Most representation techniques currently available for end-users only represent information either about structure or navigation. For example, animated menus, hyperbolic trees (e.g. http://www.inxight.com/map/) or more static pages (usually called *site maps*) were conceived to represent structure of Web sites. These techniques represent good indexes for a site but they miss relationships between nodes to provide valuable information about nodes' relevance for users.

Visualization techniques based on graphs, provide good detail about relationships but they may become confusing and overwhelming, depending on the number of nodes. Hierarchical graphs, such as SWC (Winckler et al., 2001) reduce the number of relationships to be represented, but graph manipulation is still complex, because users have to manage: a) the depth, while they navigate through the hierarchy of composite nodes; and b) the relation between nodes, when looking at transitions (hyperlinks) between them. So, SWC remains a solution only for designers.

Other techniques to show navigation, like the one by MS Frontpage™ for instance, fail to represent nodes that might end up linked to other nodes recursively as the navigation is performed. Two-dimensional and three-dimensional representations have been proposed and efforts have been made towards their evaluation (Benford et al., 1999; Risden et al., 2000).

In this paper, we discuss the importance of representing hierarchical structure and navigation information during the design of Web applications as well as to the end user. We describe a case study that highlights how structure and navigation representation can be explored in Web applications for different purposes. In Section 2, this case study is used to explain how designers can represent structure and navigation using SWC notation. In Section 3, we introduce the Bifocal Tree (Cava et al., 2002), a visual representation for displaying tree structures based on a node-edge diagram. We extend the Bifocal Tree to represent SWC designs. Section 4 discusses our approach and related techniques, and Section 5 presents our conclusions and future work.

2 Navigation Design with SWC Notation

In a very simplified way, the SWC notation is based on a finite set of states and a set of transitions between the states (Winckleret al., 2001). States are used to represent a set of objects that a user is visualizing at a specific time, such as HTML pages. A transition is typically a link that changes the set of objects being shown to the user. When selecting a link on a page, a user is in fact activating a transition, which leads to a new state that corresponds physically to a HTML page.

SWC modelling is typically a graph-based representation, in which we can compose a hierarchy of nodes by inserting nodes into other nodes. The navigation is typically represented by arrows (called *transitions*) linking different nodes, even though links can cross the boundaries of composite nodes. Figure 1 presents a partial modelling for the "The Cave of Lascaux" Web site (http://www.culture.fr/culture/arcnat/lascaux/en/) with the SWC notation. This model shows the composite states (*S0*, *S3* and *S3.1*) and leaf states (*S1, S2, S4, S5, S6, S7, S8, S9, S10, S3.1, S3.2, S3.2, S3.4* and *S3.5*).

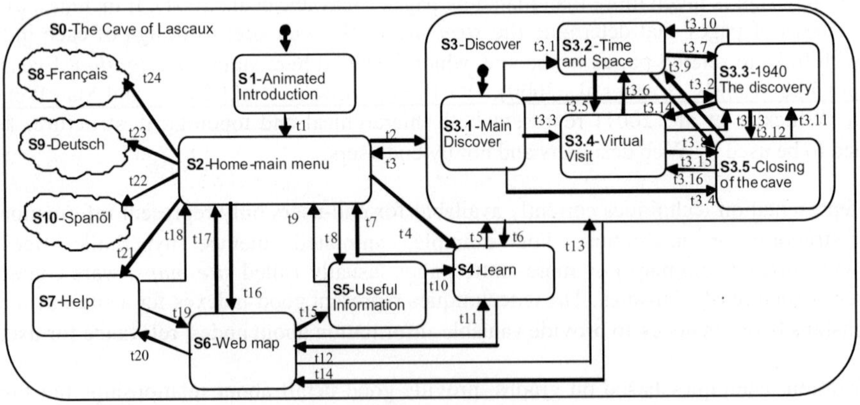

Figure 1: Partial SWC model for "The Cave of Lascaux" Web site.

The composite states represent classes of pages which share the same structure at that level. Sub-states inherit from its parent the relationships, and doing so, generic navigation can be specified once and reused for all leaf nodes in the composition. For example, in Figure 1, the transitions (*t3, t6* and *t14*) going from state *S3-Discover* to states *S2-Home-main menu*, *S4-Learn* and *S6-Web map*, respectively, are also part of sub-states in *S3* (*3.1, S3.2-, S3.3, S3.4* and *S3.5*). The states at the left are instances of classes of pages that have their own navigation. The complete model for this site includes much more levels, but due to space constraints, only two hierarchical levels are represented here: level 1, *S0-The Cave of Lascaux,* and level 2, *S3-Discover.*

Such nodes composition creates different levels of abstraction for the Web site, allowing the designer to cluster similar information, and clearly separates structural information and navigation in the model. For designers, this kind of representation contains relevant information including state types, transitions and composition with semantics meaning.

3 Representing SWC Models with the Extended Bifocal Tree

The Bifocal Tree (Cava et al., 2002) is a visual representation for displaying tree structures based on a node-edge diagram. The technique was originally conceived to display a single tree structure as a focus+context visualization. Indeed, it provides (simultaneously and in the same window) detailed and contextual views of a selected sub-tree. Given a selected node in the hierarchy (A, for example), the left part of the window displays A's parent and all the sibling sub-trees, while the right part shows A and its sub-tree. Selecting any node causes its repositioning at the focus point in the detailed area and the movement of its parent node to the focus point of the context area. The other nodes are repositioned accordingly to these two ones.

By taking advantage of this two-focus approach, we can explicitly separate information related to structure (state composition) and navigation (transitions). We extended the Bifocal Tree technique to represent Web site's hierarchy in the context area and the Web site's navigational structure in the detailed area, thus providing hierarchy and navigation information both to designers and end users. This separation would enhance user perception about both classes of information, while keeping all relevant information. This Extended Bifocal Tree representation for the SWC model of "The Cave of Lascaux" Web site (presented in Figure 1) is shown in Figure 2.

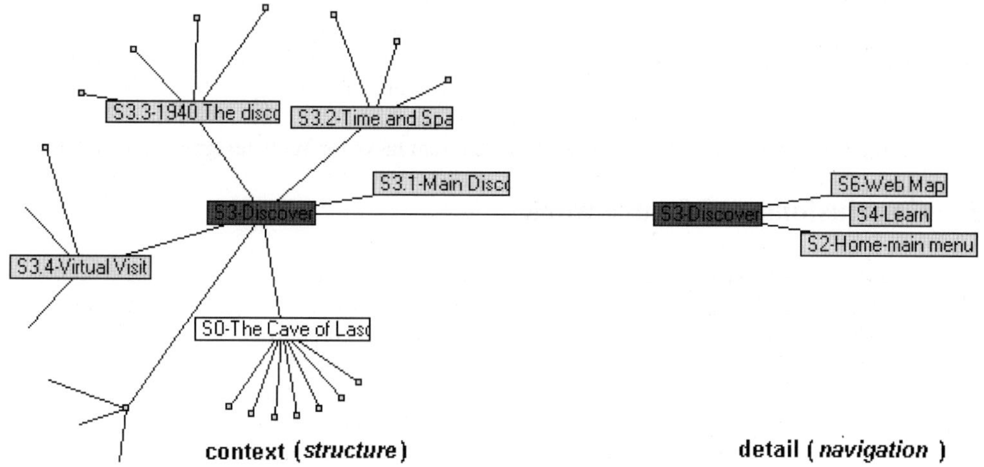

Figure 2: Extended Bifocal Tree representation of the SWC model.

The root node is drawn in white and the two foci are drawn in red. The left focus is devoted to represent the nodes hierarchy, while the right focus represents the navigation information associated to the selected node. In this example, the user would be examining the navigation options from document *S3-Discover,* having access to the whole hierarchy of documents represented in the context area. Its hierarchical composition can be seen at the left part of the diagram, while the right part shows the links that can be followed from it. The hierarchy presented by the Bifocal Tree include sub-nodes that are not represented in the SWC model (small squares going out from *S3.4, S.3.3, S3.2*) due to space constraints.

The main advantage of this visualization technique is that information about structure and navigation are separated in the presentation but the context is preserved. Users can use both foci to explore the information space. When a node in the detail area is selected the context is updated automatically to provide the appropriate view of the structure into which the selected node is located; on the other hand, the detailed view displays the navigation structure associated to it facilitating user orientation for browsing in each view.

4 Discussion

Solutions to the visualization of different aspects of Web sites have been reported in the literature along the past decade, but providing browsing aids that improve Web applications usability is still a challenge. *Site maps* that reflect the file structure of a Web site are a partial solution to this problem because they can not reflect all the possible paths obtained from the links between objects. The hyperbolic tree displays the structure of a Web site as a node-edge diagram, and allows smooth, dynamic transitions from one node to others. This representation is mainly targeted to end users. On the other hand, Munzner's 3D hyperbolic representation integrated into the XML3D browser (Risden et al., 2000) exhibits the hyperlink structure of a Web site as a single node-edge diagram in 3D space and a set of 2D lists showing parents, siblings and children of a selected node. However, the hierarchy is embedded in the hyperlink structure.

The Extended Bifocal Tree explicitly represents the hierarchical and navigation structure in an integrated view allowing a seamless transition between browsing styles. Usability tests applied to evaluate the original Bifocal Tree browser allowed us to improve the representation. Experiments with users accomplishing node finding, comparison and path navigation tasks, and heuristics evaluation employing Nielsen's usability factors as well as Bastien and Scapin's ergonomic criteria have shown that the Bifocal Tree is more efficient than the hyperbolic browser and the Windows Explorer™ in tasks like navigation through specific paths. Thus, tests with the Extended Bifocal Tree can rely on these results and target different tasks for Web designers and end users.

5 Conclusions and future work

More the information space is bigger, more easily users suffer from *lost-in-space phenomena*. Due to the complexity and huge amount of relationships between pages, large Web sites may cause users to create a wrong mental model about how information is organized. We have shown a small case study with SWC notation, demonstrating that, although specification of structure is a clear activity in the design process, the huge amount of links does not allow users to perceive the structure easily without help. Research on visualization techniques has been trying to close this gap and help users to better explore and understand such large information spaces.

In this paper, we have stressed the importance of representing structure and navigation information during the design phase of Web applications and showed a way of filtering these two classes of information to user. We believe that end users might benefit from the information they would get looking at both hierarchy and navigation structure of a Web site at the same time. The hierarchical structure view allows users seeing at a glance the entire Web site and jump directly into the information sub-space using a visual representation instead of browsing all the documents. Viewing the navigational structure allows users to better evaluate the relevance of a selected node according to the other nodes it is linked to.

Previous works have used independent techniques to represent hierarchy and navigation information. Our solution for this problem is to filter these two kinds of information during the visualization by using the Extended Bifocal Tree, which preserves the relation fromone class of information to another, by employing an integrated, but well-separated representation.

The Extended Bifocal Tree uses a specification in SWC notation (made by designers) to build a more usable representation to end users. This tool could work directly on the file system of a Web site but such approach can present problems if the directory-file hierarchy does not reflect pages classification according to users' perspective. The specification of the structure of a Web site is an arbitrary information that is hard to recover by reengineering after the Web site is launched. So, we rely on SWC specifications as a safe source for the Extended Bifocal Tree. Such notation is suitable for supporting the design process of Web applications with other advantages than the visualization of information space. The modelling with SWC allows, for example, specifying requirements for the application, testing and simulation before the implementation, and improves communication within the design team. In addition, it is important to notice that only a synergistic use of techniques that show the hierarchical and topological structures of large Web sites, at the same time, can make users benefit from this information because different browsing behaviours should be supported to improve usability of Web applications.

Next activities include users testing to improve the tool that implements the technique. Our goal is to provide the Extended Bifocal Tree as a visible tool (typically as a Java Applet), giving to users information and access about the Web site hierarchical and navigational structure.

Acknowledgments

This work has been sponsored by CAPES/COFECUB SpiderWeb project and CNPq (Brazilian Council for Research and Development). First author is also sponsored by FAPERGS.

6 References

Benford, S.; Taylor, I. Brailsford, D.; Koleva, B.; Craven, M. et al. (1999) Three dimensional visualization of the World Wide Web. *ACM Computing Surveys*, Vol. 31, No. 4, ACM Press, New York.

Cava, R.; Luzzardi, P.R. & Freitas, C.M.D.S. (2002) The Bifocal Tree: a Technique for the Visualization of Hierarchical Information Structures. *IHC200 -, V Symposium on Human Factors in Computer Systems Proceedings*. Fortaleza, Brazil. October 7-10.

Ceri, S.; Fraternali, P. & Bongio, A. (2000) Language (WebML): a modeling language for designing Web sites. *9th WWW Conference Proceedings*, Amsterdan, May 15-19.

Chi, E.H. (2002) Improving Web Usability Through Visualization. *IEEE Internet Computing*, 6(2), 65-71.

Risden, K.; Czerwinski, M.;. Munzner, T. & Cook, D.B. An initial examination of ease of use for 2D and 3D information visualizations of web content. *International Journal of Human Computer Studies*, 53(5), 695-714.

Schwabe, D.; Esmeraldo, L.; Rossi, G. & Lyardet, F. (2001) Engineering Web Applications for Reuse. *IEEE Multimedia*, 8(1), 20-31.

Winckler, M.; Farenc, C.; Palanque, P. & Bastide, R. (2001) Designing Navigation for Web Interfaces. *IHM-HCI2001 Proceedings*, Lille, França, September 10-14.

Visualizing Activity in Shared Information Spaces

Wolfgang Gräther, Wolfgang Prinz

Fraunhofer FIT
Schloss Birlinghoven
53754 Sankt Augustin
{givenname.surname}@fit.fraunhofer.de

Abstract

This paper describes Smartmaps, an applet, which provides users with an overview of all activities in shared information spaces like shared file systems, Web sites, or shared workspaces in a treemap. The treemap is constructed from the shared artifacts, links from the Smartmap to the artifacts are established, and actions on artifacts result in color coding of the respective part of the Smartmap. This presentation of activity enables activity-based navigation and helps in becoming aware of potential collaborators and advice giving experts. The Smartmap is dynamically recreated, when the underlying information space changes. Moreover Smartmaps can be integrated into the information spaces itself to build social spaces.

1 Introduction

The importance of awareness for successful use of groupware has been identified in many studies of workplaces. Several models and applications have been developed and the requirement to provide awareness about activities and actions of others in a cooperative environment is part of CSCW-engineering. Like in real world settings, situated action (Suchman 1987) requires awareness information about the working space in which the action takes place. Smartmaps provide both task-oriented and social awareness (Prinz 1999), i.e. they yield information about the state of artifacts as well as presence and activities of people. To explain the application of Smartmaps for the provision of awareness in shared information spaces, we also describe the integration of Smartmaps into BSCW shared workspaces (Appelt 1999).

The BSCW shared workspace system is a Web-based groupware tool based on the notion of shared workspaces, which may contain various kinds of objects such as documents, tables, graphics, spreadsheets or links to other Web pages. Folders are used to group the artifacts and the hierarchy of folders constitutes the shared workspace. Workspaces can be set up, members can be invited and objects can be stored, managed, edited or downloaded with any Web browser. There are many other services available, for example, discussion forums. Asynchronous as well as synchronous collaboration is supported by BSCW.

Smartmaps uses a treemap for the visualization of the objects in shared information spaces. The treemap visualization technique maps a full hierarchy onto a rectangular region in a space-filling manner that makes efficient use of space and therefore allows very large hierarchies to be displayed (Johnson and Shneiderman 1991). An overview of history and applications of treemaps is available at http://www.cs.umd.edu/hcil/treemap-history/.

2 Visualization

The Smartmap shown in Figure 1 presents activity information in an information space consisting of 42 artifacts. The 3 top-level folders have thick borders and contain 22, 9, and 11 artifacts, respectively. Folders on other levels in the hierarchy of the information space are not directly visible. All artifacts of the information space are represented as small rectangles, artifacts in the same folder are close to each other and ordered lexicographically.

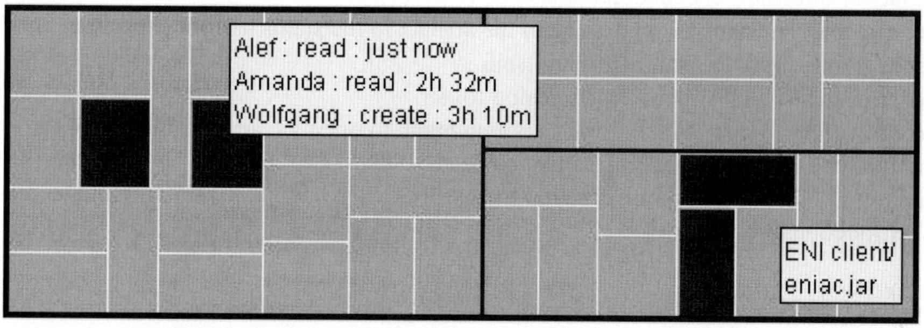

Figure 1: Smartmap with tooltips display the location of artefacts and name of persons

Highlighted rectangles, in black, indicate user activity. The default presentation mode conveys the overall activity and their distribution in the information space. Tooltips, which are activated, when the users moves the mouse over the corresponding region, indicate location and name of the artifact. When the mouse is moved over a highlighted rectangle, then the tooltip presents the names, actions, and passed times of the last 3 users who have worked with the artifact.
There are several parameters influencing the visualization. The color for highlighting can be chosen and the highlighting intensity can be specified, so that only artifacts or artifacts and enclosing folders up to a configurable level are highlighted with decreasing color intensities. In addition, the duration of highlighting could be set. Figure 2 shows the shared information space presented in Figure 1 with different highlighting parameters.

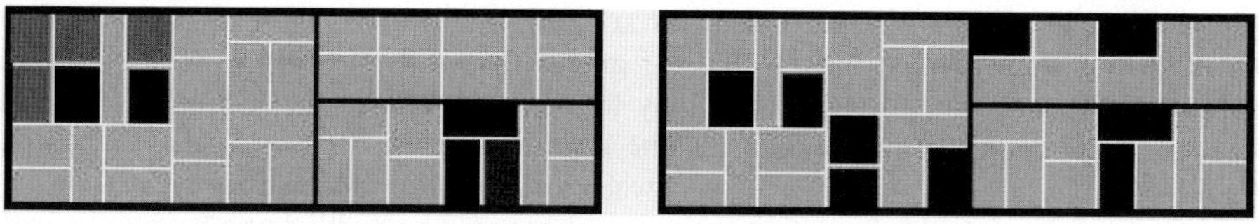

Figure 2: Smartmap with different parameters for highlighting

For the duration of highlighting, we prefer 10 minutes for Smartmaps visualizing activities on Web sites and we favor a duration of one day for BSCW shared workspaces and shared file systems with about 20 users. The latter duration enables users to see at a glance an overview of what has happened in the information space during the last 24 hours. The different parameters for visualization enables people to know that there are others present and working in the same information space, the same part or even on the same artifact.
The Smartmap applet is usually placed on Web pages. For example, it can be used as a site map, which eases navigation and presents current visitors on the Web site. In the following we will focus on the interaction possibilities of Smartmaps.

3 Interaction

Smartmaps support not only the visualization of activity information in shared information spaces, but they also ease the navigation in shared information spaces. There are a lot of interaction possibilities on Smartmaps:

- Moving the mouse over the Smartmap presents either the location and the name of the artifact or information about recent user activities,
- a mouse click presents the pathname of the artifact in the status bar of the browser window,
- a shift-mouse click opens the artifact itself,
- a control-mouse click opens the enclosing folder,
- a control-right-mouse click presents a popup menu to open the artifact, the enclosing folder, and further enclosing folders up to the top-level folder (see Figure 3),
- and a right-mouse click shows a popup menu to set visualization parameters and to access other functions.

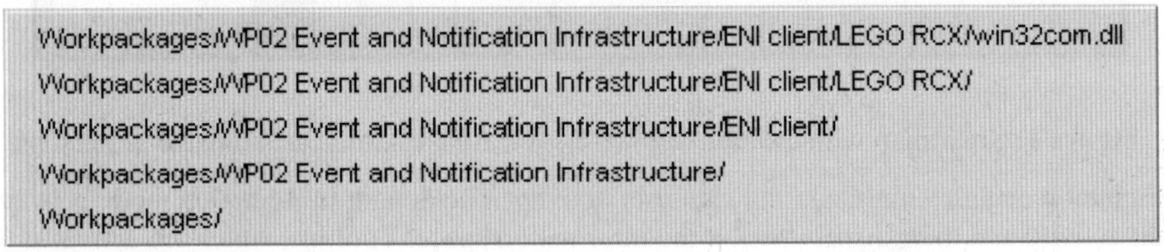

Figure 3: Popup menu to access different parts of the shared information space

Shared information spaces are often used to support cooperative work processes. The mutual orientation and information about the progress of common tasks, changes in group structure, or cues about accessibility of others is a prerequisite for successful work completion. Smartmaps peripherally inform cooperation partners about presence and ongoing activities of co-workers. The interaction possibilities enable activity-based navigation in the information space and ease traversing hierarchical structures.

A Right-mouse click shows a popup menu for setting visualization parameters. Aside from the already in section 2 mentioned parameters, the granularity of the visualization could be set to: complete (default), only enclosing folders, only the first x-levels of the hierarchy, etc. Another parameter sets the event lookup period, i.e. the time how often the Smartmap is updated. An update of the underlying treemap is necessary, when artifacts have been newly created or deleted, or the structure of the shared information space has changed. The highlighting have to be updated, when user actions on the artifacts have happened.

In addition, a filter could be set, which allows to create a Smartmap only of selected document types of the whole shared information space. A search option allows to search for artifacts and persons. When documents or persons are found, then the corresponding rectangles in the Smartmap blink. Finally, help and about information for Smartmaps is available.

4 Smartmaps as part of shared information spaces

Smartmaps complement shared information spaces with an additional mean for navigation and presentation of awareness information. In the following we will focus on the integration of Smartmaps into BSCW shared workspaces. The BSCW system offers a user customizable area, the banner, which partly changes the display of workspace and folders. Typical banners are: images or headlines according to workspace content, project logos, Web cam images, and even applets. The banner is inherited through the hierarchy of folders in the shared workspace and HTML is used to describe the content of the banner. Figure 4 shows a Smartmap integrated into a BSCW shared workspace with more than 1000 artefacts. At every place in the workspace the Smartmap is shown and augments the usual list mode presentation of the shared objects. The actual position of the user in the hierarchy of the shared workspace is indicated by an orange rectangle.

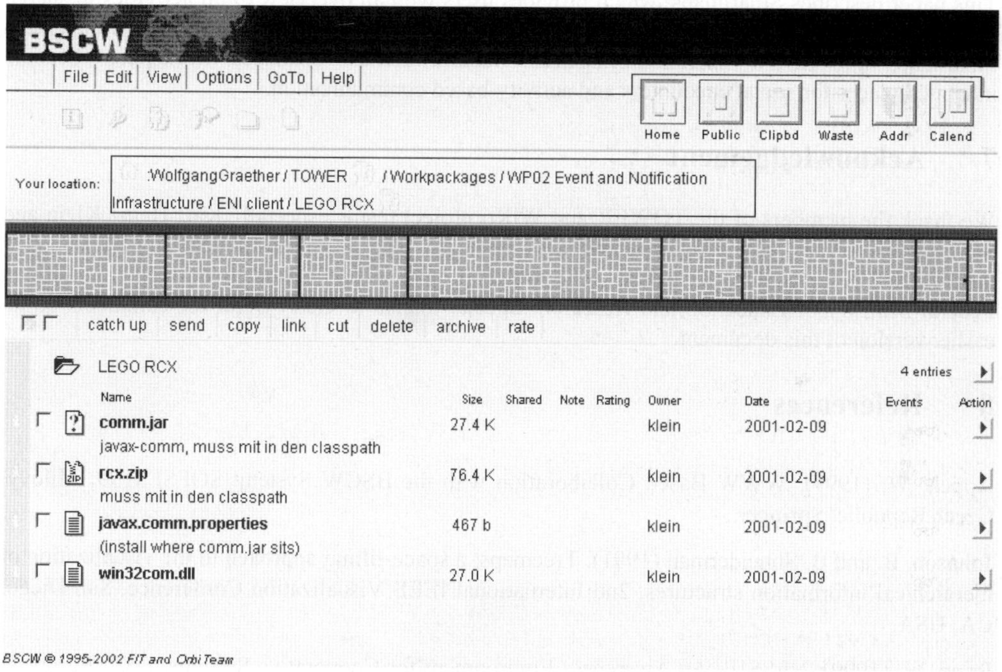

Figure 4: Smartmap in a BSCW share workspace

Smartmaps in shared workspaces complement awareness information which is already available at the shared artefacts, but visible only in the current work situation, with awareness information from other parts of the overall working context of the user. A even larger context could be built, when several Smartmaps, representing different workspaces, are put together.

We have integrated Smartmaps into several large project workspaces and we could observe that users quickly apply the Smartmap. For example, they first check the places for which activity of other co-workers is indicated. They move the mouse to the corresponding highlighted rectangle to see who acted on the artefact, what kind of action took place, an when happened the action. Then, often that link is followed and the corresponding folder or the object is opened.

5 Implementation

Smartmaps is implemented as Java applet. It employs a crawler which traverses the shared information space and delivers its structure to the applet. In case of BSCW shared workspaces the structure is updated when the user moves to another workspace. For very large workspaces the traversal could be time consuming, therefore we realized a caching mechanism.

The information about user activity in the shared information spaces is stored as events in an event server. The event server is accessible via cgi-scripts, i.e. the Smartmap applet sends queries for events using the http protocol. The events itself carry data about time, action type, actor, etc. The event server has a history of events available for certain event types.

6 Conclusion

This paper describes Smartmaps, which provides users with an overview of all activities in shared information spaces like shared file systems, Web sites, or shared workspaces. We have integrated Smartmaps into BSCW shared workspaces. Such awareness enhanced workspaces become inhabited places for social encounters and activity-based communication.

7 Acknowledgement

We thank the members of the TOWER and WiKo project teams especially Karl-Heinz Klein and Sabine Kolvenbach who supported our developments. The TOWER project is partly funded through the IST program IST-10846. The WiKo project is partly funded through the German Federal Ministry of Education and Research. Special thanks to Gerry Stahl for comments on an earlier version of this document.

8 References

Appelt, W. (1999). WWW Based Collaboration with the BSCW System. SOFSEM'99, Milovy, Czech Republic, Springer.

Johnson, B. and B. Shneiderman (1991). Treemaps: a space-filling approach to the visualization of hierarchical information structures. 2nd International IEEE Visualization Conference, San Diego, CA, USA.

Prinz, W. (1999). NESSIE: An Awareness Environment for Cooperative Settings. Sixth European Conference on Computer Supported Cooperative Work, Kopenhagen, Denmark, Kluwer Academic Publishers.

Suchmann, L.A. (1987). Plans and situated actions: The problem of human-machine communication. Cambrigde: Cambridge University Press.

The Development from Physical to Interaction Based Simulation Procedures on the Example of Virtual Cables or Hose Simulations

Elke Hergenröther

Fraunhofer Institute for Computer Graphics
Fraunhoferstrasse 5, D-64283 Darmstadt (Germany)
hergenro@igd.fhg.de

Abstract

This paper shows the development process from a physical cable procedure to a geometrical simulation procedure. The developed simulation procedures were created for interactive user manipulations. During the application presentations I learned that the physical correctness was not necessary for the visible correctness and a high simulation rate was important for the user.

1 Introduction

In car- and aircraft industry virtual environments (VE) are an important technology in the product development process. They have to replace the physical mock-ups step by step with digital mock-ups. Using digital mock-ups reduced the development time and therefore the mock-up production costs. The assumption was that digital mock-ups should have the same functionality as the physical mock-ups, but until now bendable elements like cables, tubes or rubber plugs are not integrated in the virtual prototype development process.

In the area of assembly and disassembly simulation the interaction with tubes and cables is very important. The collision detections are tested in VE, because the most VE have included sophisticated collision detection procedures. Also they provide some intuitive interaction features, but they do not have special interaction features to remove a collision between tube or cable with other elements of the scene available. Another problem are the simulation procedures itself. Many simulation procedures are not suitable for VE.

Up to now to meet the claim of physical correctness is the main effort during the development of a simulation procedure. Thus the first cable and tube simulation algorithms for VE are finite element (FE) based simulations. The classical FE-procedure provide no features to prevent the stretching and the compression of the cables length. Also the FE-simulated cable can be bent without any constraint. These deficits are very troublesome during the interaction with deformable cables or tubes. So at first I have to analyse the requirements users have on the simulation.

2 The kinematical-based cable or tube simulation procedure

The most significant demand for a cable simulation used in an assembly application is the constant cable length during the simulation. If the cable is fixed on both ends then it should not get longer while the user pushes it aside. The bending property is also important in this context. To fulfil these demands we developed a kinematics based cable simulation algorithm presented in

Hergenröther and Dähne (2000). It is a hierarchical method to ensure a real-time calculation on different processors. The kinematical chain is complemented with mass-points and rotation springs. Thereby the simulation of different materials is possible and consequently a different bending property can be calculated.

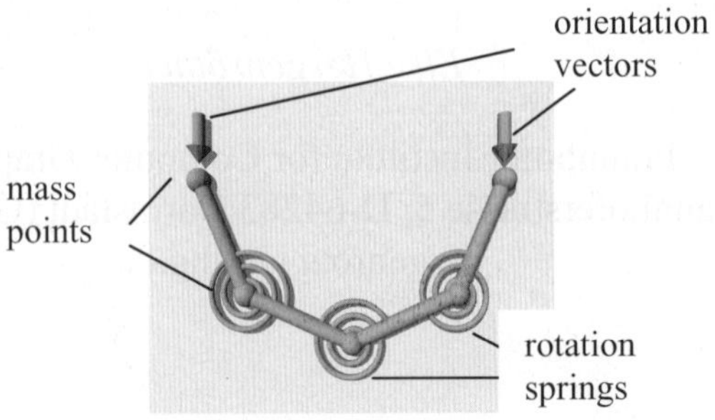

Figure 1: Cable simulation model

2.1 The comparison of a simulated with a real hose

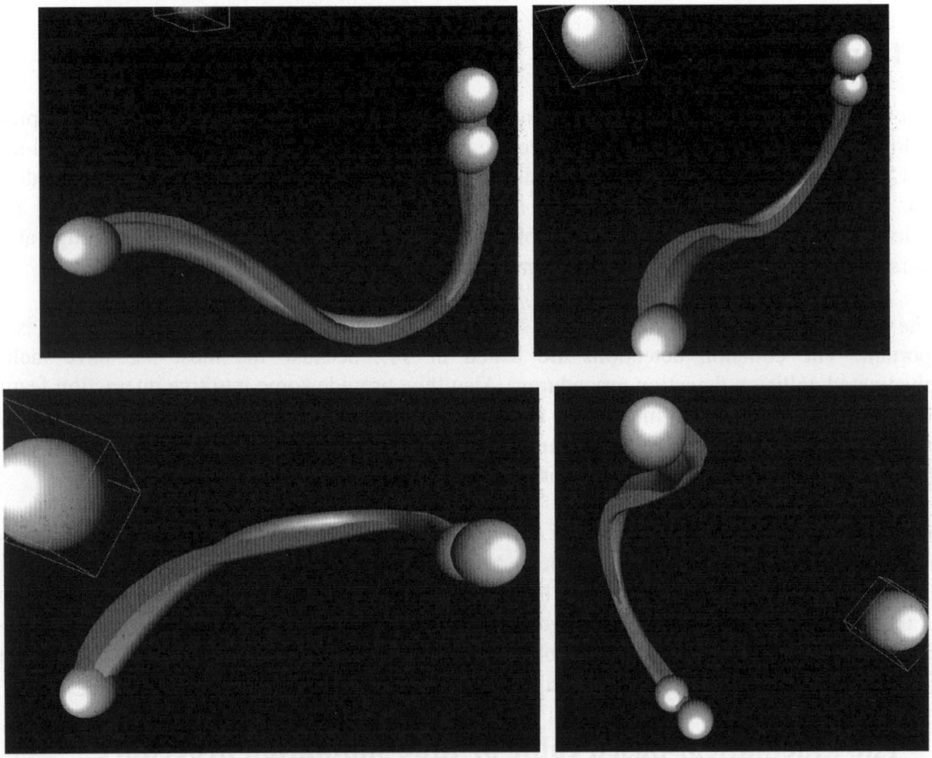

Figure 2: Different views of the comparison of a real hose (red) with the simulated hose (green)

To test the correctness of the simulation I compared the simulated results with a model of a real flexible brake hose. The results you can see in figure 2. The reasons of the difference between the real and the simulated hose is because the kinematical simulation procedure support no torsion. In figure 3 you see a real scenario. Here the hose has only a small torsion and you can see there is a high conformity.

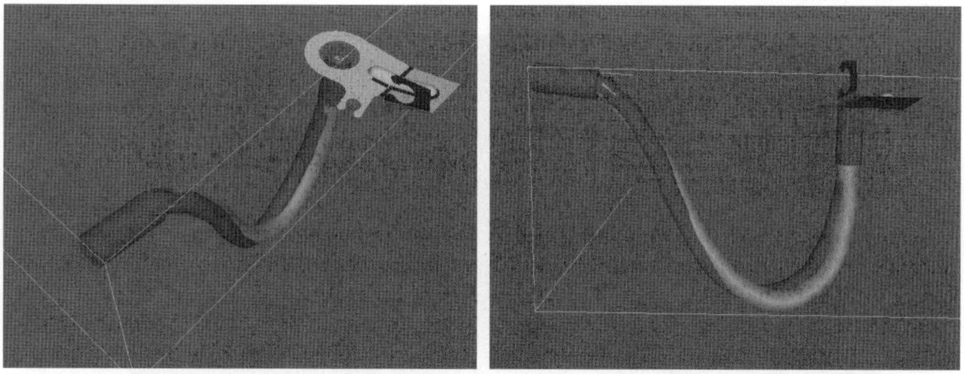

Figure 3: Comparison of a real hose (with only a few torsion) with the simulated hose

2.2 The virtual environment

This simulation procedure was included in our VE. To solve the user demands best we use a virtual laser beam. The ray is emitted from the hand in the direction the user points to. Hereby it is very intuitive to select objects, which are not too far away. We use this technique for the selection of the start and the end points of individual cables included in a harness. After the simulation the different harness pieces result in a smoothly coherent cable harness.

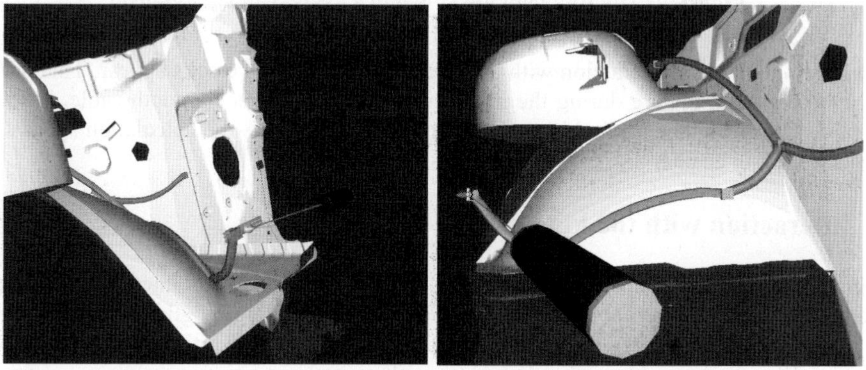

Figure 4: Cable installation in Virtual Environments from different points of view

3 The kinematical simulation

To outline the position of the cable harness or the tube it is a good solution, but to model the cables or harness run interactive this solution does not make sense. In case of modelling the shape of the cable or tube the simulation has not to be physically correct. The most problems which have to be solved are: Could the cable be bent around the corner? Or, could the cable connect object one with object two? For this purpose a physically correct simulation is not needed. All the user needs to know is the information how bendable the cable or the tube is. So a kinematically based simulation procedure where the angle of two adjacent stiff segments will be controlled by a maximal value fulfils the demands in this area.

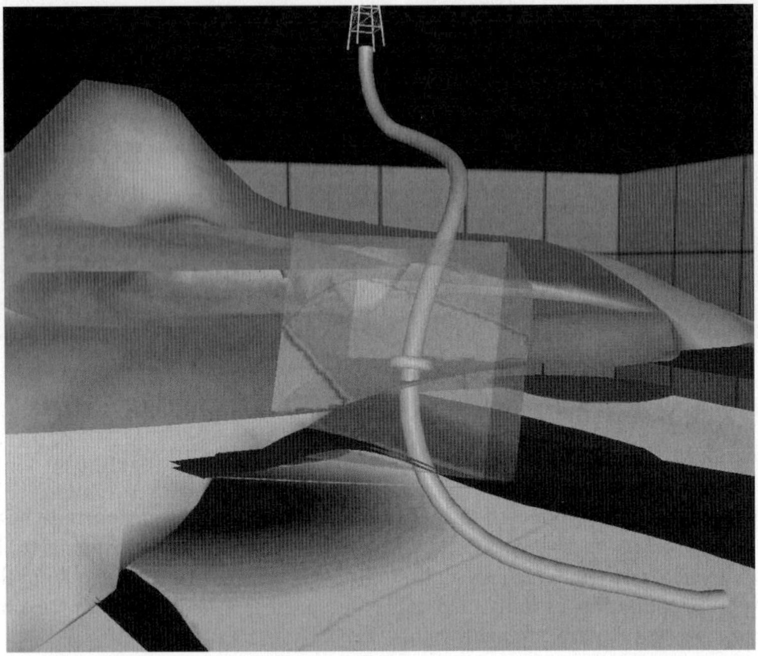

Figure 5: The well path integrated in a geological environment

Unfortunately during the interaction with a kinematical chain some cracks appear. This problem is solved with a special feature during the interaction. The solution is a smooth cable, hose or well path shape. The advantage of the kinematically based simulation is the calculation speeds, see the performance test in chapter 3.2.

3.1 Interaction with the kinematical simulation

The simulation became complete with a virtual clamp (see figure 5) as interaction feature. With the clamp you can grasp the cable or tube and move it. A modelling by gliding the clamp over the cable is also possible. We presented the application of different exhibitions and many people use this feature. Everybody could use the clamp after a short explanation.

3.2 Performance test of the kinematical simulation procedure

The test was done on an AMD Athlon (1,4 GHz). The simulated cable was composed of 40 nodes:

- pure kinematical simulation: 92.500 Hz
- includes the feature to avoid steps: 1.900 Hz

4 Conclusion and future work

The conclusion is that the user demands have to be taken much more into account during the conception of a new simulation algorithm. A detailed analysis of the user demands and expectations prevent the conception and implementation of complex and therefore costly simulation procedures.

5 References

Hergenröther, E. & Dähne, P. (2000): Real-Time Virtual Cables Based on Kinematics Simulation, *The 8th International Conference in Central Europe on Computer Graphics, Visualization an Interactive Digital Media2000* , Plzen, Czech Republic 402-409.

Hergenröther, E. & Knöpfle, C. (2001): Cable Installation in Virtual Environments, *Proceeding of the IASTED International Conference of Modeling and Simulation*, Pittsburgh, Pennsylvania, USA, pp. 276-280, May 2001

Hergenröther, E. & Knöpfle, C. (2001): Installation and Manipulation of a Cable Harness in Virtual Environments, *Proceeding of the IASTED International Conference of Robotics and Maufacturing*, Cancun, Mexico, pp. 240-245, May 2001

Hergenröther, E. & Müller, S.: Integration of Cables in the Virtual Product Development Process, *IFIP 11th International PROLMAMAT Conference on Digital Enterprise*, Budapest, Hungary, pp. 84-94, 7-10 November 2001

Visualizing Metadata: LevelTable vs. GranularityTable in the SuperTable/Scatterplot Framework

Tobias Limbach, Peter Klein, Frank Müller, Harald Reiterer

University of Konstanz
Universitätsstr. 10, Box D73, 78457 Konstanz, Germany
{Tobias.Limbach, Peter.Klein, Frank.Müller, Harald.Reiterer}@uni-konstanz.de

Abstract

This paper will describe ongoing efforts to ease the information retrieval process on metadata, using a SuperTable/Scatterplot framework recently named *VisMeB* (**Vis**ual **Me**tadata **B**rowser). Based on the combination of a SuperTable (in two design variants) and a Scatterplot (also in two design variants), users can follow different search strategies to achieve results. Usability testing has showen, that an integration of our proposed visualizations supports most user's search style.

1 Introduction

To visually explore large data bases, the generation of metadata and their meaningful visualization is common practice. Systems like FilmFinder (Ahlberg & Shneiderman 1994) and xFIND (Andrews Gütl, Moser, Sabol, & Lackner 2001) not only enable users to search with keywords, they also let the users browse through result sets and enable a user based visual assessment of the information retrieval process. Users can decide, based on their estimation of the individual relevance of each iteratively retrieved result set, if their informational need has been satisfied or not. But, to put users in the position to decide, the system has to be usable, task oriented and support the users needs in each step of the information retrieval process.
After this introduction a brief overview of the system VisMeB is given, followed in the third chapter by an overview of usability test results of the two integrated design variants, while chapter four gives an outlook to current and future work.

2 Visualizing Metadata in a SuperTable/Scatterplot Framework

Based on a redesign of the Web search system INSYDER (Reiterer, Mußler & Mann 2001), we developed a visual metadata browser with different strategies to avoid the cognitively demanding traditional representation of metadata result sets, which often leads to a long list of document attributes. We assume that users don't want to scroll through endless result lists without any possibility to filter or sort out, the way they usually would have to, following the traditional concept of search engines.
Visualizations like a ResultTable, a BarGraph, and Tilebars, which were until then used separately, were put into one embedding visualization called SuperTable complemented by a Scatterplot. This multifocal approach using focus-plus-context techniques is found in other table based visualizations, for example the TableLens (Rao & Card 1994). Users are offered different brushing and linking techniques between the two components. This was inspired by the work of North and Shneiderman (North & Shneiderman 2000). The Scatterplot, as a graphical

representation of the result set, gives a quick overview and defines own views by zooming, selecting and filtering to reduce the amount of hits, while the SuperTable follows the concept of a distortion-based table with BarGraphs to visualize document relevance (Veerasamy & Navathe 1995), and Tilebars (Hearst 1995). A RelevanceCurve is used to assess text segments, as their individual relevance is visualized through stacked columns. Documents can be analyzed in high detail. Our aim was to avoid the use of cognitively demanding result lists as long as possible. With the interaction possibilities the user is given, it should be possible to focus on the cognitive and intellectual resources only on the original document.

Following the scenario based design approach proposed by (Rosson & Carroll 2001) we developed different information retrieval process scenarios. While developing those, a second design variant for the SuperTable emerged from discussions with M. Eibl (Klein, Müller, Reiterer & Eibl 2002) and was built into the scenarios. Though it became necessary to specify them as *LevelTable* and, the newer one, as *GranularityTable*.

The first design variation, called LevelTable, was strongly influenced by lessons learned from the INSYDER approach. Metadata can be explored in four different views representing the "level of detail" for documents. We call this method of looking closer and closer at the details of a document the "focus of interest". The more one wants to know about a document, the deeper one has to look and the higher the level has to be, ranging from a mere graphical overview in the first level to the document as stored in the database in the last level. By doing so it combines several visualizations in one enclosing table to ease the interpretation and show their conceptual connections.

The second design variant follows a granularity concept, and with that further smoothens the change of the visualizations between the exploratory steps. The information retrieval process is ergonomically integrated into the semantic structure of the visualizations, as described by (Eibl 2002). Granularity is a term used in photography to describe the accuracy of pictorial presentations in a picture. The higher the granularity, the more details can be seen on a picture.

This idea is transferred to integrate the visualizations of the SuperTable smoothly: The lowest granularity results in a simple histogram, which states the overall importance of the single documents. The highest granularity would lead to a representation of the actual text. In between these margins are as many intermediate steps as are required to give the impression of a smooth transition (as far as possible technically). A visual and conceptual clue is given by a slider (known from e.g. media players, household appliances etc.). Figure 1 shows two screenshots of the html-mockups.

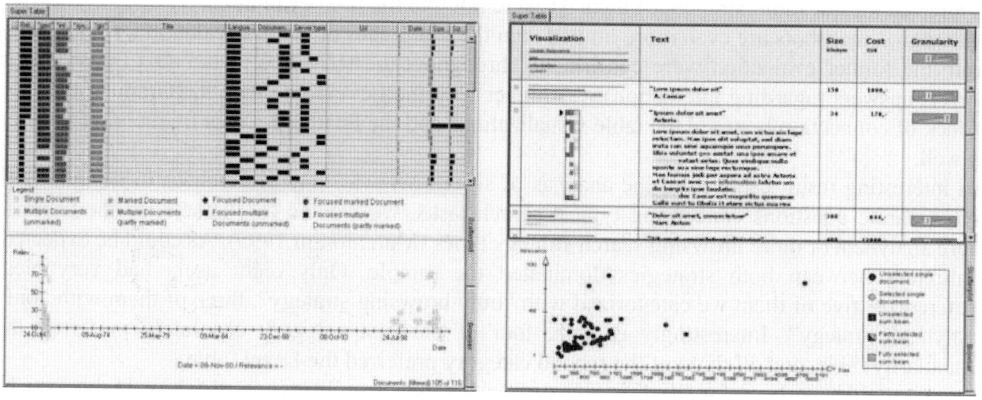

Figure 1: Html-mockups: LevelTable with overview, GranularityTable with different grades of detail

Both design variants are combined with a scatterplot. For early user tests, they were implemented as html-mockups with a limited range of interaction possibilities. Nevertheless, they fulfilled the requirements of our scenarios and were though tested.
A fundamental difference is established in the interaction concepts of the two design variants. While the LevelTable moves the result set as a whole into the next grade of detail, the GranularityTable gives the user the possibility to explore each result`s set details individually. While both start with an overview of all search results and end with a document browser, the GranularityTable can be adjusted to different grades of detail on one screen.

3 Testing GranularityTable vs. LevelTable

In October 2002 the html-mockups were tested by several users (n=8). Among individual differences in working and searching habits, the majority stated to have very high expectations concerning the work with metadata and that they expect higher efficiency with a working metadata browser.
After the pre-test questionnaire and a video introduction, the users were handed out a script with test tasks to work on. All tests were documented by minute taking, a video camera and a screen-recording.
The overall reception of the SuperTable/Scatterplot framework was good to very good. Some interactions of the Level- and GranularityTable were surprising to the users, though appreciated. Our analysis of the post-test showed, that users were in favor of the GranularityTable. Contradictory to this the results of the test sessions showed that users had more problems to fully understand the interaction concepts of the GranularityTable than of the LevelTable. We argue that, once understood, the joy-of-use is higher with the GranularityTable. Additionally it is possible that users were influenced by the clearer and more aesthetic design of the GranularityTable, and therefore gave it better ratings than the LevelTable. Our test design ruled out learning or last-item remembering effects. Nevertheless this bias in the test design has to be eradicated for the next user tests.
In parallel to the lab tests we started a web-based evaluation. Questions regarded individual search behavior, a virtual search with the two design variants, how the users would interact with them and what they would like to have differently. The participants were asked to download two short introductory videos, and several screenshots correlating to tasks. The sequence of Level-/GranularityTable was randomized to exclude learning and last item remembering effects. 35 users completed the questionnaire, out of them 31 were put into the final evaluation.
Although screenshots are even more limited than the prototypes of the lab evaluation, some results from the former evaluation were confirmed. Throughout the test the effectiveness (measured in correct answers regarding interaction) was higher with the LevelTable than the GranularityTable. A lack of connection between the table visualizations and the Scatterplot was frequently criticized as well.
An interesting result came from the analysis of search behavior and preferences in design. With five separate questions concerning typical search tasks, we wanted to characterise the users in more analytical or more browsing search strategy types (Marchionini 1995). As could be expected, a mixture between both strategies dominated the sample. Only eight users had very clear preferences, five of them we categorized with “only browsing strategy”, three of them with “only analytical strategy”. Interestingly enough, four of the first category absolutely preferred the GranularityTable, and all three of the second category preferred the LevelTable.
We assume that at least for the first steps of an iterative search process the LevelTable can be efficient in analyzing the result set as a whole, maybe find patterns or reformulate/discard the query due to unsatisfactory results. Content is not the primary goal, but filtering and reduction of

the result set is. If then the results are narrowed down to potentially interesting documents, the GranularityTable with its browsing comfort can be used. Now content is the primary goal, modalities can be changed frequently. In this manner, our initially developed scenarios were partly validated by empirical results, though our scenario characters begin the information retrieval process with analytical, very formal and sophistically formulated queries only, while during the iterative retrieval process they become more informal and data driven. Although the evidence should not be weighted too strong, we took it as a hint to handle both design variants equitable.
Using both tables integrated in one search might speed up and ease the visual information retrieval process. This and further interaction concepts are part of the ongoing redesign of VisMeB.

4 Outlook

Throughout the implementation of VisMeB, wide ranges of usability engineering efforts have been taken. With iterative expert evaluations, user tests, questionnaires and user scenarios, a certain level of security in usability questions is reached. The html-mockups were replaced by a java prototypical application, which is constantly monitored and evaluated.
The Scatterplot has now a zoom function, coupled with the result tables. Zoomed documents are placed on top of the tables, the other documents are faded out. Magic Lense Filters in the Scatterplot, inspired mainly by (Fishkin & Stone 1995), can be configured and combined with boolean operators to filter out unwanted documents. Additionally a 3D-Scatterplot is implemented, where data is displayed as transparent cubes. Rotation provides illumination from all directions, a zoom function as well as different selection mechanisms complete the equipment of the 3D-Scatterplot.

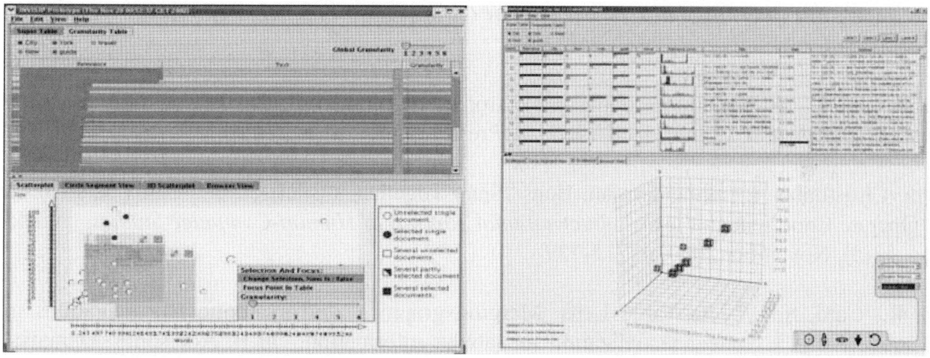

Figure 2: GranularityTable with Magic Lense Filters, Level Table with 3D-Scatterplot

Both result tables now give immediate feedback to user actions. Through a mouse-over-effect, navigational help is given, which is especially useful in both starting levels with low detail. The height of the row is changed, and the user can see this document's details from the next level. In all other levels, moving the mouse over a document highlights the corresponding data point in the Scatterplot and vice versa. Current implementation scenarios for VisMeB regard the use as a geometadata browser, a browser for a movie database and a web search. It will be enhanced by a tool for customization of the visualizations, influenced by ideas from (, Plaisant, & Shneiderman 2000). Users will be able to choose, in which level of detail they want which metadata visualized with which visualization (e.g. Barcharts, RelevanceCurves).
The next steps will be user tests on the java prototype and a comparative evaluation between a form filling and a visual query preview interface. The 2D and 3D-Scatterplots will be compared concerning their effectiveness and efficiency.

References

Ahlberg, C. & Shneiderman, B. "Visual Information Seeking: Tight Coupling of Dynamic Query Filters with Starfield Displays, In B. Adelson, S. Dumais, J.S. Olson (Eds.), *CHI 1994: Conference Proceedings Human Factors in Computing Systems,* Conference: Boston, MA, April 24-28 1994. New York (ACM Press) 1994. p. 313-317.

Andrews, K., Gütl, C., Moser, J., Sabol, V., Lackner, W.: "Search Result Visualization with xFIND".*UIDIS '01: Proceedings of the Second International Workshop on User Interfaces to Data Intensive Systems*, IEEE 2001

Eibl, M. "DEViD: a media design and software ergonomics integrating visualization for document retrieval" *Information Visualization,* 1, 2002, p.139-157

Fishkin, K. & Stone, M.C. "Enhanced Dynamic Queries via Movable Filters." In I.R. Katz; R.L. Marks et al (Eds.): *CHI 1995: Conference Proceedings Human Factors in Computing Systems.* Conference: Denver, CO, May 7-11 1995. New York (ACM Press) 1995. p. 23-29

Hearst, M. "TileBars: Visualization of Term Distribution Information in Full Text Information Access." In I. Katz; R.L. Mack; L. Marks et al. (Eds.): *CHI 1995: Conference Proceedings Human Factors in Computing Systems.* Conference: Denver, CO, May 7-11 1995. New York (ACM Press) 1995. p. 59-66.

Klein, P., Müller, F., Reiterer, H. & Eibl, M. "Visual Information Retrieval with the SuperTable + Scatterplot", *6th International Conference INFORMATION VISUALISATION,* 10-12 July 2002, London.

Marchionini, G. "Information seeking in electronic environments", *Cambridge Series on Human-Computer Interaction,* Cambridge University Press 1995

North, Christopher L.; Shneiderman, Ben: "Snap-Together Visualizations: Can Users Construct and Operate Coordinated Views". *International Journal of Human-Computer Studies,* Academic Press, 53(5), November 2000, p. 715-739

Rao, R., Card, S.K. "The Table Lens. Merging graphical and symbolic representations in an interactive focus + context visualization for tabular information." *in* B. Adelson;S. Dumais; J.S. Olson (Eds.): *CHI 1994: Conference Proceed-ings Human Factors in Computing Systems.* Conference: Boston, MA, April 24-28 1994. New York (ACM Press) 1994. p. 318-322.

Rosson, M.B., Carroll, J.M. "Usability Engineering. Scenario-Based Development of Human-Computer Interaction" San Diego (Academic Press) 2002

Veerasamy, A. , Navathe, S. B "Querying, Navigating and Visualizing a Digital Library Catalog" *Proceedings of the Second Annual Conference on the Theory and Practice of Digital Libraries*, 1995

'SnapShots' – A Household Visualisation and Planning Tool

Jon Matthews

Interact Lab
School of Cognitive and Computing Sciences
University of Sussex, Falmer
Brighton, BN1 9QH, UK
jonathlm@cogs.susx.ac.uk

Abstract

Organising a wedding is a very complex task often fraught with difficulties. A study of how couples plan a wedding highlighted key problems and obstacles in the organisational process. The findings were used to inform the design of a system to support and aid families planning such an event more effectively by facilitating collaboration and communication. This paper discusses how useful such a tool could be.

1 Introduction

The act of 'getting married' is treated as a very important life event and milestone by many cultures across the world. Weddings are a time of huge celebration for all involved including the bride and groom as well as their friends and relatives. However, getting married takes a substantial amount of time and effort in the organisational process. When planning a wedding, there are a number of details to be worked out and everyone from the 'in-laws' to friends and more distant relatives will have opinions as to how the day should look and progress. There are also many 'hazy' areas where members of the family need advice from suppliers or are at odds with each other as to how to resolve their problems. At this time, inter-personal relationships can dominate causing tension and conflict situations may arise that must be resolved through negotiation and careful management. In summary, planning a wedding can be a complex, stressful and time-consuming process. We propose that one way of reducing some of this stress and complexity is by providing a more open system that encourages collaboration and communication amongst all parties involved by giving them access to and a record of the planning process as it progresses over time and space.

2 Complex Planning

Much of the literature (for example, Stone and Veloso, 1994) regarding complex planning has been A.I.-based and involves machine learning in order to solve problems automatically such as scheduling for university timetables or other similar events. However, the type of complex planning necessary when organising an event such as a wedding is distributed across people, space and time. Currently, this is supported using a range of technologies, software applications and other artefacts that people create in order to help them solve particular problems encountered during the planning process. Also, many of the commercial applications available are temporally based in that they offer suggestions and checklists as to what tasks should be undertaken dependent on the timescale of the wedding (for instance, six months before the wedding you should consider these events…). They are also designed to be used individually and therefore, restrict who is privy to such information. The A.I. approach to complex problems and the commercially offered planning software does not offer solutions to a distributed task such as planning a wedding because, in such a case, communication is fragmented and not everyone has access to all information or is aware of the latest developments. To this end a study was undertaken to look particularly at common problems and breakdowns when planning in distributed tasks.

3 Collaborative Planning of Complex Events

The process of planning for a wedding was analysed in detail by undertaking a longitudinal study with four engaged couples, which monitored the wedding process over a number of weeks via structured interviews, and diary studies. Three other couples, who had recently been married (within 6 months), were also interviewed in order to discuss their experiences of the planning and organisation process as well as giving them opportunity to discuss if they would have done anything different or made any changes in 'hindsight'.

In terms of planning a wedding, the engaged couple often prefer a bespoke approach that reflects their individuality and personality, as well as their likes and dislikes. As one participant noted when planning her wedding, "*I actually found choosing the flowers quite difficult as I really like flowers and wanted something different...in fact I wanted every detail to reflect me*". This highlighted the fact that many suppliers require a definitive idea of what is required in order to produce quotes and possible examples. This is not always the case and many participants stated that they had an idea or theme that they wanted but did not always have the means to express it.

A number of conflict situations can arise over the course of the wedding that all have to be dealt with and resolved. For instance, a conflict of 'interest' may arise between the various suppliers and the bride and groom. Obviously, the bride and groom want their 'dream day' to look stunning but be achieved within a strict budget. On the other hand, the suppliers' motivation is to gain the business and make as much profit as possible. Conflicts of opinion and/or interest may also arise between the bride and groom and their immediate family or friends. Discussion with one of the interviewees illustrates this point perfectly, "*We decided to get married in our local football ground. My mum initially said she wouldn't come! She wanted the traditional Church wedding. I explained how much football meant to both of us and compared it to her and my dad's passion for "rock 'n' roll" dancing. It worked, and she eventually agreed to come. We all had the best days of our lives*". From one of the interviews, the bride and groom were faced with the difficult situation of the Groom's mother and father being divorced and not on speaking terms. The father had since remarried but, unfortunately, the groom's relationship with his new stepmother was strained. This is quite a common situation showing how the wedding model involves collaborative planning and distributed decision-making tasks in which areas of conflict can soon arise.

From the interviews undertaken with couples who had just got married, many of them found that planning the seating arrangement a difficult and time-consuming process. It was suggested that this problem is exacerbated by the complex inter-personal relationships that are so important during this time: "*My father was paying for the whole wedding, and when it came to the guest list, he wanted to invite all his business acquaintances. His friends did come, but I chose where they sat. I had all my friends and family near me at the top table, and they were hidden away around the edges*!" Many couples stated that they produced some sort of schematic diagram or other representation to help them with this delicate task.

Negotiation seems to be the key to the planning process and a lot of time and patience is necessary when discussing different options, requirements, budgets, problem solving and other tasks. Negotiations between the engaged couple and the venues, caterers, suppliers or other family members can get extremely complex and even if adequate time scales are available some major decisions don't actually get resolved until late in the day: "*Well, you have to choose menus and try and juggle the price you want to pay with the price of certain menus but you don't want that ingredient you want to swap it with something else. It shouldn't have been complicated but somehow it was. The venue sent a big batch of a whole load of different menus and there were different types of meal - do you want a 'champagne breakfast', a sit-down meal or hot or cold buffet? For each category there were a number of possible selections and instead of making it a bit easier by saying, we'll have that one, in the end, we decided that we'll have a bit of this one and a bit of that one and then we found out that it was now too expensive so we had to have a complete rethink and take something off. What a headache*!" In hindsight, a number of participants recommended that family and friends should know exactly what has been planned as early as possible so any disagreements can be resolved quickly in order to minimize any disappointment, conflict or resentment. For instance, "*I told my Catholic mum that I was getting married in an Anglican church a couple of days after we announced the engagement. It got a 'biggie' out of the way early on. She sulked for about three months – which was fine as we had 18 months to wait until the big day*!"

The study has highlighted that any difficulties when planning a complex distributed task are problem-based as opposed to time-based. Therefore, it is important to develop a system that helps solve this distributed problem and facilitate communication.

4 Wedding Organisation and Planning - A Normative Model

A normative model of wedding organisation and planning was developed based on the longitudinal study. Obviously, variations and exceptions from this model are possible and it is not too difficult to imagine the bride and groom wanting to plan and organise the whole day themselves without involving anyone else, or, simply, rushing off to Las Vegas and completing the whole process in about 30 minutes. However, the purpose of this model is to emphasise the proto-typical processes and tasks necessary to plan for getting married as well as to highlight the particular nuances of this special time and family activity.

At the hub of the whole process are the bride and groom and the moment they announce their engagement and decide to set a date for their subsequent wedding. Once an appropriate date has been chosen, it is now necessary to make a number of arrangements with regard to catering, florists, decoration and the venues. At this point, the process is too complicated and beyond the scope and expertise of the engaged couple to undertake by themselves, so it is then necessary for them to invite 'others' to partake in the planning at various times, taking on different roles. Thus, a small but temporary (i.e. for the duration of the planning process from the announcement of the engagement to the actual wedding day) group of people is formed around the nucleus of the bride and groom in order to achieve the objective of planning and organising their wedding. Other entities are invited to join the group, depending on what service, product, help or advice they can offer and at the discretion of the bride and groom. The products and services provided by various entities can vary enormously from tangible goods and services such as flowers, the Bride's wedding dress, catering etc., to support from friends or financial help from close relatives. Weddings generally offer opportunities for family and friends to reinforce their emotional ties with the engaged couple. It is suggested that the bride and groom are at the hub of the planning process and, therefore, the majority of interactions will involve them on some level.

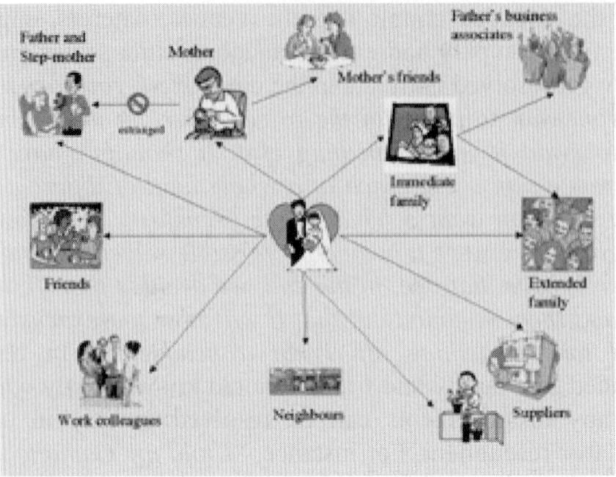

Figure 1: Wedding web: potential people involved in the wedding process

In considering this normative model, it is possible to identify the potential people involved in the wedding process (see Figure 1) and, therefore, some of the more complicated inter-personal relationships and problems that can arise. For instance, the couple may invite their neighbours, work colleagues and friends to the wedding but this may cause some tension between them if one party has more people to invite than the other. Obviously, immediate family, and, possibly, extended family will receive wedding invitations and it is easy to imagine how inter-personal relationships can cause difficulties, tensions and stresses.

5 'SnapShots' – A Visualisation and Planning Tool for Distributed Groups

'SnapShots' is a prototype broadband service designed to support a distributed group problem-solving task as well as enhance and support communication among group members. It has been designed to enable users to plan and organise major family events including weddings, family reunions, birthdays and festive celebrations. It incorporates functions to aid the planning and organisation of such events including 'to-do' lists, guest lists and various prompts based on particular timescales (for instance, 'three months before the event you should be doing the following activities…'). However, due to the higher bandwidth associated with broadband, it also allows users to create specific visualizations of the event. Initially, 'SnapShots' is based on the idea that everyone has a dream of what their ideal wedding will look like and allows them to produce this image in a digital format. This could then be shared amongst family and friends for comments and suggestions but also acts as a starting point to complement and aid the complex planning and decision-making process. It is envisaged that the new "SnapShots' broadband service will be used collaboratively amongst family members (parents of the bride and groom, extended family and friends etc) as well as potential suppliers (florists, caterers, venue management etc.). Although, 'SnapShots' will be used as a collaborative tool, it will be left to the discretion of the principal players (mainly, the bride and groom) to determine to what extent they wish to use these capabilities in terms of access rights and other privileges.

The collaborative visualisation capabilities of the 'SnapShots' tool serves three purposes but is also an integral part of the planning and organisational process. Firstly, is suggested that the process of being able to create, modify and edit collaborative representations will facilitate and support a distributed group problem-solving task, as well as aid the resolution and management of intra-group conflict and, enhance and support communication amongst distributed group members. Secondly, it could be used to develop and create a visualisation of the proposed wedding, which could then be shared amongst family, friends and other interested parties for comments and/or suggestions. More importantly, "SnapShots" would then offer the opportunity to work collaboratively and in conjunction with any potential supplier in order to design and develop the perfect day as well as automatically calculate any budgetary requirements (see Figures 2 and 3).

Figure 2: Wedding ceremony venue

Figure 3: Completed Visualization

From the interviews with couples who had just married, it was noticeable that organizing and determining the seating plan can be a very difficult and stressful process as personal relationships can dominate (for example, X cannot sit next to Y because…). It is recognized that many engaged couples planning their wedding already use some form of representation to aid them with this task. This will then be provided by the "SnapShots" service in order to aid with a solution to this difficult decision-making process. The schematic visualization, of say, the reception venue is included to help aid the organization of the seating plan but also in order to create a database of guest information. For instance, this might include personal details (names, addresses etc.) as well as wedding appropriate information.

Obviously, the bride and groom are central to the planning/decision making process and the majority of communication interactions will involve them on some level thus, creating a highly centralized network. Shaw (1981) analysed a variety of different communication networks in order to determine which was best suited for small group efficiency and productivity. He concluded that centralized networks were more efficient for routine tasks whilst decentralized networks were more efficient for tasks that require creativity and collaborative problem solving. In terms of the wedding planning process, this suggests some sort of contradiction in that the highly centralized network placing the bride and groom at the hub is unfortunately not the most productive or conducive to a collaborative task. Developing 'SnapShots' as an Internet based service allows distributed access for the collaborative creation and editing of the various visualisations. This should support a more 'open' communication network but it is still necessary for the bride and groom to have ultimate power over decision-making.

6 References

Shaw, M.E., (1981), *Group Dynamics: The Psychology of Small Group Behavior*, New York: McGraw-Hill.

Stone, P., and Veloso, M., (1994), Learning to Solve Complex Planning Problems: Finding Useful Auxiliary Problems, 1994 AAAI Fall Symposium on Planning and Learning, available online at http://www.ri.cmu.edu/pubs/pub_3176_text.html, 13/02/03.

Development of an OS Visualization System for Learning Systems Programming

Yosuke Nishino *Eiichi Hayakawa*

Takushoku University
Facluty of Engineering, Department of Computer Science
815-1 Tatemachi Hachioji Tokyo Japan
yosuken@os.cs.takushoku-u.ac.jp, hayakawa@cs.takushoku-u.ac.jp

Abstract

This paper describes about the “MINATO” project that is the environment of a software education support system and a practical environment using Visualization that is the part of the "MINATO".
In the present state, the learning of system programming proceeds with the development of the various learning environments without a specific method. The "MINATO" presents the cooperation between the hardware and the software in the learning process of a system programming.

1 Introduction

The education of system programming is indispensable for students who learn computer science or information technology (IT). An Operating System (OS) is especially important in system programming but understanding the OS is difficult. Because an OS is like a black box and a learner can't observe the behaviour of it. We did a solution by the "MINATO" project to these problems. A "MINATO" project is a system software education support environment. This research is a part of the "MINATO" project.
We proposed an OS study environment based on visualization that resolve these problems. In this environment adaptability, the learners are given according to their understanding level.
We have materialized about task management, memory management and file system and the study of the concept of multi-process programming, synchronization with processes, paging mechanism and disk allocation. Moreover, it was evaluated in the undergraduate lecture of our university for two years.
The purpose of our research is the development of the “MINATO” and visualization system that supports the education of the OS. This system can make the process of learning more easily in comparison with OS learning by the usual textbook.
Our target users are Computer science course students and the instructors.

2 The "MINATO" Project

The outline of "MINATO" is mentioned here. The reasons why system programming is difficult are mentioned in the next three points. (1) Needs the extensive knowledge of the hardware from the logic level to the architecture level. (2) Even in the software especially in the OS and the compiler's behaviour are hard to see. (3) Needs the understanding of the cooperative behaviour of the hardware and the software.

The objective of the "MINATO" project is the solution to the problems given above. Correlation figure about the research being done at present is shown in the **Figure 1**. Even if a simple substance goes respectively through the learning environment in the "MINATO" project, the processor which the thing it can think to be especially effective learns by holding the practice which general idea knowledge to learn by the visualization of premise knowledge to learn in the class college, and the OS, and this environment were used for though it comes to learn of the computer system

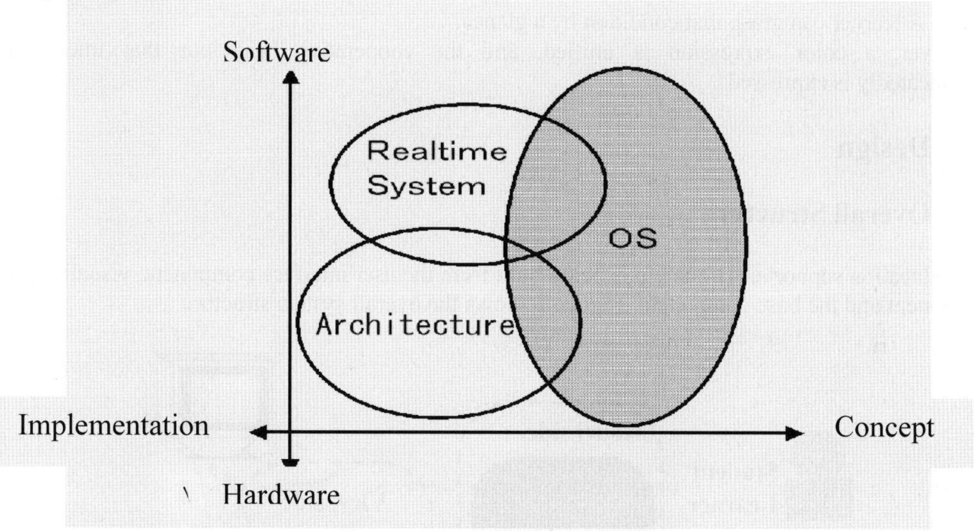

Figure 1:The "MINATO" project

3 Problem Analysis

At present many lectures are generally carried out using the textbook. But, as for the learner, the execution of the OS is hard to grasp in the learning environment that based on a textbook. It doesn't seem easy to understand because OS is like a black box and hard to see for the learners. Because of this, a student has problems to understand the concept of the OS and the state of the execution can't be imagined.

When the student learns by using the actual OS, much premise knowledge is necessary. Then, it is difficult to learn the concept of the OS from the source code and the mechanism. Furthermore, an OS gets confused because of the complex cooperative works.

4 The Design Policy of Education Support

The design policy of the visualization in the OS education is mentioned. As a Black box is a problem in the OS education. The next visualization policy is proposed in this environment to these problem elements.

(1) Abstraction of the structure.

It works in the understanding of the OS, and abstraction of the structure is the effective means of understanding promotion with being an important element. As for the OS, the behaviour for structure becomes haggy complicated. So the detailed structures are not shown to the learner.

(2) Changing the state dynamically in condition.
An OS does a dynamic condition transition and the state of that condition transition can't be followed in the commentary static figure by the textbook. The behaviour of the OS is visualized in this environment by doing the dynamic condition transfer, which contains animations.

(3) The idea of color expression.
There are conditions in the OS by each case. Condition is expressed in this environment using colors. A learner can grasp that condition by a glance.
Moreover, a color expression is unified, and the cooperative behaviour that intertwine complicatedly is expressed.

5 Design

5.1 Overall Structure

This education support environment is being built from the user interface component, visualization component and the base component. **Figure 2** shows the overall system structure.

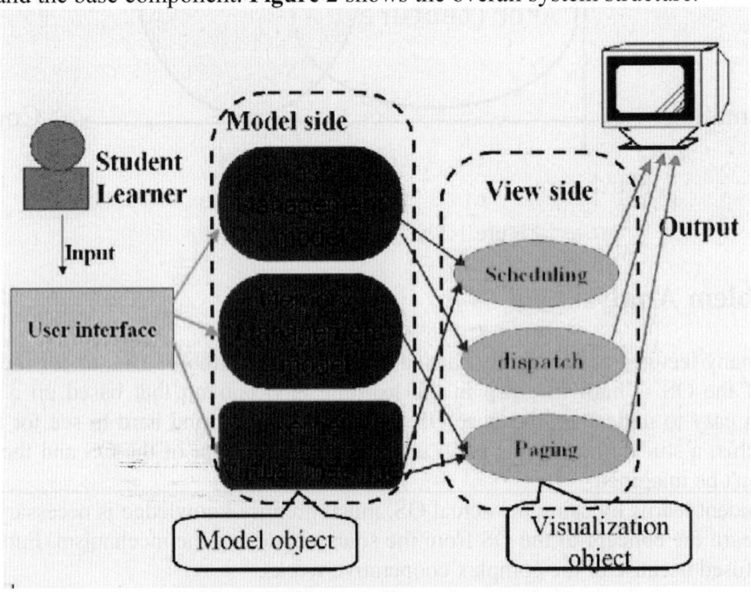

Figure 2: The overall system structure

5.2 Design of the Visualization Objects and Model Objects

Our visualization system is based on the designed by the MVC model. The View side is assembled by the visualization objects. At present, the View side that are implemented involves task scheduling, task dispatching, synchronisation, mutual exclusion and paging.
The Model side is assembled by the model objects. At present, Model side that are implemented involves task management model, memory management model and virtual machine with kernel.
The View side receives visualizing data from the model object. Then, those data are built up, and visualize it.

The virtual machine with kernel is implemented, and it learns the execution of an OS to work by making a studying applicable OS execute virtual machine actually.
As for view side, it is the important part that visualization is actually displayed to the user. As for the characteristics of visualization to do the visualization in this education support environment is as follows:

(1) Visualization by animation.
(2) Color is changed by the condition of a status in the OS.
(3) The change of execution on speed of the visualization.

We have already realized the example of (a) task scheduling, (b) task dispatching, (c) synchronization, mutual exclusion, and (d) paging. **Figure 3** shows screen shots of our OS Visualization System for learning computer systems.

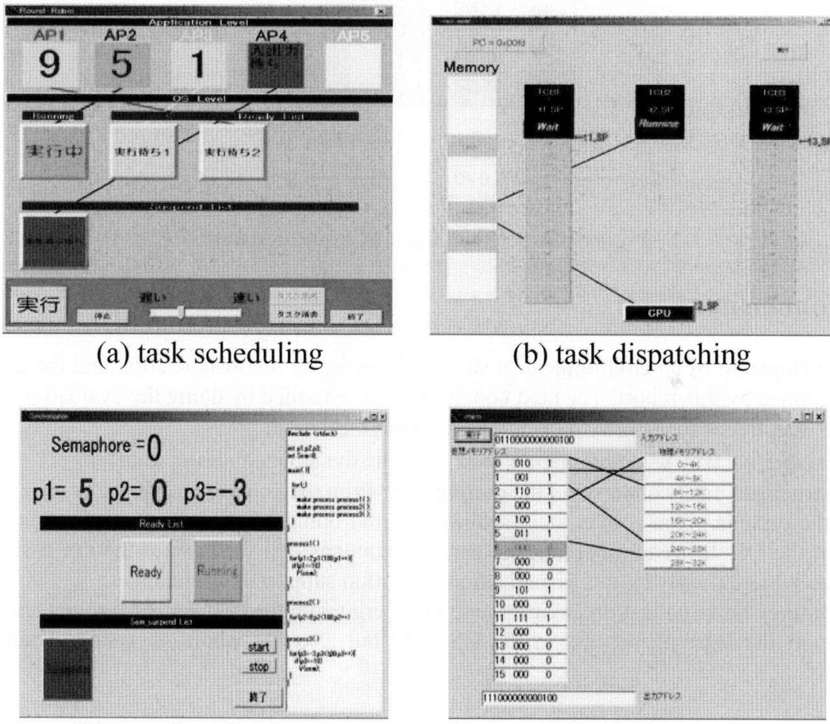

(a) task scheduling (b) task dispatching
(c) synchronization (d) paging
Figure 3: Screen shots of our OS Visualization System

6 Evaluation

For about two years, we performed the demonstration about visualization of task scheduling, a synchronization, exclusion control, and paging in the OS lecture. The third graders of the department of Takushoku University computer science were the targets that took part in the questionnaire. An official approval was done to prove the validity of the evaluation. A hypothesis could be dismissed as a result of doing the evaluation of the validity by the official approval.

Therefore, the validity of this visualization environment could be shown more than a result of an official approval.

The answer to the question that was obtained from the 100 participated students. **Figure 4** shows a detailed result.

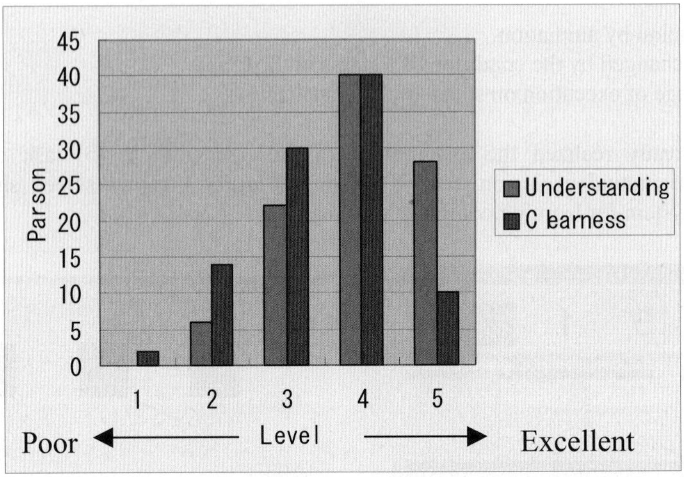

Figure4: A detailed result of a questionnaire.

7 Conclusion

The development of an environment to support learning by the visualization and the evaluation were mentioned by this report. The next conclusion was obtained by doing the evaluation.

- An effective way using visualization to support OS learning
- An effective way using animation to express the dynamic condition transfer
- Learner can easily understand the current state from the color expression

From the reasons given above, the visualization environment is helpful to support the learning process of the OS. It became possible to learn the fundamental function of the OS easily and to understand it more deeply by this visualization education support environment.

In the future, the visualization environment will merge with the "MINATO" project and will not only support the learning process in the OS field but also the whole of system programming education.

8 References

Visualization design in OS education support environment

Yosuke Nishino,Satosi Simokawa,Eiichi Hayakawa,Nobumasa Takahashi FIT (2002)

Development of an environment with FPGA for learning systems programming

Satosi Simokawa,Yosuke Nishino,Eiichi Hayakawa THE INSTITUTE OF ELECTRONICS, INFORMATION AND COMMUNICATION ENGINEERS (2003)

Lessons Learned in Designing a 3D Interface for Collaborative Inquiry in Scientific Visualization

Stephan Olbrich

RRZN / L3S / University of Hannover
Schloßwender Str. 5
D-30159 Hannover
olbrich@rrzn.uni-hannover.de

Nils Jensen

L3S / RRZN / University of Hannover
Deutscher Pavillon, 1.OG, Expo Plaza 1
D-30539 Hannover
jensen@learninglab.de

Abstract

The paper specifies the design and evaluation of DocShow-VR's user interface. DocShow-VR is a software composite of data generator, dispatcher, and interactive displays for visualizing results of computational simulations over networks by means of animated 3D graphics in real time, used by researchers. 6 researchers, developers, and students evaluated the system informally over 10 months, collaboratively and alone. Results are specified. We improve the interface for network-distributed learning groups. A formal evaluation is due June 2003. Our improvements of the software in accordance to the evaluation results will aid students of the natural and technical sciences to use complex visualization.

1 Introduction

We report the development and informal evaluation of DocShow-VR's (DSVR's) user interface. DSVR is a software that visualizes data. Users interact by use of conventional (mouse, keyboard, monitor) and Virtual Reality (VR) devices.

Users carry out virtual experiments in DSVR's synthetic 3D space. For example, researchers at the University of Hannover investigate atmospheric convections (Jensen, Olbrich, Pralle, & Raasch, 2002). The parts of DSVR are distributed on networked computers:

- a graphic generator is called by a computer simulation and forwards the result,
- a 'Server' stores and replicates graphics to workstations, and
- a 'Viewer', on each workstation, displays graphics.

Users control the scene by use of a 2D graphical user interface (GUI). Augmented 3D graphics help them to mediate the actions of collaborative workers and learners (CSCW, CSCL).

The GUI must generate user acceptance, determined by quality defined by functionality, efficiency, and usability, compare Bevan (1999). 3D interfaces for collaborative work are not standardized (Kettner, 1995), in contrast to 2D GUIs for single users (Shneiderman, 1997). We encounter questions:

1. Differ 3D and 2D interaction from each other?
2. Are 2D metaphors directly applicable to 3D?
3. Does DSVR's CSCW interface support collaboration to a sufficient extent?
4. How must the CSCL interface differ from CSCW?

Section 2 specifies DSVR and related work. Section 3 describes our evaluation method, and Section 4 the results. We suggest answers in Section 5, and state future work.

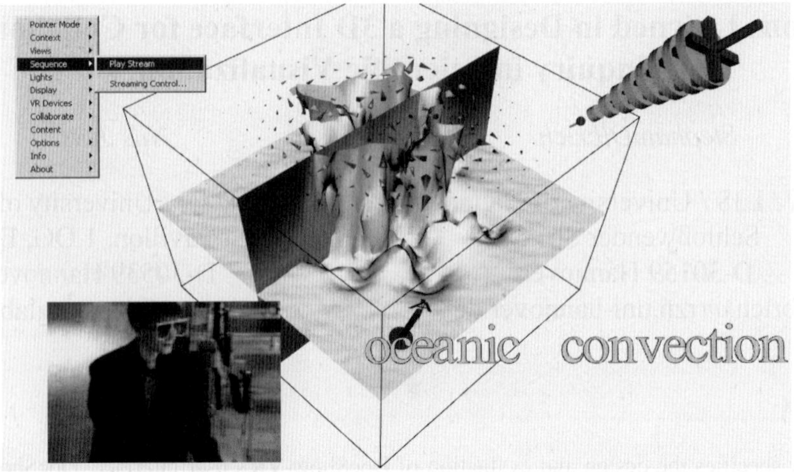

Figure 1: DocShow-VR with 2D menu and CSCW tools. Shown: oceanic convection.

2 Technology and the Evaluation of Collaborative Visualization

The Viewer is a plugin for web browsers that displays graphics in a resizable window (Figure 1). Data from remote computers are visualized by means of surfaces, geometric primitives, and compound shapes. Users control visible cues. The 2D GUI has a context menu, a mouse cursor, and windows. A workstation and conventional peripherals are used. Pushing the right mouse button opens the menu. A user selects an object and an action from the menu, for example 'move'. She moves the mouse while holding down the left button. The cursor indicates the selected action. DSVR differs to other systems because users visualize complex volumes, navigate a scene over interconnected Viewers synchronously, record ('capture') input and output, and use VR peripherals and CSCW tools:

- 3D stereoscopic display. Displays and glasses generate one view for each eye. Users reconstruct depth information that is not available through monoscopic display.
- Space mouse with 6 degrees of freedom (DOF) for 3D navigation. Mice have 2 DOF.
- Users synchronize their views and results of their actions to refer to the same things.
- A simulated video wall displays peers and their environment. The source is a teleconferencing hardware that mediates utterances over video and audio.
- A tele-pointer transports gesture, gaze, and pointing.
- Users place annotations to comment on a scene. Annotations are moved in relation to the object's coordinate system to which they are attached. Text is entered in a window.
- Clip planes extract portions of visible content. Planes are invisible and of unlimited extent. Up to six planes are displayed by means of identical symbols.

Related are 3D web browsers, visualization frameworks, and distributed virtual environments. Park, Kapoor, and Leigh (2000) evaluate CAVE6D, a tool for collaborative scientific visualization. They find students work independently over longer periods, converge their results, and synchronize their views. The number of collaborators is fixed, and their viewed data have at most 3 dimensions. Lindeman, Sibert, and Hahn (1999) study the use of 3D windows in synthetic environments. They observe: haptic feedback increases task performance, and windows must be in the same visual field as modified objects. They use 3D windows as mediators between objects and users, which we want to avoid because 3D objects can be addressed directly in a more intuitive way. A complementary strategy follow Dorohonceanu, Sletterink, and Marsic (2000) who specify an interface for collaboration in real time, similar to ours. Their system uses 3D views in 2D windows, the opposite of our approach of embedding 2D windows in 3D views. Jung, Gross, and Do (2002) describe a system, similar to our software, that helps users to annotate 3D elements in virtual space. Their ideas for improving their software could be adapted for DSVR, for example to structure discussion. They visualize architectures.

3 Method

6 people tested DSVR: 2 researchers in the natural sciences, 2 students in the engineering sciences, and 2 developers (the authors). Each had more than 5 years experience in computer use, 3 expertise in 3D graphics. 1 student used the Viewer for 3, others for 10 months. They viewed a virtual building, map, river flow, molecule set, and factory at their workplaces. 2 used Linux, 4 MS Windows. All used workstations, 2 with space mice. Developers used CSCW tools in a seminar room, and another room, each with 2 projectors and video wall. Also, we let ca. 10 colleagues and students from our department use the Phantom input device, a robotic arm to which a stylus was attached. Users translated and rotated a simulated box by means of a tele-pointer in a synthetic environment. The software that was tested in conjunction with the Phantom was delivered by the vendor.

4 Results

The 3D view attracted users. After 10 minutes they were ready to use the system.
Display: 2D elements occluded scenes. The mouse cursor appeared to hover in front of objects and generated eye strain in users that viewed stereoscopic scenes. Monoscopic display did not help us to track patterns in mostly unstructured data. Stereoscopic display mitigated, but did not eliminate the problem. We navigated the scene to display it from different viewpoints. 1 visitor could not see 3D images.
Workflow: **A researcher** suggested to keep selected windows and menus open to switch commands quickly. The researcher wanted to set which animation parts were skipped during 'fast forward' and 'fast backward'. He switched between windows and set simulation parameters manually when he started DSVR. He felt hindered in working fluently, and did not want to repeat actions.
A student tried to pick 3D objects with the mouse before he noticed the context menu. He had difficulties to rotate an object when its rotation axis was separate from the object. The student and one of us moved objects out of the visible plane and set viewpoints to locations where we lost the orientation and restarted the scene.
CSCW tools: We wanted to share selected parts of a scene with remote collaborators. For example, we tried to move a pointer simultaneously and were not aware of each other's actions. The situation required additional negotiation about who would use the object first because DSVR had no 'floor control' to regulate order of use. We wanted to show and hide tools during a session because they occluded other objects. Annoyed by that shortcoming, we often disabled tools before start.
To use the video wall we had to set the tele-conferencing hardware and video mode manually. Misconfigurations occurred. Video and audio were useful for presenting graphics and speech but did not indicate eye contact, necessary for 1-to-1 talk, between peers. A user could not determine if a peer watched her video screen and talked to her remote colleague or addressed her local physical surrounding. In one case there was no way to determine when one of us addressed his remote peer because audio transmission was distorted by a large group of visitors at the local site. The local peer switched attention to visitors without notice to his remote peer, who was unable to follow the discourse, which increased his uncertainty about when and how to reply to his partner.

Users moved tele-pointers with space mice, 2-DOF mice were too cumbersome. Users forgot or ignored the tele-pointer if they were detached from input devices or occupied by discussions with peers. One of us had difficulties to control the tele-pointer by means of a space mouse, and needed longer training (about 2 hours). The same user could move the stylus of the Phantom in a more accurate and faster way, which our evaluation of the Phantom verified for all testers.
Annotations were moved in relation to the selected camera, like tele-pointers, but after they were attached to an object they moved in relation to a different coordinate system, which made it hard for users to drag notes along the intended direction.
The state of a clip plane was not clear because its symbol, located above the plane, was symmetric around each axis. The use of identical symbols for up to six planes confused users.

5 Discussion

Differ 3D and 2D interaction from each other? Consistent with previous research, 3D interaction metaphors rely on tacit knowledge from everyday experience in reality. This is indicated by the attempts of users to grab 3D objects directly. Instant graphical feedback and VR devices helped them to carry out tasks. We assume movements in 3D must relate to the coordinate system of the viewer because actions slowed down on violation of the rule (compare our tests with annotations and space mice).
Stereoscopic viewing supplemented visual cues that eased users to perceive depth-cascaded elements. However, 2D windows interfered with stereopsis and occluded valuable space. They must be removed, which is no problem because users can control and view objects directly and instantly, as it is implemented partially in DSVR for the navigation of objects.
Are 2D metaphors directly applicable to 3D? 3D affects presentation and interaction. We observed that users got lost in complex scenes, and we noticed an increase of their arm activity while they used VR devices, which did not occur to that extent with conventional hardware. The higher burden on motor function, stamina, spatial perception, and visual memory leads to more errors of use. There are solutions:

- *Prevent*: **Ease motor function** by a context-dependent reduction of DOF, automated 'way-finding', and alignment. For example, just specify the target of an annotation by pointing and clicking at it, do not force the user to drag it to its place. **Clarify what is seen from where**. For example, make symbols indicate their orientation unambiguously, make them distinguishable from each other, and avoid abrupt changes of viewpoints and objects. **Prevent clutter**. Enforce stereopsis, and let the user navigate a scene to grasp 'what's in there'. Parameterize object transparency because clip planes may not suffice. **Indicate object states**, if they are unavoidable, to make subsequent actions predictable. When objects are manipulated, show the frame of reference (axis, coordinate, origin, boundary).
- *Guide*: Simulate landmarks that help users to navigate in 3D, e.g. haptic feedback.
- *Orient*: Provide an overview from which users recover orientation.
- *Reset*: Fix cameras, provide 'undo' and 'redo', and store settings.

Does DSVR's CSCW interface support collaboration to a sufficient extent? 1-to-1, 1-to-many, and many-to-many communication is supported. Collaboration at the same time and different times (by way of recording sessions), and at the same place and different places is possible. We must improve computer-mediated 1-to-1 talk. We must help users to select objects they want to manipulate and use together. Users may request control over shared tools to make the previous owner release the tool, and stereotyped messages may supplement the video wall to indicate eye contact and posture to add missing nonverbal signs that guide a discussion.

How must the CSCL interface differ from CSCW? The differences will be small. Pea, Gomez, Edelson, Fishman, Gordin, and O'Neill (1997) suggest that experts and novices share the same tools, and learn from each other. Tools must expose the tacit knowledge of experts to students by way of mediating communication between and among tutors and students, and direct novices in search for relevant data and actions. Helpful could be an integrated synthetic desktop for retrieving and viewing web content without disabling stereoscopic viewing.

A controlled experiment is due June '03. We will improve DSVR, test it in universities, and add multimodal interfaces. Tailored versions can attract schools and 'virtual teams' from industry.

Acknowledgements

We thank all participants for their valuable comments and suggestions. The pilot study is part of the VASE 3 project at the Learning Lab Lower Saxony.

References

Bevan, N. (1999). Design for usability. *Proc. HCI International '99*. Munich, Germany: Lawrence Erlbaum.

Dorohonceanu, B., Sletterink, B., & Marsic, I. (2000). A novel user interface for group collaboration. *Proc. 33rd Ann. Hawaii Int. Conf. on System Sciences (HICSS-33)*. Maui, Hawaii: IEEE Computer Society Press.

Jensen, N., Olbrich, S., Pralle, H., & Raasch, S. (2002). An efficient system for collaboration in tele-immersive environments. *Proc. 4th Eurographics Workshop on Parallel Graphics and Visualization* (pp. 123 – 131). Blaubeuren, Germany: ACM Press.

Jung, T., Gross, M., & Do, E. (2002). Annotating and sketching on 3D web models. *Proc. 7th Int. Conf. on Intelligent User Interfaces* (pp. 95 – 102). San Francisco, CA: ACM Press.

Kettner, L. (1995). A classification scheme of 3D interaction techniques. Technical Report B 95-05, Institute for Computer Science, Dep. of Mathematics and Computer Science, Freie Universität Berlin, Germany.

Lindeman, R., Sibert, J., & Hahn, J. (1999). Towards usable VR: an empirical study of user interfaces for immersive virtual environments. *Proc. ACM CHI '99* (pp. 64 – 71). Pittsburgh, PA: ACM Press.

Park, K., Kapoor, A., & Leigh, J. (2000). Lessons learned from employing multiple perspectives in a collaborative virtual environment for visualizing scientific data. *Proc. 3rd Int. Conf. on Collaborative Virtual Environments* (pp. 73 – 82). San Francisco, CA: ACM Press.

Pea, R., Gomez, L., Edelson, D., Fishman, B., Gordin, D., & O'Neill, D. (1997). Science education as a driver of cyberspace technology development. In K. Cohen (Ed.), *Internet Links for Science Education* (pp. 189 — 220). New York, NY: Plenum Press.

Shneiderman, B. (1997). Designing the user interface: strategies for effective human-computer interaction (3rd ed.). Reading, MA: Addison-Wesley.

Visualizing Social Navigation in Scientific Literature

Pearl Pu
Human Computer Interaction Group
Swiss Institute of Technology Lausanne
CH-1015 Ecublens, Switzerland
Pearl.pu@epfl.ch

Punit Gupta
Indian Institute of Technology
Guwahati-781039
India
Punitkg@yahoo.com

Abstract

In this paper, we investigate visualization-based social navigation systems, especially those used in scientific literature. We identify the requirements for effective visualization of reputation and recommendation, two notions that underlie most social navigation systems. First we present several designs of visual emphasis techniques addressing the dimensionality and scales of reputation, and two sets of icons for effectively distinguishing recommended items. Then we discuss the results of two user studies, which provided validation for our non-linear emphasis methods to represent reputation, and have enabled us to choose optimal icons for recommended items. Based on these findings, we present a re-design for the NEC Research Institute Research Index (Citeseer), a concrete case of reputation based recommendation systems.

1 Introduction

From maps and drawings etched on cave walls telling and sharing stories about hunting trips in ancient times, to dog-eared books in libraries marking their popularity, or smoothed trails in forests indicating paths leading somewhere, social environments have enabled humans to accomplish complex tasks based on information gathered collaboratively. Such information is often presented in the forms of objects bearing signs of how they have been used, whether as maps, tear-and-wear signs on books, or footprints [3]. Today, digital spaces are replacing traditional social environments, and employ mechanisms such as collaborative filtering and recommendation systems [5]. Unfortunately, digital spaces do not age, nor are they physical in nature. Most information concerning reputation and recommendation are text based, forcing users to "read" rather than affording them to "see" information, thus making it hard for them to quickly perceive evidence, spot and retain trends, and patterns.

We aim at developing methods to effectively communicate reputation and recommendation information to achieve the following objectives:

- provide a social space that resembles the multidimensional physical space in which we habit;
- provide distinguishable and intuitive visual signs for reputation, enabling searchers to quickly decide which items to pursue (decision making in navigation) without making decisions for them;
- provide visual representation for recommended items to enable users to "see" items in the right categories, rather than "reading" them;
- provide users with quick learning curves to understand and retain the meanings of signs in such environments.

2 REPUTATION

We adopt the definition of reputation as being the accumulated scale of opinions of products or persons from a significant population of people. Social reputation refers to such treatment of reputation within a defined social group or network for which the user belongs [1].

2.1 Dimensionality of reputation in scientific literature

In NEC Research Index (Citeseer), the dimension of reputation is limited to the number of citations and the year of publication [2], although there can be more dimensions such as the reputation of the publisher, the institute, and authors.

Table 1: Mapping from a range of citation numbers to each of the three variations of visual implementations: font size, stars, font size + weight + shading.

Citations	Font size	Stars	Font size +Weight +Shading.
(136–151]	Ex	**********	Ex
(121–136]	Ex	*********	**Ex**
(105–121]		********	Ex
(91 – 105]		*******	
(76 – 91]	Ex	******	Ex
(60 - 76]		*****	
(45 - 60]		****	Ex
(30 - 45]	Ex	***	
(15 - 30]		**	Ex
[0 - 15]	Ex	*	Ex

Table 2: User study results from the undergraduate group

Combination	No of users	Probability
(Str + De)	28	0.8
(Fs + Sh + De)	23	0.657
(Str)	15	0.428
(Fs + Sh)	10	0.286
(Fs)	2	0.057
Conditional probabilities		
(Str+De) / ((Fs) + (Str))		0.857
(Fs+Sh+De) / ((Fs) + (Fs+Sh))		0.75
(Fs+Sh+de) / ((Str) + (Str+De))		0.68
(Fs+Sh+De)/((Str+De)+(Fs+Sh))		0.636
(Str + de) / (Str)		0.56
(Str) / (Fs + Sh)		0.161

2.2 Scales of reputation

The range of values used to represent the scale of citation can vary in a wide range in scientific literature. The number goes from 0 (only self citation) to as high as 10,000 (record at Citeseer is 10473 for authors, 1198 for Documents). To understand the pattern and ranges of citations, we examined a set of 100 documents at Citeseer, and plotted a graph on the frequency of different scales that appear in the collection. It turns out a majority (60%) of papers are cited only up to 3 times, 25% are cited between 4 and 30, 10% are cited between 31-100 and 5% are cited more than 100. This graph (where the x axis represents the number of citations and the y axis the number of documents) starts with many documents cluttered around 0-30 and trails off to few documents with high citation numbers. The high density in the low ranges of citation and the wide gap between low and high number of citations thus calls for some non-linear mapping techniques between the actual citation value and what can be visually displayed. In particular, we must distinguish the scales conforming to our study of statistical frequencies of citation numbers. That is, a wide range of values must be modelled, while at the same time the mapping will result in a range of 10 distinct visual ranges. We have opted for the slow-in slow-out method [6], which magnifies the highly dense areas, and de-magnifies in the low distribution area. The method maps a given reputation value (x) to a scale as described by Equation 1.

Equation 1: Equation for slow-in-slow-out method

$$Y = \frac{(d+1)*X}{d*X+1}$$

2.3 Visually Rendering Reputation

In determining the most effective mapping between reputation scales and visual values using the slow-in and slow-out function, we consider the following choices:

- Font size– values are mapped to a scale using 5 distinct font sizes (Figure 5);
- Stars – values mapped to a scale of 10 stars linearly arranged (Figure 6);
- Font size + weight + shading –values are mapped to a scale of 7 combined effects (Figure 7).

Here bold-faced characters are heavier than non-bold faced one. The shading of characters refers to the grey scale of their fonts.

2.4 User study

Table 3: user study results for graduate students.

Combination	No of users	Probability
(Str + De)	28	0.8
(Fs + Sh + De)	23	0.657
(Str)	15	0.428
(Fs + Sh)	10	0.286
(Fs)	2	0.057
Conditional probabilities		
(Str+De) / ((Fs) + (Str))		0.857
(Fs+Sh+De) / ((Fs) + (Fs+Sh))		0.75
(Fs+Sh+de) / ((Str) + (Str+De))		0.68
(Fs+Sh+De) / ((Str+De)+ (Fs+Sh))		0.636
(Str + de) / (Str)		0.56
(Str) / (Fs + Sh)		0.161

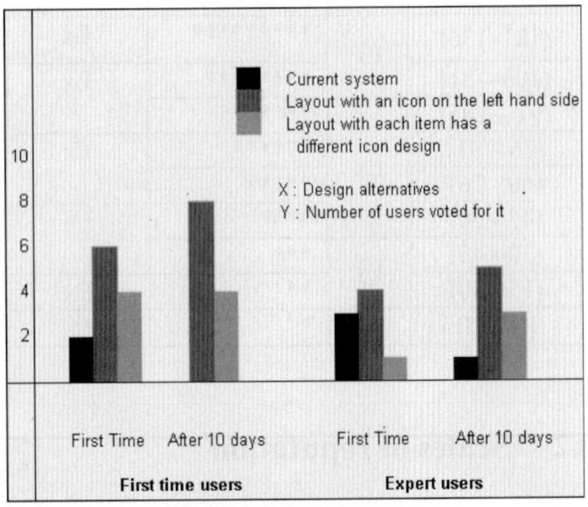

Figure 1: Results from the usability study of icon designs

The goal of this empirical study was to survey the 5 designs and determine the most effective non-linear mapping of reputation scales. A two-dimensional display is employed where the documents are ordered by the year of publication, and each document is visually rendered to show its citation number. Subjects were divided into two groups: the first group is comprised of undergraduate students (very little research experience) and the second of graduate students (research oriented). They belong to different demographic backgrounds and have different academic specialization such as computer science, communication systems, micro engineering, civil engineering and mechanical engineering. Each user receives 5 sets of interface designs where reputation information has been rendered using a combination and variation of font size, stars, and font size +

shade. Users are asked to pretend to "select 5 articles in relation to the current article to further their reading." Then they are asked to mark yes to the interface designs (marking more than once is allowed), if it allows effective scanning of the search results in order to make their selections.

Results are different for the two subject groups, which were shown in Table 2 & 3. It is clear that the undergraduate students prefer stars combined with details the most (p=0.8), while the graduate students strongly prefer the "Font sizes + weight + shade" design (p=0.933). To consolidate the difference, we computed conditional probabilities in order to account for the dependency of their voting, that is, whether subjects are univocal or ambivalent about the designs. For instance in the undergraduate group, the probability of liking 'Str + De' given the fact that they also like 'Fs' and 'Str' is quite high (0.857), indicating that many subjects have voted for these three designs simultaneously. On the other hand, the conditional probability of (Str+De) / (Fs+Sh+De) is only 0.071 in the graduate group, indicating a group more certain about their choices. Furthermore, the difference in probability for 'Str + De' (0.8) and 'Fs + Sh + De' (0.657) in the undergraduate group is rather small, compare to their difference (0.933 vs. 0.133) in the graduate group.

An explanation for the undergraduate results (stars + detail) is that stars are linear visual forms, and therefore are the easiest to compare for their relative quantities. They are also most familiar to undergraduate students, as they have seen stars used to rate music CDs for instance. On the other hand, results from the graduates showed that if users are familiar with literature seeking tasks and understand the difference between citation and the reputation of a music CD, then the visual signs based on their areas are more effective. The final conclusion is that we will use Fs + Sh + De as the emphasis technique for citations.

2.5 RECOMMENDATION

We are concerned with navigational recommendation in scientific literature (implicit recommendation), which is to provide users with additional documents based on the current document. At Citeseer, five categories of recommended documents are displayed: *cited-by* documents (those documents citing the current article), *active bibliography* (related documents at the sentence level), *Similar documents at sentence level*, *Related documents from co-citation*, and *similar documents based on text.*

Table 4. Two sets of icons for different categories of papers

Category	Icon1	Icon 2	Category	Icon1	Icon2
Cited-by documents		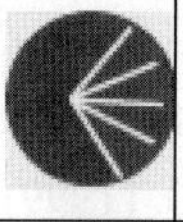	Related document from co-citation		
active bibliography		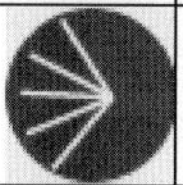	Similar document based on text		
Similar documents at sentence level					

Table 4 shows two sets of icons for each of the types of recommendations used in Citeseer. A user study followed to prove the hypotheses that icons help users

- to recognize the right category of information in subsequent web page visits much quicker (thus learning)

- to quickly recall the meaning of icons after a period of task interruption (retention)

Our experiment was set up as follows. Users were given three design layouts to examine: the current Citeseer page layout (L1), a layout where each title is marked with an icon (L2, as shown in Figure 8) and a layout where each category of items is marked by a single icon (L3, as shown in Figure 10). Study was conducted in two steps. The first step was to determine which design is more effective for users to recognize the right category of articles. The second step was conducted after 10 days, where the same experiment was repeated. This was to determine which design enables users to more effectively retain information. There were two sets of users: users from the first set were familiar with Citeseer while the second set users were new to it. The user task is defined as the speed of identifying the types of recommended items in their respective categories.

A total of 20 users were interviewed, out of which 12 were new to the Citeseer and 8 were familiar with it already. Figure 1 summarizes the user study based on their feedbacks. 16% of the new Citeseer users voted for L1, 34% for L2, and 50% of them voted for L3. On testing the same users after 10 days, we found that 66.6% of them voted for L3, and rest for L2. No user voted for L1.

In the case for expert users, we observed that 37% of them voted for L1, 13% for L2, and 50% for L3. After 10 days, our experiment showed that only 13% of the expert users voted for L1, 25% for L2, and 62% voted for L3. It is clear that the majority of users prefer L3. Even though expert users initially find L1 useful, they prefer L3 when it comes to recalling the categories.

3 Conclusion

We demonstrated how visual signs augment users perception and seeking of information in social navigation systems, as well as their retention of learned signs for intermittent navigation tasks. Our two formal user studies showed that the non-linear mapping from citation numbers to visual values using font size, shading, and detailed information is most preferred by users to distinguish items of different reputation scales, and that icon set 1 together with layout 3 is most preferred by users to recognize and recall the type of recommended items in a visual display. We thus have designed a new interface for Citeseer (see Figure 10 in [4]).

4 References

[1] Dieberger, A.., Höök, A.., Svensson, M., and Lönnqvist, P. (2001). Social Navigation Research Agenda.

[2] Gupta, P. and Pu, P. (2003). Social Cues and Awareness for Recommendation Systems. In the proceedings of Intelligent User Interfaces Conference, Miami Florida.

[3] Munro, A., Hook, K. and Benyon, D. (1999). Footprints in the snow. In A. Munro, K. Hook, and D. Benyon, editors, Social Navigation of Information Space, pages 1--14. Springer.

[4] Pu, P. and Gupta, P. (2002). Visualizing Reputation and Recommendation in Scientific Literature. Technical Report, EPFL.

[5] Resnick, P and Varian, H.R. Editors. (1997). *Special Issue on Recommender Systems*, volume 40. Communications of the ACM.

[6] Sarkar, M. and Brown, M.H. (1992). Graphical Fisheye Views of Graphs. In Proceedings of ACM Conference on Computer Human Interaction.

Visual Interfaces for Opportunistic Information Seeking

Pearl Pu, Paul Janecek

Human Computer Interaction Group
Institute of Core Computing Science
Swiss Federal Institute of Technology (EPFL)
1015 Lausanne, Switzerland
{pearl.pu, paul.janecek}@epfl.ch

Abstract

Visual interfaces play an increasingly important role in how we access, analyze, and understand information. However, most such interfaces do not support opportunistic information seeking, a type of behavior characterized by uncertainty in user's initial information needs and subsequent modification of search queries to improve on results. In this paper, we present a visual interface implemented using semantic fisheye views (SFEV) and discuss how we use a word ontology to expand search context, thus allowing users more opportunities to refine initial queries. After presenting the formalism of SFEV, we describe an application of this framework in image retrieval.

1 Introduction

Opportunistic information seeking (Bates, 1989) is characterized by the fact that users are initially unable to articulate their needs clearly. They must learn new vocabularies and domain knowledge as they interact with search results, an ontology, or both in order to refine query terms. Here we define a query as the focus of information retrieval. A context of a query is the set of semantically related terms provided by a word-ontology, or terms from the current search results. In both cases, it is desirable to visually reify (Furnas & Rauch, 1998) the current query and its context in the same interface so that users can opportunistically choose the direction of search. The more search foci they are able to see and pursue, the more chance they have for the most precise query formulation. We thus have come up with the following requirements for the visual opportunistic interface for retrieval (VOIR):

Context + focus: Users must be able to perceive other possible search goals while exploring the current one. This means that the interface should include what surrounds the focus of the user's current interest (context) as well as the focus itself. The visual emphasis of displayed items must be determined based on their *semantic distance* to the focus, thus providing a semantic context for information perception and exploration.

Dynamic selection of focus and contexts: The interface must be dynamic because the user's context shifts when their search goals change. Displayed items should follow this shift by recalculating their respective semantic distance to the new focus, and their relative visual emphasis in relation to current interest.

Multiple foci and multiple contexts: Information seeking is multi-directional with several search goals. One important user activity in exploring multi-directional search is to compare the results and detect patterns and evidences (visual inferencing). Therefore, the interface must be able to support several foci and their respective contexts. Users should be able to select any item or a group of items as a focus and any number of foci. The interface should display the corresponding contexts to allow distinction and comparison.

Flexible modelling of semantics: to explore several semantic relationships among data, a flexible mechanism must be used. That is, users must be able to define any type of semantic relationships they want, but guided by the system.

In this paper, we introduce semantic fisheye view as a general display method attempting to satisfy the above requirements for VOIR. We first describe the formalisms and various algorithms comprising a SFEV; the types of semantic relationships that can be modelled; and an applications of SFEV for opportunistic information seeking with images. We will then compare SFEV/VOIR to other visual interfaces for information retrieval (VIRI).

2 Formalisms of Semantic Fisheye View

Fisheye (also called "focus + context") views are interactive visualization techniques that balance local detail and global context by directly relating the visual emphasis of information in a representation to a measure of the user's current interest. The term "fisheye" is an analogy to the effect of a wide-angle lens, where the center of the image is in focus and objects are progressively distorted towards the periphery. Furnas introduced the concept of fisheye views as a method to interactively reduce the complexity of abstract data structures such as hierarchies and structured text, but suggested that the technique could be applied in any domain where a *degree of interest (DOI)* function could be defined (Furnas, 1986). His original equation is as follows:

$$DOI(x|fp=y)=API(x)-D(x,y)$$

Given the user's current focus, $fp=y$, the *degree of interest, DOI,* of every element x is the difference between the element's *a priori importance*, *API*, and the *distance*, *D*, between the element and the current focus. The *API* is the importance of an object independent of any focus, and may be either specified (e.g., measured from user studies), or derived algorithmically from properties of the information collection (e.g., calculated from structural metrics). Although *API* is often static, it may also change over time to reflect user interaction (Bartram, Ho, Dill, & Henigman, 1995; Lokuge & Ishizaki, 1995; Ruger, Preim, & Ritter, 1998).

The *distance* function measures the "conceptual" distance between the user's current focus and each element in the collection. We have identified several different types of metrics based on the attributes and structure of the information, the user's task, and the history of interaction (Janecek & Pu, 2002). For example, (Herman, Melancon, & Marshall, 2000) describes a range of different structural metrics that could be used to characterize "conceptual" distance in graphs.

In a previous paper we generalized the fisheye view paradigm to model multiple semantic contexts as combinations of weighted distance metrics (Janecek & Pu, 2002):

$$DOI_{context} = f(API, w_i, dist_i)$$

We also described a prototype for browsing the structural, content and task-based relationships between information in a tabular display of a flight itinerary.

3 A prototype for exploring an annotated image collection

We have developed a SFEV for information seeking in a large database of annotated images (56000 images). This prototype allows the user to explore the collection at both the image content (annotation) level and the semantic concept level. First, each image has a caption and a set of, on average, 25 keywords (a vocabulary of 28000 unique keywords over the entire collection). Second, we use WordNet (Miller, 1995), a large network of semantic and lexical relationships between words, to disambiguate and extend the keywords associated with each image.

To create a semantic fisheye view integrating these two models of the collection, we have defined a *DOI* function that combines the similarity measure of the vector space model with the path distance between senses in WordNet. We linearly scale image size to reflect *DOI*.

In the interface, the user's focus may be a keyword in the image collection, a concept in WordNet, or one of the images. The distance metric used to create the fisheye view changes depending on the type of the focus. As the focus changes, the images in the collection are dynamically resized to reflect their similarity. This visual feedback allows the user to immediately evaluate the effectiveness of their query.

Our prototype supports these basic user information-seeking tasks:

Search by key word, concept or image

Query expansion by adding terms learned from keyword list appearing in retrieved images

Query expansion by adding terms learned from related words found in WordNet

Query refinement by adding more keywords or selecting the specific sense of the keyword in WordNet.

3.1 A scenario

A user is exploring a collection of images. Initially her query is very vague: "horse". Since images usually have few words describing them, matching based on keywords can only return images with the word "horse" or some variation of that word in the annotations. In Figure 1 we see the set of images and keywords that contain "horse". The largest images directly match the animal "horse", but the results are very non-homogenous, also containing images of waterfalls ("horseshoe falls"), a monument to the Native American chieftain "Crazy Horse", and fish ("horse-eye jacks"). By navigating through the set of similar keywords, she can immediately see the related images. For example, selecting "horse shoe" reveals the images shown in Figure 2. Although she can browse through the images and see their annotations, this doesn't enable her to easily learn the different types of horses that she could have images of.

Figure 1: Images retrieved by SFEV using keyword 'horse.'

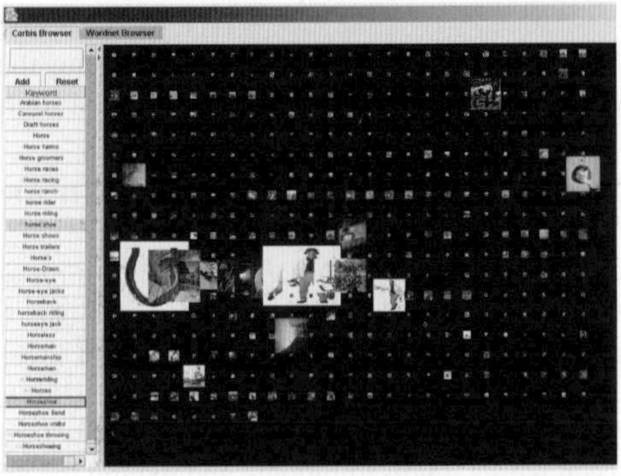

Figure 2: 'Horse shoe' is selected as the focus in SFEV.

She decides to look at the concepts related to "horse" in WordNet. Following the relationships between words, she can browse through different types of horses, such as "pintos", "Arabians", and "bays," and immediately see relevant images. Of course, searching for any of these words individually would return pictures of beans, people, and bodies of water. However, including the context of these concepts in the search for images, we clearly see results that are semantically relevant.

4 Future work

We implemented SFEV to satisfy several key requirements important to information browsing, exploring and seeking. Multiple foci are supported to strengthen SFEV's semantic emphasis by allowing related words to be used simultaneously for a query. We are currently exploring multiple foci in terms of discovering semantic relationships between two or more concepts to support opportunistic search. Furthermore, we plan to perform several user studies in order to validate our hypotheses regarding SFEV's role for providing a flexible and wide range of modelling of semantic information.

5 Related work: Visual Information Retrieval Interfaces (VIRI)

There is a long history of research on information exploration tools to help users formulate their queries and understand the relationships between collections of information, such as search results. For example, Scatter/Gather (Cutting, Karger, Pedersen, & Tukey, 1992) automatically clusters retrieved documents into categories and labels them with descriptive summaries. Similarly, Kohonen maps (Lin, Soergel, & Marchionini, 1991) cluster documents into regions of a 2-D map. The goal of these methods is to organize documents to help users more efficiently evaluate where they can find information that satisfies their needs. The problem is that when the goals of the user are ill defined and fluid, it is unlikely that any single organization will be satisfactory over time.

Semantic fisheye views (SFEV), on the other hand, are interactive techniques that modify an existing view to make semantic relationships apparent. This dynamic aspect allows the user to continuously refine or expand their goals by exploring these relationships. We propose that enhancing VIRI with SFEV techniques can effectively support the requirements of opportunistic search.

6 Conclusion

In this paper we described several important requirements for an interface to support opportunistic information seeking. We have presented a framework of semantic fisheye views, which are a type of dynamic display technique aimed at the uncovering of semantic relationships of data by exploiting users visual power for patterns. We then described an application of SFEV for finding images by exploring word semantics.

References

Bartram, L., Ho, A., Dill, J., & Henigman, F. (1995). The Continuous Zoom: A Constrained Fisheye Technique for Viewing and Navigating Large Information Spaces. In *Proceedings of the ACM Symposium on User Interface Software and Technology* (pp. 207-215).

Bates, M. J. (1989). The design of browsing and berrypicking techniques for the online search interface. *Online Review, 13*(5), 407-424.

Cutting, D. R., Karger, D. R., Pedersen, J. O., & Tukey, J. W. (1992). Scatter/gather: A cluster-based approach to browsing large document collections. In *Proceedings of ACM SIGIR '92* (pp. 318-329).

Furnas, G. W. (1986). Generalized Fisheye Views. In *Proceedings of ACM CHI'86 Conference on Human Factors in Computing Systems* (pp. 16-23).

Furnas, G. W., & Rauch, S. J. (1998). Considerations for Information Environments and the NaviQue Workspace. In *DL'98: Proceedings of the 3rd ACM International Conference on Digital Libraries* (pp. 79-88).

Herman, I., Melancon, G., & Marshall, M. S. (2000). Graph Visualization and Navigation in Information Visualization: a Survey. *IEEE Transactions on Visualization and Computer Graphics, 6*(1), 24-43.

Janecek, P., & Pu, P. (2002). A Framework for Designing Fisheye Views to Support Multiple Semantic Contexts. In *International Conference on Advanced Visual Interfaces (AVI '02)* (pp. 51-58). Trento, Italy: ACM Press.

Lin, X., Soergel, D., & Marchionini, G. (1991). A Self-Organizing Semantic Map for Information Retrieval. In *Proceedings of the 14th Annual International ACM SIGIR Conference on Research and Development in Information Retrieval* (pp. 262-269).

Lokuge, I., & Ishizaki, S. (1995). GeoSpace: An Interactive Visualization System for Exploring Complex Information Spaces. In *Proceedings of ACM CHI'95 Conference on Human Factors in Computing Systems* (Vol. 1, pp. 409-414): ACM.

Miller, G. A. (1995). WordNet: a lexical database for English. *Communications of the ACM, 38*(11), 39-41.

Ruger, M., Preim, B., & Ritter, A. (1998). Zoom Navigation. In T. Strothotte (Ed.), *Computational Visualization: Graphics, Abstraction, and Interactivity* (pp. 161-174). Berlin: Springer-Verlag.

Visualization and interaction in a SCADA throughout GIS components

M. Sordo Touza, J. A. Taboada González, J. Flores González, J. Del Rio Cumbreño

Dept. of Electronic & Computing, University of Santiago de Compostela.
Facultade de Física, Campus Universitario Sur, 15706 - Santiago de Compostela, España.
{elmaruxa, eljose, eljulian, sandman}@usc.es

Abstract

Santiago de Compostela University has developed, inside its Energetic Optimization Plan, a SCADA system for monitoring and control of its buildings which are distributed throughout the city of Santiago de Compostela area. This system management the different equipments presents in each building as the ones that imply an energetic expend (elevator, illumination, machine rooms) as the ones that have specially alertness needs (fire, intrusions, environment alarms,...). The geographical distribution of the buildings and the sensors and equipment to control give to the information a very important space component. The capabilities of a Geographic Information System has been added to the system to analyze this information throughout specific software components. The efficiency of the use of this components in the graphic interface module of this system is discussed in this paper.

1. Introduction

In the last few years the energetic optimization has proved to be more than an economic problem. The concern about ecology and the more and more obvious idea that our current energetic source is not unlimited have promoted the development of optimization and control of energetic systems. The introduction of this systems in an organization or company implies the use of control systems in their sources of energetic expenses. These control systems usually take into account the introduction of systems of remote acquisition, visualization and control called SCADA. These SCADA systems provide tools for the analysis of information. In that way we can adapt the control to the particular necessities of the company.

During these three years the Santiago de Compostela University has been involved in these struggles through the POE program (Energetic Optimization Plan). A set of goals to improved the energetic management in the buildings of this university are involved in this plan. One of this goals is the development of a SCADA system to the supervision and control of universitary installments. This installments are located in a set of 47 buildings with differents uses (faculties, university hall of residence, investigation institutes, administration buildings, etc.) which are distributed geographically throughout the city of Santiago de Compostela area.

Santiago de Compostela University has a campus in other city 70 km from Santiago, called Lugo. Nowadays it is planned to control buildings in this campus with this system.

This SCADA collect and show the different installments information inside of each building. This system implements a hierarchical architecture in which the low levels, what are near from device to control, are based in the Staefa integral AS1000 system (Siemens Building Technologies, S.A.). To communicate their different components this system implements several field buses in each building. This hardware is able to control the building according to the high level commands sent from the control center even when the communication with this center is lost. The communications in each building are centralized through one unique device which transmit all the information between the building and the control center. This communication use an ethernet management by the university.

The high level modules are been running in the control center. We focus our attention on the monitoring and control interface. This module uses to communicate with the buildings another communication module which translates the SIEMENS and USC communications protocols. The main goal of this interface is to show to the user all the information from installments like elevators, boiling rooms, air-conditionings, electric lights, etc. This interface has to allow to manipulate this equipments to the user.

Because of the geographical distribution of the buildings and the distribution of elements of consume inside the buildings, with special mention to the electric lights, we can see that our data has important distributional information on space. That is why we give to the SCADA system the capabilities from a

Geographical Information System through a specific software component for space information management. These software components (MapObjects) allow to add maps to the application and provide a set of characteristic GIS functions throughout a standard programming language.
The graphic interface has been developed using the version 4.0 of Borland Builder C++ running in a 1.6 GHz Pentium 4 under operative system Windows 2000 server. The access to ESRI MapObjects version 1.2 is through OOP C++ language.

2. Geographic Information system – MapObjects.

A geographic information system (GIS) can be defined as the set of operation over geographic information from the observation, stored and analyze, to the use of derive information for decision support[1]. GIS allows to show, through a similar symbols like a classic map, the alphanumeric information over a plan or map of the zone that we want to study. This kind of representation offers to the last user a more easy and quickly assimilation of the information.
These GIS capacities are added to our SCADA system throughout MapObjects components. The base class of this Active X (TMap) allows to add plans to the application. In this way the computer screen shows an image like a classic map.
An important characteristic of the MapObjects, and the GIS in general, is the layer stored of geographic information. In the classic map the way of the information is stored is the same as the way of representation, it means, the map is properly the set of all information. Nevertheless, the GIS stores information in a different way to its representation: the information is stored in different layers and each one is stored in independent archives. As we can see in figure 1 the information layers are structured in steps over a base layer (plan or image). This stored system allows change the map partially so, only the archive, with the layer we want change, is modified.

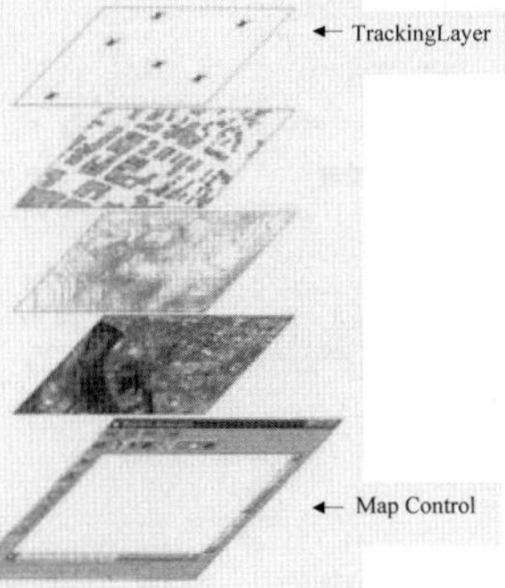

Fig. 1.- Layer stored of geographic information.

The Tracking Layer is another MapObject characteristic showed in this figure. This is an special layer usually use to monitor objects in motion. This layer has special repaint functions so any change in the object position doesn't imply the total screen repaint, instead of this, the anterior object position is deleted and the new is painted. In this way only the old and new positions areas of the object are changed, and so, these changes are management more efficiently.

3. Description of the graphic interface.

The graphic interface is the module of the system in charge of giving the operator all the necessary information for its interpretation. The data of more than 250 equipments existent inside the buildings are shown in our interface in real time. To be able to access to this 250 screens in a fast and orderly way, the equipments are classified according to the building and the system to which they belong. All the controlled buildings are shown on the main screen of the application, organized according to the geographic campus in which they are located (see Fig. 2.) After choosing the building, the different types of signals in it are shown as accesibless, classified into 11 different systems: fires, intrusions, boilers, cogeneration motors, air-conditioning systems, etc. An aditional screen opens once selected the system, already showing the signs of the required installments (see Fig. 3). If there are different zones in the system, those are shown on the left side of the screen, shaped like buttons, allowing the choice of any of them. Every equipment sign is represented on the main object of MapObjects (TMap), which allows us to visualize a map on the screen. This object is always used to present the signs coming from the different equipments of the USC buildings, although the representation used is not always the same (see Fig. 3 and 4). There are equipments with a strong component of localization of space in which the position of the different elements gives an important information, and there are others in which this might be considered of secondary importance. Thus the system uses two different representations according to the type of equipment visualized.

On the first ones, the exact spatial location of each of the devices that constitute the equipment is represented on the ground plan. A layer of points which place spatially each element is created over the ground plan, and a symbol is associated to each point, in charge of reflecting the status of the device (On/Off) through a code of colours (green for On, black for Off). Besides, different symbols are established to distinguish the status automatic (controlled by the control system) /manual (deliberately forced to take that value) of the devices. Every symbol is painted on the layer TrackingLayer, although the geographic position of each element is not going to change. The use of this layer allows the system to make good use of its partial repainting method, being able to update the status of each element each time a new value is received, without the need of refreshing all the screen.

The lightning system inside a building is shown on Figure 3, as an example of this type of representation. For this equipment is not only important to know if a certain electric panel is active or not, but is also equally important to know the zones to which it gives service. Thus when deciding to switch off a panel of illumination is necessary to know which zone is going to remain in the dark, to know which users are going to be affected by this measure, and to know if it can be carried out or not. This representation in the shape of plan is used in the equipments of illumination, elevators, fires, and intrusions.

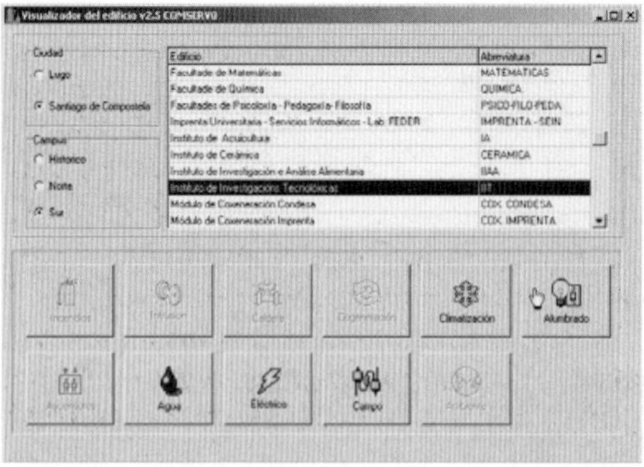

Fig. 2.- Main screen of the graphic interface: choice of the system to be visualized.

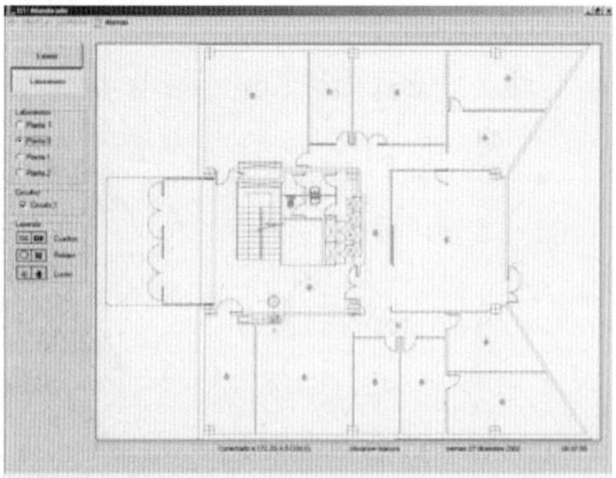
Fig. 3.- Representation on plan of the illumination of a building.

However, there are other equipments in which the information does not have a strong spatial component, and it is more useful to know its functioning status than its exact position on floor. Such is the case of boiling rooms, air- conditioning systems or other complex equipments. The high number of them (a boiling room/ an air- conditioning system or both in each building), together with the great difference of configuration existing in this equipments, makes it imposible to create a general functioning diagram which fits all the equipments, although they are of the same type. This leads us to use the independent layer organization which is used in the GIS to manage with the enormous amount of information of these comples equipments. The capacity of the GIS is used to put an image as base layer, and exactly the same structure of information as in the first representation is used. For each one of this equipments the functioning diagram is designed, and upon this, the different layers of information are created, one for each type of sign to be represented, allowing the group of layers to show the present functioning of the equipment (see Figure 4). All the elements are shown in one single screen, together with the status of each one of these in each moment and the relationship between them. Thus the operator is able to know rapidly how any action exerted upon any device is going to affect the rest of them.

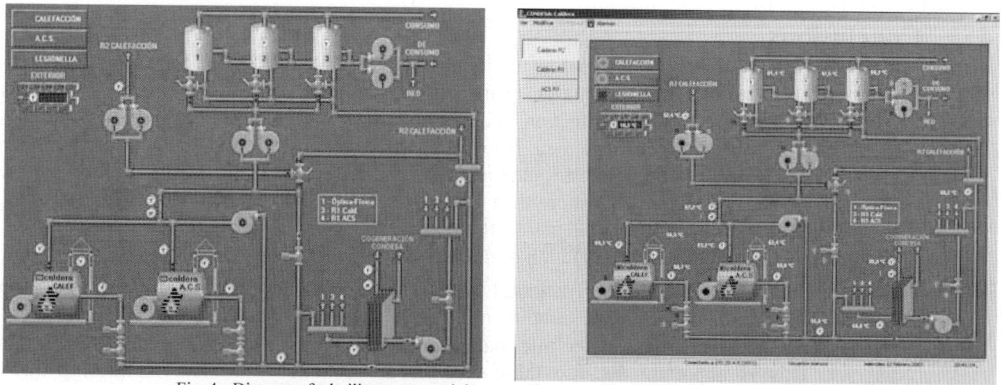

Fig. 4.- Diagram of a boiling room, and the same diagram with all the layers of information.

As in the former representation, all the digital signs are painted on the TrackingLayer, and follow the same code of colour: green for the working devices, black for the not working. The analogic values, however, do not have a visualization so direct. The dinamic data are not processed by the TrackingLayer; this causes limitations in their visualization. The TrackingLayer allows the painting of several different symbols in points belonging to the same layer, but if the symbols do not belong to this layer, as it happens in the text, they have to be the same for every point of the layer. Two points in one layer of text with different attributes (one in automatic and other in manual, for example), cannot be distinguished immediately by changing its source (size, colour, type, style, etc.). Every point belonging to this layer has to be visualized in the same way. This forces us to introduce one more layer which reflects the status manual/automatic of these points. Besides, as these symbols are not included in the TrackingLayer, its capacity of partial repainting is lost, which implies a refreshment of all the layer of data each time a new analogic value is received.

Finally, the MapObjects also makes easy the interaction of the interface with the user as they allow to recover information of the signs already monitoring. The layer structure of these GIS components allows access to the elements of a particular layer, without the need of consulting all the information of the rest of the layers. Therefore, when any point of the screen is selected, all the information of the sign associated to that point is obtained. This facilitates the creation of specific windows which are launched depending on the clicked element. When the control center commands an action, for example, the user only has to select a point (of Go/ Stop), to make the window of Light/ Switch Off show the actions that can be carried out in that device (see Figure 5).

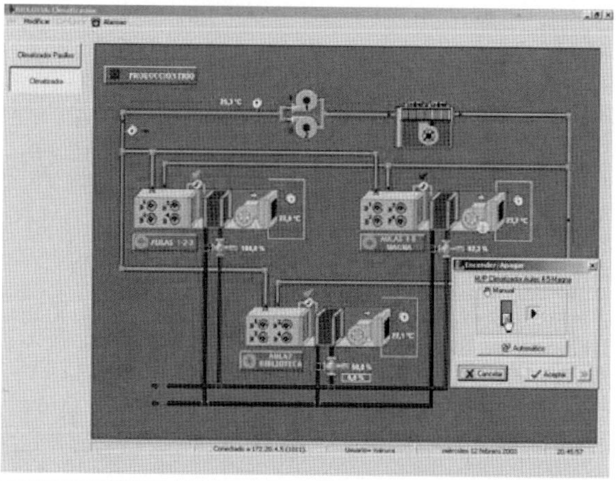

Fig. 5.- Schematic representation of an air- conditioning system and action of stopping a fan.

4. Conclusions

- The MapObject components provides an efficient way to deal with plans and diagrams. It offers us the capability to structure the information neatly using layers.
- The SCADA system has been in function since the beginning of 2002 and the number of buildings administered has been grown since then. Nowadays, the system monitoring and control more than 250 equipments of 47 buildings distributed around the three campus of the city of Santiago the Compostela. The system analyses more than 5000 parameters, 1800 of them could be configured. The system manages more than 1000 alarms too.
- We find limitations with the visualization of the text using MapObjects.
- The partial refresh methods of the Tracking Layer give us a efficient way to paint in real time a big amount of changeable values of the installments.
- The MapObjects structuration of the information in Layers allow us an easy update of plans and diagram. This is important because these changes are common when we monitoring many equipments.
- The MapObjects give us an efficient interaction with the user because it allows us to obtain information from the devices using the showed parameters.

5. Acknowledgements.

This project has been supported by:Santiago de Compostela University, “Desenvolvemento da aplicación informática do control enerxético da USC”, (1998).

Siemens Building Technologies, S. A. “Sistemas de telesupervisión de servicios técnicos y gestión del control de instalaciones”, (1999).

References.

[1] José Mª Quinteiro González, Javier Lamas Graziani, Juan D. Sandoval González: “Sistemas de control para viviendas y edificios. Domótica.”, Paraninfo Editorial, S. A., (2000).

[2] Jeffrey Star; John Estes: “Geographic Information Systems: An Introduction”, Prentice-Hall (1990).

[3] ESRI: “MapObjects-Building Applications with MapObjects”, Red Land (California, EE. UU.), Environmental Systems Research Institute (1996c).

[4] ESRI: “MapObjects-GIS and Mapping Components Programmer’s Reference”, Red Land (California, EE. UU.), Environmental Systems Research Institute (1996b).

[5] Borland C++ Builder 4: “Developer’s Guide”, Inprise Corporation (1999).

Hybrid Visualization of Manufacturing Management Information for the Shop Floor

Sascha Stowasser

ifab-Institute of Human and Industrial Engineering,
Universität Karlsruhe (TH)
Kaiserstrasse 12, 76128 Karlsruhe, Germany
Sascha.Stowasser@ifab.uni-karlsruhe.de

Abstract

Despite years of research activities concerning human-machine-interfaces in industrial manufacturing environments, innovative forms of visualization (e.g. three-dimensional or realistic visualization) have yet to be thoroughly experimentally examined from software ergonomic and cognitive psychological points of view. Even though realistic forms of visualization has been implemented in an array of fields (e.g. architecture), there are relatively few statements as to when it is recommendable to chose such a visualization form in the industrial manufacturing.
With help from a recently concluded examination (cf. Stowasser 2002), it should be methodically determined, for the first time, in which way realistic elements should be used for the presentation of manufacturing information on the shop floor and for the surveillance and controlling of shop floor processes. Such human-computer-interfaces should support the employees in performing operative tasks like the observation and monitoring of manufacturing processes, checking the conditions of machines and tools, assigning materials to orders, tracing breakdowns and so on. Incorrect decisions in shop floor management (e.g. due to insufficient presentation of the information needed for process execution) have a direct influence upon the utilization, deadline adherence, costs situation and thus the competitiveness of the manufacturing.

1 Introduction into Shop Floor Controlling Systems

1.1 Visualisation Forms for Shop Floor Controlling Systems

Shop floor controlling is defined as the short-term controlling and monitoring of shop floor processes. It is responsible for the planning compatible execution of shop floor orders while considering human, economic, quality and deadline demands. In order to execute shop floor orders in a planning compatible manner the appropriation of personnel, equipment, material and other necessary resources, as well as the execution of work tasks, are initiated, surveyed and secured by the shop floor controlling system.
The goal of the current work is to provide a contribution to the experimental analysis and configuration of computer supported, process-orientated visualization forms using the example of enterprise shop floor processes within the operative shop floor controlling. Various research studies have found that the available, shop floor controlling systems, usually window-based, have deficits with respect to their user-friendliness and information transparency (see Greenough, Kay, Fakun & Tjahjono, 2000; Kasvi & Vatiainen 2000; Stowasser 2002).

1.2 Today´s Shop Floor Controlling Systems

The window technique has been established as a standard in all enterprise application areas. In the window technique the entire screen surface is divided into individual logical groups. One differentiates between information, controlling, processing and notification parts. A perception psychologically suitable arrangement of these groups serves to ease the registration of information as well as its quick interpretation. Taking these groupings into account, the arrangement of the information occurs according to information classes. Status information and control information is required for user orientation and for the control of the dialogue. The information required for the tasks to be carried out immediately is presented in the processing part. If the user makes an error in the handling of the computer system or disregards restrictions to the input, notifications are delivered by the system. Notifications can apply to "autonomous" objects.

2 Window-based versus realistic shop floor visualization

The basis for the comparative investigation study are two forms of visualizing industrial manufacturing information on the shop floor: a traditional window-(text-)based visualization *FEWER* (Fensterbasierte Werkstattsteuerung) and an innovative realistic form of visualization called Virtual Shop Floor (*VISOR*) (see Zülch & Stowasser 2001; Stowasser 2002).

2.1 Window-based shop floor visualization *FEWER*

Within the context of this work, the shop floor system *FEWER* (cf. Stowasser 2002) bases on text formulas, masks and dialogues. Text or graphic information as well as interaction possibilities are provided to the user within the structure. Figure 1 clarifies the structure of the user interface *FEWER*.

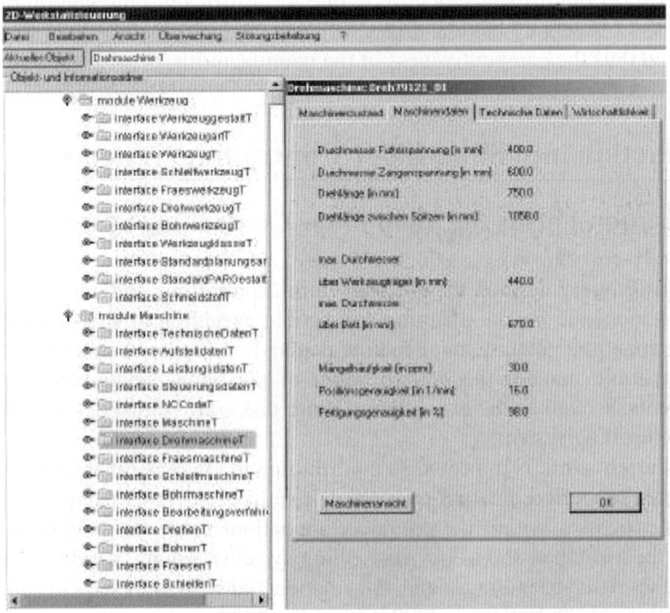

Figure 1: Window-based visualization *FEWER*

In the information part the current mask is indicated and the name of the program and the shop floor object being considered are given. With a menu selection in the control part the user is presented with a list of the functions at his disposal. The processing part of the screen mask offers a workspace for the execution of the operative shop floor controlling tasks needing to be dealt with, namely: Surveillance of preparation, quantities and deadline monitoring, quality controlling, tracing breakdowns, determining the causes for disturbances as well as intervening in the manufacturing procedure.

The processing part in *FEWER* contains a list of the shop floor objects (e.g. machines, orders) and their possible attributes. Using these lists, the user can find the objects required for the control and surveillance of the shop floor.

2.2 Virtual Shop Floor Visualization *VISOR*

VISOR (Virtual Shop Floor) is a realistic representation form based on the use of the graphic, multimedia possibilities of current computer systems. The realistic visualization is, in the context of this work, defined as (cf. Stowasser 2002) a spatial-perspective model of a three-dimensional scene, which can be considered from varying views. The model is visualized using graphic computer support. The degree of the visualization reaches an accurately detailed, objective (photographic) level. The dynamic shop floor environment is represented with *VISOR* in a spatial-perspective manner. The user can "place" himself in the realistic shop floor and move about in all three directions by using the spacemouse.

For operative shop floor observation, meaning the surveillance of appropriation (of e.g. resources, personnel), quantities, deadlines and quality, the user requires information about the current shop floor events. Such information is displayed in *VISOR* as text or graphics, whereby the information subject matter plays an important role in the conception (cf. figure 2). The data to be visualized can refer to abstract as well as to concrete circumstances.

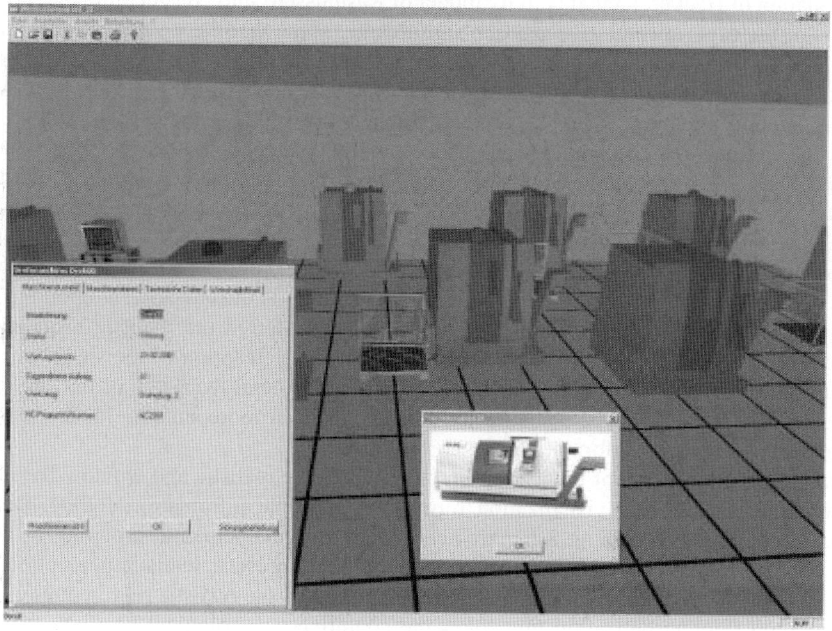

Figure 2: Realistic shop floor visualization *VISOR*

Concrete information is arranged metaphorically at the location of creation (e.g. material stocks in the warehouse) or use (e.g. NC-programs at machines). Metaphoric means that objects in an interactive system are represented in such a way that the users are familiar with them from their daily life or from their work environment. Thus, the worker can move around in the realistic shop floor and call up information at the representation of the respective object. It is for example possible to access order data (e.g. status of the current process, previous lead time, adherence to delivery dates) at the presented boxes, each representing an order. In particular fixed images (e.g. photos, sketches) or multimedia, dynamic visualization elements are used to represent current process states. Abstract information, which cannot be directly assigned to an object, is also visualized with metaphors in the representation. This functionality allows for the integration of, for example, general announcements (represented with blackboards).

3 Hybrid Visualization

The empirical methods used for the investigation of both visualization forms at the "Laboratory for Human-Machine-Interaction" of the ifab-Institute at the University of Karlsruhe are:

- eye movement registration to find out in which way the cognitive information process proceeds,
- log-file analysis to examine the interactions (e.g. mouse, keyboard) of the test persons,
- video-recording to observe the gestures of the test persons, and
- structured interviews to record the demographic data and to analyse the processing strategies used by the test persons in detail.

The study with 20 test persons (industrial shop floor experts and academic students) helped to sort out specific situations in the shop floor in which a specific visualization technique is most helpful. The design of these visualization techniques takes perceptual psychology models (e.g. theory of action regulation of Hacker, mental activity model of Rasmussen) into consideration.
The results of the experimental investigation significantly show that the visualization is markedly important for the cognitive performance of and the strain to the user during the execution of shop floor tasks. The compatible representation of mental models, upon which the shop floor worker's cognitive operation planning is based, is suitably supported by the realistic visualization. If one considers the average fixation duration as a measure for example, the test persons´ strain was approximately 25% lower when *VISOR* was used. This effect must be considered as particularly important when unforeseen operation requirements with high cognitive demand arise, in which an error-free intervention in the shop floor process is necessary.
The investigation also shows, however, that the realistic visualization is not preferred for all shop floor tasks. The choice of information coding and visualization is thus dependent upon the type of task and the information required for it. The test persons preferred the realistic visualization in particular for the execution of those tasks for which a spatial perception of the shop floor and of the object arrangement is helpful and for those for which a general overview of the shop floor is advantageous. The window-based visualization form is however more suitable for the representation of abstract management information.
The advantages of both forms of navigation, in other words the realistic and window-based forms, can be combined in a hybrid visualization form using multiple information coding. Order and organizational principles, meaning a hierarchical structuring and semantic grouping of the shop floor objects, can thereby be achieved. Additionally, already present, realistic perception experience, stored in the form of a mental model, can be supported by the realistic representation.
The hybrid shop floor visualization thus takes the cognitive psychological position, that human thought processes are marked by concentration on local detail and by a global overview of impor-

tant information, into account. This should make it possible for the user to apply the visualization form most suitable for him and to dynamically vary it depending on the task, information needs, individual preferences and personal abilities. This requirement is also fulfilled by adaptive visualizations.

Figure 3 provides, as an example, the vision of a hybrid shop floor controlling visualization. The text-based, hierarchically arranged object list allows for a quick object-orientated access to a sought shop floor element. A realistic animation of shop floor occurrences, through the effect of immersion, and the visual-spatial representation would enhance this even further. Disturbances in the shop floor process, error messages and alarms would have to be shown repeatedly and coded in various forms. (e.g. pop-up disturbance dialogue).

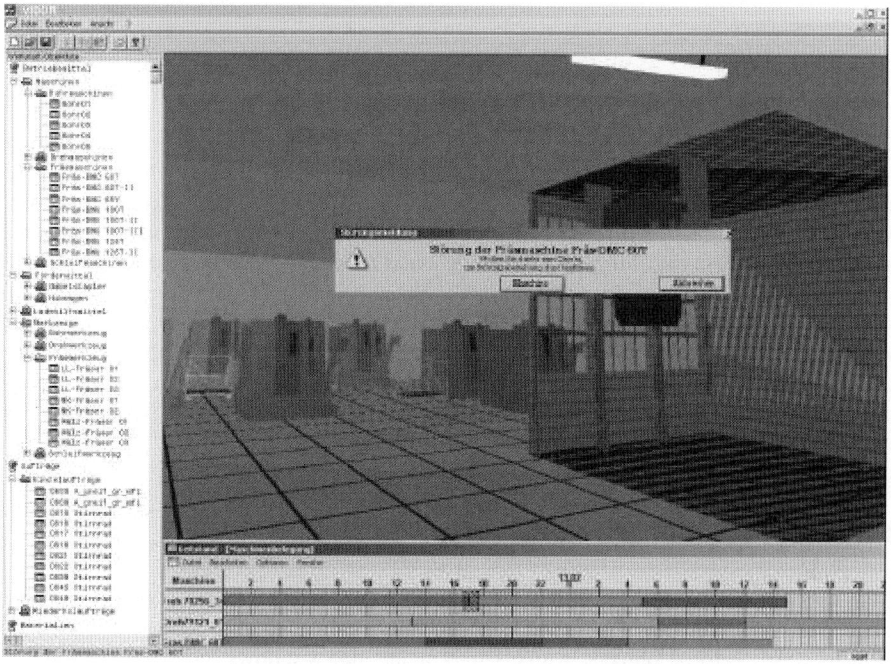

Figure 3: Hybrid visualization of the shop floor

References

Greenough, R. M.; Kay, J. M.; Fakun, & Tjahjono, B. (2000). Multimedia and Hypermedia in Manufacturing. In Commonwealth Institute (Ed.), *Technology Transfer and Innovation (tti2000)* (pp. 17-22). London UK: Commonwealth Institute.

Kasvi, J. J. J., & Vartiainen, M. (2000). Performance Support on the Shop Floor. *Performance Improvement*, 39(6), 40-46.

Stowasser, S. (2002). Vergleichende Evaluation von Visualisierungsformen zur operativen Werkstattsteuerung. Aachen: Shaker.

Zülch, G.; Stowasser, S. (2001). VISOR – Towards a Three-dimensional Shop Floor Visualisation. In M. Smith, G. Salvendy (Eds.), *Systems, Social and Internationalization Design Aspects of Human-Computer Interaction* (pp. 217-221). Mahwah, NJ: Lawrence Erlbaum Associates Publishers.

Visualizing aircraft properties: An empirical study

Monica Tavant[1]

EEC Eurocontrol Experimental Centre
Zac des Bordes BP 15 Bretigny-sur-Orge
F-91222 CEDEX France
monica.tavanti@eurocontrol.int

Geraldine Flynn

EEC Eurocontrol Experimental Centre
Zac des Bordes BP 15 Bretigny-sur-Orge
F-91222 CEDEX France
geraldine.flynn@eurocontrol.int

Abstract

The design of visualisation solutions is a significant issue whenever safety-critical environments are involved. The domain of Air Traffic Control is a safety-critical domain where design choices are constrained by users' procedural knowledge and, most of all, by their explicit need of high simplicity in the representations. These constraints compel designers to adopt poor visual solutions that, sometimes, decrease the ability to discriminate and to quickly locate visual targets. However, as the need to display higher amount of data increases, effective and creative ways to visualise data have to be found. The objective of this study is the evaluation of graphical solutions representing aircraft provided with different technical equipment. Twelve operative controllers were engaged in a location and categorization task. Despite the results of the subjective assessment, the empirical data suggest that unusual and advanced design solution do not negatively impact on controllers' performance.

1 Introduction

Novel visualization approaches are extremely important in safety critical domains such as Air Traffic Control (ATC). Nowadays, in ATC, many attempts are done to represent information in sophisticated ways: colored objects, widgets like menus and windows, etc. (Mertz et al., 2000). These systems also implement the mouse input, drag and drop interactions, animations (Athenes et al., 2000) and other highly developed solutions (Mertz & Vinot, 1999) and (Kessler & Knapen, 2000); in many cases, however, advanced interfaces remain at a simple experimental level. Frequently, operative controllers have a critical attitude towards unusual interfaces, fearing that they might have a negative impact on their activities. Controllers' main activity is maintenance of aircraft separation, that is the safety distance that has to be kept vertically and horizontally between aircraft so to avoid possible aircraft collisions.

Nowadays, in several ATC systems the digital data block, or "label", provides information about aircrafts. The label reports (at least) the aircraft identification and its altitude. Controllers rely on this information to maintain safety.

An example of what controllers have to monitor while working is shown in Figure 1, reproducing a particular area of the radar screen: aircrafts, together with their call-signs and current altitude are shown. It is common opinion within the ATC community that simple data (as the call- sign of an

[1] At the time of this study, Monica Tavanti was working for Deep Blue s.r.l, the firm hired as consultant for the TRAMS project.

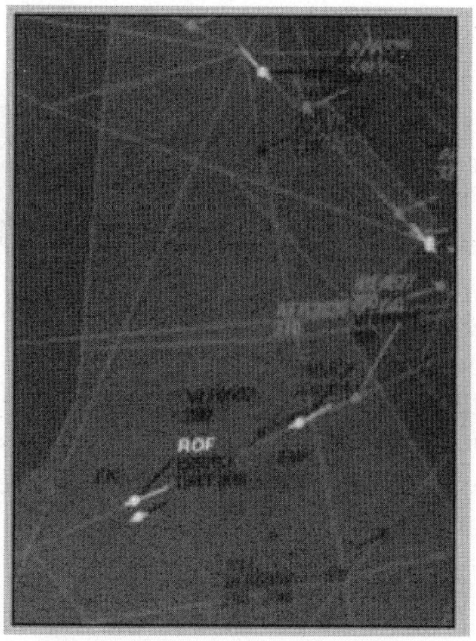

Figure 1: Traffic on a radar display

aircraft, etc…) need to be represented in a simple manner to allow faster and easier data processing and that an interface cluttered with too many objects could disturb monitoring, already a demanding activity. The design of ATC interfaces is affected by the constraint of "visualization simplicity", often misinterpreted as "poor visualization".
But the need of displaying higher amount of information increases as new technological innovations are introduced, and more creative approaches to visualization problems are required. The present study is an attempt in this sense. Three proposals of graphical representation were created and evaluated with operative controllers. Two of them were quite advanced if compared to the interfaces currently used for en-route air traffic control. The results of the empirical evaluation revealed that such visual solutions did not have a negative impact on controllers' performance. The following sections will describe the background work and the technical ATC concept lying behind the visualization problem. The design process, the different graphical solutions and the experiment are then explained. The work will be summed up in the last section.

2 TRAMS: technical concepts

The study was conducted within the European Air Traffic Management Programme MODE S (MODE Selective), during the development of the project TRAMS (TRAnsition to Mode S). The TRAMS project aimed to investigate the consequences deriving from the gradual introduction of MODE S capabilities in aircraft and control agencies. Nowadays, Mode 3/A systems are operational in European Air Traffic Management Service Providers. In these systems, controllers identify the air traffic using a dynamic SSR (Secondary Surveillance Radar) code. However, Mode 3/A implies a number of limitations (like the limited amount of SSR code to be used for identification, etc…).
For overcoming MODE 3/A limitations, MODE 3/A is going to be gradually replaced with MODE S. In MODE S systems, aircrafts will be provided with a unique individual code, which is allocated by the registering authority and does not change through the flight. The MODE S technology is distinguished in two instances, the MODE S Elementary (which will allow a "more automatic aircraft identification"), and the MODE S Enhanced, which is provided with additional downlink capabilities, so that special parameters associated to the aircraft will be down linked and visualized in real time on the RADAR screen. During the Transition to MODE S a mixed-mode condition is expected to take place. Namely, MODE A and MODE S control agencies will have to deal with MODE 3/A and MODE S Elementary or Enhanced traffic. In this mixed situation different processes and procedures have to coexist. Most importantly, if the aircraft is MODE S equipped (either Elementary or Enhanced) but the ground control position is not (a mixed-mode condition), then special identification procedures shall be applied.
Therefore, MODE S traffic has to be quickly located on the RADAR screen as soon as it enters the concerned sector. An efficient way to visualize the aircraft endowed with MODE S Elementary and Enhanced capabilities ought to be found. During the creation of visualization ideas, two controllers were involved. This was done to ensure that the users, with their knowledge and specific needs, provided feedback during the design process (Norman, 1988).

3 The symbol

A simple solution to approach the problem was to add an alphanumeric character (e.g. "S", for MODE S) either before or after the call-sign. However, when attention is lacking or in case of dense traffic, the alphanumeric character could be confused with the letters of the call-sign. A more reasonable choice was to insert a symbol. This choice was constrained because, for technical reasons, the set of available symbols was limited. Because of other technical constraints a small rectangle was chosen as the symbol. It had two versions (cf. Figure 2): a full rectangle representing the MODE S Enhanced traffic and a simple rectangle's contour for the MODE S Elementary traffic. The symbol was simple and easy to implement, but its location could hinder the legibility of the label information. In order to keep this informational area clean and uncluttered, we tried to focus on the track.

Figure 2: The rectangle

4 The track

As previously stated, the MODE S is a special technology that involves the aircraft's transponder, which spreads around information, namely the aircraft call-sign, plus the special parameters if the aircraft is MODE S Enhanced. In this solution (in Figure 3) the track, i.e., the small dot representing the aircraft, *spreads* different levels of luminance: lower for the MODE S Elementary and higher for the MODE S Enhanced aircraft (providing additional data, thus *spreading* more information). This solution seemed more intuitive than the previous and provided good visibility.

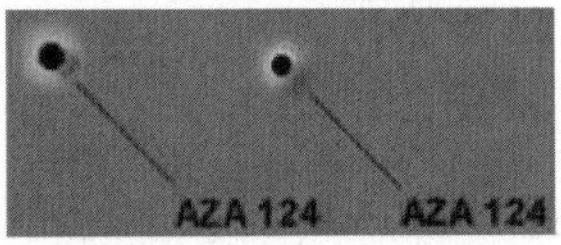

Figure 3: The track

5 Modified call-sign

The last proposal implemented a visualization solution within the area of the label. The label contains information about the aircraft; therefore it is a reasonable place to display aircraft's information. Being an important information source, the label always risks being jumbled with symbols and widgets, but with this solution, it would maintain its role without adding anything extra. The MODE S Elementary aircraft had its calls-sign slightly in relief over the background with a small luminance effect added to MODE S Enhanced aircraft.

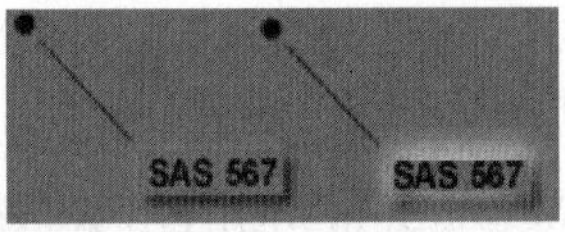

Figure 4: The call-sign

6 The experiment

The goal of the experiment was to investigate which graphical solution allowed controllers to most quickly perform the location and categorization of a target aircraft. The independent variable was the "interface type" with 3 levels ("track'', "symbol' and "call-sign" solution). The interface type was a between-subjects factor and a significance level of .05 was chosen as the decision criterion. The study was run on a portable Pentium II machine, with 64 MB of memory and a 14-inches display with 1024*768 resolution. The displays were written in Macromedia's Lingo. Twelve air traffic controllers from Ciampino CRAV (Rome, Italy) were tested. Their age average was 41.1. All of them were fully operational with 7 years of experience minimum.

6.1 The displays

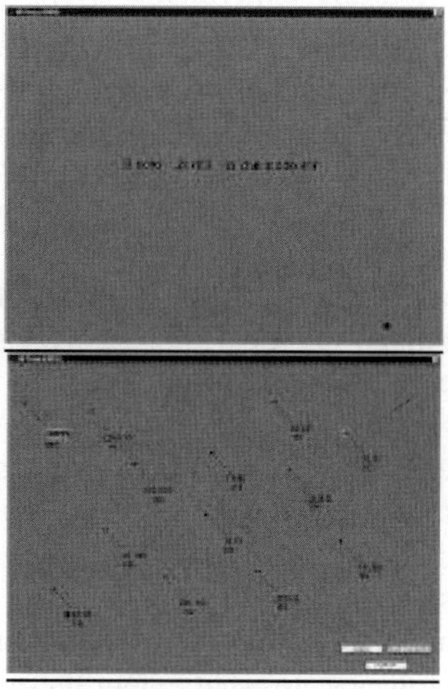

Figure 5: The displays used for the test

The displays (Figure 5) used in the test were made up of two parts. The first part of the display consisted of a window (962 pixels of width * 642 pixels of height) with grey background carrying a textual string in the very centre. The text asked the MODE of the target aircraft. After 15 seconds this first part automatically turned to the second part of the display, which comprised an equivalent window with 13 aircraft tracks. Each aircraft carried a unique call sign and the flight level. Among the 13 aircraft only 9 were used as targets. The remaining 4 aircraft were simple distracters. On the right-bottom of the window three buttons were present. They were labelled MODE S", "MODE S Advanced" and "MODE 3/A". Subjects had to click one of them click to accomplish the task.

6.2 Procedure and task

All the subjects were randomly assigned to one of the three groups (4 subjects for the track, 4 subjects for the symbol and 4 for the call-sign condition). Before the beginning of the test, each subject received a training session. A separated display with all the solutions was created. Subjects were allowed to "play with it" for around 10 minutes. Afterwards, subjects were requested to indicate the correspondence between the MODE and the graphical solutions presented. Then, the first display started. The controllers had to read the question and wait for the second display to appear. After, they had to search the target and click on the button of the corresponding MODE.

6.3 Results

The data report the time deployed by controllers to identify the aircraft's MODE and click on the appropriate button, and the number of errors made. The unit of time used for the recordings was the "tick" which, in Macromedia, corresponds to a 60th of a second. The analysis of variance revealed a significant difference among the interfaces ($F(2,105) = 30.26$, $p<.05$). A Tukey HSD test revealed a significant difference between the mean of the track and the mean of the symbol condition (difference between means = 112.05, $p<.05$). Also the difference between the track condition and the "call-sign" condition is significant (difference between means = 122.63, $p<.05$). The difference between the modified call-sign and the symbol condition is not significant (difference between means = 10.58, $p>.05$). None of the subjects made errors.

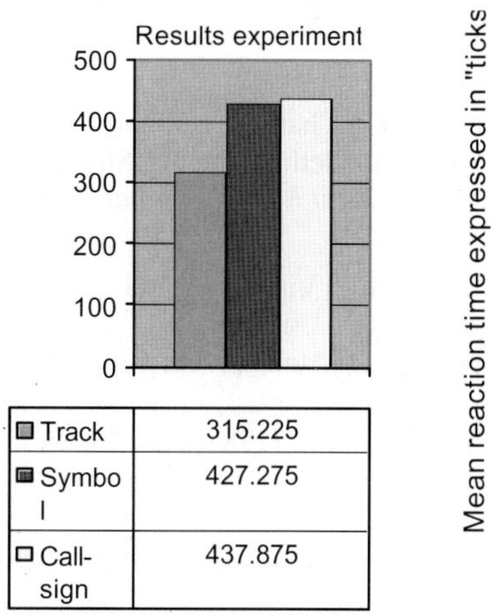

7 Subjective evaluation and discussion

After the experiment, every subject was asked to explore a new display, which showed the graphical solutions used during the tests plus other solutions not used in the experiment, such as color-coded information, a box surrounding the call-sign and a rectangle occupying several positions around the call-sign. While examining the solutions, the controllers had to comment and evaluate them according to three main criteria: visibility, clarity of display, comprehensibility.
The results of the subjective sessions revealed that controllers judged the symbol to be the best solution because it seemed the most legible, the lightest and easiest to learn. The highlighted track was judged easy to comprehend but also not clear and distracting. The modified call-sign was judged not clear, not easily visible and not easy to understand. In general, controllers showed a certain degree of unease with the more "advanced" solutions. These comments seem to confuse the results of the empirical evaluation, according to which the track solution was the most effective for the task designed for the experiment. Possibly, despite the subjective judgments, more advanced visual solutions can be effective and supportive for specific tasks. Moreover, the data collected show no difference between the symbol and the modified call-sign. It can also be speculated that more creative solutions to visualise data (e.g. like the call-sign in relief) do not adversely affect controllers' performance.
However, to what extent these empirical data can be generalized is debatable, because the number of controllers involved in the study was too small; yet, continuing the work on advanced and innovative ideas on visualization issues in safety critical domains seems a promising research area.

8 References

Athènes, S., Chatty, S., Bustico, A. (2000). Human factors in ATC alarms and notifications design: an experimental evaluation. In Proceedings of the 3rd International Air Traffic Management R&D Seminar, *ATM-2000*, Napoli, Italy.

Kesseler, E., Knapen, E.G. (2000). "Interaction: Advanced controller displays, an ATM essential". In Proceedings of the 3rd International Air Traffic Management R&D Seminar, *ATM-2000*, Napoli, Italy.

Mertz, C., Chatty, S., Vinot, J.L. (2000). Pushing the limits of ATC user interface design beyond S&M interaction: the DgiStrips Experience. In Proceedings of the *3rd USA/Europe Air Traffic Management R & D Seminar*, Napoli, Italy.

Mertz, C., Vinot, J.L. (1999). Interface gestuelle pour écran tactile, animations et design graphique: Un guide de conception. In Proceedings of *IHM'99*, Conférence AFIHM sur l'Interaction Homme-Machine, Montpellier, France.

Norman, D. A. (1988). The Psychology of everyday things. New York: Basic Books Inc.

9 Acknowledgments

Thank to the anonymous reviewers, Guglielmo Guastalla, PhD and Horst Hering for their feedback and comments.

Section 7

HCI: Application and Services

Usability Tests with Interactive Maps

Natalia Andrienko and Gennady Andrienko

Fraunhofer AIS – Institute for Autonomous Intelligent Systems
Schloss Birlinghoven, Sankt-Augustin, D-53754 Germany
E-mail: {Natalia.Andrienko, Gennady.Andrienko}@ais.fraunhofer.de

Abstract

CommonGIS is a software system for exploratory analysis of geographically referenced data by means of visualising the data on highly interactive maps and other graphical displays. We performed an experimental study for assessing the usability of five different visualisation tools implemented in CommonGIS. Usability was assessed in terms of tool learnability, memorability, and user satisfaction. We found that users were, in principle, able to understand and adopt the new ideas of map interactivity and manipulability. However, our study showed the importance of educating users, which is a particular challenge for software accessible over the Internet.

1 Introduction

Visualisation of data in a graphical form is a primary instrument for exploratory data analysis, which aims at revealing inherent features in the data and generating hypotheses (Tukey 1977). Spatially related data are traditionally visualised on maps, which are capable to convey spatial relationships. Currently, it is generally recognised that not only graphical representation of data is needed for exploration but also high degree of user interactivity, possibilities to manipulate the displays and to consider the data from various perspectives (see, for example, Cleveland & McGill 1988, Buja *et al.* 1996). In cartography, a concept of "geographic visualisation", or "geovisualisation", has recently emerged, which implies the use of interactive, dynamic maps to support hypotheses generation and decision-making (MacEachren 1994, MacEachren & Kraak 1997).

Interactive techniques and tools can support information exploration and knowledge construction only when users are able to utilise these instruments properly. This problem is addressed in the research agenda of the Commission on Visualisation and Virtual Environments of the International Cartographic Association (Slocum et al. 2001). While the Commission recognises the value of the established usability engineering principles, it warns that they may be only to a limited extent applicable in designing and evaluating geovisualisation tools. Thus, the central idea of these principles is to base upon people's previous knowledge of the world and experiences with other software and to keep consistency with people's expectations (e.g. Macintosh 1995). For such novel concepts as interactive maps and graphs, there are often no analogous situations in typical people's experience that could be utilised as metaphors. It may be also impossible to implement unconventional features using only standard user interface elements with standard behaviours. Consistency with user's expectations is also hard to keep: the very idea of a map dynamically changing its appearance conflicts with typical expectations since most people are mainly accustomed to static paper maps. Hence, most users may not be able to master novel techniques just relying on their previous experiences, and teaching may become indispensable.

In this paper we report on the testing of interactive tools implemented in our geovisualisation package CommonGIS. Our goal was to check to what degree users are able to:

- Understand the purposes of geovisualisation tools and learn how to use them;
- Retain the acquired skills after some period of not using them; and
- Develop a liking for the tools rather than being afraid of them.

We assumed that users would need some background in how to utilise the exploratory tools and wanted to see how much teaching would be required and in what form.

2 Interactive Techniques Used in the Study

Five exploratory interactive techniques were selected to test: outlier removal (focusing), visual comparison, dynamic classification, dynamic query, and dynamic linking between a map and other graphical displays (brushing). These techniques are considered in detail in Andrienko and Andrienko (1999); here we briefly describe just one of them, outlier removal, to illustrate the idea of interactive, manipulable map.

Very often, a data set contains outliers - a few very high or very low values of a numeric attribute, while the rest of the values are relatively close to each other. For example, birth rates in Europe lie between 10.56 and 15.93 births per 10,000 people, with the exception of Albania, which has a rate of 21.70. In maps as well as other graphical displays, the presence of outliers can make differences between mainstream values hard to perceive. For example, on an unclassified choropleth map representing the birth rates by proportional degrees of darkness (the higher the value, the darker the colour), Albania will be dark while all other countries will have approximately similar colouring (Figure 1, left). When outliers are removed from the view, the display represents a shorter value interval, and the differences among the mainstream values can be made more apparent (cf. Figure 1, right).

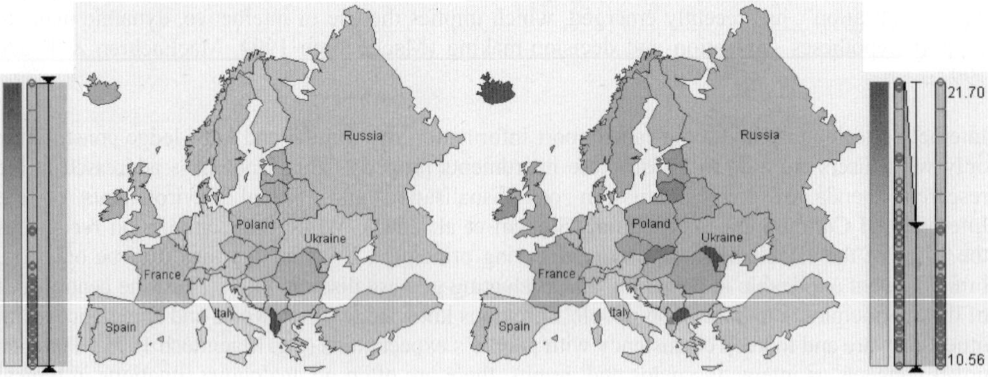

Figure 1: Effect of outlier removal (focusing). When Albania, with a very high birth rate, is taken out of consideration, differences between the other countries can be represented more distinctly.

The operation of removing outliers is performed using an interactive focusing control shown in Figure 1 beside the maps. The tool is operated by dragging the triangular delimiters along the vertical line that represents the value range of the attribute visualised on the map.

3 Test Tasks

The tasks in the test were organised into five series, each dedicated to one of the techniques selected for testing. A typical task was to perform a certain operation with an interactive tool. Each task was followed by one or more questions concerning the outcome of the operation. The questions were either closed (i.e. the user selected one or more correct responses from a list) or required a response in the form of a single number retrieved from the display. Users were also allowed to reply, "Don't know." We developed a special extension of CommonGIS that displayed the tasks and questions on the computer screen and recorded the responses. The system automatically generated all necessary thematic maps and other displays. At the beginning of each task series, the users were asked to read a short illustrated description of the tested technique.

Each task series included a variety of tasks so as to test all possible tool uses foreseen by the tool designers. The tasks differed in their complexity. Here is an example of a task requiring an overall view of a territory, which is, in general, more difficult than a consideration of particular locations:

Task: after removing the outliers from the map, consider the spatial variation of attribute values.
Question: where can you observe a cluster of districts with high values of the attribute?
Possible answers: 1) in the centre; 2) in the north-west, near the border; 3) in the south-east; 4) in the south, along the coast.

At the end of each task series, a subjective evaluation of the tested tools was performed based on the following four questions:

1. Whether it was easy or difficult to carry out the tasks in the series;
2. Whether users felt frustration or comfort when using the tool;
3. Whether users felt their performance was fast or slow; and
4. Whether the tool appeared visually appealing.

To answer each question, one could choose one of four possible positions on a "semantic differential scale" (Nielsen 1993, pp. 35-36), involving two opposites: very difficult vs. very easy, completely frustrated vs. completely comfortable, very slow vs. very fast, and very unappealing vs. very appealing.

4 Organisation of the Tests and Results Obtained

The tests were done in three rounds. The first two rounds were performed on-site using a local installation of the system while the third round was conducted over the Internet. Participants of the first round (9 persons) were given a lecture demonstrating the tools that they needed to apply in the test. The presentation took about 30 minutes. The subjects did not practice using the tools by themselves before starting the test. The second round took place one month later with participation of the same users as in the first round. The purpose was to check tool memorability, that is, whether the users could maintain the acquired knowledge and skills about the interactive techniques and re-apply them after a recess. The third round involved more than 200 subjects who did not get any prior introduction and could rely only on written explanations and illustrated examples. The purpose was to test whether people accessing CommonGIS on the web were able to master its interactive exploratory facilities without being directly taught. The participants of the first two rounds of testing were professional GIS users or software engineers while the third-round participants were students from two universities in Germany: computer science students from the Technical University of Darmstadt and geoinformatics students from the University of Münster.

The results of the first two rounds were rather encouraging. In the first round, the subjects were quite able to use the interactive tools and demonstrated solid performance in fulfilling the requested tasks. The results of the second round were even better than those of the first round. This means that people, once having understood the new tools and trained to use them, could preserve and reuse the knowledge and skills acquired. Moreover, the subjective satisfaction data showed that the subjects liked the new tools, and many very favourable comments were received. Some expressed their wish to use the system in their professional activities.

In the Internet-based tests, however, the overall performance of the subjects was much worse than in the first two rounds. The mean error rates (i.e. percentages of wrong answers to the total number of answers) were 41.67% in tasks on outlier removal, 25.29% in those on visual comparison, 37.38% in dynamic classification, 52.83% in using dynamic query, and 30.21% in tasks on dynamic linking. This differs dramatically from the results of the first two rounds: 8.89/1.67%, 8.89/0%, 9.25/2.78%, 12.50/8.33%, and 28.20/12.96%, respectively. The third round was also characterised by very high individual differences among the participants. The degree of subjective satisfaction also varied greatly and, as in the first two rounds, correlated with performance.

A probable conclusion from the Internet-based tests is that the written explanations did not adequately substitute for the introductory lecture. It was detected that many subjects did not read the explanations at all or viewed them for just a few seconds (the testing software recorded the times when an explanation page was opened, if at all, and when the next action took place). This may indicate, in particular, that some students might not be really interested in participating in the test and, therefore, not sufficiently motivated to learn the tools and achieve satisfactory results.

However, it would be wrong to say that the reluctance of the subjects to read the instructions was the only reason for the poor overall performance. Many of the test participants spent enough time for viewing the explanations but still made a lot of errors in fulfilling the tasks. At the same time, some subjects who did not read any explanations performed poorly at the beginning of the test but greatly improved by the end. This demonstrates that users may be psychologically rather diverse, and there are people who prefer and are able to master new tools by "playing" with them.

It should be taken into account that the students, in general, were less experienced in the analysis of geographic information compared to the professionals who took part in the first two rounds. This especially refers to the computer science students from Darmstadt. Indeed, these students had notably higher error rates than the geoinformatics students from Münster. One more problem was with the language: all the explanations and task formulations as well as the user interface of CommonGIS were in English while the students were native Germans. Although German university students, in general, have good knowledge of English, this probably does not apply to the particular terminology used in the test. In fact, many students complained in their comments about difficulties in understanding the explanations and task formulations.

5 Conclusion

The main overall conclusion from the usability studies is that users are, in principle, able to understand and adopt the novel ideas concerning map interactivity and manipulability. However, these ideas need to be appropriately introduced; people can hardly grasp them just from the appearance of the maps and controls. We found that an introductory demonstration may be sufficient for understanding the purposes of interactive tools, and short training enables people to use the tools. In situations when no direct teaching is possible, as in the case of first time users encountering the tools on the web, adequate substitutes need to be provided. It was seen that on-line help is not always effective. Training users by means of interactive on-line tutorials may be more appropriate but still not ideal. Many people will not invest time and effort for such training without being sufficiently convinced of the considerable benefits of the new tools. Ideally, the new techniques should be introduced to users when they perform their own data analysis, that is, each technique should be introduced exactly at the moment when it would be useful. This may be achieved by creating an intelligent (knowledge-based) software component, which could understand the users' goals, propose the appropriate instruments, and help in using them (Andrienko & Andrienko 2002).

The study also demonstrated the importance of providing the user interface in the native language of the users. The same applies to all instructions and explanations, in particular, to on-line help.

Acknowledgements

The work was done within the CommonGIS project partly funded by the European Commission (Esprit project 28983, November 1998 - June 2001). We are grateful to all partners in the project for their collaboration, to all test participants, and to Prof. Werner Kuhn and Dr. Sven Fuhrmann who organised the tool testing by students at the University of Muenster.

References

Andrienko, G., & Andrienko, N. (1999). Interactive maps for visual data exploration. *International Journal Geographical Information Science* 13(4): 355-374

Andrienko, N., & Andrienko, G. (2002). Intelligent support for geographic data analysis and decision making in the Web. *Journal of Geographic Information and Decision Analysis* 5(2: 115-28.

Buja, A., Cook, D., & Swayne, D.F. (1996). , Interactive high-dimensional data visualization. *Journal of Computational and Graphical Statistics*, 5, 78-99

Cleveland, W.S., & McGill, M.E. (1988). Dynamic graphics for statistics. Belmont: Wadsworth and Brooks

MacEachren, A. M. (1994). Visualization in modern cartography: Setting the agenda. In: *Visualisation in modern cartography*. New York, New York: Elsevier Science Inc. pp.1-12.

MacEachren, A.M., & Kraak, M.-J. (1997). Exploratory cartographic visualization: Advancing the agenda. *Computers and Geosciences* 23(4): 335-44.

Macintosh (1995). Macintosh human interface guidelines. Reading, Massachusetts: Addison-Wesley Publishing Company

Nielsen, J. (1993). Usability engineering. Boston, Massachusetts: Academic Press, Inc.

Slocum, T., Blok, C., Jiang, B., Koussoulakou, A., Montello, D., Fuhrmann, S., & Hedley, N. (2001). Cognitive and usability issues in geovisualization. *Cartography and Geographic Information Science* 28(1): 61-75.

Tukey, J.W. (1977). Exploratory data analysis. Reading, Massachusetts: Addison-Wesley Publishing Company.

Supporting the Evocation Process in Creative Design

Nathalie Bonnardel

Research center in Psychology of Cognition, Language and Emotion (PsyCLE)
University of Provence
29, av. R. Schuman
13621 Aix-en-Provence France
nathb@up.univ-mrs.fr

Evelyne Marmèche

Laboratory of Cognitive Psychology (LPC)
University of Provence
29, av. R. Schuman
13621 Aix-en-Provence France
evelyne@up.univ-mrs.fr

Abstract

In order to contribute to a better understanding of creativity in non routine design activities, we conducted an experimental study which focuses on a main cognitive mechanism involved in creative design: the re-use of aspects derived from previous sources of inspiration. Especially, our objective was to determine to which extent potential sources (intra- and inter-domain sources) are considered by designers as useful for solving a specific design problem. Since the relevancy of sources of inspiration may be differently appreciated according to the level of expertise in design, the experiment was performed with two groups of participants: students specialized in industrial design (called "experienced" designers) and students who had no training in design (called "naïve" designers). Results show differences in the number and nature of aspects derived by experienced and naïve designers as well as in the usefulness assessment of the different types of suggested sources of inspiration. On this basis, we propose ways for supporting creativity and, especially, for facilitating the progression from novice to experienced designers.

1 Creativity and design problem-solving

With regard to a cognitive approach, creativity is considered as a complex activity, belonging to problem-solving activities. However, a main characteristic of creative tasks, such as design tasks, is that the initial state is ill defined: designers have, initially, only an incomplete and imprecise mental representation of the design to be performed. The designers' mental representation evolves as the problem-solving progresses. Therefore, each designer constructs his or her own representation of the design problem and deals with a problem that has become specific to him or her. Thus, different designers, supposedly solving the same design problem, reach different solutions, due to the fact that they adopt various points of view (Bonnardel, 2002). In accordance with previous descriptions of creativity (Koestler, 1975, p. 121) creativity is explained by "the sudden interlocking of two previously unrelated skills, or matrices of thought". Creativity has also been described as consisting in the activation and re-combination in a new way of previous knowledge elements in order to generate new properties based on the previous ones (Ward, Smith, & Vaid, 1997). Especially, according to Ward's structured imagination framework, people engaged in generative cognitive activities have to extend the boundaries of a conceptual domain by mentally crafting novel instances of the concept. However, experimental results showed that people have a strong tendency to rely on exemplars (Jansson & Smith, 1991), even when they have been instructed to be the most creative as possible. But yet, the more the participants move away from the first evoked sources, the more they are creative and original (Ward, Patterson, Sifonis, Dodds, & Saunders, 2002). It therefore appears that the most successful uses of analogies may depend on the capacity to move beyond initially retrieved information to better or more refined exemplars, interpretations, and source analogs. Research on creative cognition holds the promise of identifying the most effective means of facilitating such abilities.

2 Experimental Study

2.1 Objective and Hypotheses

The objective of the study is to characterize how designers judge the relevance and usefulness of various sources of inspiration, which belong or not to the same category as the object to be designed (i.e., respectively intra- and inter-domain sources). Our hypothesis is that, depending on the designers' level of expertise, the usefulness of intra- and inter-domain sources will be judged differently. "Naïve" designers in a domain would mainly focus their interest on intra-domain sources, making use of a direct analogical reasoning based on surface similarities between the sources and the target. On the contrary, "experienced" designers would be more inclined to use remote analogies with inter-domain sources, and to adopt more diverge points of view upon each source taken into account.

2.2 Participants

The participants were differentiated according to their level of expertise in industrial design. "Experienced" students have been attending a professional training in industrial design for one or two years after the "baccalaureate" (10 participants). “Naïve” designers were students in psychology who had no experience at all in design (10 participants).

2.3 Procedure and Experimental Tasks

Firstly, the participants were provided with the description of a design problem: to design a new seat for a "cyber-café". The complete description of the problem is the following: *"The object to be designed was intended to be used in a Parisian "cyber-café". It should be a particular seat with a contemporary design in order to be attractive for young customers. Such seats should allow the user to have a good sitting position, holding the back upright. Towards this end, the users should put their knees on a support intended to this function. In addition, these seats should allow the users to relax, by offering them the possibility to rock"*. This description is deliberately vague in order to allow participants to create whatever they wish – provided that it is in accordance with the constraints expressed in the problem description. Such a design problem was new for all the participants whatever their level of expertise in design.

Then, the following instruction was read by the experimenter and given as a written text to participants: *"Professional designers already had to solve this design problem. In order to deal with it, they got inspired by various pre-existing objects. You are going to be provided with a list of objects they were inspired by. For each of these objects, we ask you to explain in what it can be useful for solving the design problem which has been described. Your answer can be as longer as you wish, and you can come back on what you wrote as frequently as you want"*.

The participants were provided with a set of 18 source-objects, presented as a list of names of objects. This list was based on objects previously evoked by professional designers while solving the problem at hand (Bonnardel & Marmèche, 2001).
The potential sources of inspiration consisted in:

- 6 "intra-domain" sources, i.e. objects which belong to the same semantic field and category as the object to design: e.g., an *"office-chair"* or a *"camping chair";*
- 6 "close inter-domain" sources, which belong to another semantic field than the object to design, but which all share a seating component: e.g. a *"swing"*, a *"sledge"*, or a *"canoe-kayak";*
- 6 "far inter-domain" sources, which also belong to another semantic field but do not include a seating aspect: e.g., a *"climbing position"*, or a *"wave"*.

Each participant had to perform two tasks.

In the first experimental phase, participants had to judge, for each source, which aspects could be re-used for designing the cyber-café seat. Towards this end, each participant was provided with a notebook comprising 18 pages (presented in a random order). On each page the following instruction was written, instantiated with the specific name of the object: *"Which aspects of [an office chair] could be useful for designing the cyber-café seat?"*. Participants had 20 minutes to perform this task.

In the second experimental phase, the whole set of source-objects was presented as a list. Participants had "*to assess the usefulness of each source-object of inspiration with regard to an utility scale in five points*" (from 1: not useful at all, to 5: very useful). Participants had 5 minutes to perform this second task.

2.4 Results

2.4.1 Number of evoked aspects

In the first experimental phase, when subjects had to comment on the different sources, the results show that, for each source-object, experienced designers evoked significantly more aspects than naïve designers, whatever the kind of the suggested sources of inspiration: in mean, respectively, 2.29 *vs* 1.38 aspects ($p < .002$). In addition, naïve designers evoked significantly more aspects for intra-domain sources than for inter-domain sources, whatever the specific nature of inter-domain sources ("close" or "far"). On the contrary, no significant effect of the nature of suggested sources was observed for experienced designers.

2.4.2 Scores of usefulness

Concerning the second experimental phase, it appears that experienced designers assigned a significantly higher score of usefulness to the different suggested sources than "naïve" designers, whatever the nature of the sources. In addition, naïve designers assigned a higher score to intra-domain sources than to inter-domain sources. Moreover, for naïve designers, "close" inter-domain sources were significantly better assessed than "far" inter-domain sources. On the contrary, no significant effect was observed for experienced designers.

2.4.3 Nature of evoked aspects

A qualitative analysis of the evoked aspects during the first experimental phase has been conducted, with regard to various categories of aspects. The aspects evoked by participants were characterized according to 6 categories:[1]

- *Functional* aspects, referring to the use of the object; for instance, a participant pointed out that a *"photomaton seat allows the user to adjust the height of the seat";*
- *Structural* aspects, including a description of parts of the object; for instance, about a *"bicycle"*, a participant proposes *"to keep the pedals in order to create a foot-rest"*;
- *Affective* aspects, reflecting sensations or feelings produced by the object; for instance, about a *"nest"*, a participant considered it as *"warm"* and another one as *"bringing a feeling of protection";*
- *Aesthetic* aspects; for instance, about an *"office-chair"*, a participant evoked its *"modern look"*;
- *General* aspects, which are too much imprecise in order to be allocated to a specific category; for instance, aspects such as *"ergonomic"* or *"rational";*
- *Other* aspects, including answers that cannot be classified otherwise.

Firstly, results showed that experienced participants evoked more various aspects than naïve participants. However, the proportions of the different kinds of aspects are close to each other for the two groups. Functional, structural and affective aspects are the most frequently evoked. Aesthetic, general and other aspects are very rarely evoked, whatever the designers' level of expertise. Nevertheless, naïve participants evoked more functional and structural aspects for intra-domain source-objects than for inter-domain sources. This is not the case for experienced participants who evoked as frequently these different kinds of aspects, whatever the kind of source, intra-, close or far inter-domain sources. When affective aspects are evoked, it is above all in the case of "far" inter-domain sources, whatever the designers' level of expertise. Interestingly, it was observed that experienced participants, but not naïve participants, frequently integrated various aspects of a source-object at a same time. For instance, about the "rocking-chair", an expert evoked a structural aspect linked to a functional one: *"to keep the roundness of the stand, so that the seat can rock".*

3 Towards supporting the evocation process in design

As it was just presented, the results we obtained showed important differences in the evocation process of the participants at different levels of expertise in design:

- naïve designers appeared to prioritize intra-domain sources of inspiration and to mainly focus on functional similarities with the target,
- whereas experienced designers were able to take advantage of a lot of various sources, semantically near or far from the target, and to adopt different points of views on the suggested sources.

Thus, contrary to naïve designers, experienced designers seem to have acquired a particular skill consisting in a strong fluency in the use of analogical reasoning. This is in line with previous findings showing that:

[1] Three judges independently categorized the aspects evoked by the participants. A high degree of agreement was obtained (.95). In case of hesitation, a short discussion allowed us to reach a complete agreement.

- spontaneously, designers engaged in the “cyber-café” design problem-solving, mainly evoke intra-domain sources, whatever their level of expertise (Bonnardel & Marmèche, 2001);
- but, it does exist a relationship between the capacity to move away from initially retrieved information and the level of creativity of the productions (Ward & al., 2002).

Therefore, a concrete objective is to support naïve and novice designers in developing more powerful evocation processes.

Based on the qualitative analysis we performed, two operational objectives can be pursued:

1. to support naïve and novice designers in really taking advantage of inter-domain sources;
2. to support them in adopting various points of views on sources, whatever they are, intra- or inter-domain.

Such objectives could be reached, at least partially, through pedagogical actions during design education (Casakin & Goldschmidt, 1999) or through specific creativity training groups (Dewulf & Baillie, 1999). In addition, the use of computational systems could be particularly useful for providing designers with large database consisting in a lot of pictures or words potentially useful for creative design tasks (Nakakoji, Yamamoto & Ohira, 2000). Such functionalities could be complemented by specific questions dealing with aspects that novices do not spontaneously consider, such as "affective", or "aesthetic" ones. Moreover, at a more general level, it could be interesting to lead the novices to become aware of the usefulness of reasoning by analogy from diverse sources, in order to become more and more creative and original.

Acknowledgments

We wish to thank the participants in this study and, particularly, René Ragueb, their professor in design. Many thanks also to Emmanuelle Aune for her precious contribution.

4 References

Bonnardel, N. (2002). Towards understanding and supporting creativity in design: Analogies in a constrained cognitive environment. *International Journal of Knowledge-Based Systems, 13*, 505-513.

Bonnardel, N., & Marmèche, E. (2001). Creative design activities: the evocation process and its evolution with regard to expertise. In J. Gero, & M.L. Maher (Eds.), *Computational and Cognitive Models of Creative Design V* (pp. 189-204). University of Sydney, Australia: Key Centre of Design Computing and Cognition.

Casakin, H. & Goldschmidt, G. (1999). Expertise and the use of visual analogy: implications for design education. *Design Studies, 20*, 153-175.

Dewulf, S. & Baillie, C. (1999). *CASE – Creativity in Art, Science and Engineering: How to foster creativity*. London: Imperial College of Science, Technology and Medicine.

Jansson, D.G., & Smith, S.M. (1991). Design fixation. *Design Studies, 12*, 3-11.

Koestler, A. (1975). *The Act of Creation*, London.

Nakakoji, K., Yamamoto, Y. & Ohira, M. (2000). A framework that supports collective creativity in design using visual images. *International Journal of Knowledge-Based Systems, 13*.

Ward, T.B., Patterson, M.J., Sifonis, C.M., Dodds, R.A., & Saunders, K.N. (2002). The role of graded category structure in imaginative thought. *Memory & Cognition, 30*, 199-216.

Ward, T.B. Smith, S.M., & Vaid, J. (1997). Conceptual structures and processes in creative thought. In T.B. Ward, S.M. Smith, & J. Vaid (Eds.), *Creative Thought: An investigation of conceptual structures and processes* (pp. 1-27). Washington, DC: American Psychological Association.

The Use of m-Commerce Services and Technologies as an Instrument of Personnel Marketing – Conceptual Considerations and Empirical Studies

Iris Bruns, Olaf Oehme, Holger Luczak

Institute of Industrial Engineering and Ergonomics, RWTH Aachen University
Bergdriesch 27, D - 52056 Aachen, Germany
{i.bruns; o.oehme; h.luczak}@iaw.rwth-aachen.de}

Abstract

Technological progress as well as modern approaches to marketing, are increasing the need to rethink the customary approaches to personnel marketing. When developing new personnel marketing procedures, it is necessary to carefully examine their potential value and appeal to the applicant. In order to compare different personnel marketing strategies, the seven well-known requirements they are to fulfill were first rated according to importance. This ranking was then used as a basis for a value analysis comparing four different application scenarios. The results indicate that personnel marketing has the highest chance of success when combined with customer relationship marketing as well as mobile devices.

1 Introduction

Nowadays, literature provides a broad consensus of the definition of personnel marketing (Scholz, 1994; Simon et al., 1995): "personnel marketing" means putting the marketing philosophy in the personnel department consistently into practice. Employees as well as potential employees are regarded as customers. Concerning the current as well as the future job situation, an effective personnel-marketing strategy as an instrument for external personnel recruitment has become indispensable for business companies of any size and sector. Obtaining qualified personnel is becoming more and more competitive. Over the past few years, the Internet has become a modern and well- established personnel marketing instrument. Especially for the target group of skilled personnel, high potentials and young professionals, e-recruiting has become a commonly used element of any personnel marketing mix. Whereas in 1999, there where 60,000 job offers in commercial German internet job centers, there are more than 200,000 at the moment (Crosswater, 2003).

However, the society of the future will be characterized by another "W" before "World Wide Web". It stands for "Wireless" and demonstrates the beginning of a completely new era of mobile communication, information and interaction. By the end of the year 2002, people possessed far more mobile communication devices than local internet accesses (Geer & Gross, 2001). No other technical device has gained as much attention as the mobile telephone. Moreover, the spectrum of handheld- and organizer technologies has developed with amazing speed. Against this background, visions of ubiquitous accesses for any kind of information are growing. Mobile devices combined with modern communication technology are expected to become very influential client-interaction-technology. Consequently this communication channel will play a leading role in an effective personnel marketing for the following reasons:

- *Speed, availability, flexibility and internationality are only some of the crucial criteria of assessment. From the company's as well as from an applicant's point of view, these aspects play an important role for the application and recruitment process in both the internal and external job market. In this respect, the mobile internet will exceed all previous technologies and possibilities.*

- *As a consequence of the previous developments and achievements of personnel marketing on the local internet, high acceptance of the mobile internet by the target groups of skilled personnel, young professionals and the so-called "high potentials" is assumed. For business companies of any size and any sector, these target groups will play an essential role, especially when looking at the current situation of the job market and the so-called "war of talents" (Giesen, 2001; Jäger, 2000).*
- *No other portable terminal has become widespread as quickly as the cell phone. It has the potential to open new dimensions concerning availability, accessibility and equal distribution of all different kinds of information.*

Besides the technological development and change, there are other factors that need to be taken into consideration within the context of mobile services. New concepts and services of the mobile market have to satisfy the expectations and needs of costumers and applicants. In this respect, characteristics such as personalizing and individualizing play an important role. An enormous amount of scientific and economic publications address the classical consumer goods marketing deals with Electronic Customer Relationship Marketing (eCRM) (Sexauer & Hagen, 2002). In contrast to former theories of consumer goods marketing, the eCRM represents a revolutionary approach. These new models and theories are based on the uniqueness of the expected mobile market compared to the ones already known. (Reischl & Sundt, 2000). "eCRM" refers to a form of relationship marketing which has developed from one-to-many marketing to a one-to-one marketing for the stationary as well as for the mobile internet. The starting point for marketing agreement and the development of services is the company's relation to its clients.

2 Scenario Definition

Within the research field of personnel marketing, there are no approaches, models or concepts which concentrate on the development from e-commerce to m-commerce, nor are there concepts for transforming the classical personnel marketing to a personalized Candidate Relationship Personnel Marketing (CRPM) as yet (Rust, 2002). However, prognoses as well as individual cases indicate that in the future one will be confronted with four varieties of electronic recruitment (see Figure 1).

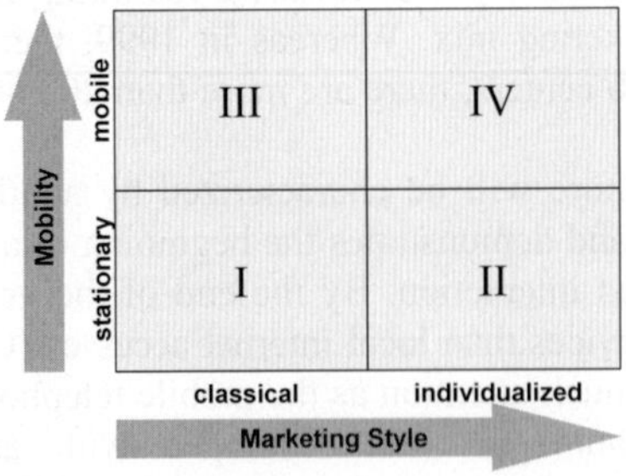

Figure 1: Mobility-Marketing Style Portfolio for e-recruiting

I) *Stationary Internet + Classical Marketing*
Alternative I) contains any current e-recruiting-measures restricted to the stationary internet.

II) *Mobile Internet + Classical Marketing*
Here, offers for the stationary Internet are supplemented with individualized and applicant-specific aspects.

III) *Stationary Internet + Personalized Marketing*
The current recruiting process is supported by the application of mobile devices (mobile telephones, PDA etc.).

IV) *Mobile Internet + Personalized Marketing*
The individual recruiting process is supported by the application of mobile devices and the transmission of individualized offers.

3 Requirements Placed on Personnel Marketing Instruments

The application process from the point of view of the applicant is a subject often described in literature (Simon et al.1995; Martin & Nowak 1997). Though the application process within an enterprise differs from company to company and from sector to sector, the requirements of the target group - which in our case are high potentials – concerning the applicant's demands on the application process remain the same. Those demands can be described by seven factors, which can be used to appreciate the quality of a personnel marketing tools (Figure 2). The extent to which these criteria are fulfilled by a given personnel marketing tool, has a reflection on the image of the company and thus the quality of the application process.

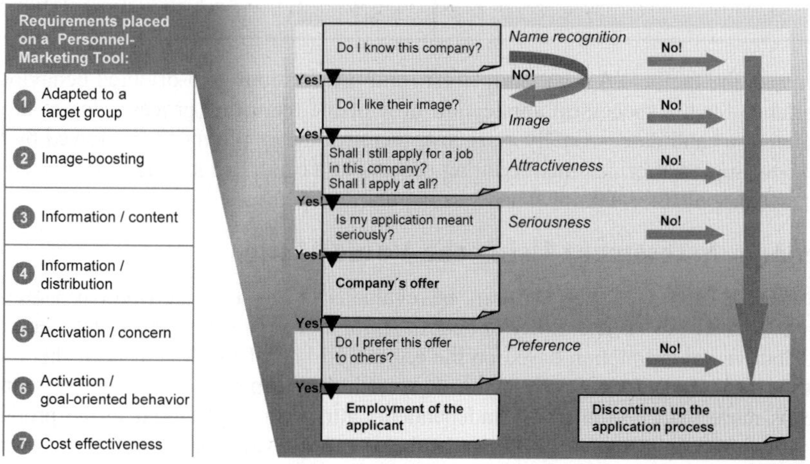

Figure 2: Recruiting process and requirements placed on personnel marketing instruments.

However, the literature cited above only emphasizes that these seven factors are of a certain importance. As yet, no empirical studies have closely examined whether the applicants do, in fact, identify and regard them as crucial factors. However, being a question that should not be ignored (e.g. to estimate the fulfillment of one factor and whether an improvement of this factor is actually necessary and worthwhile) questionnaires were sent out to high potentials. The aim was to identify applicants' experiences with and expectations of online-application offers and to determine general online-application behavior of the main target group, high potentials, skilled personnel and young professionals.

About 237 participants having the characteristics of the target group were surveyed in an investigation conducted by the Institute of Industrial Engineering and Ergonomics of RWTH Aachen University (Bruns et al., 2003). The results of the survey combined with expert interviews of twelve German Human Resource Managers, enabled us to determine the importance of each of the seven factors in a paired comparison (Table 1). The sum of the individual weights leads to a range of its importance (Pfeifer, 2001; Eversheim, 1989).

Table 1: Ranking Criteria and Weighting Factors w_i

Comparison (Evaluation from the point of view of the applicant) 2: more important 1: same important 0: lesser important is more important than	Adapted to a target group	Image-boosting	Information / content	Information / distribution	Activation / interest	Activation / goal-oriented behavior	Cost effectiveness
Adapted to a target group		1	0	0	2	2	0
Image-boosting	1		0	0	1	1	0
Information / content	2	2		1	2	2	1
Information / distribution	2	2	1		2	2	1
Activation / interest	0	1	0	0		1	0
Activation / goal-oriented behavior	0	1	0	0	1		0
Cost effectiveness	2	2	1	1	2	2	
Sum	6	9	2	3	10	10	2
Percent (w_i)	0,14	0,21	0,05	0,07	0,24	0,24	0,05

As one can see, the factors "Activation/Interest" and "Activation/ goal-oriented behaviour" are of high importance for the successful application flow of the recruiting process, due to the fact that the activation of a candidate is finally a decisive factor of the employment. Followed by the factor of "Image-boosting" that brings the recruiting process to a candidate´s mind at first. Hence, it also represents a chance also for unknown and less popular companies.

4 Scenario Assessment Using the Value Model

The scenarios defined in Section 2 can be evaluated by a multi-attributive value function (Eisenführ & Weber, 1999). The utility or benefit N_j for a given alternative j is calculated by using a multi-attributive value function based on the regression line of fulfillment v_{ji} of the alternative j for the i-th criterion and then weighting w_i these criteria. Assuming an attributive value model, which means more than two mutually independent attributes on preference, the product of the weighting and the straight line of fulfillment lead to the single value of benefit of each variable.

$$N_j = \sum_{i=1}^{n} v_{ji} w_i$$

Thus, the four scenarios were evaluated according to how well the personnel marketing tool were judged with regard to these seven factors. The benefit or utility of each tool was calculated using a linear value-model (Table 2).

Table 2: Determination of the value of benefit N_j for the four scenarios by the determined criteria

	I	II	III	IV	w_i
Adapted to a target group	2	4	2	4	0,14
Image - Boosting	2	4	2	4	0,21
Information/ Content	3	4	1	4	0,05
Information/Distribution	2	2	4	4	0,07
Activation/ Interest	2	4	2	4	0,24
Activation/ Goal-oriented behaviour	1	3	2	4	0,24
Cost effectiveness	4	4	1	1	0,05
N_j	0,48	0,9	0,51	0,96	1

Likert-Original	
strongly agree	4
agree	3
uncertain	2
disagree	1
strongly disagree	0

Simply mobilizing the application offers hardly has an effect – as demonstrated by the almost equal values of benefit (see I) and III)). In contrast, the benefit analysis produced significant values for the versions III) and IV) described above. The personalization aspect of personnel marketing analogous to the customer relationship management of the sales marketing as well as the support offered by the use of mobile communication devices will have an important influence on future personnel work.

5 Conclusions and Outlook

The efficiency analysis demonstrated that merely mobilizing the information has little effect on personnel recruitment. It is much more important to regard the applicants as clients, to attend to them personally and to be available for assistance with any questions. However, our analysis could only show which of the above-mentioned scenarios could benefit from new technology. This makes it necessary to develop special models for these technologies which meet the applicants' requirements. Therefore, nowadays one can certainly explore the potentials of state of the art technology, yet when developing these models one should not neglect the fact that this technology is rapidly changing. Developments in research fields such as Retinal Laser Projection and Ubiquitous Computing can be named in this context. Progress in these research fields can increase the standards in mobile work stations to those now common in office equipment.

6 References

Bruns, I.; Oehme, O.; Luczak, H. (2003): Schwachstellen- und Anforderungsanalyse heutiger E-Recruiting-Aktivitäten und Prognosen für die Zukunft. In: *Proceedings of the GfA/ISOES* 2003, Munich, May 07.-09. in press.

Crosswater Job Guides (eds.) (2003): Rangliste der Online-Jobbörsen in Deutschland. Retrieved February 10, 2003, from: http://www.crosswater-systems.com/ej5005ap.htm.

Eisenführ, F.; Weber, M. (1999): *Rationales Entscheiden.* 3. neubearbeitete und erweiterte Auflage. Berlin : Springer, 1999.

Eversheim, W. (1989): Organisation in der Produktionstechnik. Bd. 4 : *Fertigung und Montage.* 2. neubearb. u. erweiterte Aufl. Düsseldorf : VDI-Verlag, 1989.

Geer, R.; Gross, R. (2001): Geschäftsmodelle für das mobile Internet. Landsberg : Lech, 2001.

Martin, C.; Nowak, A. (1997): Stellenangebote im Internet oder Printmedien. In: *Personal,* (1997) 3.

Pfeifer, T. (2001): Abwicklungsqualität von Entwicklungsprozessen (Teilprojekt 5). In: *SFB 361: Modelle und Methoden zur integrierten Produkt- und Prozessgestaltung.* Aachen: RWTH Aachen, 2001. - Arbeits- und Ergebnisbericht 1999 / 2000 / 2001, S. 295-330.

Rust, U. (2002) : E-Cruiting in nationaler und internationaler Perspektive. In: *Personal,* (2002)5.

Scholz, C. (1999): Personalmarketing für High Potentials. In: Thiele, A. ; Eggers, B. (Hrsg.): *Innovatives Personalmarketing für High Potentials.* Göttingen : Verlag für angewandte Psychologie, 1999.

Sexauer, H.J. (2002): Entwicklungslinien des Costumer Relationship Management. In: *WiST* (2002) 4.

Simon, H.; Wiltinger, K. ; Sebastian, K.-H. ; Tacke, G. (1995): Effektives Personalmarketing - Strategien, Instrumente, Fallstudien. Wiesbaden : Gabler Verlag, 1995.

Developing a Framework for HCI Influences on Creativity

Winslow Burleson
MIT Media Lab
20 Ames St., Cambridge, MA, 02139
win@media.mit.edu

Abstract

This paper describes ongoing work to develop a greater understanding of the design of HCI with respect to its potential to impact users' creativity. Adaptive interfaces are being developed that are contextually aware of their users in terms of the elements of the Componential Model of Creativity: motivation; domain relevant knowledge; creative skills within the individual; and the individual's external environment. Findings of empirical studies and consensual assessment techniques are being used as the basis for developing experiments and applications with HCI interventions which strive to increase the creativity of users.

1 Introduction

Although the definition of creativity is frequently debated there is some consensus that it deals with a "process" which results in a "novel" and "useful" "product", in the most general sense of these words. Teresa Amabile developed the Componential Model of Creativity stating that individuals posses domain relevant knowledge, creative skill, and motivation which interacts with a fourth element, the external environment (work or social), to form a confluence which contributes to an individual's ability to generate creative products (Amabile, 1983). Traditional HCI methodologies have largely ignored analysis and consideration of motivational elements in the evaluation of user interaction (Burleson, 2003). Yet HCI is intrinsically a meeting point of a person and technology, which often entails interactions with domain and task relevant knowledge, the opportunity for creative skill acquisition and utilization, intrinsic and extrinsic motivation, performance, and achievement. It is also a place to connect to other people; it is a work environment and presents social opportunities. This notion of a workspace is particularly emphasized as relevant to creativity by Ben Schneiderman in his arguments for systems which facilitate users' ability to "Relate, Create, and Donate" their ongoing work and creative products (Schneiderman, 1997). According to the Componential Model of Creativity each of the many aspects of interaction has an impact on creativity. In short HCI can be conceived as a lens that overlays the entire componential model. This lens may be focused or diffuse; an interface element may have a specific intended impact or a broader and sometimes unintended impact on one or more components of creativity.

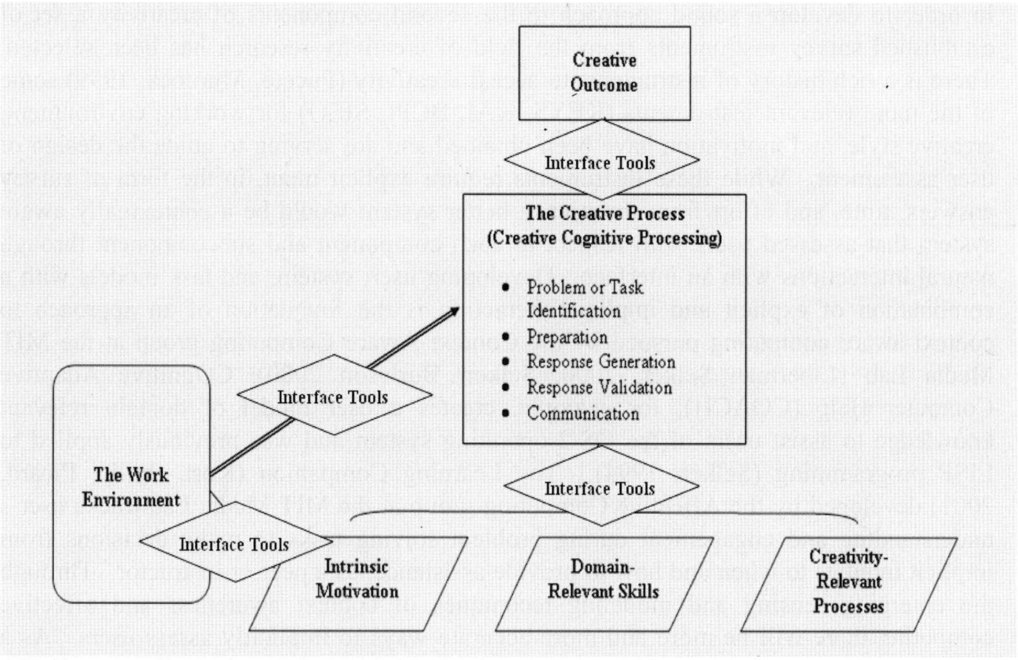

Figure 1: Componential Model of Creativity with Interface Tools as Lenses

2 Methods and Approach

There have been several elegant experiments conducted to substantiate the Componential Model of Creativity (Amabile, 1983; Amabile et al., 1990; Amabile, Mueller, 2002; Ruscio et al., 1998). A set of three diverse open ended tasks (Haiku writing, collage making, and tower building) has been used in a series of experiments with existing experimental procedures along with the development of an intrinsic motivation survey instrument (SIEQ) to asses the effects of evaluation, contract for reward, and motivation. To take an example in the external environment component, it has been shown that the perception of time pressure has a detrimental impact on the products of the creative process but paradoxically increases an individuals self-rated evaluation of having been creative (Amabile, Mueller, 2002). Drawing on the lessons of these experiments, performed in the psychological tradition, HCI user studies are being developed to explore the means and extent to which it is possible to apply and focus the HCI lens toward the various components of creativity. For example, if intrinsic motivation is effected in the interface (through affirmation, games, fun, positive affect inducement, time pressure, etc….) (Isen, 2002) or if extrinsic rewards are introduced (actualizing rewards, contract for reward, or evaluation) these interface elements and interactions should effect the creative products of those engagements in predictable ways and confirm the benefits of applying the findings of psychological creativity experiments to HCI.

In order to develop a sound approach to the several components of creativity a set of established survey instruments from the field of creativity research has been selected. There is a rich history of instruments to access creativity (Puccio, Murdock, 1999) some of the more relevant instruments (KEYS, KAI, BCPI, SIEQ) for working environment, creative style, and motivation have been obtained and are serving to guide the design of user assessment. While these instruments require explicit input, in the form of survey answers, time, and effort from the user, a better system would be a contextually aware system that assessed users with respect to each component and subcomponent through natural interactions with an interface. Developing user, system, and task models with a combination of explicit and implicit interaction is the foundation of an approach to context-aware computing pursued by the Context Aware Computing group at the MIT Media Lab (Liberman Selker, 2000; Selker, Burleson 2000). COgnitive Adaptive Computer Help (COACH), for example, creates a user model of domain relevant knowledge to assist users of the OS/2 operating system and was previously applied to LISP programming (Selker, 1994). The Learning Companion (Kort, Reilly, Picard, 2001) developed by the Affective Computing group at the MIT Media Lab tracks user's understanding and engagement during problem solving tasks to make decisions from implicit input as to when and how to provide assistance, as a peer or instructor. Through the emerging sensing and modeling techniques of context awareness and affective computing there will be more and more accurate ways to implicitly assess users. As a preliminary step existing explicit instruments, established by creativity researchers, are being correlated with users' interactions and behaviors (affect, distraction, frustration, persistence or change of task, apparent domain knowledge, etc…) to form a basis for the development of implicit assessment of users. Once an understanding of a user's state has been developed the system can select an appropriate intervention. With knowledge of a user's individual strengths and weaknesses relative to the componential model the environment can tailor its interface and interactions to present tools and opportunities to augment the user's skills, and ultimately to make the resulting products of users' efforts more creative.

3 Consensual Assessment

In order to asses the creativity of the products produced by individuals and groups under different experimental treatments the consensual assessment technique is used (Amabile, 1983). The ratings of users' products provided by this technique are the primary dependent variables for many of the studies. The technique involves judges ranking users' products, relative to each other, in the following manner. The products are review by at least four judges. Judges are recruited for domain relevant expertise. Judges are blind to the experimental conditions in which the products are created. Typical preparation of the judges has them engage in the same open ended product creation tasks as the users, under each treatment conditions. Judges are then presented sample products from pilot studies and asked to rate them on a 7-point Likert scale. The inter-judge reliability is assessed. If the reliability coefficient is acceptable the final rating for each participant's products is taken to be the mean of the judge's scores. If inter-judge reliability is unacceptable further training of the judges is conducted and/or one or more of the judges may be replaced until reliability is attained.

4 CODEX II

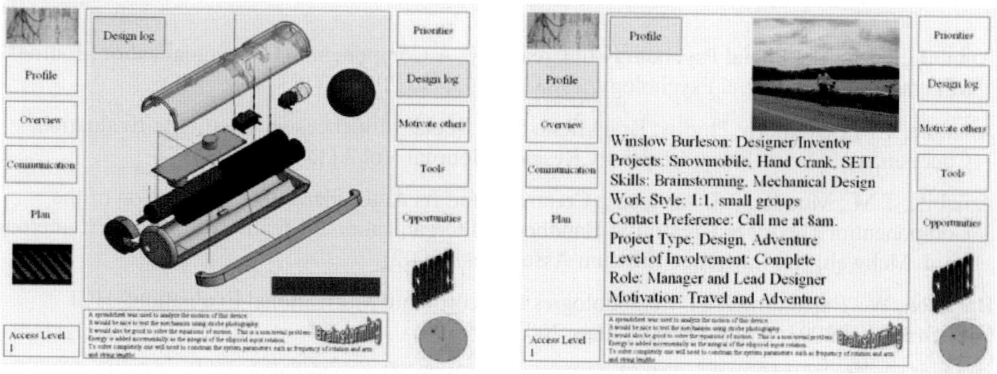

Figure 2: Screen Shots from CODEX II

Reeves and Nass report that interactions between people and computers occur in similar ways to those between people and people (Reeves, Nass, 1996). It has also been shown that negative effects of evaluation of creativity takes place when using a computer (Amabile, Mueller, 2002). On the other hand computers have used, "structured reproducible procedures that lead to ideas judged as creative (creative sparks)", (Goldenberg, 1999). In an effort to better understand the potential to impact user's creativity CODEX II (Collaborative Design and Exploration II) is being developed. It is an application which will attempt to monitor implicitly the creative process of individual users and group interactions and dynamically alter the treatment conditions; providing heuristic reminders, opportunities, agents, and manipulations; appropriately altering time pressure across algorithmic and heuristic tasks; increasing access to creativity skills; and providing positive extrinsic motivation through enabling and informational rewards. It will strive to provide actualizing rewards, which have been shown to benefit creativity. The goal is that simply by using the CODEX II environment each user will realize the facilitation of some of the components of creativity. The confluence of these components should result in the generation of more creative products.

5 Conclusions

Diverse tools are being developed to explore a framework for HCI influences to support creativity. These tools are directly related to Amabile's components of creativity. The impacts of the tools are proposed as "treatments" and are assessed through the "consensual assessment" of the products of open-ended tasks. The findings will lead to new ways to realize the benefits of incorporating analysis of motivation and creative skill, domain relevant knowledge, and social and work environments into the design and development of HCI. This effort will serve to compliment the emerging HCI approaches regarding creative and motivational elements and will increase understanding of HCI influences on users' creativity.

6 References

Amabile, T.M.; The Social Psychology of Creativity: A Componential Conceptualization; Journal of Personality and Social Psychology, 1983, Vol 45, No 2, 357-376.

Amabile, T.M.; Goldfarb, Phyllis; Brackfield, S.C.; Social Influences on Creativity: Evaluation, Coaction, and Surveillance; Creativity Research Journal Vol 3 (1) 6-12 (1990).

Amabile, T.M.; Mueller, J.S. Assessing Creativity and Its Antecedents: An Exploration of the Componential Theory of Creativity; Handbook of Organizational Creativity, edited by Cameron Ford. Mahwah, NJ: Lawrence Erlbaum Associates. (2002)

Burleson, W.; Expanding HCI Methodologies to Incorporate Motivational Evaluation; HCI International 2003; Crete, Greece, June, 2003.

Goldenberg, Jacob, Mazursky, David, Solomon, Sorin, Essays on Science and Society: Creative Sparks, Science 1999. 285: 1495-1496

Isen, A. Some Neuropsychologyical Factors in Positive Affect, Social Behavior, and Cognition; Handbook of Positive Psychology by C.R. Snyder & S. Lopez 2002.

Kort, B.; Reilly, R.; Picard, R.; 2001; An Affective Model of Interplay Between Emotions and Learning: Reengineering Educational Pedagogy—Building a *Learning Companion* ICALT-2001 (International Conference on Advanced Learning Technologies).

Leiberman, H.; Selker, T.; Out of Context: Systems That Adapt To, and Learn From, Context, IBM Systems Journal 39, Nos. 3&4, 617-632, 2000.

Puccio, G.; Murdock, M; Eds. Creativity assessment: Readings and resources; Creative Education Foundation Press; Buffalo, NY. 1999.

Reeves, B.; Nass, C. The Media Equation: how people treat computers, television, and new media like real people and places. Cambridge Univ. Press. 1996.

Ruscio, J.; Whitney, D.M.; Amabile, T.M.; Looking Inside the Fishbowl of Creativity: Verbal and Behavioral Predictors of Creative Performance; Creativity Research Journal Vol 11, No. 3, 243-263, 1998.

Selker, T. Coach: A Teaching Agent that Learns. Communications of the ACM, 37(7):pp.92—99. July 1994.

Selker, T., Burleson, W. "Context-Aware Design and Interaction in Computer Systems," IBM Systems Journal 39, Nos. 3&4. June 2000.

Shneiderman, B. Designing the user interface: strategies for effective human-computer interaction, Addison Wesley. (1997).

Direct Manipulation for E-commerce Sites: a New Approach

Giuseppe Capozzo, Mauro Mosconi, Marco Porta

Dipartimento di Informatica e Sistemistica – Università di Pavia
Via Ferrata, 1 – 27100 – Pavia - Italy
{capozzo, mauro, porta}@vision.unipv.it

Abstract

In this paper we propose a new paradigm of web interaction we devised to make e-commerce sites more "natural" and easy to use. In our model, as in real stores, users interact with the goods on the shelves through "direct manipulation"; the shopping cart is always visible, allowing customers continuously to monitor and access its contents at any time. We describe the architecture and use of our prototypal system (implemented with *Macromedia Flash®* and characterized by advanced interactivity and multimedia integration) while emphasizing the novelty and usability benefits of the approach.

1 Introduction

In the present scene of Internet technologies, concepts like *usability* and *ergonomics* of user interfaces become more and more important as greater possibilities are daily offered to developers and designers. Therefore, to make the principles of on-line interaction between users and web sites as immediate and intuitive as possible, a proper *user-centered* design approach becomes essential. This is more true when dealing with critical fields, such as e-commerce: present applications pertaining to this category introduce issues which cannot be neglected. Purchase operations are complex and far from the mental mechanisms implied by real world actions: key information (such as cart contents and purchase procedures) is not immediately available; moreover, users may experience unsatisfying feedback to their actions (long response time for server side applications and long loading times for Java ones) .

Our work was the formulation and implementation of a new approach for on-line interaction with the aim to improve the usability of e-commerce sites. Taking into account the principles of direct manipulation and choosing *drag&drop* as the main methodology of interaction, we have developed a prototype e-commerce application based on a new model of on-line purchase, more comforting and close to the real world.

After a first implementation in Javascript (Mosconi, Porta & Zanetti, 2002), which however was not fully compatible with Netscape Navigator, we chose the Macromedia Flash® platform: this way we have been able to take advantage of the great potential of a tool capable to provide "Rich Internet Applications", characterized by advanced interactivity and multimedia integration, without requiring long loading times.

2 Navigating the virtual store

The user moves inside the *virtual store* using a map built according to the principle for which the disposition of the corridors (main and secondary) reflects the organization of the store in super-categories subdivided in sub-categories (when present) of products. The virtual store, like a real one, can therefore be thought of as being subdivided into many sectors ("rooms"), each one represented by a spheric button (Figure 1).

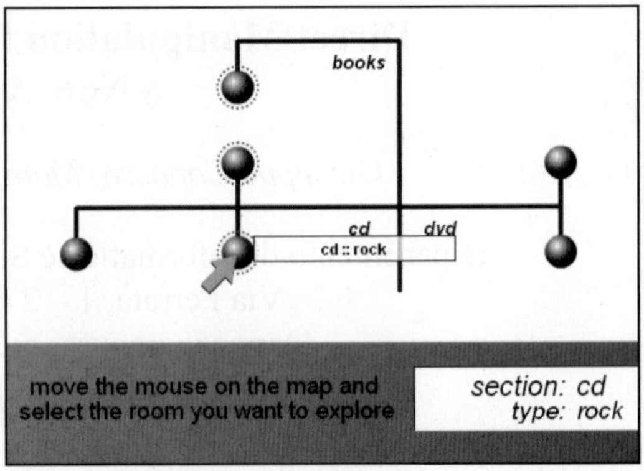

Figure 1: The interactive map

We adopted some particular strategies to support the process of exploration of the virtual store. Everytime the user presses a spheric button he/she is immediately "transported" to the correspondent sector, while on the map an arrow-shaped icon marks his/her current position (the classic "you are here"). A dotted circle will surround the same button with the aim for the user to keep track of his/her navigation: the user will avoid this way to come back to the same room if not desired. The dinamic text box at the bottom right corner of the map will show the super and sub-category of products of the present room.

3 Browsing the shelves

Each room of the virtual store has a circular perimeter coherently with the map visualization (see the spheric buttons on the map); the shelves of the room are arranged along its perimeter. Once entered in a room users can see one shelf at a time (Figure 2, left side). In order to browse all the shelves of the room, three different typologies of interaction are offered to the user: buttons, keyboard and rotational scrolling, each one providing the same effects and results. We chose to adopt this "redundancy" because of the criticalness of the action.

As can be seen in Figure 2, there are two separate control sections at the bottom of the shelf: two buttons with an arrow icon (left and right, labeled "more items") on the left end, an "explore" button on the other end. The arrow buttons implement horizontal scrolling functionalities (e.g. from the first to the second shelf and viceversa).
The same control can be achieved through the arrow keys (left and right) on the keyboard.

Our idea was to implement an angular typology of exploration by means of a proper control system: pressing the "explore" button the interface will show a semitransparent hidden device for the rotational scrolling (Figure 3b). As usual in a scrollbar, we have a draggable cursor and two buttons (CW and CCW rotation in the room) for the scrolling. Such an approach achieves two purposes: the first consists in a typology of exploration extremely immediate and immersive; the second, maybe more important, consists in a visual feedback for the user to his/her actions. In fact, looking at the rotational interface, the user gets an idea of the quantity of shelves examined and to be examined: the user will hardly be confused.

It is important to remember that there is a sharp synchronization between the three methodologies of exploration offered.

4 Examining the products

For design and *memorability* reasons, we chose to display five goods for each shelf, each one unambiguously represented by an image and described by a fixed set of general information such as name and price. Users can obtain more specific information (e.g. id code, extended textual description...) through a special "question mark" button on the left side of the image: once this button has been pressed, the system loads a draggable window displaying these details (Figure 2).

Inside this window a "demo" button allows the user to access demonstrative multimedia data, in the specific formats for the tipology of the good (*mp3, mpeg, pdf*...): the system executes the multimedia object using the default corresponding player. This way users benefit of further (and decisive for the purchase) information, along with a richer "sensory" experience.

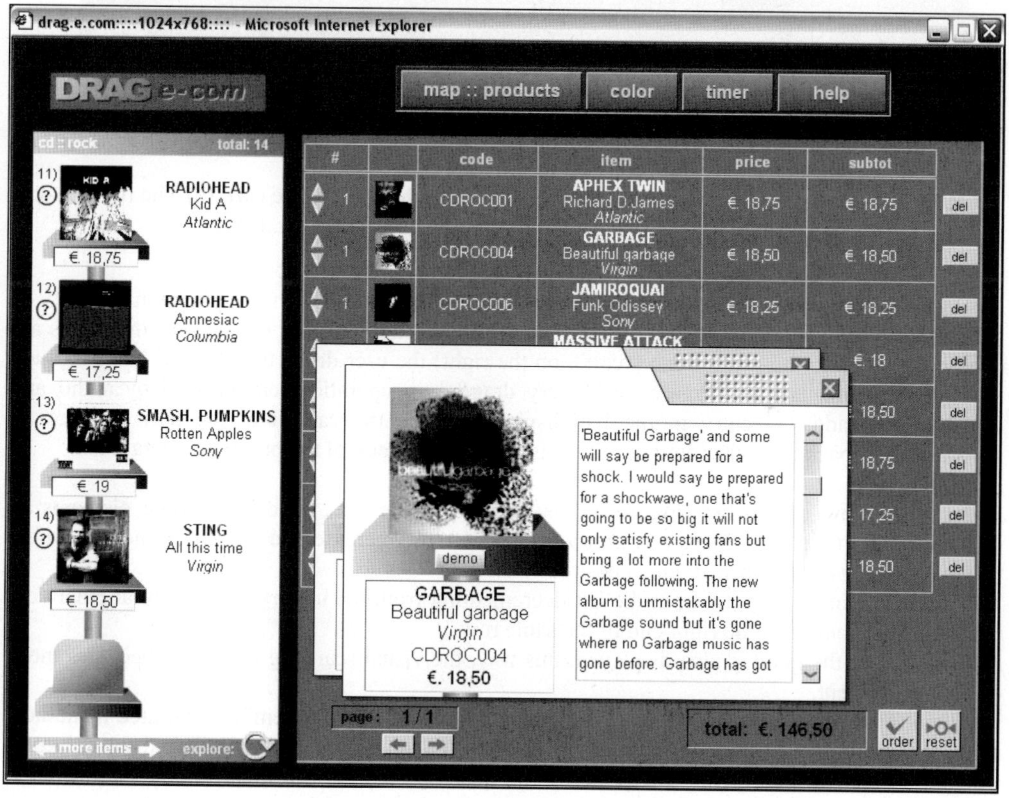

Figure 2: The main interface: shelves and cart structure and extended info windows

5 Using the shopping cart

The idea underlying our model is that, when shopping on the Web, the user should behave as much possible as the same way as he does in real situations. Hence, we imagine the potential buyer seeing the goods on the shelves, examining them and, if something meets his needs, putting it into the shopping cart...

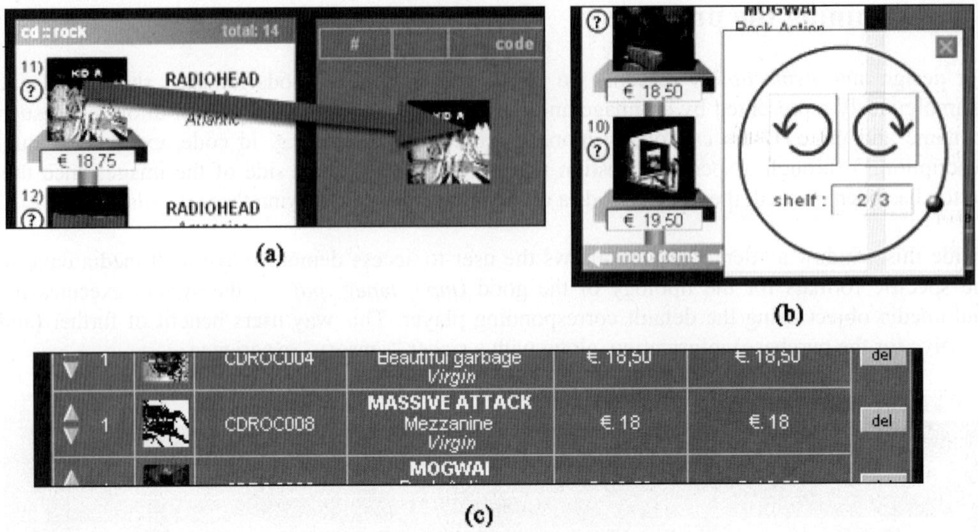

Figure 3: The drag and drop operation of an item from the shelf to the cart; (a) The rotational interface; (b) The structure of a row in the cart list. (c)

Basically, in our model (we called it "Shelves & Cart" model) the application window is subdivided into two distinct areas (Figure 2): in the first one, ("shelf", on the left) the goods are exposed, while into the second one ("cart", on the right) the user drops the selected items through a **drag&drop** mechanism (Figure 3a). Every drag&drop operation, emphasized by audio and visual effects, adds an entry to the list displayed inside the cart area: the user receives an immediate feedback of his/her actions and constantly keeps track of the purchase operations.

Every single row of the list (Figure 3c) is made of some data fields and some elements of interaction: at the left end there is a group of two buttons (increase and decrease) and a dynamic textbox showing the number of goods desired.
The user can eliminate a row from the list decrementing from the unitary quantity or alternatively using the "del" button at the right end of the same row.
A thumbnail of the good is also displayed: this way, the visual scanning of the list becomes more quick and efficient.
The typologies of data fields include an id code ("code") , a group ("item") constituted by author, name and producer of the good, the unitary price of the single object ("price") and the sub-total ("subtot") obtained multiplying the price by the quantity desired. The total (as the sum of each subtotal) is displayed in the dynamic text box at the right bottom of the cart area.

We have developed a special sorting algorithm for the list, called by every drag&drop operation, by which the order of the rows in the list reflects the order of the different typologies of goods in the store. Anyway the user can identify immediately the position in the list of the last good dropped in the cart thanks to a blinking animation of the thumbnail, lasting few seconds, which draws the user attention.
The list is organized in pages: in our prototype, every page contains ten rows. The user can flick through these pages using two arrow buttons at the left bottom of the cart area: a particular animation will stress the passage from a page to another. The user can eliminate all the rows of the

list by pressing the "reset" button: such a functionality could be very handy but dangerous at the same time (in case of an accidental pression of the button). Therefore we decided to add some *error tolerance* criteria, introducing an alert box in corrispondence with each pression of the reset button: the user, by means of the two buttons present in the box, can confirm or deny his/her intention to delete the entire list. The same strategy has been adopted for the pression of the "del" button of each row, avoiding this way any unwanted elimination.

Once the exploration of the store is completed, the user can send the cart list to the server pressing the "order" button, in order to complete the purchase procedure.

6 Usability Issues

Since the proposed approach is based on direct manipulation, it exploits a metaphor which refers explicitly to the actions we usually perform to purchase goods. This is the key point of the model, which involves many usability benefits.

Users interact directly with the system: this helps them to understand better the site functioning. They have total control over the system: being able to position objects in always-visible areas gives them a sense of "mastery". Moreover, the realistic dynamicity of dragging images between separate areas enforces the users consciousness about what they are doing. The visibility of the shopping cart is therefore a fundamental feature: as stated in (Tilson and oth., 1998) one of the most important heuristics to improve e-commerce sites usability is to provide a shopping cart that customers can see everything they have selected to purchase, and allow them to access the contents of the cart at any time.

We have carried out some *low cost* usability tests (Krug, 2000) on the prototype, which have suggested a first encouraging response to our work, highlighting interesting characteristics of *learnability* and *memorability* of the interface. The use of particular strategies of *error tolerance*, along with the presence of particularly attractive static and animated graphical elements, has allowed us to notice a certain feeling of *satisfaction* in the user during his/her permanence in the site.

Moreover we have observed that, standing a substantial difference in the behaviours of - generally speaking - "novice" and "expert" users with respect to the interface functionalities (almost completely discovered by the skilled users), even the novice ones have been able in all the cases to finalize some way their purchase procedures without any objective difficulty.

7 References

Krug, S. (2000). Don't Make Me Think: Common Sense Approach to Web Usability. New Riders Publishing.

Mac Gregor, C. (2001). Developing User Friendly Flash Content, from http://www.flazoom.com.

Mosconi, M., Porta, M., & Zanetti, F.(2002). Direct Manipulation of Pictorial Items within Web Sites: a Drag & Drop Approach to On-Line Interaction. Proceedings of the *Working Conference on Advanced Visual Interfaces* (AVI 2002), May 22-24, Trento, Italy.

Tilson, R., Dong, J., Martin, M. and Kieke, E. (1998). Factors and Principles Affecting the Usability of Four E-commerce Sites. *Proceedings of the 4th Conference on Human Factors & the Web*, June 5, Basking Ridge, NJ, USA.

A Prototype of a Graphical Guiding System

Yi-Ting Chen, Yi-Nan Lai, Zhong-You Xua, Wei-Lun Jian

Dept. of IEEM National Tsing Hua University
Kuan Fu Rd., Hsinchu, Taiwan, ROC
U880803@oz.nthu.edu.tw

Abstract

In a campus holding a broad area about 83 ㎢ with dozens of various buildings, visitors who come in the first time may spend a lot of time to find out the place they want to visit, and time is money. So we intend to design a computer program for visitors' convenience and let them get familiar with this campus more easily.

1 Problem Identification

Traditional method to find way to the particular place is to ask somebody or read the road sign and maps. However, for the first-come visitors, they have no idea about this big campus, and oral description for them is too rough to perceive. Besides, the map of this campus is too simple and lack of vividness. Since PDA and notebook computers are more and more popular nowadays, it should be workable and practical to design a computer map program to improve interaction between visitors and the map.

2 System Development

When designing this system, the first step is to take pictures of some spots in this campus as shown in Fig.1 and Fif.2. By now we have just finished partial map to be tested and evaluated.

Figure1：Campus main entrance

Figure 2: Administration building

Visual Basic was used to write the program of the guiding system. When writing this program, we also have to find out the shortest path between each point and design various directional indications as shown in Fig.3 and fig. 4. Since traveling is typically an egocentric task, ego-referenced maps are generally better choices for navigational guidance (Wickens & Hollands, 2000,p166). The directional indications as well as corresponding campus pictures will follow this principle to be shown from the visitor's view.

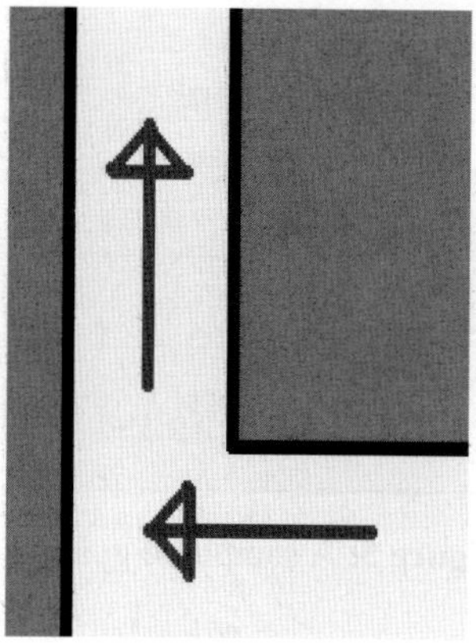

Figure 3: One sample of directional indication

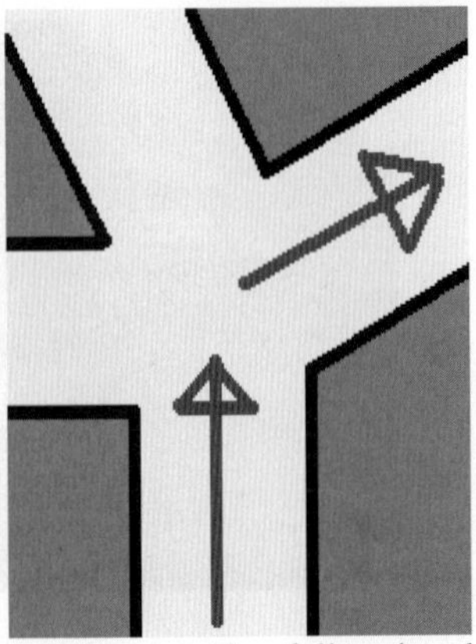

Figure 4: Another sample of directional indication

Thereinafter is the interface of our system. As shown in Fig. 5, the upper part of this graph is a map of the campus, and when users have chosen where he/she is and where he/she wants to reach, the smaller picture at southwest will show where the user is. Then, click the upper button of the three buttons, which means "next," and the picture at southwest will show the next point on the way to the destination. Additionally, the picture at southeast will be the direction between these two points The design of this computer map fits the principle of pictorial realism and principle of integration (Sanders & McCormick, 1993,p155)

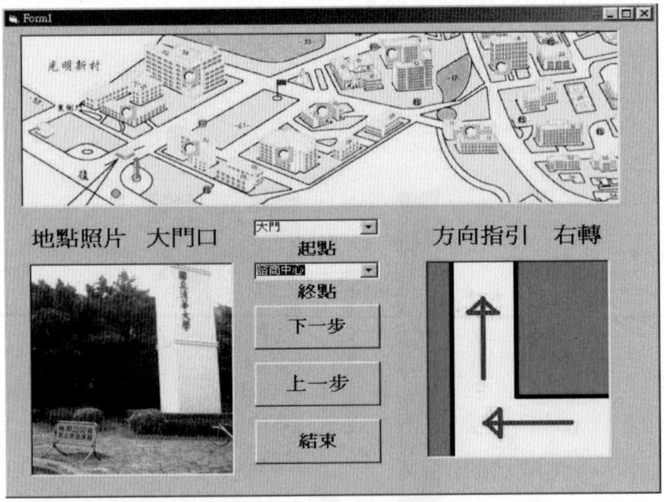

Figure 5: A sample of guiding system

3 Testing

First we chose the path from main entrance to the consulting center, and then five first-come visitors were selected at the main entrance to use this system to find out where the center is for experiment. Another five first-come visitors were asked to find out the same place without using this system. We followed them to record the time but gave neither hints nor time limit for them. Finally, the time data were analyzed.

Table1 and Table2 are the results of the experiment. Comparing the time spent by contrast group and experiment group with T-test at 0.01 significant level, we find out the fact that time spent by contrast group is statistically significant longer than that spent by experiment group.

Table 1: Contrast group

Number	Age	Gender	Time(min)
1	26	F	8.25
2	46	F	9.63
3	33	M	7.51
4	35	M	6.33
5	28	M	7.8

- Data analysis：
  - Mean of total time = 7.90（min）
  - Standard deviation = 1.19（min）

Table 2: Experiment group

Number	Age	Gender	Time(min)
1	17	F	6.16
2	20	M	5.23
3	21	F	4.98
4	39	M	4.6
5	34	M	5.21

- Data analysis：
  - Mean of total time = 5.23（min）
  - Standard deviation = 0.57（min）

So one can see that obviously first- come visitors who use this system can reach the goal more easily and quickly.

4 Discussion and Suggestions

After this experiment, we are very glad that this system is really useful and helpful to visitors. In addition, visitors who participated this experiment also gave us some precious suggestions as follows

- Adding "help" function in the program so that users can get familiar with the program more easily.
- The picture taken at some point shall be clearer so that users won't get confused.
- The program may add some voice after an action so that users can get feedback.
- The colors may be designed to be more harmonious.

5 Future Prospect

- This program may be applied not only in campuses but also in big hospitals, amusement parks and so on.
- Improving the interface of this program and let it be friendlier so that even graybeards and children can use it easily.
- Cooperate with the department of Computer Science to develop a specialized device or put this system on our school official website for visitors to download in their note- book computers or PDA before visiting.
- For the further design, we will consider the factors of cultural differences so that foreign visitors can also use this map easily.

References

Sanders, M.S. and .McCormick, E.J. (1993), Human Factors in Engineering and Design, 7th edition, McGraw –Hill, Inc. p155.

Wickens, C.D. and Hollands, J. G. (2000), Engineering Psychology and Human Performance, 3rd edition, Prentice –Hall Inc., New Jersey, p166.

Developing Interactive Art Using Visual Programming

Ernest Edmonds

Creativity and Cognition Research
Faculty of Information Technology
University of Technology, Sydney
POBOX 123
Broadway NSW
Australia
ernest@ernestedmonds.com

Linda Candy, Mark Fell, Roger Knott, Sandra Pauletto, Alastair Weakley

Creativity and Cognition Research
Department of Computer Science
Loughborough University
Leicestershire LE11 3TU
UK
{l.candy; m.j.fell; r.p.knott; s.pauletto;a.j.weakley}@lboro.ac.uk

Abstract

The paper reports on the use of a visual programming environment and discusses its impact on the creative process. It draws upon research into the collaborative interactive art development process. The context of the work is COSTART, a research project on computer systems for creative work that was undertaken at the Creativity and Cognition Research Studios, in the UK. In the process of making interactive artworks, Max/MSP is a means of implementing exploratory ideas. It is a creative tool that enables both the artist and technologist to push the boundaries of their work into previously uncharted areas.

1 Introduction

Digital technologies impact creativity in numerous ways, and may be seen as enablers of new art forms and aesthetics. It is essential that any description of this relationship closely examines how technologies reshape not only the outcome, but also art practice and the roles of those engaged in that practice. Due to the technical complexity and the expertise necessary for making such works, it is common for artists to collaborate with technologists in order to realize their ideas. Creativity and the nature of collaboration are important to the HCI community insofar as they are indicators of activity currently at the margins of this discipline. As a shift from the margins to the mainstream takes place, the development of creativity support systems provides valuable insights into future needs of the wider user community.

This paper is concerned with the development of emerging technologies for the creative arts and, specifically, the role of visual programming environments in the process of making interactive artworks. The paper describes how interactive art projects can be developed using a visual programming environment called Max/MSP (Puckette, 2002). Investigations were carried out into the development processes of an HCI team and a group of artists-in-residence. The work described arises from a UK funded research project (COSTART, 2003), the aim of which was to explore the nature of creative practice and the requirements of digital technologies for creativity. In order to achieve this, a number of artist in residency projects were put in place in order to study the process of creating new art and technology systems. In this context, whilst artists and technologists collaborated in creating digital art, at the same time, researchers observed events and gathered data for analysis. The goal of researchers is to arrive at a coherent view of events across a number of separate situations. The distinctive characteristic of this practice based research approach is that innovative art projects are developed in tandem with the research activities. The approach and previous results are described in *Explorations in Art and Technology* (Candy and Edmonds, 2002). In the following sections, an account of how the COSTART project artists' requirements were addressed by a visual programming environment called Max/MSP is provided.

2 Co-Creativity in Art and Technology: The Digital Art Projects

Initial requirements for technology solutions were identified from a set of proposals for artist-in-residency projects. The participant artists and technologists designed and developed the technical solutions in collaboration. All the artists concerned wished to create systems that would enable either the audience or artists themselves to interact with the system in such a way as to generate new works or different forms of an existing work. In order to achieve this there were a number of technical options that had to be explored. The main categories of technical requirements were:

Sensor data interaction and integration
Image capture and manipulation
Audiovisual correspondences
Live and pre-recorded data integration

Gina Czarnecki is deeply engaged in exploring the possibilities of audience interaction in a process modelled on biological evolution. The aim of her project *Interactive Evolution* was to allow the audience to control the evolution of a virtual population represented by images of people projected onto a screen. A matrix of floor pad sensors was arranged on the floor in front of the projection that allowed the audience to select a particular person in the virtual population by choosing a position in the room. The main requirement was to develop a means of controlling an interactive audiovisual installation. Technical challenges were identified arising from the requirement to identify the largest group of people in a room. Each onscreen person was a QuickTime movie. Each movie had a layer that specified the transparency value of its corresponding pixel. This enabled the compositing of a group of people where each could be individually addressed and modified. Thus the requirements in this project were concerned both with image manipulation and sensor data integration.

Adriano Abbado's interests are in audiovisual abstract correspondence and expression. The initial focus for his residency was to examine the manipulation of aural and visual objects in real-time, with a specific interest in the correspondence between them. During the residency, he turned his attention to the potential of the interactive techniques and the resulting work explores the sound visual relationships with sensor devices that enable a different kind of interactive experience from his previous works. In the piece, the focus is on the correspondences between abstract animation and synthetic sound. The user can interact with the audio and visual material by moving his/her hands closer to or away from two proximity sensors. Max/MSP was appropriate because it allowed an integrated approach and allows audiovisual manipulation. Sequences of aural and visual events were played back in real-time and the user, through the movement of his/her arms, controlled their separate density in time. Thus we can see that this project centred on audiovisual correspondences, which was where the technical challenges lay.

Yasunao Tone has been working with the conversion of calligraphic drawings into sound. The technical requirements of the project were primarily to develop a system that was capable of translating drawing into sound. For this piece the artist was drawing ancient Chinese characters and texts. Various environments for each component of the work could have been used. For example, the input could have been handled in C and the synthesis in Csound, but Max/MSP provided an integrated approach. A system was made that translated the movements of the pen into sound. A system was developed that created a vertical harmonic mapping so that sound at the far edges were tonally simple (having only the fundamental frequency present) and towards the centre band of the drawing more harmonics were introduced. Video was introduced into the system, so that the position of a pen could be used to read the brightness of a corresponding pixel. This value was used as a seed in the synthesis algorithm. The speed of movement was also calculated and this was mapped onto the volume of a sound. This parameter controlled a variation on granular density, so that slow movement produced more subtle scratchy types of sound, and fast strokes produced harsh distortion. This work required both novel sensor date interaction and novel physical and audiovisual correspondences.

George Saxon's project integrated live and pre-recorded video data, enabling real-time interaction between the viewer and the work. The end result is a display of live camera images combined with recordings of earlier input so that people see themselves and others moving around the space on large video projections. While projecting live material, the program also records sections of input, this is looped and randomly played back cross-fading it with the constantly sent live signal, creating an unusual sense of confusion about where the interacting person is. It was decided that Max/MSP was appropriate because it enabled the collaborating parties to define a set of processes and to reconfigure in a quick manner. They approached the working period with an open ended set of ideas and used the time develop a work around these. Firstly the real time video is sent straight to this display, short segments of this real time input are recorded. These typically range from half a second to two seconds. This recording is looped and played back. Finally the real time input is cross-faded with the looped playback. The requirements here, then, were particularly concerned with the integration of live and pre-recorded video data.

3 Visual Programming And Max/MSP

A useful definition is "Visual Programming (VP) refers to any system that allows the user to specify a program in a two-(or more)-dimensional fashion." (Myers, 1990). One of the best known examples is LabVIEW (National Instrument Corporation, 1990) which was developed by National Instruments and has enjoyed wide success and has been available commercially for over 10 years. It uses the dataflow paradigm and is designed for use by end-user engineers and scientists for the development of data acquisition, analysis, display and control applications. LabVIEW has many characteristics in common with Max/MSP (Puckett, 2003). Max/MSP is a dataflow development tool used widely by digital musicians. Max is a programming environment that acts as a foundation for other applications (named in honour of Max Mathews), while MSP is a program that allows Max to process, synthesize, and play back audio.

The development of the Max graphical environment began at IRCAM, a music research institute in Paris, France, by Miller Puckette in the late-1980s. It is built round the concept of patches (or objects) communicating using messages. It is object-oriented in some, but not all, senses. The Max paradigm is "a way of combining pre-defined building blocks into configurations useful for real-time computer music performance." (Puckett, 2002). However, Max is not a musical composition language as such, it is an instrument design language (lyon, 2002). It is in this instrument-building mode that it was used in the work described in this paper.

In 2002 Cycling '74 published another set of objects for Max/MSP: the Jitter objects. These objects allow processing, in interpreted, of video, images and 3D graphics. Jitter sees the visual data as matrices, and the processing of the visual material happens through the manipulation of these matrices. Jitter can be used for video processing, custom effects, 2D/3D graphics, audio/visual interaction, data visualisation and analysis (Cycling, 2003). In the work discussed in this paper, the use of a beta version of Jitter was a significant aspect of the activity.

4 Observations on the Use of Max/MSP: Multi-Purpose Environments

In many of the projects described above, use of the Max/MSP environment allowed the artists, although often not expert programmers, to take an active role in the development of the systems. Hoeben and Stappers (2001) identify this as a key requirement for computer-based tools supporting early stages of design. The following points show how Max's characteristics influence the creative process.

4.1 Immediacy and fluidity- Interpreted Programming

When using Max/MSP, changes to patches have immediate consequences. This enables the user to try options without having to stop, compile and re-start processes and creates a sense of continuous and undifferentiated action, feedback, and evaluation. This process was described as 'interpreted programming' by one technologist. This characteristic makes the process more fluent, adding a sense of enjoyment and pleasure to the process itself, and enables much more intuitive, creative and rapid exploration. One technologist describes how 'The idea got more precise in a process of continuous feedback with the technology chosen'. She also comments that using Max is much more like using a physical object. Such 'physical-world modalities of interaction' (Baroth and Hartsough, 1995) are becoming increasingly popular.

4.2 Issues in collaborative working with Max

It is clear that when working with Max the process of development becomes less rigid, and is not so clearly punctuated by a linear sequence of distinct stages. It is also apparent that these stages are largely determined by separate technological processes, and that these in turn generate a series of separate 'job descriptions'. In this case we observed that as the process of making art changes, there is a radical shifting and sharing of responsibility, where the artist was able to engage with previously technical processes, and the technologist was able to contribute to previously aesthetic or artistic content. This sharing of knowledge and crossover of skills is seen as a very positive state of affairs, one that can potentially lead to innovation in both the arts and technology.

4.3 Extendibility and Productivity

Max provides an open-ended framework for developing new objects that can be written in the C language. There are several communities of artists and programmers developing new objects, and, as new technologies and approaches to interaction arise, new objects find their way into these communities. It is in this way that Max both 'keeps up with' and contributes to a dialogue between the needs and interests of the artist-programmers and their tools. External and third party objects provide increasingly simple ways of doing complex tasks. For example, what might have been done with several more basic objects, can be achieved with one custom object. Hence the developmental process is not only streamlined, but also the general vocabulary of the user is extended.

4.4 An Integrated Approach

Max provides the facility to integrate low-level system inputs with high-level visual programming tools so a single software system can be used for the entire process. This differs from other approaches where, for example, the input processing was written in C and the sound synthesis written in Csound. Max facilitated the easy integration of separately developed components. In one residency, the technologist concentrated on developing a system for aural manipulation while the artist concentrated on visual manipulation. When the whole system needed to be evaluated, these two components were easily linked by drawn links, to establish a flow between the two sections.

5 Conclusion

During the course of the projects described here, the use of Max/MSP allowed the rapid creation both of prototype systems and methods of data analysis that proved useful in 'scoping' the projects. It is often the case when developing new systems and products that the requirements are ill defined at the start. In the process of making interactive artworks, Max/MSP is a means of implementing exploratory ideas. It is a creative tool that enables both the artist and technologist to push the boundaries of their work into previously uncharted areas. It promotes a style of working that is flexible, fluent and engaging.

References

Baroth, E. and Hartsough, C. Visual Programming in the Real World. in Burnett, M., Goldberg, A. and Lewis, T. eds. *Visual Object Oriented Programming*, Manning Publications Co, Greenwich, CT, USA, 1995, 21-42.

COSTART. http://creative.lboro.ac.uk/ccrs/COSTART2/COSTART2.html

Cycling. http://www.cycling74.com/products/jitter.html.2003

Hoeben, A. and Stappers, P.J., Tools for the Conceptual Phase of Design at the ID-Studiolab. in ACM *CHI 2001, Workshop on Tools, Conceptual Frameworks, and Empirical Studies for Early Stages of Design*, 2001.

Lyon, E. Dartmouth Symposium on the Future of Computer Music Software: a Panel Discussion. Computer Music Journal, Volume 26 (4) 2002. pp13-30.

Myers, B. A. Taxonomies of Visual programming and Program Visualization. Visual Languages and Computing. Volume 1 (1) 1990. pp 97-123.

National Instrument Corporation. LabVIEW 2: Getting started manual. Austin TX: National Instrument Corporation. 1990.

Puckette, M. Max at Seventeen. Computer Music Journal, Volume 26 (4) 2002. pp 31-43.

Persistent Cart Design: A Review of Implementations

Dena Fletcher

Columbia House
1221 6th Avenue, New York, NY 10020
dfletcher@chcmail.com

Mark E. Fletcher

Pace University
1 Pace Plaza, New York, NY 10038
mfletcther@pace.edu

Abstract

In this paper, we investigate current implementations of "persistent cart" user interface design in ecommerce websites. Our goal is to provide some clarity on this topic by identifying categories of persistent cart designs. We describe the features of the persistent carts in each of three categories: "Basic," "Itemized," and "Interactive," providing examples of current implementations within each. We discuss the advantages and disadvantages of the designs in each of these categories, as well as identify some specific user interface design issues in the examples provided.

1 Introduction

Whether it's called the shopping cart, cart, basket, or shopping bag, the order detail page is ubiquitous in ecommerce sites, and the features are by and large the same. They include: product name, price, quantity, the ability to remove items, and the ability to "checkout." The cart has been widely discussed and studied. There are various guidelines for good and "usable" cart design. The same, however, cannot be said of the persistent cart. The persistent cart is the small version of the cart that is viewable from most, if not all, pages of the site. There has been little discussion or research of this feature, though it has been touted as a feature that improves site usability and facilitates online shopping (Constantine & Lockwood, 2000; "Designing shopping," 2000).

Many of the problems with online carts have been attributed to the breakdown of the shopping cart metaphor (Constantine, 1998). The term "shopping cart" is meant to evoke in the minds of users the experience of shopping in an "offline" store. Beyond the name, however, the online cart often bears little resemblance to its real-world counterpart. The persistent cart can mitigate this situation by bringing more of the features of the offline cart to the online shopping experience. In effect, it bridges the gap between the offline and online shopping experiences.

Though the use of persistent carts is fairly uncommon, its implementation or design comes in a number of different varieties. The goal here is to review and categorize the types of persistent carts currently in use and to attempt to analyze some of the advantages and disadvantages of each.

2 Cart Link

A feature common to nearly all ecommerce sites is a link that allows users to jump to the cart at any point during the shopping process. In its simplest implementation, the cart link provides no information about the cart state and acts only as an access point.

In some cases a single piece of information about the contents of the cart is added to the cart link. Most typically this is an item count (see Figure 1 below). Examples of this abound in both horizontal and vertical layouts and occur in sites offering a wide range of products.

Web designers struggle with real estate when designing persistent carts. Real estate is, simply, the amount of space the persistent cart takes up on a single page, and typically on every page of the site. This space is significant because the ultimate goal of most ecommerce sites is clear, to sell as much product as possible. This usually means promoting as many products as the real estate will allow. An advantage of these simple cart links is the limited amount of space they take up. The item count information also has the advantage of providing feedback, confirming that an item has been added to the cart.

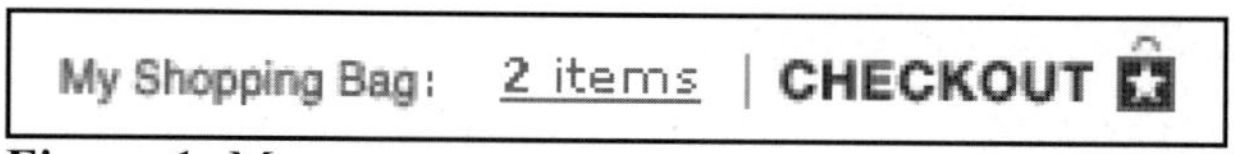

Figure 1: Macys.com

These carts do not provide much information compared to the order detail page or a real world cart. Though users know how many items are in their cart, they cannot see the items or get any sense of how much money they are spending.

3 Basic Persistent Cart

The Basic Persistent Cart provides at least one piece of information in addition to the item count. There are two subgroups within this category. The first provides users with an exact accounting of the items in the cart. This is accomplished either through a listing of items or through thumbnail images of the items (see Figure 2). The Eziba.com example uses thumbnails in a horizontal alignment. There is a clear disadvantage to this design. Since websites are intended to scroll vertically and not horizontally, the horizontal real estate is fixed. This persistent cart can display a maximum of four or five items in the persistent cart before running out of room.

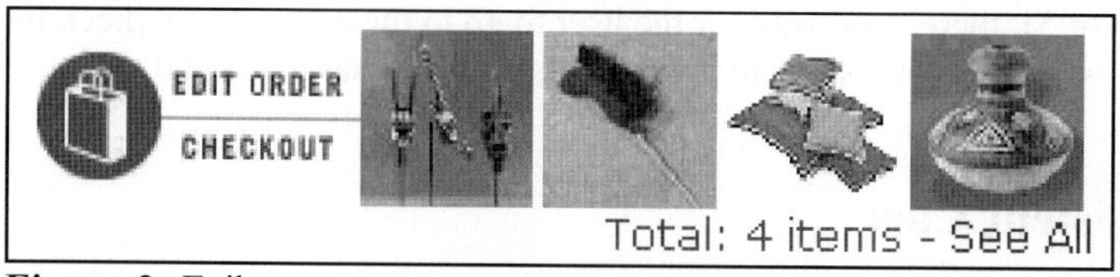

Figure 2: Eziba.com

The second type of implementation in this category provides users with the number of items and total cost of all items in the cart (see Figure 3).

Figure 3: Barnes & Noble.com

What is interesting about this second subcategory is that the running total is clearly useful information for the user, but it has no parallel in an offline environment. The addition of this kind of information is an excellent example of being sensitive to the medium, keeping in mind both its strengths and weaknesses, and designing the experience within those parameters rather than attempting to exactly duplicate an experience from a different medium.

4 Itemized Persistent Cart

The Itemized Persistent Cart is an implementation that provides users most, if not all, of the basic information of the full cart, or order detail page. In these designs, users see a listing of the items in the cart, the price per item, and typically the total cost of all items in the cart (see Figures 4 and 5).

Figure 4: Omahasteaks.com

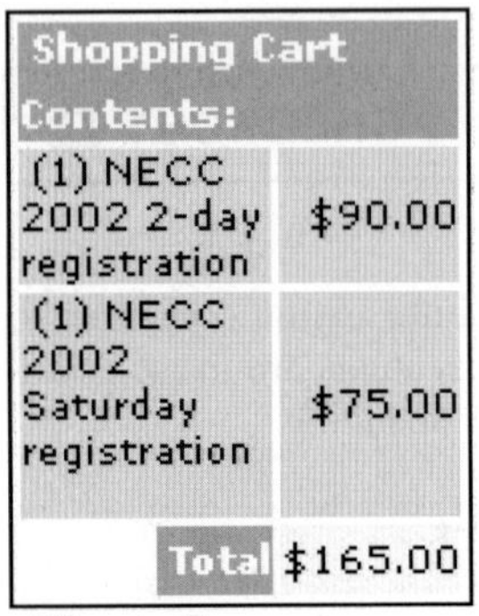

Figure 5: eco.org

The Omahasteaks.com example (Figure 4) highlights a significant usability issue. Rather than listing the cart contents by name they are listed only by product number. This is an example of designing from a systems perspective rather than from a user perspective. Particularly in a Business to Consumer environment such as this, product numbers are unlikely to hold much meaning for users.

In the eco.org example (Figure 5), there is no way for the user to go to the cart page or checkout from the persistent cart. This is a serious oversight and greatly impacts the usefulness and usability of this persistent cart.

5 Interactive Persistent Cart

The Interactive Persistent Cart is characterized by the addition of interactive elements typically reserved for the order detail page including, the ability to remove items and the ability to adjust quantities (see Figures 6 through 8).

While the carts in the other categories can be seen as snapshots of a user's cart with varying degrees of detail, the interactive cart more closely resembles the order detail page and, by extension, the real-world shopping cart experience.

Since these designs provide, in essence, a mini version of the order detail page, they typically avoid a drawback common to many ecommerce sites, taking the user to the cart page (and out of the shopping process) each time an item is added to the cart.

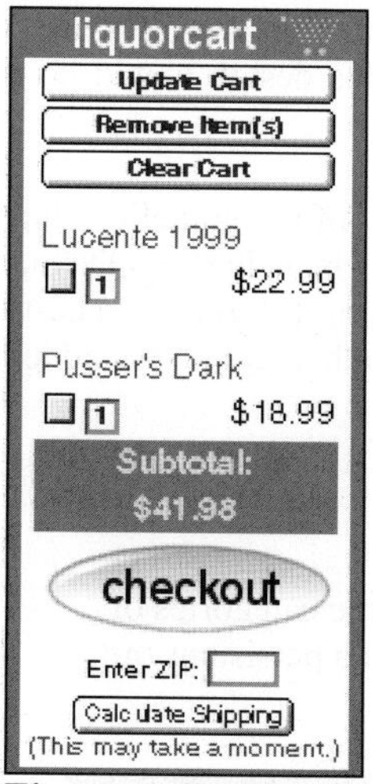

Figure 6: liquorama.net

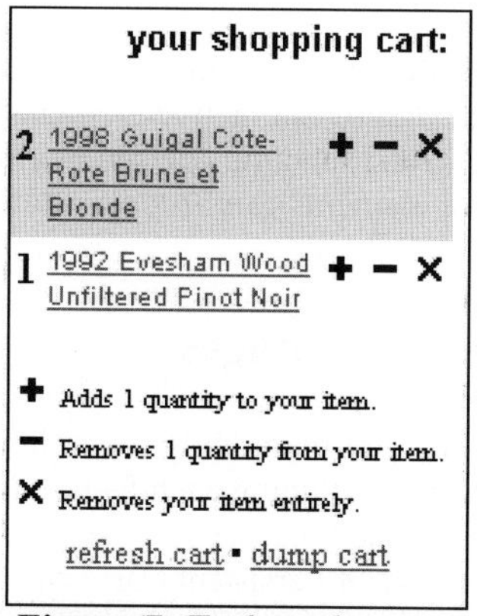

Figure 7: Taylorandnorton.com

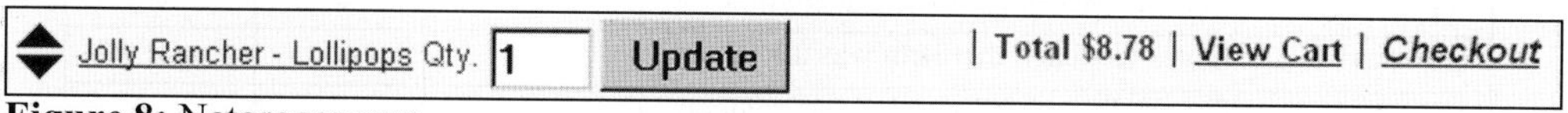

Figure 8: Netgrocer.com

The above examples of Interactive Persistent Carts illustrate the compromises that are often made in providing interactivity in a persistent cart and some of the problems that can arise from these compromises. Liquorama.net (Figure 6) is quite unique, offering virtually all the content and interactivity a user would expect to encounter on a separate cart page. Each item is listed by name; a price is listed for each item; and a subtotal of all items is provided, as is the ability to change quantities and remove items. This persistent cart even allows users to calculate their shipping costs for the order. This is truly a mini version of the order detail page. In fact, on this site, the cart page has been eliminated. The design of the interactivity in this persistent cart is somewhat confusing. It is not immediately clear what steps need to be taken to remove an item, specifically checking the box next to the item and clicking the "update cart" button. Users might expect that changing the quantity to "0" could remove the item from the cart, but this cart does not allow the quantity to be set to "0."

The Taylorandnorton.com persistent cart (Figure 7) enables users to change quantities of items or remove them from the cart. However, this cart does not include any of the pricing information found in the "Itemized" persistent carts. Further, the fact that the persistent cart includes a key that defines each of the icons indicates that the interaction tools need explaining. The inclusion of a key takes up valuable real estate that could be better used to include more information about the cart contents such as pricing information. In this cart, as in the eco.org example (Figure 5) above, there is no apparent way for the user to go to the full cart page or checkout from the persistent cart. It is also not clear what the function of the link "refresh cart" is.

The Netgrocer.com persistent cart (Figure 8) is interesting in that it is designed to consistently occupy the same small amount of real estate regardless of the number of items added to the cart. Because of the scroll mechanism within the cart, each item is viewable separately, and the overall footprint of the cart does not change. Users are provided a listing of all items in the cart, a total cost for all items, as well as the ability to change quantities. Users can remove an item from the cart by changing the quantity to "0." Though the advantage of the real estate savings is clear, users are not able to get a full view of their cart. Adding an item count to this design would help ameliorate this issue.

In adding interactivity to a persistent cart it is important to consider many of the guidelines that have been established for the design of interactive elements for cart pages including van Duyne, D. K., Landay, J. A., Hong, J. I. (2002) and the Nielsen Norman Group Report (2001). The clarity of the design of these interactive elements can become all the more critical because of the added constraint of real estate.

6 Cart Features and Advantages

Table 1 below summarizes the features commonly found in each of the three categories of persistent carts and the advantage to including each feature in the design of a persistent cart.

Table 1: Summary of Persistent Cart Features and Advantages

Feature	Advantage	Basic	Itemized	Interactive
Cart Link	Enables user to get to cart at anytime	✓	✓	✓
Item count	Gives feedback that item was added	✓	✓	✓
List of items	Provides users full view of cart contents mirroring real world experience	✓ (or total cost)	✓	✓
Total cost	Convenient information for users easily provided in web environment	✓ (or item cost)	✓	✓
Cost per item	Convenient information for users easily provided in web environment	x	✓	✓
Ability to change quantity	Eliminates need for user to add item again or leave shopping environment to go to cart page	x	x	✓
Ability to remove items	Eliminates need for user to go to cart page	x	x	✓

References

Constantine, L. (1998). Use and Misuse of Metaphor. Retrieved November 20, 2002, from http://foruse.com/articles/metaphor.pdf

Constantine & Lockwood. (2000). Shoddy Shopping Carts. Retrieved November 10, 2002, from http://www.virtuallook.com/e-comm.htm

Designing shopping cart and checkout interfaces. (2000). Retrieved November 20, 2002, from http://www.virtuallook.com/e-comm.htm

Nielsen Norman Group. (2001). E-commerce User Experience: Design Guidelines for Shopping Carts, Checkout, and Registration. Retrieved October 15, 2002, from http://www.virtuallook.com/e-comm.htm

van Duyne, D. K., Landay, J. A., Hong, J. I. (2002). The Design of Sites: Patterns, Principles, and Processes for Crafting a Customer-Centered Web Experience. Boston: Addison Wesley Professional.

Adaptive Plant Human-Machine Interface Based on State Recognition and Machine Learning

Kazuo Furuta, Ichiro Kataoka[1]*, and Keiichi Nakata*

Institute of Environmental Studies, The University of Tokyo
7-3-1 Hongo, Bunkyo-ku, Tokyo 113-0033, Japan
{furuta, nakata}@k.u-tokyo.ac.jp

Abstract

A concept of adaptive plant human-machine interface will be proposed in this work that realizes effective human-machine cooperation in process control. The interface system infers operator's state recognition process with a Bayesian network and presents the information that matches operator's requirement from the result of state recognition. The effectiveness of the proposal was demonstrated by cognitive experiment using a simple test plant of DURESS.

1 Introduction

Cooperative task execution by both humans and machines is an important issue for safety and reliability of a large and complex system. To achieve this goal both humans and machines have to understand the intention and logics each other that determine their behaviour, and the intelligence level of any operator supporting system depends on how well the system can comprehend human intention.

A concept of adaptive plant human-machine interface will be proposed in this work that realizes effective human-machine cooperation in process control. The interface understands operator's intention, provides information that the operator requires in the form that matches operator's requirement, and adapts to the personal preference of an individual operator following our previous work (Furuta, Oyama & Kondo 1999). The interface system infers operator's state recognition process with a Bayesian network and presents the information that matches operator's requirement from the result of state recognition. In addition, our interface is equipped with a learning mechanism so that it can adjust the prescribed knowledge to actual cognitive characteristics of an operator. While the interface presents information displays following the rules determined by the designer at an early stage of operation, it gradually acquires actual display rules as operation is continued and learning progresses.

2 System Architecture

2.1 Overview

The adaptive interface system of this study consists of four modules: sensor validation module, state recognition module, learning module, and information display module.

[1] Mechanical Engineering Research Laboratory, Hitachi, Ltd., Ibaraki, Japan, kataoka@gm.merl.hitachi.co.jp

The sensor validation module diagnoses anomalies of the sensing system, indicates the result on the alarm panel, and omits sending sensor data to the state recognition module. The diagnosis is performed by consensus development between agents distributed over the sensing system. The method used for this diagnosis is based on our previous work (Kataoka, Nakata & Furuta, 2000) and will not be explained in detail here.

The state recognition module obtains sensor data from the sensor validation module, and infers operator's state recognition process from system parameters that determines what the operator is going to do to the plant. The inference is performed with a Bayesian network, where the sensor data are given to symptom nodes as evidence and posterior probabilities are updated for state nodes. The state node with the highest probability determines the information display to be presented, which corresponds to the task expected in the state.

The information display module selects and presents the information display following the display rule fired by the identified plant state. The interface screen consists of the following two display areas of information channels. An overview information channel presents the global plant status information that does not depend heavily on operator's task. A display that mimics plant configuration is used in this channel. A specific information channel presents information that the operator requires in a specific situation in plant operation. The information display module controls presentation through this channel.

The learning module acquires the knowledge which information display the operator wishes to look at in each plant state in order to adjust improper display rules implemented in the initial design. Without the rule learning the information display module will repeat presenting the same display for a certain state, even though the operator did not want the presented display and called up another one before.

2.2 Method of State Recognition

Intent inference process can be categorized into two processes: inference of goals and plans from perception of the operator, and that from behaviour of the operator. The former is called state recognition, and the latter plan recognition. Since plan recognition requires observation of operator's behaviour while state recognition does not, intent inference based just on state recognition is adopted in this study. A Bayesian network is used to implement state recognition after Hatakeyama and Furuta (2000), because of its suitability to evidential reasoning under uncertainty.

It has been shown that the operator uses a hierarchical state space when he/she identifies process faults in the plant. In such a state hierarchy, since a specific state directly implies its parent state, the relation from the specific state to the general state can be interpreted as direct causality of a Bayesian network. The Bayesian network for state recognition thereby can be constructed by inverting the state hierarchy. The relation from some state to its symptom is also interpreted as direct causality in a Bayesian network.

2.3 Learning of Display Rules

An ability to learn which information display should be presented is expected, because display rules implemented by the designer may be inadequate. Individual operators may have different strategies of problem solving in plant operation. The knowledge required here is a set of rules that relate system states to information displays to be presented. It is however wise to use knowledge available in the domain rather than to acquire all the knowledge from scratch by learning, because learning from scratch is too inefficient. Weights of rules should not be initialised with random values, but display rules are to be used at the beginning that domain experts think reasonable.

In this study the profit sharing algorithm(Grefenstette, 1988) has been adopted for rule learning. Profit sharing is to learn action rules that relate perceived world states and actions to be taken. Each rule has a weight and a rule with a great weight is likely to be chosen. For fast learning, it is necessary to ensure reward gains by try-and-error in the early stage of learning and to distribute rewards to relevant rules properly. Profit sharing can be thought as a classifier system with weights in a time series, which also fires a rule with a high priority to decide the next action.

Rules to be learned are in a form Rule(State, Action). The state here means the plant state inferred by the state recognition module, and the action means an action to present an information display to the operator. If the operator overrides the display chosen by the system, the rule that presented the display is weakened and the rule to present the operator's choice is strengthened. Every time the operator calls up a new display, one episode is terminated and weights are updated. As the operator continuously calls up information displays that he/she wants to look at, the learning module learns which display should be presented in a particular plant state.

2.4 Application

The proposed adaptive interface architecture was applied to DURESS, the Dual Reservoir System Simulation (Vicente & Rasmussen, 1990). In DURESS, plant states correspond to tasks that the operator is expected to perform in operation. The task hierarchy of DURESS was constructed by goal-mean task analysis, and then it was transformed to the state hierarchy. In construction of a Bayesian network, some state nodes were added to cover all situations: normal operation, system shutdown, and system startup.

3 Experiment

3.1 Experimental Setting

An experiment was conducted to check the effectiveness of the proposed interface architecture, in particular the effectiveness of acquiring display rules by learning.

Six graduate students of engineering departments were recruited as subjects. Before experiment they were taught the structure of DURESS and the procedures of its operation, and then they practiced operation. Subjects were instructed to achieve operation goals to keep the outlet flow and the outlet temperature from Tank A and B at demanded values while keeping alarms off as long as possible.

Workload imposed on a subject during experimental runs was evaluated by NASA-TLX (Hart & Staveland, 1988). Operation performance of a subject was assessed by number of display operations, number of plant operations, total time interval with active alarms, accumulated deviation of outlet flow rate from demanded values, and that of outlet temperature.

Operation performance was compared between two conditions with and without rule learning. Influence of experimental scenarios was also studied with learning, where the same combination of events occurred in a different sequence. In order to avoid learning effect of subjects during experiment, different scenarios were used for learning and non-learning conditions. Two months later the same subjects tried the same scenario under a learning condition that was used under a non-learning condition before. Each scenario lasts 650 seconds and includes multiple events at every 100 seconds in an arbitrary order out of change in flow demand, change in temperature demand, pump failure, and valve failure.

How rule learning can absorb personal variation was studied when the all subjects operated the plant successively. The rule weights obtained by learning from a previous operator ware applied as the initial weights for the next operator. The order of subjects was determined randomly. If the rule learning works, frequency of display operations will gradually decrease even though operators change. In this successive test different scenarios were used that last 550 seconds and include multiple events of the same class at every 50 seconds.

3.2 Experimental Result

Table 1 shows each performance metric for a non-learning condition with Scenario 1, a learning condition with Scenario 1, and a learning condition with Scenario 2 The values are averages and standard deviations in parentheses over the six subjects.

The result shows that number of display operations was reduced, particularly in abnormal plant states. Once the operator calls up a particular display in a certain plant state, he/she does not need to call up the display again, because the system presents the display automatically. Rule learning resulted in reduced workload, deviation of outlet flow, and deviation of outlet temperature. The differences were significant ($p<0.05$), however, only for number of display operations in abnormal states. The differences between Scenario 1 and 2 were all insignificant.

Table 1: Experimental Result Averaged Over Subjects

Learning	no	yes	
Event Scenario	1		2
NASA-TLX	50 (22)	43 (20)	39 (15)
Deviation of Outlet Flow	180 (161)	109 (86)	105 (57)
Deviation of Outlet Temperature	3134 (2221)	3025 (2022)	3150 (1805)
Interval With Active Alarms (sec)	239 (136)	250 (176)	213 (99)
Number of Plant Operations	121 (30)	93 (26)	94 (19)
Number of Display Operations	58 (36)	26 (13)	33 (21)
in Abnormal States	29 (18)	6 (5)	12 (8)

Figure 1 shows accumulation of display operations for each operator in the successive test. The turn of operation was from Subject A through F. The first subject called up displays many times, but display call-ups were reduced for the successive operators. Display call-ups increased a little for Subject E, but decreased again for Subject F. Accumulation of display operations showed saturation for every subject, and presentation of interface displays was gradually automated.

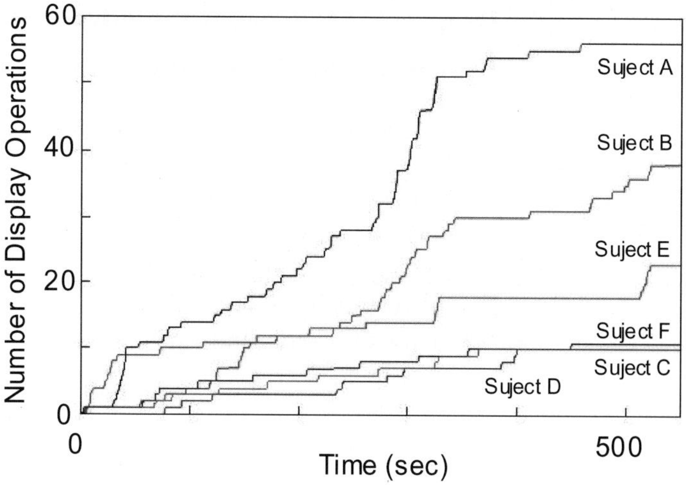

Figure 1: Accumulation of Display Operations in Successive Test

4 Conclusion

Architecture of adaptive plant human-machine interface has been proposed that aims to support cooperative task execution between the plant system and the operator. The interface infers operator's state recognition process to identify the current state of the plant and then presents an information display that best matches the task corresponding to the current state. A learning mechanism is implemented to the interface system so that the system can adjust the association rules between plant states and displays to be presented to operator's actual performance. An experiment was conducted and the result showed that the operation workload was reduced as the interface system learned appropriate information displays to be presented to the operator.

References

Furuta, K., Oyama, Y., & Kondo, S. (1999). Intelligent Plant Human-Machine Interface Based on Intent Inferencing, In M.J. Smith & G. Salvendy (Eds.), *Human-Computer Interaction* (*HCI '99*) (pp. 1192-1195). Amsterdam: Elsevier

Grefenstette, J.J. (1988). Credit Assignment in Rule Discovery Systems Based on Genetic Algorithms, *Machine Learning*, 3, 225-245.

Hatakeyama, N., & Furuta, K. (2000). Bayesian Network Modeling of Operator's State Recognition Process, In S. Kondo & K. Furuta (Eds.), *Probabilistic Safety Assessment and Management* (*PSAM 5*) (pp. 53-58). Tokyo: Universal Academy Press

Hart, S.G., & Staveland, L.E. (1988). Development of NASA-TLX (Task Load Index): Result of Empirical and Theoretical Research, In P.A. Hancock & N. Meshkai (Eds.), *Human Mental Workload* (pp. 139). Amsterdam: Elsevier

Kataoka, I., Nakata, K., & Furuta, K. (2001). A Method of Consensus Formation of Multi-Agent System in Process Control, In *Proc. Asia Pacific Symposium on Safety (APSS)*. Kyoto

Vicente, K.J., & Rasmussen, J. (1990). The Ecology of Human-Machine System II: Mediating Direct Perception in Complex Work Domains, *Ecological Psychology*, 2, 207-249.

Development of a Dynamic Operation Permission System for CRT-based operation interfaces

Akio Gofuku, Yoshihiko Ozaki**, Tadashi Ohi**, Koji Ito****

*Department of Systems Engineering, Okayama University
3-1-1, Tsushima-Naka, Okayama, 700-8530, Japan
fukuchan@sys.okayama-u.ac.jp
**Mitsubishi Electric Corporation
*** Mitsubishi Heavy Industry

Abstract

It is one of important problems to improve the usability of nuclear power plant operation without decreasing plant safety and reliability. CRT-based operation interfaces become to be introduced to newly constructed plants. Although the operation interfaces have an advantage of flexibility, operators may manipulate irrelevant software switches with their current operation. This study proposes a dynamic operation permission to reduce the commission errors in the plant operation through CRT-based operation interfaces. This paper describes a basic algorithm of the dynamic operation permission, an operation model to represent operation procedure written in operation manuals and estimation of effects and influences of an operation action by a functional model.

1 Introduction

It is one of important problems to improve the usability of nuclear plant operation without decreasing plant safety and reliability. CRT-based operation panels are widely introduced to newly constructed plants. Most conventional operation panels are replaced to CRT-based operation panels with the replacement of plant instrumentation from analogue to digital systems. A CRT-based operation panel can equip with software switches which operators can call and select operation screens and issue the commands for necessary operations by touch panel manipulation, keyboard typing, and so on. One of advantages to utilize software switches is their flexibility to design and to manipulate. However, operators may manipulate irrelevant software switches with their current operation. Therefore, a control panel with software switches needs to equip with a mechanism to avoid this kind of commission errors.

The authors proposed a conception of dynamic operation permission[Gofuku2002] to reduce the commission errors. The main idea of the dynamic operation permission is to prevent only evident commission errors and to leave operators behave as they like as far as they follow operation manuals and various operation rules. The relation between operators and a dynamic operation permission system is such that the system supervises operators in the cases of establishing suitable operations without eliminating creative ideas of operators.

The previous paper[Gofuku2002] described the conception of a dynamic operation permission system and how operations are represented as operation models. This paper describes the outline of the basic algorithm and operation models. This paper also describes an estimation method of effects and side effects of an operation for the dynamic operation permission.

2 Dynamic Operation Permission Algorithm

Figure 1 shows the flow chart of the dynamic operation permission. Depending on the plant situation, operation candidates are listed by following the descriptions of operation manuals. An operation candidate is an operation which can be made next. When operators make an operation action, the operation is first identified by the screen selected and the console button pushed.
The action is permitted if it is one of operation candidates and all prerequisites for the action are satisfied. The prerequisites of an operation action specify the plant conditions that the action can be made and the operation actions to be made before the action.
When an operator makes an action not included in operation candidates, its prerequisites are evaluated. If a prerequisite for the action is not satisfied, the operation action is not permitted. If all prerequisites are satisfied or no prerequisite is specified for the action, the permission is decided by comparing the estimated behavior after making the action with the desired and undesired behaviors of each operation candidate.
The estimation of behavior after taking an operation is made based on a functional model. The technique will be described in Section 4. If an undesired behavior is estimated to appear by the action, it is permitted but a strong warning is indicated to operators. If an undesired behavior is not estimated to appear and the estimated behaviors are coherent to desired behaviors of at least one of operation candidates, the action is permitted with a comment. If no desired behavior is estimated to appear and no undesired behavior is estimated to appear by making the action, the action is permitted with a warning. The levels of warning are temporary determined for the prototype dynamic operation permission system the authors are now developing for a steam generator tube rupture accident of a PWR (Pressurized Water Reactor) nuclear power plant.

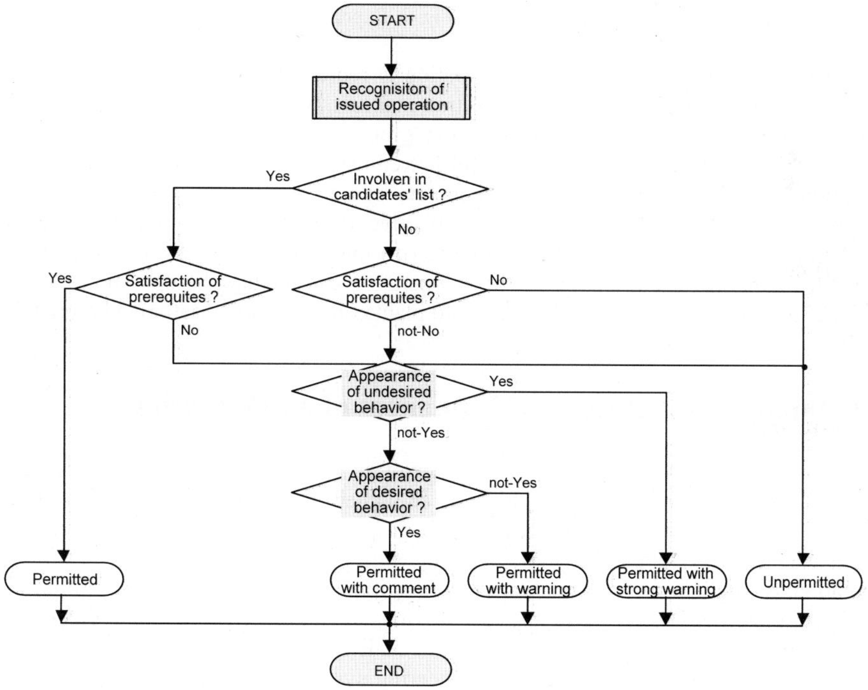

Fig. 1 Flow chart of dynamic operation permission.

3 Operation Models for Dynamic Operation Permission

To implement the algorithm of the dynamic operation permission described in the previous section as a system, the descriptions of operation manuals are needed to be modeled and to be stored as operation models. The operation manuals describe operation action sequences, necessary conditions for an operation, its concrete action, and so on. This means that operation manuals describe two types of information. The one is operation action sequences. The other is the information of an operation. These types of information are represented separately in this study.
The operation action sequences are graphically represented as operation action sequence diagrams. Some actions can be grouped into a macroscopic operation because they are taken under the same operation purposes. To deal with the part-of relations among operations, the authors define two types of operations, that is, unit operation and abstract operation. A unit operation is an action by operators. On the other hand, an abstract operation is the one composed of multiple unit operations and/or abstract operations. A detailed operation action sequence is defined for an abstract operation. In this way, operation action sequences are represented hierarchically by defining abstract operations. The initiation and termination of a unit or an abstract operation are represented by the symbol shown in Fig. 2. The operation name or identification characters are shown inside the box. The logical relations among operations in an operation action sequence are represented by the symbols of AND, OR, and XOR shown in Fig. 3.

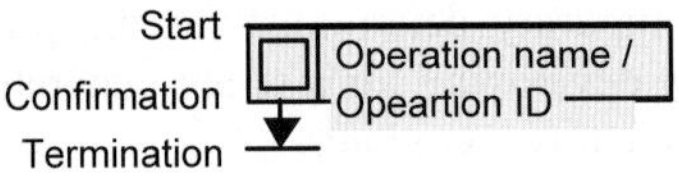

Fig. 2 Expression of unit operation.

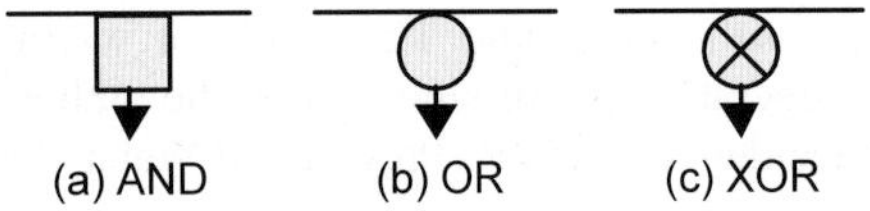

Fig. 3 Expresions of logical relations of operation.

On the other hand, the information of an operation is represented by a frame format. From our analyses of the described information in operation manuals, a unit operation is represented by (a) operation identification characters, (b) operation group identification characters, (c) operation name, (d) persons who make the operation, (e) prerequisites of operation, (f) operation purpose, (g) operation goal for control action, (h) concrete operation action, (i) qualitative functional influences, (j) method to confirm the operation effective, (k) method to evaluate the termination of operation, (l) desired behaviors by the operation, (m) undesired behaviors after the operation is made, and (n) remarks. The prerequisites of an operation can include the operation identification characters of operations necessary to be taken and plant conditions to be satisfied before taking the operation.

4 Estimation of Effects and Influences of an Operation

The estimation results of the effects and influences of an operation to plant behaviors are used in the dynamic operation permission algorithm as explained in Section 2. The authors utilize the MFM (Multi-level Flow Modelling)[Lind1994] for the framework to develop an estimation model because it can represent plant behaviors relating with goals/sub-goals of a plant although the representation is qualitative. The qualitative estimation is enough because the purpose of the dynamic operation permission is to prevent obvious commission errors and the decision of permission is basically made based on operation manuals.
The MFM is a methodology to model an engineering system from the viewpoint of means and goals. The MFM represents intentional aspects of a system from the standpoint that a system is a man-made purposeful system. The MFM models a system in two dimensions, that is, means-end

and whole-part dimensions. The relations among system goals, sub-goals, and system functions to achieve goals/sub-goals are represented by the means-end dimension. System functions are represented by a set of mass, energy, activity, and information flow substructures on several levels of abstraction.

Based on the MFM model for a target plant, the effects and influences of an operation action are estimated by the following algorithm. The algorithm is based on the influence estimation algorithm in deriving plausible counter actions for an anomalous situation of plants[Gofuku1999] although some of estimation rules are corrected. The estimation algorithm is shown in Fig. 4. First, an operation action is mapped to the corresponding function or behavior in a MFM model. To deduce the influence of the action from the corresponding function or behavior in a flow structure, the influence on each function when the function, its input, or output changes is interpreted as shown in Table 1. Propagating the influence in the flow structure, the change of function achievement at the function connecting to an upper sub-goal is known. The influence on the goal connected to the function is estimated by the qualitative causality knowledge specified beforehand between the function and the goal. The influence is propagated to the function conditioned by the goal by the qualitative causality knowledge between the goal and the function. Then, the influence in the upper flow structure is estimated. In this way, the influence of an action on the whole plant is estimated.

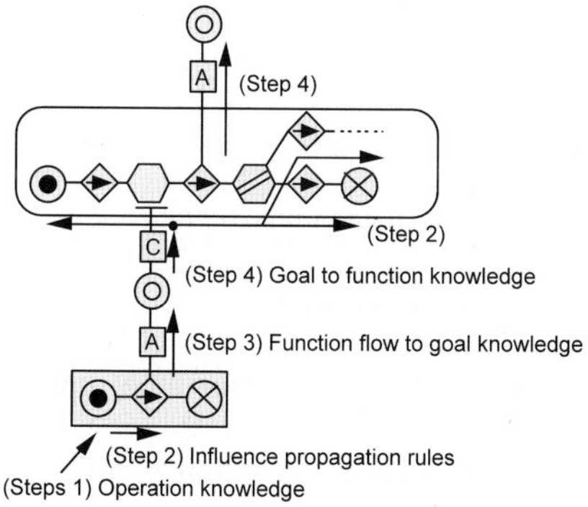

Fig. 4 Influence estimation algorithm.

Table 1 Examples of influence propagation rules

Function	Change		Influence	
Source	Function +	-	Output +	-
	Output +	-	Function +	-
Sink	Function +	-	Input +	-
	Input +	-	Function +	-
Transport	Function +	-	(Output & Input) +	-
	Input +	-	(Output & Function) +	-
	Output +	-	(Input & Function) +	-

The ultimate goal of counter actions in an accidental situation of a nuclear power plant is to safely shutdown the plant. The cooling of reactor is most important in an accident management because nuclear reaction is automatically stopped when an accidental situation is detected and the cooling of reactor is an important necessary condition to prevent any radioactive release to environment. Therefore, the authors have developed a MFM model for a prototype dynamic operation permission system of PWR plants in the case of steam generator tube rupture event by setting the cooling of reactor as the top goal of the model. Figure 5 shows the simplified version of the MFM model.

5 Conclusions

This paper describes the conception of dynamic operation permission the authors propose to reduce the commission errors by operators. The outline of the basic algorithm of the dynamic operation permission and operation models is described. The paper also describes a technique to estimate the effects and influences of an operation action used to decide the permission of the action when the action is not followed a standard operation procedure.

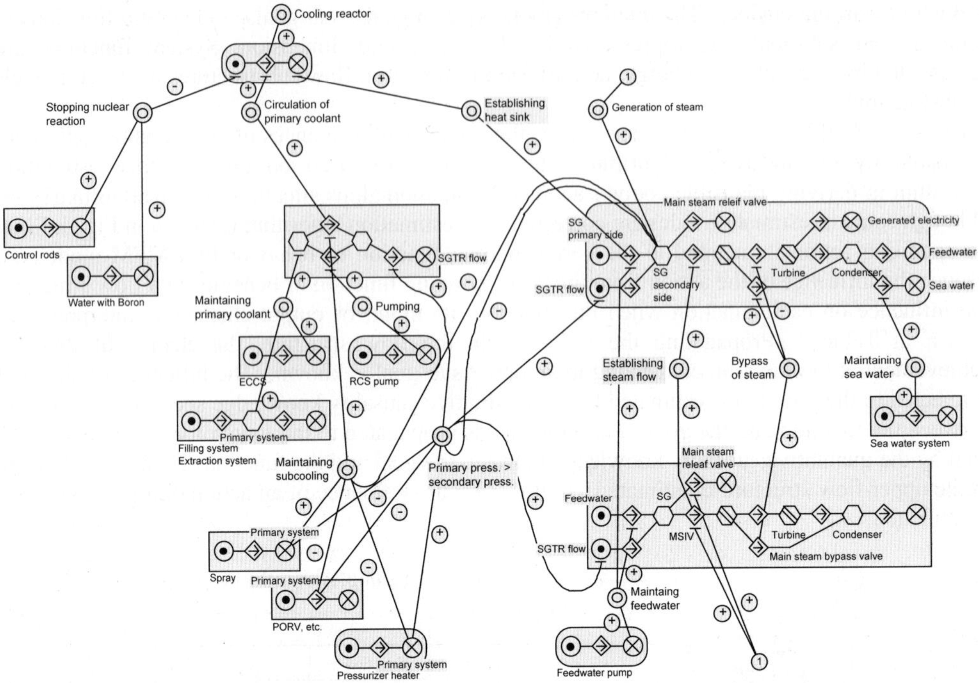

Fig. 5 Simplified version of a MFM model for a PWR plant in an steam generator tube rupture event.

The future studies related with the dynamic operation permission system include

(a) development of a prototype for an abnormal scenario of steam generator tube rupture event of a PWR plant, and
(b) evaluation of the applicability of the conception of the dynamic operation permission using the prototype.

The authors believe that dynamic operation permission systems will contribute to enhance the safety and reliability as well as the usability of engineering systems equipped with CRT-based interfaces.

References

Lind, M. (1994). Modeling Goals and Functions of Complex Industrial Plants, *Applied Artificial Intelligence*, 8 (2), 259-283.

Gofuku, A. and Tanaka, Y. (1999). Application of a Derivation Technique of Plausible Counter Actions to an Oil Refinery Plant, *Proc. IJCAI Fourth Workshop on Engineering Problems for Qualitative Reasoning*, 77-83.

Gofuku, A., Ozaki, and Y, Ito, K. (2002). A Dynamic Operation Permission System for Pressurized Water Reactor Plants, *Proc. Int. Symp. on the Future I&C for NPP*, 360-365.

Optimization of Instrument Calibration Intervals by On-line Sensor Monitoring Techniques

Nobuhiro Hayashi[1]

Mitsubishi Heavy Industries, Ltd.
nobuhiro_hayashi@kind.kobe.mhi.co.jp

Masumi Nomura[2]

Mitsubishi Heavy Industries, Ltd.
masumi_nomura@n.trdc.mhi.co.jp

Abstract

This paper introduces a new method for optimizing calibration interval of sensors by using an on-line sensor monitoring system for pressurized water reactor (PWR) plants.
An analog type technology has been applied to instrumentation & control systems for Japanese PWR plants since Mihama unit No.1, the first PWR plant in Japan, which began commercial operation in 1970. The analog system needs a lot of calibration work in every planned outage because the analog system may cause a drift and it should be kept within a required band.
In recent years, microprocessor based digital technology has been extensively applied for various control related systems in Japanese PWR plants, and this results in reduction of maintenance work-load for instrument rack and its exposure dose associated with calibration work at field. However, since a sensor part still needs calibration, there arise strong needs from utilities to eliminate sensor calibration made every planned outage or to increase calibration interval of the sensors.
We are developing an on-line monitoring system to satisfy the above-mentioned needs. This system models the characteristics of the sensors just after the planned outage and monitors sensors during a plant operation by predicting true value and evaluating just before the next planned outage sensor drifts that is difference between true value of a parameter and measured value of the sensor.
The system can detect the defect sensor, or the intact sensor exhibit a sufficiently small drift and need no calibration at the time.

1 Introduction

At the nuclear power plants, calibration work for instrumentation and control (I&C) system consists of the following three stages. (See. figure1)

Calibration of sensors
Calibration of signal processing system
Calibration of human interface (HI) & plant computer (PCCS)

[1] 1-1, Wadasaki-cho 1-Chome, Hyogo-ku, KOBE 652-8585 JAPAN
[2] 1-1, Shinhama Arai-cho 1-Chome, Takasago, Hyogo 676-8686 JAPAN

Figure2 shows share of these works in the actual PWR plants.

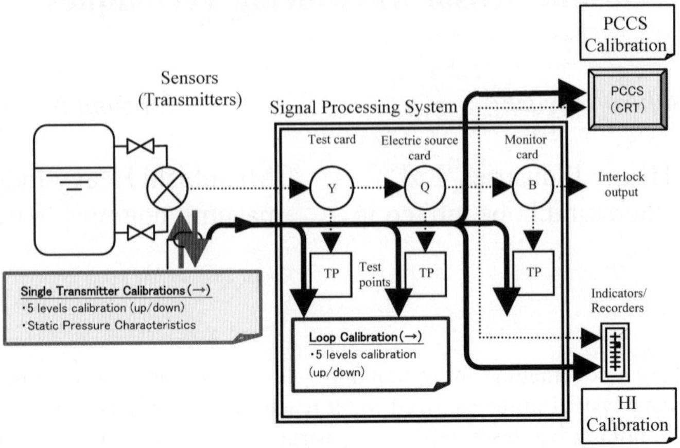

Figure 1: Calibration work at planned outage

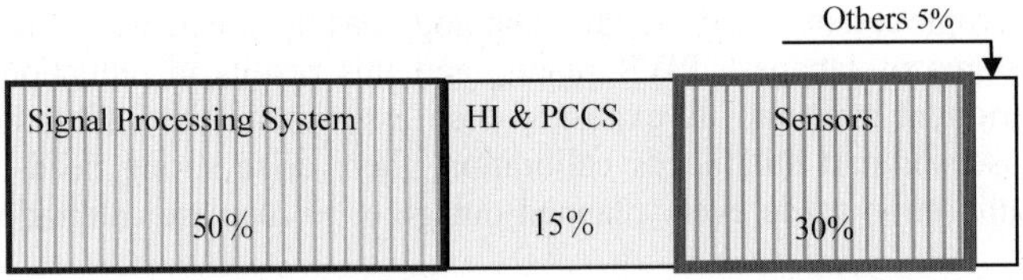

Figure 2: Share of work load for calibration

Today Japanese utilities require to reduce the maintenance work load for I&C system in order to reduce maintenance cost and to shorten the duration of planned plant outage.

We will achieve the drastic reduction of the calibration work by digitalization of the system other than sensor part. But as for sensors at present no efficient methods are realized for reducing calibration work.

Therefore, we have been developing a new system for evaluating drifts of sensors and transmitters over the full-range, judging which of them do not require calibration that now have to be made for all the sensors at every planned outage.

2 On-line sensor monitoring system

An on-line monitoring system monitors the sensors in the plants. This system models the characteristics of the sensors just after an inspection, and then monitors and evaluates sensor drifts (the difference between a true value and a measured value of the sensor). The system distinguishes sensor with drift a sufficiently small, which eliminates unnecessary calibrations.

Figure 3 shows the configuration of the on-line sensor monitoring system. The function of each block is described below.

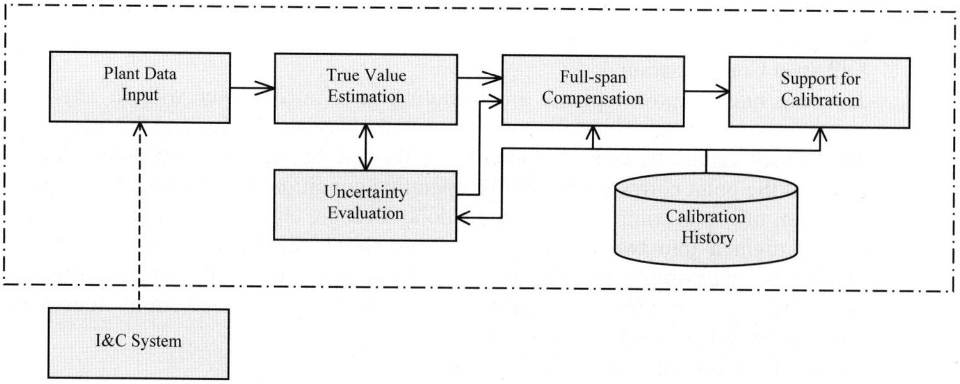

Figure 3: Configuration of the on-line sensor monitoring system

- Plant Data Input
  This part is a plant data input portion from existing I&C system, in which unit conversion, normalization and noise elimination are done if necessary.
- True Value Estimation
  True value estimation models with which the original value of the process parameter is estimated are prepared. Several kinds of estimation technique are used. One effective way is a feed forward Neural Network (NN) method, particularly an Auto-Associative Neural Network (AANN) method is used for this type of estimation (See figure4). The inputs of the AANN are set by measured values, and the AANN learns to output estimated values of the sensors corresponding to the inputs (J.W. Hines, 1998).

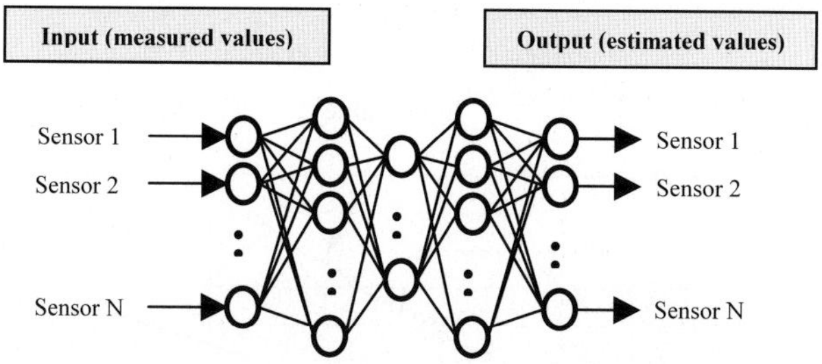

Figure 4: Structure of AANN

To cope with the sensor which reaches at 400 or 500, many sets of estimated models are provided in this part. When different estimated values for one sensor are obtained from several estimated models, joint processing of the estimated value is done by the average with the weight and so on.

- Uncertainty Evaluation
  Drift of each sensor influences the estimated value of each sensor because of using more than one correlative sensor outputs in the true value estimated model. In other words, uncertainty is contained in the estimated value which can be calculated by the model. In the uncertainty evaluation part, this uncertainty contained in estimated value

is evaluated based on the sensitivity characteristics of the estimated model and the own uncertainty of the sensor.

- Full-span Compensation
  Japanese nuclear power plants are operating at a rated power without any power change as a base load. The difference between the measured value and the true value of each sensor cannot be estimated over the full range because the measured value is in stable at the point corresponding to the operating condition. To deal with this particular situation, full-span compensation can be done focusing on the fact that the sensor drifts can be divided into two types. One is "zero-shift", the parallel shift of drifts, and another is "span-shift", the fluctuation of drifts over the sensor range. Zero-shift is estimated by true value estimation and span-shift is derived from sensor drift characteristics by statistical method (D. Asada, 2002).
- Support for Calibration Interval
  This part predicts future soundness based on the present soundness evaluated from the result over sensor full range and sensor drift characteristics. If a predicted drift range exceeds an allowance level, calibration will be necessary.

3 System Performance Evaluation

Data from a real operating plant was used to test the true value estimation methodology. Test group consists of 16 sensors, which are installed to reactor coolant system (RCS) and feed water/ main steam system (FW/MS). After learning of the estimated model, a imitative drift was intentionally added to actual process value and input to the estimated model. Figure 5 shows the test result for the main steam pressure. This graph shows that the estimated model outputs the estimated value which is almost same as the original value and that the imitative drift can be fully removed.

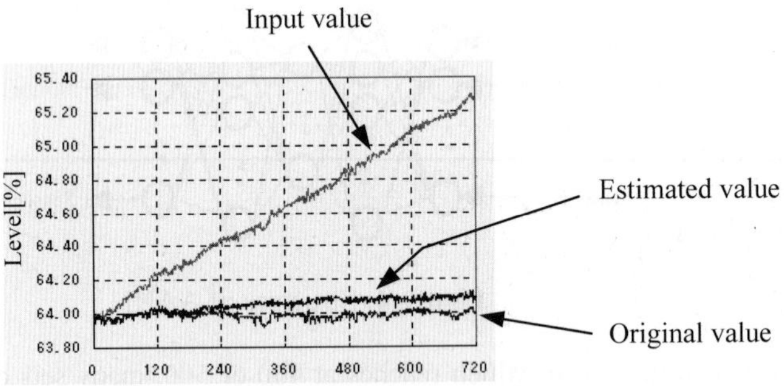

Figure 5: True value estimation

Figure 6 shows the example of support information screen which we have developed. Predicted sensor drift range is shown in two broken lines on left graph. Detailed numerical information is displayed on the right side.

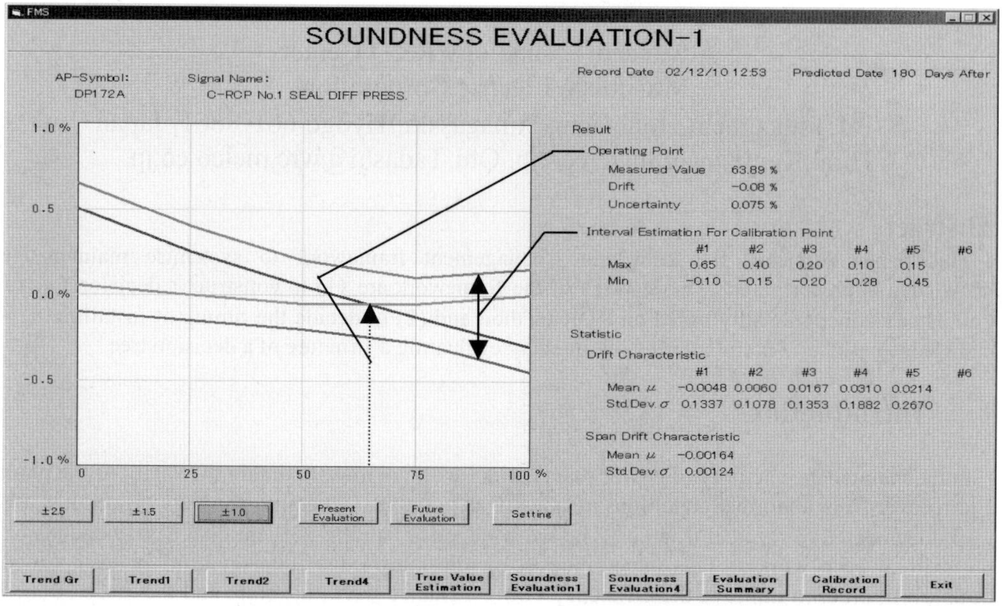

Figure 6: Example of support information display

4 Conclusion

At present we have completed the first-step development of the system, and have created a pilot system. Now we are trying to make the system more sophisticated using the real data of the various sensors gained in an actual PWR plant, and to confirm its performance. In the next phase, we plan to apply the improved system for verification of performance in the actual PWR plant. After that, also demonstrations of the feasibility of the verified system are planned for other plants. The development of the new system will be completely finished by 2004 fiscal year. After that, we will apply the developed system to the actual PWR plants.
We think it will contribute to the reduction of the maintenance cost and exposure dose.

References

D. Asada, H. Iba. (2002). Optimization of Instrument Calibration Intervals by On-line Sensor Monitoring Techniques, *ISOFIC2002 (International Symposium on the Future I&C NPP)*, pp128-130, Seoul, Korea.
J.W. Hines and R.E. Uhrig. (1998). Use of Auto Associative Neural Networks for Signal Validation. Journal of Intelligent and Robotic Systems 21, pp143-154

Development of Maintenance Knowledge Management System for Power Plant

Kenji Hirai, Tadashi Ohi

Advanced Technology R&D Center
Mitsubishi Electric Corporation
8-1-1,Tsukaguchi-Honmachi, Amagasaki, Hyogo 661-8661, Japan
Hirai.Kenji@wrc.melco.co.jp, Ohi.Tadashi@wrc.melco.co.jp

Abstract

This paper describes the knowledge management framework to systemise maintenance knowledge in power plant. The features of the framework are (1) to construct a decision tree of trouble cases incrementally based on RDR method and (2) to decide the timing of installing new knowledge based on several metrics obtained by evaluating a structure of a decision tree.

1 Introduction

In the maintenance activity of power plants, plant workers must maintain the safety of the plant and recover from abnormal conditions as quickly as possible. Generally, recovery procedures are well documented for typical troubles. The knowledge for handling exceptional cases depends on the skill of each worker. This paper focuses on the knowledge management for exceptional troubles. In order to improve the efficiency of maintenance work, the store and the reuse of such knowledge each worker has are extremely important. The knowledge sharing among workers leads to preventing occurrences of similar troubles. Additionally, knowledge acquisition from experienced workers is significant for preserving of maintenance expertise.

We are developing the knowledge management framework to systemise maintenance knowledge in power plants. The features of the framework are (1) to construct a decision tree of trouble cases incrementally based on RDR (Ripple down rules) method and (2) to decide the timing of installing new knowledge based on several metrics obtained by evaluating a structure of a decision tree.

In conventional knowledge base system, the analysis of the domain is required before constructing the knowledge base. Furthermore, it is difficult to maintain consistency of knowledge base while adding new knowledge continuously. Our approach is incremental construct of knowledge base without special analysis of the domain in advance. The addition of new cases to a decision tree is executed based on the difference list obtained by comparison with the existing similar case. The selection of appropriate distinguishing attributes is influenced by the occurrence order of cases. In addition, it has the possibility that the number of redundant nodes increases as a decision tree grows. An expert is required to check whether newly added cases are sufficiently effective to classify cases occurring in future. Therefore, the systematic approach is needed to customise the system to suit the status of accumulated knowledge. The knowledge management system provides functions of evaluating a tree structure and inducing metrics. Metrics give an expert the suggestion to decide when to reconstruct a tree.

The next section shows an overview of maintenance knowledge management system that we are developing. Section 3 describes the tree constructing method used in our system. Section 4 looks at the use of metrics for reconstructing a decision tree. Section 5 describes the main functions of the experimental system.

2 System Overview

We have designed Maintenance Knowledge Management System to support both field workers and knowledge engineers (Figure1). For field workers, the system provides services of retrieving similar cases and storing their trouble reports in the form that is reusable. Newly reports from workers are stored as a format of XML (eXtensible Markup Language) and added to a decision tree incrementally. For knowledge engineers (conventionally experienced workers), the system provides services of monitoring the performance of retrievals and checking the validation of a decision tree.

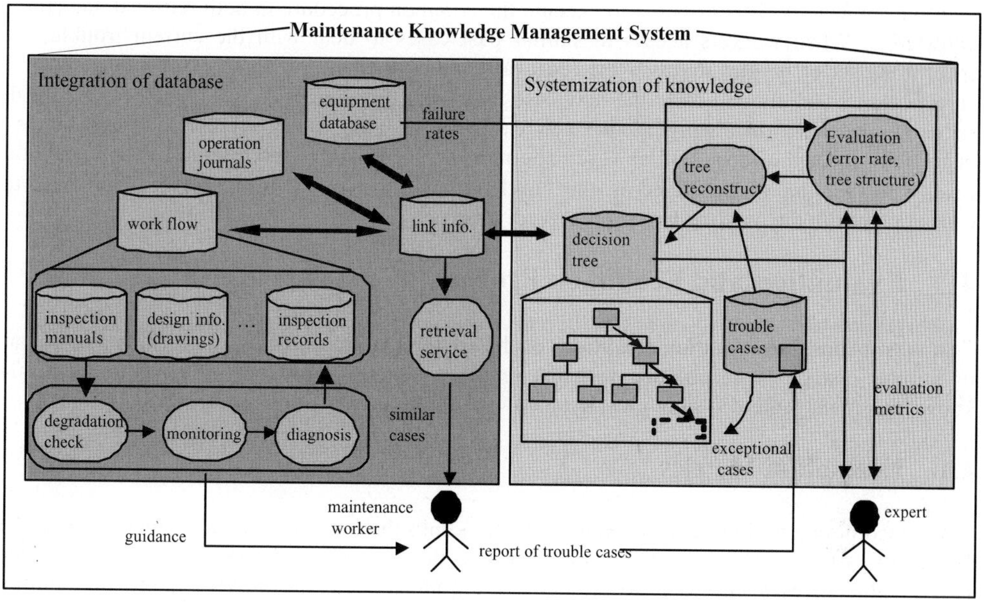

Figure 1: **System Configuration**

3 Knowledge Management based on Incremental Tree Constructing

3.1 Decision Tree Constructing

Each trouble case that is used as knowledge is described with a form of an attribute-value vector. The main attributes are the status and the cause of the trouble. The status is the attribute that can be observed externally in the occurrence of the trouble. The cause is the diagnosed result to the trouble. For example, thinking of the trouble in the controller, the status shows the color of the LED lamp on the card. The cause is the location of the wrong component that drove the breakdown of the controller.

The tree constructing method is based on RDR method ([1][2]). A decision tree is expressed in a binary tree. Each branch of a tree has the distinguishing status. Each node of a tree has the cases classified as the same class. When a new case is given, it is passed down the proper branch until a leaf node is reached. If the node selected as candidate gives the correct solution to the current case, the current case is stored to the selected node and no new nodes are added. Otherwise, the new node is added to store the current case. The distinguishing status for the new branch is calculated by the comparison with the status set of the parent node.

The accuracy of the classification and the cost of constructing a tree are in the relation of trade-off. In the viewpoint of practical use, the timing of installing a new node should be executed after recognizing the effectiveness to classify other cases. The method of deciding the timing of installation and selecting the best distinguishing status out of several candidates are described in section 4.1.

3.2 Retrieval and Store of Maintenance Knowledge

In our framework, it is possible to execute the common procedure in both retrieval and store of knowledge. When workers access to similar past cases to deal with the current trouble, they specify the set of unique status as the retrieval condition. If the retrieved result mismatches the current case, workers can store the new case by giving the correct cause. If the set of specified status is insufficient to distinguish the new case from existing cases, workers add new status to distinguish from the most similar case. The newly added case is interpreted as the exceptional case. The location to be added is the child node of the most similar case. Therefore, the consistency with the existing tree is preserved.

4 Evaluation and Feedback Process

In most plants, the operation conditions of the equipment are not fixed. The kinds and the ratio of failures change in accordance with the replacement and upgrading of the equipment. The knowledge management system must keep up with the change of such operating conditions.

4.1 Evaluation Metrics

We incorporate two kinds of evaluation metrics into the knowledge management system. The first metric value gives the statistics of usage (ex. the hit ratio of the retrieval etc). The second metric value gives a hint to make a tree more compact (ex. the metrics based on minimum description length ([3]).

At each time a case is added, metrics are recalculated. For leaf nodes and intermediate nodes that have cases, the following calculation is executed.

$$V = 1 + \log(C) + E * (\log (I * (C - 1)))$$

(C: the count of class, I: the count of instances, E: the count of exceptions)

This value shows the cost that is required to reclassify exceptional cases. For all nodes that have subsidiary trees, the following calculation is executed hierarchically.

$$V = \log (T) + V_l + V_r$$

(V_l : metrics of left side tree, V_r : metrics of right side tree, T: the count of conditions)

Figure2 shows the tree structure and metrics that are calculated to each node. First, the new data ("case1") is classified in the node ("node-B") that has the most similar cases. After this classification, node-B has exceptional cases. Therefore, the system also calculates metrics to alternative trees with a new node ("node-E") for classification. By the comparison between the metrics of the current tree with alternative trees, the system evaluates the effectiveness of reclassification and selects the better tree. Assume the lower value is the better selection.

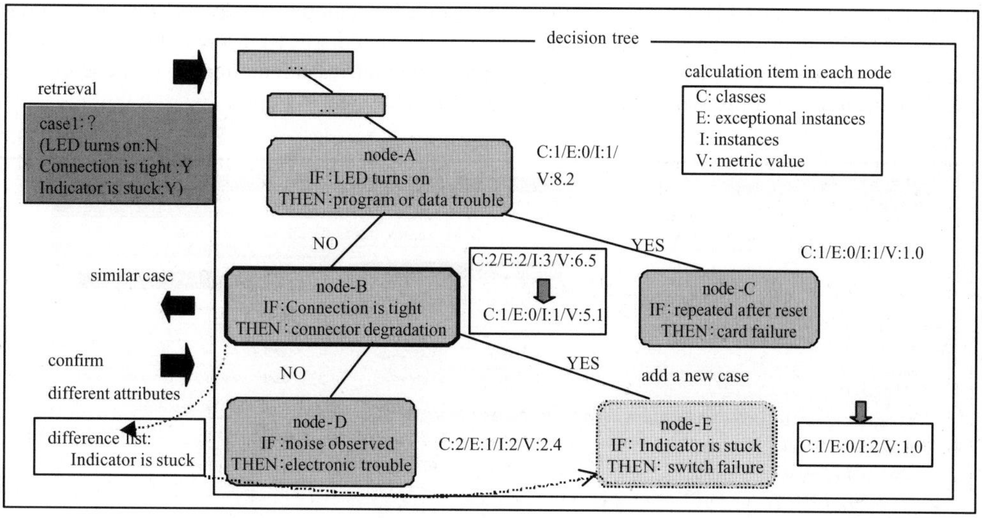

Figure 2: an example of a decision tree with metrics

4.2 Tree Reconstructing

Incremental constructing approach makes it possible to obtain a decision tree by low cost. In order to maintain the system effectively during long-term usage, it is necessary to customize the system continuously based on the appropriate evaluation.

To evaluate a tree in several points of view, we provide the function of reconstructing a decision tree. In the reconstruction phase, it is possible to simulate the process of systemisation under different set of the cases for training and testing. When the hit ratio of retrievals degraded in the system, it is possible to investigate the cause by comparing with a tree of a different structure. Such simulation can also be helpful in the investigation of the effect derived from giving cases with the different sequences and the missing value of some cases.

The procedures of reconstructing a tree are the followings.

Step1) Monitoring of the error rate in classification according to installation of new cases
The expert decides the timing of tree reconstruct.

Step2) Evaluation of the current tree structure by metrics
The expert checks the status of classification and locates sub trees that are necessary to reconstruct.

Step3) Selection of training data set in order to construct a new decision tree
Following the result of the step2, the expert selects training data set and eliminates unnecessary cases with less reference count.

5 Development of Experimental Prototype

We are developing the experimental prototype of maintenance knowledge management system. The system will be integrated with plant database (ex. the operation journals and maintenance records). The management of trouble cases is based on the format of XML ([4]). For practical purpose, each trouble case is linked with related data (ex. maintenance manuals, equipment configuration, records). Plant workers can easily search for maintenance knowledge and relevant data by specifying observed status (Figure3). On the other hand, knowledge engineers can monitor occurrence counts of exceptional cases to each node. The metrics calculation gives a cue to locate the effective node for improving the efficiency of retrieval.

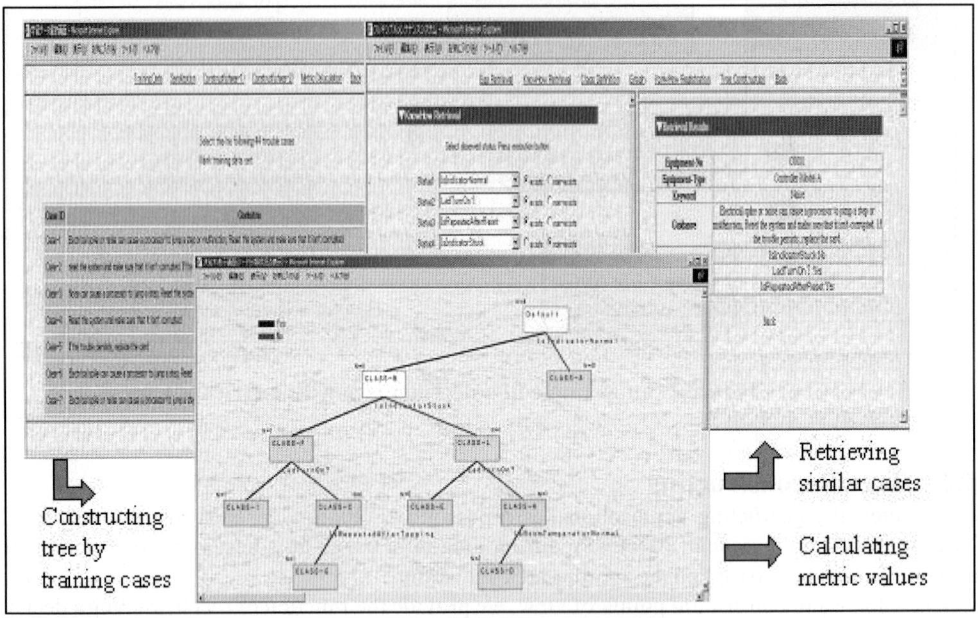

Figure 3: **screenshots of user interface**

6 Conclusion

In this paper, we presented the framework of maintenance management system for power plant. Currently, we are evaluating the functionality and the effectiveness of the system through the application to maintenance work of controller systems. We believe that this framework will be helpful to share common knowledge among plant sites as well as to manage site-specific knowledge.

References

[1] Compton, P. (1992). Ripple Down Rules: Turning Knowledge Acquisition into Knowledge Maintenance. *Artificial Intelligence in Medicine*, Vol.4, pp47-59.

[2] Davis, R. (2000). User-Centred and Driven Knowledge-Based Systems for Clinical Support using Ripple Down Rules. *Proc. of the 33rd Hawaii International Conference on System Sciences*

[3] Utgoff, P. E. (1997). Decision tree induction based on efficient tree restructuring. *Machine Learning, 29*, pp5-44.

[4] World Wide Web Consortium. Extensible Markup Language (XML). from http://www.w3.org/XML

Customer Relationship Management in E-business

Diana Horn

School of Industrial Engineering
Purdue University
1287 Grissom Hall
West Lafayette, IN 47907
dhorn2@purdue.edu

Richard Feinberg

Department of Consumer Sciences and Retailing
Purdue University
West Lafayette, IN 47907-1262
feinbergr@cfs.purdue.edu

Gavriel Salvendy

School of Industrial Engineering
Purdue University
1287 Grissom Hall
West Lafayette, IN 47907
salvendy@ecn.purdue.edu
and
Department of Industrial Engineering
Tsinghua University
Beijing P.R. China

Abstract

The shift from product based to value based competition in e-business accentuates the need for better understanding and application of customer relationship management (CRM). This paper examines the basic components of CRM and the bond between CRM and e-business (referred to as eCRM), which leads to customer acquisition, retention, and profitability. Results of a recent study on eCRM (Horn, 2002) support the overall conclusion that relational aspects of customer-business interactions play a major role in determining customer attitude. Along with CRM in e-business, future applications of CRM in mobile commerce will greatly depend on the understanding of customer-business interactions.

1 Introduction

The advance of e-business and mobile commerce heightens the need for better understanding and management of customer relationships. As competition shifts from product based to value based, customer relationship management (referred to as CRM) empowers businesses to create and nourish relationships with customers and in turn maximize customer retention, acquisition, and profitability. The ease of product and price comparisons on the Internet increases the difficulty for organizations to retain customers and stay competitive with price and product offers. CRM shifts the competition from price and product based to customer based. The Internet offers new opportunities and technologies for CRM implementation by creating additional touch-points or channels for an organization to communicate with the customer, acquire customer knowledge, and customize interactions. Therefore, the combination of CRM and e-business generates powerful tools for retaining valuable customers and increasing profits.

While there is a great deal of awareness on the need and importance of CRM in e-business (which is referred to as eCRM), not much attention has focused on clearly defining the process of eCRM from the customer's perspective. In other words, what does a customer really value in the interactions with an e-business? This purpose of this examination of eCRM is to: investigate the role of eCRM in the customer-e-business interactions; evaluate the results and implications of a recent study on eCRM (Horn, 2002); and briefly discuss the future of CRM in mobile commerce.

2 CRM in E-business

To better understand what a customer values in a relationship with an e-business, one must first define the overall goals and components of CRM in e-business. Customer relationship management is associated with various overlapping terminology such as: relationship marketing, one to one marketing, 360 degree view of the customer, back-office to front-office, knowledge management, etc. Alongside the conglomerate terminology, the definition and concept of CRM varies from person to person and boardroom to boardroom. This study defines CRM as the process of understanding and relating to the customer in order to enhance customer acquisition, retention, and profitability, and defines eCRM as the incorporation of CRM into e-business.

The eCRM goals of customer acquisition, retention, and profitability are obtained through four eCRM business components, which include *interact, connect, know, and relate* (Swift, 2001). These elements exist in an on-going cycle that begins with building customer knowledge and results in high-impact customer interactions (Swift, 2001). The first component, *interact* is defined as the series of transactions and interactions that make up a continuing dialog between a customer and the business where customer data (customer identification, behaviour, wants, and needs) are collected (Peppers & Rogers, 1999). The next component of the eCRM process is to *connect* or map and manage the interaction points between the customer and business (Swift, 2001). Because the eCRM process is based on knowing the customer and applying that knowledge, the integration and accessibility of knowledge is critical to the success of eCRM by gaining a broader view of the customer (Rollo & Garrett, 2001). The third component, *know* is the insight gained through the capture and analysis of detailed information to create continuous learning (about customers, products, channels, markets, and competitors) from the data warehouse and/or knowledge bases (Swift, 2001). A final major component of eCRM is *relate* or applying customer insight to create relevant communications between customers, channels, suppliers, and partners (Swift, 2001).

The four eCRM components operate in a cycle or process so that the e-business is continually learning and better relating to the customer. This relationship management cycle centers around interactions between the customer and the business. Interactions, whether through sales, marketing, or customer service and support departments, lay the foundation and build the relationship between the customer and the e-business.

3 eCRM Interactions and Customer Attitude

Examining business-customer interactions and customer attitude in both traditional and virtual environments provides better understanding of the eCRM process. The general human performance model suggests that optimal human performance comes from skilled individuals who perform satisfying activities in a favourable environment (Bailey, 1996). Thus the design of eCRM systems must incorporate conditions that produce satisfying interactions and favorable environments for the customer. Studies of customer-business interactions and the impacts on customer attitudes provide insight into what conditions actually contribute to the overall effectiveness of eCRM. A broader understanding of effective eCRM conditions is obtained by looking customer attitudes in both the traditional physical setting (where the customer-business interactions primarily occurs during face to face contact between the customer and the employee) and the virtual e-business environment (where interactions mainly occur between the customer and the e-business website).

In the traditional physical environment, customer attitudes are associated with various aspects of customer-business interactions. Customer satisfaction is related customer dialog (Bruhn & Grund, 2000) and the business's initiating, signaling and disclosing behaviors and frequency of interaction (Leuthesser & Kohli, 1995). Other studies show that perceived quality is related to level of service provider participation (Ennew & Binks, 1999), business tangibles (physical facilities), reliability (dependable service), responsiveness (prompt service), assurance (knowledge and courtesy of employees), and empathy (individualized attention) (Parasuraman, Zeithaml & Berry, 1988, 1991).

In the virtual environment, customer attitude is also associated with numerous e-business website features. Customer attitudes such as satisfaction, perceived quality, and trust are shown to be related to e-business website ease of use, aesthetic design, processing speed, and security (Yoo & Donthu, 2001); convenience, site design, and financial security (Szymanski & Hise, 2000); entertainment, informativeness, and organization (Chen & Wells, 1999); and comprehensive information and communication (Lee, Kim & Moon, 2000). A more complete model of CRM in e-business is gained by incorporating the key relationship aspects in the traditional retail environment with the research findings of important e-business environmental features. Thus, in order to achieve customer acquisition, retention, and profitability, eCRM must encompass the both physical and virtual aspects of interactions that impact customer attitudes.

4 An Empirical Investigation of eCRM

A study by Horn (2002) tested a model of eCRM that examined different aspects of the relationship between customers and e-businesses. The model of eCRM in this study proposed that customer-business interactions lead to customer attitudes, which in turn impacts customer retention, acquisition, and profitability. The eCRM model was tested with a 38 item eCRM survey, which was constructed based on literature findings, SITEQUAL (a website quality instrument), and SERVQUAL (a service quality instrument). Two hundred e-business customers and customer contact professionals completed the survey on the web.

The study finds three main eCRM attributes: general CRM (which accounts for 51% of the total variance); personalization (which accounts for 9% of the total variance); and privacy (which accounts for 7% of the total variance) from a factor analysis. The study also showed that these eCRM attributes significantly predict customer attitude and account for 83% of the explained variance of customer attitude. The study also emphasized that website content organization (an item within the general CRM dimension) was correlated highly with customer attitude (65% of the explained variance). This study supports the general conclusion that in the virtual environment, customer-business interactions lead to customer attitudes and these main interaction aspects are key components in eCRM.

The results of this study provide both theoretical and practical implications to the development of eCRM. Theoretically, this study provides a foundation for development of eCRM guidelines and emphasizes the need to include relational type attributes of e-business websites in further investigation of eCRM. Practically this study also reveals customer perceptions of eCRM as three major dimensions: general CRM, personalization, and privacy.

5 Future of CRM in Mobile Commerce

As with eCRM, the application of CRM in mobile commerce (m-commerce) creates more opportunities for business to provide value added services in their relationships with customers. However, the mobile commerce market space requires an even broader view of the technology, the market, and the consumers by not only considering m-commerce as a new distribution channel, but also as new aspect of consumerism (Nohria and Leestma, 2001). M-commerce enables both the customer and the employees to become mobile, thus creating new types of applications and interactions for the business-customer relationship. New interaction features generated through m-commerce include ubiquity (communication anywhere at anytime), personalization, and flexibility which result in easy, timely access to information and immediate purchase opportunities (Siau, Lim, and Shen, 2001).

The expansion of CRM into mobile commerce faces several challenges. The business must learn how to provide customer confidence through enhanced system design (Siau, Lim, and Shen, 2001). Similar to in eCRM, the application of CRM in m-commerce faces the issue of how to provide secure interactions with customers and build customer trust. The flexibility and portability of m-commerce technologies also increases the need for simplicity of use and interface design. Customer relationship management allows business, whether in e-commerce or m-commerce, to maintain valuable relationships with customers through connected and efficient interactions.

References

Bailey, R. (1996). *Human Performance Engineering: Designing High Quality, Professional User Interfaces for Computer Products, Applications, and Systems.* Upper Saddle River: Prentice Hall PTR.

Bruhn, M. & Grund, M. A. (2000). Theory, development and implementation of national customer satisfaction indices: The Swiss Index of Customer Satisfaction (SWICS). *Total Quality Management,* 11, 1017-1028.

Chen, Q. & Wells, W. (1999). Attitude toward the site. *Journal of Advertising Research*, 39, 27-37.

Ennew, C. T. & Binks, M. R. (1999). Impact of participative service relationships on quality, satisfaction, and retention: An exploratory study. *Journal of Business Research,* 46, 121-132.

Horn, D. (2002). Customer Relationship Management in E-Business, Unpublished Master's Thesis, School of Industrial Engineering, Purdue University, West Lafayette, Indiana, U.S.A.

Lee, J., Kim, J. & Moon, J.Y. (2001). *What makes Internet users visit cyber stores again? Key design factors for customer loyalty. In H. J. Bullinger & J. Ziegler (Eds.),* Human Computer Interaction, (pp. 305 - 312). *Mahwah, NJ: Lawrence Erlbaum Associates, Publishers.*

Leuthesser, L. & Kohli, A. K. (1995). Relational behavior in business markets. *Journal of Business Research,* 34, 221-233.

Nohria, N. & Leestma, M. (2001). A moving target: The mobile-commerce customer. *JMit Sloan Management Review*, 42(3), 104.

Parasuraman, A., Zeithaml, V. A. & Berry, L. (1988). SERVQUAL: A multiple-item scale for measuring customer perceptions of service quality. *Journal of Retailing*, 64, 12-40.

Parasuraman, A., Berry, L. L. & Zeithaml, V. A. (1991). Refinement and reassessment of the SERVQUAL scale. *Journal of Retailing*, 67, 420-450.

Peppers, D. & Rogers, M. (1999). *Enterprise one to one: Tools for competing in the interactive age.* NewYork: Currency Doubleday.

Rollo, C. & Garrett, H. (2001). Cheap thrills: 16 ways to cost effectively manage the ultimate customer experience part I and part II. Retrieved March 18,2001 from www.crmcommunity.com.

Siau, K., Lim, E. & Shen, Z. (2001). Mobile commerce: Promises, challenges, and research agenda. *Journal of Database Management*, 12 (3), 4-13.

Swift, R. S. (2001). *Accelerating customer relationships using CRM and relationship technologies*. Upper Saddle River, NJ: Prentice Hall PTR.

Szymanski, D. & Hise, R. (2000). e-Satisfaction: An initial examination. *Journal of Retailing*, 76, 309-322.

Yoo, B., Naveen, D. (2001). Developing a scale to measure the perceived quality of an internet shopping site (SITEQUAL). *Quarterly Journal of Electronic Commerce*, 2, 31-45.

A Computerized Graphic Interface on Emergency Operating Procedure (EOP)

Fei-hui Hwang

National Tsing Hua University
Hsinchu, Taiwan, R.O.C., 300
d907817@oz.nthu.edu.tw

Sheue-Ling Hwang

National Tsing Hua University
Hsinchu, Taiwan, R.O.C., 300
slhwang@ie.nthu.edu.tw

Abstract

Operation of a Nuclear Power Plant (NPP) is a complex task, especially in emergency operation. In an emergency, the major task of the operator is to ensure plant safety, but the complex displays and multiple messages in the control room make the supervisor sustain heavy stress in decision-making, which may cause negligence, inadequate or incorrect decision response. Thus, the objective of this paper is to develop a computerized graphic interface to assist complex operation in nuclear power plant. The experimental task involved two major procedures: Average Operating Procedure (AOP), which adopts basic watermark control procedure, and Emergency Operating Procedure (EOP), which adopts alternative watermark controls procedure. The results of experiment and subjective rating revealed that computerized graphic interface can enhance training, reduce time stress, and improve operating accuracy.

1 Introduction

During the 1980s, the commercial nuclear power industry increased its investment in automated information sharing, which has lead to more frequent use of information management tools. This growth reflects the expansion of computing power to include plant operations and maintenance personnel, who need timely system performance data. Most of the emphasis has been placed on computerized training systems for reactor operators rather than on assisting operators in their daily responsibilities. Operation of a nuclear power plant is a complex task that requires knowledge of plant systems and procedures, including system interactions and limiting conditions for operation (Thomas, 1987). In an emergency, the role of the operator is to ensure plant safety. The objective of this paper is to develop a computerized graphic interface to assist complex operations (including average operating procedure and emergency operating procedure) in a nuclear power plant.

The interface design for interactions between humans and machines or computers is becoming an important issue as automation technology and equipment have been widely used (Rook and Donnell, 1993). Using a training simulator to evaluate performance was conducted to find how human performance in emergency differs from that under normal operation conditions (Ohtsuka et al., 1994). The requirements for a computerized graphic interface are firstly developed by an existing method in NPP domains of application, and by identifying some specific functions. Then, the system is proposed, and the content of procedure and operation features are described. Finally, the experimental results of the computerized graphical interface are discussed.

2 Method

In this study, a computerized graphic interface system in a NPP is a small-scale virtual system. The procedures of basic watermark control and alternative watermark control, which are parts of Procedure 500.3 EOP-RC, are selected. The interface of watermark system was developed and then an experiment was conducted to evaluate the effect of this interface system on normal and emergency operating performance.

2.1 Applying DWCE (Dynamic Work Causality Equation) model

As a complement to the traditional reasoning models, a Dynamic Boolean Expression provides an event-driven approach for the analysis and modeling of interactive reasoning (Yang and Hwang, 2001). DWCE inherits the logical concepts of Dynamic Boolean Expression, which can analyze and predict outcomes of a dynamic decision process. The DWCE model represents a decision dynamically and indicates the changes at each step (Hwang et al., 2000; Yang and Hwang, 2001). Hwang and Hwang (in press) verified that the flowchart transformed by DWCE has improved the comprehension of operation. DWCE model was applied to transform the basic watermark control and alternative watermark control procedures into flowcharts. These flowcharts were used for the design of the computerized interface in this study.

2.2 Interface design of watermark system

Visual Basic 6.0, which is an object-oriented programming and visual program, is used to design the proposed interface. To apply the human factors principles in the DWCE model, the flowchart of the procedure is programmed. The flowchart includes the status, decision, and action components, which are represented by the green light, red light, and yellow light respectively in the interface. Hwang and Hwang (in press) have verified that applying the AOP computerized graphic interface in the training stage has significantly reduced learning time and increased operating performance. In the present experiment, both AOP and EOP computerized graphic interface are applied in the watermark system. The whole interface presents three major functions. It includes operations, reaffirming, and procedural guide.

2.3 Scenarios design

Situations have implications for system design and training (Meister, 1995). In this system, each event was presented by the combination of visual and auditory alarms. Since measurements, such as time and accuracy are good index for human performance (O'Hara and Hall, 1992), subject's operation time and number of errors during the experiment of this study will be recorded by computer program.

2.4 Experimental variables

The independent variable in the experiment includes two levels.

- *Level 1. AOP (Average Operating Procedure): basic watermark control procedure;*
- *Level 2. EOP (Emergency Operating Procedure): alternative watermark control procedure.*

There are three dependent variables in the experiment.

- *Operation time: An index of manipulation speed should be considered due to time limit for such task.*
- *Accuracy of action: The number of errors when handling events was an index of the accuracy.*
- *Open-ended subjective rating*

2.5 Subjects

Thirty-one students of industrial engineering at Tsing Hua University, Taiwan, participated in the experiment with payment. All the subjects have taken the course of Human Factors. Their ages ranged from 20 to 30. Each subject attended 2 stages in this experiment, training stage with three training steps and simulation stage. There were four events, two in basic watermark control and two in alternative watermark control, in training stage to be manipulated. In the simulation stage, there were six events, two in basic watermark control and four in alternative watermark control. The tasks include decisions and actions to deal with alarm signals.

3 Results

The performance measures, operation time and accuracy of action were analyzed, combining the results from open-ended subjective rating.

3.1 Operation time

Operation time of AOP and EOP obtained from the experiment including training 1, training 2, training 3, and simulation, were analyzed. The results of ANOVA revealed that operation time among training 1, training 2, training 3, and simulation were significantly different ($p<0.05$). The average time for each operation of AOP and EOP is compared that operation time of AOP is longer than that of EOP during each step. In addition, operation time of each procedure is decreased gradually.

3.2 Accuracy of action

The results of ANOVA revealed that accuracy of training1, training 2, and training 3 were not significantly different ($p>0.05$), but accuracy between training stage and simulation stage was significantly different ($p<0.05$). The accuracy increased as training time increased. However, the number of errors increased in the simulation stage.

3.3 Open-ended subjective rating

The subject's feedback was obtained by subjective rating after the subject finished an experiment. The purpose of the subjective rating is to collect the opinions of subjects on five topics: 1) design of auditory stimulus, 2) design of visual stimulus, 3) importance for interface design, 4) mental load and stress of subject, and 5) the reasons for making mistakes.

4 Discussion

The computerized graphic interface is designed to simulate a series of events in watermark procedure. The experimental task of the present study included operating of AOP and EOP during the training stage. Then the operating performance can be verified in simulation stage.

4.1 Training effect

The T test indicated that operation time between training 1 and training 2 [t (31)=6.1, p<0.05], between training 2 and training 3 [t (31)=2.5, p<0.05], and between training 3 and simulation [t (31)=5.6, p<0.05] were significantly different each other in the AOP. Moreover, the number of errors from training 1 to simulation decreased gradually (see Table 1). In the EOP, the operation time between training 1 and training 2 [t (31)=4.6, p<0.05], and between training 2 and training 3 [t (31)=2.2, p<0.05] were significantly different from each other. There were no significant differences between training 3 and simulation [t (31)=0.56, p>0.05], but the number of errors decreased gradually. Through experiment, it is verified that the computerized graphic interface for AOP or EOP had significant training effect on performance.

4.2 Comparison of AOP and EOP

In a nuclear power plant, the operation time is limited and the operating error is not allowable. Hwang and Hwang (in press) have verified that there exists higher learning effect for a novice when using computerized graphical interface than using non-computerized interface for AOP. The result of subjective rating indicated that subjects were less vigilant during operation of AOP, and thus the number of errors increased even though operation time was faster. Therefore, operation time and accuracy should be traded off. The designer must decide which information is important to specify and what level of detail specification is needed at certain stages of the design process (Howell et al., 1993). Standardization of visual information design can minimize the trade-offs problem.

4.3 Limitation of study

Visual information in graphical interface is a primary channel for supervisors to recognize the system status. The opinions of subjects about visual information design of the proposed system includes, 1) to give careful consideration on font type and font size of Chinese characters, 2) the numerical display should avoid too many decimal fractions, 3) use auditory feedback to reduce eye strain and increase some mood aspects positively (Rauterberg, 1998), and 4) to replace word by symbol. The computerized interface, however, has a physically limited screen size. Font type and font size should be designed as clearly as possible. Besides, the causes of errors in the proposed system may be due to individual differences, external factors, good or bad design of interface and feedback, and the level of training.

Every interaction of complex procedure possesses attributes that affect interface design. A framework for the description of procedure is needed in which computerized graphical interface can be represented as event, status, and operating patterns along with descriptive content of interfacial function in a working environment. To apply computerized graphical interface to a time-critical task improves operator vigilance, recognition ability, skill proficiency, and operation safety.

5 Conclusion and Further Study

In this study, a DWCE model was applied to develop the graphic interface of AOP and EOP. The interface provides real-time information and procedural assistance to the operator. Then, the training effect of interface was verified by conducting an experiment. The results of this experiment showed that the performance of EOP using computerized graphic interface could be significantly improved. We also observed a significant increase of operator vigilance in different procedures. Finally, one could see that there were more learning effect with computerized graphic interface. Through subjective rating, some suggestions to improve this graphic interface design were drawn. Firstly, by using different color to the numeric display, the discrimination can be significantly increased. Secondly, auditory alarm feedback helps operators to keep vigilance on the processes and to track the activity. Thirdly, some wordy statements can be replaced by simple symbols.

Since the operation of nuclear power plant is a teamwork task, for the further research, one may study the performance of teamwork via network and similar computerized graphic interface.

Acknowledgement

The authors would like to thank the Administration of Nuclear Power Plant II providing valuable information and partial of this study was supported by National Science Council of the Republic of China under contract No. NSC89-TPC-7-007-019.

References

Howell, S., Hoang, D. N., Nguyen, C. & Karangelen, N. (1993). Critical issues in the design of large-scale distributed systems. *IEEE Advance in Parallel and Distributed Systems,* 28-33.

Hwang, F.H. & Hwang, S.L. (in press). Design and evaluation of computerized operating procedures in nuclear power plants. *Ergonomics.*

Meister, D. (1995). Cognitive behavior of nuclear reactor operators. *International Journal of Industrial Ergonomics*, 16, 109-122.

Ohtsuka, T., Yoshimura, S. (1994). Nuclear power plant operator performance analysis using training simulators-operator performance under abnormal plant conditions. *Journal of Nuclear Science and Technology*, 31(11), 1184-1193.

Rauterberg, M. (1998). About the importance of auditory alarms during the operation of a plant simulator. *Interacting with Computers*, 10, 31-44.

Rook, F. W. & Donnell, M. L. (1993). Human cognition and the expert system interface: Mental models and inference explanations. *IEEE Transaction on System, Man, and Cybernetics*, 23, 1649-1661.

Thomas, J. M. (1987). Software tools for the nuclear plant operator. *Nuclear Plant Journal*, 20-22.

Yang, C.Y. and Hwang, S.L. (2001). Reappraisal of decision making models in engineering applications. *International Journal of Cognitive Ergonomics*, 5(2), 149-177.

Usability of Spatial Decision Support Tools for Collaborative Water Resource Planning

Piotr Jankowski

Department of Geography, San Diego State University
5500 Campanile Drive, San Diego, CA 92182-4493
piotr@typhoon.sdsu.edu

Abstract

The experiments involving water users from the Boise River Basin in Western Idaho, USA were conducted over the period of two years to evaluate the optimal strategy for integration of stakeholder involvement and the potential influence of technology in the decision making process. Spatial decision support tools used in experiments ranged from desktop GIS software to a prototype of spatial decision support system for collaborative water resource planning. The experiment demonstrated that spatial displays of information assist in enabling stakeholders to make informed decisions. Members of the Control Group found spatial displays to be useful and helpful in decision determinations. Members of the Test Group, while also finding spatial displays to be useful and helpful, were not able to fully utilize the enhanced technology provided for their group, and did not rate the usability of the technology as highly as the Control Group. This presents a challenge to researchers to increase the simplicity and versatility of software for future use in collaborative planning and decision-making.

1 Introduction

Recent years have witnessed a high level of interest in methods and applications of spatial information technologies for support of public participation in local planning and decision-making (Jankowski & Nyerges 2001, Craig et al., 2002). Yet despite the active research on, and practical applications of participatory decision making, our understanding of which spatial information tools are effective for support of public participation and which are not is limited by the shortage of empirical studies of their use. Especially valuable in this regard are the empirical studies offering insight into real decision problems tackled by true stakeholders. Such studies can lead to accrual of knowledge about how to effectively use geographic information (GI) in support of participatory decision-making. In order to realize this potential, empirical studies should be based on a sound theoretical framework that allows comparing results across case studies and problem domains. This paper reports about preliminary results of one such empirical study that offered to provide insights into the dynamics of collaboration involving true stakeholders in a realistic decision setting of water management planning in the Boise River Basin of southwestern Idaho. Of particular interest in this study was how 2D and 3D maps, decision models, and group work information tools influenced the dynamics of, and were influenced by a group interaction during collaborative decision process. A field experiment design with two groups – a control group and a test group was used in order to establish whether or not the use of GI tools would help to improve the decision process and outcomes. Members of a Test Group were provided with individual computers and access to spatial information technology tools, geographic data, and facilitators to assist with data exploration. Members of a Control Group conducted the identical sessions as the Test Groups, with the exception that they did not have individual computers with which to further analyze GIS data, and instead solely relied upon group information displays.

The decision task in the study involved the selection of a water management plan for the Boise River Basin in Southwestern Idaho, USA. The stakeholders taking part in two experiments, conducted between May of 2001 and September of 2002, included a mix of participants representative of common-pool resource decision problems. They were water district representatives, irrigation district representatives, municipal water provider representatives, general public representatives, and the representatives of private ground water users (irrigators and commercial water users). In the first experiment, conducted in May of 2001 the control group was exposed only to Geographic Information System- (GIS) based maps, tables, and graphs used to present the decision situation-relevant information. Lessons about the use of spatial information tools by the stakeholders during the first experiment were used in the design of prototype collaborative spatial decision support systems called *WaterGroup*, used during the second experiment in September of 2002. In the second experiment both groups used the *WaterGroup* with the only difference that the control group solely relied upon group information displays whereas the members of the test group had access to individual computers. Each group had ten participants representing different stakeholder perspectives. The reminder of the paper describes the decision task, tools used during the experiments and the results.

2 Decision Task

In the State of Idaho and in all other western states, many streams and aquifers are unable to provide sustained water supplies that fully satisfy all uses during good water years. During drought years water distribution insufficiencies are exacerbated. In order to better use the available water resources Idaho and other states have been increasingly interested in conjunctive water mangement approach. The term "conjunctive management" (CA) has often been used to refer to the interrelationship between surface and ground water. Historically, ground water distribution decisions in the western United States have often been made by water rights administrators who utilize informal weighting considerations in crafting solutions. While this informal technique has produced satisfactory results in many instances, it will not be adequate as stakeholders demand increasing participation. Hydrologists can integrate technical information mathematically in deterministic models to predict the impact of ground water withdrawals on streams. If these simulations were highly accurate, water distribution requirements could be implemented in a straightforward manner. These simulations, however, are laden with uncertainty. In addition, results of these simulations and accompanying assumptions are not widely understood by stakeholders. Accordingly, technical solutions are often not supported by constituents. Decision Support Systems must be developed to integrate data from the variety of technical information that is now available in a manner that (1) fully describes technical uncertainties, and (2) can be understood by stakeholders.

The purpose of the experiments was to receive feedback from stakeholders regarding implementation of CA in the Boise River Basin. This is potentially a broad topic, requiring some narrowing of issues to allow for focusing of the discussions. The decision space for implementation of CA covers a broad perspective including timing of implementation, scope of ground water wells to be affected, mitigation techniques and decision-making techniques. The stakeholders were asked to assist with providing information related to only a portion of the decision space in this matter. The scope of the decision task presented to the stakeholders was limited to the first two of these four areas, thus including only timing of implementation and ground water wells to be affected. Ground water wells were grouped into zones based on their proximity from the Boise River. The grouping of wells into the zones was based on the proven hydrological relationship between the volume of water withdrawn from the wells, distance of wells from the river, and the level of water flows in the river. The stakeholders were asked to consider two timing scenarios for CA. The first scenario was to select the number of years before CA should begin. The second was to select the number of years that would be required to complete CA in all zones. While the decision task was straightforward, it represented the opportunity for a fundamental shift in the distribution regime for water rights in the Boise River Basin. Diversions from ground water have never been regulated within the basin, and an affirmative timing scenario had the potential of initiating a significant layer of regulatory impact in the basin by initiating CA in the basin and requiring for the first time that ground water appropriators mitigate the impacts of their diversions on the surface water system.

3. Participatory Spatial Information Tools

Two different sets of tools were used in two experiments. In the first experiment commercial GIS-based spatial decision support software was used. In the second experiment custom-developed participatory spatial decision support software was used. Both software tools are described respectively in the following two sections.

3.1 Tools Used during the First Experiment

In the first experiment conducted in May of 2001 both groups used a spatial decision support software called GeoChoicePerspectives – GCP (Jankowski & Nyerges, 2001). The GCP supports single user and/or group-based decision making in a geographic information system (GIS) context. Stakeholders use the software to explore, evaluate and prioritise preferences on all aspects of a decision-making process involving multiple criteria and options. Options are represented as points, lines or areas with attributes. Multiple perspectives on options evaluation can be combined to provide an overall perspective. Single users can use the GCP to synthesize multiple evaluations of an option ranking. Groups can use the GCP to combine multiple perspectives on criteria and options in an iterative process for consensus building.

The GCP software is composed of three components: GeoVisual, ChoiceExplorer, and ChoicePerspectives. The GeoVisual component is used by stakeholders for exploring geographic data on maps, and presenting the results of locational options rankings for single user and/or group contexts that are generated by ChoiceExplorer or ChoicePerspectives. GeoVisual is implemented as an extension of the ArcView® GIS 3.3 platform. The ChoiceExplorer component is used by stakeholders to perform criteria selection and weighting plus options evaluation and prioritisation. ChoicePerspectives synthesizes rankings from ChoiceExplorer that are subsequently displayed as consensus maps in GeoVisual. GeoVisual and ChoiceExplorer are dynamically linked to support interactive computation and display.

During the experiment both groups used only the GeoVisual module for the visualization of wells belonging to different zones and for exploration of hydrologic characteristics of zones and wells. In addition, the test group used the ChoicePerspectives module for electronic voting on zone selection and timing of zone management.

3.2 Tools Used during the Second Experiment

In the second experiment conducted in September of 2002 both groups used a prototype spatial decision support system called WaterGroup. The design of the WaterGroup incorporated lessons learned from evaluating the software usability during the first experiments. One of the key lessons learned from the first experiment was that software functions must address values, objectives, and specific decision problem concerns of stakeholders.

Just like the GCP, the WaterGroup supports single user and/or group-based decision making. Unlike the GCP, which is a generic software tool, the WaterGroup was custom-build to specifically support group decision making activities addressing conjunctive water management. The functionalities of WaterGroup include interactive exploration functions for vector and raster data including 2D and 3D visualizations, management alternative construction tool, what-if analysis tool, option impact analysis tool, voting tool, and communication tool. The what-if analysis tool of WaterGroup allows the stakeholders to evaluate the impacts different management options have upon the ground and surface water supplies within the study area. A module was created that produced graphical year-by-year breakdowns of the number of wells and amount of water brought under management for any specified option. A second module allowed the stakeholders to view comparisons between a number of different options at once. The option impact displays consisted of multiple thumbnails of line charts, bar charts, and pie charts. Two different types of voting tools were provided: one type allowed stakeholders to vote on proposals and the other allowed them to rank their preferred management options. Other tools included in WaterGroup were the management option submission tool, which allowed individual stakeholders to submit their options for consideration by the group, the voting tool allowing the stakeholders to electronically vote for submitted management options, and the communication tool facilitating electronic communication between the stakeholders and meeting facilitator.

4. Results of Experiments

Both experiments were evaluated based on post-experiment questionnaires answered by the stakeholders in both groups (control and test group). The stakeholder responses were normalized to fit a 0-100 scale. The summary of the results of both experiments is presented in Figure 1. Three observations can be quickly discerned from the data. First, the usability of GCP software in the first experiment was rated an average 13% lower than the usability of the WaterGroup software in the second experiment. Second, the standard deviation of the ratings for both test groups (using the GCP and WaterGroup software) was on average 14% higher than the average standard deviation of both control groups (using GCP and WaterGroup software). Third, in both experiments the respective control groups rated the software's support functionality higher than the test groups.

The higher average rating of the WaterGroup software supports the premise that a customized SDSS that takes into account requirements and constraints of a specific decision problem is likely to result in a more productive decision support software than a generic SDSS. Decision support tools, in order to be perceived as useful by its users, invariably require a direct relevance to the steps of decision process requiring data processing, organizing, and synthesis capabilities – functions that computers are better at than humans. The design of relevant decision support tools requires a good understanding of decision process, people involved, their value structures, and process constraints.

The higher standard deviation of the mean ratings for the test groups reflects the high heterogeneity of groups in respect to computer skills, which is to be expected in any participatory decision making situation involving a representative sample of diverse constituencies. This fact needs a careful consideration during the design of participatory SDSS.

The higher rating given by the control groups, regardless of the software used, underscores an important constraint of using groupware tools known from other studies with group decision support systems (Jankowski and Nyerges, 2001). Heterogeneous groups of stakeholders prefer facilitated decision processes, in which a facilitator relieves them of the burden of operating the software. This was the case with both control groups. The facilitator was leading the process and simultaneously operating the software on behalf of stakeholders, who felt freed of technical tasks and could concentrate on a creative part of solving the decision task.

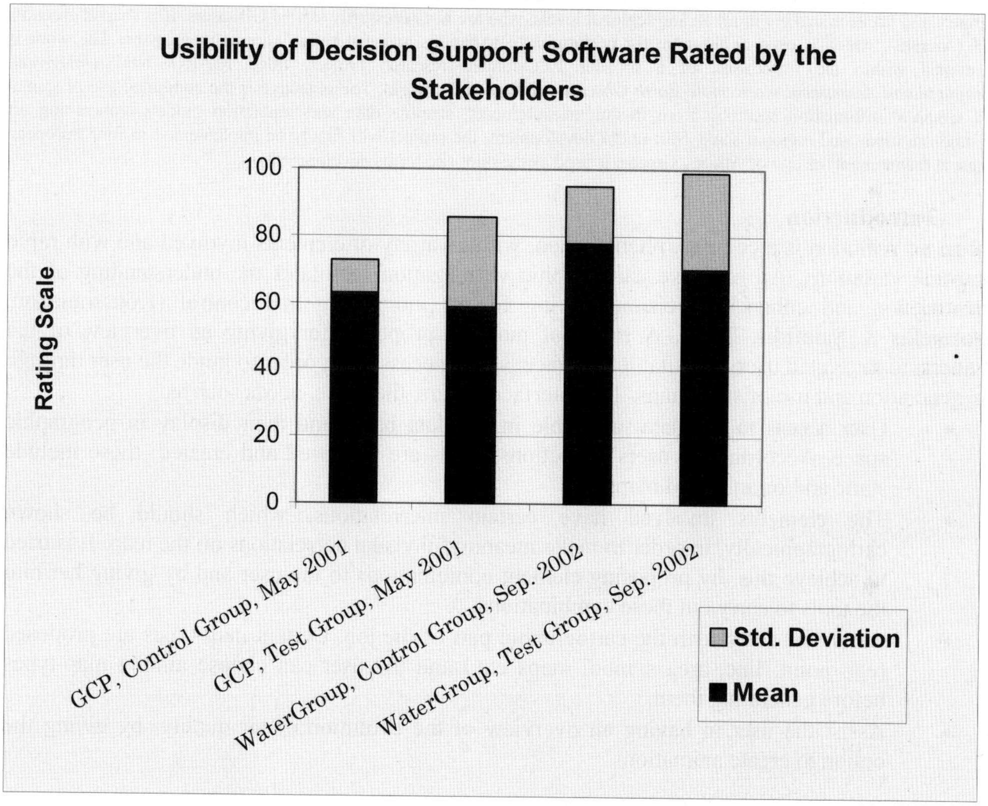

Figure 1. Summary of software ratings from both experiments.

References

Jankowski, P. and Nyerges, T. 2001. *Geographic Information Systems for Group Decision Making: Towards a Participatory, Geographic Information Science*, Taylor & Francis, London.

William J. Craig, Trevor M. Harris and Daniel Weiner (eds.). 2002. *Community Participation and Geographic Information Systems.* Taylor & Francis, London.

An Interface for Mapping Spatio-Temporal Elements of Urban Air Pollution

Alexandra Koussoulakou
Aristotle University of Thessaloniki
kusulaku@eng.auth.gr

Dimitrios Sarafidis
Aristotle University of Thessaloniki
sarafid@topo.auth.gr

Abstract: The main objective of the work is the creation of an interactive environment for mapping a variety of elements involved in urban air pollution. It is meant for users such as environmental specialists, transportation- and urban planners and decision makers in urban and regional levels, who are not necessarily skilled GIS users. It is shaped around a GIS (ArcGIS), which is used as the core for performing calculations and generating various map layers. The work is developed within the framework of a broader EEC-funded research project, titled ICAROS-NET (*Integrated Computational Assessment via Remote Earth Observation System Network*). The objective is the combined use of spatial and temporal information resulting from in-situ measurements, satellite data and simulation models, concerning air pollution in urban and regional scale. Still under development, the project will finally be implemented in four European cities; at the moment the city of Athens, Greece, is used as the pilot site for its development.

1 Introduction

Urban air pollution is a complex phenomenon, with a variety of elements involved and with rapid temporal variations. An effective cartographic visualization facilitates the understanding of the phenomenon and enhances decision making for its monitoring and control (Koussoulakou, Soulakellis & Sarafidis, 2001). A range of maps is proposed for giving an overview of the situation to interested users; a suitable interface is also necessary in order to guide the user through the generation and use of such maps. The interface covers, therefore, needs such as:

- User access to the data -available in the data base- and their display in geographic space. According to users' selections, maps are proposed and created; these include static and/or animated maps.
- The elements involved have certain interrelations, which should be shown cartographically, in order to make meaningful visual correlations on the map. It is tried to achieve this, by proposing element combinations to the user and by giving her/him the tools to carry out these combinations.
- Assist the users with the cartographic part of the job. Certain map-types are proposed (e.g. point, line, area symbol maps etc.) and the user can choose among map-types before generating them.
- Assist the user in having an overview of the evolution of air quality, by giving the option to create animations.

2 Mapping the elements involved

A prerequisite for mapping is the collection, processing and database storage of the data to be mapped. Other partners within the ICAROS-NET team carried out these tasks, together with the design and creation of the respective database. The data used and stored in the data base are provided by mainly three sources: ground measurement stations (which record a number of pollutants and of meteorological parameters), a simulation model (which calculates/predicts pollutant concentrations for any location –in practice: for a regular grid of points covering the area) and satellite images displaying air quality indicators (such as optical thickness of suspended particulates). (Soulakellis, Koussoulakou, Sarafidis, Sifakis & Sarigiannis, 2002).

The various elements involved in urban air pollution can be grouped in five general categories:

- *Topographic* e.g. 3D form of the terrain, built up area, water, green etc.
- *Meteorological / Atmospheric* e.g. wind, temperature, moisture, precipitation etc.

- *Air Quality* e.g. concentrations either at monitoring stations or at locations predicted by a simulation model, distribution of monitoring stations etc.
- *Sources and emissions* e.g. industrial, traffic, central heating.
- *Urban activities* related to air pollution e.g. land use, traffic density, population density, age of buildings, city profile etc.

These elements are both influencing and influenced by the air quality status of the area. The elements involved in urban air pollution, are therefore interrelated in certain ways. Consequently, it is meaningful to produce maps, which will allow for the superimposition of interrelated elements, so that the user can visually correlate their spatial distribution (Koussoulakou, 1994).
The user can, in principle, select any element for mapping and see the result as a map with superimposed layers. In order, however, to have meaningful combinations, according to the interrelations mentioned above, instructions have been included in the system in the form of a brief tutorial; they are explained and suggested to the user via the "*Help*" button of the user interface (see a snapshot in Figure 1). Consequently, the user can select a number of combinations, which result in the creation of the respective maps; animated maps are included, together with an interface for their generation.

3 Cartographic Visualization of Urban Air Pollution

The two key elements that distinguish Cartographic Visualization from mere mapping of geographical data are Interaction and Dynamics. Effective cartographic/geographic visualization, therefore, is the result of the combination of the maps per se, with the tools available for manipulating these maps and the data "hidden" behind them. The creation of the cartographic interface within ICAROS-NET was carried out through the development of suitable software for:

- Querying the database and selecting data for mapping (using Visual Basic).
- Cartographic visualization of the selected items (using Visual Basic and ArcObjects).

During development, it was tried to follow the general user interface design principles (see e.g. Shneiderman, 1987) and implement them within the geographic/cartographic scope. A brief description of the interface functionality is given in the following.
After the initial selection of the city in question, the user is presented with a map of the city, where *topographic* layers can be switched on and off; these layers include natural and man-made elements of importance to air pollution (e.g. elevation, land use, location of monitoring stations etc). Map contents corresponding to the main groups of elements involved and interrelated in urban air pollution are also offered to the user for display and query. These include *Air Quality* (in the form of measurements as well as model results) and *Meteorologica*l parameters (e.g. wind, temperature etc). The interface has a double function; since it is also a legend for the base map (the topographic features) and to some extend, to the thematic content (see Figure 2). At this point the user can therefore make selections to combine the above elements for mapping. When an element is selected it appears as "checked", while the "check" symbol is absent when the element is not selected. When the *Air Quality* group icon is clicked, two further selections are offered: *Measurements* (at stations) and *Model* (results). By further clicking on *Measurements* a number of pollutants appears –they are measured at the stations of the network. Each pollutant is represented by a certain, standard color. Clicking on the *Meteorology* group icon, invokes the availability for further selection of elements such as *Wind*, *Temperature* etc. At this point the geodatabase query can begin. Querying the *Air Quality* group, for example, can mean asking the concentration values for one or more pollutants. Starting from the former case, let us suppose that the value of SO_2 concentrations at the network stations is asked. When the SO_2 icon is clicked -and checked- the system requests the day of the year and the hour of the day for which to display the concentration. A menu for time selection appears: namely, Window "*Query Pollution Geodatabase*"; after this selection is made, the system queries and reads data from the database that is created in the

previously developed ICAROS-NET platform. When data are retrieved, the system asks the type of thematic map that the user would like to view: Window "*Map Types*" appears for the selection (see Figure 3). In this window the user can select among a number of continuous and discrete thematic representations: *proportional point symbols* (circles or bars), *graduated vector map*, *proportional symbol grid map*, *grid choropleth map* and *filled-in isoline map*. Depending on the number of pollutants selected previously, some of these representations might be impossible and therefore the respective buttons are inactive. In our case (i.e. one pollutant -SO_2- at the locations of the monitoring stations) it is possible to have a discrete representation with proportional point symbols (circles or bars) or/and a continuous representation with filled-in isolines; the other buttons will therefore be inactive for the moment. By then selecting, for instance, "*circles*" in the "*Map Types*" menu, a thematic layer with the values of SO_2 concentrations at monitoring stations appears; these values are drawn as circles of size proportional to their magnitude. A legend for explanation of the sizes of circles pops up at the same time. The absence of measurements at a particular station is also displayed by a suitable symbol on the maps and explained in the map-legend. In order to facilitate interaction, a bar menu on top of the map has also been created. The following items can be accessed via this bar: menu for selecting thematic *Map*-types, menu for *Time* selection, *Capture*, *Home*, *About*, *Help* and *Exit*. The first two menus are already described in the previous; via the bar menu they can be activated independently from other actions, that is, anytime the user wishes to. The "*Home*" button directs the user to the introductory screen, where a city can be selected. Guidance for meaningful combinations of air pollution elements is given through an interactive demo tool, which can be accessed via the "*Help* " button. The "*Capture*" button allows for capturing frames if the user wishes to keep a record of various dates/times/representations etc. The final purpose is to create animations from these captures. Let us suppose that the user wishes to see the same distribution by means of a continuous representation. By selecting "*Maps*" from the bar-menu the "*Map Types*" window reappears and the "*filled-in isoline map*" is clicked. The layer shown in Figure 4 appears, together with its respective legend window. As it is seen, this layer covers only partially the geographic space -this is because the interpolation considers only the stations that have recorded data; stations with no data recorded are skipped and therefore no symbolism is generated there. In case the user wishes to clear either point or area symbolism it is possible to do so by right clicking the mouse over the map area and selecting the appropriate command from the menu that appears. When the user wishes to view the distribution of (the concentrations of) more pollutants at a time he/she can select from *Air Quality>Measurements>* the pollutants of interest. It was decided to limit the maximum number of pollutants that can be selected simultaneously to five (5), in order to avoid visual confusion. Only the thematic symbolism of proportional bars is possible in this case and therefore the respective button is the only active. In the legend the absence of measurements at stations (i.e. the non- availability of data to be mapped) is represented with an empty bar of a fixed height.

5 References

Koussoulakou A. (1994). Spatial-Temporal Analysis of Urban Air Pollution. In A.M. MacEachren and D.R.F. Taylor (Eds) *Visualization in Modern Cartography* (pp. 243-267). Pergamon.

Koussoulakou, A., Soulakellis, N. & Sarafidis D. (2001). Interactive visualizations of spatial and temporal air pollution aspects for monitoring and control. In *Proceedings of the 20th International Cartographic Conference*, Beijing, China, Vol. 1, pp.367-373.

Shneiderman B. (1987). Designing the User Interface. Addison Wesley.

Soulakellis N., Koussoulakou A., Sarafidis D., Sifakis N., & Sarigiannis D. (2002). Synergistic use of GIS and satellite Remote Sensing data for mapping the status of urban air quality (in Greek). In Proceedings of the Conference *Social Practices and Spatial Information – European and Greek Experience in GIS*, held at the Aristotle University, Thessaloniki, June 2002 (Proceedings on CD-ROM).

Acknowledgements
ICAROS-NET (IST-2000-29264) is a project funded by the European Union.

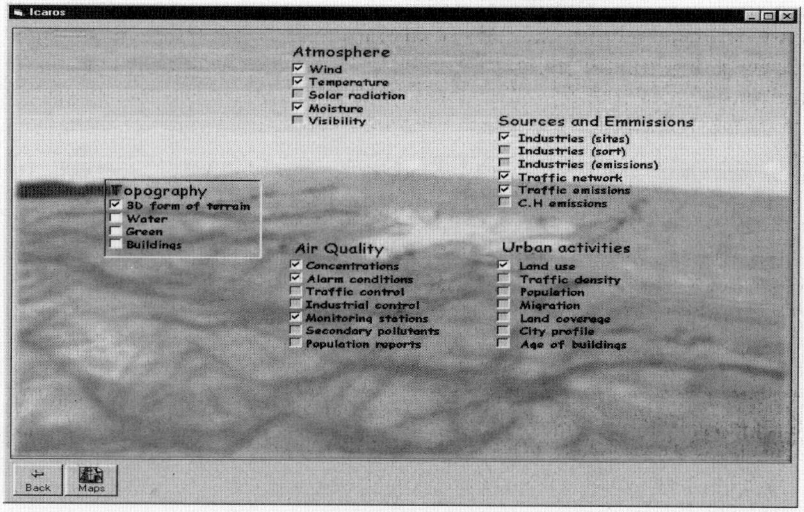

Figure 1: A snapshot from the brief introductory tutorial for showing interrelations among urban air pollution elements.

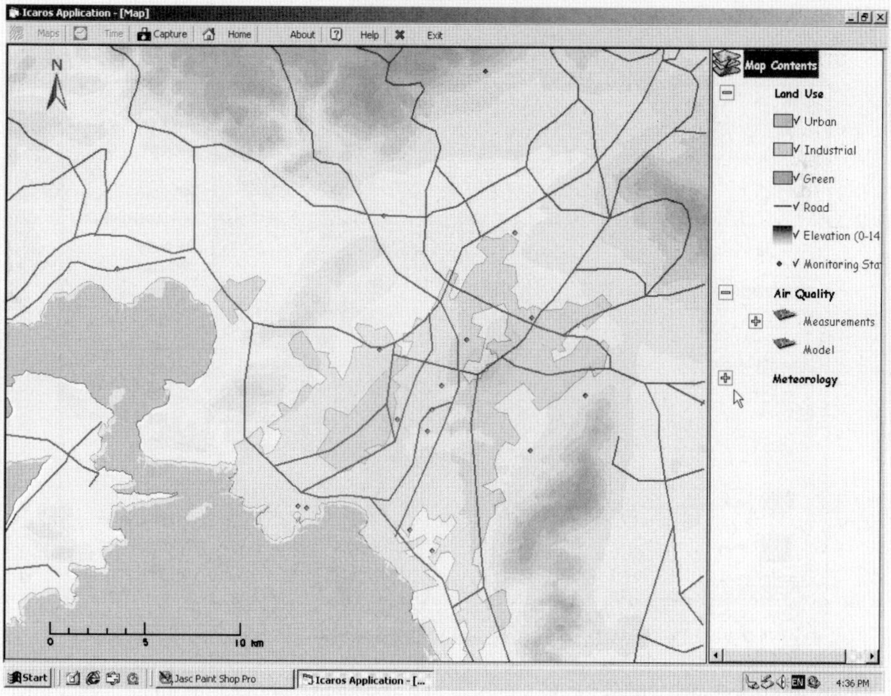

Figure 2: User interface for selecting map layers; it also functions as map legend.

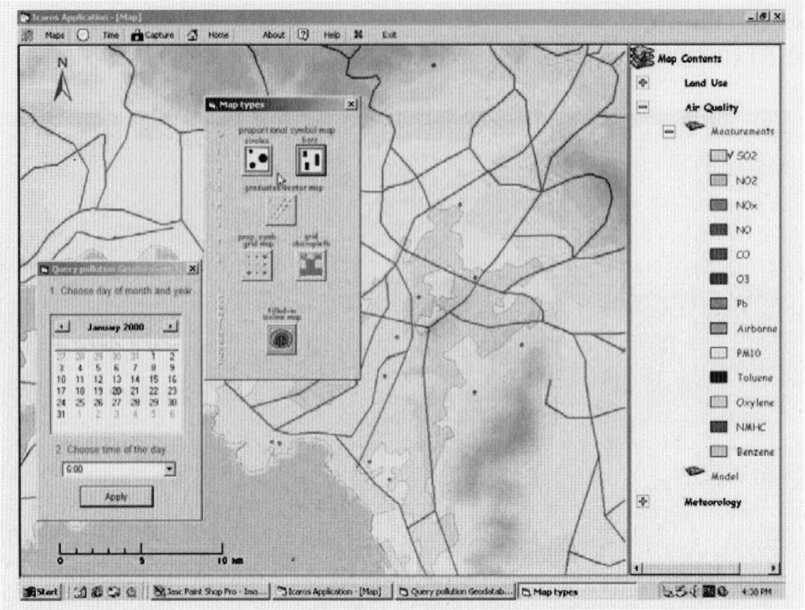

Figure 3: Menu for selecting date and time and menu for map-type selection (case of one pollutant only).

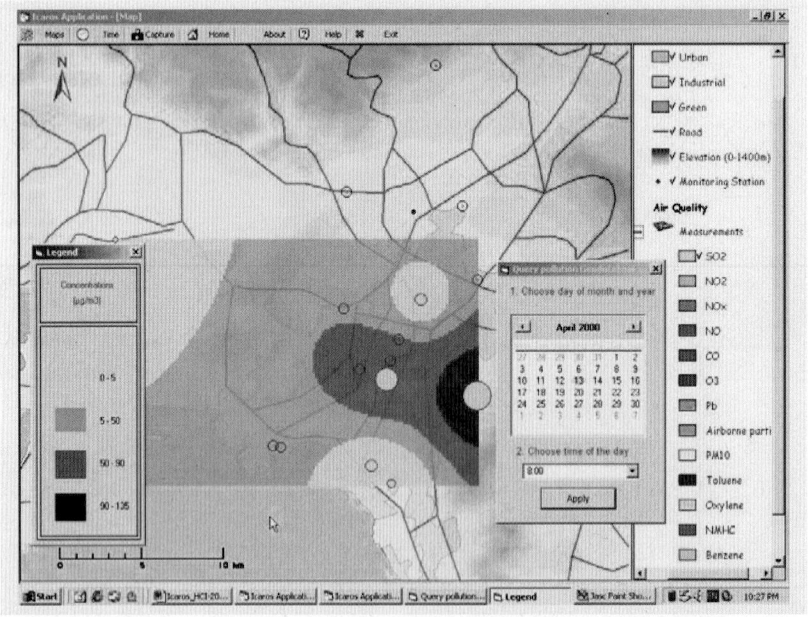

Figure 4: Proportional circles for the values of SO_2 at stations and filled-in isoline display after interpolation.

Attitudes Towards Technology Use in Public Areas: The Influence of External Factors on ATM use

Linda Little, Pam Briggs, Dave Knight

Psychology Division
Northumbria University
Newcastle upon Tyne, NE1 8ST, UK
l.little@unn.ac.uk

Lynne Coventry

NCR Financial Solutions Group Ltd
Kingsway West
Dundee, DD2 3XX, UK
lynnco@exchange.Scotland.NCR.COM

Abstract

Humans interact with technological systems in a variety of different environments. However, very little of the human-computer interaction or the psychological literature considers the interaction of people and technology in public zones. Previous research by Little, Briggs & Coventry (2002) found several factors that influence public use of a technological system such as perceived levels of privacy, and safety. This study aimed at further understanding the factors that influence the use of a technological system in a public area. Three hundred and seven participants who were either frequent or non-frequent automated teller machine (ATM) users completed a questionnaire related to attitude towards ATMs. The results of the study confirm that attitude and intention towards use are influenced by several external variables. Specifically, perceived ease of use, usefulness, privacy and personal space were found to have significant effects on attitude and intention towards use.

1 Introduction

Nearly everyone has some type of daily interaction with technology whether it is at home, work or in a public area. Take a walk down any main street and you will see people using cellular telephones, Internet kiosks and ATMs, generally without any prior training or instruction. As advances in technological systems are made and increasingly used more in public areas, a clear understanding of factors that influence system use is needed.

Several factors are already known to influence the acceptance of technology (e.g. Davis 1989) and environmental factors such as levels of privacy, social density are known to influence behaviour (see Kaya & Erkip 1999; Gifford 1987; Sundstrom 1975). However, such influences have not been previously integrated.

1.1 Technology Acceptance

The Technology Acceptance Model (TAM) proposed by Davis (1989) identifies certain system characteristics associated with adoption of technological systems. The model is based on Fishbein & Ajzen's (1975) Theory of Reasoned Action which implies there is a link between beliefs, attitudes, intentions and behaviour. In the terms of TAM the attitude towards a system drives the behavioural intention to use that system, and attitude is in turn determined by two influential beliefs: perceived usefulness (PU) and perceived ease of use (PEOU). TAM has also been extended to show the importance of other factors on adoption and use of technological systems such as social influence (Malhotra & Galletta 1999).

However, nearly all research that uses TAM to measure acceptance and future use focuses on information technology systems within the workplace and not in public zones. When considering human interaction with technology in public areas, in particular ATM's, several external factors have to be taken into account e.g. levels of privacy, social density and safety.

1.2 Environmental influences on behaviour

Several environmental factors are known to influence behaviour and are relevant to research into the use of technologies in public areas such as levels of privacy (Pedersen 1999) and social density (Kay & Erik 1999).

Privacy is a human boundary process that allows access by others according to one's own needs and situational factors. Situational factors can be either social (interaction from others) or physical (location, layout). Although there is no universal definition of privacy, the concept is highly complex and involves different perspectives and dimensions. Pedersen (1999) proposed six privacy dimensions: solitude (freedom from observation by others), reserve (not revealing personal information to others), isolation (being geographically removed from and free from observation by others), anonymity (being seen but not identified or identifiable in any way), intimacy with friends (being alone with friends) and intimacy with family (being alone with family). Pedersen suggests that the six types of privacy 'represent the basic approaches people use to satisfy their privacy needs.

However privacy is not the only factor that influences use of a system in a public area. Humans are territorial creatures and need a certain level of security and comfort from their immediate location (Valentine 2001). Therefore when a person uses technology in a public area there exist concerns over levels of safety from the actual location and other people around at that time. Situational factors often mean that a person at some point may co-exist in a space with another person with whom they don't want to interact with such as in a crowd or waiting in a queue (e.g. waiting to use a public telephone). Although waiting in a queue is a matter of social convention a person can get very irritated by thoughtless invasion into his or her space by someone else. People can also become annoyed if the user of a system takes to long. Ruback, Pope & Doriot (1989) found that people tended to take longer using a public telephone when someone else was waiting to use it. However, they suggested that different locations are confounded with different time pressures and particular types of people.

A qualitative study by Little, Briggs & Coventry (2002) found attitude and intention towards use of an ATM was influenced by several personal constructs: perceived levels of privacy, personal space, time pressure and safety. Non-personal constructs that emerged were usefulness, ease of use and location. The findings for usefulness and ease of use support TAM in its assertion that these factors are two key predictors of intention to use a system, however they were not the only important factors.

The study by Little et al supports previous research such as Lucas & Spitler (1999) in that external variables are an important influence on technological adoption and use. The study also posits that the main variables included in TAM (PU and PEOU) are not the only major determinants that influence the use of an ATM in a public area but consideration must be given to other external variables such as privacy.

Developing theory regarding use of technological systems in public areas not only adds to the literature but also provides practical future applications. The aim of the present study is to validate previous findings by Little, Briggs & Coventry that external variables such as privacy affect attitude and intention towards a technological system used in a public area such as an ATM. Also to determine how the measured variables fit with TAM.

2 Method

2.1 Design

A non-experimental independent measures design was used in this study. Variables that were measured were participant's subjective ratings of levels of privacy, personal space, safety, time pressure, ease of use, usefulness, social interaction, attitude and intention to use ATMs.

2.2 Participants

Three hundred and seven participants from the Northeast of England completed a questionnaire related to ATM use. 125 males, 182 females, age range 16–90 years (mean 49.63). All participants were chosen by opportunity sampling.

2.3 Materials

The study was carried out in two different locations in Newcastle upon Tyne, UK: Seaton Delaval and Gosforth. A self-completion questionnaire was developed based on previous findings by Little et al (2002). The questions related to participants opinions on several issues regarding ATM use: privacy, personal space, safety, time pressure, social interaction, intention to use, attitude towards use, ease of use, location and usefulness. The majority of questions in each of the subscales were explicitly linked to past research. For example, the questions for privacy were based on four of the dimensions proposed by Pedersen (1999): solitude (e.g. I do not like being observed by others when I use an ATM.), anonymity, reserve (e.g. When I use an ATM I am concerned other people can see personal information) and isolation. Questions related to PU (e.g. I am only interested in using ATM's for cash withdrawal) and PEOU (e.g. I find ATM's easy to use) were obtained from research by Davis (1989). Each of the subscales showed a high level of internal consistency, with coefficient alphas in the range of .69 to .83, mean alpha = .76. The questionnaire also included questions about the demographic characteristics (age, sex, and frequency of use) were also recorded.

2.4 Procedure

Questionnaires were delivered by hand to 1100 homes in two different areas in Newcastle. A full explanation of the research and instructions on how to complete the questionnaire were included in the delivered package. Participants were asked to read and respond to all questions by recording their responses on a continuum bipolar scale of 1-7. After completion participants were asked to return the questionnaire in the pre-paid envelope provided to Northumbria University, Newcastle by 20th November 2002. A total of 361 questionnaires were returned, 307 were fully completed this gave an overall response rate of 32.81% of which 27.9% questionnaires were used for further analysis.

3 Results

A Structural Equation Modelling approach was used to analyze the data using EQS software. Maximum likelihood estimation was used to find how the measured external variables fitted in relation to TAM. Little support was found for this model, Chi-square = 670.65, DF=15, $p < .001$; CFI = 0.25; RMSEA = 0.38.

Post hoc modifications were performed in an attempt to develop a better fit and more parsimonious model. On the basis of the Lagrange Multiplier and the Wald tests several paths were added and deleted where it made theoretical sense to do so. The revised model provided an extremely good fit to the data, Chisquare = 1.34, DF=3, p=0.72; CFI = 1.00; RMSEA = 0.00 (see Table 1).

Table 1: Comparison between the two models fits.

	TAM	NEW MODEL
CHISQUARE	670.65	1.34
CFI	0.25	1.00
RMSEA	0.38	0.00

The model provides evidence that space (0.18), privacy (0.17), usefulness (0.12) and ease of use (0.51) are all predictive of attitude towards use of a technological system used in a public area. Contrary to the TAM, the revised model also suggests that Space (0.22), Usefulness (0.16) and Ease of Use (0.25) are directly related to Intention to Use such a system. Significant relationships were also evident between several of the independent variables (see Figure 1).

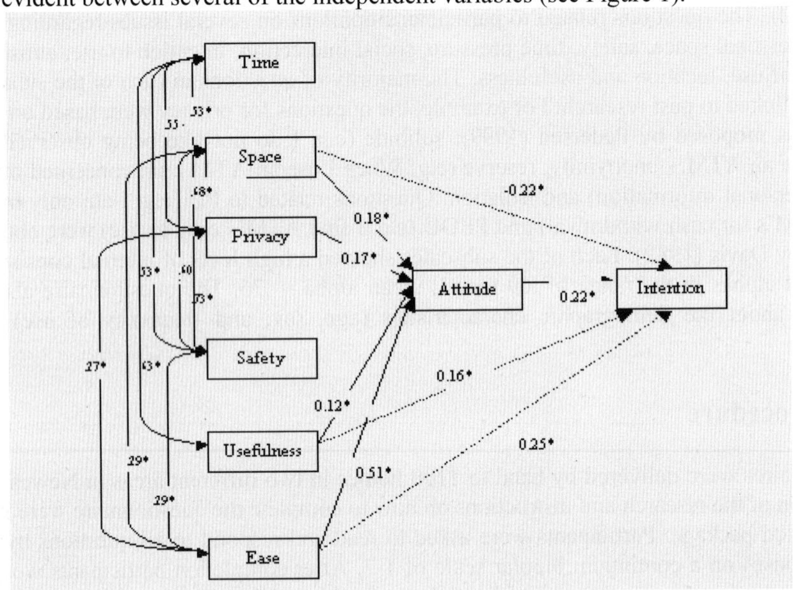

Figure 1: Revised Model of factors that influence Attitude and Intention to use an ATM.

4 Discussion

The study has provided empirical and important evidence that external variables have a direct effect on attitude towards the use of an ATM in a public area, as well as on Intention to Use such a system. The findings show perceived levels of privacy, personal space, ease of use and usefulness affect the level of use of an ATM.

This study supports previous works such as Little et al (2002), Lucas & Spitler (1999) in that external factors are an important influence on technological adoption and use. This study posits that the main variables included in TAM (PU and PEOU) are not the only major determinants of

ATM use. External variables that are both personal and non-personal influence use of an ATM in a public area. The revised and expanded TAM proposed in this study demonstrates how external variables such as personal space and privacy have a direct effect on attitude and intention to use a system. Therefore mechanisms that determine user behaviour of technological systems in public areas such as an ATM are more complex than stated in TAM. TAM has been successfully applied to specific research areas such as technology adoption in the workplace (see Davis 1989). However, the findings from this study suggest that when applying TAM to explain user behaviour or adoption of technological systems in public areas new and/or expanded theoretical models are needed.

Future research into technological system use in public zones needs to encompass these findings as the model presented here could be applied to various systems used in public areas. As technological systems are becoming smaller and more powerful, mobile personal devices are becoming more common, further research in this area is needed to see if acceptance and use in public areas and behaviours towards these types of systems is changing

In conclusion, this study has provided important evidence that attitude and intention towards use of a technological system in a public area is influenced by several external factors that either have a direct or indirect effect on use. The proposed model provides a practical tool for assessing and measuring use of technological systems in public zones. Finding the exact factors and issues that influence people's perception of use is an important area not only for current use but also in trying to predict future use of technological systems used in public areas.

References

Davis, F, D. (1989). Perceived usefulness perceived ease of use, and user acceptance of information technology. *MIS Quarterly,* 13, 319-339.

Fishbein, M., & Ajzen, I. (1975). *Belief, Attitude, Intention and Behavior: An Introduction to Theory and Research.* Reading, MA: Addison-Wesley

Gifford, R. (1987). *Environmental Psychology: Principles and Practice*. Boston: Allyn.

Kaya, N., & Erkpip, F. (1999). Invasion of personal space under the condition of short-term crowding: a case study of an automated teller machine. *Journal of Environmental Psychology. 19.* 183-189.

Little, L., Briggs, P., & Coventry, L. (2002). Attitudes and External Influences on ATM use: A Qualitative Approach. *Report for NCR Ltd. October 2002.*

Lucas, H.C., & Spitler, K. (1999). Technology Use and Performance: A Field Study of Broker Workstations. *Decision Science. 20. (2).* 291-311.

Malhotra, Y. & Galletta, D.F. (1999). Extending the Technology Acceptance Model to Account for Social Influence. 32nd Hawaii Int. Conference on System Sciences.

Pedersen, D.M. (1999). Model for types of privacy by privacy functions. *Journal of Environmental Psychology. 17.* 147-156.

Ruback, R.B., Pape, K.D. & Doriot, P. (1989). Waiting for a Phone: Intrusion on Callers Leads to Territorial Defense. Social Psychology Quarterly. 53. (3). 232-241.

Sundstrom, E. (1975). An experimental study of crowding: effects of room size, intrusion, self-disclosure, and self-reported stress. *Journal of Personality and Social Psychology. 32.* 645-654.

Valentine, G. (2001). *Social Geographies: Space and Society*. England: Pearson

The Interface Design of Alarm Signals for Improving the Performance of the Second Vigilance

Cheng-Li Liu

Department of Industrial Management, Van Nung Institute of Technology
No. 1 Van Nung Rd., Chung-Li, Tao-Yuan, Taiwan, R.O.C.
johnny@cc.vit.edu.tw

Abstract

When we study vigilance on the concept of initial signal detection, the ability of tracking and detecting follow-up situation or signals should be discussed, which is called the second vigilance. This study results show that extremely high arousal might fail to monitor performance of the second vigilance significantly, although there is alarm set. Appropriate arousal level plus adapted alarm signals could promote performance of the second vigilance efficiently. To improve the performance of the second vigilance for the safety control of complex and dynamic systems, the following information would further be needed: (1) alarms to let an operator know that some function is approaching to its boundary, and (2) artificial alarms at low arousal levels for improving arousal, especially performance of the first vigilance is poor or decrement.

1 Introduction

Research on vigilance began in World War II to determine why radar operators were failing to detect a significant number of submarine targets. As systems have become primarily one of passively monitoring displays for critical signals, vigilance research is still important. However, the most researches of vigilance are focused on the concept of signal detection. In fact, when we study vigilance on the first detecting extraordinary signals, the ability of tracking and detecting the following signals or situation should be discussed, which is called the second vigilance. In human-centered automation, we assume that a human has full authority and responsibility in decision and every operation. Then, he/she must recognize and locate a fault when it occurs, and must identify causes of the fault for taking countermeasures against it. However, as has been seen in various accidents or incidents, it is not an easy task for the operator to identify the operating condition of a large complex system correctly (Itoh & Inagaki, 1997). Previous researches have shown that the attention determines the performance of vigilance. Over the years, there have been many research studies dealing with vigilance. However, theories of vigilance are really not theories specific to vigilance, but rather are general theories applied to the situation of vigilance.

There are some researches studied on the alarm or artificial signals. Keyvan (2001) used time series analysis and regression modeling techniques for monitoring reactor components' signal condition, then to generate signals simulating the pump shaft degradation progress. Kozma & Nabeshima (1995) proposed monitoring system utilizes advanced signal processing methods based on artificial neural networks in order to achieve early detection of changes in the state of the coolant. MacRae & Falahee (1995) taked account of possible changes in response criterion as well as in sensory sensitivity to improve sensory screening by attempting to counteract vigilance decrement and the effects of relative judgement. However, all these studies are discussed on the field of the first vigilance. The cognitive situation of the second vigilance is more complex than the first one. The human operator should detect extraordinary signals on the first vigilance, and then he/she must pay attention to detect extraordinary signals on the following tracking work, but cannot ignore other condition needing the first vigilance. The purposes of this study are to discuss the characteristics of the second vigilance and the effect of alarm which warns operator to take a break and avoid the second vigilance decrement.

2 Experiment

An experiment is conducted to study the characteristics of the second vigilance and how the design of alarms can affect the second vigilance decrement. We compare the difference of conditions of alarms displayed in the control panel.

2.1 Controlled process

The experiment is done on the simulated Auxiliary Feed-Water Monitoring System and Heat Sink & Demineralized Water System which are shown in Figure 1and Figure 2 (Roth-Seefrid, & Fischer, 1988). There are four steam generators (SG1, SG2, SG3 and SG4), if any SG is under low water level, the subject must detect and inject water. These SGs are fed by the startup and shutdown system (SSS). If the SSS is failure or water level is lower, the manual feed system (MFS) should be start. When the SGs show levels below 5 m, the emergency feed-water system (EFS) should be start. There are some check points should be detected illuminated as following:

- p1: SG-level ≥ 13 m Over load, EFS and MFS should be shutdown
- p2: SG-level ≥ 15 m Over feeding, MFS, SSS and EFS all must be shutdown
- p3 :SG-level ≤ 9m Scram, SSS and MFS should be start
- p4: SG-level ≤ 5 m Emergency feeding, EFS must be start

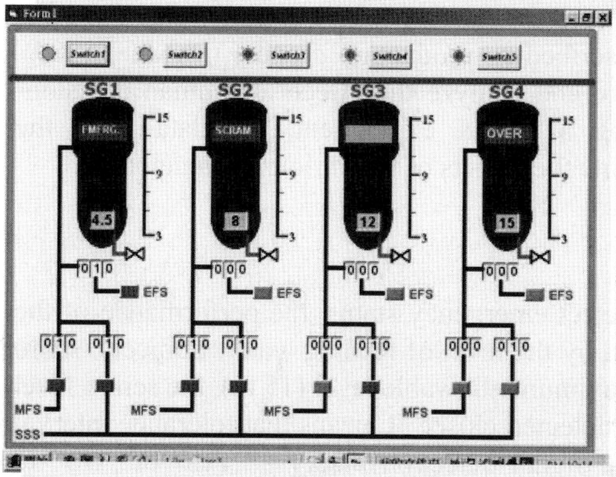

Figure 1: An simulated Auxiliary Feed-Water Monitoring System

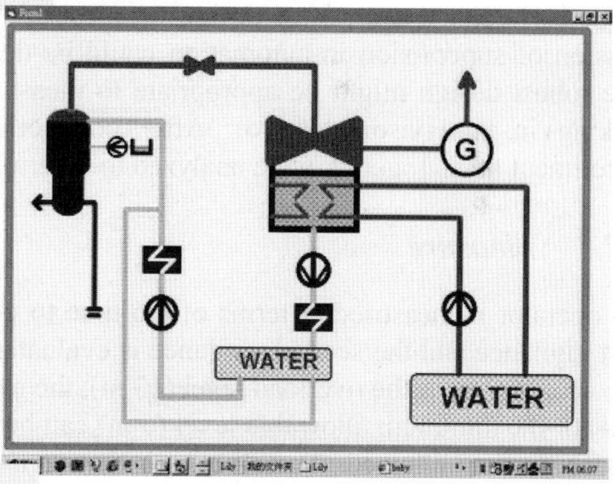

Figure 2: Heat Sink & Demineralized Water Store System

In each checkpoint, operator is alerted to imminent danger by special symbols and signal colors on the SG:

- Green: Safety condition
- Red: Over load or Scram
- Magenta: Multiple emergency as overfeeding or emergency feeding

Therefore, we can define the status at level p1 and p3 as stage of the first vigilance, and tracking and detecting at the level p2 and p4 as stage of the second vigilance.

There are two kinds of malfunction to be designed as secondary tasks:

(1) The artificial signals randomly turned on which should be turned off at the top side of Fig.1.

(2) When the system of Heat Sink & Demineralized Water Store occurs extraordinary signals, it should be recovery as shown in Fig. 2.

2.2 Method

2.2.1 Tasks.

The are 16 Subjects requested to monitor and feed cool water satisfying the SG situations, and are requested to maintain safety: If he/she detects emergency status, it should be controlled immediately. Each subject receives some training (at least 3 times) to learn how to control the simulated system in no fault conditions. They practiced and must pass a pre-experiment before a formal experiment.

2.2.2 Experimental design

Considering the experiment where we are interested in determining the effect of five parameters, which influenced the performance on decision-making. The five control factors in the study were A – *alarm status (including None signal, Alarm)*, B – *Emergency complexity (including SG1, SG1 +SG2, SG1+SG2+SG3, SG1+SG2+SG3+SG4)*, C – *Vigilance situation (including interval of five minutes between p1 and p2 or p3 and p4, interval of ten minutes between p1 and p2 or p3 and p4)*, D –*Alarm Timing* (*including Alarm at critical level, Alarm beforehand 0.5 m of critical level)*, and E –*Emergency situation (including p1+p2, p3+p4).* Simultaneously, we concern the interaction A×C, B×C, C×D and C×E. There are so many controlled factors to be designed in this experiment, and each one's effect is equally important to us. So an efficient but small matrix experiment is needed. Here, we use robust design to construct an orthogonal matrix experiment. In fact, the mission of supervision in automation could be described as production process (Phadke, 1989). The robust design might be appropriate to measure and analyze the effects of human decision-making in supervisory control. After conducting a matrix experiment, the data from the experiment taken together were analyzed to determine the effects of the various parameters.

2.2.3 Measure

An operator is measured in terms of the time to detect emergency status, the performance of the first vigilance and the second vigilance is evaluated by the level of feeding water. Subjects' score becomes higher as the overload level (13 m), the maximum allowable level (15 m), the scram level (9m) or the minimum allowable level (5 m) can be detected closer. It means that tolerance interval of detection is allowed. Therefore, we cannot use signal detection theory - "Yes" or "No" to explain and evaluate the vigilance performance at these status. We applied fuzzy logic to construct the performance measure function Λ. The membership function $\Lambda : U \rightarrow [0, 1]$ is a function with three parameters defined as

$$\Lambda(u; \alpha, \beta, \gamma) = \begin{cases} 0 & u < \alpha, \\ (u - \alpha)/(\beta - \alpha) & \alpha \leq u \leq \beta, \\ (\gamma - u)/(\gamma - \beta) & \beta \leq u \leq \gamma, \\ 0 & u > \gamma, \end{cases} \quad (1)$$

Where β is crisp value of the maximum allowable level, overload level, scram level or minimum allowable level on four status of SG, α is the infimum and γ is supremum of each status. There are four membership functions to be constructed as shown in Fig. 3.

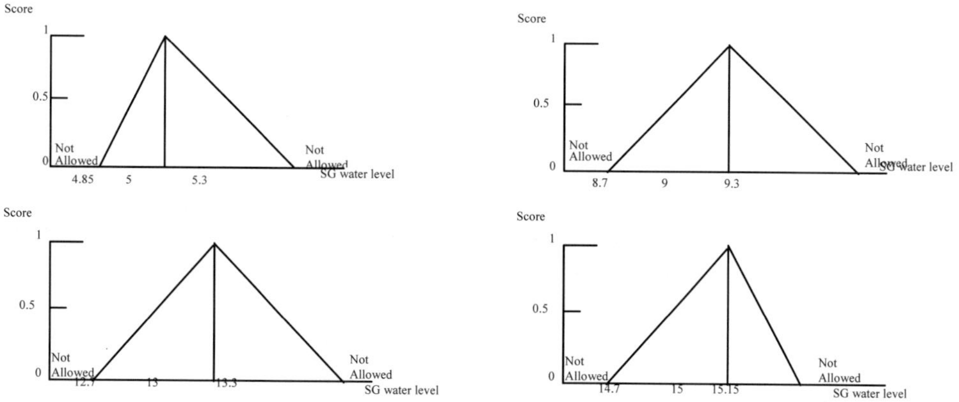

Figure 3: Membership function of performance of vigilance

There is another performance we are concerned: the integrated performance including the first vigilance and the second vigilance. The Extension Theory was used to analysis the performance as following,

$C(A_i) = a_j\, K(A_{ij})$, (2)

where $C(A_i)$ is the *i*-th trial's integrated performance, **a** is the weight coefficient vector and $\mathbf{a}_1$ is defined as weight coefficient of the performance of the first vigilance and $\mathbf{a}_2$ the second. The Analytic Hierarchy Process (AHP) was used to estimate the weight coefficient which is a powerful and flexible decision making process to help people set priorities and make the best decision when both qualitative and quantitative aspects of a decision need to be considered. By reducing complex decisions to a series of one-on-one comparisons, then synthesizing the results, AHP not only helps decision makers arrive at the best decision, but also provides a clear rationale that it is the best (Zahedi, 1986 & Abulfarej, 1994). If the weight of performance of the second vigilance was defined as the double of the first in this study, then **a** = (0.333, 0.666). $K(A_{ij})$ is membership function value of performance of the *i*-th trial and *j*-th vigilance.

3 Results and discussion

Table 1 shows sixteen individual experiment results corresponding to the sixteen rows. The entries in the first five columns of the matrix represent the levels of the five factors. An analysis of variance (ANOVA) was performed to identify significant factors. The ANOVA for the value of performance measure function (Λ) of the second vigilance shows that factor A (*alarm status*) exhibited a very significant effect ($F(1,7) = 168.0$, $p < 0.01$). It means that system designed including alarm at emergency check point could reduce the second vigilance decrement efficiently. The main effects of factor B, C and E are all significant (factor B - $F(3,11) = 29.4$, $p < 0.01$; factor C – $F(1,11) = 23.6$, $p < 0.01$; factor E – $F(1,11) = 21.1$, $p < 0.01$). It means that the performance is poor at low arousal levels, and extremely high arousal failure to monitor performance of the second vigilance significantly, although there are alarm set. If the tracking time from the vigilance to the second vigilance is too long, the performance of the second vigilance will be worst. In the trial 12, although the performance of the first vigilance is average, if it was at low arousal levels, no alarm in case of emergency, and tracking was longer, the integrated performance will be poorest.

Table 1: L_{16} orthogonal array and data summary by experiment of Feed-Water Simulation System

No.	A 2 3	B 7 8 15	C 1	D 10	E 12	A × C 3	B × C 6	B × C 9	B × C 14	C × D 11	C × E 13	Results of the first vigilance		Results of the second vigilance		Integrated perform-ance
												Response level	Perorm-ance	Response Time	Peform-ance	
1	1	1	1	1	1	1	1	1	1	1	1	13.06	0.800	15.11	0.267	0.444
2	1	4	1	2	2	1	1	2	2	2	2	8.87	0.567	4.89	0.267	0.367
3	1	2	1	1	2	1	2	1	2	1	2	8.92	0.733	5.07	0.767	0.755
4	1	3	1	2	1	1	2	2	1	2	1	13.14	0.533	15.07	0.533	0.533
5	2	2	1	2	1	2	2	1	2	2	1	13.03	0.900	15.01	0.933	0.922
6	2	3	1	1	2	2	2	2	1	1	2	8.94	0.800	4.96	0.733	0.755
7	2	1	1	2	2	2	1	1	1	2	2	8.95	0.833	4.98	0.867	0.855
8	2	4	1	1	1	2	1	2	2	1	1	13.09	0.700	15.08	0.467	0.544
9	1	4	2	1	1	2	1	2	1	2	2	13.15	0.500	15.14	0.067	0.211
10	1	2	2	2	2	2	1	1	2	1	1	8.89	0.633	4.9	0.333	0.433
11	1	3	2	1	2	2	2	2	2	2	1	8.88	0.600	4.89	0.267	0.378
12	1	1	2	2	1	2	2	1	1	1	2	13.08	0.733	15.13	0.133	0.333
13	2	4	2	2	1	1	2	2	2	1	2	13.11	0.633	15.09	0.400	0.478
14	2	1	2	1	2	1	2	1	1	2	1	8.93	0.767	5.05	0.833	0.811
15	2	2	2	2	2	1	1	2	1	1	1	8.95	0.833	4.96	0.733	0.767
16	2	3	2	1	1	1	1	1	2	2	2	12.94	0.800	14.96	0.867	0.844

4 Conclusion

The main findings of this study are : The performance of the second vigilance (1) is correlated to the first vigilance, (2) is poor at low arousal levels and (3) is worst at extremely high arousal levels, even though alarms are set. To produce artificial alarm at low arousal level but not become high arousal level, the performance would be improved. (4) In general, an alarm system makes the operator aware of system condition that requires timely assessment or action. However, more often, the time is not enough. If we can predict human reliability from previous decision behavior to fire up alarm, the operator or manager could understand current personal condition to adjust behavior or cognition, then slips or mistakes would be avoided. In this study we can find the effect of alarm in advance to keep operator in adapted second vigilance is significant.

Acknowledgements

The author would like to thank the National Science Council of the Republic of China for financially supporting this work under Contract No. NSC91-2213-E-238-005.

References

Abulfarej, W. H. (1994). Special concrete shield selection using the analytic hierarchy process. *Nuclear Technology*, 107(2), 215-226.

Itoh, M., & Inagaki, T. (1997). Design of human-interface for situation awareness II: interdependence between control panel and alarm system. *IEEE International Workshop on Robot and Human Communication*, 308-313.

Keyvan, S. (2001). Traditional signal pattern recognition versus artificial neural networks for nuclear plant diagnostics. *Progress in Nuclear Energy*, 39, 1-29.

Kozma, R., & Nabeshima, K. (1995). Studies on the detection of incipient coolant boiling in nuclear reactors using artificial neural networks. *Annals of Nuclear Energy*, 22, 483-496.

MacRae, A. W., & Falahee, M. (1995). Theoretical note on a practical problem: effective screening of drinking water for taints. *Food Quality and Preference*, 6, 69-74.

Phadke, M. S. (1989). *Quality Engineering Using Robust Design*, New Jersey, NJ: PTR Prentice-Hall.

Roth-Seefrid, H. & Fischer, H. D. (1988). Advanced information systems to enhance operational safety. *Reliability Engineering and System Safety*, 22, 91-106.

Zahedi, F., (1986). The Analytic Hierarchy Process--a Survey of the Method and its Applications. *Interfaces,* 16(4), 96-108.

Incorporating graphical interface technique into development of a training system for emergency response center

Hunszu Liu[1] *Yuan-Ming Gu*[2] *Sheue-Ling Hwang*[3]

Ming Hsin University of Science and Technology
hliu@must.edu.tw

National Tsing Hua University
toby1022@seed.net.tw

National Tsing Hua University
slhwang@ie.nthu.edu.tw

Abstract

Traditional training programs can not supply semiconductor companies with capable Emergency Response Center (ERC) supervisors. The lack of experienced instructors and the complexity of training materials have put the system under the risks of wrong judgments and decisions making by under-trained workers. Although Computer-Aided Education (CAE) system can be used to improve the workers' learning effectiveness and efficiencies, its advantage can be fully explored only through appropriate training programs and well-designed interface screen, where the symbols and icons can be easily recognized and decoded by the potential trainees and its contents represent the necessary information of the real system.

The objective of this research is to examine the relationship between different forms of training materials and the performance of trainees. The training materials are classified into three categories including documented alphanumerical training materials, documented graphical training materials and computerized graphical training materials.

The subjects in this study are 15 engineering graduate students who have the potential to work as supervisors in ERC center. Subjects were randomly divided into three different groups and trained separately by different forms of materials with the same content. Self-developed computer program, which simulate the operation of Gas Detector Alarm Informing Procedure, is used as study tool to investigate the performance of trainees. Performance measurement indexes include trainees' training time, response time, message processing time, number of errors and number of missing.

The results of study indicate that different forms of training materials have significant effects on the trainees' performance. The computerized graphical training materials require the least efforts and have the best performance than the other two. It is argued that in order to speed up the training process of the ERC trainee, the computerized graphical training materials are suggested. The results can be used to support the development of the training system for ERC personnel. Future study will focus on the development of computer-aided ERC training program.

---

[1] No.1 Hsin-Hsing Rd., Hsin-Fong, Hsinchu, Taiwan, ROC
[2] Kuang Fu Rd., Hsinchu, Taiwan
[3] Kuang Fu Rd., Hsinchu, Taiwan

1 Introduction

The manufacturing characteristics, such as utilization of various highly toxic chemicals during continue production process and long working hours, usually 12 hours per shift, of semiconductor industry required continue surveillance on the factory activities to ensure the safe operations of its production. Usually, in Taiwan, the responsibilities of monitoring the semiconductor plant activities fall upon the shoulders of working teams in Emergency Response Center (ERC) through the integration of data from automatic sensor devices, video camera and communication channels. Duty supervisor examines all these data, and appropriate instructions are initiated for related personnel to perform correct rectified actions. Computer programs are utilized to assist the data integration process, and some decision supporting systems are also installed to reduce the mental workload of monitors. As a result, various types of data are presented on several computer screens, which require experienced workers to interpret and transform them into useful information. The supervisors' knowledge of the appropriate procedures, which require lots of hands on experiences, is a key factor to the success of ERC operations.

Traditional training programs, which rely on documented training materials and senior supervisors' teaching, are established to furnish novice workers with appropriate knowledge to perform the monitoring tasks. However, the workers' jumping-ship problems due to the heavy demand of experienced supervisors from other semiconductor companies have made the capable instructors hard to find. Without instructors' help, piles of training materials often need lots of efforts and time for trainees to relate them to real working scenario. Moreover, the facilities are operated full day, which cannot be taken off-line to be used solely for training purpose. Currently, the workers can only learn the working skills by doing on site and by observing the senior worker's operations. This approach puts the system under the risks of wrong judgments and decisions making by under-trained workers.

Researchers (Su & Lin, 1998) (Gramopadhye & Bhagwat, 1998) believe that the Computer-aided education (CAE) system can be used to improve the trainees' learning effectiveness and efficiencies. However, these advantages can be fully explored only through appropriate training program and well-designed interface screen, where the symbols and icons can be easily recognized and decoded by the potential trainees, and its contents represent the necessary information of the real system(Berthelette, 1996). Other researcher (Johnson, 1996) assumed that the graphical interface technique can improve the users' usability. It is argued that incorporating graphical interface technique into ERC training system can improve training performance. This study is a pilot study for development of computer-aided ERC training program. The objective of this research is to examine the effects of different forms of training materials, including documented alphanumerical training materials, documented graphical training materials and computerized graphical training materials on the performance of monitoring staffs.

2 Background information

Although there are many expected functions that ERC can provide, currently the major task in Taiwan semiconductor industry is to monitor the status of plant facilities and inform appropriate on-site personnel for further actions. The reason for this special case is because that the concept of establishment of ERC has just emerged recently after some severe accidents which have cost Taiwan semiconductor industry more than 20 billions of losses since 1996. The industry starts to gradually installs Facilities Management Control System (FMCS) to get a tightly control of their facilities. Typical FMCS includes power system, gas supply system, water system, clean room air control system, hazard prevention system, chemical supply system, cooling water system, air

conditioning system, early warning alarm system, gas detection system and others. All these systems provide on line data through different channels of communication and interfaces.
These systems are installed by different vendors and usually come along with their own software system and manuals. Usually the vendors provide training courses after the installation. These courses, however, only illustrate how to operate the system but not the operations of ERC activities. The companies still need to develop their own standards of procedures (SOP) to teach their ERC supervisors how to perform their jobs. These SOP are written according to companies' document control system which can generate piles of technical papers and may not be suitable for directly used by trainees as training materials. More effective and efficient way of representation of training materials is needed.

3 Experiment

The gas detector alarm system informing procedure is chosen as the testing content of the training material, which is the most frequent event during three months of ERC operations based on the historical data of a local semiconductor company. This procedure consists of 18 steps, which is represented into three different forms. The documented alphanumerical representation describes the SOP through 18 sentences. The documented graphical representation describes the SOP with pictures of computer program results as well as guiding words. The computerized graphical representation provides the trainees icons with tag-along guiding words.
Since all of the supervisors currently working in ERC have master degree in engineering and the training system is targeted for new recruit, 15 engineering graduate students who have the potential to work as supervisor is chosen as study subjects. Subjects were randomly divided into three different groups (A, B and C) and trained separately by different forms of materials with the same content. A group is trained by documented alphanumerical representation. B group is trained by documented graphical representation. C group is trained by computerized graphical representation.
The environment setting is arranged according to the layout of ERC (Fig 1). The FMCS system is installed in No. 1 Display, where is the source of incident information. The Camera Control Television (CCTV) is installed in No.2 display, where is the source of on-site picture. The No.3 display is for subject to handle secondary task. Subject performs secondary task while monitoring the FMCS system.

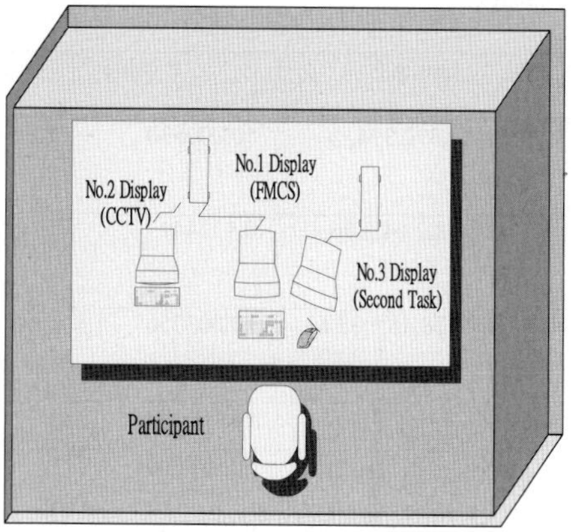

Fig. 1: Layout of experiment setting

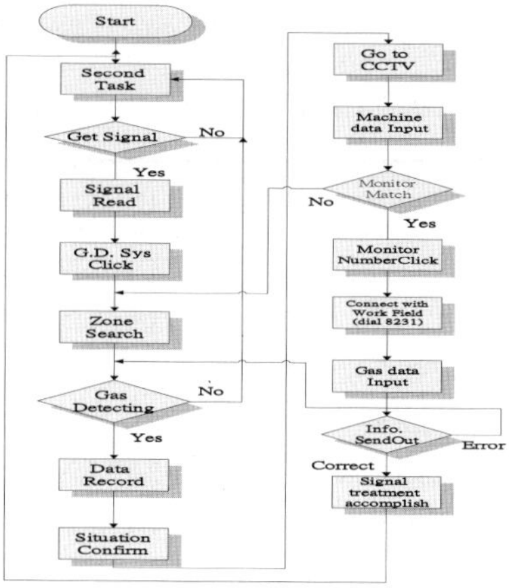

Fig. 2: Flow chart of testing program

If there is incident message appearing on screen, he/she must perform the informing procedure immediately. Once the process is completed, subject continue performing secondary task and monitoring the FMCS. Self-developed computer program, which simulate the operation of Gas Detector Alarm Informing Procedure, is used as testing tool to investigate the performance of trainees. The flow chart of this program is depicted in Fig. 2. After he/she has confidence to accept the test, each subject runs the testing program with three different frequencies of incident rate. Performance measurement index include trainees' training time, response time, message processing time, number of errors, number of missing and score from processing secondary task. The definitions for these indexes are listed as follows:

- Training time: time from trainee accessing the training material to being confident to accept the test;
- Response time: time from incident message appearing on display screen to trainee acknowledge its existence;
- Message processing time: time from trainee starts to performing the informing process to complete the process;
- Number of errors: number of wrong actions during the informing process;
- Number of missing: number of missing actions during the informing process;
- Score from processing secondary task: score from processing secondary task during the testing period.

4 Results & discussions

Experiment results indicate that different forms of training materials have significant effects on training time, processing time, number of errors, and number of missing (Table 1.). The effects on response time and secondary task score are not significant. This is because the incident message alert always appears at the same location and it can be easily identified. Further data analysis results (Table 2) indicate that subjects in group C require less training time and processing time than the other subjects in group A and B. Also they make fewer numbers of errors and missing than other subjects in group A and B. This means the performances of subjects in group C, in term of training time, processing time, number of errors and number of missing, are better than subjects in group A and B.

Table 1: Experiment results

Measurement indexes	Group A	Group B	Group C	P value
Training time (second)	480,479,677,523,607	426,452,412,427,430	387,382,372,394, 373	*0.000530
Response time (second)	2.11,4.87,2.88,4.27, 4.5	3.44,2.83,3.6,3.55, 5.55	2.05,2.38,2.04,3.06, 2.77	0.132441
Processing time (second)	52,55.62,56.93,79.77 ,58.57	43.4,54.08,35,48.92, 47.8	33.8,37.28,45.1,33.63, 43.87	*0.003756
Number of errors	4,3,3,5,3	2,3,3,3,2	3,1,2,1,1	*0.006483
Number of missing	2,3,2,4,4	1,2,2,1,4	0,0,2,1,1	*0.018305
Secondary task score	340,346,370,480,363	373,380,446,356,373	433,273,346,536,356	0.978721

*Marked effects are significant at P<0.05

Table 2: Experiment data analysis results

* Marked effects are significant at P<0.05

	Training material representation forms							
	Training time		Processing time		Number of errors		Number of missing	
Group	B	C	B	C	B	C	B	C
A	*0.002351	*0.000181	*0.14898	*0.001213	0.70262	*0.001847	0.151713	*0.005590
B		0.163919		0.196106	-	**0.070262**		0.091071

5 Conclusions

The capable ERC supervisors are hard to find. Most new recruits need to be trained to be acquainted with the unique FMCS for each company effectively and efficiently. The results of this study indicate that different forms of training materials have significant effects on the trainees' performance. The computerized graphical training materials require the least efforts and have the best performance than the other two. Therefore, it is argued that in order to speed up the training process of the ERC trainee, the computerized graphical training materials are suggested. The SOPs, currently used by semiconductor industry for guiding workers' operations, should not be used directly as the training materials for ERC supervisors. Efforts should be made to change them into computerized graphical representations. The results can be used to support the development of the training system for ERC personnel. Future study will focus on the development of computer-aided ERC training program.

Acknowledgements

The authors would like to express their gratitude to Powerchip Semiconductor Corp. and Dr. Guang-Hann Chen for their kindly support for this research. This research is funded by National Science Council in Taiwan (NSC# 91-2213-E-007-057).

References

Berthelette, D. (1996), "Evaluation of ergonomic training programs", Safety Science, Vol.23 No.2/3, pp.133-143.

Gramopadhye, A., Bhagwat, S., Kimbler, D., & Greenstein, J. (1998), "The use of advanced technology for visual inspection training", Applied Ergonomics Vol. 29, No.5, pp.361-375.

http://www.itri.org.tw/cishthome/idb/datanews/news/news.htm (in Chinese)

http://www.tsia.org.tw/Issues/IssuesESH.asp (in Chinese)

Johnson, C. W. (1996), "Integrating human factors and systems engineering to reduce the risk of operator error", Safety Science Vol. 22, pp.195-214.

Su, Y.L., & Lin, D.Y. (1998), "The impact of expert-system-based training on calibration of decision confidence in emergency management", Computers in human behaviour 14 no.1, pp.181-194.

The automated construction of relevance maps using spatial data mining

Michael May

Fraunhofer AIS
Knowledge Discovery Team
Schloss Birlinghoven,
53754 Sankt Augustin
Germany
michael.may@ais.fraunhofer.de

Abstract

An important goal of data mining is to find patterns that are understandable to a non-expert. This makes careful design of the Human-Computer-Interface an essential topic. Many data mining tasks have to deal with data that are explicitly or implicitly geo-referenced. For those tasks, embedding of data mining results in a geographic map can be a very effective visualization. We introduce the notion of a relevance map, which maps attributes and spatial features statistically relevant for a phenomenon of interest. We describe a method for automatically constructing those maps using subgroup discovery, a data mining approach searching for statistically defined deviation patterns.

1 Introduction

Data mining is the partially automated search for hidden patterns in typically large and multi-dimensional databases. It draws on results in machine learning, statistics and database theory (Klösgen & Zytkow 2002). Important goals are to find patterns that are interesting and previously unknown to the user, but also to present them in a way that is understandable to a non-expert. This makes a careful design of the Human-Computer-Interface an essential topic.

The dominant way of presenting results of a statistical or data mining analysis to the user is either in textual or numeric format (a regression equation, diagnostic summaries) or visualized in some mathematical space (e.g. as a scatter plot of *x-y* co-ordinates). However, this type of visualization is not always the best one.

It has been noted that the task of exploratory data analysis is to present the salient features of the data in a format suitable for the information processing capabilities of the human brain (Good, 1983). If the data describes some spatial phenomenon, this suggests that the result of an analysis should be presented to the user in *geographic space*, since the human visual system is well-trained for handling spatial information. In spite of this, visualization of data mining results in geographic space is rarely used, even if many data sets are implicitly or explicitly geo-referenced (e.g. using a zip-code).

In this paper we introduce the notion of a relevance map as a means to convey the results of statistical inference on spatial phenomena to the user. We show how those relevance maps can be constructed in an automated manner from a database containing both spatial and non-spatial information.

The rest of this paper is organized as follows. First we introduce the notion of a relevance map. Next we describe the algorithm used to infer those maps. Finally we discuss an example.

2 Relevance Maps

While the most common data mining algorithms such as decision trees aim at classification and prediction, there is another class whose goal it is to help the user to better understand the data and to provide an explanation of a phenomenon. This is done by identifying one or more attributes that are relevant for a phenomenon of interest according to some measure of interestingness.

We call a visualiation of such potential explanations of some phenomenon of interest a *relevance map*. It is a thematic map that displays (1) the spatial distribution of a phenomenon of interest, the *target attribute* T, and (2) another set of attributes C that is statistically relevant for T, such that the interaction between C and T becomes visible.

Objectives of constructing relevance maps are

1. to identify patterns in space;
2. to identify other objects that are potential generators of such patterns;
3. to identify information relevant for explaining the pattern;
4. to filter and hide irrelevant information.

Ideally, a pure relevance map contains everything that is relevant for a explanation of a phenomenon in a given context and nothing that is irrelevant.

There is an important difference between statistical visualizations in abstract space and visualizations in geographical space. While most visualizations of statistical data show only what is entailed by the statistical results (e.g. *x-y*-coordinates and a regression line), embedding statistical results in geographic space often adds additional information not part of the original results. Therefore, linking of statistical results to geographical data allows the user to combine the statistical information with his domain knowledge. This is especially important since embedding of a statistical result into a substantial (scientific or common-sense) theory is among the most effective ways to guard against many problems that accompany statistical inference, including discovery of outliers and of statistical artifacts (accidental findings, problems from multiple testing, multi-correlation). It can also be an effective method for spotting systematic errors in the data. Moreover, it can be instrumental in the task of forming causal hypotheses. As is well known, causal conclusions cannot be based on observational statistical data alone.

3 Automated construction of relevance maps

Geographic Information Systems (GIS) are widely used for analyzing and visualizing spatial information. The purpose of creating a thematic map in a GIS is often the manual construction of a relevance map. However these systems have limited capabilities to visualize spatial patterns on a map having more than a few dimensions. Hence, complex and unsuspected dependencies are easily overlooked. Searching great amounts of data by applying a statistical evaluation function, filtering away irrelevant and keeping relevant information is exactly what *data mining* does. For this reason, the combination of GIS and data mining technology promises great benefits. We have implemented a system for automated construction of relevance maps using the *subgroup mining* approach (Klösgen & May, 2002). The information should be presented to the user in a way that is intuitively understandable and supports further investigation. Subgroup mining will be described in this section in an informal way with the help of an example.

Assume we are interested in analysing migration rates for Stockport, a district in Greater Manchester, UK. As data sources UK 1991 census data, available on the level of enumeration districts, is used. These data are collected (and copyrighted) by the UK Office for National Statistics ONS. Also available is detailed geographical information on streets, rivers, buildings, railway lines, shopping areas. Assume we are interested in Stockport enumeration districts with a high migration rate. We ask ourselves how those enumeration districts are characterized and what might distinguish them from other enumeration districts not having a high migration rate. Spatial subgroup discovery is a data mining approach that helps to answer this kind of question. It searches the hypothesis space for interesting *deviation patterns* with respect to an attribute of interest.

Example: „Migration is high in enumeration districts with high unemployment“

The attribute of interest or target attribute T is “high migration rate”, the concept C is “enumeration districts with high unemployment” and the subgroup is the conjunction of both, i.e. “Migration is high in enumeration districts with high unemployment”. The deviation pattern is that the proportion of districts satisfying the target T is much higher in districts that satisfy pattern C than in the overall population ($p(T|C)>p(T)$).

Table 1. Terminology for subgroup discovery.

Type	**Abbreviation**	**Description**
Concept	**C**	“enumeration districts with high unemployment”
Target	**T**	“Migration rate is high”
Subgroup	**S**	„Migration rate is high in enumeration districts with high unemployment“
Attribute	**A**	Migration rate
Value	**V**	high, low, medium

The hypothesis space is searched in a top-down manner from more general to more specific; e.g. the description “area with high unemployment and a large number of medical establishments” is more specific than the description “area with high unemployment”. A *beam search* is performed so that only the best *n* hypotheses found so far are expanded at each level of search. Hypotheses are ranked by a statistical *evaluation function* based on the classical binomial test. The search terminates when the algorithm has searched the hypothesis space up to a user-defined search depth. The best *n* hypotheses are reported to the user (for details see Klösgen, May, 2002).

We have implemented the automated construction of relevance maps in a spatial data mining platform called SPIN! (http://ais.fraunhofer.de/KD/SPIN/). For visual exploratory spatial analysis the CommonGIS module for interactive manipulation of statistical maps is integrated into the system (Andrienko, Andrienko 1999). Its major strength, lies in its capabilities for interactive visual exploration of statistical data. Subgroup Mining is deeply integrated with an object-relational spatial database (Oracle Spatial 9i). The data mining search is distributed between the database, that does the counts statistics, and the search manager, that controls the search and evaluates hypotheses.

4 Example

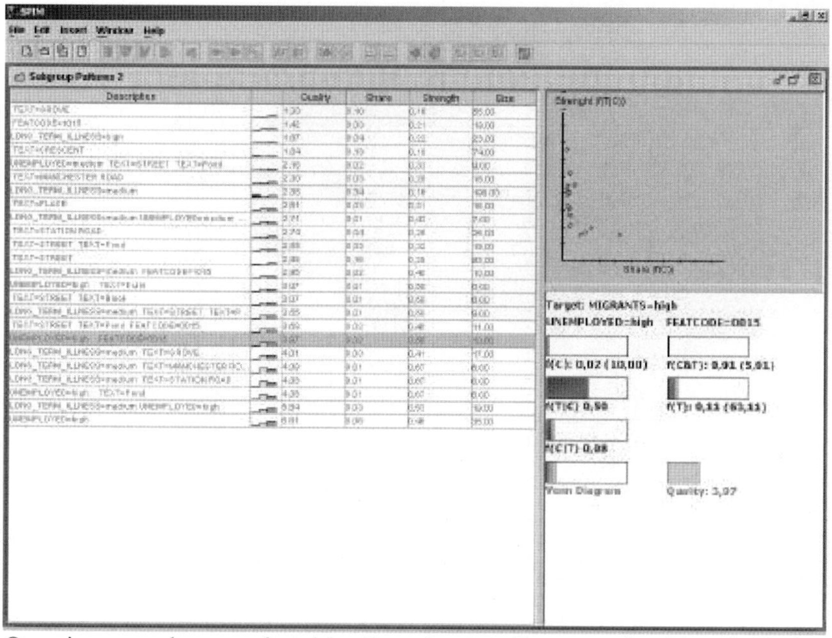

Fig 2. Overview on subgroups found by the subgroup discovery algorithm (left) showing the subgroup description. Bottom right side shows a detail view for the overlap of the concept C (e.g. located near a railway line) and the target attribute T (high migration rate). The larger the difference between p(T) and p(T|C) and the larger the subgroup C, the higher the quality of a subgroup. The window on the right top plots p(T|C) against p(C) for all subgroups.

For visualizing relevance maps, we use a combination of cartographic and non-cartographic displays linked together through simultaneous dynamic highlighting of the corresponding parts. The user navigates in the list of subgroups, which are dynamically highlighted in the map window. Figure 5 and 6 show an example for the migrant scenario, where the subgroup discovery method reports a relation between districts with high migration rate and high-unemployment. The conditional probability (p(T|C)) becomes even larger (though the subgroup C becomes smaller) if those districts are located near a railway line (shown in brown on the map). This is an example for a subgroup combining spatial features (being crossed by a railway line) and non-spatial features (unemployment rate).

Acknowledgement: Work on this paper has been partially funded by the European Commission under IST-1999-10536-SPIN!. Data on the Stockport application are providedtothe project by the project partners Manchester University and Manchester Metropolitan University.

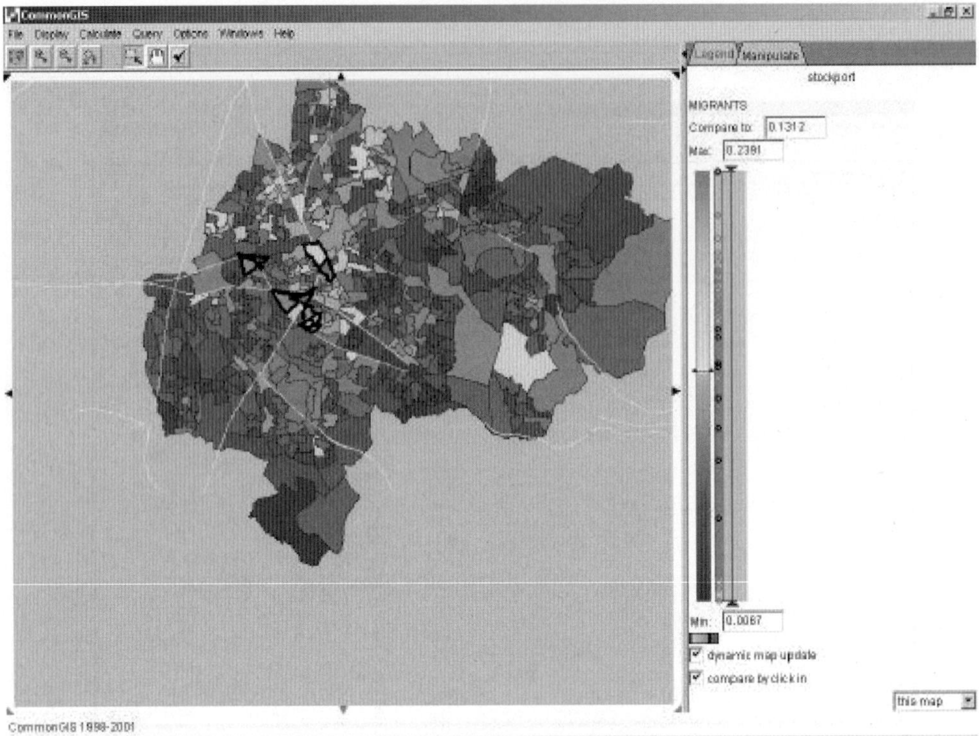

Fig. 2. Enumeration districts satisfying the subgroup description C (high unemployment rate and crossed by a railway line) are highlighted with a thicker black line. Enumeration districts also satisfying the target (high migration rate) are displayed in brown.

5 References

Andrienko, G. & Andrienko, N. (1999). Interactive Maps for Visual Data Exploration. *International Journal of Geographical Information Science* 13(5), 355-374.

Good, I. J. (1983). The Philosophy of Exploratory Data Analysis. *Philosophy of Science* 50, 283-295

Klösgen, W. & May, M. (2002) Spatial Subgroup Mining Integrated in an Object-Relational Spatial Database. In Elomaa, T.,Mannila, H. & Toivonen, H. (eds.) *Proc. Principles of Data Mining and Knowledge Discovery (PKDD),* 6th European Conference, Helsinki, Finland, August 2002, p. 275-286, Berlin: Springer.

Klösgen, W. & Zytkow, J. (eds.) (2002). *Handbook of Data Mining and Knowledge Discovery.* Oxford: Oxford University Press.

May, M., Savinov, S: An integrated platform for spatial data mining and interactive visual analysis, In *Proceedings Data Mining 2002. 3rdInternational Conference on Data Mining Methods and Databases for Engineering, Finance and Other Fields*, Bologna, Italy, 25-27. Sept, 2002.

Information Support for Annual Maintenance with Wearable Device

Takashi Nagamatsu, Tomoo Otsuji
Kobe University of mercantile marine
5-1-1, Fukae-minami,Higashi-nada,Kobe 658-0022,JAPAN
{nagamatu,otsuji} @cc.kshosen.ac.jp

Hirotake Ishii, Hiroshi Shimoda, Hidekazu Yoshikawa
Kyoto University
Gokasho,Uji,Kyoto 611-0011,JAPAN
{hirotake,shimoda,yosikawa}@energy.kyoto-u.ac.jp

Wei Wu
Mitsubishi Electric Corporation
8-1-1, Tsukaguchi Honmachi, Amagasaki, Hyogo, 661-8661, JAPAN
Wu.Wei@wrc.melco.co.jp

Abstract

One of the way in which annual maintenance is improved is to support with wearable devices connected with network. Two basic key systems for advanced plant maintenance using Head Mounted Display(HMD) have been developed, which are "a communication support system between a foreman and workers" and "a remote cooperative system utilizing gazing point information". The one enables a foreman to observe and instruct plural workers at the same time. And the other realizes a supporting function from a remote expert. The effectiveness of these prototype systems have been confirmed by basic laboratory experiments in the university.

1 Background

A concept of the satellite operation maintenance center could be a solution to decrease the total management cost of nuclear power plants. The satellite center is located apart from the control room of nuclear power plants, where the introduction of an advanced information system for operation and maintenance with the concept of augmented reality technology is the main subject of this paper. In this study we focus on improvement of annual maintenance work, which is supported with wearable devices connected with network.

The wearable devices workers will wear for conducting annual maintenance work, will need the following three functions; (1)navigation function by which maintenance worker can reach their work place through the labyrinth of nuclear power plant without straying, (2) support function which helps workers look for manual appropriate for their work, search past similar cases as the reference to their work to conduct without any mistake, and communicate smoothly with their foreman at the work place to supervise the subordinate workers by using mobile computer, and (3) support function by which experts located in the satellite center will help by giving appropriate instruction the workers to conduct their maintenance work accurately to compensate for the workers' testing skills for solving difficult problems.

As the first step to realize the three functions above, two basic key systems using Head Mounted Display(HMD) have been developed, which are "a communication support system between a foreman and workers" and "a remote cooperative system utilizing gazing point information".

2 A Communication support system between a foreman and workers

For the communication support between a foreman and workers, a new observation and instruction support system has been developed by utilizing a tablet PC and Augmented Reality technology. This support system enables a foreman to observe and instruct plural workers at the same time for machine-maintenance work. With this support system, a foreman can observe plural worker's field of views and make some instructions to the workers by superimposing the instructions over the worker's field of views.

In this system, as shown in Figure 1, the workers equip a CCD camera which captures the worker's filed of view, a mobile computer with wireless LAN, a microphone and a see-through HMD. And the foreman equips a microphone and a tablet PC with wireless LAN.

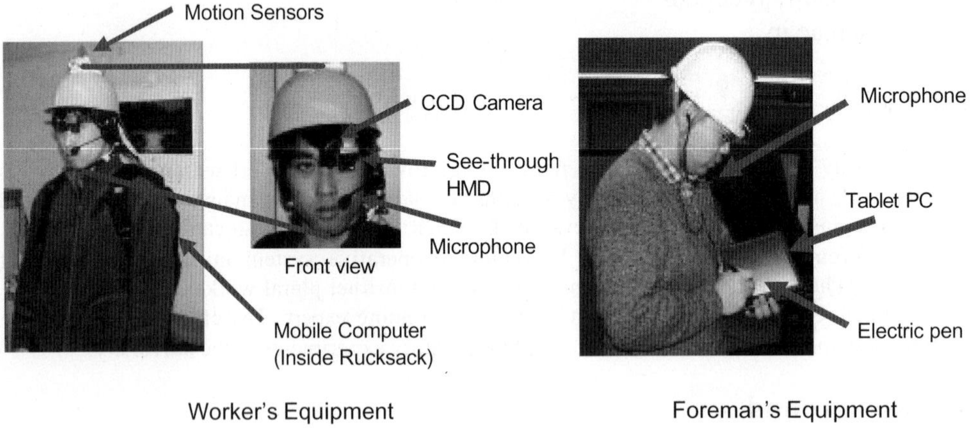

Figure1 Worker's and foreman's equipments.

The worker's filed of view images captured by the CCD camera mounted on the worker's helmet are transferred to the foreman's tablet PC via wireless LAN. As shown in Figure 2, the interface of the foreman's tablet PC consists of preview screens for observing the plural worker's field of views, an instruction screen for writing instructions to the workers and some operation buttons for controlling the support system.

The foreman can select a worker from preview screens and write some instructions such as arrows and comments over the worker's field of view with an electric pen. When the foreman pushes a send button after writing some instructions, the instructions are transferred to the workers who are equipped with a see-through HMD and superimposed upon the view of the workers.

On the worker's site, a function which estimates appropriate locations where the instructions should be superimposed has been developed in order that the locations of instructions where the foreman intends to display are properly aligned in the correct relative position even if the worker moves around the work place. With this function, the worker can understand some instructions from the foreman more intuitively such as important locations the worker should pay attention to.

In this study, in order to estimates appropriate locations, two measurement methods have been developed. One is the method that the appropriate locations are calculated based on the artificial

markers located in the work site in advance. The other is the method that the appropriate locations are calculated based on the natural markers which are acquired from the feature point of work environment such as an edge of machine apparatus (Harris et al.,1988). These methods are used together according to the situation of the work environment. For example, in the case that the accurate measurement is required, the method using the artificial markers are applied. And in the case that it is difficult to locate some artificial markers around the work environment and the accurate measurement is not required, the method using the natural markers are applied.
In addition to the above function, a function has been developed to capture images of the worker's view by using stereo-matching algorithms for measuring the three dimensional shape of equipments located in the work place. This function makes it possible to estimate the three dimensional locations of the instructions where the foreman intends to display even if the instructions written on the display of mobile computer are in two dimensions.

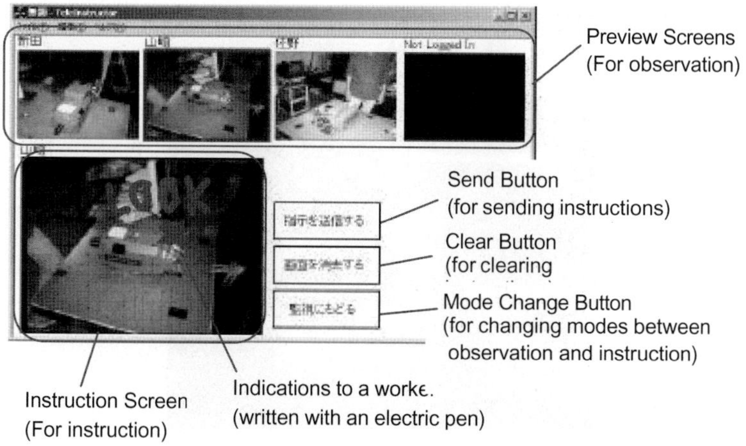

Figure 2 Foreman's Interface.

3 A remote cooperative system utilizing gazing point information

3.1 System configuration

A cooperative system utilizing gazing point has been developed. The system consists of two terminals. One is for the plant workers and the other is for the expert at the remote site such as satellite centre. The terminal for the plant worker consists of Eye-Sensing HMD (ES-HMD) (Fukushima et al.,2001), two PCs, headset, etc. as shown in Fig.3. The ES-HMD monitors worker's eyes by CMOS cameras and calculates gazing point by one PC. The other PC captures the view image of plant worker through a CCD camera which is fixed on ES-HMD. The terminal for expert receives the view image and the plant worker's gazing point via network and display where the plant worker is gazing in the view image as shown in Fig.4. The expert can know where the worker is gazing and detect what the worker thinks in combination with conversation information. And the expert direct which point the worker should pay particular attention to and what to do for supporting the worker's difficult task of the machine maintenance and testing.

Figure 3: Worker with AR devices

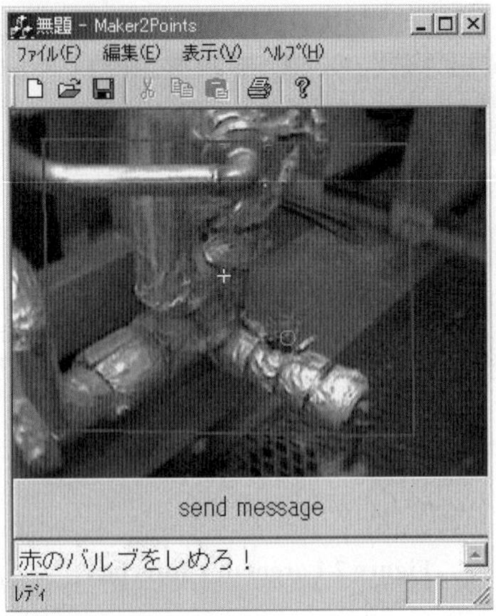

Figure 4: Interface for expert (+: worker's gazing point, O: indicating point by expert)

3.2 Experiment for System Evaluation

An experiment for system evaluation was conducted. As a test site a boiling thermal-hydraulic test loop was used. Participants are one expert and 3 workers. The expert is a researcher who use the loop and 3workers are students of university who have no particular knowledge of the loop. Task in the experiment that workers do are injection of fluid to the loop, adjustment of flow rate and draining of the fluid. The experiment were conducted 9 times (3 times by each person) as shown in table 1. In the table "AR indication" means using the function of display of superimposed indication on HMD and "gazing point" means using gazing point information for expert to understand the situation of workers.

Table 1: Experimental Condition for workers

Number of experiment	Participant A	Participant B	Participant C
1	Using AR Indication & gazing point	Using AR Indication	Not use
2	Using Indication	Using AR Indication & gazing point	Using AR Indication & gazing point
3	Not use	Not use	Using AR Indication

Table 2 shows the result of subjective evaluation of the system by questionnaire. From that result this system is bad to work, but it is useful. So to make this system useful and usable, the system must be improved to be miniaturized to wear and move easily.

Table 2: Subjective Evaluation of the System by Participants as Workers
(-2:very bad, -1:bad, 0: neutral, 1:good, 2:very good)

Division	item	Participant A	Participant B	Participant C	Ave.
Ease of work	Ease of walk	1	-1	1	0.33
	Ease of looking around work space	-1	2	-2	-0.33
	Ease of moving hands	2	2	2	2
	Ease of moving neck	0	-1	-1	-0.67
	Restriction of vsibility	-1	2	-1	0
	Brightness of view	-2	-1	-2	-1.67
Understandability of Indication	Using AR Indication	2	2	2	2
	Using Audio	2	2	-1	1
	Using Character	Not used	Not used	Not used	-
Communication of intension	Using Audio	2	2	1	1.67
	Using gazing point	0	2	2	1.33
Whole System	Usefulness of the system	2	0	2	1.33
	Usefulness of the gazing point	2	2	1	1.67
	Usefulness of Audio	1	2	2	1.67
	Usefulness of Character	Not used	Not used	Not used	-

4 Conclusion

Two basic key systems for advanced plant maintenance using Head Mounted Display(HMD) have been developed, which are "a communication support system between a foreman and workers" and "a remote cooperative system utilizing gazing point information". The effectiveness of these prototype systems have been confirmed by basic laboratory experiments. The next step will be to conduct more realistic system development to be used in the training centre of machine maintenance workers for nuclear power plant.

References

Harris, C., Stephens, M. (1988). A Combined Corner and Edge Detector, *Proceedings of Alvey Vision Conference*, 147-151.

Fukushima,S., Suzuki,K., Murakami,S., Nakajima,R. (2001). Light weight glasses for measuring pupil and eye movement equipped with display function. *Correspondences on Human Interface*,3(2),75-78.

Toward A Taxonomy of Interaction Design Techniques for Externalizing in Creative Work

Kumiyo Nakakoji *Yasuhiro Yamamoto*
RCAST, University of Tokyo and PRESTO, JST
4-6-1 Komaba, Meguro, Tokyo, 153-8904
{kumiyo, yxy}@kid.rcast.u-tokyo.ac.jp

Abstract

Externalizing, the act of creating and modifying an external representation, plays a crucial role in creative design work. Whether a tool supports externalizing in a designer's desirable manner depends not only on what the tool does but also on its interaction design; how a user interacts with the tool through what representations. As an initial attempt to the development of taxonomy of interaction design techniques for tools that support externalizing, this paper examines seven tools for externalizing to illustrate why and how fine-grained interaction influences the process of externalizing.

1 Externalizing in Creative Work

The goal of our research has been to design and develop conceptual frameworks and computational tools that support creative design work (Nakakoji et al. 2000)(Yamamoto et al. 2000)(Fischer, Nakakoji 1994). This paper focuses on a designer's externalization processes, and uses a word *externalizing* to refer to the act of creating and modifying an external representation. Tools for externalizing provide a designer a space for the act of externalizing, supporting the designer's creative process and amplifying their creative ability.

There have been a number of studies that stress the importance of externalizations for human cognition. Scaife and Rogers (1996) use the notion of *external cognition* to emphasize the importance of graphical representations in helping human cognition. Arnheim (1969) states that visual perception is not a passive recording of stimulus material but an active concern of the mind and a viewer is engaged in problem solving while perceiving the visual stimuli. Zhang (1997) shows that different external representations help people differently in solving problems, and that choosing the right representation is critical to make problem-solving more efficient.

These studies, however, mainly focus on external representations as a given entity, and not on the externalizing process. A notable exception is Schoen (1983), who argues that an important aspect of the design activity is *reflection-in-action.* Differentiating reflection-*in*-action from reflection-*on*-action, he argues that for a designer, moving his/her hand while drawing sketches is crucial in the reflective thinking. In this sense, experiential cognition is not separated from reflective cognition (Norman 1993); they are intertwined. This suggests that activity and reflection should complement and support each other (Csiksentmihalyi 1990) at a very fine-grained level of interactions.

We use the term, externalize, to mean to cause state changes in a representation resulting in a different "what you see." Resultant externalizations are not merely expressions of mental imagery (Goldschmidt 1999), but rather, externalizing, the process itself, helps construct mental imagery, affecting what the person thinks, feels, knows and understands. Sketching is a typical example of the act of externalizing. Not only sketched objects, but also the sketching process itself helps a designer identify problems and formulate a solution space.

In order for a tool to support externalizing in a desirable manner, we must be concerned not only with what functionality the tool should provide, but primarily with how the designer interacts with the tool and through what representations. While sketching is a representation, holding a pencil and moving it while pushing its lead on a sheet of paper resulting in a black line is an interaction. The selection of sketching tools affects an architect's creative process (Lawson 1994). Seemingly subtle differences in a representation and interactions have a large impact on the effect of externalizing. The design of fine-grained representation and interaction, therefore, should be a central concern in the development of tools for externalizing.

Little is known, however, about what aspects of representations and interactions with tools are important in externalizing and how they are related to the externalizing process in promoting or disturbing a creative process. The rest of this paper examines seven systems as tools for externalizing, especially in early stages of design tasks. We discuss what aspects of the externalizing process each system does and does not support to address this issue.

2 Seven Tools for Externalizing: Illustrative Examples

Ideas. The Ideas system supports pen-based sketching and helps the designer in the manipulation of pages and resultant sketched objects (Hoeben 2001). The system mimics the conventional externalizing scheme of paper and pencil on a sketchbook, while extends it by providing a natural, smooth interaction through transparent operations on the representations.

Teddy. The Teddy system takes a user's 2D drawing action as an input and produces a corresponding 3D graphic object (Igarashi 1999). For instance, while a user draws a circle, a trajectory of a mouse appears forming a circle in a window (just like a conventional pen-based sketching interface), but when the user finishes drawing the circle (by creating a closed line), the circle is converted into a sphere in the same window. When the user adds another circle on top of the sphere, it is then converted into a sphere on top of the previous sphere. Repeating this process, the user can produce a hand-drawing 3D model consisting of objects, such as a Teddy bear. With Teddy, the externalizing process takes place at two levels. The first level is through a transparent operation. While drawing a circle, the mouse trajectory appears as a curved line and gives a feeling to the user that the user is directly drawing the curved line. The second level is at a larger granularity of interactivity (Svanaes 1999). Each time a user finishes a closed shape, Teddy converts it into a 3D object and displays it in the same window replacing the original shape. This externalizing process is not as closely integrated as the first level, but gives a user a feeling of externalizing.

VR Sketchpad. The VR Sketchpad system also takes a user's hand-drawing as an input, but produces a 3D architectural object in a VRML window, such as a wall, table, chair, or a TV set (Do 2001). The user can draw a floor plan on a sketchpad window, and when requested, the system first parses the sketchpad and identifies shapes (e.g., a circle), and produces architectural objects in corresponding locations (e.g. a table). VR Sketchpad supports externalizing process in two levels similar to Teddy. The first level is the same with Ideas and Teddy. The second level is even larger granularity of interactivity than Teddy. With Sketchpad, a user has to explicitly request for conversion to produce architectural objects from sketched objects. The converted 3D objects appear on a different VRML window and therefore do not replace original hand-drawings. The user may see the sketchpad as input and the VRML window as output. In this sense, the system might provide less feeling of immediateness in externalizing.

Silk. The Silk system takes hand-drawn configuration of user interface objects as an input, and produces functional, executable user interface objects as an output in a different window when requested by a user (Landay, Myers 2001). Similar to Teddy and VR Sketchpad, Silk supports externalizing process in two levels. The first level is the transparent operation as in the same as the above systems. In the second level, hand-drawings and the generated user interface components have a little more detached relationship than VR Sketchpad because the user has to specify constraints and behavior in order to produce functional interface components.

SketchAmplifier. The SketchAmplifier is a tool that amplifies the aspect of time spent in sketching. When a designer draws strokes in a canvas window (Figure 1-a), SketchAmplifier computes the time taken to draw the lines and generates a 3D representation by taking the time as the Z-axis (Figure 1-b). This representation is used to generate an animation (Figure 1-c) that illustrates how the lines have been drawn. With SketchAmplifier, the designer is not only able to immediately reflect on what has been drawn, but also on how it has been drawn in terms of the order and the speed of drawing, which is impossible with paper and pencil.

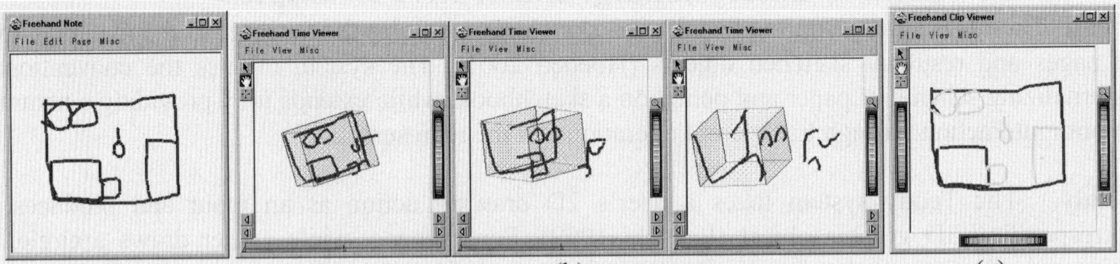

(a) (b) (c)

Figure 1: SketchAmplifier

ART#001. The ART#001 system aims at providing "sketches" for early stages of writing (Yamamoto, Nakakoji 2002). In addition to a regular way of externalizing words and sentences by typing a keyboard and applying copy-paste operations, the system allows a user to manipulate text as a chunk (element) with a 2D space view and a document view. Dragging a text element in the space view changing the relative position in the space will dynamically changes the order of the element in a document view. With ART#001, a user can get a feeling of directly manipulating text objects. Since the user is able to read elements in a document view in different orders by moving around the corresponding element in a space view without releasing the mouse (i.e., without making commitment), ART#001 allows the user to play what-if games by interacting with the representation.

SideView. The SideView tool allows a designer to compare results of applying different image manipulation filters in a GIMP-based image editor by providing previews (Terry 2002). Based on observations of graphic designers repeating a process of menu selection and undo to compare different effects of applying different image filters, the system is designed to extend each menu item with the previews of a result of applying the function to the current task. The focus of SideView is not explicitly supporting the user to externalize representations. However, how the user interacts with the system can be viewed as an externalizing process. A user takes an action (moving a mouse over a menu item), and the system shows a preview as a result of the user's action. The user then is able to reflect on what would be the result of taking the action; just like drawing a rough line in a sketching pad and examines an outcome.

3 Toward Taxonomy

All of the above seven systems support externalizing in early stages of creative tasks by employing different types of interaction design techniques. They use different representations and interactions based on their purposes, tasks, goals, algorithmic constraints and the limitation of current computational capability.

As a first step toward the development of taxonomy of interaction design techniques for externalizing, we present a list of aspects that characterizes their approaches.

- *Representational Immediateness:* The first levels of externalizing discussed above (sketching in Ideas, VR Sketchpad, Teddy, Silk and typing in ART#001) provide low-level immediate representation for a user's action. In contrast, the second level of externalizing provide domain-rich, possibly more distant representational mapping of what a user acts to what the system shows as a result of externalizing.
- *Spatial Immediateness:* Where the user's action takes place and where the resulting externalization is displayed can be the same or different. Hand-drawing interface usually draws lines where the mouse cursor is. 3D objects generated in Teddy will appear in the same window, but representations generated in VR Sketchpad and Silk, and the animation of SketchAmplifier will appear in different windows. These decisions are primarily due to computational limitations (e.g., no algorithms can draw a 2D shape on a VRML window in an effective manner).
- *Temporal Immediateness:* What unit of interaction is interpreted as a meaningful unit by the system determines temporal immediateness. While simple mappings of actions (line-drawing and direct manipulation of objects) enable temporal immediateness, algorithmic constraints and computational power are two factors that affect the temporal immediateness. Algorithmic factors affect Teddy, for instance, which requires a user to finish drawing a stroke before the system can convert it into a 3D object. Available computational power affects SketchAmplifier, for instance, with which the speed of playing animation is limited.
- *Realism toward domain:* Both VR Sketchpad and Silk produce representations that are closer to the domain the user is engaged in by taking a user's hand-drawing. By focusing on a particular domain, such systems map a user's simple action to complex domain objects, allowing the user to have the situation talks back to them more grounded in the domain.
- *Realism toward verisimilitude:* Some tools for externalizing aim at providing representations that are more realistic. They include the real practice (paper and pencil in Ideas), physical objects (3D objects in Teddy), or temporal experience (sketch animation in SketchAmplifier).
- *Allowing what-if games:* Both ART#001 and SideViews turn "expensive" operations into affordable ones, allowing a user to play what-if games during the externalizing process.

4 Discussions

In order for tools to support an externalizing process, we have to consider: *what* a designer wants to externalize (representation), and *how* the designer wants to externalize it (interaction). Whether a representation is desirable depends on what interaction is possible with the representation, and what interaction is necessary depends on what representation is to interact with.

Existing HCI approaches, such as the notion of direct manipulation, virtual reality, interaction design, are necessary but not sufficient to help us address the concern. This paper gives our initial attempt to develop a conceptual framework and a language that can be used to describe what aspect of tools support *how* and *what* aspects of externalizing processes. We will continue this effort through the development of tools for externalizing in creative work and the examination of both success and failure cases of tools developed by others.

5 References

Arnheim, R. (1969) Visual Thinking. University of California Press, Berkeley.

Csikszentmihalyi, M. (1990) Flow: The Psychology of Optimal Experience. HarperCollins Publishers, New York.

Do, E.Y-L. (2001) VR Sketchpad: Create Instant 3D Worlds by Sketching on a Transparent Window. Vries, B.d., Leeuwen, J.P.v. and Achten, H.H. eds. CAAD Futures 2001, Kluwer Academic Publishers, Eindhoven, the Netherlands, pp.161-172.

Fischer, G., Nakakoji, K. (1994) Amplifying Designers' Creativity with Domain-Oriented Design Environments, Artificial Intelligence and Creativity, in Part V, T. Dartnall (Ed.), Kluwer Academic Publishers, The Netherlands, pp. 343-364.

Goldschmidt, G. (1999) Design, in Encyclopedia of Creativity, Mark. A. Runco, Steven R. Pritzker (Eds.), Vol.1, Academic Press, San Diego, CA., pp.525-535.

Hoeben, A. & Stappers, P.J. (2001) Ideas: Concepts for a Designers' Sketching-Tool, Extended Abstract of CHI 2001, Seattle, WA.

Igarashi, T., Matsuoka, S. & Tanaka, H. (1999) Teddy: A Sketching Interface for 3D Freeform Design ACM SIGGRAPH'99, Los Angels, pp.409-416.

Landay, J.A. & Myers, B.A. (2001) Sketching Interfaces: Toward More Human Interface Design. IEEE Computer, 34 (3). pp.56-64.

Lawson, B, (1994) Design in Mind, Architectural Press, MA.

Nakakoji, K., Ohira, M., Yamamoto, Y. (2000) Computational Support for Collective Creativity, Knowledge-Based Systems Journal, Elsevier Science, Vol.13, No.7-8, pp.451-458, December.

Norman, D.A. (1993) Things That Make Us Smart. Addison-Wesley Publishing Company, Reading, MA.

Scaife, M. & Rogers, Y. (1996) External Cognition: How do Graphical Representations Work?, Int. J. Human-Computer Studies, no.45, pp.185-213, Academic Press.

Schoen, D.A. (1983) The Reflective Practitioner: How Professionals Think in Action. Basic Books, New York.

Svanaes, D. (1999) Understanding Interactivity: Steps to a Phenomenology of Human-Computer Interaction, Ph.D. Dissertation, Dept. of Computer and Information Science, Norwegian University of Science and Technology, Trondheim, Norway.

Terry, M. & Mynatt, E. (2002) Side Views: Persistent, On-Demand Previews for Open-Ended Tasks UIST 2002, pp.71-80.

Yamamoto, Y., Nakakoji, K., Takada, S. (2000) Hands-on Representations in a Two-Dimensional Space for Early Stages of Design, Knowledge-Based Systems Journal, Elsevier Science, Vol.13, No.6, pp.375-384, November.

Yamamoto, Y., Nakakoji, K. & Aoki, A. (2002) Spatial Hypertext for Linear-Information Authoring: Interaction Design and System Development Based on the ART Design Principle Proceedings of Hypertext2002, ACM Press, pp.35-44.

Zhang, J. (1997) The Nature of External Representations in Problem Solving. in Cognitive Science, pp.179-217.

Dynamic Query Choropleth Maps for Information Seeking and Decision Making

Kent L. Norman

Department of Psychology
University of Maryland
College Park, MD 20742
kent_norman@lap.umd.edu

Haixia Zhao, Ben Shneiderman, Evan Golub

Human-Computer Interaction Laboratory
University of Maryland
College Park, MD 20742
haixia@cs.umd.edu, ben@cs.umd.edu,
egolub@cs.umd.edu

Abstract

Information retrieval and visualization can be combined in dynamic query systems that allow users unparalleled access to information for decision making. In this paper, we report on the development and evaluation of a dynamic query system (YMap) that displays information on a choropleth map using double thumb sliders to select ranges of query variables. The YMap prototype is a Java-Applet that supports panning and zooming. Several usability studies were conducted on early prototypes that resulted in the current version. Applications of YMap for decision making tasks are discussed.

1 Introduction

New methods of information visualization open up rich databases to information analysts by taking advantage of the power of human visual processing in combination with complex screen displays (Card, MacKinlay, & Shneiderman, 1999). Such visualizations are particularly important in mapping geographic data (MacEachren, Hardisty, & Gahegan, 2001). Map displays can combine a number of facets of nominal and quantitative data to show vast amounts of data in a way that illuminate rather than obscure important points and trends. For example, choropleth maps combine geographic layout with quantitative information by colourizing areas using a colour gradient. Choropleth maps allow analysts to quickly scan the area and to locate highs, lows, and trends (Andrienko & Andrienko, 1999).

Similarly, new methods of dynamic database query facilitate information retrieval by creating a close coupling between the direct manipulation of query variables and the presentation of the retrieved set (Ahlberg & Shneiderman, 1993). Dynamic query applications allow the users to see the immediate result of changes in the query variables.

This paper presents a new prototype of a dynamic map query system called "YMap" based on the work of Dang, North, and Shneiderman (2001).

2 YMap Prototype

Figure 1 shows a screenshot of YMap applet. It contains a choropleth map (centre/left), a set of dynamic query double-box sliders (right/middle), a detail data panel (bottom/left), a drop-down-list-box of attributes to choose from for shading the map (above the map, with continuous shading legend), an overview map (right/bottom), and a toolbox for panning, zooming and toggling between a state view and a county view (top). The prototype follows the general interface design methodology for exploring large data sets proposed by Shneiderman (1998): overview of the data distribution via visualization (shaded choropleth map and histogram bars on the sliders), dynamic query (via double-box sliders) to filter out unwanted entries and narrow down the result, and details-on-demand to further examine individual entries of interests (mouse over a region to view the data associated with that region, select regions to compare their data in the detail information panel, and save selected data to an external file for further analysis). The screenshot shows a county view of continental United States shaded by the year 1992 furniture and home furnishing store sale volume. Counties with low number of new private housing units or with median population age out of range [26, 42] are filtered out (in grey, the same colour used for out-of-range parts of the sliders.)

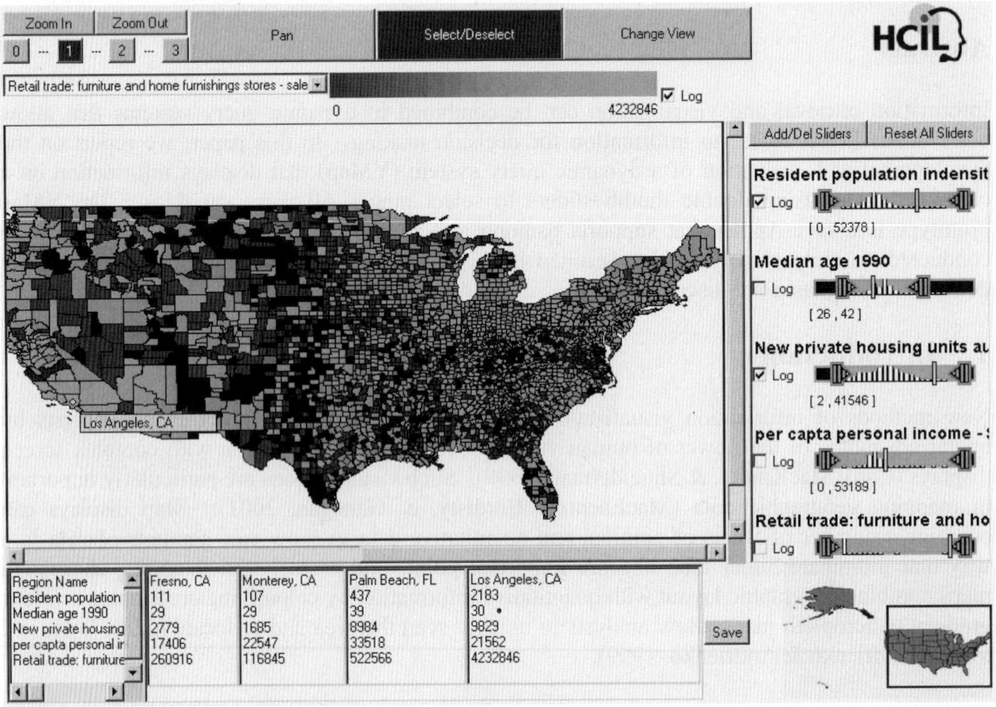

Figure 1: A screen shot of our YMap prototype

While a lot of additional features can be added to our prototype, such as scatter plots and linked brushing with the map, here we focus on three key issues in our system design:

2.1 Striving for universal usability

Our system strives to work towards universal usability (Hochheiser & Shneiderman 2001). Specifically, we aim to support occasional users with modem speed connections and low-end machines, as well as users with highspeed connections and fast machines. While long initial download time is a typical problem with most interactive Web map solutions that deliver vector-format map data to the browser for rendering, YMap employs an image-based technique that encodes geographic knowledge into raster images called base maps (Zhao & Shneiderman 2002).

This enables sub-second, fine-grained interface controls for dynamic query, dynamic classification, and geographic object data identification, with short initial system response time (to reduce the frustration of users with slow connections), near linear interaction performance scalability to the number of geographic objects (which accommodates users with low-end machines), and the potential of reducing the overall data transfer over the network.

2.2 Zooming/panning structure

With the image-based technique, zooming/panning were designed to ensure usability as well as quick the system response time. We chose to pre-generate base maps and define the number of zooming levels instead of having the server generate the base maps on-the-fly and allowing free-scale zooming. Several strategies for achieving fast zooming/panning are discussed in by Zhao & Shneiderman (2002), such as breaking a large base map into several smaller ones, two-staged zooming response using aggregated/interpolated base maps, and pre-fetching.

2.3 Slider Structure

2.3.1 Data distribution and the histogram

While it is a good idea to present each geographic object as a dot along the slider (Andrienko & Andrienko, 1999), it is often hard to judge changes when a section of the slider is overcrowded (e.g., 20 dots and 200 dots of size $1mm^2$ /ea along a 2mm slider section shows no difference to users' visual perceptions). YMap sliders use histograms (see Figure 2). The height of a histogram bar is proportional to the number of geographic objects with the attribute value falling in its range. When users mouse-over a bar, a pop-up appears showing the value range of the bar and its frequency (e.g., in Figure 2a, the highest bar shows that 8 states have resident population density within the range [62.76, 99.41] 1/SQM). The histogram serves several purposes: (a) it gives an overview of the statistical data distribution; (b) it gives users hints of how the map will be filtered when the slider is adjusted; and (c) it allow bi-directional linked identification between the map and the slider.

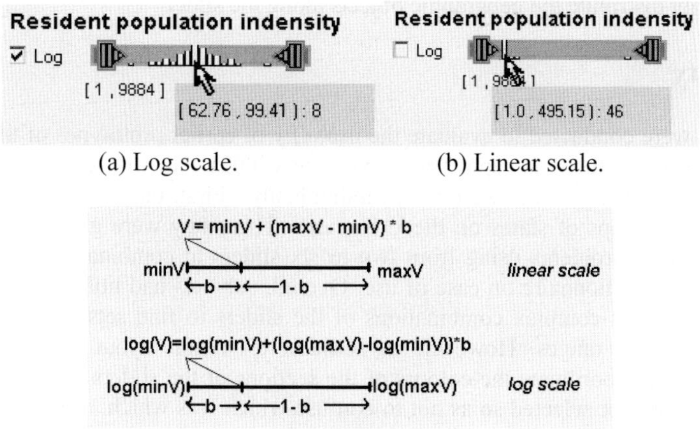

(a) Log scale. (b) Linear scale.

(c) Equations

Figure 2: Slider scales

When users mouse-over a region on the map, the value of that region is marked as a yellow bar on the slider (see Figure 3a). On the other hand, when users mouse-over a histogram bar, all the regions with the attribute values falling into that bar can be highlighted on the map (see Figure 3b) (to be implemented). In our prototype, users can move the two slider boxes to define a query width, then click a point between the two boxes, hold and move the fixed-width query range to view the change on the map (see Figure 3c).

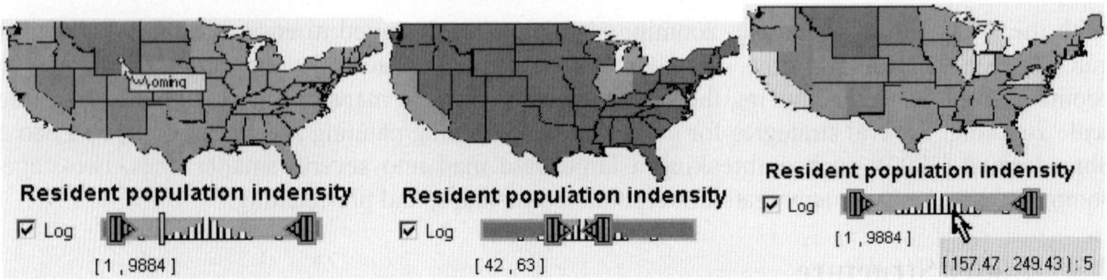

(a) (b) (c)

Figure 3: Bi-directional linked identification

2.3.2 Slider scales

Two different slider scales are supported, linear and logarithmic. In linear scale, the attribute value represented by a slider position is linearly proportional to its distance to the ends of the slider. In logarithmic scale, the values are logarithmized (see Figure 2c). The purpose of providing different scale options is to enable users to spread geographic objects along the slider, so filtering is more gradual when the slider is adjusted. As an illustration, in Figure 2b with a linear scale, 46 out of 51 states have their population density values falling into the first histogram bar. A slight movement of the slider box across the bar causes most of the states to be filtered out, hiding their difference from the exploration. When switched to logarithmic scale as in Figure 2a, states are more evenly distributed along the slider. The logarithmic scale used in our prototype helps to spread out overcrowded lower-end data. Other scale alternatives could be added, such as a percentile scale that would evenly distribute the geographic objects along the slider.

3 Usability

Several studies were conducted to evaluate the usability of earlier prototypes of the dynamic map query system. One of these is reported by Norman (2002). In this study, 43 undergraduate students from a psychology class were tested individually. First, they were instructed on how to set sliders to find groups of states on the U.S. map. Then they were given 18 problems using single sliders and 12 problems using from two to six sliders in combination. Finally, they were given a follow-up questionnaire on ease of use. Overall, subjects had little problem adjusting the single sliders and even complex combinations of the sliders to find sets of states whose values were within specified ranges. However, in addition to minor layout changes, these studies indicated the need to coordinate the colours of the sections of the sliders with the map areas that were selected versus not selected so as not to confuse which was which. Overall, the usability of YMap was rated fairly highly.

In addition, Norman (2002) found that with dynamic query maps, users quickly became aware of attributes that varied in geographic directions (e.g., north-south, east-west, coastal-central, etc.).

They also were able to detect when sliders were correlated with other sliders. Thus, dynamic query maps have the added benefit that users become more aware of the underlying structure of the database in the process of retrieval and exploration.

4 Applications

Dynamic choropleth query maps such as YMap have many applications in complex information analysis and decision making tasks. In addition to finding specific geographic regions that match a query and retrieving details, they are particularly usual in terms of sensing changes, gradients, and relationships among variables. They are particularly useful for making decisions among large sets of geographic alternatives. Decision with many alternatives generally involves a two-stage process: rapid elimination of unacceptable alternatives followed by a detailed comparison of remaining possibilities. YMap allows the user to home in on a select set of regions by adjusting sliders. For example, if one wanted to find the ideal region for a new manufacturing facility, the decision maker would eliminate many locations based on building cost or proximity to the market, select in locations with large labour forces, and select out areas with high costs of energy, etc. Thus, the first stage of the decision process involves a number of tradeoffs between the settings of the alternative sliders, adjusting one up and another down to reduce the set of alternatives to manageable, best set. The second stage involves a detailed comparison of the specific attributes of the alternatives. YMap provides the dynamic visualization interface to facilitate both stages.

References

Ahlberg, C., & Shneiderman, B. (1993). Visual information seeking: Tight coupling of dynamic query filters with starfield displays. In R. M. Baeker, G. Grudin, W. A. S. Buxton, & S. Greenberg (Eds). *Readings in human-computer interaction: Toward the year 2000* (pp. 450-456). San Francisco: Morgan Kaufmann.

Andrienko, G. L. & Andrienko, N. V., (1999). Interactive maps for visual data exploration, *International Journal of Geographical Information Science*, 13 (4), 355-374.

Card, S. K., MacKinlay, J. D., & Shneiderman, B. (Eds.) (1999). *Readings in information visualization: Using vision to think*. San Francisco: Morgan Kaufman.

Dang, G., North C. & Shneiderman B., (2001). Dynamic queries and brushing on choropleth maps, *Proc. International Conference on Information Visualization 2001*, IEEE Press, 757-764.

Hochheiser, H. & Shneiderman, B. (2001). Universal usability statements: Marking the trail for all users. *ACM Interactions*, 8 (2), 16-18.

MacEachren,A. M., Hardisty, F. & Gahegan, M. (2001, May). Supporting visual integration and analysis of geospatially-referenced data through web-deployable, cross-platform tools. Paper presented at the National Conference for Digital Government Research, Los Angeles, CA.

Norman, K. L. (2002). *Dynamic Query Maps: A Study of Performance and Usability*. (Report No. LAP-TR-2002-03). College Park, MD: Laboratory for Automation Psychology, University of Maryland.

Shneiderman, B. (1998). *Designing the user interface: Strategies for effective human-computer interaction.* Reading, Mass: Addison-Wesley.

Zhao, H. & Shneiderman, B. (2002). *Image-based Highly Interactive Web Mapping for Geo-referenced Data Publishing*. (Report No. HCIL-TR-2002-26, CS-TR-4431, UMIACS-TR-2003-2). College Park, MD: Human-Computer Interaction Laboratory, University of Maryland.

Design and Prototype Development of Building Energy Management Agent System

Fumiaki Obayashi, Yoshihiko Tokunaga, Junji Nomura

Matsushita Electric Works, Ltd.
1048 Kadoma, Kadoma-shi, Osaka, Japan
{obayashi, tokunaga, nomura}@icrl.mew.co.jp

Abstract

Energy management of buildings has many problems. Buildings have a lot of subsystems and equipment, and they have a long life and many changes in equipment or the way they operate. In energy management, characteristics of energy are different in each building, and advanced expert knowledge and experience are required to analyze energy information or optimal control. In addition, new management functions are needed. So the new Building Energy Management Agent System (BEMAS) has been designed in this research. The basic policy of BEMAS is that the agents are assigned to every component such as equipment or functions that change frequently, and their cooperation carries out the management of energy, so BEMAS can cope with the basic changes in buildings. Meanwhile, it is requested that building energy management could be carried out also by nonprofessional users, so the Literacy Free Interface, which is the user interface with user support functions such as information interpretation, has been proposed in the BEMAS. In addition, the energy management know-how is designed to be accumulated in the system, and it enables an inexperienced user to conduct energy management easily. Furthermore, a prototype system has been developed based on these designs.

1 Introduction

In recent years, buildings have become an important factor in saving energy all over the world. A reduction in energy consumption of a building contributes to energy problems. So, Building Energy Management System (BEMS) has emerged to make a mark in recent times. Buildings, however, have many different components and equipment changes in life of the buildings. A suitable management plan is different in each case. They cause an increase in maintenance costs. BEMS should have the characteristics such as open standards, high flexibility, cooperation of distributed intelligent modules; so multiagent is desirable for BEMS. This paper examines how the BEMS should be designed with multiagents, and explains the development of a prototype system.

2 Building Energy Management

The energy management of buildings have many issues to contend with. Buildings have a great many components such as energy-related apparatus, sensors, etc; and are influenced by various factors such as residents, environment, and how to operate. Buildings are getting to be more secretive, and all kind of faults go underground. Optimal control cannot be realized easily.

Buildings have a long life, and various factors change frequently. For example, change of equipment, measuring instruments, room layout, and operating way claims the modification of the management system.
Furthermore, management know-how of energy is different in each building, and advanced knowledge and experience are needed for suitable management. They increase the difficulty of energy management of buildings.

3 Building Energy Management Agent System

Considering the above issues, the new BEMS by multiagents, Building Energy Management Agent System (BEMAS) has been designed. Fig.1 shows the concept of BEMAS. BEMAS consists of 2 main parts, Energy Management Agent Unit and Literacy Free Interface.

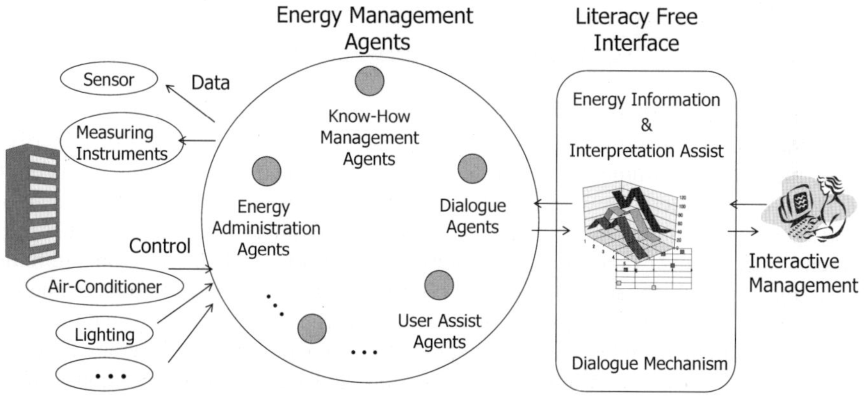

Fig.1 Concept of Building Energy Management Agent System (BEMAS)

The data of all kinds of measurement instruments and sensors in the building are sent to BEMAS. And the Energy Management Agent Unit deals with this data and conducts a variety of energy management controls. The output of Energy Management Agent Units are sent to the Literacy Free Interface. The Literacy Free Interface cooperates with Energy Management Agents, and presents all manners of information to the user, and receives user input. The user's order is sent to the Energy Management Agent Unit and controls the energy equipment of the building.
Energy Management Agent Unit has Energy Administration Agents such as Fault Detection Agents and Energy-Saving Diagnosis Agents, and User Assist Agents that interpret the data and explain the meaning, e.g., and Dialogue Agents, and the more, Know-How Management Agents that accumulate the know-how about energy management.
For the energy management, advanced expert knowledge and experience is required. However, such specialists cannot carry out the energy management of every building, so it is necessary to enable ordinary people to manage the buildings. For example, when some energy related graphs of a subsystem are indicated, it is not easy for ordinary people to analyze the meaning of these graphs and to find out faults in equipment or suitable energy control plans. So BEMAS has a Literacy Free Interface, the user interface that has user assist functions. Fig.2 shows the concept of the Literacy Free Interface. In BEMAS, Literacy Free means to relieve the burden imposed not only on the operation of IT instruments but on mastering the actual information. A Literacy Free Interface presents energy-related information, and helps to interpret the information, and provide the dialogue mechanism and the intuitive operationality. The user can manage the building and Energy Management Agent dialogically through this interface. Furthermore, this dialogue

mechanism aim is easy operation, so the user could operate BEMAS easily and dialogically when various new functions are added to BEMAS.

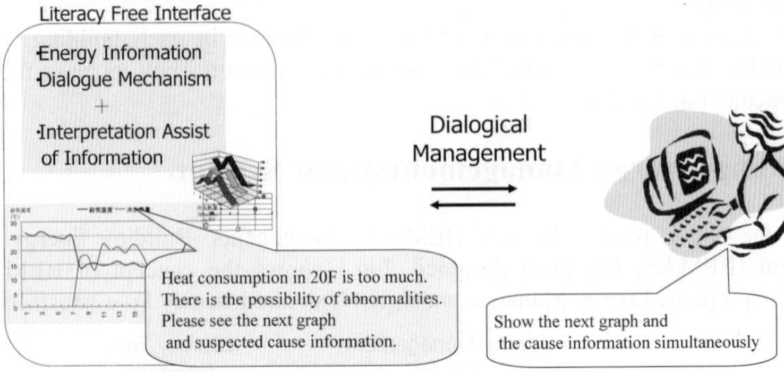

Fig.2 Concept of Literacy Free Interface

Fig.3 shows the system outline of BEMAS. BEMAS consists of 3 parts; User Interface Section, Central Administration Section, and Local Control Section. In buildings, various energy-related equipment, measuring instruments, and sensors are divided and installed into every subsystem such as each floor system, or each kind of energy system. In Local Control Sections, the embedded agents are installed in the controllers of such subsystems, and they conduct the energy management such as data acquisition or equipment control. And the Central Administration Section has agents which collect the data from each local subsystem and perform synthetic energy management, the agents are accompanied by advanced intellectual processing, and the database to store various measurements data, etc. And User Interface Section has the Literacy Free Interface, which works with the Central Administration Section, and Local Control Sections, and enables users to handle the energy management of the building.

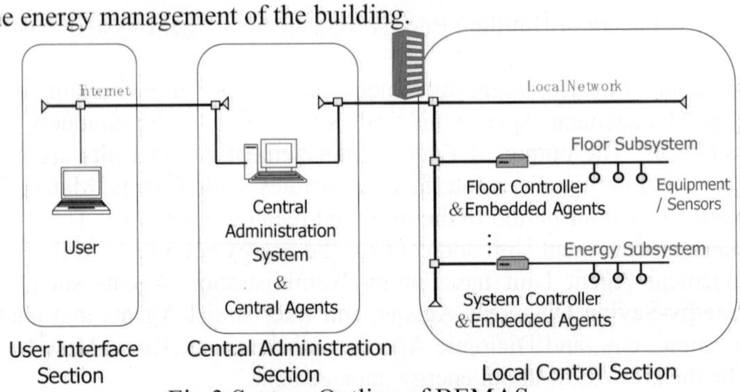

Fig.3 System Outline of BEMAS

Agent groups are assigned to installation places by the special feature of the part which it takes charge of in this way, and are also assigned for the functions to perform. For instance, the agents which take charge of its management are prepared for every apparatus, and the agents which have various functions are prepared for every field of energy management. And the management is executed by the cooperation of the "equipment agents" and the "function agents". This gives easy configuration to adapt to changes of buildings. By the change of cooperation between the agents, the cooperation of the equipment and the function can be changed easily. In the same way, it is easy to adapt to revisions, deletions, or parameter changes of the equipment or the management functions, accompanied with the changes of buildings. Furthermore, Know-How Management Agents store up the operation know-how about energy management, the arrangement method, the parameters, the fault detection method, and the cause and the solution of the fault. And the know-how is reflected in the cooperation of agents, and it enables BEMAS to easily perform the management according to the characteristic of every building. And also, the stored know-how can be applied to other systems.

By such a structure and the synergistic effect with the characteristic of multiagents, the new BEMS which cope with the problems of the conventional BEMS is realized. This BEMAS has the following features;

- Flexible adaptation to the change of the equipment or measurement items
- Easy to extend or change the management functions
- Storage of management know-how such as fault detection know-how and its application to the system
- Easy configuration or behaviour change of the system, in response to the change of operation way, etc
- Interactive management with users
- Assist non-professional user to carry out energy management

4 Prototype Development

4.1 Functions

Based on these designs, a prototype system with limited standard functions has been developed. The air-conditioning equipment is used as energy apparatus. The energy management function is limited to standard ones, the fault detection function of equipment and the function to grasp of an energy situation are chosen. In building energy systems which have numerous components and is like a black box, faults go underground. It is also a big cause of energy loss.

The contents of fault detection, however, should be made as suitable ones for the target characteristic through the actual operation, so basic contents are installed in this prototype system. Therefore, the assignment of agents is comparatively simple.

In a building, all kinds of energy data is measured and stored by measuring instruments for every minute. In this prototype system, 4 fault detections related to air-conditioning subsystems are conducted with the measured data. They detect an inefficient machine, and the diagnosis of the measurement state of water flow, the correlation state of water flow temperature, the concentration of CO_2 in the air-conditioning unit, and the interrelationship of the inverter and airflow. These diagnostic checks are performed based on the data-processing method and the judgment method which were defined beforehand. These checks are conducted periodically and are also conducted by user directions. When a fault is found, it is notified to the user and its cause and solution which were stored until now are shown. Fig.4 indicates a snapshot of the user interface showing the results of fault detection.

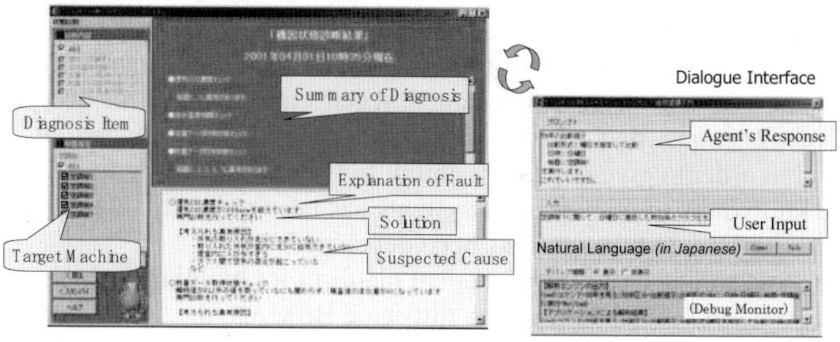

Fig.4 Showing the Results of Fault Detection / Dialogue Interface (in Japanese)

Furthermore, the graphs arranging the measured data to grasp the energy state are also presented. In order to facilitate analysis from various viewpoints, the system presents the graphs with a variety of compilations, and suggests some points of view of data interpretation.
Moreover, on configuration of the agents, the list of agents, their roles, and their configuration state is presented. The association between equipment agents and function agents, and the property of agents are set up for each agent.

On the other hand, for the purpose of easy operation of the above functions, GUI operation and the dialogical operation with natural language in Japanese are both available. Fig.4 illustrates an example of the interface. BEMAS can be handled through dialogue with the interface agent as the representative of BEMAS. And the configuration of agents can be changed through dialogue with each agent. When a user's input is an ambiguous sentence or required information is missing, the dialogue agent asks for more information and understands the user's intention correctly.

4.2 System Configuration

The composition of agents is shown in Fig.5. 3 agent groups are organized under the general manager agent, which performs administration of the agents and relay of messages among agents, and so on. They are the equipment agents which are assigned for each air-conditioning machine, the function agents which are assigned for each energy management function, and the user interface agents which cooperate with the user interface application. Now, 14 function agents, 1 to 200 equipment agents are prepared, and more can be added or deleted. In this prototype, Central Agent Section and Local Control Section are both installed in PCs. The used energy data is the data which is measured in our building and saved locally.

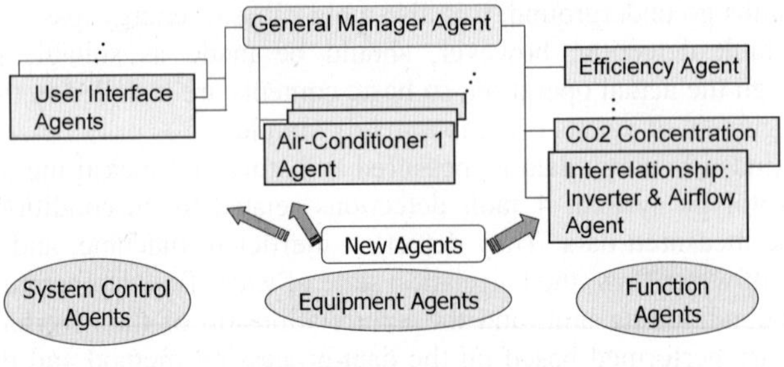

Fig.5 Composition of Agents

5 Conclusion

Energy management of buildings has a lot of issues, such as response to the change of the building or functions. In order to cope with such problems, the new building energy management system based on multiagent, BEMAS, has been designed. Agents are assigned for every component such as equipment or functions that change frequently, and their cooperation carries out the management. And in order to give maximum ability to BEMS, a Literacy Free Interface that offers user support and know-how of management systems has been proposed. Based on these designs, a prototype system with limited functions has been developed.
In the future, we are planning to develop an agent platform for buildings, which can be installed to embedded controllers / PCs. And aiming at a full-scale BEMAS, the agents' structure is due to be studied.

Development of an Advanced Human-machine Interface System to Enhance Operating Availability of Nuclear Power Plants

Tadashi Ohi[1], *Wei Wu*[1], *Yoshihiko Ozaki*[1], *Hidekazu Yoshikawa*[2], *Tetsuo Sawaragi*[2], *Masaharu Kitamura*[3], *Kazuo Furuta*[4], *Akio Gofuku*[5], *Koji Ito*[6]

[1]Mitsubishi Electric Corporation
Tsukaguchi-Honmchi 8-1-1, Amagasaki, Hyogo,661-8661,Japan
{Ohi.Tadashi, Wu.Wei, Ozaki.Yoshihiko}@wrc.melco.co.jp

Abstract

With the deregulation of the power industry worldwide and the aging of the nuclear power plants (NPP), concerns are growing over the reliability and safety of the NPP as their operating costs are to be reduced. We think that maintenance activity, plant availability, and plant managing are the sensitive fields that require more researches and improved in order to response properly to the above issues. In this paper, we propose the concept of the satellite operation maintenance center, and describe technical development plan to realize it. We describe a framework called Advanced Operation System (AOS). AOS consists of three technical aspects: dynamic operation permission system, information system using interface agent and crew performance evaluation. In the maintenance domain, we propose a framework called Ubiquitous-Computing-based Maintenance support System (UCMS). UCMS consists of three technical aspects: ubiquitous computing infrastructure, augment reality technology and intelligence information processing. We scheduled our project from its preliminary study to the final prototype for 3.5 years, starting from Dec. 2001.

1 Introduction

With the deregulation of the power industry worldwide and the aging of the nuclear power plants (NPP), concerns are growing over the reliability and safety of the NPP as their operating costs are to be reduced. The necessity to answer this issue is now felt strongly in the industrialized countries using NPP to supply enough electricity to the increasing power demand.

We think that maintenance activity, plant availability, and plant managing are the sensitive fields that require more researches and improved in order to response properly to the above issues. For that purpose we propose this project to develop an "advanced human-machine interface system to enhance the operating availability of nuclear power plants". This human-machine system applies information processing and human-interface technologies to the plant operation & maintenance domain.

In the plant-operation domain, many parts of the plant operation are done as soft operations through Visual Display Units (VDUs) in newly constructed plants and plants where existing plant-control systems have been replaced. Therefore, the development of advanced plant-operation systems that reflect a knowledge of critical human factors, such as the means of communication between crews, and the human error evaluation techniques have become important subjects. In the

[2]Kyoto University, [3]Tohoku University, [4]Tokyo University, [5]Okayama University, [6]Mitsubishi Heavy Industry

plant-maintenance domain, the development of work-support systems where portable information equipment is used by field staff doing maintenance work has also become an important subject. Furthermore, more reduction of operation / maintenance cost and more system reliability will be required for future nuclear plant. Various kinds of support function by the computer become more important.

2 Concept of the satellite operation maintenance center

As the solution system that decreases the total management cost of nuclear power plants, we propose the concept of Satellite Operation & Maintenance Center (SOMC). In the future, the SOMC will integrate the operation & maintenance activities of nuclear power plants. Figure 1 shows a conceptual scheme of SOMC. Experts of operation / maintenance will always stay at SOMC, they cooperate with field workers to manage several plants.

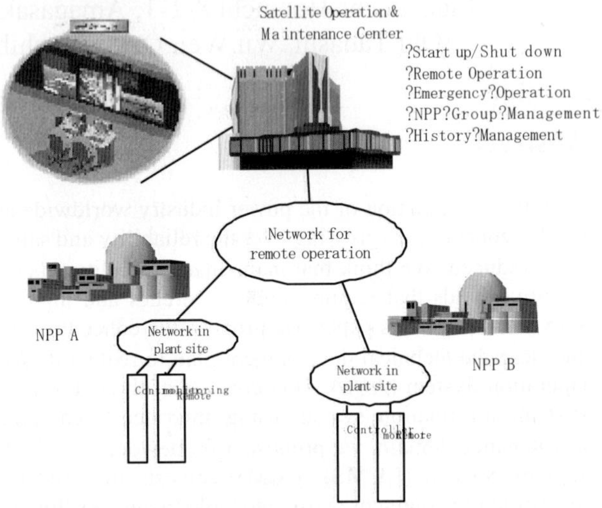

Figure 1: Conceptual scheme of SOMC

The SOMC consists of two integrated rooms, one for operation and one for maintenance. In the operation room, operation crew perform monitoring and operation (ie. startup, shutdown, steady state, emergency) of several plants. In the maintenance room, specialists for maintenance support field workers in a plant site. If required, a specialist will be dispatched to the site. The efficiency of the equipment and human resources can be increased through such organization. Considering the present regulation and social situation, it is difficult to shift to such integrated form immediately from the present form. In this project, we perform technical development so that SOMC will be realized.

From a viewpoint of a human interface, the target of this development is strengthening the following two functions.

- Situation Awareness
- Team Awareness

Details of each domain are described below.

3 Development plan

The goal of the project is to develop the advanced human machine system in a plant operation & maintenance domain, and has common subjects - improvement in situation recognition and recognition between teams. In order to clarify a subject at the beginning, first prototype system is developed in each domain. We will consider integration of operation system and a maintenance support system, in the last phase of the project.

3.1 R&D for operation domain

The composition of the human-machine system for operation is divided into the function for operator and crew. The operator's task can be divided into an operation task and a monitoring task. We have proposed a framework called advanced operation system (AOS). AOS consists of three technical aspects: dynamic operation permission system, information system using interface agent and crew performance evaluation. The target and approaches for each aspect are summarized as follows.

3.1.1 Dynamic operation permission system

This system aims at reducing the commission error in operation by CRT operation. The system infers an plant state from plant parameters, and makes usable only the operation switch corresponding to the plant state. Therefore, we will construct plant model for plant state identification and develop the sequence control system for this purpose. We are considering application of a multi-level flow model as a fundamental plant model.

3.1.2 Information system using interface agent

This system aims at improving the situation awareness of an operator. We will design the interface agent which takes both actions of monitoring and controlling a plant in collaboration with human operator. The following three points are especially important.

(1) Situation recognition model of interface agent
In emergency, cognitive resources, such as cautioning of human operators and working memory, will be restrained. We analyse the situation recognition process of skilful operators, and we model the situation recognition using decision-making theories, such as naturalistic decision-making[1]. Finally we install this function in an interface agent.

(2) Display management
We will develop a technique for presenting sufficient information reminding situation recognition, and suitable correspondence operation. We will also design the interface agent that supports the appropriate judgment of an operator using this technology.

(3) Share of intention between the human and the machine
When an interface agent and an human operator cooperate, it is necessary to communicate clearly any intention of an agent to human. For that purpose we develop the intention share technique with human operator and interface agent

3.1.3 Crew performance evaluation

The evaluation of a crew's performance and its availability within the control room is important, that is why we will develop a technique to evaluate a crew's performance, which will be included in the share of information for operating crew and in a mutual belief[2] for advanced operation system. And, we plan to apply this technology to evaluate "Dynamic Operation Permission System" and " Information System Using Interface Agent".

3.2 R&D for maintenance domain

The objectives of the NPPs maintenance is to ensure the safety of operation continuously while maintaining a high reliability and safety of various plant facilities under the constraints of low

costs. Many approaches and efforts have been done in order to reach objectives, such as the use of advanced inspection instruments, improvement of sensing system, etc. These efforts resulted in the good plant availability of Japanese NPP since 1980's. Increasing plant availability is especially important now with respect to needs of increasing the competitive ability of NPP under the electric power deregulation. But, on the other hand, the number of ageing NPP in Japan is increasing due to over 30 years have past since the first Japanese commercial NPP started operating. The maintenance works, therefore, now require more attention and technical advancement to cope with the situations mentioned above than ever before.
Using the advanced information technology to reach an innovative and practical level of maintenance works is the objective of the R&D described in this paper. We have reviewed the maintenance works in Japanese NPPs to propose appropriate support methods.

The maintenance activities have then been grouped into two categories: passive maintenance and active maintenance. Active maintenances are those activities in which a person is directed to inspect specific items of equipment, usually via written procedures. The typical active maintenance activity is the periodic maintenance inspection.
On the other hand, the passive maintenance activities are those in which a person is asked to search for deviant conditions, on periodic (daily, weekly, monthly) walk-around inspections which are scheduled inspection tour in a specified area of the plant to note any deviant conditions. Though a few written materials are used for that kind of activity, it requires both an awareness of the situation and the team in order to carry out the tasks efficiently and bye cooperating closely with the operators in the main control room.
Based on the review of the maintenance tasks in NPP, we have proposed a framework called Ubiquitous-Computing-based Maintenance support System UCMS. Ubiquitous Computing is a new approach to distributing computation in the environment. We think Ubiquitous Computing is key technology for the future I&C system. UCMS consists of three technical aspects: ubiquitous computing infrastructure, augment reality technology and intelligence information processing. The target and approaches for each aspect are summarized below.

3.2.1 Ubiquitous computing infrastructure

Based on the concept of ubiquitous computing, a small-sized device called as Ubiquitous Computing Device (UCD) will be developed to serve as an information gate for various kinds of plant equipment. UCD will be able to communicate with other devices, collect data from the equipment to which the UCD is installed, store historic data of the equipment, and process the collected and stored data to analyse the equipment's status, and other necessary functions to support the maintenance tasks. As a platform designed for to maintenance support systems, those UCDs will realize the ubiquitous computing infrastructure in NPP. The aims of the ubiquitous computing infrastructure can be summed up as below;

-To provide the staff on-site with information related to tasks easily
-To enhance the efficiency of the maintenance tasks
-To reduce human errors that might occur during the maintenance tasks

Approaches to meet these targets are currently developed and can be summarized as follows;

(1) Study on the optimal network communication method between UCD and user-side computer (such as notebook, PDA, etc) in the context of NPP maintenance tasks.
(2) Study on adaptive information processing technologies, such as the automatic information communication adapted to the tasks being carried out.

3.2.2 Augmented reality technology

Regarding the human-machine interface system, we are currently investigating a system aimed to support maintenance workers by applying augmented-reality (AR) technology[3] and head-mounted display (HMD) technology. The targets of the AR-HMD-based interface system are the following items;

-Providing a visual and easily accessible maintenance information such as plant parameters, design information and maintenance procedures, among others.

-Serving as a human-machine interface for the satellite maintenance system coming in near future.

Using the system, maintenance workers could see the live video image of the facilities around with the relevant superimposed information (text/graphics of maintenance procedures) on a lightweight HMD. Combined with UCDs installed into facility or instrument, the workers could get the necessary maintenance information about the equipment just by coming near to and looking at the specific equipment. Besides, the maintenance information could be presented in a way that worker could understand easily, e.g., temperature or flow of pipes will be displayed in digital number directly near the specific pipes on the HMD. To serve as the human machine interface for the satellite maintenance system, the related undergoing studies are;

(1) Study on the image recognition technology
(2) Study on the effective ways of presenting information
(3) Study on the see-through type HMD

3.2.3 Intelligence information processing for maintenance

The purpose of this system is to improve the working efficiency and to prevent human errors in plant maintenance. We will develop intelligent maintenance support system including effective information retrieval technique using the ubiquitous network, and the method of maintenance information presentation on ubiquitous computing system.

4 Conclusion

In operation domain, we are developing the operation model and a multi-level flow model for dynamic operation permission system, the user interface agent incorporating an expert's decision-making model, and the team intention formation model. In maintenance domain, we are developing the ubiquitous computing system, the communications processing system for maintenance work, and the adapted type maintenance information management technique.

The results so far obtained as well as the current state of advancement in our researches show that the basic technology we outlined here is feasible and some of prototype systems are constructed.
This project is supported by The Institute of Applied Energy.

References

1.Klein, G., Sources of power: How people make decisions. MIT Press, 1998.

2.Furuta, K., Kondo, S., An Approach to Assessment of Plant Man-Machine Systems by Computer Simulation of an Operator's Cognitive Behavior, Int. J. of Man-Machine Studies, 39, p473-493, 1993

3.Azuma, R.T.: A Survey of Augmented Reality. Presence : Teleoperators and Virtual Environments, Vol.6, No.4,355-385 ,1997

Supporting Creative Work

Daniel V. Oppenheim

IBM T.J. Watson Research Center
19 Skyline Drive, Hawthorne, NY, 10532, USA
music@us.ibm.com

Abstract

The work most people do every day involves ambiguity and multiple possibilities for action that can only be navigated by creativity: writing a paper for this conference, devising a business plan, or constructing a study curriculum. The flow of creative work is unstructured, not clearly understood, and therefore not well supported by computers. Improving support for creative work will significantly improve productivity across many domains and should be considered a grand challenge. This paper examines aspects of the creative process and suggests a framework of requirements, design principles, and components that will better support creative work. Significant improvement across application and domain boundaries can be made using existing technology through incremental development and conventional HCI.

1 Supporting Creativity: A Grand Challenge

Creativity is often associated with art and genius, but this is a limited viewpoint. Creativity relates to "how" we do things and not to "what" we are doing (Pirsig, 1974). In fact, most people are creative on a daily basis when doing their work: devising a study curriculum, constructing a business plan, or inventing an advertisement for a new product. In work tasks that are creative, the individuality of personality, thinking, and creativity leave a unique thumbprint on the output. A clear goal leads to the construction of a new artifact intended for a particular audience—a student body, management, or a business client. Consider, for example, the writing of this paper to be included in these proceedings.

This paper was written with a word processor. What I thought would be a simple process was not so. As I was trying to clarify my thinking on creativity this paper changed form several times. Where I began had little to do with where I ended, large sections appeared and disappeared, and the structure and logical flow refused to stabilize. I was struck by how similar this experience was to many I'd had while composing large-scale musical works. I believe the process was similar because in both cases I was trying to define, express, develop, and work out ideas. The breakdowns caused by the technology I used also seemed similar in both domains. Once I knew what I wanted to write I had no problem, but getting there was very hard. I had difficulties managing the vast amount of information, publications, references, ideas, sketches, plans, and drafts that I was constantly referring to. This was aggravated over time as I found I needed to refer to a different set of items at different times. The work itself and its structure were in constant flux, and my own thinking and planning were continuously evolving. Observing other people working in different disciplines and domains revealed similar breakdowns.

This paper examines various aspects of creative work and how technology could support them. By supporting creativity I mean the ways in which computers could be of better assistance in all aspects of our work, from the decision to do something to the production of the desired artefact. Ideally, the flow of creative thinking would remain undisturbed as we act.

I argue that the grand challenge for computer science is to figure out how to improve the support for creative work in general, regardless of domain or application. It is important because so many people routinely use computers for doing creative work in all aspects of life: business, academia, science, art, and home. If the question of the '70s and '80s was "how can we make computers easy enough for a five-year-old to use," the question today is "how can we make computers support our natural ability to be creative."

2 Workflow, Capture, Appraise, Compose, and Construct

I use the term **workflow** to describe the flow of both thought and actions that take place over time as we work on a task. Thought can be conscious or not, and can take place in the workplace or elsewhere. The flow of work may often be interrupted as we switch between different tasks or assume different roles in our lives. As a process, the workflow is as unstructured as it is elusive. No two people work the same way, and a certain person may work differently at different times. This process is interactive, iteravtive, experimental, and reflexive (Oppenheim 1986, 1996). Interaction takes place internally within the creator, e.g. his or her ideas and intentions that lead to action through the computer. Results are appraised to determine if they should be kept, modified, or discarded, and similarly, strategies are appraised and modified in light of the new context. Many additional factors constantly influence this process, such as changes in requirements or goals, new information, changes in our understanding, or changes in our frame-of-mind. There are many more factors that affect our work than we can account for. Moreover, the context in which we think is constantly evolving and affecting our work, and at the same time our work itself, as it evolves, affects the context. This is a complex and reflexsive cycle. I define four archetypical activities that comprise the workflow: *capture*, *appraise*, *compose*, and *construct* (this expands on the design rationale of QSketcher—see Abrams et. al. 2002). These activities are reflexive, feed into each other, and occur throughout the workflow in no particular order.

Capture is the process of identifying and keeping the things that will help us accomplish our task: books, requirements, goals, images, ideas, plans, sketches, or even scribbles on napkins. I refer to these as the collection of task-related objects. Capture can be done deliberately or can be triggered by seemingly random events. Deliberate capture has been well described as *collecting* where, for example, a researcher may browse the Web for relevant information (Schneiderman 2002). Random occurrences also provide ample opportunity for capture: a lecture we happen to attend, a television program we surf into, a book we read, or an overheard conversation. But notions of capture go deeper and may relate to the unconscious. Sudden realizations or formulations are common: Archimedes jumping out of the bath shouting "Eureka", Newton reacting to a falling apple, or Kekulé remembering his dream of a snake biting its own tail and predicting the ring theory of benzene. New ideas or realizations may be triggered as we work, or while working on unrelated tasks, or by seemingly random events that occur at unexpected moments away from the workplace. Unfortunately, these fragile moments can easily be lost.

Several breakdowns emanating from the word processor permeated my work. Capturing sudden ideas that could not be expressed in a few words, especially if they had several steps that built upon each other, was ineffective. By the time I wrote the first step I would often forget the next. I found that pencil and paper, or even the whiteboard, worked much better, possibly because I was able to sketch (capture) structural points elsewhere on the page as I was writing down early steps. When ideas relating to other tasks suddenly appeared I found it hard to capture them with the word processor without seriously disrupting my current flow; naming and saving a new document was also disruptive. Flashes of inspiration away from the word processor produced many notes, voice mail, and e-mail to myself that were hard to manage. Over time my desk, walls, and whiteboard became hopelessly cluttered with books, articles, e-mails, sketches, sticky notes, suggestions from colleagues, etc., and I spent too much time on their management and organization. In short, the word processor is built around the notion of a document but not around the creative task. Technology could certainly improve things by providing easy capture mechanisms that are designed with the notion of task in mind.

Appraisal is an evaluative cognitive process that intervenes between an encounter and our reaction. First we assess if an encounter is relevant or not. Then we decide what might be done. It is a rapid and intuitive process that occurs automatically, can be conscious or unconscious, and focuses on significance, categorization, and meaning (Lazarus 1984). Many factors can affect our judgment: task goals, requirements, our understanding of the problem, personality type, emotional state, historical events, peer pressure, or even social conventions. Appraisal is always done from a certain perspective, and as we appraise something we may rapidly switch between many different perspectives. For example, a new paragraph added to this paper may be appraised from the perspective of the logical flow, clarity of expression, impact on the reader, or affect on pagination and document size. Each perspective has associated with it a collection of things, or *context*, that consciously assist and influence our thought. We consciously relate to ideas, plans, to-do items, drafts, sketches, etc. Unconscious factors, or *frame-of-mind*, may relate to past life experiences, expertise, competence, and may also have to do with emotion and intuition. I found that the particular items I chose and the specific way I arranged them on my whiteboard not only captured context, but also captured frame-of-mind.

The word processor was helpful when appraising perspectives such as clarity of expression, grammar, or changes in document size. The ability to have several windows open on different sections of this document, other documents, or Web pages was also useful but became too painful to restore each time I worked. I ended using my desk and whiteboard as a large billboard on which I placed the objects that helped me through additional appraisals, contexts, and frames of mind. But over time this became disruptive as I would run out of space for supporting new needs and did not want to discard existing arrangements I found useful. At this point my work became less effective because my thinking was not getting the support needed. The burden of physically reorganizing my whiteboard, often encouraged me to work abstractly without using it. Ironically, I noticed on several occasions that the organization I had not wanted to disrupt in the past was no longer relevant to me. Had I reorganized the space when I needed it, I would have been more effective in getting to where I wanted to be. I had to develop a balance between how long I kept an arrangement and the frequency in which I updated items in a given arrangement or created new arrangements.

Composition refers broadly to the thought process that deals with strategy and structure. It is a slower, more abstract, and reflective thought process than appraisal, and has to do with analysis, categorization, organization, relationship and meaning. This is where important decisions that influence every aspect of the resulting artefact are made. Reflexively, the ongoing changes in the work context and the evolving artefact affect all aspects of this process, often leading to refinements or modifications in strategy. This is an enigmatic activity where inspiration, intuition, and spontaneity go side by side with context, discipline, logic, order, knowledge, and experience.

Composition is the activity where I felt the word processor's document metaphor fundamentally broke down as a tool that supported my needs. The page view did not provide a large enough context for me to relate the details to the larger context of the entire paper, and the outline view did not translate well to my perception of the logical flow, development, or structure. I composed by and large at the whiteboard or in my head. Several additional limitations of the whiteboard soon became apparent. I would have liked to reuse several of the items in different arrangements, such as goals, ideas, or structure. I missed an ability to relate items to other items. I would have also benefited from the ability to make changes on the whiteboard, such as restructuring the logical flow, and have those changes take effect in the actual document. Finally, I would have liked to see the changes I made in the document reflected in the appropriate items on my whiteboard.

A digital version of a whiteboard could be designed in a way that addresses many of the deficiencies I encountered. Objects on display can be connected to actual data. The written word "goal," for example, could link to a requirements document, related literature, a design, or any relevant section in the document. Thus we can hide detail when it's not needed, but always find needed information quickly. By linking the representation (view) with the data (model) we provide the possibility of directly editing anything that is on display. The visualization becomes a context that both facilitates our thinking and enables our actions. Thus, the concepts we use to think about the problem also become the concepts through which we act upon the data and manipulate the system. The environment in which we construct is also optimized for appraisal and composition. Blurring boundaries between thought and action are beneficial to the user.

Construct is the set of actions that produce the artefact or product. This activity is typically done through editors and is by-and-large domain or task specific. The word processor, for example, provides operations such as cut-copy-paste, search-replace, format, or spell-check. Two things should be considered when designing computer support. The first has to do with concepts. One should be able to operate the system using the same concepts in which he is appraising and composing (see Oppenheim 1986, 1987, 1996). The second is that this activity must integrate with the overall workflow and all the task-related objects.

3 Putting it all together

In the ideal work environment, the flow of creative thought and activity will remain undisturbed as we manipulate the environment to facilitate our personal thought process and articulate our goals. As we shift work from one activity to another, as thought leads to action, and as we appraise past actions from the multitude of perspectives we deem relevant, we facilitate and maintain *Flow* (Csikszentmihalyi, 1996) and *Engagement* (Brown, 2003). The problem is putting it all together within a simple and consistent framework that the user can manage easily. For any one aspect of the problem there is probably an existing application in some domain that supports it exceedingly well. As Riecken (2002) pointed out: "But this doesn't help me! None are on my computer. And even if they were, they would not integrate into my workflow."

It is impractical that any one application will provide good overall support for creative work. The amount of development required makes this prohibitive—imagine having to build a window system into each new application. In order to facilitate support in ways that would be useful in many domains, one must focus on components rather than applications. The right components would support the creative activities in ways that could be shared by many applications in different domains. This would also facilitate interoperability and sharing of data between applications, thus breaking down the somewhat rigid boundaries that exist today between applications. The user could then think of today's stand-alone applications as components that work in harmony within the larger conceptual framework of his task. For example, the breakdowns I encountered with the word processor that resulted from its narrow focus on the notion of document metaphor would be resolved by other components within this framework.

If this is done well, then components that facilitate capture, appraise, and compose, will be become generally available, allowing developers to focus by-and-large on the domain specific aspects of construct. Solving this problem can be accomplished through small incremental steps, using readily available technologies, while maintaining the UI conventions with which users are familiar. For the user, solving this problem may be experienced as a groundbreaking improvement.

References

Abrams, S., Bellofatto, R., Fuhrer, R., Oppenheim, D., et. al. (2002) QSketcher: An Environment for Composing Music for Film. Proceedings of *Creativity and Cognition*, Loughboro, UK.

Brown, Andrew (2003) Music Composition and the Computer. PhD Thesis. School of Music, the University of Queensland, Australia.

Csikszentmihalyi M. (1966) Creativity: Flow and the psychology of discovery and invention. New York: Harper Collins.

Lazarus, R., Folkman, S. (1984) Stress, Appraisal, and Coping. New York: Springer.

Oppenheim, D. (1986). The Need for Essential Improvements in the Machine Composer Interface used for the Composition of Electroacoustic Computer Music. Proceedings of the *International Computer Music Conference*, the Hague, Holland.

Oppenheim, D. (1987). The P-G-G environment for Music Composition. Proceedings of the *International Computer Music Conference*, Illinois, USA.

Oppenheim, D. (1991). Towards a Better Software-Design for Supporting Creative Musical Activity (CMA). *Proceedings of the ICMC*, Montreal, Canada.

Oppenheim, D. (1996) DMIX—A Multi Faceted Environment for Composing and Performing Computer Music. *Computers and Mathematics with Applications*, 32:1, pp 117-135.

Pirsig, Robert M (1974) Zen and the Art of Motorcycle Maintenance. Bantam Books.

Riecken, D. (2002) Public comment made at the 2002 Creativity and Cognition Conference, Loughboro, UK.

Schneiderman, B. (2002). Leonardo's Laptop, Human Needs and the New Computing Technologies. Cambridge, Massachusetts: MIT Press.

Cell Phone vs. Computer: A Comparison of Electronic Commerce and Mobile Commerce from the User's Perspective

A. Ant Ozok
Department of Information Systems
University of Maryland Baltimore County
1000 Hilltop Circle Baltimore, MD, 21250
E-mail: ozok@umbc.edu

Abstract

Mobile commerce is getting increasingly popular with the introduction of cell phones, PDAs and combo devices. This study discusses the main differences and the implications of these differences between classical e-commerce and mobile commerce from a user's perspective. Differences in the two businesses and the effects on the user, usability, data entry and data retrieval issues, cultural acceptance issues, and the integration of Customer Relationship Management (CRM) into mobile commerce area are discussed. It is concluded that there is a number of items to be explored regarding usability issues, and data retrieval and entry is significantly limited compared to regular PC-based e-commerce media. While it is believed that due to the worldwide popularity of cell phones different cultures will not have difficulty accepting the integration of mobile commerce, it is also concluded that CRM issues need to be integrated seamlessly into the mobile commerce area following similar steps taken in the e-commerce area, and some infrastructural steps need to be taken for this integration.

Objective & Significance

The introduction of mobile devices into the electronic commerce arena has allowed the customers to use their cell phones, Personal Digital Assistants (PDAs) or combo devices without needing the hassles of a bulky stationary desktop computer, a laptop which is still much bigger than a mobile device, and a hard-wired Internet connection. However, there also exist some substantial issues that potentially hinder mobile devices to be used as commonly as desktop and laptop computers in e-commerce. Because of its "electronic" nature, mobile commerce can be thought of part of electronic commerce. However, in order to qualify as part of mobile commerce, a transaction should be conducted using a wireless network, a mobile device, and a non-desktop operating system. Although mobile devices are presently also used in business to business (B2B) e-commerce, the objective of this study is to focus on the use of mobile devices in business to consumer (B2C) commerce. Although technical and infrastructural issues are one of the primary differences between mobile commerce and e-commerce, this study is not focused on these technical differences. The primary objective of this study is to determine the cultural, usability, and Customer Relationship Management implications of the differences between electronic commerce (e-commerce) and mobile commerce (m-commerce). The differences between the two areas and their implications from the users' perspective are discussed in the following categories: General advantages and disadvantages of m-commerce from the business point of view and their potential effects on the user, a comparison of the two media from the computation (data entry), communication (data retrieval) and general usability point of view, the reception of the two media in two different representative cultures, and a comparison of the Customer Relationship Management issues between e-commerce and m-commerce.

Business Differences and Implications

Although commerce means exchange of goods and currencies between businesses, the term Electronic Commerce is often used today for shopping of consumers using the Internet (Turban et al. 2002). With the inception of the Internet in 1994 and its rapid growth, selling and buying goods using the Internet has become highly common in the U. S. and around the world. According to Cefasoft, an on-line research firm, American consumers spent a little less than $ 40 billion on-line in 2002, with the worldwide e-commerce spending expected to exceed $3 trillion by the end of 2003 (Source: Jupiter Communications). While shopping on-line is highly popular, new alternatives are sought by consumers to make their shopping experience even more convenient. With cell phone use getting highly popular and reaching 50 million users in the U. S. according to PC magazine, a new alternative has been offered to shoppers in the form of being able to browse the Internet and doing shopping using their cell phones. Again according to PC magazine, 10 million Americans have already gone on the Internet using their cell phones. Currently, actual m-commerce customers are fairly limited in number, and no pure m-commerce companies exist, however, many traditional e-commerce companies such as Amazon.com have Wireless Application Protocol (WAP) versions of their sites where users can do shopping (Turban et al., 2002). Retail, financial applications, distance education, music on demand and office applications are the most popular applications used in relation to m-commerce (Varshney and Vetter, 2001). Users prefer mobile commerce mainly because of mobility, broad reachability, ubiquity, convenience, and localization of products and services attributes (Turban et al., 2002). Mobility refers to the ability to initiate an e-commerce transaction at any desired time. Broad reachability refers to everyone with a cell phone with minimum Internet access requirements being able to be part of m-commerce. Ubiquity can be described as the absence of geographical obstacles in m-commerce. Convenience represents the device being always at the user's disposal and increasingly easy to use. And finally, m-commerce can localize services and products by determining the user's physical location and giving the user location-based shopping capabilities. Moreover, other drivers of m-commerce include widespread availability and affordability of devices, expected to reach 1.3 to 1.4 billion worldwide by 2004, the absence of need for a computer which is on average more expensive, the mobile handset becoming a culture, and improvement of bandwidth for cell phones (with the current system in the U. S supporting Generation 2.5 infrastructure and the more capable and broader 3rd Generation (referred to as 3G) soon to be available). However, there are also some limitations from a business perspective that is preventing m-commerce from being a serious rival to e-commerce. There is no standardized security protocol, customer confidence is still low to cell phone transactions, usability issues are not yet fully explored, bandwidth is still limited and regulations for bandwidth allocation are still in the process, transmissions are more frequently interrupted due to wireless issues such as weather, many e-commerce companies don't have dedicated WAP pages, 3G networks currently have a number of infrastructural problems and the inception of 3G in the U. S. has been delayed for many times because of those problems, going on-line via cell phone significantly decreases battery life because of higher power consumption, and the fees associated with mobile Internet services are higher on average than PC-based Internet services. The positive and negative issues mentioned offer a trade-off for customers – being able to do shopping from (almost) anywhere and any time versus having limited access, limited interaction, and a hefty price to it. From a business perspective, several years are needed to have an integrated infrastructure and strong company backing for m-commerce that will allow it to become a substantial part of the e-commerce phenomenon. Consequently, for the mobile access to receive high demand by a large user population, pricing, usability, and access issues need to be improved primarily by businesses in order to make the mobile part a significant economic booster of the e-commerce business.

Computation, Communication, and General Usability Issues

Perhaps the single most important obstacle in front of mobile commerce preventing it from becoming as popular as PC-based commerce is the limited data entry and data retrieval capabilities. Sears and Arora (2002) indicated that the data entry tools for mobile environment are significantly more limited than regular computer devices and data entry performance is significantly lower with novice tasks. Additionally, the small screen size can never be a rival to computer screens unless holographic phones are invented which can project a large screen image without the need of a projection screen but on thin air. Nielsen (2001) indicated several reasons as to why cell phones are ill suited for mobile Internet access. Some of them include the form of the cell phones being not a sutiable design for data-rich interaction but more dictated by the distance between the human ear and mouth, the keypad dominating too much of the surface area, and a numeric keypad being a poor device for entering alphanumeric characters. Latest improvements include add-on and integrated keyboards to phones, PDAs and combination mobile devices which in turn result in occupation of additional space as well as in increase of the overall device size. Additionally, the screen resolution and size are presently quite far away from PC-based screens, although the recent developments in Liquid Crystal Display (LCD) technologies have allowed the mobile device makers to reach quite high levels of screen resolution and colors up to 320 by 240 pixel resolutions, which is still far away from the most common resolution of 1024 by 768 pixels in desktop computers. Additional usability difficulties with cell phone use includes one of the hands being occupied while data entry is conducted with the other hand (using a stylus pen or the keypad), and more difficulties involving retrieval of information such as graphics being to small for legibility and taking a long waiting time to download. Based on these issues, it can be concluded that although the usability issues concerning mobile commerce use are currently in the process of being explored, time is needed to find solutions to fundamental problems preventing mobile phone use to become as popular as personal computers in the e-commerce arena (for example, technological improvements allowing natural speech recognition as functional as data entry using a keyboard).

Cultural Issues

Another issue in placing the mobile commerce in a high place in the digital economy is its acceptance among the cultures. While there is nothing wrong with using a mobile device in logging on to the Internet and doing shopping, there is a considerable number of on-line shoppers who are not comfortable with giving their personal and credit card information even through commercial hard-wired lines. In the U. S., with zero liability guarantees and insurances offered by some banks and credit card companies, the possibility of on-line fraud is not at an alarming level. On an international level, people are more cautious in giving their credit card and personal information on-line. Lightner, Yenisey, Ozok and Salvendy (2002) studied the differences in on-line behaviors between Turkish and American college students and concluded that Turkish users are much more reluctant than American users in submitting their personal and credit card information on-line. Lightner et al. also indicate that conveying trustworthiness by providing a secure and private shopping environment seems paramount for online success. An additional security item that comes into the play with mobile commerce is the information traveling wirelessly in some part, which may create a lack of perceived security on the users' part. Additionally, the study by Lightner et al. indicated that users from both cultures very much prefer to touch and examine the product they are purchasing. With traditional e-commerce, people have usually detailed pictures of the products to study before they purchase it. Having a very small picture to study before buying a product undermines this particular preference even more than a picture on a PC screen. One up side regarding the expansion of mobile commerce is the high popularity of cell phones in most cultures. However, how differently people from a number of cultures perceive mobile commerce remains an area to be explored.

Customer Relationship Management Issues

Customer Relationship Management (CRM) includes actions taken in order to preserve customer retention and gain new customers. Effective CRM occurs when vendor provides effective and useful online help, product information, and tools to customers regardless of the interaction type (Turban et al., 2002). Increasing emphasis is put on CRM because it is five times as expensive to attract a new customer as it is to keep an existing one, customers who have relationships with companies are more loyal and less likely to switch when a better price is discovered, satisfied customers are more likely to refer new customers, and customer feedback can be effective (Resnick, 2001). An additional study by Ozok, Salvendy and Oldenburger (2003) indicated that customers expect consistent CRM from their on-line vendors before, during, and after their shopping transaction. A consistent treatment and offering of customer help after the purchase are no doubt items that need to be present in mobile commerce just as in classic e-commerce. Customers expect an around-the-clock service and content center they will be able to access by using their cell phone browsers or by calling the voice services. The same security concerns would logically be available with mobile CRM services as regular mobile commerce transactions. The offering of CRM services also creates additional infrastructural challenges for mobile commerce companies where all of the CRM services provided to regular e-commerce customers through their PCs and phone lines should be provided in the mobile Internet environment as well. Additional services relating to CRM may include offering customer services on the road, for example while on a plane that may require special treatment. The introduction of 3G networks is believed to also improve CRM activities in the mobile commerce arena, such as by allowing face-to-face real-time on-line conversations between the customer and the service representative, remote software diagnostics, or step-by-step interactive guidance on products. Mobile CRM can also allow customers to download software updates or receive digital newsletter or update information regarding their products, regardless of where they are. Additionally, using the mobile devices, customers can receive advertisement regarding products of their close interest as part of targeted advertisement campaigns, and these advertisements are no longer limited to text messages but audio, video and animation as well. Additionally, just as in localized product offerings, information services based on the customer's location can be delivered, such as AAA services being deployed to the location of where the customer's car broke, based on the information obtained from the Global Positioning System (GPS) included in the cell phone device. Finally, different customers who have a common attribute, such as customers having a specific product and would like to exchange information with each other regarding the product, can be connected using m-commerce CRM services.

It should be noted, however, few of the CRM items are currently available in the m-commerce arena, mainly because of a lack of necessary infrastructure to realize the CRM attributes and the business itself not being developed completely. Despite the obstacles, the author believes that the steps to be taken in order to fulfill CRM requirements in mobile commerce are not much different from those of classical e-commerce, and it is just an issue of infrastructural development and wide company acceptance and support for the m-commerce CRM level to reach the level of classic e-commerce.

Conclusions

Mobile commerce is no doubt a promising medium for electronic shopping because of its attributes of convenience superior to classical PC-based e-commerce. However, usability, cultural, business support and CRM issues need to be developed on a substantial basis before m-commerce can measure up to the popularity and technological level of e-commerce. The two are conducted through substantially different media, and especially in the usability and cultural acceptance areas, and more empirical research needs to be conducted in the process of establishing m-commerce on a broad basis. Also, technical advancements will help m-commerce applications expand to the entire population of e-commerce sites. These technical advancements can potentially include more affordable browser-capable mobile devices and all of the classical e-commerce sites having m-commerce counterparts. In conclusion, it can be said that m-commerce is a rapidly expanding phenomenon and a convenient and promising area within e-commerce.

References

Cefasoft [on-line] (2003): accessed 2. 15. 2003, available at cefasoft.co.uk

Jupiter Communications, Inc. (2003): accessed 2. 15. 2003, available at www.jup.com

Lightner, N., Yenisey, M. M., Ozok, A. A. and Salvendy, G. (2002): Shopping Behavior and Preferences in E-commerce of Turkish and American University Students: Implications from Cross-Cultural Design, *Behavior and Information Technology,* Vol. 21(6), pp. 373-385.

Nielsen (2001): useit.com Alert Box, January 2001, available at useit.com, accessed 2. 15. 2003.

Ozok A., Salvendy, G. and Oldenburger, K. (2003): Consistency of Customer Relationship Management in E-Commerce. Behavour and Information Technology, Under Review.

Resnick, M. (2001): Design and Usability Evaluation of Customer Relationship Management in Commercial Web Sites and E-Business. *Usability Solutions, Miami, Florida*, downloaded from crm2001online.com, 3.5.2002.

Sears, A. and Arora, R. (2002). Data entry for mobile devices: An empirical comparison of novice performance with Jot and Graffiti. Interacting with Computers, 14(5), 413-433.

Turban, E., King, D., Lee, J., Warkentin, M., Chung, H. M. (2002): Electronic Commerce2002: A Managerial Perspective. Prentice Hall, Upper Saddle River, New Jersey, U. S. A.

Varshney, U. and Vetter, R. (2001) : A Framework for the Emerging M-Commerce Applications. Proceedings, 34th HICSS, Hawaai, January 2001

How to Treat Your Customers: Guidelines for Consistency in E-Commerce

A. Ant Ozok[1], Gavriel Salvendy[2,3], Kristen Oldenburger[2]

[1]University of Maryland Baltimore County, Department of Information Systems, Baltimore, MD, USA
E-mail: ozok@umbc.edu

[2]Purdue University, School of Industrial Engineering, 1287 Grissom Hall, West Lafayette, IN 47907
salvendy@ecn.purdue.edu, dhorn2@purdue.edu

[3]Department of Industrial Engineering, Tsinghua University
Beijing P.R. China

Abstract

Customer Relationship Management (CRM) is one of the rapidly emerging key factors of success in the highly competitive electronic commerce arena. Fourteen items have been identified in a developed questionnaire to measure CRM factors in consistency of customer treatment. One hundred participants were presented the questionnaire and were asked to rate the importance of each of these items on a 5-point Likert Scale. The developed tool had a Cronbach's Alpha internal reliability value of 0.81. Two factors as part of a Principal Components analysis were identified: Technical and Customer Treatment Factors. The individual Cronbach's Alpha values were 0.84 for Treatment Factor and 0.74 for Technical Factor. The analysis indicated that the **technical factor** included *the consistencies of* ***shopping steps, site design, and navigation,*** while the **customer treatment factor** included *the consistencies of* ***promotions, in-stock indication, product variety, fraud protection, presented guarantees, customer fairness, and return policies***. Based on the analysis, ten guidelines for **Consistency of Customer Treatment in CRM** were produced. The new guidelines can be applied to any Business-to-Consumer (B2C) shopping site as part of the CRM policy.

1. Objective and Significance

With the ongoing rapid growth of on-line shopping in the Business-to-Customer (B2C) e-commerce arena, Customer Relationship Management (CRM) is gaining increasing importance on a daily basis. CRM deals with the issue of how to retain your existing customers, and how to obtain new customers. Consistency of customer treatment is a term including the pre-purchase, purchase, and post-purchase activities. Consistency in CRM deals with how consistent the on-line seller treats their customers before, throughout, and after the purchase. While most of the studies concerning consistency in the human-computer interaction domain deal with consistency of interface design (Ozok, Salvendy, 2001, Shneiderman, 1992), the consistency of costumer treatment in the on-line shopping arena has not been studied in the past literature. Therefore, the current study aims at determining the core elements of consistency in CRM.

2. Literature Review

The current literature on Customer Relationship Management deals with design and testing of commercial Web sites. The importance of CRM within the business-to-consumer as well as business-to-business domain is strongly emphasized by e-commerce vendors and e-tailers. Resnick (2001) indicated that CRM is essential because it is five times as expensive to attract a new customer as it is to keep one, or an estimated double the value of an average sale. Ngai and Wat (2002) indicated that policies and technological issues are also significant parts of e-commerce management, and these two items are therefore included in the current study as part of customer treatment consistency. And Anderson (2000) indicated that there are still obstacles in front of e-commerce firms largely based on false beliefs such as biases and fears, and the current study aims at improving the CRM belief in e-commerce and make on-line retailers eliminate these false beliefs. The literature search indicated that no explicit study has been conducted concerning e-commerce customer treatment consistency and its possible implications.

3. Methodology

A tool was developed in the form of a questionnaire measuring the elements of consistency in e-commerce. It was based on a pre-experiment open-ended questionnaire that asked to 29 applicants the most important factors in CRM consistency.

The Customer Relationship Management Consistency Questionnaire (CRM-CQ)

A tool was developed in the form of a questionnaire measuring the elements of consistency in e-commerce. It was based on previous literature and a pre-experiment open-ended questionnaire that asked to 29 applicants the most important factors in CRM consistency. It consisted of sixteen individual items, two of which were included for measuring the internal reliability (Cronbach's Alpha) value, which was 0.81. The factors were identified as the *consistencies of:* ***price-quoting, steps to execute a transaction, design of the shopping Web page, site navigation, promotions offered, indication of in-stock products, product variety, alternative product suggestions, fraud protection, presented guarantees involving the product, fairness of the site, help offered, return policies, and personal information.***

4. Results and Discussion

4.1 Factor Intercorrelations

In order to better determine the interrelationships between the different conceptual elements of consistency in CRM, a correlation matrix was set up containing all of the sixteen items in the questionnaire. The most significant and conceptually important correlations are discussed in this chapter. The high correlations between the duplicate questions (created for internal reliability purposes) are not discussed since these high correlations are logical because of the identical nature of the duplicate questions.

The most significant correlation in the matrix is between **Consistency of Site Navigation** and **Consistency of Site Design** ($R = 0.56$, $p<0.0001$). The logical finding indicates a strong relationship between the CRM expectations from the customer that deal with issues related to interface usability. While a consistent site design is expected, the navigation among the different pages within the same site is also expected to be consistent and these two items are strongly related to each other by the customers. Two separate results highly related to this particular result are the correlation between **Consistency of Steps to Execute a Transaction** and **Consistency of Site Design** (R = 0.44, p<0.0001) and the correlation between **Consistency of Steps to Execute a Transaction** and **Consistency of Site Navigation** (R = 0.38, p<0.0001). These findings indicate a significant relationship among elements of consistency in CRM dealing with the interface design issues. Therefore, it can be concluded by these correlation values that the consistency of navigation, site design and execution steps are all closely related within the context of Customer Relationship Management.

Since the concept of consistency in Customer Relationship Management also includes post-purchase activities, a strong correlation between **Consistency in Fraud Protection** and **Consistency of Help Offered by the Site** (R = 0.42, p<0.0001) is an indication of two post-purchase elements being strongly interrelated. The two items are parts of customer-seller trust, and the finding indicates that the expectations of the customers concerning fraud protection is strongly related to their expectations regarding on-line as well as phone help for the products they purchased in case something is or goes wrong with the product and purchase.

Another strong correlation is detected between the items **Consistency of Fairness of the Site** and **Consistency of Return Policies** (R = 0.41, p<0.0001). Additionally, a significant correlation were detected between the items **Consistency of Presented Guarantees Involving the Product** and **Consistency of Return Policies** (R = 0.41, p<0.0001). The findings lead to the conclusion that the consistency of guarantees, return policies and fairness can be viewed as belonging to the consistency of product-and post-purchase-related customer treatment in connection with Customer Relationship Management.

4.2 Principal Components Analysis and Identification of the New Factors

In order to group the factors impacting the consistency in E-commerce Customer Relationship Management, a multivariate analysis was conducted on the fourteen items of the CRM-CQ. The last two items were not included in the analysis since they were inserted for internal reliability purposes. The analysis had the primary goal of identifying the major issues of consistency in e-commerce CRM by finding large-scale clusters (similar to usability and sociability clusters in on-line communities, Preece, 2001), and also eliminating some of the items that cannot be articulated by any group. For this purpose, principal components procedure was executed. The principal components analysis was chosen because of its relative accuracy in identifying initial factors compared to the maximum likelihood procedure (Ozok, Salvendy, 2001). The results from this analysis are presented in Table 1. The principal components procedure produced loadings from each of the sixteen items on two factors. Consistent with the study by Ozok and Salvendy (2001) and Wei and Salvendy (2000), the loading threshold was chosen as 0.40.The analysis indicated that consistencies of **promotions, in-stock indication, product variety, fraud protection, presented guarantees, site fairness**, and **return policies** were clustered into Factor 1, and **consistencies of shopping steps, site design,** and **navigation** were clustered into Factor 2. The factors could easily be named as **Customer Treatment Factor** (Factor 1) and **Technical Factor** (Factor 2). The multivariate analysis indicates that there are two general items when it comes to consistency in Customer Relationship Management in e-commerce. The technical factor deals with items relating to the design of the site, such as whether the steps to execute a transaction or to navigate across the site are consistent, the designs of individual pages within the shopping site are consistent, etc. The customer treatment factor is concerned with whether all the customers are treated fairly and consistently before, during, and after the shopping process, such as whether the same promotions, product variety, fraud protection, guarantees, etc. are offered to every customer.

The clustering by the multivariate analysis was thought to result in interesting findings concurring with the conceptual factors regarding customer treatment consistency. Additionally, four items, **consistency of price quoting, consistency of alternative product suggestions, consistency of help offered by the site,** and **consistency in keeping the customer's personal information**, were eliminated by the analysis. This indicates that although these items may be important from a broader CRM point of view, they may not be as important to the customers within the context of consistency.

After the identification of the two new factors, two separate Cronbach's Alpha internal reliability analyses were conducted in order to determine whether the two factors also had high internal reliability. The Cronbach's Alpha values were 0.84 for Factor 1 and 0.75 for Factor 2, and it was therefore determined that both factors had high internal reliability values.

Table 1. Factor Loadings for the Questionnaire Items
(Numbers in bold indicate loadings greater than 0.40)

Items	Factor 1 (Treatment Factor) Loading	Factor 2 (Technical Factor) Loading
Price	0.32848	0.05914
Steps	0.15190	**0.64222**
Design	0.04865	**0.82438**
Navigation	0.18380	**0.69145**
Promotions	**0.45649**	0.12799
In-Stock	**0.69576**	-0.10108
Variety	**0.53489**	0.17493
Suggestion	0.37170	0.16374
Fraud	**0.43300**	0.16340
Guarantees	**0.51922**	0.22165
Fairness	**0.42334**	0.29647
Help	0.33000	0.24554
Returns	**0.42813**	0.10437
Personal Info	0.14931	0.31995

5. Conclusion and Recommendations

Several significant results have been obtained from the study, and they can be summarized as follows:

- A total of ten items can be identified contributing to overall consistency in e-commerce Customer Relationship Management.
- As far as mutual trust and security are concerned, users are highly expectant of consistent fraud protection and consistent guarantees from their e-commerce site.
- Relating to products they are purchasing, users expect consistent fairness, as well as consistent presentation of promotions and in-stock indications from their shopping site.
- Relating to site design, users expect consistent site design, consistent navigation options, and consistent steps to complete their transactions.
- The mentioned three factors in the previous bulletin are intercorrelated and conceptually integral to each other.

- Additionally, two items belonging to the purchased product, consistent promotions and consistent product variety for every customer, highly relate to each other which indicates the two items being integral items for the customers in order to find the best deals for their budget.
- There is also a strong positive relationship between consistencies in the site's fairness, offered guarantees, and the return policies offered. Hence it can be concluded that a site's consistent fairness can easily be related with the consistency of its return and guarantee policies.

Guidelines for Consistency In E-Commerce

Customer Treatment Guidelines

- Give all of your customers the same variety and selection of products to choose from.
- Give all of your customers the same amount of protection from on-line fraud, regardless of the value of the purchase or corporate or individual identity of the customer.
- Treat your customers fairly before, during, and after the purchase.
- Have consistent return policies for all products and manufacturers on your site.

Site Design Guidelines

- Have the same amount of steps to conduct on your shopping site to complete transactions for purchasing all of the products.
- Have a consistent site design.
- Have consistent navigation rules for the customers using your site.

References

Anderson, R. (2000): Making an E-Business Conceptualization and Design Process More "User"-Centered. *Interactions,* July+August 2000, pp. 27-30.

Ngai, E., Wat, F. (2002): A Literature Review and Classification of Electronic Commerce Research. *Information & Management,* Vol. 39 (2002), pp. 415-429.

Ozok, A. A. and Salvendy, G. (2001): How Consistent is Your Web Design? *Behavior and Information Technology*, Vol. 20(6), pp. 433-447.

Preece, J. (2001): Sociability and Usability in Online Communities: Determining and Measuring Success. *Behavior and Information Technology*, Vol. 20(5), pp. 347-356.

Resnick, M. (2001): Design and Usability Evaluation of Customer Relationship Management in Commercial Web Sites and E-Business. *Usability Solutions, Miami, Florida*, downloaded from crm2001online.com, 3.5.2002.

Shneiderman, B. (1992): Designing the User Interface: Strategies for Effective Human-Computer Interaction. New York: Addison-Wesley.

A Study of Culture Differences for Browsing Hypertext on Handheld Devices

Pei-Luen Patrick Rau

Department of Industrial Engineering
Tsinghua University
Beijing 100084, China
rpl@mail.tsinghua.edu.cn

Yun-Ju Chen

Institute of Communication Studies
Nation Chiao Tung University
Hsinchu 300, Taiwan
jillchen.ct89g@nctu.edu.tw

Abstract

The objective of this study was to investigate the effects of two cultural dimensions, time orientation and context of communication, on the browsing performance for handheld device users. An experiment was conducted with 89 participants. The experimental design was a 2x2 factorial with two independent variables: high/low context of communication, polychronic/monochronic time orientation. The dependent variables are performance time, total number of steps, enjoyment of browsing, and disorientation. The results showed that high-context of communication participants tend to feel less enjoyed and more disorientated than the low-context of communication participants.

1 Introduction

With the development of high-speed wireless networking and various types of mobile devices, mobile computing has been widely accepted and applied. However, mobile information access has to cope with the problems and restrictions (e.g., display, battery, memory size, processing power, narrow bandwidth, restricted resources) of the mobile environment. The purpose of this study was to investigate the effects of various variables on user's performance and attitude for browsing hypertext on handheld devices. The variables studied are context of communication and time orientation.

Both high and low context of communication are studied. A high context (HC) of communication is one in which most of the information is already in the person, while very little is in the coded, explicit, transmitted part of the message (Hall, 1976). In high context cultures, most of the information is in the physical context or is internalized in the people who are a part of the interaction. Very little information is actually coded in verbal message. In the opposite, information is mostly carried in verbal message, and very little is embedded in the context or within the participants in low context cultures (Porter & Samovar, 1997). Thus low context of communication users may spend more time and link more web pages. Also, low context of communication participants may feel less disorientated because they tend to pay more attention to the background information.

There are two types of time orientation studied: monochronic and polychronic. Monochronic time means paying attention to and doing only one thing at a time. Polychronic time means being involved with many things at once. In monochromic cultures, time is experienced and used in a linear way. Monochronic time is divided quite naturally into segments; it is scheduled and compartmentalized, making it possible for a person to concentrate on one thing at a time (Hall & Hall, 1990). Monochronic Time Orientation treats time in a linear manner. Monochronic users have to deal with more than one job at a time, often in a non-linear environment like hypertext systems. On the other hand, polychronic users tend to handle many things at the same time, which happens frequently in browsing.

Rau and Liang (2003) studied the effect of cultural effects on browsing a Web-based service. They found that participants with high-context of communication style were more disorientated than those with low context communication. Also, marginal significant differences were found in total steps and performance time in time orientation. Polychronic participants browse information faster and took fewer steps than monochronic participants. The limitations of handheld devices such as display size might affect the browsing performance, particularly for low context of communication users.

2 Hypotheses

Two hypotheses for the effect of culture differences on browsing hypertext on handheld devices were developed based on the literature.

Hypothesis One. For users of low context communication, the browsing performance (performance time, step, enjoyment of browsing, and disorientation) on handheld devices will be better than the browsing performance of users of high context communication.

Hypothesis Two. For users of polychronic time orientation, the browsing performance (performance time, step, enjoyment of browsing, and disorientation) on handheld devices will be better than the browsing performance of users of monochronic time orientation

3 Methodology

The purpose of this research is to empirically examine the effects of two culture differences, context of communication and time orientation, on the browsing performance of handheld devices. Eighty-nine novice mobile device users aging from 20 to 30 years were engaged in the experiment. Forty-five participants were high context of communication and forty-four participants were low context of communication. Fifty-four participants were polychronic time orientation and thirty-five were monochromic time orientation.

A 2x2 factorial design was used in this research. The independent variables were context of communication and time orientation. Context of communication has two levels: high and low. Time orientation has two levels: polychronic and monochromic.

This research used four dependent variables: performance time, step, enjoyment of browsing, and disorientation. Performance time was defined as the total time taken to complete the required browsing task. Time was recorded by the testing hypertext system to the nearest one-thousandth of a second. Step was defined as the total number of nodes that a participant visited for the tasks. Enjoyment of browsing was defined as the score obtained through a questionnaire. Disorientation was also defined as the score obtained through a satisfaction questionnaire consisting of seventy questions designed by this research and a past study (Beasley and Waugh, 1995) on the scale of 1 (*strongly disorientated)* to 7 (*never disoriented*).

Ten information search tasks on a testing hypertext system were designed for handheld devices. Participants were asked to scan and review information relevant to a task. An H3630 iPAQ Personal Digital Assistant with a touch screen of 320*240 pixels was used for the experiment. Each participant was required to perform the tasks independently with no other participants present except for the experimenter. The participants' movements throughout the system were automatically timed to the nearest .001 sec and traced by a testing hypertext system.

All participants began to fill out a general information questionnaire concerning their personal characteristics, including age, education, and computer and Internet experience. Then the participants filled out two questionnaires of cultural differences. Each participant was given a handheld device and on-screen instructions. A brief practice session was then conducted to help the participants understand the operation of the system and the tasks to be performed. Following the practice, each participant performed the information search tasks. Participants were instructed to perform the tasks as quickly as possible without sacrificing accuracy. On the completion of the tasks, each participant was given two questionnaires of attitude and disorientation.

4 Testing Hypotheses

The intention of hypothesis one was to examine how the difference in context of communication might influence browsing hypertext for handheld device users. As shown in Table 1, the result indicated that significant differences in enjoyment of browsing ($F = 4.000$, $p = 0.049$) and disorientation ($F = 4.359$, $p = 0.040$) were found. The results indicated that high context of communication participants tend to feel less enjoyed and more disorientated than the low context of communication participants. Hypothesis one is partially supported that the low-context users feel more enjoyed and less oriented when browsing hypertext on handheld devices than high-context users.

Table 1. Data for Testing Hypothesis One for Communication Context

Variables	High		Low		F	*p*
	Mean	SD	Mean	SD		
Performance time (Second)	543.9	115.19	534.3	123.10	0.031	0.862
Total steps	58.4	11.93	57.8	11.67	0.037	0.848
Enjoyment of Browsing	4.6	1.01	5.0	0.91	4.000	0.049
Disorientation	3.1	0.80	2.70	0.84	4.359	0.040

Table 2. Data for Testing Hypothesis Two for Time Orientation

Variables	Monochronic		Polychronic		F	*p*
	Mean	SD	Mean	SD		
Performance time (Second)	532.7839	126.1749	548.9015	106.8614	0.153	0.697
Total steps	57.94	11.06	58.29	12.88	0.334	0.565
User pleasure	4.7654	0.9855	4.9143	0.9679	1.678	0.199
Disorientation	2.9259	0.8545	2.8050	0.8137	1.315	0.255

The intention of hypothesis two was to examine how the difference in time orientation might influence browsing hypertext for handheld device users. As shown in Table 2, the result indicated that no significant differences in browsing performance were found. The results indicated that the polychronic time orientation participants do not perform better than the monochromic time orientation participants.

5 Discussions and Conclusions

Design of user interfaces for handheld devices requires an understanding of the characteristics of devices and users. Hypertext design and wireless Web pages design for handheld devices need attentions due to the limitations of handheld devices and characteristics of hypertext. The results of this study indicated that handheld device users' communication style affects the browsing performance in terms of enjoyment of browsing and degree of disorientation. Low-context users appear to be less disoriented compared to the users with high-context of communication style. The findings suggest that users' communication style should be considered in the design of hypertexts or Web pages for handheld devices

References

Hall, Edward T. (1976). *Beyond culture*. NY: Anchor Books.

Hall, Edward T. & Hall, Mildred Reed.(1990). *Understanding cultural differences*. Yarmouth, MA: Intercultural Press.

Kiger, J. I. (1984) .The depth/breadth tradeoff in the design of menu-driven interfaces. *International Journal of Man-Machine Studies*, 20, 201-213.

Porter, R. and Samovar L. A. (1997). An Introduction to Intercultural Communication. In Samovar, L. A. & Porter, R. (eds.), *Intercultural communication: a reader.* Belmont, CA. Wadsworth Publish Company.

Rau, P.L. P. and Liang, S.F. M. (2003). A Study of Cultural Effects on Designing User Interface for a Web-based Service, *Int. J. of Services Technology and Management,* In press.

When Computers Fade ... Pervasive Computing and Situationally-Induced Impairments and Disabilities

Andrew Sears, Min Lin

Interactive Systems Research Center, UMBC
1000 Hilltop Circle
Baltimore MD 21250
asears, mlin4@umbc.edu

Julie Jacko

Georgia Institute of Technology
765 Ferst Drive
Atlanta, GA 30332
jacko@isye.gatech.edu

Yan Xiao

University of Maryland, Baltimore
10 South Pine Street
Baltimore, MD 21201
yxiao@umaryland.edu

Abstract

Since Weiser introduced his vision of ubiquitous computing, computing devices have become lighter, smaller, cheaper, and more powerful. At the same time, pervasive computing and the concept of context have attracted significant attention, with the goal of supporting the use of computing devices anywhere, anytime, as computers fade into the environment. While there has been some success inferring the users' intensions, reliably understanding users' general goal remains a significant challenge. Using limited context information, such as location, can be useful, but the benefits are limited. Context is more than location. As computers are embedded into everyday things, the situations users encounter become more variable. As a result, situationally-induced impairment and disabilities (SIID) will become more common and user interfaces will play an even more important role. Recent understandings on context suggested the importance of applications themselves as parts of the whole context space. This article will explain and discuss the characteristics of SIID under a three-dimension (human, environment, and applications) context model. We suggest integrating information from all dimensions to have a whole picture of context. More studies are needed to understand the relationship among different dimensions, and to help design effective context-aware applications overcoming SIID.

1 Introduction

Twelve years ago, Weiser introduced his vision of ubiquitous computing as the trend for the 21st century (Weiser, 1991). He imagined a world where people used computers without noticing their existence because they were seamlessly integrated into the environment. Computers would be everywhere, would infer intent, and many would not even be noticed. As the 21st century begins, we find ourselves in a world where computers are smaller, more mobile, more powerful, and less expensive. Computers are being integrated into the environment such that they are not noticed, there has been some success inferring intent, but many computers are still quite identifiable as technological support for the activities we are engaged in. Clearly, we are making progress toward Weiser's vision. Examples include augmenting the functionality of everyday objects such as coffee cups via embedded sensors and circuits (Beigl, Gellersen, & Schmidt, 2001) and facilitating guided tours by integrating positioning devices throughout buildings (Abowd et al., 1997).

While we have made progress toward Weiser's vision, two of the key terms he used to describe this vision continue to evolve. Today, *environment* has, to a large extent, been replaced by *context*.

It seems that *environment* often caused people to focus on the physical space in which the technology was being used, while *context* is seen as expanding the focus to include not only the physical space, but also aspects of the social situation that may influence the use of technology. Currently, context-aware systems are the focus of significant research. At the same time, *pervasive computing* seems to be replacing *ubiquitous computing* to emphasize the way these devices permeate everyday activities. Everyday artifacts may look the same, but today they often include some form of computing technology which makes decisions and shapes the user's activities. Before a call is connected, recipient's phone can inform the caller if a voice message is preferred rather than a phone call (Gellersen, Schmidt, & Beigl, 2002). As Weiser envisioned, computers are beginning to vanish into our everyday lives.

Examples include multiple attempts to create smart environments within buildings. Unfortunately, difficulty inferring the user's intent can lead to inappropriate results (Bellotti & Edwards, 2001). An individual can be greeted by with personalized message when they enter an office, phone calls can be forwarded as an individual moves from room to room, and printing requests could be routed to the nearest printer for quick pick-up. While each of these scenarios could provide benefits, they can also prove problematic if the individual's goals are not adequately understood. Personalized greetings can become a disturbance if they do not provide useful information or if the person is using a mobile phone when entering the room. Automatic forwarding of telephone calls may be inappropriate if the person is going into a meeting. The "nearest" printer may not be in a convenient location when the document is finally printed.

As researchers continue to investigate techniques for inferring the user's goals, applications such as those described above will become more effective. However, focusing on environment at this level can only provide limited benefits. Many efforts to develop context-aware technologies pay little attention to the interactions that will still occur. Even as computing devices become embedded in the environment, and computers become more invisible to their users, users will still be interacting with some combination of hardware and software. These technologies may look different, and involve new interaction techniques, but they will still require input from users, generate results that are intended to benefit these users, and perhaps even present information directly to the users. As a result, information technologies will continue to play an important role in effective and efficient interactions even as technology becomes smaller, more mobile, and increasingly embedded into the environment. Interestingly, as these changes occur, the environments in which people use these technologies become more variable. This suggests that additional research is needed that focuses on understanding the difficulties users experience when interacting with information technologies in these increasingly complex environments. The concept of SIID (Sears & Young, 2003) is suggested as a framework for addressing these issues.

2 Situationally-Induced Impairments and Disabilities

According to the International Classification of Functioning, Disability and Health (ICIDH-2), impairment refers to a loss or abnormality of body structure or function and disability refers to difficulties an individual may have in executing a task or action. Although these definitions originate in the healthcare community, the concepts also apply in the context of human-computer interactions, especially when environmental factors are considered. For example, the absence of a limb is a physical impairment that may, depending on the input technologies available, result in computer-oriented disabilities. Hands-busy situations can result in similar disabilities, depending on the input technologies that are available. Similarly, a variety of health conditions, or a noisy environment, can result in disabilities associated with speech-based interactions with computers.

Both the work environment and the activities the individual is engaged in can lead to SIID. For example, a paramedic may be busy providing medical care to a patient in a moving ambulance. At the same time, the paramedic must document patient conditions and treatment so this information can be given to caregivers upon arrival at the hospital. The vibration from the moving vehicle as well as the demands of providing appropriate health care degrades the paramedic's ability to interact with physical input devices. The siren and road noise hinder the use of speech-based interactions. These SIID make interactions between users and computing devices more difficult. Other factors, including temperature, lighting, demands for the user's attention, and stress can also adversely affect an individual's ability to interact effectively with computing technologies.

While we intentionally draw an analogy to difficulties that may be experienced due to health conditions, we also explicitly acknowledge that SIIDs are different in that they tend to be temporary and dynamic. Individuals typically develop strategies to overcome health-related disabilities, but the temporary and dynamic nature of SIID makes this more difficult. A strategy may be effective at one moment and ineffective at the next. As a result, information technologies that are aware of the difficulties users may experience as a result of the current environment and tasks are increasingly important: different strategies can be presented to users, alternative interaction techniques can be supported, or the way input is processed can be adapted.

3 Defining Context (and the Role of SIID)

Schmidt et al. (1999) proposed a three dimensional model to describe the concept of "context" that explicitly separated activities, users, and the environment. More recently, Dey et al. (2001) provided a broad definition that includes "any information that can be used to characterize the situation of entities (i.e. whether a person, place, or object) that are considered relevant to the interaction between a user and an application, including the user and the application themselves". Schmidt's model provides an effective high-level representation of context, but many details were not provided leaving the relationship among different dimensions vague and largely unexplored. Dey's definition explicitly encompasses all of the information that could prove useful for context-aware technologies. It also represents the first time that applications were identified as an entity at the same level as the users.

While broad definitions have been offered, rather limited definitions of context tend to be employed in practice and the definitions that are employed tend to focus on the underlying technologies as opposed to how the difficulties users may encounter when interacting with these technologies. Location is perhaps the most studied aspect of context. Examples include the Active Badge system (Want et al., 1992), a coupon delivery service (Duri et al., 2001), and a tour guide system (Cheverst et al., 2000; Pascoe, 1997). While several groups (Brown, Bovey, & Chen, 1997; Dey, 1998) have stressed that there is more to context than location, research on context-aware has continued to emphasize the physical environment. Of particular interest is the fact that few researchers have focused on how the users and their activities factor into the definition of context. One effort that did explicitly focus on the users' activities resulted in the concept of a minimal attention user interface (MAUI) (Pascoe, Ryan, & Morse, 2000). This concept was illustrated using applications that supported an ecologist's observations of giraffes. These applications address some SIID, such as those that result from eyes-busy tasks, and represent one of the few situations where the users' interactions were the primary focus of the research. Figure 1 represents a combination of the 3D model proposed by Schmidt et al. and the definition proposed by Dey et al., that separates contextual information into three categories: human, environment, and application. The human dimension addresses the characteristics of the user (e.g., identity, preference, biophysiological conditions, emotional state), current social issues (e.g., existence of other people, social interactions), and users' current activities (e.g., walking, filling out a form, looking for a phone number, following a map). The environment dimension includes location, physical conditions (e.g., lighting, noise, temperature, speed), and the infrastructure (e.g., embedded devices, sensors, communication protocols). The application dimension includes the available functions (e.g., backlight control, font size adjustment, zoom function) and input/output channels (e.g., touch screen, voice command, gesture recognition, audio output, vibration). While most research continues to focus on a single dimension, the concept of SIID can only be addressed if the cumulative effects of all three dimensions are considered.

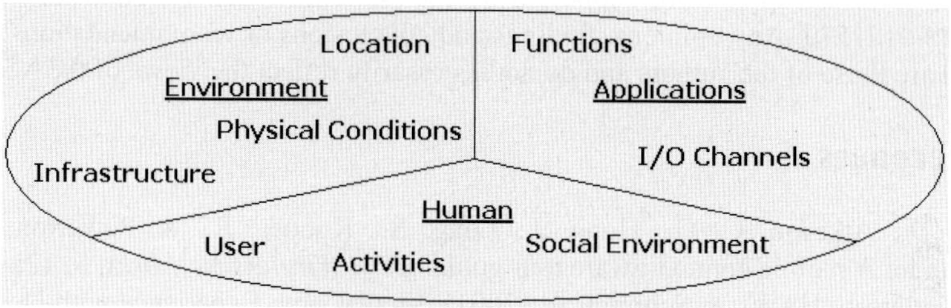

Figure 1: 3D Context Space

For example, when the user is attempting to complete a task, the application dimension defines how the task can be completed (e.g., speech input, stylus input, physical keyboard). The environmental dimension may make one or more of the alternatives more or less attractive (e.g., speech input is not effective in a noisy environment). Similarly, the human dimension can alter the efficacy of various solutions (e.g., speech impairments may preclude speech-based interactions; physical impairments may hinder the use of stylus-based input). SIID can only be addressed if all three dimensions are understood and, more importantly, the interactions among these dimensions are understood. Clearly, some of this understanding is the responsibility of the user, but the information technologies can also contribute to this process.

Users can recognize certain aspects of the context and adapt. They can stop walking if motion causes difficulties entering information, mute their phone if the conversation hinders speech-based input, or even switch to a different input device. The user may be able to turn on additional lights, turn off a radio, or move indoors if necessary. However, the context may also limit the user-based adaptations that are feasible. For example, a paramedic must continue with their primary task of providing healthcare even if this interferes to with their record keeping activities. Similarly, sensors can allow technology initiated adaptations to accommodate environmental issues and user activities (Hinckley et al., 2000). As background noise increases, speech interactions can be enhanced using noise filtering algorithms. If the background noise becomes too loud, speech recognition can be disabled and the input can be digitized for subsequent processing. If the user is moving, stylus-based input can be debounced. If the movement becomes such that debouncing is no longer effective, alternative forms of input could be suggested. If the lighting is insufficient, contrast can be increased or backlighting can be turned on.

4 Conclusion

While current definitions of context include the necessary information, the interactions among the dimensions have not been adequately explored. The concept of SIID provides a framework for this exploration within the larger framework of context-aware computing. SIID highlight the importance of understanding all three dimensions, how these dimensions interact, how users adapt, and how technology-based adaptations can facilitate more effective solutions that reduce errors and speed interactions. As computers continue to fade, users will continue to interact with these technologies, but these interactions will become more subtle, the environments will become more complex, and the interactions among the human, environment, and application dimensions will become even more important.

5 Acknowledgements

This material is based upon work supported by the National Science Foundation (NSF) under Grant No. IIS-0121570. Any opinions, findings and conclusions or recommendations expressed in this material are those of the authors and do not necessarily reflect the views of the NSF.

6 References

Abowd, G. D., Atkeson, C. G., Hong, J., Long, S., Kooper, R., & Pinkerton, M. (1997). Cyberguide: A mobile context-aware tour guide. *ACM Wireless Networks*, 3, 421-433.

Beigl, M., Gellersen, H.-W., & Schmidt, A. (2001). MediaCups: Experience with Design and Use of Computer-Augmented Everyday Artefacts. *Computer Networks*, 35 (4), 401-409.

Bellotti, V., & Edwards, K. (2001). Intelligibility and Accountability: Human Considerations in Context-Aware Systems. *Human-Computer Interaction*, 16 (2-4), 193-212.

Brown, P. J., Bovey, J. D., & Chen, X. (1997). Context-aware applications: From the laboratory to the marketplace. *IEEE Personal Communication*, 4 (5), 58-64.

Cheverst, K., Davies, N., Mitchell, K., & Friday, A. (2000). Experiences of developing and deploying a context-aware tourist guide: The GUIDE project. In *Proceedings of the MOBICOM Conference*, 20-31.

Dey, A. K. (1998). Context-aware computing: The CyberDesk project. In *Proceedings of the AAAI '98 Spring Symposium on Intelligent Environments*, 51-54.

Dey, A. K., Abowd, G. D., & Salber, D. (2001). A Conceptual Framework and a Toolkit for Supporting the Rapid Prototyping of Context-Aware Applications. *Human-Computer Interaction*, 16, 97-166.

Duri, S., Cole, A., Munson, J., & Christensen, J. (2001). An approach to providing a seamless end-user experience for location-aware applications. In *Proceedings of the WMC*, 20-25.

Gellersen, H. W., Schmidt, A., & Beigl, M. (2002). Multi-sensor context-awareness in mobile devices and smart artifacts. *Mobile Networks and Applications*, 7 (5), 341-351.

Hinckley, K., Pierce, J., Sinclair, M., & Horvitz, E. (2000). Sensing Techniques for Mobile Interaction, ACM Symposium on User Interface Software & Technology, 91-100.

Pascoe, J., Ryan, N., & Morse, D. (2000). Using while moving: HCI issues in fieldwork environments. *ACM Transactions on Computer-Human Interaction*, 7 (3), 417-437.

Schmidt, A., Aidoo, K. A., Takaluoma, A., Tuomela, U., Laerhoven, K. V., & Velde, W. V. d. (1999). Advanced Interaction in Context. In *Proceedings of the HUC*, 89-101.

Sears, A., & Young, M. (2003). Physical Disabilities and Computing Technologies: An Analysis of Impairments. In J. A. Jacko and A. Sears (Eds.), *Human-Computer Interaction Handbook* (pp. 482-503).

Want, R., Hopper, A., Falcao, V., & Gibbons, J. (1992). The Active Badge Location System. *ACM Transactions on Information Systems*, 10 (1), 91-102.

Weiser, M. (1991). The Computer for the 21st Century. *Scientific American*, 265, 94-104.

Fostering Motivation and Creativity for Computer Users

Ted Selker

MIT Media Lab
20 Ames Street Cambridge Ma 02139
Selker@media.mit.edu

Abstract

Creativity might be viewed as any process which results in a novel and useful product. People use computers for creative tasks; they flesh out ideas for text, graphics, engineering solutions, etc. Computer programming is an especially creative activity, but few tools for programming aid creativity.

Most computers are used in solitude; however, people depend on social supports for creativity. A computer could provide some of the social support and cues normally offered by humans to keep a worker motivated and help him consider useful alternatives. Computers could support and filter potentially creativity-enhancing communication with other humans. This paper develops the notion that creativity and motivation enhancement can easily be incorporated into the design of high-quality human-computer interaction.

1 Introduction

Dictionaries define "creativity" with words like originality, expressiveness and imagination (Carles, 2003). Creativity brings new ideas and improvements to people's lives. While some still hold that teaching creativity is dubious, Nickerson's work reports on teaching and measuring creativity imp rovement over an extended period of time (Sternberg, 1999). Possibly no one is more cited for his writing on creativity than Mihaly Csikszentmihalyi, the promoter of the idea of flow in creativity (Sternberg, 1999). Teaching creativity has become an industry (Heleven, 2003). Edward DeBono, best known for "lateral thinking", has published dozens of books on the topic. Most common are brainstorming prosthetics, such as outlining tools, lists of steps to go through or pictures and words to help expand ones thinking on a topic.

Computer interfaces are typically judged on many factors such as power, elegance, simplicity, ease of use and learning. Good functional user interface is taking the tool out of the task. If the user interface can avoid being the focus of attention, the user can focus on their goal; i.e., writing the paper, making the phone call, etc.. Spreadsheets, for example, allow people to compare results in what-if scenarios (Brown & Gould, 1987). The ability to start and stop secondary tasks is important to creative pursuits as well. While the ergonomics community considers fatigue important, it has not yet played a role for the human-computer interface. Much could be done to help maintain concentration and productivity, such as breaks. This could be based on recognition of the user's competencies and weaknesses, productive work pattern, signs of fatigue, and sometimes just staying out of the way.

The remainder of this paper gives examples of features that designers can add to human-computer interface that improve users' motivation, engagement and ability to be creative.

2 Setting the stage for creative work

Many factors can help set a creative context. People create rituals, special settings, go to workshops to get them in the mood to be creative. Unfortunately, people will not always be willing to stop working and start up a creativity-enhancing tool. We must search for motivational and creativity-enhancing activities that can be integrated into applications and that don't deter workflow (Burleson & Selker, 2002).

Still, some approaches to getting people to be creative just don't work. I once went to a 5-hour brainstorming session with some of the most creative people I know. It all started with drinks with the hosts the night before. We didn't brainstorm there; we would tomorrow. In the morning we went to an extended breakfast, we were asked to wait for the 20 minutes to set up the mics, 30 minutes for a group photo shot. The facilitator then tried to structure the conversation, which did not open people up. A microphone-off lunch was another time to stop thinking. In all, no one played his or her best cards. Just as the disruptions in this brainstorming meeting stalled intellectual momentum, so do complex procedures for finding, starting or changing computer applications .

An idea has to be captured to be considered. Most people have had the experience of forgetting what they wanted to say when someone else chose that moment to talk. Software should not interrupt when a person is trying to express an idea either.

I have used the Control-C command for cutting in so many text editors that it is automatic. If a person uses a command all the time, it will be executed faster and with less attention. One email system that I use requires many more selections for tasks than other email programs; a task model such as GOMS show these extra steps as distracting (Card, Moran & Newell, 1983). Recalling and entering an uncommon command is more distracting than just recognizing and selecting it (Klatzky, 1980).

Classifying and organizing things can be helpful, but where should they be put to be retrieved? Early hypertext experiments found people's focus on classifying their work distracted them from doing the work (Foss, 1989). Computers are good record keepers and potentially good indexers. Mechanisms for storing new ideas and finding them again should be predictable or easily discovered to avoid retrieval problems. Revisions are easily shown on paper with annotations. Similarly, marks on whiteboards get so important that people become reluctant to erase anything. Ever since the text editor was invented, people have been asking for infinite undo; clipboards still don't have structured undo. I even lost part of this paper due to a limited text-editor undo. Versioning text editors such as PEdit (Kruskal, 1984) have not been widely available until Microsoft Word 6. PEdit created different versions of documents for different purposes, allowing a person to make alternative versions without committing to eliminating others. While Word6 allows different people to use different colors, it does not support keeping and comparing alternative ideas.

Brainstorming is often aided by references such as dictionaries, thesaurus and prosthetics with broadening words, pictures, etc. Now people have the Internet and ask it questions at will. These tools can also cause distraction from the task at hand.

This section highlighted some standard software user interface features that help people try ideas and some of the problems that can occur. The rest of the paper will focus on how the design of human-computer interaction can be useful.

3 Cognitive science considerations

What are the precursors to motivational and creative success in a task? Perceptual-sensory, precognitive, cognitive, ergonomic, behavioural and social dimensions all impact human performance (Klatzky, 1980). Designers need to make software to take these human constraints into account.

Perceptual issues play a big role in what we notice. Time used for concentrating on perception borrows from cognitive tasks. Sensory limitations are in every interface decision from display contrast ratio to font design and even the shape of typing keys. The IBM TrackPoint (Rutledge & Selker, 1990) is at least 20% faster at selection than other joysticks because it uses a cognitive model of how fast eyes track. Even the wallpaper on graphical interfaces can distract users from finding the icons they are looking for on the screen.

Ergonomic issues affect what we can reach. The TrackPoint is placed in the middle of the keyboard as a direct consequence of literature that shows that a person spends more than a second switching from a keyboard to the mouse and back again (Rutledge & Selker, 1990). This placement saves almost a second spent reaching for a mouse or track pad, which significantly speeds activities like text editing.

Precognitive issues are recognized without even taking one's conscious attention. A blinking cursor attracts the eye. Dropped shadows can be recognized much faster than picture-frame style bezels, etc. (Enns & Rensink, 1990). Designers can use precognitive perception to help creative people attend to problems and opportunities.

Just as with physiological psychology issues, the limits of cognitive ability must be respected to allow people to learn and remember. For example, user interfaces that inadvertently prime a user with a point of view might stop other ideas from being considered (Klatzky, 1980). User interfaces that require too much short-term attention are difficult to use. The limit to how many things can be remembered at once is often covered by the legendary 7 +- 2 (Klatzky, 1980).

Word6 is a mentor for bad spellers and syntax lightweights (Heidorn, Jensen & Richardson, 1993). One study showed how formatting technology could distract people from their writing objectives (Rosson, 1983). A group with a command-line controlled text editor produced better writing than the group with a What I See Is What I Get text editor. Users must understand where the creative interest is to choose tools that allow them to focus on their goal.

Behavioural issues affect creative potential too. Most people's offices are adorned with mementos to make them feel comfortable. These belongings can help them take mini breaks and have outlets for releasing tension. A nice place to sit with a view, a beverage, a pad of paper for jotting notes or drawing can all provide needed diversion that doesn't stop the work flow. Unfortunately, solitaire, instant messaging, animated pop-ups and games that need to be finished once started are all more distracting than a sip of a beverage.

Attempting to balance our need for focus with our desire for breaks can turn into procrastination. By watching the pattern of use and mouse movements, a computer can interpret aspects of a user's intentions (Lockerd & Selker, 2002). This might be used to monitor distractions and encourage people to stop procrastinating. In any case we should make computer activities that can be non-distracting background mini-breaks.

Social issues also affect people's ability to be creative. Telling a person that they are doing well part way through a project improves their resolve to finish it (Nass & Reeves, 1996). Instant messaging and email, help a people expand their thinking and search for a solution (they can also distract and make it harder to focus). Drift Catcher (Lockerd & Selker, 2002) classifies email based on social relation and succeeds at getting people to focus on the relationships they want to improve.

Fostering motivation and creative tasks must take into account social expectations. The ability for a person to feel judged by a computer is well documented (Nass & Reeves, 1996); people are less willing to communicate something to a computer that they perceive as a judgement about it. The computer's persona could be used to help people feel creative (Burleson & Selker, 2002).

4 Discussion

Support tools could foster creativity by helping people find things, show alternatives, annotate and compare ideas and evaluate their work (Miller, 2002). Such tools should also be careful to allow a user to keep structure, communicate appropriately and notice how much time or other resources they are committing to an idea. Sometimes the most important creative inspiration will be to look for collaborators. A creativity-enhancing toolkit needs to help in the evaluation of when to encourage a person to enter a social partnership.

This paper is a call for the integration of motivation and creativity enhancing considerations in software. Developers need toolkits to help them incorporate and validate creativity and motivation enhancements into software.

The paper has presented some approaches for improving computer user motivation and focusing on the creative process. We have taken a walk through human capabilities that user interface designers must consider in their design. The emphasis has been on what systems do to distract users from productive creative work, and what could be done to enhance these processes. These approaches hold clues to the elements that should be included.

We must reduce the memory load for people to be creative. I am nostalgic for a web programming approach that didn't require keeping an alphabet soup of tools and protocols in mind to be productive.

This paper is meant to be part of the beginning of an era in which computers become partners in peoples' work. We hope to help computer application developers focus on the cognitive and emotional precursors to creativity, the highest human cognitive process of all.

References

Arroyo, E. & Selker, T. (2001). Interruptions: Which are the Less Disruptive? *Conference on Human Factors in Computing Systems*. Pittsburgh, PA.

Brown, S.P. & Gould, D.J. (1987). An Experimental Study of People Creating Spreadsheets, *Transactions on Information Systems*, 5 (3), 258 – 272.

Burleson, W. & Selker, T. (2002) Creativity and Interface. *Communications of the ACM*, 45 (10), 112-115.

Card, S., Moran, T. & Newell, A. (1983). The Psychology of Human-Computer Interaction. Hillsdale, NJ: Erlbaum.

Carles, C. Definitions of Creativity, Retrieved January 2003, from http://members.ozemail.com.au/~caveman/Creative/Basics/definitions.htm

Enns, J. T. & Rensink , R. A. (1990). Scene-based properties influence visual search. *Science*, 247, 721-723.

Foss, C.L. (1989). Detecting lost users: Empirical studies on browsing hypertext. Technical Report 972, INRIA, France.

Heidorn, G.E., Jensen, K. & Richardson S.D. (1993). Natural Language Processing: The PLNLP Approach. Boston: The Kluwer Academic Publishers.

Heleven, M. An inspirational list of creativity and innovation resources, Retrieved January 2003, from http://www.creax.com/creaxnet

Klatzky, R.L. (1980). Memory: Structures and Processes, San Franscisco: W.H. Freeman & Co.

Kruskal, V.J. (1984). Managing Multi-Version Programs with an Editor. *IBM Journal of Research and Development,* 28 (1), 74-81.

Lockerd, A. & Selker, T. (2002). DriftCatcher: Enhancing Social Networks Through Email. International Sunbelt Social Networks Conference XXII.

Miller, C. (2002). Etiquette for Human-Computer Work – Papers from the 2002 Fall Symposium, *Massachusetts Technical Report,* FS-02-02, 102.

Nass, C. & Reeves, B. (1996). The Media Equation: How People Treat Computers, Television, and New Media like Real People and Places. New York: Cambridge University Press.

Rosson, M.B. (1983). Patterns of experience in text editing. *Proceedings of Human Factors in Computing Systems,* 171-175.

Rutledge, J. D. & Selker, T. (1990). Force-to-Motion Function for Pointing. *Interact-90*, New York: Elsevier Press.

Sternberg, R. (Ed.) (1999). Hand Book of Creativity. New York: Cambridge University Press.

Inspection and Condition Monitoring Service on the Web for Nuclear Power Plants

Yukio Sonoda *Yukinori Hirose* [1]

Power and Industrial System Research and Development Center, Toshiba Corp.
8 Shinsugita, Isogo, Yokohama, Kanagawa, 235-8523 JAPAN
yukio.sonoda@toshiba.co.jp yukinori.hirose@toshiba.co.jp

Abstract

The authors developed the system that offers the inspection and condition monitoring service for a nuclear power plant on the Internet. The system is established to the maintenance center of a vendor. It receives the operation data and inspection data of the devices of the plant, executes the monitoring and inspection of aged degradation and replies the result to the maintenance personnel in the site. Aiming at the improvement of reliability of a nuclear power plant and the efficiency of maintenance work, we started the maintenance service on the Web.

1 Introduction

As for a nuclear power plant the reduction of the power generation costs is requested in addition to high reliability and stable operation. It has an effect on the cost reduction to improve the efficiency of maintenance work, which accounts for a large part of the cost. We propose remote inspection and condition monitoring as the means of the efficient maintenance. In order to support monitoring and diagnosis for ordinary plant operation and to support inspection and maintenance work in a periodical outage, it is necessary to evaluate a large amount of field data such as vibration, noise, temperature, image and so on. The remote monitoring has been tried to apply since before and real-time monitoring of important safety parameters in case of emergency situation is in practical use. On the other hand remote support system for daily monitoring and maintenance works were not going ahead because of installation cost and data security. But problem on the application of the remote monitoring is being cancelled by the progress of the wide area network and information technology in recent years.

The remote inspection and condition monitoring has following advantages;

- Offering a support of the specialists who are in the distant place and a precise evaluation by using computer analysis codes,
- Reliable deterioration detection by means of comparative evaluation with a large amount data that are gathered from many plants to the remote maintenance center,
- Making the utmost efficiency of the maintenance works with limited facilities and workforce.

---

[1] Shigeru Kanemoto, shigeru.kanemoto@toshiba.co.jp, Shun-ichi Shimizu and Tomohito Nakano, Isogo Engineering Center, Toshiba Corporation, 8 Shinsugita, Isogo, Yokohama, Kanagawa, 235-8523 JAPAN, shunich.shimizu@toshiba.co.jp and tomohito.nakano@toshiba.co.jp

Toshiba has started remote maintenance service for nuclear power plants a few years before and this paper shows our recent activities.

2 System Overview

The authors developed the system that offers inspection and condition monitoring service for a nuclear power plant on the Internet. Figure 1 shows the structure of the system. The end users, who are plant maintenance personnel, can submit various technical supports such as data evaluation, inspection, consultation, etc. to Toshiba maintenance service section, via the Internet. The Web server deals with user account management and the application server processes job control management. Each application computer carries out respective analysis such as inspection by image processing, condition-based maintenance of devices, vibration monitoring of pumps. This maintenance service on the Web has started since 2000 and has been expanding gradually.

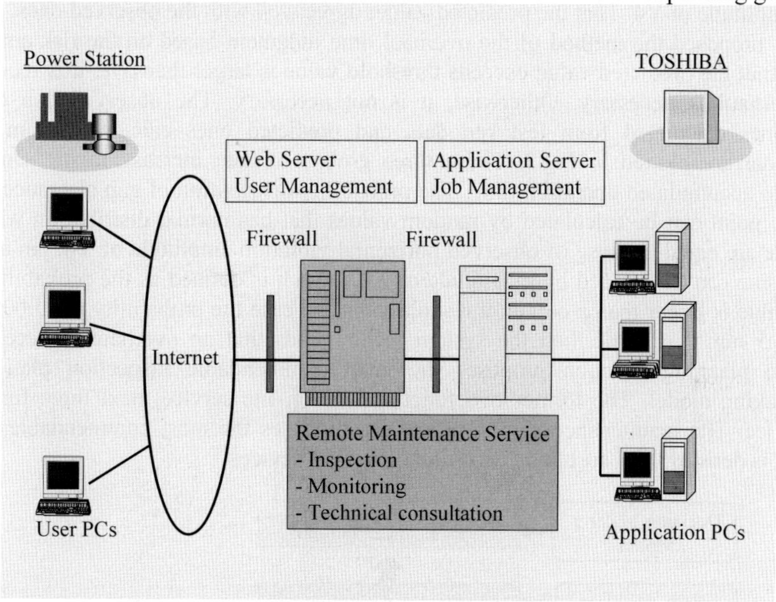

Fig. 1: The Structure of The Inspection and Condition Monitoring Service System

3 Service Contents

3.1 Device *Karte* Management System

The authors developed device *karte* management system, which supports application of condition-based maintenance to nuclear power plants. As shown in Fig.2, The system consists of five functions; database management, degradation evaluation, overhaul time judgment and maintenance planning. The database manages process data, device condition data, operation history data and maintenance history data. The process data mean plant operation parameters, such as pressure, flow rate and so on. The device condition data represents on power device conditions, such as vibration, temperature and noise. The operation history data are operation log and patrol log. The maintenance history data are overhaul records. The degradation evaluation function detects degradation symptom by monitoring field data behaviour. The function evaluates operating performance from process data, soundness from device condition data, and accumulated operating time from operation history data and aging tendency from maintenance history data. This function identifies present degradation degree according to these evaluation results. The degradation prediction function predicts long-term aging trend by statistical model using history data. This model using extrapolated device condition data and accumulated operating time calculates degradation condition in future. The overhaul time judgment function decides an appropriate overhaul time by risk analysis.

For example, the filtering pumps are overhauled in every two scheduled plant outages and gap clearances between wearing ring and pump shaft are measured at two positions. The clearance values increase from the first to the third outage and from the fourth to the sixth outage and the wearing ring was exchanged in the fourth outage, as shown in Fig.3. In the test run after overhaul, vibration at four positions and temperature at five positions were measured. The authors derived the multi-regression model, which predicts gap clearance from accumulated time and horizontal vibration amplitude of V4. That the predicted values agree well with the observed ones.

The authors proposed the method of the overhaul time judgment based on the risk analysis. The probability that the predicted value exceeds threshold value is larger than 5%; it is recommended that an overhaul is necessary. Otherwise, it is not necessary. The observed gap clearances, predicted ones calculated from test run data and predicted ones calculated from operating condition data are plotted in Fig.3. These three groups of data increase in the same way in proportion to accumulated operation time. The probability distribution of gap clearance value at a certain time point can be calculated by random values that has normal distribution whose mean and variance are equal to those of observed horizontal vibration amplitude of V4. An appropriate inspection time can be decided by magnitude of risk, which is defined as the probability that the predicted value is larger than a certain threshold value. In case the probability at 3,000 days with threshold 1.5 mm is 13.5%, then the system recommends that an overhaul is necessary. The maintenance planning function proposes the most commendable inspection plan by using decision–making model. The former two functions are on-site service; next three functions are offline service. The maintenance planning function proposes the most commendable inspection plan, which is derived from cost and risk estimation of all devices.

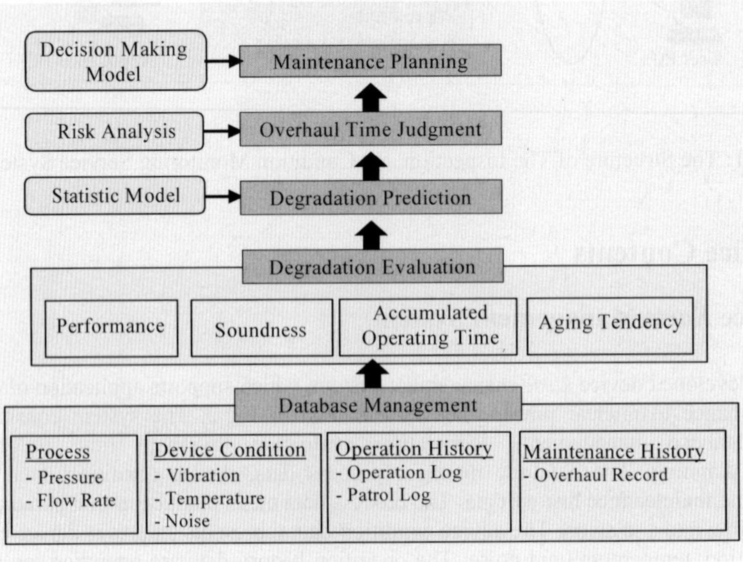

Fig.2: Functional Scheme of The Device *Karte* Management System

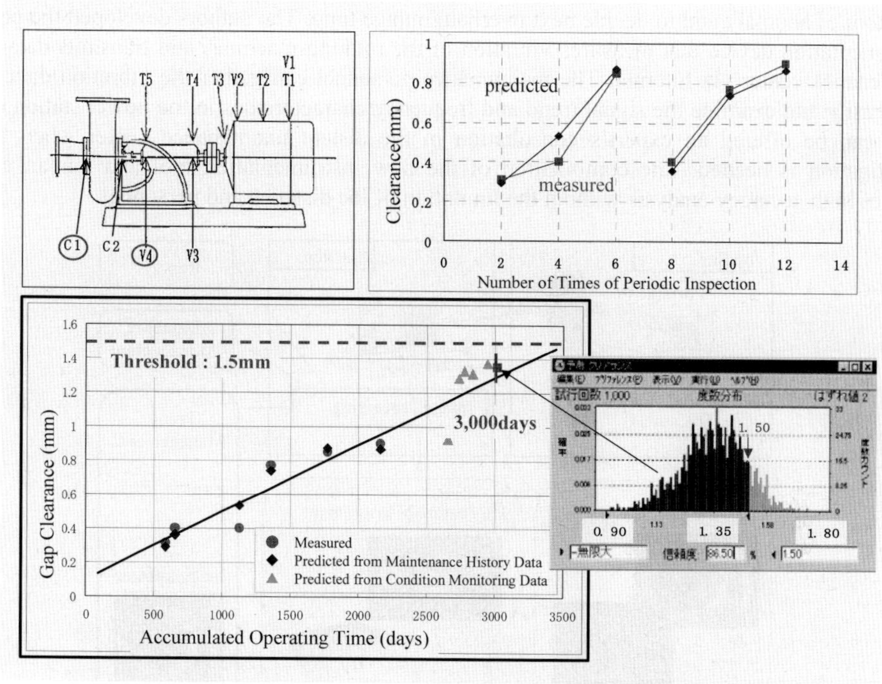

Fig.3: The Statistical Prediction Model and The Overhaul Time Judgement

3.2 Remote Inspection by Image Processing

The static devices, for example pipes, pressure tanks and pump seals, are checked with non-destructive inspection such as visual test, ultra sonic test, eddy current test and so on in periodical outage for overhaul. The visual test, which is mainly used of them, is performed by a human expert. Thus the accuracy of the test depends on human skill and there is the fear that the method is not able to keep the quality. The authors proposed automatic visual test method by using digital image processing. Figure 4 shows an example of remote inspection service by image processing technique. Photographs of the parts that should be inspected are taken by digital 2-dimensional or 3-dimensional camera. Material defects such as corrosion area or crack depth are measured by feature extraction processing. Feature data are encoded and sent to Toshiba office for further analysis. Consequently, users will get inspection results of defect extension trend, present deterioration degree, appropriate replacement time and so on. This remote inspection service has been carried out since last year.

3.3 Remote Vibration Monitoring

The maintenance of the pumps that are typical dynamic devices is very important and they require a periodical overhaul to exchange the article of consumption, such as bearing and seal. The condition-based maintenance method is adopted to optimize overhaul period, by monitoring vibration temperature and lubricating oil componential analysis. Especially, vibration monitoring of the pump is very important in order to evaluate the aged deterioration degree, to detect symptom of anomaly and to decide next overhaul outage time. The authors developed the portable instrumentation device that measures vibration of the rotating machines and transmits data to the maintenance center via Internet. The maintenance personnel can gather the vibration data in the field easily and examine the signal trend and frequency characteristics in the power station office. They can be offered an expert's consultation in the distant maintenance center when further investigation is needed. The combination of the new instrumentation tool and human experts realizes high accuracy analysis without the limitation of the distance and personnel.

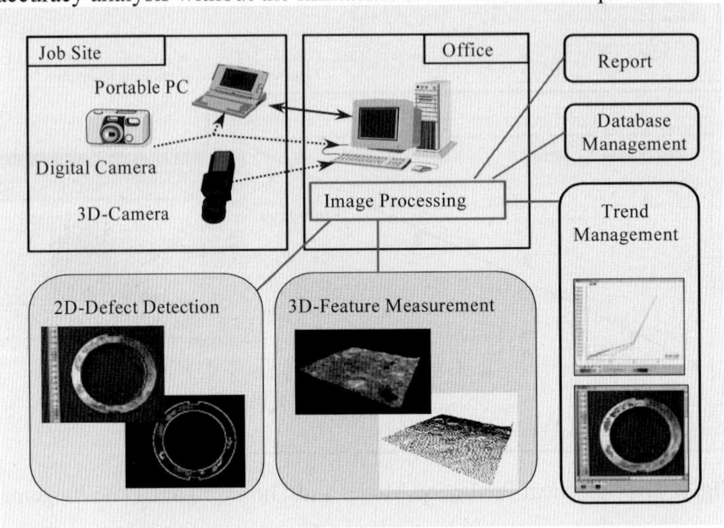

Fig. 4: Visual Test Service by Image Processing

4 Conclusions

The authors have developed the remote maintenance service system for a nuclear power plant, which offers the automatic visual test by image processing, the vibration monitoring for rotating machines and condition-based maintenance application on the Internet. This system has been being applied to several plants from a few years before. The activity of condition-based maintenance application will be continued and scope of target devices will be extended.

References

Sonoda, Y., et al., "Development of Device *Karte* Management System and Deterioration Prediction Method", 2001 Fall Meeting of the Atomic Energy Society of Japan, H22.

Sonoda, Y., et al., "Application of Condition Based Maintenance to Nuclear Power Plants", International Symposium on Artificial Intelligence, Robotics and Human Centered Technology for Nuclear Applications, 2002 Tsukuba, Japan.

Kanemoto, S., et al., "Remote Monitoring and Diagnosis System for Nuclear Plant", The Institute of Electrical Engineers of Japan, Annual Conference 2003.

Implication of Cognitive Style Questionnaire-MBTI in User Interface Design

Su, Kuo-Wei[1], *Hwang, Sheue-Ling*[2], *& Lee, Szu-Hsien*
[1]Department of Accounting, Takming College, Taipei, Taiwan, R.O.C.
[2]Department of Industrial Engineering, National Tsing-Hua University, Hsin-Chu, Taiwan, R.O.C.
kwsu@mail.takming.edu.tw

Abstract

The user interface should consider the individual difference of the users. For the complex system, the user interface should support the users' information processing style. The MBTI questionnaire had been used to indicate people styles of perceiving and judging information in many researches. We anticipate the users with different cognitive styles in sensing-intuitive dimension may influence the information processing. The computer-based procedures in nuclear power plant were taken as the experimental system. The expectative results are the different styles in information processing should be considered in user interface design.

1 Introduction

It is well recognized that the nature and quality of users' interaction with computers have been an increasing concern at the present day. Understanding of the user's individual difference is a critical component in the development of user-centered interface design. Smith and Dunckley (1998) described the effects of the "user diversity" in share interfaces design: interfaces which do not take account of different styles of cognitive decision making will lead to poor levels of learning and exploitation. Sifaqui (1999) also said: "Taste and behavior of each individual differs. This simple observation is presented in every action of each individual. Each individual, therefore, interacts in a different way, in congruency to his model of the world and to his previous experiences." Thus, a well-designed user interface shall consider the user's individual difference in interacting with computer.

Fuchs-Frohnhofen, Hartmann, Brandt, and Weydandt (1996) said: "An important aspect of human-centered system design is cognitive compatibility, which means that the structure of the human-machine interface of the computer should match the cognitive structures of the users." Benyon (1993) discussed five analysis phases that need to be considered when designing adaptive systems. One of these five analyses is user analysis. It is concerned with obtaining attributes of users that are relevant to the application such as the required intellectual capability, cognitive processing ability, and prerequisite knowledge requirements. According to these researches, a suitable interface for users should consider the traits of different users, especially the cognitive factors.

Many researches have been conducted in designing the interface to enhance the interaction between human and machine. Based on these studies (Berrais, 1997; Spence & Tsai, 1997; Doyle et al., 1998; Smith & Dunckley, 1998), we noticed the importance of individual differences for interface design and tried to use the cognitive style traits to designing suitable interfaces for different information processing styles.

Furthermore, the accuracy and the immediacy in the complex human-machine system are very important. Getting appropriate information promptly is a serious requirement for such systems.

The nuclear power plants (NPP), for example, have many operating procedures with symptom orientation and the operators must attend to the unusual parameter to decide which action should be done. In conventional NPP, the procedures were described on paperback document. There was no other way to find the procedure except finding the proper page. But in a modern NPP with computer-based procedures, the video display units (VDUs) display limited information one at a time and the operators may check the procedure in many different routes. To find the appropriate information and the corresponding procedure, the operators shall scroll the screen of visual displays to find out the usable information. Thus, the information perceiving preference of different operators will influence the performance of emergency operating procedure. According to the cognitive style researches, the information perceiving preference shall due to the difference of cognitive style.

2 The cognitive style

According to the definition of Sage (1981), the cognitive style was "the behavior that individuals exhibit in the formulation or acquisition, analysis, and interpretation of information or data of presumed value for decision making". Also, Messick (1984) defined cognitive styles as consistent individual differences in organizing information, and processing both information and experience. He described cognitive styles as generalized habits of thought.

By the discussion of cognitive style paradigm, Keen and Morton (1978) emphasized the correlation of cognitive style and problem-solving process. Scholl (2002) said, "one's cognitive style generally operates in an unconscious manner, he or she is often unaware of the mental processes used to acquire, analyze, categorize, store, and retrieve information in making decisions and solving problems". Thus, we can consider the cognitive styles as the habits or the preferences of individuals to formulate, analyze, interpret, even categorize, store and retrieve information or data for decision making and problem solving. In a simple description, cognitive styles are the ways of how individuals process information.

2.1 The Myers-Briggs Type Indicator (MBTI)

The Myers-Briggs Type Indicator (MBTI) is an instrument designed to measure four dimensions of an individual's cognitive style: introversion/extraversion, sensing/intuitive, thinking/feeling, and judging/perceiving (Myers & McCaulley, 1989). The MBTI is based on Jung's theory of psychological types. Each of the poles represents opposite preferences, and an individual prefers one pole to the other for every pair of items. Introversion/extraversion refers to a person's orientation or attitude toward the world. Sensing/intuitive refers to ways in which a person gathers or perceives information. The thinking/feeling dimension relates to functions of processing information and ways in which a person makes decisions. The judging/perceiving dimension refers to a person's orientation to handle the outer world.

2.2 The implication of cognitive style for interface design

2.2.1 Adaptive design for different cognitive styles

Smith and Dunckley (1998) used the term of "user diversity" to describe the starting point of developing shared interfaces by the Logical User Centred Interface Design (LUCID) method. The first element that should be taken into account is the different styles of cognitive decision making, otherwise it will lead to poor levels of learning and exploitation. Smith et al. (1998) have classified user factors into the following types when they used the LUCID method to design the shared interfaces.

- The objective factors, such as gender, age, and mother tongue et al.
- The subjective factors, such as cognitive style and user attitude.

2.2.2 *The influences of the four dimensions in interface design*

DiTiberio and Hammer (1993) concluded, "Although all four of these scales can be found, only two are commonly used in research, the S/N and T/F scales." Spence and Tsai (1997) used the terms, analytic and non-analytic, to split the subjects in information-seeking research. The analytic group contains subjects with a thinking type of cognitive style, whereas the non-analytic group is represented by the feeling type. Summarizing the influences of four dimensions of MBTI, the sensing/intuitive and thinking/feeling dimensions play the key roles of information perceiving and problem solving. Thus, there is no necessary to design the different processes of information display or problem solving for the users with different types in I/E and J/P dimensions.

According to the result of literatures review, the main influences of different cognitive styles on interface design are due to the differences of the S/N and T/F dimensions in MBTI. Cheng (1986) had mentioned that the cognitive styles are neither constant nor unchangeable; they may change through learning and training. The information perceiving differences due to the distinction of the sensing styles and the intuitive styles may fundamentally influence the interface design in information display ways. We will design two kinds of interfaces for the ST types and the NT types.

3 Interface design

According to the model which proposed by Furuta and Furuhama (1999), the cognitive space of the operator's knowledge may be divided into four subspaces: configuration subspace, causality subspace, state subspace, and goal subspace. Berrais (1996) proposed the user requirements and the system requirements must be defined before establishing the user model. Depending on the user model, the user interface would be designed to support the users. Considering these two proposed models, the interface design model is illustrated in Figure 1.

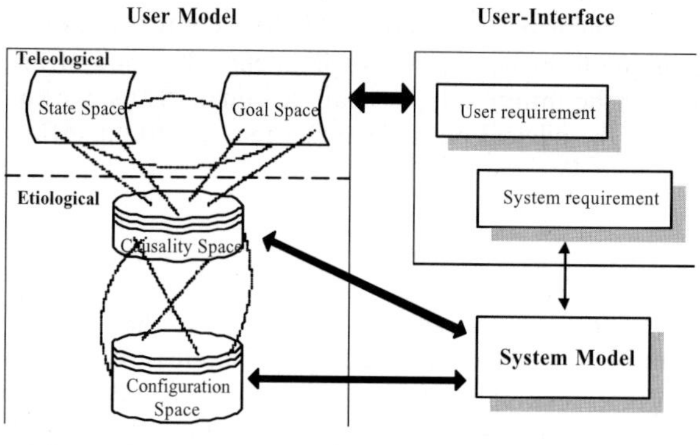

Figure 1: A Recommended User interface model.

The interface of computer-based procedure will be divided into three areas: current procedure information, parallel information and user prompts (see Table 1) . The current procedure information area indicates the status of the current step and the supporting sub-steps. Since the sensing types seeing the differences between two concepts and disassociating process with goals, we design the functional borders to distinguish different procedures for sensing types. The parallel information area is used to display information that needs to be monitored in parallel (e.g., water level, pressure). Since the intuitive types of operators prefer to view things globally and seeing the integration between parts of a system, we may only show the parallel information which alarms to intuitive types operators. The user prompt area will display the possible solutions or reasons for operators. Considering the difference of gathering and perceiving information between sensing and intuitive types (inductive and deductive), the prompts will quite different. The user prompts area may display the possible reasons for sensing types; because of the sensing types learn things experientially and may process alarm depending on experience. On the other side, the user prompts area may display the possible solutions for intuitive types because the intuitive types of operators learn things theoretically and may think too much when handling the situation.

Table 1: The different CBP design for sensing and intuitive types

	Current procedure information	**Parallel information**	**User prompts**
Sensing types	Functional borders	-	Reasons for situation
Intuitive types	-	Only showing alarm information	Solutions for situation

4 Methodology

4.1 Pilot study of MBTI

Before starting the regular experiments, we used the MBTI questionnaires as a pilot study to find out the cognitive styles of the participants. Since we want to discriminate the sensing (S) and the intuitive (N) types, we should obtain enough participants in both sensing type and intuitive type. As mentioned before, the introversion/extraversion and the judging/perceiving dimensions have few impacts in using the user interface. So, we distinguished the participants into two groups, the sensing type and the intuitive type, and the thinking/feeling dimension would be still under consideration for understanding the influence of training and balancing the performances between the sensing types and the intuitive types. Though we would not design the specific interfaces for the thinking and feeling dimension, we thought the different traits of the thinking and the feeling types will influence the performances of the experiment. The rate of the thinking/feeling types with the sensing types or the intuitive types shall be equal. Thus, we can consider the different performance of operating the nuclear power plant system due to the difference of the sensing and the intuitive.

4.2 Experiment

For examining the design principles based on the difference of the cognitive styles, we will use the experiments to find out the utilities of these design principles. We will use Visual Basic 6.0 to setup the simulate interface of computer-based procedure. According to the design principles that previously mentioned, we would design two specific interfaces for sensing and intuitive persons. Furthermore, the participants will be asked to write a questionnaire after they finish the experiment. The questionnaire includes opinions and suggestions of the participants, and more important, how they feel about the interfaces.

4.3 Expected results

The interfaces which have designed depending on the information processing styles will be used to test if the cognitive style principles are worked. We expected the user interface which designed for the specific cognitive style will make the participants to control the simulation system more easily. The results of the experiments will offer the evidence to prove the effect of these design principles by faster reaction time and lower error rate.

According to the results of the questionnaires, the interfaces may be modified by the participants' opinions and suggestions. And the design principles which based on the cognitive style traits will be resurveyed. We hope the interface design principles could be used not just in NPP area, but could be implicated extensively in any complex system.

References

Benyon, D., Adaptive systems: A solution to usability problems, J. *User Modeling and User-Adapted Interaction 3(1), 1993, p. 1-22*

Berrais, A. (1997). Knowledge-based expert systems: user interface implications. *Advances in Engineering Software 28, p. 31-41*

Cheng, Y. P. (1986). An analysis of cognitive style between designers and users in computerized information system-empirical study. *A thesis submitted to Institute of Management Science College of Management National Chiao-Tung University, Taiwan.*

DiTiberio, J. K., and Hammer, A. L. (1993). Introduction to Type in College. *Consulting Psychologists Press, Inc.: Palo Alto, CA., p.1-30*

Doyle, J. K., Radzicki, M. J., Rose, A., and Trees, W. S. (1998). Using cognitive styles typology to explain individual differences in dynamic decision making: Much ado about nothing. *Center for the Quality of Management Journal, Report No.13*

Fuchs-Frohnhofen, P., Hartmann, E. A., Brandt, D., and Weydandt, D. (1996). Designing human-machine interfaces to match the user's mental models. *Control Eng. Practice, 4, p. 13-18*

Furuta, K., and Furuhama, Y. (1999). Cognitive space of operator's knowledge. *Ergonomics, 42, p. 1429-1439*

Keen, P. G. W., and Morton, S. (1978). Decision support systems: An organizational perspective. *Addison-Wesley Reading, Mass., 1978.*

Messick, S. (1984). The nature of cognitive styles: Problems and promise in educational practice. *Educational Psychologist, 19(2), p. 59-74*

Sage, A. P. (1981). Behavioral and organizational considerations in the design of information systems and process for planning and decision support. *IEEE Transactions on Systems, Man, and Cybernetics, 11(9), p. 640-677*

Scholl, R. W. (2002). Cognitive style and the Myers-Briggs Type Inventory (MBTI). http://www.cba.uri.edu/Scholl/Notes/cognitive_style.htm (2002/08/27)

She, Y. M., & Hwang, S. L. (1996). The approach of user's mental model and cognitive type in process control system, *Taiwan, ROC: National Tsing-Hua University Press.*

Sifaqui, C. (1999). Structuring user interface with a meta-model of mental models. *Computers & Graphics, 1999 Vol.23, p. 323-330*

Smith, A. and Dunckley, L. (1998). Using the LUCID method to optimize the acceptability of shared interfaces. *Interacting with Computers 9, p. 335-345*

Spence, J. W., and Tsai, R. J. (1997). On human cognition and the design of information systems. *Information & Management 32, p. 65-73*

Information Provision for Maintenance Work with Distributed DB Framework

Makoto Takahashi , Yo Ito, Hisashi Sato and Masaharu Kitamura
Department of Quantum Science and Energy Engineering, Tohoku University
Aramaki-Aza-Aoba01, Aoba-ku, Sendai, 980-8579 Japan
makoto.takahashi@qse.tohoku.ac.jp

Wei Wu and Tadashi Ohi
Advanced Technology R&D Center
Mitsubishi Electric Corporation
8-1-1, Tsukaguchi-honmachi, Amagasaki City, Hyogo 661-8661, Japan
Wu.Wei@wrc.melco.co.jp

Abstract

The framework of intelligent support system for the maintenance of nuclear power plant is proposed in this paper with the emphasis on the combined use of a portable device and an intelligent information processing. Routine monitoring activities for working components, which are supposed to be equipped with local monitoring devices connected to in-house network, are main target of the present study. In the present framework, the monitoring activities are performed in coordination among intelligent agent with autonomous moving capability and maintenance personnel carrying portable information device. The specific features of the present study are the use of distributed database with the effective search strategy and the realization of the flexible diagnosis process through the portable device on-site. The prototype system has been developed to show the basic functionality of the proposed framework. The proposed framework is expected to contribute to increasing effectiveness of the maintenance activities for condition-based monitoring, which also lead to the higher level of safety.

1 Introduction

Significant improvements in maintenance activities are necessary in order to meet the higher level of reliability and safety required to nuclear power plant operation. In addition, the cost-effectiveness has become major issue because of the competitive situations with other energy sources. The introduction of condition-based maintenance (CBM) has become an issue of particular interest because of the reasons given below. First, social impact of a plant anomaly is becoming higher and higher irrespective of the actual severity of the anomaly. Second, prevention of the fault is generally more cost-effective than post fault recovery. These expected benefits clearly justify the employment of the CBM, which should be supported by more advanced techniques of information management. A large portion of technical activities in maintenance execution has a common nature with the operation related task such as surveillance and diagnosis activities. Several attempts have been already made to integrate these activities (Hallbert,1995:Kitamura,1996) traditionally treated more or less independently. The authors believe that much more efforts should be devoted to the improvement of the information provision for maintenance work by fully utilizing the rapidly evolving information processing paradigms such as distributed DB and mobile agent (Washio,1995).

The framework of an intelligent support system for the maintenance of nuclear power plant is proposed in this paper with the emphasis on the combined use of a portable device and an intelligent information processing. Routine monitoring activities for working components, which are supposed to be equipped with local monitoring devices connected to an in-house network, are main target of the present study. In the present framework, the monitoring activities are performed in coordination among intelligent agent and maintenance personnel carrying portable information device, which is able to communicate with the local monitoring device. The specific features of the present study are the use of distributed database with the effective search strategy and the realization of the flexible diagnosis process using the portable device on-site.

2 Basic Framework

2.1 Strategy for Information Provision

Along with the quick advance of an information technology, it is quite natural to consider the possibility of applying it to support maintenance activities in the nuclear plant. In almost all of the nuclear power stations, PCs are utilized as the tool for summarizing maintenance results and for making official documents. In some advanced cases, the portable information devices are already utilized to collect maintenance related information and the collected information can be transferred to the host computer through the in-house network. This type of off-line use of information device is effective to reduce task load and possible errors. In the present study, it is aimed to show that the combined use of mobile agents and distributed database can improve the quality and reliability of the maintenance work significantly.

Figure 1 shows the overall framework of the proposed information provision system for maintenance support. The proposed system is composed of following components:

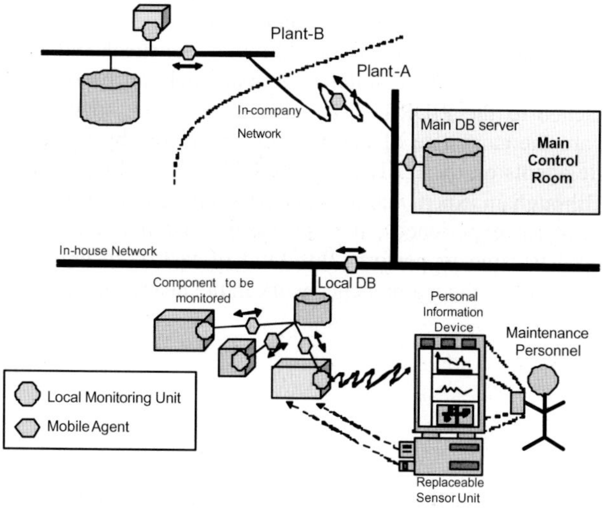

Figure 1: Overall framework of the proposed information provision system

- Central Monitoring Server(CMS)
- Local Monitoring Unit(LMU)
- Local DB server(LDB)
- Portable Information Device with Replaceable Sensor Unit(PIDRS)

The essential information to be handled in the CMS includes the records of; scheduled inspection and testing, unscheduled (anomaly-based) inspection and testing, repairs of components and devices, etc. The symptoms of experienced anomalies are also regarded to be rich sources of information and thus needed to be stored as a symptom database. The CMS is dedicated to support supervisory and higher-level decision-making. The CMS also takes the role of long-term data back-up since the memory size of the LMU cannot be sufficiently large even taking into account the rapid growth in available memory size. Each distributed LMU unit, equipped with a local processor, is assigned to take care of integrity of a predefined set of components. The LDB is used for storage of symptoms, activity records, maintenance procedural guides and related drawings. These knowledge and data are provided to local maintenance personnel upon request presented on the PIDRS. Furthermore, the maintenance personnel can make use of replaceable sensor unit to conduct supplementary measurements for better decision-making.

The advances in maintenance information management naturally necessitate more careful consideration on HMI, since increase in the amount and diversity of information is useful if and only if the information is provided in a systematic and coherent manner. In other words, special considerations must be given to the design of the information display system assigned to the PIDRS as well as the CMS. A particular emphasis is needed to visualize functional implication and contribution of each component provided by each LMU to the overall plant functionality.

The introduction of an on-demand sensing using PIDRS is another specific feature in the present framework (Takahashi, 2000). Routine monitoring of each component is supposed to be performed by a mobile agent usually residing on the LMU. It is responsible for the periodical monitoring of the objective component. When it detects considerable deviation in any of the measured signals or in the characteristic features, such as frequency spectrum, it calls for the on-site monitoring by the maintenance personnel. In addition to the check and diagnosis task through the interaction window of the PIDRS, maintenance personnel can perform on-demand sensing using the sensors attached to the PIDRS. Mechanical vibration, sound, surface temperature, atmospheric gas, etc. are the candidate of the modality measured by the additional sensors and analyzed by the mobile agents on the LMU or on the CMS, depending on the complexity of the analysis procedure. Although an expert maintenance personnel can perform this kind of diagnosis task based on his long-term experiences, it is necessary that maintenance personnel without specific expertise should be able to perform this kind of task with high reliability when less number of people are assigned to the maintenance work in the near future.

2.2 Methods

2.2.1 Mobile Agent

Mobile agents play a key role in the proposed framework of information provision. Mobile agents are basically similar to software modules which can move along the network to perform the specified task. The specific functions assigned to the mobile agents are:

(a) Periodical Monitoring of plant parameters
(b) Signal analysis for detecting anomalies
(c) Signal storing and retrieval on DB
(d) Diagnosis of malfunctions
(e) Find similar events in the DB

When any anomaly is found with the periodical monitoring by the agent residing on LMU, it calls for another agent to perform further detailed analysis. The maintenance personnel have only to send command through PIDRS in order to execute the pre-defined diagnosis procedure. Mobile agent plays another role as the search agent for the similar anomaly events stored in the database. Finding similar event previously experienced is one of the typical requirements of the personnel who face the problem to be solved. The important point here is that users don't have to indicate specific database to be searched. Instead, the mobile agent performs autonomous search starting from LDBs, ending in the database in other plant. With the help of intelligent mobile agents, the burden of the information manipulation can be drastically reduced.

2.2.2 Distributed database

In the present framework, it is assumed that the database storing various kinds of information should be distributed. For example, each components or each sub-system should be equipped with local database server storing specific information on the local component. The process variables measured and transferred by mobile agents are supposed to be stored only on the LDB unless no anomaly is found. Only when any anomaly is found or agent for data retrieval is visited, stored data is transferred to other database server. The advantage of such distributed database is twofold: one is the reduction of the amount of data on the network and the other is the tolerance to the network malfunction.

2.2.3 Search based on diverse similarity (Diantono,2000)

Finding out the similar event in the database is quite important as the first step action to solve the problem. It is rather difficult, however, to define the similarity. For example, when the noise in the monitored signal is at issue, the similar signal in terms of the time series should be found as the similar event. When the defect of the rotating machinery is at issue, similarity of the frequency spectrum should be compared to find out similar events. The similarity should be defined in the diverse characteristics and the search should be performed in terms of such diverse attributes.

3 Prototype System

The prototype system showing the basic function of the proposed framework has been developed in the PCs connecting to the local network. In each PC, the middleware called Aglet (Developed by IBM Inc.) for realizing the mobile agent has been installed. PC1 plays the role of CMS and PC 2 plays the role of LMU. It has been confirmed that the developed system is capable of transferring process parameters by using the mobile agents. Figure.2 shows the example display image on the PIDRS. As the PIDRS is now under construction, it only shows the display for the interaction performed by the mobile agents. The display window can be changed by TAB, which enables user to quickly change the content of the displayed information. It has been confirmed that the developed interface design successfully provides the user-agent interaction on the limited area.

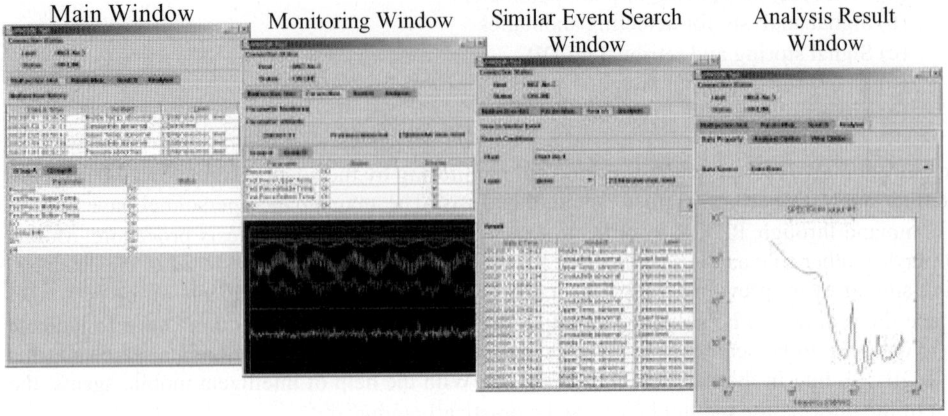

Figure 2: Example display images for PIDRS

4 Conclusion

The framework of intelligent support system for the maintenance of nuclear power plant is proposed with the emphasis on the combined use of a portable device and an intelligent information processing. The specific features of the present study are the use of distributed database with the effective search strategy and the realization of the flexible diagnosis process through the portable device on-site. The prototype system has been developed to show the basic functionality of the proposed framework. The proposed framework is expected to contribute to increasing effectiveness of the maintenance activities for condition-based monitoring, which also lead to the higher level of safety.

5 References

Hallbert,B and Meyer,P. (1995). Summary of Lessons Learned at the OECD Halden Reactor Project for the Design and Evaluation of Human-Machine Systems, Proc. Topical Meeting on Computer-Based Human Support Systems: Technology, Methods and Future, Philadelphia, 407-413.

Kitamura,M., Furukawa,H, Kozma,R & Washio,T. (1996) . Guiding Rules for Development of Intelligent Monitoring System of Nuclear Power Plants", Proceeding of SMORN VII; A Symposium on Nuclear Reactor Surveillance and Diagnosis 2, OECD, 493-501.

Washio,T, & Kitamura,M. (1995). Worm-Type Agents for Intelligent Operation of Large-Scale Man-Machine Systems, Advances in Human Factors/Ergonomics 20A (Selected papers from Proc. HCI'95; 6th International Conference on Human-Computer Interaction), 925-930.

Takahashi,M., Takei,S., Tani,H., &Kitamura,M. (2000). INTELLIGENT MOBILE SENSING FOR DYNAMIC FAILURE IDENTIFICATION, International Topical Meeting on Nuclear Plant Instrumentation, Controls, and Human-Machine Interface Technologies (NPIC&HMIT 2000), Washington, DC,November,2000. (CD-ROM), 1, 572 - 580.

Diantono,C., Takahashi,M., & Kitamura,M. (2000). MULTI-MODAL DATABASE FOR INTELLIGENT MAINTENANCE SUPPORT FOR NUCLEAR POWER PLANT", International Topical Meeting on Nuclear Plant Instrumentation, Controls, and Human-Machine Interface Technologies (NPIC&HMIT 2000), Washington,DC,November, (CD-ROM), 1, (2000), 581 - 590.

Navigating Data – Selections, Scales, Multiples

Martin Theus

Augsburg University
Universitätsstr. 14, 86135 Augsburg, Germany
martin.theus@math.uni-augsburg.de

Abstract

Traditional graphical data analysis was a tedious and static process. The construction of graphs, especially maps, was a time consuming and unique job. The rise of computers, printers and plotters and finally graphical monitors, allowed a much easier and faster generation of all kinds of graphics. But yet the simple substitute of the manual process by a machine did not reveal new insights in an analysis, though it did speed it up. New graphical representations have been developed over the last 20 years, which did benefit substantially from growing computer power.

Today graphical data exploration becomes more and more interactive. The analyst is now able to select subgroups, change scales and look at multiple plots and representations of the data at the same time. This paper investigates strategies for navigating through complex data sets in the light of advanced interactive computer human interfaces.

1 Introduction

The analysis of data can be viewed from many different angles. Statistician will usually make more use of mathematical methods and distribution assumptions. A data miner will use various computational methods "out of the box", and the domain expert might stick to traditional analysis methods, which might not really suite the problem, but are well accepted in the community. No matter from which angle we view the problem, in all cases a solid understanding of the data is necessary to get reliable results.

Without doubt, graphical analysis methods will give the broadest insight into data. As long, as the data is small, the number of possible and sensible representations is quite limited and we might be able to look at all of them. With larger data, i.e. >10k observations and >20 variables, it is getting harder to investigate the various possible relationships. One solution to overcome this problem are automated data mining methods. Unfortunately these methods usually are not able to use the domain expert's knowledge towards relevant findings. Another approach is to make the graphical analysis of the data more efficient. The next section will look at the three pillars of interactive data analysis: selection mechanisms, scale changes and multiple displays.

2 Navigating Data

In an interactive data analysis environment three important operations can be identified. Selection of data, the change of scale of a view (which is often referred to as "zoom"), and multiple views of the data. These operations are crucial for most interfaces, used to retrieve information. The reader may map the 3 operations to the problem of searching for information in the the world wide web.

2.1 Selections

Selecting data – or more specific, selecting a subgroup of interest – makes most sense, if we can link the selection information to other representations of the data.

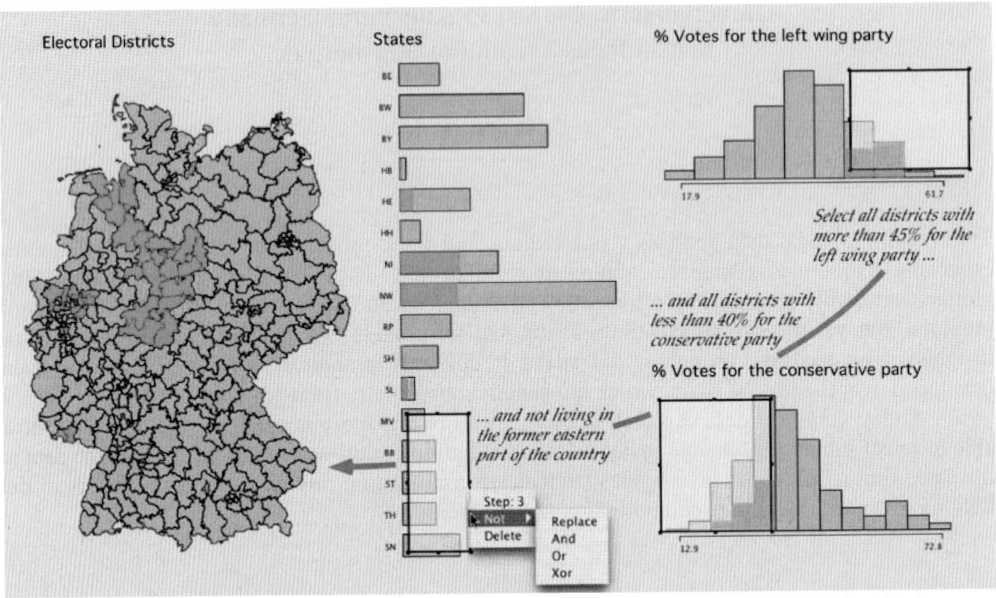

Figure 1: Data on the 2002 election in Germany. A 3-step selection is performed. Beginning with all electoral districts, having more than 45% of votes for the left wing party, the selection has been narrowed to all districts with less than 35% for the conservative party. The final selection excludes all states from the former eastern part (Map: © Statistisches Bundesamt, Wiesbaden).

Figure 1 shows an example of linked highlighting of a selected group. The selection spans over three different plots: two histograms and a barchart. The resulting subset is monitored in a map.

Two important aspects of selecting data are the flexibility of specifying a subset, and the ability to reproduce a selection. Setting up complex selections is usually achieved by stepwise refining a selected subset. This is done by combining a currently selected subset with a new selection via Boolean operators. In the example of Figure 1, the selections have been combined as *set1* **AND** *set2* **AND NOT** *set3*. To keep track of the different selection steps, the selected regions can visually be emphasized as shown in Figure 1. These *selection rectangles* (Theus et al., 1998) can be used as brush to apply dynamic selections. Each selection rectangle carries a selection mode (e.g. **AND, OR, XOR** ...) and can be altered in order to specify the selected subgroup precisely. Returning to the example of Figure 1, we could easily change the selected states, without having to redefine the selection sets from selection step one and two.

In the context of large data sets, which can only be accessed via SQL-queries, this selection mechanism can easily be translated into database queries. For the selection in Figure 1 we would yield: `SELECT districts FROM election WHERE (((SPD > 45) AND CDU < 35) AND NOT state IN (MV, BB, ST, TH, SN)).`

2.2 Scales

As long as we only look at small sized data of little complexity, there is no need to investigate the data at different scales. There are basically two situations, where an analysis must make use of scaling operations. The first situation – best known from geographical data analysis – is data, which was measured at different resolutions, building a hierarchy (e.g. census data on states, counties, districts, roads, houses, households). The second one can be found with very large data sets. Very large data sets tend to have many extreme outlier, which obscure plots like scatterplots very much. Either the plot is overwhelmed by the massive amount of data, or 99% of the data are concentrated on only 1% of the plot's real estate and vice versa.
In general we may distinguish two kinds of scale change/zoom operations[1]. The simple zoom is a pure graphical operation. Simple graphical zooms do not change the data, which is displayed, nor the representation which is used, but only rescale the plot to new bounds. More sophisticated zoom operations either switch between different resolutions of the data or different representations for the data displayed, or both. These more sophisticated zoom operations are often referred to as *logical zooms*.

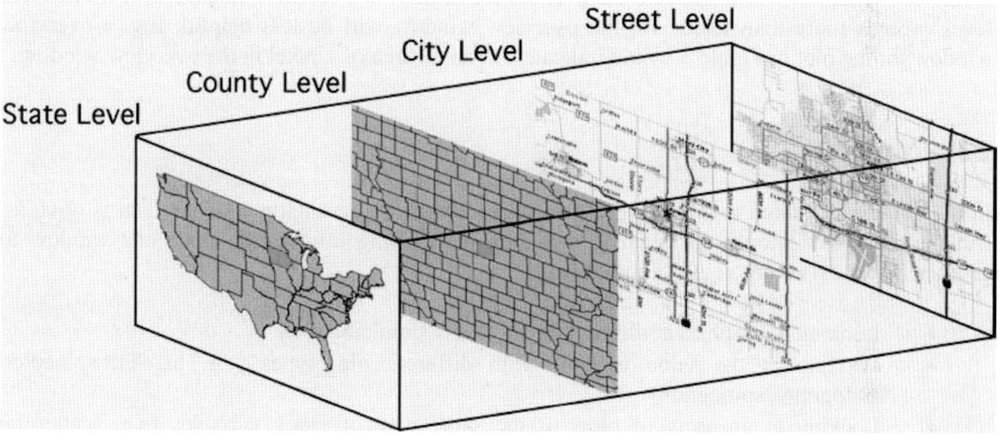

Figure 2: An example of logical zooming in maps. Starting with a map on the US states, the zoom takes us down all the way to the street level of Ames, IA.

Logical zooming usually gives a high level of guidance due to their definition. When e.g. zooming from a view on state level to a view on county level, we would expect to zoom in on a specific state. This guidance is crucial when zoom level go beyond a factor of 10-100. But even for a simple graphical zooms this guidance should be offered by the system. One method of support is the so called *hierarchical zooming*. Hierarchical zooming keeps a list of all zoom stages. If the user decides to zoom back he/she is taken to the exactly same scale of the previous zoom step. This avoids generating views, which have not been specified by the user.

Another application for logical zooming can be found in standard statistical graphics like scatterplots or mosaic plots. Scatterplots can be binned if they have to deal with hundreds of thousands of cases. Zooming in on a specific detail, will switch back to the display of raw points (compare Figure 3 for an example). Mosaic plots will only display the very first variables in the

[1] The change of scale in a plot is often referred to as zooming, which shall be regarded as synonyms.

recursion. As the user zooms in on a particular subgroup, more and more variables can be displayed without overloading the display.

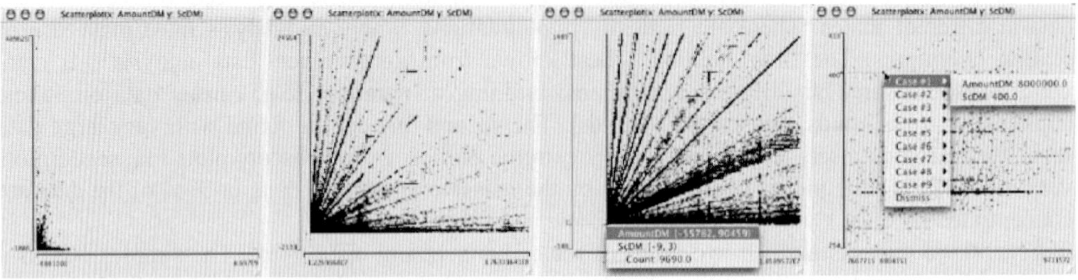

Figure 3: An example of logical zooming in scatterplots. Starting with a binned scatterplot in the leftmost plot, zooming in by a factor of more than 1,000 takes us to a detailed view of the data in a traditional scatterplot (rightmost plot).

Overview windows, which indicate the zoomed area can give a certain guidance. But if the zoom level exceeds more than factor 10, the overview window will be less helpful. E.g. a overview window for the plot in Figure 3 would indicate a zoomed area of 1 pixel in the overview window.

2.3 Multiples

In the context of linked highlighting the need of multiple representations of the data is obvious. But multiple view should go much further than just linking information from one window to another. The most common use of multiples are:

- Looking at graphics at different zoom levels simultaneously
- Looking at the same variable with different plot types (e.g. barchart/spineplot, histograms/spinogram)
- Looking at an array of plots of the same type of many variables (e.g. scatterplot matrices)
- Looking at multivariate data from different perspectives (e.g. grand tour in scatterplots and mosaic plots)

All these operations can be achieved by using standard window interface operations. The arrangement and scale of the windows needs a lot of thought.

Figure 4 gives an example of multiple views of the same data for the Titanic passenger data. In the left plot the variables gender has been conditioned upon class and age. It nicely shows the decline of survival rate from 1st class to crew members. Rates for females are always higher than for males. All children in 1st and 2nd class survived. A result we would expect from the old adage “Women and children first!”. Rearranging the plot makes a deviation from this pattern more clearly. Looking at the classes by age and gender, we see a different pattern of the survival rate for adult men than for adult women. There survival rate for adult males is increasing from 2nd class towards crew. Only 1st class males stick out with an unusual high value. A result which was not immediately visible from the left plot.

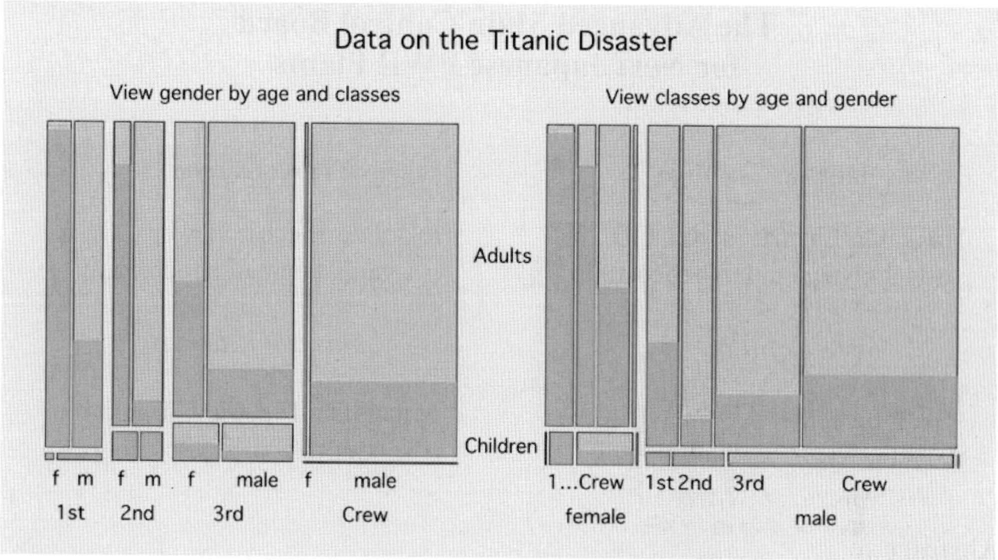

Figure 4: Two views of the Titanic data set using mosaic plots. For all 2201 passenger the variables *Class, Ages, Gender* and *Survived* have been recorded. Although the plots show exactly the same data, the left plot reveals other features than the right plot. All survivors are highlighted

3 Conclusions

Modern data analysis tools must provide various interactions to navigate through data. Today data sets are often large and of complex structure. Thus an analysis must look at very many different views on the data. The most crucial interactions are (i) to select interesting subgroups and features in the data, (ii) to look at data at different scales and aggregation and (iii) to look at multiple views simultaneously.

Unfortunately most commercial statistical packages and data mining tools still lack a lot of the described capabilities and do not support an analysis very well. Research software like *Mondrian* (Theus, 2003), *TimeSearcher* (Hochheiser & Shneiderman, 2002) and *Polaris* (Stolte et al., 2000) do implement some of the above mentioned ideas and suggestions and hopefully will influence new products in the future.

References

Chris Stolte and Pat Hanrahan. (2000) Polaris: A System for Query, Analysis and Visualization of Multi-dimensional Relational Databases, Proceedings of the Sixth IEEE Symposium on Information Visualization, October 2000. http://graphics.stanford.edu/papers/polaris/

Hochheiser, H, Shneiderman, B (2002). Visual Queries for Finding Patterns in Time Series Data, Poster Compendium of the 2002 Symposium on Information Visualization, 22-23.

Theus, M, Hofmann, H, Wilhelm; A (1998). Selection Sequences: Interactive Analysis of Massive Data Sets. *Proc. of the 29th Symposium on the Interface: Computing Science & Statistics,*

Theus, M. (2003). Mondrian – Interactive Statistical Graphics in JAVA. Retrieved February 15, 2002 from http://stats.math.uni-augsburg.de/Mondrian

The Advanced Main Control Board for Next Japanese PWR Plants

Akiyoshi Tsuchiya[1]

Hokkaido Electric Power CO., Inc
a-tsuchiya@epmail.hepco.co.jp

Takashi Yano[1]

Hokkaido Electric Power CO., Inc
t-yano@epmail.hepco.co.jp

Koji Ito[2]

Mitsubishi Heavy Industries, Ltd.
koji_ito@atom.hq.mhi.co.jp

Masashi Kitamura[3]

Mitsubishi Electric Corporation
m-kitamu@pic.melco.co.jp

Abstract

The improvement of main control board layout for the PWR (Pressurized Water Reactor) nuclear power plant is accomplished to reduce the operator's workload and potential human error through the better working condition where he could utilize his abilities in maximum.

In designing the main control board the major concerns are to create the Human-System Interface (HSI) suitable to an integrated digital I&C system so that the operator can correctly perform his tasks; diagnosis of plant problems, choice of countermeasures and implementation of control actions.

The Japanese PWR owners and Mitsubishi Group have developed an advanced main control board (console) reflecting the past studies about human factors as well as employing a state of the art electronics technology. It consists of an operator's console, a supervisor's console and a common LDP (Large Display Panel) and adapts soft switches for the manipulation through VDUs (Visual Display Unit) with touch panels.

The functional specifications have been evaluated by utility operators using a prototype main control board driven by a plant simulator.

1 Introduction

It is important to provide a HSI system with which operators can easily pick up appropriate information among a large number of plant process parameters and correctly identify the plant state.

The design of main control room in Japan has been continuously improved from conventional single large board with hard-wired indicators and switches to functionally divided boards with CRTs as major information source (Saito, 1990).

[1] 2 Higashi 1-Chome, Ohdori, Chuo-ku, SAPPORO 060-8677 JAPAN
[2] 1-1, Wadasaki-cho 1-Chome, Hyogo-ku, KOBE 652-8585 JAPAN
[3] 1-2, Wadasaki-cho 1-Chome, Hyogo-ku, KOBE 652-8555 JAPAN

A fully digital I&C system including the advanced main control board is planned for the next Japanese PWR plant with a view to achieving increased safety, reliability, operability and maintainability.
It is desired to improve the safety and efficiency and to construct the HSI system suitable to the fully digital I&C system so that operators can correctly perform their tasks; diagnosis of plant problems, choice of countermeasures and implementation of control actions.
An advanced main control board, which has the following features, has been developed to meet the above-mentioned objectives.

- Full-time sit-down-operation console for reducing monitoring areas and traffic.
- Touch operational HSI system with VDUs.
- Plant information presentation, which should be shared by the shift supervisor and operators using LDP
- Automation of high workload monitoring tasks such as plant scram, and presentation of its results.
- Suitable space allocation among large display panel, operator console and shift supervisor console for smooth communication among all the operators.

This paper describes the design concept, enhanced operability features, system configuration and evaluation results of the advanced main control board.

2 Enhancement of Operability

The recent design improvement trends of main control board are obviously directed toward the soft operation utilizing computer driven HSI devices. The benefits of the soft operation are to supply relevant process information necessary for the implementation of control as well as providing appropriate process parameters for facilitating tasks, in addition to saving spaces of instrumentation and control switches. The advanced main control board consists of operator console, supervisor console and large display panel (see Figure 1).

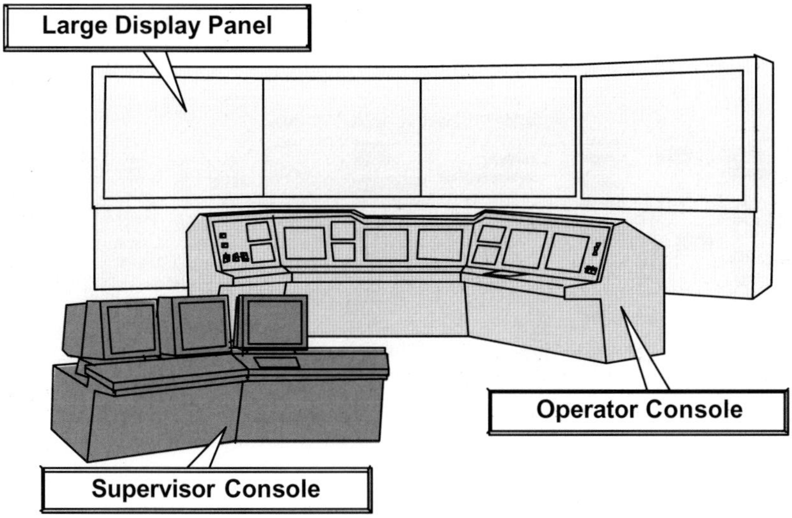

Figure 1: Configuration of the Advanced Main Control Board

The advanced main control board also enhances the operability by taking advantage of the soft-operation as described below

- Integration of monitoring and operation
  Operator's actions are composed of checking the standby-condition of equipment before his implementation, monitoring the plant parameters (the direct objects and relevant parameters) and verifying the plant behaviors after the implementation. In order to improve the operability of board all information necessary for the manipulation such as control-switch situations and relevant parameters are integrated onto the same screen. In addition, the operator's manipulations are automatically checked in comparison with the procedures in computer archive.
- Automatic verification of interlocks
  When a significant event like a plant scram or a safety injection in case of emergency (S.I.) would occur, the operator's workload and stress would become huge because he would be expected to conduct many manipulations at the same time such as collecting the safety-related information and confirming the plant conditions, etc. To put his actions in ease the interlocks between the components are automatically verified in the I&C system and the verified results are displayed on the VDUs and the LDP.
- Inter-linked display request
  All the monitoring and operation screens are inter-linked in terms of the functional and/or operational relationship. Another concerned screen can be shown in replace on the same display by requesting it in the previous one.
- Common large display panel
  The following functions are provided by the common large display panel on the back wall of the control room so as to manage the operation (see Figure 2):
  - Integrated display of important plant parameters in normal condition.
  - Effective sharing of information among the shift crew in abnormal condition
  - Easy recognition of the plant situation through the system graphic pattern view.
  - Easy recognition of important information by intelligent categorizing of the alarms.
  - Displaying timely the computer-checked results like an interlock of components and so on.

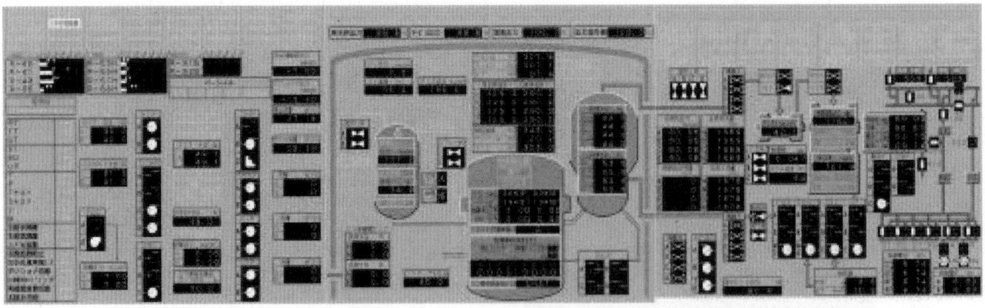

Figure 2: Large Display Panel

- Alarm handling system
  A dynamically prioritized alarm system is used for avoiding information overflow and facilitating plant state identification (Fujita, 1989). A prioritized alarm is provided in

compiling many simultaneous alarms and is displayed in VDUs and LDP with color coordination categorized by 3 levels. Moreover, the summarized alarm displayed on LDP can move to another rank depending on the importance of new-coming alarm, thereby when the more critical/important alarm activate, the plant status can be confirmed by LDP.

3 Operator Workload and Human Error Probability Evaluation

The effectiveness of improved HSI applied for the advanced main control board was quantitatively evaluated in terms of the operator's mental workload and potential human error in comparison with that of the HSI applied for conventional main control boards. The evaluation was carried out in complicated emergent situations. The result of review indicates that the integration of monitoring and operation, the automatic verification of interlocks and the inter-linked display request method could contribute greatly to the reduction of operator's load in case of these situations. It indicates that operator's load would be reduced approximately 30% and potential human error would be reduced approximately 60% in comparison with that of the HSI applied for the conventional boards (see Table 1&2). The model human processor was used to evaluate mental workload (Card, 1981) and THERP (Technique for Human Error Rate Prediction) was used in the valuation of human error rate ("NUREG-1278/CR", 1983).

Table1: Evaluation results of mental workload

Sequence	Result (Reduction)
Plant Trip	15%
S.I.	53%
B.O.	58%
Identify event (after S.I.)	0%
Cool down (after S.I.)	11%
Total	27%

Table2: Evaluation results of Human Error Rate

Sequence	Result (Reduction)
Plant Trip	28%
S.I.	96%
B.O.	76%
Steam generator tube rupture	75%
Start-up (normal condition)	18%
Total	59%

4 Validation Test by Shift Operator Crews

A prototype advanced main control board was prepared being connected to a full-scale plant simulator and the overall operability was tested by using it. This test was carried out to make sure that one shift supervisor and one operator could monitor and manipulate the whole nuclear power plant under any situations (see Figure 3).

Figure 3: Variation test facility of the advanced main control board

Operating crews from several nuclear power plants participated in and went thorough the validation under normal and accident plant conditions. The performances of the screens integrated with monitoring and operation functions, the inter-linked display request method and the alarm handling system were validated by means of checking the operator's sequential actions described in the procedures. The results showed the improvement of operability of the advanced main control board.

In addition, the operators' personal notes were gathered through interviewing them to confirm the qualitative evaluation (e.g., course of action for the recognition, easiness of monitoring and operation, etc). The notes suggested the design goals of operability improvement were achieved.

5 Conclusion

It is shown that the advanced main control board would be applicable to the next Japanese PWR. The validation results confirm the improved operability of the advanced main control board. The board is supposed to be installed in the Tomari-Nuclear Power Plant Unit 3 of Hokkaido Electric Power Company, which is expected to be in operation in 2008.

References

Card, S.K.(1981). The Model Human Processor: A Model for Making Engineering Calculations of Human Performance, Proceedings of 25th Annual Meeting of Human Factors Society, Human Factors Society, Santa Monica, U.S.A.

Fujita, Y. (1989). Improved Annunciator System for Japanese Pressurized-Water Reactors, *Nuclear Safety,* Vol.30 (pp. 209-221).

NUREG/CR-1278 (1983). Handbook of Human Reliability Analysis with Emphasis on Nuclear Power Plant Applications, U.S. Nuclear Regulatory Commission

Saito, M. (1990). Human factors Considerations Related to Design and Evaluation of PWR Plant Main Control Boards, International Symposium on Balancing Automation and Human Action in Nuclear Power Plants, IAEA -SM-315/59, Munich, Germany.

The Evaluation of the Graphical User Interfaces of Four B2C E-commerce Websites in Taiwan

Shiaw-Tsyr Uang

Department of Industrial Engineering and Management
Ming Hsin University of Science and Technology
Hsin Feng 304, Hsinchu, TAIWAN, R.O.C.
uang@must.edu.tw

Abstract

The purpose of the present research was to investigate the interface design features of four B2C e-commerce websites in Taiwan. These four websites are Pchome (shopping.pchome.com.tw), Coolbid (www.coolbid.com.tw), Gomos (gomos.sunup.net), and Bid (www.bid.com.tw). Both a videotaping retrospective method and questionnaires were used to collect the data from 15 students of Ming Hsin University of Science and Technology. Dependent measures were the averaged searching time, number of link errors, and subjective preference. Design features of the icons, including recognizability, discriminability, communicativeness, color, layout, and simplicity, were ranked across four websites. Videotaping retrospective analysis revealed that the averaged searching time for Pchome, Coolbid, Gomos, and Bid were 90.8, 142.5, 164.7, and 199.8 seconds, respectively. The number of link errors for Pchome, Coolbid, Gomos, and Bid were 7, 4, 9, and 12 during testing period. Pchome seemed to be preferred on its icon design features than the other three tested websites by the subjects.

1 Introduction

The advent of the Internet and World Wide Web technologies are radically changing the interactions between business organizations and their consumers. In order to compete in the globalized market, there is a growing interest in the use of electronic commerce (e-commerce, or EC) as a means to perform business transactions (Guo, 2002). Lots of variables seem to dominate the success of implementing e-commerce (Martinsons, 2002; Ngai and Wat, 2002). However, as indicated in one literature review paper that even though there has been an exponential growth in e-commerce research, not enough studies were focused on the human-machine interface design of the e-commerce (Ngai and Wat, 2002). Especially, the business-to-consumer (B2C or BtoC) e-commerce, which is the transaction between business organizations and end users, is significantly influenced by the efficacy of its interface design.

Past studies indicated that the usage of graphical user interface (GUI) or computer icons has significant advantages in recognition and user preference than text (Kacmar and Carey, 1991; Weidenbeck, 1999). Several design principles for computer icons also can be found in previous research (Huang, Shieh and Chi, 2002) and may be applied to the design of B2C e-commerce websites. Hence, lots of the issues need to be addressed such as how to provide attractive and clear description of products, how to arrange a convenient and user-friendly searching and ordering process, and so on. Recognizability, discriminability, communicativeness, color, layout, and simplicity are the most fundamental design features of the icons. Table 1 summaries the six design features and their brief descriptions. The purpose of the present research was to investigate the interface design features of four B2C e-commerce websites in Taiwan.

Table 1: The fundamental icon design features and their brief descriptions

Feature	Description
Recognizability	The icon should be easy to recognize
Discriminability	The icon should be easy to discriminate from other icons
Communicativeness	The icon should express the intended messages clearly
Color	Proper color design should be used
Layout	The elements in an icon should be arranged carefully
Simplicity	The icon should be as simple as possible

2 Methods

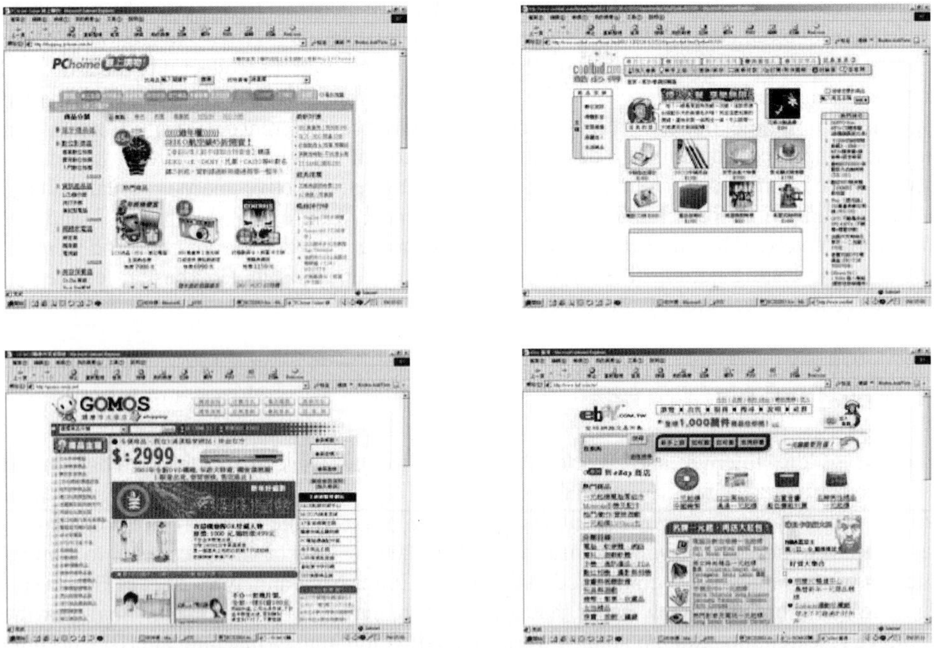

(a) Pchome	(b) Coolbid
(c) Gomos	(d) Bid

Figure 1: The averaged searching time and number of link errors of four websites

2.1 Subjects

A total of 15 students (8 males and 7 females) of Ming Hsin University of Science and Technology participated in this study. The ages of subjects were from 21 to 25 years old and with an average of 22 years old. Four B2C e-commerce websites were chosen according to popularity and product diversity. Figure 1 are the homepages of these four websites: Pchome (shopping.pchome.com.tw), Coolbid (www.coolbid.com.tw), Gomos (gomos.sunup.net), and Bid (www.bid.com.tw).

2.2 Procedures

Both a videotaping retrospective method and questionnaires were used to collect the data from 15 subjects. First, a camera was used to record the fully process of searching and ordering a product while testing each of the four websites. After the experiment, the experimenter watched the videotapes and calculated the elapsed time and errors. The videotaping retrospective method has the advantages to repetitive observe any detail operational problems; however, it takes lots of time in analyzing the tapes. Finally, a questionnaire was constructed and used to collect demographic data, consuming habits, and subjective preference toward various design features. Dependent measures were the averaged searching time, number of link errors, subjective preference, and etc. Searching time is the time from the beginning of a search till locates the description of certain items. Link error is choosing or clicking on the wrong items during the transaction process. Design features of the icons, including recognizability, discriminability, communicativeness, color, layout, and simplicity, were ranked across four websites by the subjects.

3 Results and Discussion

3.1 Videotaping retrospective analysis

Videotaping retrospective analysis revealed that the averaged searching time for Pchome, Coolbid, Gomos, and Bid were 90.8, 142.5, 164.7, and 199.8 seconds, respectively. The number of link errors for Pchome, Coolbid, Gomos, and Bid were 7, 4, 9, and 12 during testing period. Bid was worse in these two indexes. Pchome had shortest searching time and Coolbid had smallest link errors (refer to Figure 2).

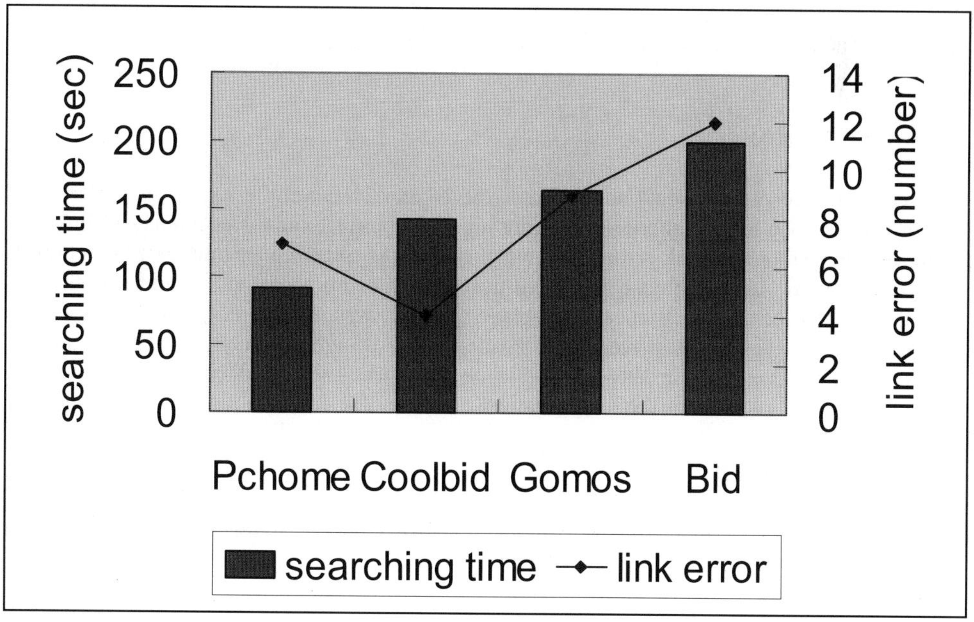

Figure 2: The averaged searching time and number of link errors of four websites

3.2 Questionnaire analysis

The GUI ranking of these four websites across different GUI features are summarized in Table 2. Pchome seemed to be preferred than the other three tested websites by the subjects. And, this finding is congruent with the previous results on searching time.

Novice and experienced users were classified according to whether a subject was familiar with a certain e-commerce website or not. It was assumed that novice and experienced users might have different system preferences and need different GUI design requirements. The results support this assumption. The experienced users emphasized more on system speed and simplicity than the novice users. If it took longer time in searching a product, then the users would rank this website worse and would not like to visit this website anymore!

Table 2: The subjective ranking of design features for four e-commerce websites

Features \ Websites	Pchome	Coolbid	Gomos	Bid
Recognizability	◎◎◎◎	◎◎◎	◎	◎◎
Discriminability	◎◎	◎◎◎◎	◎◎◎	◎
Communicativeness	◎◎◎◎	◎	◎◎	◎◎◎
Color	◎◎◎◎	◎◎◎	◎	◎◎
Layout	◎◎◎◎	◎◎◎	◎	◎◎
Simplicity	◎◎◎◎	◎◎◎	◎	◎◎
Overall	◎◎◎◎	◎◎◎	◎	◎◎

Best: ◎◎◎◎ Worse: ◎

4 Conclusions

The results of the present study showed that the e-commerce website with user preferred GUI design features also has the shortest searching time of a product and fewer number of link errors. Since searching time and link error are two critical factors affecting an Internet transaction, it is important to create icons with superior identification and comprehension. In addition, this study indicated the novice and experienced users have different system preferences and require various interface design. Hence, it increases the challenges of designing the B2C websites since the diversities of its users. It implies a flexible or intelligent user interface would be a possible solution and it still requires further investigations.

Acknowledgements

The author would like to thank the National Science Council of the Republic of China for financially supporting this work under Contract No. NSC91-2213-E-159-010.

References

Guo, C. (2002). Competing in high-growth markets: The case of e-commerce, *Business Horizons*, March-April, 77-83.

Huang, S.-M., Shieh, K.-K. and Chi, C.-F. (2002). Factors affecting the design of computer icons, *International Journal of Industrial Ergonomics*, 29, 211-218.

Kacmar, C. J. and Carey, J. M. (1991). Assessing the usability of icons in user interface, *Behavior and Information Technology*, 10, 443-457.

Martinsons, M. G. (2002). Electronic commerce in China: emerging success stories, *Information & Management*, 39, 571-579.

Ngai, E. W. T. and Wat, F. K. T. (2002). A literature review and classification of electronic commerce research, *Information & Management*, 39, 415-429.

Weidenbeck, S. (1999). The use of icons and labels in an end user application program: an empirical study of learning and retention, *Behavior and Information Technology*, 18(2), 68-82.

An evaluation of Turkish e-commerce sites according to several guidelines: An empirical study

Mehmet Mutlu Yenisey †, Cafer Erhan Bozdag ‡, Pelin Nisari ¶

Istanbul Technical University
Dept. of Industrial Engineering
80680 Macka-Istanbul Turkey
† yenisey@itu.edu.tr, ‡ bozdagc@itu.edu.tr,¶ pelin_nisari@yahoo.co.uk

Abstract

Internet is a global network that enables us to get information, communicate with others. Additionally, Internet strongly triggers E-commerce. In fact, e-commerce means that commercial activities are performed on computer media. Namely, marketplace, or store is alive in a computer. Customers, in other word users, access to the store via Internet. They start a session on server side, and begin to do shopping within their browsers. Therefore, it can be said that e-commerce is a special kind of web programming applications. Hence, everything necessary for the success of a web site is also valid for that of a commercial site. One of the most important points for a web site is its usability. Because, it is said that a customer is only one click away from a web site. If a user finds a web site is difficult to use or it is too slow to download the pages then she/he can also leave the site immediately by just one click. Therefore, some points related to usability must be taken into consideration during the design phase of the web site. These points are found in literature under the title of "Guidelines". Furthermore, if a web site serves as a commercial site than a few concepts, like security, must be added to guidelines. These guidelines also guaranty the success on the side of customer, which is measured as the customer satisfaction. Several studies comparing web sites can be found in literature. But, these studies are mostly for western countries. In this paper, an empirical study conducted for Turkish e-sellers is given. Turkey has a growing economy. Internet usage and users, the number of online customers are increasing day by day. It is necessary to determine the current status of Turkish e-commerce sites in order to supply information to the researchers. In this study, selected big and famous Turkish companies' online shopping web sites are evaluated and results are discussed.

1.Motivation

Despite the fact that the Internet was first introduced to share information among professionals, today it became an extensive source where ordinary people can satisfy all their everyday needs. 'Online shopping', which refers to the purchase of goods and services using the Internet, is one of the most important developments involved in our lives with the spread of the Internet.

There has been an increasing demand for online shopping sites as the consumers recognized the ease of use and the convenience of doing so as well as the competitive prices. Thus entrepreneurs all over the world introduced an increasing supply after they noticed the new market potential and the opportunity of starting a business with a relatively small amount of capital. As a consequence, a new and profitable way of commerce was established. Many new companies have been set up. Due to the competition and the profits involved, it became a necessity for older companies, which have not yet entered the Internet sector to get into motion.

Some companies did better on the Internet, some did not. So a new subject came into agenda: Determining the factors in relation to the effectiveness of online shopping sites and their shopping procedures as well as finding out the customer satisfaction criteria.In this context when we look at Turkey, an increasing number of firms provide online shopping on their web sites and many others planning to do so while setting up their informative company sites. It is surprising that so far no study has been made in order to measure the effectiveness of shopping sites. This paper, therefore, is a result of such an effort that aims to be an initial step for future analysis in the area. In order to base our assessments on universal norms, an extensive research was conducted on the

Internet publishing and online marketing issues and the basic form factors were gathered in a checklist. The attributes summarized from literature gave way to a questionnaire, which was applied, to 209 individuals in an effort to evaluate 63 Turkish shopping web sites: 1)The site has an efficient search engine. (Nielsen, 2002a; 2002b) 2)To view the whole page I need to use the left-right scrolling bar at the bottom. This causes a loss of time and decreases the ease of use of the site. (Nielsen, 2002a) 3)Page layouts in the site are the same. It gives the site a professional look, and users will notice easily when they get out of the site for one reason. (For example by clicking a link) (Chaplin, 2000; Hastings, 2002; Mickiewicz, 2000; 1999) 4)Pages are loading in a short period. Internet means speed, site owners have a very short time to gain and impress users. There is a huge supply, when the user gets bored of waiting (approximately 20 sec.), he quickly passes to another site; in this circumstances designing a site using new technologies and small tricks is crucial. (Hastings, 2002; Johnson, 2000b; Mickiewicz, 2000; Nielsen, 1999a; 1999b) 5)All the pages include company name and company logo. Naturally this will create a familiarity, and in the long term a reliance on the company. (Chaplin, 2000; Nielsen, 1999; Sivasubramanian, 2000; Tamsevicius, 2000) 6)There is information that the site has been updated. This increases the reliability of the site. (Mickiewicz, 1999a; Nielsen, 1999b) 7)Pages included some essential information about the contact person. In core business, in case of any need all the necessary information must be provided to customer. (Harper, 2002) 8)The site has optional versions in other languages. This prevents language from being a hindrance. (Sivasubramanian, 2000) 9)There is a link to return to home page and a menu bar to navigate through the site, on every page. (Chaplin, 2000; Hastings, 2002; Mickiewicz, 2000) 10)When I click on the links many unnecessary windows pops up. The user would like to have the control of the web site in his hands, so unexpected and numerous pages and ads will bother him/her. (Nielsen, 2002a) 11)The links in the text can be distinguished easily. (Harper, 2002; Nielsen, 1999b) 12)Some links did not work. (Mickiewicz, 2000; Sivasubramanian, 2000; Tamsevicius, 2000) 13)I can manage the number and the subjects of the e-mails that the site will send me in the future. (Nielsen, 1999) 14)When the site requested my e-mail address, I've been explained for what reason my e-mail address is being taken. (Mickiewicz, 1999b; Nielsen, 2002a; 1999) 15)The site includes adequate information about the security precautions. (Nielsen, 2002a) 16)The site includes adequate information about return and guarantee policies. (Tamsevicius, 2000) 17)There is satisfactory information that my personal information will be protected. (Nielsen, 2002a) 18)The site gives enough information about the products. Customers want to know more about what they are going to buy; a satisfactory explanation coming with a photograph and the price is essential. (Johnson, 2000a; Nielsen, 2002a) 19)The selling and delivery processes are clearly defined. (Nielsen, 2002a) 20)The photographs give a clear idea of the products. The basic element when building trust between the customer and the seller is being honest. The customer must have the option to see the photograph of the product. Site designers should avoid decreasing download times by extracting or cutting important photographs. (Nielsen, 1996)

2.Survey

The survey contains 20 questions. They ask participants to respond using a 5-point Likert scale ranging from 'strongly disagree' to 'strongly agree'. The questions are; 1)Site has an effective search engine, 2)It is necessary to use bottom scroll bar to view the entire page, 3)All pages in the site are consistent, 4)Pages are downloaded in a short time, 5)Every pages have company's name and emblem, 6)There is an information indicating that site is updated, 7)Site has names and numbers of contact persons, 8)Site's versions in other languages exist, 9)Every page has a link to return to the home page and navigation bar, 10)When I click a link, so many new windows are opened, 11)The links in the text are very easy to understand, 12)There are broken links, 13)I am able to control the number and content of e-mail to be sent by the site, 14)It is clarified the purpose of receiving my e-mail address, 15)There is sufficient information about security,

16)There is sufficient information about return policy, 17)There is sufficient information about privacy, 18) There is sufficient information about products, 19)Sales and shipment methods are clarified, 20)Images are reflecting products sufficiently. The general evaluation of each question is given in Table.1.

Table.1.Evaluation of Questions

Question No.	Point	Question No.	Point	Question No.	Point	Question No.	Point
1	3.62	6	2.26	11	3.87	16	3.62
2	4.30	7	3.81	12	4.25	17	3.49
3	4.02	8	1.75	13	2.82	18	3.94
4	3.89	9	4.29	14	3.07	19	3.90
5	4.46	10	4.40	15	3.75	20	3.81

The overall average is 3.67. Totally, 63 Turkish sites are evaluated. 15 sites get above 4, 46 sites get between 3 and 3.99, and only 2 sites get below 2. 24% of all sites meet the general rules. This shows that online shopping is being considered seriously and started to develop in Turkey. The worst answers are 'Site's versions in other languages exist', 'There is an information indicating that site is updated', and 'I am able to control the number and content of e-mail to be sent by the site'. These questions get 1.75, 2.26 and 2.82 respectively. The sites and their average scores are given in Table.2.

Table.2.Average scores of Turkish Sites

Site	Puan	Site	Puan	Site	Puan
www.lufthansa.com.tr	4.85	www.genpatech.com	3.85	www.deppo.com	3.50
www.thy.com.tr	4.65	www.pandora.com.tr	3.78	www.orjinsolar.com	3.47
www.sonfiyat.com	4.55	www.baskuda.com	3.77	www.carsi.com.tr	3.40
www.dr.com.tr	4.40	www.arena.com.tr	3.75	www.sahibinden.com	3.40
www.sanalmagaza.com	4.30	www.gittigidiyor.com	3.75	www.alogurme.com	3.39
www.office1.com.tr	4.20	www.ykm.com.tr	3.75	www.rotaline.com	3.38
www.fenerium.com	4.15	www.yemeksepeti.com	3.75	www.zureyfa.com	3.38
www.biletix.com	4.15	www.kangurum.com.tr	3.75	online.migros.com.tr	3.30
www.bookinturkey.com	4.10	www.kitapyurdu.com	3.72	www.2a-engineering.com	3.25
www.genpa.com.tr	4.05	www.hepsiburada.com	3.71	www.serisonu.com	3.23
www.turkcell.com	4.05	www.siberalem.com	3.70	www.pet.gen.tr	3.21
www.ideefixe.com.tr	4.03	www.iccamasiri.com	3.67	www.iskender.com	3.15
www.kvk.com.tr	4.00	www.dod.com.tr	3.66	www.ihturkey.com	3.10
www.mezun.com/flower	4.00	www.gima.com.tr	3.60	www.tansas.com.tr	3.10
www.rehberatlas.com	4.00	www.kariyer.net	3.60	www.kontor.com.tr	3.06
www.akcag.com.tr	3.95	www.hobievi.com	3.59	www.armada.com.tr	3.05
www.teba.com.tr	3.95	www.kozmetikpazar.com	3.56	www.psikologum.net	3.03
www.estore.com.tr	3.90	www.sarayhali.com.tr	3.55	www.ozlemanten.com	3.02
www.garantialisveris.com	3.89	www.arcelik.com.tr	3.54	www.poweralsat.com	3.00
www.afrodit.com.tr	3.85	www.evdeyiz.com	3.50	www.dijitalsatis.com	2.90
www.bekoonline.com	3.85	www.nanclimon.com.tr	3.50	www.sanalcarsi.com	2.70

In the second part of the statistical analysis, questions are grouped and each group is individually evaluated. The groups, their questions, and average scores are given in Table.3.

Table.3.Questions and groups

Group	Questions	Average
Usability	1, 2, 4, 8, 9, 10, 11, 12, 13	3.69
Page Layout	2, 3, 5, 6, 11, 12	3.86
Security	7, 14, 15, 17	3.60
Sales	7, 16, 18, 19, 20	3.82
Costumer Relations	7, 13, 14, 16, 17, 19	3.45

2.1.Usability

This is the only group that there is not any site scores full point. Usability is an important feature for online shopping sites. Turkish sites get average 3.69 point for this feature. 15 sites get above 4. The sites scores above 4 are given in Table.4.

Table.4.Usability evaluation of the sites

Site	Point	Site	Point	Site	Point
www.lufthansa.com.tr	4.89	www.biletix.com	4.30	www.sanalmagaza.com	4.11
www.thy.com.tr	4.78	www.rehberatlas.com	4.22	www.turkcell.com	4.11
www.bookinturkey.com	4.67	www.akcag.com.tr	4.11	www.fenerium.com	4.00
www.sonfiyat.com	4.44	www.bekoonline.com	4.11	www.genpatech.com	4.00
www.dr.com.tr	4.33	www.office1.com.tr	4.11	www.nanclimon.com.tr	4.00

2.2.Page Layout

This group is the one, which has the highest average. 21 sites out of 63 are above 4. Overall average is 3.86 for this category. The sites score above 4 are given in Table.5.

Table.5.Page layout evaluation of the sites

Site	Point	Site	Point	Site	Point
www.lufthansa.com.tr	5.00	www.genpatech.com	4.33	www.kitapyurdu.com	4.02
www.sonfiyat.com	5.00	www.mezun.com/flower	4.30	www.carsi.com.tr	4.00
www.thy.com.tr	4.83	www.office1.com.tr	4.22	www.dijitalsatis.com	4.00
www.teba.com.tr	4.58	www.sanalmagaza.com	4.11	www.genpa.com.tr	4.00
www.estore.com.tr	4.42	www.sarayhali.com.tr	4.11	www.sahibinden.com	4.00
www.bookinturkey.com	4.33	www.fenerium.com	4.17	www.siberalem.com	4.00
www.dr.com.tr	4.33	www.ideefixe.com.tr	4.15	www.turkcell.com	4.00

2.3.Security

Average is 3.60 for security issues. It is the third one in the rank. Two sites get 5 while two athers 1.25. Scores for the sites above 4 is given Table.6.

Table.6.Security evaluation of the sites

Site	Point	Site	Point	Site	Point
www.lufthansa.com.tr	5.00	www.gittigidiyor.com	4.25	www.genpa.com.tr	4.00
www.rehberatlas.com	5.00	www.kvk.com.tr	4.25	www.nanclimon.com.tr	4.00
www.dr.com.tr	4.50	www.sonfiyat.com	4.25	www.office1.com.tr	4.00
www.mezun.com/flower	4.50	www.fenerium.com	4.13	www.sanalcarsi.com	4.00
www.sanalmagaza.com	4.50	www.ideefixe.com.tr	4.06	www.turkcell.com	4.00
www.siberalem.com	4.50	www.afrodit.com.tr	4.00	www.ykm.com.tr	4.00
www.thy.com.tr	4.50	www.akcag.com.tr	4.00		

2.4.Sales

This is the second highest average. Average is 3.82. This means that Turkish online sites are careful about the issues related sales. Sites above 4 are given in Table.7.

Table.7.Sales evaluation of the sites

Site	Point	Site	Point	Site	Point
www.dr.com.tr	5.00	www.mezun.com/flower	4.60	www.teba.com.tr	4.20
www.office1.com.tr	5.00	www.thy.com.tr	4.60	www.biletix.com	4.14
www.sanalmagaza.com	5.00	www.fenerium.com	4.50	www.estore.com.tr	4.10
www.kvk.com.tr	4.90	www.kangurum.com.tr	4.33	www.baskuda.com	4.04
www.sonfiyat.com	4.80	www.garantialisveris.com	4.27	www.akcag.com.tr	4.00
www.arena.com.tr	4.60	www.ideefixe.com.tr	4.22	www.bekoonline.com	4.00

www.genpa.com.tr	4.60	www.afrodit.com.tr	4.20	www.gittigidiyor.com	4.00
www.lufthansa.com.tr	4.60	www.iskender.com	4.20	www.turkcell.com	4.00

2.5.Costumer relations

Finally, Turkish sites are evaluated according to costumer relations. This is the lowest category. Turkish sites must be more careful about their customers' e-mail addresses. The sites get more than 4 are given Table.8.

Table.8.Costumer relation evaluation of the sites

Site	Point	Site	Point	Site	Point
www.lufthansa.com.tr	5.00	www.thy.com.tr	4.33	www.kvk.com.tr	4.00
www.bekoonline.com	4.33	www.fenerium.com	4.25	www.nanclimon.com.tr	4.00
www.dr.com.tr	4.33	www.genpa.com.tr	4.17	www.rehberatlas.com	4.00
www.siberalem.com	4.33	www.gittigidiyor.com	4.17	www.sanalmagaza.com	4.00
www.sonfiyat.com	4.33	www.ideefixe.com.tr	4.02	www.turkcell.com	4.00

3.Conclusion

In this research, our objective is to take a photo of Turkish web sites according to general rules. Turkish online shopping sites get fair scores from the survey. But, it is necessary to extend this survey as include the non-Turkish sites. A benchmarking must be conducted for future research. Hence, the place of Turkish sites can be found.

4.References

Chaplin, T. (2000). Consistency: is your website two-faced? Retrieved January 5, 2003, from http://www.webmasterbase.com/article/123

Ellis, A. (2002). Foster the feel good factor. Retrieved January 5, 2003, from http://www.webmasterbase.com/article/787

Fortin, Ph.D. M. (2000). Overcoming the front-page hurdle. Retrieved January 5, 2003, from http://www.webmasterbase.com/article/94

Harper, M. (2002). Diary of a Webmaster part 1 - my site design checklist. Retrieved January 5, 2003, from http://www.webmasterbase.com/article/655

Hastings, S. (2002). Design effective navigation in 10 steps. Retrieved January 5, 2003, from http://www.webmasterbase.com/article/904

Johnson, J. (2000a). Designing a user-friendly site. Retrieved January 5, 2003, from http://www.webmasterbase.com/article/75

Johnson, J. (2000b). Give your site a facelift. Retrieved January 5, 2003, from http://www.webmasterbase.com/article/110

Mickiewicz, M. (1999a). Design guidelines. Retrieved January 5, 2003, from http://www.webmasterbase.com/article/20

Mickiewicz, M. (1999b) . Design guidelines. Retrieved January 5, 2003, from http://www.webmasterbase.com/article/15

Mickiewicz, M. (2000). Design guidelines. Retrieved January 5, 2003, from http://www.webmasterbase.com/article/205

Nielsen, J. (1996). Marginalia of web design. Retrieved January 5, 2003, from http://www.useit.com/alertbox/9611.html

Nielsen, J. (1999a). Trust or bust: communicating trustworthiness in web design. Retrieved January 5, 2003, from http://www.useit.com/alertbox/990307.html

Nielsen, J. (1999b). "Top ten mistakes"revisited three years later. Retrieved January 5, 2003, from http://www.useit.com/alertbox/990502.html

Nielsen, J. (2002a). Top ten web-design mistakes of 2002. Retrieved January 5, 2003, from http://www.useit.com/alertbox/20021223.html

Nielsen, J. (2002b). Top ten guidelines for homepage usability. Retrieved January 5, 2003, from http://www.useit.com/alertbox/20020512.html

Sivasubramanian, Dr. M. (2000). 'ReDesign' is not a dirty word. Retrieved January 5, 2003, from http://www.webmasterbase.com/article/291

Tamsevicius, K. (2000). Dazzle your visitors with a dynamic home page. Retrieved January 3, 2003, from http://www.webmasterbase.com/article/304

Perception of E-commerce: A View from an Industrial Engineer's Perspective

Mehmet Mutlu Yenisey

Istanbul Technical University
Industrial Engineering Department
80680 Macka-Istanbul Turkey
yenisey@itu.edu.tr

Abstract

E-commerce is a day-by-day expanding commercial media and mainly is to perform whole commercial activities on a virtual media. Today, when we talk about e-commerce we understand getting information from the Internet. But, it is clear that e-commerce is not limited to the Internet, and didn't begin with Internet era. The total amount of shopping, in terms of both money and quantity, is also rapidly increasing. Additionally, these augmentations are valid for not only B2B but also B2C type e-commerce. Several definitions for the taxonomy of e-commerce can be found in literature. The main objective of this paper is not to give taxonomies. It is contented with a short reminder of taxonomies. E-commerce can be classified according to several aspects like parties, functions, types of activities etc. The most important thing is to define why we do, or need, e-commerce. This definition will give us the clues for a successful e-commerce. Hence, we can obtain the revenues, in fact profits, expected from e-commerce. There are a lot of issues to be taken into consideration for the success of the commercial activities in the virtual world, i.e. content management, usability, security, privacy, availability, cross-cultural aspects etc. These issues will lead to the formation of the customer loyalty that is the item of utmost importance we expect to create for the success. The classical approach cannot be sufficiently adopted to e-commerce because of new technologies. Technological developments involving e-commerce can be examined in two ways: Innovations in hardware and software. The biggest improvement in e-commerce physical infrastructure is the direction towards mobility. Which allows conducting shopping transactions on the go. Another important development happens in the back-end of e-commerce. Some intelligent agents like data warehousing and data mining are already part of e-commerce. In fact, today we must comprehend that e-commerce is not only a kind of reconstruction of commercial activities in a digital environment. It must be taken into consideration that e-commerce has, or must have, a knowledge management logic.

1. Introduction

E-commerce has existed since 60s. The primitive type of e-commerce was only in the form of simply EDI and for business-to-business (B2B) applications. But, especially after the Internet era has begun and new information technologies have been seen, its importance has improved considerably. Additionally, business-to-consumer (B2C) applications also have risen with the growth of Internet (Trepper, 2000). But e-commerce, or generally e-everything, is not simply to put a letter of "e" in front of the word. Embedding e-technologies relates to people, structure and culture. It has critical issues of implementation and innovation (Stace, Holtham and Courtney, 2001).

2. Properties of E-Commerce

Perhaps, people do not think about e-commercial activities during performing them. But, when we take a closer look at e-commerce, we can see that there are many Industrial Engineering applications.

2.1. E-commerce is a knowledge management system

E-commerce should be considered as a knowledge management system. There are several forces driving the knowledge management. When we think about these forces, we easily can see that they are mostly overlapped. First of all, the society of today is, with its all components, turning into a knowledge society. This means that organizations, people, products and services are becoming more intelligent. It is obvious that the same development is also valid for e-commercial activities, and of course for e-commerce itself. The second force behind knowledge management is globalization. We have no doubt that e-commerce is dramatically becoming global day-by-day. Especially, after the big revolution called Internet, new information technologies are driving e-commerce. Additionally, companies wanting to have a secure place and make profit in globalizing economy should take going global seriously into consideration. Also, we can talk about the need for speed and reduction of cycle times. In today's more global environment, we need faster processes to catch the speed of the information age, hence, short transaction and cycle times. Another important driving force is the need for organizational growth. The organizations in e-business need more growth abilities in order to adopt the today's more competitive environment. Other factors are the abilities of organizational redesign and employee mobility for adopting to the new corporate world. Knowledge management approach has features for these abilities. Lastly and perhaps most importantly, e-commerce is a cooperative kind of business bringing several partners like members of a value chain together. This feature requires good organization and coordination for the success. One of the basic promises of knowledge management is cooperation among the society, organization, people and employees. It provides both an infrastructure for value chain and a supra-structure for regulatory environment. Recently developed information and communication technologies enable parties of e-commerce to link at relatively low costs. Moreover, these technologies provide a good and a very usable media to exchange the knowledge. Especially, the last feature is very important for accessing the end-users, meaning the customers (Maier, 2002).

2.2. E-commerce has cultural aspects

Another important point for the success of e-commerce is the cross-cultural issues. Because every society, or community, has its own special behaviors for daily shopping. These behaviors are also being reflected on on-line shopping. For example, a survey conducted in 2001 shows that Turkish university students wish bidding possibilities more than the U.S. students (Lightner, Yenisey, Ozok & Salvendy, 2002). Chau et al. (2002) claim that on-line consumers in different countries with different ethnic origins use the Web for different purposes while on-line costumers' patterns for purchasing are gradually becoming standardized. But, these different purposes lead them to have different impressions of the same web sites. Different impressions do not arise only from cultural differences but also from personal ones. This aspect leads to provide different interfaces to the users. Hence, adaptive user interface approach comes into play, since only a person can determine what interface is the best one and the most usable for her/his comfort. Helander and Khalid (2000) emphasize the importance of customizing the web site individually according to the user's needs.

3. Views From an Industrial Engineer's Perspective

First of all, e-commerce is an interdisciplinary approach and an inter-organizational business. Therefore, several service and business branches are involved with e-commerce. It requires the coordination of several professions. Naturally, most applications of industrial engineering can be used for solving the problems of e-commerce, since industrial engineering is also an interdisciplinary branch of the engineering. Moreover, industrial engineers usually play a role of coordination, or being mortar.

E-commerce is a project and it requires the project planning techniques to be applied. The whole process of e-commerce must be planned carefully. Every activity in the process must be determined and defined. A detailed analysis will be very useful for definitions. It is clear that an e-commerce project differs from classical construction projects because of having high-technological properties. The activities of e-commerce projects are mostly made of either software development or hardware procurement. Procurement activities are deterministic and easy to plan. But, software development activities are highly probabilistic due to their nature. Therefore, they require a specific attention for both planning and control. Even, sometimes it may be necessary to be involved with more than one project or the project may be divided into several sub-projects. Additionally, some of them may be outsourced. In this case, the problem becomes multi-project management and has additional variables (Archibald, 1976).

Another important issue is that e-commerce is an investment. Moreover, it can be easily said that e-commerce is an expensive investment since it is very much technology-oriented. Therefore, e-commerce projects must be evaluated with the methods of engineering economics. There are difficulties for the definition of costs and revenues. It is necessary that the investment analyzer must deal with some social costs and benefits that are hard to define. Another difficulty arises from uncertainties of them. This aspect makes the decision-maker to deal with high risks. Therefore, deterministic models may not provide a suitable solution. Specifying a suitable probability distribution for each input variable can overcome this problem. E-commerce has a cost structure that must be managed very well. Examples for the IT part of e-commerce costs are hardware, software, installation, training and maintenance costs. Other costs directly related to the commercial activities, like shipment, financial, procurement costs must be taken into consideration. Hence, a good cost structure can be defined (Remeyni et al., 2000).

Österle (1995) discusses the importance of workflow and flow planning, and gives examples for business in information age. It is obvious that e-commerce requires well-defined workflow. Workflow definitions must be done for every process in e-commercial activities. These activities consist of order acquisition, shipment and logistics management, payment and other financial transactions. Every step in every transaction must be defined and sequenced in correct order.

E-commerce is human-oriented. Therefore, human factors issues must be taken into consideration. Besides web-specific studies, those for software development may be applied to e-commerce activities. The most important point is to keep in mind the accommodation between human and computer in terms of both software and hardware. Additionally, it is also important to make users feel comfortable. More usable web sites means more costumers, and of course more profit.

E-commerce is a technology-oriented application, and it has expert staff that requires a good human resources management system. IT staff's knowledge must be kept up-to-date for the success of e-commerce because of the day-by-day improvements in IT. Well-planned training programs will help us to catch up with the current status. Additionally, a well-organized career planning and development must be performed. These will increase the motivation, and of course performance of our staff. Another important point is the organization of company. Clarkson's (2001) study mentions four organizational approaches; traditional, project-based, matrix and hybrid. The study suggests matrix and hybrid structures for medium-to large-sized IT departments.

Today, the application of intelligent agents like AI, neural networks etc. is the new direction for the development of e-commerce. The success of e-commerce goes through the prediction of customer behavior. This will develop costumer loyalty that will lead to increase in sales and profit. The logic behind this is simple; satisfied costumer do not shop elsewhere (Walsh and Godfrey, 2002). If we are able to foresee the reasons of costumers' behaviors then we can easily adopt ourselves according to them. Therefore, applying intelligent agents is very important. Examples of the application of intelligent agents are interface design, estimation of costumer demands, security etc.

Today, development direction of IT is towards mobility. Mobility started by using cellular phones. People experienced the pleasure of being mobile while communicating each other. As time passes, mobile devices got cheaper and most people became enterable to the wireless world. Today, even a very cheap phone has features for connecting to the Internet. This development forces e-commerce to transform towards m-commerce. But, at the beginning there were problems arising from the limitations on the communication speed. Today, these limitations are beingve gradually overcome.

4.Conclusion

The future is e-world, or another word is e-globe. Today, we have relatively a smaller world because of new IT technologies and recent developments at their disposal. But the question is how people make profit from them or whether they make any profit at all. As industrial engineers, we assist people to use them better by finding and implementing good models to fit the virtu-real world of e-commerce. This will increase the productivity, which is important for obtaining high profits by low costs. Applying our optimization logic to every part of e-commerce will help us to achieve this goal.

5.References

Archibald, R.D., (1976). Managing High-Technology Programs and Projects, John Wiley & Sons Inc.

Chau, P.Y.K., Cole, M., Massey, A.P., Montoya-Weiss, M., & O'Keefe, R.M. (2002). Cultural differences in the online behavior of consumers. Communications of the ACM, 45 (10), 138-143

Clarkson, M. (2001). Developing IT Staff: A Practical Approach. Springer-Verlag London

Helander, M. G., Khalid, H.M. (2000). Modeling the customer in electronic commerce. Applied Ergonomics, 31, 609-619

Lightner, N., Yenisey, M.M., Ozok, A.A., & Salvendy, G. (2002). Shopping Behaviour and Preferences in E-commerce of Turkish and American University Students: Implications from Cross-Cultural Design. Behaviour & Information Technology, 21 (6), 373-385

Maier, R. (2002). Knowledge Management Systems: Information and Communication Technologies for Knowledge Management, Springer Verlag Berlin

Österle, H. (1995). Business in the Information Age: Heading for New Processes, Springer-Verlag

Remeyni, D., Money, A., Sherwood-Smith, M. (2000). The effective measurement and management of IT costs and benefits: 2nd edition, Butterworth-Heinemann

Stace, D., Holtham, C., and Courtney, N. (2001). E-change: Charting a path towards sustainable e-strategies. Strategic Change, 10, 403-418

Trepper, C. (2000). E-commerce Strategies, Microsoft Press, Washington

Walsh, J., Godfrey, S. (2000). The Internet: A new era in Costumer Service. European Management Journal, 18 (1), 85-92

Section 8

Human Factors and Ergonomics

Ergonomic Analysis of a Distributed System

Ahmet E. Çakir

ERGONOMIC Institute for Work Sciences
Soldauer Platz 3 – D-14055 Berlin
Ahmet.Cakir@ergonomic.de

Abstract

A software as a part of an application developed centrally but used in a widely distributed company was evaluated for finding the sources of reported "ergonomic" problems. The analyses have revealed that the higher proportion of the problems have not been caused by the software to be analyzed but by the organizational structure of the company and the distribution of the responsibilities for the application over many organizational units. The analysis of the history of the application showed that the basic assumptions in software ergonomics, e.g. the existence of a target user population or of a work task for the system, are not valid for real work systems. During the life span of a software within a company, not only the structure of users may change, but also the structure of that organization. The common notion in ergonomics proves too static for real working life.

1 Introduction

In textbooks for ergonomists, work systems are described as a result of a design process that starts with the definition of the work task followed by the assignment of functions to humans and machinery. Later, the required quality of technology and interaction with users is considered. The procedure ends with the last step of designing the hardware needed. The way in which people think while planning a complex work system does not differ very much from designing a truck or similar. One of the best illustrations of this is displayed in Fig. 1.

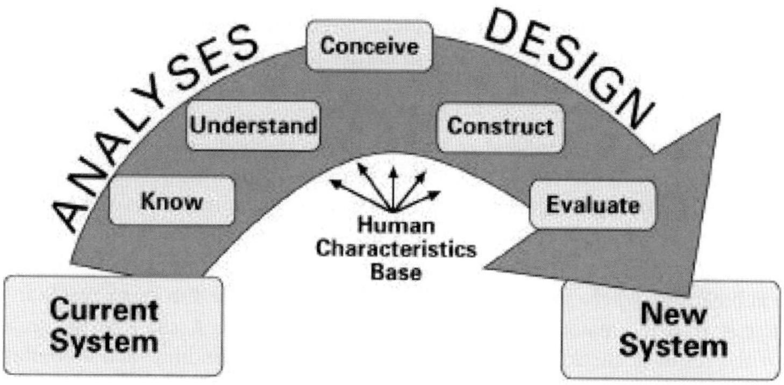

Fig. 1 Illustration of "System Ergonomics" as an integrative approach (after TNO http://www.tm.tno.nl/organisation/group/distributed/system.html)

The notion is that designing a system has a goal from the very beginning, and the analysis of the existing system would result in information meaningful for the target system. In order to apply the information from the human characteristics base the designer needs a clear view of the "user" or user population. The model as described in the original source of the figure is an application of the models of "Systems Engineering" on ergonomic design of work systems.

This model can be successfully applied for different design tasks under the condition that not too many relevant changes do occur between the analysis ("Know") and the introduction of the new system. The analyses described in this paper were performed in a big administration deemed stable by most people. The task was uncovering the source of an apparent failure of a software used by some thousand employees in different locations.

2 Short description of the project

2.1 Problem description

Before the researchers entered the scene, various persons and groups within the company had discussed the objective of the project for at least two years. The unofficial diagnosis was that the software had some important bugs that made working with it very unpleasant, difficult and error prone. The most important problem reported was that the software would force the users to change the order of the steps while discussing an issue with a customer. And this task was the most important. Some important facts had been listed and sent to higher ranks of the company. The complainers claimed that the software had been designed without considering user needs and introduced after an insufficient training. The organisation resp onsible for maintaining the system would not respond to fault reports.

All together, the problems reported matched the common prejudice that software people would not care much for the users. Thus, the primary task was finding in which direction they had failed while analysing the user needs and converting them into user requirements and software.

2.2 Methods used

Since the users were distributed over many cities it was decided to involve a sufficient number of locations but not all. The company has suggested inviting the required number of users to a certain office where the tests could be performed. Instead, we have decided to go to the relevant locations and consider the entire social environment at the workplace and not the software only. The reason for this was that in earlier projects, other types of failures with the same software in different locations occurred although it was running on the same machine.

The overall method we have applied is the coherency method, mostly applied for complex investigations (e.g. in murder cases) rather than in normal scientific research. The meaning of coherent in the sense of this method is "holding together as a harmonious and credible whole". The starting point of the method is considering each and every argument put forth as valid. If all arguments are valid any logical combinations of them are valid, too. If one manages finding combinations where arguments are piled up like single bricks to result in a logical statement (e.g. if B is greater than A and C is greater than B, A shall be the smallest), the stability of the result is checked until the pile collapses. After finding such a combination, all arguments are checked until the reason for the inconsistency with the expectations is found, i.e. the point of inco herence.

For this project, a scenario was developed including a series of everyday tasks. If all complaints were considered true the software would fail regardless of the context of use (e.g. user population, customer behaviour, training, hardware etc.). If this assumption were true, any user would experience problems. To check this assumption, the researchers completed the same formal training as a user so that they were able to perform all the tasks of the scenario. After the training, they worked on real tasks with the help of an expert who was responsible for the correctness of the decisions taken.

The next step was examining the software using a standardized checklist based on ISO 9241-12 to ISO 9241-17. In addition, the entire functionality as described in the user manual was checked for validity. In both cases, only such features were considered that could be evaluated without involving the users.

The final step was running the scenario at different locations with real users under real working conditions (about 50 workplaces). All sessions were video taped with the permission of the user representatives and of each user. The users were asked to speak loud what they did. For each task, the entire procedure was compared with the intended way of performing the task. For this purpose, the designer responsible for programming and maintaining the relevant part of the software was asked to work in the same way as she or he had assumed for the users.

After performing the task, the user was interviewed using the version of the questionnaire ISOMETRICS (Gediga et al., 1999) for walkthrough evaluation. The questionnaire consists of 72 items formulated after ISO 9241-10. The user can either give a fully positive rating for each item or show up to three examples explaining why the rating differs from fully positive. Since the examples need to be demonstrated with the software under investigation, each detected problem can be shown to the group responsible for running the software. The entire list of the features causing complaints was validated after a walkthrough with the designers.

3 Some results of the project

3.1 General findings

The major outcome of the project was that the design of the specific software was not responsible for about the half of the problems claimed before or discovered during the project. The severest software related problems were caused by other software that was to be used in the same context. The designers of the software have been aware of some potential problems and had tried to influence the other applications. However, they have not been able to take the hurdles caused by the distribution of responsibilities within the organisation.

For the users, any features of their work disliked or misunderstood seemed to be an ergonomic problem for which the designers of the software were held responsible including parts of the task required by the "employer". However, it was not clear for them who the employer really was, the company or the local plant or both of them. This distinction was important because of some important differences in the interest. E.g., while the company defines the policy for which the application was created the local plant has to employ the users. This means in practice, e.g. that while the software was designed for all plants the work is being organized locally. If the local plant employs people with insufficient skills or refuses to train them properly the users tend to blame the software while real usability problems caused by the software may constitute noticeable cost and performance problems for the plant.

While trying to trace the source of the problems, it was detected that the goal set for the design of the software, matching the needs of a certain task, had been changed during the development process, and the organisation of work and the tasks of the users had been changed after the introduction of the software. In addition, the "users" did not match the initial target population because many of them had joined the company after a merger with another organization that itself was the result of another merger. All these developments had taken place within the life span of the software. To make things even worse, the age of other software to be used in the same context ranged from some months to more than 30 years. Queries with other companies from the same business areas revealed that this situation was not even considered remarkable. It seems to be normal.

The analyses of the organizational and social environment performed after learning these facts have shown that many of the problems that had motivated the research work are caused outside the domain where ergonomists interested in hard- and software (named dialogue interface or tool interface in Fig. 2). We call these "Non technical interfaces" because these domains are normally not accessible for ergonomists. In case of this project, the software and application designers had limited access, and only few influence on procedures in the higher levels where the tasks are distributed between humans. Thus, they rely on relevant targets determined there, and suffer from later decisions changing the targets.

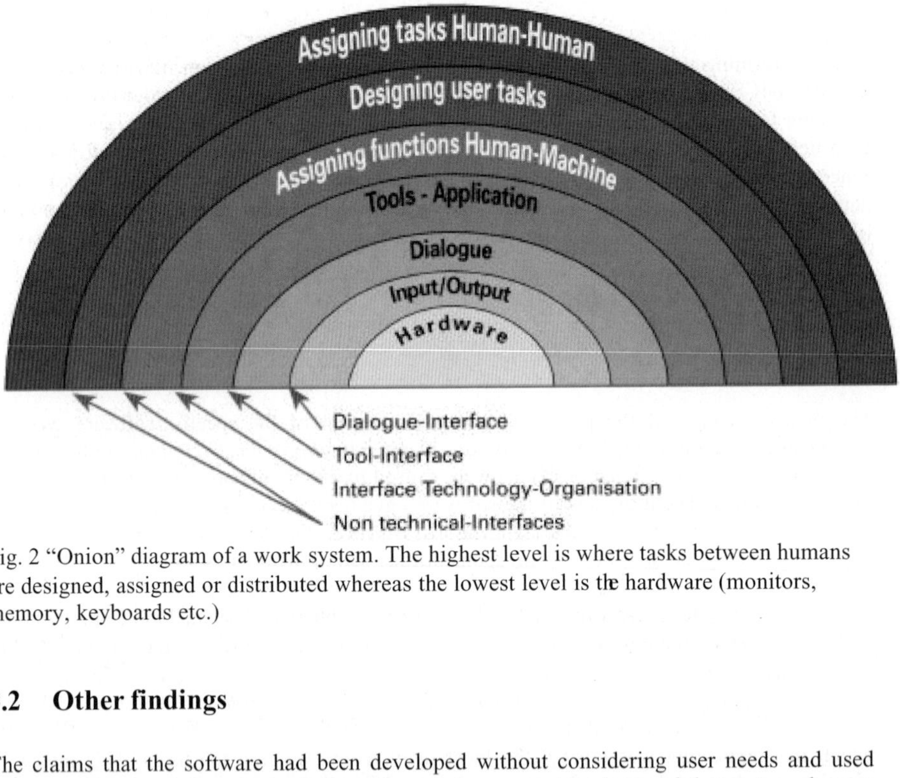

Fig. 2 "Onion" diagram of a work system. The highest level is where tasks between humans are designed, assigned or distributed whereas the lowest level is the hardware (monitors, memory, keyboards etc.)

3.2 Other findings

The claims that the software had been developed without considering user needs and used without taking care of error reports turned out to have a real background, but not much more than a tale. In reality, the software designers were obliged to involve users in the development and they had done so, however, without a sound methodology. Their chances in developing an acceptable method have not been very high because "the users" have been employees of organisational units feeling free whether or not to send users to the project. Those who had done so had sent persons with very different backgrounds. Some of them were "end users" whereas others were representatives of the organizational unit. In some case, highly motivated and skilled people were sent, and in other cases those believed dispensable.

It was not even nearly true that the software people did not care for user problems reported to them. They have been continuously working on the further development of the system, not only debugging it. The real problem causing the users to believe they would not be taken seriously had occurred after the introduction, when the users were asked to report their problems. They did, and the software developers were overwhelmed by the submissions. Later, when there were only few reports, the software team believed the problems had been mostly solved. The real reason was the disappointment of the users who did not receive adequate response.

The comparison between the anticipated work procedure for a given task and the real way how the users worked revealed that almost no user worked as expected, sometimes for surprising reasons having nothing to do with the software. E.g., they kept feeding a database with new records instead of checking the validity of the existing corresponding record because creating a new record was awarded with bonus points but not checking an existing one. When the software was being planned the bonus system was not existent.

This system had also generated the problem the users hated most. They complained that transactions initiated and numbered by the system would be too difficult to work on whereas self-initiated transactions would cause less or no problems. The description of the problem gave the impression of a major software problem. However, this did not exist at all. The real cause was very human. In fact, working on a case initiated by the computer took longer and required more work to achieve the same goal but the cause was not the computer initiating the task. It had to start a transaction when the responsible worker for a given task was not present. In this case, somebody else had to collect the data needed to process the transaction and hand it over to the responsible person. There were two major human problems associated to this. The first is that collecting data for somebody else, mostly an unknown colleague may always be difficult because of different working styles. The second problem generated by the bonus system was that bonus points were given for starting a transaction, not for completeness and quality of the work done. The responsible user had to check the entire record and all what had been done before taking further steps. And only checking required more work than creating the entire transaction. Since everybody in the company played the role of the responsible person for a part of the day, all of them disliked the feature and, in consequence, also the software. Fact is the feature and the problem created by it had nothing to do with the software. The problem will exactly be the same if somebody takes records for a colleague during her of his absence.

3.3 Lessons learned

The analyses performed during this project showed that an application in a widely distributed company is a living entity rather than a product developed following a plan to reach a target. Unfortunately, this living entity is not single a well-shaped body with organs living in synergy, it is also no organized network of groups, but rather a tangle of networks or clumps of different organisms like in a Portuguese man of war. In difference to the groups in a Portuguese man of war, parts of the entity may disappear suddenly or new parts may join the party.

Technology is only part of the entity with some extremely vivid parts but also with other antiquated elements. If severe problems with technology occur their analysis should always include the entire entity to be able to find the real causes. Else, one may end with symptoms.

References

Gediga , G., Hamborg, K.-C., & Düntsch: The IsoMetrics Usability Inventory An operationalisation of ISO 9241-10 supporting summative and formative evaluation of software systems, Behaviour and Information Technology, 18 , 151 - 164

A Training System for Maintenance Personnel in Nuclear Power Plants

Mitsuko Fukuda *Yukiharu Ohga*

Hitachi, Ltd. Power & Industrial Systems R & D Laboratory
7-2-1, Omika, Hitachi, Ibaraki, Japan
fukuda@erl.hitachi.co.jp ohga@erl.hitachi.co.jp

Abstract

To realize effective and efficient training of maintenance personnel in nuclear power plants, a training system for the operation of the instrumentation and control equipment is being developed. The system consists of virtual training devices, a plant simulator, and human simulators. The virtual training devices provide various training scenarios involving various controllers and field devices. Human simulators take over the tasks of the trainee's coworkers and enable training for cooperative maintenance tasks.

1 Introduction

In nuclear power plants, there are many instrumentation and control systems. At the present time, the training of maintenance personnel is performed by using actual devices or their mock-ups. To train personnel in maintenance of a control system, it is necessary to prepare a set of controllers for the system that is almost the same as the actual set used in a real plant. There are many different control systems in nuclear power plant, so that it is difficult to prepare all the control systems required for training. As a result, the scope of training is restricted to a limited number of control systems.

Instrumentation and control equipment connected functionally is installed separately in different locations in power plants. The maintenance and inspection of equipment are usually performed cooperatively by a maintenance team, the members of which communicate with each other by telephone. For example, one member covers the operating console in the control room, another covers the controllers at the rear of the console area in the control room, viz., the back panels, and the other covers the field devices, such as valves, installed in the plant local area.

In the current training program, an instructor is required to operate equipment in order to make changes in the trainee's environment and to communicate with the trainee. Therefore, the trainee requires the assistance of an instructor to carry out training in cooperative work.

A training system is now being developed to cope with these two problems; the restriction placed on the number and type of systems that can be used in training and the requirement for an instructor. The system simulates the various situations encountered in maintenance operations by applying virtual reality techniques. It provides a self-instructional environment for cooperative work by using a human simulator.

2 System Composition

The composition of the training system is shown in figure 1. The training system consists of three groups. One is a group of training devices that provide the corresponding training environment for

each position, such as the console in the control room, back panels, and plant local area. These three training environments can be created by three-dimensional computer graphics, so that various training scenarios with various controllers and field devices can be provided. During training on console operation, a trainee uses a virtual operating console. To conduct training in the operation of the back panels, a trainee uses virtual back panels.
For the field devices, the trainee can see the status of devices and change it, but physical exercises, such as operations to open valves, are not included. Commercial haptic devices have been successfully applied to medical training, etc. (Salisbury & Srinivasan, 1997), but it is difficult to realize physical feedbacks for the operation of field devices, such as valves, using the commercial haptic devices. The current training program using real devices is suitable for physical exercises of field device operations. The training system is focused on the training of cooperative maintenance task on instrumentation and control equipment, so that we set aside the physical training for field devices.

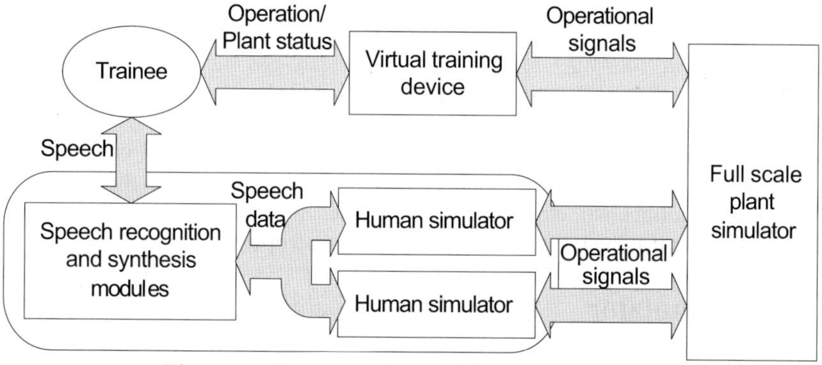

Figure 1: The composition of the training system

The training system includes a full-scale simulator of a boiling water reactor, including controllers. It receives operational signals, calculates the influences of the signals on the plant status, and changes the conditions of the console, the back panels, and the devices in the local area.
One or more human simulators are used according to the training situation. One human simulator takes over as a co-worker at a certain position. It executes the required operations of the maintenance task, and communicates with the trainee by means of a speech recognition module and a speech synthesis module. So that the trainee can carry out training in cooperative tasks without the assistance of an instructor.

3 Virtual reality training environment

A virtual training device consists of display and input devices. To conduct training in the operation of the back panels, the trainee uses a touch-sensitive display as shown in figure 2. The trainee performs an operation, such as jumper setting and lifting of cables, on the virtual controller. A touch-sensitive display that displays a controller unit in real size has been selected as the device, because every component can be placed at its proper position and the trainee can practise in the actual posture. Also by using a touch-sensitive display, the trainee can easily perform the operations with the small switches and terminals.
For a virtual training device, the controller data are prepared for each control system. The controller data include three-dimensional graphic data for appearance, and functional data for motion of the controller. The virtual training device with the neutron monitoring controller data

provides a training environment for neutron monitoring system maintenance. With the feedwater controller data, a virtual training device provides the training environment for feedwater system maintenance.

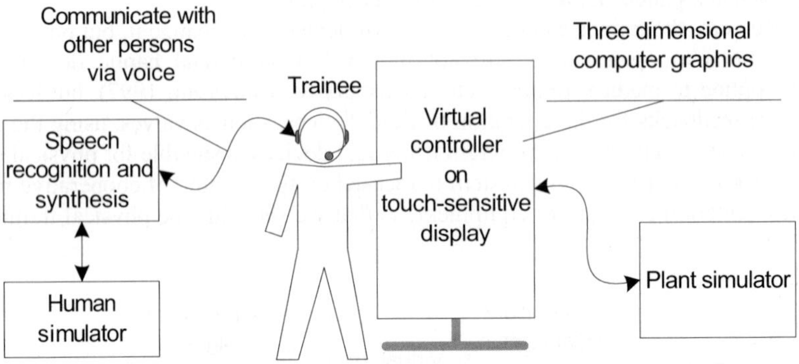

Figure 2: A virtual training device for back panels

During training on console operation in the control room, the trainee uses a virtual console. To do this, the trainee can choose a head-mounted display. Every switch on the virtual console is operated by a mouse, a trackball, etc. For the field devices, the trainee can see the status of devices, such as a valve status, on a display and change it by a mouse operation.

4 Human simulator

The human simulator takes over as the trainee's co-worker. The simulator consists of speech recognition/synthesis modules and an action module. Using these modules, the human simulator recognizes the words spoken by the trainee, operates devices and switches and confirms their status, and speaks to the trainee. These modules are activated in the order specified in action procedure data, in which procedures of operation and communication with the trainee are prepared for each unit of action, such as operation of a switch, etc.

Figure 3 shows the functional structure of a human simulator. When the trainee speaks, a speech recognition module generates the recognized word sequence. A keyword matching unit refers to the words and the action procedure data, and judges the activate condition in a current action statement, for example '(wait "Valve-lock condition is set")'. When the recognized words agree with the words in the statement, the succeeding statements are executed. If the statement is on an operation, the human simulator generates the corresponding operational signals and sends them to the plant simulator. If it is a confirmation of the plant status, the human simulator looks up the plant status and reports the result to the trainee.

Preparing the action procedure data manually for various training situations is a very laborious task because it must include the detailed action sequence that enables a human simulator to execute step-by-step maintenance actions. To cope with this problem, a program tool has been developed to generate the action procedure data automatically from the existing documents on maintenance procedures. Figure 4 outlines the generation of action procedure data. The document described in a certain style, such as a table, is parsed and analyzed to extract the data that specify maintenance tasks. Usually the document only describes the essential operations and the related plant status without the information about executers and timings. The developed tool refers to the knowledge about the typical team configuration and the standard execution process of each type of maintenance task, such as interlock test. The tool derives the skeleton of the action procedure from the document data. The skeleton describes when and who executes the specified task. Next, the detailed action procedure data are derived by using the production rules to add the communication data necessary for cooperation and the complementary actions to execute the task. By using the tool, the repertoire of training scenarios can be added to easily.

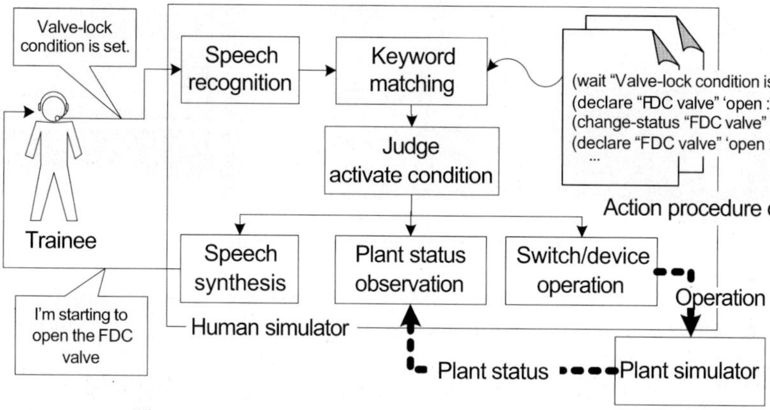

Figure 3: Functional structure of a human simulator

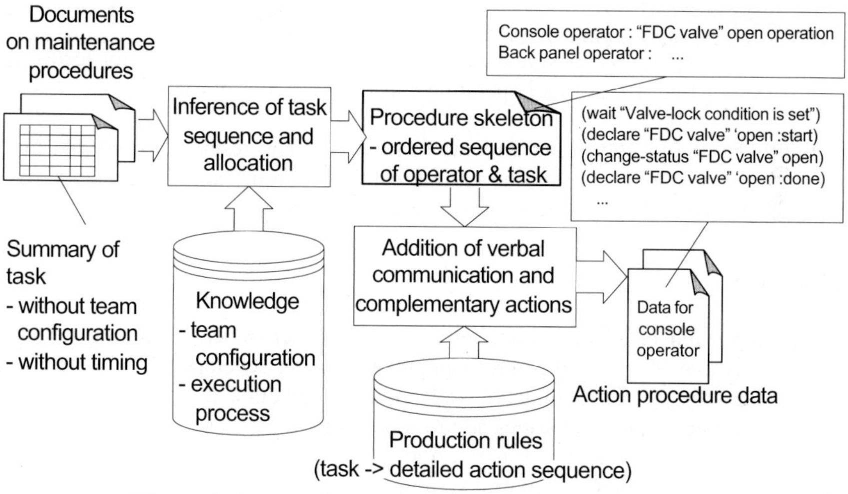

Figure 4: Automatic generation of action procedure data

5 Prototype system for feedwater control system maintenance

A prototype of the training system has been developed. It includes the control room console, the back panels for the feedwater control system, and a boiling water reactor plant. Figure 5 shows example graphics of the virtual feedwater controller. A trainee can watch the virtual controller and open the door of the unit. Each module rack can be opened. Every module on the rack can be pulled out and inserted, and each switch and terminal of the modules can be touched.

Figure 6 shows the operation consoles in the virtual control room. During training on console operation, a trainee walks through the virtual control room to operate the switches, and to look at monitors and indicators.

(1) Overview (2) Module rack (3) CPU modules

Figure 5: Virtual feedwater controller

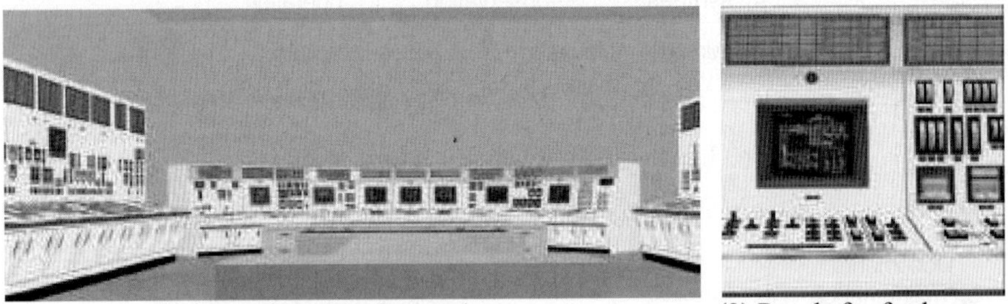

(1) Operation console panels (2) Panels for feedwater

Figure 6: Virtual operating consoles in a control room

An evaluation using the prototype system was carried out. The results showed that the training environment can be prepared by using the virtual reality techniques and the human simulator can correctly simulate a maintenance worker and can communicates with the trainee.

6 Summary

To realize effective and efficient training of maintenance personnel, a training system for the operation of the instrumentation and control equipment is being developed. The system consists of virtual training devices, a plant simulator, and human simulators. It provides various types of training for cooperative work with various controllers and field devices.

The training system will facilitate the effective and efficient training of the maintenance personnel in operations on instrumentation and control systems.

Reference

Salisbury, J.K. and Srinivasan, M. A. (1997). Phantom-Based Haptic Interaction with Virtual Objects, IEEE Computer Graphics and Applications, 17-5, 6-10

Evaluation of Workplace for People with Alternative Abilities

Ewa Górska, Jerzy Lewandowski

Institute of Production Systems Organization
Narbutta 85, 02-524 Warsaw Poland
e.gorska@wip.pw.edu .pl; j.lewandowski@wip.pw.edu.pl

Abstract
In light of the present law, workplaces should be organized in such a way as to enable people with alternative abilities to be employed and work there. It is self evident that in many cases people with alternative abilities can only perform their work if workstations and rooms have been properly adapted – that is involving all the architectural requirements to their needs. This research paper presents the concept of tasks selection and organization of workstations in accordance with the kinetic and perception abilities of a person with alternative abilities with limb impairment.

1 Introduction

A steady increase in disabilities makes different countries and international organizations take wide range of actions to help people with alternative abilities and limit disabilities threats.

In Poland, as GUS (The Main Statistics Office) revealed, in 1988 there were more than 3,258 people with alternative abilities that is about 10% of the whole population, 85% of which have no chance to be employed. Expertise carried out by The Committee of Rehabilitation and Social Adaptation of PAN (National Academy of Science) showed that in 1999 that number grew to 4,538 and it is estimated that in 2000 it is bound to reach about 12-15% of the general population. The number of handicapped is growing dramatically, for example annually there are 1600 new cases of people who have to use a wheelchair permanently.

2 Formulation of the Problem

Legal regulations, both in the country and worldwide, oblige employers to create new workplaces for people with alternative abilities. However they do not specify how to proceed with this serious problem in order to make people with alternative abilities use the opportunity and take the employers' job offer.

Hence in many scientific centers there are new initiatives and actions being taken in hope to increase chances for employment of the disabled. This is done by learning about needs and potentialities of people with alternative abilities (Kurkus-Rozowska, 1997), (Nevala-Puranen, Louhevaara, Itäkannas, Alaranta, 2000), (Nevala-Puranen, Seuri, Simola, Elo, 1999), developing methods which allow work and workstation analysis (Tortosa, Ferreras, Garcia-Molina, Diez, Lazaro, Cerezo, Page, 1999), (Välima, Mikkonen, Kantolahti, Palukka, Mattila, Rautjärvi, Kivi, 1999) and introducing methods which concern adaptation of workstation to the limited abilities of people with alternative abilities (GUS, 1997).

In spite of many attempts, there are no efficient tools allowing precise and effective adaptation of workstation to the needs of people with alternative abilities. This situation is brought about by both great variety of diseases on one hand and workstations on the other.

Amongst many areas, which require close analysis, this research focuses on the issue of the spatial design of workstation. First of all, it is a question of designing workstations in accordance with the needs of people with alternative abilities with limbs impairment.

The analysis, as it has been mentioned before, is limited to movement disabilities as the most numerous and at the same time most complex area of dysfunctions connected with human movement system.

It has been assumed that in order to choose a good workplace for a person with alternative abilities it is necessary to:

- carry out complex analysis and evaluation of a person with alternative abilities paying special attention to his natural skills and abilities, and not disabilities,
- carry out analysis of work and workstations requirements,
- compare the results of both analyses and define relations between skills and abilities of a person with alternative abilities on the one hand and workstation requirements on the other and on the basis of both generate an adequate workplace for a given person.

The first step was intended to systematize many different types of movement dysfunctions for the needs of engineering design of workplace spatial structure.

The second step was meant to systematize and classify tasks and activities performed at workstations according to needs and requirements of people with alternative abilities with movement dysfunction.

Physical adaptation of workstations can be resolved into introducing changes or modifications which consist in establishing adequate measure proportions of workstations, adequate arrangement of equipment, installing additional elements or removing those which hinder performance of some given professional activities. If need be, an employer should provide a disabled employee with rehabilitation equipment which is adapted to the type of disability and which enables successful performance of professional activities.

The research has been carried out with the assumption that a workstation should be designed in such a way as to guarantee autonomy and independence of a disabled employee. What is more a workstation should not be recognized as designed specially for a person with alternative abilities – it should be universal enough to make it equally good for a regular, not disabled employee.

The research has been finalized with the elaboration of computer program which supports decision taking in conditions where there are given two sets of elements interacting in some specific situations. These interacting sets or modal teams are: an employee with perception and movement dysfunctions and workstation, which can be adapted to needs and abilities.

3 Main Assumptions on the Workstation Designing System for People with Alternative Abilities

With the use of the above mentioned questionnaires a lot of information has been collected on perception and movement abilities of people with alternative abilities and requirements posed by workstations. However we analyzed only workstations situated in the companies of work protection. Work results served to build database and database of system knowledge allowing to:

- collect information on disabilities among certain population
- collect information on working environment where people with alternative abilities could be possibly employed
- search both databases in order to find functional correspondences in the other database. The search can be carried out using equivocal criteria. Functioning of such tool enables finding for example not concrete and ready workstations for people with alternative abilities but a group of workstations together with a list of their possible modifications. The role of decision-maker consists in establishing which of the changes proposed by the system are to be actually realized
- indicate proper matches at a workstation in order to employ adequate people with alternative abilities at a given workstation
- suggest possible modification of a workstation in a given range so that modifications could meet the requirements of an employed person

What is more the elaborated system provides opportunity for creating a general form endowed with specific dysfunctions. It also allows creating a suitable workstation for a disabled employee, fully adapted to his/her needs. Finally it offers a simulation of human – workstation interaction system.

Project of shape modeling has been divided into the following stages:

- model of database including information on anthropometrical model forms,
- model of database on diseases and disturbances according to medical terminology,
- model of knowledge base enabling translation of "medical base" into parameters required by the base of models' parameters,
- interface: base of parameters – Anthropos or Catia

An introductory classification has been carried out for the method's sake. It includes:

- damages and losses in the anatomic structure of movement organs
- disturbances of motor functions
- deformations

Designed system has to serve:

- selection of workstations possessing suitable parameters and advantages for development of a given type of dysfunction
- finding among people with alternative abilities those who can work at the workstation possessing specific parameters

In both cases we establish the priority: a human can only work at such a workstation where basic functional and ergonomic parameters are at least on the level which fulfills the requirements (mainly healthy ones) of a potential employee. This is the basis criterion deciding about selection of workstations for people and people for workstations.

The system will in the first case search workstations database for such places where functional parameters are superior to the minimum requirements of a person with alternative abilities.

In the second case the working of the system is similar and basically conducted in the same way but due to the output of database, the base of persons will be searched in order to find people whose requirements are lowers to those accessible at the analyzed workstation. In both cases a human with his dysfunctions is the objective and most important factor.

Information on workstations and people with movement dysfunctions is introduced into the system from specially designed data questionnaires.

The system has computer implementation and because of this it is very accessible and does not require from the user any special knowledge.

It is designed for two groups of users:

- "trainer" – user whose task is to teach system how to make associations according to the given criteria. The user has right to change the rules of drawing conclusions and to adapt source database. Source data serves in turn the process of verifying hypotheses and conclusions, which enable efficient functioning of the system.
- "client" – user using the system in order to select and match successfully people with alternative abilities with workstations

4 Results

Issues presented and discussed in the article have put into view complexity of the problem concerning adapting working conditions to the requirements of people with alternative abilities. Employing people with alternative abilities in normal companies where there are work conditions suitable to their psychophysical abilities and needs helps abolish barriers and creates better chances for active participation in social life. It also gives people with alternative abilities equal opportunities to function as fully valuable members of our society.

5 References

Kurkus-Rozowska B, (1997). Evaluation of Psycho-Physical Needs of Disabled People for Their Most Favorable Employment, Problems of Social and Professional Rehabilitation No 1, CEBRON Warszawa

Nevala-Puranen N., Louhevaara V., Itäkannas E., Alaranta H., (2000). ERGODIS: A Method for Evaluating Ergonomics of Workplaces for Employees with Physical Disabilities, Proceedings of ERGON'AXIA 2000, May 19-21, 2000, Warsaw, 237-239

Nevala-Puranen N., Seuri M., Simola A., Elo J., (1999). Physically Disabled at Work: Need for Ergonomic Interventions, Journal of Occupational Rehabilitation, Vol. 9, No. 4, 215-225

Health Condition and Needs of Disabled People in Poland in 1997, (1997). GUS, Warszawa

Tortosa l., Ferreras A., Garcia-Molina C., Diez R., Lazaro A., Cerezo C., Page A., 1999, Ergowork-IBV: Method for ergonomic adaptation of workplace for disabled people w: Proceedings of CAES'99 International Conference on Computer-Aided Ergonomics and Safety, May 19-21, 1999, Barcelona

Välima M., Mikkonen P., Kantolahti T., Palukka P., Mattila M., Rautjärvi L., Kivi P., (1999). The Model for Analysis of Work Ability at SMEs, w: Proceeding of the International Conference on Computer Aided Ergonomics and Safety CAES'99, May 19-21 1999, Barcelona.

6 Permissions

The research was supported by a grant from the National Science Committee No 7 053 16 supervised by Phd. Ewa Górska

Symptoms of depression in the VDT – operators.

Anna Janocha[1], *Ewa Salomon*[1], *Ludmiła Borodulin-Nadzieja*[1], *Robert Skalik*[1], *Małgorzata Sobieszczańska*[2]

Department of Physiology[1] and Department of Pathophysiology[2] Medical University of Wrocław
Chałubińskiego 10 [1], and Marcinkowskiego 1 [2]; 50-368 Wrocław, Poland
ajanocha@fizjo.am.wroc.pl

Abstract

Everyday long-term contact with computer screens is an occupational risk, which induces various undesirable health consequences, including subjective complaints of central nervous system (CNS) e.g., fatigue or loss of energy, headache, insomnia, diminished ability to think or concentrate and depressed mood. The same symptoms cause clinically significant distress during depressive episode.

The present investigations were undertaken in order to evaluate usefulness of the self-assessment charts like **B**eck **D**epression **I**nventory (BDI) for detecting depressive symptoms and **K**imberly **Y**oung **T**est (KYT) for detecting Net-addiction in the VDT – operators.

The examinations were carried out in the 60 persons. The study population comprised two groups. One group, was constituted by the 30 students of Technical University, working with computer from 5 to 8 hours every day (VDT-operators). The second group, serving as a control one, comprised the 30 students of Medical University, who worked with computer occasionally.

The obtained results suggest that everyday, long-time work with computer may lead to depressive symptoms. Our observations are supposed to be the direct implications of the exposure of the VDT-operators to the harmful occupational factors resulting in functional disorders within the CNS.

1 Introduction

Computer is a flabbergasting and full of genius invention of 20th century, which made human work more efficient. In spite of undeniable benefits, computer revolution is assumed to endanger human health. Sight exposure to electromagnetic irradiation is the most frequently mentioned negative factor. However, it is often forgotten that there is a whole gamut of negative computer-related factors like forced slouching position in front of the computer screen and the others, which may impair human muscle-skeletal and circulatory system. What is more, stress and addiction to computer is often ignored (Bednarski & Zieliński, 2000).

Internet is a medium, which gives a brand-new dimension to human communication. More and more people take advantage of web site for professional, education and distraction reasons. Anonymity, comfort, easy access and low usage cost add to the steadfast continuously grow demand for computer technology (Matuszczyk, 2000).

In spite of conspicuous advantage from the market expansion of web site, one should remember the negative aspects of this phenomenon, namely **I**nternet **A**ddiction **D**isorder (IAD). A psychologist, Kimberly Young from University of Pittsburgh-Brandford (USA) established the following IAD subtypes: cyber-sexual and cyber-relationship addiction, net compulsions, information overload and computer addiction (Świerzbin, 2002).

Computers and Internet attracted millions of people from all over the world. According to the statistical data, the number of computer addicts aged from 6 to 70 years reaches approximately 5 million in the USA. The official European computer addict statistics lags behind. However, the data are supposed to be underestimated, because the addicted persons rarely look for specialist counseling.

2 Methodology

2.1 Material

The 30 students of Technical University, VDT – operators (20 male and 10 female, age range: 20-25 years, mean age: 22,3 ± 1,47 years) were included into the study. A 7-hour task of the dialogue type was performed by these operators. A control group (group II) comprised 30 students of Medical University, working with computer occasionally, was also studied. The both groups were matched for sex and age. In the control group mean age was 23,5 ± 1,42.

The VDT operators performed their work using monochromatic monitors with a graphic card „Hercules". A „light on dark" mode of presentation was used on the VDT screen (green and amber characters against dark background). No protective filters were used. The distance between the eyes of VDT operators and the monitor was about 60 cm and the magnitude of the presented characters was about 4,5 mm. Monitors were situated on the ordinary desks with their sides facing the windows. It was not possible to regulate the height of different elements used during VDT operations, such as monitors, documents and keyboards nor different parts of chairs used at the work places.

2.2 Methods

Self-assessment charts, like, e.g., **B**eck **D**epression **I**nventory (BDI) used in the present study, are considered to be of great usefulness in establishing a diagnosis of depression. It is very easy to apply such a tool. The examined subject is requested to complete the questionnaire which consists of 22 issues; for each of the questions you can choose only one answer (yielding 0-3 scores).

It is assumed that the outcome of the whole test exceeding 10 scores indicates depression. This test concerns the previous day and enables to assess the psychological condition of the patient in the more objective manner. Although this scale is used as a self-assessment test, the reliable interpretation of the results can be performed exclusively by the person highly qualified in psychopathology. In our study, BDI was evaluated by a psychiatrist.

For detecting IAD self-assessment charts, like, e.g., **K**imberly **Y**oung **T**est (KYD) used in the present study, were undertaken in the VDT – operators. It is very easy to apply such a tool. The examined subject is requested to complete the questionnaire, which consists of 8 issues; for each of the questions you can answer only yes or no (1 score). It is assumed that the outcome of the whole test exceeding 5 scores indicates IAD.

2.2.1 Statistical analysis.

Non-parametric data was given as the absolute numbers or percentages. A comparison analysis was performed using the Chi-square method.

3 Results

The analysis of the BDI outcomes was the base for dividing the VDT – operators group into the three subgroups depending on the BDI scores obtained:
Ia – students obtained borderline results = 10 scores,
Ib – students obtained ≥ 12 scores,
Ic – students obtained normal results < 10 scores.
Average result for the group I was 9,4 ± 1,38 scores. The BDI outcomes were correlated with KYT results evaluating Net- addiction. Result of this correlation is depicted in the table 1.

Table 1. Characteristics of group I and group II.

Subgroups	Number of subjects	BDI scores	KYD skores	Male	Female	Symptoms of depression
			GROUP II			
Ia	15	10	5 – 7	10	5	+
Ib	3	≥12	8	0	3	++
Ic	12	<10	<5	10	2	–
			GROUP II			
IIa	5	10	<5	2	3	+
IIb	25	<10	<5	18	7	–

Referring to KYT, IAD was recognized in all persons from Ia and Ib subgroups. However, persons from the Ib subgroup reached maximal score. It must be also stressed that persons from the Ib subgroup spent their whole time for browsing the Internet. Probably, the addicted persons did not reveal the factual time they browsed through Internet web sites. We assume that the Web browsing time was longer than it was stated by the examined persons on the questionnares. Net-addiction was not found in the Ic subgroup.
Among addicted persons typical depressive states of low spirit are accompanied by the additional symptoms such as fatigue or lack of energy, inability of joy experiencing, low self-esteem, headache, insomia and diminished ability to think or concentrate.
Completely different results were obtained in the group II, serving as a control. The analysis of the BDI outcomes permitted to divide this group into 2 subgroups:
IIa – students obtained borderline results = 10 scores,
IIb – students obtained normal results < 10 scores.
In the control group average result was 6,16 ± 1,80 scores. The average result of BDI revealed statistically significant differences ($p<0,001$) in the both groups. The control group was Net-addiction free. Comparison of group I and group II is also depicted in the table I.
The comparative analysis with relation to the male and female percentage contribution demonstrating various states of Net-addiction showed the evident distinctions – figure 1.

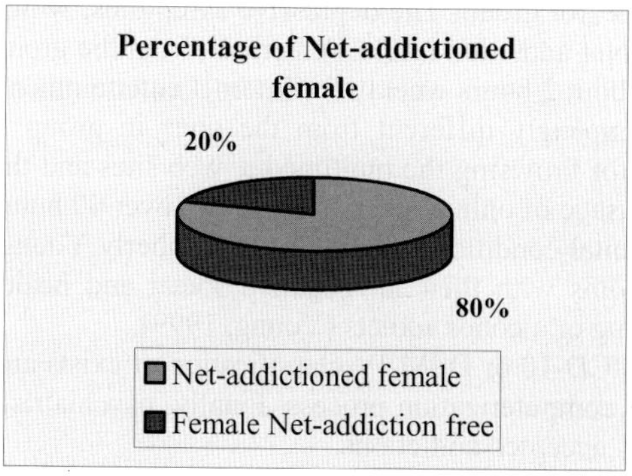

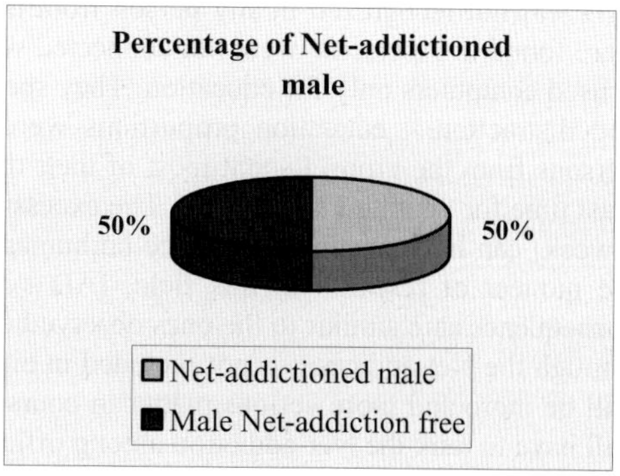

Figure 1. Percentage of Net-addicted male and female in the VDT – operators group.

The statistically significant differences (p<0,001) were found in the subjects depicted in the figure 1.

4 Conclusions

A lot of monographs on detrimental influence of computers on human organism, and especially visual organ, have been written as yet. However, there are few reports on influence of computerization on human mentality. The question of Net-addiction is more and more frequently raised nowadays. According to American specialists, multi-hour Internet Web-browsing can make computer users addicted (Kościelniak, 1997). Psychologists state that only 1-hour Web-browsing a day is safe for computer users. Persons, who spend about 40 hours weekly in front of computer screen, are deemed to be addicted (Stępień, 2002).

The authors of this study took an effort to evaluate influence of computers and Internet on mentality of the examined persons by means of simple, but generally used psychological tests. The results in the group of VDT – operators, obtained by means of BDI, confirmed the preliminary diagnosis of depressive syndrome on the basis of VDT – operators complaint report, especially depressed mood, lack of energy and inability of joy experiencing (of, at least 2-week duration). BDI results correlated proportionately with KYT score – the more pronounced symptoms of depression the more pronounced symptoms of IAD (table 1). At this stage of the research, it is extremely difficult to find out which of these phenomena is primary, because persons with diagnosed affective disorders seem to be more inclined to lapse into addiction, but on the other hand every addiction can be a cause of depressive states (Matuszczyk, 2000). It must be also stressed that women are more inclined to be addicted to Internet than men, which is not generally known fact. IAD was recognized in 80% of women and 50% of men in our study (figure 1). The same results were obtained by H.Petrie and D.Gunn from Hertfordshire University (Wojtasiński, 1998).

As it results from the table 1, not all persons from the group I are prone to be addicted, though the average Internet-browsing time for all members of the group I was approximately the same (7 ± 1,5 hours a day). The essential difference concerned the model of contact with computer. Persons from the subgroup Ic used computer for work and education and they spent only 4 hours weekly for Internet distraction. On the contrary, persons from the subgroup Ib used computer exclusively for entertainment and especially spent time in so called Chat Rooms, which had negative influence on their work and education.

IAD was not recognized in any person from the control group. The depressive symptoms, which were found in 5 persons, were not connected with Net-addiction (table 1). Persons from the group II used computers only for education. They spent about 2 hours weekly for Internet entertainment, and distraction – education proportions were completely different from the ones in group I. Persons from the group I spent most of their time for browsing the multimedia, web sites and the least time for visiting Chat Rooms. The excessive usage of online computer service (over 40 hours a week) can have negative influence on human mental condition. According to Kimberly Young, the pioneer of research in this field, IAD symptoms with their all negative social and health consequences are similar to the ones observed in drug or alcohol addicts (Young, 1999).

Though the Net-addiction is not included in either ICD-10 or DSM-IV classification, it exists and will be more and more serious matter in course of computerization process. Finally, psychiatrists will have to rank the Net-addiction among officially accepted addictions.

References

1. Bednarski T., Zieliński T. (2000). Komputer a zdrowie. Retrieved December 9, 2002, from www.infoholizm.prv.pl
2. Matuszczyk M. (2000). Uzależnienie od Internetu. Biuletyn Elektroniczny Psychiatrii Online, 9/2000.
3. Świerzbin K.M. (2002). Uzależnienie od Internetu jako nowa jednostka chorobowa. Szkic problematyki. Retrieved December 9, 2002, from www.infoholizm.prv.pl
4. Kościelniak P. (1997). W sieci nałogu. Archiwum Rzeczpospolitej Online, 8/1997.
5. Stępień A. (2002). Komputer czy wróg. Retrieved December 9, 2002, from www.infoholizm.e-sfera.pl
6. Wojtasiński Z. (1998). Kobiety uzależniają się od Internetu. Archiwum Rzeczpospolitej Online, 12/1998.
7. Young K. (1999) Internet addiction: symptoms, evaluation and treatment. W: Vande Creek L., Jackson T. (Red.). Innovations in Clinical Practice: A Source Book. Profesional Resourse Press 1999, 17, 19-31.

Human Factors as a Determinant of Quality of Work

Aleksandra Kawecka-Endler

Poznan University of Technology, Institute of Management Engineering
Strzelecka Str. 11 60-965 Poznan Poland
Aleksandra.Kawecka-Endler@put.poznan.pl

Abstract

The achievement of high results in the way of quality is possible by assurance of industrial production. This production is the integrate process, in which the most essential and most decisive factor is a man. In the paper has been present the analysis of many factors and problems participated in creating of quality of work, according to the principles of humanizing work.

1. Introduction

Poland is approaching the date of entering the European Union and the fact is accompanied both by expectations and future responsibilities. One of the priority obligations is for the Polish regulations and solutions (in many aspects of life, in economic, legal and social fields) to conform to the Union requirements.

One of the most important factor of enterprise competitiveness is the quality of products and services offered on a market (national and international). Quality determines the effectiveness and productiveness of an enterprise which are necessary to succeed.

Major goal of common European market of products and services is unification of a system of requirements and the possibility of their precise determination and exaction. Achieving this goal guarantees several benefits:

- Improvement of quality and competitiveness of products and services to customers (price, modernity, attractiveness, security, variety).
- Systematic and consequent withdrawing of technologies harmful to men and environment, which will force implementation of solutions compatible with ergonomics and environment protection standards.
- Development and spread of ergonomic design methods (or ergonomic verification) of products and processes, as well as technical objects, machines, equipments and tools.
- Gaining detailed knowledge about the influence of production processes on work conditions (micro scale) and on environment (macro scale) in order to efficiently diminish them (to meet the requirements stated in norms) or totally eliminate them.
- Improvement of general awareness among the enterprise workers (and entire society) of quality aspects, the meaning of good and safe work conditions, which should be the base of all integrated actions aiming at environment protection.

The research run in both polish and foreign enterprises prove the existence of immense reserves connected with several aspects of humanization of work environment, being hidden in so called human factor. However, in order to free this potential, it is necessary to use particular solutions which take into consideration the humanization aspects in designing the work processes and which are integrated with the whole production (or service) activity of an enterprise (Kawecka-Endler, 2002).

2. Condition of polish enterprises

In the last ten years Poland has experienced transformations which have brought about fundamental changes in the company structure and organization and in its surroundings. To meet various demands and to be competitive on national and international markets, Polish companies must cope with current tasks in a rational and methodical way. They must also analyze their execution and adjust it appropriately, taking into account organizational capabilities and surrounding hazards.

In countries characterized by stabilized free market economy such requirements are best met by small and average enterprises, which lead in creating the GDP and constitute around 70% of all enterprises (in western countries, USA, Japan). It should be assumed that its share in country's industry should be close to those figures.

In Poland, the number of small and average enterprises has been systematically increasing since 1990. According to GUS (Central Statistical Office) in 2000 its share in GDP was around 49,4% and its number is estimated at 1,7 – 1,8 millions. This relation is still far from assumed.

An existing state is influenced by a variety of factors, among which the most important is insufficient help in: financing, crediting, advising, informing. Such help is necessary in order to create the organizational structures of small and average enterprises adequate to EU requirements.

In Poland it is crucial to work out and implement the effective solutions for small and average enterprises as they form a flexible and effective organizational structure able to quickly transform and adjust to changing market conditions.

An important role in this transformation is also played by innovation processes, and among them the implementation of systems of quality management and environment protection and work safety. Innovation processes are very complex and require an individual approach in each and every enterprise, depending on the character of realized processes. Financial means available in small and average enterprises are not sufficient to hire the advisors (experts).

Financing such undertakings is the domain of average and big enterprises, since they posses certain means and funds necessary to realize them.

Assuming that in production (and services) the quality is a priority, both in home market and in international exchange (EU and world standards), it is necessary to work out solutions which can help small and average enterprises in quick and successful achieving of good quality of work and production.

3. Human factors of work processes

The 60's and 70's popularized the sociological approach and anthropocentrism and started the systematical actions for humanization of work environment in industry. The core of modern organizational, ergonomic and quality solutions is a paradigm which states that *"a worker is the most important and most valuable factor of production processes"*.

Major differences in design of work processes in both approaches are shown in Table 1.

Table 1: Changes in an approach to work shaping (Bullinger 1994, p.20)

HUMAN in work process	TECHNOCENTRIC approach (technology controls human)	ANTHROPOCENTRIC approach (human controls technology)
WORK CONTENT	• Small • Deep work division • without decision making • without responsibility	• Large • Integration of tasks • Oriented at making important decisions • With responsibility
QUALIFICATIONS	• Technical	• Technical (expert) • Organizational • Social
WORK HOURS	• Inflexible	• Flexible
WORK SAFETY	• Protections	• Conceptual (design) • Prevention
ERGONOMICS	• Adjusting human to work conditions	• Adjusting work conditions to human

The requirements of **work humanization** point out the existing range of a problem, which in reality requires detailed solutions for particular line of business or even for particular work group. The following are the most important of those problems:

- Health losses caused by accidents at work and work related illnesses (they can be caused by noise, working with harmful materials and dangerous tools etc.) Most frequent work related illnesses are: loss of hearing, respiratory system illnesses, skin diseases.
- Work conditions, not necessarily harmful, can be unpleasant and hardly acceptable (can be caused by high or low temperature, noise, unpleasant smells, working outside regardless to season and in extreme climate, etc.)
- Activities requiring hard physical labor (ex. Assembly of big and heavy elements, loading and unloading), constant concentration (ex. Product control, sight quality control) or uncomfortable and not physiological body position (ex. Assembly or over-head welding)
- Monotony, especially while actions are repeated in short cycles (ex. Manual assembly and disassembly), activities in ordered sequence and beat (forced work beat) and activities without necessary work space and without possible cooperation (planning and shaping own work)
- Social isolation and (or) little communication during work as an effect of separation of work stands, which require certain conditions (ex. Materials examining with UV beams, stands creating so called protection screen) or working at home (new, developing forms of work in western countries)
- Organizational conditions regarding social relations after work, work time-offs and unfavorable for human time of work (nights, weekends, work shifts). Apart from those social life domains in which the unfavorable work time is stipulated by the nature of the work (ex. Public transport, health care, power plants) there are also situation where shifted work and weekend work are used for economical reasons, allowing better use of very expensive means of production (ex. engineers work at manager’s work stands) (Luczak, 1991, p.4).

Every problem regarding human work should be solved systematically, by analyzing all connections and relations existing in particular system. Elements of work system are shown at Figure 1.

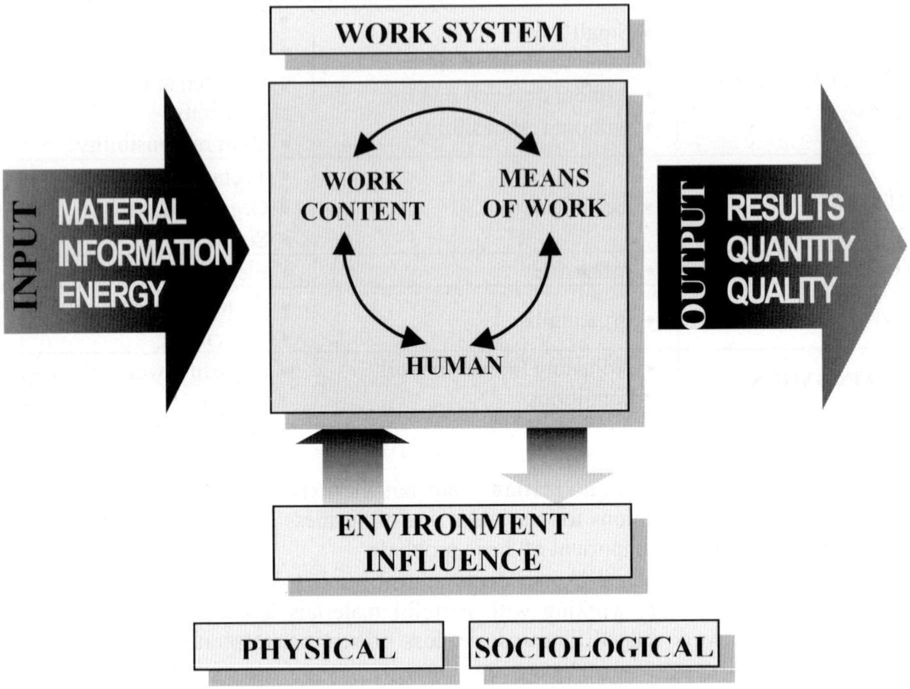

Figure 1: Elements of work system (based on Bullinger 1994, p.2)

4. Human factors as a determinant of quality of work

Quality at every stage its formation through contact with man, who directly or indirectly takes part in this process, is inseparably connected with ergonomics. It is a result of exact connection and correlation, which appear between forms of work organization, ergonomics of work condition and quality, which is received in these conditions. The real part in formation of high level of quality has system, which controls the production quality. The system contains all stages of quality formation (search at development of construction and technology, realization, product service).
Formation of work conditions should be understood as protecting the workers as direct executors of work processes in human-work system. That is why all actions in the area of work safety and security require systematic solutions. Safety and hygiene of work is the core of each and every organization and directly affects variety aspects of its activity, including the economic results. Design of work in technological processes needs integration of the accepted rules of work organization, work humanization and work production with overall activities of the entire company. According to these rules it is possible to identify some psychosocial criteria which would allow for optimization both: labor division and the resultant quality of work. Some of the criteria are as follows:

- compliance of professional qualifications of workers with the qualifications required to perform particular activities,
- degree of standard work time,
- safety and hygiene of work and work conditions at stands and at productions department,
- motivation system,
- modern methods of work organization (Kawecka-Endler, 2001).

During the work process, in which most of labor and operations is carried out at stands, work partition offers most of the possibilities to improve good work conditions. When the above mentioned principles and criteria are taken into account in the design of work structure, safe working conditions are secured in processes, which guarantees high quality of work and final production.

Quality could be systematically improvements in each phase it's formatting and it is possible to prevent mistakes by discovering them quite early. Disseminating such a activity for constant improvement quality process, products and organization.

5. Summary

The complexity and variety of quality aspects (ex. Technical, economical, social, ecological etc.) is a reason of an increase of difficulties with finding good solutions. Also from the experience of enterprises in many countries it is well known that achieving good quality results and competitiveness is not an easy task. This issue requires constant solutions searching for dynamically changing technological and organizational conditions set by various production tasks, various technologies, changing (often lowered) investment funds correlated with market fluctuation which is not easy to foresee.

Such solutions can be optimal only for specifically described technological and organizational conditions of particular enterprise. Each time they require individual, systematical and complex analysis of many factors connected with work humanization.

The great prospects on scale of implicated of work quality and products issue from wide understanding of humanizing work, especially according to work conditions, creation the division of labor and formation the compact system of motive impulses, which integrate good work with suitable wages and economic results of enterprise.

6. References

Bullinger H.-J., (1994). *Ergonomie. Produkt- und Arbeitsgestaltung,*: B.G.Teubner, Stuttgart.

Kawecka-Endler A., (2001). Ergonomic and Quality Aspects in assembly,. [in:] *Systems, Social and Internalization Design Aspects of Human-Computer-Interaction*, Volume 2, Edited by Michael J. Smith, G. Salvendy, Lawrence Erlbaum Associates, Publishers Mahwah, New Jersey – London, IEA'2001, s. 410-413.

Kawecka-Endler A., (2002). Work safety in assembly, [in:] *Occupational risk in didactic, learning and training of ergonomics, work safety and labour protection,* Monograph, University of Technology, Poznan, s. 195-203.

Luczak H., (1991). Arbeitswissenschaft (Konzepte, Arbeitspersonnen), T. 2, Technische Universität Berlin, 3. Auflage, Berlin.

Technostress, Quality of Work Life, and Locus of Control.

Khairunnisa Khan & Prof. James Fisher

University of Witwtersrand
Johannesburg, South Africa
k_khan_000@hotmail.com

Abstract

The aim of this study was to explore the associations between the proposed variables (technostress (TS), quality of work life (QWL) and locus of control (LOC)). Cluster analysis was firstly applied to all three variable instruments to establish groups of individuals with similar profiles. Tests of association were then applied between these groupings searching for qualitative structural relationships. Results reveal eight distinct clusters and significant associations between TS and QWL as well as between QWL and LOC.

1 Introduction

Brod (1982, p16) defines TS as "a modern disease of adaptation caused by an inability to cope with new computer technologies in a healthy manner" and depicts TS as manifesting in two separate ways. First is through a struggle to accept technology. Second is in an over identification with technology (Genco, 2000). This definition alludes to a paradox in the way TS unfolds. On the one hand are those who struggle to allow technology into their lives, and on the other are those who embrace technology to such a degree that it is impossible to imagine life without it. Weil & Rosen (1997) propose a more direct attitudinal influence of TS on QWL when they point out that "the omnipresence of technology, time wasting computer glitches, rapid changes in systems and machines, worries about job security and constant techno scrutiny has resulted in generalised technological angst." This overall anxiety in turn interferes with people's ability to concentrate and results in workers being less productive, less efficient and generally less satisfied with jobs. LOC orientation (Rotter, 1966) refers to a person's perception of personal control of the environment; people attribute the cause of events either to themselves or to their environments. Internal control refers to a belief that reinforcements are contingent on the individual's own behaviour, attributes or capacities. External control on the other hand is based on the belief that reinforcements are not under the individuals personal control but rather depends on fate, powerful others, chance, luck etc. (Rotter,

1966). Crable et al (1994) hypothesised that there would be link between locus of control and computer anxiety since previous research had indicated that people with a strong locus of control experience less life stress and anxiety than externals. However, their results did not confirm this hypothesis and was attributed to a lack of variance between I-E orientations in the sample. Igbaria & Parasuraman (1989) however did find a moderate relationship between computer anxiety and locus of control.

2. Methodology

This study adopts a non-experimental; cross sectional, quantitative, ex post facto research design. Various sources of TS were identified and assessed using Weil & Rosen's (1997) TS checklist as a guide to formulate a comprehensive checklist. It consists of a total of six items, (e.g. I feel as though I am overwhelmed with information) followed by a five point Likert option to respond to. QWL was assessed using the Leiden Quality of Work questionnaire (1999). Rotter's (1966) Internal –External scale was used to assess LOC orientation. A volunteer sample of (N=119) employees from a single organisation in Johannesburg (55 female: 64 male) completed these questionnaires. Cluster analysis and chi-squared analysis were applied to form groupings and test associations between groups. This provides a different insight into the similarity and differences between groups and allows a more qualitative interpretation than can be gleaned through the use of scale total scores or sub scale scores.

3. Results and Discussion

Cluster analysis revealed eight distinct clusters. Wards minimum variance was applied firstly to the profiles for TS variables. Four clusters emerged. For the QWL and LOC profiles two clusters emerged.

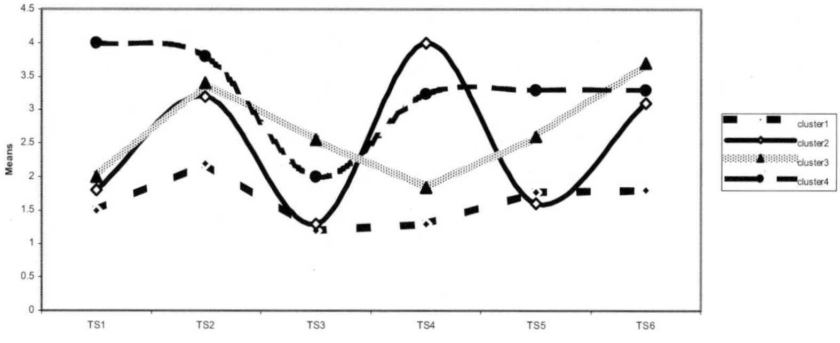

Figure 1: Pattern of responses (TS)

3.1 Comparison of TS clusters according to patterns of responses.

From figure 1 the first TS cluster reflects relatively low TS overall. The second TS cluster displayed a highly fluctuating pattern of responses. E.g. the highest mean on item four could indicate that this cluster of individuals often experience feelings of anxiety at the thought of having to operate a new technological device. The lowest peak on the third item shows that they have not been overcome by the desire to physically destroy equipment. This cluster also does not feel overwhelmed with information but they do feel frustrated at how much time it takes to use technology. The third cluster felt most frustrated at how much time it took to use technology. The second highest peak was observed at item 2 indicating that this cluster of individuals felt frustrated at the huge amount of e-mails and/ or telephone calls they received daily. Similar to the second cluster, they did not feel overwhelmed with information, but in contrast while the second cluster experienced anxiety at the thought of learning to use new technology this cluster did not. The fourth cluster stands out. The highest mean at item one shows that this cluster felt overwhelmed with information and also became frustrated at the huge amount of e-mails or telephone calls they received daily. The lowest peak was at item three indicating an occasional desire to physically destroy equipment. Items four, five and six indicate moderate feelings of anxiety at the thought of having to learn to operate new technology, feelings of intrusion on personal time and space by telephones and pagers, as well as moderate feelings of frustration at the time taken to use new technology.

3.2 Comparison of QOW clusters according to patterns of responses

QWL cluster one experienced low levels of quality of work on item two showing that this cluster do not perceive themselves as having a lot of responsibility in their jobs. The second cluster however illustrated a perception of more responsibility. The first cluster displayed the highest mean at item 18 indicating that their jobs required them to work very hard as well as a high mean at item 8 indicating their jobs to involve a lot of repetitive work. Similarly, the second cluster also perceived their jobs to involve a high degree of repetitive work. Overall the first cluster generally illustrated lower levels of quality of work with the majority of responses ranging from 1.5 to 2 while the second cluster showed higher levels of quality of work with the majority of responses ranging between 2 and 4. While the results of the cluster analysis clearly revealed a split between an intrinsically orientated group of respondents and an externally orientated group within the sample, in contrast to Anderson (1977), there were no significant associations between LOC and TS.

3.3 Is there an association between TS and QWL?

Results revealed a significant association between TS and QWL. This confirms both Bradley's (1989) and Weil & Rosen's (1997) studies that suggest an association between the experience of TS and ones perception of QWL. Most of this association was explained by the TS cluster one and the QWL cluster 2 (p = 6.26). This result seems to support Levine et al's (1984) point that QWL takes on different meanings for different segments of the workforce and thus while there is evidence to support an overall association between the two constructs, closer observation via the clustering process illustrates how certain groups tend to support such an association more strongly than the rest of the sample.

	TS clusters				
QWL	1	2	3	4	Total
1	21	19	17	18	75
2	1	6	14	23	44
Total	22	25	31	41	119

*Chi squared (df= 3) 19.06: p.0003

3.4 Is there an association between QWL and LOC?

A significant association exists between QWL and LOC. The LOC cluster one was strongly associated with the QWL cluster two (p = 1.97). This demonstrates an association between internal orientation and high perceptions of QWL and more specifically a perception of high responsibility in the job.

	LOC clusters		
QWL	1	2	Total
1	25	50	75
2	7	37	44
Total	32	87	119

*Chi squared (df= 1) 4.28: p.04

4. Implications

The implication of finding an association between internal LOC and QWL is that it represents a direct response to Spector's (1982) concern for a need to research the direction of the difference in perception of job characteristics between internals and externals. The significant result found in this study serves to establish an association between internal orientation and QWL and paves the way for more conclusive research in this area. The significant association between TS and QWL also illustrates room for research into the effects of TS and its components in the workplace. Theoretically then this research has implications for the way in which people's experience of work is measured in future. Specifically, the method of using scale totals to establish relationships between constructs and hence applying this to entire populations seems to offer a lot less in terms of explaining the qualitative differences that people may experience on these same variables.

4. Conclusion

Individual's experience of work and work related variables is not a simple linear process. Results indicated an association between low levels of TS and high levels of QWL, as well as between internal LOC and QWL.

References

- Anderson, C. R. (1977) Locus of Control, Coping Behaviors, and Performance in a Stress Setting: A Longitudinal Study *Journal of Applied Psychology, 62,* 446-451
- Brod, C. (1982). Managing Technostress: Optimising the Use of Computer Technology. *Personnel Journal, Oct,* 753-757
- Bradley, G (1989) Computers and the Psychosocial work environment Taylor & Francis: NY
- Crable, E.A; Brodzinski, J.D; Schrerer, R.F; Jones, P.D. (1994). The Impact of Cognitive Appraisal, Locus of Control, and Level of Exposure on the Computer Anxiety of Novice Computer Users. *Journal of Educational Computing Research, 10 (4),* 329-340
- Genco, P. (2000). Technostress in our schools and lives. *Book Report, 19(2),* 42-44 Retrieved May 11, 2002 from http//ehostvgwl18.epnet.com/delivery.asp? deliveryoption
- Igbaria, M & Parauraman, S. (1989) A Path Analytic Study of Individual Characteristics, Computer Anxiety, and Attitudes toward Microcomputers. *Journal of Management, 15,* 129-138
- Weil, M.M & Rosen, L.D. (1997) *Coping with Technology: At Home, At Work, At Play.* John Wiley & Sons, Inc.: New York
- Weiten, W (1992) *Psychology; Themes and variations (2nd edition)* Pacific grove: Brooks/Cole
- Levine, M. F; Taylor, J.C. & Davis, L. E. (1984) Defining Quality of Working Life *Human Relations, 37,*81-104
- Spector, P. E. (1982) Behavior in Organizations as a Function of Employee's Locus of Control *Psychological Bulletin, 91,* 482-497

The Framework for Indirect Management Features of Process Control User Interfaces

Toni Koskinen, Marko Nieminen

Software Business and Eng. Institute,
Helsinki University of Technology
P.O.Box 9600, FIN-02015, FINLAND
Firstname.Lastname@hut.fi

Hannu Paunonen, Jaakko Oksanen

Metso Automation Inc.
P.O.Box 237, FIN-33101, FINLAND
Firstname.Lastname@metso.com

Abstract

The dominant interaction metaphor in process control user interfaces has been direct manipulation meaning that users have to initiate all tasks and monitor all events. Since the amount of employees' tasks and responsibilities has increased, new instruments should be designed so that they cope with the situation. The objective of the study is to define the framework of indirect management features of process control user interfaces. The framework could be utilized when considering and designing the indirect management functionality in control room systems.

1 Introduction

Nowadays one operator in process control environment may have thousands of measurements, and actuators in his control. To cope with this responsibility, an operator has a control system at his disposal. The control system has several means to support the operator's work and, more often, decision making. The control system 1) automates the routine tasks allowing the operator to concentrate on exceptions and more goal oriented tasks; 2) it informs the organization by showing the process status in a versatile presentation on different abstraction levels (Zuboff, 1988). Since the number of tasks to take care of and the number of events to monitor is continuously increasing, operators need complementary techniques and new instruments to manage the situation. Indirect management is one of the complementary techniques. Kay (1990) defined indirect management as a group of autonomous agents operating in the background in order to satisfy users' goals. Maes (1994) further complemented the definition stating that in indirect management the user is engaged in a cooperative process in which human and computer agents initiate communication, monitor events and perform tasks. Maes (1994) presented indirect management as a metaphor of a personal assistant who is collaborating with the user in the same work environment.

During recent years several researchers have pointed out the possible usability threats that agent technologies may cause, e.g. losing of control over the agents or misunderstanding the functionality of the agents. In process control the usability threats should be taken into account carefully, since the work domain is complex; situation needs to be interpreted correctly; decisions must be made in real-time; and consequences of misjudgements are often safety-critical.

2 Objective of the Study

The objective of this study is to define the framework of indirect management features in process control user interfaces (UI). The framework is defined on the basis of usefulness (Nielsen, 1993).

3 Definition of the Framework Model

Usefulness is a key issue when deciding whether a system can be used to achieve some desired goal. Usefulness can be divided into the two categories: utility and usability. Utility should give answer to the question of what is needed that the functionality of the system is sufficient. Usability should give answer to the question of how well users can use that sufficient functionality (Nielsen, 1993).

3.1 Utility Aspect: Indirect Management in Process Control

The nature of process control work is quite different from office work for example. The process itself generates exceptions through equipment malfunctions, blockages in pipes, quality problems caused by raw materials etc. The operator has to monitor the process and sometimes he has to react rapidly to changes in order to save the valuable production. The causes of problems are not always obvious and solutions to these problems are hard to discover. On the other hand in problematic situations the operator can contact the rest of the organization or utilize the knowledge contained by the control system. (Paunonen, 1997)

The tasks of the production organization including operators can be divided into classes: monitoring, predefined tasks, disturbance control, information exchange, knowledge management, learning, and development (Paunonen, 1997). These are also categories of roles for the system to support the process control work. In this framework the utility of indirect management is evaluated on the basis of different tasks and roles. This will lead to agent categories like: monitoring, task support, disturbance control, information exchange, and knowledge and information management. Learning and development can probably be included to some of the categories above.

Monitoring is traditionally automated through alarms and messages which are predefined during the implementation and installation of the control system. Using indirect management the operator could set an agent to monitor whether some expected situation described for the agent occurs. This allows the operator to concentrate on some other tasks instead of monitoring the situation. Compared to system alarms indirect management brings more flexibility to choose the situations which may sometimes be unique (e.g. maintenance and repair situations).

In a predefined task the operator usually has to perform setting of values, starts, stops and different kinds of checkings in a predefined order. Applying indirect management could mean to set the agent to do the task or part of the task for him. These kinds of tasks tend to vary depending on the situation and they are known perfectly only after the plant have operated a while.

Disturbance control puts emphasis on figuring out what has originally caused the problem and what has happened after that. For that purpose the operator could, for instance, set an agent to find possible sources of poor quality. The agent could study the history of given measurements and compare them to the history of quality data and present the results as potential causes of the disturbance.

Support for *information exchange* in the organization is required because distances at plants are usually long and operators work in shifts. Today the production organizations use diary, bulletin boards and e-mails to transfer information. Looking at process information of some other department is also used for purposes of knowing what decisions are made elsewhere. Agents could be used for instance to write an automatic entry to the diary or send an e-mail to some other department if a given situation in the process occurs.

Process control calls for *knowledge*. Part of the knowledge is stored in the control system as guides, manuals and descriptions. Also diary entries describing past incidents can be regarded as knowledge because they can be used to solve present problems at hand. In process control environment the relevant knowledge should be immediately available (Paunonen & Oksanen, 1999). However, sometimes a situation in which knowledge support is needed is persistent and not so clear. In these situations, agents defined by operators are useful tools for searching help from different knowledge sources. The agents can give hints to solve the problem.

3.2 Usability Aspect: Design Suggestions

When considering the design criteria of above presented tasks there are some work domain specific issues that should be taken into account. Work domain is complex; part of the decision making is both time- and safety-critical; and some process situations are unique. Therefore it is essential that operators are always aware of what their actions can produce. This includes that operators have to be aware of when agents actually interfere and steer the process and when they are carrying out tasks that are not directly safety-critical. For instance to apply indirect management features in predefined tasks is safety-critical. In other tasks the agents will work as supporting and/or monitoring roles. The generic design suggestions for indirect management emphasize that the user should feel in control through providing a sense of predictability and transparency to the system, and thereby gain the user's trust (Höök, 1999).

In process control the utility of indirect management lies in the categories of monitoring, predefined tasks, disturbance control, information exchange, and knowledge and information management. In the following list we address the issue of how this utility should be introduced.

- As a consequence of work domain complexity and unique situations the agents should be end-user configurable.
- Users have to be both in charge and responsible in all situations. Therefore the agents particularly in predefined tasks should be designed to restrict user autonomy to protect users making misjudgments (Friedman & Nissenbaum 1997). In other words users should not be able to assign tasks that could potentially cause significant harm to process.
- Users have to be trained to configure the agents so that they have always a clear presumption about its expected behavior.
- In hasty on-line decision-making situations the configuration of the agent should be possible within few seconds. This may mean a compromise between the expected results and the configuration time set for the agent. However, in most situations incomplete advice is better than no advice at all (St. Amant & Dulberg, 1998). For instance in decision making situations it is important to receive information quickly even if it is somewhat inaccurate. The web search engines have proven that people can form quickly an outlook of an issue by combining contents of several somewhat inaccurate documents.
- The agents should keep user informed about their actions (feedback). Process control user interfaces consist of significant amount of functionality and information. The agents should be considered as features of the user interface and not separate applications. The user interface should have a functionality to show active agents and their status.
- The most of the traditional usability heuristics (Nielsen, 1993) can be adapted to agent applications, regardless the fact that the heuristics were specified for direct manipulation.

4 The Case Study: Early Evaluation

4.1 Implementation of the Study

To illustrate the concept of indirect management with agents, we performed a constructive empirical study. The empirical study consisted of specifying, constructing and evaluating the indirect management prototype (The monitoring agent, MA). The task of the MA is to monitor the process measurements given by the user and inform the user when the given conditions are fulfilled (Figure 1). The prototype was designed and constructed according to the principles presented in the framework. Additionally, key user-centred design activities were followed (ISO 13407, 1999). The MA prototype was integrated into the MetsoDNA operator desktop user interface. The structure and the functionality of the MA was kept simple and informative.

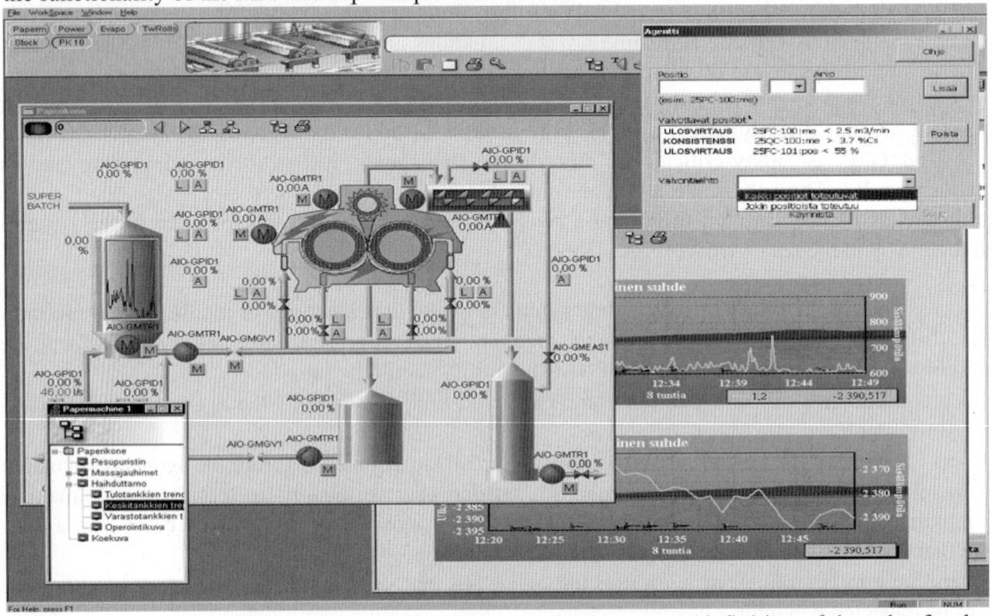

Figure 1: The Metso DNA operator desktop, the monitoring agent (definition of the rules for the agent) can be seen at the top right corner. Users could, for instance, "drag and drop" for selecting the desired measures from the desktop.

The prototype was evaluated in real settings in a power plant control room. Employees (N=5) were given a realistic scenario describing a typical routine monitoring task. However, instead of carrying out the task traditionally they were asked to use the MA. The evaluation was observed, recorded, and the users were interviewed after the evaluation.

4.2 Results

All users (N=5) were able to perform the task according to the given scenario. However, some users had problems in understanding the capability of the agent and sometimes they thought that the agent actually steered the simulated process. However, recalling the role of the agent helped them understand the functionality of the agent.

The users thought of several different tasks in which the MA could be utilized. One operator stated: *"this would be a significant help in dozens of situations. When you're busy, it's so easy to forget to monitor some values. You don't wake up until the process control system alarms."* All the users claimed that they would use the MA regularly for predefined monitoring purposes.

The users considered the monitoring agent was easy to use. Especially they liked the drag-and-drop feature and the simple dialogue structure. These issues were considered important, since they made it possible to define a new monitoring task within a few seconds. Also the feedback in the form of sounds and visual cues was considered useful.

5 Conclusions and Future Work

The framework for indirect management features of process control user interfaces was defined on the basis of usefulness, which includes utility and usability. The utility of indirect management was categorized into the different tasks. The results from the case study indicate that this categorization helps employees to understand the roles of the agents. The advantage of categorizing the indirect management features according to the tasks of the operators is that the operators can understand the functionality of the agent through considering their own tasks. The usability aspect is essential as well, when considering the usefulness of indirect management. The case study results supported well the design suggestions. The design of agents should be kept simple. The agents should be possible to define and start within few seconds.

In the future, we aim to design the indirect management features according to the principles presented in the framework. The framework will be focused and revised on the basis of forthcoming experiences.

References

Friedman, B., & Nissenbaum, H., (1997). Software Agents and User Autonomy. Proceedings of the First International Conference on Autonomous Agents. Feb., 1997 New York: ACM.

Höök, K., (1999). Designing and evaluating intelligent user interfaces. In IUI '99: Proceedings of the 1999 International Conference on Intelligent User Interfaces, Los Angeles, California, January 5 - 8, 1999, Redondo Beach, CA. ACM.

ISO/IEC, (1999) 13407 Human-Centred Design Processes for Interactive Systems. ISO/IEC 13407: 1999.

Kay, A., (1990). User Interface: A Personal View, in The Art of Human Computer Interface Design, B. Laurel [ed], Addison-Wesley, pp.191-207.

Maes, P., (1994). Agents that reduce work and information overload. *Communications of the ACM, 37(7):31-40.*

Nielsen, J., (1993). Usability engineering. Academinc Press Inc., Boston.

Paunonen, H., (1997). Roles of informating process control systems. Tampere. University of Technology, Publication 225, Tampere. 166 p., ISBN 951-722-916-X.

Paunonen H. & Oksanen J., (1999). Usability criteria for knowledge support in process control (in Finnish). Prosessinohjausjärjestelmän tietämystuen käytettävyyskriteerit. Automaatiopäivät 99, 14.-16.9.1999. Helsinki, Finland.

St. Amant, R. & Dulberg, M., (1998) Experimental evaluation of intelligent assistance for navigation Knowledge-Based Systems, 11(1): 61-70.

Zuboff, S., (1988). In the age of the smart machine. Basic books, New York.

The Assessment of the Working Computer Systems in the Enterprise Made by Computer Operators

Katarzyna Lis *Jerzy Olszewski* *Mariusz Szczubelek*

University of Economics Department of Labour and Social Policy
61-875 Poznan, Al. Niepodleglosci 10, Poland
lis@novci1.ae.poznan.pl olszewsk@novc1.ae.poznan.pl

Abstract

The paper presents the results of research of computer operators who use computer systems in several companies. This paper consist of six parts: introduction, four chapters and conclusions. The purpose of research is analysis of different attitude of the computer systems under implementation in the enterprises. The introduction presents this general problem. The second part of this article describes employment conditions of computer operators. In this part the computer is often identify as a tool who reduce the employment. This point of view causes reluctance to use the informatization. The third part of this article presents problems of the work organization concern some computer positions in the researching enterprises. Implementation of computers is often connected with changes in organization structure of the enterprises. This fact causes series relations of psychological nature. The fourth part of this article presents results of the research in the field of requirements in attitude to computer operators. The special attention was paid to wrong activities of employers in relation to subordinates. The next part of this article includes the assessment of the computerisation influence on the interperson relations in the enterprise. Very important and observe problem are effects of the implementation of computer systems in the enterprise, they are psychological nature. It appears in accumulation of information on workers (confidential data) and in assurance of its access. The workers mistrust the enterprise and they are frightened of consequences of little errors, reluctance to take the responsibility, etc. The last part of this article contains conclusions of the empirical research.

1 Introduction

Nowadays we can observe that the enterprises, both big corporations and small companies, function in the conditions of constant change of outer surrounding as well as inner factors. The result of these changes are the noticeable modifications in the functioning of the enterprise as well as the reorientation in the behavior of employees. The factors which condition significant changes of the employees` behavior are strong competition, growing consumer expectations, changing law and society and computerization. These factors cause changes not only in the organization of the working process, communication processes, logistic, financial etc. but also in the psychological sphere of man which reflects his or her attitude towards these processes and their evaluation.
The assessment of employees by their employers, co-workers, clients and even subordinates is an important issue. Self - evaluation is becoming more and more popular in many companies. These forms of assessment serve administrative, informative and motivational purposes (Pocztowski, 2002). The aim of this article is to show the importance of these purposes, especially motivational

purpose, which can be used by the employee as reversible information and which should motivate employees for self development and professional development as well as give information (informative purpose) about their strong and weak points.

2 The Employment Conditions of Computer Operators

The employment conditions include such elements as working conditions, which include such elements as financial condition, equipment, organization of work, motivational system as well as the attitude of the employee towards his or her duties, which can also result in the form of stress. Stress is connected with working at a computer and is the result of psychological overpressure which as the computer system develops, succumbs to evolution and limitations. One of the causes of psychological overpressure connected with the use of computer is the necessity to memorize a lot of instructions, as well as the demand of coming to terms with efficient computer work, the demand of error free work and effectiveness of this work. Also, it is well known that people who work at a computer, work in solitude. Computer work creates the feeling of artificiality, unreality, dehumanization of relationship between people as well as desocialization of man (Cellary, 2002).
In the first period of working with a computer a lot of employees show the unwillingness to work, which is caused by the fear of the unknown and failed attempts to work with the computer. The source of psychosocial barriers can be found in the fear of destroying an expensive equipment, in the feeling of disproportion between our abilities to use a computer and potential possibilities that the computer creates, in the fear of becoming redundant as well as unhealthy and cumbersome factors which occur when working in front of monitors. There is then, a strong need of pragmatic actions that will limit the anxiety, psychological and physical danger caused by this process. Psychological difficulties connected with the use of computer can be softened by the establishment and implementation of successful social-technical recommendations.
In order to investigate the working conditions of computer operators, a questionnaire research has been carried out in 164 companies. The questionnaire has been designed on the basis of problem list based on Dortmund list.

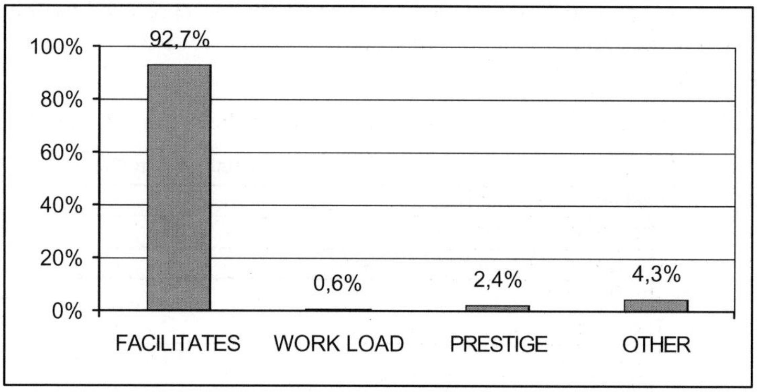

Figure 1: The assessment of the functioning of computer system

The authors made an attempt to investigate the working conditions focusing on such issues as the attitude of the employees towards computerization and their opinion on the functioning of the computer system in the company. Data presented (see Figure 1) show that of computer operators think that the existing computer system makes their work easier (92,7%), however, few of the respondents say that it is an additional work load for them (0,6%).

Table 1: The attitude of employees towards the computerization

	Observation [%]
The Increase of Work Load	7,3
Becoming Redundant	9,8
Transfer to Other Work	4,9
Current Skills Will Become Useless	14,1
The Lack of Understanding of the Aim of the Computerization	18,0
Unwillingness to Get Used to New Conditions	22,0
The Loss of Relationships	0,5
The Lack of Possibility to Participate in Taking Decisions	0,0
The Conviction that Only The Enterprise Will Benefit from the Computerization	2,0
Other	2,0
There Are No Obstacles	19,5

Table 1 shows the results of the attitude of computer operators towards computerization of the company. The major problem is the unwillingness of the employee to get used to new working conditions (22,0%). Some of the respondents show the lack of understanding of the aim of the computerization (18,0%). Another group does not notice any obstacles in the computerization of companies (19,5%).
There is a significant disproportion in the opinion of the respondents about the computerization of the companies. This can be explained by the differentiation of the investigated population as far as the preparation for the profession is concerned such as qualifications, professional experience and cultural background.

3 The Organization of Work

The organization of work place of computer operator consists of many technical, organizational, economical activities which aim at rational synthesis of work potential with computer system. Work place design of computer operator is connected with the choice of the rational working method i.e. designing the appropriate processes, activities and movements.

Table 2: The influence of the computerization on the organization of work

	Observation [%]
Fast Access to Information	60,7
The Precision of Information	6,2
The Speed of Document Circulation	8,4
The Reliability	4,7
The Rhythm	2,3
The Efficiency	16,8
Other	0,9

The article presents the results of research on the influence of the computerization on the fast access to information (see Table 2). Most respondents (60,7%) said that computerization has positive influence. The efficiency of computer system was mentioned in the second place (16,8%). Such elements of work place design as the rhythm of work and reliability were given the fewest number of positive opinions (4,7%). Gathered information shows that the computerization has both positive and negative influence on the organization of working process. This type of information is especially important for all sorts of managers. The results of the analysis show that measures must be taken in order to improve such areas of computer system as the rhythm of work, reliability, the precision of data and the speed of document circulation.

4 Requirements Towards Computer Operators

The important element in the assessment of employees apart from such factors as behavior, personality and effectiveness is competence. The competence of computer operator consists of such features as education, professional experience, qualifications, effectiveness and psychological predisposition.

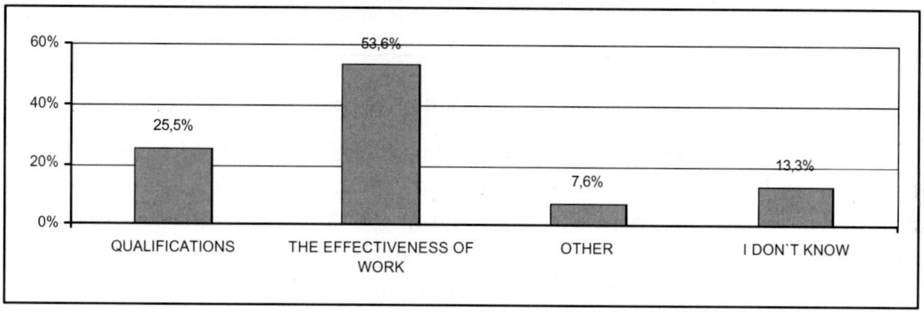

Figure 2: The influences computer operators` payment

The investigation of the relationship between the level of competence of computer operators and payment is also important (see Figure 2). Motivational system influences computer operators` effectiveness most (53,6%). Qualifications are less valued (25,5%). Some respondents are not aware of the range of motivational system (13,3%), others (7,6%) think that motivational system includes other assessment elements then effectiveness and qualifications. It means that about 20% of respondents do not have the knowledge about the motivational system.

5 The Assessment of the Influence of Computerization on the Interpersonal Relationship in the Enterprise

The interpersonal relationships are an important element in shaping working conditions. These relationships are an important factor in shaping general working conditions of computer operators. Because of that, interpersonal relationships have been investigated (between computer operators and managers).

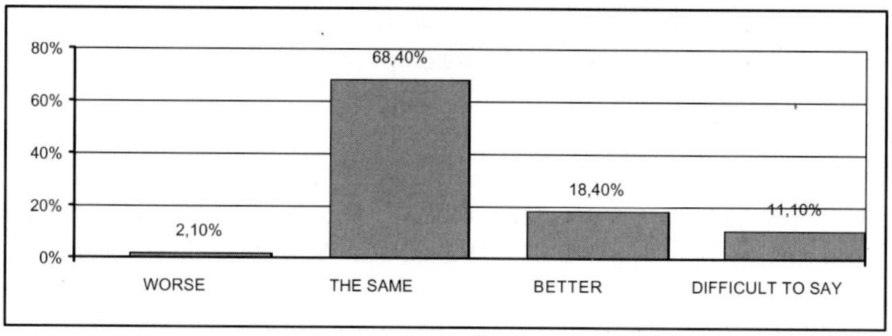

Figure 3: The treatment of computer engineers by managers

The research shows (see Figure 3) that computer operators are treated in the same way as other employees (68,4%). Some of the respondents think that computer operators are treated better than other employees (18,4%) and only a small group of the respondents think that computer operators are treated worse (2,1%). Other respondents (11,1%) did not have any opinion about it.

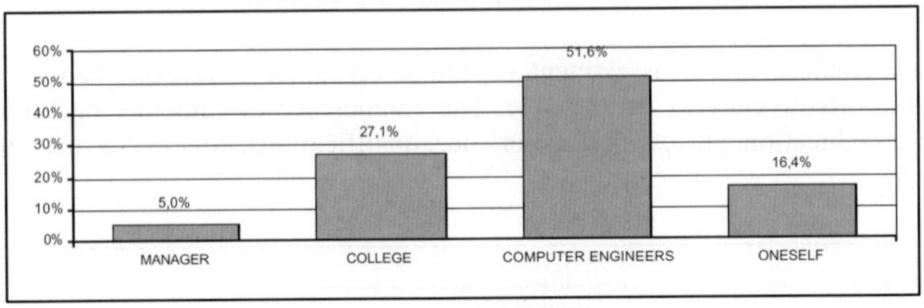

Figure 4: The assistance with which the employees are provided

The access to assistance at work is an important element of interpersonal relationships. Empirical data show that computer engineers (51,6%), also often colleges (27,1%) or even the very employees (16,4%) solve the problem (see Figure 4). Managers are less likely to provide employees with such assistance. In conclusion it should be highlighted that the emp loyees are most often assisted by computer engineers. This can be explained by the fact that, that group of people is best prepared as far as theory and practice is concerned.

6 Conclusion

The research method that has been used enabled us to investigate the conditions of work of computer operators in 164 enterprises. The majority of computer operators state that computer systems facilitate the working process. Only about 0,6% of respondents claim the opposite. The other problem is the attitude of employees towards computerization. About 1/5 of respondents expressed the negative attitude towards computerization but another 1/5 of respondents do not have see any negative attitude towards computerization. To sum up there is a considerable differentiation in the opinions about computerization. This can explained by the different level of respondents` competence.

The results of research on the organization of work are also interesting. Most respondents - 60% claim that computerization speeds up the access to information. However, there are certain problems that occur in the organization of work (rhythm, reliability). Computer operators were also subject of the research. Most of them claim that there is a relationship between competence and payment. The last part of the article has been devoted to the investigation of correlation between computerization and interpersonal relationships. The research shows that computer engineers are treated in the same way as other employee. This equal treatment of the investigated groups positively influences interpersonal relationships at work.

References

Cellary, W. (2002). Rynek pracy w gospodarce elektronicznej, w. Polska w drodze do globalnego spoleczenstwa informacyjnego, (pp. 62-64). Warszawa: Wyd. Program Narodów Zjednoczonych ds. Rozwoju.

Pocztowski, A. (2002). Zarzadzanie zasobami ludzkimi, Zarys problematyki i metod. Kraków: Oficyna Ekonomiczna.

Operator's Contribution to the Success of Control Board Renewal Project of Genkai 1&2 of Japanese PWR Nuclear Power Plant

Shuuji Miyanari[1], Kazuhide Tomita[2], Kenji Hattori[3]

[1] Kyushu Electric Power Co., Inc. :Imamura, Genkai-Cho, Higashimatsuura-Gun, Saga Pref., 847-1411, Japan
Shuuji_Miyanari@kyuden.co.jp
[2] Mitsubishi Heavy Industries, Ltd.: 1-1-1, Wadasaki-Cho, Hyogo-Ku, Kobe, Hyogo Pref., 652-8555, Japan
kazuhide_tomita@mhi.co.jp
[3] Mitsubishi Electric Corporation: 1-1-2, Wadasaki-Cho, Hyogo-Ku, Kobe, Hyogo Pref., 652-8555, Japan
hattori@pic.melco.co.jp

Abstract

This paper describes how operators contributed to a control board renewal (CBR) project at Genkai units No.1&2 which was completed in 2001 successfully. The renewal work was the total modernization of the control room consisting of the replacement of the control board and the relevant instrumentation and control (I&C) panels, the relocation of them and the replacing of old cables connecting them each other with new ones. Operators took part in all kinds of the renewal work, such as conceptual planning, designing a replica simulator, verifying the operability of the new control board, training at the simulator and exerting complementary work at field. They were positive to the renewal work and solve the problems reflecting their own opinions to it, which could lead the project to the success.

1 Introduction

Genkai unit No.1 started commercial operation in October 1975 and unit No.2 in March 1981. These units are categorized as the first generation of the pressurized water reactor (PWR) of nuclear power plants in Japan.

After the start of commercial operation, the control board had been modified several times such as to reflect lessons learned from TMI or to retrofit to meet new regulations etc. and the control board were desirable to be improved from operability or monitoring function viewpoint. Operators had adapted them to the new circumstances so that both units had achieved safe operation without any problems.

On the other hand, the control board became annoyed with component's obsolescence or lack of space for future modifications. It was considered the control board to be renewed. Additionally the units No.3&4 which adopt the improved control board utilizing many CRTs as an integrated information device started commercial operation in 1994 and 1997 respectively.

A top management of Kyushu Electric Power Co. has a policy that the working conditions for the operators who are directly in charge of power production should be preferentially considered. Then the study on the comprehensive control board renewal for units No.1&2 started in 1994.

2 Features of Renewal Project

The renewal work aimed to enhance the safety, reliability, maintainability, and future margin of the control board and the instrumentation and control panels, adopting state-of-the-art technology regarding computers or digital control equipment as much as possible.

In case of the construction of a new plant, engineers lead the whole project. All engineering related works are done, and then handed over to the operating personnel to start commissioning. It is mainly by the engineers to deal with planning, designing, verifying the I&C systems in shop and at field. After commissioning, the plant is handed over to a client, the utility. After that the utility's operators are in charge of power production and the maintenance personnel are in charge of equipment maintenance.

But the renewal case might be different. It should start with grasping problems for the existing systems and operators would be suitable to clarify the current issues to be resolved. Under these circumstances the management of Kyushu Electric Power Co. thought the operators grasped those issues and decided to incorporate them into a task force of this project. Figure 1 delineates the feature of the renewal work focusing on the operators' roles.

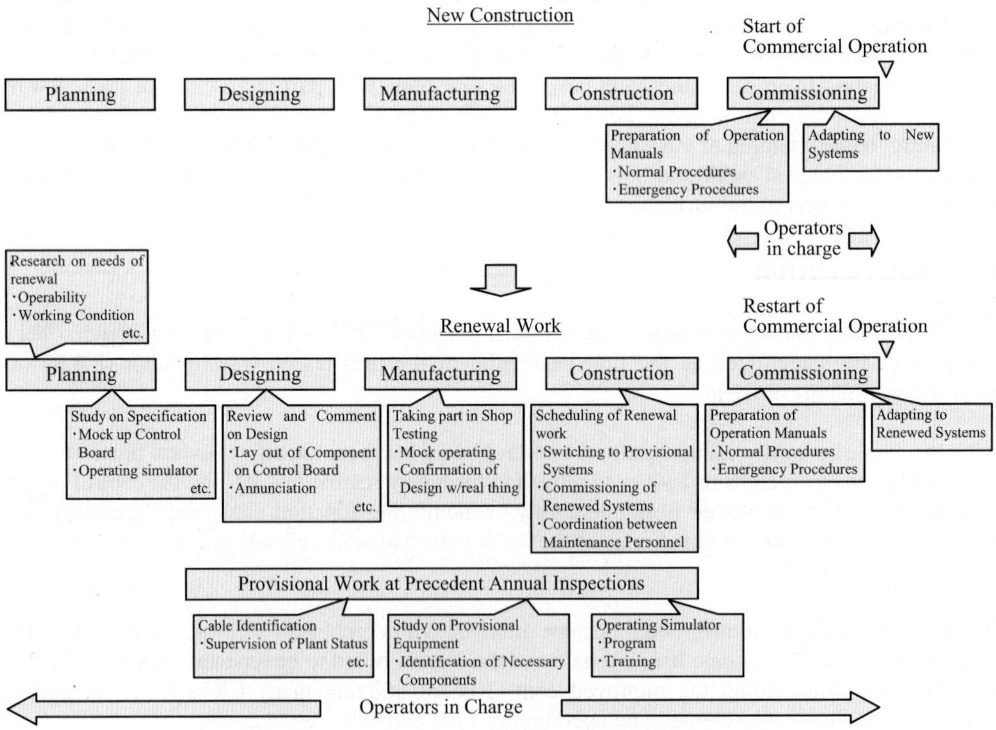

Figure1: Operators in charge at Renewal Work

In Japan the applicants are required to inspect a nuclear power plant nearly once every year. The renewal work should be done within this annual inspection time. It is necessary the renewal work to be completed in this annual inspection time, and also important to make the optimized schedule not only to inspect the plant thoroughly but also to perform the renewal work.

3 Operator's Characteristics

Operators are considered to be conservative to the renewal. They have accustomed themselves to the present control board by training or studying the relevant knowledge. If the system is revised, they have to exert the same amount of training and studying for the new system.

Crew members of the operator shift are normally organized from wide generations from twenties to fifties, sometimes called a "family". The different generations use the same control device so that the maximum adaptability to all generations is required to the renewed system. It should also be taken into consideration that the renewed system will be inherited to the next generation. Therefore it is also important the renewal system is to be easily adapted to the next generation.

The above situation was one of the reasons why the operators were incorporated into the task force of this project from the early stage. They had exerted themselves to reflect the opinions of other colleagues the majority of whom were in charge of power production as shift members.

4 Operator's Contribution to Renewal Work

It took nearly eight years to complete the project. The first year was used for feasibility study and the next two years for establishing basic specifications, followed by four years for engineering and manufacturing the equipment and the final year for installation and implementation of the renewed systems at site.

A plot plan focusing on the renewal concept was investigated intensively, considering such as the result of inquiries to the operators or the current technologies. Through these inquiries it became evident that improvement of the operability from the physical layout was especially requested. The comprehensive renewal plan was settled, not only the control board but the instrumentation and control systems, its layout and cabling system between them.

The basic concept of the renewal had been discussed extensively for nearly two years and finally the specifications were settled. In the course of these studies, operators had reviewed the design specifications proposed by engineers and presented their comments from the viewpoint of the long term user after the renewal. Especially to what extent or how much the hard devices to be computerized were studied with the collaboration of engineers.

Computerized soft operation technique was selectively adopted for the non-safety functions, such as an electrical system for switchyard equipment and some auxiliary systems of water treatment in reactor and turbine systems, being the latter system also centralised from localised control. But the soft operation was not selected for the safety systems owing to the time schedule because the soft operation for the safety system had not been adopted in the real use in Japanese PWR plants until that time.

Computerized monitoring function was utilized as much as possible. The related hard devices were replaced by flat display panels (FDP). Ten units of FDP were installed on the new control board. Each FDP unit could present all pieces of information as long as it is input to the plant computer.

Alarm window style was not changed but the actuation logic was sophisticated. A dynamic alarm suppression system was adopted by changing the inputs to digitalized signal conditioning cards from the hard wired windows. If a window alarm comprised of several elements, an individual element which activated the window was easily identified by a telop message shown on the FDP screen because those signals are also input to the plant computer.

These detail specifications were studied through communication with engineers and confirmed by the mock-ups, then reflected on the design documents.

A replica simulator was installed in the training center at the site considering the importance of pre-operational training. The operators exerted the preparation of training programs and the documentation of new operation manuals and the confirmation of them using the plant simulator.

Operators also played an active role in the detail design. Genkai units 1&2 are twin unit type, but the name plates on the control board, such as for switches or annunciators, were not always identical because of the time gap in construction time. The renewed design had to be coordinated in this practical areas and the operator took effort to list up all deviations and decide the appropriate names.

The operators also exerted complementary works at field. Many control boards or panels were removed and replaced or relocated. To reconnect cables between them, field termination cabinets were provided adjacent to these panels in order to marshal the cables and install new multi-conductor cables to reduce the volume of cables. Each cable coming into the cable spreading area should be correctly identified one by one so that the re-cabling could be safely done. For this purpose, the new technology which could identify a cable at any arbitrary point was developed by the engineers. This technique was such that a coded signal was injected to the drain wire and the signal was detected at a sensor winded around the cable. Although this was done after the plant was shut down, the operators collaborated in judging that the signal injection did not affect the plant and confirming their judgement was correct by monitoring the plant status.

5 Operator's Evaluation on Renewal

Inquiries on operators after the CBR reveal the majority of them are satisfied. Figure 2 explains this fact.

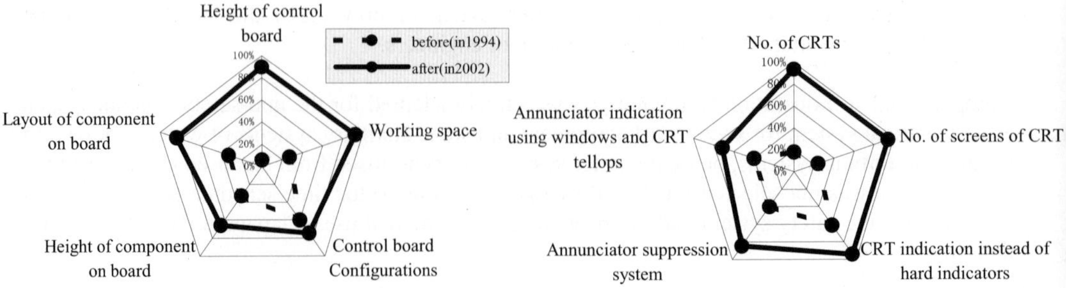

Figure2: Operator satisfaction chart

A figure of the renewed control room is shown in figure 3.

Figure3: Renewed Control Room

6 Conclusion

The computer technology is getting innovative, but this is not a target but a tool for the user. It can be meaningful only after the human machine interface is correctly accomplished. In this sense the operators' role achieved through the Genkai CBR project is very instructive.

The operators at Genkai 1&2 were positive to the renewal work and to reflect their own opinions on the design of the renewed systems. If the designer compels his self-righteous design to the user, the result would be worse. If the designer develops his idea in collaboration with the user, the result could be better.

It is considered that these kinds of computerized modernization project are under way or will be planned in the industrial fields widely all over the world. We are confident in suggesting that the key to the success of the renewal project is whether the end user, namely, the operator is effectively incorporated or not.

Definition and prototyping of ErgoMonitor - an online monitoring system for ergonomic evaluation of human-computer interaction in a *Web* environment

Marcelo Morandini

Department of Computer Sciences – State University of Maringa

morandin@din.uem.br

Walter de Abreu Cybis

Department of Computer Sciences – Federal University of Santa Catarina

cybis@inf.ufsc.br

Abstract

This paper presents the ErgoMonitor as an environment to help web sites usability evaluation tasks. The ErgoMonitor represents an approximation between general log files analyzers in web servers and the systems employed to provide usability evaluation based on observation of the interactions between users and computational systems. The ErgoMonitor is being developed as a way to quantitatively validate the qualitative diagnoses concerning usability problems on specific interactions between consumers (users) and e-commerce web sites. Once the parameters and data needed for the analysis (ie, *urls* of the tasks to monitor and their associated log files) are provided, the ErgoMonitor will analyze the log files to determine metrics of interactions′ productivity, in an objective and fast way. These metrics shall be able to validate the diagnoses related to the specific usability problems.

Introduction

Web Site designers are always dealing with daily challenges: their products must attract user's attention, provide good usability and guarantee satisfaction during the interaction. These requirements are even more important for companies that use e-commerce, once their financial health is strongly dependent on the quality of the user interfaces developed (W3C, 2002). The usability evaluation approach becomes here a strategic partner (Nilesen, 2000).

The ErgoColn (Cybis et alli, 2002) is an environment developed to support co-evaluative inspection of ergonomic qualities that are spread by web interfaces components. Through the use of this technique, users and designers task perspectives and remarks on the web site usability are taken under consideration to the definition of scenarios. Having the description of web pages concerned to these scenarios as initial parameters, the ErgoColn Support Tool (available at www.labiutil.inf.ufsc.br/ergocoin) automatically assembles specific sets of usability checklists to be used by the evaluator.

The qualitative diagnoses issued from the ErgoColn application refers to "possible" usability problems within a specific, and some times supposed, context of use. In this approach, the evaluators produce usability diagnoses just having some information concerned to the web site operation conditions all over the world-wide-web. These contexts of use may, in fact, be related to unexpected users, computers, browsers and even environment conditions. So, web site's usability validation diagnoses are expected and necessary.

On the other hand, a log monitor system can gather data from real interaction and is able to identify the existence of problems in the real context of use, despite unknowing how, in fact, this context is featured. It produces results on quantitative basis that are independent of the context condition, but can validate usability problems diagnoses in qualitative basis.

This paper presents the definition and architecture of the ErgoMonitor, a system that collects and treats specific data from real users' interactions with the web site. The ErgoMonitor is designed to validate qualitative diagnosis issued from the ErgoColn application. The interaction data gathering is done in total absence of evaluator and evaluation apparatus and data is monitored on large scale. It will permit that the scope of data concerned to the usability diagnoses validation be increased while the costs involved with this process be decreased.

ErgoMonitor conceptual basis

The ErgoMonitor represents an approximation between general log files analyzers in web servers and usability evaluation support systems based on observation of the interactions between users and the computational systems.

The log files analyzers (Lift, 2002; Mach-5, 2002, Bobby, 2002) treat data from these files to produce information about users/consumers geographical localization and also about frequency and access time of the web pages visited in specific periods of time. This information could support marketing analysis and also help the identification of web sites' working and usability problems. In this kind of analysis, it is not possible to identify the interactions context of use (related to the user, environment, equipment, browser, etc.), but the data collected are free of "contamination" risks that can be produced by unreal testing conditions. It is important to mention that this approach produces quantitative results in a very fast, easy and objective way.

The systems designed to support evaluation based on observation of the user's interaction (Noldus, 2003; Paterno, 2003) support qualitative analysis, done by the evaluators, and perform quantitative treatments of the interactions data collected in usability tests, in which, the contexts of use (user, task, environment features, etc) are well known. Particularly, these systems allow the marking of interaction parts as important or, otherwise, irrelevant to further qualitative and/or quantitative analysis. The information type collected by these systems allows richer and more detailed qualitative and quantitative analysis, but may require, and normally do, an increased workload. So, the data collected may be "affected" by the testing conditions in artificial environments that are not representative of the real context of use.

Description of the ErgoMonitor operation

The ErgoMonitor will analyze traces of some tasks and produce metrics associated with individual behaviors. For achieving this goal, the web site designer and/or usability staff will initially inform the parameters needed plus the way to access the server log files. He/she will inform the *urls* of pages associated with the tasks (interactions chunks) which had presented diagnoses of usability problems and the period of time to be considered in these analyses. Specifically, the initial and the final *urls* corresponding to the tasks compliment must be informed, as well as the specific *urls* related to error and help messages. In the sequence, the ErgoMonitor will analyze the log files of the web site under evaluation and quickly present reports concerning the productivity of marked interactions.

Ideally, the tasks to be followed are very precise, with beginning and ending points clearly determined, which easily allow tasks marking. Examples of this kind of tasks are:

- user login;
- user register;
- search;
- download.

The data collected by following these tasks will produce very objective interaction metrics that include: task effectiveness, error occurrences, help solicitations, time consuming, and retasking.

Some metrics processing depends on special markings in the log files data to include specific points (*urls*) for tasks beginning and ending, and also the error messages and help web pages. These markings can be done by the evaluator responsible staff, which must inform and submit these parameters to the ErgoMonitor. This process defines a typical customization of the ErgoMonitor.

These metrics will be processed to produce interaction Verified Factors (the average of real interaction metrics) considering all users (discarding the distribution outliers). For example, considering the login task, can be defined:

- verified factor of task effectiveness;
- verified factor of error occurrences;
- verified factor of help asking;
- verified factor of time consuming;
- verified factor of retasking;

The rules of an expert system (that is also part of the ErgoMonitor environment) will compare Verified Factors with the Allowable Factors that characterize good and bad user-computer interactions for similar tasks. In fact, these allowable behaviors are essential components of the ErgoMonitor´s logical processing, and will be pre-determined by its authors, through observing good and bad interactions, or by some other common sense determination.

The knowledge of disturbs on these tasks (issues of triggering the rules comparing the real interaction metrics with those associated with good and bad interactions), will allow designers the possibility of creating new versions for the site, fixing or preventing problems in specific forms, fields, pages, icons, figures, files, etc; and the adoption of management activities, like a search about users, tasks and environments.

For validating this methodology, a "manual" version of this environment was projected based on filling specific *Interaction Descriptive Cards*, which collect real interaction data. The Real Metrics and the Verified Factors are manually processed, as well as the inference rules that are faced with the Allowable Factors.

In the future, we can examine the utilization of the ErgoMonitor to generate problems indications, independently of any previous diagnoses, just considering the interaction data collected. Nevertheless, this approach still asks for further studies and analysis. To summarize, it is necessary to reinforce that the ErgoMonitor environment is basically proposed to be a support tool for validating e-commerce web sites usability evaluations and DOES NOT have intention to exclude the use of the well-known usability evaluation techniques and methodologies.

Bibliographic References

(Bobby, 2002) Bobby – A Tool to Support Web Site Evaluation, available at www.cast.org, accessed in may 2002.

(Cybis et alli, 2002) Cybis, W. A., Scapin, D. L., Morandini, M.; Conception and Definition of Ergo-CoIn, A Tool to Support Usability Evaluation of Web Sites; *Rapport Technique, March 2002*, INRIA – Rocquencourt, France.

(Nielsen 2000) Nielsen, J.; *Designing Web Usability*, Campus Editors, 2000.

(Lift, 2002) WebSAT – An Usability Support Tool for Web Sites Usability Evaluation, developed by *UsableNet Company*, available at **www.usablenet.com**, accessed in may, 2002.

(Mach-5, 2002) Mach 5 Enterprises Corporation, *Fast Stats Analyzer: Lightining Web Site Statistics*; available at www.mach5.com, accessed in april 2002.

(Noldus, 2003) Noldus Information Technology Co.; *The Observer 4.1 Tool*; available at http://www.noldus.com/, accessed in January, 2003.

(Paterno, 2003) Paterno, F.; *Remote User Interface Evaluator Tool*, available at http://giove.cnuce.cnr.it/~fabio/remusine.html, accessed in January, 2003.

(W3C, 2002) World Wide Web Consortium's (W3C) - *Web Accessibility Initiative*, disponível em http://www.w3c.com, acessado em maio de 2002.

The State of Ergonomic Consciousness of Employees as An Index of Safety Quality of the Organisation

Musioł Teresa

Silesian Technical University
Faculty of Organisation & Management
ul. Roosevelta 26-28, 41- 800 Zabrze
tessa2002@wp.pl

Abstract

The paper tries to get assessment of relation between the state of ergonomic consciousness of employees and safety of organisation in which they are participating. The quality of the culture of the organisation is measurable value in the relation between real state and expected one depending on the human behaviour in human work and life environment. There is a need of the transpersonality communication (human potential), whose an essence is empathy category of the phenomenon. This is base of benchmarking tools in safety management.

1 Introductory remarks – What this is consciousness?

The feeling of human's safety state is his emotional one we must take into consideration the consciousness role into subjective and objective assessment of this state. It is related to the degree of the consciousness during the undertaking the activities protecting human in the work and life environment. These activities are characterised by the definition of the state of thermodynamical equilibrium (it concern the human's physiology) and also of the area of cultural conditions in everyone of his aspect. (Musioł & Karłaszewski, 1990)

A human regulates his own relations with environment not only by the adaptation to it, but also throughout its active transformation coming into various effects, not always positive ones. A goal of human activity is also the performance of the tasks that are determined by human environment, first of all of the subject of a community in social system consisting first of all from such elements like task, activity and conditions. (Golubińska, 1974)

A human in this system manifests himself the various forms of activity, in which the economic activities and the process related to it is their existence fundament. The human activity in the work process includes rational and irrational elements. The irrational subjects of this activity are rational but the rational methods of their achievements are sure. The rationality of activity methods in the achievement of irrational subjects can be measurable and immeasurable. The measurability can be assessed by the use of formalised mathematics and economic methods. There is worse situation with reference to the immeasurability. It is subjected to the specific evaluating judgements basing on many criteria related to the assessment of attitude and pronouncement of events participant, their activity effects that just exist and that can appear in the future. Because a human is the participant of the collective life this activities standard is determined by cultural standards. (Musioł, 2002)

2 A consciousness and safe behaviour

Regardless of the culture researchers opinion in its scope depending on antropological depiction the culture is always the way of the organisation of the collective life. It creates the logical and psychological profile of the individual life in this community. This culture is collection of learned phenomena that are passed on during the teaching process and the practical. Norbert Sillamy in his own dictionary of psychology (Sillamy, 1994) defines a consciousness as direct knowledge about the existence, acts and external world, which everybody possesses. To organise data delivered into consciousness by the senses and the memory our consciousness makes that we can place ourselves in time and space. A consciousness does not exist as separate brain function. It is not anything external or internal one, this is our attitude of perceptive world. Søren Kiergegaard (Kiergegaard, 1995), who was existentialist claims that consciousness is self-knowledge, which determines human's relation to himself. The lack of the personality and the self-knowledge loose brain control and therefore this is proof of the lack of consciousness. (Kiergegaard, 1995) In the other words the consciousnesses missing is always the reason of negative activities that is to say - evil. The human activity first of all in work process without the ergonomic consciousness will be always the reason of destructive activities and it proves the lack of safety consciousness of all work process participants.

The essence of the process of learning the ergonomics cognitive is the growth of the ergonomical consciousness degree related to the environment of human life and work. For this purpose the knowledge about the all threats around us and methods of their identification helps us first of all. It is influenced by the ergonomic cognitive, which stands a human as a subject in his environment and it is natural cognitive act performed in the relation to the environment.

3 The raise of behaviours consciousness state

Because the area of the culture represents the collection of skilled phenomena (Nowicka, 1998) it means that the culture is not transmitted throughout the genes activity but throughout the learning process. The learning may be the result of the imitation (benchmarking) or the active training. As it was mentioned above the essence of learning process of the ergonomics cognitive is the growth of ergonomic consciousness. It means capturing the theoretical and practical information about manners of the behaviour in many kinds of situations related to many different phenomena. A result of a lack of these of information is a conflict focus between the conceptional and perceptual assessment of the given situation and it is the primitive threats. These threats have an exogenous character and they are the reason of destructive behaviours. On the Figure 1 the above mentioned relations are presented in the graphical form.

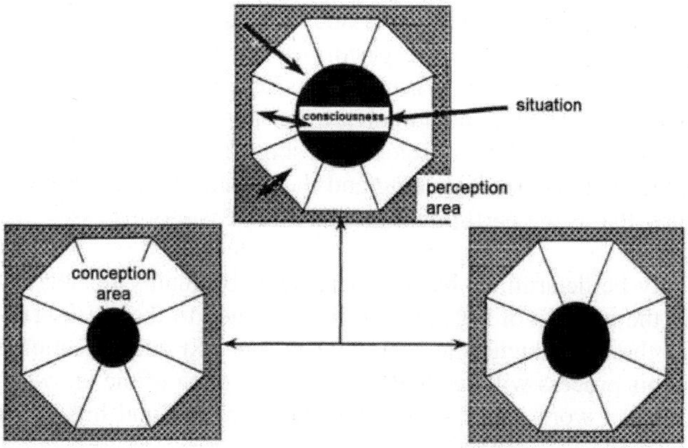

Figure 1: The relations between the consciousness and the behaviour. (Musioł, 2002)

The imperative of recognition for behaviours consciousness are exogenous and endogenous factors that change human work and therefore the state changes for safety of the environment. Continuing the question it may be worth to ask how large is the evil, which is the result of consciousness missing in every society that the safety is an element of the cultures in their psychological dimension. It is also everyday and basic obligation of the ethical duty not only in relation to oneself but also to other people. So it is necessary to change in the thinking process basing on the "Newton's paradigm", which is going into the thinking basing on the "paradigm of the imagination".

And there is a question: why? The answer this question is: because a human as a unique individual is a phenomenon, in which the psychophysical and emotional processes determine the satisfaction of the need of capturing a self-knowledge about itself and surrounding external world by category as like empathy, in other words of understanding situation of another human. (Stein, 1998)

4 A consciousness state and organisation safety as a relation kind

As it was just mentioned earlier the culture is related to a human on many manners.
A human is its creator and at the same time is formed by it. A culture as system possesses its own tools to serve the human adaptation to his specific behaves in his material and psychological environment.

Taking notice, that organisation culture quality is the relation between real state and expected one:

$$Q_k = \frac{F_k}{O_k} \prec 1 \quad \text{(Equition 1)}$$

Q_k – quality of organisation culture;
F_k – real state;
O_k – expected state; where 1 is ideal state, in which $P(t) = 0$ when P- probability and t - time.

An instrument to obtain the estimations of the above-mentioned states, are work stations audit results. (Stein, 1998) These estimations are performed basing on the check list. The check list structure must be created basing on the existence of measurable and immeasurable criteria of the threats in the work process. This estimation should be realised by the workers with the use of ergonomic self-knowledge. A worker consciousness should enable to estimate the threat level in an accepted scale. It needs the correct training and the organisation culture level, which includes the work safety culture, that is the performance culture in the work process.

These assessments may be determined by the coefficients of quality related to the consciousness state about threats in the process of organisation work related to their risk. The knowledge about the methods of these threats reduction and the threats risk is a quantification of state quality of organisation safety. This process will demand continuous growth of the knowledge level about the human in the organisation work process, in other words: higher and higher ergonomic human's consciousness.

Quickly changing work and life environment, particularly on the level of the inter-human communication demands bigger and bigger self – knowledge from us. This environment is changing in the way being dependent and independent of us. In the field of these changes the following sciences may be considered:

- the prakseology as a science considering the effective activities in the learning the logic and the technological activities,
- the ergonomics, which is accompanied to prakseology as a base in ergological methods use in designing and diagnosing system: human – environment. (Musioł, 1997)

Of course the purpose of all activities is endeavour to well-being of a human during his activity in every organisation.

5 Conclusions

In order to raise and monitoring of worker consciousness state and the observation of the consciousness indicators it is necessary:

- to implement work safety management systems, which will force the activities directed to ergonomic consciousness raise (Charytonowicz, 1996);
- to intensify professional training delivered for obtaining an ergonomic knowledge (Studenski, 1996);
- to plan tasks assignments adapted to the qualifications and the individual psychophysical predisposition of worker;
- to give reasons to safe behaviours in work process;
- to react on all signals of lowering of consciousness state of the threat on work place;
- to introduce corrective activities as an reaction on detected destructive users' behaviours. (Musioł & Grzesiek, 2001)

6 References

Charytonowicz, J. (1998). Perspectives of ergonomic education of the Technical University of Wrocław. In L. M. Pacholski, J. S. Marcinkowski (Ed.) *Proceedings of International Seminar of Ergonomics Teachers*, (pp. 205-215). Poznań: Pub. Poznan University of Technology.

Golubińska, K. (1974). Rozwój cywilizacji a psychiczne obciążenie człowieka (Development of civilisation and psychical charge of human). Warszawa: PWN, (in Polish).

Kiergegaard, S. (1995). Choroba na śmierć (Sickness on death). Poznań: Wydawnictwo Zysk i s-ka, (in Polish).

Musioł, T. (1997). Audit bezpieczeństwa (Audit of safety). In J. Kicki, M. Mazurkiewicz, Z. Pilecki, (Ed.), *Proceedings of the School of underground Mining 97*, (pp. 621-631), Warszawa: Wydawnictwo Polskiej Akademii Nauk, (in Polish).

Musioł, T. (2002). Kultura bezpieczeństwa pracy a jakość kultury organizacji (The culture of the safety of work and the quality of the culture of organisation). Gliwice: Wydawnictwo Politechniki Śląskiej, (in Polish).

Musioł, T., & Grzesiek, J. (2001). The simulation of catastrophic event as tools for identification of threat. In L. M. Pacholski, J. S. Marcinkowski.(Ed.), *Proceedings of the International Seminar of Ergonomics Teachers,* (pp. 210 - 221) Poznań: Pub. Poznan University of Technology.

Musioł, T., & Karłaszewski, B. (1990). Psychic load reduction of a power plan operator. In L.K.Noro and O. Brown (Ed), *Proceedings of the 3rd International Symposium on Human Factors in Organizational Design & Management. Human Factors in Organisational design and Management,* (pp. 7 - 11) Nord Holland: Eliservier Science Publishers B.V.T.

Nowicka, E. (1998). Inkulturyzacja i personalizacja. Podstawowe pojęcia z antropologii – Wprowadzenie do pedagogiki wybór tekstów. (The culturisation and personalisation. The basic knowledge in antropology). Kraków: Wydawnictwo Impuls, (in Polish).

Sillamy, N. (Ed.) (1994) *Słownik psychologii (Psychological Dictionary - translation of Libraire Larousse 1989)*. Katowice: Wydawnictwo Książnica (in Polish).

Stein, E. (1998): O zagadnieniu wczucia (About the empathy problem). Kraków: Wydawnictwo Znak, (in Polish).

Studenski, R. (1996). Organizacja bezpiecznej pracy w przedsiębiorstwie (Organisation of safety of work in enterprise). Gliwice: Wydawnictwo Politechniki Śląskiej, (in Polish).

The Adaptation to Main Control Room of a New Human Machine Interface Design

Yuji Niwa

Institute of Nuclear Safety System, Inc.
64 Sada, Mikata-gun, Mihama-cho,
Fukui 919-1205 Japan
niwa@inss.co.jp

Hidekazu Yoshikawa

Kyoto University
Gokashou, Uji, Kyoto,
Japan
yosikawa@uji.energy.kyoto-u.ac.jp

Abstract

Several innovative interface designs have been proposed especially after the catastrophic accident in Three Mile Island No.2. The Safety Parameter Display System (SPDS) (Woods et al., 1982) has actually introduced in nuclear power plants. However, nevertheless subsequent efforts of human machine interface (HMI) designers, operators in nuclear power plants are conservative to more convenient and advanced HMI designs. As a result, such interface designs have rarely implemented in commercial plants. Recently, the criticism that researches on HMI design might not be necessary has been arisen. The paper suggests HMI designers should focus on comprehensive designs for average operators rather than highly abstracted ones. Even when such comprehensive and preferred HMI are shown in demonstration, operators never agreed with the real implementation and use in main control room during accidents. From this experience, the paper also concerns the issues to be resolved to adapt a new HMI design.

1 Introduction

Almost of commercial HMI designs as well as new ones, they employ the graphical representation to express plant status/situation. The typical and conservative designs are presentation of each significant plant parameters with relevant piping and instrument diagram (P&ID), since operators are very familiar with plant structure. In addition, trend graphs are frequently displayed in HMI screen to monitor plant behaviour. Viewing current HMI designs, information presented is effective for preventive maintenance and operation, which means current HMIs are not necessarily designed assuming the use of post-accident stage. As obviously seen in the TMI accident, operators extremely upset and they could not proper recovery actions. Besides the SPDS, many HMI designs have been proposed. Some of them are quite attractive for HMI researchers' eyes. However, almost of such new designs forget subject to use HMI is operators who are not necessarily familiar with high academic outcomes. Moreover, HMI designers never know the ultimate degradation of thinking capability just after reactor trip. As the famous Human Factors research proposed, the information process of in human is observe, interpret, decision-making and scheduling. Via the process, human can act in several situations. Interpretation requires operators' calmness, which can be commonly accepted. Graphical representation frequently requires interpretation even applied elements of information are familiar with operators and operators are trained well by full-scale simulator. If the cognitive workload of interpretation is mitigated by some ways, then operators are expected to use HMIs. The syntactical presentation, i.e., the presentation of emergency operating procedure (EOP) will mitigates or bypasses the interpretation process even if operators can read documents. If the syntactical instruction is presented by appropriate way considering human factors aspects, it is worthwhile taking as a HMI of post-accident support. It can be called as "direct support" in the sense that the workload for interpretation is not required or extremely light. Whilst direct support, many newly developed and traditional HMIs using graphics can be called as "indirect support". The schematic diagram of direct and indirect support is depicted in Fig. 1.

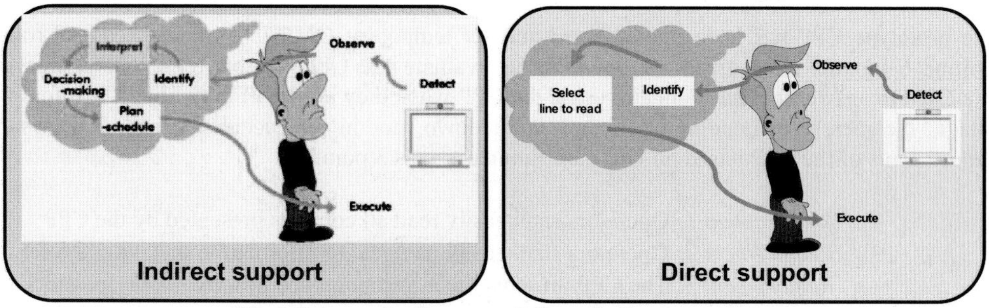

Fig.1 Schematic diagram of direct/indirect support

Needless to say that both have advantages and disadvantages. Conclusively from the research, combined use of both supports will improve reliability of recovery operations in post-trip stage when operators upset. Obviously the direct support must be quite comprehensive for operators. Thus procedure presentation will assist operators to guide what they should do even in upset condition as a direct support.

Even when such comprehensive HMI are shown in demonstration, operators never agreed with the real implementation and use in main control room during accidents although they appreciate the design principles. From this experience, the paper also concerns the issues to be resolved to adapt a new HMI design. First, user requirement should be decided from operators' voice. Statistical treatment is NOT the solution to make user requirement document (URD). Considering strong correlation of operators tastes, it has only to be analyse/integrate a few operators' opinions. In such way, the deep contribution of user in HMI design is a one of significant solution of adaptation whilst the stepped extension of application is the other one. The former requires the collaboration of utility, vendor and occasionally researchers in university. Thus utility-vendor-university union may lead to real application and adaptation of a new HMI that is basically required to be comprehensive. In this paper, a field evaluation of only direct support is concerned and makes suggestions for adaptation.

2 The CPP (computerized procedure presentation) study

The sentences in EOP are as follows:

[Operation]::=[Condition]||[Action], (1)

and additional information is taken as a comment. If [Condition] is true then the specified [Action] is taken. The design principle is that all texts in an EOP are decomposed into these 3 elements and assigned to *tiled* format screen. The instruction to be read by operator is highlighted and check box is blinking. Insertion of check mark is easily made by clicking the corresponding check box or merely by "Return/Enter" key. The CPP system automatically highlights the next EOP sentence to be followed. For mitigating the workload to search appropriate EOP to be followed, the CPP has comprehensive menu (navigation) screen and for better orientation, it has "operations map" in the presented screen. A typical screen image is depicted in Figure 2. (Niwa et al, 1996)

3 The evaluation of the CPP by site operators

3.1 Basic questionnaire and result

The prototype CPP has been implemented and set temporary *OUTSIDE* of main control room. Around 6 months were given for operators to evaluate the CPP. In this term, operators were requested to answer the questionnaire using the CPP when they are in off duty time. Therefore the data is the merely evaluation when they calm down, not in real accident stage. The system requirements of the CPP based on URD are summarised as 3 points.

- *Requirement 1:* Operators shall easily read all contents presented by the CPP
- *Requirement 2:* Operators shall easily navigate to information they would refer to
- *Requirement 3:* The CPP presents the information necessary without operators' frustration

The requirements (concerns) of major design principle are decomposed hierarchically so that the specific questions. The resultant questions that were queried to operators are shown in **Table 1 together with mean scores (see below)**. As a questionnaire style, "direct method" that means operators responds using numerical value to questions given *vis-à-vis* traditional questionnaire adopts some scale (categorised into interval scale or ordered scale). Operators can insert arbitrary number from 0 (quite disagreeable) to 5 (full mark, perfectly agree). The reason such unique questionnaire was given is we are interested in the item that 0 is given, which might bring significant information to improve the CPP. As is shown in Table 1, a respondent has given score 0 to the query numbered 2-1 and 3-1. However, due to re-interview, this respondent did not understand the meaning of queries. Figure 3 shows a selection of distribution of scores from respondents. Apparently all distributions tend to shift to high score and all means are over 2.5. The evaluation by users are considerably well from the viewpoint of the CPP designer.

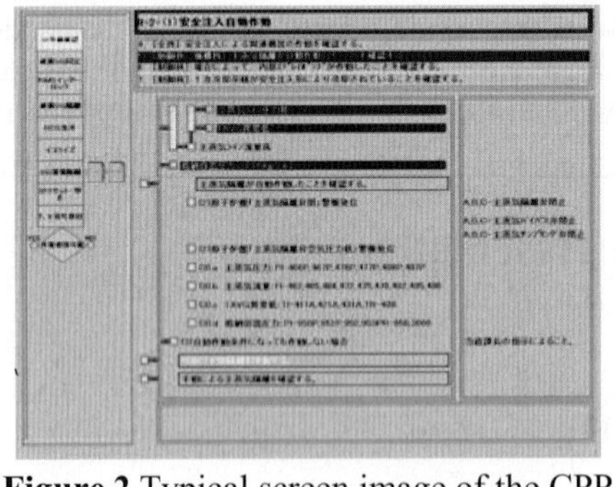

Figure 2 Typical screen image of the CPP

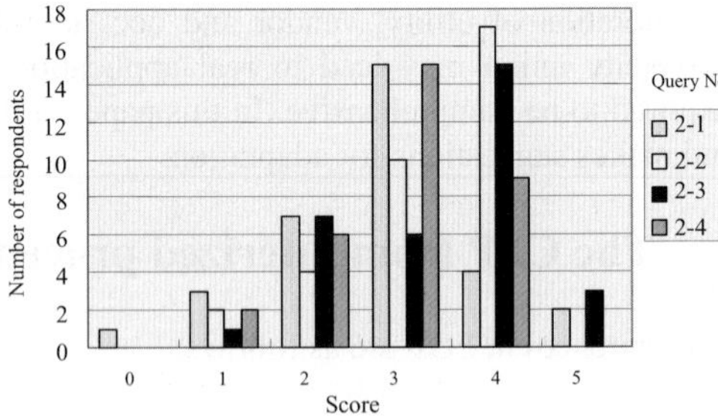

Figure-3 Distribution of score

Table-1 Structured Interview Design and specific questions

Requirement No.	Decomposed requirement	Specific question	Query No.	Mean
1 (Reformat technology)	The CPP shall present all information in EOPs	**The CPP presentation is any feeling of disorder from EOP?**	1-1	3.0
	What are presented shall be sufficiently comprehensive	**The CPP makes easy to search EOP you would like to refer to?**	2-1	2.9
		Icons in menu screen are sufficiently understandable?	2-2	3.5
		Menu screen is easy to use?	2-3	3.4
		System messages in dialogue box are sufficiently understandable?	2-4	3.1
2 (Navigation)	Easy to navigate to where they would like to refer to	**Easy to navigate to procedure STEP you would like to refer to?**	3-1	3.2
		Logical condition is sufficiently understandable?	3-2	3.4
		Procedure STEP presented by the CPP is sufficiently understandable?	3-3	3.5
		Comments are presented clearly?	3-4	3.4
		Screen layout is easy to read?	3-5	3.2
	Easily to grasp orientation	**CPP makes clear where you are making operation?**	4-1	3.3
		Check boxes are easy to use?	4-2	3.2
		CPP makes clear operations you have already done?	4-3	3.5
3 (System response)	System response shall be sufficiently fast considering real use	**Response time for presentation of the CPP is sufficiently fast?**	5-1	4.1

3.2 Basic questionnaire and discussion to operators for adaptation

In many previous HMI researches, they have been finalised at this stage, which means they concluded the HMI is successful in the sense that the designed HMI is accepted well by users. However, considering the better adaptation, it is a good idea to give the further question if operators can use the CPP in real main control room. Contrary to the basic design evaluation, most operators are against the introduction of the CPP in main control room. Why they refuse the CPP as a HMI in real use is summarised as:

- Manual insertion to check box will increase workload during an accident
- At least an operator is occupied by The CPP, although operators are suffering from the lack of human resources during an accident
- The CPP will be monitored by each operators in main control room (operator would like to check own operation made)
- The CPP might present incorrect operations due to system failure

These requirements for adaptation in control room created new researches for site adaptation. First, it has been realised operators will never accept the system that accompanies additional workload even if it is very light, which will lead to the research on an "automated scrolling CPP." For this purpose, machine has to recognise and monitor human action. Second, operators would like to communise the information on plant status and to assist supervisor's decision-making. The research on *decentirised* CPP presented by PDA (portable digital assistant) would be required.

Of course, Verification and validation including software of the CPP is the most important as the last complaint indicates.

Further discussion has been made with operators and why they have hesitation to introduce the CPP has been clarified conclusively. The reason is operators have anxiety to use the CPP in main control room, since operators have never trained the recovery actions referencing the CPP using full-scale simulator. Vendors always make such training before introduction and operators require same steps even to such simple system.

4 Proposal and conclusion

As this experience indicates, operators are ultimately conservative to HMI. In this research, the quite comprehensive HMI has been composed and accepted as education use rather than real service during accident. Operators only accept HMI supplied by only a vendor whole plant constructed. Unfortunately vendors cannot necessarily grasp user requirements during service time for power generation. Moreover vendors are requested to make profit to introduce new technologies prior to the development of design methodology for the best convenience during an accident or preventive monitoring. The critical gap to design and evaluate HMI is that operators cannot necessarily organise the user requirements before construction. Average utility staffs can evaluate HMI and discuss requirement that will never lead to major change of system design of HMI once the prototype system has been demonstrated. Universities arte proposing new design principles that seem to be hard to understand for average operators. They tend to finalise a research on HMI when evaluation, whose respondent is merely students, has been completed. Many innovative HMI design have been proposed and most of them remain not to introduce to commercial plant. However, the recent deregulation market of electricity has to be taken into account also in Japan. A new HMI design is strongly required to introduce in commercial plant. All nuclear industries have to make clear profit to implement easy to use HMI and extend the market outside of nuclear field. For this purpose, the composition of university-utility-vendor research union is strongly recommended. In this paper, such necessity has been shown quoting the real experience.

References

Woods, D. Wise J., & Hanses L (1982): Evaluation of safety parameter display concept", EPRI final report (1982)

Niwa, Y, Hollnagel, E & Green, M, (1996): Guidelines for computerised procedure for emergency operating procedure", *Nuclear Engineering and Design*, 167, (pp114-127). Elvesier (1996)

A Phenomenon of Computer in Human Life

Przemyslaw Nowakowski *Jerzy Charytonowicz*

Department of Architecture, Wroclaw University of Technology
Prusa str 53, 50-317 Wroclaw, Poland
pn@arch.pwr.wroc.pl georgech@novell.arch.pwr.wroc.pl

Abstract

Computer development is a special phenomenon in the field of production of various artifacts – each year brings changes that contribute to rapid (moral and physical) aging of computer models manufactured before. Computers are manufactured for all groups of consumers, including children and the disabled. Increasingly, they are integral parts of technical products like phones, cars, etc. Such technical factors as ease of operation, data processing speed, etc. (user 'friendliness') as well as their aesthetical qualities combine into ergonomic quality of computer systems (devices and software). Computers have already become everyday use devices. They are used in virtually all areas of social life, especially in the environment of work. They have become an important element of changes related to organization of work and human well-being (health effects, sociological and social changes). Universal computerization is characterized by many measurable qualities in the context of social evolution, such as development of new types of services, access to information, contact unhindered by any barriers, etc. However, computerization creates also many dangers in the social field, such as new kinds of crime, lack of privacy, alienation. Recognizing and coping with those dangers becomes an important task for ergonomics in the future.

1 Initial remarks

Technical development has revolutionary character, i.e. rapid change, contrary to social changes, which usually occur in an evolutionary way, i.e. benign, calm and natural. This creates a problem of reconciling the rapid pace of technical development with a process of social acceptation of new technical achievements that change existing habits and lifestyles in society. Technical development has always outpaced development of social consciousness and this aspect seems crucial in introducing new inventions because it is mainly their social acceptation which determines effectiveness and pace of technical development in particular social environments. Social acceptation of all types of technical innovations depends to a large extent on their main purpose and the degree of their effect on human imagination in the aspect of potential dangers. Sociologists, psychiatrists and psychologists more and more often indicate that an untamed technical development destroys origins and foundations connected with the shaping of human personality and proper formation of social relations. The creation and acceptation of technical development should be therefore accompanied by introduction of mechanisms enabling human control and power over new technologies with attention to the wide scope of their social effects, which consequently should constitute another task of future ergonomics. Technical development besides unquestionable improvements carries with it also different kinds of dangers to both the proper development of the area of human personality and the sustaining of equilibrium in the natural ecosystem of which humans are an active element.

Perception, analysis and data processing as well as reactions to stimuli are attributes of almost all living organisms. Since the beginning of ages humans strived to find a convenient place for their existence as well as predict and protect themselves from changes disturbing the arranged order. Innate human limitations in data processing and desire to increase the comfort of life founded a basis for invention and employment of increasingly advanced calculating machines in everyday activities.

2 Evolution of calculating machines

The foundation for constructing of "calculating machines" was provided by the formulation of numerical system, mathematic laws and ability to quantify material elements of the environment (measurement and weight). A calculating machine that has been in the longest use (for centuries) may be considered **abacus**, which continues to be successfully used in many countries today. The 19th century saw popularization of **arithmometers**, small mechanical calculating machines, whose electronic versions can still be seen on pay-desks in shops. Almost to the end of 1970s engineers had used **slide rules** that were then replaced by **calculators**.

The first digital calculating machine was ENIAC, which was constructed in the U.S.A. at the University of Pennsylvania in 1940s. It consisted of over 18,000 lamps, weighted 30 tons and occupied a large chamber. The lamps were relatively big, hence the dimensions of that type of devices were enormous and required strong cooling. As early as in mid 1960s computers became a permanent element of office work. The archetype of modern personal computers was "Alto" with a half-tone screen which was constructed by Xerox and the first microcomputer was based on Intel 8080 microprocessor by IBM in 1975. In mid 1960s Douglas Engelbert invented the "mouse" which although brick-sized immediately became readily used as a replacement for keys in the keyboard (Gelernter, 1999).

The last two decades have witnessed dynamic development of successive computer generations of ever increasing operational capacities, faster and increasingly miniaturized. The scope of their application in the material environment of human life grows dynamically and covers practically all areas of human activity – from the environment of work to the environment of rest, from children's toys to military and space industries.

3 Evolution of software

The foundation for modern software was provided by **binary system of "0" and „1"** that was used for the first time in the **Turing's calculating machine** of 1936. The method of **object programming** ("Algol 60" and "Simula 67") invented in mid 1960s enabled modeling of software upon the structure of real environment thus making creation of complex applications possible. An idea of the simultaneous use of a single application by many users was born in that period. The principle of **shared-time work** consisted in a program quickly contacting with consecutive users. The first screen windows and light pen that enabled drawing on a screen were invented by Ivan Sutherland from Xerox in 1960s ("Sketchpad" application). Not much later Alan Kay programmed overlapping windows and at the same time eliminated space limitations of software related to the size of monitor screen. The above inventions, however, were abandoned for about 20 years. Xerox' inventions were used and improved by Apple only after 1977, which in 1984 created its „MAC" application in which the screen with windows imitated a traditional desk with piles of documents. "MAC" turned out to be an easy-to-use „picture application". Simultaneously, Microsoft promoted its "DOS" application which required, however, introducing commands with the use of often complicated algorithms. Typing errors rendered that system less effective than „MAC". Only in 1990 did Microsoft replace „DOS" with „Windows 3.0" application modeled after „MAC". The

above mentioned principle of shared-time work provided the basis for creation of the first computer nets (cyberstructures) in 1960s. The archetype of the modern **Internet** was „Linda" application that was used for the first time at the Yale University in the U.S.A. That application could be simultaneously used by many terminals (Gelernter, 1999). Modern computer applications have complicated, extended structure and are therefore often referred to as **virtual engines** – machines without actual body.
Apprehensions related to difficulties in using complicated applications have become a factor impeding full acceptation of the computer as a tool assumed to be helpful for humans, especially among the elderly. Modern directions of ergonomic activities indicate growing tendencies towards designing simpler and more user-friendly software.

4 Computerization of everyday life

Computers have already become everyday use devices. They find application in virtually all areas of social life. Their use promotes, among other things, development of new types of services, access to information and contact over all barriers. Almost all devices contain electronic components and software, which is a result of miniaturization and increasing operating rates of digital components. The use of devices "armed" with electronic components promotes in turn improvements in technical comfort of life. The diversity of software enables meeting nearly all needs. The complex structure of software frequently limits user's intervention and its volume makes it prone to errors. Unergonomic and overdeveloped applications with unnecessary functions make technology live its own life while the actual needs of users are becoming unimportant. Most people uses but a small part of what software offers and at the same time is nervous about its failure frequency. Studies show that development of an average application last over 50% longer than planned and every forth one is abandoned at all (Gelernter, 1999). A certain shortcoming of software is also the imitating of its "paper" counterparts (files, folders, etc.).
Introduction and development of computer technologies result in new forms of work. Flexible working hours, workplace changeability, independent decision-making, responsibility, task programming, team work and frequent necessity of interdisciplinary cooperation determine specifics of the modern environment of work. Owing to computer systems, new qualities increasingly appear in the way office work is done. All necessary data are easy to acquire and process further while videoconferences, digital image processing, mobile computers and phones replace many time-consuming activities. In the opinion of experts, flexible working hours, constant availability of workers and office computerization *(Satellite Office, Telework Center, Virtual Office)* promote rationalization and reduction of the costs of work and increases in its effectiveness. At the same time there are warnings against social phenomena accompanying those processes, such as mental stress resulting from the feeling of continuous and undisclosed control, constant availability, responsibility, competitiveness and fear of job loss as well as egoism, disorientation, alienation and social disintegration within family or colleagues (Charytonowicz, 1999).
Computer is becoming the basic tool of work and entertainment. It is assumed that the use of paper as the basic medium of information will become significantly reduced in time to come. Effective work with a computer screen, however, requires good sight and constantly raised concentration. Linking several computer screens within a network renders constant presence at the workplace unnecessary. Moreover, a computer screen is a universal tool, hence one workplace can be used by different people in different time (*Desk Sharing).* This principle may be advantageous especially to the disabled, in the first place through easy access to the workplace and possibility of working in flexible hours or part-time. Modern computer workplaces distributed all over the office space are not only more and more ofter interconnected by a computer network but they are also integrated with multimedia as well as data acquisition, processing and access systems (*DMS-Document Management System*).

Sensor-controlled screens („*Dyna Wall*"), control panels („*Interac Table*") and rest-furniture plugged into a network („*Com Chairs*") can change the „official" character of office space and create a home ambience, which may in turn significantly increase effectiveness.
Experiments with „work by telephone" (*Telework*) have been conducted for more than ten years now thus abandoning traditional work in office centers in favour of work at home (*Home Office*). However, separation of a professional office workplace within home space and work characterized by lack of personal contact with colleagues are accepted by only some few occupational groups (e.g., by freelance professionals or people preferring flexible working hours, including contract workers) (Wehr, 1996). It should be also kept in mind that professional work done at home tends to negatively influence the lifestyle of household members and social relations between them.
Enthusiasts of the Internet and virtual reality believe that one of the main advantages of computer techniques is the fact of people staying at home where the outside world (e.g., work environment) is brought in by means of electronic devices, which in consequence disrupts the basic functions of the environment of rest that includes also home environment. It collides also with the principles of ergonomic shaping of two fundamental environments of human life, i.e. the environment of work and broadly conceived environment of rest where the basic precondition for mental relaxation is leaving of professional problems outside home. Computer, which is becoming a ubiquitous element of the human habitat and allows people to stay home and keep in touch with the outside world at the same time, significantly influences reduction in mutual contacts between people which is a condition for proper social development of individuals through, among other things, formation of skills of coexistence and cooperation with other members of the society. At the same time, transfer of work to the environment of rest results in disappearance of fixed working hours and clear distinction between private and professional lives. The argument of computer manufacturers thay new technologies will for instance enable women to work at home and thus be with their children more frequently is irrational since effective work requires concentration and focus. Work at home in connection with looking after other household members means its poor quality for both the employer and family of the worker. Under the disguise of concern with comfort of the working man, manufacturers of computer systems and software try in the first place to increase demand for their products. At the same time, owners of businesses switching to the „virtual office" system save the costs of office space rental by shifting their functions to the homes of their workers. Moreover, it offers possibility of making a demand for flexible working hours and control of workers in the environment of their rest. Experts observe that anonymity of work at home *(Home Office)* or occasional work at a random office workplace *(Time Sharing)* may lead to reduced motivation to activities and consequently to decrease in their effectiveness.
Doing professional work at home, however, offers a chance for employment of the disabled on equal conditions. It is particularly true with regard to people with impaired motion, sight and audition but intellectually sound. In such cases it is sufficient to supplement a computer workplace with suitable technical devices individually adjusted to the disfunction (Charytonowicz, 1998). Development of the concept of multimedia and virtual office may influence change in the forms of office buildings and structure of cities. Deconcentration of the environment of work and its more even distribution will occur which will also cover peripheral areas of cities. Up to several dozens people may be employed in paricular objects. Smaller, several stories high buildings are more open to transformation and even to radical change of their function than hitherto preferred high-rise buildings. This will make workplaces „come" closer to the home environment. Such a perspective seems to be an advantageous alternative to the concept of "Home Office".
Computers came earliest into common use in the U.S.A. and it was also there that results of their negative effects on the human organism were earliest recognized. They were systematized within the frames of so-called **RSI symptoms** understood as an internal reaction of the operator's organism to stress caused by work with computer, which actually means a group of injuries resulting from repeated strains. In the late 1980s RSI symptoms were added to the list of occupational diseases in the U.S.A. American studies show that over 20%

people working with a computer keyboard for more than ten years suffer from typical RSI symptoms. Effects of those symptoms frequently resemble unnatural degenerations resulting from intensive practicing of some sports (Fritsch, 1996).

Revolutionary changes, like steam engine in the age of Industrial Revolution, nowadays were brought by computer. Development of digital technology and microprocessors began a new stage of technical development called *computer revolution* or *information age* in the late 1980s, which initiated another race of man and machine, this time not a steam engine but a digital device which unlike the former one influences human mind as well, thus contributing to formation of **stresses** whose causes include fears of human work being replaced by the work of computerized robots and not being able to follow the machine in term of working speed. Protection of the human being from stresses resulting from contacts with technology seems to be a priority direction of activities for ergonomics of the future especially because various somatic disorders connected with longtime computer operation have already been well recognized (e.g., wrist, spine, cardiovascular diseases, etc.) contrary to mental and social effects of universal computerization, which have been neglected in hitherto deliberations of ergonomists. Contemporary computer science offers such splendid means of creating synthetic and mobile images interactively reacting to the presence of spectator (*virtual reality*) that it is a source of concern for psychiatrists and psychologists. In their opinion, frequent exposure to virtual reality may discourage participation in the real world and its problems.

5 Summary

Universal use of computer technologies is characteristic of the last forty years only. The dynamics of qualitative changes (miniaturization, data processing rates) and quantitative changes (popularization in technology and everyday life) is an unusual phenomenon in the entire history of technique. Computerization is perceived by the majority of users as a remedy for almost all existing problems. Too little attention is paid to shortcomings of computer systems and negative social effects that slowly become visible in strongly technicized countries. The pace of life and information overload increasingly undermine traditional values and social order established through centuries. However, computer progress will continue. Elimination of negative effects of computerization can be helped by their universal awareness and systematic counteraction.

6 References

Charytonowicz, J. (1998). Social effects of integrating design. Procedings of the Global Ergonomics Conference. Cape Town: Elsevier. p.p. 203-206

Charytonowicz, J. (1999). Social aspects of of universal computerisation. Procedings of HCI International '99. Munich: LEA Publ. Vol. 2. p.p. 573-577

Gelernter, D. (1999). Machine Beauty (Polish edition). Warsaw: Publisher W.A.B.

Fritsch, M. (1996). Handbuch gesundes Bauen und Wohnen. Munich: Deut. Taschenbuch Verlag

Wehr, W. (1996). Gegen die Proletariesierung. AIT 10/96. Leinfenden-Echterdingen: Verlaganstalt Alexander Koch GmbH. p.p. 78-81

Have Operators to Forget the Old System in Order to Acquire the New One? A Case Study in the Health Care Context

Francesca Rizzo, Oronzo Parlangeli

University of Siena
Department of Communication Sciences
Via dei Termini 653100 Siena
francy@media.unisi.it
parlangeli@unisi.it

Sebastiano Bagnara

Politecnico of Milan
Via Durando 10 20131 Milan
Sebastiano.bagnara@polimi.it

Abstract

During the 2001 the top management of the Reservation Center of Hospital Services and Treatments (RCHST) of Florence Health Department was projecting to substitute the old information system with a new one. Managers also thought that the system in use had to undergo a usability evaluation: results so obtained were expected to be a good basis on which implementing the new software. A study was therefore conducted using an integrated series of both qualitative and quantitative techniques. Results showed that the system usability was quite poor and that strongly affected the general level of performance that the RCHST was able to guarantee. The performance offered by the front office reservation services negatively affected the degree of satisfaction expressed by customers. However, contrary to researchers expectations, the low level of usability of the information system did not affect the degree of satisfaction of front desk operators. An innovative hypothesis of training program, based on the forgetting of the system in use in order to allow the acquisition of the new one without affecting the level of satisfaction, has been consequently put forward.

1 Introduction

In this paper we report a research activity conducted by the Department of Communication Science of the University of Siena, from December 2001 to May 2002. Research was initially planned to evaluate the usability level of the software human operators use to make reservations and it was expected to release, as final results, data and suggestions to inform the implementation process of the new information system. At the time of the investigation the information system that allowed operators to control the private and public health center agendas in order to make a reservation was not the only channel to access this service. It was also possible to directly contact, by phone or in presence, one of the Florence health centers. This resulted in a variety of internal and external communication problems that required the top management to introduce a new software and impose it as the only tool to be used to reserve health services and treatments. By and large, it was clear that the RCHST performance depended on a series of factors related to:

- Operators
- Customers
- Software system
- Work organisation (environment and practices)

all together interacting and having a role in delivering reservation services. This suggested researchers to adopt a systemic view: the assessment of the information system usability level used by the operators also had to imply a deep knowledge of the RCHST components and of their interactions (Fig 1).

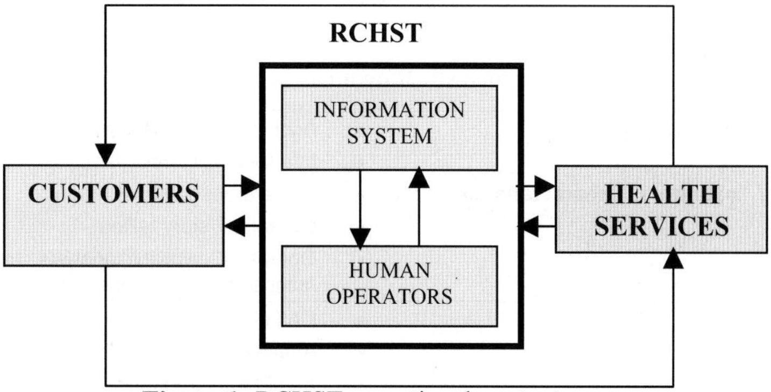

Figure 1: RCHST operational context

Some sub-goals were thus considered:

- knowing information system operators' characteristics;
- knowing the information system in use;
- knowing the activities performed by operators;
- knowing the working context in which the information system was used.

The RCHST was a large net composed by a series of front desks located in some public hospitals of Florence and in private health centers (private hospitals, chemist's shops). In addition to reservation desks, customers had the possibility to make a reservation by phone contacting a call center specifically structured to this aim. Synthesising the reservation center was composed by: i)Front desks located in public and private hospitals; ii)Front desks located in chemist's shops; and iii) Call center.

2 Methodology

A series of qualitative and quantitative techniques coming from the etnographics approach (Suchman, 1987) and successfully used in user center design methods (Hutchins, 1995) were adopted. In table 1 the three investigation areas pertaining to research activity are summarised and characterised by some of the chosen technique.

The reservation Center employed 100 operators working to deliver reservation services. They were:

- Front office operators; people with some experience in health care services and the longest experience with the software.
- Chemists; people usually engaged in many others activities. They had a poor confidence with the software. When a chemist's shop customer required a health care reservation the operator accessed a workstation, opened an intranet connection and entered the Information System.
- Call center operators. This group is composed by operators with 3-6 months of experience with the system.

During their activity researchers visited 8 front desks inside 6 private and public Florence health care structures, the call center, and 4 chemist's shops.

Front desks were small offices where two operators worked separately on different workstations. Operator had to perform a series of tasks: identifying customer, understanding which health service he/she needs, identifying the correct code for that health service, opening the corresponding electronic agendas, identifying possible dates, asking customer about his/her agenda, fixing, changing, deleting or finding a reservation. Many times in a day operators helped each other in interacting with the information system. They had a personal notebook where they registered all the information that the system did not provide. On field observation revealed that operators memorised most of the information needed to interact with the software.

Table 1: Summary of the research investigation areas in relation to the inquiring techniques used

1. Knowing operators' characteristics and activities	• Acquisition of data provided by the organisation • Interviews (management, operators) • On field observations • Operators' story telling • Operators' activity analysis (number of tasks, number of errors, time to deliver a service etc.)
2. Knowing working context	• On field observations • Analysis of the front desk physical environment • Semi-structured interviews with customers
3. Knowing Information System	• System analysis (instructions handbook, conceptual structure, information architecture) • System observation in use • Heuristic evaluation

Call center operators worked individually in a box. Their workstation was composed by a computer, a phone, and a headphone. The operational sequence they usually performed was the same of that described for front desk operators. Call center operators used notes, post-it, city map to access information not provided by the software. They often helped each other to solve a problem; otherwise they asked to the supervisor.

Chemists were operators usually working in a chemist's shop but sometime they could also deliver the reservation service. The operational sequence they usually performed was the same of that described for front desk operators. The workstation to put reservations was differently located in the chemist's shops we visited. On field observations showed that also these operators, in order to accomplish their tasks, often used notes, post-it instructions and so on. In some cases researchers observed operator calling, by phone, a front desk point to require support in completing a complex task.

The information system was based on a database storing information about people that can access the national health care assistance and living in the Florence town district. System was based on a command line interface. Operators could not use the mouse to move on the video, this being possible only by the keyboard. Software allowed the execution of the following functions: making a reservation, deleting a reservation, creating a new profile, modifying a reservation, finding a reservation.

3 Results and discussion

In the table 2 main results are referred to research investigation areas.

Results showed a low level of usability of the system. This was clearly indicated by data collected through the on field observations during which researchers analysed the software system, the operators' activity, and the tools and procedures characterising work practices. Also the heuristic evaluation activity (Nielsen, 1994) underlined a series of interactive breakdowns and violations of the user centered design principles (Norman, 1989). Additional problems were observed due to the possibility to differently access to the reservation service, since these accesses were not coordinated each other. All these factors together co-occur in weakening the RCHST capability to foster a satisfactory level of performance in the customers' view.

Table 2: Results obtained according to research investigation areas

1. Knowing operators' characteristics and activities	• Operators had to memorise most of the information needed to interact with the system • Operators were not involved in management decisions and organisational changes • Operators were scarcely trained on information system • Operators were scarcely trained on customer interaction techniques • Front desk operators judged the system adequate to support their activity
2. Knowing working context	• Ineffective External communication • Customers expressed a low level of satisfaction in relation to the service • Inefficient Internal communication • In order to perform their activity Operators used many tools which were not provided by RCHST (personal book, notes, post-it, black board) • The call center was located in another town • Front desk operators worked in pair • Call center operators worked alone • Chemists worked alone
3. Knowing Information System	• System did not provide any information to support task execution • System did not suggested any of its function • System interface did not allow any direct object manipulation • System did not provide any feedback to operators' actions • System did not support navigation • There were no mapping between command and functions • System functions were managed by keyboard • System language was not clear • System information architecture was poor • System did not allow error management

However, front desk operators did not complain a low satisfaction level about the information system. This in contrast with the poor system effectiveness and with its difficulty of use observed during the study. This result can be explained in relation to two factors: i) the long time spent by front desk operators with the system; ii) the low level of information technology expertise manifested by operators. It can be reasonably argued that the familiarity with the tool, the development of informal procedures an practices to perform everyday activities, the creation of external tools (personal book, blackboard, paper information repositories etc.) could seriously prevent front desk operators from accepting a new information system. One of the solutions the RCHST management could apply to avoid this negative perspective could be a planned intervention of a deliberate forgetting strategy. Recent literature on information processing mechanisms underlines the positive role of forgetting in different situations in which loosing, temporarily or definitively, some piece of information can improve the cognitive performance under complex circumstances as, for example, circumstances where a human operator has to interact with unknown systems (Parlangeli, Rizzo, Bagnara, 2003). These considerations provide the logical basis for conceptualising, in this context, forgetting of the current information system as a process that can be managed in order to obtain its rejection and to facilitate the introduction of a new information system. In this case, the intervention that can be considered as more effective is referred to the interference hypothesis, this being focused on a competition mechanism occurring among pieces of knowledge sharing some cognitive resources (Underwood, 1957; Bower and Mann, 1992). Forgetting can thus be related to the interference mechanisms intervening when new information is stored in memory, thus damaging old information already acquired (retroactive interference). More practically front desk operators could be exposed to a training program in which they use and learn new unknown information systems. The acquisition of new knowledge should result in the improvement of operators' information technologies skills and in the weakening of the knowledge about the old information system. After this first training phase a second one on the new information system, the one that will be actually used, should be implemented.

Finally, results about the relation between the RCHST level of performance and the degree of satisfaction expressed by the customers showed how the first variable negatively affect the second one. Less clear are the mechanisms by which the low performance level offered by the RCHST influence the perception of the organisation identity by customers. At the state of the art it is necessary to consider that the RCHST is not the only channel by which to obtain a reservation. Consequently it is almost impossible to investigate the nature of the relation reported above. Anyway top management has to consider that when the RCHST will be the only contact channel to access health care services this will become the most important way through which organisational identity and values (Mok, 1996) are communicated.

4 References

Bower, G.H., and Mann, T. (1992). Improving recall by recoding interfering material at the time of retrieval, *Journal of Experimental Psychology: Learning, Memory and Cognition*, 18, 1310-1320.

Hutchins, E. (1995). Cognition in the wild. Cambridge, MA: MIT Press.

Mok, C. (1996). Designing Business. Multiple Media, Multiple Disciplines. San Jose, CA, Adobe Press.

Nielsen, J. Mack, R.L. (1994); Usability Inspection Methods. New York: John Wiley.

Norman, D. A. (1988). The psychology of everyday things. New York. Basic Book.

Parlangeli, O., Rizzo, F. & Bagnara, S. (2003). Delete memories learning through deliberate forgetting, *Tecnology & Cognition*, in press.

Suchman, L. (1987). Plans and Situated Actions. The problem of human machine communication. Cambridge, University Press Cambridge.

Underwood, B.J. (1957). Interference and forgetting, *Psychological Review*, 64, 49-60.

An Operator Training System based on Man Machine Simulator

Kunihide Sasou, Kenichi Takano, Mitsuhiro Ebisu

Central Research Institute of Electric Power Industry
2-11-1, Iwado-kita, Komae, Tokyo, 201-8511, Japan
sasou@criepi.denken.or.jp

Abstract

This paper describes an operator training system based on Man Machine Simulator (MMS) developed by CRIEPI, which simulates the behavior of an operating team under abnormal operating conditions at a nuclear power plant. This training system shows how an operating team under abnormal operating conditions behaves as results of computer simulation of them. This system uses 3D computer graphics to show the behavior of the simulated team and the control panels. This system uses Boston Dynamics Inc.'s PeopleShop for the 3D images of operators' behavior. On the 3D CG control panels, indications of indicators move and lights of alarms go on and out, reflecting the simulated plant behavior. This system shows how to use research results of cognitive models of operators and also shows a new approach to operators' training with the cognitive model and computer graphics.

1 Introduction

The group of authors has developed a system called "Man Machine Simulator (MMS)"(Takano, Sunaoshi & Suzuki, 1999). This system simulates behavior of a team of operators facing with abnormal operating conditions. The simulated behavior of the team includes operators' reading of indicators on the control panel, decision-makings, operations through the control panel and communications between members. The simulated behavior of individual operators is based on operator's mental model that shows the predicted behavior of a plant. For example, when the first warning sounds off, the mental model depicts the plant behavior under the assumption that no operations are done against the condition. The next step is, the model finds the most important plant parameter and executes actions to mitigate deterioration of the parameter with using knowledge stored in the Knowledge Base of MMS. This decision-making process simulates the way that real operators go ahead of the plant and execute actions. Man Machine Simulator is confirmed to be simulate the behavior of the operation team facing with each of 40 abnormal operating conditions simulated by the plant simulator. The authors developed a training system named - Operator Training System (MMS-OTS) (Takano et al., 2001) based on the technique of MMS and 3 D computer graphics.

2 Points to be Improved

Man Machine Simulator is short to be used for training of real operators. One is because the simulated operations and observations are limited due to the simplifications of the plant simulator in MMS. For example, the plant simulator simulates only one train of a system that actually has two trains. The other is that presentation of the simulated results such as movements of each operator is not so good. Considering the system to be used for the training, the quality of the plant

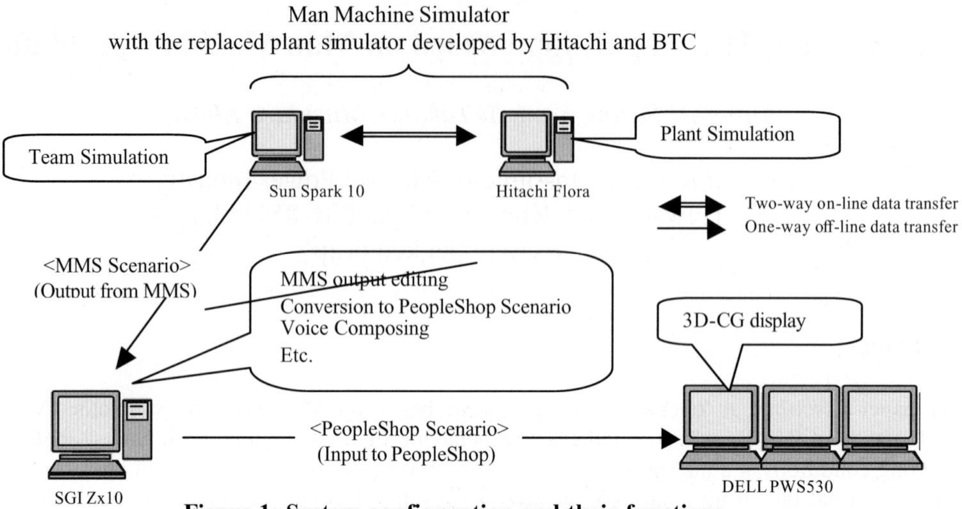

Figure 1: System configuration and their functions

simulation and presentation of the simulated behavior should be improved. Therefore, the authors replaced the plant simulator with another one running on a personal computer developed by Hitachi, Ltd. and BWR Operation Training Center Corp. or BTC in Japan. This simulator simulates the plant behavior of a real nuclear power plant in Japan. The CG software introduced for this project is Boston Dynamics Inc.' PeopleShop also running on a MS-Windows PC.

3 System Configuration and Functions

Figure 1 shows the system configuration and their functions.

3.1 MMS Scenario

MMS scenario, the output of Man Machine Simulator, is generated by Sun Spark 10 and Hitachi Flora with two-way on-line data transfer. Sun Spark 10 calculates the behavior of a team (actions, communications) and Flora calculates the plant operating conditions. MMS scenario is the combined results of the outputs of each computer.

3.2 MMS edit, Conversion and Voice-composing

3.2.1 MMS edit

This MMS scenario would be converted to the data fitted to PeopleShop, or PeopleShop scenario. However, if the original MMS scenario is be converted to PeopleShop scenario, 3D computer graphics shows lots of deficiencies in the motion of CG operators such as extremely quick walk, turn, operation, overlapped utterances by the same operator, unnatural contents of utterances, etc. Unnatural quick motions come from that PeopleShop is showing the motion-captured behavior in the period of the time given by the MMS scenario. For example, now consider that an operator turns 2 switches in order. Real human moves his/her hand to the second switch from the first one directly. However, CG operator moves his/her hand back to the side of his/her body, its initial position, and moves the hand again to the second switch. Inevitably, the time required to finish the all actions is longer than the real one. In addition, PeopleShop is finishing the whole actions at the designated time by MMS scenario. As a result, the motion of the CG operator is much faster than the real one. The other deficiencies, such as overlapped utterances and unnatural contents are due to the MMS's modeling. MMS does not consider the time required to pronounce sentences nor natural utterance.

Therefore, the authors edited the MMS scenario in order to make the motion of CG operators and communications more natural and realistic. For example, repeated actions such as described above are combined to one action with longer necessary time. A CG operator keeps his motion longer as if he/she repeats the actions. The authors changed the timing of utterances and their contents, sometime deleted utterances or combined two utterances into one.

3.2.2 Conversion and Voice-composing

Post-edited MMS scenario is converted to PeopleShop scenario by a converter program developed for this project. Each content of an utterance is converted to a WAVE file by a Toshiba's "LaLa voice" based application also developed for this project.

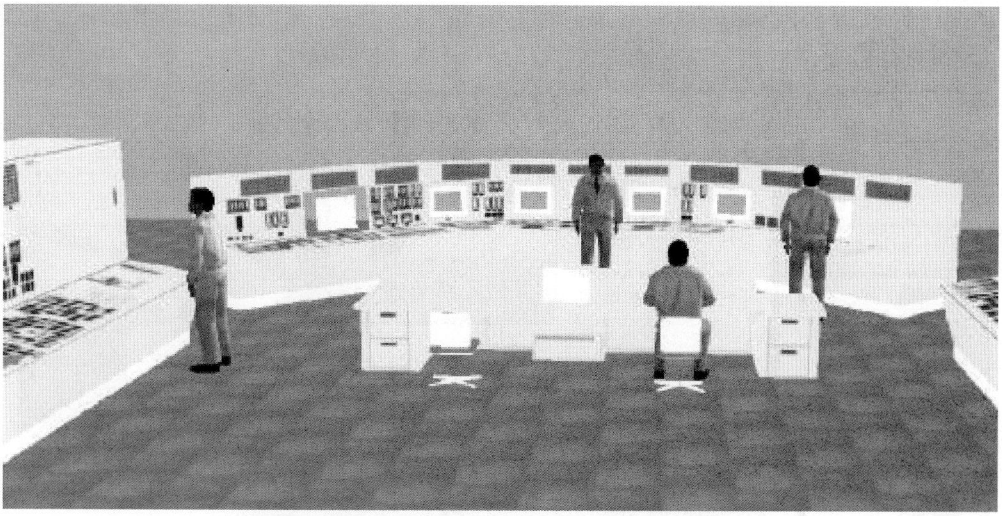

Figure 2: Example of 3D computer graphics

3.2.3 3D Display

The system uses 3 displays. The center display shows the overview of the simulated team. Camera angle and position is changeable. The control panel has 2 modes of LOD (Level of Detail) and LOD between these modes is automatically filled by the graphic board (Wildcat4210). Quick response to the change of camera angle/position is realized in the low LOD (40 frame/sec). Also detailed graphics such as showing the movement of the indication in an indicator is also realized in high LOD when the camera position is close to the control panel (10 frame/sec). Figure 2 shows an example of CG on this display. The left display shows the CG of indicators or switches each operator watches and turns. On the right display, several windows show warnings, each operator's utterances. Other options are possible such as displaying trend graph of plant parameters, etc. but those were not developed for financial reasons.

4 MMS-OTS presentation

4.1 Example

The abnormal operating conditions simulated in MMS-OTS are (1) alterations of System Frequency and steam leak on Main Steam Line and (2) steam leak in Reactor Containment, stuck open on Condenser Vacuum Brake Valve and mechanical failure of Containment Spray Injection System. With referencing the real operation manuals, data are installed into the knowledge base in MMS on procedures to stop/start equipments, theoretical relations between plant parameters, relations of parameters and warnings, relations of operations and their effects, etc. Through repeated simulations, contents of the knowledge base are modified to find the best fitting to these two abnormal operating conditions.

Figure 3 shows an example of the simulation and related computer graphics. The operating conditions are "alterations of System Frequency and steam leak on Main Steam Line". The figure shows the scene of the period of 68 seconds that they are decreasing the plant power by deceleration of PLR (Primarily Re-circulation) pump speed and then manually shut down (scram)

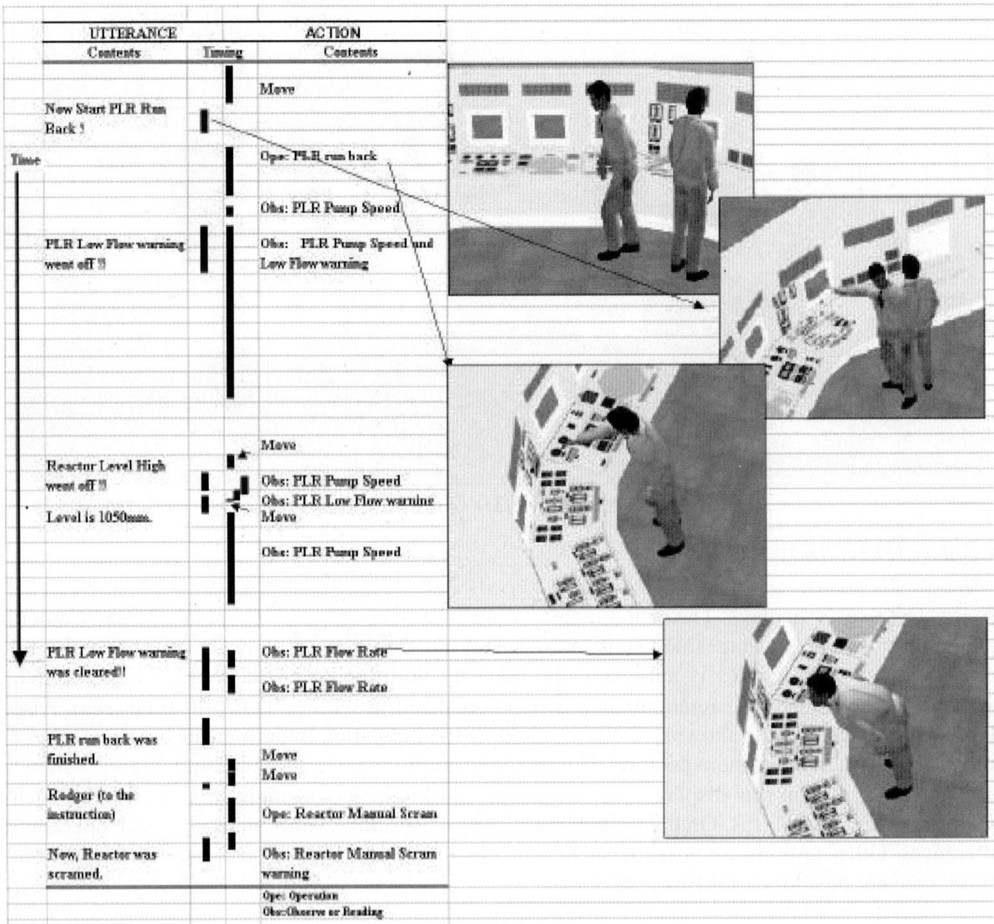

Figure 3: Example of the simulation

the reactor. The figure descries only the utterance and actions by the reactor operator who is in charge of these operations.

4.2 Modification on MMS output

Table 1: Modification on MMS output

Pre-edit	44 actions or utterances	
↓	modification in 33	
	(non modification in 11)	
Post-edit	42 actions or utterances	
contents of modification		
	- timming	7
	- long time action	4
	- non CG presentation	17
	- sentence	5
	- combining utterances	2
	- edit-in one behavior	1
	- edit-out one behavior	1

Actions that are not shown in CG are not listed in Figure 3. The input data to PeopleShop, or post-edited data of the original MMS output, of the reactor operator's behaviours for this period consists of 42 simulated actions or utterances, including those unlisted behaviours in Figure 3. Table 1 shows contents of the modification. The original had 44 actions and utterances on which several modifications were done. The modification in the largest number is making actions not presented in CG. For example, several readings showing repeated looking into a indicator (looks like vowing to the panel several times) were combined into one reading with long observation time so that it looked like reading indicators continuously. "Combining utterances" in Table 1 means that separate utterances were combined into one utterance. For example, the original were "PLR (A) Low Flow warning went off" and "PLR (B) Low Flow warning went off". That is because an utterance is designed in Man Machine Simulator to be done after each reading of indicator/warning is done. Then, these were combined into "PLR Low Flow warning went off" in Figure 3. This table indicates that it is necessary to make many modifications in order to make CG character' actions and utterances more realistic and the modifications do not deny the MMS output itself.

5 Conclusion

The authors developed MMS-OTS, an operator training system based on Man Machine Simulator (MMS) developed by CRIEPI. This system uses 3D computer graphics to show the dynamic behaviours of the simulated team and the control panels. Many modifications are necessary on MMS output to make the actions or utterances of CG operators smoother. However, this system shows how to use research results of cognitive models of operators and also shows a new approach to operators' training with the cognitive model and computer graphics.

This is the result of the joint research project with BTC.

References

Takano,K., Sunaoshi,W. & Suzuki,K. (1999). Intellectual Simulation of Operating Team Behavior in Coping with Anomalies Occurring at Commercial Nuclear Power Plants, In H.J.Bullinger & J.Ziegler (Ed.), Human-Computer Interaction: Communication, Cooperation, and Application Design, Volume 2 of the Proceedings of the 8th International Conference on Human-Computer Interaction, (pp.1201-1205). London, LEA

Takano,K. et al. (2001). Development of the Operator Training System Using Computer Graphics (Part 1). CRIEPI Report S00003. (Japanese)

Experimental Studies of Computerized Support System from Human-Centered Aspect

Hidekazu Yoshikawa

Graduate School of Energy Science,
Kyoto University
Gokasho, Uji, Kyoto 611-0011 Japan
yosikawa@energy.kyoto-u.ac.jp

Takahisa Ozawa

Research Institute for Advanced Technology,
Matsushita Electric Works, Ltd.
Kadoma 1048, Kadoma, Osaka 571-8686 Japan
ozawa@ai.mew.co.jp

Abstract

Issues on human-centered approach for computer usage in process control is discussed from the experimental studies for introducing computerized operator support systems. One is the problem in the human cognitive characteristics for diagnosing the modern automatic process systems and the other is the use of eye tracking function in the computerized training system with expert's teaching to the trainees for effective transfer of meta-knowledge for process diagnosis.

1 Introduction

It has been commonly said in the field of process control that the introduction of computers into human-machine system has brought about the change of human operator's role from traditional manual controller to supervisory control (Sheridan, 1992), and according to Sheridan, the level of automation is classified into ten levels from complete manual control to full automation in accordance with how to introduce computer in the human-machine system. But his view is mainly based on a technology-centered automation that it classifies the relation of machine to human when introducing computer into the human-machine system, where no scope is given to think over how the human sees the reality of the complicated machine system caused by the addition of automatic functions.

In this paper, the introduction of computerized operator support system in human-machine system will be discussed not from technology-centered but from human-centered viewpoint. Concretely, the two issues of (i) what kind of problem will give rise to the human who deals with the machine systems which have equipped with more automatic functions and (ii) how to cope with the raised problem during the process of education and training for human operators, will be discussed on the basis of two experimental studies.

2 First Step Study

What kind of problem will be brought to human who will deal with the machine systems which have more automatic function? In order to understand the raised issue, the experimental studies have been conducted by the following three order; (i) experimental evaluation on how the human operator would constitute his/her mental model for the automated machine system during his/her process of education and training to master the knowledge and skill of the plant anomaly diagnosis, (ii) the construction of a human operator simulator based on the mental model derived from the experimental evaluation in (i), and (iii) the conductance of computer simulation experiment on human-machine interaction in order to know what kinds of problem are foreseen in the diagnostic practice on the basis of his/her conformed mental model (i.e., the evaluation of the accuracy of the human diagnosis model).

2.1 Derivation of human operator model

A laboratory experiment using a nuclear power plant simulator was conducted for observing human operators' cognitive behaviour for diagnosing plant anomaly with the participation of both the novice operator and the experienced ones (Takahashi, et al., 1996; Wu et al, 2000). The reduced human model for plant diagnosis from the experimental analysis could be interpreted as the combination of the three models, (i) the cause-consequence diagram to describe the event-chain propagation in the plant parameters on the human-machine interface which represents the plant dynamics including automatic control system, (ii) the qualitative increase/decrease relationship of the connected parameter pairs in the cause-consequence diagram, and (iii) a set of fuzzy logics to judge that individual parameters would exceed the normal range (Yoshikawa, et al., 1999; Ozawa, et al., 1999; Wu, et al, 2000). In fact, it can be said that the basic knowledge of human's diagnosis for plant anomaly conformed by his/her education/training process is basically the assembly of piece-wise knowledge of monotonous reasoning of event propagation.

2.2 Conductance of human-machine interaction simulation

The derived three models were utilized to organize a computerized human operator model simulator, and it was connected to a real-time plant dynamic simulator of a nuclear power plant to conduct versatile human-machine interaction simulation to simulate the diagnostic behaviour of human operator to find the root cause of anomaly (i.e., the root parameter seen in the disturbance propagation of the plant parameters in the cause-consequence diagram) for various types of accidents in nuclear power plant. In the human-machine interaction simulation, the accuracy of the human operator's diagnosis was estimated by adopting Dempster-Shafer's theory as the evaluation basis of upper and lower bounds of reliability for the diagnosed root causes used by the human operator model simulator. The example results of the human-machine interaction simulation are summarized in Table 1, with respect to the root cause symptoms, and upper and lower bounds of reliability for each anomaly type. In Table 1, the anomalies that have the same values of 1.0 for both the upper and lower bounds are the ones that can be easily diagnosed by the human operator model, while the other anomalies like zero or almost zero reliability are difficult cases for which human operator model can diagnose but with low confidence. And it is seen in Table 1 that the latter anomaly cases often happen when the compensating function by the automated system works so that the transient behaviour of plant be more difficult to interpret by the human model simulator.

2.3 Implication for human factors issues related with automated system

The result of the above first step experimental study can be summarized by the following way: "The basic knowledge of human's diagnosis for plant anomaly conformed by his/her education/training process is basically the assembly of piece-wise knowledge of monotonous reasoning of event propagation. This brings two issues. One is that the human diagnosis has basic weakness to deal with the process system response with automatic control system, and the other is the danger of constructing automatic diagnosis system simply by human diagnostic knowledge, what we call expert system.

Table 1: Example results of diagnosis by human operator model

Anomaly type	Symptom for root cause	Reliability	
		Upper bound	Lower bound
RCS leakage small size	Containment gas monitor high	1	1
	Containment dust monitor high	1	1
Steam generator tube rupture	Condenser gas monitor high	1	0
	SG blow down water monitor high	1	0
	Feed water flow control valve control signal low	1	0
Pressurizer spray valve failed open (large)	Spray valve openness high	0	0
	Main feed water flow control valve openness low	0	0
	Main steam flow low	0	0
	T-avg high	0.1	0
	Feed water flow control valve control signal low	1	0
Main feed water flow sensor fail high	Main feed water flow high	1	0
	SG water level low	1	0
	Pressurizer pressure high	1	0

3 The Second Step Study

Then, how should we do for human-centered approach? A new experimental study was conducted on the development of a computerized education and training system by the participation of a skilled expert for effective transfer of plant diagnosis skill. The target training system is a sort of interactive CAI equipped with a plant dynamic simulator, by which novice trainee will build up his/her skill by man-to-man training with a real expert as instructor (Ozawa, et al., 2001). The whole configuration of this training system is illustrated in Figure 1.

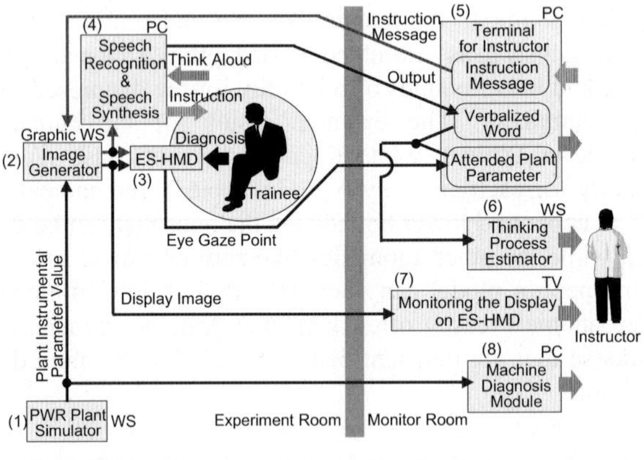

Figure 1: Whole configuration of training system

There are two interfaces in the CAI, one for trainee and the other for instructor. Concerning the interface for trainee, the authors' developed special head-mounted display (HMD) called as Eye-Sensing HMD (ES-HMD) (Fukushima, et al, 1999) was utilized in order to track the trainee's eye focusing point while showing him/her the graphic interface where the plant parameters change from the plant simulator are exhibited. The appearance of ES-HMS is shown in Figure 2 and the interface display for the trainee is shown in Figure 3. During the experiment, the trainee was asked to diagnose the root cause of various accidental transient by looking into the graphic display. As for the interface system to the instructor to teach trainee during the experiment, the same graphic display to the trainee through ES-HMD is also given to the instructor's interface where the trainee's eye focal point is superimposed to make the instructor more easily understand where the trainee attended in the display. The communication means from the instructor to the trainee were both the voice and text message put in through the instructor's interface to be given as pop-up display in the trainee's display any time during training. The interface for the instructor is shown in Figure 4.

Figure 2: Appearance of eye-sensing HMD

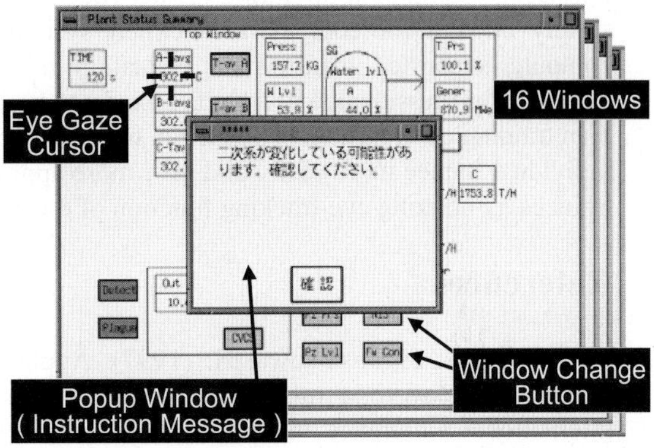

Figure 3: Interface display for trainee

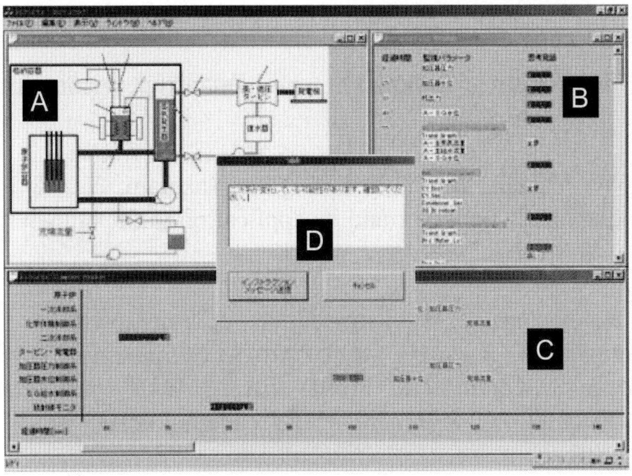

Figure 4: Interface display for instructor

Although the results of versatile experimental analysis of this second step experiment were presented in (Ozawa, et al, 2001), the most important findings from the view point of human-centered support can be summarized by the following statement: "By the afore-mentioned computerized support environment for the instructor, the instructor can properly recognize the failure of individual trainee's situation awareness, and give appropriate instructions in accordance with individual trainee's level of knowledge/skill such as where to attend in the interface display, how to apply the knowledge learned on-the-desk education." It was found that the instructor conveyed this kind of implicit "meta-knowledge" properly during the teaching process. In fact, those kind of meta-knowledge conveyed to the individual trainees by the instructor who deals with many trainees with varieties of ability and versatile personalities seems very difficult to implement into computerized CAI successfully, even by the use of advanced AI technology.

4 Conclusion

In this paper, with the basic assumption of the present practice of introduction of automatic control in the machine system as being given a priori, it was first pointed out that the human diagnosis knowledge being built up during his/her education process would not be necessarily a reliable one simply utilized for computerized expert system. And in the latter part of the paper, a new computer-supported man-to-man education training environment was proposed for effective transfer of professional skill from the expert to trainee. The essential point in the proposed man-to-man training environment is that it can help enhancing the instructor's "situation awareness" of the individual trainee more easily and give more appropriate support to meet with levels of individual trainees by utilizing eye-tracking function of the eye-sensing HMD.

References

Fukushima, S. & Yoshikawa, H. (1997). Application of a newly developed eye sensing head-mounted display to a mutual adaptive CAI for plant diagnosis. Proceedings of the seventh international conference on human-computer interaction, 2, 225-228.

Ozawa, T., Shimoda, H., Yoshikawa, H., & Hollnagel, E. (2001). An experimental study on an adaptive CAI system for training of diagnosing nuclear power plant anomalies, Proceedings of CSAPC2001, 319-328.

Sheridan, T.B. (1992). Telerobotics, automation, and human supervisory control. Cambridge: The MIT Press.

Takahashi, M., Wu, W., Yasuta, A., Yoshikawa, H. Nakatani, Y., & Nakagawa, T. (1996). Analysis of operator's diagnostic behavior and its application to the human modeling, Proceedings CSEPC96, 26-33.

Wu, W., Yoshikawa, H., Nakagawa, T., Kameda, A., & Fumizawa, M. (2000). Human model simulation of plant anomaly diagnosis (HUMOS-PAD) to estimate time cognitive reliability curve for HRA/PSA practice. In S. Kondo & K. Furuta (Ed.), *PSAM5-Probabilistic safety assessment and management* (pp.1001-1007). Tokyo: Universal Academic Press.

Yoshikawa, H. & Wu, W. (1999). An experimental study on estimating human error probability (HEP) parameters for PSA/HRA by using human model simulation. *Ergonomics,* 42(11), 1588-1595.

Author Index

Subject Index